Toxoplasma gondii

THE MODEL APICOMPLEXAN - PERSPECTIVES AND METHODS

Toxoplasma gondii

The Model Apicomplexan - Perspectives and Methods

Edited by

LOUIS M. WEISS AND KAMI KIM

AMSTERDAM • BOSTON • HEIDELBERG • LONDON
NEW YORK • OXFORD • PARIS • SAN DIEGO
SAN FRANCISCO • SINGAPORE • SYDNEY • TOKYO
Academic Press is an Imprint of Elsevier

Acquiring Editor: Linda Versteeg-Buschman
Development Editor: Halima Williams
Project Managers: Karen East and Kirsty Halterman
Designer: Matthew Limbert

Academic Press is an Imprint of Elsevier
32 Jamestown Road, London NW1 7BY, UK
225 Wyman Street, Waltham, MA 02451, USA
525 B Street, Suite 1800, San Diego, CA 92101-4495, USA

Second edition 2014

British Library Cataloguing-in-Publication Data
A catalogue record for this book is available from the British Library

Library of Congress Cataloging-in-Publication Data
A catalog record for this book is available from the Library of Congress

ISBN: 978-0-12-396481-6

For information on all Academic Press publications
visit our website at elsevierdirect.com

Typeset by TNQ Books and Journals
www.tnq.co.in

Printed and bound by CPI Group (UK) Ltd, Croydon, CRo 4YY

Cover photograph: Tissue cyst of ME49 (Type II strain) *T. gondii* isolated from mouse brain tissue. Electron microscopy performed at the Albert Einstein College of Medicine, Bronx, New York, USA in 1995. Photomicrograph courtesy of Dr. Sandra Halonen, Montana State University

Dedication

We would like to thank our families (Lisa, Hannah, Talia and Oren; Tom, Clayton and Vaughan) for their patience, understanding and tolerance during the completion of this book. In addition, we want to thank the *Toxoplasma* research community for their enthusiasm and contributions to this project. The *Toxoplasma* research community is legendary for its generosity toward colleagues and new investigators. It has been a unique pleasure to be involved with such a welcoming and intellectually stimulating group of researchers.

There have been many key research groups and individual researchers who have contributed to the development of the critical knowledge base required for progress on this pathogen. This book is a testament to these researchers.

We would like to dedicate this book to Elmer Pfefferkorn PhD, Dartmouth College. Elmer's work paved the way for the explosion in molecular biology, cell biology and genomic research associated with this organism. Elmer's intellectual rigour and deep thinking have had a significant influence on current researchers on *Toxoplasma gondii*, and we are all indebted to his generosity of spirit and profound insights into this pathogen.

LMW, KK
Bronx, New York, USA, 2007

Once again, we are deeply grateful for the support of our families, especially Tom and Lisa, who again displayed patience, understanding and tolerance during the completion of this revised edition.

The second edition of this book is dedicated to the *Toxoplasma* scientific community, whose dedication and productivity have made this updated volume a necessity.

LMW, KK
Bronx, New York, USA, 2013

Contents

19. ToxoDB: An Integrated Functional Genomic Resource for *Toxoplasma* and Other Sarcocystidae

OMAR S. HARB, DAVID S. ROOS ON BEHALF OF THE EUPATHDB GROUP

20. Comparative Aspects of Nucleotide and Amino Acid Metabolism in *Toxoplasma gondii* and Other Apicomplexa

KSHITIZ CHAUDHARY, BARBARA A. FOX, DAVID J. BZIK

21. *Toxoplasma gondii* Chemical Biology

MATTHEW BOGYO, GARY WARD

22. Proteomics of *Toxoplasma gondii*

JONATHAN M. WASTLING, DONG XIA

23. Cerebral Toxoplasmosis: Pathogenesis, Host Resistance and Behavioural Consequences

YASUHIRO SUZUKI, QILA SA, ERI OCHIAI, JEREMI MULLINS, ROBERT YOLKEN, SANDRA K. HALONEN

Preface to the First Edition

Toxoplasma gondii is a ubiquitous, Apicomplexan parasite of warm-blooded animals and is one of the most common parasitic infections of humans. Infection can result in encephalitis in immune compromised hosts, chorioretinitis in immune competent hosts or congenital transmission with fetopathy if a seronegative pregnant women becomes infected. It has been estimated that, in the absence of effective antiretroviral therapy and immune reconstitution, the risk for development of toxoplasmosis in a patient with AIDS with positive serologic findings for *Toxoplasma* is as high as 30%. Waterborne outbreaks of acute infection with chorioretinitis and an association of infection with increased mortality rates in California sea otter are emerging epidemiologic trends due to *T. gondii* infection.

The Apicomplexa are parasites that cause a wide variety of diseases in animals. *T. gondii* has become a model organism for the study of the Apicomplexa, as it is the most experimentally tractable organism in this important group of intracellular parasites that includes *Plasmodium, Eimeria, Cryptosporidium, Neospora* and *Theileria*. Currently, *T. gondii* remains the Apicomplexan species most readily amenable to genetic manipulation with refined protocols for both classic and reverse genetics. Transient transfection efficiency is high (routinely over 50%), and expression of epitope tags, reporter constructs and heterologous proteins is relatively uncomplicated. Due to the difficulties in genetic manipulation of most Apicomplexa, *T. gondii* has been used as an expression system for these parasites. *T. gondii* has also been used for testing the biological or biochemical function of proteins that, for one reason or other, cannot be readily expressed in other organisms. The pathogenic stages of *T. gondii* are easily propagated and quantified in the laboratory; the mouse animal model is well-established and reagents for study of the host response as well as basic biology of the parasite are widely available. Because of these experimental advantages, *T. gondii* has emerged as a major model organism for the study of Apicomplexan biology.

Immunity to *T. gondii* is a complex process involving innate and adaptive immune responses. *T. gondii* has been a useful model system pathogen for understanding the immune response to an intracellular pathogen, including studies on macrophage function, cell mediated immunity, dendritic cells and the gut associated immune response. The ease with which it can be cultured *in vitro,* availability of reporter parasite lines and its pathogenicity in mouse models has facilitated genetic studies of the immune response to this organism.

The availability of genome sequences has revolutionized the study of microbial pathogens. Genome sequences for the Apicomplexa are in various stages of completion. The *T. gondii* genome is ≈65 Mb and has been sequenced for a type 2 strain (ME49) at 12X (http://www.toxodb.org; http://www.apidb.org). These data have been integrated with genetic mapping data. Plans are in place for sequencing other strains of interest, as well.

In general, *T. gondii* genes are much more intron rich than those of *Plasmodium* or *Cryptosporidium*. This has made gene prediction more problematic, but recent gene models have been devised which address these problems and have permitted proteomic studies on *T. gondii*. Both genetic and proteomic studies have resulted in rapid advances in our understanding of the composition of the various organelles in this organism and how these specialized structures interact to allow successful intracellular parasitism by this organism.

This book was an outgrowth of discussions held at the Seventh International Congress on Toxoplasmosis in Tarrytown, New York in 2003 and the publication of papers and review articles from this congress in March 2004 in the International Journal for Parasitology (volume 34, number 3, pages 249–432). It was evident at this congress and confirmed at the Eighth International Congress in 2005 (Corsica, France) that the field of study of this pathogen had matured considerable since the First International Congress occurred in 1990 at Dartmouth University (Hanover, New Hampshire, USA). This has been paralleled by attendance at this congress, which has grown from an initial group of 26 investigators (Fig. 1) to over 150 participants and the increasing number of laboratories working on this organism.

There has been no recently published unified source for information on the biology, ultrastructure, genetics, immunology and animal models of this pathogen. We believe the current book fills this unmet need. Authors were encouraged to review older literature comprehensively so that their chapters could serve as free-standing reference articles. Many chapters include summary tables and provide key reference material,

FIGURE 1 **Participants in the First International Congress on Toxoplasmosis held in 1990 at Dartmouth College, Minary Conference Center, Squam Lake, Holderness, New Hampshire, USA**
Bottom: Lloyd Kasper, Rima McLeod, Jean Francois Dubremetz, John Boothroyd, Abbott Laboratories (unknown)
Middle: Joe Schwartzman, Elmer Pfefferkorn, Takuro Endo, Louis M. Weiss, Francoise Darcy, Philippe Thuiliez, Marie France Cesbron, Judy Smith, Alan Johnson, Yasu Suzuki, Takashi Asai, J. P. Dubey
Top: Greg Felice, James L. Fishback, David Sibley, Jack Remington, Alan Sher, Jack Frenkel, David Roos, Keith Joiner, Ben Luft, Bill Current

including photomicrographs and data from the literature. We hope that the chapters can serve as summaries of the current state of the literature providing an easy access point for studies on this organism.

The enthusiastic participation of the research community was critical in making this project a reality. We hope that the final product will serve as a key reference material for researchers who want to study *T. gondii* or use it as a model eukaryotic pathogen.

LMW, KK
Bronx, New York, USA, 2007

Preface to the Second Edition

Since the publication of the first edition of this book in 2007, the 'omics' revolution has influenced all aspects of the *Toxoplasma* field. Analysis of the genome and comparative genomics has enabled identification of novel gene families linked to pathogenesis and virulence, as well as facilitating the identification of new genes associated with novel organelles such as the apicoplast. Further study of the population biology of *T. gondii* using more sophisticated molecular tools has resulted in a more complete understanding of the worldwide distribution and evolution of *T. gondii* strains. Large transcriptomics, proteomics and epigenomics datasets are available and led to new insights into gene regulation, including the identification of the APETELA or AP2 family of plant-like transcription factors. Unique families of secreted proteins have been identified as modulators of host signalling and gene expression. Among the most intriguing are the ROP2 kinase and pseudokinase family members that are secreted into the host, but this family is only one of many sets of *Toxoplasma* genes that manipulate the host. Further studies suggest that other rhoptry and dense granule proteins are implicated in a complex modulation of host metabolism and gene expression leading to alterations of host signalling and immune response, and we anticipate that over the next years, new efforts will continue to illuminate our understanding of the host-pathogen interplay.

Next generation sequencing has been applied to the resequencing and reannotation of the genome of *Toxoplasma gondii.* As part of a community driven 'white paper' conceived at the Tenth International *Toxoplasma* meeting in 2009, high coverage sequencing of DNA and RNA from prototypic representatives of the 16 major groups was completed. These data are complemented by lower coverage sequences of 35 other representative strains. Transcriptomic and proteomic data are available for sporozoite and bradyzoite stages, enabling generation of new hypotheses about these important and experimentally inaccessible developmental stages. The reannotation of the genome has been a community effort that also incorporates data from proteomics and epigenomics studies as well as functional studies by many groups. Numerous advances in technology, including new methods to regulate gene or protein expression, have also facilitated state of the art molecular genetics and chemical biology approaches that permit investigation of aspects of *T. gondii* biology previously intractable to experimental inquiry.

Preparations are being made for the Twelfth International Congress on Toxoplasmosis and many young investigators are entering the field. With the many new technologies, enormous new datasets and new groups now contributing to the field, there continues to be a need for a comprehensive text that synthesizes and summarizes the literature for beginners as well as experienced investigators. We are fortunate that so many members of the *Toxoplasma* research community have generously donated their time to update this book and lend their expertise. We anticipate the next wave of research will enable us to use new and old tools to more thoroughly understand the parasite and its relationship with its animal and human hosts giving us further insights into how *T. gondii* has evolved into such a successful intracellular parasite.

LMW, KK
Bronx, New York, USA, 2013

List of Contributors

James W. Ajioka Department of Pathology, Cambridge University, Cambridge, England, United Kingdom

Daniel Ajzenberg Department of Parasitology and Mycology, Toxoplasma Biological Resource Center, Faculté de Médecine, Université de Limoges, Limoges, France

James Alexander Strathclyde Institute of Pharmacy and Biomedical Sciences, University of Strathclyde, Glasgow, Scotland, United Kingdom

Takashi Asai Department of Tropical Medicine and Parasitology—Infectious Diseases, Keio University School of Medicine, Tokyo, Japan

Matthew Bogyo Department of Pathology, Stanford University School of Medicine, Stanford, California, USA

John C. Boothroyd Department of Microbiology and Immunology, Stanford University School of Medicine, Stanford, California, USA

Barbara A. Butcher Department of Microbiology and Immunology, Cornell University College of Veterinary Medicine, Ithaca, New York, USA

David J. Bzik Department of Microbiology and Immunology, Geisel School of Medicine at Dartmouth, Lebanon, New Hampshire, USA

Vern B. Carruthers Department of Microbiology and Immunology, University of Michigan Medical School, Ann Arbor, Michigan, USA

Marie-France Cesbron-Delauw UMR5163-CNRS-UJF, Jean-Roget Institute, Grenoble, France

Kshitiz Chaudhary Vaccines Research, Pfizer, Pearl River, New York, USA

Isabelle Coppens Department of Molecular Microbiology and Immunology, Johns Hopkins University Bloomberg School of Public Health, Baltimore, Maryland, USA

Marie-Laure Dardé Toxoplasma Biological Resource Center and Faculté de Médecine, Université de Limoges, Limoges, France

Eric Y. Denkers Department of Microbiology and Immunology, Cornell University College of Veterinary Medicine, Ithaca, New York, USA

Jitender P. Dubey Animal Parasitic Diseases Laboratory, Beltsville Agricultural Research Center, Agricultural Research Service, United States Department of Agriculture, Beltsville, Maryland, USA

Jean-François Dubremetz UMR CNRS 5235, Université de Montpellier 2, Montpellier, France

Jean E. Feagin Emerging and Neglected Diseases Programe, Seattle Biomedical Research Institute, Seattle, Washington, USA; Departments of Global Health and Pharmacy, University of Washington, Seattle, Washington, USA

David J. P. Ferguson Nuffield Department of Clinical Laboratory Science, University of Oxford, John Radcliffe Hospital, Oxford, UK

Barbara A. Fox Department of Microbiology and Immunology, Geisel School of Medicine at Dartmouth, Lebanon, New Hampshire, USA

Ricardo T. Gazzinelli Federal University of Minas Gerais, Minas Gerais, Brazil and University of Massachusetts, Worcester, Massachusetts, USA

Uwe Gross Institute for Medical Microbiology, University Medical Centre, Georg-August-University, Göttingen, Germany

Marc-Jan Gubbels Department of Biology, Boston College, Chestnut Hill, Massachusetts, USA

Sandra K. Halonen Department of Microbiology, Montana State University, Bozeman, Montana, USA

Omar S. Harb Department of Biology, University of Pennsylvania, Philadelphia, Pennsylvania, USA

Fiona L. Henriquez Institute of Biomedical and Environmental Health Research, School of Science, University of the West of Scotland, Paisley, Scotland, United Kingdom

Andrea Honda New York Eye and Ear Infirmary, New York, USA

Christopher A. Hunter Department of Pathobiology, School of Veterinary Medicine, University of Pennsylvania, Philadelphia, Pennsylvania, USA

Damien Jacot Department of Microbiology and Molecular Medicine, Faculty of Medicine, University of Geneva, Geneva, Switzerland

Farzana Khaliq Strathclyde Institute of Pharmacy & Biomedical Sciences, University of Strathclyde, Glasgow, Scotland, UK

Imtiaz A. Khan Department of Microbiology, Immunology and Tropical Medicine, George Washington University, Washington DC, USA

Kami Kim Departments of Medicine, Pathology, and Microbiology & Immunology, Albert Einstein College of Medicine, Bronx, New York, USA

Laura J. Knoll Department of Medical Microbiology and Immunology, University of Wisconsin-Madison, Madison, Wisconsin, USA

Paul Latkany New York Eye and Ear Infirmary, New York, USA

Maryse Lebrun Laboratoire Dynamique des interactions membranaires normales et pathologiques, Université de Montpellier, Montpellier, France

David S. Lindsay Department of Biomedical Sciences and Pathobiology, Virginia-Maryland Regional College of Veterinary Medicine, Virginia Tech, Blacksburg, Virginia, USA

Carsten G.K. Lüder Institute for Medical Microbiology, University Medical Centre, Georg-August-University, Göttingen, Germany

Rima McLeod Divisions of Ophthalmology and Visual Sciences (Surgery), Infectious Diseases (Pediatrics), Committees on Immunology, Molecular Medicine, Fellow Institute Genetics, Genomics and Systems Biology, The University of Chicago, Chicago, Illinois, USA

Markus Meissner Wellcome Trust Centre for Molecular Parasitology, Institute of Infection, Immunity and Inflammation, College of Medical, Veterinary and Life Sciences, University of Glasgow, Glasgow, Scotland, United Kingdom

José G. Montoya Department of Medicine, Palo Alto Medical Foundation, Stanford University, Palo Alto, California, USA

Dana G. Mordue Department of Microbiology and Immunology, New York Medical College, Valhalla, New York, USA

Silvia N.J. Moreno Department of Cellular Biology and Center for Tropical and Emerging Global Diseases, University of Georgia, Athens, Georgia, USA

Naomi Morrissette Department of Molecular Biology and Biochemistry, University of California Irvine, California, USA

Jeremi Mullins Department of Microbiology, Immunology and Molecular Genetics, University of Kentucky College of Medicine, Lexington, Kentucky, USA

Eri Ochiai Department of Microbiology, Immunology and Molecular Genetics, University of Kentucky College of Medicine, Lexington, Kentucky, USA

Douglas A. Pace Department of Biological Sciences, California State University, Long Beach, California, USA

Marilyn Parsons Emerging and Neglected Diseases Program, Seattle Biomedical Research Institute, Seattle, Washington, USA; Department of Global Health, University of Washington, Seattle, Washington, USA

Lucas Borges Pereira Department of Parasitology, Institute of Biomedical Sciences, University of São Paulo, São Paulo, Brazil

Eskild Petersen Department of Infectious Diseases, Aarhus University Hospital, Skejby, Aarhus, Denmark

Sheela Prasad The University of Chicago, Chicago, Illinois, USA

Joshua B. Radke Departments of Molecular Medicine and Global Health and the Florida Center for Drug Discovery and Innovation, University of South Florida, Tampa, Florida, USA

Michael L. Reese Department of Microbiology and Immunology, Stanford University School of Medicine, Stanford, California, USA

Utz Reichard Institute for Medical Microbiology, University Medical Centre, Georg-August-University, Göttingen, Germany

Craig W. Roberts Strathclyde Institute of Pharmacy and Biomedical Sciences, University of Strathclyde, Glasgow, Scotland, United Kingdom

David S. Roos Department of Biology, University of Pennsylvania, Philadelphia, Pennsylvania, USA

Qila Sa Department of Microbiology, Immunology and Molecular Genetics, University of Kentucky College of Medicine, Lexington, Kentucky, USA

Frank Seeber FG16, Parasitology, Robert Koch-Institute, Berlin, Germany

Lilach Sheiner Center for Tropical and Emerging Global Diseases, University of Georgia, Athens, Georgia, USA

L. David Sibley Department of Molecular Microbiology, Washington University School of Medicine, St Louis, Missouri, USA

Anthony P. Sinai Department of Microbiology, Immunology and Molecular Genetics, University of Kentucky College of Medicine, Lexington, Kentucky, USA

Dominique Soldati-Favre Department of Microbiology and Molecular Medicine, Faculty of Medicine, University of Geneva, Geneva, Switzerland

Boris Striepen Center for Tropical and Emerging Global Diseases, University of Georgia, Athens, Georgia, USA

Emily Su St Luke's Roosevelt Hospital, New York, USA

Chunlei Su Department of Microbiology, The University of Tennessee, USA

William J. Sullivan, Jr. Departments of Pharmacology & Toxicology, and Microbiology & Immunology, Indiana University School of Medicine, Indianapolis, Indiana, USA

Yasuhiro Suzuki Department of Microbiology, Immunology and Molecular Genetics, University of Kentucky College of Medicine, Lexington, Kentucky, USA

Stanislas Tomavo Centre for Infection and Immunity of Lille, Institut Pasteur de Lille, Université Lille Nord de France, France

Tadakimi Tomita Department of Pathology and Medicine, Albert Einstein College of Medicine, Bronx, New York, USA

Christine Van Tubbergen The University of Chicago, Chicago, Illinois, USA

Gary Ward Department of Microbiology and Molecular Genetics, University of Vermont College of Medicine, Burlington, Vermont, USA

Jonathan M. Wastling Department of Infection Biology, Institute of Infection and Global Health, University of Liverpool, Liverpool, England, United Kingdom

Louis M. Weiss Departments of Pathology and Medicine, Albert Einstein College of Medicine, Bronx, New York, USA

Michael W. White Departments of Molecular Medicine and Global Health and the Florida Center for Drug Discovery and Innovation, University of South Florida, Tampa, Florida, USA

Dong Xia Department of Infection Biology, Institute of Infection and Global Health, University of Liverpool, Liverpool, England, UK

Robert Yolken Stanley Neurology Laboratory, Johns Hopkins University, Baltimore, Maryland, USA

CHAPTER

1

The History and Life Cycle of *Toxoplasma gondii*

Jitender P. Dubey

Animal Parasitic Diseases Laboratory, Beltsville Agricultural Research Center, Agricultural Research Service, United States Department of Agriculture, Beltsville, Maryland, USA

OUTLINE

Toxoplasma gondii, Second Edition
http://dx.doi.org/10.1016/B978-0-12-396481-6.00001-5

2014 Published by Elsevier Ltd

1.1 INTRODUCTION

Infections by the protozoan parasite *Toxoplasma* gondii are widely prevalent in humans and other animals on all continents. There are many thousands of references to this parasite in the literature and it is not possible to give equal treatment to all authors and discoveries (Dubey, 2008). The objective of this chapter is, rather, to provide a history of the milestones in our acquisition of knowledge of the biology of this parasite (Fig. 1.1).

1.2 THE ETIOLOGICAL AGENT

Nicolle and Manceaux (1908) found a protozoan in tissues of a hamster-like rodent, the gundi, *Ctenodactylus gundi,* which was being used for leishmaniasis research in the laboratory of Charles Nicolle at the Pasteur Institute in Tunis. They initially believed the parasite to be *Leishmania,* but soon realized that they had discovered a new organism and named it *Toxoplasma gondii* based on the morphology (Modern Latin *Toxo* = arc or bow, *plasma* = life) and

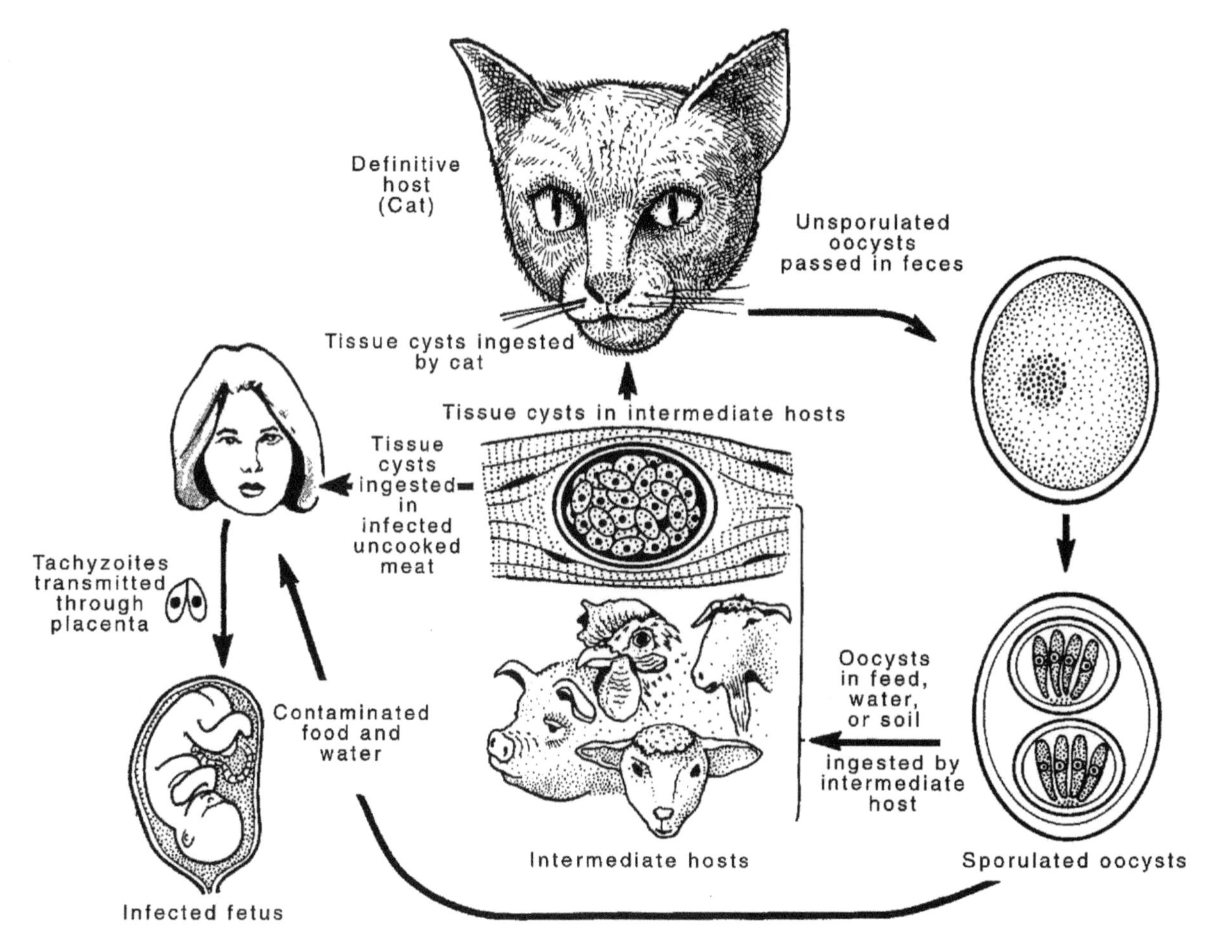

FIGURE 1.1 **Life cycle of *T. gondii*.**

the host (Nicolle and Manceaux, 1909). Thus, its complete designation is *Toxoplasma gondii* (Nicolle and Manceaux, 1908). In retrospect, the correct name for the parasite should have been *Toxoplasma gundii*; Nicolle and Manceaux (1908) had incorrectly identified the host as *Ctenodactylus gondi*. Splendore (1908, translated in 2009 into English) discovered the same parasite in a rabbit in Brazil, also erroneously identifying it as *Leishmania*, but he did not name it. It is a remarkable coincidence that this disease was first recognized in laboratory animals and was first thought to be *Leishmania* by both groups of investigators.

1.3 PARASITE MORPHOLOGY AND LIFE CYCLE

1.3.1 Tachyzoites

The tachyzoite (Frenkel, 1973) is lunate and is the stage that Nicolle and Manceaux (1909) found in the gundi (Fig. 1.2A). This stage has also been called trophozoite, the proliferative form, the feeding form, and endozoite. It can infect virtually any cell in the body. It divides by a specialized process called endodyogeny, first described by Goldman *et al.* (1958). Gustafson *et al.* (1954) first studied the ultrastructure of the

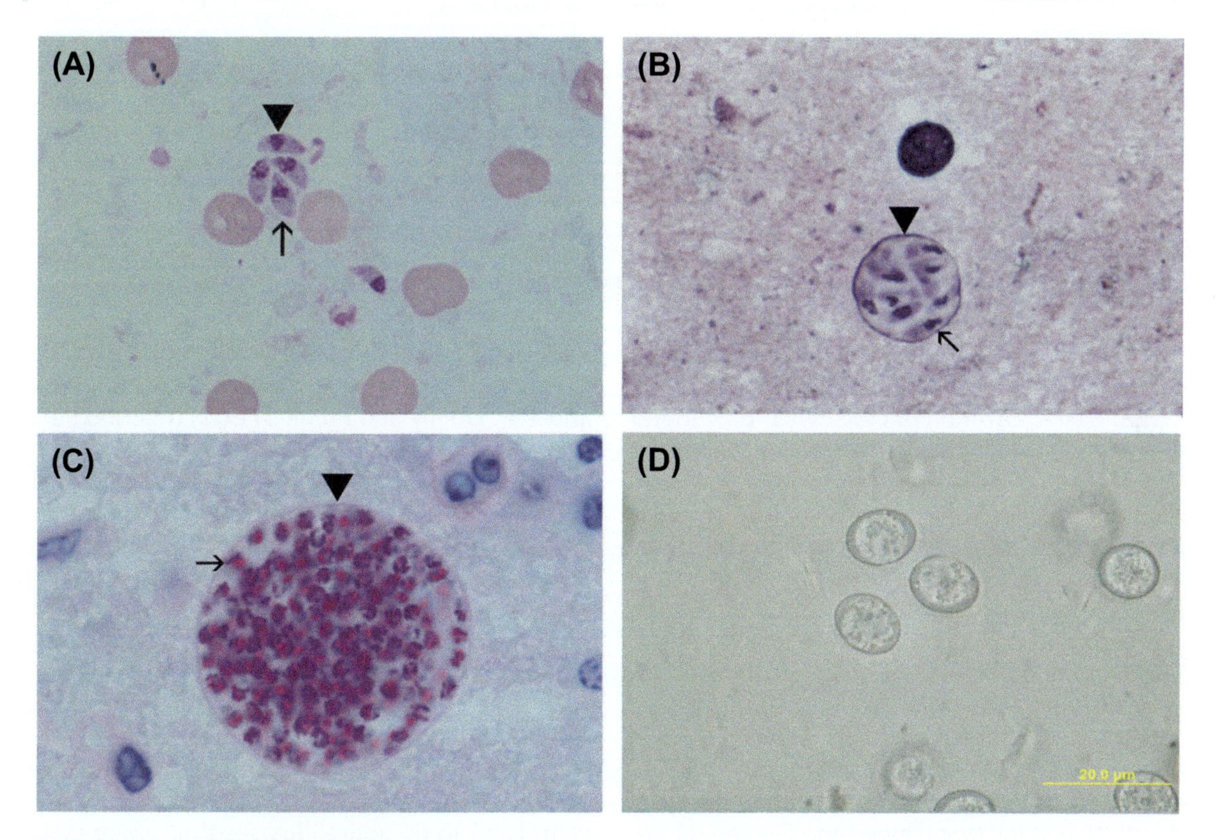

FIGURE 1.2 **Life cycle stages of *T. gondii*.**
A) Tachyzoites (arrowhead) in smear. Giemsa stain. Note nucleus dividing into two nuclei (arrow).
B) A small tissue cyst in smear stained with Giemsa and a silver stain. Note the silver-positive tissue cyst wall (arrow head) enclosing bradyzoites that have a terminal nucleus (arrow).
C) Tissue cyst in section, PAS. Note PAS-positive bradyzoites (arrow) enclosed in a thin PAS-negative cyst wall (arrowhead).
D) Unsporulated oocysts in cat faeces. Unstained.

tachyzoite. Sheffield and Melton (1968) provided a complete description of endodyogeny when they fully described its ultrastructure.

1.3.2 Bradyzoite and Tissue Cysts

The term 'bradyzoite' (Greek brady = slow) was proposed by Frenkel (1973) to describe the stage encysted in tissues. Bradyzoites are also called cystozoites. Dubey and Beattie (1988) proposed that cysts should be called tissue cysts to avoid confusion with oocysts (Figs 1.2B,C). It is difficult to determine from the early literature who first identified the encysted stage of the parasite (Lainson, 1958). Levaditi *et al.* (1928) were apparently the first to report that *T. gondii* may persist in tissues for many months as 'cysts'; however, considerable confusion between the term 'pseudocysts' (a group of tachyzoites) and tissue cysts existed for many years. Frenkel and Friedlander (1951) and Frenkel (1956) characterized cyst cytologically as containing organisms with a subterminal nucleus and periodic acid Schiff (PAS)-positive granules surrounded by an argyrophilic cyst wall (Fig. 1.2B,C). Wanko *et al.* (1962) first described the ultrastructure of the *T. gondii* cyst and its contents. Jacobs *et al.* (1960a) first provided a biological characterization of cysts when they found that the cyst wall was destroyed by pepsin or trypsin, but the cystic organisms were resistant to digestion by gastric juices (pepsin-HCl), whereas tachyzoites were destroyed immediately. Thus, tissue cysts were shown to be important in the life cycle of *T. gondii* because carnivorous hosts can become infected by ingesting infected meat. Jacobs *et al.* (1960b) used the pepsin digestion procedure to isolate viable *T. gondii* from tissues of chronically infected animals. When *T. gondii* oocysts were discovered in cat faeces in 1970, oocyst shedding was added to the biological description of the cyst (Dubey and Frenkel, 1976).

Dubey and Frenkel (1976) performed the first in-depth study of the development of tissue cysts and bradyzoites and described their ontogeny and morphology. They found that tissue cysts formed in mice as early as three days after their inoculation with tachyzoites. Cats shed oocysts with a short prepatent period (three to 10 days) after ingesting tissue cysts or bradyzoites, whereas after they ingested tachyzoites or oocysts the prepatent period was longer (≥18 days), irrespective of the number of organisms in the inocula (Dubey and Frenkel, 1976; Dubey, 1996, 2001, 2006) (Fig. 1.2D). Prepatent periods of 11–17 days are thought to result from the ingestion of transitional stages between tachyzoite and bradyzoite (Dubey, 2002, 2005).

Wanko *et al.* (1962), and Ferguson and Hutchison (1987) reported on the ultrastructural development of *T. gondii* tissue cysts. The biology of bradyzoites including morphology, development in cell culture *in vivo*, conversion of tachyzoites to bradyzoites, and *vice versa*, tissue cyst rupture, and distribution of tissue cysts in various hosts and tissues was reviewed critically by Dubey *et al.* (1998).

1.3.3 Enteroepithelial Asexual and Sexual Stages

Asexual and sexual stages were reported in the intestine of cats in 1970 (Frenkel, 1970) (Figs 1.3 and 1.4). Dubey and Frenkel (1972) described the asexual and sexual development of *T. gondii* in enterocytes of the cat and designated the asexual enteroepithelial stages as types A through E rather than as generations conventionally known as schizonts in other coccidian parasites (Figs 1.3 and 1.4). These stages were distinguished morphologically from tachyzoites and bradyzoites, which also occur in cat intestine (Fig. 1.3D). The challenge was to distinguish different stages in the cat intestine because there was profuse multiplication of *T. gondii* three days post-infection (Fig. 1.4A). The entire cycle was completed in 66 hours after feeding tissue cysts to cats (Dubey and Frenkel, 1972). There are reports on the ultrastructure of schizonts (Sheffield, 1970; Piekarski *et al.*, 1971; Ferguson *et al.*, 1974), gamonts (Ferguson *et al.*, 1974, 1975;

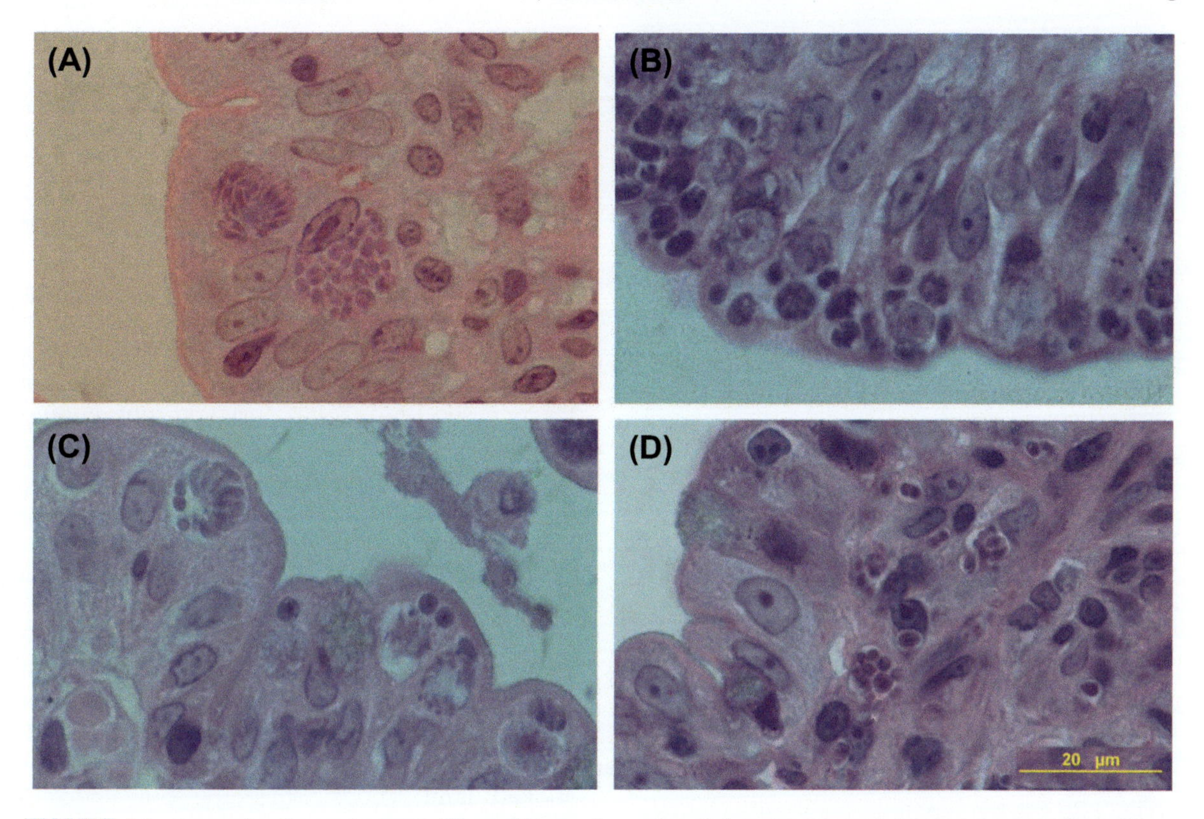

FIGURE 1.3 **Asexual and sexual stages of *T. gondii* in sections of small intestine of cats fed tissue cysts. H and E stain.** A) Type C (arrow) schizont with a residual body and a Type B schizont with a hypertrophied host cell nucleus (arrowhead). 52 hours post-infection.
B) Heavily infected small intestine with schizonts and gamonts in the epithelium. Five days post-infection.
C) Types D and E schizonts (a), a mature female gamont (e), young female gamont (b), and two male gamonts (c) in the epithelium.
D) Tachyzoites in the lamina propria (arrows). Types B and D schizonts are below the enterocyte nucleus and often cause hypertrophy of the parasitized cell whereas Types D and E schizonts are always above the enterocyte nucleus and do not cause hypertrophy of the host cell even in hyperparasitized cases. Tachyzoites are found in the lamina propria of the cat intestine.

Speer and Dubey, 2005), oocysts and sporozoites (Christie *et al.*, 1978; Ferguson *et al.*, 1979a,b; Speer *et al.*, 1998). In 2005, Speer and Dubey described the ultrastructure of asexual enteroepithelial types B through E and distinguished their merozoites.

1.4 TRANSMISSION

1.4.1 Congenital

The mechanism of transmission of *T. gondii* remained a mystery until its life cycle was discovered in 1970. Soon after the initial discovery of the organism it was found that the *C. gundi* were not infected in the wild and had acquired *T. gondii* infection in the laboratory. Initially, transmission by arthropods was suspected, but this was never proven (Frenkel, 1970, 1973). Congenital *T. gondii* infection in a human child was initially described by Wolf *et al.* (1939a,b) and later found to occur in many species of animals, particularly sheep, goats, and rodents. Congenital infections can be repeated in some strains of mice (Beverley, 1959), with infected mice producing congenitally infected offspring for at least 10 generations.

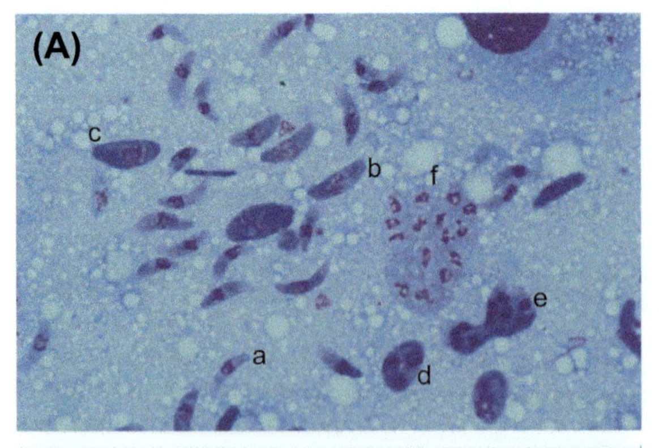

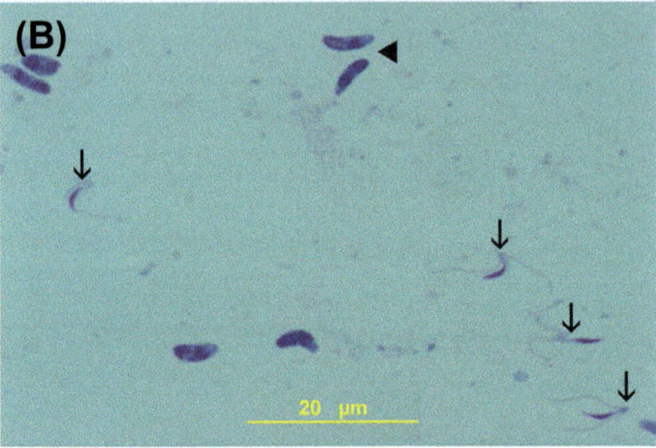

FIGURE 1.4 **Smears of intestinal epithelium of a cat seven days after feeding tissue cysts (Giemsa stain).** A) Note different sizes of merozoites (a–c), schizont with three nuclei (d), schizont with six or more nuclei and merozoites are budding from the surface (e), and a multinucleated schizont (f).
B) Four biflagellated microgametes (arrows) and merozoites (arrowhead) for size comparison.

Beverley discontinued his experiments because of high mortality in some lines of congenitally infected mice and because the progeny from the last generation of infected mice were seronegative and presumed not to be infected with *T. gondii*. Jacobs (1964) repeated these experiments and found that congenitally infected mice may be infected but not develop antibodies because of immune tolerance. Dubey *et al.* (1995a) isolated viable *T. gondii* from seronegative, naturally-infected mice. These findings are of epidemiological significance.

1.4.2 Carnivorism

Congenital transmission occurs too rarely to explain widespread infection in man and animals worldwide. Weinman and Chandler (1954) suggested that transmission might occur through the ingestion of undercooked meat. Jacobs *et al.* (1960a) provided evidence to support this idea by demonstrating the resistance to proteolytic enzymes of *T. gondii* derived from cysts. They found that the cyst wall was immediately dissolved by such enzymes but the released bradyzoites survived long enough to infect the host. This hypothesis of transmission through the ingestion of infected meat was experimentally tested by Desmonts *et al.* (1965) in an experiment with children in a Paris sanatorium. They compared the acquisition rates of *T. gondii* infection in children before and after admission to the sanatorium. The 10% yearly acquisition rate of *T. gondii* antibody rose to 50% after adding two portions of undercooked beef or horse meat to the daily diet and to a 100% yearly rate after the addition of undercooked lamb chops. Since the prevalence of *T. gondii* is much higher in sheep than in horses or cattle this illustrated the importance of carnivorism in transmission of *T. gondii*. Epidemiological evidence indicates it is common in humans in some localities where raw meat is routinely eaten. In a survey in Paris, Desmonts *et al.* (1965) found that over 80% of the adult population sampled had antibodies to *T. gondii*. Kean *et al.* (1969) described toxoplasmosis in a group of medical students who had eaten undercooked hamburgers.

1.4.3 Faecal–Oral

While congenital transmission and carnivorism can explain some of the transmission of *T. gondii* it does not explain the widespread infection in vegetarians and herbivores. A study in Bombay, India, found the prevalence of *T. gondii* in strict vegetarians to be similar to that in non-vegetarians (Rawal, 1959). Hutchison (1965), a biologist at Strathclyde University in Glasgow, first discovered *T. gondii* infectivity associated with cat faeces. In a preliminary experiment, Hutchison (1965) fed *T. gondii* cysts to a cat infected with the nematode *Toxocara cati* and collected faeces containing nematode ova. Faeces floated in 33% zinc sulphate solution and stored in tap water for 12 months induced toxoplasmosis in mice. This discovery was a breakthrough because, until then, both known forms of *T. gondii* (i.e. tachyzoites and bradyzoites) were killed by water. Microscopic examination of faeces revealed only *T. cati* eggs and *Isospora* oocysts. In Hutchison's report, *T. gondii* infectivity was not attributed to oocysts or *T. cati* eggs. He repeated the experiment with two *T. cati*-infected and two *T. cati*-free cats. *T. gondii* was transmitted only in association with *T. cati* infection. On this basis, Hutchison (1967) hypothesized that *T. gondii* was transmitted through nematode ova. He suspected transmission of *T. gondii* through the eggs of the nematode *Toxocara* similar to the transmission of the fragile flagellate *Histomonas* through *Heterakis* eggs. Hutchison initially wanted to test the nematode theory using *Toxocara canis* and *T. gondii* transmission in the dog but decided on the cat and *Toxocara cati* model because there was no place to house dogs (J.P. Dubey, 1965, personal communication). Transmission of *T. gondii* by *Toxocara canis* eggs made more sense because of the known zoonotic potential of *T. canis*; *Toxocara cati* was not, at that time, known to infect humans, but *T. canis* was. Discovery of the life cycle of *T. gondii* would have been delayed if Hutchison had worked with dogs instead of cats.

Hutchison's 1965 report stimulated other investigators to examine faecal transmission of *T. gondii* through *T. cati* eggs (Dubey, 1966, 1968; Jacobs, 1967; Hutchison *et al.*, 1968; Frenkel *et al.*, 1969; Sheffield and Melton, 1969). The nematode egg theory of transmission was discarded after *Toxoplasma* infectivity was dissociated from *T. cati* eggs (Frenkel *et al.*, 1969) and *Toxoplasma* infectivity was found in faeces of worm-free cats fed *T. gondii* (Frenkel *et al.*, 1969; Sheffield and Melton, 1969). Finally, in 1970, knowledge of the *T. gondii* life cycle was completed by discovery of the sexual phase of the parasite in the small intestine of the cat (Frenkel *et al.*, 1970). *T. gondii* oocysts, the product of schizogony and gametogony, were found in cat faeces and characterized morphologically and biologically (Dubey *et al.*, 1970a, b).

Several groups of workers, independently and at about the same time, found *T. gondii* oocysts in cat faeces (Hutchison *et al.*, 1969, 1970, 1971; Frenkel *et al.*, 1970; Dubey *et al.*, 1970a,b; Sheffield and Melton, 1970; Overdulve, 1970; Weiland and Kűhn, 1970; Witte and Piekarski, 1970). The discovery of the *T. gondii* oocyst in cat faeces and its implications have been reviewed by Frenkel (1970, 1973) and Garnham (1971).

In retrospect the discovery and characterization of the *T. gondii* oocyst in cat faeces was delayed because (1) *T. gondii* oocysts were morphologically identical to oocysts of the previously described coccidian parasite of cats and dogs (Dubey *et al.*, 1970a) and (2) until 1970 coccidian oocysts were sporulated in 2.5% potassium dichromate. Chromation of the oocyst's wall interfered with excystation of the sporozoites when oocysts were fed to mice and, thus, the mouse infectivity titre of the oocysts was lower than expected from the number of oocysts administered (Dubey *et al.*, 1970a). These findings led to the use of 2% sulphuric acid as the best medium for sporulation and storage of *T. gondii* oocysts. Unlike dichromate, which was difficult to wash off the oocysts, sulphuric acid could be easily neutralized and the oocysts could be injected without washing into mice (Dubey *et al.*, 1972). Unlike other coccidians, *T. gondii* oocysts were found to excyst efficiently when inoculated parenterally into mice and thus alleviated the need for oral inoculation for bioassay of oocysts (Dubey and Frenkel, 1973, 2006).

Ben Rachid (1970) fed *T. gondii* oocysts to gundis which died 6–7 days later from toxoplasmosis. This knowledge about the life cycle of *T. gondii* probably explains how gundis became infected in the laboratory of Nicolle. At least one cat was present in the Pasteur laboratory in Tunis (Dubey, 1977).

Of the many species of animals experimentally infected with *T. gondii,* only felids shed *T. gondii* oocysts (Miller *et al.*, 1972; Jewell *et al.*, 1972; Janitschke and Werner, 1972; Polomoshnov, 1979; Dubey, 2010). Oocysts shed into the environment have caused several outbreaks of disease in humans (Teutsch *et al.*, 1979; Benenson *et al.*, 1982; Bowie *et al.*, 1997; de Moura *et al.*, 2006). *T. gondii* oocysts found in the faeces of naturally-infected cougars (Aramini *et al.*, 1998) were epidemiologically linked to the largest recorded waterborne outbreak of toxoplasmosis in humans (Bowie *et al.*, 1997). Seroepidemiological studies on isolated islands in the Pacific (Wallace, 1969), Australia (Munday, 1972), and the USA (Dubey *et al.*, 1997) have shown an absence of *Toxoplasma* on islands without cats, confirming the important role of the cat in the natural transmission of *T. gondii*. Vaccination of cats with a live mutant strain of *T. gondii* on eight pig farms in the USA reduced the transmission of *T. gondii* infection in mice and pigs (Mateus-Pinilla *et al.*, 1999), thus supporting the role of the cat in natural transmission of *T. gondii*.

Historically, before the discovery of the coccidian cycle of *T. gondii,* coccidian parasites were considered host- and site-specific and to be transmitted by the faecal–oral route. After the discovery of the sexual cycle of *T. gondii*, several other genera (e.g. *Sarcocystis, Besnoitia*) were found to be coccidian. Although *T. gondii* has a wide host range it has retained the definitive-host specificity restricted to felids. Dr. J.K. Frenkel deserves the credit for initiating testing of many species of animals, including wild felids, for oocysts shedding under difficult housing conditions (it was not easy handling bob cats and ocelots in cages) (Dubey, 2009). Only the felids were found to shed *T. gondii* oocysts (Frenkel *et al.*, 1970; Miller *et al.*, 1972). Although *T. gondii* can be transmitted in several ways, it has adapted to be transmitted most efficiently by carnivorism in the cat and by the faecal–oral (oocysts) route in other hosts. Pigs and mice (and presumably humans) can be infected by ingesting even one oocyst (Dubey *et al.*, 1996), whereas 100 oocysts may not infect cats (Dubey, 1996). Cats can shed millions of oocysts after ingesting only one bradyzoite, while ingestion of 100 bradyzoites may not infect mice orally (Dubey, 2001, 2006). This information has proved very useful in conducting epidemiological studies and for the detection by feeding to cats of low numbers of *T. gondii* in large samples of meat (Dubey *et al.*, 2005).

After the discovery of the life cycle of *T. gondii* in the cat it became clear why Australasian marsupials and New World monkeys are highly susceptible to clinical toxoplasmosis. The former evolved in the apparent absence of the cat (there were few or no cats in Australia and New Zealand before settlement by Europeans) and the latter live on tree tops, not exposed to cat faeces. In contrast, marsupials in America and Old World monkeys are resistant to clinical toxoplasmosis (Dubey and Beattie, 1988).

1.5 TOXOPLASMOSIS IN HUMANS

1.5.1 Congenital Toxoplasmosis

Three pathologists, Wolf, Cowen, and Paige from New York, USA, first conclusively identified *T. gondii* in an infant girl who was delivered full term by caesarean section on May 23, 1938 at Babies Hospital, New York (Wolf *et al.*, 1939a,b). The girl developed convulsive seizures at three days of age and lesions were noted in the maculae of both eyes through an ophthalmoscope. She died when she was a month old and an autopsy was performed. At post mortem, brain, spinal cord, and right eye were removed for examination. Free and intracellular *T. gondii* were found in lesions of encephalomyelitis and retinitis of the girl. Portions of cerebral cortex and spinal cord were homogenized in saline and inoculated intracerebrally into rabbits and mice. These animals developed encephalitis, *T. gondii* was demonstrated in their neural lesions and *T. gondii* from these animals was successfully passaged into other mice.

Wolf, Cowen and Paige reviewed in detail their own cases and those reported by others, particularly Janků (1923) and Torres (1927) of *T. gondii*-like encephalomyelitis and chorioretinitis in infants (Wolf and Cowen, 1937, 1938; Wolf *et al.*, 1939a,b, 1940; Cowen *et al.*, 1942; Paige *et al.*, 1942). Joseph Janků (1923), an ophthalmologist from Czechoslovakia, was credited earlier with finding a *T. gondii*-like parasite in a human eye (Janků, 1923). The following description of the case of Janků is taken from the English translation published by Wolf and Cowen (1937):

> The patient was born with left microphthalmus and became blind at the age of three months, and had hydrocephalus. The child died when 11 months old. The eyes and brain were removed at autopsy. Grossly, the child had internal hydrocephalus but the brain was not available for histopathological examination. Chorioretinitis was present in both eyes and cysts-like structures *(termed sporocysts by Janků)* were seen in the right eye.

Janků (1923, reprinted 1959) thought that this parasite was *Encephalitozoon* (a microsporidia). The material from this case is thought to have been destroyed in World War bombing and so confirmation of these findings is not possible. Torres (1927) found protozoa in lesions of encephalitis in a two day old girl in Rio de Janeiro, Brazil. Numerous organisms were seen but these were thought to be a new species of *Encephalitozoon*. This patient also had myocarditis and myositis. In the Netherlands, de Lange (1929) found protozoa in sections of the brain of a four month old child that was born with hydrocephalus. These sections were reexamined by Wolf and Cowen and a full account was reviewed by Sabin (1942).

Sabin (1942) summarized all that was known of congenital toxoplasmosis in 1942 and proposed typical clinical signs of congenital toxoplasmosis: hydrocephalus or microcephalus, intracerebral calcification, and chorioretinitis. These signs helped in the clinical recognition of congenital toxoplasmosis. Frenkel and Friedlander (1951) published a detailed account of five fatal cases of toxoplasmosis in infants that were born with hydrocephalus; *T. gondii* was isolated from two. They described the pathogenesis of internal hydrocephalus as a blockage of the aqueduct of Sylvius due to ventriculitis resulting from a *T. gondii* antigen–antibody reaction. This lesion is unique to human congenital toxoplasmosis and has never been verified in other animals (J.P. Dubey, unpublished). This report was the first in-depth description of lesions of congenital toxoplasmosis not only in the central nervous system but also other organs. Hogan (1951) also provided the first detailed clinical description of ocular toxoplasmosis.

1.5.2 Acquired Toxoplasmosis

1.5.2.1 Children

Sabin (1941) reported toxoplasmosis in a six year old boy from Cincinnati, Ohio. An asymptomatic child with the initials R.H. was hit with a baseball bat on October 22, 1937. He developed a headache two days later and convulsions the day after that. He was admitted to the hospital on the seventh day but without obvious clinical signs. Except for lymphadenopathy and an enlarged spleen, nothing abnormal was found. He then developed neurological signs and died on the thirtieth day of illness. The brain and spinal cord were removed for histopathological examination and bioassay. Because of the suspicion of polio virus infection a homogenate of cerebral cortex was inoculated into mice. *T. gondii* was isolated from the inoculated mice and this isolate was given the initials of the child and became the famous RH strain. Only small lesions of non-suppurative encephalitis were found microscopically in the brain of this child. Neither gross lesions and nor any viral or bacterial infections were found. This child most likely had acquired *T. gondii* infection recently and the blow to the head was coincidental and unrelated to the onset of symptoms. It is noteworthy that some mice infected with the original RH strain did not die until day 21 post-inoculation. By the third passage mice died in three to five days after inoculation. The RH strain of *T. gondii* has, since 1938, been passaged in mice in many laboratories. After this prolonged passage its pathogenicity for mice has been stabilized (Dubey, 1977) and it has lost the capacity to produce oocysts in cats (Frenkel *et al.*, 1976). Additional details of history of toxoplasmosis in humans were given by Dubey (2008) and Weiss and Dubey (2009).

1.5.2.2 *Toxoplasmosis in Adults*

Pinkerton and Weinman (1940) identified *T. gondii* in the heart, spleen and other tissues of a 22 year old patient who died in 1937 in Lima, Peru. The patient exhibited fever and concomitant *Bartonella* spp. infection. Pinkerton and Henderson (1941) isolated *T. gondii* from blood and tissues of two individuals (aged 50 and 43 years) who died in St. Louis, Missouri. Recorded symptoms included rash, fever and malaise. These were the first reports of acute toxoplasmosis in adults without neurological signs.

1.5.2.2.1 LYMPHADENOPATHY

Siim (1956) drew attention to the fact that lymphadenopathy is a frequent sign of acquired toxoplasmosis in adults and these findings were confirmed by Beverley and Beattie (1958) who reported on the cases of 30 patients. A full appreciation of the clinical symptoms of acquired toxoplasmosis was achieved when outbreaks of acute toxoplasmosis were reported in adults in the USA (Teutsch *et al.*, 1979) and in Canada (Bowie *et al.*, 1997).

1.5.2.2.2 OCULAR DISEASE

Before 1950, virtually all cases of ocular toxoplasmosis were considered to result from congenital transmission (Perkins, 1961). Wilder (1952) identified *T. gondii* in eyes that had been enucleated. The significance of this finding lies in the way this discovery was made. These eyes were suspected of being syphilitic, tubercular, or of having tumours. Wilder was a technician in the registry of Ophthalmic Pathology at the Armed Forces Institute for Pathology (AFIP) and she routinely examined microscopically the sections that she prepared. She put enormous effort into identifying microbes in these 'tuberculous' eyes, but never identified bacteria or spirochetes by special staining. Then she found *T. gondii* in the retinas of these eyes. She subsequently collaborated with Jacobs and Cook and found most of these patients with histologically confirmed *T. gondii* infection had low levels of dye test antibodies (a titre of 1:16) and in one patient antibodies were demonstrable only in undiluted serum (Jacobs *et al.*, 1954a). Jacobs *et al.* (1954b) made the first isolation of *T. gondii* from an eye of a 30 year old male hospitalized at the Walter Reed Army Hospital. The eye had been enucleated because of pain associated with elevated intraocular pressure. A group of ophthalmologists from southern Brazil initially discovered ocular toxoplasmosis in siblings. Among patients with postnatally-acquired toxoplasmosis who did not have retinochoidal scars before, 8.3% developed retinal lesions during a seven year follow-up (Silveira *et al.*, 1988, 2001). Ocular toxoplasmosis was diagnosed in 20 of 95 patients with acute toxoplasmosis associated with the Canadian waterborne outbreak of toxoplasmosis in 1995 (Burnett *et al.*, 1998; also see Holland, 2003).

1.5.2.2.3 AIDS EPIDEMIC

Before the epidemic of the acquired immunodeficiency syndrome (AIDS) in adults in the 1980s, neurological toxoplasmosis in adults was rarely reported and limited essentially to patients treated for tumours or those given transplants. Luft and Remington (1983) reported acute toxoplasmosis induced encephalitis that was fatal if not treated. In almost all cases clinical disease occurred as a result of reactivation of chronic infection initiated by the depression of intracellular immunity due to HIV infection. Initially, many of these cases of toxoplasmosis in AIDS patients were thought to be lymphoma.

1.6 TOXOPLASMOSIS IN OTHER ANIMALS

Mello (1910) in Turin, Italy, first reported fatal toxoplasmosis in a domestic animal (a four month old dog) that died of acute visceral toxoplasmosis. Over the next 30 years canine toxoplasmosis was reported in Cuba, France, Germany, India, Iraq, Tunisia, the USSR, and the USA (Dubey and Beattie, 1988). Campbell *et al.* (1955) found that most cases of clinical toxoplasmosis were in dogs infected with the Canine Distemper Virus (CDV) infection. Even vaccination with live attenuated CDV vaccine can trigger clinical toxoplasmosis in dogs (Dubey *et al.*, 2003). The incidence of clinical toxoplasmosis in dogs has decreased dramatically after vaccination with CDV vaccine became a routine practice.

Strangely enough, the first case of toxoplasmosis was not reported in a cat until 1942 when Olafson and Monlux found it in a cat from Middletown, New York, USA. In the 1950s and 1960s, Galuzo and Zasukhin published in Russian their own studies and those of other researchers on many species of animals from the former USSR. This information was made available to scientists in other countries when their book was translated into English by Plous Jr. and edited by Fitzgerald (1970). Jirá and Kozojed (1970, 1983) published the most comprehensive bibliography of toxoplasmosis, listing more than 12,000 references and categorizing them by hosts and topics. This work proved useful for literature searches before electronic databases.

Toxoplasmosis in sheep deserves special attention because of its economic impact. William Hartley, a well-known veterinary pathologist from New Zealand, and his associates J.L. Jebson and D. McFarlane, discovered *T. gondii*-like organisms in the placentas and foetuses of several unexplained abortions in ewes in New Zealand. They called it New Zealand type II abortion. Hartley and Marshall (1957) finally isolated *T. gondii* from aborted foetuses. Hartley (1961) and Jacobs and Hartley (1964) experimentally induced toxoplasmic abortion in ewes. The identification of *T. gondii* abortion in ewes was a landmark discovery in veterinary medicine. Prior to that, protozoa were not recognized as a cause of epidemic abortion in livestock. Subsequently, Jack Beverley and Bill Watson recognized epidemics of ovine abortion in the UK (Beverley and Watson, 1961). Dubey and Towle (1986) and Dubey and Beattie (1988) summarized all that was known about toxoplasmosis in sheep and its impact on agriculture. Millions of lambs are still lost worldwide due to this infectious disease.

Sanger and Cole (1955) were first to isolate *T. gondii* from a food animal. Dubey and Beattie (1988) and Dubey (2010) reviewed the international literature on toxoplasmosis in humans and other animals. The discovery and naming of two new organisms, *Neospora caninum* (Dubey *et al.*, 1988) and *Sarcocystis neurona* (Dubey *et al.*, 1991), which were previously thought to be *T. gondii*, resulted in new information on the host distribution of *T. gondii*. We now know that cattle and horses are resistant to clinical *T. gondii*, that *N. caninum* is a common cause of abortion in cattle worldwide (Dubey, 2003), and that *S. neurona* is a common cause of fatal encephalomyelitis in horses in the Americas (Dubey *et al.*, 2001). There have been no confirmed cases of clinical toxoplasmosis in either cattle or horses (J.P. Dubey, unpublished).

The finding of *T. gondii* in marine mammals deserves special mention. Before the discovery of the *T. gondii* oocyst no one would have suspected that the marine environment would be contaminated with *T. gondii* and that fish-eating marine mammals would be found infected with *T. gondii* (Dubey *et al.*, 2003; Conrad *et al.*, 2005; Thomas *et al.*, 2007; Miller *et al.*, 2008; Dubey 2010). Thomas and Cole (1996) and Cole *et al.* (2000) isolated viable *T. gondii* from sea otters in the USA. Several reports have now appeared that confirm that *T. gondii* can occur in many species of marine mammals (Dubey, 2010).

1.7 DIAGNOSIS

1.7.1 Sabin–Feldman Dye Test

Development of a novel serologic test, the dye test, in 1948 by Albert Sabin and Harry Feldman was perhaps the greatest advancement in the field of toxoplasmosis (Sabin and Feldman, 1948). The dye test is highly sensitive and specific with no evidence for false results in humans. Even titres as low as 1:2 are meaningful for the diagnosis of ocular disease. The ability to identify *T. gondii* infections based on a simple serological test opened the door for extensive epidemiological studies on the incidence of infection. It became clear that *T. gondii* infections are widely prevalent in humans in many countries. It also demonstrated that the so-called tetrad of clinical signs considered indicative of clinical congenital toxoplasmosis occurred in other diseases and assisted in the differential diagnosis (Sabin and Feldman, 1949; Feldman and Miller, 1956).

1.7.2 Detection of IgM Antibodies

Remington *et al.* (1968) first proposed the usefulness of the detection of IgM antibodies in cord blood or infant serum for the diagnosis of congenital toxoplasmosis since IgM antibodies do not cross the placenta, whereas IgG antibodies do. Remington (1969) modified the indirect fluorescent antibody test (IFAT) and the enzyme-linked immunosorbent assay (ELISA) (Naot and Remington, 1980) to detect IgM in cord blood. Desmonts *et al.* (1981) developed a modification of IgM–ELISA, combining it with the agglutination test (IgM–ISAGA) to eliminate the necessity for an enzyme conjugate. Although IgM tests are not perfect, they have proved useful for screening programmes (Remington *et al.*, 2001)

1.7.3 Direct Agglutination Test

The development of a simple direct agglutination test has aided tremendously in the serological diagnosis of toxoplasmosis in humans and other animals. In this test no special equipment or conjugates are needed. This test was initially developed by Fulton (1965) and improved by Desmonts and Remington (1980) and Dubey and Desmonts (1987) who called it the modified agglutination test (MAT). The MAT has been used extensively for the diagnosis of toxoplasmosis in animals. The sensitivity and specificity of MAT has been validated by comparing serologic data and isolation of the parasite from naturally- and experimentally-infected pigs (Dubey, 1997; Dubey *et al.*, 1995a,b) and naturally-infected chickens (J.P. Dubey, unpublished).

1.7.4 Detection of *T. gondii* DNA

Burg *et al.* (1989) first reported detection of *T. gondii* DNA from a single tachyzoite using the B1 gene in a polymerase chain reaction (PCR). Several subsequent PCR tests have been developed using different gene targets. Overall, this technique has proven very useful in the diagnosis of clinical toxoplasmosis (Remington *et al.*, 2011).

1.8 TREATMENT

Sabin and Warren (1942) reported the effectiveness of sulphonamides against murine toxoplasmosis and Eyles and Coleman (1953) discovered the synergistic effect of combined therapy with sulphonamides and pyrimethamine; the latter is the standard therapy for toxoplasmosis in humans (Remington *et al.*, 2001). Garin and Eyles (1958) found spiramycin to have antitoxoplasmic activity in mice. Since spiramycin is nontoxic and does not cross the placenta; it has been used prophylactically in women during pregnancy to reduce transmission of the parasite from mother to foetus (Desmonts and Couvreur, 1974b).

1.9 PREVENTION AND CONTROL

1.9.1 Serologic Screening During Pregnancy

Georges Desmonts initiated studies in Paris, France, in the 1960s looking at seroconversion in women during pregnancy and the transmission of *T. gondii* to the foetus. Blood was obtained at the first visit, at 7 months, and at the time of parturition. Desmonts initiated prophylactic treatment of women who seroconverted during pregnancy. Results of the 15 year study demonstrated that: (1) infection acquired during the first two trimesters was most damaging to the foetus; (2) not all women that acquired infection transmitted it to the foetus; (3) women seropositive prior to pregnancy did not transmit infection to the foetus; and (4) treatment with spiramycin reduced congenital transmission, but not clinical disease in infants (Desmonts and Couvreur, 1974a,b). At about the same time, Otto Thalhammer initiated a similar screening programme for pregnant women in Austria (Thalhammer, 1973, 1978). In addition to scientific knowledge, these screening programmes have helped to disseminate information for the prevention of toxoplasmosis.

A neonatal serological screening and early treatment for congenital *T. gondii* infection was initiated in Massachusetts, USA, in the 1980s (Guerina *et al.*, 1994). The efficacy of treatment of *T. gondii* infection in the foetus and newborn is not fully delineated, and many issues related to the cost and benefit of screening and treatment in pregnancy and in newborns remain to be examined.

1.9.2 Hygiene Measures

After the discovery of the life cycle of *T. gondii* in 1970 it became possible to advise pregnant women and other susceptible populations on avoiding contact with oocysts (Frenkel and Dubey, 1972). Studies were conducted to construct thermal curves showing temperatures required to kill *T. gondii* in infected meat by freezing (Kotula *et al.*, 1991), cooking (Dubey *et al.*, 1990), and by gamma irradiation (Dubey *et al.*, 1986). These data are now used by regulatory agencies to advise consumers about the safety of meat. Freezing of meat overnight in a household freezer before human or animal consumption remains the easiest and most economical method of reducing transmission of *T. gondii* through meat.

1.9.3 Animal Production Practices

Extensive epidemiological studies on pig farms in the USA in the 1990s concluded that keeping cats out of pig barns and raising pigs indoors can reduce *T. gondii* infection in pigs (Dubey *et al.*, 1995a,b; Weigel *et al.*, 1995). As a result of changes in pig husbandry, prevalence of viable *T. gondii* in pigs is reduced to less than 1% (Dubey *et al.*, 2005). Because ingestion of infected pork is considered the main meat source of *T. gondii* for humans (at least in the USA), it will hopefully reduce prevalence of *T. gondii* in humans.

1.9.4 Vaccination

Vaccination of sheep with a live cystless strain of *T. gondii* reduces neonatal mortality in lambs and this vaccine is available commercially (Wilkins and O'Connel, 1983; Buxton and Innes, 1995). To date there is no vaccine suitable for human use.

Acknowledgements

I would like to thank Drs. Georges Desmonts (now deceased), David Ferguson, Jack Frenkel, H.R. Gamble, Garry Holland, Jeff Jones, and Jack Remington for their helpful discussions in the preparation of this manuscript.

References

Aramini, J.J., Stephen, C., Dubey, J.P., 1998. *Toxoplasma gondii* in Vancouver Island cougars (*Felis concolor vancouverensis*): serology and oocyst shedding. J. Parasitol. 84, 438–440.

Ben Rachid, M.S., 1970. Contribution à l'étude de la toxoplasmose du gondi. II. Comportement du *Ctenodactylus gundi* vis-à-vis de *Isospora bigemina*. Arch. Inst. Pasteur Tunis 47, 33–35.

Benenson, M.W., Takafuji, E.T., Lemon, S.M., Greenup, R.L., Sulzer, A.J., 1982. Oocyst-transmitted toxoplasmosis associated with ingestion of contaminated water. N. Engl. J. Med. 307, 666–669.

Beverley, J.K.A., 1959. Congenital transmission of toxoplasmosis through successive generations of mice. Nature 183, 1348–1349.

Beverley, J.K.A., Beattie, C.P., 1958. Glandular toxoplasmosis. A survey of 30 cases. Lancet 23, 379–384.

Beverley, J.K.A., Watson, W.A., 1961. Ovine abortion and toxoplasmosis in Yorkshire. Vet. Rec. 73, 6–11.

Bowie, W.R., King, A.S., Werker, D.H., Isaac-Renton, J.L., Bell, A., Eng, S.B., Marion, S.A., 1997. Outbreak of toxoplasmosis associated with municipal drinking water. Lancet 350, 173–177.

Burg, J.L., Grover, C.M., Pouletty, P., Boothroyd, J.C., 1989. Direct and sensitive detection of a pathogenic protozoan, *Toxoplasma gondii*, by polymerase chain-reaction. J. Clin. Microbiol. 27, 1787–1792.

Burnett, A.J., Shortt, S.G., Isaac-Renton, J., King, A., Werker, D., Bowie, W.R., 1998. Multiple cases of acquired toxoplasmosis retinitis presenting in an outbreak. Ophthalmology 105, 1032–1037.

Buxton, D., Innes, E.A., 1995. A commercial vaccine for ovine toxoplasmosis. Parasitology 110, S11–S16.

Campbell, R.S.F., Martin, W.B., Gordon, E.D., 1955. Toxoplasmosis as a complication of canine distemper. Vet. Rec. 67, 708–716.

Christie, E., Pappas, P.W., Dubey, J.P., 1978. Ultrastructure of excystment of *Toxoplasma gondii* oocysts. J. Protozool. 25, 438–443.

Cole, R.A., Lindsay, D.S., Howe, D.K., Roderick, C.L., Dubey, J.P., Thomas, N.J., Baeten, L.A., 2000. Biological and molecular characterizations of *Toxoplasma gondii* strains obtained from southern sea otters (*Enhydra lutris nereis*). J. Parasitol. 86, 526–530.

Conrad, P.A., Miller, M.A., Kreuder, C., James, E.R., Mazet, J., Dabritz, H., Jessup, D.A., Gulland, F., Grigg, M.E., 2005. Transmission of *Toxoplasma*: clues from the study of sea otters as sentinels of *Toxoplasma gondii* flow into the marine environment. Int. J. Parasitol. 35, 1155–1168.

Cowen, D., Wolf, A., Paige, B.H., 1942. Toxoplasmic encephalomyelitis. VI. Clinical diagnosis of infantile or congenital toxoplasmosis; survival beyond infancy. Arch. Neurol. Psychiat. 48, 689–739.

de Lange, C., 1929. Klinische und pathologisch-anatomische Mitteilungen über Hydrocephalus chronicus congenitus und acquisitus. Ztschr. f. d. ges. Neurol. u. Psychiat. 120, 433–500.

de Moura, L., Bahia-Oliveira, L.M.G., Wada, M.Y., Jones, J.L., Tuboi, S.H., Carmo, E.H., Ramalho, W.M., Camargo, N.J., Trevisan, R., Graça, R.M.T., da Silva, A.J., Moura, I., Dubey, J.P., Garrett, D.O., 2006. Waterborne outbreak of toxoplasmosis in Brazil, from field to gene. Emerg. Infect. Dis. 12, 326–329.

Desmonts, G., Couvreur, J., Alison, F., Baudelot, J., Gerbeaux, J., Lelong, M., 1965. Étude épidémiologique sur la toxoplasmose: de l'influence de la cuisson des viandes de boucherie sur la fréquence de l'infection humaine. Rev. Fr. Études Clin. Biol. 10, 952–958.

Desmonts, G., Couvreur, J., 1974a. Congenital toxoplasmosis. A propective study of 378 pregnancies. N. Engl. J. Med. 290, 1110–1116.

Desmonts, G., Couvreur, J., 1974b. Toxoplasmosis in pregnancy and its transmission to the foetus. Bull. N. Y. Acad. Med. 50, 146–159.

Desmonts, G., Remington, J.S., 1980. Direct agglutination test for diagnosis of Toxoplasma infection: method for increasing sensitivity and specificity. J. Clin. Microbiol. 11, 562–568.

Desmonts, G., Naot, Y., Remington, J.S., 1981. Immunoglobulin M immunosorbent agglutination assay for diagnosis of infectious diseases. Diagnosis of acute congenital and acquired *Toxopalsma* infections. J. Clin. Microbiol. 14, 544–549.

Dubey, J.P., 1966. Toxoplasmosis and its transmission in cats with special reference to associated *Toxocara cati* infestations. Ph.D. Thesis. University of Sheffield, England, pp. 1–169.

Dubey, J.P., 1968. Isolation of *Toxoplasma gondii* from the faeces of a helminth free cat. J. Protozool. 15, 773–775.

Dubey, J.P., 1977. *Toxoplasma, Hammondia, Besnoitia, Sarcocystis*, and other tissue cyst-forming coccidia of man and animals. In: Kreier, J.P. (Ed.), Parasitic protozoa, vol. 3. Academic Press, New York, pp. 101–237.

Dubey, J.P., 1996. Infectivity and pathogenicity of *Toxoplasma gondii* oocysts for cats. J. Parasitol. 82, 957–960.

Dubey, J.P., 1997. Validation of the specificity of the modified agglutination test for toxoplasmosis in pigs. Vet. Parasitol. 71, 307–310.

Dubey, J.P., 1998. Refinement of pepsin digestion method for isolation of *Toxoplasma gondii* from infected tissues. Vet. Parasitol. 74, 75–77.

Dubey, J.P., 2001. Oocyst shedding by cats fed isolated bradyzoites and comparision of infectivity of bradyzoites of the VEG strain *Toxoplasma gondii* to cats and mice. J. Parasitol. 87, 215–219.

Dubey, J.P., 2002. Tachyzoite-induced life cycle of *Toxoplasma gondii* in cats. J. Parasitol. 88, 713–717.

Dubey, J.P., 2003. Neosporosis in cattle. J. Parasitol. 89 (suppl.), S42–S46.

Dubey, J.P., 2005. Unexpected oocyst shedding by cats fed *Toxoplasma gondii* tachyzoites: *in vivo* stage conversion and strain variation. Vet. Parasitol. 133, 289–298.

Dubey, J.P., 2006. Comparative infectivity of oocysts and bradyzoites of *Toxoplasma gondii* for intermediate (mice) and definitive (cats) hosts. Vet. Parasitol. 140, 69–75.

Dubey, J.P., 2008. The history of *Toxoplasma gondii* – the first 100 years. J. Eukaryot. Microbiol. 55, 467–475.

Dubey, J.P., 2009. History of the discovery of the life cycle of *Toxoplasma gondii*. Int. J. Parasitol. 39, 877–882.

Dubey, J.P., 2010. Toxoplasmosis of animals and humans, second ed. CRC Press, Boca Raton, Florida, pp. 1–330.

Dubey, J.P., Beattie, C.P., 1988. Toxoplasmosis of animals and humans. CRC Press, Boca Raton, Florida, pp. 1–220.

Dubey, J.P., Desmonts, G., 1987. Serological responses of equids fed *Toxoplasma gondii* oocysts. Equine Vet. J. 19, 337–339.

Dubey, J.P., Frenkel, J.K., 1972. Cyst-induced toxoplasmosis in cats. J. Protozool. 19, 155–177.

Dubey, J.P., Swan, G.V., Frenkel, J.K., 1972. A simplified method for isolation of *Toxoplasma gondii* from faeces of cats. J. Parasitol. 58, 1005–1006.

Dubey, J.P., Frenkel, J.K., 1973. Experimental *Toxoplasma* infections in mice with strains producing oocysts. J. Parasitol. 59, 505–512.

Dubey, J.P., Frenkel, J.K., 1976. Feline toxoplasmosis from acutely infected mice and the development of *Toxoplasma* cysts. J. Protozool. 23, 537–546.

Dubey, J.P., Towle, A., 1986. Toxoplasmosis in sheep: a review and annotated bibliography. Commonwealth Institute of Parasitology, Herts, U.K. pp. 1–152.

Dubey, J.P., Miller, N.L., Frenkel, J.K., 1970a. The *Toxoplasma gondii* oocyst from cat faeces. J. Exp. Med. 132, 636–662.

Dubey, J.P., Miller, N.L., Frenkel, J.K., 1970b. Characterization of the new faecal form of *Toxoplasma gondii*. J. Parasitol. 56, 447–456.

Dubey, J.P., Brake, R.J., Murrell, K.D., Fayer, R., 1986. Effects of irradiation on the viability of *Toxoplasma gondii* cysts in tissues of mice and pigs. Am. J. Vet. Res. 47, 518–522.

Dubey, J.P., Carpenter, J.L., Speer, C.A., Topper, M.J., Uggla, A., 1988. Newly recognized fatal protozoan disease of dogs. J. Am. Vet. Med. Assoc. 192, 1269–1285.

Dubey, J.P., Kotula, A.W., Sharar, A., Andrews, C.D., Lindsay, D.S., 1990. Effect of high temperature on infectivity of *Toxoplasma gondii* tissue cysts in pork. J. Parasitol. 76, 201–204.

Dubey, J.P., Davis, S.W., Speer, C.A., Bowman, D.D., de Lahunta, A., Granstrom, D.E., Topper, M.J., Hamir, A.N., Cummings, J.F., Suter, M.M., 1991. *Sarcocystis neurona* n. sp. (Protozoa: Apicomplexa), the etiologic agent of equine protozoal myeloencephalitis. J. Parasitol. 77, 212–218.

Dubey, J.P., Weigel, R.M., Siegel, A.M., Thulliez, P., Kitron, U.D., Mitchell, M.A., Mannelli, A., Mateus-Pinilla, N.E., Shen, S.K., Kwok, O.C.H., Todd, K.S., 1995a. Sources and reservoirs of *Toxoplasma gondii* infection on 47 swine farms in Illinois. J. Parasitol. 81, 723–729.

Dubey, J.P., Thulliez, P., Weigel, R.M., Andrews, C.D., Lind, P., Powell, E.C., 1995b. Sensitivity and specificity of various serologic tests for detection of *Toxoplasma gondii* infection in naturally infected sows. Am. J. Vet. Res. 56, 1030–1036.

Dubey, J.P., Lunney, J.K., Shen, S.K., Kwok, O.C.H., Ashford, D.A., Thulliez, P., 1996. Infectivity of low numbers of *Toxoplasma gondii* oocysts to pigs. J. Parasitol. 82, 438–443.

Dubey, J.P., Rollor, E.A., Smith, K., Kwok, O.C.H., Thulliez, P., 1997. Low seroprevalence of *Toxoplasma gondii* in feral pigs from a remote island lacking cats. J. Parasitol. 83, 839–841.

Dubey, J.P., Lindsay, D.S., Speer, C.A., 1998. Structure of *Toxoplasma gondii* tachyzoites, bradyzoites and sporozoites, and biology and development of tissue cysts. Clin. Microbiol. Rev. 11, 267–299.

Dubey, J.P., Lindsay, D.S., Saville, W.J.A., Reed, S.M., Granstrom, D.E., Speer, C. A., 2001. A review of *Sarcocystis neurona* and equine protozoal myeloencephalitis (EPM). Vet. Parasitol. 95, 89–131.

Dubey, J.P., Zarnke, R., Thomas, N.J., Wong, S.K., Van Bonn, W., Briggs, M., Davis, J.W., Ewing, R., Mensea, M., Kwok, O.C.H., Romand, S., Thulliez, P., 2003. *Toxoplasma gondii, Neospora caninum, Sarcocystis neurona,* and *Sarcocystis canis*-like infections in marine mammals. Vet. Parasitol. 116, 275–296.

Dubey, J.P., Ross, A.D., Fritz, D., 2003. Clinical *Toxoplasma gondii, Hammondia heydorni,* and *Sarcocystis* sp. infections in dogs. Parassitologia 45, 141–146.

Dubey, J.P., Hill, D.E., Jones, J.L., Hightower, A.W., Kirkland, E., Roberts, J.M., Marcet, P.L., Lehmann, T., Vianna, M.C.B., Miska, K., Sreekumar, C., Kwok, O.C.H., Shen, S.K., Gamble, H.R., 2005. Prevalence of viable *Toxoplasma gondii* in beef, chicken and pork from retail meat stores in the United States: risk assessment to consumers. J. Parasitol. 91, 1082–1093.

Eyles, D.E., Coleman, N., 1953. Synergistic effect of sulfadiazine and daraprim against experimental toxoplasmosis in the mouse. Antibiot. Chemother. 3, 483–490.

Feldman, H.A., Miller, L.T., 1956. Serological study of toxoplasmosis prevalence. Am. J. Hygiene 64, 320–335.

Ferguson, D.J.P., Hutchison, W.M., 1987. An ultrastructural study of the early development and tissue cyst formation of *Toxoplasma gondii* in the brains of mice. Parasitol. Res. 73, 483–491.

Ferguson, D.J.P., Hutchison, W.M., Dunachie, J.F., Siim, J.C., 1974. Ultrastructural study of early stages of asexual multiplication and microgametogony of *Toxoplasma gondii* in the small intestine of the cat. Acta. Pathol. Microbiol. Scand. B 82, 167–181.

Ferguson, D.J.P., Hutchison, W.M., Siim, J.C., 1975. The ultrastructural development of the macrogamete and formation of the oocyst wall of *Toxoplasma gondii.* Acta. Pathol. Microbiol. Scand. B 83, 491–505.

Ferguson, D.J.P., Birch-Andersen, A., Siim, J.C., Hutchison, W.M., 1979a. Ultrastructural studies on the sporulation of oocysts of *Toxoplasma gondii.* I. Development of the zygote and formation of sporoblasts. Acta. Pathol. Microbiol. Scand. B 87, 171–181.

Ferguson, D.J.P., Birch-Andersen, A., Siim, J.C., Hutchison, W.M., 1979b. An ultrastructural study on the excystation of the sporozoites of *Toxoplasma gondii.* Acta. Pathol. Microbiol. Scand. B 87, 277–283.

Fitzgerald, P.R., 1970. Toxoplasmosis of animals. University of Illinois, Urbana, Illinois, pp. 1–472.

Frenkel, J.K., 1956. Pathogenesis of toxoplasmosis and of infections with organisms resembling *Toxoplasma.* Ann. N. Y. Acad. Sci. 64, 215–251.

Frenkel, J.K., 1970. Pursuing *Toxoplasma.* J. Infect. Dis. 122, 553–559.

Frenkel, J.K., 1973. *Toxoplasma* in and around us. BioScience 23, 343–352.

Frenkel, J.K., Dubey, J.P., 1972. Toxoplasmosis and its prevention in cats and man. J. Infect. Dis. 126, 664–673.

Frenkel, J.K., Friedlander, S., 1951. Toxoplasmosis. Pathology of neonatal disease, pathogenesis, diagnosis, and treatment. 1–150. Public Health Service Publication No. 141. United States Government Printing Office, Washington, D.C., USA.

Frenkel, J.K., Dubey, J.P., Miller, N.L., 1969. *Toxoplasma gondii*: faecal forms separated from eggs of the nematode *Toxocara cati.* Science 164, 432–433.

Frenkel, J.K., Dubey, J.P., Miller, N.L., 1970. *Toxoplasma gondii* in cats: faecal stages identified as coccidian oocysts. Science 167, 893–896.

Frenkel, J.K., Dubey, J.P., Hoff, R.L., 1976. Loss of stages after continuous passage of *Toxoplasma gondii* and *Besnoitia jellisoni.* J. Protozool. 23, 421–424.

Fulton, J.D., 1965. Studies on agglutination of *Toxoplasma gondii.* Trans. Roy. Soc. Trop. Med. Hyg. 59, 694–704.

Garin, J.P., Eyles, D.E., 1958. Le traitement de la toxoplasmose experimentale de la souris par la spiramycine. La Presse Médicale 66, 957–958.

Garnham, P.C.C., 1971. Progress in parasitology. University of London, The Athlone Press, London, pp. 1–223.

Goldman, M., Carver, R.K., Sulzer, A.J., 1958. Reproduction of *Toxoplasma gondii* by internal budding. J. Parasitol. 44, 161–171.

Guerina, N.G., Hsu, H.W., Meissner, H.C., Maguire, J.H., Lynfield, R., Stechenberg, B., Abroms, I., Pasternack, M.S., Hoff, R., Eaton, R.B., Grady, G. F., Cheeseman, S.H., McIntosh, K., Medearis, D.N., Robb, R., Weiblen, B.J., 1994. Neonatal serologic screening and early treatment for congenital *Toxoplasma gondii* infection. N. Engl. J. Med. 330, 1858–1863.

Gustafson, P.V., Agar, H.D., Cramer, D.I., 1954. An electron microscope study of *Toxoplasma.* Am. J. Trop. Med. Hyg. 3, 1008–1021.

Hartley, W.J., 1961. Experimental transmission of toxoplasmosis in sheep. N. Z. Vet. J. 9, 1–6.

Hartley, W.J., Marshall, S.C., 1957. Toxoplasmosis as a cause of ovine perinatal mortality. N. Z. Vet. J. 5, 119–124.

Hogan, M.J., 1951. Ocular toxoplasmosis. Columbia Unviersity Press, New York, pp. 1–86.

Holland, G.N., 2003. Ocular toxoplasmosis: a global reassessment. Part I: epidemiology and course of disease. Am. J. Ophthalmol. 136, 973–988.

Hutchison, W.M., 1965. Experimental transmission of *Toxoplasma gondii.* Nature 206, 961–962.

Hutchison, W.M., 1967. The nematode transmission of *Toxoplasma gondii.* Trans. Roy. Soc. Trop. Med. Hyg. 61, 80–89.

Hutchison, W.M., Dunachie, J.F., Work, K., 1968. The faecal transmission of *Toxoplasma gondii.* Acta. Pathol. Microbiol. Scand. 74, 462–464.

Hutchison, W.M., Dunachie, J.F., Siim, J.C., Work, K., 1969. Life cycle of *Toxoplasma gondii.* Br. Med. J. 4, 806.

Hutchison, W.M., Dunachie, J.F., Siim, J.C., Work, K., 1970. Coccidian-like nature of *Toxoplasma gondii.* Br. Med. J. 1, 142–144.

Hutchison, W.M., Dunachie, J.F., Work, K., Siim, J.C., 1971. The life cycle of the coccidian parasite *Toxoplasma gondii,* in the domestic cat. Trans. Roy. Soc. Trop. Med. Hyg. 65, 380–399.

Jacobs, L., 1964. The occurrence of *Toxoplasma* infection in the absence of demonstrable antibodies. Proc. First Int. Congr. Parasitol. 1, 176–177.

Jacobs, L., 1967. *Toxoplasma* and toxoplasmosis. Adv. Parasitol. 5, 1–45.

Jacobs, L., Hartley, W.J., 1964. Ovine toxoplasmosis: studies on parasitaemia, tissue infection, and congenital transmission in ewes infected by various routes. Br. Vet. J. 120, 347–364.

Jacobs, L., Cook, M.K., Wilder, H.C., 1954a. Serologic data on adults with histologically diagnosed toxoplasmic chorioretinitis. Trans. Am. Acad. Ophthalmol. Otolaryngol. 58, 193–200.

Jacobs, L., Fair, J.R., Bickerton, J.H., 1954b. Adult ocular toxoplasmosis. Report of a parasitologically proved case. A. M. A. Arch. Ophthalmol. 52, 63–71.

Jacobs, L., Remington, J.S., Melton, M.L., 1960a. The resistance of the encysted form of *Toxoplasma gondii*. J. Parasitol. 46, 11–21.

Jacobs, L., Remington, J.S., Melton, M.L., 1960b. A survey of meat samples from swine, cattle, and sheep for the presence of encysted *Toxoplasma*. J. Parasitol. 46, 23–28.

Janitschke, K., Werner, H., 1972. Untersuchungen uber die Wirtsspezifitat des geschlechtlichen Entwicklungszyklus von *Toxoplasma gondii*. Z. Parasitenk. 39, 247–254.

Janků, J., 1923. Pathogenesa a pathologická anatomie tak nazvaneho vrozeného kolombu žluté skvrany voku normálné velikem a microphthalmickém s nalezem parasitu v sítnici. Časopis lékařeů českŷck 62, 1021, 1052, 1081, 1111 and 1138.

Janků, J., 1959. Die Pathogenese und pathologische Anatomie des sogenannten angeborenen Koloboms des gelben Flecks in normal grossen sowie im mikrophtalmischen Auge mit Parasitenbefund in der Netzhaut. Česk. Parasitol. 6, 9–57.

Jewell, M.L., Frenkel, J.K., Johnson, K.M., Reed, V., Ruiz, A., 1972. Development of *Toxoplasma* oocysts in neotropical felidae. Am. J. Trop. Med. Hyg. 21, 512–517.

Jíra, J., Kozojed, V., 1970. Toxoplasmosis 1908–1967. Gustav Fischer Verlag, Stuttgart, pp. 1–464.

Jíra, J., Kozojed, V., 1983. Toxoplasmosis 1968–1975. Gustav Fischer Verlag, Stuttgart, pp. 1–394.

Kean, B.H., Kimball, A.C., Christenson, W.N., 1969. An epidemic of acute toxoplasmosis. J. Am. Med. Assoc. 208, 1002–1004.

Kotula, A.W., Dubey, J.P., Sharar, A.K., Andrew, C.D., Shen, S.K., Lindsay, D.S., 1991. Effect of freezing on infectivity of *Toxoplasma gondii* tissue cysts in pork. J. Food Protection 54, 687–690.

Lainson, R., 1958. Observations on the development and nature of pseudocysts and cysts of *Toxoplasma gondii*. Trans. Roy. Soc. Trop. Med. Hyg. 52, 396–407.

Levaditi, C., Schoen, R., Sanchis Bayarri, V., 1928. L'encéphalo-myélite toxoplasmique chronique du lapin et de la souris. C. R. Soc. Biol. 99, 37–40.

Luft, B.J., Conley, F.K., Remington, J.S., Laverdiere, M., Wagner, K.F., Levine, J.F., Craven, P.C., Strandberg, D.A., File, T.M., Rice, N., Meunier-Carpenter, F., 1983. Outbreak of central-nervous-system toxoplasmosis in Western Europe and North America. Lancet I, 781–784.

Mateus-Pinilla, N.E., Dubey, J.P., Choromanski, L., Weigel, R.M., 1999. A field trial of the effectiveness of of a feline *Toxoplasma gondii* vaccine in reducing *T. gondii* exposure for swine. J. Parasitol. 85, 855–860.

Mello, U., 1910. Un cas de toxoplasmose du chien observé à Turin (2). Bull. Soc. Pathol. Exot. Fil. 3, 359–363.

Miller, M.A., Miller, W.A., Conrad, P.A., James, E.R., Melli, A.C., Leutenegger, C. M., Dabritz, H.A., Packham, A.E., Paradies, D., Harris, M., Ames, J., Jessup, D.A., Worcester, K., Grigg, M.E., 2008. Type X *Toxoplasma gondii* in a wild mussel and terrestrial carnivores from coastal California: new linkages between terrestrial mammals, runoff and toxoplasmosis of sea otters. Int. J. Parasitol. 38, 1319–1328.

Miller, N.L., Frenkel, J.K., Dubey, J.P., 1972. Oral infections with *Toxoplasma* cysts and oocysts in felines, other mammals, and in birds. J. Parasitol. 58, 928–937.

Munday, B.L., 1972. Serological evidence of *Toxoplasma* infection in isolated groups of sheep. Res. Vet. Sci. 13, 100–102.

Naot, Y., Remington, J.S., 1980. An enzyme-linked immunosorbent assay for detection of IgM antibodies to *Toxoplasma gondii*: use for diagnosis of acute acquired toxoplasmosis. J. Infect. Dis. 142, 757–766.

Nicolle, C., Manceaux, L., 1908. Sur une infection à corps de Leishman (ou organismes voisins) du gondi. C. R. Seances Acad. Sci. 147, 763–766.

Nicolle, C., Manceaux, L., 1909. Sur un protozoaire nouveau du gondi. C. R. Seances Acad. Sci. 148, 369–372.

Olafson, P., Monux, W.S., 1942. Toxoplasma infection in animals. Cornell Vet. 32, 176–190.

Overdulve, J.P., 1970. The identity of *Toxoplasma* Nicolle & Manceaux, 1909 with *Isospora* Schneider, 1881 (I). Proc. Kon. Ned. Akad. Wet. C 73, 129–151.

Paige, B.H., Cowen, D., Wolf, A., 1942. Toxoplasmic encephalomyelitis. V. Further observations of infantile toxoplasmosis; intrauterine inception of the disease; visceral manifestations. Am. J. Dis. Child 63, 474–514.

Perkins, E.S., 1961. Uveitis and toxoplasmosis. J. & A. Churchill Ltd., London, pp. 1–142.

Piekarski, G., Pelster, B., Witte, H.M., 1971. Endopolygenie bei *Toxoplasma gondii*. Z. Parasitenk. 36, 122–130.

Pinkerton, H., Weinman, D., 1940. *Toxoplasma* infection in man. Arch. Pathol. 30, 374–392.

Pinkerton, H., Henderson, R.G., 1941. Adult toxoplasmosis. A previously unrecognized disease entity simulating the typhus-spotted fever group. J. Am. Med. Assoc. 116, 807–814.

Polomoshnov, A.P., 1979. Definitive hosts of *Toxoplasma*. Problems of Natural Nidality of Diseases. Institute of Zoology, Kazakh Academy of Sciences, Alma Ata. 10, 68–72. [In Russian].

Rawal, B.D., 1959. Toxoplasmosis. A dye-test on sera from vegetarians and meat eaters in Bombay. Trans. Roy. Soc. Trop. Med. Hyg. 53, 61–63.

Remington, J.S., 1969. The present status of the IgM fluorescent antibody technique in the diagnosis of congenital toxoplasmosis. J. Pediatr. 75, 1116–1124.

Remington, J.S., Miller, M.J., Brownlee, I.E., 1968. IgM antibodies in acute toxoplasmosis. II. Prevalence and significance in acquired cases. J. Lab. Clin. Med. 71, 855–866.

Remington, J.S., McLeod, R., Thulliez, P., Desmonts, G., 2001. Toxoplasmosis. In: Remington, J.S., Klein, J. (Eds.), Infectious diseases of the foetus and newborn infant, fifth ed. W. B. Saunders, Philadelphia, pp. 205–346.

Remington, J.S., McLeod, R., Wilson, C.B., Desmonts, G., 2011. Toxoplasmosis. In: Remington, J.S., Klein, J. (Eds.), Infectious diseases of the foetus and newborn infant, seventh ed. W. B. Saunders, Philadelphia, pp. 918–1041.

Sabin, A.B., 1941. Toxoplasmic encephalitis in children. J. Am. Med. Assoc. 116, 801–807.

Sabin, A.B., 1942. Toxoplasmosis. A recently recognized disease of human beings. Adv. Pediatr. 1, 1–53.

Sabin, A.B., Feldman, H.A., 1948. Dyes as microchemical indicators of a new immunity phenomenon affecting a protozoon parasite (*Toxoplasma*). Science 108, 660–663.

Sabin, A.B., Feldman, H.A., 1949. Persistence of placentally transmitted toxoplasmic antibodies in normal children in relation to diagnosis of congenital toxoplasmosis. Pediatrics 4, 660–664.

Sabin, A.B., Warren, J., 1942. Therapeutic effectiveness of certain sulfonamide on infection by an intracellular protozoon (*Toxoplasma*). Proc. Soc. Exp. Biol. Med. 51, 19–23.

Sanger, V.L., Cole, C.R., 1955. Toxoplasmosis. VI. Isolation of *Toxoplasma* from milk, placentas, and newborn pigs of asymptomatic carrier sow. Am. J. Vet. Res. 16, 536–539.

Sheffield, H.G., 1970. Schizogony in *Toxoplasma gondii*: an electron microscope study. Proc. Helminthol. Soc. Wash. 37, 237–242.

Sheffield, H.G., Melton, M.L., 1968. The fine structure and reproduction of *Toxoplasma gondii*. J. Parasitol. 54, 209–226.

Sheffield, H.G., Melton, M.L., 1969. *Toxoplama gondii*: transmission through faeces in absence of *Toxocara cati* eggs. Science 164, 431–432.

Sheffield, H.G., Melton, M.L., 1970. *Toxoplasma gondii*: the oocyst, sporozoite, and infection of cultured cells. Science 167, 892–893.

Siim, J.C., 1956. Toxoplasmosis acquisita lymphonodosa; clinical and pathological aspects. Ann. N. Y. Acad. Sci. 64, 185–206.

Silveira, C., Belfort, R., Burnier, M., Nussenblatt, R., 1988. Acquired toxoplasmic infection as the cause of toxoplasmic retinochoroiditis in families. Am. J. Ophthalmol. 106, 362–364.

Silveira, C., Belfort, R., Muccioli, C., Abreu, M.T., Martins, M.C., Victora, C., Nussenblatt, R.B., Holland, G.N., 2001. A follow-up study of *Toxoplasma gondii* infection in Southern Brazil. Am. J. Ophthalmol. 131, 351–354.

Speer, C.A., Dubey, J.P., 2005. Ultrastructural differentiation of *Toxoplasma gondii* schizonts (types B to E) and gamonts in the intestines of cats fed bradyzoites. Int. J. Parasitol. 35, 193–206.

Speer, C.A., Clark, S., Dubey, J.P., 1998. Ultrastructure of the oocysts, sporocysts and sporozoites of *Toxoplasma gondii*. J. Parasitol. 84, 505–512.

Splendore, A., 1908. Un nuovo protozoa parassita de' conigli. incontrato nelle lesioni anatomiche d'une malattia che ricorda in molti punti il Kala-azar dell' uomo. Nota preliminare pel. Rev. Soc. Scient, Sao Paulo 3, 109–112.

Splendore, A., 2009. A new protozoan parasite in rabbits. Int. J. Parasitol. 39, 861–862.

Teutsch, S.M., Juranek, D.D., Sulzer, A., Dubey, J.P., Sikes, R.K., 1979. Epidemic toxoplasmosis associated with infected cats. N. Engl. J. Med. 300, 695–699.

Thalhammer, O., 1973. Prevention of congenital toxoplasmosis. Neuropadiatrie 4, 233–237.

Thalhammer, O., 1978. Prevention of congenital infections. In: Perinatal Medicine, 6th European Congress. Perinatal Medicine, Vienna, pp. 44–51.

Thomas, N.J., Cole, R.A., 1996. The risk of disease and threats to the wild population. Endangered Species Update, Conservation and Management of the Southern Sea Otter (Special Issue 13), 23–27.

Thomas, N.J., Dubey, J.P., Lindsay, D.S., Cole, R.A., Meteyer, C.U., 2007. Protozoal meningoencephalitis in sea otters (*Enhydra lutris*): a histopathological and immunohistochemical study of naturally occurring cases. J. Comp. Pathol. 137, 102–121.

Torres, C.M., 1927. Morphologie d'un nouveau parasite de l'homme, *Encephalitozoon chagasi*, N. sp., observe dans un cas de meningo-encephalomyelite congenitale avec myosite et myocardite. C. R. Soc. Biol. 97, 1787–1790.

Wallace, G.D., 1969. Serologic and epidemiologic observations on toxoplasmosis on three. Pacific Atolls. Am. J. Epidemiol. 90, 103–111.

Wanko, T., Jacobs, L., Gavin, M.A., 1962. Electron microscope study of *Toxoplasma* cysts in mouse brain. J. Protozool. 9, 235–242.

Weigel, R.M., Dubey, J.P., Siegel, A.M., Kitron, U.D., Mannelli, A., Mitchell, M.A., Mateus-Pinilla, N.E., Thulliez, P., Shen, S.K., Kwok, O.C.H., Todd, K.S., 1995. Risk factors for transmission of *Toxoplasma gondii* on swine farms in Illinois. J. Parasitol. 81, 736–741.

Weiland, G., Kühn, D., 1970. Experimentelle *Toxoplasma*-Infektionen bei der Katze. II. Entwicklungsstadien des Parasiten im Darm. Berl. Münch. Tierärztl. Wochenschr. 83, 128–132.

Weinman, D., Chandler, A.H., 1954. Toxoplasmosis in swine and rodents. Reciprocal oral infection and potential human hazard. Proc. Soc. Exp. Biol. Med. 87, 211–216.

Weiss, L.M., Dubey, J.P., 2009. Toxoplasmosis: a history of clinical observations. Int. J. Parasitol. 39, 895–901.

Wilder, H.C., 1952. Toxoplasma chorioretinitis in adults. A. M. A. Arch. Ophthalmol. 48, 127–136.

Wilkins, M.F., O'Connell, E., 1983. Effect on lambing percentgage of vaccinating ewes with Pathogenesis and pathologic anatomy of coloboma of the macula lutea in an eye of normal dimensions and in a microphthalmic eye with parasites in the retina. N. Z. Vet. J. 31, 181–182.

Witte, H.M., Piekarski, G., 1970. Die Oocysten-Ausscheidung bei experimentell infizierten Katzen in Abhängigkeit vom Toxoplasma-Stamm. Z. Parasitenk. 33, 358–360.

Wolf, A., Cowen, D., 1937. Granulomatous encephalomyelitis due to an encephalitozoon (encephalitozoic encephalomyelitis). A new protozoan disease of man. Bull. Neur. Inst. NY 6, 306–371.

Wolf, A., Cowen, D., 1938. Granulomatous encephalomyelitis due to a protozoan (*Toxoplasma* or *Encephalitozoon*). II. identification of a case from the literature. Bull. Neur. Inst. NY 7, 266–290.

Wolf, A., Cowen, D., Paige, B., 1939a. Human toxoplasmosis: occurrence in infants as an encephalomyelitis verification by transmission to animals. Science 89, 226–227.

Wolf, A., Cowen, D., Paige, B.H., 1939b. Toxoplasmic encephalomyelitis. III. A new case of granulomatous encephalomyelitis due to a protozoon. Am. J. Pathol. 15, 657–694.

Wolf, A., Cowen, D., Paige, B.H., 1940. Toxoplasmic encephalomyelitis. IV. Experimental transmission of the infection to animals from a human infant. J. Exp. Med. 71, 187–214.

CHAPTER

2

The Ultrastructure of *Toxoplasma gondii*

David J. P. Ferguson and Jean-François Dubremetz***

*Nuffield Department of Clinical Laboratory Science, University of Oxford, John Radcliffe Hospital, Oxford, UK and **UMR CNRS 5235, Université de Montpellier 2, Montpellier, France

OUTLINE

Toxoplasma gondii, Second Edition
http://dx.doi.org/10.1016/B978-0-12-396481-6.00002-7

2.1 INVASIVE STAGE ULTRASTRUCTURE AND GENESIS

This chapter reviews the electron microscopic data on *Toxoplasma gondii* and its life cycle stages. Corresponding light microscopy of these stages can be found in Chapter 1: The History and Life Cycle of *Toxoplasma gondii*.

2.1.1 Basic Ultrastructural Morphology

There are four invasive forms of *T. gondii*: the tachyzoite, bradyzoite, merozoite and sporozoite. Tachyzoites and bradyzoites are associated with the intermediate host and merozoites and sporozoites with the definitive host. Tachyzoites and merozoites are responsible for the expansion of the population within a host while the bradyzoites and sporozoites are capable of environmental transmission to new hosts.

All of the infectious stages have the same basic morphology with only minor variations. The standard features will be described in this section and are based mainly on observations of tachyzoites. The tachyzoite is the most extensively studied stage in the *T. gondii* life cycle because of the ease with which large numbers can be obtained both *in vivo* and *in vitro*. These invasive stages are crescent shaped cells (2×7 μm approximately) with a slightly more pointed anterior end (the anterior being defined by the direction of motility) (Fig. 2.1A). They are comprised of a unique cytoskeleton, secretory organelles (rhoptries, micronemes, dense granules), endosymbiotic derived organelles (mitochondrion, apicoplast), eukaryotic universal organelles (nucleus, endoplasmic reticulum, Golgi apparatus, ribosomes), specific structures (acidocalcisomes, plant-like vacuoles) all enclosed by a complex membranous structure termed the pellicle.

The cytoskeleton comprises:

- Two apical rings located beneath the plasma membrane at the apical tip of the parasite. They are uncharacterized at the molecular level but both are made of a thin ring of electron dense material, the upper one is 160 nm, the posterior one 200 nm in diameter (200 nmD).
- The conoid is a hollow, truncated cone consisting of fibres wound into a spiral, like a compressed spring, 400 nm in diameter at the base and 250 nm high, made of tubulin organized in a unique fashion, consisting of asymmetrical filaments of about 9 protofilaments, very different from typical microtubules (Hu *et al.*, 2002b).
- Two polar rings encircle the top of the resting conoid (Fig. 2.2A). The outer ring consists of dense material covering the anterior rim of the inner membrane complex (IMC, see below). The inner ring is formed of material which anchors the 22 subpellicular microtubules that extend underneath the IMC for approximately 2/3 of the body length (Nichols and Chiappino, 1987). These microtubules are classic 22 nm diameter hollow tubes, comprising 13 protofilaments made of tubulin (Hu *et al.*, 2002b).
- A pair of adjacent intraconoidal microtubules is also found, extending for a short distance (less than 1 μm) into the apical cytoplasm and ending anteriorly next to an apical vesicle of 40 nm that adheres to the plasma membrane (Hu *et al.*, 2002b).

The pellicle is a distinctive membrane complex that encloses the infectious stages. It consists of an outer unit membrane (plasmalemma) that completely encloses the organism and an inner layer of two closely applied unit membranes found at a fixed distance (approx 15 nm) from the plasmalemma. The inner membrane complex (IMC) consists of fused plates formed from flattened vesicles derived from the ER–Golgi system (Vivier and Petitprez, 1969). The inner layer is interrupted by circular apertures at the anterior end (outer polar ring), where the conoid protrudes and at the posterior end. The organization of the IMC has been essentially unravelled by electron

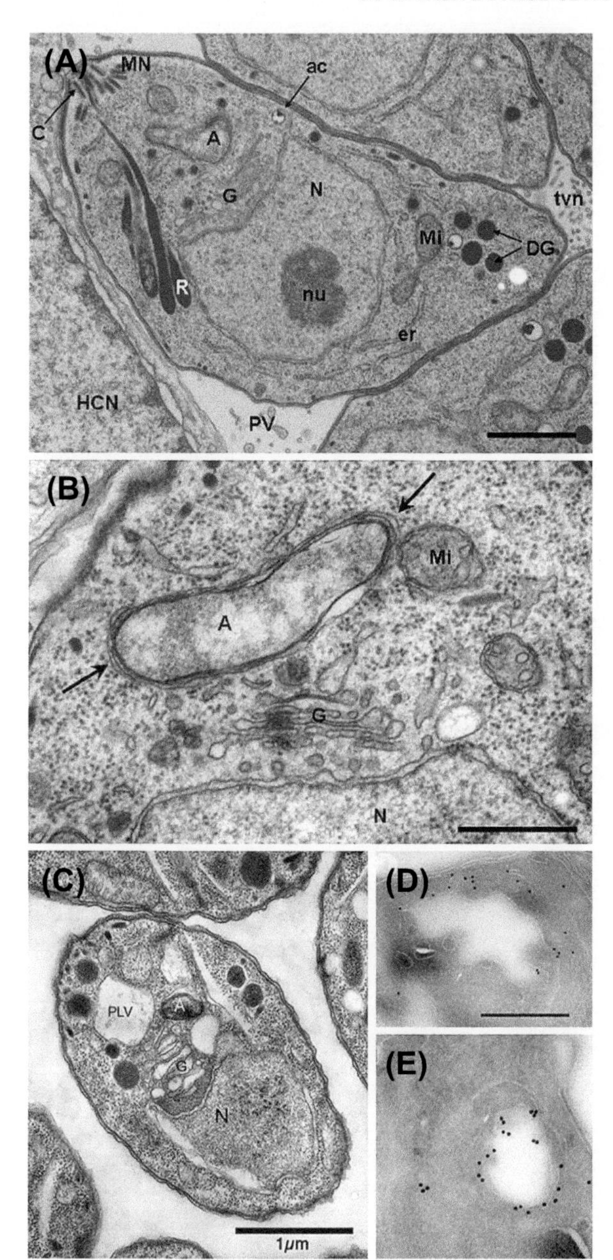

FIGURE 2.1 ***Toxoplasma gondii* tachyzoite ultrastructure.** A) Sagittal section of an intravacuolar tachyzoite. A, apicoplast; C, conoid; DG, dense granule; ER, endoplasmic reticulum; G, Golgi body; HCN, host cell nucleus; MN, micronemes; Mi, mitochondria; N, nucleus; NU, nucleolus; PV, parasitophorous vacuole; R, rhoptry. Bar is 1 μm.

microscopy (EM) freeze fracture (Morrissette *et al.*, 1997; Porchet and Torpier, 1977). It is made of an apical plate which is a single, truncated cone approximately 1 μm high, on which 6 longitudinal rows of rectangular plates are attached. The rows end at the posterior end of the tachyzoite by triangular plates joined in a turban-like fashion. The rows can extend straight or be twisted helicoidally. The protoplasmic faces on both sides of the IMC are covered with lines of intramembranous particles (IMPs), with 22 lines of higher density corresponding to the underlying subpellicular microtubules (Fig. 2.2C). IMPs have a 32 nm longitudinal periodicity and are lined approximately 30 nm apart (Morrissette *et al.*, 1997). The organization of IMPs in the apical plate is distinct from that in the other plates, suggesting a distinct molecular structure in this apical area. An additional organized structure associated with the inner side of the IMC has been described by negative staining after detergent extraction, as a network of 8–10 nm filaments containing two novel proteins with extended coiled–coiled domains that may play a role in determining cell shape (Mann and Beckers, 2001). The precise correlation between this network and the IMP alignments has not been defined.

The pellicle has an additional adaptation termed the micropore, which is located in the apical half of the cell normally just anterior to

B) Enlargement of the Golgi area showing the apicoplast (A) surrounded by four membranes (arrows). Bar is 0.5 μm.

C) Thin sections of whole tachyzoites showing a large empty vacuole: PLV/VAC. The conditions used to prepare the cells were the same as the ones used for Fig. 2 of Miranda *et al.*, 2010. A, apicoplast; G, Golgi; N, nucleus. Bar is 200 nm. Image courtesy of Wandy Beatty, Microbiology Imaging Facility, Washington University.

D) and E) Immunogold electron microscopy labelling with anti-TgVP1 antibody (1:100) of a large, empty vacuole. The antibody used was an affinity purified rabbit anti-serum generated against the peptide: SGKNEYGMSEDDPRN. The conditions used to prepare the cells were the same as the ones used for Fig. 1 of Miranda *et al.*, 2010. Bar is 200 nm. Images courtesy of Wandy Beatty, Microbiology Imaging Facility, Washington University.

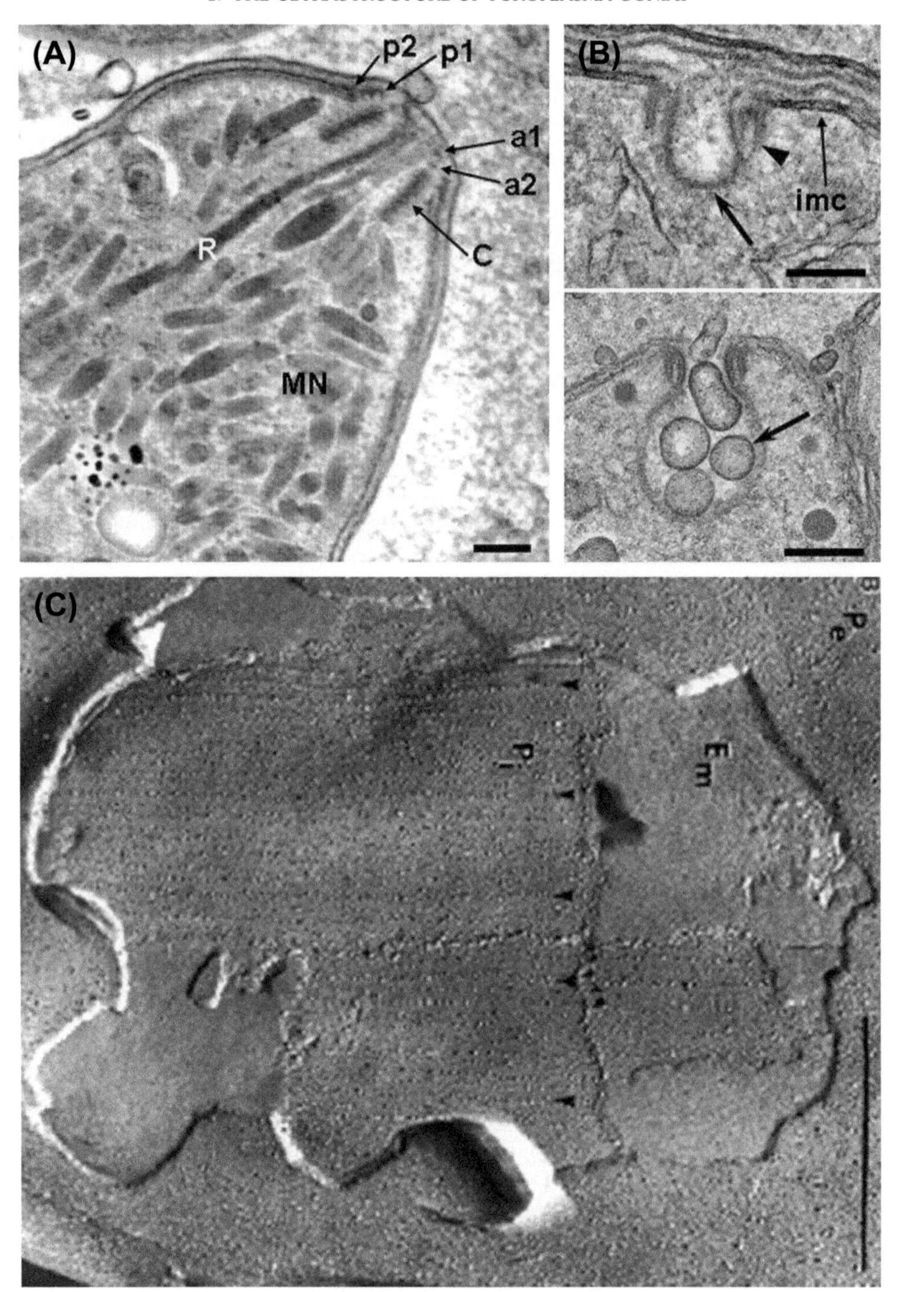

FIGURE 2.2 **Details of bradyzoite (A, B) and tachyzoite (C) ultrastructure.**
A) Apical area of a bradyzoite showing the two apical rings (a1, a2), the two polar rings (p1, p2) above and around the conoid. C, conoid; MN, micronemes; R, rhoptry. Lower picture: uptake of PV vesicular material through the micropore. Bar is 0.1 μm.
B) Upper picture: micropore showing the invagination of the zoite plasmalemma (arrow) through an opening and indentation (arrowhead) of the inner membrane complex (IMC). Lower picture: uptake of PV vesicular material through the micropore (arrow). Bar is 0.1 μm.
C) Freeze fracture image of the pellicle of a tachyzoite (from Morrissette *et al.*, 1997). The three successive membranes are shown: Pe, protoplasmic face of the plasmalemma; Em, exoplasmic face of the external layer of the inner membrane complex; Pi, protoplasmic face of the inner layer of the inner membrane complex. Bar is 0.2 μm.

the nucleus. The single micropore consists of a circular (approximately 150 nm in diameter) invagination of the plasmalemma through a break in the inner membrane complex. The latter infolds to form an electron dense collar around the invagination (Fig. 2.2B). These structures are present throughout development and increase in number during endopolygeny and gametogony. They are thought to act as a cytostome-like structure with extensions of the invaginated plasmalemma budding off resulting in the uptake of material (Nichols *et al.*, 1994) (Fig. 2.2B). This process has been clearly shown to be important in the malaria parasite where the micropore is responsible for the ingestion the erythrocyte haemoglobin (Aikawa *et al.*, 1966).

Three distinct secretory organelles have been identified, which can vary in numbers and shape between the invasive stages (see below) (Figs. 2.1A and 2.2A). First are small rod-shaped micronemes (250 × 50 nm), located in the most apical area of the parasite, behind the conoid. They are homogeneously electron dense. Second are the rhoptries, organized as a group of elongated, club-shaped organelles that extend from within the conoid toward the nucleus. They show a long, narrow neck up to 2.5 μm long and a sac-like body about 0.25 × 1 μm in the posterior area. The contents are electron dense, except in the widened part, where the structure can be either labyrinthine or electron dense in appearance depending of the specific stage. A third type of organelle, found throughout the cell but mostly in the posterior part of the parasite, are spherical-shaped (0.3 μm diameter) structures with electron dense contents, which have been termed the dense granules.

Immunoelectron microscopy has played an important role in our understanding of the functions of these organelles. With the development of antibodies to specific proteins, it has been possible to begin to identify proteins specifically located in the different organelles. It has been possible to identify proteins (MIC proteins) that are only present in the micronemes or proteins located in the dense granules (GRA proteins). Indeed in the case of the rhoptries it has been possible to identify proteins not just located in the rhoptries but to differentiate between those located in the bulbous region (the ROP proteins) and those specific to the neck region (the RON proteins) (Bradley *et al.*, 2005).

The nucleus occupies a central or basal location depending on the invasive stage (see below). It is often flattened on the upper side, where the Golgi apparatus is located. It contains a central nucleolus and small clumps of electron dense heterochromatin scattered throughout the nucleoplasm. The nuclear envelope has numerous nuclear pores and is covered on its external side with ribosomes, except on the upper face, where the Golgi apparatus is located (Figs. 2.1A, B and 2.23C). The nuclear envelope is in continuity with sheets of rough endoplasmic reticulum that extend into the cytoplasm of the tachyzoite.

On the upper surface of the nucleus, a layer of clear vesicles of 70 nm diameter, some of which can be seen budding from the nuclear envelope, is topped by three to four flat Golgi cisternae, on top of which numerous vesicles of various contents and size can be observed (Figs. 2.1A,B and 2.23B,C).

Using certain preparative techniques, one or two vesicles of c. 200 nm containing one or several electron dense droplets or crystals of various sizes in a clear background are found near the nucleus or in the posterior part of tachyzoites (Fig. 2.1A). These have been termed the acidocalcisomes and the dark content is believed to be a calcium phosphate crystal (Moreno and Zhong, 1996). In addition, under other preparative techniques, a vesicular compartment termed the plant-like vacuole (PLV) can be seen (Miranda *et al.*, 2010) (Fig. 2.1C). This compartment contains single-membrane-bounded vesicles of diverse size and appearance and is occupied by a less electron dense material than that present in the cytosol. The PLV has been demonstrated to have immunoreactivity to antisera to proton pyrophosphatase (TgVP1) as well as antisera to cathepsin L-like enzyme (TgCPL) and an aquaporin water-channel (TgAQP1) (Miranda *et al.* 2010) (Fig. 2.1D,E).

Several mitochondrial profiles of 0.5 μM width and various lengths can usually be observed at various locations above and below the nucleus (Fig. 2.1A,B). These represent sections through a single-branched and elongated mitochondrion. They show the typical apicomplexan structure, with bulbous cristae.

Above the Golgi is the apicoplast (Fig. 2.1B,C). This organelle, limited by multiple membranes, has been identified morphologically since the early 1960s (Ogino and Yoneda, 1966; Sheffield and Melton, 1968; Vivier and Petitprez, 1969) but was only recently shown to be a typical plastid (Kohler *et al.*, 1997). Since it appears to be a feature of all members of the *Apicomplexa*, with the exception of *Cryptosporidium* spp., the term apicoplast was proposed. In the infectious stage it is relatively uniform in shape, up to 500 nm in diameter, bounded by possibly four membranes and filled with granular and filamentous content, in which ribosomes can be observed. The origin of the four membranes is still a matter of debate but could result from a secondary phagocytosis (Kohler *et al.* 1997). It has recently been proposed that the four-membrane structure could result from the extensive invagination of the inner membrane of a double membraned organelle, but this requires confirmation (Kohler, 2005).

2.1.2 Comparison of the Invasive Stages

The infectious stages consisting of the tachyzoite, bradyzoite, merozoite and sporozoite differ from each other in the number of the apical organelles, the shape and electron density of the rhoptries, the location of the nucleus and the presence or absence of polysaccharide granules. The nucleus is more centrally located in the tachyzoite (Fig. 2.1A) and merozoite (Fig. 2.11B) and more basally located in the bradyzoite (Fig. 2.23B) and sporozoite (Fig. 2.21B). An additional cytoplasmic structure not described above is the polysaccharide granule. The polysaccharide granules are ovoid structures (250–180 nm) of variable electron density located in both the apical and basal cytoplasm. They contain an unusual form of carbohydrate which is biochemically more similar to plant amylopectin than animal glycogen (Coppin *et al.*, 2005). These granules are rarely found in tachyzoites or merozoites, but are present in large numbers in bradyzoites and sporozoites (Figs. 2.21B and 2.23B). The granules appear to represent a stored energy source which would be consistent with a possible requirement for the long-term survival of the bradyzoites and sporozoites or the extra energy needed during transmission between hosts. The most marked variations are in the apical organelles (see review by (Dubey *et al.*, 1998). There are relatively few micronemes in the tachyzoite and merozoites but these are more numerous in the bradyzoite and sporozoite. In the case of the dense granules, these are numerous (5–12) in the tachyzoite and sporozoites, with fewer in the bradyzoite and very few (2–3) in the merozoite. In the case of the rhoptries there are differences in the number, shape and electron density between stages. The number of rhoptries is relatively similar (5–12) for the tachyzoite, bradyzoite and sporozoite with fewer in the merozoite (2–4). The shape of the rhoptries in the tachyzoite and sporozoites appears to have an elongated swelling with a labyrinthine appearance in comparison to the more bulbous and electron dense swelling in the merozoites and bradyzoites. These differences are summarised in Table 2.1.

2.1.3 Host Cell Invasion

Observations on *T. gondii* invasion have been performed in cell cultures and on red blood cells (which appear to be a possible, although unusual, abortive host cell for this parasite) (Michel *et al.*, 1979). Invasion is operated by

TABLE 2.1 Summary of the Morphological Differences between Stages of *T. gondii*

Life Cycle Stage	Nucleus	Micronemes	Rhoptries		Dense Granules	Polysaccharide Granules
			Number	Appearance		
Tachyzoite	Central	Few	5–12	Labyrinthine	Numerous	**Few**
Bradyzoite	Basal	Numerous	5–10	Solid	Numerous	**Numerous**
Merozoite	Central	Few	3–5	Solid	Few	**Absent**
Sporozoite	Basal	Numerous	5–10	Labyrinthine	Numerous	Numerous

a moving junction, which has the same morphological features as the one described for *Plasmodium knowlesi* (Aikawa *et al.*, 1981), both in thin section and in freeze fracture. Interestingly, *T. gondii* makes the same junction with nucleated cells and red blood cells (Porchet-Hennere and Torpier, 1983). It is a very close apposition of the parasite and host plasma membrane (Figs. 2.3B and 2.4B), with thickening of the host side and an accumulation of rhomboidally organized intramembranous particles on the protoplasmic face of the host plasma membrane (Fig. 2.3C). This forms a very tight junction which excludes small electron dense tracers such as Ruthenium Red. The molecular organization of the moving junction is still not fully elucidated, but data have shown that it involves proteins derived from the rhoptry neck in association with the microneme protein AMA1 (Alexander *et al.*, 2005; Lebrun *et al.*, 2005).

Microneme exocytosis has never been clearly visualized, although it is thought to occur both during gliding motility and invasion. What has been shown is the accumulation of alignment of small clear vesicles inside the conoid in conditions of chemically triggered microneme exocytosis, as if these dense rod-like organelles gave rise to these small vesicles before or after exocytosis (Carruthers and Sibley, 1999). The docking site for microneme exocytosis is not known.

Rhoptry exocytosis is easily documented upon invasion, as an apical opening in continuity with the parasite plasma membrane, facing the developing parasitophorous vacuole (PVM) (Nichols *et al.* 1983). Freeze fracture shows an open pit in the PVM at that location suggesting continuity between rhoptry contents and PVM or even host cell cytoplasm (Fig. 2.3A). The role of the apical vesicle and of the apical rosette of intramembranous particles located at the rhoptry exocytosis site have never been elucidated, but the rhoptries open precisely at this location and the IMP rosette disappears, just as in trichocyst exocytosis in *Paramecium* spp. (Beisson *et al.*, 1976).

At very early stages of invasion, when the moving junction forms, small vesicles can be seen budding from the developing vacuole or laying in the host cell cytoplasm (Figs. 2.3B and 2.4A). At this stage, empty rhoptries are already observed. Therefore these vesicles correspond to the physiological counterpart of the vacuoles, which are the product of frustrated rhoptry exocytosis in the host cell cytoplasm when invasion is blocked by cytochalasin D (Hakansson *et al.*, 2001).

The membrane of the developing vacuole is completely devoid of intramembranous particles (Fig. 2.4A) (Dubremetz *et al.*, 1993), reflecting the selective exclusion of the intra-membranous host cell proteins at the moving junction. However, it will acquire IMPs during the first hour of development (Porchet-Hennere and Torpier, 1983), likely due to parasite contribution, especially from dense granule protein translocation in the PVM (Dubremetz *et al.*, 1993). Progression of the moving junction along the zoite is

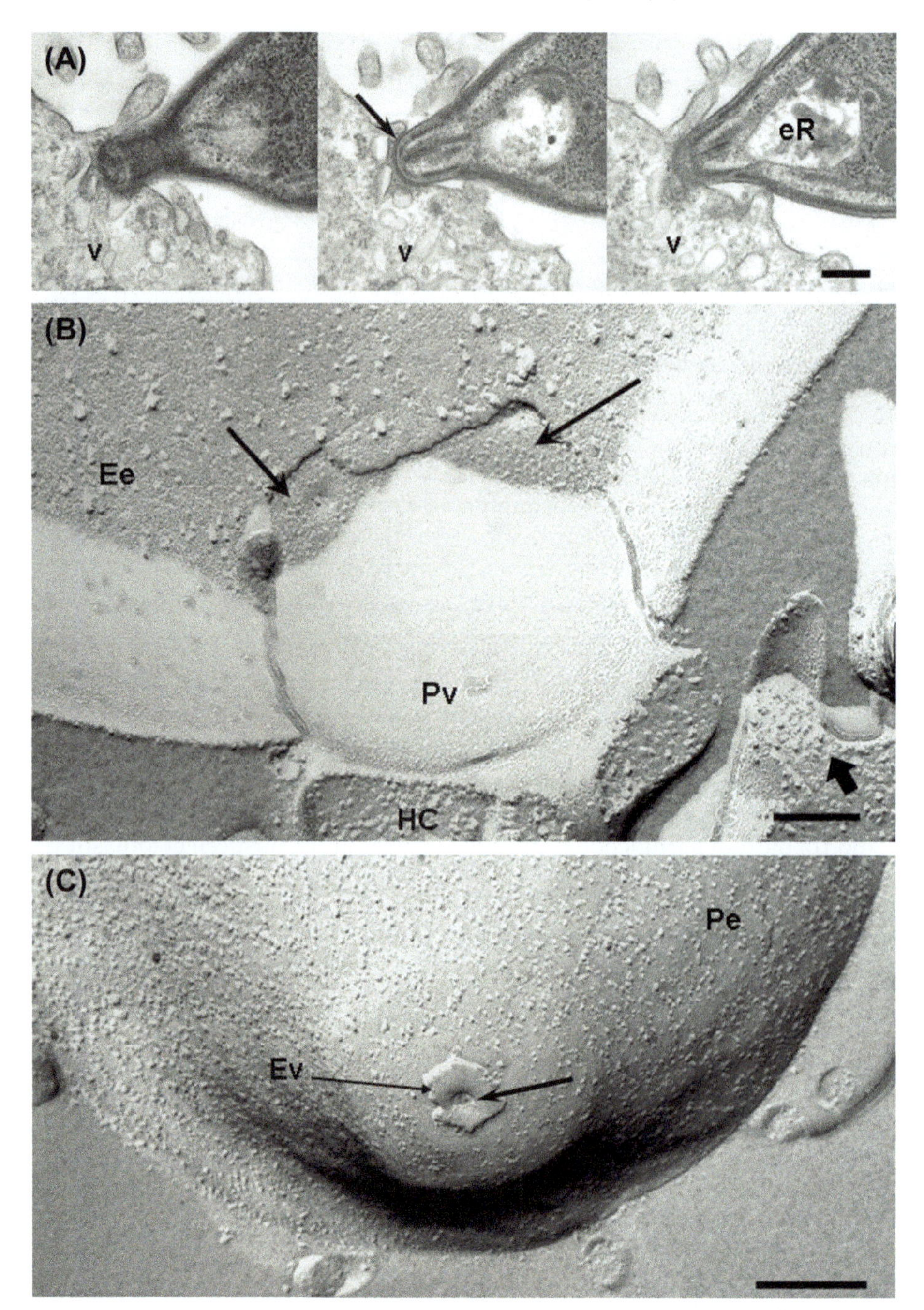

FIGURE 2.3 **Host cell invasion *in vitro*.**
A) Serial section though the apical area of a tachyzoite at an early stage of Hela cell invasion. The moving junction is covering the apex of the tachyzoite (arrow); an empty rhoptry has exocytosed its contents in the neighbouring host cell cytoplasm as small vesicles (v). eR, empty rhoptry. Bar is 0.1 μm.

sometimes, but not always, correlated with parasite constriction.

2.1.4 Parasitophorous Vacuole, Intracellular Development

Within minutes after closure of the parasitophorous vacuole (PV), the posterior part of the parasite invaginates and the tubulovesicular network (TVN) starts developing in this invagination (Sibley *et al.*, 1995). The origin of the TVN is not fully understood. It contains dense granule proteins that are exocytosed from the anterior end of the parasite; this exocytosis begins before the completion of the invasion process (Dubremetz *et al.*, 1993). But the origin of the tubular material itself, which is likely made of phospholipids, has never been elucidated; what is known is that the GRA2 protein is required to organize this network (Mercier *et al.*, 2002), and that these tubules are in direct continuity with the PVM, although these two structures contain distinct dense granules-derived proteins (Cesbron-Delauw, 1994).

Immediately after invasion, host cell mitochondria and endoplasmic reticulum (ER) surround the PV and persist throughout the intracellular development (Fig. 2.8A,B). The ER is devoid of ribosomes on the side facing the vacuole. The distance between these organelles and the PV is highly conserved and is about 12 and 18 nm for mitochondria and ER respectively (Sinai *et al.*, 1997). The PVM-associated mitochondria may look normal but sometimes show morphological changes, with the cristae becoming larger and irregular in shape and the stroma becoming electron dense.

The parasitophorous vacuole described above is formed by actively invading parasites and is characterized by the absence of the fusion of the host cell lysosomes, thus protecting the parasite during intracellular development (Jones and Hirsch, 1972; Jones *et al.*, 1972). This probably relates to the exclusion of the host cell intramembranous proteins from the parasitophorous vacuole membrane during invasion. In contrast, parasites within vacuoles formed by host cell phagocytosis exhibit lysosome fusion and the parasites are broken down in typical phagolysosomes (Jones and Hirsch, 1972).

2.1.5 Endodyogeny

The tachyzoite is unique in its ability to undergo indefinite proliferation by a distinctive process termed endodyogeny, which involves parasite growth and division to form two daughters. Despite grossly resembling binary fission, endodyogeny is a highly complex event, related to the structural complexity of the formation of polarized daughters. In addition, in the tachyzoite, contrasting with the canonical asexual division mode of most *Apicomplexa* and even the coccidian stages of *T. gondii*, the apical complex and invasion related organelles of the mother persist until the end of the endodyogeny process.

Although both events occur simultaneously, we will describe the mitosis and daughter tachyzoite genesis successively.

B) Freeze fracture image of the apical area of an invading tachyzoite at a stage corresponding to part A above. The typical structure of the moving junction in the protoplasmic face (Pv) of the host cell (HC) plasmalemma (which will turn into the parasitophorous vacuole membrane) is visible (arrows), below the parasite apical exoplasmic plasmalemmal face. Bar is 0.1 μm.

C) Freeze fracture image of the apex of an invading tachyzoite at a similar stage of invasion as part B, but corresponding to the complementary fracture faces, showing the pit (arrow) in the parasitophorous vacuole membrane (Ev, exoplasmic face) covering the site of rhoptry exocytosis in the tachyzoite plasmalemma. Pe, protoplasmic face. Bar is 0.2 μm.

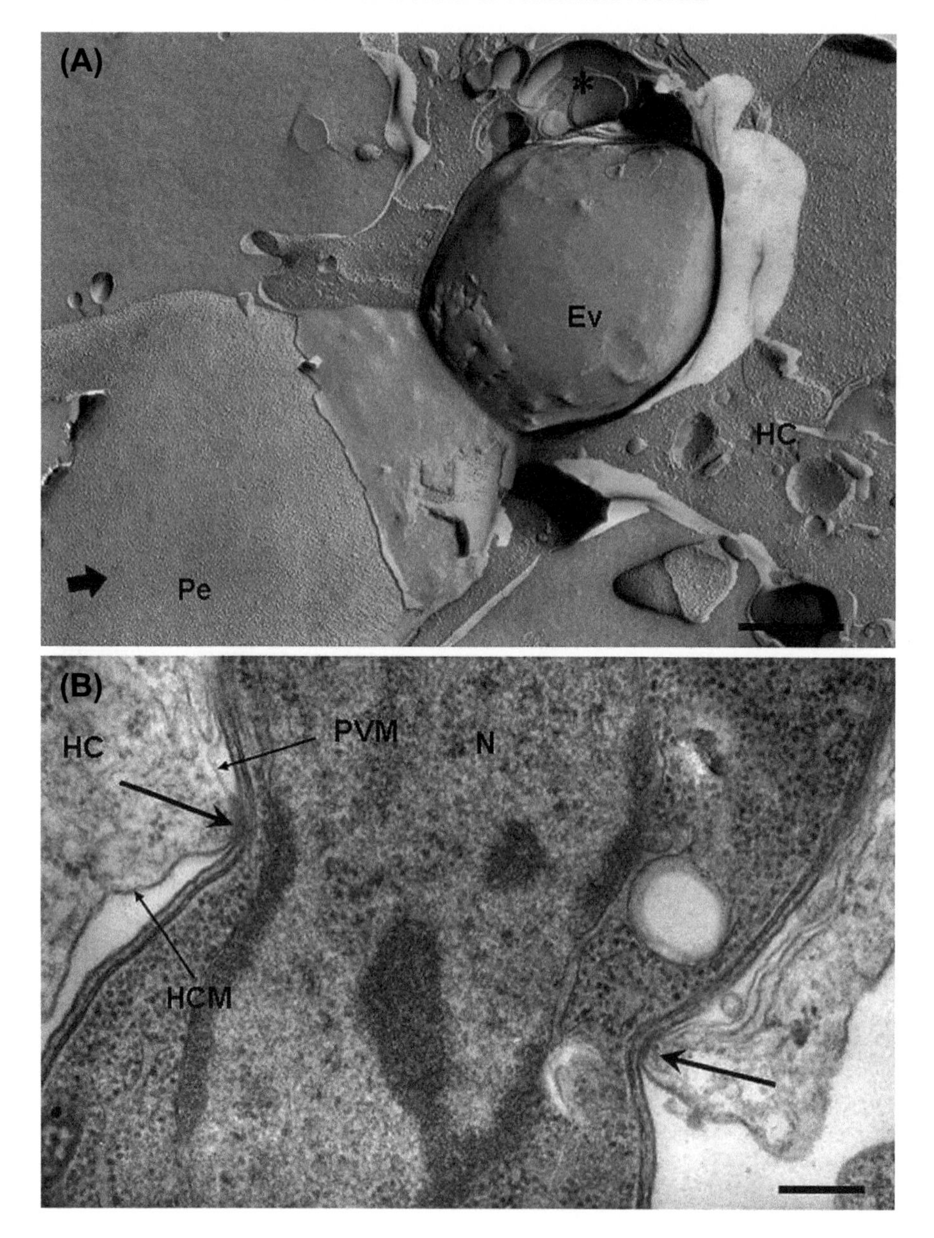

FIGURE 2.4 **Host cell invasion *in vitro*.**
A) Freeze fracture of an invading tachyzoite showing the parasitophorous membrane (Ev), a clump of membrane whorls that may correspond to material exocytosed from the rhoptries (asterisk), and the plasmalemma of the tachyzoite. HC, host cell; Pe, protoplasmic face. Bar is 0.5 μm.
B) Section through an invading tachyzoite showing the moving junction (arrows) and the continuity and the difference in electron density between the host cell plasmalemma (HCM) and the parasitophorous vacuole membrane (PVM). N, nucleus. Bar is 0.2 μm.

2.1.5.1 Mitosis

There have been few descriptions of *T. gondii* mitosis at the ultrastructural level and what has been observed can be interpreted by comparison with more detailed studies in related *Apicomplexa*, especially *Eimeria* spp. (Dubremetz, 1973). One unique feature of apicomplexan mitosis is the retention of an intact nuclear membrane throughout the process of division. Coccidian type centrioles (150 nm diameter) consist of nine short tubules (100 nm long) centred on a central tubule. Centrosomes, or spindle pole bodies, are made of two centrioles oriented in parallel (Fig. 2.5B). Centrosomes are always

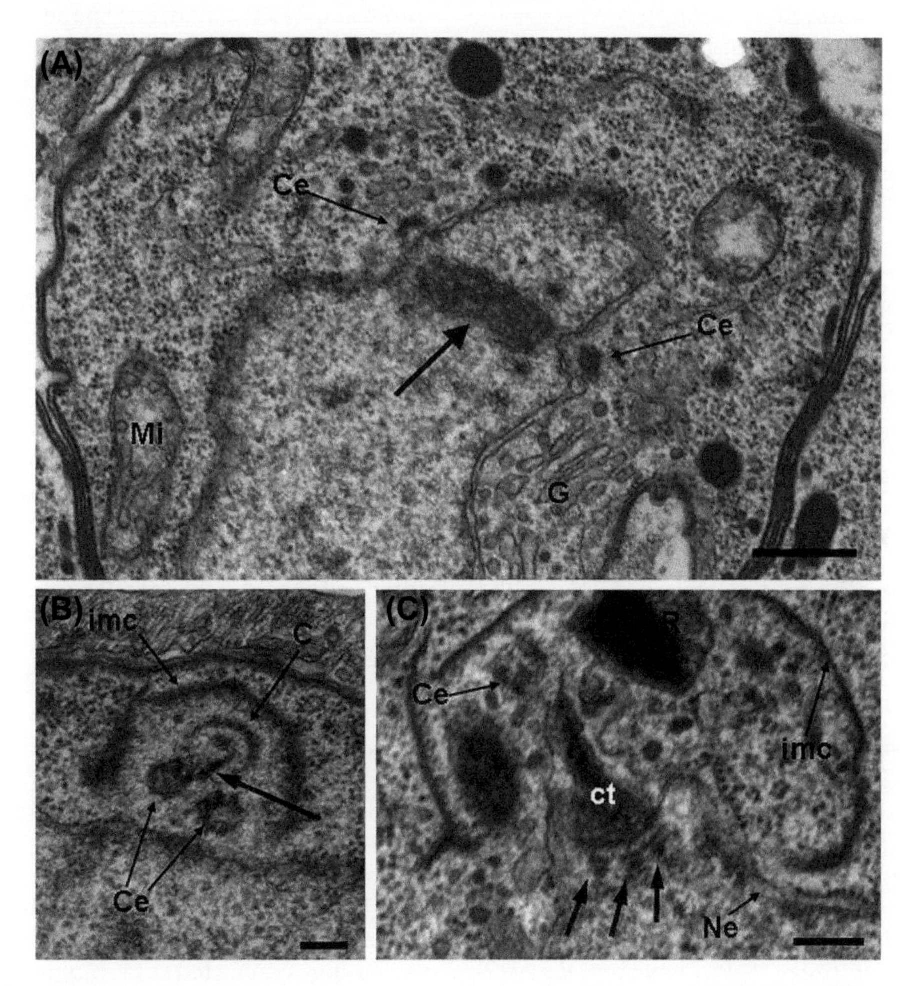

FIGURE 2.5 **Tachyzoite endodyogeny.**
A) Early stage of mitosis: the mitotic spindle elongates into a cytoplasmic funnel through the nucleus (arrow), between the centrioles (Ce). G, Golgi body; Mi, mitochondria. Bar is 0.5 μm.
B) Early stage of daughter zoite genesis where a dense fibre (arrow) extends between the centrosome (Ce, centriole) and the newly formed conoid (C). The apical part of the inner membrane complex (IMC) and subpellicular microtubules have started developing. Bar is 0.2 μm.
C) Centrocone (ct) budding off the nuclear envelope (Ne) between a centriole (Ce) and three kinetochores (arrows) in an early stage of daughter zoite formation. R, rhoptry. Bar is 0.2 μm.

found associated with centrocones, or mitotic spindle poles, usually on the apical side of the nucleus. The earliest stage of mitosis is a transnuclear funnel containing fibrous material, corresponding to an invagination of the nuclear envelope opened on both sides towards the cytoplasm (Fig. 2.5A). The mitotic spindle most likely polymerizes in this funnel which then opens in the nucleoplasm in its middle part, whereas the poles give rise to the centrocones. These latter are at first subspherical invaginations of the nuclear envelope opened towards the centrosomes and through which the spindle microtubules extend. The intranuclear spindle is usually very short and transient and has rarely been described. What occurs most likely is that the kinetochores are separated immediately after the funnel opening and assemble on the nucleoplasmic side of the centrocones. Indeed, in Coccidia, caryokinesis does not depend on mitotic spindle elongation. Centrocones soon become conical evaginations of the nuclear envelope, opened on the centrosomes and covered on the nucleoplasmic side with a layer of multilayered structures corresponding to the kinetochores (Fig. 2.5C). What is specific to this stage is that each round of mitosis occurs simultaneously with the development of two daughter individuals.

2.1.5.2 Zoite Biogenesis

Soon after the centrosomes separate and centrocones are formed, the future apical complex of each daughter tachyzoite starts to develop adjacent to each centrosome. The details of this biogenesis have not been studied as thoroughly as in *Eimeria* spp. (Dubremetz, 1975), but follow the same scheme (Hu *et al.*, 2002a; Vivier, 1970; Vivier and Petitprez, 1969; Francia *et al.*, 2012). A very early stage of development shows a bent fibre originating between the pair of centrioles and joining an area where the conoid is being assembled (Fig. 2.5B). The inner membrane complex and underlying subpellicular microtubules array appears to form around the conoid and, in a coordinated manner, starts to grow posteriorly (Figs. 2.5C and 2.6A). This occurs within the mother cell cytoplasm rather than in association with the mother cell plasmalemma that is characteristic of daughter formation in classic schizogony undergone by most apicomplexans. Early stages are short, flattened cones above the centrocones (Fig. 2.6A), which later elongate into the grossly cylindrical shape that will eventually surround the mature organism (Fig. 2.6B). The Golgi apparatus divides concomitantly with spindle formation with each newly formed dictyosome being found on the upper nuclear envelope, near each centrocone (Pelletier *et al.*, 2002) (Figs. 2.5A and 2.6A). The apicoplast elongates and appears to divide during daughter formation with a portion entering each daughter. Rhoptry precursors are observed at this time as heterogeneous, irregularly-shaped vesicles of about 0.3 μm located near the dictyosomes, inside the inner membrane complex (Fig. 2.6A). As development proceeds, the nucleus becomes U-shaped and the developing inner complex elongates and engulfs the daughter nuclei (Fig. 2.6B) while additional organelles (rhoptry precursors and then micronemes) are formed anterior to the dictyosome. The rhoptry contents condense to eventually acquire their mature labyrinthine appearance, while the rhoptry ducts appear and elongate towards the conoid. As the daughters grow, the inner membrane complex of the mother cell breaks down along with the anterior organelles. The fully formed daughters fill much of the mother cell cytoplasm and their inner membrane complex comes in contact with the mother cell plasmalemma to form the daughter pellicle (Fig. 2.7A,B). This is initiated at the anterior end and results in the daughters remaining connected via a small portion of residual cytoplasm before finally separating. Repeated rounds of division lead to accumulation of tachyzoites within the vacuole, which may adopt a typical rosette appearance when grown in flat cell such as human foreskin fibroblasts (Fig. 2.8A).

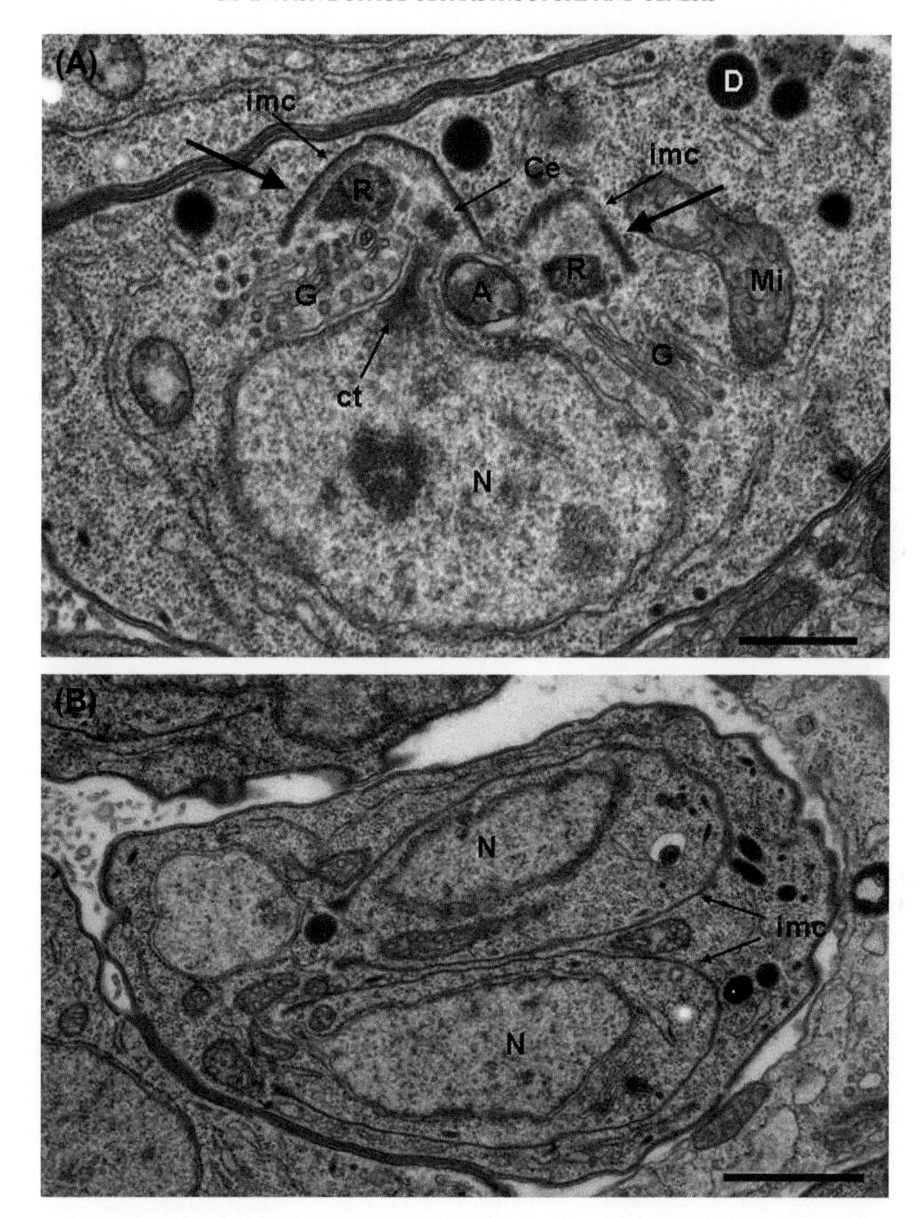

FIGURE 2.6 **Tachyzoite endodyogeny.**
A) Early stage of endodyogeny showing two developing daughters (arrows), with early rhoptries. The Golgi body has divided. Only one nuclear pole (ct, Ce) is in the section plane. A, apicoplast; Ce, centrioles; ct, centrocone; G, Golgi body; imc, inner membrane complex; Mi, mitochondria; N, nucleus; R, rhoptry. Bar is 0.5 μm.
B) Later stage of endodyogeny where the daughter nuclei (N) have entered the developing zoites. Bar is 1 μm.

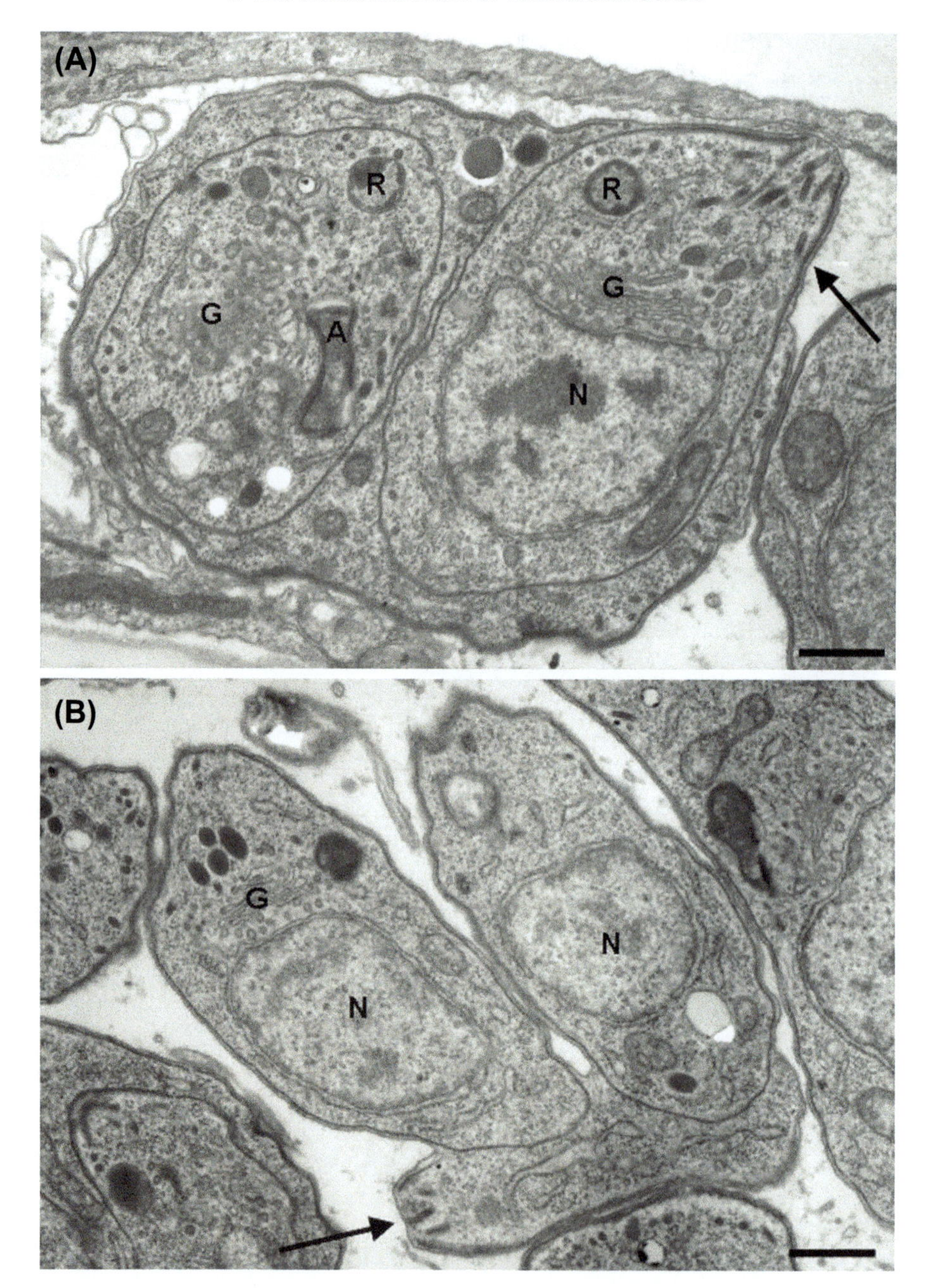

FIGURE 2.7 **Tachyzoite endodyogeny.**
A) Early budding stage where one of the daughter tachyzoites is protruding out of the mother cell by getting wrapped in the mother plasmalemma (arrow). A, apicoplast; G, Golgi body; N, nucleus; R, rhoptry. Bar is 0.5 μm.
B) Late stage of daughter budding at a stage where the remnants of the mother cell apical complex are still visible (arrow) while the two daughters are almost completely formed. Bar is 0.5 μm.

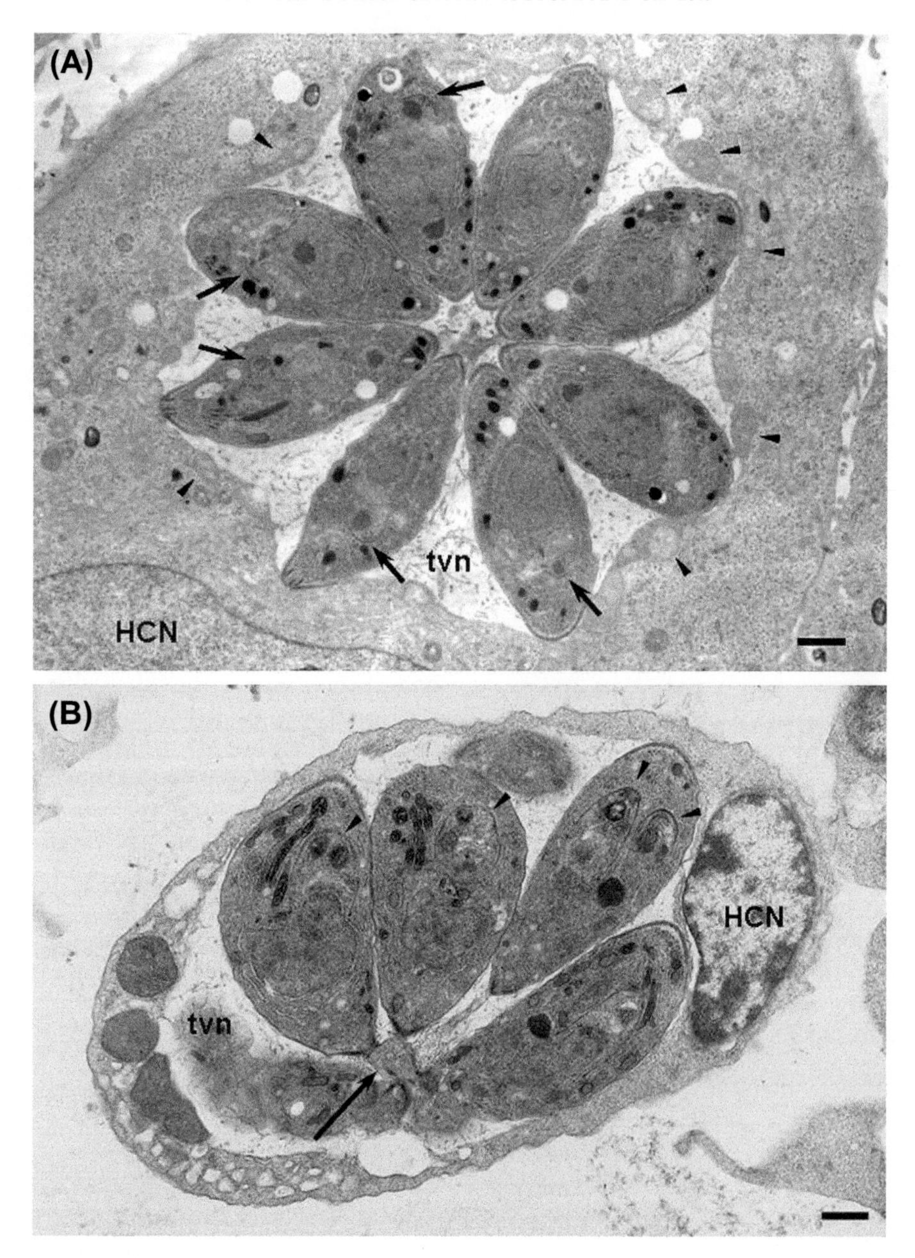

FIGURE 2.8 **Intracellular rosette of tachyzoites.**
A) Typical figure of intracellular tachyzoite multiplication in adherent cells grown *in vitro*, where divisions occur in one single plane. The vacuole and tubulovesicular network (tvn) surround the parasites, all of which are in an early stage of endodyogeny (arrows). Host cell mitochondria (arrowheads) surround the parasitophorous vacuole membrane. HCN, host cell nucleus. Bar is 1 μm.
B) Repeated endodyogeny showing the synchronized initiation of a new round of daughter formation (arrowheads) while the original daughters are still connected at the posterior end (arrow). HCN, host cell nucleus. Bar is 1 μm.

In certain cases, each of the daughters, while remaining attached by their posterior ends, can undergo new cycles of endodyogeny (Fig. 2.8B). There is evidence for the synchronized division of the tachyzoites within a vacuole (Fig. 2.8A, B). In quantitative studies, it was observed that this was more common *in vivo* for avirulent parasites compared to virulent parasites (Ferguson and Hutchison, 1981). This was only observed for tachyzoites and not seen during division of bradyzoites in tissue cysts.

Endodyogeny is the exclusive form of asexual division undergone within the intermediate host (during tachyzoite and bradyzoite formation) and differs from the processes undergone in the definitive host (merozoite formation) or within the oocyst (sporozoite formation) (see later sections).

2.2 COCCIDIAN DEVELOPMENT IN THE DEFINITIVE HOST

2.2.1 Host–Parasite Relationship

Coccidian development is limited to the epithelial cells of the small intestine of the cat (the definitive host). In all stages undergoing coccidian development the parasites are located within a tight fitting, thick walled parasitophorous vacuole (Fig. 2.9A) (Ferguson, 2004; Ferguson *et al.*, 1974). At higher power, the wall has a laminated appearance, which in certain areas can be seen to consist of three closely-applied unit membranes (Fig. 2.9B, D). In addition, there are a number of conical-shaped dense structures impinging on the luminal surface of the PV (Ferguson, 2004) which, in certain cases, appeared to connect the surface of the parasite to the parasitophorous vacuole membrane (Fig. 2.9C, D). In contrast to the host/parasite relationship of the tachyzoite, there was no evidence of formation of the tubular structure within the PV or the congregation of the host cell mitochondrion or strands of rough endoplasmic reticulum (rER) around the periphery of the PV (Fig. 2.9A). These structural differences correlated with the lack of expression of the majority of dense granule proteins. Of the GRAs 1–8 and NTPase identified in the tachyzoite, only GRA7 and NTPase are expressed by the gut stages (Ferguson, 2004; Ferguson *et al.*, 1999a, 1999b). This laminated, thick-walled PV is similar to that observed for certain *Isospora* species (Ferguson *et al.*, 1980) to which *T. gondii* is closely related, but differs from those of the Genus *Eimeria* which are limited by a single-unit membrane (Ferguson *et al.*, 1976).

2.2.2 Asexual Development

During coccidian development only a single asexual process has been observed, which has unique structural features and has been termed endopolygeny (Piekarski *et al.*, 1971). This term had been used previously to describe an abnormal type of development observed for the tachyzoite (Vivier, 1970). However, the abnormal tachyzoite development described did not represent an internal budding process. Therefore, because of the accuracy of the description and its usage over the years, it would appear appropriate to retain the term for the description of the asexual multiplication of the coccidian stages. In studies of coccidian development of both types 2 and 3 strains of *T. gondii* occurring between four and 10 days post-infection (PI), only a single process was observed, although there were marked variations in the number of daughters formed. The process involved growth of the parasite and repeated nuclear divisions (Fig. 2.9E) employing an eccentric intranuclear spindle, as described during endodyogeny. There is also a marked increase in the size of the mitochondria-which are located predominantly around the periphery. In addition, it was possible to observe multiple profiles of the apicoplast (limited by four membranes) but these were more centrally located and, from immunocytochemistry, appeared to consist of a single branched structure

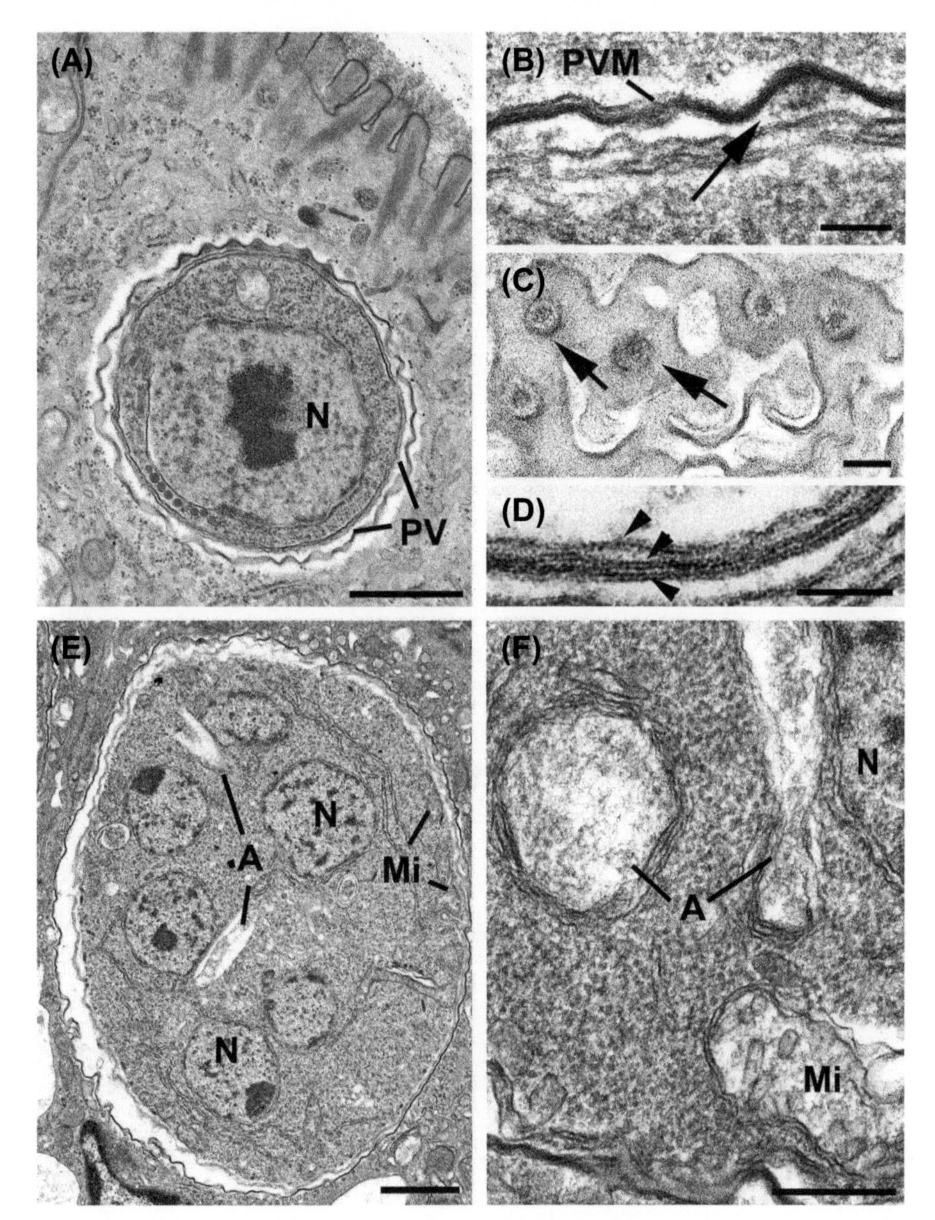

FIGURE 2.9 ***Toxoplasma gondii*** **developing in enterocytes of the cat intestine.**
A) Early developmental stage located in a thick walled, tight fitting parasitophorous vacuole (PV). N, nucleus. Bar is 1 μm.
B) Enlargement showing the laminated structure of the electron dense membrane limiting the parasitophorous vacuole (PVM). Note a conical structure protruding into the membrane (arrow). Bar is 100 nm.
C) Tangential section through the membrane of the parasitophorous vacuole illustrating the circular shape of the conical protrusion into the parasitophorous vacuole membrane (arrows). Bar is 100 nm.
D) Detail in which the membrane of the vacuole can be resolved into three unit membranes (arrowheads). Bar is 100 nm.
E) Mid stage schizont with a number of nuclei (N) and a centrally located elongated apicoplast (A). Bar is 1 μm.
F) Detail showing the double membranes enclosing the mitochondrion (Mi) and nucleus (N) compared to the multiple membranes enclosing the apicoplast (A). Bar is 0.5 μm.

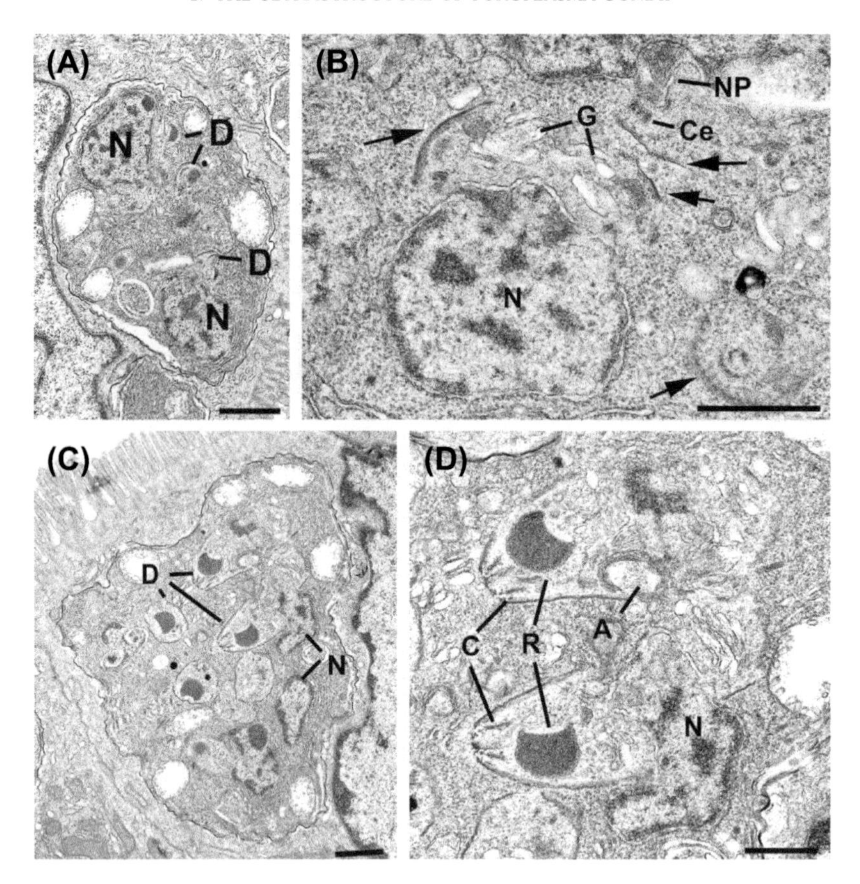

FIGURE 2.10 **Early stages of endopolygeny.**
A) Small schizont with few nuclei (N) but showing the initiation of daughter formation (D). Bar is 1 μM.
B) Detail of a schizont showing the plate-like structures of the inner membrane complex representing the initiation of daughter formation (arrows). Ce, centriole; N, nucleus; NP, nuclear pole; G, Golgi body. Bar is 0.5 μm.
C) Low power of a large schizont with a number of nuclei (N) showing the formation of a larger number of daughters (D). Bar is 1 μm.
D) Detail showing the posterior growth on the inner membrane complex of the daughter to partially enclose the apicoplast (A) and nucleus (N). Note the anlagen of the rhoptry (R) and the conoid (C) in the apex of the daughter. Bar is 0.5 μm.

(Ferguson and Hutchison, 1981). These three organelles could be differentiated by their ultrastructural features (Fig. 2.9F). The number of nuclear divisions varied between parasites which had a direct effect on the number of daughters formed. It is the presence of this proliferative phase prior to daughter formation that distinguishes endopolygeny from endodyogeny. It is not clear how the number of nuclear divisions is controlled but it does not appear to relate to parasite size or a given number of nuclear divisions since these can vary markedly between parasites (Figs. 2.10A, C).

The trigger for the end of the proliferative phase and the initiation of the differentiation phase (daughter formation) is unclear. It is at the end of the proliferative phase that the elongated apicoplast divides simultaneously into a number of fragments equal to the number of nuclei. Daughter formation can occur in parasites containing between four and twenty nuclei and is initiated during or just after the final nuclear division (Fig. 2.10A, B). The first evidence of the initiation of daughter formation was the appearance of a conical structure formed by a number of flattened vesicles each with underlying longitudinally-running microtubules and with the conoid in the apex (Fig. 2.10B). The initiation of daughter formation is synchronized with all daughters forming at the same time (Fig. 2.10A, C). The mechanism of daughter formation is similar to that observed for the two daughters formed during endodyogeny of the tachyzoite or bradyzoite. Since this occurs at a multi-nucleated stage, numerous daughters are formed, thus the appropriateness of the term endopolygeny. The simultaneous formation of a large number of daughters requires an extremely well coordinated process to ensure that all daughters receive a full complement of organelles and are therefore viable. As daughter formation progresses by the posterior growth of the inner membrane complex, it encloses a nucleus, apicoplast and mitochondrion (Fig. 2.10D). In the apical cytoplasm, one or two electron dense spherical structures representing the nascent rhoptries and a number of cigar shaped micronemes appear (Fig. 2.10D). The merozoites have relatively few dense granules and these appear to form late in daughter development. This posterior growth continues until the merozoite is fully formed and contains the full complement of organelles. In the apical cytoplasm there is maturation of the rhoptries with the development of the duct leading to the conoid. In contrast to the tachyzoite and bradyzoites, the bulbous end of the rhoptry remains spherical. At this point the daughters fill the mother cell cytoplasm but are still enclosed in the schizont plasmalemma. The final stage is the invagination of the mother cell plasmalemma, starting at the anterior of the daughter and progressing posteriorly to form the outer membrane of the pellicle of each daughter (Fig. 2.11A). A single micropore is formed in the pellicle just anterior to the nucleus. The merozoites often remain attached to a small amount of residual cytoplasm at the posterior end. These banana-shaped daughters can be seen forming fan-like structures (Fig. 2.11B, C). The merozoites are released from the host cell into the lumen, where they can reinvade enterocytes. However, this process has not been observed by electron microscopy.

Unlike many *Eimeria* species, there does not appear to be the distinct sequential generations of schizogony, which differ from each other in their size and number of daughters formed. However, in studies of the early stages of infection (one to three days), additional asexual processes have been described (Speer and Dubey, 2005). It has been reported that certain developing parasites have a similar host/parasite relationship and undergo endodyogeny and repeated endodyogeny (type B schizonts) while others appeared intermediate (type C schizonts). The type B schizonts have a similar relationship to that described for parasites invading the small intestine of the intermediate host (Dubey, 1997; Speer and Dubey, 1998). These stages appear to be rare and could represent examples where the initial invading bradyzoite failed to undergo conversion to coccidian development but underwent conversion to tachyzoite development as seen in the intermediate host. It is known that the cat can act as an intermediate host as well as the definitive host.

2.2.3 Sexual Development

After an unknown number of asexual cycles, certain merozoites on entering a new enterocyte can develop into either male (microgametocyte) or female (macrogametocyte) gametocytes. In microgametogony, this results in the formation of multiple (15–30) male (microgametes) and, in macrogametogony, the formation of a single female (macrogamete). The trigger for the conversion from asexual to sexual development is unknown. Nor is it known what is responsible for deciding whether an invading merozoite

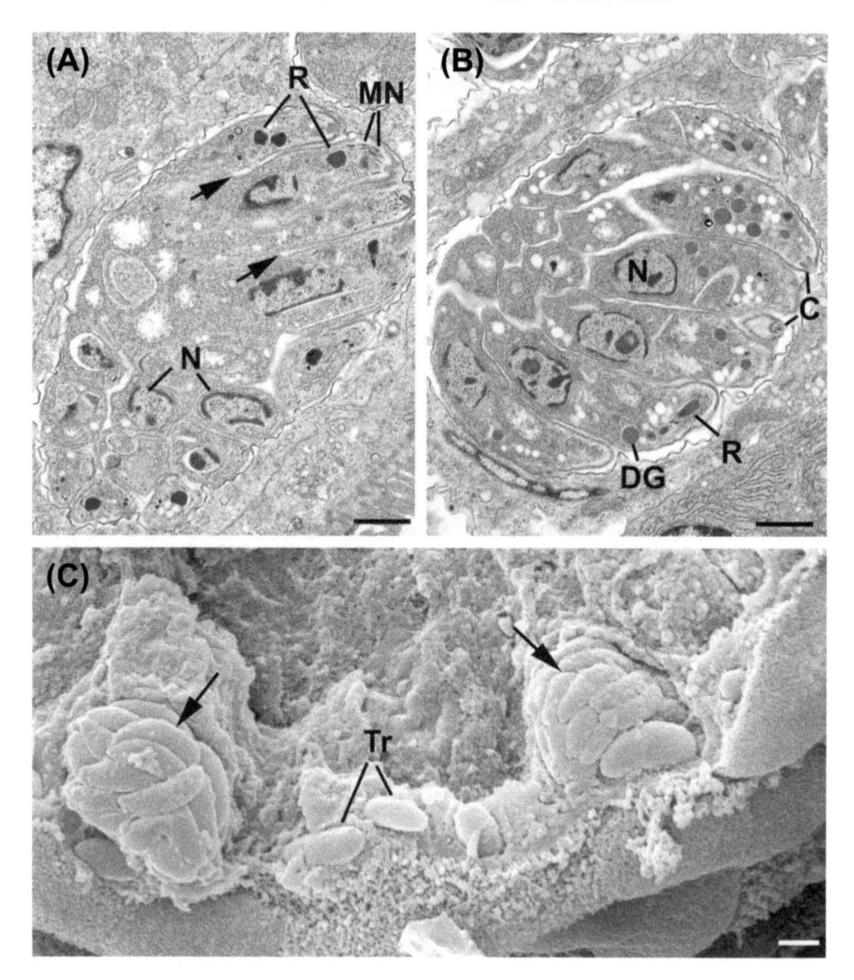

FIGURE 2.11 **Late stages of endopolygeny.**
A) Late schizont showing the daughters filling the mother cell cytoplasm and the invagination of the plasmalemma around the daughters (arrows). N, nucleus; R, rhoptry; MN, microneme. Bar is 1 μm.
B) Mature schizont with fully formed merozoites with the characteristic apical organelles. N, nucleus; C, conoid; R, rhoptry; DG, dense granule. Bar is 1 μm.
C) Scanning electron micrograph of a fracture through the epithelial cells of a villus. A number of small trophozoites (Tr) and two mature schizonts with crecentic shaped merozoites (arrows) can be seen. Bar is 2 μm.

develops into either a microgametocyte or macrogametocyte.

The initiation of gametocyte formation appears to be less controlled in *T. gondii* than other species of Coccidia. In the majority of *Eimeria* spp., there are a fixed number of asexual cycles followed by the vast majority of merozoites simultaneously develop into sexual stages. In *T. gondii*, there does not appear to be a distinct conversion with a mixture of both asexual and sexual stages observed throughout enteric development. There were no ultrastructural features that could identify the merozoite that would develop into sexual stages, nor were there any differences in the host/parasite relationship or parasitophorous vacuole.

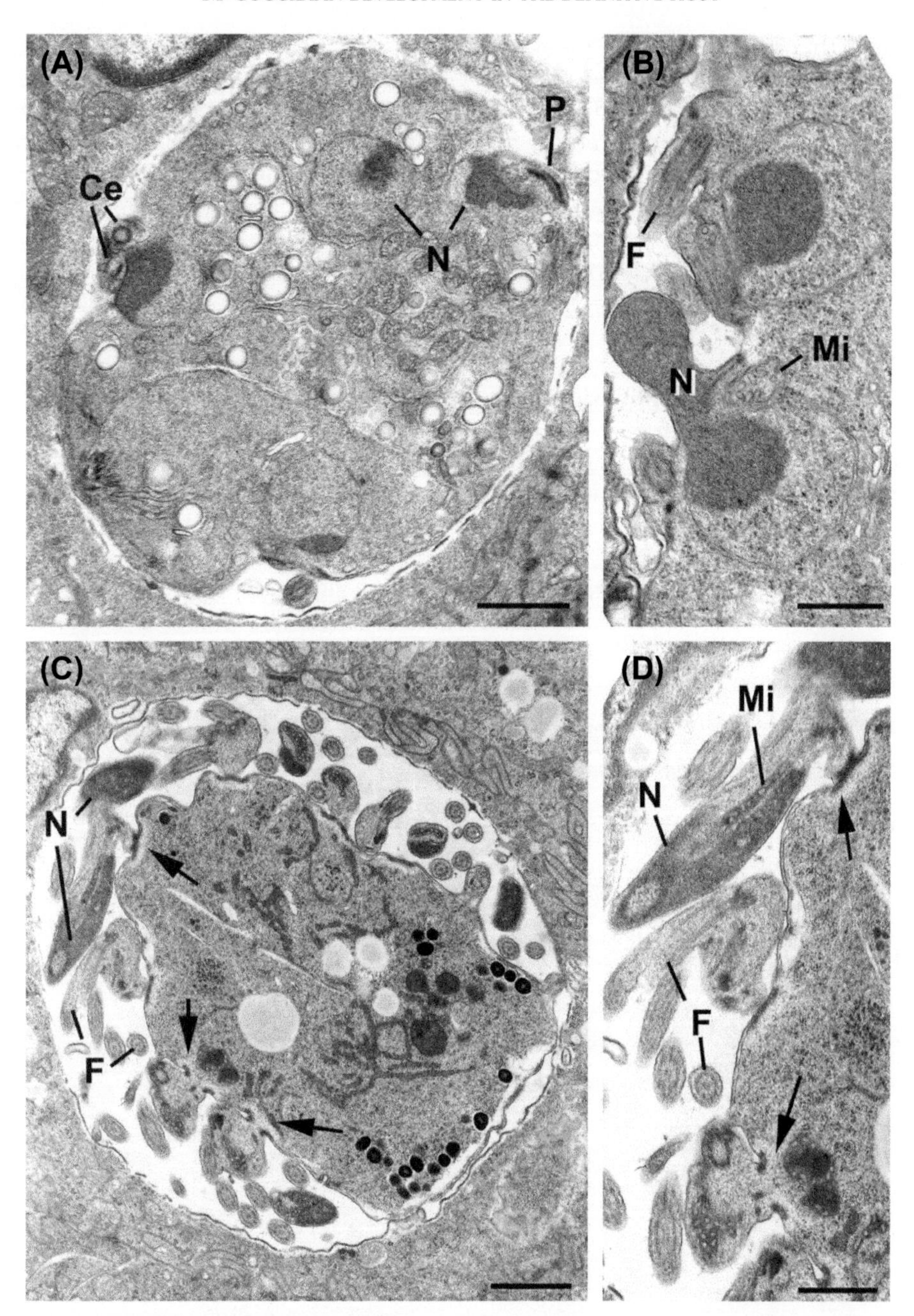

FIGURE 2.12 **Electron micrographs of various stages in the process of microgametogony.**
A) Mid stage microgametocyte showing the peripherally located nuclei with areas of condensed chromatin. Centriole (Ce) and the plate-like perforatorium (P) can be seen located between the nuclei (N) and the plasmalemma. Bar is 1 μm.
B) Detail showing the protrusion the flagella (F) plus the dense portion of the nucleus and a mitochondrion (Mi) into the PV. Bar is 0.5 μm.
C) Late stage showing a number of microgametes forming in the PV while still attached to the mother cell (arrows). N, nucleus; F, flagellum. Bar is 1 μm.
D) Detail from part c showing elongating nucleus (N) and mitochondrion (Mi) of the microgamete still connected to the mother cell (arrows). Bar is 0.5 μm.

On entering the host cell, the merozoite becomes more spherical and loses the majority of its apical organelles such as the rhoptries and dense granules, although the conoid and a few micronemes remain. This stage (trophozoite) begins to grow and there appears to be an increase in the size of the mitochondrion/mitochondria, which are located around the periphery.

2.2.3.1 *Microgametogony and the Microgamete*

There are relatively few descriptions of microgametogony (Colley and Zaman, 1970; Dubey *et al.*, 1998; Ferguson *et al.*, 1974; Pelster and Piekarski, 1971). Initially it is impossible to differentiate between the proliferative phase of endopolygeny and microgametogony with both processes involving continued growth and repeated nuclear divisions. It has been reported that the earliest stage allowing differentiation between schizogony and microgametogony is based on the difference in the distribution of the nuclear chromatin (Ferguson *et al.*, 1974). During schizogony, the electron dense heterochromatin remains dispersed throughout the nuclei (Fig. 2.9E) whereas in the later stages of microgametogony the heterochromatin condenses into electron dense masses at the periphery of the nucleus (Fig. 2.12A). In microgametogony, the nuclei move to the periphery of the cell with two centrioles and a dense plaque (perforatorium) located between the nuclei and the plasmalemma (Fig. 2.12A). The centrioles become the basal bodies for the developing flagella, which begin to grow by protruding into the parasitophorous vacuole (Fig. 2.12B). Interestingly, although the centrioles differ from

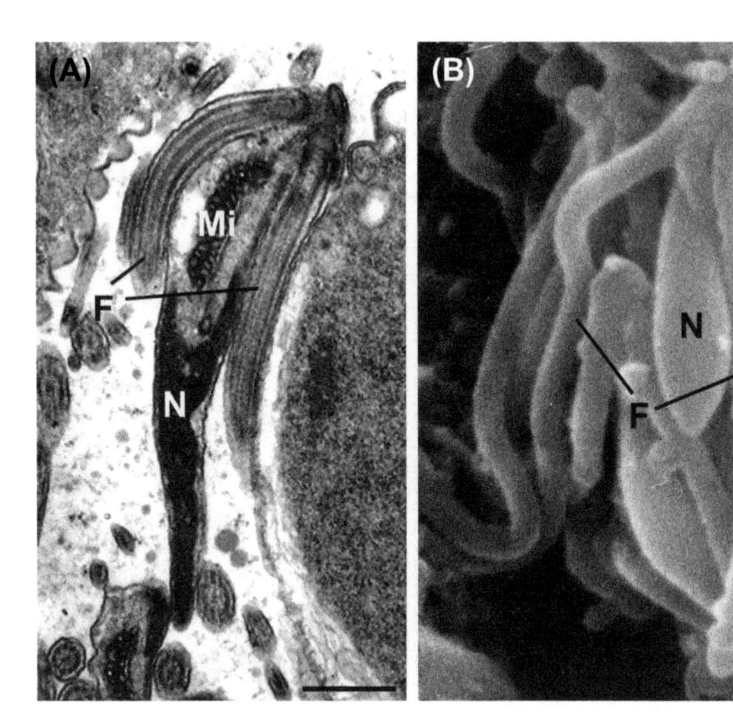

FIGURE 2.13 **The structure of the mature microgamete.**
A) Longitudinal TEM section through a microgamete showing the dense nucleus (N) and the anterior mitochondrion (Mi) and the basal bodies of the two flagella (F). Bar is 1 μm.
B) SEM of a microgamete illustrating the nucleus (N) and the two very long posteriorly pointing flagella (F). Bar is 1 μm.

metazoan centrioles, the flagella have the typical nine peripheral duplet tubules with the two central microtubules. As this flagellar growth occurs, the chromatin condensation continues at the side of the nucleus closest to the centrioles with the other part of the nucleus having a more electron lucent appearance. In addition, a mitochondrion is located adjacent to each nucleus (Fig. 2.12B). There is no significant development in the apicoplast during this process (Ferguson *et al.*, 2005). Microgamete development continues with flagellar growth and protrusion of a portion of cytoplasm containing the basal bodies, the electron dense portion of the nucleus and a mitochondrion into the lumen of the PV (Fig. 2.12B, C, D). As this occurs there is division of the nucleus with the electron dense portion separating from the electron lucent portion. The electron dense portion enters the developing microgamete and the lucent portion remains within the mother cell as a residual nucleus. The microgametocyte of *T. gondii* produces relatively few (15–30) microgametes (Fig. 2.12C). The immature microgametes are still attached to the mother cell by a narrow cytoplasmic isthmus (Fig. 2.12C, D). Maturation continues with each microgamete becoming elongated in appearance and consisting of an electron dense nucleus with a mitochondrion located between the nucleus and the basal bodies from which the two very long flagella run toward the posterior (Fig. 2.13A, B). In addition, there is an electron dense plate termed the perforatorium in the apex and four longitudinally running microtubules (Ferguson *et al.*, 1974). Once fully formed, the microgametes detach from the microgametocyte leaving a large residual cytoplasmic body.

2.2.3.2 Macrogametogony and the Macrogamete

The development of the macrogametocyte has been described in a few studies (Colley and Zaman, 1970; Ferguson *et al.*, 1975; Pelster and Piekarski, 1972). It is associated with growth of the trophozoite and the appearance of a large nucleus with dispersed chromatin and a large nucleolus but no nuclear division. As the macrogametocyte grows there is a marked increase in the size of the peripherally located mitochondrion and the centrally located apicoplast (Fig. 2.14A). In addition, a number of Golgi bodies are distributed throughout the cytoplasm. The first distinct organelle of macrogametogony is the appearance of flocculent material condensed within dilatations of the rER (Fig. 2.14B, D). This material represents the initiation of formation of the wall forming body type 2 (WFB2) so called because of their role in the formation of the oocyst walls (see below). A Golgi body is often associated with the membrane of ER surrounding the WFB2. As maturation continues, there is an increase in size and number of the WFB2 and a number of electron dense membrane bound granules appear to form from vesicles produced by the Golgi bodies (Fig. 2.14C). These were of various sizes and were termed the wall forming body type 1 (WFB1) (Fig. 2.14D). However, it has been possible using immunoelectron microscopy to identify two populations of membrane bound electron dense granules (Ferguson *et al.*, 2000). One population which appears to be involved in the formation of the outer veil is termed the veil forming bodies (VFB). These were originally termed WFB type 1a, but, with the identification of similar granules in the macrogametocyte of *Eimeria maxima,* it was proposed that VFB may be a more appropriate term (Ferguson *et al.*, 2003). The WFB1 appear slightly larger than the VFB. As the veil and wall forming bodies are being synthesized there is also the synthesis of numerous polysaccharide granules and lipid droplets and an expansion of the apicoplast (Fig. 2.14C). When fully developed, the macrogametocyte can be considered to be a mature macrogamete (Fig. 2.14C). This is not a sharp division but one of convenience to differentiate the developing stage from the mature gamete.

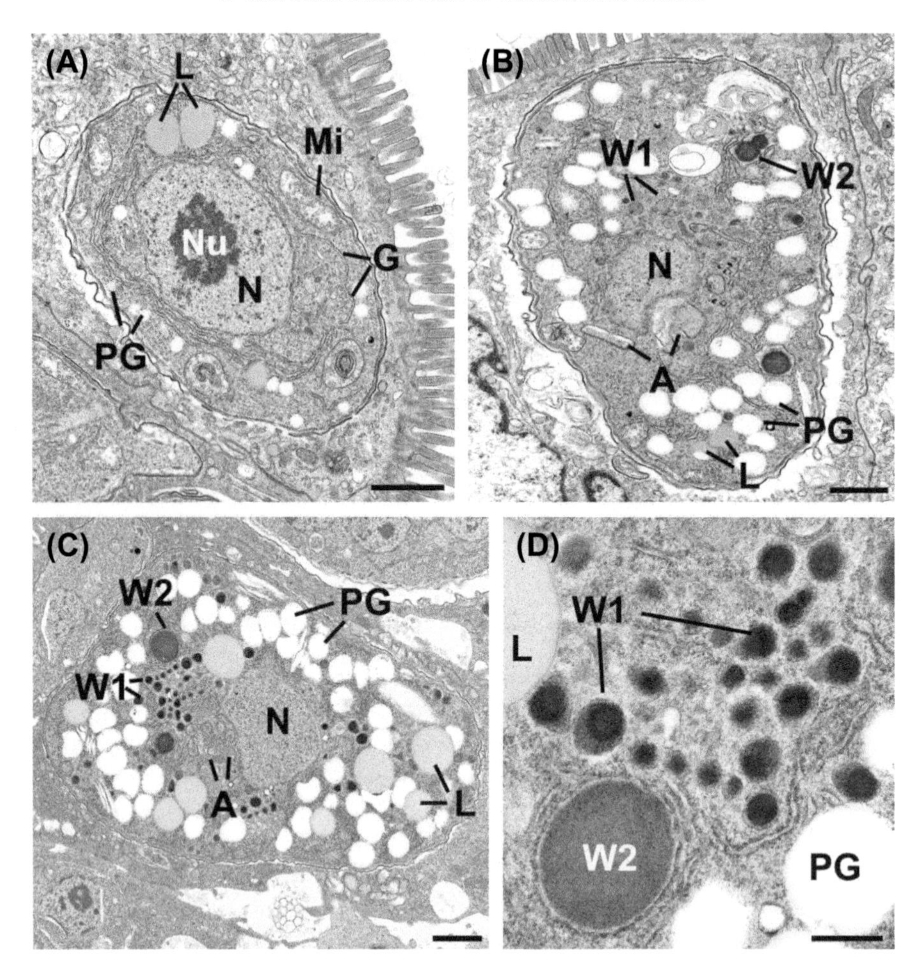

FIGURE 2.14 **Various stages in the development of the macrogametocyte.**
A) Early macrogametocyte characterized by the central nucleus (N) with a large nucleolus (Nu). The cytoplasm contains a number of profiles of an enlarged mitochondrion (Mi) and an enlarged Golgi body (G). A few polysaccharide granules (PG) and lipid droplets (L) were present in the cytoplasm. Bar is 1 μm.
B) Mid stage macrogametocyte showing increasing numbers of polysaccharide granules (PG) and lipid droplets (L) and the appearance of wall forming bodies type 1 (W1) and type 2 (W2) in the cytoplasm and an increase in size of the apicoplast (A). N, nucleus. Bar is 1 μm.
C) Mature macrogamete showing the centrally located nucleus (N) with adjacent apicoplast (A). The cytoplasm contains numerous wall forming bodies of type 1 (W1) and a few type 2 (W2) plus numerous polysaccharide granules (PG) and lipid droplets (L). Bar is 2 μm.
D) Detail of the cytoplasm of a mature macrogamete showing the numerous dense granules representing the wall forming bodies type 1 (W1). The wall forming body type 2 (W2) is located within the rough endoplasmic reticulum. PG, polysaccharide granule; L, lipid droplet. Bar is 0.5 μm.

2.2.4 Oocyst Wall Formation

The oocyst wall is a multi-layer structure, which is extremely resistant to physical and chemical insults. As such it is fundamental to the survival of the parasite. Without this wall the parasite could not survive in the external environment for the extended periods required for transmission between hosts by faecal contamination. The oocyst wall is a complex structure consisting of distinct layers (Ferguson *et al.*, 1975; Speer *et al.*, 1998). The oocyst wall is synthesized while the macrogamete is still within the host cell. In reviewing these data with reference to later observations for both *T. gondii* and *Eimeria* spp., the wall can be divided into three zones. The first is the formation of a loose outer veil consisting of 2/3 membranes (termed layer 1–3 (Ferguson *et al.*, 1975)) which is formed by the fusion of the veil forming bodies with the macrogamete plasmalemma and release of their contents (Ferguson *et al.*, 2000). This occurs during the maturation of the macrogamete. This is followed by the triggered secretion of the WFB1, which occurs simultaneously in the mature macrogamete to form the outer layer of the oocyst wall (termed layer 4 (Ferguson *et al.*, 1975)) (Fig. 2.15A). This initially forms a thick layer that undergoes polymerization to form a 30–70 nm electron dense layer. Finally, the contents of the WFB2 are released and coalesce to form the electron lucent inner layer of the oocyst wall (termed layer 5 (Ferguson *et al.*, 1975)) (Fig. 2.15B, D). The cytoplasmic mass loses the WFBs during oocyst wall formation and is characterized by a central electron lucent nucleus and cytoplasm packed with polysaccharide granules and lipid droplets (Fig. 2.15B). This process is identical to that described for the closely related genus *Eimeria* (Ferguson *et al.*, 2003). For correct formation of the oocyst wall there is a requirement for tight control and sequential secretion of the VFB and the WFB1 and 2. From the available data for the Coccidia it would be most accurate to consider the outer veil as part of the early development as it is lost by the time oocysts are released with the faeces. The oocyst proper can be considered as a double-layered structure (reviewed by Belli *et al.* 2006). The outer electron dense layer is thinner in the *T. gondii* oocyst (Fig. 2.15D) than those of *Eimeria* spp. (Belli *et al.*, 2006). The formation and polymerization of the inner layer has a dramatic effect on the ability to process the oocyst for ultrastructural examination. To date no technique has been developed that will allow the oocysts of *T. gondii* or any other Coccidian oocyst to be examined by electron microscopy. Over the past 30 years numerous attempts, using many electron microscopic fixatives and embedding protocols, have resulted in failure. The two layers provide different structural and chemical protection. The outer layer contains mostly proteins and carbohydrate and appears to provide structural strength. In contrast, the inner layer has high lipid content and appears to provide the protection from chemical insult by its impervious nature (even to electron microscopy reagents). Work on the properties of the oocyst is continuing in the closely related genus *Eimeria* (Belli *et al.*, 2006).

2.2.5 Fertilization

It would appear logical that, if sexual development takes place, there will be fusion between a microgamete and a macrogamete to form a fertilized zygote. However, this process has never been visualized. It could be expected that the mature microgametes and macrogametes are released from the host cells and fertilization takes place in the lumen. Indeed, macrogametes/oocysts with attached microgametes have been observed on rare occasions (Ferguson, 2002) (Fig. 2.15C). However, as has been described above, oocyst wall formation is initiated prior to release of the macrogamete from the host cell. An additional anomaly in *T. gondii* is the formation of very few microgametocytes producing relatively few

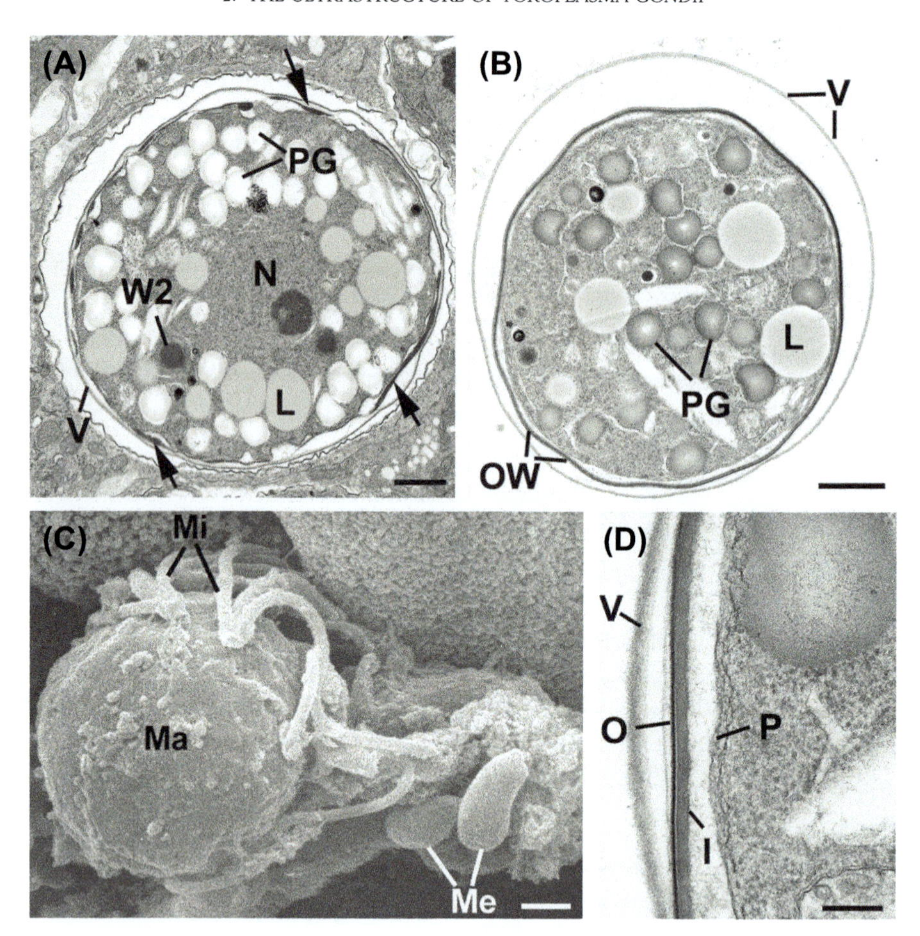

FIGURE 2.15 **Early stages of oocyst formation.**
A) Early stage of oocyst wall showing the formation of the outer veil (V) and partial formation of the outer layer of the oocyst wall (arrows). Note this is associated with the loss of the VFB and the WFB1 from the macrogamete cytoplasm while the WFB2 (W2) remains. L, lipid droplet; PG, polysaccharide granule; N, nucleus. Bar is 1 μm.
B) Newly released oocyst showing the outer veil (V) and fully formed oocyst wall (OW) enclosing a cytoplasmic mass containing polysaccharide granules (PG) and lipid droplets (L). Bar is 1 μm.
C) SEM showing a number of microgametes (Mi) apparently attach to a macrogamete/oocyst (Ma) with two adjacent merozoites (Me). Bar is 1 μm.
D) Detail of the oocyst wall consisting of the outer veil (V) plus the thin electron dense outer layer (O) and the thicker inner layer (I) which separates from the plasmalemma (P) of the cytoplasmic mass. Bar is 100 nm.

microgametes in relation to the number of macrogametes. It is a universal feature of plants and animals that there is a vast excess of male gamete formation because of the importance of ensuring maximum fertilization of the female gametes. That fertilization can occur has been proven from the identification of cross fertilized parasites (Pfefferkorn and Pfefferkorn, 1980). However, *T. gondii,* unlike most other metazoan, is normally haploid and whether this will affect the necessity for fertilization is open to question (Ferguson, 2002).

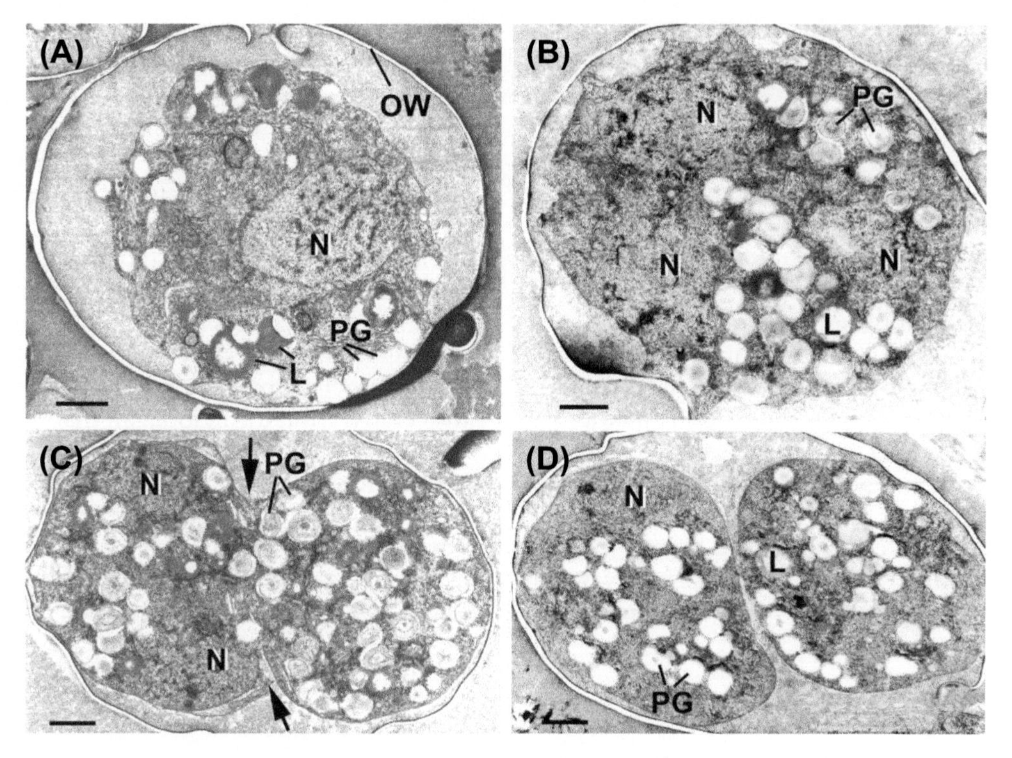

FIGURE 2.16 **Early oocyst sporulation in the external environment.**
A) Section through an unsporulated oocyst (zero hours) showing the central nucleus (N) with cytoplasm containing a Golgi body (G), mitochondria (Mi) and a number of polysaccharide granules (PG) and lipid droplets (L). Bar is 1 μm.
B) Early stage in sporulation with the cytoplasmic mass containing a number of nuclei (N). PG, polysaccharide granules; L, lipid droplets. Bar is 1 μm.
C) Section through an oocyst in which the cytoplasmic mass has started to divide (arrows) to form the two secondary sporoblasts. Note the two nuclei (N) in one of the forming sporoblasts. PG, polysaccharide granules. Bar is 1 μm.
D) Section through the two secondary sporoblasts. N, nucleus; L, lipid droplet; PG, polysaccharide granules. Bar is 1 μm.

2.2.6 Oocyst and Extracellular Sporulation

The oocyst is the only stage of *T. gondii* that is capable of undergoing extracellular development – all other development processes can only occur within viable host cells. The oocysts are excreted in an unsporulated form with a single undifferentiated cytoplasmic mass: the primary sporoblast (Fig. 2.16A). In the external environment, asexual development (sporulation) occurs, which finally results in the formation of two sporocysts, each of which contains four sporozoites. Initial attempts to study this process were unsuccessful because of our inability to process the oocyst for ultrastructural examination. It was only possible to overcome this problem by developing a technique that involved freezing and cryosectioning of the oocysts prior to processing for electron microscopy (Birch-Andersen *et al.*, 1976). The aim was to fracture the oocyst wall without destroying the cytoplasmic mass within. This technique was inefficient with destruction of a large proportion of oocysts. However, a few oocysts remained intact and these were used to examine the ultrastructural changes associated with sporulation. Due to the difficulties these studies have been limited to a series of papers on the sporulation of *E. brunetti* (Ferguson *et al.*, 1978a,b) as model for the genus *Eimeria* and *T. gondii* (Ferguson *et al.*, a–c). The quality of the ultrastructural observations is limited due to the technical difficulties, but the developmental process could be followed.

The original central cytoplasmic mass termed the primary sporoblast was similar to that of the macrogamete (Figs. 2.15B and 2.16A). The cytoplasm containing a single large nucleus plus numerous polysaccharide and lipid droplets admixed with mitochondria and rER is enclosed by a unit membrane (Fig. 2.16A). In *T. gondii,* the nucleus underwent two rounds of division giving rise to four nuclei (Fig. 2.16B). The cytoplasmic mass then underwent elongation and became limited by two additional membranes. This was followed by cytokinesis of the cytoplasmic mass with the formation of centrally located infoldings of the limiting membranes (Fig. 2.16C), which finally fused, thus dividing the primary sporoblast into two spherical secondary sporoblasts, each with two nuclei (Fig. 2.16D). This process is summarized diagrammatically in Fig. 2.17. As each secondary sporoblast develops it becomes more elongated or cigar-shaped and develops into the sporocyst, which is characterized by the formation of the sporocyst wall (Fig. 2.18A,D,E). In *T. gondii,* the wall of the sporocyst appears to be formed by material secreted from the cytoplasm that condenses on one of the limiting membranes. It forms a distinctive structure comprised of four plates (Fig. 2.18 D, E) joined by specialized sutures with an overlaying thin layer of electron dense material (Figs. 2.18B, 2.19A). This wall has various banded striations which could be

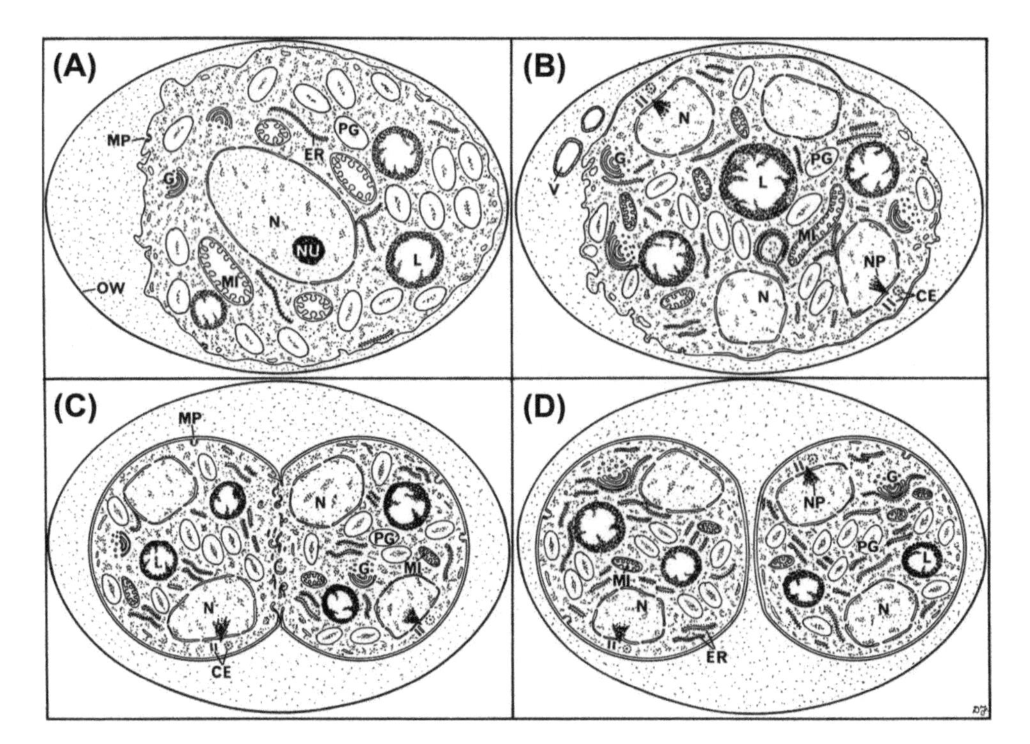

FIGURE 2.17 **A diagrammatic representation of the changes observed during the development of the zygote and formation of the sporoblasts.**

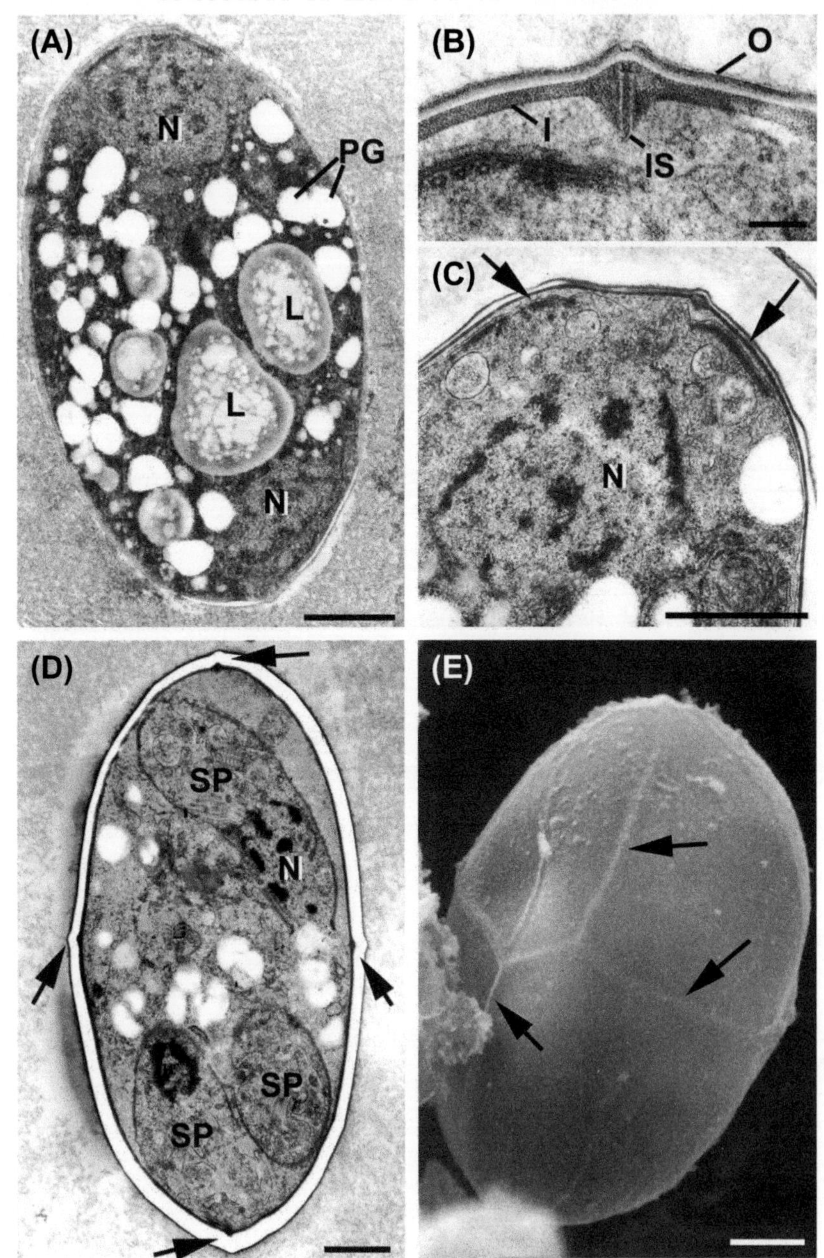

FIGURE 2.18 **Development of the sporocyst with sporozoite formation.**
A) Early development of the sporocyst showing the elongated appearance with a nucleus (N) located at either end, and the cytoplasm containing polysaccharide granules (PG) and lipid droplets (L). Bar is 1 μm.
B) Cross section through the sporocyst wall at the junction between plates showing the outer (O) and inner (I) layers. There is a swelling of the plates of the inner layer at the junction where they are joined by an intermediate strip (IS) of material. Bar is 100 nm.
C) Enlargement of part of a sporoblast showing the nucleus (N) and the two dense plaques (arrows) representing the initiation of daughter formation. Bar is 1 μm.
D) Advanced stage of sporozoite formation (SP) showing the nucleus becoming enclosed by the inner membrane complex of the daughters. The junction between the four plates of the sporocyst wall can be seen (arrows). Bar is 1 μm.
E) Scanning electron micrograph illustrating the raised junctions between the plates (arrows). Bar is 1 μm.

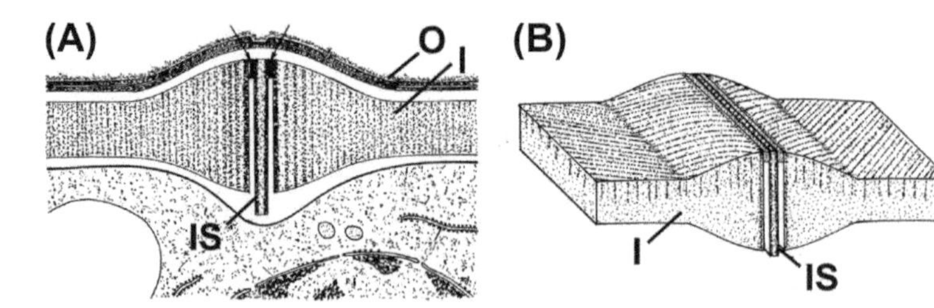

FIGURE 2.19 A diagrammatic representation of a cross section sectional (A) and the 3D (B) appearances of the sporocyst wall at the junction between the four plates which form the sporocyst. I, inner layer; IS, intermediate strip.

consistent with organized repetitive protein structures which probably provide structural strength and increase resistance to external insult (Fig. 2.19B).

Within the cytoplasm of the developing sporocyst a nucleus was observed at either end of the elongated sporocyst with the majority of the cytoplasm containing polysaccharide granules and lipid droplets (Fig. 2.18A). It was observed that the anlagen of two daughters formed adjacent to the plasmalemma above each nucleus at either end of the sporocyst (Fig. 2.18C). The process of daughter formation was similar to that observed during endodyogeny with the nucleus appearing to divide during the posterior growth of the inner membrane complex of each daughter with two daughters forming at either end of the sporocyst (Fig. 2.18C). This inner membrane growth continued until the daughters were fully formed and enclosed a nucleus, apicoplast and mitochondrion and the apical organelles (micronemes, rhoptries and dense granules). This formation of the daughters adjacent to the sporocyst plasmalemma differs from the internal formation associated with endodyogeny or endopolygeny. In this situation it was observed that the plasmalemma invaginated with the growth of the inner membrane complex to form the sporozoite pellicle and in this respect is similar to classical schizogony. This resulted in the formation of four daughters (two from each end). A small residual cytoplasmic mass remains within each sporocyst. The process of sporulation is represented diagrammatically in Fig. 2.20.

2.2.7 Excystation

There have been few ultrastructural studies on the process of excystation (Christie *et al.*, 1978; Ferguson *et al.*, 1979d; Speer *et al.*, 1998). In certain studies, the oocyst wall was broken mechanically by grinding although it has been reported that reasonable excystation can occur without this process (Speer *et al.*, 1998). Excystation is stimulated by incubation in a mixture of trypsin and bile salts (sodium taurocholate). This excystation fluid appears to act on the sporocyst wall causing increased tension, which results in an infolding of the edges of the plates along the suture lines (Fig. 2.21A, B). At the sutures, there is a separation of the inner aspect of the inner layer of the sporocyst wall which initially remained attached at the outer edge (Fig. 2.21C). This connection eventually ruptures and with it the outer membrane of the sporocyst wall (Fig. 2.21D). There appears to be rapid separation of the plates and infolding to form scroll-like structures that allow the sporozoites to escape (Fig. 2.21E).

2.3 DEVELOPMENT IN THE INTERMEDIATE HOST

2.3.1 Tachyzoite Development

When an intermediate host is infected by ingestion of tissue cysts or oocysts, the bradyzoites and sporozoites are released into the lumen of the small intestine. These invade the enterocyte or intraepithelial lymphocytes of the small intestine or pass through into the

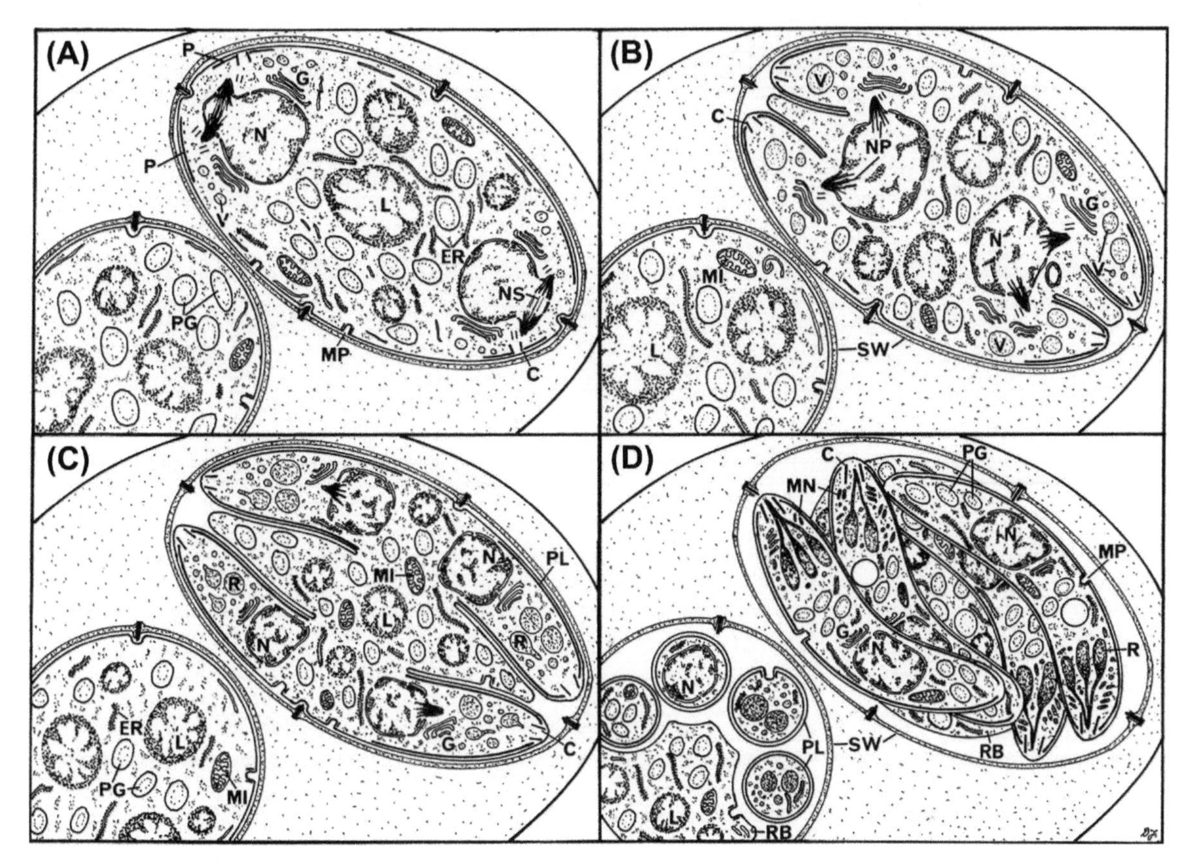

FIGURE 2.20 **A diagrammatic representation of the changes occurring during development of the sporocysts and formation of the sporozoites.**

lamina propria and invade cells there. The process of invasion has not been observed but is likely to be similar to that described previously for tachyzoites (Section 2.1.3). In either case the parasite (bradyzoite or sporozoite) defaults to tachyzoite development with formation of the characteristic parasitophorous vacuole and undergoes multiplication by endodyogeny (see Section 1.5). This process has been described in detail by Speer and Dubey (Dubey *et al.*, 1998; Speer and Dubey, 1998). From *in vitro* studies it was originally proposed that the sporozoite entered a host cell and formed an enlarged PV, which it then left to form a second vacuole within which it underwent tachyzoite development (Speer *et al.*, 1995); however, this was not observed in their *in vivo* studies and could represent an *in vitro* artefact (Dubey *et al.*, 1997). Thereafter, the tachyzoites disseminated systemically via the vascular system to all organs of the body. In the various organs they undergo proliferation by endodyogeny in many cell types and were present initially in large numbers in the lungs and spleen but by 6–10 days had invaded all organs, including the brain. However, in immunocompetent mice (genetically resistant to *Toxoplasma*), the number of lesions and tachyzoites peaked at about 12 days and by 21 days it

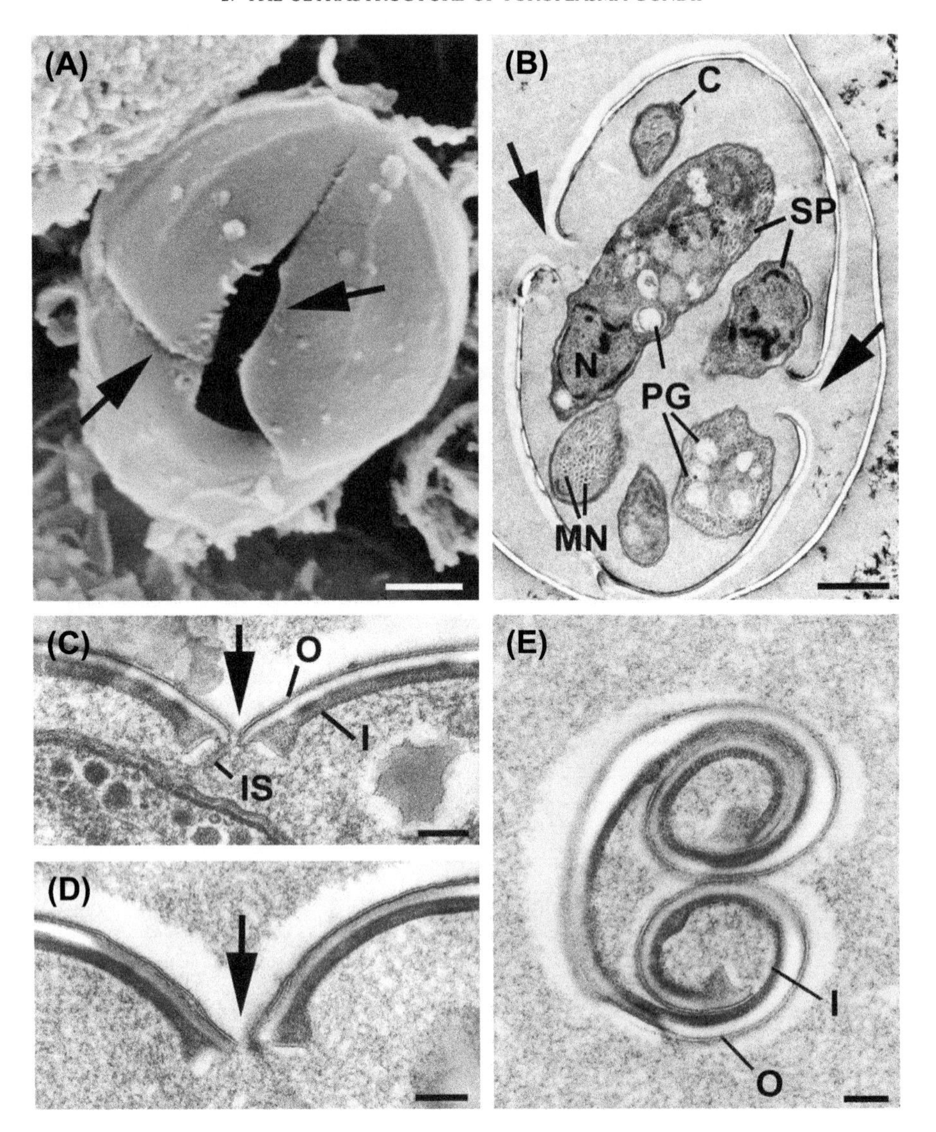

FIGURE 2.21 **Changes in the sporocyst associated with excystation.**
A) Scanning electron micrograph of a sporocyst undergoing excystation showing the separation of the plates of the sporocyst wall (arrows). Bar is 1 μm.
B) Transmission electron micrograph through an excysting sporocyst showing separation and infolding of the plates of the sporocyst wall (arrows). The sporozoites (SP) contain a posteriorly located nucleus (N) and numerous micronemes (MN) and polysaccharide granules (PG). C, conoid. Bar is 1 μm.
C) Early stage in excystation showing inward curling (arrow) of the sporocyst wall at the junction of the plates of the inner layer (I) resulting in a separation of the plates and the intermediate strip (IS). O, outer layer. Bar is 100 nm.
D) More advance stage of excystation showing separation of the inner plates and rupture of the outer layer (arrow). Bar is 200 nm.
E) Example of the continued curling of the plates of the sporocyst wall to form tightly wound scroll-like structures. Bar is 100 nm.

was difficult to identify tachyzoites in any organ, including the brain, even using immunocytochemistry.

2.3.2 Stage Conversion: Tachyzoite to Bradyzoite

There appears to be very marked tissue tropism in relation to the organs where the majority of tissue cysts are formed. The two tissues where the majority of tissue cysts are observed are the striated muscles, including the heart and the central nervous system. However, this can vary between species. For example, the majority of tissue cysts are located in the musculature in pigs (Dubey, 1986) but predominantly in the brains of mice. This again contrasts with the *in vitro* situation where almost any cell type can act as a host cell during stage conversion. This again emphasizes the need for caution when extrapolating from *in vitro* results.

Stage conversion was examined in detail in mice. At 12–15 days post-oral infection, lesions were observed in the brain which consisted of numbers of parasites undergoing tachyzoite development admixed with parasites forming early tissue cysts (Fig. 2.22A). It was observed that only a small sub-population of the tachyzoites underwent conversion. The lesions consisted of numbers of extracellular tachyzoites plus a few intracellular organisms located in typical tachyzoite-like PVs (Fig. 2.22B, D). These could often be identified as inflammatory cells, which formed part of the lesion. However, it was also possible to identify a number of early tissue cysts (Fig. 2.22C, E). A number of cysts were seen within the lesion and indeed it was possible to observe two cysts forming within the one host cell. These could be differentiated from tachyzoite-like vacuoles on the distinctive structure of the PV and PMV enclosing parasites that could be identified at the 1–2 cell stage (Fig. 2.22E). The distinctive PV appears to form at the time of invasion. This is also consistent with the immunocytochemical results using the stage specific antibodies (SAG1 and ENO2 for tachyzoites and BAG1 and ENO1 for bradyzoites). The parasitophorous vacuoles of early tissue cyst were characterized by their tight fit and being limited by a membrane with numerous irregular, shallow invaginations (Fig. 2.22E). These vacuoles lacked the tubular network but possessed a thin layer of amorphous material. In addition, there was no congregating of the host cell mitochondrion or rER around the vacuole. When examined by immunoelectron microscopy, it was found that the material beneath the membrane reacted positively to the cyst wall protein recognized by the antibody CC2. It was reported that a small sub-population of the tachyzoites from peritoneal exudates contained lucent cytoplasmic granules which were positive with CC2 (Gross *et al.*, 1996).

It is interesting to speculate that there is a sub-population of tachyzoites which, on reaching the correct environment, may be pre-programmed and are able to initiate tissue cyst formation directly. These tissue cysts continue to enlarge over the next three weeks with the bradyzoites dividing by endodyogeny (Fig. 2.23A). Initially, a large number of the parasites are undergoing endodyogeny but the proportion of dividing organism reduces during the first 28 days post-infection and from three months onward very few dividing parasites are seen (Ferguson and Hutchison, 1987a). During this early development the cyst enlarges and the depth of the invaginations increases and a more distinct underlying amorphous layer forms (Fig. 2.23E). Within the early tissue cyst the bradyzoites still have similar ultrastructural features to the tachyzoites, particularly rhoptries, which have a labyrinthine appearance (Ferguson and Hutchison, 1987b). This shows that although the structure is that of a tissue cyst and the expression of marker molecules (BAG1, ENO1, LDH2) are those of bradyzoites, the specific ultrastructural features lag behind. It was often 21–28 days before typical bradyzoites could be identified. In addition, many of the dividing

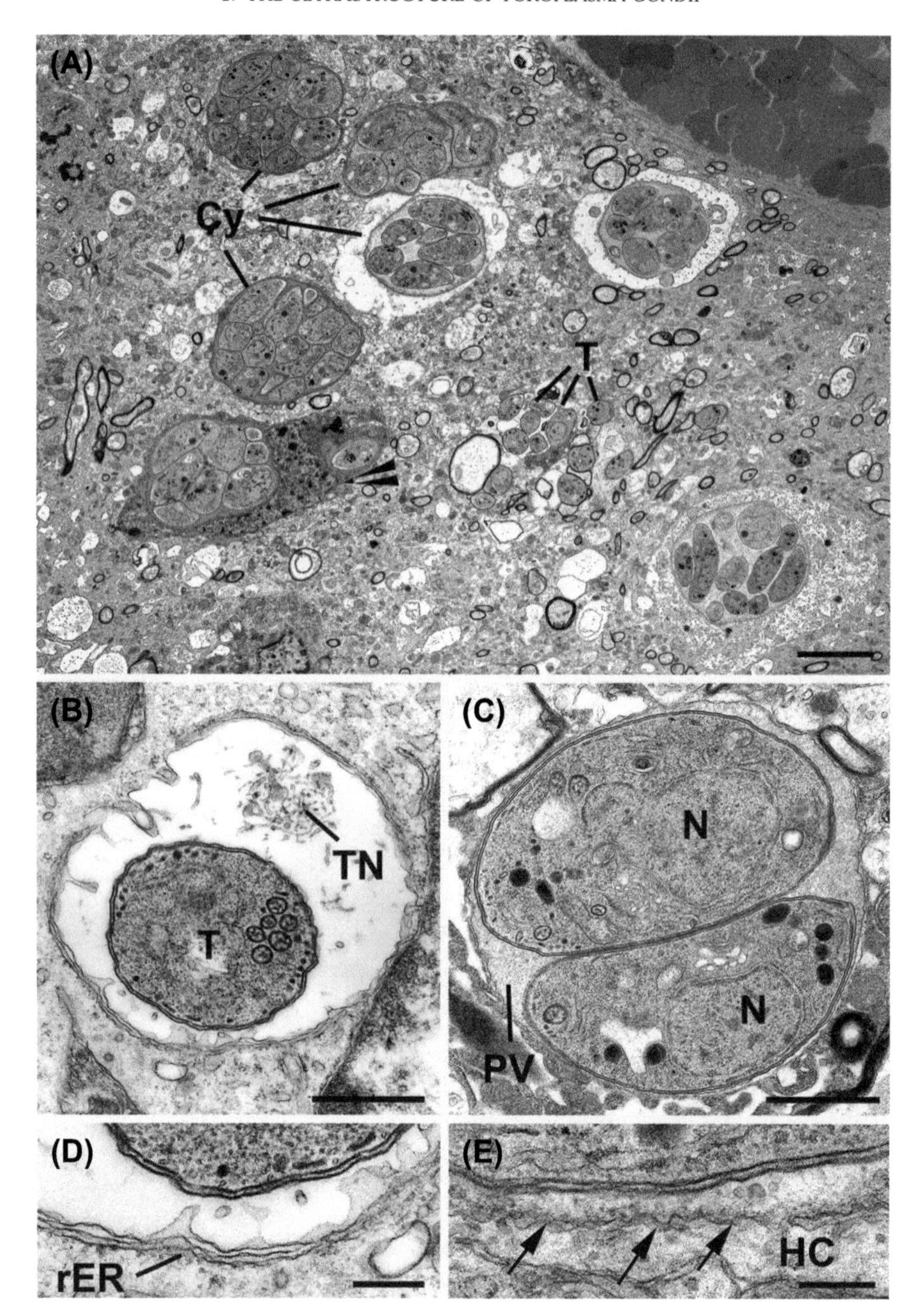

FIGURE 2.22 **Early stages of development in the mouse brain.**
A) Low magnification of a section through the brain of a mouse at 15 days post-infection showing lesion undergoing stage conversion. Note the formation of a number of early tissue cysts (Cy) and the group of tachyzoite-like organisms (T). Bar is 5 μm. B) Part of a cell in the brain showing a tachyzoite (T) within a typical tachyzoite parasitophorous vacuole with its tubular network (TN). Bar is 1 μm.

parasites had more electron lucent cytoplasm with few organelles (Fig. 2.23A).

2.3.3 Structure of the Tissue Cyst and Bradyzoite

The structure of the mature tissue cyst observed from three to 24 months post-infection (approximately the life span of the mouse) remained relatively unchanged (Fig. 2.23D). The first important observation was that throughout this period the tissue cysts were retained within a viable host cell (Fig. 2.23D, E). It was originally thought that the mature cysts were extracellular. However, on ultrastructural examination, a thin rim of host cell cytoplasm could be observed enclosing the tissue cysts (Ferguson and Hutchison, 1987B). This may explain the lack of an immune response to the tissue cysts; they are masked by the host cell. With the limited host cytoplasm available it is difficult to identify the cell type. However, in the majority of cases, the host cells could be identified as neurones because of the defining presence of synapses (Fig. 2.23E). It is not possible to identify the other host cells although their features are consistent with neurones. There are variations in the thickness of the cyst wall between tissue cysts with some showing deep invaginations of the limiting membrane forming a complex network of interconnecting channels all embedded in the homogeneous granular material. There is also some vesicle formation on the inner aspect (Fig. 2.23E) (Ferguson and Hutchison, 1987a). In the mature cysts, the bradyzoites appeared more elongated than the tachyzoites with a posteriorly located nucleus (Fig. 2.23B). There were numerous micronemes and few dense granules although this was variable. The rhoptries had more bulbous ends and were electron dense. The major difference was the presence of numerous polysaccharide granules (Fig. 2.23B, d).

The majority of tissue cysts appear as a single structure but it is possible to find small groups of tissue cysts of different sizes. It has been suggested, from the very early studies (Lainson, 1958), that this could represent daughter cyst formation resulting from the escape of individual bradyzoites to form new cysts. However, extensive immunocytochemical and ultrastructural studies fail to find supporting evidence. It is possible to observe the formation of groups of cysts from as early as 14 days post-infection (Fig. 2.22A). In addition, when examined by electron microscopy, the cysts, although of different sizes, appear to be of similar maturity. The bradyzoites in all the cysts appeared as mature organisms with no evidence of the feature described above for immature cysts. In addition, given the immunological response of the host to exposed parasite antigens, it would be expected that any escaped bradyzoite would invoke an inflammatory response (see Section 2.3.5 below).

2.3.4 Inflammatory Changes in the Brains of Infected Mice

Toxoplasma infection of the brain was associated with inflammatory changes in around the meninges and certain blood vessels within the brain (Ferguson *et al.*, 1991). During the acute phase many of the inflammatory cells were

C) Example of a very early tissue cyst containing two parasites contained within a tight fitting vacuole (PV). N, nucleus. Bar is 1 μm.

D) Detail from part B showing the strand of rough endoplasmic reticulum (rER) associated with the parasitophorous vacuole membrane characteristic of a tachyzoite containing vacuole. Bar is 100 nm.

E) Detail from part C showing the undulating parasitophorous vacuole membrane (arrows) and the absence of host cell organelles associated with the membrane. HC, host cell. Bar is 100 nm.

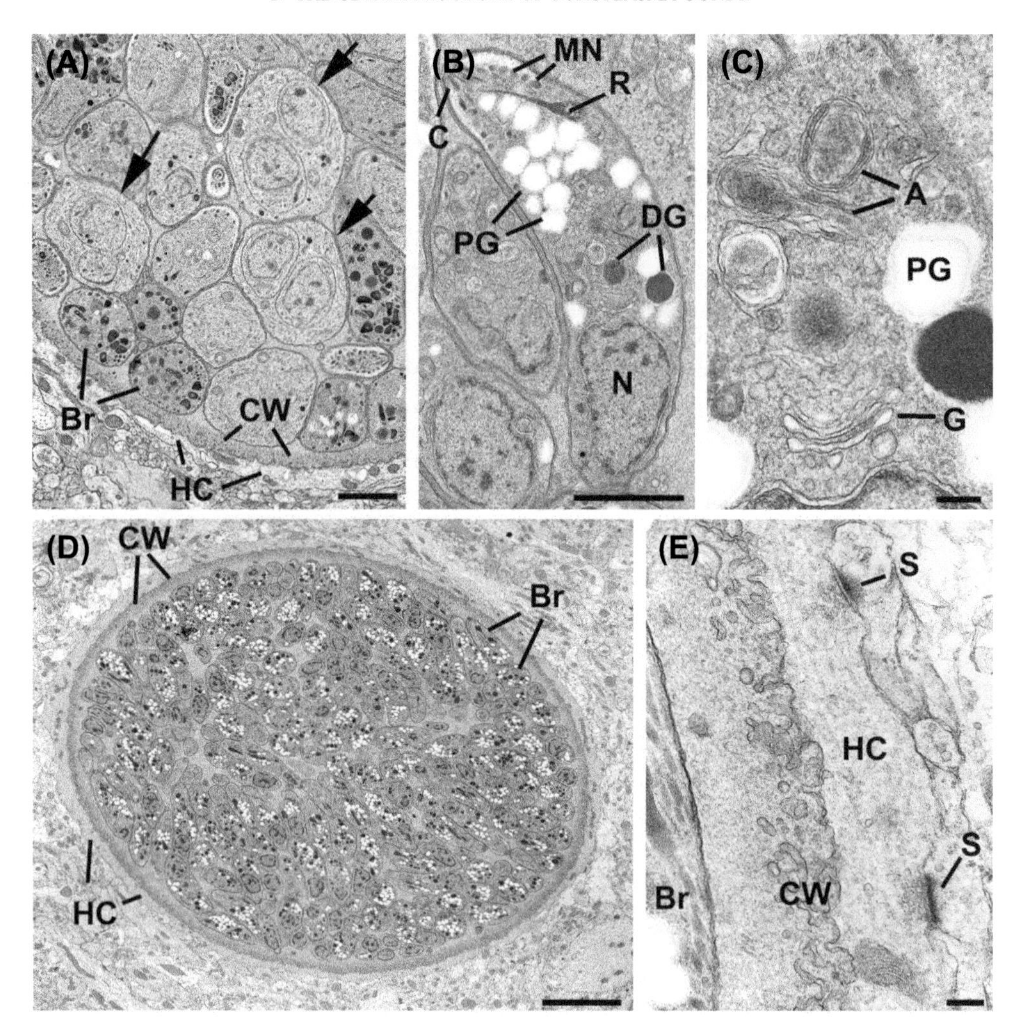

FIGURE 2.23 **Mature tissue cysts and bradyzoites in mouse brain.**
A) Part of the periphery of an early tissue cyst (21 days post-infection) showing the cyst wall (CW) and the thin rim of host cell (HC) cytoplasm. The cyst contains a number of bradyzoites (Br) and electron lucent organisms undergoing endodyogeny (arrows). Bar is 1 μm.
B) Longitudinal sectioned bradyzoite showing the posteriorly located nucleus, numerous polysaccharide granules (PG) plus dense granules (DG), rhoptries (R), micronemes (MN) and conoid (C). Bar is 1 μm.
C) Detail of the region just anterior to the nucleus of the bradyzoite in part B showing the Golgi body (G) and the apicoplast (A). PG, polysaccharide granule. Bar is 100 nm.
D) Low power of a mature tissue cyst (1 year post-infection) showing a large number of bradyzoites (Br) enclosed by a cyst wall retained within a host cell (HC). Bar is 10 μm.
E) Detail from the periphery of the cyst in part d showing the cyst wall (CW) with deep invaginations of the limiting membrane of the cyst into the underlying granular material. Note the host cell (HC) can be definitively identified as a neurone because of the presence of synapses (S). Bar is 100 nm.

lymphocytes and monocytes with a few polymorphic leucocytes. In chronically infected mice, with numerous tissue cysts in the brain, there was the continuous presence of inflammatory cells which cuff the small blood vessels within the neuropil and the vessels of the meninges (Fig. 2.24A, B, C). However, it should be noted that these inflammatory cells show no tropism toward the tissue cysts (Fig. 2.24B). In the chronic infections (three to 24 months post-infection), the majority of inflammatory cells were plasma cells or monocytes/macrophages (Fig. 2.24C). In addition, it is was possible to observe numerous plasma cells around the small vessels and free in the neuropil (Fig. 2.24A).

2.3.5 Cyst Rupture in Immunocompetent Hosts

One of the clinical problems in immunocompromised hosts was recrudescence of the infection resulting in stage conversion back to actively proliferating and tissue destructive tachyzoites. To examine the situation in an

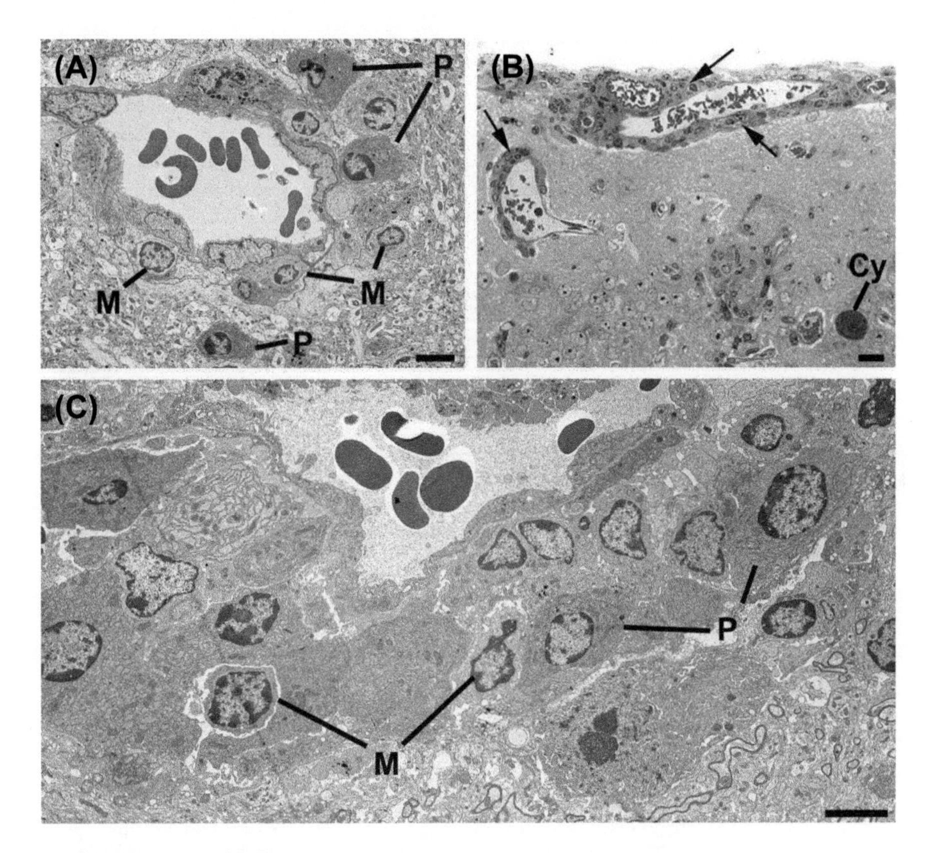

FIGURE 2.24 **Inflammatory cell infiltration of chronically infected mouse brain.**
A) Small blood vessel from the brain of a chronically infected mouse showing a number of monocytes (M) and plasma cells (P) cuffing the vessel and also a plasma cell in the neuropil. Bar is 10 μm.
B) Light micrograph of the brain of a chronically infected mouse showing numerous inflammatory cells within the meninges and cuffing the blood vessels (arrows). Note the tissue cyst (Cy) invokes no reaction. Bar is 20 μm.
C) Electron micrograph of a large vessel close to the meninges showing the large number of cuffing monocytes (M) and plasma cells (P). Bar is 10 μm.

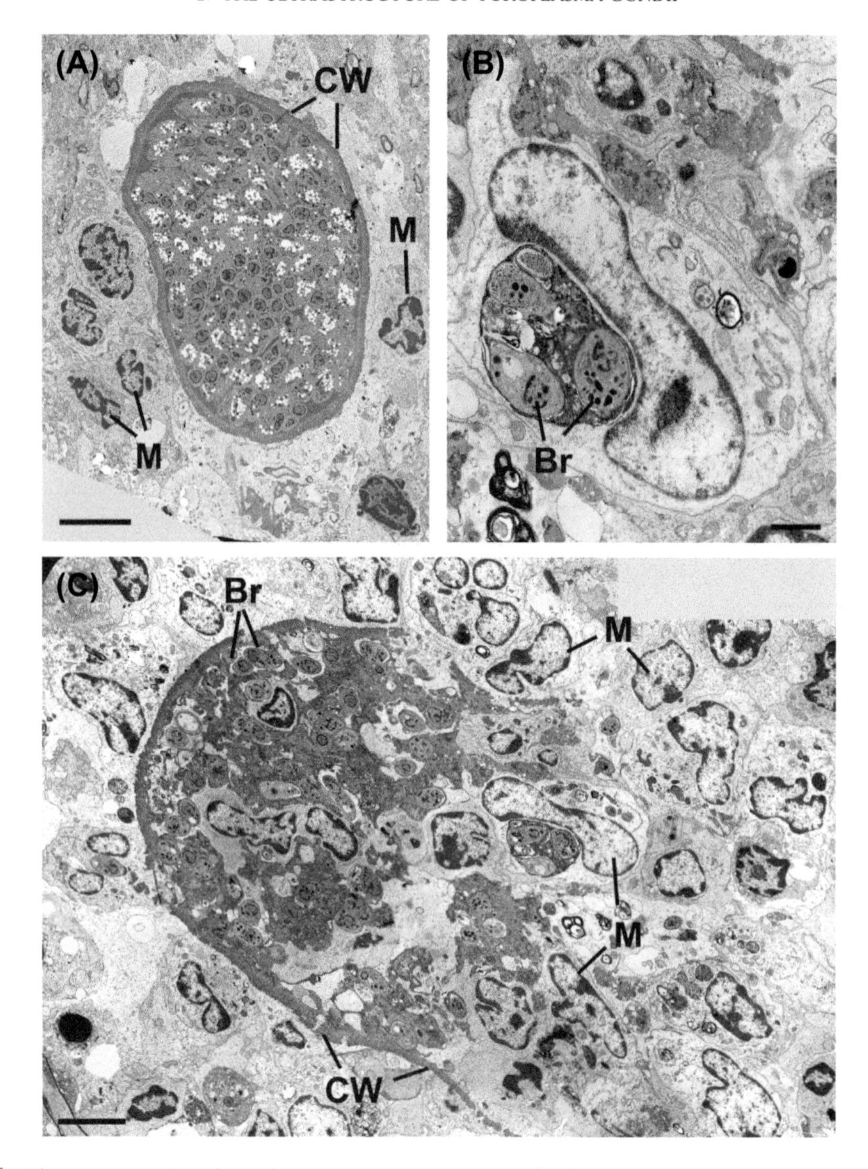

FIGURE 2.25 **Tissue cyst rupture in an immune competent mouse brain.**
A) Low power image of a tissue cyst with an intact cyst wall (CW) but with loss of the host cell. Note the number of inflammatory cells, monocytes (M), surrounding the cyst. Bar is 10 μm.
B) Detail from the tissue cyst in part C showing a macrophage with a phagocytic vacuole containing degenerating bradyzoites (Br). Bar is 1 μm.
C) Section through a ruptured tissue cyst in an immunocompetent host showing the fractured cyst wall (CW) partially enclosing the bradyzoites (Br). Note the numerous macrophages (M) surrounding and invading into the tissue cyst and phagocytizing the bradyzoites. Bar is 5 μm.

immunocompetent host, the brains of immunocompetent, chronically infected mice were examined. It was found that, indeed, a very small percentage of tissue cysts were rupturing at any given time during chronic infections (Ferguson *et al.*, 1989). The initial change appeared to be death of the host cell. With the exposure of the parasite antigens in the cyst wall there was evidence for a rapid and massive cell mediated immune response involving numerous inflammatory cells (monocytes and even neutrophils). These were observed around the still apparently intact cyst (Fig. 2.25A). With the rupture of the cyst wall there was further influx of macrophages into the cyst (Fig. 2.25C). The macrophages phagocytosed the bradyzoites where heterophagosomes were formed, resulting in destruction of the parasites (Fig. 2.25B). This resulted in the formation of small inflammatory lesions (microglia nodules) with some evidence of host cell apoptosis but limited host tissue damage. The bradyzoites appeared to be destroyed before they could undergo replication or stage conversion to tachyzoites. These observations were confirmed by immunocytochemical examination of the lesions.

2.3.6 Development *in Vitro*

It is often stated that *T. gondii* is very easy to culture and this has made it very popular as a molecular model. However, it needs to be emphasized that normally only the tachyzoite and tachyzoite development occurs in cell cultures. More recently it has been possible to trigger tissue cyst formation but, to date, there has not been a single successful attempt to reproduce development undergone by the coccidian stages.

2.3.6.1 *Tachyzoite Development* in Vitro

In vitro, tachyzoites undergo similar development to that described above (Sections 1.4 and 1.5) irrespective of the type of host cell used. The host/parasite relationship and the proliferation by endodyogeny and repeated endodyogeny as identical to that described above.

2.3.6.2 *Bradyzoite Development* in Vitro

This was first described in astrocytes in the 1980s (Jones *et al.*, 1986) and techniques for inducing this stage conversion were identified in the 1990s. It was observed that factors (pH changes, oxygen tension) that induce stress in the culture system appear to stimulate conversion to cyst formation. Unlike the *in vivo* situation where there is marked selection of the type of host cell (neurones, muscle cells), *in vitro* almost any cell type could act as a host cell for cyst formation. In certain *in vitro* studies (Ferguson, unpublished), the process of cyst formation from the time the parasite entered the host cell was similar to that observed *in vivo*. However, other studies suggest that there may be some tachyzoite-like development before the vacuole takes on the features of the tissue cyst (Soete *et al.*, 1994).

References

Aikawa, M., Hepler, P.K., Huff, C.G., Sprinz, H., 1966. The feeding mechanism of avian malarial parasites. J. Cell Biol. 28, 355–373.

Aikawa, M., Miller, L.H., Rabbege, J.R., Epstein, N., 1981. Freeze-fracture study on the erythrocyte membrane during malarial parasite invasion. J. Cell Biol. 91, 55–62.

Alexander, D.L., Mital, J., Ward, G.E., Bradley, P.J., Boothroyd, J.C., 2005. Identification of the moving junction complex of *Toxoplasma gondii*: a collaboration between distinct secretory organelles. PLoS pathog. 1 (2), e17.

Beisson, J., Lefort-Tran, M., Pouphile, M., Rossignol, M., Satir, B., 1976. Genetic analysis of membrane differentiation in *Paramecium*. Freeze-fracture study of the trichocyst cycle in wild-type and mutant strains. J. Cell Biol. 69, 126–143.

Belli, S.I., Smith, N.C., Ferguson, D.J.P., 2006. The coccidian oocyts – a tough nut to crack. Trends Parasitol. 22 (9), 416–423.

Birch-Andersen, A., Ferguson, D.J., Pontefract, R.D., 1976. A technique for obtaining thin sections of coccidian oocysts. Acta. Pathol. Microbiol. Scand. [B] 84, 235–239.

Bradley, P.J., Ward, C., Cheng, S.J., Alexander, D.L., Coller, S., Coombs, G.H., Dunn, J.D., Ferguson, D.J., Sanderson, S.J., Wastling, J.M., Boothroyd, J.C., 2005. Proteomic analysis of rhoptry organelles reveals many novel constituents for host-parasite interactions in *Toxoplasma gondii*. J. Biol. Chem. 280, 34245–34258.

Carruthers, V.B., Sibley, L.D., 1999. Mobilization of intracellular calcium stimulates microneme discharge in *Toxoplasma gondii*. Mol. Microbiol. 31, 421–428.

Cesbron-Delauw, M.F., 1994. Dense-granule organelles of *Toxoplasma gondii*: their role in the host-parasite relationship. Parasitol. Today 10, 293–296.

Christie, E., Pappas, P.W., Dubey, J.P., 1978. Ultrastructure of excystment of *Toxoplasma gondii* oocysts. J. Protozool. 25, 438–443.

Colley, F.C., Zaman, V., 1970. Observations on the endogenous stages of *Toxoplasma gondii* in the cat ileum. II Electron microscope study. South East Asian J. Trop. Med. Pub. Health 1, 456–480.

Coppin, A., Varre, J.S., Lienard, L., Dauvillee, D., Guerardel, Y., Soyer-Gobillard, M.O., Buleon, A., Ball, S., Tomavo, S., 2005. Evolution of plant-like crystalline storage polysaccharide in the protozoan parasite *Toxoplasma gondii* argues for a red alga ancestry. J. Mol. Evol. 60, 257–267.

Dubey, J.P., 1986. A review of toxoplasmosis in pigs. Vet. Parasitol. 19, 181–223.

Dubey, J.P., 1997. Bradyzoite-induced murine toxoplasmosis: stage conversion, pathogenesis and tissue cyst formation in mice fed bradyzoites of different strains of *Toxoplasma gondii*. J. Eukaryot Microbiol. 44, 592–602.

Dubey, J.P., Lindsay, D.S., Speer, C.A., 1998. Structures of *Toxoplasma gondii* tachyzoites, bradyzoites and sporozoites and biology and development of tissue cysts. Clin. Microbiol. Rev. 11, 267–299.

Dubey, J.P., Speer, C.A., Shen, S.K., Kwok, O.C., Blixt, J.A., 1997. Oocyst-induced murine toxoplasmosis: life cycle, pathogenicity and stage conversion in mice fed *Toxoplasma gondii* oocysts. J. Parasitol. 83, 870–882.

Dubremetz, J.F., 1973. Etude ultrastructurale de la mitose schizogonique chez la coccidie *Eimeria necatrix* (Johnson 1930). J. Ultrastr. Res. 42, 354–376.

Dubremetz, J.F., 1975. La genèse des Mérozoïtes chez la coccidie *Eimeria necatrix*. Etude Ultrastructurale. J. Protozool. 22, 71–84.

Dubremetz, J.F., Achbarou, A., Bermudes, D., Joiner, K.A., 1993. Kinetics and pattern of organelle exocytosis during *Toxoplasma gondii*/host-cell interaction. Parasitol. Res. 79, 402–408.

Ferguson, D., 2002. Toxoplasma gondii and sex: essential or optional extra? Trends Parasitol. 18, 351.

Ferguson, D.J., 2004. Use of molecular and ultrastructural markers to evaluate stage conversion of *Toxoplasma gondii* in both the intermediate and definitive host. Int. J. Parasitol. 34, 347–360.

Ferguson, D.J., Belli, S.I., Smith, N.C., Wallach, M.G., 2003. The development of the macrogamete and oocyst wall in *Eimeria maxima*: immuno-light and electron microscopy. Int. J. Parasitol. 33, 1329–1340.

Ferguson, D.J., Birch-Andersen, A., Hutchinson, W.M., Siim, J.C., 1980. Ultrastructural observations showing enteric multiplication of *Cystoisospora (Isospora) felis* by endodyogeny. Z. Parasitenkd. 63, 289–291.

Ferguson, D.J., Birch-Andersen, A., Hutchison, W.M., Siim, J.C., 1976. Ultrastructural studies on the endogenous development of *Eimeria brunetti*. Acta. Pathol. Microbiol. Scand. [B] 84B, 401–413.

Ferguson, D.J., Birch-Andersen, A., Hutchison, W.M., Siim, J.C., 1978a. Light and electron microscopy on the sporulation of the oocysts of *Eimeria brunetti*. I. Development of the zygote and formation of the sporoblasts. Acta. Pathol. Microbiol. Scand. [B] 86, 1–11.

Ferguson, D.J., Birch-Andersen, A., Hutchison, W.M., Siim, J.C., 1978b. Light and electron microscopy on the sporulation of the oocysts of *Eimeria brunetti*. II. Development into the sporocyst and formation of the sporozoite. Acta. Pathol. Microbiol. Scand. [B] 86, 13–24.

Ferguson, D.J., Birch-Andersen, A., Siim, J.C., Hutchison, W.M., 1979a. Ultrastructural studies on the sporulation of oocysts of *Toxoplasma gondii*. I. Development of the zygote and formation of the sporoblasts. Acta. Pathol. Microbiol. Scand. [B] 87B, 171–181.

Ferguson, D.J., Birch-Andersen, A., Siim, J.C., Hutchison, W.M., 1979b. Ultrastructural studies on the sporulation of oocysts of *Toxoplasma gondii*. II. Formation of the sporocyst and structure of the sporocyst wall. Acta. Pathol. Microbiol. Scand. [B] 87B, 183–190.

Ferguson, D.J., Birch-Andersen, A., Siim, J.C., Hutchison, W.M., 1979c. Ultrastructural studies on the sporulation of oocysts of *Toxoplasma gondii*. III. Formation of the sporozoites within the sporocysts. Acta. Pathol. Microbiol. Scand. [B] 87, 253–260.

Ferguson, D.J., Birch-Andersen, A., Siim, J.C., Hutchison, W.M., 1979d. An ultrastructural study on the excystation of the sporozoites of *Toxoplasma gondii*. Acta. Pathol. Microbiol. Scand. [B] 87, 277–283.

Ferguson, D.J., Brecht, S., Soldati, D., 2000. The microneme protein MIC4, or an MIC4-like protein, is expressed within the macrogamete and associated with oocyst wall formation in *Toxoplasma gondii*. Int. J. Parasitol. 30, 1203–1209.

Ferguson, D.J., Cesbron-Delauw, M.F., Dubremetz, J.F., Sibley, L.D., Joiner, K.A., Wright, S., 1999a. The expression and distribution of dense granule proteins in the enteric (Coccidian) forms of *Toxoplasma gondii* in the small intestine of the cat. Exp. Parasitol. 91, 203–211.

Ferguson, D.J., Graham, D.I., Hutchison, W.M., 1991. Pathological changes in the brains of mice infected with *Toxoplasma gondii*: a histological, immunocytochemical and ultrastructural study. Int. J. Exp. Pathol. 72, 463–474.

Ferguson, D.J., Henriquez, F.L., Kirisits, M.J., Muench, S.P., Prigge, S.T., Rice, D.W., Roberts, C.W., McLeod, R.L., 2005. Maternal inheritance and stage-specific variation of the apicoplast in *Toxoplasma gondii* during development in the intermediate and definitive host. Eukaryot Cell 4, 814–826.

Ferguson, D.J., Hutchison, W.M., 1981. Comparison of the development of avirulent and virulent strains of *Toxoplasma gondii* in the peritoneal exudate of mice. Ann. Trop. Med. Parasitol. 75, 539–546.

Ferguson, D.J., Hutchison, W.M., 1987a. The host-parasite relationship of *Toxoplasma gondii* in the brains of chronically infected mice. Virchows Arch. A Pathol. Anat. Histopathol. 411, 39–43.

Ferguson, D.J., Hutchison, W.M., 1987b. An ultrastructural study of the early development and tissue cyst formation of *Toxoplasma gondii* in the brains of mice. Parasitol. Res. 73, 483–491.

Ferguson, D.J., Hutchison, W.M., Dunachie, J.F., Siim, J.C., 1974. Ultrastructural study of early stages of asexual multiplication and microgametogony of *Toxoplasma gondii* in the small intestine of the cat. Acta. Pathol. Microbiol. Scand [B] Microbiol. Immunol. 82, 167–181.

Ferguson, D.J., Hutchison, W.M., Pettersen, E., 1989. Tissue cyst rupture in mice chronically infected with *Toxoplasma gondii*. An immunocytochemical and ultrastructural study. Parasitol. Res. 75, 599–603.

Ferguson, D.J., Hutchison, W.M., Siim, J.C., 1975. The ultrastructural development of the macrogamete and formation of the oocyst wall of *Toxoplasma gondii*. Acta. Pathol. Microbiol. Scand. [B] 83, 491–505.

Ferguson, D.J., Jacobs, D., Saman, E., Dubremetz, J.F., Wright, S.E., 1999b. In vivo expression and distribution of dense granule protein 7 (GRA7) in the exoenteric (tachyzoite, bradyzoite) and enteric (coccidian) forms of *Toxoplasma gondii*. Parasitology 119 (Pt 3), 259–265.

Francia, M.E., Jordan, C.N., Patel, J.D., Sheiner, L., Demerly, J.L., Fellows, J.D., de Leon, J.C., Morrissette, N.S., Dubremetz, J-F., Striepen, B., 2012. Cell division in apicomplexan parasites is organized by a homolog of the striated fiber of algal flagella. PLoS 10 (12), e1001444. http://dx.doi.org/10.1371/Journal.pbio.1001444.

Gross, U., Bohne, W., Soete, M., Dubremetz, J.F., 1996. Developmental differentiation between tachyzoites and bradyzoites of *Toxoplasma gondii*. Parasitol. Today 12, 30–33.

Hakansson, S., Charron, A.J., Sibley, L.D., 2001. *Toxoplasma* evacuoles: a two-step process of secretion and fusion forms the parasitophorous vacuole. Embo. J. 20, 3132–3144.

Hu, K., Mann, T., Striepen, B., Beckers, C.J., Roos, D.S., Murray, J.M., 2002a. Daughter cell assembly in the protozoan parasite *Toxoplasma gondii*. Mol. Biol. Cell. 13, 593–606.

Hu, K., Roos, D.S., Murray, J.M., 2002b. A novel polymer of tubulin forms the conoid of *Toxoplasma gondii*. J. Cell Biol. 156, 1039–1050.

Jones, T.C., Bienz, K.A., Erb, P., 1986. In vitro cultivation of *Toxoplasma gondii* cysts in astrocytes in the presence of gamma interferon. Infect. Immun. 51, 147–156.

Jones, T.C., Hirsch, J.G., 1972. The interaction between *Toxoplasma gondii* and mammalian cells. II. The absence of lysosomal fusion with phagocytic vacuoles containing living parasites. J. Exp. Med. 136, 1173–1194.

Jones, T.C., Yeh, S., Hirsch, J.G., 1972. The interaction between *Toxoplasma gondii* and mammalian cells. I. Mechanism of entry and intracellular fate of the parasite. J. Exp. Med. 136, 1157–1172.

Kohler, S., 2005. Multi-membrane-bound structures of Apicomplexa: I. The architecture of the *Toxoplasma gondii* apicoplast. Parasitol. Res. 96, 258–272.

Kohler, S., Delwiche, C.F., Denny, P.W., Tilney, L.G., Webster, P., Wilson, R.J., Palmer, J.D., Roos, D.S., 1997. A plastid of probable green algal origin in Apicomplexan parasites. Science 275, 1485–1489.

Lainson, R., 1958. Observations on the development and nature of pseudocysts and cysts of *Toxoplasma gondii*. Trans. R. Soc. Trop. Med. Hyg. 52, 396–407.

Lebrun, M., Michelin, A., El Hajj, H., Poncet, J., Bradley, P.J., Vial, H., Dubremetz, J.F., 2005. The rhoptry neck protein RON4 relocalizes at the moving junction during *Toxoplasma gondii* invasion. Cell. Microbiol. 7, 1823–1833.

Mann, T., Beckers, C., 2001. Characterization of the subpellicular network, a filamentous membrane skeletal component in the parasite *Toxoplasma gondii*. Mol. Biochem. Parasitol. 115, 257–268.

Mercier, C., Dubremetz, J.F., Rauscher, B., Lecordier, L., Sibley, L.D., Cesbron-Delauw, M.F., 2002. Biogenesis of nanotubular network in *Toxoplasma* parasitophorous vacuole induced by parasite proteins. Mol. Biol. Cell. 13, 2397–2409.

Michel, R., Schupp, K., Raether, W., Bierther, F.W., 1979. Formation of a close junction during invasion of erythrocytes by *Toxoplasma gondii* in Vitro. Int. J. Parasitol. 10, 309–313.

Miranda, K., Pace, D.A., Cintron, R., Rodrigues, J.C., Fang, J., Smith, A., Rohloff, P., Coelho, E., de Haas, F., de Souza, W., Coppens, I., Sibley, L.D., Moreno, S.N., 2010. Characterization of a novel organelle in Toxoplasma gondii with similar composition and function to the plant vacuole. Mol. Microbiol. 76 (6), 1358–1375. http://dx.doi.org/10.1111/j.1365-2958.2010.07165.x.

Moreno, S.N.J., Zhong, L., 1996. Acidocalcisomes in *Toxoplasma gondii* tachyzoites. Biochem. J. 313, 655–659.

Morrissette, N.S., Murray, J.M., Roos, D.S., 1997. Subpellicular microtubules associate with an intramembranous particle lattice in the protozoan parasite *Toxoplasma gondii*. J. Cell Sci. 110 (Pt 1), 35–42.

Nichols, B.A., Chiappino, M.L., 1987. Cytoskeleton of *Toxoplasma gondii*. J. Protozool. 34, 217–226.

Nichols, B.A., Chiappino, M.L., O'Connor, G.R., 1983. Secretion from the rhoptries of *Toxoplasma gondii* during host-cell invasion. J. Ultrastruct. Res. 83, 85–98.

Nichols, B.A., Chiappino, M.L., Pavesio, C.E., 1994. Endocytosis at the micropore of *Toxoplasma gondii*. Parasitol. Res. 80, 91–98.

Ogino, N., Yoneda, C., 1966. The fine structure and mode of division of *Toxoplasma gondii*. Arch. Ophthalmol. 75, 218–227.

Pelletier, L., Stern, C.A., Pypaert, M., Sheff, D., Ngo, H.M., Roper, N., He, C.Y., Hu, K., Toomre, D., Coppens, I., Roos, D.S., Joiner, K.A., Warren, G., 2002. Golgi biogenesis in *Toxoplasma gondii*. Nature 418, 548–552.

Pelster, B., Piekarski, G., 1971. Electron microscopical studies on the microgametogeny of *Toxoplasma gondii*. Z. Parasitenkd. 37, 267–277.

Pelster, B., Piekarski, G., 1972. Ultrastructure of the macrogametes in *Toxoplasma gondii*. Z. Parasitenkd. 39, 225–232.

Pfefferkorn, L.C., Pfefferkorn, E.R., 1980. *Toxoplasma gondii*: genetic recombination between drug resistant mutants. Exp. Parasitol. 50, 305–316.

Piekarski, G., Pelster, B., Witte, H.M., 1971. Endopolygeny in *Toxoplasma gondii*. Z. Parasitenkd. 36, 122–130.

Porchet, E., Torpier, G., 1977. Etude du germe infectieux de *Sarcocystis tenella* et *Toxoplasma gondii* par la technique du cryodecapage. Z. Parasitenkd. 54, 101–124.

Porchet-Hennere, E., Torpier, G., 1983. Relations entre *Toxoplasma* et sa cellule-hôte. Protistologica 19, 357–370.

Sheffield, H.G., Melton, M.L., 1968. The fine structure and reproduction of *Toxoplasma gondii*. J. Parasitol. 54, 209–226.

Sibley, L.D., Niesman, I.R., Parmley, S.F., Cesbron-Delauw, M.F., 1995. Regulated secretion of multi-lamellar vesicles leads to formation of a tubulovesicular network in host-cell vacuoles occupied by *Toxoplasma gondii*. J. Cell Sci. 108, 1669–1677.

Sinai, A.P., Webster, P., Joiner, K.A., 1997. Association of host cell endoplasmic reticulum and mitochondria with the *Toxoplasma gondii* parasitophorous vacuole membrane: a high affinity interaction. J Cell. Sci 110 (Pt 17), 2117–2128.

Soete, M., Camus, D., Dubremetz, J.F., 1994. Experimental induction of bradyzoite-specific antigen expression and cyst formation by the RH strain of *Toxoplasma gondii* in vitro. Exp. Parasitol. 78, 361–370.

Speer, C.A., Clark, S., Dubey, J.P., 1998. Ultrastructure of the oocysts, sporocysts and sporozoites of *Toxoplasma gondii*. J. Parasitol. 84, 505–512.

Speer, C.A., Dubey, J.P., 1998. Ultrastructure of early stages of infections in mice fed *Toxoplasma gondii* oocysts. Parasitology 116 (Pt 1), 35–42.

Speer, C.A., Dubey, J.P., 2005. Ultrastructural differentiation of *Toxoplasma gondii* schizonts (types B to E) and gamonts in the intestines of cats fed bradyzoites. Int. J. Parasitol. 35, 193–206.

Speer, C.A., Tilley, M., Temple, M.E., Blixt, J.A., Dubey, J.P., White, M.W., 1995. Sporozoites of *Toxoplasma gondii* lack dense-granule protein GRA3 and form a unique parasitophorous vacuole. Mol. Biochem. Parasitol. 75, 75–86.

Vivier, E., 1970. Observations nouvelles sur la reproduction asexuée de *Toxoplasma gondii* et considérations sur la notion d'endogenèse. C. R. Acad. Sc. Paris 271, 2123–2126.

Vivier, E., Petitprez, A., 1969. Le complexe membranaire et son évolution lors de l'élaboration des individus-fils de *Toxoplasma gondii*. J. Cell. Biol. 43, 329–342.

CHAPTER

3

Molecular Epidemiology and Population Structure of *Toxoplasma gondii*

Marie-Laure Dardé, Daniel Ajzenberg†, Chunlei Su***

***Toxoplasma Biological Resource Center and Faculté de Médecine, Université de Limoges, Limoges, France †Department of Parasitology and Mycology, Toxoplasma Biological Resource Center, Faculté de Médecine, Université de Limoges, Limoges, France **Department of Microbiology, The University of Tennessee, USA**

OUTLINE

Toxoplasma gondii, Second Edition
http://dx.doi.org/10.1016/B978-0-12-396481-6.00003-9

3.1 INTRODUCTION

Toxoplasma gondii is the most successful protozoan parasite that infects mammals and birds worldwide. It has been recognized as an important food- and water-borne pathogen in humans due to frequent outbreaks of infections. One third of the human population worldwide is chronically infected (Dubey, 2010). Infections in healthy pregnant women may cause blindness, mental retardation or even death of the foetus. Reactivation of latent infection in immunocompromised patients such as those with AIDS can cause life-threatening encephalitis (Montoya and Liesenfeld, 2004). In a mouse model, there are marked differences of virulence among distinct *T. gondii* strains. Most strains are avirulent and lead to chronic infection without symptoms. However, a few strains are highly virulent and result in fatal toxoplasmosis (Sibley and Boothroyd, 1992). The cases of severe acute disseminated toxoplasmosis in immunocompetent human patients are often reported in South America due to infection with atypical isolates (Bossi and Bricaire, 2004; Delhaes *et al.*, 2010). Given the biological and epidemiological diversity of the parasite, it is expected that the parasite is genetically diverse. Therefore, understanding genetic diversity and evolutionary history of *T. gondii* is of great importance to identify major reservoirs and transmission modes through which the parasite spreads among different hosts. In the past two decades, a large number of genetic markers have been developed to distinguish *T. gondii* strains from a variety of hosts in different geographical regions, providing essential tools for the study of molecular epidemiology, genetic diversity and population structure of *T. gondii*.

3.2 GENETIC MARKERS

Genetic markers can be divided into two categories including diagnostic and genotyping markers. Diagnostic markers rely on Polymerase Chain Reaction (PCR) amplification of *T. gondii* specific DNA in the clinical specimen. To achieve high sensitivity, the repetitive DNA elements are usually the favourable targets of choice. The most often used methods target at the 35-copy B1 gene (Burg *et al.*, 1989), the 300-copy 529 bp repeat element (Homan *et al.*, 2000), and the 110-copy internal transcribed spacer (ITS-1) or 18S rDNA gene sequences (Calderaro *et al.*, 2006; Hurtado *et al.*, 2001; Jauregui *et al.*, 2001). These methods have been modified and used in different laboratories since their initial development. The target sequences can be amplified by conventional PCR, nested PCR or quantitative real-time PCR. The most sensitive assay reported so far is quantitative real-time PCR of the 529 bp element with the sensitivity of detecting 1/50 of a genome equivalent (Kasper *et al.*, 2009; Su *et al.*, 2010).

Genotyping markers have been developed to distinguish *T. gondii* isolates. They are suitable for the studies of molecular epidemiology and population genetics of the parasite. These markers include multilocus enzyme electrophoresis (MLEE) (Darde *et al.*, 1992), mobile genetic elements PCR (MGE–PCR) (Terry *et al.*, 2001), random amplified polymorphic DNA polymerase chain reaction (RAPD–PCR) (Ferreira *et al.*, 2004; Guo and Johnson, 1995), PCR-restriction fragment length polymorphism (PCR–RFLP) (Cristina *et al.*, 1995; Ferreira *et al.*, 2006; Grigg *et al.*, 2001b; Howe and Sibley, 1995; Khan *et al.*, 2005; Sibley and Boothroyd, 1992; Su *et al.*, 2006), microsatellite analysis (Ajzenberg *et al.*, 2002b, 2004, 2010; Blackston *et al.*, 2001; Lehmann *et al.*, 2004; Lehmann *et al.*, 2006), multilocus DNA sequencing (Frazão-Teixeira *et al.*, 2011; Khan *et al.*, 2011a; Khan *et al.*, 2007; Khan *et al.*, 2006; Khan *et al.*, 2011b; Lehmann *et al.*, 2000; Miller *et al.*, 2008b; Pan *et al.*, 2012) and serotyping (Kong *et al.*, 2003). Serotyping is a very different approach for strain typing and for population genetic study. It is based on the use of synthetic peptides derived from polymorphic sites of the genes coding for *T. gondii* antigens including GRA6 and GRA7 (Kong *et al.*, 2003). The detection of antibodies against these peptides would allow identification of the parasite strain with no need for strain isolation or DNA extraction. However, due to very limited number of markers available, this method has been used in only a few studies and is not particularly powerful in population genetic studies (Sousa *et al.*, 2009; Sousa *et al.*, 2008; Sousa *et al.*, 2010).

Of the many different genotyping methods reported in the literature, four of them, including multilocus MLEE, PCR–RFLP, microsatellites and DNA sequencing, are more often used and have generated significant data to our understanding of *T. gondii* genetic diversity and population genetics. The advantage of multilocus markers is that they can readily capture genetic diversity and genetic recombination with good resolution. In addition, they are able to detect mixed infections. As *T. gondii* is haploid, only one allele is expected for a given locus. More than one allele for a locus of a given sample is indicative of mixed infections (Ajzenberg *et al.*, 2002b).

3.2.1 Genotyping Methods for Epidemiological and Population Genetics Studies

3.2.1.1 MLEE

The first series of studies on strain diversity in *T. gondii* relied on MLEE analysis. Six polymorphic enzymatic systems were characterized in *T. gondii* including aspartate aminotransferase, glutathione reductase, amylase, glucose phosphate isomerase, acid phosphatase, and propionyl esterase. These enzymes exhibited two to three isoforms in a collection of 83 *T. gondii* stocks (Ajzenberg, *et al.*, 2002a; Dardé *et al.*, 1992; Dardé *et al.*, 1988). This allowed the description of 12 zymodemes within this collection. The majority of stocks are clustered into three main zymodemes Z1, Z2 and Z3. Zymodeme 4 was closely related to zymodeme 2, differing by only one allozyme. The designation of zymodemes Z1, the Z2–4 cluster and Z3 were later demonstrated to be consistent with the three main lineages, type 1, type 2, and type 3 respectively, defined by multilocus PCR–RFLP analysis (Howe and Sibley, 1995). Other zymodemes (from Z5–12) found were represented by single isolates, each of which was later defined with additional markers to be different strains.

The MLEE revealed relatively low resolution in *T. gondii* genetic studies. The main disadvantage is that it requires large numbers of purified parasites (approximately 7×10^6 tachyzoites per enzyme). This is particularly difficult for slow growing isolates and it may take up to 2 months of repeated passage to obtain sufficient quantities of tachyzoites for analysis (Dardé *et al.*, 1992). Nevertheless, the findings by MLEE analysis stimulated a great deal of interest in studying the parasite's genetic diversity and its potential association with disease manifestations.

3.2.1.2 Microsatellites

Microsatellite sequences are short nucleotide tandem repeats of two to six nucleotides that occur two to 20 times (Ajzenberg, *et al.*, 2002a; Ajzenberg *et al.*, 2004; Ajzenberg *et al.*, 2010; Blackston *et al.*, 2001; Lehmann *et al.*, 2004; Lehmann *et al.*, 2006). They are hyper variable due to fast accumulation of length polymorphisms by intra-allelic polymerase slippage on microsatellite sequence during replication. In general, the mutation rate for microsatellites is 10^{-2} to 10^{-5} per locus per replication, which is several orders of magnitude faster than that of single nucleotide polymorphisms (SNPs) (Goldstein and Schlotterer, 1999). The fast mutation rate makes microsatellites well suited for individual identification of *T. gondii* isolates, especially informative for molecular epidemiological studies. To reveal microsatellite polymorphisms, the target sequences are first amplified by PCR using fluorescent primers and the products resolved in an automated sequencer. Several sets of microsatellite markers have been used in different studies with five to 15 markers, and there is no consensus on a uniform set of markers for population genetic studies (Ajzenberg, *et al.*, 2002a; Ajzenberg *et al.*, 2010; Blackston *et al.*, 2001; Lehmann *et al.*, 2004; Lehmann *et al.*, 2006; Sreekumar *et al.*, 2003). Recently, a multiplex PCR for 15 microsatellites was developed for multilocus analysis of isolates in a single PCR amplification, providing a simple method with high resolution in genotyping *T. gondii* strains (Ajzenberg *et al.*, 2010).

3.2.1.3 PCR–RFLP

Multilocus PCR–RFLP typing of *T. gondii* has been widely used for population genetic studies due to its ease of use and cost effectiveness (Cristina *et al.*, 1995; Ferreira *et al.*, 2006; Grigg *et al.*, 2001a; Howe and Sibley, 1995; Khan *et al.*, 2005; Pena *et al.*, 2008; Sibley and Boothroyd, 1992; Su *et al.*, 2006). This method relies on endonucleases to recognize SNPs among DNA sequences, digest PCR products and reveal distinct DNA banding patterns by agarose gel electrophoresis. The mutation rate for SNP in *T. gondii* is not known, but expected to be close to 10^{-9} to 10^{-10} per nucleotide position per replication for eukaryotes in general (Goldstein and Schlotterer, 1999), making it an excellent tool to study distantly related parasite strains. Sibley and Boothroyd were the first to use PCR–RFLP markers in studying *T. gondii* genetic diversity and its association with virulence in mice (Sibley and Boothroyd, 1992). When analysing 28 strains from a variety of hosts on five continents using three PCR–RFLP markers, they demonstrated that the 10 virulent strains have an identical genotype, whereas the 18 avirulent strains were moderately polymorphic, suggesting that virulent strains originated from a single lineage. Since then, many sets of PCR–RFLP markers were developed and applied to distinguish *T. gondii* strains collected from a variety of hosts in different regions. Here we have no intention to make an exhaustive list of all published articles on this subject, but to include a few representatives in the references (Cristina *et al.*, 1995; Ferreira *et al.*, 2006; Grigg *et al.*, 2001a; Howe and Sibley, 1995; Khan *et al.*, 2005; Pena *et al.*, 2008; Su *et al.*, 2006). Two of these studies made significant contribution to our understanding of *T. gondii* genetic diversity and population structure. By analysing 106 strains from a variety of hosts (including human) from North America and Europe using five PCR–RFLP markers, Howe and Sibley revealed the dominance of three clonal lineages (types 1, 2 and 3) of *T. gondii* and concluded that the parasite was clonal with the type 2 lineage accounting for 50% of isolates, and the population diversity was very limited in the regions (Howe and Sibley, 1995). However, when analysing 125 isolates from chickens, dogs and cats in Brazil using 10 PCR–RFLP markers, 48 different genotypes were identified, most were different from strains in North America and

Europe. Furthermore, there were no major dominant lineages and complete lack of the type 2 strains (Pena *et al.*, 2008), indicating different population structures among the continents.

3.2.1.4 DNA Sequencing Typing (MLST)

Multilocus DNA sequencing typing has been increasingly used to study *T. gondii* genetic diversity due to its high resolution and decreased cost for sequencing (Frazão-Teixeira *et al.*, 2011; Grigg *et al.*, 2001; Khan *et al.*, 2011a; Khan *et al.*, 2007; Khan 2011b; Lehmann *et al.*, 2000; Miller *et al.*, 2008b; Pan *et al.*, 2012). The advantage of MLST is its high resolution in distinguishing *T. gondii* strains. However, at present, most studies have used only a limited number of loci to investigate a small collection of strains, therefore the information from these studies is limited. Most recently, whole genome sequencing has been used to study *T. gondii* (Minot *et al.*, 2012). This approach will provide greatest resolution for the study of genetic diversity and evolution of the genome when applied to a large number of parasite strains.

3.3 GENOTYPE DESIGNATION

Currently, there is no standard for genotype designation. Given that different methods (microsatellites, PCR–RFLP, sequencing, etc.) are used to type a variety of isolates, and each method has its own scheme of classification, genotype designation has been confusing. A brief comparison of designation schemes is summarized in Table 3.1. Among these methods, the conventional designation of genotypes assumed the clonal population structure with the three dominant lineages, namely Types I, II and III (also known as types 1, 2 and 3). All other genotypes were considered rare and denoted as 'atypical' or 'exotic'. The ToxoDB PCR–RFLP genotype naming system applies to isolates typed by 10 genetic markers, including *SAG1*, *SAG2* (5′-3′*SAG2* and alt.-*SAG2*), *SAG3*, *BTUB*, *GRA6*, *L358*, *c22-8*, *c29-2*, *PK1* and *Apico* (Su *et al.*, 2010). In literature, it is often counted as 11 markers, since 5′-3′*SAG2* and alt.*SAG2* are considered two markers instead of one. These markers can further divide the type 2 lineage into 'type 2 clonal' and 'type 2 variant'. The type 2 clonal has a type 2 allele at locus Apico, whereas the type 2 variant has a type 1 allele. The Haplogroup naming system is based on DNA sequencing of five introns, including *UPRT*, *MIC*, *BTUB*, *HP*, and *EF* (Khan *et al.*, 2007; Khan *et al.*, 2011b). At the time of this writing, 15 haplogroups have been identified (Khan *et al.*, 2011b; Su *et al.*, 2012). The *Toxoplasma* BRC code naming system is based on 15 microsatellite markers, including *TUB2*, *W35*, *TgM-A*, *B18*, *B17*, *M33*, *MIV.1*, *MXI.1*, *M48*, *M102*, *N60*, *N82*, *AA*, *N61*, and *N83* (Ajzenberg *et al.*, 2010). A list of 138 *T. gondii* genotypes identified from 956 isolates by the combination of these three methods is summarized in Table 3.2.

3.4 MOLECULAR EPIDEMIOLOGICAL AND POPULATION STUDIES

Molecular tools may be used (1) to investigate an outbreak, (2) to understand the circulation of *T. gondii* strains at a local level, between different environments and hosts, and (3) to determine geographical distribution of genotypes and global population structure.

3.4.1 Outbreak Investigations

To investigate a *T. gondii* outbreak requires intensive studies which employ polymorphic markers to track strains in contiguous space and time. High-resolution genotyping by microsatellite markers represents the best solution for these studies (Carme, *et al.*, 2009a; Demar *et al.*, 2007; Wendte *et al.*, 2010). If the strains are

TABLE 3.1 Comparison of Genotype Designation Schemes

ToxoDB PCR-RFLP Genotype #	Haplo-Groups	*Toxoplasma* BRC code of representative isolate	Conventional Designation	Representative Isolate	Comment
1	2	TgA00001	Type II, type 2	PTG	type 2 clonal
2	3	TgH00005	Type III, type 3	VEG	
3	2	TgH00001	Type II, type 2	PRU	type 2 variant
4	12	N/A	Type 12, atypical, exotic	B41	
5	12	N/A	Type 12, atypical, exotic	ARI	
6	6	TgH00007	Type BrI, atypical, exotic	FOU	identical to *Africa 1*
7	3	N/A	atypical, exotic	G622M	
8	9	TgA00005	Type BrIII, Atypical, exotic	P89 (TgPgUs15)	
9	13	N/A	*Chinese 1*, atypical, exotic	TgCtPRC04	
10	1	TgA00004	Type I, type 1	GT1	
11	4	N/A	Type BrII, atypical, exotic	TgCatBr01	
12	3	N/A	atypical, exotic	TgCkGy02	
13	3	N/A	atypical, exotic	TgCatStk07a	
14	9	N/A	atypical, exotic	TgCatBr15	
15	4	N/A	atypical, exotic	CASTELLS	
16	N/A	N/A	atypical, exotic	TgCkNi01	
17	4	TgH00006	Type BrIV, atypical, exotic	MAS	
18	N/A	N/A	atypical, exotic	TgCatPRC1	
19	8	TgA00007	atypical, exotic	TgCatBr05	
20	N/A	N/A	atypical, exotic	TgDgSL4	
28	7	TgH00008	atypical, exotic	CAST	
60	5	TgH18002	atypical, exotic	GUY-KOE	
60	10	TgH00009	atypical, exotic	GUY-VAND	
66	11	N/A	Cougar, atypical, exotic	TgCgCa01	
111	15	N/A	atypical, exotic	TgCatBr64	
203	14	TgA15004	atypical, exotic	GAB2-2007-GAL-DOM2	identical to Africa 3

N/A, no designation assigned.

TABLE 3.2 List of 138 *Toxoplasma gondii* Genotypes (Su *et al.*, 2012)

Isolate (*Toxoplasma* BRC code)	ToxoDB PCR-RFLP Genotype #	Haplogroups	Geographical origin	Host	Microsatellite Markers															RFLP Markers										
					TUB2	*W35*	*TgM-A*	*B18*	*B17*	*M33*	*MIV.1*	*MXI.1*	*M48*	*M102*	*N60*	*N82*	*AA*	*N61*	*N83*	*SAG1*	*5-3SAG2*	*alt.SAG2*	*SAG3*	*BTUB*	*GRA6*	*c22-8*	*c29-2*	*L358*	*PK1*	*Apico*
GT1 (TgA00004)	10	1	America/ Northern/ USA	Animal/ Goat	291	248	209	160	342	169	274	358	209	168	145	119	265	87	306	I	I	I	I	I	I	I	I	I	I	I
TgCatBr80	55	1	America/ Southern/ Brazil	Animal/ Cat	289	248	209	160	342	169	272	356	233	166	142	123	265	87	306	I	I	I	I	III	I	u-1	I	I	I	I
TgCkCr04	35	1	America/ Central/ Costa Rica	Animal/ Chicken	291	248	205	160	348	167	274	358	209	166	nd	123	279	89	304	I	I	I	I	I	I	I	III	I	I	III
TgCkNi09	27	1	America/ Central/ Nicaragua	Animal/ Chicken	291	242	205	160	346	167	274	358	211	166	147	127	265	85	304	I	I	I	I	I	I	I	III	I	I	I
TgDgCo17	38	1	America/ Southern/ Colombia	Animal/ Dog	291	244	209	160	336	165	278	358	213	166	147	119	263	91	306	I	III	III	III	I	I	I	III	I	I	III
B73	127	2	America/ Northern/ USA	Animal/ Bear	289	242	207	160	336	169	274	356	213	190	142	111	267	101	310	II or III	II	II	II	II	II	II	II	III	III	III
PRU (TgH00001)	3	2	Europe/ Western/ France	Human/ Congenital	289	242	207	158	336	169	274	356	209	176	142	117	265	123	310	II or III	II	II	II	II	II	II	II	II	II	I
PTG (TgA00001)	1	2	America/ Northern/ USA	Animal/ Sheep	289	242	207	158	336	169	274	356	215	174	142	111	265	91	310	II or III	II	II	II	II	II	II	II	II	II	II
TgCkBr168	129	2	America/ Southern/ Brazil	Animal/ Chicken	291	242	207	158	336	169	274	356	235	176	138	119	261	93	310	II or III	II	II	II	III	II	II	II	II	II	II

(*Continued*)

TABLE 3.2 List of 138 *Toxoplasma gondii* Genotypes (Su *et al.*, 2012) (*cont'd*)

Isolate (*Toxoplasma* BRC code)	ToxoDB PCR-RFLP Genotype #	Haplogroups	Geographical origin	Host	Microsatellite Markers															RFLP Markers										
					TUB2	*W35*	*TgM-A*	*B18*	*B17*	*M33*	*MIV.1*	*MXI.1*	*M48*	*M102*	*N60*	*N82*	*AA*	*N61*	*N83*	*SAG1*	*5-3SAG2*	*alt.SAG2*	*SAG3*	*BTUB*	*GRA6*	*c22-8*	*c29-2*	*L358*	*PK1*	*Apico*
TgCtCo08	128	2	America/ Southern/ Colombia	Animal/ Cat	289	242	207	158	336	169	274	356	211	174	142	115	267	103	308	II or III	II	II	II	II	II	II	III	II	II	I
ENVL-MAC	83	3	America/ Caribbean/ Barbade	Animal/ New World monkey	291	242	205	160	336	165	278	356	213	190	142	111	277	87	312	I	I	I	III	I	III	II	III	III	I	III
G622M	7	3	America/ Central/ Panama	Animal/ Dove	289	242	205	160	336	165	278	356	221	190	145	111	279	87	314	I	III	III	III	III	III	III	III	III	III	I
M7741	133	3	America/ Northern/ USA	Animal/ Sheep	289	242	205	160	336	165	278	356	215	190	147	111	267	91	312	II or III	III	III	III	III	I	III	III	III	III	III
ROD-US	72	3	America/ Northern/ USA	Human/ Transplant	289	242	205	160	336	165	278	356	213	190	147	111	267	89	314	I	III	III	III	III	III	III	III	III	u-2	III
TgBBeCa01	90	3	America/ Northern/ Canada	Animal/ Bear	289	248	209	160	336	165	278	356	213	166	149	107	265	87	306	I	I	I	III	III	III	III	III	III	I	III
TgCatBr66	26	3	America/ Southern/ Brazil	Animal/ Cat	289	248	209	160	336	165	278	356	211	190	147	111	267	89	306	II or III	III	III	III	III	III	I	III	III	I	III
TgCatCa01	130	3	America/ Northern/ Canada	Animal/ Cat	289	248	209	160	336	165	278	358	213	190	142	113	263	93	306	II or III	III	III	I	III	III	II	I	III	III	III
TgCatPr08	118	3	America/ Caribbean/ Puerto Rico	Animal/ Cat	289	250	209	160	336	165	278	358	213	166	142	111	265	87	310	I	III	III	III	III	I	I	III	III	I	nd
TgCatPr09	115	3	America/ Caribbean/ Puerto Rico	Animal/ Cat	289	242	205	164	342	165	278	356	213	172	147	111	267	91	306	I	III	III	I	III	III	III	III	III	III	I

TgCatStK01	141	3	America/ Caribbean/ Saint Kitts	Animal/ Cat	289	242	205	162	336	165	278	nd	213	164	147	111	265	89	310	II or III	III	III	III	III	III	III	III	III	I	III
TgCatStk07a	13	3	America/ Caribbean/ Saint Kitts	Animal/ Cat	291	242	205	162	342	165	278	356	211	164	142	109	277	87	312	I	I	I	I	I	III	II	III	III	I	III
TgCkBr008	125	3	America/ Southern/ Brazil	Animal/ Chicken	289	242	203	156	342	169	278	358	229	170	147	105	265	91	312	I	III	III	III	III	III	II	III	III	u-2	III
TgCkGy01	31	3	America/ Southern/ Guyana	Animal/ Chicken	291	242	205	162	336	165	278	356	213	164	149	111	281	87	312	I	III	III	III	I	III	III	III	III	I	III
TgCkGy02	12	3	America/ Southern/ Guyana	Animal/ Chicken	291	242	205	162	336	165	278	nd	213	164	142	109	265	103	312	I	III	III	I	I	III	II	III	III	I	III
TgCkNi27	140	3	America/ Central/ Nicaragua	Animal/ Chicken	289	242	205	160	336	165	278	356	213	188	142	125	277	87	312	II or III	III	III	III	III	III	II	III	III	I	III
TgCkNi45	50	3	America/ Central/ Nicaragua	Animal/ Chicken	289	242	205	160	336	165	278	356	213	190	147	111	273	87	304	II or III	III	III	I	III	I	III	III	III	III	III
TgDgCo13	79	3	America/ Southern/ Colombia	Animal/ Dog	289	248	205	160	342	169	274	358	209	166	140	125	281	87	312	I	I	I	I	III	I	I	III	III	I	III
VEG (TgH00005)	2	3	America/ Northern/ USA	Human/ AIDS	289	242	205	160	336	165	278	356	213	188	153	111	267	89	312	II or III	III	III	III	III	III	III	III	III	III	III
TgCkGy08	25	3	America/ Southern/ Guyana	Animal/ Chicken	289	242	205	162	336	165	278	356	237	164	149	109	269	87	312	I	III	III	I	III	III	III	III	III	I	I
CASTELLS	15	4	America/ Southern/ Uruguay	Animal/ Sheep	287	242	207	158	358	169	274	356	239	164	138	109	283	87	324	u-1	I	II	III	III	III	III	I	I	III	I
IPP-URB (TgH20005)	17	4	Europe/ Western/ France	Human/ Congenital	287	242	207	160	354	169	274	356	221	168	147	109	289	87	324	u-1	I	II	III	III	III	u-1	I	I	III	I
LGE-CUV (TgH21016)	202	4	Europe/ Western/ France	Human/ Congenital	287	242	207	158	362	169	274	356	217	164	138	109	273	97	326	u-1	I	II	III	III	II	III	I	I	III	I
MAS (TgH00006)	17	4	Europe/ Western/ France	Human/ Congenital	291	242	205	162	362	169	272	358	221	166	142	111	332	95	338	u-1	I	II	III	III	III	u-1	I	I	III	I

(Continued)

TABLE 3.2 List of 138 *Toxoplasma gondii* Genotypes (Su *et al.*, 2012) (*cont'd*)

Isolate (*Toxoplasma* BRC code)	ToxoDB PCR-RFLP Genotype #	Haplogroups	Geographical origin	Host	Microsatellite Markers															RFLP Markers										
					TUB2	*W35*	*TgM-A*	*B18*	*B17*	*M33*	*MIV.1*	*MXI.1*	*M48*	*M102*	*N60*	*N82*	*AA*	*N61*	*N83*	*SAG1*	*5-3SAG2*	*alt.SAG2*	*SAG3*	*BTUB*	*GRA6*	*c22-8*	*c29-2*	*L358*	*PK1*	*Apico*
TgCatBr01	11	4	America/ Southern/ Brazil	Animal/ Cat	289	242	205	160	342	165	278	358	233	164	147	111	316	89	308	I	I	II	III	III	III	I	III	I	II	III
TgCatBr06-20	126	4	America/ Southern/ Brazil	Animal/ Cat	291	242	205	162	342	169	272	356	237	164	145	111	265	89	314	I	nd	I	III	III	II	u-1	I	I	u-1	I
TgCatBr18	119	4	America/ Southern/ Brazil	Animal/ Cat	291	242	207	160	338	169	272	358	229	164	142	111	263	89	308	I	III	III	III	III	II	u-1	I	I	u-1	I
TgCatBr25	47	4	America/ Southern/ Brazil	Animal/ Cat	291	242	207	160	338	169	272	358	229	164	142	111	263	89	308	I	III	III	III	III	II	u-1	I	I	II	I
TgCatBr34	104	4	America/ Southern/ Brazil	Animal/ Cat	291	248	205	160	338	169	272	356	245	164	136	111	316	87	314	I	I	II	III	I	III	u-1	I	I	III	I
TgCatBr40	92	4	America/ Southern/ Brazil	Animal/ Cat	291	242	207	162	338	169	272	358	223	176	145	113	316	91	308	I	I	II	I	III	II	II	I	I	II	I
TgCatBr57	108	4	America/ Southern/ Brazil	Animal/ Cat	289	242	203	164	338	169	272	356	219	180	142	105	289	81	332	I	I	II	III	III	III	II	I	I	III	I
TgCatBr81	124	4	America/ Southern/ Brazil	Animal/ Cat	291	242	207	156	338	165	276	358	225	166	145	107	265	91	316	I	III	III	III	III	III	II	III	I	u-1	I
TgCkBr037	107	4	America/ Southern/ Brazil	Animal/ Chicken	291	242	205	162	338	165	276	356	231	164	138	105	289	105	330	I	I	II	III	III	II	u-1	I	I	III	I
TgCkBr061	93	4	America/ Southern/ Brazil	Animal/ Chicken	291	242	205	162	338	165	276	356	235	176	140	119	289	99	330	I	I	II	I	III	II	u-1	I	I	III	I
TgCkBr147	17	4	America/ Southern/ Brazil	Animal/ Chicken	289	242	205	162	334	169	272	356	227	164	142	117	334	87	308	u-1	I	II	III	III	III	u-1	I	I	III	I

TgCkBr155	76	4	America/ Southern/ Brazil	Animal/ Chicken	289	242	205	160	334	167	276	356	213	190	140	119	267	87	308	u-1	III	III	III	III	III	u-1	I	I	III	I
TgDgBr18	106	4	America/ Southern/ Brazil	Animal/ Dog	291	242	205	162	338	169	276	358	229	164	142	111	324	89	316	I	I	II	III	III	II	u-1	I	I	II	I
TOU-ALI (TgH38021)	99	4	Africa/ Eastern/ Réunion	Human/ Congenital	289	242	205	160	338	169	272	356	231	178	142	119	301	89	308	I	I	II	I	III	III	u-1	I	I	III	I
GUY-BAS1 (TgH18006)	100	5	America/ Southern/ French Guiana	Human/ Amazonian toxoplasmosis	289	242	203	160	344	165	276	356	211	168	142	109	271	87	320	I	I	II	I	III	III	u-1	I	III	III	I
GUY-KOE (TgH18002)	60	5	America/ Southern/ French Guiana	Human/ Amazonian toxoplasmosis	289	246	203	160	337	165	274	356	209	172	136	111	251	109	310	I	I	II	I	III	III	III	I	III	III	I
GUY-MAT (TgH18003)	95	5	America/ Southern/ French Guiana	Human/ Amazonian toxoplasmosis	291	242	203	160	339	165	272	358	221	174	138	107	277	95	312	I	I	II	I	III	III	II	I	I	III	I
GUY-RUB (TgH00002)	98	5	America/ Southern/ French Guiana	Human/ Amazonian toxoplasmosis	289	242	205	170	360	167	274	356	223	190	142	109	259	85	312	I	I	II	I	III	III	III	III	I	III	III
TgCkBr036	37	5	America/ Southern/ Brazil	Animal/ Chicken	291	246	207	162	358	169	276	358	221	164	142	105	263	91	314	I	II	II	III	III	III	u-1	I	I	III	I
TgCkBr038	22	5	America/ Southern/ Brazil	Animal/ Chicken	291	246	207	162	362	169	272	358	233	164	142	105	322	87	314	u-1	I	II	III	III	III	u-1	I	III	III	III
TgCkBr089	65	5	America/ Southern/ Brazil	Animal/ Chicken	291	246	207	162	362	169	278	358	227	164	142	107	326	87	314	I	I	II	III	III	III	u-1	I	I	III	I
TgRsCr01	52	5	America/ Central/ Costa Rica	Animal/ Toucan	291	248	205	160	364	165	274	356	209	192	140	115	263	97	304	u-1	I	II	III	I	III	u-2	I	I	III	I
FOU (TgH00007)	6	6	Europe/ Western/ France	Human/ Transplant	291	248	205	160	342	165	274	354	227	166	147	111	281	89	306	I	I	I	III	I	II	u-1	I	I	I	I

(Continued)

TABLE 3.2 List of 138 *Toxoplasma gondii* Genotypes (Su *et al.*, 2012) (*cont'd*)

Isolate (*Toxoplasma* BRC code)	ToxoDB PCR-RFLP Genotype #	Haplogroups	Geographical origin	Host	Microsatellite Markers															RFLP Markers										
					TUB2	*W35*	*TgM-A*	*B18*	*B17*	*M33*	*MIV.1*	*MXI.1*	*M48*	*M102*	*N60*	*N82*	*AA*	*N61*	*N83*	*SAG1*	*5-3SAG2*	*alt.SAG2*	*SAG3*	*BTUB*	*GRA6*	*c22-8*	*c29-2*	*L358*	*PK1*	*Apico*
PBr	84	6	America/ Southern/ Brazil	Animal	291	248	205	160	334	165	274	354	213	190	140	119	271	87	308	I	I	I	III	I	III	u-1	I	I	III	I
TgCatBr09 (TgA00008)	42	6	America/ Southern/ Brazil	Animal/ Cat	291	242	205	160	362	165	278	354	227	174	140	111	269	89	308	I	I	I	III	III	II	I	I	I	u-1	I
TgCatBr26	80	6	America/ Southern/ Brazil	Animal/ Cat	291	248	205	160	362	165	278	354	229	174	140	111	271	89	308	I	I	I	III	I	II	I	I	I	u-1	I
TgCatBr41	117	6	America/ Southern/ Brazil	Animal/ Cat	291	242	205	160	342	165	278	354	231	166	149	111	263	91	308	I	III	III	III	I	III	u-1	I	I	u-1	III
TgCatBr45	56	6	America/ Southern/ Brazil	Animal/ Cat	291	242	205	160	348	169	272	354	237	166	147	111	269	89	306	I	I	I	III	I	II	u-1	I	III	I	I
TgCatBr50	86	6	America/ Southern/ Brazil	Animal/ Cat	291	248	205	160	342	165	274	354	231	166	147	111	265	89	310	I	I	I	III	III	II	u-1	I	I	III	I
TgCatBr72	85	6	America/ Southern/ Brazil	Animal/ Cat	289	248	205	160	342	165	274	354	231	164	147	111	271	89	308	I	I	I	III	III	II	u-1	I	I	II	I
TgCkBr041	33	6	America/ Southern/ Brazil	Animal/ Chicken	291	248	207	162	366	169	272	358	233	164	142	111	269	91	314	u-1	I	II	III	I	III	u-1	I	I	I	I
TgCkBr054	82	6	America/ Southern/ Brazil	Animal/ Chicken	291	248	203	160	342	165	274	354	221	166	142	111	269	89	330	I	I	I	III	I	III	II	I	I	I	III
TgCkBr107	70	6	America/ Southern/ Brazil	Animal/ Chicken	289	242	205	162	340	165	278	358	243	164	145	111	267	93	306	I	III	III	III	III	II	u-1	I	I	I	III
TgCkBr136	41	6	America/ Southern/ Brazil	Animal/ Chicken	291	242	205	160	342	165	278	354	229	166	140	111	271	91	308	I	I	I	III	I	II	I	I	I	I	I

TgCkBr143	105	6	America/ Southern/ Brazil	Animal/ Chicken	289	242	205	162	340	169	278	358	237	164	147	111	293	95	308	I	I	II	III	III	II	u-1	III	III	III	I
TgDgBr06	51	6	America/ Southern/ Brazil	Animal/ Dog	291	248	205	160	362	165	278	354	231	166	151	111	273	91	342	u-1	I	II	III	I	III	II	I	I	I	I
CAST (TgH00008)	28	7	America/ Northern/ USA	Human/ AIDS	291	242	205	158	342	167	276	356	211	168	147	119	279	87	306	I	I	I	I	I	I	II	I	III	I	III
TgCkBr141	77	7	America/ Southern/ Brazil	Animal/ Chicken	291	248	209	160	344	165	278	356	209	166	142	113	281	95	306	I	I	I	I	I	I	u-1	I	I	III	III
TgCatBr05 (TgA00007)	19	8	America/ Southern/ Brazil	Animal/ Cat	291	242	205	160	362	165	278	356	237	174	140	111	265	89	314	I	III	III	III	III	III	I	I	I	u-1	I
TgCatBr67	121	8	America/ Southern/ Brazil	Animal/ Cat	289	242	205	160	348	165	278	358	213	164	145	111	263	123	308	I	III	III	III	III	III	I	III	I	III	III
TgCkBr016	94	8	America/ Southern/ Brazil	Animal/ Chicken	291	242	205	160	362	165	278	356	229	164	136	113	265	91	314	I	I	II	I	III	III	I	I	I	II	I
TgCkBr019	64	8	America/ Southern/ Brazil	Animal/ Chicken	289	242	207	160	342	165	278	358	223	164	136	111	318	87	314	I	I	II	III	III	III	u-1	I	I	u-2	I
TgCkBr026	71	8	America/ Southern/ Brazil	Animal/ Chicken	291	242	207	162	342	165	278	358	229	164	151	105	265	91	310	I	III	III	III	III	III	II	I	III	III	I
TgCkBr040	59	8	America/ Southern/ Brazil	Animal/ Chicken	291	248	205	160	342	165	274	356	235	166	147	111	269	89	308	I	I	I	III	III	II	u-1	I	I	I	I
TgCkBr048	75	8	America/ Southern/ Brazil	Animal/ Chicken	289	242	205	164	362	165	278	356	225	164	157	111	297	91	342	u-1	I	II	III	III	III	II	I	I	III	III
TgCkBr075	40	8	America/ Southern/ Brazil	Animal/ Chicken	291	242	207	162	342	165	278	358	231	164	155	107	295	95	308	u-1	I	II	III	III	III	III	III	I	III	I
TgCkBr093	69	8	America/ Southern/ Brazil	Animal/ Chicken	291	242	207	160	342	165	278	358	235	174	140	111	265	91	308	I	III	III	III	III	II	I	III	I	II	I

(Continued)

TABLE 3.2 List of 138 *Toxoplasma gondii* Genotypes (Su *et al.*, 2012) (*cont'd*)

Isolate (*Toxoplasma* BRC code)	ToxoDB PCR-RFLP Genotype #	Haplogroups	Geographical origin	Host	Microsatellite Markers															RFLP Markers										
					TUB2	*W35*	*TgM-A*	*B18*	*B17*	*M33*	*MIV.1*	*MXI.1*	*M48*	*M102*	*N60*	*N82*	*AA*	*N61*	*N83*	*SAG1*	*5-3SAG2*	*alt.SAG2*	*SAG3*	*BTUB*	*GRA6*	*c22-8*	*c29-2*	*L358*	*PK1*	*Apico*
TgDgBr15	53	8	America/ Southern/ Brazil	Animal/ Dog	291	242	205	162	362	165	278	356	233	166	149	111	295	91	342	u-1	I	II	III	III	III	II	I	I	III	I
TgRaW03	32	8	America/ Northern/ USA	Animal/ Racoon	291	242	205	160	362	165	278	354	229	174	140	111	265	91	308	I	III	III	III	III	II	I	I	I	I	I
P89 (TgA00005)	8	9	America/ Northern/ USA	Animal/ Pig	291	242	205	160	348	165	278	356	213	190	142	111	261	87	314	I	III	III	III	III	III	II	III	III	III	III
TgCatBr10	21	9	America/ Southern/ Brazil	Animal/ Cat	291	242	207	160	360	165	278	356	229	174	140	105	263	91	314	I	III	III	III	III	III	I	I	I	III	III
TgCatBr15	14	9	America/ Southern/ Brazil	Animal/ Cat	289	242	205	162	344	165	278	358	225	164	142	111	263	105	312	I	III	III	III	III	III	III	I	III	III	III
TgCatBr20	120	9	America/ Southern/ Brazil	Animal/ Cat	289	242	205	160	360	165	278	356	213	174	140	105	265	105	314	I	III	III	III	III	III	I	I	III	III	III
TgCatBr76	67	9	America/ Southern/ Brazil	Animal/ Cat	291	242	205	160	342	165	278	356	237	176	140	111	265	93	314	I	III	III	III	I	III	I	III	III	u-1	III
TgCkBr074	138	9	America/ Southern/ Brazil	Animal/ Chicken	291	242	207	162	340	169	272	358	227	164	138	111	263	91	314	u-1	III	III	III	III	III	III	I	III	III	III
TgCkBr126	45	9	America/ Southern/ Brazil	Animal/ Chicken	289	248	205	160	342	169	278	354	213	166	142	111	269	105	310	I	III	III	III	I	II	II	III	I	I	III
TgCkBr130	116	9	America/ Southern/ Brazil	Animal/ Chicken	289	242	205	160	348	169	278	356	213	192	142	111	265	103	314	I	III	III	III	I	III	II	III	III	III	III
TgCkBr166	114	9	America/ Southern/ Brazil	Animal/ Chicken	289	242	207	160	342	165	278	356	211	190	149	105	263	101	306	I	III	III	I	III	III	III	I	III	I	I

TgCkBr169	78	9	America/ Southern/ Brazil	Animal/ Chicken	291	242	205	162	342	165	278	356	227	164	142	109	277	91	312	I	I	I	I	I	III	II	I	III	I	III
TgCkGy34	123	9	America/ Southern/ Guyana	Animal/ Chicken	289	242	205	160	336	165	278	356	213	174	140	111	265	89	316	I	III	III	III	III	III	I	III	III	III	III
TgDgCo16	46	9	America/ Southern/ Colombia	Animal/ Dog	291	242	205	160	358	165	274	356	221	190	142	121	265	115	304	I	III	III	III	I	III	II	I	III	III	I
GUY-DOS (TgH18001)	97	10	America/ Southern/ French Guiana	Human/ Amazonian toxoplasmosis	289	246	203	160	344	167	272	356	229	176	142	113	263	85	312	I	I	II	I	III	III	II	III	I	III	I
GUY-VAND (TgH00009)	60	10	America/ Southern/ French Guiana	Human/ Amazonian toxoplasmosis	291	242	203	162	344	167	276	356	217	170	142	113	277	91	308	I	I	II	I	III	III	III	I	III	III	I
TgCatBr44	34	10	America/ Southern/ Brazil	Animal/ Cat	291	242	205	162	342	165	278	356	231	166	153	111	265	91	344	u-1	I	II	III	III	III	II	I	I	u-1	I
TgCkBr109	96	10	America/ Southern/ Brazil	Animal/ Chicken	291	246	203	156	336	167	276	356	215	174	145	111	265	101	316	I	I	II	I	III	III	II	III	I	III	III
ARI	5	12	America/ Northern/ USA	Human/ Transplant	289	242	209	158	336	169	274	362	215	170	147	131	295	89	316	u-1	II	II	II	II	II	II	II	I	II	I
B41	4	12	America/ Northern/ USA	Animal/ Bear	289	242	207	162	336	169	274	356	213	170	142	111	287	107	316	II or III	II	II	II	II	II	II	II	I	II	I
RAY	5	12	America/ Northern/ USA	Human/ Congenital	289	242	211	160	336	169	274	362	233	176	151	113	283	99	320	u-1	II	II	II	II	II	II	II	I	II	I
TgCatPr05	49	12	America/ Caribbean/ Puerto Rico	Animal/ Cat	289	246	207	162	336	169	274	356	233	166	147	113	267	93	312	II or III	II	II	I	II	II	II	II	III	II	nd
TgSoUs01	39	12	America/ Northern/ USA	Animal/ Sea otter	289	242	207	162	336	169	274	356	215	170	142	111	267	97	310	II or III	II	II	II	II	II	II	II	I	II	II

(Continued)

TABLE 3.2 List of 138 *Toxoplasma gondii* Genotypes (Su *et al.*, 2012) (*cont'd*)

Isolate (*Toxoplasma* BRC code)	ToxoDB PCR-RFLP Genotype #	Haplogroups	Geographical origin	Host	Microsatellite Markers															RFLP Markers										
					TUB2	*W35*	*TgM-A*	*B18*	*B17*	*M33*	*MIV.1*	*MXI.1*	*M48*	*M102*	*N60*	*N82*	*AA*	*N61*	*N83*	*SAG1*	*5-3SAG2*	*alt.SAG2*	*SAG3*	*BTUB*	*GRA6*	*c22-8*	*c29-2*	*L358*	*PK1*	*Apico*
TgWtdUs08	74	12	America/ Northern/ USA	Animal/ White tailed deer	289	242	207	160	336	169	274	356	227	190	142	111	269	103	310	II or III	III	III	III	II	II	II	III	II	II	I
GAB2-2007-GAL-DOM2 (TgA15004)	203	14	Africa/ Central/ Gabon	Animal/ Chicken	291	242	207	160	342	165	278	354	223	166	142	111	277	97	310	I	I	I	III	I	II	II	III	III	I	III
TgCkBr059	36	14	America/ Southern/ Brazil	Animal/ Chicken	291	248	205	160	342	165	278	354	227	166	145	105	261	93	306	I	I	I	III	I	III	II	I	III	I	III
TgCkBr186	88	14	America/ Southern/ Brazil	Animal/ Chicken	291	248	205	160	342	165	278	354	237	166	142	105	263	89	306	I	I	I	III	III	III	II	I	III	I	III
TgCatBr38	136	15	America/ Southern/ Brazil	Animal/ Cat	291	242	205	162	362	165	278	356	221	166	149	111	261	93	336	u-1	I	u-1	III	III	III	II	I	I	III	I
TgCatBr64	111	15	America/ Southern/ Brazil	Animal/ Cat	289	242	207	160	338	165	278	356	225	190	136	105	263	97	310	I	I	u-1	III	III	III	u-1	I	III	III	I
TgCkBr013	63	15	America/ Southern/ Brazil	Animal/ Chicken	289	242	205	160	338	169	276	356	213	164	145	111	316	89	312	I	I	II	III	III	III	I	III	I	II	I
TgCkBr045	135	15	America/ Southern/ Brazil	Animal/ Chicken	291	242	207	164	342	169	272	356	229	172	142	105	320	81	310	u-1	I	II	III	III	III	II	I	III	III	I
TgCkBr173	81	15	America/ Southern/ Brazil	Animal/ Chicken	289	248	205	160	348	169	278	354	213	194	149	111	275	89	308	I	I	I	III	I	II	u-1	I	I	III	III
TgCkBr177	109	15	America/ Southern/ Brazil	Animal/ Chicken	291	242	207	166	334	165	278	356	239	168	147	105	nd	93	310	I	I	II	III	III	III	III	I	III	III	III
TgCkBr178	134	15	America/ Southern/ Brazil	Animal/ Chicken	291	242	205	160	336	165	278	356	213	nd	142	109	299	101	312	u-1	I	II	III	I	III	II	III	III	I	III

TgCkNi04	23	15	America/ Central/ Nicaragua	Animal/ Chicken	291	242	205	160	336	165	274	356	223	194	142	121	281	89	304	I	I	II	III	I	III	II	I	III	u-1	I
TgCtCo05	61	15	America/ Southern/ Colombia	Animal/ Cat	291	242	205	160	336	165	276	356	223	166	142	121	279	87	304	I	I	II	III	I	III	II	III	I	u-2	I
TgCtCo15	101	15	America/ Southern/ Colombia	Animal/ Cat	291	242	205	160	336	165	276	356	223	166	142	121	279	87	304	I	I	II	III	I	III	II	I	I	u-2	I
TgDgCo06	44	15	America/ Southern/ Colombia	Animal/ Dog	291	242	205	160	336	165	276	356	223	166	142	130	287	91	304	I	I	II	III	I	III	II	I	I	u-3	I
GUY-JAG1 (TgA18001)	197	11	America/ Southern/ French Guiana	Animal/ Jaguar	289	242	203	160	336	169	274	358	209	176	142	111	257	107	340	I	II	II	II	II	II	II	II	I	II	I
TgCgCa01	66	11	America/ Northern/ Canada	Animal/ Cougar	289	242	205	158	336	169	274	354	219	174	151	119	259	79	332	I	II	II	III	II	II	II	u-1	I	u-2	I
TgCkGh01	137	13	Africa/ Western/ Ghana	Animal/ Chicken	293	242	203	156	336	165	nd	nd	215	176	130	109	281	nd	306	u-1	II	II	III	III	II	II	III	II	III	I
TgCtPRC04	9	13	Asia/ Eastern/ China	Animal/ Cat	293	242	211	160	336	169	274	354	nd	172	145	nd	281	nd	308	u-1	II	II	III	III	II	II	III	II	II	I
SOU	139	ND	America/ Northern/ USA	Human/ AIDS	289	242	205	158	336	165	278	356	225	174	142	111	259	89	312	II or III	II	II	III	III	III	II	II	III	III	I
TgCatPr06	112	ND	America/ Caribbean/ Puerto Rico	Animal/ Cat	291	246	209	162	336	169	274	356	213	168	142	113	267	87	306	I	II	II	I	I	I	III	II	III	I	nd
TgCkBr114	29	ND	America/ Southern/ Brazil	Animal/ Chicken	291	242	205	160	358	165	274	356	233	194	142	121	293	nd	304	I	I	II	III	I	III	II	I	III	III	I
TgCkCr01	91	ND	America/ Central/ Costa Rica	Animal/ Chicken	291	248	209	160	348	nd	276	358	209	166	142	125	259	89	306	I	I	II	I	I	I	II	I	I	I	I
TgCkCr07	43	ND	America/ Central/ Costa Rica	Animal/ Chicken	293	246	209	160	352	nd	278	356	209	166	140	105	259	95	304	I	I	II	I	II	I	I	I	I	u-1	I

(Continued)

TABLE 3.2 List of 138 *Toxoplasma gondii* Genotypes (Su *et al.*, 2012) (*cont'd*)

Isolate (*Toxoplasma* BRC code)	ToxoDB PCR-RFLP Genotype #	Haplogroups	Geographical origin	Host	Microsatellite Markers															RFLP Markers										
					TUB2	*W35*	*TgM-A*	*B18*	*B17*	*M33*	*MIV.1*	*MXI.1*	*M48*	*M102*	*N60*	*N82*	*AA*	*N61*	*N83*	*SAG1*	*5-3SAG2*	*alt.SAG2*	*SAG3*	*BTUB*	*GRA6*	*c22-8*	*c29-2*	*L358*	*PK1*	*Apico*
TgCkGh02	132	ND	Africa/ Western/ Ghana	Animal/ Chicken	289	242	205	160	336	165	278	356	213	190	140	111	267	93	312	II or III	III	III	III	II	II	II	III	III	II	III
TgCkNi01	16	ND	America/ Central/ Nicaragua	Animal/ Chicken	289	242	209	160	348	167	276	356	209	166	147	111	259	nd	304	I	I	II	III	III	I	III	I	III	III	I
TgCkNi35	102	ND	America/ Central/ Nicaragua	Animal/ Chicken	291	242	205	160	336	165	274	356	223	190	142	111	279	91	304	I	I	II	III	I	III	II	I	III	u-1	III
TgCtCo03	62	ND	America/ Southern/ Colombia	Animal/ Cat	291	242	205	162	358	nd	274	356	233	194	142	123	297	nd	304	I	I	II	III	I	III	II	III	III	III	I
TgWtdUs10	54	ND	America/ Northern/ USA	Animal/ White tailed deer	289	242	205	160	336	165	278	356	215	190	151	111	259	93	312	II or III	II	II	III	III	III	III	III	III	III	II
TgCkBr171	57	ND	America/ Southern/ Brazil	Animal/ Chicken	289	242	207	160	348	169	278	356	213	196	147	111	265	87	308	I	I	I	III	I	II	u-1	I	III	II	III
TgCkGy07	30	ND	America/ Southern/ Guyana	Animal/ Chicken	289	242	209	162	336	165	278	356	233	164	147	109	267	87	306	I	III	III	I	III	III	III	III	III	I	III
TgCkGy22	48	ND	America/ Southern/ Guyana	Animal/ Chicken	289	242	205	162	336	165	278	356	233	164	151	111	277	87	306	I	III	III	III	III	III	III	III	III	III	III
TgDgCo07	122	ND	America/ Southern/ Colombia	Animal/ Dog	289	242	209	162	336	165	278	356	213	164	140	111	263	97	306	I	III	III	III	III	III	I	III	III	I	III
TgShUs28	73	ND	America/ Northern/ USA	Animal/ Sheep	291	248	209	160	336	165	278	354	213	166	147	111	277	87	304	II or III	III	III	I	I	I	III	III	III	I	I
TgShUs32	131	ND	America/ Northern/ USA	Animal/ Sheep	289	242	209	160	336	169	274	354	213	166	147	111	277	87	304	II or III	III	III	II	II	I	III	III	II	II	I

nd, allele not determined
ND, haplogroup not defined
u-1, u-2, and u-3 are different alleles from conventional I, II and III alleles

available and provide enough DNA, sequencing may be added to microsatellite markers.

A microsatellite-based typing scheme was applied to determine the molecular genotypes of *T. gondii* isolates associated with a human water-borne outbreak in Brazil which occurred over a short time span in 2001 (Vaudaux *et al.*, 2010; Wendte *et al.*, 2010). This outbreak was linked to oocyst-contamination of a municipal water supply in the town of Santa Isabel do Ivai and resulted in infection and symptomatic disease in hundreds of people. A previous study utilizing a limited set of PCR–RFLP markers concluded to the identity of the two strains isolated from the water cistern implicated as the source of the outbreak and of 4/11 chicken isolates collected in the surrounding environment immediately following the outbreak event. So, the outbreak genotype seemed to be highly prevalent (Vaudaux *et al.*, 2010). When applying microsatellite and sequencing markers on the same set of isolates, the two strains isolated from the water cistern were confirmed as identical, but genetically different from chicken strains found in the neighbouring environment except for one chicken isolate. The higher level of resolution provided by microsatellite and sequence typing imparts a higher level of confidence to the conclusion that only one strain was associated with this clonal epidemic.

In another outbreak of human toxoplasmosis concerning 11 patients in a small village of 33 inhabitants in Suriname, microsatellite genotyping demonstrated that all five patients from whom parasites were isolated were infected with the same, previously undiscovered genotype (Demar *et al.*, 2007). Similarly, microsatellite markers were used to analyse two fatal outbreaks of toxoplasmosis that occurred in 2001 and 2006 in an outdoor captive breeding colony of squirrel monkeys (*Saimiri sciureus*) at the Institut Pasteur in French Guiana (Carme, *et al.*, 2009a). The 2001 and 2006 outbreaks were shown to be due to two different *T. gondii* strains. The 2001 strain exhibited a type 2 genotype whereas the 2006 strain was an atypical strain. In 2006, the outbreak comported two successive episodes, three weeks apart. The second event leading to the death of 20 squirrel monkeys was believed to be related to direct contamination by tachyzoites of bronchopulmonary origin from dying monkeys of the first event. All samples collected during both episodes of the 2006 outbreak had the same multilocus genotype with 12 microsatellite markers, which here again strongly suggests that only one *T. gondii* strain was involved in that outbreak.

3.4.2 Dynamics of Transmission Between Different Environments or Hosts

The use of an ecological approach for *T. gondii* strain epidemiology is still rarely performed. For analysing this phenomenon which can occur over a relatively short period of time, the rapidly evolving microsatellites are the best suited markers. Microsatellites were able to show the interpenetration of strains from anthropized and from wild areas in French Guiana, with the possibility of hybrid strains, or the influence of human activities on clusterization of *T. gondii* population (Mercier *et al.*, 2011). Genetic studies using PCR–RFLP markers and DNA sequencing were also essential for understanding transmission networks for *T. gondii* in Californian marine mammals. The same strains were found in Californian sea otters, in the terrestrial mammals, notably in wild felid hosts, from the adjacent coastal areas, as well as in a filter-feeding invertebrate collected near the shoreline. This study provided evidence for a mechanism through which this terrestrial parasite could infiltrate the marine environment via land-to-sea run-off and bio-concentration of oocysts in prey species of sea otters and other marine mammals (Miller *et al.*, 2008b). Further discussion of the dynamics of transmission in the environment is found in the paragraph below about the environmental factors influencing genetic population structure of *T. gondii* strains.

3.4.3 Geographical Distribution

Numerous studies have been conducted in diverse countries for analyzing *T. gondii* genotypes of isolates circulating in diverse countries and hosts. Most of them use a multilocus and multi-chromosome genotyping by PCR–RFLP markers (Su *et al.*, 2006), less frequently by microsatellite markers (Ajzenberg *et al.*, 2010) or gene sequencing. Studies are most often carried out in free-range chickens, as sentinels of the environmental contamination with *T. gondii* oocysts (Lehmann *et al.*, 2006). Genotypes infecting wild species or other domestic animals are less frequently studied. Although epidemiological studies presenting genotypes infecting humans may be biased due to travel or to consumption of imported food, they usually reflect genotypes circulating in animals and environment (Ajzenberg *et al.*, 2002a).

3.4.3.1 *Europe*

In Europe, type 2 is largely predominant in published studies, from the extreme north (the Arctic archipelago of Svalbard, Finland) (Prestrud *et al.*, 2008; Jokelainen *et al.*, 2011, 2012) to the Mediterranean countries (Italy, Portugal, Greece). However, although this has to be confirmed by more studies, there may be a gradient in the prevalence of this type from north to south, type 2 being more prevalent in Northern and Western Europe, than in southern countries, closer to Africa or the Middle East. In France, it is found in more than 90% of human congenital toxoplasmosis and also in nearly all isolates originating from a large variety of animals (Ajzenberg, *et al.*, 2002b; Dardé, 2008; Richomme *et al.*, 2009; Aubert *et al.*, 2010; Dumètre *et al.*, 2006; Halos *et al.*, 2010). It was also predominant in Germany (Herrmann *et al.*, 2013; Herrmann *et al.*, 2012b), Switzerland (Frey *et al.*, 2012) and in Poland (Nowakowska *et al.*, 2006). In southern countries, although type 2 is still present, type 3 seems to be more frequently isolated (Messaritakis *et al.*, 2008; Dubey *et al.*, 2006; de Sousa *et al.*, 2006). Recent studies from Germany detected genotypes different from clonal lineages type 2 or 3 (Herrmann *et al.*, 2010; Herrmann, *et al.*, 2012b). An oocyst sample shed by a single cat was further demonstrated as being a mixture of different genotypes resulting from a single cross between type 2 and 3 isolates, suggesting that naturally recombinant isolates may be circulating in Europe (Herrmann *et al.*, 2010; Herrmann *et al.*, 2012a). Natural recombinant of type 2 and type 3 isolates were also detected in foxes (Herrmann *et al.*, 2012b). Taken together, there are two clonal lineages of *T. gondii* circulating in Europe, with the type 2 overwhelmingly dominant, followed by the type 3 lineage. The previously reported clonal type 1 lineage is rarely isolated from animals as well as from humans in Europe.

3.4.3.2 *Africa*

Several genetic studies on *T. gondii* in Africa also suggested clonal population structure with a few dominant lineages, including type 2, type 3, *Africa 1*, and *Africa 3* genotypes. Multilocus PCR–RFLP genotyping studies of 154 samples – including 19 chicken isolates from six African countries (Egypt, Nigeria, Congo, Mali, Burkina Faso and Kenya) (Velmurugan *et al.*, 2008), 115 isolates in feral cats from Egypt (Al-Kappany *et al.*, 2010), and 20 samples from Ugandan chickens (Lindström *et al.*, 2008) – collectively identified 74 type 2, 56 type 3, six type 1, four ToxoDB genotype #20, five belonging to four other genotypes, and nine to mixed infection. However, a recent study of 69 chicken isolates from Gabon in West Africa using 13 microsatellite markers revealed 35 type 3, 19 *Africa 3*, 11 *Africa 1*, three type 3-like, and one other genotype (Mercier *et al.*, 2010). *Africa 3* and *Africa 1* were also found in African patients from other West and Central African countries (Ajzenberg *et al.*, 2009), suggesting that these lineages may be prevalent over large areas in Africa. It is notable that the *Africa 1* genotype is identical to ToxoDB genotype #6 (type BrI) which has been frequently identified in Brazil, South America (Pena *et al.*, 2008). Given the geographical diversity in Africa, a broader sampling of *T. gondii* is needed to better understand its population structure.

3.4.3.3 *Asia*

Existing studies indicate that *T. gondii* has a clonal population structure in eastern China with the dominance of two genotypes – including ToxoDB PCR–RFLP genotype #9 (*Chinese 1*, Haplogroup 13) and type 1 lineages. PCR–RFLP genotyping of 75 samples from China, including 17 isolates from cats in Guangdong province (Dubey *et al.*, 2007b), 11 from cats in Beijing (Qian *et al.*, 2012), 14 from cats in four different provinces (Chen *et al.*, 2011), 17 from cats, pigs, sheep and humans (Zhou *et al.*, 2009), 16 from pigs in two provinces (Zhou *et al.*, 2010) – collectively identified 59 ToxoDB #9, 13 type 2, two ToxoDB PCR–RFLP #18, and one type 2 variant strains. In two provinces of western China (Xinjiang and Qinghai), 1600 km distant from each other, the isolation of type 2 in four wild birds and in one sheep may indicate that this clonal type 2 lineage is prevalent in western regions of China (Zhou *et al.*, 2009; Huang *et al.*, 2012). A study of eight dog isolates from Vietnam identified four ToxoDB #9 and four ToxoDB #18 strains (Dubey *et al.*, 2007a). In Middle East countries (Iran, Egypt, United Arab Emirates, Qatar), type 2 and 3 isolates were also identified (Zia-Ali *et al.*, 2007; Dubey *et al.*, 2010). Overall, available data indicated low genetic diversity of *T. gondii* in Asia and the ToxoDB PCR–RFLP genotype #9 is the dominant genotype.

3.4.3.4 *Australia*

Data on *T. gondii* diversity in Australia is limited. A type 2 strain was reported from a domestic dog (Al-Qassab *et al.*, 2009). Different genotypes and recombinant strains were isolated from Australian wildlife (a wombat, a wallaby, two woylies, a mouse, a meerkat, and eight kangaroo samples) and from domestic animals (a horse, a cat, and two goats) (Parameswaran *et al.*, 2010). These genotypes were characterized by atypical alleles that were interpreted as a result of genetic drift following the isolation of archetypal strains (types 1, 2 or 3) imported into Australia during early European settlement. This possible genetic drift was further confirmed by sequencing of only three loci (B1, SAG2 and SAG3) performed on isolates from 16 macropods (Pan *et al.*, 2012).The observed allelic diversity suggested that most genotypes exist as minor variants of established archetypal lineages by the accumulation of new mutations. Recombination of these variants was responsible for the diversification of strains on this continent. The presence of multiple infections in these macropods will favour recombination in feral cats, enhancing diversity (Pan *et al.*, 2012). Overall, the number of *T. gondii* isolates studied in Australia is still very limited, and more studies are needed to have a better view of *T. gondii* genetic diversity and population structure.

3.4.3.5 *North America*

Genetic studies of *T. gondii* in North America revealed higher diversity than in Europe, Africa and Asia. However, a few genotypes including type 2, type 12 and type 3 are dominant in this region. Type 1 is rarely encountered. Type 12 has been recently identified as the major genotype in wildlife in North America (Dubey *et al.*, 2011b; Khan *et al.*, 2011a). It includes previously reported Type A and X strains from sea otters (Miller *et al.*, 2004; Miller *et al.*, 2008a; Sundar *et al.*, 2008). Typing results from a collection of 169 wildlife isolates in the USA and Canada comprised of 22 genotypes, including type 12, type 2, type 3 and 19 others. The type 12 accounted for 46.7% (79/169) of isolates and was the dominant lineage, followed by type 2 which accounted for 27.8% (47/169), and the type 3 accounted for 10.1% (17/169). Together, these three genotypes accounted for 84.6% (143/169) of total isolates (Dubey *et al.*, 2011b). Studies of 182 isolates from domestic pigs in the USA revealed nine genotypes, including type 2 (clonal and variant), type 3, type 12 (include ToxoDB PCR–RFLP genotypes #4 and #5), and four others. Among these isolates, 36.3% (66/182) were type 2 clonal, 19.8% (36/182) type 2 variant, 26.9% (49/182) type 3, 9.9% (18/182) type 12, and 7.1% (13/182) belong to four other genotypes. The top three lineages (types 2, 3 and 12) accounted for 94.5% (172/182) of these isolates, indicating low genetic diversity in domestic pigs (Velmurugan *et al.*, 2009). Comparison of *T. gondii* isolates from domestic animals and wildlife indicates a lower genetic diversity in the former. This may suggest the partition of domestic versus sylvatic transmission routes for *T. gondii*. Genotypes different from type 2 or 3 were also found in wildlife from Canada (Dubey 2008a; Lehmann *et al.*, 2000) and in marine mammals living near the North Pacific coast (Gibson *et al.*, 2011). Non-type 2 isolates may also be spread in humans as suggested by a serotyping study performed on cases of congenital toxoplasmosis acquired in the USA. Human congenital infections due to non-type 2 serotype appeared to be more frequent in the southern and western parts of the country (McLeod *et al.*, 2012).

3.4.3.6 Central and South America

After the initial discovery of clonal population structure of *T. gondii* in Europe and North America (Darde *et al.*, 1992; Howe and Sibley, 1995; Sibley and Boothroyd, 1992), it was speculated that this general rule applies worldwide. However, a decade later, several studies on *T. gondii* diversity in South America revealed distinct genotypes and higher genetic diversity than expected (Ajzenberg *et al.*, 2004; Ferreira *et al.*, 2006; Khan *et al.*, 2006; Lehmann *et al.*, 2004). These findings challenged the dogma of clonal population structure and stimulated a wave of population genetics studies on *T. gondii*. Up to date, numerous studies have confirmed the high genetic and genotypic diversity among *T. gondii* strains circulating in Central and South America. But the genetic diversity and population structure are not homogeneous in this subcontinent.

Genotyping analysis with PCR–RFLP markers performed on 164 isolates from three countries in Central America (Guatemala, Nicaragua, Costa Rica), from one country in Caribbean (Grenada) and five countries from the northern and western parts of South America (Venezuela, Colombia, Peru, Chile, and Argentina) showed a moderate diversity but with the presence of the type 2 lineage, and the high predominance of type 3 and related genotypes belonging to haplogroup 3 (Rajendran *et al.*, 2012; Su *et al.*, 2012).

In Brazil, a large number of isolates from a variety of animals from different areas were intensively studied revealing a high genotypic diversity. Pena *et al.* genotyped 125 isolates from chickens, dogs and cats in Brazil and identified 48 genotypes with 26 of these genotypes having single isolates (Pena *et al.*, 2008). Four of the 48 genotypes with multiple isolates from different hosts and locations were designated as types BrI, BrII, BrIII and BrIV, and they were considered the common clonal lineages in Brazil. These results indicated that the *T. gondii* population in Brazil is highly diverse with a few successful clonal lineages expanding into wide geographical areas. The genotyping results (10 PCR–RFLP markers) were recently summarized for 363 isolates from Brazil (Dubey, *et al.*, 2012). The three most common genotypes designated in this study as ToxoDB PCR–RFLP genotype #6, #8 and #11, which were previously designated as type Br1, BrIII, and BrII (Pena *et al.*, 2008) may represent common clonal lineages. The same genotypes were also found in Argentina (Moré *et al.*, 2012). In the globalizing study by Su *et al.* (2012), Brazilian isolates were found in different haplogroups including 6, 14, 4, 8, 9 and 15, emphasizing their diversity. Types 2 and 3 isolates are infrequently encountered in this part of the continent (Moré *et al.*, 2012). The high genetic diversity observed in South America is maximal in the Amazonian area, with the highly divergent Amazonian strains observed in the rainforest of French Guiana (Ajzenberg *et al.*,

2004; Mercier *et al.*, 2011). Some of them were clustered in the study by Su *et al.* (2012) in the haplogroups 5 and 10. In the nearby coastal area of the same country, genotypes similar to those found in Caribbean islands are circulating.

3.4.3.7 Global Genetic Diversity and Population Structure of T. gondii

Genetic diversity and population structure are strongly influenced by the intensity of sexual recombination and evolutionary selection of the recombinants. Reproduction without genetic recombination leads to a clonal population. With such population structure, it is expected to see identical genotypes over large geographic areas and at intervals of many years. If a background level of genetic recombination is coupled with clonal expansion of a few genotypes, then an epidemic population structure is expected. In this scenario, a number of uncommon multilocus genotypes co-exist with an overrepresentation of a few major genotypes. Random mating without selective expansion of any progeny will lead to a panmictic population structure. Existing data suggested that *T. gondii* has a clonal population structure in Europe, Africa, Asia and North America, but an epidemic structure in Central and South America.

Lehmann *et al.* studied 275 chicken isolates worldwide using five microsatellite, one minisatellite and *SAG2* RFLP markers and revealed four major populations of *T. gondii* (Lehmann *et al.*, 2006). One population (type 2 related) was found in all continents except South America, one population (type 3 related) was found worldwide and two populations were largely confined in South America. In addition, the southern populations had the highest diversity. It was suggested that South American and Eurasian populations had evolved separately until the last several hundred years when transatlantic trading transmigrated a few lineages which rapidly expanded in the north. It is further concluded that, given the higher genetic diversity observed in South America, modern-day *T. gondii* populations were originated in this region (Lehmann *et al.*, 2006).

Khan *et al.* did structural analysis of 46 representative *T. gondii* isolates by sequencing eight introns from five unlinked loci (Khan *et al.*, 2007). These authors concluded that modern day *T. gondii* populations are derived from genetic recombination of four ancestral lineages. The first ancestral lineage was previously denoted as the α strain which crossed with an ancestral type 2 strain (theoretical type II_1) and gave rise to the modern day type 1 strains (Boyle *et al.*, 2006). This ancestral lineage α most closely resembles modern day strains GPHT, BOF, FOU and TgCatBr2, which are also known as type BrI or Africa 1 (Pena *et al.*, 2008; Mercier *et al.*, 2010). The second ancestral lineage (theoretical type II_2) is most closely related to modern day strains PTG, DEG and PIH, which are also known as the type 2 lineage. The third ancestral lineage was previously denoted as the β strain which crossed with an ancestral type 2 strain (theoretical type II_3) and gave rise to the modern day type 3 strains (Boyle *et al.*, 2006). The third ancestral lineage is closely related to modern day strains P89 and TgCatBr3, which are also known as type BrIII (Pena *et al.*, 2008). The fourth ancestral lineage most closely resembles modern day strains MAS, TgCatBr18 and TgCatBr25, which are also known as type BrIV (Pena *et al.*, 2008).

The recent analysis of a large collection of 956 *T. gondii* isolates worldwide made a significant contribution to our understanding of the parasite's global genetic diversity and population structure (Su *et al.*, 2012). This study integrated the information generated by different types of genetic markers including PCR–RFLP, microsatellite and intron sequencing, and for the first time made it possible to bridge and compare the data generated by each individual method. From the 956 isolates, 138 unique genotypes were identified based the combined markers of PCR–RFLP, microsatellite and intron sequencing (Fig. 3.1). Fifteen haplogroups were defined by intron sequencing data. The 138

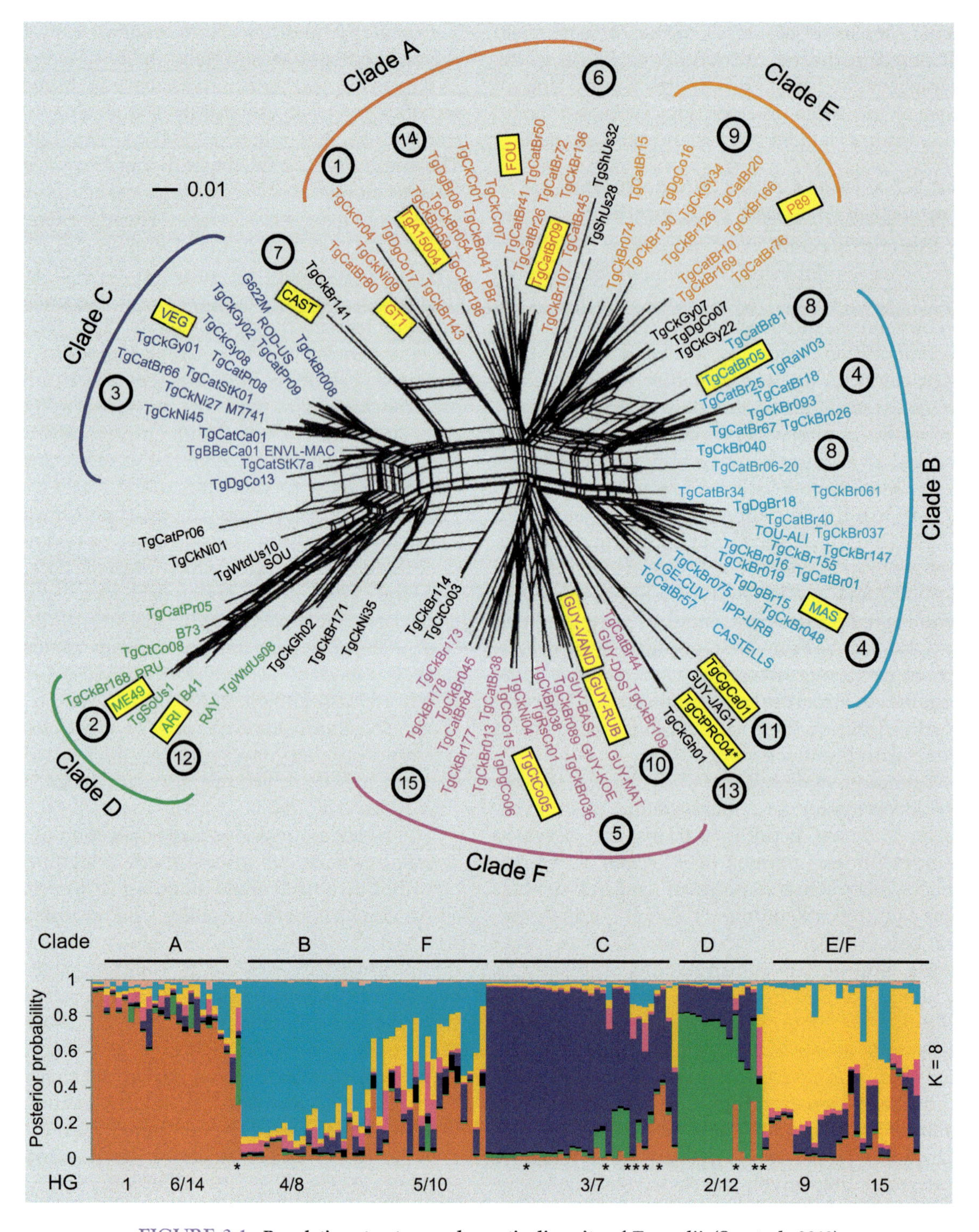

FIGURE 3.1 **Population structure and genetic diversity of *T. gondii*. (Su *et al.*, 2012)**

genotypes were clustered into six major clades (A through F). The conventional types 1, 2 and 3 lineages are grouped into clades A, D and C, respectively. The overall population consisted of clusters of highly abundant, overrepresented clonal genotypes along with more diverse groups that may be derived from genetic crosses. The distribution of genotypes showed a strong geographic separation, with widespread clonal genotypes in the northern hemisphere and highly diverse genotypes in South America. Population structure analysis of these genotypes indicated that a small number of ancestral lineages gave rise to the existing diversity through a process of limited admixture (Su *et al.*, 2012).

3.5 FACTORS AFFECTING TRANSMISSION AND GENETIC EXCHANGE

Toxoplasma gondii is a heterogamous parasite with sexual and asexual life stages in definitive and intermediate hosts, respectively. However, population structure studies reveal a high proportion of clonality in some parts of the world, suggesting a restricted use of this sexual cycle. The limited genetic exchange in a given population may result in an upstream inhibition of recombination or a downstream elimination of recombinant genotypes by natural selection (Tibayrenc and Ayala, 2002). Both biological and environmental factors influence the proportion of genetic exchanges.

3.5.1 Biological Factors

Several of the biological properties of *T. gondii* may account for the limited genetic exchanges:

Horizontal transmission through carnivory among intermediate hosts: T. gondii can effectively bypass the sexual stage by cycling, presumably indefinitely, among intermediate hosts (Su *et al.*, 2003; Grigg *et al.*, 2009).

Vertical, transplacental transmission: there is evidence that, in some host species, *T. gondii* may be serially, vertically transmitted (Williams *et al.*, 2005; Hide *et al.*, 2009; Innes *et al.*, 2009; Miller *et al.*, 2008a). This would lead to a local clonal expansion of the parasite.

Self-fertilization or self-mating: successful mating of male and female gametes originating from a single parasite clone (i.e. with identical genetic background) has been demonstrated experimentally (Cornelissen and Overdulve, 1985; Pfefferkorn *et al.*, 1977). This results in a clonal expansion via sex and meiosis (Wendte *et al.*, 2010). The sexual stage infection in the cat is relatively transient and it is thus likely that the majority of infections involve only a single *T. gondii* isolate derived from a single prey source. This means that self-fertilization or 'selfing' would be common and would limit the flow of genes between strains (Howe and Sibley, 1995). This has also been suggested as being the main factor in the emergence of epidemic clones (Wendte *et al.*, 2010).

Parthenogenesis: many macrogametes of the parasite remain unfertilized but are capable of forming oocysts in the small intestine of cats by parthenogenesis (Ferguson, 2002).

Immunity against reinfection: the presence of only one strain in a prey species is a key to the transmission dynamics of the parasite, conditioning clonal expansion both by selfing and by carnivory transmission. Multiple infections may be due either to simultaneous ingestion of different preys or of an oocyst sample with different genotypes or to reinfection. *Toxoplasma* is classically known to induce a strong immune response both in intermediate and in definitive hosts, preventing infection by a new isolate. In murine models, reinfection was possible only when the prime and the challenge strains belonged to two different types (Dao *et al.*, 2001; Elbez-Rubinstein *et al.*, 2009). In a context of a highly predominant type circulating in the environment, such as type 2 in Europe, reinfection should be the exception and infection with a single isolate the rule, reinforcing the clonal structure. A mixed infection with two different isolates both belonging to type 2 has only been described once, with highly polymorphic microsatellite markers (Ajzenberg, *et al.*, 2002b). Ingestion by a cat of such a prey infected by two different isolates belonging to the same type will not result in the emergence of a new clonal type. Mixed infections, possibly resulting from reinfections, were more often described in countries where numerous different genotypes are already circulating and where recombination events occur more frequently (Dubey *et al.*, 2007c; Dubey *et al.*, 2009; Pan *et al.*, 2012).

Fitness: the expansion of types 2 and 3 was suggested to be facilitated by their ability to effectively outcompete other genotypes (Grigg and Sundar, 2009; Sibley and Aijoka, 2008). These superfit strains were supposed to possess the right mix of alleles for key proteins such as ROP or GRA proteins (Boothroyd, 2009). Some recombinant genotypes resulting from a genetic outcrossing in a definitive host may be unable to develop in some intermediate hosts. But, at present, this has not been demonstrated. The notion of fitness was also hypothesized to explain the higher diversity of *T. gondii* strains paralleling the host range diversity (see below, influence of environmental factors) (Mercier *et al.*, 2011).

3.5.2 Environmental Factors

The analysis of a genetic population structure must consider the opportunities for transmission between hosts in different environments. For *T. gondii,* important factors to take into account are the felid population (density, diversity, hunting areas and dietary intake), the richness of prey species and the possibility of circulation of these species. The variations in the dynamics of *T. gondii* transmission in an urban/rural/non-anthropized gradient, in temperate and in tropical environments, have recently been reviewed in Gilot-Fromont *et al.* (2012). They have consequences in terms of *T. gondii* genetic population structure.

In domestic areas, only a limited number of host species such as cats, a few meat-producing animals and peridomestic mammals and birds are involved in the *T. gondii* domestic cycle. Among many genotypes, the three clonal lineages seemed to be most successfully adapted to these domestic hosts (Lehmann *et al.*, 2003). They would have diverged about 10,000 years ago which coincides with the domestication of companion and agricultural animals. In Europe or North America, intensive breeding of a narrow range of domestic meat-producing animals together with cat domestication offered a major niche to these lineages. In rural areas, farm buildings which shelter both cats and small intermediate hosts (small peridomestic rodents and birds) are reservoirs of infection, from which transmission of clonal types can radiate to the surrounding wild environment (Lehmann *et al.*, 2003), leading to an impoverishment of genetic diversity even in wildlife. In such a context, strains found in the neighbouring wildlife are usually similar to strains found in domestic animals (De Craeye *et al.* 2011; Aubert *et al.*, 2010; Halos *et al.*, 2010). Human activities lead to an impoverishment of diversity limiting recombination and thus gene flow in *T. gondii*.

Remote tropical areas, like the rainforest, are species-rich habitats which harbour many species of mammals and birds. This could sustain a greater diversity of parasite genotypes in order to colonize the maximum of ecological niches (Ajzenberg *et al.*, 2004; Mercier *et al.*, 2011). The different behavior of wild felids compared to that of domestic cats (larger hunting areas and dietary intake) and the number of possible preys influence hybridization patterns and gene flow of the parasite and thus its genetic structure of populations. Definitive hosts are more frequently infected by multiple *T. gondii* genotypes, which then cross and recombine before transmission to a new intermediate host (Gilot-Fromont *et al.*, 2012; Pan *et al.*, 2012). The possibility of reinfection by different strains may be another source of increasing diversity (Elbez-Rubinstein *et al.*, 2009). All these ecological factors lead to a high genetic diversity. For instance, in the French Guiana rainforest, one of the most important hotspots of diversity with at least 183 mammal species – including eight of 39 known wild felid species, and 718 bird species – 'wild' strains exhibited a remarkably higher genetic diversity than strains from the adjacent anthropized environment (Mercier *et al.*, 2011). Besides, strains from the anthropized environment were clustered into a few widespread lineages whereas the 'wild' population of strains does not exhibit any clear genetic clustering structure or any linkage disequilibrium, supporting the hypothesis of an important mixing in this natural, undisturbed ecosystem.

An intermediate situation may be present in countries such as the USA or Canada where large territories are still non-anthropized. Here, a genotypic diversity of *T. gondii* in the wild animals is present, co-existing with clonal lineages in anthropized areas (Dubey *et al.*, 2008b, 2011; Wendte *et al.*, 2011). At the confluence between both environments, wild animals may penetrate in anthropized areas and domestic animals come in contact with the wild through wild game, soil or running water. The consequences of this interpenetration in terms of *T. gondii* genotypes are diverse: (1) detection of *T. gondii* strains with 'hybrid' genotypes between the 'wild' population and the anthropized population reflecting genetic exchanges, (2) strains from the wild environment found in domestic animals, or (3) on the opposite strains from the anthropized environment found in wild animals. This was shown in the French Guianan context (Mercier *et al.*, 2011), but we can also detect the influence of wild strains in North America by the presence of diverse genotypes in domestic animals (Dubey *et al.*, 2011a; Dubey *et al.*, 2008b).

Human activities, such as urbanization, fragmentation of landscape, deforested areas, farming, domestication of cats and other animals, modify *T. gondii* ecology by clustering the parasite population; meanwhile, transportation of these strains through large distances by human trade exchange and transportation of animals leads occasionally to expansion of clonal lineages. This mechanism was suggested as an explanation for the large distribution of the same lineages across countries or continents (Lehman *et al.*, 2006; Mercier *et al.*, 2011).

3.6 *TOXOPLASMA* GENOTYPE AND BIOLOGICAL CHARACTERISTICS

Virulence in the mouse model is the most recognized phenotype and has been well described for the classical clonal lineages, types 1, 2 and 3. Type 1 strain led to a widespread parasite dissemination and death of mice less than 10 days after inoculation of less than 10 tachyzoites; in contrast, mice survived to infection with a type 2 or 3 strain (LD_{50} ranging from 10^2 to 10^5) and tachyzoite dissemination was much less extensive. Type 3 is also generally considered as avirulent in mice, although progressive deterioration and death of mice, notably with neurological symptoms, can occur a few weeks or months after inoculation (Dardé *et al.*, 1988). The genetic differences between these three strains elicit a different modulation in the host cells and different immune response in the host that could in part explain the different patterns of virulence (Gavrilescu and Denkers, 2001; Fux *et al.*, 2003; Nguyen *et al.*, 2003; Diana *et al.*, 2004; Robben *et al.*, 2004; Dubremetz and Lebrun, 2012; Niedelman *et al.*, 2012; Rosowski *et al.*, 2011).

The higher virulence of type 1 in mice compared to type 2 or 3 has been correlated with *in vitro* biological properties: type 1 displays enhanced migration under soft agarose plates, as well as enhanced transmigration across polarized epithelia or across extracellular matrix. They also show a higher rate of *ex vivo* penetration of lamina propria and submucosa (Barragan and Sibley, 2002, 2003). This ability to cross epithelial barriers rapidly and reach the bloodstream within hours post-infection might be an important predeterminant of parasite dissemination *in vivo* in susceptible host species. In cell culture, type 1 grows faster than type 2 or 3 and has a lower rate of interconversion from tachyzoite to bradyzoite than type 2 strains (Soete *et al.*, 1993). The higher growth rate of type 1 parasites due either to a higher reinvasion rate or to a shorter doubling time may also explain the higher tissue burden observed in mice infected with virulent strains (Saeij *et al.*, 2005).

Other genotypes are usually more virulent in mice than type 2 or 3. They ranged from highly virulent to intermediate or non-virulent phenotype according to the differences in the precise combination of alleles they have inherited (Grigg and Suzuki, 2003; Niedelmann *et al.*, 2012). Experimental crosses between types 1, 2 or 3 strains proved useful in identifying genes that determine mouse virulence (Saeij *et al.*, 2006; Taylor *et al.*, 2006; Behnke *et al.*, 2011; Reese *et al.*, 2011; Niedelmann *et al.*, 2012). They also showed that mouse mortality caused by the progeny clones of these crosses ranged from low levels to 100%, consistent with a multilocus trait. The same finding was observed in naturally recombinant isolates. Among the five recombinant (2×3) genotypes identified in an oocyst sample, two were of low virulence in mice, one showed an intermediate virulence (LD_{50} greater than or equal to 10^2 but less than 10^4 tachyzoites) or high (LD_{50} less than 10^2 tachyzoites) virulence phenotype in mice, and two were highly virulent (LD_{50} less than 10^2 tachyzoites) in mice (Herrmann *et al.*, 2012a).

Although *in vitro* studies demonstrate different intrinsic properties of the different strains, the expression of this virulence in a given host species is a more complex trait which depends on several host and parasite characteristics. The definition of *T. gondii* virulence with respect to mouse infection led to much ambiguity when defining virulence factors, since other hosts and especially man may behave quite differently from mice (Dubremetz and Lebrun, 2012). The most divergent strains, Amazonian strains, which have virulence traits in mice but also in immunocompetent humans, are useful for fundamental studies of pathogenicity.

3.7 *TOXOPLASMA GONDII* GENOTYPE AND HUMAN DISEASE

3.7.1 Circumstances of Isolation and Genetic Typing

The circumstances during which *T. gondii* strains can be isolated from human cases of toxoplasmosis are relatively rare. They necessitate the retrieval of tachyzoites from pathological samples collected for diagnosis purposes (blood, amniotic fluid, bronchoalveolar lavage fluid, aqueous humour, etc.) or of cysts in tissues collected via biopsy or necropsy. Isolates from human toxoplasmosis come predominantly from cases of congenital toxoplasmosis or from immunodeficient patients, and much less frequently from symptomatic acquired toxoplasmosis in immunocompetent patients. As a diagnostic assay, strain isolation via mouse inoculation is not easy to perform and tends to be abandoned in favour of PCR-based assays which are more sensitive and standardized, give faster results, and bypass the need of animal care facilities.

The sensitivity of the PCR-based methods theoretically allows direct analysis of the parasite genotype from primary clinical sample. This would enable genotyping to be performed on small quantities of pathological products such as ocular fluids or on formaldehyde fixed tissues. In reality, the low number of parasites present in some samples and the frequent presence of PCR inhibitors often led to negative results with single-copy genotyping markers. Increasing the sensitivity of genotyping is theoretically possible with nested PCR assays, but this procedure greatly amplifies the risk of DNA contamination. Genetic typing of clinical samples must rely on multilocus markers and be suitable for screening of a large number of isolates (Ajzenberg, 2010; Su *et al.*, 2010). Myriad monolocus studies with the SAG2 marker were reported in the literature, but they will not be cited in this chapter because they failed to capture the true genotypes of *T. gondii*. The most widespread techniques currently used for genotyping *T. gondii* in human toxoplasmosis rely on PCR–RFLP and microsatellite markers (Su *et al.*, 2006; Ajzenberg *et al.*, 2010).

Strain isolation or direct typing on pathological products can only be performed during symptomatic toxoplasmosis. To fully interpret the influence of *T. gondii* genotype on the different clinical aspects of human toxoplasmosis, we should also genotype the vast majority of human infections that are asymptomatic. Serological typing, a non-invasive method dependent on recognition of variant antigenic determinants (Kong *et al.*, 2003), may be useful in generating data on all the strains circulating in the human population.

3.7.2 Congenital Toxoplasmosis

Congenital toxoplasmosis is the main source of *T. gondii* isolation in humans from samples of amniotic fluid, placenta, cord blood and tissues of aborted foetuses. Amniotic fluids or even placenta are available mainly in countries, such as France, where a systematic prenatal screening of congenital toxoplasmosis is performed. In this case, nearly all the isolates responsible for congenital toxoplasmosis, including the majority of asymptomatic cases at birth, can be submitted for genotyping. Otherwise, strain isolation comes mainly from the more severe cases of symptomatic congenital toxoplasmosis, introducing a clinical bias.

The largest series of congenital cases with genotyping data was conducted in a multicentre study in France (Ajzenberg, *et al.*, 2002a). Typing performed on all the strains consecutively isolated in congenital cases in French laboratories revealed that they almost all belonged to type 2. Type 2 isolates were found in all the different aspects of congenital disease from lethal infection and severe neuro-ocular involvement in early maternal infections to isolated chorioretinitis or latent toxoplasmosis in late maternal infections. Though type 2 strains can sometimes prove highly pathogenic to the human foetus (Kieffer *et al.*, 2011), the severity of foetal infection caused by type 2 strains is clearly linked to the period of gestation at which the mother becomes infected with a decreasing risk of clinical disease when gestational age increases.

The question is to know whether non-type 2 strains are more virulent for the foetus than type 2 strains. According to the data available, type 3 strains seem to behave roughly like type 2 strains in congenital toxoplasmosis and true type 1 strains are too rarely isolated to understand how they are interacting with the human host (Ajzenberg, 2010). However, there is a growing body of evidence to link atypical strains to a higher burden in congenital toxoplasmosis (Demar *et al.*, 2007; Elbez-Rubinstein *et al.*, 2009; Delhaes *et al.*, 2010). The higher pathogenicity of atypical strains has been suggested to explain the more severe ocular disease in congenitally infected children in Brazil when compared to those infected in Europe (Gilbert *et al.*, 2008). It is now clear that Europe and Brazil have totally different population structures of *T. gondii*. If diversity is low in Europe with a large predominance of type 2 strains, it is not the case in Brazil where countless diverse and atypical genotypes have been characterized in animals. One can assume that this diversity of strains is also true in humans and could explain the higher burden of the disease in infants congenitally infected with some virulent *T. gondii* strains in Brazil. Unfortunately, strain collection from human cases is lacking in this area to fully support the hypothesis (Ferreira *et al.*, 2011). In fact, reports of atypical strains in congenital toxoplasmosis are scarce but most of these cases were severe even after late maternal infection during pregnancy (Ajzenberg, *et al.*, 2002a; Demar *et al.*, 2007; Elbez-Rubinstein *et al.*, 2009; Delhaes *et al.*, 2010). Furthermore, the fact that atypical strains are able to reinfect women with a past infection to *T. gondii* is another clue to suggest that these strains with a different genetic background trigger an inappropriate immune response in the host which increases the risk of severe damages for the foetus (Elbez-Rubinstein *et al.*, 2009). Because isolation of strains is a difficult task from clinical samples, a partial answer to the question of the higher pathogenicity of non-type 2 strains versus type 2 strains may come from serotyping instead of genotyping even though serotyping assays have limitations (Ajzenberg, 2012). In a recent study conducted in congenitally infected infants in the USA with a serotyping assay, severe disease at birth was more common in infants with non-type 2 serotypes than in those with type 2 serotypes (MacLeod *et al.*, 2012). These non-type 2 serotypes could belong to the haplogroup 12 which was also named fourth clonal lineage (Khan *et al.*, 2011a). Strains belonging to the haplogroup 12 are overabundant and endemic in the USA and appear more pathogenic in congenital toxoplasmosis than the type 2 strains commonly found in Europe (Ajzenberg, 2012).

Whilst severity of disease is influenced by trimester in which infection is acquired by the mother and likely by the genetic background of strains, other factors including genetic predisposition may contribute. Studies suggest that polymorphisms in genes affecting immune response, including HLA (Mack *et al.*, 1999), or in genes playing a role in developmental processes, such as ABCA4 and COL2A1, could influence clinical outcome in congenital toxoplasmosis (Jamieson *et al.*, 2008).

3.7.3 Postnatally Acquired Toxoplasmosis in Immunocompetent Patients

3.7.3.1 Ocular Toxoplasmosis

The influence of *T. gondii* genotype in ocular toxoplasmosis (OT) has been poorly studied because of the scarcity of ocular fluid samples for genotyping analysis. These low-volume samples (usually less than 200 μL) require invasive diagnostic procedures, and most of the collected volume is dedicated for the biological diagnosis of OT. Isolation of strains from ocular samples in mice or by cell culture is usually never attempted because of these volume limitations. When PCR-based assays are performed for detecting *T. gondii*, the extracted DNA may be used for direct genotyping analyses but the amount of parasitic DNA is often too low to permit amplification of single-copy genetic markers leading to unsuccessful or incomplete typing results. Moreover, genotyping analyses performed directly on aqueous and vitreous humour samples may lead to a bias toward an under representation of the common cases of OT because these ocular fluid samples are usually collected when the clinical presentation is severe and/or atypical (Subauste *et al.*, 2011).

Ocular toxoplasmosis (OT) is the most common identifiable cause of posterior uveitis in many parts of the world, but the burden of postnatal OT varies significantly across geographical areas. For example, it was estimated that 0.6% of the general population of Alabama and Maryland in the USA had retinochoroidal scars consistent with *T. gondii* infection (Holland, 2003). Much higher figures of prevalence of toxoplasmic retinal scars have been reported in population-based studies in South America, ranging from 6% in Armenia, Colombia (de la Torre *et al.*, 2007), to 21% in Erechim, southern Brazil (Glasner *et al.*, 1992). Among the different hypotheses to explain this variation in the prevalence of OT between the northern and the southern parts of the Americas, the well-known genetic divergence between strains from both areas has been regarded as the most plausible (Subauste *et al.*, 2011).

The available multilocus genotyping data of OT in Brazil were mainly collected in the southern states of the country where the prevalence of the disease is known to be extremely high. *T. gondii* genotypes were retrieved directly from patients' samples in the city of Erechim, Rio Grande do Sul state (Khan *et al.*, 2006) and in diverse cities of Saõ Paulo state (Ferreira *et al.*, 2011), whereas one strain was collected indirectly from environmental samples implicated as the source of a large waterborne toxoplasmosis outbreak that caused a high prevalence of eye involvement in Santa Isabel do Ivai, Paraná state (de Moura *et al.*, 2006; Vaudaux *et al.*, 2010). It is hard to draw any definite conclusion from these studies because the number of cases is very limited and additional studies are needed to rule out cross-contamination of DNA samples, but some unexpected findings deserve attention. The genetic diversity of *T. gondii* strains in OT appears limited when compared to the genetic data of strains collected in animals in Brazil. In Saõ Paulo state, for example, it is striking to see that only one genotype (ToxoDB genotype #65) was collected from seven patients with OT (Ferreira *et al.*, 2011) whereas, in the same state of Saõ Paulo, 20 different genotypes were collected from 46 cats (Pena *et al.*, 2008). The ToxoDB genotype #65 seems to be common in Brazil since it has also been described in immunosuppressed patients with cerebral toxoplasmosis (Ferreira *et al.*, 2011) and in animals elsewhere in this country (Pena *et al.*, 2008). Similarly, most of the 11 patients from Erechim were infected by only one genotype and the strain involved in the outbreak of Santa Isabel do Ivai had the BrI genotype which is considered as a major clonal lineage in Brazil because it has been sampled in numerous animals from different areas. All these data suggest that, among the numerous genotypes circulating in Brazil, only a few are responsible for the high burden of OT in Brazil and are common in the environment. A major argument for the high virulence of strains collected in patients from Erechim and Saõ Paulo state is indirectly highlighted by the fact that they were not characterized from ocular fluid samples but from peripheral blood (Khan *et al.*, 2006; Ferreira *et al.*, 2011). This prolonged parasitaemia in ocular disease in Brazil has been confirmed by direct microscopic observation of tachyzoites in blood samples (Silveira *et al.*, 2011).

Studies in the USA and Europe have reported contradictory results. In France and in several European countries, OT involves predominantly type 2 strains (Morisset *et al.*, 2008; Fekkar *et al.*, 2011). This result is not surprising since type 2 strains are the most common strains in the European environment. According to one survey that investigated the genetic background of strains causing atypical and/or severe OT in the USA, most cases in immunocompetent patients were not due to the common type 2 or haplogroup 12 strains but involved atypical genotypes (Grigg *et al.*, 2001b). Because epidemiological information is lacking, it is not possible to know if these atypical strains were collected in patients originating from South America where such atypical genotypes are common. Given the unclear population structure in North America, it is however possible that these atypical genotypes circulate in the

USA and may be more virulent than type 2 strains. Notably, in Canada, the largest outbreak of toxoplasmosis ever reported was likely associated to an atypical strain shed by a cougar in one of two main reservoirs of municipal drinking water serving the Greater Victoria area (Bowie *et al.*, 1997; Burnett *et al.*, 1998; Dubey, *et al.*, 2008a).

3.7.3.2 Disseminated Toxoplasmosis

Atypical strains with highly divergent genotypes have been involved in the rare cases of disseminated toxoplasmosis observed in otherwise healthy people. The disease is characterized by high and prolonged fever associated with lung involvement which may be life threatening and need intensive care management. The parasitic DNA is usually genotyped from peripheral blood or bronchoalveolar lavage fluid samples. Most of these cases have been described in French Guiana (Carme *et al.*, 2002, 2009b; Demar *et al.*, 2007, 2012; Groh *et al.*, 2012), but some cases may occur elsewhere like in Europe (De Salvador-Guillouët *et al.*, 2006; Pomares *et al.*, 2011). In French Guiana this severe infection is acquired after consumption of wild game or drinking unfiltered water from the Amazonian rainforest of French Guiana. In Europe, these cases are thought to be acquired after consumption of imported meat, especially horsemeat, from South America.

The determinants of strain-specific differences in virulence for human infection remain unknown. The *ROP18*, *ROP5*, *ROP16*, and *GRA15* loci of *T. gondii* have been associated with virulence in mice, but it cannot be assumed that these four virulence loci in mice similarly affect survival in human cells (Niedelman *et al.*, 2012). An outbreak in Suriname showed that the same atypical strain could lead to a broad spectrum of clinical manifestations ranging from mild illness to life threatening disease (Demar *et al.*, 2007). These data strongly suggest that the inoculum size of *T. gondii* related to different dietary practices or hygiene habits is of paramount importance to explain the severity of symptoms in toxoplasmosis.

3.7.4 Postnatally Acquired Toxoplasmosis in Immunocompromised Patients

In AIDS patients, the most frequent clinical presentation of toxoplasmosis is encephalitis. The material for isolating and genotyping strains is rarely available in these patients because cerebral biopsies are almost never performed and PCR assays usually test negative in blood samples. In patients with severe toxoplasmosis whose immunosuppression is not caused by HIV infection, such as those who undergo allogeneic haematopoietic stem cell transplantation (HSCT), *T. gondii* is often detected in blood and BAL samples and therefore could be genotyped, but the incidence of this opportunistic infection in those patients is much lower than in AIDS patients.

Immunosuppressed patients usually reactivate the strain they acquired asymptomatically years before when they were immunocompetent. It is then expected that the genetic background of strains in these patients reflects the one observed in the general population in a given area. For example, an immunocompromised patient with a past immunity to *T. gondii* and who has always lived in France will have a high probability to reactivate a type 2 strain because type 2 strains account for greater than 95% of strains in the French environment. In other areas with a higher diversity of strains, the overrepresentation of a given genotype in immunocompromised patients is likely to reflect a disproportionately high infection with this genotype in meat-producing animals and environment of this area.

The largest collection of multilocus *T. gondii* genotypes with epidemiological, clinical, and outcome data in immunocompromised people was obtained from a French multi-centre study (Ajzenberg *et al.*, 2009). The genotype of *T. gondii* strains was strongly linked to the presumed geographical origin of infection. Type 2 strains were predominant in patients that acquired the infection in Europe whereas non-type 2 strains were more commonly recovered from patients with infections acquired outside Europe. Two main atypical genotypes were recovered from patients who acquired the infection in several sub-Saharan African countries (genotype *Africa 1*) and in the French West Indies (genotype *Caribbean 1*) suggesting that these atypical genotypes are common in these areas. The distribution of *T. gondii* genotypes (type 2 vs. non-type 2) was not significantly different between patients with AIDS and non-HIV patients, nor was it significantly different for different sites of infection and outcome. These data suggested that the genotype of the strain do not play a major role in the pathophysiology and severity of toxoplasmosis in immunocompromised patients.

3.7.5 Conclusion and Perspective on *Toxoplasma* Genotype and Human Disease

The role of the strain in the severity of human toxoplasmosis is likely to be a reality but is difficult to assess because the scenario of pathogenicity is complex and involves several factors such as genetic background of the strain, inoculum dose, suppression or immaturity of the host immune system and genetic predisposition. Because of their association with a high burden of ocular disease in Brazil and their ability to cause life threatening symptoms in otherwise healthy people in the Amazonian forest, the atypical strains circulating in tropical South America appear to be more pathogenic for humans than those circulating elsewhere in the world. However, it is unclear whether the higher pathogenicity of the South American strain is associated with true strain-specific differences in virulence or lack of adaptive host response. Future research should answer this question. Paradoxically, strains from this area were mainly collected from animals and genotyping data on human isolates are very limited. The collection of more strains from human toxoplasmosis with accurate clinical and epidemiological data should be encouraged in South America. The genotypes of these strains were classified as atypical because their genetic polymorphisms did not fit within the three clonal lineages initially described in Europe and the USA. In fact, the term 'atypical' includes a big variety of divergent genotypes, and efforts should be conducted for clarifying their classification in a simple, consensual and universal manner. Generation of whole genome sequences for some South American strains is ongoing and will greatly expand our understanding of genetic diversity of *T. gondii*.

References

Ajzenberg, D., Cogné, N., Paris, L., et al., 2002a. Genotype of 86 *Toxoplasma gondii* isolates associated with human congenital toxoplasmosis, and correlation with clinical findings. J. Infect. Dis. 186, 684–689.

Ajzenberg, D., Banuls, A.L., Tibayrenc, M., Darde, M.L., 2002b. Microsatellite analysis of *Toxoplasma gondii* shows considerable polymorphism structured into two main clonal groups. Int. J. Parasitol. 32, 27–38.

Ajzenberg, D., Banuls, A.L., Su, C., et al., 2004. Genetic diversity, clonality and sexuality in. Toxoplasma gondii. Int. J. Parasitol. 34, 1185–1196.

Ajzenberg, D., Year, H., Marty, P., et al., 2009. Genotype of 88 *Toxoplasma gondii* isolates associated with toxoplasmosis in immunocompromised patients and correlation with clinical findings. J. Infect. Dis. 199, 1155–1167.

Ajzenberg, D., 2010. Type I strains in human toxoplasmosis: myth or reality? Future Microbiol. 5, 841–843.

Ajzenberg, D., Collinet, F., Mercier, A., Vignoles, P., Darde, M.L., 2010. Genotyping of *Toxoplasma gondii* isolates with 15 microsatellite markers in a single multiplex PCR assay. J. Clin. Microbiol. 48, 4641–4645.

Ajzenberg, D., 2012. High burden of congenital toxoplasmosis in the United States: the strain hypothesis? Clin. Infect. Dis. 54, 1606–1607.

Al-Qassab, S., Reichel, M.P., Su, C., et al., 2009. Isolation of *Toxoplasma gondii* from the brain of a dog in Australia and its biological and molecular characterization. Vet. Parasitol. 164, 335–339.

Al-Kappany, Y.M., Rajendran, C., Abu-Elwafa, S.A., Hilali, M., Su, C., Dubey, J. P., 2010. Genetic diversity of *Toxoplasma gondii* isolates in Egyptian feral cats reveals new genotypes. J. Parasitol. 96, 1112–1114.

Aubert, D., Ajzenberg, D., Richomme, C., et al., 2010. Molecular and biological characteristics of *Toxoplasma gondii* isolates from wildlife in France. Vet. Parasitol. 171, 346–349.

Barragan, A., Sibley, L.D., 2002. Transepithelial migration of *Toxoplasma gondii* is linked to parasite motility and virulence. J. Exp. Med. 195, 1625–1633.

Barragan, A., Sibley, L.D., 2003. Migration of *Toxoplasma gondii* across biological barriers. Trends Microbiol. 11, 426–430.

Behnke, M.S., Khan, A., Wootton, J.C., Dubey, J.P., Tang, K., Sibley, L.D., 2011. Virulence differences in *Toxoplasma* mediated by amplification of a family of polymorphic pseudokinases. Proc. Natl. Acad. Sci. USA 108, 9631–9636.

Blackston, C.R., Dubey, J.P., Dotson, E., et al., 2001. High-resolution typing of *Toxoplasma gondii* using microsatellite loci. J. Parasitol. 87, 1472–1475.

Boothroyd, J.C., 2009. Expansion of host range as a driving force in the evolution of Toxoplasma. Mem. Inst. Oswaldo Cruz 104, 179–184.

Bossi, P., Bricaire, F., 2004. Severe acute disseminated toxoplasmosis. Lancet 364, 579.

Bowie, W.R., King, A.S., Werker, D.H., et al., 1997. Outbreak of toxoplasmosis associated with municipal drinking water. Lancet 350, 173–177.

Boyle, J.P., Rajasekar, B., Saeij, J.P., et al., 2006. Just one cross appears capable of dramatically altering the population biology of a eukaryotic pathogen like Toxoplasma gondii. Proc. Natl. Acad. Sci. USA. 103, 10514–10519.

Burg, J.L., Grover, C.M., Pouletty, P., Boothroyd, J.C., 1989. Direct and sensitive detection of a pathogenic protozoan *Toxoplasma gondii* by polymerase chain reaction. J. Clin. Microbiol. 27, 1787–1792.

Burnett, A.J., Shortt, S.G., Isaac-Renton, J., King, A., Werker, D., Bowie, W.R., 1998. Multiple cases of acquired toxoplasmosis retinitis presenting in an outbreak. Ophthalmology 105, 1032–1037.

Carme, B., Bissuel, F., Ajzenberg, D., et al., 2002. Severe acquired toxoplasmosis in immunocompetent adult patients in French Guiana. J. Clin. Microbiol. 40, 4037–4044.

Carme, B., Ajzenberg, D., Demar, M., et al., 2009a. Outbreaks of toxoplasmosis in a captive breeding colony of squirrel monkeys. Vet. Parasitol. 163, 132–135.

Carme, B., Demar, M., Ajzenberg, D., Dardé, M.L., 2009b. Severe acquired toxoplasmosis caused by wild cycle of *Toxoplasma gondii*, French Guiana. Emerg. Infect. Dis. 15, 656–658.

Calderaro, A., Piccolo, G., Gorrini, C., et al., 2006. Comparison between two real-time PCR assays and a nested-PCR for the detection of *Toxoplasma gondii*. Acta Biomed. 77, 75–80.

Chen, Z.W., Gao, J.M., Huo, X.X., et al., 2011. Genotyping of *Toxoplasma gondii* isolates from cats in different geographic regions of China. Vet. Parasitol. 183, 166–170.

Cornelissen, A.W., Overdulve, J.P., 1985. Sex determination and sex differentiation in coccidia: gametogony and oocyst production after monoclonal infection of cats with free-living and intermediate host stages of *Isospora (Toxoplasma) gondii*. Parasitology 90, 35–44.

Cristina, N., Dardé, M.L., Boudin, C., Tavernier, G., Pestre-Alexandre, M., Ambroise-Thomas, P., 1995. A DNA fingerprinting method for individual characterization of *Toxoplasma gondii* strains: combination with isoenzymatic characters for determination of linkage groups. Parasitol. Res. 81, 32–37.

Dao, A., Fortier, B., Soete, M., Plenat, F., Dubremetz, J.F., 2001. Successful reinfection of chronically infected mice by a different *Toxoplasma gondii* genotype. Int. J. Parasitol. 31, 63–65.

Dardé, M.L., Bouteille, B., Pestre-Alexandre, M., 1988. Isoenzymic characterization of seven strains of *Toxoplasma gondii* by isoelectrofocusing in polyacrylamide gels. Am. J. Trop. Med. Hyg. 39, 551–558.

Darde, M.L., Bouteille, B., Pestre-Alexandre, M., 1992. Isoenzyme analysis of 35 *Toxoplasma gondii* isolates and the biological and epidemiological implications. J. Parasitol. 78, 786–794.

Dardé, M.L., 2008. Toxoplasma gondii, 'new' genotypes and virulence. Parasite 15, 366–371.

De Craeye, S., Speybroeck, N., Ajzenberg, D., et al., 2011. *Toxoplasma gondii* and Neospora caninum in wildlife: common parasites in Belgian foxes and Cervidae? Vet. Parasitol. 178, 64–69.

de Sousa, S., Ajzenberg, D., Canada, N., et al., 2006. Biologic and molecular characterization of *Toxoplasma gondii* isolates from pigs from Portugal. Vet. Parasitol. 135, 133–136.

Delhaes, L., Ajzenberg, D., Sicot, B., et al., 2010. Severe congenital toxoplasmosis due to a *Toxoplasma gondii* strain with an atypical genotype: case report and review. Prenatal Diagn. 30, 902–905.

Demar, M., Ajzenberg, D., Maubon, D., et al., 2007. Fatal outbreak of human toxoplasmosis along the Maroni River: epidemiological, clinical, and parasitological aspects. Clin. Infect. Dis. 45, e88–e95.

Demar, M., Hommel, D., Djossou, F., et al., 2012. Acute toxoplasmoses in immunocompetent patients hospitalized in an intensive care unit in French Guiana. Clin. Microbiol. Infect. 18, E221–E231. http://dx.doi.org/10.1111/j.1469-0691.2011.03648.x.

de Moura, L., Bahia-Oliveira, L.M., Wada, M.Y., et al., 2006. Waterborne toxoplasmosis, Brazil, from field to gene. Emerg. Infect. Dis. 12, 326–329.

De Salvador-Guillouët, F., Ajzenberg, D., Chaillou-Opitz, S., et al., 2006. Severe pneumonia during primary infection with an atypical strain of *Toxoplasma gondii* in an immunocompetent young man. J. Infect. 53, e47–e50.

Diana, J., Persat, F., Staquet, M.J., et al., 2004. Migration and maturation of human dendritic cells infected with *Toxoplasma gondii* depend on parasite strain type. FEMS. Immunol. Med. Microbiol. 42, 321–331.

Dubey, J.P., Vianna, M.C., Sousa, S., et al., 2006. Characterization of *Toxoplasma gondii* isolates in free-range chickens from Portugal. J. Parasitol. 92, 184–186.

Dubey, J.P., Huong, L.T.T., Sundar, N., Su, C., 2007a. Genetic characterization of *Toxoplasma gondii* isolates in dogs from Vietnam suggests their South American origin. Vet. Parasitol. 146, 347–351.

Dubey, J.P., Zhu, X.Q., Sundar, N., Zhang, H., Kwok, O.C.H., Su, C., 2007b. Genetic and biologic characterization of *Toxoplasma gondii* isolates of cats from China. Vet. Parasitol. 145, 352–356.

Dubey, J.P., Lopez-Torres, H.Y., Sundar, N., et al., 2007c. Mouse-virulent *Toxoplasma gondii* isolated from feral cats on Mona Island, Puerto Rico. J. Parasitol. 93, 1365–1369.

Dubey, J.P., Quirk, T., Pitt, J.A., et al., 2008a. Isolation and genetic characterization of *Toxoplasma gondii* from raccoons (*Procyon lotor*), cats (*Felis domesticus*), striped skunk (*Mephitis mephitis*), black bear (*Ursus americanus*), and cougar (*Puma concolor*). J. Parasitol. 94, 42–45.

Dubey, J.P., Sundar, N., Hill, D., et al., 2008b. High prevalence and abundant atypical genotypes of *Toxoplasma gondii* isolated from lambs destined for human consumption in the USA. Int. J. Parasitol. 38, 999–1006.

Dubey, J.P., Moura, L., Majumdar, D., et al., 2009. Isolation and characterization of viable *Toxoplasma gondii* isolates revealed possible high frequency of mixed infection in feral cats (Felis domesticus) from St Kitts, West Indies. Parasitology 136, 589–594.

Dubey, J.P., Pas, A., Rajendran, C., et al., 2010. Toxoplasmosis in Sand cats (Felis margarita) and other animals in the Breeding Centre for Endangered Arabian Wildlife in the United Arab Emirates and Al Wabra Wildlife Preservation, the State of Qatar. Vet. Parasitol. 172, 195–203.

Dubey, J.P., 2010. Toxoplasmosis of animals and humans, Second ed. CRC Press, Boca Raton, FL., 336 p.

Dubey, J.P., Rajendran, C., Ferreira, L.R., et al., 2011a. High prevalence and genotypes of *Toxoplasma gondii* isolated from goats, from a retail meat store, destined for human consumption in the USA. Int. J. Parasitol. 41, 827–833.

Dubey, J.P., Velmurugan, G.V., Rajendran, C., et al., 2011b. Genetic characterisation of *Toxoplasma gondii* in wildlife from North America revealed widespread and high prevalence of the fourth clonal type. Int. J. Parasitol. 41, 1139–1147.

Dubey, J.P., Lago, E.G., Gennari, S.M., Su, C., Jones, J.L., 2012. Toxoplasmosis in humans and animals in Brazil: high prevalence, high burden of disease, and epidemiology. Parasitology 139, 1375–1424.

Dubremetz, J.F., Lebrun, M., 2012. Virulence factors of *Toxoplasma gondii*. Microb. Infect. Sep 21. http://dx.doi.org/10.1016/j.micinf.2012.09.005. [Epub ahead of print].

Dumètre, A., Ajzenberg, D., Rozette, L., Mercier, A., Dardé, M.L., 2006. *Toxoplasma gondii* infection in sheep from Haute-Vienne, France: seroprevalence and isolate genotyping by microsatellite analysis. Vet. Parasitol. 142, 376–379.

Elbez-Rubinstein, A., Ajzenberg, D., Darde, M.L., et al., 2009. Congenital toxoplasmosis and reinfection during pregnancy: case report, strain characterization, experimental model of reinfection, and review. J. Infect. Dis. 199, 280–285.

Fekkar, A., Ajzenberg, D., Bodaghi, B., et al., 2011. Direct genotyping of *Toxoplasma gondii* in ocular fluid samples from 20 patients with ocular toxoplasmosis: predominance of type II in France. J. Clin. Microbiol. 49, 1513–1517.

Ferreira, Ade M., Vitor, R.W., Carneiro, A.C., Brandao, G.P., Melo, M.N., 2004. Genetic variability of Brazilian *Toxoplasma gondii* strains detected by random amplified polymorphic DNA-polymerase chain reaction (RAPD-PCR) and simple sequence repeat anchored-PCR (SSR-PCR). Infect. Genet. Evol. 4, 131–142.

Ferreira, Ade M., Vitor, R.W., Gazzinelli, R.T., Melo, M.N., 2006. Genetic analysis of natural recombinant Brazilian *Toxoplasma gondii* strains by multilocus PCR-RFLP. Infect. Genet. Evol. 6, 22–31.

Ferreira, I.M., Vidal, J.E., de Mattos, C., et al., 2011. *Toxoplasma gondii* isolates: multilocus RFLP-PCR genotyping from human patients in Sao Paulo State, Brazil identified distinct genotypes. Exp. Parasitol. 129, 190–195.

Frazão-Teixeira, E., Sundar, N., Dubey, J.P., Grigg, M.E., de Oliveira, F.C.R., 2011. Multi-locus DNA sequencing of *Toxoplasma gondii* isolated from Brazilian pigs identifies genetically divergent strains. Vet. Parasitol. 175, 33–39.

Frey, C., Berger-Schoch, A., Herrmann, D., et al., 2012. Incidence and genotypes of *Toxoplasma gondii* in the muscle of sheep, cattle, pigs as well as in cat feces in Switzerland. Schweiz Arch Tierhei. 154, 251–255.

Ferguson, D.J., 2002. Toxoplasma gondii and sex: essential or optional extra? Trends Parasitol. 18, 355–359.

Fux, B., Rodrigues, C.V., Portela, R.W., et al., 2003. Role of cytokines and major histocompatibility complex restriction in mouse resistance to infection with a natural recombinant strain (type I-III) of *Toxoplasma gondii*. Infect. Immun. 71, 6392–6401.

Gavrilescu, L.C., Denkers, E.Y., 2001. IFN-gamma overproduction and high level apoptosis are associated with high but not low virulence *Toxoplasma gondii* infection. J. Immunol. 167, 902–909.

Gilbert, R.E., Freeman, K., Lago, E.G., et al., 2008. European Multicentre Study on Congenital Toxoplasmosis (EMSCOT). Ocular sequelae of congenital toxoplasmosis in Brazil compared with Europe. PLoS. Negl. Trop. Dis. 13, e277.

Gibson, A.K., Raverty, S., Lambourn, D.M., Huggins, J., Magargal, S.L., Grigg, M.E., 2011. Polyparasitism is associated with increased disease severity in *Toxoplasma gondii*-infected marine sentinel species. PLoS. Negl. Trop. Dis. 5, e1142.

Gilot-Fromont, E., Lélu, M., Dardé, M.L., et al., 2012. The Life Cycle of *Toxoplasma gondii* in the Natural Environment. 'Toxoplasmosis - Recent Advances' O. Djurković Djaković (Eds.), ISBN 978-953-51-0746-0, InTech, September 9, 2012. Available from: http://www.intechopen.com/books/toxoplasmosis-recent-advances/the-life-cycle-of-toxoplasma-gondii-in-the-natural-environment.

Glasner, P.D., Silveira, C., Kruszon-Moran, D., et al., 1992. An unusually high prevalence of ocular toxoplasmosis in southern Brazil. Am. J. Ophthalmol. 114, 136–144.

Goldstein, D.B., Schlotterer, C., 1999. Microsatellites: Evolution and Applications. Oxford University Press, Oxford, 352 p.

Grigg, M.E., Bonnefoy, S., Hehl, A.B., Suzuki, Y., Boothroyd, J.C., 2001a. Success and virulence in *Toxoplasma* as the result of sexual recombination between two distinct ancestries. Science 294, 161–165.

Grigg, M.E., Ganatra, J., Boothroyd, J.C., Margolis, T.P., 2001b. Unusual abundance of atypical strains associated with human ocular toxoplasmosis. J. Infect. Dis. 184, 633–639.

Grigg, M.E., Suzuki, Y., 2003. Sexual recombination and clonal evolution of virulence in. Toxoplasma. Microbes Infect. 5, 685–690.

Grigg, M.E., Sundar, N., 2009. Sexual recombination punctuated by outbreaks and clonal expansions predicts *Toxoplasma gondii* population genetics. Int. J. Parasitol. 39, 925–933.

Groh, M., Faussart, A., Villena, I., et al., 2012. Acute lung, heart, liver, and pancreatic involvements with hyponatremia and retinochoroiditis in a 33-year-old French guianan patient. PLoS. Negl. Trop. Dis. 6, e1802.

Guo, Z.G., Johnson, A.M., 1995. Genetic characterization of *Toxoplasma gondii* strains by random amplified polymorphic DNA polymerase chain reaction. Parasitology 111, 127–132.

Halos, L., Thébault, A., Aubert, D., et al., 2010. An innovative survey underlining the significant level of contamination by *Toxoplasma gondii* of ovine meat consumed in France. Int. J. Parasitol. 40, 193–200.

Herrmann, D.C., Pantchev, N., Vrhovec, M.G., et al., 2010. Atypical *Toxoplasma gondii* genotypes identified in oocysts shed by cats in Germany. Int. J. Parasitol. 40, 285–292.

Herrmann, D.C., Bärwald, A., Maksimov, A., et al., 2012a. *Toxoplasma gondii* sexual cross in a single naturally infected feline host: Generation of highly mouse-virulent and avirulent clones, genotypically different from clonal types I, II and III. Vet. Res. 43, 39.

Herrmann, D.C., Maksimov, P., Maksimov, A., et al., 2012b. *Toxoplasma gondii* in foxes and rodents from the German Federal States of Brandenburg and Saxony-Anhalt: seroprevalence and genotypes. Vet. Parasitol. 185, 78–85.

Herrmann, D.C., Wibbelt, G., Götz, M., Conraths, F.J., Schares, G., 2013. Genetic characterisation of *Toxoplasma gondii* isolates from European beavers (Castor fiber) and European wildcats (Felis silvestris silvestris). Vet. Parasitol. 191, 108–111.

Hide, G., Morley, E.K., Hughes, J.M., et al., 2009. Evidence for high levels of vertical transmission in *Toxoplasma gondii*. Parasitology 136, 1877–1885.

Holland, G.N., 2003. Ocular toxoplasmosis: a global reassessment. Part I: epidemiology and course of disease. Am. J. Ophthalmol. 136, 973–988.

Homan, W.L., Vercammen, M., De Braekeleer, J., Verschueren, H., 2000. Identification of a 200- to 300-fold repetitive 529 bp DNA fragment in *Toxoplasma gondii*, and its use for diagnostic and quantitative PCR. Int. J. Parasitol. 30, 69–75.

Howe, D.K., Sibley, L.D., 1995. *Toxoplasma gondii* comprises three clonal lineages: correlation of parasite genotype with human disease. J. Infect. Dis. 172, 1561–1566.

Huang, S.Y., Cong, W., Zhou, P., et al., 2012. First report of genotyping of *Toxoplasma gondii* isolates from wild birds in China. J. Parasitol. 98, 681–682.

Hurtado, A., Aduriz, G., Moreno, B., Barandika, J., Garcia-Perez, A.L., 2001. Single tube nested PCR for the detection of *Toxoplasma gondii* in fetal tissues from naturally aborted ewes. Vet. Parasitol. 102, 17–27.

Innes, E.A., Bartley, P.M., Buxton, D., Katzer, F., 2009. Ovine toxoplasmosis. Parasitology 136, 1887–1894.

Jamieson, S.E., de Roubaix, L.A., Cortina-Borja, M., et al., 2008. Genetic and epigenetic factors at COL2A1 and ABCA4 influence clinical outcome in congenital toxoplasmosis. PLoS One 3, e2285.

Jauregui, L.H., Higgins, J., Zarlenga, D., Dubey, J.P., Lunney, J.K., 2001. Development of a real-time PCR assay for detection of *Toxoplasma gondii* in pig and mouse tissues. J. Clin. Microbiol. 39, 2065–2071.

Jokelainen, P., Isomursu, M., Näreaho, A., Oksanen, A., 2011. Natural *Toxoplasma gondii* infections in European brown hares and mountain hares in Finland: proportional mortality rate, antibody prevalence, and genetic characterization. J. Wildl. Dis. 47, 154–163.

Jokelainen, P., Nylund, M., 2012. Acute fatal toxoplasmosis in three Eurasian red squirrels (*Sciurus vulgaris*) caused by genotype II of *Toxoplasma gondii*. J. Wildl. Dis. 48, 454–457.

Kasper, D.C., Sadeghi, K., Prusa, A.R., et al., 2009. Quantitative real-time polymerase chain reaction for the accurate detection of *Toxoplasma gondii* in amniotic fluid. Diagnostic Microbiol. Infect. Dis. 63, 10–15.

Khan, A., Su, C., German, M., Storch, G.A., Clifford, D.B., Sibley, L.D., 2005. Genotyping of *Toxoplasma gondii* strains from immunocompromised patients reveals high prevalence of type I strains. J. Clin. Microbiol. 43, 5881–5887.

Khan, A., Jordan, C., Muccioli, C., et al., 2006. Genetic divergence of *Toxoplasma gondii* strains associated with ocular toxoplasmosis. Brazil. Emerg. Infect. Dis. 12, 942–949.

Khan, A., Fux, B., Su, C., et al., 2007. Recent transcontinental sweep of *Toxoplasma gondii* driven by a single monomorphic chromosome. Proc. Natl. Acad. Sci. USA 104, 14872–14877.

Khan, A., Dubey, J.P., Su, C., Ajioka, J.W., Rosenthal, B.M., Sibley, L.D., 2011a. Genetic analyses of atypical *Toxoplasma gondii* strains reveal a fourth clonal lineage in North America. Int. J. Parasitol. 41, 645–655.

Khan, A., Miller, N., Roos, D.S., et al., 2011b. A monomorphic haplotype of chromosome Ia is associated with widespread success in clonal and non-clonal populations of *Toxoplasma gondii*. MBio. 2, e00228–00211.

Kieffer, F., Rigourd, V., Ikounga, P., Bessieres, B., Magny, J.F., Thulliez, P., 2011. Disseminated congenital toxoplasma infection with a type II strain. Pediatr. Infect. Dis. J. 30, 813–815.

Kong, J.T., Grigg, M.E., Uyetake, L., Parmley, S.F., Boothroyd, J.C., 2003. Serotyping of *Toxoplasma gondii* infections in humans using synthetic peptides. J. Infect. Dis. 187, 1484–1495.

Lehmann, T., Blackstone, C.R., Parmley, S.F., Remington, J.S., Dubey, J.P., 2000. Strain typing of *Toxoplasma gondii*: comparison of antigen-coding and housekeeping genes. J. Parasitol. 86, 960–971.

Lehmann, T., Graham, D.H., Dahl, E., et al., 2003. Transmission dynamics of *Toxoplasma gondii* on a pig farm. Infect. Genet. Evol. 3, 135–141.

Lehmann, T., Graham, D.H., Dahl, E.R., Bahia-Oliveira, L.M.G., Gennari, S.M., Dubey, J.P., 2004. Variation in the structure of *Toxoplasma gondii* and the roles of selfing, drift, and epistatic selection in maintaining linkage disequilibria. Infect. Genet. Evol. 4, 107–114.

Lehmann, T., Marcet, P.L., Graham, D.H., Dahl, E.R., Dubey, J.P., 2006. Globalization and the population structure of *Toxoplasma gondii*. Proc. Natl. Acad. Sci. USA 103, 11423–11428.

Lindström, I., Sundar, N., Lindh, J., et al., 2008. Isolation and genotyping of *Toxoplasma gondii* from Ugandan chickens reveals frequent multiple infections. Parasitology 135, 39–45.

Mack, D.G., Johnson, J.J., Roberts, F., et al., 1999. HLA-class II genes modify outcome of *Toxoplasma gondii* infection. Int. J. Parasitol. 29, 1351–1358.

McLeod, R., Boyer, K.M., Lee, D., et al., 2012. Prematurity and severity are associated with *Toxoplasma gondii* alleles (NCCCTS, 1981–2009). Clin. Infect. Dis. 54, 1595–1605.

Mercier, A., Devillard, S., Ngoubangoye, B., et al., 2010. Additional haplogroups of *Toxoplasma gondii* out of Africa: population structure and mouse-virulence of strains from Gabon. PLoS Negl. Trop. Dis. 4, e876.

Mercier, A., Ajzenberg, D., Devillard, S., et al., 2011. Human impact on genetic diversity of *Toxoplasma gondii*: example of the anthropized environment from French Guiana. Infect. Genet. Evol. 11, 1378–1387.

Messaritakis, I., Detsika, M., Koliou, M., Sifakis, S., Antoniou, M., 2008. Prevalent genotypes of *Toxoplasma gondii* in pregnant women and patients from Crete and Cyprus. Am. J. Trop. Med. Hyg. 79, 205–209.

Miller, M.A., Grigg, M.E., Kreuder, C., et al., 2004. An unusual genotype of *Toxoplasma gondii* is common in California sea otters (Enhydra lutris nereis) and is a cause of mortality. Int. J. Parasitol. 34, 275–284.

Miller, M., Conrad, P., James, E.R., et al., 2008a. Transplacental toxoplasmosis in a wild southern sea otter (Enhydra lutris nereis). Vet. Parasitol. 153, 12–18.

Miller, M.A., Miller, W.A., Conrad, P.A., et al., 2008b. Type X *Toxoplasma gondii* in a wild mussel and terrestrial carnivores from coastal California: new linkages between terrestrial mammals, runoff and toxoplasmosis of sea otters. Int. J. Parasitol. 38, 1319–1328.

Minot, S., Melo, M.B., Li, F., et al., 2012. Admixture and recombination among *Toxoplasma gondii* lineages explain global genome diversity. Proc. Natl. Acad. Sci. USA 109, 13458–13463.

Montoya, J.G., Liesenfeld, O., 2004. Toxoplasmosis. Lancet 363, 1965–1976.

Moré, G., Maksimov, P., Pardini, L., et al., 2012. *Toxoplasma gondii* infection in sentinel and free-range chickens from Argentina. Vet. Parasitol. 184, 116–121.

Morisset, S., Peyron, F., Lobry, J.R., et al., 2008. Serotyping of *Toxoplasma gondii*: striking homogeneous pattern between symptomatic and asymptomatic infections within Europe and South America. Microbes Infect. 10, 742–747.

Nguyen, T.D., Bigaignon, G., Markine-Goriaynoff, D., et al., 2003. Virulent *Toxoplasma gondii* strain RH promotes T-cell-independent overproduction of proinflammatory cytokines IL12 and gamma-interferon. J. Med. Microbiol. 52, 869–887.

Niedelman, W., Gold, D.A., Rosowski, E.E., et al., 2012. The rhoptry proteins ROP18 and ROP5 mediate *Toxoplasma gondii* evasion of the murine, but not the human, interferon-gamma response. PLoS Pathog. 8, e1002784.

Nowakowska, D., Colón, I., Remington, J.S., et al., 2006. Genotyping of *Toxoplasma gondii* by multiplex PCR and peptide-based serological testing of samples from infants in Poland diagnosed with congenital toxoplasmosis. J. Clin. Microbiol. 44, 1382–1389.

Pan, S., Thompson, R.C.A., Grigg, M.E., Sundar, N., Smith, A., Lymbery, A.J., 2012. Western Australian marsupials are multiply infected with genetically diverse strains of *Toxoplasma gondii*. PLoS One 7, e45147.

Parameswaran, N., Thompson, R.C., Sundar, N., et al., 2010. Non-archetypal Type II-like and atypical strains of *Toxoplasma gondii* infecting marsupials of Australia. Int. J. Parasitol. 40, 635–640.

Pena, H.F.J., Gennari, S.M., Dubey, J.P., Su, C., 2008. Population structure and mouse-virulence of *Toxoplasma gondii* in Brazil. Int. J. Parasitol. 38, 561–569.

Pfefferkorn, E.R., Pfefferkorn, L.C., Colby, E.D., 1977. Development of gametes and oocysts in cats fed cysts derived from cloned trophozoites of *Toxoplasma gondii*. J. Parasitol. 63, 158–159.

Pomares, C., Ajzenberg, D., Bornard, L., et al., 2011. Toxoplasmosis and horse meat, France. Emerg. Infect. Dis. 17, 1327–1328.

Prestrud, K.W., Åsbakk, K., Mørk, T., Fuglei, E., Tryland, M., Su, C., 2008. Direct high-resolution genotyping of *Toxoplasma gondii* in arctic foxes (*Vulpes lagopus*) in the remote arctic Svalbard archipelago reveals widespread clonal Type II lineage. Vet. Parasitol. 158, 121–128.

Qian, W., Wang, H., Su, C., et al., 2012. Isolation and characterization of *Toxoplasma gondii* strains from stray cats revealed a single genotype in Beijing, China. Vet. Parasitol. 187, 408–413.

Rajendran, C., Su, C., Dubey, J.P., 2012. Molecular genotyping of *Toxoplasma gondii* from Central and South America revealed high diversity within and between populations. Infect. Genet. Evol. 12, 359–368.

Reese, M.L., Zeiner, G.M., Saeij, J.P., Boothroyd, J.C., Boyle, J.P., 2011. Polymorphic family of injected pseudokinases is paramount in *Toxoplasma* virulence. Proc. Natl. Acad. Sci. USA 108, 9625–9630.

Richomme, C., Aubert, D., Gilot-Fromont, E., et al., 2009. Genetic characterization of *Toxoplasma gondii* from wild boar (*Sus scrofa*) in France. Vet. Parasitol. 164, 296–300.

Robben, P.M., Mordue, D.G., Truscott, S.M., Takeda, K., Akira, S., Sibley, L. D., 2004. Production of IL-12 by macrophages infected with *Toxoplasma gondii* depends on the parasite genotype. J. Immunol. 172, 3686–3694.

Rosowski, E.E., Lu, D., Julien, L., et al., 2011. Strain-specific activation of the NF-kappaB pathway by GRA15, a novel *Toxoplasma gondii* dense granule protein. J. Exp. Med. 17, 195–212.

Saeij, J.P., Boyle, J.P., Coller, S., et al., 2006. Polymorphic secreted kinases are key virulence factors in toxoplasmosis. Science 314, 1780–1783.

Saeij, J.P., Boyle, J.P., Boothroyd, J.C., 2005. Differences among the three major strains of *Toxoplasma gondii* and their specific interactions with the infected host. Trends Parasitol. 21, 476–481.

Sibley, L.D., Boothroyd, J.C., 1992. Virulent strains of *Toxoplasma gondii* comprise a single clonal lineage. Nature 359, 82–85.

Sibley, L.D., Ajioka, J.W., 2008. Population structure of *Toxoplasma gondii*: clonal expansion driven by infrequent recombination and selective sweeps. Annu. Rev. Microbiol. 62, 329–351.

Silveira, C., Vallochi, A.L., Rodrigues da Silva, U., et al., 2011. *Toxoplasma gondii* in the peripheral blood of patients with acute and chronic toxoplasmosis. Br. J. Ophthalmol. 95, 396–400.

Soete, M., Fortier, B., Camus, D., Dubremetz, J.F., 1993. *Toxoplasma gondii*: kinetics of bradyzoite-tachyzoite interconversion in vitro. Exp. Parasitol. 76, 259–264.

Sousa, S., Ajzenberg, D., Marle, M., et al., 2009. Selection of polymorphic peptides from GRA6 and GRA7 sequences of *Toxoplasma gondii* strains to be used in serotyping. Clin. Vac. Immunol. 16, 1158–1169.

Sousa, S., Ajzenberg, D., Vilanova, M., Costa, J., Darde, M.-L., 2008. Use of GRA6-derived synthetic polymorphic peptides in an immunoenzymatic assay to serotype *Toxoplasma gondii* in human serum samples collected from three continents. Clin. Vac. Immunol. 15, 1380–1386.

Sousa, S., Canada, N., Correia da Costa, J.M., Dardé, M.-L., 2010. Serotyping of naturally *Toxoplasma gondii* infected meat-producing animals. Vet. Parasitol. 169, 24–28.

Sreekumar, C., Graham, D.H., Dahl, E., et al., 2003. Genotyping of *Toxoplasma gondii* isolates from chickens from India. Vet. Parasitol. 118, 187–194.

Su, C., Evans, D., Cole, R.H., et al., 2003. Recent expansion of *Toxoplasma* through enhanced oral transmission. Science 299, 414–416.

Su, C., Zhang, X., Dubey, J.P., 2006. Genotyping of *Toxoplasma gondii* by multilocus PCR-RFLP markers: A high resolution and simple method for identification of parasites. Int. J. Parasitol. 36, 841–848.

Su, C., Shwab, E.K., Zhou, P., Zhu, X.Q., Dubey, J.P., 2010. Moving towards an integrated approach to molecular detection and identification of *Toxoplasma gondii*. Parasitology 137, 1–11.

Su, C., Khan, A., Zhou, P., et al., 2012. Globally diverse *Toxoplasma gondii* isolates comprise six major clades originating from a small number of distinct ancestral lineages. Proc. Natl. Acad. Sci. USA 109, 5844–5849.

Subauste, C.S., Ajzenberg, D., Kijlstra, A., 2011. Review of the series 'Disease of the year 2011: toxoplasmosis' pathophysiology of toxoplasmosis. Ocul. Immunol. Inflamm. 19, 297–306.

Sundar, N., Cole, R.A., Thomas, N.J., Majumdar, D., Dubey, J.P., Su, C., 2008. Genetic diversity among sea otter isolates of *Toxoplasma gondii*. Vet. Parasitol. 151, 125–132.

Taylor, S., Barragan, A., Su, C., et al., 2006. A secreted serine-threonine kinase determines virulence in the eukaryotic pathogen *Toxoplasma gondii*. Science 314, 1776–1780.

Terry, R.S., Smith, J.E., Duncanson, P., Hide, G., 2001. MGE-PCR: a novel approach to the analysis of *Toxoplasma gondii* strain differentiation using mobile genetic elements. Int. J. Parasitol. 31, 155–161.

Tibayrenc, M., Ayala, F.J., 2002. The clonal theory of parasitic protozoa: 12 years on. Trends Parasitol. 18, 405–410.

Vaudaux, J.D., Muccioli, C., James, E.R., et al., 2010. Identification of an atypical strain of *Toxoplasma gondii* as the cause of a waterborne outbreak of toxoplasmosis in Santa Isabel do Ivai, Brazil. J. Infect. Dis. 202, 1226–1233.

Velmurugan, G.V., Dubey, J.P., Su, C., 2008. Genotyping studies of *Toxoplasma gondii* isolates from Africa revealed that the archetypal clonal lineages predominate as in North America and Europe. Vet. Parasitol. 155, 314–318.

Velmurugan, G.V., Su, C., Dubey, J.P., 2009. Isolate designation and characterization of *Toxoplasma gondii* isolates from pigs in the United States. J. Parasitol. 95, 95–99.

Wendte, J.M., Miller, M.A., Lambourn, D.M., et al., 2010. Self-mating in the definitive host potentiates clonal outbreaks of the Apicomplexan parasites *Sarcocystis neurona* and *Toxoplasma gondii*. PLoS Genet. 6, e1001261.

Wendte, J.M., Gibson, A.K., Grigg, M.E., 2011. Population genetics of *Toxoplasma gondii*: new perspectives from parasite genotypes in wildlife. Vet. Parasitol. 182, 96–111.

Williams, R.H., Morley, E.K., Hughes, J.M., et al., 2005. High levels of congenital transmission of *Toxoplasma gondii* in longitudinal and cross-sectional studies on sheep farms provides evidence of vertical transmission in ovine hosts. Parasitology 130, 301–307.

Zhou, P., Zhang, H., Lin, R.-Q., et al., 2009. Genetic characterization of *Toxoplasma gondii* isolates from China. Parasitol. Int. 58, 193–195.

Zhou, P., Nie, H., Zhang, L.-X., et al., 2010. Genetic characterization of *Toxoplasma gondii* isolates from pigs in China. J. Parasitol. 96, 1027–1029.

Zia-Ali, N., Fazaeli, A., Khoramizadeh, M., Ajzenberg, D., Dardé, M., Keshavarz-Valian, H., 2007. Isolation and molecular characterization of *Toxoplasma gondii* strains from different hosts in Iran. Parasitol. Res. 101, 111–115.

CHAPTER

4

Human *Toxoplasma* Infection

Rima McLeod, Christine Van Tubbergen*, José G. Montoya†, Eskild Petersen***

***The University of Chicago, Chicago, Illinois, USA**

†Department of Medicine, Palo Alto Medical Foundation, Stanford University, Palo Alto, California, USA

****Department of Infectious Diseases, Aarhus University Hospital, Skejby, Aarhus, Denmark**

OUTLINE

Toxoplasma gondii, *Second Edition*
http://dx.doi.org/10.1016/B978-0-12-396481-6.00004-0

4.1 CLINICAL MANIFESTATIONS AND COURSE

4.1.1 Introduction and History

Toxoplasmosis refers to disease caused by *Toxoplasma gondii* (Remington *et al.*, 2011; Delair *et al.*, 2011; McLeod, *et al.*, 2012; Dubey *et al.*, 2012c). As reviewed in Chapter 1 the first record of a human case ascribed to infection with *T. gondii* was Janku's report of a child with hydrocephalus in 1923 (Janku, 1923). In 1939, Wolf, Cowen, and Paige described a case of toxoplasmic encephalitis (Wolf *et al.*, 1939). Systemic symptoms of infection with *T. gondii* were first reported in adults in 1940 (Pinkerton and Weinman, 1940). Sabin reported the first case of encephalitis due to *T. gondii* in infants (Sabin, 1941) and, with Feldman, developed a serologic test, the Sabin-Feldman dye test, to diagnose infection (Sabin and Feldman, 1948). Lymphadenopathy was recognized as a symptom in older children and adults by Siim (1951) and Gard and Magnusson (1951). Encephalitis due to *T. gondii* in immune compromised patients was first reported from patients with Hodgkin's disease during immunosuppressive treatment (Flament-Durand *et al.*, 1967) and shortly after by Vietzke *et al.* (1968). Hogan recognized *T. gondii* as a cause of ocular infection, retinitis, and scarring (Hogan *et al.*, 1957) and O'Connor and Perkins confirmed this could be effectively treated with pyrimethamine and sulphadiazine (O'Connor, 1974; Perkins, 1956). Several authors and groups described different aspects of congenital infections noting symptomatic congenital infections at birth in persons living in Europe and North America and/or late, progressive neurological and ophthalmologic manifestations in untreated children, even in those with subclinical infections at birth, and improved outcomes with prompt diagnosis and initiation of treatment prenatally as well as during infancy (Frenkel and Friedlander, 1951; Eichenwald, 1957, 1960; Couvreur and Desmonts, 1962, 1964; Kimball *et al.*, 1971; Saxon *et al.*, 1973; Alford *et al.*, 1974; Thalhammer, 1975; Couvreur *et al.*, 1976; Remington and Desmonts, 1976; Stagno *et al.*, 1977; Desmonts and Remington, 1980; Wilson *et al.*, 1980a, 1980b; Aspock and Pollack, 1982; Desmonts, 1982; Koppe and Kloosterman, 1982; Couvreur *et al.*, 1955, 1979, 1984; Desmonts and Couvrer, 1979, 1984; McLeod *et al.*, 1985, 1992, 2000, 2006a, 2009, 2012; Moncada and Montoya, 2012; Morin *et al.*, 2012; Koppe *et al.*, 1986; Daffos *et al.*, 1988; Hohlfeld *et al*, 1989, 1994; Forestier, 1991; McAuley *et al.*, 1994; Swisher *et al.*, 1994; Roizen *et al.*, 1995, 2006; Mets *et al.*, 1996; Patel *et al.*, 1996; Thulliez, 2001a; Beghetto *et al.*, 2003; Romand *et al.*, 2004; Boyer *et al.*, 2005, 2011; Kieffer *et al.*, 2008; Phan *et al.*, 2008a, 2008b; Weiss and Dubey, 2009; Cortina-Borja *et al.*, 2010; Noble *et al.*, 2010; Villena et al., 2010; Gilbert *et al.*, 2011; Peyron *et al.*, 2011; Remington *et al.*, 2011; Ajzenberg, 2012; Andrade *et al.*, 2012; Dubey *et al.*, 2012b; Olariu *et al.*, 2011; Wallon *et al.*, 2013). Manifestations of congenital infection may be prevented or reduced and there is clear benefit from early treatment as demonstrated by several large studies (see Section 4.3). Also, such treatment appears to be effective for infections with parasites with differing genotypes in the USA (McLeod *et al.*, 2012).

Toxoplasma can cause neurological and ophthalmologic disease when acquired postnatally in those without apparent immune compromise, as well as for patients with lymphomas and other immune compromise (Townsend *et al.*, 1975; Couvreur and Thulliez, 1996). This postnatal infection has since been recognized as the most common infectious cause of retinitis and uveitis in the world (Silveira *et al.*, 1988; Glasner *et al.*, 1992; Kortbeek *et al.*, 2009; Delair *et al.*, 2011). Acute acquired toxoplasmosis causing a mononucleosis-like illness during postnatally acquired infection in persons without known immune compromise was reported by Remington *et al.* (1962). *T. gondii* causing destruction and/or inflammation of the heart, pericardium, skeletal muscle, lung, skin (dermatomyositis), as well as brain and eye, were reported subsequently (Palma *et al.*, 1984; Pollock, 1979; Montoya *et al.*, 1997; Cunningham, 1982; Paspalaki *et al.*, 2001; Prado *et al.*, 1978; Greenlee *et al*, 1975; McCabe *et al.*, 1987; Montoya and Remington, 2008). Clinicians in Brazil (Silveira *et al.*, 1988, 2001, 2002; Glasner *et al.*, 1992; Carmé *et al.*, 2002, 2009; Amorim Garcia *et al.*, 2004; Lehmann *et al.*, 2006; Gilbert *et al.*, 2008; Vasconcelos-Santos *et al.*, 2009, 2011, 2012; Vaudaux *et al.*, 2010; Ajzenberg *et al.*, 2012; Andrade *et al.*, 2012; reviewed in Dubey *et al.*, 2012c; Lavinsky *et al.*, 2012; Orefice *et al.*, 2012) and in tropical areas such as Panama and regions with Amazon tributaries (Darde *et al.*, 1998; Carmé *et al.*, 2002, 2009; Demar *et al.*, 2007, 2012) noted severe, frequent and recurrent eye and other substantial disease manifestations, including pneumonias requiring respiratory support, and death, with acute acquired toxoplasmosis (Demar *et al.*, 2007, 2012). It appears that the genotype of the infecting strain of *T. gondii* might sometimes in humans contribute significantly to differences in virulence of parasites as it does in other animals (see Chapter 3 for a discussion of genotypes) (Sibley *et al.*, 1992; Grigg *et al.*, 2001; Vallochi *et al.*, 2005; Lehmann *et al.*, 2006; Darde, 2008; Gilbert *et al.*, 2008; Ajzenberg *et al.*, 2012; Demar *et al.*, 2012, reviewed in Dubey *et al.*, 2012c; McLeod *et al.*, 2012).

Toxoplasma infection, transmitted into seronegative heart transplant recipients by hearts from donors with *T. gondii* infection, in one case presenting as a brain abscess, was reported in 1979 (Ryning *et al.*, 1979; McLeod *et al.*, 1979). It later was confirmed in several series of cases that a heart from a seropositive donor transplanted into a seronegative recipient could lead to reactivation of latent *T. gondii* infection in the donor heart and present significant risk of life-threatening disease for the recipient. Pyrimethamine prophylaxis now effectively prevents disease in such patients. Cases with disseminated or organ damaging toxoplasmosis with each type of solid organ, bone marrow and stem cell transplantation, and with many different kinds of immunosuppressive treatments, were noted in subsequent years (Remington, 1974; Ryning *et al.*, 1979; Conley *et al.*, 1981; Luft *et al.*, 1983; Wreghitt *et al.*, 1992, 1995; Singer *et al.*, 1993; Slavin *et al.*, 1994; Bretagne *et al.*, 1995;

Aubert, 1996; Renoult *et al.*, 1997; Sing *et al.*, 1999; Martino *et al.*, 2000, 2005; Abgrall *et al.*, 2001; Montoya *et al.*, 2001, 2010; Botterel *et al.*, 2002; Mele *et al.*, 2002; Lassoued *et al.*, 2007; Matsuo *et al.*, 2007; Derouin and Pelloux, 2008; Bautista *et al.*, 2012; Bories *et al.*, 2012; Busemann *et al.*, 2012; Caselli *et al.*, 2012; Fernàndez-Sabé, 2012; Osthoff *et al.*, 2012; Strabelli *et al.*, 2012; Vaughan *et al.*, 2012).

One of the earliest documented cases of toxoplasmic encephalitis in persons with AIDS occurred in a homosexual man in the late 1970s, who lived in Chicago but had travelled to Haiti. He also developed *Pneumocystis* pneumonia and tuberculosis and his HIV infection was recognized later. He was one of the first patients in the USA AIDS epidemic (Levin *et al.*, 1983), before HIV was identified as a cause of the concomitant opportunistic infections in AIDS. It soon became clear that toxoplasmic encephalitis was the most common of the CNS infections that were presenting manifestations during the HIV epidemic, particularly in areas of higher seroprevalence of *T. gondii* infection (Luft *et al.*, 1983; Post *et al.*, 1983; Luft and Remington, 1984, 1992; Levy *et al.*, 1986, 1988, 1990; Haverkos *et al.*, 1987; Kovacs, 1995; Leport *et al.*, 1996; Abgrall *et al.*, 2001; Furco *et al.*, 2008; Arslan *et al.*, 2012; Basavaprabhu *et al.*, 2012; Guevara-Silva *et al.*, 2012; Kim *et al.*, 2012).

Carefully designed and implemented prenatal serologic screening programmes to diagnose primary *T. gondii* infections acquired in gestation were mandated by law in France in 1972 (Desmonts and Couvreur, 1974a, 1974b; Daffos *et al.*, 1988). This was combined with an approach for diagnosis and treatment of the infection *in utero*, including diminishing intervals to each month for obtaining serum samples for gestational screening, and continuous treatment with pyrimethamine and sulphadiazine with leucovorin when congenital infection was likely. There has been an almost complete eradication of severe symptomatic congenital toxoplasmosis in France since the implementation of these programmes (Kieffer *et al.*, 2008; Cortina-Borja *et al.*, 2010; McLeod *et al.*, 2009; Peyron *et al.*, 2011; Stillwaggon *et al.*, 2011; Wallon *et al.*, 2013). Serotype II predominance in France was documented, indicating that the analyses of outcomes applied to serotype II infections (Ajzenberg 2012; Darde *et al.*, 2008; Peyron *et al.*, 2011). An alternative approach was developed in Austria (Aspock, 1982) and Germany (Hotop *et al.*, 2012), where any acutely infected pregnant women receive pyrimethamine, sulphadiazine and leucovorin beginning at the eighteenth week of gestation or beginning after that time if the infection is acquired later in gestation. A European group, EMSCOT, analysed pooled data over time from many countries that had different approaches (Dunn *et al.*, 1999; Naessens *et al.*, 1999; Gilbert *et al.*, 1999). There is an improvement of neurological findings at birth with shorter intervals between diagnosis and treatment in gestation (Cortina-Borja *et al.*, 2010; SYROCOT, 2007). The EMSCOT–SYROCOT studies included evaluations that were combined from centres in Europe. There were considerable variations in their standard of care, diagnostic measures, treatments, and timing of implementation of treatment. In the analyses, when data from these differing centres were pooled, these varied methods did not allow the investigators to identify significant effects on neurological signs and symptoms when durations from diagnosis to initiation of varied treatments was greater than four weeks, or to identify effectiveness of treatment in reducing retinal disease. Effectiveness of treatment in preventing subsequent retinal disease could be identified, however, when data from the more uniform individual centres were analysed separately (Kieffer *et al.*, 2008). Studies of additional modifications to try to identify whether alternative, abbreviated treatment durations might be effective, but less costly, are currently ongoing in France.

Feldman, Eichenwald, Stagno, Alford, Wilson and Remington, and the NCCCTS in earlier reports, and in one manuscript, the NCCCTS with the Palo Alto Research Foundation Remington and Montoya USA serology reference laboratory group, have emphasized how severe congenital infection can be at birth without prenatal screening and subsequent treatment. This is often the case in regions without screening programmes, such as in North America (Feldman, 1953; Eichenwald, 1957, 1960; Alford *et al.*, 1974; Stagno *et al.*, 1977; Wilson *et al.*, 1980a, 1980b; McLeod *et al.*, 1985, 1992, 2000, 2006a; 2012, 2013a, b; McAuley *et al.*, 1994; Montoya and Remington, 2008; Olariu *et al.*, 2011; Remington *et al.*, 2011; Moncado and Montoya, 2012). The mixture of parasite genetic serotypes during this time in the USA was defined by the NCCCTS with the Grigg group at the National Institutes of Health (NIH) (McLeod *et al.*, 2012). These investigators demonstrated that, in the USA, congenital toxoplasmosis is not caused predominantly by serotype II parasites as in Europe; only approximately a third of those infected in the North American NCCCTS have type II parasites.

The oocyst life cycle stages were discovered in cat intestine contemporaneously by Hutchison and Frenkel (Hutchison *et al.*, 1970; Frenkel *et al.*, 1970; reviewed in Dubey, 1998). Proteins specific to the oocyst and sporozoite life cycle stages have been identified (Kasper and Ware, 1989; Hill *et al.*, 2011). The frequency of unrecognized exposure to oocysts in mothers of congenitally infected infants (referred to the NCCCTS) was studied by Hill *et al.* (2011) and Boyer *et al.* (2011). Using a serologic assay, to detect antibody to a sporozoite protein present only for less than eight months after acquisition, family clusters and recreational riding stable, work, and waterborne epidemics were identified that were caused by unrecognized exposure to oocysts in North America. This demonstrated that education alone about avoiding risk factors in the USA, although it may be helpful in limiting infections, is not likely to be sufficient to eliminate this infection (Gollub *et al.*, 2008; Boyer *et al.*, 2011). A mathematical model and algorithm to predict the cost and benefit of gestational serologic screening, diagnoses, and prenatal and postnatal treatment has been developed (Stillwaggon *et al.*, 2011) and, when applied to data from the USA, suggested that prenatal screening and *in utero* treatment would be cost effective (Stillwaggon *et al.*, 2011), while another study estimated the healthcare costs of this disease (Collier *et al.*, 2012). A newborn screening programme for *T. gondii* infections has documented occult infections with brain and eye damage in infants born in Massachusetts and New Hampshire (Guerina, 1994). A small number of obstetrical practices in the USA have implemented screening and treatment for primary acquisition of *T. gondii* infection during gestation.

The development of serologic and molecular tests for diagnosis involved many groups (Sabin and Feldman, 1948; Araujo *et al.*, 1971; Desmonts and Couvreur, 1974a, 1974b; Walls and Bullock, 1977; Balsari *et al.*, 1980; Desmonts and Remington, 1980; Naot *et al.*, 1981a, 1981b; Naot and Remington, 1980; Remington *et al.*, 1985; Thulliez *et al.*, 1986, 1989; Hedman *et al.*, 1989, 1993; Dannemann *et al.*, 1990; Grover *et al.*, 1990; Stepick-Biek *et al.*, 1990; Wong *et al.*, 1993; Petithory *et al.*, 1996; Pinon *et al.*, 1996; Liesenfeld *et al.*, 1997, 2001a, 2001b; Montoya, 2002; Montoya *et al.*, 2002; Romand *et al.*, 2004; Beghetto *et al.*, 1995; Press *et al.*, 2005; Bricker-Hidalgo *et al.*, 2007; Montoya and Remington, 2008; Remington *et al.*, 2011; L'Ollivier *et al.*, 2012; Pinto *et al.*, 2012; Villard *et al.*, 2012). This has occurred through the latter half of the previous century and continues to the present. These advances in serologic methods defined and changed the way in which physicians diagnose the various syndromes, with varying manifestations, caused by this infection (Remington *et al.*, 1962, 2011; Greenlee *et al*, 1975; Prado *et al.*, 1978; Pollock, 1979; Cunningham, 1982; Palma *et al.*, 1984; McCabe *et al.*, 1987; Montoya *et al.*, 1997; Paspalaki *et al.*, 2001; Binquet *et al.*, 2003, Wallon *et al.*, 2004, Garweg *et al.*, 2005; Kodjikian *et al.*, 2006; Montoya *et al.*, 2010; Delair *et al.*, 2011; Bories *et al.*, 2012; Brownback *et al.*, 2012; Burrowes *et al.*, 2012; Karanis *et al.*, 2012; McLeod *et al.*, 2012; Toporowski *et al.*, 2012). These serologic tests facilitated a large number and variety of sero-epidemiologic studies (e.g. Desmonts *et al.*, 1965a, 1965b; Ambroise-Thomas *et al.*, 1966; Benensen *et al.*, 1982; Luft and Remington, 1984; Papoz *et al.*, 1986; Cook *et al.*, 2000; Jones *et al.*, 2001, 2007, 2009, 2012a, 2012b; Alvarado-Esquivel *et al.*, 2002, 2011; Bahia-Oliveira *et al.*, 2003; Lindsay *et al.*, 2003; Conrad *et al.*, 2005; Wainwright *et al.*, 2007; Aptouramani *et al.*, 2012; Garabedian *et al.*, 2012; Hill and Dubey, 2012; Mosti *et al.*, 2012). Definition of differences in parasite genetics and serotyping for a few of these differences have been developed (Sibley and Boothroyd, 1992; Kong *et al.*, 2003; Lehmann *et al.*, 2006; Darde, 2008; Dubey *et al.*, 2012a; Maksimov *et al.*, 2012; Minot *et al.*, 2012; Carneiro *et al.*, 2013).

Pyrimethamine (with folinic acid) and sulphadiazine can treat the infection synergistically *in vitro* and in animal models (Eyles and Coleman, 1955) and was effective in treating human retinochoroiditis and uveitis including in placebo controlled studies (Perkins *et al.*, 1956; Garin *et al.*, 1985; Mets *et al.*, 1996; Brezin *et al.*, 2003; Holland *et al.*, 2008; Delair *et al.*, 2011). Folinic acid was found to rescue mammalian bone marrow from anti-folate toxicity without affecting the inhibitory effect of pyrimethamine on the parasite *in vitro* at the concentrations achieved with current treatment regimens (Frenkel and Hitchings, 1957; McLeod *et al.*, 1990). It was also found that pyrimethamine (with folinic acid) and sulphadiazine can effectively treat the infection in immune compromised persons, including transplant recipients, persons with AIDS, and those with other immune suppression (Weiss *et al.*, 1988; LePort *et al.*, 1996; Delair *et al.*, 2011, Ryning *et al.*, 1979; Ruskin and Remington, 1976). In pregnant women, in an algorithm including spiramycin in some cases and pyrimethamine and sulphadiazine in others, treatment appeared to block transmission to the foetus and attenuated or eliminated disease at birth (Desmonts and Couvreur, 1974a, 1974b; Aspock and Pollack, 1982; Couvreur *et al.*, 1988; Hohlfeld *et al.*, 1989; Brezin *et al.*, 2003; McLeod *et al.*, 2006a, 2009, 2012; Kieffer *et al.*, 2008; Cortina-Borja *et al.*, 2010; Wallon *et al.*, 2013; Peyron *et al.*, 2011; Hotop *et al.*, 2012). In infants, treatment has been shown to attenuate or eliminate signs of infection and illness (Couvreur and Desmont, 1962; McAuley *et al.*, 1994; McLeod *et al.*, 2006a, 2012). Further, it was demonstrated that antimicrobial prophylaxis against recurrence of retinal disease (e.g. with TMP-SMX; Silveira *et al.*, 2002) is effective. Alternative medicines for sulphadiazine hypersensitive persons (McLeod *et al.*, 2006b) also have been identified.

A study identifying clinical findings across human lifetimes combined with characterization of parasite and host genetics is being carried out. This is proving to be a powerful tool for understanding pathogenesis and consequences of this infection and the diseases it causes. It is, as well, providing fundamental insights into pathogenic mechanisms for humans (Mack *et al.*, 1999; Jamieson *et al.*, 2008, 2009, 2010; Lees *et al.*, 2010; Peixoto *et al.*, 2010; Tan *et al.*, 2010; Cong *et al.*, 2011a, 2011b, 2012; Witola *et al.*, 2011; Béla *et al.*, 2012; McLeod *et al.*, 2012; Dutra *et al.*, 2013). It allows the human infection to be studied in a limited way consistent with appropriate protections for those in these studies. Experiments of nature in which mutations have occurred in human genes, or where diseases profoundly modify a specific immune function, also are informative; e.g. toxoplasmosis in the setting of disruption of the function of CD4 T cells, CD40 ligands, or TNFα.

T. gondii infection is very common, present in 1/3 to 1/2 of humans in the world. This infection may be asymptomatic, have presenting symptoms that range from mild to severe, or possibly have symptoms not attributed to the infection. The possibility that there may be neurobehavioural consequences from chronic infection in some persons has been proposed, with findings associated with infection in animal models, transcriptome studies with neuronal cells, and with seroprevalence association studies providing a foundation raising questions about causality (Ferguson *et al.*, 1991; Brown *et al.*, 2005; Hermes *et al.*, 2008; Niebuhr *et al.*, 2008; Bech-Nielsen, 2012; Goodwin *et al.*, 2012; Guenter *et al.*, 2012; Horacek *et al.*, 2012; Hamdani *et al.*, 2012; Holub *et al.*, 2012). The hypothesis is that treatment of underlying toxoplasmosis might provide adjunctive or even definitive therapy for these neurobehavioural diseases (Torres *et al*, 1993; Kankova *et al.*, 2012; Kaushik *et al.*, 2012; Lester, 2012; Pearce *et al.*, 2012; Pederson *et al.*, 2012; Torrey *et al.*, 2012; Xiao *et al.*, 2012a, 2012b; Flegr, 2013). The fact that the parasite resides in the brain across lifetimes has raised the possibility that it might even shape language and cultures (Lafferty, 2006). Cause and effect have not been proven for humans for any of these associations. The well-recognized, proven, treatable, and long-standing problems that congenital and ocular infection cause, and disease in immune compromised persons that impact on quality of life, cause deaths, morbidity, and costs for care (Stillwaggon *et al.*, 2011) provide reason enough to develop definitive, better treatments and preventive measures. These include effective application of prenatal serologic screening and treatment programmes, vaccines and treatments to prevent infection and/or sequelae, which are challenges for the future.

4.1.2 Postnatally Acquired Infection in Children and Adults

4.1.2.1 *Adults and Older Children with Primary, Acute Acquired* T. gondii *Infection*

T. gondii can be acquired by inadvertent and unrecognized ingestion of meat which has not been cooked to 'well done', which contains encysted *T. gondii* (Dubey and Beattie, 1998; Dubey *et al.*, 2012b; Fig. 4.1). Desmonts *et al.* documented that infections from undercooked meat in an orphanage in France could lead to infection (Desmonts *et al.*, 1965a, 1965b). Ingestion of shellfish was also defined as a risk factor in the USA in epidemiologic surveys (Jones *et al.*, 2009). Mussels can become infected when they filter sea water contaminated with oocysts from surrounding rivers. Ingestion of contaminated shellfish caused mortality in sea otters (Conrad *et al.*, 2005; Miller *et al.*, 2002). Infection can also be acquired from oocysts contaminating fresh water or food, including fruits and vegetables, and likely from inhaling oocyst contaminated dust/soil (Teutsch *et al.*, 1979; Benenson *et al.*, 1982; Isaac-Renton *et al.*, 1998; Wainwright *et al.*, 2007; Shapiro *et al.*, 2010; Boyer *et al.*, 2011; Hill *et al.*, 2012; Lass *et al.*, 2012; Gallas-Lindemann *et al.*, 2013; Dubey *et al.*, 2012c). Dog fur can be contaminated with oocysts and serve as a transport vector (Frenkel and Parker, 1996), as can insects such as cockroaches and flies. Oocysts can persist in water and warm, moist soil for up to a year, even one oocyst is infectious, and an acutely infected member of the cat family can excrete up to 5 million oocysts during a period of two weeks after acquisition (Dubey and Beattie, 1998; Lindsay *et al.*, 2003; Fritz *et al.*, 2012b).

T. gondii can cause isolated family (Luft and Remington, 1984), educational facility (dos Santos *et al.*, 2010; Morris *et al.*, 2012), or work (Magaldi *et al.*, 1969) clusters and epidemics of infection with multiple exposures to the same source of infection (Isaac-Renton *et al.*, 1998; Bowie *et al.*, 1997; Burnett *et al.*, 1998; Dubey and Beattie, 1998; Jones *et al.*, 2009, 2012a; Vaudaux *et al.*, 2010; Karanis *et al.*, 2012; Lass *et al.*, 2012). In the NCCCTS, a source of infection was not often recognized or identified correctly by mothers of congenitally infected children (Boyer *et al.*, 2011). Seventy-eight percent of the mothers whose serum was tested in the perinatal period had an antibody to an 11 kDa *T. gondii* sporozoite protein, present exclusively in oocyst acquired infections in the six to eight months after infection is acquired (Boyer *et al.*, 2011; Hill *et al.*, 2011).

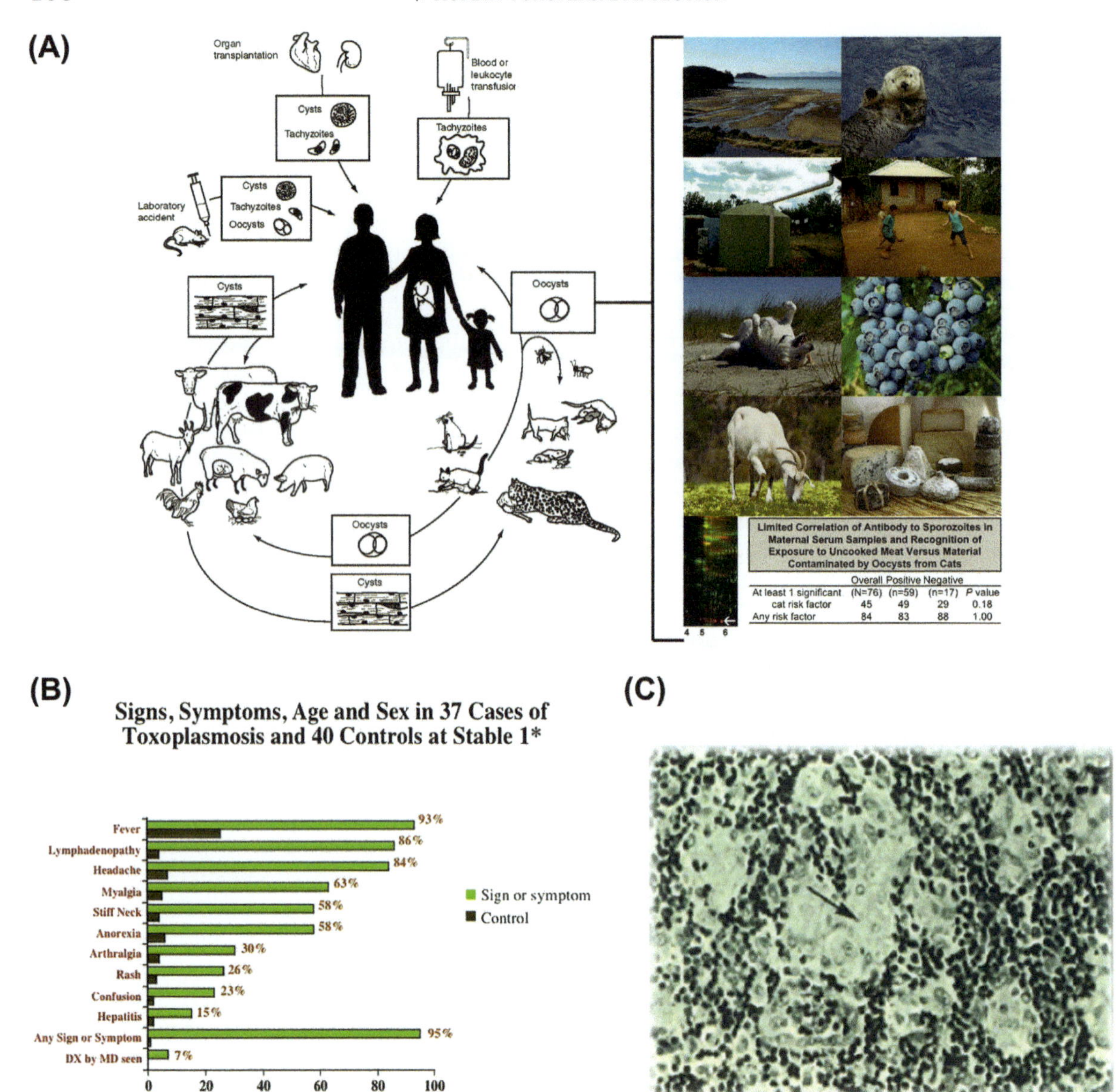

Limited Correlation of Antibody to Sporozoites in Maternal Serum Samples and Recognition of Exposure to Uncooked Meat Versus Material Contaminated by Oocysts from Cats

	Overall (N=76)	Positive (n=59)	Negative (n=17)	*P* value
At least 1 significant cat risk factor	45	49	29	0.18
Any risk factor	84	83	88	1.00

FIGURE 4.1 **Acute acquired *Toxoplasma gondii* infection.**
A) Life cycle of *Toxoplasma gondii* and modes of acquisition.
B) The definitive hosts of *T. gondii* are members of the felidae family (Gerhold and Jessup, 2012). Colour photos depict recently documented modes of oocyst acquisition, including contaminated mussels, drinking water and schoolyards in Brazil, the fur of dogs, unwashed fruits and vegetables, and goat cheese. 2D gel depicts a protein specific to *T. gondii* sporozoites (shown in red and indicated by white arrow). Table shows limits of correlation between presence of antibodies to oocysts and self-determined risk factors for oocyst exposure during pregnancy of mothers in NCCCTS, 1981–1998 (Hill *et al.*, 2011 with permission; Boyer *et al.*, 2011 with permission). *Image from: Remington JS, McLeod R. Toxoplasmosis. In: Braude AI, editor.* International Textbook of Medicine, Medical Microbiology and Infectious Disease, *vol. II. WB Saunders, Philadelphia. 1981; 1818, with permission.*

Frequency and severity of disease manifestation may vary by geographic location or be associated with parasite genetics (McLeod *et al.*, 2012). Dose, route of infection, and host genetics likely influence manifestations and outcomes for people (Fig. 4.1). Acute acquired infection in older children and adults in Europe is believed to most often be asymptomatic. Desmonts reported that ~10% of mothers of congenitally infected children reported symptoms or signs. In an epidemic in the USA in Atlanta, Georgia (Teutsch *et al.*, 1979; Dubey *et al.*, 1981), 95% of the 37 infected persons reported symptoms including headache, myalgias and fever, but only 7% were correctly diagnosed by their primary physician (Fig. 4.1B). These associated symptoms of headache and myalgia in patients with lymphadenopathy were emphasized in the USA (Wong *et al.*, 2012) and Brazilian cases (reviewed in Dubey *et al.*, 2012c). In the USA, illness and small peripheral retinal scars appear to be relatively common (~10%) in mothers of infants with *T. gondii* infection (Noble *et al.*, 2010). Symptoms are, in part, associated with parasite serotype (McLeod *et al.*, 2012). Non-type II infections more often, but not exclusively, were more symptomatic. Pregnancy may also result in recrudescence of retinal disease (Andrade *et al.*, 2012).

Toxoplasmic lymphadenopathy can involve any nodes, but most often involves cervical nodes which present as non-tender, discrete, firm, not matted or fixed to contiguous tissues, and do not suppurate. Mesenteric nodes, that cause pain, with fever may be mistaken for appendicitis and a pectoral node may be mistaken for breast cancer. Lymphadenopathy may be either self-limited or associated with prolonged symptoms, such as fatigue, for more than a year. Toxoplasmosis may account for 5% of clinically significant lymphadenopathy cases (Dorfman and Remington, 1973; Luft *et al.*, 1984; McCabe *et al.*, 1987; Montoya *et al.*, 2010; Natella *et al.*, 2012; Rollins-Ravel *et al.*, 2012). Fever, malaise, night sweats, myalgias, sore throat, hepatosplenomegaly or small numbers of circulating, atypical lymphocytes may accompany the adenopathy. Adenopathy and symptoms usually resolve within a few months to a year. The nodes have characteristic, distinctive histopathology with epithelioid histiocytes and monocytoid cells that encroach on and blur the margins of germinal centres (Dorfman and Remington, 1973; Luft and Remington,

Photographs from:
Estuary: http://www.travelmezze.com/images/estuary-abel-tasman-national-park-new-zealand1.JPG
Otter: http://carinbondar.com/wp-content/uploads/2010/11/seaotter2.jpg
Water supply: http://www.sswm.info/sites/default/files/toolbox/DOLMAN%20and%20LUNDQUIST%202008%20Roof%20Water%20Harvesting%20for%20a%20low%20Impact%20Water%20Supply%20Brazil.jpg
Schoolyard: http://thisismyhappiness.com/wp-content/uploads/2011/10/Playing-soccer-with-local-kids.jpg
Dog: http://images.cpcache.com/merchandise/514_400x400_NoPeel.jpg?region=name:FrontCentre,id:70011053,w:16
Blueberries: http://www.thehomesteadgarden.com/wp-content/uploads/2012/10/Blueberry_Cluster.jpg
Goat: http://www.formaggiokitchen.com/shop/images/goat%20sampler2.jpg
Goat cheese: http://upload.wikimedia.org/wikipedia/commons/a/a1/Domestic_goat_feeding_on_capeweed.jpg
2D gel from: Hill *et al.*, 2011. *Data in table from: Teutsch* et al., *1979, with permission.*
Symptoms and signs in persons with acute acquired toxoplasmosis from a riding stable epidemic and in uninfected controls in Atlanta, Georgia, USA .
C) Oocysts were the source of infection in the riding stable epidemic. Light green bars represent persons with acute acquired toxoplasmosis. Dark green bars represent uninfected controls. Percentages refer to persons with acute acquired toxoplasmosis displaying a particular symptom or sign of infection. Persons with acute acquired toxoplasmosis are more likely to exhibit symptoms, such as fever, lymphadenopathy, headache, myalgia, stiff neck, anorexia, arthralgia, rash, confusion and hepatitis, than those uninfected. Most persons (95%) with acute acquired toxoplasmosis in this epidemic displayed a sign or symptom of infection, but only a few (7%) were correctly diagnosed by their physician. *Data from Teutsch* et al., *1979, with permission.*
Histopathology of lymph node from a person with acute toxoplasmosis.
Black arrow indicates cluster of epithelioid histiocytes. *Image from: Dorfman* et al., *1973, with permission.*

1984; Natella *et al.*, 2012; Rollins-Ravel *et al.*, 2012; Fig. 4.1C). Plasma dendritic cells are identified in the nodes but are not in tightly packed clusters as in certain malignancies (Rollins-Raval *et al.*, 2012). Fine-needle aspirate of lymph nodes has been used to establish the diagnosis of toxoplasmic lymphadenopathy (Natella *et al.*, 2012).

Occasionally, acquired infection may be associated with myositis or a sepsis-like syndrome (Demar *et al* 2012). Townsend *et al.* (1975) described toxoplasmic encephalitis, meningoencephalitis, and meningitis with 50% of the cases occurring in persons without known immune compromise. Polymyositis, dermatomyositis, (Greenlee *et al.*, 1975; Pollock, 1979; Palma *et al.*, 1984), myocarditis, pericarditis, pancarditis, (Prado *et al.*, 1978; Cunningham, 1982; Montoya *et al.*, 1997), pneumonia, mesenteric lymphadenopathy mimicking appendicitis, gastrointestinal symptoms (Schreiner and Liesenfeld, 2009), hepatocellular abnormalities, Guillain–Barre syndrome, headaches, fever, and visual symptoms have all been reported with acute acquired infection (Remington *et al.*, 1962). Couvreur and Thulliez described neurological and retinal disease in acute acquired infection (Couvreur and Thulliez, 1996). There is evidence in animal models that gastrointestinal involvement occurs in some instances (Schreiner and Liesenfeld, 2009; Benson *et al.*, 2012; Mennechet *et al.*, 2004; please see chapters on genetics and immunity). How frequently this infection causes these symptoms and whether they are dependent on inoculum size, life cycle stage, parasite strain, unrecognized immune deficiencies, and variability in immune responses of the person remain to be determined.

Host genetics can influence toxoplasmic chorioretinitis (Jamieson *et al.*, 2008, 2009; Dutra *et al.*, 2013) and parasite strain, as defined by Kong *et al.* (2003), may be associated with eye disease in adults (Holland *et al.*, 1996; Grigg *et al.*, 2001). Symptoms have ranged from short and self-limited to severe symptoms with prolonged fever, chronic lymphadenopathy, fatigue and progressive, recurrent retinochoroiditis (Masur *et al.*, 1978; Teutsch *et al.*, 1979; Benenson *et al.*, 1982; Luft and Remington, 1984; Silveira *et al.*, 1988, 2001; Glasner *et al.*, 1992; Couvreur and Thulliez, 1996; Bowie *et al.*,1997; Burnett *et al.*, 1998; Delair *et al.*, 2008; Montoya and Remington, 2008; Holland *et al.*, 2008).

Severe acute infection associated with interstitial pneumonia and death have been reported in epidemics in the Amazon region, and along its tributaries, including the Maroni River in Guyana (Carme *et al.*, 2002, 2009; Demar *et al.*, 2007, 2012). Other common symptoms included fever, weight loss, increased liver enzymes, lymphadenopathy, headache, rash, retinochoroiditis, myocarditis, myositis and neurological disorders. Severe manifestations, such as pneumonia, in adults also have been described from Brazil (Leal *et al.*, 2007; Dubey *et al.*, 2012c). In some cases, the severe clinical syndromes have been associated with infection with atypical *T. gondii* genotypes (Grigg *et al.*, 2001; Vallochi *et al.*, 2005; Khan *et al.*, 2006; Vaudaux *et al.*, 2010; Carneiro *et al.*, 2013). The severe *T. gondii* infections that have been reported from South America (Carme *et al.*, 2002, 2009; Demar *et al.*, 2007), with diffuse interstitial pneumonia as a key symptom, often required ventilator support and intensive care modalities and even then fatalities occurred in some cases. Severe retinal involvement also characterizes both acute acquired, chronic recurrent and congenital infections, particularly in Brazil, but also elsewhere (Silveira *et al.*, 1988; Glasner *et al.*, 1992; Burnett *et al.*, 1998; Delair *et al.*, 2008; Gilbert *et al.*, 2008; Montoya and Remington, 2008; Vasconcelos-Santos and Andrade, 2011; Brownback *et al.*, 2012; Denes *et al.*, 2012; Dubey *et al.*, 2012c; Toporovski *et al.*, 2012).

4.1.2.2 The Special Problem of Primary Infection During Gestation

Acute *T. gondii* infection in pregnant women does not differ from *T. gondii* infection in other immune competent individuals. The infection may be asymptomatic, although cervical lymphadenopathy may occur, and chorioretinal lesions may develop. Due to an asymptomatic course, acute infection may go unnoticed, or symptoms may not be attributed to *T. gondii* infection. Thus, the infection may be transmitted to the foetus without the pregnant woman realizing she is acutely infected. A programme for serodiagnosis and treatment of primary infection during gestation and PCR diagnosis of congenital toxoplasmosis in the foetus was developed in France and Belgium (Fig. 4.2). Treatment is used to prevent transmission to the foetus and to eliminate, or reduce, sequelae if transmission occurs. A separate approach has been developed and utilized in Germany (Hotop *et al.*, 2012).

4.1.2.3 Postnatally Acquired Infection in Older Children and Adults – the Chronic Infection

There is now extensive literature attributing many behavioural and neurological diseases to *T. gondii* infections (Brown *et al.*, 2005; Niebuhr *et al.*, 2008; Bech-Nielsen, 2012; Fabiani *et al.*, 2012; Goodwin *et al.*, 2012; Guenter *et al.*, 2012; Hamdani *et al.*, 2012; Horacek *et al.*, 2012; Holub *et al.*, 2012). This grew from observations that chronic infection in rodents is associated with ongoing neurobehavioural, neuroimaging, histologic, dopaminergic, cytokine and transcriptomic abnormalities in brain in both infected and contiguous cells (Ferguson *et al.*, 1991; Webster *et al.*, 2006; Hermes *et al.*, 2008; Vyas *et al.*, 2010; Mitra *et al.*, 2012). Differing animal species have varied manifestations of disease based on their genetics, e.g. new and old world primates (Araujo *et al.*, 1973) and various strains of mice. There is a literature that has ascribed causal associations between an increased seroprevalence of *T. gondii* with a wide variety of illnesses including seizures (Stommel *et al.*, 2001), bipolar disease and schizophrenia (Webster *et al.*, 2006, 2013; Henriquez *et al.*, 2009; Torrey *et al.*, 2012; Webster *et al.*, 2013), motor vehicle accidents, slowed reaction times which become progressively worse with duration of infection, sex of offspring, curmudgeonly behaviour in men and 'sex kitten' behaviour in women (Flegr, 2013), suicide (Godwin, 2012; Okusaga and Postolache, 2012; Pedersen *et al.*, 2012), memory loss, depression, gluten intolerance (Severance *et al.*, 2012) and diminished verbal memory measured on RBANS test in young urban professionals (Torrey and Yolken, abstract, Toxoplasmosis International Meeting, The Netherlands, 2009), among others (Ferguson *et al.*, 1991; Brown *et al.*, 2005; Hermes *et al.*, 2008; Niebuhr *et al.*, 2008; Bech-Nielsen, 2012; Fabiani *et al.*, 2012; Godwin *et al.*, 2012; Goodwin *et al.*, 2012; Guenter *et al.*, 2012; Hamdani *et al.*, 2012; Horacek *et al.*, 2012; Holub *et al.*, 2012). At present, there is no definitive proof that *T. gondii* causes or contributes significantly to any of these conditions, other than seizures, in humans. Proposed mechanisms include parasite production of dopamine, tyrosine hydroxylase, and the inflammatory immune response elicited by infection. Webster *et al.* (2013) have proposed how rat behavioural changes with a dopamine connection might be studied in humans. If there is a cause-and-effect relationship between *T. gondii* infection and neurological and behavioural diseases, such as schizophrenia, bipolar disease, memory loss and dementia, for some persons, it must be uncommon, at least in the first few decades after infection. This is because, in carefully studied cohorts of mothers and their congenitally infected children, there have been only rare persons with these neuropsychiatric or other diseases (McLeod, unpublished data). This indicates that if there is a cause-and-effect relationship between *T. gondii* infection and neurobehavioural disease, other cofactors must play a substantial role.

4.1.3 Congenital Infection

4.1.3.1 The Foetus, Infant and Older Child

During the 1940s an understanding developed that acute acquired maternal infection resulted in congenital toxoplasmosis in the foetus and newborn infant. Holmdahl and Holmdahl (1955) found that two out of 23,260

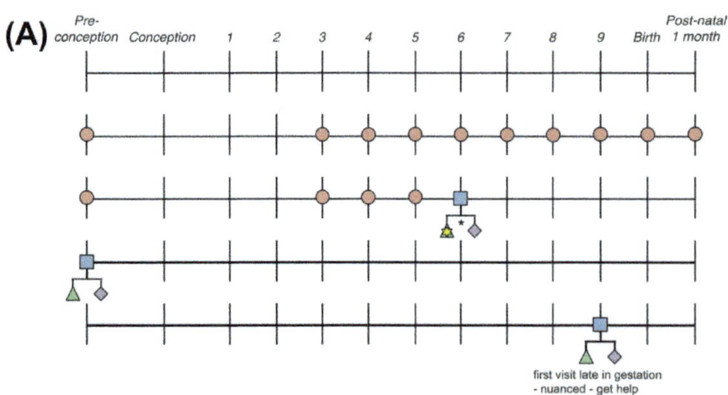

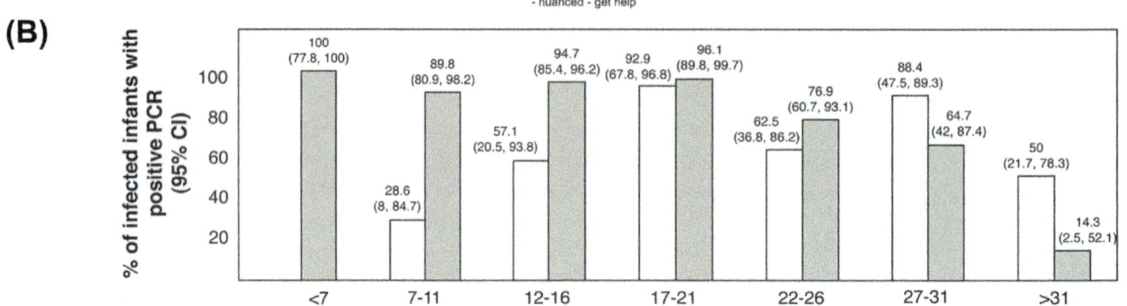

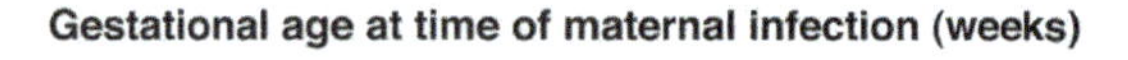

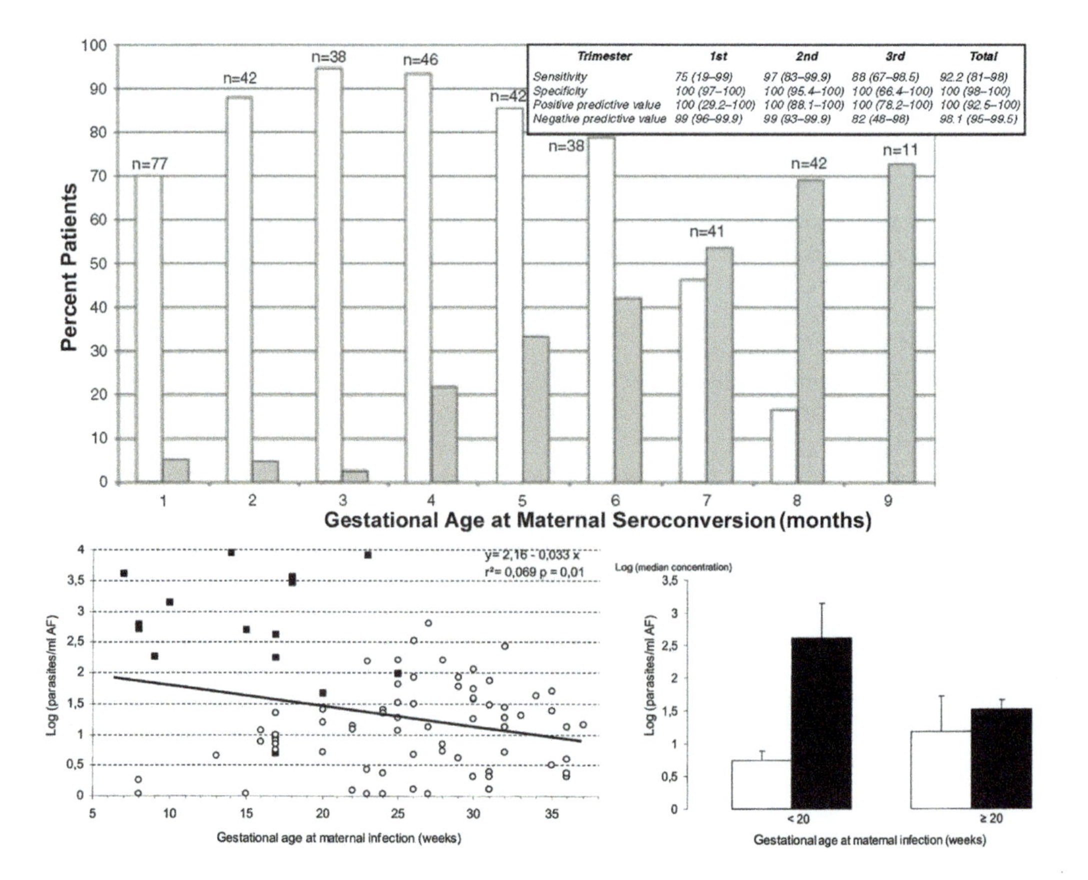

Trimester	1st	2nd	3rd	Total
Sensitivity	75 (19–99)	97 (83–99.9)	88 (67–98.5)	92.2 (81–98)
Specificity	100 (97–100)	100 (95.4–100)	100 (66.4–100)	100 (98–100)
Positive predictive value	100 (29.2–100)	100 (88.1–100)	100 (78.2–100)	100 (92.5–100)
Negative predictive value	99 (96–99.9)	99 (93–99.9)	82 (48–98)	98.1 (95–99.5)

children had clinical toxoplasmosis in a study performed from 1948 to 1951. In 1953, Feldman reported a series of 103 congenitally infected children. In this series, 99% had eye lesions, 63% had intracranial calcifications and 56% had psychomotor retardation (Feldman, 1953). Examples of some of these manifestations are shown in Figs. 4.3, 4.4 and 4.5. These observations initiated interest in congenital infection among scientists in Europe (Couvreur, 1955). A study from Austria reported frequent, similar symptoms in children with congenital toxoplasmosis (Aspock and Pollack, 1982). A study from France demonstrated that the seroprevalence in pregnant women in Paris was 85%, with a relatively high risk of acquisition of *Toxoplasma* infection in seronegative women (Desmonts *et al.*, 1965a, 1965b). The incidence of seropositivity among non-immigrant women in Stockholm, Sweden ranged from 47.7% in 1957 to 21.1% in 1987 (Forsgren *et al.*, 1991). Another group demonstrated that, in pregnant women, seropositivity was 14% in Stockholm, Sweden and 26% in southern Sweden (Evengard *et al.*, 2001). Seropositivity in Poland is nearly 60% and increases with age (Paul *et al.*, 2000). Eichenwald described the presence of severe disease in a cohort of children referred to him in Texas with a poor prognosis for vision, cognition, and motor function with seizures and hearing loss in two of four children who did not have symptoms at birth (Eichenwald, 1960; Fig. 4.2). The earlier work of Couvreur (1955) and Desmonts (1962, 1964) was followed by a larger study from France of 374 pregnancies (Desmonts and Couvreur, 1974a, b). Koppe (1982, 1989) and Stagno, Wilson, Remington, and Alford (1974) described substantial ophthalmologic and neurological sequelae in untreated children with congenital toxoplasmosis, both when symptomatic and asymptomatic in the perinatal period (Desmonts and Couvreur, 1984; McLeod *et al.*, 2009; Thulliez *et al.*, 2001a; Remington *et al* 2011). *T. gondii* in North America still often presents as a severe disease when diagnosed at birth in the absence of systematic screening programmes (McLeod *et al.*, 1990, 2006a, 2012; McAuley, 1994; Olariu *et al.*, 2011). The severity of symptomatic infection as well as sequelae of infections thought to be mild or asymptomatic at birth was detailed by Eichenwald in 1957. Wilson, Remington, Reynolds and Stagno (1980a) noted later neurological and ophthalmologic sequelae even for those thought to be asymptomatic at birth.

FIGURE 4.2 **Primary *T. gondii* infection during gestation: the French approach and data enable prenatal diagnosis of congenital infection**

A) Schematic diagram of screening algorithm for acquisition of primary *Toxoplasma* infection during gestation utilized in France and some USA obstetrical practices. In France, IgG and IgM are tested simultaneously in screening programs (i.e. pink circle = IgG- and IgM-). *Image from: McLeod* et al., *2013b, with permission.*

B) Top: Diagnosis of infants with PCR in amniotic fluid (AF) according to gestational age, in weeks, at maternal infection. Open bars indicate PCR sensitivity. Shaded bars indicate negative predictive values of PCR assay. 95% confidence interval (CI) shown in parenthesis above bars.

Middle: Percentage of infants undergoing amniocentesis and cases of congenital toxoplasmosis according to gestational age, in months, at maternal infection. Open bars indicate amniocentesis. Shaded bars indicate cases of congenital toxoplasmosis.

Bottom Left: Correlation between concentration of *Toxoplasma* in amniotic fluid and gestational age, in weeks, at maternal infection Shaded squares indicate severe signs of infection. Open circles indicate mild or no signs of infection. Higher concentrations of *T. gondii* in amniotic fluid are more likely earlier in gestation and correspond to severe disease in infants. Lower concentrations of *T. gondii* are more likely later in gestation and correspond to mild or no disease in infants. n = sample size.

Bottom Right: Comparison of median parasite concentration in amniotic fluid between infants with mild/no infection and infants with severe infection for maternal infections acquired before or after 20 weeks' gestation. Open bars indicate infants with subclinical infection. Shaded bars represent infants with infectious sequelae. Severe infections in infants are more likely if maternal infection is acquired before 20 weeks' gestation. *Image from: McLeod* et al., *2013a; Romand* et al., *2004; Wallon* et al., *2011, with permission.*

(A)

Eichenwald Study Outcome for 101 Patients at ≥ 4 Years Old

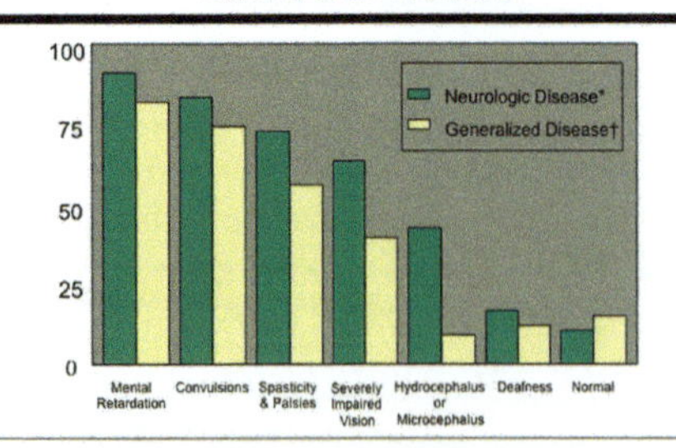

* *Patients with neurologic disease had otherwise undiagnosed central nervous system disease in the first year of life (n=70).*
† *Patients with generalized disease had otherwise undiagnosed nonneurologic disease in the first 2 months of life (n=31).*

(B)

Koppe Study Visual Outcome for 12 Children who were Asymptomatic at Birth, Untreated or Treated less than Once a Month and Evaluated When They Were 6 and 20 Years Old

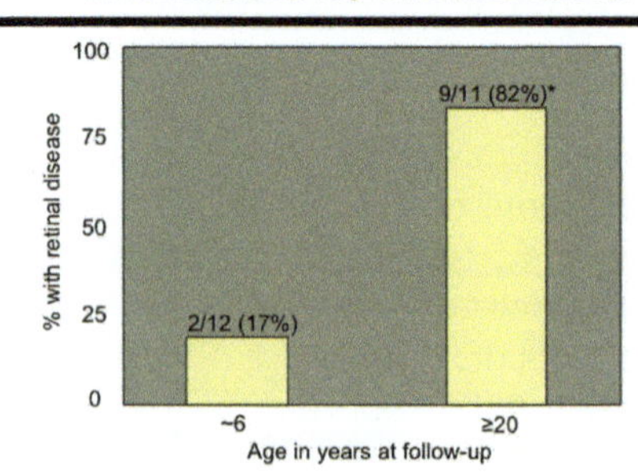

* *One child lost to follow-up*

(C)

Percentage with this Manifestation at or Near Birth[a]

100
80
60
40
20
0

Premature[b] · R/O sepsis · Rash · Hepatomegaly · Splenomegaly · Microphthalmia · Chorioretinitis · Hydrocephalus · Microcephalus · Seizures · CNS calcifications · Thrombocytopenia · Anaemia

(D)

Disease Manifestations at Birth

Splenomegaly: enlarged spleen
Hepatomegaly: enlarged liver
Skin rash: petechiae, blueberry muffin rash

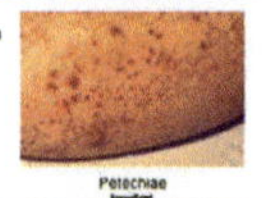

Petechiae

Blueberry muffin rash

Chorioretinal disease: scar, active recurring disease

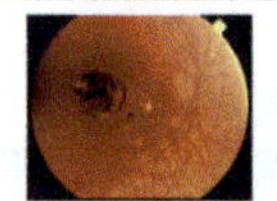

Patient at age 4.5 before new lesion.

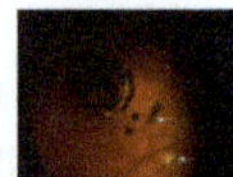

Patient at age 7 with active disease.

CNS Disease: Hydrocephalus; CNS calcifications

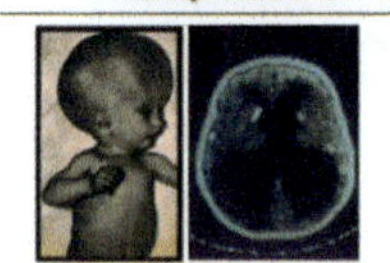

Hydrocephalus: ventriculomegaly

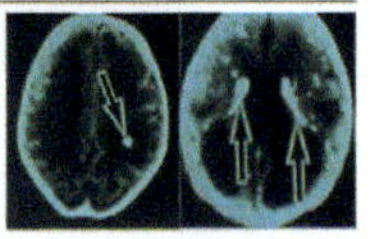

CNS Calcifications

(E)

OPHTHALMOLOGIC MANIFESTATIONS OF CONGENITAL TOXOPLASMOSIS

	NO. OF TREATED PATIENTS WITH FINDING (%) (N = 76)	NO. OF HISTORICAL PATIENTS WITH FINDING (%) (N = 18)	TOTAL NO. OF PATIENTS WITH FINDING (%) (N = 94)
Strabismus	26 (34)	5 (28)	31 (33)
Nystagmus	20 (26)	5 (28)	25 (27)
Microphthalmia	10 (13)	2 (11)	12 (13)
Phthisis	4 (5)	0 (0)	4 (4)
Microcornea	15 (20)	3 (17)	18 (19)
Cataract	7 (9)	2 (11)	9 (10)
Vitritis (active)	3 (4)*	2 (11)	5 (5)
Retinitis (active)	6 (8)	4 (22)	10 (11)
Chorioretinal scars	56 (74)	18 (100)	74 (79)
Macular	39/72 (54)†	13/17 (76)	52/89 (58)
Juxtapapillary	37/72 (51)	9/17 (53)	46/89 (52)
Peripheral	43/72 (58)	14/17 (82)	57/89 (64)
Retinal detachment	7 (9)	2 (11)	9 (10)
Optic atrophy	14 (18)	5 (28)	19 (20)

*Two additional patients, not included in this table, were receiving treatment and retinochoroiditis had resolved, but vitreous cells and veils persisted at time of examination.
†Numerator represents number of patients with finding; denominator is the total number, unless otherwise specified. Number in parentheses is percentage. Patients with bilateral retinal detachment in whom the location of scars was not possible were excluded from the denominator.

(F)

Incidence of Central Nervous System and Retinal Abnormalities Among Infants in Whom *T. gondii* Infection was Diagnosed Solely by Neonatal Screening

Site of Abnormality	Incidence*
Central nervous system	14/48 (29)
Increased CSF protein (≥100 mg/dl)	8/32 (25)
Intracranial calcifications on CT scan	9/46 (20)
Enlarged lateral ventricles on CT or US	1/47 (2)
Retina	9/48 (19)
Active chorioretinitis	2/48 (4)
Retinal scars without active inflammation	7/48 (15)
Either site	19/48 (40)

**Number affected/number examined (percent).*
From Guerina et al, NEJM, 330:1861, 1994.

Several studies have documented that prompt treatment can prevent transmission from mother to child and reduce clinical symptoms in children (Couvreur and Desmonts, 1962; Roux *et al.*, 1976; Desmonts, 1982; Couvreur *et al.*, 1984, 1993; Daffos *et al.*, 1988; Hohlfeld *et al.*, 1989, 1994; Forestier 1991; Foulon *et al.*, 1994, 1999; Gilbert *et al.*, 2001; Thulliez, 2001a; Peyron *et al.*, 2011; Berrebi *et al.*, 2007; Kieffer *et al.*, 2008; Montoya *et al.*, 2010; Wallon *et al.*, 2013).

Congenital infection results when acute acquired *T. gondii* infection occurs in previously seronegative pregnant women. Women who are seropositive before conception, as a rule, do not transmit infection to the foetus. There have, however, been rare cases reported of congenital infection where the mother acquired infection one to two months before conception (e.g. a previously seropositive woman from Brazil but living in Switzerland transmitted to her foetus after a trip to Brazil during her pregnancy (Kodjikian *et al.*, 2004) and a woman in Brazil with chronicinfection and active retinal disease who transmitted to her foetus (Andrade *et al.*, 2012). Immune compromised women (e.g. those treated with corticosteroids) may occasionally transmit a chronic infection during pregnancy resulting in congenital infection (Desmonts and Couvreur, 1984; Mitchell *et al.*, 1990). Persons with HIV in the context of available retroviral treatment may, but in one series did not, commonly transmit their chronic *T. gondii* infection to their foetus when pregnant (Dunn *et al.*, 1996). The clinical manifestations of congenital toxoplasmosis vary depending on the trimester when the infection was acquired. Congenital infection of the foetus in women infected just before conception is extremely rare (Vogel *et al.*, 1996; Remington *et al.*, 2011) and even during the first few weeks of pregnancy, the maternal–foetal transmission rate is only a small percent. There is an inverse relationship between the rate of transmission and the severity of the infection (Desmonts and Couvreur; 1984; Remington *et al.*, 2011; Wallon *et al.*, 2013; Fig. 4.7D). Without treatment, infections acquired in the first trimester

FIGURE 4.3 ***T. gondii*** **congenital infection causes disease when untreated.**
A) Data from Eichenwald study of patient outcome at 4 years of age or older. n = sample size. *Image from: McLeod* et al., *2009.*
B) Data from Koppe study of visual outcome for children who were asymptomatic at birth, untreated or treated for less than a month, and evaluated at ages six and 20 years. Percentages are those children with retinal disease. There are negative outcomes for children with congenital toxoplasmosis who were untreated or treated for only one month. *Image from: McLeod* et al., *2009; based on data from Eichenwald, 1957, 1960; Koppe and Kloosterman, 1982; Koppe* et al., *1986, with permission.*
C) Prevalence of various manifestations of *Toxoplasma* infection in children at or near birth referred to the NCCCTS (1981–2009). These include prematurity, r/o sepsis, skin rash, hepatomegaly, splenomegaly, microphthalmia, chorioretinitis, hydrocephalus, microcephalus, seizures, CNS calcifications, thrombocytopenia and anaemia. *Image from: McLeod et al., 2013b, with permission.*
D) Images depicting manifestations of congenital toxoplasmosis. Manifestations include splenomegaly and hepatomegaly. Manifestations of the skin can include petechiae and/or blueberry muffin rash. The infant depicted here has a blueberry muffin rash secondary to CMV infection. In *Toxoplasma* infection, appearance of blueberry muffin rash appears similar. Manifestations involving the eye include chorioretinitis which can result in scarring. Involvement of the brain can manifest in hydrocephalus and/or calcifications.
Petechiae image from: http://dermatology.about.com/library/blpetechiaephoto.htm
Blueberry muffin rash image from: Mehta, V., Balachandran, C., Lonikar, V. (2008). Blueberry muffin baby: a pictorial differential diagnosis. *Dermatol Online J.* 14(2):8.
Fundus photographs from: Phan *et al.*, 2008a, with permission.
Hydrocephalus image from: Swisher *et al.*, 1994, with permission.
Calcification images from: Patel *et al.*, 1996, with permission.
E) Incidence of ophthalmologic manifestations of congenital toxoplasmosis. n = sample size. *Data from Mets* et al., *1996, with permission.*
F) Incidence of central nervous system and retinal manifestations of congenital toxoplasmosis as documented by the New England Screening Program.

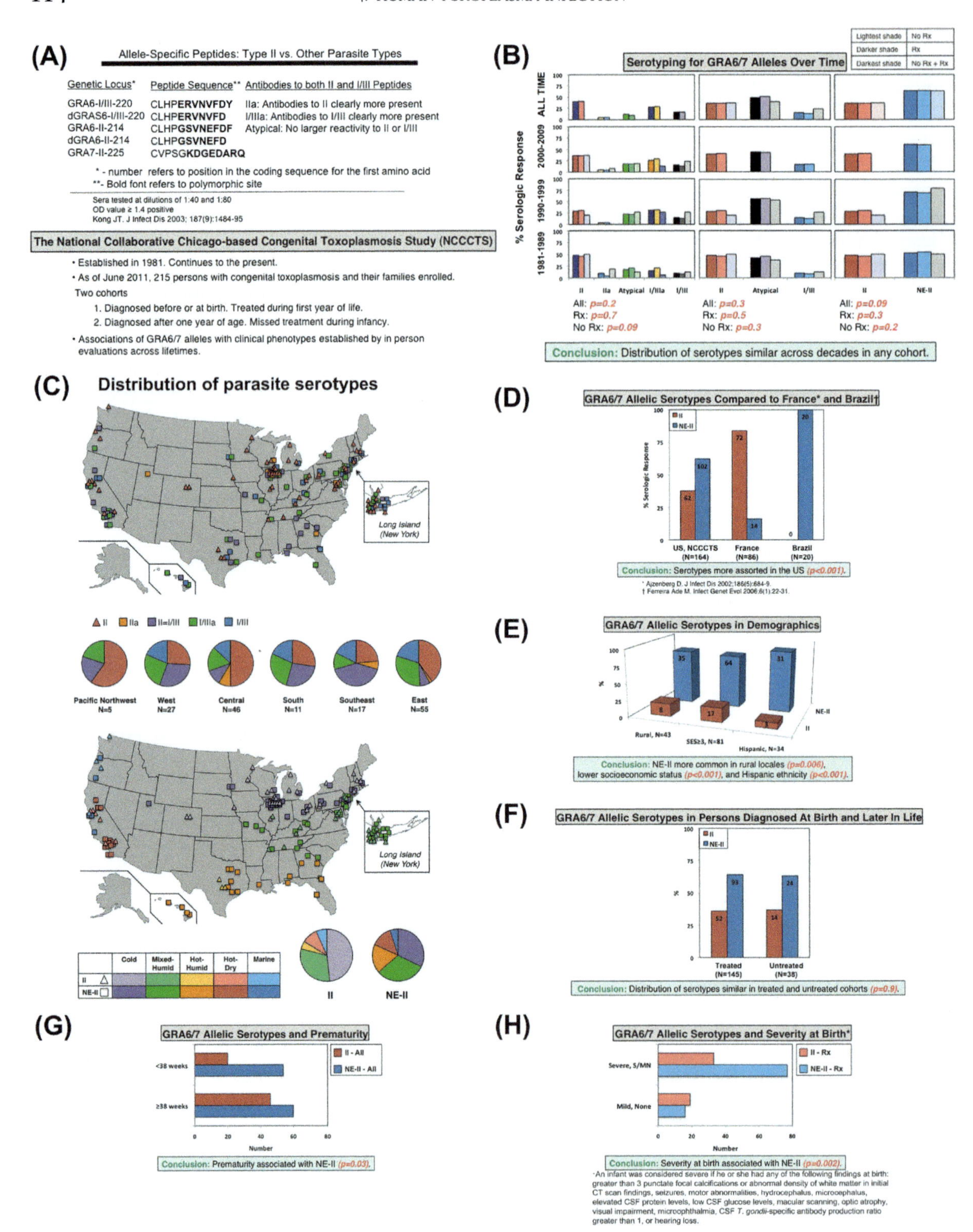
(A)
Allele-Specific Peptides: Type II vs. Other Parasite Types
Genetic Locus* Peptide Sequence** Antibodies to both II and I/III Peptides
GRA6-I/III-220 CLHPERVNVFDY IIa: Antibodies to II clearly more present
dGRAS6-I/III-220 CLHPERVNVFD I/IIIa: Antibodies to I/III clearly more present
GRA6-II-214 CLHPGSVNEFDF Atypical: No larger reactivity to II or I/III
dGRA6-II-214 CLHPGSVNEFD
GRA7-II-225 CVPSGKDGEDARQ
* - number refers to position in the coding sequence for the first amino acid
**- Bold font refers to polymorphic site
Sera tested at dilutions of 1:40 and 1:80
OD value ≥ 1.4 positive
Kong JT. J Infect Dis 2003; 187(9):1484-95
The National Collaborative Chicago-based Congenital Toxoplasmosis Study (NCCCTS)
• Established in 1981. Continues to the present.
• As of June 2011, 215 persons with congenital toxoplasmosis and their families enrolled.
Two cohorts
1. Diagnosed before or at birth. Treated during first year of life.
2. Diagnosed after one year of age. Missed treatment during infancy.
• Associations of GRA6/7 alleles with clinical phenotypes established by in person evaluations across lifetimes.
(B)
Serotyping for GRA6/7 Alleles Over Time
Lightest shade No Rx
Darker shade Rx
Darkest shade No Rx + Rx
% Serologic Response
ALL TIME
2000-2009
1990-1999
1981-1989
II IIa Atypical I/IIIa I/III
II Atypical I/III
II NE-II
All: p=0.2
Rx: p=0.7
No Rx: p=0.09
All: p=0.3
Rx: p=0.5
No Rx: p=0.3
All: p=0.09
Rx: p=0.3
No Rx: p=0.2
Conclusion: Distribution of serotypes similar across decades in any cohort.
(C)
Distribution of parasite serotypes
Long Island (New York)
II IIa II=I/III I/IIIa I/III
Pacific Northwest N=5
West N=27
Central N=46
South N=11
Southeast N=17
East N=55
Long Island (New York)
Cold
Mixed-Humid
Hot-Humid
Hot-Dry
Marine
II
NE-II
II
NE-II
(D)
GRA6/7 Allelic Serotypes Compared to France* and Brazil†
II
NE-II
% Serologic Response
62
102
72
14
0
20
US, NCCCTS (N=164)
France (N=86)
Brazil (N=20)
Conclusion: Serotypes more assorted in the US (p<0.001).
* Ajzenberg D. J Infect Dis 2002;186(5):684-9.
† Ferreira Ade M. Infect Genet Evol 2006;6(1):22-31.
(E)
GRA6/7 Allelic Serotypes in Demographics
%
35
64
31
8
17
3
Rural, N=43
SES≥3, N=81
Hispanic, N=34
NE-II
II
Conclusion: NE-II more common in rural locales (p=0.006), lower socioeconomic status (p<0.001), and Hispanic ethnicity (p<0.001).
(F)
GRA6/7 Allelic Serotypes in Persons Diagnosed At Birth and Later In Life
II
NE-II
%
52
93
14
24
Treated (N=145)
Untreated (N=38)
Conclusion: Distribution of serotypes similar in treated and untreated cohorts (p=0.9).
(G)
GRA6/7 Allelic Serotypes and Prematurity
II - All
NE-II - All
<38 weeks
≥38 weeks
Number
Conclusion: Prematurity associated with NE-II (p=0.03).
(H)
GRA6/7 Allelic Serotypes and Severity at Birth*
II - Rx
NE-II - Rx
Severe, S/MN
Mild, None
Number
Conclusion: Severity at birth associated with NE-II (p=0.002).
*An infant was considered severe if he or she had any of the following findings at birth: greater than 3 punctate focal calcifications or abnormal density of white matter in initial CT scan findings, seizures, motor abnormalities, hydrocephalus, microcephalus, elevated CSF protein levels, low CSF glucose levels, macular scarring, optic atrophy, visual impairment, microophthalmia, CSF T. gondii-specific antibody production ratio greater than 1, or hearing loss.

result in congenital infection in 10% to 25% of foetuses (Couvreur and Desmonts, 1962, 1964; Stillwaggon *et al.*, 2011). The rate of transmission rises to 30% to 50% in those women infected during the second trimester and 60% to 70% for those infected during the third trimester. These studies on risk and severity of transmission were done in areas with a predominance of type II strains and not in areas (e.g. the USA or South America) where other, differing genotypes predominate. Several studies have demonstrated that treatment of the pregnant woman with anti-*T. gondii* medicines reduces the incidence and severity of manifestations of congenital infection (Desmonts and Couvreur, 1984; Thulliez 1992, 2001a; Kieffer *et al.*, 2008; McLeod *et al.*, 2009, 2012; Cortina-Borja *et al.*, 2010; Hotop *et al.*, 2012; Remington *et al.*, 2011; Stillwaggon *et al.*, 2011; Wallon *et al.*, 2013; Figs. 4.2, 4.3, 4.4, 4.5, 4.7).

The clinical manifestations of congenital toxoplasmosis usually are most severe if infection is acquired before week 26 of gestation. In these cases, the retina and central nervous system are commonly affected with non-specific signs including retinochoroiditis, strabismus, blindness, epilepsy, psychomotor or mental retardation, encephalitis, microcephaly, intracranial calcification, hydrocephalus anaemia, jaundice, rash, and petechiae due to thrombocytopenia (Remington *et al.*, 2001, 2011; Fig. 4.3). Newborns infected in the third trimester may be asymptomatic at birth but, with careful examination, signs such as meningitis and retinitis are frequently noted at birth. Sequelae, such as chorioretinitis, often develop later in life without treatment (Koppe *et al.*, 1982, 1986; Phan *et al.*, 2008a, 2008b; Kieffer *et al.*, 2008; Wallon *et al.*, 2010; Peyron *et al.*, 2011). Desmonts and Couvreur (1984), Stagno *et al.* (1977), and The New England Screening Program (Guerina *et al.*, 1994) found that up to 50% or more of those infants believed to have no signs at birth actually did have clinical findings due to their *T. gondii* infection when carefully examined in the perinatal period (Figs. 4.3 to 4.8).

FIGURE 4.4 **Parasite serotype and congenital infection in the USA.**
A) Allele-specific peptides. Top: The peptide sequence at the GRA6/7 locus determines parasite serotype II or NE-II (not exclusively II). Bottom: Summary of the NCCCTS.
B) NCCCTS patients with antibodies to GRA6/7 peptides to type II or type I/III parasites, or both, based on decade of birth. Right: Distribution of patients with serologic responses to parasite serotype II or NE-II. Left: Distribution of patients with serologic responses designated as serotype II, IIa, atypical, I/IIIa, or I/III. Persons with responses to both serotype II and NE-II peptides are designated as IIa if the response was greater to serotype II, designated as I/IIIa if the response was greater to serotype I/III, or designated as atypical if the response was equal to serotypes II and I/III. Rx indicates the persons in the NCCCTS cohort who were diagnosed in the perinatal period. No Rx indicates persons in the NCCCTS cohort who missed being treated during the first year of life. All includes Rx plus No Rx. p = p-value. *Image from: McLeod* et al., *2012, with permission.*
C) Distribution of parasite serotypes. Top map shows distribution of parasite serotype in the USA by birthplace and USA region. Bottom map shows distribution of parasite serotypes by birthplace and climate. N = sample size. *Image from: McLeod* et al., *2012, with permission.*
D) Prevalence of parasite serotype by country. Type II serotype toxoplasmic infections predominate in France. NE-II toxoplasmic infections predominate in Brazil. Serotypes of toxoplasmic infections are more varied in the USA. N = sample size. p = p-value.
E) Associations between parasite serotype and demographics. Toxoplasmic infection with NE-II serotype associated with residing in rural locales, lower socioeconomic status (SES) and Hispanic ethnicity. N = sample size. p = p-value.
F) Associations between parasite serotype and treated/untreated cohorts. Serotypes are similarly distributed between treated and untreated cohorts. N = sample size. p = p-value.
G) Associations between parasite serotype and prematurity. Toxoplasmic infection with NE-II serotype associated, but not exclusively, with prematurity. p = p-value.
H) Associations between parasite serotype and disease severity at birth. Toxoplasmic infection with NE-II serotype is associated, but not exclusively, with severe disease at birth. *Images from: McLeod* et al., *2013a; data from McLeod* et al., *2012, with permission.*

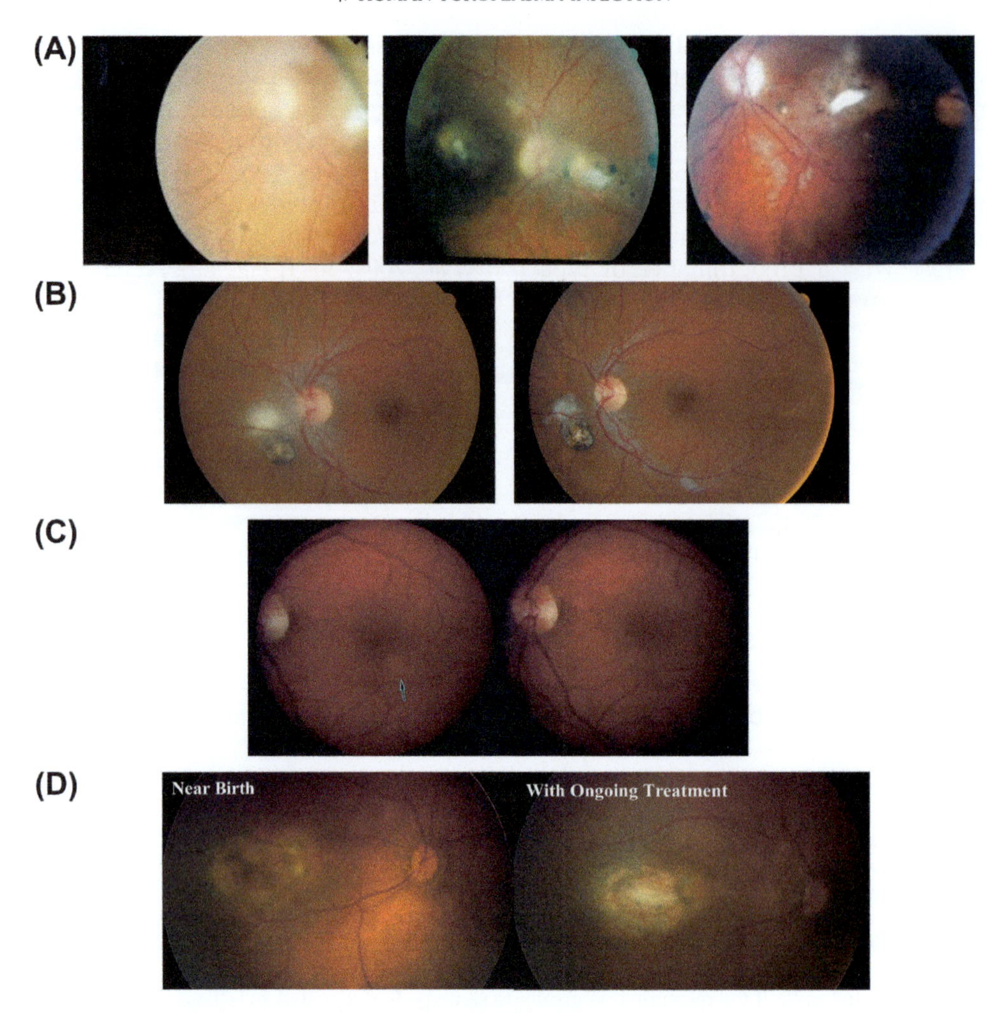

FIGURE 4.5 **Manifestations of ocular toxoplasmosis improve with treatment.**
A) Fundus photographs of follow-up of vitritis. Left panel depicts severe vitritis. Middle panel depicts resolving vitritis with active lesion. Right panel depicts resolved vitritis and healed lesion.
B) Fundus photographs of follow-up of toxoplasmic lesion. Left image depicts active lesion. Right image depicts healed scar.
C) Fundus photographs of follow-up of toxoplasmic lesion. Left image depicts active retinal lesion (arrow). Right panel depicts completely resolved lesion within one month of initiating treatment.
D) Fundus photographs of follow-up of toxoplasmic lesion in an infant. Left panel, 'near birth', depicts active vitritis. Right panel, 'with ongoing treatment', depicts clearing of vitritis.
E) Ocular toxoplasmosis with submacular neovascular membrane. Fundus photographs shown in top row, A–D. Indocyanin green angiograph images of choroidal neovascular membrane with hemorrhage depicted in middle row, A–C. Optical coherence tomography (OCT) images of resolution of choroiditis, depicted in middle row, with pyrimethamine, sulphadiazine, and intraocular injection of Lucentis shown in bottom row, A–D.
F) New eye lesions in children who had less than eight weeks to eight weeks or more delay from diagnosis *in utero* to treatment. Left, Kaplan–Meier plot shows the age at diagnosis of retinochoroiditis. (Solid line: delay of less than four weeks; dashed line: delay of four to eight weeks; dotted line: delay of more than eight weeks). Right, Kaplan–Meier plot estimates age at diagnosis of retinochoroiditis during first two years of life among 300 infants.
G) Recurrent retinal disease and new lesions in children. Those who missed treatment in the first year of life (left panel) and those who were treated in the first year of life with pyrimethamine and sulphadiazine (right panel). Blue shaded area represents confidence interval. *Images A–D from: Delair* et al., *2011, with permission; Image E from: Benevento* et al., *2008, with permission; Image F from: Kieffer* et al., *2008, with permission; Image G from: McLeod* et al., *2009, with permission.*

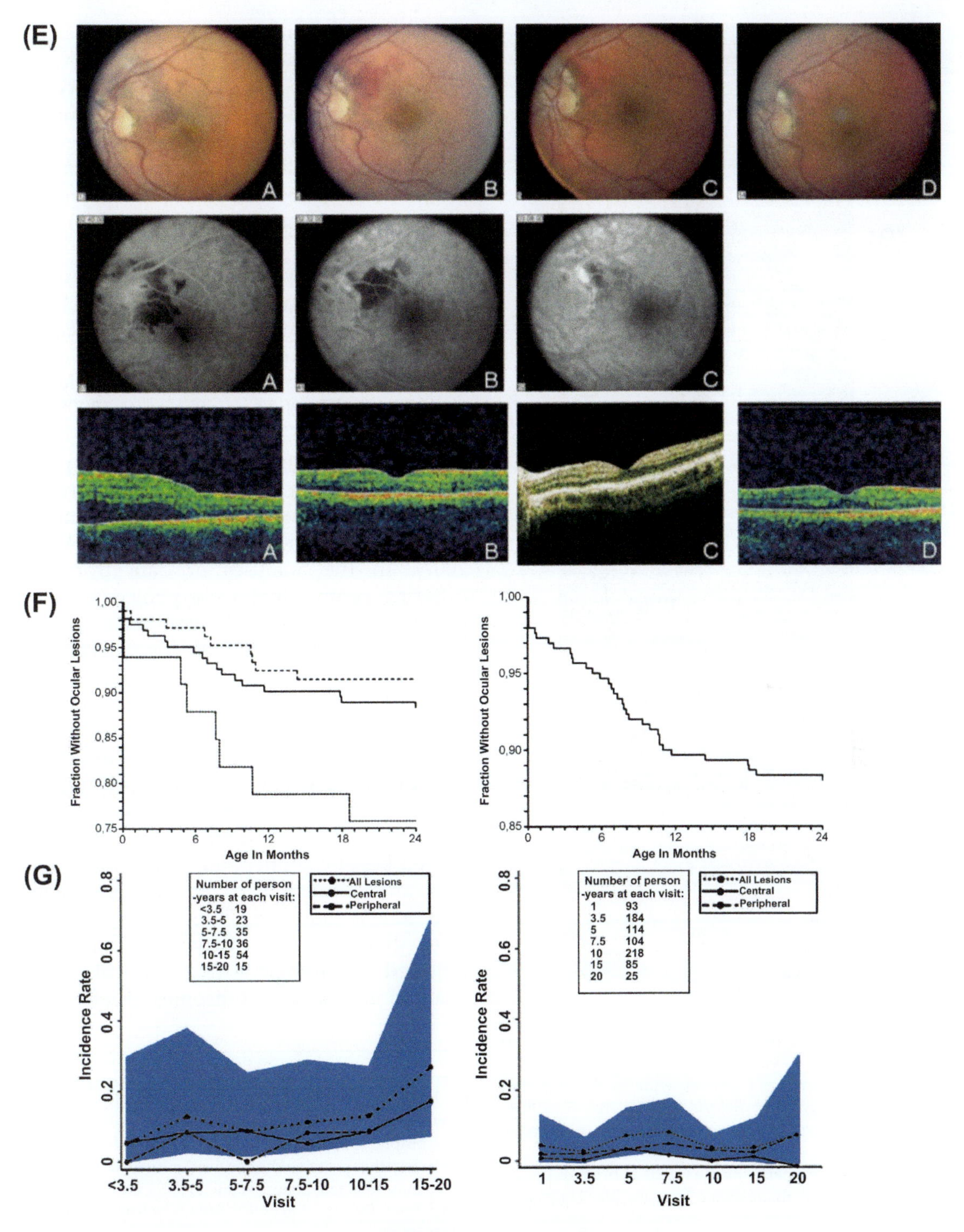

FIGURE 4.5 (*continued*).

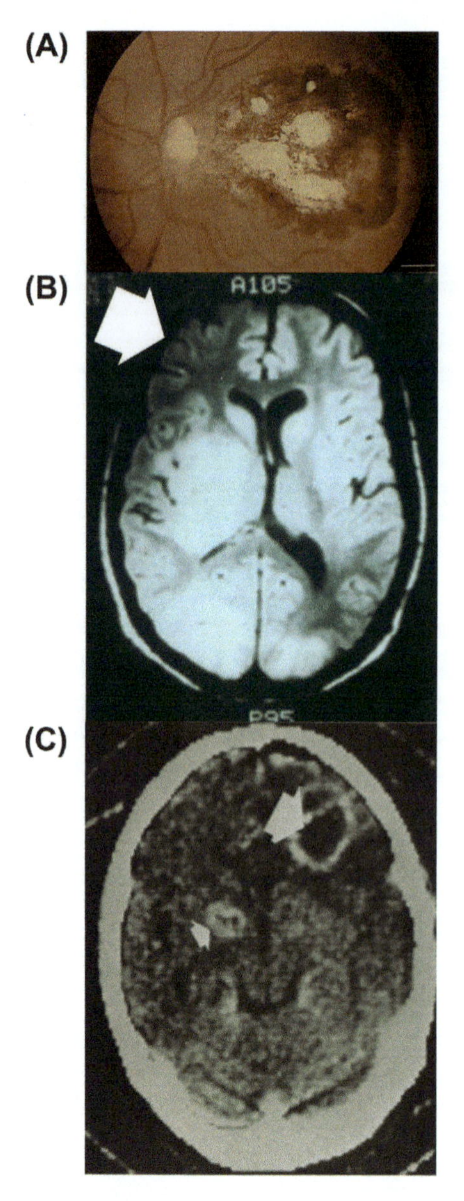

FIGURE 4.6 **Disease due to *T. gondii* infection in immune compromised persons can involve any organ, but frequently affects the eye and brain.**

A) Fundus photograph of person with HIV/AIDS and toxoplasmosis. *Image from: Roberts, F., McLeod, R. (1999). Pathogenesis of toxoplasmic retinochoroiditis. Parasitol Today. 15(2): 51–57.*

B) Brain magnetic resonance image (MRI) of brain of person with HIV/AIDS and toxoplasmic encephalitis. Light gray area indicated by the white arrow is the only normal area of the brain in this person. *Image from: Levin* et al., *1983, with permission.*

C) Computed tomography (CT) image of the brain of a seronegative person who received a heart transplant from an acutely infected donor. White arrows indicate abscesses. *Image from: Ryning* et al., *1979, with permission.*

A recent European multi-centre study reported that 19 (8%) out of 244 newborns with congenital toxoplasmosis had cerebral calcifications and that treatment within four weeks of prenatal diagnosis reduced the risk of neurological findings (Adjusted Odds Ratio 0.28; CI 0.08–0.75) (Gras *et al.*, 2005). The same study found chorioretinitis in 30 (12%) of 255 newborns with congenital toxoplasmosis; however, they reported that treatment did not reduce the risk of chorioretinitis (Gras *et al.*, 2005). More recently it has become clear that the more rapidly treatment is initiated, the less risk there is of eye disease due to *T. gondii* infection (EMSCOT, 2003; Kieffer *et al.*, 2008). The EMSCOT–SYROCOT studies included evaluations combined from centres in Europe that had variations in their standard of care diagnostic measures, treatments, and timing of implementation (Kieffer *et al.*, 2008; SYROCOT, 2007). In the analyses of data from these differing centres that were pooled, effects on neurological signs and symptoms were not significant when durations from diagnosis to initiation of varied treatments was greater than four weeks. However, when data from individual centres were deconvoluted and reanalysed separately by those centres in Paris and Lyon, beneficial effect in preventing eye disease was demonstrated (Kieffer *et al.*, 2008). Similarly, a recent study from Lyon has demonstrated beneficial effect on prevention of infection and reduction of adverse sequelae of infection (Wallon *et al.*, 2013). A shorter interval from diagnosis to initiation of treatment is reflected in serologic testing protocols and has resulted in the best outcomes (Kieffer *et al.*,

(A) French Approach to Prenatal Prevention, Dx, and Rx

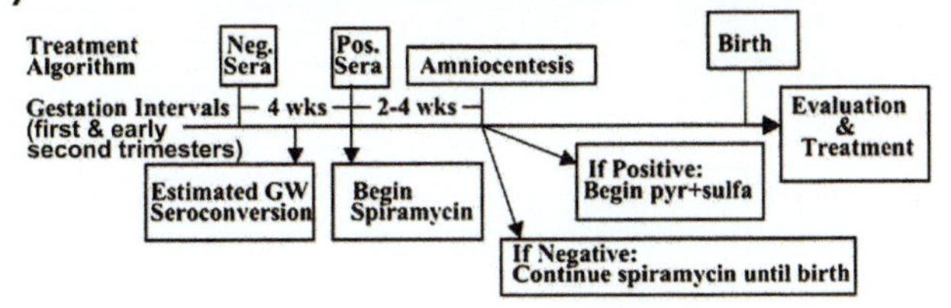

(B)

Outcomes Using the French Algorithm

- **Diagnose Mother:** Systematic serologic screening before conception, intrapartum and postpartum.
- **Treat Mother:** If acute serology, spiramycin reduces transmission of *T. gondii* across placenta to foetus, but does not treat the foetus.
  - Untreated 94 (60%) of 154 vs. treated 91 (23%) of 388[a].
- **Diagnose Fetus**: Ultrasounds, amniocentesis, PCR at ≥ 18 wk gestation[b].
  - Sensitivity 37 (97%) of 38; specificity 301 of 301[c]. For mid-gestation infections, please see Figure 2.
- **Treat Fetus:** Treat mother with pyrimethamine and sulfadiazine, which cross the placenta to treat the foetus.
  - **Hohlfeld**[c,d]: N=54 live births; 34 terminations; All 54 normal development. 19% subtle findings: 7 (13%) intracranial calcifications, 3 (6%) chorioretinal scars. Follow-up of 18 children (median age 4.5 yr; range 1-11 yr): 39% retinal scars, most scars were peripheral.
  - **Couveur**[e]: Compared to a spiramycin regimen, adding pyrimethamine and sulfadiazine to a spiramycin regimen reduce the number of isolates from placenta from 77% to 42%, reduced *T. gondii*-specific immunoglobulin load at birth from 139 IU/g IgG to 86 IU/g IgG and at 6 mo. from 137 IU/g IgG to 70 IU/g IgG, reduced *T. gondii*-specific IgM prevalence in the neonate from 69% to 17%, and increased subclinical infections, presumably by delaying transmission, from 17% to 30%.
  - **Kieffer**[f]: Shorter interval between diagnosis and treatment reduces subsequent retinal disease.
  - **Syrocot**[g]: Shorter interval between diagnosis and treatment reduces subsequent neurologic disease.
- Conclusion: Favourable outcomes with treatment *in utero* in France and as French Algorithm applied in U.S.[h,i].

(C)

Ocular and Intracranial Lesions in Treated Infants with Congenital Toxoplasmosis in European Countries with Systematic Prenatal Screening and Treatment Programs for Toxoplasmosis During Gestation[j]

Cohort Region	Recruitment Period	Infected Live-Born Children	Clinical Manifestations		
			Any	Ocular Lesions	Intracranial Lesions
France					
Nice	1996-2000	15	1	1	0
Grenoble	1996-2000	6	2	1	1
Lyon	1996-2000	43	10	9	3
Marseille	1996-2000	20	2	2	0
Nice	1996-2000	8	4	2	2
Paris	1996-2000	65	8	8	1
Reims	1996-2000	8	2	8	0
Toulouse	1996-2000	22	3	8	1
Austria	1992-2005	33	3	3	2
	1996-2000	24	5	5	2
Italy					
Naples	1996-2000	11	3	3	3
Milan	1996-2000	4	0	0	0
Slovenia					
Ljubljana	1996	3	3	0	0
Total (%)		262	43 (16.4)	38 (14.5)	15 (5.7)

Conclusions:

- Gestational age of *T. gondii* infection acquisition predicts maternal-fetal transmission
- Delay in prenatal treatment increases the risk of clinical signs in infected children
- Prenatal treatment results in:
  - Decreased number of cases with severe infection
  - Decreased number of cases with mild infection
  - Decreased incidence of sequelae at birth
  - Decreased number of late sequelae
  - Decreased incidence of vertical transmission

FIGURE 4.7 *(continued)*

(D)

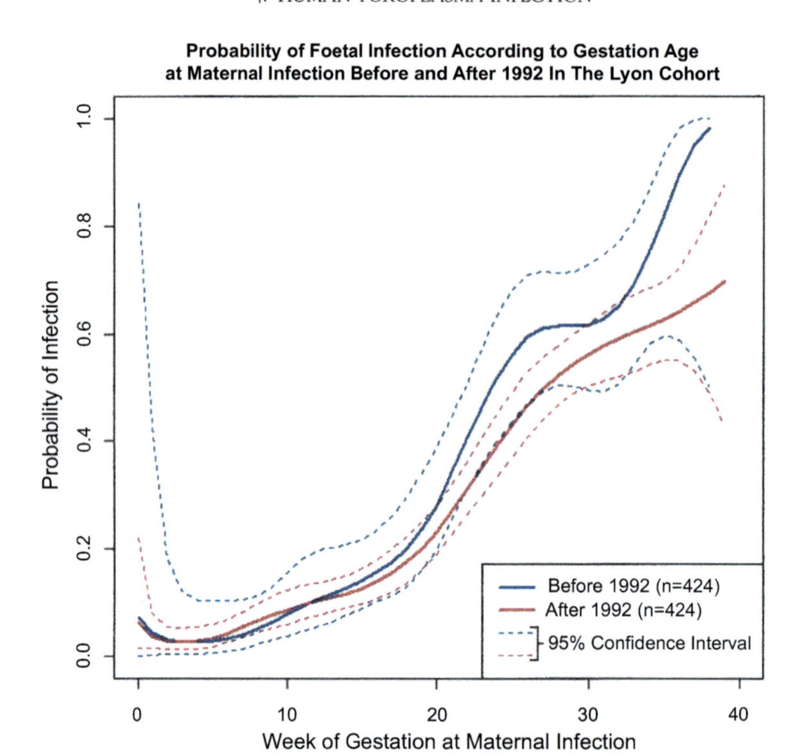

Reduction In The Risk of Infection and of Clinical Signs Following Changes In The Retesting Policy (1992) and In Antenatal Diagnosis and Treatment Procedures (1995) (Lyon-Cohort)

Periods	1987-1991	1992-1995	1996-2008	
Risk of Infection				
Retesting Policy for Women Identified as Susceptible at The First Prenatal Test Implemented In 1985	Recommended Without Specific Frequency	Mandatory and Monthly		
Infected Children/Mothers	125/424	388/1,624		P<0.018
Risk of Clinical Signs at Age 3 Years				
PCR Availability on Amniotic Fluid	No		Yes	
PS Antenatal Treatment	Alternating with Spiramycin for 3-Week Periods		Continuous	
Clinical Signs/Infected Children	87/794		46/1,150	P<10^{-4}

FIGURE 4.7 *(continued)*

2008; SYROCOT, 2007), where testing is done monthly starting at week 11 of gestation for seronegative pregnant women. Outcomes appear to improve with treatment *in utero* and postnatally (Thulliez *et al* 2001a, 2001b; Peyron *et al.*, 2003, 2011; McLeod *et al.*, 2006a, 2009, 2012; Kieffer *et al* 2008; Cortina-Borja *et al.*, 2010; Remington *et al* 2011; Wallon *et al.*, 2011).

The genetic type of parasite is associated with manifestations of the infant at birth (McLeod *et al.*, 2012), although not absolutely (Fig. 4.4). There appears to be a favourable response to treatment for both type II or non-type II infections (McLeod *et al.*, 2012; see Section 4.3). Congenital transmission from mother to foetus with chorioretinitis is more common in Brazil after primary infection in pregnancy, compared to Europe, and manifestations are more severe (Gilbert *et al.*, 2008). In Minas Girais, one in ~770 infants born has had congenital toxoplasmosis and half these infants had active retinal disease at birth (Vasconcelos Santos and Andrade, 2011). This might reflect the different recombinant I/III genotypes of *T. gondii* that are prevalent in Minas Girais, Brazil, which is different from Europe and North America. About a third of the NCCCTS participants in North America have a serotype II parasitic infection. In animals in North America there are type I, II, III, IV (haplogroup X11) and other *T. gondii* serotypes, but the type I, III parasites are not, for the most part, the same as those found in Brazil.

4.1.3.2 Congenital Toxoplasmosis in Different Countries

1.3.2.1 FRANCE AND BELGIUM

The pioneering work of Desmonts, Couvreur, Thulliez, Romand, Costa, Peyron, Wallon, Daffos, Foulon, Kieffer, McLeod and many others (summarized above) has defined much of what we understand about congenital toxoplasmosis and how to prevent, diagnose and treat it as well as the natural history of both treated and untreated congenital infection. This experience was derived predominantly from infections with type II *T. gondii* in Europe. The information about severity being greater and transmission less frequent in early gestation with decreasing severity later in gestation, but

FIGURE 4.7 **The french approach to treatment of congenital toxoplasmosis during gestation and infancy and outcomes.** A) The French approach to prenatal prevention, diagnosis and prescribed treatment. *Image from: McLeod* et al., *2009, with permission.*
B) Treatment and diagnosis regimens to follow throughout gestation.
These include diagnosing and treating the pregnant women as well as the foetus. Table adapted from Remington *et al.*, 2011.
[a] Data from Desmonts *et al.*, 1984.
[b] Data from Hohlfeld *et al.*, 1994.
[c] Data from Daffos *et al.*, 1988.
[d] Data from Hohlfeld *et al.*, 1989.
[e] From Couveur J *et al.*, 1993.
[f] Data from Kieffer *et al.*, 2008.
[g] Data from SYROCOT, 2007.
[h] Data from Peyron *et al.*, 2011.
[i] Data from Stillwaggon *et al.*, 2011.
[j] Data from SYROCOT, 2007. Table from Olariu *et al.*, 2011.
C) Clinical manifestations in treated infants with congenital toxoplasmosis in European countries with screening and treatment programmes and summary of effects. *Data from SYROCOT, 2007. Table from Olariu* et al., *2011, with permission.*
D) Top: Probability of foetal infection according to gestational age at the time of maternal infection before (n = 451) and after (n = 1,624) mid-1992 – Lyon Cohort 1987–2008
Bottom: Reduction in the risk of infection and of clinical signs following changes in the retesting policy (1992) and in antenatal diagnosis and treatment procedures (1995) – Lyon Cohort 1987–2008. *Images from: Wallon* et al., *2013, with permission.*

with higher transmission rates, is derived from data in these studies. Studies in these countries (many in collaboration with the Remington laboratory in the USA) defined the diagnostic approaches to this infection. The universal prenatal screening in France was instrumental in defining this disease and approaches to its treatment. Overall, this disease has changed from one with a poor prognosis to one in which the prognosis was for a favourable outcome (McAuley *et al.*, 1994; Kieffer *et al.*, 2008; Peyron *et al.*, 2011; McLeod *et al.*, 1992; 2000; 2006; 2009; 2012; Wallon *et al.*, 2013).

4.1.3.2.2 AUSTRIA, GERMANY, THE NETHERLANDS AND ITALY

Prenatal screening has also occurred in several other European countries. In Italy, there is serodiagnosis during gestation, treatment before and after birth throughout infancy and follow-up in a centralized manner, often in a programme headed by Buffolano *et al.* (2013, submitted) in Naples. Koppe and Kloosterman (1982), in the Netherlands, described recurrent, new eye disease in almost all children by adolescence even though they had no signs or symptoms of disease at birth. Aspock and Pollak (1982) described favourable outcomes for infants born to acutely infected mothers following the treatment of primary infections in pregnant women with pyrimethamine and sulphadiazine. Hotop *et al.* (2012) described the data of an approach used in Germany for treatment of *T. gondii* infection acquired during pregnancy. In this approach, spiramycin is administered until the sixteenth week of gestation, followed by at least four weeks of treatment of the pregnant woman with pyrimethamine, sulphadiazine and leucovorin (folinic acid) (abbreviated PSL). This was independent of the timing of the primary infection during gestation. If infection of the foetus was confirmed by PCR or if foetal ultrasound indicated severe symptoms (e.g. hydrocephalus or ventricular dilation), treatment with PSL was continued until and after delivery. In France, only spiramycin is given unless infection of the foetus is proven or if primary infection of the pregnant woman occurs after 21 weeks' gestation. Hotop *et al.*'s (2012) retrospective analysis of 685 women who had a serologic profile consistent with primary infection in pregnancy and their children noted an overall transmission rate (4.8%) and rate of clinical manifestations in newborns (1.6%) that was lower than that reported in other countries.

4.1.3.2.3 THE USA

The NCCCTS was established in 1981 and performed a phase 1 study to determine early safety and efficacy of continued use of pyrimethamine and sulphadiazine throughout the first year of life and the influence of such treatment on later outcomes. A phase 2 randomized clinical trial followed with an additional observational study of historical untreated persons and those who were treated in the first year of life. These evaluations were performed in Chicago at pre-specified time intervals (near birth, 1 year, 3.5 years, 5 years, 7.5 years, 10 years and then continuing at 5-year intervals) for pre-specified endpoints, evidence for toxicity as well as other findings. This phase 2 randomized trial uses a higher or lower dose of pyrimethamine (two months or six months of daily pyrimethamine, 1 mg/kg, followed by 1 mg/kg each Monday, Wednesday, Friday for the remainder of the year of treatment) plus sulphadiazine and leucovorin begun at diagnosis near birth. Occasionally, there was also treatment of the foetus by treatment of the mother prenatally. This has allowed studies of pathogenesis, genetics and immune function in families, pharmacokinetics of medicines, outcomes as well as other studies (Sibley and Boothroyd, 1992; Silveira *et al.*, 2002; Vallochi *et al.*, 2005; Khan *et al.*, 2006; Lehmann *et al.*, 2006; Dubey *et al.*, 2012c). This study defined parasite types in those participating in the NCCCTS in North America (II or non-II) (McLeod *et al.*, 2012), the high frequency of infections with oocysts (Boyer *et al.*, 2011),

(A)

Design

Patients
- Referred <2.5m, usually with clinical signs. Dx confirmed (by isolation PCR, antibody [birth, 1 yr], lymphocyte blastogenesis). Willing to be evaluated in Chicago, no concomitant immunosuppressive conditions

Treatment
- High (6 mos) vs.low (2 mos) pyr 1mg/kg/d
- Pyrimethamine (Pyr), sulfadiazine, folinic acid, 1 yr
- No concomitant untreated controls for ethical reasons
- Feasibility then randomized phase

Evaluations
- Uniform, longitudinal, at ≤ 0.25, 1, 3.5, 5, 7.5, 10, 15, and 20 yrs
- Collection of history, demographics
- Evaluation by specialists in pediatric infectious diseases, ophthalmology, neurology, audiology, and development
- Review of CT, laboratory tests, compliance with medications

Outcome
- Contrast Rx 1 yr this study vs. no Rx or Rx ≤ 1 mos (literature)
- Compare toxicity & outcome with high vs. low pyr

(B)

Phase I, 1981-1991-Feasibility, Pyrimethamine pharmacokinetics & toxicity

Pyrimethamine, sulfadiazine, and leukovorin treatment for one year (pyrimethamine and sulfadiazine adjusted weekly for infant's weight)

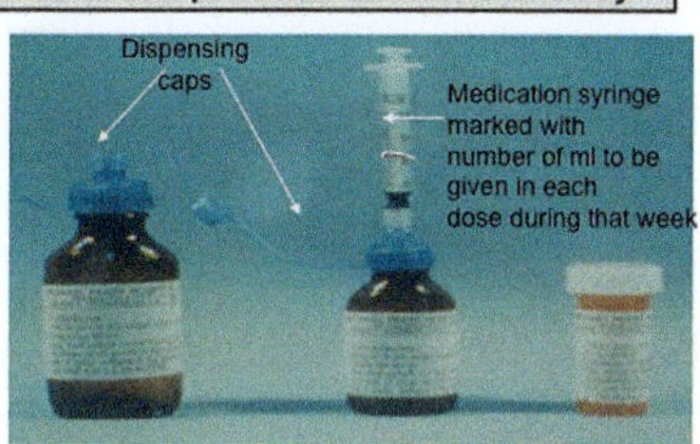

Serum pyrimethamine levels obtained for infants with congenital Toxoplasma infection

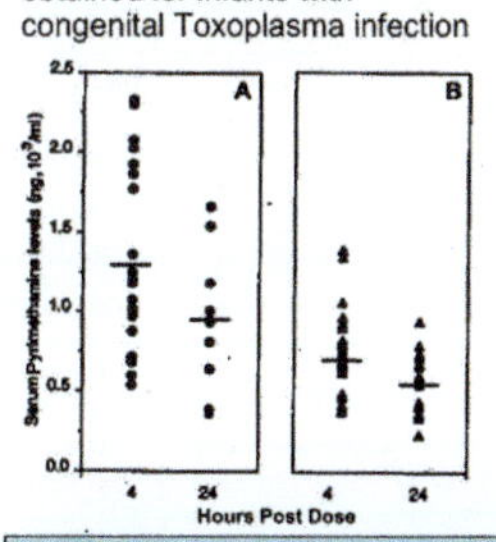

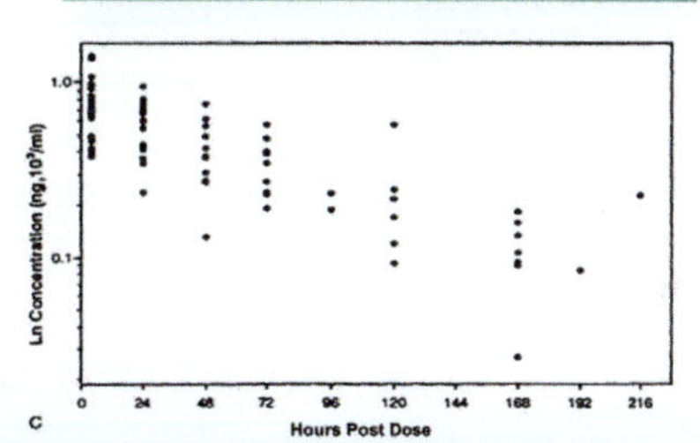

Conclusions:
- Medicines (PSL) could be formulated for infants, appeared to be well-tolerated, and could be administered safely throughout the first year of life (N=14).
- Signs of active infection, such as thrombocytopenia, hepatitis, and chorioretinitis, resolved within weeks after initiating treatment
- Pyrimethamine pharmakinetics defined for infants. Pyrimethamine levels in serum and CSF were sufficient to inhibit the parasite *in vitro*.
- ~70% normal development and functional.
- Active retinal disease treatable with no loss of visual acuity.
- No hearing loss.
- Two children tolerated higher dose which formed the basis for comparison of dosages in the randomized controlled trial (RCT).

(C)

Phase II,1991-2012 –RCT: Pre-established Statistical Guidelines, Power Analysis

Detectable Treatment Effects and Relative Effects at 10 Years for Asymptomatic and Mild, Moderate Systemic Groups/Moderate Neurologic, Severe Group*

Endpoint	Control Incidence (p_c%)	Detectable treatment Incidence (p_t%)[a,b]	Detectable relative efficacy (%)[c]
Visual acuity <20/20	50/70	19.1/13.1	62/56
New retinal lesions	85/90	55.2/55.3	35/39
Motor abnormality	10/60	–[d]/21.8	–[d]/64
IQ<70	25/95	2.5/63.4	90/33
ΔIQ≥15	50/95	19.0/63.4	62/33
Hearing loss >30dB	30/30	5.2/1.6	83/95

Comparisons between treatment groups involve sample sizes as follows: for asymptomatic ("none"), mild, moderate systemic - treatment A = 33, treatment C = 33; for moderate neurologic, severe- treatment A = 22; treatment C =22.

[b] *The calculation of "detectable treatment incidence" is carried out by solving the equation: $2.8\ SE\ (p_c-p_t) = p_c-p_t$ where $SE\ (p_c-p_t) = \sqrt{[(p_c q_c/n_c) = (p_t q_t/n)]}$; $q_c = 100-p_c$; $q_t = 100-p_t$.*

[c] *Detectable relative efficacy = $([p_c-p_t]/p_c) \times 100$.*

[d] *With as low as 10% control incidence, there is a reasonable probability that no control cases with this deficit will be observed. Therefore, there is no reduction that could be detected with assurance.*

* *Methods for calculation, Snedocor GW and Cochran WG. 6.4 Sample size and comparative experiments. Statistical methods, 7th Edition, Iowa State University Press, Ames Iowa. 1980, p. 102-5.*

FIGURE 4.8 *(continued)*

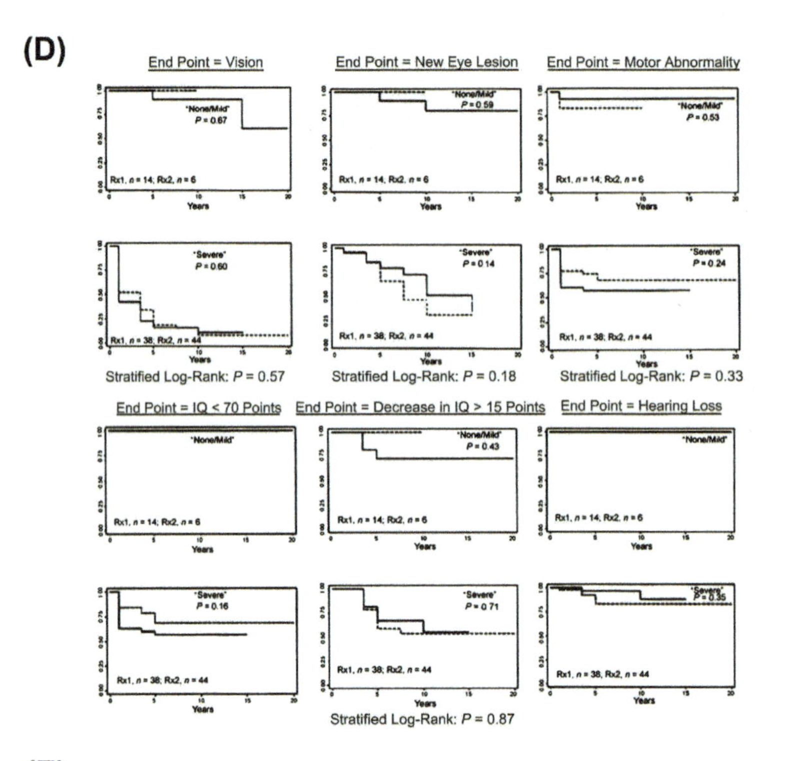

(E)

Endpoint Outcomes Compared to Literature Data

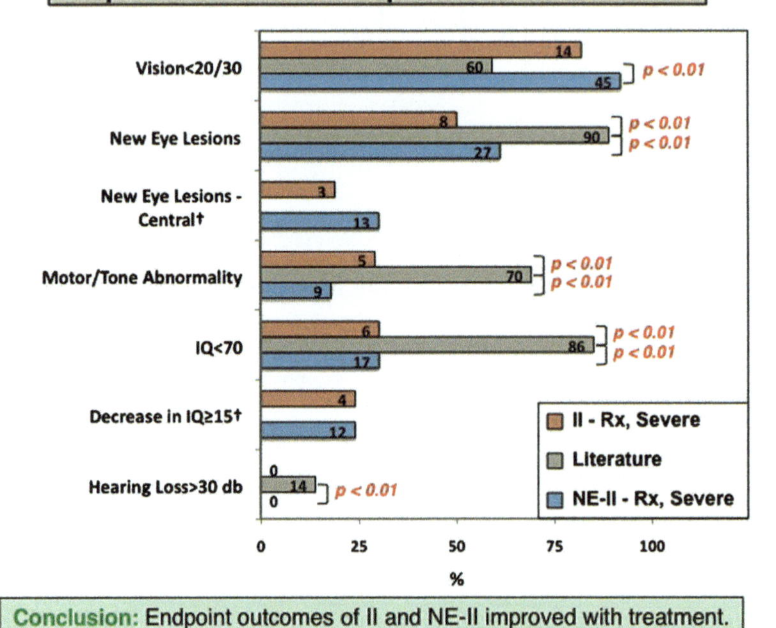

* Eichenwald HF. Human toxoplasmosis. Copenhagen: Munksgaard;1960. p. 41-49.
† Endpoint not measured in literature data

FIGURE 4.8 (*continued*)

(F)

Manifestations at Birth for Groups with Gestational Treatment

	In Utero Rx		No *In Utero* Rx		p value‡
	Type II	Type NE-II	Type II	Type NE-II	
Gestational Age <38 weeks	3*/15† (20%)	3/13 (23%)	15/37 (41%)	40/79 (51%)	0.82
Severe, S/MN	8/15 (53%)	6/13 (46%)	25/37 (68%)	71/80 (89%)	0.08
Splenomegaly	1/15 (7%)	1/13 (8%)	8/36 (22%)	36/78 (46%)	0.54
Hepatomegaly	4/15 (27%)	2/13 (15%)	10/36 (28%)	40/78 (51%)	0.11
Skin Rash	1/15 (7%)	0/13 (0%)	7/36 (19%)	29/78 (37%)	Not Estimable
Chorioretinal Scars	8/15 (53%)	5/13 (38%)	24/37 (65%)	67/80 (84%)	0.07

Conclusion: Without gestational Rx, prematurity, severity, hepatomegaly, splenomegaly, skin rash, chorioretinal scars more prevalent with NE-II.
With gestational Rx, associations of NE-II with manifestations at birth no longer significant, less prevalent with NE-II.

Not Estimable = due to 0 cell.
* Number of patients with manifestation at birth in each *in utero* treatment and parasite serotype cohort.
† Total number of patients in each *in utero* treatment and parasite serotype cohort; table only includes those diagnosed at birth and treated during the first year of life.
‡ Based on a test of the *in utero* treatment group by parasite serotype interaction from a logistic regression model. A statistically significant interaction would indicate that the effect of parasite serotype on disease manifestations depends on whether *in utero* treatment was received.

(G)

Endpoint Outcomes Based On Treatment Group

	Type II		Type NE-II		p-value‡	Type II	Type NE-II	p-value§
	A	C	A	C		A+C	A+C	
Vision<20/30	10*/19† (53%)	5/10 (50%)	19/27 (70%)	28/31 (90%)	0.17	15/29 (52%)	47/58 (81%)	<0.01
New Eye Lesions	5/17 (29%)	4/7 (57%)	9/24 (38%)	19/27 (70%)	0.85	9/24 (38%)	28/51 (55%)	0.22
New Eye Lesions - Central	2/17 (12%)	2/7 (29%)	6/24 (25%)	7/27 (26%)	0.42	4/24 (17%)	13/51 (25%)	0.56
Motor/Tone Abnormality	3/19 (16%)	2/10 (20%)	6/27 (22%)	3/31 (10%)	0.32	5/29 (17%)	9/58 (16%)	1.00
IQ<70	2/19 (11%)	4/13 (31%)	10/30 (33%)	8/37 (22%)	0.08	6/32 (19%)	18/67 (27%)	0.46
Decrease In IQ ≥15	3/19 (16%)	2/10 (20%)	6/27 (22%)	6/31 (19%)	0.70	5/29 (17%)	12/58 (21%)	0.78
Hearing Loss >30 db	There Was No Hearing Loss In Any Group.				-	-	-	-

Conclusion: Treatment dosage and endpoint outcomes not associated.
Serotype and endpoint outcomes not associated.

* Number of persons with endpoint in each parasite serotype and treatment group cohort.
† Total number of persons in each parasite serotype and treatment group cohort.
‡ Based on a test of the treatment group by parasite serotype interaction from a logistic regression model. A statistically significant interaction would indicate that the effect of treatment on outcome depends on parasite serotype.
§ P values are from two-sided Fisher's exact tests.

A = treatment of daily pyrimethamine and sulfadiazine for two months, followed by pyrimethamine on Monday, Wednesday, and Friday, and continued daily sulfadiazine for the remainder of the year of therapy. Both randomized and feasibility cohorts included.
C = treatment of daily pyrimethamine and sulfadiazine for six months, followed by pyrimethamine on Monday, Wednesday, and Friday, and continued daily sulfadiazine for the remainder of the year. Both randomized and feasibility cohorts included.

Conclusions:
- Endpoint outcomes for II and NE-II serotypes improve with treatment.
- Associations of NE-II manifestations at birth are not significant with gestational treatment.
- Associations are not absolute.
- **Prompt prenatal diagnosis and treatment is likely to be especially important for those with NE-II serotype, but critical for all infected persons.**

FIGURE 4.8 (*continued*)

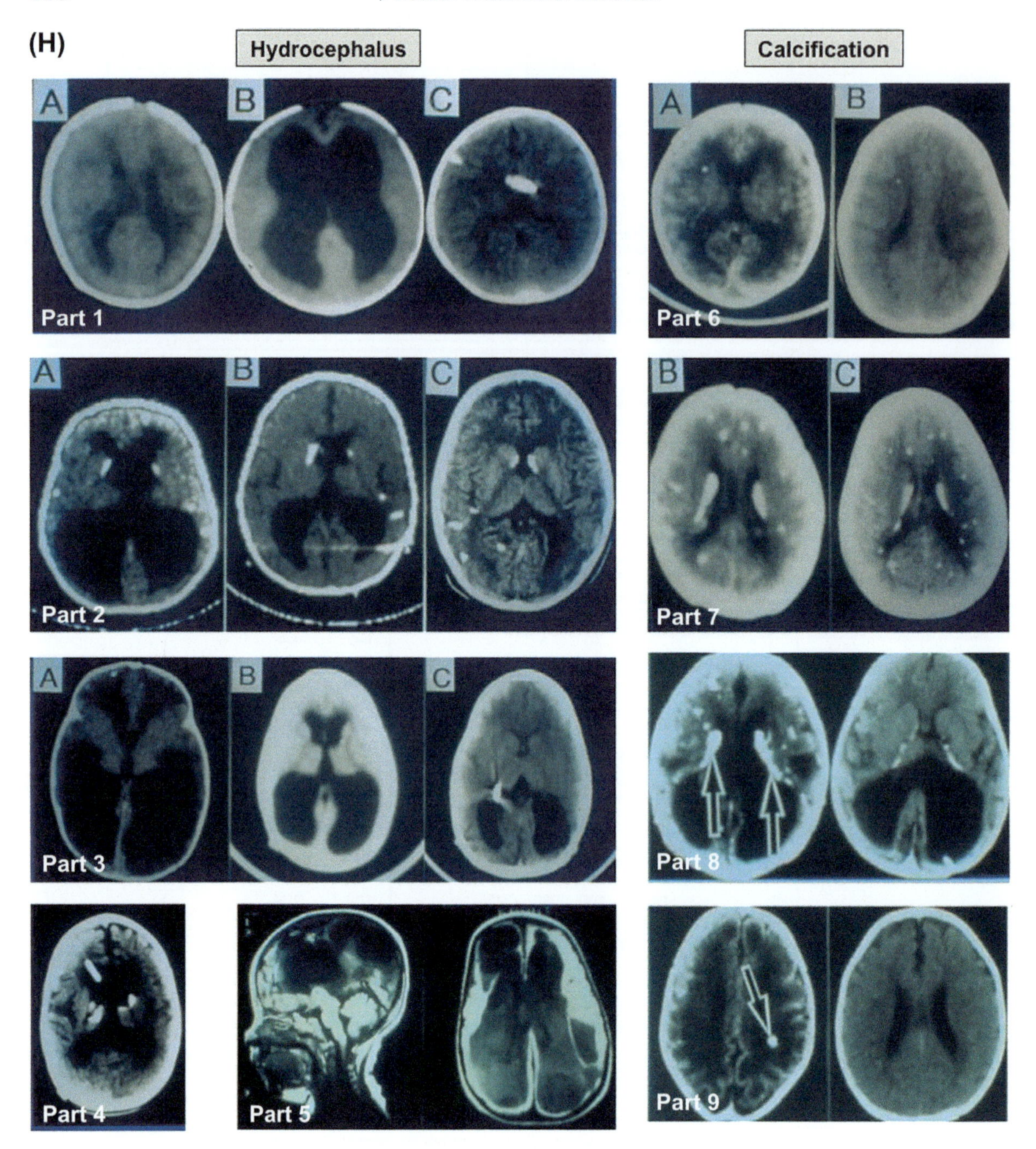

FIGURE 4.8 (*continued*)

and demonstrated that outcomes with treatment are improved (McLeod *et al.*, 2000, 2006a, 2009, 2012) with the best outcomes for those persons treated *in utero* (Fig. 4.4). A cost benefit analysis suggested that prenatal screening and treatment would likely reduce morbidity and mortality and be cost beneficial in the USA (Stillwaggon *et al.*, 2011). Practice guidelines regarding management of congenital toxoplasmosis and prenatal screening for the USA are being considered at present.

4.1.3.2.4 BRAZIL

The first case of what was later diagnosed as congenital toxoplasmosis was recognized in Brazil by Carlos Bastos Magarinos Torres in 1927 during an autopsy of a two day old infant. Dubey, Lagos and Jones *et al.* recently summarized many aspects of *T. gondii* infections in Brazil (reviewed in Dubey *et al.*, 2012c). Gilbert, Lagos *et al.* described the clinical manifestations of congenital toxoplasmosis in a cohort in Brazil (Gilbert *et al.*, 2008). In

FIGURE 4.8 **Treatment of congenital toxoplasmosis in the USA and outcomes.**
A) Design of the NCCCTS.
B) Treatment for toxoplasmosis. Top: Toxoplasmosis medications for infants. Suspended in 2% sugar solution. Suspension at usual concentration must be made up each week. Store refrigerated. *Image from: McAuley* et al., *1994, with permission.* Middle: Pyrimethamine serum levels (4 and 24 hours after a dose) of children given 1 mg of pyrimethamine per kg daily. *Image from: McLeod* et al., *1992, with permission.* Bottom: Conclusions concerning administering medication to treat congenital toxoplasmosis.
C) Detectable effects of treatment for congenital toxoplasmosis in the NCCCTS Phase II RCT, 1991–2012. *Data from: McLeod* et al., *unpublished, with permission.*
D) Kaplan–Meier plots showing the outcomes for each endpoint for NCCCTS patients in the pooled feasibility/observational phase and the randomized phase. Patients received either treatment 1 (solid line) or treatment 2 (dotted line). There is no visible trend for superiority or statistically significant superiority at this time. IQ, intelligence quotient; Rx1, treatment arm 1; Rx2, treatment arm 2. *Data from: McLeod* et al., *2006a, with permission.*
Image and caption verbatim from: McLeod et al., *2009, with permission.*
E) Endpoint outcomes and parasite serotype of persons with congenital toxoplasmosis treated in the NCCCTS cohort and untreated in literature data. *Image from: McLeod* et al., *2012, with permission.*
F) Manifestations at birth for newborns with *in utero* treatment, NCCCTS. *Table and caption from: McLeod* et al., *2012, with permission.*
G) Endpoint outcomes based on treatment group, NCCCTS. *Table and caption from: McLeod* et al., *2012, with permission.*
H) Resolution of hydrocephalus and calcifications is associated with shunting and treatment, NCCCTS. Part 1: CT of the brain at birth (A), showing development of hydrocephalus at three months of age with treatment (B), and at one year of age (C). This child had normal development. Part 2: CT of the brain at three months of age, showing hydrocephalus (A), at four months of age, after shunt placement (B), and at eight years of age (C). This child had normal development. Part 3: CT of brain at two months of age before shunt placement (A), at four months of age after shunt placement (B), and at 14 months of age after shunt placement (C). Note the marked increase in the size of cortical mantle. This child had normal development. Part 4: CT of the brain at one year of age. This patient appeared normal at birth but meningoencephalitis developed and was untreated until three months of age. At this age, hydrocephalus and bilateral macular chorioretinitis led to the diagnosis of congenital toxoplasmosis and initiation of treatment. Note the significant residual atrophy and calcifications. This child experienced substantial motor dysfunction, developmental delays and visual impairment. Part 5: MRI of patient at nine months of age demonstrating changes likely due to perinatal anoxia and hypoglycaemia. Such complications of toxoplasmosis and delays in shunting have been associated with the most severe sequelae. Part 6: CT of the brain at birth (A) with areas of hypolucency, mildly dilated ventricles and small calcifications, and at one year of age (B) with normal findings except for two small calcifications. This child had normal development. Part 7: CT of brain at birth (B) and at one year of age (C). Note the growth of cortex and resolution of encephalomalacia. There were no new calcifications nor increase in size of calcifications. Part 8: CT of the brain of a treated infant (left) and at one year follow-up (right). Note the diminution and/or resolution of the calcifications indicated by the arrows. Part 9: CT of the brain of a treated infant (left) and at one year follow-up (right). Note the diminution and/or resolution of the calcification indicated by the arrow. *Images of Pt 1–7 and descriptions from: McAuley* et al., *1994, with permission. Images of Pt 8–9 and descriptions from: Patel* et al., *1996, with permission.*

Brazil, the state of Minas Gerais has very recently implemented a prenatal *T. gondii* screening programme that will likely be using the German (Hotop *et al.*, 2012) method for prenatal treatment.

Studies surveying the seroprevalence of *T. gondii* in the general population of Brazil dating back to 1962 indicate that seroprevalence was, and remains, higher in Brazil than in the USA. Seroprevalence in children and pregnant women is one of the highest worldwide. Toxoplasmosis, especially congenital toxoplasmosis and eye disease, have been described as very prevalent with severe morbidity in Brazil. The variability of presentation of Brazilian *T. gondii* eye disease was first noted in 1992 when Glasner *et al.* observed a remarkably high incidence of retinal disease caused by toxoplasmosis in the Brazilian city of Erechim (Glasner *et al.*, 1992). Congenital toxoplasmosis in the Brazilian state of Minas Gerais had a prevalence of one in 770 births with approximately 50% of infants having active eye disease at birth (Vasconcelos-Santos *et al.*, 2009) and up to 35% having neurological findings (Dubey *et al.*, 2012c). A similar prevalence of 1/1000 births or more has been documented in other regions of Brazil as well. There is also a remarkably high incidence of hearing loss, 40% in some series (Dubey *et al.*, 2012c), as compared to treated congenital infection in the USA. Studies of the genotypes of *T. gondii* isolates has indicated that the predominant strains are different than those seen in Europe and North America (Sibley and Boothroyd, 1992; Silveira *et al.*, 2001, 2002; Vallochi *et al.*, 2005; Lehmann *et al.*, 2006; Dubey *et al.*, 2012c) and this may account for the difference in clinical presentations.

The high incidence of toxoplasmosis in Brazil can partially be attributed to the high level of environmental contamination with oocysts and there have been large epidemics involving hundreds of people with substantial illness. For example, in one survey of 31 soil samples from different school yards, *T. gondii* was isolated from seven (23%) locations (dos Santos *et al.*, 2010). Other epidemics have been documented to be due to contaminated drinking water (e.g. Santa Isabel do Ivai (Vaudaux *et al.*, 2010)). Another epidemic occurred among university students in São-José-dos Campos (Magaldi *et al.*, 1969).

In addition to congenital toxoplasmosis, postnatally acquired toxoplasmosis also poses a serious threat and causes severe ocular disease with recurrences of retinochoroiditis (Dubey *et al.*, 2012c). An unusual finding, which appears to be idiosyncratic, was reported recently from Brazil (Vasconcelos-Santos, 2012). This group reported a chronically infected pregnant woman who transmitted *T. gondii* to her foetus. Whether this was due to previous exposure or re-exposure to a more virulent strain of *T. gondii* endogenous to Brazil remains to be determined through genetic studies of both parasite and host.

4.1.4 The Special Problem of Ocular Disease

Toxoplasmic retinochoroiditis may be the result of congenitally or postnatally acquired infection (Figs. 4.5 and 4.6). Involvement of the eye occurs during the acute stage of infection and reactivation during the chronic stage of the infection may occur (Fig. 4.4; Delair *et al.*, 2011). Appearance of eye lesions may be the same in both congenital and postnatally acquired infections. Symptoms include change in visual acuity, blurred vision, scotoma, pain, photophobia and epiphora. There may be complete or partial loss of vision, increased intraocular pressure or glaucoma (Delair *et al.*, 2011). There appears to be a predilection for involvement of the macula, especially the fovea, which results in impairment or a loss of visual acuity. Chorioretinitis due to *T. gondii* may be a relapsing disease due to reactivation of latent infection.

In a study in Brazil (Holland *et al.*, 1996), the relapses with acute, postnatally acquired infection occurred most frequently near the time of acquisition and diminished over time. This may be due to the existing genotype in the area although higher inoculation rates and infection from oocysts in the environment could be alternative explanations. Especially severe disease has been noted in the elderly. Reactivation of congenital toxoplasmosis appears to occur at school-entry age, adolescence and other times of considerable stress (Fig. 4.4). During reactivation, bradyzoites in cysts apparently transform to tachyzoites that proliferate and cause active chorioretinal inflammation that is associated with vitritis causing the above symptoms. Impaired vision in adult patients may be due to congenital infection with *T. gondii* or postnatally acquired infection. One study in the United Kingdom found a life time risk of symptomatic *T. gondii* eye disease of approximately two per 10,000 and a 100-fold higher risk in persons born in West Africa, living in the UK (Gilbert *et al.*, 1999).

T. gondii chorioretinitis is more common in South America compared to North America and Europe, and differences in *T. gondii* genetics may be responsible (Khan *et al.*, 2006; Vallochi *et al.*, 2005). Eye disease due to *T. gondii* is highly prevalent in Brazil, and a recent study found a prevalence of chorioretinitis in adults of 1.2% (Amorim Garcia *et al.*, 2004). In Erechim, Southern Brazil, 17.7% of the adult population has ocular toxoplasmosis (Glasner *et al.*, 1992; Silveira *et al.*, 2001). In some regions of Brazil, 80% of persons are infected, with 50% of those over the age of 50 years and 20% of the total numbers of persons having retinal disease that is recurrent and debilitating (Glasner *et al.*, 1992; Silveira *et al.*, 1988, 2001, 2002; Vasconcelos-Santos *et al.*, 2009; Vasconcelos-Santos, 2012; Lavinsky *et al.*, 2012; Oréfice *et al.*, 2012; Vaudaux *et al.*, 2010; Carmé *et al.*, 2002, 2009; Gilbert 2008; Lehmann *et al.*, 2006; Demar *et al.*, 2007; Andrade, 2012; Ajzenberg *et al.*, 2012; reviewed in Dubey *et al.*, 2012c; Darde *et al.*, 1998; Amorim Garcia, 2004). Ocular disease also appears to be quite severe with recurrences in persons in Colombia and Mexico (London *et al.*, 2011). Retinal scars due to ocular toxoplasmosis were reported in 6% of adults in Quindio, Colombia (de la Torre *et al.*, 2007). Separately, branch retinal artery occlusion attributed to toxoplasmosis was recently reported to have occurred in an adolescent (Chiang *et al.*, 2012).

One complication of *T. gondii* infection for which there is a specific treatment is choroidal neovascular membranes (CNVMs). An example of a CNVM is shown in fundus photographs and optical coherence tomography (OCT) images in Fig. 4.5. Activity is associated with loss of visual acuity, oedema and haemorrhage. Pathogenesis appears to involve hypoxia induction factor including VegF because anti-VegF treatment results in rapid resolution of haemorrhages and other CNVM.

4.1.5 Immune Compromised Patients

4.1.5.1 HIV-Infected Patients

The increased frequency of *Toxoplasma* encephalitis in patients with AIDS was reported soon after the start of the HIV epidemic (Levin *et al.*, 1983; Roue *et al.*, 1984; Enzensberger *et al.*, 1985; Suzuki *et al.*, 1988a) and *Toxoplasma* encephalitis was an important end-stage cause of death in HIV patients before the introduction of combination Anti-Retroviral Therapy (cART) in the USA and Europe (Luft *et al.*, 1983, 1993; Basavaprabhu *et al.*, 2012; Luft and Remington, 1984, 1992; Kim *et al.*, 2012; Kovacs, 1995; Guevara-Silva *et al.*, 2012; Leport *et al.*, 1996; Abgrall *et al.*, 2001; Post *et al.*, 1983; Haverkos *et al.*, 1987; Levy *et al.*, 1986, 1988, 1990) (Fig. 4.6). Incidence has diminished in the era of cART. Toxoplasmosis in patients with AIDS is most often the result of reactivation of latent disease in patients with a low CD4+ T-cell count. Patients with reactivated toxoplasmosis often present with signs and symptoms of encephalitis (Luft and Remington, 1992) and/or much less frequently with eye disease (Liesenfeld *et al.*, 1999). Acute acquired infection in AIDS patients has been reported and may involve multiple organs. The disease most often presents with subacute, focal deficits including hemiparesis (39%–52%), altered mental state (30%–42%), seizures (15%–29%), cranial nerve disturbances (7%–28%), abnormalities of speech (6%–26%), cerebellar signs (9%–30%), meningismus (10%–16%) and behavioural/psychomotor manifestations (30%–42%) including psychosis, dementia and anxiety. Pulmonary toxoplasmosis presenting as febrile illness with cough and dyspnea has also been reported (Rabaud *et al.*, 1996). Almost all organs have been reported as being involved, although the predominance is in the brain (Luft *et al.*, 1983, 1993; Basavaprabhu *et al.*, 2012; Luft and Remington, 1984, 1992; Kim *et al.*, 2012; Kovacs, 1995; Guevara-Silva *et al.*, 2012; Leport *et al.*, 1996; Abgrall *et al.*, 2001; Haverkos *et al.*, 1987; Levy *et al.*, 1986, 1988, 1990; Post *et al.*, 1983; summarized in Montoya *et al.*, 2010).

4.1.5.2 Persons with Cardiac and Renal Transplants

Toxoplasma infection has been described after heart, kidney and liver transplantation (Aubert *et al.*, 1996; Giordano *et al.*, 2002; Renoult *et al.*, 1997, Wulf *et al.*, 2005). In most cases, infection manifests within three months after transplantation. Patients with organ transplants or malignancy may develop central nervous system, retina, myocardial or pulmonary involvement due to reactivation (Fernàndez-Sabé, 2012; Aubert, 1996; Ryning *et al.*, 1979; Mele *et al.*, 2002; Slavin *et al.*, 1994; Abgrall *et al.*, 2001; Remington, 1974; Conley *et al.*, 1981; Lassoued *et al.*, 2007; Wreghitt *et al.*, 1992, 1995; Renoult *et al.*, 1997; Singer *et al.*, 1993; Botterel *et al.*, 2002; Martino *et al.*, 2000, 2005; Matsuo *et al.*, 2007; Sing *et al.*, 1999; Luft *et al.*, 1983; Bretagne *et al.*, 1995; Bautista *et al.*, 2012; Bories *et al.*, 2012; Busemann *et al.*, 2012; Caselli *et al.*, 2012; Ostoff *et al.*, 2012; Vaughan *et al.*, 2012; Strabelli *et al.*, 2012; Montoya *et al.*, 2001, 2010) (Fig. 4.6). Clinical signs of infection are similar to those in patients with AIDS and involve fever and encephalitis. The parasite may also be present in eyes, liver, heart, lungs, pancreas, adrenal and kidney.

Seronegative persons may become infected with *T. gondii* through transplanted organs from seropositive donors. More rarely, seropositive transplant recipients may reactivate their latent infection due to the transplant-related immune-suppression, although typically if a seropositive person receives a transplant from a seropositive donor, there is a rise in IgG antibody titre and development of IgM antibody specific for *T. gondii* without illness ascribed to the parasite. Thus, the frequency of transplant-related infection depends on the seroprevalence of infection with *T. gondii* in a population. Gallino *et al.* (1996) reported that 14 (87%) of 16 infections in *T. gondii* naïve seronegative recipients seroconverted after receiving a cardiac transplant from a seropositive donor. The use of antiparasitic prophylaxis also impacts the rate of infection. A review of 257 cases of heart transplants between 1985 and 1993 and 33 cases of heart–lung transplants found that 4.5% (13) were between a seropositive donor and a seronegative recipient. Nine of these cases were followed up and only one patient was documented as seroconverting. All these patients received trimethoprim/sulphamethoxazole prophylaxis for *Pneumocystis* (Orr *et al.*, 1994). In patients receiving a cardiac transplant, six weeks pyrimethamine prophylaxis reduced infection from 57% (4/7) to 14% (5/37) (Wreghitt *et al.*, 1992). When a seropositive person receives a heart transplant from a donor who is seropositive, there is an increase in *T. gondii*-specific IgG antibody titre and a new presence of *T. gondii*-specific IgM antibody but usually this occurs without causing illness.

4.1.5.3 Bone Marrow and Hematopoietic Stem Cell Transplantation

The prevalence of *T. gondii* in bone marrow and hematopoietic stem cell transplantation

(BMT) patients also varies with the seroprevalence in the population: 0.5% in the USA to 5% in France (Fernàndez-Sabé, 2012; Aubert, 1996; Ryning *et al.*, 1979; Mele *et al.*, 2002; Slavin *et al.*, 1994; Remington, 1974; Conley *et al.*, 1981; Martino *et al.*, 2000, 2005; Matsuo *et al.*, 2007; Sing *et al.*, 1999; Bretagne *et al.*, 1995; Bautista *et al.*, 2012; Bories *et al.*, 2012; Busemann *et al.*, 2012; Caselli *et al.*, 2012; Ostoff *et al.*, 2012; Vaughan *et al.*, 2012; Strabelli *et al.*, 2012; Montoya *et al.*, 2010). Most patients are seropositive for *T. gondii* before transplantation and reactivate their latent infection. Symptoms of *T. gondii* infections in bone marrow transplant patients include fever (43%), seizures (14%), headache (13%), confusion (13%) and pulmonary symptoms (12%). Ninety-two percent had more than one symptom and the average onset of symptoms was 62 days post BMT (range 1−689 days) (Mele *et al.*, 2002). Disseminated infection that is rapidly progressive may occur. Mortality rates are high (Chandrasekar *et al.*, 1997). The European Group for Blood and Bone Marrow Transplantation reported on 106 allogenic stem cell transplants of which 55% of the recipients were *Toxoplasma* IgG positive. All received prophylaxis with trimethoprim and sulphamethoxazole for six months and 15% (16/106; 95% CI: 8%−21%) had at least one *T. gondii* PCR positive blood sample and 6% (6/106; 95% CI: 1%−10%) experienced clinical disease due to *T. gondii* infection (Martino *et al.*, 2005). The median days to diagnosis from onset of symptoms was 42 days (range 1−178 days) and the presenting symptoms were localized encephalitis in four patients, pulmonary toxoplasmosis in one patient and one patient presenting acute disseminated disease (Martino *et al.*, 2005).

4.2 DIAGNOSIS OF INFECTION WITH *TOXOPLASMA GONDII*

Strategies for control and prevention of congenital toxoplasmosis vary between countries, as discussed above, and the diagnostic challenges are different in pre- and neonatal screening programmes. Systematic, prenatal screening is performed in Austria, France, Slovenia, Minas Girais, Brazil, and widespread on-demand screening takes place in Belgium, Germany, Italy, and Spain. Samples are obtained during pregnancy and analysed for *Toxoplasma*-specific IgM and IgG antibodies (Table 4.1; Figs. 4.2 and 4.7). When seroconversion is detected, this confirms that the mother is infected, and treatment is usually started.

Neonatal screening for congenital toxoplasmosis is performed in New England, was formerly performed in Denmark and is performed in parts of Brazil by analysing the blood samples obtained on filter paper (Guthrie cards) in the days immediately postpartum (Guerina *el al.*, 1994; Lebech *et al.*, 1999; Neto *et al.*, 2004). Detection of *Toxoplasma*-specific IgM antibodies eluted from the PKU-filter paper is followed by a request of a blood sample from both the mother and infant for confirmatory testing (Sorensen *et al.*, 2002). Fifteen to twenty percent of these sera are found to be negative for *Toxoplasma*-specific IgM (Decoster *et al.*, 1992; Lebech *et al.*, 1999; Naot *et al.*, 1981). Low levels of *Toxoplasma*-specific IgM antibodies may be found for up to several years after acute infection, and the mere demonstration of low levels of *Toxoplasma*-specific-IgM antibodies is therefore not regarded as a sign by itself of acute infection with *T. gondii* (Liesenfeld *et al.*, 1997, 2001a, 2001b).

One diagnostic challenge is the situation when *Toxoplasma*-specific IgG and IgM antibodies are found in the first sample after conception, where the time of infection is the key to estimate whether the foetus is at risk or not (Ades, 1991). The measurement of the avidity of IgG antibodies was first demonstrated for *T. gondii* infections in 1989 (Hedman *et al.*, 1989, 1993; Lecolier and Pucheu, 1993) and has since then been further developed (Dannemann *et al.*, 1990; Marcolino *et al.*, 2000; Beghetto *et al.*, 2003; Petersen *et al.*, 2005) but is not reliable

TABLE 4.1 *Diagnosis of Toxoplasmosis*

Summary of Laboratory Findings in the Diagnosis of Toxoplasmosis
• **Acute Infection (e.g. lymphadenitis):** IgM and IgA serology are positive. Avidity is usually low. If avidity is high, this indicates an infection occurred > 12-16 wk earlier. A high avidity can be useful to date onset of the infection to longer than 12-16 wk. Serial serum specimens at a 3 wk interval can demonstrate seroconversion or a rising titer.
• **Acute Toxoplasmosis in Pregnancy:** The serologic response is similar to that seen for other persons with acute *T. gondii* infection above; however, the issue for assessing the risk of disease transmission is determining if the infection occurred during or before pregnancy, as *T. gondii*-specific IgM can persist for months or years after the acute infection. Risk for transmission is considered to be present if the infection occurred during pregnancy. The use of IgG avidity (high in most chronic infections, see above), differential agglutination (AC/HS), and the presence of IgA and/or IgE antibodies (which disappear more quickly than IgM) can help determine the timing of the infection, whether an amniocentesis is indicated to identify congenital toxoplasmosis, or if use of medicines would be helpful.
• **Chronic Infection:** IgM is negative and IgG is present. Antibody levels do not change with serial specimens. Avidity is high and AC/HS has a chronic pattern.
• **Reactivation of Disease During Immune-Suppression (e.g. Toxoplasma encephalitis in HIV/AIDS):** IgM is negative and IgG may be present. In some cases, there is no detectable serologic response to *T. gondii*. However, if clinical presentation is highly suspicious of infection in the absence of positive serologic results, CSF, blood, and possibly tissue samples should be obtained for diagnosis as indicated. While the sensitivity of PCR has been variable in this setting, if positive, PCR can be useful for diagnosis. In some clinical circumstances, presumptive treatment may be warranted.
• **Congenital Toxoplasmosis**: For diagnosis *in utero* for a fetus of an acutely infected pregnant woman who appears to have acquired the infection during gestation, PCR of amniotic fluid and ultrasound imaging of the fetus are used to establish fetal infection. Newborns will be IgG positive, due to passage of maternal antibody across the placenta. Compatible clinical findings in an infant of an acutely infected mother or the presence of *T. gondii*-specific IgM or IgA in a newborn confirms the diagnosis of congenital infection. Serial serology with a stable or rising IgG titer can also confirm the diagnosis of congenital infection. An example of fall in *T. gondii*-specific antibody load is as follows:

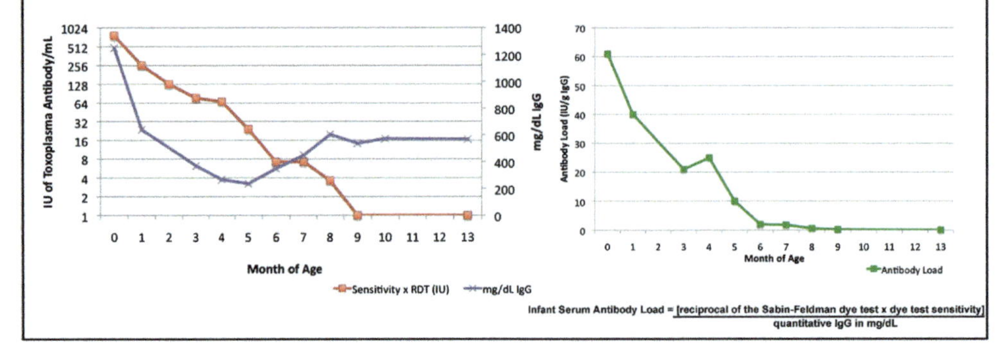

Images from: McLeod, *et al.*, unpublished data.
Equation from: McLeod, *et al.*, 2013a, with permission.

in the beginning of the infection when there are low levels of IgG antibodies (Press *et al.*, 2005). Since previous studies have shown that some individuals have low avidity IgG antibodies many months after infection (Petersen *et al.*, 2005), Petersen *et al.* investigated whether treatment influences the maturation of IgG antibodies and, based on twelve untreated patients, it seems that treatment and/or pregnancy may delay IgG maturation (Petersen *et al.*, 2005). A study of the value of different diagnostic tests for acute infection with *T. gondii*, including *Toxoplasma*-specific IgG, IgM, IgA antibodies, and the IgG avidity index, demonstrated that the combination of a sensitive test for *Toxoplasma*-specific IgM antibodies and a *Toxoplasma*-specific IgG avidity assay has the highest predictive value of the time of infection (Robert *et al.*, 2001).

4.2.1 *Toxoplasma* Antigens and Diagnostic Assays

Diagnosis of active, acute infection can be made on the basis of serologic tests or by histopathology, isolation of *T. gondii* or PCR from a variety of samples (e.g. CSF, placenta, foetal tissues, or peripheral blood). *T. gondii* has three distinct life-stages, each with stage-specific expression of antigens (Kasper and Ware, 1989; Singh *et al.*, 2002; Hill, *et al.*, 2011; Boyer *et al.*, 2011). An important immunodominant antigen is the tachyzoite-specific Surface Antigen 1, SAG1 (p30, SRS29B) which comprises up to 5% of the protein of the tachyzoite (Burg *et al.*, 1988). The antigen expressed in *E. coli* has been shown to be recognized by natural SAG1 antibodies (Harning *et al.*, 1996) and SAG1 is considered a prime candidate antigen in diagnostic tests because of its immunodominance and lack of known cross-reactivity to antigens from other microorganisms. Other surface antigens, SAG2 (SRS34A), SAG3 (SRS57) and SAG4 have been identified, SAG2 and SAG3 being tachyzoite specific and SAG4 being bradyzoite specific (Cesbron-Delauw, 1994, 1995; Howe and Sibley, 1994; Odberg-Ferragut *et al.*, 1996). Two other groups of *T. gondii* antigens have been studied for use in diagnostic assays, the dense granule antigens, GRAs, and in particular GRA1, GRA6 (Lecordier *et al.*, 2000) and GRA7 (Fischer *et al.*, 1998) and microneme antigens, MICs (Garcia-Règuet *et al.*, 2000; Lourenco *et al.*, 2001; Cerede *et al.*, 2002). The MIC antigens have also been shown to be important in the induction of protective immunity (Beghetto *et al.*, 2005). Bradyzoite-specific antigens like the bradyzoite antigen 1, BAG1 (Bohne *et al.*, 1993) and matrix antigen 1, MAG1 (Parmley *et al.*, 1994) should, in theory, be important in the antibody repertoire in infections past the acute stage, but they still remain to find their place in future diagnostic assays and studies have suggested their utility (Yolken *et al.*, 2012). The diagnostic value of oocyst-specific antigens has been studied in a single study of *T. gondii* oocyst-infected cats (Dubey *et al.*, 1995). Hill *et al.* (2011, 2012) developed a test for defining infection with sporozoites using response to an 11 kDa sporozoite protein.

4.2.2 The Development of Diagnostic Assays

The complement fixation assay (CFA) was the first diagnostic test for *Toxoplasma*-specific antibodies (Warren and Sabin, 1941; Steen and Kaas, 1951). The dye-test described by Sabin and Feldman (1948) is based on antibody-mediated killing of live *T. gondii* parasites in the presence of complement. Methylene blue is a vital dye. Thus, if antibodies are present parasites are lysed in the presence of complement. They then appear thin and blue as opposed to plump and robust when they are alive. The dye-test (DT) has proved a very sensitive assay, but the requirement of live, *T. gondii* parasites makes it more complicated and expensive to perform (Reiter-Owona *et al.*, 1999). The DT is not included in reference panels circulated as part of external quality control programmes, and multi-centre studies show a considerable variability

(Pethithory *el al.*, 1996; Reiter-Owona *et al.*, 1999; Rigsby *et al.*, 2004). Immunofluorescence assays (IFA) were introduced in the 1960s (Ambroise-Thomas *et al.*, 1966) and proved specific, but with a lower sensitivity compared to the DT. The IFA for *Toxoplasma*-specific IgM antibodies is still used by some centres because it is highly specific, but it has a low sensitivity (Robert *et al.*, 2001).

The Enzyme Immuno Assay technique (EIA) became available in 1972 (Engvall and Perlmann, 1972). The first *Toxoplasma*-specific IgM assay was developed by Remington and Miller (1968) and the first EIA based assay by Naot and Remington (1980). By the end of the 1980s, the direct EIA measuring *Toxoplasma*-specific IgG antibodies and the μ-capture EIA measuring *Toxoplasma*-specific IgM antibodies were well established in reference centres and the first commercial test introduced (Schaefer *et al.*1989). The μ-capture IgM assays were an improvement over the direct EIA IgM assays, but continued to have problems with false positive results (Liesenfeld *et al.*, 1997). The development of the Immunosorbent Agglutination assays (ISAGA) solved this by using whole cell formalin fixed *T. gondii* (Beghetto *et al.*, 2003) and tests based on this technique are regarded as highly sensitive and specific for *Toxoplasma*-specific IgM and IgA antibodies (LeFichoux *et al.*, 1984; Pouletty *et al.*, 1985). Immunoblot using single antigens has also been tested as a means to improve diagnostic sensitivity (Gross *et al.*, 1992).

A method to measure the maturation of *Toxoplasma*-specific IgG antibodies to determine the time of infection was described by Hedman *et al.* (1989). The test explores the increasing avidity (sum of all affinities) by the specific IgG antibodies with the maturation of the immune response and it was shown that the time of infection could be determined within a three month window after infection. The test has been adapted for automated systems (Petersen *et al.*, 2005). Newer IgG avidity tests allow one to exclude an acute infection within the last three to four months in patients with high avidity antibodies. In contrast, the presence of low or intermediate avidity IgG antibodies does not necessarily allow one to diagnose an acute infection, since the maturation of IgG antibodies may show marked differences between individuals. The same principle is used in the differential agglutination test (Thulliez *et al.*, 1989). Recently, Villard *et al.* (2012) compared four commercially available avidity assays and found greater than 95% sensitivity. The bioMerieux test had the greatest sensitivity and efficacy (99%). Recent studies to develop improved assays include peptides, recombinant proteins, and PCR (Raymond *et al.*, 1990; Fuentes *et al.*, 1996; Reischl *et al.*, 2003; Flori *et al.*, 2004; Lefevre-Pettazzoni *et al.*, 2007; Meroni *et al.*, 2009; Chapey *et al.*, 2010; Dai *et al.*, 2012a, 2012b; Faucher *et al.*, 2012; Fayyaz *et al.*, 2012; Holec-Gasior *et al.*, 2012a, 2012b; Kotresha *et al.*, 2012; Morelle *et al.*, 2012; Murat *et al.*, 2012; Prusa *et al.*, 2012; Saadatnia *et al.*, 2012).

4.2.3 Diagnosis of *Toxoplasma gondii* Infection in Pregnant Women

In countries where prenatal screening programmes are in place, a test of the first blood sample from pregnant women for *Toxoplasma*-specific IgG and IgM antibodies in the first trimester is performed. A high avidity assay can help to date acquisition of the infection to greater than 12 weeks earlier, but a low avidity test does not prove a recent infection has not occurred. Approximately 5% of seropositive women in the first trimester have *Toxoplasma*-specific IgM antibodies, but only approximately 4% of these give birth to a child with congenital *Toxoplasma* infection. It is, therefore, a considerable problem to diagnose whether women with specific IgM antibodies are infected before or after conception. This is particularly a problem in countries where testing of pregnant women in the beginning of pregnancy is common. This problem has been partly solved by obtaining two samples from pregnant women to see if there is any development of the specific immune response. It is generally agreed that there is a development of the *Toxoplasma*-specific IgG antibody response within the first six to eight weeks after infection after which the IgG levels are maintained at a high and stable level, with or without declining IgM antibodies (Jenum *et al.*, 1997, 1998).

The question of too many low-level *Toxoplasma*-specific IgM-positive patients and whether the diagnostic performance could be improved by not merely repeating the same tests two weeks apart was investigated in a European multi-centre study (Robert *et al.*, 2001). All highly sensitive assays were found to have a low specificity, and single tests were unable to reliably distinguish between acute and latent infections. Only the sequential analysis of sera by a highly sensitive IgM assay in combination with IgG avidity testing gave excellent diagnostic performances. In contrast, IgA or IgM assays were less useful to diagnose acute infections by confirming positive IgM results.

4.2.3.1 IgG Avidity Index

In a European multi-centre study, many laboratories contributed samples from patients in whom the time of infection was known. This panel was used to determine the proportion of sera showing specific IgM and IgA antibodies to *T. gondii* (Beghetto *et al.*, 2003) as well as the IgG avidity index within one to three months, three to twelve months or more than twelve months after seroconversion. These data were used to propose a two level strategy for diagnosis (Robert *et al.*, 2001). Treatment may delay the development of high-avidity *Toxoplasma*-specific IgG antibodies (Meroni *et al.*, 2009).

4.2.3.2 Combined, Two-Test Strategies

Robert *et al.* (2001) demonstrated that the best strategy for diagnosing acute and recent infection with *T. gondii* was a two-test strategy with a sensitive IgM test first followed by an IgG avidity test. Thus, the study confirmed the need of the *Toxoplasma*-specific IgG avidity index assay in the diagnosis of acute and recent infection. The increased use of the *T. gondii* IgG avidity test has highlighted an inherent problem, which is that many pregnant women have long-lasting, low IgG avidity antibodies and the IgG avidity assay needs further development (Beghetto *et al.*, 2003).

4.2.4 Improvement of EIA tests for *Toxoplasma*-specific IgG and IgM Antibodies

The problems with IgM based diagnostics in *T. gondii* infections have resulted in attempts to improve the tests. The accepted reference test is the ISAGA, but most analyses are performed with an EIA capture test. The assays use whole-cell, lysed *T. gondii* as the antigen, and attempts have been made to improve the test by using recombinant antigens (Ferrandiz *et al.*, 2004). The ISAGA IgM, EIA IgM and IgM immunofluorescence tests were evaluated in a prospective, European cohort study (EMSCOT) of women diagnosed with primary *T. gondii* infection during pregnancy with newborns identified through neonatal screening. The EMSCOT study provided data on sensitivities for diagnosing congenital infection in the newborn of four *Toxoplasma*-specific IgM antibody assays and three *Toxoplasma*-specific IgA antibody assays. The study also provided data on the sensitivity of neonatal testing related to the estimated gestational age of infection with *T. gondii*, showing that the IgM seropositivity at birth only detects infections in the second half of pregnancy. The study included 5223 samples from 996 children of which 3742 were tested with an EIA system, 2011 with an ISAGA IgM, and 316 with the IgM immunofluorescence assay. The children were followed for one year to ascertain the diagnosis by demonstrating the presence of *Toxoplasma*-specific IgG antibodies at 12 months of age, which is the gold standard for confirming congenital infection with *T. gondii*. The sensitivity for EIA test was low, 29.3%, and demonstrated clearly the need for better tests. The study was the first study to provide data on the sensitivity of *Toxoplasma*-specific IgM antibodies related to gestational age at infection. It was demonstrated clearly that a sensitivity of 50% is not reached until after the thirtieth gestational week (Petersen *et al.*, 2012).

4.2.5 Recombinant IgG Assays – Adults

Conventional assays have so far used whole-cell, lysed, *T. gondii* antigens, which have batch variations. With increasing emphasis on the need for reproducibility, the use of recombinant antigens in diagnostics assays provides a theoretical advantage. Previous studies have shown that the GRA1, GRA7 and SAG1 molecules are immunodominant (Aubert *et al.*, 2000; Harning *et al.*, 1996; Jacobs *et al.*, 1999; Johnson *et al.*, 1992; Li *et al.*, 2000; Suzuki *et al.*, 1988b). Pietkiewicz *et al.* (2004) demonstrated that recombinant antigens, including a mixture of GRA1, GRA7 and SAG1, were not as sensitive as whole-cell, lysed, antigen if sera had an IgG titre of less than 1:1600 using an EIA test and less than 1:512 in an IgG immunofluorescence test. The test did, however have a 100% sensitivity in a panel of sera from individuals who had *Toxoplasma*-specific IgM and/or IgA antibodies (i.e. those in whom the infection was recent) (Pietkiewicz *et al.*, 2004). Future assays for *Toxoplasma*-specific IgG antibodies relying on recombinant antigens need to include a panel of antigens but this test has not yet been optimized to the same sensitivity as the whole-cell, lysed antigen assay.

4.2.6 Recombinant IgM and IgG Assays – Newborns

Diagnostic assays based on recombinant antigens for measuring the *Toxoplasma*-specific IgM antibodies were evaluated in infants with or without congenital toxoplasmosis born to mothers with toxoplasmosis acquired during pregnancy (Petersen *et al.*, 2012). Antigen fragments from the MIC2, MIC3, GRA3, GRA7, M2AP and SAG1 proteins were tested in an EIA test (RecEIA) on 104 serum samples from newborns born to mothers infected with *T. gondii* during pregnancy (Beghetto *et al.*, 2006). Thirty-five were congenitally infected and 34 of 35 (97%) serum samples from the congenitally infected patients reacted with at least one of the recombinant antigens (Buffolano *et al.*, submitted). Remarkably, all sera from the 22 *Toxoplasma*-infected newborns who were clinically and serologically undiagnosed at birth were reactive using the IgM Rec–ELISA analysis, allowing the confirmation of congenital toxoplasmosis as soon as two months after birth. The presence of *T. gondii*-specific IgM antibodies against recombinant MIC2, MIC3, M2AP and SAG1 antigens may be used for the early postnatal diagnosis of congenital toxoplasmosis.

Buffolano *et al.* found that the newborn *Toxoplasma*-infected child produces primarily IgG_2 and IgG_3 against recombinant *Toxoplasma* antigens, whereas the maternally transferred antibodies primarily were IgG_1 (Buffolano *et al.*, 2013, submitted for publication). This subclass analysis of serum samples from mother and child against defined recombinant antigens may further improve diagnosis of congenital *Toxoplasma* infection in newborns. Peyron adapted this approach and found interferon gamma (INFγ) production in response to *T. gondii* antigens can also been used to establish diagnosis of congenital toxoplasmosis.

4.2.7 The *Toxoplasma*-Specific IgG Avidity Index

The maturation of the IgG response varies considerably between individuals. In the study of Lappalainen *et al.* (1993) two seroconverting mothers already had an IgG avidity index above

20% at the time of diagnosis, but most patients had developed an IgG avidity index above 15% after 180 days (Lappalainen *et al.*, 1993). A study from France found an average IgG avidity index of 0.2 in pregnant women infected within five months (Leolier and Pucheu, 1993). The original method developed by Hedman *et al.* (1989) used serial dilutions tested in EIA with and without 6M urea, but automated assays today calculate the IgG avidity index from two single measurements of the sample with and without urea. This introduces an uncertainty, although experiments with only two serum sample dilutions showed an excellent agreement with IgG avidity measurements using four serial serum sample dilutions (Korhonen *et al.* 1999). Prince and Wilson (2001) evaluated the IgG avidity assay by using single dilution assays with and without urea and showed that it makes a difference whether the *Toxoplasma* IgG avidity index is calculated from the OD (optical density) values or the activity measured in International Units of *Toxoplasma*-specific IgG antibodies per ml because the signal obtained in an EIA system is not linear (Prince and Wilson, 2001).

The IgG avidity results found in one study (Petersen *et al.*, 2005) demonstrated that a persistent, low IgG avidity index poses a diagnostic problem, at least in pregnant women receiving treatment during pregnancy (Petersen *et al.*, 2005). Up to half of the patients with acute infections may show a low or borderline IgG avidity index six months after the infection (Montoya *et al.* 2002; Rossi, 1998), which is in concordance with the results reported in this study (Petersen *et al.*, 2005). The LIAISON® avidity results were compared with the semi-automated VIDAS system for measuring the *Toxoplasma*-specific IgG avidity index and there was a good correlation between the results from the two systems showing that persistence of low-level *Toxoplasma*-specific IgG avidity antibodies is an inherent problem of measuring the *Toxoplasma*-specific IgG avidity index unrelated to the assay system.

The cut-off value defining a low IgG avidity index differs markedly between different studies and one study found that patients infected within the past three months had IgG avidity index below 0.45 (Holliman *et al.*, 1994). A comparison between the VIDAS and the Labsystems IgG avidity index showed a correlation coefficient of 0.6 in pregnant women but 0.88 in other patients (Alvarado-Esquivel *et al.*, 2002), but the difference was not further discussed. Improvement of the IgG avidity assay using Western blot technique has been attempted and revealed differences in the maturation of the specific IgG to different antigens (Villavedra *et al.*, 1999).

The IgG response matures rapidly in some individuals and this has been reported in several studies. For instance, the cut-off of the *Toxoplasma*-specific IgG avidity index using the VIDAS system (bioMérieux) was defined as 0.3 to ensure that all sera from acute infections had a low-avidity index (Pelloux *et al.*, 1998). The same study showed that, at least in pregnant women, a low IgG avidity index persisted up to nine months post-infection and all women were treated. In a study of *T. gondii*-infected pregnant women identified prospectively through prenatal screening, Jenum *et al.* (1997) found that two out of 73 women had an IgG avidity index above 0.2 before 20 weeks' gestation, but many continued to have a low IgG avidity index even a year after infection. It is assumed that all women were treated during pregnancy. The IgG avidity results found in one study (Petersen *et al.*, 2005) demonstrated that long lasting low IgG avidity was a common finding in pregnant women. The study compared the IgG avidity maturation in treated, pregnant women with samples from patients with acute infection with *T. gondii* who were not pregnant and were not treated, and found a significantly more rapid IgG maturation during the first four months after infection in subjects who were not treated and not pregnant (Petersen *et al.*, 2005).

The observation that the *Toxoplasma gondii*-specific IgG-maturation is delayed in treated pregnant women compared to non-treated, non-pregnant individuals has been reported in one previous study, which found a significantly delayed IgG-maturation in treated individuals (Sensini *et al.*, 1996). The finding that treatment may influence the IgG avidity maturation underlines the need for further studies to better clarify the avidity maturation process in pregnant women under therapy in comparison with untreated individuals. If confirmed, different cut-off values will have to be defined for treated and untreated and/or pregnant and non-pregnant individuals.

4.2.8 Molecular and Other Diagnostic Techniques

The diagnosis of acute toxoplasmosis may be established by the detection of anti-*T. gondii* antibodies by serological tests or the detection of tachyzoites or *T. gondii*-specific DNA in body fluids or tissue samples. In most cases of toxoplasmosis in immune-competent individuals, diagnosis is established by serological tests; however, molecular (i.e. PCR) diagnostic tests have proven useful in the diagnosis of infection *in utero* as well as in immune compromised hosts. The detection of *T. gondii* tachyzoite DNA in body fluids and tissues by isolation and PCR amplification is effective to diagnose congenital (Grover *et al.*, 1990), ocular (Montoya *et al.*, 1999), and cerebral toxoplasmosis (Holliman *et al.*, 1990). PCR is a key part of the diagnosis of *in utero* infection. Sensitivity in initial reports was 100%, but subsequent studies have indicated this is dependent on gestational age of infection (Montoya, 2002; Thalib *et al.*, 2005; Switaj *et al.*, 2005). Sensitivity also varies with gene target (e.g. the B1 gene is present at 35 copies and AF146527 is present at 300 copies). For amniotic fluid, sensitivity appears best with PCR assay to detect the 300 copy 592bp gene (Romand *et al.*, 2004) (Fig. 4.2). In a French study of 2000 consecutive amniotic fluid samples, it was confirmed that a positive PCR correlates with disease and that PCR is more sensitive than any other available test (Thulliez, 2001b).

Isolation of *T. gondii* from blood or body fluids (e.g. CSF, amniotic, or BAL fluids) also establishes diagnosis of the acute infection. Isolation can be performed by inoculation of the samples in mice or in tissue cultures. The demonstration of tachyzoites in histological sections or smears of body fluids by immunoperoxidase staining with anti-*T. gondii* antibodies also establishes the diagnosis (Conley *et al.*, 1981). This technique has been very useful in the diagnosis of CNS mass lesions in the setting of HIV/AIDS.

4.2.9 Diagnosis of *Toxoplasma gondii* Infection in Newborn Infants

Diagnosis of congenital infection with *T. gondii* may be difficult at birth if *Toxoplasma*-specific IgM and/or IgA antibodies are not present, because present diagnostic methods can only, with difficulty, distinguish between maternal and foetal IgG. If the child has been treated continuously with sulphadiazine and pyrimethamine, the synthesis of *Toxoplasma*-specific IgG antibodies often is suppressed and the serological confirmation of the infection can sometimes not be made with certainty before the second year of life (Wallon *et al.*, 2001). This situation is found when *T. gondii* infection is suspected, but *Toxoplasma*-specific IgM and/or IgA antibodies cannot be demonstrated in the child and parasitological investigations, such as PCR analysis for *Toxoplasma*-specific nucleic acid, are negative or not appropriate. IgM and IgA antibodies do not cross the placenta, and neonatal screening programmes for congenital toxoplasmosis are based on detection of *Toxoplasma*-specific IgM antibodies eluted from blood spots from PKU filter papers (Guthrie cards) (Guerina *et al.*, 1994; Lebech *et al.*, 1999; Neto *et al.*, 2004; Sorensen *et al.*, 2002). Different cut-offs for maternal and newborn *Toxoplasma*-specific IgM antibodies have been proposed (Candolfi *et al.*, 1993). Treatment of acute toxoplasmosis during pregnancy reduced the magnitude and duration of the *Toxoplasma*-specific IgM response in a number of earlier studies (Desmonts and Couvreur, 1984; Holfeld *et al.*, 1989). Two studies did not report such an effect (Gras *et al.*, 2004; Petersen *et al.*, 2012). Presumably this was due to differences in methods and cohorts.

Demonstration of *Toxoplasma*-specific IgG antibodies with different specificities in sera from the mother and child shows that the child synthesizes her/his own IgG antibodies, confirming that the child is infected with *T. gondii.* Previous studies have shown that transferred maternal and neo-synthesized *T. gondii*-specific IgG antibodies can be differentiated by immunoblot or immunocomplexing (Chumpitazi *et al.*, 1995; Gross *et al.*, 2000; Remington *et al.*, 2004; Robert-Gagneaux *et al.*, 1999). Differentiation of the specificities of IgG antibodies in the mother and child can also be done by comparing *T. gondii* antigen precipitated with maternal or child sera before performing an electrophoresis of the antigen–antibody complex (Pinon *et al.*, 1996, 2001; Robert-Gangneux *et al.*, 1999).

The immunoblot and immunocomplexing techniques were compared in a double-blind study and found to be equally sensitive (Pinon *et al.*, 2001). The immunoblot technique identifies newborns with congenital toxoplasmosis with a sensitivity of approximately 70% (Rilling *et al.*, 2003; Tissot-Dupont *et al.*, 2003) increasing to 85% within the first three months of life (Gross *et al.*, 2000; Rilling *et al.*, 2003; Tissot Dupont *et al.*, 2003). These results still leave 15%–30% of congenitally infected newborns without a confirmed diagnosis. To improve the diagnosis of congenital toxoplasmosis, a two-dimensional immunoblot (2DIB) assay was developed that is capable of distinguishing between maternal and neonate *Toxoplasma*-specific IgG with a better sensitivity than previous assays (Nielsen *et al.*, 2005). The 2DIB methodology greatly increased the resolution of the antibody response by allowing identification of up to a thousand spots where the most sensitive immunoblots do not allow distinction of less than fifty bands or often considerably less (Nielsen *et al.*, 2005).

4.2.10 Immune Compromised Patients

Because reactivation of latent *Toxoplasma* infection is the most common cause of toxoplasmosis in immune compromised patients, detection of *T. gondii* IgG antibodies is indicated. Patients with a positive result are at risk of reactivation of the infection; patients with a negative result should be instructed on how they can prevent becoming infected, recognizing that the mode of acquisition often goes unnoticed or unrecognized. The most important factor in the management of the seropositive immune-suppressed patient is to consider *T. gondii* as a potential causative agent in patients presenting with non-specific symptoms including focal symptoms from CNS, heart, lungs and liver. Since immune compromised patients do not reliably produce antibodies, serology has been complemented by direct detection methods such as PCR analysis of *T. gondii*-specific nucleic acid. The definitive diagnosis of toxoplasmosis relies on detection of *T. gondii* DNA by PCR or on histologic demonstration of the parasite. Whereas tachyzoites are diagnostic of the active infection, *T. gondii* tissue cysts may indicate latent infection. In patients with toxoplasmic encephalitis, blood, CSF and brain tissue may be used to detect *T. gondii*-specific DNA. Sensitivities of PCR range between 25% to 80% for blood and 35% to 100% for CSF samples (Colombo *et al.*, 2005; Vidal *et al.*, 2004; Parmley *et al.* 1992). For brain tissue, tachyzoites and histopathology suggest active infection, but cysts will also result in a positive PCR.

Pulmonary toxoplasmosis occurs in immune compromised persons, such as those with stem cell transplantation or in HIV infected patients with low CD4+ T cell counts (Rabaud *et al.*, 1996). A study of bronchoalveolar lavage (BAL) samples from 332 Danish HIV infected patients found 2.1% (7/332) positive samples using a new, sensitive real-time PCR method (Petersen *et al.*, BAL). The patients were in an advanced stage of immune-suppression with a mean $CD4^+$ T cell count of 39×10^6 per litre (range 0 and 161×10^6 per litre; normal values greater than 650×10^6 per litre). Monitoring bone marrow transplant patients by PCR on peripheral blood, BAL fluid and CSF (according to local symptoms) and treatment with pyrimethamine for positive PCR results reduced the mortality to the same levels as for *T. gondii* negative BMT patients. The same strategy could be applied to other immune-suppressed patients at risk of developing *T. gondii* infection, including those patients with HIV infection.

4.3 TREATMENT

It is useful to separate considerations about therapy of toxoplasmosis into several categories (Table 4.2). The decision to treat is based on who has the infection or manifestations, i.e. the location, activity of infection, symptomatology of the infection, severity, immune status of the patient, age and whether or not a woman with acute acquired *T. gondii* infection is pregnant. There have been several studies of prophylaxis and treatment for toxoplasmosis in the setting of AIDS, congenital disease, pregnancy and ophthalmologic disease. The recommended therapies are based on extrapolations from *in vitro* studies and animal models (mostly murine) and the clinical experience and practice of physicians experienced in the treatment of *T. gondii* infection. The standard therapeutic agents for the treatment of toxoplasmosis are the combination of pyrimethamine (administered with leucovorin) and sulphadiazine, or in the case of sulphonamide allergy, clindamycin or azithromycin or clarithromycin. Atovaquone targets cytochrome c and inhibitors of cytochrome c appear to affect cysts in animal models, at least for short times, although resistance or lack of efficacy have been described for other organisms and recurrences have been described in persons with AIDS during atovaquone treatment. In general, the medicines to treat toxoplasmosis are active against the rapidly replicating tachyzoite stage and have no demonstrated, or little, efficacy against established tissue cysts; therefore, patients treated for toxoplasmosis will have a latent infection at the conclusion of treatment of their active disease.

Pyrimethamine is a substituted phenylpyrimidine that is an inhibitor of dihydrofolate reductase. The serum half-life of pyrimethamine is 35 to 175 hours and serum levels on a dose of 1 mg/kg per day in infants (~25 to 75 mg for normal adult) range from 1000 to 4000 ng/mL (McLeod *et al.*, 1992). Serum levels for an individual are not completely predictable due to the wide variation in absorption and serum half-life perhaps reflecting differences in metabolism. Phenobarbital induces the enzymes which degrade pyrimethamine and thus have been associated with lower levels. Other medicines which can alter metabolism include theophylline. CSF levels of pyrimethamine are 10% to 25% of the corresponding serum levels (Fig. 4.8). Dose related bone marrow suppression primarily with neutropenia may develop. Anaemia and thrombocytopenia have only been rarely reported. Leucovorin (folinic acid) is routinely given at a dose of 10 mg/d, Monday, Wednesday, Friday (MWF) or more often, orally, to prevent these effects. Folinic acid does not inhibit the action of pyrimethamine on *T. gondii*, as the parasite cannot take up folinic acid at the concentrations achieved in serum. The parasite can take up folate (folic acid) which bypasses the effect of inhibitors of DHFR synthesis or other upstream enzymes.

Sulphonamides inhibit dihydropteroate synthetase, which is another enzyme involved in folate synthesis. Thus, sulphonamides are synergistic with pyrimethamine. Pyrimethamine and sulphadiazine, together, are eight-fold more active than either compound alone. These medicines are well absorbed with good penetration into cerebrospinal fluid. Adverse reactions such as hypersensitivity (McLeod *et al.*, 2006b) to sulphonamides are relatively common, particularly in AIDS patients. Bone marrow suppression is seen and this responds to folinic acid or, when severe, withholding medicines for short periods of time. Hypersensitivity reactions with rash, or less frequently, Stevens–Johnson syndrome as well as renal stones have also been reported. Sulphamethazine, sulphamerazine (along with sulphadiazine) in triple sulphonamides are highly active. Other sulphonamides are less active and, when they are used in established combinations, the ratios are suboptimal for both the DHFR and dihydropteroate synthetase component, although trimethoprim-sulfamethoxazole (TMP-SMX) was useful for suppressing recurrent retinal disease, but is less effective in tissue culture and murine models than pyrimethamine and sulphadiazine and is not the treatment of choice for active disease. (Silveira *et al.*, 2002).

4.3.1 Asymptomatic Infection or Latent Infection

There are currently no effective treatments tested in humans that can eliminate chronic, latent infection. Thus, immune competent individuals with latent toxoplasmosis, as evidenced by positive serology, cannot be treated to completely eradicate infection. For persons with AIDS who are seropositive for *T. gondii*, the risk of developing encephalitis has been estimated at 10% to 50%. TMP-SMX, or dapsone plus pyrimethamine, are effective in preventing *Toxoplasma* encephalitis (Bozzette *et al.*, 1995; Torres *et al.*, 1993). In cardiac transplantation, prophylaxis with pyrimethamine for six weeks is used for *T. gondii* seronegative recipients receiving hearts from seropositive donors (Wreghitt *et al.*, 1992). This was recently demonstrated to be effective in a small series from Brazil (Strabelli *et al.*, 2012).

4.3.2 Acute/Acquired Toxoplasmosis

In immune competent individuals, treatment is rarely considered standard of care. However, in a rare patient whose symptoms are persistent and disabling, treatment should be as described for disseminated disease. Myocarditis, encephalitis, a sepsis-syndrome with shock and hepatitis, and pneumonia are occasionally seen. This is especially notable in South American toxoplasmosis. The standard dose for a normal sized adult would be approximately as follows: pyrimethamine (100 mg loading dose divided BID for two days and, then beginning on the third day, 50 mg per day) and sulphadiazine (3 to 4 grams per day BID, i.e. 2 grams BID). For immune competent persons, this is continued for one week after resolution of signs and symptoms. Dosages for children are adjusted accordingly by weight. Leucovorin (10 mg daily or administered less frequently, e.g. on Monday, Wednesday, Friday) should also be given while pyrimethamine is administered and in the week after it is discontinued, due to the long half-life of this drug. Infections acquired through a laboratory accident or blood transfusion may be more severe and should also be treated as described above. As in all cases in which pyrimethamine is given, complete blood counts are monitored bi-weekly during treatment and in the week after it is discontinued when leucovorin is continued. Table 4.2 summarizes the current consensus on the treatment of toxoplasmosis.

4.3.3 Acute/Acquired Toxoplasmosis During Pregnancy

There are two approaches and regimens that have been reported to be successful in reducing infection and/or manifestations of infection in newborn infants (Figs. 4.2, 4.7 and Table 4.2). In the French approach, women are screened pre-pregnancy and/or by the eleventh week of gestation. Acutely infected pregnant women

TABLE 4.2 *Treatment of Toxoplasmosis in Different Clinical Settings*

Clinical Setting and Manifestation	Treatment
• Acute, asymptomatic infection	• The current standard of care is no treatment
• Acute infection with self-limited adenopathy, fever, or malaise in immune-competent persons	
• Latent, asymptomatic infection detected by positive serologic test	
• Severely symptomatic disease in immune-competent adults	• Pyrimethamine[b], Sulfadiazine[d], and Leukovorin (folinic acid)[b]
• Laboratory infection with *T. gondii* tachyzoites	
• Active disease in immune-compromised persons	
• In pregnant women infected during gestation:	
• First 18 wk gestation, or until term, if fetus found not to be infected by amniocentesis at 18 wk gestation and to have no clinical findings	• Spiramycin[a]
• If fetal infection confirmed or if infection acquired after 24 wk gestation	• Pyrimethamine[b] *, Sulfadiazine[d], and Leukovorin (folinic acid)[b] * ***Do NOT** use pyrimethamine in the first 14 wk of gestation*
• Congenital *T. gondii* infection in infant	• Pyrimethamine[b, c], Sulfadiazine[d], and Leukovorin (folinic acid)[b] • Occasionally corticosteroids (prednisone)[e] have been used when CSF protein is ≥ 1 g/dL or when active chorioretinitis threatens vision
• Active chorioretinitis in older children and adults	• Pyrimethamine[b], Sulfadiazine[d], and Leukovorin (folinic acid)[b] • Corticosteriods (prednisone)[e] if macula or posterior pole is involved or vitritis threatens vision
• Active choroidal neovascular membrane due to *T. gondii* infection[f]	• Pyrimethamine[b], Sulfadiazine[d], and Leukovorin (folinic acid)[b] • Lucentis (antibody to VegF) used as in algorithm[f] below 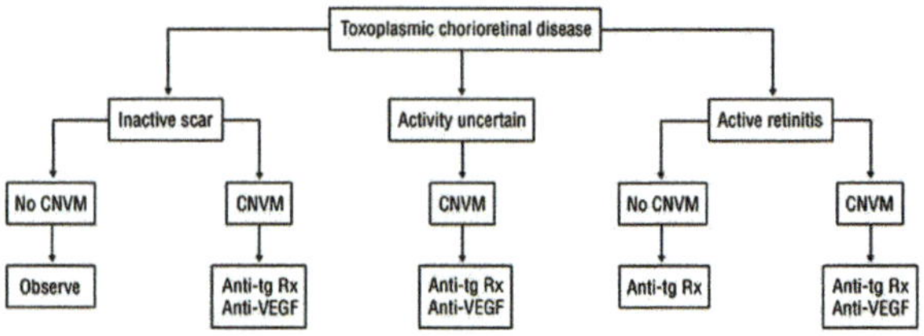 Toxoplasmic chorioretinal disease Inactive scar → No CNVM → Observe; CNVM → Anti-tg Rx, Anti-VEGF Activity uncertain → CNVM → Anti-tg Rx, Anti-VEGF Active retinitis → No CNVM → Anti-tg Rx; CNVM → Anti-tg Rx, Anti-VEGF

are given 3 g/d of spiramycin divided three times a day once maternal infection is suspected or diagnosed to decrease transmission prior to the eighteenth week of gestation if there is no evidence of foetal infection (Remington *et al.*, 2011). Spiramycin is a macrolide antibiotic that has activity against *T. gondii* due to its ability to inhibit apicoplast function but does not reach sufficient levels in foetal tissue to treat the foetus. Amniocentesis by 18 weeks' gestation and foetal ultrasonography every two weeks should be used to assess infection in the foetus and treatment should be changed to PSL if there is evidence consistent with foetal infection. Ultrasonography should be performed every two weeks as ventricular dilation, intracerebral calcifications, necroses, and/or hepatic calcifications (echogenic areas) may develop in as little as 10 days (Remington *et al.*, 2011). Foetal toxoplasmosis may be diagnosed by subinoculation of amniotic fluid into mice, PCR of amniotic fluid to identify presence of *T. gondii* genes, or ultrasonographic evidence of *T. gondii* infection including ventricular dilatation, hepatosplenomegaly, intrauterine growth retardation, or intracranial calcifications. If this occurs, then specific therapy with pyrimethamine given to the pregnant women (50 mg/d following a loading dose of 50 mg BID for two days), sulphadiazine (3 g/d divided BID) and folinic acid (10 mg/d) should be administered to the pregnant woman (Daffos *et al.*, 1988; Remington *et al.*, 2011) until delivery. Pyrimethamine is not used in the first 14 weeks of pregnancy due to concerns about teratogenicity. The majority of infants born to women treated with this regimen had subclinical disease at birth (Fig. 4.7). No group was randomized to receive a placebo because of equipoise.

An alternative approach was used by Hotop *et al.* in Germany and Austria and is beginning to be used in Minas Gerais, Brazil. In this approach (Hotop *et al.*, 2012), spiramycin was given until 16 weeks' gestation, followed by pyrimethamine and sulphadiazine with leucovorin for all patients for four weeks and if PCR of amniotic fluid was positive for *T. gondii* DNA or ultrasound abnormal, then pyrimethamine sulphadiazine is used until birth of the infant and during the first year of life (Table 4.2).

Outcomes are reported to appear to be favourable with both the French and German treatment approaches (Figs. 4.7 and 4.8). The more rapidly the treatment is initiated, the better the ocular and neurological outcomes (Kieffer *et al.*, 2008; Cortina-Borja *et al.*, 2010).

4.3.3 Congenital Toxoplasmosis

Treatment *in utero*, as described above, and in infancy has benefit for congenital infection (McLeod *et al.*, 1992, 2000, 2006a, 2009, 2012). Pathology in this infection demonstrates active infection and inflammation that is treated during

[a]In the USA, available only on request from the USA Food and Drug Administration (telephone number 301-443-5680), and then with this approval by the physician's request to Aventis (908-231-3365).
[b]Adjusted for granulocytopenia; complete blood counts, including platelets, should be monitored each Monday and Thursday.
[c]Both regimens, a higher and a lower dose, appear to be feasible and relatively safe. The duration of therapy is unknown for infants and children, especially those with AIDS.
[d]Alternative medicines for patients with atopy or severe intolerance of sulphonamides have included pyrimethamine and leucovorin with clindamycin, or azithromycin, or atovaquone, with standard dosages as recommended according to weight. In the unusual circumstance that medicines cannot be administered orally or by intraintestinal tube feeding, trimethoprim, sulphamethoxazole, and clindamycin have been administered intravenously.
[e]Corticosteroids should be used only in conjunction with pyrimethamine, sulphadiazine and leucovorin treatment and should be continued until signs of inflammation (high CSF protein, $\geq$1 g/dL) or active chorioretinitis that threatens vision have subsided, usually ~10–14 days; dosage can then be tapered and the steroids discontinued.
[f]*fImage figure from: Benevento et al., 2008, with permission. Table and caption adapted from: Remington* et al., *2011, with permission.*

foetal life and infancy (Ferguson, 2013, summarized in Remington *et al.*, 2011) when the immune system is not mature and able to effectively limit this infection. This improvement is apparent when outcomes in those treated are compared to historical controls (McLeod *et al.*, 1990, 2000, 2006a, 2012; McAuley *et al.*, 1994; Figs. 4.5 and 4.8). It has been demonstrated that neonates who appear normal at birth (subclinical disease) may later demonstrate serious sequelae (primarily retinitis) if untreated (Fig. 4.3).

Congenital toxoplasmosis is treated in the U.S.A. with a loading dose of pyrimethamine of (PYR) 2 mg/kg/d for two days followed by 1 mg/kg/d beginning on the third day and continued for 2 or 6 months. Dose is then decreased to 1 mg/kg every Monday, Wednesday and Friday (MWF) for the remainder of the first year of life. In addition, sulphadiazine at 100 mg/kg/day in 2 divided doses and folinic acid 10 mg daily or MWF are administered throughout the year. Corticosteroids (1 mg/kg/d) are added for patients who have active macular chorioretinitis or vision threatening chorioretinitis involving the posterior pole of the eye or CSF protein greater than 1 gm/dl. Table 4.2 summarizes treatment regimens. The formulations of medicines are summarized in Remington *et al.* (2011) and McLeod *et al.* (2012, 2013a in press, 2013b in press). As no paediatric solutions are made, a suspension is prepared for each medicine using the available tablets (Fig. 4.7). Most children have an absolute neutrophil count of ~1000 neutrophils/mm^3 throughout the year of treatment. In the USA, CBCs are monitored biweekly using heel stick to obtain the 0.5 mL blood samples required for this test.

Fansidar is not used in the USA because of concerns about serious adverse reactions with the long half-life of the sulphadioxine, and a suboptimal ratio of pyrimethamine and sulphadioxine. However, reports indicate good long-term outcomes and apparent lack of toxicity with treatment with this drug after a pre- and post-natal course of pyrimethamine and sulphadiazine for several months (Peyron *et al.*, 2011; Wallon *et al.*, 2013). In Paris, pyrimethamine and sulphadiazine have been utilized as in the USA (Kieffer *et al.*, 2008; McLeod *et al.*, 2009). This regimen is now also used in Italy. Rapid diagnosis and initiation of treatment appears to result in better ocular and central nervous system outcomes in infected children. With prenatal treatment followed by postnatal treatment in Paris, rapid diagnosis and initiation of medicines was associated with less retinal disease later (Kieffer *et al.*, 2008).

Shunting for hydrocephalus is an important adjunctive measure and it should be performed expeditiously. Normalization of brain parenchyma may occur but it is not possible to predict whether this will occur based on initial appearance, as expansion of cortex can occur with less than 1 mm of cortex initially. The approach to aqueductal obstruction should not be a third ventriculostomy (Cinalli *et al.*, 1999), but rather ventriculoperitoneal shunt. Third ventriculostomy often fails because of the associated inflammatory process (Renier *et al.*, 1988).

4.3.4 Ocular Toxoplasmosis

Pyrimethamine (with leucovorin: folinic acid) and sulphadiazine (PSL) are the gold standard treatment found to be effective in decreasing retinitis (its time course, subsequent retinal destruction, and the inflammatory response). This was shown in a placebo trial by Perkins published in 1956 (Perkins, 1956). It has been confirmed by direct observation of rapid response of lesions in persons studied by the NCCCTS (e.g. Fig. 4.5). *In vitro* and in murine models TMP–SMX is less effective than PSL (Grossman and Remington, 1979). This also appears anecdotally to be the case in treating recurrent retinal activity within the NCCCTS. For example, following one year of recurrences with severe vitritis and active lesions treated with repeated courses of TMP–SMX, a 60 year old patient was treated with PSL and tapering course of prednisone. In contrast to the previous year with worsening symptoms and signs while being treated with TMP–SMX, within four weeks her vitritis resolved and her active lesions had become quiescent. Nonetheless, in Brazil, suppression with TMP–SMX following treatment and resolution of active lesions was demonstrated to be effective (Silveira *et al.*, 2002) although associated with hypersensitivity in a substantial percentage of patients. Since TMP–SMX appears less effective for treatment but poses a similar risk for hypersensitivity as sulphadiazine, pyrimethamine and sulphadiazine are the preferred treatment. Other medicines such as pyrimethamine with azithromycin also appear to be effective. Suppression with azithromycin or clindamycin following treatment of active disease also anecdotally appears to be associated with lack of recurrence.

There appear to be fewer recurrences of retinal disease in children treated *in utero* and/or in the first year of life with PSL than in a cohort of persons diagnosed postnatally after the first years of life and not treated during infancy, even though these untreated children presumably had milder disease at birth that had not been recognized (Fig. 4.5). With postnatal treatment for one year, ~10% for those with mild disease at birth and ~30% for those with moderate/severe disease at birth had reactivation of retinal disease that responded quickly to re-treatment. Reactivations occur most often around the age of school entry, adolescence and times of considerable stress (Phan *et al.*, 2008a, 2008b; Fig. 4.5).

Corticosteroids (prednisone 1 mg/kg/day for an infant or child usually for ~10 to 14 days) are indicated if the macula, optic nerve head or papillomacular bundle are involved. When prednisone is given it is tapered when vitritis or macular oedema resolves.

Azithromycin, clindamycin, and atovaquone (Derouin, 2001; Meneceur *et al.*, 2008) are second-line medicines used when there is sulphonamide hypersensitivity. Azithromycin has been found to be synergistic with pyrimethamine (Derouin *et al.*, 1992), but atovaquone has been found to be antagonistic with pyrimethamine (Romand *et al.*, 1993). The azithromycin dose is 250 mg/d (first day's loading dose is 500 mg/d) for adults and dose adjusted according to weight for children. Clindamycin (1200 mg/d) has been used as an alternative drug to sulphadiazine, but it was inferior to PSL (Tabbara and O'Connor, 1980). In the NCCCTS's experience, lesions responded more slowly to second line medicines than PSL and a recent comparative study confirmed this observation.

A monthly, intraocular injection of toxoplasmic antibody to VegF (Lucentis®) has been associated with prompt resolution of choroidal neovascular membranes secondary to *T. gondii* infection (Benevento *et al.*, 2008; Fig. 4.5; Table 4.2). Choroidal neovascular membranes due to *T. gondii* respond to Lucentis® (antibody to VegF) (Fig. 4.5; Table 4.2). Use of anti-VegF for other retinal diseases in infants has been described (Capone *et al.*, 2006).

4.3.5 *Toxoplasma* Infection in Immune Compromised Persons

The same PSL treatment is used for *T. gondii* encephalitis, or other significant manifestations of disease, in the setting of immune dysfunction. The major difference is the duration of treatment. In the absence of immune reconstitution, treatment is continued to prevent relapse of infection, but if immune reconstitution occurs (i.e. in AIDS patients a CD4+ T cell count above 200) then therapy can be discontinued as it would be for an immune competent host. For a normal sized adult, the dose of pyrimethamine is 100 mg as a loading dose (divided BID for two days) followed by 50 mg/d beginning on the third day with sulphadiazine 3 to 4 grams BID (e.g. 1.5 to 2 gm every 12 hours) and folinic acid 10 mg/d (Liesenfeld *et al.*, 1999). Therapy is often started empirically and response is expected within 14 days. If no resolution occurs, then brain biopsy may be required for diagnosis. In patients intolerant to sulphadiazine, clindamycin 600 to 1200 mg every 6 hours can be used with pyrimethamine (Remington *et al.*, 1991). Alternative combinations with reported efficacy in case reports include pyrimethamine with one of the following: clarithromycin or azithromycin 1000 mg every day in an average size adult or dapsone. Desensitization to sulphadiazine has been reported to be successful. Corticosteroids are often used to control elevated intracranial pressure.

In about 30% of patients relapse if encephalitis occurs when treatment is stopped if the reason for immune-suppression remains, although relapse may not be evident for several weeks. Treatment is continued for several weeks after resolution of abnormalities. Patients may need to be maintained on pyrimethamine 25-50 mg/d, sulphadiazine 3-4 gm/d (divided BID), and folinic acid 10 mg/d after they have completed a treatment course if the cause of immune-suppression remains. With the use of antiretroviral therapy, if the CD4+ is restored to over 200, secondary prophylaxes can be discontinued.

4.3.6 Future Development of Newer Improved Anti-*T. gondii* Agents

Newer potential anti-*T. gondii* compounds have been described (Harbut *et al.*, 2012; Kavitha *et al.*, 2012; Mahamed *et al.*, 2012; Johnson *et al.*, 2012; Martins-Duarte *et al.*, 2012; Munro and Siliva, 2012; Chew *et al.*, 2012, Andrews *et al.*, 2012; Lee *et al.*, 2011; Barbosa, 2012; Lai *et al.*, 2012; Camara *et al.*, 2012; Doggett *et al.*, 2012; Fomovska *et al.*, 2012). Newer agents on the horizon may include a new echinoquinone that targets cytochrome b/c, as atovaquone (which is not synergistic with pyrimethamine) does, and eliminates cysts in an animal model (Dogget *et al.*, 2012) and a new triazine, as active alone as the synergetic combination of pyrimethamine and sulphadiazine (Mui *et al.*, 2008). Anti-sense treatments with molecular transporters and small molecule inhibitors of other essential molecular targets also are being developed (Lai *et al.*, 2012; Samuel *et al.*, 2003). Rothbard, Wender and Kumar found that molecular transporters can bring molecular cargo from the outside of the eye to the retina and across the blood–brain barrier, and McLeod *et al.* found that they carried molecular cargoes into encysted bradyzoites (Samuel *et al.*, 2003), but they have not yet been utilized for this infection in humans. Among the most promising findings that might lead to new antimicrobial agents include: PPMO targeting enoyl reductase, the Apetela 2 proteins, echinoquinone targeting cytochrome c, more highly active derivatives of azithromycin and a triazine which is, as a single agent, as active against the parasite as the synergistic combination of pyrimethamine and sulphadiazine.

4.4 PREVENTION

Methods of avoiding this infection include cooking meat to 'well done' before ingesting, washing fruits and vegetables before they are consumed (Koletzko *et al.*, 2012), wearing gloves while gardening, avoiding contact with materials excreted by cats and avoiding raw mussels. There are studies concerning detection (Gallas-LIndemann *et al.*, 2013) and disinfection of oocysts (Dumetre *et al.*, 2008) and irradiation of meat to inactivate bradyzoites in cysts. Screening to determine whether a pregnant woman is seronegative or positive for *T. gondii* infection during gestation can be used to determine risk of infection as well as to diagnosis infections with a potential for congenital transmission. Prenatal screening and prompt treatment of the infected foetus and infant have improved outcomes in a number of studies. Prophylaxis of seropositive patients with immune dysfunction or recurrent active retinal disease can prevent reactivation of latent infection. There are currently no vaccines to prevent infection or disease in humans.

4.5 CONCLUSIONS

T. gondii is a common infection of humans with 1/3 to 1/2 of all people infected. Manifestations of *T. gondii* infections vary in each clinical setting. This variation depends on genetics of the human host, genetics of the parasite, immune status of the host and probably inoculum size and parasite stage acquired, although these factors influencing pathogenesis are only partially characterized. Clinical manifestations, methods for diagnosis of primary and chronic infections in the immune competent person aid in management of this infection. This active infection (e.g. in the foetus and infant, severely ill older child or adult or immune compromised person) can be effectively treated with available medicines when treatment is begun expeditiously. Carefully performed serologic screening during gestation to diagnose primary infection in the pregnant woman, in order to facilitate treatment to eliminate disease in the foetus and infant, non-toxic medicines to eliminate encysted bradyzoites, as well as tachyzoites, and a vaccine to prevent the infection in humans are future needs.

Acknowledgements

We gratefully acknowledge patients, their families and physicians in the NCCCTS for working with us to help us to understand the clinical manifestations and pathogenesis of this infection. We also gratefully acknowledge the participants in both EMSCOT and SYROCOT.

Rima McLeod also thanks many colleagues for helpful discussions; she especially gratefully acknowledges Phillip Thulliez, Jack Remington, Jack Frenkel, Francois Kieffer, Francois Peyron, Martine Wallon, Fernand Daffos, Stephan Romand, John Costa, Herve-Pelloux, Wilma Buffolano, Kenneth Boyer, Paul Meier, Charles Swisher, Peter Heydemann, Gwen Noble, Paul Latkany, Peter Rabiah, Kristen Wroblewski, Theodore Karrison and all the members of the NCCCTS who have generously shared their experience, skills and insights and guided the development of the information included herein. She also especially acknowledges George Desmonts and Jacques Couvreur whose insights and conversations over many years helped define concepts in this chapter. Eskild Petersen especially acknowledges and thanks Ruth Gilbert who helped with the direction of the EMSCOT and SYROCOT studies.

This work was supported by: National Institute of Allergy and Infectious Diseases (Al 16945, Al 27530 and Al 014717); March of Dimes (6–528); the USA Food and Drug Administration (FD–R–000192); the United Airlines foundation; Angel Flight; United to Save Children, Gerico and Hyatt Hotel Corporation foundations; and the Michael Reese Institute Council, the Rooney Alden, Engel, Taub, Harris, Pritzker, Donelly, Cornwall and Mann families.

References

Abgrall, S., Rabaud, C., Costagliola, D., 2001. Incidence and risk factors for toxoplasmic encephalitis in human immunodeficiency virus-infected patients before and during the high active anti-retroviral therapy era. Clin. Infect. Dis. 33, 1747–1755.

Ades, A.E., 1991. Evaluating the sensitivity and predictive value of tests of recent infection: toxoplasmosis in pregnancy. Epidemiol. Infect. 107, 527–535.

Ajzenberg, D., 2012. High burden of congenital toxoplasmosis in the USA: the strain hypothesis? Clin. Infect. Dis. 54 (11), 1606–1607.

Alford Jr., C., Stagno, S., Reynolds, D.W., 1974. Congenital toxoplasmosis: clinical, laboratory, and therapeutic considerations, with special reference to subclinical disease. Bull. N. Y. Acad. Med. 50, 160–181.

Alvarado-Esquivel, C., Sethi, S., Janitschke, K., Hahn, H., Liesenfeld, O., 2002. Comparison of two commercially available avidity tests for *Toxoplasma*-specific IgG antibodies. Arch. Med. Res. 33, 520–523.

Alvarado-Esquivel, C., Estrada-Martinez, S., Liesenfeld, O., 2011. *Toxoplasma gondii* infection in workers occupationally exposed to unwashed raw fruits and vegetables: a case control seroprevalence study. Parasit. Vectors 4 (1), 235.

Ambroise-Thomas, P., Garin, J.P., Rigaud, A., 1966. Improvement of the immunofluorescence technique by the use of counterdyes. Application to *Toxoplasma gondii*. Presse Med. 74, 2215–2216.

Amorim Garcia, C.A., de, Orefice, F., de Oliveira Lyra, C., Gomes, A.B., Franca, M., Amorim Garcia Filho, C.A., de., 2004. Socioeconomic conditions as determining factors in the prevalence of systemic and ocular toxoplasmosis in Northeastern Brazil. Ophthalmic Epidemiol. 11, 301–317.

Andrade, G.M., Vasconcelos-Santos, D.V., Carellos, E.V., Romanelli, R.M., Vitor, R.W., Carneiro, A.C., Januario, J.N., 2012. Congenital toxoplasmosis from a chronically infected woman with reactivation of retinochoroiditis during pregnancy. J. Pediatr. (Rio. J.) 86 (1), 85–88.

Andrews, K.T., Haque, A., Jones, M.K., 2012. HDAC inhibitors in parasitic diseases. Immunol. Cell. Biol. 90 (1), 66–77.

Aptouramani, M., Theodoridou, M., Syrogiannopoulos, G., Mentis, A., Papaevangelou, V., Gaitana, K., Daponte, A., Hadjichristodoulou, C., the Toxoplasmosis Study Group of the Greece-Cyprus Pediatric Surveillance Unit, 2012. A dedicated surveillance network for congenital toxoplasmosis in Greece, 2006–2009: assessment of the results. BMC Public Health 12 (1), 1019.

Araujo, F.G., Barnett, E.V., Gentry, L.O., *et al.*, 1971. False-positive anti-*Toxoplasma* fluorescent-antibody tests in patients with antinuclear antibodies. Appl. Microbiol. 22, 270–275.

Araujo, F.G., Wong, M.M., Theis, J., Remington, J.S., 1973. Experimental *Toxoplasm gondii* infection in a nonhuman primate. Am. J. Trop. Med. Hyg. 22 (4), 465–472.

Arslan, F., Batirel, A., Ramazan, M., Ozer, S., Mert, A., 2012. Macrophage activation syndrome triggered by primary disseminated toxoplasmosis. Scand. J. Infect. Dis. 44 (12), 1001–1004.

Aspöck, H., Pollak, A., 1982. Prevention of prenatal toxoplasmosis by serological screening of pregnant women in Austria. Scand. J. Infect. Dis. 84 (Suppl.), 43–45.

Aubert, D., Foudrinier, F., Villena, I., Pinon, J.M., Biava, M.F., Renoult, E., 1996. PCR diagnosis and follow up of two cases of disseminated toxoplasmosis after kidney grafting. J. Clin. Microbiol. 34, 1347.

Aubert, G., Maine, G.T., Villena, I., Hunt, J.C., Howard, L., Sheu, M., Brojanac, S., Chovan, L.E., Nowlan, S.F., Pinon, J.M., 2000. Recombinant antigens to detect *Toxoplasma gondii*-specific immunoglobulin G and immunoglobulin M in human sera by enzyme immunoassay. J. Clin. Microbiol. 38, 1144–1150.

Bahia-Oliverira, L.M., Jones, J.L., Azevedo-Silva, J., *et al.*, 2003. Highly endemic, waterborne transmission in north Rio de Janeiro state. Brazil Emerg. Infect. Dis. 9, 55–62.

Balsari, A., Poli, G., Molina, V., *et al.*, 1980. ELISA for *Toxoplasma* antibody detection: a comparison with other serodiagnostic tests. J. Clin. Pathol. 33, 640–643.

Barbosa, B.F., Gomes, A.O., Ferro, E.A., Napolitano, D.R., Mineo, J.R., Silva, N.M., 2012. Enrofloxacin is able to control *Toxoplasma gondii* infection in both *in vitro* and *in vivo* experimental models. Vet. Parasitol. 187 (1–2), 44–52.

Basavaprabhu, A., Soundarya, M., Deepak, M., Satish, R., 2012. CNS toxoplasmosis presenting with obstructive hydrocephalus in patients of retroviral disease – a case series. Med. J. Malaysia 67 (2), 214–216.

Bautista, G., Ramos, A., Forés, R., Regidor, C., Ruiz, E., de Laiglesia, A., Navarro, B., Bravo, J., Portero, F., Sanjuan, I., Fernández, M.N., Cabrera, R., 2012. Toxoplasmosis in cord blood transplantation recipients. Transpl. Infect. Dis. 14 (5), 496–501.

Bech-Nielsen, S., 2012. *Toxoplasma gondii* associated behavioural changes in mice, rats and humans: evidence from current research. Prev. Vet. Med. 103 (1), 78–79.

Beghetto, E., Buffolano, W., Spadoni, A., del Pezza, M., di Cristina, M., Minenkova, O., Petersen, E., Felici, F., Gargano, N., 2003. Use of an immunoglobulin G avidity assay based on recombinant antigens for diagnosis of primary *Toxoplasma gondii* infection during pregnancy. J. Clin. Microbiol. 41, 5414–5418.

Beghetto, E., Nielsen, H.V., Del Porto, P., Buffolano, W., Guglietta, S., Felici, F., Petersen, E., Gargano, N., 2005. A combination of antigenic regions of *T. gondii* microneme proteins induce protective immunity against oral infections with parasite cysts. J. Infect. Dis. 191, 637–645.

Beghetto, E., Spadoni, A., Bruno, L., Buffolano, W., Gargano, N., 2006. Chimeric antigens of *Toxoplasma gondii*: toward standardization of toxoplasmosis serodiagnosis using recombinant proteins. J. Clin. Microbiol. 44 (6), 2133–2140.

Bela, S.R., Dutra, M.S., Mui, E., Montpetit, A., Oliveira, F.S., Oliveira, S.C., Arantes, R.M., Antonelli, L.R., McLeod, R., Gazzinelli, R.T., 2012. Impaired innate immunity in mice deficient in interleukin-1 receptor-associated kinase 4 leads to defective type 1 T cell responses, B cell expansion, and enhanced susceptibility to infection with *Toxoplasma gondii*. Infect. Immun. 80 (12), 4298–4308.

Benenson, M.W., Takafuji, E.T., Lemon, S.M., Greenup, R.L., Sulzer, A.J., 1982. Oocyst-transmitted toxoplasmosis associated with ingestion of contaminated water. N. Engl. J. Med. 3007, 666–669.

Benevento, J.D., Jager, R.D., Noble, A.G., *et al.*, 2008. Toxoplasmosis-associated neovascular lesions treated successfully with ranibizumab and antiparasitic therapy. Arch. Ophthalmol. 126, 1152–1156.

Benson, A., Pifer, R., Behrendt, C.L., Hooper, L.V., Yarovinsky, F., 2012. Gut commensal bacteria direct a protective immune resposne against *Toxoplasma gondii*. Cell. Host Microbe. 6 (2), 187–196.

Berrebi, A., Bardou, M., Bessieres, M., *et al.*, 2007. Outcome for children infected with congenital toxoplasmosis in the first trimester and with normal ultrasound findings: A study of 36 cases. Eur. J. Obstet. Gynaecol. Reprod. Bio. 135, 53–57.

Binquet, C., Wallon, M., Quantin, C., Kodjikian, L., Garweg, J., Fleury, J., Peyron, F., Abrahomawicz, M., 2003. Prognostic factors for the long-term development of ocular lesions in 327 children with congenital toxoplasmosis. Epidemiol. Infect. 131, 1157–1168.

Bohne, W., Heesemann, J., Gross, U., 1993. Induction of bradyzoite-specific *T. gondii* antigens in gamma Interferon-treated macrophages. Infect. Immun. 61, 1141–1145.

Bories, P., Zink, E., Mattern, J.F., Villard, O., Berceanu, A., Bilger, K., Candolfi, E., Herbrecht, R., Abou-Bacar, A., Lioure, B., 2012. Febrile pancytopenia as uncommon presentation of disseminated toxoplasmosis after BMT. Bone Marrow Transplant 47 (2), 301–303.

Botterel, F., Ichai, P., Feray, C., *et al.*, 2002. Disseminated toxoplasmosis, resulting from infection of allograft, after orthopedic liver transplantation: usefulness of quantitative PCR. J. Clin. Microbiol. 40, 1648–1650.

Boyer, K.M., Holfels, E., Roizen, N., *et al.*, 2005. Risk factors for *Toxoplasma gondii* infection in mothers of infants with congenital toxoplasmosis: Implications for prenatal management and screening. Am. J. Obstet. Gynecol. 192, 564–571.

Boyer, K., Hill, D., Mui, E., *et al.*, 2011. Unrecognized ingestion of *Toxoplasma gondii* oocysts leads to congenital toxoplasmosis and causes epidemics in North America. Clin. Infect. Dis. 53 (11), 1081.

Bowie, W.R., King, A.S., Werker, D.H., Isaac-Renton, J.L., Bell, A., Eng, S.B., Marion, S.A., 1997. Outbreak of toxoplasmosis associated with municipal drinking water. The BC Toxoplasma Investigation Team. Lancet 350, 173–177.

Bozzette, S., Findelstein, D., Spector, S., Frame, P., Powderly, W.G., He, W., Phillips, L., Craven, D., van der Horst, C., Feinberg, J., 1995. A randomized trial of three antipneumocystis agents in patients with advanced human immunodeficiency virus infection. New England Journal of Medicine 332, 693–699.

Bretagne, S., Costa, J.M., Keuntz, M., *et al.*, 1995. Late toxoplasmosis evidenced by PCR in marrow transplant recipient. Bone Marrow Transplant 15, 809–811.

Brezin, A.P., Thulliez, P., Couvreur, J., *et al.*, 2003. Ophthalmic outcome after prenatal and postnatal treatment of congenital toxoplasmosis. Am. J. Ophthalmol. 138, 779–784.

Bricker-Hildalgo, H., Brenier-Pinchart, M.P., Schaal, J.P., *et al.*, 2007. Value of *Toxoplasma gondii* detection in one hundred thirty-three placentas for the diagnosis of congenital toxoplasmosis. Pediatr. Infect. Dis. J. 26, 845–846.

Brown, A.S., Schaefer, C.A., Wuesenberry Jr., C.P., *et al.*, 2005. Maternal exposure to toxoplasmosis and risk of schizophrenia in adult offspring. Am. J. Psychiatry 162, 767–773.

Brownback, K.R., Crosser, M.S., Simpson, S.Q., 2012. A 49-year-old man with chest pain and fever after returning from France. Chest 141 (6), 1618–1621.

Buffolano, W., Beghetto, E., Del Pezzo, M., Spadoni, A., Di Cristina, M., Petersen, E., Gargano, N. (Submitted). Neonatal diagnosis of congenital toxoplasmosis using an enzyme-linked IgM capture assay with a combination of recombinant antigens.

Burg, J.L., Perelman, L., Kasper, L.H., Ware, P.L., Boothroyd, J.C., 1988. Molecular analysis of the gene encoding the major surface antigen of *Toxoplasma gondii*. J. Immunol. 141, 3584–3591.

Burnett, A.J., Shortt, S.G., Isaac-Renton, J., *et al.*, 1998. Multiple cases of acquired toxoplasmosis retinitis presenting in an outbreak. Ophthalmology 105, 1032–1037.

Burrowes, D., Boyer, K., Swisher, C.N., *et al.*, 2012. Spinal cord lesions in Congenital Toxoplasmosis demonstrated with neuroimaging, including their successful treatment in an adult. J. Neuroparasitology.

Busemann, C., Ribback, S., Zimmermann, K., Sailer, V., Kiefer, T., Schmidt, C.A., Schulz, K., Steinmetz, I., Dombrowski, F., Dölken, G., Krüger, W.H., 2012. Toxoplasmosis after allogeneic stem cell transplantation – a single centre experience. Ann. Hematol. 91 (7), 1081–1089.

Camara, D., Bisanz, C., Barette, C., Van Daele, J., Human, E., Barnard, B., Van der Straeten, D., Stove, C.P., Lambert, W.E., Douce, R., Maréchal, E., Birkholtz, L.M., Cesbron-Delauw, M.F., Dumas, R., Rébeillé, F., 2012. Inhibition of p-aminobenzoate and folate syntheses in plants and apicomplexan parasites by natural product rubreserine. J. Biol. Chem. 287 (26), 22367–22376.

Candolfi, E., Bessieres, M.H., Mart, P., Cimon, B., Gandilhon, F., Pelloux, H., Thulliez, P., 1993. Determination of a new cut-off value for the diagnosis of congenital toxoplasmosis by detection of specific IgM in an enzyme immunoassay. Eur. J. Clin. Microbiol. Infect. Dis. 12, 396–398.

Carme, B., Bissuel, F., Ajzenberg, D., Bouyne, R., Aznar, C., Demar, M., Bichat, S., Louvel, D., Bourbigot, A.M., Peneau, C., Neron, P., Dardé, M.L., 2002. Severe acquired toxoplasmosis in immunocompetent adult patients in French Guiana. J. Clin. Microbiol. 2002 (40), 4037–4044.

Carme, B., Demar, M., Ajzenberg, D., Dardé, M.L., 2009. Severe acquired toxoplasmosis caused by wild cycle of *Toxoplasma gondii*, French Guiana. Emerg. Infect. Dis. 15, 656–658.

Carneiro, A.C., Andrade, G.M., Costa, J.G., Pinheiro, B.V., Vasconcelos-Santos, D.V., Ferreira, A.D., Su, C., Januário, J.N., Vitor, R.W., 2013. Genetic characterization of *Toxoplasma gondii* revealed highly diverse genotypes from human congenital toxoplasmosis in southeastern Brazil. J. Clin. Microbiol.

Capone Jr., A., 2006. Treatment of neovscularization in infants with retinopathy of prematurity with anti-VEGF: vitreous surgery for retinopathy of prematurity. Paper presented at: Annual Meeting of the American Academy of Ophthalmology, Vegas, NV.

Caselli, D., Andreoli, E., Paolicchi, O., Savelli, S., Guidi, S., Pecile, P., Aricò, M., 2012. Acute encephalopathy in the immune-compromised child: never forget toxoplasmosis. J. Pediatr. Hematol. Oncol. 34 (5), 383–386.

Cerede, O., Dubremetz, J.F., Bout, D., Lebrun, M., 2002. The *T. gondii* protein MIC3 requires pro-peptide cleavage and dimerization to function as adhesion. EMBO J. 21, 2526–2536.

Cesbron-Delauw, M.F., Tomavo, S., Beauchamps, P., Fourmaux, M.P., Camus, D., Capron, A., Dubremetz, J.F., 1994. Similarities between the primary structure of two distinct major surface proteins of *Toxoplasma gondii*. J. Biol. Chem. 269, 16217–16222.

Cesbron-Delauw, M.F., 1995. The SAG2 antigen of *T. gondii* and the 31-kDa surface antigen of sarcocystis muris share similar sequence features. Parasitol. Res. 81, 444–445.

Chandrasekar, P.H., Momin, F., the Bone Marrow Transplant Team, 1997. Disseminated toxoplasmosis in marrow recipients: a report of three cases and a review of the litreature. Bone Marrow Transpl. 19, 685–689.

Chapey, E., Wallon, M., Debize, G., Rabilloud, M., Peyron, F., 2010. Diagnosis of congenital toxoplasmosis by using a whole-blood gamma interferon release assay. J. Clin. Microbiol. 48 (1), 41–45.

Chew, W.K., Segarra, I., Ambu, S., Mak, J.W., 2012. Significant reduction of brain cysts caused by *Toxoplasma gondii* after treatment with spiramycin coadministered with metronidazole in a mouse model of chronic toxoplasmosis. Antimicrob. Agents Chemother. 56 (4), 1762–1768.

Chiang, E., Goldstein, D.A., Shapiro, M.J., Mets, M.B., 2012. Branch retinal artery occlusion caused by toxoplasmosis in an adolescent. Case Report Ophthalmol. 3 (3), 333–338.

Chumpitazi, B.F., Boussaid, A., Pelloux, H., Racinet, C., Bost, M., Goullier-Fleuret, A., 1995. Diagnosis of congenital toxoplasmosis by immunoblotting and relationship with other methods. J. Clin. Microbiol. 33, 1479–1485.

Cinalli, G., Sainte-Rose, C., Chumas, P., *et al.*, 1999. Failure of third ventriculostomy in the treatment of aqueductal stenosis in children. Neurosurg. Focus 6 (4) e3.

Collier, S.A., Stockman, L.J., Hicks, L.A., Garrison, L.E., Zhou, F.J., Beach, M.J., 2012. Direct healthcare costs of selected diseases primarily or partially transmitted by water. Epidemiol. Infect. 140 (11), 2003–2013.

Colombo, F.A., Vidal, J.E., Penalva de Oliveira, A.C., Hernandez, A.V., Bonasser-Filho, F., Nogueira, R.S., Focaccia, R., Pereira-Chioccola, V.L., 2005. Diagnosis of cerebral toxoplasmosis in AIDS patients in Brazil: importance of molecular and immunological methods using peripheral blood samples. J. Clin. Microbiol. 43, 5044–5047.

Cong, H., Mui, E.J., Witola, W.H., Sidney, J., Alexander, J., Sette, A., Maewal, A., McLeod, R., 2011a. Towards an immunosense vaccine to prevent toxoplasmosis: protetive *Toxoplasma gondii* epitopes restricted by HLA-A*0201. Vaccine 29 (4), 754–762.

Cong, H., Mui, E.J., Witola, W.H., Sidney, J., Alexander, J., Sette, A., Maewal, A., McLeod, R., 2011b. Human immunome, bioinformatic analyses using HLA supermotifs and the parasite genome, binding assays, studies of human T cell responses, and immunization of HLA-A*1101 transgenic mice including novel adjuvants provide a foundation for HLA-A03 restricted CD8+T cell epitope based, adjuvanted vaccine protective against Toxoplasma gondii 6, 12.

Cong, H., Mui, E.J., Witola, W.H., Sidney, J., Alexander, J., Sette, A., Maewal, A., El Bissati, K., Zhou, Y., Suzuki, Y., Lee, D., Woods, S., Sommerville, C., Henriquez, F.L., Roberts, C.W., McLeod, R., 2012. *Toxoplasma gondii* HLA-B*702-restricted GRA7(20-28) peptide with adjuvants and universal helper T cell epitope elicits CD8(+) T cells producing interferon-γ and reduces parasite burden in HLA-B*0702 mice. Hum. Immunol. 73 (1), 1–10.

Conley, F.K., Jenkins, K.A., Remington, J.S., 1981. Toxoplasma gondii infection of the central nervous system: use of the peroxidase-antiperoxidase method to demonstrate *Toxoplasma* in formalin-fixed paraffin embedded tissue section. Human Pathology 12, 690–698.

Conrad, P.A., Miller, M.A., Kreuder, C., *et al.*, 2005. Transmission of *Toxoplasma*: clues from the study of sea otters as sentinels of *Toxoplasma gondii* flow into the marine environment. Int. J. Parasitol. 35 (11–12), 1155–1168.

Cook, A.J.C., Gilbert, R., Dunn, D., Buffolano, W., Zuffrey, J., Reymondin, M., Hohfeld, P., Foulon, W., Jenum, P., Petersen, E., Lebech, M., Semprini, E., Masters, J., 2000. Sources of Toxoplasma-infection in pregnant women: a European multicentre case-control study. Br. Med. J. 312, 142–147.

Cortina-Borja, M., Tan, H.K., Wallon, M., *et al.*, 2010. Prenatal treatment for serious neurological sequelae of congenital toxoplasmosis: an observational prospective cohort study. PLoS Med. 7 (10), 1–11.

Couvreur, J., 1955. Étude de la toxoplasmose congenitale à propo de 20 observation These. Paris.

Couvreur, J., Desmonts, G., 1962. Congenital and maternal toxoplasmosis. A review of 300 cases. Develop. Med. Chld. Neurol. 4, 519–530.

Couvreur, J., Desmonts, G., 1964. Late evolution outbreaks of congenital toxoplasmosis. Cah. Coll. Med. Hop. Paris 115, 752–758. French.

Couvreur, J., Desmonts, G., Girre, J.Y., 1976. Congenital toxoplasmosis in twins: a series of 14 pairs of twins: absence of infection in one twin in two pairs. J. Pediatr. 89 (2), 235–240.

Couvreur, J., Desmonts, G., Aron-Rosa, D., 1984. Le prognostic oculare de la toxoplasmose congenitale: role du traitement. Ann. Pediatr. 31.

Couvreur, J., Desmonts, G., Thulliez, P., 1988. Prophylaxis of congenital toxoplasmosis. effects of spiramycin on placental infection. J. Antimicrob. Chemother. 22, 193–200.

Couvreur, J., Thulliez, P., Daffos, F., Aufrant, C., Bompard, Y., Gesquiere, A., Desmonts, G., 1993. *In utero* treatment of toxoplasmic fetopathy with the combination pyrimethamine-sulfadiazine. Foetal. Diagn. Ther. 8 (1), 45–50.

Couvreur, J., Thulliez, P., 1996. Acquired toxoplasmosis with ocular or neurologic involvement. Presse Med. 25, 438–442.

Cunningham, T., 1982. Pancarditis in acute toxoplasmosis. Am. J. Clin. Pathol. 78, 403–405.

Daffos, F., Forestier, F., Capella-Pavlovsky, M., *et al.*, 1988. Prenatal management of 746 pregnancies at risk for congenital toxoplasmosis. N. Engl. J. Medicine 318, 271–275.

Dai, J., Jiang, M., Wang, Y., Qu, L., Gong, R., Si, J., 2012a. Evaluation of a recombinant multiepitope peptide for serodiagnosis of *Toxoplasma gondii* infection. Clin. Vaccine Immunol. 19 (3), 338–342.

Dai, J.F., Jiang, M., Qu, L.L., Sun, L., Wang, Y.Y., Gong, L.L., Gong, R.J., Si, J., 2012b. *Toxoplasma gondii*: Enzyme-linked immunosorbent assay based on a recombinant multi-epitope peptide for distinguishing recent from past infection in human sera. Exp. Parasitol. 133 (1), 95–100.

Dannemann, B.R., Vaughan, W.C., Thulliez, P., Remington, J.S., 1990. Differential agglutination test for diagnosis of recently acquired infection with *Toxoplasma gondii*. J. Clin. Microbiol. 28, 1928–1933.

Darde, M.L., Villena, I., Pinon, J.M., Beguinot, I., 1998. Severe toxoplasmosis causd by a *Toxoplasma gondii* strain with a new isoenzyme type acquired in French Guyana. J. Clin. Microbiol. 1, 324.

Darde, M.L., 2008. *Toxoplasma gondii*, 'new' genotypes and virulence. Parasite 15, 366–371.

Decoster, A., Darcy, F., Caron, A., Vinatier, D., Houze de L'Aulnoit, D., Vittu, G., Niel, G., Heyer, F., Lecolier, B., Delcroix, M., 1992. Anti-P30 IgA antibodies as prenatal markers of congenital toxoplasma infection. Clin. Exp. Immunol. 87, 310–315.

Delair, E., Monnet, D., Grabar, S., Dupouy-Camet, J., Yera, H., Brezin, A.P., 2008. Respective roles of acquired and congenital infections in presumed ocular toxoplasmosis. Am. J. Ophthalmol. 146 (6), 851–855.

Delair, E., Latkany, P., Noble, A.G., *et al.*, 2011. Clinical manifestations of ocular toxoplasmosis. Ocul. Immunol. Inflamm. 19 (2), 91–102.

Demar, M., Ajzenberg, D., Maubon, D., Djossou, F., Panchoe, D., Punwasi, W., Valery, N., Peneau, C., Daigre, J.L., Aznar, C., Cottrelle, B., Terzan, L., Dardé, M.L., Carme, B., 2007. Fatal outbreak of human toxoplasmosis along the Maroni River: epidemiological, clinical, and parasitological aspects. Clin. Infect. Dis. 45, e88–e95.

Demar, M., Hommel, D., Djossou, F., Peneau, C., Boukhari, R., Louvel, D., Bourbigot, A.M., Nasser, V., Ajzenberg, D., Darde, M.L., Carme, B., 2012.

Acute toxoplasmoses in immunocompetent patients hospitalized in an intensive care unit in French Guiana. Clin. Microbiol. Infect. 18, E221–E231.

Denes, E., Vidal, J., Monteil, J., 2012. Spinal cord toxoplasmosis. Infection.

Derouin, F., Almadany, R., Chau, F., Rouveix, B., Pocidalo, J.J., 1992. Synergistic activity of azithromycin and pyrimethamine or sulfadiazine in acute experimental toxoplasmosis. Antimicrob. Agents Chemother. 36 (5), 997–1001.

Derouin, F., 2001. Anti-toxoplasmosis drugs. Curr. Opin. Investig. Drugs 2 (10), 1368–1374.

Derouin, F., Pelloux, H., ESCMID Study Group on Clinical Parasitology, 2008. Prevention of toxoplasmosis in transplant patients. Clin. Microbiol. Infect. 14 (12), 1089–1101.

Desmonts, G., Couvreur, J., Ben Rachid, M.S., 1965a. Le toxoplasme, la mère et l'enfant. Arch. Franç. Pédiatr. 22, 1183–1200.

Desmonts, G., Couvreur, J., Alison, F., *et al.*, 1965b. Etude épidémiologique sur la toxoplasmose: de l'Influence de la cuisson des viandes de boucherie sur la fréquence de l'Infection Humaine. Rev. Franc. Etud. Clin. Biol. 10, 952–958.

Desmonts, G., Couvreur, J., 1974a. Congenital toxoplasmosis: a prospective study of 378 pregnancies. N. Engl. J. Med. 290, 1110–1116.

Desmonts, G., Couvreur, J., 1974b. Toxoplasmosis in pregnancy and its transmission to the foetus. Bull. N. Y. Acad. Med.: Journal of Urban Health 50, 146–159.

Desmonts, G., Couvreur, J., 1979. Congenital toxoplasmosis: a prospective study of the offspring of 542 women who acquired toxoplasmosis during pregnancy. In: Thalhammer, O., Pollak, A., Baumgarten, K. (Eds.), Pathophysiology of Congenital Disease: Perinatal Medicine (6th European Congress, Vienna). Stuttgart Georg Thieme Publishers, pp. 51–60.

Desmonts, G., Remington, J.S., 1980. Direct agglutination test for diagnosis of *Toxoplasma* infection: method for increasing sensitivity. J. Clin. Microbiol. 11, 562–568.

Desmonts, G., 1982. Acquired toxoplasmosis in pregnant women. Evaluation of the frequency of transmission of *Toxoplasma* and of congenital toxoplasmosis. Lyon Medical. 248, 115–123.

Desmonts, G., Couvreur, J., 1984. Congenital toxoplasmosis. Prospective study of the outcome of pregnancy in 542 women with toxoplasmosis acquired during pregnancy. Ann. Pediatr. (Paris) 31, 805–809.

Doggett, J.S., Nilsen, A., Forquer, I., Wegmann, K.W., Jones-Brando, L., Yolken, R.H., Bordón, C., Charman, S.A., Katneni, K., Schultz, T., Burrows, J.N., Hinrichs, D.J., Meunier, B., Carruthers, V.B., Riscoe, M.K., 2012. Endochin-like quinolones are highly efficacious against acute and latent experimental toxoplasmosis. Proc. Natl. Acad. Sci. U. S. A. 109 (39), 15936–15941.

Dorfman, R.F., Remington, J.S., 1973. Value of lymph-node biopsy in the diagnosis of acute toxoplasmosis. N. Engl. J. Med. 289, 878–881.

Dubey, J.P., Sharma, S.P., Juranek, D.D., Sulzer, A.J., Teutsch, S.M., 1981. Characterization *of Toxoplasma gondii* isolates from an outbreak of toxoplasmosis in Atlanta, Georgia. Am. J. Vet. Res. 42 (6), 1007–1010.

Dubey, J.P., Lappin, M.R., Thulliez, P., 1995. Long-term antibody responses of cats fed *T. gondii* tissue cysts. J. Parasitol. 81, 887–893.

Dubey, J.P., 1998. Advances in the life cycle of *Toxoplasma gondii*. Int. J. Parasitol. 28 (7), 1019–1024.

Dubey, J.P., Beattie, C.P., 1998. Toxoplasmosis of animals and man. CRC Press, Boca Raton, FL.

Dubey, J.P., Ferreira, L.R., Martins, J., McLeod, R., 2012a. Oral oocyst-induced mouse model of toxoplasmosis: effect of infection with *Toxoplasma gondii* strains of different genotypes, dose, and mouse strains (transgenic, out-bred, in-bred) on pathogenesis and mortality. Parasitology 139 (1), 1–13.

Dubey, J.P., Hill, D.E., Rozeboom, D.W., Rajendran, C., Choudhary, S., Ferreira, L.R., Kwok, O.C., Su, C., 2012b. High prevalence and genotypes of *Toxoplasma gondii* isolated from organic pigs in northern USA. Vet. Parasitol. 188 (1-2), 14–18.

Dubey, J.P., Lago, E.G., Gennari, S.M., Su, C., Jones, J.L., 2012c. Toxoplasmosis in humans and animals in Brazil: high prevalence, high burden of disease, and epidemiology. Parasitology 139 (11), 1375–1424.

Dumetre, A., Le Bras, C., Baffet, M., Meneceur, P., Dubey, J.P., Derouin, F., Dugeut, J.P., Joyeux, M., Moulin, L., 2008. Effects of ozone and ultraviolet radiation treatments on the infectivity of *Toxoplasma gondii* oocysts. Vet. Parasitol. 153 (3-4), 209–213.

Dunn, D., Wallon, M., Peyron, F., Petersen, E., Peckham, C.S., Gilbert, R.E., 1999. Mother to child transmission of toxoplasmosis: risk estimates for clinical counselling. Lancet 353, 1829–1833.

Dunn, D., Newell, M.L., Gilbert, R., Mok, J., Petersen, E., Peckham., C., 1996. European Collaborative Study and Research Network on Congenital Toxoplasmosis. Low incidence of congenital toxoplasmosis in children born to women infected with human immunodeficiency virus. Eur. J. Obstet. Gyn. Reprod. Hlth. 68 (1–2), 93–96.

Dutra, M.S., Bela, S.R., Peixoto-Rangel, A.L., Fakiola, M., Cruz, A.G., Gazzinelli, A., Quites, H.F., Bahia-Oliveira, L.M., Peixe, R.G., Campos, W.R., Higino-Rocha, A.C., Miller, N.E., Blackwell, J.M., Antonelli, L.R., Gazzinelli, R.T., 2013. Association of a NOD2 gene polymorphism and T-helper 17 cells with presumed ocular toxoplasmosis. J. Infect. Dis. 207 (1), 152–163.

Eichenwald, H.F., 1957. Congenital toxoplasmosis: a study of 150 cases. Am. J. Dis. Chld. 94, 411–412.

Eichenwald, H.F., 1960. A study of congenital toxoplasmosis, with particular emphasis on clinical manifestations, sequelae and therapy. In: Siim, J. (Ed.), Human toxoplasmosis. Munksgaard, Copenhagen, pp. 41–49.

Engvall, E., Perlmann, P., 1972. Enzyme-linked immunosorbent assay, Elisa. 3. Quantitation of specific antibodies by enzyme-labeled anti-immunoglobulin in antigen-coated tubes. J. Immunol. 109, 129–135.

Enzensberger, W., Helm, E.B., Hopp, G., Stille, W., Fischer, P.A., 1985. Toxoplasma encephalitis in patients with AIDS. Dtsch. Med. Wochenschr. 110, 83–87.

European Multicentre Study on Congenital Toxoplasmosis. Effect of timing and type of treatment on the risk of mother to child transmission of *Toxoplasma gondii*. BJOG 2003;110:112–20.

Evengård, B., Petterson, K., Engman, M.-L., Wiklund, S., Ivarsson, S.A., Tear-Fahnehjelm, K., Forsgren, M., Gilbert, R., Malm, G., 2001. Low incidence of toxoplasma infection during pregnancy and in newborns in Sweden. Epidemiol. Infect. 127, 121–127.

Eyles, D.E., Coleman, N., 1955. An evaluation of the curative effects of pyrimethamine and sulfadiazine, alone and in combination, on experimental mouse toxoplasmosis. Antibiot. Chemother. 5, 529–539.

Fabiani, S., Pinto, B., Bruschi, F., 2012. Toxoplasmosis and neuropsychiatric diseases: can serological studies establish a clear relationship? Neurol. Sci..

Faucher, B., Miermont, F., Ranque, S., Franck, J., Piarroux, R., 2012. Optimization of *Toxoplasma gondii* DNA extraction from amniotic fluid using NucliSENS easyMAG and comparison with QIAamp DNA minikit. Eur. J. Clin. Microbiol. Infect. Dis. 31 (6), 1035–1039.

Fayyaz, H., Rafi, J., 2012. TORCH screening in polyhydramnios: an observational study. J. Matern. Foetal Neonatal Med. 25 (7), 1069–1072.

Feldman, H.A., 1953. Congenital toxoplasmosis – a study of 103 cases. Am. J. Dis. Child. 86, 487.

Ferguson, D.J., Graham, D.I., Hutchison, W.M., 1991. Pathological changes in the brains of mice infected with *Toxoplasma gondii*: a histological, immunocytochemical and ultrastructural study. Int. J. Exp. Pathol. 72 (4), 463–474.

Ferguson, D.J., Bowker, C., Jeffery, K.J., Chamberlain, P., Squier, W., 2013. Congenital toxoplasmosis: continued parasite proliferation in the foetal brain despite maternal immunological control in other tissues. Clin. Infect. Dis. 56 (2), 204–208.

Fernàndez-Sabé, N., Cervera, C., Fariñas, M.C., Bodro, M., Muñoz, P., Gurguí, M., Torre-Cisneros, J., Martín-Dávila, P., Noblejas, A., Len, O., García-Reyne, A., Del Pozo, J.L., Carratalà, J., 2012. Risk factors, clinical features, and outcomes of toxoplasmosis in solid-organ transplant recipients: a matched case-control study. Clin. Infect. Dis. 54 (3), 355–361.

Ferrandiz, J., Mercier, C., Wallon, M., Picot, S., Cesbron-Delauw, M.F., Peyron, F. 2004. Limited value of assays using detection of immunoglobulin G antibodies to the two recombinant dense granule antigens, GRA1 and GRA6 Nt of *Toxoplasma gondii*, for distinguishing between acute and chronic infections in pregnant women. Clin. Diagn. Lab. Immunol. 11:1016–1021

Fischer, H.G., Stachelhaus, S., Sahm, M., Meyer, H.M., Reichmann, G., 1998. GRA7, an excretory 29 kDa *T. gondii* dense granule antigen released by infected host cells. Mol. Biochem. Parasitol. 91, 251–262.

Flament-Durand, J., Coers, C., Waelbroeck, C., van Geertruyden, J., Tousaint, C., 1967. Toxoplasmic encephalitis and myositis during treatment with immunodepressive drugs. Acta. Clin. Belg. 22, 44–54.

Flegr, J., 2013. Influence of latent Toxoplasma infection on human personality, physiology and morphology: pros and cons of the Toxoplasma-human model in studying the manipulation hypothesis. J. Exp. Biol. 216 (Pt 1), 127–133.

Flori, P., Tardy, L., Patural, H., *et al.*, 2004. Reliability of immunoglobulin G anti-*Toxoplasma* avidity test and effects of treatment on avidity indexes of infants and pregnant women. Clin. Diagn. Lab. Immunol. 11 (4), 669–674.

Fomovska, A., Wood, R.D., Mui, E., Dubey, J.P., Ferreira, L.R., Hickman, M.R., Lee, P.J., Leed, S.E., Auschwitz, J.M., Welsh, W.J., Sommerville, C., Woods, S., Roberts, C.W., McLeod, R., 2012. Salicylanilide inhibitors of *Toxoplasma gondii*. J. Med. Chem. 55 (19), 8375–8391.

Forestier, F., 1991. Foetal disease, prenatal diagnosis and practical measures. Presse Med., 1448–1454.

Forsgren, M., Gille, E., Ljungstrom, I., 1991. *Toxoplasma* antibodies in pregnant women in Sweden in 1969, 1979 and 1987. Lancet 337, 1413–1414.

Foulon, W., Naessens, A., Derde, M.P., 1994. Evaluation of the possibilities for preventing congenital toxoplasmosis. Am. J. Perinatol. 11, 57–62.

Foulon, W., Villena, I., Stray-Pedersen, B., *et al.*, 1999. Treatment of toxoplasmosis during pregnancy: A multicentre study of impact on foetal transmission and children's sequelae at age 1 year. Am. J. Obstet. Gynecol. 180, 410–415.

Frenkel, J.K., Friedlander, S., 1951. Toxoplasmosis: pathology of neonatal disease, pathogenesis, diagnosis and treatment. USA Government Printing Office, Washington, Publication n 141. US Public Health Service, p. 107.

Frenkel, J.K., Hitchings, G.H., 1957. Relative reversal by vitamins (p-Aminobenzoic, folic and folinic acid) of the effects of sulfadiazine and pyrimethamine on *Toxoplasma*, mouse and man. Antibiot. Chemother. 7, 630–638.

Frenkel, J.K., Dubey, J.P., Miller, N.L., 1970. *Toxoplasma gondii* in cats: fecal stages identified as coccidian oocysts. Science 167 (3919), 893–896.

Frenkel, J.K., Parker, B.B., 1996. An apparent role of dogs in the transmission of *Toxoplasma gondii*. The probable importance of xenosmophilia. Ann. N. Y. Acad. Sci. 791, 402–407.

Fritz, H., Barr, B., Packham, A., Melli, A., Conrad, P.A., 2012a. Methods to produce and safely work with large numbers of *Toxoplasma gondii* oocysts and bradyzoite cysts. J. Microbiol. Methods 88 (1), 47–52.

Fritz, H.M., Bowyer, P.W., Bogyo, M., Conrad, P.A., Boothroyd, J.C., 2012b. Proteomic analysis of fractionated *Toxoplasma* oocysts reveals clues to their environmental resistance. PLoS One 7 (1), e29955.

Fuentes, I., Rodriguez, M., Domingo, C.J., *et al.*, 1996. Urine sample used for congenital toxoplasmosis diagnosis by PCR. J. Clin. Microbiol. 34 (10), 2368–2371.

Furco, A., Carmagnat, M., Chevret, S., Garin, Y.J., Pavie, J., De Castro, N., Charron, D., Derouin, F., Rabian, C., Molina, J.M., 2008. Restoration of *Toxoplasma gondii*-specific immune responses in patients with AIDS staring HAART. AIDS 22 (16), 2087–2096.

Gallas-Lindemann, C., Sotiriadou, I., Mahmoodi, M.R., Karanis, P., 2013. Detection of *Toxoplasma gondii* oocysts in different water resources by Loop Mediated Isothermal Amplification (LAMP). Acta. Trop. 125 (2), 231–236.

Gallino, A., Maggiorini, M., Kiowski, W., Martin, X., Wunderli, W., Schneider, J., Turina, M., Follath, F., 1996. Toxoplasmosis in heart transplant recipients. Eur. J. Clin. Microbiol. Infect. Dis. 15, 389–393.

Garabedian, C., Le Goarant, J., Delhaes, L., Rouland, V., Vaast, P., Valat, A.S., Subtil, D., Houfflin-Debarge, V., 2012. [Periconceptional toxoplasmic seroconversion: about 79 cases]. J. Gynecol. Obstet. Biol. Reprod. (Paris) 41 (6), 546–552.

Garcia-Règuet, N., Lebrun, M., Fourmaux, M.N., Mercereau-Puijalon, O., Mann, T., Beckers, J.M., Samin, B., Van Beeumen, J., Bout, D., Dubremetz, J.F., 2000. The microneme protein MIC3 of *T. gondii* is a secretory adhesion that blnds to both the surface of the host cells and the surface of the parasite. Cell. Microbiol. 2, 353–364.

Gard, S., Magnusson, J.H., 1951. A glandular form of toxoplasmosis in connection with pregnancy. Acta. Med. Scand. 141, 59–64.

Garin, J.P., Brossier, N., Sung, R.T.M., Moyne, T., 1985. Effect of pyrimethamine sulfadioxine (Fansidar) on an avirulent cystogenic strain of *Toxoplasma gondii* (Prugniaud strain) in white mice. Bull. Soc. Pathol. Exot. Filiales 78, 821–824.

Garweg, J.G., Kodjikian, L., Peyron, F., Binquet, C., Fleury, J., Grange, J.D., Quantin, C., Wallon, M., 2005. [Congential ocular toxoplasmosis – ocular manifestations and prognosis after early diagnosis of infection] 222 (9), 721–727.

Gerhold, R.W., Jessup, D.A., 2012. Zoonotic Diseases Associated with Free-Roaming Cats. Zoonoses Public Health.

Gilbert, R.E., Dunn, D.T., Lightman, S., Murray, P.I., Pavesio, C.E., Gormley, P.D., Masters, J., Parker, S.P., Stanford, M.R., 1999. Incidence of symptomatic toxoplasma eye disease: aetiology and public health implications. Epidemiol. Infect. 123, 283–289.

Gilbert, R.E., Gras, L., Wallon, M., Peyron, F., Ades, A.E., Dunn, D.T., 2001. Effect of prenatal treatment on mother to child transmission of *Toxoplasma gondii*: retrospective cohort study of 554 mother-child pairs in Lyon, France. Int. J. Epidemiol. 30 (6), 1303–1308.

Gilbert, R.E., Freeman, K., Lago, E.G., Bahia-Oliveira, L.M., Tan, H.K., Wallon, M., Buffolano, W., Stanford, M.R., Petersen, E., European Multicentre Study on Congenital Toxoplasmosis (EMSCOT), 2008. Ocular sequelae of congenital toxoplasmosis in Brazil compared with Europe. PLoS Negl. Trop. Dis. 2, e277.

Giordano, L.F.C.M., Lasmar, E.P., Tavora, E.R.F., Lasmar, M.F., 2002. Toxoplasmosis transmitted via kidney allograft: case report and review. Transplant Proc. 34, 498–499.

Glasner, P.D., Silveira, C., Kruszon-Moran, D., Martins, M.C., Burnier Júnior, M., Silveira, S., Camargo, M.E., Nussenblatt, R.B., Kaslow, R.A., Belfort Júnior, R., 1992. An unusually high prevalence of ocular toxoplasmosis in southern Brazil. Am. J. Ophthalmol. 114, 136–144.

Godwin, R., 2012. *Toxoplasma gondii* and elevated suicide risk. Vet. Rec. 171 (9), 225.

Gollub, E.L., Leroy, V., Gilbert, R., Chêne, G., Wallon, M., 2008. European Toxoprevention Study Group (EUROTOXO): Effectiveness of health education on *Toxoplasma*-related knowledge, behaviour, and risk of seroconversion in pregnancy. Eur. J. Obstet. Gynecol. Reprod. Biol. 136 (2), 137–145.

Goodwin, D., Hrubec, T.C., Klein, B.G., Strobl, J.S., Werre, S.R., Han, Q., Zajac, A. M., Lindsay, D.S., 2012. Congenital infection of mice with *Toxoplasma gondii* induces minimal change in behaviour and no change in neurotransmitter concentrations. J. Parasitol. 98 (4), 706–712.

Gras, L., Gilbert, R.E., Wallon, M., Peyron, F., Cortina-Borja, M., 2004. Duration of the IgM response in women acquiring *T. gondii* during pregnancy: implications for clinical practices and cross sectional incidence studies. Epidemiol. Infect. 132, 541–548.

Gras, L., Wallon, M., Pollak, A., Cortina-Borja, M., Evengard, B., Hayde, M., Petersen, E., Ruth Gilbert, R., Multicentre, The European, Study on Congenital Toxoplasmosis, 2005. Association between prenatal treatment and clinical manifestations of congenital toxoplasmosis in infancy: a cohort study in 13 European centres. Acta. Paediatrica. 94, 1721–1731.

Greenlee, J.E., Johnson Jr., W.D., Campa, J.F., *et al.*, 1975. Adult toxoplasmosis presenting as polymyositis and cerebellar ataxia. Ann. Intern. Med. 82, 367–371.

Grigg, M.E., Ganatra, J., Boothroyd, J.C., Margolis, T.P., 2001. Unusual abundance of atypical strains associated with human ocular toxoplasmosis. J. Infect. Dis. 184, 633–639.

Gross, U., Roos, T., Heesemann, J., 1992. Improved serological diagnosis of *T. gondii* infection by detection of immunoglobulin A (IgA) and IgM antibodies against P30 by using the immunoblot technique. J. Clin. Microbiol. 30, 1436–1441.

Gross, U., Luder, C.G., Hendgen, V., Heeg, C., Sauer, I., Weidner, A., Krczal, D., Enders, G., 2000. Comparative immunoglobulin G antibody profiles between mother and child (CGMC test) for early diagnosis of congenital toxoplasmosis. J. Clin. Microbiol. 38, 3619–3622.

Grossman, P.L., Remington, J.S., 1979. The effect of trimethoprim and sulfamethoxazole on *Toxoplasma gondii in vitro* and *in vivo*. Am. J. Trop. Med. Hyg. 28 (3), 445–455.

Grover, C.M., Thulliez, P., Remington, J.S., Boothroyd, J.D., 1990. Rapid prenatal diagnosis of congenital Toxoplasma infection by using polymerase chain reaction and amniotic fluid. Journal of Clinical Microbiology 28, 2297–2301.

Guenter, W., Bieliński, M., Deptuła, A., Zalas-Wiecek, P., Piskunowicz, M., Szwed, K., Buciński, A., Gospodarek, E., Borkowska, A., 2012. Does *Toxoplasma gondii* infection affect cognitive function? A case control study. Folia Parasitol. (Praha) 59 (2), 93–98.

Guerina, N.G., Hsu, H.W., Meissner, H.C., Maguire, J.H., Lynfield, R., Stechenberg, B., Abroms, I., Pasternack, M.S., Hoff, R., Eaton, R.B., 1994.

Neonatal serologic screening and early treatment for congenital *T. gondii* infection. The New England Regional Toxoplasma Working Group. N. Engl. J. Med. 330, 1858–1863.

Guevara-Silva, E.A., Ramírez-Crescencio, M.A., Soto-Hernández, J.L., Cárdenas, G., 2012. Central nervous system immune reconstitution inflammatory syndrome in AIDS: experience of a Mexican neurological centre. Clin. Neurol. Neurosurg. 114 (7), 852–861.

Hamdani, N., Daban-Huard, C., Lajnef, M., Richard, J.R., Delavest, M., Godin, O., Guen, E.L., Vederine, F.E., Lépine, J.P., Jamain, S., Houenou, J., Corvoisier, P.L., Aoki, M., Moins-Teisserenc, H., Charron, D., Krishnamoorthy, R., Yolken, R., Dickerson, F., Tamouza, R., Leboyer, M., 2012. Relationship between *Toxoplasma gondii* infection and bipolar disorder in a French sample. J. Affect Disord.

Harning, D., Spenter, H., Metsis, A., Vuust, J., Petersen, E., 1996. Recombinant *T. gondii* SAG1 (P30) expressed in *Eschericia coli* is recognized by human *Toxoplasma*-specific IgM and IgG antibodies. Clin. Diag. Lab. Immunol. 3, 355–357.

Harbut, M.B., Patel, B.A., Yeung, B.K., McNamara, C.W., Bright, A.T., Ballard, J., Supek, F., Golde, T.E., Winzeler, E.A., Diagana, T.T., Greenbaum, D.C., 2012. Targeting the ERAD pathway via inhibition of signal peptide peptidase for antiparasitic therapeutic design. Proc. Natl. Acad. Sci. U. S. A. 109 (52), 21486–21491.

Haverkos, H.W., Remington, J.S., Chan, J.C., 1987. Assessment of *Toxoplasma* encephalitis (TE) therapy: a cooperative study. Am. J. Med. 82, 907–914.

Hedman, K., Lappalainen, M., Seppala, I., Makela, O., 1989. Recent primary *Toxoplasma* infection indicated by a low avidity of specific IgG. J. Infect. Dis. 159, 726–739.

Hedman, K., Lappalainen, M., Söderlund, M., Hedman, L., 1993. Avidity of IgG in serodiagnosis of infectious diseases. Rev. Med. Microbiol. 4, 123–129.

Henriquez, S.A., Brett, R., Alexander, J., Pratt, J., Roberts, C.W., 2009. Neuropsychiatric disease and *Toxoplasma gondii* infection. Neuroimmunomodulation 16 (2), 122–133.

Hermes, G., Ajioka, J.W., Kelly, K.A., *et al.*, 2008. Neurological and behavioural abnormalities, ventricular dilatation, altered cellular functions, inflammation, and neuronal injury in brains of mice due to common, persistent, parasitic infection. J. Neuroinflammation 5, 48.

Hill, D., Coss, C., Dubey, J.P., *et al.*, 2011. Identification of a sporozoite-specific antigen from *Toxoplasma gondii*. J. Parasitol. 97 (2), 328–337.

Hill, D.E., Dubey, J.P., 2012. *Toxoplasma gondii* prevalence in farm animals in the USA. Int. J. Parasitol.

Hogan, M.J., Kimura, S.J., Lewis, A., Zweigart, P.A., 1957. Early and delayed ocular manifestations of congenital toxoplasmosis. Trans. Am. Ophthalmol. Soc. 55, 275–296.

Hohlfeld, P., Daffos, F., Thulliez, P., *et al.*, 1989. Foetal toxoplasmosis: outcome of pregnancy and infant follow-up after *in utero* treatment. J. Pediatr. 115 (5 Pt 1), 765–769.

Hohlfeld, P., Daffos, F., Costa, J.M., *et al.*, 1994. Prenatal diagnosis of congenital toxoplasmosis with a polymerase-chain-reaction test on amniotic fluid. N. Engl. J. Med., 695–699.

Holec-Gasior, L., Ferra, B., Drapala, D., 2012a. MIC1-MAG1-SAG1 Chimeric Protein, a Most Effective Antigen for Detection of Human Toxoplasmosis. Clin. Vaccine Immunol. 19 (12), 1977–1979.

Holec-Gasior, L., Ferra, B., Drapała, D., Lautenbach, D., Kur, J., 2012b. A new MIC1-MAG1 recombinant chimeric antigen can be used instead of the *Toxoplasma gondii* lysate antigen in serodiagnosis of human toxoplasmosis. Clin. Vaccine Immunol. 19 (1), 57–63.

Holland, G.N., O'Connor, G.R., Belfort Jr., R., Remington, J.S., 1996. Toxoplasmosis. In: Pepose, J.S., Holland, G.N., Wilhelmus, K.R. (Eds.), Ocular Infection and Immunity. Mosby Year Book, St Louis, pp. 1183–1223.

Holland, G.N., Crespi, C.M., ten Dam-van Loon, N., *et al.*, 2008. Analysis of recurrence patterns associated with toxoplasmic retinochoroiditis. Am. J. Ophthalmol. 145, 1007–1013.

Holliman, R.E., Johnsos, J.D., Savva, D., 1990. Diagnosis of cerebral toxoplasmosis in association with AIDS using the polymerase chain reaction. Scandinavian Journal of Infectious Diseases 22, 243–244.

Holliman, R.E., Raymond, R., Renton, N., Johnson, J.D., 1994. The diagnosis of toxoplasmosis using IgG avidity. Epidemiol. Infect. 112, 399–408.

Holmdahl, S.C., Holmdahl, K., 1955. The frequency of congenital toxoplasmosis and some viewpoints on the diagnosis. Acta. Paediatr. 44, 322–329.

Holub, D., Flegr, J., Dragomirecká, E., Rodriguez, M., Preiss, M., Novák, T., Cermák, J., Horáček, J., Kodym, P., Libiger, J., Höschl, C., Motlová, L.B., 2012. Differences in onset of disease and severity of psychopathology between toxoplasmosis-related and toxoplasmosis-unrelated schizophrenia. Acta. Psychiatr. Scand.

Horacek, J., Flegr, J., Tintera, J., Verebova, K., Spaniel, F., Novak, T., Brunovsky, M., Bubenikova-Valesova, V., Holub, D., Palenicek, T., Höschl, C., 2012. Latent toxoplasmosis reduces gray matter density in schizophrenia but not in controls: voxel-based-morphometry (VBM) study. World J. Biol. Psychiatry 13 (7), 501–509.

Hotop, A., Hlobil, H., Gross, U., 2012. Efficacy of rapid treatment initiation following primary *Toxoplasma gondii* infection during pregnancy. Clin. Infect. Dis. 54 (11), 1545–1552.

Howe, D.K., Sibley, L.D., 1994. *T. gondii* – analysis of different laboratory stocks of the RH strain reveals genetic heterogeneity. Exp. Parasitol. 78, 242–245.

Hutchison, W.M., Dunachie, J.F., Siim, J.C., Work, K., 1970. Coccidian-like nature of *Toxoplasma gondii*. Br. Med. J. 1, 142–144.

Isaac-Renton, J., Bowie, W.R., King, A., *et al.*, 1998. Detection of *Toxoplasma gondii* Oocysts in Drinking Water. Applied and Enviromental Microbiology 64, 2278–2280.

Jacobs, D., Vercammen, M., Saman, E., 1999. Evaluation of recombinant dense granule antigen 7 (GRA7) of *T. gondii* for detection of immunoglobulin G antibodies and analysis of a major antigenic domain. Clin. Diag. Lab. Immunol. 6, 24–29.

Jamieson, S.E., de Roubaix, L.A., Cortina-Borja, M., *et al.*, 2008. Genetic and epigenetic factors at COL2A1 and ABCA4 influence clinical outcome in congenital toxoplasmosis. PLoS One 3(6):e2285.

Jamieson, S.E., Cordell, H., Petersen, E., *et al.*, 2009. Host genetic and epigenetic factors in toxoplasmosis. Mem. Inst. Oswaldo Cruz 104 (2), 162–169.

Jamieson, S.E., Peixoto-Rangel, A.L., Hargrave, A.C., *et al.*, 2010. Evidence for associations between the purinergic receptor P2X(7) (P2RX7) and toxoplasmosis. Genes. Immun. 11 (5), 374–383.

Janku, J., 1923. Pathogenesa a patologicka anatomie tak nazvaneho vrozenehonalezem parasitu v sitnici. Cas. Lek. Ces. 62, 1021–1027.

Jenum, P.A., Stray-Pedersen, B., Gundersen, A.G., 1997. Improved diagnosis of primary *T. gondii* infection in early pregnancy by determination of anti-toxoplasma immunoglobulin G avidity. J. Clin. Microbiol. 35, 1972–1977.

Jenum, P.A., Stray-Pedersen, B., 1998. Development of specific immunoglobulins G, M and A following primary *T. gondii* infection in pregnant women. J. Clin. Microbiol. 36, 2907–2913.

Johnson, A.M., Roberts, H., Tenter, A.M., 1992. Evaluation of a recombinant antigen ELISA for the diagnosis of acute toxoplasmosis and comparison with traditional ELISAs. J. Med. Microbiol. 37, 404–409.

Johnson, S.M., Murphy, R.C., Geiger, J.A., DeRocher, A.E., Zhang, Z., Ojo, K.K., Larson, E.T., Perera, B.G., Dale, E.J., He, P., Reid, M.C., Fox, A.M., Mueller, N.R., Merritt, E.A., Fan, E., Parsons, M., Van Voorhis, W.C., Maly, D.J., 2012. Development of *Toxoplasma gondii* calcium-dependent protein kinase 1 (TgCDPK1) inhibitors with potent anti-*toxoplasma* activity. J. Med. Chem. 55 (5), 2416–2426.

Jones, J.L., Kruszon-Moran, D., Wilson, M., *et al.*, 2001. *Toxoplasma gondii* Infection in the USA: Seroprevalence and Risk Factors. Am. J. Epidemiol. 154, 357–365.

Jones, J.L., Kruszon-Moran, D., Sanders-Lewis, K., *et al.*, 2007. *Toxoplasma gondii* infection in the USA, 1999-2004, decline from the prior decade. Am. J. Trop. Med. Hyg. 77, 405–410.

Jones, J.L., Dargelas, V., Roberts, J., *et al.*, 2009. Risk factors for *Toxoplasma gondii* infection in the USA. Clin. Infect. Dis. 49 (6), 878–884.

Jones, J.L., Dubey, J.P., 2012a. Foodborne toxoplasmosis. Clin. Infect. Dis. 55 (6), 845–851.

Jones, J.L., Roberts, J.M., 2012b. Toxoplasmosis hospitalizations in the USA, 2008, and trends, 1993-2008. Clin. Infect. Dis. 54 (7), e58–e61.

Kaňková, S., Sulc, J., Křivohlavá, R., Kuběna, A., Flegr, J., 2012. Slower postnatal motor development in infants of mothers with latent toxoplasmosis during the first 18 months of life. Early Hum. Dev. 88 (11), 879–884.

Karanis, P., Aldeyarbi, H.M., Mirhashemi, M.E., Khalil, K.M., 2012. The impact of the waterborne transmission of *Toxoplasma gondii* and analysis efforts for water detection: an overview and update. Environ. Sci. Pollut. Res. Int.

Kasper, L.H., Ware, P.L., 1989. Identification of stage specific antigens of *Toxoplasma gondii*. Infect. Immun. 57, 688–692.

Kaushik, M., Lamberton, P.H., Webster, J.P., 2012. The role of parasites and pathogens in influencing generalised anxiety and predation-related fear in the mammalian central nervous system. Horm. Behav. 62 (3), 191–201.

Kavitha, N., Noordin, R., Chan, K.L., Sasidharan, S., 2012. In vitro anti-*Toxoplasma gondii* activity of root extract/fractions of Eurycoma longifolia Jack. BMC Complement Altern. Med. 12, 91.

Khan, A., Jordan, C., Muccioli, C., Vallochi, A.L., Rizzo, L.V., Belfort Jr., R., Vitor, R.W., Silveira, C., Sibley, L.D., 2006. Genetic divergence of *Toxoplasma gondii* strains associated with ocular toxoplasmosis. Brazil Emerg. Infect. Dis. 12, 942–949.

Kimball, A.C., Kean, B.H., Fuchs, F., 1971. toxoplasmosis: a prospective study of 4,048 obstetric patients. Am. J. Obstet. Gynaecol. 111, 211–218.

Kieffer, F., Wallon, M., Garcia, P., *et al.*, 2008. Risk factors for retinochoroiditis during the first 2 years of life in infants with treated congenital toxoplasmosis. Pediatr. Infect. Dis. J., 27–32.

Kim, J.H., Psevdos Jr., G., Gonzalez, E., Singh, S., Kilayko, M.C., Sharp, V., 2012. All-cause mortality in hospitalized HIV-infected patients at an acute tertiary care hospital with a comprehensive outpatient HIV care program in New York City in the era of highly active antiretroviral therapy (HAART). Infection.

Kimball, A., Kean, B.H., Fuchs, F., 1971. Congenital toxoplasmosis: a prospective study of 4,048 obstetric patients. Am. J. Obstet. Gynecol. 111, 211–218.

Kodjikian, L., Hoigne, I., Adam, O., Jacquier, P., Aebi-Ochsner, C., Aebi, C., Garweg, J.G., 2004. Vertical transmission of toxoplasmosis from a chronically infected immunocompetent woman. Pediatr. Infect. Dis. 23 (3), 272–274.

Kodjikian, L., Wallon, M., Fleury, M., *et al.*, 2006. Ocular manifestations in congenital toxoplasmosis. Graefes Arch. Clin. Exp. Ophthalmol. 244, 14–21.

Koletzko, B., Bauer, C.P., Bung, P., Cremer, M., Flothkötter, M., Hellmers, C., Kersting, M., Krawinkel, M., Przyrembel, H., Rasenack, R., Schäfer, T., Vetter, K., Wahn, U., Weißenborn, A., Wöckel, A.; Gesund ins Leben [Nutrition in pregnancy - Practice recommendations of the Network 'Healthy Start - Young Family Network']. (2012). Dtsch Med Wochenschr. 137(25-26):1366–72.

Kong, J.T., Grigg, M.E., Uyetake, L., Parmley, S., Boothroyd, J.C., 2003. Serotyping of *Toxoplasma gondii* infections in humans using synthetic peptides 187 (9), 1484–1495.

Koppe, J.G., Kloosterman, G.J., 1982. Congenital toxoplasmosis: long-term follow-up. Padiatr. Padol. 17, 171–179.

Koppe, J.G., Loewer-Sieger, D.H., de Roever-Bonnet, H., 1986. Results of 20-year follow-up of congenital toxoplasmosis. Lancet 1 (8475), 254–256.

Kortbeek, L.M., Hofhuis, A., Nijhuis, C.D., Havelaar, A.H., 2009. Congenital toxoplasmosis and DALYs in the Netherlands. Mem. Inst. Oswaldo Cruz 104 (2), 370–373.

Korhonen, M.H., Brunstein, J., Haario, H., Katnikov, A., Rescaldani, R., Hedman, K., 1999. A new method with general diagnostic utility for the calculation of immuglobulin G avidity. Clin. Diag. Lab. Immunol. 6, 725–728.

Kotresha, D., Poonam, D., Muhammad Hafiznur, Y., Saadatnia, G., Nurulhasanah, O., Sabariah, O., Tan, S.Y., Izzati Zahidah, A.K., Rahmah, N., 2012. Recombinant proteins from new constructs of SAG1 and GRA7 sequences and their usefulness to detect acute toxoplasmosis. Trop. Biomed. 29 (1), 129–137.

Kovacs, J.A., 1995. Toxoplasmosis in AIDS: keeping the lid on. Ann. Intern. Med. 123 (3), 230–231.

Lafferty, K.D., 2006. Can the common brain parasite, *Toxoplasma gondii*, influence human culture? Proc. Biol. Sci. 273 (1602), 2749–2755.

Lai, B.S., Witola, W.H., El Bissati, K., Zhou, Y., Mui, E., Fomovska, A., McLeod, R., 2012. Molecular target validation, antimicrobial delivery, and potential treatment of *Toxoplasma gondii* infections. Proc. Natl. Acad. Sci. U. S. A. 109 (35), 14182–14187.

Lappalainen, M., Koskela, P., Koskiniemi, M., Ämmälä, P., Hiilesmaa, V., Teramo, K., Raivio, K.O., Remington, J.S., Hedman, K., 1993. Toxoplasmosis acquired during pregnancy: improved serodiagnosis based on avidity of IgG. J. Infect. Dis. 167, 691–697.

Lass, A., Pietkiewicz, H., Szostakowska, B., Myjak, P., 2012. The first detection of *Toxoplasma gondii* DNA in environmental fruits and vegetables samples. Eur. J. Clin. Microbiol. Infect. Dis. 31 (6), 1101–1108.

Lassoued, S., Zabraniecki, L., Marin, F., *et al.*, 2007. Toxoplasmic chorioretinitis and antitumor necrosis factor treatment in rheumatoid arthritis. Semin. Arthritis Rheum. 36, 262–263.

Lavinsky, D., Romano, A., Muccioli, C., Belfort Jr., R., 2012. Imaging in ocular toxoplasmosis. Int. Ophthalmol. Clin. 52 (4), 131–143.

LeFichoux, Y., Marty, P., Chan, H., Doucet, J., 1984. Détection des IgG anti-toxoplasmiques par I.s.ag.A. A propos de 3,786 sérologies. Bull. Soc. Fr. Parasitol. 3, 13–18.

Leal, F.E., Cavazzana, C.L., de Andrade Jr., H.F., Galisteo Jr., A.J., de Mendonça, J.S., Kallas, E.G., 2007. *Toxoplasma gondii* pneumonia in immunocompetent subjects: case report and review. Clin. Infect. Dis. 15, e62–e66.

Lebech, M., Andersen, O., Christensen, N.C., Hertel, J., Nielsen, H.E., Peitersen, B., Rechnitzer, C., Larsen, S.O., Norgaard-Pedersen, B., Petersen, E., 1999. Feasibility of neonatal screening for toxoplasma infection in the absence of prenatal treatment. Lancet 353, 1834–1837.

Lecordier, L., Fourmaux, M.P., Mercier, C., Dehecq, E., Masy, E., Cesbron-Delauw, M.F., 2000. Enzyme-linked immunosorbent assays using the recombinant dense granule antigens GRA6 and GRA1 of *T. gondii* for detection of immunoglobulin G antibodies. Clin. Diagn. Lab. Immunol. 7, 607–611.

Lecolier, B., Pucheu, B., 1993. Intérêt de l'étude de l'avidité des IgG pour le diagnostic de la toxoplasmose. Pathol. Biol. Paris 41, 155–158.

Lee, Y., Choi, J.Y., Fu, H., Harvey, C., Ravindran, S., Roush, W.R., Boothroyd, J.C., Khosia, C., 2011. Chemistry and biology of macrolide antiparasitic agents. J. Med. Chem. 54 (8), 2792–2804.

Lees, M.P., Fuller, S.J., McLeod, R., *et al.*, 2010. P2X7 receptor-mediated killing of an intracellular parasite, *Toxoplasma gondii*, by human and murine macrophages. J. Immunol. 184 (12), 7040–7046.

Lefevre-Pettazzoni, M., Bissery, A., Wallon, M., *et al.*, 2007. Impact of spiramycin treatment and gestational age on maturation of *Toxoplasma gondii* immunoglobulin G avidity in pregnant women. Clin. Vaccine Immunol. 14 (3), 239–243.

Lehmann, T., Marcet, P.L., Graham, D.H., Dahl, E.R., Dubey, J.P., 2006. Globalization and the population structure of *Toxoplasma gondii*. Proc. Natl. Acad. Sci. USA 103 (30), 11423–11428.

LePort, C., Chene, G., Morlat, P., *et al.*, 1996. Pyrimethamine for primary prophylaxis of toxoplasmic encephalitis in patients with human immunodeficiency virus infection: a double-blind, randomized trial. J. Infect. Dis. 173, 91–97.

Lester, D., 2012. *Toxoplasma gondii* and homicide. Psychol. Rep. 111 (1), 196–197.

Levin, M., McLeod, R., Young, Q., *et al.*, 1983. Pneumocystis pneumonia: importance of gallium scan for early diagnosis and description of a new immunoperoxidase technique to demonstrate *Pneumocystis carinii*. Am. Rev. Respir. Dis. 128, 182–185.

Levy, R.M., Rosenbloom, S., Perrett, L.V., 1986. Neuroradiologic findings in AIDS: a review of 200 cases. AJNR Am. J. Neurodaiol. 147, 977–983.

Levy, R.M., Bredesen, D.E., Rosenblum, M.I., 1988. Opportunistic central nervous system pathology in patients with AIDS. Ann. Neurol. 23, S7–S12.

Levy, R.M., Mills, C.M., Posin, J.P., *et al.*, 1990. The efficacy and clinical impact of brain imaging in neurologically symptomatic AIDS patients: a prospective CT/MRI study. J. Acquir. Immune. Defic. Syndr. 3, 461–471.

Li, S., Galvan, G., Araujo, F.G., Suzuki, Y., Remington, J.S., Parmley, S., 2000. Serodiagnosis of recently acquired *T. gondii* infection. Clin. Diagn. Lab. Immunol. 7, 781–787.

Liesenfeld, O., Press, C., Montoya, J.G., Gill, R., Isaac-Renton, J.L., Hedman, K., Remington, J.S., 1997. False-positive results in immunoglobulin M (IgM) toxoplasma antibody tests and importance of confirmatory testing: the Platelia Toxo IgM test. J. Clin. Microbiol. 1997 (35), 174–178.

Liesenfeld, O., Wong, S.Y., Remington, J.S., 1999. Toxoplasmosis in the setting of AIDS. In: Bartlett, J.G., Merigan, T.C., Bolognesi, D. (Eds.), Textbook of AIDS Medicine. Williams & Wilkins, Baltimore, pp. 225–259.

Liesenfeld, O., Montoya, J.G., Kinney, S., Press, C., Remington, J.S., 2001a. Effect of testing for IgG avidity in the diagnosis of *T. gondii* infection in pregnant women: experience in a USA reference laboratory. J. Infect. Dis. 183, 1248–1253.

Liesenfeld, O., Montoya, J.G., Kinney, S., Press, C., Remington, J.S., 2001b. Confirmatory serological testing for acute toxoplasmosis and rate of induced abortions among women reported to have positive *Toxoplasma* immunoglobulin M antibody titres. Am. J. Obstet. Gynecol. 184, 140–145.

Lindsay, D.S., Collins, M.V., Mitchell, S.M., *et al.*, 2003. Sporulation and survival of *Toxoplasmosis gondii* oocysts in seawater. J. Eukaryot. Microbiol. 50, 687–688.

L'Ollivier, C., Wallon, M., Faucher, B., Piarroux, R., Peyron, F., Franck, J., 2012. Comparison of mother and child antibodies that target high-molecular-mass *Toxoplasma gondii* antigens by immunoblotting improves neonatal diagnosis of congenital toxoplasmosis. Clin. Vaccine Immunol. 19 (8), 1326–1328.

London, N.J., Hovakimyan, A., Cubillan, L.D., Siverio Jr., C.D., Cunningham Jr., E.T., 2011. Prevalence, clinical characteristics, and causes of vision loss in patients with ocular toxoplasmosis. Eur. J. Ophthalmol. 21, 811–819.

Lourenco, E.V., Pereira, S.R., Faca, V.M., Coelho-Castelo, A.A.M., Mineo, J.R., Roque-Barreira, M.C., Greene, L.J., Panunto-Castelo, A., 2001. *T. gondii* micronemal protein MIC1 is a lactose-binding lectin. Glycobiology 11, 541–547.

Luft, B.J., Conley, F., Remington, J.S., *et al.*, 1983. Outbreak of central-nervous-system toxoplasmosis in western Europe and North America. Lancet 1 (8328), 781–784.

Luft, B.J., Remington, J.S., 1984. Acute Toxoplasma infection among family members of patients with acute lymphadenopathic toxoplasmosis. Arch. Intern. Med. 144, 53–56.

Luft, B.J., Remington, J.S., 1992. Toxoplasmic encephalitis in AIDS (AIDS commentary). Clin. Infect. Dis. 15, 211–222.

Luft, B.J., Hafner, R., Korzun, A.H., *et al.*, 1993. Toxoplasmic encephalitis in patients with the acquired immunodeficiency syndrome. N. Engl. J. Med. 329, 995–1000.

Mack, D.G., Johnson, J.J., Roberts, F., *et al.*, 1999. HLA-class II genes modify outcome of *Toxoplasma gondii* infection. Int. J. Parasitol. 29 (9), 1351–1358.

Magaldi, C., Elkis, H., Pattoli, D., Coscina, A.L., 1969. Epidemic of toxoplasmosis at a university in São-José-dos Campos, S.P. Brazil. 1. Clinical and serologic data. Rev. Latinoam Microbiol. Parasitol. (Mex.) 11 (1), 5–13.

Mahamed, D.A., Mills, J.H., Egan, C.E., Denkers, E.Y., Bynoe, M.S., 2012. CD73-generated adenosine facilitates *Toxoplasma gondii* differentiation to long-lived tissue cysts in the central nervous system. Proc. Natl. Acad. Sci. U. S. A. 109 (40), 16312–16317.

Maksimov, P., Zerweck, J., Maksimov, A., Hotop, A., Gross, U., Spekker, K., Däubener, W., Werdermann, S., Niederstrasser, O., Petri, E., Mertens, M., Ulrich, R.G., Conraths, F.J., Schares, G., 2012. Analysis of clonal type-specific antibody reactions in *Toxoplasma gondii* seropositive humans from Germany by peptide-microarray. PLoS One 7 (3), e34212.

Marcolino, P.T., Silva, D.A.O., Camargo, M.E., Mineo, J.R., 2000. Molecular markers in acute and chronic phases of human Toxoplasmosis: Determination of immunoglobulin G avidity by Western blotting. Clin. Diag. Lab. Immunol. 7, 384–389.

Martino, R., Bretagne, S., Einsele, H., *et al.*, 2000. Toxoplasmosis after hematopoietic stem cell transplantation. Clin. Infect. Dis. 31, 1188–1195.

Martino, R., Bretagne, S., Einsele, H., Maertens, J., Ullmann, A.J., Parody, R., Schumacher, U., Pautas, C., Theunissen, K., Schindel, C., Munoz, C., Margall, N., 2005. Cordonnier C and the Infectious Disease Working Party of the European Group for Blood and Marrow Transplantation. Early detection of *Toxoplasma* infection by molecular monitoring of *T. gondii* in peripheral blood samples after allogenic stem cell transplantation. Clin. Infect. Dis. 40, 67–78.

Martins-Duarte, E.S., Souza, W.D., Vommaro, R.C., 2012. Toxoplasma gondii: The effect of Fluconazole combined with Sulfadiazine and Pyrimethamine against acute toxoplasmosis in murine model. Exp. Parasitol.

Masur, H., Jones, T.C., Lempert, J.A., *et al.*, 1978. Outbreak of toxoplasmosis in a family and documentation of acquired retinochoroiditis. Am. J. Med. 64, 396–402.

Matsuo, Y., Takeishi, S., Miyamoto, T., *et al.*, 2007. Toxoplasmosis encephalitis following severe graft-vs.-host disease after allogeneic hematopoietic stem cell transplantation: 17 yr experience in Fukuoka BMT group. Eur. J. Haematol. 79, 317–321.

McAuley, J., Boyer, K.M., Patel, D., Mets, M., Swisher, C., Roizen, N., Wolters, C., Stein, L., Stein, M., Schey, W., 1994. Early and longitudinal evaluations of treated infants and children and untreated historical patients with congenital toxoplasmosis: The Chicago Collaborative Treatment Trial. Clinical Infectious Diseases 18, 38–72.

McCabe, R.E., Brooks, R.G., Dorfman, R.F., Remington, J.S., 1987. Clinical spectrum in 107 cases of toxoplasmic lymphadenopathy. Rev. Infect. Dis. 9, 754–774.

McLeod, R., Berry, P.F., Marshall Jr., W.H., *et al.*, 1979. Toxoplasmosis presenting as brain abscesses. Diagnosis by computerized tomography and cytology of aspirated purulent material. Am. J. Med. 67 (4), 711–714.

McLeod, R., Beem, M.O., Estes, R.G., 1985. Lymphocyte anergy specific to *Toxoplasma gondii* antigens in a baby with congenital toxoplasmosis. J. Clin. Lab. Immunol. 17, 149–153.

McLeod, R., Mack, D.G., Boyer, K., *et al.*, 1990. Phenotypes and functions of lymphocytes in congenital toxoplasmosis. J. Lab. Clin. Med. 116 (5), 623–635.

McLeod, R., Mack, D., Foss, R., *et al.*, 1992. Levels of pyrimethamine in sera and cerebrospinal and ventricular fluids from infants treated for congenital toxoplasmosis. Antimicrob. Agents Chemother. 36, 1040–1048.

McLeod, R., Boyer, K., Roizen, N., Stein, L., Swisher, C., Holfels, E., Hopkins, J., Mack, D., Karrison, T., Patel, D., Pfiffner, L., Remington, J., Withers, S., Meyers, S., Aitchison, V., Mets, M., Rabiah, P., Meier, P., 2000. The child with congenital toxoplasmosis. Curr. Clin. Top Infect. Dis. 20, 189–208.

McLeod, R., Boyer, K., Karrison, T., Kasza, K., Swisher, C., Roizen, N., Jalbrzikowski, J., Remington, J., Heydemann, P., Noble, A.G., Mets, M., Holfels, E., Withers, S., Latkany, P., Meier, P., Toxoplasmosis Study Group, 2006a. Outcome of treatment for congenital toxoplasmosis, 1981-2004: the National Collaborative Chicago-Based, Congenital Toxoplasmosis Study. Clin. Infect. Dis. 42, 1383–1394. Epub 2006 Apr 11.

McLeod, R., Khan, A.R., Noble, G.A., *et al.*, 2006b. Severe sulfadiazine hypersensitivity in a child with reactivated congenital toxoplasmic chorioretinitis. Pediatr. Infect. Dis. J. 25 (3), 270–272.

McLeod, R., Kieffer, F., Sautter, M., Hosten, T., Pelloux, H., 2009. Why prevent, diagnose and treat congenital toxoplasmosis? Mem. Inst. Oswaldo Cruz 104, 320–344.

McLeod, R., Boyer, K.M., Lee, D., Mui, E., Wroblewski, K., Karrison, T., Noble, A.G., Withers, S., Swisher, C.N., Heydemann, P.T., Sautter, M., Babiarz, J., Rabiah, P., Meier, P., Grigg, M.E., 2012. Toxoplasmosis Study Group. Prematurity and severity are associated with Toxoplasma gondii alleles (NCCCTS, 1981–2009). Clin. Infect. Dis. 54 (11), 1595–1605.

McLeod, R., Lee, D., Boyer, K. (2013a in press). Toxoplasmosis in the foetus and newborn infant.

McLeod, R., Lee, D., Boyer, K. (2013b in press). Diagnosis of congenital toxoplasmosis: a practical procedural atlas.

Mele, A., Paterson, P.J., Prentice, H.G., Leoni, P., Kibbler, C.C., 2002. Toxoplasmosis in bone marrow transplantation: a report of two cases and systematic review of the litreature. Bone Marrow Transplant 29, 691–698.

Meneceur, P., Bouldouyre, M.A., Aubert, D., Villena, I., Menotti, J., Sauvage, V., Garin, J.F., Derouin, F., 2008. *In vitro* susceptibility of various genotypic strains of *Toxoplasma gondii* in pyrimethamine, sulfadiazine, and atovaquone. Antimicrob. Agents Chemother. 52 (4), 1269–1277.

Mennechet, F.J., Kasper, L.H., Rachinel, N., Minns, L.A., Luangsay, S., Vandewalle, A., Buzoni-Gatel, D., 2004. Intestinal intraepithelial lymphocytes prevent pathogen-driven inflammation and regulate the Smad/T-bet pathway of lamina propria CD4+ T cells. Eur. J. Immunol. 34 (4), 1059–1067.

Meroni, V., Genco, F., Tinelli, C., *et al.*, 2009. Spiramycin treatment of *Toxoplasma gondii* infection in pregnant women impairs the production and the avidity maturation of *T. gondii*-specific immunoglobulin G antibodies. Clin. Vaccine Immunol. 16 (10), 1517–1520.

Mets, M., Holfels, E., Boyer, K.M., *et al.*, 1996. Eye manifestations of congenital toxoplasmosis. Am. J. Ophthalmol. 122, 309–324.

Miller, M.A., Gardner, J.A., Kreuder, C., *et al.*, 2002. Coastal freshwater runoff is a risk factor for *Toxoplasmosis gondii* infection of southern sea otters (*Enhydra lutris nereis*). Int. J. Parasitol. 32, 997–1006.

Minot, S., Melo, M.B., Li, F., Lu, D., Niedelman, W., Levine, S.S., Saeij, J.P., 2012. Admixture and recombination among *Toxoplasma gondii* lineages explain global genome diversity. Proc. Natl. Acad. Sci. U. S. A. 109 (33), 13458–13463.

Mitchell, C.D., Erlich, S.S., Mastrucci, M.T., Hutto, S.C., Parks, W.P., Scott, G.B., 1990. Congential toxoplasmosis occurring in infants perinatally infected with human immunodeficiency virus 1. Pediatric Infectious Diseases Journal 9, 512–518.

Mitra, R., Sapolsky, R.M., Vyas, A., 2012. *Toxoplasma gondii* infection induces dendritic retraction in basolateral amygdala accompanied by reduced corticosterone secretion. Dis. Model Mech.

Moncada, P.A., Montoya, J.G., 2012. Toxoplasmosis in the foetus and newborn: an update on prevalence, diagnosis and treatment. Expert Rev. Anti. Infect. Ther. 10 (7), 815–828.

Montoya, J.G., Jordan, R., Lingamneni, S., *et al.*, 1997. Toxoplasmic myocarditis and polymyositis in patients with acute acquired toxoplasmosis diagnosed during life. Clin. Infect. Dis. 24, 676–683.

Montoya, J.G., Parmley, S., Liesenfeld, O., Jaffe, O., Remington, J.S., 1999. Use of the poluymerase chain reaction for diagnosis of ocular toxoplasmosis. Ophthalmology 106, 1554–1563.

Montoya, J.G., Giraldo, L.F., Efron, B., *et al.*, 2001. Infectious complications among 620 consecutive heart transplant patients at Stanford University Medical Centre. Clin. Infect. Dis. 33, 629–640.

Montoya, J.G., Liesenfeld, O., Kinney, S., Press, C., Remington, J.S., 2002. VIDAS test for avidity of *Toxoplasma*-specific immunoglobulin G for confirmatory testing of pregnant women. J. Clin. Microbiol. 40, 2504–2508.

Montoya, J.G., 2002. Laboratory diagnosis of Toxoplasma gondii infection and toxoplasmosis. J. Infect. Dis. 2002 Feb. 15 185 (Suppl. 1), S73–S82.

Montoya, J.G., Remington, J.S., 2008. Management of *Toxoplasma gondii* infection during pregnancy. Clin. Infect. Dis. 47 (4), 554–566.

Montoya, J.G., Boothroyd, J.C., Kovacs, J.A., 2010. *Toxoplasma gondii.* Mandell, Douglas, and Bennett's Principles and Practice of Infectious Diseases. Jama: the Journal of the American Medical Association 274 (8), 3495–3526.

Morelle, C., Varlet-Marie, E., Brenier-Pinchart, M.P., Cassaing, S., Pelloux, H., Bastien, P., Sterkers, Y., 2012. Comparative assessment of a commercial kit and two laboratory-developed PCR assays for molecular diagnosis of congenital toxoplasmosis. J. Clin. Microbiol. 50 (12), 3977–3982.

Morin, L., Lobry, J.R., Peyron, F., Wallon, M., 2012. Seasonal variations in acute toxoplasmosis in pregnant women in the Rhône-Alpes region (France). Clin. Microbiol. Infect. 18 (10), E401–E403.

Morris Jr., J.G., Greenspan, A., Howell, K., Gargano, L.M., Mitchell, J., Jones, J.L., Potter, M., Isakov, A., Woods, C., Hughes, J.M., 2012. Southeastern Centre for Emerging Biologic Threats tabletop exercise: foodborne toxoplasmosis outbreak on college campuses. Biosecur. Bioterror. 10 (1), 89–97.

Mosti, M., Pinto, B., Giromella, A., Fabiani, S., Cristofani, R., Panichi, M., Bruschi, F., 2012. A 4-year evaluation of toxoplasmosis seroprevalence in the general population and in women of reproductive age in central Italy. Epidemiol. Infect. 11, 1–4.

Mui, E.J., Schiehser, G.A., Milhous, W.K., Hsu, H., Roberts, C.W., Kirisits, M., Muench, S., Rice, D., Dubey, J.P., Fowble, J.W., Rathod, P.K., Queener, S.F., Liu, S.R., Jacobus, D.P., McLeod, R., 2008. Novel triazine JPC-2067-B inhibits *Toxoplasma gondii in vitro* and *in vivo*. PLoS Negl. Trop. Dis. 2 (3), e190.

Munro, J.B., Silva, J.C., 2012. Ribonucleotide reductase as a target to control apicomplexan diseases. Curr. Issues Mol. Biol. 14 (1), 9–26.

Murat, J.B., L'Ollivier, C., Fricker Hidalgo, H., Franck, J., Pelloux, H., Piarroux, R., 2012. Evaluation of the new Elecsys Toxo IgG avidity assay for toxoplasmosis and new insights into the interpretation of avidity results. Clin. Vaccine Immunol. 19 (11), 1838–1843.

Naessens, A., Jenum, P.A., Pollak, A., *et al.*, 1999. Diagnosis of congenital toxoplasmosis in the neonatal period: A multicentre evaluation. J. Pediatr. 135, 714–719.

Naot, Y., Remington, J.S., 1980. An enzyme-linked immunosorbent assay for detection of IgM antibodies to *Toxoplasma gondii* use for diagnosis of acute acquired toxoplasmosis. J. Infect. Dis. 142, 757–766.

Naot, Y., Desmonts, G., Remington, J.S., 1981a. IgM enzyme-linked immunosorbent assay test for the diagnosis of congenital *Toxoplasma* infection. J. Pediatr. 98, 32–36.

Naot, Y., Barnett, E.V., Remington, J.S., 1981b. Method for avoiding false-positive results occurring in immunoglobulin M enzyme-linked immunosorbent assays due to presence of both rheumatoid factor and antinuclear antibodes. J. Clin. Microbiol. 14, 73–78.

Natella, V., Cozzolino, I., Sosa Fernandez, L.V., Vigliar, E., 2012. Lymph nodes fine needle cytology in the diagnosis of infectious diseases: clinical settings. Infez. Med. 20 (Suppl. 3), 12–15.

Neto, E.C., Rubin, R., Schulte, J., Giugliani, R., 2004. Newborn Screening for Congenital Infectious Diseases. Emer. Infect. Dis. 10, 1069–1073.

Niebuhr, D.W., Millikan, A.M., Cowan, D.N., *et al.*, 2008. Selected infectious agents and risk of schizophrenia among USA military personnel. Am. J. Psychiatry 165, 99–106.

Nielsen, H.V., Schmidt, D.R., Petersen, E., 2005. Diagnosis of congenital toxoplasmosis by two-dimensional immunoblot differentiation of mother and child IgG-profiles. J. Clin. Microbiol. 2005 (43), 711–715.

Noble, A.G., Latkany, P., Kusmierczyk, *et al.*, 2010. Chorioretinal lesions in mothers of children with congenital toxoplasmosis in the National Collaborative Chicago-based, Congenital Toxoplasmosis Study. Sci. Med. 20 (1), 20–26.

Norrby, R., Eilard, T., Svedhem, A., Lycke, E., 1975. Treatment of toxoplasmosis with trimethoprim-sulphamethoxazole. Scand. J. Infect. Dis. 7, 72–75.

O'Connor, G.R., 1974. Manifestations and management of ocular toxoplasmosis. Bull. NY Acad. Med. 50, 192–210.

Odberg-Ferragut, C., Soete, M., Engels, A., Samyn, B., Loyens, A., van Beeumen, J., Camus, D., Dubremetz, J.F., 1996. Molecular cloning of the *T. gondii* SAG4 gene encoding an 18 kDa bradyzoite specific surface protein. Mol. Biochem. Parasitol. 82, 237–244.

Okusaga, O., Postolache, T.T., 2012. *Toxoplasma gondii*, the immune system, and suicidal behaviour. In: Dwivdei, Y. (Ed.), The Neurological Basis of Suicide. CRC Press, Boca Raton (FL) Chapter 19. Frontiers in Neuroscience.

Olariu, T., Remington, S., McLeod, R., Alam, A., Montoya, J., 2011. Severe Congenital Toxoplasmosis in the United State: Clinical and Serologic Findings in Untreated Infants. Pediatr. Infect. Dis. J. 30 (12), 1056–1061.

Oréfice, J.L., Costa, R.A., Scott, I.U., Calucci, D., Oréfice, F.; on behalf of the Grupo Mineiro de Pesquisa em Doenças Oculares Inflamatórias (MINAS). (2012). Spectral optical coherence tomography findings in patients with ocular toxoplasmosis and active satellite lesions (MINAS Report 1). Acta Ophthalmol.

Orr, K.E., Gould, F.K., Short, G., Dark, J.H., Hilton, C.J., Corris, P.A., Freeman, R., 1994. Outcome of *T. gondii* mismatches in heart transplant recipients over a period of 8 years. J. Infect. 29, 249–253.

Osthoff, M., Chew, E., Bajel, A., Kelsey, G., Panek-Hudson, Y., Mason, K., Szer, J., Ritchie, D., Slavin, M., 2012. Disseminated toxoplasmosis after allogeneic stem cell transplantation in a seronegative recipient. Transpl. Infect. Dis.

Palma, S., Reyes, H., Guzman, L., *et al.*, 1984. Dermatomyositis and toxoplasmosis. Rev. Med. Chil. 111, 164–167.

Papoz, L., Simondon, F., Saurin, W., Sarmini, H., 1986. A simple model relevant to toxoplasmosis applied to epidemiologic results in France. Am. J. Epidemiol. 154–161.

Parmley, S.F., Goebel, F.D., Remington, J.S., 1992. Detection of *Toxoplasma gondii* in cerebrospinal fluid from AIDS patients by polymerase chain reaction. J. Clin. Microbiol. 30, 3000–3002.

Parmley, S.F., Yang, S.M., Harth, G., Sibley, L.D., Sucharczuk, A., Remington, J.S., 1994. Molecular characterization of a 65-kDa *T. gondii* antigen expressed abundantly in the matrix of tissue cysts. Mol. Biochem. Parasitol. 66, 283–296.

Paspalaki, P.K., Mihailidou, E.P., Bitsori, M., *et al.*, 2001. Polymyositis and myocarditis associated with acquired toxoplasmosis in an immunocompetent girl. BMC Musculoskelet Disord. 2, 8.

Patel, D.V., Holfels, E.M., Vogel, N.P., *et al.*, 1996. Resolution of intracranial calcifications in infants with treated congenital toxoplasmosis. Radiology 199 (2), 433–440. 1996.

Paul, M., Petersen, E., Pawlowski, Z.S., Szczapa, J., 2000. Neonatal screening for congenital toxoplasmosis in the Poznan region of Poland by analysis of *Toxoplasma gondii*-specific IgM antibodies eluted from filter paper blood spots. Pediatr. Infect. Dis. J. 19, 30–36.

Pearce, B.D., Kruszon-Moran, D., Jones, J.L., 2012. The relationship between *Toxoplasma gondii* infection and mood disorders in the third National Health and Nutrition Survey. Biol. Psychiatry 72 (4), 290–295.

Pedersen, M.G., Mortensen, P.B., Norgaard-Pedersen, B., Postolache, T.T., 2012. *Toxoplasma gondii* infection and self-directed violence in mothers. Arch. Gen. Psychiatry 69 (11), 1123–1130.

Peixoto, L., Chen, F., Harb, O.S., Davis, P.H., Beiting, D.P., Brownback, C.S., Ouloguem, D., Roos, D.S., 2010. Integrative genomic approaches highlight a family of parasite-specific kinases that regulate host responses. Cell. Host Microbe 8 (2), 208–218.

Pelloux, H., Brun, E., Vernet, G., Marcillat, S., Jolivet, M., Guergour, D., Fricker-Hidalgo, H., Goullier-Fleuret, A., Ambroise-Thomas, P., 1998. Determination of anti-*T. gondii* immunoglobulin G avidity: Adaption to the Vidas system (bioMérieux). Diag. Microbiol. Infect. Dis. 32, 69–73.

Perkins, E.S., Smith, C.H., Schofield, P.B., 1956. Treatment of uveitis with pyrimethamine (Daraprim). Br. J. Ophthalmol. 40, 577–586.

Petersen, E., Dubey, J.P., 2001. Biology of Toxoplasmosis. In: Joynson, D.H.M., Wreghitt, T.G. (Eds.), Clinical Toxoplasmosis: Prevention and Management. Cambridge University Press, Cambridge, pp. 1–42.

Petersen, E., Borobio, M.V., Guy, E., Liesenfeld, O., Meroni, V., Naessens, A., Spranzi, E., Thulliez, P., 2005. European multicentre study of the LIAISON[a] automated diagnostic system for determination of specific IgG, IgM and IgG-avidity index in toxoplasmosis. J. Clin. Microbiol. 43, 1570–1574.

Petersen, E., Kijlstra, A., Stanford, M., 2012 Apr. Epidemiology of ocular toxoplasmosis. Ocul Immunol Inflamm 20 (2), 68–75. http://dx.doi.org/10.3109/09273948.2012.661115.

Petithory, J.C., Reiter-Owona, I., Berthelot, F., Milgram, M., De Loye, J., Petersen, E., 1996. Performance of European Laboratories testing serum samples for *Toxoplasma gondii*. Eur. J. Clin. Microbiol. Infect. Dis. 15, 45–49.

Peyron, F., Ateba, A.B., Wallon, M., *et al.*, 2003. Congenital toxoplasmosis in twins: a report of fourteen consecutive cases and a comparison with published data. Pediatr. Infect. Dis. J. 22, 695–701.

Peyron, F., Garweg, J.G., Wallon, M., *et al.*, 2011. Long-term Impact of Treated Congenital Toxoplasmosis on Quality of Life and Visual Performance. Pediatr. Infect. Dis. J. 30 (7), 597–600.

Phan, L., Kasza, K., Jalbrzikowski, J., *et al.*, 2008a. Longitudinal study of new eye lesions in treated congenital toxoplasmosis. Ophthalmology 115.

Phan, L., Kasza, K., Jalbrzikowski, J., *et al.*, 2008b. Longitudinal study of new eye lesions in children with toxoplasmosis who were not treated during the first year of life. Am. J. Ophthalmol. 146.

Pietkiewicz, H., Hiszczyńska-Sawicka, E., Kur, J., Petersen, E., Nielsen, H.V., Stankiewicz, M., Andrzejewska, I., Myjak, P., 2004. Usefulness of *T. gondii* recombinant antigens in serodiagnosis of human toxoplasmosis. J. Clin. Microbiol. 42, 1779–1781.

Pinkerton, H., Weinman, D., 1940. Toxoplasma infection in man. Arch. Pathol. 30, 374–392.

Pinon, J.M., Chemla, C., Villena, I., Foudrinier, F., Aubert, D., Puygauthier-Toubas, D., Leroux, B., Dupouy, D., Quereux, C., Talmud, M., Trenque, T., Potron, G., Pluot, M., Remy, G., Bonhomme, A., 1996. Early neonatal diagnosis of congenital toxoplasmosis: value of comparative enzyme-linked immunofiltration assay immunological profiles and anti-*T. gondii* immunoglobulin M (IgM) or IgA immunocapture and implications for postnatal therapeutic strategies. J. Clin. Microbiol. 34, 579–583.

Pinon, J.M., Dumon, H., Chemla, C., Franck, J., Petersen, E., Lebech, M., Zufferey, J., Bessieres, M.H., Marty, P., Holliman, R., Johnson, J., Luyasu, V., Lecolier, B., Guy, E., Joynson, D.H., Decoster, A., Enders, G., Pelloux, H., Candolfi, E., 2001. Strategy for diagnosis of congenital toxoplasmosis: evaluation of methods comparing mothers and newborns and standard methods for postnatal detection of immunoglobulin G, M, and A antibodies. J. Clin. Microbiol. 39, 2267–2271.

Pinto, B., Castagna, B., Mattei, R., Bruzzi, R., Chiumiento, L., Cristofani, R., Buffolano, W., Bruschi, F., 2012. Seroprevalence for toxoplasmosis in individuals living in north west Tuscany: access to Toxo-test in central Italy 31 (6), 1151–1156.

Pollock, J.L., 1979. Toxoplasmosis appearing to be dermatomyositis. Arch. Dermatol. 115, 736–737.

Post, M.J., Chan, J.C., Hensley, G.T., *et al.*, 1983. *Toxoplasma* encephalitis in Haitian adults with acquired immunodeficiency syndrome: a clinical-pathologic-CT correlation. AJR Am. J. Roentgenol. 140, 861–868.

Pouletty, P., Kadouche, J., Garcia-Gonzalez, M., Mihaesco, E., Desmonts, G., Thulliez, P., Thoannes, H., Pinon, J.M., 1985. An anti-human chain monoclonal antibody: use for detection of IgM antibodies to *T. gondii* by reverse immunosorbent assay. J. Immunol. Meth. 76, 289–298.

Prado, S.P., Pacheco, V.C., Noemi, I.H., *et al.*, 1978. [Toxoplasma pericarditis and myocarditis]. Rev. Chil. Pediatr. 49, 179–185.

Press, C., Montoya, J.G., Remington, J.S., 2005. Use of a single serum sample for diagnosis of acute toxoplasmosis in pregnant women and other adults. J. Clin. Microbiol. 43, 3481–3483.

Prince, H.E., Wilson, M., 2001. Simplified assay for measuring *T. gondii* immunoglobulin G avidity. Clin. Diag. Lab. Immunol. 8, 904–908.

Prusa, A.R., Hayde, M., Pollak, A., Herkner, K.R., Kasper, D.C., 2012. Evaluation of the liaison automated testing system for diagnosis of congenital toxoplasmosis. Clin. Vaccine Immunol. 19 (11), 1859–1863.

Rabaud, C., May, T., Lucet, J.C., Leport, C., Ambroise-Thomas, P., Canton, P., 1996. Pulmonary toxoplasmosis in patients infected with human immunodeficiency virus: a French National Survey. Clin. Infect. Dis. 23, 1249–1254.

Raymond, J., Poissonnier, M.H., Thulliez, P.H., *et al.*, 1990. Presence of gamma interferon in human acute and congenital toxoplasmosis. J. Clin. Microbiol. 28 (6), 1434–1437.

Reischl, U., Bretagne, S., Krüger, D., Ernault, P., Costa, J.M., 2003. Comparison of two DNA targets for the diagnosis of Toxoplasmosis by real-time PCR using fluorescence resonance energy transfer hybridization probes. BMC Infect. Dis. 3, 7.

Reiter-Owona, I., Petersen, E., Joynson, D., Aspöck, H., Dardé, M.L., Disko, R., Dreazen, O., Dumon, H., Grillo, R., Gross, U., Hayde, M., Holliman, R., Ho-Yen, D.O., Janitschke, K., Jenum, P., Naser, K., Olszewski, M., Thulliez, P., Seitz, H.M., 1999. Looking back on half a century of the Sabin-Feldman dye-test: its past and present role in the serodiagnosis of toxoplasmosis. Results of an European multicentre study, 77. W.H.O. Bull 929–935.

Remington, J.S., Barnett, C.G., Meikel, M., *et al.*, 1962. Toxoplasmosis and infectious mononucleosis. Arch. Intern. Med. 110, 744–753.

Remington, J.S., Miller, M.J., 1966. 19S and 7S anti-toxoplasma antibodies in diagnosis of acute congenital and acquired toxoplasmosis. Proc. Soc. Exp. Biol. Med. 121, 357–363.

Remington, J.S., Miller, M.J., Brownlee, I., 1968. IgM antibodies in acute toxoplasmosis. I. Diagnostic significance in congenital cases and a method for their rapid demonstration. Pediatr. 41, 1082–1091.

Remington, J.S., 1974. Toxoplasmosis in the adult. Bull. N. Y. Acad. Med. 50, 211–227.

Remington, J.S., Desmonts, G., 1976. Congenital Toxoplasmosis. In: Remington, J.S., Klein, J.O. (Eds.), Infectious diseases of the foetus and newborn infant, first ed. WB Saunders, Philadelphia, PA, pp. 297–299.

Remington, J.S., Araujo, F.G., Desmonts, G., 1985. Recognition of different *Toxoplasma gondii* antigens by IgM and IgG antibodies in mothers and their congenitally infected newborns. J. Infect. Dis. 152, 1020–1024.

Remington, J.B., Wilder Antunes, F., *et al.*, 1991. Clindamycin for toxoplasmosis encephalitis in AIDS (letter). Lancet 338, 1142–1143.

Remington, J.S., McLeod, R., Thulliez, P., 2001. Toxoplasmosis. In: Remington, J.S., Klein, J. (Eds.), Infectious Diseases of the Foetus and Newborn Infant. WB Saunders, Philadelphia, pp. 205–346.

Remington, J.S., Thulliez, P., Montoya, J.G., 2004. Recent developments for diagnosis of toxoplasmosis. J. Clin. Microbiol. 42 (3), 941–945.

Remington, J.S., McLeod, R., Thulliez, P., Desmonts, G., 2011. Toxoplasmosis. In: Remington, J., Klein, J. (Eds.), Infectious Diseases of the Foetus and Newborn Infant, seventh ed. WB Saunders, Philadelphia, PA.

Renier, D., Sainte-Rose, C., Pierre-Kahn, A., Hirsch, J.F., 1988. Prenatal hydrocephalus: outcome and prognosis. Childs Nerv. Syst. 4 (4), 213–222.

Renoult, E., Georges, E., Biava, M.F., Hulin, L., Frimat, D., Hestin, D., Kessler, M., 1997. Toxoplasmosis in kidney transplant recipients: report of six cases and review. Clin. Infect. Dis. 24, 625–634.

Rigsby, P., Rijpkema, S., Guy, E.C., Francis, J., Das, R.G., 2004. Evaluation of a candidate international standard preparation for human anti-*Toxoplasma* immunoglobulin G. J. Clin. Microbiol. 42, 5133–5138.

Rilling, V., Dietz, K., Krczal, D., Knotek, F., Enders, G., 2003. Evaluation of a commercial IgG/IgM Western blot assay for early postnatal diagnosis of congenital toxoplasmosis. Eur. J. Clin. Microbiol. Infect. Dis. 22, 174–180.

Robert, A., Luyasu, V., Zuffrey, J., Hedman, K., Petersen, E., European Network on Congenital Toxoplasmosis, 2001. Potential of the specific markers in the early diagnosis of Toxoplasma-infection: A multicentre study using combination of isotype IgG, IgM, IgA and IgE with values of avidity assay. Eur. J. Clin. Microbiol. Infect. Dis. 20, 467–474.

Robert-Gangneux, F., Commerce, V., Tourte-Schaefer, C., Dupouy-Camet, J., 1999. Performance of a Western blot assay to compare mother and newborn anti-*Toxoplasma* antibodies for the early neonatal diagnosis of congenital toxoplasmosis. Eur. J. Clin. Microbiol. Infect. Dis. 18, 648–654.

Roizen, N., Swisher, C.N., Stein, M.A., *et al.*, 1995. Neurologic and developmental outcome in treated congenital toxoplasmosis. Pediatrics 95, 11–20.

Roizen, N., Kasza, K., Karrison, T., *et al.*, 2006. Impact of visual impairment on measures of cognitive function for children with congenital toxoplasmosis: implications for compensatory intervention strategies. Pediatrics 118, e379–e390.

Rollins-Raval, M.A., Marafioti, T., Swerdlow, S.H., Roth, C.G., 2012. The number and growth pattern of plasmacytoid dendritic cells vary in different types of reactive lymph nodes: an immunohistochemical study. Hum. Pathol.

Romand, S., Pudney, M., Derouin, F., 1993. *In vitro* and *in vivo* activities of the hydroxynaphthoquinone atovaquone alone or combined with pyrimethamine, sulfadiazine, clarithromycin, or minocycline against *Toxoplasma gondii*. Antimicrob. Agents Chemother. 37 (11), 2371–2378.

Romand, S., Chosson, M., Franck, J., *et al.*, 2004. Usefulness of quantitative polymerase chain reaction in amniotic fluid as early prognostic marker of foetal infection with *Toxoplasma gondii*. Am. J. Obstet. Gynecol. 190, 797–802.

Rossi, C.L., 1998. A simple, rapid enzyme-linked immunosorbent assay for evaluating immunoglobulin G antibody avidity in Toxoplasmosis. Diag. Microbiol. Infect. Dis. 30, 25–30.

Roue, R., Debord, T., Denamur, E., Ferry, M., Dormont, D., Barre-Sinoussi, F., Rouzioux, C., 1984. Diagnosis of Toxoplasma encephalitis in absence of neeurological signs by early computerised tomography scanning in patients with AIDS. Lancet ii. 1472.

Roux, C., Desmont, G., Mulliez, N., *et al.*, 1976. [Toxoplasmosis and pregnancy. Evaluation of 2 years of prevention of congenital toxoplasmosis in the maternity ward of Hôpital Saint-Antoine (1973–1974)]. J. Gynecol. Obstet. Biol. Reprod. (Paris) 5 (2), 249–264.

Ruskin, J., Remington, J.S., 1976. Toxoplasmosis in the compromised host. Ann. Intern. Med. 84, 193–199.

Ryning, F.W., McLeod, R., Maddox, J.C., *et al.*, 1979. Probable transmission of *Toxoplasma gondii* by organ transplantation. Ann. Intern. Med. 90, 47–49.

Saadatnia, G., Mohamed, Z., Ghaffarifar, F., Osman, E., Moghadam, Z.K., Noordin, R., 2012. *Toxoplasma gondii* excretory secretory antigenic proteins of diagnostic potential. APMIS 120 (1), 47–55.

Sabin, A.B., 1941. Toxoplasmic encephalitis in children. J. Am. Med. Assoc. 116, 801–807.

Sabin, A.B., Feldman, H.A., 1948. Dyes as microchemical indicators of a new immunity phenomenon affecting a protozoan parasite (Toxoplasma). Science 108, 660–663.

Samuel, B.U., Hearn, B., Mack, D., Wender, P., Rothbard, J., Kirisits, M.J., Mui, E., Wernimont, S., Roberts, C.W., Muench, S.P., Rice, D.W., Prigge, S.T., Law, A.B., McLeod, R., 2003. Delivery of antimicrobials into parasites. Proc. Natl. Acd. Sci. USA 100 (24), 14281–14286.

dos Santos, T.R., Nunes, C.M., Luvizotto, M.C., *et al.*, 2010. Detection of *Toxoplasma gondii* oocysts in environmental samples from public schools. Vet. Parasitol. 171 (1-2), 53–57.

Saxon, S.A., Knight, W., Reynolds, D.W., Stagno, S., Alford, C.A., 1973. Intellectual deficits in children born with subclinical congenital toxoplasmosis: a preliminary report. J. Pediatr. 85 (5), 792–797.

Schaefer, L.E., Dyke, J.W., Meglio, F.D., Murray, P.R., Crafts, W., Niles, A.C., 1989. Evaluation of microparticle enzyme immunoassays for immunoglobulins G and M to rubella virus and *Toxoplasma gondii* on the Abbott IMx automated analyzer. J. Clin. Microbiol. 27, 2410–2413.

Schreiner, M., Liesenfeld, O., 2009. Small intestinal inflammation following oral infection with *Toxoplasma gondii* does not occur exclusively in C57BL/6 mice: review of 70 reports from the litreature. Mem. Inst. Oswaldo Cruz 104 (2), 221–233.

Sensini, A., Pascoli, S., Marchetti, D., Castronari, A., Marangi, M., Sbaraglia, G., Cimmino, C., Favero, A., Castelletto, M., Mottola, A., 1996. IgG avidity in the serodiagnosis of acute *T. gondii* infection: a multicentre study. Clin. Microbiol. Infect. 2, 25–29.

Severance, E.G., Kannan, G., Gressitt, K.L., Xiao, J., Alaedini, A., Pletnikov, M.V., Yolken, R.H., 2012. Anti-gluten immune response following *Toxoplasma gondii* infection in mice 7 (11) E50991.

Shapiro, K., Conrad, P.A., Mazet, J.A., *et al.*, 2010. Effect of estuarine wetland degradation on transport of *Toxoplasma gondii* surrogates from land to sea. Appl. Environ. Microbiol. 76 (20), 6821–6828.

Sibley, L.D., Boothroyd, J.C., 1992. Virulent strains of *Toxoplasma gondii* comprise a single clonal lineage. Nature 359, 82–85.

Siegel, S.E., Lunde, M.N., Gelderman, A.H., *et al.*, 1971. Transmission of toxoplasmosis by leukocyte transfusion. Blood 37, 388–394.

Siim, J.C., 1951. Acquired toxoplasmosis: report of seven cases with strongly positive serologic reactions. J. Am. Med. Assoc. 147, 1641–1645.

Silveira, C., Belfort Jr., R., Burnier Jr., M., *et al.*, 1988. Acquired toxoplasmic infection as the cause of toxoplasmic retinochoroiditis in families. Am. J. Ophthalmol. 106, 362–364.

Silveira, C., Belfort Jr., R., Muccioli, C., Abreu, M.T., Martins, M.C., Victora, C., Nussenblatt, R.B., Holland, G.N., 2001. A follow-up study of *Toxoplasma gondii* infection in southern Brazil. Am. J. Ophthalmol. 131, 351–354.

Silveira, C., Belfort Jr., R., Muccioli, C., Holland, G.N., Victora, C.G., Horta, B.L., Yu, F., Nussenblatt, R.B., 2002. The effect of long-term intermittent trimethoprim/sulfamethoxazole treatment on recurrences of toxoplasmic retinochoroiditis. Am. J. Ophthalmol. 134 (1), 41–46.

Sing, A., Leitritz, L., Roggenkamp, A., *et al.*, 1999. Pulmonary toxoplasmosis in bone marrow transplant recipients: report of two cases and review. Clin. Infect. Dis. 29, 429–433.

Singer, M.A., Hagler, W.S., Grossniklaus, H.E., 1993. *Toxoplasma gondii* retinochoroiditis after liver transplantation. Retina 13, 40–45.

Singh, U., Brewer, J.L., Boothroyd, J.C., 2002. Genetic analysis of tachyzoite to bradyzoite differentiation mutants in *T. gondii* reveals a hierarchy of gene induction. Mol. Microbiol. 44, 721–733.

Slavin, M.A., Meyers, J.D., Remington, J.S., *et al.*, 1994. *Toxoplasma gondii* infection in marrow transplant recipients: a 20 year experience. Bone Marrow Transplant 13, 549–557.

Sørensen, T., Spenter, J., Jaliashvili, I., Christiansen, M., Nørgarrd-Pedersen, B., Petersen, E., 2002. An automated time-resolved immunofluometric assay for detection of *T. gondii* -specific IgM and IgA antibodies in filterpaper samples from newborns. Clin. Chemistry 48, 1981–1986.

Stagno, S., Reynolds, D.W., Amos, C.S., *et al.*, 1977. Auditory and visual defects resulting from symptomatic and subclinical congenital cytomegaloviral and *Toxoplasma* infections. Pediatrics, 669–678.

Steen, E., Kåss, E., 1951. A new *Toxoplasma* antigen for complement fixation test. Acta Path. Microbiol. Scand. 28, 36–39.

Stepick-Biek, P., Thulliez, P., Araujo, F.G., Remington, J.S., 1990. IgA antibodies for diagnosis of acute congenital and acquired toxoplasmosis. J. Infect. Dis. 162 (1), 270–273.

Stillwaggon, E., Carrier, C.S., Sautter, M., McLeod, R., 2011. Maternal serologic screening to prevent congenital toxoplasmosis: a decision-analytic economic model. PLoS Negl. Trop. Dis. 5 (9), e1333.

Stommel, E.W., Seguin, R., Thadani, V.M., Schwartzman, J.D., Gilbert, K., Ryan, K.A., Tosteson, T.D., Kasper, L.H., 2001. Cryptogenic epilepsy: an infectious etiology? Epilepsia 42 (3), 436–438.

Strabelli, T.M., Siciliano, R.F., Vidal Campos, S., Bianchi Castelli, J., Bacal, F., Bocchi, E.A., Uip, D.E., 2012. *Toxoplasma gondii* Myocarditis after Adult Heart Transplantation: Successful Prophylaxis with Pyrimethamine. J. Trop. Med. 2012, 853562.

Suzuki, Y., Israelski, D.M., Dannemann, B.R., Stepick-Biek, P., Thulliez, P., Remington, J.S., 1988a. Diagnosis of toxoplasmic encephalitis in patients with acquired immunodeficiency syndrome by using a new serological method. J. Clin. Microbiol. 26, 2541–2543.

Suzuki, Y., Thulliez, P., Desmonts, G., Remington, J.S., 1988b. Antigen(s) responsible for immunoglobulin G responses specific for the acute stage of *Toxoplasma* infection in humans. J. Clin. Microbiol. 26 (5), 901–905.

Swisher, C.N., Boyer, K., McLeod, R., The Toxoplasmosis Study Group, 1994. Congenital toxoplasmosis. Semin. Pediatr. Neurol., 4–25.

Switaj, K., Master, A., Skrzypczak, M., Zaborowski, P., 2005 Mar. Recent trends in molecular diagnostics for Toxoplasma gondii infections. Clin. Microbiol. Infect. 11 (3), 170–176.

SYROCOT – Systemic Review on Congenital Toxoplasmosis Study Group, Thiebaut, R., Leproust, S., Chene, G., Gilbert, R., 2007. Effectiveness of prenatal treatment for congenital toxoplasmosis: a meta-analysis of individual patients' data. Lancet 369, 115–122.

Tabbara, K.F., O'Connor, G.R., 1980. Treatment of ocular toxoplasmosis with clindamycin and sulfadiazine. Opthalmology 87, 129–134.

Tan, T.G., Mui, E., Cong, H., *et al.*, 2010. Identification of *T. gondii* epitopes, adjuvants, and host genetic factors that influence protection of mice and humans. Vaccine 28 (23), 3977–3989.

Teutsch, S.M., Juranek, D.D., Sulzer, A., Dubey, J.P., Sikes, R.K., 1979. Epidemic toxoplasmosis associated with infected cats. N. Engl. J. Med. 300, 695–699.

Thalhammer, O., 1975. Die Toxoplasmose-Untersuchung von Schwangeren und neugeborenen. Wien Klein. Wochenschr. 87, 676–681.

Thalib, L., Gras, L., Romand, S., Prusa, A., Bessieres, M.H., Petersen, E., Gilbert, R.E., 2005 May. Prediction of congenital toxoplasmosis by polymerase chain reaction analysis of amniotic fluid. BJOG 112 (5), 567–574.

Thulliez, P., Remington, J.S., Santoro, F., *et al.*, 1986. A new agglutination test for the diagnosis of acute and chronic *Toxoplasma* infection. Pathol. Biol. (Paris) 34, 173–177.

Thulliez, P., Remington, J.S., Santoro, F., Ovlaque, G., Sharma, S., Desmonts, G., 1989. A new agglutination reaction for the diagnosis of the development stage of acquired toxoplasmosis. Pathol. Biol. Paris 34, 173–177.

Thulliez, P., 1992. Screening programme for congenital toxoplasmosis in France. Scand. J. Infect. Dis. 84 (Suppl.), 43–45.

Thulliez, P., 2001a. Commentary: Efficacy of prenatal treatment for toxoplasmosis: a possibility that cannot be ruled out. Int. J. Epidemiol. 30, 1315–1316.

Thulliez, P., 2001b. Maternal and foetal infection. In: Joynson, D.H.M., Wreghitt, T.G. (Eds.), Toxoplasmosis: A comprehensive clinical guide. Cambridge University Press, pp. 193–213.

Tissot-Dupont, D., Fricker-Hidalgo, H., Brenier-Pinchart, M.P., Bost-Bru, C., Ambroise-Thomas, P., Pelloux, H., 2003. Usefulness of Western blot in serological follow-up of newborns suspected of congenital toxoplasmosis. Eur. J. Clin. Microbiol. Infect. Dis. 22, 122–125.

de-la-Torre, A., González, G., Díaz-Ramirez, J., Gómez-Marín, J.E., 2007. Screening by ophthalmoscopy for Toxoplasma retinochoroiditis in Colombia. Am. J. Ophthalmol. 143, 354–356.

Toporovski, J., Romano, S., Hartmann, S., Benini, W., Chieffi, P.P., 2012. Nephrotic syndrome associated with toxoplasmosis: report of seven cases. Rev. Inst. Med. Trop. Sao Paulo 54 (2), 61–64.

Torres, R., Barr, M., Thorn, M., Gregory, G., Kiely, S., Chanin, E., Carlo, C., Martin, M., Thornton, J., 1993. Randomized trial of dapsone and aerosolized pentamidine for the prophylaxis of *Pneumocystis carinii* pneumonia and toxoplasmic encephalitis. American Journal of Medicine 95, 573–583.

Torrey, E.F., Bartko, J.J., Yolken, R.H., 2012. *Toxoplasma gondii* and other risk factors for schizophrenia: an update. Schizophr. Bull. 38 (3), 642–647.

Townsend, J.J., Wolinsky, J.S., Baringer, J.R., Johnson, P.C., 1975. Acquired toxoplasmosis. A neglected cause of treatable nervous system disease 32 (5), 335–343.

Vallochi, A.L., Muccioli, C., Martins, M.C., Silveira, C., Belfort Jr., R., Rizzo, L.V., 2005. The genotype of *Toxoplasma gondii* strains causing ocular toxoplasmosis in humans in Brazil. Am. J. Ophthalmol. 139, 350–351.

Vasconcelos-Santos, D.V., Machado Azevedo, D.O., Campos, W.R., Oréfice, F., Queiroz-Andrade, G.M., *et al.*, 2009. Congenital toxoplasmosis in southeastern Brazil: results of early ophthalmologic examination of a large cohort of neonates. Ophthalmology 116, 2199–2205.

Vasconcelos-Santos, D.V., Queiroz Andrade, G.M., 2011. Geographic differences in outcomes of congenital toxoplasmosis. Pediatr. Infect. Dis. J. 30 (9), 816–817.

Vasconcelos-Santos, D.V., 2012. Ocular manifestations of systemic disease: toxoplasmosis. Curr. Opin. Ophthalmol. 23 (6), 543–550.

Vaudaux, J.D., Muccioli, C., James, E.R., *et al.*, 2010. Identification of an Atypical Strain of *Toxoplasma gondii* as the Cause of a Waterborne Outbreak of Toxoplasmosis in Santa Isabel do Ivai. Brazil J. Infect. Dis. 202, 1226–1233.

Vaughan, L.B., Wenzel, R.P., 2012. Disseminated toxoplasmosis presenting as septic shock five weeks after renal transplantation. Transpl. Infect. Dis.

Vidal, J.E., Colombo, F.A., de Oliveira, A.C., Focaccia, R., Pereira-Chioccola, V.L., 2004. PCR assay using cerebrospinal fluid for diagnosis of cerebral toxoplasmosis in Brazilian AIDS patients. J. Clin. Microbiol. 42, 4765–4768.

Vietzke, W.M., Geldermann, A.H., Grimley, P.M., *et al.*, 1968. Toxoplasmosis complicating malignancy: experience at the National Cancer Institute. Cancer 21, 816–827.

Villard, O., Breit, L., Cimon, B., Franck, J., Fricker-Hidalgo, H., Godineau, N., Houze, S., Paris, L., Pelloux, H., Villena, I., Candolfi, E., the French National Reference Centre for Toxoplasmosis Network, 2012. Comparison of four commercially available avidity tests for *Toxoplasma*-specific IgG antibodies. Clin. Vaccine Immunol.

Villena, I., Ancelle, T., Delmas, C., *et al.*, 2010. Congenital toxoplasmosis in France in 2007: first results from a national surveillance system. Euro. Surveill. 15 pii-19600.

Vogel, N., Kirisits, M., Michael, E., Bach, H., Hostetter, M., Boyer, K., Simpson, R., Holfels, E., Hopkins, J., Mack, D., Mets, M.B., Swisher, C.N., Patel, D., Roizen, N., Stein, L., Stein, M., Withers, S., Mui, E., Egwuagu, C., Remington, J., Dorfman, R., McLeod, R., 1996. Congenital toxoplasmosis transmitted from an immunologically competent mother infected before conception. Clinical Infectious Diseases 23, 1055–1060.

Villavedra, M., Battistoni, J., Nieto, A., 1999. IgG recognizing 21–24 kDa and 30–33 kDa tachyzoite antigens show maximum avidity maturation during natural and accidental human toxoplasmosis. Rev. Inst. Med. Trop. Sao Paulo 41, 297–303.

Vyas, A., Sapolsky, R., 2010. Manipulation of host behaviour by *Toxoplasma gondii*: what is the minimum a proposed proximate mechanism should explain? Folia Parasitol. (Praha) 57 (2), 88–94.

Wainwright, K.E., Miller, M.A., Kreuder, C., *et al.*, 2007. Chemical inactivation of *Toxoplasma gondii* oocysts in water. J. Parasitol. 93, 925–931.

Wallon, M., Cozon, G., Ecochard, R., Lewin, P., Peyron, F., 2001. Serological rebound in congenital toxoplasmosis: long-term follow-up of 133 children. Eur. J. Pediatr. 160, 534–540.

Wallon, M., Kodjikian, L., Binquet, C., *et al.*, 2004. Long-term ocular prognosis in 327 children with congenital toxoplasmosis. Pediatr. 113, 1567–1572.

Wallon, M., Franck, J., Thulliez, P., *et al.*, 2010. Accuracy of real-time polymerase chain reaction for Toxoplasma gondii in amniotic fluid. Obstet. Gynecol. 115, 727–733.

Wallon, M., Kieffer, F., Binquet, C., Thulliez, P., Garcia-Meric, P., Dureau, P., Franck, J., Peyron, F., Bonnin, A., Villena, I., Bonithon-Kopp, C., Gouyon, J.B., Masson, S., Felin, A., Cornu, C., 2011. Congenital toxoplasmosis: randomized comparison of strategies for retinochoroiditis prevention. Therapie 66 (6), 473–480.

Wallon, M., Peyron, F., Cornu, C., Vinault, S., Abrahamowicz, M., Bonithon Kopp, C., Binquet, C., 2013. Congenital *Toxoplasma* infection: monthly prenatal screening decreases transmission rate and improves clinical outcome at 3 years. Clin. Infect. Dis. Epub ahead of print.

Walls, K.W., Bullock, S.L., English, D.K., 1977. Use of the enzyme-linked immunosorbent assay (ELISA) and its microadaptation for the serodiagnosis of toxoplasmosis. J. Clin. Microbiol. 5, 273–277.

Warren, J., Sabin, A.B., 1942. The complement fixation reaction in *Toxoplasma* infection. Proc. Soc. Exp. Biol. Med. 51, 11–14.

Webster, J.P., Lamberton, P.H., Donnelly, C.A., Torrey, E.F., 2006. Parasites as causative agents of human affective disorders? The impact of anti-psychotic, mood-stabilizer and anti-parasite medication on *Toxoplasma gondii*'s ability to alter host behaviour. Proc. Biol. Sci. 273 (1589), 1023–1030.

Webster, J.P., Kaushik, M., Bristow, G.C., McConkey, G.A., 2013. *Toxoplasma gondii* infection, from predation to schizophrenia: can animal behaviour help us understand human behaviour? J. Exp. Biol. 216 (Pt 1), 99–112.

Weiss, L.M., Harris, C., Berger, M., Tanowitz, H.B., Wittner, M., 1988. Pyrimethamine concentration in serum and cerebrospinal fluid during treatment of Toxoplasma encephalitis in patients with AIDS. Journal of Infectious Diseases 157, 580–583.

Weiss, L.M., Dubey, J.P., 2009. Toxoplasmosis: A history of clinical observations. Int. J. Parasitol. 39 (8), 895–901.

Wilson, C., Remington, J.S., Stagno, S., Reynolds, D.W., 1980a. Development of adverse sequelae in children born with subclinical congenital *Toxoplasma* infection. Pediatrics 66, 767–774.

Wilson, C.B., Desmonts, G., Couvreur, J., Remington, J.S., 1980b. Lymphocyte transformation in the diagnosis of congenital toxoplasma infection. N. Engl. J. Med. 302 (14), 785–788.

Witola, W.H., Mui, E., Hargrave, A., *et al.*, 2011. NALP1 influences susceptibility to human congenital toxoplasmosis, proinflammatory cytokine response, and fate of *Toxoplasma gondii*-infected monocytic cells. Infect. Immun. 79 (2), 756–766.

Wolf, A., Cowen, D., Paige, B.H., 1939. Toxoplasmic encephalomyelitis III. A new case of ganulomatous encephalitis due to a protozoon. Am. J. Pathol. 15, 657–694.

Wong, W.K., Upton, A., Thomas, M.G., 2012. Neuropsychiatric symptoms are common in immunocompetent adult patients with *Toxoplasma gondii* acute lymphadenitis. Scand. J. Infect. Dis.

Wong, S.Y., Hajdu, M.P., Ramirez, R., *et al.*, 1993. Role of specific immunoglobulin E in diagnosis of acute *Toxoplasma* infection and toxoplasmosis. J. Clin. Microbiol. 31 (11), 2952–2959.

Wreghitt, T.G., Gray, J.J., Pavel, P., Balfour, A., Fabbri, A., Sharples, L.D., Wallwork, J., 1992. Efficacy of pyrimethamine for the prevention of donor-acquired *T. gondii* infection in heart and heart-lung transplant patients. Transplant Int. 5, 197–200.

Wreghitt, T.G., McNeil, K., Roth, C., *et al.*, 1995. Antibiotic prophylaxis for the prevention of donor-acquired *Toxoplasma gondii* infection in transplant patients. [letter, comment]. J. Infect. 31, 253–254.

Wulf, M.W.H., van Crevel, R., Portier, R., Meulen, C.G., Melchers, W.J.G., van der Ven, A., Galama, J.M.D., 2005. Toxoplasmosis after renal transplantation: Implications of a missed diagnosis. J. Clin. Microbiol. 43, 3544–3547.

Xiao, J., Kannan, G., Jones-Brando, L., Brannock, C., Krasnova, I.N., Cadet, J.L., Pletnikov, M., Yolken, R.H., 2012a. Sex-specific changes in gene expression and behaviour induced by chronic *Toxoplasma* infection in mice. Neuroscience 206, 39–48.

Xiao, J., Viscidi, R.P., Kannan, G., Pletnikov, M.V., Li, Y., Severance, E.G., Yolken, R.H., Delhaes, L., 2012b. The *Toxoplasma* MAG1 peptides induce sex-based humoral immune response in mice and distinguish active from chronic human infection. Microbes. Infect.

CHAPTER

5

Ocular Disease due to *Toxoplasma gondii*

*Emily Su**, Andrea Honda*[†], Paul Latkany*[†]*

*St Luke's Roosevelt Hospital, New York USA

[†]New York Eye and Ear Infirmary, New York, USA

OUTLINE

Toxoplasma gondii, Second Edition
http://dx.doi.org/10.1016/B978-0-12-396481-6.00005-2

5.1 INTRODUCTION

Even though ocular toxoplasmosis is the most common etiology of posterior uveitis in the USA and the world, it remains a poorly understood disease. For instance, there is limited understanding as to why macular lesions are common in congenitally infected individuals, there is limited understanding as to the details of ocular recurrences, and there is no agreement on best treatment. There is no regimen that can eliminate the bradyzoite stage of infection; therefore, once an individual is infected by *Toxoplasma gondii*, the retina is randomly, undetectably 'seeded' resulting in a chance for local recurrences in the future. There is irreversible damage to the involved retina where a recurrence occurs. If a recurrence occurs within the central macula, the consequence is severe visual morbidity.

T. gondii is one of the most common parasitic infections in the world, with as many as one third of all humans being infected (Jackson and Hutchison, 1989; Holliman, 1997). Seroprevalence appears to be diminishing in some regions and is less common in colder climates; the seropositivity is 22.5% in the USA and 74% in El Salvador (Montoya and Liesenfeld, 2004). Acquired toxoplasmosis is usually an asymptomatic systemic infection (Bowie *et al.*, 1997), however, severe acquired disease has been reported (Carme *et al.*, 2002). Several features including size of inoculum (Liesenfeld, 1999), sex (Roberts *et al.*, 1995), immune status (Suzuki *et al.*, 1996) and virulence (Su *et al.*, 2002) influence the outcome of infection. Lymphadenopathy is the most common manifestation of acute systemic infection.

Contrary to previous belief, it appears that most people with ocular toxoplasmosis are not congenitally infected, but have been infected postnatally. The rate of ocular involvement in acquired toxoplasmosis was previously estimated to be no more than 3% (Perkins, 1973) but data suggest that it is as high as 20% (Burnett *et al.*, 1998). Congenital infection with *Toxoplasma* can cause a range of effects from foetal demise or foetal abnormalities to minimal symptomatic sequelae. The classic triad of signs suggestive of congenital toxoplasmosis is chorioretinitis, intracranial calcifications, and hydrocephalus (Jones *et al.*, 2001b).

Of the myriad signs and symptoms associated with congenital *Toxoplasma* infection, *Toxoplasma* chorioretinitis is the most common, present in 70 to 90% of patients (Koppe *et al.*, 1986; Stagno *et al.*, 1977; McAuley *et al.*, 1994; Couvreur *et al.*, 1984). Indeed chorioretinitis may be the sole symptomatic manifestation of congenital infection (Alford *et al.*, 1974; Fair, 1958). Congenital infection is serious and should be managed by experts; estimates of the rate of occurrence range from 400 to 4000 births in the USA each year (Lopez *et al.*, 2000). Almost one quarter of eyes affected by ocular toxoplasmosis in congenital disease are legally blind, and congenital toxoplasmosis can cause central and even complete vision loss (Bosch-Driessen *et al.*, 2000). No well-designed recent randomized controlled trials have been conducted of the treatment of *Toxoplasma* retinochoroiditis and there is no widespread agreement regarding optimal therapeutic agents or the duration of treatment (St Georgiev, 1993).

Currently, there is considerable controversy concerning the treatment of ocular toxoplasmosis. Some clinicians reserve treatment for active disease unless there is imminent threat to a patient's vision, e.g. the presence of lesions within the macula. There has even been controversy regarding the efficacy of medicines during acute illness (Gilbert *et al.*, 2002). One meta-analysis of three randomized controlled treatment trials found little evidence supporting the use of routine antibiotic treatment for acute *Toxoplasma* chorioretinitis (Stanford *et al.*, 2003). In addition, long-term treatment of patients with chronic recurrent *Toxoplasma* chorioretinitis remains controversial (Kopec *et al.*, 2003). There are confounding areas of chronic suppressive treatment that remain unexplored. For instance it is not known if the outcome in the treatment of chronic ocular disease is dependent on the strain of *T. gondii*, i.e. type 1 virulent strain or type 2 avirulent strains, causing the infection.

5.2 HISTORICAL FEATURES OF OCULAR TOXOPLASMOSIS

The first description of ocular disease as a result of *T. gondii* infection was in 1923 by J. Jankû (Jankû, 1923). C. Levaditi (Levaditi, 1928) subsequently identified the parasite observed by Jankû as *T. gondii*. At the time, ocular involvement was noted as a manifestation of congenital disease in newborns, but it was not until 1939 that it became widely appreciated (Wolf, 1939). Rieger is credited with the origin of the concept of postnatally acquired *T. gondii* as well as the theory that recurrence may be related to immunocompromised states (Rieger, 1951). These ideas were revolutionary in 1951 and remain topics of inquiry today.

In 1952, the link between *T. gondii* and eyes with chorioretinal lesions was confirmed by Helenor Campbell Wilder Foerster (Holland *et al.*, 2002). In the years between the Sabin–Feldman dye test's introduction and this discovery, the potential role of *T. gondii* as an unrecognized cause of ocular disease in adults was under suspicion. D. Vail (Vail, 1943), J. Frenkel (Frenkel, 1949), and H. Rieger (Rieger, 1951) all described series of adult patients with chorioretinal lesions and positive *T. gondii* antibody tests. Frenkel noted positive *Toxoplasma* skin test on patients with chorioretinal lesions (Frenkel, 1949).

Clear histological evidence was not given until Wilder presented her landmark case series. This case series of 53 eyes that had been enucleated due to pain and blindness unequivocally established a strong relationship between *T. gondii* and the particular characteristics of its ocular manifestations. Through careful laboratory techniques and persistent investigation she not only provided some answers to the enigma of *Toxoplasma* chorioretinitis, but put forth central questions that remain unanswered… Each eye in Wilder's cohort had lesions that were granulomatous with central necrosis, and *T. gondii* was consistently found in the necrotic areas. Serologic testing on these patients revealed all of them to test positive for *T. gondii* antibodies. As a result of Wilder's work, ocular toxoplasmosis resulting from congenital infection became accepted as the leading cause of posterior uveitis in otherwise healthy adults. This work solidified the hypothesis that toxoplasmosis, not tuberculosis, causes ocular disease characterized by retinochoroidal lesions. Prior to Wilder's report, tuberculosis was routinely, erroneously ascribed as the source of what Wilder ultimately demonstrated was ocular toxoplasmosis.

Postnatally acquired infection with ocular involvement, as well as ocular manifestations of congenital disease, were fully characterized by M. J. Hogan in 1958 (Hogan, 1958); however, during the 1960s, Hogan and his associates made the incorrect assumption that ocular symptoms of toxoplasmosis occur largely in the presence of systemic symptoms and rarely alone (Hogan *et al.*, 1964). They recognized that a large population of patients with postnatally acquired *T. gondii* infection is asymptomatic and deduced that ocular involvement in patients with acquired infection was uncommon. The classic teaching that most, if not all, *Toxoplasma* chorioretinitis is congenital was given further support in 1973 when Perkins concluded that nearly all cases of *Toxoplasma* ocular involvement in the UK resulted from congenital infection (Perkins, 1973). Based on the belief that ocular involvement only occurs immediately after infection, episodes of recurrent chorioretinitis in children and adults were attributed to congenital infection that went undetected at birth (Hogan, 1961). Hogan, Perkins and others based their studies on a set of assumptions that are now known to be incorrect, and their conclusions are now being challenged.

Later studies have refuted many of the earlier assumptions about the nature and course of *Toxoplasma* retinochoroiditis. For example, it is now understood that ocular disease is often the only manifestation of recent, postnatally acquired infection (Ongkosuwito *et al.*, 1999). Also, retinal lesions have been known to develop long after initial infection (Silveira *et al.*, 2001). These new data have led to a shift in some basic beliefs about this disease and the consideration of the possibility that most ocular toxoplasmosis is acquired (Holland, 1999; Gilbert and Stanford, 2000).

Treatment of ocular toxoplasmosis with antimicrobial drugs began in the early 1950s. In 1953, Eyles and Coleman described the effects of pyrimethamine used in conjunction with sulphonamides (Eyles and Coleman, 1953). Hogan was the first to note that therapy seemed to elicit resolution in adult patients with *Toxoplasma* chorioretinitis (Hogan, 1958). A combination of antimalarial drugs and sulphonamides specifically pyrimethamine and sulphadiazine, was used by Hogan in 1958 and remains widely used today (Holland and Lewis, 2002). Currently, drug therapy for ocular toxoplasmosis is usually administered only if there is reactivation of the infection. Some clinicians do not administer specific drug therapy when a peripheral *Toxoplasma* chorioretinitis recurrence occurs in an immunocompetent person; however, we believe that treatment should be administered to these patients.

A survey of the members of the American Uveitis Society highlights the lack of uniformity regarding therapy. The most common regimen used in the 1991 published survey was pyrimethamine, sulphadiazine, prednisone and folinic acid in 32% of respondents and an additional 27% added clindamycin to the most common regimen (Engstrom *et al.*, 1991). Other agents with activity against *Toxoplasma* include quinolones and macrolides. Adjunctive therapies such as laser treatment or cryotherapy (Jacklin, 1975) within and adjacent to chorioretinal scars are today rarely employed.

5.3 EPIDEMIOLOGY

Prior to the mid-twentieth century, *T. gondii* was not widely recognized as a cause of ocular disease in general or of chorioretinal lesions in specific. With the introduction of the Sabin–Feldman dye test (Sabin, 1948), it became apparent that *Toxoplasma* infection was not only widespread but largely asymptomatic. There is wide variation of seroprevalence among different countries, geographic areas, and ethnic backgrounds with as many as 22.5% of adults in the USA (Jones *et al.*, 2001a) and 90% in Panama (Sousa *et al.*, 1988) seropositive for antibodies against *Toxoplasma*. At least 0.6% of residents of Alabama (Maetz *et al.*, 1987) and Maryland (Smith and Ganley, 1972) have chorioretinal scars consistent with prior *Toxoplasma* chorioretinitis. Without clusters of lesions, it is unknown whether a particular pigmentary lesion is a normal variant or secondary to *Toxoplasma* infection. Recent studies suggest that while *T. gondii* is still the leading cause of posterior uveitis, acquired disease occurs more frequently than congenital infection (Gilbert and Stanford, 2000).

Congenital toxoplasmosis appears to be the highest risk of a systemic infection for the development of ocular lesions. The risk of retinochoroiditis from intrauterine infection is 20% in the early childhood years and can rise to as high as 80% in adolescence. Congenital infection of ocular toxoplasmosis has been estimated to affect 3,000 infants born in the USA each year with a resultant annual cost of between 400 million and 8.8 billion dollars each year (Roberts and Frenkel, 1990; Roberts *et al.*, 1994; Wilson and Remington, 1980). *Toxoplasma* chorioretinitis is present in 70% to 90% of patients with congenital *Toxoplasma* infection, and it is the most common manifestation of disease (Alford *et al.*, 1974). Although 85% of congenitally infected infants appear normal at birth, studies indicate that if these patients are not treated, approximately 85% of them will go on to develop chorioretinal lesions, some resulting in vision loss, by adulthood (Koppe, 1974; Wilson *et al.*, 1980).

While the classic teaching was that most if not all ocular toxoplasmosis is secondary to intrauterine exposure, it is clear that acquired disease is more common than previously thought (Holland, 1999) – and indeed postnatal infection may account for the majority of *Toxoplasma* retinochoroiditis (Gilbert and Stanford, 2000). The origin of the classic teaching that ocular toxoplasmosis results from congenital infection was from studies showing that in South Pacific islands there is a high seroprevalence of exposure to toxoplasmosis prior to pregnancy and there is also a low rate of ocular disease (Darrell *et al.*, 1964). In addition, supporting the classic teaching is the rarity in earlier studies of multiple siblings having ocular toxoplasmosis. Additionally, there was no increase in ocular toxoplasmosis prevalence with age not paralleling the concomitant increase of seroprevalence of toxoplasmosis with age – both arguing against acquired disease as a source of ocular toxoplasmosis.

Recent studies have reported ocular toxoplasmosis in non-twin siblings (Glasner *et al.*, 1992) and ocular infection in acquired disease. Serologic testing at the Palo Alto Medical Foundation suggests that many patients displaying only ocular symptoms of toxoplasmosis may have recently acquired infection, as opposed to recurring congenital infection. It is plausible that previous lesions in these patients may be due to earlier, undetected episodes of acquired infection as chorioretinal lesions acquired during childhood or in the periphery of the eye may go unreported and be incorrectly diagnosed as congenital.

Several serologic studies of specific populations suggest that ocular toxoplasmosis is mostly acquired. Seropositivity for *T. gondii* is relatively high in the region of Erechim in southern Brazil. This is presumed to be due to the cultural practice of ingesting raw pork. One study found that the prevalence in Brazil of ocular disease increases with age (Glasner *et al.*, 1992). Congenital toxoplasmosis is rare, since the majority of women are exposed to infection prior to their first pregnancy. Cases of non-twin siblings both having ocular toxoplasmosis (Silveira *et al.*, 1988) along with the high rate of ocular disease there (17.7%) (Glasner *et al.*, 1992) suggest that previous notions about the epidemiology of this disease are worth reconsideration. Higher rates of disease due to acquired infection in Brazil may be related to any number of factors. Various factors concerning the host, the parasite and the environment have been hypothesized to affect the course of disease: differences in the genetics or age of the host upon exposure, intensity, frequency and duration of exposure, manner of transmission, and clonal type, life stage and virulence of the parasite.

Acquired ocular toxoplasmosis has been reported in France (Couvreur and Thulliez, 1996), Canada (Burnett *et al.*, 1998), Brazil (Glasner *et al.*, 1992; Silveira *et al.*, 2001) and the USA

(McCannel *et al.*, 1996; Montoya and Remington, 1996). A fascinating population based epidemiologic study of a rural area of Brazil revealed household correlated with severity of infection (Portela *et al.*, 2004). No ocular disease was found in patients younger than 10 years of age even though there was in 47% evidence of serologic infection. Of all age groups, 12.9% of seropositive patients had ocular lesions, and patients older than 55 years of age had the highest prevalence of ocular lesions. Similar households had slightly increased risk factor for ocular toxoplasmosis; however, age was the strongest risk for ocular lesions. High values for GIPL specific IgA were associated with larger eye lesions. The IgA appeared to be directed against GIPL derived from tachyzoites. Whereas in the USA, it appears that Type 2 strains of *Toxoplasma* are the most common strains associated with human infection, in southeast Brazil Type 1, Type 3 or Type 1/3 hybrids with a relative absence of Type 2 have been reported (Dubey *et al.*, 2002; Dubey *et al.*, 2003; Fux *et al.*, 2003).

There are several modes of transmission of *Toxoplasma* to humans, and the exact hierarchy of importance of transmission is unknown and will likely vary depending on region. It is usually difficult to identify the source of exposure to *T. gondii* in a particular patient. Most transmission is assumed to occur by ingestion whether it is from contaminated water, undercooked meat or contaminated produce. An example of an identified food source (deer), which appears to be associated with five cases of acquired toxoplasmosis from handling or eating venison, is compelling (Ross *et al.*, 2001). There are likely several as yet unidentified food sources within different cultures and regions that are a common source of transmission of toxoplasmosis. Unfiltered drinking water was identified as a major source for the high prevalence of infection in a rural region of northern Brazil (Bahia-Oliveira *et al.*, 2003). There was one report of aerosolization in a horse stable as a source of transmission (Teutsch *et al.*, 1979).

Toxoplasmosis has been shown to occur as a result of contaminated municipal drinking water (Bowie *et al.*, 1997) and even with a strict vegetarian diet (Hall *et al.*, 1999). Contamination of a municipal reservoir was the source of a 1995 outbreak of toxoplasmosis in Victoria, British Columbia (Bowie *et al.*, 1997) and this same problem has been associated with high risk for infection in Brazil (Bahia-Oliveira *et al.*, 2003). Though the high rates of seropositivity in Brazil are traditionally attributed to the practice of ingesting raw or undercooked pork (Silveira, 1987), there is evidence that drinking contaminated water is linked with higher rates of ocular toxoplasmosis (Silveira, 2002). Infection from multiple sources could be part of the reason for the very high rates of postnatally acquired ocular toxoplasmosis. The details of the 1995 Victoria outbreak have been recounted by Bowie and associates (Bowie *et al.*, 1997). One hundred people were found to be acutely infected with *T. gondii* caused by the outbreak, 19 of whom had infection-related retinitis. The range of ages in the entire cohort was six to 83 years, but the eye disease cases were significantly older than other cases of symptomatic disease, such as lymphadenopathy. A large outbreak of ocular toxoplasmosis affecting patients residing in Tamil Nadu in southern India in 2004 also pointed to contaminated municipal drinking water as the most plausible source (Balasundaram *et al.*, 2010). There were 198 out of 248 patients with active retinochoroiditis who were seropositive for IgM anti-toxoplasma, suggesting recently acquired infections. Interestingly, only 35 patients (14.1%) had prodrome fever prior to onset of retinitis.

Depending on the study, anywhere from 2 to 20% of infected individuals develop ocular toxoplasmosis (McCannel *et al.*, 1996; Couvreur and Thulliez, 1996; Burnett *et al.*, 1998; Glasner *et al.*, 1992; Silveira *et al.*, 2001; Montoya and Remington, 1996; Holland, 1999).

Jones and Holland estimated that 14% of the population in the USA is infected with *T. gondii* by age 40, and roughly 2% of these patients have ocular involvement (Jones and Holland, 2010). Based on the 2009 US census estimate, 21,000 people will develop toxoplasmic ocular lesions and 4,800 will be symptomatic. Other regions of the world, such as southern Brazil, have much higher disease burden due to both higher disease prevalence (56%) and rate of ocular manifestation (Dubey *et al.*, 2012). Further studies on disease burden are warranted for determining appropriate public health programmes on disease prevention, and treatment for affected individuals.

5.4 THE MECHANISM OF TISSUE DAMAGE IN OCULAR TOXOPLASMOSIS

There are several theories as to why ocular toxoplasmosis results in its inflammatory process with resultant damage.

Cyst rupture with subsequent infection
: Initially cysts generate little inflammatory response, perhaps as a result of incorporation of host cell components into the cyst wall (Dutton *et al.*, 1986a). Although there is no direct evidence of cyst rupture inducing inflammation, there is a related organism, *Besnoita jellisoni,* in which a cyst can be observed ophthalmoscopically and can undergo a spontaneous rupture which induces an inflammatory response (Frenkel, 1955). However, there does appear to be cyst degeneration by electron micrographic studies in retinochoroidal lesions (Rao and Font, 1977). In addition, from animal models it appears that the number of tachyzoites released during cyst rupture determines the presence of an inflammatory response as 500 tachyzoites cause minimal reaction while 5,000 tachyzoites induce more of an inflammatory reaction (O'Connor and Nozik, 1971; Culbertson *et al.*, 1982).

Toxic mediators released from parasite
: A dialysate from the peritoneal exudates of *Toxoplasma* infected mice was injected intravitreally into rabbit eyes and found to result in severe retina damage (Hogan *et al.*, 1971).

Damage by inflammatory mediators
: Lysis of inflammatory cells within the retina results in surrounding tissue damage (Dutton *et al.*, 1986b).

Hypersensitivity to *T. gondii* antigens
: Histopathologic demonstration of granulomas have been reported in *Toxoplasma* retinochoroiditis. Patients previously exposed to *T. gondii* develop a delayed type hypersensitivity reaction to subcutaneously administered antigen (Frenkel, 1949). Cyst rupture would result in exposure of sequestered antigen which results in an intraocular delayed type hypersensitivity reaction. Injection of *T. gondii* antigen intraocularly did not result in an inflammatory response (Culbertson *et al.*, 1982). From these experiments it is unlikely that *T. gondii* initiates the inflammatory cascade, though hypersensitivity may prolong inflammation once initiated.

Auto-antigens
: Several assays on peripheral blood from patients indicate that humoral (Abrahams and Gregerson, 1982) and cell mediated immunity (Nussenblatt *et al.*, 1989) against retinal antigens are present. Animal studies demonstrate selective destruction of outer retina similar to that observed in experimental induced uveitis with S-antigen. However, auto-antigens are unlikely to be the primary impetus behind visual morbidity as, empirically, patients who are immunocompromised do worse secondary to parasite replication. *In vitro* lymphocyte proliferation was found to be S-antigen in 16 of 40 patients with ocular

toxoplasmosis (Nussenblatt *et al.*, 1989). In addition, 34 of 36 patients had photoreceptor activity as detected by indirect immunofluoresence.

5.5 HOST FACTORS IN OCULAR TOXOPLASMOSIS

The eye is an unusual immunologic environment designed to reduce inflammation, in that transforming growth factor beta is constitutively expressed, and a high concentration of Fas ligand exists on ocular cells (Streilein and Stein-Streilein, 2000). In addition, Class I MHC molecules are down regulated (Streilein *et al.*, 1997). The unusual immunologic microenvironment of the eye may decrease clearing of *T. gondii* infection. Transforming growth factor® has been shown to reduce interleukin-12 production of interferon© (Hunter *et al.*, 1995). Nitric oxide also plays an important role in interferon gamma mediated host reaction against *T. gondii* in ocular toxoplasmosis (Langermans *et al.*, 1992; Roberts *et al.*, 2000). TNF-alpha, iNOS, IL-1, and IL-6 are all up regulated in ocular toxoplasmosis (Gazzinelli *et al.*, 1994; Hayashi *et al.*, 1996). Five genes, including one within the region of the H-2 gene, have been associated with an impact on survival after infection with *T. gondii* (McLeod *et al.*, 1989). Class I MHC and CD8+ fraction of T cells determine cyst load after infection (Brown and McLeod, 1990). Reactivation of *Toxoplasma* in the murine model of ocular toxoplasmosis has been demonstrated to be influenced by down regulation of interferon-gamma and tumour necrosis factor-alpha. Interleukin-6 enhanced intracellular reproduction of *T. gondii* (Beaman *et al.*, 1994) but this contradicts the finding of an Il-6 knockout model (Lyons *et al.*, 2001). Of note, the retinal pigmented epithelium produces a large amount of IL-6.

Vitreous fluid mitogenically stimulated cell lines of ocular toxoplasmosis patients did not show any reactivity to retinal antigens but showed a strong response to *T. gondii* antigens (Feron *et al.*, 2001). In a human retinal pigment epithelial (HRPE) cell culture assay, interferon results in L-tryptophan starvation through induction of indoleamine 2,3-dioxygenase, an enzyme that converts tryptophan to N-formylkynurenine; however, the administration of exogenous tryptophan did not result in complete reversal of the inhibitory effect. Interferon© was the most potent cytokine in HRPE cells and indeed by itself inhibited growth of parasite (Nagineni *et al.*, 1996). Interleukin 10 in an intraocular inoculum route of infection in the mouse appears to play a role in limiting inflammation, and because knockout IL-10 mice have similar gamma interferon levels to controls, it appears to have only a partial role in regulation gamma interferon production (Lu *et al.*, 2003).

A clue to the complexity of host factors in toxoplasmosis is that even though in animal models inhibition of tumor necrosis factor results in worse ocular disease, inhibition of TNF rarely results in ocular toxoplasma disease recurrence in humans. Infliximab has been confirmed by biopsy to cause CNS toxoplasmosis (Young and McGwire, 2005), but despite hundreds of thousands of patients having received infliximab, no concerning increase in ocular toxoplasmosis has occurred.

5.6 PARASITE FACTORS IN OCULAR INFECTION

Three clonal types of *T. gondii* predominate in nature (Howe and Sibley, 1995). Virulent strains of *T. gondii* appear to have their origin in a single, genetically homogeneous lineage. This is despite the parasite's pervasive nature and ability to reproduce sexually (Sibley and Boothroyd, 1992). Recently, a composite genome map of the 14 chromosomes of *T. gondii* was reported (Khan *et al.*, 2005). There is at most a 1% difference between the three strains, and reports examining only one locus are likely to misclassify *Toxoplasma* because of shared alleles between strains (Saeij *et al.*, 2005a). The most surprising development regarding strains and ocular disease was the finding of several novel genotypes from clinical ocular samples (Grigg *et al.*, 2001b). These novel strains had combinations of alleles from the three classic strains, and rather than having unique virulent genes, rather it is likely the combination of genes that results in increased virulence. It has been demonstrated experimentally that a combination of non-virulent strains can lead to more virulent progeny (Grigg *et al.*, 2001a).

The identification of *Toxoplasma* genotype using peripheral blood samples of patients with ocular disease may have clinical and epidemiologic value (Bou *et al.*, 1999). Most immunocompromised individuals do not have infection with the more virulent Type 1 strain, but patients who are immunocompetent appear to have Type 1 as a commonly identified strain associated with ocular disease (Fardeau *et al.*, 2002). Type 2 strains appear to be the most common cause of human disease in the USA (Darde, 2004). In a study limited by the sample size of 12 clinical isolates, patients with a higher proportion of atypical toxoplasmosis were also found to have clonal types associated with more virulence (eight out of 12) (Grigg *et al.*, 2001b). In addition, all six patients who were immunocompetent with severe ocular toxoplasmosis had a more virulent *SAG1* gene. When the samples were analysed by PCR RFLP assay for *SAG3, SAG2,* and *B1,* five of 12 isolates were identified to have new recombinant strains. A larger cohort is required to validate whether Type 1 associated infection is in fact more likely associated with more severe ocular disease.

A recent validated multiplex PCR has been developed allowing rapid multilocus strain typing from five microsatellite marker locations (Ajzenberg *et al.*, 2005). Type 2 strain was present in 85% of isolates of a cohort of congenital toxoplasmosis from France (Ajzenberg *et al.*, 2005; Bowie *et al.*, 1997). Severe acquired ocular toxoplasmosis disease has been documented to occur from atypical strains (Grigg *et al.*, 2001b; Burnett *et al.*, 1998). In the domestic cycle of toxoplasmosis, there are three clonal strains widely prevalent; however, in other areas where there are a wide variety of intermediate hosts (such as in French Guiana) atypical strains are more common.

While there is limited understanding as to why ocular disease remains one of the most frequent causes of recurrent morbidity, there is a suggestion from endothelial cell culture as to why ocular disease is common in systemic infection (Smith *et al.*, 2004). It appears that *T. gondii* grows better in retinal endothelial cells than in aortic, dermal or umbilical endothelial cell lines and even 2.8-fold higher than in human foreskin fibroblasts (Smith *et al.*, 2004). There are two theories as to how *T. gondii* crosses the blood brain barrier. One is that the tachyzoites cross the blood brain barrier when circulating lymphocytes are arrested within the ocular vasculature and lyse, permitting free tachyzoites to infect adjacent retinal endothelium (Roberts and McLeod, 1999). An additional theory, which has yet to be substantiated, is that infected lymphocytes cross the blood brain barrier present within the eye and transport *T. gondii* into the eye.

5.7 ANIMAL MODELS

5.7.1 Animal Models of Ocular Toxoplasmosis

Several intraocular models of toxoplasmosis have been developed. The primate (Culbertson *et al.*, 1982), mouse and rabbit (Nozik and O'Connor, 1968) models suffer from the lack of similarity to the human toxoplasmosis where ocular seeding occurs from systemic endogenous infection. Both marsupials and New World monkeys appear to be unusually susceptible to

Toxoplasma infection, which is believed to be due to the absence of feline populations in their environment (Innes, 1997; Gustafsson *et al.*, 1997; Epiphanio *et al.*, 2003). There are no ideal small animal models of human ocular toxoplasmosis. The macula is the central anatomical segment of retina within the visual axis that gives humans their most fine acuity. Only primates have maculas, and macula disease is an important feature of congenital toxoplasmosis. Since the mouse does not have a macula, it cannot demonstrate macula disease. In addition, the proportional volume of lens is much greater and the proportional vitreous volume much less in the mouse than in the human eye. Despite these shortcomings, mice have been used to mimic human retinal disease. If mice can reliably be manipulated to develop lesions similar to human ocular toxoplasmosis, it is helpful as such models provide a springboard to investigate human ocular toxoplasmosis.

5.7.2 Murine Models of Ocular Toxoplasmosis

One example of a murine model of ocular toxoplasmosis is the injection of tachyzoites into the anterior chamber of the mouse (Hu *et al.*, 1999). The first step of the technique is to remove some aqueous humour by paracentesis. (The eye is a closed environment and removing fluid reduces the risk of raised intraocular pressure.) This is followed by an injection of five microlitres of parasite suspension in Dulbecco's modified Eagle medium. It should be noted, however, that the most significant route of transmission of ocular toxoplasmosis in humans is endogenous through systemic infection and not directly through an exogenous route. The benefit of the model is that intraocular replicating parasites can be readily identified. The one difference from the human model is that the initial infection may permanently alter the normal blood brain barrier of the eye due to initial inflammatory response and result in atypical ocular environment. This intraocular inoculation model has yielded interesting results. For instance, the model has demonstrated an extraordinary susceptibility of C57bl mice to mortality within two weeks as compared to BALB/c mice which have almost no mortality from intraocular inoculation. A recent study reports that in murine models, genetic factors of the host mouse as well as the parasite strain are significant in determining susceptibility to experimental ocular toxoplasmosis (Lu *et al.*, 2005).

An alternative to avoid the damage from intraocular injection is the simple topical application of a parasite suspension of 10^3 tachyzoites RH strain in RPMI (Gil *et al.*, 2002). It is a surprisingly useful way to obtain a detectable intraocular parasite load with preservation of the intraocular architecture and no alteration of the lens architecture (Tedesco *et al.*, 2005). Surprisingly, there appears to be no detectable difference in measured intraocular infection by either the topical or intravitreal route of infection in mice. Both routes had detectable parasites within intraocular vessels, glial reaction of the inner plexiform layer by day 7 and disruption of the retinal pigmented epithelium. Intravitreal injection of phosphate buffered saline alone also resulted in glial changes within the inner plexiform layer (Tedesco *et al.*, 2005). Similar to rabbit models of ocular infection, a high inoculum dose given intraocularly can result in ocular disease even in previously primed mice (Hu *et al.*, 1999; Dao *et al.*, 2001). It has been demonstrated that even without the overwhelming of host defenses likely from intraocular inoculums, systemic re-infection of mice with different strains after previous seroconversion is possible (Dao *et al.*, 2001).

A congenital model of ocular toxoplasmosis has been reported in which infected dams are inoculated during gestation (Hay *et al.*, 1984). Unfortunately, a wide range of clinical disease occurred in this model. Murine congenital toxoplasmosis differs significantly from human disease. In the murine model, there is significant opaque cataract formation, even with lens autolysis. This can be seen by examination using a 35 dioptre lens and documented with a Zeiss operating microscope (Dutton *et al.*, 1986a). Even though ME49 strain infection of mice is a reliable model for *Toxoplasma* cysts in the brain, it is unusual to find evidence of intraocular toxoplasmosis even with PCR amplification. Antibodies in the same model against interferon©, tumour necrosis factor registered sign, or CD4 and CD8 cells results in frequent demonstration of parasite in association with more severe ocular lesions (Gazzinelli *et al.*, 1994). L-NAME through its inhibition of NO can cause disease reactivation in chronically infected mice; however, the cyst load was not robust enough to identify on histopathology (Roberts *et al.*, 2000).

In vivo imaging of the mouse fundus is possible and is greatly facilitated by the appropriate imaging set-up (Hawes *et al.*, 1999). The digital Kowa Genesis Small Animal Fundus camera was an excellent imaging device for mice that was available with a fluorescein angiographic mode. Digital capture helped to instantly confirm capture of appropriate image. Kowa has plans to release an updated model. Regular clinical fundus cameras can also take images of the posterior pole of mice eyes, but is not as easeful and usually requires manipulation of the mouse eye to achieve the best image. Recent observations document systemic infection using *in vivo* imaging employing the IVIS system (Xenogen, Alameda, California) (Saeij *et al.*, 2005b). This paper demonstrates some bioluminescence due to luciferase transgenic parasites with pixels correlating with ocular involvement; however, *ex vivo* imaging of eyes confirming this localization was not reported (Saeij *et al.*, 2005b).

5.7.3 Hamster Models of Ocular Toxoplasmosis

Frenkel first reported hamsters as an ocular model for toxoplasmosis (Frenkel, 1955). The RH strain he had used required therapy to prevent mortality and the CJ strain did not consistently produce ocular lesions. Reliable hamster models of ocular disease with oral infection (Gormley *et al.*, 1999) and intraperitoneal injection (Pavesio *et al.*, 1995) have subsequently been reported using 100 cysts of the ME49 *T. gondii* strain for infection. Although the ocular disease does not exactly mimic human disease, the models are attractive in that hamsters reliably develop ocular lesions with little systemic disease and resolve spontaneously without treatment. Unlike humans, ocular toxoplasmosis in the hamsters does not result in pigmentation, the overlying retina alone appears atrophic and the disease is bilateral.

5.7.4 Rabbit Models of Ocular Toxoplasmosis

Hogan was the first to create a published animal model of ocular toxoplasmosis by injecting tachyzoites into the carotid arteries (Garweg *et al.*, 1998; Hogan, 1951). However, the RH strain used in this model frequently resulted in meningoencephalitis and rapid mortality. Beverley and others injected the inoculum into the anterior chamber of rabbits (Beverley *et al.*, 1954). The BK strain in PBS inoculated intravitreally near the retinal wing of the posterior pole (the entry site was through the pars plana 4–5 mm behind the limbus at the superior rectus muscle insertion into the globe) into previously primed Burgundy rabbits results in an inflammatory response that mimics inflammation of human disease (Garweg *et al.*, 1998). The caveats previously mentioned apply here, as any intraocular inoculum route results in disruption of the normal intraocular architecture. Inoculation of parasites into the suprachoroidal space differs from human infection in that there is a limited local infection followed quickly by a systemic infection and there is also a risk of retinal defects (Nozik and O'Connor, 1968). The suprachoroidal route does not induce an intraocular inflammatory response in primed rabbits in contrast to the intravitreal route. In addition, high inoculum can result in ocular disease even in previously primed rabbits (Garweg *et al.*, 1998). Cyclosporine A (Friedrich *et al.*, 1992) and total body irradiation (O'Connor, 1983a) induce reactivation of ocular disease in previously infected rabbits, but local trauma does not induce reactivation in the rabbit (Nozik and O'Connor, 1970).

5.7.5 Feline Models of Ocular Toxoplasmosis

A feline model of intracarotid inoculation of 5,000 tachyzoites was successful in producing a reliable model of lesions of ocular toxoplasmosis (Davidson *et al.*, 1993). Because the feline model has primarily choroidal involvement, it differs from human ocular infection. Usually, similar to human infection, initial infection with *Toxoplasma* in cats is subclinical (Dubey, 1988). The organism has been found on histopathology throughout the eye (Dubey and Carpenter, 1993). Since over 50% of cats may be seropositive for *T. gondii*, antibody assays of ocular fluid have been developed to help diagnose *Toxoplasma* as an etiology of feline uveitis (Chavkin *et al.*, 1994). The infrequent identification of organisms by histopathology (Peiffer and Wilcock, 1991) led to theories (Davidson and English, 1998) hypothesizing an indirect *T. gondii* antigen etiology to feline *Toxoplasma* related uveitis. PCR assays of intraocular fluid have demonstrated, however, that direct infection due to *T. gondii* is the source of feline uveitis (Burney *et al.*, 1998; Lappin *et al.*, 1996). Less frequent ocular lesions occurred with other routes of infection. Oral administration of tissue cysts or oocysts resulted in infrequent ocular lesions (Dubey and Frenkel, 1974; Dubey and Frenkel, 1972; Dubey, 1977). In a model of neonatal infection of ocular toxoplasmosis, it appeared that the Mozart strain (initially isolated from intraocular fluid) resulted in more frequent eye lesions than either Maggie or ME49 (Powell and Lappin, 2001). No areas of nontapetal retina were infected in these cats and neonatal infection can result in systemic infection with apparent involvement confined to the eye even without evidence of serologic seroconversion (Powell and Lappin, 2001). Necropsy has identified frequent intraocular lesions by transplacentally infected cats (Dubey *et al.*, 1996).

5.7.6 Non-Human Primate Models of Ocular Toxoplasmosis

As expected, non-human primates have the most similar ocular environment mimicking the human eye. In addition, the published models mimic current sensibility regarding human ocular toxoplasmosis. For example, the non-human primate models support the theory that recurrent ocular disease is from the direct presence of *T. gondii* and not from indirect antigenic immunogenicity. The attempt to induce a necrotizing retinochoroiditis from either intravascular or intraocular injection of *T. gondii* antigens was unsuccessful in non-human primates (Newman *et al.*, 1982; Holland *et al.*, 1988b). In addition, no necrotizing retinitis developed in previously immunized primates after injection of live parasites; however, as expected a local inflammatory response occurred when injected intraocularly in previously sensitized animals (Webb *et al.*, 1984). Despite total lymphoid irradiation of 2,000 centigrays none of six cynomolgus monkeys developed recurrent ocular disease even with intraocular injection of Beverly strain or the presence of pre-existing chorioretinal lesions.

5.8 CLINICAL CHARACTERISTICS

5.8.1 The Clinical Features of Ocular Toxoplasmosis

The classic symptoms of ocular toxoplasmosis are similar to the classic symptoms of uveitis in general. When ocular toxoplasmosis is active usually there is pain, redness, photophobia and decreased vision. There are variations of the above classic four symptoms with some patients, for instance, having only decreased vision. The pain and photophobia are minimal unless there is severe iridocyclitis. It can be difficult to detect early onset of any of these symptoms in children as they cannot articulate their symptoms appropriately. As the congenital *Toxoplasma* study cohort (Mets *et al.*, 1997) demonstrated, if children are instructed to promptly report any change in their vision to their caregivers, this can increase the detection of active disease. Ocular toxoplasmosis can present with unusual manifestations of retina pathology. Instead of the classic severely involved focal chorioretinitis, it has also been reported to resemble unilateral acute idiopathic maculopathy (Lieb *et al.*, 2004).

The most common clinical presentation of ocular infection due to *T. gondii* is a unilateral chorioretinitis associated with a pre-existing chorioretinal scar and an overlying vitritis. In addition, the clinical ophthalmic diagnosis of retinal vasculitis, of both arterioles and veins, is commonly made in active disease secondary to interaction between antibodies and antigens (O'Connor, 1974). *Toxoplasma gondii* accounts for greater than one quarter of all cases of posterior uveitis. Lesions can occur in any part of the fundus, but in patients with congenital infection, severe macula lesions appear more common than in acquired infection. One study (Mets *et al.*, 1997) found that macular lesions were present in 58% of a cohort of 94 children with congenital toxoplasmosis, 76 of whom had 1 year of treatment with pyrimethamine and sulphadiazine, peripheral scars were present in 64%. This could be due to the early vascularization of the posterior pole during foetal development, or to the unique vascularization of the foetal macula, which contains end arterioles or to the higher concentration of cells. Evidence of bilateral infection, without simultaneous bilateral active disease because lack of bilateral inflammation but with presence of bilateral scars, was found in 46% of eyes from one study (Hogan *et al.*, 1964)

After resolution of an active lesion, patients will have decreased vision in the area of retinochoroiditis. If the lesion is small and in the periphery then the patient will probably be asymptomatic. If, however, the lesion is small but within the macula then the patient will probably be symptomatic. It is useful to have patients check their vision one eye at a time on a daily basis. *Toxoplasma* lesions within one disc diameter of the optic disc result in very significant 'downstream' visual field defects. This means that an entire region of retina away from the actual lesion, but whose communicating nerve fibres pass over the lesion, can have loss of input as measured by visual field testing (Stanford *et al.*, 2005).

As overlapping visual fields and the fact that ocular toxoplasmosis is usually active unilaterally, not bilaterally, a change in vision may not be initially detected unless unilateral daily screening occurs. An Amsler Grid, a graph paper-like grid of boxes, mounted to a flat surface may help compliance with self-vision screening. In addition to decreased vision in the area of reactivation, patients will likely have floaters or other media opacity related complaints that will vary depending on the degree of inflammation when the lesion was active and may persist after resolution of the underlying reactivation secondary to inflammatory debris being trapped within the vitreous.

Vision loss can be caused by many of the complications associated with ocular toxoplasmosis. Involvement of the macula or optic nerve can directly decrease central vision. Complications secondary to inflammation can indirectly affect vision. These include macular edema, vitreous opacity, epiretinal membrane and retinal detachment (Bosch-Driessen *et al.*, 2000; Mets *et al.*, 1997; Friedmann and Knox, 1969). Under normal circumstances, peripheral scars can affect the visual field but do not impair central vision. However, in rare cases, peripheral scarring may lead to central vision loss. An example of a rare manifestation of ocular toxoplasmosis is one case wherein central vision loss occurred due to a giant macular hole (Blaise *et al.*, 2005). This macular hole was the result of vitreous traction caused by peripheral ocular toxoplasmosis. Subretinal neovascularization in ocular toxoplasmosis has been a reported as an unusual cause of vision loss (Cotliar and Friedman, 1982).

Ocular involvement has been shown to occur long after the time of infection, either acquired or congenital infection. New lesions are likely to occur near the borders of existing lesions, and a larger lesion surrounded by smaller, satellite lesions has been considered the hallmark of a recurrence of both congenital or postnatally acquired disease. Little is known about what influences the rate of recurrence of ocular toxoplasmosis. Though Bosch-Driessen and associates report a cumulative increase in the prevalence of chorioretinal recurrence, (Bosch-Driessen *et al.*, 2002) it cannot be assumed that the risk of recurrence in an individual patient increases over time. Holland's impression was that the pattern of recurrence decreased over time (Holland, 2003). One possible explanation for this pattern is that tissue cysts in human hosts have a limited lifespan and lose their ability to reactivate. There is no evidence that short-term therapy at the time of infection has any effect on the pattern of recurrence (Bosch-Driessen *et al.*, 2002).

5.8.2 Recurrence of Ocular Toxoplasmosis

The classic description of recurrent active *Toxoplasma* chorioretinitis is a focus of retinitis appearing at the border of a retinochoroidal scar; however, there are several reports illustrating the variance in the clinical features of this disease. Active chorioretinitis does resolve without treatment and leaves a hyperpigmented scar, and recurrences develop as 'satellite' lesions. A recurrence is usually symptomatic with the redness, pain, light sensitivity and decreased vision that occur with any generalized panuveitis. Recurrence of chorioretinitis can lead to vision loss (Friedmann and Knox, 1969) and blindness.

There is limited understanding between the many factors that exist between infection and recurrence of disease. There is a strong pattern of recurrence during the teenage and adult years. Women with ocular toxoplasmosis are at a higher risk of recurrence during pregnancy (O'Connor, 1983b), though the foetus (except for rare reports; Silveira *et al.*, 2003) appears to be at risk only during a mother's initial infection.

Initially it was unclear what recurrence actually was, and it certainly remains unclear why recurrence occurs. Prior to the AIDS epidemic there was controversy as to whether recurrence represents an autoimmune process alone without the presence of active parasites. Histopathology on patients with ocular toxoplasmosis and HIV infection has demonstrated parasites in areas of involved retina. In addition, eyes that have received corticosteroid treatment alone have had very poor outcomes associated with parasites demonstrated on tissue biopsy (Sabates *et al.*, 1981). Frenkel's theory that recurrence represents a hypersensitivity reaction appears unlikely to be the central cause of recurrent ocular toxoplasmosis (Frenkel, 1974). Release of *T. gondii* antigen into the retina is not associated with a hypersensitivity reaction. Reactivation is thought to represent a shift of *T. gondii* from the dormant phase known as bradyzoites to the more active phase as a tachyzoites. There is not a clear understanding of how this shift from bradyzoite to tachyzoite occurs within the eye or what factors influence or cause this shift.

Evidence of prior recurrence is the presence of inactive satellite lesions, which are local areas of chorioretinal scars. Recurrent lesions usually occur close in proximity to prior areas of chorioretinitis as is evident in the usual clusters of scars that exist (Fig. 5.1). Recurrent disease occurs when new areas of retina are involved in an infectious inflammatory process that results in permanent destruction of involved tissue. Resolution of the inflammatory process will usually occur spontaneously after several weeks. Although the general eye inflammation will resolve, the area of retina with focal chorioretinitis is irreversibly impaired. If the lesion or recurrence is in the

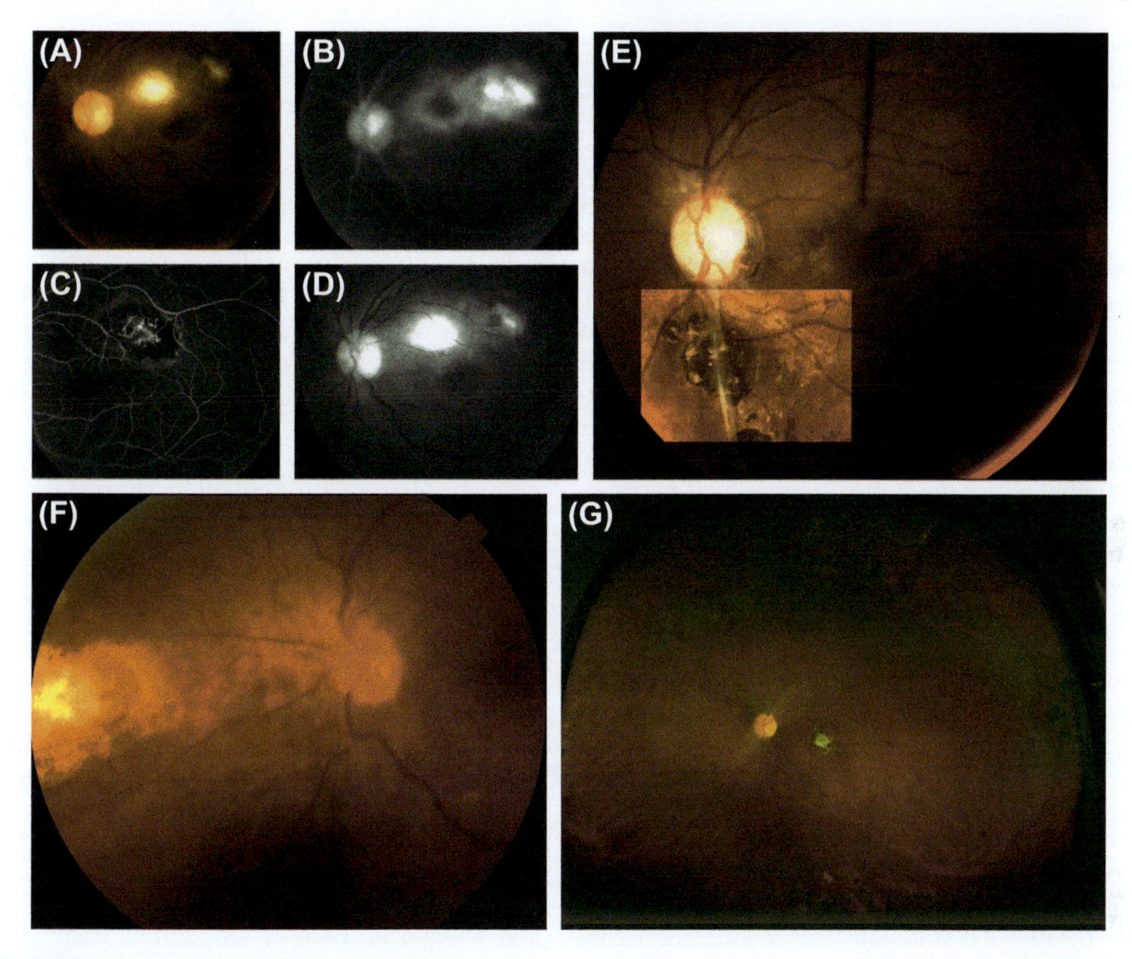

FIGURE 5.1 ***Toxoplasma gondii*** **chorioretinitis** A. Color photo of left macula with old scar superior temporal to new active lesion.

B. Late fluorescein angiogram of color photo A demonstrating incompetence of the blood brain barrier of both the active and old lesion.

C. Hypofluorescent lesion where the active lesion is on color photo A and vascular remodeling of superior temporal old lesion.

D. Red free photograph showing active lesion in part A within two millimeters of the fovea.

E. Old chorioretinal lesion inferior to the optic nerve with overlying vitreal condensation.

F. Nasal to optic nerve; there is an active lesion adjacent to old chorioretinal scar.

G. Wide-angle fundus image of central macula inactive ocular toxoplasmosis lesion (there are also peripheral retinal laser scars). This image represents an important new modality that captures most of the fundus with one image.

peripheral retina, the impact of the recurrence on the infected individual's vision can be minimal, or even asymptomatic, because the impaired vision exists in a small area of the peripheral field. Usually, the new lesion is a focal chorioretinitis; however, a more generalized vitritis and anterior uveitis usually develops secondarily as a generalized intraocular inflammatory process. This more generalized intraocular inflammatory process is responsible for the complaint of decreased vision in patients with ocular toxoplasmosis. The secondary inflammatory process can lead to retinal detachment or other ocular morbidity such as epiretinal membranes.

5.8.3 The Clinical Features of Congenital Ocular Toxoplasmosis

Congenital infection is more common later in gestation, but disease manifestations are worse if acquired earlier in gestation (Dunn *et al.*, 1999). Classically, congenital disease is associated with bilateral macular scarring, but acquired infection can also result in macular disease and rarely bilateral scarring as well (Glasner *et al.*, 1992). Other manifestations include optic neuritis, iritis, neuroretinitis, retinal vasculitis, acute retinal necrosis, recurrent iridocyclitis and persistent vitritis. Long term follow-up of congenitally infected children results in identification of further ocular sequelae not present at birth, e.g. in one study four of six untreated congenitally infected children developed scars subsequent to birth during the next twenty years (Koppe *et al.*, 1986). It is estimated that 85% of infants untreated and without ocular lesions at birth will subsequently develop ocular toxoplasmosis (Wilson *et al.*, 1980; Koppe, 1974). Microphthalmos and microcornea can occur as a consequence of severe congenital eye disease (Suhardjo *et al.*, 2003). Nystagmus and strabismus and amblyopia secondary to congenital toxoplasmosis are more complex than even most expert ophthalmologists realize (O'Neill, 1998). There is a tendency for clinicians not to struggle with the complex care involved in trying to achieve optimal visual outcome in congenital infection. Less initially severe but still disabling disease such as anterior uveitis (Cano-Parra *et al.*, 2000) secondary to *T. gondii* is likely underdiagnosed because of the limitations of current non-invasive diagnostic tests.

Exposure to *Toxoplasma* six months prior to conception is thought to eliminate the possibility of congenital transmission secondary to lifelong immunity in immunocompetent individuals. Rarely, reactivation of toxoplasmosis in previously infected immunodeficient women can result in congenital transmission of toxoplasmosis (Mitchell *et al.*, 1990). There is one recent case report of treated acquired ocular toxoplasmosis during pregnancy occurring in the mother without any subsequent foetal disease (Ramchandani *et al.*, 2002). The exact mechanism of transmission is not yet understood but is thought to be secondary to transplacental transmission of the parasite. The severity of ocular manifestations parallels the severity of CNS disease in congenital infection (Roberts *et al.*, 2001).

Mets *et al.* highlighted the ophthalmic findings of congenital *Toxoplasma* infection in treated and untreated individuals (Mets *et al.*, 1996). Seventy nine percent of children had retinochoroidal scars. Twenty eight percent of individuals had significant unilateral vision loss. Twenty nine percent of children had bilateral vision loss. The presence of inactive chorioretinal lesions in congenitally infected newborns indicates that the complete cycle of infection, activation and resolution of chorioretinal lesions may occur *in utero* (Guerina *et al.*, 1994; Mets *et al.*, 1996). The New England Regional *Toxoplasma* Working Group detected 100 of 635,000 infants who were seropositive for IgG and IgM against *Toxoplasma*. Four of 39 treated children observed for as long as six years had new postnatally developed retinal scars, and a separate nine of 48 patients had retinal lesions at birth (Guerina *et al.*, 1994). In a separate study from England, after 20 years of follow-up, nine of 11 patients with congenital toxoplasmosis had evidence of chorioretinitis and four had severe impairment (Koppe *et al.*, 1986).

The largest report of congenital toxoplasmosis in twins highlights that multiple factors beyond time of exposure during gestation influence ocular outcome in congenital infection (Peyron *et al.*, 2003). Although there are possible confounding issues of shared placentas and mortality, as is true of any infectious congenital disease involving twins, if concordance of the disease is more common among monozygotic twins than among dizygotic ones, then genetic susceptibility is likely more important than environmental influence in disease outcome. While there is no rigorous protocol that has been published focusing on a cohort of ocular outcome in twins (Rieger, 1959), it appears there is a lack of identical outcome between eyes and between twins. It is, therefore, not time nor inoculum alone that leads to presence or absence of ocular disease, size of lesions, or location of lesions (Couvreur *et al.*, 1976). There are differences in specific ocular outcomes in both dizygotic and monozygotic twins. The general disease impact with respect to symptomatic involvement and eventual ocular involvement appears more concordant in dizygotic than in monozygotic twins. It is clear that to definitively answer patterns of ocular toxoplasmosis in twins a long rigorous follow-up report remains to be published.

It is unclear why macular lesions commonly occur in congenital infection. Other frequently involved areas in the brain are the periaqueductal, periventricular, and basal ganglia regions. One theory is that secondary to a high affinity transport protein for putrescein *T. gondii* thrives in the putrescein-rich foetal retina (Seabra *et al.*, 2004). A separate theory is that the macula is the first part of the retina that is vascularized as the vasculogenesis spreads peripherally from central posterior retina to the far periphery. The macula is, therefore, affected because it is the region that is vascularized longest and more likely to be infected than the peripheral retina.

5.8.4 The Clinical Features of Ocular Presentation in the Elderly

Until recently, it was thought that the elderly were an uncommon risk group for severe ocular involvement. In a publication describing an epidemic of toxoplasmosis, ocular involvement in the elderly was identified as being more common than in other age groups (Bowie *et al.*, 1997). There was a statistically significant older mean age of ocular involvement of 56.1 years of age, with a range of 15–83 years of age in the 19 reported ocular cases (Bowie *et al.*, 1997). It is hard to imagine significant confounders to this finding, as young and older patients are usually symptomatic with ocular toxoplasmosis.

5.8.5 Atypical Presentations Common in Immunocompromised Patients

Typical severe inflammation when extensive areas of retina are involved from active ocular toxoplasmosis can be absent in immunocompromised patients. In addition, the disease is clearly associated with replication of active parasites as demonstrated by histopathology (Nicholson and Wolchok, 1976; Yeo *et al.*, 1983). Although severe *Toxoplasma* chorioretinitis does develop in patients with AIDS (Parke and Font, 1986), it is uncommon (Newsome *et al.*, 1984).

5.8.6 Classification Systems for Uveitis and Chorioretinitis

Recently, an attempt has been made to develop consensus regarding uveitis nomenclature within the field of uveitis (Jabs *et al.*, 2005). The three main subdivisions of uveitis are broken down by anatomic location – anterior, posterior or intermediate uveitis – *Toxoplasma* may cause inflammation in any subdivision (either primarily or secondarily). Anterior uveitis refers to inflammation in the front of the eye anterior to the vitreous. Intermediate uveitis refers to inflammation in the vitreous or in the pars plana (tissue located just anterior to the retina). Posterior uveitis refers to inflammation within the retina or choroids.

Friedmann and Knox (Friedmann and Knox, 1969) described three types of chorioretinal lesions found in a cohort of patients with no other clinical manifestations of *T. gondii* infection. Fifty-six percent of these patients had 'large destructive lesions'. This type of lesion is characterized by its size; a large destructive lesion is usually larger than the optic disc and is likely to be associated with vision loss and/or complications such as retinal detachment, cataracts, cystoid macular edema, glaucoma and chronic intraocular inflammatory reactions. 'Punctate inner lesions' accounted for 27% of the patients. These were described as smaller and less likely to be associated with vitreous traction. 'Punctate deep lesions' appeared in 17% of the patients. This type of lesion is always located in the macula or peripapillary area. Because the infected tissue is separate from the vitreous by uninvolved retinal layers, punctate deep lesions are usually not associated with vitreous inflammation.

5.8.7 Punctate Outer Retinal Toxoplasmosis

Other reports have described a distinctly different category of *T. gondii* lesions known as 'punctuate outer retinal toxoplasmosis' (McAuley *et al.*, 1994; Doft and Gass, 1985). This condition consists of multifocal lesions that are grey to white in color and less than 1000 μM in size. These lesions appear at the deep level of the retina and retinal pigment epithelium. There is little inflammatory reaction in the vitreous, and involvement is sometimes bilateral. Though these lesions appear as a cluster, there is generally only one focus of active disease at any point in time.

5.8.8 Atypical *Toxoplasma* Chorioretinitis

Unusual manifestations of ocular toxoplasmosis exist. Bilateral, multifocal and extensive ocular toxoplasmosis can rarely occur and these cases have usually been in immunocompromised individuals. Toxoplasmosis can mimic acute retinal necrosis (ARN), an explosively blinding chorioretinitis usually caused by herpes viruses. Rarely, toxoplasmosis can mimic viral acute retinal necrosis (Moshfeghi *et al.*, 2004). Since diagnosis is usually made by clinical appearance alone, therapy is usually directed first against the herpes virus and subsequently a broader differential is entertained if instituted therapy fails to stop progression of disease. An anti-*Toxoplasma* regimen may be instituted in severe posterior uveitis as empiric therapy. A dramatic and unusual periarteritis, eponymously named Kyrielieis-type periarteritis with deposition of focal plaque-like deposits, can occur along the major arcades and may persist after resolution of active disease (Schwartz, 1977). Rarely, a very opaque form called frosted branch angiitis (Ysasaga and Davis, 1999) may occur without retinochoroidal lesions (Holland *et al.*, 1999). A neuroretinitis can also occur with its classic stellate exudative-like appearance within the macula as a consequence of *T. gondii* infection (Perrotta *et al.*, 2003; Kucukerdonmez *et al.*, 2002).

5.8.9 Optic Nerve Involvement in Ocular Toxoplasmosis

The optic nerve is populated by approximately 1.3 million nerves that originate in the retina ganglion cell layer and connect the eye to the brain. In congenital toxoplasmosis, because eyes available for autopsy are from individuals with severe CNS involvement, it is impossible to definitively determine whether optic nerve changes are secondary to direct infection or secondarily to active CNS processes such as *Toxoplasma* encephalitis. Optic nerve atrophy was present in 20% of individuals in one congenital toxoplasmosis cohort (Mets *et al.*, 1996). 'Jensen juxtapapillary retinitis' is an occasionally used eponym associated label of peripapillary *Toxoplasma* retinitis and highlights that posterior lesions are more likely to be symptomatic.

5.8.10 *Toxoplasma* and Glaucoma

Ocular toxoplasmosis appears to have a weak, but important, association with glaucoma. However, there is no prospective study that has examined the relationship between toxoplasmosis and glaucoma. Unlike herpetic uveitis which is frequently associated with glaucoma, ocular toxoplasmosis in a retrospective study found the highest incidence, 38%, of elevated intraocular pressure in patients with active ocular toxoplasmosis (Westfall *et al.*, 2005). There was, however, no associated glaucomatous nerve damage. The elevated intraocular pressure appeared to resolve with resolution of the retinochoroiditis. A prospective study will be required to further examine any possible association.

5.9 DIAGNOSTIC TESTS AND PATHOLOGY

5.9.1 Histopathology

Several reports characterizing the destructive retinochoroiditis exist (Wilder, 1952; Zimmerman, 1961; Rao and Font, 1977; Hogan, 1951). Organisms in immunocompetent individuals are identified in the retina and optic nerve, but not in the choroids. *Toxoplasma* cysts have also been demonstrated in the retinal pigmented epithelium (Nicholson and Wolchok, 1976). Tachyzoites stain well by both Wright and Giemsa stains, and bradyzoites stain well with periodic acid Schiff stain. Granulomatous choroidal inflammation and scleral thickening can occur adjacent to the retinal lesions. After resolution of active disease, the involved retina shows severe destruction with retinal atrophy and chorioretinal adhesions. In a murine congenital model of ocular toxoplasmosis, surprisingly, the photoreceptors (not the *Toxoplasma* cysts) appeared to be the focal activity of the mononuclear intraocular infiltrate (Dutton *et al.*, 1986b). Wilder's case series highlighted the value of thicker sections in an attempt to identify tachyzoites. She used celloidin which permits 18 micron sections vs. the eight microns performed with paraffin thin sections (Holland *et al.*, 2002); however, celloidin requires six weeks before sectioning and paraffin thin sections have subsequently been shown (Holland *et al.*, 1988a) to be sensitive enough to detect parasites. Wilder noted in her report of severe disease that active disease can extend into the sclera.

The original identification of *T. gondii* as an etiology of ocular disease came through histopathologic examination of ocular tissues in congenital infection which revealed organisms in the retina and adjacent choroids (Wolf, 1939). Parasites were identified in 10 of 18 eyes of infants and foetuses in a recent report (Roberts *et al.*, 2001). In the report eight of 15 eyes had focal retinal lesions some with retinal necrosis. A lesion 7 mm large was identified in a 22 week old foetus. Only one of four cases had bilateral disease. In addition, 10 of 15 eyes had clinical lesions in the peripheral retina only. A vitritis was identified in seven of the 18 eyes. In five of eight eyes where there was optic nerve present in sections, a leptomeningitis was identified and three of the eyes had disruption of the optic nerve architecture. A diffuse choroiditis was identified in all cases. Eyes from seven and five day old infants showed similar findings to the foetuses with more evidence of organization and a retinal detachment. A two year old child demonstrated areas of an end stage continuum compared to the foetal eyes with evidence of retinal atrophy, retinal pigmentary epithelial changes and overlying gliosis. The inflammatory cells present in the eyes consisted of lymphocytes, plasma cells and macrophages. Immunohistochemical staining showed both T and B cells with the B cells fewer in number and confined primarily to the choroid. The T cell population contained both CD4+ and CD8+ cells. No tissue cysts were identified in the group of eyes. However, immunohistochemical staining confirmed the presence of parasites in 10 of 15 eyes. Parasites were not identified in the choroid or substance of the optic nerve of any eyes. A key point in understanding the complexity of ocular toxoplasmosis is that normal appearing retina can harbour *Toxoplasma* (McMenamin *et al.*, 1986).

5.9.2 Ocular Biopsies

There is a risk of irreversible blindness with any type of intraocular procedure from possible complications that include bacterial endophthalmitis and retinal detachment. Since most cases of ocular toxoplasmosis can be diagnosed on clinical grounds, the need for an intraocular procedure to make the diagnosis is highly unusual. Although the risks are severe, the actual likelihood of risk of blindness is low (less than 1/100). There are two intraocular compartments that can be sampled. The anterior chamber is in the front of the eye and the aqueous humour is fluid alone without a gel matrix being present. There are several slightly different approaches to performing an anterior chamber paracentesis (Van der Lelij and Rothova, 1997). There is still a risk of complications from sampling this fluid; however, there is less risk than doing a vitreous biopsy.

Vitreous biopsy is performed by sampling the posterior chamber or vitreous cavity. Even though the vitreous is 99% water by weight, it is a collagen gel matrix, and sampling the vitreous runs the risk of other complications such as retinal detachment. This is in contrast to an anterior chamber paracentesis which would not be likely to cause a retinal detachment. Based on recent widely used medicines, for macular degeneration that require intraocular injections, a consensus on using topical povidone and a lid speculum appear to be important steps in any office based intraocular procedure to prevent bacterial endophthalmitis (Aiello *et al.*, 2004). A vitreal biopsy can be performed as an office based procedure, depending on the local office setting, operating room access and clinical scenario as to where the vitreal biopsy should be performed. The vitreal biopsy has more risks than an anterior chamber paracentesis because of the entry into the vitreous which is a gel-like substance and can cause traction on the retina. There are a wide variety of approaches to perform a vitreal biopsy.

Analysis of T cells recovered after vitrectomy in 10 patients with active suspected *Toxoplasma* chorioretinitis demonstrated the presence of *T. gondii* associated T cell clones (Feron *et al.*, 2001). In addition, there was an absence of T cell clones against retinal antigens. Three of eight patients were positive for *T. gondii* by PCR.

5.9.3 Serology

There is currently a limited role for serologic testing in recurrent ocular toxoplasmosis clinically. The author always obtains a confirmatory IgG ELISA in suspected recurrent disease; however, there is no test available that can confirm that ocular inflammation is in fact from ocular toxoplasmosis. High avidity (greater than 40%) antibodies are associated with infections that are over six months old (Liesenfeld *et al.*, 2001; Paul, 1999). Documented seroconversion is a scenario where serologic testing is useful in suspected cases of acquired ocular disease. Patients concomitantly infected with HIV and *T. gondii* can have a negative serology (Moshfeghi *et al.*, 2004).

Antibodies against toxoplasmosis may be present in the eye because of damage to the blood brain barrier leading to passive accumulation of antibody. The Goldmann–Wittner coefficient is an attempt to overcome the false positive possibility by determining the relative local concentration of antibody compared to systemic levels. While any coefficient greater than one would indicate increased local production, most authors use coefficients greater than three as indicative of a more valid and specific threshold value (de Boer *et al.*, 1994; de Boer *et al.*, 1996; Baarsma *et al.*, 1991). A Goldmann–Wittner coefficient greater than 3.0 is seen in from 50% to 70% of patients with ocular toxoplasmosis. This coefficient provides evidence of local antibody production against *T. gondii* compared to systemic production and provides support for a diagnosis of active local ocular disease. The Goldmann–Wittner coefficient is the ratio of (anti-*Toxoplasma* IgG in aqueous humour/anti-*Toxoplasma* IgG in serum)/(total IgG in serum/total IgG in aqueous humour). Ophthalmologists should understand that any infusion solution routinely used to prepare instruments immediately prior to surgery may result in a false negative result from its resultant dilution. Ophthalmologists will need to tell the operating staff not to prime equipment prior to sampling of vitreous.

Surprisingly, most cases with a positive Goldmann–Wittner coefficient have a negative PCR assay for *T. gondii*. Also, most cases with a positive PCR for *T. gondii* have a negative Goldmann–Wittner coefficient. From the above, we can conclude that the local humoral response appears not to coincide with the local proliferation of parasite. In addition, just as peripheral serology may be negative in immunocompromised patients local antibody production measurement has been reported to be negative. In one series, seven of nine immunocompromised patients had negative local antibody production but had positive PCR (Fardeau *et al.*, 2002).

5.9.4 Immunoblotting

No universal pattern exists based on immunoblotting to aid the diagnosis of ocular toxoplasmosis. One paper highlighted the antibody binding to molecular sizes below 16 kDa and above 116 kD (De Marco *et al.*, 2003); a different report highlighted binding to antibodies to a 28-kDa antibody was revealed (Klaren *et al.*, 1998), and two further papers did not detail findings (Riss *et al.*, 1995; Villard *et al.*, 2003). There is a broad range of data regarding reported immunoglobulin subtypes in ocular toxoplasmosis. IgA in the aqueous humour has been demonstrated in from 26% to 63% of patients and IgM has been reported from less than 1% to 11% in the aqueous humour (Garweg *et al.*, 2000; Ronday *et al.*, 1999). IgE has been reported from 0% to 14% in the vitreous fluid (Liotet *et al.*, 1992; Gomez-Marin *et al.*, 2000).

5.9.5 Polymerase Chain Reaction (PCR)

Since Polymerase Chain Reaction (PCR) testing requires small sample volume, it is well suited for the small samples available from most invasive ocular biopsies (Burg *et al.*, 1989; Brezin *et al.*, 1990). Unfortunately, the vitreous being 99% by weight water does not have the dense cellular substrate ideally suited for the presence of *T. gondii* (Brezin *et al.*, 1991). Large atypical lesions which would be more likely to necessitate biopsy to help substantiate diagnosis also appear to be more likely to have positive vitreous PCR results. In one series, five of seven patients with severe toxoplasmosis had positive vitreous PCR results (Montoya *et al.*, 1999).

Should a sensitive specific test requiring intraocular sampling be developed for the eye, sampling the aqueous humour would be safest. However, currently PCR testing of the aqueous has low yield in ocular toxoplasmosis (Aouizerate *et al.*, 1993; Montoya *et al.*, 1999; Bou *et al.*, 1999; Figueroa *et al.*, 2000). The yield of positive PCR testing has been reported to be from 17% in patients with chorioretinitis (Brezin *et al.*, 1991) to 100% of patients with large lesions (Fardeau *et al.*, 2002). In elderly patients larger lesions had higher yield of aqueous PCR sampling compared to smaller lesions (60% vs. 25%) (Labalette *et al.*, 2002).

When a fulminant vitritis transforms the normally transparent vitreous to an opaque structure, it can be impossible to examine the retina. Without the availability of detecting clinical features on fundus examination more dependence on ancillary tests is necessary. A multiplex-PCR has been established to help differentiate between the more common causes of retinitis (Dabil *et al.*, 2001). The *T. gondii* based primer is based on the repetitive B1 gene, which has been found to have sensitivity of 60%–70% (Montoya *et al.*, 1999). Fifteen minutes of boiling of the vitreous is required to remove PCR reaction inhibitors. As compared to lower sensitivity initially reported, modifications to this PCR, including more recent primers, have been reported with a sensitivity approaching one tachyzoite (Jones *et al.*, 2000).

The most common use of PCR in ocular disease is to differentiate *T. gondii* from herpes family viruses (Van Gelder, 2003). Short tandem multiple pathogen PCR analysis has been validated as a possible approach given the inherent limited sample from ocular tissues (Dabil *et al.*, 2001). While PCR can be less sensitive for ocular toxoplasmosis, it appears to be very sensitive for herpetic ocular disease with reported sensitivities as high as 97% (Abe *et al.*, 1998; Ganatra *et al.*, 2000).

5.9.6 Clinical Tissue Culture Systems

Tissue culture systems developed for viral culture are widely available in most eye centres and have been used to culture *T. gondii* (Miller *et al.*, 2000). Because it can take several weeks before a result becomes positive, therapy should be initiated empirically while awaiting results.

5.9.7 Ocular Imaging

The ultra-wide-field scanning laser ophthalmoscope produces ultra-wide angle fundus photography and fluorescent angiography with satisfactory image resolution down to the level of retinal capillaries (Manivannan *et al.*, 2005). A traditional fundus exam is usually sufficient for clinical diagnosis of ocular toxoplasmosis. Occasionally, retinochoroidal lesions that lie in the far peripheral retina can be missed even by the most experienced retina and uveitis specialists on carefully conducted clinical exams. Ultra-wide field retinal photography and fluorescein angiography can aid in the detection and documentation of such lesions in uveitis (Kaines *et al.*, 2009) including ocular toxoplasmosis. As the technology has only been made available commercially in recent years, there are few reports investigating the advantages of ultra-wide-field fundus photography for other retinal diseases such as diabetic retinopathy, retinal vessel occlusion and sickle cell retinopathy (Spaide, 2011; Cho and Kiss, 2011; Wessel *et al.*, 2012). In general, ultra-wide-field fundus photography provides images of similar resolution as conventional retinal photography, with additional benefits of higher sensitivity in detecting pathologies in the peripheral retina and capability of imaging through small pupils. Disadvantages of ultra-wide-field retinal photography include higher cost and limited access to the technology in some geographic areas. It is our opinion that the ultra-wide-field scanning laser ophthalmoscope is the ocular imaging modality of choice for detection and documentation of ocular toxoplasmosis where available.

Optical coherent tomography (OCT) is a noninvasive imaging technique that allows investigation of retinal morphology and pathologies. Advances in recent years have significantly improved the resolution of OCT images. Certain newer devices and software also allow for overlaying images from fundus photography, angiography and perimetry data on retinal tomography (Kiernan *et al.*, 2010). OCT can be helpful in differentiating old retinochoroidal scars from currently active lesions, detecting small serous retinal detachment and neovascular membrane formation in ocular toxoplasmosis (Diniz *et al.*, 2011).

5.9.8 Differential Diagnosis

Congenital toxoplasmosis must be differentiated from other possible causes of the classic clinical acronym 'TORCH' for a series of etiologies that share similar signs and symptoms. The acronym includes *Toxoplasma*, rubella, cytomegalovirus, syphilis and herpes simplex virus. However, emerging pathogens such as West Nile virus must be considered as part of any differential in known congenital infection (Alpert *et al.*, 2003). Recurrent toxoplasmosis with its unilateral active lesions associated with multiple adjacent chorioretinal scars with the appropriate clinical history is virtually pathognomonic. However, clinical syndromes such as serpiginous chorioretinitis and other infectious etiologies such as cytomegalovirus may occasionally be considered. For the many other possible and unusual manifestations of ocular toxoplasmosis such as pars planitis the differential diagnosis is even broader and includes autoimmune disorders such as multiple sclerosis and infections such as Lyme disease.

Importantly, there are likely to be many cases of unusual manifestations of ocular toxoplasmosis that remain undiagnosed because the limits of non-invasive assays.

5.10 THE TREATMENT AND MANAGEMENT OF OCULAR TOXOPLASMOSIS

Despite *Toxoplasma* being the most common cause of posterior uveitis in the USA and the world there is no consensus on best treatment and controversy exists as to whether treatment should be initiated. In addition, there is no treatment that prevents recurrences. Part of the lack of evidence for a best treatment is that the inflamed active component of ocular toxoplasmosis resolves in immunocompetent individuals without treatment. In addition, there is a variable course depending on host factors, environment and parasite. A very controversial review of the literature highlighted that only three designed, prospective, randomized, placebo-controlled studies exist (Stanford *et al.*, 2003) and much of the literature was deemed inappropriate for their analysis because of a lack of placebo. The conclusion of this meta-analysis went against what most would consider standard of care. Given the known clinical experience with this disease, most clinical scientists would not have sufficient ethical equipoise to design a placebo trial for the management of ocular toxoplasmosis.

A recent survey highlights the uncertainty around the treatment and understanding of toxoplasmosis (Lum *et al.*, 2005). The formal survey completed in the year 2000 was conducted of one thousand ophthalmologists in the USA and 48% of surveyed ophthalmologists responded. During the two years of 1999 and 2000, there were an estimated 253,000 visits to ophthalmologists in the USA for ocular toxoplasmosis, 24,000 of which were for active disease. There was surprising lack of understanding among surveyed respondents regarding importance of acquired disease (50%), elderly as a high-risk group (16%), and the unlikelihood of transmission to foetus from recurrence of ocular toxoplasmosis during pregnancy (30%). Only 19% of respondents compared to 15% of uveitis subspecialists treated all patients with ocular toxoplasmosis (Holland and Lewis, 2002). Surprisingly, a zone of recurrence called the papillomacular bundle, which is a vital area for vision, would only merit treatment by 51% of the respondents. In this author's opinion, this area of recurrence should always warrant treatment.

There are many different regimens that are used in the treatment of ocular toxoplasmosis. A recent survey of uveitis subspecialists reported that nine different commercially available drugs were used in 24 different possible combinations as the treatment of choice in the treatment of typical ocular toxoplasmosis (Holland and Lewis, 2002). The following 17 different oral drugs had been used by the 80 respondents as treatment in all types of ocular toxoplasmosis in descending order of frequency: clindamycin (74 [94%]), pyrimethamine (71 [90%]), sulphadiazine (64 [81%]), trimethoprim/sulphamethoxazole (64 [81%]), sulphadiazine/sulphamerazine/sulphamethazine ('triple sulpha,' 37 [47%]), doxycycline (27 [34%]), atovaquone (26 [33%]), tetracycline (25 [32%]), minocycline (20 [25%]), azithromycin (15 [19%]), sulphasoxazole (14 [18%]), pyrimethamine/sulphadoxine, clarithromycin (six [8%]), spiramycin (six [8%]), trimethoprim (six [8%]), dapsone (five [6%]), and trimetrexate (one [1%]).

Comparing results between a 1991 and 2001 survey of uveitis specialists indicates a trend toward more aggressive treatment of uveitis among respondents. There was decreased use of clindamycin between the two surveys. The initial enthusiasm for clindamycin was the finding that clindamycin appeared both to achieve good intraocular concentrations and enter into cysts well (Tabbara and O'Connor, 1975). However, the decrease in use of clindamycin is presumed to be the lack of evidence of improved outcome and the fear of side effects with its use such as pseudomembranous enterocolitis. The most commonly used treatment regimen is a combination of sulphadiazine, pyrimethamine, corticosteroids and folinic acid (Holland and Lewis, 2002). This regimen has demonstrated *in vitro* synergy for its activity against *T. gondii*. The plasma half-life of pyrimethamine in adults is 100 hours and in children is about 60 hours (McLeod *et al.*, 1992). The author and most experts consider the combination of pyrimethamine and sulphadiazine to be the gold standard for the treatment of ocular toxoplasmosis.

In a recent study from France where serologic testing for *T. gondii* is a routine part of prenatal care, 18 of 24 consecutive congenitally infected patients were examined for ocular outcome with treatment (Brezin *et al.*, 2003). An oral regimen was used to treat mothers prenatally. Pyrimethamine (50 mg per day) was alternated throughout gestation with four weeks of concomitant sulphadiazine (3 g per day) and folinic acid followed by a two week cycle of spiramycin (9 million IU per day). Postnatal treatment was continued for one year with a regimen of 1 mg/kg/day of pyrimethamine, 50 mg/kg/day of sulphadiazine, and 50 mg/week of folinic acid. The ocular outcome was that 61% had no lesions, peripheral lesions were seen in nine eyes of five children (four eyes also had posterior pole lesions), and posterior pole lesions were detected in six eyes of five children (all of which had good visual acuity). Only one patient had severe visual impairment which was associated with sensory deprivation nystagmus.

In a different study where 15 of 39 cases of congenital *Toxoplasma* infection did not result in termination of gestation, the treatment regimen was less aggressive: three grams of spiramycin per day was administered when infection was suspected and pyrimethamine plus sulphonamides was added when diagnosis in the foetus was confirmed; in a shorter median follow-up of 12 months only two patients had eye lesions (Daffos *et al.*, 1988).

For infants, the pyrimethamine dose is usually 1 mg/kg/day and for sulphadiazine 100 mg/kg/day in two equal doses. Folinic acid is given 10 mg TIW with apple juice. This infant regimen is derived from the Chicago Collaborative Treatment Trial from which there is a helpful dispensing aid based on weekly weight assessment (McAuley *et al.*, 1994). In addition, treatment is not dictated by presence or absence of eye involvement alone in congenital toxoplasmosis as extended treatment appears to be indicated to provide optimal outcome for the multiple systemic complications (Remington *et al.*, 2001). The regimen can result in prompt resolution of active ocular toxoplasmosis in newborns (Mets *et al.*, 1996).

The most common side effect from use of pyrimethamine is bone marrow toxicity. A peripheral complete blood count should be performed as much as twice a week. Folinic acid is commonly used to help ward off the toxicity associated with pyrimethamine therapy (80% of respondents; Holland and Lewis, 2002). Folinic acid can be used at doses of 5 mg (two to seven times daily per week). Folinic acid does not cross the cellular membrane of *T. gondii* and therefore has no impact on pyrimethamine efficacy (Allegra *et al.*, 1987). Sulphadiazine can cause a crystalluria which usually promptly responds to alkalinization of the urine, and there is one report in the ophthalmic literature of acute ureteric obstruction soon after initiation of therapy for ocular toxoplasmosis (Smith *et al.*, 2001).

5.10.1 Chronic Suppressive Therapy

One paper details success of chronic long term therapy in the treatment of ocular toxoplasmosis (Silveira *et al.*, 2002); however, there are no large scale multi-centre studies that validate the use of chronic suppressive therapy in ocular toxoplasmosis, although much data exist of improved outcome with chronic suppressive therapy in central nervous system toxoplasmosis in patients with AIDS. Extrapolation based on the anatomic similarities between the eye and the brain suggests that suppressive therapy could be useful; however, further study will be required to validate chronic suppressive therapy for routine care. If an individual has serious visual morbidity from frequent recurrences or very visually threatening lesions, chronic suppressive therapy should be instituted with azithromycin or clarithromycin. The duration and optimal regimen of chronic suppressive therapy remains unknown. In a recent survey, 15 out of 32 uveitis specialists indicate that they would initiate prophylactic treatment in patients with high number of recurrences (Wakefield *et al.*, 2011). Other indications suggested by survey respondents include immunocompromised patients (eight out of 32), vision threatening eye lesions (11 out of 32), monocular patients (one out of 32) and prior to cataract or vitrectomy surgery (three out of 32).

5.10.2 Corticosteroids

Intravitreal, topical, oral, and periocular corticosteroids have been used as part of the regimen to treat ocular toxoplasmosis. Topical corticosteroids were used by 80% of respondents presumably to prevent presumed complications of anterior segment inflammation such as posterior synechiae (scarring of the iris to the underlying lens). Only 17% of respondents used corticosteroids in all patients regardless of severity of inflammation. Seventy-one percent of uveitis specialist respondents would use therapy in severe vitreous inflammation. Highlighting how ocular toxoplasmosis is secondary to actively replicating parasites are several reports about poor outcome, with patients treated with corticosteroids alone without a concomitant antiparasitic regimen (O'Connor and Frenkel, 1976; Nozik, 1977; Sabates *et al.*, 1981).

5.10.3 Laser Treatment

In the past, laser treatment was given to the retina surrounding *Toxoplasma* scars (Spalter *et al.*, 1966). The theory was that the destruction of retina by local laser treatment would dramatically decrease if not eliminate the risk of recurrence. Since *T. gondii* bradyzoites have been demonstrated in normal appearing retina, the practice of laser treatment of retina as a means of prophylaxis against recurrence is rarely, if ever, employed today.

5.10.4 Subconjunctival Therapy

Clindamycin has been administered subconjuctivally with an every-other-day injection of 50 mg for fifteen injections over thirty days (Ferguson, 1981). The benefit of subconjunctival injection as a route of local medicine administration is that if the medicine were able to penetrate the sclera and get sufficient intraocular concentrations, it would be very unlikely to develop a serious route of administration complications with subconjunctival injection as compared to an intravitreal injection. Of note, a larger amount of subconjunctival clindamycin (150 mg) resulted in corneal edema (Tabbara and O'Connor, 1975).

5.10.5 Ocular Toxoplasmosis CNVM

Separate from antiparasitic therapy, complications of retinochoroiditis may require different types of clinical management and therapy. Some of these complications may require surgical interventions, such as retinal detachment, epiretinal membranes, and neovascularization (Delair *et al.*, 2011). There are recent case reports of successfully treating choroidal neovascular membrane (CNVM) with intravitreal injection of medication. CNVM is a rare but potentially vision threatening complication of ocular toxoplasmosis. Off-label use of intravitreal ranibizumab (an anti-VEGF agent) along with antiparasitic therapy demonstrated satisfactory results in treating patients with CNVM secondary to ocular toxoplasmosis (Benevento *et al.*, 2008; Ben Yahia *et al.*, 2008). Treating toxoplasmosis associated CNVM borrows from the success of treating other retinal diseases with anti-VEGF medication

(Rosenfeld *et al.*, 2006). One common factor that has been found to play a major role in the pathophysiology in the diseases that are associated with retinochoroidal ischemia is elevated intraocular vascular endothelial growth factor (VEGF) (Admis and Shima, 2005). Elevated intraocular VEGF can lead to increased vascular permeability and growth of abnormal blood vessels, which disrupt regular retinal anatomy and decrease vision. For example, intravitreal ranibizumab or bevacizumab injections provided favourable preservation of vision in patients with neovascular age-related macular degeneration (CATT Group, 2012) in an ongoing, large multi-centre randomized trial. Based on clinical exam every four weeks 1,185 patients are randomized to receive either ranibizumab or bevacizumab intravitreal injections regularly or as needed. In both drug groups, most patients show improvement of choroidal neovascularization and decrease in subretinal fluids. Sixty percent of the patients achieved 20/40 or better visual acuity at two years after initiation of treatments. If left untreated, fewer than 10% of these patients would have been able to maintain the same level visual acuity. Intravitreal therapy has dramatically improved outcomes in more common diseases, and the optimal regimen and safety in large clinical trials has not been investigated.

5.10.6 Intravitreal Therapy

Each intravitreal injection has the risk of irreversible blindness as well as other potential complications. However, recently, intravitreal therapy has been the mode of administration of medicines for a variety of eye diseases. The benefit of intravitreal therapy is that it has excellent bioavailability and has almost no risk of systemic side effects. Intravitreal clindamycin/dexamethasone, clindamycin/triamcinolone acetonide with systemic anti-*Toxoplasma* therapy (Aggio *et al.*, 2005), and liposomal encapsulated clindamycin (Peyman *et al.*, 1988) have all been used in ocular toxoplasmosis.

Intravitreal administration of clindamycin was used as an adjunct therapy in a retrospective case series of six patients, who were either intolerant or unresponsive to systemic therapy (Sobrin *et al.*, 2007). All six patients, whether with or without concomitant pars plana vitrectomy, demonstrated resolution of active toxoplasmic retinochoroiditis. Another small case series recruited twelve patients with vision threatening disease, with active retinochoroidal lesions located within 3,000 microns from fovea, or 1500 microns from the optic disc. Damage within the central retina often leads to permanent visual impairment or distortion. Three of the patients had perceived contraindication to systemic anti-toxoplasma therapy secondary to pregnancy, and the remaining either showed lack of response after at least thirty days of systemic therapy or intolerance (Lasave *et al.*, 2010). These patients received intravitreal injections of a combination of 1.5 mg of clindamycin and 400 microgram of dexamethasone. Five patients also continued with concurrent systemic therapy to minimize damage to the fovea or optic nerve. All twelve patients showed resolution of active retinochoroiditis with a mean of 3.6 intravitreal injections, with range of two to five injections. Visual acuity either improved or stabilized in almost all of the patients except one with a lesion at the fovea.

In a more recent randomized, single masked clinical trial that included 68 patients with active ocular toxoplasmosis, the efficacy of intravitreal clindamycin/dexamethasone was studied against a regimen of the more classic systemic regimen consisting of pyrimethamine, sulphadiazine and prednisolone (Soheilian *et al.*, 2011). Patients that were randomized to receive intravitreal 1 mg of clindamycin and 400 μg of dexamethasone injections had comparable reduction of active retinal lesions, visual acuity improvement and resolution of vitreous inflammation. The mean number of injections required for resolution of active retinal lesions was 1.6 with the range of one to three injections. During the two year follow-up period, both intravitreal and systemic groups had the same disease recurrence rate (5.9%). The limited number of studies thus far suggests that intravitreal injection of clindamycin/dexamethasone may be as effective as systemic therapy for recurrent ocular toxoplasmosis, with much less adverse side effects, and less demand for patient compliance. An international survey of leading uveitic experts in 2011 reports that only nine out of 32 respondents had experience with intravitreal clindamycin (Wakefield *et al.*, 2011). A larger trial with a longer follow-up period will need to be conducted to better know how intravitreal therapy fits into standard therapy.

5.11 CONCLUSION

The devastation of ocular toxoplasmosis, even though widespread in our societies, remains without appropriate attention. Perhaps this is because ocular toxoplasmosis is a disease that crosses several disciplines: epidemiology, infectious disease, ophthalmology, pediatrics, internal medicine, pathology and parasitology. Also, ocular toxoplasmosis is a disease whose active component will resolve; however, visual morbidity remains and is often permanent after inflammation resolves. With the recent identification of unique intracellular targets (Roberts *et al.*, 1998) to effectively kill *T. gondii* with little likelihood of any impact on our own cellular machinery, the future of care of patients with ocular toxoplasmosis will likely change dramatically. Indeed, it is possible in the future that regimens may not only treat active disease, but could also, by effectively killing all stages of the parasite, eliminate the frustrating possibility of recurrence as well.

Acknowledgments

We would like to thank Louis Weiss, MD, Kami Kim, MD and Rima McLeod, MD for their unusual combination of empathy, intelligence, support and enthusiasm. We would also like to thank Jessica Coyne for her assistance with early drafts of this chapter.

References

Abe, T., Sato, M., Tamai, M., 1998. Correlation of varicella-zoster virus copies and final visual acuities of acute retinal necrosis syndrome. Graefes Arch. Clin. Exp. Ophthalmol. 236, 747–752.

Abrahams, I.W., Gregerson, D.S., 1982. Longitudinal study of serum antibody responses to retinal antigens in acute ocular toxoplasmosis. Am. J. Ophthalmol. 93, 224–231.

Admis, P., Shima, T., 2005. The role of vascular endothelial growth factor in ocular health and disease. Retina 25, 11–118.

Aggio, F.B., Muccioli, C., Belfort, R., 2005. Intravitreal triamcinolone acetonide as an adjunct in the treatment of severe ocular toxoplasmosis. Eye.

Aiello, L.P., Brucker, A.J., Chang, S., Cunningham Jr., E.T., D'Amico, D.J., Flynn Jr., H.W., Grillone, L.R., Hutcherson, S., Liebmann, J.M., O'brien, T.P., Scott, I.U., Spaide, R.F., Ta, C., Trese, M.T., 2004. Evolving guidelines for intravitreous injections. Retina 24, S3–19.

Ajzenberg, D., Dumetre, A., Darde, M.L., 2005. Multiplex PCR for typing strains of Toxoplasma gondii. J. Clin. Microbiol. 43, 1940–1943.

Alford Jr., C.A., Stagno, S., Reynolds, D.W., 1974. Congenital toxoplasmosis: clinical, laboratory, and therapeutic considerations, with special reference to subclinical disease. Bull. N. Y. Acad. Med. 50, 160–181.

Allegra, C.J., Kovacs, J.A., Drake, J.C., Swan, J.C., Chabner, B.A., Masur, H., 1987. Potent in vitro and in vivo antitoxoplasma activity of the lipid-soluble antifolate trimetrexate. J. Clin. Invest. 79, 478–482.

Alpert, S.G., Fergerson, J., Noel, L.P., 2003. Intrauterine West Nile virus: ocular and systemic findings. Am. J. Ophthalmol. 136, 733–735.

Aouizerate, F., Cazenave, J., Poirier, L., Verin, P., Cheyrou, A., Begueret, J., Lagoutte, F., 1993. Detection of Toxoplasma gondii in aqueous humour by the polymerase chain reaction. Br. J. Ophthalmol. 77, 107–109.

Baarsma, G.S., Luyendijk, L., Kijlstra, A., De Vries, J., Peperkamp, E., Mertens, D. A., Van Meurs, J.C., 1991. Analysis of local antibody production in the vitreous humor of patients with severe uveitis. Am. J. Ophthalmol. 112, 147–150.

Bahia-Oliveira, L.M., Jones, J.L., Azevedo-Silva, J., Alves, C.C., Orefice, F., Addiss, D.G., 2003. Highly endemic, waterborne toxoplasmosis in north Rio de Janeiro state. Brazil. Emerg. Infect. Dis. 9, 55–62.

Balasundaram, M., Andavar, R., Palaniswamy, M., Venkatapathy, N., 2010. Outbreak of acquired ocular toxoplasmosis involving 248 patients. Arch. Ophthalmol. 128, 28–32.

Beaman, M.H., Hunter, C.A., Remington, J.S., 1994. Enhancement of intracellular replication of Toxoplasma gondii by IL-6. Interactions with IFN-gamma and TNF-alpha. J. Immunol. 153, 4583–4587.

Benevento, D., Jager, D., Noble, G., Latkany, P., Lieler, F., Sautter, M., Meyers, S., Mets, M., Grassi, A., Rabiah, P., Boyer, K., Swisher, C., McLeod, R., Toxoplasmosis Study Group, 2008. Toxoplasmosis-associated neovascular lesions treated successfully with ranibizumab and antiparasitic therapy. Arch. Ophthalmol. 126, 1152–1156.

Beverley, J.K., Beattie, C.P., Fry, B.A., 1954. Experimental toxoplasmosis of the uveal tract. Br. J. Ophthalmol. 38, 489–496.

Blaise, P., Comhaire, Y., Rakic, J.M., 2005. Giant macular hole as an atypical consequence of a toxoplasmic chorioretinitis. Arch. Ophthalmol. 123, 863–864.

Bosch-Driessen, L.E., Berendschot, T.T., Ongkosuwito, J.V., Rothova, A., 2002. Ocular toxoplasmosis: clinical features and prognosis of 154 patients. Ophthalmology 109, 869–878.

Bosch-Driessen, L.H., Karimi, S., Stilma, J.S., Rothova, A., 2000. Retinal detachment in ocular toxoplasmosis. Ophthalmology 107, 36–40.

Bou, G., Figueroa, M.S., Marti-Belda, P., Navas, E., Guerrero, A., 1999. Value of PCR for detection of Toxoplasma gondii in aqueous humor and blood samples from immunocompetent patients with ocular toxoplasmosis. J. Clin. Microbiol. 37, 3465–3468.

Bowie, W.R., King, A.S., Werker, D.H., Isaac-Renton, J.L., Bell, A., Eng, S.B., Marion, S.A., 1997. Outbreak of toxoplasmosis associated with municipal drinking water. The BC Toxoplasma Investigation Team. Lancet 350, 173–177.

Brezin, A.P., Egwuagu, C.E., Burnier Jr., M., Silveira, C., Mahdi, R.M., Gazzinelli, R.T., Belfort Jr., R., Nussenblatt, R.B., 1990. Identification of Toxoplasma gondii in paraffin-embedded sections by the polymerase chain reaction. Am. J. Ophthalmol. 110, 599–604.

Brezin, A.P., Eqwuagu, C.E., Silveira, C., Thulliez, P., Martins, M.C., Mahdi, R.M., Belfort Jr., R., Nussenblatt, R.B., 1991. Analysis of aqueous humor in ocular toxoplasmosis. N. Engl. J. Med. 324, 699.

Brezin, A.P., Thulliez, P., Couvreur, J., Nobre, R., McLeod, R., Mets, M.B., 2003. Ophthalmic outcomes after prenatal and postnatal treatment of congenital toxoplasmosis. Am. J. Ophthalmol. 135, 779–784.

Brown, C.R., McLeod, R., 1990. Class I MHC genes and CD8+ T cells determine cyst number in Toxoplasma gondii infection. J. Immunol. 145, 3438–3441.

Burg, J.L., Grover, C.M., Pouletty, P., Boothroyd, J.C., 1989. Direct and sensitive detection of a pathogenic protozoan, Toxoplasma gondii, by polymerase chain reaction. J. Clin. Microbiol. 27, 1787–1792.

Burnett, A.J., Shortt, S.G., Isaac-Renton, J., King, A., Werker, D., Bowie, W.R., 1998. Multiple cases of acquired toxoplasmosis retinitis presenting in an outbreak. Ophthalmology 105, 1032–1037.

Burney, D.P., Chavkin, M.J., Dow, S.W., Potter, T.A., Lappin, M.R., 1998. Polymerase chain reaction for the detection of Toxoplasma gondii within aqueous humor of experimentally-inoculated cats. Vet. Parasitol. 79, 181–186.

Cano-Parra, J.L., Diaz, L.M.L., Cordoba, J.L., Gobernado, M.L., Navea, A.L., Menezo, J.L., 2000. Acute iridocyclitis in a patient with AIDS diagnosed as toxoplasmosis by PCR. Ocul. Immunol. Inflamm. 8, 127–130.

Carme, B., Bissuel, F., Ajzenberg, D., Bouyne, R., Aznar, C., Demar, M., Bichat, S., Louvel, D., Bourbigot, A.M., Peneau, C., Neron, P., Darde, M.L., 2002. Severe acquired toxoplasmosis in immunocompetent adult patients in French Guiana. J. Clin. Microbiol. 40, 4037–4044.

Chavkin, M.J., Lappin, M.R., Powell, C.C., Cooper, C.M., Munana, K.R., Howard, L.H., 1994. Toxoplasma gondii-specific antibodies in the aqueous humor of cats with toxoplasmosis. Am. J. Vet. Res. 55, 1244–1249.

Cho, M., Kiss, S., 2011. Detection and monitoring of sickle cell retinopathy using ultra-wide field color photography and fluorescein angiography. Retina 31, 728–747.

Comparison of Age-Related Macular Degeneration Treatments Trials (CATT) Research Group, Martin, F., Maquire, G., Fine, L., Ying, S., Jaffe, J., Grunwal, E., Toth, C., Redford, M., Ferris, L., 2012. Ranibizumab and bevacizumab for treatments of neovascular age-related macular degeneration: two-year results. Ophthalmology 119, 1388–1398.

Cotliar, A.M., Friedman, A.H., 1982. Subretinal neovascularisation in ocular toxoplasmosis. Br. J. Ophthalmol. 66, 524–529.

Couvreur, J., Desmonts, G., Girre, J.Y., 1976. Congenital toxoplasmosis in twins: a series of 14 pairs of twins: absence of infection in one twin in two pairs. J. Pediatr. 89, 235–240.

Couvreur, J., Desmonts, G., Tournier, G., Szusterkac, M., 1984. [A homogeneous series of 210 cases of congenital toxoplasmosis in 0 to 11-month-old infants detected prospectively]. Ann. Pediatr. (Paris) 31, 815–819.

Couvreur, J., Thulliez, P., 1996. Acquired toxoplasmosis of ocular or neurologic site: 49 cases. Presse Med. 25, 438–442.

Culbertson, W.W., Tabbara, K.F., O'Connor, R., 1982. Experimental ocular toxoplasmosis in primates. Arch. Ophthalmol. 100, 321–323.

Dabil, H., Boley, M.L., Schmitz, T.M., Van Gelder, R.N., 2001. Validation of a diagnostic multiplex polymerase chain reaction assay for infectious posterior uveitis. Arch. Ophthalmol. 119, 1315–1322.

Daffos, F., Forestier, F., Capella-Pavlovsky, M., Thulliez, P., Aufrant, C., Valenti, D., Cox, W.L., 1988. Prenatal management of 746 pregnancies at risk for congenital toxoplasmosis. N. Engl. J. Med. 318, 271–275.

Dao, A., Fortier, B., Soete, M., Plenat, F., Dubremetz, J.F., 2001. Successful reinfection of chronically infected mice by a different Toxoplasma gondii genotype. Int. J. Parasitol. 31, 63–65.

Darde, M.L., 2004. Genetic analysis of the diversity in Toxoplasma gondii. Ann. Ist Super Sanita 40, 57–63.

Darrell, R.W., Pieper Jr., S., Kurland, L.T., Jacobs, L., 1964. Chorioretinopathy and Toxoplasmosis, an Epidemiologic Study on a South Pacific Island. Arch. Ophthalmol. 71, 63–68.

Davidson, M.G., English, R.V., 1998. Feline ocular toxoplasmosis. Vet. Ophthalmol. 1, 71–80.

Davidson, M.G., Lappin, M.R., English, R.V., Tompkins, M.B., 1993. A feline model of ocular toxoplasmosis. Invest Ophthalmol. Vis. Sci. 34, 3653–3660.

de Boer, J.H., Luyendijk, L., Rothova, A., Baarsma, G.S., De Jong, P.T., Bollemeijer, J.G., Rademakers, A.J., Van Der Lelij, A., Zaal, M.J., Kijlstra, A., 1994. Detection of intraocular antibody production to herpesviruses in acute retinal necrosis syndrome. Am. J. Ophthalmol. 117, 201–210.

de Boer, J.H., Verhagen, C., Bruinenberg, M., Rothova, A., De Jong, P.T., Baarsma, G.S., Van Der Lelij, A., Ooyman, F.M., Bollemeijer, J.G., Derhaag, P.J., Kijlstra, A., 1996. Serologic and polymerase chain reaction analysis of intraocular fluids in the diagnosis of infectious uveitis. Am. J. Ophthalmol. 121, 650–658.

de Marco, R., Ceccarelli, R., Frulio, R., Palmero, C., Vittone, P., 2003. Retinochoroiditis associated with congenital toxoplasmosis in children: IgG antibody profiles demonstrating the synthesis of local antibodies. Eur. J. Ophthalmol. 13, 74–79.

Delair, E., Latkany, P., Noble, G., Rabiah, P., McLeod, R., Brezin, A., 2011. Clinical manifestations of ocular toxoplasmosis. Ocul. Immu. Infl. 19, 91–102.

Diniz, B., Regatieri, C., Andrade, R., Maia, A., 2011. Evaluation of spectral domain and time domain optical coherence tomography findings in toxoplasmic retinochoroiditis. Clin. Ophthalomol. 5, 645–650.

Doft, B.H., Gass, D.M., 1985. Punctate outer retinal toxoplasmosis. Arch. Ophthalmol. 103, 1332–1336.

Dubey, J., Beattie, C.P., 1988. Toxoplasmosis in cats. Toxoplasmosis of Animals and Man. Boca Raton, CRC Press Inc.

Dubey, J., Lago, E., Gennari, S., Su, C., Jones, J., 2012. Toxoplasmosis in humans and animals in Brazil: high prevalence, high burden of disease, and epidemiology. Parasitology 139, 1375–1424.

Dubey, J.P., 1977. Persistence of Toxoplasma gondii in the tissues of chronically infected cats. J. Parasitol. 63, 156–157.

Dubey, J.P., Carpenter, J.L., 1993. Histologically confirmed clinical toxoplasmosis in cats: 100 cases (1952-1990). J. Am. Vet. Med. Assoc. 203, 1556–1566.

Dubey, J.P., Frenkel, J.K., 1972. Cyst-induced toxoplasmosis in cats. J. Protozool. 19, 155–177.

Dubey, J.P., Frenkel, J.K., 1974. Immunity to feline toxoplasmosis: modification by administration of corticosteroids. Vet. Pathol. 11, 350–379.

Dubey, J.P., Graham, D.H., Blackston, C.R., Lehmann, T., Gennari, S.M., Ragozo, A.M., Nishi, S.M., Shen, S.K., Kwok, O.C., Hill, D.E., Thulliez, P., 2002. Biological and genetic characterisation of Toxoplasma gondii isolates from chickens (Gallus domesticus) from Sao Paulo, Brazil: unexpected findings. Int. J. Parasitol. 32, 99–105.

Dubey, J.P., Graham, D.H., Da Silva, D.S., Lehmann, T., Bahia-Oliveira, L.M., 2003. Toxoplasma gondii isolates of free-ranging chickens from Rio de Janeiro, Brazil: mouse mortality, genotype, and oocyst shedding by cats. J. Parasitol. 89, 851–853.

Dubey, J.P., Mattix, M.E., Lipscomb, T.P., 1996. Lesions of neonatally induced toxoplasmosis in cats. Vet. Pathol. 33, 290–295.

Dunn, D., Wallon, M., Peyron, F., Petersen, E., Peckham, C., Gilbert, R., 1999. Mother-to-child transmission of toxoplasmosis: risk estimates for clinical counselling. Lancet 353, 1829–1833.

Dutton, G.N., Hay, J., Hair, D.M., Ralston, J., 1986a. Clinicopathological features of a congenital murine model of ocular toxoplasmosis. Graefes Arch. Clin. Exp. Ophthalmol. 224, 256–264.

Dutton, G.N., McMenamin, P.G., Hay, J., Cameron, S., 1986b. The ultrastructural pathology of congenital murine toxoplasmic retinochoroiditis. Part II: The morphology of the inflammatory changes. Exp. Eye Res. 43, 545–560.

Engstrom Jr., R.E., Holland, G.N., Nussenblatt, R.B., Jabs, D.A., 1991. Current practices in the management of ocular toxoplasmosis. Am. J. Ophthalmol. 111, 601–610.

Epiphanio, S., Sinhorini, I.L., Catao-Dias, J.L., 2003. Pathology of toxoplasmosis in captive new world primates. J. Comp. Pathol. 129, 196–204.

Eyles, D., Coleman, N., 1953. Synergistic effect of sulphadiazine and daraprim against experimental toxoplasmosis in the mouse. Antibiot. Chemother. 3, 483–490.

Fair, J.R., 1958. Congenital toxoplasmosis; chorioretinitis as the only manifestation of the disease. Am. J. Ophthalmol. 46, 135–154.

Fardeau, C., Romand, S., Rao, N.A., Cassoux, N., Bettembourg, O., Thulliez, P., Lehoang, P., 2002. Diagnosis of toxoplasmic retinochoroiditis with atypical clinical features. Am. J. Ophthalmol. 134, 196–203.

Ferguson Jr., J.G., 1981. Clindamycin therapy for toxoplasmosis. Ann. Ophthalmol. 13, 95–100.

Feron, E.J., Klaren, V.N., Wierenga, E.A., Verjans, G.M., Kijlstra, A., 2001. Characterization of Toxoplasma gondii-specific T cells recovered from vitreous fluid of patients with ocular toxoplasmosis. Invest Ophthalmol. Vis. Sci. 42, 3228–3232.

Figueroa, M.S., Bou, G., Marti-Belda, P., Lopez-Velez, R., Guerrero, A., 2000. Diagnostic value of polymerase chain reaction in blood and aqueous humor in immunocompetent patients with ocular toxoplasmosis. Retina 20, 614–619.

Frenkel, J., 1949. Uveitis and toxoplasmin sensitivity. Am. J. Ophthalmol. 32, 127–135.

Frenkel, J., 1955. Ocular lesions in hamsters; with chronic Toxoplasma and Besnoitia infection. Am. J. Ophthalmol. 39, 203–225.

Frenkel, J., 1974. Pathology and pathogenesis of congenital toxoplasmosis. Bull. N. Y. Acad. Med. 50, 182–191.

Friedmann, C.T., Knox, D.L., 1969. Variations in recurrent active toxoplasmic retinochoroiditis. Arch. Ophthalmol. 81, 481–493.

Friedrich, R., Simon, H.U., Muller, W.A., Sych, F.J., 1992. Ocular toxoplasmosis: the role of cellular immune defense in the development of recurrences. Results from animal experiments. Ger. J. Ophthalmol. 1, 54–57.

Fux, B., Rodrigues, C.V., Portela, R.W., Silva, N.M., Su, C., Sibley, D., Vitor, R.W., Gazzinelli, R.T., 2003. Role of cytokines and major histocompatibility complex restriction in mouse resistance to infection with a natural recombinant strain (type 1-3) of Toxoplasma gondii. Infect. Immun. 71, 6392–6401.

Ganatra, J.B., Chandler, D., Santos, C., Kuppermann, B., Margolis, T.P., 2000. Viral causes of the acute retinal necrosis syndrome. Am. J. Ophthalmol. 129, 166–172.

Garweg, J.G., Jacquier, P., Boehnke, M., 2000. Early aqueous humor analysis in patients with human ocular toxoplasmosis. J. Clin. Microbiol. 38, 996–1001.

Garweg, J.G., Kuenzli, H., Boehnke, M., 1998. Experimental ocular toxoplasmosis in naive and primed rabbits. Ophthalmologica 212, 136–141.

Gazzinelli, R.T., Brezin, A., Li, Q., Nussenblatt, R.B., Chan, C.C., 1994. Toxoplasma gondii: acquired ocular toxoplasmosis in the murine model, protective role of TNF-alpha and IFN-gamma. Exp. Parasitol. 78, 217–229.

Gil, C.D., Mineo, J.R., Smith, R.L., Oliani, S.M., 2002. Mast cells in the eyes of Calomys callosus (Rodentia: Cricetidae) infected by Toxoplasma gondii. Parasitol. Res. 88, 557–562.

Gilbert, R.E., See, S.E., Jones, L.V., Stanford, M.S., 2002. Antibiotics versus control for Toxoplasma retinochoroiditis. Cochrane Database Syst. Rev. CD002218.

Gilbert, R.E., Stanford, M.R., 2000. Is ocular toxoplasmosis caused by prenatal or postnatal infection? Br. J. Ophthalmol. 84, 224–226.

Glasner, P.D., Silveira, C., Kruszon-Moran, D., Martins, M.C., Burnier Junior, M., Silveira, S., Camargo, M.E., Nussenblatt, R.B., Kaslow, R.A., Belfort Junior, R., 1992. An unusually high prevalence of ocular toxoplasmosis in southern Brazil. Am. J. Ophthalmol. 114, 136–144.

Gomez-Marin, J.E., Montoya-De-Londono, M.T., Castano-Osorio, J.C., Heine, F. A., Duque, A.M., Chemla, C., Aubert, D., Bonhomme, A., Pinon, J.M., 2000. Frequency of specific anti-Toxoplasma gondii IgM, IgA and IgE in colombian patients with acute and chronic ocular toxoplasmosis. Mem. Inst. Oswaldo Cruz 95, 89–94.

Gormley, P.D., Pavesio, C.E., Luthert, P., Lightman, S., 1999. Retinochoroiditis is induced by oral administration of Toxoplasma gondii cysts in the hamster model. Exp. Eye Res. 68, 657–661.

Grigg, M.E., Bonnefoy, S., Hehl, A.B., Suzuki, Y., Boothroyd, J.C., 2001a. Success and virulence in Toxoplasma as the result of sexual recombination between two distinct ancestries. Science 294, 161–165.

Grigg, M.E., Ganatra, J., Boothroyd, J.C., Margolis, T.P., 2001b. Unusual abundance of atypical strains associated with human ocular toxoplasmosis. J. Infect. Dis. 184, 633–639.

Guerina, N.G., Hsu, H.W., Meissner, H.C., Maguire, J.H., Lynfield, R., Stechenberg, B., Abroms, I., Pasternack, M.S., Hoff, R., Eaton, R.B., et al., 1994. Neonatal serologic screening and early treatment for congenital Toxoplasma gondii infection. The New England Regional Toxoplasma Working Group. N. Engl. J. Med. 330, 1858–1863.

Gustafsson, K., Wattrang, E., Fossum, C., Heegaard, P.M., Lind, P., Uggla, A., 1997. Toxoplasma gondii infection in the mountain hare (Lepus timidus) and domestic rabbit (Oryctolagus cuniculus). II. Early immune reactions. J. Comp. Pathol. 117, 361–369.

Hall, S.M., Pandit, A., Golwilkar, A., Williams, T.S., 1999. How do Jains get Toxoplasma infection? Lancet 354, 486–487.

Hawes, N.L., Smith, R.S., Chang, B., Davisson, M., Heckenlively, J.R., John, S.W., 1999. Mouse fundus photography and angiography: a catalogue of normal and mutant phenotypes. Mol. Vis. 5, 22.

Hay, J., Lee, W.R., Dutton, G.N., Hutchison, W.M., Siim, J.C., 1984. Congenital toxoplasmic retinochoroiditis in a mouse model. Ann. Trop. Med. Parasitol. 78, 109–116.

Hayashi, S., Chan, C.C., Gazzinelli, R.T., Pham, N.T., Cheung, M.K., Roberge, F.G., 1996. Protective role of nitric oxide in ocular toxoplasmosis. Br. J. Ophthalmol. 80, 644–648.

Hogan, M., 1951. Ocular Toxoplasmosis. Columbia University Press, New York.

Hogan, M.J., 1958. Ocular toxoplasmosis. Am. J. Ophthalmol. 46, 467–494.

Hogan, M.J., 1961. Ocular toxoplasmosis in adult patients. Surv. Ophthalmol. 6, 835–851.

Hogan, M.J., Kimura, S.J., O'Connor, G.R., 1964. Ocular Toxoplasmosis. Arch. Ophthalmol. 72, 592–600.

Hogan, M.J., Moschini, G.B., Zardi, O., 1971. Effects of Toxoplasma gondii toxin on the rabbit eye. Am. J. Ophthalmol. 72, 733–742.

Holland, G.N., 1999. Reconsidering the pathogenesis of ocular toxoplasmosis. Am. J. Ophthalmol. 128, 502–505.

Holland, G.N., 2003. Ocular toxoplasmosis: a global reassessment. Part I: epidemiology and course of disease. Am. J. Ophthalmol. 136, 973–988.

Holland, G.N., Engstrom Jr., R.E., Glasgow, B.J., Berger, B.B., Daniels, S.A., Sidikaro, Y., Harmon, J.A., Fischer, D.H., Boyer, D.S., Rao, N.A., et al., 1988a. Ocular toxoplasmosis in patients with the acquired immunodeficiency syndrome. Am. J. Ophthalmol. 106, 653–667.

Holland, G.N., Lewis, K.G., 2002. An update on current practices in the management of ocular toxoplasmosis. Am. J. Ophthalmol. 134, 102–114.

Holland, G.N., Lewis, K.G., O'Connor, G.R., 2002. Ocular toxoplasmosis: a 50th anniversary tribute to the contributions of Helenor Campbell Wilder Foerster. Arch. Ophthalmol. 120, 1081–1084.

Holland, G.N., Muccioli, C., Silveira, C., Weisz, J.M., Belfort Jr., R., O'Connor, G.R., 1999. Intraocular inflammatory reactions without focal necrotizing retinochoroiditis in patients with acquired systemic toxoplasmosis. Am. J. Ophthalmol. 128, 413–420.

Holland, G.N., O'Connor, G.R., Diaz, R.F., Minasi, P., Wara, W.M., 1988b. Ocular toxoplasmosis in immunosuppressed nonhuman primates. Invest. Ophthalmol. Vis. Sci. 29, 835–842.

Holliman, R.E., 1997. Toxoplasmosis, behaviour and personality. J. Infect. 35, 105–110.

Howe, D.K., Sibley, L.D., 1995. Toxoplasma gondii comprises three clonal lineages: correlation of parasite genotype with human disease. J. Infect. Dis. 172, 1561–1566.

Hu, M.S., Schwartzman, J.D., Lepage, A.C., Khan, I.A., Kasper, L.H., 1999. Experimental ocular toxoplasmosis induced in naive and preinfected mice by intracameral inoculation. Ocul. Immunol. Inflamm. 7, 17–26.

Hunter, C.A., Bermudez, L., Beernink, H., Waegell, W., Remington, J.S., 1995. Transforming growth factor-beta inhibits interleukin-12-induced production of interferon-gamma by natural killer cells: a role for transforming growth factor-beta in the regulation of T cell-independent resistance to Toxoplasma gondii. Eur. J. Immunol. 25, 994–1000.

Innes, E.A., 1997. Toxoplasmosis: comparative species susceptibility and host immune response. Comp. Immunol. Microbiol. Infect. Dis. 20, 131–138.

Jabs, D.A., Nussenblatt, R.B., Rosenbaum, J.T., 2005. Standardization of uveitis nomenclature for reporting clinical data. Results of the First International Workshop. Am. J. Ophthalmol. 140, 509–516.

Jacklin, H.N., 1975. Cryotreatment of toxoplasmosis retinochoroiditis. Ann. Ophthalmol. 7, 853–855.

Jackson, M.H., Hutchison, W.M., 1989. The prevalence and source of Toxoplasma infection in the environment. Adv. Parasitol. 28, 55–105.

Jankû, J., 1923. Pathogenesa a Pathologická Anatomie T. Zv. Vrozeneho Kolobomu Žlute Škurny Oku Normál ně Velikém de Mikrophtailmickěm s Nălezem Parasitu v Sitnici. Čăs Lék Čes 62, 1021–1027.

Jones, J.L., Holland, GN, 2010. Annual burden of ocular toxoplasmosis in the US. Am. J. Trop. Med. Hyg. 82, 464–465.

Jones, C.D., Okhravi, N., Adamson, P., Tasker, S., Lightman, S., 2000. Comparison of PCR detection methods for B1, P30, and 18S rDNA genes of T. gondii in aqueous humor. Invest. Ophthalmol. Vis. Sci. 41, 634–644.

Jones, J.L., Kruszon-Moran, D., Wilson, M., McQuillan, G., Navin, T., McAuley, J.B., 2001a. Toxoplasma gondii infection in the USA: seroprevalence and risk factors. Am. J. Epidemiol. 154, 357–365.

Jones, J.L., Lopez, A., Wilson, M., Schulkin, J., Gibbs, R., 2001b. Congenital toxoplasmosis: a review. Obstet Gynecol. Surv. 56, 296–305.

Kaines, A., Tsui, I., Sarraf, D., Schwartz, S., 2009. The use of ultra-wide-field fluorescein angiography in evaluation and management of uveitis. Semin. Ophthalmol. 24, 19–24.

Khan, A., Taylor, S., Su, C., Mackey, A.J., Boyle, J., Cole, R., Glover, D., Tang, K., Paulsen, I.T., Berriman, M., Boothroyd, J.C., Pfefferkorn, E.R., Dubey, J.P., Ajioka, J.W., Roos, D.S., Wootton, J.C., Sibley, L.D., 2005. Composite genome map and recombination parameters derived from three archetypal lineages of Toxoplasma gondii. Nucleic Acids Res. 33, 2980–2992.

Kiernan, D., Mieler, W., Hariprasad, S., 2010. Spectral-Domain Opital Coherence Tomography: A comparison of modern high resolution retinal imaging system. Am. J. Ophthalmol. 149, 18–31.

Klaren, V.N., Van Doornik, C.E., Ongkosuwito, J.V., Feron, E.J., Kijlstra, A., 1998. Differences between intraocular and serum antibody responses in patients with ocular toxoplasmosis. Am. J. Ophthalmol. 126, 698–706.

Kopec, R., De Caro, G., Chapnick, E., Ghitan, M., Saffra, N., 2003. Prophylaxis for ocular toxoplasmosis. Clin. Infect. Dis. 37, e147–e148.

Koppe, J., Klooterman, G.J., De Roever-Bonnet, H., et al., 1974. Toxoplasmosis and pregnancy, with a long-term follow-up of the children. Eur. J. Obstet. Gynecol. Reprod. Biol. 4, 101–110.

Koppe, J.G., Loewer-Sieger, D.H., De Roever-Bonnet, H., 1986. Results of 20-year follow-up of congenital toxoplasmosis. Lancet 1, 254–256.

Kucukerdonmez, C., Akova, Y.A., Yilmaz, G., 2002. Ocular toxoplasmosis presenting as neuroretinitis: report of two cases. Ocul. Immunol. Inflamm. 10, 229–234.

Labalette, P., Delhaes, L., Margaron, F., Fortier, B., Rouland, J.F., 2002. Ocular toxoplasmosis after the fifth decade. Am. J. Ophthalmol. 133, 506–515.

Langermans, J.A., Van Der Hulst, M.E., Nibbering, P.H., Hiemstra, P.S., Fransen, L., Van Furth, R., 1992. IFN-gamma-induced L-arginine-dependent Toxoplasmastatic activity in murine peritoneal macrophages is mediated by endogenous tumor necrosis factor-alpha. J. Immunol. 148, 568–574.

Lappin, M.R., Burney, D.P., Dow, S.W., Potter, T.A., 1996. Polymerase chain reaction for the detection of Toxoplasma gondii in aqueous humor of cats. Am. J. Vet. Res. 57, 1589–1593.

Lasave, A., Diaz-Llopis, M., Muccioli, C., Belfort, R., Arevalo, F., 2010. Intravitreal clindamycin and dexamethasone for zone 1 toxoplasmic retinochoroiditis at twenty-four months. Ophthalmology 117, 1831–1838.

Levaditi, C., 1928. Au sujet de certaines protozooses Hereditaires humaines a localization oculaires et nerveuses. C. R. Soc. Biol. (Paris) 98, 297–200.

Lieb, D.F., Scott, I.U., Flynn Jr., H.W., Davis, J.L., Demming, S.M., 2004. Acute acquired Toxoplasma retinitis may present similarly to unilateral acute idiopathic maculopathy. Am. J. Ophthalmol. 137, 940–942.

Liesenfeld, O., 1999. Immune responses to Toxoplasma gondii in the gut. Immunobiology 201, 229–239.

Liesenfeld, O., Montoya, J.G., Kinney, S., Press, C., Remington, J.S., 2001. Effect of testing for IgG avidity in the diagnosis of Toxoplasma gondii infection in pregnant women: experience in a US reference laboratory. J. Infect. Dis. 183, 1248–1253.

Liotet, S., Bloch-Michel, E., Petithory, J.C., Batellier, L., Chaumeil, C., 1992. Biological modifications of the vitreous in intraocular parasitosis: preliminary study. Int. Ophthalmol. 16, 75–80.

Lopez, A., Dietz, V.J., Wilson, M., Navin, T.R., Jones, J.L., 2000. Preventing congenital toxoplasmosis. MMWR Recomm. Rep. 49, 59–68.

Lu, F., Huang, S., Hu, M.S., Kasper, L.H., 2005. Experimental ocular toxoplasmosis in genetically susceptible and resistant mice. Infect. Immun. 73, 5160–5165.

Lu, F., Huang, S., Kasper, L.H., 2003. Interleukin-10 and pathogenesis of murine ocular toxoplasmosis. Infect. Immun. 71, 7159–7163.

Lum, F., Jones, J.L., Holland, G.N., Liesegang, T.J., 2005. Survey of ophthalmologists about ocular toxoplasmosis. Am. J. Ophthalmol. 140, 724–726.

Lyons, R.E., Anthony, J.P., Ferguson, D.J., Byrne, N., Alexander, J., Roberts, F., Roberts, C.W., 2001. Immunological studies of chronic ocular toxoplasmosis: up-regulation of major histocompatibility complex class I and transforming growth factor beta and a protective role for interleukin-6. Infect. Immun. 69, 2589–2595.

Maetz, H.M., Kleinstein, R.N., Federico, D., Wayne, J., 1987. Estimated prevalence of ocular toxoplasmosis and toxocariasis in Alabama. J. Infect. Dis. 156, 414.

Manivannan, A., Plskova, J., Farrow, A., McKay, S., Sharp, F., Forrester, V., 2005. Ultra-wide-field fluorescein angiography of the ocular fundus. Am. J. Ophthalmol. 140, 525–527.

Mcauley, J., Boyer, K.M., Patel, D., Mets, M., Swisher, C., Roizen, N., Wolters, C., Stein, L., Stein, M., Schey, W., et al., 1994. Early and longitudinal evaluations of treated infants and children and untreated historical patients with congenital toxoplasmosis: the Chicago Collaborative Treatment Trial. Clin. Infect. Dis. 18, 38–72.

Mccannel, C.A., Holland, G.N., Helm, C.J., Cornell, P.J., Winston, J.V., Rimmer, T.G., 1996. Causes of uveitis in the general practice of ophthalmology. UCLA Community-Based Uveitis Study Group. Am. J. Ophthalmol. 121, 35–46.

McLeod, R., Mack, D., Foss, R., Boyer, K., Withers, S., Levin, S., Hubbell, J., 1992. Levels of pyrimethamine in sera and cerebrospinal and ventricular fluids from infants treated for congenital toxoplasmosis. Toxoplasmosis Study Group. Antimicrob. Agents Chemother. 36, 1040–1048.

McLeod, R., Skamene, E., Brown, C.R., Eisenhauer, P.B., Mack, D.G., 1989. Genetic regulation of early survival and cyst number after peroral Toxoplasma gondii infection of A x B/B x A recombinant inbred and B10 congenic mice. J. Immunol. 143, 3031–3034.

Mcmenamin, P.G., Dutton, G.N., Hay, J., Cameron, S., 1986. The ultrastructural pathology of congenital murine toxoplasmic retinochoroiditis. Part I: The localization and morphology of Toxoplasma cysts in the retina. Exp. Eye Res. 43, 529–543.

Mets, M.B., Holfels, E., Boyer, K.M., Swisher, C.N., Roizen, N., Stein, L., Stein, M., Hopkins, J., Withers, S., Mack, D., Luciano, R., Patel, D., Remington, J.S., Meier, P., McLeod, R., 1996. Eye manifestations of congenital toxoplasmosis. Am. J. Ophthalmol. 122, 309–324.

Mets, M.B., Holfels, E., Boyer, K.M., Swisher, C.N., Roizen, N., Stein, L., Stein, M., Hopkins, J., Withers, S., Mack, D., Luciano, R., Patel, D., Remington, J.S., Meier, P., McLeod, R., 1997. Eye manifestations of congenital toxoplasmosis. Am. J. Ophthalmol. 123, 1–16.

Miller, D., Davis, J., Rosa, R., Diaz, M., Perez, E., 2000. Utility of tissue culture for detection of Toxoplasma gondii in vitreous humor of patients diagnosed with toxoplasmic retinochoroiditis. J. Clin. Microbiol. 38, 3840–3842.

Mitchell, C.D., Erlich, S.S., Mastrucci, M.T., Hutto, S.C., Parks, W.P., Scott, G.B., 1990. Congenital toxoplasmosis occurring in infants perinatally infected with human immunodeficiency virus 1. Pediatr. Infect. Dis. J. 9, 512–518.

Montoya, J.G., Liesenfeld, O., 2004. Toxoplasmosis. Lancet 363, 1965–1976.

Montoya, J.G., Parmley, S., Liesenfeld, O., Jaffe, G.J., Remington, J.S., 1999. Use of the polymerase chain reaction for diagnosis of ocular toxoplasmosis. Ophthalmology 106, 1554–1563.

Montoya, J.G., Remington, J.S., 1996. Toxoplasmic chorioretinitis in the setting of acute acquired toxoplasmosis. Clin. Infect. Dis. 23, 277–282.

Moshfeghi, D.M., Dodds, E.M., Couto, C.A., Santos, C.I., Nicholson, D.H., Lowder, C.Y., Davis, J.L., 2004. Diagnostic approaches to severe, atypical toxoplasmosis mimicking acute retinal necrosis. Ophthalmology 111, 716–725.

Nagineni, C.N., Pardhasaradhi, K., Martins, M.C., Detrick, B., Hooks, J.J., 1996. Mechanisms of interferon-induced inhibition of Toxoplasma gondii replication in human retinal pigment epithelial cells. Infect. Immun. 64, 4188–4196.

Newman, P.E., Ghosheh, R., Tabbara, K.F., O'Connor, G.R., Stern, W., 1982. The role of hypersensitivity reactions to Toxoplasma antigens in experimental ocular toxoplasmosis in nonhuman primates. Am. J. Ophthalmol. 94, 159–164.

Newsome, D.A., Green, W.R., Miller, E.D., Kiessling, L.A., Morgan, B., Jabs, D.A., Polk, B.F., 1984. Microvascular aspects of acquired immune deficiency syndrome retinopathy. Am. J. Ophthalmol. 98, 590–601.

Nicholson, D.H., Wolchok, E.B., 1976. Ocular toxoplasmosis in an adult receiving long-term corticosteroid therapy. Arch. Ophthalmol. 94, 248–254.

Nozik, R.A., 1977. Results of treatment of ocular toxoplasmosis with injectable corticosteroids. Trans. Sect. Ophthalmol. Am. Acad. Ophthalmol. Otolaryngol. 83, 811–818.

Nozik, R.A., O'Connor, G.R., 1968. Experimental toxoplasmic retinochoroiditis. Arch. Ophthalmol. 79, 485–489.

Nozik, R.A., O'Connor, G.R., 1970. Studies on experimental ocular toxoplasmosis in the rabbit. II. Attempts to stimulate recurrences by local trauma, epinephrine, and corticosteroids. Arch. Ophthalmol. 84, 788–791.

Nussenblatt, R.B., Mittal, K.K., Fuhrman, S., Sharma, S.D., Palestine, A.G., 1989. Lymphocyte proliferative responses of patients with ocular toxoplasmosis to parasite and retinal antigens. Am. J. Ophthalmol. 107, 632–641.

O'Connor, G.R., 1974. Manifestations and management of ocular toxoplasmosis. Bull. N. Y. Acad. Med. 50, 192–210.

O'Connor, G.R., 1983a. Factors related to the initiation and recurrence of uveitis. XL Edward Jackson memorial lecture. Am. J. Ophthalmol. 96, 577–599.

O'Connor, G., 1983b. The roles of parasite invasion and of hypersensitivity in the pathogenesis of toxoplasmic retinochoroiditis. Ocular Inflammation Therapy 1, 37.

O'Connor, G.R., Frenkel, J.K., 1976. Editorial: Dangers of steroid treatment in toxoplasmosis. Periocular injections and systemic therapy. Arch. Ophthalmol. 94, 213.

O'Connor, G.R., Nozik, R.A., 1971. Studies on experimental ocular toxoplasmosis in the rabbit. 3. Recurrent inflammation stimulated by systemic administration of antilymphocyte serum and normal horse serum. Arch. Ophthalmol. 85, 718–722.

O'Neill, J.F., 1998. The ocular manifestations of congenital infection: a study of the early effect and long-term outcome of maternally transmitted rubella and toxoplasmosis. Trans. Am. Ophthalmol. Soc. 96, 813–879.

Ongkosuwito, J.V., Bosch-Driessen, E.H., Kijlstra, A., Rothova, A., 1999. Serologic evaluation of patients with primary and recurrent ocular toxoplasmosis for evidence of recent infection. Am. J. Ophthalmol. 128, 407–412.

Parke, D.W., Font, R.L., 1986. Diffuse toxoplasmic retinochoroiditis in a patient with AIDS. Arch. Ophthalmol. 104, 571–575.

Paul, M., 1999. Immunoglobulin G avidity in diagnosis of toxoplasmic lymphadenopathy and ocular toxoplasmosis. Clin. Diagn. Lab Immunol. 6, 514–518.

Pavesio, C.E., Chiappino, M.L., Gormley, P., Setzer, P.Y., Nichols, B.A., 1995. Acquired retinochoroiditis in hamsters inoculated with ME 49 strain Toxoplasma. Invest. Ophthalmol. Vis. Sci. 36, 2166–2175.

Peiffer Jr., R.L., Wilcock, B.P., 1991. Histopathologic study of uveitis in cats: 139 cases (1978–1988). J. Am. Vet. Med. Assoc. 198, 135–138.

Perkins, E.S., 1973. Ocular toxoplasmosis. Br. J. Ophthalmol. 57, 1–17.

Perrotta, S., Nobili, B., Grassia, C., Sebastiani, A., Parmeggiani, F., Costagliola, C., 2003. Bilateral neuroretinitis in a 6-year-old boy with acquired toxoplasmosis. Arch. Ophthalmol. 121, 1493–1496.

Peyman, G.A., Charles, H.C., Liu, K.R., Khoobehi, B., Niesman, M., 1988. Intravitreal liposome-encapsulated drugs: a preliminary human report. Int. Ophthalmol. 12, 175–182.

Peyron, F., Ateba, A.B., Wallon, M., Kodjikian, L., Binquet, C., Fleury, J., Garweg, J.G., 2003. Congenital toxoplasmosis in twins: a report of fourteen consecutive cases and a comparison with published data. Pediatr. Infect. Dis. J. 22, 695–701.

Portela, R.W., Bethony, J., Costa, M.I., Gazzinelli, A., Vitor, R.W., Hermeto, F.M., Correa-Oliveira, R., Gazzinelli, R.T., 2004. A multihousehold study reveals a positive correlation between age, severity of ocular toxoplasmosis, and levels of glycoinositolphospholipid-specific immunoglobulin A. J. Infect. Dis. 190, 175–183.

Powell, C.C., Lappin, M.R., 2001. Clinical ocular toxoplasmosis in neonatal kittens. Vet. Ophthalmol. 4, 87–92.

Ramchandani, M., Weaver, J.B., Joynson, D.H., Murray, P.I., 2002. Acquired ocular toxoplasmosis in pregnancy. Br. J. Ophthalmol. 86, 938–939.

Rao, N.A., Font, R.L., 1977. Toxoplasmic retinochoroiditis: electron-microscopic and immunofluorescence studies of formalin-fixed tissue. Arch. Ophthalmol. 95, 273–277.

Remington, J.S., McLeod, R., Thulliez, P., Desmonts., G., 2001. Toxoplasmosis. In: Remington, J.A.K.J.O. (Ed.), Infectious Diseases of the Fetus and Newborn Infant, fifth ed. W.B. Saunders Company, Philadephia.

Rieger, H., 1951. Zur Klinik de rim Erwachsenenalter erworbenen Toxoplasmose. Klin Monatsbl Augenheilkd 119, 459–476.

Rieger, H., 1959. Toxoplasmosis congenita und Zwillingsschwangerschaft. Klin Monatsbl Augenheilkd 134, 862–871.

Riss, J.M., Carboni, M.E., Franck, J.Y., Mary, C.J., Dumon, H., Ridings, B., 1995. Ocular toxoplasmosis: value of immunoblotting for the determination of an intra-ocular synthesis of antibodies. Pathol. Biol. (Paris) 43, 772–778.

Roberts, C.W., Cruickshank, S.M., Alexander, J., 1995. Sex-determined resistance to Toxoplasma gondii is associated with temporal differences in cytokine production. Infect. Immun. 63, 2549–2555.

Roberts, F., McLeod, R., 1999. Pathogenesis of toxoplasmic retinochoroiditis. Parasitol. Today 15, 51–57.

Roberts, F., Mets, M.B., Ferguson, D.J., O'Grady, R., O'Grady, C., Thulliez, P., Brezin, A.P., McLeod, R., 2001. Histopathological features of ocular toxoplasmosis in the foetus and infant. Arch. Ophthalmol. 119, 51–58.

Roberts, F., Roberts, C.W., Ferguson, D.J., McLeod, R., 2000. Inhibition of nitric oxide production exacerbates chronic ocular toxoplasmosis. Parasite Immunol. 22, 1–5.

Roberts, F., Roberts, C.W., Johnson, J.J., Kyle, D.E., Krell, T., Coggins, J.R., Coombs, G.H., Milhous, W.K., Tzipori, S., Ferguson, D.J., Chakrabarti, D., McLeod, R., 1998. Evidence for the shikimate pathway in apicomplexan parasites. Nature 393, 801–805.

Roberts, T., Frenkel, J.K., 1990. Estimating income losses and other preventable costs caused by congenital toxoplasmosis in people in the USA. J. Am. Vet. Med. Assoc. 196, 249–256.

Roberts, T., Murrell, K.D., Marks, S., 1994. Economic losses caused by foodborne parasitic diseases. Parasitol. Today 10, 419–423.

Ronday, M.J., Ongkosuwito, J.V., Rothova, A., Kijlstra, A., 1999. Intraocular anti-Toxoplasma gondii IgA antibody production in patients with ocular toxoplasmosis. Am. J. Ophthalmol. 127, 294–300.

Rosenfeld, J., Brown, M., Heier, S., Boyer, S., Kaiser, K., Chung, Y., Kim, Y., Marina Study Group, 2006. Ranibizumab for neovascular age-related macular degeneration. N. Engl. J. Med. 355, 1419–1431.

Ross, R.D., Stec, L.A., Werner, J.C., Blumenkranz, M.S., Glazer, L., Williams, G.A., 2001. Presumed acquired ocular toxoplasmosis in deer hunters. Retina 21, 226–229.

Sabates, R., Pruett, R.C., Brockhurst, R.J., 1981. Fulminant ocular toxoplasmosis. Am. J. Ophthalmol. 92, 497–503.

Sabin, H.A., Feldman, H.A., 1948. Dyes as microchemical indicators of a new immunity phenomenon affecting a protozoon parasite (Toxoplasma). Science 108, 660–663.

Saeij, J.P., Boyle, J.P., Boothroyd, J.C., 2005a. Differences among the three major strains of Toxoplasma gondii and their specific interactions with the infected host. Trends Parasitol. 21, 476–481.

Saeij, J.P., Boyle, J.P., Grigg, M.E., Arrizabalaga, G., Boothroyd, J.C., 2005b. Bioluminescence imaging of Toxoplasma gondii infection in living mice reveals dramatic differences between strains. Infect. Immun. 73, 695–702.

Schwartz, P.L., 1977. Segmental retinal periarteritis as a complication of toxoplasmosis. Ann. Ophthalmol. 9, 157–162.

Seabra, S.H., Damatta, R.A., De Mello, F.G., De Souza, W., 2004. Endogenous polyamine levels in macrophages is sufficient to support growth of Toxoplasma gondii. J. Parasitol. 90, 455–460.

Sibley, L.D., Boothroyd, J.C., 1992. Virulent strains of Toxoplasma gondii comprise a single clonal lineage. Nature 359, 82–85.

Silveira, C., 2002. Estudo dos Fatores de Risco. Toxoplasmose: Dúvidas e Controvérsias. Erechrim, Rs, Brazil: EdiFAPES 60.

Silveira, C., Belfort Jr., R., Burnier, M.J., 1987. Toxoplasmose ocular: Identificacao de cistos de Toxoplasma gondii na retina de irmaos gemeos com diagnositco ed toxoplasmose ocular recidivante. Primeiro caso mundial. Arq. Bras. Oftalmol. 50, 215.

Silveira, C., Belfort Jr., R., Burnier Jr., M., Nussenblatt, R., 1988. Acquired toxoplasmic infection as the cause of toxoplasmic retinochoroiditis in families. Am. J. Ophthalmol. 106, 362–364.

Silveira, C., Belfort Jr., R., Muccioli, C., Abreu, M.T., Martins, M.C., Victora, C., Nussenblatt, R.B., Holland, G.N., 2001. A follow-up study of Toxoplasma gondii infection in southern Brazil. Am. J. Ophthalmol. 131, 351–354.

Silveira, C., Belfort Jr., R., Muccioli, C., Holland, G.N., Victora, C.G., Horta, B.L., Yu, F., Nussenblatt, R.B., 2002. The effect of long-term intermittent trimethoprim/sulfamethoxazole treatment on recurrences of toxoplasmic retinochoroiditis. Am. J. Ophthalmol. 134, 41–46.

Silveira, C., Ferreira, R., Muccioli, C., Nussenblatt, R., Belfort Jr., R., 2003. Toxoplasmosis transmitted to a newborn from the mother infected 20 years earlier. Am. J. Ophthalmol. 136, 370–371.

Smith, J.M., Curi, A.L., Pavesio, C.E., 2001. Crystalluria with sulphadiazine. Br. J. Ophthalmol. 85, 1265.

Smith, J.R., Franc, D.T., Carter, N.S., Zamora, D., Planck, S.R., Rosenbaum, J.T., 2004. Susceptibility of retinal vascular endothelium to infection with Toxoplasma gondii tachyzoites. Invest Ophthalmol. Vis. Sci. 45, 1157–1161.

Smith, R.E., Ganley, J.P., 1972. Ophthalmic survey of a community. 1. Abnormalities of the ocular fundus. Am. J. Ophthalmol. 74, 1126–1130.

Sobrin, L., Kump, L., Foster, C., 2007. Intravitreal clindamycin for toxoplasmic retinochoroiditis. Retina 27, 952–957.

Soheilian, M., Ramezani, A., Azimzadeh, A., Sadoughi, M., Dehghan, M., Shahghadami, R., Yaseri, M., Peyman, G., 2011. Randomized trial of intravitreal clindamycin and dexamehtasone versus pyrimethamine, sulfadiazine, and prednisolone in treatment of ocular toxoplamosis. Ophthalmology 118, 134–141.

Sousa, O.E., Saenz, R.E., Frenkel, J.K., 1988. Toxoplasmosis in Panama: a 10-year study. Am. J. Trop. Med. Hyg. 38, 315–322.

Spaide, F., 2011. Peripheral areas of nonperfusion in treated central retinal vein occlusion as imaged by wide-field fluorescein angiography. Retina 31, 829–837.

Spalter, H.F., Campbell, C.J., Noyori, K.S., Rittler, M.C., Koester, C.J., 1966. Prophylactic photocoagulation of recurrent toxoplasmic retinochoroiditis. A preliminary report. Arch. Ophthalmol. 75, 21–31.

St Georgiev, V., 1993. Opportunistic/nosocomial infections. Treatment and developmental therapeutics. Toxoplasmosis Med. Res. Rev. 13, 529–568.

Stagno, S., Reynolds, D.W., Amos, C.S., Dahle, A.J., McCollister, F.P., Mohindra, I., Ermocilla, R., Alford, C.A., 1977. Auditory and visual defects resulting from symptomatic and subclinical congenital cytomegaloviral and Toxoplasma infections. Pediatrics 59, 669–678.

Stanford, M.R., See, S.E., Jones, L.V., Gilbert, R.E., 2003. Antibiotics for toxoplasmic retinochoroiditis: an evidence-based systematic review. Ophthalmology 110, 926–931. quiz 931–2.

Stanford, M.R., Tomlin, E.A., Comyn, O., Holland, K., Pavesio, C., 2005. The visual field in toxoplasmic retinochoroiditis. Br. J. Ophthalmol. 89, 812–814.

Streilein, J.W., Ksander, B.R., Taylor, A.W., 1997. Immune deviation in relation to ocular immune privilege. J. Immunol. 158, 3557–3560.

Streilein, J.W., Stein-Streilein, J., 2000. Does innate immune privilege exist? J. Leuk. Biol. 67, 479–487.

Su, C., Howe, D.K., Dubey, J.P., Ajioka, J.W., Sibley, L.D., 2002. Identification of quantitative trait loci controlling acute virulence in Toxoplasma gondii. Proc. Natl. Acad. Sci. U. S. A. 99, 10753–10758.

Utomo Suhardjo, P.T., Agni, A.N., 2003. Clinical manifestations of ocular toxoplasmosis in Yogyakarta, Indonesia: a clinical review of 173 cases. Southeast Asian J. Trop. Med. Public Health 34, 291–297.

Suzuki, Y., Wong, S.Y., Grumet, F.C., Fessel, J., Montoya, J.G., Zolopa, A.R., Portmore, A., Schumacher-Perdreau, F., Schrappe, M., Koppen, S., Ruf, B., Brown, B.W., Remington, J.S., 1996. Evidence for genetic regulation of susceptibility to toxoplasmic encephalitis in AIDS patients. J. Infect. Dis. 173, 265–268.

Tabbara, K.F., O'Connor, G.R., 1975. Ocular tissue absorption of clindamycin phosphate. Arch. Ophthalmol. 93, 1180–1185.

Tedesco, R.C., Smith, R.L., Corte-Real, S., Calabrese, K.S., 2005. Ocular toxoplasmosis in mice: comparison of two routes of infection. Parasitology 131, 303–307.

Teutsch, S.M., Juranek, D.D., Sulzer, A., Dubey, J.P., Sikes, R.K., 1979. Epidemic toxoplasmosis associated with infected cats. N. Engl. J. Med. 300, 695–699.

Vail, D., Strong, J.C., Stephenson, W.V., 1943. Chorioretinitis associated with positive serologic tests for Toxoplasma in older children and adults. Am. J. Ophthalmol. 26, 133–141.

Van Der Lelij, A., Rothova, A., 1997. Diagnostic anterior chamber paracentesis in uveitis: a safe procedure? Br. J. Ophthalmol. 81, 976–979.

Van Gelder, R.N., 2003. Cme review: polymerase chain reaction diagnostics for posterior segment disease. Retina 23, 445–452.

Villard, O., Filisetti, D., Roch-Deries, F., Garweg, J., Flament, J., Candolfi, E., 2003. Comparison of enzyme-linked immunosorbent assay, immunoblotting, and PCR for diagnosis of toxoplasmic chorioretinitis. J. Clin. Microbiol. 41, 3537–3541.

Wakefield, D., Cunningham, E., Pavesio, C., Garweg, J., Zierhut, M., 2011. Controversies in ocular toxoplasmosis. Ocul. Immun. Inflamm. 19, 2–9.

Webb, R.M., Tabbara, K.F., O'Connor, G.R., 1984. Retinal vasculitis in ocular toxoplasmosis in nonhuman primates. Retina 4, 182–188.

Wessel, M., Aaker, D., Parlitsis, F., Cho, M., D'Amico, J., Kiss, S., 2012. Ultra-wide-field angiography improves the detection and classification of diabetic retinopathy. Retina 32, 785–791.

Westfall, A.C., Lauer, A.K., Suhler, E.B., Rosenbaum, J.T., 2005. Toxoplasmosis retinochoroiditis and elevated intraocular pressure: a retrospective study. J. Glaucoma 14, 3–10.

Wilder, H.C., 1952. Toxoplasma chorioretinitis in adults. AMA Arch. Ophthalmol. 48, 127–136.

Wilson, C.B., Remington, J.S., 1980. What can be done to prevent congenital toxoplasmosis? Am. J. Obstet Gynecol. 138, 357–363.

Wilson, C.B., Remington, J.S., Stagno, S., Reynolds, D.W., 1980. Development of adverse sequelae in children born with subclinical congenital Toxoplasma infection. Pediatrics 66, 767–774.

Wolf, A., Cowen, D., Paige, B.H., 1939. Human toxoplasmosis: occurrence in infants as an encephalomyelitis: Verification by transmission to animals. Science 89, 226–227.

Yeo, J.H., Jakobiec, F.A., Iwamoto, T., Richard, G., Kreissig, I., 1983. Opportunistic toxoplasmic retinochoroiditis following chemotherapy for systemic lymphoma. A light and electron microscopic study. Ophthalmology 90, 885–898.

Young, J.D., Mcgwire, B.S., 2005. Infliximab and reactivation of cerebral toxoplasmosis. N. Engl. J. Med. 353, 1530–1531. discussion 1530-1.

Ysasaga, J.E., Davis, J., 1999. Frosted branch angiitis with ocular toxoplasmosis. Arch. Ophthalmol. 117, 1260–1261.

Zimmerman, L., 1961. Ocular pathology of toxoplasmosis. Surv. Ophthalmol. 6, 832–838.

CHAPTER

6

Toxoplasmosis in Wild and Domestic Animals

David S. Lindsay, Jitender P. Dubey*[†]

*Department of Biomedical Sciences and Pathobiology, Virginia-Maryland Regional College of Veterinary Medicine, Virginia Tech, Blacksburg, Virginia, USA [†]Animal Parasitic Diseases Laboratory, Beltsville Agricultural Research Center, Agricultural Research Service, United States Department of Agriculture, Beltsville, Maryland, USA

OUTLINE

Toxoplasma gondii, *Second Edition*
http://dx.doi.org/10.1016/B978-0-12-396481-6.00006-4

6.1 INTRODUCTION

Toxoplasma gondii is widely distributed in wild and domestic animals, thus, the present chapter reviews toxoplasmosis in wild and domestic animals. Coverage in wild animal species is limited to confirmed cases of toxoplasmosis, cases with parasite isolation, cases with parasite detection by PCR, and experimental infection studies (Fig 6.1-6.3). Studies concerning serological prevalence have not been included for the majority of host species. This was done because many serological tests (latex agglutination, indirect hemagglutination) have been demonstrated to underestimate the prevalence of *T. gondii*.

6.2 TOXOPLASMOSIS IN WILD LIFE

6.2.1 Felids

Congenital toxoplasmosis has been reported in bobcats (*Felis rufus*) kits (Dubey *et al.*, 1987). Toxoplasmic meningoencephalitis has been observed in a six month old bobcat (Smith *et al.*, 1995). *Toxoplasma gondii* has been isolated from the tissues of adult bobcats (Lindsay *et al.*, 1997b; Dubey *et al.*, 2004b). Bobcats are important in maintaining *T. gondii* in wild herbivores in many areas of the USA (Fig. 6.1). Oocysts excreted by cougars (*Felis concolor*) were thought to be the source of a large water borne outbreak of human toxoplasmosis in Victoria, British Columbia, Canada, and oocysts were isolated from the faeces of cougars collected around the water shed (Aramini *et al.*, 1998). *T. gondii* has been isolated by bioassay in mice from a jaguarundi (*Puma yagouaroundi*) (Pena *et al.*, 2011), a jaguar (*Panthera onca*) (Demar *et al.*, 2008), and sand cat (*Felis margarita*) (Dubey *et al.*, 2010). Experimental infections resulting in oocyst excretion have been demonstrated in jaguarundi (*Puma yagouaroundi*), ocelot (*F. pardalis*), bobcats (*Lynx rufus*), and cheetah (*Acinonyx jubatus*) (Jewell *et al.*, 1972; Miller *et al.*, 1972). In general, these felids are not as efficient at producing oocysts as are domestic cats. Congenital toxoplasmosis is a major factor hindering breeding programmes for endangered Pallas's cats (*Otocolobus manul)* and sand cats (*Felis margarita*) in zoos worldwide (*see below*).

6.2.2 Canids

Acute toxoplasmosis has been reported in arctic foxes (*Alopex lagopus*) (Sorensen *et al.*, 2005), fennec foxes (*Fennecus zerda*) (Kottwitz *et al.*, 2004), grey foxes (*Urocyon cinereoargenteus*) (Davidson *et al.*, 1992; Dubey and Lin, 1994; Kelly and Sleeman, 2003), red foxes (*Vulpes vulpes*) (Reed and Turek, 1985; Dubey *et al.*, 1990; Kelly and Sleeman, 2003) and sand foxes (*Vulpes rueppelli*) (Pas and Dubey, 2008c). Co-infection with canine distemper virus is often associated with clinical toxoplasmosis in grey foxes (Davidson *et al.*, 1992; Kelly and Sleeman, 2003) and red foxes (Reed and Turek, 1985). Clinical toxoplasmosis has not been documented in wolves, coyotes, hyenas or dingoes. *T. gondii* has been isolated from arctic foxes (Dubey *et al.*, 2011b), red foxes (Smith and Frenkel, 1995; Dubey *et al.*, 2004b, 2011b), grey foxes (Dubey *et al.*, 2004b), and coyotes (Lindsay *et al.*, 1997b; Dubey *et al.*, 2004b). Aubert *et al.* (2010) found MAT

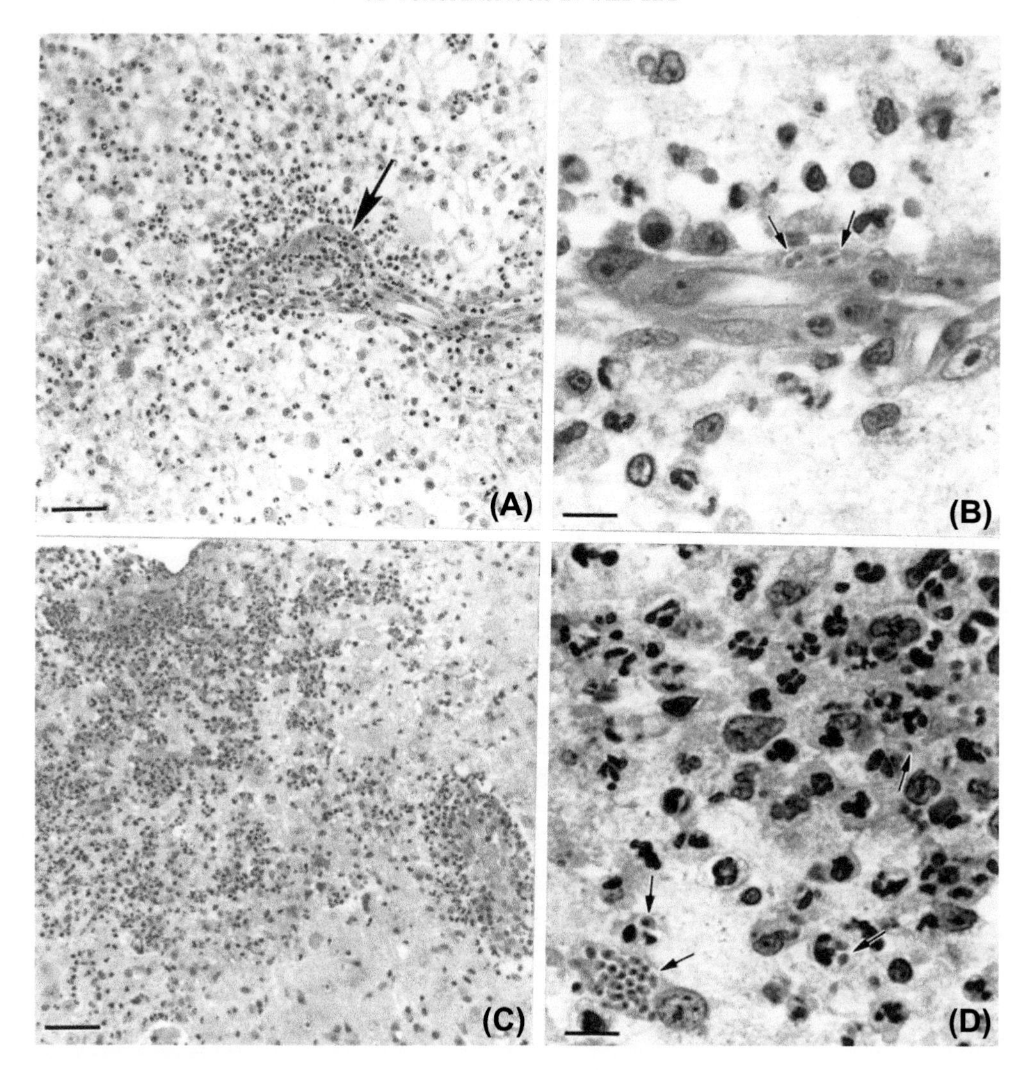

FIGURE 6.1 **Fatal toxoplasmic encephalitis in a naturally infected bobcat. H and E stain.** A) Necrosis and inflammation of a blood vessel (arrow). Bar is 50 μm.
B) Tachyzoites (arrows) in a capillary. Bar is 10 μm.
C) Vasculitis and suppurative encephalitis. Bar is 100 μm.
D) An abscess with degenerating neutrophils and tachyzoites (arrows). Bar is 10 μm.

antibodies in 14 of 19 (74%) red foxes from France and isolated *T. gondii* from the hearts of nine (69%) of 13 seropositive red foxes. The isolates were all genotype Type 2.

Herrmann *et al.* (2012) used serology (immunoblot) and PCR to examine the prevalence of *T. gondii* in red foxes and rodents from the German Federal States of Brandenburg and Saxony-Anhalt. They found 152/204 (74.5%) and 149/176 (84.7%) of red foxes in Brandenburg and Saxony-Anhalt were immunoblot positive, respectively, but none of 72 rodents (69 common

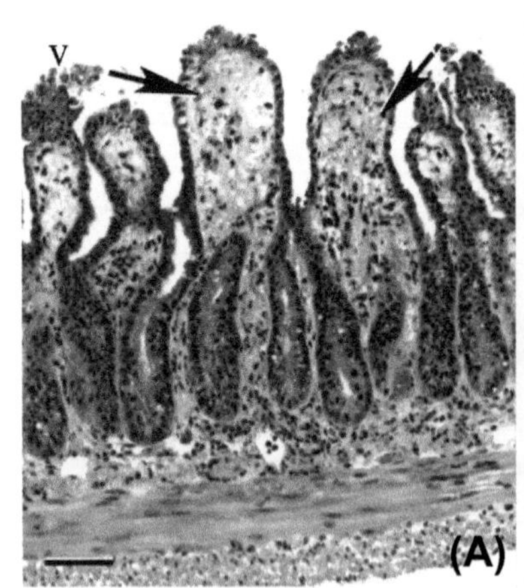

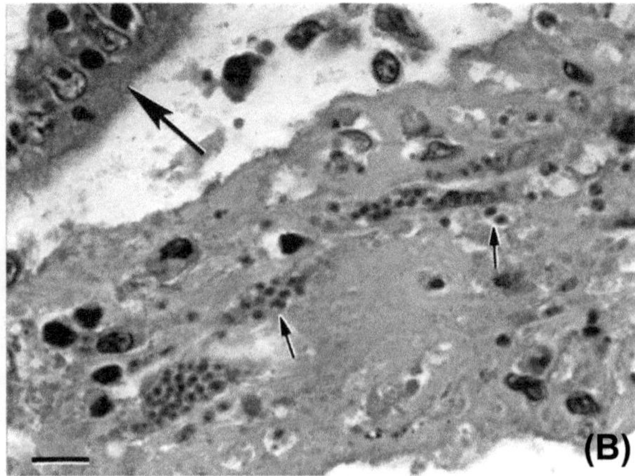

FIGURE 6.2 **Necrosis associated with *T. gondii* in small intestine.** H and E stain.
A) Necrosis of lamina propria (arrows) of villi seven days after feeding oocysts to a mouse. The surface epithelium is not affected. Numerous tachyzoites are present in lesions but are not visible at this magnification. Bar is 100 μm.
B) Necrosis of the lamina propria cells including blood vessels in a naturally infected Pallas's cat. Numerous tachyzoites (small arrows) are present. The surface epithelium (large arrow) was not affected. Bar is 10 μm.

voles *Microtus arvalis*, two Mediterranean water shrews *Neomys anomalus* and one striped field mouse *Apodemus agrarius*) had antibodies to *T. gondii*. PCR was conducted on heart tissue from seropositive red fox tissues and 28/152 (18%) and 20/149 (13%) of seropositive foxes from Brandenburg and Saxony-Anhalt, respectively, were positive (Herrmann *et al.*, 2012). PCR was done on heart and lung samples from the 72 rodents and none tested positive (Herrmann *et al.*, 2012).

6.2.3 Bears

Clinical toxoplasmosis has not been reported from bears. Viable *T. gondii* has been isolated from black bears (*Ursus americanus*) (Dubey *et al.*, 1995a) and brown bears (*Ursus arctos horribis*) (Dubey *et al.*, 2011b). Serological surveys indicate *T. gondii* infections occur in polar bears (*Ursus maritimus)* (Rah *et al.*, 2005) and in grizzly bears (*Ursus arctos*) (Chomel *et al.*, 1995). Meat from any species of bear should be considered a potential source of *T. gondii*.

6.2.4 Raccoons

Many serosurveys indicate that *T. gondii* is highly prevalent in raccoons (*Procyon lotor*) (reviewed by Hancock *et al.*, 2005). Encysted *T. gondii* has been isolated from the tissues on naturally infected raccoons (Lindsay *et al.*, 1997b; Dubey *et al.*, 2004c, 2011b). Clinical toxoplasmosis has not been reported from raccoons and they are resistant to experimental infection (Dubey *et al.*, 1993b).

6.2.5 Squirrels

Acute toxoplasmosis has been reported in grey squirrels (*Sciurus carolinensis*) (Dubey *et al.*, 2006a), American red squirrels (*Tamiasciurus hudsonicus*) (Bangari *et al.*, 2007), 13-lined ground squirrels (*Citellus tridecemlineatus*) (Van Pelt and Dieterich, 1972), Eurasian red squirrels (*Sciurus vulgaris*) (Jokelainen and Nylund, 2012), and Korean squirrels (*Tanias sibericus*) (Carrasco *et al.*, 2006). *T. gondii* has been isolated from grey squirrels (Smith and Frenkel, 1995)

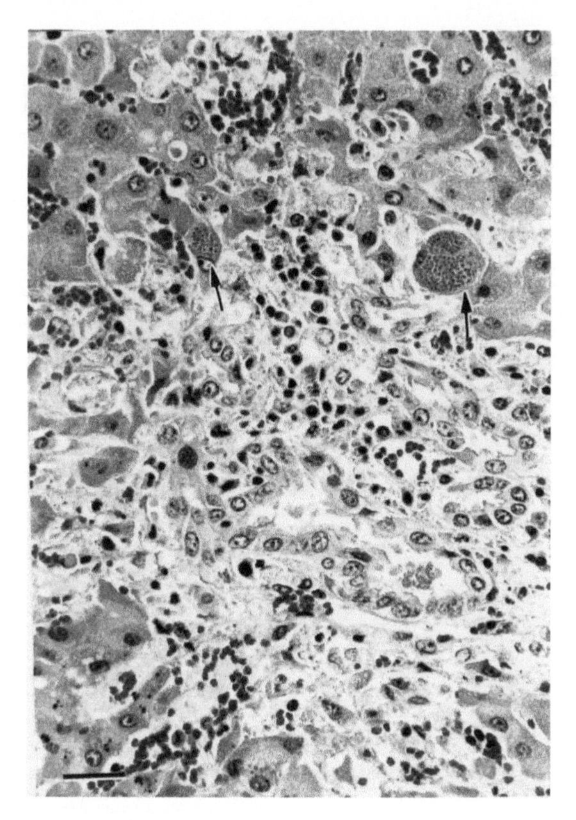

FIGURE 6.3 **Section of liver from a gazelle with toxoplasmosis showing a central area of hepatitis.** Note T. gondii (arrows) in hepatocytes at the periphery of the lesion. H and E stain. Bar is 25 μm.

and Formosan giant flying squirrels (*Petaurista petaurista grandis*) (Cross *et al.*, 1969).

6.2.6 Rabbits and Hares

Fatal toxoplasmosis has been reported from three domestic (*Oryctolagus cuniculus*) rabbits from two different sources in the USA (Dubey *et al.*, 1992a). The rabbits died after an acute illness characterized by fever, lethargy and diarrhea in one rabbit and no clinical signs in the other two rabbits. The most striking lesion in all three rabbits was foci of necrosis of the spleen and liver associated with massive presence of multiplying tachyzoites (Dubey *et al.*, 1992a). Similar findings were present in two to 18 month old domestic rabbits from 15 flocks in Germany. Necropsy examinations of 49 rabbits revealed lesions of a generalized granulomatous-necrotizing toxoplasmosis in the spleen, liver, lungs and lymph nodes (Bergmann *et al.*, 1980). Both authors of the current chapter have inoculated domestic rabbits orally and subcutaneously with *T. gondii* oocysts (usually 10,000/rabbit) to generate immune serum for immunohistochemistry. All inoculated rabbits have or would have developed fatal toxoplasmosis had they not been put down for humane reasons prior to dying. Viable *T. gondii* has been isolated from domestic rabbits (*Oryctologus cuniculus*) from Brazil (Dubey *et al.*, 2011a).

Brown hares (*Lepus europaeus*) develop fatal toxoplasmosis after experimental infection with as few as 10 oocysts and all inoculated hares died within eight to 19 days after ingesting oocysts (Sedlak *et al.*, 2000). The typical pathological finding in hares is hemorrhagic enteritis, enlargement and hyperemia of mesenteric lymph nodes, splenomegaly and multiple necrotic lesions in the parenchyma of the liver and other organs (Sedlak *et al.*, 2000). Mountain hare (*Lepus timidus*) experimentally inoculated with 50 *T. gondii* oocysts and examined 7 days later had gross lesions in the mesenteric lymph nodes and liver (Gustafsson *et al.*, 1997). Histologically, the hares had extensive necrotic areas in the small intestine, mesenteric lymph nodes and liver and less prominent foci of necrosis in various other organs (Gustafsson *et al.*, 1997). Recent retrospective studies in Finland (Jokelainen *et al.*, 2011) have documented natural toxoplasmosis in hares similar to these experimental reports. Acute generalized toxoplasmosis was demonstrated immunohistochemically and *T. gondii* was confirmed as the cause of death in 14 (8%) of 173 European brown hares (*Lepus europaeus*) and four (3%) of 148 mountain hares (*Lepus timidus*) from Finland (Jokelainen *et al.*, 2011). Aubert *et al.* (2010) demonstrated that three (13%) of 23 European brown hares

(*Lepus europaeus*) from France were positive in the MAT but were not able to isolate *T. gondii* from the hearts of two seropositive animals.

6.2.7 Skunks and Fisher

T. gondii genotype 3 was isolated from three of six asymptomatic striped skunks (*Mephitis mephitis*) from Mississippi (Dubey *et al.*, 2004d). Two of the three isolated were mouse pathogenic even though they were molecularly consistent with the mouse avirulent genotype 3. Lesions of toxoplasmosis and *T. gondii* parasites were not observed at necropsy of 37 striped skunks from Illinois (Gehrt *et al.*, 2010). This population was serologically 60% positive for exposure to *T. gondii* (Gehrt *et al.*, 2010).

T. gondii was detected by PCR from brain and skeletal muscle of a free-ranging juvenile fisher (*Martes pennanti*) from Maryland (Gerhold *et al.*, 2005). Clinically this animal had encephalitis, but it was not associated with the *T. gondii* infection.

6.2.8 Beavers

Toxoplasma gondii has been isolated from beaver (*Castor canadensis*) tissue (Dubey, 1983; Smith and Frenkel, 1995). Fatal systematic toxoplasmosis was seen in a five month old beaver that was in a rehabilitation centre in Connecticut (Forzán and Frasca, 2004). Histologic lesions contained *T. gondii* positive stages by immunohistochemistry and consisted of lymphohistiocytic encephalitis, myocarditis and interstitial pneumonia with multinucleated cells (Forzán and Frasca, 2004).

6.2.9 Woodchucks and Other Large Rodents

Central nervous system toxoplasmosis has been observed in a woodchuck (*Marmota monax*) (Bangari *et al.*, 2007) from New York. The woodchuck was euthanized because of progressive clinical signs of head tilt, circling, and rapid weight loss. The brain and heart were positive for *T. gondii* by immunohistochemistry and PCR (Bangari *et al.*, 2007).

Clinical toxoplasmosis has not been reported in capybara (*Hydrochaeris hydrochaeris*) or nutria (*Myocastor coypus*). However, the parasite has been isolated from capybara from Brazil (Yai *et al.*, 2009) and *T. gondii* DNA has been detected by PCR in nutria from Italy (Nardoni *et al.*, 2011).

6.2.10 Insectivores

Little is known about toxoplasmosis in insectivores. The prevalence of *T. gondii* using the Sabin–Feldman dye test was less than 1% in 578 insectivores from the Czech Republic (Hejlicek *et al.*, 1997). Fatal toxoplasmosis was diagnosed in a juvenile male common mole (*Talpa europaea*) from Germany (Geisel *et al.*, 1995). The brains and/or hearts from three of 22 white-toothed shrews (*Crocidura russula*) from organic pig farms in the Netherlands were positive for *T. gondii* by PCR (Kijlstra *et al.*, 2008). In another study from organic pig farms from the Netherlands, none of the brains from nine common shrews (*Sorex araneus*) and two (2%) brains from 102 white-toothed shrews (*Crocidura russula*) were positive by PCR for *T. gondii* (Meerburg *et al.*, 2012). None of two Mediterranean water shrews (*Neomys anomalus*) from Germany were positive by serology or PCR (Herrmann *et al.*, 2012).

6.2.11 Bats

Acute toxoplasmosis has been observed in a juvenile spectacled flying fox (*Pteropus conspicillatus*) and a juvenile little red flying fox (*Pteropus scapulatus*) from Australia (Sangster *et al.*, 2012). One was a captive born member of a colony and the other was undergoing rehabilitation at a wildlife hospital. Severe, acute interstitial pneumonia with varying combinations of neutrophils, large foamy macrophages and fibrin present within alveoli were seen in the lungs and *T. gondii* confirmed using immunohistochemistry. Lesions in the CNS consisted of multiple foci of gliosis, including gemistocytic astrocytes, at all levels of the cerebrum, cerebellum and brainstem of the bats (Sangster *et al.*, 2012). The bats are arboreal in nature and it was suggested that the *T. gondii* infections might have been acquired in captivity by food accidentally contaminated with oocysts (Sangster *et al.*, 2012).

Isolation of *T. gondii* was reported from pipistrelle bats (*Vespertilio pipistrellus*) and the red night bat *Nyctalus noctula* from Alma-Ata, Kazakhstan, USSR (Galuzo *et al.*, 1970). Inoculation of RH *T. gondii* did not induce clinical disease in red night bats in these studies (Galuzo *et al.*, 1970).

6.2.12 White-Tailed and Mule Deer

T. gondii is prevalent in deer from North America. Consumption of venison has been linked with clinical toxoplasmosis in humans (Sacks *et al.*, 1983; Ross *et al.*, 2001). Clinical toxoplasmosis has not been described from naturally infected deer in North America. *T. gondii* has been isolated from the tissues of white-tailed deer (*Odocoileus virginianus*) (Lindsay *et al.*, 1991b, 1997b; Dubey *et al.*, 2004b) and mule deer (*Odocoileus hemionus*) (Dubey, 1982). Viable *T. gondii* was isolated from six of 61 white-tailed deer foetuses whose dams were in early pregnancy (45–85 days of gestation) from Iowa and nine of 27 white-tailed deer foetuses from Minnesota whose dams were in mid gestation (130–150 days) of a gestational period of seven months (Dubey *et al.*, 2008b). The foetuses from *T. gondii* positive white-tailed and mule deer were negative for *T. gondii* antibodies in one study, suggesting that seropositive dams do not transmit the infection to their foetuses (Lindsay *et al.*, 2005) unless an acute infection occurs during pregnancy. Acute toxoplasmosis and death can occur in mule deer experimentally inoculated with *T. gondii* oocysts (Dubey *et al.*, 1982).

6.2.13 Other Deer

Congenital toxoplasmosis has been observed in reindeer (*Rangifer tarandus*) from a private collection in the USA (Dubey *et al.*, 2002a). Yearling reindeer may develop enteritis and die after experimental oral infection with *T. gondii* oocysts (Oksanen *et al.*, 1996).

Aubert *et al.* (2010) demonstrated that 36 (60%) of 60 roe deer (*Capreolus capreolus*), from France were positive in the MAT and obtained 12 isolates from the hearts (38%) of 33 MAT positive roe deer. They also reported that *T. gondii* antibodies were present in four (17%) of 24 red deer (*Cervus elaphus*) (Aubert *et al.*, 2010). One (25%) of four fallow deer (*Dama dama*) from France examined by Aubert *et al.* (2010) was positive (MAT titre 1:25) but attempts to isolate *T. gondii* by bioassay in mice were not successful.

6.2.14 Other Wild Ruminants

Elk (*Cervus canadensis*) are resistant to clinical disease following oral infection with oocysts but *T. gondii* can be isolated from many of their tissues indicating that elk are a potential source of infection for humans (Dubey *et al.*, 1980). *T. gondii* was been isolated by bioassay in mice from one of 21 naturally infected pronghorn antelope (*Antilocapra americana*) from Montana (Dubey, 1981). The isolate was not pathogenic for mice. Acute toxoplasmosis and death can occur in pronghorn antelopes experimentally inoculated with *T. gondii* oocysts (Dubey *et al.*, 1982). Toxoplasmic encephalitis has been observed in a four month old Rocky Mountain bighorn sheep (*Ovis canadensis canadensis*) (Baszler *et al.*, 2000). Aubert *et al.* (2010) demonstrated MAT antibodies in seven (23%) of 31 mouflons (*Ovis gmelini musimon*) from France and isolated the parasite from one (25%) of four hearts from seropositive mouflons. *T. gondii* was isolated by bioassay in mice from one of seven naturally infected moose (*Alces alces*) from Montana (Dubey, 1981). The isolate was not pathogenic for mice.

6.2.15 Sea Otters and Other Marine Mammals

Toxoplasmosis was recognized as a significant cause of mortality in southern sea otters (*Enhydra lutris nereis*) in the early 1990s (Cole *et al.*, 2000). Encephalitis is the primary cause of *T. gondii* associated death in these sea otters (Cole *et al.*, 2000). This was unexpected as sea otters do not ingest the usual intermediate hosts of *T. gondii* and their location in seawater keeps them segregated from cats. Definitive proof that *T. gondii* was killing the sea otters came when viable *T. gondii* was isolated from the tissues of sea otters (Cole *et al.*, 2000; Lindsay *et al.*, 2001a) and isolated parasites from sea otters were shown to retain the ability to make oocysts when fed to cats (Cole *et al.*, 2000). Initial isolates were all type 2 genotypes of *T. gondii* (Cole *et al.*, 2000). It has been postulated that *T. gondii* oocysts excreted in the faeces of feral cats living along the Pacific coast enter the marine environment and are ingested by sea otters when they feed on paratenic hosts (Cole *et al.*, 2000) and this is supported by the fact that coastal freshwater run-off is a risk factor for *T. gondii* infection in southern sea otters (Miller *et al.*, 2002). *T. gondii* oocysts will sporulate in seawater (Lindsay *et al.*, 2003) and remain infectious for 1.5 hours at room temperature and for at least two years at 4°C (Lindsay and Dubey, 2009) and viable *T. gondii* and *T. gondii* DNA can be recovered from experimentally inoculated bivalves (Lindsay *et al.*, 2001b, 2004; Arkush *et al.*, 2003), further supporting these assumptions. Additionally, two species of filter feeding fish, northern anchovies (*Engraulis mordax*) and Pacific sardines (*Sardinops sagax*), have been shown to be able to remove *T. gondii* oocysts from sea water and can potentially serve as biotic vectors for *T. gondii* within the marine environment (Massie *et al.*, 2010). *T. gondii* also causes deaths in other marine mammals off the Pacific coast of the USA, often in the same areas as the sea otters (Miller *et al.*, 2001) and it has also been isolated from Pacific harbour seal (*Phoca vitulina*) and California sea lion (*Zalophus californianus*).

Toxoplasmosis is frequently reported from dolphins worldwide. Congenital toxoplasmosis has been reported in bottlenose dolphins (*Tursiops aduncus*) (Jardine and Dubey, 2002). Disseminated toxoplasmosis with transplacental foetal infection has been seen in a pregnant Risso's dolphin (*Grampus griseus*) (Resendes *et al.*, 2002). Acute cases of toxoplasmosis have been seen in humpbacked dolphins (*Sousa chinensis*) (Bowater *et al.*, 2003), spinner dolphins (*Stenella longirostris*) (Migaki *et al.*, 1990), striped dolphins (*Stenella coeruleoalba*) (Di Guardo *et al.*, 2010) and Atlantic bottle-nosed dolphins (*Tursiops truncatus*) (Inskeep *et al.*, 1990). *T. gondii* has been isolated from the hearts of three of 52 bottlenose dolphins (*Tursiops aduncus*) from the eastern USA by mouse bioassay (Dubey *et al.*, 2008a). *Toxoplasma gondii* was isolated from the brain of a stranded female striped dolphin (*Stenella coeruleoalba*) from Costa Rica that died from non-*T. gondii* related causes (Dubey *et al.*, 2007a).

Toxoplasmosis has been reported from several additional species of marine mammals such as beluga whales (*Delphinapterus leucas*) (Mikaelian *et al.*, 2000), Mediterranean fin whale (*Balaenoptera physalus*) (Mazzariol *et al.*, 2012), California sea lion (*Zalophus californianus*) (Migaki *et al.*, 1977), northern fur seal (*Callorhinus ursinus*) (Holshuh *et al.*, 1985), elephant seal (*Mirounga angustirostris*) (Dubey *et al.*, 2004a), Hawaiian monk seal (*Monachus schauinslandi*) (Honnold *et al.*, 2005), Antillean manatee (*Trichechus manatus manatus*) (Dubey *et al.*, 2003; Bossart *et al.*, 2012) and West Indian manatee (*Trichechus manatus*) (Buergelt and Bonde, 1983). Experimental infection of grey seals (*Halichoerus grypus*) with up to 10,000 *T. gondii* oocysts did not induce overt clinical disease (Gajadhar *et al.*, 2004). Mild behavioral changes were the only adverse effects and *T. gondii* was isolated from brain and muscles of the experimentally infected seals.

6.2.16 New World Monkeys

Toxoplasmosis can be a problem in exhibited new world monkeys (Table 6.1). Many reports of acute disease have come from squirrel monkeys (*Saimiri sciureus*) (Cedillo-Peláez *et al.*, 2011; Epiphanio *et al.*, 2003) and golden lion tamarins (*Leontopithecus rosalia*) (Dietz *et al.*, 1997; Pertz *et al.*, 1997; Juan-Salles *et al.*, 1998; Epiphanio *et al.*, 2003). Squirrel monkeys and Panamanian night monkeys (*Aotus lemurinus*) are highly susceptible to oral tissue cyst inoculation and develop acute fatal disease (Harper *et al.*, 1985; Escajadillo and Frenkel, 1991; Furuta *et al.*, 2001). Pena *et al.* (2011) isolated *T. gondii* by bioassay in mice of the heart and brain of a young male red-handed howler monkey (*Alouatta belzebul*) with suspected toxoplasmosis from a zoo in Brazil.

TABLE 6.1 Summary of Host Species Reports of Clinical Toxoplasmosis in New World Primates

Cotton-top tamarin (*Saguinus oedipus*)
Yellow-handed marmoset (*Saguinus midas midas*)
Black marmoset (*Saguinus midas niger*)
Emperor marmoset (*Saguinus imperator*)
Red-bellied white-lipped tamarin (*Saguinus labiatus*)
Black lion tamarin (*Leontopithecus chrysopygus*)
Golden-headed lion tamarins (*Leontopithecus chrysomelas*)
Golden lion tamarins (*Leontopithecus rosalia*)
Squirrel monkeys (*Saimiri sciureus*)
Pygmy marmoset (*Callithrix pygmaea*)
Common marmoset (*Callithrix jacchus*)
Black ear-tufted marmoset (*Callithrix penicllata*)
Pale-headed saki (*Pithecia pithecia*)
Night monkey (*Aotus trivirgatus*)
Howler monkey (*Alouatta fusca*)
Woolly monkey (*Lagothrix lagotricha*)

Sources: Dietz et al., *1997; Bouer* et al., *1999; Epiphanio* et al., *2001, 2003; Dubey, 2010; Cedillo-Peláez* et al., *2011; Pena* et al., *2011.*

6.2.17 Old World Monkeys

Toxoplasmosis is reported infrequently in Old World monkeys. A case of concurrent central nervous system toxoplasmosis and simian immunodeficiency virus-induced AIDS encephalomyelitis was seen in a Barbary macaque (*Macaca sylvana*) (Sasseville *et al.*, 1995). Rhesus monkeys (*Macaca mulatta*) and stump-tailed macaques (*Macaca arctoides*) have been used as experimental models for human congenital toxoplasmosis (Wong *et al.*, 1979; Schoondermark-Van de Ven *et al.*, 1993) and cynomolgus monkeys (*Macaca fascicularis*) have been used as a model for recurrent toxoplasmic retinochoroiditis (Holland *et al.*, 1988).

6.2.18 American Marsupials

Clinical toxoplasmosis has not been reported from marsupials from the Americas. The parasite has been isolated from opossums (*Didelphis virginiana*) from Georgia (Dubey *et al.*, 2011b) and Kansas (Smith and Frenkel, 1995) in the USA and black-eared opossums (*Didelphis aurita*) from Brazil (Pena *et al.*, 2011) have been examined and proven to be positive by bioassay in mice.

6.2.19 Australian Marsupials

T. gondii infection is usually life ending in marsupials from Australia or New Zealand. Outbreaks of toxoplasmosis often occur in these animals when housed in zoos (see below). These animals evolved in the absence of cats and *T. gondii* and this is maybe why they are so highly susceptible.

Canfield *et al.* (1990) summarized clinical signs, necropsy findings and histopathological changes for 43 macropods (species not given), two common wombats (*Vombatus ursinus*), two koalas (*Phascolarctos cinereus*), six possums (species not given), 15 dasyurids (species not given), two numbats (*Myrmecobius fasciatus*), eight bandicoots (species not given) and one bilby (*Macrotis lagotis*). The animals either died suddenly without clinical signs or exhibited signs associated with respiratory, neurological or enteric disease. At necropsy, many had no visible lesions. Common necropsy findings included pulmonary congestion, oedema and consolidation, adrenal enlargement and reddening, haemorrhage and ulceration of stomach and small intestine, and lymphadenomegaly and splenomegaly (Canfield *et al.*, 1990). Congenital toxoplasmosis apparently occurs in black-faced kangaroos (*Macropus fuliginosus melanops*) based on the finding of *T. gondii* in the tissues of a 82 day old joey that died from toxoplasmosis (Dubey *et al.*, 1988b). *T. gondii* was seen in the heart, kidney, liver, lung, lymph node, spleen, small intestine and stomach of two koalas (*Phascolarctos cinereus*) that died suddenly in a fauna park in Sydney, Australia (Hartley *et al.*, 1990).

Experimental studies support the assumption that Australian marsupials are highly susceptible to toxoplasmosis. Eastern barred bandicoots (*Perameles gunnii*) developed acute toxoplasmosis when fed 100 *T. gondii* oocysts and died 15 and 17 days post infection (Bettiol *et al.*, 2000). Lesions consistent with acute toxoplasmosis were present in their tissues. The authors indicated that *T. gondii* may be a cause for a reduction in wild populations of eastern barred bandicoots. Tammar wallabies (*Macropus eugenii*) fed 500, 1000 or 10,000 *T. gondii* oocysts died of acute toxoplasmosis nine to 15 days after challenge (Reddacliff *et al.*, 1993). The lesions consisted of foci of necrosis and inflammation in the intestines, lymphoid tissue, adrenal cortex, heart, skeletal muscle and brain, and severe generalized pulmonary congestion and oedema.

6.2.20 African Wildlife

Surprisingly, little is known about *T. gondii* and toxoplasmosis from African mammals. Clinical disease has not been reported from free ranging elephants, hippopotami, giraffes, gazelles, wildebeests, impalas, chimpanzees, baboons, orangutans and gorillas nor has the parasite been isolated from the tissues of these mammals. *T. gondii* has been isolated from domestic chickens and ducks from Africa (Dubey, 2010).

6.2.21 Wild Rodents

A detailed discussion of *T. gondii* prevalence in wild rodents (mice and rats) is beyond the scope of this chapter. *T. gondii* has been isolated from the tissues from wild rodents worldwide (Dubey, 2010). Genotypes of these isolates are similar to isolates from other animals in the same geographic area.

Dabritz *et al.* (2008) has recently reviewed what was known about the global serological prevalence of *T. gondii* in wild rodents. Until large scale studies are conducted using bioassay or molecular detection methods the role of wild rodents in maintaining *T. gondii* in the environment will not be fully understood. Properly conducted serological studies usually indicate that few (less than 10%) wild mice or rats are usually found to be seropositive. For example, two (0.8%) of 238 rats (*Rattus norvegicus*) from Grenada, West Indies, were found to be serologically positive using the MAT (Dubey *et al.*, 2006b). When the brains and hearts of all 238 rats were examined by bioassay in mice *T. gondii* was isolated from only one of the 238 rats with the positive rat being one of the two serologically positive animals. This clearly demonstrates a low prevalence of *T. gondii* infection in this rat population.

6.2.22 Wild Birds

Table 6.2 lists the wild avian hosts from which viable *T. gondii* has been isolated and Table 6.3 lists the avian species which have been reported to suffer from clinical toxoplasmosis.

TABLE 6.2 Host Records for *Toxoplasma gondii* Isolation from Wild Birds

ANSERIFORMES
Mallards (*Anas platyrhynchos*)
Pochard (*Aythya ferrina*)
Tufted ducks (*Aythya fuligula*)
Pintail (*Anas acuta*)
Gadwall (*Anas strepera*)
Canada goose (*Branta canadensis*)
ACCIPITRIFORMES
Goshawk (*Accipiter gentilis*)
Cooper's hawk (*Accipiter cooperi*)
Common buzzard (*Buteo buteo*)
Kestrel (*Falco tinnunculus*)
American kestrel (*Falco sparverius*)
Pallid harrier (*Circus macrourus*)
Bald eagle (*Haliaeetus leucocephalus*)
Black vulture (*Aegypius monachus*)
Red-tailed hawk (*Buteo jamaicensis*)
Red-shouldered hawk (*Buteo lineatus*)
GALLIFORMES
Partridge (*Perdix perdix*)
Pheasant (*Phasianus colchicus*)
Wild turkey (*Meleagris gallopavo*)
GRUIFORMES
Coot (*Fulica atra*)
CHARADRIFORMES
Blackheaded gull (*Larus ridibundus*)
Common tern (*Sterna hirundo*)
COLUMBIFORMES
Collared dove (*Streptopelia decaocto*)
Laughing dove (*Streptopelia senegalensis*)
Woodpigeon (*Columba palumbus*)
Pigeon (*Columba livia*)
Ruddy ground dove (*Columbina talpacoti*)
STRIGIFORMES
Ferruginious pygmy owl (*Glaucidium brasilianum*)
Little owl (*Athene noctua*)
Great horned owl (*Bubo virginianus*)
Barred owl (*Strix varia*)
PASSERIFORMES
Great grey shrike (*Lanius excubitor*)
Yellowhammer (*Emberiza citrinella*)

(*Continued*)

TABLE 6.2 Host Records for *Toxoplasma gondii* Isolation from Wild Birds (*cont'd*)

PASSERIFORMES
Chaffinch (*Fringilla coelebs*)
House sparrow (*Passer domesticus*)
Tree sparrow (*Passer montanus*)
Jay (*Garrulus glandarius*)
Starling (*Sturnus vulgaris*)
Palm tanager (*Thraupis palmarum*)
Blackbird (*Turdus merula*)
Mistle thrush (*Turdus viscivorus*)
Song thrush (*Turdus philomelos*)
Robin (*Erithacus rubecula*)
Great tit (*Parus major*)
Nuthatch (*Sitta europea*)
Treecreeper (*Certhia familiaris*)
Greenfinch (*Chloris chloris*)
American crow (*Corvus brachyrhynchos*)
Carrion crow (*Corvus corone*)
Jackdaw (*Corvus monedula*)
Rook (*Corvus frugilegus*)

Sources: Dubey, 2002, 2010; Lindsay et al., *1993.*

T. gondii is readily isolated from the hearts and breast muscles of raptors (Lindsay *et al.*, 1993; Dubey *et al.*, 2011b, Table 2). Necrotizing myocarditis caused by *T. gondii* has been observed in a bald eagle (*Haliaeetus leucocephalus*) from New Hampshire (Szabo *et al.*, 2004). Severe toxoplasmic hepatitis was seen in an adult barred owl (*Strix varia*) from Quebec, Canada (Mikaelian *et al.*, 1997). No clinical signs were seen in three red-tailed hawks (*Buteo jamaicensis*) fed *T. gondii* tissue cysts (Lindsay *et al.*, 1991a). *T. gondii* was isolated from all three red-tailed hawks. No clinical signs were seen in great horned owls (*Bubo virginianus*), barred owls (*Strix varia*) or screech owls (*Asio otus*) fed *T. gondii* tissue cysts (Dubey *et al.*, 1992b), but parasites were isolated from the tissues of the owls at necropsy. *T. gondii* was not re-isolated from a sparrow hawk (*Falco sparverius*) that had been experimentally infected (Miller *et al.*, 1972).

Viable *T. gondii* was isolated from the hearts of eight of 16 wild turkeys (*Meleagris gallopavo*)

TABLE 6.3 List of Avian Species in Which Clinical Toxoplasmosis Has Been Reported

COLUMBIFORMES
Common pigeon (*Columba livia*)
Crown pigeons (*Goura sp.*)
Torres Strait pigeon (*Ducula spilorrhoa*)
Wonga pigeon (*Leucosarcia melanoleuca*)
Bleeding-heart dove (*Gallicolumba luzonica*)
Nicobar pigeon (*Caloenas nicobarica*)
Luzon bleeding-heart pigeon (*Gallicoluba luzonica*)
Orange-breasted green pigeon (*Treron bicinta*)
Crested wood partridge (*Rolulus roul roul*)
Yellow-headed rockfowl (*Picathartes gymnocephaus*)
PASSERIFORMES
Canaries (*Serinus canarius*)
Greenfinches (*Carduelis chloris*)
Goldfinches (*Carduelis carduelis*)
Sirkins (*Carduelis spinus*)
Linnets (*Carduelis cannabina*)
Bullfinches (*Pyrrhula pyrrhula*)
Hawaiian crow (*Corvus hawaiiensis*)
Satin bowerbird (*Ptilornorhyncus violaceus*)
Regent bowerbird (*Sericulus chrysocephalus*)
Red-whiskered bulbul (*Pycnonotus jocosus*)
PSITTACIFORMES
Budgerigars (*Melopsittacus undulatus*)
Regent parrot (*Polytelis anthopeplus*)
Superb parrot (*Polytelis swansonii*)
Red lory (*Eos bornea*)
Swainson's lorikeet (*Trichologlossus moluccanus*)
Crimson rosella (*Platycercus elegans*)
STRIGIFORMES
Barred owl (*Strix varia*)
GALLIFORMES
Wild turkeys (*Meleagris gallapavo*)
Partridges (*Perdix perdix*)
Capercaillie (*Tetrao urogallus*)
Erckel's francolin (*Francolinus erckelii*)
Guinea fowl (*Numida meleagris*)
ANSERIFORMES
Magpie geese (*Anseranas semipalmata*)
Hawaiian nene goose (*Nesochen sandicensis*)
SPHENISCIFORMES
Humboldt penguin (*Spheniscus humboldti*)
Megellanic penguin (*Spheniscus magellanicus*)

TABLE 6.3 List of Avian Species in Which Clinical Toxoplasmosis Has Been Reported (*cont'd*)

SPHENISCIFORMES
Black-footed penguin (*Spheniscus demersus*)
Little penguin (*Eudyptula minor*)
Indian pangolin *Manis crassi caudato*)
PELECANIFORMES
Red-footed booby (*Sula sula*)

Sources: Dubey 2002; Dubey, 2010.

from Alabama (Lindsay *et al.*, 1994). Fatal systemic toxoplasmosis has been seen in wild turkeys from Georgia (Howerth and Rodenroth, 1985) and West Virginia (Quist *et al.*, 1995).

6.3 TOXOPLASMOSIS IN ZOOS

Toxoplasmosis is a zoo management problem because wild felids can excrete *T. gondii* oocysts in their faeces (Jewell *et al.*, 1972; Miller *et al.*, 1972; Lukesova and Literak, 1998) and the occurrence of feral cats in zoos (Gorman *et al.*, 1986). Oocysts excreted by these felids can make their way into highly susceptible species.

Mammalian species which frequently develop toxoplasmosis in zoos include Australian marsupials (Portas, 2010), New World and arboreal monkeys (Cunningham *et al.*, 1992; Dietz *et al.*, 1997; Pertz *et al.*, 1997; Juan-Salles *et al.*, 1998; Epiphanio *et al.*, 2000), lemurs (Dubey *et al.*, 1985), and Pallas's cats (*Otocolobus manul*) (Riemann *et al.*, 1974; Dubey *et al.*, 1988a, 2002; Basso *et al.*, 2005) (Fig. 6.2). Lesions in these animals are consistent with acute toxoplasmosis and are usually most severe in visceral tissues such as the lungs, liver and spleen.

Toxoplasmosis is common in lemurs exhibited in zoos worldwide (Dubey *et al.*, 1985). A female ring-tailed lemur (*Lemur catta*) died of toxoplasmosis in a zoo in Spain one week after the delivery of four stillborn offspring which all had disseminated toxoplasmosis (Juan-Sallés *et al.*, 2011). *T. gondii* was isolated from the tissues of a three year old secundiparous female ring-tailed lemur from a zoo in Alabama that died of acute toxoplasmosis (Spencer *et al.*, 2004). The isolate was not pathogenic for mice and was genetically a type 2 isolate. This case points out the difficulty in preventing toxoplasmosis in highly susceptible animals because this lemur was housed in a group on an island in the zoo (Spencer *et al.*, 2004) making it easier to prevent contact with feral cats. Oocysts on the lemur's food (fruit, etc.) or carried in by black birds were considered likely sources of infection in this case (Spencer *et al.*, 2004).

Sporadic cases of acute toxoplasmosis have been reported in exhibited dik-dik (*Madoqua guentheri smithi*) (Dubey *et al.*, 2002b), slender-tailed meerkats (*Suricata suricatta*) (Juan-Salles *et al.*, 1997), African crested porcupines (*Hystrix cristata*) (Harrison *et al.* 2007) and Brazilian prehensile-tailed porcupines (*Coendou mexicanus*) (Morales *et al.*, 1996). Fatal disseminated toxoplasmosis in three captive slender-tailed meerkats (*Suricata suricatta*) in a zoo in La Plata, Argentina, was found to be caused by the normally nonpathogenic genotype type 3 isolate of the parasite suggesting that meerkats are highly susceptible to infection (Basso *et al.*, 2009). A case of abortion due to *T. gondii* has been reported in a Greenland musk ox (*Ovibos moshatus wardi*) (Crawford *et al.*, 2000).

Abortion and neonatal death have been observed in captive nilgais (*Boselaphus tragocamelus*). *T. gondii* DNA was demonstrated in the tissues of the nilgais using PCR (Sedlak *et al.*, 2004). Fatal toxoplasmosis was diagnosed in a captive, adult female saiga antelope (*Saiga tatarica*). *T. gondii* was detected in the liver, lung, spleen, kidney and intestine and confirmed by PCR (Sedlak *et al.*, 2004). Acute toxoplasmosis has been seen in captive Cuvier's gazelle (*Gazella cuvieri*), slender-horned gazelle (*G. leptoceros*), dama gazelle (*G. dama*) and gerenuk (*Litocranius walleri*) housed in North American zoos (Stover *et al.*, 1990; Junge *et al.*, 1992). These infections are disseminated and most lesions are in the liver, lungs, lymph nodes, adrenal glands, spleen, intestines and brain (Fig. 6.3).

Outbreaks of toxoplasmosis also occur in avian species exhibited in zoos (Poelma *et al.*, 1972; Hubbard *et al.*, 1986). Toxoplasmosis in canaries has been reported from aviaries worldwide (reviewed by Dubey, 2002). *T. gondii* genotype 3 was isolated from five of five black-winged lories (*Eos cyanogenia*) from an acute toxoplasmosis outbreak in an aviary in South Carolina (Dubey *et al.*, 2004d). Acute systemic toxoplasmosis was reported to be the cause of death of three of 10 Nicobar pigeons (*Caloenas nicobaria*) in an aviary collection in South Africa (Las and Shivaprasad, 2008). Feral cats were a known management problem and lesions were consistent with oocyst acquired infection. Three one to three month old black-footed penguin chicks (*Spheniscus demersus*) died from acute toxoplasmosis within 24 hours of showing central nervous signs (Ploeg *et al.*, 2011). The birds were housed in a baby penguin crèche in a zoo in the Netherlands. A cat with a litter of kittens had recently been observed feeding on fish intended for the penguins in the zoo and the cat was suspected as the source of infection (Ploeg *et al.*, 2011).

Management and husbandry programmes can be designed to help achieve prevention of toxoplasmosis in highly susceptible species in zoos and aviaries. Felids should never be fed fresh unfrozen meats because of the possibility of contamination with *T. gondii* tissue cysts. Meat that has been frozen solid and then thawed can be safely fed because freezing kills *T. gondii* tissue cysts (Kotula *et al.*, 1991). Feral cats should be actively controlled in zoos to prevent them from shedding oocysts. Highly susceptible species should not be housed near felids.

Outdoor aviaries are at risk because of oocysts excreted by domestic cats. Aviaries should be designed to exclude cat faeces and transport hosts (flies, roaches, etc.) that may bring in *T. gondii* on or in their bodies.

6.4 *TOXOPLASMA GONDII* AND ENDANGERED SPECIES

Toxoplasmosis can adversely affect endangered avian and mammalian species. The 'Alala (Hawaiian crow, *Corvus hawaiiensis*) is an endangered species and only about 25 were left in captivity and the wild in the year 2000 (Work *et al.*, 2000). Tragically, these birds are highly susceptible to fatal toxoplasmosis and develop disease after being introduced back into the wild. Toxoplasmosis appears to pose a significant threat and management challenge to reintroduction programmes for 'Alala in Hawaii (Work *et al.*, 2000).

Captive breeding groups of golden lion tamarins (*Leontopithecus rosalia*) have developed acute toxoplasmosis and suffered many fatalities both in North American and European zoos (Pertz *et al.*, 1997; Juan-Salles *et al.*, 1998). These arboreal monkeys are endangered and attempts to breed them in captivity for eventual release in the wild are hampered because it is difficult to keep them from being exposed to *T. gondii*.

Repeated transplacental transmission of *T. gondii* by Pallas's cats may be responsible for the high rate of impact of this disease on the Pallas's cat population in zoos. Efforts by North American zoos to establish genetically viable captive populations of Pallas's cats (*Otocolobus manul*) have been compromised by high newborn kitten mortality due to toxoplasmosis (Brown *et al.*, 2005). In their natural environment, Pallas's cats generally have little exposure to *T. gondii* and it is believed that they acquire *T. gondii* infection after captivity (Brown *et al.*, 2005). The mortality rate for toxoplasmosis of Pallas's cat kittens born in zoos in the USA is 35% to 60% (Kenny *et al.*, 2002; Brown *et al.*, 2005).

Sand cats (*Felis margarita*) housed at the Breeding Centre for Endangered Arabian Wildlife in the United Arab Emirates and Al Wabra Wildlife Preservation, Qatar, have been reported to suffer from congenital (Pas and Dubey, 2008a) and acquired toxoplasmosis (Dubey *et al.*, 2010). Serological examination of endangered Gordon's wildcat (*Felis silvestris gordoni*) kept at the same institution (Pas and Dubey, 2008b) indicated that seropositive Gordon's wildcats were present but no clinical history consistent with toxoplasmosis has been reported in these animals. Unlike domestic cats, Sand cat (*Felis margarita*) queens will repeatedly infect litters of kittens making it very difficult to keep up numbers of healthy kittens in breeding programmes. Fortunately, Gordon's wildcats appear to behave like domestic cats in their responses to *T. gondii* infection.

6.5 TOXOPLASMOSIS IN PETS

6.5.1 Cats

Most cats are asymptomatic during a primary *T. gondii* infection. Fever (40.0°C–41.7°C) is present in many cats with clinical toxoplasmosis. Clinical signs of dyspnea, polypnea, icterus and signs of abdominal discomfort were the most frequent findings in 100 cats with histologically confirmed toxoplasmosis (Dubey and Carpenter, 1993). Uveitis and retinochoroiditis are also common clinical signs in cats with toxoplasmosis. Gross and microscopic lesions are found in many organs but are most common in the lungs. Gross lesions in the lungs consist of oedema and congestion, failure to collapse, and multifocal areas of firm, white to yellow, discoloration. Pericardial and abdominal effusions may be present. The liver is the most frequently affected abdominal organ and diffuse necrotizing hepatitis may be visible grossly. Gross lesions associated with necrosis can also be observed in the mesenteric lymph nodes and pancreas.

All ages, sexes, and breeds of domestic cats are susceptible to *T. gondii* infection (Dubey *et al.*, 1977). Transplacentally or lactogenically infected kittens will excrete oocysts but the prepatent period is usually three weeks or more because the kittens are infected with tachyzoites (Dubey *et al.*, 1995b). Domestic cats under one year of age produce the most numbers of *T. gondii* oocysts. Cats that are born and raised outdoors usually become infected with *T. gondii* shortly after they are weaned and begin to hunt. *T. gondii* naive adult domestic cats will excrete oocysts if fed tissue cysts but they usually will excrete fewer numbers of oocysts and excrete oocysts for a shorter period of time than recently weaned kittens.

Intestinal immunity to *T. gondii* is strong in cats that have excreted oocysts (Dubey, 1995). Primary *T. gondii* infection in cats does not cause immunosuppression (Lappin *et al.*, 1992; Davis and Dubey, 1995). Serum antibody does not play a significant role in resistance to intestinal infection and intestinal immunity is most likely cell mediated. Oocysts begin to be excreted in the faeces before IgM, IgG or IgA antibodies are present in the serum (Lappin *et al.*, 1989). Partial development of the enteroepithelial stages occur in the intestines of immune cats but oocyst production is prevented (Davis and Dubey, 1995). Most cats that have excreted oocysts once do not re-excrete oocysts if challenged within six months to one year. Intestinal immunity will last up to six years in about 55% of cats (Dubey, 1995).

Vaccination of cats against intestinal *T. gondii* infection has been successfully achieved using a mutant strain (T-263) of the parasite (Frenkel *et al.*, 1991; Freyre *et al.*, 1993). Oral administration of strain T-263 bradyzoites results in intestinal infection but does not result in oocyst production in cats. These vaccinated cats do not excrete oocysts when challenged with oocyst producing strains of *T. gondii*. The T-263 strain is safe to use in healthy cats. It is not recommended for use in pregnant cats or FeLV positive cats or immunocompromised cats (Choromanski *et al.*, 1994, 1995). It has only limited ability to persist in the tissues of cats and cannot survive more than three back passages in cats. No reversion to oocyst excretion or increase in virulence has been observed in over 200 inoculated cats. The T-263 strain is rapidly cleared from the mouth of inoculated cats.

It is logical to assume that cat owners and veterinarians would be at a greater risk for developing toxoplasmosis; however, serological studies do not confirm this assumption. In one study in AIDS patients it was conclusively shown that owning cats did not increase the risk of developing toxoplasmosis (Wallace *et al.*, 1993). The role of cat ownership and exposure to *T. gondii* is, however, not completely clear at present. Many studies have been conducted to determine the association between cat ownership or cat exposure and the prevalence of *T. gondii* infection in humans. Many studies do not find a positive relationship while many find a positive relationship. It must be stressed that preventing exposure to cats is not the same as preventing exposure to *T. gondii* oocysts. Pregnant women or immunocompromised individuals should not change the cat's litter box. If faeces are removed daily this will also help prevent exposure by removing oocysts before they can sporulate. *T. gondii* oocyst can survive in the soil for years and can be disseminated from the original site of deposition by erosion, other mechanical means, and by phoretic vectors. Inhalation of oocysts stirred up in the dust by horses has been associated with an outbreak of human toxoplasmosis at a riding stable (Teutsch *et al.*, 1979). Oocysts are not likely to remain in the air for extended periods of time. Washing fruits and vegetables and wearing gloves while gardening are means of preventing exposure to oocysts.

T. gondii oocysts were not isolated from the fur of oocyst excreting cats (Dubey, 1995). Therefore, it is unlikely that infection can be obtained by petting a cat. Tachyzoites are not likely to be present in the oral cavity of cats with active *T. gondii* infection and none would be in a chronic infection; therefore, it is unlikely that a cat bite would transmit *T. gondii* infection. Cat scratches are also unlikely to transmit *T. gondii* infection.

T. gondii has been isolated from the tissues from domestic cats worldwide (Dubey, 2010). Genotypes of feline isolates are similar to isolates from other animals in the same geographic area.

6.5.2 Dogs

T. gondii was once confused with *Neospora caninum* as a cause of disease in dogs and many reports of toxoplasmosis in dogs are actually neosporosis (Dubey and Lindsay, 1996; Lindsay and Dubey, 2000). True toxoplasmosis does occur in dogs (Dubey *et al.*, 1989). Clinical toxoplasmosis in dogs is often associated with immunosuppression induced by canine distemper virus infection. Clinical signs are usually most apparent in the respiratory and hepatic systems and probably result from reactivation of latent infections (Dubey *et al.* 1989). Transplacental infections have not yet been confirmed in naturally infected dogs. Dogs are resistant to experimental infection with tissue cysts and oocysts (Lindsay *et al.*, 1996, 1997a).

A role for dogs in the transmission of *T. gondii* to humans has been postulated based on serological surveys and observations that dogs ingest cat faeces and often roll in cat faeces and other foul smelling substances (Frenkel *et al.*, 2003). It is believed that dogs can bring oocysts to a home after ingesting them and deposit them in or around the home when they defecate. Experimentally infective *T. gondii* oocysts can be found in dog faeces for up to two days after they ingest oocysts (Lindsay *et al.*, 1997a). *T. gondii* oocysts will not sporulate when placed on dog fur (Lindsay *et al.*, 1997a). Recently, Schares *et al.* (2005) found viable *T. gondii* oocysts in two of 24,089 dogs in Germany. The role of dogs as potential transport hosts for *T. gondii* needs further examination.

T. gondii has been isolated from the tissues from domestic dogs worldwide (Dubey, 2010). Genotypes of feline isolates are similar to isolates from other animals in the same geographic area. A related Apicomplexan parasitic protozoa *Neospora caninum* is present in dogs.

6.5.3 Ferrets

Congenital toxoplasmosis has been observed in farmed raised ferrets (*Mustela putorius furo*) from New Zealand (Thornton and Cook, 1986). Thirty percent of the kits on the farm died acutely and had lesions of disseminated toxoplasmosis. An epizootic of toxoplasmosis occurred among a population of endangered black-footed ferrets (*Mustela nigripes*) at a zoo in the USA (Burns *et al.*, 2003). Twenty-two adults and 30 kits died from acute toxoplasmosis and an additional 13 adults died from chronic toxoplasmosis after the initial outbreak.

6.6 DOMESTIC FARM ANIMALS

6.6.1 Mink

Acute toxoplasmosis with abortions has been reported in farmed mink (*Mustela vison*) from Europe and the USA (Frank, 2001; Smielewska-Los and Turniak, 2004). The practice of feeding non-frozen slaughter offal was blamed for acute toxoplasmosis in one report (Smielewska-Los and Turniak, 2004). Toxoplasmosis was diagnosed using PCR and immunohistochemistry in a young, free ranging mink (*Mustela vison*) that had signs of left hind limb lameness, ataxia, head tremors, and bilateral blindness and was found on a college campus in Michigan (Jones *et al.*, 2006). *T. gondii* has been isolated from wild mink from the USA (Smith and Frenkel, 1995).

6.6.2 Horses

Horses are resistant to experimental infection with 1×10^4 or 1×10^5 oocysts. *T. gondii* can persist in edible tissues of horses for up to 476 days (Dubey, 1985). Although *T. gondii* has been isolated from tissues of horses, there is no confirmed report of clinical toxoplasmosis in horses (Al-Khalidi and Dubey, 1979). A related Apicomplexan parasitic protozoa *Neospora hughesi* is present in horses.

6.6.3 Swine

Abortion can occur in sows. Sows only abort once. Abortions are rare in most pork producing regions of the world with the exception of Taiwan. Pigs raised on dirt are more likely to have *T. gondii* in their tissues. Diagnosis of *T. gondii* abortion in sows is best done by examining foetal fluids for antibodies using the modified agglutination test. Undercooked pork is a source of human infection and viable tissue cysts can remain in pork for up to 865 days (Dubey, 1988).

T. gondii has been isolated from the tissues from domestic pigs worldwide (Dubey, 2010). Genotypes of pig isolates are similar to isolates from other animals in the same geographic area.

6.6.4 Cattle

Clinical toxoplasmosis in cattle is rare and abortions are uncommon. Many reports of bovine abortion due to *T. gondii* are actually due to *N. caninum* (Dubey and Lindsay, 1996). Attempts to isolate *T. gondii* from seropositive cattle are often unsuccessful indicating that beef may not be a significant source of human infection in the USA (Dubey *et al.*, 2005). For example, no *T. gondii* was isolated from 2,094 samples of beef obtained from retail markets in the USA (Dubey *et al.*, 2005). However, viable tissue cysts can remain in cattle for up to 1,191 days (Dubey and Thulliez, 1993). Additional studies are needed to fully document these experimental findings.

6.6.5 Sheep

T. gondii is a common cause of abortion in ewes and an important production problem. Multiple abortions can occur in a flock indicating a common oocyst source for ewes. Ewes develop solid immunity after aborting *T. gondii* infected foetuses. A vaccine to prevent abortion in ewes is available in several countries (Buxton and Innes, 1995) Diagnosis of *T. gondii* abortion in ewes is best done by examining foetal fluids for antibodies using the modified agglutination test. Undercooked lamb and mutton is a source of human infection. *T. gondii* has been isolated from the tissues from domestic sheep worldwide (Dubey, 2010). Genotypes of sheep isolates are similar to isolates from other animals in the same geographic area.

6.6.6 Goats

T. gondii is a common cause of abortion in does. Multiple abortions can occur in a flock indicating a common oocyst source for does. Does develop immunity after aborting *T. gondii* infected foetuses but a repeat abort can occur. Diagnosis of *T. gondii* abortion in goats is best done by examining foetal fluids for antibodies using the modified agglutination test. Undercooked goat meat is a source of human infection. *T. gondii* has been isolated from the tissues from domestic goats worldwide (Dubey, 2010). Genotypes of goat isolates are similar to isolates from other animals in the same geographic area. Drinking raw, unpasteurized goats' milk is a potential source of *T. gondii* for humans and the parasite can survive for up to seven days in refrigerated goat milk (Walsh *et al.*, 1999).

6.6.7 Buffalos

Naturally occurring clinical toxoplasmosis has not been observed in buffalos (*Bison bison, Bubalus bubalis, Syncerus caffer*) and viable *T. gondii* has not been isolated from buffalos.

Serological surveys indicate that buffalos are exposed to the parasite.

6.6.8 Camels

Acute toxoplasmosis was observed in a six year old camel (*Camelus dromedarius*) (Hagemoser *et al.*, 1990). Dyspnea was the main clinical sign and many tachyzoites were found in its lungs and plural exudates. *T. gondii* has been isolated from camel meat using cat bioassays (Hilali *et al.*, 1995).

6.6.9 Llamas, Alpaca and Vicunas

Experimental studies indicate that llamas (*Lama glama*) are resistant to clinical toxoplasmosis even if challenged during pregnancy (Jarvinen *et al.*, 1999). Naturally occurring toxoplasmosis has not been reported in llamas, alpacas (*Lama pacos*) or vicunas (*Lama vicugna*).

6.6.10 Chickens

Chickens (*Gallus domesticus*) that are raised on the ground are a potential source of *T. gondii* due to high level of exposure to oocysts. Chickens do not usually develop clinical signs even after oral inoculation of large numbers of oocysts (Dubey *et al.*, 1993c; Kaneto *et al.*, 1997). Egg production may be affected adversely in laying hens fed large numbers of oocysts but *T. gondii* is not readily transmitted to the eggs of these hens (Biancifiori *et al.*, 1986). Clinical toxoplasmosis does not occur on modern chicken farms where birds are raised indoors. Chickens raised in modern production facilities in confinement indoors are not likely to have viable *T. gondii* in their tissues. None of 2,094 samples from commercial chickens in retail markets from the USA contained viable *T. gondii* in a survey from the USA (Dubey *et al.*, 2005).

T. gondii has been isolated from the tissues from chickens worldwide (Dubey, 2010). Genotypes of chicken isolates are similar to isolates from other animals in the same geographic area. The prevalence of isolation is dependent on the methods used to raise the chickens with chickens raised outside having a higher prevalence of infection.

6.6.11 Turkeys

Domestic turkeys (*Meleagris gallopavo*) fed *T. gondii* oocysts remained clinically normal except a few that developed pneumonia associated with *Aspergillus*-like fungi (Dubey *et al.*, 1993a). Tissue cysts are present in breast and leg muscles of inoculated turkeys. Clinical toxoplasmosis does not occur on modern turkey farms. *T. gondii* has been isolated from the tissues from turkeys worldwide (Dubey, 2010). Genotypes of turkey isolates are similar to isolates from other animals in the same geographic area. The prevalence of isolation is dependent on the methods used to raise the turkeys, with turkeys raised outside having a higher prevalence of infection.

6.6.12 Ducks and Geese

Domestic ducks (*Anas platyrhynchos*) fed *T. gondii* oocysts do not develop clinical toxoplasmosis (Sedlack *et al.*, 2004). Viable *T. gondii* has been isolated from the tissues of a naturally infected domestic ducks (Dubey *et al.*, 2003) and from a domestic goose (*Asner asner*) (Dubey *et al.*, 2007b).

6.7 FISH, REPTILES AND AMPHIBIANS

Toxoplasmosis does not occur in fish, reptiles or amphibians. Reports of infections in these animals in nature are erroneous. Reptiles can be manipulated to make them susceptible to *T. gondii* but they have to be experimentally infected and kept at temperatures of around 37°C–40°C.

References

Al-Khalidi, N.W., Dubey, J.P., 1979. Prevalence of *Toxoplasma gondii* infection in horses. J. Parasitol. 65, 331–334.

Aramini, J.J., Stephen, C., Dubey, J.P., 1998. *Toxoplasma gondii* in Vancouver Island cougars (*Felis concolor vancouverensis*): serology and oocyst shedding. J. Parasitol. 84, 438–440.

Aubert, D., Ajzenberg, D., Richomme, C., Gilot-Fromont, E., Terrier, M.E., de Gevigney, C., Game, Y., Maillard, D., Gibert, P., Dardé, M.L., Villena, I., 2010. Molecular and biological characteristics of *Toxoplasma gondii* isolates from wildlife in France. Vet. Parasitol. 171, 346–349.

Arkush, K.D., Miller, M.A., Leutenegger, C.M., Gardner, I.A., Packham, A.E., Heckeroth, A.R., Tenter, A.M., Barr, B.C., Conrad, P.A., 2003. Molecular and bioassay-based detection of *Toxoplasma gondii* oocyst uptake by mussels (*Mytilus galloprovincialis*). Int. J. Parasitol. 33, 1087–1097.

Bangari, D.S., Mouser, P., Miller, M.A., Stevenson, G.W., Vemulapalli, R., Thacker, H.L., 2007. Toxoplasmosis in a woodchuck (*Marmota monax*) and two American red squirrels (*Tamiasciurus hudsonicus*). J. Vet. Diagn. Invest. 19, 705–709.

Basso, W., Edelhofer, R., Zenker, W., Mostl, K., Kubber-Heiss, A., Prosl, H., 2005. Toxoplasmosis in Pallas' cats (*Otocolobus manul*) raised in captivity. Parasitology 130, 293–299.

Basso, W., Moré, G., Quiroga, M.A., Pardini, L., Bacigalupe, D., Venturini, L., Valenzuela, M.C., Balducchi, D., Maksimov, P., Schares, G., Venturini, M.C., 2009. Isolation and molecular characterization of *Toxoplasma gondii* from captive slender-tailed meerkats (*Suricata suricatta*) with fatal toxoplasmosis in Argentina. Vet. Parasitol. 161, 201–206.

Baszler, T.V., Dubey, J.P., Lohr, C.V., Foreyt, W.J., 2000. Toxoplasmic encephalitis in a free-ranging Rocky Mountain bighorn sheep from Washington. J. Wildl. Dis. 36, 752–754.

Bergmann, V., Heidrich, R., Kiupel, H., 1980. Acute toxoplasmosis outbreak in rabbit flocks. Angew Parasitol. 21, 1–6.

Bettiol, S.S., Obendorf, D.L., Nowarkowski, M., Goldsmid, J.M., 2000. Pathology of experimental toxoplasmosis in eastern barred bandicoots in Tasmania. J. Wildl. Dis. 36, 141–144.

Biancifiori, F., Rondini, C., Grelloni, V., Frescura, T., 1986. Avian toxoplasmosis: experimental infection of chicken and pigeon. Comp. Immunol. Microbiol. Infect. Dis. 9, 337–346.

Bouer, A., Werther, K., Catao-Dias, J.L., Nunes, A.L., 1999. Outbreak of toxoplasmosis in *Lagothrix lagotricha*. Folia Primatol. (Basel) 70, 282–285.

Bossart, G.D., Mignucci-Giannoni, A.A., Rivera-Guzman, A.L., Jimenez-Marrero, N.M., Camus, A., Bonde, R., Dubey, J.P., Reif, J.S., 2012. Disseminated toxoplasmosis in Antillean manatees (*Trichechus manatus manatus*) from Puerto Rico. Dis. Aqua. Org. in press.

Bowater, R.O., Norton, J., Johnson, S., Hill, B., O'Donoghue, P., Prior, H., 2003. Toxoplasmosis in Indo-Pacific humpbacked dolphins (*Sousa chinensis*), from Queensland. Aust. Vet J. 81, 627–632.

Brown, M., Lappin, M.R., Brown, J.L., Munkhtsog, B., Swanson, W.F., 2005. Exploring the ecologic basis for extreme susceptibility of Pallas' cats (*Otocolobus manul*) to fatal toxoplasmosis. J. Wildl. Dis. 41, 691–700.

Burns, R., Williams, E.S., O'Toole, D., Dubey, J.P., 2003. *Toxoplasma gondii* infections in captive black-footed ferrets (*Mustela nigripes*), 1992–1998: clinical signs, serology, pathology, and prevention. J. Wildl. Dis. 39, 787–797.

Buergelt, C.D., Bonde, R.K., 1983. Toxoplasmic meningoencephalitis in a West Indian manatee. J. Am. Vet. Med. Assoc. 183, 1294–1296.

Buxton, D., Innes, E.A., 1995. A commercial vaccine for ovine toxoplasmosis. Parasitology 110, S11–S16.

Canfield, P.J., Hartley, W.J., Dubey, J.P., 1990. Lesions of toxoplasmosis in Australian marsupials. J. Comp. Pathol. 103, 159–167.

Chomel, B.B., Zarnke, R.L., Kasten, R.W., Kass, P.H., Mendes, E., 1995. Serologic survey of *Toxoplasma gondii* in grizzly bears (*Ursus arctos*) and black bears (*Ursus americanus*), from Alaska, 1988 to 1991. J. Wildl. Dis. 31, 472–479.

Canfield, P.J., Hartley, W.J., Dubey, J.P., 1990. Lesions of toxoplasmosis in Australian marsupials. J. Comp. Pathol. 103, 159–167.

Carrasco, L., Raya, A.I., Núñez, A., Gómez-Laguna, J., Hernández, S., Dubey, J.P., 2006. Fatal toxoplasmosis and concurrent (*Calodium hepaticum*) infection in Korean squirrels (*Tanias sibericus*). Vet. Parasitol. 137, 180–183.

Cedillo-Peláez, C., Rico-Torres, C.P., Sales-Garrido, C.G., Correa, D., 2011. Acute toxoplasmosis in squirrel monkeys (*Saimiri sciureus*) in Mexico. Vet. Parasitol. 180, 368–371.

Cole, R.A., Lindsay, D.S., Howe, D.K., Roderick, C.L., Dubey, J.P., Thomas, N.J., Baeten, L.A., 2000. Biological and molecular characterizations of *Toxoplasma gondii* strains obtained from southern sea otters (*Enhydra lutris nereis*). J. Parasitol. 86, 526–530.

Choromanski, L., Freyre, A., Brown, K., Popiel, I., Shibley, G., 1994. Safety aspects of a vaccine for cats containing a *Toxoplasma gondii* mutant strain. J. Eukaryot. Microbiol. 41, 8S.

Choromanski, L., Freyre, A., Popiel, R., Brown, K., Grieve, R., Shibley, G., 1995. Safety and efficacy of modified live feline *Toxoplasma gondii* vaccine. Dev. Biol. Stand. 84, 269–281.

Crawford, G.C., Dunker, F.H., Dubey, J.P., 2000. Toxoplasmosis as a suspected cause of abortion in a Greenland muskox (*Ovibos moshatus wardi*). J. Zoo. Wildl. Med. 31, 247–250.

Cross, J., Lein, J., Hsu, M., 1969. *Toxoplasma* isolated from the Formosan giant flying squirrel. Taiwan Yi Xue Hui Za Zhi 68, 678–683.

Dabritz, H.A., Miller, M.M., Gardner, I.A., Packham, A.E., Atwill, E.R., Conrad, P.A., 2008. Risk factors for *Toxoplasma gondii* infection in wild rodents from Central Coastal California and a review of *T. gondii* prevalence in rodents. J. Parasitol. 94, 675–683.

Davidson, W.R., Nettles, V.F., Hayes, L.E., Howerth, E.W., Couvillion, C.E., 1992. Diseases diagnosed in gray foxes (*Urocyon cinereoargenteus*) from the southeastern USA. J. Wildl. Dis. 28, 28–33.

Davis, S.W., Dubey, J.P., 1995. Mediation of immunity to *Toxoplasma gondii* oocyst shedding in cats. J. Parasitol. 81, 882–886.

Demar, M., Ajzenberg, D., Serrurier, B., Dardé, M.L., Carme, B., 2008. Case Report: Atypical *Toxoplasma gondii* strain from a free-living Jaguar (*Panthera onca*) in French Guiana. Am. J. Trop. Med. Hyg. 78, 195–197.

Dietz, H.H., Henriksen, P., Bille-Hansen, V., Henriksen, S.A., 1997. Toxoplasmosis in a colony of New World monkeys. Vet. Parasitol. 68, 299–304.

Di Guardo, G., Agrimi, U., Morelli, L., Cardeti, G., Terracciano, G., Kennedy, S., 1995. Post mortem investigations on cetaceans found stranded on the coasts of Italy between 1990 and 1993. Vet. Rec. 136, 439–442.

Di Guardo, G., Proietto, U., Di Francesco, C.E., Marsilio, F., Zaccaroni, A., Scaravelli, D., Mignone, W., Garibaldi, F., Kennedy, S., Forster, F., Iulini, B., Bozzetta, E., Casalone, C., 2010 Mar. Cerebral toxoplasmosis in striped dolphins (Stenella coeruleoalba) stranded along the Ligurian Sea coast of Italy. Vet Pathol 47 (2), 245–253.

Dubey, J.P., 1981. Isolation of encysted *Toxoplasma gondii* from musculature of moose and pronghorn in Montana. Am. J. Vet. Res. 42, 126–127.

Dubey, J.P., 1982. Isolation of encysted *Toxoplasma gondii* from muscles of mule deer in Montana. J. Am. Vet. Med. Assoc. 181, 1535.

Dubey, J.P., 1983. *Toxoplasma gondii* infection in rodents and insectivores from Montana. J. Wildl. Dis. 19, 149–150.

Dubey, J.P., 1985. Persistence of encysted *Toxoplasma gondii* in tissues of equids fed oocysts. Am. J. Vet. Res. 46, 1753–1754.

Dubey, J.P., 1988. Long-term persistence of *Toxoplasma gondii* in tissues of pigs inoculated with *T. gondii* oocysts and effect of freezing on viability of tissue cysts in pork. Am. J. Vet. Res. 49, 910–913.

Dubey, J.P., 1995. Duration of immunity to shedding of *Toxoplasma gondii* oocysts by cats. J. Parasitol. 81, 410–415.

Dubey, J.P., 2002. A review of toxoplasmosis in wild birds. Vet. Parasitol. 106, 121–153.

Dubey, J.P., 2010. Toxoplasmosis of Animals and Humans, second ed. CRC Press, Boca Raton. FL, USA, 1–313.

Dubey, J.P., Thorne, E.T., Sharma, S.P., 1980. Experimental toxoplasmosis in elk (*Cervus canadensis*). Am. J. Vet. Res. 41, 792–793.

Dubey, J.P., Thorne, E.T., Williams, E.S., 1982. Induced toxoplasmosis in pronghorns and mule deer. J. Am. Vet. Med. Assoc. 181, 1263–1267.

Dubey, J.P., Carpenter, J.L., 1993. Histologically confirmed clinical toxoplasmosis in cats: 100 cases (1952–1990). J. Am. Vet. Med. Assoc. 203, 1556–1566.

Dubey, J.P., Thulliez, P., 1993. Persistence of tissue cysts in edible tissues of cattle fed *Toxoplasma gondii* oocysts. Am. J. Vet. Res. 54, 270–273.

Dubey, J.P., Lin, T.L., 1994. Acute toxoplasmosis in a gray fox (*Urocyon cinereoargenteus*). Vet. Parasitol. 51, 321–325.

Dubey, J.P., Lindsay, D.S., 1996. A review of *Neospora caninum* and neosporosis. Vet. Parasitol. 67, 1–59.

Dubey, J.P., Hoover, E.A., Walls, K.W., 1977. Effect of age and sex on the acquisition of immunity to toxoplasmosis in cats. J. Protozool. 24, 184–186.

Dubey, J.P., Kramer, L.W., Weisbrode, S.E., 1985. Acute death associated with *Toxoplasma gondii* in ring-tailed lemurs. J. Am. Vet. Med. Assoc. 187, 1272–1273.

Dubey, J.P., Quinn, W.J., Weinandy, D., 1987. Fatal neonatal toxoplasmosis in a bobcat (*Lynx rufus*). J. Wildl. Dis. 23, 324–327.

Dubey, J.P., Gendron-Fitzpatrick, A.P., Lenhard, A.L., Bowman, D., 1988a. Fatal toxoplasmosis and enteroepithelial stages of *Toxoplasma gondii* in a Pallas cat (*Felis manul*). J. Protozool. 35, 528–530.

Dubey, J.P., Ott-Joslin, J., Torgerson, R.W., Topper, M.J., Sundberg, J.P., 1988b. Toxoplasmosis in black-faced kangaroos (*Macropus fuliginosus melanops*). Vet. Parasitol. 30, 97–105.

Dubey, J.P., Carpenter, J.L., Topper, M.L., Uggla, A., 1989. Fatal toxoplasmosis in dogs. J. Am. Anim. Hosp. Assoc. 25, 659–664.

Dubey, J.P., Hamir, A.N., Rupprecht, C.E., 1990. Acute disseminated toxoplasmosis in a red fox (*Vulpes vulpes*). J. Wildl. Dis. 26, 286–290.

Dubey, J.P., Brown, C.A., Carpenter, J.L., Moore 3rd, J.J., 1992a. Fatal toxoplasmosis in domestic rabbits in the USA. Vet. Parasitol. 44, 305–309.

Dubey, J.P., Porteer, S.I., Tseng, F., Shen, S.K., Thulliez, P., 1992b. Induced toxoplasmosis in owls. J. Zoo. Wildl. Med. 23, 98–102.

Dubey, J.P., Camargo, M.E., Ruff, M.D., Wilkins, G.C., Shen, S.K., Kwok, O.C., Thulliez, P., 1993a. Experimental toxoplasmosis in turkeys. J. Parasitol. 79, 949–952.

Dubey, J.P., Hamir, A.N., Shen, S.K., Thulliez, P., Rupprecht, C.E., 1993b. Experimental *Toxoplasma gondii* infection in raccoons (*Procyon lotor*). J. Parasitol. 79, 548–552.

Dubey, J.P., Ruff, M.D., Camargo, M.E., Shen, S.K., Wilkins, G.L., Kwok, O.C., Thulliez, P., 1993c. Serologic and parasitologic responses of domestic chickens after oral inoculation with *Toxoplasma gondii* oocysts. Am. J. Vet. Res. 54, 1668–1672.

Dubey, J.P., Humphreys, J.G., Thulliez, P., 1995a. Prevalence of viable *Toxoplasma gondii* tissue cysts and antibodies to *T. gondii* by various serologic tests in black bears (*Ursus americanus*) from Pennsylvania. J. Parasitol. 81, 109–112.

Dubey, J.P., Lappin, M.R., Thulliez, P., 1995b. Diagnosis of induced toxoplasmosis in neonatal cats. J. Am. Vet. Med. Assoc. 207, 179–185.

Dubey, J.P., Lewis, B., Beam, K., Abbitt, B., 2002a. Transplacental toxoplasmosis in a reindeer (*Rangifer tarandus*) foetus. Vet. Parasitol. 110, 131–135.

Dubey, J.P., Tocidlowski, M.E., Abbitt, B., Llizo, S.Y., 2002b. Acute visceral toxoplasmosis in captive dik-dik (*Madoqua guentheri smithi*). J. Parasitol. 88, 638–641.

Dubey, J.P., Zarnke, R., Thomas, N.J., Wong, S.K., Van Bonn, W., Briggs, M., Davis, J.W., Ewing, R., Mense, M., Kwok, O.C., Romand, S., Thulliez, P., 2003. *Toxoplasma gondii, Neospora caninum, Sarcocystis neurona*, and *Sarcocystis canis*-like infections in marine mammals. Vet. Parasitol. 116, 275–296.

Dubey, J.P., Lipscomb, T.P., Mense, M., 2004a. Toxoplasmosis in an elephant seal (*Mirounga angustirostris*). J. Parasitol. 90, 410–411.

Dubey, J.P., Graham, D.H., De Young, R.W., Dahl, E., Eberhard, M.L., Nace, E.K., Won, K., Bishop, H., Punkosdy, G., Sreekumar, C., Vianna, M.C., Shen, S.K., Kwok, O.C., Sumners, J.A., Demarais, S., Humphreys, J.G., Lehmann, T., 2004b. Molecular and biologic characteristics of *Toxoplasma gondii* isolates from wildlife in the USA. J. Parasitol. 90, 67–71.

Dubey, J.P., Navarro, I.T., Sreekumar, C., Dahl, E., Freire, R.L., Kawabata, H.H., Vianna, M.C., Kwok, O.C., Shen, S.K., Thulliez, P., Lehmann, T., 2004c. *Toxoplasma gondii* infections in cats from Parana, Brazil: seroprevalence, tissue distribution, and biologic and genetic characterization of isolates. J. Parasitol. 90, 721–726.

Dubey, J.P., Parnell, P.G., Sreekumar, C., Vianna, M.C., De Young, R.W., Dahl, E., Lehmann, T., 2004d. Biologic and molecular characteristics of *Toxoplasma gondii* isolates from striped skunk (*Mephitis mephitis*), Canada goose (*Branta canadensis*), black-winged lory (*Eos cyanogenia*), and cats (*Felis catus*). J. Parasitol. 90, 1171–1174.

Dubey, J.P., Hill, D.E., Jones, J.L., Hightower, A.W., Kirkland, E., Roberts, J.M., Marcet, P.L., Lehmann, T., Vianna, M.C.B., Miska, K., Sreekumar, C., Kwok, O.C.H., Shen, S.K., Gamble, H.R., 2005. Prevalence of viable *Toxoplasma gondii* in beef, chicken and pork from retail meat stores in the USA; risk assesment to consumers. J. Parasitol. 91, 1082–1093.

Dubey, J.P., Hodgin, E.C., Hamir, A.N., 2006a. Acute fatal toxoplasmosis in squirrels (*Sciurus carolensis*) with bradyzoites in visceral tissues. J. Parasitol. 92, 658–659.

Dubey, J.P., Bhaiyat, M.I., Macpherson, C.N.L., de Allie, C., Chikweto, A., Kwok, O.C.H., Sharma, R.N., 2006b. Prevalence of *Toxoplasma gondii* in rats (*Rattus norvegicus*) in Grenada, West Indies. J. Parasitol. 92, 1107–1108.

Dubey, J.P., Morales, J.A., Sundar, N., Velmurugan, G.V., González-Barrientos, C.R., Hernández-Mora, G., Su, C., 2007a. Isolation and genetic characterization of *Toxoplasma gondii* from striped dolphin (*Stenella coeruleoalba*) from Costa Rica. J. Parasitol. 93, 710–711.

Dubey, J.P., Webb, D.M., Sundar, N., Velmurugan, G.V., Bandini, L.A., Kwok, O.C.H., Su, C., 2007b. Endemic avian toxoplasmosis on a farm in Illinois: Clinical disease, diagnosis, biologic and genetic characteristics of *Toxoplasma gondii* isolates from chickens (*Gallus domesticus*), and a goose (*Anser anser*). Vet. Parasitol. 148, 207–212.

Dubey, J.P., Fair, P.A., Sundar, N., Velmurugan, G., Kwok, O.C.H., McFee, W.E., Majumdar, D., Su, C., 2008a. Isolation of *Toxoplasma gondii* from bottlenose dolphins (*Tursiops truncatus*). J. Parasitol. 94, 821–823.

Dubey, J.P., Velmurugan, G.V., Ulrich, V., Gill, J., Carstensen, M., Sundar, N., Kwok, O.C.H., Thulliez, P., Majumdar, D., Su, C., 2008b. Transplacental toxoplasmosis in naturally-infected white-tailed deer: Isolation and genetic characterisation of *Toxoplasma gondii* from foetuses of different gestational ages. Int. J. Parasitol. 38, 1057–1063.

Dubey, J.P., Pas, A., Rajendran, C., Kwok, O.C.H., Ferreira, L.R., Martinc, J., Hebel, C., Hammer, S., Su, C., 2010. Toxoplasmosis in Sand cats (*Felis margarita*) and other animals in the Breeding Centre for Endangered Arabian Wildlife in the United Arab Emirates and Al Wabra Wildlife Preservation, the State of Qatar. Vet. Parasitol. 172, 195–203.

Dubey, J.P., Passos, I.M.F., Rajendran, C., Ferreira, I.R., Gennari, S.M., Su, C., 2011a. Isolation of viable *Toxoplasma gondii* from feral guinea fowl (*Numida meleagris*) and domestic rabbits (*Oryctolagus cuniculus*) from Brazil. J. Parasitol. 97, 842–845.

Dubey, J.P., Velmurugan, G.V., Rajendran, C., Yabsley, M.J., Thomas, N.J., Beckmen, K.B., Sinnett, D., Ruid, D., Hart, J., Fair, P.A., McFee, W.E., Shearn-Bochsler, V., Kwok, O.C., Ferreira, L.R., Choudhary, S., Faria, E.B., Zhou, H., Felix, T.A., Su, C., 2011b. Genetic characterisation of *Toxoplasma gondii* in wildlife from North America revealed widespread and high prevalence of the fourth clonal type. Int. J. Parasitol. 41, 1139–1147.

Forzán, M.J., Frasca, S., 2004. Systemic toxoplasmosis in a five-month-old beaver, (*Castor canadensis*). J. Zoo. Wildl. Med. 35, 113–115.

Epiphanio, S., Guimaraes, M.A., Fedullo, D.L., Correa, S.H., Catao-Dias, J.L., 2000. Toxoplasmosis in golden-headed lion tamarins (*Leontopithecus chrysomelas*) and emperor marmosets (*Saguinus imperator*) in captivity. J. Zoo Wildl. Med. 31, 231–235.

Epiphanio, S., Sa, L.R., Teixeira, R.H., Catao-Dias, J.L., 2001. Toxoplasmosis in a wild-caught black lion tamarin (*Leontopithecus chrysopygus*). Vet. Rec. 149, 627–628.

Epiphanio, S., Sinhorini, I.L., Catao-Dias, J.L., 2003. Pathology of toxoplasmosis in captive new world primates. J. Comp. Pathol. 129, 196–204.

Escajadillo, A., Frenkel, J.K., 1991. Experimental toxoplasmosis and vaccine tests in *Aotus* monkeys. Am. J. Trop. Med. Hyg. 44, 382–389.

Frank, R.K., 2001. An outbreak of toxoplasmosis in farmed mink (*Mustela vison* S.). J. Vet. Diagn. Invest. 13, 245–249.

Frenkel, J.K., Pfefferkorn, E.R., Smith, D.D., Fishback, J.L., 1991. Prospective vaccine prepared from a new mutant of *Toxoplasma gondii* for use in cats. Am. J. Vet. Res. 52, 759–763.

Frenkel, J.K., Lindsay, D.S., Parker, B.B., Dobesh, M., 2003. Dogs as possible mechanical carriers of *Toxoplasma* and dog fur as source of infection for young children. Int. J. Infect. Dis. 7, 292–293.

Freyre, A., Choromanski, L., Fishback, J.L., Popiel, I., 1993. Immunization of cats with tissue cysts, bradyzoites, and tachyzoites of the T-263 strain of *Toxoplasma gondii*. J. Parasitol. 79, 716–719.

Furuta, T., Une, Y., Omura, M., Matsutani, N., Nomura, Y., Kikuchi, T., Hattori, S., Yoshikawa, Y., 2001. Horizontal transmission of *Toxoplasma gondii* in squirrel monkeys (*Saimiri sciureus*). Exp. Anim. 50, 299–306.

Gajadhar, A.A., Measures, L., Forbes, L.B., Kapel, C., Dubey, J.P., 2004. Experimental *Toxoplasma gondii* infection in grey seals (*Halichoerus grypus*). J. Parasitol. 90, 255–259.

Galuzo, I.G., Vysokova, L.A., Krivkova, A.M., 1970. Toxoplasmosis in bats. In: Fitzgerald, P.R. (Ed.), Toxoplasmosis in Animals. University of Illinois, College of Veterinary Medicine, Urbana, IL, pp. 190–194.

Gehrt, S.D., Michael, J., Kinsel, M.J., Anchor, C., 2010. Pathogen dynamics and morbidity of striped skunks in the absence of rabies. J. Wildl. Dis. 46, 335–347.

Geisel, O., Breuer, W., Minkus, G., Hermanns, W., 1995. Toxoplasmosis causing death in a mole (*Talpa europaea*). Berl. Munch. Tierarztl. Wochenschr. 108, 241–243.

Gerhold, R.W., Howerth, E.W., Lindsay, D.S., 2005. *Sarcocystis neurona*-associated meningoencephalitis and description of intramuscular sarcocysts in a fisher (*Martes pennanti*). J. Wildl. Dis. 41, 224–230.

Gorman, T.R., Riveros, V., Alcaino, H.A., Salas, D.R., Thiermann, E.R., 1986. Helminthiasis and toxoplasmosis among exotic mammals at the Santiago National Zoo. J. Am. Vet. Med. Assoc. 189, 1068–1070.

Gustafsson, K., Uggla, A., Jarplid, B., 1997. *Toxoplasma gondii* infection in the mountain hare (*Lepus timidus*) and domestic rabbit (*Oryctolagus cuniculus*). I. Pathology. J. Comp. Pathol. 117, 351–360.

Hagemoser, W.A., Dubey, J.P., Thompson, J.R., 1990. Acute toxoplasmosis in a camel. J. Am. Vet. Med. Assoc. 196, 347.

Hancock, K., Thiele, L.A., Zajac, A.M., Elvinger, F., Lindsay, D.S., 2005. Prevalence of antibodies to *Toxoplasma gondii* in raccoons (*Procyon lotor*) from an urban area of Northern Virginia. J. Parasitol. 91, 694–695.

Harper 3rd, J.S., London, W.T., Sever, J.L., 1985. Five drug regimens for treatment of acute toxoplasmosis in squirrel monkeys. Am. J. Trop. Med. Hyg. 34, 50–57.

Harrison, T.M., Moorman, J.B., Bolin, S.R., Grosjean, N.L., Lim, A., Fitzgerald, S. D., 2007. *Toxoplasma gondii* in an African crested porcupine (*Hystrix cristata*). J. Vet. Diagn. Invest. 19, 191–194.

Hartley, W.J., Dubey, J.P., Spielman, D.S., 1990. Fatal toxoplasmosis in koalas (*Phascolarctos cinereus*). J. Parasitol. 76, 271–272.

Hejlicek, K., Literak, I., Nezval, J., 1997. Toxoplasmosis in wild mammals from the Czech Republic. J. Wildl. Dis. 33, 480–485.

Herrmann, D.C., Maksimov, P., Maksimov, A., Sutor, A., Schwarz, S., Jaschke, W., Schliephake, A., Denzin, N., Conraths, F.J., Schares, G., 2012. *Toxoplasma gondii* in foxes and rodents from the German Federal States of Brandenburg and Saxony-Anhalt: Seroprevalence and genotypes. Vet. Parasitol. 185, 78–85.

Hilali, M., Fatani, A., al-Atiya, S., 1995. Isolation of tissue cysts of *Toxoplasma, Isospora, Hammondia* and *Sarcocystis* from camel (*Camelus dromedarius*) meat in Saudi Arabia. Vet. Parasitol. 58, 353–356.

Holland, G.N., O'Connor, G.R., Diaz, R.F., Minasi, P., Wara, W.M., 1988. Ocular toxoplasmosis in immunosuppressed nonhuman primates. Invest. Ophthalmol. Vis. Sci. 29, 835–842.

Holshuh, H.J., Sherrod, A.E., Taylor, C.R., Andrews, B.F., Howard, E.B., 1985. Toxoplasmosis in a feral northern fur seal. J. Am. Vet. Med. Assoc. 187, 1229–1230.

Howerth, E.W., Rodenroth, N., 1985. Fatal systemic toxoplasmosis in a wild turkey. J. Wildl. Dis. 21, 446–449.

Hubbard, G., Witt, W., Healy, M., Schmidt, R., 1986. An outbreak of toxoplasmosis in zoo birds. Vet. Pathol. 23, 639–641.

Honnold, S.P., Braun, R., Scott, D.P., Sreekumar, C., Dubey, J.P., 2005. Toxoplasmosis in a Hawaiian monk seal (*Monachus schauinslandi*). J. Parasitol. 91, 695–697.

Inskeep 2nd, W., Gardiner, C.H., Harris, R.K., Dubey, J.P., Goldston, R.T., 1990. Toxoplasmosis in Atlantic bottle-nosed dolphins (*Tursiops truncatus*). J. Wildl. Dis. 26, 377–382.

Jardine, J.E., Dubey, J.P., 2002. Congenital toxoplasmosis in an Indo-Pacific bottlenose dolphin (*Tursiops aduncus*). J. Parasitol. 88, 197–199.

Jarvinen, J.A., Dubey, J.P., Althouse, G.C., 1999. Clinical and serologic evaluation of two llamas (*Lama glama*) infected with *Toxoplasma gondii* during gestation. J. Parasitol. 85, 142–144.

Jewell, M.L., Frenkel, J.K., Johnson, K.M., Reed, V., Ruiz, A., 1972. Development of *Toxoplasma* oocysts in neotropical felidae. Am. J. Trop. Med. Hyg. 21, 512–517.

Jokelainen, P., Nylund, M., 2012. Acute fatal toxoplasmosis in three Eurasian Red Squirrels (*Sciurus vulgaris*) caused by genotype II of *Toxoplasma gondii*. J. Wildl. Dis. 48, 454–457.

Jokelainen, P., Isomursu, M., Näreaho, A., Oksanen, A., 2011. Natural *Toxoplasma gondii* infections in European brown hares and mountain hares in Finland: proportional mortality rate, antibody prevalence, and genetic characterization. J. Wildl. Dis. 47, 154–163.

Jones, Y.L., Fitzgerald, S.D., Sikarske, J.G., Murphy, A., Grosjean, N., Kiupel, M., 2006. Toxoplasmosis in a free-ranging mink. J. Wildl. Dis. 42, 865–869.

Juan-Salles, C., Prats, N., Lopez, S., Domingo, M., Marco, A.J., Moran, J.F., 1997. Epizootic disseminated toxoplasmosis in captive slender-tailed meerkats (*Suricata suricatta*). Vet. Pathol. 34, 1–7.

Juan-Salles, C., Prats, N., Marco, A.J., Ramos-Vara, J.A., Borras, D., Fernandez, J., 1998. Fatal acute toxoplasmosis in three golden lion tamarins (*Leontopithecus rosalia*). J. Zoo Wildl. Med. 29, 55–60.

Juan-Sallés, C., Mainez, M., Marco, A., Sanchís, A.M.M., 2011. Localized toxoplasmosis in a ring-tailed lemur (*Lemur catta*) causing placentitis, stillbirths, and disseminated foetal infection. J. Vet. Diag. Invest. 23, 1041–1045.

Junge, R.E., Fischer, J.R., Dubey, J.P., 1992. Fatal disseminated toxoplasmosis in a captive Cuvier's gazelle (*Gazella cuvieri*). J. Zoo Wildl. Med. 23, 342–345.

Kaneto, C.N., Costa, A.J., Paulillo, A.C., Moraes, F.R., Murakami, T.O., Meireles, M.V., 1997. Experimental toxoplasmosis in broiler chicks. Vet. Parasitol. 69, 203–210.

Kelly, T.R., Sleeman, J.M., 2003. Morbidity and mortality of red foxes (*Vulpes vulpes*) and gray foxes (*Urocyon cinereoargenteus*) admitted to the Wildlife Center of Virginia, 1993–2001. J. Wildl. Dis. 39, 467–469.

Kenny, D.E., Lappin, M.R., Knightly, F., Baler, J., Brewer, M., Getzy, D.M., 2002. Toxoplasmosis in Pallas' cats (*Otocolobus felis manul*) at the Denver Zoological Gardens. J. Zoo Wildl. Med. 33, 131–138.

Kijlstra, A., Meerburg, B., Cornelissen, J., De Craeye, S., Vereijken, P., Jongert, E., 2008. The role of rodents and shrews in the transmission of *Toxoplasma gondii* to pigs. Vet. Parasitol. 156, 183–190.

Kottwitz, J.J., Preziosi, D.E., Miller, M.A., Ramos-Vara, J.A., Maggs, D.J., Bonagura, J.D., 2004. Heart failure caused by toxoplasmosis in a Fennec fox (*Fennecus zerda*). J. Am. Anim. Hosp. Assoc. 40, 501–507.

Kotula, A.W., Dubey, J.P., Sharar, A.K., Andrews, C.D., Shen, S.K., Lindsay, D.S., 1991. Effect of freezing on infectivity of *Toxoplasma gondii* tissue cysts in pork. J. Food Protect. 54, 687–690.

Lappin, M.R., Greene, C.E., Prestwood, A.K., Dawe, D.L., Tarleton, R.L., 1989. Diagnosis of recent *Toxoplasma gondii* infection in cats by use of an enzyme-linked immunosorbent assay for immunoglobulin M. Am. J. Vet. Res. 50, 1580–1585.

Lappin, M.R., Dawe, D.L., Lindl, P., Greene, C.E., Prestwood, A.K., 1992. Mitogen and antigen-specific induction of lymphoblast transformation in cats with subclinical toxoplasmosis. Vet. Immunol. Immunopathol. 30, 207–220.

Las, R.D., Shivaprasad, H.L., 2008. An outbreak of toxoplasmosis in an aviary collection of Nicobar pigeons (*Caloenas nicobaria*). J. S. Afr. Vet. Assoc. 79, 149–152.

Lindsay, D.S., Dubey, J.P., 2000. Canine neosporsis. J. Vet. Parasitol. 14, 4–14.

Lindsay, D.S., Dubey, J.P., Blagburn, B.L., 1991a. *Toxoplasma gondii* infections in red-tailed hawks inoculated orally with tissue cysts. J. Parasitol. 77, 322–325.

Lindsay, D.S., Blagburn, B.L., Dubey, J.P., Mason, W.H., 1991b. Prevalence and isolation of *Toxoplasma gondii* from white-tailed deer in Alabama. J. Parasitol. 77, 62–64.

Lindsay, D.S., Smith, P.C., Hoerr, F.J., Blagburn, B.L., 1993. Prevalence of encysted *Toxoplasma gondii* in raptors from Alabama. J. Parasitol. 79, 870–873.

Lindsay, D.S., Smith, P.C., Blagburn, B.L., 1994. Prevalence and isolation of *Toxoplasma gondii* from wild turkeys in Alabama. J. Helminthol. Soc. Wash. 61, 115–117.

Lindsay, D.S., Dubey, J.P., Butler, J.M., Blagburn, B.L., 1996. Experimental tissue cyst induced *Toxoplasma gondii* infections in dogs. J. Eukaryot. Microbiol. 43, 113S.

Lindsay, D.S., Dubey, J.P., Butler, J.M., Blagburn, B.L., 1997a. Mechanical transmission of *Toxoplasma gondii* oocysts by dogs. Vet. Parasitol. 73, 27–33.

Lindsay, D.S., Sundermann, C.A., Dubey, J.P., Blagburn, B.L., 1997b. Update on *Toxoplasma gondii* infections in wildlife and exotic animals from Alabama. J. Ala. Acad. Sci. 68, 246–254.

Lindsay, D.S., Thomas, N.J., Rosypal, A.C., Dubey, J.P., 2001a. Dual *Sarcocystis neurona* and *Toxoplasma gondii* infection in a Northern sea otter from Washington state, USA. Vet. Parasitol. 97, 319–327.

Lindsay, D.S., Phelps, K.K., Smith, S.A., Flick, G., Sumner, S.S., Dubey, J.P., 2001b. Removal of *Toxoplasma gondii* oocysts from sea water by eastern oysters (*Crassostrea virginica*). J. Eukaryot. Microbiol. (Suppl.), 197S–198S.

Lindsay, D.S., Collins, M.V., Mitchell, S.M., Cole, R.A., Flick, G.J., Wetch, C.N., Lindquist, A., Dubey, J.P., 2003. Sporulation and survival of *Toxoplasma gondii* oocysts in seawater. J. Eukaryot Microbiol. 50 (Suppl.), 687–688.

Lindsay, D.S., Collins, M.V., Mitchell, S.M., Wetch, C.N., Rosypal, A.C., Flick, G.J., Zajac, A.M., Lindquist, A., Dubey, J.P., 2004. Survival of *Toxoplasma gondii* oocysts in Eastern oysters (*Crassostrea virginica*). J. Parasitol. 90, 1054–1057.

Lindsay, D.S., McKown, R.D., DiCristina, J., Jordan, C.N., Mitchell, S.M., Oates, D.W., Sterner, M.C., 2005. Prevalence of agglutinating antibodies to *Toxoplasma gondii* in adult and foetal mule deer (*Odocoileus hemionus*) from Nebraska. J. Parasitol. 91 (6), 1490–1491.

Lindsay, D.S., Dubey, J.P., 2009. Long-term survival of *Toxoplasma gondii* sporulated oocysts in seawater. J. Parasitol. 95, 1019–1020.

Lukesova, D., Literak, I., 1998. Shedding of *Toxoplasma gondii* oocysts by Felidae in zoos in the Czech Republic. Vet. Parasitol. 74, 1–7.

Massie, G.N., Ware, M.N., Villegas, E.N., Black, M.W., 2010. Uptake and transmission of Toxoplasma *gondii* oocysts by migratory, filter-feeding fish. Vet. Parasitol. 169, 296–303.

Mazzariol, S., Federica Marcer, F., Mignone, W., 2012. *Toxoplasma gondii* coinfection in a Mediterranean fin whale (*Balaenoptera physalus*). BMC Vet. Res. 8, 20.

Meerburg, B.G., De Craeye, S., Dierick, K.A., Kijlstra, A., 2012. *Neospora caninum* and *Toxoplasma gondii* in brain tissue of feral rodents and insectivores caught on farms in the Netherlands. Vet. Parasitol. 184, 317–320.

Migaki, G., Allen, J.F., Casey, H.W., 1977. Toxoplasmosis in a California sea lion (*Zalophus californianus*). Am. J. Vet. Res. 38, 135–136.

Migaki, G., Sawa, T.R., Dubey, J.P., 1990. Fatal disseminated toxoplasmosis in a spinner dolphin (*Stenella longirostris*). Vet. Pathol. 27, 463–464.

Mikaelian, I., Dubey, J.P., Martineau, D., 1997. Severe hepatitis resulting from toxoplasmosis in a barred owl (*Strix varia*) from Quebec, Canada. Avian Dis. 41, 738–740.

Mikaelian, I., Boisclair, J., Dubey, J.P., Kenned, Y.S., Martineau, D., 2000. Toxoplasmosis in beluga whales (*Delphinapterus leucas*) from the St Lawrence estuary: two case reports and a serological survey. J. Comp. Pathol. 122, 73–76.

Miller, M.A., Sverlow, K., Crosbie, P.R., Barr, B.C., Lowenstine, L.J., Gulland, F.M., Packham, A., Conrad, P.A., 2001. Isolation and characterization of two parasitic protozoa from a Pacific harbor seal (*Phoca vitulina richardsi*) with meningoencephalomyelitis. J. Parasitol. 87, 816–822.

Miller, M.A., Gardner, I.A., Kreuder, C., Paradies, D.M., Worcester, K.R., Jessup, D.A., Dodd, E., Harris, M.D., Ames, J.A., Packham, A.E., Conrad, P.A., 2002. Coastal freshwater runoff is a risk factor for *Toxoplasma gondii* infection of southern sea otters (*Enhydra lutris nereis*). Int. J. Parasitol. 32, 997–1006.

Miller, N.L., Frenkel, J.K., Dubey, J.P., 1972. Oral infections with *Toxoplasma* cysts and oocysts in felines, other mammals, and in birds. J. Parasitol. 58, 928–937.

Morales, J.A., Pena, M.A., Dubey, J.P., 1996. Disseminated toxoplasmosis in a captive porcupine (*Coendou mexicanus*) from Costa Rica. J. Parasitol. 82, 185–186.

Nardoni, S., Angelici, M.C., Mugnaini, L., Mancianti, F., 2011. Prevalence of Toxoplasma gondii infection in *Myocastor coypus* in a protected Italian wetland. Parasit. Vect. 4, 240.

Oksanen, A., Gustafsson, K., Lunden, A., Dubey, J.P., Thulliez, P., Uggla, A., 1996. Experimental *Toxoplasma gondii* infection leading to fatal enteritis in reindeer (*Rangifer tarandus*). J. Parasitol. 82, 843–845.

Pas, A., Dubey, J.P., 2008a. Fatal toxoplasmosis in sand cats (*Felis margarita*). J. Zoo Wildl. Med. 39, 362–369.

Pas, A., Dubey, J.P., 2008b. Seroprevalence of antibodies to *Toxoplasma gondii* in Gordon's wild cat (*Felis silvestris gordoni*) in the Middle East. J. Parasitol. 94, 1169.

Pas, A., Dubey, J.P., 2008c. Toxoplasmosis in Sand fox (*Vulpes rueppelli*). J. Parasitol. 94, 976–977.

Pena, H.F.J., Marvulo, M.F.V., Horta, M.C., Silva, M.A., Silva, J.C.R., Siqueira, D.B., Lima, P.A.C.P., Vitaliano, S.N., Gennari, S.M., 2011. Isolation and genetic characterisation of *Toxoplasma gondii* from a red-handed howler monkey (Alouatta belzebul), a jaguarundi (*Puma yagouaroundi*), and a black-eared opossum (*Didelphis aurita*) from Brazil. J. Parasitol. 175, 377–381.

Pertz, C., Dubielzig, R.R., Lindsay, D.S., 1997. Fatal *Toxoplasma gondii* infection in golden lion tamarins (*Leontopithecus rosalia rosalia*). J. Zoo Wildl. Med. 28, 491–493.

Ploeg, M., Ultee, T., Marja Kik, M., 2011. Disseminated toxoplasmosis in black-footed penguins (*Spheniscus demersus*). Avian Dis. 55, 701–703.

Poelma, F.G., Zwart, P., 1972. Toxoplasmosis in crowned pigeons and other birds at the Royal Zoo, Blijdorp, in Rotterdam. Acta Zool. Pathol. Antverp. 55, 29–40.

Portas, T.J., 2010. Toxoplasmosis in macropodids: a review. J. Zoo Wildl. Med. 41, 1–6.

Quist, C.F., Dubey, J.P., Luttrell, M.P., Davidson, W.R., 1995. Toxoplasmosis in wild turkeys: a case report and serologic survey. J. Wildl. Dis. 31, 255–258.

Rah, H., Chomel, B.B., Follmann, E.H., Kasten, R.W., Hew, C.H., Farver, T.B., Garner, G.W., Amstrup, S.C., 2005. Serosurvey of selected zoonotic agents in polar bears (*Ursus maritimus*). Vet. Rec. 156, 7–13.

Reddacliff, G.L., Hartley, W.J., Dubey, J.P., Cooper, D.W., 1993. Pathology of experimentally-induced, acute toxoplasmosis in macropods. Aust. Vet. J. 70, 4–6.

Reed, W.M., Turek, J.J., 1985. Concurrent distemper and disseminated toxoplasmosis in a red fox. J. Am. Vet. Med. Assoc. 187, 1264–1265.

Resendes, A.R., Almeria, S., Dubey, J.P., Obon, E., Juan-Salles, C., Degollada, E., Alegre, F., Cabezon, O., Pont, S., Domingo, M., 2002. Disseminated toxoplasmosis in a Mediterranean pregnant Risso's dolphin (*Grampus griseus*) with transplacental foetal infection. J. Parasitol. 88, 1029–1032.

Riemann, H.P., Fowler, M.E., Schulz, T., Lock, A., Thilsted, J., Pulley, L.T., Hendrickson, R.V., Henness, A.M., Franti, C.E., Behymer, D.E., 1974. Toxoplasmosis in Pallas's cats. J. Wildl. Dis. 10, 471–477.

Ross, R.D., Stec, L.A., Werner, J.C., Blumenkranz, M.S., Glazer, L., Williams, G.A., 2001. Presumed acquired ocular toxoplasmosis in deer hunters. Retina 21, 226–229.

Sacks, J.J., Delgado, D.G., Lobel, H.O., Parker, R.L., 1983. Toxoplasmosis infection associated with eating undercooked venison. Am. J. Epidemiol. 118, 832–838.

Sangster, C.R., Gordon, A.N., Hayes, D., 2012. Systemic toxoplasmosis in captive flying-foxes. Aust. Vet. J. 90, 140–142.

Sasseville, V.G., Pauley, D.R., MacKey, J.J., Simon, M.A., 1995. Concurrent central nervous system toxoplasmosis and simian immunodeficiency virus-induced AIDS encephalomyelitis in a Barbary macaque (*Macaca sylvana*). Vet. Pathol. 32, 81–83.

Schares, G., Pantchev, N., Barutzki, D., Heydorn, A.O., Bauer, C., Conraths, F.J., 2005. Oocysts of *Neospora caninum, Hammondia heydorni, Toxoplasma gondii* and *Hammondia hammondi* in faeces collected from dogs in Germany. Int. J. Parasitol. 35 (14), 1525–1537.

Schoondermark-Van de Ven, E., Melchers, W., Galama, J., Camps, W., Eskes, T., Meuwissen, J., 1993. Congenital toxoplasmosis: an experimental study in rhesus monkeys for transmission and prenatal diagnosis. Exp. Parasitol. 77, 200–211.

Sedlak, K., Literak, I., Faldyna, M., Toman, M., Benak, J., 2000. Fatal toxoplasmosis in brown hares (*Lepus europaeus*): possible reasons of their high susceptibility to the infection. Vet. Parasitol. 93, 13–28.

Sedlak, K., Bartova, E., Literak, I., Vodicka, R., Dubey, J.P., 2004. Toxoplasmosis in nilgais (*Boselaphus tragocamelus*) and a saiga antelope (*Saiga tatarica*). J. Zoo Wildl. Med. 35, 530–533.

Smielewska-Los, E., Turniak, W., 2004. *Toxoplasma gondii* infection in Polish farmed mink. Vet. Parasitol. 122, 201–206.

Smith, K.E., Fisher, J.R., Dubey, J.P., 1995. Toxoplasmosis in a bobcat (*Felis rufus*). J. Wildl. Dis. 31, 555–557.

Smith, D.D., Frenkel, J.K., 1995. Prevalence of antibodies to *Toxoplasma gondii* in wild mammals of Missouri and east central Kansas: biologic and ecologic considerations of transmission. J. Wildl. Dis. 31, 15–21.

Spencer, J.A., Joiner, K.S., Hilton, C.D., Dubey, J.P., Toivio-Kinnucan, M., Minc, J.K., Blagburn, B.L., 2004. Disseminated toxoplasmosis in a captive ring-tailed lemur (*Lemur catta*). J. Parasitol. 90, 904–906.

Sorensen, K.K., Mork, T., Sigurdardottir, O.G., Asbakk, K., Akerstedt, J., Bergsjo, B., Fuglei, E., 2005. Acute toxoplasmosis in three wild arctic foxes (*Alopex lagopus*) from Svalbard; one with co-infections of *Salmonella Enteritidis* PT1 and *Yersinia pseudotuberculosis* serotype 2b. Res. Vet. Sci. 78, 161–167.

Sreekumar, C., Graham, D.H., Dahl, E., Lehmann, T., Raman, M., Bhalerao, D.P., Vianna, M.C., Dubey, J.P., 2003. Genotyping of *Toxoplasma gondii* isolates from chickens from India. Vet. Parasitol. 118, 187–194.

Szabo, K.A., Mense, M.G., Lipscomb, T.P., Felix, K.J., Dubey, J.P., 2004. Fatal toxoplasmosis in a bald eagle (*Haliaeetus leucocephalus*). J. Parasitol. 90, 907–908.

Teutsch, S.M., Juranek, D.D., Sulzer, A., Dubey, J.P., Sikes, R.K., 1979. Epidemic toxoplasmosis associated with infected cats. N. Engl. J. Med. 300, 695–699.

Thornton, R.N., Cook, T.G., 1986. A congenital *Toxoplasma*-like disease in ferrets (*Mustela putorius furo*). N.Z. Vet. J. 34, 31–33.

van Pelt, R.W., Dieterich, R.A., 1972. Toxoplasmosis in thirteen-lined ground squirrels. J. Am. Vet. Med. Assoc. 161, 643–647.

Wallace, M.R., Rossetti, R.J., Olson, P.E., 1993. Cats and toxoplasmosis risk in HIV-infected adults. J. Am. Med. Assoc. 269, 76–77.

Walsh, C.P., Hammond, S.E., Zajac, A.M., Lindsay, D.S., 1999. Survival of *Toxoplasma gondii* tachyzoites in goat milk: Potential source of human toxoplasmosis. J. Eukart. Microbiol., 73s–74s.

Wong, M.M., Kozek, W.J., Karr Jr., S.L., Brayton, M.A., Theis, J.H., Hendrickx, A.G., 1979. Experimental congenital infection of *Toxoplasma gondii* in *Macaca arctoides*. Asian J. Infect. Dis. 3, 61–67.

Work, T.M., Massey, J.G., Rideout, B.A., Gardiner, C.H., Ledig, D.B., Kwok, O.C., Dubey, J.P., 2000. Fatal toxoplasmosis in free-ranging endangered 'Alala from Hawaii. J. Wildl. Dis. 36, 205–212.

Yai, L.E.O., Ragozo, A.M.A., Soares, R.M., Pena, H.F.J., Su, C., Gennari, S.M., 2009. Genetic diversity among capybara (*Hydrochaeris hydrochaeris*) isolates of *Toxoplasma gondii* from Brazil. Vet. Parasitol. 162, 332–337.

CHAPTER

7

Toxoplasma Animal Models and Therapeutics

Carsten G.K. Lüder, Utz Reichard, Uwe Gross

Institute for Medical Microbiology, University Medical Centre, Georg-August-University, Göttingen, Germany

OUTLINE

Toxoplasma gondii, Second Edition
http://dx.doi.org/10.1016/B978-0-12-396481-6.00007-6

7.1 INTRODUCTION

This chapter will discuss animal models on toxoplasmosis, with special regard to pharmacological applications, and thereby try to update existing reviews (Darcy and Zenner, 1993; Derouin *et al.*, 1995; Reichard and Gross, 2007).

Virtually all mammals can be infected with *Toxoplasma gondii*. However, different animal species differ markedly in their resistance to *Toxoplasma* infection. For example, rats are usually resistant against symptomatic toxoplasmosis but most mouse strains in general are susceptible (Fujii *et al.*, 1983; Dubey and Frenkel, 1998; Zenner *et al.*, 1998). In addition, the outcome of infection is not only dependent on the animal species but also on the animal strain. The genetic background seems to be of importance since, after infection with *Toxoplasma*, striking differences in susceptibility of various strains of inbred and outbred mice and rats were observed (McLeod *et al.*, 1984, 1989; Araujo *et al.*, 1976; Williams *et al.*, 1978; Suzuki *et al.*, 1991, 1994; Brown *et al.*, 1995; Kempf *et al.*, 1999; Sergent *et al.*, 2005). We know today that, at least in part, such animal strain-dependent variation in susceptibility may be attributed to certain chromosomal regions and even to the presence of certain single genes that may influence the parasite burden dramatically (Brown and McLeod, 1990; Deckert-Schlüter *et al.*, 1994; Brown *et al.*, 1995; Johnson *et al.*, 2002; Cavailles *et al.*, 2006; Hunn *et al.*, 2011). Nevertheless, many questions in this respect still remain to be solved, which may be exemplarily underlined by the intriguing finding that the control of host genetic resistance against acute infection itself differs according to virulence and genotype of the *T. gondii* strain used (Suzuki *et al.*, 1995).

The situation becomes even more complex since these differences are not uniform with respect to the strains but are also a function of the mode of inoculation. This means that, for instance, a mouse strain that is highly susceptible to intraperitoneal (i.p.) infection compared to another mouse strain may not necessarily be highly susceptible to oral infection and *vice versa*. Indeed, almost mirror-image susceptibility between oral and i.p. challenge has been found with inbred mouse strains (Johnson, 1984), suggesting at least partly independent modes of the defence involved in each infection route. In this context one has to be aware that, mostly, two infection modes are used: (i) injection of tachyzoites grown in culture or in mice i.p., s.c. or i.v. and (ii) oral administration of tissue cysts of *T. gondii* obtained mostly from brains of chronically infected mice (McLeod *et al.*, 1984). Since oocysts are only shed by the definitive host, i.e. the cat, and their production is more difficult, they are less frequently used for oral infection although this mode of infection represents one of the two major natural routes of parasite transmission. Oral administration of tissue cysts is inherently less reproducible due to variable size and parasite content of the cysts, but has the advantage that it is a natural route of infection and that tissue cysts can be easily isolated from chronically infected mice. As tachyzoites are not resistant to the acidic pH in the stomach, they are only less infective when given orally (Dubey, 1998).

In addition to host factors, the outcome of a challenge with *Toxoplasma* is largely influenced by the nature of the infectious agent itself and one of the most common characteristics of many *Toxoplasma* strains is their variation in virulence. Depending on the parasite dose which is lethal in mice, on the time before animals succumb to infection, or the percentage of animals that do succumb, highly virulent, moderately virulent and less virulent strains have been characterized (Kaufman *et al.*, 1958; Sibley and Boothroyd, 1992; Sibley and Ajioka, 2008). Recently, a growing number of excreted/secreted effector molecules including rhoptry and dense granule proteins have been identified as key determinants of virulence in mice (Taylor *et al.*, 2006; Saeij *et al.*, 2006; Behnke *et al.*, 2011; Rosowski *et al.*, 2011). However, strain-dependent virulence does not appear as a static feature as it can be enhanced by the continuous passage of the parasite in laboratory animals. For example, when the RH strain was initially isolated, mice succumbed to infection 17 to 21 days in the first passage, seven to eight days in the second, and three to five days in the third passage and thereafter (Sabin, 1941). In any case, the virulence of *T. gondii* strains is commonly assessed according to the outcome of a systemic i.p. infection in mice and studies of the population genetic structure have defined that the most commonly used laboratory strains of *T. gondii* belong to three clonal lineages, of which type 1 includes mouse-virulent strains, and types 2 and 3, mouse-avirulent strains (Sibley and Boothroyd, 1992; Howe and Sibley, 1995; Sibley and Ajioka, 2008). In fact, type 1 strains may be lethal in mice when a single infectious organism is injected (Howe *et al.*, 1996).

Figure 7.1 summarizes the factors that influence the outcome of *Toxoplasma* infection in animals.

7.2 CONGENITAL TOXOPLASMOSIS

An ideal animal model for congenital toxoplasmosis, useful for drug studies as well as for immunological studies and vaccine design, would be one that mimics the conditions of the infection in the human host. In addition, the outcome of the infection should be clearly assessable. In practice, the fulfilment of these requirements seems to be dependent on several conditions: (i) the nature of the animal placenta as the organ where transmission takes place, (ii) the duration of pregnancy, (iii) the duration of parasitaemia and (iv) the size of the animal (for review see Derouin *et al.*, 1995). However, the latter point is not as important as it used to be since technical developments during the last few decades, such as quantitative real-time PCR and advances in imaging techniques, have greatly enhanced the sensitivity of assessment of infection in small animal models (for examples, see Flori *et al.*, 2002, 2003).

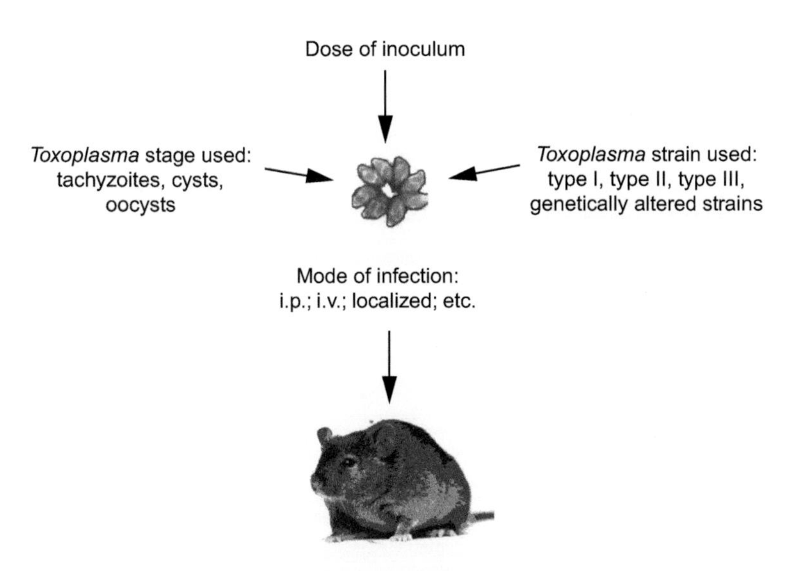

FIGURE. 7.1 Factors which influence the outcome of *Toxoplasma* animal infection.

Unlike most other organs, the placenta shows a wide variation in structure among different mammalian species and may be classified according to the number of maternal and foetal cell layers into different classes (Loke, 1982). While the placentas of carnivores and ruminants in general have four to six layers, the placenta of humans consists of only three, a foetal trophoblastic, mesenchymal and endothelial cell layer. Such a relatively thin interface which may facilitate parasite transmission is called a haemochorial placenta type, and this type is also present in primates and rodents (Loke, 1982; Darcy and Zenner, 1993). Indeed, most studies on congenital toxoplasmosis were conducted using rodents. Of course, they were mostly not chosen because of the nature of their placenta, but rather for practical reasons connected with the ease of keeping and handling small laboratory animals. However, major limitations working with small animals lie in the assessment of the foetal infection itself and in most cases are confined to the evaluation of the foetal transmission rate by direct observation or mouse subinoculation. Except for simple models for congenital chorioretinitis, studies involving the anatomical assessment of the foetal infection often require larger animals, such as pigs, sheep or even primates. In the following, different animal species and their use in models of congenital toxoplasmosis are discussed. A summary is provided in Tables 7.1 and 7.2.

7.2.1 Mouse

As indicated by numerous articles that have appeared during the recent decades, mice are well studied animals in congenital toxoplasmosis (cf. Beverley, 1959; Remington *et al.*, 1961; Beverley and Henry, 1970; Hay *et al.*, 1981, 1984; Roberts and Alexander, 1992; Roberts

TABLE 7.1 Pharmacological Studies on Congenital Toxoplasmosis

Animal Species	Specific Comments	Pharmacological Studies (Literature Examples)
Mouse	Haemotrichorial placenta; vertical transmission during chronic infection possible (depending on mouse strain); vertical transmission in inbred BALB mice during acute infection common	Nguyen and Stadtsbaeder, 1985; Fux *et al.*, 2000
Rat	Haemotrichorial placenta; rare vertical transmission during chronic infection; strain-dependent variation in transmission rates	Usmanova, 1965
Calomys callosus	Haemotrichorial placenta; transmission rates during acute infection higher than in humans; no vertical transmission during chronic infection	Costa *et al.*, 2009
Guinea pig	Haemomonochorial placenta; vertical transmission during chronic infection observed; 3-fold longer gestation time than mice and rats	Youssef *et al.*, 1985
Primate	Haemomonochorial placenta; transmission rates resemble those in humans	Schoondermark-van de Ven *et al.*, 1994a, 1994b, 1995

TABLE 7.2 Models of Congenital Toxoplasmosis

Species and Strain of Animal Used*	*Toxoplasma* Strain Administered*	*Toxoplasma* Stage and Route of Infection	Administration Time Point and *Toxoplasma* Dose in Relation to Pregnancy	Outcome of Animal Infection I (With Regard to Foetal Transmission in Acutely Infected Animals)	Outcome of Animal Infection II (With Regard to Transmission in Chronically Infected Animals)	Publication**
Mouse: NIH strain	RH, 113-CE, Beverley, M7727	Tachyzoites (RH) and cysts (113-CE, Beverley, M7727) i.p. or s.c.	Undefined numbers for acute and chronic infections; acute (RH) on day 16 of pregnancy; chronic (113-CE, Beverley, M7727) 2–15 months prior delivery	RH: No infection of young if delivery before the fifth day after infection; 56% infection on day 7 after i.p. inoculation	Transmission of chronically infected mice occurred, most frequently with the Beverley strain	Remington *et al.*, 1961
Mouse: NIH strain	Prugniaud	Cysts orally	Five cysts on day 5, 10 or 15 of gestation	Overall transmission rate greater than 90%; clinical outcome in offspring more severe if infected early during pregnancy	Not determined	Wang *et al.*, 2011
Mouse: NMRI	S93, K8, 558/72, Witting, Gail, KSU, 1070, 162/74, 248/70, MO, ALT	Cysts i.p. prior to pregnancy; cysts orally during pregnancy only in a primary infection trial with *Toxoplasma* ALT strain	20 cysts 4–8 weeks before mating; 20 cysts on day 10 of pregnancy	Primary infection during pregnancy (ALT strain only): Transmission rate 28%	Infection prior to pregnancy: S93, K8, 558/72, Witting, Gail: No transmission; KSU, 1070, 162/74, 248/70, MO, ALT: Transmission rate 1–3%	Werner *et al.*, 1977

(*Continued*)

TABLE 7.2 Models of Congenital Toxoplasmosis (*cont'd*)

Species and Strain of Animal Used*	*Toxoplasma* Strain Administered*	*Toxoplasma* Stage and Route of Infection	Administration Time Point and *Toxoplasma* Dose in Relation to Pregnancy	Outcome of Animal Infection I (With Regard to Foetal Transmission in Acutely Infected Animals)	Outcome of Animal Infection II (With Regard to Transmission in Chronically Infected Animals)	Publication**
Mouse: NMRI	Beverley	Cysts s.c.	Undefined number of cysts 5–8 days after mating	75% offspring with positive serology; offspring survival 40%; number of offspring approximately 20% compared to non-infected control; 30% mortality of infected mothers during gestation	Cotrimoxazole treatment during pregnancy almost normalized offspring number and survival	Nguyen and Stadtsbaeder, 1985
Mouse: BALB/c, BALB/K (inbred)	Beverley	Cysts orally	8 weeks before mating 5 cysts; day 12 of pregnancy 20 cysts	Transmission rate of approximately 50% (cysts for first time in pregnancy only)	Congenital infection occurred only if the mother was infected for the first time during pregnancy	Roberts and Alexander, 1992
Mouse: BALB/c (inbred)	P	Cysts orally	Two cysts between days 6–15 of pregnancy	Foetal transmission rate between 50% and 60%	Minocycline-treated mice showed transmission in only 3.6%	Fux *et al.*, 2000
Mouse: BALB/c (inbred)	ME49, Prugniaud, M7741, M3	Bradyzoites and oocysts orally	Chronically infected mice, reinfection during pregnancy	Cross-protection of chronic infection against homologous and heterologous reinfection	Transmission in two of 10 chronically infected mice or no transmission	Freyre *et al.*, 2006a; Pezerico *et al.*, 2009
Rat: Sprague–Dawley	RH	Tachyzoites i.p.	1×10^7 and 2×10^7 tachyzoites 6–8 weeks before mating		Virtually no foetal transmission (three offspring from 140 in total, all in a single litter)	Remington *et al.*, 1958

Rat: Sprague–Dawley, Osborne–Mendel, Black rat, Holtzman rat	RH, S-6, Beverley	Tachyzoites (RH, S-6) and cysts (Beverley) i.p.	1×10^4 to 1×10^7 tachyzoites and undefined number of cysts 2–8 months prior gestation		No foetal transmission from chronically infected rats with RH or S-6; Beverley: 5% transmission	Remington *et al.*, 1961
Rat: Sprague–Dawley	CT-1	Oocysts orally or s.c.; Bradyzoites s.c.	1×10^4 oocysts at 7–15 days of pregnancy; 1×10^4 bradyzoites at 10–14 days of pregnancy	Transmission rate of 82.1% (oocysts orally), 90.9% (oocysts s.c.), 43.8 (bradyzoites s.c.)	No transmission in chronically infected rats	Dubey and Shen, 1991
Rat: Fischer	RH, 76K, Prugniaud	RH: tachyzoites i.p.; Prugniaud and 76K: cysts orally	Between days 8–12 of pregnancy: 8 $\times 10^6$ RH tachyzoites; 1200 Prugniaud or 76K cysts	Transmission rates of 58% (RH), 63% (Prugniaud), 35% (76K)	No transmission to foetuses of chronically infected rats, even if rats were reinfected during pregnancy	Zenner *et al.*, 1993, 1999a
Rat: Sprague–Dawley	VEG	Oocysts orally	1×10^4 oocysts on day 6, 9, 12 or 15 of gestation	Transmission rates: 33% (day 6), 55% (day 9), 83% (day 12) and 57% (day 15)	Virtually no transmission to the next generation	Dubey *et al.*, 1997
Rat: Wistar and Long Evans	12 different strains of low to high pathogenicity for mice (e.g. M7741, Beverley, M49)	Cysts orally	2×10^2 to 3.4×10^3 cysts at 6–8 and 15 days of pregnancy	Overall transmission of 44% with a range of 0%–90% attributed to genetically based susceptibility of outbred Wistar rats; frequency of transmission not affected by the strain or dose of *Toxoplasma* nor by the time point of infection	Transmission more frequent in Long Evans than in Wistar rats	Freyre *et al.*, 2001a

(*Continued*)

TABLE 7.2 *Models of Congenital Toxoplasmosis (cont'd)*

Species and Strain of Animal Used*	*Toxoplasma* Strain Administered*	*Toxoplasma* Stage and Route of Infection	Administration Time Point and *Toxoplasma* Dose in Relation to Pregnancy	Outcome of Animal Infection I (With Regard to Foetal Transmission in Acutely Infected Animals)	Outcome of Animal Infection II (With Regard to Transmission in Chronically Infected Animals)	Publication**
Rat: Wistar	Six different strains of low to high pathogenicity for mice	Oocysts orally	1×10^4 oocysts at 15 days of pregnancy	Transmission rates of 10 to 80%; higher transmission with strains more pathogenic for mice	No statistically significant differences in the rate of transmission in rats fed with cysts (previous work)	Freyre *et al.*, 2003b
Rat: Sprague–Dawley, Fischer	RH, Prugniaud, M3	Tachyzoites (RH), cysts and oocysts orally	Chronic infection with RH, Prugniaud or M3 2 months before gestation, reinfection during pregnancy	Complete protection against vertical transmission after reinfection with same strain and stage; partial protection after heterologous reinfection	Not determined	Freyre *et al.*, 2006b
Rat: Fischer	Prugniaud	Bradyzoites and oocysts orally	1×10^4 bradyzoites or 100 oocysts	Transmission rate greater than 50%	Not determined	Freyre *et al.*, 2008
Calomys callosus: Canabrava strain	ME49	Cysts orally	20 cysts at days −50, −30, −10 of pregnancy or during pregnancy	Transmission rate 100% if infected at day −10 or during pregnancy	No transmission during chronic infection (infection 50–30 days before pregnancy)	Pereira Mde *et al.*, 1999; Barbosa *et al.*, 2007; Costa *et al.*, 2009
Hamster	ME49 for chronic infection; Prugniaud, M7741, M3 for acute infection	Bradyzoites orally, oocysts orally	10^3–10^4 bradyzoites; 10^2–10^4 oocysts; time of administration not specified	Transmission rates of 25% to 100%	Transmission rate of 9%	Freyre *et al.*, 2009, 2012

Guinea pig: not specified	Beverley	Cysts i.p.	60 or 100 cysts at 4–54 days of pregnancy; 100–200 cysts 2–6 months before mating	Transmission rate of 100% if infection during pregnancy	Transmission rate of 30% if infection before pregnancy; overall high number of stillborn or severely ill pups	Wright, 1972
Guinea pig: Dunkin–Hartley	C56	Tachyzoites intradermally	5 × 10^5 tachyzoites after 7 weeks of gestation	Transmission in more than 80%	SAG1-immunized animals with lower transmission rates	Haumont *et al.*, 2000
Guinea pig: Dunkin–Hartley	RH, 76K, Prugniaud	Tachyzoites (RH) i.p.; cysts (76K, Prugniaud) i.p. or orally	Various time points from 90 days before pregnancy to day 40 of pregnancy: 100 tachyzoites (i.p.); 100 cysts (i.p. or orally)	Transmission rates if infection during pregnancy: 54% (RH); 84% (Prugniaud); 86% (76K)	Transmission rate of 17% if infection 30–90 days before mating; overall high number of stillborn and non-viable foetuses	Flori *et al.*, 2002
Guinea pig: Dunkin–Hartley	76K	Cysts orally	100 cysts at days 20 or 40 of pregnancy	Transmission rates of 84.6% and 100% after inoculation on days 20 and 40, respectively	Infection assessed by real time PCR	Flori *et al.*, 2003
Rabbit: not specified	Witting/ALT	Cysts i.p. prior pregnancy; cysts orally during pregnancy	200 cysts 267–19 days before mating; 200 cysts during the first or second trimester to preinfected or non-preinfected animals	Foetal transmission rate up to 79%	No congenital transmission when first infection was placed at least 35 days prior mating	Werner *et al.*, 1977
Primate: *Macaca mulatta* (rhesus)	RH	Tachyzoites i.v.	5 × 10^6 tachyzoites on days 90 or 130 of pregnancy	Overall transmission rate of 61% which is comparable to that found in humans		Schoondermark-van de Ven *et al.*, 1993

* *For definition of animal or* Toxoplasma *strains please see respective articles*

** *Due to numerous publications on this topic, only few were chosen exemplarily for this table*

et al., 1994; Wang *et al.*, 2011). This may be somewhat astonishing as early studies indicated that under natural conditions, vertical transmission occurs during chronic infection and through successive generations of mice (Remington *et al.*, 1961; Beverley, 1959). Thus, at first glance, the mouse model may not be best suited to mimic the situation in humans, where vertical transmission usually occurs only during primary infection in pregnancy. However, other findings prove that whether transmission to the foetus during a chronic infection of the mother occurs is largely dependent on the parasite strain as well as on the mouse strain used; for example, in mice that were latently infected with eleven different *Toxoplasma* strains, placental transmission succeeded with only six strains (Werner *et al.*, 1977). Furthermore, chronically infected BALB/c mice normally do not allow vertical transmission whereas acutely infected BALB/c mice transmit the parasite to approximately 50% of the littermates (Roberts and Alexander, 1992). In addition, no congenital transmission was detected in the litters of chronically infected BALB/K mice, although the mothers themselves were found to have extremely high cyst counts (Roberts and Alexander, 1992). The explanation for this contradictive outcome of vertical transmission of chronically infected mice most likely lies in their genetic background (as known for low susceptibility of BALB/c mice in general), and this view may be supported by the fact that in contrast to the inbred BALB mice, most earlier studies used outbred animals or NMRI mice. Another critical factor that can impact vertical transmission of *T. gondii* is whether chronically infected hosts are reinfected during pregnancy and which parasite strain is used for reinfection (Elbez-Rubinstein *et al.*, 2009). At least in BALB/c mice, a chronic infection seems to protect against vertical transmission after reinfection with a homologous or even a heterologous parasite strain (Freyre *et al.*, 2006a; Pezerico *et al.*, 2009). Since, in principle, reinfection of mice with heterologous *Toxoplasma* strains is possible (Elbez-Rubinstein *et al.*, 2009), further studies using other mouse and parasite strains are required in order to verify this issue in the congenital infection model. Recent investigations on congenital toxoplasmosis tend to use a BALB/c mouse model rather than models with other inbred or outbred mouse strains (Thouvenin *et al.*, 1997; Fux *et al.*, 2000; Elsaid *et al.*, 2001; Abou-Bacar *et al.*, 2004a, 2004b; Beghetto *et al.*, 2005; Freyre *et al.*, 2006a). On the other hand, the mouse model was rarely used as a model for evaluating therapeutics (Nguyen and Stadtsbaeder, 1985; Fux *et al.*, 2000) but rather as a model for studying pathogenetic mechanisms and, more recently, for an evaluation of vaccine approaches (McLeod *et al.*, 1988; Roberts *et al.*, 1994; Elsaid *et al.*, 2001; Couper *et al.*, 2003; Letscher-Bru *et al.*, 2003; Ali *et al.*, 2003; Mevelec *et al.*, 2005; Ismael *et al.*, 2006). However, Fux *et al.* were able to show positive effects of a minoxycycline treatment on congenital *Toxoplasma* transmission (Fux *et al.*, 2000). Considering the advantages of easy handling of mice as laboratory animals and the availability of new sensitive PCR-based detection methods that allow better diagnostics also in small animals, a BALB/c mouse model, if further developed, may have its place as a first line screening model for testing new chemotherapeutic agents against congenital toxoplasmosis.

7.2.2 Rat

Like humans and primates, rats are relatively resistant to *T. gondii* with respect to a clinical manifestation of the infection. While transmission during acute infection of maternal rats induced by intracerebral or i.v. infection was first reported during the early 1950s (Schultz and Bauer, 1952; Hellbrügge and Dahme, 1953; Hellbrügge *et al.*, 1953, 1955), it has been shown more recently that also oral ingestion of oocysts or tissue cysts leads to foetal transmission (Dubey and Shen, 1991; Zenner *et al.*, 1993; Freyre *et al.*, 2001a, 2003b, 2008). In general, transmission rates seem to be high and were reported to lie mostly between 30% and 90%. However, there was great variation attributed to the *Toxoplasma* strains and to the rat strains used (Zenner *et al.*, 1993; Freyre *et al.*, 2003b). Indeed, also a wide variability for the formation of *Toxoplasma* cysts in rats of the same outbred strain and age, inoculated with the same strain, stage and dose of *Toxoplasma*, using the same infection route was observed (Freyre *et al.*, 2001a). Such an individual resistance of rats belonging to the same outbred strain may be attributed to the individual genetic background of the infected rat (Freyre *et al.*, 2001a; Cavailles *et al.*, 2006). In an experimental study design for drug efficacy testings, for example, such a lack of individual reproducibility may be overcome either by a comparatively high number of animals per group or by the use of inbred animals.

Except in rare instances, when unnaturally high doses of several million organisms were used for infection (Hellbrügge, 1955), *T. gondii* parasites were either not transmitted at all or extremely rarely from chronically infected rats to their littermates, irrespective of the route of inoculation, stage, strain or size of inoculum (Remington *et al.*, 1958, 1961; Dubey and Shen, 1991; Zenner *et al.*, 1993; Dubey *et al.*, 1997). Recently, partial protection of chronically infected rats against vertical transmission after reinfection with parasites of a different clonotype during pregnancy was reported (Freyre *et al.*, 2006b). In contrast to the situation in some, but not all, mouse strains in which the organism is transmitted repeatedly during chronic infection, vertical transmission of chronically infected rats requires reinfection with a heterologous parasite strain during pregnancy (Freyre *et al.*, 2006b), as also reported for humans (Elbez-Rubinstein *et al.*, 2009). Thus, with respect to clinical course and *in utero* transmission, toxoplasmosis in rats and humans is similar and the infection in rats may serve as a proper model especially for human congenital toxoplasmosis (for review, see also Dubey and Frenkel, 1998).

In spite of the obvious analogies concerning transmission, transmission rates and rates of clinical manifestation, rat models, with rare exceptions (Usmanova, 1965), have so far not been used for drug testing in congenital toxoplasmosis. This may be due to the fact that infected litters usually appear healthy. However, regarding the *T. gondii* strains, stage and routes of inoculation and probably also the rat strains, rats may serve as excellent models, especially when emphasis lies on placental transmission. In any case, as total protection against congenital toxoplasmosis can be achieved regardless of the *Toxoplasma* strain, rats may be attractive models for evaluating future vaccine candidates against the disease (Zenner *et al.*, 1993, 1999a).

7.2.3 *Calomys callosus*

More recently, *Calomys callosus*, a cricetid rodent from South America has been evaluated as a model for human congenital toxoplasmosis (Pereira *et al.*, 1999; Ferro *et al.*, 2002; Franco *et al.*, 2011). Results show that in *C. callosus*, *T. gondii* is efficiently transmitted during acute infection to their foetuses with rates up to 100% (Pereira *et al.*, 1999; Costa *et al.*, 2009). In contrast, vertical transmission does not seem to occur during chronic infection or even after reinfection with a heterologous *Toxoplasma* strain during pregnancy (Barbosa *et al.*, 2007; Franco *et al.*, 2011). Due to the high vertical transmission rates and the obviously low variability, *C. callosus* may represent a valuable model to test the efficacy of novel drug and vaccine regimes during congenital toxoplasmosis. A drawback certainly is represented by the limited availability of immunological reagents in order to investigate pathogenic issues. However, using this model, a high efficacy of azithromycin as compared to spiramycin, a combination of sulphadiazine, pyrimethamine and folinic acid, or *Artemisia annua* infusion in preventing vertical transmission has been reported (Costa *et al.*, 2009).

7.2.4 Hamster

Although it is known for several decades that the hamster supports vertical transmission of *Toxoplasma* (De Roever-Bonnet, 1960), it has only recently been reevaluated as another model for congenital toxoplasmosis (Freyre *et al.*, 2009, 2012). Depending on the parasite strain and stage, the rate of vertical transmission during acute infection varies between 25% and 100% (Freyre *et al.*, 2009). Female hamsters also transmit *T. gondii* to their foetuses during chronic infection, although with lower frequency (less than 10%) than during acute infection (Freyre *et al.*, 2009). Chemotherapeutic drugs have not yet been evaluated using the hamster model.

7.2.5 Guinea Pig

A guinea pig model for congenital toxoplasmosis has been described in various studies (Adams *et al.*, 1949; Huldt, 1960; Wright, 1972; Haumont *et al.*, 2000; Flori *et al.*, 2002, 2003; Berard-Badier *et al.*, 1968). As in humans, the guinea pig placenta is of the haemomonochorial type (Darcy and Zenner, 1993) suggestive of similar modes of transmission. However, transmission rate and sensitivity of guinea pigs to *T. gondii* after i.p. or oral infection are much higher and about halfway between the rat and mouse. In addition, congenital transmission during pregnancy of chronically infected females was observed (Wright, 1972; Flori *et al.*, 2002). As a possible advantage in comparison to mice and rats, guinea pigs have approximately three-fold longer gestation periods (with a duration of 65 days), which are long enough to enable comparative studies with different inoculation times and comparative chemotherapy studies (Flori *et al.*, 2002). For this application, the guinea pig model may be best suited; however, except for rare instances (Youssef *et al.*, 1985), in spite of its potential advantages, a guinea pig model for congenital toxoplasmosis has not yet been employed for drug testing.

7.2.6 Primate

With respect to haemochorial placentation (Ramsey and Harris, 1966; Darcy and Zenner, 1993), foetal blood sampling and assessment of foetal infection, a primate model should actually best meet the requirements for the study of the effect of medication in the infected foetus. One of the first such studies was conducted with *Macaca arctoides* as a model for primates (Wong *et al.*, 1979). Data obtained with this model suggested that although certain developmental stages of the *Toxoplasma* organism and of the foetus may favour the occurrence of congenital infection, transmission rate in general seems to be low and very little neonatal disease results (Wong *et al.*, 1979). In contrast, a more applicable model was established by Schoondermark *et al.* with rhesus monkeys (*Macaca mulatto*) (Schoondermark-van de Ven *et al.*, 1993). Herein, the frequencies of transmission which were found in the rhesus monkey after maternal infection in the second and third trimester of gestation equal those observed in humans (Schoondermark-van de Ven *et al.*, 1993). The rhesus model was then used to prove the efficacy of spiramycin or the combination of pyrimethamine and sulphadiazine for the treatment of congenital *T. gondii* infection (Schoondermark-van de Ven *et al.*, 1994b, 1995). The results showed that both regimens were clearly effective in reducing the number of parasites in the infected foetus as proven by PCR and mouse inoculation with amniotic fluid. However, spiramycin was less active (Schoondermark-van de Ven *et al.*, 1994b) and was not found anywhere in the foetal brain (Schoondermark-van de Ven *et al.*, 1994a).

The rhesus monkey model is perhaps theoretically the best animal model to prove the efficacy of a drug against human congenital toxoplasmosis, especially with regard to placental transmission. However, housing and handling of monkeys require special facilities and trained employees. It is also time-consuming and expensive which may limit the numbers of animals used for the studies. Considering this fact and considering that the outcome of the congenital infection, as in humans, often seems to be subclinical or at least shows a high degree of variation (Schoondermark-van de Ven *et al.*, 1993), no direct drug evaluation concerned with its influence on the clinical course of a congenital toxoplasmosis seems practical so far. Thus, drug efficacy must be extrapolated indirectly from the demonstration of the parasite or parasite DNA in the amniotic fluid or foetal tissue and by pharmacokinetic data of the drug that may also be obtained from the monkey foetus. In summary, the rhesus monkey model might have its place as a last evaluation step for a new drug before it is admitted to clinical trials.

7.2.7 Rabbit

Surprisingly, rabbit congenital toxoplasmosis has not been extensively studied although rabbits are widely used laboratory animals, and transmission from the mother to the foetus has been demonstrated (Uhlikova and Hubner, 1973). In addition, foetuses from chronically infected mothers were protected, thus a rabbit model would share common features with the infection in humans (Werner *et al.*, 1977). A rabbit model may be of particular interest when the small size of other common laboratory animals such as mice and rats hampers experiments, for instance, when larger volumes or subsequent blood samples are needed for examination of an antibody response (Araujo and Remington, 1975). To our knowledge no pharmaceutical studies on congenital toxoplasmosis have been conducted using a rabbit model.

7.2.8 Other Animals

Various other animals have been suggested as a model for congenital toxoplasmosis but have never been broadly used. As an example, although pigs are well known to acquire toxoplasmosis and to play a decisive role in transmitting the disease to humans via their meat (for review see Tenter *et al.*, 2000; Montoya *et al.*, 2004), their use as animal models has never been thoroughly investigated. However, when infecting pregnant miniature pigs with strains of different virulence, even a strain that was considered apathogenic to both pigs and mice, resulted in significant numbers of congenitally infected piglets (Jungersen *et al.*, 2001). Such a pig model of congenital toxoplasmosis may therefore be of value for situations where big animals are needed and they may be relevant animal models for certain *Toxoplasma* strains with low virulence that are obviously transmitted to the foetus (Jungersen *et al.*, 2001).

Congenital disease due to *T. gondii* is a major cause of abortion and neonatal mortality in sheep. In addition, undercooked meat from infected sheep is an important source of infection for man. Although congenital transmission is well described and, in fact, is known as a major cause of abortion in this species (Dubey and Rommel, 1992; Anderson *et al.*, 1994), an animal model mimicking congenital infection in humans has, to our knowledge, not been established. Recent studies with sheep indicate that reactivation of chronic infection during pregnancy may be a major cause for the congenital infection (Duncanson *et al.*, 2001; Williams *et al.*, 2005). In this respect, congenital toxoplasmosis in sheep seems more to resemble a common mode of congenital transmission in mice than in humans.

Recently, Que *et al.* reported a novel chicken embryo model which had been adopted from a chicken model that had been developed for the study of metastic diseases (Que *et al.*, 2004). Basically, tachyzoites were injected directly into the chorioallantoic veins of 12 days old chickens and after an incubation period of three to six days, the degree of infection was assessed by histopathological examination and quantitative real-time PCR on the *Toxoplasma* DNA. As this model also provides the possibility of injecting drugs via the chorioallantoic vein, it may prove precious for an initial drug screening setup with a course of infection that is shorter than that in mice (Que *et al.*, 2004).

7.3 OCULAR TOXOPLASMOSIS

Ocular toxoplasmosis may result from an *in utero* infection of a foetus via a mother whose primary infection was acquired during pregnancy or from postnatal infection. Nowadays, postnatally acquired infection is considered to account for the majority of cases of human ocular toxoplasmosis. The number of cases and severity of disease are higher in South and Central America, the Caribbean and parts of tropical Africa than in Europe and North America; this may relate to the high prevalence of atypical genotypes of the parasite (for review, see Holland, 2003; Petersen *et al.*, 2012). Different modes of infection and pathogenesis may lead to different outcomes in ocular toxoplasmosis. The majority of animal models for human ocular toxoplasmosis have been developed to mimic a primary infection of adults, although there have been also efforts which have successfully established an ocular disease as a consequence of a transplacental *Toxoplasma* transmission (Lee *et al.*, 1983; Hay *et al.*, 1984; Dutton *et al.*, 1986; Hutchison *et al.*, 1982; Hay and Dutton, 1996; Lopes *et al.*, 2009; Lahmar *et al.*, 2010).

In patients with underlying immunosuppression or immune defects such as bone marrow transplantation or HIV, toxoplasmic chorioretinitis is often associated with concurrent toxoplasmic encephalitis (TE) or disseminated infection (for review, see Montoya *et al.*, 2004). However, even in individuals with AIDS, toxoplasmic chorioretinitis is encountered relatively infrequently (Holland *et al.*, 1988a), so that most of the human cases of ocular toxoplasmosis are found in immunocompetent patients. Indeed, *T. gondii* is one of the most frequently identified cause of uveitis and the most commonly identified pathogen infecting the retina of otherwise immunocompetent individuals (Holland, 1999). Regardless of whether ocular toxoplasmosis is due to a reactivated congenital infection or to an infection that is acquired after birth, it usually presents in the immunocompetent host as a more or less localized eye disease (Montoya *et al.*, 2004). An animal model, particularly one that may be suited to the evaluation of the efficacy of a given drug, should ideally be characterized by a localized eye infection or at least predominantly by a localized eye infection, rather than by a generalized or CNS infection where the eye is just one disease location among others – an experimental challenge that is not easy to fulfil. Current models of postnatally acquired ocular toxoplasmosis are either based on primary local inoculation of *T. gondii* into the animal's eye, on a semilocal infection via the carotid artery, or on a primary systemic infection which then affects the eyes as their predominant organ of manifestation (see Tables 7.3 and 7.4).

7.3.1 Models Based on Local Eye Infection

In order to meet the above mentioned requirements, the localized infection of animal eyes was the first choice which, for technical reasons, required the use of larger animals. Thus, until 1982, the rabbit served as the most important existing experimental model for ocular toxoplasmosis and morphological lesions of acute experimental *Toxoplasma* chorioretinitis were produced by injection of parasites intravitreally (Kaufman, 1960), into the anterior chamber (Beverley *et al.*, 1954, Beverley, 1958, 1961; Jacobs *et al.*, 1964) or by transscleral inoculation into the suprachoroidal space (Nozik and O'Connor, 1968; Tabbara *et al.*, 1974; Rollins *et al.*, 1982). In fact, the anterior chamber model was used to show efficacy of pyrimethamine and sulphadiazine (Jacobs *et al.*, 1964), whereas the latter model was used to demonstrate the efficacy of clindamycin and minocycline on toxoplasmic chorioretinitis (Tabbara *et al.*, 1974; Rollins *et al.*, 1982).

TABLE 7.3 Pharmacological Studies on Ocular Toxoplasmosis

Infection Model	Specific Comments	Animal Species	Pharmacological Studies (Literature Examples)
Localized eye infection models	Best for confining infection to eyes but with the principal drawback of tissue needle damage	Rabbit	Beverley, 1958; Kaufman, 1960; Jacobs *et al.*, 1964; Tabbara *et al.*, 1974; Rollins *et al.*, 1982
		Primate	–
		Mouse	–
		Guinea pig	Jacobs *et al.*, 1964
Infection via the carotid artery	Cat is definitive natural host which may influence, for example, immunoreactions	Cat	Davidson *et al.*, 1996
Systemic infection models	Particularly hamsters and *C. callosus* produce consistent ocular disease following i.p. or oral infection	Mouse	Olle *et al.*, 1996; Norose et al., 2006
		Hamster	Gormley et al., 1998
		Calomys callosus	Lopes *et al.*, 2009

Intraocular injection of *T. gondii* tachyzoites in naive or chronically infected rabbits has recently been employed to monitor antibody levels in serum and aqueous humour (Garweg and Boehnke, 2006).

In 1982 a primate model was established that reliably produced acute toxoplasmic chorioretinitis by injection of viable RH strain *Toxoplasma* organisms (Culbertson *et al.*, 1982; Newman *et al.*, 1982). Whereas in the rabbit model the injections were made transsclerally into the suprachoroidal space at the posterior pole, in monkeys it was not possible to expose the posterior part of the sclera for direct injection. Therefore, the retinal injections were made through the pars plana across the vitreous cavity, directly into the superficial part of the retina at the posterior pole (Culbertson *et al.*, 1982). Non-human primates as well as rabbits infected via a transvitreal approach were later also used in other studies (Webb *et al.*, 1984; Holland *et al.*, 1988b; Garweg *et al.*, 1998). These transvitreal inoculation models have the principal drawback that the integrity of the vitreous cavity is disrupted and that they produce some mechanical damage to the retina (Culbertson *et al.*, 1982), but on the other hand, the blood-retinal barrier may be better maintained than by using the suprachoroidal approach, at least during the initial phase of the infection (Garweg *et al.*, 1998). Hence, the disease profile in the suprachoroidal model involves an initial local infection followed by a very early systemic one and, as such, the situation is not immunologically comparable with that evinced in humans (Friedrich and Müller, 1989). In any case, apart from the shorter time course, the transvitreal inoculation primate model represents clinical and histopathological conditions resembling those of the natural disease in humans (Culbertson *et al.*, 1982). It may also circumvent a disadvantage of the rabbit model which is the anatomic dissimilarity of the retina compared to humans (O'Connor, 1984). However, to our knowledge, this model has not yet been used for assessing drugs but rather to elucidate the pathogenesis of ocular toxoplasmosis.

In addition to the local eye infection in larger animals, a more recent small animal model uses C57BL/6 as well as MRL mice and injection of *T. gondii* (PLK strain, a clone of ME49) into the anterior eye chamber (Hu *et al.*, 1999a, 1999b).

TABLE 7.4 Models of Ocular Toxoplasmosis

Species and Strain of Animal Used*	*Toxoplasma* Strain Administered*	*Toxoplasma* Dose, Stage and Route of Infection	Outcome of Animal Infection	Remarks	Publication**
Mouse: strain A albino	Beverley	10 cysts s.c. on day 12 of pregnancy	Approximately 50% of offspring infected, 5.3% of these developed cataract; acute uveitis in a small proportion of eyes	Model of congenital ocular toxoplasmosis	Hutchison *et al.*, 1982; Dutton *et al.*, 1986
Mouse: C57BL/6	ME49	10–20 cysts i.p.	Mild uveitis and retinal vasculitis in all infected animals at 15 days postinfection	In most ocular lesions, the parasite could not be demonstrated even with PCR	Gazzinelli *et al.*, 1994
Mouse: Swiss Webster	Beverley	Cysts (undefined number) i.p.; immunosuppression by injecting polyclonal antibodies against γ-interferon	26% of mice developed chorioretinitis on day 13–15	*T. gondii* was revealed by routine cellular cultures in all immunosuppressed mice with ocular lesions	Olle *et al.*, 1994
Mouse: Swiss Webster	PRU	10 cysts orally on day 12 of pregnancy	Mortality in littermates 60%; ocular lesions in eyes from less than 20% of surviving littermates until 4 weeks after birth; *T. gondii* DNA in 75% of eyes from surviving mice	Congenital ocular toxoplasmosis model	Lahmar *et al.*, 2010
Mouse: Swiss Webster	PRU	Five cysts s.c. on day 7 after birth	100% of mice developed retinitis and retinal vasculitis until 4 weeks p.i.; *T. gondii* DNA present in all eyes	Neonatal ocular toxoplasmosis model	Lahmar *et al.*, 2010
Mouse: C57BL/6, MRL–MpJ	PLK (derived from ME49)	50–5×10^4 tachyzoites injected into the anterior chamber	Dose-dependent intraocular inflammation: 50 (none), 5×10^2 and 5×10^3 (moderate to severe), 5×10^4 tachyzoites (severe, early mortality)	Some protection if mice were pre-infected before challenge	Hu *et al.*, 1999a

Mouse: 129/SVJ (WT), IL-6-deficient strain with same genetic background	Beverley	10 cysts i.p.	Regular mild to moderate retinochoroiditis 4 weeks after challenge, severe inflammation in IL-6 deficient mice	Cytokine expression study	Lyons *et al.*, 2001
Mouse: C57BL/6, B6MRL/Ipr and B6MRL/gld (defective Fas or FasL expression, respectively)	ME49	20–30 cysts i.p.	Regular mild retinochoroiditis and moderate encephalitis after 14 days, becoming worse after 28 days	No significant difference in the degree of ocular inflammation between wild type and Fas or FasL mutant mice	Shen *et al.*, 2001
Mouse: C57BL/6, BALB/c and IFN-γ knockout mice (GKO) of both wildtype backgrounds	Fukaya	5 cysts perorally	Evidence of eye inflammation in GKO mice only; assessment via PCR, histopathology and fluorescein angiography	Toxoplasmic eye vasculitis model for GKO mice; GKO mice died 11–12 days after infection; improved outcome after sulphamethoxazole treatment	Norose *et al.*, 2003, 2006
Mouse: C57BL/6	ME49	5×10^3 bradyzoites injected intravitreally or via conjunctival instillation	Regular retinochoroiditis with both infection routes 7 days after infection	Additional eye damage caused by the intravitreal injection: instillation route preferable	Tedesco *et al.*, 2005
Mouse: C57BL/6, BALB/c, CBA/J	RH, PLK, SAG1 (P30)-deficient RH derived mutant strain	100 tachyzoites injected into the anterior chamber (Lu *et al.*, 2005); 10^3–10^4 tachyzoites intravitreally (Charles *et al.*, 2007)	C57BL/6: severe eye inflammation and 100% mortality with all *T. gondii* strains; BALB/c and CBA/J: mild to medium eye inflammation most pronounced with RH (all mice survived)	All strains of mice were protected after i.p. vaccination with temperature-sensitive mutant tachyzoites (ts-4)	Lu *et al.*, 2005; Charles *et al.*, 2007
Mouse: CD4-, CD8-, B cell-, IL-10-deficient C57BL/6	Temperature-sensitive strain ts-4 (RH background), RH	100 tachyzoites injected into the anterior chamber	Ocular lesions but no host death, most severe eye lesions in CD8 KO and IL-10 KO	Partial protection by intraocular ts-4 immunization against intraocular RH challenge	Lu *et al.*, 2009

(*Continued*)

TABLE 7.4 Models of Ocular Toxoplasmosis (*cont'd*)

Species and Strain of Animal Used*	*Toxoplasma* Strain Administered*	*Toxoplasma* Dose, Stage and Route of Infection	Outcome of Animal Infection	Remarks	Publication**
Rabbit: not specified	113-CE	5×10^3 or 1×10^4 tachyzoites into the anterior chamber	Uveitis developed after a few days	Pyrimethamine-sulphadiazine treatment	Jacobs *et al.*, 1964
Rabbit: pigmented Dutch rabbits, New Zealand white rabbits, pigmented California rabbits	RH, Beverley	Transscleral inoculation of 1000–2000 tachyzoites into the suprachoroidal space	RH: animal death from encephalitis; Beverley: retinochoroiditis in most animals after 7 days, resolving after 3 weeks	California rabbits best for technical handling	Nozik and O'Connor, 1968
Rabbit: pigmented California rabbits	Beverley	Transscleral inoculation of 400 tachyzoites into the suprachoroidal space	Constant induction of retinochoroiditis	Clinical improvement of retinochoroiditis in clindamycin treated rabbits	Tabbara *et al.*, 1974
Rabbit: pigmented California rabbits	RH	Transscleral inoculation of 400 tachyzoites into the suprachoroidal space	Non-treated animals developed retinitis but all died from toxoplasmic encephalitis	Minocycline prevented death from toxoplasmic encephalitis in 75% of animals	Rollins *et al.*, 1982
Rabbit: Burgundy	BK	Injection of 5×10^3 tachyzoites across the vitreous cavity into the superficial part of the retina	All animals developed retinochoroiditis after 7 days; 22% of naive rabbits died from generalized infection	Primed animals also with high incidence of retinochoroiditis (greater than 90%); Model used to monitor humoral response intraocularly	Garweg *et al.*, 1998; Garweg and Boehnke, 2006
Syrian Golden Hamster	ME49	10–25 cysts i.p.	All animals developed bilateral eye disease peaking after 4–5 weeks	No animal developed signs of systemic disease	Pavesio *et al.*, 1995

Syrian Golden Hamster	ME49	100 cysts orally	All animals developed unilateral or bilateral eye disease 4–8 weeks after infection	No animal developed signs of systemic disease	Gormley *et al.*, 1999
Guinea pig: not specified	RH	5×10^3 tachyzoites injected into the posterior chamber	Most animals developed acute chorioretinitis within 1–3 weeks	Recovery of *Toxoplasma* from eyes even after 10 months	Hogan *et al.*, 1956
Guinea pig: not specified	RH	5×10^3 tachyzoites injected into the posterior chamber	Not specified	Pyrimethamine-sulphadiazine treatment	Jacobs *et al.*, 1964
Calomys callosus: Canabrava strain	ME49	20 cysts orally at day 5–7 of pregnant or non-pregnant animals	40% of foetuses presented ocular lesions; 50%–75% of adult animals presented unilateral ocular cysts	Model for congenital as well as for acquired ocular toxoplasmosis; Congenital ocular toxoplasmosis prevented by azithromycin	Pereira Mde *et al.*, 1999 Lopes *et al.*, 2009
Domestic cat	ME49	5×10^3 tachyzoites into the common carotid artery	Progressive multifocal retinal and choroidal inflammatory foci (mostly bilateral) beginning 5–8 days postinoculation in all cats tested	Minimal to no clinical signs of generalized toxoplasmosis; resolution of lesions 21–70 days postinoculation	Davidson *et al.*, 1993
Primate: *Macaca fascicularis*, *Cercopithecus aethiops*, *Macaca mulatta* (rhesus)	RH	Injection of 5×10^3 to 1×10^5 tachyzoites across the vitreous cavity into the superficial part of the retina	Retinitis was reliably produced in all monkeys' eyes injected with 1×10^4 or more living tachyzoites	After 20 days the lesions began to resolve	Culbertson *et al.*, 1982

* *For definition of animal or* Toxoplasma *strains please see respective articles*

** *Due to numerous publications on this topic, only few were chosen exemplarily for this table*

Pathological and histopathological features of this model resemble in part acute ocular toxoplasmosis in humans, particularly when mice have been primed (preinfected perorally) (Hu *et al.*, 1999a). As local infection models with larger animals and especially primates are rather difficult and costly, this mouse model may offer a rational alternative, at least for larger-scaled controlled studies with therapeutics to be screened. In addition, the disadvantage of the small infection area that has to be investigated may be overcome partly by the use of current, highly sensitive detection methods such as PCR or quantitative real-time PCR. The other disadvantage of potentially extensive needle damage, particularly when such small animals as mice are used, may perhaps be circumvented when instead of intraocular injection, local instillation of the parasite is used. A recent investigation has shown that this, in principle, leads to infection of the retinal vessels with glial reactions (Tedesco *et al.*, 2005). However, to our knowledge, local eye infection models in mice have not yet been used for pharmacological drug testing but rather for immunological and pathogenetic studies (Hu *et al.*, 1999b, 2001; Lu *et al.*, 2004, 2005, 2009; Charles *et al.*, 2007).

Local infection of guinea pig eyes was also reliably achieved as early as in 1956 by injecting the RH strain into the vitreous humour (Hogan *et al.*, 1956). This animal was selected because of its relative resistance to toxoplasmic infection, and because the size of the eye did allow ophthalmoscopic examination. In fact, as early as in 1964, a guinea pig model with posterior chamber inoculation of a low virulent *T. gondii* strain, together with a rabbit model mentioned above, was successfully used to show a therapeutic effect of sulphadiazine and pyrimethamine in the treatment of ocular eye disease (Jacobs *et al.*, 1964). However, to our knowledge, local infection models using guinea pigs have not been used for pharmacological studies during the last few decades.

7.3.2 Models Based on Infection via the Carotid Artery

An intermediate model situated between the localized eye infection and the eye infection as a consequence of a generalized challenge was established in cats by Davidson *et al.* who used intracarotid inoculation to direct the parasites to ocular tissues to gain more predictable experimental ocular lesions with fewer systemic side effects (Davidson *et al.*, 1993). Indeed, all eight cats infected with a relatively small number of the ME49 strain developed the ocular disease but showed no signs or only mild signs of a generalized infection (three cats developed an increase in temperature). The multifocal areas of choroidal and retinal inflammation exhibited many similarities to ocular toxoplasmosis in humans; however, it differed from human ocular toxoplasmosis in its primary choroidal versus retinal nature (Davidson *et al.*, 1993).

The cat model has been used once to examine the effect of clindamycin in the treatment of ocular toxoplasmosis (Davidson *et al.*, 1996). Paradoxically, clindamycin administration was associated with increased morbidity and mortality from hepatitis and interstitial pneumonia, which are characteristic of generalized toxoplasmosis. The reasons for this outcome were unclear and may have been due to various aspects of the experimental setting (Davidson *et al.*, 1996). As the natural definitive host, the cat may also differ from humans in some undefined manner with regard to its immunologic response to the parasite, which leads us to believe that it is not the ideal laboratory animal for drug testing.

7.3.3 Models Based on Systemic Infection

The most frequently used animal for systemic infection models is the mouse. Essentially, two different methods have been employed to establish the disease: (i) infection of pregnant mice to induce the development of ocular lesions in the pups (Hay *et al.*, 1981, 1984; Lee *et al.*, 1983; Dutton *et al.*, 1986; Hutchison *et al.*, 1982; Hay and Dutton, 1996; Lahmar *et al.*, 2010) and (ii) systemic infection of mice which then predominantly develop ocular manifestations (Gazzinelli *et al.*, 1994; Olle *et al.*, 1994, 1996; Lyons *et al.*, 2001; Shen *et al.*, 2001, Norose *et al.*, 2003, 2005). Lahmar and colleagues have proposed an alternative model which may be viewed as an intermediate model of congenital and 'classical' (i.e. during adulthood) postnatally acquired ocular toxoplasmosis: the neonatal systemic infection of mice (Lahmar *et al.*, 2010). Such an infection leads to a similar pathophysiology as compared to the congenital ocular toxoplasmosis model but at higher and more reproducible rates of ocular involvement.

In the congenital model, the infection is not established via direct inoculation of vital parasites into the eye but via infection of gestating female mice, for example, with the Beverley strain (Hay *et al.*, 1981; Hay and Kerrigan, 1982; Hutchison *et al.*, 1982). This model has the advantage that its aetiology is probably analogous to a substantial extent to human ocular toxoplasmosis in that the foetus becomes infected *in utero* via a mother whose primary infection is acquired during pregnancy (Hay *et al.*, 1984). Interestingly, it was found that the ocular lesions in this model resemble those that have been described in experimental allergic uveitis (EAU) (Lee *et al.*, 1983; Hay *et al.*, 1984; Dutton *et al.*, 1986) and, in fact, a mouse model (the adult, not the congenital) was thereafter used to further clarify the pathogenesis and particularly the nature and influence of the immune response involved in the ocular disease (Gazzinelli *et al.*, 1994; Olle *et al.*, 1996). However, for pharmacological screening studies, the congenital model as described does not seem to be appropriate because, in spite of low postnatal mortality, it has the disadvantage that ocular morbidity, discovered by cataract manifestation, is only present in 5% of the pups (Hutchison *et al.*, 1982). Nevertheless, the use of different mice and *Toxoplasma* strains as well as new sensitive screening methods may further develop such a model to be also suitable for drug testing. Alternatively, the neonatal infection model (see above; Lahmar *et al.*, 2010) may lead to ocular pathology in a sufficiently high percentage of animals that warrants evaluating novel drug regimes.

Gazzinelli and coworkers reported that C57BL/6 (B6) mice develop mild intraocular inflammation commonly observed 15 days after intraperitoneal injection of cysts of the ME49 strain, demonstrating the possible usefulness of adult mice for an eye model (Gazzinelli *et al.*, 1994). In most of the ocular lesions, the presence of the parasite could not be demonstrated even with the PCR technique, but parasite load did increase after treatment of mice with antibodies directed against lymphocytes or cytokines (IFN-γ or TNF-α) (Gazzinelli *et al.*, 1994). Treatment with anti-IFN-γ also ended with clinical eye lesions, including single foci of chorioretinitis, and multifocal lesions of diffuse areas of retinal necrosis in an experimental model of chronically infected Swiss Webster mice (avirulent Beverley strain) (Olle *et al.*, 1996). Using also the *T. gondii* Beverley strain for i.p. infection, Lyons *et al.* found retinal inflammation most marked in the inner retinal layers of wild type 129/SVJ mice and more severe in corresponding IL-6 knockout mice (Lyons *et al.*, 2001).

Recently, Norose *et al.* established a mouse model for the ocular disease that followed the natural peroral route of infection with five cysts of a *T. gondii* avirulent strain (Norose *et al.*, 2003). The model was established for immune competent mice using resistant BALB/c or susceptible C57BL/6 and for the immune deficient mice using IFN-γ knockout (GKO) mice. Whereas all GKO mice died after 11 to 12 days demonstrating disseminated infection, both strains of WT mice survived for more than one month. In contrast to the knockout mice, there was no histopathological evidence for inflammation in the eyes and brains of wild type mice, and no characteristic findings using fluorescein angiography and documentation with a fundus camera (Norose *et al.*, 2003). Electroretinograms, as shown later, were also only changed in GKO mice (Norose *et al.*, 2005). However, the authors were able to show differential parasite distribution in the eyes of WT mice using a quantitative competitive polymerase chain reaction (QC–PCR). The GKO ocular toxoplasmosis model has recently been used to show a beneficial effect of sulphamethoxazole on the parasite load in different ocular tissues (Norose *et al.*, 2006). However, due to the severe immune defect, the improvement in the outcome of infection was not specific to the eyes, but rather a general decrease in systemic infection. In addition, the authors observed an increased percentage of bradyzoites indicating that the drug, besides killing the parasite, may also favour stage conversion from the fast replicating tachyzoite to the latent bradyzoite stage. Altogether, this suggests that the GKO model may not be best suited to mimic a localized eye infection that is commonly observed in human ocular toxoplasmosis. Nevertheless, it may have its place in evaluating ocular manifestations and their therapeutic prevention in immunocompromised individuals.

In summary, adult mouse models may provide reasonable tools for investigation of various pathological or pathogenetic aspects of ocular toxoplasmosis particularly in the immunocompromised host; they may also provide a system for screening drugs in the immunocompromised host when using various knockout mice as, for example, IFN-γ knockouts (Belal *et al.*, 2004; Norose *et al.*, 2006). Because of the mild effects on eyes using avirulent *T. gondii* strains or overwhelming systemic infection using virulent strains, adult WT mice have not yet been used and seem to be not well suited for drug testing in particular respect to ocular toxoplasmosis. At any rate, QC–PCR as described before (Kobayashi *et al.*, 1999) combined with DNA extraction from different eye parts (cornea, iris/ciliary body, lens, posterior retina, peripheral retina, choroids, sclera, optical nerve and brain) (Norose *et al.*, 2003) allows measurement of the parasite load in the eyes of small animals such as mice and may prove valuable in other models of ocular toxoplasmosis.

Based on the observations of Frenkel, who found frequent but sporadic ocular disease in Syrian golden hamsters several months after inoculation with the RH or CJ strains of *Toxoplasma* (Frenkel, 1953, 1955), reliable models with this animal and the ME49 strain were developed that show ocular disease following the i.p. or oral infection route (Pavesio *et al.*, 1995; Gormley *et al.*, 1999). The main advantage of these models is that they consistently produced ocular disease with a short incubation time but without artificial breaching of the ocular barrier and also without causing any clinical signs of systemic disease. In addition, hamster eyes are large enough to allow fundus photography to document the progression of the infection and, as usually encountered in humans when immunity is not impaired, the disease resolves spontaneously with time, without treatment (Pavesio *et al.*, 1995; Gormley *et al.*, 1999). There are, however, marked differences to the human disease as, for example, vitritis, which in hamsters was usually not significant (Pavesio *et al.*, 1995). In conclusion, the hamster model may be an option as a drug screening model mainly because of its good reproducibility and monitoring possibilities. However, one study compared conventional therapies (pyrimethamine plus sulphadiazine, clindamycin, spiramycin) with atovaquone and did not show any drug effects on the acute disease but only on the number of cerebral *Toxoplasma* cysts (Gormley *et al.*, 1998).

In addition to conventional laboratory animals, an acquired, as well as a congenital, model of ocular toxoplasmosis has been established in *Calomys callosus* (Pereira Mde *et al.*, 1999). Following oral infection with the ME49 strain, 40% of foetuses presented ocular lesions (examined after laparotomic removal of the foetus) while 75% of females and 50% of males presented ocular lesions in the acquired study setting. Adult animals survived the infection for several months without treatment and demonstrated no clinical signs of systemic disease (Pereira Mde *et al.*, 1999). Recently, treatment of acutely infected pregnant females with azithromycin (300 mg/kg) has been shown to efficiently prevent ocular toxoplasmosis in the littermates and to be superior to a treatment with pyrimethamine, sulphadiazine and folinic acid (Lopes *et al.*, 2009). Whether azithromycin also reduces acquired ocular toxoplasmosis in adult *C. callosus* remains to be established.

7.4 CEREBRAL TOXOPLASMOSIS

This section will discuss animal models available for toxoplasmic encephalitis (TE) as the predominant manifestation of the disease in the immunocompromised host. It will also include acute systemic models where disseminated infection is prominent and where the brain is usually involved as a part of the systematic infection. The most often used animal for acute or cerebral toxoplasmosis, particularly with respect to pharmacological testing, is the mouse (for examples see Perea and Daza, 1976; Grossman and Remington, 1979; Hofflin and Remington, 1987a, 1987b; Chang and Pechere, 1987; Chang *et al.*, 1988, 1991, 1994; Israelski and Remington, 1990; Derouin *et al.*, 1991, 1992; Araujo *et al.*, 1991b, 1992a, 1992b, 1996, 1998; Weiss *et al.*, 1992; Romand *et al.*, 1993, 1995, 1996; Rodriguez-Diaz *et al.*, 1993; Dumas *et al.*, 1994, 1999; Olliaro *et al.*, 1994; Alder *et al.*, 1994; Martinez *et al.*, 1996; Khan *et al.*, 1996, 2000; Aguirre-Cruz and Sotelo, 1998; Sordet *et al.*, 1998; Aguirre-Cruz *et al.*, 1998; Djurkovic-Djakovic *et al.*, 1999, 2002, 2005; Moshkani and Dalimi, 2000; Schöler *et al.*, 2001; Ferreira *et al.*, 2002; Degerli *et al.*, 2003; Belal *et al.*, 2004; Dunay *et al.*, 2004, 2009; Lescano *et al.*, 2004; Grujic *et al.*, 2005; Ling *et al.*, 2005; Mitchell *et al.*, 2006; Jost *et al.*, 2007; Shubar *et al.*, 2008, 2009, 2011; Bajohr *et al.*, 2010; Martins-Duarte *et al.*, 2010). For some purposes, hamsters (Frenkel *et al.*, 1975; Gormley *et al.*, 1998) or rats (Foulet *et al.*, 1994; Dubey, 1996; De Champs *et al.*, 1997; Kempf *et al.*, 1999; Zenner *et al.*, 1999b; Freyre *et al.*, 2001b, 2003a, 2004) have been used but not usually for drug evaluation. Models are based either on a primary acute infection of the animals, on direct inoculation of the parasite into the animal brain with or without immune suppression, or on a chronic infection of the animal that may be immune suppressed (for example, with immunosuppressive drugs or radiation, antibodies directed against lymphocytes or cytokines, or concomitant infections with viruses that modulate the immune response) (see Tables 7.5 and 7.6). For certain applications, the use of genetically modified mice with various defects in their immune system may also be an option. In general, the acute and strictly localized models have most often been used to evaluate the treatment efficacy of anti-parasitic drugs whereas the chronic infection models have been used to study their influence on cyst formation and/or prevention of a relapsing disease.

7.4.1 Acute Infection Models

Acute infection models are usually associated with a high level of mortality (often up to 100%) of laboratory animals within eight to 10 days and survival in particular treatment groups, which are estimated by the Kaplan–Meier or product limit survival analysis. So far they have been the overall preferred models for drug testing mainly because of their consistent reproducibility. Most often, between 2×10^2 and 2×10^4 tachyzoites of the virulent RH strain are injected i.p. into female Swiss Webster mice (for examples see Khan *et al.*, 1996, 1998; Araujo *et al.*, 1997; Djurkovic-Djakovic *et al.*, 1999), but

TABLE 7.5 Pharmacological Studies on Cerebral and Acute Toxoplasmosis

Infection Model (Mouse)	Specific Comments	Pharmacological Studies (Literature Examples)
Acute infection models	Standard models for drug tests; i.p. infection with up to 100% mortality within 8–10 days (type 1 strains); oral infection with cysts of low virulent strains (type 2) and quantification of parasite numbers at different time points	Khan *et al.*, 1996, 1998; Araujo *et al.*, 1997; Mui *et al.*, 2005; Shubar *et al.*, 2009, 2011; Dunay *et al.*, 2009
Localized brain infection models	Direct inoculation of parasites into the brain; does not follow the natural route of infection	Hofflin and Remington, 1987a, 1987b; Arribas *et al.*, 1995
Progressive *Toxoplasma* encephalitis models	For example i.p. or oral infection with brain cysts of type 2 strains; assessment, e.g. via brain cysts counting	Araujo *et al.*, 1991a, 1996; Dumas *et al.*, 1994, 1999
Chronic relapsing infection models (reactivated toxoplasmosis)	Infection with type 2 or 3 strains; drug-induced immunosuppression of mice or use of genetically altered immunodeficient mice	Djurkovic-Djakovic *et al.*, 2002; Schöler *et al.*, 2001; Dunay *et al.*, 2004, 2009; Mitchell *et al.*, 2006

sometimes animals are infected i.p. or orally with, for example, 10 cysts of *T. gondii* ME49 or C56 (McLeod *et al.*, 1989; Araujo *et al.*, 1997; Khan *et al.*, 1998; Yardley *et al.*, 2002; Shubar *et al.*, 2009, 2011; Martins-Duarte *et al.*, 2010). Mice are then often observed for up to 30 days from the date of infection. Surviving mice are examined for residual infection by microscopy of brain tissue for the presence of *T. gondii* cysts or by i.p. subinoculation of suspensions of portions of various organs into healthy mice (bioassay). The probability of untreated mice to succumb during acute infection following peroral infection critically depends on the genetic background of the mouse strain used with C3H/HeJ being highly resistant, BALB/c being intermediately resistant, and C57BL/6J being highly susceptible (Araujo *et al.*, 1976; Williams *et al.*, 1978; McLeod *et al.*, 1989). In addition, when choosing a mouse model, one has to consider that the route of infection (for example, i.p. or oral infection) also largely influences the outcome of infection (Johnson, 1984).

In general, acute infection models are not very close to most *Toxoplasma*-induced human diseases of the immunocompromised host, where the infection takes a more clinically localized course often, though not always, confined to the brain. In contrast, acute primary toxoplasmosis in animal models is a generalized infection, substantially involving organs other than the brain, such as the lungs or the liver. Thus, there are inherent difficulties in these models for the projection of drug efficacy deduced from a Kaplan–Meier diagram which is ultimately dependent not only on direct drug action but also on organ-specific pharmacokinetics and drug metabolism. This means, as a consequence, that a drug successfully tested in acute infection models may fail in the treatment of localized brain infection and *vice versa*. At any rate, this model is the standard model for the first *in vivo* screening of a new drug.

TABLE 7.6 Models of Cerebral Toxoplasmosis

Chronic Relapsing Infection Models (Reactivated Toxoplasmosis)						
Species and Strain of Mouse Used*	***Toxoplasma* Strain Administered***	***Toxoplasma* Dose, Stage and Route of Infection**	**Means for Reactivation**	**Outcome of Animal Infection**	**Remarks**	**Publication****
Porton	M3	30 cysts i.p.	Dexamethasone 6 weeks after infection	40% of mice developed clinical signs of toxoplasmosis, most of them with brain inflammation	Brain cysts observed in only 30–40% of mice with suspected toxoplasmosis	Nicoll *et al.*, 1997
Swiss–Webster	ME49	10 cysts orally	Dexamethasone (DXM) alone or combined with cortisone acetate (CA) 6 weeks after infection	Mortality after 7 weeks with immunosuppression: Untreated 0%, DXM 61.1%, DXM + CA 85%, uninfected + DXM 33%	Mean cyst number 2–9-fold increased compared to untreated control; 14.2% developed clinical TE (both treatment regimens)	Djurkovic-Djakovic and Milenkovic, 2001
B6	C56 (or ME49)	1×10^5 tachyzoites i.p. of C56 followed by 2 weeks treatment with sulphadiazine (20 cysts ME49 i.p. for mice preinfected with virus)	Coinfection with LP–BM5 murine leukemia virus 8 weeks after challenge with C56 (coinfection 12, 8, 4 or 2 weeks before ME49 challenge)	Chronic infection with C56: 30–40% mortality by 80 days following viral coinfection; mice with encephalitic lesions	No effects if challenged (ME49) 4 or 2 weeks after viral infection; all mice died if challenged 12 or 8 weeks after viral infection (pneumonitis, only occasional necrotic areas in brain)	Gazzinelli *et al.*, 1992
C57BL/6	Fukaya	10 cysts i.p.	Coinfection with LP–BM5 murine leukemia virus 6 weeks after *Toxoplasma* infection	All mice infected with both *T. gondii* and LP–BM5 MuLV died 9–14 weeks after the virus infection due to severe encephalitis	In contrast to other studies with the LP–BM5 virus, in this study other organs than brain seem to be less affected	Watanabe *et al.*, 1993
C57BL/6	C-strain	10 cysts orally	Coinfection with LP–BM5 murine leukemia virus 30 days before or 20, 30 and 60 days after challenge	None of the animals developed *Toxoplasma* encephalitis	Increase in *Toxoplasma* lung counts	Lacroix *et al.*, 1994

(*Continued*)

TABLE 7.6 Models of Cerebral Toxoplasmosis (*cont'd*)

Chronic Relapsing Infection Models (Reactivated Toxoplasmosis)						
Species and Strain of Mouse Used*	***Toxoplasma* Strain Administered***	***Toxoplasma* Dose, Stage and Route of Infection**	**Means for Reactivation**	**Outcome of Animal Infection**	**Remarks**	**Publication****
C57BL/6	ME49	15 cysts orally	Coinfection with LP–BM5 2 weeks after challenge	70%–80% animals succumbed to disseminated infection (including brain but also lung, spleen and liver) by 12 weeks after LP–BM5 challenge	Transfer of immune CD8 + T-cells prevented reactivation	Khan *et al.*, 1999
SCID (on C.B-17/Smn background)	ME49	20 cysts i.p.	Sulphadiazine treatment started 10 days after infection for 18 days, then discontinuation of therapy	All SCID mice died with TE 6–9 days after sulphadiazine treatment was stopped	No other organs except the brain with cysts, tachyzoites or inflammation foci	Johnson, 1992
SCID	ME49	10 cysts orally	Sulphadiazine treatment started 2 days after infection for 3 weeks, then discontinuation of therapy	Mortality 100% within 2 weeks (TE); splenocyte transfer from immune syngenic donors prevented reactivation	Reactivation started from liver spreading into other organs (including the brain)	Beaman *et al.*, 1994
BALB/c IFN-$\gamma^{-/-}$ (interferon-γ-deficient)	ME49	10 cysts i.p or orally	Sulphadiazine treatment started 4 days after infection for 3 weeks, then discontinuation of therapy	Mortality due to TE 100% within 1 week independent from infection mode; control WT BALB/c mice survived for more than 3 months	Treatment with recombinant IFN-γ prevented TE	Suzuki *et al.*, 2000
C57BL/6 IFN-$\gamma^{-/-}$ (interferon-γ-deficient)	PRU-Luc-GFP (transgenic type 2 strain)	20 cysts orally	Sulphadiazine treatment from 2 days p.i. onwards for 3 weeks, then discontinuation of therapy	Mortality 100% within 10 days of discontinuation of sulphadiazine	Treament with artemiside or artemisone reduced mortality to 20% and 40%, respectively	Dunay *et al.*, 2009

BALB/c IFN-$\gamma^{-/-}$ (interferon-γ-deficient)	Type 3 strain isolated from chicken	10 cysts orally	Sulphadiazine treatment started 4 days after infection for 11 weeks, then discontinuation of therapy	100% mortality within 10 days after discontinuation of sulphadiazine treatment	Treatment with ponazuril prevented reactivated TE in 5 out of 7 mice; no mice died	Mitchell *et al.*, 2006
C57BL/6 ICSBP/ IRF-8$^{-/-}$ (Interferon regulatory factor 8-deficient)	ME49	5–10 cysts orally	Sulphadiazine treatment started 2–7 days after infection for 2–4 weeks, then discontinuation of therapy	Mortality and time to death depend on the time of sulphadiazine treatment; synchronized development of TE	Treatment as well as maintenance therapy studies possible	Schöler *et al.*, 2001; Dunay *et al.*, 2004 ; Jost *et al.*, 2007 ; Shubar *et al.*, 2009, 2011

Models Based on Localized Brain Infection Instead of Reactivation

Species and Strain of Mouse Used*	*Toxoplasma* Strain Administered*	*Toxoplasma* Dose, Stage and Route of Infection	Outcome of Animal Infection	Remarks	Publication**
Swiss Webster	C56	1×10^4 tachyzoites injected intracerebrally	Normal mice survived but immunosuppressed died from progressive disease (no rates specified); immunosuppression with either cortisone, cyclophosphamide or cyclosporine	Cerebral lesions: inflammation intensity, tachyzoite and cyst number dependent on type of immunosuppression	Hofflin *et al.*, 1987
Swiss Webster	C56	1×10^4 tachyzoites injected intracerebrally	Mortality of 40% in normal mice and 100% in cortisone-treated animals	Drug evaluation study (efficacy of clindamycine was shown)	Hofflin and Remington, 1987a
Swiss Ico	RH	1×10^3 tachyzoites injected intracerebrally	100% of animals died within 5 days after infection due to necrotizing *Toxoplasma* meningoencephalitis	Pharmaceutical study: evaluation of highly active drugs possible	Arribas *et al.*, 1995

(Continued)

TABLE 7.6 Models of Cerebral Toxoplasmosis (*cont'd*)

Progressive *Toxoplasma* Encephalitis Models					
Species and Strain of Mouse Used*	***Toxoplasma* Strain Administered***	***Toxoplasma* Dose, Stage and Route of Infection**	**Outcome of Animal Infection**	**Remarks**	**Publication****
A/J, CBA/J, C57BL/6, C3/H, C57BL/10, DBA/2,BALB/c, different B10 mice	ME49	10 cysts i.p. or 100 cysts orally	Mice with $H\text{-}2^b$ and $H\text{-}2^k$ haplotype developed TE, mice of $H\text{-}2^d$ and $H\text{-}2^a$ haplotype did not	Histological evaluation of the number of cysts in brain sections and inflammation in the brain	Brown and McLeod, 1990; Suzuki *et al.*, 1991
B10 and BALB congenic mice	DX (type 2)	10 cysts orally	Outcome of infection differed between mouse strains: B10 generally more susceptible than BALB, important impact of MHC haplotype; mortality due to TE in some strains	Assessment of mouse survival, number of tachyzoites and cysts in the brain	Deckert-Schlüter *et al.*, 1994

C57BL/6 WT and various knockout strains including cell type-specific knockouts (cre/lox system)	DX	5 cysts orally	WT mice generally survived until 40–60 days; thereafter 30%–50% of the mice may have died due to TE depending on the individual experiment	Mainly used for pathogenicity studies; evaluation of parasites/cysts by histology and PCR	Schlüter *et al.*, 2003; Drögemüller *et al.*, 2008; Händel *et al.*, 2012
CBA/Ca (inbred)	ME49	20 cysts i.p. or orally	Development of a chronic progressive encephalitis and mice began to die approximately 6 weeks after infection	Assessment by cyst counting (light microscopy) or brain histopathology scoring	Araujo *et al.*, 1996
C57BL/6J	PRU	10 cysts i.p.	Development of a chronic progressive encephalitis characterized by brain cysts and inflammation with mortality of 60%–80% within the following months	Assessment by brain histopathology scoring	Dumas *et al.*, 1999

* *For definition of animal or* Toxoplasma *strains please see respective articles*
** *Due to numerous publications on this topic, only few were chosen exemplarily for this table*

In addition to survival curves, the count or titration of cysts of succumbed and/or surviving animal brains, as well as subinoculation into naive mice, may be performed. Alternatively, tissue-culture methods or detection methods such as quantitative PCR from organs of survivors may be used to assess the residual parasite load (Miedouge *et al.*, 1997; Weiss *et al.*, 1991; Djurkovic-Djakovic *et al.*, 1999; Bajohr *et al.*, 2010; Martins-Duarte *et al.*, 2010). An important new development is the possibility to use bioluminescence imaging of mice infected with luciferase (*luc*)-transgenic *T. gondii* parasites (Saeij *et al.*, 2005). *Luc*-transgenic parasites of type 1 and 2 background are available, and after injection of the substrate (e.g. firefly luciferin), bioluminescence intensity can be regularly monitored, e.g. every two days. This enables one to monitor the course of an acute disease in single living animals quantitatively, although the imaging output is not directly related to the parasite number, but is a relative measurement for the intensity of an infection (Saeij *et al.*, 2005; Dunay *et al.*, 2009). This system has been used to show the efficacy of artemisone and artemiside against an acute i.p. infection of CD1 outbred mice with 10^6 of a type 2 *Toxoplasma* strain (PRU background) (Dunay *et al.*, 2009).

In addition to primary evaluation of drug efficacy based on a fatal outcome of acute toxoplasmosis, mouse models have been developed that do not have the major disadvantage of the possibly severe suffering of mice before they die from overwhelming *T. gondii* infection (Samuel *et al.*, 2003; Mui *et al.*, 2005; Shubar *et al.*, 2008, 2009, 2011). In one of these models, mice are inoculated i.p. with the RH strain but on the fourth day after infection, 1.5 ml of phosphate buffered saline (PBS) is injected i.p. and then withdrawn together with all peritoneal fluid. Total numbers of parasites and concentrations of parasites are quantified microscopically as the basis for subsequent statistical analysis (Samuel *et al.*, 2003). In fact, this model has been proven by the addition of sulphadiazine to the drinking water of infected mice, which significantly reduced the parasite burden (Samuel *et al.*, 2003), and thus may well be suited to replacing the survival based acute infection models in the future. A relatively new and interesting alternative to evaluate the efficacy of a drug regime is the use of transgenic parasites expressing fluorescent proteins, e.g. the green fluorescent protein (GFP). It allows determining the number of infected host cells isolated from the peritoneal cavity or other tissues for instance on day 5 of infection using flow cytometry (Shubar *et al.*, 2008). This read-out system avoids the time-consuming and error-prone microscopic histochemical evaluation of the number of parasites and enables gathering of additional information as, for example, the types of the parasite-infected host cells or their viability. Infection of NMRI mice with 10^5 GFP-expressing type 1 RH parasites has been applied to prove an excellent activity of quinolones and distinct bisphosphonates against *T. gondii* (Shubar *et al.*, 2008; Bajohr *et al.*, 2010). In another model, mice are orally infected with ME49 tissue cysts, and the outcome of disease is examined 16 days p.i. by evaluating the number of tachyzoites, tissue cysts and inflammatory foci in sections of the brain (Shubar *et al.*, 2009, 2011). This model was successfully applied to prove a beneficial effect of atovaquone nanosuspensions in acute murine toxoplasmosis (Shubar *et al.*, 2009, 2011).

7.4.2 Localized Brain Infection Models

The direct inoculation of tachyzoites into the frontal lobe of mice may give the most reproducible results with lesions histologically resembling those that could be observed in immunocompromised humans (Hofflin *et al.*, 1987). The model was successfully used to demonstrate the efficacy of clindamycin, roxithromycin and gamma-interferon on *Toxoplasma* encephalitis (Hofflin and Remington, 1987a, 1987b). Subsequently, in another study using a localized brain infection model with the highly virulent RH strain, clindamycin showed no detectable effect; instead pyrimethamine, sulphadiazine and their combination were useful in terms of reducing mortality and histopathology (Arribas *et al.*, 1995). The procedure of local brain inoculation requires high technical expertise and thus may not be suitable for investigations on a larger scale in average laboratories. In addition, the main disadvantage of this model is that TE does not follow the natural history of *T. gondii* infection, as it is not a consequence of recrudescence of a previously established infection. It therefore may be not ideal for pharmacological trials related to these aspects (e.g. for investigating drug interference with mechanisms of recrudescence) but rather for studying certain features of pathogenesis.

7.4.3 Progressive *Toxoplasma* Encephalitis Models

Subacute or chronic infection models may be suited to evaluating the efficacy of drugs against the cyst form of *T. gondii*. Indeed, the cyst form is critical within the life cycle of the parasite with regard to pathogenesis of toxoplasmosis in immunocompromised individuals, especially with regard to development of toxoplasmic encephalitis (Frenkel and Escajadillo, 1987; Frenkel *et al.*, 1975; Hofflin *et al.*, 1987; Ferguson *et al.*, 1989). Two different types of murine models with prolonged infection times have been widely used, mostly for studying the pathogenesis of chronic toxoplasmosis but also for evaluating drug treatments: (i) progressive TE and (ii) reactivation of asymptomatic chronic infections. It is important to note that the development of disease in both models involves recrudescence of the tachyzoite stage of *T. gondii*. Hence, an improvement in the outcome of the disease after drug treatment cannot be taken as an indication for a microbicidal activity against the cyst form of the parasite.

In progressive TE models, mice are usually infected i.p. or orally with brain cysts of a mildly virulent typical type 2 *Toxoplasma* strain such as ME49 or DX. Mice can then develop a chronic progressive toxoplasmic encephalitis (type 2 strains such as the ME49 strain tend to give rise to new cyst formation during the first weeks of infection, presumably preceded by cyst rupture and proliferation of tachyzoites that are then again converted into bradyzoites) and, unless treated, begin to die within the following weeks or months. Morbidity and mortality rates due to progressive TE vary considerably depending on the *Toxoplasma* strain, the dose of infection, and particularly the mouse strain used (Araujo *et al.*, 1991a; Suzuki *et al.*, 1991; Dumas *et al.*, 1994; Deckert-Schlüter *et al.*, 1994; Djurkovic-Djakovic *et al.*, 2002). Genes present within the major histocompatibility complex (MHC) and genes outside of the MHC are decisive for the development of TE and for the time until the majority of mice succumb due to TE (Deckert-Schlüter *et al.*, 1994). In order to evaluate drug efficacies in these models, treated mice and untreated controls are usually sacrificed at determined time points and the activity of drugs is assessed by light microscopic examination and counting of cysts from brains previously ground with a pestle and mortar, homogenized by needle passage or by using glass beads and suspended in a defined volume of PBS (Araujo *et al.*, 1991a; Sarciron *et al.*, 1997; Lescano *et al.*, 2004; Djurkovic-Djakovic *et al.*, 2002). Alternatively, brains of mice may be examined histopathologically by scoring the severity of inflammatory lesions (Araujo *et al.*, 1996) and/or Kaplan–Meier curves may be obtained (Dumas *et al.*, 1999; Djurkovic-Djakovic *et al.*, 2002). A well established model of chronic TE is the infection of C57BL/6 WT mice or those with deficiencies in various immunity-related proteins with five cysts of the low virulent type 2 strain DX (for an example, see Schlüter *et al.*, 2003). Survival of WT mice is 100% until at least 40 to 50 days p.i. with increasing mortality due to TE thereafter. This model has been extensively used to define immune factors which prevent the development of TE (Schlüter *et al.*, 2003; Drögemüller *et al.*, 2008; Händel *et al.*, 2012). It may also be useful for evaluating the efficacy of drug therapies against a mild course of the chronic stage of toxoplasmosis.

7.4.4 Chronic Relapsing Infection Models (Reactivated Toxoplasmosis)

In order to describe and mimic the natural course of reactivation of *T. gondii* infection in humans, animal models based on reactivation of a chronic infection have been attempted to be developed. It was first observed in 1966 that when infecting splenectomized mice or mice treated with cortisone or 6-mercaptopurine (6-MP) with the *T. gondii* Beverley strain that the course of the disease was greatly altered in the experimental animals, distinguished by signs of severe neurological involvement and meningoencephalitis (Stahl *et al.*, 1966). Reactivation of an otherwise chronic infection had succeeded in hamsters infected with the RH strain by the administration of cortisone, cyclophosphamide or whole body irradiation (Frenkel *et al.*, 1975). On the other hand, in mice, only a little reactivation using cortisol acetate, azathioprine or cyclosporine was observed (Sumyuen *et al.*, 1996), but reactivation was obtained with dexamethasone (DMX) (Nicoll *et al.*, 1997). However, brain cysts were only demonstrated in a few of the animals.

Perhaps the most promising mouse model based on immunosuppressive drugs is that of Djurkovic-Djakovic *et al.* who used a mildly virulent type 2 *T. gondii* strain (ME49) to assess the efficacy of atovaquone combined with clindamycin treatment (Djurkovic-Djakovic *et al.*, 2002). Indeed, type 2 strains are also responsible for most cases of human TE in Europe and North America (Howe and Sibley, 1995). Reactivation was achieved in those animals that had been previously orally infected with 10 tissue cysts of the ME49 strain by immunosuppression with DXM alone and more efficiently by combined hydrocortisone-21-acetate (CA) treatment (Djurkovic-Djakovic and Milenkovic, 2001; Djurkovic-Djakovic *et al.*, 2002).

In addition to the above mentioned models, reactivation may be induced in dual infection models with *T. gondii* and viruses such as CMV (Pomeroy *et al.*, 1989) or LP—BM5 murine leukaemia virus which is responsible for murine AIDS (Gazzinelli *et al.*, 1992; Watanabe *et al.*, 1993; Lacroix *et al.*, 1994; Khan *et al.*, 1999). Such models are most precious for investigation of the immunopathogenesis of analogous disease in humans; however, due to their complexity and obstacles in reproducibility, they may not be well suited to pharmacological investigations (Lacroix *et al.*, 1994).

Another strategy besides suppressing the host immune system of immune competent mice is based on the use of severely immune compromised mice, such as those with severe combined immunodeficiency (SCID) mice lacking T- and B-lymphocytes, or athymic (nude mice) which lack functional T-cells. An otherwise fatal infection in such mice may be made chronic by administering sulphadiazine treatment and withdrawal of sulphadiazine then leads to relapse of infection (Johnson, 1992; Beaman *et al.*, 1994). To date, however, such models have been mostly used to investigate the immunopathogenesis of relapsing or acute toxoplasmosis rather than to assess drug regimens for treatment or prevention of the reactivated disease (for review see Denkers and Gazzinelli, 1998). In addition, severely immune compromised animals, such as SCID or athymic mice, are difficult to work with because of their impaired immune systems and the requirement for them to be kept in sterile conditions to prevent opportunistic infections.

A promising alternative to SCID mouse models that has been presented by Suzuki *et al.* is BALB/c mice deficient for interferon-γ (Suzuki *et al.*, 2000). BALB/c mice are genetically relatively resistant to *T. gondii* infection (Suzuki *et al.*, 1991, 1994; Brown *et al.*, 1995). IFN-γ, but not TNF-α or iNOS, is crucial for resistance against the development of TE, and mice deficient for IFN-γ die after an infection with the ME49 *T. gondii* strain when treatment with sulphadiazine is discontinued (Suzuki *et al.*, 2000). IFN-γ knockout mice treated with sulphadiazine in order to establish a chronic latent infection were successfully employed to show that ponazuril, i.e. an anticoccidial triazine used in the poultry industry, or artemisinin analogues efficiently prevented TE after the discontinuation of the sulphadiazine treatment (Mitchell *et al.*, 2006; Dunay *et al.*, 2009).

Based on the findings with the IFN-γ knockout model, reactivated TE in chronically infected mice (*T. gondii* strain ME49) deficient for the interferon regulatory factor 8 (interferon consensus sequence binding protein ICSBP/IRF-8$^{-/-}$ on a C57BL/6 background) was achieved by withdrawal of sulphadiazine treatment (Schöler *et al.*, 2001). This model, which relies on an impairment of the IL-12 dependent IFN-γ production (Scharton-Kersten *et al.*, 1997; Holtschke *et al.*, 1996), was then used to show the efficacy of different formulations of atovaquone nanosuspensions in the treatment of TE (Schöler *et al.*, 2001; Shubar *et al.*, 2009, 2011) and, thereafter, to prevent the disease from relapsing by an atovaquone maintenance therapy (Dunay *et al.*, 2004). The results obtained with this murine model of reactivated toxoplasmosis mimicked the signs of reactivated toxoplasmosis in immunocompromised patients, including the presence of parasite-associated focal necrotic lesions in the brain parenchyma and meningeal inflammation (Schöler *et al.*, 2001; Dunay *et al.*, 2004). However, it appears to be important to start the sulphadiazine treatment within two to three days after infection because later treatment increases the mortality rate considerably and leads to TE due to continuous invasion of the brain by tachyzoites rather than by reactivation of latent cysts (Jost *et al.*, 2007). Mice with impaired IFN-γ production, in contrast to SCID or nude mice, may not be as prone to common infection; however, they have to be kept under specific-pathogen-free conditions, reducing the number of facilities where such experiments can be performed. Advantages of these models in immunodeficient mice as compared to those involving reactivation by administration of immunosuppressive drugs are (i) the easiness of reactivation simply by discontinuation of sulphadiazine and (ii) a relatively synchronized development of TE within days (Schöler *et al.*, 2001).

References

Abou-Bacar, A., Pfaff, A.W., Georges, S., Letscher-Bru, V., Filisetti, D., Villard, O., Antoni, E., Klein, J.P., Candolfi, E., 2004a. Role of NK cells and gamma interferon in transplacental passage of *Toxoplasma gondii* in a mouse model of primary infection. Infect. Immun. 72, 1397–1401.

Abou-Bacar, A., Pfaff, A.W., Letscher-Bru, V., Filisetti, D., Rajapakse, R., Antoni, E., Villard, O., Klein, J.P., Candolfi, E., 2004b. Role of gamma interferon and T cells in congenital *Toxoplasma* transmission. Parasite Immunol. 26, 315–318.

Adams, F.H., Cooney, M., Adams, J.M., Kabler, P., 1949. Experimental toxoplasmosis. Proc. Soc. Exp. Biol. Med. 70, 258–260.

Aguirre-Cruz, L., Sotelo, J., 1998. Lack of therapeutic effect of colchicine on murine toxoplasmosis. J. Parasitol. 84, 163–164.

Aguirre-Cruz, L., Velasco, O., Sotelo, J., 1998. Nifurtimox plus pyrimethamine for treatment of murine toxoplasmosis. J. Parasitol. 84, 1032–1033.

Alder, J., Hutch, T., Meulbroek, J.A., Clement, J.C., 1994. Treatment of experimental *Toxoplasma gondii* infection by clarithromycin-based combination therapy with minocycline or pyrimethamine. J. Acquir. Immune Defic. Syndr. 7, 1141–1148.

Ali, S.M., Allam, S.R., Negm, A.Y., El Zawawy, L.A., 2003. Vaccination against congenital toxoplasmosis. J. Egypt. Soc. Parasitol. 33, 863–874.

Anderson, M.L., Barr, B.C., Conrad, P.A., 1994. Protozoal causes of reproductive failure in domestic ruminants. Vet. Clin. North Am. Food Anim. Pract. 10, 439–461.

Araujo, F.G., Remington, J.S., 1975. IgG antibody suppression of the IgM antibody response to *Toxoplasma gondii* in newborn rabbits. J. Immunol. 115, 335–338.

Araujo, F.G., Williams, D.M., Grumet, F.C., Remington, J.S., 1976. Strain-dependent differences in murine susceptibility to *Toxoplasma*. Infect. Immun. 13, 1528–1530.

Araujo, F.G., Huskinson, J., Remington, J.S., 1991a. Remarkable *in vitro* and *in vivo* activities of the hydroxynaphthoquinone 566C80 against tachyzoites and tissue cysts of *Toxoplasma gondii*. Antimicrob. Agents Chemother. 35, 293–299.

Araujo, F.G., Shepard, R.M., Remington, J.S., 1991b. *In vivo* activity of the macrolide antibiotics azithromycin, roxithromycin and spiramycin against *Toxoplasma gondii*. Eur. J. Clin. Microbiol. Infect. Dis. 10, 519–524.

Araujo, F.G., Lin, T., Remington, J.S., 1992a. Synergistic combination of azithromycin and sulfadiazine for treatment of toxoplasmosis in mice. Eur. J. Clin. Microbiol. Infect. Dis. 11, 71–73.

Araujo, F.G., Prokocimer, P., Remington, J.S., 1992b. Clarithromycin-minocycline is synergistic in a murine model of toxoplasmosis. J. Infect. Dis. 165, 788.

Araujo, F.G., Suzuki, Y., Remington, J.S., 1996. Use of rifabutin in combination with atovaquone, clindamycin, pyrimethamine, or sulfadiazine for treatment of toxoplasmic encephalitis in mice. Eur. J. Clin. Microbiol. Infect. Dis. 15, 394–397.

Araujo, F.G., Khan, A.A., Slifer, T.L., Bryskier, A., Remington, J.S., 1997. The ketolide antibiotics HMR 3647 and HMR 3004 are active against *Toxoplasma gondii in vitro* and in murine models of infection. Antimicrob. Agents Chemother. 41, 2137–2140.

Araujo, F.G., Khan, A.A., Bryskier, A., Remington, J.S., 1998. Use of ketolides in combination with other drugs to treat experimental toxoplasmosis. J. Antimicrob. Chemother. 42, 665–667.

Arribas, J.R., de Diego, J.A., Gamallo, C., Vazquez, J.J., 1995. A new murine model of severe acute *Toxoplasma* encephalitis. J. Antimicrob. Chemother. 36, 503–512.

Bajohr, L.L., Ma, L., Platte, C., Liesenfeld, O., Tietze, L.F., Groß, U., Bohne, W., 2010. In vitro and in vivo activities of 1-hydroxy-2-alkyl-4(1H)quinolone derivates against *Toxoplasma gondii*. Antimicrob. Agents Chemother. 54, 517–521.

Barbosa, B.F., Silva, D.A.O., Costa, I.N., Pena, J.D.O., Mineo, J.R., Ferro, E.A.V., 2007. Susceptibility to vertical transmission of *Toxoplasma gondii* in temporally dependent on the preconceptional infection in *Calomys callosus*. Plancenta 28, 624–630.

Beaman, M.H., Araujo, F.G., Remington, J.S., 1994. Protective reconstitution of the SCID mouse against reactivation of toxoplasmic encephalitis. J. Infect. Dis. 169, 375–383.

Beghetto, E., Nielsen, H.V., Del Porto, P., Buffolano, W., Guglietta, S., Felici, F., Petersen, E., Gargano, N., 2005. A combination of antigenic regions of *Toxoplasma gondii* microneme proteins induces protective immunity against oral infection with parasite cysts. J. Infect. Dis. 191, 637–645.

Behnke, M.S., Khan, A., Wootton, J.C., Dubey, J.P., Tang, K., Sibley, L.D., 2011. Virulence differences in *Toxoplasma* mediated by amplification of a family of polymorphic pseudokinases. Proc. Natl. Acad. Sci. USA 108, 9631–9636.

Belal, U.S., Norose, K., Aosai, F., Mun, H.S., Ahmed, A.K., Chen, M., Mohamed, R.M., Piao, L.X., Iwakura, Y., Yano, A., 2004. Evaluation of the effects of sulfamethoxazole on *Toxoplasma gondii* loads and stage conversion in IFN-gamma knockout mice using QC-PCR. Microbiol. Immunol. 48, 185–193.

Berard-Badier, M., Laugier, M., Louchet, E., Payan, H., 1968. The placenta in experimental acute congenital toxoplasmosis in guinea pigs. Pathol. Biol. 16, 829–835.

Beverley, J.K.A., Beattie, C.P., Fry, B.A., 1954. Experimental toxoplasmosis of the uveal tract. Br. J. Ophthalmol. 38, 489–496.

Beverley, J.K.A., 1958. A rational approach to the treatment of toxoplasmic uveitis. Trans. Opthal. Soc. UK 78, 109–121.

Beverley, J.K.A., 1959. Congenital transmission of toxoplasmosis through successive generation of mice. Nature 183, 1348–1349.

Beverley, J.K.A., 1961. Experimental ocular toxoplasmosis. Surv. Ophthalmol. 6, 897–923.

Beverley, J.K.A., Henry, L., 1970. Congenital toxoplasmosis in mice. J. Pathol. 101, xxi.

Brown, C.R., McLeod, R., 1990. Class I MHC genes and CD8+ T cells determine cyst number in *Toxoplasma gondii* infection. J. Immunol. 145, 3438–3441.

Brown, C.R., Hunter, C.A., Estes, R.G., Beckmann, E., Forman, J., David, C., Remington, J.S., McLeod, R., 1995. Definitive identification of a gene that confers resistance against *Toxoplasma* cyst burden and encephalitis. Immunology 85, 419–428.

Cavailles, P., Sergent, V., Bisanz, C., Papapietro, O., Colacios, C., Mas, M., Subra, J.F., Lagrange, D., Calise, M., Appolinaire, S., Faraut, T., Druet, P., Saoudi, A., Bessieres, M.H., Pipy, B., Cesbron-Delauw, M.F., Fournie, G.J., 2006. The rat Toxo1 locus directs toxoplasmosis outcome and controls parasite proliferation and spreading by macrophage-dependent mechanisms. Proc. Natl. Acad. Sci. USA 103, 744–749.

Chang, H.R., Pechere, J.C., 1987. Effect of roxithromycin on acute toxoplasmosis in mice. Antimicrob. Agents Chemother. 31, 1147–1149.

Chang, H.R., Rudareanu, F.C., Pechere, J.C., 1988. Activity of A-56268 (TE-031), a new macrolide, against *Toxoplasma gondii* in mice. J. Antimicrob. Chemother. 22, 359–361.

Chang, H.R., Comte, R., Piguet, P.F., Pechere, J.C., 1991. Activity of minocycline against *Toxoplasma gondii* infection in mice. J. Antimicrob. Chemother. 27, 639–645.

Chang, H.R., Arsenijevic, D., Comte, R., Polak, A., Then, R.L., Pechere, J.C., 1994. Activity of epiroprim (Ro 11-8958), a dihydrofolate reductase inhibitor, alone and in combination with dapsone against *Toxoplasma gondii*. Antimicrob. Agents Chemother. 38, 1803–1807.

Charles, E., Callegan, M.C., Blader, I.J., 2007. The SAG1 *Toxoplasma gondii* surface protein is not required for acute ocular toxoplasmosis in mice. Infect. Immun. 75, 2079–2083.

Costa, I.N., Angeloni, M.B., Santana, L.A., Barbosa, B.F., Silva, M.C.P., Rodrigues, A.A., Rostkowsa, C., Magalhaes, P.M., Pena, J.D.O., Silva, D.A. O., Mineo, J.R., Ferro, E.A.V., 2009. Azithromycin inhibits vertical transmission of *Toxoplasma gondii* in *Calomys callosus* (Rodentia: Cricetidae). Placenta 30, 884–890.

Couper, K.N., Nielsen, H.V., Petersen, E., Roberts, F., Roberts, C.W., Alexander, J., 2003. DNA vaccination with the immunodominant tachyzoite surface antigen (SAG-1) protects against adult acquired *Toxoplasma gondii* infection but does not prevent maternofoetal transmission. Vaccine 21, 2813–2820.

Culbertson, W.W., Tabbara, K.F., O'Connor, R., 1982. Experimental ocular toxoplasmosis in primates. Arch. Ophthalmol. 100, 321–323.

Darcy, F., Zenner, L., 1993. Experimental models of toxoplasmosis. Res. Immunol. 144, 16–23.

Davidson, M.G., Lappin, M.R., English, R.V., Tompkins, M.B., 1993. A feline model of ocular toxoplasmosis. Invest. Ophthalmol. Vis. Sci. 34, 3653–3660.

Davidson, M.G., Lappin, M.R., Rottman, J.R., Tompkins, M.B., English, R.V., Bruce, A.T., Jayawickrama, J., 1996. Paradoxical effect of clindamycin in experimental, acute toxoplasmosis in cats. Antimicrob. Agents Chemother. 40, 1352–1359.

De Champs, C., Imbert-Bernard, C., Belmeguenai, A., Ricard, J., Pelloux, H., Brambilla, E., Ambroise-Thomas, P., 1997. *Toxoplasma gondii: in vivo* and *in vitro* cystogenesis of the virulent RH strain. J. Parasitol. 83, 152–155.

Deckert-Schlüter, M., Schlüter, D., Schmidt, D., Schwendemann, G., Wiestler, O.D., Hof, H., 1994. *Toxoplasma* encephalitis in congenic B10 and BALB mice: impact of genetic factors on the immune response. Infect. Immun. 62, 221–228.

Degerli, K., Kilimcioglu, A.A., Kurt, O., Tamay, A.T., Ozbilgin, A., 2003. Efficacy of azithromycin in a murine toxoplasmosis model, employing a *Toxoplasma gondii* strain from Turkey. Acta. Trop. 88, 45–50.

Denkers, E.Y., Gazzinelli, R.T., 1998. Regulation and function of T-cell-mediated immunity during *Toxoplasma gondii* infection. Clin. Microbiol. Rev. 11, 569–588.

De Roever-Bonnet, H., 1960. Congenital *Toxoplasma* infections in mice and hamsters with avirulent strain. Trop. Geogr. Med. 21, 443–450.

Derouin, F., Piketty, C., Chastang, C., Chau, F., Rouveix, B., Pocidalo, J.J., 1991. Anti-*Toxoplasma* effects of dapsone alone and combined with pyrimethamine. Antimicrob. Agents Chemother. 35, 252–255.

Derouin, F., Almadany, R., Chau, F., Rouveix, B., Pocidalo, J.J., 1992. Synergistic activity of azithromycin and pyrimethamine or sulfadiazine in acute experimental toxoplasmosis. Antimicrob. Agents Chemother. 36, 997–1001.

Derouin, F., Lacroix, C., Sumyuen, M.H., Romand, S., Garin, Y.J., 1995. Experimental models of toxoplasmosis. Pharmacological applications. Parasite 2, 243–256.

Djurkovic-Djakovic, O., Nikolic, T., Robert-Gangneux, F., Bobic, B., Nikolic, A., 1999. Synergistic effect of clindamycin and atovaquone in acute murine toxoplasmosis. Antimicrob. Agents Chemother 43, 2240–2244.

Djurkovic-Djakovic, O., Milenkovic, V., 2001. Murine model of drug-induced reactivation of *Toxoplasma gondii*. Acta Protozoologica 40, 99–106.

Djurkovic-Djakovic, O., Milenkovic, V., Nikolic, A., Bobic, B., Grujic, J., 2002. Efficacy of atovaquone combined with clindamycin against murine infection with a cystogenic (Me49) strain of *Toxoplasma gondii*. J. Antimicrob. Chemother. 50, 981–987.

Djurkovic-Djakovic, O., Nikolic, A., Bobic, B., Klun, I., Aleksic, A., 2005. Stage conversion of *Toxoplasma gondii* RH parasites in mice by treatment with atovaquone and pyrrolidine dithiocarbamate. Microbes. Infect. 7, 49–54.

Drögemüller, K., Helmuth, U., Brunn, A., Sakowicz-Burkiewicz, M., Gutmann, D.H., Mueller, W., Deckert, M., Schlüter, D., 2008. Astrocyte gp130 expression is critical for the control of *Toxoplasma* encephalitis. J. Immunol. 181, 2683–2693.

Dubey, J.P., 1996. Pathogenicity and infectivity of *Toxoplasma gondii* oocysts for rats. J. Parasitol. 82, 951–956.

Dubey, J.P., 1998. Re-examination of resistance of *Toxoplasma gondii* tachyzoites and bradyzoites to pepsin and trypsin digestion. Parasitology 116 (Pt 1), 43–50.

Dubey, J.P., Shen, S.K., 1991. Rat model of congenital toxoplasmosis. Infect. Immun. 59, 3301–3302.

Dubey, J.P., Rommel, M., 1992. [Abortions caused by protozoa in agricultural animals]. Dtsch. Tierarztl. Wochenschr. 99, 355–362.

Dubey, J.P., Shen, S.K., Kwok, O.C., Thulliez, P., 1997. Toxoplasmosis in rats (*Rattus norvegicus*): congenital transmission to first and second generation offspring and isolation of *Toxoplasma gondii* from seronegative rats. Parasitology 115 (Pt 1), 9–14.

Dubey, J.P., Frenkel, J.K., 1998. Toxoplasmosis of rats: a review, with considerations of their value as an animal model and their possible role in epidemiology. Vet. Parasitol. 77, 1–32.

Dumas, J.L., Chang, R., Mermillod, B., Piguet, P.F., Comte, R., Pechere, J.C., 1994. Evaluation of the efficacy of prolonged administration of azithromycin in a murine model of chronic toxoplasmosis. J. Antimicrob. Chemother. 34, 111–118.

Dumas, J.L., Pizzolato, G., Pechere, J.C., 1999. Evaluation of trimethoprim and sulphamethoxazole as monotherapy or in combination in the management of toxoplasmosis in murine models. Int. J. Antimicrob. Agents 13, 35–39.

Dunay, I.R., Heimesaat, M.M., Bushrab, F.N., Muller, R.H., Stocker, H., Arasteh, K., Kurowski, M., Fitzner, R., Borner, K., Liesenfeld, O., 2004. Atovaquone maintenance therapy prevents reactivation of toxoplasmic encephalitis in a murine model of reactivated toxoplasmosis. Antimicrob. Agents Chemother 48, 4848–4854.

Dunay, I.R., Chan, W.C., Haynes, R.K., Sibley, L.D., 2009. Artemisone and artemiside control acute and reactivated toxoplasmosis in a murine model. Antimicrob. Agents Chemother. 53, 4450–4456.

Duncanson, P., Terry, R.S., Smith, J.E., Hide, G., 2001. High levels of congenital transmission of *Toxoplasma gondii* in a commercial sheep flock. Int. J. Parasitol. 31, 1699–1703.

Dutton, G.N., Hay, J., Hair, D.M., Ralston, J., 1986. Clinicopathological features of a congenital murine model of ocular toxoplasmosis. Graefes Arch. Clin. Exp. Ophthalmol. 224, 256–264.

Elbez-Rubinstein, A., Ajzenberg, D., Dardé, M.L., Cohen, R., Dumètre, A., Yera, H., Gordon, E., Janaud, J.C., Thulliez, P., 2009. Congentital toxoplasmosis and reinfection during pregnancy: case report, strain characterization, experimental model of reinfection, and review. J. Infect. Dis. 199, 280–285.

Elsaid, M.M., Martins, M.S., Frezard, F., Braga, E.M., Vitor, R.W., 2001. Vertical toxoplasmosis in a murine model. Protection after immunization with antigens of *Toxoplasma gondii* incorporated into liposomes. Mem. Inst. Oswaldo Cruz. 96, 99–104.

Ferguson, D.J., Hutchison, W.M., Pettersen, E., 1989. Tissue cyst rupture in mice chronically infected with *Toxoplasma gondii*. An immunocytochemical and ultrastructural study. Parasitol. Res. 75, 599–603.

Ferreira, R.A., Oliveira, A.B., Gualberto, S.A., Vitor, R.W., 2002. Activity of natural and synthetic naphthoquinones against *Toxoplasma gondii, in vitro* and in murine models of infection. Parasite 9, 261–269.

Ferro, E.A.V., Silva, D.A.O., Bevilacqua, E., Mineo, J.R., 2002. Effect of *Toxoplasma gondii* infection kinetics on trophoblast cell population in *Calomys callosus*, a model of congenital toxoplasmosis. Infect. Immun. 70, 7089–7094.

Flori, P., Hafid, J., Bourlet, T., Raberin, H., Genin, C., Sung, R.T., 2002. Experimental model of congenital toxoplasmosis in guinea-pigs: use of quantitative and qualitative PCR for the study of maternofoetal transmission. J. Med. Microbiol. 51, 871–878.

Flori, P., Hafid, J., Thonier, V., Bellete, B., Raberin, H., Tran Manh Sung, R., 2003. Parasite load in guinea pig foetus with real time PCR after maternofoetal transmission of *Toxoplasma gondii*. Parasite 10, 133–140.

Foulet, A., Zenner, L., Darcy, F., Cesbron-Delauw, M.F., Capron, A., Gosselin, B., 1994. Pathology of *Toxoplasma gondii* infection in the nude rat. An experimental model of toxoplasmosis in the immunocompromised host? Pathol. Res. Pract. 190, 775–781.

Franco, P.S., Silva, D.A.O., Costa, I.N., Gomes, A.O., Silva, A.L.N., Pena, J.D.O., Mineo, J.R., Ferro, E.A.V., 2011. Evaluation of vertical transmission of *Toxoplasma gondii* in *Calomys callosus* model after reinfection with heterologous and virulent strain. Placenta 32, 116–120.

Frenkel, J.K., 1953. Host, strain and treatment variation as factors in the pathogenesis of toxoplasmosis. Am. J. Trop. Med. Hyg. 2, 390–415.

Frenkel, J.K., 1955. Ocular lesions in hamsters; with chronic *Toxoplasma* and *Besnoitia* infection. Am. J. Ophthalmol. 39, 203–225.

Frenkel, J.K., Nelson, B.M., Arias-Stella, J., 1975. Immunosuppression and toxoplasmic encephalitis: clinical and experimental aspects. Hum. Pathol. 6, 97–111.

Frenkel, J.K., Escajadillo, A., 1987. Cyst rupture as a pathogenic mechanism of toxoplasmic encephalitis. Am. J. Trop. Med. Hyg. 36, 517–522.

Freyre, A., Correa, O., Falcon, J., Mendez, J., Gonzalez, M., Venzal, J.M., 2001a. Some factors influencing transmission of *Toxoplasma* in pregnant rats fed cysts. Parasitol. Res. 87, 941–944.

Freyre, A., Falcon, J., Correa, O., Mendez, J., Gonzalez, M., Venzal, J.M., 2001b. Residual infection of 15 *Toxoplasma* strains in the brain of rats fed cysts. Parasitol. Res. 87, 915–918.

Freyre, A., Falcon, J., Correa, O., Mendez, J., Gonzalez, M., Venzal, J.M., Morgades, D., 2003a. Cyst burden in the brains of Wistar rats fed *Toxoplasma* oocysts. Parasitol. Res. 89, 342–344.

Freyre, A., Falcon, J., Mendez, J., Gonzalez, M., Venzal, J.M., Morgades, D., 2003b. Foetal *Toxoplasma* infection after oocyst inoculation of pregnant rats. Parasitol. Res. 89, 352–353.

Freyre, A., Falcon, J., Mendez, J., Correa, O., Morgades, D., Rodriguez, A., 2004. An investigation of sterile immunity against toxoplasmosis in rats. Exp. Parasitol. 107, 14–19.

Freyre, A., Falcon, J., Mendez, J., Rodreguez, A., Correa, L., Gonzalez, M., 2006a. Refinement of the mouse model of congenital toxoplasmosis. Exp. Parasitol. 113, 154–160.

Freyre, A., Falcon, J., Mendez, J., Rodriguez, A., Correa, L., Gonzalez, M., 2006b. *Toxoplasma gondii*: partial cross-protection among several strains of the parasite against congenital transmission in a rat model. Exp. Parasitol. 112, 8–12.

Freyre, A., Falcon, J., Mendez, J., Gonzalez, M., 2008. *Toxoplasma gondii*: an improved rat model of congenital infection. Exp. Parasitol. 120, 142–146.

Freyre, A., Fialho, C.G., Bigatti, L.E., Araujo, F.A., Falcon, J.D., Mendez, J., Gonzalez, M., 2009. *Toxoplasma gondii*: congenital transmission in a hamster model. Exp. Parasitol. 122, 140–144.

Freyre, A., Araujo, F.A.P., Fialho, C.G., Bigatti, L.E., Falcon, J.D., 2012. Protection in a hamster model of congenital toxoplasmosis. Vet. Parasitol. 183, 359–363.

Friedrich, R., Müller, W.A., 1989. The effect of a subretinal injection of *Toxoplasma gondii* on the serum antibody titer in a rabbit model of ocular toxoplasmosis. Angew. Parasitol. 30, 15–17.

Fujii, H., Kamiyama, T., Hagiwara, T., 1983. Species and strain differences in sensitivity to *Toxoplasma* infection among laboratory rodents. Jpn. J. Med. Sci. Biol. 36, 343–346.

Fux, B., Ferreira, A., Cassali, G., Tafuri, W.L., Vitor, R.W., 2000. Experimental toxoplasmosis in Balb/c mice. Prevention of vertical disease transmission by treatment and reproductive failure in chronic infection. Mem. Inst. Oswaldo Cruz. 95, 121–126.

Garweg, J.G., Kuenzli, H., Boehnke, M., 1998. Experimental ocular toxoplasmosis in naive and primed rabbits. Ophthalmologica 212, 136–141.

Garweg, J.G., Boehnke, M., 2006. The antibody response in experimental ocular toxoplasmosis. Graefes Arch. Clin. Exp. Ophthalmol. 244, 1668–1679.

Gazzinelli, R.T., Hartley, J.W., Fredrickson, T.N., Chattopadhyay, S.K., Sher, A., Morse 3rd, H.C., 1992. Opportunistic infections and retrovirus-induced immunodeficiency: studies of acute and chronic infections with *Toxoplasma gondii* in mice infected with LP-BM5 murine leukemia viruses. Infect. Immun. 60, 4394–4401.

Gazzinelli, R.T., Brezin, A., Li, Q., Nussenblatt, R.B., Chan, C.C., 1994. *Toxoplasma gondii*: acquired ocular toxoplasmosis in the murine model, protective role of TNF-alpha and IFN-gamma. Exp. Parasitol. 78, 217–229.

Gormley, P.D., Pavesio, C.E., Minnasian, D., Lightman, S., 1998. Effects of drug therapy on *Toxoplasma* cysts in an animal model of acute and chronic disease. Invest. Ophthalmol. Vis. Sci. 39, 1171–1175.

Gormley, P.D., Pavesio, C.E., Luthert, P., Lightman, S., 1999. Retinochoroiditis is induced by oral administration of *Toxoplasma gondii* cysts in the hamster model. Exp. Eye Res. 68, 657–661.

Grossman, P.L., Remington, J.S., 1979. The effect of trimethoprim and sulfamethoxazole on *Toxoplasma gondii in vitro* and *in vivo*. Am. J. Trop. Med. Hyg. 28, 445–455.

Grujic, J., Djurkovic-Djakovic, O., Nikolic, A., Klun, I., Bobic, B., 2005. Effectiveness of spiramycin in murine models of acute and chronic toxoplasmosis. Int. J. Antimicrob. Agents 25, 226–230.

Händel, U., Brunn, A., Drögemüller, K., Müller, W., Deckert, M., Schlüter, D., 2012. Neuronal gp130 expression is crucial to prevent neuronal loss, hyperinflammation, and lethal course of murine *Toxoplasma* encephalitis. Am. J. Pathol. 181, 163–173.

Haumont, M., Delhaye, L., Garcia, L., Jurado, M., Mazzu, P., Daminet, V., Verlant, V., Bollen, A., Biemans, R., Jacquet, A., 2000. Protective immunity against congenital toxoplasmosis with recombinant SAG1 protein in a guinea pig model. Infect. Immun. 68, 4948–4953.

Hay, J., Hutchison, W.M., Lee, W.R., Siim, J.C., 1981. Cataract in mice congenitally infected with *Toxoplasma gondii*. Ann. Trop. Med. Parasitol. 75, 455–457.

Hay, J., Kerrigan, P., 1982. Photography of cataract in mice congenitally infected with *Toxoplasma gondii*. Ann. Trop. Med. Parasitol. 76, 363–365.

Hay, J., Lee, W.R., Dutton, G.N., Hutchison, W.M., Siim, J.C., 1984. Congenital toxoplasmic retinochoroiditis in a mouse model. Ann. Trop. Med. Parasitol. 78, 109–116.

Hay, J., Dutton, G.N., 1996. Fundal with dots: observations from a murine model of congenital ocular toxoplasmosis. Br. J. Ophthalmol. 80, 189.

Hellbrügge, T.F., Dahme, E., 1953. Experimental toxoplasmosis. Klin. Wochenschr. 31, 789–791.

Hellbrügge, T.F., Dahme, E., Hellbrügge, F.K., 1953. Experimental animal observations on transplacental infection with *Toxoplasma*. Z. Tropenmed. Parasitol. 4, 312–322.

Hellbrügge, T.F., 1955. Foetal infection during the acute and chronic phase of latent toxoplasmosis in rats. Arch. Gynakol. 186, 384–388.

Hofflin, J.M., Remington, J.S., 1987a. Clindamycin in a murine model of toxoplasmic encephalitis. Antimicrob. Agents Chemother. 31, 492–496.

Hofflin, J.M., Remington, J.S., 1987b. *In vivo* synergism of roxithromycin (RU 965) and interferon against *Toxoplasma gondii*. Antimicrob. Agents Chemother. 31, 346–348.

Hofflin, J.M., Conley, F.K., Remington, J.S., 1987. Murine model of intracerebral toxoplasmosis. J. Infect. Dis. 155, 550–557.

Hogan, M.J., Lewis, A., Zweigart, P.A., 1956. Persistence of *Toxoplasma gondii* in ocular tissues. I. Am. J. Ophthalmol. 42, 84–89.

Holland, G.N., 1999. Reconsidering the pathogenesis of ocular toxoplasmosis. Am. J. Ophthalmol. 128, 502–505.

Holland, G.N., 2003. Ocular toxoplasmosis: a global reassessment. Part I: epidemiology and course of disease. Am. J. Ophthalmol. 136, 973–988.

Holland, G.N., Engstrom Jr., R.E., Glasgow, B.J., Berger, B.B., Daniels, S.A., Sidikaro, Y., Harmon, J.A., Fischer, D.H., Boyer, D.S., Rao, N.A., et al., 1988a. Ocular toxoplasmosis in patients with the acquired immunodeficiency syndrome. Am. J. Ophthalmol. 106, 653–667.

Holland, G.N., O'Connor, G.R., Diaz, R.F., Minasi, P., Wara, W.M., 1988b. Ocular toxoplasmosis in immunosuppressed nonhuman primates. Invest. Ophthalmol. Vis. Sci. 29, 835–842.

Holtschke, T., Lohler, J., Kanno, Y., Fehr, T., Giese, N., Rosenbauer, F., Lou, J., Knobeloch, K.P., Gabriele, L., Waring, J.F., Bachmann, M.F., Zinkernagel, R. M., Morse 3rd, H.C., Ozato, K., Horak, I., 1996. Immunodeficiency and chronic myelogenous leukemia-like syndrome in mice with a targeted mutation of the ICSBP gene. Cell 87, 307–317.

Howe, D.K., Sibley, L.D., 1995. *Toxoplasma gondii* comprises three clonal lineages: correlation of parasite genotype with human disease. J. Infect. Dis. 172, 1561–1566.

Howe, D.K., Summers, B.C., Sibley, L.D., 1996. Acute virulence in mice is associated with markers on chromosome VIII in *Toxoplasma gondii*. Infect. Immun. 64, 5193–5198.

Hu, M.S., Schwartzman, J.D., Lepage, A.C., Khan, I.A., Kasper, L.H., 1999a. Experimental ocular toxoplasmosis induced in naive and preinfected mice by intracameral inoculation. Ocul. Immunol. Inflamm. 7, 17–26.

Hu, M.S., Schwartzman, J.D., Yeaman, G.R., Collins, J., Seguin, R., Khan, I.A., Kasper, L.H., 1999b. Fas-FasL interaction involved in pathogenesis of ocular toxoplasmosis in mice. Infect. Immun. 67, 928–935.

Hu, M.S., Schwartzman, J.D., Kasper, L.H., 2001. Apoptosis within mouse eye induced by *Toxoplasma gondii*. Chin. Med. J. (Engl.) 114, 640–644.

Huldt, G., 1960. Experimental toxoplasmosis transplacental transmission in guinea pigs. Acta Pathol. Microbiol. Scand. 49, 176–188.

Hunn, J.P., Feng, C.G., Sher, A., Howard, J.C., 2011. The immunity-related GTPases in mammals: a fast-evolving cell-autonomous resistance system against intracellular pathogens. Mamm. Genome 22, 43–54.

Hutchison, W.M., Hay, J., Lee, W.R., Siim, J.C., 1982. A study of cataract in murine congenital toxoplasmosis. Ann. Trop. Med. Parasitol. 76, 53–70.

Ismael, A.B., Dimier-Poisson, I., Lebrun, M., Dubremetz, J.F., Bout, D., Mevelec, M.N., 2006. Mic1-3 knockout of *Toxoplasma gondii* is a successful vaccine against chronic and congenital toxoplasmosis in mice. J. Infect. Dis. 194, 1176–1183.

Israelski, D., Remington, J., 1990. Activity of gamma interferon in combination with pyrimethamine or clindamycin in treatment of murine toxoplasmosis. Eur. J. Clin. Microbiol. Infect. Dis. 9, 358–360.

Jacobs, L., Melton, M.L., Kaufman, H.E., 1964. Treatment of experimental ocular toxoplasmosis. Arch. Ophthalmol. 71, 111–118.

Johnson, A.M., 1984. Strain dependent, route of challenge-dependent susceptibility to toxoplasmosis. Z. Parasitenk. 70, 303–309.

Johnson, J., Suzuki, Y., Mack, D., Mui, E., Estes, R., David, C., Skamene, E., Forman, J., McLeod, R., 2002. Genetic analysis of influences on survival following *Toxoplasma gondii* infection. Int. J. Parasitol. 32, 179–185.

Johnson, L.L., 1992. SCID mouse models of acute and relapsing chronic *Toxoplasma gondii* infections. Infect. Immun. 60, 3719–3724.

Jost, C., Reiter-Owona, I., Liesenfeld, O., 2007. The timing of sulfadiazine therapy impacts the reactivation of latent *Toxoplasma* infection in IRF-8$^{-/-}$ mice. Parasitol. Res. 101, 1603–1609.

Jungersen, G., Bille-Hansen, V., Jensen, L., Lind, P., 2001. Transplacental transmission of *Toxoplasma gondii* in minipigs infected with strains of different virulence. J. Parasitol. 87, 108–113.

Kaufman, H.E., Remington, J.S., Jacobs, L., 1958. Toxoplasmosis: the nature of virulence. Am. J. Ophthalmol. 46, 255–261.

Kaufman, H.E., 1960. The effect of corticosteroids on experimental ocular toxoplasmosis. Am. J. Ophthalmol. 50, 919–926.

Kempf, M.C., Cesbron-Delauw, M.F., Deslee, D., Gross, U., Herrmann, T., Sutton, P., 1999. Different manifestations of *Toxoplasma gondii* infection in F344 and LEW rats. Med. Microbiol. Immunol. (Berl.) 187, 137–142.

Khan, A.A., Slifer, T., Araujo, F.G., Remington, J.S., 1996. Trovafloxacin is active against *Toxoplasma gondii*. Antimicrob. Agents Chemother. 40, 1855–1859.

Khan, A.A., Nasr, M., Araujo, F.G., 1998. Two 2-hydroxy-3-alkyl-1,4-naphthoquinones with *in vitro* and *in vivo* activities against *Toxoplasma gondii*. Antimicrob. Agents Chemother. 42, 2284–2289.

Khan, I.A., Green, W.R., Kasper, L.H., Green, K.A., Schwartzman, J.D., 1999. Immune CD8(+) T cells prevent reactivation of *Toxoplasma gondii* infection in the immunocompromised host. Infect. Immun. 67, 5869–5876.

Khan, A.A., Araujo, F.G., Craft, J.C., Remington, J.S., 2000. Ketolide ABT-773 is active against *Toxoplasma gondii*. J. Antimicrob. Chemother. 46, 489–492.

Kobayashi, M., Aosai, F., Hata, H., Mun, H.S., Tagawa, Y., Iwakura, Y., Yano, A., 1999. *Toxoplasma gondii*: difference of invasion into tissue of digestive organs between susceptible and resistant strain and influence of IFN-gamma in mice inoculated with the cysts perorally. J. Parasitol. 85, 973–975.

Lacroix, C., Levacher-Clergeot, M., Chau, F., Sumuyen, M.H., Sinet, M., Pocidalo, J.J., Derouin, F., 1994. Interactions between murine AIDS (MAIDS) and toxoplasmosis in co-infected mice. Clin. Exp. Immunol. 98, 190–195.

Lahmar, I., Guinard, M., Sauer, A., Marcellin, L., Abdelrahman, T., Roux, M., Mousli, M., Moussa, A., Babba, H., Pfaff, A.W., Candolfi, E., 2010. Murine neonatal infection provides an efficient model for congenital ocular toxoplasmosis. Exp. Parasitol. 124, 190–196.

Lee, W.R., Hay, J., Hutchison, W.M., Dutton, G.N., Siim, J.C., 1983. A murine model of congenital toxoplasmic retinochoroiditis. Acta. Ophthalmol. (Copenh.) 61, 818–830.

Lescano, S.A., Amato Neto, V., Chieffi, P.P., Bezerra, R.C., Gakiya, E., Ferreira, C.S., Braz, L.M., 2004. [Evaluation of the efficacy of azithromycin and pyrimethamine, for treatment of experimental infection of mice with *Toxoplasma gondii* cystogenic strain]. Rev. Soc. Bras. Med. Trop. 37, 460–462.

Letscher-Bru, V., Pfaff, A.W., Abou-Bacar, A., Filisetti, D., Antoni, E., Villard, O., Klein, J.P., Candolfi, E., 2003. Vaccination with *Toxoplasma gondii* SAG-1 protein is protective against congenital toxoplasmosis in BALB/c mice but not in CBA/J mice. Infect. Immun. 71, 6615–6619.

Ling, Y., Sahota, G., Odeh, S., Chan, J.M., Araujo, F.G., Moreno, S.N., Oldfield, E., 2005. Bisphosphonate inhibitors of *Toxoplasma gondi* growth: in vitro, QSAR, and in vivo investigations. J. Med. Chem. 48, 3130–3140.

Loke, Y.W., 1982. Transmission of parasites across the placenta. Adv. Parasitol. 21, 155–228.

Lopes, C.D., Silva, M.N., Ferro, E.A., Sousa, R.A., Firminot, M.L., Bernardes, E.S., Roque-Barreira, M.C., Pena, J.D., 2009. Azithromycin reduces ocular infection during congenital transmission of toxoplasmosis in the *Calomys callosus* model. J. Parasitol. 95, 1005–1010.

Lu, F., Huang, S., Kasper, L.H., 2004. CD4+ T cells in the pathogenesis of murine ocular toxoplasmosis. Infect. Immun. 72, 4966–4972.

Lu, F., Huang, S., Hu, M.S., Kasper, L.H., 2005. Experimental ocular toxoplasmosis in genetically susceptible and resistant mice. Infect. Immun. 73, 5160–5165.

Lu, F., Huang, S., Kasper, L.H., 2009. The temperature-sensitive mutants of *Toxoplasma gondii* and ocular toxoplasmosis. Vaccine 27, 573–580.

Lyons, R.E., Anthony, J.P., Ferguson, D.J., Byrne, N., Alexander, J., Roberts, F., Roberts, C.W., 2001. Immunological studies of chronic ocular toxoplasmosis: up-regulation of major histocompatibility complex class I and transforming growth factor beta and a protective role for interleukin-6. Infect. Immun. 69, 2589–2595.

Martinez, A., Allegra, C.J., Kovacs, J.A., 1996. Efficacy of epiroprim (Ro11-8958), a new dihydrofolate reductase inhibitor, in the treatment of acute *Toxoplasma* infection in mice. Am. J. Trop. Med. Hyg. 54, 249–252.

Martins-Duarte, E.S., Lemgruber, L., de Souza, W., Vommaro, R.C., 2010. *Toxoplasma gondii*: fluconazole and itraconalzole activity against toxoplasmosis in a murine model. Exp. Parasitol. 124, 466–469.

McLeod, R., Estes, R.G., Mack, D.G., Cohen, H., 1984. Immune response of mice to ingested *Toxoplasma gondii*: a model of *Toxoplasma* infection acquired by ingestion. J. Infect. Dis. 149, 234–244.

McLeod, R., Frenkel, J.K., Estes, R.G., Mack, D.G., Eisenhauer, P.B., Gibori, G., 1988. Subcutaneous and intestinal vaccination with tachyzoites of *Toxoplasma gondii* and acquisition of immunity to peroral and congenital *Toxoplasma* challenge. J. Immunol. 140, 1632–1637.

McLeod, R., Eisenhauer, P., Mack, D., Brown, C., Filice, G., Spitalny, G., 1989. Immune responses associated with early survival after peroral infection with *Toxoplasma gondii*. J. Immunol. 142, 3247–3255.

Mevelec, M.N., Bout, D., Desolme, B., Marchand, H., Magne, R., Bruneel, O., Buzoni-Gatel, D., 2005. Evaluation of protective effect of DNA vaccination with genes encoding antigens GRA4 and SAG1 associated with GM-CSF plasmid, against acute, chronical and congenital toxoplasmosis in mice. Vaccine 23, 4489–4499.

Miedouge, M., Bessieres, M.H., Cassaing, S., Swierczynski, B., Seguela, J.P., 1997. Parasitemia and parasitic loads in acute infection and after anti-gamma-interferon treatment in a toxoplasmic mouse model. Parasitol. Res. 83, 339–344.

Mitchell, S.M., Zajac, A.M., Kennedy, T., Davis, W., Dubey, J.P., Lindsay, D.S., 2006. Prevention of recrudescent toxoplasmic encephalitis using ponazuril in an immunodeficient mouse model. J. Eucaryot. Microbiol. 53, S164–S165.

Montoya, J.G., Kovacs, J.A., Remington, J.S., 2004. Toxoplasma gondii. In: Mandell, G.L., Bennett, J.E., Dolin, R. (Eds.), Principles and Practice of Infectious Diseases, vol. 2. Churchill Livingston, Philadelphia – London, pp. 3170–3198.

Moshkani, S.K., Dalimi, A., 2000. Evaluation of the efficacy of atovaquone alone or in combination with azithromycin against acute murine toxoplasmosis. Vet. Res. Commun. 24, 169–177.

Mui, E.J., Jacobus, D., Milhous, W.K., Schiehser, G., Hsu, H., Roberts, C.W., Kirisits, M.J., McLeod, R., 2005. Triazine inhibits *Toxoplasma gondii* tachyzoites *in vitro* and *in vivo*. Antimicrob. Agents Chemother. 49, 3463–3467.

Newman, P.E., Ghosheh, R., Tabbara, K.F., O'Connor, G.R., Stern, W., 1982. The role of hypersensitivity reactions to *Toxoplasma* antigens in experimental ocular toxoplasmosis in nonhuman primates. Am. J. Ophthalmol. 94, 159–164.

Nguyen, B.T., Stadtsbaeder, S., 1985. Comparative effects of cotrimoxazole (trimethoprim-sulphamethoxazole) and spiramycin in pregnant mice infected with *Toxoplasma gondii* (Beverley strain). Br. J. Pharmacol. 85, 713–716.

Nicoll, S., Wright, S., Maley, S.W., Burns, S., Buxton, D., 1997. A mouse model of recrudescence of *Toxoplasma gondii* infection. J. Med. Microbiol. 46, 263–266.

Norose, K., Mun, H.S., Aosai, F., Chen, M., Piao, L.X., Kobayashi, M., Iwakura, Y., Yano, A., 2003. IFN-gamma-regulated *Toxoplasma gondii* distribution and load in the murine eye. Invest. Ophthalmol. Vis. Sci. 44, 4375–4381.

Norose, K., Aosai, F., Mizota, A., Yamamoto, S., Mun, H.S., Yano, A., 2005. Deterioration of visual function as examined by electroretinograms in *Toxoplasma gondii*-infected IFN-gamma-knockout mice. Invest. Ophthalmol. Vis. Sci. 46, 317–321.

Norose, K., Aosai, F., Mun, H.S., Yano, A., 2006. Effects of sulfamethoxazole on murine ocular toxoplasmosis in interferon-gamma knockout mice. Invest. Ophthalmol. Vis. Sci. 47, 265–271.

Nozik, R.A., O'Connor, G.R., 1968. Experimental toxoplasmic retinochoroiditis. Arch. Ophthalmol. 79, 485–489.

O'Connor, G.R., 1984. Experimental models of toxoplasmosis. In: Tabbara, K.F., Cello, R.M. (Eds.), Animal Models of Ocular Disease. William Thomas Publishing, Springfield, MA, pp. 97–104.

Olle, P., Bessieres, M.H., Cassaing, S., Esteve, T., Cazabonne, P., Seguela, J.P., 1994. Experimental murine toxoplasmic retinochoroiditis. J. Eukaryot. Microbiol. 41, 16S.

Olle, P., Bessieres, M.H., Malecaze, F., Seguela, J.P., 1996. The evolution of ocular toxoplasmosis in anti-interferon gamma treated mice. Curr. Eye Res. 15, 701–707.

Olliaro, P., Gorini, G., Jabes, D., Regazzetti, A., Rossi, R., Marchetti, A., Tinelli, C., Della Bruna, C., 1994. *In vitro* and *in vivo* activity of rifabutin against *Toxoplasma gondii*. J. Antimicrob. Chemother. 34, 649–657.

Pavesio, C.E., Chiappino, M.L., Gormley, P., Setzer, P.Y., Nichols, B.A., 1995. Acquired retinochoroiditis in hamsters inoculated with ME 49 strain *Toxoplasma*. Invest. Ophthalmol. Vis. Sci. 36, 2166–2175.

Perea, E.J., Daza, R.M., 1976. The effect of minocycline, doxycycline and oxytetracycline on experimental mouse toxoplasmosis. Bull. Soc. Pathol. Exot. Filiales 69, 367–372.

Pereira Mde, F., Silva, D.A., Ferro, E.A., Mineo, J.R., 1999. Acquired and congenital ocular toxoplasmosis experimentally induced in *Calomys callosus* (Rodentia, Cricetidae). Mem. Inst. Oswaldo Cruz. 94, 103–114.

Petersen, E., Kijlstra, A., Stanford, M., 2012. Epidemiology of ocular toxoplasmosis. Ocul. Immunol. Inflamm. 20, 68–75.

Pezerico, S.B., Langoni, H., Da Silva, A.V., Da Silva, R.C., 2009. Evaluation of *Toxoplasma gondii* placental transmission in BALB/c mice model. Exp. Parasitol. 123, 168–172.

Pomeroy, C., Kline, S., Jordan, M.C., Filice, G.A., 1989. Reactivation of *Toxoplasma gondii* by cytomegalovirus disease in mice: antimicrobial activities of macrophages. J. Infect. Dis. 160, 305–311.

Que, X., Wunderlich, A., Joiner, K.A., Reed, S.L., 2004. Toxopain-1 is critical for infection in a novel chicken embryo model of congenital toxoplasmosis. Infect. Immun. 72, 2915–2921.

Ramsey, E.M., Harris, J.W.S., 1966. Comparison of uteroplacental vasculature and circulation in the rhesus monkey and man. Contributions to Embryology/Carnegie Institution of Washington 38, 59–70.

Reichard, U., Gross, U., 2007. *Toxoplasma* animal models and therapeutics. In: Weiss, L., Kim, K. (Eds.), Toxoplasma gondii. The model apicomplexan – perspectives and methods. Elsevier Academic Press, San Diego CA, London UK, pp. 153–184.

Remington, J.S., Jacobs, L., Kaufman, H.E., 1958. Studies on chronic toxoplasmosis; the relation of infective dose to residual infection and to the possibility of congenital transmission. Am. J. Ophthalmol. 46, 261–267. discussion 268.

Remington, J.S., Jacobs, L., Melton, M.L., 1961. Congenital transmission of toxoplasmosis from mother animals with acute and chronic infections. J. Infect. Dis. 108, 163–173.

Roberts, C.W., Alexander, J., 1992. Studies on a murine model of congenital toxoplasmosis: vertical disease transmission only occurs in BALB/c mice infected for the first time during pregnancy. Parasitology 104 (Pt 1), 19–23.

Roberts, C.W., Brewer, J.M., Alexander, J., 1994. Congenital toxoplasmosis in the Balb/c mouse: prevention of vertical disease transmission and foetal death by vaccination. Vaccine 12, 1389–1394.

Rodriguez-Diaz, J.C., Martinez-Grueiro, M.M., Martinez-Fernandez, A.R., 1993. Comparative activity of several antibiotics against *Toxoplasma gondii* in a mouse model. Enferm. Infecc. Microbiol. Clin. 11, 543–546.

Rollins, D.F., Tabbara, K.F., Ghosheh, R., Nozik, R.A., 1982. Minocycline in experimental ocular toxoplasmosis in the rabbit. Am. J. Ophthalmol. 93, 361–365.

Romand, S., Pudney, M., Derouin, F., 1993. *In vitro* and *in vivo* activities of the hydroxynaphthoquinone atovaquone alone or combined with pyrimethamine, sulfadiazine, clarithromycin, or minocycline against *Toxoplasma gondii*. Antimicrob. Agents Chemother. 37, 2371–2378.

Romand, S., Bryskier, A., Moutot, M., Derouin, F., 1995. *In vitro* and *in vivo* activities of roxithromycin in combination with pyrimethamine or sulphadiazine against *Toxoplasma gondii*. J. Antimicrob. Chemother. 35, 821–832.

Romand, S., Della Bruna, C., Farinotti, R., Derouin, F., 1996. *In vitro* and *in vivo* effects of rifabutin alone or combined with atovaquone against *Toxoplasma gondii*. Antimicrob. Agents Chemother. 40, 2015–2020.

Rosowski, E.E., Lu, D., Julien, L., Rodda, L., Gaiser, R.A., Jensen, K.D.C., Saeij, J. P.J., 2011. Strain-specific activation of the NF-kB pathway by GRA15, a novel *Toxoplasma gondii* dense granule protein. J. Exp. Med. 208, 195–212.

Sabin, A.B., 1941. Toxoplasmic encephalitis in children. JAMA 116, 801–807.

Saeij, J.P.J., Boyle, J.P., Grigg, M., Arrizabalaga, G., Boothroyd, J.C., 2005. Bioluminescence imaging of *Toxoplasma gondii* infection in living mice reveals dramatic differences between strains. Infect. Immun. 73, 695–702.

Saeij, J.P.J., Boyle, J.P., Coller, S., Taylor, S., Sibley, L.D., Brooke-Powell, E.T., Ajioka, J.W., Boothroyd, J.C., 2006. Polymorphic secreted kinases are key virulence factors in toxoplasmosis. Science 314, 1780–1783.

Samuel, B.U., Hearn, B., Mack, D., Wender, P., Rothbard, J., Kirisits, M.J., Mui, E., Wernimont, S., Roberts, C.W., Muench, S.P., Rice, D.W., Prigge, S.T., Law, A.B., McLeod, R., 2003. Delivery of antimicrobials into parasites. Proc. Natl. Acad. Sci. U. S. A. 100, 14281–14286.

Sarciron, M.E., Lawton, P., Saccharin, C., Petavy, A.F., Peyron, F., 1997. Effects of 2′,3′-dideoxyinosine on *Toxoplasma gondii* cysts in mice. Antimicrob. Agents Chemother. 41, 1531–1536.

Scharton-Kersten, T., Contursi, C., Masumi, A., Sher, A., Ozato, K., 1997. Interferon consensus sequence binding protein-deficient mice display impaired resistance to intracellular infection due to a primary defect in interleukin 12 p40 induction. J. Exp. Med. 186, 1523–1534.

Schlüter, D., Kwok, L.Y., Lütjen, S., Soltek, S., Hoffmann, S., Körner, H., Deckert, M., 2003. Both lymphotoxin-alpha and TNF are crucial for control of *Toxoplasma gondii* in the central nervous system. J. Immunol. 170, 6172–6182.

Schöler, N., Krause, K., Kayser, O., Muller, R.H., Borner, K., Hahn, H., Liesenfeld, O., 2001. Atovaquone nanosuspensions show excellent therapeutic effect in a new murine model of reactivated toxoplasmosis. Antimicrob. Agents Chemother. 45, 1771–1779.

Schoondermark-van de Ven, E., Melchers, W., Galama, J., Camps, W., Eskes, T., Meuwissen, J., 1993. Congenital toxoplasmosis: an experimental study in rhesus monkeys for transmission and prenatal diagnosis. Exp. Parasitol. 77, 200–211.

Schoondermark-van de Ven, E., Galama, J., Camps, W., Vree, T., Russel, F., Meuwissen, J., Melchers, W., 1994a. Pharmacokinetics of spiramycin in the rhesus monkey: transplacental passage and distribution in tissue in the foetus. Antimicrob. Agents Chemother. 38, 1922–1929.

Schoondermark-van de Ven, E., Melchers, W., Camps, W., Eskes, T., Meuwissen, J., Galama, J., 1994b. Effectiveness of spiramycin for treatment of congenital *Toxoplasma gondii* infection in rhesus monkeys. Antimicrob. Agents Chemother. 38, 1930–1936.

Schoondermark-van de Ven, E., Galama, J., Vree, T., Camps, W., Baars, I., Eskes, T., Meuwissen, J., Melchers, W., 1995. Study of treatment of congenital *Toxoplasma gondii* infection in rhesus monkeys with pyrimethamine and sulfadiazine. Antimicrob. Agents Chemother. 39, 137–144.

Schultz, W., Bauer, H., 1952. Experimentelle Toxoplasmose. Klin. Wochenschr. 30, 850–851.

Sergent, V., Cautain, B., Khalife, J., Deslee, D., Bastien, P., Dao, A., Dubremetz, J. F., Fournie, G.J., Saoudi, A., Cesbron-Delauw, M.F., 2005. Innate refractoriness of the Lewis rat to toxoplasmosis is a dominant trait that is intrinsic to bone marrow-derived cells. Infect. Immun. 73, 6990–6997.

Shen, D.F., Matteson, D.M., Tuaillon, N., Suedekum, B.K., Buggage, R.R., Chan, C.C., 2001. Involvement of apoptosis and interferon-gamma in murine toxoplasmosis. Invest. Ophthalmol. Vis. Sci. 42, 2031–2036.

Shubar, H.M., Mayer, J.P., Hopfenmüller, W., Liesenfeld, O., 2008. A new combined flow-cytometry-based assay reveals excellent acivity against *Toxoplasma gondii* and low toxicity of new bisphosphonates in vitro and in vivo. J. Antimicrob. Chemother. 61, 1110–1119.

Shubar, H.M., Dunay, I.R., Lachenmaier, S., Dathe, M., Bushrab, F.N., Mauludin, R., Müller, R.H., Fitzner, R., Borner, K., Liesenfeld, O., 2009. The role of apolipoprotein E in uptake of atovaquone into the brain in murine acute and reactivated toxoplasmosis. J. Drug Target 17, 257–267.

Shubar, H.M., Lachenmaier, S., Heimesaat, M.M., Lohmann, U., Maudulin, R., Müller, R.H., Fitzner, R., Borner, K., Liesenfeld, O., 2011. SDS-coated atovaquone nanosuspensions show improved therapeutic efficacy against experimental acquired and reactivated toxoplasmosis by improving passage of gastrointestinal and blood-brain barriers. J. Drug Target 19, 114–124.

Sibley, L.D., Boothroyd, J.C., 1992. Virulent strains of *Toxoplasma gondii* comprise a single clonal lineage. Nature 359, 82–85.

Sibley, L.D., Ajioka, J.W., 2008. Population structure of *Toxoplasma gondii*: clonal expansion driven by infrequent recombination and selective sweeps. Ann. Rev. Microbiol. 62, 329–351.

Sordet, F., Aumjaud, Y., Fessi, H., Derouin, F., 1998. Assessment of the activity of atovaquone-loaded nanocapsules in the treatment of acute and chronic murine toxoplasmosis. Parasite 5, 223–229.

Stahl, W., Matsubayashi, H., Akao, S., 1966. Modification of subclinical toxoplasmosis in mice by cortisone, 6-mercaptopurine and splenectomy. Am. J. Trop. Med. Hyg. 15, 869–874.

Sumyuen, M.H., Garin, Y.J., Derouin, F., 1996. Effect of immunosuppressive drug regimens on acute and chronic murine toxoplasmosis. Parasitol. Res. 82, 681–686.

Suzuki, Y., Joh, K., Orellana, M.A., Conley, F.K., Remington, J.S., 1991. A gene(s) within the H-2D region determines the development of toxoplasmic encephalitis in mice. Immunology 74, 732–739.

Suzuki, Y., Joh, K., Kwon, O.C., Yang, Q., Conley, F.K., Remington, J.S., 1994. MHC class I gene(s) in the D/L region but not the TNF-alpha gene determines development of toxoplasmic encephalitis in mice. J. Immunol. 153, 4649–4654.

Suzuki, Y., Yang, Q., Remington, J.S., 1995. Genetic resistance against acute toxoplasmosis depends on the strain of *Toxoplasma gondii*. J. Parasitol. 81, 1032–1034.

Suzuki, Y., Kang, H., Parmley, S., Lim, S., Park, D., 2000. Induction of tumor necrosis factor-alpha and inducible nitric oxide synthase fails to prevent toxoplasmic encephalitis in the absence of interferon-gamma in genetically resistant BALB/c mice. Microbes. Infect. 2, 455–462.

Tabbara, K.F., Nozik, R.A., O'Connor, G.R., 1974. Clindamycin effects on experimental ocular toxoplasmosis in the rabbit. Arch. Ophthalmol. 92, 244–247.

Taylor, S., Barragan, A., Su, C., Fux, B., Fentress, S.J., Tang, K., Beatty, W.L., El Hajj, H., Jerome, M., Behnke, M.S., White, M., Wootton, J.C., Sibley, L.D., 2006. A secreted serine-threonine kinase determines virulence in the eukaryotic pathogen *Toxoplasma gondii*. Science 314, 1776–1780.

Tedesco, R.C., Smith, R.L., Corte-Real, S., Calabrese, K.S., 2005. Ocular toxoplasmosis in mice: comparison of two routes of infection. Parasitology 131, 303–307.

Tenter, A.M., Heckeroth, A.R., Weiss, L.M., 2000. *Toxoplasma gondii*: from animals to humans. Int. J. Parasitol. 30, 1217–1258.

Thouvenin, M., Candolfi, E., Villard, O., Klein, J.P., Kien, T., 1997. Immune response in a murine model of congenital toxoplasmosis: increased susceptibility of pregnant mice and transplacental passage of *Toxoplasma gondii* are type 2-dependent. Parasitologia 39, 279–283.

Uhlikova, M., Hubner, J., 1973. Congenital transmission of toxoplasmosis in domestic rabbits. Folia Parasitol. (Praha) 20, 285–291.

Usmanova, A.F., 1965. Apropos of the effect of chloridine and sulfadimezin on the course of pregnancy. (Experimental study). Sov. Zdravookhr. Kirg. 2, 7–10.

Wang, T., Liu, M., Gao, X.J., Zhao, Z.J., Chen, X.G., Lun, Z.R., 2011. *Toxoplasma gondii*: the effects of infection at different stages of pregnancy on the offspring of mice. Exp. Parasitol. 127, 107–112.

Watanabe, H., Suzuki, Y., Makino, M., Fujiwara, M., 1993. *Toxoplasma gondii*: induction of toxoplasmic encephalitis in mice with chronic infection by inoculation of a murine leukemia virus inducing immunodeficiency. Exp. Parasitol. 76, 39–45.

Webb, R.M., Tabbara, K.F., O'Connor, G.R., 1984. Retinal vasculitis in ocular toxoplasmosis in nonhuman primates. Retina 4, 182–188.

Weiss, L.M., Udem, S.A., Salgo, M., Tanowitz, H.B., Wittner, M., 1991. Sensitive and specific detection of *Toxoplasma* DNA in an experimental murine model: use of *Toxoplasma gondii*-specific cDNA and the polymerase chain reaction. J. Infect. Dis. 163, 180–186.

Weiss, L.M., Luft, B.J., Tanowitz, H.B., Wittner, M., 1992. Pyrimethamine concentrations in serum during treatment of acute murine experimental toxoplasmosis. Am. J. Trop. Med. Hyg. 46, 288–291.

Werner, H., Janitschke, K., Masihi, M., Adusu, E., 1977. The effect of *Toxoplasma* antibodies after reinfection with *T. gondii*. III. Communication: investigations on the question of placental transmission of *Toxoplasma* in immunised pregnant animals. Zentralbl. Bakteriol. [Orig A] 238, 128–142.

Williams, D.M., Grumet, F.C., Remington, J.S., 1978. Genetic control of murine resistance to *Toxoplasma gondii*. Infect. Immun. 19, 416–420.

Williams, R.H., Morley, E.K., Hughes, J.M., Duncanson, P., Terry, R.S., Smith, J.E., Hide, G., 2005. High levels of congenital transmission of *Toxoplasma gondii* in longitudinal and cross-sectional studies on sheep farms provides evidence of vertical transmission in ovine hosts. Parasitology 130, 301–307.

Wong, M.M., Kozek, W.J., Karr Jr., S.L., Brayton, M.A., Theis, J.H., Hendrickx, A.G., 1979. Experimental congenital infection of *Toxoplasma gondii* in *Macaca arctoides*. Asian J. Infect. Dis. 3, 61–67.

Wright, I., 1972. Transmission of *Toxoplasma gondii* across the guinea-pig placenta. Lab. Anim. 6, 169–180.

Yardley, V., Khan, A.A., Martin, M.B., Slifer, T.R., Araujo, F.G., Moreno, S.N.J., Docampo, R., Croft, S.L., Oldfield, E., 2002. *In vivo* activities of farnesyl pyrophosphate synthase inhibitors against *Leishmania donovani* and *Toxoplasma gondii*. Antimicrob. Agents Chemother. 46, 929–931.

Youssef, M.Y., el-Ridi, A.M., Arafa, M.S., el-Sawy, M.F., el-Sayed, W.M., 1985. Effect of levamisole on toxoplasmosis during pregnancy in guinea-pigs. J. Egypt. Soc. Parasitol. 15, 41–48.

Zenner, L., Darcy, F., Cesbron-Delauw, M.F., Capron, A., 1993. Rat model of congenital toxoplasmosis: rate of transmission of three *Toxoplasma gondii* strains to foetuses and protective effect of a chronic infection. Infect. Immun. 61, 360–363.

Zenner, L., Darcy, F., Capron, A., Cesbron-Delauw, M.F., 1998. *Toxoplasma gondii*: kinetics of the dissemination in the host tissues during acute phase of infection of mice and rats. Exp. Parasitol. 90, 86–94.

Zenner, L., Estaquier, J., Darcy, F., Maes, P., Capron, A., Cesbron-Delauw, M.F., 1999a. Protective immunity in the rat model of congenital toxoplasmosis and the potential of excreted-secreted antigens as vaccine components. Parasite Immunol. 21, 261–272.

Zenner, L., Foulet, A., Caudrelier, Y., Darcy, F., Gosselin, B., Capron, A., Cesbron-Delauw, M.F., 1999b. Infection with *Toxoplasma gondii* RH and Prugniaud strains in mice, rats and nude rats: kinetics of infection in blood and tissues related to pathology in acute and chronic infection. Pathol. Res. Pract. 195, 475–485.

CHAPTER

8

Biochemistry and Metabolism of *Toxoplasma gondii*: Carbohydrates, Lipids and Nucleotides

Isabelle Coppens, Takashi Asai†, Stanislas Tomavo***

*Department of Molecular Microbiology and Immunology, Johns Hopkins University Bloomberg School of Public Health, Baltimore, Maryland, USA †Department of Tropical Medicine and Parasitology—Infectious Diseases, Keio University School of Medicine, Tokyo, Japan **Centre for Infection and Immunity of Lille, Institut Pasteur de Lille, Université Lille Nord de France, France

OUTLINE

Toxoplasma gondii, Second Edition
http://dx.doi.org/10.1016/B978-0-12-396481-6.00008-8

8.1 INTRODUCTION

Toxoplasma gondii and all other parasites of the phylum Apicomplexa replicate exclusively within eukaryotic cells indicating that these parasites may depend on metabolic activities of their hosts and they have evolved metabolic pathways reflecting their intracellular life style. These parasites display specific or alternate metabolic networks as documented by the recent discovery of unique metabolic pathways in the vestigial plastid (apicoplast) found in many Apicomplexa. Evidence for distinctive parasite metabolic pathways has direct implications for understanding parasite's requirements for intracellular growth. Further, these pathways may provide unique targets for compounds designed to inhibit and eradicate infection caused by these important human and animal pathogens.

While targeting unique parasite pathways is an attractive strategy, in practice, procurement of enough pure parasites for biochemical study or purification of parasite components such as enzymes is extremely difficult. Biochemical and metabolic studies of *T. gondii* have also been difficult to execute because of continuous problems with contamination of parasite preparations with host cell components. Accordingly, only a few studies of biochemistry and metabolism of *T. gondii* have been reported. Recent advances in gene engineering technology and molecular biology, including the genome database projects, have eliminated many of the barriers to our understanding of the biochemistry of these parasites. For instance, one can now easily obtain a predicted amino acid sequence of an enzyme or a transporter of *T. gondii* from the genome database (http://www.ToxoDB.org/). These data can be used to clone or synthesize the gene and to produce a recombinant enzyme to characterize the enzymatic activity. Bioinformatic schemes for several metabolic and salvage pathways of *T. gondii* have been compiled and are available at http://www.ToxoDB.org.

This chapter discusses the carbohydrate, lipid and nucleotide metabolism during the life cycle of *T. gondii*. It will also focus on new insights into distinct metabolic pathways, their evolutionary roots and their contribution to *T. gondii* survival during intracellular development and differentiation. In addition to this chapter, metabolic networks in place in the apicoplast and mitochondrion are discussed in Chapter 9 as well as the role of the apicoplast in fatty acid synthesis and salvage. Chapter 20 encompasses a comprehensive discussion of amino acid and nucleotide pathways with comparison to other Apicomplexa.

8.2 CARBOHYDRATE METABOLISM

8.2.1 Developmentally Regulated Expression of Amylopectin in *T. gondii*

One evolutionary hallmark of the close relationship between the photosynthetic microorganisms and apicomplexan parasites is defined by the existence of a storage polysaccharide named amylopectin in the cytoplasm of some Apicomplexa. Another reflection of this relationship is the presence of a vestigial plastid 'apicoplast', discussed in Chapter 9. In contrast to plants, which contain starch defined as a branched amylopectin and amylose in the chloroplasts, the encysted bradyzoite and sporozoite forms of Apicomplexa such as *T. gondii* (Coppin *et al.*, 2005), *Eimeria* (Karkhanis *et al.*, 1993) and *Cryptosporidium* (Harris *et al.*, 2004) accumulate amylopectin, a polymer of linear glucose that is not present in *Plasmodia* and *Babesia* species. It has been speculated that the disappearance of amylopectin in *Eimeria* sporozoites resulted in the inability of the parasite to establish successful infection in mice (Augustine, 1980; Nakai and Ogimoto, 1983). In *T. gondii*, the bradyzoites accumulate abundant amylopectin granules and numerous micronemes. In contrast, the tachyzoites lack amylopectin, and fewer micronemes are present (Fig. 8.1).

Amylopectin is probably consumed when the encysted and dormant bradyzoites switch into the rapidly replicating tachyzoites. It is postulated that amylopectin provides an energy source as its degradation to glucose can provide metabolic intermediates or substrates for glycolysis or mitochondrial oxidative phosphorylation.

8.2.2 *T. gondii* Displays a Uniquely Simple Pathway for Amylopectin Synthesis

Using bioinformatic searches, several gene candidates encoding enzymes that are probably involved in amylopectin biosynthesis were identified (Coppin *et al.*, 2005). These putative enzymes can be grouped in two classes: (i) enzymes that are involved in amylopectin synthesis such as amylopectin synthase, branching enzymes, UDP–glucose pyrophosphorylase, isoamylase, indirect debranching enzyme, α1-4-glucanotransferase and glycogenin and (ii) enzymes for amylopectin degradation like α-amylase, dikinase or R1 protein, phosphorylase and α-glucosidase. Based on the presence of these enzymes, metabolic

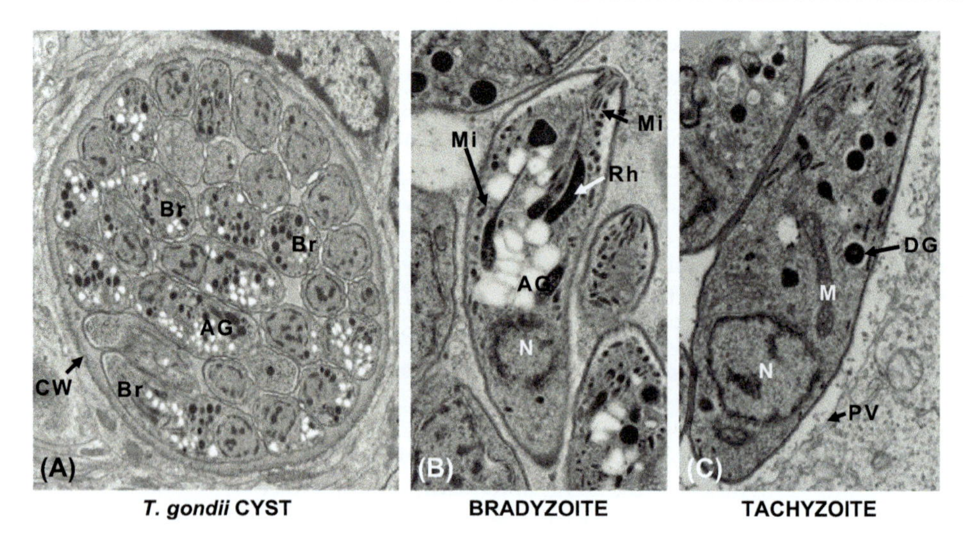

FIGURE 8.1 A) Transmission electron micrographs of bradyzoites (Br) within a tissue cyst. Note the presence of the cyst wall (CW) and numerous amylopectin granules (AG) in the cytoplasm of the bradyzoites.
B) and C) show a higher magnification of ultrastructural morphology of bradyzoite and tachyzoite which lacks amylopectin granules. CW, cyst wall; DG, dense granules; M, mitochondrion; Mi, micronemes; N, nucleus; PV, parasitophorous vacuole; Rh, rhoptry.

pathways and enzymes involved in amylopectin synthesis in *T. gondii* are probably similar to that of starch synthesis in the unicellular green algae *Chlamydomonas reinhardtii* (Fig. 8.2).

Surprisingly, all of these genes are present in the *Toxoplasma* genome as a unique copy, suggesting that redundant genes are not required for the synthesis of a genuine crystalline amylopectin in this protozoan parasite (Coppin *et al.*, 2005; Ball and Morell, 2003). This is in violation of the current dogma that suggests that redundancy of genes is required to build a crystalline starch in plants. Even in the simplest unicellular picophytoplanktonic green algae, *Ostreococcus tauri*, there is a multiplicity of genes and redundancy of isoenzymes involved in starch synthesis (Ral *et al.*, 2004).

Only UDP−glucose pyrophosphorylase and UDP−glucose utilizing amylopectin synthase are found in *T. gondii*. Comparative genomic analyses involving the unicellular red alga *Cyanidioschyzon merolae* (Matsuzaki *et al.*, 2004), the unicellular green alga *Chlamydomonas reinhardtii*, the yeast *Saccharomyces cerevisae* and the bacteria *Escherichia coli* revealed that both *C. merolae* and *T. gondii* contain a UDP−glucose utilizing glycogen (starch) synthase-like sequences and glycogenins. These enzymes are specific for the eukaryote UDP−glucose based pathway. In addition, UDP−glucose utilizing glycogen synthase activity has been detected in the crude extract from *T. gondii* while only ADP−glucose dependent activity is present in *Chlamydomonas* lysates (Coppin *et al.*, 2005). *T. gondii* also contains an indirect debranching enzyme, a bifunctional enzyme that carries both α-1,4-glucano-transferase and amylo-1,6-glucosidase activities in fungi and animals (Fig. 8.2).

However, the characteristic most typifying the amylopectin biosynthetic pathway in *T. gondii* is the presence of genes that are of plant origin. Among the genes that distinguish plant starch metabolism from those of the animal, fungal and bacterial glycogen pathways are the isoamylase and R1 (glucan water dikinase activity)-like sequences in *T. gondii*. This suggests that both plant and animal-like amylopectin biosynthetic pathways are required for the synthesis of crystalline amylopectin in the parasite (Fig. 8.2).

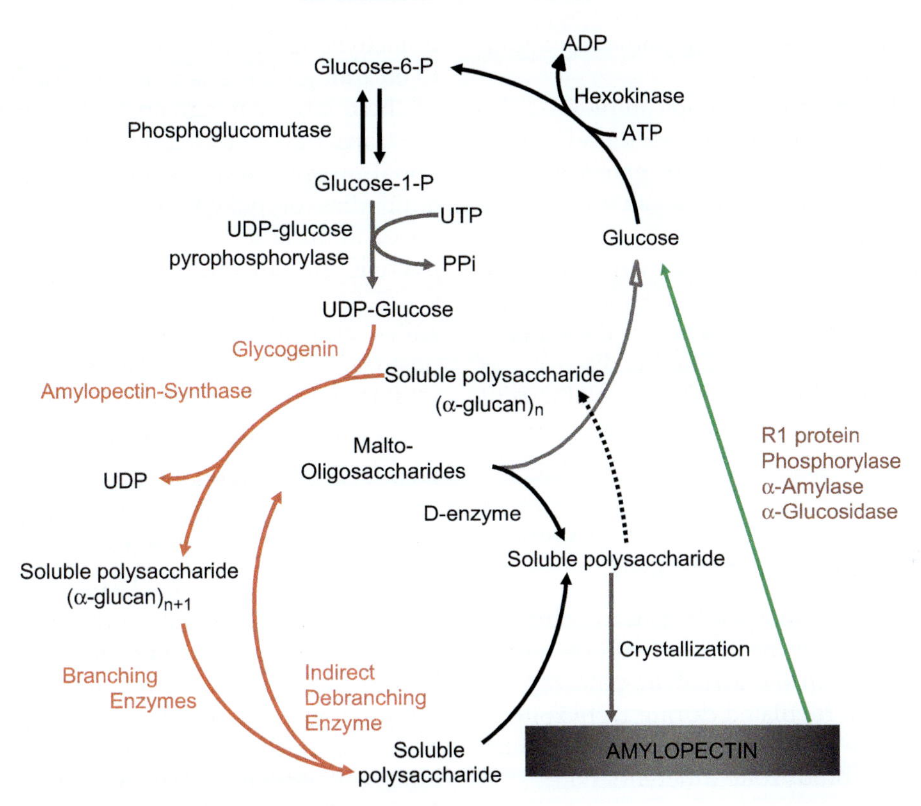

FIGURE 8.2 **Proposed metabolic pathways involved in the biosynthesis of semi-crystalline polysaccharide storage, amylopectin of *T. gondii*.** The biosynthetic pathway and enzymes involved therein (glycogenin, UDP–glucose pyrophosphorylase, amylopectin synthase, branching enzymes, indirect debranching enzymes) are shown in black and red arrows, respectively. The enzymes (α-amylase, R1 protein, phosphorylase and α-glucosidase) involved in the degradation pathway (green arrow) are indicated in brown. The putative genes encoding these enzymes have been identified in the genome sequence of *T. gondii* (http://www.ToxoDB.org/). The presence of active amylopectin synthase using UDP–glucose as a substrate has been assayed in tachyzoite and bradyzoite crude extracts.

8.2.3 Evolutionary Origins of Enzymes of Amylopectin Biosynthesis

Both plant and animal-like metabolism are probably involved in amylopectin biosynthesis in *T. gondii*. This is likely a signature of the evolutionary origin of apicomplexan parasites. These parasites contain a vestigial plastid 'apicoplast' that is derived from a secondary endosymbiosis with the engulfment of a unicellular algae (McFadden *et al.*, 1996; Köhler *et al.*, 1997; Cai, *et al.*, 2003; Waller *et al.*, 2003). Phylogenetic analyses performed with two key enzymes (amylopectin synthase and the R1 protein) demonstrate that the *T. gondii* amylopectin-synthetic pathway has evolved from the red algal starch synthesizing machinery through a secondary endosymbiotic event (Coppin *et al.*, 2005). These phylogenetic data, together with the presence of the enzymes and the enzymatic activities described above, establish that apicomplexans and red algae such as *C. merolae*, use a UDP–glucose pathway to build water-insoluble amylopectin. This is the pathway used by all floridean starch accumulating organisms (Nyvall *et al.*, 1999).

It is equally apparent that apicomplexans also contain plant-like genes that are not found in yeast and mammals. These consist of genes that are required in plants and green algae and encode enzymes (water dikinase or R1 protein, α-1,4-glucanotransferase or D-enzyme and isoamylase) that are involved in breakdown and synthesis of starch. The presence of these plant-like enzymes could be useful for the discovery of inhibitors that can interfere with the synthesis or degradation of amylopectin in apicomplexan parasites.

8.2.4 Stage-Specific Expression of Genes Involved in Glucose Catabolism

It is noteworthy that some genes coding for enzymes involved in the biosynthesis of amylopectin and others in the glycolytic pathway are developmentally regulated during tachyzoite to bradyzoite stage conversion. See Chapter 15 for a discussion on bradyzoite differentiation.

8.2.4.1 Stage-Specific Expression of Genes Involved in Amylopectin Degradation

The expression pattern of the genes involved in amylopectin biosynthesis in tachyzoites and bradyzoites isolated from mouse brain cysts has been investigated by RT−PCR. Transcripts coding for enzymes known to be involved in the catabolic functions, such as the R1 protein, α-glucan phosphorylase, α-glucosidase and α-amylase, are preferentially expressed in bradyzoites (Coppin *et al.*, 2005). In constrast, transcripts coding for enzymes known to be involved in amylopectin synthesis (glycogenin, amylopectin synthase, branching enzyme) are preferentially expressed in tachyzoites but can also be detected at lower expression levels in bradyzoites.

This pattern is consistent with the production of amylopectin during differentiation of tachyzoites into bradyzoites and with the mobilization of the glucose stores during bradyzoite-to-tachyzoite interconversion (Tomavo, 2001). Even though there is evidence for transcription of these genes, it remains to be determined if the transcripts detected are translated into functional proteins and enzymes. Therefore, specific antibodies or enzymatic activities need to be tested in order to demonstrate protein synthesis. Even if the enzymatic activity of amylopectin synthase is been demonstrated in both tachyzoites and bradyzoites (Coppin *et al.*, 2005), post-transcriptional regulation may also occur on some of the stage-specific transcripts detected.

8.2.4.2 Stage-Specific Expression of Genes Involved in Glycolysis

GLUCOSE 6-PHOSPHATE ISOMERASES AND LACTATE DEHYDROGENASES

The glycolytic enzyme lactate dehydrogenase (LDH, EC 1.1.1.27) is a glycolytic enzyme that catalyses the interconversion of pyruvate to lactate using NAD^+ as a co-enzyme (Fig. 8.3). Two stage-specific LDH genes have been identified, the tachyzoite-specific *LDH1* and the bradyzoite-specific *LDH2* (Yang and Parmley, 1995, 1997). The transcript of LDH2 was only detected in the bradyzoite stage while mRNA of LDH1 was found in both bradyzoite and tachyzoite stages. The absence of LDH2 mRNA in the tachyzoite suggests that the transcription of LDH2 is suppressed during transition from the bradyzoite to the tachyzoite stage. Conversely, the data indicate that LDH1 is the only isoenzyme produced by the tachyzoites. The level of LDH2 mRNA increased markedly *in vitro* during bradyzoite induction, suggesting that transcription activation and/or mRNA stability may explain the stage-specific expression of LDH2 gene in *T. gondii* (Yang and Parmley, 1997). As the predicted isoelectric point of the two LDHs is different, two-dimensional electrophoresis has been used to demonstrate that only one LDH protein is expressed in each developmental parasitic stage.

It has also been demonstrated that LDH1 and LDH2 share a unique structural feature with

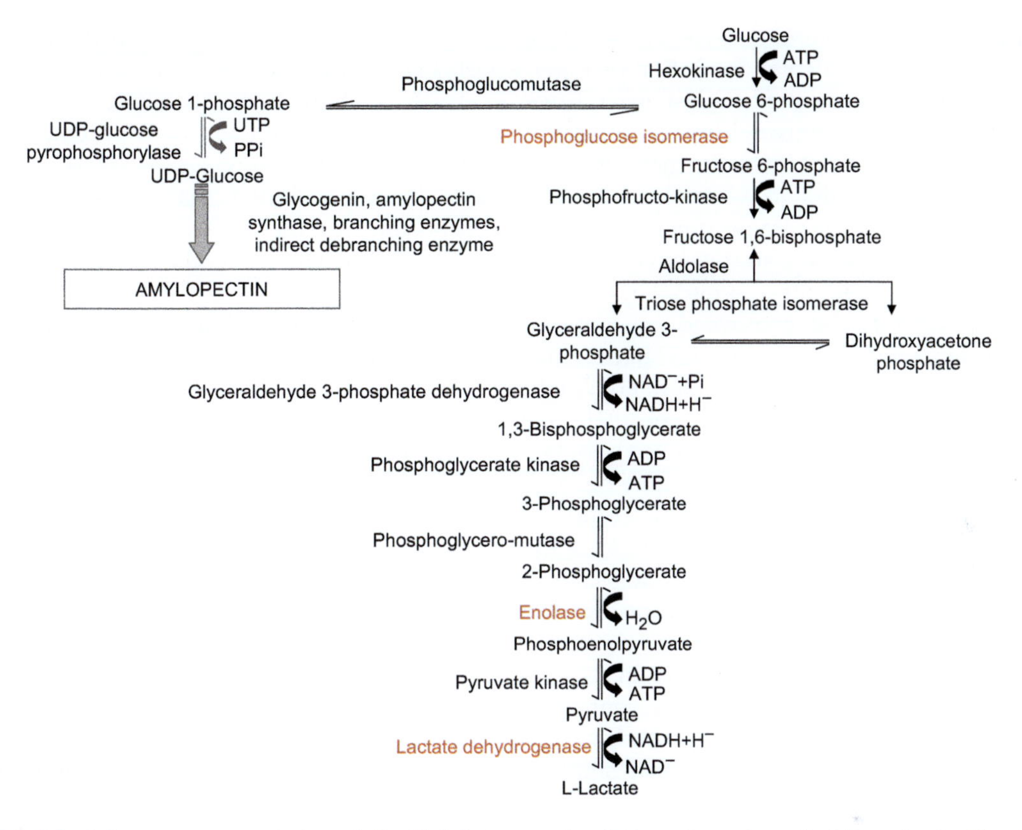

FIGURE 8.3 **Schematic representation of the link between glycolysis and amylopectin biosynthesis.** The classical pathway of glycolysis is shown in the right panel. Phosphoglucose isomerase, enolase and lactate dehydrogenase (shown in red) are found as two iso-enzymes, each enzyme is stage-specifically expressed in the tachyzoites or in the bradyzoites (see more details in the text).

LDH from the malarial parasite *Plasmodium falciparum* (pLDH), namely a five-amino acid insertion into the substrate specificity loop. This insertion has only been observed in pLDH, LDH1 and LDH2 (Bzik *et al.*, 1993; Yang and Parmley, 1997). All other LDH enzymes described so far do not contain this insertion. The insertion in LDH2 is identical to the insertion in pLDH (KSDKE) but differs slightly from the insertion in LDH1 (KPDSE).

Comparative studies on the kinetic properties of *T. gondii* LDH1 and LDH2 and *P. falciparum* LDH showed that LDH1 and LDH2 exhibit broader substrate specificity than pLDH. For both LDH1 and LDH2, 3-phenylpyruvate is an excellent substrate, even better than pyruvate when LDH2 was tested with both substrates. In contrast, pLDH does not utilize 3-phenylpyruvate (Dando *et al.*, 2001). In addition, both LDH1 and LDH2 can utilize the NAD analogue 3-acetylpyridine adenine dinucleotide (APAD) efficiently, similarly to pLDH. A range of inhibitors including gossypol and derivatives inhibit LDH1, LDH2 and pLDH, but in general LDH2 is more sensitive than LDH1. LDH1 also shows substrate inhibition despite the substitution in both LDH1 and LDH2 of a methionine for serine163, a residue that is thought to be critical for production of substrate inhibition (Dando *et al.*, 2001). Most importantly, gossypol and gossylic iminolactone have been shown to display inhibition of *T. gondii* tachyzoite growth in fibroblast in cultures. The differences in sensitivities to inhibitors between LDH1 and LDH2 further illustrate how these enzymes may have evolved to serve separate roles during stage development. It should be noted that these studies have been performed with recombinant enzymes. These observations remain to be confirmed with native purified enzymes since other co-factors may also be involved in the modulation of enzymatic properties. Nevertheless, one can speculate that the apparent greater sensitivity of recombinant LDH2 to gossypol and derivatives may be leads for the design of inhibitors that could be exploited as chemotherapeutic agents to eliminate cysts from chronically infected hosts.

In the design of specific inhibitors, crystal structures of LDH may be required. Kavanagh *et al.* (2004) have crystallized LDH1 in its apo-form and in its ternary complexes containing NAD+ or the NAD+-analogue 3-acetylpyridine adenine dinucleotide (APAD(+)) and sulphate or the inhibitor oxalate. Superimposition of LDH1 with human muscle- and heart-specific LDH isoforms reveals differences in residues that line the active site. This increases the hydrophobicity of LDH1. It was concluded that these differences would aid in designing inhibitors specific for LDH1 that may be useful in treating toxoplasmic encephalitis and other complications that arise in immune-compromised patients.

To investigate the significance of *T. gondii* lactate dehydrogenases, LDH1 and LDH2 in the control of a metabolic flux during parasite differentiation, the expression of these two isoenzymes was knocked down in a stage-specific manner (Al-Anouti *et al.*, 2004). These LDH knockdown parasites exhibited variable growth rates in either the tachyzoite or the bradyzoite stage when compared with the wild type parasites. Their differentiation processes were impaired *in vitro* and they were unable to form tissue cysts in a murine model system. In addition, all mice infected with the knockdown of LDH1 and LDH2 expression gave rise to virulence-attenuated parasites, and survived a subsequent challenge with parental parasites at a dose that usually causes 100% mortality. It has been concluded that LDH expression is important for the cell cycle and differentiation of *T. gondii*. However, the precise mechanisms by which LDH knockdown impairs parasite growth and differentiation remain to be elucidated. Another glycolytic enzyme that is stage-specific is glucose 6-phosphate isomerase (G6-PI, EC 5.3.1.9) which catalyses the interconversion of glucose 6-phosphate to fructose 6-phosphate (Fig. 8.3). A cDNA fragment encoding G6-PI was isolated from a bradyzoite-specific subtractive library and the full-length cDNA was used to complement an *E. coli* mutant lacking G6−PI (Yahiaoui *et al.*, 1999; Dzierszinski *et al.*, 1999). RT−PCR data have demonstrated that the transcript coding G6−PI is preferentially present in bradyzoites while a minute amount can be detected in tachyzoites. Immunoblot analysis performed with specific polyclonal antibodies revealed G6−PI only in encysted bradyzoites, demonstrating the stage-specific expression of G6−PI in *T. gondii*. It remains to be determined, however, if the other putative G6−PI coding genes presently described in the *Toxoplasma* genome correspond to the tachyzoite-specific G6−PI. Here, only the *T. gondii* cDNA coding G6−PI has been tested in *E. coli* complementation; its enzymatic activity with the purified enzyme has not been directly assayed.

ENOLASES

Within the glycolytic pathway, enolase or ENO (2-phospho-D-glycerate hydrolase, EC 4.2.1.11) catalyses the conversion of 2-phosphoglycerate to phosphoenol pyruvate (Fig. 8.3). As for LDH, two stage-specific enolase-coding genes have been described (Dzierszinski *et al.*, 1999). The two genes are located on the same chromosome and separated only by an intergenic sequence of 1.6 kilobases. Both transcript and protein corresponding to the product of the gene named ENO1 are only detected in bradyzoite while ENO2's transcript and protein are found in tachyzoites. The amino acid identity between ENO1 and ENO2 was found to be 73%.

Interestingly, when compared with human and other mammalian enolases, both enolases contain a pentapeptide insertion: EWGYC in ENO2 and the almost identical EWGWS motif in ENO1 (Dzierszinski *et al.*, 1999) and enolase of *Plasmodium falciparum* (Read *et al.*, 1994), respectively. In addition, another dipeptide EK/DK insertion was also found in ENO1 and ENO2 of *T. gondii* and in *P. falciparum* enolase. Superimposition of the model tridimensional structure of ENO1 or ENO2 with that of human enolase revealed a perfect match between their 3D models except for the presence of two extra-loops corresponding to the pentapeptide EWGWC and the dipeptide EK insertion, respectively. The presence of these two loops was also evident in *P. falciparum* and, surprisingly, in plant enolases (Dzierszinski *et al.*, 1999). The functions of these two loops were investigated by site-directed mutagenesis of the pentapeptide, the dipeptide and both loops in the ENO1 recombinant enzymes (Dzierszinski *et al.*, 2001). The enzymatic properties of these mutated enzymes, and that of the wild type enzyme demonstrated that the deletion of a single EK loop does not affect the Km of the enzyme but the deletion of both loops causes a 13-fold increase of the enzyme Km. Deletion of the pentapeptide EWGWC gave a 5-fold increase of the Km compared to the values of the wild type enzyme.

The Km, Vmax, and temperature stability of pure recombinant ENO1 and ENO2 enzymes have been compared. While the Km values are identical, ENO1 and ENO2 display distinct Vmax with a value 3-fold higher for ENO2 than that of ENO1, suggesting that the two isoenzymes have the same affinity for the substrate 2-PGA but exhibit different rates of substrate consumption. The denaturation temperature of ENO1 was also found to be higher than that of ENO2, indicating that the tachyzoite ENO2 is more thermolabile than the bradyzoite ENO1.

The enzymatic properties of the two stage-specific enolases seem to be in good agreement with the metabolic and physiological adaptation required during *T. gondii* differentiation and encystation. It can be postulated that these enzymes play discrete biological functions that most probably involve profound carbohydrate metabolism modifications such as the biosynthesis or degradation of amylopectin that occurs during the stage conversion of *T. gondii*.

Extensive studies performed on the stage-specific expression of enolases of *T. gondii* using the polyclonal antibodies specific to ENO1 and ENO2 revealed that both enolase isoenzymes can be detected in the nucleus of the parasite (Ferguson *et al.*, 2002). The accumulation of enolase signal in the nucleus is observed in both tachyzoites and bradyzoites but only in dividing zoites. The nuclear signal of ENO1 in the brain of 12 day-infected mice is detected early in tachyzoites that are differentiating into bradyzoites. The expression of ENO1 in these intermediate zoites appears earlier than that of the classical bradyzoite surface marker, P36 or SAG4, and these ENO1 expressing intermediate zoites are still expressing the tachyzoite SAG1 surface marker. In addition, it has been found that only the tachyzoite specific ENO2 is expressed in the dividing sexual forms of *T. gondii* examined in infected cells derived from the cat. Taken together, these data suggest that this novel subcellular localization can be ascribed to novel nuclear activity displayed by ENO1 and ENO2.

It should be noticed that the nuclear localization of enolase was first described in human cells where the enzyme binds to the c-Myc promoter and acts as transcriptional repressor in cancer cells (Feo *et al.*, 2000). The binding to DNA target and the domain in enolase that is involved in the transcriptional regulation has been identified

(Subramanian *et al.*, 2000). Interestingly, the factor isolated from cold-resistant mutants of *Arabidopsis thaliana* using genetic approaches was identified as *A. thaliana* enolase which binds to a DNA motif similar to that described in human cells (Lee *et al.*, 2002). Thus, one could postulate that *T. gondii* enolases might display similar transcriptional or other nuclear functions involved in the intracellular growth of the parasite. However, further experimental support needs to be provided for the precise nuclear activity of enolases in *T. gondii*.

8.2.5 Parasite Glycolytic Enzymes Involved in Other Biological Functions

In eukaryotic cells, many kinds of multifunctional regulatory proteins have been identified that perform distinct biochemical functions in the nucleus, the cytoplasm, or both. Recent studies establish that metabolic enzymes display biological roles distinct from their cognate functions. Perhaps the best-studied examples are enzymes that double as eye lens proteins essential for normal vision: lactate dehydrogenase (crystalline in ducks and crocodiles), α-enolase (crystalline in lamprey and turtles) and argininosuccinate lyase (crystalline in birds and reptiles) (Piatigorsky, 2003).

Another example of a multifunctional protein is glyceraldehyde 3-phosphate dehydrogenase (GAPDH), which interacts with a wide variety of RNAs, including ribosomal RNA, tRNA, hammerhead ribozymes and the 3'untranslated region of many mRNAs. Proposed functions for GAPDH include regulation of helicase activity, tRNA and mRNA export, RNA unfolding, translational regulation, regulation of viral gene expression and regulation of mRNA stability (Sirover, 2005). In *P. falciparum*, it has been shown that the N-terminal domain of GAPDH mediates GTPase Rab2-dependent recruitment to membranes, suggesting that GAPDH exerts non-glycolytic function(s) in the parasite, possibly including a role in vesicular transport and biogenesis of apical organelles (Daubenberger *et al.*, 2003).

In *T. gondii*, several studies have described several glycolytic enzymes with functions distinct from their primary functions. Using pull-down assays, Jewett and Sibley (2003) reported that the parasite aldolase is the molecular link between micronemal thrombospondin-anonymous related proteins (TRAP family) and the actomyosin motor involved in apicomplexan parasite gliding and host cell invasion. Parafusin related protein 1 (PRP1) is an orthologue of phosphoglucomutase that is associated with micronemes (Matthiesen *et al.*, 2001, 2003).

8.3 N-GLYCOSYLATION IN *TOXOPLASMA GONDII*

T. gondii recognizes, binds, and penetrates virtually any kind of mammalian cells using a repertoire of proteins released from late secretory organelles and a unique form of gliding motility (also named glideosome) that critically depends on actin filaments and myosin. How *T. gondii* glycosylated proteins mediate host–parasite interactions remains elusive. To date, only limited evidence is available concerning *N*-glycosylation in Apicomplexa. Three independent studies have demonstrated that *T. gondii* tachyzoites are capable of modifying proteins by *N*-glycans than previously thought (Luk *et al.*, 2008; Fauquenoy *et al.*, 2008; Luo *et al.*, 2011). Comprehensive proteomics and glycomics analyses show that several key components required for host cell–*T. gondii* interactions are *N*-glycosylated (Fauquenoy *et al.*, 2008; Luo *et al.*, 2011). Detailed structural characterization confirmed that *N*-glycans from *T. gondii* total protein extracts consist of oligomannosidic (Man(5-8)(GlcNAc)2) and paucimannosidic (Man(3-4)(GlcNAc)2) sugars, which are rarely present on mature eukaryotic glycoproteins (Fauquenoy *et al.*, 2008). *In situ* fluorescence using concanavalin A and *Pisum sativum* agglutinin predominantly stained the parasite. Visualization of *Toxoplasma* glycoproteins purified by affinity chromatography followed by detailed proteomics and glycan analyses identified components involved in gliding motility, moving junction, and other additional functions implicated in intracellular development. Importantly tunicamycin-treated parasites were considerably reduced in motility, host cell invasion and growth. Thus, these data suggest that *N*-glycosylation probably participates in modifying key proteins that are essential for host cell invasion by *T. gondii*. Data have also been obtained that suggest that some post translational glycosylation on *T. gondii* proteins may be due to host cell enzymes that are taken up by and then used by the parasite to modify its own proteins (Eller *et al.*, 2012).

A study using combined glycomic and proteomic approaches has confirmed that all three potential *N*-glycosylated sites of GAP50 are specifically occupied by unusual *N*-glycan structures that are rarely found on mature mammalian glycoproteins (Fauquenoy *et al.*, 2011). GAP50 is a key component of the multiprotein complex of *T. gondii* motility, which is essential for host cell entry, migration through host tissues and invasion. It is powered by a motor complex mainly composed of actin, myosin heavy chain A, myosin light chain 1, gliding associated proteins GAP45, and GAP50, the only integral membrane anchor so far described. Using site-directed mutagenesis, it has been demonstrated that *N*-glycosylation is a prerequisite for GAP50 transport from the endoplasmic reticulum to the Golgi apparatus and for its subsequent delivery into the inner complex membrane. Assembly of key partners into the gliding complex, and parasite motility are severely impaired in the unglycosylated GAP50 mutants (Fauquenoy *et al.*, 2011). Moreover, comparative affinity purification using *N*-glycosylated and unglycosylated GAP50 as bait identified three novel hypothetical proteins including the recently described gliding associated protein GAP40, and it has been shown that *N*-glycans are required for efficient binding to gliding partners. Taken together, these data provide the detailed analyses of *T. gondii N*-glycosylation functions that are vital for parasite motility and host cell entry.

8.4 GLYCOLIPID ANCHORS

8.4.1 Structure of *Toxoplasma* Glycosylphosphatidylinositol and its Role in Membrane Anchoring

Extensive studies have demonstrated that many proteins are attached to the eukaryote cell membranes via inositol-containing glycophospholipids (GPIs). In *T. gondii*, the major surface proteins are anchored to the parasite surface by a GPI-membrane anchor (Nagel *et al.*, 1989; Tomavo *et al.*, 1989). This type of anchor seems to be more frequently used in *Toxoplasma* and other protozoa than in higher eukaryotes.

When it became clear that numerous surface antigens of *T. gondii* are GPI-anchored, several studies established the structure and biosynthesis of these membrane anchors and, in particular, of their putative precursors. The structure of GPI anchors of these proteins has been determined by combining metabolic labelling (tritiated glucosamine, mannose, galactose, palmitic and myristic acids, and inositol) and their sensitivity to a lipase named phosphatidylinositol phospholipase C (PI–PLC) (Tomavo *et al.*, 1989). The release of these proteins from the surface of live parasites, which causes a cross-reacting determinant (CRD) of the soluble forms to be accessible to anti-CRD serum of trypanosomes, also confirms that the major surface proteins (SAGs) of *Toxoplasma* possess GPI anchors.

Further investigations led to the determination of the *Toxoplasma* GPI anchor structure that is composed of the evolutionarily conserved linear

GPI core, ethanolamine-PO_4-6Manα1-2Manα1-6Manα1-4GlcNAcα1-6-inositol on SAG1 (P30) and P23 (Fig. 8.3, Tomavo *et al.*, 1992a, 1993). Candidate glycolipid precursors that are probably transferred *en bloc* to the nascent membrane proteins of *T. gondii* have been identified and isolated either in living parasites or in parasite extracts.

Four mature glycolipids and other intermediate forms have been characterized (Tomavo *et al.*, 1992b). These four major glycolipids have the same GPI core structure and can serve as preassembled precursors of GPI anchors linked to these proteins. In addition, a lipophilic 'low molecular weight' antigen (4.6 kDa), identified using human patient sera and monoclonal antibodies (Sharma *et al.*, 1983; Tomavo *et al.*, 1994), was shown to be glycophosphoinositols that are not linked to proteins (free GPIs) and localize in the plasma membrane of the parasites. These free-GPIs or low molecular weight antigens were shown to elicit an early immunoglobulin M response in humans.

The detailed structures of these free GPIs have been determined using metabolic labelling, enzymatic digestion, followed by classical chromatographic analysis, nuclear magnetic resonance and fast-atom bombardment-mass spectroscopy (Striepen *et al.*, 1997). The following two GPI-structures were elucidated: the classical structure (ethanolamine-PO4)-Manα1-2Manα1-6(GalNacβ1-4)Manα1-4GlcNα-inositol-PO4-lipid and a novel structure (ethanolamine-PO4)-Manα1-2Manα1-6(Glcα1-4GalNacβ1-4)Manα1-4Glc Nα-inositol-PO4-lipid both with and without terminal ethanolamine phosphate. Only *T. gondii* GPIs bearing the unique glucose-N-acetylgalactosamine side branch are immunogenic in humans (Striepen *et al.*, 1997).

8.4.2 Role of GPIs in Cell Signalling and Host Immune Response

Except for their role in membrane insertion of surface proteins in *T. gondii,* the biological functions of GPIs are presently unknown. In other eukaryotic systems, GPIs can display functions involved in signal transduction. One possible function of the GPI anchor might be to allow a closer association of the proteins with themselves and other surface proteins in the membrane (Tomavo, 1996). Consistent with this idea, genetically engineered transmembrane-anchored SAG1 does not show the usual observed association of GPI-anchored SAG1 with itself and/or other proteins (Seeber *et al.*, 1998).

Toxoplasma-free GPIs elicit strong and early immunogenic responses during host infection. Data from other protozoa suggest that other functions of GPIs in host immune response are possible. In *P. falciparum,* the GPI moiety, free or associated with protein, induces tumour necrosis factor and interleukin-1 production by macrophages and regulates metabolism in adipocytes (Schofield and Hackett, 1993). Deacylation with specific phospholipases abolishes cytokine induction. When administered to mice *in vivo*, the malaria parasite GPI induces cytokine release, a transient pyrexia and hypoglycemia and profound and lethal cachexia in presence of sensitizing agents. The data suggest that the GPI of *Plasmodium* is a potent glycolipid toxin that may be responsible for a novel pathogenic process. It has been further demonstrated that *Plasmodium* GPI directly and specifically increases cell adhesion molecule expression in HUVECs and parasite cytoadherence (Schofield *et al.*, 1996). These parasite GPIs induce rapid activation of a tyrosine kinase in macrophages.

The minimal structure requirement for tyrosine kinase activation is the evolutionarily conserved core glycan sequence Manα1,2-Manα1,6Manα1-4GlcN1-6myo-inositol. The GPI alone appears sufficient to mimic the activities of malaria parasite extracts in the signalling pathway leading to TNF expression (Tachado *et al.*, 1997). Thus, GPIs of intraerythrocytic *P. falciparum* induce proinflammatory cytokine responses. It was also reported that adults who have resistance to clinical malaria contain high levels of pertinent anti-GPI antibodies, whereas susceptible children lack or have low levels of short-lived antibody response. Individuals who were not exposed to *P. falciparum* lack anti-GPI antibodies completely. Absence of a pertinent anti-GPI antibody response correlated with malaria-specific anemia and fever, suggesting that anti-GPI antibodies provide protection against clinical malaria (Naik *et al.*, 2000). These results could be evaluated in studies aimed at defining the activity of chemically defined structures for toxicity and results would have implications for the development of GPI-based therapies or vaccines.

The *P. falciparum* GPI glycan consisting of the sequence NH(2)-CH(2)-CH(2)-PO(4)-(Manα1-2) 6Manα1-2Manα1-6Manα1-4GlcNH(2) α1-6myo-inositol-1,2-cyclic-phosphate was chemically synthesized, conjugated to carriers, and used to immunize mice infected with *Plasmodium berghei*, a rodent model of severe malaria. The recipients were substantially protected against malarial acidosis, pulmonary oedema, cerebral syndrome and fatality (Schofield *et al.*, 2002). Altogether, the above data suggest that GPI is a significant proinflammatory endotoxin of parasitic origin and it may contribute to pathogenesis and fatalities in humans. In addition, GPI may also be used as a prototype carbohydrate anti-toxin vaccine against malaria. It remains to be seen whether GPI has a similar role in clinical toxoplasmosis.

8.5 LIPID METABOLISM

Lipid metabolic activities in *T. gondii* are essential for the production of infectious progeny and the persistence of the parasites in their mammalian hosts. Bioinformatic tools, e.g. Kyoto Encyclopedia of Genes and Genomes framework, have delineated the metabolic maps for lipid syntheses in *T. gondii*, and have concomitantly revealed auxotrophies for several lipid species (http://www.ToxodB.org). Much recent work has indeed documented that *T. gondii* acquires the necessary lipids through intricate and complex networks of synthesis and uptake. *T. gondii* has sophisticated lipid homeostatic pathways in that it has retained the genetic capacity to express redundant lipid biosynthetic pathways, and it has developed efficient mechanisms to scavenge several host lipids or lipidic precursors. Lipid inventory in *T. gondii* reveals the presence of neutral and polar lipids as found in any eukaryotic organisms. In addition, the parasites have uncommon lipid molecular species that can serve as signatures of these pathogens. This indicates that unique biosynthetic enzymes and lipid translocators/transporters are operational in these parasites and raises the prospect that lipid metabolic pathways in *Toxoplasma* may abound in valid drug targets. As a prime example, fatty acid biosynthetic pathways are being successfully exploited as antimicrobial targets in *Toxoplasma* (reviewed in Goodman and McFadden, 2007).

The parasite shows amazingly diverse features in lipid metabolic pathways, as some of them share close similarities to mammalian pathways whereas some are more evolutionarily related to bacteria and plant ones. From a cell biological viewpoint, no doubt exists that the lipid metabolism and great plasticity of *T. gondii* will likely reveal many more metabolic surprises in the future. In the context of toxoplasmosis chemotherapy, characterization of more lipid-based target molecules and knowledge about mechanisms promoting host lipid delivery to the PV will hold considerable potential. To this end, rationally designed lipid synthesis/uptake inhibitors would represent exciting prospects for the next generation of anti-*Toxoplasma* agents.

8.5.1 Fatty Acids

8.5.1.1 Fatty Acid Biosynthetic Pathways – Generalities

Fatty acid synthesis is a critical anabolic pathway in most organisms. In addition to being the major component of membranes, fatty acids are important energy storage molecules, and fatty acyl derivatives possess a variety of physiological functions, including post-translational modification of numerous proteins. The fundamental process of fatty acid biosynthesis is highly conserved among species. The key feature is the sequential extension of an alkanoic chain, two carbons at a time, by a series of decarboxylative condensation reactions. This process is generally initiated with the carboxylation of acetyl-CoA to yield malonyl-CoA (Smith *et al.*, 2003). The malonate group of malonyl-CoA is transferred to the phosphopantetheine prosthetic group of a small, acidic protein or protein domain, called the acyl carrier protein (ACP). Malonyl-ACP is then condensed with acetyl-CoA, reduced, dehydrated and reduced once again yielding an acyl-ACP. The elongation of the chain occurs by condensing another malonyl-ACP with the acyl-ACP and repeating the reaction cycle.

In nature, there are two basic types of fatty acid synthesis (FAS) architectures. The prototypical FASI is found in vertebrates and fungi. This pathway is an associated system since it consists of a single gene that produces a multifunctional protein, which contains all of the reaction centres required to produce a fatty acid molecule (Smith *et al.*, 2003). By contrast, plants, bacteria and lower eukaryotes such as yeast and some protozoa, contain two genes that are implicated in fatty acid production, and whose polypeptide products coalesce to form a multifunctional complex (White *et al.*, 2005). This dissociated system named FASII is characterized by the encoding of each component by a separate gene that produces a unique protein, which catalyses a single step in the pathway.

The Type I FAS is thought to have evolved by the fusion of a Type II complex into a single protein. The multifunctional protein of FASI is localized in the cytosol. In plants, FASII takes place in the plastid (chloroplast) that is derived from a cyanobacterial endosymbiont. The genes for these enzymes are all encoded in the nuclear genome, and the proteins are post-translationally targeted to the plastid as is common with plastid enzymes in plants and algae (McFadden, 1999). FASI is usually considered more efficient than FAII because the enzymatic activities are fused into a single polypeptide template and the intermediates do not diffuse from the complex. However, FASI produces only palmitate whereas FASII is capable of producing a large diversity of fatty acids with different chain lengths. Unsaturated fatty acids, iso- and anteiso-branched-chain fatty acids, and hydroxy fatty acids are generated by FASII. In addition, some FASII intermediates are used in the synthesis of key cellular constituents, such as lipoic acid and quorum-sensing molecules. This enormous diversity of products is possible because the ACP intermediates in the type II pathway are diffusible entities that can be diverted into other biosynthetic pathways.

Finally, significant amounts of the fatty acids can further be elongated into very long chain fatty acids by individual membrane-bound enzymes, named elongases located in the endoplasmic reticulum (ER). The synthesis of very long chain fatty acids is a ubiquitous system found in different organisms and cell types (reviewed in Jakobsson *et al.*, 2006; Uttaro *et al.*, 2006). These specific fatty acids serve commonly as building blocks of sphingolipids, but they are also important constituents of glycerophospholipids, triacylglycerols, steryl- and wax-esters.

8.5.1.2 Fatty Acid Synthesis in Toxoplasma

Toxoplasma has the great ability to infect various mammals and birds and to multiply in any type of cells in their hosts. In these different environments, the parasite must encounter different nutritional challenges. It is no surprise then if *T. gondii* has evolved to express a broad set of fatty acid-related genes for *de novo* synthesis (Mazumdar and Striepen, 2007). Bioinformatic, genetic, and biochemical studies document that *T. gondii* express three fatty acid synthetic pathways localized to different cellular compartments (Waller *et al.*, 1998; Seeber *et al.*, 2003; Ramakrishnan *et al.*, 2012; D. R. Roos, personal communication): a cytosolic FASI-like pathway producing palmitic acid; a FASII present in the apicoplast producing myristic and palmitic acids, in addition to lipoic acid; and an elaborate fatty acid elongation pathway compartmentalized into the ER and responsible for the production of very long chain monounsaturated fatty acids (Fig. 8.4).

Concerning the FASI pathway, gene structures reveal the existence of a single large polypeptide that harbours the ACP, FabD, FabH, FabG, FabZ and FabI activities. FASI seems to primarily provide bulk products (mainly palmitic acid) to apicoplast FASII, implicating that the FASI and FASII pathways act in concert to satisfy parasite fatty acid metabolism.

The apicoplast, a secondary chloroplast organelle, contains enzymes of the type II complex, ascribing an important role of this organelle for *de novo* fatty acid production. Acetyl-CoA is carboxylated to form malonyl-CoA by an acetyl coenzyme A carboxylase (ACC) using bicarbonate as a source of the carboxyl group, biotin as a cofactor, and ATP as a source of energy (Jelenska *et al.*, 2001). Indeed, ACC consists of three major functional domains: the biotin carboxylase domain, the carboxyltransferase domain, and the biotin carboxyl carrier domain containing covalently attached biotin. The first step of the

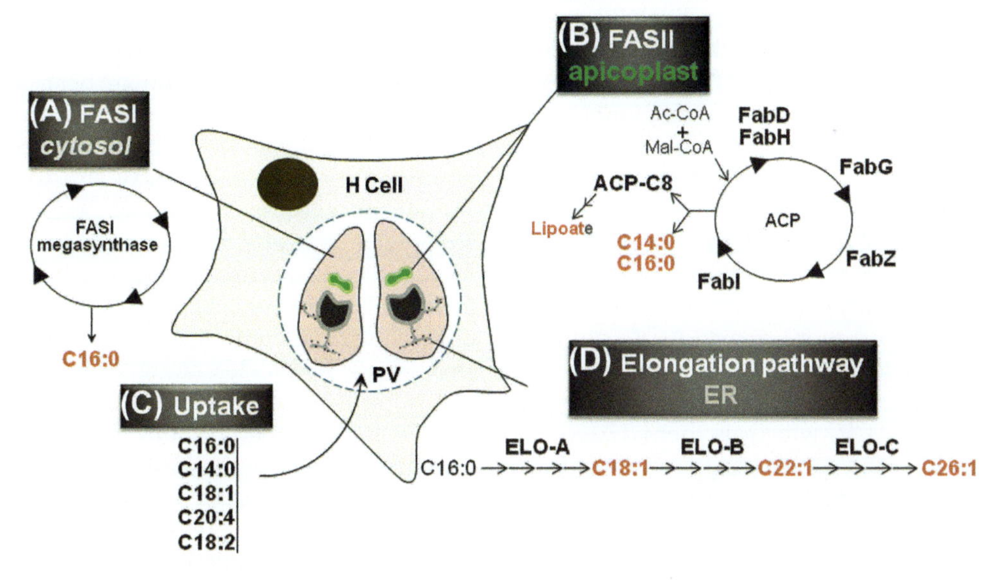

FIGURE 8.4 **Fatty acid salvage and biosynthetic pathways in *Toxoplasma*.** The parasite harbours three fatty acid synthetic pathways that are localized to different cellular compartments.
A) Cytosolic-located FASI.
B) Apicoplast-localized FASII producing myristic and palmitic acid in addition to lipoic acid.
C) ER-associated elongase system that synthesizes very long chain monounsaturated fatty acids using the activity of elongases (ELO). Major products are highlighted in red. The parasite is also able to scavenge several fatty acids from the environment. Ac, acetate; H Cell, host cell; Mal, malonate; PV, parasitophorous vacuole *(adapted from Ramakrishnan et al., 2012).*

ACC-catalysed reaction is an ATP-dependent transfer of the carboxyl group from bicarbonate to the biotin residue (first half-reaction). The carboxyl group is then transferred to acetyl-CoA producing malonyl-CoA (second half-reaction). Malonyl-CoA is used for *de novo* fatty acid biosynthesis as well as in fatty acid elongation. Clearly, incubation of radiolabelled malonyl-CoA with *T. gondii* extracts results in the production of palmitate.

ACC is encoded in the nucleus, synthesized in the cytosol before transport to the plastid through a 25- to 125-aa N-terminal transit peptide (Jelenska *et al.*, 2001). The ACC compartmentalization in the apicoplast raises the question of the source of acetyl-CoA used for FASII. Since acetyl-CoA cannot cross membranes, this molecule has to be produced in the apicoplast or has to be transported as free acetate or as another metabolite that could easily be converted into acetyl-CoA. Intriguingly, *T. gondii* expresses a second ACC, most probably cytosolic although having the multidomain type as the ACC prototype found in the cytoplasm of eukaryotes and in plastids of some plants.

Subsequent to the ACC activity, acetyl-CoA and malonyl-CoA are transferred to an acyl carrier protein (ACP) by the actions of acetyl-CoA-ACP transacylase and malonyl-CoA:ACP transacylase (FabD), respectively. β-Ketoacyl:-ACP is then synthesized from acetyl-ACP and malonyl-ACP by a β-ketoacyl:ACP synthase (FabH). β-Ketoacyl-ACP is reduced by β-ketoacyl:ACP reductase (FabG) to form β-hydroxyacyl-ACP which is dehydrated by β-hydroxyacyl-ACP dehydrase (FabZ) to form α,β-trans enoyl-ACP. This is further reduced to butyryl-ACP by the action of enoyl-ACP reductase (FabI). This cycle occurs up to seven times in *T. gondii*. ACP plays a central role in fatty acid biosynthesis by holding the forming acyl chain, whereas FabH and FabZ are involved in the condensation and dehydration steps, respectively, of acetyl addition during acyl chain elongation. Similarly to ACC, ACP and FabH, FabF, and FabI are localized to the apicoplast despite the presence of an unusual 'internal' signal peptide in FabH.

Loss of FASII severely compromises the growth and the virulence of *Toxoplasma* (Mazumdar *et al.*, 2006). In particular, genetic disruption of ACP leads to defects in apicoplast biogenesis and a consequent loss of the organelle. In addition to providing fatty acids, FASII is also required for the *de novo* synthesis of lipoic acid. Lipoic acid is an essential cofactor for oxidative decarboxylases and is involved in the response to oxidative stress. ACP-knockdown parasites are impaired in protein lipoylation of the apicoplast pyruvate dehydrogenase complex, an important source of the metabolic precursor acetyl-CoA. In summary, a major function of FASII is related to apicoplast maintenance and biogenesis, likely because of a requirement for the synthesis of lipids required for the growth and division of plastid membranes or organellar protein import.

Fundamentally different from the cytosolic Type I pathway of the human host, apicoplast FASII has tremendous potential for the development of parasite-specific inhibitors. Many components of this pathway are already the target for existing antibiotics and herbicides (reviewed in Roberts *et al.*, 2003; Sonda and Hehl, 2006; Goodman and McFadden, 2007; Martins-Duarte *et al.*, 2009). For example, *in vitro* and *in vivo* tests with selected aryloxyphenoxypropionate herbicides show that the carboxyltransferase domain of the apicoplast *T. gondii* ACC is the binding target for this class of inhibitors (Jelenska *et al.*, 2002). Expectedly, the cytosolic form of *T. gondii* ACC and human ACC are resistant to aryloxyphenoxypropionates. Triclosan is also a potent inhibitor of type II FabI (Baldock *et al.*, 1996). This compound restricts the growth of *T. gondii in vitro* (McLeod *et al.*, 2001). Triclosan blocks the incorporation of radioactive acetate into the fatty acids of *Toxoplasma* and specifically inhibits the FASII pathway. Thiolactomycin, a fungal secondary metabolite (Oishi *et al.*, 1982) selectively inhibits type II FabH of *T. gondii* (Martins-Duarte, 2009). Thiolactomycin decreases rapidly the growth of this parasite. Cerulenin, a metabolite of *Cephalosporium caelurens*, is an inhibitor of both types I and II FabH (Heath *et al.*, 2001). Cerulenin is found to act synergistically with triclosan in inhibiting FASII in the related malaria parasite. Thiolactomycin and cerulenin represent potential drugs that may also affect FAS pathways in *T. gondii*, and therefore growth.

If the apicoplast represents a significant source of fatty acids, its products can further be modified in the ER- of *Toxoplasma* (Ramakrishnan *et al.*, 2012). Of interest, the existence of membrane contact sites between the apicoplast outermost membrane and the ER as shown at the ultrastructural level suggests a non-vesicular trafficking of lipid between the two organelles (Tomova *et al.*, 2009). The synthesis of very long chain fatty acids is primarily dependent on a fatty acid elongation system comprising three elongases, two reductases and a dehydratase expressed by *T. gondii* (Ramakrishnan *et al.*, 2012). Metabolic labelling studies with radioactive glucose show that intracellular parasites synthesize a range of long and very long chain fatty acids (C14:0–26:1). The enzymatic steps involved in the FA elongation process are similar to those in FASI and FASII, but the growing chain is held by CoA instead of ACP. Genetic ablation of each individual FA elongase in *T. gondii* does not result in global growth defect despite observations of decreased amounts of very long fatty acids for most single elongase-deficient parasites.

8.5.1.3 *Fatty Acid Salvage by* Toxoplasma

In addition to the three fatty acid biosynthetic pathways existing in *Toxoplasma*, mechanisms of import of selected fatty acids from the host cell have been observed in this parasite (Tomavo *et al.*, 1989; Quittnat *et al.*, 2004; Polonais and Soldati-Favre, 2010; Ramakrishnan *et al.*, 2012). Exogenous fatty acids are used for incorporation into complex lipids or degraded all the way to acetyl-CoA via mitochondrial β-oxidation. After transport to other compartments, these acetyl-CoA molecules can be used for *de novo* fatty acid biosynthesis. Intravacuolar *T. gondii* scavenges the fluorescent fatty acid analogue 5-butyl-4,4-difluoro-4-bora-3a,4a-diaza-s-indacene-3-nonanoic acid (C4-BODIPY-C9) from the medium and delivers this lipid to the parasitophorous vacuole (PV) membrane, Golgi/ER (Charron and Sibley, 2002) and thin tubules filling the lumen forming the intravacuolar network (Caffaro and Boothroyd, 2011). This selective compartmentalization of diverted lipids reflects sorting activities mediated by the parasites to adeptly distribute exogenous lipids into proper organelles. Intracellular *T. gondii* and host-free parasites are competent to accrue various free fatty acids from their environment (Quittnat *et al.*, 2004), including palmitic, oleic, stearic, linoleic, and arachidonic acids with a preferential internalization of palmitic acid. Inside parasites, exogenous fatty acids are manufactured into triacylglycerols (Quittnat *et al.*, 2004) and cholesteryl esters (Nishikawa *et al.*, 2005). Extracellular parasites incorporate palmitic or myristic acids into GPI anchors, such as that of SAG1 (Tomavo *et al.*, 1989). After uptake of butyric acid by the parasites, the lipid is anabolized into phosphatidylcholine (Charron and Sibley, 2002). Interestingly, the reduced growth of *T. gondii* consecutive to ablation of type II fatty acid synthase activity is partially restored by addition of long chain fatty acids in the medium (Ramakrishnan *et al.*, 2012), suggestive of compensatory activities between fatty acid synthesis and uptake of *T. gondii*. Obviously, the diversity and redundancy of the fatty acid pathways might be taken as an indication that the availability of the right fatty acids is an essential determinant for successful adaptation of the parasite to various host cells. If fatty acid uptake is as essential as fatty acid synthesis for the parasite growth and pathogenesis, fatty acid homeostatic pathways hold promise for much more *Toxoplasma*-specific drug targets.

8.5.2 Glycerophospholipids

8.5.2.1 Phospholipid Biosynthetic Pathways – Generalities

Glycerophospholipids, known as phospholipids, are key molecules that contribute to the structural definition of cells and that participate in the regulation of many cellular processes. Phospholipid metabolism is a major activity that cells engage in throughout their growth (Carman and Zeimetz, 1996). These amphiphilic lipids insert in cell membranes and form into a sheet two molecules thick with the fat-soluble portions inside, shielded on both sides by the water-soluble portions. This stable structure provides the cell membrane with its integrity. In mammalian cells, the most important glycerophospholipids are phosphatidylcholine, phosphatidylethanolamine and phosphatidylinositol. Phosphatidylcholine, phosphatidylethanolamine and phosphatidylserine are synthesized from phosphatidate by the CDP–diacylglycerol pathway, while phosphatidylcholine and phosphatidylethanolamine are also synthesized by the Kennedy (CDP–choline and CDP–ethanolamine) pathway (Fig. 8.5). CDP–diacylglycerol is also used for the synthesis of other phospholipids, including phosphatidylinositol (Carman and Zeimetz, 1996; Vance and Vance, 2004).

8.5.2.2 Phospholipid Composition and Physiological Relevance in Toxoplasma

Quantification of the phospholipid profile of *Toxoplasma* reveals that phosphatidylcholine is the most prevalent lipid, accounting for about 75% of total phospholipids (Gupta *et al.*, 2005). The next most abundant lipids are phosphatidylethanolamine (10%), phosphatidylinositol (7.5%), phosphatidylserine (6%) and phosphatidic acid (1%). Compared with the human cells, *T. gondii* has higher levels of phosphatidylcholine but lower levels of sphingomyelin and phosphatidylserine (Welti *et al.*, 2007). Mass spectrometric analyses on polar lipids reveal unique lipid composition and unusual abundance in *Toxoplasma*. For example, the parasite contains a relatively high level of ceramide phosphoethanolamine (~2% of the total polar lipids) with a fatty amide profile with only 16- and 18-carbon species. *T. gondii* has also greater amounts of shorter-chain fatty acid in its polar lipids compared to their hosts. Diacyl phosphatidylcholine with two saturated acyl chains with 12, 14 or 16 carbons makes up over 11% of parasite phosphatidylcholine but less than 3% of the host phosphatidylcholine molecular species.

The distinctive *T. gondii* phospholipid profile may be particularly suited to the function of parasitic membranes and the interaction of the parasite with the host cell and the host's immune system. Present at outer leaflet of the plasma membrane of eukaryotic cells, phosphatidylserine is a major ligand involved in the uptake of apoptotic cells (Fadock *et al.*, 2001). Phagocytosis of apoptotic cells by macrophages induces a noninflammatory response based on the exposure of phosphatidylserine that leads to TGF-β1 secretion (Fadock *et al.*, 1998). Phosphatidylserine exposure on the cell surface has also been related to evasion mechanisms of parasites, a concept known as apoptotic mimicry. *T. gondii* mimics apoptotic cells by exposing phosphatidylserine, inducing secretion of TGF-β1 by infected activated macrophages which leads to the degradation of inducible nitric oxide synthase and inhibition of nitric oxide production, and consequently parasite persistence in macrophages (Santos *et al.*, 2011). A phosphatidylserine-negative subpopulation of *Toxoplasma* enters macrophages by phagocytosis and is unable to inhibit nitric oxide synthesis whereas a phosphatidylserine-positive subpopulation invades macrophages by active penetration and no sign of inflammation is detected in mouse. This indicates that the escape mechanism

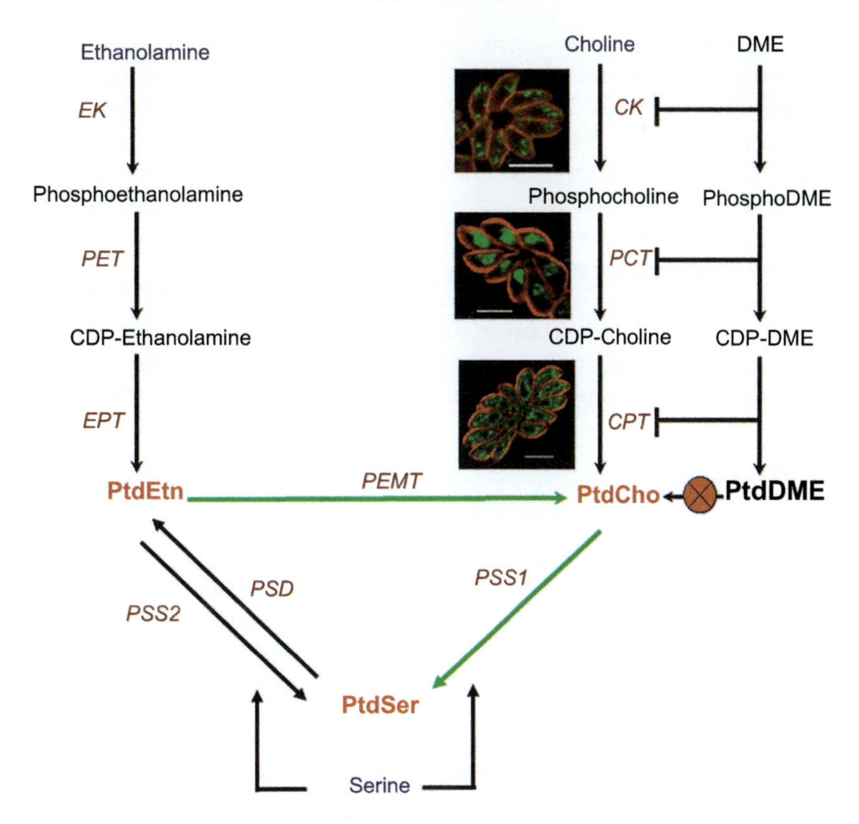

FIGURE 8.5 **Biosynthetic pathways of three major phospholipids in human and *Toxoplasma*.** The *H. sapiens* pathways are adapted from literature and, of *T. gondii*, are constructed based on the reported enzyme activities and annotations in the parasite database (http://www.ToxoDB.org). The common pathways are shown in black; those specific to human are depicted in green. Initial precursors are shown in blue; the intermediates of lipid synthesis are in black; phospholipids are in red, and the enzymes are in brown colour. Dimethylethanolamine (DME) is metabolized via the CDP–choline route and produces phosphatidyldimethylethanolamine (PtdDME), which is not methylated to phosphatidylcholine (PtdCho) in *T. gondii* causing disruption of membrane biogenesis. CK, choline kinase (forming clusters in the cytosol in green); PCT, phosphocholine cytidylyltransferase (localized to the nucleus in green); CPT, CDP–choline phosphotransferase (localized to the ER in green); EK, ethanolamine kinase; PET, phosphoethanolamine cytidylyltransferase; EPT, CDP–ethanolamine phosphotransferase; PtdEtn, phosphatidylethanolamine; PEMT, phosphatidylethanolamine methyltransferase; PtdSer, phosphatidylserine; SD, serine decarboxylase; PSS, phosphatidylserine synthase; PSD, phosphatidylserine decarboxylase *(adapted from Sampels* et al., *2012).*

of *T. gondii* is dependent on the exposure of phosphatidylserine, making this lipid essential for a successful infection and survival. Interfering with the phosphatidylserine biosynthetic pathways would dramatically reduce the parasite burden in their hosts.

Pharmacological studies have revealed that phosphatidylcholine is also a central phospholipid for *Toxoplasma* physiology. The choline analogue *N,N*-dimethylethanolamine is taken up by intracellular parasites as efficiency as choline (Gupta *et al.*, 2005). As a result, *T. gondii* growth is progressively arrested, probably due to dramatic phosphatidylcholine depletion and/or toxic phosphatidyldimethylethanolamine amassing in parasite membranes. In mammalian cells, phosphatidyldimethylethanolamine is normally produced as a short-lived intermediate in the conversion of phosphatidylethanolamine to phosphatidylcholine (Vance and Vance, 2004). Clearly, dimethylethanolamine interferes with choline uptake and metabolism to phosphatidylcholine, resulting in selective alteration in parasite membrane morphology at concentrations non-toxic for the host cell. This indicates that the dominance of phosphatidylcholine as a major lipid in *T. gondii* membranes offers great potentialities to disrupt the membrane biogenesis of the parasite.

8.5.2.3 Phospholipid Synthesis in **Toxoplasma**

Toxoplasma is enzymatically equipped to synthesize *de novo* all its aminoglycerophospholipids via the Kennedy pathway (Fig. 8.5). *T. gondii* incorporates choline, ethanolamine and serine into phosphatidylcholine, phosphatidylethanolamine and phosphatidylserine, respectively (see Section 8.5.2.4 below). Unlike its mammalian host (Vance and Vance, 2004), *Toxoplasma* does not possess activity for a serine decarboxylase and phosphatidylethanolamine or phosphatidylethanolamine methyltransferase, and thus appears incompetent in making phosphatidylcholine from serine and/or ethanolamine (Gupta *et al.*, 2005; Sampels *et al.*, 2012). Conversely, a parasite choline kinase has potential to counteract the loss of ethanolamine kinase and to sustain *de novo* synthesis of phosphatidylethanolamine in *T. gondii*. The parasite also harbours a phosphatidylserine decarboxylase route to produce phosphatidylethanolamine from phosphatidylserine. The parasite resilience to a perturbation of choline kinase, compositional flexibility of its membranes, and likely redundant routes of phosphatidylethanolamine synthesis can confer an adjustable membrane biogenesis to *T. gondii* in dissimilar nutrient environments in host cells. The observed metabolic plasticity might allow *T. gondii* to fine-tune membrane biogenesis according to the intracellular niche and contribute to its evolution as a promiscuous pathogen.

Diphosphatidylglycerol, a mitochondrial cardiolipin representative and phosphatidylglycerol are also produced by *Toxoplasma* as monitored by metabolic experiments using radioactive acetate (Bisanz *et al.*, 2006).

8.5.2.4 Phospholipid Salvage by **Toxoplasma**

Quantitative data on the rates of phospholipid syntheses reveal that *T. gondii* has an adequate synthetic capacity to produce all of the phosphatidylethanolamine species, but only 50% of phosphatidylserine and ~5–10% of phosphatidylcholine as required for a parasite doubling. This indicates that *T. gondii* must be auxotrophic for phosphatidylserine and phosphatidylcholine – or their precursors to acquire sufficient amounts of all phospholipids (Gupta *et al.*, 2005; Charron and Sibley, 2002). Activities of phospholipid uptake by *T. gondii* have been exemplified by using fluorescent glycerophospholipid analogues 2-(4,4-difluoro-5-(4-phenyl-1,3-butadienyl)-4-bora-3a,4a-diaza-s-indacene-3-pentanoyl)-1-hexadecanoyl-*sn*-glycero-3-phosphocholine (BODIPY-phosphatidylcholine) and 2-(4,4-difluoro-5,7-dimethyl-4-bora-3a,4a-diaza-s-indacene-3-pentanoyl)-1-hexadecanoyl-*sn*-glycero-3 phosphate (BODIPY-phosphatidic acid) that label parasite compartments after internalization in host cells (Charron and Sibley, 2002). BODIPY-phosphatidylcholine is mobilized to plasma membrane and dispersed small vesicles while BODIPY-phosphatidic acid moves to the parasite compartments similar to those containing fluorescent C4–BODIPY–C9, corresponding presumably to the Golgi/ER and the PV membrane. Interestingly, diversion of BODIPY-phosphatidic acid loaded in host cells prior to infection shows a bright fluorescent labelling in the PV membrane. By contrast, BODIPY-phosphatidylcholine is completely excluded from parasites that have invaded prelabelled hosts.

Intracellular parasites and host-free parasites are also competent to take up L-α-phosphatidic acid but only parasites inside host cells further metabolize the scavenged lipid into phosphatidylcholine (Charron and Sibley, 2002). *T. gondii* can acquire the phospholipid head group precursors from its environment and use them for the synthesis of major lipids (Gupta *et al.*, 2005). Labelled serine internalized by free parasites is metabolized into phosphatidylserine and phosphatidylethanolamine after phosphatidylserine decarboxylation, as well as in minor sphingolipids. Phosphatidylethanolamine is the main polar lipid generated after uptake of ethanolamine. Like serine, the metabolism of ethanolamine shows a time-dependent increase in lipid synthesis that progressively slows over a 6 hour period. No significant radioactive phosphatidylcholine is detected in *Toxoplasma* membranes after incubation in the presence of either tritiated serine or ethanolamine, suggesting that the parasites, at least in axenical conditions, have a negligible phosphatidylethanolamine methyltransferase activity (Gupta *et al.*, 2005). However, another study conducted on parasites grown in host cells fed with labelled L-serine or ethanolamine shows that *T. gondii* can synthesize phosphatidylcholine as a major resultant end product using these labelled precursors (Charron and Sibley, 2002). This discrepancy is probably ascribed to differences in metabolic requirements between parasites released from cells or growing inside cells. When host cells are loaded with labelled L-serine or ethanolamine prior to infection, no subsequent metabolization of radioactive serine or ethanolamine is observed (Charron and Sibley, 2002). This may be linked to the rapid conversion of serine and ethanolamine into phosphatidylcholine by mammalian cells and the inability of the parasite to scavenge intact phosphatidylcholine from host cells, as corroborated in fluorescence studies (see above).

Exposure of parasites to labelled choline or methylcholine results in the uptake and metabolization of these compounds into phosphatidylcholine (Charron and Sibley, 2002; Gupta *et al.*, 2005). Choline preaccumulated into host cells before infection is also further metabolized into phosphatidylcholine by the parasites, in accordance to the competence of the intravacuolar *T. gondii* to readily taken up choline. Various forms of radioactive choline-containing lipids are only observed in intravacuolar parasites. This parallels the observation showing that the metabolism of choline is increased by about 2-fold in host-free parasites incubated in an intracellular-type medium compared to parasites maintained in an extracellular-type medium. This leads to the assumption that choline metabolism and phosphatidylcholine synthesis are stimulated in response to parasitic invasion and replication within host cells.

8.5.3 Glycerolipids

8.5.3.1 Glycerolipid Biosynthetic Pathways – Generalities

Bacteria, yeast, plants and animals all have the ability to synthesize glycerolipids, mainly triacylglycerols and diacylglycerols, a critical function during periods of nutritional excess and/or nutritional stress (Coleman and Lee, 2004). In higher eukaryotes, triacylglycerols are packaged in circulating lipoproteins for distribution to peripheral tissues where they can be used immediately or stored in cytosolic lipid droplets. Such energy-dense triacylglycerol stores can free organisms temporally and spatially from the need for an immediate energy supply and provide a reserve depot that can be used when local resources fail or when specific kinds of fatty acids or lipid precursors are required. Triacylglycerol stores can also be partially hydrolysed to form diacylglycerols, which perform two distinct roles: supporting the biosynthesis/degradation of phospholipids and regulating the protein kinase C activity that controls cell growth. In animals, triacylglycerols are energy stores, repositories of fatty acids and precursors for phospholipid biosynthesis, and depots of signalling molecules (van Blitterswijk and Houssa, 1999). By contrast, in bacteria and lower eukaryotes, triacylglycerols are solely synthesized during times of stress or resource depletion, and they are used primarily for phospholipid synthesis. In this view it is not surprising that higher organisms have developed several pathways for triacylglycerol synthesis and regulation, as compared to unicellular organisms. Commonly, the formation of triacylglycerols is catalysed by the activity of microsomal acyl-CoA:diacylglycerol acyltransferases (DGAT).

8.5.3.2 *Glycerolipid Synthesis in* Toxoplasma

In *T. gondii*, triacylglycerol synthesis occurs via the glycerol-3-phosphate pathway and involves a DGAT, named *Tg*DGAT1 (Quittnat *et al.*, 2004). Fatty acid can be incorporated into *Toxoplasma* diacylglycerol, revealing that this latter lipid is the acyl acceptor. *Tg*DGAT1 contains signature motifs characteristic of the DGAT1 family. *Tg*DGAT1 is an integral membrane protein localized to the ER. When a *Saccharomyces cerevisiae* mutant strain lacking neutral lipid production is transformed with *Tg*DGAT1, a significant DGAT activity is reconstituted, resulting in the biogenesis of cytosolic lipid inclusions. In contrast to human DGAT1 lacking fatty acid specificity, *Tg*DGAT1 preferentially incorporates palmitate into triacylglycerols. Triacylglycerols are stored in parasite cytosolic lipid bodies. Stored triacylglycerols may be a reservoir of fatty acids utilizable for phospholipid biosynthesis and/or exploitable as respiratory substrates in *Toxoplasma*. *T. gondii* plastids seem to contain the complete pathway to synthesize galactosylglycerolipids from fatty acids and glycerol-phosphate (Marechal *et al.*, 2002; Botte *et al.*, 2008). These lipids are generally found in plant chloroplast membranes and might be therefore synthesized in the parasite apicoplast.

8.5.4 Sterols and Steryl Esters

8.5.4.1 *Sterol Lipid Biosynthetic Pathways – Generalities*

Cholesterol is the major sterol molecule ubiquitously present in mammalian cells. This lipid has been selected in the long natural evolution process for its ability to maintain a delicate balance between membrane rigidity (e.g. to allow large cell volumes) and membrane fluidity (e.g. to allow membrane-embedded proteins to function properly) (Bretscher and Munro, 1993). Mammalian cells obtain cholesterol both by internalization of plasma low density lipoprotein particles (LDL) or by *de novo* synthesis via the mevalonate pathway in the ER (Goldstein and Brown, 1990). The cholesterol molecule is formed from acetate units. The acetate units are joined in a series of reactions to form farnesyl pyrophosphate, a branch point for the biosynthesis of other isoprenoid compounds such as ubiquinone, dolichol and farnesylated proteins. Hydroxy-3-methylglutaryl-CoA reductase from the mevalonate pathway is the rate determining enzyme for the entire pathway from acetate to cholesterol.

Deposition of excess cellular cholesterol in the form of cholesteryl esters is catalysed by acyl-CoA:cholesterol acyltransferases (ACAT), ER resident enzymes. Native and exogenous cholesterol has several possible fates: incorporation into membranes, efflux to extracellular acceptors, conversion into cholesteryl esters or, depending on the cell type, metabolism into bile acids or steroid hormones. Rates of cholesterol biosynthesis, LDL internalization and cholesterol esterification are exquisitely sensitive to cellular levels of free cholesterol. Three possible mechanisms of cholesterol movement include aqueous diffusion, vesicle-mediated transport, and soluble carriers, which may work together or separately to mobilize cholesterol within the cell (Liscum and Underwood, 1995). Evidence has accrued that biological membranes are made of a mosaic of lipids domains. Maintenance of domain structure is critical for cell function. Cholesterol plays a key role in organizing signalling lipids and proteins within these membrane domains (Anderson and Jacobson, 2002).

8.5.4.2 Sterol Salvage and Transport in **Toxoplasma**

Toxoplasma membranes contain β-hydroxysterols, as probed using the polyene antibiotic filipin routinely used to reveal the steady-state distribution of sterols by fluorescence microscopy (Coppens *et al.*, 2000). A predominantly parasite staining is located to the plasma membrane and the rhoptries, apical secretory organelles whose content is discharged upon parasite invasion (Coppens and Joiner, 2003). *T. gondii* diverts LDL-derived cholesterol that has transitted through host lysosomes (Coppens et al., 2000) as demonstrated after incubation of infected cells with labelled cholesterol incorporated into LDL. Cholesterol movement from lysosomes to the PV requires temperatures permissive for vesicular transport, metabolic energy and host microtubules, but no fusion (Sehgal *et al.*, 2005) *Toxoplasma* sequesters cholesterol-filled host lysosomes within invaginations of the PV membrane (Coppens *et al.*, 2005). Cholesterol delivery into the PV requires functional host Niemann–Pick type C proteins (Sehgal *et al.*, 2005) that mediate cholesterol egress across the endo-lysosomal membranes (Sleat *et al.*, 2004). Additionally, a host cell P-glycoprotein transporter, a membrane-bound efflux pump, is required for cholesterol transport to the PV (Bottova *et al.*, 2009). Cholesterol incorporation into the parasite is abolished after protease treatment on parasite plasma membrane (Sehgal *et al.*, 2005). A lipid-translocating importer of the ATP-binding cassette (ABC) transporter G subfamily (ABCG family) located to the PV and parasite plasma membrane delivers cholesterol to the parasite interior (Ehrenman *et al.*, 2010). Inside the parasite, a D-bifunctional protein containing two sterol-carrier protein-2 domains promotes the circulation of cholesterol, phospholipids and fatty acids between parasite organelles and the plasma membrane. Other parasite ABCG transporters are involved in cholesterol and phospholipid movement from the parasite to the PV and may contribute to the expansion of PV membrane size (Ehrenman *et al.*, 2010).

The availability of host cholesterol has a direct impact on *Toxoplasma* development inside its PV. LDL deprivation impairs parasite growth whereas overabundance of these lipoproteins stimulates parasite replication (Coppens *et al.*, 2000; Nishikawa *et al.*, 2011). Delivery of sterol analogues incorporated into LDL could be an efficient strategy to substitute the indispensable cholesterol by structural-related compound with growth-reducing activity. The sterol analogues 22,26-azasterol and 24,25-(R,S)-epiminolanosterol, inhibitors of sterol-24-methyl transferase producing 24-alkyl sterols, have potent and selective antiproliferative activity against *T. gondii* (Dantas-Leite *et al.*, 2004; Martins-Duarte *et al.*, 2011). The molecular mechanism of these lipids is unclear since 24-alkyl sterols are not detected in this parasite. It is observed, however, that the rapid accumulation of these lipid analogues in diverse membranes alters maintenance, fusogenicity and function of the parasite organelles. It is reported that selected sterol analogues (e.g. cholesteryl chloride, cholestanone or thiocholesterol) can affect the growth of various cholesterol-auxotroph organisms (Clayton, 1964). Their antiproliferative properties represent new therapeutic approaches for the treatment of *Toxoplasma*.

8.5.4.3 Sterol Storage in **Toxoplasma**

Nile red that strongly fluoresces in the presence of steryl esters detects the presence of cytosolic lipid bodies in *Toxoplasma* (Sonda *et al.*, 2001; Charron and Sibley, 2002; Quittnat *et al.*, 2004), indicating the ability of cholesterol storage by the parasites (Nishikawa *et al.*, 2005). Of the 21 molecular species detected in *Toxoplasma*, cholesteryl oleate C18:1 (42%) and palmitate C16:0

(26%) are the main esters for mammalian cells but the parasite also has uniquely large amounts of cholesteryl eicosanoate C20:1 (7%). In addition, the parasite contains, to a lesser extent, cholesteryl palmitoleate C16:1, stearate C18:0, linoleate C18:2, arachidonate C20:4 and some polyunsaturated C22 fatty acids. Other cholesteryl ester fatty acid species represented 3.5% of the total species (Lige *et al.*, 2011). *Toxoplasma* is competent to synthesize cholesteryl esters by two ACAT-related enzymes located to the ER (Nishikawa *et al.*, 2005; Lige *et al.*, 2013). Parasite ACATs present a broad sterol substrate affinity but preferentially use palmitate to form cholesteryl esters. The loss of individual ACAT can be relatively tolerated by the parasite, though a slower growth compared to wild-type parasites has been observed. This suggests that the parasites can endure the resulting slight differences in their cholesteryl ester pools and that the ACAT enzymes partially complement each other. Loss of both ACAT results in synthetic lethality, indicating the cholesterol storage is an important function for *T. gondii*. Host LDL and fatty acids scavenged by *Toxoplasma* serve as ACAT activators by stimulating cholesteryl ester synthesis and lipid droplet biogenesis in the parasite. Lipoprotein depletion causes a progressive consumption of material stored in the parasite's lipid bodies. Under conditions of excess LDL, the activity of cholesterol esterification is significantly increased, entailing that the parasites adeptly control the massive supply of cholesterol by producing the storage form of cholesterol. A Niemann–Pick, type C1-related protein in *Toxoplasma* controls the intracellular levels of several lipids (Lige *et al.*, 2011). Parasites lacking the Niemann–Pick-related protein accumulate lipid bodies enriched in cholesteryl esters.

The replication rate of intracellular *T. gondii* correlates to the LDL concentration in the medium. Excess cholesterol diverted by the parasite is rapidly neutralized and stored in lipid bodies. Blockade of cholesteryl ester synthesis is deleterious for the parasite, leading to rapid induction of free cholesterol crystallization in parasite membranes and rupture of the plasma membrane (Nishikawa *et al.*, 2005). ACAT-deficient parasites are particularly sensitive to ACAT inhibitors. The higher vulnerability of *T. gondii* towards ACAT inhibitors compared with mammalian cells is probably linked to the absence of cholesterol acceptors (mainly lipoproteins) in the PV, which are known to desorb excess cholesterol from membranes.

8.5.5 Sphingolipids

8.5.5.1 Sphingolipid Biosynthetic Pathways – Generalities

Glycosylphosphatidylinositols (GPIs) are a class of glycolipids that are used by a wide variety of eukaryotic cells to anchor proteins, polysaccharides, and small oligosaccharides to the plasma membrane through covalent linkages (Ferguson and Williams, 1988). Comparison of the chemical structures of GPI membrane anchors of different organisms indicates that the anchors contain a remarkably conserved core glycan structure, suggesting that a common biosynthetic pathway may have been conserved throughout eukaryotic evolution. The transfer of GPI anchors to proteins occurs in the ER with concurrent displacement of a C-terminal hydrophobic peptide, followed by the rapid substitution of the peptide tail by the GPI anchor. Ceramides are the principal lipid components present in sphingomyelin, complex glycolipids, cerebrosides and gangliosides (Sharma and Shi, 1999). Ceramides are broadly recognized as vital second messengers in the signal transduction process mediated by receptors of many cytokines and growth factors.

8.5.5.2 Sphingolipid Synthesis in **Toxoplasma**

Toxoplasma can readily incorporate sugars and amino acids as precursors of sphingolipids. *De novo* synthesis of ceramide, glycosylated ceramide and sphingomyelin in *T. gondii* has been demonstrated by metabolic labelling studies using tritiated serine and galactose. After internalization, labelled galactose is metabolized in various glycosphingolipids (e.g. di- and triglycosylated ceramide; Azzouz *et al.*, 2002). Parasite incubation with labelled serine leads to the production of ceramide. After uptake, labelled glucosamine serves as a GPI glycolipid precursor and is associated with the dominant surface protein, SAG1 (Striepen *et al.*, 1997; Zinecker *et al.*, 2001).

GPI in *T. gondii* serve as membrane anchors for a large number of plasma membrane proteins (Tomavo *et al.*, 1992; Striepen *et al.*, 1997; Zinecker *et al.*, 2001). Their biosynthetic pathway is initiated on the parasite ER with the transfer of *N*-acetylglucosamine to phosphatidylinositol involving a phosphatidylinositol-glycan class A (PIGA)-like protein (Wichroski and Ward, 2003). The GPI core glycan is then assembled via sequential glycosylation of phosphatidylinositol (Tomavo *et al.*, 1992). *Toxoplasma* PIGA sequence contains a potential transmembrane domain followed by a stretch of mostly hydrophilic residues extending to the C-terminus. A functional copy of PIGA is required for viability, demonstrating that GPI biosynthesis is an essential process in *T. gondii*.

Glycosphingolipids, e.g. inositol phosphorylceramide, are synthesized *de novo* via the 3-ketosphinganine pathway from serine and palmitoyl-CoA (Azzouz et al., 2002; Sonda et al., 2005) with ceramide as an intermediate. Metabolic studies show that *T. gondii* readily incorporates radioactive acetate into glycosylcerebroside, lactosylcerebroside and globotriosylcerebroside while only intracellular parasites produce globoside (Bisanz *et al.*, 2005).

GPI-anchored proteins dominate the surface of *T. gondii* and are implicated in both host cell attachment and modulation of the host immune response (Lekutis *et al.*, 2001). Although the GPI core glycan is conserved in all organisms, some differences in additional modifications to GPI structures and biosynthetic pathways have been reported for *T. gondii* (de Macedo *et al.*, 2003). This indicates that the GPI biosynthetic pathway is a potential target for the development of new chemotherapeutics against this parasite. Indeed the lethal consequences of PIGA disruption in *T. gondii* may result from a deficiency in GPI-anchored proteins, free GPI, or both (Wichroski and Ward, 2003). *In vitro* and *in vivo* studies reveal that sugars and amino acid analogues, synthetic mannoside acceptor substrates and natural compounds specifically interfere with GPI biosynthesis in many different pathogenic organisms (de Macedo *et al.*, 2003). Synthesis of parasite ceramide is dramatically decreased after incubation of intracellular *Toxoplasma* with either threo-phenyl-2-palmitoylamino-3-morpholino-1-propanol, a specific inhibitor of glucosylceramide synthesis, or L-cycloserine that blocks the serine palmitoyltransferase activity (Azzouz *et al.*, 2002). The antibiotic aureobasidin A, a potent inhibitor of inositol phosphorylceramide that is absent from mammalian cells, abrogates *T. gondii* replication by the severe reduction of total complex sphingolipids' synthesis without noticeable host cell alterations (Sonda *et al.*, 2005).

8.5.5.3 Sphingolipid Salvage by **Toxoplasma**

Intracellular *T. gondii* is also able to retrieve sphingolipids intact from the culture and accumulate the scavenged lipids into the Golgi apparatus. In mammalian cells, exogenous ceramides concentrate the Golgi complex to be further metabolized into major sphingolipids and glucosylceramides. After incubation with NBD-C6-ceramide, the Golgi of intravacuolar *T. gondii* is stained, suggesting that the parasite can intercept the ceramide pathway of the host cell to acquire exogenous ceramides or other sphingolipids manufactured in the host Golgi (de Melo and de Souza, 1996; Coppens, unpublished).

8.5.6 Isoprenoid Derivatives

8.5.6.1 Isoprenoid Biosynthetic Pathways – Generalities

The post-translational modification of proteins by isoprenoid residues such as farnesyl and geranylgeranyl is a major mechanism by which cytosolic proteins interact with cellular membranes (Swiezewskaa and Danikiewicz, 2005). Isoprenylation is also required for the proper membrane localization and the biological activity of several cellular proteins implicated in the regulation of DNA replication and cell cycling, therefore having important roles in the regulation of cell proliferation. Two coexisting isoprenoid pathways exist in organisms, the mevalonate pathway present in the cytosol of mammalian cells (see above) and the recently described 1-deoxy-*d*-xylulose-5-phosphate pathway. This latter pathway seems to be restricted so far to bacteria, plastids in plants and apicoplast in Apicomplexa (Jomaa *et al.*, 1999).

8.5.6.2 Isoprenoid Synthesis in Toxoplasma

T. gondii membranes contain both farnesylated and geranylgeralynated proteins (Ibrahim *et al.*, 2001; Ling *et al.*, 2005). Enzymes of the 1-deoxy-*d*-xylulose-5-phosphate pathway seem to have an apicoplast origin and may contribute to the production of farnesyl and geranylgeranyl molecules for the parasite (Seeber, 2003). Two farnesyl-diphosphate synthase homologues have been identified in *Toxoplasma* and localize to the mitochondria; this enzyme is bifunctional, catalysing the formation of both farnesyl diphosphate and geranylgeranyl diphosphate (Ling *et al.*, 2007).

Prenylated proteins are ubiquitously important for cell proliferation regulation. Two categories of protein farnesyltransferase inhibitors have been described so far (Qian *et al.*, 1997): isoprene analogues and peptidomimetics based on the consensus CAAX motif that is required for isoprenylation. Peptidomimetics act as alternative substrates *in vitro*, and thereby competitively block protein farnesylation. Clearly, specific inhibition of *T. gondii* protein farnesyltransferase activity is observed using selected modified heptapeptides (Ibrahim *et al.*, 2001). Bisphosphonates, diphosphate analogues in which a carbon atom replaces the oxygen atom bridge between the two phosphorus atoms of the diphosphate, are potent inhibitors of farnesyl-diphosphate synthase and inhibit the growth of *Toxoplasma* (Martin *et al.*, 2001). Moreover, *in vivo* testing of bisphosphonates against *T. gondii* in mice has shown that one of the nitrogen-containing bisphosphonates, risedronate, significantly increases the survival of mice infected by *T. gondii* (Yardley *et al.*, 2002). All these results indicate that bisphosphonates are promising candidate drugs to treat infections caused by *T. gondii* as well as some other protozoan parasites.

8.5.6.3 Isoprenoid Salvage by Toxoplasma

Alternatively, the parasite can salvage isoprenoids such as labelled trans,trans farnesol and labelled trans,trans,cis geranylgeraniol, from the medium to produce its prenylated proteins (Ibrahim *et al.*, 2001). This implies the presence of functional protein farnesyltransferase and geranylgeranyl transferase in *T. gondii*. Indeed, a protein farnesyltransferase activity has been detected in the parasite, responsible for the catalysis of isoprene lipid modifications.

8.6 NUCLEOTIDE BIOSYNTHESIS

The most extensively studied metabolic pathways in *T. gondii* are those of pyrimidine and purine nucleotide biosynthesis. These pathways provide the substrates for DNA/RNA biosynthesis and are commonly targeted for chemotherapy. They are logical areas to study in the rapidly multiplying tachyzoite form. Due to limited material nearly all studies have been performed exclusively in tachyzoites. Illustrations of the present overall knowledge of nucleotide biosynthesis in the tachyzoite form of *T. gondii* are shown in Fig 8.6–8.9. Comparative analyses of genome sequences for several Apicomplexa have revealed surprising differences among the Apicomplexa in nucleotide metabolism that are discussed further in Chapter 20.

Early works determined that *T. gondii* has both *de novo* (Hill *et al.*, 1981; O'Sullivan *et al.*, 1981; Schwartzman and Pfefferkorn, 1981; Asai *et al.*, 1983a) and salvage (O'Sullivan *et al.*, 1981; Pfefferkorn, 1978; Pfefferkorn and Pfefferkorn, 1977, 1980; Iltzsch, 1993) pyrimidine nucleotide biosynthetic pathways. The *de novo* pyrimidine biosynthetic pathway is more important than the salvage pathway and is essential for *T. gondii* growth and virulence. Pyrimidine auxotrophic mutants of *T. gondii* with disruption of the single copy carbamyl phosphate synthetase II gene (CPSII, E.C. 6.3.5.5) are avirulent in mice (Fox and Bzik, 2002). CPSII catalyses the first step of the *de novo* pyrimidine biosynthetic pathway. Furthermore, pyrimidine starvation is one of the conditions that cause stage conversion from the tachyzoite to the bradyzoite (Bohne and Roos, 1997). The *de novo* pyrimidine biosynthetic pathway is shown in Fig. 8.6.

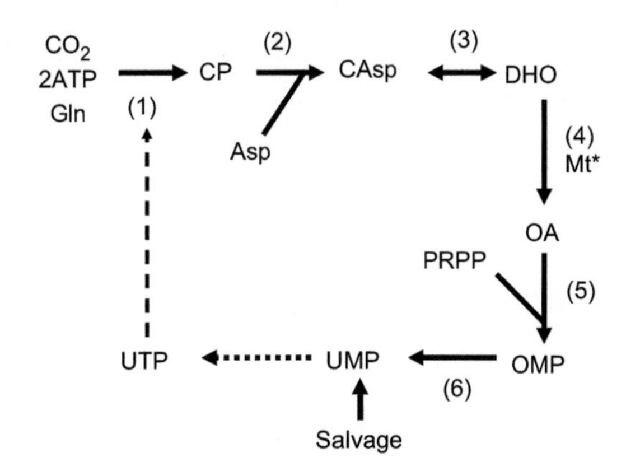

FIGURE 8.6 ***T. gondii* pyrimidine *de novo* biosynthetic pathways.** Solid lines represent activities that were detected in *T. gondii*, a short dashed line represents activities of uridylate kinase and nucleoside diphosphate kinase that are not considered to be *de novo* enzymes and a long dashed line represents an inhibitory effect on CPSII by UTP. CP; carbamyl phosphate, CAsp; aspartate asparatate, DHO; dihydroorotate, OA; orotic acid, OMP; orotidine 5′-phosphate, PRPP; 5-phosphoribosyl-1-pyrophosphate, Mt; mitochondrion. Enzyme activities are numbered as follows: (1) carbamyl phosphate synthetase II (CPSII), (2) aspartate carbamyltransferase (ATCase), (3) dihydroorotase (DHOase), (4) dihydroorotate dehydrogenase (DHO–DHase), (5) orotate phosphoribosyltransferase (OPRTase), (6) orotidylate decarboxylase (ODCase). *DHO–DHase activity is present in membranous fraction and may link to the respiratory chain of mitochondrion.

8.6.1 Pyrimidine *De Novo* Biosynthetic Pathway

The preliminary characterization of all six enzymes of the *de novo* pyrimidine biosynthetic pathway (Asai *et al.*, 1983a) indicated some distinctions between *T. gondii* enzyme and host enzyme activities. The mammalian host CPSII is part of a large multifunctional protein (CAD) composed of three enzymes: CPSII, aspartate carbamyltransferase (ATCase, E.C. 2.1.3.2), and dihydroorotase (DHOase, E.C. 3.5.2.3) (Mori and Tatibana, 1978; Davidson *et al.*, 1993). These three enzymes comprise the first three enzymes in the pathway. In contrast to mammalian CAD, *T. gondii* CPSII is a cytosolic protein that is an independent enzyme with approximate molecular mass of 540 kDa (Asai *et al.*, 1983a). *T. gondii* has a single CPSII gene which is interrupted by 36 introns. The predicted protein encoded by the 37 CPSII exons is a 1687 amino acid polypeptide with approximate molecular mass of 186 kDa (Fox and Bzik, 2003). Consequently, the native *T. gondii* CPSII appears to be a trimer of identical subunits. This large CPSII is a common characteristic of other apicomplexan parasites (Flores *et al.*, 1994, Chansiri and Bagnara, 1995). Bacteria and plants also possess the independent CPS proteins (Jones, 1980; Zhou *et al.*, 2000), however, the structural organization of these enzymes is different from those of mammalian and parasitic protozoan CPSII including *T. gondii* CPSII.

The enzyme reaction of glutamine-dependent CPSII consists of two reactions, the reaction of glutamine amidotransferase (GATase, E.C. 2.4.2.14) and the reaction of CPS. *T. gondii* and other parasitic protozoa including *Trypanosoma*, *Babesia* and *Plasmodium* express a bifunctional glutamine-dependent CPSII composed of an N-terminal GATase domain fused with C-terminal CPS domains (Aoki *et al.*, 1994; Flores *et al.*, 1994; Chansiri and Bagnara, 1995; Nara *et al.*, 1998; Gao *et al.*, 1999; Fox and Bzik, 2003). Bacteria and plants express a monofunctional GATase as well as a monofunctional CPS (Jones, 1980; Zhou *et al.*, 2000).

Mammalian CPSII is characteristically an allosterically controlled enzyme. Forward activation is by 5-phosphoribosyl-1-pyrophosphate (PRPP), and backward inhibition is by UTP (Jones, 1980). PRPP is a substrate of nucleotide biosynthesis and provides the sugar phosphate moiety of nucleotides. *T. gondii* CPSII activity is inhibited by UTP but no activation by PRPP is reported (Asai *et al.*, 1983a). The regulatory domain for the allosteric control of mammalian CPSII is the C-terminal ~150 amino acid domain (Liu *et al.*, 1994). *T. gondii* CPSII contains significant amino acid insertions in this expected regulatory domain (Fox and Bzik, 2003). Enlarged C-terminal regulatory domains are also reported for other apicomplexan parasites (Flores *et al.*, 1994; Chansiri and Bagnara, 1995). The lack of activation of *T. gondii* CPSII by PRPP may be due to this enlarged C-terminal domain. It has been demonstrated that deletion of the GATase region as well as the C-terminal regulatory region completely abolished complementation activity of CPSII cDNA minigenes for the uracil auxotrophy of CPSII-deficient mutants (Fox *et al.*, 2009).

T. gondii ATCase is a cytosolic monofunctional enzyme protein with approximate molecular mass of 140 kDa (Asai *et al.*, 1983a). The open reading frame of cloned *T. gondii* ATCase cDNA encodes a putative 423 amino acid polypeptide with a predicted molecular mass of 46.8 kDa. Recombinant *T. gondii* ATCase with catalytic activity exhibits a molecular mass of 144 kDa (Mejias-Torres and Zimmermann, 2002). Native *T. gondii* ATCase is likely to be a trimer of identical subunits. Plant ATCase (Khan *et al.*, 1999) is inhibited by UMP and *E. coli* ATCase (Wales *et al.*, 1999) is inhibited by CTP and UTP (*E. coli*). However, no significant effect by any nucleotide has been reported on the activity of *T. gondii* ATCase (Asai *et al.*, 1983a, Mejias-Torres and Zimmermann, 2002).

T. gondii DHOase is a cytosolic monofunctional enzyme with approximate molecular mass of 70 kDa whose activity is not regulated by nucleotides (Asai *et al.*, 1983a). No characterization of the *T. gondii* DHOase gene has been reported.

Dihydroorotate dehydrogenase (DHO-DHase, E.C. 1.3.3.1) is the fourth enzyme of *de novo* pyrimidine biosynthetic pathway. DHO-DHases of living organisms are classified into two families. Found in some bacteria and some lower eukaryotes, family 1 enzymes are cytosolic proteins. On the other hand, family 2 enzymes are membrane associated and link to the respiratory chain in mitochondria for their catalytic redox force (Bjornberg *et al.*, 1997). The *T. gondii* DHO-DHase is exclusively recovered in the particulate fraction of tachyzoite extract and inhibited by respiratory chain inhibitors (Asai *et al.*, 1983a). The predicted amino acid sequence, with approximate molecular mass of 65 kDa from cloned cDNA, is most similar to family 2 DHO-DHases (Sierra-Pagan and Zimmermann, 2003).

Orotate phosphribosyltransferase (OPRTase, E.C. 2.4.2.10) is the fifth enzyme of the pathway, and orotidylate decarboxylase (ODCase, E.C. 4.1.1.23) is the sixth enzyme of the pathway. These are cytosolic enzymes and cosediment by sucrose gradient centrifugation at a position corresponding to a molecular mass of approximately 70 kDa (Asai *et al.*, 1983a). In some higher eukaryotes, these two enzymes are on the same polypeptide (Jones, 1980). In the apicomplexan parasite *Plasmodium falciparum,* the two enzymes exist as a multienzyme complex containing two subunits each of 33 kDa OPRTase and 38 kDa ODCase (Krungkrai *et al.,* 2005). It is not clear if *T. gondii* OPRTase and ODCase are probably similar to the *P. falciparum* type.

8.6.2 Pyrimidine Salvage Biosynthetic Pathway

The *T. gondii* tachyzoite pyrimidine salvage pathway is illustrated in Fig. 8.7. *T. gondii* has five enzyme activities that are involved in the salvage of pyrimidine nucleosides and nucleobases: cytidine deaminase (E.C. 3.5.4.5); deoxycytidine deaminase (E.C. 3.5.4.14); uridine phosphorylase (E.C. 2.4.2.3); deoxyuridine phosphorylase (E.C. 2.4.2.23); and uracil phosphoribosyltransferase (UPRTase, E.C. 2.4.2.19) (Iltsch, 1993). No detailed studies for the properties of pyrimidine nucleoside deaminases have

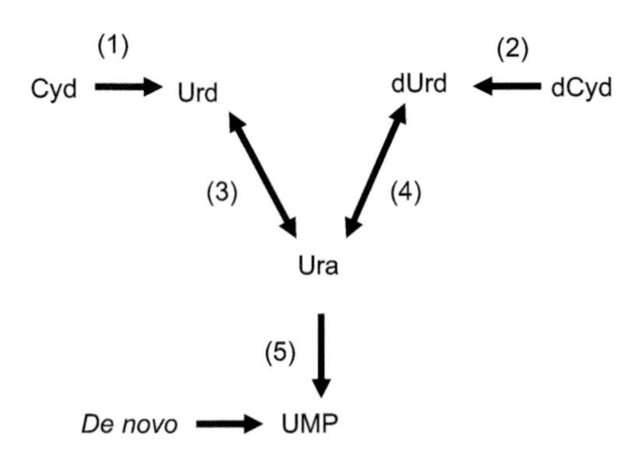

FIGURE 8.7 ***T. gondii*** **pyrimidine salvage biosynthetic pathways.** Solid lines represent activities that were detected in *T. gondii*. Cyd, cytidine; dCyd, deoxycytidine; dUrd, deoxyuridine; Ura, uracil; Urd, uridine. Enzyme activities are numbered as follows: (1) cytidine deaminase, (2) deoxycytidine deaminase, (3) uridine phosphorylase (4) deoxyuridine phosphorylase, (5) uracil phosphoribosyltransferase (UPRTase). It may be highly possible that deaminase activities are catalysed by one enzyme and that phosphorylase activities are also catalysed by one enzyme.

been reported. Uridine phosphorylase has been characterized and utilizes thymidine poorly (Chaudhary, Ting, Roos and Kim, personal communication).

It appears that all pyrimidine salvage in *T. gondii* proceeds through uracil, which is salvaged to the nucleotide level by UPRTase. However, the salvage pathway is not essential for tachyzoite viability and growth, since UPRTase-deficient mutants exhibit the same growth rate as normal tachyzoites (Donald and Roos, 1995). Although UPRTase may not be important for tachyzoite growth, UPRTase is thought to be a possible therapeutic target and is one of the most extensively studied enzymes in *T. gondii* at molecular level including crystal structure analysis (Carter *et al.,* 1997; Schumacher *et al.,* 1998, 2002). *T. gondii* UPRTase recognizes uracil only and no other naturally occurring pyrimidine and purine bases (Carter *et al.,* 1997). Without substrates or its activator GTP, recombinant *T. gondii* UPRTase behaves as a homodimer composed of two identical subunits (27 kDa). The physiologically active UPRTase is a tetramer, and this tetrameric structure is stabilized by binding of GTP leading to a more active state (Schumacher *et al.,* 2002). The other native protozoan UPRTases from *Crithidia luciliae* (Asai *et al.,* 1990) and *Giardia intestinalis* (Dai *et al.,* 1995) also behave as homodimers. Microarray-based RNA analysis by biosynthetic tagging (RABT) in which 4-thiouracil was used as an UPRTase substrate has been developed (Cleary, 2008; Zeiner *et al.,* 2008).

8.6.3 Purine Salvage Biosynthetic Pathway

Like all parasitic protozoa and many intracellular pathogens, *T. gondii* is incapable of *de novo* purine nucleotide biosynthesis. It relies on salvage pathways for purine nucleotides that are essential for parasite growth and survival (Perrotto *et al.,* 1971; Schwartzman and Pfefferkorn, 1982; Krug *et al.,* 1989; Ullman and Carter,

1995). An illustration of the *T. gondii* tachyzoite purine salvage pathway is shown in Fig. 8.8. *T. gondii* is thought to have six enzyme activities that are involved in the salvage of purine nucleosides and nucleobases: adenine deaminase (E.C. 3.5.4.2); adenosine deaminase (E.C. 3.5.4.4); inosine phosphorylase (E.C. 2.4.2.1); guanosine phosphorylase (E.C. 2.4.2.15); adenosine kinase (AK, E.C. 2.7.1.20); and hypoxanthine-xanthine-guanine phosphoribosyltransferase (HXGPRTase, E.C. 2.4.2.8) (Chaudhary *et al.*, 2004). No detailed studies for the properties of purine nucleoside deaminase or purine base deaminase have been reported. A gene for adenine deaminase, but not for adenosine deaminase, has been identified in the *T. gondii* genome (Chaudhary *et al.*, 2004). Purine nucleoside phosphorylase has been characterized activity against inosine and guanosine but not adenosine (Chaudhary, Ting, Kim, Roos; submitted).

AK and HXGPRTase have been extensively studied at molecular level including crystal structure analyses (Darling *et al.*, 1999; Schumacher *et al.*, 2000, 1996; Donald *et al.*, 1996; White *et al.*, 2000; Heroux *et al.*, 1999a, 1999b). Neither AK nor HXGPRTase is essential for *T. gondii* survival, since tachyzoites can survive elimination of either activity alone (Donald *et al.*, 1996; Sullivan *et al.*, 1999; Chaudhary *et al.*, 2004). It is reported that double stranded RNAs targeting both AK and HXGPRTase were effective at eliciting suppression of the corresponding gene in cultured tachyzoites (Yu *et al.*, 2009).

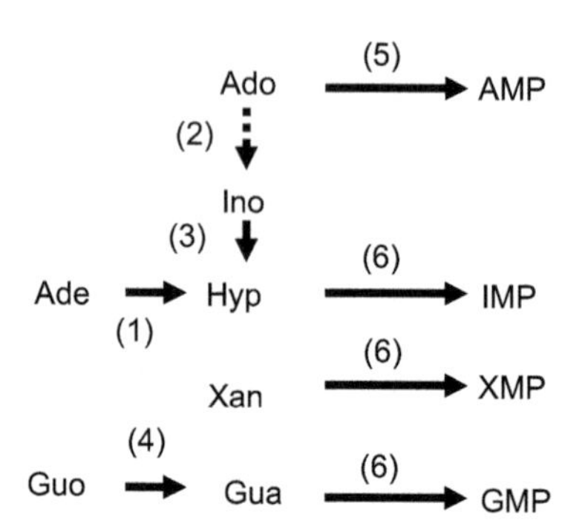

FIGURE 8.8 ***T. gondii* purine salvage biosynthetic pathways.** Solid lines represent activities or enzyme genomes that were detected in *T. gondii*, a short dashed line represents activity of adenosine deaminase whose existence is highly suspected. Ade, adenine; Ado, adenosine; Guo, guanosine; Gua, guanine; Hyp, hypoxanthine; Ino, inosine; Xan, xanthine. Enzyme activities are numbered as follows: (1) adenine deaminase, (2) adenosine deaminase, (3) inosine phosphorylase, (4) guanosine phosphorylase, (5) adenosine kinase (AK), (6) hypoxanthine-xanthine-guanine phosphoribosyltransferase (HXGPRTase).

The generation of a double knockout mutant of both AK and HXGPRTase is impossible, suggesting that *T. gondii* accommodates only these two routes of purine salvage (Chaudhary *et al.*, 2004). This situation is not common to all apicomplexan parasites, as discussed in Chapter 20. For instance, *Cryptosporidium* has AK but lacks HXGPRTase (Striepen *et al.*, 2004), and *Plasmodium* has HXGPRTase (Vasanthakumar *et al.*, 1990) but lacks AK (Chaudhary *et al.*, 2004).

The *T. gondii* AK is a monomeric protein with molecular mass of 39.3 kDa and shares less than 30% sequence identity with the AKs of other organisms (Sullivan *et al.*, 1999). The *T. gondii* AK shows a strict specificity for adenosine among naturally occurring nucleosides, such as inosine and guanosine (Darling *et al.*, 1999).

Two isozymes of *T. gondii* HXGPRTase have been identified as the predicted translation products of differentially spliced mRNAs transcribed from a single gene (Donald *et al.*, 1996). The crystal structure of isozyme-I is a tetramer composed of four identical subunits (26.4 kDa) (Schumacher *et al.*, 1996, Heroux *et al.*, 1999a). The subunit of isozyme-II possesses an extra 49 amino acids that are inserted seven amino acids downstream of the N-terminus, and another sequence of isozyme-II is identical to that of isozyme-I (Donald *et al.*, 1996). The isozyme-I is predominantly cytosolic, whereas the isozyme-II localizes to the tachyzoite inner membrane complex and the extra 49 amino acids sequence of isozyme-II contains a membrane-targeting signal (Chaudhary *et al.*, 2005). The two isozymes form heterotetramers when co-expressed in *E. coli* (White *et al.*, 2000) or in tachyzoite *in vivo* (Chaudhary *et al.*, 2005). For the membrane association of enzyme, at least two isozyme-II subunits in the tetramer are necessary (Chaudhary *et al.*, 2005). There is no significant difference in kinetic properties between isozymes and it is not understood why *T. gondii* possesses two HXGPRTase isozymes.

The adenosine transporter in *T. gondii* has been identified and a presence of additional permeation pathways for other purine nucleosides and purine bases of *T. gondii* have been suggested (Schwab *et al.*, 1995; De Kong *et al.*, 2003). The gene of *T. gondii* adenosine transporter is cloned and expressed in *Xenopus laevis* oocytes (Chiang *et al.*, 1999). Adenosine uptake by the expressed adenosine transporter is inhibited by various nucleosides, nucleoside analogues, hypoxanthine, guanine, and dipyridamole (Chiang *et al.*, 1999), suggesting that one transporter may play a role of all the transportations of nucleosides and nucleobases into the tachyzoite. Recent studies using nitrobenzylthioinosine, an inhibitor of nucleoside transporter in mammalian cells, and various nonphysiological β-L-enantiomers of purine nucleosides demonstrate that the *T. gondii* adenosine/purine nucleoside transporter(s) lacks stereospecificity and substrate specificity in the transportation of purine nucleosides (Al Safarialani *et al.*, 2003). There is still no consensus of how many transport systems for purine nucleosides/nucleobases are present in the tachyzoite.

8.7 NUCLEOSIDE TRIPHOSPHATE HYDROLASE (NTPASE)

8.7.1 Distinctive Features and Uniqueness

The tachyzoite form of *T. gondii* has a novel nucleoside triphosphate hydrolase (NTPase; EC 3.6.1.3) described for the first time in 1983 (Asai *et al.*, 1983b). The properties of the *T. gondii* NTPase, such as substrate specificity and divalent cation requirements, are most similar to those of ecto(E)-type ATPases (Plesner, 1995). E-type ATPases are ubiquitous in eukaryotic cells and a number of parasitic protozoan E-type ATPases have been reported (see reviews; Plesner, 1995, Meyer-Fernandes, 2002). In the past half century, there have been several suggestions regarding the function of E-type ATPase. The physiological function of *T. gondii* NTPase is undoubtedly different from those of ubiquitous common E-type ATPases for the reasons discussed below. X-ray crystallography has confirmed a unique structure for this NTPase molecule (Matoba *et al.*, 2010; Krung *et al.*, 2012). Although its function is not yet understood, *T. gondii* NTPase must play an extraordinary and unique role for tachyzoite replication and survival.

The most striking feature of *T. gondii* NTPase is its great abundance in the tachyzoite cell. The *T. gondii* NTPase is one of the main proteins, and the calculated amount of NTPase protein is about 2%–8% of the total tachyzoite proteins (Asai *et al.*, 1983b, 1987; Nakaar *et al.*, 1998). In contrast, the common E-type ATPases are extremely low abundance. Even the most highly purified E-type ATPases ever reported would not be visible as a band in electrophoretic gels (Plesner, 1995). The next striking feature of *T. gondii* NTPase is that the enzyme seems to be a dormant enzyme under ordinary conditions. Dithiol compounds, such as dithiothreitol (DTT), are essential for activation of NTPase activity *in vitro* (Asai *et al.*, 1983b). No common E-type ATPases that require activation by DTT have been reported. This feature with the great abundance of NTPase in the tachyzoite cell leads to the surprising conclusion that *T. gondii* conceals an enormous potency for ATP hydrolysis. In fact, the specific activity of ATP hydrolysis (U/mg protein), assayed in the presence of DTT, in the whole tachyzoite cell is about five thousands times higher than that of mouse spleen cell (Asai *et al.*, 1986).

The presence of DTT-activated NTPase in closely related apicomplexan parasite, *Neospora caninum*, has been reported also (Asai *et al.*, 1998). However, DTT-activated NTPase enzymes have not been identified in other protozoa including the apicomplexan parasites, *Plasmodium berghei* (Asai *et al.*, 1986) and *Eimeria tenella* (Asai *et al.*, 1998). DTT-activated NTPase enzymes are thought to be restricted to *T. gondii* and *N. caninum*.

Another critical difference between the *T. gondii* NTPase and ubiquitous common E-type ATPases is that *T. gondii* NTPase seems to fulfil its function from outside of the tachyzoite cell. Tachyzoite replication occurs in the host cell within the parasitophorous vacuole. During formation of the parasitophorous vacuole, the tachyzoite secretes NTPase into the vacuole from the dense-granule, one of the secretory organelles of tachyzoite (Bermudes *et al.*, 1994; Sibley *et al.*, 1994). The secreted soluble NTPase gradually associates with the intravacuolar network that is constructed within the vacuole, and it changes to a membrane-associated insoluble enzyme (Carruthers and Sibley, 1997). The ubiquitous common E-type ATPases including other protozoan E-type enzymes are not secreted from the cells (Plesner, 1995; Meyer-Fernandes, 2002). An illustration showing NTPase in the tachyzoite-infected cell is shown in Fig. 8.9.

8.7.2 NTPase Isoforms and Their Molecular Properties

The type I strains of *T. gondii*, which are acutely virulent in mice (Sibley and Boothroyd, 1992), contain two isoforms of NTPase (Bermudes *et al.*, 1994, Asai *et al.*, 1995). One of the isoforms (NTPase-I in Asai *et al.*, 1995; NTP3 in Bermudes *et al.*, 1994) preferentially hydrolyses triphophate nucleosides, while another isoform (NTPase-II in Asai *et al.*, 1995; NTP1 in Bermudes *et al.*, 1994) hydrolyses tri- and diphosphate nucleosides at approximately equal rates. The NTPase-I isoform appears to be present only in the type I virulent strains, while NTPase-II is universally present in all

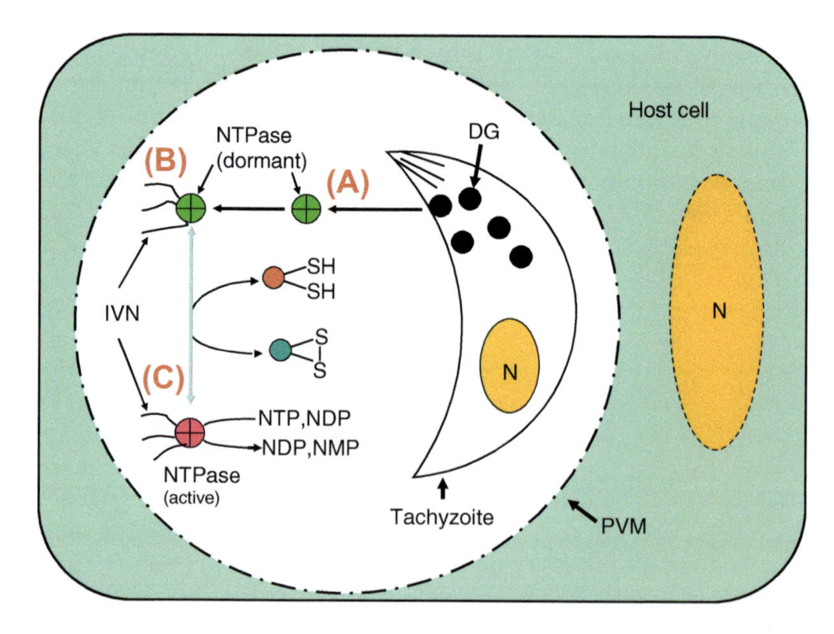

FIGURE 8.9 Behaviour of NTPase in the tachyzoite-infected cell. NTPase is secreted from the dense granule of tachyzoite into the parasitophorous vacuolar space and associates to the intravacuolar network. Then, NTPase activity may be regulated by oxido-reduction change in its molecule affected by dithiol compound or unknown dithiol-disulfide oxidoreductase within parasitophorous vacuole. N-terminal α-helix participates for forming tetramer soluble enzyme (A). N-terminal amino acid residues probably combine with IVN (B). Opening an S-S bond in amino acid residues at 258 and 268 is necessary for activation of enzyme (C). DG, dense granule; IVN, intravacuolar network; N, nucleus; PVM, parasitophorous vacuole membrane.

T. gondii strains (Asai *et al.*, 1995). Although the presence of NTPase-I seems to be one of the primary factors of virulence in mice, no direct evidence for a role in virulence has been proven. *Neospora caninum*, which is avirulent in mice, has only NTPase-I type enzyme, and no NTPase-II type enzyme has been detected (Asai *et al.*, 1998).

The complete cDNAs for NTPase-I and NTPase-II encode predicted open reading frames of the identical size that differ in 16 of 628 amino acids between the two isoforms (Asai *et al.*, 1995). The molecular mass of native NTPases are approximately 260 kDa, composed of four identical subunits with predicted molecular mass of 67 kDa (Asai *et al.*, 1995). Both isoforms of the NTPase contain an N-terminal hydrophobic signal peptide (25 amino acids) that is absent in native NTPase purified from the tachyzoites (Asai *et al.*, 1995). It is supposed that this signal peptide serves as a signal for transport of NTPase to the dense granule. Other dense granule proteins have similar N-terminal hydrophobic signal peptides (Cesbron-Delauw *et al.*, 1996).

As mentioned above, the primary difference between these NTPase isoforms lies in their ability to hydrolyse nucleoside triphosphate versus diphosphate substrates. While NTPase-II hydrolyses ATP to ADP and ADP to AMP at almost the same rate, both native and recombinant NTPase-I hydrolyse ADP to AMP at a much slower rate, less than 1% of the rate for ATP (Asai *et al.*, 1995, 1998). This suggests that *T. gondii* NTPase should be classified as NTP diphosphohydrolase (apyrase; EC 3.6.1.5), a new gene family of E-type ATPases. *T. gondii* NTPase has homology to apyrase, and an antibody against *T. gondii* NTPase recognizes apyrases of potato and *Schistosoma mansoni* (Vasconcelos *et al.*, 1996), and *Trypanosoma cruzi* (Fietto *et al.*, 2004). The abundant DTT-activated NTPase in *N. caninum* has no apyrase activity, and its substrate specificity is restricted to nucleoside triphosphate (Asai *et al.*, 1998).

8.7.3 Possible Physiological Function for NTPase

Protozoan parasites are purine auxotrophs (Berens *et al.*, 1995). It has been speculated that the NTPases of *T. gondii* are involved in the salvage of purine nucleosides from the host cell (Bermudes *et al.*, 1994). The *T. gondii* NTPase is secreted from the tachyzoite into the parasitophorous vacuole; therefore, it contains higher potency of hydrolyzing ATP to ADP and AMP (Fig. 8.7). As mentioned earlier, purines are salvaged through adenosine kinase or HXGPRTase. However, the tachyzoite lacks an ecto-5′-nucleotidase activity that would dephosphorylate AMP to adenosine, the substrate of adenosine kinase (Ngo *et al.*, 2000). Furthermore, it is clear that the primary role of the NTPase of *N. caninum* tachyzoites is not likely to involve purine acquisition, since that parasite lacks enzymatic activities for progressively cleaving nucleotides to their monophosphate form (Asai *et al.*, 1998). It is therefore doubtful whether the *T. gondii* NTPases are involved in purine salvage.

Exposure of the tachyzoites to DTT can activate egress of previously nonmotile intravacuolar tachyzoites within 60 seconds with a concurrent Ca^{2+} flux (Stommel *et al.*, 1997). Furthermore, *T. gondii* NTPase is found largely in an inactive-oxidized form in the parasitophorous vacuole and readily activated by DTT. This leads to subsequent rapid depletion of host ATP and egress of tachyzoites (Silverman *et al.*, 1998). Whether or not the *T. gondii* NTPase is involved in naturally occurring tachyzoite egress from host cell, these phenomena are quite suggestive for the physiological function of *T. gondii* NTPase.

Since DTT is not present in nature, thioredoxin, the most abundant cellular reducing dithiol catalyst, was tested as an activator. The reduced form of thioredoxin can activate NTPase and the oxidized form of thioredoxin has a reversible effect on NTPase activity *in vitro*

(Asai and Kim, 1987). This suggests that thioredoxin may regulate NTPase activity *in vivo*. However, host cell thioredoxin may be excluded from the parasitophorous vacuole by the parasitophorous vacuolar membrane (Schwab *et al.*, 1994). Although glutathione itself has no activation effect on NTPase *in vitro* (Asai *et al.*, 1983b), it is reported that glutathione promoters activate a Ca^{2+} flux and decrease ATP levels in tachyzoite-infected fibroblasts (Stommel *et al.*, 2001). These investigators further postulate that the NTPase is activated by glutaredoxin, a small protein with an active centre disulphide. Glutaredoxin may be reduced by glutathione, leading ultimately to tachyzoite egress. Finally, the tachyzoites control their egress by secreting glutaredoxin from their bodies into the parasitophorous vacuole for activation of NTPase (Stommel *et al.*, 2001). This model requires further testing but it is possible that NTPase may be activated by unknown dithiol compound derived from the host cell or tachyzoites. Alternatively NTPase activity may be regulated by an unknown protein disulphide oxidoreductase within the vacuole.

There is still no consensus for the physiological function of *T. gondii* NTPase. Despite this, it is thought that the *T. gondii* NTPase plays an extremely important function in tachyzoite replication and survival, since the enzyme is expressed mainly in tachyzoite form (Ferguson *et al.*, 1999). Attempts to disrupt the genes have failed and antisense depletion studies suggest that NTPase is essential for parasite replication (Nakaar *et al.*, 1999). Furthermore, it is reported that monoclonal antibodies against NTPase can reduce the replication of *T. gondii* (Tan *et al.*, 2010).

References

Al-Anouti, F., Tomavo, S., Parmley, S., Ananvoranich, S., 2004. The expression of lactate dehydrogenase is important for the cell cycle of *Toxoplasma gondii*. J. Biol. Chem. 279, 52300–52311.

Al Safarialani, O.N., Naguib, F.N.M., El Kouni, M.H., 2003. Uptake of nitrobenzylthioinosine and purine beta-L-nucleosides by intracellular *Toxoplasma gondii*. Antimicrob. Agents Chemother. 47, 3247–3251.

Anderson, R.G., Jacobson, K., 2002. A role for lipid shells in targeting proteins to caveolae, rafts, and other lipid domains. Science 296, 1821–1825.

Aoki, T., Shimogawara, R., Ochiai, K., Yamasaki, H., Shimada, J., 1994. Molecular characterization of a carbamoyl-phosphate synthetase II (CPSII) gene from *Trypanosoma cruzi*. Adv. Exp. Med. Biol. 370, 513–516.

Asai, T., O'Sullivan, W.J., Kobayashi, M., Gero, A.M., Yokogawa, M., Tatibana, M., 1983a. Enzymes of the *de novo* pyrimidine biosynthetic pathway in *Toxoplasma gondii*. Mol. Biochem. Parasitol. 7, 89–100.

Asai, T., O'Sullivan, W.J., Tatibana, M., 1983b. A potent nucleoside triphosphate hydrolase from the parasitic protozoan *Toxoplasma gondii*. J. Biol. Chem. 258, 6816–6822.

Asai, T., Kanazawa, T., Kobayashi, S., Takeuchi, T., Kim, T., 1986. Do protozoa conceal a high potency of nucleoside triphosphate hydrolysis present in *Toxoplasma gondii*? Comp. Biochem. Physiol. B 85, 365–368.

Asai, T., Kim, T., Kobayashi, M., Kojima, S., 1987. Detection of nucleoside triphosphate hydrolase as a circulating antigen in sera of mice infected with *Toxoplasma gondii*. Infect. Immun. 55, 1332–1335.

Asai, T., Kim, T., 1987. Possible regulation mechanism of potent nucleoside triphosphate hydrolase in *Toxoplasma gondii*. Zbl. Bakt. Hyg. A 264, 464–467.

Asai, T., Lee, C.S., Chandler, A., O'Sullivan, W.J., 1990. Purification and characterization of uracil phosphoribosyltransferase from *Crithidia luciliae*. Comp. Biochem. Physiol. B 95, 159–163.

Asai, T., Miura, S., Sibley, L.D., Okabayashi, H., Takeuchi, T., 1995. Biochemical and molecular characterization of nucleoside triphosphate hydrolase isozymes from the parasitic protozoan *Toxoplasma gondii*. J. Biol. Chem. 270, 11391–11397.

Asai, T., Howe, D.K., Nakajima, K., Nozaki, T., Takeuchi, T., Sibley, L.D., 1998. *Neospora caninum*: Tachyzoites express a potent type-I nucleoside triphosphate hydrolase, but lack nucleoside diphosphate hydrolase activity. Exp. Parasitol. 90, 277–285.

Augustine, P.C., 1980. Effects of storage time and temperature on amylopectin levels and oocysts production of *Eimeria meleagrimitis* oocysts. Parasitology 81, 519–524.

Azzouz, N., Rauscher, B., Gerold, P., Cesbron-Delauw, M.F., Dubremetz, J.F., Schwarz, R.T., 2002. Evidence for de novo sphingolipid biosynthesis in *Toxoplasma gondii*. Int. J. Parasitol. 32, 677–6784.

Baldock, C., Rafferty, J.B., Sedelnikova, S.E., Baker, P.J., Stuitje, A.R., Slabas, A.R., Hawkes, T.R., Rice, D.W., 1996. A mechanism of drug action revealed by structural studies of enoyl reductase. Science 274, 2107–2110.

Ball, S.G., Morell, M.K., 2003. From bacterial glycogen to starch: understanding the biogenesis of the plant starch granule. Annu. Rev. Plant. Biol. 54, 207–233.

Berens, R.L., Krug, E.C., Marr, J.J., 1995. Purine and pyrimidine metabolism. In: Mar, J.J., Muller, M. (Eds.), Biochemistry and Molecular Biology of Parasites. Academic Press, London, pp. 89–117.

Bermudes, D., Peck, K.R., Afifi-Afifi, M., Beckers, C.J.M., Joiner, K.A., 1994. Tandemly repeated genes encode nucleoside triphosphate hydrolase isoforms secreted into the parasitophorous vacuole of *Toxoplasma gondii*. J. Biol. Chem. 269, 29252–29260.

Bisanz, C., Bastien, O., Grando, D., Jouhet, J., Marechal, E., Cesbron-Delauw, M.F., 2006. *Toxoplasma gondii* acyl-lipid metabolism: de novo synthesis from apicoplast generated fatty acids versus scavenging of host cell precursors. Biochem. J. 394, 297–305.

Bjornberg, O., Rowland, P., Larsen, S., Jensen, K.F., 1997. Active site of dihydroorotate dehydrogenase A from *Lactococcus lactis* investigated by chemical modification and mutagenesis. Biochemistry 36, 16197–16205.

Bohne, W., Roos, D.S., 1997. Stage-specific expression of a selectable marker in *Toxoplasma gondii* permits selective inhibition of either tachyzoites or bradyzoites. Mol. Biochem. Parasitol. 88, 115–126.

Botté, C., Saïdani, N., Mondragon, R., Mondragón, M., Isaac, G., Mui, E., McLeod, R., Dubremetz, J.F., Vial, H., Welti, R., Cesbron-Delauw, M.F., Mercie, C., Maréchal, E., 2008. Subcellular localization and dynamics of a digalactolipid-like epitope in *Toxoplasma gondii*. J. Lipid Res. 49, 746–762.

Bottova, I., Hehl, A.B., Stefanić, S., Fabriàs, G., Casas, J., Schraner, E., Pieters, J., Sonda, S., 2009. Host cell P-glycoprotein is essential for cholesterol uptake and replication of *Toxoplasma gondii*. J. Biol. Chem. 284, 17438–17448.

Bretscher, M.S., Munro, S., 1993. Cholesterol and the Golgi apparatus. Science 261, 1280–1281.

Bzik, D.J., Fox, B.A., Gonyer, K., 1993. Expression of *Plasmodium falciparum* lactate dehydrogenase in *Escherichia coli*. Mol. Biochem. Parasitol. 59, 155–166.

Caffaro, C.E., Boothroyd, J.C., 2011. Evidence for host cells as the major contributor of lipids in the intravacuolar network of *Toxoplasma*-infected cells. Eukaryo. Cell. 10, 1095–1099.

Cai, X., Fuller, A.L., McDougald, L.R., Zhu, G., 2003. Apicoplast genome of the coccidian *Eimeria tenella*. Gene 321, 39–46.

Carman, G.M., Zeimetz, G.M., 1996. Regulation of phospholipid biosynthesis in the yeast *Saccharomyces cerevisiae*. J. Biol. Chem. 271, 13293–13296.

Carruthers, V.B., Sibley, L.D., 1997. Sequential protein secretion from three distinct organelles of *Toxoplasma gondii* accompanies invasion of human fibroblasts. Eur. J. Cell Biol. 73, 114–123.

Carter, D., Donald, R.G., Roos, D., Ullman, B., 1997. Expression, purification, and characterization of uracil phosphoribosyltransferase from *Toxoplasma gondii*. Mol. Biochem. Parasitol. 87, 137–144.

Chansiri, K., Bagnara, A.S., 1995. The structural gene for carbamyl phosphate synthetase from the protozoan parasite *Babesia bovis*. Mol. Biochem. Parasitol. 74, 239–234.

Charron, A.J., Sibley, L.D., 2002. Host cells: mobilizable lipid resources for the intracellular parasite *Toxoplasma gondii*. J. Cell Sci. 115, 3049–3059.

Chaudhary, K., Darling, J.A., Fohl, L.M., Sullivan Jr., W.J., Donald, R.K.G., Pfefferkorn, E.R., Ullman, B., Roos, D.S., 2004. Purine salvage pathways in the apicomplexan parasite *Toxoplasma gondii*. J. Biol. Chem. 279, 31221–31227.

Chaudhary, K., Donald, R.K.G., Nishi, M., Carter, D., Ullman, B., Roos, D.S., 2005. Differential localization of alternatively spliced hypoxanthine-xanthine-guanine phosphoribosyltransferase isoforms in *Toxoplasma gondii*. J. Biol. Chem. 270, 22053–22059.

Chiang, C.W., Carter, N., Sullivan Jr., W.J., Donald, R.G.K., Roos, D.S., Naguib, F.N.M., El Kouni, M.H., Ullman, B., Wilson, C.M., 1999. The adenosine transporter of *Toxoplasma gondii*. Identification by insertional mutagenesis, cloning, and recombinant expression. J. Biol. Chem. 274, 35255–35261.

Cesbron-Delauw, M.F., Lecordier, L., Mercier, C., 1996. Role of secretory dense granule organelles in the pathogenesis of toxoplasmosis. Curr. Trop. Microbiol. Immunol. 219, 59–65.

Clayton, R.B., 1964. The utilization of sterols by insects. J. Lipid Res. 15, 3–19.

Cleary, M.D., 2008. Cell type-specific analysis of mRNA synthesis and decay in vivo with uracil phosphoribosyltransferase and 4-thiouracil. Methods Enzymol. 448, 379–406.

Coleman, R.A., Lee, D.P., 2004. Enzymes of triacylglycerol synthesis and their regulation. Prog. Lipid Res. 43, 134–176.

Coppens, I., Joiner, K.A., 2003. Host but not parasite cholesterol controls *Toxoplasma* cell entry by modulating organelle discharge. Mol. Biol. Cell 14, 3804–3820.

Coppens, I., Sinai, A.P., Joiner, K.A., 2000. *Toxoplasma gondii* exploits host low-density lipoprotein receptor-mediated endocytosis for cholesterol acquisition. J. Cell Biol. 149, 167–180.

Coppens, I., Vielemeyer, O., 2005. Insights into unique physiological features of neutral lipids in Apicomplexa: from storage to potential mediation in parasite metabolic activities. Int. J. Parasitol. 35, 597–615.

Coppin, A., Varre, J.S., Lienard, L., Dauvillee, D., Guerardel, Y., Soyer-Gobillard, M.O., Buleon, A., Ball, S., Tomavo, S., 2005. Evolution of plant-like crystalline storage polysaccharide in the protozoan parasite *Toxoplasma gondii* argues for a red alga ancestry. J. Mol. Evol. 60, 257–267.

Dai, Y.P., Lee, C.S., O'Sullivan, W.J., 1995. Properties of uracil phosphoribosyltransferase from *Giardia intestinalis*. Int. J. Parasitol. 25, 207–214.

Dando, C., Schroeder, E.R., Hunsaker, L.A., Deck, L.M., Royer, R.E., Zhou, X., Parmley, S.F., Vander Jagt, D.L., 2001. The kinetic properties and sensitivities to inhibitors of lactate dehydrogenases (LDH1 and LDH2) from *Toxoplasma gondii* comparisons with pLDH from *Plasmodium falciparum*. Mol. Biochem. Parasitol. 118, 23–32.

Dantas-Leite, L., Urbina, J.A., de Souza, W., Vommaro, R.C., 2004. Selective anti-*Toxoplasma* gondii activities of azasterols. Int. J. Antimicrob. Agents 23, 620–626.

Darling, J.A., Sullivan Jr., W.J., Carter, D., Ullman, B., Roos, D.S., 1999. Recombinant expression, purification, and characterization of *Toxoplasma gondii* adenosine kinase. Mol. Biochem. Parasitol. 103, 15–23.

Daubenberger, C.A., Tisdale, E.J., Curcic, M., Diaz, D., Silvie, O., Mazier, D., Eling, W., Bohrmann, B., Matile, H., Pluschke, G., 2003. The N'-terminal domain of glyceraldehyde-3-phosphate dehydrogenase of the apicomplexan *Plasmodium falciparum* mediates GTPase Rab2-dependent recruitment to membranes. Biol. Chem. 384, 1227–1237.

Davidson, J.N., Chen, K.C., Jamison, R.S., Musmanno, L.A., Kern, C.B., 1993. The evolutionary history of the first three enzymes in pyrimidine biosynthesis. Bioessays 15, 157–163.

De Kong, H.P., Al-Salabi, M.I., Cohen, A.M., Coombs, G.H., Wastling, J.M., 2003. Identification and characterisation of high affinity nucleoside and nucleobase transporters in *Toxoplasma gondii*. Int. J. Parasitol. 33, 821–831.

de Macedo, C.S., Shams-Eldin, H., Smith, T.K., Schwarz, R.T., Azzouz, N., 2003. Inhibitors of glycosyl-phosphatidylinositol anchor biosynthesis. Biochimie 85, 465–472.

de Melo, E.J., de Souza, W., 1996. Pathway of C6-NBD-Ceramide on the host cell infected with *Toxoplasma gondii*. Cell Struct. Funct. 21, 47–52.

Donald, R.G.K., Roos, D.S., 1995. Insertional mutagenesis and marker rescue in a protozoan parasite: cloning of the uracil phosphoribosyltransferase locus from *Toxoplasma gondii*. Proc. Natl. Acad. Sci. USA 92, 5749–5753.

Donald, R.G.K., Carter, D., Ullman, B., Roos, D.S., 1996. Insertional tagging, cloning, and expression of the *Toxoplasma gondii* hypoxanthine-xanthine-guanine phosphoribosyltransferase gene. J. Biol. Chem. 271, 14010–14019.

Dzierszinski, F., Popescu, O., Toursel, C., Slomianny, C., Yahiaoui, B., Tomavo, S., 1999. The protozoan parasite *Toxoplasma gondii* expresses two functional plant-like glycolytic enolases. Implications for evolutionary origin of apicomplexans. J. Biol. Chem. 274, 24888–24895.

Dzierszinski, F., Mortuaire, M., Dendouga, N., Popescu, O., Tomavo, S., 2001. Differential expression of two plant-like enolases with distinct enzymatic and antigenic properties during stage conversion of the protozoan parasite *Toxoplasma gondii*. J. Mol. Biol. 309, 1017–1027.

Ehrenman, K., Sehgal, A., Lige, B., Stedman, T.T., Joiner, K.A., Coppens, I., 2010. Novel roles for ATP-binding cassette G transporters in lipid redistribution in *Toxoplasma*. Mol. Microbiol. 76, 1232–1249.

Eller, B., Datta, A., Lynn, B.C., Sinai, A.P., 2012. A case for host involvement in the glycosylation of *Toxoplasma tissue cysts*. 12th International Workshops on Opportunistic Pathogens August 2012, Tarrytown, NY. Abstract T46, page 64.

Fadok, V.A., Bratton, D.L., Konowal, A., Freed, P.W., Westcott, J.Y., Henson, P.M., 1998. Macrophages that have ingested apoptotic cells in vitro inhibit proinflammatory cytokine production through autocrine/paracrine mechanisms involving TGF-beta, PGE2, and PAF. J. Clin. Invest. 101, 890–898.

Fadok, V.A., Bratton, D.L., Henson, P.M., 2001. Phagocyte receptors for apoptotic cells: recognition, uptake, and consequences. J. Clin. Invest. 108, 957–962.

Fauquenoy, S., Morelle, W., Hovasse, A., Bednarczyk, A., Slomianny, C., Schaeffer, C., Van Dorsselaer, A., Tomavo, S., 2008. Proteomics and glycomics analyses of *N*-glycosylated structures involved in *Toxoplasma gondii*-host cell interactions. Mol. Cell. Proteomics 7, 891–910.

Fauquenoy, S., Hovasse, A., Sloves, P.-J., Morelle, W., Alayi, D.T., Slomianny, C., Werkmeister, E., Schaeffer, C., Van Dorsselaer, A., Tomavo, S., 2011. Unusual N-glycan structures required for trafficking *Toxoplasma gondii* GAP50 to the inner membrane complex regulate host cell entry through parasite motility. Mol. Cell. Proteomics 10. M111.008953.

Feo, S., Arcuri, D., Piddini, E., Passantino, R., Giallongo, A., 2000. ENO1 gene product binds to the c-myc promoter and acts as a transcriptional repressor: relationship with Myc promoter-binding protein 1 (MBP-1). FEBS Lett. 472, 47–52.

Fietto, J.L., DeMarco, R., Nascimento, I.P., Castro, I.M., Carvalho, T.M., de Souza, W., Bahia, T., Alves, M.J., Verjovski-Almeida, S., 2004. Characterization and immunolocalization of an NTP diphosphohyrolase of *Trypanosoma cruzi*. Biochem. Biophys. Res. Commun. 316, 454–460.

Flores, M.V., O'Sullivan, W.J., Stewart, T.S., 1994. Characterisation of the carbamoyl phosphate synthetase gene from *Plasmodium falciparum*. Mol. Biochem. Parasitol. 68, 315–318.

Fox, B.A., Bzik, D.J., 2002. *De novo* pyrimidine biosynthesis is required for virulence of *Toxoplasma gondii*. Nature 415, 85–88.

Fox, B.A., Bzik, D.J., 2003. Organization and sequence determination of glutamine-dependent cabamoyl phosphate synthestase II in *Toxoplasma gondii*. Int. J. Parasitol. 33, 89–96.

Fox, B.A., Ristuccia, J.G., Bzik, D.J., 2009. Genetic identification of essential indels and domains in carbamoyl phosphate synthetase II of *Toxoplasma gondii*. Int. J. Parasitol. 39, 533–539.

Ferguson, M.A., Williams, A.F., 1988. Cell-surface anchoring of proteins via glycosyl-phosphatidylinositol structures. Annu. Rev. Biochem. 57, 285–320.

Ferguson, D.J., Cesbron-Delauw, M.F., Dubremetz, J.F., Sibley, L.D., Joiner, K.A., Wright, S., 1999. The expression and distribution of dense granule proteins in the enteric (Coccidian) forms of *Toxoplasma gondii* in the small intestine of the cat. Exp. Parasitol. 91, 203–211.

Ferguson, D.J., Parmley, S.F., Tomavo, S., 2002. Evidence for nuclear localisation of two stage-specific isoenzymes of enolase in *Toxoplasma gondii* correlates with active parasite replication. Int. J. Parasitol. 32, 1399–1410.

Gao, G., Nara, T., Nakajima-Shimada, J., Aoki, T., 1999. Novel organization and sequences of five genes encoding all six enzymes for *de novo* pyrimidine biosynthesis in *Trypanosoma cruzi*. J. Mol. Biol. 285, 149–161.

Goldstein, J.L., Brown, M.S., 1990. Regulation of the mevalonate pathway. Nature 343, 425–430.

Goodman, C.D., McFadden, G.I., 2007. Fatty acid biosynthesis as a drug target in apicomplexan parasites. Curr. Drug Targets 5, 15–30.

Gupta, N., Zahn, M.M., Coppens, I., Joiner, K.A., Voelker, D.R., 2005. Selective disruption of phosphatidylcholine metabolism of the intracellular parasite *Toxoplasma gondii* arrests its growth. J. Biol. Chem. 280, 16345–16353.

Harris, J.R., Adrian, M., Petry, F., 2004. Amylopectin: a major component of the residual body in *Cryptosporidium parvum* oocysts. Parasitology 128, 269–282.

Heath, R.J., White, S.W., Rock, C.O., 2001. Lipid biosynthesis as a target for antibacterial agents. Prog. Lipid Res. 40, 467–497.

Heroux, A., White, E.L., Ross, L.J., Borhani, D.W., 1999a. Crystal structures of the *Toxoplasma gondii* hypoxanthine-guanine phosphoribosyltransferase-GMP and -IMP complexes: comparison of purine binding interactions with the XMP complex. Biochemistry 38, 14485–14494.

Heroux, A., White, E.L., Ross, L.J., Davis, R.L., Borhani, D.W., 1999b. Crystal structure of the *Toxoplasma gondii* hypoxanthine-guanine phosphoribosyltransferase with XMP, pyrophosphate, and two Mg(2+) ions bound: insights into the catalytic mechanism. Biochemistry 38, 14495–14506.

Hill, B., Kilsby, J., Rogerson, G.W., McIntosh, R.T., Ginger, C.D., 1981. The enzymes of pyrimidine biosynthesis in a range of parasitic protozoa and helminths. Mol. Biochem. Parasitol. 2, 123–134.

Ibrahim, M., Azzouz, N., Gerold, P., Schwarz, R.T., 2001. Identification and characterisation of *Toxoplasma gondii* protein farnesyltransferase. Int. J. Parasitol. 31, 1489–1497.

Iltzsch, M.H., 1993. Pyrimidine salvage pathways in *Toxoplasma gondii*. J. Euk. Microbiol. 40, 24–28.

Jakobsson, A., Westerberg, R., Jacobsson, A., 2006. Fatty acid elongases in mammals: their regulation and roles in metabolism. Prog. Lipid Res. 45, 237–249.

Jelenska, J., Crawford, M.J., Harb, O.S., Zuther, E., Haselkorn, R., Roos, D.S., Gornicki, P., 2001. Subcellular localization of acetyl-CoA carboxylase in the apicomplexan parasite *Toxoplasma gondii*. Proc. Natl. Acad. Sci. USA 98, 2723–2728.

Jelenska, J., Sirikhachornkit, A., Haselkorn, R., Gornicki, P., 2002. The carboxyltransferase activity of the apicoplast acetyl-CoA carboxylase of Toxoplasma gondii is the target of aryloxyphenoxypropionate inhibitors. J. Biol. Chem. 277, 23208–23215.

Jewett, T.J., Sibley, L.D., 2003. Aldolase forms a bridge between cell surface adhesions and the actin cytoskeleton in apicomplexan parasites. Mol. Cell. 11, 885–894.

Jones, M.E., 1980. Pyrimidine nucleotide biosynthesis in animals: genes, enzymes, and regulation of UMP biosynthesis. Annu. Rev. Biochem. 49, 253–279.

Jomaa, H., Wiesner, J., Sanderbrand, S., Altincicek, B., Weidemeyer, C., Hintz, M., Turbachova, I., Eberl, M., Zeidler, J., Lichtenthaler, H.K., Soldati, D., Beck, E., 1999. Inhibitors of the nonmevalonate pathway of isoprenoid biosynthesis as antimalarial drugs. Science 285, 1573–1576.

Karkhanis, Y.D., Allocco, J.J., Schmatz, D.M., 1993. Amylopectin synthase of *Eimeria tenella*: identification and kinetic characterization. J. Euk. Microbiol. 40, 594–598.

Kavanagh, K.L., Elling, R.A., Wilson, D.K., 2004. Structure of *Toxoplasma gondii* LDH1: active-site differences from human lactate dehydrogenases and the structural basis for efficient APAD+ use. Biochemistry 43, 879–889.

Khan, A.I., Chowdhry, B.Z., Yon, R.J., 1999. Wheat-germ aspartate transcarbamoylase: revised purification, stability and re-evaluation of regulatory kinetics in terms of the Monod-Wyman-Changeux model. Eur. J. Biochem. 259, 71–78.

Köhler, S., Delwiche, C.F., Denny, P.W., Tilney, L.G., Webster, P., Wilson, R.J., Palmer, J.D., Roos, D.S., 1997. A plastid of probable green algal origin in apicomplexan parasites. Science 275, 1485–1489.

Krug, E.C., Mar, J.J., Berens, R.L., 1989. Purine metabolism in *Toxoplasma gondii*. J. Biol. Chem. 264, 10601–10607.

Krung, U., Zebisch, M., Krauss, M., Sträter, N., 2012. Structural insight into activation mechanism of *Toxoplasma gondii* nucleoside triphosphate diphosphohydrolases by disulfide reduction. J. Biol. Chem. 287, 3051–3066.

Krungkrai, S.R., DelFraino, B.J., Smiley, J.A., Prapunwattana, P., Mitamura, T., Horii, T., Krungkrai, J., 2005. A novel enzyme complex of orotate phosphoribosyltransferase and orotidine 5′-monophosphate decarboxylase in human malaria parasite *Plasmodium falciparum*: physical association, kinetics, and inhibition characterization. Biochemistry 44, 1643–1652.

Lee, H., Guo, Y., Ohta, M., Xiong, L., Stevenson, B., Zhu, J.K., 2002. LOS2, a genetic locus required for cold-responsive gene transcription encodes a bi-functional enolase. EMBO J. 21, 2692–2702.

Liscum, L., Underwood, K.W., 1995. Intracellular cholesterol transport and compartmentation. J. Biol. Chem. 270, 15443–15446.

Lekutis, C., Ferguson, D.J., Grigg, M.E., Camps, M., Boothroyd, J.C., 2001. Surface antigens of *Toxoplasma gondii*: variations on a theme. Int. J. Parasitol. 31, 1285–1292.

Lige, B., Romano, J.D., Bandaru, V.V., Ehrenman, K., Levitskaya, J., Sampels, V., Haughey, N.J., Coppens, I., 2011. Deficiency of a Niemann-Pick, type C1-related protein in *Toxoplasma* is associated with multiple lipidoses and increased pathogenicity. PLoS Pathog. 7, e1002410.

Lige, B., Sampels, V., Coppens, I., 2013. Characterization of a second sterol-esterifying enzyme in Toxoplasma highlights the importance of cholesterol storage pathways for the parasite. Mol. Microbiol. 87, 951–967.

Ling, Y., Sahota, G., Odeh, S., Chan, J.M., Araujo, F.G., Moreno, S.N., Oldfield, E., 2005. Bisphosphonate inhibitors of *Toxoplasma gondii* growth: *In vitro*, QSAR, and in vivo investigations. J. Med. Chem. 48, 3130–3140.

Liu, X., Guy, H.I., Evans, D.R., 1994. Identification of the regulatory domain of the mammalian multifunctional protein CAD by the construction of an *Escherichia coli* hamster hybrid carbamoyl-phosphate synthetase. J. Biol. Chem. 269, 27747–22775.

Luk, F.C., Johnson, T.M., Beckers, C.J., 2008. N-linked glycosylation of proteins in the protozoan parasite *Toxoplasma gondii*. Mol. Biochem. Parasitol. 157, 169–178.

Luo, Q., Upadhya, R., Zhang, H., Madrid-Aliste, C., Nieves, E., Kim, K., Angeletti, R.H., Weiss, L.M., 2011. Analysis of the glycoproteome of *Toxoplasma gondii* using lectin affinity chromatography and tandem mass spectrometry. Microbes. Infect. 13, 1199–1210.

Marechal, E., Azzouz, N., de Macedo, C.S., Block, M.A., Feagin, J.E., Schwarz, R.T., Joyard, J., 2002. Synthesis of chloroplast galactolipids in apicomplexan parasites. Eukaryot. Cell 1, 653–656.

Martin, M.B., Grimley, J.S., Lewis, J.C., Heath 3rd, H.T., Bailey, B.N., Kendrick, H., Yardley, V., Caldera, A., Lira, R., Urbina, J.A., Moreno, S.N., Docampo, R., Croft, S.L., Oldfield, E., 2001. Bisphosphonates inhibit the growth of *Trypanosoma brucei*, *Trypanosoma cruzi*, *Leishmania donovani*, *Toxoplasma gondii*, and *Plasmodium falciparum*: a potential route to chemotherapy. J. Med. Chem. 44, 909–916.

Martins-Duarte, E.S., Jones, S.M., Gilbert, I.H., Atella, G.C., de Souza, W., Vommaro, R.C., 2009. Thiolactomycin analogues as potential anti-*Toxoplasma gondii* agents. Parasitol. Int. 58, 411–415.

Martins-Duarte, E.S., Lemgruber, L., Lorente, S.O., Gros, L., Magaraci, F., Gilbert, I.H., de Souza, W., Vommaro, R.C., 2011. Evaluation of three novel azasterols against Toxoplasma gondii. Vet. Parasitol. 177, 157–161.

Matoba, K., Shiba, T., Takeuchi, T., Sibley, L.D., Seki, M., Kikyo, F., Horiuchi, T., Asai, T., Harada, S., 2010. Crystallization and preliminary X-ray structural analysis of nucleoside triphosphate hydrolase from Neospora caninum and *Toxoplasma gondii*. Acta. Cryst. F66, 1445–1448.

Matsuzaki, M., Misumi, O., Shin, I.T., Maruyama, S., et al., 2004. Genome sequence of the ultrasmall unicellular red alga *Cyanidioschyzon merolae* 10D. Nature 428, 653–657.

Matthiesen, S.H., Shenoy, S.M., Kim, K., Singer, R.H., Satir, B.H., 2001. A parafusin related *Toxoplasma* protein in Ca^{2+} regulated secretory organelles. Eur. J. Cell. Biol. 80, 775–783.

Matthiesen, S.H., Shenoy, S.M., Kim, K., Singer, R.H., Satir, B.H., 2003. Role of the parafusin ortholog, PRP1, in microneme exocytosis and cell invasion in *Toxoplasma gondii*. Cell Microbiol. 5, 613–662.

Mazumdar, J., Striepen, B., 2007. Make it or take it: fatty acid metabolism of apicomplexan parasites. Eukaryot. Cell 6, 1727–1735.

Mazumdar, J.H., Wilson, E., Masek, K.A., Hunter, C., Striepen, B., 2006. Apicoplast fatty acid synthesis is essential for organelle biogenesis and parasite survival in *Toxoplasma gondii*. Proc. Natl. Acad. Sci. USA 103, 13192–13197.

McFadden, G.I., Reith, M.E., Munholland, J., Lang-Unnasch, N., 1996. Plastid in human parasites. Nature 381, 482.

McFadden, D.C., Boothroyd, J.C., 1999 Sep-Oct. Cytochrome b mutation identified in a decoquinate-resistant mutant of Toxoplasma gondii. J Eukaryot Microbiol. 46 (5), 81S–82S.

McLeod, R., Muench, S.P., Rafferty, J.B., Kyle, D.E., Mui, E.J., Kirisits, M.J., Mack, D.G., Roberts, C.W., Samuel, B.U., Lyons, R.E., Dorris, M., Milhous, W.K., Rice, D.W., 2001. Triclosan inhibits the growth of *Plasmodium falciparum* and *Toxoplasma gondii* by inhibition of apicomplexan Fab I. Int. J. Parasitol. 31, 109–113.

Mejias-Torres, I.A., Zimmermann, B.H., 2002. Molecular cloning, recombinant expression and partial characterization of the asparatate transcarbamoylase from *Toxoplasma gondii*. Mol. Biochem. Parasitol. 119, 191–201.

Meyer-Fernandes, J.R., 2002. Ecto-ATPases in protozoa parasites: looking for a function. Parasitol. Int. 51, 299–303.

Mori, M., Tatibana, M., 1978. Multi-enzyme complex of glutamine-dependent carbamyl phosphate synthetase with asparatate carbamyltransferase and dihydroorotase from rat ascites-hepatoma cells. Eur. J. Biochem. 86, 381–388.

Nagel, S.D., Boothroyd, J.C., 1989. The major surface antigen, P30, of *Toxoplasma gondii* is anchored by a glycolipid. J. Biol. Chem. 264, 5569–5574.

Naik, R.S., Branch, O.H., Woods, A.S., Vijaykumar, M., Perkins, D.J., Nahlen, B.L., Lal, A.A., Cotter, R.J., Costello, C.E., Ockenhouse, C.F., Davidson, E.A., Gowda, D.C., 2000. Glycosylphosphatidylinositol anchors of *Plasmodium falciparum*: molecular characterization and naturally elicited antibody response that may provide immunity to malaria pathogenesis. J. Exp. Med. 192, 1563–1576.

Nakaar, V., Bermudes, D., Peck, K.R., Joiner, K.A., 1998. Upstream elements required for expression of nucleoside triphosphate hydrolase genes of *Toxoplasma gondii*. Mol. Biochem. Parasitol. 92, 229–239.

Nakaar, V., Samuel, B.U., Ngo, E.O., Joiner, K.A., 1999. Target reduction of nucleoside triphosphate hydrolase by antisense RNA inhibitors *Toxoplasma gondii* proliferation. J. Biol. Chem. 274, 5083–5087.

Nakai, Y., Ogimoto, K., 1983. Relationship between amylopectin and viability of *Eimeria tenella* sporozoite. Jpn. J. Vet. Sci. 45, 127–129.

Nara, T., Gao, G., Yamasaki, H., Nakajima-Shimada, J., Aoki, T., 1998. Carbamoyl-phosphate synthetase II in kinetoplastids. Biochim. Biophys. Acta. 1387, 462–468.

Ngo, H.M., Ngo, E.O., Bzik, D.J., Joiner, K.A., 2000. *Toxoplasma gondii*: Are host cell adenosine nucleotides a direct source for purine salvage? Exp. Parasitol. 95, 48–153.

Nishikawa, Y., Ibrahim, H.M., Kameyama, K., Shiga, I., Hiasa, J., Xuan, X., 2011. Host cholesterol synthesis contributes to growth of intracellular Toxoplasma gondii in macrophages. J. Vet. Med. Sci. 73, 633–639.

Nishikawa, Y., Quittnat, F., Stedman, T.T., Voelker, D.R., Choi, J.Y., Zahn, M., Yang, M., Pypaert, M., Joiner, K.A., Coppens, I., 2005. Host cell lipids control cholesteryl ester synthesis and storage in intracellular *Toxoplasma*. Cell. Microbiol. 7, 849–867.

Nyvall, P., Pelloux, J., Davies, H.V., Pedersen, M., Viola, R., 1999. Purification and characterization of a novel starch synthase selective for uridine 5′-diphosphate glucose from the red alga *Gracilaria tenuistipitata*. Planta 209, 143–152.

Oishi, H., Noto, T., Sasaki, H., Suzuki, K., Hayashi, T., Okazaki, H., Ando, K., Sawada, M., 1982. Thiolactomycin, a new antibiotic. I. Taxonomy of the producing organism, fermentation and biological properties. J. Antibiot. 35, 391–395.

O'Sullivan, W.J., Johnson, A.M., Finney, K.G., Gero, A.M., Hagon, E., Holland, J.W., Smithers, G.W., 1981. Pyrimidine and purine enzymes in *Toxoplasma gondii*. Aust. J. Exp. Biol. Med. Sci. 59, 763–767.

Perrotto, J., Keister, D.B., Gelderman, A.H., 1971. Incorporation of precursors into *Toxoplasma* DNA. J. Protozool. 18, 470–473.

Pfefferkorn, E.R., Pfefferkorn, L.C., 1977. Specific labeling of intracellular *Toxoplasma* with uracil. J. Protozool. 24, 449–453.

Pfefferkorn, E.R., 1978. *Toxoplasma gondii*: the enzymic defect of a mutant resistant to 5-fluorodeoxyuridine. Exp. Parasitol. 44, 26–35.

Pfefferkorn, E.R., Pfefferkorn, L.C., 1980. *Toxoplasma gondii*: genetic recombination between drug resistant mutants. Exp. Parasitol. 50, 305–316.

Piatigorsky, J., 2003. Crystallin genes: specialization by changes in gene regulation may precede gene duplication. J. Struct. Funct. Genomics 3, 131–137.

Plesner, L., 1995. Ecto-ATPase: identities and function. Int. Rev. Cytol. 151, 141–214.

Polonais, V., Soldati-Favre, D., 2010. Versatility in the acquisition of energy and carbon sources by the Apicomplexa. Biol. Cell. 102, 435–445.

Qian, Y., Sebti, S.M., Hamilton, A.D., 1997. Farnesyltransferase as a target for anticancer drug design. Biopolymers 43, 25–41.

Quittnat, F., Nishikawa, Y., Stedman, T.T., Voelker, D.R., Choi, J.Y., Zahn, M.M., Murphy, R.C., Barkley, R.M., Pypaert, M., Joiner, K.A., Coppens, I., 2004. On the biogenesis of lipid bodies in ancient eukaryotes: synthesis of triacylglycerols by a *Toxoplasma* DGAT1-related enzyme. Mol. Biochem. Parasitol. 138, 107–122.

Ral, J.P., Derelle, E., Ferraz, C., Wattebled, F., Farinas, B., Corellou, F., Buléon, A., Slomianny, M.C., Delvalle, D., d'Hulst, S., Rombauts, S., Moreau, H., Ball, S., 2004. Starch division and partitioning a mechanism for granule propagation and maintenance in the picophytoplanktonic green alga *Ostreococcus tauri*. Plant. Physiol. 36, 3333–3340.

Ramakrishnan, S., Docampo, M.D., Macrae, J.I., Pujol, F.M., Brooks, C.F., van Dooren, G.G., Hiltunen, J.K., Kastaniotis, A.J., McConville, M.J., Striepen, B., 2012. Apicoplast and endoplasmic reticulum cooperate in fatty acid biosynthesis in apicomplexan parasite *Toxoplasma gondii*. J. Biol. Chem. 287, 4957–4971.

Read, M., Hicks, K.E., Sims, P.F., Hyde, J.E., 1994. Molecular characterization of the enolase gene from the human malaria parasite *Plasmodium falciparum*. Evidence for ancestry within a photosynthetic lineage. Eur. J. Biochem. 267, 513–520.

Roberts, C.W., McLeod, R., Rice, D.W., Ginger, M., Chance, M.L., Goad, L.J., 2003. Fatty acid and sterol metabolism: potential antimicrobial targets in apicomplexan and trypanosomatid parasitic protozoa. Mol. Biochem. Parasitol. 126, 129–142.

Sampels, V., Hartmann, A., Dietrich, I., Coppens, I., Sheiner, L., Striepen, B., Herrmann, A., Lucius, R., Gupta, N., 2012. Conditional mutagenesis of a novel choline kinase demonstrates plasticity of phosphatidylcholine biogenesis and gene expression in *Toxoplasma gondii*. J. Biol. Chem. 287, 16289–16299.

Santos, T.A., Portes, Jde A., Damasceno-Sá, J.C., Caldas, L.A., Souza, Wd., Damatta, R.A., Seabra, S.H., 2011. Phosphatidylserine exposure by *Toxoplasma gondii* is fundamental to balance the immune response granting survival of the parasite and of the host. PLoS One 6, e27867.

Schofield, L., Hackett, F., 1993. Signal transduction in host cells by a glycosylphosphatidylinositol toxin of malaria parasites. J. Exp. Med. 177, 145–153.

Schofield, L., Novakovic, S., Gerold, P., Schwarz, R.T., McConville, M.J., Tachado, S.D., 1996. Glyccosylphosphatidylinositol toxin of Plasmodium up-regulates intercellular adhesion molecule-1, vascular cell adhesion molecule-1, and E-selectin expression in vascular endothelial cells and increases leukocyte and parasite cytoadherence via tyrosine kinase-dependent signal transduction. J. Immunol. 156, 1886–1896.

Schofield, L., Hewitt, M.C., Evans, K., Siomos, M.A., Seeberger, P.H., 2002. Synthetic GPI as a candidate anti-toxic vaccine in a model of malaria. Nature 418, 785–789.

Schumacher, M.A., Carter, D., Scott, D.M., Roos, D.S., Ullman, B., 1998. Crystal structures of *Toxoplasma gondii* uracil phosphoribosyltransferase reveal the atomic basis of pyrimidine discrimination and prodrug binding. EMBO J. 17, 3219–3233.

Schumacher, M.A., Carter, D., Roos, D.S., Ullman, B., Brennan, R.G., 1996. Crystal structures of *Toxoplasma gondii* HGXPRTase reveal the catalytic role of a long flexible loop. Nat. Struct. Biol. 3, 881–887.

Schumacher, M.A., Carter, D., Scott, D.M., Mathews, I.I., Ealick, S.E., Roos, D.S., Ullman, B., Brennan, R.G., 2000. Crystal structures of *Toxoplasma gondii* adenosine kinase reveal a novel catalytic mechanism and prodrug binding. J. Mol. Biol. 298, 875–893.

Schumacher, M.A., Bashor, C.J., Song, M.H., Otsu, K., Zhu, S., Parry, R.J., Ullman, B., Brennan, R.G., 2002. The structural mechanism of GTP stabilized oligomerization and catalytic activation of the *Toxoplasma gondii* uracil phosphoribosyltransferase. Proc. Natl. Acad. Sci. USA 99, 78–83.

Schwab, J.C., Beckers, C.J.M., Joiner, K.A., 1994. The parasitophrous vacuole membrane surrounding intracellular *Toxoplasma gondii* functions as a molecular sieve. Proc. Natl. Acad. Sci. USA 91, 509–513.

Schwab, J.C., Afifi-Afifi, M., Pizzorno, G., Handscumacher, R.E., Joiner, K.A., 1995. *Toxoplasma gondii* tachyzoites possess an unusual plasma membrane adenosine transporter. Mol. Biochem. Parasitol. 70, 59–69.

Schwartzman, J.D., Pfefferkorn, E.R., 1981. Pyrimidine synthesis by intracellular *Toxoplasma gondii*. J. Prasitol. 67, 150–158.

Schwartzman, J.D., Pfefferkorn, E.R., 1982. *Toxoplasma gondii*: purine synthesis and salvage in mutant host cells and parasites. Exp. Parasitol. 53, 77–86.

Seeber, F., 2003. Biosynthetic pathways of plastid-derived organelles as potential drug targets against parasitic apicomplexa. Curr. Drug Targets Immune Endocr. Metabol. Disord. 3, 99–109.

Seeber, F., Dubremetz, J.F., Boothroyd, J.C., 1998. Analysis of *Toxoplasma gondii* stably transfected with a transmembrane variant of its major surface protein, SAG1. J. Cell Sci. 111, 23–29.

Sehgal, A., Bettiol, S., Wenk, M.R., Pypaert, M., Kaasch, A., Blader, I., Joiner, K.A., Coppens, I., 2005. Peculiarities of host cholesterol transport to the unique intracellular compartment containing *Toxoplasma gondii*. Traffic 6, 1–17.

Sharma, S.D., Mullenax, J., Araujo, F.G., Erlich, H.A., Remington, J.S., 1983. Western blot analysis of the antigens of *Toxoplasma gondii* recognized by human IgM and IgG antibodies. J. Immunol. 131, 977–983.

Sharma, K., Shi, Y., 1999. The yins and yangs of ceramide. Cell Res. 9, 1–10.

Sibley, L.D., Boothroyd, J.C., 1992. Construction of a molecular karyotype for *Toxoplasma gondii*. Mol. Biochem. Parasitol. 51, 291–300.

Sibley, L.D., Niesman, I.R., Asai, T., Takeuchi, T., 1994. *Toxoplasma gondii*: Secretion of a potent nucleoside triphosphate hydrolase into parasitophorous vacuole. Exp. Parasitol. 79, 301–311.

Sierra-Pagan, M.L., Zimmermann, B.H., 2003. Cloning and expression of the dihydroorotate dehydrogenase from *Toxoplasma gondii*. Biochim. Biophys. Acta. 1637, 178–181.

Silverman, J.A., Qi, H., Riehl, A., Becker, C., Nakaar, V., Joiner, K.A., 1998. Induced activation of the *Toxoplasma gondii* nucleoside triphosphate hydrolase leads to depletion of host cell ATP levels and rapid exit of intracellular parasites from infected cells. J. Biol. Chem. 273, 12352–12359.

Sirover, M.A., 2005. New nuclear functions of the glycolytic protein, glyceraldehyde-3-phosphate dehydrogenase, in mammalian cells. J. Cell. Biochem. 95, 45–52.

Sleat, D.E., Wiseman, J.A., El-Banna, M., Price, S.M., Verot, L., Shen, M.M., Zhao, Q., Passini, M.A., Davidson, B.L., Stewart, G.R., Lobel, P., 2004. Genetic evidence for nonredundant functional cooperativity between NPC1 and NPC2 in lipid transport. Proc. Natl. Acad. Sci. USA 101, 5886–5891.

Smith, S., Witkowski, A., Joshi, A.K., 2003. Structural and functional organization of the animal fatty acid synthase. Prog. Lipid Res. 42, 289–317.

Sonda, S., Hehl, A.B., 2006. Lipid biology of Apicomplexa: perspectives for new drug targets, particularly for *Toxoplasma gondii*. Trends Parasitol. 22, 41–47.

Sonda, S., Sala, G., Ghidoni, R., Hemphill, A., Pieters, J., 2005. Inhibitory effect of aureobasidin A on *Toxoplasma gondii*. Antimicrob. Agents Chemother. 49, 1794–1801.

Sonda, S., Ting, L.M., Novak, S., Kim, K., Maher, J.J., Farese Jr., R.V., Ernst, J.D., 2001. Cholesterol esterification by host and parasite is essential for optimal proliferation of Toxoplasma gondii. J. Biol. Chem. 276, 34434–34440.

Stommel, E.W., Ely, K.H., Schwarzman, J.D., Kasper, L.H., 1997. *Toxoplasma gondii*: dithiol-induced Ca2+ flux causes egress of parasites from the parasitophorous vacuole. Exp. Parasitol. 87, 88–97.

Stommel, E.W., Cho, E., Steide, J.A., Seguin, R., Barchowsky, A., Schwarzman, J.D., Kasper, L.H., 2001. Identification and role of thiols in *Toxoplasma gondii* egress. Exp. Biol. Med. 226, 229–236.

Striepen, B., Zinecker, C.F., Damm, J.B., Melgers, P.A., Gerwig, G.J., Koolen, M., Vliegenthart, J.F., Dubremetz, J.F., Schwarz, R.T., 1997. Molecular structure of the 'low molecular weight antigen' of *Toxoplasma gondii*: a glucose alpha1-4 N-acetylgalactosamine makes free glycosyl-phosphatidylinositols highly immunogenic. J. Mol. Biol. 266, 797–813.

Striepen, B., Pruijssers, A.J., Huang, J., Li, C., Gubbels, M.J., Umejiego, N.H., Hedstrom, L., Kissinger, J.C., 2004. Gene transfer in the evolution of parasite nucleotide biosynthesis. Proc. Natl. Acad. Sci. USA 101, 3154–3159.

Subramanian, A., Miller, D.M., 2000. Structural analysis of alpha-enolase. Mapping the functional domains involved in down-regulation of the c-myc protooncogene. J. Biol. Chem. 275, 5958–5965.

Sullivan Jr., W.J., Chiang, C.W., Wilson, C.M., Naguib, F.N., El-Kouni, M.H., Donald, R.G.K., Roos, D.S., 1999. Insertional tagging of at least two loci associated with resistance to adenine arabinoside in *Toxoplasma gondii*, and cloning of the adenosine kinase locus. Mol. Biochem. Parasitol. 103, 1–14.

Swiezewskaa, E., Danikiewiczb, W., 2005. Polyisoprenoids: structure, biosynthesis and function. Prog. Lipid Res. 44, 235–258.

Tachado, S.D., Gerold, P., Schwarz, R., Novakonic, S., McConville, M., Schofield, L., 1997. Signal transduction in macrophages by glycosylphosphatidylinositols of *Plasmodium, Trypanosoma,* and *Lesihmania*: activation of protein tyrosine kinases and protein kinase C by inositolglycan and diacylglycerol moieties. Proc. Natl. Acad. Sci. USA 94, 4022–4027.

Tan, F., Hu, X., Pan, C.-W., Ding, J.-Q., Chen, X.-G., 2010. Monoclonal antibodies against nucleoside triphosphate hydrolase-II can reduce the replication of *Toxoplasma gondii*. Parasitol. Int. 59, 141–146.

Tomavo, S., Schwarz, R.T., Dubremetz, J.F., 1989. Evidence for glycosyl-phosphatidylinositol anchor of *Toxoplasma gondii* major surface antigens. Mol. Cell. Biol. 9, 4576–4580.

Tomavo, S., Dubremetz, J.F., Schwarz, R.T., 1992a. A family of glycolipids from *Toxoplasma gondii*: identification of candidate glycolipid precursor(s) for *Toxoplasma gondii* glycosylphosphatidylinositol membrane anchors. J. Biol. Chem. 267, 11721–11728.

Tomavo, S., Dubremetz, J.F., Schwarz, R.T., 1992b. Biosynthesis of glycolipid precursors for glycosyl-phosphatidylinositol membrane anchors in a *Toxoplasma gondii* cell-free system. J. Biol. Chem. 267, 21446–21458.

Tomavo, S., Dubremetz, J.F., Schwarz, R.T., 1993. Structural analysis of glycosyl-phosphatidylinositol membrane anchor of the *Toxoplasma gondii* tachyzoite surface glycoprotein gp23. Biol. Cell. 78, 155–162.

Tomavo, S., Couvreur, G., Leriche, M.A., Sadak, A., Achbarou, A., Fortier, B., Dubremetz, J.F., 1994. Immunolocalization and characterization of the 4.6 kD antigen *of Toxoplasma gondii* that elicits an early IgM response upon primary infection. Parasitology 108, 139–145.

Tomavo, S., 1996. The major surface proteins of *Toxoplasma gondii*: structures and functions. Curr. Top. Microbiol. Immunol. 219, 45–54.

Tomavo, S., 2001. The differential expression of multiple isoenzyme forms during stage conversion of *Toxoplasma gondii*: an adaptive developmental strategy. Int. J. Parasitol. 31, 1023–1031.

Tomova, C., Humbel, B.M., Geerts, W.J., Entzeroth, R., Holthuis, J.C., Verkleij, A.J., 2009. Membrane contact sites between apicoplast and ER in *Toxoplasma gondii* revealed by electron tomography. Traffic 10, 1471–1480.

Ullman, B., Carter, D., 1995. Hypoxanthine-guanine phosphoribosyltransferase as a therapeutic target in protozoal infections. Infect. Agents Dis. 4, 29–40.

Uttaro, A.D., 2006. Biosynthesis of polyunsaturated fatty acids in lower eukaryotes. IUBMB 58, 563–571.

van Blitterswijk, W.J., Houssa, B., 1999. Diacylglycerol kinases in signal transduction. Chem. Phys. Lipids 98, 95–108.

Vance, J.E., Vance, D.E., 2004. Phospholipid biosynthesis in mammalian cells. Biochem. Cell. Biol. 82, 113–128.

Vasanthakumar, G., Davis Jr., R.L., Sullivan, M.A., Donahue, J.P., 1990. Cloning and expression in Escherichia coli of a hypoxanthine-guanine phosphoribosyltransferase-encoding cDNA from *Plasmodium falciparum*. Gene 91, 63–69.

Vasconcelos, E.G., Ferreira, S.T., Carvalho, T.M., Souza, W., Kettlun, A.M., Mancilla, M., Valenzuela, M.A., Verjovski-Almeida, S., 1996. Partial purification and immunohistochemical localization of ATP diphosphhydrolase from *Schistosoma mansoni*. Immunological cross-reactitivities with potato apyrase and *Toxoplasma gondii* nucleoside triphosphate hydrolase. J. Biol. Chem. 271, 22139–22145.

Waller, R.F., Keeling, P.J., Donald, R.G., Striepen, B., Handman, E., Lang-Unnasch, N., Cowman, A.F., Besra, G.S., Roos, D.S., McFadden, G.I., 1998. Nuclear-encoded proteins target to the plastid in *Toxoplasma gondii* and *Plasmodium falciparum*. Proc. Natl. Acad. Sci. USA 95, 12352–12357.

Waller, R.F., Keeling, P.J., Van Dooren, G.G., McFadden, G.I., 2003. Comment on a green algal apicoplast ancestor. Science 301, 49.

Wales, M.E., Madison, L.L., Glaser, S.S., Wild, J.R., 1999. Divergent allosteric patterns verify the regulatory paradigm for asparatate transcarbamylase. J. Mol. Biol. 294, 1387–1400.

Welti, R., Mui, E., Sparks, A., Wernimont, S., Isaac, G., Kirisits, M., Roth, M., Roberts, C.W., Botté, C., Maréchal, E., McLeod, R., 2007. Lipidomic analysis of *Toxoplasma gondii* reveals unusual polar lipids. Biochemistry 46, 13882–13890.

Wichroski, M.J., Ward, G.E., 2003. Biosynthesis of glycosylphosphatidylinositol is essential to the survival of the protozoan parasite *Toxoplasma gondii*. Eukaryot. Cell. 2, 1132–1136.

White, E.L., Ross, L.J., Davis, R.L., Ginkel, S.Z., Vasanthakumar, G., Borhani, D.W., 2000. The two *Toxoplasma gondii* hypoxanthine-guanine phosphoribosyltransferase isozymes form heterotetramers. J. Biol. Chem. 275, 19218–19223.

White, S.W., Zheng, J., Zhang, Y.M., Rock, C.O., 2005. The structural biology of type II fatty acid biosynthesis. Annu. Rev. Biochem. 74, 791–831.

Yahiaoui, B., Dzierszinski, F., Bernigaud, A., Slomianny, C., Camus, D., Tomavo, S., 1999. Isolation and characterization of a subtractive library enriched for developmentally regulated transcripts expressed during encystations of *Toxoplasma gondii*. Mol. Biochem. Parasitol. 99, 223–235.

Yang, S., Parmley, S.F., 1995. A bradyzoite stage-specifically expressed gene of *Toxoplasma gondii* encodes a polypeptide homologous to lactate dehydrogenase. Mol. Biochem. Parasitol. 73, 291–294.

Yang, S., Parmley, S.F., 1997. *Toxoplasma gondii* expresses two distinct lactate dehydrogenase homologous genes during its life cycle in intermediate hosts. Gene 184, 1–12.

Yardley, V., Khan, A.A., Martin, M.B., Slifer, T.R., Araujo, F.G., Moreno, S.N., Docampo, R., Croft, S.L., Oldfield, E., 2002. In vivo activities of farnesyl pyrophosphate synthase inhibitors against *Leishmania donovani* and *Toxoplasma gondii*. Antimicrob. Agents Chemother. 46, 929–931.

Yu, L., Gao, Y.-F., Li, X., Qiao, Z.-P., Shen, J.-L., 2009. Double-stranded RNA specific to adenosine kinase and hypoxanthine-xanthine-guanine-phosphoribosyltransferase retards growth of *Toxoplasma gondii*. Parasitol. Res. 104, 377–383.

Zeiner, G.M., Cleary, M.D., Foute, A.E., Meiring, C.D., Mocarski, E.S., Boothroyd, J.C., 2008. RNA analysis by biosynthetic tagging using 4-thiouracil and uracil phosphoribosyltransferase. Meth. Mol. Biol. 419, 135–145.

Zinecker, C.F., Striepen, B., Geyer, H., Geyer, R., Dubremetz, J.F., Schwarz, R.T., 2001. Two glycoforms are present in the GPI-membrane anchor of the surface antigen 1 (P30) of *Toxoplasma gondii*. Mol. Biochem. Parasitol. 116, 127–135.

Zhou, Z., Metacalf, A.E., Lovatt, C.J., Hyman, B.C., 2000. Alfalfa (*Medicago sativa*) carbamoylphosphate synthetase gene structure records the deep linage of plant. Gene 243, 105–114.

CHAPTER

9

The Apicoplast and Mitochondrion of *Toxoplasma gondii*

Frank Seeber, Jean E. Feagin[†,**,‡], Marilyn Parsons[†,**]*

*FG16, Parasitology, Robert Koch-Institute, Berlin, Germany [†] Emerging and Neglected Diseases Program, Seattle Biomedical Research Institute, Seattle, Washington, USA

**Department of Global Health, University of Washington, Seattle, Washington, USA

[‡]Department of Pharmacy, University of Washington, Seattle, Washington, USA

OUTLINE

Toxoplasma gondii, Second Edition
http://dx.doi.org/10.1016/B978-0-12-396481-6.00009-X

9.1 INTRODUCTION

The hallmark of a eukaryotic cell is the division of the cell contents into specialized membrane-bounded compartments. These organelles provide multiple benefits to the cell, including protecting the rest of the cell from dangerous reaction products, generating gradients that can be exploited for biological processes, and separating potentially interfering pathways.

The partitioning of the eukaryotic cell led Wallin to suggest, in the 1920s, that it is a collection of symbiotic microorganisms (Wallin, 1927). Of the numerous organelles in eukaryotic cells, two provide evidence of endosymbiosis since they contain small genomes and are bounded by double instead of single membranes: the mitochondrion and the chloroplast (or more generically, plastids). Their genomes encode key proteins needed for the specialized function of these organelles, including some components of a separate translation system, and a variable phalanx of other genes. By 1970, these oddities spurred Margulis to propose the endosymbiont theory (Margulis, 1970), which postulates that the present-day eukaryote originated from multiple, interacting organisms and more specifically that these organelles are the remnants of engulfed prokaryotic cells. Initially viewed as outlandish, today this idea is firmly entrenched in biological doctrine. Along with the endosymbiont theory, students learn that the mitochondrion is the powerhouse of the cell and that photosynthesis occurs in chloroplasts, but generally know little regarding the other vital metabolic activities performed by these organelles.

In the origins of these organelles lie new possibilities for intervention in diseases caused by apicomplexan parasites, including birth defects, blindness, and encephalitis due to *T. gondii*, malaria due to *Plasmodium* species, and numerous veterinary diseases. The intense interest in the endosymbiont organelles of apicomplexans lies in the realization that these parasites have not one but two extrachromosomal DNAs, each residing in its own organelle. The single mitochondrion has a minimal genome with unique rRNA genes and presumably unique ribosomes. The second organelle is the apicoplast (apicomplexan plastid), a leftover from a photosynthetic past. There is only one apicoplast per cell and the genome it contains is a remnant chloroplast DNA. This novel and totally unanticipated finding suggested the presence of plant-like metabolic pathways quite different from those in the vertebrate hosts. Such pathways would provide a variety of new chemotherapeutic targets. Indeed, while *T. gondii* can survive temporarily without an apicoplast, such cells are incapable of proliferation following invasion of new host cells (see Section 9.2.8). The unique characteristics of apicomplexan mitochondria also present possibilities for intervention. Studying the origins and activities of the DNA-containing organelles of *T. gondii* and of the malaria parasite *Plasmodium falciparum* has paid big dividends already and there are undoubtedly more to come.

Work on the apicoplast and mitochondrion of *T. gondii* is inextricably intertwined with studies of these organelles in *Plasmodium*. *Plasmodium falciparum* in particular has been intensively studied due to its major health relevance. While the sections that follow will focus on *T. gondii*, work with other apicomplexans will be noted. For some topics, work on *P. falciparum* predominates and will be described in greater detail to provide a framework for understanding the *T. gondii* organelles. Despite the many similarities, there are also some surprising differences between *T. gondii* and *P. falciparum*. Consequently, predictions of apicoplast or mitochondrial functions based on data from just one organism must be tempered with caution. Topics include a brief history of the identification and origins of the organelles, genome content and gene expression, replication, and trafficking of proteins to the organelles. We also discuss insights from antibiotic sensitivity studies, organelle metabolism and the potential for further drug development. For those seeking additional detail, there have been several recent reviews on topics in this chapter, most focusing on the apicoplast (Ralph *et al.*, 2004b; Sato and Wilson, 2005; Wilson, 2005; Waller and McFadden, 2005; Wiesner and Seeber, 2005; Parsons *et al.*, 2007; Obornik *et al.*, 2008; Lim and McFadden, 2010; Seeber and Soldati-Favre, 2010; McFadden, 2011).

9.2 THE APICOPLAST

9.2.1 History

Electron micrographs provided the first indication of the variety of subcellular organelles in *T. gondii* (Fig. 9.1). Some appear quite conventional, such as the ER, Golgi, and a mitochondrion bounded by two membranes. Others are novel to apicomplexans. Prominent examples include rhoptries, micronemes, dense granules, and the conoid, comprising the apical complex from which the phylum takes its name (described in Chapters 12 and 13). The parasites also possess an unusual structure called the inner membrane complex, which lies under the plasma membrane in the apical half of the cell. Among the novel organelles described early in the study of apicomplexans was a small structure surrounded by multiple membranes, but which had no obvious function. It did not even have the same name in different apicomplexans, being known as the spherical body in *Plasmodium*, the 'Lamellärer Körper', 'Hohlzylinder', 'Golgi adjunct', and 'vésicule plurimembranaire' in *T. gondii*, and the 'grosse Vakuole mit kräftiger Wandung' in *Eimeria*, to name a few (cited in Siddall, 1992). The multiple membranes provided the first clue to the organelle's unusual identity, but the possibility that this might be relevant to the mystery organelle was missed for decades even though secondary endosymbiosis had already been invoked as the origin of some chloroplasts (Lee, 1977; Greenwood *et al.*, 1977; Gibbs, 1978; Whatley *et al.*, 1979).

The next clue came from studies of parasite genomes. In 1984, Borst and colleagues (Borst *et al.*, 1984) reported 12 micron and 23 micron circular extrachromosomal DNAs from *T. gondii*, the latter being head-to-tail dimers of the former. When spread for electron microscopy, many of the molecules adopted a cruciform structure. These data echoed earlier reports of similarly sized circular extrachromosomal DNAs in *Plasmodium lophurae* (Kilejian, 1975) and *Plasmodium berghei* (Dore *et al.*, 1983). These molecules matched the size range and conformation expected for mitochondrial genomes of unicellular eukaryotes, so were immediately labelled as such. It was, of course, the logical conclusion. It was also wrong. Who would have suspected these were remnant chloroplast genomes? The only clue was the cruciform structure, typical of chloroplast but not mitochondrial genomes. Certainly no one connected the circular genomes with the multi-membraned organelle.

Research on organelle DNA in apicomplexans was initially pursued exclusively in *Plasmodium*. Williamson, Wilson, and colleagues identified three bands in isopycnic sucrose density gradients of *Plasmodium knowlesi* (Williamson *et al.*, 1985) and *P. falciparum* (Gardner *et al.*, 1988) lysates. One was lighter than the main band of nuclear DNA, as is usually the case for mitochondrial genomes. It proved to be a ~35 kb circular DNA (Williamson *et al.*, 1985; Gardner *et al.*, 1988) and when spread for electron microscopy, it demonstrated a cruciform structure (Williamson *et al.*, 1985). It thus displayed the characteristics of the previously reported 'mitochondrial' DNAs of apicomplexans.

The lowest band on the *P. falciparum* gradient, described as 'diffuse and weakly fluorescent' (Feagin *et al.*, 1992), migrated just below the nuclear DNA band and proved to contain tandem repeats of a 6 kb DNA sequence.

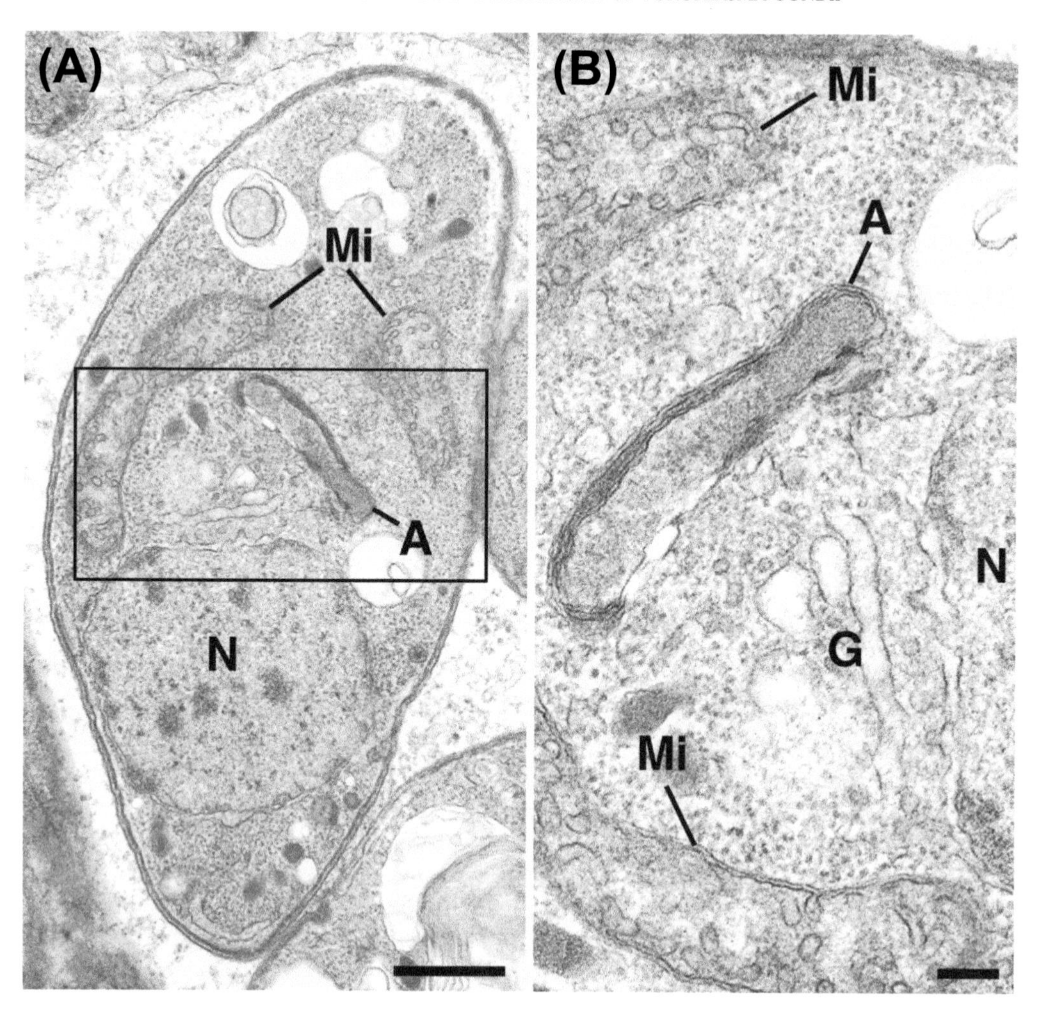

FIGURE 9.1 **Ultrastructural appearances of the apicoplast and mitochondrion in *T. gondii*.** Part A is an electron micrograph of a section through intracellular tachyzoite showing the apicoplast (A) and two regions of the elongated mitochondrion (Mi) anterior to the nucleus (N). Bar is 500 nm. Part B is an enlargement of the boxed area from panel A, rotated 90° left. Note the multiple membranes bounding the apicoplast, the finger-like cristae of the mitochondrion, and the Golgi body (G). Bar is 100 nm. Image courtesy of David Ferguson. *Reproduced with permission from Ferguson* et al. *(2005) Euk. Cell 4, 818, with slight modification (rotation).*

A similar repeated sequence had been identified in *P. yoelii* (Vaidya and Arasu, 1987; Suplick *et al.*, 1988; Vaidya *et al.*, 1989) and *P. gallinaceum* (Aldritt *et al.*, 1989; Joseph, 1990). Upon sequencing, the '6 kb element' in all three species was found to encode classic mitochondrial proteins (apocytochrome *b* (*cob*); cytochrome *c* oxidase subunit I (*cox*1); and subunit III (*cox*3)), and small, fragmented rRNAs (Aldritt *et al.*, 1989; Vaidya *et al.*, 1989; Joseph, 1990; Feagin *et al.*, 1992; Feagin, 1992; Feagin *et al.*, 2012) (see Section 9.3.2 The Mitochondrial Genome). Despite its minute size, this repeated element has the requisite minimum of genes expected in mitochondrial genomes (Gillham, 1994). But if the 6 kb element was the mitochondrial genome, what was the 35 kb DNA?

Analysis of the 35 kb DNA revealed that it contains a large inverted repeat composed of two copies of small and large subunit rRNAs, arranged tail to tail (Gardner *et al.*, 1988; 1991a; 1993). The rRNAs are similar to those of prokaryotes, as expected for both mitochondrial and plastid rRNAs. But mitochondrial genomes do not typically have duplicated rRNAs, while those of chloroplasts do (Gillham, 1994). Further sequencing showed that the 35 kb DNA also encodes subunits of a eubacterial-like RNA polymerase (Gardner *et al.*, 1991b). This is unequivocally a plastid characteristic; all plastid genomes studied thus far encode and are transcribed by such RNA polymerases (although some plastids import additional RNA polymerases (Liere *et al.*, 2011)). Indeed, the subunits encoded by *RPOB* (Gardner *et al.*, 1994b) and *RPOC1* (Howe, 1992) were determined to be more like chloroplast counterparts than bacterial ones. In contrast, almost all mitochondria employ a single subunit RNA polymerase most closely related to phage RNA polymerases (Gray and Lang, 1998). Further analysis from the *P. falciparum* 35 kb DNA identified genes encoding components of an organelle translation system, but no photosynthesis-related genes.

Serendipitously, the plastid genomes of *Epifagus virginiana*, a non-green plant, and *Astasia longis*, a non-green alga, were under analysis at the same time. These are both much reduced in size compared to those of green plants. While plastid-encoded genes related to photosynthesis were missing, those needed for expression were present (reviewed in dePamphilis and Palmer, 1989). The parallels with the *P. falciparum* 35 kb DNA are striking (Fig. 9.2). With the accumulating data, the formerly implausible explanation that the 35 kb DNA was derived from chloroplast DNA became increasingly believable (Wilson *et al.*, 1991; Palmer, 1992). It is now well

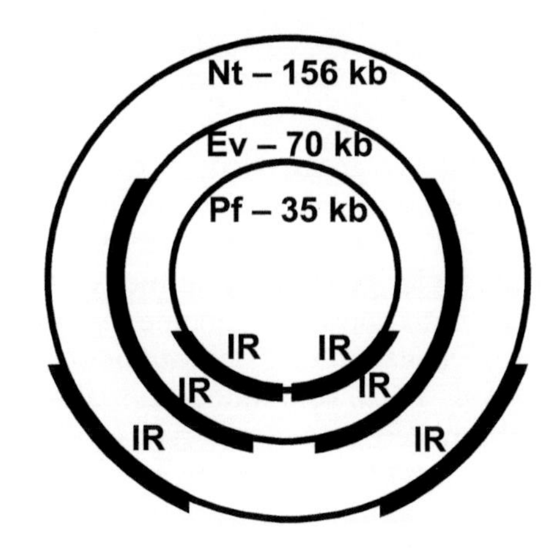

FIGURE 9.2 **Plastid genome structure.** Schematic depiction of the plastid genomes of *Nicotiana tabacum* (Nt), a green plant, *Epifagus virginiana* (Et), a non-green plant, and *P. falciparum* (Pf). Genome sizes are indicated. The salient feature of each genome is an inverted repeat, producing two IR regions per genome (thickened lines). The IRs include rRNAs, tRNAs and, except for *P. falciparum*, some protein-coding genes. IR size for *N. tabacum*, *E. virginiana*, and *P. falciparum*: 25 kb, 23 kb, and 5 kb, respectively.

established that apicomplexans have algal ancestors (see Section 9.2.2 Evolution below).

The *T. gondii* plastid genome was sequenced by the year 1997 (GenBank accession U87145, RefSeq NC001799). It is strikingly similar to its *P. falciparum* counterpart in gene content and organization (see Section 9.2.3 The Apicoplast Genome below). Complete or near-complete apicoplast genome sequences are now available for multiple species of *Plasmodium* (Sato *et al.*, 2000; Carlton *et al.*, 2008) and for multiple other apicomplexans, including the chicken pathogen *Eimeria tenella* (Cai *et al.*, 2003) and the bovine pathogens *Theileria parva* (Gardner *et al.*, 2005) and *Babesia bovis* (Brayton *et al.*, 2007). Conservation of gene content and genome organization is strong; the principal difference is that the piroplasms *Babesia* and *Theileria* have only one copy of the rDNA transcription unit (Gardner *et al.*, 2005; Brayton *et al.*, 2007). This high degree of genome similarity is matched by functional conservation that is largely dependent on the import of nuclearly encoded proteins (see Section 2.7 Protein Trafficking to the Apicoplast and Section 2.9 Apicoplast Metabolism). This makes *Toxoplasma* an excellent model for study of the apicomplexan plastid.

A lingering question was the subcellular location of the 35 kb genome. It did not co-localize with the mitochondrial genome in sucrose gradient fractionation of *P. falciparum* organelles (Wilson *et al.*, 1992), so a mitochondrial location appeared unlikely. As a remnant chloroplast genome, it should reside in an organelle with more than one bounding membrane. The 35 kb DNA was a genome without a home, and the spherical body was an organelle without a role – might they intersect? The well-defined subcellular structure of *T. gondii* makes it more amenable for cell biological studies than *P. falciparum* so it is unsurprising that the localization question was first answered for *T. gondii*. In the mid-1990s, *in situ* hybridization studies using probes derived from the *T. gondii* 35 kb DNA showed that this genome resides in an organelle located just apical to the nucleus, the hitherto mysterious multi-membraned organelle (McFadden *et al.*, 1996; Köhler *et al.*, 1997). The *P. falciparum* 35 kb DNA has also been demonstrated to reside in the corresponding organelle (Wilson *et al.*, 1996). The plethora of names for the organelle has been replaced with a single term: the apicoplast, for apicomplexan plastid. Solving the initial mysteries of the homeless genome and the unexplained organelle has generated a number of fascinating questions. How does a group of obligate intracellular parasites get a plastid? Why has it been maintained in non-photosynthetic organisms? What role does it play for the cell? What possibilities for disease intervention result from the presence of 'plant' genes in protozoan pathogens?

9.2.2 Evolution

Evidence points to an α-proteobacterial ancestor for mitochondria, with some components of the mitochondrial metabolome otherwise derived (reviewed in Gray, 2012). Despite earlier suggestions that there might be more than one origin of chloroplasts, the most recent data now point to a single endosymbiotic event there as well, with the engulfed organism being a photosynthetic cyanobacterium (reviewed in McFadden and van Dooren, 2004; Keeling, 2010). The characteristic double membranes of both mitochondria and chloroplasts are believed to reflect the endosymbiotic event that produced them. As the relationship evolved to symbiosis, many genes in the engulfed partner were transferred to the nucleus, and those gene products essential to organellar function were translated in the cytosol and subsequently imported across two membranes into the organelle. But a number of organisms have not just two, but three or four bounding membranes around their chloroplasts. To explain this, the endosymbiont hypothesis was proposed (reviewed in Archibald and Keeling, 2002). All cases known to date involve engulfment of a photosynthetic alga that already bears a chloroplast (Fig. 9.3). The two inner membranes are believed to be the original chloroplast membranes, the third membrane representing the plasma membrane of the algal cell and the outermost membrane deriving from the host endomembrane system. Prominent examples of organisms with secondary chloroplasts with four membranes are chlorarachniophyte algae and cryptomonad algae (the outermost membrane appears fused to the host rough ER). Loss of one membrane has occurred in some organisms as in euglenoids (three membranes) and dinoflagellates (usually three membranes). The abundance of other membranous structures in *T. gondii* and *P. falciparum* cells, coupled with the small size of the apicoplast (~300 nm in diameter in *T. gondii*) and the close apposition of its membranes, made it challenging to determine the number of membranes surrounding the apicoplast. Numerous electron micrographs show four

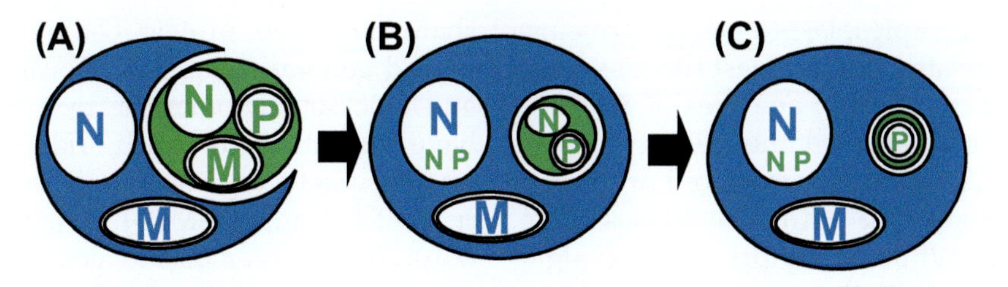

FIGURE 9.3 **Secondary endosymbiosis.** Secondary endosymbiosis entails engulfment of an algal cell by a eukaryote (A); loss of genes from the algal cell, with some transferred to the host nucleus (B); and finally loss of the algal nucleus, leaving behind an organelle bounded by four membranes (C). Further steps may reduce the number of bounding membranes to three. P, plastid; M, mitochondrion; N, nucleus. Blue, host cell; green, algal cell.

membranes surrounding the *T. gondii* apicoplast. That number is now generally accepted as well for *Plasmodium* (Köhler *et al.*, 1997; McFadden and Roos, 1999; McFadden, 2011).

Cryptomonad and chlorarachniophyte algae, which also bear secondary plastids, provide firm proof of secondary endosymbioses involving a red and green alga, respectively. Not only are their chloroplasts surrounded by four membranes but a remnant nucleus of the algal symbiont, the nucleomorph, nestles between the second and third membranes. Nucleomorph genomes are much reduced, with just three small chromosomes totalling a few hundred kb of DNA (Douglas *et al.*, 2001; Gilson *et al.*, 2006; Moore and Archibald, 2009). Despite their small size, these genomes encode components for their own perpetuation and expression plus a few other proteins, including some destined for import to the chloroplast (Gilson and McFadden, 1996, 1997; Zauner *et al.*, 2000; Loffelhardt, 2011; Hirakawa *et al.*, 2011) and even some that participate in the import process (Douglas *et al.*, 2001; Hirakawa *et al.*, 2012). Most of the genes needed for photosynthesis have been relocated to the nucleus and their protein products are trafficked across all four membranes. In contrast to these algae, apicomplexans have lost the endosymbiont nucleus and photosynthesis-related genes, which they no longer need as intracellular parasites. However, they continue to employ the apicoplast as a synthetic compartment (see Section 9.2.9 Apicoplast Metabolism). In many lineages, it appears that secondary chloroplasts were acquired by a common ancestor and then lost by some extant representatives (Keeling, 2010). For example, the gregarine apicomplexans lack an apicoplast as well as the enzymes typically found in the organelle, but it is presumed that the ancestral apicomplexan possessed a chloroplast.

The movement of foreign genes into a host nucleus is called lateral gene transfer (reviewed in Doolittle *et al.*, 2003; Bapteste *et al.*, 2004). As noted above, some transferred genes encode products that are trafficked back to the organelle. Others may provide the host with new synthetic capabilities in the cytosol or other compartments. Still others may be redundant, allowing for the evolution of new functions, while some may simply replace the corresponding endogenous genes. The rapidly increasing pool of genome sequences has greatly accelerated our understanding of lateral gene transfer, showing it to be widespread and of considerable scope. Secondary endosymbiosis permits additional lateral gene transfer, from both the nucleus and the chloroplast of the endosymbiont to the nucleus of the host. Importantly, that means that genes from the endosymbiont nucleus, which may have no relationship to plastid function, may still provide insights useful for deciphering the history of secondary endosymbionts. Phylogenetic analyses of apicomplexans have considered genes encoded by the apicoplast genome, nuclear genes encoding apicoplast-targeted proteins, and genes not related to apicoplast function.

A recurring question in analyses of apicomplexan evolution has been the identity of the secondary endosymbiont: was it a red or a green alga? Data supporting both contentions have been gathered. While the imprint of the algal ancestor could potentially be seen in the nuclear genome, this provides little clarity because some genes have been proposed to be of red algal origin (Fast *et al.*, 2001; Coppin *et al.*, 2005) and others to be of green algal origin (Funes *et al.*, 2002). Phylogenetic studies of numerous nuclear genes unrelated to the endosymbiont show that apicomplexans cluster most closely with dinoflagellates; together, these comprise the alveolate clade (Gajadhar *et al.*, 1991; Wolters, 1991). Since the secondary plastids of dinoflagellates are widely considered to derive from a red algal lineage, this relationship suggests a red algal lineage for apicomplexan plastids. A red algal origin is also favored by the arrangement of plastid-encoded ribosomal protein genes, a characteristic often considered to discriminate red from green algae (Stoebe and Kowallik, 1999), although this conclusion is tempered by significant rearrangement of plastid genomes in non-photosynthetic algae and plants as compared to their close photosynthetic relatives (Stoebe and Kowallik, 1999). Gene loss comparisons also favor a red algal ancestry for apicoplasts, since their genomes encode seven proteins common in the red algal lineage but missing in the green algal lineage (Blanchard and Hicks, 1999; Janouskovec *et al.*, 2010).

Given the above findings, it was surprising that phylogenetic analyses of apicoplast-encoded genes, considered singly or in groups, usually favoured a green algal origin (Köhler *et al.*, 1997; Cai *et al.*, 2003; Lau *et al.*, 2009). This may reflect the fact that direct comparisons of apicoplast genes with other red algal-derived plastid genomes mostly involved distantly related organisms since dinoflagellates and Apicomplexa share only two protein coding genes in their plastid genomes. Recently, however, the plastid and nuclear genomes of two alveolate algae have been obtained which bridge the gap. These photosynthetic organisms, *Chromera velia* (Moore *et al.*, 2008) and CCMP3155 (Janouskovec *et al.*, 2010), have larger plastid genomes with highly overlapping gene content. Together they encode orthologues of all protein-coding genes on the plastid genomes of both dinoflagellates and apicomplexans. Phylogenetic analysis using these sequences supports a close relationship of the apicomplexan genes to these red alga-derived plastids (Moore *et al.*, 2008; Janouskovec *et al.*, 2010). However, this point is not firmly settled (Woehle *et al.*, 2011; Burki *et al.*, 2012). In other plastid and nuclear genomes, TGA of course specifies a stop codon. Interestingly, a novel codon for tryptophan is shared by *T. gondii* and *C. velia* plastid genomes: both use TGA as well as the canonical TGG (Moore *et al.*, 2008). Taken together, it seems that the evidence now favours a red algal origin for the apicoplast. However, the debate is not yet over: analysis of nuclear genes provides evidence of both red and green algal ancestors in these organisms as was seen in apicomplexans (see above).

9.2.3 The Apicoplast Genome

The number of plastid genomes per cell is contested. Based on nucleic acid hybridization, Köhler *et al.* (1997) reported five to six copies per cell for the *T. gondii* plastid genome and a single plastid genome per cell for *P. falciparum*. This is consistent with a prior estimate of one to two copies for *P. falciparum* (Wilson *et al.*, 1993). Matsuzaki *et al.* (2001) have examined genome copy number by measurement of apicoplast-localized fluorescence following DAPI staining. Their analysis suggests 25 copies of the apicoplast genome in *T. gondii* and 15 in *P. falciparum*, a considerable increase in both cases. Multiple copies of a genome would facilitate repair of mutations by gene conversion, making the higher numbers attractive from a functional point of view, but the matter remains unresolved.

When photosynthetic capability is not needed, the loss of photosynthesis-related genes from the chloroplast genome is profound (reviewed in dePamphilis and Palmer, 1989). Chloroplast genomes average 150–200 kb in size but those of non-photosynthetic plants are ~70 kb and the *T. gondii* plastid genome is ~35 kb (Fig. 9.2). Apicoplast genomes are quite similar, suggesting that much of the reduction in coding capacity happened prior to splitting the apicomplexan lineages. The apicoplast genomes of *P. falciparum* (Wilson *et al.*, 1996), *T. gondii* (ToxoDB, Gajria *et al.*, 2008), and *E. tenella* (Cai *et al.*, 2003) are all ~35 kb in size, with a large inverted repeat. The repeat unit consists of small subunit (SSU) and large subunit (LSU) rRNAs encoded head to head and separated by seven tRNA genes. A single tRNA gene is found at the 3′ ends of both rRNAs. As noted above, this organization is highly reminiscent of chloroplast genomes. Curiously, the 39.4 kb and 33 kb apicoplast genomes of the piroplasms *T. parva* (Gardner *et al.*, 2005) and *B. bovis* (Brayton *et al.*, 2007), respectively, lack the repeat rRNA structure and have only a single set of rRNAs with no intervening tRNAs. Both species also differ from the *P. falciparum*, *T. gondii* and *E. tenella* apicoplast genomes by being unidirectionally transcribed.

The gene content of apicoplast genomes is highly conserved although the relative location of genes can differ. The similarities of these genomes to each other are greatest within each taxonomic group, as anticipated. On the other hand, *Cryptosporidium parvum*, which had once been considered a coccidian (like *Toxoplasma* and *Eimeria*), lacks an apicoplast genome as well as the organelle and its metabolic pathways (Riordan *et al.*, 1999, 2003; Zhu *et al.*, 2000b). Consideration of *C. parvum*'s genome sequence further supports its considerable differences from other coccidians (Keithly *et al.*, 1997; Zhu *et al.*, 2000a). It is now considered more similar to the gregarines, apicomplexan parasites of insects and molluscs.

Most apicoplast genes encode components needed for expression of the apicoplast genome (Fig. 9.4). All of the tRNAs needed for apicoplast

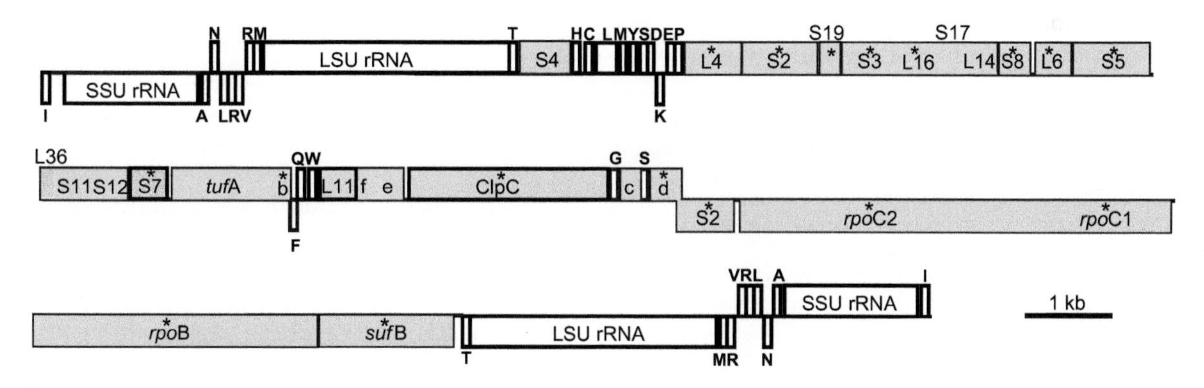

FIGURE 9.4 **The *T. gondii* apicoplast genome.** A schematic map of the *T. gondii* apicoplast genome is shown, with genes above or below the line depending on direction of transcription (left to right above the line). Protein coding genes are grey; those marked with asterisks contain internal TGA codons. ORFs of unknown significance are identified with a single lower case letter. Large and small subunit ribosomal protein genes (*rpl* and *rps*) are designated L and S. Other protein-coding genes are named as in the text. Non-coding RNA genes are white; tRNA genes are identified by the single letter code above or below their location. LSU rRNA and SSU rRNA, large and small subunit rRNAs, respectively. Data derived from GenBank reference sequence NC_001799.

protein synthesis are encoded by the apicoplast genome, although they are clustered differently in piroplasms than in coccidians and sporozoans (Wilson *et al.*, 1996; Gardner *et al.*, 2005; Brayton *et al.*, 2007, ToxoDB). The protein coding genes include three subunits of a multi-subunit, eubacterial-like RNA polymerase: *RPOB*, *RPOC1*, and *RPOC2* (Wilson *et al.*, 1996), as is the case for chloroplast genomes. Chloroplast genomes also generally encode *RPOA*, the remaining major subunit of the RNA polymerases (Gillham, 1994). In *P. falciparum*, that gene is found in the nuclear genome and has a predicted apicoplast targeting sequence (PlasmoDB). Highly scoring sequences can be detected in the *T. gondii* genome by BLAST query (J. E. Feagin, unpublished results). The other protein-coding genes include 17 ribosomal proteins, and the translation elongation factor Tu (Wilson *et al.*, 1996). There are only two apicoplast-encoded genes that have predicted functions other than gene expression, *CLPC* and *SUFB* (see Section 9.2.9 Apicoplast Metabolism). Finally, a handful of unidentified open reading frames are modestly conserved between *T. gondii* and *P. falciparum* (see Bahl *et al.*, 2010 and http://roos.bio.upenn.edu/~rooslab/jkissing/TgPfcomparison.html). They predict small basic proteins which may be additional ribosomal proteins, some of which are difficult to identify due to limited conservation. This possibility is particularly attractive, given that the total number of identified plastid ribosomal proteins, including those predicted to be imported from the cytosol, is less than that needed for functional ribosomes (Gardner *et al.*, 2002).

A tiling microarray has been used to search for single nucleotide polymorphisms along the apicoplast genome of three prototype *T. gondii* strains as compared to the originally sequenced apicoplast genome from the type I strain RH. The apicoplast genome sequences of RH and a second type I strain (GT1) were identical. However, 12 differences were noted in a type II strain (Prugniaud) and 48 in a type III strain (CTG) (Bahl *et al.*, 2010).

9.2.4 Expression of the Apicoplast Genome

Apicoplast gene expression data derive primarily from *P. falciparum*. Individual transcripts exist for the rRNAs (Gardner *et al.*, 1988, 1991a, 1993) and tRNAs (Preiser *et al.*, 1995) but transcription appears to be polycistronic (Feagin and Drew, 1995; Wilson *et al.*, 1996). Although a precursor/mature RNA relationship has not been definitively shown, primer extension analysis of the plastid rRNAs, corroborated by RNase protection experiments, provided evidence for longer, less abundant RNAs (Gardner *et al.*, 1991a, 1993). Serendipitously, two RNase protection products at the 3′ end of the small subunit rRNA differed in size by 70 nt, and the larger product later was shown to encompass a tRNA (Gardner *et al.*, 1994a).

In contrast to the rapidly processed rRNA/tRNA transcripts, transcripts from *RPOB* and *RPOC1* remain on long polycistronic transcripts, 10–12 kb in length (Feagin and Drew, 1995). Wilson *et al.* (1996) detected long transcripts for additional *P. falciparum* apicoplast genes. Their attempts to map the 5′ ends of individual genes were often unsuccessful. Together, these points suggest extensive polycistronic transcription, with little accompanying cleavage. Supporting evidence comes from RT-PCR studies showing that transcripts spanning multiple genes can be detected (Wilson *et al.*, 1996). In *P. falciparum*, the abundance of many RNAs encoded by the apicoplast genome appears to be co-regulated during the erythrocytic cycle. They are most abundant in the schizont phase (Bozdech *et al.*, 2003), which is consistent with earlier data assessing individual transcripts (Feagin and Drew, 1995). The near identity of the expression patterns for adjacent genes is likely strongly affected by polycistronic transcription.

A substantial proportion of the apicoplast genome is devoted to genes for translation and several lines of indirect evidence point to functional protein synthesis in the organelle. The rRNAs and tRNAs encoded by the *P. falciparum* apicoplast genome are conventional in size and predicted structure and have the expected conservation of sequence (Gardner *et al.*, 1991, 1993, 1994a). The apicoplast rRNAs and tRNAs of *T. gondii*, while not analysed in depth, appear similar. Electron micrographs of the *P. falciparum* (Hopkins *et al.*, 1999) and *T. gondii* (McFadden *et al.*, 1996) apicoplasts show granular structures of a size consistent with organelle ribosomes. Finally, a number of compounds known to inhibit protein synthesis on eubacterial and organelle ribosomes have negative effects on growth of *T. gondii* and *P. falciparum* in culture (see Section 9.2.8). Direct evidence for apicoplast protein synthesis is scant. Roy *et al.* (1999) reported data consistent with apicoplast polysomes but these have not been pursued. Chaubey *et al.* (2005) used antibodies raised against heterologously expressed apicoplast EF-Tu to detect its synthesis in *P. falciparum*. However, there are other elongation factors and related proteins in the *P. falciparum* genome and the authors did not report the specificity of their antisera.

The *T. gondii* apicoplast genome sequence has 33 TGA stop codons embedded in 17 of the 28 predicted protein coding genes. *E. tenella* (Cai *et al.*, 2003), *Neospora caninum* (Lang-Unnasch and Aiello, 1999), *T. parva* (Gardner *et al.*, 2005), and *B. bovis* (Brayton *et al.*, 2007) have also been found to have numerous TGA stop codons in different positions in apicoplast genes. It has long been known that many mitochondrial genomes decode UGA as a tryptophan codon rather than a stop codon (Gillham, 1994) although this alternate codon usage has not been reported for chloroplast genomes (M.W. Gray, personal communication). The apicoplast genomes appear to be the exception since many of the TGA codons occur at sites of conserved tryptophans. Use of TGA for tryptophan means that it is unavailable as a stop codon so *T. gondii* apicoplast genes end with TAA or TAG. Given the apparent ubiquity of TGA use for tryptophan in apicoplast genomes, it is interesting to note that there are no TGA codons in *Plasmodium* apicoplast genes, either internally or terminally. This suggests that *Plasmodium* lost the UGA decoding mechanism or that the mechanism developed after *Plasmodium* branched from other apicomplexans. The tryptophan tRNA encoded by the *T. gondii* plastid genome has an anticodon (CCA) that is incapable of pairing with UGA. Other organisms have developed several mechanisms to overcome this difficulty (reviewed in Gray *et al.*, 2004), including RNA editing. The mechanism employed by *T. gondii* is not yet known.

Some apicoplast genes have internal TAA or TAG stop codons. *T. gondii RPOC2* has one of each, both located toward the middle of the gene, as well as a TAA within *RPS8*. It is unclear whether these are pseudogenes or whether a mechanism exists to allow translation through the internal stop codons. Both the *T. gondii* and *P. falciparum* apicoplast genomes have only a single gene with an intron: a conserved $tRNA^{leu}$ (see www.ToxoDB.org).

9.2.5 Apicoplast Genome Replication

Although the *T. gondii* and *P. falciparum* apicoplast DNAs are strikingly similar in gene content and organization, they differ in an important element of physical structure: about 90% of the *P. falciparum* plastid genomes are circular while the *T. gondii* apicoplast DNA is in linear tandem arrays (Williamson *et al.*, 2001, 2002). This observation has immediate implications for replication of plastid DNA.

Several lines of evidence point to the linearity of *T. gondii* plastid DNA. The movement of circular molecules in gels is sensitive to electrophoresis conditions, revealed typically by comparing pulsed field gels electrophoresed using different pulse times and intensities. Williamson *et al.* (2001) found that Southern blots of such gels showed only a minor proportion of the *T. gondii* plastid DNA migrating as if circular. Instead, over 90% corresponded to a ladder consistent with differing numbers of the unit plastid genome, suggesting a tandem array. Restriction analysis of tandemly repeated sequences and circular molecules both can generate a circularly permuted map, but the ends of a linear array produce additional restriction fragments. This is the result obtained for the *T. gondii* apicoplast genome. Further analysis confirmed that it is linearized in the middle of the inverted repeat and exists largely as tandemly repeated linear arrays. These data strongly suggest that the genome is replicated via a rolling circle mechanism. The progression of DNA polymerase around a circular genome can produce a displaced linear DNA. If it is not cleaved and recircularized, a linear concatamer of genomes is produced.

The proportion of linear to circular molecules is reversed in *P. falciparum*, with circular DNAs predominating. Using two-dimensional gel analysis to track branch points, Williamson and colleagues (Williamson *et al.*, 2002) showed that most *P. falciparum* plastid DNA replication initiates at two sites mirrored in the inverted repeat and then proceeds via D-loop intermediates, eventually yielding a circular replicate. The remaining DNA appears to replicate via a rolling circle mechanism, as has been predicted for *T. gondii*. By hybridization of nascent DNA to apicoplast sequences, Singh and colleagues initially identified two, then later three replication initiation sites in each copy of the *P. falciparum* inverted repeat (Singh *et al.*, 2003, 2005). One is located at the tRNA cluster between the small and large subunit rRNAs, consistent with prior data (Williamson *et al.*, 2002), and appears to be substantially more active than the other two. Replication in other apicomplexans has not been studied so it is unclear whether D-loop or rolling circle replication will be most common in the apicomplexans.

The importance of apicoplast functions as potential drugs targets (see Section 9.2.8 Drug Sensitivities) has fostered interest in replication of its genome. The DNA replication machinery for the apicoplast appears largely conventional – for a chloroplast genome – but with some twists. Several DNA binding proteins have been identified. A *P. falciparum* DnaJ homologue has been reported to bind at least one of the apicoplast DNA replication initiation sites (Kumar *et al.*, 2010). Another DNA-binding protein has characteristics similar to bacterial single-stranded binding (SSB) proteins, localizes to the apicoplast and preferentially binds apicoplast DNA (Prusty *et al.*, 2010). *P. falciparum* SSB binds DNA as a tetramer, like *E. coli* SSB, but the polarity of binding is reversed (Antony *et al.*, 2012). An apicoplast-localized *P. falciparum* orthologue of the bacterial histone-like DNA binding protein HU has been shown to bind apicoplast DNA in a non-sequence specific manner (Ram *et al.*, 2008; Sasaki *et al.*, 2009), to promote DNA condensation (Ram *et al.*, 2008), and to complement an *E. coli* HU mutant (Sasaki *et al.*, 2009). Efforts to delete the gene in *P. berghei* were unsuccessful, suggesting it is essential (Sasaki *et al.*, 2009). *T. gondii* HU is also apicoplast-localized and complemented an *E. coli* HU mutant. A knock-out mutant was achieved in *T. gondii* but it showed greatly reduced growth kinetics and was associated with missegregation of the apicoplast genome and loss of the apicoplast from many cells (Reiff *et al.*, 2012), as previously reported for ciprofloxacin-treated *T. gondii* (Fichera and Roos, 1997).

Treatment of *P. falciparum* with topoisomerase II inhibitors results in cleavage of apicoplast as well as nuclear DNA (Weissig *et al.*, 1997). Consistent with that observation, genes similar to the A and B subunits of DNA gyrase, a bacterial topoisomerase, are found in the nuclear genome (Khor *et al.*, 2005), and their protein products are apicoplast-localized (Dar *et al.*, 2007). A combination of inhibitor studies and characterization of expressed proteins, primarily focused on the B subunit, indicate that it has expected functions and is involved in apicoplast DNA replication (Khor *et al.*, 2005; Dar *et al.*,

2007, 2009; Raghu Ram *et al.*, 2007). Selective cleavage of the A subunit mRNA inhibits parasite growth and development (Ahmed and Sharma, 2008; Augagneur *et al.*, 2012). *T. gondii* encodes putative A and B subunits of gyrase (Sheiner *et al.*, 2011) but these have not yet been studied functionally. Recently, Sheiner *et al.* (2011) have identified additional *T. gondii* proteins expected to function in plastid genome repair and replication. One is a putative ATP-dependent helicase similar to a bacterial DNA replication and repair enzyme, and another has helicase-like domains. The third has a domain similar to phage integrases and DNA break-rejoining enzymes.

The *PREX* gene, most studied in *Plasmodium* but present in many apicomplexans, encodes a multifunctional, apicoplast-localized protein with primase, helicase, exonuclease, and polymerase domains, the last being similar to bacterial DNA polymerase I. The primase/helicase and the exonuclease/polymerase domains (Seow *et al.*, 2005), and the primase domain with an adjacent zinc-binding domain (Lindner *et al.*, 2011), have been expressed recombinantly and the expected activities confirmed. Antibodies raised against each of these domains detect band sizes consistent with proteolytic processing to yield individual primase and exonuclease/polymerase proteins. The *P. falciparum* PREX targeting sequence directs proteins to the apicoplast (Seow *et al.*, 2005; Lindner *et al.*, 2011); fluorescently tagged primase associates with distinct foci within the apicoplast during replication (Lindner *et al.*, 2011). Efforts to knock out *PREX* in *P. yoelii* were unsuccessful, suggesting the gene is essential. However, because the entire gene was targeted, it is unclear whether all the activities are essential. Recently, the activity of recombinantly expressed *P. falciparum* PREX has been further characterized, showing that its fidelity, when the adjacent 3′–5′ exonuclease is included, is on par with other high fidelity DNA polymerases (Kennedy *et al.*, 2011). A *T. gondii PREX* orthologue has been identified and its polymerase domain shown to be active (Mukhopadhyay *et al.*, 2009).

9.2.6 Apicoplast Division

To ensure perpetuation of their genomes, mitochondria and plastids must be present at all times, even if inactive. That means that mitochondria and plastids must divide and be partitioned during each cell cycle to provide organelles for the daughter cells. In unicellular eukaryotes, the events of organelle division tend to be coordinated with the cell cycle. This coordination is especially important since only a single mitochondrion and apicoplast are present in Apicomplexa, as opposed to the multiple chloroplasts and mitochondria found in each cell of multicellular eukaryotes.

Until recently, methods for synchronizing *T. gondii* have been lacking, leading to a paucity of data on some aspects of analysis of coordination of events with the cell cycle. For example, it is unclear whether replication of plastid DNA is coordinated with nuclear DNA replication in *T. gondii*, as it is in *P. falciparum* (Shaw *et al.*, 2001; Williamson *et al.*, 2002). However, analysis of events in single *T. gondii* cells has produced insights into organelle division. The use of fluorescent reporters targeted to different cellular structures has expanded the tools available for cell cycle analysis, permitting experiments to follow the partitioning of the apicoplast into daughter cells.

Division of the apicoplast is coordinated with the cell cycle. In an elegant study of apicoplast division in *T. gondii* (Striepen *et al.*, 2000), a number of important observations were made (Fig. 9.5). In apicomplexans, the centrosome is extranuclear and located close to the apicoplast, which is itself apical to the nucleus. During daughter cell formation, as the duplicated centrosomes move apart, the ends of the apicoplast follow. Apicoplast DNA is localized in a nucleoid and as the organelle lengthens into a dumbbell, then a U-shape, two nucleoids

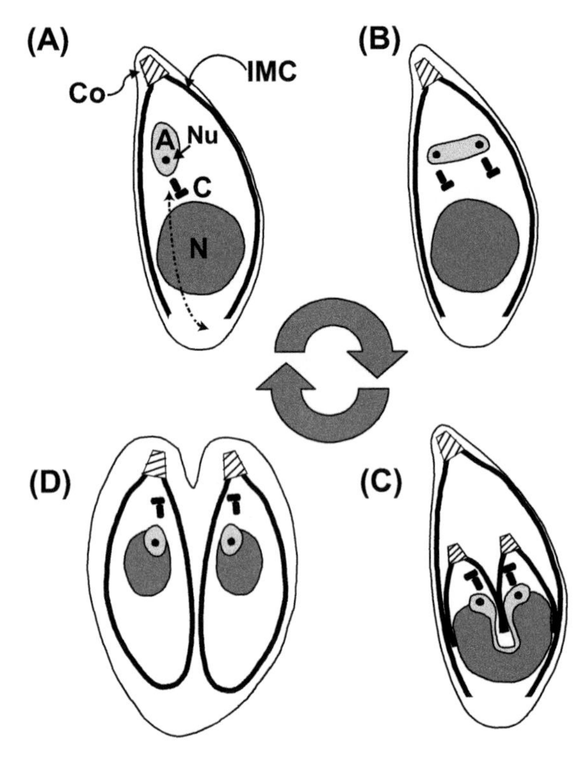

FIGURE 9.5 **Apicoplast division during endodyogeny.** A schematic depiction of endodyogeny is shown. Cell components are labelled in Part A: A, apicoplast; C, centrosome; Co, conoid; IMC, inner membrane complex; N, nucleus; Nu, nucleoid. Prior to duplication, the centrosome migrates from its position near the apicoplast to the basal end of the cell and back (dashed line). In Part B, the nucleoid and centrosome have both been duplicated and the apicoplast lengthens as the centrosomes and nucleoids move outward. In Part C, new conoids have formed and new inner membrane complex is developing. The nucleus and apicoplast are beginning to move into the forming daughter cells, both becoming U-shaped. The relative position of the centrosomes and apicoplast has reversed. In Part D, daughter cell formation is nearing completion. The organelles will return to the positions shown in panel A as cytokinesis is completed. See (Striepen *et al.* (2000) and Nishi *et al.* (2008) for additional details.

can be seen, one at each end and each adjacent to a centrosome. Coordinately, the parasite nucleus is repositioned to the basal end of the cell and develops two arms that move toward the centrosomes in the forming daughter cells. The timing of apicoplast growth, division, and segregation was further placed into the context of other subcellular organelles in later studies which took advantage of parasites expressing multiple tagged proteins (Nishi *et al.*, 2008). The Golgi body duplicates around the time that apicoplast begins to elongate as the inner membrane complex forms. Next, the endoplasmic reticulum (ER) begins to segregate with the nucleus. Mitochondrial division and partitioning to the daughter cells is completed well after apicoplast segregation, just prior to daughter cell emergence. The secretory organelles begin to form at this late stage. Matsuzaki and colleagues (Matsuzaki *et al.*, 2001) analysed plastid division employing DAPI staining to follow DNA and transmission electron microscopy to assess structural changes during apicoplast division. They describe similar changes to apicoplast shape during division as well as other features: thickening at the ends of the elongating apicoplast, a 'scratched' appearance at its central constriction, and a dark-stained structure associated with the constriction. They hypothesize that these may be, respectively, sites for attachments of centriole microtubules, a plastid division ring, and a structure involved in organelle scission.

The close juxtaposition of the centrosomes (as visualized by the centriole marker centrin) and apicoplast at the time of division led to the hypothesis that centrosome movement elongates the apicoplast and that linkages with the newly forming inner membrane complex are responsible for the division, mechanically pulling the centrosomes and the linked apicoplasts into the daughter cell. However, subsequent studies showed that the centrioles exhibit a curious behaviour prior to their duplication, migrating to the basal side of the nucleus and back, indicating that the apparent linkage with the apicoplast is not consistent through the cell cycle (Nishi *et al.*, 2008). Later work showed that treatment of parasites with 3-methyladenine blocked centrosome duplication and plastid division but not apicoplast elongation (Wang *et al.*, 2010), demonstrating that plastid elongation can proceed in the absence of a daughter centrosome.

Organelle division in many other organisms involves a division ring that constricts around the middle of an elongated organelle until it splits in two. However, genes for the responsible proteins of the ring or its formation, such as FTSZ (a tubulin-like GTPase derived from the prokaryotic ancestor) and the dynamin-related protein ARC5 (a eukaryotic invention), have not been detected in apicomplexan genomes. Further studies showed that a novel dynamin-related protein, DRPA, localized to the apicoplast of *T. gondii* (Breinich *et al.*, 2009). The dynamin family of GTPases is involved in the fission of various organelles including mitochondria. Mutations that abrogate their activities have been studied extensively, allowing the generation of a dominant negative mutant of *DRPA*. The mutant protein was fused to a domain that promotes degradation unless stabilized by a cell-penetrant ligand and then expressed in *T. gondii*. In the presence of the stabilizing ligand, the mutant drpa protein blocked division of the plastid even though elongation occurred. As a result many daughter cells lacked an apicoplast and were non-viable for subsequent rounds of invasion and proliferation. Interestingly, even though the only known function of *DRPA* is in apicoplast scission, phylogenetic analysis indicates that *DRPA* is present in *Cryptosporidium* species, which lack an apicoplast (Breinich *et al.*, 2009). The two membranes of the chloroplast have separate but presumably interacting division machineries, so this work represents the start of understanding the division of the apicoplast with its four membranes.

Another parasite protein essential for apicoplast division is MORN1, a protein bearing multiple 'membrane occupation and recognition nexus' domains (Lorestani *et al.*, 2010). Using a system that allows tetracycline-regulated expression of a target gene, a conditional knockout of MORN1 was developed. As MORN1 levels decreased, defects in apicoplast division and segregation increased dramatically. In contrast to the highly specific phenotype of the *drpa* mutant, the *morn1* mutant also had defects in the assembly of a basal complex and cytokinesis, implying a potential linkage between each of these processes, which may be critical to their coordination within the cell cycle.

Apicoplast structure has also been examined in bradyzoites and in sexual stage *T. gondii* (Dzierszinski *et al.*, 2004; Ferguson *et al.*, 2005). Tachyzoites are rapidly replicating stages that are principally responsible for the pathology of toxoplasmosis. Bradyzoites are encysted parasites that are only slowly replicative, but they can switch to the tachyzoite stage. It is bradyzoites that linger in the host and are responsible for recurrence of toxoplasmosis in previously infected individuals. Tachyzoites can be induced to switch to bradyzoites in culture; these share many characteristics with bradyzoites isolated from animals. Using fluorescent reporters localized to the apicoplast, Dzierszinski *et al.* (2004) found that 10%–20% of *in vitro*-induced bradyzoites apparently lack plastids. Both mis-segregation and loss of signal without cell division were observed. In contrast, fluorescent tags showed no loss of mitochondria. Consistent with these observations, the authors noted parts of a mature *in vivo* bradyzoite cyst did not stain with antibodies to the apicoplast protein ACP. Complementing this observation are data from a serendipitous mis-segregating apicoplast mutant in tachyzoites (He *et al.*, 2001a). Analysis of its phenotype demonstrated that *T. gondii* without apicoplasts are viable as long as they are within the original vacuole. They are, however, unable to replicate in a new vacuole (see Fig. 9.10). Thus, any bradyzoites lacking an apicoplast would not be able to initiate a productive infection upon reactivation.

Ferguson *et al.* (2005) performed a comprehensive examination of the apicoplast during *in vivo* infection, including in bradyzoites and in the asexual and sexual forms that occupy the small intestinal villi of the cat. Electron micrographs and immunofluorescence assays were used to examine each stage. For bradyzoites, the authors report that there is an apicoplast adjacent to each nucleus in mature cysts, in contrast to previous studies (Dzierszinski *et al.*, 2004), and attribute this discrepancy to possible differences between *in vivo* and *in vitro* cysts. The life cycle stages of *T. gondii* that occur in the feline host include the coccidian phase, which has some similarities to the asexual erythrocytic cycle of *Plasmodium*. When resident in villi of the cat small intestine, the plastid has the expected small ovoid shape in trophozoites, but in schizonts it has elongated and developed branches. This is quite similar to changes observed in *P. falciparum* apicoplast morphology in the corresponding stages (Waller *et al.*, 2000). The plastid in *T. gondii* microgametocytes, macrogametocytes and macrogametes appears condensed and almost globular. In contrast, microgametes lack apicoplasts (Ferguson *et al.*, 2005). This finding implies that the apicoplast is maternally inherited in *T. gondii*, as it is in *P. falciparum* (Vaidya *et al.*, 1993; Creasey *et al.*, 1994).

9.2.7 Protein Trafficking to the Apicoplast

9.2.7.1 Targeting Sequences

With the identification of the apicoplast genome, it became apparent that the transcription and translation of its resident genes (including many ribosomal proteins), as well as any apicoplast-specific functions, would require the collaboration of additional proteins. These proteins are encoded in the nucleus and hence are called nucleus-encoded apicoplast-targeted (NEAT) proteins. Prior to the availability of the complete genome sequence, a search of the *T. gondii* EST databases for proteins homologous to chloroplast proteins led to the identification of several candidates for apicoplast ribosomal proteins, as well as some enzymes involved in fatty acid biosynthesis. Initially, two such proteins, acyl carrier protein (ACP) and ribosomal protein S9 (S9), were confirmed to be localized to the apicoplast by microscopic analysis using specific antisera (Waller *et al.*, 1998). These genes, in turn, provided tools for dissecting the manner in which proteins are targeted to the apicoplast.

Sequence analysis clearly showed that these proteins, all of which were predicted to reside within the lumen of the apicoplast, possessed N-terminal extensions as compared to their bacterial orthologues. In eukaryotic organisms, such extensions often contain topogenic information. For example, proteins localized to primary chloroplasts are usually targeted via an N-terminal transit peptide directly from the cytosol. Localization to the secretory system usually involves an N-terminal signal sequence. These presequences are rapidly removed by specific processing enzymes upon import (Bruce, 2001).

Interestingly, in those organisms with secondary chloroplasts, N-terminal extensions appear to contain a signal sequence followed by a transit peptide (Nassoury *et al.*, 2005). This organization is exactly what is observed for both *T. gondii* and *P. falciparum* NEAT proteins predicted to reside in the apicoplast lumen (Waller *et al.*, 1998). Using gene fusions, Waller *et al.* (Waller *et al.*, 1998) showed that the N-terminal extension of *T. gondii* ACP was able to target green fluorescent protein (GFP) to the apicoplast. Furthermore, presequences of *P. falciparum* predicted NEAT proteins were able to target GFP to the *T. gondii* apicoplast. These data suggest that at least some mechanisms of targeting are conserved across the Apicomplexa (Waller *et al.*, 1998; Jomaa *et al.*, 1999).

Both the signal and transit regions of the N-terminal extension of NEAT proteins are required to target a reporter to the apicoplast; deletion of either region results in mis-targeting (Fig. 9.6). Without a transit sequence, the S9 signal sequence targets GFP for secretion

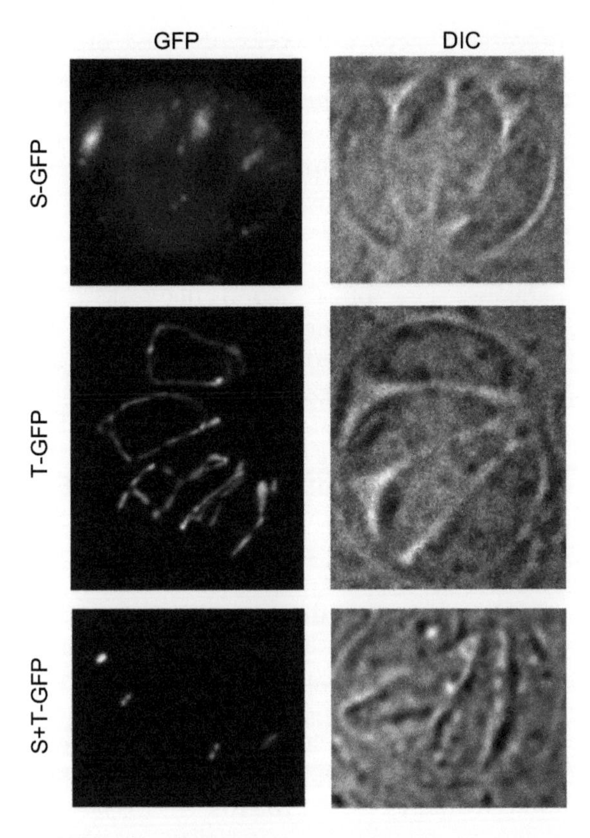

FIGURE 9.6 **Both domains of the N-terminal extension are required for targeting of NEAT proteins.** GFP fusions containing the entire N-terminal extension of ribosomal protein S9 (aa 1–159, S+T-GFP), its signal sequence (aa 1–42, S-GFP) or its transit sequence (aa 33–159, T-GFP) were expressed in *T. gondii*. The left hand panels show GFP fluorescence, while the right hand panels show DIC images of the same cells, which are residing within a parasitophorous vacuole in host fibroblasts. Co-localization with apicoplast markers (DNA or acetyl CoA carboxylase) demonstrated that the single dot observed upon expression S+T-GFP corresponds to the apicoplast. Co-localization with mitochondrial markers HSP60 and Mitotracker showed that the T-GFP protein is found in the mitochondrion. S-GFP is found primarily in the parasitophorous vacuole, although some material can be seen within the endomembrane system of the parasite. Image courtesy of Dr. Amy DeRocher.

(DeRocher *et al.*, 2000; Yung *et al.*, 2001), indicating that the reporter protein had entered the secretory system. Similar work in *P. falciparum* showed that the N-terminal region of ACP is able to target GFP for secretion (Waller *et al.*, 2000). The signal sequences for these proteins do not appear any different from those of proteins targeted to other destinations in the secretory system. Indeed, replacing the endogenous signal sequence with one from a heterologous secretory protein did not alter targeting to the apicoplast (Tonkin *et al.*, 2006). Without a signal sequence, the S9 transit peptide directs GFP to the mitochondrion, while a cytosolic localization is seen with GFP fusions to the transit peptide of ferredoxin-NADP$^+$ reductase (FNR) in *T. gondii* (Harb *et al.*, 2004) (Fig. 9.6). Similar findings were seen for the ACP transit peptide in malaria parasites (Waller *et al.*, 2000). Taken together, these studies indicate that the first step in protein targeting to the apicoplast lumen is entry into the secretory system.

The transit peptides of NEAT proteins vary in length, from about 50–200 aa and are very diverse in sequence. Like chloroplast transit peptides, these peptides have few acidic or hydrophobic residues (Foth *et al.*, 2003). In fact, the transit peptide of *T. gondii* S9, when fused to GFP, allows GFP to be imported into isolated pea chloroplasts (DeRocher *et al.*, 2005). Apicoplast transit peptides have a net positive charge. *T. gondii* and *P. falciparum* transit sequences show different amino acid biases; these appear to result from the different nucleotide composition of their genomes (that of *P. falciparum* is very AT-rich) (Ralph *et al.*, 2004a). The *T. gondii* transit sequences are enriched for serine and threonine, amino acids shown to be important in plant transit peptides (Bruce, 2001). *P. falciparum* transit peptides are enriched for asparagine and lysine residues (Foth *et al.*, 2003) and mutational analysis showed that hydroxylated residues are not crucial for targeting in that species (Waller *et al.*, 2000). Site-directed mutagenesis of the *P. falciparum* ACP transit peptide indicated that while basic residues at positions 2 and 6 were not essential, an acidic residue at position 2 prevented apicoplast targeting (Foth *et al.*,

2003). A predicted HSP70 binding site in the transit peptide was also found to be important (Foth *et al.*, 2003), suggesting that an unfolded structure could be important in apicoplast targeting. Indeed, recent studies have demonstrated that *P. falciparum* transit peptides exist in a disordered state (Gallagher *et al.*, 2011). Several studies have shown that *T. gondii* apicoplast transit peptides contain redundant information, since non-overlapping segments can still mediate targeting (DeRocher *et al.*, 2000; Yung *et al.*, 2001, 2003; Harb *et al.*, 2004). Detailed mapping of transit peptide functions of *T. gondii* FNR suggests that release from the ER, localization to the apicoplast, binding to chaperones, and processing are specified by discrete domains (Harb *et al.*, 2004).

Processing of chloroplast transit peptides is rapid, such that only the mature protein is seen under steady state conditions (for example, see Shanklin *et al.*, 1995). In contrast, both the mature form and the precursor protein containing the transit peptide are observed for NEAT proteins, whether native proteins or artificial gene fusions (Waller *et al.*, 1998; DeRocher *et al.*, 2005). Pulse–chase studies indicate that little processing is seen until 45–120 minutes after synthesis for GFP fused to signal and transit sequences of ACP in both *T. gondii* and *P. falciparum* (van Dooren *et al.*, 2002; DeRocher *et al.*, 2005). Whether this delay (as compared to chloroplast protein processing) reflects the time required for complete import or relative inefficiency of processing is not yet clear. A *P. falciparum* putative orthologue of the chloroplast stromal processing peptidase has been identified (van Dooren *et al.*, 2002). The predicted protein includes a bipartite targeting sequence, suggesting that it lies within the apicoplast. A putative *T. gondii* orthologue can be detected in the gene models of ToxoDB v7.3 (E values of $\sim 10^{-56}$). A functional apicoplast is required for processing of the transit peptide in *T. gondii* (He *et al.*, 2001b). However, analysis of S9 and FNR transit peptide cleavage indicates that multiple processing steps can occur, the first of which may happen before import is complete (Yung *et al.*, 2001; Harb *et al.*, 2004).

The identification of NEAT proteins and characterization of the bipartite apicoplast targeting sequence have allowed development of bioinformatic models that predict which proteins may be localized to the apicoplast of malaria parasites. Use of a neural network (PATS: Zuegge *et al.*, 2001) and a later rule-based algorithm (PlasmoAP: Foth and McFadden, 2003) allowed the identification of candidate *Plasmodium* NEAT proteins, which has been useful in predicting apicoplast metabolic pathways. Both programmes are available through a web-based interface at http://www.PlasmoDB.org. Differences in the amino acid bias of *P. falciparum* and *T. gondii* transit peptides mean that these programmes are not directly applicable to *T. gondii*. However, recently a rule-based algorithm was devised that is based on the known characteristics of the bipartite targeting sequence and is generally applicable to different apicomplexans (Cilingir *et al.*, 2012). After training on a set of 35 known apicoplast proteins of *T. gondii* (and a similar number of non-apicoplast proteins bearing a signal sequence), approximately 400 proteins were predicted to be localized to the apicoplast. While the identification of these candidates represent a good starting point, it will be important to verify individual proteins of interest, and furthermore to use experimental methods to identify proteins that may not have the types of targeting sequences detected by this algorithm.

Interestingly, not all proteins that reside in the apicoplast lumen bear bipartite targeting sequences. A screen for apicoplast proteins based on phylogenetic distribution and timing of expression during the cell cycle identified four genes encoding apicoplast luminal proteins that apparently lack a signal sequence or a transmembrane domain (Sheiner *et al.*, 2011). This discrepancy could reflect misidentification of the start codon or perhaps biological phenomena such as processing to reveal a recessed bipartite sequence or import by piggybacking with another protein, although such a process has not been described for apicoplast proteins as yet. Yet another twist that has emerged in recent work is that several proteins are dually targeted to the apicoplast and mitochondrion; the mechanisms underlying this phenomenon are discussed in Section 9.3.5 Protein Trafficking to the Mitochondrion. Additionally, Apicomplexans lack enough genes to encode individual tRNA synthetases for all translational compartments, so some of these enzymes are dually targeted to the cytosol and apicoplast in both *T. gondii* and *P. falciparum* (Pino *et al.*, 2010; Jackson *et al.*, 2012). The cytosolic form predominates, as expected, and may arise from use of an alternative initiation codon that eliminates the targeting information.

The above studies concern the localization of soluble matrix proteins to the apicoplast. It can be presumed that a substantial membrane proteome is required for the import of proteins and substrates and for the export of products, with membranes and intermembrane spaces having a different constellation of protein constituents. Several such proteins have now been identified in apicomplexans, including some functioning in import (see below) and others presumably functioning in metabolism. The proteins show a characteristic ring-like staining surrounding the luminal marker upon immunofluorescence analysis by deconvolution microscopy and localization to the outer compartments of the apicoplast by immunoelectron microscopy (Fig. 9.7A). The precedent of chloroplasts suggests that different targeting sequences may be utilized for different proteins, in part related to the protein's localization within the plastid. Targeting of proteins to the inner membrane of the chloroplast is often mediated by a transit peptide (Silva-Filho *et al.*, 1997; Roth *et al.*, 2004) whereas the picture with outer membrane proteins is more mixed

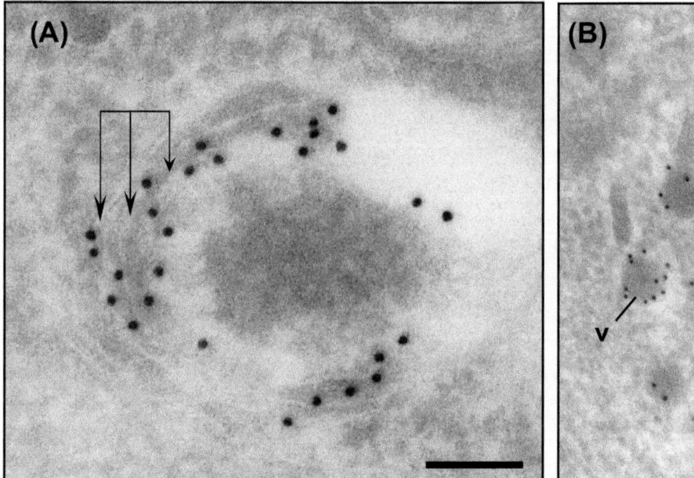

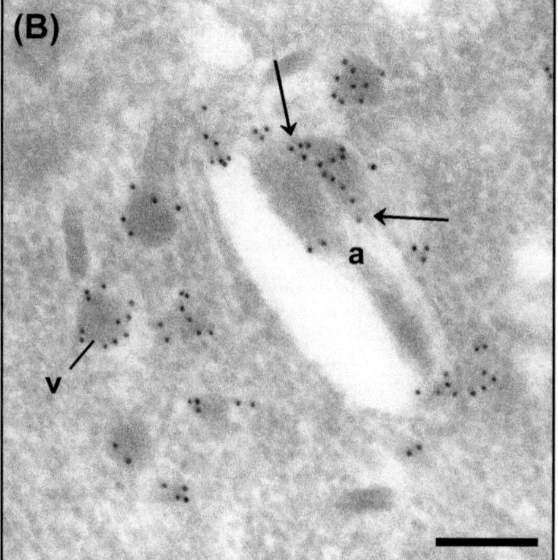

FIGURE 9.7 **Localization of the ATRX1 thioredoxin to vesicles and the outer compartments of the apicoplast.** Immunogold electron microscopy was used to detect epitope-tagged ATRX1 in *T. gondii*.
A) Close view of the apicoplast and its multiple membranes. Triple arrows mark the apparent localization of the molecule to different compartments. Bar is 100 nm.
B) Presence of ATRX1 at the apicoplast (a) and abundant small vesicles (v). Note the apparent fusion of a vesicle with the apicoplast, marked by arrows. Bar is 200 nm.

with some bearing transit sequences (such as Toc75), others having an N-terminal region resembling a signal sequence (Hofmann and Theg, 2005), and yet others apparently lacking either type of sequence (Funes *et al.*, 2004). The findings regarding the apicoplast are somewhat reminiscent of that, with the caveat that the tight spacing of apicoplast membranes makes it impossible to assign proteins to individual membranes via microscopy (even immunoelectron microscopy).

Several proteins targeted to the outer compartments of the apicoplast have a recognizable bipartite sequence, indicating entry into the ER followed by sorting to the apicoplast. Among these are *T. gondii* TIC20 and *P. falciparum* TIC22, which are likely localized to the innermost membrane and the adjacent intermembrane space (Kouranov *et al.*, 1999), and PfDer1-1 (Spork *et al.*, 2009), which is likely localized to the periplastid membrane. In contrast, other membrane proteins lack the canonical bipartite extension, although all appear to possess sequences that mediate entry into the ER. The transmembrane protease FTSH1 utilizes a signal anchor sequence to enter the endomembrane system (Karnataki *et al.*, 2007b), as does the peripheral membrane thioredoxin ATRX1 (DeRocher *et al.*, 2008). Notably, neither FTSH1 nor ATRX1 possess an obvious transit peptide. The *T. gondii* apicoplast sugar phosphate transporter APT1 (Karnataki *et al.*, 2007a) and its *P. falciparum* orthologue that resides in the outer membrane (Mullin *et al.*, 2006) lack any recognizable signal or transit sequence although it is likely that one of the transmembrane domains functions in localization to the ER. Interestingly, the paralogous *P. falciparum* iTPT, thought to reside in the innermost membrane, does have a bipartite sequence (Mullin *et al.*, 2006).

The molecular determinants that mark membrane proteins lacking transit peptides for localization to the apicoplast have not been well studied. In addition to the signal anchor sequence, ATRX1 requires some information in the 200 amino acids following the anchor for proper localization (DeRocher *et al.*, 2008). However, the requisite characteristics of this region have not been identified. More information has been obtained with respect to APT1. There, an N-terminal segment prior to the first of several transmembrane domains is essential for the protein's localization to the apicoplast (DeRocher *et al.*, 2012). The region was further narrowed by alanine scanning to a tyrosine-glycine motif that faces the cytosol. Mutation of this tyrosine to any other amino acid led to APT1 localization to the Golgi body, whereas mutation of the glycine to alanine (or several other amino acids) led to partial mislocalization. The YG motif could be situated as far as 64 aa from the transmembrane domain and still function, but when placed within seven aa, mislocalization occurred even though the YG motif was intact. Tyrosine-based motifs (with the consensus of YXXΦ, where Φ represents a hydrophobic amino acid) function in other pathways within eukaryotic secretory systems (Bonifacino and Traub, 2003). While it might seem surprising that specificity is provided by such a minimal motif, tyrosine offers multiple opportunities for interactions. Indeed, crystallographic analysis of the cargo binding subunit of the human AP2 adaptor complex with target tyrosine-based cargo peptides showed a combination of hydrophobic interactions with the tyrosine aromatic ring and hydrogen bonding with its hydroxyl moiety, providing specificity and binding affinity (Owen and Evans, 1998). Thus one model consistent with the observed *T. gondii* data is that the cytosolic YG motif interacts with a protein that selects cargo for transport to the apicoplast. Although similar YG motifs are present in the N-terminal region of *Plasmodium* APT1 orthologues, such motifs have not been identified in other apicoplast membrane proteins. Thus it appears likely that other proteins may employ distinct determinants for targeting to apicoplast membranes, possibly related to their final destination.

9.2.7.2 Trafficking Mechanisms

The path that NEAT proteins follow from the ER to the apicoplast lumen is not yet clear, although several models have been proposed (Fig. 9.8). One hypothesis is that the apicoplast lies literally within the secretory system with its outer membrane contiguous with ER. All proteins would pass it on their way to other destinations and NEAT proteins would be grabbed by the apicoplast by virtue of their transit peptide. They would then be imported through the next three membranes. Some organisms with secondary plastids have ribosomes studding the outer membrane, suggesting continuity with the ER (Gibbs, 1981). There is not any specific evidence supporting this mechanism in Apicomplexa: ribosomes have not been detected on the apicoplast surface and although the apicoplast lies very close to the nuclear envelope (Tomova *et al.*, 2009), which is an extension of the ER in *T. gondii* (Hager *et al.*, 1999), there is no evidence for fusion of the membranes (Tomova *et al.*, 2009). In *P. falciparum* sporozoites, the apicoplast has been seen within 10 nm of the nuclear envelope (Kudryashev *et al.*, 2010). Another possibility is that proteins move by vesicular trafficking from the ER to apicoplast, either directly or indirectly. In this case, the transit peptide would be responsible for packaging the proteins into appropriate vesicles. Specificity of vesicle trafficking would likely be conveyed by an additional component, possibly proteins such as SNAREs at the vesicle and apicoplast surfaces, as occurs for targeting to other destinations in the secretory system (Hong, 2005). However, none of the RAB family GTPases, nor any other molecules that typically function in vesicular trafficking examined thus far, appear to be uniquely required for protein targeting to the apicoplast. There is some evidence that the apparatus for selecting proteins from the ER for apicoplast import is regulated during the erythrocytic cycle of *P. falciparum*. Expression of mRNAs for known NEAT proteins is temporally regulated during the erythrocytic cycle, peaking in the late trophozoite/early schizont stages (Bozdech *et al.*, 2003; Le Roch *et al.*, 2003). When a *P. falciparum* NEAT GFP fusion protein was expressed earlier than normal (by driving transcription using a promoter active earlier in the erythrocytic cycle), it trafficked to the parasitophorous vacuole (Cheresh *et al.*, 2002). The general significance of this finding and whether such regulation exists in *T. gondii* is not yet clear.

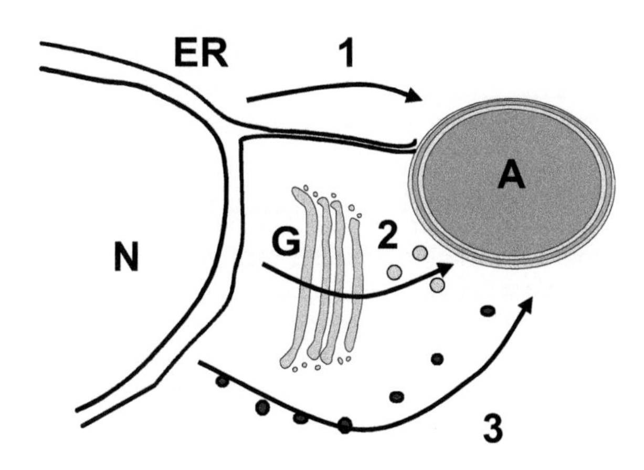

FIGURE 9.8 **Three models for protein targeting to the apicoplast.** Model 1 shows direct import from the ER. Model 2 postulates vesicular trafficking through the Golgi. Model 3 proposes vesicular trafficking from the ER, bypassing the Golgi. Organelles indicated are: A, apicoplast; ER; G, Golgi; N, nucleus.

It is unlikely that *T. gondii* apicoplast luminal proteins traffic through the Golgi. Exposure of *T. gondii* to brefeldin A (BFA) or low temperature block (15°C), which inhibit Golgi trafficking showed no effect on marker protein localization to the apicoplast (DeRocher *et al.*, 2005). To circumvent potential confusion arising from pre-existing marker proteins in the apicoplast, the exit of apicoplast-targeted GFP from the ER was blocked by tagging it with a conditional aggregation domain (CAD) (Rollins *et al.*, 2000; DeRocher *et al.*, 2005). In the absence of a synthetic ligand, the CAD domains aggregate, blocking trafficking of the fusion protein, as seen

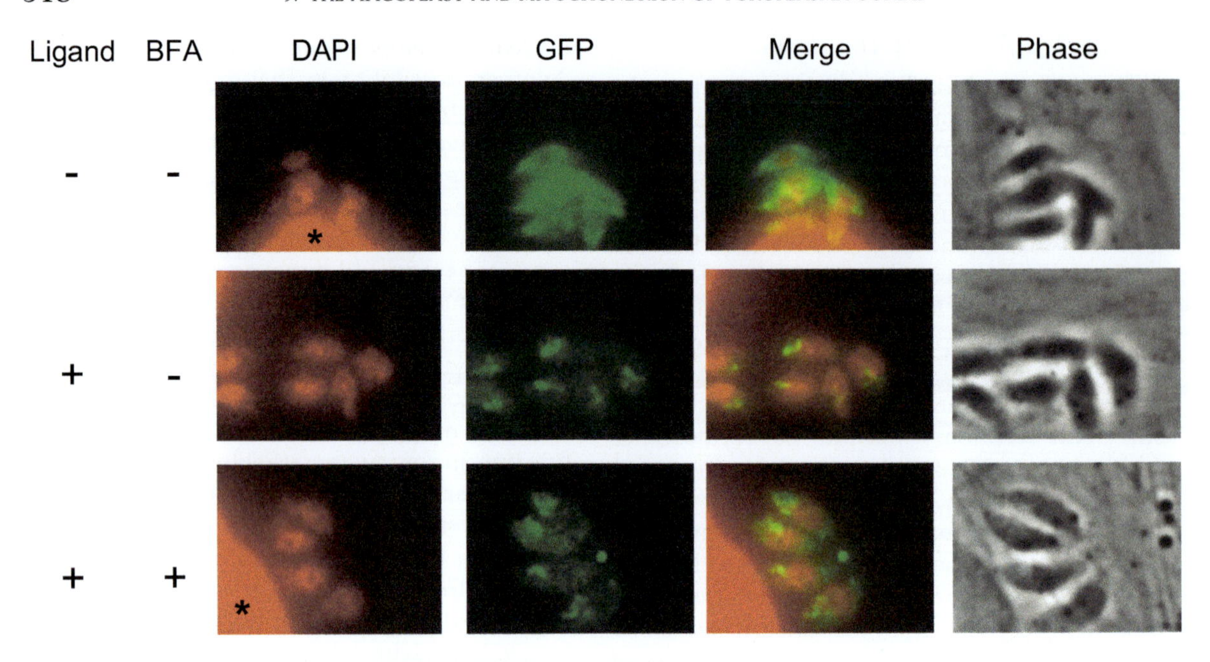

FIGURE 9.9 **Apicoplast protein targeting studied by conditional aggregation.** A GFP fusion protein bearing the bipartite extension of ribosomal protein S9, plus a tandem array of four CAD domains, was expressed in *T. gondii*. Removal of ligand causes aggregation of the CAD domains, while addition of ligand yields monomerization. When ligand is removed from the stable transfectants, the GFP is detected in the ER. When ligand is added for four hours, the protein traffics to the apicoplast. This tracking occurs even in the presence of BFA. GFP was detected using anti-GFP antibodies. DAPI staining reveals the DNA in the parasite nucleus and apicoplast, and in the upper and lower images, a portion of the host cell nucleus is visible (asterisk). Image courtesy of Dr. Amy DeRocher.

in other systems (Rivera *et al.*, 2000). After two days, the apicoplast was depleted of the marker protein, which instead was seen in the ER (Fig. 9.9). Subsequent addition of ligand released the fusion protein, which rapidly localized to the plastid region (DeRocher *et al.*, 2005). This localization was not blocked by BFA or by incubation at 15°C. These results were corroborated by studies in *P. falciparum*, in which BFA resistance was also seen (Tonkin *et al.*, 2006). Furthermore, addition of an ER-retrieval sequence to an apicoplast targeted marker protein did not affect targeting. Since the ER retrieval system connects with mislocalized proteins in the Golgi body, this finding further indicated that the modified protein did not transit through the Golgi (DeRocher *et al.*, 2005; Tonkin *et al.*, 2006). Interestingly, although no effect of BFA was seen on localization at the light microscopy level, a BFA-sensitive step is involved in protein maturation or localization within the organelle in *T. gondii*, since no transit peptide cleavage was observed in pulse–chase experiments in the presence of BFA (DeRocher *et al.*, 2005). However, BFA does not inhibit transit peptide cleavage in *P. falciparum* (Tonkin *et al.*, 2006).

While vesicles transporting luminal proteins have not been described, large vesicles bearing certain proteins that are also present in the outer apicoplast compartments have been detected in *T. gondii* (Karnataki *et al.*, 2007a, b; DeRocher *et al.*, 2008; Tawk *et al.*, 2011). In addition to residing at the apicoplast, tagged versions of APT1, FTSH1, and ATRX1, were seen in 'dots' and 'tubules' upon immunofluorescence. These structures were resolved as electron-dense vesicles upon immunoelectron microscopy (Fig. 9.7B). The vesicles are most abundant at the time of apicoplast enlargement prior to division, and are lacking in newly formed daughter cells. Some images show what appears to be recent or ongoing fusion of vesicles with the outer membrane of the apicoplast.

The vesicles themselves appear to bear phosphatidylinositol 3-phosphate (PI3P), as does the apicoplast (Tawk *et al.*, 2011). Interestingly, over-expression of a PI3P-binding protein led to the eventual loss of the apicoplast in many cells, as well as accumulation of electron lucent vesicles (Tawk *et al.*, 2011). Thus PI3P is important for the biogenesis and maintenance of the apicoplast but the functional role it plays has not been dissected. In other vesicular trafficking systems, phosphatidylinositol phosphates appear to function in enhancing the interaction of molecules with the vesicle membrane (reviewed by Mayinger, 2012). It appears unlikely that the vesicles arise from the apicoplast, because they are visible in parasites that have lost their apicoplast due to interference with PI3P (Tawk *et al.*, 2011) or over-expression of certain toxic constructs (unpublished results, DeRocher and Parsons). In the studies of mutant APT1 proteins described above, the presence of specific mutant proteins at the apicoplast was always accompanied by their presence in vesicles, implying a similar mode of recognition of these proteins as cargo. It is tempting to view these vesicles as the carriers of membrane proteins to the apicoplast, but direct evidence for this is lacking. An alternative could be that they represent relatively stable structures that are lost upon daughter cell formation and regenerated each cell cycle. Little data are available on whether trafficking of membrane proteins from the ER involves the Golgi body. Our preliminary data indicate that the large vesicles bearing the membrane proteins are still present following treatment with BFA under conditions in which the Golgi body is disrupted (Bouchut and Parsons, unpublished). However, it remains to be seen whether these vesicles were pre-existing or arose following administration of the Golgi body inhibitor.

Insights into the import apparatus of the apicoplast have increased in recent years. It is assumed, based on phylogenetic origins and similarity of targeting sequences, that the translocon spanning the inner membranes of apicoplasts resembles that of chloroplasts. The chloroplast translocon is composed of multiple components in the inner (Tic) and outer (Toc) membranes (Jarvis and Soll, 2002; Nassoury *et al.*, 2005). Among the functions provided in the complex are specific binding to the transit sequence, channel formation, energy generation, and chaperone activity. Two components of the apicoplast inner translocon *T. gondii* TIC20 (van Dooren *et al.*, 2008) and *P. falciparum* TIC22 (Kouranov *et al.*, 1999) have been identified. Conventional TOC proteins have not been demonstrated in Apicomplexa. However, a diverged candidate TOC protein, derived from the cyanobacterial outer membrane protein OMP85, has been identified in the diatom *Phaeodactylum tricornatum* (Bullmann *et al.*, 2010). It localizes to the plastid membranes and is able to generate a pore when incorporated into liposomes. A putative orthologue of this protein has been proposed in *T. gondii* (Bullmann *et al.*, 2010), but expression levels of the corresponding mRNA are quite low and further experiments are needed to confirm apicoplast localization. Presumably, other orthologous Tic and Toc functions are present but the machinery is substantially diverged with respect to protein sequence. Isolated secondary plastids (red-algal derived) from the cryptomonad *Guillardia theta*, described by the authors as having two membranes (the outer membranes being stripped), were capable of importing proteins bearing transit peptides from nucleomorph-encoded plastid proteins but not those bearing transit peptides from nuclearly-encoded proteins (Wastl and Maier, 2000). These data raise the possibility of multiple mechanisms of protein import.

Insights into the means of protein translocation across the periplastid membrane came from studies of *G. theta* in which the nucleomorph genome was scanned for candidate proteins that might mediate protein translocation. These analyses identified two proteins likely functioning in the import pathway for the inner membrane: Tic22 and Iap100 (Douglas *et al.*, 2001). Additionally, paralogues of several proteins associated with the ER associated degradation (ERAD) pathway were detected (Sommer *et al.*, 2007). The ERAD pathway extrudes misfolded proteins from the ER through a multicomponent translocon into the cytosol for degradation. Thus by analogy, the ERAD-like machinery was proposed to function in importing proteins from the outermost plastid space across the periplastid membrane. One of these components, *G. theta* Der1p, was shown to complement *S. cerevisiae der1* mutants, demonstrating functionality. This nucleomorph-encoded ERAD-like machinery exists in parallel with canonical ERAD machinery which is nuclearly encoded. Extending these studies to *P. falciparum* (Sommer *et al.*, 2007; Spork *et al.*, 2009; Kalanon *et al.*, 2009) and later *T. gondii* (Agrawal *et al.*, 2009) also demonstrated dual ERAD-like genes, with one set more closely related to genes present only in organisms with plastids derived from red algal endosymbionts. In these organisms, which lack a nucleomorph, all components are nuclearly encoded. Key features of the ERAD machinery are the pore (possibly formed by DER1), ubiquitination machinery (important for translocation as well as degradation in canonical ERAD), and an AAA ATPase mechanoprotein CDC48 that assists in pulling the ubiquitinated proteins through the pore. The symbiont-derived ERAD-like machinery (dubbed SELMA) identified thus far in *T. gondii* (Agrawal *et al.*, 2009) and *P. falciparum* (Sommer *et al.*, 2007; Spork *et al.*, 2009; Kalanon *et al.*, 2009) includes two Der1 isoforms, a CDC48 protein, and paralogues several components of the ubiquitination machinery. Interestingly, SELMA is not the only example in which the ERAD machinery has been repurposed to drive protein import. Multiple similar components also function in the import of peroxisomal proteins, including AAA proteins and a specialized mono-ubiquitination machinery (Ma *et al.*, 2011). Protein unfolding is not required for import into peroxisomes, but the role of unfolding in protein localization to the various compartments of the apicoplast has not been studied.

With four bounding membranes, the apicoplast offers multiple options for localization. The mechanisms by which membrane proteins are shuttled from one membrane to another are not understood nor are the determinants that specify their final localization. Membrane composition may differ between them and may convey some specificity to the targeting of membrane proteins. The lipid composition of chloroplast membranes is very distinct from other cellular membranes, with high levels of galactoglycerolipids (Block *et al.*, 1983). Although little is known about apicoplast membranes directly, galactoglycerolipids have been identified in *P. falciparum* and *T. gondii* (Marechal *et al.*, 2002; Bisanz *et al.*, 2006). Their relative distribution across the multiple membranes remains to be determined.

9.2.8 Drug Sensitivities

The potential of the apicoplast as a drug target reflects its algal, and thus also its cyanobacterial, origin, with many proteins and pathways not shared by the human host. Furthermore, many apicoplast proteins are enzymes, which bind small molecules and hence have a higher possibility of being druggable (Hopkins and Groom, 2002). Indeed, several of the prokaryotic-like features of apicoplast functions can be inhibited by existing antimicrobials. These have been important research tools, and some are important clinically as well (Wiesner and Seeber, 2005; Fleige *et al.*, 2007; Dahl and Rosenthal, 2008). The ability of inhibitors of organellar

DNA replication, transcription, and translation to kill *T. gondii* and *P. falciparum* indicate that organellar functions are essential. For example, the rifamycin S antibiotics, such as rifampicin and rifabutin, inhibit eubacterial RNA polymerases like those encoded by the apicoplast genome and are active against both *T. gondii* and *Plasmodium* species (Alger *et al.*, 1970; Divo *et al.*, 1985; Strath *et al.*, 1993; Araujo *et al.*, 1994; Pukrittayakamee *et al.*, 1994; Olliaro *et al.*, 1994). Both the A and B subunits of *P. falciparum* DNA gyrase bear apicoplast targeting sequences (Khor *et al.*, 2005), and inhibitors of DNA gyrase are toxic to *T. gondii* (Fichera and Roos, 1997).

T. gondii is sensitive to the macrolide antibiotic clindamycin, an inhibitor of prokaryotic-like translation, and a *T. gondii* mutant resistant to clindamycin was cross-resistant to the macrolide antibiotics azithromycin and spiramycin (Pfefferkorn and Borotz, 1994). Two other clindamycin-resistant *T. gondii* mutants were found to have a point mutation in the plastid large subunit rRNA that mapped close to known clindamycin specificity determinants (Camps *et al.*, 2002). Macrolide interaction sites are known to be restricted to a small region of the ribosome's peptidyl transferase domain and co-crystallizations of these drugs with ribosomes have shown that they all block the peptide exit tunnel (Hermann, 2005). Consequently, the observed cross-resistance is unsurprising.

Other inhibitors of prokaryotic-like translation include thiostrepton, which acts against *P. falciparum* (McConkey *et al.*, 1997; Rogers *et al.*, 1997; Clough *et al.*, 1997) but not *T. gondii*. The differential sensitivity is likely due to an alternate nucleotide at the critical position in the *T. gondii* apicoplast large subunit rRNA (Clough *et al.*, 1997). The antibiotic actinonin inhibits the peptide deformylase that removes the formyl group from the initiator methionine in eubacterial proteins. It shows some activity against *P. falciparum in vitro* (Wiesner *et al.*, 2001), likely acting against only the apicoplast since targeting experiments in *T. gondii* have indicated that the product of the single peptide deformylase gene locates exclusively to this organelle (Pino *et al.*, 2010). Tetracyclines, which inhibit translation by prokaryotic and prokaryotic-like ribosomes, decrease growth of both *T. gondii* (Tabbara *et al.*, 1982; Chang *et al.*, 1990, 1991) and malaria parasites (Tabbara *et al.*, 1982; Geary and Jensen, 1983; Divo *et al.*, 1985). These antibiotics bind to the small subunit of the ribosome (Brodersen *et al.*, 2000; Anokhina *et al.*, 2004) and it had been suggested initially that, in apicomplexans, the mitochondrial ribosome was the main site of action. This was consistent with antibiotic effects on mitochondrial morphology, functions, and the pattern of protein synthesis inhibition in *T. gondii* and *P. falciparum* (Kiatfuengfoo *et al.*, 1989; Beckers *et al.*, 1995; Goodman *et al.*, 2007). However, Camps *et al.*, (2002) reported that the kinetics of *T. gondii* growth inhibition by clindamycin and tetracycline were quite similar, including a delayed death phenotype (see below). Studies with *P. falciparum* (Dahl *et al.*, 2006; Goodman *et al.*, 2007) have substantiated the apicoplast as a site for protein translation inhibition by tetracyclines, in addition to – or instead of – the mitochondrion (see below).

Although an effect on the mitochondrion often cannot be ruled out, many of the drugs discussed above clearly affect the apicoplast. They show an intriguing phenotype, called delayed death (Fichera *et al.*, 1995), that is easily detected in *T. gondii* because it replicates by endodyogeny. An example is the case of clindamycin, but other antibiotics show a similar phenotype (Dahl and Rosenthal, 2008). After drug is added, parasite multiplication continues vigorously within the first vacuole. However, when establishing the second vacuole, the parasites fail to replicate. This effect occurs even if the drug is removed at the second cycle. As a consequence and as a characteristic indication of a delayed death effect by a compound, much less drug is necessary to kill 50% of parasites (IC_{50}) when the growth assay is run for longer times (72–96 hours) than when the drug effect is evaluated after 48 hours of growth (Dahl and Rosenthal, 2008). The delayed death phenomenon was further explored using an unusual system for generating apicoplast-deficient cells (He *et al.*, 2001a). Transient transfection of a construct containing ACP–GFP fused to the rhoptry targeting sequences of ROP1 was found to interfere with division of the apicoplast. After a series of divisions, the parasitophorous vacuole contained many parasites but only one with an apicoplast. Within the original vacuole, the cells remained healthy whether they contained an apicoplast or not and all were capable of invading a new host cell (Fig. 9.10). However, only those with an apicoplast could proliferate in the new host. Hence, it appears that some molecules produced directly or indirectly by the apicoplast in the preceding cycle are required in the next round of infection. One proposed mechanism to explain delayed death is based on the projected need for a formylated methionine–tRNA for translation initiation in the mitochondrion of *T. gondii*. The apicoplast genome, but not the mitochondrial genome, encodes a methionine tRNA and there is only a single gene each for the methionine–tRNA formyltransferase and deformylase in *T. gondii*, the products of which localize to the apicoplast only (Pino *et al.*, 2010). Howe and Purton (2007) invoke a requirement for transfer of formylated $tRNA^{Met}$ from the site of its synthesis within the apicoplast to the closely associated mitochondrion to enable mitochondrial translation. However, the lack of both genes in the genomes of *Babesia* sp. as well as *Theileria* sp. argues against a requirement of formylated methionine–tRNA for mitochondrial translation initiation in *T. gondii* (Pino *et al.*, 2010) and thus also as an explanation for the delayed death phenotype. In this regard, it is important to remember that two of the three mitochondrially-encoded protein genes in *P. falciparum* lack an AUG initiation codon (Feagin, 1992), as do a variable subset of those genes in other apicomplexans (J.E. Feagin, unpublished results). Alternate initiation codons for mitochondrial genes are well known.

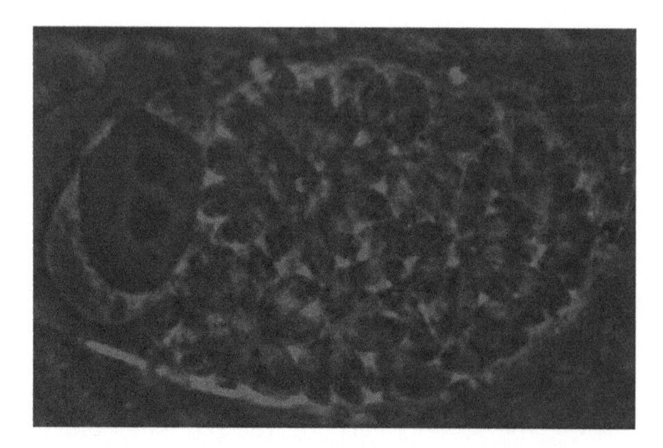

FIGURE 9.10 **Replication without apicoplast division.** *T. gondii* were transiently transfected with ACP–GFP–m-ROP1 and inoculated into an HFF cell monolayer. This image, taken 48 hours after transfection, shows a vacuole with about 64 parasites, only one of which contains an apicoplast as shown by the fluorescence of the GFP reporter. Image courtesy of C.Y. He and D.S. Roos. Reprinted by permission from EMBO J. 20: 330, 2001.

A recent *in vitro* study in *P. falciparum* blood stages revealed a surprising aspect of the delayed death caused by clindamycin, chloramphenicol, and doxycycline (a tetracycline). Yeh and Derisi (2011) showed that supplementation of treated cultures with 200 μM isopentenyl pyrophosphate (IPP) neutralized the growth inhibitory effect of these drugs at the late time point of culture (96 hours) so that higher concentrations of drugs (similar to those required to kill at 48 hours) were required to kill those parasites. As will be detailed below (Section 9.2.9 Apicoplast Metabolism), IPP is the end product of the isoprenoid biosynthesis pathway in the apicoplast. These observations indicate that, at least in *P. falciparum* blood stages, the isoprenoid pathway is a central player for understanding the basis of the delayed death phenotype. However, the data do not mean that these drugs only target the apicoplast since they still kill at low micromolar levels in the presence of exogenous IPP (Yeh and DeRisi, 2011).

At least one molecular defect related to delayed death caused by transcription/translation inhibitors in *T. gondii* has been identified. The level of apicoplast DNA is dramatically reduced upon clindamycin treatment (Fichera *et al.*, 1995; Fichera and Roos, 1997), as it is by similar inhibitors. Evidently the inability to translate certain proteins encoded by the apicoplast genome prevents the proper replication of the apicoplast DNA, either directly or indirectly, potentially causing delayed death. One of the few apicoplast-encoded proteins not involved in transcription/translation is the putative chaperone CLPC (an Hsp93 homologue), which may be required for import of NEAT proteins necessary for apicoplast DNA replication. Failure to translate it could negatively affect import of the proteins needed for apicoplast functions. Another hypothesis builds upon the growing list of DNA and RNA processing enzymes containing iron-sulphur clusters (ISC), which are important for their respective functions (Netz *et al.*, 2011; White and Dillingham, 2012; Wu and Brosh, 2012). Since import of NEAT proteins likely requires protein unfolding and concomitant loss of ISC, the apicoplast has kept its own ISC synthesis and assembly machinery (see Section 9.2.9 Apicoplast Metabolism). One component of it is the apicoplast-encoded SUFB, and it thus seems possible that impairment in ISC synthesis and/or assembly could result in an inactive primase or polymerase, thereby inhibiting replication. However, so far no data are available on ISC-containing DNA or RNA processing enzymes in the organelle.

Several of the antibiotics that target apicoplast functions are in clinical use, especially against malaria. The antibiotic clindamycin serves as a second line drug for toxoplasmosis while spiramycin is chosen for toxoplasmosis during early pregnancy. However, some of the agents, while effective *in vitro*, are not sufficiently antiparasitic for use in treatment of disease, e.g., ciprofloxacin. A number of comprehensive reviews discuss various aspects of the apicoplast as a drug target in apicomplexan diseases (Wiesner and Seeber, 2005; Dahl and Rosenthal, 2008; Fleige and Soldati-Favre, 2008; Lizundia *et al.*, 2008; Wiesner *et al.*, 2008; Botté *et al.*, 2012), and the topic is also covered in Chapter 21 of this volume.

9.2.9 Apicoplast Metabolism

What functions are potentially localized to the apicoplast? As noted above, the apicoplast genome does not encode any proteins that provide clues to its metabolic role. Most of the genes encode proteins required for transcription or translation. A potential orthologue of CLPC (a chaperone component of the Tic complex) is encoded on the apicoplast genome. Apicoplast-encoded SUFB was first thought to encode an ABC transporter, but is now believed to be involved in ISC metabolism (Rangachari *et al.*, 2002; Wilson *et al.*, 2003) (see below). Hence, most of the clues to apicoplast function come from the predicted NEAT proteins (see Section 9.2.7 Protein Trafficking to the Apicoplast). Using PlasmoAP, Ralph *et al.* (2004b) identified more than 500 *P. falciparum* proteins as potentially targeted to the apicoplast. The identification of the constellation of potential apicoplast proteins provided clues as to why the organelle is essential. Nonetheless, about 70% of the predicted NEAT proteins are of unknown function, being annotated as hypothetical proteins (Ralph *et al.*, 2004b). Further analysis will be required to determine which of these predicted proteins are expressed and are targeted to the apicoplast. A systematic comparison of the *P. falciparum* predicted NEAT proteins (in particular the hypothetical proteins) with all the *T. gondii* predicted proteins has not been reported to date. However, due to the cyanobacterial origin of the apicoplast, it has been possible to map several known metabolic pathways to NEAT proteins, and these have recently been

TABLE 9.1 Comparison of genome size, predicted protein coding sequences and known metabolic pathways of the apicoplast of some apicomplexan parasites

	Genome size (Mb)	No. protein coding sequences	FD/FNR*	[Fe-S]	DOXP	LIPA/B	PDH	FAS II	Haem
T. gondii (ME49)	63.01	7993	✓	✓	✓	✓	✓	✓	✓
Neospora caninum	61.04	7082	✓	✓	✓	✓	✓	✓	✓
P. falciparum (3D7)	23.33	5538	✓	✓	✓	✓	✓	✓	✓
B. bovis	8.14	3706	✓	✓	✓				
T. parva	8.31	4082	✓	✓	✓				

** FD/FNR, ferredoxin redox system; [Fe-S], iron-sulphur cluster biosynthesis; DOXP, isoprenoid biosynthesis; LIPA/B, lipoic acid metabolism; PDH; pyruvate dehydrogenase complex; FAS II, fatty acid biosynthesis type II; Haem, haem biosynthesis. Checks indicate the presence of the pathway. Assembled from data obtained from EuPathDB Gene Metrics (http://eupathdb.org); August 2012.*

compared between several sequenced apicomplexans (see Seeber and Soldati-Favre, 2010 for a comprehensive review). The data indicate that *T. gondii* is the most 'complete' apicomplexan parasite with regard to the pathways discussed below, giving it a higher metabolic flexibility and thus presumably also a less restricted host cell range compared to *Plasmodium* sp. and the piroplasms (Table 9.1).

As apicoplast pathways have an evolutionary history that does not overlap that of the human host, the organelle has been considered to harbour promising potential drug targets from its initial discovery. Some of the functions mapped to the apicoplast have already been experimentally demonstrated to be essential through the use of inhibitors or conditional gene knock-outs, or have a high probability of being essential given their biological importance. A summary of metabolic pathways currently predicted or demonstrated to reside in the apicoplast is given in Figure 9.11, together with their interrelationship with those of the mitochondrion.

Acyl carrier protein, part of the type II fatty acid biosynthesis pathway (FAS II), was one of the first proteins recognized to be targeted to the apicoplast (Waller *et al.*, 1998). Fatty acids are of great importance for a variety of cellular functions, with their role as precursors for lipid synthesis being amongst the most important. The type II pathway, typical of bacteria and algae, is mediated by a series of individual enzymes, all of which have been identified in the sequenced apicomplexan genomes except those of piroplasms *Babesia* sp. and *Theileria* sp. (Table 9.1). Gene knock-out experiments have clearly shown that the type II pathway is essential in *T. gondii* (Mazumdar *et al.*, 2006). The situation is different in *Plasmodium* sp.: the erythrocytic stages can grow in the absence of a functional FAS II pathway presumably because they can scavenge sufficient amounts of fatty acids/lipids from the host cell (Yu *et al.*, 2008; Tarun *et al.*, 2009; Vaughan *et al.*, 2009). Liver stage parasites, however, have to synthesize fatty acids via the FAS II pathway in order to supply the amount of lipids required by the massive accumulation of merozoites during late stage differentiation. Consequently, FAS II gene knock-outs are detrimental in liver stage parasites (Yu *et al.*, 2008; Vaughan *et al.*, 2009).

In contrast to the dissociated bacterial-type FAS II, the type I pathway is mediated by a single, large multidomain protein complex

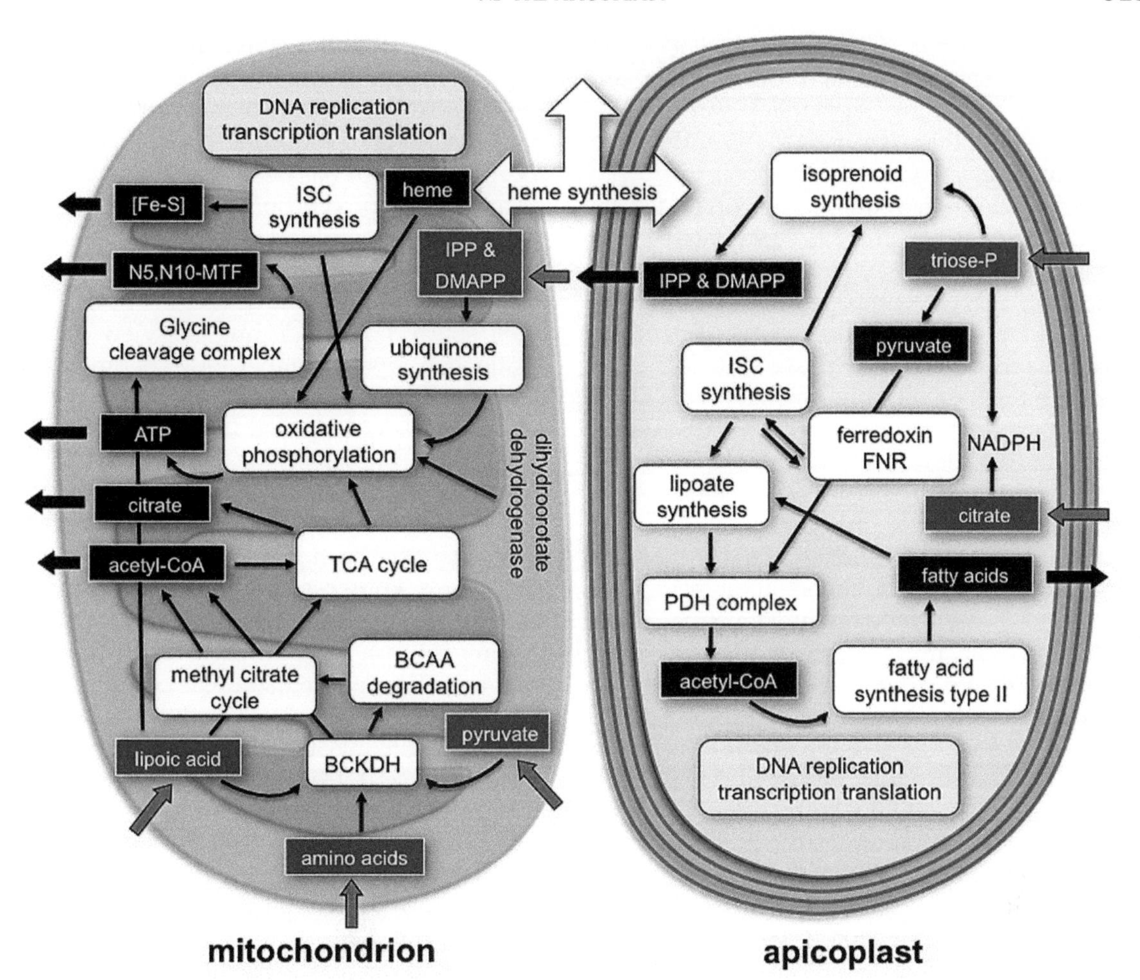

FIGURE 9.11 **Metabolism of the apicoplast and mitochondrion.** Metabolic pathways or important reactions in both compartments are shown (white boxes) and their interactions indicated by arrows. Compounds that are imported into the organelle (grey boxes) or produced inside it (black boxes) are also drawn. Their import (large grey arrows) or export (large black arrows) next to the compound indicates flux and consumption of the metabolites. Haem synthesis takes place in three places, apicoplast, mitochondrion and cytosol. For more details see text and cited reviews. BCAA, branched-chain amino acids; BCKDH, branched-chain keto acid dehydrogenase; DMAPP, dimethylallyl diphosphate; FNR, ferredoxin-NADP$^+$ reductase; IPP, isopentenyl pyrophosphate; ISC, iron sulphur cluster; N5,N10-MTF, N5,N10-methylenetetrahydrofolate.

which is not plastid-localized. *T. gondii*, *N. caninum*, and *E. tenella* all encode a protein related to FAS I (Mazumdar and Striepen, 2007). *T. gondii* can take up fatty acids from the host cell and perhaps utilize the type I enzyme to synthesize additional long-chain fatty acids. An ER-resident fatty acid elongation system supports the provision of long-chain FA (Ramakrishnan *et al.*, 2012 and references therein). Interestingly, *Plasmodium* sp. lacks the type I pathway. Conversely, *Cryptosporidium parvum* has only the type I pathway (Zhu *et al.*, 2004). This observation correlates with the apparent lack of an apicoplast in *C. parvum* (Riordan *et al.*, 1999; Zhu *et al.*, 2000b; Keithly *et al.*, 2005) and the absence of genes (both nuclear and apicoplast) encoding apicoplast proteins in *C. parvum* (Abrahamsen *et al.*, 2004) and *Cryptosporidium hominis* (Xu *et al.*, 2004).

Acetyl-CoA is a required precursor for the FAS II pathway and is generated from glycolysis-derived pyruvate by the action of the large multienzyme complex pyruvate dehydrogenase (PDH). Several studies have shown that *T. gondii* and *P. falciparum* possess nuclear genes encoding the four subunits needed for an apicoplast PDH even though they lack the typical mitochondrial form (McMillan *et al.*, 2005; Foth *et al.*, 2005; Crawford *et al.*, 2006; Fleige *et al.*, 2007; Pino *et al.*, 2007). To be active, subunit E2 of PDH must be post-translationally modified by a dithiol-fatty acid, called lipoic acid (LA). LA synthesis itself requires two enzymes (LIPA and LIPB) that are localized to the apicoplast (Thomsen-Zieger *et al.*, 2003), as well as its substrate octanoyl-ACP, itself a product of the FAS II pathway. Thus this pathway indirectly requires its own product. The free LA is presumably attached to PDH-E2 via a lipoate ligase (Seeber and Soldati-Favre, 2010). In *P. falciparum*, this ligase is dually targeted to the mitochondrion and the apicoplast (Günther *et al.*, 2007). Recent data suggest that FAS II-synthesized fatty acids are also required outside the apicoplast (Ramakrishnan *et al.*, 2011) but it is not known how and in what form they leave the organelle.

The first step in production of fatty acids from acetyl-CoA is catalysed by acetyl-CoA carboxylase (ACC) which contains a covalently attached biotin prosthetic group. The biotin group can be visualized with FITC-labelled streptavidin and co-localizes with apicoplast markers, confirming the targeting prediction for one of two *T. gondii* ACCs; the other is cytoplasmic (Jelenska *et al.*, 2001). An enzyme required for the ligation of this co-factor, biotin ligase, is expected to be apicoplast localized (Seeber and Soldati-Favre, 2010). Plastid ACCs are inhibited by aryloxyphenoxyproprionate herbicides and *T. gondii* growth is inhibited by these compounds (Zuther *et al.*, 1999). Recombinant ACC corresponding to the apicoplast-localized protein has been confirmed as a target of aryloxyphenoxyproprionates (Jelenska *et al.*, 2002). The compound triclosan was initially presumed to be a specific FAS II inhibitor, acting against enoyl-ACP reductase, in both *Plasmodium* sp. and *T. gondii* but this view has been challenged by a number of experiments that suggest severe off-target effects of triclosan and its derivatives (discussed in Seeber and Soldati, 2007; Botté *et al.*, 2012).

Isoprenoid synthesis is perhaps the most important pathway localized to the apicoplast. It has recently been recognized via genomics to be the only known anabolic pathway present in all apicoplast-bearing parasites (Table 9.1), suggesting it has a vital function for these cells. In fact, recent experiments in *P. falciparum* erythrocytic stages have demonstrated that it is the only essential apicoplast pathway in this parasite stage (Yeh and DeRisi, 2011). Isoprenoids are a large class of natural compounds and fulfil important cellular functions in signalling processes and protein modifications like prenylation, but are also required for synthesis of the cofactor ubiquinone (coenzyme Q) and for tRNA modifications. The enzymes mediating this pathway in the apicoplast are distinct from those of the mevalonate pathway found in the human host. The non-mevalonate parasite pathway, also called the DOXP pathway for its early intermediate 1-deoxy-D-xylulose 5′ phosphate, consists of seven enzymatic steps, at the end of which stand the two isomeric products isopentenyl diphosphate (IPP) and dimethylallyl diphosphate (DMAPP) (for review see Wiesner and Jomaa, 2007; Seeber and Soldati-Favre, 2010). Fosmidomycin, an antibiotic which inhibits the second enzyme, DOXP reductase, was shown to kill the malaria parasite (Jomaa *et al.*, 1999). Although fosmidomycin and derivatives had shown little activity against *T. gondii* and other apicomplexans like *Eimeria* sp. and *Theileria* sp. (Ling *et al.*, 2005; Clastre *et al.*, 2007; Lizundia *et al.*, 2008), it is now clear that *T. gondii* has a fully functional DOXP pathway that is essential for its survival (Baumeister *et al.*, 2011; Nair *et al.*, 2011). The low activity of fosmidomycin reflects poor bioavailability in *T. gondii* (Baumeister *et al.*, 2011; Nair *et al.*, 2011), but additional factors may also play a role. These are still open questions regarding the transport of the highly charged drug through the various cellular membranes and compartments between susceptible and non-susceptible apicomplexans (Seeber and Soldati-Favre, 2010; Baumeister *et al.*, 2011; Nair *et al.*, 2011).

How do the initial substrates for these pathways (phosphoenolpyruvate and dihydroxyacetone phosphate) enter the apicoplast? By analogy with the chloroplast and due to the charged nature of the compounds, it is likely that specific carriers are required, and that such proteins could be identified in the genomes. The malaria parasite possesses two sugar phosphate transporters that are differentially localized to the inner and the outer membrane of the apicoplast, respectively (Mullin *et al.*, 2006). They were shown in a cell-free assay system to mediate the uptake of the substrates into the plastid (Lim *et al.*, 2010). In *T. gondii* the same task seems to be fulfilled by a single sugar phosphate translocator, APT1 (Fleige *et al.*, 2007; Karnataki *et al.*, 2007a), that likely localizes to multiple membranes of the apicoplast and that is able to transport different compounds like triose phosphates and phosphoenolpyruvate (Mullin *et al.*, 2006; Karnataki *et al.*, 2007a; Brooks *et al.*, 2010; Lim *et al.*, 2010). In contrast, very little is known about how the products, IPP and DMAPP, leave the apicoplast, but it is possible that the aforementioned sugar phosphate transporters could also assist in the export of the products. Moreover, the mitochondrion is a likely 'consumer' of the isoprenoids from the apicoplast, e.g., for condensation of IPP and DMAPP by mitochondrial farnesyl-pyrophosphate synthase (Ling *et al.*, 2007) as a precursor for ubiquinone modification. The close association of both organelles might simplify metabolite exchange through membrane contact sites like those described between the apicoplast and ER (Tomova *et al.*, 2009). In the plastid itself, tRNA modification by isoprenylation via an isopentenylpyrophosphate transferase (MIAA) is a likely activity that requires IPP/DMAPP. Such modification is presumably required to suppress stop codons and frameshifts known to occur frequently in apicoplast genomes (see Section 2.4 Apicoplast Genome).

Connected to the DOXP pathway are two accessory systems required for the functions of individual enzymes that also reside in the apicoplast. One is a redox system comprised of the small iron-sulfur protein ferredoxin (FD) and its partner ferredoxin-NADP$^+$ reductase (FNR) (Vollmer *et al.*, 2001). The ferredoxin redox system is most similar to those of non-photosynthetic tissues of plants and donates electrons derived from NADPH to acceptor proteins via protein–protein interactions (Seeber *et al.*, 2005; Aliverti *et al.*, 2008). NADPH, in turn, can potentially be generated via apicoplast-resident aconitase and isocitrate dehydrogenase and/or by glyceraldehyde 3-phosphate dehydrogenase isoenzymes, respectively (Pino *et al.*, 2007). The last two enzymes of the DOXP pathway (ISPH/LYTB and ISPG/GCPE) are known to be FD-interacting proteins in *Plasmodium* sp., plants, and cyanobacteria, and thus presumably obtain their electrons from this redox system (Okada and Hase, 2005; Röhrich *et al.*, 2005; Seemann *et al.*, 2006). Further putative electron acceptors for FD are the LIPA protein involved in LA synthesis (see above) and a putative haem oxygenase described in *Plasmodium* (Okada, 2009) but with poorly defined function and localization in apicomplexans (Seeber and Soldati-Favre, 2010).

Ferredoxin is a small protein with a two iron–two sulphur cluster (2Fe-2S), and as such, pointed very early to a requirement for the biosynthesis of these essential prosthetic groups in the apicoplast (Seeber, 2002). The labile nature (both chemically and structurally) of pre-assembled ISC proteins would likely not survive the unfolding thought to be required for import of NEAT proteins into the plastid. ISCs fulfil a variety of functions in proteins, most notably facilitating electron transfer processes. Three of the four currently known ISC proteins in the apicoplast, besides FD, are the enzymes noted above to interact with FD (i.e. LIPA, LYTB and GCPE). The fourth is MIAB, which is presumably involved in tRNA modification (Ralph *et al.*, 2004b).

Synthesis of ISCs commences through a series of reactions that are basically conserved in prokaryotes as well as eukaryotes, although with different players in different cellular compartments (see Table 9.2, and Balk and Pilon, 2011; Chahal and Outten, 2012). The generation of elemental sulphur from cysteine is mediated by the action of cysteine desulphurase, and its reduction is coupled to the oxidation of iron (imported by still undefined iron transporters) on a specific protein scaffold complex where the clusters are assembled (Balk and Pilon, 2011; Chahal and Outten, 2012). Additional proteins are required for the release of ISCs from the scaffold and for their transfer to apoproteins and coordination with specific amino acid side chains, usually cysteines.

In eukaryotes the main site of ISC synthesis is the mitochondrion, although in plastid-bearing organisms plastids are a second site for ISC synthesis (Balk and Pilon, 2011). The compartmentation of these reactions is likely beneficial, segregating the generation of reactive intermediates and thereby avoiding damage to other cellular components. The first hint that ISC synthesis also occurs in the apicoplast (Ellis *et al.*, 2001) was the finding that the apicoplast-encoded gene *SUFB* (formerly ORF470 or *ycf24*) is related to bacterial *suf*B. That gene lies in the same operon as other genes that are involved in ISC biosynthesis via the so-called SUF pathway (Nachin *et al.*, 2001, 2003). Subsequently, genes encoding all the other homologous proteins of the SUF pathway, in addition to other accessory proteins, have been identified in *T. gondii* (see Table 9.2) and other Apicomplexa, and several of them have predicted apicoplast targeting sequences (Ellis *et al.*, 2001; Seeber, 2002; Ralph *et al.*, 2004b; Fleige *et al.*, 2010; Seeber and Soldati-Favre, 2010). Few additional experimental data have been published to date, except for the apicoplast localization of *Plasmodium* SUFC and its interaction with SUFB (Kumar *et al.*, 2011), and of the *T. gondii* NFU-like scaffold protein (Sheiner *et al.*, 2011). Interestingly, this latter protein seems to be non-essential since a gene knock-out could be obtained, although no further data were reported (Sheiner *et al.*, 2011). In contrast, point mutations in human mitochondrial NFU have severe effects on the ISC assembly of lipoate synthase and thus on the TCA cycle (Cameron *et al.*, 2011; Navarro-Sastre *et al.*, 2011). ISC synthesis in plastids basically follows the bacterial SUF system (Wollers *et al.*, 2010; Balk and Pilon, 2011; Chahal and Outten, 2012). Less is known about apicoplast ISC synthesis; it possibly has features of both mitochondrial and plant plastid systems. For instance, while sequences similar to the plant plastid scaffold proteins HCF101 (Schwenkert *et al.*, 2009) and GRXS14/16 (Rouhier *et al.*, 2010) can be identified in the genome of several apicomplexans, a prediction of their possible involvement in ISC synthesis also in the apicoplast was not possible by analysing sequence features alone (Seeber and Soldati-Favre, 2010). More interest in this essential pathway is clearly necessary.

Haem biosynthesis is another organellar process in eukaryotes, occurring in the mitochondrion of the human host. The cyclic tetrapyrrole haem is an essential iron-containing prosthetic group for proteins such as cytochromes and cytochrome *c* oxidase, which are also present in *T. gondii*. Once again compartmentation may serve to protect the cell from potentially damaging

TABLE 9.2 Putative Proteins Involved in the Apicoplast ISC Synthesis of *T. gondii*

Designation (ToxoDB acc. no.)	Putative Function	Apicoplast Localization*				
		SignalP	TargetP	ChloroP	Pprowler	Experimental
SUFA (TGME49_097930)	transfer protein; transfer of [Fe–S] to apo-proteins like FD or LipA		mit	chl	chl	
SUFB (TogoCp26)	part of SUF BCD scaffold complex; sulphide acceptor	apicoplast-encoded				
SUFC (TGME49_025800)	part of SUF BCD scaffold complex; ATPase activity	sp	sp			api (Pf)
SUFD (TGME49_073450)	part of SUF BCD scaffold complex; ATPase activity					
SUFE (TGME49_077010)	SUFS activator and sulphide 'transferase'; part of SUF ES complex		mit	chl	sp	
SUFS (TGME49_016170)	cysteine desulfurase; generates sulphide from L-cysteine		sp	chl	sp	
HCF101 (TGME49_118590)	scaffold protein for [4Fe–4S]?					
CPN60 (TGME49_040600)	chaperonin 60; help in [Fe–S] transfer?		sp	chl	sp	api (Tg,Pf)
CPN20 (TGME49_073960)	chaperonin 20; help in [Fe–S] transfer?		mit	chl		api (Pf)
GRX14/16-like (TGME49_047580)	glutaredoxin-like scaffold protein for [2Fe–2S]?					
NFU (TGME49_021920)	scaffold and targeting protein; maturation of lipoate synthase and FD?					api (Tg)
FD (TGME49_015070)	apicopast ferredoxin; redox system with FNR for electron transfer?	sp		chl	chl	api (Tg)
FNR (TGME49_098990)	apicoplast ferredoxin–NADP+ reductase; redox system with FD for electron transfer	sp	sp		sp	api (Tg)

* *Predictions were performed at the following servers:*
SignalP 4.0, http://www.cbs.dtu.dk/services/SignalP/; TargetP 1.1, http://www.cbs.dtu.dk/services/TargetP/; ChloroP 1.1, http://www.cbs.dtu.dk/services/ChloroP/; PProwler Plant; http://pprowler.itee.uq.edu.au/pprowler_webapp_1–2/.
Abbreviations: sp, signal peptide; mit, mitochondrion; chl, chloroplast; api, apicoplast in either *T. gondii* (Tg) or *P. falciparum* (Pf); blank, no prediction

intermediates or by-products. In contrast to its host, in *T. gondii* and *P. falciparum*, haem biosynthesis is accomplished through a metabolic cooperation between the apicoplast, the cytosol and the mitochondrion. All enzymes required for the *de novo* synthesis of haem are present in the genomes of *P. falciparum* (with one exception) and *T. gondii* while absent from the piroplasms (Table 9.1; Seeber and Soldati-Favre, 2010). The localization of the different proteins mentioned below has been determined in both parasites by either GFP targeting experiments and/or by antibody staining.

The enzyme mediating the first step in the pathway is the mitochondrial δ-aminolevulinic acid synthase (Varadharajan *et al.*, 2002; Sato *et al.*, 2004; Wu, 2006), generating 5-aminolevulinate from glycine and succinyl–CoA (as in animals). The next three enzymes in the pathway, porphobilinogen synthase, porphobilinogen deaminase, and uroporphyrinogen III synthase, are considered apicoplast-resident in *T. gondii* (Wu, 2006) and *P. falciparum*, although in the malaria parasite the latter two enzymatic activities are reported to reside in one protein, porphobilinogen deaminase (Nagaraj *et al.*, 2008). This would explain the apparent absence of a gene with homology to uroporphyrinogen III synthases. The next enzyme, uroporphyrinogen decarboxylase, is thought to be cytosolic in *T. gondii* (Wu, 2006) and apicoplast-localized in *Plasmodium* (Nagaraj *et al.*, 2009a), whereas coproporphyrinogen III oxidase is cytosolic in both parasites (Wu, 2006; Nagaraj *et al.*, 2010). Finally, the last two enzymes (protoporphyrinogen oxidase and ferrochelatase) are reported to be mitochondrial again for both organisms (van Dooren *et al.*, 2006; Wu, 2006; Nagaraj *et al.*, 2009b), allowing the prosthetic haem group to be finally attached to cytochromes *c* and c_1. The complex mosaic nature of the pathway, both in terms of phylogenetic origin (Obornik and Green, 2005; Koreny *et al.*, 2011) and cellular localization, raises several issues. The most obvious one is how intermediates, which are considered to be membrane-impermeable, are transported between the compartments (Seeber and Soldati-Favre, 2010). More work is required to shed light on this interesting aspect.

9.3 THE MITOCHONDRION

In surprising contradiction to the apicoplast, comparatively little is known about the *Toxoplasma* mitochondrion. This relative neglect in part reflects the novelty of the apicoplast, perhaps drawing attention away from the more mundane endosymbiotic organelle in the parasite. An even larger damper on mitochondrial research in *T. gondii* has been the unexpected difficulty of defining the mitochondrial genome in this parasite. However, ultrastructurally, *T. gondii* has a conventional protozoan mitochondrion: a single, double-membraned organelle per cell, with tubular rather than plate-like cristae (Fig. 9.1).

9.3.1 Evolution

Phylogenetic analyses of the apicomplexan mitochondrial genomes have focused on relationships within genera and the association of parasite and host speciation (Rathore *et al.*, 2001; Joy *et al.*, 2003; Jongwutiwes *et al.*, 2005; Mu *et al.*, 2005). Most of these studies have compared *Plasmodium* species but the number and diversity of apicomplexan mitochondrial genome sequences has significantly expanded in recent years, and continues to grow. Complete or near-complete mitochondrial genome sequences are now available for 25 species of *Plasmodium*, and for eight other species of hemosporidians, representing four genera. Complete mitochondrial genome sequences are also available for *E. tenella* and multiple species of *Babesia* and *Theileria*. For some species, especially *P. falciparum*, sequence coverage is deep, with greater than 100 independent isolates having been sequenced. Still, the limited gene content of these genomes – only three protein genes and fragmented ribosomal RNAs – restricts what questions can be addressed and what inferences can be drawn. Consequently, the deep evolutionary roots of the apicomplexan mitochondrion have been relatively neglected.

In contrast to data on *Plasmodium*, little has been published concerning the *T. gondii* mitochondrial genome. Here we will draw from other apicomplexans to illustrate the likely characteristics of the *T. gondii* mitochondrial genome and its expression.

9.3.2 The Mitochondrial Genome

Despite considerable variation in size and gene content among species, all mitochondrial genomes known to date encode apocytochrome b (*cob*) and cytochrome c oxidase I (*cox1*). Cytochrome c oxidase III (*cox3*) is nearly invariant, missing from the mitochondrial genome only in some green algae. Mitochondrial genomes also invariably encode eubacterial-like large and small subunit rRNAs (reviewed in Gillham, 1994; Lang *et al.*, 1999; Gray *et al.*, 2004). The *T. gondii* mitochondrial genome has proven very difficult to identify, due to the presence of *cob* and *cox1* pseudogenes scattered throughout the nuclear genome (Ossorio *et al.*, 1991). These pseudogene sequences and the neighbouring nuclear DNA in *T. gondii* are flanked by inverted repeats, suggesting a retroposon-like mechanism for insertion and intragenomic proliferation of the mitochondrial sequence (Ossorio *et al.*, 1991). Attempts to differentiate the *bona fide* mitochondrial genes from nuclear gene copies by first isolating the mitochondrion, then its DNA, have thus far been unsuccessful (S. Tomavo, personal communication). However, other apicomplexan mitochondrial DNAs provided clues for the likely characteristics of the *T. gondii* mitochondrial DNA.

Apicomplexan mitochondrial genomes are the smallest ever reported, having gone to extremes in reduction of coding capacity. Like the plastid genomes, they are conserved in gene content but not in organization or topology. Only three protein coding genes remain in the ~6 kb mitochondrial genomes, compared to 13 in the 16 kb mammalian mitochondrial genomes (reviewed in Feagin, 2000). However, those three are the hallmark mitochondrial genes: *cob*, *cox1* and *cox3*. Apicomplexan mitochondrial genomes also contain short sequences similar to highly conserved portions of eubacterial-like large and small subunit rRNAs (reviewed in Feagin, 2000; see also Feagin *et al.*, 2012). Unexpectedly, these sequences are not contiguous but are fragmented and scattered out of order across the mitochondrial genome. Fragmented rRNAs have been identified in a number of organisms (reviewed in Gray and Schnare, 1990) but the fragments encoded by apicomplexan mtDNAs are by far the smallest and their cumulative size is far less than conventional rRNAs.

The *P. falciparum* mitochondrial rRNA genes and the corresponding transcripts have been studied in greatest detail (Feagin *et al.*, 1992, 1997, 2012). Thirty-four small RNAs, ranging in size from 23 to 190 nt, have been mapped to the *P. falciparum* mitochondrial genome. Of these, 27 specify regions of large and small subunit rRNA. They are well conserved across *Plasmodium* species and other haemosporidians, and corresponding sequences have been found in the *T. parva* mitochondrial genome for 23 of them. They encode highly conserved and functionally important parts of conventional rRNAs, with expected base-pairing, long-distance interactions, and sequence elements known to be associated with ribosomal function. When mapped to a 3-D model of the ribosome, they cluster at the interface between the large and small ribosomal subunits (Feagin *et al.*, 2012). While their function in mitochondrial protein synthesis has not been formally demonstrated, all evidence suggests these tiny transcripts comprise functional rRNAs which, when assembled into mitochondrial ribosomes, translate the three mitochondrial mRNAs.

Small RNAs corresponding to the rRNA genes have also been reported for *P. yoelii* (Suplick *et al.*, 1990) and *T. parva* (Kairo *et al.*, 1994). More recently, Raabe *et al.* (2010) prepared from size-selected small RNAs of *P. falciparum*. Among these were many that matched the mitochondrial genome, including the 34 transcripts noted above. Dinoflagellates, considered the closest relatives of apicomplexans, also have fragmented mitochondrial rRNAs (Norman, 2001; Slamovits *et al.*, 2007; Nash *et al.*, 2007; Jackson *et al.*, 2012), so this characteristic appears to be quite ancient.

Mitochondrial genome content is quite similar among apicomplexans, but the organization of genome elements differs. The mitochondrial genomes of *Plasmodium* species and other haemosporidians (*Haemoproteus, Hepatocystis, Leucocytozoon*, and *Parahemoproteus* species) have the same gene order (Feagin *et al.*, 2012) but the order of elements in other apicomplexans is quite variable. The mitochondrial genomes of most *Babesia* species share a common organization, with a single set of genes flanked by inverted repeats. *B. microti* is a surprising exception. Its mitochondrial genome is an 11.1 kb circular molecule containing a pair of 2.5 kb inverted repeats and two unique regions which encode the expected content of an apicomplexan mitochondrial genome (Cornillot *et al.*, 2012). The mitochondrial genomes of *Theileria* and *Eimeria* species share a common gene order and are flanked by inverted repeats; however, the gene order differs from that of *Babesia*.

Gene loss to the nucleus is a classic feature of endosymbiont organelles and recent data show that the mitochondrion can survive loss of its genome, most typically in modified form. The *Trichomonas* hydrogenosome has long been suggested to be a remnant mitochondrion (reviewed in Dyall and Johnson, 2000). More recently, other 'amitochondriate' protozoa, including *Entamoeba histolytica, Giardia intestinalis* and *Encephalitozoon cuniculi,* have been shown to possess a membrane-bound compartment that contains mitochondrial markers such as mitochondrial HSP70, cpn60, and proteins needed for mitochondrial ISC biosynthesis (Tovar *et al.*, 1999; Williams *et al.*, 2002; Vavra, 2005; Regoes *et al.*, 2005; Goldberg *et al.*, 2008). The apicomplexan *C. parvum* provides another example of organelle retention following organelle genome loss. It lacks both extrachromosomal genomes but has a multi-membraned organelle. That it is a relict mitochondrion and not an apicoplast became clear when the markers noted above were shown to localize to the organelle (LaGier *et al.*, 2003; Riordan *et al.*, 2003; Slapeta and Keithly, 2004). Gregarines also appear to lack mitochondrial genomes (Zhu *et al.*, 2000a).

The mitochondrial protein import machinery in these protozoa is much reduced but components have been shown to function in heterologous mitochondrial import assays (Burri *et al.*, 2006; Alcock *et al.* 2012). An ADP/ATP transporter has been identified in several of the 'amitochondriate' protozoa (Chan *et al.*, 2005; Williams *et al.*, 2008; Tsaousis *et al.*, 2008) and the proteome of the *Giardia* mitosome has been evaluated, demonstrating the presence of the ISC assembly pathway and a simplified protein import system (Jedelsky *et al.*, 2011). On a more recent evolutionary time scale, certain subspecies of African trypanosomes, *Trypanosoma evansi* and *T. equiperdum*, lack most or all mitochondrial DNA, depending on the isolate (Lun *et al.*, 2010; Schnaufer, 2010). The parasites retain their mitochondrion but in at least some cases have compensatory mutations in ATP synthase (Schnaufer *et al.*, 2005). Might the peppering of mitochondrial pseudogenes in the *T. gondii* nucleus reflect a next step in gene loss from an already tiny mitochondrial genome?

The *T. gondii* mitochondrion is clearly functional (see Section 3.6 Energy Metabolism) but it is equally clear that some mitochondrial functions do not depend on having a mitochondrial genome. So is there a genome in the *T. gondii* mitochondrion? The answer is likely yes. Matsuzaki *et al.* (2001) describe DAPI-stained nucleoids in the *T. gondii* mitochondrion, and Williamson *et al.* (2001) noted unspecified mitochondrial sequences in a band from an isopycnic CsCl gradient of *T. gondii* DNA. A preliminary *T. gondii* mitochondrial genome sequence has been assembled as an outgrowth of the parasite genome project, and confirmed by amplification and mapping studies (D. Shanmugam, L. Peixoto, and D.S. Roos, personal communication and as noted in Bahl *et al.*, 2010). Its characteristics are quite similar to other apicomplexan mitochondrial genomes, as shown in Fig. 9.12. The draft sequence is ~6 kb in length, and includes the three expected protein coding genes, although all are encoded on the same DNA strand (in contrast to the situation in other apicomplexans). The *T. gondii* mitochondrial genome also includes sequences similar to highly conserved regions of rRNAs (J.E. Feagin, unpublished results). The mitochondrial genome is flanked by repeated sequences and has some internal repeats. This is reminiscent of *Theileria* mitochondrial DNAs, which are single units with terminal inverted repeats (Hall *et al.*, 1990; Kairo *et al.*, 1994) but contrasts with the tandemly repeated mitochondrial DNAs of *Plasmodium* species (Vaidya and Arasu, 1987; Joseph *et al.*, 1989). Verification that this DNA resides within the mitochondrion awaits further analysis.

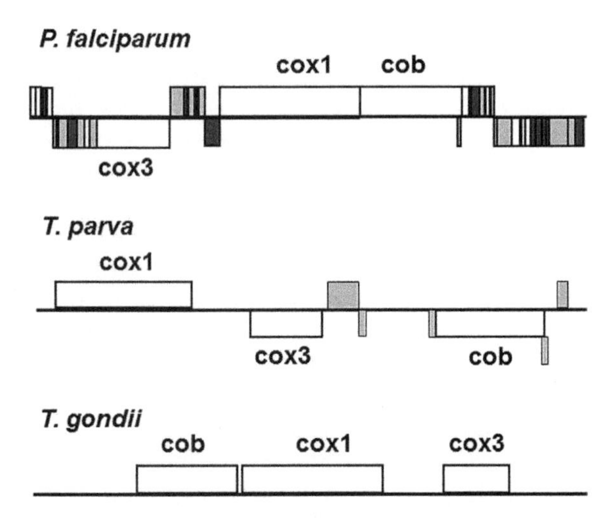

FIGURE 9.12 **Comparisons of apicomplexan mitochondrial genomes.** Schematic maps of the mitochondrial genomes of *P. falciparum*, *T. parva*, and *T. gondii* are shown with genes above or below the line depending on direction of transcription (left to right above the line). Protein-coding genes are white, fragments of large and small subunit rRNAs are shown in light grey and dark grey boxes, respectively. Open boxes indicate small transcripts with characteristics similar to the rRNA fragments; these may correspond to less-conserved regions of rRNA. The *P. falciparum* mitochondrial genome has been most thoroughly studied, with fewer rRNA fragments identified for *T. parva* and none identified as yet for *T. gondii*. Preliminary analyses show the presence of further rRNA sequence on both these mitochondrial genomes, in small scattered fragments as in *P. falciparum*.

9.3.3 Mitochondrial Gene Expression

Few studies have addressed expression of mitochondrial genes in *T. gondii*. The *cob* gene is the principal exception. COB is a component of complex 3 of the electron transport chain, which is the site of action of atovaquone. This drug is effective against *T. gondii* and *P. falciparum* but resistance develops rapidly in both parasites (Chiodini *et al.*, 1995; McFadden *et al.*, 2000). To evaluate the mechanism of resistance in *T. gondii*, McFadden *et al.* (2000) cloned and sequenced *cob* cDNA from atovaquone-sensitive and -resistant *T. gondii*. Sequence comparisons showed changes at two positions important for electron transfer, matching data from atovaquone-resistant lines of *Plasmodium* (Srivastava *et al.*, 1999; McFadden *et al.*, 2000). Despite the facile development of resistance (Chiodini *et al.*, 1995), atovaquone has been successfully used for malaria treatment in combination with other anti-malarials (Patel and Kain, 2005). In combination with other agents, atovaquone is recommended for prevention or treatment of toxoplasmosis when first line agents have failed or are not tolerated by HIV-infected individuals.

The predicted C-terminus of *T. gondii* cob is hydrophobic and would likely be buried in the mitochondrial membrane, whereas in other organisms it is usually hydrophilic. Extending past the stop codon, the predicted amino acid sequence would be a good match for *Plasmodium cob* genes (McFadden *et al.*, 2000). It is unclear whether translation of the cob mRNA somehow bypasses the stop codon or if *T. gondii* COB simply has an unexpected topology. Regardless of this consideration, the *cob* gene is expressed; *cob* mRNA (McFadden *et al.*, 2000; Pino *et al.*, 2010) and COB were detected in both tachyzoites and bradyzoites (McFadden *et al.*, 2000). Tags for both *cob* and *cox1* are found in recently reported SAGE libraries. Based on relative abundance of SAGE tags, mitochondrial transcripts appear to be more abundant than those derived from the apicoplast genome (Radke *et al.*, 2005).

Given the presence of *cob* DNA in the nuclear genome, it is formally possible that the mRNA and protein being detected were made from nuclear genes. However, evidence for mitochondrial transcription in *T. gondii* was obtained recently. Following partial solubilization with digitonin, PCR amplification of cDNA provided evidence for *cob* mRNA in the total and organellar fractions but not the cytosolic fraction (Pino *et al.*, 2010). While transcription from the *T. gondii* mitochondrial genome remains minimally investigated, some predictions may reasonably be made. It is common for mitochondrial genomes to be polycistronically transcribed, with the precursor RNA then cleaved to produce individual RNAs (Gillham, 1994). Mitochondrial transcripts have been reported for *P. falciparum*, *P. yoelii*, *P. gallinaceum*, and *T. parva* (reviewed in Feagin, 2000) and data from *P. falciparum* support polycistronic transcription and post-transcriptional processing (Ji *et al.*, 1996; Gillespie *et al.*, 1999; Rehkopf *et al.*, 2000; Feagin *et al.*, 2012). It is likely that mitochondrial transcription in *T. gondii* will have similar characteristics. The three *P. falciparum* mitochondrial mRNAs are coordinately regulated and are most abundant in the late trophozoite and schizont stages (Feagin and Drew, 1995). Nuclearly encoded mitochondrial proteins also show a general up-regulation in trophozoite and schizont stages (Bozdech *et al.*, 2003; Le Roch *et al.*, 2003).

As yet, mitochondrial protein synthesis in *T. gondii* remains largely unexplored. Localization of the protein elongation factor EF-Tu to the mitochondrion suggests it is needed there for protein synthesis (Pino *et al.*, 2010). Given the content and expression of mitochondrial genomes in other apicomplexans, it is likely that rRNAs will be encoded as small fragments. Ribosomal proteins and tRNAs will need to be imported. Indeed, several tRNAs have already been shown to be imported into the *T. gondii* mitochondrion (Esseiva *et al.*, 2004). To be functional, these tRNAs must be charged by tRNA synthetases. Some amino acids correspond to a single tRNA synthetase gene in the *T. gondii* genome. Surprisingly, these synthetases have been shown to be present only in the cytosol and apicoplast (Pino *et al.*, 2010). These authors suggest that the tRNAs are imported into the mitochondrion fully charged with the corresponding amino acid. A challenge with this hypothesis is that once the amino acid is transferred from the tRNA to the growing polypeptide, the tRNA would not be able to be recharged within the organelle. Another curious feature is that unlike mitochondria of other organisms, protein synthesis does not apparently start with an N-formylated methionine, since both the methionyl tRNA formyl transferase and the peptide deformylase appear to be present only in the apicoplast (Pino *et al.*, 2010). Mitochondrial protein synthesis, like that of the plastid, is likely to be sensitive to inhibitors of eubacterial and organelle ribosomes. Analyses of such drug sensitivities are discussed above (see Section 9.2.8 Drug Sensitivities).

9.3.4 Mitochondrial Genome Replication

Mitochondrial genome replication has been examined in *P. falciparum* and appears to involve recombination as well as replication. Linear concatamers, likely produced by rolling circle replication from a few circular templates, appear to be capable of priming further rounds of replication on the templates and each other (Preiser *et al.*, 1996). Preliminary indications suggest that the *T. gondii* mitochondrial genome is not tandemly repeated. Thus even if it is replicated via the rolling circle method, it is unlikely that complex networks of mitochondrial DNA will be produced, as reported for *P. falciparum* (Preiser *et al.*, 1996).

As noted earlier, mitochondrial division and segregation is one of the last events of the tachyzoite cell cycle. The means by which the mitochondrion is partitioned to daughter cells remains unclear. In other organisms, the movement of mitochondria requires cytoskeletal elements and their associated motors. An association with the inner membrane complex via the intermediate filament protein IMC1 may be important, as proposed by Nishi *et al.* (2008)

9.3.5 Protein Trafficking to the Mitochondrion

As described above, proteins destined for secondary plastids traffic first to the secretory system and then enter the plastid. In contrast, entry into primary endosymbionts – both mitochondria and plastids – typically does not involve the secretory system. Rather, proteins are translated on free cytosolic ribosomes, or on ribosomes associated with the organelle's outer membrane, and enter the organelles directly by virtue of translocation machinery. This is a largely understudied topic for apicomplexans but a few points have been investigated, mostly in concert with plastid targeting studies (see Section 9.2.7 Protein Trafficking to the Apicoplast). Paradoxically, one of the first sequences shown to function as a mitochondrial targeting sequence in *T. gondii* was derived from the NEAT protein S9. At its N-terminal, *T. gondii* S9 gene encodes a signal sequence and a predicted plastid targeting sequence, which together is able to target GFP to the apicoplast. However, when only the transit sequence was present, GFP was targeted to the mitochondrion (DeRocher *et al.*, 2000). Plastid transit peptides do resemble mitochondrial targeting sequences. Both types of topogenic signals show considerable flexibility in sequence and are rich in basic amino acids, but mitochondrial targeting sequences generally adopt an amphipathic structure, which is not essential for transit peptides (reviewed in Haucke and Schatz, 1997). Delivery of proteins to the correct location is aided by diversion of *T. gondii* apicoplast proteins into the secretory system. Hence cryptic mitochondrial targeting sequences may not be revealed until the signal sequence is cleaved, at which point mitochondrial targeting can no longer occur.

Brydges and Carruthers (2003) reported a similarly unexpected result when assessing the subcellular location of a *T. gondii* iron superoxide dismutase (FeSOD20). This gene has a predicted apicoplast targeting sequence consisting of an apparent N-terminal signal sequence followed by a transit sequence. Unexpectedly, when the targeting sequence was fused to GFP and transfected into *T. gondii*, it was the mitochondrion and not the apicoplast that fluoresced. Changing a single amino acid in the 'signal' sequence from arginine to alanine was sufficient to render it functional, diverting GFP to the apicoplast (Brydges and Carruthers, 2003). More recently, Soldati and colleagues further investigated targeting of *T. gondii* FeSOD2, this time looking at the full protein fused to a C-terminal epitope tag instead of just the targeting sequence fused to GFP. The protein encoded by this construct localized to both the apicoplast and the mitochondrion (Pino *et al.*, 2007). Because the construct employed a cDNA sequence, alternative splicing was ruled out, and because the presequence lacked additional methionines after the start codon, alternative translation initiation was also ruled out.

Dual targeting allows a gene to do double duty, encoding products that function in more than one subcellular location. Examples of dual-targeted proteins are growing more abundant, especially in *Arabidopsis* (Duchene *et al.*, 2005). Most often, the proteins involved – tRNA synthetases, RNA polymerases, and various enzymes, to name a few – are needed for both mitochondrial and plastid functions (reviewed in Peeters and Small, 2001; Silva-Filho, 2003; Karniely and Pines, 2005). However, there are also examples of dual targeting to mitochondria and peroxisomes (Petrova *et al.*, 2004), to mitochondria and the ER (Levitan *et al.*, 2005), to chloroplasts and the ER (Kiessling *et al.*, 2004; Levitan *et al.*, 2005) and triple targeting to mitochondria, nuclei, and the cytosol (Martin and Hopper, 1994). In many instances, alternative splicing or use of alternate initiation codons produces sufficient differences to direct the slightly different gene products to different parts of the cell. In other instances, post-translational modifications affect localization (Silva-Filho, 2003).

Based on the precedents from other organisms, it is not surprising that multiple examples of dual targeting in *T. gondii* have been discovered. Use of alternative start codons is another means by which dual localization to the apicoplast and mitochondrion can be achieved (Saito *et al.*, 2008). Here, the predicted presequence of pyruvate kinase II contains multiple methionines. When the third methionine was mutated to alanine, the protein was no longer found in the mitochondrion. Conversely, when the N-terminal region containing the signal sequence was deleted so that translation could only start at the third methionine, the protein was prevented from being diverted to the ER and was routed to the mitochondrion. One of the more complex examples is a thioredoxin-dependent peroxidase which is alternatively spliced to yield different 5′ transcript ends (Pino *et al.*, 2007). cDNAs corresponding to these splicing variants, when transfected into *T. gondii*, yield proteins localized to the cytosol (shorter cDNA) or apicoplast and mitochondrion (longer cDNA). Several other examples of proteins dually targeted to the apicoplast and mitochondrion have been identified in *T. gondii* (Pino *et al.*, 2007) and *P. falciparum* (Günther *et al.*, 2007; Kehr *et al.*, 2010). These experiments suggest that the context of the N-terminal targeting sequence also affects localization and point to the need for caution in ascribing cellular localizations on the basis of predicted trafficking sequences and experiments with tagged proteins. They also indicate that dual targeting of nuclearly encoded proteins is a factor in *T. gondii* organelle function.

Comparatively little progress has been made concerning the identification of components of the *T. gondii* mitochondrial import system. The *T. gondii* genome encodes orthologues of at least four putative mitochondrial inner membrane translocases: Tim17, Tim22, Tim23, Tim44 and possibly Tim9. Candidate outer membrane translocon components have been proposed: Tom40 in both *T. gondii* and *P. falciparum* and Tom22 in *P. falciparum* (Macasev *et al.*, 2004). Only CPN60 (HSP60), a classical mitochondrial chaperone involved in import, has been characterized (Toursel *et al.*, 2000). The HSP60 mRNA is alternatively spliced and both of the mRNAs are more abundant in bradyzoites than tachyzoites, as is the HSP60 protein. Immunofluorescence analysis with anti-*T. gondii* HSP60 antibodies confirms mitochondrial localization in tachyzoites but, curiously, early in the bradyzoite to tachyzoite conversion, the same antibodies detect two unknown bodies within or below the nucleus. This transitory localization is lost by 18 hours and its significance is unknown (Toursel *et al.*, 2000).

9.3.6 Energy Metabolism

The mitochondrion is commonly described in introductory biology classes as the powerhouse of the cell. It is the site of the tricarboxylic acid (TCA) cycle and the electron transport chain, which act in concert to efficiently convert the products of glucose catabolism to ATP. The other functions of the organelle often get little mention. However, while mitochondrial respiration is key for aerobically respiring cells, many parasite cells are found in environments with low oxygen tension and rely at least partly on alternate means for energy generation. This can result in modification of the energy generating pathways by loss or replacement of some enzymes or steps. Evaluation of the enzyme content of the *T. gondii* mitochondrion has been largely done by annotation of the *T. gondii* genome based on known metabolic pathways from other organisms (Fig. 9.11) (Seeber and Soldati, 2007; Mather *et al.*, 2007; Seeber *et al.*, 2008; Polonais and Soldati-Favre, 2010) and additional effort will be needed to validate predictions and fully assess their mitochondrial roles.

The *T. gondii* mitochondrion can be stained with fluorescent dyes like rhodamine 123, MitoTracker or $DiOC_6$, which accumulate in organelles that have a high transmembrane potential (Tanabe, 1985; Lin *et al.*, 2009). Another classical way of marking active mitochondria is by detecting diaphorase activity, which uses NADH or NADPH as substrate to reduce nitroblue tetrazolium, producing a dense black precipitate. In the presence of NADPH, *T. gondii* tachyzoites stain darkly while mitochondria in the host cell do not. The reverse is true for NADH (Seeber *et al.*, 1998). These data point to generation of an electrochemical gradient but the relative importance of glycolysis and mitochondrial respiration in *T. gondii* energy generation is still under investigation (see below).

Tachyzoites have cristate mitochondria, typically associated with mitochondrial respiration, but they rely heavily on glycolysis for energy production. Bradyzoites do as well, but there are differences in enzyme activities between these stages. Activity of the glycolytic enzyme phosphofructokinase is 2–3-fold lower in bradyzoites than tachyzoites (Denton *et al.*, 1996). Two other glycolytic enzymes, glucose-6-phosphate isomerase and enolase, are up-regulated in bradyzoites (Dzierszinski *et al.*, 1999). On the other hand, isocitrate dehydrogenase activity, a TCA cycle indicator, was low and unchanged between the stages (Denton *et al.*, 1996). Together, these findings suggest that glycolysis is necessary for energy production in both tachyzoites and bradyzoites, although perhaps not to the same degree. Glycolysis-derived pyruvate is converted to lactate by lactate dehydrogenase. *T. gondii* has two lactate dehydrogenase isoenzymes (LDH1 and LDH2), which are stage-specifically expressed: LDH1 in tachyzoites and LDH2 in bradyzoites (Ferguson *et al.*, 2005). Bradyzoites have about three times the LDH activity of tachyzoites and it was hypothesized that bradyzoites need LDH for catabolism of amylopectin, a storage polysaccharide which is abundant in the encysted parasites (Denton *et al.*, 1996). Pyruvate is also usually funnelled to the mitochondrion where the pyruvate dehydrogenase complex (PDH) converts it to acetyl-CoA, the entry point into the TCA cycle. But both *T. gondii* and *P. falciparum* have only one PDH, which resides in the plastid rather than the mitochondrion (see Section 9.2.9 Apicoplast Metabolism). In the absence of a mitochondrial PDH, a key question thus is: what is the source for mitochondrial acetyl-CoA?

Based on *in silico* analysis, a complete pathway for the degradation of branched-chain amino acids (BCAA) has been annotated, including a downstream methylcitrate cycle that serves to detoxify propionyl-CoA derived from BCAA degradation (Seeber *et al.*, 2008). In this pathway the amino acids Ile, Leu and Val are first deaminated by BCAA aminotransferase (BCAT), giving rise to ketoacids that are then decarboxylated by the four subunit encompassing branched-chain keto acid dehydrogenase (BCKDH) complex. In four enzyme-catalyzed further steps the final products are acetyl-CoA plus propionyl-CoA in the case of Ile; acetoacetate plus acetyl-CoA in the case of Leu; and propionyl-CoA only in the case of Val. The latter is further metabolized by the five-enzyme methyl citrate cycle and fed into the TCA cycle as succinate or as oxaloacetate after conversion of pyruvate by mitochondrial pyruvate carboxylase (Limenitakis, 2013). The BCKDH complex is similar to the PDH complex, at the amino acid sequence level as well as enzymatically and structurally (Perham, 2000). Based on these data, it has been hypothesized that the BCKDH complex in *T. gondii* and *P. falciparum* (despite lacking BCAT as well as the four enzymes subsequent to the BCKDH complex) can substitute for the missing mitochondrial PDH complex, accepting pyruvate as substrate and thereby directly producing acetyl-CoA (Seeber *et al.*, 2008). Recent analysis of *T. gondii* BCKDH−E1a knock-out strains by metabolomics provide strong evidence that this is indeed the case − disruption of this enzyme leads to accumulation of pyruvate and a severe drop in acetyl-CoA (D. Soldati-Favre, personal communication). The presence of clear homologues of the recently reported mitochondrial pyruvate transporter subunits MPC-1/-2 (Bricker *et al.*, 2012; Herzig *et al.*, 2012) in the genomes of *T. gondii* and *P. falciparum* (F. Seeber, unpublished results) lends further support for a direct connection between cytosolic glycolysis and mitochondrial acetyl-CoA generation via pyruvate. This view is supported by similar results from recent metabolomics experiments with the early dinoflagellate *Perkinsus marinus*, which apparently lacks a mitochondrial PDH but has a BCKDH (Danne *et al.*, 2013). In *P. falciparum* it remains to be seen whether the previously reported half canonical, branched TCA cycle that was shown to utilize glutamine as fuel (Olszewski *et al.*, 2010) also has the capacity to run 'full circle' by using the BCKDH complex to stand in for the missing mitochondrial PDH.

Another important question is the extent to which the electrochemical gradient formed by the electron transport chain (ETC) via the TCA cycle is used in *T. gondii* to generate ATP (oxidative phosphorylation). An almost complete inner membrane resident ETC can be compiled from genome data (with a type II NAD(P)H oxidoreductase, NDH2, as an alternative to complex I (for details see Seeber *et al.*, 2008; Vaidya and Mather, 2009); and should be able to collect reducing equivalents from the TCA cycle in the matrix. However, the actual ATP-generating system, ATP synthase (complex V), is apparently incomplete. While the five F1 subunits are found in *T. gondii* and *P. falciparum*, a homologue can be found for only one of the three F0 subunits, the c subunit (Seeber *et al.*, 2008; Vaidya and Mather, 2009 and references therein). This is in contradiction to the observed inhibition of O_2 consumption and ATP levels by the ATP-synthase inhibitor oligomycin in *T. gondii* (Vercesi *et al.*, 1998; Lin *et al.*, 2009). A plausible explanation would be that parasite sequences for subunits a and b evolved to a degree that prevents their identification by BLAST searches. This is consistent with recent data from *P. falciparum* that confirmed the mitochondrial localization of all but one predicted ATP synthase subunit and the assembly of the complexes into dimers, thus providing evidence for the presence of unknown and presumably unconventional subunits without which dimer formation would seem unlikely (Balabaskaran *et al.*, 2011). Attempts to knock out the known subunit genes were unsuccessful, suggesting an essential function for the complex.

To further complicate matters, it is unclear whether the potent growth inhibitory action of drugs like atovaquone and other ETC inhibitors on *T. gondii* and *P. falciparum* can be regarded as an indication for ETC involvement in ATP generation (Painter *et al.*, 2007; Fisher *et al.*, 2008). Experiments in *P. falciparum* suggest that in blood stage parasites, the ETC is only essential for pyrimidine biosynthesis (Painter *et al.*, 2007). Dihydroorotate dehydrogenase (DHODH) catalyses the conversion of dihydroorotate to orotate in the *de novo* pyrimidine biosynthesis pathway and is an essential mitochondrial enzyme also in *T. gondii* (Hortua Triana *et al.*, 2012). Since the expression of a ubiquinone-independent (and thus atovaquone-insensitive) yeast DHODH in transgenic parasites does not affect their growth in the presence of many ETC inhibitors, the conclusion of Painter *et al.* (2007) was that NADH dehydrogenase (NDH2) activity is dispensable in the erythrocytic stages *in vitro*. Gene knock-outs of *NDH2* in *P. falciparum* confirmed this notion but also showed that NDH2 was essential in the mosquito stage where ATP provision by glycolysis is limited (Boysen and Matuschewski, 2011). Other work on ETC inhibitors still maintains NDH2 to be a vital component of the ETC in blood stage parasites (Fisher *et al.*, 2008; Biagini *et al.*, 2012). In contrast, attempts to knock down the two NDH2 isoforms simultaneously in *T. gondii* were unsuccessful (Lin *et al.*, 2011). This, together with a moderate decline in ATP levels (c. 25%) of the individual knock-outs, may indicate a different role for the ETC in this parasite. Obviously, more work is required to elucidate the importance of ATP synthesis via oxidative phosphorylation in *T. gondii* and *P. falciparum*.

9.3.7 Other Metabolic Pathways

The importance of oxidative phosphorylation is so great in many cells that other functions of the mitochondrion are sometimes overlooked (Fig. 9.11). Among those functions in *T. gondii* are formation of ISC complexes, haem biosynthesis (see above), ubiquinone synthesis, and a key step in pyrimidine synthesis, i.e. the reduction of dihydroorotate to orotate catalysed by dihydroorotate dehydrogenase (DHODH; see above and Chapter 20). Several recent reviews on apicomplexan mitochondrial metabolism are available (van Dooren *et al.*, 2006; Seeber *et al.*, 2008; Vaidya and Mather, 2009; Polonais and Soldati-Favre, 2010).

The importance of ISCs has already been pointed out above (see Section Section 9.2.9 Apicoplast Metabolism). Mitochondrial ISC synthesis not only serves that organelle with these co-factors but also provides them for the cytosol and nucleus (Lill and Muhlenhoff, 2008; Lill, 2009; Lill *et al.*, 2012). Some authors consider this to be the most crucial function of mitochondria overall. Bioinformatic analysis has allowed *in silico* assembly of a mitochondrial ISC synthesis pathway in *T. gondii* (Seeber, 2002; Seeber and Soldati, 2007; Seeber *et al.*, 2008) that does not appear to deviate much from that known in yeast and higher eukaryotes. One notable exception is the apparent absence of a homologue of the protein frataxin which is implicated in iron transport and assembly of ISC and present in most eukaryotes (Busi and Gomez-Casati, 2012), including even *T. parva* and *C. parvum* (Seeber *et al.*, 2008). A parasite homologue to the small bacterial protein YfhJ has been proposed but not demonstrated to have taken over the function of frataxin in *Plasmodium* (Pastore *et al.*, 2006). However, a clear sequence homologue even to the putative *Plasmodium YfhJ* cannot be found in *T. gondii* (F. Seeber, unpublished results). No experimental verification has been reported on any of these components with the exception of the localization of two ISC scaffold proteins, ISCU and ISCA, to mitochondria (Pino *et al.*, 2007).

The ETC depends heavily on the co-factor ubiquinone (coenzyme Q). Based on sequence data and localization predictions, synthesis of this compound in the mitochondrion of *T. gondii*, starting from the ubiquinone precursor 4-hydroxybenzoic acid (4-HB), appears plausible (Seeber *et al.*, 2008). However, the source of 4-HB is currently unknown since there is no evidence for a chorismate lyase that could convert chorismate (derived from the shikimate pathway; McConkey *et al.*, 2004) to 4-HB. Whether this is again a failure of bioinformatic detection of divergent enzymes or a real 'metabolic gap' needs to be determined. In the latter case, it could mean that *T. gondii* relies on host cell 4-HB for ubiquinone biosynthesis. It should be noted that the modification of 4-HB by a specific prenyltransferase to yield ubiquinone requires isoprenoids of different complexities and lengths, and the precursors for those, IMPP and DMAPP, are synthesized in the apicoplast (see Section 9.2.9 Apicoplast Metabolism). In the mitochondrion these are then condensed by the enzyme farnesyl pyrophosphate synthase into more complex isoprenoids (Ling *et al.*, 2007).

Some metabolic activities that might have been expected in *T. gondii*'s mitochondrion are apparently absent and, on the other hand, pathways that are only present *in T. gondii* but not in *P. falciparum* shed some light on the evolutionary forces that shaped the organellar metabolic pathways (Seeber and Soldati, 2007). Most eukaryotes have a dissociated type II-like mitochondrial FAS pathway that is mainly required to synthesize lipoic acid (see Section 9.2.9 Apicoplast Metabolism) (Hiltunen *et al.*, 2010), but *T. gondii* and other Apicomplexa have lost this entire synthetic route. The consequences are that LA, which is required in the mitochondrion by three proteins for enzymatic activity (E2 subunit of the 2-oxoglutarate dehydrogenase of the TCA cycle; E2 subunit of the BCKDH; and the H-protein of the glycine cleavage complex) has to be scavenged from the host cell mitochondria (Crawford *et al.*, 2006). This LA is then attached to the mentioned enzymes by an ATP- or GTP-dependent mitochondrial lipoate ligase. Experiments in both *T. gondii* (Crawford *et al.*, 2006), as well as in the liver stages of *P. berghei* (Deschermeier *et al.*, 2012), suggest that the apicoplast is not able to contribute substantial amounts of LA to the mitochondrion, thereby making the parasites totally dependent on LA supplied by the host cell. The benefits for this would be that a number of proteins became superfluous. Why there is no metabolic crosstalk between the otherwise highly connected parasite organelles with respect to LA exchange is not known.

The above mentioned glycine cleavage system (GCS) is involved in the conversion of glycine into N5,N10-methylenetetrahydrofolate, a precursor required in pyrimidine biosynthesis and folate metabolism (Kikuchi *et al.*, 2008). Three of the four genes that encode the GCS enzymes (so-called T-, H- and L-proteins) are found in the genomes of *T. gondii* and *P. falciparum* and are predicted to be mitochondrial, whereas a gene for the P-protein could not be identified (Seeber *et al.*, 2008). No data are available that have addressed the functional role of GCS in apicomplexans, but *T. gondii* is the only one of these parasites that seems to be able to produce the co-factor pantothenate, the precursor for coenzyme A (Seeber *et al.*, 2008; Spry *et al.*, 2008). As explained below, this provides a connection between the GCS and the BCAA degradation pathway and may be a further reason why BCAA degradation has been retained in *T. gondii*. Pantothenate synthesis is initiated by the conversion of 3-methyl-2-oxobutanoate to 2-dehydropantoate. The former is the transamination product of valine, produced by the transaminase BCAT of the BCAA degradation pathway. The reaction is catalysed by the enzyme 3-methyl-2-oxobutanoate hydroxymethyltransferase, which requires GCS-derived N5,N10-methylenetetrahydrofolate and presumably takes place in the cytosol, like the other steps that lead to pantothenate. No experimental data have so far been reported on this pathway.

9.3.8 Mitochondrial Function and the Tachyzoite to Bradyzoite Switch

The transition of active tachyzoites to encysted bradyzoites is a bit of a black box but changes in certain environmental conditions stimulate the transition *in vitro* (Sullivan and Jeffers, 2012). Among these are changes in pH, alteration in nutrient accessibility, change in O_2 tension, and exposure to compounds that limit or block mitochondrial function. It is this latter point that is the focus here. *T. gondii* tachyzoites transition quickly to *in vitro* bradyzoites following exposure to atovaquone and other mitochondrial inhibitors (Bohne *et al.*, 1994; Tomavo and Boothroyd, 1995). Bradyzoites appear to be less reliant on mitochondrial respiration than intracellular tachyzoites (reviewed in Gross *et al.*, 1996). It is not clear if the switch from active growth to an encysted stage reflects a drop in ATP availability or interference with the mitochondrial redox machinery. In either case, identifying the molecular correlates that connect mitochondrial function to stage conversion provides an avenue to understanding this critical step.

9.4 PERSPECTIVES

The endosymbiotic organelles of apicomplexans have proven much more exciting than imagined years ago. The mitochondrion still searches for delineation of function, but its fragmented rRNAs provide a look at a minimal ribosome. From a multi-membraned organelle of unknown significance, we now have a relict plastid with pathways that are yielding targets for new drugs against the diseases caused by these parasites. Much research on both organelles has previously focused on *P. falciparum*, for good reasons: the organelle genomes were first identified in *P. falciparum*; its nuclear genome sequence was completed years earlier than any other apicomplexan; and the magnitude of morbidity and mortality it causes is dramatically greater than for *T. gondii*. But now, with many tools in place and more being developed, it is likely, and indeed highly desirable, that there will be a rapid expansion of knowledge about the apicoplast and mitochondrion of *T. gondii*, with implications for treatment and control for diseases caused by numerous apicomplexans.

Acknowledgements

We thank Dr. Michael Gray for advice on the evolution of endosymbiosis. Searches noted in the text were carried out in Release 1.0 of OrthoMCL (Li *et al.*, 2003), Release 4.4 of PlasmoDB (Kissinger *et al.*, 2002), or Releases 3.0 and 7.3 of the ToxoDB database (Gajria *et al.*, 2008). Genomic and/or cDNA sequence data was accessed via http://ToxoDB.org and/or http://www.tigr.org/tdb/t_gondii/. Genomic data were provided by the Institute for Genomic Research (supported by the NIH grant #AI05093), and by the Sanger Center (Wellcome Trust). EST sequences were generated by Washington University (NIH grant #1R01AI045806-01A1). Support for work in the authors' labs was provided by NIH grants R01 AI50506 to MP, by the Deutsche Forschungsgemeinschaft to FS (Se 622/4-3).

References

Abrahamsen, M.S., Templeton, T.J., Enomoto, S., Abrahante, J.E., Zhu, G., Lancto, C.A., Deng, M., Liu, C., Widmer, G., Tzipori, S., Buck, G.A., Xu, P., Bankier, A.T., Dear, P.H., Konfortov, B.A., Spriggs, H.F., Iyer, L., Anantharaman, V., Aravind, L., Kapur, V., 2004. Complete genome sequence of the apicomplexan, *Cryptosporidium parvum*. Science 304, 441–445.

Agrawal, S., van Dooren, G.G., Beatty, W.L., Striepen, B., 2009. Genetic evidence that an endosymbiont-derived ERAD system functions in import of apicoplast proteins. J. Biol. Chem. 284, 33683–33691.

Ahmed, A., Sharma, Y.D., 2008. Ribozyme cleavage of *Plasmodium falciparum* gyrase A gene transcript affects the parasite growth. Parasitol. Res. 103, 751–763.

Alcock, F., Webb, C.T., Dolezal, P., Hewitt, V., Shingu-Vasquez, M., Likic, V.A., Traven, A., Lithgow, T., 2012. A small Tim homohexamer in the relict mitochondrion of *Cryptosporidium*. Mol Biol Evol. 29, 113–122.

Aldritt, S.M., Joseph, J.T., Wirth, D.F., 1989. Sequence identification of cytochrome *b* in *Plasmodium gallinaceum*. Mol. Cell. Biol. 9, 3614–3620.

Alger, N.E., Spira, D.T., Silverman, P.H., 1970. Inhibition of rodent malaria in mice by rifampicin. Nature 227, 381–382.

Aliverti, A., Pandini, V., Pennati, A., de, R.M., Zanetti, G., 2008. Structural and functional diversity of ferredoxin-NADP(+) reductases. Arch. Biochem. Biophys. 474, 283–291.

Anokhina, M.M., Barta, A., Nierhaus, K.H., Spiridonova, V.A., Kopylov, A.M., 2004. Mapping of the second tetracycline binding site on the ribosomal small subunit of *E. coli*. Nucleic. Acids Res. 32, 2594–2597.

Antony, E., Weiland, E.A., Korolev, S., Lohman, T.M., 2012. *Plasmodium falciparum* SSB tetramer wraps single-stranded DNA with similar topology but opposite polarity to E. coli SSB. J. Mol. Biol. 420, 269–283.

Araujo, F.G., Slifer, T., Remington, J.S., 1994. Rifabutin is active in murine models of toxoplasmosis. Antimicrob. Agents Chemother. 38, 570–575.

Archibald, J.M., Keeling, P.J., 2002. Recycled plastids: a 'green movement' in eukaryotic evolution. Trends Genet. 18, 577–584.

Augagneur, Y., Wesolowski, D., Tae, H.S., Altman, S., Ben, M.C., 2012. Gene selective mRNA cleavage inhibits the development of *Plasmodium falciparum*. Proc. Natl. Acad. Sci. USA 109, 6235–6240.

Bahl, A., Davis, P.H., Behnke, M., Dzierszinski, F., Jagalur, M., Chen, F., Shanmugam, D., White, M.W., Kulp, D., Roos, D.S., 2010. A novel multifunctional oligonucleotide microarray for *Toxoplasma gondii*. BMC Genomics 11, 603.

Balabaskaran, N.P., Morrisey, J.M., Ganesan, S.M., Ke, H., Pershing, A.M., Mather, M.W., Vaidya, A.B., 2011. ATP synthase complex of *Plasmodium falciparum*: dimeric assembly in mitochondrial membranes and resistance to genetic disruption. J. Biol. Chem. 286, 41312–41322.

Balk, J., Pilon, M., 2011. Ancient and essential: the assembly of iron-sulfur clusters in plants. Trends Plant Sci. 16, 218–226.

Bapteste, E., Boucher, Y., Leigh, J., Doolittle, W.F., 2004. Phylogenetic reconstruction and lateral gene transfer. Trends Microbiol. 12, 406–411.

Baumeister, S., Wiesner, J., Reichenberg, A., Hintz, M., Bietz, S., Harb, O.S., Roos, D.S., Kordes, M., Friesen, J., Matuschewski, K., Lingelbach, K., Jomaa, H., Seeber, F., 2011. Fosmidomycin uptake into *Plasmodium* and *Babesia*-infected erythrocytes is facilitated by parasite-induced new permeability pathways. PLoS ONE 6, e19334.

Beckers, C.J.M., Roos, D.S., Donald, R.G., Luft, B.J., Schwab, J.C., Cao, Y., Joiner, K.A., 1995. Inhibition of cytoplasmic and organellar protein synthesis in *Toxoplasma gondii*. Implications for the target of macrolide antibiotics. J. Clin. Invest. 95, 367–376.

Biagini, G.A., Fisher, N., Shone, A.E., Mubaraki, M.A., Srivastava, A., Hill, A., Antoine, T., Warman, A.J., Davies, J., Pidathala, C., Amewu, R.K., Leung, S.C., Sharma, R., Gibbons, P., Hong, D.W., Pacorel, B., Lawrenson, A.S., Charoensutthivarakul, S., Taylor, L., Berger, O., Mbekeani, A., Stocks, P.A., Nixon, G.L., Chadwick, J., Hemingway, J., Delves, M.J., Sinden, R.E., Zeeman, A.M., Kocken, C.H., Berry, N.G., O'Neill, P.M., Ward, S.A., 2012. Generation of quinolone antimalarials targeting the *Plasmodium falciparum* mitochondrial respiratory chain for the treatment and prophylaxis of malaria. Proc. Natl. Acad. Sci. USA 109, 8298–8303.

Bisanz, C., Bastien, O., Grando, D., Jouhet, J., Marechal, E., Cesbron-Delauw, M.F., 2006. *Toxoplasma gondii* acyl-lipid metabolism: de novo synthesis from apicoplast generated fatty acids versus scavenging of host cell precursors. Biochem. J. 394, 197–205.

Blanchard, J.L., Hicks, J.S., 1999. The non-photosynthetic plastid in malarial parasites and other apicomplexans is derived from outside the green plastid lineage. J. Euk. Microbiol. 46, 367–375.

Block, M.A., Dorne, A.J., Joyard, J., Douce, R., 1983. Preparation and characterization of membrane fractions enriched in outer and inner envelope membranes from spinach chloroplasts. II. Biochemical characterization. J. Biol. Chem. 258, 13281–13286.

Bohne, W., Heesemann, J., Gross, U., 1994. Reduced replication of *Toxoplasma gondii* is necessary for induction of bradyzoite-specific antigens: A possible role for nitric oxide in triggering stage conversion. Infect. Immun. 62, 1761–1767.

Bonifacino, J.S., Traub, L.M., 2003. Signals for sorting of transmembrane proteins to endosomes and lysosomes. Annu. Rev. Biochem. 72, 395–447.

Borst, P., Overdulve, J.P., Weijers, P.J., Fase-Fowler, F., Van den Berg, M., 1984. DNA circles with cruciforms from *Isospora (Toxoplasma) gondii*. Biochim. Biophys. Acta 781, 100–111.

Botté, C.Y., Dubar, F., McFadden, G.I., Marechal, E., Biot, C., 2012. *Plasmodium falciparum* apicoplast drugs: targets or off-targets? Chem. Rev. 112, 1269–1283.

Boysen, K.E., Matuschewski, K., 2011. Arrested oocyst maturation in *Plasmodium* parasites lacking type II NADH:ubiquinone dehydrogenase. J. Biol. Chem. 286, 32661–32671.

Bozdech, Z., Llinás, M., Pulliam, B.L., Wong, E.D., Zhu, J., DeRisi, J.L., 2003. The transcriptome of the intraerythrocytic developmental cycle of *Plasmodium falciparum*. PLoS Biol. 1, E5.

Brayton, K.A., Lau, A.O., Herndon, D.R., Hannick, L., Kappmeyer, L.S., Berens, S.J., Bidwell, S.L., Brown, W.C., Crabtree, J., Fadrosh, D., Feldblum, T., Forberger, H.A., Haas, B.J., Howell, J.M., Khouri, H., Koo, H., Mann, D.J., Norimine, J., Paulsen, I.T., Radune, D., Ren, Q., Smith Jr., R.K., Suarez, C.E., White, O., Wortman, J.R., Knowles Jr., D.P., McElwain, T.F., Nene, V.M., 2007. Genome sequence of *Babesia bovis* and comparative analysis of apicomplexan hemoprotozoa. PLoS Pathog. 3, 1401–1413.

Breinich, M.S., Ferguson, D.J., Foth, B.J., van Dooren, G.G., Lebrun, M., Quon, D.V., Striepen, B., Bradley, P.J., Frischknecht, F., Carruthers, V.B., Meissner, M., 2009. A dynamin is required for the biogenesis of secretory organelles in *Toxoplasma gondii*. Curr. Biol. 19, 277–286.

Bricker, D.K., Taylor, E.B., Schell, J.C., Orsak, T., Boutron, A., Chen, Y.C., Cox, J.E., Cardon, C.M., Van Vranken, J.G., Dephoure, N., Redin, C., Boudina, S., Gygi, S.P., Brivet, M., Thummel, C.S., Rutter, J., 2012. A mitochondrial pyruvate carrier required for pyruvate uptake in yeast, *Drosophila*, and humans. Science. 337, 96–100.

Brodersen, D.E., Clemons, W.M., Carter, A.P., Morgan-Warren, R.J., Wimberly, B.T., Ramakrishnan, V., 2000. The structural basis for the action of the antibiotics tetracycline, pactamycin, and hygromycin B on the 30S ribosomal subunit. Cell 103, 1143–1154.

Brooks, C.F., Johnsen, H., van Dooren, G.G., Muthalagi, M., Lin, S.S., Bohne, W., Fischer, K., Striepen, B., 2010. The Toxoplasma apicoplast phosphate translocator links cytosolic and apicoplast metabolism and is essential for parasite survival. Cell Host Microbe 7, 62–73.

Bruce, B.D., 2001. The paradox of plastid transit peptides: conservation of function despite divergence in primary structure. Biochim. Biophys. Acta 1541, 2–21.

Brydges, S.D., Carruthers, V.B., 2003. Mutation of an unusual mitochondrial targeting sequence of SODB2 produces multiple targeting fates in *Toxoplasma gondii*. J. Cell Sci. 116, 4675–4685.

Bullmann, L., Haarmann, R., Mirus, O., Bredemeier, R., Hempel, F., Maier, U.G., Schleiff, E., 2010. Filling the gap, evolutionarily conserved Omp85 in plastids of chromalveolates. J. Biol. Chem. 285, 6848–6856.

Burki, F., Flegontov, P., Obornik, M., Cihlar, J., Pain, A., Lukes, J., Keeling, P.J., 2012. Re-evaluating the green versus red signal in eukaryotes with secondary plastid of red algal origin. Genome Biol. Evol. 4, 626–635.

Burri, L., Williams, B.A., Bursac, D., Lithgow, T., Keeling, P.J., 2006. Microsporidian mitosomes retain elements of the general mitochondrial targeting system. Proc. Natl. Acad. Sci. USA 103, 15916–15920.

Busi, M.V., Gomez-Casati, D.F., 2012. Exploring frataxin function. IUBMB Life 64, 56–63.

Cai, X., Fuller, A.L., McDougald, L.R., Zhu, G., 2003. Apicoplast genome of the coccidian *Eimeria tenella*. Gene 321, 39–46.

Cameron, J.M., Janer, A., Levandovskiy, V., Mackay, N., Rouault, T.A., Tong, W.H., Ogilvie, I., Shoubridge, E.A., Robinson, B.H., 2011. Mutations in iron-sulfur cluster scaffold genes NFU1 and BOLA3 cause a fatal deficiency of multiple respiratory chain and 2-oxoacid dehydrogenase enzymes. Am. J. Hum. Genet. 89, 486–495.

Camps, M., Arrizabalaga, G., Boothroyd, J., 2002. An rRNA mutation identifies the apicoplast as the target for clindamycin in *Toxoplasma gondii*. Mol. Microbiol. 43, 1309–1318.

Carlton, J.M., Adams, J.H., Silva, J.C., Bidwell, S.L., Lorenzi, H., Caler, E., Crabtree, J., Angiuoli, S.V., Merino, E.F., Amedeo, P., Cheng, Q., Coulson, R. M., Crabb, B.S., Del Portillo, H.A., Essien, K., Feldblyum, T.V., Fernandez-Becerra, C., Gilson, P.R., Gueye, A.H., Guo, X., Kang'a, S., Kooij, T.W., Korsinczky, M., Meyer, E.V., Nene, V., Paulsen, I., White, O., Ralph, S.A., Ren, Q., Sargeant, T.J., Salzberg, S.L., Stoeckert, C.J., Sullivan, S.A., Yamamoto, M.M., Hoffman, S.L., Wortman, J.R., Gardner, M.J., Galinski, M. R., Barnwell, J.W., Fraser-Liggett, C.M., 2008. Comparative genomics of the neglected human malaria parasite *Plasmodium vivax*. Nature 455, 757–763.

Chahal, H.K., Outten, F.W., 2012. Separate Fe-S scaffold and carrier functions for SufB2C2 and SufA during in vitro maturation of [2Fe-2S] Fdx. J. Inorg. Biochem. 116, 126–134.

Chan, K.W., Slotboom, D.J., Cox, S., Embley, T.M., Fabre, O., van der Giezen, M., Harding, M., Horner, D.S., Kunji, E.R., León-Avila, G., Tovar, J., 2005. A novel ADP/ATP transporter in the mitosome of the microaerophilic human parasite *Entamoeba histolytica*. Curr. Biol. 15, 737–742.

Chang, H.R., Comte, R., Pechere, J.C., 1990. In vitro and in vivo effects of *doxycycline* on *Toxoplasma gondii*. Antimicrob. Agents Chemother. 34, 775–780.

Chang, H.R., Comte, R., Piguet, P.F., Pechere, J.C., 1991. Activity of minocycline against *Toxoplasma gondii* infection in mice. J. Antimicrob. Chemother. 27, 639–645.

Chaubey, S., Kumar, A., Singh, D., Habib, S., 2005. The apicoplast of *Plasmodium falciparum* is translationally active. Mol. Microbiol. 56, 81–89.

Cheresh, P., Harrison, T., Fujioka, H., Haldar, K., 2002. Targeting the malarial plastid via the parasitophorous vacuole. J. Biol. Chem. 277, 16265–16277.

Chiodini, P.L., Conlon, C.P., Hutchinson, D.B.A., Farquhar, J.A., Hall, A.P., Peto, T.E.A., Birley, H., Warrell, D.A., 1995. Evaluation of atovaquone in the treatment of patients with uncomplicated *Plasmodium falciparum* malaria. J. Antimicrob. Chemother. 36, 1073–1078.

Cilingir, G., Broschat, S.L., Lau, A.O., 2012. ApicoAP: The first computational model for identifying apicoplast-targeted proteins in multiple species of Apicomplexa. PLoS ONE 7, e36598.

Clastre, M., Goubard, A., Prel, A., Mincheva, Z., Viaud-Massuart, M.C., Bout, D., Rideau, M., Velge-Roussel, F., Laurent, F., 2007. The methylerythritol phosphate pathway for isoprenoid biosynthesis in coccidia: presence and sensitivity to fosmidomycin. Exp. Parasitol. 116, 375–384.

Clough, B., Strath, M., Preiser, P., Denny, P., Wilson, I., 1997. Thiostrepton binds to malarial plastid rRNA. FEBS Letters 406, 123–125.

Coppin, A., Varre, J.S., Lienard, L., Dauvillee, D., Guerardel, Y., Soyer-Gobillard, M.O., Buleon, A., Ball, S., Tomavo, S., 2005. Evolution of plant-like crystalline storage polysaccharide in the protozoan parasite *Toxoplasma gondii* argues for a red alga ancestry. J. Mol. Evol. 60, 257–267.

Cornillot, E., Hadj-Kaddour, K., Dassouli, A., Noel, B., Ranwez, V., Vacherie, B., Augagneur, Y., Bres, V., Duclos, A., Randazzo, S., Carcy, B., Debierre-Grockiego, F., Delbecq, S., Moubri-Ménage, K., Shams-Eldin, H., Usmani-Brown, S., Bringaud, F., Wincker, P., Vivares, C.P., Schwarz, R.T., Schetters, T.P., Krause, P.J., Gorenflot, A., Berry, V., Barbe, V., Ben, M.C., 2012. Sequencing of the smallest Apicomplexan genome from the human pathogen *Babesia microti*. Nucleic Acids Res. 40, 9102–9114.

Crawford, M.J., Thomsen-Zieger, N., Ray, M., Schachtner, J., Roos, D.S., Seeber, F., 2006. *Toxoplasma gondii* scavenges host-derived lipoic acid despite its de novo synthesis in the apicoplast. EMBO J. 25, 3214–3222.

Creasey, A., Mendis, K., Carlton, J., Williamson, D., Wilson, I., Carter, R., 1994. Maternal inheritance of extrachromosomal DNA in malaria parasites. Mol. Biochem. Parasitol. 65, 95–98.

Dahl, E.L., Rosenthal, P.J., 2008. Apicoplast translation, transcription and genome replication: targets for antimalarial antibiotics. Trends Parasitol. 24, 279–284.

Dahl, E.L., Shock, J.L., Shenai, B.R., Gut, J., DeRisi, J.L., Rosenthal, P.J., 2006. Tetracyclines specifically target the apicoplast of the malaria parasite *Plasmodium falciparum*. Antimicrob. Agents Chemother. 50, 3124–3131.

Danne, J.C., Gornik, S.G., Macrae, J.I., McConville, M.J., Waller, R.F., 2013. Alveolate mitochondrial metabolic evolution: dinoflagellates force reassessment of the role of parasitism as a driver of change in apicomplexans. Mol. Biol. Evol. 30, 123–139.

Dar, A., Prusty, D., Mondal, N., Dhar, S.K., 2009. A unique 45-amino-acid region in the toprim domain of *Plasmodium falciparum* gyrase B is essential for its activity. Eukaryot. Cell 8, 1759–1769.

Dar, M.A., Sharma, A., Mondal, N., Dhar, S.K., 2007. Molecular cloning of apicoplast-targeted *Plasmodium falciparum* DNA gyrase genes: unique intrinsic ATPase activity and ATP-independent dimerization of PfGyrB subunit. Eukaryot. Cell 6, 398–412.

Denton, H., Roberts, C.W., Alexander, J., Thong, K.W., Coombs, G.H., 1996. Enzymes of energy metabolism in the bradyzoites and tachyzoites of *Toxoplasma gondii*. FEMS Microbiol. Lett. 137, 103–108.

dePamphilis, C.W., Palmer, J.D., 1989. Evolution and function of plastid DNA: A review with special reference to nonphotosynthetic plants. In: Boyer, C. D., Shannon, J.C., Hardison, R.C. (Eds.), Physiology, Biochemistry, and Genetics of Nongreen Plants. The American Society of Plant Physiologists, pp. 182–202.

DeRocher, A., Gilbert, B., Feagin, J.E., Parsons, M., 2005. Dissection of brefeldin A-sensitive and -insensitive steps in apicoplast protein targeting. J. Cell Sci. 118, 565–574.

DeRocher, A., Hagen, C.B., Froehlich, J.E., Feagin, J.E., Parsons, M., 2000. Analysis of targeting sequences demonstrates that trafficking to the *Toxoplasma gondii* plastid branches off the secretory system. J. Cell Sci. 113, 3969–3977.

DeRocher, A.E., Coppens, I., Karnataki, A., Gilbert, L.A., Rome, M.E., Feagin, J.E., Bradley, P.J., Parsons, M., 2008. A thioredoxin family protein of the apicoplast periphery identifies abundant candidate transport vesicles in *Toxoplasma gondii*. Eukaryot. Cell 7, 1518–1529.

DeRocher, A.E., Karnataki, A., Vaney, P., Parsons, M., 2012. Apicoplast targeting of a *T. gondii* transmembrane protein requires a cytosolic tyrosine-based motif. Traffic 13, 694–704.

Deschermeier, C., Hecht, L.S., Bach, F., Rutzel, K., Stanway, R.R., Nagel, A., Seeber, F., Heussler, V.T., 2012. Mitochondrial lipoic acid scavenging is essential for *Plasmodium berghei* liver stage development. Cell. Microbiol. 14, 416–430.

Divo, A.A., Geary, T.G., Jensen, J.B., 1985. Oxygen- and time-dependent effects of antibiotics and selected mitochondrial inhibitors on *Plasmodium falciparum* in culture. Antimicrob. Agents Chemother. 27, 21–27.

Doolittle, W.F., Boucher, Y., Nesbo, C.L., Douady, C.J., Andersson, J.O., Roger, A.J., 2003. How big is the iceberg of which organellar genes in nuclear genomes are but the tip? Philos. Trans. R. Soc. B. 358, 39–57.

Dore, E., Frontali, C., Forte, T., Fratarcangeli, S., 1983. Further studies and electron microscopic characterization of *Plasmodium berghei* DNA. Mol. Biochem. Parasitol. 8, 339–352.

Douglas, S., Zauner, S., Fraunholz, M., Beaton, M., Penny, S., Deng, L.T., Wu, X., Reith, M., Cavalier-Smith, T., Maier, U.G., 2001. The highly reduced genome of an enslaved algal nucleus. Nature 410, 1091–1096.

Duchene, A.M., Giritch, A., Hoffmann, B., Cognat, V., Lancelin, D., Peeters, N.M., Zaepfel, M., Marechal-Drouard, L., Small, I.D., 2005. Dual targeting is the rule for organellar aminoacyl-tRNA synthetases in *Arabidopsis thaliana*. Proc. Natl. Acad. Sci. USA 102, 16484–16489.

Dyall, S.D., Johnson, P.J., 2000. Origins of hydrogenosomes and mitochondria: evolution and organelle biogenesis. Curr. Opin. Microbiol. 3, 404–411.

Dzierszinski, F., Nishi, M., Ouko, L., Roos, D.S., 2004. Dynamics of *Toxoplasma gondii* differentiation. Eukaryot. Cell 3, 992–1003.

Dzierszinski, F., Popescu, O., Toursel, C., Slomianny, C., Yahiaoui, B., Tomavo, S., 1999. The protozoan parasite *Toxoplasma gondii* expresses two functional plant-like glycolytic enzymes. Implications for evolutionary origin of apicomplexans. J. Biol. Chem. 274, 24888–24895.

Ellis, K.E., Clough, B., Saldanha, J.W., Wilson, R.J., 2001. Nifs and Sufs in malaria. Mol. Microbiol. 41, 973–981.

Esseiva, A.C., Naguleswaran, A., Hemphill, A., Schneider, A., 2004. Mitochondrial tRNA import in *Toxoplasma gondii*. J. Biol. Chem. 279, 42363–42368.

Fast, N.M., Kissinger, J.C., Roos, D.S., Keeling, P.J., 2001. Nuclear-encoded, plastid-targeted genes suggest a single common origin for apicomplexan and dinoflagellate plastids. Mol. Biol. Evol. 18, 418–426.

Feagin, J.E., 1992. The 6 kb element of *Plasmodium falciparum* encodes mitochondrial cytochrome genes. Mol. Biochem. Parasitol. 52, 145–148.

Feagin, J.E., 2000. Mitochondrial genome diversity in parasites. Int. J. Parasitol. 30, 371–390.

Feagin, J.E., Drew, M.E., 1995. *Plasmodium falciparum*: Alterations in organelle transcript abundance during the erythrocytic cycle. Exp. Parasitol. 80, 430–440.

Feagin, J.E., Harrell, M.I., Lee, J.C., Coe, K.J., Sands, B.H., Cannone, J.J., Tami, G., Schnare, M.N., Gutell, R.R., 2012. The fragmented mitochondrial ribosomal RNAs of *Plasmodium falciparum*. PLoS One 7, e38320.

Feagin, J.E., Mericle, B.L., Werner, E., Morris, M., 1997. Identification of additional rRNA fragments encoded by the *Plasmodium falciparum* 6 kb element. Nucleic. Acids Res. 25, 438–446.

Feagin, J.E., Werner, E., Gardner, M.J., Williamson, D.H., Wilson, R.J.M., 1992. Homologies between the contiguous and fragmented rRNAs of the two *Plasmodium falciparum* extrachromosomal DNAs are limited to core sequences. Nucleic. Acids Res. 20, 879–887.

Ferguson, D.J., Henriquez, F.L., Kirisits, M.J., Muench, S.P., Prigge, S.T., Rice, D.W., Roberts, C.W., McLeod, R.L., 2005. Maternal inheritance and stage-specific variation of the apicoplast in *Toxoplasma gondii* during development in the intermediate and definitive host. Eukaryot. Cell 4, 814–826.

Fichera, M.E., Bhopale, M.K., Roos, D.S., 1995. In vitro assays elucidate peculiar kinetics of clindamycin action against *Toxoplasma gondii*. Antimicrob. Agents Chemother. 39, 1530–1537.

Fichera, M.E., Roos, D.S., 1997. A plastid organelle as a drug target in apicomplexan parasites. Nature 390, 407–409.

Fisher, N., Bray, P.G., Ward, S.A., Biagini, G.A., 2008. Malaria-parasite mitochondrial dehydrogenases as drug targets: too early to write the obituary. Trends Parasitol. 24, 9–10.

Fleige, T., Fischer, K., Ferguson, D.J., Gross, U., Bohne, W., 2007. Carbohydrate metabolism in the *Toxoplasma gondii* apicoplast: localization of three glycolytic isoenzymes, the single pyruvate dehydrogenase complex and a plastid phosphate translocator. Eukaryot. Cell 6, 984–996.

Fleige, T., Limenitakis, J., Soldati-Favre, D., 2010. Apicoplast: keep it or leave it. Microbes Infect. 12, 253–262.

Fleige, T., Soldati-Favre, D., 2008. Targeting the transcriptional and translational machinery of the endosymbiotic organelle in apicomplexans. Curr. Drug Targets 9, 948–956.

Foth, B.J., McFadden, G.I., 2003. The apicoplast: a plastid in *Plasmodium falciparum* and other Apicomplexan parasites. Int. Rev. Cytol. 224, 57–110.

Foth, B.J., Ralph, S.A., Tonkin, C.J., Struck, N.S., Fraunholz, M., Roos, D.S., Cowman, A.F., McFadden, G.I., 2003. Dissecting apicoplast targeting in the malaria parasite *Plasmodium falciparum*. Science 299, 705–708.

Foth, B.J., Stimmler, L.M., Handman, E., Crabb, B.S., Hodder, A.N., McFadden, G.I., 2005. The malaria parasite *Plasmodium falciparum* has only one pyruvate dehydrogenase complex, which is located in the apicoplast. Mol. Microbiol. 55, 39–53.

Funes, S., Davidson, E., Reyes-Prieto, A., Magallon, S., Herion, P., King, M.P., Gonzalez-Halphen, D., 2002. A green algal apicoplast ancestor. Science 298, 2155.

Funes, S., Reyes-Prieto, A., Perez-Martinez, X., Gonzalez-Halphen, D., 2004. On the evolutionary origins of apicoplasts: revisiting the rhodophyte vs. chlorophyte controversy. Microbes Infect. 6, 305–311.

Gajadhar, A.A., Marquardt, W.C., Hall, R., Gunderson, J., Ariztia-Carmona, E.V., Sogin, M.L., 1991. Ribosomal RNA sequences of *Sarcocystis muris, Theileria annulata* and *Crypthecodinium cohnii* reveal evolutionary relationships among apicomplexans, dinoflagellates, and ciliates. Mol. Biochem. Parasitol. 45, 147–154.

Gajria, B., Bahl, A., Brestelli, J., Dommer, J., Fischer, S., Gao, X., Heiges, M., Iodice, J., Kissinger, J.C., Mackey, A.J., Pinney, D.F., Roos, D.S., Stoeckert Jr., C.J., Wang, H., Brunk, B.P., 2008. ToxoDB: an integrated *Toxoplasma gondii* database resource. Nucleic. Acids Res. 36, D553–D556.

Gallagher, J.R., Matthews, K.A., Prigge, S.T., 2011. *Plasmodium falciparum* apicoplast transit peptides are unstructured in vitro and during apicoplast import. Traffic 12, 1124–1138.

Gardner, M.J., Preiser, P., Rangachari, K., Moore, D., Feagin, J.E., Williamson, D.H., Wilson, R.J.M., 1994a. Nine duplicated tRNA genes on the plastid-like DNA of the malaria parasite *Plasmodium falciparum*. Gene 144, 307–308.

Gardner, M.J., Goldman, N., Barnett, P., Moore, P.W., Rangachari, K., Strath, M., Whyte, A., Williamson, D.H., Wilson, R.J.M., 1994b. Phylogenetic analysis of the *rpo*B gene from the plastid-like DNA of *Plasmodium falciparum*. Mol. Biochem. Parasitol. 66, 221–231.

Gardner, M.J., Bates, P.A., Ling, I.T., Moore, D.J., McCready, S., Gunasekera, M. B.R., Wilson, R.J.M., Williamson, D.H., 1988. Mitochondrial DNA of the human malarial parasite *Plasmodium falciparum*. Mol. Biochem. Parasitol. 31, 11–17.

Gardner, M.J., Bishop, R., Shah, T., de Villiers, E.P., Carlton, J.M., Hall, N., Ren, Q., Paulsen, I.T., Pain, A., Berriman, M., Wilson, R.J., Sato, S., Ralph, S.A., Mann, D.J., Xiong, Z., Shallom, S.J., Weidman, J., Jiang, L., Lynn, J., Weaver, B., Shoaibi, A., Domingo, A.R., Wasawo, D., Crabtree, J., Wortman, J.R., Haas, B., Angiuoli, S.V., Creasy, T.H., Lu, C., Suh, B., Silva, J.C., Utterback, T.R., Feldblyum, T.V., Pertea, M., Allen, J., Nierman, W.C., Taracha, E.L., Salzberg, S.L., White, O.R., Fitzhugh, H.A., Morzaria, S., Venter, J.C., Fraser, C.M., Nene, V., 2005. Genome sequence of *Theileria parva*, a bovine pathogen that transforms lymphocytes. Science 309, 134–137.

Gardner, M.J., Feagin, J.E., Moore, D.J., Rangachari, K., Williamson, D.H., Wilson, R.J.M., 1993. Sequence and organization of large subunit rRNA genes from the extrachromosomal 35 kb circular DNA of the malaria parasite *Plasmodium falciparum*. Nucleic. Acids Res. 21, 1067–1071.

Gardner, M.J., Feagin, J.E., Moore, D.J., Spencer, D.F., Gray, M.W., Williamson, D.H., Wilson, R.J.M., 1991a. Organization and expression of small subunit ribosomal RNA genes encoded by a 35-kb circular DNA in *Plasmodium falciparum*. Mol. Biochem. Parasitol. 48, 77–88.

Gardner, M.J., Williamson, D.H., Wilson, R.J.M., 1991b. A circular DNA in malaria parasites encodes an RNA polymerase like that of prokaryotes and chloroplasts. Mol. Biochem. Parasitol. 44, 115–124.

Gardner, M.J., Hall, N., Fung, E., White, O., Berriman, M., Hyman, R.W., Carlton, J.M., Pain, A., Nelson, K.E., Bowman, S., Paulsen, I.T., James, K., Eisen, J.A., Rutherford, K., Salzberg, S.L., Craig, A., Kyes, S., Chan, M.S., Nene, V., Shallom, S.J., Suh, B., Peterson, J., Angiuoli, S., Pertea, M., Allen, J., Selengut, J., Haft, D., Mather, M.W., Vaidya, A.B., Martin, D.M., Fairlamb, A.H., Fraunholz, M.J., Roos, D.S., Ralph, S.A., McFadden, G.I., Cummings, L.M., Subramanian, G.M., Mungall, C., Venter, J.C., Carucci, D.J., Hoffman, S.L., Newbold, C., Davis, R.W., Fraser, C.M., Barrell, B., 2002. Genome sequence of the human malaria parasite *Plasmodium falciparum*. Nature 419, 498–511.

Geary, T.G., Jensen, J.B., 1983. Effects of antibiotics on *Plasmodium falciparum* in vitro. Am. J. Trop. Med. Hyg. 32, 221–225.

Gibbs, S.P., 1978. The chloroplasts of *Euglena* may have evolved from symbiotic green algae. Can. J. Bot. 56, 2883–2889.

Gibbs, S.P., 1981. The chloroplast endoplasmic reticulum: structure, function, and evolutionary signfiicance. Int. Rev. Cytol. 72, 49–99.

Gillespie, D.E., Salazar, N.A., Rehkopf, D.H., Feagin, J.E., 1999. The fragmented mitochondrial ribosomal RNAs of *Plasmodium falciparum* have short A tails. Nucleic Acids Res. 27, 2416–2422.

Gillham, N.W., 1994. Organelle Genes and Genomes. Oxford University Press, New York.

Gilson, P.R., McFadden, G.I., 1996. The miniaturized nuclear genome of eukaryotic endosymbiont contains genes that overlap, genes that are cotranscribed, and the smallest known spliceosomal introns. Proc. Natl. Acad. Sci. USA 93, 7737–7742.

Gilson, P.R., McFadden, G.I., 1997. Good things in small packages: the tiny genomes of chlorarachniophyte endosymbionts. BioEssays 19, 167–173.

Gilson, P.R., Su, V., Slamovits, C.H., Reith, M.E., Keeling, P.J., McFadden, G.I., 2006. Complete nucleotide sequence of the chlorarachniophyte nucleomorph: nature's smallest nucleus. Proc. Natl. Acad. Sci. USA 103, 9566–9571.

Goldberg, A.V., Molik, S., Tsaousis, A.D., Neumann, K., Kuhnke, G., Delbac, F., Vivares, C.P., Hirt, R.P., Lill, R., Embley, T.M., 2008. Localization and functionality of microsporidian iron-sulphur cluster assembly proteins. Nature 452, 624–628.

Goodman, C.D., Su, V., McFadden, G.I., 2007. The effects of anti-bacterials on the malaria parasite *Plasmodium falciparum*. Mol. Biochem. Parasitol. 152, 181–191.

Gray, M.W., 2012. Mitochondrial evolution. Cold Spring Harb. Perspect. Biol. 4.

Gray, M.W., Lang, B.F., 1998. Transcription in chloroplasts and mitochondria: a tale of two polymerases. Trends Microbiol. 6, 1–3.

Gray, M.W., Lang, B.F., Burger, G., 2004. Mitochondria of protists. Annu. Rev. Genet. 38, 477–524.

Gray, M.W., Schnare, M.N., 1990. Evolution of the modular structure of rRNA. In: Hill, W.E., Dahlberg, A., Garrett, R.A., Moore, P.B., Schlessinger, D., Warner, J.R. (Eds.), The Ribosome: Structure, Function, & Evolution. American Society for Microbiology, Washington, D.C., pp. 589–597.

Greenwood, A.D., Griffiths, H.B., Santore, U.U., 1977. Chloroplasts and cell compartments in Cryptophyceae. Br. Phycol. J. 12, 119.

Gross, U., Bohne, W., Soête, M., Dubremetz, J.F., 1996. Developmental differentiation between tachyzoites and bradyzoites of *Toxoplasma gondii*. Parasitol. Today 12, 30–33.

Günther, S., Wallace, L., Patzewitz, E.M., McMillan, P.J., Storm, J., Wrenger, C., Bissett, R., Smith, T.K., Muller, S., 2007. Apicoplast lipoic acid protein ligase B is not essential for *Plasmodium falciparum*. PLoS Pathog. 3, e189.

Hager, K.M., Striepen, B., Tilney, L.G., Roos, D.S., 1999. The nuclear envelope serves as an intermediary between the ER and Golgi complex in the intracellular parasite *Toxoplasma gondii*. J. Cell Sci. 112, 2631–2638.

Hall, R., Coggins, L., McKellar, S., Shiels, B., Tait, A., 1990. Characterisation of an extrachromosomal DNA element from *Theileria annulata*. Mol. Biochem. Parasitol. 38, 253–260.

Harb, O.S., Chatterjee, B., Fraunholz, M.J., Crawford, M.J., Nishi, M., Roos, D.S., 2004. Multiple functionally redundant signals mediate targeting to the apicoplast in the apicomplexan parasite *Toxoplasma gondii*. Eukaryot. Cell 3, 663–674.

Haucke, V., Schatz, G., 1997. Import of proteins into mitochondria and chloroplasts. Trends Cell Biol. 7, 103–106.

He, C.Y., Shaw, M.K., Pletcher, C.H., Striepen, B., Tilney, L.G., Roos, D.S., 2001a. A plastid segregation defect in the protozoan parasite *Toxoplasma gondii*. EMBO J. 20, 330–339.

He, C.Y., Striepen, B., Pletcher, C.H., Murray, J.M., Roos, D.S., 2001b. Targeting and processing of nuclear-encoded apicoplast proteins in plastid segregation mutants of *Toxoplasma gondii*. J. Biol. Chem. 276, 28436–28442.

Hermann, T., 2005. Drugs targeting the ribosome. Curr. Opin. Struct. Biol. 15, 355–366.

Herzig, S., Raemy, E., Montessuit, S., Veuthey, J.L., Zamboni, N., Westermann, B., Kunji, E.R., Martinou, J.C., 2012. Identification and functional expression of the mitochondrial pyruvate carrier. Science 337, 93–96.

Hiltunen, J.K., Autio, K.J., Schonauer, M.S., Kursu, V.A., Dieckmann, C.L., Kastaniotis, A.J., 2010. Mitochondrial fatty acid synthesis and respiration. Biochim. Biophys. Acta 1797, 1195–1202.

Hirakawa, Y., Burki, F., Keeling, P.J., 2011. Nucleus- and nucleomorph-targeted histone proteins in a chlorarachniophyte alga. Mol. Microbiol. 80, 1439–1449.

Hirakawa, Y., Burki, F., Keeling, P.J., 2012. Genome-based reconstruction of the protein import machinery in the secondary plastid of a chlorarachniophyte alga. Eukaryot. Cell 11, 324–333.

Hofmann, N.R., Theg, S.M., 2005. Chloroplast outer membrane protein targeting and insertion. Trends Plant Sci. 10, 450–457.

Hong, W., 2005. SNAREs and traffic. Biochim. Biophys. Acta 1744, 493–517.

Hopkins, A.L., Groom, C.R., 2002. The druggable genome. Nat. Rev. Drug Discov. 1, 727–730.

Hopkins, J., Fowler, R., Krishna, S., Wilson, I., Mitchell, G., Bannister, L., 1999. The plastid in *Plasmodium falciparum* asexual blood stages: a three-dimensional ultrastructural analysis. Protist 150, 283–295.

Hortua Triana, M.A., Huynh, M.H., Garavito, M.F., Fox, B.A., Bzik, D.J., Carruthers, V.B., Loffler, M., Zimmermann, B.H., 2012. Biochemical and molecular characterization of the pyrimidine biosynthetic enzyme dihydroorotate dehydrogenase from *Toxoplasma gondii*. Mol. Biochem. Parasitol. 184, 71–81.

Howe, C.J., 1992. Plastid origin of an extrachromosomal DNA molecule from *Plasmodium*, the causative agent of malaria. J. Theor. Biol. 158, 199–205.

Howe, C.J., Purton, S., 2007. The little genome of apicomplexan plastids: its raison d'etre and a possible explanation for the 'delayed death' phenomenon. Protist 158, 121–133.

Jackson, K.E., Pham, J.S., Kwek, M., De Silva, N.S., Allen, S.M., Goodman, C.D., McFadden, G.I., de Pouplana, L.R., Ralph, S.A., 2012. Dual targeting of aminoacyl-tRNA synthetases to the apicoplast and cytosol in *Plasmodium falciparum*. Int. J. Parasitol. 42, 177–186.

Janouskovec, J., Horak, A., Obornik, M., Lukes, J., Keeling, P.J., 2010. A common red algal origin of the apicomplexan, dinoflagellate, and heterokont plastids. Proc. Natl. Acad. Sci. USA 107, 10949–10954.

Jarvis, P., Soll, J., 2002. Toc, tic, and chloroplast protein import. Biochim. Biophys. Acta 1590, 177–189.

Jedelský, P.L., Doležal, P., Rada, P., Pyrih, J., Smíd, O., Hrdý, I., Sedinová, M., Marcinčiková, M., Voleman, L., Perry, A.J., Beltrán, N.C., Lithgow, T., Tachezy, J., 2011. The minimal proteome in the reduced mitochondrion of the parasitic protist *Giardia intestinalis*. PLoS ONE 6, e17285.

Jelenska, J., Crawford, M.J., Harb, O.S., Zuther, E., Haselkorn, R., Roos, D.S., Gornicki, P., 2001. Subcellular localization of acetyl-CoA carboxylase in the apicomplexan parasite *Toxoplasma gondii*. Proc. Natl. Acad. Sci. USA 98, 2723–2728.

Jelenska, J., Sirikhachornkit, A., Haselkorn, R., Gornicki, P., 2002. The carboxyltransferase activity of the apicoplast acetyl-CoA carboxylase of *Toxoplasma gondii* is the target of aryloxyphenoxypropionate inhibitors. J. Biol. Chem. 277, 23208–23215.

Ji, Y., Mericle, B.L., Rehkopf, D.H., Anderson, J.D., Feagin, J.E., 1996. The 6 kb element of *Plasmodium falciparum* is polycistronically transcribed. Mol. Biochem. Parasitol. 81, 211–223.

Jomaa, H., Wiesner, J., Sanderbrand, S., Altincicek, B., Weidermeyer, C., Hintz, M., Turbachova, I., Eberl, M., Zeidler, J., Lichtenthaler, H.K., Soldati, D., Beck, E., 1999. Inhibitors of the nonmevalonate pathway of isoprenoid biosynthesis as antimalarial drugs. Science 285, 1573–1578.

Jongwutiwes, S., Putaporntip, C., Iwasaki, T., Ferreira, M.U., Kanbara, H., Hughes, A.L., 2005. Mitochondrial genome sequences support ancient population expansion in *Plasmodium vivax*. Mol. Biol. Evol. 22, 1733–1739.

Joseph, J.T., 1990. The mitochondrial microgenome of malaria parasites. Harvard Medical School, Boston. Ph.D. thesis.

Joseph, J.T., Aldritt, S.M., Unnasch, T., Puijalon, O., Wirth, D.F., 1989. Characterization of a conserved extrachromosomal element isolated from the avian malarial parasite *Plasmodium gallinaceum*. Mol. Cell. Biol. 9, 3621–3629.

Joy, D.A., Feng, X., Mu, J., Furuya, T., Chotivanich, K., Krettli, A.U., Ho, M., Wang, A., White, N.J., Suh, E., Beerli, P., Su, X.Z., 2003. Early origin and recent expansion of *Plasmodium falciparum*. Science 300, 318–321.

Kairo, A., Fairlamb, A.H., Gobright, E., Nene, V., 1994. A 7.1 kb linear DNA molecule of *Theileria parva* has scrambled rDNA sequences and open reading frames for mitochondrially-encoded proteins. EMBO J. 13, 898–905.

Kalanon, M., Tonkin, C.J., McFadden, G.I., 2009. Characterization of two putative protein translocation components in the apicoplast of *Plasmodium falciparum*. Eukaryot. Cell 8, 1146–1154.

Karnataki, A., DeRocher, A., Coppens, I., Nash, C., Feagin, J.E., Parsons, M., 2007a. Cell cycle-regulated vesicular trafficking of *Toxoplasma* APT1, a protein localized to multiple apicoplast membranes. Mol. Microbiol. 63, 1653–1668.

Karnataki, A., DeRocher, A.E., Coppens, I., Feagin, J.E., Parsons, M., 2007b. A membrane protease is targeted to the relict plastid of *toxoplasma* via an internal signal sequence. Traffic 8, 1543–1553.

Karniely, S., Pines, O., 2005. Single translation–dual destination: mechanisms of dual protein targeting in eukaryotes. EMBO Rep. 6, 420–425.

Keeling, P.J., 2010. The endosymbiotic origin, diversification and fate of plastids. Philos. Trans. R. Soc. B 365, 729–748.

Kehr, S., Sturm, N., Rahlfs, S., Przyborski, J.M., Becker, K., 2010. Compartmentation of redox metabolism in malaria parasites. PLoS Pathog. 6, e1001242.

Keithly, J.S., Langreth, S.G., Buttle, K.F., Mannella, C.A., 2005. Electron tomographic and ultrastructural analysis of the *Cryptosporidium parvum* relict mitochondrion, its associated membranes, and organelles. J. Eukaryot. Microbiol. 52, 132–140.

Keithly, J.S., Zhu, G., Upton, S.J., Woods, K.M., Martinez, M.P., Yarlett, N., 1997. Polyamine biosynthesis in *Cryptosporidium parvum* and its implications for chemotherapy. Mol. Biochem. Parasitol. 88, 35–42.

Kennedy, S.R., Chen, C.Y., Schmitt, M.W., Bower, C.N., Loeb, L.A., 2011. The biochemistry and fidelity of synthesis by the apicoplast genome replication DNA polymerase Pfprex from the malaria parasite *Plasmodium falciparum*. J. Mol. Biol. 410, 27–38.

Khor, V., Yowell, C., Dame, J.B., Rowe, T.C., 2005. Expression and characterization of the ATP-binding domain of a malarial *Plasmodium vivax* gene homologous to the B-subunit of the bacterial topoisomerase DNA gyrase. Mol. Biochem. Parasitol. 140, 107–117.

Kiatfuengfoo, R., Suthiphongchai, T., Prapunwattana, P., Yuthavong, Y., 1989. Mitochondria as the site of action of tetracycline on *Plasmodium falciparum*. Mol. Biochem. Parasitol. 34, 109–115.

Kiessling, J., Martin, A., Gremillon, L., Rensing, S.A., Nick, P., Sarnighausen, E., Decker, E.L., Reski, R., 2004. Dual targeting of plastid division protein FtsZ to chloroplasts and the cytoplasm. EMBO Rep. 5, 889–894.

Kikuchi, G., Motokawa, Y., Yoshida, T., Hiraga, K., 2008. Glycine cleavage system: reaction mechanism, physiological significance, and hyperglycinemia. Proc. Jpn. Acad. Ser. B Phys. Biol. Sci. 84, 246–263.

Kilejian, A., 1975. Circular mitochondrial DNA from the avian malarial parasite *Plasmodium lophurae*. Biochim. Biophys. Acta 390, 276–284.

Kissinger, J.C., Brunk, B.P., Crabtree, J., Fraunholz, M.J., Gajria, B., Milgram, A.J., Pearson, D.S., Schug, J., Bahl, A., Diskin, S.J., Ginsburg, H., Grant, G.R., Gupta, D., Labo, P., Li, L., Mailman, M.D., McWeeney, S.K., Whetzel, P., Stoeckert, C.J., Roos, D.S., 2002. The *Plasmodium* genome database. Nature 419, 490–492.

Köhler, S., Delwiche, C.F., Denny, P.W., Tilney, L.G., Webster, P., Wilson, R.J., Palmer, J.D., Roos, D.S., 1997. A plastid of probable green algal origin in Apicomplexan parasites. Science 275, 1485–1489.

Koreny, L., Sobotka, R., Janouskovec, J., Keeling, P.J., Obornik, M., 2011. Tetrapyrrole synthesis of photosynthetic chromerids is likely homologous to the unusual pathway of apicomplexan parasites. Plant Cell. 23, 3454–3462.

Kouranov, A., Wang, H., Schnell, D.J., 1999. Tic22 is targeted to the intermembrane space of chloroplasts by a novel pathway. J. Biol. Chem. 274, 25181–25186.

Kudryashev, M., Lepper, S., Stanway, R., Bohn, S., Baumeister, W., Cyrklaff, M., Frischknecht, F., 2010. Positioning of large organelles by a membrane-associated cytoskeleton in Plasmodium sporozoites. Cell. Microbiol. 12, 362–371.

Kumar, A., Tanveer, A., Biswas, S., Ram, E.V., Gupta, A., Kumar, B., Habib, S., 2010. Nuclear-encoded DnaJ homologue of *Plasmodium falciparum* interacts with replication *ori* of the apicoplast genome. Mol. Microbiol. 75, 942–956.

Kumar, B., Chaubey, S., Shah, P., Tanveer, A., Charan, M., Siddiqi, M.I., Habib, S., 2011. Interaction between sulphur mobilisation proteins SufB and SufC: evidence for an iron-sulphur cluster biogenesis pathway in the apicoplast of *Plasmodium falciparum*. Int. J. Parasitol. 41, 991–999.

LaGier, M.J., Tachezy, J., Stejskal, F., Kutisova, K., Keithly, J.S., 2003. Mitochondrial-type iron-sulfur cluster biosynthesis genes (IscS and IscU) in the apicomplexan *Cryptosporidium parvum*. Microbiology 149, 3519–3530.

Lang, B.F., Gray, M.W., Burger, G., 1999. Mitochondrial genome evolution and the origin of eukaryotes. Annu. Rev. Genet. 33, 351–397.

Lang-Unnasch, N., Aiello, D.P., 1999. Sequence evidence for an altered genetic code in the *Neospora caninum* plastid. Int. J. Parasitol. 29, 1557–1562.

Lau, A.O., McElwain, T.F., Brayton, K.A., Knowles, D.P., Roalson, E.H., 2009. *Babesia bovis*: A comprehensive phylogenetic analysis of plastid-encoded genes supports green algal origin of apicoplasts. Exp. Parasitol. 123, 236–243.

Le Roch, K.G., Zhou, Y., Blair, P.L., Grainger, M., Moch, J.K., Haynes, J.D., de la Vega, P., Holder, A.A., Batalov, S., Carucci, D.J., Winzeler, E.A., 2003. Discovery of gene function by expression profiling of the malaria parasite life cycle. Science 301, 1503–1508.

Lee, R.E., 1977. Evolution of algal flagellates with chloroplast endoplasmic reticulum from the ciliates. S. Afr. J. Sci. 73, 179–182.

Levitan, A., Trebitsh, T., Kiss, V., Pereg, Y., Dangoor, I., Danon, A., 2005. Dual targeting of the protein disulfide isomerase RB60 to the chloroplast and the endoplasmic reticulum. Proc. Natl. Acad. Sci. USA 102, 6225–6230.

Li, L., Stoeckert Jr., C.J., Roos, D.S., 2003. OrthoMCL: identification of ortholog groups for eukaryotic genomes. Genome Res. 13, 2178–2189.

Liere, K., Weihe, A., Borner, T., 2011. The transcription machineries of plant mitochondria and chloroplasts: Composition, function, and regulation. J. Plant Physiol. 168, 1345–1360.

Lill, R., 2009. Function and biogenesis of iron-sulphur proteins. Nature 460, 831–838.

Lill, R., Hoffmann, B., Molik, S., Pierik, A.J., Rietzschel, N., Stehling, O., Uzarska, M.A., Webert, H., Wilbrecht, C., Muhlenhoff, U., 2012. The role of mitochondria in cellular iron-sulfur protein biogenesis and iron metabolism. Biochim. Biophys. Acta 1823, 1491–1508.

Lill, R., Muhlenhoff, U., 2008. Maturation of iron-sulfur proteins in eukaryotes: mechanisms, connected processes, and diseases. Annu. Rev. Biochem. 77, 669–700, 669-700.

Lim, L., Linka, M., Mullin, K.A., Weber, A.P., McFadden, G.I., 2010. The carbon and energy sources of the non-photosynthetic plastid in the malaria parasite. FEBS Lett. 584, 549–554.

Lim, L., McFadden, G.I., 2010. The evolution, metabolism and functions of the apicoplast. Philos. Trans. R. Soc. B. 365, 749–763.

Limenitakis, J., Oppenheim, R.D., Creek, D.J., Foth, B.J., Barrett, M.P., Soldati-Favre, D., 2013. The 2-methylcitrate cycle is implicated in the detoxification of propionate in *Toxoplasma gondii*. Mol Microbiol. 87, 894–908.

Lin, S.S., Gross, U., Bohne, W., 2009. Type II NADH dehydrogenase inhibitor 1-hydroxy-2-dodecyl-4(1H)quinolone leads to collapse of mitochondrial inner-membrane potential and ATP depletion in *Toxoplasma gondii*. Eukaryot. Cell 8, 877–887.

Lin, S.S., Gross, U., Bohne, W., 2011. Two internal type II NADH dehydrogenases of *Toxoplasma gondii* are both required for optimal tachyzoite growth. Mol. Microbiol. 82, 209–221.

Lindner, S.E., Llinas, M., Keck, J.L., Kappe, S.H., 2011. The primase domain of PfPrex is a proteolytically matured, essential enzyme of the apicoplast. Mol. Biochem. Parasitol. 180, 69–75.

Ling, Y., Li, Z.H., Miranda, K., Oldfield, E., Moreno, S.N., 2007. The farnesyl-diphosphate/geranylgeranyl-diphosphate synthase of *Toxoplasma gondii* is a bifunctional enzyme and a molecular target of bisphosphonates. J. Biol. Chem. 282, 30804–30816.

Ling, Y., Sahota, G., Odeh, S., Chan, J.M., Araujo, F.G., Moreno, S.N., Oldfield, E., 2005. Bisphosphonate inhibitors of *Toxoplasma gondi* growth: in vitro, QSAR, and in vivo investigations. J. Med. Chem. 48, 3130–3140.

Lizundia, R., Werling, D., Langsley, G., Ralph, S.A., 2008. The Theileria apicoplast as a target for chemotherapy. Antimicrob. Agents Chemother. 53, 1213–1217.

Loffelhardt, W., 2011. The chlorarachniophyte nucleomorph is supplemented with host cell nucleus-encoded histones. Mol. Microbiol. 80, 1413–1416.

Lorestani, A., Sheiner, L., Yang, K., Robertson, S.D., Sahoo, N., Brooks, C.F., Ferguson, D.J., Striepen, B., Gubbels, M.J., 2010. A *Toxoplasma* MORN1 null mutant undergoes repeated divisions but is defective in basal assembly, apicoplast division and cytokinesis. PLoS ONE 5, e12302.

Lun, Z.R., Lai, D.H., Li, F.J., Lukes, J., Ayala, F.J., 2010. *Trypanosoma brucei*: two steps to spread out from Africa. Trends Parasitol. 26, 424–427.

Ma, C., Agrawal, G., Subramani, S., 2011. Peroxisome assembly: matrix and membrane protein biogenesis. J. Cell. Biol. 193, 7–16.

Mácašev, D., Whelan, J., Newbigin, E., Silva-Filho, M.C., Mulhern, T.D., Lithgow, T., 2004. Tom22′, an 8-kDa trans-site receptor in plants and protozoans, is a conserved feature of the TOM complex that appeared early in the evolution of eukaryotes. Mol. Biol. Evol. 21, 1557–1564.

Marechal, E., Azzouz, N., de Macedo, C.S., Block, M.A., Feagin, J.E., Schwarz, R. T., Joyard, J., 2002. Synthesis of chloroplast galactolipids in apicomplexan parasites. Eukaryot. Cell 1, 653–656.

Margulis, L., 1970. Origin of Eukaryotic Cells. Yale University Press, New Haven.

Martin, N.C., Hopper, A.K., 1994. How single genes provide tRNA processing enzymes to mitochondria, nuclei and the cytosol. Biochimie 76, 1161–1167.

Mather, M.W., Henry, K.W., Vaidya, A.B., 2007. Mitochondrial drug targets in apicomplexan parasites. Curr. Drug Targets 8, 49–60.

Matsuzaki, M., Kikuchi, T., Kita, K., Kojima, S., Kuroiwa, T., 2001. Large amounts of apicoplast nucleoid DNA and its segregation in *Toxoplasma gondii*. Protoplasma 218, 180–191.

Mayinger, P., 2012. Phosphoinositides and vesicular membrane traffic. Biochim. Biophys. Acta 1821, 1104–1113.

Mazumdar, J., Striepen, B., 2007. Make it or take it: fatty acid metabolism of apicomplexan parasites. Eukaryot. Cell 6, 1727–1735.

Mazumdar, J., Wilson, H., Masek, K., Hunter, A., Striepen, B., 2006. Apicoplast fatty acid synthesis is essential for organelle biogenesis and parasite survival in *Toxoplasma gondii*. Proc. Natl. Acad. Sci. USA 103, 13192–13197.

McConkey, G.A., Rogers, M.J., McCutchan, T.F., 1997. Inhibition of protein synthesis. Targeting the plastid-like organelle with thiostrepton. J. Biol. Chem. 272, 2046–2049.

McConkey, G.A., Pinney, J.W., Westhead, D.R., Plueckhahn, K., Fitzpatrick, T.B., Macheroux, P., Kappes, B., 2004. Annotating the *Plasmodium* genome and the enigma of the shikimate pathway. Trends Parasitol. 20, 60–65.

McFadden, D.C., Tomavo, S., Berry, E.A., Boothroyd, J.C., 2000. Characterization of cytochrome *b* from *Toxoplasma gondii* and Q(o) domain mutations as a mechanism of atovaquone-resistance. Mol. Biochem. Parasitol. 108, 1–12.

McFadden, G.I., 2011. The apicoplast. Protoplasma 248, 641–650.

McFadden, G.I., Reith, M.E., Munholland, J., Lang-Unnasch, N., 1996. Plastid in human parasites. Nature 381, 482.

McFadden, G.I., Roos, D.S., 1999. Apicomplexan plastids as drug targets. Trends Microbiol. 7, 328–333.

McFadden, G.I., van Dooren, G.G., 2004. Evolution: red algal genome affirms a common origin of all plastids. Curr. Biol. 14, R514–R516.

McMillan, P.J., Stimmler, L.M., Foth, B.J., McFadden, G.I., Muller, S., 2005. The human malaria parasite *Plasmodium falciparum* possesses two distinct dihydrolipoamide dehydrogenases. Mol. Microbiol. 55, 27–38.

Moore, C.E., Archibald, J.M., 2009. Nucleomorph genomes. Annu. Rev. Genet. 43, 251–264.

Moore, R.B., Obornik, M., Janouskovec, J., Chrudimsky, T., Vancova, M., Green, D.H., Wright, S.W., Davies, N.W., Bolch, C.J., Heimann, K., Slapeta, J., Hoegh-Guldberg, O., Logsdon, J.M., Carter, D.A., 2008. A photosynthetic alveolate closely related to apicomplexan parasites. Nature. 451, 959–963.

Mu, J., Joy, D.A., Duan, J., Huang, Y., Carlton, J., Walker, J., Barnwell, J., Beerli, P., Charleston, M.A., Pybus, O.G., Su, X.Z., 2005. Host switch leads to emergence of *Plasmodium vivax* malaria in humans. Mol. Biol. Evol. 22, 1686–1693.

Mukhopadhyay, A., Chen, C.Y., Doerig, C., Henriquez, F.L., Roberts, C.W., Barrett, M.P., 2009. The *Toxoplasma gondii* plastid replication and repair enzyme complex, PREX. Parasitology 136, 747–755.

Mullin, K.A., Lim, L., Ralph, S.A., Spurck, T.P., Handman, E., McFadden, G.I., 2006. Membrane transporters in the relict plastid of malaria parasites. Proc. Natl. Acad. Sci. USA 103, 9572–9577.

Nachin, L., El Hassouni, M., Loiseau, L., Expert, D., Barras, F., 2001. SoxR-dependent response to oxidative stress and virulence of *Erwinia chrysanthemi*: the key role of SufC, an orphan ABC ATPase. Mol. Microbiol. 39, 960–972.

Nachin, L., Loiseau, L., Expert, D., Barras, F., 2003. SufC: an unorthodox cytoplasmic ABC/ATPase required for [Fe-S] biogenesis under oxidative stress. EMBO J. 22, 427–437.

Nagaraj, V.A., Arumugam, R., Chandra, N.R., Prasad, D., Rangarajan, P.N., Padmanaban, G., 2009a. Localisation of *Plasmodium falciparum* uroporphyrinogen III decarboxylase of the heme-biosynthetic pathway in the apicoplast and characterisation of its catalytic properties. Int. J. Parasitol. 39, 559–568.

Nagaraj, V.A., Arumugam, R., Gopalakrishnan, B., Jyothsna, Y.S., Rangarajan, P.N., Padmanaban, G., 2008. Unique properties of *Plasmodium falciparum* porphobilinogen deaminase. J. Biol. Chem. 283, 437–444.

Nagaraj, V.A., Prasad, D., Arumugam, R., Rangarajan, P.N., Padmanaban, G., 2010. Characterization of coproporphyrinogen III oxidase in *Plasmodium falciparum* cytosol. Parasitol. Int. 59, 121–127.

Nagaraj, V.A., Prasad, D., Rangarajan, P.N., Padmanaban, G., 2009b. Mitochondrial localization of functional ferrochelatase from *Plasmodium falciparum*. Mol. Biochem. Parasitol. 168, 109–112.

Nair, S.C., Brooks, C.F., Goodman, C.D., Strurm, A., McFadden, G.I., Sundriyal, S., Anglin, J.L., Song, Y., Moreno, S.N., Striepen, B., 2011. Apicoplast isoprenoid precursor synthesis and the molecular basis of fosmidomycin resistance in *Toxoplasma gondii*. J. Exp. Med. 208, 1547–1559.

Nash, E.A., Barbrook, A.C., Edwards-Stuart, R.K., Bernhardt, K., Howe, C.J., Nisbet, R.E., 2007. Organization of the mitochondrial genome in the dinoflagellate *Amphidinium carterae*. Mol. Biol. Evol. 24, 1528–1536.

Nassoury, N., Wang, Y., Morse, D., 2005. Brefeldin a inhibits circadian remodeling of chloroplast structure in the dinoflagellate *Gonyaulax*. Traffic 6, 548–561.

Navarro-Sastre, A., Tort, F., Stehling, O., Uzarska, M.A., Arranz, J.A., Del, T.M., Labayru, M.T., Landa, J., Font, A., Garcia-Villoria, J., Merinero, B., Ugarte, M., Gutierrez-Solana, L.G., Campistol, J., Garcia-Cazorla, A., Vaquerizo, J., Riudor, E., Briones, P., Elpeleg, O., Ribes, A., Lill, R., 2011. A fatal mitochondrial disease is associated with defective NFU1 function in the maturation of a subset of mitochondrial Fe-S proteins. Am. J. Hum. Genet. 89, 656–667.

Netz, D.J., Stith, C.M., Stumpfig, M., Kopf, G., Vogel, D., Genau, H.M., Stodola, J.L., Lill, R., Burgers, P.M., Pierik, A.J., 2011. Eukaryotic DNA polymerases require an iron-sulfur cluster for the formation of active complexes. Nat. Chem. Biol. 8, 125–132.

Nishi, M., Hu, K., Murray, J.M., Roos, D.S., 2008. Organellar dynamics during the cell cycle of *Toxoplasma gondii*. J. Cell Sci. 121, 1559–1568.

Norman, J.E., 2001. Mitochondrial genome organization, expression and evolution in the dinoflagellate *Crypthecodinium cohnii*. Dalhousie University. Dissertation.

Obornik, M., Green, B.R., 2005. Mosaic origin of the heme biosynthesis pathway in photosynthetic eukaryotes. Mol. Biol. Evol. 22, 2343–2353.

Obornik, M., Janouskovec, J., Chrudimsky, T., Lukes, J., 2008. Evolution of the apicoplast and its hosts: From heterotrophy to autotrophy and back again. Int. J. Parasitol. 39, 1–12.

Okada, K., 2009. The novel heme oxygenase-like protein from *Plasmodium falciparum* converts heme to bilirubin IXα in the apicoplast. FEBS Lett. 583, 313–319.

Okada, K., Hase, T., 2005. Cyanobacterial non-mevalonate pathway: (E)-4-hydroxy-3-methylbut-2-enyl diphosphate synthase interacts with ferredoxin in *Thermosynechococcus elongatus* BP-1. J. Biol. Chem. 280, 20672–20679.

Olliaro, P., Gorini, G., Jabes, D., Regazzetti, A., Rossi, R., Marchetti, A., Tinelli, C., Della, B.C., 1994. In-vitro and in-vivo activity of rifabutin against *Toxoplasma gondii*. J. Antimicrob. Chemother. 34, 649–657.

Olszewski, K.L., Mather, M.W., Morrisey, J.M., Garcia, B.A., Vaidya, A.B., Rabinowitz, J.D., Llinás, M., 2010. Branched tricarboxylic acid metabolism in *Plasmodium falciparum*. Nature 466, 774–778.

Ossorio, P.N., Sibley, L.D., Boothroyd, J.C., 1991. Mitochondrial-like DNA sequences flanked by direct and inverted repeats in the nuclear genome of *Toxoplasma gondii*. J. Mol. Biol. 222, 525–536.

Owen, D.J., Evans, P.R., 1998. A structural explanation for the recognition of tyrosine-based endocytotic signals. Science 282, 1327–1332.

Painter, H.J., Morrisey, J.M., Mather, M.W., Vaidya, A.B., 2007. Specific role of mitochondrial electron transport in blood-stage *Plasmodium falciparum*. Nature 446, 88–91.

Palmer, J.D., 1992. A degenerate plastid genome in malaria parasites? Curr. Biol. 2, 318–320.

Parsons, M., Karnataki, A., Feagin, J.E., DeRocher, A., 2007. Protein trafficking to the apicoplast: deciphering the apicomplexan solution to secondary endosymbiosis. Eukaryot. Cell 6, 1081–1088.

Pastore, C., Adinolfi, S., Huynen, M.A., Rybin, V., Martin, S., Mayer, M., Bukau, B., Pastore, A., 2006. YfhJ, a molecular adaptor in iron-sulfur cluster formation or a frataxin-like protein? Structure 14, 857–867.

Patel, S.N., Kain, K.C., 2005. Atovaquone/proguanil for the prophylaxis and treatment of malaria. Expert. Rev. Anti. Infect. Ther. 3, 849–861.

Peeters, N., Small, I., 2001. Dual targeting to mitochondria and chloroplasts. Biochim. Biophys. Acta 1541, 54–63.

Perham, R.N., 2000. Swinging arms and swinging domains in multifunctional enzymes: catalytic machines for multistep reactions. Annu. Rev. Biochem. 69, 961–1004, 961-1004.

Petrova, V.Y., Drescher, D., Kujumdzieva, A.V., Schmitt, M.J., 2004. Dual targeting of yeast catalase A to peroxisomes and mitochondria. Biochem. J. 380, 393–400.

Pfefferkorn, E.R., Borotz, S.E., 1994. Comparison of mutants of *Toxoplasma gondii* selected for resistance to azithromycin, spiramycin, or clindamycin. Antimicrob. Agents Chemother. 38, 31–37.

Pino, P., Aeby, E., Foth, B.J., Sheiner, L., Soldati, T., Schneider, A., Soldati-Favre, D., 2010. Mitochondrial translation in absence of local tRNA aminoacylation and methionyl tRNA Met formylation in Apicomplexa. Mol. Microbiol. 76, 706–718.

Pino, P., Foth, B.J., Kwok, L.Y., Sheiner, L., Schepers, R., Soldati, T., Soldati-Favre, D., 2007. Dual targeting of antioxidant and metabolic enzymes to the mitochondrion and the apicoplast of *Toxoplasma gondii*. PLoS Pathog. 3, e115.

Polonais, V., Soldati-Favre, D., 2010. Versatility in the acquisition of energy and carbon sources by the Apicomplexa. Biol. Cell 102, 435–445.

Preiser, P., Williamson, D.H., Wilson, R.J., 1995. tRNA genes transcribed from the plastid-like DNA of *Plasmodium falciparum*. Nucleic Acids Res. 23, 4329–4336.

Preiser, P.R., Wilson, R.J., Moore, P.W., McCready, S., Hajibagheri, M.A., Blight, K.J., Strath, M., Williamson, D.H., 1996. Recombination associated with replication of malarial mitochondrial DNA. EMBO J. 15, 684–693.

Prusty, D., Dar, A., Priya, R., Sharma, A., Dana, S., Choudhury, N.R., Rao, N.S., Dhar, S.K., 2010. Single-stranded DNA binding protein from human malarial parasite *Plasmodium falciparum* is encoded in the nucleus and targeted to the apicoplast. Nucleic Acids Res. 38, 7037–7053.

Pukrittayakamee, S., Viravan, C., Charoenlarp, P., Yeamput, C., Wilson, R.J., White, N.J., 1994. Antimalarial effects of rifampin in *Plasmodium vivax* malaria. Antimicrob. Agents Chemother. 38, 511–514.

Raabe, C.A., Sanchez, C.P., Randau, G., Robeck, T., Skryabin, B.V., Chinni, S.V., Kube, M., Reinhardt, R., Ng, G.H., Manickam, R., Kuryshev, V.Y., Lanzer, M., Brosius, J., Tang, T.H., Rozhdestvensky, T.S., 2010. A global view of the nonprotein-coding transcriptome in *Plasmodium falciparum*. Nucleic Acids Res. 38, 608–617.

Radke, J.R., Behnke, M.S., Mackey, A.J., Radke, J.B., Roos, D.S., White, M.W., 2005. The transcriptome of *Toxoplasma gondii*. BMC Biol. 3, 26.

Raghu Ram, E.V., Kumar, A., Biswas, S., Kumar, A., Chaubey, S., Siddiqi, M.I., Habib, S., 2007. Nuclear gyrB encodes a functional subunit of the *Plasmodium falciparum* gyrase that is involved in apicoplast DNA replication. Mol. Biochem. Parasitol. 154, 30–39.

Ralph, S.A., Foth, B.J., Hall, N., McFadden, G.I., 2004a. Evolutionary pressures on apicoplast transit peptides. Mol. Biol. Evol. 21, 2183–2194.

Ralph, S.A., van Dooren, G.G., Waller, R.F., Crawford, M.J., Fraunholz, M.J., Foth, B.J., Tonkin, C.J., Roos, D.S., McFadden, G.I., 2004b. Tropical infectious diseases: metabolic maps and functions of the *Plasmodium falciparum* apicoplast. Nat. Rev. Microbiol. 2, 203–216.

Ram, E.V., Naik, R., Ganguli, M., Habib, S., 2008. DNA organization by the apicoplast-targeted bacterial histone-like protein of *Plasmodium falciparum*. Nucleic Acids Res. 36, 5061–5073.

Ramakrishnan, S., Docampo, M.D., Macrae, J.I., Pujol, F.M., Brooks, C.F., van Dooren, G.G., Hiltunen, J.K., Kastaniotis, A.J., McConville, M.J., Striepen, B., 2012. The apicoplast and endoplasmic reticulum cooperate in fatty acid biosynthesis in the apicomplexan parasite *Toxoplasma gondii*. J. Biol. Chem. 287, 4957–4971.

Rangachari, K., Davis, C.T., Eccleston, J.F., Hirst, E.M., Saldanha, J.W., Strath, M., Wilson, R.J., 2002. SufC hydrolyzes ATP and interacts with SufB from *Thermotoga maritima*. FEBS Lett. 514, 225–228.

Rathore, D., Wahl, A.M., Sullivan, M., McCutchan, T.F., 2001. A phylogenetic comparison of gene trees constructed from plastid, mitochondrial and genomic DNA of *Plasmodium* species. Mol. Biochem. Parasitol. 114, 89–94.

Regoes, A., Zourmpanou, D., Leon-Avila, G., van der, G.M., Tovar, J., Hehl, A.B., 2005. Protein import, replication, and inheritance of a vestigial mitochondrion. J. Biol. Chem. 280, 30557–30563.

Rehkopf, D.H., Gillespie, D.E., Harrell, M.I., Feagin, J.E., 2000. Transcriptional mapping and RNA processing of the *Plasmodium falciparum* mitochondrial mRNAs. Mol. Biochem. Parasitol. 105, 91–103.

Reiff, S.B., Vaishnava, S., Striepen, B., 2012. The HU protein is important for apicoplast genome maintenance and inheritance in *Toxoplasma gondii*. Eukaryot. Cell 11, 905–915.

Riordan, C.E., Ault, J.G., Langreth, S.G., Keithly, J.S., 2003. *Cryptosporidium parvum* Cpn60 targets a relict organelle. Curr. Genet. 44, 138–147.

Riordan, C.E., Langreth, S.G., Sanchez, L.B., Kayser, O., Keithly, J.S., 1999. Preliminary evidence for a mitochondrion in *Cryptosporidium parvum*: Phylogenetic and therapeutic implications. J. Euk. Microbiol. 46, 52S–55S.

Rivera, V.M., Wang, X., Wardwell, S., Courage, N.L., Volchuk, A., Keenan, T., Holt, D.A., Gilman, M., Orci, L., Cerasoli Jr., F., Rothman, J.E., Clackson, T., 2000. Regulation of protein secretion through controlled aggregation in the endoplasmic reticulum. Science 287, 826–830.

Rogers, M.J., Bukhman, Y.V., McCutchan, T.F., Draper, D.E., 1997. Interaction of thiostrepton with an RNA fragment derived from the plastid-encoded ribosomal RNA of the malaria parasite. RNA 3, 815–820.

Röhrich, R.C., Englert, N., Troschke, K., Reichenberg, A., Hintz, M., Seeber, F., Balconi, E., Aliverti, A., Zanetti, G., Kohler, U., Pfeiffer, M., Beck, E., Jomaa, H., Wiesner, J., 2005. Reconstitution of an apicoplast-localised electron transfer pathway involved in the isoprenoid biosynthesis of *Plasmodium falciparum*. FEBS Lett. 579, 6433–6438.

Rollins, C.T., Rivera, V.M., Woolfson, D.N., Keenan, T., Hatada, M., Adams, S.E., Andrade, L.J., Yaeger, D., van Schravendijk, M.R., Holt, D. A., Gilman, M., Clackson, T., 2000. A ligand-reversible dimerization system for controlling protein-protein interactions. Proc. Natl. Acad. Sci. USA 97, 7096–7101.

Roth, C., Menzel, G., Petetot, J.M., Rochat-Hacker, S., Poirier, Y., 2004. Characterization of a protein of the plastid inner envelope having homology to animal inorganic phosphate, chloride and organic-anion transporters. Planta 218, 406–416.

Rouhier, N., Couturier, J., Johnson, M.K., Jacquot, J.P., 2010. Glutaredoxins: roles in iron homeostasis. Trends Biochem. Sci. 35, 43–52.

Roy, A., Cox, R.A., Williamson, D.H., Wilson, R.J., 1999. Protein synthesis in the plastid of *Plasmodium falciparum*. Protist 150, 183–188.

Saito, T., Nishi, M., Lim, M.I., Wu, B., Maeda, T., Hashimoto, H., Takeuchi, T., Roos, D.S., Asai, T., 2008. A novel GDP-dependent pyruvate kinase isozyme from *Toxoplasma gondii* localizes to both the apicoplast and the mitochondrion. J. Biol. Chem. 283, 14041–14052.

Sasaki, N., Hirai, M., Maeda, K., Yui, R., Itoh, K., Namiki, S., Morita, T., Hata, M., Murakami-Murofushi, K., Matsuoka, H., Kita, K., Sato, S., 2009. The *Plasmodium* HU homolog, which binds the plastid DNA sequence-independent manner, is essential for the parasite's survival. FEBS Lett. 583, 1446–1450.

Sato, S., Clough, B., Coates, L., Wilson, R.J., 2004. Enzymes for heme biosynthesis are found in both the mitochondrion and plastid of the malaria parasite *Plasmodium falciparum*. Protist 155, 117–125.

Sato, S., Tews, I., Wilson, R.J.M., 2000. Impact of a plastid-bearing endocytobiont on apicomplexan genomes. Int. J. Parasitol. 30, 427–439.

Sato, S., Wilson, R.J., 2005. The plastid of *Plasmodium* spp.: a target for inhibitors. Curr. Top. Microbiol. Immunol. 295, 251–273.

Schnaufer, A., 2010. Evolution of dyskinetoplastic trypanosomes: how, and how often? Trends Parasitol. 26, 557–558.

Schnaufer, A., Clark-Walker, G.D., Steinberg, A.G., Stuart, K., 2005. The F1-ATP synthase complex in bloodstream stage trypanosomes has an unusual and essential function. EMBO J. 24, 4029–4040.

Schwenkert, S., Netz, D.J., Frazzon, J., Pierik, A.J., Bill, E., Gross, J., Lill, R., Meurer, J., 2009. Chloroplast HCF101 is a scaffold protein for [4Fe-4S] cluster assembly. Biochem. J. 425, 207–214.

Seeber, F., 2002. Biogenesis of iron-sulphur clusters in amitochondriate and apicomplexan protists. Int. J. Parasitol. 32, 1207–1217.

Seeber, F., Aliverti, A., Zanetti, G., 2005. The plant-type ferredoxin-NADP+ reductase/ferredoxin redox system as a possible drug target against apicomplexan human parasites. Curr. Pharm. Des. 11, 3159–3172.

Seeber, F., Ferguson, D.J., Gross, U., 1998. *Toxoplasma gondii*: a paraformaldehyde-insensitive diaphorase activity acts as a specific histochemical marker for the single mitochondrion. Exp. Parasitol. 89, 137–139.

Seeber, F., Limenitakis, J., Soldati-Favre, D., 2008. Apicomplexan mitochondrial metabolism: a story of gains, losses and retentions. Trends Parasitol. 24, 468–478.

Seeber, F., Soldati, D., 2007. The metabolic functions of the mitochondrion and the apicoplast. In: Ajioka, J., Soldati, D. (Eds.), *Toxoplasma*: molecular and cellular biology. Norfolk: Horizon Bioscience, pp. 409–435.

Seeber, F., Soldati-Favre, D., 2010. Metabolic pathways in the apicoplast of apicomplexa. Int. Rev. Cell. Mol. Biol. 281, 161–228.

Seemann, M., Tse Sum, B.B., Wolff, M., Miginiac-Maslow, M., Rohmer, M., 2006. Isoprenoid biosynthesis in plant chloroplasts via the MEP pathway: direct thylakoid/ferredoxin-dependent photoreduction of GcpE/IspG. FEBS Lett. 580, 1547–1552.

Seow, F., Sato, S., Janssen, C.S., Riehle, M.O., Mukhopadhyay, A., Phillips, R.S., Wilson, R.J., Barrett, M.P., 2005. The plastidic DNA replication enzyme complex of *Plasmodium falciparum*. Mol. Biochem. Parasitol. 141, 145–153.

Shanklin, J., DeWitt, N.D., Flanagan, J.M., 1995. The stroma of higher plant plastids contain ClpP and ClpC, functional homologs of *Escherichia coli* ClpP and ClpA: an archetypal two-component ATP-dependent protease. Plant Cell 7, 1713–1722.

Shaw, M.K., Roos, D.S., Tilney, L.G., 2001. DNA replication and daughter cell budding are not tightly linked in the protozoan parasite *Toxoplasma gondii*. Microbes Infect. 3, 351–362.

Sheiner, L., Demerly, J.L., Poulsen, N., Beatty, W.L., Lucas, O., Behnke, M.S., White, M.W., Striepen, B., 2011. A systematic screen to discover and analyze apicoplast proteins identifies a conserved and essential protein import factor. PLoS Pathog. 7, e1002392.

Siddall, M.E., 1992. Hohlzylinders. Parasitol. Today 8, 90–91.

Silva-Filho, M.C., 2003. One ticket for multiple destinations: dual targeting of proteins to distinct subcellular locations. Curr. Opin. Plant Biol. 6, 589–595.

Silva-Filho, M.D., Wieers, M.C., Flugge, U.I., Chaumont, F., Boutry, M., 1997. Different in vitro and in vivo targeting properties of the transit peptide of a chloroplast envelope inner membrane protein. J. Biol. Chem. 272, 15264–15269.

Singh, D., Chaubey, S., Habib, S., 2003. Replication of the *Plasmodium falciparum* apicoplast DNA initiates within the inverted repeat region. Mol. Biochem. Parasitol. 126, 9–14.

Singh, D., Kumar, A., Raghu Ram, E.V., Habib, S., 2005. Multiple replication origins within the inverted repeat region of the *Plasmodium falciparum* apicoplast genome are differentially activated. Mol. Biochem. Parasitol. 139, 99–106.

Slamovits, C.H., Saldarriaga, J.F., Larocque, A., Keeling, P.J., 2007. The highly reduced and fragmented mitochondrial genome of the early-branching dinoflagellate *Oxyrrhis marina* shares characteristics with both apicomplexan and dinoflagellate mitochondrial genomes. J. Mol. Biol. 372, 356–368.

Slapeta, J., Keithly, J.S., 2004. *Cryptosporidium parvum* mitochondrial-type HSP70 targets homologous and heterologous mitochondria. Eukaryot. Cell 3, 483–494.

Sommer, M.S., Gould, S.B., Lehmann, P., Gruber, A., Przyborski, J.M., Maier, U.G., 2007. Der1-mediated pre-protein import into the periplastid compartment of chromalveolates? Mol. Biol. Evol. 24, 918–928.

Spork, S., Hiss, J.A., Mandel, K., Sommer, M., Kooij, T.W., Chu, T., Schneider, G., Maier, U.G., Przyborski, J.M., 2009. An unusual ERAD-like complex is targeted to the apicoplast of *Plasmodium falciparum*. Eukaryot. Cell 8, 1134–1145.

Spry, C., Kirk, K., Saliba, K.J., 2008. Coenzyme A biosynthesis: an antimicrobial drug target. FEMS Microbiol. Rev. 32, 56–106.

Srivastava, I.K., Morrisey, J.M., Darrouzet, E., Daldal, F., Vaidya, A.B., 1999. Resistance mutations reveal the atovaquone-binding domain of cytochrome *b* in malaria parasites. Mol. Microbiol. 33, 704–711.

Stoebe, B., Kowallik, K.V., 1999. Gene-cluster analysis in chloroplast genomics. Trends Genet. 15, 344–347.

Strath, M., Scott-Finnigan, T., Gardner, M., Williamson, D., Wilson, I., 1993. Antimalarial activity of rifampicin in vitro and in rodent models. Trans. R. Soc. Trop. Med. Hyg. 87, 211–216.

Striepen, B., Crawford, M.J., Shaw, M.K., Tilney, L.G., Seeber, F., Roos, D.S., 2000. The plastid of *Toxoplasma gondii* is divided by association with the centrosomes. J. Cell. Biol. 151, 1423–1434.

Sullivan Jr., W.J., Jeffers, V., 2012. Mechanisms of *Toxoplasma gondii* persistence and latency. FEMS Microbiol. Rev. 36, 717–733.

Suplick, K., Akella, R., Saul, A., Vaidya, A.B., 1988. Molecular cloning and partial sequence of a 5.8 kilobase pair repetitive DNA from *Plasmodium falciparum*. Mol. Biochem. Parasitol. 30, 289–290.

Suplick, K., Morrisey, J., Vaidya, A.B., 1990. Complex transcription from the extrachromosomal DNA encoding mitochondrial functions of *Plasmodium yoelii*. Mol. Cell. Biol. 10, 6381–6388.

Tabbara, K.F., Sakuragi, S., O'Connor, G.R., 1982. Minocycline in the chemotherapy of murine toxoplasmosis. Parasitology 84, 297–302.

Tanabe, K., 1985. Visualization of the mitochondria of *Toxoplasma gondii*-infected mouse fibroblasts by the cationic permeant fluorescent dye rhodamine 123. Experientia 41, 101–102.

Tarun, A.S., Vaughan, A.M., Kappe, S.H., 2009. Redefining the role of de novo fatty acid synthesis in *Plasmodium* parasites. Trends Parasitol. 25, 545–550.

Tawk, L., Dubremetz, J.F., Montcourrier, P., Chicanne, G., Merezegue, F., Richard, V., Payrastre, B., Meissner, M., Vial, H.J., Roy, C., Wengelnik, K., Lebrun, M., 2011. Phosphatidylinositol 3-monophosphate is involved in *Toxoplasma* apicoplast biogenesis. PLoS Pathog. 7, e1001286.

Thomsen-Zieger, N., Schachtner, J., Seeber, F., 2003. Apicomplexan parasites contain a single lipoic acid synthase located in the plastid. FEBS Lett. 547, 80–86.

Tomavo, S., Boothroyd, J.C., 1995. Interconnection between organellar functions, development and drug resistance in the protozoan parasite, *Toxoplasma gondii*. Int. J. Parasitol. 25, 1293–1299.

Tomova, C., Humbel, B.M., Geerts, W.J., Entzeroth, R., Holthuis, J.C., Verkleij, A.J., 2009. Membrane contact sites between apicoplast and ER in *Toxoplasma gondii* revealed by electron tomography. Traffic 10, 1471–1480.

Tonkin, C.J., Struck, N.S., Mullin, K.A., Stimmler, L.M., McFadden, G.I., 2006. Evidence for Golgi-independent transport from the early secretory pathway to the plastid in malaria parasites. Mol. Microbiol. 61, 614–630.

Toursel, C., Dzierszinski, F., Bernigaud, A., Mortuaire, M., Tomavo, S., 2000. Molecular cloning, organellar targeting and developmental expression of mitochondrial chaperone HSP60 in *Toxoplasma gondii*. Mol. Biochem. Parasitol. 111, 319–332.

Tovar, J., Fischer, A., Clark, C.G., 1999. The mitosome, a novel organelle related to mitochondria in the amitochondrial parasite *Entamoeba histolytica*. Mol. Microbiol. 32, 1013–1021.

Tsaousis, A.D., Kunji, E.R., Goldberg, A.V., Lucocq, J.M., Hirt, R.P., Embley, T.M., 2008. A novel route for ATP acquisition by the remnant mitochondria of *Encephalitozoon cuniculi*. Nature 453, 553–556.

Vaidya, A.B., Akella, R., Suplick, K., 1989. Sequences similar to genes for two mitochondrial proteins and portions of ribosomal RNA in tandemly arrayed 6-kilobase-pair DNA of a malarial parasite. Mol. Biochem. Parasitol. 35, 97–108.

Vaidya, A.B., Arasu, P., 1987. Tandemly arranged gene clusters of malarial parasites that are highly conserved and transcribed. Mol. Biochem. Parasitol. 22, 249–257.

Vaidya, A.B., Mather, M.W., 2009. Mitochondrial evolution and functions in malaria parasites. Annu. Rev. Microbiol. 63, 249–267, 249-267.

Vaidya, A.B., Morrisey, J., Plowe, C.V., Kaslow, D.C., Wellems, T.E., 1993. Unidirectional dominance of cytoplasmic inheritance in two genetic crosses of *Plasmodium falciparum*. Mol. Cell. Biol. 13, 7349–7357.

van Dooren, G.G., Stimmler, L.M., McFadden, G.I., 2006. Metabolic maps and functions of the *Plasmodium* mitochondrion. FEMS Microbiol. Rev. 30, 596–630.

van Dooren, G.G., Su, V., D'Ombrain, M.C., McFadden, G.I., 2002. Processing of an apicoplast leader sequence in *Plasmodium falciparum* and the identification of a putative leader cleavage enzyme. J. Biol. Chem. 277, 23612–23619.

van Dooren, G.G., Tomova, C., Agrawal, S., Humbel, B.M., Striepen, B., 2008. *Toxoplasma gondii* Tic20 is essential for apicoplast protein import. Proc. Natl. Acad. Sci. USA 105, 13574–13579.

Varadharajan, S., Dhanasekaran, S., Bonday, Z.Q., Rangarajan, P.N., Padmanaban, G., 2002. Involvement of delta-aminolaevulinate synthase encoded by the parasite gene in de novo haem synthesis by *Plasmodium falciparum*. Biochem. J. 367, 321–327.

Vaughan, A.M., O'Neill, M.T., Tarun, A.S., Camargo, N., Phuong, T.M., Aly, A.S., Cowman, A.F., Kappe, S.H., 2009. Type II fatty acid synthesis is essential only for malaria parasite late liver stage development. Cell. Microbiol. 11, 506–520.

Vavra, J., 2005. 'Polar vesicles' of microsporidia are mitochondrial remnants ('mitosomes')? Folia Parasitol. (Praha.) 52, 193–195.

Vercesi, A.E., Rodrigues, C.O., Uyemura, S.A., Zhong, L., Moreno, S.N., 1998. Respiration and oxidative phosphorylation in the apicomplexan parasite *Toxoplasma gondii*. J. Biol. Chem. 273, 31040–31047.

Vollmer, M., Thomsen, N., Wiek, S., Seeber, F., 2001. Apicomplexan parasites possess distinct nuclear-encoded, but apicoplast-localized, plant-type ferredoxin-NADP+ reductase and ferredoxin. J. Biol. Chem. 276, 5483–5490.

Waller, R.F., Keeling, P.J., Donald, R.G., Striepen, B., Handman, E., Lang-Unnasch, N., Cowman, A.F., Besra, G.S., Roos, D.S., McFadden, G.I., 1998. Nuclear-encoded proteins target to the plastid in *Toxoplasma gondii* and *Plasmodium falciparum*. Proc. Natl. Acad. Sci. USA 95, 12352–12357.

Waller, R.F., McFadden, G.I., 2005. The apicoplast: a review of the derived plastid of apicomplexan parasites. Curr. Issues Mol. Biol. 7, 57–79.

Waller, R.F., Reed, M.B., Cowman, A.F., McFadden, G.I., 2000. Protein trafficking to the plastid of *Plasmodium falciparum* is via the secretory pathway. EMBO J. 19, 1794–1802.

Wallin, I.E., 1927. Symbionticism and the Origin of Species. Williams and Wilkins, Baltimore.

Wang, Y., Karnataki, A., Parsons, M., Weiss, L.M., Orlofsky, A., 2010. 3-Methyladenine blocks *Toxoplasma gondii* division prior to centrosome replication. Mol. Biochem. Parasitol. 173, 142–153.

Wastl, J., Maier, U.G., 2000. Transport of proteins into cryptomonads complex plastids. J. Biol. Chem. 275, 23194–23198.

Weissig, V., Vetro-Widenhouse, T.S., Rowe, T.C., 1997. Topoisomerase II inhibitors induce cleavage of nuclear and 35-kb plastid DNAs in the malarial parasite *Plasmodium falciparum*. DNA Cell Biol. 16, 1483–1492.

Whatley, J.M., John, P., Whatley, F.R., 1979. From extracellular to intracellular: the establishment of mitochondria and chloroplasts. Proc. R. Soc. London Ser. B 204, 165–187.

White, M.F., Dillingham, M.S., 2012. Iron-sulphur clusters in nucleic acid processing enzymes. Curr. Opin. Struct. Biol. 22, 94–100.

Wiesner, J., Jomaa, H., 2007. Isoprenoid biosynthesis of the apicoplast as drug target. Curr. Drug Targets 8, 3–13.

Wiesner, J., Reichenberg, A., Heinrich, S., Schlitzer, M., Jomaa, H., 2008. The plastid-like organelle of apicomplexan parasites as drug target. Curr. Pharm. Des. 14, 855–871.

Wiesner, J., Sanderbrand, S., Altincicek, B., Beck, E., Jomaa, H., 2001. Seeking new targets for antiparasitic agents. Trends Parasitol. 17, 7–8.

Wiesner, J., Seeber, F., 2005. The plastid-derived organelle of protozoan human parasites as a target of established and emerging drugs. Expert Opin. Ther. Targets 9, 23–44.

Williams, B.A., Hirt, R.P., Lucocq, J.M., Embley, T.M., 2002. A mitochondrial remnant in the microsporidian *Trachipleistophora hominis*. Nature 418, 865–869.

Williams, B.A., Haferkamp, I., Keeling, P.J., 2008. An ADP/ATP-specific mitochondrial carrier protein in the microsporidian *Antonospora locustae*. J. Mol. Biol. 375, 1249–1257.

Williamson, D.H., Denny, P.W., Moore, P.W., Sato, S., McCready, S., Wilson, R.J., 2001. The in vivo conformation of the plastid DNA of *Toxoplasma gondii*: implications for replication. J. Mol. Biol. 306, 159–168.

Williamson, D.H., Preiser, P.R., Moore, P.W., McCready, S., Strath, M., Wilson, R.J., 2002. The plastid DNA of the malaria parasite *Plasmodium falciparum* is replicated by two mechanisms. Mol. Microbiol. 45, 533–542.

Williamson, D.H., Wilson, R.J.M., Bates, P.A., McCready, S., Perler, F., Qiang, B., 1985. Nuclear and mitochondrial DNA of the primate malarial parasite *Plasmodium knowlesi*. Mol. Biochem. Parasitol. 14, 199–209.

Wilson, I., Gardner, M., Rangachari, K., Williamson, D., 1993. Extrachromosomal DNA in the Apicomplexa. In: Smith, J.E. (Ed.), Toxoplasmosis. NATO ASI series H. Springer-Verlag, Heidelberg, pp. 51–62.

Wilson, R.J., 2005. Parasite plastids: approaching the endgame. Biol. Rev. Cambridge. Phil. Soc. 80, 129–153.

Wilson, R.J., Rangachari, K., Saldanha, J.W., Rickman, L., Buxton, R.S., Eccleston, J.F., 2003. Parasite plastids: maintenance and functions. Philos. Trans. R. Soc. B. 358, 155–162.

Wilson, R.J.M., Denny, P.W., Preiser, P.R., Rangachari, K., Roberts, K., Roy, A., Whyte, A., Strath, M., Moore, D.J., Moore, P.W., Williamson, D.H., 1996. Complete gene map of the plastid-like DNA of the malaria parasite *Plasmodium falciparum*. J. Mol. Biol. 261, 155–172.

Wilson, R.J.M., Fry, M., Gardner, M.J., Feagin, J.E., Williamson, D.H., 1992. Subcellular fractionation of the two organelle DNAs of malaria parasites. Curr. Genet. 21, 405–408.

Wilson, R.J.M., Gardner, M.J., Williamson, D.H., Feagin, J.E., 1991. Have malaria parasites three genomes? Parasitol. Today 7, 134–136.

Woehle, C., Dagan, T., Martin, W.F., Gould, S.B., 2011. Red and problematic green phylogenetic signals among thousands of nuclear genes from the photosynthetic and apicomplexa-related Chromera velia. Genome Biol. Evol. 3, 1220–1230.

Wollers, S., Layer, G., Garcia-Serres, R., Signor, L., Clemancey, M., Latour, J.M., Fontecave, M., de Ollagnier, C.S., 2010. Iron-sulfur (Fe-S) cluster assembly: the SufBCD complex is a new type of Fe-S scaffold with a flavin redox cofactor. J. Biol. Chem. 285, 23331–23341.

Wolters, J., 1991. The troublesome parasites - molecular and morphological evidence that Apicomplexa belong to the dinoflagellate-ciliate clade. BioSystems 25, 75–83.

Wu, B., 2006. Heme biosynthetic pathway in apicomplexan parasites. University of Pennsylvania. Ref Type: Thesis/Dissertation.

Wu, Y., Brosh Jr., R.M., 2012. DNA helicase and helicase-nuclease enzymes with a conserved iron-sulfur cluster. Nucleic Acids Res. 40, 4247–4260.

Xu, P., Widmer, G., Wang, Y., Ozaki, L.S., Alves, J.M., Serrano, M.G., Puiu, D., Manque, P., Akiyoshi, D., Mackey, A.J., Pearson, W.R., Dear, P.H., Bankier, A.T., Peterson, D.L., Abrahamsen, M.S., Kapur, V., Tzipori, S., Buck, G.A., 2004. The genome of *Cryptosporidium hominis*. Nature 431, 1107–1112.

Yeh, E., DeRisi, J.L., 2011. Chemical rescue of malaria parasites lacking an apicoplast defines organelle function in blood-stage *Plasmodium falciparum*. PLoS Biol. 9, e1001138.

Yu, M., Kumar, T.R., Nkrumah, L.J., Coppi, A., Retzlaff, S., Li, C.D., Kelly, B.J., Moura, P.A., Lakshmanan, V., Freundlich, J.S., Valderramos, J.C., Vilcheze, C., Siedner, M., Tsai, J.H., Falkard, B., Sidhu, A.B., Purcell, L.A., Gratraud, P., Kremer, L., Waters, A.P., Schiehser, G., Jacobus, D.P., Janse, C.J., Ager, A., Jacobs Jr., W.R., Sacchettini, J.C., Heussler, V., Sinnis, P., Fidock, D.A., 2008. The fatty acid biosynthesis enzyme FabI plays a key role in the development of liver-stage malarial parasites. Cell Host Microbe 4, 567–578.

Yung, S., Unnasch, T.R., Lang-Unnasch, N., 2001. Analysis of apicoplast targeting and transit peptide processing in *Toxoplasma gondii* by deletional and insertional mutagenesis. Mol. Biochem. Parasitol. 118, 11–21.

Yung, S.C., Unnasch, T.R., Lang-Unnasch, N., 2003. Cis and trans factors involved in apicoplast targeting in *Toxoplasma gondii*. J. Parasitol. 89, 767–776.

Zauner, S., Fraunholz, M., Wastl, J., Penny, S., Beaton, M., Cavalier-Smith, T., Maier, U.-G., Douglas, S., 2000. Chloroplast protein and centrosomal genes, a tRNA intron, and odd telomeres in an unusually compact eukaryotic genome, the cryptomonad nucleomorph. Proc. Natl. Acad. Sci. USA 97, 200–205.

Zhu, G., Keithly, J.S., Philippe, H., 2000a. What is the phylogenetic position of *Cryptosporidium?* Int. J. Syst. Evol. Microbiol. 50 (Pt 4), 1673–1681.

Zhu, G., Li, Y., Cai, X., Millership, J.J., Marchewka, M.J., Keithly, J.S., 2004. Expression and functional characterization of a giant Type I fatty acid synthase (CpFAS1) gene from *Cryptosporidium parvum*. Mol. Biochem. Parasitol. 134, 127–135.

Zhu, G., Marchewka, M.J., Keithly, J.S., 2000b. *Cryptosporidium parvum* appears to lack a plastid genome. Microbiology U.K. 146 (part 2), 315–321.

Zuegge, J., Ralph, S., Schmuker, M., McFadden, G.I., Schneider, G., 2001. Deciphering apicoplast targeting signals—feature extraction from nuclear-encoded precursors of *Plasmodium falciparum* apicoplast proteins. Gene 280, 19–26.

Zuther, E., Johnson, J.J., Haselkorn, R., McLeod, R., Gornicki, P., 1999. Growth of *Toxoplasma gondii* is inhibited by aryloxyphenoxypropionate herbicides targeting acetyl-CoA carboxylase. Proc. Natl. Acad. Sci. USA 96, 13387–13392.

CHAPTER

10

Calcium Storage and Homeostasis in *Toxoplasma gondii*

*Silvia N.J. Moreno**, *Lucas Borges Pereira*†, *Douglas A. Pace***

***Department of Cellular Biology and Center for Tropical and Emerging Global Diseases, University of Georgia, Athens, Georgia, USA †Department of Parasitology, Institute of Biomedical Sciences, University of São Paulo, São Paulo, Brazil **Department of Biological Sciences, California State University, Long Beach, California, USA**

OUTLINE

Toxoplasma gondii, Second Edition
http://dx.doi.org/10.1016/B978-0-12-396481-6.00010-6

10.1 INTRODUCTION

The calcium ion (Ca^{2+}) is a universal signalling molecule and plays a crucial role as a second messenger for the control of a variety of cell functions in eukaryotes, including contraction, secretion, cell division and differentiation, and sodium and potassium permeability. In *Toxoplasma gondii*, Ca^{2+} plays a critical role in several parasite-specific functions including host cell invasion, motility, differentiation and egress.

The cytosolic concentration of Ca^{2+} is maintained at 70–100 nm, which is likely the result of the concerted operation of a number of transporters and Ca^{2+}-binding proteins, many of them still to be identified and characterized. The endoplasmic reticulum (ER) and acidocalcisomes have been identified as major calcium stores. Other potential calcium storage organelles include the Golgi, and the recently described plant-like vacuole/vacuolar compartment (PLV/VAC) (a lysosomal compartment). The molecular players involved in the control and regulation of *T. gondii* intracellular calcium are partially identified and the available information is, at best, fragmentary. In this review, we will try to summarize the information available about the specialized systems for uptake and release of Ca^{2+} in *T. gondii.* We will also analyse the available methods to study Ca^{2+} storage and homeostasis.

10.2 FLUORESCENT METHODS TO STUDY CALCIUM HOMEOSTASIS IN *TOXOPLASMA GONDII*

Because of the importance of Ca^{2+} in biology, numerous methodological approaches are currently in use for analysing the mechanisms of cellular and/or subcellular Ca^{2+} homeostasis in eukaryotic cells. Although each method for analysing Ca^{2+} fluctuations has certain advantages over the others, each also suffers drawbacks. Investigation of the intracellular Ca^{2+} concentration ($[Ca^{2+}]_i$) was boosted dramatically with the development of fluorescent dyes of high sensitivity by Tsien (Grynkiewicz *et al.*, 1985; Tsien, 1989). Many other techniques exist for measuring $[Ca^{2+}]_i$ fluxes, including $^{45}Ca^{2+}$, microelectrodes, metallochromic indicators, and Ca^{2+}-sensitive photoproteins. For *T. gondii*, the most widely used methods have been those involving chemical fluorescent probes.

10.2.1 Fluorescent Probes

One of the most important and accessible techniques is the employment of chemical probes that exhibit significant changes in structural properties upon binding to calcium. These changes result in spectral shifts that can be monitored and quantified in real time. In general, these indicator probes possess a modular design in which there is a calcium-binding site (similar in structure to EGTA) combined to a fluorescent indicator. There has been a rapid evolution of fluorescent probes over the past two decades and there is now a wide range of commercially available probes for a variety of purposes (e.g. see Life Technologies/Invitrogen, www.invitrogen.com). It is important to keep in mind that there is no single probe that will be a good fit for all situations. As will be discussed, the selection of a fluorescent probe is dependent on a number of factors including the wavelengths to be used, cell permeability, cellular location of interest, regulatory timescale of interest, and dynamic concentration range of calcium to be measured.

The work of Tsien and colleagues resulted in the first generation of highly sensitive fluorescent probes that could be used for measuring changes in cellular Ca^{2+} concentration (reviewed in Tsien, 1989). Fluorescent probes have very low background light emission due to their wavelength specificity and exhibit dramatic changes upon Ca^{2+} binding. Perhaps the most important development in fluorescent probes was the addition of an acetoxymethyl (AM) ester group to the indicators making them cell permeable and Ca^{2+} insensitive. The AM-fluorescent probe is able to freely diffuse through the plasma membrane due to the masking of charged portions by the AM group (Bruton *et al.*, 2012). Upon diffusing through the plasma membrane the AM group is cleaved off by cytosolic esterases, effectively trapping the probe in the cytosol (Tsien, 1983). A significant issue when using AM probes is their capacity to become 'compartmentalized' within organelles. Due to the ester group, the probes can diffuse into organelles before being made membrane impermeable. Additionally, the cell impermeant version can become enriched in certain organelles or expelled from the cell through organic anion channels (Tsien, 1989). To overcome these issues it is critical to load cells with the minimum effective concentration of fluorescent probe. This has the additional benefit of decreasing the ability of the probe from interfering with Ca^{2+} signalling due to buffering effects from binding to free calcium. The use of organic anion channel blockers is recommended to prevent organellar compartmentalization of the probe (Tsien, 1989). The organic anion channel blocker, probenecid, has been shown to be specific and effective in reducing the sequestration of probe to organelles or back to the extracellular environment (Di Virgilio *et al.*, 1988). For *T. gondii*, the general protocol in our laboratory is to load with 5 μM of the ratiometric fluorescent probe, Fura2-AM. We load the parasites for 27 minutes at 27 °C in the presence of 40 μM probenecid. Additionally, it is useful to keep the parasites on ice after washing until their use for fluorescent measurements as this helps to slow the compartmentalization of the probe.

Calcium probes fall into two general categories: non-ratiometric (single wavelength) and ratiometric probes. The non-ratiometric probes have a fixed wavelength for excitation and emission. Upon Ca^{2+} binding they display an increase in fluorescent intensity with very little spectral shift. Ratiometric probes are analysed by simultaneously determining the fluorescent output at either two excitation wavelengths (e.g. Fura2) or two emission wavelengths (e.g. Indo1). The data are collected in the form of a ratio showing fluorescence for the probe in its bound and unbound states. The use of Fura2 involves exciting the probe at 340 nm and 380 nm while monitoring the emission at 510 nm. When Fura2 is unbound to Ca^{2+} its maximum fluorescence occurs when excited at 380 nm, however the bound state of Fura2 has a maximum fluorescence at 340 nm. As such, the amount of bound and unbound probe is determined simultaneously. The primary advantage of ratiometric probes is that interfering factors such as differences in cell density, cell volume, loading variation, photobleaching, and differences in path-length are negated (Bright *et al.*, 1987). Given the batch-to-batch variability that is typically encountered when loading *T. gondii* tachyzoites in terms of total amount of Fura2-AM in the cytosol, this is an important benefit. Importantly, the calibration of ratiometric probes is mathematically straightforward, so that the proportion of bound to unbound probe can be converted into relevant units of concentration (e.g. see Grynkiewicz *et al.*, 1985). Figure 10.1 shows a typical experiment with *T. gondii* tachyzoites loaded with Fura2-AM and cytosolic calcium fluctuations are followed after transforming the ratio values to concentrations. These experiments were published in Moreno and Zhong (1996) and Miranda *et al.* (2010) (Fig. 10.1A, B).

Non-ratiometric and ratiometric probes can be purchased through commercial vendors (e.g. Invitrogen) and are available as free acid (non-AM) and cell permeable (AM esters) versions. Free acid, cell impermeable versions of Ca^{2+} probes are typically used when measuring Ca^{2+} in permeabilize cells (e.g. digitonin treatment) or when Ca^{2+} needs to be monitored in the extracellular environment. Table 10.1 shows a list of popular chemical Ca^{2+} probes for *T. gondii* research and their relevant characteristics as well as studies employing their use. For a fuller comparison of

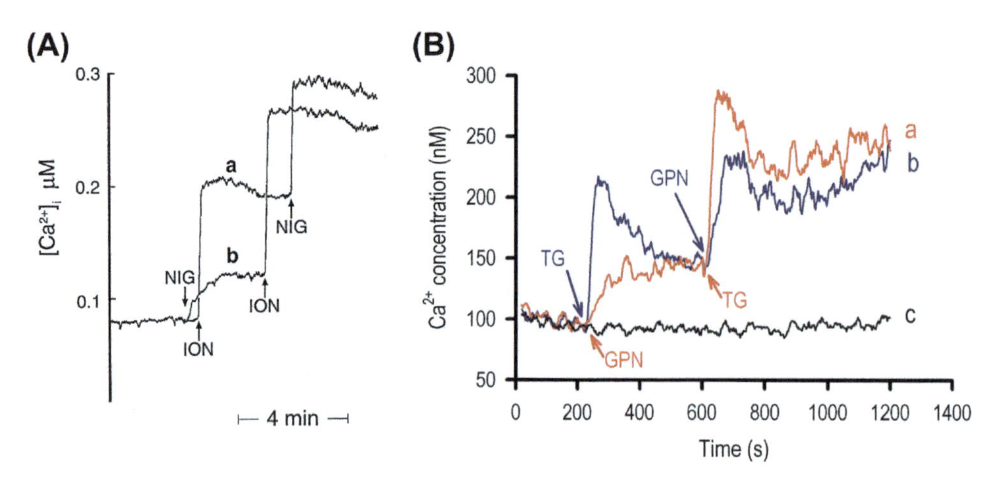

FIGURE 10.1 **Acidic calcium pools from *T. gondii* tachyzoites.**
A) Tachyzoites were loaded with Fura2-AM. Ionomycin (ION: 1 µM) or nigericin (NIG: 1 µM) was added where indicated. *Reproduced from Moreno and Zhong, Biochem. J. 313, 655–659, 1996.*
B) GPN, a cathepsin C substrate, is hydrolysed in an acidic compartment and its product produces swelling of a lysosomal compartment and calcium release into the cytoplasm. Thapsigargin (TG) treatment causes calcium release from the ER and is independent of GPN-induced calcium release. *Reproduced from Miranda* et al. *Mol. Microbiol., 76:1358, 2010.*

available Ca^{2+} probes the paper by Takahashi (Takahashi *et al.*, 1999) is recommended.

The dissociation constant (K_d) of a fluorescent Ca^{2+} probe (whether non-ratiometric or ratiometric) is a critical variable to consider when choosing it for a given experiment. The K_d of a probe should be similar to the concentration of Ca^{2+} in the system being measured. Because the cytosolic calcium of *T. gondii* is ~100 nm (Moreno and Zhong, 1996), Fura2, with a K_d of 140 nm, represents a good choice. However, dramatic Ca^{2+} responses, such as those evoked by the Ca^{2+} ionophores ionomycin or A23187, can result in cytosolic Ca^{2+} concentrations in excess of 1 µM, well beyond the saturation point of the probe. High affinity Ca^{2+} probes such as Fura2 also suffer from slow release times of bound Ca^{2+}, an important consideration if Ca^{2+} transients (i.e. waves and spikes) are to be observed. A remedy for these situations is the use of a similar probe Fura4F (K_d = 770 nM), which is able to quantify higher concentrations of Ca^{2+} and has the temporal resolution necessary to capture Ca^{2+} transients.

The choice of wavelengths is important for both logistical and biological reasons. Probes such as Fluo3 and Fluo4, which are excited at 488 nm, are most commonly used for microscopic analysis of Ca^{2+} signalling. Emission and excitation wavelengths are compatible with most epifluorescent microscopes and the increase upon Ca^{2+} binding is very large (greater than 100 times) making visual detection easy to observe. Calcium Green-5N is a probe with similarities to Fluo3 but has the benefit of being more fluorescent in resting cells (due to higher affinity than Fluo3). We have used it to measure Ca^{2+} uptake and release from Ca^{2+} stores in *T. brucei* permeabilized cells and it showed higher sensitivity than Arsenazo III (Huang, 2013). Preliminary experiments indicate that it will also work for permeabilized tachyzoites (Moreno and Vercesi, unpublished). In this regard, there are many probes available that have visible emission/excitation wavelengths in a variety of colours, making them useful for dual labelling studies (see Life Technologies/www.invitrogen.com). Ratiometric probes, on the other hand, require specialized

TABLE 10.1 Calcium Probes and Selected References

Probe	Probe Type*	Wavelengths (nm)**	K_d (nM)	Refs
Fura2	R (Ex)	Em = 380/340 Ex = 510	145	1
FuraPE3	R (Ex)	Em = 380/340 Ex = 510	146	2
Fura4F	R (Ex)	Em = 380/340 Ex = 510	770	†
Indo-1	R (Em)	Ex = 338 Em = 485/405	230	3
Fluo3	S	Ex = 496 Em = 507	390	4
Fluo4	S	Ex = 494 Em = 516	345	5
Calcium Green-1	S	Ex = 506 Em = 531	190‡‡	‡‡‡
Calcium Crimson	S	Ex = 590 Em = 615	185	‡

* *R: Ratiometric, S: Single wavelength, Abs: absorbance probe (non-fluorometric). For ratiometric probes mode of ratio (excitation or emission) given in parenthesis.*
** *Em: emission wavelength, Ex: excitation wavelength. Dual wavelengths for ratiometric dyes are listed with regards to calcium binding state (unbound/bound).*
† *Unpublished studies on* T. gondii *that used Fura4F-AM were found, however our lab has used a similar discontinued probe, Fura5F-AM, to measure ionomycin release with good results (Moreno, unpublished data).*
‡ *Unpublished studies on* T. gondii. *However, our lab has observed cytotoxic effects and not appropriate for live cell analysis.*
‡‡ *Also available in a lower calcium affinity version (Invitrogen, Life Sciences): Calcium green-2, K_d = 550 nM. Suitable for calcium spikes up to 25 μM.*
‡‡‡ *Published study* T. brucei *(Huang* et al., *2013). We have used Calcium Green-5N successfully in our lab on* T. gondii *using permeabilization strategies (unpublished results).*
1, *(Carruthers and Sibley, 1999; Vieira and Moreno, 2000; Lovett* et al., *2002; Luo* et al., *2005; Nagamune* et al., *2007; Miranda* et al., *2010)*
2, *(Moudy* et al., *2001)*
3, *(Pingret* et al., *1996; Stommel* et al., *1997)*
4, *(Schwab* et al., *1994)*
5, *(Lovett and Sibley, 2003; Wetzel* et al., *2004)*

setups for rapid switching of the two wavelengths in use as well as a way of quantifying fluorescent signal. Ratiometric probes are most useful in plate reader formats or large fluorometers using cuvettes and cell suspensions (e.g. Hitachi F-4500). In this format the lower fluorescent signals are easily quantified and calibration is straightforward, resulting in real-time analysis of Ca^{2+} regulatory events. Plate readers are especially useful for high throughput, large sample size screening using ratiometric protocols that have already been validated on a real time fluorometer.

From a biological standpoint, measurements using probes such as Fura2 can be problematic due to their excitation wavelengths, which are in the UV range. Excitation at shorter wavelengths can cause cellular damage and negatively influence the process being observed. While longer wavelength probes like Fluo4 can get around this issue there are still instances when visible wavelength excitation probes result in high background fluorescence (Mbatia and Burdette, 2012). The recent advent of near-IR probes has successfully avoided both of these issues in mammalian cells. Probes such as KFCA and CaSIR-1 are examples of near-IR probes that are just being used in mammalian Ca^{2+} regulation studies (e.g. Matsui *et al.*, 2011). However, these probes have not been used in *T. gondii* or other apicomplexans thus far, so their appropriateness still requires testing. When using new probes caution is required in setting up optimal loading and washing conditions, as well as checking for cytotoxic effects. For example, our laboratory has used the long wavelength (visible), non-ratiometric probe Calcium Crimson. Although this probe should avoid the pitfalls of near-UV cytotoxicity, we still found that it was toxic to parasites and therefore not useful for observing real-time Ca^{2+} regulation.

10.2.2 Manipulation of Ca^{2+}

Calcium buffers and ionophores are frequently used to lower or raise intracellular or extracellular Ca^{2+}. The purpose of these protocols is to study Ca^{2+}-dependent cellular processes (Kao, 1994).

EGTA is highly selective for binding Ca^{2+} over Mg^{2+} and because of this, it is the most commonly used Ca^{2+} buffer. However, the Ca^{2+}-binding activity by EGTA is very pH-dependent when used at physiological pH. This is because at these pH values EGTA exists primarily as protonated species (H_2EGTA^{2-}) (highest pK_as 8.90 and 9.52). Upon binding Ca^{2+}, 2 H^+ will be released suggesting that this reaction should have very steep pH dependence. In fact a drop in pH from 7.2 to 7.1 changes the K_d(Ca) of EGTA by a factor of approximately 1.6. Acidification of the medium by high levels of EGTA was proposed to be responsible for the postulated (Pezzella *et al.*, 1997) requirement of extracellular Ca^{2+} for invasion of host cells by tachyzoites (Lovett and Sibley, 2003; Pezzella *et al.*, 1997)

Tsien developed an analogue of EGTA in which the methylene links between oxygen and nitrogen atoms were replaced with benzene rings to yield a compound called BAPTA (Tsien, 1980). This compound has a considerably lower pH sensitivity than EGTA at physiological pH values since its highest pK_as are 5.47 and 6.36 (Tsien, 1980). Because of this, BAPTA is a less troublesome Ca^{2+} buffer to use, although it is significantly more costly than EGTA.

To study the correlation of a biological process with changes in $[Ca^{2+}]_i$, it is useful to be able to block the change in $[Ca^{2+}]_i$ with a Ca^{2+} chelator. The easiest way to introduce extra Ca^{2+} buffering capacity into cells is by loading them with BAPTA/AM, the ester form of BAPTA (Vieira and Moreno, 2000). Cells can be loaded with the AM ester of BAPTA and a Ca^{2+} indicator simultaneously since similar conditions can be used (Kao, 1994). This BAPTA buffering method has been widely used in *T. gondii* to understand the role of Ca^{2+} in microneme secretion by *T. gondii* (Carruthers *et al.*, 1999a), conoid extrusion (Mondragon and Frixione, 1996), gliding motility (Wetzel *et al.*, 2004), and invasion (Vieira and Moreno, 2000). It is important to load the cells with BAPTA analogues unable to chelate Ca^{2+} as a control to ensure that BAPTA-dependent results are specific to its chelating ability and not a result of nonspecific cytotoxicity. Analogues such as 'half-BAPTA' (Vieira and Moreno, 2000) or D-BAPTA (Saoudi *et al.*, 2004) can be used, although half-BAPTA is not currently available. It is important to know that BAPTA can display side effects. It has a potent microtubule depolymerizing effect and decreases the ATP pool of the cells (Saoudi *et al.*, 2004). It is also important to provide controls showing that the concentrations of BAPTA/AM used are able to chelate intracellular Ca^{2+} (Vieira and Moreno, 2000).

The ionophores A23187 and ionomycin form lipid-soluble complexes with divalent metal cations and increase the permeability of biological membranes to Ca^{2+}. There are significant differences in the properties of both ionophores that should be considered when using them in an experiment. Ca^{2+} transport by both ionophores is pH-dependent, and this pH-dependence differs (Liu and Hermann, 1978). Transport of Ca^{2+} by A23187 is best at pH 7.5 whereas Ca^{2+} transport by ionomycin does not reach a maximum until pH 9.5. In addition, ionomycin has better selectivity for Ca^{2+} over Mg^{2+} whereas A23187 shows no preference between the two (Liu and Hermann, 1978). Both ionophores are inefficient in mediating Ca^{2+} transport at low Ca^{2+} concentrations. Due to its fluorescent properties, A23187 should not be used for fluorescence microscopy, whereas ionomycin has very little fluorescent interference (or alternatively, the related compound, Br-A32187 can be used).

Figure 10.1A shows a typical tracing with Fura 2-loaded *T. gondii* tachyzoites in suspension in a buffer containing 1 mM EGTA. Under these conditions, fluorescent changes reflect Ca^{2+} movements from intracellular Ca^{2+} stores. The addition of ionomycin shows a large increase in intracellular Ca^{2+} indicating that Ca^{2+} is released from an intracellular compartment with neutral pH into the cytosol (endoplasmic reticulum). When nigericin is added after ionomycin, a second increase in cytosolic Ca^{2+} occurs due to its independent release from an acidic compartment. Similar results are observed if the order of additions is reversed (Fig. 10.1A). Nigericin is a K^+/H^+ exchanger and because of this property it alkalinizes acidic compartments by allowing protons to be released in exchange for potassium (Fig. 10.1A). A postulated H^+/Ca^{2+} exchanger takes protons back and releases Ca^{2+} into the cytosol.

An important consideration when working with *T. gondii* tachyzoites and Ca^{2+} measurements is their appropriate treatment and preparation for analyses. *T. gondii* has a heightened sensitivity to Ca^{2+} even at low contaminating concentrations. Pace *et al.* (submitted for publication) observed increased microneme secretion in parasites after the addition of 0.4 μM Ca^{2+}. It is therefore important to control the parasites' exposure to Ca^{2+} during collection, loading, washing and other types of preparation. The use of a Ca^{2+}-defined buffer containing ~100 nm free calcium (representative of the parasite and host cytosolic free calcium) is critical for obtaining parasites in an appropriate physiological state with regards to Ca^{2+} responsiveness. Buffers with Ca^{2+} at blood plasma levels (1.8 mM) or even buffers with only contaminating calcium (5–10 μM) run the risk of causing 'calcium fatigue' in which due to the long-term exposure to elevated Ca^{2+} (e.g. 30–60 minutes of preparation time), the parasites have already responded to the increased Ca^{2+} environment. Invasion-linked behaviour such as microneme secretion or conoid extension can therefore be diminished when measured under conditions not employing a Ca^{2+}-defined buffer. There are several web-based applications (http://maxchelator.stanford.edu/) that can be used for determining how much EGTA and calcium are required to reach different free calcium concentrations.

10.2.3 Genetic Indicators

Genetically encoded Ca^{2+} indicators (GECIs) include Ca^{2+} sensors that can be expressed inside cells by transgenic approaches. There are three classes of GECIs that have been developed: i) the bioluminescent reporters based on the aequorin photoprotein which generates light by a chemical reaction; ii) the single fluorescent proteins like the camgaroos, G-Camps, pericams, 'case sensors', and grafted EF-hands and iii) the chameleon-type reporters in which a Ca^{2+}-responsive element is inserted between two fluorescent proteins so that upon Ca^{2+} binding, an alteration in the efficiency of fluorescence resonance energy transfer (FRET) occurs. The single fluorescent GECIs contain a circularly permutated GFP, a Ca^{2+} binding domain (CaM) and an M13 domain. Upon Ca^{2+} binding, conformational changes in the CaM–M13 complex induce fluorescence changes in the circularly permutated enhanced GFP (Tian *et al.*, 2012).

There are no reports on the use of these indicators in *T. gondii* but our laboratory has successfully been able to isolate parasites expressing GCamp3, which can be used to measure Ca^{2+} during the lytic cycle of the parasite (Borges, Liu *et al.*, manuscript in preparation).

10.3 REGULATION OF $[Ca^{2+}]_I$ IN *TOXOPLASMA GONDII*

Calcium plays an essential role in several critical processes of the lytic cycle of *Toxoplasma gondii*. This unicellular organism needs to regulate the cytosolic Ca^{2+} within a narrow physiological range. Ca^{2+} signalling, which possibly triggers downstream effects leading to motility, invasion, and egress, is probably initiated by the mobilization of Ca^{2+} from intracellular stores. In mammalian cells the second messenger inositol 1,4,5-trisphosphate (IP_3) gates an intracellular calcium channel (IP3R), which will open and release Ca^{2+} from the ER into the cytoplasm of the cell. The information for this signalling pathway in *T. gondii* is fragmentary. A phosphoinositide phospholipase C has been biochemically characterized (Fang *et al.*, 2006) although it is not known how it is regulated or its mode of activation. There is pharmacological evidence for the presence of channels responsive to IP_3 (Lovett and Sibley, 2003) although there is no gene annotated in ToxoDB that resembles any of the mammalian IP_3R and/or ryanodine receptors (RyRs). There is an ER calcium pool, sensitive to thapsigargin, and the Ca^{2+}-ATPase involved in the pumping of Ca^{2+} from the cytosol into the ER has been characterized (Nagamune *et al.*, 2007). The role of the mitochondria in Ca^{2+} uptake is not clear. No orthologue of the mitochondrial Ca^{2+} uniporter has been found and no Ca^{2+} uptake has been detected in energized mitochondria (Vercesi *et al.*, unpublished results). In addition, *T. gondii* contains several unique organellar compartments that potentially could contribute to diverse Ca^{2+} transients necessary for vital functions within the parasites. These include the apicoplast, a remnant plastid derived from a secondary endosymbiotic event, and various acidic calcium stores (Patel and Docampo, 2010) including the acidocalcisomes (Rohloff *et al.*, 2011) and the recently described plant-like vacuole (PLV) (Miranda *et al.*, 2010).

10.3.1 Ca^{2+} Transport Across the Plasma Membrane

The $[Ca^{2+}]_i$ in tachyzoites is maintained at 70 $\pm$ 6 nm in the absence of extracellular Ca^{2+} (with the Ca^{2+} chelator EGTA added to the medium) and 100 $\pm$ 9 nm in the presence of 1 mM extracellular Ca^{2+}, as detected in Fura2-loaded cells (Moreno and Zhong, 1996). These concentrations are in the range observed in many studies with eukaryotic cells (Grynkiewicz *et al.*, 1985).

Ca^{2+} enters eukaryotic cells across the plasma membrane through a number of channels, some of which are under the control of receptors (receptor-operated Ca^{2+} channels), the potential across the plasma membrane (voltage-gated Ca^{2+} channels), and the content of intracellular Ca^{2+} stores (store-operated Ca^{2+} channels), whereas others appear to be nonselective leak channels (Tsien and Tsien, 1990). Voltage-dependent Ca^{2+} channels have been detected in free-living protists (Plattner *et al.*, 2012; Ogura and Takahashi, 1976), and the most recent annotation of ToxoDB indicates the presence of a cation channel family protein (TGME49_205265) with 20 predicted TM domains, which had previously been predicted to be a voltage gated Ca^{2+}-channel (Nagamune and Sibley, 2006; Prole and Taylor, 2011). Further research is needed to characterize this protein. Other potential Ca^{2+} channels in *T. gondii* are proposed in both publications (Nagamune and Sibley, 2006; Prole and Taylor, 2011).

The active export of Ca^{2+} from eukaryotic cells is accomplished by the action of a Na^+/Ca^{2+} exchanger and a Plasma Membrane Ca^{2+}-ATPase (PMCA). There is no biochemical evidence for the presence of a Na^+/Ca^{2+} exchanger in *T. gondii*, which is likely the case because there are no reports of the presence of Na^+/Ca^{2+} exchangers in early eukaryotes (Pozos *et al.*, 1996). In contrast, a PMCA-type Ca^{2+}-ATPase (*Tg*A1) has been characterized in these parasites, which is located in the plasma membrane and acidocalcisomes (Luo *et al.*, 2001) (Section 10.4.3). Mammalian PMCAs are activated by the Ca^{2+} binding protein calmodulin (CaM), and biochemical evidence for CaM stimulation has been reported for a Ca^{2+}-ATPase activity from *T. gondii* (Bouchot *et al.*, 2001). However, *Tg*A1 appears to lack a typical CaM-binding domain, which might suggest the presence of a different domain able to bind CaM. A second gene encoding for a potential PMCA has been found in the *T. gondii* genome (TGME49_233770) (Gajria *et al.*, 2008). The deduced amino acid sequence (1448 aa) shows 45% identity with *Tg*A1 (Luo *et al.*, 2001). Three other genes are annotated (TGME49_278660, TGME49_318460 and TGME49_247690) that are or were predicted to encode for Ca^{2+}-ATPases and await further research. We have localized TGME49_278660 to the plasma membrane using a C-terminal tagging strategy (Chasen *et al.*, manuscript in preparation) but we have no biochemical data on the function of this protein.

10.3.2 Ca^{2+}-Binding Proteins

Calcium binding proteins contain a highly conserved helix–loop–helix structure or EF hand motif. Typically, EF hand motifs occur in pairs (EF hand domain) and facilitate the cooperative binding of two Ca^{2+} ions per domain (Luan *et al.*, 2002). Calcium binding proteins containing a single or odd number of EF hand motifs have been reported in both bacteria and eukaryotes and are thought to function via dimerization mechanisms (Grabarek, 2006). Some of these Ca^{2+} binding proteins such as calmodulin (CaM) act as Ca^{2+} receptors. Other proteins appear to act as Ca^{2+} storing devices (e.g. calsequestrin, calreticulin families). A large number of EF hand domain-containing proteins are found in the *T. gondii* genome (ToxoDB) (Gajria *et al.*, 2008) and very few of these proteins have been characterized.

Structurally, calmodulins are acidic proteins comprised of two globular domains (each with a pair of EF hands) linked to each other by a flexible helical region (Gifford *et al.*, 2007). *T. gondii* CaM is a small (16 kDa) acidic calcium-binding protein with four Ca^{2+}-binding sites (EF hands) and a high level of identity (92.5%) with human CaM (Seeber *et al.*, 1999). By immunofluorescence analysis using monoclonal antibodies reactive against CaM from different species, *T. gondii* CaM was found in the apical end of released tachyzoites (together with actin and myosin), and also beneath the membrane in intracellular parasites (Pezzella-D'Alessandro *et al.*, 2001). Immunogold electron microscopy using monoclonal antibodies against mammalian CaM confirmed the localization of CaM in the anterior region of tachyzoites, together with actin (Song *et al.*, 2004).

The CaM inhibitors calmidazolium and trifluoperazine significantly reduced parasite invasion *in vitro* and caused changes in the tachyzoites' shape (Pezzella *et al.*, 1997) although, in the case of calmidazolium, this effect could also be due to its ability to increase $[Ca^{2+}]_i$ and stimulate microneme secretion (Wetzel *et al.*, 2004).

Subcellular localization studies using green fluorescence protein (GFP)-tagged chimeras have also localized three calmodulin-like proteins to the conoid (TGME49_046930 and TGME49_062010) (Hu *et al.*, 2006) and plasma membrane (TGME49_305050) (Moreno *et al.*, unpublished observations). These findings suggest a speciallization of the different CaMs to distinct signalling processes in apicomplexan parasites.

Calcineurin B-like (CBL) proteins have been identified only in higher plants and in some protist genomes (Batistic and Kudla, 2009). These EF hand motif containing proteins interact with a group of serine/threonine protein kinases called CBL-interacting protein kinases (CIPK). The *Toxoplasma* genome identifies one gene that could encode a functional calcineurin B-like protein (TGME49_013800). However, no CIPKs were identified to suggest the existence of a functional CBL–CIPK Ca^{2+} decoding system in these parasites.

An important group of Ca^{2+} binding proteins is comprised of the many enzymes that are modulated through direct interactions with Ca^{2+}. These include kinases (e.g. Ca^{2+}-dependent protein kinases (CDPKs) and calcium-calmodulin-dependent kinases (CCaMKs)), proteases, phosphatases, synthases, nucleoside triphosphatases (NTPases), etc. (Moreau, 1987; Kretsinger, 1976). An investigation of EF hand motif-containing proteins in *T. gondii* identified several CDPKs, Ser/Thr protein phosphatases, tRNA synthases, ubiquitin C-terminal hydrolases and cathepsins that potentially can undergo Ca^{2+}-induced effector responses.

In addition, the *Toxoplasma* genome contains several hypothetical proteins, heat-shock proteins, centrin/troponin C-like proteins, which likely exhibit vital functions in these parasites such as buffering and species-specific signalling processes (Gajria *et al.*, 2008).

The CDPK family constitutes a group of kinases that are only found in plants and protists. In plants, CDPKs mediate Ca^{2+} signals that regulate a diverse number of pathways including cell cycle progression and stress responses. The canonical CDPK is composed of an amino-terminal serine/threonine kinase domain, followed by a junction domain (also known as the autoinhibitory domain) that connects to the carboxy-terminal calmodulin-like domain (Klimecka and Muszynska, 2007). The calmodulin-like domain typically consists of four EF hand domains for Ca^{2+} binding. The autoinhibitory domain apparently regulates CDPK by interacting with the kinase domain and acting as a pseudosubstrate. Binding of Ca^{2+} to the EF hand domains relieves the autoinhibition (Harmon *et al.*, 2000).

Phylogenetic analyses of CDPKs in *T. gondii* indicate the presence of 12 CDPKs (Billker *et al.*, 2009; Nagamune *et al.*, 2008b). CDPKs have been implicated as a mechanistic link between Ca^{2+} signalling and virulence-related traits including differentiation, motility, invasion and egress. The protein kinase inhibitor KT5926 was shown to block microneme secretion in *T. gondii* (Dobrowolski *et al.*, 1997). Subsequent work demonstrated that KT5926 was targeting a CDPK and that inhibition of Ca^{2+}-dependent kinase activity correlated very well with inhibition of microneme secretion (Kieschnick *et al.*, 2001). Genetic ablation of CDPK1, the orthologue of CDPK4 in *P. berghei*, demonstrated that its activity is essential for motility, invasion and egress of *T. gondii* (Lourido *et al.*, 2010). TgCDPK2 may not be expressed in *T. gondii* tachyzoites. The activity of TgCDPK3, an orthologue of PfCDPK1, has been characterized recently using a CaMK as peptide substrate. Ca^{2+} stimulates the enzyme and its mechanism of activation was characterized (Wernimont *et al.*, 2010). More recent studies have revealed a role for this kinase in egress (discussed below) (Garrison *et al.*, 2012; McCoy *et al.*, 2012; Lourido *et al.*, 2012).

Future work on CDPKs will help understand the role of Ca^{2+} signalling in the lytic cycle of *T. gondii* (see below). One important aspect would be to find the specific substrates of CDPKs, which will reveal critical information about their role in invasion and differentiation. Additionally, Ca^{2+}-dependent protein kinases exhibit a significant level of crosstalk with other protein kinases, most notably the cyclic nucleotide-dependent kinases (Klimecka and Muszynska, 2007; Billker *et al.*, 2009). Taking this into account along with the large number and structural diversity of Ca^{2+}-dependent kinases, it is apparent that further research on the CDPKs in apicomplexans is likely to be very rewarding in terms of cell biology and development of potential drug targets for clinical use.

10.4 CALCIUM SOURCES

10.4.1 Endoplasmic Reticulum

The largest Ca^{2+} store in cells is typically the endoplasmic reticulum, with local concentrations reaching millimolar levels. The endoplasmic reticulum also possesses two independent pathways for Ca^{2+} influx and efflux. The influx is catalysed by the very well known sarco-endoplasmic reticulum Ca^{2+}-ATPase (SERCA), which actively translocates 2 Ca^{2+} for the hydrolysis of 1 ATP molecule. Evidence for the presence of a SERCA-type Ca^{2+}-ATPase in *T. gondii* was first provided by experiments using Fura2-loaded tachyzoites in which thapsigargin, a very specific inhibitor of this pump when used at low concentrations (Thastrup *et al.*, 1990), was shown to increase $[Ca^{2+}]_i$ in tachyzoites (Moreno and Zhong, 1996). Molecular evidence for the presence of a SERCA-type Ca^{2+}-ATPase in *T. gondii* has been reported (Nagamune *et al.*, 2007). The gene encoding this pump was able to complement yeast deficient in Ca^{2+} pumps providing evidence of its function as a Ca^{2+} pump, and the encoded protein has an apparent molecular mass of 120 kDa. Known inhibitors of this pump, such as thapsigargin (Thastrup *et al.*, 1990), or artemisinin (Eckstein-Ludwig *et al.*, 2003), were able to stimulate microneme secretion, a process that relies on elevated $[Ca^{2+}]_i$ (see below, under 'Ca^{2+}- signaling').

Ca^{2+} release from the endoplasmic reticulum of eukaryotic cells is mediated by ryanodine (RyR) and inositol 1,4,5-trisphosphate (IP_3R) channels. RyR are activated by a rise in $[Ca^{2+}]_i$ (Ca^{2+}-induced Ca^{2+} release, CICR). In addition there are RyR-like channels activated by cyclic ADP-ribose (cADPR), sphingosine and a distinct Ca^{2+}-release pathway activated by nicotinic acid adenine dinucleotide phosphate (NAADP). The *T. gondii* phosphoinositide-specific phospholipase C (TgPI-PLC), the enzyme that generates the second messengers inositol 1,4,5-trisphosphate (IP_3) and diacylglycerol (DAG) was characterized biochemically (Fang *et al.*, 2006). IP_3/ryanodine-sensitive stores had been postulated to be present in *T. gondii* on the basis of pharmacological studies (Lovett *et al.*, 2002). Treatment with ethanol increased IP_3 and $[Ca^{2+}]_i$ and this pathway was sensitive to inhibitors of IP_3R channels. *T. gondii* also responded to agonists of cADPR-gated channels such as ryanodine and caffeine (Lovett *et al.*, 2002). Evidence for the presence of cADPR cyclase and hydrolase activities, the two enzymes that control cADPR levels, was also found (Chini *et al.*, 2005). *T. gondii* microsomes that were loaded with $^{45}Ca^{2+}$ released Ca^{2+} when treated with cADPR, and the RyR antagonists 8-bromo-cADPR and ruthenium red blocked this response. Although *T. gondii* microsomes also responded to IP_3, the inhibition profiles of these Ca^{2+}-release channels were mutually exclusive (Chini *et al.*, 2005). IP_3Rs are present in unicellular organisms like ciliates (Ladenburger and Plattner, 2011) and trypanosomes (Huang *et al.*, 2013) but the molecular entities of the IP_3R or RyR channels in *T. gondi* remain elusive. This could be due to lack of homology with the channels of animal cells, as occurs in plants that otherwise respond to the same second messengers (Nagata *et al.*, 2004).

10.4.2 Mitochondria

Mitochondria possess a high capacity to sequester Ca^{2+} although under physiologic conditions, total mitochondrial Ca^{2+} levels and free Ca^{2+} reflect and parallel cytosolic Ca^{2+}. Mammalian cells and trypanosomes possess a mitochondrial calcium uniporter (MCU), which utilizes the electrochemical gradient generated by respiration or ATP hydrolysis promoting the entry of Ca^{2+} into the mitochondria (Docampo and Lukes, 2012). Calcium efflux, on the other hand, takes place by a different pathway, which appears to catalyse the electroneutral exchange of internal Ca^{2+} by external sodium or protons. Biochemical evidence for mitochondrial Ca^{2+} uptake is available in some malaria parasites (Gazarini and Garcia, 2004) but experimental data from our own laboratory showed that the *T. gondii* mitochondria does not take up Ca^{2+} upon energization (Vercesi and Moreno, unpublished observations). Unlike the mammalian mitochondria, where intracellular Ca^{2+} regulates the activity of several dehydrogenases, no such Ca^{2+}-regulated dehydrogenases have been reported in *T. gondii.*

10.4.3 Acidocalcisomes

T. gondii tachyzoites possess a significant amount of Ca^{2+} stored in an acidic compartment which can be released into the cell cytoplasm with nigericin (a K^+/H^+ exchanger) or the weak base NH_4Cl (Fig. 10.1A) (Moreno and Zhong, 1996). Further characterization of this compartment led to the discovery of acidocalcisomes, which were first described and characterized in *Trypanosoma brucei* (Vercesi *et al.*, 1994) and *T. cruzi* (Docampo *et al.*, 2005) (Fig. 10.2). These organelles are characterized by their acidic

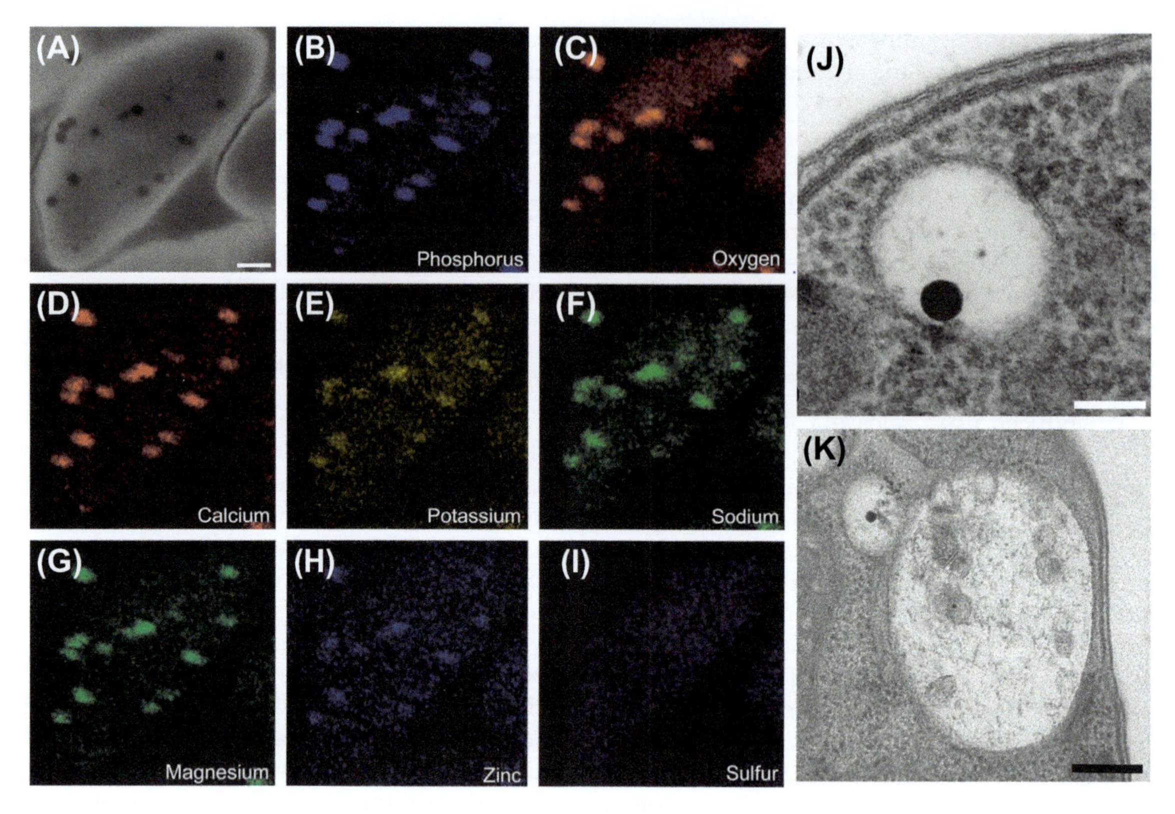

FIGURE 10.2 **Acidocalcisomes of *T. gondii*.** Transmission electron microscopy (TEM) of a *T. gondii* tachyzoite. Visualization of acidocalcisomes and spatial mapping of elemental distribution by whole-cell electron microscopy. Acidocalcisomes are clearly identified as round, electron dense structures seen throughout the cell (A). Specific elements are mapped from the cells identified in A (B–I). Scale bar is 0.5 μM. *A–I were published as supplemental material in Rohloff et al., PloS One. 6:e18390, 2011.* Thin section of a *T. gondii* tachyzoite showing one empty acidocalcisome (J). Note the small amount of electron dense material accumulated in the matrix of the organelle. Scale bar is 0.1 μM. Interaction between acidocalcisome and the PLV (K). Scale bar is 0.5 μm. *Reproduced from Miranda* et al. *Mol. Microbiol., 76:1358, 2010.*

nature, high density (both in weight and by electron microscopy), and high content of pyrophosphate, polyphosphate (polyP), calcium, magnesium, and other elements (Fig. 10.2A–I) (Docampo *et al.*, 2005; Rodrigues *et al.*, 2002). Acidocalcisomes are similar to the volutin or metachromatic granules first described more than a hundred years ago (Kunze, 1907) in Coccidia, and detected in 1966 in *T. gondii* for their ability to stain red when treated with toluidine blue (metachromasia) (Mira Gutierrez, 1966). They were also named 'black granules' (Bonhomme *et al.*, 1993) and more recent work in *Leishmania* (Besteiro *et al.*, 2008) and *T. brucei* (Huang *et al.*, 2011) suggested that acidocalcisomes are lysosome-related organelles.

In thin sections, the acidocalcisomes of *T. gondii* appear as empty vesicles occasionally bearing an electron dense material that sticks to the inner face of the membrane (Fig. 10.2J) (Miranda *et al.*, 2008). Tachyzoites show acidocalcisomes with different degrees of preservation of the electron dense material, from an almost totally empty vacuole to others containing a considerable amount of electron dense material (Fig. 10.2A, J). Occasionally, acidocalcisomes of

Toxoplasma are observed fusing with each other or with larger organelles (Fig. 10.2K).

A large number of acidocalcisomes can be seen in electron spectroscopic images of whole cells directly dried on Formvar-coated grids (Fig. 10.2A) (Miranda *et al.*, 2008). The advantage of this type of preparation is the observation of the whole parasite (and whole organelles) without the addition of fixatives and other chemicals used in the routine procedures for transmission electron microscopy. This procedure reduces significantly the extraction of material from the acidocalcisomes and, therefore, allows the observation of the organelle in its 'native' state (Luo *et al.*, 2001). In these preparations, acidocalcisomes appear as spherical electron dense organelles dispersed throughout the cell body (Fig. 10.2A). X-ray microanalysis (Luo *et al.*, 2001) reveals considerable amounts of oxygen, sodium, magnesium, phosphorus, chlorine, potassium, calcium and zinc concentrated in these compartments, similar to what has been reported previously in the acidocalcisomes of trypanosomatids (Fig. 10.2A–I) (Scott *et al.*, 1997; Rodrigues *et al.*, 1999; Miranda *et al.*, 2004; LeFurgey *et al.*, 2001).

A Ca^{2+}-ATPase from *T. gondii* (TgA1) was characterized and localized to acidocalcisomes (Luo *et al.*, 2001). The TgA1 gene was able to complement yeasts deficient in the vacuolar Ca^{2+}-ATPase gene *PMC1*, providing genetic evidence for its function (Luo *et al.*, 2001). The protein product is closely related to the family of plasma membrane calcium ATPases (PMCA). In addition, Ca^{2+} transport activity was recently shown in an acidocalcisome fraction, which was insensitive to thapsigargin and sensitive to bafilomycin A_1 (Rohloff *et al.*, 2011). An analysis of conserved core sequences of all PMCA-type Ca^{2+}-ATPases has identified a subcluster within these sequences that is formed by the acidocalcisomal Ca^{2+}-ATPase of *T. cruzi, T. brucei, T. gondii* and *Dictyostelium discoideum*, and the vacuolar Ca^{2+}-ATPases of yeast and *Entamoeba histolytica* (Luo *et al.*, 2001). Mutants deficient in TgA1 were shown to have decreased virulence *in vitro* and *in vivo* due to their deficient invasion of host cells (Luo *et al.*, 2005). Biochemical analysis revealed that the tachyzoite polyP content was drastically reduced, and that the basal Ca^{2+} levels were increased and unstable. Ionophore-induced microneme secretion was also impaired. Complementation of *null* mutants with TgA1 restored most functions (Luo *et al.*, 2005).

The acidification of acidocalcisomes is due to the activity of two proton pumps: the *T. gondii* V-H^+-PPase (TgVP1) (Drozdowicz *et al.*, 2003) and a V-H^+-ATPase (Rodrigues *et al.*, 2002). The *TgVP1* gene was cloned and sequenced and a truncated version of the enzyme (without the *N*-terminal) could be functionally expressed in yeast (Drozdowicz *et al.*, 2003). TgVP1 belongs to the K^+-stimulated group of V-H^+-PPases (type I) and its biochemical characterization was done using *T. gondii* purified fractions enriched in the enzyme (Miranda *et al.*, 2010; Rohloff *et al.*, 2011). The activity of the enzyme has been successfully used as a marker for acidocalcisome purification (Rohloff *et al.*, 2011). TgVP1 also localizes to the recently described PLV (see below) (Miranda *et al.*, 2010; Parussini *et al.*, 2010). The V-H^+-ATPase was first identified in *T. gondii* by its sensitivity to bafilomycin A_1, a specific inhibitor of this proton pump when used at low concentrations (Bowman *et al.*, 1988). In experiments using intact tachyzoites loaded with the fluorescent Ca^{2+} indicator Fura2, bafilomycin A_1 was able to release Ca^{2+} from an intracellular compartment of *T. gondii* (Moreno and Zhong, 1996). The V-H^+-ATPase was also shown to play a role in intracellular pH homeostasis (Moreno *et al.*, 1998).

There is also experimental evidence for the presence of Na^+/H^+ and Ca^{2+}/H^+ exchangers in acidocalcisomes of *T. gondii* (Rohloff *et al.*, 2011). In subcellular fractions it was possible to measure a collapse of the proton gradient upon addition of Ca^{2+} in acidocalcisome fractions supporting the participation of a Ca^{2+}/H^+ exchanger

(Fig. 10.3A, trace a). A gene annotated as Manganese resistance 1 protein (TGME49_207910), which encodes for a protein of 501 amino acids was recently cloned and localized to a novel vacuolar compartment proposed to be the PLV. Further studies are needed (Guttery et al., 2013).

Although the Ca^{2+} content of acidocalcisomes is very high (probably in the molar range), most of it is bound to polyP and can be released only upon alkalinization (Moreno and Zhong, 1996) or after polyP hydrolysis (Rodrigues *et al.*, 2002). All acidocalcisomes described so far have been found to have high levels of phosphorus in the form of inorganic pyrophosphate (PP_i) and polyphosphate (polyP). *T. gondii* acidocalcisomes are especially rich in short chain polyP such as $polyP_3$ (Rodrigues *et al.*, 2000; Moreno *et al.*, 2001).

PP_i is a byproduct of many biosynthetic reactions (synthesis of nucleic acids, coenzymes, proteins, activation of fatty acids, and isoprenoid synthesis) and its hydrolysis by inorganic pyrophosphatases makes these reactions thermodynamically favourable. None of these pathways have been found in *T. gondii* acidocalcisomes. One possibility is that PP_i is there as a byproduct of the hydrolysis of polyP or as an intermediate for its synthesis. Only three reactions are known to use PP_i in *T. gondii*, one catalysed by the phosphofructokinase (Peng *et al.*, 1995) another catalysed by the V-H^+-PPase responsible for acidification of acidocalcisomes (Rodrigues *et al.*, 2000; Drozdowicz *et al.*, 2003) and PLV (Miranda *et al.*, 2010) and a third one catalysed by an inorganic pyrophosphatase (Pace *et al.*, 2011).

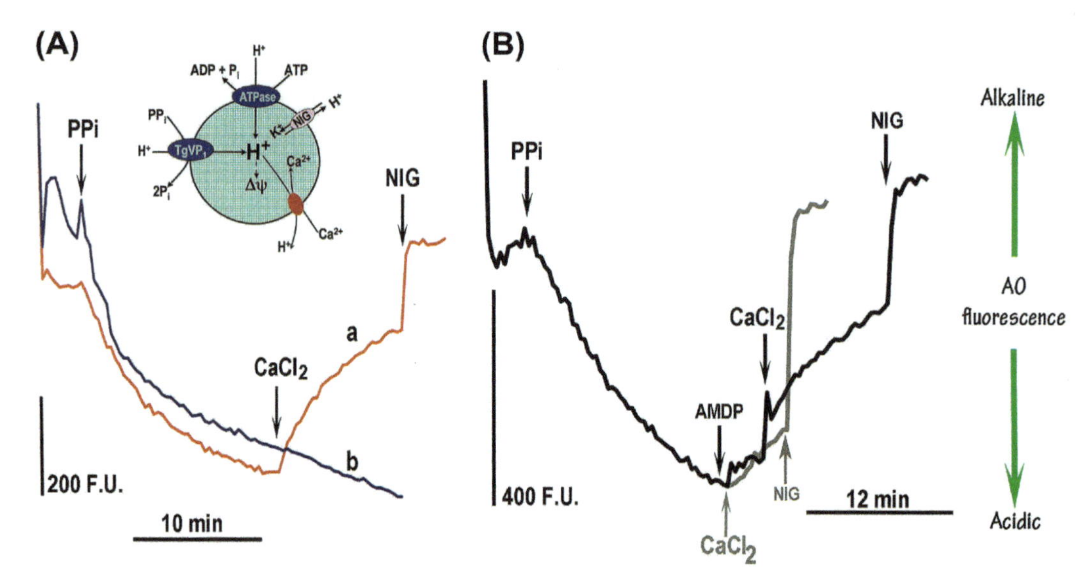

FIGURE 10.3 **Calcium proton exchange activities in enriched fractions for the PLV and acidocalcisomes.** Addition of phyrophosphate (PP_i) leads to acidification by a V-H^+-PPase, which results in a decrease in green fluorescence (526 nm) of Acridine Orange. Addition of $CaCl_2$ collapses the proton gradient leading to alkalynization of the organelle and subsequent increase in green and decrease in orange fluorescence. This proton release is because of an exchange mechanism present in the membrane of the organelle, which exchanges protons for calcium. The cartoon at the top shows an illustration of the Ca^{2+}/H^+ exchange activity (A). *This tracing is part of a figure published in Miranda* et al. *Mol. Microbiol., 76:1358, 2010.* Similar activity to that in 'A' was also observed with a purified acidocalcisome fraction (B). AMDP is a specific inhibitor of the V-H^+-PPase and upon addition of $CaCl_2$ there is a collapse of the proton gradient due to the same mechanism explained in panel 'A'. Nigericin (NIG) exchanges protons for potassium ions leading to full collapse of the proton gradient. *This figure was published in Rohloff* et al., *PloS One. 6:e18390, 2011.*

The low sulphur content detected by elemental analysis suggested a low content of proteins within acidocalcisomes (Fig. 10.2I). However, in addition to the proton and calcium pumps, another enzymatic activity has been detected. Acidocalcisomes from *T. gondii* were shown to contain a polyphosphatase activity (Rodrigues *et al.*, 2002). A recent study localized a p-glycoprotein (P-gp) to the acidocalcisome by IFA and ImmunoEM using a commercial antibody (Bottova *et al.*, 2010). The specific P-gp inhibitor GF120918 had an effect on parasite motility, microneme secretion and egress from the host cell, all cellular processes known to depend on Ca^{2+} signalling in the parasite.

10.4.3.1 Methods to Study Acidocalcisomes

In addition to the light and electron microscopy techniques used to visualize acidocalcisomes described above, spectrofluorometric techniques using Fura2-loaded parasites have been used to detect changes in acidocalcisome Ca^{2+} (Moreno and Zhong, 1996; Rodrigues *et al.*, 2002). Acidocalcisome Ca^{2+} is released when the cells are submitted to alkalinizing agents (NH_4Cl), inhibitors of the V-H^+-ATPase (bafilomycin A_1), and Na^+/H^+ (monensin), and K^+/H^+ exchangers (nigericin), or when incubated with ionophores (ionomycin) after their alkalinization (Moreno and Zhong, 1996). Phosphorous compounds were investigated by ^{31}P-NMR spectroscopy (Rodrigues *et al.*, 2000; Moreno *et al.*, 2001), or by biochemical techniques (Rodrigues *et al.*, 2002). Isolation of acidocalcisomes has been done using iodixanol gradient centrifugation using enzymatic markers (Rodrigues *et al.*, 2002; Rohloff *et al.*, 2011). This purification protocol was much improved allowing the isolation of acidocalcisomes with proton transport and Ca^{2+} uptake activities and almost no mitochondrial contamination (Rohloff *et al.*, 2011). In contrast to procedures used before to isolate acidocalcisomes from other cells, this method used lower concentrations of iodixanol in the gradient steps but with the sample added in the middle of the gradient in a 20% iodixanol layer rather than applied to the top of the gradient without added iodixanol. This strategy allowed a better separation of acidocalcisomes from the fragmented and empty membranes ('ghosts') present in the intermediate fractions and a better recovery of the pellet fraction at the base of the gradient (Rodrigues *et al.*, 2002).

10.4.4 Plant-Like Vacuole

The Plant-Like Vacuole (PLV), also named as Vacuolar Compartment (VAC) is a novel organelle in the tachyzoite stage of *T. gondii* visible by light microscopy and with broad similarity to the plant vacuole (Miranda *et al.*, 2010; Parussini *et al.*, 2010). This organelle contains a plant-like vacuolar proton pyrophosphatase (TgVP1), a vacuolar proton ATPase, several cathepsin proteases (TgCPL, TgCPB, others), an aquaporin (TgAQP1), a Na^+/H^+ exchanger (Francia *et al.*, 2011) as well as Ca^{2+}/H^+ exchange activity supporting its similarity to the plant vacuole (Fig. 10.4). It is a dynamic structure that undergoes significant changes as the extracellular tachyzoite invades a host cell and begins replication within the parasitophorous vacuole (Parussini *et al.*, 2010). In the extracellular tachyzoite stage the PLV occupies a large portion of the parasite and is multivesicular (Miranda *et al.*, 2010). Physiological experiments in intact cells loaded with the Ca^{2+} indicator Fura2-AM, showed release of Ca^{2+} from an intracellular store stimulated by glycyl-L-phenylalanine-naphthylamide (GPN), which was different from other Ca^{2+} stores such as the ER and acidocalcisomes (Fig. 10.1B). GPN is specifically hydrolysed in the lysosomes of a variety of different cell types by a cathepsin C protease, and had been used before to demonstrate the presence of Ca^{2+} in lysosome-like compartments (Haller *et al.*, 1996; Srinivas *et al.*, 2002). The PLV contains a cathepsin C as shown by proteomic data of enriched fractions and IFA analysis of cells expressing

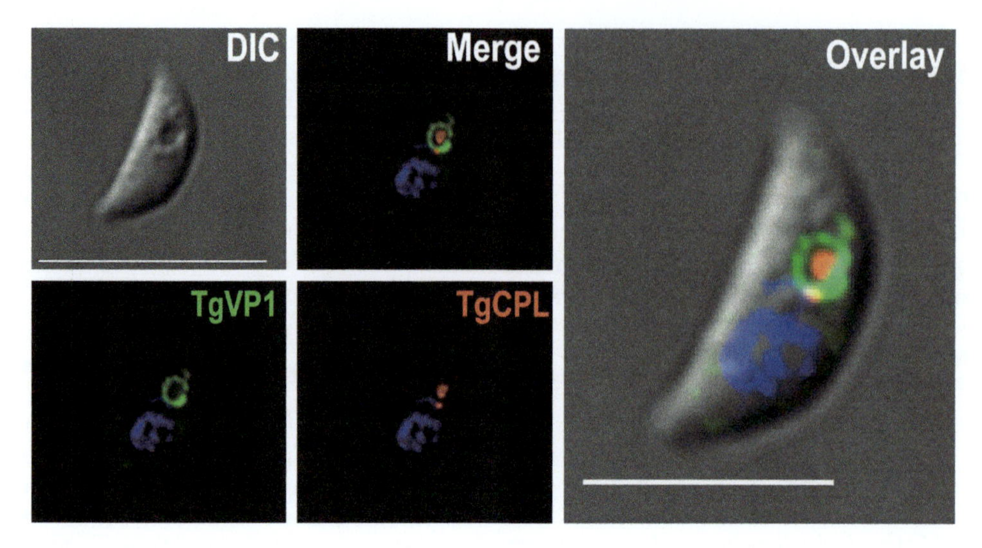

FIGURE 10.4 **Co-localization of TgVP1 with TgCPL to the PLV.** IFA of a *T. gondii* tachyzoite with a rabbit polyclonal antibody generated against a peptide of TgVP1 (1:4000) (green) and mouse anti-TgCPL (1:400) (red) and DAPI (blue). Both antibodies localize to the same compartment (merge and overlay). TgVP1 labels the membrane and TgCPL appears to label the lumen of the organelle. *Reproduced from Miranda* et al. *Mol. Microbiol., 76:1358, 2010.*

a C-terminal tagged CPC gene (Moreno and Carruthers, unpublished). This would indicate that the Ca^{2+}-containing compartment responsible for the cytosolic Ca^{2+} increase in the presence of GPN is the PLV. This Ca^{2+} release has also been implicated in store-operated Ca^{2+} entry (see below). In addition, proton transport measurements in enriched fractions showed Ca^{2+} exchange activity after pyrophosphate-driven acidification (Fig. 10.3A–B). Furthermore, proteomic data and immunofluorescence assays showed the localization of TgA1 to this organelle (Moreno *et al.*, unpublished observations). The localization of both TgVP1 and TgA1 to acidic organelles, acidocalcisomes and PLV, supports our hypothesis that both organelles interact or that the PLV has a role in the biogenesis of the acidocalcisome. The co-localization of TgAQP1 to both PLV and to acidocalcisomes also supports this hypothesis (Fig. 10.5). Biochemical experiments of enriched PLV fractions have shown the presence of several ion exchange mechanisms, one of which being a calcium-proton exchanger (Ca^{2+}/H^+-exchanger). The genome of *T. gondii* does possess an annotated Ca^{2+}/H^+-exchanger (TGME49_207910), however confirmation that this gene encodes for the protein that localizes to the PLV is still lacking. The significance of the PLV as a Ca^{2+} store is being investigated, but due to its large size and dynamic behaviour, it seems likely that the PLV contributes directly to many of the Ca^{2+}-related behaviours that have been documented for *T. gondii*.

10.4.5 Extracellular Calcium and Store-Operated Ca^{2+} Entry

Recent research from our laboratory has demonstrated that the extracellular environment is an important Ca^{2+} reservoir for extracellular tachyzoites. As described above, the intracellular stores of Ca^{2+} are critical for the initiation of many virulence-related traits such as conoid extension (Mondragon and Frixione, 1996), gliding (Wetzel *et al.*, 2004), microneme secretion (Carruthers *et al.*, 1999a; Lovett *et al.*,

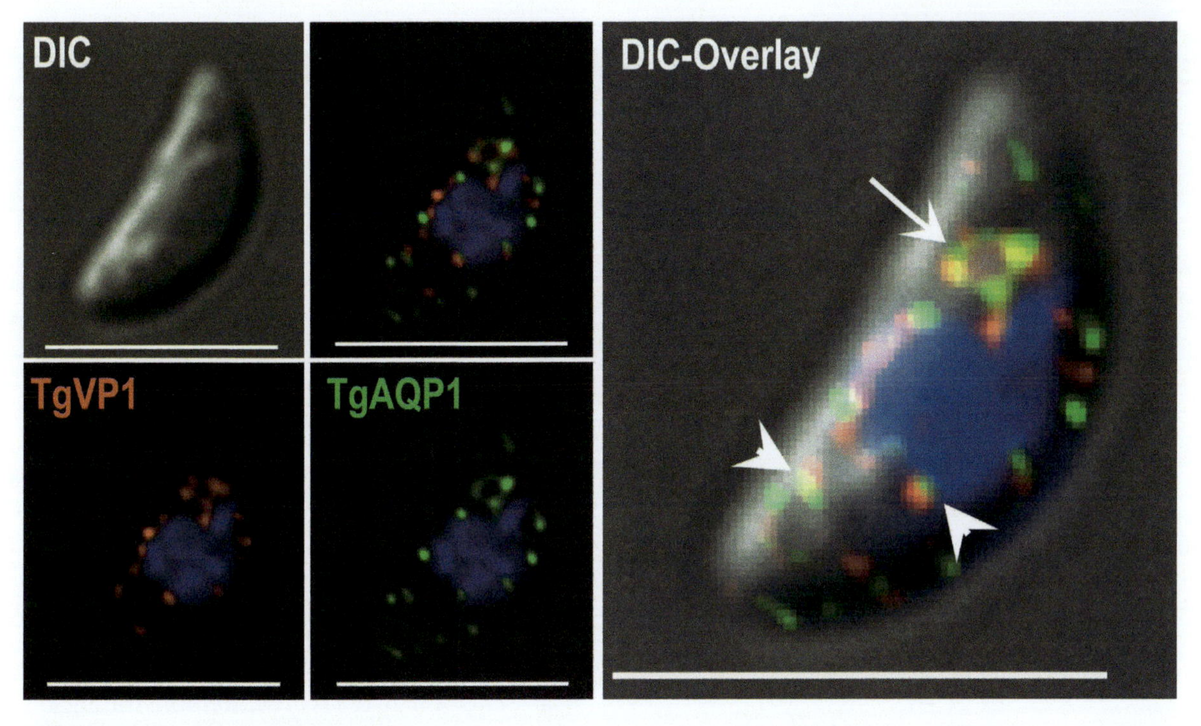

FIGURE 10.5 **TgVP1 and TgAQP1 co-localize in the PLV and in acidocalcisomes of extracellular tachyzoites of *T. gondii*.** Parasites over-expressing TgAQP1 were simultaneously analysed for localization of TgAQP1 using an affinity purified antibody and endogenous TgVP1 using an affinity purified antibody, both described previously (Miranda *et al.*, 2010). Green, TgAQP1 localization; Red, TgVP1 localization. Arrows show areas of TgVP1 and TgAQP1 co-localization to acidocalcisomes. Larger ring structure of co-localization is the plant-like vacuole (PLV).

2002), and invasion (Vieira and Moreno, 2000). However, these traits can experience significant enhancement when parasites are within a Ca^{2+} replete environment. Pace *et al.* (submitted for publication) showed the physiological existence of store-operated Ca^{2+} entry in extracellular tachyzoites (Fig. 10.6). Store-operated calcium entry (SOCE) is defined as the rapid influx of extracellular Ca^{2+} that is prompted by the release of Ca^{2+} from intracellular stores (Bird *et al.*, 2008). In *T. gondii*, SOCE was observed after the release of Ca^{2+} from the ER using the SERCA inhibitor thapsigargin. SOCE was also observed after release of Ca^{2+} from the PLV using GPN (Pace *et al.*, submitted for publication). It seems likely that the intracellular Ca^{2+} stores are sufficient to initiate invasion in tachyzoites, but the access to extracellular Ca^{2+} may play a significant role in the ability to sustain invasion over longer periods of time, a situation that may be very important for *in vivo* infection conditions. The molecular participants for SOCE in mammalian cells have been the centre of many significant discoveries recently (reviewed in Parekh, 2006; Schindl *et al.*, 2009). Interestingly, the genes encoding the proteins involved in these mechanisms in mammalian cells for SOCE (e.g. STIM1 and Orai1), are not present in the *T. gondii* genome (Prole and Taylor, 2011). This opens up a very exciting aspect of Ca^{2+} regulation in *T. gondii*. The use of *T. gondii* as a model system for understanding basal eukaryotic regulatory strategies of Ca^{2+} may prove to be very fruitful.

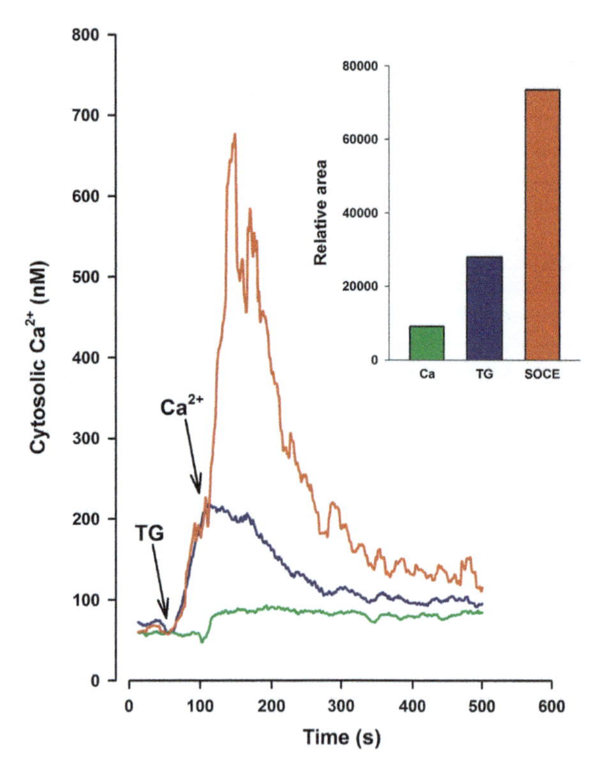

FIGURE 10.6 **Store-operated calcium entry in extracellular tachyzoites of *T. gondii*.** Tachyzoites were loaded with Fura2-AM and fluorescence was monitored on a Hitachi F-4500 fluorometer. Each tracing represents an independent experiment (2.5 ml cuvette volume with 5×10^7 parasites in extracellular buffer conditions). Experiments were as follows: green tracing, 2 mM Ca^{2+} addition at 100 s; blue tracing, 1 μM thapsigargin (TG) at 50 s; red tracing, TG (50 s) followed by Ca^{2+} (100 s). Red tracing shows store-operated calcium entry where TG causes depletion of ER calcium because of its release to the cytoplasm resulting in enhanced calcium entry upon addition of extracellular calcium (compare green and red tracings when calcium added). Inset represents quantification of total calcium entry by integrating area under each tracing shown.

10.5 Ca^{2+} AND CELL FUNCTION IN *TOXOPLASMA GONDII*

T. gondii is an obligate intracellular parasite and relies on cell invasion as a strategy to avoid host response and clearance. Invasion also initiates the parasite lytic cycle leading to intracellular replication and egress ultimately destroying the infected cell (Black and Boothroyd, 2000). A large amount of experimental evidence has shown that fluctuations in the parasite intracellular Ca^{2+} play a key role in several processes of its lytic cycle (Arrizabalaga and Boothroyd, 2004; Moreno *et al.*, 2011). *T. gondii* tachyzoites contain micronemes, specialized apically distributed secretory organelles, which appear to play an important role in the early phase of the invasion process. Increases in intracellular Ca^{2+} mediate microneme secretion even in the absence of host cells, as shown when cells are treated with Ca^{2+} ionophores like ionomycin and A23187, the ATPase inhibitor thapsigargin (Carruthers *et al.*, 1999a), ethanol (Carruthers *et al.*, 1999b; Matthiesen *et al.*, 2003), the anti-calmodulin agent calmidazolium (Wetzel *et al.*, 2004) or the plant hormone abscisic acid (ABA) (Nagamune *et al.*, 2008a). This effect can be blocked with the intracellular Ca^{2+} chelator BAPTA/AM demonstrating that the secretion of micronemes is triggered by an increase in intracellular Ca^{2+}. Studies with parasites pre-loaded with BAPTA/AM have also revealed the role of calcium in conoid extrusion (Mondragon and Frixione, 1996; Del Carmen *et al.*, 2009), gliding motility (Wetzel *et al.*, 2004) and invasion (Vieira and Moreno, 2000; Bouchot *et al.*, 1999).

Alcohols also stimulate microneme secretion, intracellular Ca^{2+} and IP_3 levels (Carruthers *et al.*, 1999b; Lovett *et al.*, 2002). In spite of the lack of genetic evidence for the presence of both ryanodine receptors (RyR) and IP_3R in *T. gondii*, previous studies showed that ryanodine and caffeine enhance Ca^{2+} release and microneme secretion, and that ethanol, which is known to cause Ca^{2+} release, stimulates an increase in inositol 1,4,5-trisphosphate (IP_3) (Lovett *et al.*, 2002). It has been postulated that the effect of alcohol on Ca^{2+} may occur through the stimulation of a phosphoinositide phospholipase C (PI-PLC) (Fang *et al.*, 2006), which would lead to an increase in the levels of IP_3 and release

Ca^{2+} from the ER. In addition, xestospongin C, an IP_3 receptor antagonist, inhibited microneme secretion and blocked parasite attachment and invasion of host cells (Lovett *et al.*, 2002). These studies indicated that IP_3 is acting as a second messenger and *T. gondii* possesses an intracellular Ca^{2+} release channel with properties of the IP_3/ryanodine receptors (RyR) superfamily. There is also evidence showing intracellular Ca^{2+} release by cyclic ADP ribose (cADPR). Inhibition of this pathway with the non-hydrolysable analogue 8-Br-cADPR or the RyR inhibitor dantrolene decreased microneme secretion and gliding motility (Chini *et al.*, 2005). Taken together, these results indicate that both IP_3 and RyR channels are important for efficient motility and cellular invasion, suggesting that they may work cooperatively, as in other cells (Chini *et al.*, 2005).

Changes in the intracellular Ca^{2+} concentration of *T. gondii* during their interaction with host cells have been directly demonstrated by digital fluorescence microscopy of parasites loaded with Fura2 (Vieira and Moreno, 2000) and by time-lapse microscopy of parasites loaded with Fluo4 (Lovett and Sibley, 2003). Loading tachyzoites with BAPTA-AM, at concentrations able to chelate intracellular Ca^{2+}, but not with the chemical analogue half-BAPTA-AM, which does not chelate Ca^{2+}, prevents Ca^{2+} transients and decreases host invasion (Vieira and Moreno, 2000). Inhibition of microneme release by chelation of intracellular Ca^{2+} with BAPTA-AM also inhibits parasite invasion of host cells (Carruthers *et al.*, 1999a). In summary, these results indicate that a Ca^{2+} increase in the parasite precedes the events associated with host cell invasion like conoid extrusion, microneme secretion, and invasion. An informative mutant is the TgA1 *null* parasites, which have an invasion defect, which correlates with a reduced virulence *in vivo* (Luo *et al.*, 2005). These cells have altered intracellular Ca^{2+} levels, and are unable to maintain physiological Ca^{2+} levels under experimental conditions, and as a result, are deficient in microneme secretion (Luo *et al.*, 2005).

Changes in intracellular Ca^{2+} have also been involved in *T. gondii* egress from the host cells on the basis of the use of Ca^{2+} ionophores (Endo *et al.*, 1982; Black *et al.*, 2000; Pingret *et al.*, 1996) reviewed in Arrizabalaga and Boothroyd (2004). Addition of the Ca^{2+} ionophore A23187 to infected macrophages stimulated the movement and egress of tachyzoites resulting in host cell lysis (Endo *et al.*, 1982). Most of the parasite population exited from infected human foreskin fibroblasts after only a 2 min exposure to 1 μM A23187 and this event was temperature dependent (Black *et al.*, 2000). The importance of Ca^{2+} for *T. gondii* exit from the host cell was confirmed by experiments showing that microinjection of intracellular Ca^{2+} also stimulated the exit of the parasites (Schwab *et al.*, 1994). A *T. gondii* mutant that is altered in its response to the Ca^{2+} ionophore A23187 was found to be defective in a Na^+/H^+ exchanger located on the parasite's plasma membrane. These mutant cells have increased levels of intracellular Ca^{2+}, which explains its decreased sensitivity to A23187 (Arrizabalaga *et al.*, 2004).

A role for a *T. gondii* PI-PLC in parasite egress has also been postulated on the basis of studies with the PI-PLC inhibitor U73122 (Moudy *et al.*, 2001). It was shown that permeabilized *Toxoplasma*-infected cells preincubated with U73122 but not with the inactive analogue U73343, prevented parasite egress in the presence of extracellular buffer, and it was proposed that parasite egress depended on the intracellular Ca^{2+} increase stimulated by the decrease in the external K^+ concentration (Moudy *et al.*, 2001). Since the inhibitor U73122 apparently affects PI-PLC activity in mammalian cells through its effects on heterotrimeric G proteins (Thompson *et al.*, 1991), and these have not been described in *T. gondii*, direct measurements of the products of *Tg*PI-PLC activity (IP_3, DAG) will be necessary to confirm this proposal.

Exposure of tachyzoites to 5 mM dithiotretitol (DTT) was also shown to activate egress of previously nonmotile intravacuolar parasites within 60 seconds. This was accompanied by an increase in intra-parasitophorous vacuole (PV) fluorescence ratio of Indo1-loaded infected human fibroblasts. The parasite activation and Ca^{2+} increase were prevented by chelation of extracellular Ca^{2+} by EGTA and BAPTA-AM, although ionomycin was still able to increase Ca^{2+} in the PV, and motility and egress of the parasite (Stommel *et al.*, 1997). Since DTT was known to activate the nucleoside triphosphate hydrolase (NTPase) of the parasite, it was proposed that the hydrolysis of ATP would render the SERCA-Ca^{2+}-ATPase non-functional leading to Ca^{2+} leaking into the cytoplasm which would start motility and egress (Stommel *et al.*, 2001). Thus, infected cells treated with DTT exhibit a Ca^{2+} increase resulting in egress and host cell lysis (Stommel *et al.*, 1997).

Previous work by Black *et al.* (Black *et al.*, 2000) hypothesized that the origin of the Ca^{2+} required for *T. gondii* egress could come from the extraparasitic environment. In this work it was shown that treating the infected host cell with BAPTA-AM, expected not to enter the parasite, inhibited egress. In addition, selectively permeabilizing the host cell with 0.005% saponin was able to induce egress, which was blocked by chelating extracellular Ca^{2+} with EGTA (Black *et al.*, 2000; Arrizabalaga and Boothroyd, 2004). However, studies by Moudy *et al.* (2001) argued that the intraparasitic Ca^{2+} is sufficient to induce *T. gondii* exit, since the absence of Ca^{2+} in the media did not inhibit egress of parasites after artificially permeabilizing host cells with alpha-toxin. In addition, this study showed that the loss of K^+ by the host cell was able to trigger parasite egress suggesting that Ca^{2+} mobilization is a downstream event (Moudy *et al.*, 2001; Arrizabalaga and Boothroyd, 2004).

Calcium dependent protein kinases (CDPKs) also play a role in *T. gondii* egress. Specifically, the work of Moudy *et al.* (2001) showed that KT5926, a CDPK1 inhibitor, blocked parasite egress. The ionophore A23187 did not restore egress inhibition indicating that the effect of this compound is because of the inhibition of Ca^{2+}-dependent effectors (Moudy *et al.*, 2001). The laboratory of Arrizabalaga (Garrison *et al.*, 2012) performed genetic screens for egress mutants. Four mutants were isolated, all of them harbouring a missense mutation in the same gene, TgCDPK3, encoding a Ca^{2+}-dependent protein kinase. All four mutations were predicted to alter key regions of the *TgCDPK3* gene and this was confirmed by biochemical studies of each mutant recombinant form. Complementation with a wild type copy of CDPK3 coding sequence was able to restore the rapid egress response to A23187 at levels similar to those produced by the wild type strain. Thus, a role of CDPK3 in regulating the *T. gondii* egress is suggested, although its mechanism of action needs to be precisely elucidated (Garrison *et al.*, 2012). Recent work postulates that the role of TgCDPK3 is to rapidly respond to Ca^{2+} signalling in specific ionic environments resulting in up-regulating processes required for gliding motility (McCoy *et al.*, 2012). In addition, Lourido *et al.* (2012) demonstrated that both invasion and egress are controlled by different kinases: both CDPK1 and CDPK3 were required for egress but only CDPK1 was required during invasion.

Secretion of microneme proteins is also important for motility of *T. gondii*. This is the result of the binding of the adhesion proteins to the substratum, and the link of the membrane to the cytoskeleton (Wetzel *et al.*, 2004). *T. gondii* parasites loaded with the Ca^{2+} indicator Fluo4 were analysed by live imaging and periodic oscillation in the intracellular Ca^{2+} levels were found to be linked to gliding of the parasites. Caffeine treatment blocked motility because it causes a prolonged elevation in intracellular Ca^{2+} while calmidazolium treatment prolonged motility, stimulated microneme secretion and increased the frequency of Ca^{2+} oscillations (Wetzel *et al.*, 2004).

10.6 CONCLUSIONS

Ca^{2+} homeostasis in *T. gondii* differs in several aspects from the processes that occur in other eukaryotic cells, providing great opportunities for targeting them for new therapies (Fig. 10.7). Acidocalcisomes are distinct acidic calcium stores present in *T. gondii*, in which Ca^{2+} is mostly bound to polyP, although no information is available on second messengers involved in Ca^{2+} release from these organelles. Further studies are necessary to understand the biogenesis and function of acidocalcisomes in *T. gondii*. PP_i, polyP, and cations are accumulated in large amounts in acidocalcisomes but both their transport mechanism and functions in *T. gondii* are largely unknown. Ca^{2+}-ATPases are present but apparently different from their mammalian counterparts. SERCA-type Ca^{2+}-ATPases are sensitive to artemisinin, which does not inhibit the mammalian pump, while the PMCA-type Ca^{2+}-ATPase, which also localizes in acidocalcisomes, does not possess a typical calmodulin-binding domain.

A large compartment, which contains Ca^{2+} released by cathepsin C substrate GPN, has

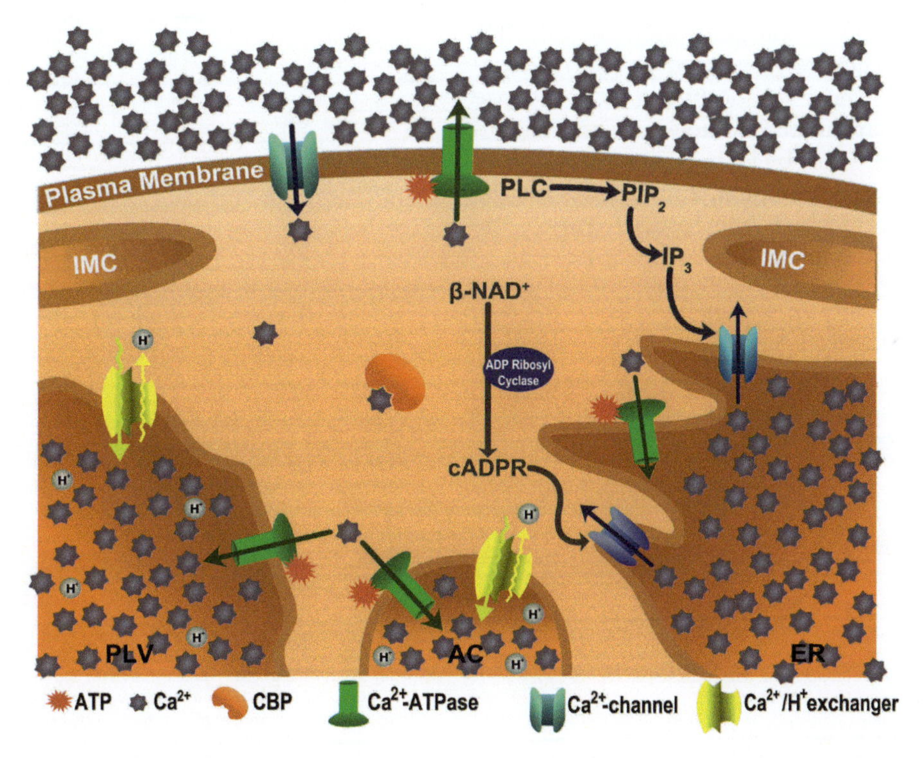

FIGURE 10.7 **Calcium homeostasis in *T. gondii*.** Ca^{2+} entry is through a Ca^{2+} channel gated by an unknown mechanism. Once inside the cells, Ca^{2+} can be translocated back to the extracellular environment by a PMCA (plasma membrane-type Ca^{2+}-ATPase). In addition, Ca^{2+} will interact with Ca^{2+}-binding proteins (CBP) or become sequestered by the endoplasmic reticulum (ER), acidocalcisome (AC), or the PLV. The endoplasmic reticulum (ER) contains the SERCA (sarcoplasmic–endoplasmic reticulum) and acidocalcisomes and PLV a Ca^{2+}-ATPase (TgA1). Both the AC and the PLV contain Ca^{2+}/H^+ exchanger(s). A PI-PLC becomes active by an unknown mechanism producing IP_3, which will act on a Ca^{2+}-release channel in the ER. An ADPR cyclase forms cADPribose which acts on a Ca^{2+}-release channel in the ER. Illustration by Christina Moore.

been described but its role in regulating cytosolic Ca^{2+} is not known. The molecular players for the store operated calcium entry pathway are a mystery as is the channel(s) involved in Ca^{2+} release from the ER (Fig. 10.7). Intracellular Ca^{2+} increase has been linked to critical elements of the *T. gondii* lytic cycle such as conoid extrusion, microneme secretion, invasion of host cells and egress from host cells. Understanding the direct regulation of these processes by calcium signalling, as well as other functions that are causally linked to calcium release, will continue to be aided by new information provided by microbial genome sequencing as well as reagent and protocol innovations. An example of one such innovation is the generation of parasites expressing genetically encoded calcium indicators (GECIs). These developments will provide new insight on calcium homeostasis and signalling in *T. gondii* and be instrumental in the creation of specific novel therapeutic agents for the disruption of calcium homeostasis during the lytic cycle of *T. gondii*.

Acknowledgements

Work in our laboratories was funded by the National Institutes of Health grants R21AI079625 and R01AI096836 to SNJM. DAP was partially supported by an NIH T32 training grant AI-60546 to the Center for Tropical and Emerging Global Diseases. LBP was supported by a fellowship from the University of São Pablo. Christina Moore developed and designed the cartoon for Figure 10.7.

References

Arrizabalaga, G., Boothroyd, J.C., 2004. Role of calcium during *Toxoplasma gondii* invasion and egress. Int. J. Parasitol. 34, 361–368.

Arrizabalaga, G., Ruiz, F., Moreno, S., Boothroyd, J.C., 2004. Ionophore-resistant mutant of *Toxoplasma gondii* reveals involvement of a sodium/hydrogen exchanger in calcium regulation. J. Cell Biol. 165, 653–662.

Batistic, O., Kudla, J., 2009. Plant calcineurin B-like proteins and their interacting protein kinases. Biochim. Biophys. Acta 1793, 985–992.

Besteiro, S., Tonn, D., Tetley, L., Coombs, G.H., Mottram, J.C., 2008. The AP3 adaptor is involved in the transport of membrane proteins to acidocalcisomes of Leishmania. J. Cell Sci. 121, 561–570.

Billker, O., Lourido, S., Sibley, L.D., 2009. Calcium-dependent signaling and kinases in apicomplexan parasites. Cell host & Microbe 5, 612–622.

Bird, G.S., Dehaven, W.I., Smyth, J.T., Putney Jr., J.W., 2008. Methods for studying store-operated calcium entry. Methods 46, 204–212.

Black, M.W., Arrizabalaga, G., Boothroyd, J.C., 2000. Ionophore-resistant mutants of *Toxoplasma gondii* reveal host cell permeabilization as an early event in egress. Mol. Cell. Biol. 20, 9399–9408.

Black, M.W., Boothroyd, J.C., 2000. Lytic cycle of *Toxoplasma gondii*. Microbiol. Mol. Biol. Rev. 64, 607–623.

Bonhomme, A., Pingret, L., Bonhomme, P., Michel, J., Balossier, G., Lhotel, M., Pluot, M., Pinon, J.M., 1993. Subcellular calcium localization in *Toxoplasma gondii* by electron microscopy and by X-ray and electron energy loss spectroscopies. Microsc. Res. Tech. 25, 276–285.

Bottova, I., Sauder, U., Olivieri, V., Hehl, A.B., Sonda, S., 2010. The P-glycoprotein inhibitor GF120918 modulates Ca^{2+}-dependent processes and lipid metabolism in *Toxoplasma gondii*. PloS One 5. e10062.

Bouchot, A., Jaillet, J.D., Bonhomme, A., Alessandro, N.P., Laquerriere, P., Kilian, L., Burlet, H., Gomez-Marin, J.E., Pluot, M., Bonhomme, P., Pinon, J.M., 2001. Detection and localization of a Ca^{2+}-ATPase activity in *Toxoplasma gondii*. Cell Struct. Funct. 26, 49–60.

Bouchot, A., Zierold, K., Bonhomme, A., Kilian, L., Belloni, A., Balossier, G., Pinon, J.M., Bonhomme, P., 1999. Tachyzoite calcium changes during cell invasion by *Toxoplasma gondii*. Parasitol. Res. 85, 809–818.

Bowman, E.J., Siebers, A., Altendorf, K., 1988. Bafilomycins: a class of inhibitors of membrane ATPases from microorganisms, animal cells, and plant cells. Proc. Natl. Acad. Sci. U. S. A. 85, 7972–7976.

Bright, G.R., Fisher, G.W., Rogowska, J., Taylor, D.L., 1987. Fluorescence ratio imaging microscopy: temporal and spatial measurements of cytoplasmic pH. The J. Cell Biol. 104, 1019–1033.

Bruton, J.D., Cheng, A.J., Westerblad, H., 2012. Methods to detect Ca^{2+} in living cells. Adv. Exper. Med. Biol. 740, 27–43.

Carruthers, V.B., Giddings, O.K., Sibley, L.D., 1999a. Secretion of micronemal proteins is associated with toxoplasma invasion of host cells. Cell Microbiol. 1, 225–235.

Carruthers, V.B., Moreno, S.N., Sibley, L.D., 1999b. Ethanol and acetaldehyde elevate intracellular $[Ca^{2+}]$ and stimulate microneme discharge in *Toxoplasma gondii*. Biochem. J. 342 (Pt 2), 379–386.

Carruthers, V.B., Sibley, L.D., 1999. Mobilization of intracellular calcium stimulates microneme discharge in *Toxoplasma gondii*. Mol. Microbiol. 31, 421–428.

Chini, E.N., Nagamune, K., Wetzel, D.M., Sibley, L.D., 2005. Evidence that the cADPR signalling pathway controls calcium-mediated microneme secretion in *Toxoplasma gondii*. Biochem. J. 389, 269–277.

Del carmen, M.G., Mondragon, M., Gonzalez, S., Mondragon, R., 2009. Induction and regulation of conoid extrusion in *Toxoplasma gondii*. Cell Microbiol. 11, 967–982.

Di Virgilio, F., Steinberg, T.H., Swanson, J.A., Silverstein, S.C., 1988. Fura-2 secretion and sequestration in macrophages. A blocker of organic anion transport reveals that these processes occur via a membrane transport system for organic anions. J. Immunol. 140, 915–920.

Dobrowolski, J.M., Carruthers, V.B., Sibley, L.D., 1997. Participation of myosin in gliding motility and host cell invasion by *Toxoplasma gondii*. Mol. Microbiol. 26, 163–173.

Docampo, R., De Souza, W., Miranda, K., Rohloff, P., Moreno, S.N., 2005. Acidocalcisomes – conserved from bacteria to man. Nat. Rev. Microbiol. 3, 251–261.

Docampo, R., lukes, J., 2012. Trypanosomes and the solution to a 50-year mitochondrial calcium mystery. Trends Parasitol. 28, 31–37.

Drozdowicz, Y.M., Shaw, M., Nishi, M., Striepen, B., Liwinski, H.A., Roos, D.S., Rea, P.A., 2003. Isolation and characterization of TgVP1, a type I vacuolar H+-translocating pyrophosphatase from *Toxoplasma gondii*. The dynamics of its subcellular localization and the cellular effects of a diphosphonate inhibitor. J. Biol. Chem. 278, 1075–1085.

Eckstein-Ludwig, U., Webb, R.J., Van Goethem, I.D., East, J.M., Lee, A.G., Kimura, M., O'Neill, P.M., Bray, P.G., Ward, S.A., Krishna, S., 2003. Artemisinins target the SERCA of *Plasmodium falciparum*. Nature 424, 957–961.

Endo, T., Sethi, K.K., Piekarski, G., 1982. *Toxoplasma gondii*: calcium ionophore A23187-mediated exit of trophozoites from infected murine macrophages. Exp. Parasitol. 53, 179–188.

Fang, J., Marchesini, N., Moreno, S.N., 2006. A *Toxoplasma gondii* phosphoinositide phospholipase C (TgPI-PLC) with high affinity for phosphatidylinositol. Biochem. J. 394, 417–425.

Francia, M.E., Wicher, S., Pace, D.A., Sullivan, J., Moreno, S.N., Arrizabalaga, G., 2011. A *Toxoplasma gondii* protein with homology to intracellular type Na(+)/H(+) exchangers is important for osmoregulation and invasion. Exp. Cell Res. 317, 1382–1396.

Gajria, B., Bahl, A., Brestelli, J., Dommer, J., Fischer, S., Gao, X., Heiges, M., Iodice, J., Kissinger, J.C., Mackey, A.J., Pinney, D.F., Roos, D.S., Stoeckert Jr., C.J., Wang, H., Brunk, B.P., 2008. ToxoDB: an integrated *Toxoplasma gondii* database resource. Nuc. Acids Res. 36, D553–D556.

Garrison, E., Treeck, M., Ehret, E., Butz, H., Garbuz, T., Oswald, B.P., Settles, M., Boothroyd, J., Arrizabalaga, G., 2012. A Forward Genetic Screen Reveals that Calcium-dependent Protein Kinase 3 Regulates Egress in Toxoplasma. PLoS Path. 8, e1003049.

Gazarini, M.L., Garcia, C.R., 2004. The malaria parasite mitochondrion senses cytosolic Ca^{2+} fluctuations. Biochem. Biophys. Res. Comm. 321, 138–144.

Gifford, J.L., Walsh, M.P., Vogel, H.J., 2007. Structures and metal-ion-binding properties of the Ca^{2+}-binding helix-loop-helix EF-hand motifs. Biochem. J. 405, 199–221.

Grabarek, Z., 2006. Structural basis for diversity of the EF-hand calcium-binding proteins. J. Mol. Biol. 359, 509–525.

Grynkiewicz, G., Poenie, M., Tsien, R.Y., 1985. A new generation of Ca^{2+} indicators with greatly improved fluorescence properties. J. Biol. Chem. 260, 3440–3450.

Guttery, D.S., Pittman, J.K., Frénal, K., Poulin, B., McFarlane, L.R., Slavic, K., Wheatley, S.P., Soldati-Favre, D., Krishna, S., Tewari, R., Staines, H.M., 2013. The Plasmodium berghei Ca^{2+}/H^{+} exchanger, PbCAX, is essential for tolerance to environmental Ca^{2+} during sexual development. PloS Path. 9, e1003191.

Haller, T., Volkl, H., Deetjen, P., Dietl, P., 1996. The lysosomal Ca^{2+} pool in MDCK cells can be released by ins(1,4,5)P3-dependent hormones or thapsigargin but does not activate store-operated Ca^{2+} entry. Biochem. J. 319 (Pt 3), 909–912.

Harmon, A.C., Gribskov, M., Harper, J.F., 2000. CDPKs – a kinase for every Ca^{2+} signal? Trends Plant Sci. 5, 154–159.

Hu, K., Johnson, J., Florens, L., Fraunholz, M., Suravajjala, S., Dilullo, C., Yates, J., Roos, D. S., Murray, J.M., 2006. Cytoskeletal components of an invasion machine – the apical complex of *Toxoplasma gondii*. PLoS Pathog. 2. e13.

Huang, G., Barlett, P., Thomas, A.P., Moreno, S.N., Docampo, R., 2012. Acidocalcisomes of Trypanosoma brucei have an inositol 1,4,5-trisphosphate receptor that is required for growth and infectivity. Proc. Natl. Acad. Sci. U. S. A. 110, 1887–1892.

Huang, G., Fang, J., Sant'anna, C., Li, Z.H., Wellems, D.L., Rohloff, P., Docampo, R., 2011. Adaptor protein-3 (AP-3) complex mediates the biogenesis of acidocalcisomes and is essential for growth and virulence of *Trypanosoma brucei*. The Journal of Biological Chemistry 286, 36619–36630.

Kao, J.P., 1994. Practical aspects of measuring [Ca^{2+}] with fluorescent indicators. Methods Cell Biol. 40, 155–181.

Kieschnick, H., Wakefield, T., Narducci, C.A., Beckers, C., 2001. *Toxoplasma gondii* attachment to host cells is regulated by a calmodulin-like domain protein kinase. J. Biol. Chem. 276, 12369–12377.

Klimecka, M., Muszynska, G., 2007. Structure and functions of plant calcium-dependent protein kinases. Acta Biochim. Pol. 54, 219–233.

Kretsinger, R.H., 1976. Calcium-binding proteins. Annu. Rev. Biochem. 45, 239–266.

Kunze, W., 1907. Uber Ocheobius herpobdellae schuberg et kunze. Arch. Protistenk. 9, 383–390.

Ladenburger, E.M., Plattner, H., 2011. Calcium-release channels in paramecium. Genomic expansion, differential positioning and partial transcriptional elimination. PloS One 6, e27111.

Lefurgey, A., Ingram, P., Blum, J.J., 2001. Compartmental responses to acute osmotic stress in Leishmania major result in rapid loss of Na^{+} and Cl^{-}. Comp. Biochem. Physiol. A Mol. Integr Physiol. 128, 385–394.

Liu, C., Hermann, T.E., 1978. Characterization of ionomycin as a calcium ionophore. J. Biol. Chem. 253, 5892–5894.

Lourido, S., Shuman, J., Zhang, C., Shokat, K.M., Hui, R., Sibley, L.D., 2010. Calcium-dependent protein kinase 1 is an essential regulator of exocytosis in Toxoplasma. Nature 465, 359–362.

Lourido, S., Tang, K., Sibley, L.D., 2012. Distinct signalling pathways control Toxoplasma egress and host-cell invasion. Embo J. 31, 4524–4534.

Lovett, J.L., Marchesini, N., Moreno, S.N., Sibley, L.D., 2002. *Toxoplasma gondii* microneme secretion involves intracellular Ca(2+) release from inositol 1,4,5-triphosphate (IP(3))/ryanodine-sensitive stores. J. Biol. Chem. 277, 25870–25876.

Lovett, J.L., Sibley, L.D., 2003. Intracellular calcium stores in *Toxoplasma gondii* govern invasion of host cells. J. Cell Sci. 116, 3009–3016.

Luan, S., Kudla, J., Rodriguez-Concepcion, M., Yalovsky, S., Gruissem, W., 2002. Calmodulins and calcineurin B-like proteins: calcium sensors for specific signal response coupling in plants. The Plant Cell 14 (Suppl.), S389–S400.

Luo, S., Ruiz, F.A., Moreno, S.N., 2005. The acidocalcisome Ca^{2+}-ATPase (TgA1) of *Toxoplasma gondii* is required for polyphosphate storage, intracellular calcium homeostasis and virulence. Mol. Microbiol. 55, 1034–1045.

Luo, S., Vieira, M., Graves, J., Zhong, L., Moreno, S.N., 2001. A plasma membrane-type Ca^{2+}-ATPase co-localizes with a vacuolar H^{+}-pyrophosphatase to acidocalcisomes of *Toxoplasma gondii*. Embo J. 20, 55–64.

Matsui, A., Umezawa, K., Shindo, Y., Fujii, T., Citterio, D., Oka, K., Suzuki, K., 2011. A near-infrared fluorescent calcium probe: a new tool for intracellular multicolour Ca^{2+} imaging. Chem. Commun. 47, 10407–10409.

Matthiesen, S.H., Shenoy, S.M., Kim, K., Singer, R.H., Satir, B.H., 2003. Role of the parafusin orthologue, PRP1, in microneme exocytosis and cell invasion in *Toxoplasma gondii*. Cell Microbiol. 5, 613–624.

Mbatia, H.W., Burdette, S.C., 2012. Photochemical tools for studying metal ion signaling and homeostasis. Biochem. 51, 7212–7224.

Mccoy, J.M., Whitehead, L., Van Dooren, G.G., Tonkin, C.J., 2012. TgCDPK3 Regulates Calcium-Dependent Egress of *Toxoplasma gondii* from Host Cells. PLoS Pathog. 8, e1003066.

Mira Gutierrez, J., Del Rey Calero, J., 1966. Volutina in *Toxoplasma gondii*. Med. Trop. (Madr.) 42, 20–29.

Miranda, K., De souza, W., Plattner, H., Hentschel, J., Kawazoe, U., Fang, J., Moreno, S.N., 2008. Acidocalcisomes in Apicomplexan parasites. Exper. Parasitol. 118, 2–9.

Miranda, K., Pace, D.A., Cintron, R., Rodrigues, J.C., Fang, J., Smith, A., Rohloff, P., Coelho, E., De Haas, F., De souza, W., Coppens, I., Sibley, L.D., Moreno, S.N., 2010. Characterization of a novel organelle in *Toxoplasma gondii* with similar composition and function to the plant vacuole. Mol. Microbiol. 76, 1358–1375.

Miranda, K., Rodrigues, C.O., Hentchel, J., Vercesi, A., Plattner, H., De Souza, W., Docampo, R., 2004. Acidocalcisomes of *Phytomonas francai* possess distinct morphological characteristics and contain iron. Microsc. Microanal. 10, 647–655.

Mondragon, R., Frixione, E., 1996. Ca^{2+}-dependence of conoid extrusion in *Toxoplasma gondii* tachyzoites. J. Eukaryot. Microbiol. 43, 120–127.

Moreau, R.A., 1987. Calcium-binding proteins in fungi and higher plants. J. Dairy Sci. 70, 1504–1512.

Moreno, B., Bailey, B.N., Luo, S., Martin, M.B., Kuhlenschmid, T.M., Moreno, S.N., Docampo, R., Oldfield, E., 2001. (31)P NMR of apicomplexans and the effects of risedronate on *Cryptosporidium parvum* growth. Biochem. Biophys. Res. Commun. 284, 632–637.

Moreno, S.N., Ayong, L., Pace, D.A., 2011. Calcium storage and function in apicomplexan parasites. Essays. Biochem. 51, 97–110.

Moreno, S.N., Zhong, L., 1996. Acidocalcisomes in *Toxoplasma gondii* tachyzoites. Biochem. J. 313 (Pt 2), 655–659.

Moreno, S.N., Zhong, L., Lu, H.G., Souza, W.D., Benchimol, M., 1998. Vacuolar-type H^{+}-ATPase regulates cytoplasmic pH in *Toxoplasma gondii* tachyzoites. Biochem. J. 330 (Pt 2), 853–860.

Moudy, R., Manning, T.J., Beckers, C.J., 2001. The loss of cytoplasmic potassium upon host cell breakdown triggers egress of *Toxoplasma gondii*. J. Biol. Chem. 276, 41492–41501.

Nagamune, K., Beatty, W.L., Sibley, L.D., 2007. Artemisinin induces calcium-dependent protein secretion in the protozoan parasite *Toxoplasma gondii*. Eukaryotic Cell 6, 2147–2156.

Nagamune, K., Hicks, L.M., Fux, B., Brossier, F., Chini, E.N., Sibley, L.D., 2008a. Abscisic acid controls calcium-dependent egress and development in *Toxoplasma gondii*. Nature 451, 207–210.

Nagamune, K., Moreno, S.N., Chini, E.N., Sibley, L.D., 2008b. Calcium regulation and signaling in apicomplexan parasites. Sub-cellular Biochemistry 47, 70–81.

Nagamune, K., Sibley, L.D., 2006. Comparative genomic and phylogenetic analyses of calcium ATPases and calcium-regulated proteins in the apicomplexa. Mol. Biol. Evol. 23, 1613–1627.

Nagata, T., Iizumi, S., Satoh, K., Ooka, H., Kawai, J., Carninci, P., Hayashizaki, Y., Otomo, Y., Murakami, K., Matsubara, K., Kikuchi, S., 2004. Comparative analysis of plant and animal calcium signal transduction element using plant full-length cDNA data. Mol. Biol. Evol. 21, 1855–1870.

Ogura, A., Takahashi, K., 1976. Artificial deciliation causes loss of calcium-dependent responses in Paramecium. Nature 264, 170–172.

Pace, D.A., Fang, J., Cintron, R., Docampo, M.D., Moreno, S.N., 2011. Over-expression of a cytosolic pyrophosphatase (TgPPase) reveals a regulatory role of PP(i) in glycolysis for *Toxoplasma gondii*. Biochem. J. 440, 229–240.

Parekh, A.B., 2006. Cell biology: cracking the calcium entry code. Nature 441, 163–165.

Parussini, F., Coppens, I., Shah, P.P., Diamond, S.L., Carruthers, V.B., 2010. Cathepsin L occupies a vacuolar compartment and is a protein maturase within the endo/exocytic system of *Toxoplasma gondii*. Mol. Microbiol. 76, 1340–1357.

Patel, S., Docampo, R., 2010. Acidic calcium stores open for business: expanding the potential for intracellular Ca^{2+} signaling. Trends Cell Biol. 20, 277–286.

Peng, Z.Y., Mansour, J.M., Araujo, F., Ju, J.Y., Mckenna, C.E., Mansour, T.E., 1995. Some phosphonic acid analogs as inhibitors of pyrophosphate-dependent phosphofructokinase, a novel target in *Toxoplasma gondii*. Biochem. Pharmacol. 49, 105–113.

Pezzella, N., Bouchot, A., Bonhomme, A., Pingret, L., Klein, C., Burlet, H., Balossier, G., Bonhomme, P., Pinon, J.M., 1997. Involvement of calcium and calmodulin in *Toxoplasma gondii* tachyzoite invasion. Eur. J. Cell Biol. 74, 92–101.

Pezzella-D'Alessandro, N., Le Moal, H., Bonhomme, A., Valere, A., Klein, C., Gomez-Marin, J., Pinon, J.M., 2001. Calmodulin distribution and the actomyosin cytoskeleton in *Toxoplasma gondii*. J. Histochem. Cytochem. 49, 445–454.

Pingret, L., Millot, J.M., Sharonov, S., Bonhomme, A., Manfait, M., Pinon, J.M., 1996. Relationship between intracellular free calcium concentrations and the intracellular development of *Toxoplasma gondii*. J. Histochem. Cytochem. 44, 1123–1129.

Plattner, H., Sehring, I.M., Mohamed, I.K., Miranda, K., De Souza, W., Billington, R., Genazzani, A., Ladenburger, E.M., 2012. Calcium signaling in closely related protozoan groups (Alveolata): non-parasitic ciliates (Paramecium, Tetrahymena) vs. parasitic Apicomplexa (Plasmodium, Toxoplasma). Cell Calcium 51, 351–382.

Pozos, T.C., Sekler, I., Cyert, M.S., 1996. The product of HUM1, a novel yeast gene, is required for vacuolar Ca^{2+}/H^{+} exchange and is related to mammalian Na^{+}/Ca^{2+} exchangers. Mol. Cell Biol. 16, 3730–3741.

Prole, D.L., Taylor, C.W., 2011. Identification of intracellular and plasma membrane calcium channel homologues in pathogenic parasites. PloS One 6, e26218.

Rodrigues, C.O., Ruiz, F.A., Rohloff, P., Scott, D.A., Moreno, S.N., 2002. Characterization of isolated acidocalcisomes from *Toxoplasma gondii* tachyzoites reveals a novel pool of hydrolyzable polyphosphate. J. Biol. Chem. 277, 48650–48656.

Rodrigues, C.O., Scott, D.A., Bailey, B.N., De Souza, W., Benchimol, M., Moreno, B., Urbina, J.A., Oldfield, E., Moreno, S.N., 2000. Vacuolar proton pyrophosphatase activity and pyrophosphate (PPi) in *Toxoplasma gondii* as possible chemotherapeutic targets. Biochem. J. 349 (Pt 3), 737–745.

Rodrigues, C.O., Scott, D.A., Docampo, R., 1999. Presence of a vacuolar H^{+}-pyrophosphatase in promastigotes of *Leishmania donovani* and its localization to a different compartment from the vacuolar H^{+}-ATPase. Biochem. J. 340 (Pt 3), 759–766.

Rohloff, P., Miranda, K., Rodrigues, J.C., Fang, J., Galizzi, M., Plattner, H., Hentschel, J., Moreno, S.N., 2011. Calcium uptake and proton transport by acidocalcisomes of *Toxoplasma gondii*. PloS One 6, e18390.

Saoudi, Y., Rousseau, B., Doussiere, J., Charrasse, S., Gauthier-Rouviere, C., Morin, N., Sautet-Laugier, C., Denarier, E., Scaife, R., Mioskowski, C., Job, D., 2004. Calcium-independent cytoskeleton disassembly induced by BAPTA. Eur. J. Biochem. 271, 3255–3264.

Schindl, R., Muik, M., Fahrner, M., Derler, I., Fritsch, R., Bergsmann, J., Romanin, C., 2009. Recent progress on STIM1 domains controlling Orai activation. Cell Cal. 46, 227–232.

Schwab, J.C., Beckers, C.J., Joiner, K.A., 1994. The parasitophorous vacuole membrane surrounding intracellular *Toxoplasma gondii* functions as a molecular sieve. Proc. Natl. Acad. Sci. U. S. A. 91, 509–513.

Scott, D.A., Docampo, R., Dvorak, J.A., SHI, S., Leapman, R.D., 1997. In situ compositional analysis of acidocalcisomes in *Trypanosoma cruzi*. J. Biol. Chem. 272, 28020–28029.

Seeber, F., Beuerle, B., Schmidt, H.H., 1999. Cloning and functional expression of the calmodulin gene from *Toxoplasma gondii*. Mol. Biochem. Parasitol. 99, 295–299.

Song, H.O., Ahn, M.H., Ryu, J.S., Min, D.Y., Joo, K.H., Lee, Y.H., 2004. Influence of calcium ion on host cell invasion and intracellular replication by *Toxoplasma gondii*. Korean. J. Parasitol. 42, 185–193.

Srinivas, S.P., Ong, A., Goon, L., Bonanno, J.A., 2002. Lysosomal Ca(2+) stores in bovine corneal endothelium. Investigative Ophthalmology & Visual Science 43, 2341–2350.

Stommel, E.W., CHO, E., Steide, J.A., Seguin, R., Barchowsky, A., Schwartzman, J.D., Kasper, L.H., 2001. Identification and role of thiols in *Toxoplasma gondii* egress. Exp. Biol. Med. (Maywood) 226, 229–236.

Stommel, E.W., Ely, K.H., Schwartzman, J.D., Kasper, L.H., 1997. *Toxoplasma gondii*: dithiol-induced Ca^{2+} flux causes egress of parasites from the parasitophorous vacuole. Exper. Parasitol. 87, 88–97.

Takahashi, A., Camacho, P., Lechleiter, J.D., Herman, B., 1999. Measurement of intracellular calcium. Physiol. Rev. 79, 1089–1125.

Thastrup, O., Cullen, P.J., Drobak, B.K., Hanley, M.R., Dawson, A.P., 1990. Thapsigargin, a tumor promoter, discharges intracellular Ca^{2+} stores by specific inhibition of the endoplasmic reticulum Ca^{2+}-ATPase. Proc. Natl. Acad. Sci. U. S. A. 87, 2466–2470.

Thompson, A.K., Mostafapour, S.P., Denlinger, L.C., Bleasdale, J.E., Fisher, S.K., 1991. The aminosteroid U-73122 inhibits muscarinic receptor sequestration and phosphoinositide hydrolysis in SK-N-SH neuroblastoma cells. A role for Gp in receptor compartmentation. J. Biol. Chem. 266, 23856–23862.

Tian, L., Hires, S.A., Looger, L.L., 2012. Imaging neuronal activity with genetically encoded calcium indicators. Cold Spring Harbor Protocols 2012, 647–656.

Tsien, R.W., Tsien, R.Y., 1990. Calcium channels, stores, and oscillations. Annu. Rev. Cell Biol. 6, 715–760.

Tsien, R.Y., 1980. New calcium indicators and buffers with high selectivity against magnesium and protons: design, synthesis, and properties of prototype structures. Biochem. 19, 2396–2404.

Tsien, R.Y., 1983. Intracellular measurements of ion activities. Annu. Rev. Biophys. Bioeng. 12, 91–116.

Tsien, R.Y., 1989. Fluorescent indicators of ion concentrations. Methods Cell Biol. 30, 127–156.

Vercesi, A.E., Moreno, S.N., Docampo, R., 1994. Ca^{2+}/H^{+} exchange in acidic vacuoles of *Trypanosoma brucei*. Biochem. J. 304 (Pt 1), 227–233.

Vieira, M.C., Moreno, S.N., 2000. Mobilization of intracellular calcium upon attachment of *Toxoplasma gondii* tachyzoites to human fibroblasts is required for invasion. Mol. Biochem. Parasitol. 106, 157–162.

Wernimont, A.K., Artz, J.D., Finerty Jr., P., Lin, Y.H., Amani, M., Allali-hassani, A., Senisterra, G., Vedadi, M., Tempel, W., Mackenzie, F., Chau, I., Lourido, S., Sibley, L.D., Hui, R., 2010. Structures of apicomplexan calcium-dependent protein kinases reveal mechanism of activation by calcium. Nature Struct. Mol. Biol. 17, 596–601.

Wetzel, D.M., Chen, L.A., Ruiz, F.A., Moreno, S.N., Sibley, L.D., 2004. Calcium-mediated protein secretion potentiates motility in *Toxoplasma gondii*. J. Cell Sci. 117, 5739–5748.

CHAPTER

11

The *Toxoplasma gondii* Parasitophorous Vacuole Membrane: A Multifunctional Organelle in the Infected Cell

Anthony P. Sinai

Department of Microbiology, Immunology and Molecular Genetics, University of Kentucky College of Medicine, Lexington, Kentucky, USA

OUTLINE

11.1 INTRODUCTION

In contrast to most pathogens, Apicomplexan parasites like *Toxoplasma gondii* actively invade target host cells using an actin–myosin based active invasion mechanism (Dobrowolski and Sibley, 1996; Carruthers, 2002a). This results in the invading parasites establishing themselves,

Toxoplasma gondii, Second Edition
http://dx.doi.org/10.1016/B978-0-12-396481-6.00011-8

in some cases transiently (e.g. *Theileria* spp. (Shaw, 1997)), in a specialized vacuolar compartment termed the parasitophorous vacuole (PV). The *T. gondii* PV is a highly dynamic compartment defining the replication permissive niche for the actively growing tachyzoite form of the parasite (Sinai and Joiner, 1997; Mercier *et al.*, 2005). The delimiting membrane defining the parasitophorous vacuole, the parasitophorous vacuole membrane, or PVM, is increasingly being recognized as a specialized 'organelle' that is unique as an organelle given that it resides outside the corpus of the organism itself (Martin *et al.*, 2007) (Figure 11.1). In recent years, the complexity and diversity of activities associated with the PVM are becoming evident, thus renewing interest in the biology of the PVM.

A systematic study of this enigmatic organelle has been severely limited by several issues. Primary among these is the fact that it is formed only in the context of the infected cell thereby limiting the amount of material for detailed biochemical analysis. Secondly, unlike other cellular organelles that can often be purified by conventional approaches, the PVM cannot be purified away from host cell organelles (Sinai *et al.*, 1997) (see below) (Figure 11.1). Studies demonstrate that the PVM, on its own and by virtue of its interactions with cellular components, plays critical functions in the structural integrity of the vacuole, nutrient acquisition and the manipulation of cellular functions (Sinai and Joiner, 1997; Lingelbach and Joiner, 1998). In addition it appears that the repertoire of activities at the PVM is likely to be plastic reflecting temporal changes associated with the replicative phase of parasite growth (Lingelbach and Joiner, 1998). Finally, the PVM likely forms the foundation for the cyst wall as the parasite differentiates in the establishment of latent infection (Weiss and Kim, 2000). As the critical border crossing between the parasite and invaded cell the study of the PVM provides a fertile area for new investigation aided by the decoding of the *Toxoplasma* genome (available at http://www.toxoDB.org) and the application of proteomic analyses to basic questions in parasite biology. This chapter discusses the emerging themes associated with the functional dissection of the PVM that indicate activities associated with nutrient acquisition, manipulation of host signalling affecting both immune and non-immune functions and immunity itself, at the level of both the innate and acquired arms of the immune system. The emerging data points to the PVM being much more than a passive demarcation between the parasite and host and

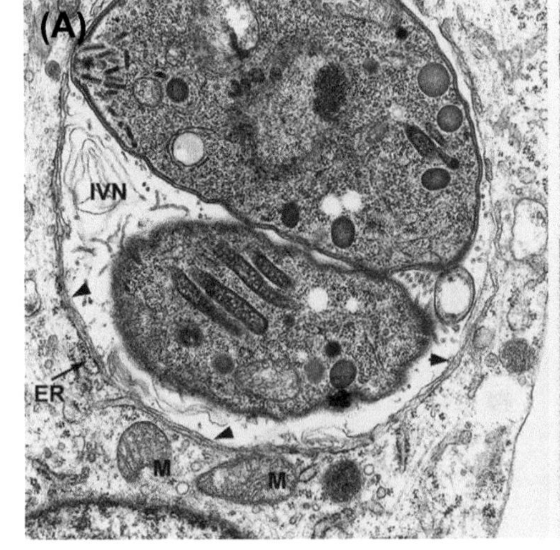

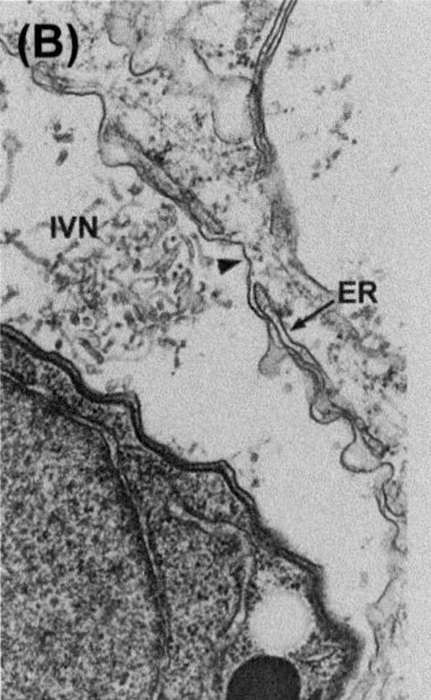

FIGURE 11.1 Transmission electron micrographs (Panels A and B) of *T. gondii* ME49 strain in primary murine astrocytes demonstrating flattened endoplasmic reticulum (ER) in close proximity around the parasitophorous vacuole membrane (indicated by arrowheads) and associated mitochrondria (M). The membranous network (IVN) can be demonstrated within the parasitophorous vacuole. *Images courtesy of Dr. Sandra Halonen (Montana State University) were obtained at the Albert Einstein College of Medicine Ultrastructure Facility.*

to the PVM being a dynamic entity affecting every aspect of intracellular residence. What is also evident is how little we know about this enigmatic organelle, thus opening new trails for the discovery of fundamental knowledge concerning host parasite interactions.

11.2 BIOGENESIS OF THE PVM

Unlike a typical phagosome membrane formed by conventional phagocytosis or macropinocytosis (Jutras and Desjardins, 2005; Desjardins *et al.*, 1994) the *T. gondii* PVM is formed during the active invasion of the host cell by the parasite (Dobrowolski and Sibley, 1996). *Toxoplasma* invasion is rapid and temporally staged. Productive attachment via the apical end to the plasma membrane results from the secretion of the micronemes (Carruthers, 2002b; Carruthers *et al.*, 1999). The formation of the moving junction accompanied by the secretion of rhoptries and the engagement of the actin–myosin mediated invasion complex results in the entry of the parasite (Besteiro *et al.*, 2011). These events are rapid (5–30 seconds) and force the host plasma membrane around the site of invasion to invaginate at the point of entry thus contributing the bulk of the membrane associated with the newly formed vacuole (Nichols *et al.*, 1983; Coppens and Joiner, 2003). Additional lipids in the nascent vacuole may be contributed to by the discharge of the contents of the rhoptries that have been captured as a 'cloud' in electron micrographs of invading parasites (Nichols *et al.*, 1983; Coppens and Joiner, 2003). The physical force exerted by the parasite drives the parasite completely within the confines of the host cell and the entry scar at the plasma membrane is sealed.

Several features of this invasion process directly impact the protein composition of the vacuolar membrane. The early freeze-etch microscopic studies of Porchet-Hennere demonstrated that the early vacuolar membrane was devoid of intramembrane particles suggesting the selective exclusion of membrane proteins in the course of invasion (Porchet-Hennere *et al.*, 1985, Porchet-Hennere and Torpier, 1983). A significant and important refinement to the model came from the studies of Mordue *et al.* (1999) who demonstrated that the exclusion of membrane proteins at the time invasion was in fact selective, by excluding proteins containing Type I-single spanning transmembrane domains while permitting inclusion of lipid-anchored proteins (Mordue *et al.*, 1999). Signals for exclusion from the PVM were not encoded in the transmembrane domain itself but rather the cytoplasmic tail suggesting sorting occurs at the level of the cytoskeleton (Mordue *et al.*, 1999). A refinement of this original study revealed that the lipid environment did not contribute to the sorting of proteins into the nascent vacuole as both lipid raft associated and non-raft Type I transmembrane domain containing proteins were generally excluded from the vacuole (Charron and Sibley, 2004). Interestingly, while proteins with single transmembrane domains were excluded, polytopic membrane proteins in the raft fractions gained access to the PVM (Charron and Sibley, 2004). These studies demonstrated that the *T. gondii* PVM is not in fact devoid of host proteins as originally surmised Porchet-Hennere *et al.*, 1985; Porchet-Hennere and Torpier, suggesting the vacuole is not entirely segregated from cellular functions.

The function of stripping the nascent vacuole of excluded proteins is mediated by the moving junction at the site of parasite invasion. First characterized morphologically in the context of *Plasmodium* and *Toxoplasma* entry, the 'moving junction' appeared as an area of increased electron density at the interface between the parasite surface and the host plasma membrane (Aikawa *et al.*, 1978; Michel *et al.*, 1980). Additional studies revealed the moving junction to be critical in parasite invasion resulting in the significant deformation of the transitting organism (Aikawa *et al.*, 1977; Nichols and O'Connor, 1981; Michel *et al.*, 1980; de Souza, 2005), exclusion of host markers (Suss-Toby *et al.*, 1996; Mordue *et al.*, 1999) as well as antibodies opsonizing the parasite surface (Dubremetz *et al.*, 1985). The first insights into the molecular composition of the moving junction emerged from the Boothroyd (Alexander *et al.*, 2005) and Dubremetz (Lebrun *et al.*, 2005) laboratories. The molecular dissection of the moving junction complex has provided considerable insight (Besteiro *et al.*, 2011) into its function although aspects governing the molecular sieving properties remain elusive. The potential that parasite proteins interacting with the host cytoskeleton at the moving junction may play some role has been suggested (Straub *et al.*, 2009, 2011).

Stripped of most host proteins, the lipid bilayer forming the nascent vacuole appears to be derived predominantly from the host plasma membrane (Suss-Toby *et al.*, 1996). This was elegantly demonstrated by the Ward laboratory using real-time electrophysiological analyses of invading parasites (Suss-Toby *et al.*, 1996). Their studies convincingly showed that the parasite contributes little if any lipid to the new vacuole (Suss-Toby *et al.*, 1996); this is despite the discharge of lipid by the rhoptries during invasion (Coppens and Vielemeyer, 2005; Coppens and Joiner, 2003; Foussard *et al.*, 1991). The newly formed PVM is rapidly modified by the temporally and spatially staged release of proteins from the parasite rhoptries and dense granules (Carruthers *et al.*, 1999; Carruthers and Sibley, 1997; de Souza, 2005). The first proteins to appear in the PVM are derived from the rhoptries (Carruthers and Sibley, 1997; Dubremetz, 2007).

Within minutes of complete internalization, a large scale secretion event at the posterior end of the parasite results in the release of a number of dense granule proteins (Sibley *et al.*, 1986). Several of these GRA proteins, including GRA3, -5, -7, -8, -9 and -10, appear at the PVM (Bermudes *et al.*, 1994; Henriquez *et al.*, 2005; Ahn *et al.*, 2005; Lecordier *et al.*, 1999; Jacobs *et al.*, 1998; Carey *et al.*, 2000; Adjogble *et al.*, 2004) (reviewed in Mercier *et al.*, 2005). Unlike rhoptries, where the prevailing wisdom holds that discharge is limited to parasite invasion, dense granules form the basis of the constitutive and default secretory pathway (Karsten *et al.*, 1998; Mercier *et al.*, 2005). What is not known is whether the relative concentrations of GRA proteins loaded into the secretory granules varies at different phases in the parasites' intracellular residence – including differentiation into the tissue cyst (see below).

11.3 THE PHYSICAL ORGANIZATION OF THE PV AND PVM

The PVM is a dynamic entity. To better understand the organization of the PVM it is important to consider the structural basis of the PV. Electron microscopic studies indicate that the vacuolar space contains a proliferation of membrane tubules referred to as the PV network (reviewed in Mercier *et al.*, 2005). The network is structurally (Magno *et al.*, 2005a; Sibley *et al.*, 1995) and antigenically (Mercier *et al.*, 1993, 2005; Labruyere *et al.*, 1999; Sibley *et al.*, 1995; Lecordier *et al.*, 1995) distinct from the PVM but does exhibit some degree of connectivity to it. Studies examining the organization of the *T. gondii* PV using scanning and transmission electron microscopy of thick preparations reveal a remarkably detailed picture of vacuolar organization (Schatten and Ris, 2004; Magno *et al.*, 2005a). In these stunning images the parasites appear to be decorated with the network structure which forms a mesh-like pattern with connections to the PVM (Schatten and Ris, 2002, 2004; Magno *et al.*, 2005a). These structures may in fact be elements of the host microtubule cytoskeleton that have been shown to project into the vacuole (Coppens *et al.*, 2006). On the outer leaflet of the PVM are associated organelles (Sinai *et al.*, 1997; Magno *et al.*, 2005b; de Souza, 2005; Melo and de Souza, 1997; de Melo *et al.*, 1992; Schatten and Ris, 2004) and elements of the cytoskeleton (Sehgal *et al.*, 2005; Andrade *et al.*, 2001; Melo *et al.*, 2001;

Halonen and Weidner, 1994; Cintra and De Souza, 1985).

The PVM itself is not a uniform membrane. This heterogeneity is evident on a gross level by the differences in the distribution and extent of PVM–organelle association (Sinai *et al.*, 1997; Magno *et al.*, 2005b). In addition there may be functional subdomains within the vacuolar membrane as well. Another 'modification' of the PVM is the membranous extensions emanating from many, but not all, vacuoles. These PVM-extensions were first described for a number of dense granules proteins including GRA3 (Bermudes *et al.*, 1994), GRA5 (Lecordier *et al.*, 1999), GRA7 (Coppens *et al.*, 2006; Sehgal *et al.*, 2005) and GRA8 (Carey *et al.*, 2000). How the level of organization in the PVM impacts function remains an open question.

11.4 ACTIVITIES ASSOCIATED WITH THE EARLY AND DEVELOPING PVM

11.4.1 Organelle Association

Morphologically, a feature defining the *T. gondii* PVM is the intimate association of host mitochondria and the endoplasmic reticulum (Sinai *et al.*, 1997; Magno *et al.*, 2005b; de Souza, 2005; Melo and de Souza, 1997; de Melo *et al.*, 1992; Schatten and Ris, 2004). A key element of this interaction is the tight apposition of the host organelle membranes with the PVM but the absence of membrane fusion. The extent of the PVM–organelle association and the relative distribution have been quantified morphometrically and appear to vary, particularly for the extent of PVM–mitochondrial association depending on the cell type and approach to quantification (Sinai *et al.*, 1997; Magno *et al.*, 2005b). This variation very likely reflects the relative concentration of these organelles in the cell, particularly in the juxtanuclear area where the vacuoles tend to reside (Sinai *et al.*, 1997). All indications point to the PVM–organelle association being a highly stable and in effect irreversible interaction (Sinai *et al.*, 1997).

The kinetics of PVM–organelle association are rapid and detectable within a minute of parasite entry (Sinai and Joiner, 2001). These properties correlate with those for the discharge of rhoptries at the time of parasite invasion (Carruthers and Sibley, 1997; Carruthers, 1999). The temporal connection to rhoptry discharge implicates a rhoptry-derived factor. This view was supported by the exposure of a functional mitochondrial localization signal in the processed form of rhoptry ROP2. The view was further supported by the high affinity of this N-terminal domain to host mitochondrial and other membranes in general (Sinai and Joiner, 2001) that was subsequently found to be mediated by specific amphipathic helices encoded in the protein (Reese and Boothroyd, 2009). While fitting the temporal and functional characteristics associated with a mediator of PVM–mitochondrial association, the involvement of the ROP2 locus alone in mediating this activity has been challenged by the fact that targeted disruption of the gene does not affect the recruitment of mitochondria to the PVM (Pernas and Boothroyd, 2010). While redundant activities in the ROP2 family of proteins, representing both active and pseudokinases (Peixoto *et al.*, 2010), may complement the defect, the possibility of an entirely new mechanism for mitochondrial recruitment is now a viable option.

The molecular mechanism for PVM–endoplasmic reticulum (ER) association remains elusive. However, recent data suggest that the recruitment of the ER to the PVM may in fact impact antigen presentation thus affecting the development of adaptive immunity (Goldszmid *et al.*, 2009). It is notable that while overt fusion between the PVM and associated ER was not detected, the clear delivery of host ER resident proteins into the lumen of the vacuole is highly suggestive of such events occurring (Goldszmid *et al.*, 2009). This presents an additional level of interaction between the parasite and host cell and could provide a mechanism for the bi-directional transport of macromolecules. We are currently exploring this possibility as a mechanism by which host glycosyltransferases may be delivered to the vacuole as a means to target and modify parasite proteins, particularly in the context of differentiation to the tissue cyst forms (see below).

11.4.2 Resistance to Lysosome Fusion

The seminal studies of Jones and Hirsch (Jones *et al.*, 1972; Jones and Hirsch, 1972) were the earliest to demonstrate that the *T. gondii* vacuole failed to fuse with lysosomes. This observation confirmed and refined in multiple subsequent studies indicate that the *T. gondii* PVM is non-fusogenic with regard to components of the endocytic and exocytotic pathways of membrane traffic (De Carvalho and deSouza, 1989; Joiner *et al.*, 1990; Mordue and Sibley, 1997; DeMelo and Souza, 1996) and fails to acidify (Sibley *et al.*, 1985; Mordue and Sibley, 1997). The ability to prevent lysosomal fusion is critically dependent on active parasite invasion as internalization by phagocytosis results in rapid lysosome fusion and the killing of the parasite (Joiner *et al.*, 1990). The property of resistance to lysosome fusion is established at the time of invasion by the exclusion of signals for entry into the endocytic cascade by the moving junction (Sinai and Joiner, 1997; Mordue *et al.*, 1999).

The capacity to establish and maintain the non-fusogenic state appears to be intrinsic to the PVM as demonstrated by the elegant experiments of Hakansson and colleagues (Hakansson *et al.*, 2001). In this study the authors used the microfilament depolymerizing agent cytochalasin D to block the entry of partially invaded parasites. While the parasites were frozen at the host plasma membrane in the process of invasion, rhoptry discharge was not affected (Hakansson *et al.*, 2001). PVM vesicles defined by the presence of rhoptry proteins and termed E-vacuoles (empty vacuoles) were found to have retained their non-fusogenic properties indicating that the PVM and PVM alone is responsible for this property (Hakansson *et al.*, 2001).

Despite the absence of lysosomal fusion, lysosomal contents, most notably lipoproteins and sterols, are actively sequestered by the parasite in a host microtubule dependent transport mechanism (Coppens *et al.*, 2006). This reveals that the parasite, once inside the host cell specifically reorganizes the structural organization of the host cell to facilitate its own growth and potentially its developmental decisions.

11.4.3 The PVM as a Signalling Platform in the Neutralization of Host Defences

The ability to resist lysosomal fusion and acidification were long considered the primary defence mechanism against cellular defences (Sinai and Joiner, 1997). The identification in interferon stimulated cells of an additional defence mechanism mediated by the direct deposition of murine p47-resistance GTPases on the PVM resulting in their disruption (Martens *et al.*, 2005; Halonen *et al.*, 2001; Melzer *et al.*, 2008; Khaminets *et al.*, 2010) presented another later of host defence. The class of compounds involved are the immunity-related GTPases (IRGs) that represent a family of related proteins that are extensively expanded in mice but not humans (Bekpen *et al.*, 2005, 2009), resulting in this class of molecules contributing to host defence selectively in mice (Konen-Waisman and Howard, 2007).

The mechanisms underlying the destruction of the PVM and subsequent clearance of the now cytoplasmic parasites is the coordinated loading of the IRG proteins in a manner akin to the membrane attack complex associated with the activation of complement (Martens *et al.*, 2005; Khaminets *et al.*, 2010). Not surprisingly, the expansion of host defence mechanism is met by the development of parasite countermeasures. In *Toxoplasma* this role is mediated by the ROP–kinase family members ROP18 (an active kinase) and ROP5 (an inactive pseudokinase).

Both proteins emerged to prominence from studies that used forward genetic approaches (reviewed in Sibley *et al.*, 2009) aimed at identifying virulence determinants by the examination of genetic crosses between the prototypical *Toxoplasma* strains (Reese *et al.*, 2011; Saeij *et al.*, 2005, 2006; Behnke *et al.*, 2012; Taylor *et al.*, 2006). ROP18 was identified independently as a PVM-localized kinase required for intracellular replication (El Hajj *et al.*, 2007). At the time the substrate was not known.

The identification of substrate as being members of the murine IRG family of proteins demonstrated that phosphorylation of this molecule resulted in its inactivation, thus affording protection to the parasites within the vacuole (Steinfeldt *et al.*, 2010). Interestingly, the protective effects of ROP18 (and ROP5) in the context of the interferon response are restricted to murine cells and not seen in human cells consistent with the IRG pathway being largely absent in humans (Niedelman *et al.*, 2012; Bekpen *et al.*, 2005, 2009). While the virulence impact is most pronounced with murine IRGs as the primary target, additional host proteins and pathways have also been implicated as substrates of ROP18 (Cheng *et al.*, 2012; Yamamoto and Takeda, 2012).

In addition to ROP18, which is a bona fide kinase, the genetic evidence points to a vital contribution of ROP5, a pseudokinase (Peixoto *et al.*, 2010). Exciting recent data point to the fact that, despite lacking kinase activity, ROP5 regulates ROP18 activity at the PVM in a strain dependent manner (Behnke *et al.*, 2012; Fleckenstein *et al.*, 2012; Reese *et al.*, 2011). This points to how host-specific pressures may influence the evolution of parasite countermeasures to ensure effective parasitism.

11.4.4 The PVM as a Signalling Platform in the Modulation of Host Activities

The *Toxoplasma* infected cell is fundamentally remodelled at the level of transcription (Blader *et al.*, 2001) and by extension the proteome (Wastling *et al.*, 2005; Nelson *et al.*, 2008). Many of these changes are due to the parasite-mediated subversion of critical cellular transcriptional machineries. Work in our laboratory (Molestina *et al.*, 2003; Molestina and Sinai, 2005b) and that of others (Blader *et al.*, 2001; Brenier-Pinchart *et al.*, 2000; Denney *et al.*, 1999; Kim *et al.*, 2001; Jensen *et al.*, 2011) has found that the activation of the host transcription factor NFκB accompanies infection in several cell types. The activation of NFκB is associated to varying degrees with the identification of the phosphorylated form of the inhibitor I κB at the PVM. The activation of NFκB is mediated by the phosphorylation of its inhibitor I κB at two critical serine residues by the I κB kinase (IKK) complex and its subsequent degradation (reviewed in Ghosh and Karin, 2002). Remarkably, in infected cells, this event happens at the PVM and has been linked to a parasite encoded I κB kinase activity (TgIKK) (Molestina and Sinai, 2005a, b). The identity of the TgIKK, however, has remained elusive as bioinformatic and purification attempts have been unsuccessful in isolating the candidate gene.

Phospho-IκBα localization at the PVM is pronounced in Type I parasites (Molestina *et al.*, 2003) and much reduced to absent in Type II organisms (Jensen *et al.*, 2011). Interestingly, the extent of NFκB activation does not appear to correlate with levels of Phospho-IκB at the PVM suggesting that the regulation of this pathway is more complex than a single determinant being involved (Jensen *et al.*, 2011; Molestina and Sinai, 2005b). Indeed, work from the Saeij lab identified the dense granule secreted GRA15 as a potent activator of NFκB (Jensen *et al.*, 2011), while our studies suggest that the elusive TgIKK activity is incapable of directly activating NFκB in the absence of the host I κB kinase (IKK) but is involved in sustaining the activation of the pathway during the course of the infection (Molestina and Sinai, 2005b).

While a subset of parasite effectors are secreted into the host cell, thereby impacting signalling events, it is likely that additional signalling activities are likely to be found at the PVM. These could play functions in both the defence of intracellular niche as well as the manipulation of host cell activities.

11.5 STRUCTURAL MODIFICATIONS IN THE HOST CELL

Several studies have examined the effect on *T. gondii* development on the host cytoskeleton. Studies have focused on microtubules (Sehgal *et al.*, 2005; Melo *et al.*, 2001; Cintra and De Souza, 1985) and intermediate filaments (Halonen and Weidner, 1994; Cintra and De Souza, 1985), both of which appear to form a scaffold or overcoating of the vacuole. The distortion of the host cytoskeleton in infected cells is actually dynamic and not merely a function of spatial constraints in the infected cell (Sehgal *et al.*, 2005; Coppens *et al.*, 2006).

The extent of manipulation of the microtubule cytoskeleton by *T. gondii* was revealed in a groundbreaking study from the Coppens laboratory (Coppens *et al.*, 2006). As noted above, the fact *T. gondii* PVM resists lysosomal fusion was taken as evidence for the complete lack of interaction with this organelle system (reviewed in Sinai and Joiner, 1997). This is no longer believed to be true. In studies examining cholesterol utilization by *T. gondii* it was noted that the bulk of parasite cholesterol comes from exogenous sources along the lysosomal LDL pathway (Coppens *et al.*, 2000; Sehgal *et al.*, 2005). Perturbation of this pathway resulted in the sequestration of cholesterol and the inhibition of parasite growth (Coppens *et al.*, 2000). In this recent study Coppens' colleagues executed painstakingly detailed electron microscopic analyses to reveal that intact lysosomes are delivered to the lumen of the vacuole by microtubule supported invaginations of the PVM (Coppens *et al.*, 2006). Thus, while the lysosomes themselves do not fuse with the PVM, their contents including cholesterol and hydrolytic enzymes are safely delivered to the parasite (Coppens *et al.*, 2006). The presence of secreted proteinase inhibitors in the lumen of the vacuole (Morris *et al.*, 2002), and the pore activity of the PVM (Schwab *et al.*, 1994b) (see below) that would prevent acidified environment likely contribute to the neutralization of the lysosomes degradative power.

In addition to the microtubules involved in the delivery of lysosomes a far more profound reorganization of the microtubule cytoskeleton is observed. The microtubules overcoating the PVM are short and disordered (Coppens *et al.*, 2006). Remarkably, the microtubule organizing centre (MTOC) relocated to a location adjacent to the vacuole, potentially drawn in by the thread-like PVM-extension that appears to extend along microtubules (Coppens *et al.*, 2006; Romano *et al.*, 2008).

11.6 ROLE OF THE PVM IN NUTRIENT ACQUISITION

The central tenet of parasitism is the acquisition of nutrients from the host. The *T. gondii* PVM appears to be adapted to this scavenging function allowing for both building blocks of intermediary metabolism and lipids transit. The *T. gondii* PVM is porous by virtue of a non-specific pore activity allowing free bidirectional access to compounds under 1300 daltons (Schwab *et al.*, 1994b). While described almost two decades ago, the molecular basis of this activity remains elusive. With regard to lipidic nutrients, PVM-associated organelles are the likely source at the site of membrane–membrane contact (Sinai and Joiner, 1997; Sinai *et al.*, 1997) as has been described in diverse systems for inter-organellar lipid transfer (Vance and Shiao, 1996; Vance and Vance, 2004). Sterols, on the other hand, are actively scavenged from the lysosomal pathway as described above (Coppens *et al.*, 2000; Sehgal *et al.*, 2005). Additionally, lipids are also scavenged from lipid storage droplets in the host cell (Charron and Sibley, 2002; Coppens and Vielemeyer, 2005) presumably using the cytoskeleton as conduits for delivery to the PVM. Other sources of host lipids across the PVM have been suggested by studies using lipid tracers (Caffaro and Boothroyd, 2011) which, taken together, indicate that the PVM may serve as a sorting station for the active scavenging of lipids from diverse host sources.

11.7 THE PVM AS THE SUBSTRATE FOR THE DEVELOPING TISSUE CYST WALL

The tachyzoite PVM is typically a transient structure that is destroyed once the parasites egress from the infected host cell. However, its fate in the differentiation to the tissue cyst containing bradyzoites remains unknown. Our understanding of this dramatic transformation is very poor and can be summarized in the conversion of the unit membrane PVM into a dense, heavily glycosylated tissue cyst wall (Scholtyseck *et al.*, 1974; Ferguson and Hutchison, 1987). Glycosylation of the tissue cyst wall is readily detected using several lectins, the most well characterized of which is Dolichos lectin (DBA) that recognizes terminal α-GalNAc residues (Scholtyseck *et al.*, 1974; de Carvalho *et al.*, 1991; Sethi *et al.*, 1977; Fauquenoy *et al.*, 2008; Derouin *et al.*, 1981). Interestingly, DBA labels the rhoptries of intracellular tachyzoites but not the PVM (de Carvalho *et al.*, 1991), despite many rhoptry proteins localizing to this compartment. These data suggest that gross changes in glycosylation account for the modification of the PVM to the dense glycocalyx that is the cyst wall. Of note, lectin labelling studies indicate additional glycan reactivity profiles that predominate in the tissue cyst and are largely absent in the PVM (Scholtyseck *et al.*, 1974; Sethi *et al.*, 1977). This begs several questions regarding the origins of the cyst wall including: (i) does the PVM serve as the physical substrate for the cyst wall? (ii) What is the fate of tachyzoite PVM in a mature tissue cyst? (iii) Does the cyst wall form on the luminal or host cell cytoplasm facing aspect of the PVM? and (iv) What is the fate of PVM-associated organelles?

The bulk of the data regarding these questions comes from early ultrastructural studies on mature tissue cysts (Scholtyseck *et al.*, 1974; Ferguson, 2004; Ferguson and Hutchison, 1987). Here, the dense and heavily glycosylated tissue cyst wall is found to contain membranous elements within it – whether these derive from the PVM or from host organelles is not clear (Ferguson, 2004; Ferguson and Hutchison, 1987). The limited molecular evidence, defined by the detection predominantly of dense granule (GRA) proteins secreted by tachyzoites within the tissue cyst wall and matrix (Ferguson, 2004) suggest that at least the protein components in cyst wall formation are in fact delivered to the PVM during tachyzoite growth. The fact that knock-outs of several GRA proteins have no clear effect on tachyzoite growth but do impact the extent and efficiency of tissue cyst formation suggest the primary function of many GRA proteins may in fact be the assembly of the tissue cyst wall should that represent a developmental course for a given vacuole (Mercier *et al.*, 1998, 2005; Craver *et al.*, 2010; Pszenny *et al.*, 2012). The fact that most of the GRA proteins are in effect orphans without homologues outside of the tissue cyst forming coccidia suggests their functions may be linked to this critical activity.

An additional enigma surrounds the conversion of the poorly or unglycosylated PVM into the highly glycosylated tissue cyst wall. The question regarding the origin of glycosylation has been one that has persisted, without serious investigation, for decades (Sethi *et al.*, 1977; Derouin *et al.*, 1981). The sources of glycosylation could be the parasite, the host or both. In fact, studies put forth by the Schwarz laboratory suggest that both host and parasite glycans decorate parasite proteins (Garenaux *et al.*, 2008). Currently being addressed is the cell biological basis of this finding driven by the observation that the *Toxoplasma* PVM exhibits low level fusion activity with PVM-associated ER (Goldszmid *et al.*, 2009; Cebrian *et al.*, 2011) while possibly redirecting Golgi traffic (Coppens, personal communication). As the primary host cell organelles are involved in protein glycosylation, these activities, which might be accelerated during tissue cyst formation, could allow for the recruitment of host enzymes into the lumen of the vacuole to which the substrate nucleotide sugars could gain access from the host cytoplasm via the nutrient pore (Schwab *et al.*, 1994a). Data from the author's laboratory group using host cells with defects in the glycosylation pathways suggests that some post-translational glycosylation on *T. gondii* proteins is indeed due to host cell enzymes that are taken up by and then used by the parasite to modify its own proteins (Eller *et al.*, 2012). Regardless of the mechanisms employed, the dissection of the molecular basis underlying the conversion of the unit membrane PVM to the heavily glycosylated and dense tissue cyst wall present some fascinating biological questions.

11.8 IDENTIFICATION OF NOVEL ACTIVITIES AT THE PVM

Over the last few years there has been a burst of interest in the study of the PVM, addressing its structural, nutrient scavenging and signalling interactions with the host cell. As noted throughout the chapter, many questions remain to be answered, providing a fertile area of investigation. The completion of the *Toxoplasma* genome (http://www.toxodb.org/) as well as proteomic studies defining the secretory organelles (Bradley *et al.*, 2005; Zhou *et al.*, 2005) have and will continue to advance our understanding of PVM. A comprehensive dissection of PVM biology will likely emerge from the study of activities localizing to this enigmatic organelle. The development of efficient gene tagging strategies in both Type I (Fox *et al.*, 2009; Huynh and Carruthers, 2009) and cyst forming Type II parasites (Fox *et al.*, 2011) provides an efficient means to systematically assign localization and/or function from knockout studies. In this regard, the author's laboratory undertook a proteomic analysis of the PVM by the generation of poly-specific antisera directed against the PVM fraction. While this strategy proved too 'noisy', resulting in the identification of targets localizing to the cytoplasm and mitochondrion of the parasite, many known and several novel dense granule and rhoptry targeting proteins were also identified (Sinai, in preparation). Those of most significant interest are activities without known homology where assignment of localization and/or function can be assessed using the high efficiency gene targeting approaches. The unique nature of the PVM suggests most activities will likely represent hypothetical open reading frames or annotated genes without function. This presents challenges, but also opportunities, as many will provide new insights into the biology of both the PVM and the parasite itself.

References

Adjogble, K.D., Mercier, C., Dubremetz, J.F., Hucke, C., Mackenzie, C.R., Cesbron-Delauw, M.F., Daubener, W., 2004. GRA9, a new Toxoplasma gondii dense granule protein associated with the intravacuolar network of tubular membranes. Int. J. Parasitol. 34, 1255–1264.

Ahn, H.J., Kim, S., Nam, H.W., 2005. Host cell binding of GRA10, a novel, constitutively secreted dense granular protein from Toxoplasma gondii. Biochem. Biophys. Res. Commun. 331, 614–620.

Aikawa, M., Komata, Y., Asai, T., Midorikawa, O., 1977. Transmission and scanning electron microscopy of host cell entry by Toxoplasma. Am. J. Pathol. 87, 285–296.

Aikawa, M., Miller, L.H., Johnson, J., Rabbege, J., 1978. Erythrocyte entry by malaria parasites. A moving junction between erythrocyte and parasite. J. Cell Biol. 77, 77–82.

Alexander, D.L., Mital, J., Ward, G.E., Bradley, P., Boothroyd, J.C., 2005. Identification of the Moving Junction Complex of Toxoplasma gondii: A Collaboration between Distinct Secretory Organelles. PLoS Pathog. 1, e17.

Andrade, E.F., Stumbo, A.C., Monteiro-Leal, L.H., Carvalho, L., Barbosa, H.S., 2001. Do microtubules around the Toxoplasma gondii-containing parasitophorous vacuole in skeletal muscle cells form a barrier for the phagolysosomal fusion? J. Submicrosc. Cytol. Pathol. 33, 337–341.

Behnke, M.S., Fentress, S.J., Mashayekhi, M., Li, L.X., Taylor, G.A., Sibley, L.D., 2012. The Polymorphic Pseudokinase ROP5 Controls Virulence in Toxoplasma gondii by Regulating the Active Kinase ROP18. PLoS Pathogens 8, e1002992.

Bekpen, C., Hunn, J.P., Rohde, C., Parvanova, I., Guethlein, L., Dunn, D.M., Glowalla, E., Leptin, M., Howard, J.C., 2005. The interferon-inducible p47 (IRG) GTPases in vertebrates: loss of the cell autonomous resistance mechanism in the human lineage. Genome Biology 6, R92.

Bekpen, C., Marques-Bonet, T., Alkan, C., Antonacci, F., Leogrande, M.B., Ventura, M., Kidd, J.M., Siswara, P., Howard, J.C., Eichler, E.E., 2009. Death and resurrection of the human IRGM gene. PLoS Genetics 5, e1000403.

Bermudes, D., Dubremetz, J.F., Joiner, K.A., 1994. Molecular characterization of the dense granule protein GRA3 from *Toxoplasma gondii*. Mol. Biochem. Parasit. 68, 247–257.

Besteiro, S., Dubremetz, J.F., Lebrun, M., 2011. The moving junction of apicomplexan parasites: a key structure for invasion. Cellular Microbiology 13, 797–805.

Blader, I.J., Manger, I.D., Boothroyd, J.C., 2001. Microarray analysis reveals previously unknown changes in Toxoplasma gondii-infected human cells. J. Biol. Chem. 276, 24223–24231.

Bradley, P.J., Ward, C., Cheng, S.J., Alexander, D.L., Coller, S., Coombs, G.H., Dunn, J.D., Ferguson, D.J., Sanderson, S.J., Wastling, J.M., Boothroyd, J.C., 2005. Proteomic analysis of rhoptry organelles reveals many novel constituents for host-parasite interactions in Toxoplasma gondii. J. Biol. Chem. 280, 34245–34258.

Brenier-Pinchart, M.P., Pelloux, H., Simon, J., Ricard, J., Bosson, J.L., Ambroise-Thomas, P., 2000. Toxoplasma gondii induces the secretion of monocyte chemotactic protein-1 in human fibroblasts, in vitro. Mol. Cell Biochem. 209, 79–87.

Caffaro, C.E., Boothroyd, J.C., 2011. Evidence for host cells as the major contributor of lipids in the intravacuolar network of Toxoplasma-infected cells. Eukaryotic Cell 10, 1095–1099.

Carey, K.L., Donahue, C.G., Ward, G.E., 2000. Identification and molecular characterization of GRA8, a novel, proline-rich, dense granule protein of Toxoplasma gondii. Mol. Biochem. Parasitol. 105, 25–37.

Carruthers, V., Sibley, L., 1997. Sequential protein secretion from three distinct organelles of Toxoplasma gondii accompanies invasion of human fibroblasts. Eur. J. Cell Biol. 73, 114–123.

Carruthers, V.B., 1999. Armed and dangerous: Toxoplasma gondii uses an arsenal of secretory proteins to infect host cells. Parasitol. Int. 48, 1–10.

Carruthers, V.B., 2002a. Host cell invasion by the opportunistic pathogen Toxoplasma gondii. Acta Tropica 81, 111–122.

Carruthers, V.B., 2002b. Host cell invasion by the opportunistic pathogen Toxoplasma gondii. Acta. Trop. 81, 111–122.

Carruthers, V.B., Giddings, O.K., Sibley, L.D., 1999. Secretion of micronemal proteins is associated with toxoplasma invasion of host cells. Cell Microbiol. 1, 225–235.

Cebrian, I., Visentin, G., Blanchard, N., Jouve, M., Bobard, A., Moita, C., Enninga, J., Moita, L.F., Amigorena, S., Savina, A., 2011. Sec22b regulates phagosomal maturation and antigen crosspresentation by dendritic cells. Cell 147, 1355–1368.

Charron, A.J., Sibley, L.D., 2002. Host cells: mobilizable lipid resources for the intracellular parasite Toxoplasma gondii. J. Cell Sci. 115, 3049–3059.

Charron, A.J., Sibley, L.D., 2004. Molecular partitioning during host cell penetration by Toxoplasma gondii. Traffic 5, 855–867.

Cheng, L., Chen, Y., Chen, L., Shen, Y., Shen, J., An, R., Luo, Q., Du, J., 2012. Interactions between the ROP18 kinase and host cell proteins that aid in the parasitism of Toxoplasma gondii. Acta Tropica 122, 255–260.

Cintra, W.M., De Souza, W., 1985. Immunocytochemical localization of cytoskeletal proteins and electron microscopy of detergent extracted tachyzoites of Toxoplasma gondii. J. Submicrosc. Cytol. 17, 503–508.

Coppens, I., Dunn, J.D., Romano, J.D., Pypaert, M., Zhang, H., Boothroyd, J.C., Joiner, K.A., 2006. Toxoplasma gondii sequesters lysosomes from mammalian hosts in the vacuolar space. Cell 125, 261–274.

Coppens, I., Joiner, K.A., 2003. Host but not parasite cholesterol controls Toxoplasma cell entry by modulating organelle discharge. Mol. Biol. Cell 14, 3804–3820.

Coppens, I., Sinai, A.P., Joiner, K.A., 2000. Toxoplasma gondii exploits host low-density lipoprotein receptor-mediated endocytosis for cholesterol acquisition. J. Cell Biol. 149, 167–180.

Coppens, I., Vielemeyer, O., 2005. Insights into unique physiological features of neutral lipids in Apicomplexa: from storage to potential mediation in parasite metabolic activities. Int. J. Parasitol. 35, 597–615.

Craver, M.P., Rooney, P.J., Knoll, L.J., 2010. Isolation of Toxoplasma gondii development mutants identifies a potential proteophosphogylcan that enhances cyst wall formation. Mol. Biochem. Parasitol. 169, 120–123.

De Carvalho, L., Desouza, W., 1989. Cytochemical localization of plasma membrane enzyme markers during interiorization of tachyzoites of *Toxoplasma gondii* by macrophages. J. Protozol. 36, 164–170.

De Carvalho, L., Souto-Padron, T., De Souza, W., 1991. Localization of lectin-binding sites and sugar-binding proteins in tachyzoites of Toxoplasma gondii. J. Parasitol. 77, 156–161.

De Melo, E.J., De Carvalho, T.U., De Souza, W., 1992. Penetration of Toxoplasma gondii into host cells induces changes in the distribution of the mitochondria and the endoplasmic reticulum. Cell Struct. Funct. 17, 311–317.

De Souza, W., 2005. Microscopy and cytochemistry of the biogenesis of the parasitophorous vacuole. Histochem. Cell Biol. 123, 1–18.

Demelo, E.J.T., Souza, W.D., 1996. Pathway for C6-NBD-ceramide on the host cell infected with *Toxoplasma gondii*. Cell Struct. Funct. 21, 47–52.

Denney, C.F., Eckmann, L., Reed, S.L., 1999. Chemokine secretion of human cells in response to Toxoplasma gondii infection. Infect. Immun. 67, 1547–1552.

Derouin, F., Beauvais, B., Lariviere, M., Guillot, J., 1981. Binding of fluorescein-labelled lectins on trophozoites and cysts of 3 strains of Toxoplasma gondii. C R Seances Soc. Biol. Fil. 175, 761–768.

Desjardins, M., Huber, L., Parton, R., Griffiths, G., 1994. Biogenesis of phagolysosomes proceeds through a sequential series of interactions with the endocytic apparatus. J. Cell Biol. 124, 677–688.

Dobrowolski, J.M., Sibley, L.D., 1996. Toxoplasma invasion of mammalian cells is powered by the actin cytoskeleton of the parasite. Cell 84, 933–939.

Dubremetz, J.F., 2007. Rhoptries are major players in Toxoplasma gondii invasion and host cell interaction. Cell Microbiol. 9, 841–848.

Dubremetz, J.F., Rodriguez, C., Ferreira, E., 1985. Toxoplasma gondii: redistribution of monoclonal antibodies on tachyzoites during host cell invasion. Exp. Parasitol. 59, 24–32.

El Hajj, H., Lebrun, M., Arold, S.T., Vial, H., Labesse, G., Dubremetz, J.F., 2007. ROP18 is a rhoptry kinase controlling the intracellular proliferation of Toxoplasma gondii. PLoS Pathog. 3, e14.

Eller, B., Datta, A., Lynn, B.C., Sinai, A.P., 2012. A case for host involvement in the glycosylation of Toxoplasma tissue cysts. 12th International Workshops on Opportunistic Pathogens August 2012, Tarrytown, NY. Abstract T46, page 64.

Fauquenoy, S., Morelle, W., Hovasse, A., Bednarczyk, A., Slomianny, C., Schaeffer, C., Van Dorsselaer, A., Tomavo, S., 2008. Proteomics and glycomics analyses of N-glycosylated structures involved in Toxoplasma gondii—host cell interactions. Mol. Cell Proteomics 7, 891–910.

Ferguson, D.J., 2004. Use of molecular and ultrastructural markers to evaluate stage conversion of Toxoplasma gondii in both the intermediate and definitive host. Int. J. Parasitol. 34, 347–360.

Ferguson, D.J., Hutchison, W.M., 1987. An ultrastructural study of the early development and tissue cyst formation of Toxoplasma gondii in the brains of mice. Parasitol. Res. 73, 483–491.

Fleckenstein, M.C., Reese, M.L., Konen-Waisman, S., Boothroyd, J.C., Howard, J.C., Steinfeldt, T., 2012. A Toxoplasma gondii pseudokinase inhibits host IRG resistance proteins. PLoS Biology 10, e1001358.

Foussard, F., Leriche, M.A., Dubremetz, J.F., 1991. Characterization of the lipid content of Toxoplasma gondii rhoptries. Parasitology 102, 367–370.

Fox, B.A., Falla, A., Rommereim, L.M., Tomita, T., Gigley, J.P., Mercier, C., Cesbron-Delauw, M.F., Weiss, L.M., Bzik, D.J., 2011. Type II Toxoplasma gondii KU80 Knockout Strains Enable Functional Analysis of Genes Required for Cyst Development and Latent Infection. Eukaryot. Cell.

Fox, B.A., Ristuccia, J.G., Gigley, J.P., Bzik, D.J., 2009. Efficient gene replacements in Toxoplasma gondii strains deficient for nonhomologous end joining. Eukaryot. Cell 8, 520–529.

Garenaux, E., Shams-Eldin, H., Chirat, F., Bieker, U., Schmidt, J., Michalski, J.C., Cacan, R., Guerardel, Y., Schwarz, R.T., 2008. The dual origin of Toxoplasma gondii N-glycans. Biochemistry 47, 12270–12276.

Ghosh, S., Karin, M., 2002. Missing pieces in the NF-kappaB puzzle. Cell 109 (Suppl.), S81–S96.

Goldszmid, R.S., Coppens, I., Lev, A., Caspar, P., Mellman, I., Sher, A., 2009. Host ER-parasitophorous vacuole interaction provides a route of entry for antigen cross-presentation in Toxoplasma gondii-infected dendritic cells. J. Exp. Med. 206, 399–410.

Hakansson, S., Charron, A.J., Sibley, L.D., 2001. Toxoplasma evacuoles: a two-step process of secretion and fusion forms the parasitophorous vacuole. Embo. J. 20, 3132–3144.

Halonen, S.K., Weidner, E., 1994. Overcoating of Toxoplasma parasitophorous vacuoles with host cell vimentin type intermediate filaments. J. Euk. Microbiol. 41, 65–71.

Halonen, S.K., Taylor, G.A., Weiss, L.M., 2001. Gamma interferon-induced inhibition of Toxoplasma gondii in astrocytes is mediated by IGTP. Infect. Immun. 69 (9), 5573–5576.

Henriquez, F.L., Nickdel, M.B., Mcleod, R., Lyons, R.E., Lyons, K., Dubremetz, J.F., Grigg, M.E., Samuel, B.U., Roberts, C.W., 2005. Toxoplasma gondii dense granule protein 3 (GRA3) is a type I transmembrane protein that possesses a cytoplasmic dilysine (KKXX) endoplasmic reticulum (ER) retrieval motif. Parasitology 131, 169–179.

Huynh, M.H., Carruthers, V.B., 2009. Tagging of endogenous genes in a Toxoplasma gondii strain lacking Ku80. Eukaryot. Cell 8, 530–539.

Jacobs, D., Dubremetz, J.-F., Loyens, A., Bosman, F., Saman, E., 1998. Identification and heterologous expression of a new dense granule protein (GRA7) from *Toxoplasma gondii*. Molec. Biochem. Parasitol. 91, 237–249.

Jensen, K.D., Wang, Y., Wojno, E.D., Shastri, A.J., Hu, K., Cornel, L., Boedec, E., Ong, Y.C., Chien, Y.H., Hunter, C.A., Boothroyd, J.C., Saeij, J.P., 2011. Toxoplasma polymorphic effectors determine macrophage polarization and intestinal inflammation. Cell Host & Microbe 9, 472–483.

Joiner, K.A., Fuhrman, S.A., Mietinnen, H., Kasper, L.L., Mellman, I., 1990. *Toxoplasma gondii*: Fusion competence of parasitophorous vacuoles in Fc receptor transfected fibroblasts. Science 249, 641–646.

Jones, T.C., Hirsch, J.G., 1972. The interaction between *Toxoplasma gondii* and mammalian cells. II. The absence of lysosomal fusion with phagocytic vacuoles containing living parasites. J. Exp. Med. 136, 1173.

Jones, T.C., Veh, S., Hirsch, J.G., 1972. The interaction between *Toxoplasma gondii* and mammalian cells. I. Mechanism of entry and intracelluar fate of the parasite. J. Exp. Med. 136, 1157–1172.

Jutras, I., Desjardins, M., 2005. PHAGOCYTOSIS: At the Crossroads of Innate and Adaptive Immunity. Annu. Rev. Cell Dev. Biol. 21, 511–527.

Karsten, V., Qi, H., Beckers, C.J.M., Reddy, A., Dubremetz, J.F., Webster, P., Joiner, K.A., 1998. The protozoan parasite Toxoplasma gondii targets proteins to dense granules and vacuolar space using both conserved and unusual mechanisms. J. Cell Biol. 141, 911–914.

Khaminets, A., Hunn, J.P., Konen-Waisman, S., Zhao, Y.O., Preukschat, D., Coers, J., Boyle, J.P., Ong, Y.C., Boothroyd, J.C., Reichmann, G., Howard, J.C., 2010. Coordinated loading of IRG resistance GTPases on to the Toxoplasma gondii parasitophorous vacuole. Cellular Microbiology 12, 939–961.

Kim, J.M., Oh, Y.K., Kim, Y.J., Cho, S.J., Ahn, M.H., Cho, Y.J., 2001. Nuclear factor-kappa B plays a major role in the regulation of chemokine expression of HeLa cells in response to Toxoplasma gondii infection. Parasitol. Res. 87, 758–763.

Konen-Waisman, S., Howard, J.C., 2007. Cell-autonomous immunity to Toxoplasma gondii in mouse and man. Microbes and Infection/Institut. Pasteur. 9, 1652–1661.

Labruyere, E., Lingnau, M., Mercier, C., Sibley, L.D., 1999. Differential membrane targeting of the secretory proteins GRA4 and GRA6 within the parasitophorous vacuole formed by Toxoplasma gondii. Mol. Biochem. Parasitol. 102, 311–324.

Lebrun, M., Michelin, A., El Hajj, H., Poncet, J., Bradley, P.J., Vial, H., Dubremetz, J.F., 2005. The rhoptry neck protein RON4 re-localizes at the moving junction during Toxoplasma gondii invasion. Cell Microbiol. 7, 1823–1833.

Lecordier, L., Meleon-Borodowski, I., Dubremetz, J.F., Tourvielle, B., Mercier, C., Deslee, D., Capron, A., Cesbron-Delauw, M.F., 1995. Characterization of a dense granule antigen of Toxoplasma gondii (GRA6) associated to the network of the parasitophorous vacuole. Mol. Biochem. Parasitol. 70, 85–94.

Lecordier, L., Mercier, C., Sibley, L.D., Cesbron-Delauw, M.F., 1999. Transmembrane insertion of the Toxoplasma gondii GRA5 protein occurs after soluble secretion into the host cell. Mol. Biol. Cell 10, 1277–1287.

Lingelbach, K., Joiner, K., 1998. The parasitophorous vacuole membrane surrounding Plasmodium and Toxoplasma: an unusual compartment in infected cells. J. Cell Sci. 111, 1467–1475.

Magno, R.C., Lemgruber, L., Vommaro, R.C., De Souza, W., Attias, M., 2005a. Intravacuolar network may act as a mechanical support for Toxoplasma gondii inside the parasitophorous vacuole. Microsc. Res. Tech. 67, 45–52.

Magno, R.C., Straker, L.C., De Souza, W., Attias, M., 2005b. Interrelations between the parasitophorous vacuole of Toxoplasma gondii and host cell organelles. Microsc. Microanal. 11, 166–174.

Martens, S., Parvanova, I., Zerrahn, J., Griffiths, G., Schell, G., Reichmann, G., Howard, J.C., 2005. Disruption of Toxoplasma gondii Parasitophorous Vacuoles by the Mouse p47-Resistance GTPases. PLoS Pathog. 1, e24.

Martin, A.M., Liu, T., Lynn, B.C., Sinai, A.P., 2007. The Toxoplasma gondii parasitophorous vacuole membrane: transactions across the border. J. Eukaryot. Microbiol. 54, 25–28.

Melo, E.J., Carvalho, T.M., De Souza, W., 2001. Behaviour of microtubules in cells infected with Toxoplasma gondii. Biocell 25, 53–59.

Melo, E.J., De Souza, W., 1997. Relationship between the host cell endoplasmic reticulum and the parasitophorous vacuole containing Toxoplasma gondii. Cell Struct. Funct. 22, 317–323.

Melzer, T., Duffy, A., Weiss, L.M., Halonen, S.K., 2008. The gamma interferon (IFN-gamma)-inducible GTP-binding protein IGTP is necessary for toxoplasma vacuolar disruption and induces parasite egression in IFN-gamma-stimulated astrocytes. Infect. Immun. 76 (11), 4883–4894.

Mercier, C., Adjogble, K.D., Daubener, W., Delauw, M.F., 2005. Dense granules: are they key organelles to help understand the parasitophorous vacuole of all apicomplexa parasites? Int. J. Parasitol. 35, 829–849.

Mercier, C., Howe, D., Mordue, D., Lingnau, M., Sibley, L., 1998. Targeted disruption of the GRA2 locus in *Toxoplasma gondii* decreases acute virulence in mice. Infect. Immun. 66, 4176–4182.

Mercier, C., Lecordier, L., Darcy, F., Deslee, D., Murray, A., Tourvieille, B., Maes, P., Capron, A., Cesbron-Delauw, M.-F., 1993. Molecular characterization of a dense granule antigen (GRA2) associated with the network of the parasitophorous vacuole in *Toxoplasma gondii*. Mol. Biochem. Parasitol. 58, 71–82.

Michel, R., Schupp, S., Raether, W., Bierther, F.W., 1980. Formation of a close junction during invasion of eythrocytes by *Toxoplasma gondii* in vitro. Int. J. Parasitol. 10, 309–313.

Molestina, R.E., Payne, T.M., Coppens, I., Sinai, A.P., 2003. Activation of NF-{kappa}B by Toxoplasma gondii correlates with increased expression of antiapoptotic genes and localization of phosphorylated I{kappa}B to the parasitophorous vacuole membrane. J. Cell Sci. 116, 4359–4371.

Molestina, R.E., Sinai, A.P., 2005a. Detection of a novel parasite kinase activity at the *Toxoplasma gondii* parasitophorous vacuole membrane capable of phosphorylating host IkBa. Cell Microbiol. 7, 351–362.

Molestina, R.E., Sinai, A.P., 2005b. Host and parasite-derived IKK activities direct distinct temporal phases of NF-{kappa}B activation and target gene expression following Toxoplasma gondii infection. J. Cell Sci. 118, 5785–5796.

Mordue, D., Sibley, L., 1997. Intracellular fate of vacuoles containing Toxoplasma gondii is determined at the time of formation and depends on the mechanism of entry. J. immunol. 159, 4452–4459.

Mordue, D.G., Desai, N., Dustin, M., Sibley, L.D., 1999. Invasion by Toxoplasma gondii establishes a moving junction that selectively excludes host cell plasma membrane proteins on the basis of their membrane anchoring. J. Exp. Med. 190, 1783–1792.

Morris, M.T., Coppin, A., Tomavo, S., Carruthers, V.B., 2002. Functional analysis of Toxoplasma gondii protease inhibitor 1. J. Biol. Chem. 277, 45259–45266.

Nelson, M.M., Jones, A.R., Carmen, J.C., Sinai, A.P., Burchmore, R., Wastling, J. M., 2008. Modulation of the host cell proteome by the intracellular apicomplexan parasite Toxoplasma gondii. Infect. Immun. 76, 828–844.

Nichols, B.A., Chiappino, M.L., O'Connor, G.R., 1983. Secretion from the rhoptries of Toxoplasma gondii during host-cell invasion. Journal of Ultrastructure Research 83, 85–98.

Nichols, B.A., O'Connor, G.R., 1981. Penetration of mouse peritoneal macrophages by the protozoon Toxoplasma gondii. Lab. Invest. 44, 324–334.

Niedelman, W., Gold, D.A., Rosowski, E.E., Sprokholt, J.K., Lim, D., Farid Arenas, A., Melo, M.B., Spooner, E., Yaffe, M.B., Saeij, J.P., 2012. The rhoptry proteins ROP18 and ROP5 mediate Toxoplasma gondii evasion of the murine, but not the human, interferon-gamma response. PLoS Pathogens 8, e1002784.

Peixoto, L., Chen, F., Harb, O.S., Davis, P.H., Beiting, D.P., Brownback, C.S., Ouloguem, D., Roos, D.S., 2010. Integrative genomic approaches highlight a family of parasite-specific kinases that regulate host responses. Cell Host & Microbe 8, 208–218.

Pernas, L., Boothroyd, J.C., 2010. Association of host mitochondria with the parasitophorous vacuole during Toxoplasma infection is not dependent on rhoptry proteins ROP2/8. International Journal for Parasitology 40, 1367–1371.

Porchet-Hennere, E., G., T, 1983. Relations entre Toxoplasma et sa cellule-hote. Protistologica 19, 357–370.

Porchet-Hennere, E., Vivier, E., Torpier, G., 1985. Origine des membranes de la paroi chez Toxoplasma. Annales de parasitologie humaine et compar 60, 101–110.

Pszenny, V., Davis, P.H., Zhou, X.W., Hunter, C.A., Carruthers, V.B., Roos, D.S., 2012. Targeted disruption of Toxoplasma gondii serine protease inhibitor 1 increases bradyzoite cyst formation in vitro and parasite tissue burden in mice. Infection and Immunity 80, 1156–1165.

Reese, M.L., Boothroyd, J.C., 2009. A helical membrane-binding domain targets the Toxoplasma ROP2 family to the parasitophorous vacuole. Traffic 10, 1458–1470.

Reese, M.L., Zeiner, G.M., Saeij, J.P., Boothroyd, J.C., Boyle, J.P., 2011. Polymorphic family of injected pseudokinases is paramount in Toxoplasma virulence. Proceedings of the National Academy of Sciences of the United States of America 108, 9625–9630.

Romano, J.D., Bano, N., Coppens, I., 2008. New host nuclear functions are not required for the modifications of the parasitophorous vacuole of Toxoplasma. Cellular Microbiology 10, 465–476.

Saeij, J.P., Boyle, J.P., Boothroyd, J.C., 2005. Differences among the three major strains of Toxoplasma gondii and their specific interactions with the infected host. Trends Parasitol. 21, 476–481.

Saeij, J.P., Boyle, J.P., Coller, S., Taylor, S., Sibley, L.D., Brooke-Powell, E.T., Ajioka, J.W., Boothroyd, J.C., 2006. Polymorphic secreted kinases are key virulence factors in toxoplasmosis. Science 314, 1780–1783.

Schatten, H., Ris, H., 2002. Unconventional specimen preparation techniques using high resolution low voltage field emission scanning electron microscopy to study cell motility, host cell invasion, and internal cell structures in Toxoplasma gondii. Microsc. Microanal. 8, 94–103.

Schatten, H., Ris, H., 2004. Three-dimensional imaging of Toxoplasma gondii-host cell interactions within the parasitophorous vacuole. Microsc. Microanal. 10, 580–585.

Scholtyseck, E., Mehlhorn, H., Muller, B.E., 1974. Fine structure of cyst and cyst wall of Sarcocystis tenella, Besnoitia jellisoni, Frenkelia sp. and Toxoplasma gondii. J. Protozool. 21, 284–294.

Schwab, J.C., Beckers, C.J., Joiner, K.A., 1994a. The parasitophorous vacuole membrane surrounding intracellular Toxoplasma gondii functions as a molecular sieve. Proc. Natl. Acad. Sci. U. S. A. 91, 509–513.

Schwab, J.C., Beckers, C.J.M., Joiner, K.A., 1994b. The parasitophorous vacuole membrane surrounding intracellular *Toxoplasma gondii* functions as a molecular sieve. Proc. Natl. Acad. Sci. 91, 509–513.

Sehgal, A., Bettiol, S., Pypaert, M., Wenk, M.R., Kaasch, A., Blader, I.J., Joiner, K.A., Coppens, I., 2005. Peculiarities of host cholesterol transport to the unique intracellular vacuole containing toxoplasma. Traffic 6, 1125–1141.

Sethi, K.K., Rahman, A., Pelster, B., Brandis, H., 1977. Search for the presence of lectin-binding sites on Toxoplasma gondii. J. Parasitol. 63, 1076–1080.

Shaw, M.K., 1997. The same but different: the biology of Theileria sporozoite entry into bovine cells. Int. J. Parasitol. 27, 457–474.

Sibley, L.D., Krahenbuhl, J.L., Adams, G.M.W., Weidner, E., 1986. Toxoplasma modifies macrophage phagosomes by secretion of a vesicular network rich in surface proteins. J. Cell Biol. 103, 867–874.

Sibley, L.D., Niesman, I.R., Parmley, S.F., Cesbron-Delauw, M.-F., 1995. Regulated secretion of multi-lamellar vesicles leads to formation of a tubulo-vesicular network in host-cell vacuoles occupied by Toxoplasma gondii. J. Cell Sci. 108, 1669–1677.

Sibley, L.D., Qiu, W., Fentress, S., Taylor, S.J., Khan, A., Hui, R., 2009. Forward genetics in Toxoplasma gondii reveals a family of rhoptry kinases that mediates pathogenesis. Eukaryot. Cell 8, 1085–1093.

Sibley, L.D., Weidner, E., Krahenbuhl, J.L., 1985. Phagosome acidification blocked by intracellular *Toxoplasma gondii*. Nature 315, 416–419.

Sinai, A.P., Joiner, K.A., 1997. Safe haven: the cell biology of nonfusogenic pathogen vacuoles. Annu. Rev. Microbiol. 51, 415–462.

Sinai, A.P., Joiner, K.A., 2001. The Toxoplasma gondii protein ROP2 mediates host organelle association with the parasitophorous vacuole membrane. J. Cell Biol. 154, 95–108.

Sinai, A.P., Webster, P., Joiner, K.A., 1997. Association of host cell endoplasmic reticulum and mitochondria with the *Toxoplasma gondii* parasitophorous vacuole membrane: a high affinity interaction. J. Cell Sci. 110, 2117–2128.

Steinfeldt, T., Konen-Waisman, S., Tong, L., Pawlowski, N., Lamkemeyer, T., Sibley, L.D., Hunn, J.P., Howard, J.C., 2010. Phosphorylation of mouse immunity-related GTPase (IRG) resistance proteins is an evasion strategy for virulent Toxoplasma gondii. PLoS Biology 8, e1000576.

Straub, K.W., Cheng, S.J., Sohn, C.S., Bradley, P.J., 2009. Novel components of the Apicomplexan moving junction reveal conserved and coccidia-restricted elements. Cellular Microbiology 11, 590–603.

Straub, K.W., Peng, E.D., Hajagos, B.E., Tyler, J.S., Bradley, P.J., 2011. The moving junction protein RON8 facilitates firm attachment and host cell invasion in Toxoplasma gondii. PLoS Pathogens 7, e1002007.

Suss-Toby, E., Zimmerberg, J., Ward, G.E., 1996. Toxoplasma invasion: The parasitophorous vacuole is formed from host cell plasma membrane and pinches off via a fission pore. Proc. Natl. Acad. Sci. 93, 8413–8418.

Taylor, S., Barragan, A., Su, C., Fux, B., Fentress, S.J., Tang, K., Beatty, W.L., Hajj, H.E., Jerome, M., Behnke, M.S., White, M., Wootton, J.C., Sibley, L.D., 2006. A secreted serine-threonine kinase determines virulence in the eukaryotic pathogen Toxoplasma gondii. Science 314, 1776–1780.

Vance, J.E., Shiao, Y.-I., 1996. Intracellular trafficking of phospholipids: Import of phosphatidylserine into mitochondria. Anticancer Res. 16, 1333–1340.

Vance, J.E., Vance, D.E., 2004. Phospholipid biosynthesis in mammalian cells. Biochem. Cell Biol. 82, 113–128.

Wastling, J.M., Burchmore, R., Nelson, M., May 27–31 2005. Manipualtion of the host cell proteome by *Toxoplasma gondii*. 8th international Congress on Toxoplasmosis. Porticcio, Corsica, France.

Weiss, L.M., Kim, K., 2000. The development and biology of bradyzoites of Toxoplasma gondii. Front Biosci. 5, D391–D405.

Yamamoto, M., Takeda, K., 2012. Inhibition of ATF6beta-dependent host adaptive immune response by a Toxoplasma virulence factor ROP18. Virulence 3, 77–80.

Zhou, X.W., Kafsack, B.F., Cole, R.N., Beckett, P., Shen, R.F., Carruthers, V.B., 2005. The opportunistic pathogen Toxoplasma gondii deploys a diverse legion of invasion and survival proteins. J. Biol. Chem. 280, 34233–34244.

CHAPTER

12

Toxoplasma Secretory Proteins and Their Roles in Cell Invasion and Intracellular Survival

*Maryse Lebrun**, *Vern B. Carruthers*[†], *Marie-France Cesbron-Delauw***

Laboratoire Dynamique des interactions membranaires normales et pathologiques, Université de Montpellier, Montpellier, France [†]Department of Microbiology and Immunology, University of Michigan Medical School, Ann Arbor, Michigan, USA **UMR5163-CNRS-UJF, Jean-Roget Institute, Grenoble, France

OUTLINE

Toxoplasma gondii, Second Edition
http://dx.doi.org/10.1016/B978-0-12-396481-6.00012-X

12.1 INTRODUCTION

Being an obligate intracellular parasite, *Toxoplasma gondii* has a limited capacity to survive outside of a cell during infection. Invading a new host cell is therefore crucial for its survival and the expansion of infection. Invasion mechanisms seem to be highly conserved in the Apicomplexa, meaning that results obtained with one organism can often be extrapolated to other members of the phylum. Accordingly, in places *T. gondii* invasion will be described using supplemental findings obtained from other Apicomplexa. Almost all of our knowledge on *T. gondii* host cell interactions is derived from *in vitro* studies using tachyzoites, the stage that is the most amenable to experimentation.

T. gondii shows two major peculiarities when compared to other members of the phylum. First, it has almost no host specificity, being able to invade all cell types, from mammalian to fish and even insect cells. Only plant protoplasts have proven refractory to invasion (Werk and Fischer, 1982). Second, contrasting with most other Apicomplexa that multiply by schizogony, *T. gondii* tachyzoites proliferate by endodyogeny, bypassing the dedifferentiation stage that occurs in schizonts. Tachyzoites are thus invasive at any stage of their cell cycle, except during the short period of cytokinesis.

Host cell invasion by *T. gondii* is an active, parasite-driven process (Morisaki *et al.*, 1995). It leads to the formation of a new subcellular compartment in which the parasites settle and multiply. The organelles of the apical complex – micronemes, rhoptries, and dense granules – play successively their part in the process by exocytosing their contents (Carruthers and Sibley, 1997; Dubremetz *et al.*, 1993).

This chapter will highlight the key elements involved in invasion and emphasize the sequential exocytosis of secretory organelles, including some peripheral aspects such as protein trafficking to organelles and biogenesis of parasitophorous vacuole.

12.2 INVASION: A RAPID AND ACTIVE PROCESS UNIQUE TO APICOMPLEXAN PARASITES

Host cell invasion by *T. gondii* is fundamentally different from phagocytosis or endocytosis induced by intracellular pathogens such as viruses, bacteria or *Trypanosoma cruzi* (Antoine *et al.*, 1998; Finlay and Cossart, 1997; Sibley and Andrews, 2000). Invasion is completed in less than 20 seconds and occurs with an apparent passivity of the host cell (i.e. without inducing host cell membrane ruffling or tyrosine phosphorylation of host cell) (Morisaki *et al.*, 1995). Entry mechanisms are largely driven by the parasite itself, although the host cell may contribute to some extent in the invasion of *T. gondii* tachyzoite (Caldas *et al.*, 2009; Gonzalez *et al.*, 2009; Sweeney *et al.*, 2010). Therefore Apicomplexa invasion is termed 'active invasion' instead of 'induced invasion' used for entry of bacteria. The parasite penetrates into a tight-fitting vacuole formed by invagination of the host plasma membrane. This active invasion has been observed in many different cell types, including professional phagocytes, and the parasite then resides within a specialized compartment called the parasitophorous vacuole (PV) that does not acidify (Sibley *et al.*, 1985) or fuse with lysosomes (Jones and Hirsch, 1972). The PV also does not intersect with the host endocytic or exocytic systems (Mordue *et al.*, 1999b).

Contrasting with this active invasion, opsonized parasites are internalized by macrophages over a period of two to four minutes in a spacious vacuole involving a series of profound changes of the host cell similar to those triggered when a bacterium is captured (Morisaki *et al.*, 1995). The vacuole surrounding the parasite is then a traditional phagosome that quickly acquires the markers of endocytosis, fuses with endosomes/lysosomes, and rapidly acidifies (Mordue and Sibley, 1997; Morisaki *et al.*, 1995; Sibley *et al.*, 1985). The parasite is then destroyed by lysosomal enzymes. Opsonized internalization does not involve reorientation of the parasite to bring the apical end into contact with the host cell as observed for active invasion and does not involve secretion of apical proteins stored in secretory organelles (Morisaki *et al.*, 1995). The fate of the parasite is therefore largely dependent on its ability to actively invade.

Like other apicomplexan zoites, the invasive stages of *T. gondii* are motile. Motility and invasion are closely linked and the molecular mechanisms driving these two phenomena are considered as closely related. Non-motile parasites are unable to penetrate a cell. *Toxoplasma* cells do not have specialized organelles for motility such as cilia or flagella, yet they are able to move along a substrate or a host cell surface by gliding, with antero-posterior polarity. This gliding motility, which is specific for Apicomplexa and highly conserved within the phylum, is defined by the absence of shape change, in contrast to amoebae and others' vertebrate cells that use crawling motility accompanied by the emission of pseudopods and lamellipods in the direction of the motion. Apicomplexan locomotion is exceptionally fast, reaching speeds of up to 10 μm/s *in vitro*, i.e. 10 to 50 times faster than the rates of motility of the majority of the other cells (keratinocytes, amoebas, etc.).

Videomicroscopy of extracellular *T. gondii* tachyzoites reveals that this gliding motility consists of a succession of several stereotyped behaviours: (i) circular gliding, which commences while the crescent-shaped parasite lies on its right side, from where it moves in a counter-clockwise manner; (ii) upright twirling, which occurs when the parasite is attached to the substrate by its posterior end, producing clockwise spinning; and (iii) helical rotation, which is a horizontal twirling movement resulting in forward displacement (Hakansson *et al.*, 1999). Helical gliding is the only long-distance productive movement observed *in vitro*, but it should be noted that *Plasmodium* sporozoites exhibit only circular gliding *in vitro* yet they are able to migrate considerable linear distances during liver invasion (Frevert *et al.*, 2005). The directional and twisting nature of helical motility strongly suggests a connection between the driving system of the parasite and the helical organization of the subpellicular microtubules. This connection could be established indirectly by the intramembrane particles (IMP) present on the protoplasmic faces of the inner membrane complex (see Chapter 2).

During helical gliding, *Toxoplasma* comes into direct contact with the host cell by its apical end. A counter-clockwise torsion of the body of the parasite is sometimes visible during entry (Nichols, 1985). An apical attachment step is difficult to visualize, as invasion is a rapid process and both gliding locomotion and invasion appear continuous except for a ~5-fold reduction in speed during entry. However, there is some evidence of attachment with apical reorientation, since treatment of invading parasites with cytochalasin D (cytD), an actin-disrupting drug which blocks motility and invasion, does not affect attachment (Dobrowolski *et al.*, 1997; Miller *et al.*, 1979), suggesting that attachment and active invasion are uncoupled.

The conoid, a thimble-shaped cytoskeletal structure at the extreme apex of the parasite, extends and retracts repeatedly as the parasite moves along the host cell surface. A compound (inhibitor 6) identified in a small molecule screen selectively blocks conoid extrusion and parasite invasion (Carey *et al.*, 2004b). However, conoid extrusion is not involved in motility since this same inhibitor has no effect on either parasite motility or microneme secretion. Although its precise role in invasion is unknown, the movements of the conoid may contribute to bringing the apex of the parasite in close proximity to the host cell surface (Schwartzman and Saffer, 1992). Pretreatment with cytochalasin D inhibits conoid extrusion (Mondragon and Frixione, 1996) but as previously mentioned does not affect attachment, indicating that a direct role of conoid in intimate attachment can be excluded. The projection of this structure during motility or invasion is evident in *Toxoplasma*, *Eimeria* and *Sarcocystis*, apicomplexan species that initiate infection through the gut. The absence of this organelle in *Plasmodium* sp. suggests it may be an accessory apparatus used to penetrate particularly robust barriers such as the intestinal epithelium (Mondragon and Frixione, 1996).

The area of close contact between the cell and the extended conoid invaginates as a small depression of the host cell membrane to create the moving junction (MJ) through which apicomplexan zoites propel themselves into the nascent PV (Aikawa *et al.*, 1978). The MJ is a region of tight apposition, but not fusion, between parasite and host cell membranes, the latter being markedly thickened at this level (Aikawa *et al.*, 1978). Freeze-fracture of invading tachyzoites (Porchet and Torpier, 1977) (Dubremetz and Ferguson, Chapter 2) shows that the MJ contains rhomboidally arrayed particles and is identical in appearance to that formed by invading *Plasmodium knowlesi* merozoites (Aikawa *et al.*, 1981). Although this interface has been called the 'moving junction', this is a fixed point of reference between the host cell and the parasite membrane and the parasite glides through this junction as it enters the cell. The MJ gives the illusion of moving as a circumferential ring around the parasite as invasion proceeds (see Dubremetz and Ferguson, Chapter 2). It coincides generally but not systematically with a prominent constriction of the parasite at the site of penetration. This constriction may result from physical constraints imposed by the host cytoskeleton. Analogous to an iris diaphragm, the MJ expands as it slides over the midsection of the parasite then closes at the posterior end when entry is completed by fusion of host membrane behind the parasite (Aikawa *et al.*, 1981; Lebrun *et al.*, 2005). The parasite is now located within the PV, the membrane of which originates mainly from host plasmalemma (Suss-Toby *et al.*, 1996). A striking difference in the density of IMP between the P face of the host cell and P face of the PV is observed (Aikawa *et al.*, 1981; Dubremetz *et al.*, 1993), illustrating the diffusion barrier that excludes transmembrane proteins at the MJ (Mordue *et al.*, 1999a).

The parasite is then isolated from the cytoplasm of the host cell in a PV, which increases in size thereafter to allow its development. Sequestered within the PV, the parasite salvages essential nutrients from the host cell and multiplies rapidly. Schwab *et al.* (1994) demonstrated that the PV membrane (PVM) surrounding *T. gondii* permits free bi-directional diffusion of small charged and uncharged molecules (less than 1300 Da) between the host cell cytoplasm and the vacuolar space. This finding is consistent with a pore of protein origin crossing the PVM. Since most of the integral membrane proteins from the host are excluded from the PVM (Mordue *et al.*, 1999a), the pore is probably parasite derived. Parasite proteins released from two different secretory organelles, the rhoptries and dense granules, are associated with the PVM (Lecordier *et al.*, 1999; Saffer *et al.*, 1992) and could contribute to pore formation.

In contrast to the tight-fitting PV formed by tachyzoites, sporozoite entry *in vitro* is characterized by a large primary vacuole containing a single parasite and lacking molecular pores (Speer *et al.*, 1997; Tilley *et al.*, 1997). The parasite does not replicate in this vacuole and instead within 24 hours it forms and exits into a secondary vacuole that is accompanied by the secretion of dense granule proteins (Tilley *et al.*, 1997). The significance of this two-step invasion process is unknown but since the parasite replicates only within the secondary vacuole (Speer *et al.*, 1995), specific secretory modification of the compartment is probably crucial for parasite intracellular survival and multiplication.

12.3 INVASION: A TIGHTLY COUPLED SECRETION MACHINERY

As described above, the successive steps of invasion are well characterized morphologically. The molecular mechanisms are far less understood, in particular those leading to building of the PV. The involvement of the apical organelle contents in the invasion process has been clearly demonstrated. Microneme proteins (MICs), followed by rhoptry proteins, are sequentially secreted to play distinct functions coordinated in time and space (Carruthers and Sibley, 1997). Dense granule secretion has also been described (Carruthers and Sibley, 1997; Dubremetz *et al.*, 1993; Leriche and Dubremetz, 1991), but it occurs mainly at the end of the invasion process, after PV formation, and dense granule proteins are therefore most likely involved in PV remodelling.

12.3.1 Invasion and Motility: Central Role of MICs

The first insight into understanding gliding motility and invasion by apicomplexan zoites came from observing relocalization of surface bound molecules. Vanderberg described the

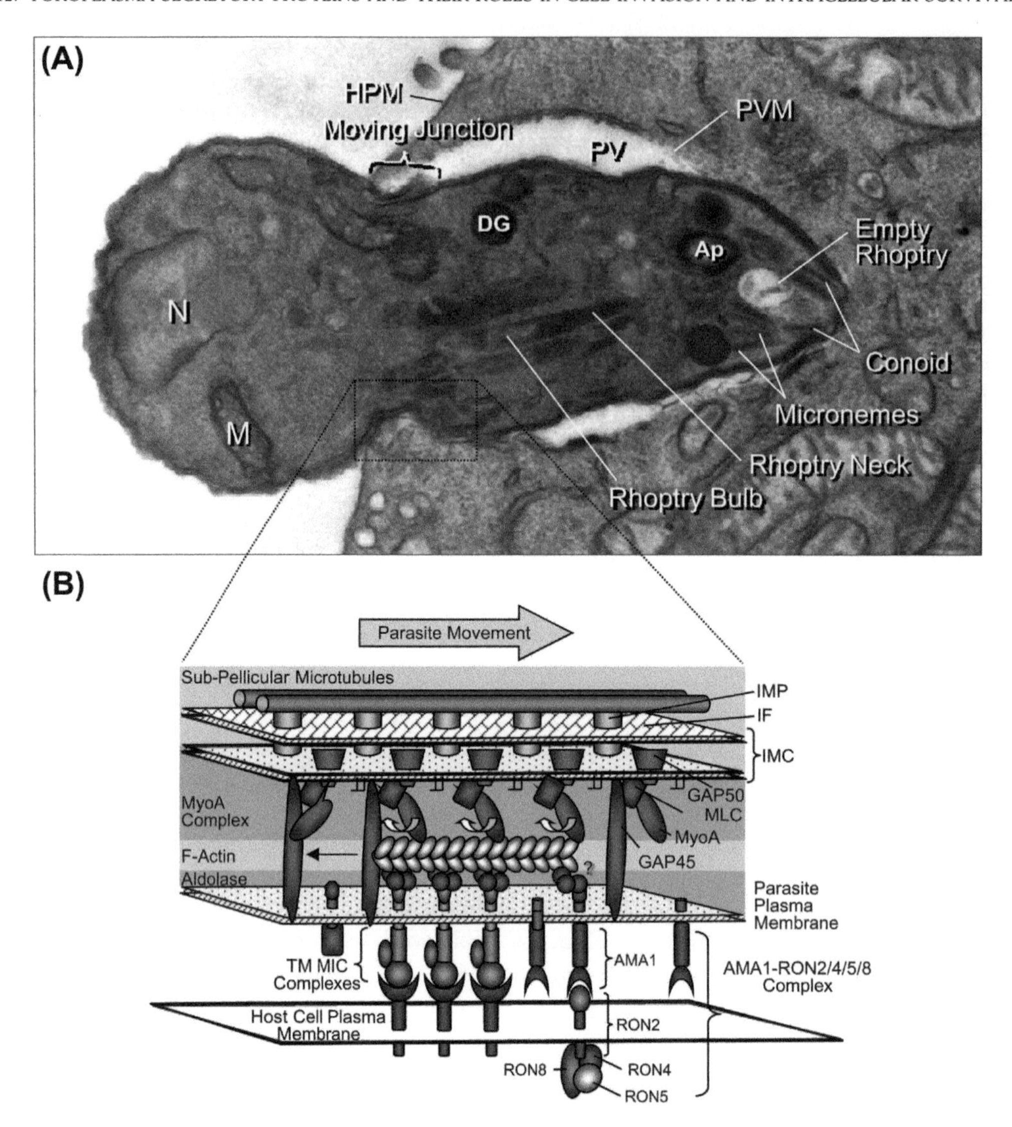

FIGURE 12.1 ***Toxoplasma*** **ultrastructure and molecular components of the moving junction during invasion.** A) Transmission electron micrograph of a *Toxoplasma* tachyzoite actively invading a Caco2 cell. Relevant structures of the parasite and host cell are labelled including the parasitophorous vacuole (PV), which is created by invaginating the host plasma membrane (HPM) to form the parasitophorous vacuolar membrane (PVM). Parasite internal structures including the nucleus (N), mitochondrion (M), dense granule (DG), and apicoplast (Ap) are labelled. Also shown is the moving junction (bracket and rectangle), the site of intimate contact between parasite's plasma membrane and the HPM where receptor engagement occurs. *Photomicrograph courtesy of David A. Elliott.*

B) Model of the glideosome system at the moving junction. As discussed in the text, TM MIC complexes are thought to bind host receptors and transmit the glideosome's mechanical force by connecting through aldolase to F-actin, which is translocated posteriorly by the ATP driven myoA complex anchored into the inner membrane complex (IMC). Intramembrane particles (IMP) consisting of TM proteins may connect the IMC with sub-pellicular microtubules and intermediate filament-like (IF) structures on the cytosolic face of the IMC. Also shown is the AMA1−RON2/4/5/8 complex, which is proposed to form the MJ. RON2 is predicted to be an integral membrane protein anchored into the host plasma membrane (as depicted here). RON2 interacts directly with AMA1 exposed on the parasite surface, providing a bridge between the parasite and its circumsporozoite reaction whereby antibodies were capped on the trailing end of gliding *Plasmodium berghei* sporozoites (Vanderberg, 1974). Then, Dubremetz and Ferreira showed that cationized ferritin bound the *Eimeria* sporozoite surface and was rapidly capped posteriorly (Dubremetz and Ferreira, 1978). The velocity and susceptibility to cytochalasin D and low temperature of this capping phenomenon and of gliding motility were identical (Dubremetz and Ferreira, 1978; Russell and Sinden, 1981).

Further studies eventually led to a model in which parasite gliding results from antero-posterior translocation of surface proteins interacting with cell surface or substrate receptors while being coupled to a submembranous actomyosin motor (Daher and Soldati-Favre, 2009; Menard, 2001; Opitz and Soldati, 2002; Sibley, 2010). Several studies have shown that these proteins were not resident at the parasite surface, but came from microneme exocytosis. Microneme exocytosis is an inducible process that has been first observed to occur at the apical attachment site between parasite and host cell before invasion. Inhibition of microneme exocytosis is correlated to inhibition of invasion (Carruthers and Sibley, 1999). After apical exocytosis, MIC proteins are translocated distally and released posteriorly as the parasite glides on a substrate or enters a cell (Carruthers and Sibley, 1997; Garcia-Reguet *et al.*, 2000; Kappe *et al.*, 1999; Opitz *et al.*, 2002; Asai *et al.*, 1995). This antero-posterior relocalization matches exactly the progression of the invasion. In addition, it is inhibited by cytochalasin D, suggesting a link between these proteins and the inner actin cytoskeleton.

The molecular characterization of microneme proteins (MICs) has shown that adhesive motifs usually found in higher eukaryote proteins are present in these proteins (Tomley *et al.*, 2001). A functional part played by these adhesins in host cell attachment, motility, invasion, and of a synergistic role of MICs in the infectious process (Cerede *et al.*, 2005; Huynh *et al.*, 2003; Marchant *et al.*, 2012; Mital *et al.*, 2005; Huynh and Carruthers, 2006) has been demonstrated, as well as a critical role of the C-terminal cytosolic domain (CD) of transmembrane MICs in motility and invasion (Jewett and Sibley, 2003; Kappe *et al.*, 1999; Sheiner *et al.*, 2008). This domain, which in many cases is highly conserved in Apicomplexa, is indeed the link between the extracellular domains of MICs interacting with host cell receptors and the submembranous actomyosin motor.

In the molecular model of invasion that has emerged, secreted MICs serve as ligands recognizing both the host cell surface and the submembranous cytoskeleton associated motor, called the glideosome (Fig. 12.1). The parasite myosin contractile system is anchored to the inner membrane complex (IMC) and allows anterior to posterior translocation of MICs, the outermost membrane of the IMC providing a conveyor belt-like system. When a transmembrane MIC protein is tethered to a fixed extracellular receptor, the parasite is propelled forward. The posterior release of the interaction needed for efficient motion or invasion involves specific proteolytic activities. In this model, although experimental proof is still lacking, the MJ is thought to provide anchorage to the host cell while the parasite's actin–myosin motor powers invasion. In this context, gliding machinery plays the central part of the invasion process, which would be also used for egress from the host cells. The molecular actors of gliding and invasion are described in detail in Chapter 13.

host cell. AMA1 also resides on either side of the MJ without RONs. AMA1 binds to aldolase (*via* its CD) suggesting that it serves as a bridging protein that physically connects the glideosome to other components of the MJ complex (*via* its ectodomain) and thus plays a critical role in the posterior translocation of the MJ complex during invasion; but this model awaits further demonstration. The other members of the RON complex (RON4/5/8) are tethered to RON2 and exposed to the cytosolic face of the host cell membrane, suggesting a role in anchoring the MJ to the host cell cytoskeleton.

12.3.2 Parasitophorous Vacuole Formation and Host Cell Reprogramming: Role of Rhoptries

There is no formal proof for a role of rhoptries in PV formation. However, rhoptry exocytosis is visualized at the beginning of invasion and rhoptry proteins (ROPs) are found associated with the PVM at early stages of biogenesis. The rhoptry contents sometimes appear as multi-lamellar vesicles both before and after exocytosis in *T. gondii* and in *Plasmodium* (Bannister *et al.*, 1986; Nichols *et al.*, 1983; Stewart *et al.*, 1986). When invasion is interrupted with cytD, the intracellular accumulation of this membrane-like material called evacuoles is particulary conspicuous even though the PV is not formed (Hakansson *et al.*, 2001). Invagination of the host cell membrane leads to PV formation, this latter being modified by fusing with these rhoptry derived secretory vesicles (Hakansson *et al.*, 2001). Consistent with this, rhoptry discharge is closely tied to the abrupt spike of conductance that is detected upon initial binding of the parasite to the host cell (Suss-Toby *et al.*, 1996). How these vesicles assume a multi-lamellar structure is not known, but their morphology suggests they contain lipids organized in sheets or bilayers. Their lipid contents are strictly parasite derived (Hakansson *et al.*, 2001). Subcellular fractionation of rhoptries has also shown an abundance of lipids in these organelles (Foussard *et al.*, 1991) and especially an enrichment in cholesterol, but this latter was not confirmed in a subsequent study (Besteiro *et al.*, 2008). Therefore, rhoptry exocytosis serves to export proteins and lipids modifying the membrane of the developing PV.

A growing number of studies have indicated that ROPs may also manipulate the host cell to the parasite's benefit. Indeed, *Toxoplasma* ROP kinases or pseudokinases targeted to the PVM or relocated in the host nucleus early after invasion hijack the host cell. These emerging roles will be discussed below and in Chapter 14.

12.3.3 Moving Junction Formation: Cooperative Role between Micronemes and Rhoptries

The molecular structure of the MJ has long been an enigma. MIC protein complexes associated with host cell receptors were believed to build up the junction, although no convincing data supported this hypothesis at the time. The first protein associated with a MJ was the merozoite cap protein-1 (MCP-1), described 15 years ago in *Plasmodium falciparum* (Hudson-Taylor *et al.*, 1995; Klotz *et al.*, 1989). This protein had been located under the merozoite plasmalemma suggesting it is not directly involved in the parasite–red cell interaction, but re-examination of its location demonstrated that MCP-1 is a chromatin associated nuclear peroxiredoxin, renamed PfnPrx (Richard *et al.*, 2011).

Two independant studies in *T. gondii* have led to the characterization of the first components of the MJ, an important step toward solving the molecular architecture of this structure central to apicomplexan invasion (Alexander *et al.*, 2005; Lebrun *et al.*, 2005). Surprisingly, among these MJ proteins are hypothetical proteins restricted to Apicomplexa and derived from the anterior part of the rhoptries, known as rhoptry neck proteins (RONs) and the microneme protein apical membrane antigen 1 (AMA1) suggesting that the MJ derives from collaboration between MICs and RONs. These findings will be detailed below (Fig. 12.1).

12.3.4 Maturation of the Parasitophorous Vacuole: A Prominent Role of Dense Granules

Toxoplasma and *Plasmodium* have evolved two distinct types of parasitophorous vacuoles (PV), linked to differential strategies of intracellular parasitism. While initial formation of both types of vacuoles is driven by the same active invasion process, maturation of the nascent PV is different in terms of both architecture and metabolism and leads to two distinct modes of parasite replication: endodyogeny *versus* schizogony.

The fate of the *Toxoplasma*-containing vacuole strictly correlates with the intravacuolar release of dense granule proteins (GRAs) and the formation of multi-lamellar structures at a posterior invaginated pocket of the parasite formed 10–20 minutes post-invasion (Sibley *et al.*, 1995). These structures extend into the vacuolar lumen as a membranous nanotubular network (MNN) that connects with the PVM (Sibley *et al.*, 1986; 1995). GRA proteins are segregated to either the MNN or PVM (Mercier *et al.*, 2005). The role of the MNN remains unclear but based on its spatial importance within the vacuole, current hypotheses favour a role in supporting metabolic exchange between the parasite and the host cell and/or an architectural role in the spatial organization of the vacuole. Both GRA proteins and the network persist during replication of the parasite within the PV. Later, when parasite stage conversion takes place, the vacuolar structures redistribute and contribute to the formation of an intracellular cyst (Ferguson, 2004; Torpier *et al.*, 1993). Hence, an essential function of some GRAs might be at the encystation stage for the construction and maintenance of the cyst wall. Emerging roles of GRAs in modulation of the host immune response will be also discussed in Chapter 14.

By contrast, the *Plasmodium*-containing vacuolar space remains tightly apposed to the parasite plasma membrane. The parasite expands its membranes beyond the limits of the initial vacuole, thus developing a prominent tubovesicular network (TVN). The *Plasmodium* PV behaves as a sorting compartment for the secretory machinery, leading to the export of parasite proteins of major importance for host–parasite interactions. Although the presence of electron dense organelles distinct from both micronemes and rhoptries was reported (Blackman and Bannister, 2001), there is no evidence that these are functionally similar to the *Toxoplasma* dense granules. In particular, so far no GRA orthologues have been found in the *Plasmodium* databases. The dense granules might thus reflect an evolutionary divergence required for specialized intracellular development and host–parasite relationship.

12.4 MICRONEMES

12.4.1 Trafficking of MICs through the Secretory Pathway

Fifty to one hundred micronemes (from Greek for 'small threads') populate the apical portion of invasive *Toxoplasma* zoites. Micronemes are often seen in an arc-like pattern by their association with the cytoplasmic face of the inner membrane complex in the apical region. This association is likely mediated by binding to microtubules, as was shown in *P. falciparum* (Bannister *et al.*, 2003; Schrevel *et al.*, 2008).

The first evidence of microneme-specific forward targeting elements was reported by Di Cristina *et al.* (2000) (Di Cristina *et al.*, 2000). Using a chimera consisting of the major surface antigen SAG1 ectodomain, CD46 transmembrane anchor (TM), and the MIC2 cytosolic domain (CD), they showed that the MIC2 CD was sufficient for microneme targeting. The key targeting elements were narrowed down to two motifs: YHYY, located immediately adjacent to the cytosolic side of the transmembrane (TM) anchor; and EIEYE, located further downstream. Although it was proposed that these motifs function as tyrosine-based sorting signals that interact with medium chain (μ-chain) within adaptor protein (AP) complexes for sorting into clathrin-coated vesicles, the YHYY motif is probably too close to the TM anchor to be accessible for μ-chain binding. Also, the EIEYE(AD…) element does not conform to the YXXΦ consensus μ-chain binding sequence.

Nonetheless, further support for tyrosine-based motifs in targeting to the micronemes came from showing that the CD of the lysosomal protein LAMP was sufficient for microneme targeting, and that the tyrosine in the sequence GYQTI was crucial since a mutant expressing GAQTI was retained in a Golgi-associated compartment (Hoppe *et al.*, 2000). The GYQTI conforms to the YXXΦ consensus sequence suggesting that *Toxoplasma* expresses microneme directive APs that can recognize this canonical element. It is unclear why mutation of the MIC2 CD sorting sequences (YHYY and EIEYE) resulted in secretion into the PV (i.e. a targeting defect) whereas altering the LAMP tyrosine-based motif led to retention (i.e. a trafficking defect), but this presumably reflects the fine specificity of APs functioning at different steps in the pathway. Mutational analysis of tyrosine-based motifs in the context of full-length MIC proteins expressed in a corresponding knockout strain may further clarify their role in targeting to the micronemes.

Reiss *et al.* (2001) revealed the interdependence of microneme proteins for trafficking and targeting by showing that the TM microneme protein MIC6 serves as an escorter for MIC1 and MIC4. Targeted deletion of *MIC6* resulted in secretion of MIC1 and MIC4 into the PV. Together with their finding that the MIC6 CD was sufficient to confer microneme targeting, these results strongly suggested that MIC6 uses elements in its CD for targeting of MIC1 and MIC4 to the micronemes. Targeting interdependence was further revealed by the discoveries that MIC2 is an escorter for M2AP (Huynh *et al.*, 2004). These studies support the concept that soluble MICs are escorted to the micronemes by binding to TM MICs containing CD forward targeting information.

Although the CDs of MIC2 and MIC6 are sufficient for microneme targeting when isolated from the rest of the protein, several studies show that 'soluble' partner MICs are also necessary for correct trafficking. For example, the absence of MIC1 causes arrest of MIC4 and MIC6 in the ER/Golgi and failure to reach the micronemes (Reiss *et al.*, 2001). Similarly, MIC2 accumulates in the ER/Golgi of M2AP deficient parasites (Huynh *et al.*, 2003). These findings suggest that signals in the CD of TM MICs are sufficient to reach the Golgi but that additional signals in partner MICs are necessary to complete the journey to the micronemes. Supporting this notion, several recent studies have shown that soluble MICs that are initially made as preproproteins contain trafficking information in their respective propeptides. For example, a propeptide deletion mutant (M2APΔpro) expressed in M2AP knockout parasites accumulated in post-Golgi compartment associated with the endosomal system, despite correctly oligomerizing with MIC2 (Harper *et al.*, 2006). Similarly, expression of MIC3 or MIC5 propeptide deletion mutants in their respective gene knockout backgrounds resulted in diversion to the PV or vesicular retention within the parasite, respectively (Brydges *et al.*, 2008; El Hajj *et al.*, 2008). More recently the key residues of several MIC propropeptides were mapped to the extreme N-terminus of the propeptide (Gaji *et al.*, 2011). The principal residues were shown to be aliphatic amino acids, particularly leucine or valine, positioned within the first three amino acids of the propeptide. These findings indicate that the propeptides of several MICs contain forward targeting information that supplements the tyrosine-based sorting signals expressed in the CDs of transmembrane MICs. Nevertheless, studies with MIC3 (El Hajj *et al.*, 2008) and M2AP (Harper and Carruthers, unpublished) suggest that the propeptide is not sufficient for microneme targeting when appended to a reporter protein. Proper targeting of MIC3 required both the propeptide and at least one of the three EGF domains, but neither the N-terminal chitin-binding domain nor dimerization of the protein was necessary to access the micromenes (El Hajj *et al.*, 2008). The subtilase SUB1 appears to be an exception to the requirement for a domain from the main body of the protein since its propeptide was shown to be both necessary and sufficient for trafficking to the micronemes (Binder *et al.*, 2008). Also, the propeptide of MIC6 is not necessary for microneme targeting (Meissner *et al.*, 2002).

Although it remains unclear how soluble MICs aid in the trafficking of adhesive complexes, the recent identification of the transmembrane protein TgSortilin (TgSORTLR) as a cargo receptor for MIC and ROP proteins offers a potential mechanism (Sloves *et al.*, 2012). In other systems sortilin is well known for ushering proteins from the Golgi to the endocytic system and lysosomes in a mannose-6-phosphate independent manner. Consistent with a similar function, the N-terminal luminal domain of TgSORTLR was shown to bind several MICs and ROPs including MIC1, MIC4, MIC5, M2AP, ROP1, ROP2, ROP4 and ROP5, whereas the CD bound to clathrin-associated trafficking complexes responsible for anterograde Golgi to endosome trafficking and retrograde retrieval. Parasites conditionally deficient in TgSORTLR were devoid of micronemes and rhoptries, establishing a crucial role for TgSORTLR in the biogenesis of apical secretory organelles. MICs and ROPs were either absent altogether or mislocalized to the intracellular vesicles, the parasite surface, the PV or the host cytoplasm due to leakage of the PV. That TgSORTLR resides mainly in the Golgi and post-Golgi endosome-like vesicles concurs with a role in shuttling proteins through the endosomal system *en route* to the micronemes and rhoptries. Apicomplexan parasites lack mannose-6-phosphate as an alternative endosomal trafficking signal and as such they appear to solely rely on TgSORTLR.

The studies described above support a model in which forward targeting signals from the CD of a TM MIC license it to escort soluble MICs through the ER to the Golgi where additional signals in soluble MICs are recognized by TgSORTLR to navigate the complexes from the Golgi through some portion of the endosomal system to the micronemes. After transporting its cargo to the micronemes, TgSORTLR likely returns to the Golgi to reload for another delivery cycle. The role of TgSORTLR in ROP trafficking and rhoptry biogenesis is discussed in greater detail below in Section 12.5.1.

12.4.2 Microneme Proteins

T. gondii expresses a large, diverse array of MICs including TM and soluble proteins. Several approaches have been used to identify MICs. Before the genome sequencing era, MICs were discovered using monoclonal antibodies (MIC1, MIC2, MIC3) followed by immunoscreening of cDNA libraries (Achbarou *et al.*, 1991a; Fourmaux *et al.*, 1996b; Garcia-Reguet *et al.*, 2000) or immunoaffinity purification (Donahue *et al.*, 2000). Genome sequencing of *T. gondii* has allowed the characterization of additional MICs, either by similarities to MICs of other Apicomplexa (AMA1, MIC2, SUB1) (Hehl *et al.*, 2000; Miller *et al.*, 2001; Wan *et al.*, 1997), by searching transmembrane domains and CDs with homologies to that of TRAP/MIC2 (MIC6, MIC7, MIC8, MIC9, MIC12) (Meissner *et al.*, 2002; Opitz *et al.*, 2002) or by similarities to already described *Toxoplasma* MICs (MIC13, MIC16) (Friedrich *et al.*, 2010a; Sheiner *et al.*, 2010). Novel MICs were identified by cell fractionation and calcium-modulating compounds that enhance or block microneme secretion, and N-terminal sequencing (MIC4, MIC5, MIC10, MIC11) (Brecht *et al.*, 2001; Brydges *et al.*, 2000; Harper *et al.*, 2004b; Hoff *et al.*, 2001). Co-precipitation with MIC2 antibodies was used to identify M2AP (Rabenau *et al.*, 2001). Highly sensitive and complementary technologies such as two-dimensional electrophoresis/mass spectroscopy (2-DE/MS) and liquid chromatography/electrospray ionization MS-MS (LC/ESI MS-MS) have added new candidates to the already long list of MICs previously identified (Zhou *et al.*, 2005; Kawase *et al.*, 2007). New bioinformatic tools allowing the identification of proteins showing similar cell cycle dependent expression profiles promises to add more candidates to the growing list.

12.4.2.1 MICs Sharing Homologies with Structural Domains of Eukaryotic Proteins Involved in Protein–Protein or Protein–Carbohydrate Interactions

The molecular characterization of MICs has revealed a striking conservation of structural domains known in higher eukaryotic cells (see Table 12.1). These domains are found on both soluble and TM MICs, in single or multiple copies and in a variety of combinations, so that every MIC protein is structurally unique. The repertoire of MICs in *T. gondii* constitutes a patchwork of proteins, some of them sharing orthologues in the other Apicomplexa. Several of these conserved structural domains are known in higher eukaryote to be responsible for protein–protein or carbohydrate–protein interactions. The cell surface binding properties of MICs have been demonstrated (MIC1, MIC2, MIC3, MIC4) (Carruthers and Sibley, 1999; Fourmaux *et al.*, 1996a; Garcia-Reguet *et al.*, 2000; Marchant *et al.*, 2012), but the presence of such domains has not been systematically associated with an attachment function (e.g. MIC6, which does not directly bind host receptors) (Saouros *et al.*, 2005). In this section, we will review the various structural domains found in MICs.

12.4.2.1.1 I- OR A-DOMAIN

MIC2 is the only characterized *Toxoplasma* protein known to possess this domain (Wan *et al.*, 1997). MIC2 is a member of the TRAP family, highly conserved among the phylum (Clarke *et al.*, 1990; Robson *et al.*, 1988; Spano *et al.*, 1998). Consistent with its conserved nature, conditional ablation of MIC2 expression markedly decreased parasite gliding motility and parasite attachment, resulting in diminished invasion and loss of virulence in a mouse model of acute toxoplasmosis (Huynh and Carruthers, 2006). MIC2 possesses a single integrin-like A-domain. This domain is present in the α-chain of some integrins, which are type I integral membrane proteins that promote cell–cell and cell–extracellular matrix (ECM) contacts (Larson *et al.*, 2009; Pytela, 1988). A similar domain is also found in various plasma proteins (e.g. von Willbrand factor), in soluble matrix proteins or in proteins involved in cell–cell and cell–ECM matrix interactions during homeostasis, inflammation or cell migration (Whittaker and Hynes, 2002). These domains bind to various ligands including collagens, laminin, fibronectin, ICAM-1 and the complement product iC3b. A unique feature of A-domains is that they possess a metal ion dependent adhesion site (MIDAS) motif composed of five non-contiguous amino acids (Lee *et al.*, 1995). *In vitro* assays of binding to several different putative receptors revealed that the MIC2 A-domain binds specifically to heparin (Harper *et al.*, 2004a), a ubiquitous sulphated proteoglycan found in the extracellular matrix (ECM). This binding is not dependent on the MIDAS site, a property also observed for binding of α2β1 integrin to laminin (Dickeson *et al.*, 1998) and of the *P. falciparum* TRAP domain with heparin (McCormick *et al.*, 1999). Homology modelling-based structural analysis of the TRAP A-domain identified a cluster of basic residues on the surface that potentially confers heparin-binding ability (Akhouri *et al.*, 2004). A similar small basic patch was later identified on the surface of the MIC2 A-domain upon solving its crystal structure (Tonkin *et al.*, 2010). These findings suggest spatial separation of the putative heparin-binding site from the MIDAS region, indicating that these distinct molecular surfaces may be involved in the recognition of multiple receptors (Akhouri *et al.*, 2004). Although heparin binding is MIDAS independent, a considerable loss of infectivity and host cell invasiveness was observed for *P. berghei* sporozoite expressing TRAP mutated in the MIDAS site (Matuschewski *et al.*, 2002), showing the importance of the motif in these functions, and suggesting that its function is at least partly MIDAS dependent. It was also shown that the TRAP A-domain interacts with fetuin-A on hepatocyte membranes and that this interaction enhances the parasite's ability to invade hepatocytes (Jethwaney *et al.*, 2005). In *T. gondii*, MIC2 binds to ICAM-1 and this

TABLE 12.1 Properties of *Toxoplasma* Secretory Proteins-Microneme Proteins

Location/ Protein	Calculated MW (kDa)[1]	Domains(#)[2]	Interacting Partners	Mutant Phenotypes	Function	Post-Secretory Trafficking	References
Microneme							
MIC1	49	MAR(2), galectin-like domain(1)	MIC4, MIC6	Non-essential protein KO mutant is less invasive and less virulent in mice	Adhesion, binding to sialylated oligosaccharides, folding, assembly of the MIC1/4/6 complex	Secreted and posterior capping	(Fourmaux *et al.*, 1996; Saouros *et al.*, 2005; Cerede *et al.*, 2005; Blumenschein *et al.*, 2007; Sawmynaden *et al.*, 2008; Garnett *et al.*, 2009)
MIC2	83	A/I-domain(1), TSR(5), degenerate TSR(1), TM(1)	M2AP	Key protein. C-terminal truncation mutant was not recoverable. Substitution with *Eimeria* MIC1 is viable but less invasive. Conditional knockdown mutant showed markedly reduced gliding and invasion	Escorter, adhesion	Secreted and posterior capping	(Wan *et al.*, 1997; Jewett & Sibley, 2004; Carruthers *et al.*, 1999; Huynh *et al.*, 2004; Starnes *et al.*, 2006; Huynh & Carruthers, 2006; Tonkin *et al.*, 2010)
MIC3	38 (90 dimer)	CBL(1), EGF(5)	MIC8	Non-essential protein KO mutant has reduced virulence in mice	Adhesion	Secreted and posterior capping	(Garcia-Reguet *et al.*, 2000; Cerede *et al.*, 2002, 2005)
MIC4	63	Pan/Apple(6)	MIC1, MIC6	Non-essential protein	Adhesion, binding to galactose	Secreted and posterior capping	(Brecht *et al.*, 2001; Reiss *et al.*, 2001; Marchant *et al.*, 2012)
MIC5	20	Protease prodomain		Non-essential protein KO mutant shows elevated SUB1 proteolysis of other secretory proteins	Subtilisin protease inhibitor	Secreted and posterior capping	(Brydges *et al.*, 2000; Saouros *et al.*, 2012)
MIC6	37	EGF(3), acidic	MIC1, MIC4	Non-essential protein	Escorter		(Reiss *et al.*, 2001)
MIC7	36	EGF(5), TM(1)					(Meissner *et al.*, 2002)
MIC8	75	CBL(1), EGF(10), TM(1)	MIC3				(Meissner *et al.*, 2002)
MIC9	32	EGF(3), TM(1)					(Meissner *et al.*, 2002)
MIC10	23					Secreted	(Hoff *et al.*, 2001)

(Continued)

TABLE 12.1 Properties of *Toxoplasma* Secretory Proteins-Microneme Proteins (*cont'd*)

Location/ Protein	Calculated MW (kDa)[1]	Domains(#)[2]	Interacting Partners	Mutant Phenotypes	Function	Post-Secretory Trafficking	References
MIC11	22	Strong charge asymmetry				Secreted	(Harper *et al.*, 2004)
MIC12 (TGME49_ 267680)	234	EGF(31), TSR(4), TM(1)				Secreted	(Opitz *et al.*, 2002)
MIC13 (MCP2)	51	MAR(3)			Adhesion, binding to sialylated oligosaccharides		(Friedrich *et al.*, 2010, Fritz *et al.*, 2012)
MIC14 (TGME49_ 218310)	106	TSR(2), TM(1)					Unpublished
MIC15 (TGME49_ 247195)	320	TSR(3), TM(1)					Unpublished
MIC16 (TGME49_ 289630)	75	TSR(6), TM(1)					Unpublished
MIC17A,B,C	~39	Pan/Apple(4)					(Chen *et al.*, 2008; Sohn *et al.*, 2011)
AMA1	60	Pan/Apple(2), TM(1)	RON2, RON4, RON5	Key protein Conditional knock-down mutant can attach but cannot penetrate. Fails to organize the MJ. Expression of a ROM4 dominant negative mutant interferes with AMA1 shedding post-invasion and reduces the initiation of parasite replication. A shedding resistant mutant is defective in invasion	MJ organization, binding to RON2	Secreted Detected on both the external and internal regions of invading parasites. Visible at the MJ in conditional knockdown mutant	(Hehl *et al.*, 2000; Alexander *et al.*, 2005; Mital *et al.*, 2005; Donahue *et al.*, 2000; Howell *et al.*, 2005; Besteiro *et al.*, 2009; Crawford *et al.*, 2010; Santos *et al.*, 2011; Lamarque *et al.*, 2011; Tyler & Boothroyd, 2011; Parussini *et al.*, 2012; Tonkin *et al.*, 2011)

M2AP	35	Galectin-like(1), coil(1)	MIC2	Non-essential protein KO mutant shows defects in MIC2 trafficking; deficient attachment and invasion; virulence defect		Secreted and posterior capping	(Rabenau *et al.*, 2001; Huynh *et al.*, 2003)
SUB1	85	Subtilase, GPI		Non-essential protein KO mutant shows defects the surface processing of several secretory proteins; deficient in gliding motility and invasion; virulence defect	Proteolysis	Secreted and posterior capping	(Miller *et al.*, 2001; Binder & Kim, 2004; Lagal *et al.*, 2010)
ROM1	28	Rhomboid, TM(7)		Non-essential protein KO mutant shows a fitness defect due to modest deficiencies in invasion and replication	Proteolysis		(Brossier *et al.*, 2005; Dowse *et al.*, 2005; Brossier *et al.*, 2008)
SPATR	58	EGF(1), TSR(2)		Non-essential protein KO is partially defective in invasion and virulence		Secreted	(Kawase *et al.*, 2007; Huynh and Carruthers, unpublished)
Toxolysin 4 (TLN4)	247	Metalloprotease (M16 family)		Non-essential protein Difficult to KO in RH but KO in Δku80 is viable	Proteolysis	Secreted	(Laliberte & Carruthers, 2011; Agrawal and Carruthers, unpublished)

[1] *Based on the complete open reading frame including signal sequence or GPI anchor signal, if present*

[2] *Abbreviations: CBL, Chitin binding like domain; CCP, domain abundant in complement control proteins; CobT, Cobalamin biosynthesis protein; DUF1222, domain of unknown function; EGF, epidermal growth factor; TSR, thrombospondin type-1 repeat; TM, transmembrane*

interaction facilitates migration across polarized epithelial cells (Barragan *et al.*, 2005). Interestingly, the MIC2 A-domain structure revealed the absence of a divalent metal in the MIDAS site, a finding that is consistent with the divergent spatial positioning of MIDAS residue D188. This feature, which appears to be unique to the MIC2 A-domain, might dictate its receptor specificity thus influencing target cell preference.

12.4.2.1.2 THROMBOSPONDIN TYPE 1 REPEAT DOMAIN: TSR

This domain is found in MIC2, MIC14, MIC15, MIC16 and SPATR (Kawase *et al.*, 2010; Opitz *et al.*, 2002; Sheiner *et al.*, 2010; Wan *et al.*, 1997), with one degenerated motif in each case. The TSR domain is present in many proteins from distantly related organisms including vertebrates. TSR proteins are usually involved in cell–cell and cell–matrix interactions (Lawler, 1986) in clotting or innate immunity (Haefliger *et al.*, 1989). The TSR domain allows thrombospondin and properdin to bind sulphated sugars and especially glycosaminoglycans (GAGs) (Chen *et al.*, 2000; Holt *et al.*, 1990). It shows also a high affinity for heparin (Guo *et al.*, 1992).

There are no reports of a function for the TSR domain of MIC2. However, in *Plasmodium*, the TRAP TSR domain binds heparan sulphate proteoglycans on the hepatocyte surface *in vivo* and *in vitro* (Muller *et al.*, 1993). The importance of the TRAP TSR–heparan sulfate interaction in the host cell invasion by *Plasmodium* sporozoite was demonstrated by Matuschewski and coworkers (Matuschewski *et al.*, 2002). Structural data on the TSP-1 of thrombospondin (Tan *et al.*, 2002) suggests that the MIC2 TSR likely forms an extended 'stalk' on which the A-domain sits, optimally positioned for interaction with host receptors. Although the TSR domain appears to participate in the interaction with M2AP (M. Huynh, J. Harper and V.B. Carruthers, unpublished) additional studies will be required to determine whether it also binds to a host cell receptor.

MIC14 and MIC15 have not been characterized but are annotated in http://www.Toxodb.org. MIC16 is a TM protein that displays six TSRs in its ectodomain and is conserved amongst coccidian parasites (Sheiner *et al.*, 2010). Although the function of MIC16 remains to be determined, it appears to be shed from the parasite surface by intramembrane cleavage similar to, for example, MIC2 and AMA1. The CD of MIC16 lacks targeting information for the micronemes, implying that MIC16 interacts with another MIC protein(s) that guides it to the micronemes.

Stephen Hoffman and colleagues implicated a *P. falciparum* surface protein called Secreted Protein with Altered Thrombospondin Repeat (PfSPATR) in sporozoite invasion (Chattopadhyay *et al.*, 2003). PfSPATR is expressed in merozoites, gametocytes and sporozoites, making it an attractive multistage vaccine candidate. Recombinant PfSPATR bound to HepG2 cells and an antibody to PfSPATR reduced sporozoites invasion, suggesting a role in adhesion. Similar findings were reported for *P. knowlesi* SPATR (Mahajan *et al.*, 2005). More recently a *Toxoplasma* orthologue, TgSPATR, was identified in a proteomics analysis of secretory proteins (Kawase *et al.*, 2007). Further characterization showed that TgSPATR is secreted from the micronemes and is on the surface of extracellular tachyzoites (Kawase *et al.*, 2010). Like its *Plasmodium* counterpart, TgSPATR contains an N-terminal EGF domain and one (Kawase *et al.*, 2007) or more likely two (V.B. Carruthers, unpublished) C-terminal TSR repeats. TgSPATR knockout parasites are deficient in invasion and virulence (M. Huynh and V.B. Carruthers, unpublished), thus providing genetic confirmation that the conserved SPATR family is an important player in cell invasion by apicomplexan parasites.

12.4.2.1.3 EPIDERMAL GROWTH FACTOR-LIKE DOMAIN

EGF-like domains are found in a large variety of proteins, mainly of animal origin (growth factors, lipoprotein receptors, selectins, clotting factors, extracellular matrix proteins, etc.) and are frequently seen in tandem repeats with various degrees of conservation. The functional significance of these domains is not understood yet. A common feature of these repeated domains is that they are found in the extracellular portion of TM proteins or in secreted proteins engaged in protein-protein interactions (Appella *et al.*, 1988; Davis, 1990). EGF domains typically contain six cysteine residues that form disulphide bridges. The sub-domain lengths between cysteines vary extensively. Some EGF-like domains have the capacity to bind calcium and are therefore termed calcium binding EGF (cbEGF). Calcium binding EGF domains form a more rigid, protease resistant structure upon binding calcium, as shown for the *Eimeria* protein EtMIC4 (Periz *et al.*, 2005), a TM protein containing 31 EGF-like and 12 TSR domains (Tomley *et al.*, 2001). This property may help form an elongated stalk-like structure to maximally project cbEGF-containing proteins from the cell surface.

EGF-like domains are found in apicomplexan MICs and on resident surface proteins of *Plasmodium*. Five of the EGF containing proteins in *T. gondii* are TM (MIC6, MIC7, MIC8, MIC9 and MIC12) (Meissner *et al.*, 2002; Opitz *et al.*, 2002; Reiss *et al.*, 2001) and two are soluble (MIC3 and SPATR) (Garcia-Reguet *et al.*, 2000; Kawase *et al.*, 2010). MIC3, MIC6, MIC8 and SPATR are expressed by tachyzoites and bradyzoites (Garcia-Reguet *et al.*, 2000; Meissner *et al.*, 2002) whereas MIC7 and MIC9 are predominantly expressed by bradyzoites (Meissner *et al.*, 2002). A *Toxoplasma* orthologue of EtMIC4 exists in the ToxoDB database. This protein likely corresponds to *T. gondii* MIC12 since the deduced amino acid sequence is virtually identical to the partial sequence of MIC12 published by Opitz *et al.* (Opitz *et al.*, 2002). At least two of the EGF-containing MICs contribute to tachyzoite cell invasion. MIC6 knockout parasites show mislocalization of the carbohydrate binding adhesins MIC1 and MIC4, resulting in a ~50% loss of cell invasion (Sawmynaden *et al.*, 2008). Conditional ablation of MIC8 revealed an essential role in invasion characterized by a block in rhoptry secretion but normal gliding motility and apical attachment (Kessler *et al.*, 2008), suggesting a potential role for MIC8 in signalling the discharge of rhoptries for parasite invasion (see below).

The involvement of MIC EGF-like domains in the interaction of these proteins with the host cell has not been shown. The MIC3 EGF-like domains are not involved in MIC3 binding to cells. They might, however, be involved in the appropriate exposure of the MIC3 binding site located in the chitin-binding-like domain (Cerede *et al.*, 2002). MIC6 contains three EGF-like domains, but does not bind to cells directly (Saouros *et al.*, 2005). EGF-like domains allow the formation of heteromeric MIC protein complexes. For instance, the second and third EGF domains of MIC6 associate with the MIC1 galectin like domain (GLD), thus increasing the valancy of MIC1 on the parasite surface during invasion (Reiss *et al.*, 2001). The structure of the MIC6–EGF2/MIC3–GLD revealed a novel interaction mode for an EGF domain involving an intimate hydrophobic surface (Sawmynaden *et al.*, 2008).

12.4.2.1.4 PAN/APPLE MODULE

The Apple module contains a conserved core of three disulphide bridges. In some members of the family an additional disulphide bridge links the N- and C-termini of the domain. This later type is commonly seen in tandem repeats (Tordai *et al.*, 1999) and mediates protein–protein or protein–carbohydrate interactions. It is found in the N-terminal domain of members of the plasminogen/hepatocyte growth factor family, in the plasma prekallikrein/coagulation factor XI family (Tordai *et al.*, 1999), in various nematode proteins and recently in apicomplexan MIC proteins (Brown *et al.*, 2001). The Apple domains of plasma prekallikrein are known to mediate its binding to high molecular weight kininogen (Herwald *et al.*, 1996), and the Apple domains of factor XI bind to factor XIIa, platelets, kininogen, factor IX, and heparin (Ho *et al.*, 1998). This domain was described for the first time in MIC4 through searching sequence homologies (Brecht *et al.*, 2001). It is also found in the micronemes of two other Apicomplexa: *Eimeria tenella* MIC5 containing multiple copies of apple domains (Brown *et al.*, 2000), and the *Sarcocystis muris* lectin (SML) possessing two of these domains (Klein *et al.*, 1998). Brecht and coworkers suggested that MIC4, which contains six Apple domains, could be an adhesin (Brecht *et al.*, 2001). However, subsequent studies have shown that MIC4 does not directly bind host cells and that one of its partner proteins, MIC1, is responsible for receptor recognition (Lourenco *et al.*, 2001; Saouros *et al.*, 2005). More recently the fifth Apple domain of MIC4 was shown to bind specifically to galactose terminating oligosaacharides (Marchant *et al.*, 2012). The authors suggested that the proteolytic release of the fifth and sixth Apple domains from the parasite surface could mimic the function of galectins involved in innate immunity, thus modulating the host immune response.

The Apple domain is a subfamily of the PAN (Plasminogen, Apple, Nematode) module that has been detected in the AMA1 (Pizarro *et al.*, 2005), a microneme protein found in all *Plasmodium* species and in at least six other Apicomplexa including *T. gondii*, *Neospora caninum*, *Sarcocystis neurona*, *Eimeria tenella*, *Babesia bovis* and *Theileria*. The structural characterization of the AMA1 ectodomains from *P. falciparum* (Bai *et al.*, 2005), *P. vivax* (Pizarro *et al.*, 2005), *T. gondii* (Crawford *et al.*, 2010), *N. caninum* and *B. divergens* (Tonkin *et al.*, 2012), has revealed a three domain architecture, originally proposed based on the disulphide bonding pattern (Hodder *et al.*, 1996). Domains I (DI) and II (DII) are structurally homologous whereas domain III (DIII) is highly divergent. The seminal study of Pizarro *et al.* (Pizarro *et al.*, 2005) established that DI and DII of *Plasmodium vivax* AMA1 adopted a PAN motif. Intriguingly, however, the DI and DII domains of TgAMA1, BdAMA1 and NcAMA1 are not recognized as PAN module containing proteins (Crawford *et al.*, 2010; Tonkin *et al.*, 2012), as was identified for Pf/PvAMA1.

The membrane proximal DIII domain displays the most structural divergence of the three ectoplasmic AMA1 domains. In TgAMA1, NcAMA1 and PvAMA1, it adopts the structurally ultra-stable cystine knot motif that probably stabilizes and orients the ectodomain. In contrast, BdAMA1 DIII is noticeably less compact, forming a more extended layer across the base of DI and DII. The DIII domain has been shown to function as an adhesin for *Plasmodium* attachment to erythrocytes (Kato *et al.*, 2005), based on its expression on CHO cells being sufficient to bind to the Kx membrane protein on trypsin-treated erythrocytes. Erythrocyte binding activity was also conferred by a construct comprising domains I and II of *P. yoelii* AMA1 (Fraser *et al.*, 2001). However, other investigations failed to demonstrate binding of PfAMA1 (shed from cultured supernatent, and encompassing most of the PfAMA1 ectodomain) to host cells (Howell *et al.*, 2001).

Of particular interest is a conserved apical hydrophobic groove in the AMA1 DI domain, which is surrounded by a series of polymorphic loops from domain I and by an extended non-polymorphic DII loop. This structural feature, conserved across the AMA1 molecules of all apicomplexans examined, is crucial for the invasion. Known to be the target for invasion-inhibitory monoclonal antibodies (Coley *et al.*, 2006; 2007; Collins *et al.*, 2009; Pizarro *et al.*, 2005) and a peptide identified from a phage-display library (Richard *et al.*, 2010), this trough has been recently shown to bind the rhoptry neck protein RON2 (Tonkin *et al.*, 2011; Vulliez-Le Normand *et al.*, 2012a), a component of a multiprotein complex that is crucial for host cell invasion (Alexander *et al.*, 2006; Besteiro *et al.*, 2009; Lamarque *et al.*, 2011; Tyler and Boothroyd, 2011). These and other recent studies discerning the role of the AMA1 ectodomain during invasion are described in greater detail below.

Proteomic analysis of secretory products in *T. gondii* revealed at least three additional proteins containing PAN/Apple domains (Zhou *et al.*, 2005). The same series of paralogous genes were identified in an *in silico* screen for putative secretory proteins (Chen *et al.*, 2008). The *N. caninum* orthologue of one member of this family was recently identified by a monoclonal antibody that stains the micronemes, providing strong evidence of its localization to these organelles (Sohn *et al.*, 2011). These proteins were termed NcMIC17A, NcMIC17B and NcMIC17C, according to their order on the chromosome. NcMIC17B was the family member identified as being expressed in *N. caninum*, which is consistent with transcriptomic data indicating robust expression of NcMIC17B, but little or no expression of NcMIC17A or NcMIC17C. Interestingly, based on proteomic and transcriptomic data in http://www.Toxodb.org it appears that TgMIC17C is the principally expressed member of the family in *T. gondii*. Whether this genus specific expression pattern contributes to the distinct biology of *N. caninum* and *T. gondii* awaits further study.

A novel PAN/Apple domain-containing protein (P104) was recently described (Gong *et al.*, 2012). It is composed of 10 putative PAN/Apple domains and displays affinity for chondroitin sulphate (CS). It localizes to the apical pole of the parasite, with partial co-localization with micronemes.

12.4.2.1.5 THE CHITIN-BINDING-LIKE DOMAIN: CBL

MIC3 and MIC8 contain in their N-terminal region a domain homologous to one found in chitin binding proteins (hevein domain or chitin-binding motif), followed by several EGF-like domains (Garcia-Reguet *et al.*, 2000; Meissner *et al.*, 2002). This chitin-binding-like domain (CBL) is also found in an *E. tenella* protein (ETH_00017540, http://www.Toxodb.org) possessing a signal peptide and two EGF-like domains. The chitin-binding domain is a well conserved 30–40 AA stretch found in plants and fungi (Wright *et al.*, 1991). It binds specifically to N-acetyl glucosamine oligosacharides and is involved in the cross-linking of chitin subunits. One of the best-known plant lectins is the wheat germ agglutinin (WGA), a homodimeric protein. Each subunit contains four repeats of the CBL domain (Wright and Kellogg, 1996), which features eight conserved cysteine residues implicated in four disulphide bridges that are responsible for structural conformation. A conserved serine residue forms a hydrogen bond with the non-reduced end of chitin polymers and stabilizes the interaction. This serine residue is followed by an aromatic AA pocket responsible for sugar binding (Wright *et al.*, 1991). The CBL domain has been shown to be responsible for cell surface interaction in MIC3 (Cerede *et al.*, 2002) and in MIC8 (O. Cérède and M. Lebrun, unpublished). This domain could allow binding to a large array of cell types (Cerede *et al.*, 2002). Similar to WGA, the binding ability of MIC3 involves aromatic amino acids (Cerede *et al.*, 2005).

The CBL domain is also found in three additional ORFs of the *T. gondii* genome (TgME49_086740, TgME49_044180 and TgME49_038220). TgME49_086740 has also been termed MIC8-like1 (Kessler *et al.*, 2008) or MIC8.2 (Sheiner *et al.*, 2010). The CD of this TM protein can bind to aldolase and is functionally interchangeable with the CD of MIC8 (Kessler *et al.*, 2008; Sheiner *et al.*, 2010).

12.4.2.1.6 GALECTIN-LIKE DOMAIN

A galectin-like domain is present in the C-terminal part of MIC1 (Saouros *et al.*, 2005). Galectins are a unique family of soluble carbohydrate binding proteins (Barondes *et al.*, 1994). The carbohydrate recognition involves critical AAs in the central region of the six-stranded beta-sheet and comprises an array of hydrophilic residues and a key aromatic side-chain (Seetharaman *et al.*, 1998; Umemoto *et al.*, 2003). These positions are not conserved in the equivalent locations within MIC1, which shares a hydrophobic environment in this region, reminiscent of protein–protein interface (Saouros *et al.*, 2005). Carbohydrate experiments showed no detectable carbohydrate binding except for 100 mM lactose which substantiates an altered binding motif (Saouros *et al.*, 2005). A role in protein–protein interaction has been confirmed. Indeed, the galectin domain of MIC1 is involved in the interaction with MIC6, assisting the correct folding of MIC6 and stabilizing the C-terminal fragment encompassing the third EGF and the acidic region (Saouros *et al.*, 2005). This allows the MIC1/4/6 complex to exit the ER (Reiss *et al.*, 2001).

12.4.2.1.7 MICRONEME ADHESIVE REPEAT DOMAIN

The N-terminal region of MIC1 was initially thought to contain degenerate TSR domains (Fourmaux *et al.*, 1996a). However, the crystal structure of this region revealed that it consists of a new fold termed the Microneme Adhesive Repeat (MAR) (Blumenschein *et al.*, 2007). MIC1 has two MAR domains that bind specifically to sialylated oligosacharides in a novel-binding mode involving a key threonine in the binding pocket rather than conventional basic AAs (arginine or lysine). Additional studies revealed that this binding site is found in several parasite, bacterial and viral proteins (Garnett *et al.*, 2009). The tandem MAR domains bind to branched sialylated oligosacharides. Active site mutation of either domain abolishes binding to host cells, confirming the importance of its multivalent binding interaction. The significance of parasite recognizing sialic acids as receptors is underscored by the ~30%–90% diminished invasion of sialic acid deficient host cells (Blumenschein *et al.*, 2007; Monteiro *et al.*, 1998) and the ~50% reduced invasion by MIC1 knockout parasites (Cerede *et al.*, 2005).

Three additional MAR containing proteins, initially termed MCP2, MCP3 and MCP4, are encoded in the *Toxoplasma* genome (Friedrich *et al.*, 2010a). MCP2, now termed MIC13, is poorly expressed in tachyzoites but is abundantly expressed in bradyzoites and sporozoites (Fritz *et al.*, 2012a). Like MIC1, MIC13 binds sialylated oligosacharides but with some unique preferences that suggest a role for parasite binding to sialylated receptors in the gut. Interestingly, a similar situation appears to exist for the *Eimeria* MAR protein EtMIC4, which preferentially recognizes the non-N-glycolylated sialylated oligosaccharides that are expressed in the gut of its chicken host (Lai *et al.*, 2011).

MCP3 and MCP4 do not have the key active site threonine, a finding that is consistent with the absence of detectable host cell binding activity (Friedrich *et al.*, 2010a). The subcellular location of MCP3 has not been determined, but it appears to be moderately expressed in tachyzoites and highly expressed in bradyzoites. MCP4 is mainly expressed in bradyzoites where it is a component of the cyst lumen and wall where it might play a structural role (Buchholz *et al.*, 2011).

12.4.2.2 Other MICs

Although many MICs contain structural motifs suggestive of a role in protein–protein interactions and parasite attachment, a growing subset of microneme proteins (MICs) do not possess obvious adhesive features suggesting they may play alternative functions (MIC5, MIC10, MIC11). MIC5 encodes a previously identified immunodominant antigen called H4 (Brydges *et al.*, 2000; Johnson and Illana, 1991). Although MIC5 possesses a consensus sequence characteristic of members of the parvulin family of peptidyl–prolyl cis-trans isomerases (PPIases), no PPIase activity was detected with recombinant MIC5 (Brydges *et al.*, 2006). Targeted deletion of *MIC5* resulted in hyperproteolysis of MICs on the parasite surface, suggesting that MIC5 regulates protease activity. Recent NMR analysis of MIC5 revealed that it mimics the structure of the prodomain of subtilisin proteases (Saouros *et al.*, 2012). Like many proteases, subtilisins contain an N-terminal prodomain that suppresses activity by occluding the active site. Consistent with it mimicking this feature, MIC5 inhibits the activity of the micronemal protease SUB1 via a flexible C-terminal peptide. The flexible peptide contains conserved amino acids that are thought to mediate interactions with the SUB1 active site. Together the findings suggest that MIC5 regulates SUB1 activity on the parasite surface to preclude over processing of specific substrates or inappropriate processing of other surface or secretory proteins.

MIC10 is an 18 kDa protein lacking a putative TM domain. Although MIC10 is secreted during invasion it does not remain associated with the parasite membrane after microneme discharge like other MICs and does not bind to host cells (Hoff *et al.*, 2001). The role of MIC10 during infection remains to be determined. MIC11 is also a small soluble protein of unknown function that displays a pronounced charge asymmetry with an acidic N-terminal region (pI 4.3) and a basic C-terminal region (pI 8.2) (Harper *et al.*, 2004b), reminiscent of the rhoptry protein ROP1 (Ossorio *et al.*, 1992). It is hypothesized that MIC11 and ROP1 proteins act as molecular organizers within rhoptries and micronemes, respectively, using their charge asymmetry to form an organized scaffold based on homotypic ionic interactions (Hoff *et al.*, 2001).

Perforin-like Protein 1, which was discovered in a proteomics screen of secretory products (Zhou *et al.*, 2005), contains a membrane attack complex (MAC)/perforin (PF) domain implicated in membrane disruption. The MACPF domain is the main functional entity allowing the immune effector proteins perforin and terminal components of the MAC to form large oligomeric ring-like pores in target membranes. Consistent with a similar membranolytic function, PLP1 expression is necessary for the formation of large membrane lesions in the PV membrane during induced parasite egress (Kafsack *et al.*, 2009). Further support for a role in egress came from showing that PLP1 deficient parasites show a significant delay in escape from the cell, with some parasites failing to leave host cells altogether (Kafsack *et al.*, 2009). PLP1 null parasites are profoundly attenuated in a mouse model of acute toxoplasmosis, revealing this secreted microneme protein as a key virulence factor (Kafsack *et al.*, 2009). These findings also highlight the importance of rapid egress for progression of lethal infection by a type I parasite strain and they established that micronemes contribute to parasite egress in addition to invasion.

At least three proteases are found in micronemes. SUB1 is a subtilisin-type serine protease that is recognized by an anti-*Plasmodium* SUB1 antibody, suggesting that the two subtilases share areas of conserved antigenic structure (Miller *et al.*, 2001). SUB1 trims other MIC proteins on the parasite surface during invasion and this appears to modulate their adhesive activity (see Section 4.4.2 below).

Another serine protease belonging to the rhomboid family has been identified in micronemes, named ROM1 (Brossier *et al.*, 2005). Conditional disruption of ROM1 resulted a moderate reduction of invasion and intracellular replication (Brossier *et al.*, 2008), thus confering a parasite fitness defect.

TLN4 is a metalloproteinase of the insulinase family that was recently identified in the micronemes (Laliberte and Carruthers, 2011). Although the substrates for ROM1 and TLN4 are not known, these proteases are likely to play a role in MIC processing before or after secretion of the organelles. This will be discussed further in Section 4.4.2.

12.4.2.3 MICs Assemble in Complexes

Another characteristic feature of *Toxoplasma* MICs is to assemble into protein complexes that work in concert. The first demonstration of the occurrence of such a complex was obtained by *MIC6* gene deletion (Reiss *et al.*, 2001). Indeed, the latter led to MIC1 and MIC4 missorting to the PV instead of microneme. In addition, artificial expression of MIC6 at the parasite plasma membrane induced the same localization for MIC1 and MIC4. Co-precipitation confirmed the association of the three proteins into a stable complex (Reiss *et al.*, 2001). Then, the MIC2/M2AP complex was also found (Rabenau *et al.*, 2001). In this case, a 450 kDa hexameric complex was found to form in the ER, to be targeted in microneme and to persist on the surface during invasion (Jewett and Sibley, 2004; Rabenau *et al.*, 2001). Last, the existence of a MIC3/MIC8 complex has been suggested by the expression of MIC3 on the surface of the parasite expressing artificially a GPI anchored version of MIC8 (Meissner *et al.*, 2002).

The density of MICs in complexes may serve two functions. They are likely to be important for proper trafficking in the secretory pathway. A role in quality control has been shown for MIC1 (Reiss *et al.*, 2001; Saouros *et al.*, 2005). Indeed, soluble MIC1 and MIC4 require interaction with the membrane bound protein MIC6 to ensure proper targeting to the micronemes. As mentioned previously, the galectin domain of MIC1 binds MIC6 and ensures proper folding of MIC6 to pass the ER quality control. Then, MIC6 ensures targeting of the complex to micronemes. A similar role has been postulated for the soluble protein M2AP (Huynh *et al.*, 2003; 2004). It stabilizes the complex formed with MIC2 and influences the correct targeting to micronemes. In the absence of M2AP, MIC2 appears to trimerize correctly but does not target efficiently to the micronemes and as a consequence undergoes degradation. Interestingly though, unlike the other confirmed MIC complexes MIC3 and MIC8 are not interdependent for microneme targeting since targeted disruption of either protein does not alter the localization of the other protein (Cerede *et al.*, 2005; Kessler *et al.*, 2008).

A second function of MICs complexes may be in cooperative binding to enhance cell recognition and facilitate cell entry. Indeed, oligomerization is often critical for enhancing affinity, and, in addition, to cluster receptors and other components within complexes to induce signalling. Multimerization of MIC2 has been shown to increase the number of interactions with host cell receptors, thereby forming a multivalent adhesive protein (Harper *et al.*, 2004a). MIC3 is a homodimer and this oligomerization is required for host cell binding (Cerede *et al.*, 2002). The MIC3 homodimer contains two CBL domains and the TM protein MIC8 possesses one, the complex therefore expressing three CBL domains. CBL domain oligomerization has been shown to increase its avidity (Cerede *et al.*, 2002). CBL domain oligomerization in the MIC3/MIC8 complex would therefore be used to increase the affinity with the host cell surface. MIC1 has affinity for host cell surfaces when dimerized (Fourmaux *et al.*, 1996a) and evidence of MIC1 as a trimer has also been reported (Marchant *et al.*, 2012). Mature MIC6 has two binding sites for MIC1 and each copy of MIC1 appears to be associated with one copy of MIC4 (Marchant *et al.*, 2012, Sawmynaden *et al.*, 2008). Accordingly, each molecule of MIC6 can theoretically support the display of six copies of MIC1 and six copies of MIC4 on the parasite surface during invasion. To summarize, all microneme proteins described so far as showing affinity for host cells work as oligomers and this increased valency likely supports a robust interface for apical attachment to host cells prior to and during cell invasion.

12.4.2.4 Cytosolic Domain of TM MICs

The cytosolic domain of several TM MICs is highly conserved in Apicomplexa. Replacing the cytosolic domain of TRAP with that of TgMIC2 in *P. berghei* sporozoites resulted in normal motility, invasive capability, and infectivity *in vivo* (Kappe *et al.*, 1999), underscoring the functional conservation of the invasive mechanism in Apicomplexa. This domain contains three essential motifs: (i) two of them including first a stretch of acidic residues and second tyrosine residues are involved in membrane trafficking (Di Cristina *et al.*, 2000; Sheiner *et al.*, 2010), (ii) the last C-terminal amino acids and especially a tryptophan are involved in the interactions of MICs with aldolase, a component of the glideosome (see Chapter 13) (Kappe *et al.*, 1999; Sheiner *et al.*, 2010; Jewett and Sibley, 2003). Aldolase is a glycolytic enzyme that binds actin filaments *in vitro* and *in vivo* (Schindler *et al.*, 2001; Wang *et al.*, 1997). In Apicomplexa, it links the TM MICs interacting extracellularly with their receptors and the submembranous actomyosin motor of the parasite (see also below) (Jewett and Sibley, 2003; Sheiner *et al.*, 2010; Zheng *et al.*, 2009). In *Toxoplasma* the TM domains of MIC2, MIC6, MIC12 and AMA1 bind to aldolase. In PfTRAP, MIC2 and MIC6, the C-terminal tryptophan is required for binding aldolase (Jewett and Sibley, 2003; Sheiner *et al.*, 2010) as is a second acid cluster of residues near the critical tryptophan (Starnes *et al.*, 2006). AMA1 does not contain a tryptophan at this position, but two aromatic residues (WF), more distal to the C-terminus are crucial for binding to aldolase and contribute to the function of AMA1 in invasion (Sheiner *et al.*, 2010). These residues are present in a highly conserved motif in all apicomplexan AMA1 proteins that precede a stretch of acidic residues in TgAMA1. This result suggests that TgAMA1 serves as a bridging protein that physically connects the glideosome (via its CD) to other components of the MJ complex (via its ectodomain) and thus plays a critical role in the posterior translocation of the MJ complex during invasion (Sheiner *et al.*, 2010). This model awaits further demonstration.

12.4.3 Microneme Secretion

Microneme secretion is a regulated process (also known as stimulus-coupled secretion). Sensing a low potassium environment upon host cell rupture enables *T. gondii* to determine extracellularity, and is important in regulating parasite cytoplasmic calcium levels (Moudy *et al.*, 2001), which in turn activates microneme secretion (Carruthers and Sibley, 1999). Accordingly, fluorescent imaging studies have revealed that parasites in association with host cells show elevated levels of cytoplasmic calcium (Vieira and Moreno, 2000). Chelation of extracellular calcium with EGTA or BAPTA or addition of excess calcium has little effect on microneme secretion, parasite motility, or cell invasion (Lovett and Sibley, 2003). Chelation of host cell intracellular calcium prior to invasion with the cell permeable chelator BAPTA-AM also has no effect on attachment and invasion (Lovett and Sibley, 2003). Host cell calcium levels also remain relatively constant during invasion (Lovett and Sibley, 2003). These observations indicate that parasite invasion occurs independently of host calcium. By contrast, secretion of micronemes is inhibited by chelation of parasite intracellular calcium (by pre-loading of parasites with BAPTA-AM), indicating that parasite calcium is critical to the process of invasion (Carruthers and Sibley, 1999; Lovett and Sibley, 2003). Microneme secretion can be induced in the absence of host cells by incubating *T. gondii* parasites with ethanol to elevate intracellular calcium (Carruthers *et al.*, 1999b), which occurs through IP_3 generation (Lovett *et al.*, 2002). This calcium transient can be triggered naturally by the accumulation of abscisic acid produced by parasites (Nagamune *et al.*, 2008), or exposure to a low potassium environment (Moudy *et al.*, 2001).

Apicomplexans have multiple calcium stores including the endoplasmic reticulum (ER), mitochondria, and acidocalcisomes (Moreno and Docampo, 2003). The intracellular origin of the releasable intracellular calcium and how the release of calcium is regulated are discussed in Chapter 10.

The kinase inhibitors staurosporine, KT5926, and compound 1 block microneme secretion and attachment/invasion when added to parasites before infecting HFF cells (Carruthers *et al.*, 1999a; Dobrowolski *et al.*, 1997; Kieschnick *et al.*, 2001; Wiersma *et al.*, 2004) supporting a role for parasite protein kinases in these processes. The effect of staurosporine is not reversed by treatment with the calcium ionophore A23187 (Carruthers *et al.*, 1999a), indicating that the staurosporine block is downstream of the calcium step in the microneme secretion signal transduction pathway. A calcium-dependent calmodulin domain protein kinase sensitive to KT5926 (CDPK1) that phosphorylates three unknown parasite proteins was identified in tachyzoites (Kieschnick *et al.*, 2001). CDPK1 belongs to the calcium-dependent protein kinases (CDPKs) that represent the dominant calcium-responsive kinases in Apicomplexa (Billker *et al.*, 2009). The role of TgCDPK1 in calcium-dependent microneme secretion, invasion and egress was confirmed recently by generating a conditional knockout of TgCDPK1 and using chemical genetics (Lourido *et al.*, 2010). The authors exploited the unique ATP-binding pocket of TgCDPK1, which has a glycine at a key position termed as the 'gatekeeper', to specifically inhibit the kinase using bulky pyrazolo [3,4-*d*] pyrimidine (PP) derivatives (Lourido *et al.*, 2010).

Also, the activity of a cyclic GMP-dependent protein kinase (PKG) sensitive to compound 1 is required for microneme secretion, gliding motility, and invasion (Wiersma *et al.*, 2004; Donald and Liberator, 2002). Inhibition of PKG activity by compound 1 was not reversed by calcium agonist suggesting that it functions downstream of the calcium dependent events in the signalling pathway. Precisely how PKG regulates microneme secretion and motility awaits identification of its substrates.

Chemical genetic screens have recently identified two proteins involved in microneme secretion (Farrell *et al.*, 2012; Hall *et al.*, 2011). By screening small molecules that block the process of host cell invasion by *T. gondii* and using click-chemistry approach, Hall *et al.* identified a poorly characterized protein homologous to the human protein DJ-1 (Hall *et al.*, 2011). Modification of Cys127 on TgDJ-1 resulted in a block of microneme secretion and motility, even in the presence of direct stimulators of calcium release, suggesting that TgDJ-1 plays a role downstream of the calcium flux required for microneme secretion, parasite motility, and subsequent invasion of host cells. TgDJ-1 has a papain-like protease domain, but the potential proteolytic activity of this domain and the mechanism of DJ-1 function remains to be determined.

Mutational profiling has also identified TgDOC2, a protein conserved in Apicomplexa and involved in calcium-dependent secretion of micronemes (Farrell *et al.*, 2012). DOC2 contains a tandem C2 domain that is known to bind Ca^{2+} and to be involved in Ca^{2+}-mediated exocytosis (for example, neurotransmitter release), facilitating membrane fusion of secretory vesicles with the plasma membrane (Friedrich *et al.*, 2010b; Groffen *et al.*, 2010). Therefore, TgDOC2.1 constitutes a second level of Ca^{2+}-dependent control of *Toxoplasma* microneme secretion that probably facilitates membrane fusion, and likely acts downstream of the CDPKs.

Finally, an homologue of parafusin, a protein involved in calcium regulated exocytosis in *Paramecium*, named PRP1 (parafusin related protein, a member of the phosphoglucomutase family), is associated with a subset of micronemes in *T. gondii*, particularly those situated near the apical end (Matthiesen *et al.*, 2003). Its involvement in calcium regulated microneme exocytosis has been suggested based on its dynamic redistribution to the cytoplasm during microneme discharge (Matthiesen *et al.*, 2001; 2003). Together, these studies suggest a multilayered Ca^{2+}-mediated control of microneme secretion in *T. gondii* and underscore the tightly temporal regulation of this process.

12.4.4 Post-Secretory Traffic of MICs

12.4.4.1 Parasite Surface Exposition and Posterior Capping of MICs

As the parasite penetrates the host cell, most MICs are excluded from entering the vacuole and are progressively capped behind the MJ, remaining confined to the portion of the parasite that still protrudes out of the host cell (Carruthers *et al.*, 1999a; Carruthers and Sibley, 1997; Garcia-Reguet *et al.*, 2000). As a consequence of MIC protein capping, binding to a fixed substrate would lead to forward locomotion, and binding to cell surface receptors would lead to penetration into the cell. The backward capping of *T. gondii* MICs is an actin-dependent process, implying that either actin polymerization or actin filaments are required (Jensen and Edgar, 1976; Miller *et al.*, 1979; Russell and Sinden, 1981; Ryning and Remington, 1978).

The current view is that the actomyosin motor, located beneath the plasma membrane of the parasite, interacts indirectly with the cytoplasmic tail of the TM MICs including MIC2 and AMA1 and translocates them toward the posterior end of the parasite (Keeley and Soldati, 2004; Sheiner *et al.*, 2010; Sibley, 2003; Soldati and Meissner, 2004). Thus as an immobilized myosin walks along the actin filament, the MICs—cell receptor complexes are capped backwards and the parasite propels itself on the substrate or into the host cell. Substantial evidence of this capping model has been obtained by reverse genetic approaches in *Plasmodium*. Deletion of the TRAP C-terminus abolishes gliding motility and cell invasion (Kappe *et al.*, 1999).

Directional invasion and helical gliding motility are sustained by connection of the actomyosin motor with the IMC. This additional membrane layer is supported by microtubules that provide the pitch for the spiral gliding action and serves as the 'tramline' for the capping reaction. The myosin is indeed anchored in the inner membrane complex of the pellicle and generates the mechanochemical force for translocating the actin filaments (Gaskins *et al.*, 2004). The different players of the invasion apparatus are called glideosome and described in Chapter 13.

12.4.4.2 Proteolytic Cleavages Before and During Invasion

As mentioned in Section 4.1, many MICs are initially synthesized as preproproteins that undergo proteolytic processing (maturation) during transit to the micronemes. An endosomal cathepsin protease L (CPL) was recently identified as one of the proteases responsible for proMIC maturation (Parussini *et al.*, 2010). CPL resides in the vacuolar compartment (VAC) (Parussini *et al.*, 2010), which is also known as the plant-like vacuole (PLV) (Miranda *et al.*, 2010). CPL also exists in endosome-like vesicles where it discretely encounters its proMIC substrates (Parussini *et al.*, 2010). Targeted disruption of *CPL* resulted in the delayed maturation of M2AP and MIC3 but not MIC6 or AMA1. Thus, CPL contributes to the maturation of some but not all proMICs, indicating the existence of additional maturase within the microneme trafficking system. Recombinant CPL cleaved the propeptide from recombinant M2AP *in vitro* at the same cleavage site used *in vivo*. This cleavage site contains amino acid residues that are favourable for CPL cleavage based on testing a library of peptide substrates. The contribution of an endosomal protease to proMIC maturation is consistent with the trafficking of these substrates through part of the endosomal system *en route* to the micronemes.

MIC proteins are also extensively processed on the parasite surface. These post-exocytosis processing events likely regulate adhesion and permit the disruption of the adhesive MIC complexes and the dissociation of the parasite—host interaction at the end of the invasion process. Post-exocytosis proteolysis is several distinct protease activities that were initially termed microneme protein protease 1, 2 and 3 (MMP1, MMP2 and MMP3) (Carruthers *et al.*, 2000; Zhou *et al.*, 2004). As detailed below, ROM4 and possibly ROM5 are responsible for MMP1 activity, which sheds TM MIC proteins from the parasite surface by intramembrane cleavage near the C-terminus. SUB1 conveys MMP2 and possibly MMP3 activity on the parasite surface, resulting in the trimming of terminal peptides from MICs.

The post-exocytosis C-terminal cleavage of MIC2, MIC6, MIC8, MIC12 and AMA1 by MMP1 sheds these TM MICs from the parasite surface (Carruthers *et al.*, 2000; Donahue *et al.*, 2000; Meissner *et al.*, 2002; Opitz *et al.*, 2002; Reiss *et al.*, 2001). This event in MIC2 constitutes an essential step during host cell invasion, since mutation-based impairment of MIC2 C-terminal proteolysis impairs host cell entry (Brossier *et al.*, 2003). The essentiality of this cleavage event for AMA1 is disputed. In one study, dominant negative interference with cleavage did not impair invasion but prevented post-invasion initiation of replication (Santos *et al.*, 2011). However, another study that expressed non-cleavable mutants of AMA1 in the conditional knockout strain concluded that AMA1 C-terminal cleavage is dispensible for replication but it facilitates cell invasion (Parussini *et al.*, 2012). The disparate outcomes are likely due to the different approaches used to interfere with AMA1 cleavage and possibly also to compensatory mechanisms in the parasite.

Genetic and biochemical evidence shows that C-terminal cleavage occurs within the TM domain by regulated intramembrane proteolysis (RIP) (Howell *et al.*, 2005; Opitz *et al.*, 2002; Zhou *et al.*, 2004). The intramembrane cleavage site is conserved between TM MICs of *Toxoplasma* and other Apicomplexa (Dowse *et al.*, 2005). These cleavage sites resemble the recognition sequence for rhomboid-like proteases, whose activity is sensitive to the serine protease inhibitor 3,4-dichloroisocoumarin (DIC) (Urban and Freeman, 2003). DIC inhibits intramembrane cleavage of TM MICs (Howell *et al.*, 2005) and host cell invasion by *T. gondii* (Conseil *et al.*, 1999), consistent with the participation of a rhomboid-like protease.

Rhomboids are multipass membrane serine proteases that cleave within the TM of their substrate. First evidence that a parasite rhomboid may cleave TM MICs came from a study by Urban and Freeman (2003) showing that Rhomboid-1 from Drosophila and RHBDL2 from humans cleave chimeric proteins containing the TM domain of MIC2, MIC6 or MIC12 (Urban and Freeman, 2003).

Rhomboid-like genes are present in the genome of all apicomplexan parasites currently sequenced. *T. gondii* contains six rhomboid genes (*ROM1−6*) (Dowse and Soldati, 2005). ROM2 and ROM3 are mainly expressed in sporozoites, ruling out their role as MMP1 activities in all stages of the parasite (Brossier *et al.*, 2005). MMP1 activity is predicted to be constitutive on the parasite surface (Opitz *et al.*, 2002). ROM1 localizes to micronemes and ROM4 and ROM5 are expressed at the plasma membrane (Brossier *et al.*, 2005; Dowse *et al.*, 2005). Whereas ROM4 is evenly distributed along the parasite surface, ROM5 localizes at the posterior pole of the parasite where the cleavage of MICs is thought to occur (Brossier *et al.*, 2005). Definitive evidence of ROM4's involvement in intramembrane cleavage of MIC2 came from conditional ablation of ROM4 expression (Buguliskis *et al.*, 2010). ROM4 deficient parasites showed markedly reduced shedding of MIC2 into the culture supernatant, resulting in the accumulation of MIC2 on the parasite surface. Mutant parasites showed increased adhesion to host cells but reduced invasion thereof and defective gliding motility, supporting the model that intramembrane shedding of MIC2 is important parasite active entry entry into host cells. Although the role of ROM5 in intramembrane shedding of TM MICs during invasion remains to be determined, the pronounced phenotype of ROM4 deficient parasites suggests that ROM5 alone is not capable of supporting normal MIC shedding from the parasite surface.

MMP2 trims off a short N-terminal extension upstream of the A-domain from MIC2 and it cleaves the M2AP C-terminal domain at several sites (Zhou *et al.*, 2004). This protease also processes MIC4 at its N-terminus and C-terminus, releasing the last two Apple domains (Brecht *et al.*, 2001; Zhou *et al.*, 2004). MMP2 activity is markedly enhanced by treatment with cytD, which blocks the capping of MICs toward the posterior end of the parasite. This finding was interpreted as evidence that MMP2 is a resident apical surface protein or microneme protein since cytD restricts MICs to the apical surface, thereby facilitating processing (Carruthers *et al.*, 2000; Zhou *et al.*, 2004). Based on partial inhibition by chymostatin and PMSF and nearly complete inhibition by ALLN and ALLM, MMP2 was predicted to be a serine or cysteine protease.

The likely identity of MMP2 was revealed upon finding that SUB1 deficient parasites failed to process any of the MMP2 substrates including MIC2, M2AP and MIC4 (Lagal *et al.*, 2010). As mentioned above, SUB1 is a subtilisin-like protease that is secreted from the micronemes onto the parasite apical surface during invasion, placing it in the correct location for MMP2 activity. SUB1 activity is also sensitive to PMSF (Miller *et al.*, 2001), which is consistent with the inhibitor profile of MMP2. Interestingly, ablation of SUB1 also extinguished MMP3 activity, which was identified previously by its ability to cleave the C-terminal most peptide of M2AP in an ALLN insensitive manner (Zhou *et al.*, 2004). SUB1 cleavage of the C-terminal most M2AP peptide might occur rapidly upon secretion of both the protease and substrate, thus potentially explaining its refractivity to ALLN. The N-terminal processing of MIC2 now attributed to SUB1 has been shown to facilitate MIC2 interaction with ICAM-1, presumably by optimal exposure of the MIC2 A-domain (Barragan *et al.*, 2005). SUB1 deficient parasites are moderately deficient in gliding motility and parasite attachment to host cells, consistent with a general role of surface trimming in the activation of MIC adhesins. The lack of surface trimming and activation of MIC adhesins also results in partial attenuation of virulence in mice, confirming the *in vivo* significance of this form of proteolysis (Lagal *et al.*, 2010).

12.4.5 Why Does *T. gondii* Exhibit this Patchwork of MICs?

The vast MIC repertoire of *T. gondii* MICs may be correlated with the broad host cell specificity *in vitro* and the spreading of infection in all organs in toxoplasmosis, contrasting with the high cell and organ specificity found in *Plasmodium*. Apicomplexa show variable cell specificity, particularly at different stages of infection: specificity may be related to the MIC repertoire, quite different from one genus to the other, and depend upon the stage in the life cycle.

MIC gene deletion in *T. gondii* has shown that at least some of the MICs are not essential for invasion *in vitro* (M2AP, MIC1, MIC3, MIC4, MIC5, MIC6, SUB1, ROM1, PLP1) (Brossier *et al.*, 2008; Cerede *et al.*, 2005; Huynh *et al.*, 2003; Lagal *et al.*, 2010; Reiss *et al.*, 2001; Kafsack *et al.*, 2009). However, *in vitro* findings using a limited number of often non-physiologically relevant cell types might not reflect the biological roles of MICs *in vivo*. Indeed *MIC3* gene disruption does not modify fibroblast invasion, but induces a death delay in mice (Cerede *et al.*, 2005). Consistent with this, MIC3 could be essential for invading specific cell types other than those that have been tested thus far *in vitro*. This hypothesis is supported by the demonstration that the binding ability of MIC3 to host cells is crucial for parasite virulence *in vivo* (Cerede *et al.*, 2005). This shows also that the importance of MICs is better evaluated *in vivo* than *in vitro*. Indeed, motility, adhesion and invasion, which are functions postulated for MIC proteins, are multifactorial phenomena, the expression of which is likely to differ *in vivo* from what occurs in the static environment found in a culture dish.

Deletion of the *MIC1* gene results in defective targeting of both MIC4 and MIC6 in micronemes (Reiss *et al.*, 2001) and *mic1KO* parasites are 50%

impaired in invasion (Cerede *et al.*, 2005). This defect in invasion can thus be assigned to the absence of the MIC1/4/6 complex, without further precision. As previously mentioned, only MIC1 binds the host cell *in vitro* (Saouros *et al.*, 2005). Therefore, MIC6 anchors the two other proteins and interacts with the underlying motor, whereas MIC1 would establish specific interactions with host cell receptors necessary for host cell invasion.

The disruption of M2AP gives also a partial phenotype. M2AP knockout parasites were more than 80% impaired in host cell entry and show delayed death in mice (Harper *et al.*, 2006; Huynh *et al.*, 2003). In these parasites, MIC2 partially accumulates in the parasite ER/Golgi apparatus and is poorly secreted. This invasion defect is likely due to defective expression of MIC2/M2AP complex.

The importance of MIC protein diversity has been further stressed by simultaneous disruption of MICs (Cerede *et al.*, 2005). Indeed, a spectacular decrease in virulence *in vivo* was observed by simultaneous disruption of the *MIC1* and *MIC3* genes (which corresponds in fact to a MIC1/3/4/6 functional KO). This result demonstrates that MICs have a synergistic effect on infection *in vivo*.

Repeated attempts to directly disrupt the gene encoding MIC2 were unsuccessful (L.D. Sibley, personal communication (Huynh *et al.*, 2004)), suggesting that, as for TRAP in *Plasmodium* sporozoites, MIC2 is a key protein. Moreover, although parasites bearing a targeted deletion of the MIC2 C-domain could be initially detected by flow cytometry in a transfected population, they were rapidly lost upon further growth (Jewett and Sibley, 2004). Consistent with a crucial role, conditional ablation of MIC2 reduced expression to approximately 5% of normal levels resulted in more than 80% impaired in invasion because of a defect in attachment (Huynh and Carruthers, 2006). Interestingly, MIC2 deficient parasites display almost exclusively one type of motion, circular gliding. These parasites are also markedly attenuated in mice, requiring a greater than 600-fold higher inoculation for lethal infection. No defect in intracellular replication or non-induced egress was seen. Whether MIC2 is involved in the MJ formation has not been established, although it has been shown to occupy the interface between the parasite and host membranes during invasion (Carruthers *et al.*, 1999a). In *Plasmodium*, the two adhesive motifs of TRAP have been suggested to be involved in MJ formation. Indeed, independent or simultaneous mutation in *Plasmodium* TRAP A-domain and TSP does not alter sporozoite motility but specifically decreases, or abolishes in the case of simultaneous mutations, host cell invasion *in vitro* and *in vivo* (Matuschewski *et al.*, 2002). Therefore, gliding motility and host cell invasion likely involve distinct extracellular associations, perhaps with invasion requiring stronger binding to host receptors than what gliding motility needs.

Like TRAP, AMA1 is an integral membrane protein widely conserved in apicomplexan parasites. Considerable evidence points toward a role of AMA1 in invasion. The first demonstration of such a role came from the inhibitory effects of anti-AMA1 antibodies on invasion of *P. knowlesi* into erythrocytes (Thomas *et al.*, 1984), then confirmed in other apicomlexan parasite with anti-*Toxoplasma* AMA1 antibodies (Hehl *et al.*, 2000) or anti-*Babesia* AMA1 (Gaffar *et al.*, 2004). Finally, anti-AMA1 antibodies inhibit sporozoite invasion, suggesting that the protein is also involved in invasion of hepatocytes (Silvie *et al.*, 2004). AMA1 was proposed to be involved in the formation of a tight binding interface between merozoite and erythrocyte surfaces (Mitchell *et al.*, 2004). Indeed, the initial random surface attachment of merozoites to red blood cells is not affected by the presence of inhibitory antibodies, but the normal apical reorientation of merozoites does not occur and the close junctional contact (6 nM) is absent. In this study, the existence of membranous blebs inside some red blood cells suggested that some transient apical contacts occur in presence of anti-AMA1 and are sufficient to induce the secretion of the rhoptries, although insufficient to create the prolonged intimate contact between the red blood cell membrane and merozoite apex which is required to allow completion of invasion.

Consistent with this, Mital *et al.* (2005) generated a conditional knockdown of *T. gondii* AMA1 and showed that AMA1 is not involved in gliding motility, or in the initial step of attachment, or in microneme release, but fails to attach intimately to host cells (Mital *et al.*, 2005). Altogether, these data suggested a role for AMA1 in building the MJ. This has been supported by recent studies, which showed that secreted AMA1 is associated at the parasite surface with a complex of secreted rhoptries neck proteins that relocalize at the MJ (Alexander *et al.*, 2005; Besteiro *et al.*, 2009; Lamarque *et al.*, 2011; Tyler and Boothroyd, 2011). During invasion, AMA1 displays a surface localization with a high steady-state level (Howell *et al.*, 2005). AMA1 is present at the MJ but the majority of AMA1 is clearly on both sides of this adhesion zone. But in knock-down parasites expressing low level of AMA1, this protein precisely colocalizes with RON4 at the MJ (Alexander *et al.*, 2005). These studies suggest a model in which AMA1, with the cooperation of RONs, is involved in the formation of the MJ, being a major player in invasion (see more details in Section 5.4.1).

Conditional ablation of *Toxoplasma* AMA1 reduced expression to an undetectable level of AMA1 and resulted in more than 80% impaired in invasion (Mital *et al.*, 2005), showing a key role for AMA1 during invasion but suggesting also compensatory mechanisms in the *Toxopplasma*. In AMA1 deficient parasites, ROP secretion into host cells in the form of evacuoles is reduced, suggesting a role of AMA1 in rhoptry secretion (Mital *et al.*, 2005).

A universal and essential role of the AMA1–RON complex in MJ formation is disputed in a recent study (Giovannini *et al.*, 2011) that showed when AMA1 is reduced to undetectable levels in *P. berghei* sporozoites, the entry of hepatocytes is not affected, while the parasites fail to invade erythrocytes. This illustrates that the invasion machinery components may be different at different stages of the parasite life cycle. In contrast to PbAMA1, *Pb*RON4 is required for sporozoite invasion of hepatocytes (Giovannini *et al.*, 2011), which implies that AMA1 works independently of the RON proteins at the hepatic stage. The authors also re-examine the conditional *Toxoplasma* AMA1 and show that (i) the ring staining of RON4 at the MJ appears normal in the absence of any detectable AMA1, (ii) in the absence of surface AMA1, tachyzoites fail to adhere throughout their length and instead bind only via their anterior portion, and finally (iii) the parasites invade at the same speed as wild-type parasite. The authors proposed that the AMA1–RON complex does not fulfil the force-transducing role of the MJ, and moreover that AMA1 is not involved in MJ formation per se but contributes to an independent step before MJ formation that helps the parasite to bind host cell surfaces through its entire length. This controversy clearly illustrates that a better understanding of the role of AMA1 is needed.

Conditional ablation of MIC8 revealed a key role in tachyzoite invasion (Kessler *et al.*, 2008). Although MIC8 deficient parasites showed normal egress, gliding motility, and apical attachment, they failed to form a MJ for invasion of host cells. The arrest in invasion appears to correspond to a block in rhoptry secretion characterized by the absence of RON secretion at the MJ and ROP secretion into host cells in the form of evacuoles. That the MIC8 CD does not interact with aldolase and the glideosome (Sheiner *et al.*, 2010) is consistent with the absense of a role in gliding motility and active penetration. The findings point toward a role for MIC8 in the assembly of the MJ and/or rhoptry discharge.

12.5 RHOPTRIES

12.5.1 Biogenesis of Rhoptries and Traffic of ROPs in the Secretory Pathway

Rhoptries are present at a multiplicity of 6–12 per cell in *T. gondii*, while *Plasmodium* zoites contain only two. Rhoptries are club shaped organelles with a bulbous base and an extended duct that opens at the anterior pole of parasite. There is a clear segregation of rhoptry proteins between the neck and the bulb (Roger *et al.*, 1988), although no membrane delimits these sub-compartments. The denomination of ROP is for those proteins located in the bulb while RON is for the proteins associated with the neck.

Rhoptry proteins traffic along the classic secretory pathway from the ER, through the Golgi and then to the rhoptry. The rhoptries are formed first as immature rhoptries (or pre-rhoptries), which are trans-Golgi derived vesicles (see Dubremetz and Ferguson, Chapter 2). The underlying molecular mechanism that enables trans-Golgi derived vesicle formation is still unknown. A recent study in *Toxoplasma* showed a central role of the alveolate specific dynamin related protein B (DrpB) in the biogenesis of secretory organelles (Breinich *et al.*, 2009). DrpB is a mechanoenzyme expressed transiently during replication in a region close to the Golgi. Conditional expression of a dominant-negative of DrpB leads to parasites that are able to replicate but do not form micronemes and rhoptries *de novo*. In the absence of rhoptries, ROPs and RONs are mistargeted to the constitutive secretory pathway. It has been suggested that DrpB is required in a very dynamic process to form vesicles for the regulatory secretory pathway.

Rhoptries and pre-rhoptries are thought to be the only acidified organelles in the parasite (Shaw *et al.*, 1998). Rhoptries are hypothesized to be derived from both the secretory and endocytic pathways (Ngo *et al.*, 2004; Yang *et al.*, 2004), although endocytosis has never been formally proven in *T. gondii*. Delivery of members of the *ROP2* gene family (e.g. ROP2, ROP4, ROP7; see below) was proposed to be dependent on a direct binding of adaptins with the tyrosine-based (*YXXΦ*) or dileucine-based (LL) motifs contained in the C-terminal end of ROP2 family proteins (Hoppe and Joiner, 2000; Ngo *et al.*, 2003), which were proposed to be type 1 TM proteins (Sinai and Joiner, 2001). Mutation of either of these signals was shown to diminish the delivery of ROP2 or ROP4 to mature rhoptries and proteins accumulated in multivesicular body (MVB). This compartment was distinct from the dense granules, micronemes and Golgi/TGN, and colocalized with a multivesicular endosomal compartment marked by the endosomal marker VPS4 (vacuolar protein sorting 4). These results suggested that rhoptry targeting occurs along the endocytic pathway and is mediated by adaptins. Consistent with this, the YXXΦ motif has been shown to facilitate protein sorting by direct binding to the *T. gondii* μ-chain component (Tgμ1) of the AP-1 clathrin adaptor (Hoppe and Joiner, 2000). In mammalian cells, AP-1 is an essential regulator of membrane trafficking by mediating budding of clathrin-coated vesicles from the TGN, immature secretory granules and endosomes. Tgμ1 localized to Golgi/TGN, to juxtaposed coated Golgi-associated vesicles, and both the rhoptry membrane and membranous lumen. Dominant-negative expression of Tgμ1 disrupted rhoptry sorting and rhoptry biogenesis (Ngo *et al.*, 2003). Large tubular and multivesicular structures were mostly observed and ROP2 accumulated in MBV ressembling the vesicles to which ROP2 with cytoplasmic targeting signal mutations localized. There was no significant alteration of dense granules and micronemes. These suggested that adaptins are involved in sorting of ROPs to TGN to an endosomal compartment.

The way this sorting occurs was recently questioned (Labesse *et al.*, 2009; Qiu *et al.*, 2009), as the resolution of the structure of the

ROP2 protein ruled out the previous structural model (Sinai and Joiner, 2001) implying that these proteins exposed the adaptin-binding motifs outside the rhoptry. The only hydrophobic stretch, once considered to be the TM, was found buried inside the crystal structure of ROP2, demonstrating that these proteins could not adopt a type 1 TM insertion and that a direct interaction with adaptins was unlikely. In addition, in the course of identifying signals responsible for sorting into specific compartments with a green fluorescent protein (GFP)-based motif-trap, Bradley *et al.* showed that a ROP4–GFP fusion lacking the two previously described targeting signals (a dileucine and YXXΦ motif) is correctly delivered to rhoptries (Bradley *et al.*, 2004). Similarly, ROP17 (El Hajj *et al.*, unpublished) and ROP18 (El Hajj *et al.*, 2007a), two others members of the ROP2 family localize to rhoptries although they do not possess these motifs.

However, the involvement of an endolysosomal-like traffic for rhoptries remains relevant, since Sortilin, also known as VPS10 in yeast, a cargo receptor that functions in mannose-6-phosphate independent sorting to the endolysosomal system, is essential for ROP trafficking and rhoptry biogenesis (Sloves *et al.*, 2012). *Toxoplasma* sortilin-like receptor, TgSORTLR, is a type I single pass TM protein, residing within Golgi–endosomal related compartments. It interacts specifically with rhoptry and microneme proteins through its lumenal domain, while through its cytoplasmic tail, it recruits the cytosolic sorting machinery involved in anterograde and retrograde protein transport, including clathrin heavy chain and the β-, γ- and μ-chain components of the AP-clathrin adaptors. It also pulls down the coat complex transport proteins Sec23/Sec24 that ensure the directionality of anterograde membrane flow from the ER to the Golgi apparatus (Lord *et al.*, 2011), as well as three homologues of retromer-associated vacuolar sorting proteins Vsp9, Vps26 and Vsp35, that mediates retrograde transport from endosomes to the trans-Golgi network (Kim *et al.*, 2010). Ectopic expression of the N-terminal TgSORTLR lumenal domain resulted in dominant negative effects with the mislocalization of endogenous TgSORTLR as well as rhoptry, microneme proteins and the retromer–protein Vps26 protein. Conditional ablation of TgSORTLR disrupted rhoptry and microneme biogenesis, inhibited parasite motility, and blocked both invasion into and egress from host cells. Altogether, these findings suggest that TgSORTLR binds to rhoptry or microneme proteins in the Golgi lumen and recruits cytosolic cargo sorting proteins to form a sorting complex, which is presumably associated with vesicles that exit from the external Golgi cisternae. TgSORTLR then guides its cargo through prerhoptries, or immature micronemes before releasing its payload. The 'empty' cargo receptor then recruits components of the retromer complex for retrograde translocation and recycling to the Golgi to reload with new cargo. As rhoptries are formed prior to micronemes during daughter cell formation (Nishi *et al.*, 2008), this recycling model would allow the biogenesis of two distinct compartments using the same molecular machinery.

How TgSORTLR recognizes different cargo proteins from rhoptries and micronemes remains to be precisely elucidated. One possibility is that sorting within the Golgi occurs via a clustering mechanism whereby proteins en route to a particular destination aggregate into distinct sub-domains with only a subset of proteins forming direct contacts with the cargo receptor. An escorter process has been described for rhoptries in *Plasmodium*, in which soluble rhoptry proteins RAP2 and RAP3 form an oligomeric complex with the TM RAP1 protein (Baldi *et al.*, 2000), which is escorted to the rhoptry via a transient interaction of the N-terminal of RAP1 with the GPI-anchored rhoptry protein RAMA (Topolska *et al.*, 2004; Richard *et al.*, 2009). Mutation of aromatic amino acids that abolishes the RAMA–RAP1 interaction also interferes with RAP1 targeting. A second possibility is that the timing of expression dictates specificity. Rhoptries are made before micronemes during and following daughter cell formation (Besteiro *et al.*, 2009; Nishi *et al.*, 2008). Thus, TgSORTLR might facility rhoptry biogenesis prior to turning its attention to packaging micronemes.

Rhoptry proteins complexes also exist in *T. gondii*. A recent study reports the characterization of two novel rhoptry neck proteins, RON9 and RON10, which form a new complex independent of the MJ complex (Lamarque *et al.*, 2012). Disruption of either *RON9* or *RON10* leads to the retention of the partner in the ER followed by subsequent degradation, suggesting that the RON9/RON10 complex formation is required for proper sorting to the rhoptries. The correct targeting of the moving junctional complex RON2/RON4/RON5 to the rhoptry is dependent on the expression of RON2, while RON8 travels alone to its terminal compartment (M. Lebrun, unpublished results) and is dispensable for the correct targeting of the rest of the complex (Straub *et al.*, 2011).

The mechanism by which proteins are partitioned within a single membrane bound organelle is not understood. Data from *P. falciparum* argues for the presence of a bulb-retention signal, as RAP1 appears to possess a distinct signal that avoids re-localization of the protein from the bulb of the rhoptries to the neck (Richard *et al.*, 2009). These findings are consistent with the observation that the prodomain of the *T. gondii* rhoptry bulb ROP1 protein directs trafficking of a reporter more prominently to the rhoptry neck, whereas full-length ROP1 is enriched in bulb (Bradley and Boothroyd, 2001). Mapping of the domains involved in targeting have also involved the propeptides of RON8 and Toxolysin-1 as sorting signals (Straub *et al.*, 2009; Hajagos *et al.*, 2012). However, these sequences lack apparent similarities at the primary sequence level, precluding the identification of a region or a consensus sequence sufficient for rhoptry targeting.

12.5.2 ROPs and RONs Processing

Most characterized *T. gondii* rhoptry proteins are proteolytically cleaved during transit in the secretory pathway. Although only three prodomain cleavage sites have been confirmed experimentally at a conserved SΦXE site, first characterized in ROP1 (Bradley *et al.*, 2002; Miller *et al.*, 2003; Turetzky *et al.*, 2010), the presence of this 'ROP1-like' site in a processed rhoptry protein has become a predictive tool and many putative cleavage sites have been reported without experimental confirmation in ROP4, ROP13, ROP18, RON2, RON5, RON8 (Besteiro *et al.*, 2009; Carey *et al.*, 2004a; El Hajj *et al.*, 2007a; Turetzky *et al.*, 2010). Interestingly, a degenerated site has been shown to be cleaved in Toxolysin-1 (SΦXD) (Hajagos *et al.*, 2012), allowing refinement of the ROP1-like cleavage to SΦX(E/D).

The importance of this cleavage site is unknown. It does not seem to play a role in trafficking since targeting of ROP1, ROP13 and Toxolysin-1 to rhoptries does not depend on their proteolytic processing (Bradley *et al.*, 2002; Hajagos *et al.*, 2012; Turetzky *et al.*, 2010). It has been postulated that it allows the removal of organellar targeting sequences once the protein is *en route* to the rhoptries, preventing premature activation of the rhoptry protein where it might be detrimental to the parasite. This removal seems indeed regulated in time and space. The processing of ROP1 and of the RON complexes (see below) are blocked by brefeldin A (Besteiro *et al.*, 2009; Soldati *et al.*, 1998) and in the case of ROP1 is not inhibited by the dominant negative Tgμ1 (Ngo *et al.*, 2003) or by the dominant negative DrpB construct in the case of RON8 (Breinich *et al.*, 2009), indicating that the processing event occurs post-Golgi and before the intervention of AP-1 complex and DrpB. Alternatively, the proteases responsible for processing could be diverted along with the substrates and the AP-1 and DrpB mutant strains.

Once cleaved, the pro-domains are likely degraded, as a specific antiserum generated against the pro-domain labelled only the pre-rhoptries which correspond to the nascent rhoptry of daughter parasites during endodyogeny and did not label the mature organelle in S phase parasites (Besteiro *et al.*, 2009; Carey *et al.*, 2004a; Hajagos *et al.*, 2012; Turetzky *et al.*, 2010), indicating that cleavage should take place within immature rhoptries by a resident protease. The maturase(s) involved in processing at the SΦXE site is not yet clearly defined but potential candidates are discussed later.

A second cleavage site SXL↓Q for rhoptry protein Toxolysin-1 has been recently identified (Hajagos *et al.*, 2012), where mutation of the P3 serine and P1 leucine inhibit cleavage. This cleavage occurs at the C-terminal end of Toxolysin-1 prior to rhoptry commitment, suggesting the presence of an additional rhoptry maturase in the secretory pathway.

12.5.3 Rhoptry Content

Identification of the major components of the rhoptries in *T. gondii* has been obtained by subcellular fractionation and generation of monoclonal antibodies (see Table 12.2). Proteomics analysis of purified rhoptries has further allowed the characterization of 38 previously unidentified proteins (Bradley *et al.*, 2005) of which some have been confirmed to be localized in the organelles. Most of the ROPs contain a signal peptide, and many have at least one predicted TM domain or a GPI anchor, suggesting an association with membranes. Proteins devoid of signal peptide have been also shown to be associated with the cytosolic face of the rhoptry (Herm-Gotz *et al.*, 2007; Cabrera *et al.*, 2012). More specifically, their association to the membrane of the rhoptry is mediated by myristoylation and palmitoylation motifs (Cabrera *et al.*, 2012). In addition, rhoptry proteomic analysis led to the characterization of proteins specifically localized in the rhoptry neck, i.e. RONs. Many ROPs and RONs contain repeated motifs that may be involved in protein–protein interactions.

Virtually all the information about the contents or functions of the rhoptries has come from the tachyzoite stage of *T. gondii*. Only a few stage specific ROP and RON have been described (Fritz *et al.*, 2012a,b; Schwarz *et al.*, 2005). The bradyzoite specific rhoptry protein 1 (BRP1) was identified by a bioinformatic analysis of previously identified genes that are highly expressed during bradyzoite development and prediction of genes encoding secretory proteins (Cleary *et al.*, 2002; Schwarz *et al.*, 2005). BRP1 is also expressed in the merozoite stages in the gut of infected cats. The only homologue known is in the closely related parasite *N. caninum. In vitro* and *in vivo* analysis of *BRP1* knock-out parasites show that BRP1 does not play an essential role in development of the bradyzoite stage, development of brain cysts, or oral infection of new hosts. The quantitative comparison of the transcriptomes of three major developmental stages of *Toxoplasma*: tachyzoites, bradyzoites and oocysts, has revealed a specific sporozoite–rhoptry neck protein 2 (sporoRON2) (Fritz *et al.*, 2012b), which is the paralogue of RON2 (see below). The expression data for this paralogue in both tachyzoites and bradyzoites was close to background levels, and proteomic evidence supports the expression of sporoRON2 in oocysts (Fritz *et al.*, 2012a).

12.5.3.1 ROPs

To date, most of the proteins contained in the bulb are restricted to *Toxoplasma*. As they are exported to the host cell and function as effectors to modulate the host cell machinery, one can therefore suggest that these proteins have been subjected to evolutionary pressure and may have derived from some specialized property of these organisms, such as their ability to form tissues cysts or from their complex heteroxenous life cycle.

TABLE 12.2 Properties of *Toxoplasma* Secretory Proteins – Rhoptry Proteins

Location/ Protein	Calculated MW (kDa)[1]	Domains(#)[2]	Interacting Partners	Mutant Phenotypes	Function	Post-Secretory Trafficking	References
Rhoptry							
Rhoptry Body							
ROP1	46	Strong charge asymmetry		Non-essential protein KO shows normal growth, invasion and virulence, but abnormal rhoptry morphology.		Associated with the PVM	(Saffer *et al.*, 1992; Ossorio *et al.*, 1992; Soldati *et al.*, 1995)
ROP2	64	Degenerate kinase(1)		Non-essential protein *in vitro*		Associated with the PVM	(Sadak *et al.*, 1988, Beckers *et al.*, 1994; Sinai and Joiner, 2001; Sadak *et al.*, 2003; Pernas and Boothroyd, 2010)
ROP4 (ROP2 family)	64	Degenerate kinase(1)				Associated with the PVM and phosphorylated	(Carey *et al.*, 2004)
ROP5 (ROP2 family)	61	Degenerate kinase(1)		Non-essential protein *in vitro* Virulence determinant in mouse	Control host immune response	Associated with the PVM	(Bradley *et al.*, 2005; Leriche and Dubremetz, 1991; El Hajj *et al.*, 2007b; Behnke *et al.*, 2011; Reese *et al.*, 2011)
ROP7 (ROP2 family)	63	Degenerate kinase(1)				Associated with the PVM	(Bradley et al., 2005; Hajj *et al.*, 2005)
ROP8 (ROP2 family)	64	Degenerate kinase(1)		Non-essential protein *in vitro*			(Beckers *et al.*, 1997, Pernas and Boothroyd, 2010)
ROP9 (P36)	39						(Reichmann *et al.*, 2002)
ROP10	61						(Bradley *et al.*, 2005)
ROP11 (ROP2 family)	57	Degenerated kinase(1)					(Bradley *et al.*, 2005)
ROP12	25						(Bradley *et al.*, 2005)

ROP13	45		Non-essential protein *in vitro* Null mutant shows normal virulence *in vivo*		Host cytoplasm	(Bradley et al., 2005; Turetzky *et al.*, 2010)
ROP14	122	TM(2), DUF1222(1)				(Bradley *et al.*, 2005)
ROP15	34					(Bradley *et al.*, 2005)
ROP16 (ROP2 family)	76	Serine/ threonine kinase(1)	Non-essential protein *in vitro* Virulence determinant in mouse	Control host immune response	Associated with the host cell nucleus	(Bradley et al., 2005; Coller *et al.*, 2005; Butcher *et al.*, 2011; Ong *et al.*, 2010; Saeij *et al.*, 2007; Yamamoto *et al.*, 2009)
ROP17 (ROP2 family)	70	Putative kinase(1)			Associated with PVM	(Peixoto *et al.*, 2010)
ROP18 (ROP2 family)	62	Kinase(1)	Non-essential protein *in vitro* Virulence determinant in mouse	Control host immune response	Associated with PVM	(Bradley et al., 2005; El Hajj *et al.*, 2007; Fentress *et al.*, 2010; Steinfeld *et al.*, 2010; Yamamoto *et al.*, 2001)
ROP19 (ROP2 family)	61	Putative kinase(1)			Associated with PVM	(Peixoto et al., 2010)
ROP20 (ROP2 family)	61	Putative kinase(1)			Associated with PVM	(Peixoto et al., 2010)
ROP23 (ROP2 family)	61	Degenerate kinase(1)			Associated with PVM	(Peixoto et al., 2010)
ROP24 (ROP2 family)	61	Degenerate kinase(1)			Associated with PVM	(Peixoto et al., 2010)
ROP25 (ROP2 family)	120	Putative kinase(1)			Associated with PVM	(Peixoto et al., 2010)
ROP26 (ROP2 family)	48	Degenerate kinase				{Peixoto, 2010 #1340
ROP31 (ROP2 family)	59	Putative kinase(1)				(Peixoto et al., 2010)
ROP38 (ROP2 family)	60	Putative kinase(1)		Control host cell response	Associated with PVM	(Peixoto et al., 2010)

(*Continued*)

TABLE 12.2 Properties of *Toxoplasma* Secretory Proteins – Rhoptry Proteins (*cont'd*)

Location/ Protein	Calculated MW (kDa)[1]	Domains(#)[2]	Interacting Partners	Mutant Phenotypes	Function	Post-Secretory Trafficking	References
ROP39 (ROP2 family)	65	Putative kinase(1)				Associated with PVM	(Peixoto et al., 2010)
ROP40 (ROP2 family)	58	Degenerate kinase(1)				Associated with PVM	(Peixoto et al., 2010)
SUB2	142	Subtilase(1), TM(1)		Essential protein Abnormal rhoptry formation in SUB2 antisense knockdown	Rhoptry protein maturase		(Miller *et al.*, 2003; Binder and Kim, 2004)
Toxolysin 1 (TLN1)	177	Metalloprotease (M16 family)		Non-essential protein	Proteolysis	Associated with the PVM	(Bradley *et al.*, 2005), (Hajagos *et al.*, 2012)
Toxofilin	27		Host actin	Non-essential protein Facilitates parasite invasion by disrupting host cortical actin cytoskeleton	Actin binding	Host cytoplasm	(Bradley et al., 2005; Poupel *et al.*, 2000; Lodoen *et al.*, 2010; Delorme-Walker *et al.*, 2012)
Rab11a	25	Small GTPase		Control assembly of the inner membrane complex	Vesicular trafficking?	Associated with the cytoplasmic side of the rhoptry membrane	(Bradley et al., 2005; Herm-Gotz *et al.*, 2007; Agop-Nersesian *et al.*, 2009)
NHE2	91	sodium hygrogen exchanger		Non-essential protein KO is defective in calcium-dependent egress	Ion Homeostasis?		(Karasov *et al.*, 2005)
BRP1 (Bradyzoite specific Rhoptry Protein 1)	18						(Schwarz *et al.*, 2005)
Armadillo Repeats-Only[3] (ARO)	31	Armadillo-like repeats(2)					(Cabrera *et al.*, 2012)
Rhoptry neck							

RON1	127	TM(1), CobT(1) CCP(1)				(Bradley *et al.*, 2005)
RON2	155	TM(2)	In complex with RON4, RON5, RON8 Interacts with AMA1		MJ localisation and posterior translocation	(Bradley *et al.*, 2005; Lebrun *et al.*, 2005; Besteiro *et al.*, 2009; Lamarque *et al.*, 2011; Tonkin *et al.*, 2011; Tyler 2011)
RON3	223	TM(1)				(Bradley *et al.*, 2005)
RON4	107		In complex with RON2, RON5, RON8		MJ localization and posterior translocation	(Alexander *et al.*, 2005; Bradley *et al.*, 2005; Lebrun *et al.*, 2005)
RON5	179		In complex with RON2, RON4, RON8		MJ localization and posterior translocation	(Bradley *et al.*, 2005; Lebrun *et al.*, 2005; Besteiro *et al.*, 2009; Straub *et al.*, 2009)
RON6	239					Bradley, unpublished
RON8	329		In complex with RON4, RON5, RON2	Non-essential protein KO is defective in attachment, invasion and virulence	MJ localisation and posterior translocation	(Besteiro *et al.*, 2009; Straub *et al.*, 2009; Straub *et al.*, 2011)
RON9	204		RON10	Non-essential protein KO has no discernible phenotype apart from mislocalization of RON10		(Lamarque *et al.*, 2012)
RON10	140		RON9	Non-essential protein KO has no discernible phenotype apart from mislocalization of RON9		(Lamarque *et al.*, 2012)

[1] *Based on the complete open reading frame including signal sequence or GPI anchor signal, if present*
[2] *Abbreviations: CCP, domain abundant in complement control proteins; CobT, Cobalamin biosynthesis protein; DUF1222, domain of unknown function; TM, transmembrane*
[3] *Also associated with rhoptry neck*

12.5.3.1.1 ROP2 FAMILY, SO CALLED RHOPTRY KINASES (ROPKS)

The ROP2 family was first described as three rhoptry proteins recognized by a single monoclonal antibody (Sadak *et al.*, 1988). Additional members of the family have been successively identified using proteomics, *in silico* searches, integrative genomic analysis and gene cloning to obtain a comprehensive view of the family that comprises at least 44 members. These proteins share several common features such as the presence in the C-terminus of a protein kinase-like domain, a similar size (50 kDa range), a basic amino acid-rich N-terminal area (El Hajj *et al.*, 2006; El Hajj *et al.*, 2007a,b; Saeij *et al.*, 2006; Saeij *et al.*, 2007; Taylor *et al.*, 2006; Peixoto *et al.*, 2010). This family is now called the 'rhoptry kinases' (ROPKs). Twenty-one of the ROPKs were localized. Nineteen of those were found in the rhoptry (ROP2, 4, 5, 7, 8, 11, 16, 17, 18, 19, 20, 23, 24, 25, 26, 31, 38, 39, 40) and two were designated apical but did not co-localize with rhoptry markers (ROP21 and 22) (Peixoto *et al.*, 2010).

Seventeen of the ~44 ROPK family members are predicted to be active based on the putative presence of a complete catalytic triad (ROP18, 35, 31, 17, 21, 27, 30, 16, 25, 28, 20, 39, 38, 29, 19, 41, 42) (Peixoto *et al.*, 2010). Among them, the role of ROP16, ROP18 and ROP38 has been elucidated (see below and Chapter 14).

A high degree of divergence (average 16% pairwise identity) is present in the ROPK family. The N-terminal portion of the kinase domain encompassing the activation loop and substrate-binding site is the most conserved part. Most members of the ROPK family lack the glycine loops responsible for stabilization of the αβ-phosphate of ATP and the conserved aspartic acid in the catalytic loop critical for phosphotransferase activity. These degenerate members, including ROP2, ROP4, ROP5, ROP7, ROP8, are not expected to be enzymatically active and are refered to as pseudokinases. Crystallographic analysis reveals that ROP2, ROP5 and ROP18 maintain a conserved kinase fold with a unique regulatory domain, suggesting that these pseudokinases may function in scaffolding and/or sequestering substrates (Labesse et al., 2009; Qiu et al., 2009; Reese and Boothroyd, 2011). The role of the pseudokinase ROP5 is discussed below.

Although the ROPK family forms a single clade distinct from previously characterized kinase families, phylogenetic analysis defines two main clades among ROPKs (Peixoto *et al.*, 2010). One contains most previously identified rhoptry proteins (ROP2, ROP4, ROP5, ROP7, ROP8 and ROP18), including many recent duplications. Half of the members of this clade have degenerated into pseudogenes; ROP18 being the only active member of this clade. The second clade contains most of the active kinases, including ROP16 and ROP38.

The array of roles played by these kinases and pseudokinases has considerably expanded in recent years, making them the major players of the modulation of the host cell biology to promote a suitable environment for growth and proliferation (see details in Chapter 14). Comparative genomic analysis shows that these coccidian-specific secreted kinases are differentially expressed between strains, and/or differentiate stage (Peixoto *et al.*, 2010). Some of these proteins are strongly expressed in mature oocysts/sporozoites (Fritz *et al.*, 2012a,b; Peixoto *et al.*, 2010) suggesting that they may serve similar function in other stages.

12.5.3.1.2 PROTEASES

Rhoptries contain several kinds of proteases. A subtilisin-like serine protease or subtilase (SUB2) has been identified in *T. gondii* rhoptries by homology to *P. falciparum* SUB2 (Miller *et al.*, 2003). Another rhoptry serine protease (TGME49_062920), closer to chymotrysin has been identified in the proteome study (Bradley *et al.*, 2005). SUB2 is predicted to be a type I TM protein with a conserved catalytic domain. It is autocatalytically processed at the N-terminus. The mapping of the endogenous cleavage site of SUB2, harbouring a similar cleavage site to ROP cleavage sites, supports the proposal that TgSUB2 is the maturase of rhoptries proteins at SΦXE/D consensus (Bradley and Boothroyd, 1999). Consistent with this hypothesis, SUB2 co-immunoprecipitates with ROP1 (Miller *et al.*, 2003) and also with ROP2 and ROP4 (Binder and Kim, 2004). Interestingly, ROP5, a member of the ROP2 family, does not possess the SΦXE/D motif and is not cleaved (El Hajj *et al.*, 2007b). A knockout of SUB2 could not be obtained (Miller *et al.*, 2003) but SUB2 antisense treatment of parasites induced abnormal rhoptry formation (Binder and Kim, 2004), accumulation of vesicular structures and impaired replication, leading to the hypothesis that SUB2 is involved in organelle biogenesis.

Based on homology with eukaryotic cysteine proteinases, a cathepsin B-like protein named toxopain-1 or cathepsin protease B (CPB) was identified in *T. gondii* rhoptries (Que *et al.*, 2002). CPB is produced as pro-protein that is autocatalytically cleaved to generate an active enzyme. A role of CPB in rhoptry protein processing and rhoptry biogenesis was suggested based on showing that a protease inhibitor delays ROP2 family protein processing and causes abnormal rhoptry biogenesis and reduced invasion (Que *et al.*, 2002). On the other hand, more recent findings (Dou *et al.*, 2012) suggest that CPB is principally located in the lysosome-like VAC (Parussini *et al.*, 2010). Targeted deletion of CPB did not affect rhoptry ultrastructure or the processing of ROP2 based on metabolic pulse—chase experiments. Also, that CPB recognition of its substrate cleavage site is mainly via the P2 residue (Parussini *et al.*, 2010) is inconsistent with mutational studies of ROP pro-domain cleavage sites (SΦXE/D), which suggest that the P3 S and P1 E/D are the key residues.

An insulinase-like protein has been identified in the rhoptry fraction (Bradley *et al.*, 2005) and named Toxolysin-1 (Hajagos *et al.*, 2012). It belongs to the M16 family of metalloproteases that generally depend on divalent cations for their activity (Rawlings *et al.*, 1991). The function of this insulinase is unknown, but it is not essential for the parasite (Hajagos *et al.*, 2012). Toxolysin-1 is subject to two different processing events (see above), one of them separating the functional protease domain in two portions. Interestingly, a tight association between the two portions is observed, probably reconstituting a mature enzyme.

12.5.3.1.3 NA$^+$/H$^+$ EXCHANGER

In searches of the *Toxoplasma* genome, three homologues of sodium—hydrogen exchangers have been found. These proteins catalyze the Na^+/H^+ exchange and are involved in regulation of internal pH and cell volume. One of them, NHE2, is associated with rhoptries (Karasov *et al.*, 2005). The pH of rhoptries changes from acidic (estimated to be ~3.5—5.5) for nascent rhoptries to more neutral (~5.0—7.0) in mature organelles (Shaw *et al.*, 1998). It is therefore hypothesized that NHE2 may be involved in pH regulation during the course of rhoptry biogenesis and rhoptry protein processing. However, the disruption of NHE2 does not affect targeting of several rhoptry proteins, including ROP1 and ROP2/3/4, nor does it alter egress or rate of growth (Karasov *et al.*, 2005). In addition, *in vivo* studies of virulence yielded no difference compared to a wild-type strain.

12.5.3.1.4 PHOSPHATASE

The presence of a phosphatase 2C (PP2C) in rhoptries is very attractive (Bradley *et al.*, 2005; Gilbert *et al.*, 2007). PP2C-type protein phosphatases are monomeric enzymes present in both prokaryotes and eukaryotes. Members of this family of phosphoprotein phosphatases are involved in the regulation of several signalling pathways including regulation of the cell cycle, adaptation and cell recovery after DNA double-strand breaks or environmental stress response (Schweighofer *et al.*, 2004). Interestingly, *Toxoplasma* PP2C contains a NLS sequence and is targeted to the host cell nucleus (see below). *PP2C* is not an essential gene, and parasites knocked out for this phosphatase show only a mild growth defect (Gilbert *et al.*, 2007).

12.5.3.1.5 TOXOFILIN

Toxofilin is a 27-kDa protein that has been shown to bind mammalian G-actin (Poupel *et al.*, 2000). It is an actin sequestering protein that caps actin filaments. Toxofilin was first suggested to be present in the cytosol of the apical end of the parasite and involved in control of parasite actin polymerization during invasion and motility (Poupel *et al.*, 2000). Then, toxofilin was localized in rhoptries (Bradley *et al.*, 2005) and shown to be secreted into the host cell (Lodoen *et al.*, 2010), where it upregulates the host cortical actin cytoskeleton dynamics facilitating *Toxoplasma* invasion (Delorme-Walker *et al.*, 2012). *In vitro*, the control of actin dynamics by Toxofilin has been shown to depend on a casein kinase II and a phosphatase 2C (Delorme *et al.*, 2003).

12.5.3.1.6 ROP1

ROP1 was the first ROP protein identified in *Toxoplasma* (Schwartzman, 1986), encoded by the first rhoptry gene cloned and sequenced (Ossorio *et al.*, 1992). It is a soluble protein whose function is as yet unknown. Although it has no TM anchor, it is associated with the PVM after invasion (Saffer *et al.*, 1992) and is completely degraded within 24 hours suggesting that it had fulfilled its role by then. ROP1 KO parasites are not impaired in growth, invasion or virulence but do show altered rhoptry ultrastructure (Kim *et al.*, 1993, Soldati *et al.*, 1995).

12.5.3.1.7 ROPS CONSERVED IN APICOMPLEXA

Until the proteomic work by Bradley *et al.*, there were no homologies described for rhoptry proteins within Apicomplexa, except for *T. gondii* ROP9 (P36) (Reichmann *et al.*, 2002), which has a homologue in *Plasmodium*. This was fairly surprising, since these organelles are likely to perform the same function. Indeed, both in *Plasmodium* and *Toxoplasma*, membranous structures containing protein and lipids are secreted from the rhoptries during invasion and are incorporated into the expanding PVM (Bannister *et al.*, 1986; Hakansson *et al.*, 2001). The proteomic study by Bradley *et al.* identified ROPs (ROP9 and ROP14) that have homologues in other Apicomplexa (Bradley *et al.*, 2005). ROP14 also shows homology with a hypothetical membrane protein of mammalian cells. Other hypothetical proteins found in the *T. gondii* rhoptry fraction have apicomplexan homologues and may serve conserved roles; however, their localization in the organelle has not been confirmed.

12.5.3.2 RONs

Most of the proteins located in the rhoptry neck, RONs show homologies restricted to Apicomplexa (Bradley *et al.*, 2005).

12.5.3.2.1 MOVING JUNCTIONAL RONS COMPLEX

Independent studies, one looking for AMA1 associated proteins and the other searching for antigens recognized by an antibody that stained the MJ, isolated a complex of RON proteins (Alexander *et al.*, 2005; Besteiro *et al.*, 2009; Lebrun *et al.*, 2005; Straub *et al.*, 2009). This complex is composed of RON2, RON4, RON5 and RON8, conserved in other Apicomplexa, except RON8, which is specific to coccidian parasites (Besteiro *et al.*, 2011; Besteiro *et al.*, 2009; Straub *et al.*, 2009). Once secreted, these proteins are associated to the moving junction and involved in invasion (see below). *Cryptosporidium,* whose assignment as *Coccidia* is being reevaluated and which remains extracytoplasmic, does not possess any orthologue of the MJ RONs complex in its genome.

These proteins do not bear recognizable domains or motifs that could suggest a particular molecular interaction. They are subjected to proteolytic maturation (Besteiro *et al.*, 2009; Straub *et al.*, 2009) in the pre-rhoptry compartments, which is not a prerequisite to their interaction, as immature proteins were found to be interacting *in vitro* (Besteiro *et al.*, 2009). Bioinformatic analysis indicates that RON4, RON5 and RON8 are putative soluble proteins, while RON2 is predicted to have two TM domains. How the complex is organized and its stochiometry is not yet known.

Two paralogues of RON2 are found in *T. gondii,* one specific for the sporozoite stage (SporoRON2 or RON2L2) (Fritz *et al.*, 2012a,b). *Tg*RON4 has one additional copy, which is presumably expressed in all stages according to mass spectrometry and EST datasets. No additional orthologues for RON5 or RON8 are present in the *Toxoplasma* genome. Database searches also reveal additional paralogues of MJ complex members in *Coccidia*, such as *Neospora* and *Eimeria*, while they are absent in *Plasmodium* genomes. This could highlight a more specialized function for these proteins in the invasion of *Coccidia*.

12.5.3.2.2 RON1

RON1 is a Sushi-containing protein (also called CCP for complement control protein), found in the neck of the rhoptry (Bradley *et al.*, 2005). RON1 is conserved in all Apicomplexa parasites. Its orthologue in *Plasmodium* is called PfASP1 (O'Keeffe *et al.*, 2005; Srivastava *et al.*, 2010). The Sushi domain spans approximatively 60 residues and contains four invariant cysteine residues. It is present in various complement regulator proteins found in mammals. The role played by RON1 is currently unknown.

12.5.3.2.3 RON9/RON10 COMPLEX

A highly stable hetero-complex formed by the association of RON9 and RON10 was recently identified in the rhoptry neck of tachyzoites. This complex is distinct from the MJ-complex (Lamarque *et al.*, 2012). Both RON9 and RON10 are conserved in *Coccidia* and *Cryptosporidia*. RON10 does not display known domains, while RON9 contains a Sushi domain, repetitions of ankyrin motifs and a set of repetitions enriched in proline (P), glutamic acid (D), aspartic acid (E), and serine (S) or threonine (T) typical of PEST sequences, which are targets for rapid degradation. The ankyrin repeat is a common motif in nature, predominantly found in eukaryotic proteins and involved in protein–protein interactions. Genetic disruption of *RON9* or *RON10* leads to loss of the entire complex, but does not result in development defect in HFF *in vitro* or virulence in mice (Lamarque *et al.*, 2012). The conservation of RON9/RON10 in the genome of *Cryptosporidia*, which does not form a MJ, argues against a participation of this complex in MJ formation. As the primary site of infection for *Coccidia*, as well as *Cryptosporidium* spp., is the epithelial cells of the gastrointestinal tract, further studies are needed to determine if the function of the RON9–10 complex might be linked to the interaction with a brush border membrane.

12.5.3.3 Proteins Associated with the Cytosolic Face of the Rhoptry

12.5.3.3.1 RAB11A

Rab11a was found in the purified rhoptry proteome (Bradley *et al.*, 2005). Rab11a is highly dynamic and can be also observed in an endosomal-like compartment and at the IMC in *T. gondii* (Agop-Nersesian *et al.*, 2009). In higher eukaryotes, Rab11 belongs to the family of small GTPases involved in the regulation of vesicular traffic. It is usually localized at early endosomes, perinuclear recycling endosomes, as well as at the trans-Golgi network, and it is considered as the Rab that controls slow endosomal recycling, as well as traffic to the Golgi apparatus (Chen *et al.*, 1998; Ullrich *et al.*, 1996). *Toxoplasma* Rab11a, which does not contain any signal peptide, is likely associated with the cytoplasmic side of the rhoptry membrane through geranyl–geranyl modification of the two cysteines of the CCXX site (Bradley *et al.*, 2005), a usual feature of Rab11. It was therefore proposed that Rab11a may act as a regulator of trafficking to the rhoptries (Bradley *et al.*, 2005).

Parasites expressing dominant negative Rab11a show a severe growth defect (Herm-Gotz *et al.*, 2007). However, no defects in rhoptries are observed in this mutant. The fate of other organelles, including micronemes, dense granules, Golgi, apicoplast and mitochondria and nucleus are not changed and the formation and early elongation of the IMC appeared normal (Agop-Nersesian *et al.*, 2009). Ablation of Rab11a function results in daughter parasites having an incompletely formed IMC that leads to a block at a late stage of division. In fact, Rab11a appears to regulate the assembly of the motor complex at the IMC, an essential step in parasite development. Why Rab11a is associated with rhoptries remains unknown.

12.5.3.3.2 TgARO

A conserved apicomplexan protein termed Armadillo Repeats-Only (ARO) protein is localized to the cytosolic face of *P. falciparum* and *T. gondii* rhoptries (Cabrera *et al.*, 2012). Although ARO does not possess a signal peptide for entering the secretory pathway it is nevertheless targeted exclusively to the rhoptries, by attachment to the cytosolic face of the membrane by N-terminal acyl-modifications. Gene disruption experiments are consequently required for deciphering the precise function of ARO in apicomplexan parasites.

12.5.3.4 Lipids

T. gondii rhoptries not only contain proteins but also lipids, which may form the membranous structures sometimes observed in the organelles. Only few studies have attempted to determine the lipid composition of rhoptries. The first one, using TLC and enzymatic assays on a rhoptry-enriched fraction, showed that these organelles seemed to be particularly enriched in cholesterol, with a cholesterol to phospholipid ratio greater than one (Foussard *et al.*, 1991) and phosphatidylcholine being the major rhoptry phospholipid. Significant amounts of phosphatidic acid and lysophospholipids were also found, but not phosphatidylserine, phosphatidylinositol or sphingomyelin (Foussard *et al.*, 1991). It was suggested that secretion of phosphatidic acid and lysophospholipids may facilitate initial vacuole membrane formation (Foussard *et al.*, 1991).

Re-evaluation of the rhoptry-lipid content using HPLC and capillary GLC shows that cholesterol is present in lower proportions (Besteiro *et al.*, 2008). Consistent with this, cholesterol in the rhoptries does not appear to be essential for invasion, as parasites depleted of rhoptry cholesterol (16%–23%) are still able to invade cells (Besteiro *et al.*, 2008; Coppens and Joiner, 2003).

An enrichment of saturated fatty acids has been also observed using these improved technological tools (Besteiro *et al.*, 2008), suggesting an elevated rigidity for membranes derived from rhoptries. This was supported by a lower fluidity of rhoptry-derived membranes than membranes from whole *T. gondii* cells by fluorescence anisotropy (Besteiro *et al.*, 2008). Overall, total membranes from rhoptries (possibly including internal membranous structure) appear to have high membrane rigidity.

12.5.4 Secretion of Rhoptries and Post-Secretion Trafficking of ROPs and RONs

Rhoptry discharge occurs early during the invasion process. Electron microscopic observations show empty rhoptries as early as the initial apical contact creating the MJ. How many rhoptries are secreted during invasion is unknown, but not all rhoptries are discharged in tachyzoite (see Fig. 12.1A). In contrast, in *P. falciparum,* the RAP1 rhoptry signal disappears from the parasite just after invasion, indicating that this parasite exocytoses the content of its two rhoptries (Riglar *et al.*, 2011). Interestingly, electron micrographs show the fusion between the neck of two rhoptries at the initial stage of invasion (Aikawa *et al.*, 1978), implying a fusion machinery at this step. A single burst of rhoptry secretion occurs during the invasion process, as proteins associated with the MJ in *T. gondii* are found exclusively at the MJ and not in the vacuole.

The signals and molecular mechanisms defining how rhoptries fuse to the parasite membrane are almost entirely unknown. Recently, an inhibitor of phospholipase A2s called 4-bromophenacyl bromide has been shown to interfere with rhoptry secretion, while it did not affect micronemes secretion (Ravindran *et al.*, 2009). *T. gondii* encodes in its genome many predicted phospholipases that might be direct targets of this inhibitor.

The importance of microneme proteins in rhoptry secretion has been recently documented. Indeed, ROP protein discharge is reduced in *Toxoplasma* in the absence of AMA1 (Mital *et al.*, 2005) and MIC8 (Kessler *et al.*, 2008), and these parasites display a major defect of invasion. In the case of MIC8, the deletion of the cytoplasmic domain is sufficient to impair secretion of rhoptry proteins during invasion. Altogether, these results indicate that the attachment of the parasite to the host cell is a prerequisite to efficient rhoptry secretion, and that interaction of MICs with a ligand probably triggers signal transduction pathways through their cytosolic tails. This was recently supported by an elegant study in *P. falciparum* (Singh *et al.*, 2010), which demonstrates that the interaction of EBA175 with glycophorin A, its receptor on erythrocytes, diminishes the elevated cytosolic calcium levels necessary for microneme secretion (see above) and, importantly, triggers release of rhoptry proteins. These observations were also observed with EBA140–glycophorin C receptor engagement (Singh *et al.*, 2010).

Whether RON and ROP protein discharge is simultaneous or sequential is not known, although some evidence suggests simultaneity (Lebrun *et al.*, 2005). How ROPs are introduced in the cell is unknown. Details of rhoptry structure during red cell invasion by *Plasmodium* show a dome-like structure of the red cell membrane that enters the tip of the rhoptry duct which is difficult to interpret (Bannister and Mitchell, 1989). Concomitant to the intimate attachment of the apical tip of the parasite with the host cell and rhoptry secretion is a transient rise in the conductance of the host cell membrane (Suss-Toby *et al.*, 1996). This has been interpreted as a transient break in the host membrane allowing rhoptry proteins to enter the host cell. Whether it corresponds to the 40 nM pore observed by freeze fracture during the invasion process (Dubremetz, 2007), which seems to connect the rhoptry contents with the host cell cytoplasm, remains to be demonstrated.

Once secreted, rhoptry proteins associate with different compartments and most likely perform a wide variety of roles: (i) some RONs remain associated with the MJ, (ii) some ROPs associate with the PVM, and (iii) other ROPs are found in the host cell nucleus (Gilbert *et al.*, 2007; Saeij *et al.*, 2007).

12.5.4.1 Moving Junction-Associated Proteins

The discovery of RON proteins at the MJ a few years ago has greatly improved our knowledge of the molecular organization of the MJ. The RON2/RON4/RON5/RON8 complex is secreted during invasion and then associated with the MJ during the entire invasion process, giving the characteristic ring-shaped labelling by IFA. This complex is inserted into the host cell membrane, where the transmembrane protein RON2 spans the host cell membrane (Lamarque *et al.*, 2011), and RON4, RON5 and RON8, which do not contain TM domains, are exposed to the cytosolic face of the host membrane (Besteiro *et al.*, 2009; Lamarque *et al.*, 2011), The N-terminal part of RON2, which is inside the host cell (Lamarque *et al.*, 2011), is presumably responsible for maintaining the rest of the complex in this location, while some direct interaction of RONs with membranous components are also likely. For example, when expressed in mammalian cells, RON8 localizes at the the host cell membrane in a manner dependent on the C-terminal portion of the protein (Straub *et al.*, 2009). Fractionation experiments of these RON8-expressing cells support a loose association of RON8 with the cortical cytoskeleton, a host TM protein or host lipids.

The MJ holds the parasite and the host plasma membranes in close apposition. It is static on the host cell membrane side but the parasite membrane flows along this link, hence the name moving junction. Yet, according to this organization, the MJ RON proteins are ideally positioned to perform the anchoring on the host cell side. On the parasite side, the discovery that RON2 binds directly to the microneme protein AMA1 (Besteiro *et al.*, 2009), which is secreted just prior to rhoptry secretion at the tip of the parasite and exposed on the parasite surface, allows us to propose a model where the parasite would insert its own receptor (RON2 and associated RON proteins) for the ligand AMA1 to form the close apposition of the host cell and parasite plasma membranes that constitute the MJ. How this junction then moves along the parasite surface will be discussed below. The AMA1−RON2 interaction has been detected by co-immunopurification *in vitro*, but the functional relevance of this interaction *in vivo* has been further validated by the inhibition of the invasion process by incubating tachyzoites with a recombinant protein corresponding to a C-terminal portion of RON2 (Lamarque *et al.*, 2011; Tyler and Boothroyd, 2011). Interestingly, although the AMA1 and RON2 primary sequences differ among Apicomplexa, the AMA1−RON2 interaction is evolutionarily conserved in *Plasmodium* sp., with the same invasion-inhibitory effects of the orthologous region of the *Pf*RON2 domain (Lamarque *et al.*, 2011; Vulliez−Le Normand *et al.*, 2012b), or of a peptide (R1 peptide) or an antibody which inhibits the *Pf*AMA1−*Pf*RON complex formation (Coley *et al.*, 2006; 2007; Collins *et al.*, 2009; Pizarro *et al.*, 2005; Richard *et al.*, 2010). No interspecies or intergeneric cross-binding was seen, highlighting the separate co-evolution of the AMA1−RON2 pair in Apicomplexa and the importance of maintaining this interaction for efficient invasion (Lamarque *et al.*, 2011).

Deletions and mutations in the RON2 ectodomain define the essential AMA1 binding domain of RON2 to a U-shaped polypeptide loop containing a pair of cysteine residues in a disulfide bridge (Tonkin *et al.*, 2011). Co-crystallization of the AMA1 with this peptide reveals that the RON2 peptide is deeply anchored within the hydrophobic grove of AMA1 to form an embedded structure, which likely withstands the mechanic forces that occur during the invasion process (Tonkin *et al.*, 2011). These atomic details provide insight into the molecular basis of the intimate attachment, and are consistent with the 6 nM distance between the host cell and parasite plasma membranes at the MJ. Accordingly, conditional depletion of AMA1 results in parasites that are largely impaired in invasion; the early steps of invasion (motility, distant attachment) are not affected but TgAMA1-depleted tachyzoites examined by transmission electron microscopy show that parasites have less frequently progressed beyond the distant attachment step (Mital *et al.*, 2005).

By solving the co-structure in both *Toxoplasma* and *P. falciparum* (Tonkin *et al.*, 2011; Vulliez-Le Normand *et al.*, 2012b), researchers have defined the molecular basis of species specificity of the AMA1−RON2 pair. While the overall U-shaped architecture of RON2 in complex with AMA1 appears to be remarkably well maintained between the two parasites, clear specific features playing an influential role in species selectivity are visible in the cystine loop, which is the most divergent region within the RON2s. For example, the *T. gondii* cystine loop is two residues shorter which mirrors the narrower groove of *Tg*AMA1. Or, an Arg2041 residue, specific to the *P. falciparum* species, located at the tip of the β-hairpin with its guanidyl group, fits snugly into a preformed cavity of *Pf*AMA1, which is also occupied by an arginine of the invasion-inhibitory immunoglobulin IgNAR141-1 (Henderson *et al.*, 2007), a lysine of the inhibitory monoclonal antibody 1F9 (Coley *et al.*, 2007) or of the R1 peptide (Vulliez-Le Normand *et al.*, 2012b). These inhibitors form an extensive interface with this region and impair the binding of the cystine loop of RON2.

Apart from being located at the MJ, the precise role of the RON4/5/8 proteins during invasion remains elusive. At the MJ, host cell proteins are segregated depending on their anchoring in the membrane, leading to the exclusion of most of the TM proteins but not GPI-anchored proteins from the forming parasitophorous vacuole (Mordue *et al.*, 1999a). Whether these RONs participate in the selective segregation is yet unknown. Likewise, whether the junctional complex facilitates PV closure by PVM fusion remains to be investigated.

12.5.4.2 PVM-Associated ROPs

Rhoptry-derived vesicles (evacuoles; Hakansson *et al.*, 2001) are secreted in host cell cytoplasm. These vesicles contain lipid components presumably derived from the rhoptry lipids and some rhoptry proteins. These vesicles fuse with the nascent PVM (Hakansson *et al.*, 2001) and ROPs proteins become exposed to the host cell side (Beckers *et al.*, 1994; Carey *et al.*, 2004a; El Hajj *et al.*, 2007a,b; Hajj *et al.*, 2005).

While ROP2, ROP4 and ROP5 were first described as a PVM-integral TM proteins (Beckers *et al.*, 1994; El Hajj *et al.*, 2007b), it has been recently shown that ROPKs are associated with membrane through arginine-rich-amphipathic helices (RAH domain) conserved in the N-terminal part of the ROP2 family (Labesse *et al.*, 2009; Reese and Boothroyd, 2009). This domain displays a preferential tropism for the PVM rather than host cell membranes and appears to be a good predictor of a *Toxoplasma* protein's localization to the PVM. When expressed in the host cell, this domain does not associate to any specific membrane but is targeted to the PVM in cells infected by *T. gondii*. As the PVM is composed largely from membrane derived from the host cell, the specificity of the RAH domain is probably due to an interaction with an unidentified parasite dependent factor, such as another ROP or a specific lipid anchored to, or integral within, the PVM. It is also possible that the RAH domain recognizes a specific physical property of the PVM that the host membranes lack. The RAH membrane association is resistant to extraction with a basic carbonate solution, which disrupts charge interactions. Moreover, elimination of the positive charge on RAH had no effect on PVM association, while mutations of the basic residues decrease specificity for the PVM, suggesting that an integral rather than a peripheral association with membrane is involved (Reese and Boothroyd, 2009). The amphipathic helices are proposed to be integrated through their hydrophobic faces into the hydrocarbon core of a bilayer.

In any intracellular parasite surrounded by a vacuolar membrane, the membrane serves as a critical functional interface between the parasite and the host cell cytoplasm. For *T. gondii*, nutrient import occurs across the PVM; host mitochondria and endoplasmic reticulum are recruited and physically tethered to the PVM (Sinai *et al.*, 1997). The host cell microtubules (Melo *et al.*, 2001; Coppens *et al.*, 2006) and the intermediate filament network are reorganized around the PVM (Halonen and Weidner, 1994), and the intracellular parasite is capable of inhibiting host cell apoptosis from within the PVM. Rhoptry proteins of the PVM are well situated to play an important role(s) in PVM function. ROP2 was been first suggested to be essential (Nakaar *et al.*, 2003) and one of the mediators of the PVM–mitochondrion association (Sinai and Joiner, 2001; Sinai *et al.*, 1997). However, a mutant with a deletion of the entire ROP2 locus (comprising three related genes, ROP2a, ROP2b and ROP8) is viable and retains the ability to recruit host mitochondria in a manner that is indistinguishable from the parental strain, re-opening the question of which molecules mediate this association (Pernas and Boothroyd, 2010).

The role of some ROPKs has been extensively studied during the past years. ROP18 can act as a virulence factor for *T. gondii* by phosphorylating and thereby inactivating the interferon-gamma-inducible immunity-related GTPases (IRG proteins) responsible for the successful early resistance of mice to *T. gondii* strains (Fentress *et al.*, 2010; Steinfeldt *et al.*, 2010) (for a review see Fentress and Sibley, 2011). ROP18 has also been reported to act on another target, ATF6β, a host endoplasmic reticulum-localizing transcription factor that regulates genes as part of the unfolded protein response (UPR) (Yamamoto *et al.*, 2011). During stress, such as invasion by an intracellular parasite, unfolded proteins can accumulate in the ER. When this occurs, ATF6β is cleaved and the cytoplasmic tail translocates to the nucleus where it activates immune related genes. The ATF6β level was reduced in cells infected with WT and the kinase activity of ROP18 is essential for this proteasome-dependent degradation of ATF6β and for parasite virulence. Consistent with a key role for ATF6β in resistance against this intracellular pathogen, ATF6β-deficient mice exhibit a high susceptibility to infection by ROP18-deficient parasites (Yamamoto *et al.*, 2011). The N-terminal extension of ROP18 binds to ATF6β, and the subsequent threonine phosphorylation of ATF6β is believed to be the signal to send the protein for degradation mediating downregulated CD8 T-cell-mediated type I adaptive immune responses. Interestingly, the ROP18 associates with a luminal domain of ATF6β. How ROP18 gains access to the lumen of the host ER awaits further study.

The other kinase, ROP38, not only appears to influence the expression of a large number of genes, but is also induced itself during differentiation of the parasite from one stage to another (Peixoto *et al.*, 2010). Specifically, ROP38 downregulates host genes associated with MAPK signalling and the control of apoptosis and proliferation, but how these changes impact the virulence *in vivo* in unknown.

Although ROP5 codes for a pseudokinase it is still a major virulence determinant for *T. gondii* (Behnke *et al.*, 2011; Reese *et al.*, 2011). *ROP5* was shown to exist as a tandem cluster of nearly identical genes, rather than a single gene. All members are considered to be pseudokinases, as the catalytic Asp in the HRD motif is replaced with either an Arg or His depending on the ROP5 isoform. Genetic disruption of the entire ROP5 locus in the type I lineage led to complete attenuation of acute virulence, and complementation with ROP5 restored lethality to WT levels (Behnke *et al.*, 2011; Reese *et al.*, 2011). These findings reveal that a locus of catalytically inactive kinases plays an important role in pathogenesis of toxoplasmosis in the mouse model. Moreover, the activity of ROP5 is dependent upon the non-canonical motif in the ROP5 pseudoactive site, as replacement of the arginine by the canonical aspartic residue did not completely restore virulence (Reese and Boothroyd, 2011). The mechanism of action of the pseudokinases ROP5 appears to occur in large part by the ability of ROP5 to activate ROP18 kinase activity, through complex formation with the IRG substrates of ROP18 (Behnke *et al.*, 2012; Fleckenstein *et al.* 2012; Niedelman *et al.*, 2012).

12.5.4.3 ROPs Targeting to the Host Nucleus

Cells infected with *T. gondii* undergo up- or down-regulation of a subset of genes and, depending of the virulence, different *T. gondii* strains show dramatic differences in the ability to modulate host cell genes (Blader *et al.*, 2001; Gail *et al.*, 2001). These genes encode proteins that function in pathways needed to help the parasite satisfy its nutritional needs, as well as proteins involved in inflammation, apoptosis, cell growth and differentiation. Recently, it has been evident that depending on the parasite genotype, the host modulation differs profoundly. The impact of the genotype on *Toxoplasma* host signalling is discussed in more detail in Chapters 14 and 16.

By virtue of their direct access to the host cell cytosol or host nucleus, ROPs are uniquely positioned to modulate changes in host transcription, but direct evidence has been obtained only for ROP16. ROP16 was identified as a virulence determinant, distinguishing how type I/III and II strains activate STAT3/6 during parasite infection (Saeij *et al.*, 2007). One function of STAT transcription factors is to regulate the expression of genes involved in the immune response to intracellular pathogens. ROP16 is a bona fide serine/threonine kinase that binds and directly phosphorylates STAT3 and STAT6 (Yamamoto *et al.*, 2009; Ong *et al.*, 2010) resulting in downregulation of proinflammatory cytokine signalling (Ong *et al.*, 2010; Saeij *et al.*, 2007; Yamamoto *et al.*, 2009) and upregulation of host cell arginase-1 synthesis (Butcher *et al.*, 2011), which in turn limits the availability of arginine, an amino acid that is required for parasite growth and host-inducible nitric oxide production. By secreting a protein that interacts with STAT pathways, the parasite can limit the host immune response and protect itself from destruction (Denkers *et al.*, 2012).

The early growth response 1 (EGR1) and EGR2 transcription factors, as well as their downstream targets, are rapidly up-regulated in *Toxoplasma*-infected cells (Blader *et al.*, 2001). Although the nature of the parasite factor involved in this upregulation is unknown, upregulation of EGR2 required direct contact between the host cell and parasite and appears to involve a rhoptry-derived factor (Phelps *et al.*, 2008).

12.6 DENSE GRANULES

12.6.1 The Dense Granule Proteins: GRAs and Others

Characterization of dense granule molecules started with the production of monoclonal antibodies against *in vitro* excreted–secreted antigens (ESA) and subcellular fractionation of tachyzoites (Cesbron-Delauw *et al.*, 1989; Charif *et al.*, 1990; Leriche and Dubremetz, 1991). The list of dense granule proteins that have been identified and characterized has grown steadily over the past 15 years since Cesbron-Delauw *et al.* first described P23 (Cesbron-Delauw *et al.*, 1989), which was later renamed GRA1 according to the nomenclature proposed by Sibley *et al.* (1991) (see Table 12.3). Dense granule proteins constitute a group of relatively small proteins, all presenting an N-terminal hydrophobic sequence which fits the characteristics of a signal peptide (for review, see Mercier *et al.*, 2005). Except for a few examples where a specific enzymatic or regulatory function has been defined (see below), the majority of dense granule proteins do not present any significant similarity to each other nor to proteins with known function. These were named GRA proteins solely on the basis of their subcellular localization within the dense granules of the tachyzoite stage (Sibley *et al.*, 1991). Up to now fifteen proteins of 21–58 kDa given the GRA designation have been characterized.

GRA1 is an abundant protein (~2% of the total ESTs derived from an RH strain cDNA library (Ajioka *et al.*, 1998) that is soluble and characterized by two predicted EF–Hand domains (AA 149–180 and 197–223, respectively) whose calcium binding property was confirmed experimentally (Cesbron-Delauw *et al.*, 1989). With the exception of GRA1, all GRA proteins described so far are predicted to encode membrane-associated domains on secondary structure analysis. These are either a single hydrophobic α-helix (GRA3, GRA4, GRA5, GRA6, GRA7, GRA8, GRA12, GRA14 and GRA15) or amphipathic α-helices (GRA2 and GRA9). Indeed, all these GRA proteins were detected associated with membranous systems of the PV (i.e. the MNN or PVM).

Another common feature of GRA proteins is the difference observed between their theoretical molecular weight, calculated from the amino acid sequence, and the molecular weight of the

TABLE 12.3 Properties of *Toxoplasma* Secretory Proteins – Dense Granule Proteins

Location/ Protein	Calculated MW (kDa)[1]	Domains(#)[2]	Interacting Partners	Mutant Phenotypes	Function	Post-Secretory Trafficking	References
Dense Granule							
GRA1	20	EF-Hand(2)			Calcium binding	Lumen of the PV or loosely associated with MNN	(Cesbron-Delauw *et al.*, 1989)
GRA2	20	Amphipathic α-helix(2)	GRA4, GRA6	Non-essential protein KO mutant shows complete disruption of the tubular architecture of MNN; less virulent in mice	MNN architecture	Associated with MNN	(Mercier *et al.*, 1998a,b,2002; Labruyere *et al.*, 1999)
GRA3	24	TM(1)		Reduced growth rate under starving condition in culture		Associated with the MNN and PVM	(Henriquez *et al.*, 2005; Craver & Knoll, 2007)
GRA4	36	TM(1), ATP/GTP binding domain(1)	GRA2, GRA6			Associated with the MNN	(Mevelec *et al.*, 1992; Labruyere *et al.*, 1999)
GRA5	13	TM(1)		Secreted soluble products from the KO mutants display a reduced capacity to induced dendritic cells migration		Associated with the PVM	(Lecordier *et al.*, 1993, 1999; Mercier *et al.*, 2001; Persat *et al.*, 2012)
GRA6	24	TM(1)	GRA2, GRA4	KO mutant results in vesiculation of the MNN; less virulent in mice	MNN architecture	Associated with the MNN	(Lecordier *et al.*, 1995; Labruyere *et al.*, 1999; Mercier *et al.*, unpublished)
GRA7	26	TM(1)		KO mutant results in both the loss of HOST structure and reduced growth rate under starving condition in culture	HOST tubulation	Associated with the MNN and PVM	(Jacobs *et al.*, 1998; Coppens *et al.*, 2006)

GRA8	29	TM(1)			Associated with the PVM	(Carey *et al.*, 2000)
GRA9	35	Amphipathic α-helix(1), RGD(1)			Associated with the MNN	(Adjogble *et al.*, 2004)
GRA10	36	TM(1)			Associated with the PVM	
GRA11						
GRA12	47.9	TM(1)			Associated with the MNN	
GRA 14	44.7	TM(1)	Reduced growth rate under starvation conditions *in vitro*		Associated with the PVM	(Rome *et al.*, 2008)
GRA15		NLR(1)	Effect on *in vitro* and *in vivo* parasite growth and cytokine production (IL−12)	Activation of the NF−κB pathway	Associated with the PVM	(Rosowski *et al.*, 2011)
NTPase I	69		Targeted reduction by antisense RNA inhibits parasite proliferation	Nucleotide triphosphatase, egress	Lumen of the PV or loosely associated with MNN	(Asai *et al.*, 1995; Bermudes *et al.*, 1994)
NTPase II	70			Nucleotide triphosphatase, egress	Lumen of the PV or loosely associated with MNN	(Asai *et al.*, 1995)
PI-1	33	Kazal(4)		Serine protease inhibitor	Lumen of the PV	(Morris *et al.*, 2002; Pszenny *et al.*, 2000, 2002)
PI-2	27	Kazal(4)		Serine protease inhibitor	Lumen of the PV	(Morris and Carruthers, 2003)
Cyp18	20	Cyclophilin		Protein folding		(High *et al.*, 1994; Carey *et al.*, 2000)

[1] *Based on the complete open reading frame including signal sequence or GPI anchor signal, if present*
[2] *TM, transmembrane*

native protein, as detected on SDS-PAGE of tachyzoite lysates. This suggests potential post-translational modifications (for review, Mercier et al., 2005). Although numerous N- and O-glycosylation sites are predicted within the GRA amino acid sequences, only a few GRA proteins were shown to be O-glycosylated: GRA2 (Zinecker *et al.*, 1998), GRA4 (Achbarou *et al.*, 1991b) and GRA6 (Travier *et al.*, unpublished). The relative richness of GRA proteins in charged amino acids and in proline residues could also account for the differences between the theoretical and apparent molecular weights. Cy-18, present within DGs as well as within the PV (High *et al.*, 1994), through its potential peptidylprolylcis–isomerase activity, might help in the folding of proline-rich proteins like GRA4 and GRA8, for better trafficking (Carey *et al.*, 2000).

The second group comprises a few soluble dense granule proteins with known function as cyclophilins (Carey *et al.*, 2000), nucleotide triphosphatases (NTPases) (Asai *et al.*, 1995; Bermudes *et al.*, 1994), and serine protease inhibitors of the Kazal family (TgPIs) (Morris and Carruthers, 2003; Morris *et al.*, 2002; Pszenny *et al.*, 2000, 2002). Two NTPase isoforms, TgNTPase-I and TgNTPase-II (E.C. 3.6.1.3), slightly different at the genetic level (97% of identity), were described in *Toxoplasma* tachyzoites and one in *Neospora caninum* (Asai *et al.*, 1998). The gene encoding TgNTPase-II is found in all strains of *T. gondii*, while the gene encoding TgNTPase-I is confined to virulent strains (Asai *et al.*, 1995; Bermudes *et al.*, 1994). *In vitro* activity of these enzymes was extensively characterized but their vacuolar function remains unclear (see below).

12.6.2 Biogenesis of Dense Granule Organelles: Dual Features of Both Constitutive and Regulated Secretory Pathways

Dense granules (DG) are homogeneous spherical electron dense vesicles ~200 nM in diameter and enclosed by a unit membrane. The existence of subpopulations of dense granules storing specific GRAs was examined by double or triple labellings with specific antibodies. All dense granules exhibited multiple labelling, showing localization of different GRAs within the same granules (Ferguson *et al.*, 1999a; Labruyere *et al.*, 1999; Sibley *et al.*, 1995). The number of DGs varies between the different infectious stages. The largest numbers (~15) have been observed in the tachyzoites and sporozoites with intermediate numbers (8–10) in the bradyzoites and few (3–6) in the merozoites. This may correlate with the number of GRAs expressed and the type of PV formed (discussed below).

There are still several unresolved questions regarding both dense granule formation and the sorting of GRA proteins into these organelles. In particular, whether dense granules are functionally analogous to constitutive *versus* regulated secretory vesicles is not fully established. In other eukaryotic cells, sorting of proteins to constitutive *versus* regulated secretory pathways occurs in the trans-Golgi network (TGN). Proteins destined for regulated secretion aggregate and are packaged into immature secretory granules (ISG). Clathrin coated vesicles bud from these ISGs and recycle proteins back to the TGN or endosomes, resulting in further concentration of the secretory proteins. The specific coat proteins, identified at the TGN, include both the adaptor proteins AP1 and AP3, and adaptor related proteins GGAs (for a review, see Arvan and Castle, 1998).

In *Toxoplasma*, immature dense granules have never been observed. Moreover, transient and stable expression of several soluble reporter proteins in *Toxoplasma* showed that any soluble protein, provided it possesses a signal peptide, is delivered to dense granules and later secreted into the PV. By contrast, addition of a GPI signal anchor targets the same reporter protein to the plasma membrane through transport vesicles (Karsten *et al.*, 1998). Conversely, the TgSAG1 protein for which the GPI signal anchor domain has been deleted is routed to the PV *via* the dense granules (Striepen *et al.*, 1998). Thus, dense granules constitute the default constitutive pathway for soluble proteins in *Toxoplasma.*

However, morphologically, dense granules are similar to the dense core granules involved in regulated secretion in mammalian cells, indicating that retention and condensation of secretory products may occur during dense granule formation. The prevailing model of sorting by retention in higher organisms is the selective aggregation of regulated, but not constitutive secretory proteins, which limits the ability of the former to escape from maturing granules during the process of constitutive vesicle budding (Arvan and Castle, 1998). This condensation may be due to an inherent property of the regulated secretory proteins to aggregate *via* subtle changes in the forming granule, while trafficking through the different compartments of the secretory pathway. These changes include mild acidification or an increasing concentration of bivalent cations such as calcium (Chanat and Huttner, 1991).

Most of the GRA proteins are not intrinsically soluble and are predicted as transmembrane proteins, a state which in turn occurs following their secretion into the PV (Lecordier *et al.*, 1999; Gendrin *et al.*, 2010). While transmembrane surface proteins are targeted to the parasite plasma membrane *via* an alternative vesicular route (Gendrin *et al.*, 2008), the TM domain-bearing GRA proteins are trafficked to the DG as solubilized aggregates (Sibley *et al.*, 1995; Lecordier *et al.*, 1999; Labruyere *et al.*, 1999) (Ruffiot and Cesbron-Delauw, unpublished data). These aggregates include high molecular weight complexes, in which GRA proteins interact together (Braun *et al.*, 2008). The mechanism by which these proteins remain excluded from the endomembranous system is still not fully understood. For the proteins GRA5 and GRA6, it is mediated by their N-terminal hydrophilic domain (Gendrin *et al.*, 2008, 2010). Whether this mechanism relies on properties of these highly charged N-terminal sequences to spontaneously aggregate, i.e. without any additional interaction of other co-factor, remains to be investigated. Since dense granules have never been reported to constitute an acidic compartment, protein condensation is unlikely to result from a substantial decrease in pH. An AP-1 orthologue was localized at the trans-most cisternae of the Golgi apparatus (Ngo *et al.*, 2003) and despite the fact that YXXΦ motifs were localized in the cytoplasmic tails of two GRA proteins (GRA4 and GRA7), these were not recognized by the *Toxoplasma* AP-1 (Ngo *et al.*, 2003). Given that GRA1, the most abundant dense granule product, is a soluble calcium binding protein (Cesbron-Delauw *et al.*, 1989), a role for Ca^{2+} in regulating aggregation remains a possibility.

12.6.3 Exocytosis of Dense Granules

The secretion of the dense granules has been difficult to capture: fusion of the dense granule membrane with the parasite plasma membrane (PPM) takes place sub-apically, at supposed gaps between the plates forming the inner membrane complex (Leriche and Dubremetz, 1990; Dubremetz *et al.*, 1993). Dense granule secretion appears to respond to signals associated with both constitutive and regulated pathways of secretion.

In favour of constitutive secretion, dense granule fusion with the target PPM is assisted by small GTPases of the Rab family and by soluble accessory factors (*N*-ethylmaleimide Soluble Factor (NSF), Soluble NSF-Associated Protein REceptor/Soluble NSF Associated Protein machinery (SNARE/SNAP) (Chaturvedi *et al.*, 1999). Also, an increase in intracellular Ca^{2+}, which usually triggers fusion of mammalian dense core granules with the plasmalemma, has no effect on dense granules exocytosis (Chaturvedi *et al.*, 1999; Liendo and Joiner, 2000) but does result in secretion of the micronemes (Carruthers *et al.*, 1999a).

In contrast, several features are consistent with regulated secretion including: (i) the burst of dense granule secretion into the PV occurring shortly after its formation (Dubremetz *et al.*, 1993), (ii) the fact that brefeldin A has no effect on the release of pre-stored GRAs (Coppens *et al.*, 1999) and (iii) that dense granule secretion was shown to be quantitatively and specifically induced by heat-inactivated serum (Coppens *et al.*, 1999). Two mechanisms driving dense granule secretion are thus hypothesized. Since two distinct populations of dense granules were never observed, the type of secretion might be related to the compaction stage of dense granules.

12.6.4 Post-Secretory Trafficking of GRAs

The involvement of GRAs in the maturation of the PV has only been examined in detail during tachyzoite development. Shortly after invasion, the burst of dense granule secretion into the PV coincides with the specific structural changes of the PV (see Section 3.4). At this stage, only GRA1, TgPIs and NTPases remain primarily in the lumen of the vacuole (Morris and Carruthers, 2003; Pszenny *et al.*, 2002; Sibley *et al.*, 1994, 1995). By immunoelectron microscopy, most of the GRAs are detected associated with the membranous system of the PV that includes the MNN, the HOST and the PVM; GRA2, GRA4, GRA6, GRA9 and GRA12 were found associated with the MNN (Adjogble *et al.*, 2004; Bonhomme *et al.*, 1998; Charif *et al.*, 1990; Dubremetz *et al.*, 1993; Labruyere *et al.*, 1999; Lecordier *et al.*, 1995; Michelin *et al.*, 2008; Sibley *et al.*, 1994). In these membranes, GRA2, GRA4 and GRA6 are components of a multimeric protein complex (Labruyere *et al.*, 1999). GRA3, GRA5, GRA7, GRA8, GRA10, GRA14 and GRA15 were preferentially detected at both the PVM and its membranous extensions within the host cell cytoplasm (Achbarou *et al.*, 1991b; Bonhomme *et al.*, 1998; Carey *et al.*, 2000; Dubremetz *et al.*, 1993; Lecordier *et al.*, 1993; Sinai *et al.*, 1997; Rome *et al.*, 2008; Rosowski *et al.*, 2011). Up to now, while GRA7 is the sole DG protein described as specifically associated to the HOSTs (Coppens *et al.*, 2006), GRA15 is the only one which is known to be associated with the evacuoles and then targeted to host cell nucleus (Rosowski *et al.*, 2011).

The PV-targeted GRAs exhibit various types of membrane association. While both GRA1 and NTPases exhibit a very loose association to the MNN (Sibley *et al.*, 1994, 1995), GRA4 is only displaced by urea treatment (Labruyere *et al.*, 1999), suggesting an association based mainly on hydrogen bonds. In contrast, GRA2, GRA3 and GRA5–GRA14 are only quantitatively displaced from their respective membranes by non-ionic detergents, indicating membrane spanning domains stabilized by hydrophobic interactions. For both GRA2 (Mercier *et al.*, 1998a) and GRA5 (Lecordier *et al.*, 1999), the respective putative membrane domains (the GRA2 amphipathic α-helices and the GRA5 transmembrane domain) were shown to be responsible for membrane association. Furthermore, topologically, the N-terminal domain of GRA5 is exposed to the host cell cytoplasm whereas its C-terminal faces the lumen of the PV (Lecordier *et al.*, 1999). This mechanism of post-translational membrane insertion is unconventional and is becoming elucidated. In comparison to what is observed within the DGs, both the compaction state and the interactor number of GRA complexes decreases following their secretion into the PV. This structural change is proposed to favour membrane association (Mercier and Cesbron-Delauw, unpublished; Braun *et al.*, 2008). Moreover, the targeting of either GRA6 to the MNN or GRA5 to the PVM is essentially mediated by their respective N-terminal hydrophilic sequence (Gendrin *et al.*, 2008, 2010) This led to the model that the selective targeting to PVM or MNN relies on specific protein–protein or protein–lipid interactions between the N-terminal domain of GRAs and specific components of these membranes (Cesbron-Delauw *et al.*, 2008). Even if the lipid composition of the different membranous compartments found within the PV remains unknown, their width differences (Magno *et al.*, 2005) could imply specificities in their lipid composition. One might thus suggest that specific interactions established between GRA proteins and a particular class of lipids could contribute to the specific protein targeting. Preliminary results, such as the specific interaction of GRA7 with phosphoinositides (Coppens *et al.*, 2006) and of the GRA6 N-terminal domain with negatively charged lipids (Gendrin *et al.*, 2010) are in agreement with this hypothesis.

12.6.5 Dense Granule Protein Function

At the tachyzoite stage, the burst of DG proteins secretion into the PV following host cell invasion and their selective targeting within the PV compartment suggest that they might contribute significantly to the structural organization of this new compartment and/or have important functions in the PV metabolism (Carruthers and Sibley, 1997; Cesbron-Delauw *et al.*, 1994; Dubremetz *et al.*, 1993).

12.6.5.1 GRAs

BLAST searches performed on GRA sequences did not reveal any significant similarity between the GRAs and other proteins in the databases. However, motif searches indicated some potential biochemical properties: two EF–Hands predicted in GRA1 (AA 149–180 and 197–223) suggested calcium-binding activity, which was subsequently demonstrated (Cesbron-Delauw *et al.*, 1989), an ATP/GTP binding site in GRA4 (AA 307–314), and an RGD adhesion motif in both GRA9 (AA 170–172) and GRA10 (AA 294–296) (Mercier *et al.*, 2005).

The construction of genetic knock-out (KO) parasites from the virulent RH background has provided clues to the function of several GRAs. The analysis of *GRA2*, *GRA6* and *GRA2–GRA6* RH KOs showed that these proteins contribute to the formation of the MNN (Mercier *et al.*, 2002). GRA2, *via* its amphipathic α-helices, induces tubulation of the vesicular material observed at the posterior end of the parasite shortly after invasion. GRA6 further stabilizes these pre-formed membranous tubules. Despite such a dramatic disruption of the MNN architecture, both the *GRA2* and *GRA6* KO parasites displayed normal *in vitro* growth rate (Mercier *et al.*, 1998b), indicating that the network organization is dispensable for parasite intracellular replication. *In vivo*, however, both the *GRA2* and *GRA6* KO parasites are less virulent than their parental RH strain, allowing mice to survive to acute infection and to develop brain cyst-like structures (Mercier *et al.*, 1998b; Mercier *et al.*, unpublished). Whether this avirulent phenotype is linked to the absence of intravacuolar tubular structures remains to be investigated.

The KOs of several PVM-associated GRAs are also available in the virulent RH background and their study has provided interesting findings in terms of host–parasite crosstalk. While the *GRA5* KO mutant did not exhibit any obvious phenotypic change in either *in vitro* parasite proliferation or virulence in mice (Mercier *et al.*, 2001), studies of the *GRA5* KO highlighted an unexpected role of GRA5 as an inducer of human CD34–dendritic cell migration towards CCL19 (Persat *et al.*, 2012). The *in vitro* phenotypic analysis of the *GRA7* KO mutant, in host cells maintained in culture medium depleted in lipids, revealed the disappearance of HOSTs from the PV associated with the alteration of parasite morphology and the decrease of parasite growth rate. In addition, recombinant GRA7 was demonstrated to induce liposome tabulation. These results led to the conclusion that GRA7 is the principal actor in the formation of PV HOSTs (Coppens *et al.*, 2006). Reduced *in vitro* growth rates under starvation conditions were also observed for both a cultured *GRA3KO* (Craver and Knoll, 2007) as well as a cultured *GRA14* KO (Rome *et al.*, 2008).

Up to now, despite numerous attempts, KO of *GRA1* has been unsuccessful, suggesting that it may be essential for tachyzoite intracellular survival (Braun *et al.*, unpublished). GRA1 is the most abundant EST (2% of the total EST derived from the RH strain, http://toxodb.org/News/Item-3.0-newESTdata-Mf.shtml) and its Ca^{2+} binding properties suggest an important role in the calcium homeostasis of the PV or in assisting the packaging of other GRA proteins as proposed. Further investigations of the function of these GRA proteins may require the construction of conditional KO mutants.

12.6.5.2 Other Dense Granule Proteins

Despite their homology with well-characterized proteins, the function of the other dense granule proteins is not clearly established. The abundant NTPases are essential and display apyrase activity (Asai *et al.*, 1995; Nakaar *et al.*, 1999). Hence, as *Toxoplasma* is auxotroph for purines, it was initially postulated that the presence of vacuolar NTPases, as well as of a 5′ nucleotidase would allow stepwise degradation of ATP into ADP, AMP and adenosine, with the latter being eventually transported across the parasite plasma membrane *via* a low affinity adenosine transporter (Stedman and Joiner, 1999). However, it was also demonstrated that *in vitro*, maximal activation of NTPases requires dithiols (Asai *et al.*, 1983; Bermudes *et al.*, 1994) and this enzyme is minimally active in the PV (Silverman *et al.*, 1998). Moreover, no 5′ nucleotidase activity was detected on the parasite surface suggesting that the parasite is incapable of converting AMP to adenosine for transport across the parasite plasma membrane (Ngo *et al.*, 2000). Thus, the initial hypothesis that NTPases are involved in purine salvage may be incorrect.

An alternative role for NTPases in parasite egress has been proposed. With the PVM becoming more permeable as parasites develop, intravacuolar activation of NTPases would occur just before parasite egress (Silverman *et al.*, 1998; Stommel *et al.*, 1997), so that the ATP pumped from the host cell could be degraded and provide the energy necessary for parasite motility. According to this model, and given their abundance, NTPase activity would be tightly regulated to avoid rapid depletion of the ATP vacuolar stock, which would trigger premature egress of the parasite from the vacuole (Silverman *et al.*, 1998).

Among the serine protein inhibitors detected within the PV, TgPI-1 is a broad-spectrum inhibitor that is capable of neutralizing trypsin, chymotrypsin and elastase *in vitro*, whereas TgPI-2 appears to be specific for trypsin (Morris and Carruthers, 2003; Morris *et al.*, 2002). The intravacuolar function of these inhibitors is unclear. Because both are active against digestive enzymes, TgPI-1 and TgPI-2, possibly released into the extracellular environment during the host cell lysis, may protect the parasite from proteases as it traverses the gastrointestinal tract (Morris *et al.*, 2002).

Cyclophilins (CyPs) are highly conserved proteins associated with an *in vitro* peptidyl–prolyl *cis/trans* isomerase (PPIase or rotamase) activity and are implicated more broadly in mediating protein–protein interactions within large protein complexes. As such, TgCyP-18 could be involved in regulating the assembly of protein complexes within the dense granules and/or the PV. In particular, GRA4 and/or GRA8 which are rich in proline residues, could be potential substrates (Carey *et al.*, 2000).

12.6.6 Stage Specific Expression of Dense Granule Proteins

Based on the few studies available, bradyzoites and sporozoites express nearly the full tachyzoite repertoire of dense granule proteins (Mercier *et al.*, 2005; Ferguson, 2004). This is consistent with the 'tachyzoite fate' of both bradyzoites and sporozoites into the host. The dense granule repertoire would be required to adapt the PV for optimal tachyzoite development and facilitate parasite proliferation in many cell types. In contrast, merozoites which undergo limited proliferation and rapidly differentiate into the sexual stages in the enterocytes of the cat small intestine, express a very limited repertoire of dense granule proteins (Ferguson, 2004).

12.6.6.1 Bradyzoite/Tissue Cyst

Bradyzoites are quiescent parasites formed in intracellular tissue cysts found within muscle cells and within cells of the central nervous system, predominantly neurons. The bradyzoite PV is limited by a unit membrane with numerous shallow invaginations (see Chapter 15). An underlying layer of moderately electron dense fine granular material contributes to the wall of the tissue cyst (Ferguson *et al.*, 1989). At the bradyzoite stage, the dense granules contain all the GRAs identified in the tachyzoite (GRA1–8) although there is evidence for reduced expression of NTPase (Ferguson *et al.*, 1999a,b; Nakaar *et al.*, 1998). Furthermore, GRA4, GRA8 and NTPases were not detected within the cyst wall, which may be a consequence of their degradation or modification during encystment. All the other GRAs were shown to be present in the cyst wall, with a location reminiscent of that observed within the tachyzoite PV (Ferguson, 2004; Torpier *et al.*, 1993). Evidence for dense granule proteins trafficking beyond the cyst wall has only been reported for GRA7 (Ferguson *et al.*, 1999a; Fischer *et al.*, 1998). The importance of the GRA proteins for the establishment of cyst burden has been demonstrated in the case of both GRA4 and GRA6. The isolation of a KU80 KO mutant (deficient in non-homologous end joining) in an avirulent, type II cyst-forming strain allowed the subsequent deletion of the *GRA4* and/or the *GRA6* gene(s). Noticeably, while no significant difference was observed in the *in vitro* growth rate between the mutants and the parental strain, dramatic reductions in cyst burdens (reduction by 91% in the case of single deletions and by 99% in the case of the double deletion) were observed three weeks post-infection in mice (Fox *et al.*, 2011). Whether the lack in cyst burden of the mutants relies on a default in cyst formation (potential structural role of GRA4 and/or GRA6) or on an interference with the host response, which would dramatically prevent parasite dissemination, remains to be investigated.

12.6.6.2 Merozoite

During the coccidian development of *T. gondii* in the enterocytes of the cat, the ultrastructural appearance of the PV is very different from that observed for the tachyzoite PV (Ferguson, 2004). The parasites are located in a tightly fitting PV limited by a thickened membrane with a laminated appearance consisting of three closely applied unit membranes. The PV lacks the MNN and there is no association of the host cell rough ER or mitochondria (Ferguson, 2004). At this stage, only two dense granule proteins, GRA7 and TgNTPase, are detected within the dense granules. Both are released into the PV shortly after invasion but their level of staining drops as the parasites mature (Ferguson *et al.*, 1999a,b).

12.6.6.3 Sporozoite

The dense granules of the sporozoite appear to contain at least GRA1–8 with the exception of TgGRA3 and TgNTPase (Tilley *et al.*, 1997). *In vitro*, sporozoites entering into a host cell form an unusual large vacuole (PV1) devoid of dense granule proteins (Tilley *et al.*, 1997) but leave this compartment to enter a new vacuole (PV2) that more closely is similar to a tachyzoite PV. Formation of PV2 is correlated with the secretion of the tachyzoite repertoire of dense granule proteins (Tilley *et al.*, 1997). In fact, GRA3, GRA5 and NTPases are the first to be detected, followed by GRA1, GRA2, GRA4 and GRA6 (Tilley *et al.*, 1997).

12.7 CONCLUSIONS

Toxoplasma pathogenesis is intimately associated with the parasite ability to invade host cells, an active process that has no counterpart outside the Apicomplexa phylum. The mechanisms involved are likely to be similar throughout the phylum, as first exemplified by the conservation of a family of microneme proteins containing thrombospondin-related domains first identified in *Plasmodium* spp. The development of modern genetic analysis in *T. gondii* (complete genetic map, efficient genetic transformation and knock-out or knock-down of essential genes) has greatly facilitated improving our knowledge on the invasion process of *T. gondii* and has helped to identify the key molecular actors of the invasion machinery of *Plasmodium*, a parasite that is less amenable to genetic manipulation. Molecules conserved across the phylum have been recently extended to the motor complex, proteases of the rhomboid family, and proteins translocated at the moving junction. This has highlighted a number of novel putative targets for therapeutic intervention to treat major apicomplexan diseases of humans and livestock.

Acknowledgements

We are grateful to Jean-François Dubremetz for critical review of the manuscript and to David Elliott for providing the electron micrograph shown in Figure 12.1.

References

Achbarou, A., Mercereau-Puijalon, O., Autheman, J.M., Fortier, B., Camus, D., Dubremetz, J.F., 1991a. Characterization of microneme proteins of *Toxoplasma gondii*. Mol. Biochem. Parasitol. 47, 223–233.

Achbarou, A., Mercereau-Puijalon, O., Sadak, A., Fortier, B., Leriche, M.A., Camus, D., Dubremetz, J.F., 1991b. Differential targeting of dense granule proteins in the parasitophorous vacuole of *Toxoplasma gondii*. Parasitology 3, 321–329.

Adjogble, K.D., Mercier, C., Dubremetz, J.F., Hucke, C., Mackenzie, C.R., Cesbron-Delauw, M.F., Daubener, W., 2004. GRA9, a new *Toxoplasma gondii* dense granule protein associated with the intravacuolar network of tubular membranes. Int. J. Parasitol. 34, 1255–1264.

Agop-Nersesian, C., Naissant, B., Ben Rached, F., Rauch, M., Kretzschmar, A., Thiberge, S., Menard, R., Ferguson, D.J., Meissner, M., Langsley, G., 2009. Rab11A-controlled assembly of the inner membrane complex is required for completion of apicomplexan cytokinesis. PLoS Pathog. 5, e1000270.

Aikawa, M., Miller, L.H., Johnson, J., Rabbege, J., 1978. Erythrocyte entry by malarial parasites. A moving junction between erythrocyte and parasite. J. Cell Biol. 77, 72–82.

Aikawa, M., Miller, L.H., Rabbege, J.R., Epstein, N., 1981. Freeze-fracture study on the erythrocyte membrane during malarial parasite invasion. J. Cell Biol. 91, 55–62.

Ajioka, J.W., Boothroyd, J.C., Brunk, B.P., Hehl, A., Hillier, L., Manger, I.D., Marra, M., Overton, G.C., Roos, D.S., Wan, K.L., Waterston, R., Sibley, L.D., 1998. Gene discovery by EST sequencing in *Toxoplasma gondii* reveals sequences restricted to the Apicomplexa. Genome Res. 8, 18–28.

Akhouri, R.R., Bhattacharyya, A., Pattnaik, P., Malhotra, P., Sharma, A., 2004. Structural and functional dissection of the adhesive domains of *Plasmodium falciparum* thrombospondin-related anonymous protein (TRAP). Biochem. J. 379, 815–822.

Alexander, D.L., Kapur, S.A., Dubremetz, J.F., Boothroyd, J.C., 2006. *Plasmodium falciparum* AMA1 (PfAMA1) binds a rhoptry neck protein homologous to TgRON4, a component of the moving junction in *Toxoplasma*. Eukaryot. Cell 5, 1169–1173.

Alexander, D.L., Mital, J., Ward, G.E., Bradley, P., Boothroyd, J.C., 2005. Identification of the Moving Junction Complex of *Toxoplasma gondii*: A Collaboration between Distinct Secretory Organelles. PLoS Pathog. 1, e17.

Antoine, J.C., Prina, E., Lang, T., Courret, N., 1998. The biogenesis and properties of the parasitophorous vacuoles that harbour *Leishmania* in murine macrophages. Trends Microbiol. 6, 392–401.

Appella, E., Weber, I.T., Blasi, F., 1988. Structure and function of epidermal growth factor-like regions in proteins. FEBS Lett. 231, 1–4.

Arvan, P., Castle, D., 1998. Sorting and storage during secretory granule biogenesis: looking backward and looking forward. Biochem. J. 332, 593–610.

Asai, T., Howe, D.K., Nakajima, K., Nozaki, T., Takeuchi, T., Sibley, L.D., 1998. Neospora caninum: tachyzoites express a potent type-I nucleoside triphosphate hydrolase. Exp. Parasitol. 90, 277–285.

Asai, T., Miura, S., Sibley, L.D., Okabayashi, H., Takeuchi, T., 1995. Biochemical and molecular characterization of nucleoside triphosphate hydrolase isozymes from the parasitic protozoan *Toxoplasma gondii*. J. Biol. Chem. 270, 11391–11397.

Asai, T., O'Sullivan, W.J., Tatibana, M., 1983. A potent nucleoside triphosphate hydrolase from the parasitic protozoan *Toxoplasma gondii*. Purification, some properties, and activation by thiol compounds. J. Biol. Chem. 258, 6816–6822.

Bai, T., Becker, M., Gupta, A., Strike, P., Murphy, V.J., Anders, R.F., Batchelor, A.H., 2005. Structure of AMA1 from *Plasmodium falciparum* reveals a clustering of polymorphisms that surround a conserved hydrophobic pocket. Proc. Natl. Acad. Sci. U. S. A. 102, 12736–12741.

Baldi, D.L., Andrews, K.T., Waller, R.F., Roos, D.S., Howard, R.F., Crabb, B.S., Cowman, A.F., 2000. RAP1 controls rhoptry targeting of RAP2 in the malaria parasite *Plasmodium falciparum*. Embo. J. 19, 2435–2443.

Bannister, L.H., Hopkins, J.M., Dluzewski, A.R., Margos, G., Williams, I.T., Blackman, M.J., Kocken, C.H., Thomas, A.W., Mitchell, G.H., 2003. *Plasmodium falciparum* apical membrane antigen 1 (PfAMA-1) is translocated within micronemes along subpellicular microtubules during merozoite development. J. Cell Sci. 116, 3825–3834.

Bannister, L.H., Mitchell, G.H., 1989. The fine structure of secretion by *Plasmodium* knowlesi merozoites during red cell invasion. J. Protozool. 36, 362–367.

Bannister, L.H., Mitchell, G.H., Butcher, G.A., Dennis, E.D., 1986. Lamellar membranes associated with rhoptries in erythrocytic merozoites of *Plasmodium knowlesi*, a clue to the mechanism of invasion. Parasitology 92 (Pt 2), 291–303.

Barondes, S.H., Castronovo, V., Cooper, D.N., Cummings, R.D., Drickamer, K., Feizi, T., Gitt, M.A., Hirabayashi, J., Hughes, C., Kasai, K., et al., 1994. Galectins: a family of animal beta-galactoside-binding lectins. Cell 76, 597–598.

Barragan, A., Brossier, F., Sibley, L.D., 2005. Transepithelial migration of *Toxoplasma gondii* involves an interaction of intercellular adhesion molecule 1 (ICAM-1) with the parasite adhesin MIC2. Cell Microbiol. 7, 561–568.

Beckers, C.J., Dubremetz, J.F., Mercereau-Puijalon, O., Joiner, K.A., 1994. The *Toxoplasma gondii* rhoptry protein ROP 2 is inserted into the parasitophorous vacuole membrane, surrounding the intracellular parasite, and is exposed to the host cell cytoplasm. J. Cell Biol. 127, 947–961.

Behnke, M.S., Fentress, S.J., Mashayekhi, M., Li, L.X., Taylor, G.A., Sibley, L.D., 2012. The Polymorphic Pseudokinase ROP5 Controls Virulence in *Toxoplasma gondii* by Regulating the Active Kinase ROP18. PLoS Pathog. 8, e1002992.

Behnke, M.S., Khan, A., Wootton, J.C., Dubey, J.P., Tang, K., Sibley, L.D., 2011. Virulence differences in *Toxoplasma* mediated by amplification of a family of polymorphic pseudokinases. Proc. Natl. Acad. Sci. U. S. A. 108, 9631–9636.

Bermudes, D., Peck, K.R., Afifi, M.A., Beckers, C.J., Joiner, K.A., 1994. Tandemly repeated genes encode nucleoside triphosphate hydrolase isoforms secreted into the parasitophorous vacuole of *Toxoplasma gondii*. J. Biol. Chem. 269, 29252–29260.

Besteiro, S., Bertrand-Michel, J., Lebrun, M., Vial, H., Dubremetz, J.F., 2008. Lipidomic analysis of *Toxoplasma gondii* tachyzoites rhoptries: further insights into the role of cholesterol. Biochem. J 415, 87–96.

Besteiro, S., Dubremetz, J.F., Lebrun, M., 2011. The moving junction of apicomplexan parasites: a key structure for invasion. Cell Microbiol. 13, 797–805.

Besteiro, S., Michelin, A., Poncet, J., Dubremetz, J.F., Lebrun, M., 2009. Export of a *Toxoplasma gondii* rhoptry neck protein complex at the host cell membrane to form the moving junction during invasion. PLoS Pathog. 5, e1000309.

Billker, O., Lourido, S., Sibley, L.D., 2009. Calcium-dependent signaling and kinases in apicomplexan parasites. Cell Host Microbe 5, 612–622.

Binder, E.M., Kim, K., 2004. Location, location, location: trafficking and function of secreted proteases of *Toxoplasma* and Plasmodium. Traffic 5, 914–924.

Binder, E.M., Lagal, V., Kim, K., 2008. The prodomain of *Toxoplasma gondii* GPI-anchored subtilase TgSUB1 mediates its targeting to micronemes. Traffic 9, 1485–1496.

Blackman, M.J., Bannister, L.H., 2001. Apical organelles of Apicomplexa: biology and isolation by subcellular fractionation. Mol. Biochem. Parasitol. 117, 11–25.

Blader, I.J., Manger, I.D., Boothroyd, J.C., 2001. Microarray analysis reveals previously unknown changes in *Toxoplasma gondii*-infected human cells. J. Biol. Chem. 276, 24223–24231.

Blumenschein, T.M., Friedrich, N., Childs, R.A., Saouros, S., Carpenter, E.P., Campanero-Rhodes, M.A., Simpson, P., Chai, W., Koutroukides, T., Blackman, M.J., Feizi, T., Soldati-Favre, D., Matthews, S., 2007. Atomic resolution insight into host cell recognition by *Toxoplasma gondii*. EMBO J. 26, 2808–2820.

Bonhomme, A., Maine, G.T., Beorchia, A., Burlet, H., Aubert, D., Villena, I., Hunt, J., Chovan, L., Howard, L., Brojanac, S., Sheu, M., Tyner, J., Pluot, M., Pinon, J.M., 1998. Quantitative immunolocalization of a P29 protein (GRA7), a new antigen of *Toxoplasma gondii*. J. Histochem. Cytochem. 46, 1411–1422.

Bradley, P.J., Boothroyd, J.C., 1999. Identification of the pro-mature processing site of *Toxoplasma* ROP1 by mass spectrometry. Mol. Biochem. Parasitol. 100, 103–109.

Bradley, P.J., Boothroyd, J.C., 2001. The pro region of *Toxoplasma* ROP1 is a rhoptry-targeting signal. Int. J. Parasitol. 31, 1177–1186.

Bradley, P.J., Hsieh, C.L., Boothroyd, J.C., 2002. Unprocessed *Toxoplasma* ROP1 is effectively targeted and secreted into the nascent parasitophorous vacuole. Mol. Biochem. Parasitol. 125, 189–193.

Bradley, P.J., Li, N., Boothroyd, J.C., 2004. A GFP-based motif-trap reveals a novel mechanism of targeting for the *Toxoplasma* ROP4 protein. Mol. Biochem. Parasitol. 137, 111–120.

Bradley, P.J., Ward, C., Cheng, S.J., Alexander, D.L., Coller, S., Coombs, G.H., Dunn, J.D., Ferguson, D.J., Sanderson, S.J., Wastling, J.M., Boothroyd, J.C., 2005. Proteomic analysis of rhoptry organelles reveals many novel constituents for host-parasite interactions in *Toxoplasma gondii*. J. Biol. Chem. 280, 34245–34258.

Braun, L., Travier, L., Kieffer, S., Musset, K., Garin, J., Mercier, C., Cesbron-Delauw, M.F., 2008. Purification of *Toxoplasma* dense granule proteins reveals that they are in complexes throughout the secretory pathway. Mol. Biochem. Parasitol. 157, 13–21.

Brecht, S., Carruthers, V.B., Ferguson, D.J., Giddings, O.K., Wang, G., Jakle, U., Harper, J.M., Sibley, L.D., Soldati, D., 2001. The *Toxoplasma* micronemal protein MIC4 is an adhesin composed of six conserved apple domains. J. Biol. Chem. 276, 4119–4127.

Breinich, M.S., Ferguson, D.J., Foth, B.J., van Dooren, G.G., Lebrun, M., Quon, D.V., Striepen, B., Bradley, P.J., Frischknecht, F., Carruthers, V.B., Meissner, M., 2009. A dynamin is required for the biogenesis of secretory organelles in *Toxoplasma gondii*. Curr. Biol. 19, 277–286.

Brossier, F., Jewett, T.J., Lovett, J.L., Sibley, L.D., 2003. C-terminal processing of the *Toxoplasma* protein MIC2 is essential for invasion into host cells. J. Biol. Chem. 278, 6229–6234.

Brossier, F., Jewett, T.J., Sibley, L.D., Urban, S., 2005. A spatially localized rhomboid protease cleaves cell surface adhesins essential for invasion by *Toxoplasma*. Proc. Natl. Acad. Sci. U. S. A. 102, 4146–4151.

Brossier, F., Starnes, G.L., Beatty, W.L., Sibley, L.D., 2008. Microneme rhomboid protease TgROM1 is required for efficient intracellular growth of *Toxoplasma gondii*. Eukaryot. Cell 7, 664–674.

Brown, P.J., Billington, K.J., Bumste ad, J.M., Clark, J.D., Tomley, F.M., 2000. A microneme protein from *Eimeriatenella* with homology to the Apple domains of coagulation factor XI and plasma pre-kallikrein. Mol. Biochem. Parasitol. 107, 91–102.

Brown, P.J., Gill, A.C., Nugent, P.G., McVey, J.H., Tomley, F.M., 2001. Domains of invasion organelle proteins from apicomplexan parasites are homologous with the Apple domains of blood coagulation factor XI and plasma pre-kallikrein and are members of the PAN module superfamily. FEBS Lett. 497, 31–38.

Brydges, S.D., Harper, J.M., Parussini, F., Coppens, I., Carruthers, V.B., 2008. A transient forward-targeting element for microneme-regulated secretion in *Toxoplasma gondii*. Biol. Cell 100, 253–264.

Brydges, S.D., Sherman, G.D., Nockemann, S., Loyens, A., Daubener, W., Dubremetz, J.F., Carruthers, V.B., 2000. Molecular characterization of TgMIC5, a proteolytically processed antigen secreted from the micronemes of *Toxoplasma gondii*. Mol. Biochem. Parasitol. 111, 51–66.

Brydges, S.D., Zhou, X.W., Huynh, M.H., Harper, J.M., Mital, J., Adjogble, K.D., Daubener, W., Ward, G.E., Carruthers, V.B., 2006. Targeted deletion of MIC5 enhances trimming proteolysis of *Toxoplasma* invasion proteins. Eukaryot. Cell 5, 2174–2183.

Buchholz, K.R., Fritz, H.M., Chen, X., Durbin-Johnson, B., Rocke, D.M., Ferguson, D.J., Conrad, P.A., Boothroyd, J.C., 2011. Identification of tissue cyst wall components by transcriptome analysis of in vivo and in vitro *Toxoplasma gondii* bradyzoites. Eukaryot. Cell 10, 1637–1647.

Buguliskis, J.S., Brossier, F., Shuman, J., Sibley, L.D., 2010. Rhomboid 4 (ROM4) affects the processing of surface adhesins and facilitates host cell invasion by *Toxoplasma gondii*. PLoS Pathog. 6, e1000858.

Butcher, B.A., Fox, B.A., Rommereim, L.M., Kim, S.G., Maurer, K.J., Yarovinsky, F., Herbert, D.R., Bzik, D.J., Denkers, E.Y., 2011. *Toxoplasma gondii* rhoptry kinase ROP16 activates STAT3 and STAT6 resulting in cytokine inhibition and arginase-1-dependent growth control. PLoS Pathog. 7, e1002236.

Cabrera, A., Herrmann, S., Warszta, D., Santos, J.M., John Peter, A.T., Kono, M., Debrouver, S., Jacobs, T., Spielmann, T., Ungermann, C., Soldati-Favre, D., Gilberger, T.W., 2012. Dissection of minimal sequence requirements for rhoptry membrane targeting in the malaria parasite. Traffic 13, 1335–1350.

Caldas, L.A., Attias, M., de Souza, W., 2009. Dynamin inhibitor impairs *Toxoplasma gondii* invasion. FEMS Microbiol. Lett. 301, 103–108.

Carey, K.L., Donahue, C.G., Ward, G.E., 2000. Identification and molecular characterization of GRA8, a novel, proline-rich, dense granule protein of *Toxoplasma gondii*. Mol. Biochem. Parasitol. 105, 25–37.

Carey, K.L., Jongco, A.M., Kim, K., Ward, G.E., 2004a. The *Toxoplasma gondii* rhoptry protein ROP4 is secreted into the parasitophorous vacuole and becomes phosphorylated in infected cells. Eukaryot. Cell 3, 1320–1330.

Carey, K.L., Westwood, N.J., Mitchison, T.J., Ward, G.E., 2004b. A small-molecule approach to studying invasive mechanisms of *Toxoplasma gondii*. Proc. Natl. Acad. Sci. U. S. A. 101, 7433–7438.

Carruthers, V.B., Giddings, O.K., Sibley, L.D., 1999a. Secretion of micronemal proteins is associated with *Toxoplasma* invasion of host cells. Cell Microbiol. 1, 225–235.

Carruthers, V.B., Moreno, S.N., Sibley, L.D., 1999b. Ethanol and acetaldehyde elevate intracellular [Ca2+] and stimulate microneme discharge in *Toxoplasma gondii*. Biochem. J. 342 (Pt 2), 379–386.

Carruthers, V.B., Sherman, G.D., Sibley, L.D., 2000. The *Toxoplasma* adhesive protein MIC2 is proteolytically processed at multiple sites by two parasite-derived proteases. J. Biol. Chem. 275, 14346–14353.

Carruthers, V.B., Sibley, L.D., 1997. Sequential protein secretion from three distinct organelles of *Toxoplasma gondii* accompanies invasion of human fibroblasts. Eur. J. Cell Biol. 73, 114–123.

Carruthers, V.B., Sibley, L.D., 1999. Mobilization of intracellular calcium stimulates microneme discharge in *Toxoplasma gondii*. Mol. Microbiol. 31, 421–428.

Cerede, O., Dubremetz, J.F., Bout, D., Lebrun, M., 2002. The *Toxoplasma gondii* protein MIC3 requires pro-peptide cleavage and dimerization to function as adhesin. Embo. J. 21, 2526–2536.

Cerede, O., Dubremetz, J.F., Soete, M., Deslee, D., Vial, H., Bout, D., Lebrun, M., 2005. Synergistic role of micronemal proteins in *Toxoplasma gondii* virulence. J. Exp. Med. 201, 453–463.

Cesbron-Delauw, M.F., Gendrin, C., Travier, L., Ruffiot, P., Mercier, C., 2008. Apicomplexa in mammalian cells: trafficking to the parasitophorous vacuole. Traffic 9, 657–664.

Cesbron-Delauw, M.F., Guy, B., Torpier, G., Pierce, R.J., Lenzen, G., Cesbron, J.Y., Charif, H., Lepage, P., Darcy, F., Lecocq, J.P., et al., 1989. Molecular characterization of a 23-kilodalton major antigen secreted by *Toxoplasma gondii*. Proc. Natl. Acad. Sci. U. S. A. 86, 7537–7541.

Cesbron-Delauw, M.F., Tomavo, S., Beauchamps, P., Fourmaux, M.P., Camus, D., Capron, A., Dubremetz, J.F., 1994. Similarities between the primary structures of two distinct major surface proteins of *Toxoplasma gondii*. J. Biol. Chem. 269, 16217–16222.

Chanat, E., Huttner, W.B., 1991. Milieu-induced, selective aggregation of regulated secretory proteins in the trans-Golgi network. J. Cell Biol. 115, 1505–1519.

Charif, H., Darcy, F., Torpier, G., Cesbron-Delauw, M.F., Capron, A., 1990. *Toxoplasma gondii*: characterization and localization of antigens secreted from tachyzoites. Exp. Parasitol. 71, 114–124.

Chattopadhyay, R., Rathore, D., Fujioka, H., Kumar, S., de la Vega, P., Haynes, D., Moch, K., Fryauff, D., Wang, R., Carucci, D.J., Hoffman, S.L., 2003. PfSPATR, a *Plasmodium falciparum* protein containing an altered thrombospondin type I repeat domain is expressed at several stages of the parasite life cycle and is the target of inhibitory antibodies. J. Biol. Chem. 278, 25977–25981.

Chaturvedi, S., Qi, H., Coleman, D., Rodriguez, A., Hanson, P.I., Striepen, B., Roos, D.S., Joiner, K.A., 1999. Constitutive calcium-independent release of *Toxoplasma gondii* dense granules occurs through the NSF/SNAP/SNARE/Rab machinery. J. Biol. Chem. 274, 2424–2431.

Chen, H., Herndon, M.E., Lawler, J., 2000. The cell biology of thrombospondin-1. Matrix Biol. 19, 597–614.

Chen, W., Feng, Y., Chen, D., Wandinger-Ness, A., 1998. Rab11 is required for trans-golgi network-to-plasma membrane transport and a preferential target for GDP dissociation inhibitor. Mol. Biol. Cell 9, 3241–3257.

Chen, Z., Harb, O.S., Roos, D.S., 2008. In silico identification of specialized secretory-organelle proteins in apicomplexan parasites and in vivo validation in *Toxoplasma gondii*. PLoS One 3, e3611.

Clarke, L.E., Tomley, F.M., Wisher, M.H., Foulds, I.J., Boursnell, M.E., 1990. Regions of an *Eimeria tenella* antigen contain sequences which are conserved in circumsporozoite proteins from *Plasmodium* spp. and which are related to the thrombospondin gene family. Mol. Biochem. Parasitol. 41, 269–279.

Cleary, M.D., Singh, U., Blader, I.J., Brewer, J.L., Boothroyd, J.C., 2002. *Toxoplasma gondii* asexual development: identification of developmentally regulated genes and distinct patterns of gene expression. Eukaryot. Cell 1, 329–340.

Coley, A.M., Gupta, A., Murphy, V.J., Bai, T., Kim, H., Foley, M., Anders, R.F., Batchelor, A.H., 2007. Structure of the malaria antigen AMA1 in complex with a growth-inhibitory antibody. PLoS Pathog. 3, 1308–1319.

Coley, A.M., Parisi, K., Masciantonio, R., Hoeck, J., Casey, J.L., Murphy, V.J., Harris, K.S., Batchelor, A.H., Anders, R.F., Foley, M., 2006. The most polymorphic residue on *Plasmodium falciparum* apical membrane antigen 1 determines binding of an invasion-inhibitory antibody. Infect. Immun. 74, 2628–2636.

Collins, C.R., Withers-Martinez, C., Hackett, F., Blackman, M.J., 2009. An inhibitory antibody blocks interactions between components of the malarial invasion machinery. PLoS Pathog. 5, e1000273.

Conseil, V., Soete, M., Dubremetz, J.F., 1999. Serine protease inhibitors block invasion of host cells by Toxoplasma gondii. Antimicrob. Agents Chemother. 43, 1358–1361.

Coppens, I., Andries, M., Liu, J.L., Cesbron-Delauw, M.F., 1999. Intracellular trafficking of dense granule proteins in *Toxoplasma gondii* and experimental evidences for a regulated exocytosis. Eur. J. Cell Biol. 78, 463–472.

Coppens, I., Dunn, J.D., Romano, J.D., Pypaert, M., Zhang, H., Boothroyd, J.C., Joiner, K.A., 2006. *Toxoplasma gondii* sequesters lysosomes from mammalian hosts in the vacuolar space. Cell 125, 261–274.

Coppens, I., Joiner, K.A., 2003. Host but not parasite cholesterol controls *Toxoplasma* cell entry by modulating organelle discharge. Mol. Biol. Cell 14, 3804–3820.

Craver, M.P., Knoll, L.J., 2007. Increased efficiency of homologous recombination in *Toxoplasma gondii* dense granule protein 3 demonstrates that GRA3 is not necessary in cell culture but does contribute to virulence. Mol. Biochem. Parasitol. 153, 149–157.

Crawford, J., Tonkin, M.L., Grujic, O., Boulanger, M.J., 2010. Structural characterization of apical membrane antigen 1 (AMA1) from *Toxoplasma gondii*. J. Biol. Chem. 285, 15644–15652.

Daher, W., Soldati-Favre, D., 2009. Mechanisms controlling glideosome function in apicomplexans. Curr. Opin. Microbiol. 12, 408–414.

Davis, C.G., 1990. The many faces of epidermal growth factor repeats. New Biol. 2, 410–419.

Delorme-Walker, V., Abrivard, M., Lagal, V., Anderson, K., Perazzi, A., Gonzalez, V., Page, C., Chauvet, J., Ochoa, W., Volkmann, N., Hanein, D., Tardieux, I., 2012. Toxofilin upregulates the host cortical actin cytoskeleton dynamics, facilitating *Toxoplasma* invasion. J. Cell Sci. 125, 4333–4342.

Delorme, V., Cayla, X., Faure, G., Garcia, A., Tardieux, I., 2003. Actin dynamics is controlled by a casein kinase II and phosphatase 2C interplay on *Toxoplasma gondii* Toxofilin. Mol. Biol. Cell 14, 1900–1912.

Denkers, E.Y., Bzik, D.J., Fox, B.A., Butcher, B.A., 2012. An inside job: hacking into Janus kinase/signal transducer and activator of transcription signaling cascades by the intracellular protozoan *Toxoplasma gondii*. Infect. Immun. 80, 476–482.

Di Cristina, M., Spaccapelo, R., Soldati, D., Bistoni, F., Crisanti, A., 2000. Two conserved amino acid motifs mediate protein targeting to the micronemes of the apicomplexan parasite *Toxoplasma gondii*. Mol. Cell Biol. 20, 7332–7341.

Dickeson, S.K., Walsh, J.J., Santoro, S.A., 1998. Binding of the alpha 2 integrin I domain to extracellular matrix ligands: structural and mechanistic differences between collagen and laminin binding. Cell Adhes. Commun. 5, 273–281.

Dobrowolski, J.M., Carruthers, V.B., Sibley, L.D., 1997. Participation of myosin in gliding motility and host cell invasion by *Toxoplasma gondii*. Mol. Microbiol. 26, 163–173.

Donahue, C.G., Carruthers, V.B., Gilk, S.D., Ward, G.E., 2000. The *Toxoplasma* homologue of *Plasmodium* apical membrane antigen-1 (AMA-1) is a microneme protein secreted in response to elevated intracellular calcium levels. Mol. Biochem. Parasitol. 111, 15–30.

Donald, R.G., Liberator, P.A., 2002. Molecular characterization of a coccidian parasite cGMP dependent protein kinase. Mol. Biochem. Parasitol. 120, 165–175.

Dou, Z., Coppens, I., Carruthers, V.B., 2012. Non-canonical maturation of two papain-family proteases in *Toxoplasma gondii*. J. Biol. Chem 288, 3523–3534.

Dowse, T.J., Pascall, J.C., Brown, K.D., Soldati, D., 2005. Apicomplexan rhomboids have a potential role in microneme protein cleavage during host cell invasion. Int. J. Parasitol. 35, 747–756.

Dowse, T.J., Soldati, D., 2005. Rhomboid-like proteins in Apicomplexa: phylogeny and nomenclature. Trends Parasitol. 21, 254–258.

Dubremetz, J.F., 2007. Rhoptries are major players in *Toxoplasma gondii* invasion and host cell interaction. Cell Microbiol. 9, 841–848.

Dubremetz, J.F., Achbarou, A., Bermudes, D., Joiner, K.A., 1993. Kinetics and pattern of organelle exocytosis during *Toxoplasma gondii*/host cell interaction. Parasitol. Res. 79, 402–408.

Dubremetz, J.F., Ferreira, E., 1978. Capping of cationized ferritin by coccidian zoites. J. Protozool. 25, 9B.

El Hajj, H., Demey, E., Poncet, J., Lebrun, M., Wu, B., Galeotti, N., Fourmaux, M.N., Mercereau-Puijalon, O., Vial, H., Labesse, G., Dubremetz, J.F., 2006. The ROP2 family of *Toxoplasma gondii* rhoptry proteins: proteomic and genomic characterization and molecular modeling. Proteomics 6, 5773–5784.

El Hajj, H., Lebrun, M., Arold, S.T., Vial, H., Labesse, G., Dubremetz, J.F., 2007a. ROP18 is a rhoptry kinase controlling the intracellular proliferation of *Toxoplasma gondii*. PLoS Pathog. 3, e14.

El Hajj, H., Lebrun, M., Fourmaux, M.N., Vial, H., Dubremetz, J.F., 2007b. Inverted topology of the *Toxoplasma gondii* ROP5 rhoptry protein provides new insights into the association of the ROP2 protein family with the parasitophorous vacuole membrane. Cell Microbiol. 9, 54–64.

El Hajj, H., Papoin, J., Cerede, O., Garcia-Reguet, N., Soete, M., Dubremetz, J.F., Lebrun, M., 2008. Molecular signals in the trafficking of *Toxoplasma gondii* protein MIC3 to the micronemes. Eukaryot. Cell 7, 1019–1028.

Farrell, A., Thirugnanam, S., Lorestani, A., Dvorin, J.D., Eidell, K.P., Ferguson, D.J., Anderson-White, B.R., Duraisingh, M.T., Marth, G.T., Gubbels, M.J., 2012. A DOC2 protein identified by mutational profiling is essential for apicomplexan parasite exocytosis. Science 335, 218–221.

Fentress, S.J., Behnke, M.S., Dunay, I.R., Mashayekhi, M., Rommereim, L.M., Fox, B.A., Bzik, D.J., Taylor, G.A., Turk, B.E., Lichti, C.F., Townsend, R.R., Qiu, W., Hui, R., Beatty, W.L., Sibley, L.D., 2010. Phosphorylation of immunity-related GTPases by a *Toxoplasma gondii*-secreted kinase promotes macrophage survival and virulence. Cell Host Microbe 8, 484–495.

Fentress, S.J., Sibley, L.D., 2011. The secreted kinase ROP18 defends *Toxoplasma's* border. Bioessays 33, 693–700.

Ferguson, D.J., 2004. Use of molecular and ultrastructural markers to evaluate stage conversion of *Toxoplasma gondii* in both the intermediate and definitive host. Int. J. Parasitol. 34, 347–360.

Ferguson, D.J., Cesbron-Delauw, M.F., Dubremetz, J.F., Sibley, L.D., Joiner, K.A., Wright, S., 1999a. The expression and distribution of dense granule proteins in the enteric (coccidian) forms of *Toxoplasma gondii* in the small intestine of the cat. Exp. Parasitol. 91, 203–211.

Ferguson, D.J., Hutchison, W.M., Pettersen, E., 1989. Tissue cyst rupture in mice chronically infected with *Toxoplasma gondii*. An immunocytochemical and ultrastructural study. Parasitol. Res. 75, 599–603.

Ferguson, D.J., Jacobs, D., Saman, E., Dubremetz, J.F., Wright, S.E., 1999b. In vivo expression and distribution of dense granule protein 7 (GRA7) in the exoenteric (tachyzoite, bradyzoite) and enteric (coccidian) forms of *Toxoplasma gondii*. Parasitology 119, 259–265.

Finlay, B.B., Cossart, P., 1997. Exploitation of mammalian host cell functions by bacterial pathogens. Science 276, 718–725.

Fischer, H.G., Stachelhaus, S., Sahm, M., Meyer, H.E., Reichmann, G., 1998. GRA7, an excretory 29 kDa *Toxoplasma gondii* dense granule antigen released by infected host cells. Mol. Biochem. Parasitol. 91, 251–262.

Fleckenstein, M.C., Reese, M.L., Konen-Waisman, S., Boothroyd, J.C., Howard, J.C., Steinfeldt, T., 2012. A *Toxoplasma gondii* pseudokinase inhibits host IRG resistance proteins. PLoS Biol. 10, e1001358.

Fourmaux, M.N., Achbarou, A., Mercereau-Puijalon, O., Biderre, C., Briche, I., Loyens, A., Odberg-Ferragut, C., Camus, D., Dubremetz, J.F., 1996a. The MIC1 microneme protein of *Toxoplasma gondii* contains a duplicated receptor-like domain and binds to host cell surface. Mol. Biochem. Parasitol. 83, 201–210.

Fourmaux, M.N., Garcia-Reguet, N., Mercereau-Puijalon, O., Dubremetz, J.F., 1996b. *Toxoplasma gondii* microneme proteins: gene cloning and possible function. Curr. Top. Microbiol. Immunol. 219, 55–58.

Foussard, F., Leriche, M.A., Dubremetz, J.F., 1991. Characterization of the lipid content of *Toxoplasma gondii* rhoptries. Parasitology 102 (Pt 3), 367–370.

Fox, B.A., Falla, A., Rommereim, L.M., Tomita, T., Gigley, J.P., Mercier, C., Cesbron-Delauw, M.F., Weiss, L.M., Bzik, D.J., 2011. Type II *Toxoplasma gondii* KU80 knockout strains enable functional analysis of genes required for cyst development and latent infection. Eukaryot Cell 10, 1193–1206.

Fraser, T.S., Kappe, S.H., Narum, D.L., VanBuskirk, K.M., Adams, J.H., 2001. Erythrocyte-binding activity of *Plasmodium* yoelii apical membrane antigen-1 expressed on the surface of transfected COS-7 cells. Mol. Biochem. Parasitol. 117, 49–59.

Frevert, U., Engelmann, S., Zougbede, S., Stange, J., Ng, B., Matuschewski, K., Liebes, L., Yee, H., 2005. Intravital observation of *Plasmodium berghei* sporozoite infection of the liver. PLoS Biol. 3, e192.

Friedrich, N., Santos, J.M., Liu, Y., Palma, A.S., Leon, E., Saouros, S., Kiso, M., Blackman, M.J., Matthews, S., Feizi, T., Soldati-Favre, D., 2010a. Members of a novel protein family containing microneme adhesive repeat domains act as sialic acid-binding lectins during host cell invasion by apicomplexan parasites. J. Biol. Chem. 285, 2064–2076.

Friedrich, R., Yeheskel, A., Ashery, U., 2010b. DOC2B, C2 domains, and calcium: A tale of intricate interactions. Mol. Neurobiol. 41, 42–51.

Fritz, H.M., Bowyer, P.W., Bogyo, M., Conrad, P.A., Boothroyd, J.C., 2012a. Proteomic analysis of fractionated *Toxoplasma* oocysts reveals clues to their environmental resistance. PLoS One 7, e29955.

Fritz, H.M., Buchholz, K.R., Chen, X., Durbin-Johnson, B., Rocke, D.M., Conrad, P.A., Boothroyd, J.C., 2012b. Transcriptomic analysis of *Toxoplasma* development reveals many novel functions and structures specific to sporozoites and oocysts. PLoS One 7, e29998.

Gaffar, F.R., Yatsuda, A.P., Franssen, F.F., de Vries, E., 2004. Erythrocyte invasion by Babesia bovis merozoites is inhibited by polyclonal antisera directed against peptides derived from a homologue of *Plasmodium falciparum* apical membrane antigen 1. Infect. Immun. 72, 2947–2955.

Gail, M., Gross, U., Bohne, W., 2001. Transcriptional profile of *Toxoplasma gondii*-infected human fibroblasts as revealed by gene-array hybridization. Mol. Genet. Genomics 265, 905–912.

Gaji, R.Y., Flammer, H.P., Carruthers, V.B., 2011. Forward targeting of *Toxoplasma gondii* proproteins to the micronemes involves conserved aliphatic amino acids. Traffic 12, 840–853.

Garcia-Reguet, N., Lebrun, M., Fourmaux, M.N., Mercereau-Puijalon, O., Mann, T., Beckers, C.J., Samyn, B., Van Beeumen, J., Bout, D., Dubremetz, J.F., 2000. The microneme protein MIC3 of *Toxoplasma gondii* is a secretory adhesin that binds to both the surface of the host cells and the surface of the parasite. Cell Microbiol. 2, 353–364.

Garnett, J.A., Liu, Y., Leon, E., Allman, S.A., Friedrich, N., Saouros, S., Curry, S., Soldati-Favre, D., Davis, B.G., Feizi, T., Matthews, S., 2009. Detailed insights from microarray and crystallographic studies into carbohydrate recognition by microneme protein 1 (MIC1) of *Toxoplasma gondii*. Protein Sci. 18, 1935–1947.

Gaskins, E., Gilk, S., DeVore, N., Mann, T., Ward, G., Beckers, C., 2004. Identification of the membrane receptor of a class XIV myosin in *Toxoplasma gondii*. J. Cell Biol. 165, 383–393.

Gendrin, C., Bittame, A., Mercier, C., Cesbron-Delauw, M.F., 2010. Post-translational membrane sorting of the *Toxoplasma gondii* GRA6 protein into the parasite-containing vacuole is driven by its N-terminal domain. Int. J. Parasitol. 40, 1325–1334.

Gendrin, C., Mercier, C., Braun, L., Musset, K., Dubremetz, J.F., Cesbron-Delauw, M.F., 2008. *Toxoplasma gondii* uses unusual sorting mechanisms to deliver transmembrane proteins into the host cell vacuole. Traffic 9, 1665–1680.

Gilbert, L.A., Ravindran, S., Turetzky, J.M., Boothroyd, J.C., Bradley, P.J., 2007. *Toxoplasma gondii* targets a protein phosphatase 2C to the nuclei of infected host cells. Eukaryot Cell. 6, 73–83.

Giovannini, D., Spath, S., Lacroix, C., Perazzi, A., Bargieri, D., Lagal, V., Lebugle, C., Combe, A., Thiberge, S., Baldacci, P., Tardieux, I., Menard, R., 2011. Independent roles of apical membrane antigen 1 and rhoptry neck proteins during host cell invasion by apicomplexa. Cell Host Microbe 10, 591–602.

Gong, H., Kobayashi, K., Sugi, T., Takemae, H., Kurokawa, H., Horimoto, T., Akashi, H., Kato, K., 2012. A novel PAN/apple domain-containing protein from *Toxoplasma gondii*: characterization and receptor identification. PLoS One 7, e30169.

Gonzalez, V., Combe, A., David, V., Malmquist, N.A., Delorme, V., Leroy, C., Blazquez, S., Menard, R., Tardieux, I., 2009. Host cell entry by apicomplexa parasites requires actin polymerization in the host cell. Cell Host Microbe 5, 259–272.

Groffen, A.J., Martens, S., Diez Arazola, R., Cornelisse, L.N., Lozovaya, N., de Jong, A.P., Goriounova, N.A., Habets, R.L., Takai, Y., Borst, J.G., Brose, N., McMahon, H.T., Verhage, M., 2010. Doc2b is a high-affinity Ca2+ sensor for spontaneous neurotransmitter release. Science 327, 1614–1618.

Guo, N.H., Krutzsch, H.C., Negre, E., Vogel, T., Blake, D.A., Roberts, D.D., 1992. Heparin- and sulfatide-binding peptides from the type I repeats of human thrombospondin promote melanoma cell adhesion. Proc. Natl. Acad. Sci. U. S. A. 89, 3040–3044.

Haefliger, J.A., Tschopp, J., Vial, N., Jenne, D.E., 1989. Complete primary structure and functional characterization of the sixth component of the human complement system. Identification of the C5b-binding domain in complement C6. J. Biol. Chem. 264, 18041–18051.

Hajagos, B.E., Turetzky, J.M., Peng, E.D., Cheng, S.J., Ryan, C.M., Souda, P., Whitelegge, J.P., Lebrun, M., Dubremetz, J.F., Bradley, P.J., 2012. Molecular dissection of novel trafficking and processing of the *Toxoplasma gondii* rhoptry metalloprotease toxolysin-1. Traffic 13, 292–304.

Hajj, H.E., Lebrun, M., Fourmaux, M.N., Vial, H., Dubremetz, J.F., 2005. Characterization, biosynthesis and fate of ROP7, a ROP2 related rhoptry protein of *Toxoplasma gondii*. Mol. Biochem. Parasitol 146, 98–100.

Hakansson, S., Charron, A.J., Sibley, L.D., 2001. *Toxoplasma* evacuoles: a two-step process of secretion and fusion forms the parasitophorous vacuole. Embo. J. 20, 3132–3144.

Hakansson, S., Morisaki, H., Heuser, J., Sibley, L.D., 1999. Time-lapse video microscopy of gliding motility in *Toxoplasma gondii* reveals a novel, biphasic mechanism of cell locomotion. Mol. Biol. Cell 10, 3539–3547.

Hall, C.I., Reese, M.L., Weerapana, E., Child, M.A., Bowyer, P.W., Albrow, V.E., Haraldsen, J.D., Phillips, M.R., Sandoval, E.D., Ward, G.E., Cravatt, B.F., Boothroyd, J.C., Bogyo, M., 2011. Chemical genetic screen identifies *Toxoplasma* DJ-1 as a regulator of parasite secretion, attachment, and invasion. Proc. Natl. Acad. Sci. U. S. A. 108, 10568–10573.

Halonen, S.K., Weidner, E., 1994. Overcoating of *Toxoplasma* Parasitophorous Vacuoles with Host Cell Vimentin Type Intermediate Filaments. J. Eukaryot. Microbiol. 41, 65–71.

Harper, J.M., Hoff, E.F., Carruthers, V.B., 2004a. Multimerization of the *Toxoplasma gondii* MIC2 integrin-like A-domain is required for binding to heparin and human cells. Mol. Biochem. Parasitol. 134, 201–212.

Harper, J.M., Huynh, M.H., Coppens, I., Parussini, F., Moreno, S., Carruthers, V.B., 2006. A cleavable propeptide influences *Toxoplasma* infection by facilitating the trafficking and secretion of the TgMIC2-M2AP invasion complex. Mol. Biol. Cell 17, 4551–4563.

Harper, J.M., Zhou, X.W., Pszenny, V., Kafsack, B.F., Carruthers, V.B., 2004b. The novel coccidian micronemal protein MIC11 undergoes proteolytic maturation by sequential cleavage to remove an internal propeptide. Int. J. Parasitol. 34, 1047–1058.

Hehl, A.B., Lekutis, C., Grigg, M.E., Bradley, P.J., Dubremetz, J.F., Ortega-Barria, E., Boothroyd, J.C., 2000. *Toxoplasma gondii* homologue of *Plasmodium* apical membrane antigen 1 is involved in invasion of host cells. Infect. Immun. 68, 7078–7086.

Henderson, K.A., Streltsov, V.A., Coley, A.M., Dolezal, O., Hudson, P.J., Batchelor, A. H., Gupta, A., Bai, T., Murphy, V.J., Anders, R.F., Foley, M., Nuttall, S.D., 2007. Structure of an IgNAR-AMA1 complex: targeting a conserved hydrophobic cleft broadens malarial strain recognition. Structure 15, 1452–1466.

Herm-Gotz, A., Agop-Nersesian, C., Munter, S., Grimley, J.S., Wandless, T.J., Frischknecht, F., Meissner, M., 2007. Rapid control of protein level in the apicomplexan *Toxoplasma gondii*. Nat. Methods 4, 1003–1005.

Herwald, H., Renne, T., Meijers, J.C., Chung, D.W., Page, J.D., Colman, R.W., Muller-Esterl, W., 1996. Mapping of the discontinuous kininogen binding site of prekallikrein. A distal binding segment is located in the heavy chain domain A4. J. Biol. Chem. 271, 13061–13067.

High, K.P., Joiner, K.A., Handschumacher, R.E., 1994. Isolation, cDNA sequences, and biochemical characterization of the major cyclosporin-binding proteins of *Toxoplasma gondii*. J. Biol. Chem. 269, 9105–9112.

Ho, D.H., Badellino, K., Baglia, F.A., Walsh, P.N., 1998. A binding site for heparin in the apple 3 domain of factor XI. J. Biol. Chem. 273, 16382–16390.

Hodder, A.N., Crewther, P.E., Matthew, M.L., Reid, G.E., Moritz, R.L., Simpson, R.J., Anders, R.F., 1996. The disulfide bond structure of *Plasmodium* apical membrane antigen-1. J. Biol. Chem. 271, 29446–29452.

Hoff, E.F., Cook, S.H., Sherman, G.D., Harper, J.M., Ferguson, D.J., Dubremetz, J.F., Carruthers, V.B., 2001. *Toxoplasma gondii*: molecular cloning and characterization of a novel 18-kDa secretory antigen, TgMIC10. Exp. Parasitol. 97, 77–88.

Holt, G.D., Pangburn, M.K., Ginsburg, V., 1990. Properdin binds to sulfatide [Gal(3-SO4)beta 1-1 Cer] and has a sequence homology with other proteins that bind sulfated glycoconjugates. J. Biol. Chem. 265, 2852–2855.

Hoppe, H.C., Joiner, K.A., 2000. Cytoplasmic tail motifs mediate endoplasmic reticulum localization and export of transmembrane reporters in the protozoan parasite *Toxoplasma gondii*. Cell Microbiol. 2, 569–578.

Hoppe, H.C., Ngo, H.M., Yang, M., Joiner, K.A., 2000. Targeting to rhoptry organelles of *Toxoplasma gondii* involves evolutionarily conserved mechanisms. Nat. Cell Biol. 2, 449–456.

Howell, S.A., Hackett, F., Jongco, A.M., Withers-Martinez, C., Kim, K., Carruthers, V.B., Blackman, M.J., 2005. Distinct mechanisms govern proteolytic shedding of a key invasion protein in apicomplexan pathogens. Mol. Microbiol. 57, 1342–1356.

Howell, S.A., Withers-Martinez, C., Kocken, C.H., Thomas, A.W., Blackman, M.J. , 2001. Proteolytic processing and primary structure of *Plasmodium falciparum* apical membrane antigen-1. J. Biol. Chem. 276, 31311–31320.

Hudson-Taylor, D.E., Dolan, S.A., Klotz, F.W., Fujioka, H., Aikawa, M., Koonin, E.V., Miller, L.H., 1995. *Plasmodium falciparum* protein associated with the invasion junction contains a conserved oxidoreductase domain. Mol. Microbiol. 15, 463–471.

Huynh, M.H., Carruthers, V.B., 2006. *Toxoplasma* MIC2 is a major determinant of invasion and virulence. PLoS Pathog. 2, e84.

Huynh, M.H., Opitz, C., Kwok, L.Y., Tomley, F.M., Carruthers, V.B., Soldati, D., 2004. Trans-genera reconstitution and complementation of an adhesion complex in *Toxoplasma gondii*. Cell Microbiol. 6, 771–782.

Huynh, M.H., Rabenau, K.E., Harper, J.M., Beatty, W.L., Sibley, L.D., Carruthers, V.B., 2003. Rapid invasion of host cells by *Toxoplasma* requires secretion of the MIC2-M2AP adhesive protein complex. Embo. J. 22, 2082–2090.

Jensen, J.B., Edgar, S.A., 1976. Effects of antiphagocytic agents on penetration of *Eimeria* magna sporozoites into cultured cells. J. Parasitol. 62, 203–206.

Jethwaney, D., Lepore, T., Hassan, S., Mello, K., Rangarajan, R., Jahnen-Dechent, W., Wirth, D., Sultan, A.A., 2005. Fetuin-A, a hepatocyte-specific protein that binds *Plasmodium berghei* thrombospondin-related adhesive protein: a potential role in infectivity. Infect. Immun. 73, 5883–5891.

Jewett, T.J., Sibley, L.D., 2003. Aldolase forms a bridge between cell surface adhesins and the actin cytoskeleton in apicomplexan parasites. Mol. Cell 11, 885–894.

Jewett, T.J., Sibley, L.D., 2004. The *Toxoplasma* proteins MIC2 and M2AP form a hexameric complex necessary for intracellular survival. J. Biol. Chem. 279, 9362–9369.

Johnson, A.M., Illana, S., 1991. Cloning of *Toxoplasma gondii* gene fragments encoding diagnostic antigens. Gene 99, 127–132.

Jones, T.C., Hirsch, J.G., 1972. The interaction between *Toxoplasma gondii* and mammalian cells. II. The absence of lysosomal fusion with phagocytic vacuoles containing living parasites. J. Exp. Med. 136, 1173–1194.

Kafsack, B.F., Pena, J.D., Coppens, I., Ravindran, S., Boothroyd, J.C., Carruthers, V.B., 2009. Rapid membrane disruption by a perforin-like protein facilitates parasite exit from host cells. Science 323, 530–533.

Kappe, S., Bruderer, T., Gantt, S., Fujioka, H., Nussenzweig, V., Menard, R., 1999. Conservation of a gliding motility and cell invasion machinery in Apicomplexan parasites. J. Cell Biol. 147, 937–944.

Karasov, A.O., Boothroyd, J.C., Arrizabalaga, G., 2005. Identification and disruption of a rhoptry-localized homologue of sodium hydrogen exchangers in *Toxoplasma gondii*. Int. J. Parasitol. 35, 285–291.

Karsten, V., Qi, H., Beckers, C.J., Reddy, A., Dubremetz, J.F., Webster, P., Joiner, K.A., 1998. The protozoan parasite *Toxoplasma gondii* targets proteins to dense granules and the vacuolar space using both conserved and unusual mechanisms. J. Cell Biol. 141, 1323–1333.

Kato, K., Mayer, D.C., Singh, S., Reid, M., Miller, L.H., 2005. Domain III of *Plasmodium falciparum* apical membrane antigen 1 binds to the erythrocyte membrane protein Kx. Proc. Natl. Acad. Sci. U. S. A. 102, 5552–5557.

Kawase, O., Nishikawa, Y., Bannai, H., Igarashi, M., Matsuo, T., Xuan, X., 2010. Characterization of a novel thrombospondin-related protein in *Toxoplasma gondii*. Parasitol. Int. 59, 211–216.

Kawase, O., Nishikawa, Y., Bannai, H., Zhang, H., Zhang, G., Jin, S., Lee, E.G., Xuan, X., 2007. Proteomic analysis of calcium-dependent secretion in *Toxoplasma gondii*. Proteomics 7, 3718–3725.

Keeley, A., Soldati, D., 2004. The glideosome: a molecular machine powering motility and host cell invasion by Apicomplexa. Trends Cell Biol. 14, 528–532.

Kessler, H., Herm-Gotz, A., Hegge, S., Rauch, M., Soldati-Favre, D., Frischknecht, F., Meissner, M., 2008. Microneme protein 8—a new essential invasion factor in *Toxoplasma gondii*. J. Cell Sci. 121, 947—956.

Kieschnick, H., Wakefield, T., Narducci, C.A., Beckers, C., 2001. *Toxoplasma gondii* attachment to host cells is regulated by a calmodulin-like domain protein kinase. J. Biol. Chem. 276, 12369—12377.

Kim, E., Lee, Y., Lee, H.J., Kim, J.S., Song, B.S., Huh, J.W., Lee, S.R., Kim, S.U., Kim, S.H., Hong, Y., Shim, I., Chang, K.T., 2010. Implication of mouse Vps26b-Vps29-Vps35 retromer complex in sortilin trafficking. Biochem. Biophys. Res. Commun. 403, 167—171.

Kim, K., Soldati, D., Boothroyd, J.C., 1993. Gene replacement in *Toxoplasma gondii* with chloramphenicol acetyltransferase as selectable marker. Science 262, 911—914.

Klein, H., Loschner, B., Zyto, N., Portner, M., Montag, T., 1998. Expression, purification, and biochemical characterization of a recombinant lectin of Sarcocystis muris (Apicomplexa) cyst merozoites. Glycoconj. J. 15, 147—153.

Klotz, F.W., Hadley, T.J., Aikawa, M., Leech, J., Howard, R.J., Miller, L.H., 1989. A 60-kDa *Plasmodium falciparum* protein at the moving junction formed between merozoite and erythrocyte during invasion. Mol. Biochem. Parasitol. 36, 177—185.

Labesse, G., Gelin, M., Bessin, Y., Lebrun, M., Papoin, J., Cerdan, R., Arold, S.T., Dubremetz, J.F., 2009. ROP2 from *Toxoplasma gondii*: A Virulence Factor with a Protein-Kinase Fold and No Enzymatic Activity. Structure 17, 139—146.

Labruyere, E., Lingnau, M., Mercier, C., Sibley, L.D., 1999. Differential membrane targeting of the secretory proteins GRA4 and GRA6 within the parasitophorous vacuole formed by *Toxoplasma gondii*. Mol. Biochem. Parasitol. 102, 311—324.

Lagal, V., Binder, E.M., Huynh, M.H., Kafsack, B.F., Harris, P.K., Diez, R., Chen, D., Cole, R.N., Carruthers, V.B., Kim, K., 2010. *Toxoplasma gondii* protease TgSUB1 is required for cell surface processing of micronemal adhesive complexes and efficient adhesion of tachyzoites. Cell Microbiol. 12, 1792—1808.

Lai, L., Bumstead, J., Liu, Y., Garnett, J., Campanero-Rhodes, M.A., Blake, D.P., Palma, A.S., Chai, W., Ferguson, D.J., Simpson, P., Feizi, T., Tomley, F.M., Matthews, S., 2011. The role of sialyl glycan recognition in host tissue tropism of the avian parasite *Eimeria tenella*. PLoS Pathog. 7, e1002296.

Laliberte, J., Carruthers, V.B., 2011. *Toxoplasma gondii* toxolysin 4 is an extensively processed putative metalloproteinase secreted from micronemes. Mol. Biochem. Parasitol. 177, 49—56.

Lamarque, M., Besteiro, S., Papoin, J., Roques, M., Vulliez-Le Normand, B., Morlon-Guyot, J., Dubremetz, J.F., Fauquenoy, S., Tomavo, S., Faber, B.W., Kocken, C.H., Thomas, A.W., Boulanger, M.J., Bentley, G.A., Lebrun, M., 2011. The RON2-AMA1 interaction is a critical step in moving junction-dependent invasion by apicomplexan parasites. PLoS Pathog. 7, e1001276.

Lamarque, M.H., Papoin, J., Finizio, A.L., Lentini, G., Pfaff, A.W., Candolfi, E., Dubremetz, J.F., Lebrun, M., 2012. Identification of a new rhoptry neck complex RON9/RON10 in the Apicomplexa parasite *Toxoplasma gondii*. PLoS One 7, e32457.

Larson, E.T., Parussini, F., Huynh, M.H., Giebel, J.D., Kelley, A.M., Zhang, L., Bogyo, M., Merritt, E.A., Carruthers, V.B., 2009. *Toxoplasma gondii* cathepsin L is the primary target of the invasion-inhibitory compound morpholinurea-leucyl-homophenyl-vinyl sulfone phenyl. J. Biol. Chem. 284, 26839—26850.

Lawler, J., 1986. The structural and functional properties of thrombospondin. Blood 67, 1197—1209.

Lebrun, M., Michelin, A., El Hajj, H., Poncet, J., Bradley, P.J., Vial, H., Dubremetz, J.F., 2005. The rhoptry neck protein RON4 relocalizes at the moving junction during *Toxoplasma gondii* invasion. Cell Microbiol. 7, 1823—1833.

Lecordier, L., Mercier, C., Sibley, L.D., Cesbron-Delauw, M.F., 1999. Transmembrane insertion of the *Toxoplasma gondii* GRA5 protein occurs after soluble secretion into the host cell. Mol. Biol. Cell 10, 1277—1287.

Lecordier, L., Mercier, C., Torpier, G., Tourvieille, B., Darcy, F., Liu, J.L., Maes, P., Tartar, A., Capron, A., Cesbron-Delauw, M.F., 1993. Molecular structure of a *Toxoplasma gondii* dense granule antigen (GRA 5) associated with the parasitophorous vacuole membrane. Mol. Biochem. Parasitol. 59, 143—153.

Lecordier, L., Moleon-Borodowsky, I., Dubremetz, J.F., Tourvieille, B., Mercier, C., Deslee, D., Capron, A., Cesbron-Delauw, M.F., 1995. Characterization of a dense granule antigen of *Toxoplasma gondii* (GRA6) associated to the network of the parasitophorous vacuole. Mol. Biochem. Parasitol. 70, 85—94.

Lee, J.O., Rieu, P., Arnaout, M.A., Liddington, R., 1995. Crystal structure of the A domain from the alpha subunit of integrin CR3 (CD11b/CD18). Cell 80, 631—638.

Leriche, M.A., Dubremetz, J.F., 1990. Exocytosis of *Toxoplasma gondii* dense granules into the parasitophorous vacuole after host cell invasion. Parasitol. Res. 76, 559—562.

Leriche, M.A., Dubremetz, J.F., 1991. Characterization of the protein contents of rhoptries and dense granules of *Toxoplasma gondii* tachyzoites by subcellular fractionation and monoclonal antibodies. Mol. Biochem. Parasitol. 45, 249—259.

Liendo, A., Joiner, K.A., 2000. *Toxoplasma gondii*: conserved protein machinery in an unusual secretory pathway? Microbes Infect. 2, 137—144.

Lodoen, M.B., Gerke, C., Boothroyd, J.C., 2010. A highly sensitive FRET-based approach reveals secretion of the actin-binding protein toxofilin during *Toxoplasma gondii* infection. Cell Microbiol. 12, 55—66.

Lord, C., Bhandari, D., Menon, S., Ghassemian, M., Nycz, D., Hay, J., Ghosh, P., Ferro-Novick, S., 2011. Sequential interactions with Sec23 control the direction of vesicle traffic. Nature 473, 181—186.

Lourenco, E.V., Pereira, S.R., Faca, V.M., Coelho-Castelo, A.A., Mineo, J.R., Roque-Barreira, M.C., Greene, L.J., Panunto-Castelo, A., 2001. *Toxoplasma gondii* micronemal protein MIC1 is a lactose-binding lectin. Glycobiology 11, 541—547.

Lourido, S., Shuman, J., Zhang, C., Shokat, K.M., Hui, R., Sibley, L.D., 2010. Calcium-dependent protein kinase 1 is an essential regulator of exocytosis in *Toxoplasma*. Nature 465, 359—362.

Lovett, J.L., Marchesini, N., Moreno, S.N., Sibley, L.D., 2002. *Toxoplasma gondii* microneme secretion involves intracellular Ca(2+) release from inositol 1,4,5-triphosphate (IP(3))/ryanodine-sensitive stores. J. Biol. Chem. 277, 25870—25876.

Lovett, J.L., Sibley, L.D., 2003. Intracellular calcium stores in *Toxoplasma gondii* govern invasion of host cells. J. Cell Sci. 116, 3009—3016.

Magno, R.C., Lemgruber, L., Vommaro, R.C., De Souza, W., Attias, M., 2005. Intravacuolar network may act as a mechanical support for *Toxoplasma gondii* inside the parasitophorous vacuole. Microsc. Res. Tech. 67, 45—52.

Mahajan, B., Jani, D., Chattopadhyay, R., Nagarkatti, R., Zheng, H., Majam, V., Weiss, W., Kumar, S., Rathore, D., 2005. Identification, cloning, expression, and characterization of the gene for *Plasmodium* knowlesi surface protein containing an altered thrombospondin repeat domain. Infect. Immun. 73, 5402—5409.

Marchant, J., Cowper, B., Liu, Y., Lai, L., Pinzan, C., Marq, J.B., Friedrich, N., Sawmynaden, K., Liew, L., Chai, W., Childs, R.A., Saouros, S., Simpson, P., Roque Barreira, M.C., Feizi, T., Soldati-Favre, D., Matthews, S., 2012. Galactose recognition by the apicomplexan parasite *Toxoplasma gondii*. J. Biol. Chem. 287, 16720—16733.

Matthiesen, S.H., Shenoy, S.M., Kim, K., Singer, R.H., Satir, B.H., 2001. A parafusin-related *Toxoplasma* protein in Ca2+-regulated secretory organelles. Eur. J. Cell Biol. 80, 775—783.

Matthiesen, S.H., Shenoy, S.M., Kim, K., Singer, R.H., Satir, B.H., 2003. Role of the parafusin orthologue, PRP1, in microneme exocytosis and cell invasion in *Toxoplasma gondii*. Cell Microbiol. 5, 613—624.

Matuschewski, K., Nunes, A.C., Nussenzweig, V., Menard, R., 2002. *Plasmodium* sporozoite invasion into insect and mammalian cells is directed by the same dual binding system. Embo. J. 21, 1597—1606.

McCormick, C.J., Tuckwell, D.S., Crisanti, A., Humphries, M.J., Hollingdale, M.R., 1999. Identification of heparin as a ligand for the A-domain of *Plasmodium falciparum* thrombospondin-related adhesion protein. Mol. Biochem. Parasitol. 100, 111—124.

Meissner, M., Reiss, M., Viebig, N., Carruthers, V.B., Toursel, C., Tomavo, S., Ajioka, J.W., Soldati, D., 2002. A family of transmembrane microneme proteins of *Toxoplasma gondii* contain EGF-like domains and function as escorters. J. Cell Sci. 115, 563—574.

Melo, E.J., Carvalho, T.M., De Souza, W., 2001. Behaviour of microtubules in cells infected with *Toxoplasma gondii*. Biocell 25, 53—59.

Menard, R., 2001. Gliding motility and cell invasion by Apicomplexa: insights from the *Plasmodium* sporozoite. Cell Microbiol. 3, 63–73.

Mercier, C., Adjogble, K.D., Daubener, W., Delauw, M.F., 2005. Dense granules: Are they key organelles to help understand the parasitophorous vacuole of all apicomplexa parasites? Int. J. Parasitol. 35, 829–849.

Mercier, C., Cesbron-Delauw, M.F., Sibley, L.D., 1998a. The amphipathic alpha helices of the *Toxoplasma* protein GRA2 mediate post-secretory membrane association. J. Cell Sci. 111, 2171–2180.

Mercier, C., Dubremetz, J.F., Rauscher, B., Lecordier, L., Sibley, L.D., Cesbron-Delauw, M.F., 2002. Biogenesis of nanotubular network in *Toxoplasma* parasitophorous vacuole induced by parasite proteins. Mol. Biol. Cell 13, 2397–2409.

Mercier, C., Howe, D.K., Mordue, D., Lingnau, M., Sibley, L.D., 1998b. Targeted disruption of the GRA2 locus in *Toxoplasma gondii* decreases acute virulence in mice. Infect. Immun. 66, 4176–4182.

Mercier, C., Rauscher, B., Lecordier, L., Deslee, D., Dubremetz, J.F., Cesbron-Delauw, M.F., 2001. Lack of expression of the dense granule protein GRA5 does not affect the development of *Toxoplasma* tachyzoites. Mol. Biochem. Parasitol. 116, 247–251.

Michelin, A., Bittame, A., Bordat, Y., Travier, L., Mercier, C., Dubremetz, J.F., Lebrun, M., 2008. GRA12, a *Toxoplasma* dense granule protein associated with the intravacuolar 1 membranous 2 nanotubular network. Int. J. Parasitol. 39, 299–306.

Miller, L.H., Aikawa, M., Johnson, J.G., Shiroishi, T., 1979. Interaction between cytochalasin B-treated malarial parasites and erythrocytes. Attachment and junction formation. J. Exp. Med. 149, 172–184.

Miller, S.A., Binder, E.M., Blackman, M.J., Carruthers, V.B., Kim, K., 2001. A conserved subtilisin-like protein TgSUB1 in microneme organelles of *Toxoplasma gondii*. J. Biol. Chem. 276, 45341–45348.

Miller, S.A., Thathy, V., Ajioka, J.W., Blackman, M.J., Kim, K., 2003. TgSUB2 is a *Toxoplasma gondii* rhoptry organelle processing proteinase. Mol. Microbiol. 49, 883–894.

Miranda, K., Pace, D.A., Cintron, R., Rodrigues, J.C., Fang, J., Smith, A., Rohloff, P., Coelho, E., de Haas, F., de Souza, W., Coppens, I., Sibley, L.D., Moreno, S.N., 2010. Characterization of a novel organelle in *Toxoplasma gondii* with similar composition and function to the plant vacuole. Mol. Microbiol. 76, 1358–1375.

Mital, J., Meissner, M., Soldati, D., Ward, G.E., 2005. Conditional expression of *Toxoplasma gondii* apical membrane antigen-1 (TgAMA1) demonstrates that TgAMA1 plays a critical role in host cell invasion. Mol. Biol. Cell 16, 4341–4349.

Mitchell, G.H., Thomas, A.W., Margos, G., Dluzewski, A.R., Bannister, L.H., 2004. Apical membrane antigen 1, a major malaria vaccine candidate, mediates the close attachment of invasive merozoites to host red blood cells. Infect. Immun. 72, 154–158.

Mondragon, R., Frixione, E., 1996. Ca(2+)-dependence of conoid extrusion in *Toxoplasma gondii* tachyzoites. J. Eukaryot Microbiol. 43, 120–127.

Monteiro, V.G., Soares, C.P., de Souza, W., 1998. Host cell surface sialic acid residues are involved on the process of penetration of *Toxoplasma gondii* into mammalian cells. FEMS Microbiol. Lett. 164, 323–327.

Mordue, D.G., Desai, N., Dustin, M., Sibley, L.D., 1999a. Invasion by *Toxoplasma gondii* establishes a moving junction that selectively excludes host cell plasma membrane proteins on the basis of their membrane anchoring. J. Exp. Med. 190, 1783–1792.

Mordue, D.G., Hakansson, S., Niesman, I., Sibley, L.D., 1999b. *Toxoplasma gondii* resides in a vacuole that avoids fusion with host cell endocytic and exocytic vesicular trafficking pathways. Exp. Parasitol. 92, 87–99.

Mordue, D.G., Sibley, L.D., 1997. Intracellular fate of vacuoles containing *Toxoplasma gondii* is determined at the time of formation and depends on the mechanism of entry. J. Immunol. 159, 4452–4459.

Moreno, S.N., Docampo, R., 2003. Calcium regulation in protozoan parasites. Curr. Opin. Microbiol. 6, 359–364.

Morisaki, J.H., Heuser, J.E., Sibley, L.D., 1995. Invasion of *Toxoplasma gondii* occurs by active penetration of the host cell. J. Cell Sci. 108 (Pt 6), 2457–2464.

Morris, M.T., Carruthers, V.B., 2003. Identification and partial characterization of a second Kazal inhibitor in *Toxoplasma gondii*. Mol. Biochem. Parasitol. 128, 119–122.

Morris, M.T., Coppin, A., Tomavo, S., Carruthers, V.B., 2002. Functional analysis of *Toxoplasma gondii* protease inhibitor 1. J. Biol. Chem. 277, 45259–45266.

Moudy, R., Manning, T.J., Beckers, C.J., 2001. The loss of cytoplasmic potassium upon host cell breakdown triggers egress of *Toxoplasma gondii*. J. Biol. Chem. 276, 41492–41501.

Muller, H.M., Reckmann, I., Hollingdale, M.R., Bujard, H., Robson, K.J., Crisanti, A., 1993. Thrombospondin related anonymous protein (TRAP) of *Plasmodium falciparum* binds specifically to sulfated glycoconjugates and to HepG2 hepatoma cells suggesting a role for this molecule in sporozoite invasion of hepatocytes. Embo. J. 12, 2881–2889.

Nagamune, K., Hicks, L.M., Fux, B., Brossier, F., Chini, E.N., Sibley, L.D., 2008. Abscisic acid controls calcium-dependent egress and development in *Toxoplasma gondii*. Nature 451, 207–210.

Nakaar, V., Bermudes, D., Peck, K.R., Joiner, K.A., 1998. Upstream elements required for expression of nucleoside triphosphate hydrolase genes of *Toxoplasma gondii*. Mol. Biochem. Parasitol. 92, 229–239.

Nakaar, V., Ngo, H.M., Aaronson, E.P., Coppens, I., Stedman, T.T., Joiner, K.A., 2003. Pleiotropic effect due to targeted depletion of secretory rhoptry protein ROP2 in *Toxoplasma gondii*. J. Cell Sci. 116, 2311–2320.

Nakaar, V., Samuel, B.U., Ngo, E.O., Joiner, K.A., 1999. Targeted reduction of nucleoside triphosphate hydrolase by antisense RNA inhibits *Toxoplasma gondii* proliferation. J. Biol. Chem. 274, 5083–5087.

Ngo, H.M., Ngo, E.O., Bzik, D.J., Joiner, K.A., 2000. *Toxoplasma gondii*: are host cell adenosine nucleotides a direct source for purine salvage? Exp. Parasitol. 95, 148–153.

Ngo, H.M., Yang, M., Joiner, K.A., 2004. Are rhoptries in Apicomplexan parasites secretory granules or secretory lysosomal granules? Mol. Microbiol. 52, 1531–1541.

Ngo, H.M., Yang, M., Paprotka, K., Pypaert, M., Hoppe, H., Joiner, K.A., 2003. AP-1 in *Toxoplasma gondii* mediates biogenesis of the rhoptry secretory organelle from a post-Golgi compartment. J. Biol. Chem. 278, 5343–5352.

Nichols, B.A., 1985. Interactions between *Toxoplasma* and host phagocytes. Int. Ophthalmol. Clin. 25, 71–80.

Nichols, B.A., Chiappino, M.L., O'Connor, G.R., 1983. Secretion from the rhoptries of *Toxoplasma gondii* during host cell invasion. J. Ultrastruct. Res. 83, 85–98.

Niedelman, W., Gold, D.A., Rosowski, E.E., Sprokholt, J.K., Lim, D., Farid Arenas, A., Melo, M.B., Spooner, E., Yaffe, M.B., Saeij, J.P., 2012. The rhoptry proteins ROP18 and ROP5 mediate *Toxoplasma gondii* evasion of the murine, but not the human, interferon-gamma response. PLoS Pathog. 8, e1002784.

Nishi, M., Hu, K., Murray, J.M., Roos, D.S., 2008. Organellar dynamics during the cell cycle of *Toxoplasma gondii*. J. Cell Sci. 121, 1559–1568.

O'Keeffe, A.H., Green, J.L., Grainger, M., Holder, A.A., 2005. A novel Sushi domain-containing protein of *Plasmodium falciparum*. Mol. Biochem. Parasitol. 140, 61–68.

Ong, Y.C., Reese, M.L., Boothroyd, J.C., 2010. *Toxoplasma* rhoptry protein 16 (ROP16) subverts host function by direct tyrosine phosphorylation of STAT6. J. Biol. Chem. 285, 28731–28740.

Opitz, C., Di Cristina, M., Reiss, M., Ruppert, T., Crisanti, A., Soldati, D., 2002. Intramembrane cleavage of microneme proteins at the surface of the apicomplexan parasite *Toxoplasma gondii*. Embo. J. 21, 1577–1585.

Opitz, C., Soldati, D., 2002. 'The glideosome': a dynamic complex powering gliding motion and host cell invasion by *Toxoplasma gondii*. Mol. Microbiol. 45, 597–604.

Ossorio, P.N., Schwartzman, J.D., Boothroyd, J.C., 1992. A *Toxoplasma gondii* rhoptry protein associated with host cell penetration has unusual charge asymmetry. Mol. Biochem. Parasitol. 50, 1–15.

Parussini, F., Coppens, I., Shah, P.P., Diamond, S.L., Carruthers, V.B., 2010. Cathepsin L occupies a vacuolar compartment and is a protein maturase within the endo/exocytic system of *Toxoplasma gondii*. Mol. Microbiol. 76, 1340–1357.

Parussini, F., Tang, Q., Moin, S.M., Mital, J., Urban, S., Ward, G.E., 2012. Intramembrane proteolysis of *Toxoplasma* apical membrane antigen 1 facilitates host cell invasion but is dispensable for replication. Proc. Natl. Acad. Sci. U. S. A. 109, 7463–7468.

Peixoto, L., Chen, F., Harb, O.S., Davis, P.H., Beiting, D.P., Brownback, C.S., Ouloguem, D., Roos, D.S., 2010. Integrative genomic approaches highlight a family of parasite-specific kinases that regulate host responses. Cell Host Microbe 8, 208–218.

Periz, J., Gill, A.C., Knott, V., Handford, P.A., Tomley, F.M., 2005. Calcium binding activity of the epidermal growth factor-like domains of the apicomplexan microneme protein EtMIC4. Mol. Biochem. Parasitol. 143, 192–199.

Pernas, L., Boothroyd, J.C., 2010. Association of host mitochondria with the parasitophorous vacuole during *Toxoplasma* infection is not dependent on rhoptry proteins ROP2/8. Int. J. Parasitol. 40, 1367–1371.

Persat, F., Mercier, C., Ficheux, D., Colomb, E., Trouillet, S., Bendridi, N., Musset, K., Loeuillet, C., Cesbron-Delauw, M.F., Vincent, C., 2012. A synthetic peptide derived from the parasite *Toxoplasma gondii* triggers human dendritic cells' migration. J. Leukoc. Biol. 92, 1241–1250.

Phelps, E.D., Sweeney, K.R., Blader, I.J., 2008. *Toxoplasma gondii* rhoptry discharge correlates with activation of the early growth response 2 host cell transcription factor. Infect. Immun. 76, 4703–4712.

Pizarro, J.C., Vulliez-Le Normand, B., Chesne-Seck, M.L., Collins, C.R., Withers-Martinez, C., Hackett, F., Blackman, M.J., Faber, B.W., Remarque, E.J., Kocken, C.H., Thomas, A.W., Bentley, G.A., 2005. Crystal structure of the malaria vaccine candidate apical membrane antigen 1. Science 308, 408–411.

Porchet, E., Torpier, G., 1977. Freeze fracture study of *Toxoplasma* and Sarcocystis infective stages (author's transl). Z. Parasitenkd 54, 101–124.

Poupel, O., Boleti, H., Axisa, S., Couture-Tosi, E., Tardieux, I., 2000. Toxofilin, a novel actin-binding protein from *Toxoplasma gondii*, sequesters actin monomers and caps actin filaments. Mol. Biol. Cell 11, 355–368.

Pszenny, V., Angel, S.O., Duschak, V.G., Paulino, M., Ledesma, B., Yabo, M.I., Guarnera, E., Ruiz, A.M., Bontempi, E.J., 2000. Molecular cloning, sequencing and expression of a serine proteinase inhibitor gene from *Toxoplasma gondii*. Mol. Biochem. Parasitol. 107, 241–249.

Pszenny, V., Ledesma, B.E., Matrajt, M., Duschak, V.G., Bontempi, E.J., Dubremetz, J.F., Angel, S.O., 2002. Subcellular localization and post-secretory targeting of TgPI, a serine proteinase inhibitor from *Toxoplasma gondii*. Mol. Biochem. Parasitol. 121, 283–286.

Pytela, R., 1988. Amino acid sequence of the murine Mac-1 alpha chain reveals homology with the integrin family and an additional domain related to von Willebrand factor. Embo. J. 7, 1371–1378.

Qiu, W., Wernimont, A., Tang, K., Taylor, S., Lunin, V., Schapira, M., Fentress, S., Hui, R., Sibley, L.D., 2009. Novel structural and regulatory features of rhoptry secretory kinases in *Toxoplasma gondii*. EMBO J. 28, 969–979.

Que, X., Ngo, H., Lawton, J., Gray, M., Liu, Q., Engel, J., Brinen, L., Ghosh, P., Joiner, K.A., Reed, S.L., 2002. The cathepsin B of *Toxoplasma gondii*, toxopain-1, is critical for parasite invasion and rhoptry protein processing. J. Biol. Chem. 277, 25791–25797.

Rabenau, K.E., Sohrabi, A., Tripathy, A., Reitter, C., Ajioka, J.W., Tomley, F.M., Carruthers, V.B., 2001. TgM2AP participates in *Toxoplasma gondii* invasion of host cells and is tightly associated with the adhesive protein TgMIC2. Mol. Microbiol. 41, 537–547.

Ravindran, S., Lodoen, M.B., Verhelst, S.H., Bogyo, M., Boothroyd, J.C., 2009. 4-Bromophenacyl bromide specifically inhibits rhoptry secretion during *Toxoplasma* invasion. PLoS One 4, e8143.

Rawlings, N.D., Polgar, L., Barrett, A.J., 1991. A new family of serine-type peptidases related to prolyl oligopeptidase. Biochem. J. 279 (Pt 3), 907–908.

Reese, M.L., Boothroyd, J.C., 2009. A helical membrane-binding domain targets the *Toxoplasma* ROP2 family to the parasitophorous vacuole. Traffic 10, 1458–1470.

Reese, M.L., Boothroyd, J.C., 2011. A conserved non-canonical motif in the pseudoactive site of the ROP5 pseudokinase domain mediates its effect on *Toxoplasma* virulence. J. Biol. Chem. 286, 29366–29375.

Reese, M.L., Zeiner, G.M., Saeij, J.P., Boothroyd, J.C., Boyle, J.P., 2011. Polymorphic family of injected pseudokinases is paramount in *Toxoplasma* virulence. Proc. Natl. Acad. Sci. U. S. A. 108, 9625–9630.

Reichmann, G., Dlugonska, H., Fischer, H.G., 2002. Characterization of TgROP9 (p36), a novel rhoptry protein of *Toxoplasma gondii* tachyzoites identified by T cell clone. Mol. Biochem. Parasitol. 119, 43–54.

Reiss, M., Viebig, N., Brecht, S., Fourmaux, M.N., Soete, M., Di Cristina, M., Dubremetz, J.F., Soldati, D., 2001. Identification and characterization of an escorter for two secretory adhesins in *Toxoplasma gondii*. J. Cell Biol. 152, 563–578.

Richard, D., Bartfai, R., Volz, J., Ralph, S.A., Muller, S., Stunnenberg, H.G., Cowman, A.F., 2011. A genome-wide chromatin-associated nuclear peroxiredoxin from the malaria parasite *Plasmodium falciparum*. J. Biol. Chem. 286, 11746–11755.

Richard, D., Kats, L.M., Langer, C., Black, C.G., Mitri, K., Boddey, J.A., Cowman, A.F., Coppel, R.L., 2009. Identification of rhoptry trafficking determinants and evidence for a novel sorting mechanism in the malaria parasite *Plasmodium falciparum*. PLoS Pathog. 5, e1000328.

Richard, D., MacRaild, C.A., Riglar, D.T., Chan, J.A., Foley, M., Baum, J., Ralph, S.A., Norton, R.S., Cowman, A.F., 2010. Interaction between *Plasmodium falciparum* apical membrane antigen 1 and the rhoptry neck protein complex defines a key step in the erythrocyte invasion process of malaria parasites. J. Biol. Chem. 285, 14815–14822.

Riglar, D.T., Richard, D., Wilson, D.W., Boyle, M.J., Dekiwadia, C., Turnbull, L., Angrisano, F., Marapana, D.S., Rogers, K.L., Whitchurch, C.B., Beeson, J.G., Cowman, A.F., Ralph, S.A., Baum, J., 2011. Super-resolution dissection of coordinated events during malaria parasite invasion of the human erythrocyte. Cell Host Microbe 9, 9–20.

Robson, K.J., Hall, J.R., Jennings, M.W., Harris, T.J., Marsh, K., Newbold, C.I., Tate, V.E., Weatherall, D.J., 1988. A highly conserved amino-acid sequence in thrombospondin, properdin and in proteins from sporozoites and blood stages of a human malaria parasite. Nature 335, 79–82.

Roger, N., Dubremetz, J.F., Delplace, P., Fortier, B., Tronchin, G., Vernes, A., 1988. Characterization of a 225 kilodalton rhoptry protein of *Plasmodium falciparum*. Mol. Biochem. Parasitol. 27, 135–141.

Rome, M.E., Beck, J.R., Turetzky, J.M., Webster, P., Bradley, P.J., 2008. Intervacuolar transport and unique topology of GRA14, a novel dense granule protein in *Toxoplasma gondii*. Infect. Immun. 76, 4865–4875.

Rosowski, E.E., Lu, D., Julien, L., Rodda, L., Gaiser, R.A., Jensen, K.D., Saeij, J.P., 2011. Strain-specific activation of the NF-kappaB pathway by GRA15, a novel *Toxoplasma gondii* dense granule protein. J. Exp. Med. 208, 195–212.

Russell, D.G., Sinden, R.E., 1981. The role of the cytoskeleton in the motility of coccidian sporozoites. J. Cell Sci. 50, 345–359.

Ryning, F.W., Remington, J.S., 1978. Effect of cytochalasin D on *Toxoplasma gondii* cell entry. Infect. Immun. 20, 739–743.

Sadak, A., Taghy, Z., Fortier, B., Dubremetz, J.F., 1988. Characterization of a family of rhoptry proteins of *Toxoplasma gondii*. Mol. Biochem. Parasitol. 29, 203–211.

Saeij, J.P., Boyle, J.P., Coller, S., Taylor, S., Sibley, L.D., Brooke-Powell, E.T., Ajioka, J.W., Boothroyd, J.C., 2006. Polymorphic secreted kinases are key virulence factors in toxoplasmosis. Science 314, 1780–1783.

Saeij, J.P., Coller, S., Boyle, J.P., Jerome, M.E., White, M.W., Boothroyd, J.C., 2007. *Toxoplasma* co-opts host gene expression by injection of a polymorphic kinase homologue. Nature 445, 324–327.

Saffer, L.D., Mercereau-Puijalon, O., Dubremetz, J.F., Schwartzman, J.D., 1992. Localization of a *Toxoplasma gondii* rhoptry protein by immunoelectron microscopy during and after host cell penetration. J. Protozool. 39, 526–530.

Santos, J.M., Ferguson, D.J., Blackman, M.J., Soldati-Favre, D., 2011. Intramembrane cleavage of AMA1 triggers *Toxoplasma* to switch from an invasive to a replicative mode. Science 331, 473–477.

Saouros, S., Dou, Z., Henry, M., Marchant, J., Carruthers, V.B., Matthews, S., 2012. Microneme protein 5 regulates the activity of *Toxoplasma* subtilisin 1 by mimicking a subtilisin prodomain. J. Biol. Chem. 287, 36029–36040.

Saouros, S., Edwards-Jones, B., Reiss, M., Sawmynaden, K., Cota, E., Simpson, P., Dowse, T.J., Jakle, U., Ramboarina, S., Shivarattan, T., Matthews, S., Soldati-Favre, D., 2005. A novel galectin-like domain from *Toxoplasma gondii* micronemal protein 1 assists the folding, assembly, and transport of a cell adhesion complex. J. Biol. Chem. 280, 38583–38591.

Sawmynaden, K., Saouros, S., Friedrich, N., Marchant, J., Simpson, P., Bleijlevens, B., Blackman, M.J., Soldati-Favre, D., Matthews, S., 2008. Structural insights into microneme protein assembly reveal a new mode of EGF domain recognition. EMBO Rep. 9, 1149–1155.

Schindler, R., Weichselsdorfer, E., Wagner, O., Bereiter-Hahn, J., 2001. Aldolase-localization in cultured cells: cell-type and substrate-specific regulation of cytoskeletal associations. Biochem. Cell Biol. 79, 719–728.

Schrevel, J., Asfaux-Foucher, G., Hopkins, J.M., Robert, V., Bourgouin, C., Prensier, G., Bannister, L.H., 2008. Vesicle trafficking during sporozoite development in *Plasmodium berghei*: ultrastructural evidence for a novel trafficking mechanism. Parasitology 135, 1–12.

Schwab, J.C., Beckers, C.J., Joiner, K.A., 1994. The parasitophorous vacuole membrane surrounding intracellular *Toxoplasma gondii* functions as a molecular sieve. Proc. Natl. Acad. Sci. U. S. A. 91, 509–513.

Schwartzman, J.D., 1986. Inhibition of a penetration-enhancing factor of *Toxoplasma gondii* by monoclonal antibodies specific for rhoptries. Infect. Immun. 51, 760–764.

Schwartzman, J.D., Saffer, L.D., 1992. How *Toxoplasma gondii* gets in and out of host cells. Subcell Biochem. 18, 333–364.

Schwarz, J.A., Fouts, A.E., Cummings, C.A., Ferguson, D.J., Boothroyd, J.C., 2005. A novel rhoptry protein in *Toxoplasma gondii* bradyzoites and merozoites. Mol. Biochem. Parasitol 144, 159–166.

Schweighofer, A., Hirt, H., Meskiene, I., 2004. Plant PP2C phosphatases: emerging functions in stress signaling. Trends Plant. Sci. 9, 236–243.

Seetharaman, J., Kanigsberg, A., Slaaby, R., Leffler, H., Barondes, S.H., Rini, J.M., 1998. X-ray crystal structure of the human galectin-3 carbohydrate recognition domain at 2.1-A resolution. J. Biol. Chem. 273, 13047–13052.

Shaw, M.K., Roos, D.S., Tilney, L.G., 1998. Acidic compartments and rhoptry formation in *Toxoplasma gondii*. Parasitology 117 (Pt 5), 435–443.

Sheiner, L., Dowse, T.J., Soldati-Favre, D., 2008. Identification of trafficking determinants for polytopic rhomboid proteases in *Toxoplasma gondii*. Traffic 9, 665–677.

Sheiner, L., Santos, J.M., Klages, N., Parussini, F., Jemmely, N., Friedrich, N., Ward, G.E., Soldati-Favre, D., 2010. *Toxoplasma gondii* transmembrane microneme proteins and their modular design. Mol. Microbiol. 77, 912–929.

Sibley, L.D., 2003. *Toxoplasma gondii*: perfecting an intracellular life style. Traffic 4, 581–586.

Sibley, L.D., 2010. How apicomplexan parasites move in and out of cells. Curr. Opin. Biotechnol. 21, 592–598.

Sibley, L.D., Andrews, N.W., 2000. Cell invasion by un-palatable parasites. Traffic 1, 100–106.

Sibley, L.D., Krahenbuhl, J.L., Adams, G.M., Weidner, E., 1986. *Toxoplasma* modifies macrophage phagosomes by secretion of a vesicular network rich in surface proteins. J. Cell. Biol. 103, 867–874.

Sibley, L.D., Niesman, I.R., Asai, T., Takeuchi, T., 1994. *Toxoplasma gondii*: secretion of a potent nucleoside triphosphate hydrolase into the parasitophorous vacuole. Exp. Parasitol. 79, 301–311.

Sibley, L.D., Niesman, I.R., Parmley, S.F., Cesbron-Delauw, M.F., 1995. Regulated secretion of multi-lamellar vesicles leads to formation of a tubulo-vesicular network in host cell vacuoles occupied by *Toxoplasma gondii*. J. Cell Sci. 108, 1669–1677.

Sibley, L.D., Pfefferkorn, E.R., Boothroyd, J.C., 1991. Proposal for a uniform genetic nomenclature in *Toxoplasma gondii*. Parasitol. Today 7, 327–328.

Sibley, L.D., Weidner, E., Krahenbuhl, J.L., 1985. Phagosome acidification blocked by intracellular *Toxoplasma gondii*. Nature 315, 416–419.

Silverman, J.A., Qi, H., Riehl, A., Beckers, C., Nakaar, V., Joiner, K.A., 1998. Induced activation of the *Toxoplasma gondii* nucleoside triphosphate hydrolase leads to depletion of host cell ATP levels and rapid exit of intracellular parasites from infected cells. J. Biol. Chem. 273, 12352–12359.

Silvie, O., Franetich, J.F., Charrin, S., Mueller, M.S., Siau, A., Bodescot, M., Rubinstein, E., Hannoun, L., Charoenvit, Y., Kocken, C.H., Thomas, A.W., Van Gemert, G.J., Sauerwein, R.W., Blackman, M.J., Anders, R.F., Pluschke, G., Mazier, D., 2004. A role for apical membrane antigen 1 during invasion of hepatocytes by *Plasmodium falciparum* sporozoites. J. Biol. Chem. 279, 9490–9496.

Sinai, A.P., Joiner, K.A., 2001. The *Toxoplasma gondii* protein ROP2 mediates host organelle association with the parasitophorous vacuole membrane. J. Cell Biol. 154, 95–108.

Sinai, A.P., Webster, P., Joiner, K.A., 1997. Association of host cell endoplasmic reticulum and mitochondria with the *Toxoplasma gondii* parasitophorous vacuole membrane: a high affinity interaction. J. Cell Sci. 110 (Pt 17), 2117–2128.

Singh, S., Alam, M.M., Pal-Bhowmick, I., Brzostowski, J.A., Chitnis, C.E., 2010. Distinct external signals trigger sequential release of apical organelles during erythrocyte invasion by malaria parasites. PLoS Pathog. 6, e1000746.

Sloves, P.J., Delhaye, S., Mouveaux, T., Werkmeister, E., Slomianny, C., Hovasse, A., Dilezitoko Alayi, T., Callebaut, I., Gaji, R.Y., Schaeffer-Reiss, C., Van Dorsselear, A., Carruthers, V.B., Tomavo, S., 2012. *Toxoplasma* sortilin-like receptor regulates protein transport and is essential for apical secretory organelle biogenesis and host infection. Cell Host Microbe 11, 515–527.

Sohn, C.S., Cheng, T.T., Drummond, M.L., Peng, E.D., Vermont, S.J., Xia, D., Cheng, S.J., Wastling, J.M., Bradley, P.J., 2011. Identification of novel proteins in Neospora caninum using an organelle purification and monoclonal antibody approach. PLoS One 6, e18383.

Soldati, D., Kim, K., Kampmeier, J., Dubremetz, J.F., Boothroyd, J.C., 1995. Complementation of a *Toxoplasma gondii* ROP1 knock-out mutant using phleomycin selection. Mol. Biochem. Parasitol. 74, 87–97.

Soldati, D., Lassen, A., Dubremetz, J.F., Boothroyd, J.C., 1998. Processing of *Toxoplasma* ROP1 protein in nascent rhoptries. Mol. Biochem. Parasitol. 96, 37–48.

Soldati, D., Meissner, M., 2004. *Toxoplasma* as a novel system for motility. Curr. Opin. Cell Biol. 16, 32–40.

Spano, F., Putignani, L., Naitza, S., Puri, C., Wright, S., Crisanti, A., 1998. Molecular cloning and expression analysis of a Cryptosporidium parvum gene encoding a new member of the thrombospondin family. Mol. Biochem. Parasitol. 92, 147–162.

Speer, C.A., Dubey, J.P., Blixt, J.A., Prokop, K., 1997. Time lapse video microscopy and ultrastructure of penetrating sporozoites, types 1 and 2 parasitophorous vacuoles, and the transformation of sporozoites to tachyzoites of the VEG strain of *Toxoplasma gondii*. J. Parasitol. 83, 565–574.

Speer, C.A., Tilley, M., Temple, M.E., Blixt, J.A., Dubey, J.P., White, M.W., 1995. Sporozoites of *Toxoplasma gondii* lack dense-granule protein GRA3 and form a unique parasitophorous vacuole. Mol. Biochem. Parasitol. 75, 75–86.

Srivastava, A., Singh, S., Dhawan, S., Mahmood Alam, M., Mohmmed, A., Chitnis, C.E., 2010. Localization of apical sushi protein in *Plasmodium falciparum* merozoites. Mol. Biochem. Parasitol. 174, 66–69.

Starnes, G.L., Jewett, T.J., Carruthers, V.B., Sibley, L.D., 2006. Two separate, conserved acidic amino acid domains within the *Toxoplasma gondii* MIC2 cytoplasmic tail are required for parasite survival. J. Biol. Chem. 281, 30745–30754.

Stedman, T.T., Joiner, K.A., 1999. En route to the vacuole. Tracing the secretory pathway of *Toxoplasma gondii*. In: G. S. (Ed.), Phagocytosis: microbial invasion. JAI Press Inc., Stamford, Connecticut, pp. 233–261.

Steinfeldt, T., Konen-Waisman, S., Tong, L., Pawlowski, N., Lamkemeyer, T., Sibley, L.D., Hunn, J.P., Howard, J.C., 2010. Phosphorylation of mouse immunity-related GTPase (IRG) resistance proteins is an evasion strategy for virulent *Toxoplasma gondii*. PLoS Biol. 8, e1000576.

Stewart, M.J., Schulman, S., Vanderberg, J.P., 1986. Rhoptry secretion of membranous whorls by *Plasmodium falciparum* merozoites. Am. J. Trop. Med. Hyg. 35, 37–44.

Stommel, E.W., Ely, K.H., Schwartzman, J.D., Kasper, L.H., 1997. *Toxoplasma gondii*: dithiol-induced Ca2+ flux causes egress of parasites from the parasitophorous vacuole. Exp. Parasitol. 87, 88–97.

Straub, K.W., Cheng, S.J., Sohn, C.S., Bradley, P.J., 2009. Novel components of the Apicomplexan moving junction reveal conserved and coccidia-restricted elements. Cell Microbiol. 11, 590–603.

Straub, K.W., Peng, E.D., Hajagos, B.E., Tyler, J.S., Bradley, P.J., 2011. The moving junction protein RON8 facilitates firm attachment and host cell invasion in *Toxoplasma gondii*. PLoS Pathog. 7, e1002007.

Striepen, B., He, C.Y., Matrajt, M., Soldati, D., Roos, D.S., 1998. Expression, selection, and organellar targeting of the green fluorescent protein in *Toxoplasma gondii*. Mol. Biochem. Parasitol. 92, 325–338.

Suss-Toby, E., Zimmerberg, J., Ward, G.E., 1996. *Toxoplasma* invasion: the parasitophorous vacuole is formed from host cell plasma membrane and pinches off via a fission pore. Proc. Natl. Acad. Sci. U. S. A. 93, 8413–8418.

Sweeney, K.R., Morrissette, N.S., LaChapelle, S., Blader, I.J., 2010. Host cell invasion by *Toxoplasma gondii* is temporally regulated by the host microtubule cytoskeleton. Eukaryot. Cell 9, 1680–1689.

Tan, K., Duquette, M., Liu, J.H., Dong, Y., Zhang, R., Joachimiak, A., Lawler, J., Wang, J.H., 2002. Crystal structure of the TSP-1 type 1 repeats: a novel layered fold and its biological implication. J. Cell Biol. 159, 373–382.

Taylor, S., Barragan, A., Su, C., Fux, B., Fentress, S.J., Tang, K., Beatty, W.L., Hajj, H.E., Jerome, M., Behnke, M.S., White, M., Wootton, J.C., Sibley, L.D., 2006. A secreted serine-threonine kinase determines virulence in the eukaryotic pathogen *Toxoplasma gondii*. Science 314, 1776–1780.

Thomas, A.W., Deans, J.A., Mitchell, G.H., Alderson, T., Cohen, S., 1984. The Fab fragments of monoclonal IgG to a merozoite surface antigen inhibit *Plasmodium* knowlesi invasion of erythrocytes. Mol. Biochem. Parasitol. 13, 187–199.

Tilley, M., Fichera, M.E., Jerome, M.E., Roos, D.S., White, M.W., 1997. *Toxoplasma gondii* sporozoites form a transient parasitophorous vacuole that is impermeable and contains only a subset of dense-granule proteins. Infect. Immun. 65, 4598–4605.

Tomley, F.M., Billington, K.J., Bumstead, J.M., Clark, J.D., Monaghan, P., 2001. EtMIC4: a microneme protein from *Eimeria tenella* that contains tandem arrays of epidermal growth factor-like repeats and thrombospondin type-I repeats. Int. J. Parasitol. 31, 1303–1310.

Tonkin, M.L., Crawford, J., Lebrun, M.L., Boulanger, M.J., 2012. Babesia divergens and Neospora caninum Apical Membrane Antigen 1 (AMA1) structures reveal selectivity and plasticity in apicomplexan parasite host cell invasion. Protein Sci. 22, 114–127.

Tonkin, M.L., Grujic, O., Pearce, M., Crawford, J., Boulanger, M.J., 2010. Structure of the micronemal protein 2 A/I domain from *Toxoplasma gondii*. Protein Sci. 19, 1985–1990.

Tonkin, M.L., Roques, M., Lamarque, M.H., Pugniere, M., Douguet, D., Crawford, J., Lebrun, M., Boulanger, M.J., 2011. Host cell invasion by apicomplexan parasites: insights from the co-structure of AMA1 with a RON2 peptide. Science 333, 463–467.

Topolska, A.E., Lidgett, A., Truman, D., Fujioka, H., Coppel, R.L., 2004. Characterization of a membrane-associated rhoptry protein of *Plasmodium falciparum*. J. Biol. Chem. 279, 4648–4656.

Tordai, H., Banyai, L., Patthy, L., 1999. The PAN module: the N-terminal domains of plasminogen and hepatocyte growth factor are homologous with the apple domains of the prekallikrein family and with a novel domain found in numerous nematode proteins. FEBS Lett. 461, 63–67.

Torpier, G., Charif, H., Darcy, F., Liu, J., Darde, M.L., Capron, A., 1993. *Toxoplasma gondii*: differential location of antigens secreted from encysted bradyzoites. Exp. Parasitol. 77, 13–22.

Turetzky, J.M., Chu, D.K., Hajagos, B.E., Bradley, P.J., 2010. Processing and secretion of ROP13: A unique *Toxoplasma* effector protein. Int. J. Parasitol. 40, 1037–1044.

Tyler, J.S., Boothroyd, J.C., 2011. The C-Terminus of *Toxoplasma* RON2 Provides the Crucial Link between AMA1 and the Host-Associated Invasion Complex. PLoS Pathog. 7, e1001282.

Ullrich, O., Reinsch, S., Urbe, S., Zerial, M., Parton, R.G., 1996. Rab11 regulates recycling through the pericentriolar recycling endosome. J. Cell Biol. 135, 913–924.

Umemoto, K., Leffler, H., Venot, A., Valafar, H., Prestegard, J.H., 2003. Conformational differences in liganded and unliganded states of Galectin-3. Biochemistry 42, 3688–3695.

Urban, S., Freeman, M., 2003. Substrate specificity of rhomboid intramembrane proteases is governed by helix-breaking residues in the substrate transmembrane domain. Mol. Cell 11, 1425–1434.

Vanderberg, J.P., 1974. Studies on the motility of *Plasmodium* sporozoites. J. Protozool. 21, 527–537.

Vieira, M.C., Moreno, S.N., 2000. Mobilization of intracellular calcium upon attachment of *Toxoplasma gondii* tachyzoites to human fibroblasts is required for invasion. Mol. Biochem. Parasitol. 106, 157–162.

Vulliez-Le Normand, B., Tonkin, M.L., Lamarque, M.H., Langer, S., Hoos, S., Roques, M., Saul, F.A., Faber, B.W., Bentley, G.A., Boulanger, M.J., Lebrun, M., 2012a,b. Structural and functional insights into the malaria parasite moving junction complex. PLoS Pathog. 8, e1002755.

Vulliez-Le Normand, B., Tonkin, M.L., Lamarque, M.H., Langer, S., Hoos, S., Roques, M., Saul, F.A., Faber, B.W., Bentley, G.A., Boulanger, M.J., Lebrun, M., 2012b. Structural and functional insights into the malaria parasite moving junction complex. PLoS pathogens 8, e1002755.

Wan, K.L., Carruthers, V.B., Sibley, L.D., Ajioka, J.W., 1997. Molecular characterisation of an expressed sequence tag locus of *Toxoplasma gondii* encoding the micronemal protein MIC2. Mol. Biochem. Parasitol. 84, 203–214.

Wang, J., Tolan, D.R., Pagliaro, L., 1997. Metabolic compartmentation in living cells: structural association of aldolase. Exp. Cell Res. 237, 445–451.

Werk, R., Fischer, S., 1982. Attempts to infect plant protoplasts with *Toxoplasma gondii*. J. Gen. Microbiol. 128, 211–213.

Whittaker, C.A., Hynes, R.O., 2002. Distribution and evolution of von Willebrand/integrin A domains: widely dispersed domains with roles in cell adhesion and elsewhere. Mol. Biol. Cell 13, 3369–3387.

Wiersma, H.I., Galuska, S.E., Tomley, F.M., Sibley, L.D., Liberator, P.A., Donald, R.G., 2004. A role for coccidian cGMP-dependent protein kinase in motility and invasion. Int. J. Parasitol. 34, 369–380.

Wright, C.S., Kellogg, G.E., 1996. Differences in hydropathic properties of ligand binding at four independent sites in wheat germ agglutinin-oligosaccharide crystal complexes. Protein Sci. 5, 1466–1476.

Wright, H.T., Sandrasegaram, G., Wright, C.S., 1991. Evolution of a family of N-acetylglucosamine binding proteins containing the disulfide-rich domain of wheat germ agglutinin. J. Mol. Evol. 33, 283–294.

Yamamoto, M., Ma, J.S., Mueller, C., Kamiyama, N., Saiga, H., Kubo, E., Kimura, T., Okamoto, T., Okuyama, M., Kayama, H., Nagamune, K., Takashima, S., Matsuura, Y., Soldati-Favre, D., Takeda, K., 2011. ATF6beta is a host cellular target of the *Toxoplasma gondii* virulence factor ROP18. J. Exp. Med. 208, 1533–1546.

Yamamoto, M., Standley, D.M., Takashima, S., Saiga, H., Okuyama, M., Kayama, H., Kubo, E., Ito, H., Takaura, M., Matsuda, T., Soldati-Favre, D., Takeda, K., 2009. A single polymorphic amino acid on *Toxoplasma gondii* kinase ROP16 determines the direct and strain-specific activation of Stat3. J. Exp. Med. 206, 2747–2760.

Yang, M., Coppens, I., Wormsley, S., Baevova, P., Hoppe, H.C., Joiner, K.A., 2004. The *Plasmodium falciparum* Vps4 homologue mediates multivesicular body formation. J. Cell Sci. 117, 3831–3838.

Zheng, B., He, A., Gan, M., Li, Z., He, H., Zhan, X., 2009. MIC6 associates with aldolase in host cell invasion by *Toxoplasma gondii*. Parasitol. Res. 105, 441–445.

Zhou, X.W., Blackman, M.J., Howell, S.A., Carruthers, V.B., 2004. Proteomic analysis of cleavage events reveals a dynamic two-step mechanism for proteolysis of a key parasite adhesive complex. Mol. Cell Proteomics 3, 565–576.

Zhou, X.W., Kafsack, B.F., Cole, R.N., Beckett, P., Shen, R.F., Carruthers, V.B., 2005. The opportunistic pathogen *Toxoplasma gondii* deploys a diverse legion of invasion and survival proteins. J. Biol. Chem. 280, 34233–34244.

Zinecker, C.F., Striepen, B., Tomavo, S., Dubremetz, J.F., Schwarz, R.T., 1998. The dense granule antigen, GRA2 of *Toxoplasma gondii* is a glycoprotein containing O-linked oligosaccharides. Mol. Biochem. Parasitol. 97, 241–246.

CHAPTER

13

The *Toxoplasma* Cytoskeleton: Structures, Proteins and Processes

*Naomi Morrissette**, *Marc-Jan Gubbels*[†]

*Department of Molecular Biology and Biochemistry, University of California Irvine, California, USA

[†]Department of Biology, Boston College, Chestnut Hill, Massachusetts, USA

OUTLINE

Toxoplasma gondii, Second Edition
http://dx.doi.org/10.1016/B978-0-12-396481-6.00013-1

13.1 MORPHOLOGY

13.1.1 Life Cycle and Parasite Appearance

Toxoplasma gondii is an obligate intracellular parasite that infects diverse nucleated cells in a wide variety of vertebrates, including humans. Like many other apicomplexans, *Toxoplasma* has a complex life cycle that involves differentiation into distinct forms which occupy discrete niches (Chapters 1 and 2). The three asexual invasive 'zoite' forms of *Toxoplasma* (sporozoites, tachyzoites and bradyzoites) have similar morphological properties and are likely to share many cytoskeletal features (Chobotar and Scholtyseck, 1982). Genetic recombination in *Toxoplasma* is achieved by a sexual cycle that only occurs in the intestinal epithelium of felids and consequently has not been exhaustively studied (Pelster and Piekarski, 1971; Scholtyseck *et al.*, 1971a, 1971b; Ferguson *et al.*, 1974, 1975). During this process, *Toxoplasma* macrogametes are fertilized by microgametes, which are the only developmental stage that constructs flagella to power motility (Scholtyseck *et al.*, 1972; Ferguson *et al.*, 1974). Microgametes contain apical basal bodies that template two flagellar axonemes which extend past the cell body away from the apex. Although asexual zoite forms lack flagella, they are still motile, using a characteristic actin and myosin-based gliding motility to invade and cause the acute symptoms of disease (Frixione *et al.*, 1996; Hakansson *et al.*, 1999). Since the tachyzoite form is most amenable to experimental manipulation, nearly all studies of the cytoskeleton have characterized this stage. Although it is likely that many aspects of the tachyzoite cytoskeleton will be conserved with bradyzoites and sporozoites, some cytoskeletal components are expressed or essential in a developmentally specific fashion. This includes the bradyzoite-expressed myosin TgMyoD (Herm-Gotz *et al.*, 2006; Polonais *et al.*, 2011; Delbac *et al.*, 2001) as well as proteins that are used to build the flagellar axoneme (Scholtyseck *et al.*, 1972; Ferguson *et al.*, 1974; Hodges *et al.*, 2010).

13.1.2 IMC and Pellicle-Associated Structures

Apicomplexans, along with ciliates and dinoflagellates, are classified as alveolate organisms (Cavalier-Smith and Chao, 2004; Leander and Keeling, 2003). Alveolates have a system of flattened vesicles (alveoli) that closely underlie the plasma membrane, creating a pellicle structure that is composed of three unit membranes. In apicomplexans, the patchwork of plasma membrane-associated alveoli is called the inner membrane complex (IMC). This structure is integral to *Toxoplasma* replication, motility and invasion of host cells. The apical polar ring (APR) marks the site where the IMC begins, an arrangement that leaves the extreme apical region of the parasite enclosed by only plasma membrane, perhaps to facilitate secretion (Fig. 13.1A and B). The IMC extends to the posterior of *Toxoplasma*, where the individual vesicle plates join in a turbine-shaped structure (Porchet and Torpier, 1977; Morrissette *et al.*, 1997). The elongated shape of *Toxoplasma* zoites is maintained by association of 22 evenly spaced subpellicular microtubules which interact with the cytosolic face of the pellicle (Nichols and Chiappino, 1987). These microtubules extend in a gentle spiral from the APR to end in the region beyond the nucleus. Subpellicular microtubules impose both an elongated serpentine shape and characteristic apical polarity in *Toxoplasma* and other apicomplexan parasites (Stokkermans *et al.*, 1996; Morrissette and Sibley, 2002b). The minus ends of subpellicular microtubules are inserted into the APR with attachment supported by projections of the APR which resemble a cogwheel in transverse views (Russell and Burns, 1984). The APR is therefore an unusual circular microtubule organizing centre (MTOC) that is unique to apicomplexan parasites.

The apical and basal regions of the pellicle have distinct properties and protein components from the central (lateral) region of the pellicle (Fig. 13.1C). The regular spacing between the IMC and plasma membrane is wider in an apical cap region than in the body of the parasite and is dictated in part by the length of a coiled-coil domain in the related TgGAP70 and TgGAP45 proteins which have an apical and central localization, respectively (Frenal *et al.*, 2010). In the related coccidian *Eimeria nieschulzi*, investigators have speculated that the apical cap region is more flexible in order to elongate and contract during motility and host cell invasion (Dubremetz and Torpier, 1978). Discrete pellicle domains are also defined by the ISPs (a family of novel myristoylated and palmitoylated IMC Sub-compartment Proteins) that localize to the apical (ISP1), lateral (ISP2, ISP3 and ISP4) and basal (ISP3) regions of the IMC (Beck *et al.*, 2010; Fung *et al.*, 2012). The cytoskeletal proteins TgCAP and TgDLC localize to the cap region (Hu *et al.*, 2006; Lorestani *et al.*, 2012) and TgCen2 labels a ring of approximately six annuli which mark the lower boundary of the apical cap region (Hu *et al.*, 2006). The detergent-insoluble TgPhIL1 protein, which associates with both the apical and basal portions of the IMC, is also a likely cytoskeletal component. Lastly, the basal (posterior) region of the pellicle contains a number of cytoskeletal proteins including TgMORN1, TgMyoB and C, Tg14-3-3, TgDLC, TgMSC1a and the alveolin motif containing proteins IMC5, IMC8, IMC9 and IMC13 (Lorestani *et al.*, 2012;

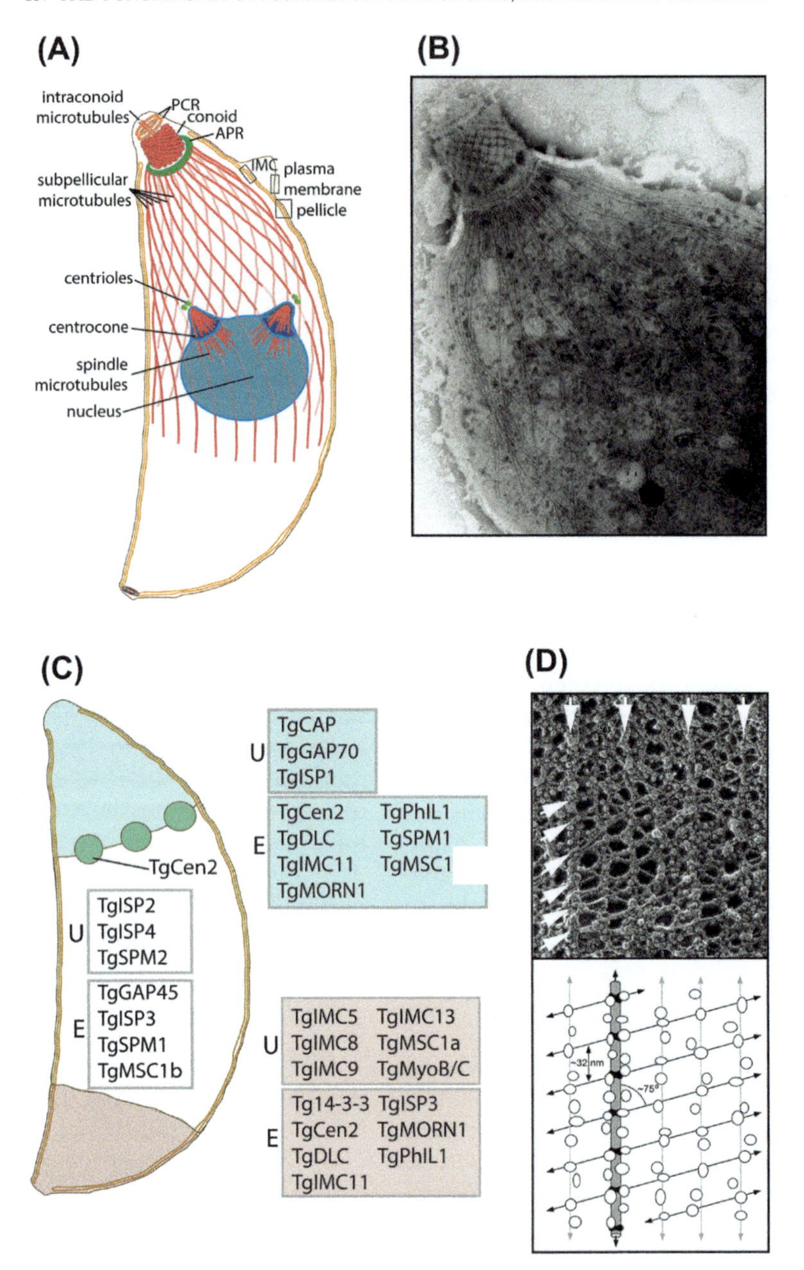

FIGURE 13.1 **Cytoskeletal elements in *Toxoplasma* tachyzoites.**
A) The *Toxoplasma* pellicle consists of the plasma membrane and an underlying patchwork of associated vesicular alveoli termed the IMC. The apical polar ring (APR) marks the site where the IMC begins, leaving the extreme apical region of the parasite only enclosed by plasma membrane. The APR is also a MTOC: twenty-two subpellicular microtubules extend from this site along the cytoplasmic face of the pellicle to a region just beyond the parasite nucleus. The conoid is a small motile organelle that can be extended from or retracted into the APR. Two fibrous preconoidal rings (PCR) associate with the conoid while two closely apposed microtubules transit the conoid lumen. Spindle microtubules originate at poles located in the

Anderson-White *et al.*, 2011; Hu, 2008; Ferguson *et al.*, 2008; Hu *et al.*, 2006; Gubbels *et al.*, 2006; Delbac *et al.*, 2001).

Freeze fracture studies of *Toxoplasma* and other apicomplexans have shown that IMC membranes are characterized by regular arrays of uniformly sized intramembranous particles (IMPs) which likely represent the transmembrane domains of integral membrane proteins (Porchet and Torpier, 1977; Morrissette *et al.*, 1997; Dubremetz and Torpier, 1978). The IMPs are organized into a highly regular two dimensional lattice of IMP rows (Fig. 13.1D). The apical cap region of the IMC contains twenty-two rows of IMPs radiating out from a ring, corresponding to the twenty-two underlying subpellicular microtubules. Each row consists of two parallel lines of IMPs separated by a small distance. Below the apical cap region, the IMC contains similar IMP rows which are immediately adjacent to each other in double particle columns. In addition, unlike in the cap region, many single particle rows are interspersed between the double particle rows. In order to accommodate increasing parasite diameter along the long axis of the parasite, the distance between double particle rows increases and the number of intervening single particle rows increases to maintain a constant spacing between rows. Moreover, the continuity of IMP rows is maintained across IMC plates, which represent topologically distinct vesicles. The organization of the particle rows suggests an intimate association with both the subpellicular microtubules and a second, non-microfilament cytoskeletal network. Fourier analysis of isolated subpellicular microtubules reveals that they are coated with a microtubule associated protein (MAP) that binds with a 32 nm periodicity (Morrissette *et al.*, 1997). Similarly, the double IMP rows overlying the microtubules also exhibit a 32 nm periodicity that is revealed by Fourier analysis of freeze-fracture images. This suggests that the MAP coordinates the close interaction of subpellicular microtubules with the IMC. The 32 nm longitudinal repeat extends to the single rows of IMPs, creating a two-dimensional lattice. The organization of the IMP rows persists to the posterior of the parasite, past the region associated with subpellicular microtubules. The integrity of the particle lattice is independent of microtubules and actin filaments, suggesting that other

cytoplasm and penetrate to the interior of the nucleus through pores in a specialized region of the nuclear membrane termed the centrocone. A centriole pair, organized in parallel configuration, is located at each spindle pole. During replication, daughter parasite buds emerge above the spindle poles but are omitted from this image for reasons of clarity. This image is based on previously published figures (Nichols and Chiappino; 1987; Hammond *et al.*, 1973).

B) A detergent-extracted, negative stained *Toxoplasma* tachyzoite cytoskeleton showing the conoid, preconoidal rings, APR and twenty-two subpellicular microtubules. This image is reproduced from Morrissette *et al.*, (1997).

C) The apical and basal regions of the tachyzoite pellicle have distinct properties and protein components from the central (lateral) region of the pellicle. The apical cap is defined by the TgISP1 marker and is also uniquely (U) the location of the cytoskeletal proteins TgGAP70 and TgCAP. Other listed cytoskeletal components are enriched (E) in this region. The lower boundary of the apical cap is defined by a ring of approximately five to six TgCen2-containing annuli. Conversely, several other proteins are located in the lateral and/or basal regions of mature parasites and are excluded from the cap region. The lateral region is uniquely (U) labelled by TgISP2 and the microtubule associated protein SPM2, while several proteins are also found in the apical (TgSPM1) or basal (TgISP3, TgGAP45 and TgMSC1b) regions. The basal (posterior) region of the pellicle is associated with a specific set of cytoskeletal proteins which mediate cytokinesis and remain assembled at the parasite posterior. While most components are found in the basal complex structures of developing daughter buds, TgMSC1a is only a marker of the mature basal complex. A number of proteins are uniquely found in this pellicle compartment (U) while others are enriched here but also found in an apical (TgPhIL1, TgDLC) or lateral (TgISP3) location.

D) A proposed filament system that links the underlying subpellicular microtubules to a lattice of intramembranous particles (Morrissette *et al.*, 1997) appears quite similar to the organization of presumptive alveolin proteins revealed by freeze-drying a glycerol and detergent-extracted replica for electron microscopy, image courtesy of John Heuser, reproduced from Morrissette and Sibley (2002a).

IMC-associated cytoskeletal filaments organize this structure. A filamentous network with a similar organization to the IMP lattice can be observed after glycerol or deoxycholate-extraction of *Toxoplasma* or other apicomplexans (Morrissette and Sibley, 2002a; D'Haese *et al.*, 1977). The filaments have a diameter of 8–10 nm and the extracted network maintains the general shape of the parasite (Fig. 13.1D).

Initial studies of detergent-insoluble pellicle proteins characterized four detergent-insoluble IMC proteins (IMC1, 2, 3 and 4) that associate with the IMC of nascent and mature parasites (Mann and Beckers, 2001; Hu *et al.*, 2006; Gubbels *et al.*, 2004). IMC1, 3 and 4 are members of a large family of proteins that contain 'alveolin' repeats, found in diverse alveolate organisms. The alveolin repeat has conserved 'EKIVEVP' and 'EVVR' or 'VPV' sub-repeats in valine- and proline-rich domains (Gould *et al.*, 2008; Anderson-White *et al.*, 2011). These proteins likely function to provide tensile strength to parasites, similar to intermediate filament proteins found in metazoan organisms. The *Toxoplasma* genome contains genes for a large family of alveolin IMC proteins (Anderson-White *et al.*, 2011). Some alveolin proteins are found in a wide variety of apicomplexans, likely indicating that these proteins have conserved roles in building the IMC-associated cytoskeleton. However, the alveolin protein family is expanded in *Toxoplasma*, allowing non-conserved IMC proteins to coordinate *Toxoplasma*-specific functions (such as endodyogeny) that are not shared by all members of the Apicomplexa (see below).

In addition to IMC proteins and microtubules, actin and the class XIV myosins TgMyoA and TgMyoD localize to the space between the plasma membrane and the IMC (Meissner *et al.*, 2002; Polonais *et al.*, 2011; Herm-Gotz *et al.*, 2006; Gaskins *et al.*, 2004). TgMyoA and actin filaments power a gliding motility that is critically important to invasion of host cells (Dobrowolski and Sibley, 1996; Dobrowolski *et al.*, 1997a; Skillman *et al.*, 2011; Meissner *et al.*, 2002). TgMyoA associates with the IMC via interactions with a group of glideosome associated proteins (GAP40, GAP45/GAP70 and GAP50). Microfilaments indirectly associate with the plasma membrane: the cytoplasmic tail of the secreted MIC2 adhesin binds to the glycolytic enzyme aldolase which in turn interacts with actin (Jewett and Sibley, 2003; Starnes *et al.*, 2006; Starnes *et al.*, 2009).

13.1.3 Apical Structures

Toxoplasma and other members of the phylum Apicomplexa are named for their distinctive polarized apex, which contains organelles that coordinate interaction with host cells. Apical secretory organelles (micronemes and rhoptries) release components that mediate gliding motility, host cell invasion and establishment of the parasitophorous vacuole. The gregarine and coccidian subsets of apicomplexans (the Conoidasida) have an additional apical structure termed the conoid (Leander and Keeling, 2003). The *Toxoplasma* conoid is composed of approximately ten to 14 ~430 nm long fibres that follow a left-handed helical path to create a 380 nm diameter funnel-shaped structure that is thought to facilitate invasion of host cells (Nichols and Chiappino, 1987; Hu *et al.*, 2002b; Morrissette *et al.*, 1997) (Fig. 13.1A and B). Although conoid filaments contain tubulin, cross-sections through the fibres appear as a nonsymmetrical 'comma' shape with nine subunits rather than closed tubules (Hu *et al.*, 2002b). The conoid is surmounted by two preconoidal rings and two ~400 nm long microtubules are located within the conoid (Nichols and Chiappino, 1987). The intraconoid microtubules are closely apposed and appear connected to the preconoidal rings by filaments. A complex composed of the conoid, intraconoid microtubules and preconoidal rings can extend beyond the APR or retract within it to be surrounded by the corset of subpellicular microtubules. Conoid extension and retraction is visible during invasion and can be triggered in extracellular parasites by ionomycin treatment (Mondragon and Frixione, 1996; Del Carmen *et al.*, 2009).

The origins of the conoid structure are enigmatic. Conoids are found in apicomplexan gregarines such as *Selenidium* and in coccidian parasites including *Toxoplasma* (Leander and Keeling, 2003). Other apicomplexans such as *Plasmodium* and *Cryptosporidium* lack conoids, most likely because this organelle was lost from these lineages. Remarkably, some closely related non-apicomplexan alveolates (*Colpodella, Chromera, Cryptophagus* and *Perkinsus*) have similar apical structures termed 'incomplete conoids' or 'pseudoconoids' and apical secretory organelles that resemble micronemes and rhoptries. The 'incomplete conoid' or 'pseudoconoid' structures consist of an open-sided cone built from a set of nearly vertical apical microtubules which are separate from adjacent subpellicular microtubules. It is therefore likely that the common ancestor of apicomplexans and dinoflagellates had an open-sided conoid that was lost from dinoflagellates and modified into a closed conoid in apicomplexans (Leander and Keeling, 2003). The biology of alveolates that contain incomplete conoids or pseudoconoids may provide insight into how apicomplexan parasites evolved. For example, *Colpodella vorax* and *Rastrimonas subtilis* (previously *Cryptophagus subtilis*) use apical secretory organelles and a pseudoconoid to partially (*Colpodella*) or fully (*Rastrimonas*) invade other free-living aquatic protozoa in order to aspirate their cytoplasmic contents or to grow intracellularly (Brugerolle, 2002a, b; Leander *et al.*, 2003). Notably, in these organisms, incomplete conoids or pseudoconoids are adjacent to apical basal bodies with associated flagella while subpellicular microtubules are not organized by an APR. In fact, the morphology of these organisms is most like the organization of apicomplexan microgametes which have a set of short microtubules adjacent to apical basal bodies with associated flagellar axonemes (Ferguson *et al.*, 1974; Scholtyseck *et al.*, 1971b). In contrast, the 'closed conoid' of asexual stage coccidians can be extended through or retracted through the adjacent APR into a basket of associated subpellicular microtubules and flagella are not present. *Rastrimonas* absorbs its flagella during intracellular growth and this feature may have become more pronounced during the evolution of apicomplexan parasites, ultimately leading to the loss of flagella in all stages other than microgametes. Perhaps as a consequence, the centrioles (basal bodies without an associated transition zone) reside above the nucleus rather than at the apical end of the parasite. Recent studies indicate that at least three proteins that associate with the base of flagella (TgSAS6L, TgSFA2 and TgSFA3) are retained in *Toxoplasma* tachyzoites (see below) (de Leon *et al.*, 2013 Francia *et al.*, 2012). Notably, centrioles have only been documented in the asexual stages of coccidian apicomplexans and other apicomplexan species (*Plasmodium, Cryptosporidium*) apparently lack both conoids and centrioles (see below).

13.1.4 Basal Structures

The posterior end of *Toxoplasma* contains a novel cytoskeletal assembly, the basal complex, which functions in place of a contractile ring to complete cytokinesis at the conclusion of parasite replication (Anderson-White *et al.*, 2011; Beck *et al.*, 2010; Hu, 2008; Hu *et al.*, 2006). Daughter basal complexes emerge proximal to duplicated centrioles and extend at the leading edge of the IMC of lengthening daughter buds. TgMORN1 (TGGT1_086660) is a scaffolding protein that appears early during the emergence of daughter buds. Other basal complex proteins are recruited when the leading edge is at its widest point. These include the myosins TgMyoB and C (TGGT1_076590), as well as Tg14-3-3 (TGGT1_043200), TgCen2 (TGGT1_101950), TgDLC1 (TGGT1_066540) and the alveolin motif containing proteins IMC5 (TGGT1_079150), IMC8 (TGGT1_079140), IMC9

(TGGT1_081370) and IMC13 (TGGT1_002640) (Lorestani *et al.*, 2010; Anderson-White *et al.*, 2011; Hu, 2008; Delbac *et al.*, 2001). The assembly of basal complexes is independent of the structural integrity of the daughter cortical cytoskeleton but requires the action of TgMORN1, as loss of this protein is associated with organelle segregation defects and the formation of conjoined daughters (Gubbels *et al.*, 2006; Heaslip *et al.*, 2010a; Lorestani *et al.*, 2010). Moreover, although microtubules are not evident in this region by immunofluorescence or electron microscopy, an antibody that recognizes detyrosinated tubulin brightly labels the basal region (Xiao *et al.*, 2010). Once daughter cell buds have fully elongated to enclose a complete complement of cellular organelles, the basal complex constricts to close off the IMC at the posterior of the parasite. In contrast to the contractile ring, which disassembles at the completion of cytokinesis, the mature basal complex (the basal cap) is a permanent cytoskeletal structure located at the posterior of tachyzoites, suggesting that it continues to play a role in maintenance of IMC closure (Fig. 13.1C). TgMSC1a (TGGT1_098490) is a specific marker of mature basal caps; it does not localize to basal complexes in developing daughter buds (Lorestani *et al.*, 2012).

13.1.5 The Nucleus

Replication in *Toxoplasma* can occur by endodyogeny (Sheffield and Melton, 1968), which creates two daughter parasites per replication cycle, or by endopolygeny, where many daughters synchronously bud from a mother parasite (Piekarski *et al.*, 1971; Ferguson *et al.*, 1974; Sheffield and Melton, 1968; Gubbels *et al.*, 2008). In both processes, the nuclear membrane remains intact during mitosis and chromosome segregation occurs without chromosome condensation (Fig. 13.3). TgMORN1 (TGGT1_086660) is a marker of the centrocone, a specialized apical region of the nuclear envelope associated with the spindle poles (Ferguson *et al.*, 2008; Gubbels *et al.*, 2006; Heaslip *et al.*, 2010a; Lorestani *et al.*, 2010). During mitosis, spindle microtubules originate in the cytoplasm, proximal to asexual stage centrioles, and pass through nuclear pores located in the region of the centrocone to mediate chromosome segregation (Morrissette and Sibley, 2002b; Stokkermans *et al.*, 1996; Swedlow *et al.*, 2002; Striepen *et al.*, 2000) (Fig. 13.1A). The chromosome centromeres appear to be durably associated with the centrocone throughout the cell cycle, although it is not clear whether this association depends upon permanent spindle microtubules or anchoring via other nuclear membrane components (Brooks *et al.*, 2011).

13.1.6 Centrioles, Centrosomes and Basal Bodies

Centrioles and basal bodies are highly ordered structures that are conserved in organisms ranging from protozoa to vertebrates. Typical centrioles are barrel-shaped, 100–250 nm in diameter and 100–400 nm in length (Dutcher, 2004; Preble *et al.*, 2000). Most centrioles have a '9+0' structure of triplet microtubules, exist in pairs and are arranged orthogonally after duplication. Atypical centriole organization occurs in some organisms: centrioles in *Caenorhabditis elegans* have nine singlet microtubules while those in *Drosophila melanogaster* embryos have nine doublets. *Toxoplasma* tachyzoite centrioles have an unconventional form consisting of a central tubule surrounded by nine singlet microtubules (Morrissette and Sibley, 2002b). This organization also occurs in other coccidian parasites such as *Besnoitia jellisoni* and *Eimeria bovis*; however, non-coccidian apicomplexans such as *Plasmodium* species appear to lack asexual stage centrioles altogether (Sheffield, 1966; Dubremetz and Elsner, 1979; Morrissette and Sibley, 2002b; Sinden, 1982, 1983, 2010; Sinden and Smalley, 1979), although they encode several centrin proteins of which PfCentrin3 localizes to a discrete perinuclear region (Mahajan *et al.*, 2008). The available transmission electron microscopy images of thin sections through coccidian centrioles indicate that they are quite short and are arranged in a parallel rather than orthogonal configuration. To date, there are no available images of procentriole forms during centriole duplication (Naomi Morrissette, personal observation). In contrast to centrioles, centrosomes consist of a centriole pair surrounded by pericentriolar matrix (PCM) which contains γ-tubulin, pericentrin and ninein (Rieder *et al.*, 2001). This term was coined to define the MTOC of metazoan cells which contains centrioles and PCM. In the current *Toxoplasma* literature, some investigators describe the behaviour of 'centrioles' to reflect the apparent lack of PCM components (pericentrin and ninein) and characteristic morphology while others use the term 'centrosome' to convey the microtubule organizing capacity of this centriole structure.

Basal bodies are structurally identical to centrioles, but are continuous with a transition zone and associated flagellar axoneme. The typical flagellar axoneme has a central pair of microtubules surrounded by nine doublet microtubules (9+2 organization). The conversion from 9+0 triplet microtubules in the basal body to 9+2 doublets in the axoneme occurs in the transition zone, which contains the machinery for intraflagellar transport required to assemble the flagellum. *Toxoplasma* microgametes are the only stage of this organism that builds flagella for motility (Mehlhorn, 1972; Ferguson *et al.*, 1974). It is possible that in addition to building a transition zone, the singlet microtubule centrioles found in asexual stages must be converted into a structure with triplet blades in order to template the flagellar axoneme (Fig. 13.2). In *Plasmodium* spp., the asexual stages appear to lack centrioles and basal bodies are apparently formed *de novo* during microgametogenesis (Sinden, 1982, 1983; Sinden *et al.*, 2010). Transverse sections through the barrel of the basal body structure are rare, and some images show a form consisting of nine singlet microtubules surrounding a central tubule. Since these basal bodies are believed to appear *de novo*, this may represent an incomplete or intermediate structure during biogenesis. A conventional '9+0' triplet basal body has been observed in a rare image of an *Eimeria* microgamete (David Ferguson, personal communication). However, it is unclear if the asexual stage *Toxoplasma* centrioles mature to form a basal body with triplet microtubules or whether microgamete basal bodies are formed *de novo* as in *Plasmodium* spp.

13.2 CYTOSKELETAL ELEMENTS

13.2.1 Microtubules, MAPs, Motors and MTOC

Microtubules in *Toxoplasma* and other apicomplexans are critically important components of the mitotic spindle used for chromosome segregation during cell division (Morrissette and Sibley, 2002b; Shaw *et al.*, 2000; Stokkermans *et al.*, 1996; Swedlow *et al.*, 2002). They are also important structural components of the membrane cytoskeleton, intimately associating with the overlying pellicle to confer apical polarity and a rigid elongated cell shape. The microtubule-containing flagellar axoneme is constructed in a developmentally regulated fashion in order to drive microgamete motility leading to fertilization of the macrogamete (Ferguson *et al.*, 1974, 1975).

13.2.1.1 α- and β-Tubulins

Microtubules are built by polymerization of α-β tubulin heterodimer subunits and typically contain 13 protofilaments. Many organisms have more than one gene for α- and/or β-tubulin subunits. These tubulin isotypes have slightly different amino acid sequences and cells can modulate the properties of a microtubule population by changing relative isotype expression

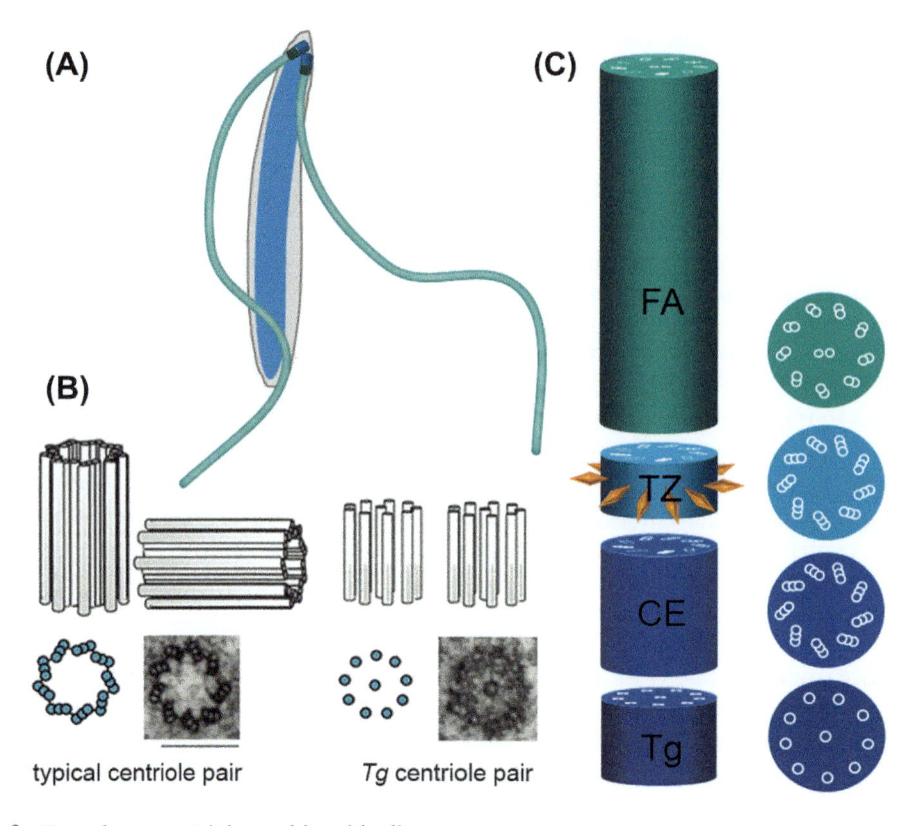

FIGURE 13.2 ***Toxoplasma* centrioles and basal bodies.**
A) A diagram of a *Toxoplasma* microgamete, which is the only stage to build flagella for motility: microgametes contain apical basal bodies that template two flagellar axonemes which extend past the cell body away from the apex.
B) In organisms ranging from *Chlamydomonas* to humans centrioles are made of triplet microtubules and are organized in an orthogonal configuration. In *Toxoplasma* tachyzoites, two short centrioles consisting of nine singlet microtubules are found in parallel at spindle poles. Electron micrographs of human (left) and tachyzoite (right) centriole structure: scale bar is 200 nm.
C) Studies from diverse organisms have established that in order for centrioles (CE) to become basal bodies, cells must build a centriole extension termed the transition zone (TZ) which templates the associated 9+2 flagellar axoneme (FA). If the singlet microtubule centrioles found in asexual stage *Toxoplasma* (Tg) are used to template flagella in microgametes, this function may require that they first be converted into a structure with triplet blades.

levels. In some organisms specific isotypes are required for construction of specialized structures, such as the flagellar axoneme. The apicomplexan parasites *P. falciparum and P. berghei* have two unlinked α-tubulin genes: α-tubulin-II is especially highly expressed in male gametes, where it localizes to the axoneme whereas α-tubulin-I appears to be a largely a housekeeping protein (Holloway *et al.*, 1989, 1990; Rawlings *et al.*, 1992; Kooij *et al.*, 2005; Schwank *et al.*, 2010). The *Toxoplasma* genome contains genes for three α-tubulin isotypes and three β-tubulin isotypes (Xiao *et al.*, 2010; Nagel and Boothroyd, 1988). The α1-tubulin isotype (TGGT1_094210) is abundantly represented in both tachyzoite and oocyst proteomes and is mutated in all identified cases of dinitroaniline resistance (Morrissette *et al.*, 2004; Ma *et al.*, 2007). Mass spectrometry and RNA-seq data deposited at http://ToxoDB.org suggest that the α2-tubulin gene (TGGT1_116000) may be expressed at low levels in tachyzoites and is clearly expressed in oocysts

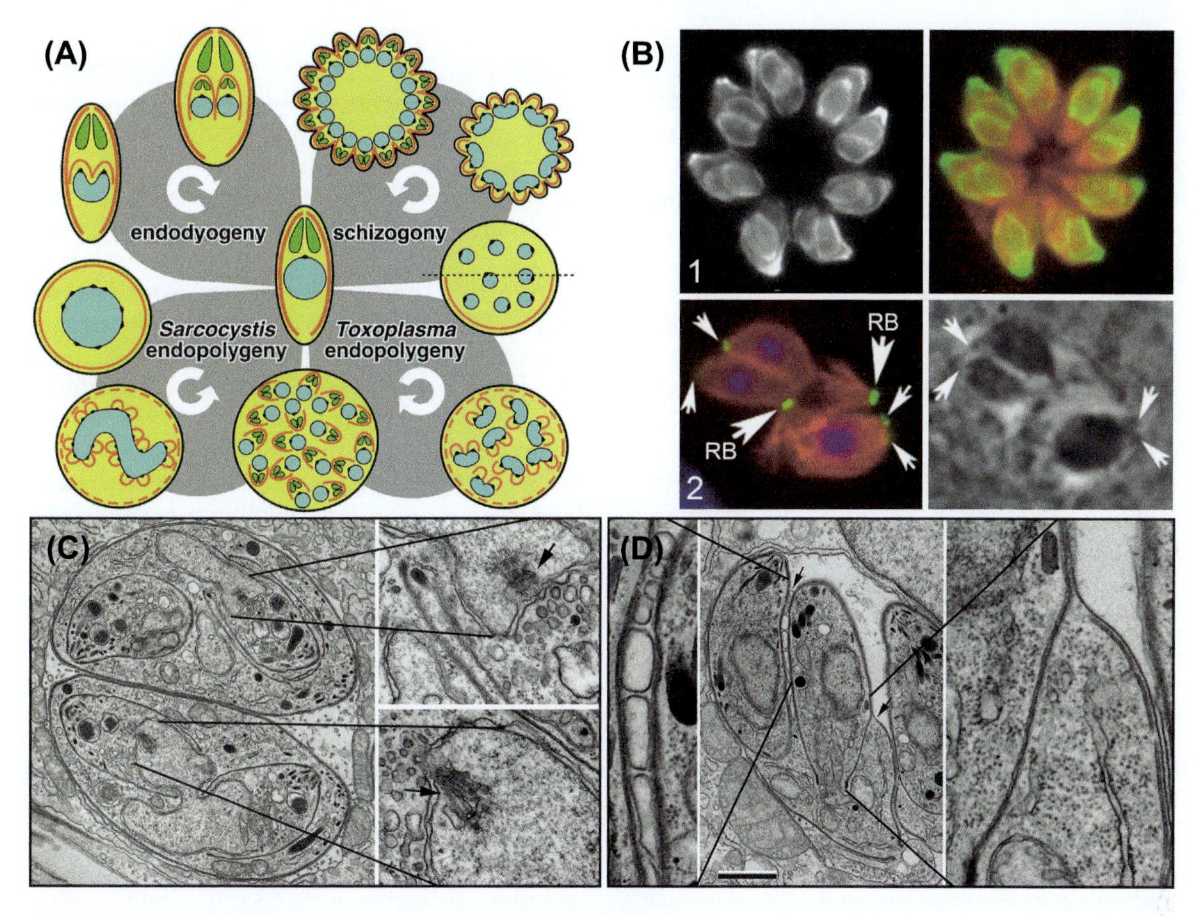

FIGURE 13.3 **Replication in *Toxoplasma* and other apicomplexans.**

A) Apicomplexans exhibit cell division flexibility. The beginning and end product of division is the host cell invasion-competent 'zoite' (centre). Endodyogeny is the formation of two daughters through internal budding. In schizogony, parasites first go through multiple rounds of DNA replication and mitosis before budding. It is set apart from other strategies in that budding takes place at the plasma membrane, and that the multi-nucleated schizont does not contain an IMC (marked by the black dotted line, central right panel). Division by endopolygeny as carried out by *Toxoplasma* in the cat gut also involves multiple rounds of DNA replication and mitosis but budding takes place internally within the cytoplasm. Endopolygeny in *Sarcocystis neurona* goes through several rounds of DNA replication but omits karyokineses and budding takes place internally within the cytoplasm. Regardless of the division mode, the assembly of daughters is linked to a final round of DNA replication and mitosis. In this image, the nucleus is light blue; the IMC cytoskeleton is red and apical rhoptry organelles are green. The black knobs on the nucleus represent the spindle pole/centriolar complex. This image was modified from Ferguson *et al.* (2008).

B1) Immunofluorescence image of mature daughter buds defined by subpellicular microtubules (green) inside replicating tachyzoites (red).

B2) Newly emerging daughters defined by subpellicular microtubules (red), RNG1 (green) and DNA (blue) discard the maternal peripheral cytoskeleton into the residual body (RB). This image is reproduced from Tran *et al.* (2010).

C) Electron microscopy captures the spindle poles (enlarged) within developing daughter buds during the process of endodyogeny. This image was originally published in Morrissette and Sibley (2002b).

D) Daughter buds emerge by annealing their IMC membranes to the maternal plasma membrane as the maternal IMC is discarded along with associated cytoskeletal elements. The scale bar is 1 μm and this image was originally published in Morrissette and Sibley (2002b).

while the α3-tubulin gene (TGGT1_116270) is expressed in both tachyzoites and oocysts. The β1-tubulin gene (TGGT1_075980) is mutated in a subset of parasite lines that have acquired compensatory mutations to modulate fitness defects conferred by dinitroaniline resistance mutations in the α1-tubulin gene (Ma *et al.*, 2008). Since similar mutations were not identified in β2- or β3-tubulin genes, this suggests that the α1- and β1-tubulin proteins form a prominent population of tubulin heterodimers. Mass spectrometry and RNA-seq data deposited at http://ToxoDB.org suggest that both β2- (TGGT1_065490) and β3- (TGGT1_071990) isotypes are expressed in tachyzoites and oocysts. Although the three β-tubulin isotypes are quite similar (~98% identity), the α-tubulin isotypes have dramatically distinct amino acid sequences (40–70% identity). Protein alignments indicate that the three α-tubulins are conserved in the regions of the tubulin signature motif (GTP binding) and in α-helices 11 and 12 which interact with motors and MAPs (Naomi Morrissette, personal observation). However, other significant domains, such as the H1–S2 and M loops that coordinate interactions between protofilaments in the microtubule lattice are surprisingly diverse. For example, of the 39 amino acids that define the H1–S2 loop, only ten are conserved between α1- and α3-tubulins. Similarly, there are 16 residues in the M loop, and only six are conserved between α1- and α3-tubulins. For comparison, there are only two differences in the H1–S2 loop and one difference in the M loop when *Toxoplasma* α1-tubulin is compared to bovine tubulin. The most striking difference between α1- and α2-tubulins is that α2-tubulin has a four amino acid insert in the H1–S2 loop which may increase the flexibility and/or strength of α2-containing microtubules. Notably, α2-tubulin residues P165, V238 and M378 and the α3-tubulin residues F136 and L235 correspond to point mutations in α1-tubulin that confer dinitroaniline resistance, perhaps indicating that α2- and α3-tubulins do not bind to dinitroanilines (Ma *et al.*, 2007; Morrissette *et al.*, 2004). The α3-tubulin subunit also has an uncharacteristically long C-terminal tail that contains a 35 amino acid extension. In contrast to the globular nature of most of the tubulin molecule, the C-terminal tails of α- and β-tubulins are unstructured and extend from the microtubule surface to promote interactions with MAPs and motors. The extended α3-tubulin tail is much less acidic than the portion that is closer to the microtubule surface and is also likely to be unstructured because there are four prolines along its length.

Toxoplasma microtubule polymerization is inhibited by dinitroanilines which selectively bind to plant and protozoan tubulin but not vertebrate tubulins (Morrissette and Sept, 2008). Computational modelling suggests that dinitroanilines bind to the *Toxoplasma* α1-tubulin subunit but not to vertebrate α-tubulin, consistent with the selective binding of these compounds to sensitive tubulins (Mitra and Sept, 2006; Morrissette *et al.*, 2004). The proposed dinitroaniline binding site is situated beneath the H1–S2 (N) loop and binding is likely to inhibit protofilament–protofilament interactions, causing microtubule disruption. Diverse single point mutations to *Toxoplasma* α1--tubulin can confer resistance to dinitroanilines (Ma *et al.*, 2007; Morrissette *et al.*, 2004). The individual substitutions are clustered in a core region and in areas of subunit contact within the microtubule lattice. Biochemical studies using resistant α-tubulins expressed in *Tetrahymena* indicate that resistance mutations can reduce tubulin affinity for dinitroanilines and/or increase subunit–subunit affinity within the microtubule lattice to compensate for the destabilizing effect of dinitroanilines (Lyons-Abbott *et al.*, 2010). Some of the mutations cause tachyzoites to have longer subpellicular microtubules, suggesting that increased microtubule stability shifts the dimer–polymer equilibrium. *Toxoplasma* lines with resistance mutations have increased incidence of replication defects due to the failure of spindle microtubules and subpellicular microtubules to work in synchrony (Ma *et al.*, 2007). In the absence of continuous dinitroaniline selection, spontaneous suppressor mutations appear and reduce both replication defects and oryzalin resistance. A larger study of the relationship of the effect of compensatory mutations on fitness and resistance identified 29 novel α1-tubulin and 17 novel β1--tubulin mutations that reduce fitness defects in dinitroaniline resistant lines (Ma *et al.*, 2008). In the absence of dinitroanilines, lines with compensatory mutations have increased fitness relative to resistant parental strains, but are still less fit than wild type parasites. Moreover, compensatory mutations lower the level of resistance conferred by the primary α1-tubulin mutations, indicating that dinitroaniline resistance is associated with a fitness disadvantage and increased fitness with compensatory mutations occurs at the cost of resistance.

Tubulin dimers can be altered by post-translational modifications that differentially mark distinct microtubule subpopulations within a cell. The α-tubulin subunit can be acetylated on K40 and its C-terminal tail can be reversibly modified by deletion of the terminal tyrosine (de-tyrosination) or irreversibly modified by deletion of the last two residues (Δ2). The C-terminal tails of both α- and β-tubulin can be reversibly modified by glutamylation and/or glycylation. The recent identification of enzymes that modify tubulin has allowed researchers to investigate the effects of specific post-translational modifications. Deletion of enzymes that acetylate or glycylate tubulin in related alveolate *Tetrahymena* is associated with increased lability of microtubules indicated by increased resistance to Taxol and increased sensitivity to dinitroanilines (Akella *et al.*, 2010; Wloga and Gaertig, 2010; Wloga *et al.*, 2008). Although it is possible that post-translational modifications directly influence microtubule stability, most evidence indicates that post-translational modifications alter how microtubules interact with associated proteins and this in turn influences the relative sensitivity of microtubules to stabilizing or destabilizing drugs. A recent study used mass spectrometry (MS) and modification-specific antibodies to document diverse post-translational modifications to *Toxoplasma* tubulin (Xiao *et al.*, 2010). Post-translational modifications to α1--tubulin include acetylation of K40, detyrosination (removal of the C-terminal Y453) and truncation of the last five amino acids (ΔYGDEY). The last alteration is apparently a novel modification of tubulin in *Toxoplasma*. Notably, only α1-tubulin is susceptible to detyrosination, as α2-tubulin ends with a valine and α3-tubulin ends with an asparagine. Similarly, although α1-tubulin has a K40 which is acetylated, α2-tubulin lacks a lysine in this region and the α3-tubulin K39 is in a distinct local environment that may not permit acetylation. Both α1- and β1-tubulins are polyglutamylated, a modification observed on axoneme, centriole and some spindle microtubules. Glycylation, a modification that is restricted to ciliated cell types where it is enriched on axonemes and basal bodies, may be reserved for flagellated male gametes and was not reported in this study. Significantly, this study uncovered a novel tubulin modification (methylation) through the fortuitous cross-reactivity of an antibody that detects methylated histones. This modification was confirmed by mass spectrometry, which detected methylation of the C-terminal tails of α1- and β1-tubulin.

13.2.1.2 Microtubule Associated Proteins (TgICMAP1, SPM1 and SPM2)

Toxoplasma contains at least five distinct tubulin-containing structures: spindle, subpellicular and intra-conoid microtubules, the centrioles and the conoid. Each of these is part of a complex structure, which likely contains a unique set of microtubule associated proteins (MAPs). A mass spectrometry survey of proteins found in a detergent-extracted sample enriched for the APR, conoid and subpellicular microtubules identified a number of peptides that potentially interact with the *Toxoplasma* tachyzoite membrane skeleton (Hu *et al.*, 2006). A number of hits from this screen have been validated as cytoskeletal proteins. These include TgICMAP1,

TgIMC4, TgSPM1, TgSPM2, TgCen2, TgCen3 and a dynein light chain (TgDLC1) (Hu *et al.*, 2006; Tran *et al.*, 2012; Heaslip *et al.*, 2009).

13.2.1.2.1 TgICMAP1

TgICMAP1 (TGGT1_051670) is a 135 kDa protein that localizes to the intra-conoid microtubules in *Toxoplasma* tachyzoites (Heaslip *et al.*, 2009). It contains a SMC-like domain and tagged truncation mutants indicate that this motif is required for localization to the intraconoid microtubules. Both *in vitro* assays and ectopic expression in mammalian cells indicate that TgICMAP1 directly binds to microtubules. Although there are *Neospora* and *Eimeria* homologues of ICMAP1, there are no homologues in *Plasmodium* or *Cryptosporidium*. This is consistent with its localization to intraconoid microtubules, which do not exist in non-coccidian apicomplexans.

13.2.1.2.2 TgSPM1

TgSPM1 (subpellicular microtubule 1; TGGT1_043740) is localized along the full length of subpellicular microtubules in *Toxoplasma* tachyzoites, but does not localize to other tubulin containing structures (Tran *et al.*, 2012). TgSPM1 protein has six copies of a 32 amino acid repeat. SPM1 homologues are found in other apicomplexans: there are five (considerably more degenerate) copies of the 32 amino acid repeat in *Cryptosporidium* SPM1 and in different *Plasmodium* species SPM1 homologues vary in size because the repeat has expanded to different degrees. Interestingly, the *Plasmodium* repeats show the highest degree of conservation, which may be the consequence of successive gene recombination events that homogenized the sequence while increasing the motif copy number. The repeats are critical for correct localization of the *Toxoplasma* SPM1 protein and increased deletion of repeats is associated with reduced and ultimately eliminated localization. A TgSPM1 null line has reduced fitness and the subpellicular microtubules have increased susceptibility to DOC-extraction relative to matched control parasites. When the null line is complemented with SPM1−YFP, tachyzoite subpellicular microtubules are again detergent-stable.

13.2.1.2.3 TgSPM2

TgSPM2 (subpellicular microtubule 2; TGGT1_038020) has a more restricted localization pattern than SPM1: it is associated with the middle portion of the subpellicular microtubules (Tran *et al.*, 2012). SPM2 lacks characterized protein motifs and there are apparent homologues of this protein in *Neospora* and *Plasmodium* but not *Cryptosporidium*. Protein alignments indicate that conservation of *Plasmodium* and *Toxoplasma* SPM2 proteins is restricted to residues 49−60 and 189−268 of the *Toxoplasma* protein. Surprisingly, although the majority of conservation between *Toxoplasma* and *Plasmodium* SPM2 proteins is in the C-terminus, the N-terminal 63 residues are required for localization to the apical periphery of mother and daughter parasites. Deletion of SPM2 indicates that the protein is not essential. Competition assays show that the *spm2* null line grows comparably to control lines, indicating that the SPM2 protein is not required for overall tachyzoite fitness *in vitro*.

13.2.1.2.4 EB1

EB1 is found at the plus ends of dynamic microtubules in organisms ranging from land plants to budding and fission yeast to vertebrates (Lansbergen and Akhmanova, 2006). An alignment of the *Toxoplasma* homologue of EB1 with human and *Chlamydomonas* EB1 proteins indicates that TgEB1 (TGGT1_083050) has an extension of ~70 amino acids at the C-terminus and contains an insert in the central portion of the protein. TgEB1−YFP localizes diffusely in the cytoplasm early in daughter replication. Upon centriole duplication, it is found between but not coincident with duplicated centrioles, consistent with localization to the ends of mitotic microtubules. As growth of daughter buds continues, TgEB1−YFP is located adjacent to each duplicated centriole as well as diffusely in the cytoplasm of the growing daughter. This persists until replication begins again. Notably, TgEB1—YFP does not localize to the tips of extending subpellicular microtubules of daughter parasites, consistent with the non-dynamic nature of these microtubules (J. de Leon and N. Morrissette, unpublished observations).

13.2.1.3 Microtubule Motors

Microtubule motors are multi-subunit complexes that harness energy released by ATP hydrolysis to move along the length of microtubules. Kinesins typically move toward the microtubule plus end while dyneins move toward the microtubule minus end. The *Toxoplasma* genome has a number of genes that encode kinesin and dynein subunits. With the exception of one dynein light chain (TgDLC1), which localizes to the spindle poles, centrioles, basal complex and conoid region of tachyzoites (Hu *et al.*, 2006), nothing is known of the localization or putative functions of these motors.

13.2.1.3.1 KINESINS

In many eukaryotic organisms, kinesins have been shown to function in mitosis and meiosis and in transport of cellular cargo. Typically, kinesins are composed of two heavy and two light chains (Verhey *et al.*, 2011; Verhey and Hammond, 2009; Miki *et al.*, 2005). The heavy chain has a globular head containing the motor domain connected to a stalk region. The two intertwined stalks interact with kinesin light chains in a tail region. The light chains coordinate motor association with cargo. The *Toxoplasma* genome encodes 19 kinesin heavy chain homologues, but lacks annotated kinesin light chains. Motor domains may be in the centre of the kinesin heavy chain (Kin I motors), or at the N-terminal (Kin N) or C-terminal (Kin C) regions. Kin N kinesins are plus-end directed motors, Kin C kinesins are minus-end directed motors and Kin I kinesins typically disrupt microtubules rather than move on their surface (Verhey *et al.*, 2011; Desai and Walczak, 2001; Kardon and Vale, 2009; Miki *et al.*, 2005; Reed *et al.*, 2006; Verhey and Hammond, 2009). *Toxoplasma* has an apparent orthologue (TGGT1_038260) of a Kin I motor from *P. falciparum* that has been used to model microtubule interactions for this class of kinesins (Mulder *et al.*, 2009). Based on the location of motor domains within the other *Toxoplasma* kinesin heavy chains, it appears that there are 15 (+) end directed Kin N kinesins and 3 (−) end directed Kin C motors. Although the kinesin heavy chains components have not been characterized to date, mass spectrometry data in ToxoDB suggests nine of the 19 are expressed in tachyzoites (Naomi Morrissette, personal observation).

13.2.1.3.2 DYNEINS

Dyneins function in two capacities in eukaryotic cells: some function in the axoneme to power flagellar beating while others, located in the cytoplasm, have important roles in organelle transport, spindle function and centrosome assembly (Pfister *et al.*, 2006; Hook and Vallee, 2006; Asai and Wilkes, 2004). Dynein motors are substantially larger than kinesins and typically contain 12 subunits. Although some subunits are shared by cytoplasmic and axonemal dyneins, other components are unique to each motor type. Dynein heavy chains (DHCs) contain a motor domain. In the case of cytoplasmic dyneins, two identical cytoplasmic DHCs associate with two intermediate chains, which anchor dynein to its cargo. The functions of four other intermediate chains and several light chains associated with cytoplasmic dyneins are not well understood. Axonemal dyneins may contain one, two or three distinct axonemal DHCs which bridge two microtubules to coordinate the sliding that underlies flagellar motility. The *Toxoplasma* GT1 genome contains ten annotated DHCs and a number of types of intermediate and light chains. Of the ten DHC homologues, seven appear most like axonemal subunits while three are likely cytoplasmic dyneins (Naomi Morrissette, personal observations). There are two light chain subunits that are specific for cytoplasmic dyneins present in the *Toxoplasma* genome: light intermediate chain (one homologue) and dynein light chain type 1 (three homologues). One of the dynein light chain type 1 homologues (TgDLC1; TGGT1_066540) has been tagged and localizes to the spindle poles, centrioles, basal complex and conoid region of tachyzoites (Hu *et al.*, 2006). Two light chain subunits shared by flagellar and cytoplasmic dyneins are represented in the *Toxoplasma* genome: Tctex-1 (three homologues) and Roadblock/LC7 (two homologues). There are three accessory subunit types that are specific for flagellar dyneins represented in the *Toxoplasma* genome: the WD40 repeat intermediate chain, axonemal assembly factor and axonemal light chain. The *Toxoplasma* genome encodes four axonemal intermediate chains. The axonemal light chain interacts directly with the N-terminal half of the heavy chains and plays an important role in flagellar motility: there are two *Toxoplasma* axonemal light chain homologues. Lastly, *Toxoplasma* has two leucine rich repeat-containing axonemal assembly factor subunits that are required for cytoplasmic pre-assembly of axonemal dyneins. It is likely that a subset of the dyneins, particularly the cytoplasmic dyneins, will be constitutively expressed in all stages while other dynein subunits (particularly flagellar dynein components) are likely to be expressed in a developmentally regulated fashion to participate in microgamete motility. The activity of cytoplasmic dynein is regulated by the multi-subunit dynactin complex. The *Toxoplasma* genome encodes homologues of a number of dynactin complex proteins including Arp1 (see Section 13.2.7.2.1) (Gordon and Sibley, 2005).

13.2.1.3.3 DYNACTIN

Dynactin is a multi-subunit complex that binds to dynein and the kinesin-2 subset of kinesins (Kardon and Vale, 2009), where it links motors to organelle or vesicle cargo, is often essential for motor activity and enhances motor processivity. It has established roles in chromosome alignment, spindle organization and centriole function. *Toxoplasma*, *Plasmodium* and *Cryptosporidium* have genes that encode obvious orthologues of the Arp1 (TGGT1_024220), p25 (TGGT1_090320), p27 (TGGT1_028520), and p62 (TgTwinScan_5099) subunits of the dynactin complex (Gordon and Sibley, 2005). Other subunits have a lower degree of conservation and have not been identified by homology searches.

13.2.2 Centriole and Basal Body Components

Centrioles and basal bodies are highly ordered organelles with conserved structural features and protein components. Previous work has shown that *Chlamydomonas* basal bodies are composed of more than 150 polypeptides and comparative genomics suggests that there are greater than 600 genes encoding specific basal body and flagellar components (Dutcher, 2004). Although some structures and components are shared by all centrioles, different species show diversity of organization and protein content. *Toxoplasma* tachyzoite centrioles have a simplified form consisting of a central tubule surrounded by nine singlet microtubules (Morrissette and Sibley, 2002b) and comparative bioinformatics suggest that a number of centriole and basal body components are missing from the genomes of *Toxoplasma* and *Plasmodium* (Hodges *et al.*, 2010). Both observations suggest that apicomplexan centrioles have reduced complexity. Although the *Toxoplasma* genome contains homologues of a number of conserved centriolar components (δ- and ε-tubulin, centrin, WDR16, SAS-4 and SAS-6), it also lacks genes for a considerable number of widely conserved basal body proteins such as CEP192/SPD-2, PCM1, ninein, rootletin, and asterless.

13.2.2.1 γ-, δ- and ε-Tubulin

In addition to α-β tubulin dimers, other members of the tubulin superfamily are critically important centriole components. Centrioles and alternative MTOCs (such as yeast spindle pole bodies) contain γ-tubulin which is found in the centriole lumen and is tightly bound to the microtubules of the centriolar cylinder. In some cellular models, loss of γ-tubulin (silencing) blocks centriole duplication. The δ- and ε-tubulins are found in protozoa and vertebrates (Dutcher, 2001). *Chlamydomonas* δ-tubulin mutants fail to correctly assemble doublet and triplet microtubules and the ε-tubulin null is lethal although *Drosophila* lacks both δ- and ε-tubulins but is capable of building triplet microtubule containing basal bodies in male gametes. There is a single copy gene for *Toxoplasma* γ-tubulin (TGGT1_082260). Alignment of γ-tubulin protein sequences from *Homo sapiens*, *Toxoplasma* and *Chlamydomonas* indicates a high degree of conservation throughout most of the protein. There is sequence divergence between the homologues in the C-terminal 50 residues. When the endogenous copy of *Toxoplasma* γ-tubulin is tagged with an in-frame fusion to YFP, γ-tubulin−YFP localizes to the cytoplasm and centrioles (J. de Leon and Naomi Morrissette, unpublished observations). In contrast, γ-tubulin−YFP is not a component of the APR which organizes the subpellicular microtubules in *Toxoplasma* and other apicomplexans. An alignment of human, *Toxoplasma* and *Chlamydomas* δ-tubulin indicates that *Toxoplasma* δ-tubulin possesses several insertions relative to *Chlamydomonas* and human δ-tubulin. These include a 51 N-terminal residue extension, followed by 34, 58, and 20 residue inserts, and a subsequent 29 C-terminal residue extension. There is a higher degree of conservation between human, *Toxoplasma* and *Chlamydomonas* ε-tubulin, although *Toxoplasma* ε-tubulin has a 16 amino acid insert at the N-terminus of the protein. Interestingly, mutations to δ- and ε-tubulin are associated with loss of triplet microtubule blades in *Chlamydomonas*. When the endogenous copies of *Toxoplasma* δ- and ε-tubulin (TGGT1_020490 and TGGT1_000470) were tagged with YFP knock-in, fluorescence was not detected, suggesting that these tubulins are either expressed at extremely low levels or not expressed at all (J. de Leon and Naomi Morrissette, unpublished observations). Consistent with this, although γ-tubulin mRNA and protein is detected in tachyzoite data deposited at http://ToxoDB.org, a lack of corresponding data for δ- and ε-tubulin may indicate that these proteins are not expressed in tachyzoites.

13.2.2.2 Centrins

Centrins are highly conserved proteins that contain four EF hands and were initially identified as a calcium-sensing regulator of centriole structure, orientation, and position in unicellular green algae (Salisbury *et al.*, 1984). In metazoans, centrin is a ubiquitous centrosomal protein located in the pericentriolar matrix, in fibres linking centrioles, and within the transition zone at the base of flagella (Baron and Salisbury, 1988; Moudjou *et al.*, 1991). There are two categories of centrins: centriolar centrins typically localize to centriolar spokes while centrosomal centrins function in centrosome duplication. Three isoforms of centrin have been identified in *Toxoplasma* (Hu, 2008; Striepen *et al.*, 2000; Hartmann *et al.*, 2006). TgCen1 (TGGT1_009500) and TgCen3 (TGGT1_009500) have the hallmarks of centrosomal centrins (a regular distribution of lysine and arginine residues and hydrophobic amino acids in the N-terminal 20 residues) while TgCen2 (TGGT1_101950) lacks this property. In tachyzoites, TgCen1 exclusively localizes to the centrioles/centrosomes, while TgCen3 is predominantly present in the centrosome and to a much lesser extent at the conoid. In contrast, TgCen2 is found at both the apical and basal ends of the parasite in addition to the centrioles and 5−6 apical annuli of unknown function (Hu *et al.*, 2006). It has been speculated that TgCen2 in the basal complex may contract at the end of endodyogeny, although its specific role at this site or in the apical cap annuli has yet to be elucidated (Hu *et al.*, 2008).

Centrins can be regulated by phosphorylation (Lukasiewicz *et al.*, 2011; Lutz *et al.*, 2001; Yang *et al.*, 2010). The mitotic serine/threonine

Aurora A (AurA) kinase phosphorylates the KKTSLY consensus sequence found at the C-terminus of centriolar centrins. Aberrant phosphorylation of this site is associated with centrosome over-duplication and consequent aneuploidy (Lukasiewicz *et al.*, 2011). Several putative aurora kinases are present within the *Toxoplasma* genome but have not been characterized to date and although TgCen3 contains a KQTSLY sequence, there is no experimental evidence that this site is phosphorylated in *Toxoplasma* (Treeck *et al.*, 2011; Nebl *et al.*, 2011). However, the available phosphoproteome profiles focused on host cell egress rather than intracellular replication and therefore may have missed an essential window of time when this modification appears. The phosphoproteome did capture TgCen1 phosphorylation on Ser9, a protein kinase A (PKA) phosphorylation consensus site, and there is a characterized PKA (TGGT1_081170) in *Toxoplasma* (Kurokawa *et al.*, 2011). For TgCen2, a phosphorylation site at Ser180 is evident, but this site does not fit any consensus for known kinases.

13.2.2.3 Assemblins/SFA

Assemblin (striated fibre assemblin, SFA) is a major component of striated microtubule associated fibres, a non-contractile fibre that associates with basal bodies and microtubule rootlets in green algae (Lechtreck *et al.*, 1996; Lechtreck and Melkonian, 1991). SFA may function in basal body orientation as well as in assembly and stability of rootlet microtubules. Genes for SFA are also found in the genomes of a variety of alveolates, including apicomplexans (Harper *et al.*, 2009; Lechtreck, 2003). The *Toxoplasma* genome encodes three homologues (TGGT1_042960, TGGT1_063240, and TGGT1_029850) with TgSFA2 and TgSFA3 expressed in tachyzoites. Studies using a cross-reacting anti-SFA antibody showed co-localization with the centriole marker centrin in *Toxoplasma* tachyzoites: the signal appears shortly after centriole duplication and is degraded later in the cell cycle (Lechtreck, 2003). More recently, specific TgSFA3 antiserum and TgSFA2 and TgSFA3 tagged lines were used to establish that TgSFA2 and TgSFA3 co-localize to fibre-like structures in tachyzoites during mitosis and budding while parasites in interphase do not have detectable protein. SFA fibres emerge from the centrioles to tether them to the emerging apical daughter buds at the APR and conoid. The loss of either SFA2 or SFA3 blocks initiation of daughter APRs and consequently creation of daughter buds. Since nuclear division is not inhibited, multiple nuclei accumulate akin to replication by schizogony in other apicomplexans. These observations suggest that although the asexual stages of apicomplexans have dispensed with building flagellar structures, they retain SFAs as a means to ensure reliable inheritance of apical complex organelles (Francia *et al.*, 2012).

13.2.2.4 SAS-6

SAS-6 is a ubiquitous centriolar protein that has been shown to be involved in forming the cartwheel during early biogenesis of the daughter centriole in many organisms (van Breugel *et al.*, 2011; Kitagawa *et al.*, 2011; Leidel *et al.*, 2005; Nakazawa *et al.*, 2007). There is a single copy gene for the *Toxoplasma* homologue of SAS-6 (TGGT1_040920) as well as a SAS-6 related protein that is not a true homologue (see Section 13.2.3.1). The structural properties of SAS-6 elegantly dovetail with its ability to template a centriole. SAS-6 assembles into homodimers that contain a globular head domain at the N-terminus and an extended coiled-coil rod. *In vitro*, SAS-6 homodimers self-assemble into a structure with a central hub and nine rods, akin to the cartwheel observed in structural studies of centrioles. Relative to *Chlamydomonas* and human SAS-6, *Toxoplasma* SAS-6 has an N-terminal extension: these additional 270 amino acids may account for the robust tubule 'hub' observed at the centre of *Toxoplasma* centrioles. *Toxoplasma* SAS-6—YFP is found in the cytoplasm and at the centrioles during parasite replication (de Leon *et al.*, 2013).

13.2.2.5 IMC15

The alveolin-motif containing IMC family of proteins localize to the inner membrane complex (see below). IMC15 (TGGT1_000660) cycles between the centrioles and IMC. It first associates with duplicated centrioles at the beginning of the cell cycle, then relocates to the IMC of emerging daughters. This behaviour suggests that IMC15 may function to regulate the number of daughter buds and to ensure that their construction proceeds with the correct timing (Anderson-White *et al.*, 2011).

13.2.2.6 Tg14-3-3

Tg14-3-3 (TGGT1_043200) is one of four 14-3-3 proteins in the *Toxoplasma* genome. The 14-3-3 protein family is a group of conserved regulatory molecules that bind to diverse signalling proteins such as kinases and phosphatases (Morrison, 2009). Tg14-3-3 localizes to daughter buds, the basal complex and centrioles (Lorestani *et al.*, 2012).

13.2.3 The Conoid

13.2.3.1 SAS6L

In addition to a SAS-6 homologue, which localizes to the tachyzoite centriole, the genome of *Toxoplasma* contains a related SAS-6-like (SAS6L) protein which localizes above the conoid in the region of the preconoidal rings (de Leon *et al.*, 2013). Loss of SAS6L causes reduced fitness in *Toxoplasma* tachyzoites whereas elevated SAS6L–YFP expression induces formation of prominent cytoplasmic filaments. SAS6L homologues are found in simple eukaryotes that contain SAS-6 and centrioles, but have been lost from the holozoan lineage. In *Trypanosoma brucei*, SAS6L localizes to the region of the basal plate, the site in the axoneme where the central pair microtubules are nucleated. It is likely that SAS6L also localizes to this region in apicomplexan microgametes. SAS6L proteins consist of a conserved domain that is located at the N-terminal head of SAS-6. The localization of SAS6L to the flagellar plate and the tip of the conoid may indicate that the closed conoid evolved from a flagellar axoneme or flagellar components. Alternatively, SAS6L may have been retained along with the basal body-associated SFA proteins in order to coordinate apical organelle inheritance in aflagellar tachyzoites (Francia *et al.*, 2012).

13.2.3.2 Centrins

The *Toxoplasma* genome encodes three isoforms of centrin (TgCen1, -2, and -3). As described above, tagged TgCen1 (TGGT1_009500) and tagged TgCen3 (TGGT1_009500) localize to the centrioles/centrosome (Hu *et al.*, 2006; Hu, 2008; Hartmann *et al.*, 2006). TgCen3 may also localize to the conoid and tagged TgCen2 (TGGT1_101950) is found at both the apical and basal ends of the parasite in addition to the centrioles and 5–6 apical annuli of unknown function (Hu *et al.*, 2006).

13.2.3.4 Dynein Light Chain

The TgDLC1 (TGGT1_066540) subunit has been mRFP-tagged and localizes to the spindle poles, centrioles, basal complex and conoid region of tachyzoites (Hu *et al.*, 2006). This subunit is characteristic of cytoplasmic dyneins.

13.2.4 The Apical Polar Ring (APR)

13.2.4.1 RNG1

TgRNG1 is a small, low-complexity, detergent-insoluble protein that localizes to the *Toxoplasma* APR (Tran *et al.*, 2010). Although *Neospora caninum* and *Sarcocystis neurona* have related proteins, there are no obvious homologues in other apicomplexan organisms. Although this may be explained by the small size (79 amino acids) and low complexity of the RNG1 protein, it is also possible that RNG1 was lost from some lineages, perhaps in tandem with loss of the conoid. This latter possibility might imply that RNG1 serves to coordinate interactions between the APR and the conoid. TgRNG1 is a late component of the developing APR, appearing only after nuclear division is complete and daughter buds are quite mature. When TgRNG1−YFP is ectopically overexpressed to produce TgRNG1−YFP throughout the cell cycle, excess protein aggregates in the cell cytoplasm but RNG1−YFP is not prematurely incorporated into the APR and does not drive formation of additional APR structures. A second protein (TGGT1_001950) that localizes to this region also shares the same late cell cycle kinetics (Gould *et al.*, 2011). It was identified as a conserved protein found in alveolate pellicles from *Tetrahymena* and *Toxoplasma* and we propose that it be designated as RNG2.

13.2.4.2 *Tetrahymena* Pellicle Protein Homologues

The ciliate *Tetrahymena thermophila* is an alveolate organism with a trilaminar pellicle structure that is similar to the *Toxoplasma* pellicle. A proteomic survey of the components of a *Tetrahymena* pellicle fraction identified 529 novel proteins (Gould *et al.*, 2011). Many of the pellicle proteins have similarity to the A-type inclusion protein of Pox viruses. Nearly a quarter of these proteins have repetitive regions with an amino acid bias for K, E, Q, L, I and V, and were termed charged repeat motif proteins (CRMPs). This data set was used to identify related CRMPs in the *Toxoplasma* genome. When two putative CRMP proteins were tagged in tachyzoites, they localized to the pellicle in the region of the APR. The TGGT1_045460 protein is found at one or two rings in the apical region of *Toxoplasma* tachyzoites. These rings are apparent at a very early stage of daughter cell formation. In contrast, although the TGGT1_001950 protein is also located at the tachyzoite apex, it forms a smaller spot and only appears in mature cells. The behaviour of the latter protein is similar to RNG1 since both appear in mature tachyzoites but are not present in developing daughter cells (Tran *et al.*, 2010).

13.2.5 Alveolins and Other Pellicle Proteins

The *Toxoplasma* membrane skeleton is composed of a detergent-resistant network of 10 nm filaments that maintains the general shape of the parasite after extraction. Initial characterization of detergent-insoluble IMC-associated proteins identified the first member of a family of proteins that contain 'alveolin' repeats (TgIMC1; TGGT1_116030) as well as a distinct IMC component (TgIMC2; TGGT1_083640) that is unrelated to the alveolin family proteins (Mann and Beckers, 2001; Hu *et al.*, 2002a). Reanalysis of the ToxoDB annotated IMC2 gene suggests that it may encode a larger protein than initially reported. In addition there is an IMC2 paralogue, which we tentatively name IMC2b (TGGT1_083630) located next to IMC2a. BLASTp search against the NCBI NR database and SMART analysis indicates that both predicted proteins contain metallophosphatase superfamily and N-terminal transmembrane domains. The IMC network of immature parasites is detergent-soluble and its conversion to a detergent-resistant network coincides with proteolytic release of the C-terminus of TgIMC1 (Hu *et al.*, 2002a). Subsequently, IMC3 was identified by a morphology-based screen of YFP tagged genomic DNA (Gubbels *et al.*, 2004) and IMC4 was identified in a mass spectrometry screen for components of the apical membrane cytoskeleton (Hu *et al.*, 2006).

When the alveolin motif is used to identify other members in the *Toxoplasma* genome, the number of alveolin family members is increased to 14 proteins (IMC5−15) (Anderson-White *et al.*, 2011). While orthologues of some alveolin motif proteins are found in a wide variety of alveolates, the alveolin family is expanded in *Toxoplasma*, perhaps allowing non-conserved IMC proteins to coordinate *Toxoplasma*-specific functions such as replication by endodyogeny. With the exception of TgIMC7 (TGGT1_066030), TgIMC12 (TGGT1_023950) and TgIMC14 (TGGT1_009650), the *Toxoplasma* alveolin genes exhibit coordinated expression that peaks at the time of daughter budding (between mitosis and early G1-phase). Maximal TgIMC14 expression occurs an hour after budding and TgIMC7 and TgIMC12 peak between late G1- and S-phase. Individual alveolin proteins exhibit distinct patterns of localization to the cytoskeletons of mother and daughter parasites. Both TgIMC1 and TgIMC4 (TGGT1_116040) have equivalent levels of localization to mothers and developing daughters. In contrast, TgIMC3 (TGGT1_099250), TgIMC6 (TGGT1_014960) and TgIMC10 (TGGT1_117680) distribution is skewed toward developing daughter parasites and as daughter parasites emerge and mature, these proteins diminish in concentration. Consistent with their distinct expression profiles, TgIMC7, TgIMC12 and TgIMC14 exclusively associate with the mature cortical cytoskeleton in G1-phase. Since the cytoskeleton is fully formed at this time, these proteins may function to distinguish the mother cytoskeleton from that of maturing daughters in order to permit selective disassembly of maternal structures during late endodyogeny. This speculation is supported by the observation that other apicomplexans that replicate by schizogony (which does not require that mother and daughter components be distinguished) lack orthologues for TgIMC7, TgIMC12 and TgIMC14. TgIMC11 (TGGT1_051060) is found at the apical cap and basal regions. TgIMC5 (TGGT1_079150), TgIMC8 (TGGT1_079140), TgIMC9 (TGGT1_081370) and TgIMC13 (TGGT1_002640) are reorganized during replication: during the first portion of budding they are found along the entire daughter bud, but over time shift to localize to the basal complex of mature parasites. Pairwise comparisons indicate differences in the association of individual IMC proteins at the basal complex: the largest region is covered by TgIMC9, TgIMC13 and TgIMC15 (TGGT1_000660), as well as MORN1. TgIMC5 and TgIMC8 encompass a smaller region and the basal tip is delimited by TgCen2. The shift of TgIMC5, TgIMC8, TgIMC9 and TgIMC13 from small buds to the basal complex occurs in parallel with assembly of TgCen2 on the basal complex, suggesting that they may work with TgCen2 and other basal complex components to establish the tapered basal end of the cytoskeleton. Perhaps the most exciting observation is that TgIMC15 associates with duplicated centrioles and transitions to the cytoskeleton of budding daughters. Localization of TgIMC15 to early daughter buds precedes the localization of TgMORN1. Ultimately, TgIMC15 localizes to an extreme apical region consistent with the conoid. Association of TgIMC15 with duplicated centrioles may provide a cue to initiate daughter budding by coordinating recruitment of additional critical factors. The dynamic association of TgIMC15 with structures in developing daughter parasites illuminates a previously unappreciated connection between the cell cycle, mitosis and cytokinesis by budding.

13.2.5.1 TgILP1

TgILP1 (TGGT1_089780) is a detergent-insoluble, IMC-localized protein that is widely conserved across the Apicomplexa. It has been designated *Toxoplasma* IMC localizing protein 1 in order to distinguish it from the alveolin repeat containing IMC proteins. TgILP1−YFP localizes to the IMC of both mother and daughter parasites, although it is preferentially targeted to daughter buds (Lorestani *et al.*, 2012). It appears to be a single copy gene. Overexpression of TgILP1−YFP causes parasites to have a malformed cytoskeleton and irregular nuclei, suggesting that nuclear division and cytokinesis have been uncoupled. The fusion protein colocalizes with TgIMC3 in the mature malformed

IMC and forms small TgIMC3-independent ring structures in the cytoplasm.

13.2.5.2 TgPhIL1

TgPhIL1 (TGGT1_011800) is a 22-kDa protein that was first identified using a photoactivatable reagent to selectively label integral membrane proteins within the tachyzoite pellicle (Gilk *et al.*, 2006). Surprisingly, TgPhIL1 (photosensitized INA-labeled protein 1) is not predicted to have a transmembrane domain; rather, it has a peripheral localization and resistance to detergent extraction that is consistent with it being a component of the peripheral (IMC-associated) cytoskeleton. TgPhIL1 localizes to the tachyzoite pellicle with increased distribution at the apical and basal caps. Separation of the plasma membrane and IMC by α-toxin treatment reveals that TgPhIL1 associates with the IMC. Moreover, in DOC-extracted samples TgPhIL1 sometimes appears in spiralling 'stripes' that may represent the longitudinal 'sutures' between the flattened plates of the IMC. During tachyzoite replication, TgPhIL1 is localized to the apical IMC and the elongating posterior edge of daughter parasite buds. TgPhIL1 does not contain any identifiable protein motifs and homologous proteins are only found in other apicomplexans. These homologues have divergent N-terminal regions but a high degree of sequence identity in the C-terminal regions. Deletion analysis indicates that the C-terminal half of TgPhIL1 is sufficient for correct localization.

13.2.6 TgMORN1

Membrane occupation and recognition nexus (MORN) repeats are 23 amino acids in length and are found in a variety of bacterial and eukaryotic proteins. Eukaryotic MORN proteins coordinate membrane–membrane or membrane–cytoskeleton interactions (Shimada *et al.*, 2004; Takeshima *et al.*, 2000). Two MORN-repeat proteins have been reported in *Toxoplasma*. TgMORN2 (TGGT1_031450) was first identified in a morphology-based screen for novel components of distinct subcellular structures (Gubbels *et al.*, 2004). Although the initial localization suggested that TgMORN2–YFP localized to the plasma membrane, once the full-length sequence was tagged in various ways, its localization shifted to other membranous structures depending on expression level, kind and location of the tag on either the N- or C-terminus (Marc-Jan Gubbels, unpublished data). As such its exact subcellular localization is still elusive. TgMORN2 does not have clear orthologues in other apicomplexans, whereas TgMORN1 (TGGT1_086660) is strongly conserved among the Apicomplexa (Gubbels *et al.*, 2006; Heaslip *et al.*, 2010a). TgMORN1 and TgMORN2 share 24.5% identity. TgMORN1 is dynamically associated with several subcellular locations within replicating tachyzoites. It is found at the centrocone, a specialized region of the nuclear envelope and in two rings that represent the earliest appearance of daughter buds. Following chromosome duplication, the TgMORN1 spot elongates and splits into two, consistent with localization at the poles of the bipolar mitotic spindle. The two 'daughter' rings increase in size and each again splits to resolve into an apical (conoidal) ring and a posterior ring. These rings define the ends of the nascent IMC in the two daughter parasites. The posterior TgMORN1 rings define a leading edge for the lengthening IMC structures and extend to enclose the two arms of a 'U-shaped' nucleus during endodyogeny. Within the cat intestine, *Toxoplasma* replicates by endopolygeny, a process in which repeated nuclear divisions are followed by synchronous formation of many daughter parasites within the mother cell cytoplasm. Under these conditions, multiple MORN1--containing rings are observed within the mother cell cytoplasm (Ferguson *et al.*, 2008). Similarly, when MORN1 is localized in a diverse set of apicomplexan parasites, it is found in different patterns, but consistently identifies regions of emerging IMC during replication. Deletion of the TgMORN1 gene or conditional depletion of TgMORN1 protein inhibits assembly of the basal complex leading to aberrant replication (see below) (Lorestani *et al.*, 2010; Heaslip *et al.*, 2010a). More recently, putative TgMORN1 interacting proteins have been identified by co-purification and mass spectrometry. These include TgMSC1a, Tg14-3-3, and TgILP1 (Lorestani *et al.*, 2012).

13.2.6.1 *TgMSC1a*

TgMSC1a (TGGT1_098490) was identified in a screen for proteins that interact with TgMORN1 (Lorestani *et al.*, 2012). It does not contain any detectable motifs. TgMSC1a co-localizes with TgIMC5 and TgIMC8 in the posterior ring and is largely non-overlapping with the TgMORN1 signal in the basal cap. Significantly, TgMSC1a is the first unique marker of mature basal caps, as it does not label the basal complex of developing daughter parasites. It has a paralogue in *Toxoplasma*, TgMSC1b (TGGT1_018690).

13.2.6.2 *TgMSC1b*

TgMSC1b (TGGT1_018690) was identified as a paralogue of TgMSC1a: the proteins share 36% identity. TgMSC1b localizes to the mature pellicle (Lorestani *et al.*, 2012). This pattern is similar to TgGAP45, which also is only detectable in the maternal cytoskeleton. Non-coccidian Apicomplexa only encode a single orthologue of the TgMSC1a/TgMSC1b proteins, which correlates with the lack of maintenance of a mature basal complex in these species.

13.2.6.3 *Tg14-3-3*

As described above, Tg14-3-3 (TGGT1_043200) is one of four 14-3-3 proteins in *Toxoplasma*. It labels developing daughter buds, as well as the basal complex and the centriole (Lorestani *et al.*, 2012). Since 14-3-3 proteins bind diverse signalling proteins, they may regulate daughter cell development, akin to their roles in cell cycle control in other eukaryotic organisms.

13.2.7 Actin, Actin-Like Proteins and Actin Binding Proteins

The *Toxoplasma* genome has genes for a single actin isoform (TgACT1), and a number of actin-like and actin-related proteins (ALPs and Arps). Although the actin-related proteins Arp2 and Arp3 regulate actin polymerization in many eukaryotes, *Toxoplasma* and other apicomplexans lack Arp2 and Arp3 orthologues as well as other Arp2/3 complex subunits (Gordon and Sibley, 2005). Formins, a second category of conserved proteins that promote actin polymerization, are represented in apicomplexan genomes and play a critical role in initiating actin polymerization (Daher *et al.*, 2010, 2012; Skillman *et al.*, 2012). Homology searches indicate that Apicomplexan genomes encode a few conserved actin-regulatory proteins (formins, profilin, ADF/cofilin, capping protein, cyclase-associated protein, and coronin) (Schuler and Matuschewski, 2006b). This may indicate that a reduced set of actin binding proteins is sufficient to regulate polymerization. Alternatively, it may indicate that we have yet to identify components of the actin machinery that are uniquely found in this phylum. Indeed, apicomplexan myosins interact with GAP40, GAP45 and GAP70 proteins that are unique to this phylum (Frenal *et al.*, 2010; Bullen *et al.*, 2009; Gaskins *et al.*, 2004; Baum *et al.*, 2006; Johnson *et al.*, 2007). Moreover, biochemical studies indicate that the apicomplexan orthologues of actin binding proteins have unusual features relative to the properties of these proteins in other eukaryotes. Nonetheless, disruption of genes for proteins that promote polymerization such as the formins (Daher *et al.*, 2010) or depolymerization such as ADF (Mehta and Sibley, 2011) and profilin (Plattner *et al.*, 2008) impair motility and invasion, indicating that regulation of actin assembly is key to these processes.

13.2.7.1 Actin

Actin is an essential and conserved component of nearly all eukaryotic cells. This ~42 kDa protein can exist as a free subunit (globular, G-actin) or polymerize to form filamentous microfilaments (F-actin). Polymerized actin is essential to a number of conserved eukaryotic cell processes including motility, shape and cytokinesis. Inhibitor studies using the microfilament-disrupting drug cytochalasin D first suggested that assembled *Toxoplasma* actin is essential for motility and invasion of host cells (Schwartzman and Pfefferkorn, 1983; Dobrowolski and Sibley, 1996). The *Toxoplasma* genome contains a single actin gene (TgACT1; TGGT1_021970) and mutation of alanine 136 to glycine confers cytochalasin-resistance (Dobrowolski and Sibley, 1996). Resistant parasites, but not wild-type tachyzoites, are capable of invasion into host cells in the presence of cytochalasin. In striking contrast to other eukaryotes, where the bulk of actin is assembled into F-actin, actin is largely unassembled in tachyzoites (Skillman *et al.*, 2011; Holzinger, 2009; Sahoo *et al.*, 2006; Dobrowolski *et al.*, 1997b; Dobrowolski and Sibley, 1996). *Toxoplasma* F-actin is stabilized by treatment with phalloidin or jasplakinolide (Skillman *et al.*, 2011; Shaw and Tilney, 1999; Shaw *et al.*, 2000; Poupel and Tardieux, 1999; Wetzel *et al.*, 2003; Holzinger, 2009) and although jasplakinolide treatment increases the rate of gliding, it disrupts normal motility and reversibly inhibits host cell invasion (Wetzel *et al.*, 2003; Skillman *et al.*, 2011; Poupel and Tardieux, 1999). Jasplakinolide-treated parasites contain bundles of stabilized actin filaments, most often observed as an extended protrusion of the plasma membrane from the apical end (Shaw and Tilney, 1999). Treatment can also enhance F-actin at the posterior end, which appears as spiralling filaments (Wetzel *et al.*, 2003); similar structures can sometimes be observed in untreated parasites (Dobrowolski *et al.*, 1997b).

Actin from *Toxoplasma* or other apicomplexans is divergent from actin in other eukaryotes (Gordon and Sibley, 2005; Skillman *et al.*, 2011). Baculovirus-expressed recombinant actin has been used to establish that differences in the amino acid sequence account for the intrinsically unstable nature of apicomplexan F-actin. Although recombinant actin from *Toxoplasma* (TgACT1) and *Plasmodium falciparum* (PfACT1 and PfACT2) can polymerize, the resulting filaments are short and unstable (Skillman *et al.*, 2011; Sahoo *et al.*, 2006). Molecular modelling suggested that unusual residues in TgACT1 (G200 and K270) may contribute to the intrinsic instability of *Toxoplasma* actin by influencing monomer–monomer interactions in actin filaments. When these residues are mutated to restore conserved amino acids, the K270M substitution modestly enhances assembly and the G200S substitution significantly increases polymerization (Skillman *et al.*, 2011). Since creating the reverse substitutions in yeast actin does not significantly alter its assembly properties, it is likely that other amino acid differences also contribute to the reduced assembly of apicomplexan actin. When K270M or G200S substituted TgACT1 genes are expressed in tachyzoites, they impair parasite proliferation and cause aberrant motility, indicating that intrinsically unstable microfilaments are essential for correct actin function.

13.2.7.2 Actin-Related and Actin-Like Proteins (Arps and ALPs)

In addition to a single gene for conventional actin (TgACT1), the *Toxoplasma* genome contains genes for a number of actin-like and actin-related proteins (Gordon and Sibley, 2005). Phylogenetic analysis indicates that there are ten distinct groups of apicomplexan proteins that contain a conserved actin domain (pfam00022). While some of these proteins are orthologues of Arps (actin-related proteins), a widely conserved group of eukaryotic proteins with defined roles in modulating the cytoskeleton and regulating chromatin remodelling, others (termed actin-like proteins, ALPs) are unique to apicomplexans.

13.2.7.2.1 ACTIN-RELATED PROTEINS (ARPS)

Actin-related proteins share a common actin-fold and an overall sequence similarity with actin (Schroer *et al.*, 1994). They are central to cytoplasmic and nuclear functions, including regulation of microtubule motor activity (Arp1, Arp10 and Arp11), actin polymerization (Arp2 and Arp3) and chromatin remodelling (Arp4, Arp5, Arp6, Arp7, Arp8 and Arp9). It is notable that although diverse eukaryotes use the conserved Arp2/3 complex to regulate the actin cytoskeleton, apicomplexans lack genes for Arp2 and Arp3 proteins as well as for the p34, p21 and p16 subunits of the complex (Gordon and Sibley, 2005). The presence of remnant p41 and p20 homologues in some apicomplexans along with the observation that the related alveolate *Tetrahymena thermophila* encodes a canonical Arp2/3 complex suggests that the Arp2/3 complex was lost or has diverged to become unrecognizable.

Three of the ten distinct actin-related protein groups in apicomplexan parasites are Arp1, Arp4 and Arp6 orthologues (Gordon and Sibley, 2005). Arp1 (TGGT1_024220) is a conserved component of the dynactin complex which activates microtubule motors (dyneins and kinesin II motors) and coordinates interactions with cargo. Other dynactin complex subunits (p25, p27 and p62) are also represented in apicomplexan genomes. The *Toxoplasma* genome also contains a gene for Arp6 (TGGT1_012980) and two genes for Arp4 orthologues: Arp4a (TGGT1_002020) and Arp4b (TGGT1_012160). Both Arp4 and Arp6 are conserved nuclear proteins. Arp4 is a component of chromatin remodelling and histone acetyltransferase complexes. An I621T mutation in TgArp4a causes temperature-sensitive growth arrest of tachyzoites as a consequence of destabilized and mislocalized TgArp4a protein leading to chromosome loss during nuclear division (Suvorova *et al.*, 2012).

13.2.7.2.2 ACTIN-LIKE PROTEINS (ALPS)

Seven of the ten distinct actin-related protein groups in apicomplexan organisms represent actin-like proteins (ALPs) that are unique to apicomplexans and may serve in parasite-specific cellular roles (Gordon and Sibley, 2005). The ALPs do not group with Arp clades. While some ALPs are highly conserved (ALP1, ALP2, and ALP3), others are organism-specific (ALP5, ALP7, and ALP8). The *Toxoplasma* genome contains genes for ALP1 (TGGT1_030250), ALP2a (TGGT1_100560), ALP3 (TGGT1_065260), ALP8 (TGGT1_075190) and two forms of ALP9: ALP9a (TGME49_034660) and ALP9b (TGME49_069240). TgALP1 is most closely related to conventional actin and is likely to have a role in formation of daughter cell membranes during cell division in apicomplexan parasites (Gordon *et al.*, 2008). TgALP1 readily interconverts between diffusible and complex-bound forms (Gordon *et al.*, 2010) and localizes to a discrete region of the nuclear envelope, on transport vesicles, and on the early emerging IMC of daughter cells prior to the arrival of the alveolin family member IMC1. Overexpression of ALP1 disrupts daughter cell IMC formation and causes defects in nuclear and apicoplast segregation.

13.2.7.3 Formins (TgFRM1, TgFRM2 and TgFRM3)

Formins are large, multi-domain regulatory proteins that associate with the fast-growing (plus/barbed) end of actin filaments to promote polymerization (Goode and Eck, 2007). In contrast to a characteristic branched filament polymerization driven by the Arp2/3 complex, formins promote linear filament assembly. All formins contain a FH2 (formin homology) domain which is required for self-association (dimerization), actin nucleation and barbed end binding. Formins may also contain FH1 and FH3 domains. The FH1 domain contains polyproline stretches that recruit profilin and mediate interactions with SH3 or WW domain proteins. The FH3 domain is less conserved and directs the subcellular localization of formins. Formins promote filament nucleation and remain associated with the barbed end to prevent association of capping proteins.

The documented lack of Arp2/3 complex homologues in apicomplexans including *Toxoplasma* (Gordon and Sibley, 2005) illustrates the critical importance of formins for driving actin polymerization. The *Toxoplasma* genome contains three formin genes (Daher *et al.*, 2010, 2012). TgFRM1 (TGGT1_062340) and TgFRM2 (TGGT1_062180) are conserved in other apicomplexan organisms but TgFRM3 (TGGT1_039280) is restricted to the coccidian subset of parasites. TgFRM1 has a predicted molecular weight of 552 kDa and a FH2 domain located at the very C-terminal end of the protein. Although it lacks a canonical FH1 domain, it has a proline-rich region adjacent to the N-terminal side of the FH2 domain, which may function as a FH1 domain. TgFRM1 also has four tetratricopeptide repeats in the N-terminal half of the protein. TgFRM2 is slightly smaller than TgFRM1 with a predicted molecular weight of 492 kDa. Although the TgFRM2 FH2 domain is also located in the C-terminal portion of the protein, it is followed by an additional 666 residues. TgFRM3 has a predicted molecular weight of 299 kDa and also contains a FH2 domain in the C-terminal half of the protein sequence, between residues 1480 and 2181. Although TgFRM2 and TgFRM3 lack tetratricopeptide repeats, like TgFRM1, TgFRM2 and TgFRM3 have proline-rich domains proximal to the FH2 domain. It is unclear if these domains are sufficient for profilin binding, as co-immunoprecipitation assays fail to show an interaction between *Toxoplasma* profilin and the *Toxoplasma* formins (see below). Although both formins localize to the pellicle, separation of the plasma membrane from the IMC reveals that TgFRM1 remains associated with the plasma membrane while TgFRM2 is retained by the IMC. Although TgFRM3 is nonessential in tachyzoites (a knockout line does not show motility or invasion defects), overexpression of TgFRM3 is deleterious and the protein could only be localized in a line with regulated overexpression of a tagged construct. Over-expressed TgFRM3 localizes to patches at the apex and posterior end of tachyzoites. TgFRM1 and TgFRM2 are essential for tachyzoite survival and only TgFRM1 was amenable to investigation using a tetracycline regulatable promoter to control protein levels. Although loss of TgFRM1 protein only influences motility and invasion modestly, biochemical experiments indicate that TgFRM1 and TgFRM2 promote F-actin assembly (Skillman *et al.*, 2012).

13.2.7.4 *Profilin (TgPRF)*

In other eukaryotes, profilins have been shown to play a complex role in regulating actin polymerization. Although they sequester actin monomers to promote depolymerization, they also enhance nucleotide exchange to increase the pool of ATP–actin that is available for polymerization (Kovar, 2006). Moreover, profilins are typically recruited to the barbed ends of actin filaments by polyproline stretches in formins and in this context promote filament polymerization. Apicomplexan profilins are divergent from other profilins and *Toxoplasma* profilin (TgPRF; TGGT1_074080) does not complement a *Saccaromyces cerevisiae* conditional profilin knockout (Plattner *et al.*, 2008). Experiments using either homologous (baculovirus expressed TgACT1) or heterologous (rabbit skeletal muscle) actin indicate that TgPRF has unusual properties relative to other profilins (Plattner *et al.*, 2008; Skillman *et al.*, 2012). TgPRF weakly inhibits both actin nucleotide exchange and the ability of FH1 and FH2 domains from TgFRM1 and TgFRM2 to stimulate polymerization of TgACT1, suggesting that it predominantly functions as a monomer sequestering protein. This is consistent with the observation that the bulk of actin in tachyzoites is in an unpolymerized state. Moreover, as described above, although *Toxoplasma* formins have polyproline stretches proximal to the FH2 domain, co-immunoprecipitation assays fail to show an interaction between *Toxoplasma* formins and TgPRF (Daher *et al.*, 2010). TgPRF is also a dominant antigen that signals 'danger' through Toll-like receptor-11 and induces IL-12 in mouse splenic dendritic cells (Yarovinsky and Sher, 2006). Conditional loss of profilin impairs motility and invasion and PRF-depleted parasites are avirulent and do not induce IL-12. Complementation with the *Plasmodium falciparum* profilin homologue (PfPRF) restores motility but not IL-12 induction, indicating that influences of PRF on virulence can be separated into motility-dependent and motility-independent functions (Plattner *et al.*, 2008). Surprisingly, these studies also established that depletion of TgPRF does not alter the pool of pelletable actin recovered after treatment with drugs that stabilize (jasplakinolide) or destabilize (cytochalasin D) actin filaments.

13.2.7.5 ADF/Cofilin

ADF/cofilin family proteins regulate actin-driven processes (polarity, migration, and division) in diverse eukaryotes (Poukkula *et al.*, 2011). They increase actin filament turnover by severing F-actin and may also influence actin dynamics by sequestering G-actin monomers. Most apicomplexan parasites have a single ADF homologue and alignment of apicomplexan ADFs with other homologues indicates that the bulk of residues involved in G-actin binding are conserved, while amino acids required for F-actin interaction are missing (Mehta and Sibley, 2010, 2011; Allen *et al.*, 1997). Recombinant TgADF (TGGT1_015110) has a high affinity for ADP–actin, a limited ability to sever microfilaments, and unlike other ADFs does not interact with PIP_2 which typically inhibits severing activity (Yadav *et al.*, 2011; Mehta and Sibley, 2010). Although TgADF does not significantly co-sediment with rabbit microfilaments, it is capable of inducing F-actin disassembly in a dose-dependent manner. This suggests that TgADF-induced depolymerization is a consequence of a strong G-actin sequestering activity. A conditional depletion of TgADF causes F-actin accumulation and induces slowed, non-productive serpentine movements and rocking, akin to the effect of the F-actin stabilizing drug jasplakinolide (Mehta and Sibley, 2011). As a consequence of inappropriately stabilized actin filaments, tachyzoite motility, invasion and egress are inhibited. The activity of TgADF may be modulated in bradyzoites by interaction with deoxyribose phosphate aldolase-like protein (TgDPA) since recombinant TgDPA enhances the ability of TgADF to disassemble microfilaments (Ueno *et al.*, 2010). The unusual requirements for regulation of apicomplexan actin polymerization have likely re-shaped TgADF to function in a G-actin rich environment by reducing its capacity for filament severing and maximizing its capacity to sequester monomers.

13.2.7.6 Aldolase

Toxoplasma tachyzoites and other asexual stage apicomplexans exhibit an unusual substrate-dependent gliding motility which also powers host cell invasion (Dobrowolski *et al.*, 1997a; Schwartzman, 1998; Sibley *et al.*, 1998; Heintzelman, 2003; Keeley and Soldati, 2004; Baum *et al.*, 2006). During this process, transmembrane adhesins such as MIC2 are secreted from the parasite apex and translocated along the parasite pellicle to the parasite posterior by actin and myosin-based machinery (Soldati *et al.*, 2001). Adhesins are linked to the cytoskeleton by the glycolytic enzyme aldolase (TGGT1_069700). Although this may seem unusual, aldolase is also recognized to interact with the actin cytoskeleton in a number of cellular contexts such as in 3T3L1 cells where it has been shown to mediate the association of actin filaments with the insulin-regulated glucose transporter GLUT4 (Kao *et al.*,

1999). In *Toxoplasma*, the C-termini of adhesins such as MIC2 interact with aldolase, which binds to F-actin to link adhesins to microfilaments (Jewett and Sibley, 2003; Starnes *et al.*, 2006, 2009). In turn, the actin filament–aldolase–adhesin complex is translocated towards the parasite posterior by the action of a myosin motor complex (MyoA and the GAP proteins) which remains fixed in the IMC. Aldolase is a homotetrameric enzyme with a basic groove that surrounds the substrate-binding pocket. This region is also important to F-actin and MIC2 binding which occurs by electrostatic interactions. Directed mutations to both MIC2 and aldolase demonstrate that acidic residues in the C-terminal tail of MIC2 interact with the aldolase basic groove. Although many of the charged residues in aldolase that contribute to MIC2 tail interactions are also required for enzyme activity, it is possible to identify mutations (K41A and R42A) that selectively reduce MIC2 tail binding without inactivating enzyme activity (Starnes *et al.*, 2009).

13.2.7.7 Capping Protein

The F-actin capping protein is a well-conserved heterodimer of α- (CPA) and β- (CPB) subunits that binds to the plus (barbed) end of actin filaments to block subunit exchange at this site (Cooper and Sept, 2008). The *Toxoplasma* genome has a clear homologue of the β-subunit (TGGT1_030260) and a less conserved putative homologue of the α-subunit (TGGT1_021400) that contains the cap A superfamily motif. The role of capping protein has been investigated in *Plasmodium berghei* by creating a gene knockout for the β-subunit (Ganter *et al.*, 2009). Although null parasites develop normally during the erythrocytic cycle, reduced ookinete motility leads to fewer oocysts and sporozoites in *Anopheles* and inhibition of transmission.

13.2.7.8 Cyclase-Associated Protein

Cyclase-associated proteins (CAPs) are a family of conserved actin-binding proteins distinct from CAPZ/capping protein (above). They contain a conserved C-terminal actin-binding domain that regulates actin remodelling in response to cellular signals (Hubberstey and Mottillo, 2002). CAPs bind to G-actin to control the availability of monomeric actin. Although CAP homologues exist in *Toxoplasma*, *Cryptosporidium parvum* and *Plasmodium* spp., they are considerably smaller than CAPs in yeast and vertebrates and lack the eponymous adenylate cyclase binding domain (Hliscs *et al.*, 2010). Previous work has shown that *C. parvum* CAP protein binds G-actin *in vitro* and *P. berghei* CAP protein is essential for oocysts development in the mosquito vector, although it is dispensable for blood stage parasites. When the endogenous copy of TgCAP (TGGT1_086070) is tagged with a C-terminal YFP fusion, it localizes to the apical sub-compartment of the pellicle (Lorestani *et al.*, 2012). TgCAP is incorporated into the apical region of daughter buds late in endodyogeny, prior to their emergence from the mother: this behaviour is similar to the kinetics displayed by the APR marker TgRNG1 (Tran *et al.*, 2010). The localization of TgCAP depends upon environmental cues: in intracellular tachyzoites TgCAP colocalizes with TgISP1, an established marker of the apical cap region, but in extracellular parasites TgCAP, but not TgISP1, moves to the tachyzoite cytoplasm. This suggests that TgCAP may be sequestered to prevent motility in intracellular parasites but in the presence of extracellular cues is released to the cytoplasm so that it may function to regulate actin-associated motility.

13.2.7.9 Coronin

Coronins are WD-repeat containing actin binding proteins found in a wide variety of eukaryotes. *Toxoplasma* (TGGT1_098160) and other apicomplexans have apparent coronin homologues (Figueroa *et al.*, 2004; Tardieux *et al.*, 1998). Relative to other coronins, the *Toxoplasma* and *Neospora* proteins have a C-terminal extension that contains a SMC domain and a potential microtubule binding domain. Peptides corresponding to coronin are identified in a preparation of soluble tachyzoite antigen (STAg) (Ma *et al.*,

2009) and an anti-coronin antibody directed against *Dictyostelium* coronin cross-reacts with *Plasmodium* coronin on immunoblots (Tardieux *et al.*, 1998). To date, the subcellular distribution of coronin has not been determined in apicomplexan parasites.

13.2.8 Myosins

Myosins are molecular motors that move on F-actin polymers and are essential for diverse processes in nearly all eukaryotes (O'Connell *et al.*, 2007). Myosins are used for cytokinesis, vesicle transport, phagocytosis and organelle movement. They consist of motor domain-containing heavy chains that power motility and light chains which have regulatory roles. Myosin heavy chains typically contain head, neck, and tail domains: the head domain coordinates actin binding, ATPase activity and generation of movement; the neck domain typically interacts with myosin light chains that regulate activity and the variable tail region binds the cargo to define the motor specificity. Myosin heavy chains have been grouped into 18 phylogenetic classes based on their head domains. The *Toxoplasma* genome contains 11 myosin heavy chains (TgMyoA to TgMyoK) many of which are members of the class XIV subset (Heintzelman and Schwartzman, 1999, 2001; Heintzelman, 2003; Foth *et al.*, 2006). Class XIV myosins are found in apicomplexans and ciliates, and share some domain architecture. Phylogenetic analysis has identified four sub-groups within class XIV myosins. TgMyoA (TGGT1_070410) and TgMyoD (TGGT1_043280) are members of subgroup XIVa, while TgMyoB/C (TGGT1_076590) and TgMyoE (TGGT1_051300) are placed in XIVb. TgMyoH (TGGT1_051300) is in subgroup XIVc and a number of ciliate myosins are in XIVd. Myosins in groups XIVa–b lack tails while those in XIVc have an ATS1 motif, a feature of proteins that may act as guanine nucleotide exchange factors. Strikingly, TgMyoH has 6–8 IQ motifs; in other myosins, IQ motifs bind calmodulin in the absence of calcium. TgMyoF (TGGT1_103490; class XXII) has 3–6 IQ motifs and a WD40 domain. TgMyoG (TGGT1_092070; class XXIII) has a MyTH4 domain; in other systems, myosins with MyTH4 and FERM domains bind to microtubules (Foth *et al.*, 2006). It is unclear whether class XXIII myosins, which possess a MyTH4 but lack a FERM domain, interact with microtubules. TgMyoI (TGGT1_092070; class XXIV) and TgMyoK (TGGT1_062370; class VI) myosins are characterized by two IQ motifs and coiled-coil domains, while TgMyoJ (TGGT1_013230; class VI) lacks these features. Surprisingly, mRNAs for TgMyoB, TgMyoC, and TgMyoD are more abundant in bradyzoite rather than tachyzoite stage organisms (Delbac *et al.*, 2001). This may reflect either increased transcription rates and/or increased transcript stability in bradyzoites.

13.2.8.1 TgMyoA, TgMLC1 and TgELC1

The class XIVa myosin TgMyoA is located between the plasma membrane and the IMC where it is critically important for tachyzoite motility (Hettmann *et al.*, 2000; Herm-Gotz *et al.*, 2002; Meissner *et al.*, 2002). It is a fast, single-headed myosin that moves in unitary steps towards the barbed (plus) ends of actin filaments with biochemical properties that are most similar to fast skeletal muscle myosins (Herm-Gotz *et al.*, 2002). Although the head of MyoA interacts with F-actin in an ATP-dependent fashion, the C-terminal region is sufficient for pellicle association (Hettmann *et al.*, 2000). Two arginine residues within this domain (R814 and R815) are critically important to localization: replacing these amino acids with alanines displaces TgMyoA from the pellicle to the cytoplasm. To date, TgMyoA has been shown to associate with two light chains: TgMLC1 (TGGT1_087590) (Gaskins *et al.*, 2004; Heaslip *et al.*, 2010b; Herm-Gotz *et al.*, 2002) and TgELC1 (TGGT1_107770) (Nebl *et al.*, 2011). TgMyoA lacks a tail which serves to localize the motor to cargo. This function is supplied in *trans* by a novel N-terminal extension of TgMLC1.

A degenerate C-terminal IQ motif in TgMyoA binds to a calmodulin-like domain in TgMLC1 while the N-terminal domain of TgMLC1 associates with the C-terminus of TgGAP45 (see below) to localize the motor at the IMC. Although the interaction between the *Plasmodium* MyoA and MLC1 orthologues has been shown to be calcium-independent (Green *et al.*, 2006), TgMyoA association with TgELC1 (essential light chain) is likely regulated by both direct Ca^{2+} binding and by phosphorylation in order to initiate motility (Nebl *et al.*, 2011). The *Toxoplasma* genome also encodes a TgELC1 paralogue (TGGT1_042450) with nearly identical expression throughout tachyzoite development (Marc-Jan Gubbels, personal observation). Two lines of evidence demonstrate that TgMyoA is essential to gliding motility and host cell invasion. Parasites depleted for TgMyoA (a conditional knockout) are non-motile, exhibit reduced invasion and egress and cannot form plaques in culture (Meissner *et al.*, 2002). Moreover, tachypleginA, a small molecule inhibitor of *Toxoplasma* tachyzoite invasion and motility irreversibly alters the electrophoretic mobility of TgMLC1 and TgMyoA containing tachypleginA-modified TgMLC1 has decreased motor activity (Heaslip *et al.*, 2010b).

13.2.8.2 TgMyoD and MLC2

TgMyoD is a second class XIVa myosin that is smaller, but closely related to TgMyoA (Hettmann *et al.*, 2000; Polonais *et al.*, 2011; Herm-Gotz *et al.*, 2006). Although TgMyoD has *in vitro* biochemical properties that are similar TgMyoA (Herm-Gotz *et al.*, 2006), it is unique to the coccidian subset of apicomplexans and is primarily expressed in bradyzoites, suggesting that it has a specialized role during tissue cyst stages. TgMyoD and its coccidia-specific associated light chain, TgMLC2 (TGGT1_096730) localize to the tachyzoite pellicle. Like TgMLC1, TgMLC2 has an N-terminal extension relative to the calmodulin domain which is necessary and sufficient for pellicle targeting (Polonais *et al.*, 2011). However, in contrast to TgMLC1, it does not appear to interact with additional proteins (such as the GAP proteins) and likely associates with the plasma membrane, by direct palmitoylation of TgMLC2. TgMLC2 is not modified by the invasion inhibitor tachypleginA (Heaslip *et al.*, 2010b). A conditional knockout of TgMyoD demonstrates that it is not essential for tachyzoite growth and gliding motility *in vitro* or virulence in mice (Herm-Gotz *et al.*, 2006). The TgMLC2 protein disappears when TgMyoD is knocked out, indicating that it is stabilized by association with the TgMyoD heavy chain.

13.2.8.3 Other Myosin Components

TgMyoB and TgMyoC are myosin heavy chains encoded by a single gene which can be alternatively spliced to create distinct C-terminal tails (Delbac *et al.*, 2001). The low expression levels of endogenous TgMyoB and TgMyoC led researchers to localize these motors in lines that transiently express tagged constructs over endogenous protein. These studies suggest that TgMyoB and TgMyoC localize to the tachyzoite cytoplasm, often at its periphery. Especially strong labelling is present at the posterior end and is evident in developing daughter parasites during endodyogeny. Although the roles of TgMyoB and TgMyoC have not been reinvestigated in light of more recently identified components of the basal complex, these motors may play a role in daughter cell scission at the completion of endodyogeny. In addition to the existing data on myosin heavy chains in *Toxoplasma*, seven myosin light chains (MLCs) have been annotated (Polonais *et al.*, 2011). Of these, TgMLC6 is identical to TgCAM2, which localizes to the conoid (Hu *et al.*, 2006).

13.2.8.4 The Glideosome Complex (GAP40, GAP45/70 and GAP50)

While TgMyoD likely associates with the plasma membrane via direct palmitoylation of its associated light chain TgMLC2 (Polonais *et al.*, 2011), TgMyoA and TgMLC1 localize to the pellicle of mature parasites in a stable complex with the gliding associated proteins

TgGAP40, TgGAP50, and TgGAP45/TgGAP70 (Gaskins *et al.*, 2004; Frenal *et al.*, 2010). TgGAP40 and TgGAP50 are integral IMC proteins that stably anchor the motor complex to the IMC and TgGAP45 and TgGAP70 are related acylated proteins which are tethered to the pellicle. Other apicomplexans have homologues of the GAP proteins, indicating that the glideosome is a highly conserved structure (Baum *et al.*, 2006). TgGAP40 (TGGT1_101540) is an integral membrane protein that spans the IMC nine times and has an apparent molecular weight of 37 kDa (Frenal *et al.*, 2010). The other GAP proteins (GAP50, GAP45 and GAP70) have been characterized in greater detail and are described below.

TgGAP45 (TGGT1_078320) is a novel, likely essential 27.3 kDa protein (Gaskins *et al.*, 2004). It lacks homology to characterized proteins, although there are orthologous proteins in the genomes of other apicomplexan parasites. TgGAP45 is predicted to contain an N-terminal coiled-coil domain and a C-terminal globular domain and is modified by N- and C-terminal acylations (Gaskins *et al.*, 2004; Johnson *et al.*, 2007; Rees-Channer *et al.*, 2006) as well as by serine phosphorylation (Gilk *et al.*, 2009) to regulate targeting and assembly into the mature glideosome structure (Fig. 13.4). The C-terminus of TgGAP45 interacts with the N-terminal domain of TgMLC1 to correctly localize TgMyoA in the glideosome (Frenal *et al.*, 2010). Phosphorylation of TgGAP45 creates several electrophoretically distinct TgGAP45 species (Gilk *et al.*, 2009) and likely regulates assembly of a mature GAP complex, although it does not alter targeting to the IMC. Depletion of TgGAP45 causes TgMLC1 to be redistributed to the cytoplasm and disrupts the intimate association of the plasma membrane and underlying IMC. Tachyzoites lacking TgGAP45 have dramatically reduced gliding motility, invasion, and egress (Frenal *et al.*, 2010). Defects associated with the loss of TgGAP45 can be complemented with the *P. falciparum* homologue of GAP45 or with a TgGAP45 mutant protein that is incapable of C-terminal acylation, indicating that this modification is dispensable for functional association of TgGAP45 with the motor complex

In addition to GAP45, the coccidian subset of apicomplexans has a GAP45-related protein, GAP70 (TGGT1_114490) (Frenal *et al.*, 2010). GAP70 has a conserved N-terminal region with predicted acylation sites, an extended coiled-coil domain (that is longer than the corresponding region of GAP45) and a conserved C-terminus. Although GAP70 has a predicted molecular mass of 37.6 kDa, it migrates at ~70 kDa, perhaps due to a high percentage of negatively charged amino acids. In contrast to the lateral location of TgGAP45 at the pellicle, TgGAP70 localizes at the apical region of the pellicle. Deletion analysis indicates that it is dually anchored to the plasma membrane through N-terminal acylations and to the apical cap through its C-terminal domain. TgGAP70 is not essential and null tachyzoites do not have discernible defects. Over-expression of TgGAP70 can partially complement the depletion of TgGAP45. Interestingly, the spacing between IMC and plasma membrane is increased when TgGAP70 is used to complement a TgGAP45 depleted strain. This is consistent with the extended coiled-coil domain of TgGAP70 relative to TgGAP45. TgGAP70 may facilitate the looser cohesion of IMC and plasma membrane in the apical region to permit greater pellicle flexibility required for conoid extension and retraction.

TgGAP50 (TGGT1_030300) was first identified as a protein that co-immunoprecipitates with TgMyoA, TgMLC1 and TgGAP45 (Gaskins *et al.*, 2004). Orthologous proteins are found in other apicomplexans, and these share 22%–26% identity with members of the calcineurin-like/purple acid phosphatase superfamily although they lack key enzymatic residues and detectable activity (Gaskins *et al.*, 2004; Bosch *et al.*, 2012). In other systems, calcineurin-like phosphatases have been shown to interact with myosins in a genetic screen (Fujita *et al.*, 2002), and regulate muscle contraction (Galler *et al.*, 2010). TgGAP50 has transmembrane domains at both termini. The N-terminal 50 amino acids are a signal peptide required for ER targeting which must be removed for maturation and localization to the IMC while the C-terminal transmembrane helix serves to anchor this protein in the IMC. Mature TgGAP50 contains a 351-residue lumenal domain with three N-linked glycans followed by a transmembrane domain, and a six-residue cytoplasmic domain at the C-terminus. Photobleaching experiments indicate that TgGAP50–YFP is non-mobile, a likely consequence of its localization to cholesterol-containing detergent-resistant domains of the IMC (Johnson *et al.*, 2007). Since the bulk of TgGAP50 resides in the IMC lumen, interactions with other components of the glideosome machinery are likely limited to the C-terminal membrane-spanning helix and short cytoplasmic domain. However, the strong conservation of surface-exposed GAP residues among various apicomplexan homologues may indicate that there are other lumen-localized components of the invasion machinery, which have yet to be identified (Bosch *et al.*, 2012).

13.3 PUTTING IT ALL TOGETHER: PROCESSES

13.3.1 Replication

13.3.1.1 Endodyogeny and Endopolygeny

Replication in *Toxoplasma* typically occurs by endodyogeny (Sheffield and Melton, 1968), although endopolygeny has also been observed, particularly in the context of the feline intestine (Ferguson *et al.*, 1974). In endodyogeny a single nuclear division is coupled to formation of two daughter parasites per replication cycle while in endopolygeny, creation of a multinucleated mother cell precedes the synchronous emergence of many daughter parasites (Fig. 13.3A). In both processes, mitosis proceeds without nuclear membrane breakdown (closed mitosis). Moreover, both endodyogeny and endopolygeny employ the formation of internal buds to package a complete set of organelles for each daughter within IMC compartments that emerge from the mother by adopting her plasma membrane. Other apicomplexans, by comparison, replicate by schizogony, a process that drives the synchronous emergence of thousands of daughter parasites from a multi-nucleated cell. The related but distinct modes of replication by endodyogeny, endopolygeny and schizogony illustrate that apicomplexan parasites can independently regulate DNA replication and chromosome segregation, nuclear division and daughter cell budding in contrast to many other eukaryotic lineages where these processes are inextricably linked (Striepen *et al.*, 2007; Gubbels *et al.*, 2008).

13.3.1.2 Nuclear Division

Toxoplasma mitosis has a number of features that distinguish it from mitosis in metazoan cells. The nuclear membrane remains intact and chromosome segregation occurs without chromosome condensation. Studies using YFP-tagged α1-tubulin to measure tubulin content are consistent with the association of each chromosome with a single $\sim$1 μm long spindle microtubule (Swedlow *et al.*, 2002). Spindle poles pass through an electron-dense invagination of the nuclear membrane termed the centrocone which is marked by TgMORN1 (Gubbels *et al.*, 2006; Ferguson *et al.*, 2008; Heaslip *et al.*, 2010a; Lorestani *et al.*, 2010) to end in the cytoplasm, proximal to asexual stage centrioles. Since centromeres are retained in a specialized apical region of the nuclear envelope throughout the cell cycle (Brooks *et al.*, 2011), either spindle microtubules or some other stable association anchors chromosomes to the centrocone region of the nuclear envelope. Dinitroanilines such as oryzalin selectively disrupt microtubules in intracellular tachyzoites but do not affect microtubules in host cells (Shaw *et al.*, 2000; Morrissette and

Sibley, 2002b; Stokkermans *et al.*, 1996). Although tachyzoites cannot undergo nuclear division in the absence of microtubules, centriole duplication, assembly of the apical membrane structures, protein synthesis and DNA replication continue unchecked (Stokkermans *et al.*, 1996; Morrissette and Sibley, 2002b; Tran *et al.*, 2010; Beck *et al.*, 2010).

13.3.1.3 Organelle Segregation and Development of Daughter Buds

Since nuclear division and cytokinesis are not closely coupled during apicomplexan replication, these organisms must have mechanisms in place to correctly segregate complete sets of organelles (Striepen *et al.*, 2007; Gubbels *et al.*, 2008). During *Toxoplasma* endodyogeny, two daughter parasites are enclosed by nascent IMC with associated APR, subpellicular microtubules and an alveolin/IMC protein cytoskeleton (Hu *et al.*, 2002a). Each daughter contains a complete set of apical complex organelles as well as a mitochondrion, Golgi apparatus, apicoplast and nucleus which are acquired with characteristic timing during replication (Nishi *et al.*, 2008). In emerging daughter parasites TgSFA2 and TgSFA3 tether the juxtanuclear centrioles to the apical conoid and associated apical secretory organelles (Lechtreck, 2003; Francia *et al.*, 2012). Conditional depletion of either TgSFA2 or TgSFA3 blocks daughter bud formation by inhibiting construction of the daughter APRs, although nuclear division proceeds and multiple nuclei accumulate in the cytoplasm. Like TgSFA2 and TgSFA3, TgIMC15 associates with duplicated centrioles at the beginning of the cell cycle, then relocates to the IMC of emerging daughters (Anderson-White *et al.*, 2011). This suggests that it may function to link centriole duplication with formation of the correct number of daughter buds and to ensure that nuclear division and cytokinesis (daughter cell budding) proceed with the correct timing. Later in replication, the plastid is segregated into daughter buds by association with centrioles (Striepen *et al.*, 2000).

The development of daughter buds occurs in a series of orderly steps that are intimately associated with the cytoskeleton. Some of the earliest events include the appearance of TgIMC15 on newly duplicated centrioles/centrosomes and localization of TgALP1 to emergent daughter buds (Anderson-White *et al.*, 2011; Gordon *et al.*, 2008). This is followed by localization of TgMORN1 first at the centrocone, then in rings which define the apical and posterior ends of the nascent IMC (Gubbels *et al.*, 2006; Heaslip *et al.*, 2010a; Lorestani *et al.*, 2010). As the lengthening IMC buds envelop the two halves of the horse-shoe shaped nucleus, subpellicular microtubules and their associated MAPs (TgSPM1 and TgSPM2) extend along the growing daughter IMC structures (Tran *et al.*, 2012; Morrissette and Sibley, 2002b) as do the alveolin family proteins TgIMC1, 3–6, 8–11, 13, and 15 (Anderson-White *et al.*, 2011). In particular, TgIMC3, 6 and 10 are enriched in daughter buds relative to their apparent concentration in the mother IMC. A regulated proteolysis removes the C-terminus of the major network component, TgIMC1, which coincides with conversion of the network from a detergent-labile to a detergent-resistant state late in daughter cell development (Hu *et al.*, 2002a). Once daughter cells are mature, the maternal apical complex is disassembled and daughter parasites bud from the mother, adopting her plasma membrane. Remnants of the maternal IMC and apical organelles are deposited in a structure at the posterior of the emerging parasites termed a residual body. While many cytoskeletal components are put into place during the development of internal daughter buds (including TgGAP40, TgGAP50, microtubules, TgSPM1, TgSPM2 and TgIMC1, 3–6, 8–11, 13, and 15), others (TgMyoA, TgMLC1, TgGAP45, TgGAP70, TgMSC1b, TgRNG1, and TGGT1_001950/RNG2) do not appear until daughter

IMC associates with the maternal plasma membrane to form a mature pellicle.

13.3.1.4 Emergence of Daughter Parasites

Once daughter buds contain a complete set of organelles, their emergence requires several coordinated processes in order to substitute daughter IMC for mother IMC at the plasma membrane, to separate the plasma membrane at the interface between daughters and to detach the posterior of daughters from remnants of the mother cell. Differential localization of cytoskeletal components distinguishes maternal structures from daughter buds and ensures that the glideosome—TgMyoA complex is only active in mature parasites. Changes in the composition of alveolin proteins may permit the regulated dissociation of the maternal IMC from the plasma membrane: TgIMC3, TgIMC6 and TgIMC10 are most concentrated in daughter buds but diminish after daughter cell emergence while levels of TgIMC7, TgIMC12 and TgIMC14 increase after daughter emergence (Anderson-White *et al.*, 2011). The association of daughter IMC with maternal plasma membrane triggers the recruitment of additional cytoskeletal components (TgGAP45, TgGAP70, TgMyoA, TgMLC1, TgMSC1b, TgRNG1, and TGGT1_001950/RNG2) (Gaskins *et al.*, 2004; Gilk *et al.*, 2009; Frenal *et al.*, 2010; Tran *et al.*, 2010; Gould *et al.*, 2011; Lorestani *et al.*, 2012) and Rab11A GTPase activity mediates vesicle fusion to separate the lateral surfaces of daughter buds (Agop-Nersesian *et al.*, 2009). Conditional depletion of GAP45 disrupts the intimate association of the plasma membrane and IMC suggesting that it has a role in pellicle formation as well as in assembly of the myosin motor complex (Frenal *et al.*, 2010).

Daughters remain attached to a residual body containing remnants of the mother cell. Scission requires the formation of a basal complex that constricts the posterior ends of daughters much like a contractile ring in order to complete cytokinesis (Anderson-White *et al.*, 2011; Heaslip *et al.*, 2010a; Hu, 2008; Lorestani *et al.*, 2010). TgMORN1, TgCen2, TgMyoB/C, actin, detyrosinated tubulin, and TgDLC1 have all been localized to the posterior region of emerging daughters (Ferguson *et al.*, 2008; Gubbels *et al.*, 2006; Heaslip *et al.*, 2010a; Lorestani *et al.*, 2010; Delbac *et al.*, 2001; Dobrowolski *et al.*, 1997b; Xiao *et al.*, 2010; Hu *et al.*, 2006, 2008) and after nuclear division TgIMC5, TgIMC8, TgIMC9 and TgIMC13 move to the basal complex from the IMC (Anderson-White *et al.*, 2011). Loss of TgMORN1 is associated with an aberrant basal complex and defects in cytokinesis leading to conjoined daughter parasites (Lorestani *et al.*, 2010; Heaslip *et al.*, 2010a). Moreover, none of the known basal complex components localize correctly without TgMORN1. It is unclear what proteins drive closure of the basal complex, although recruitment of TgCen2 correlates with the constriction and this is enhanced by increased cellular calcium (Hu, 2008). In contrast to the contractile ring, which disassembles at the completion of cytokinesis, the mature basal complex is a permanent cytoskeletal structure located at the posterior of parasites and TgMSC1a is a specific marker of the mature basal complex (Lorestani *et al.*, 2012).

In addition to its critical role in organelle packaging during *Toxoplasma* replication, the IMC serves as a rigid substrate to tether glideosome-associated components directly beneath the plasma membrane. The regulated maturation of this motor complex during pellicle formation likely prevents the premature activation of this machine in developing daughters. TgGAP40 and TgGAP50 are integral membrane proteins that are embedded in the IMC of both immature (daughter buds) and mature (mother) parasites (Frenal *et al.*, 2010; Gaskins *et al.*, 2004; Johnson *et al.*, 2007; Gilk *et al.*, 2009). In contrast, TgGAP45, TgGAP70, TgMyoA and TgMLC1 only associate with the outer face of the IMC in mature parasites (Fig. 13.4B). TgMyoA, TgMLC1, and TgGAP45 associate in a soluble 'proto-glideosome' complex which must interact with TgGAP50 at the IMC to

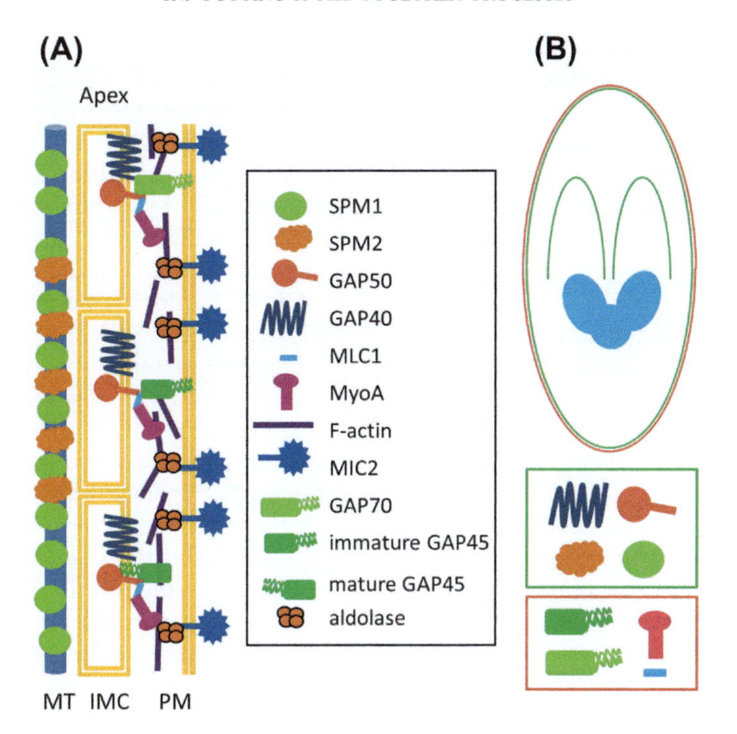

FIGURE 13.4 **Organization of glideosome components and other pellicle associated cytoskeletal elements in *Toxoplasma* tachyzoites.**
A) The pellicle is associated with diverse cytoskeletal elements, including subpellicular microtubules and alveolin family members which provide an essential rigidity to the IMC. Elements of the actin and myosin gliding motility machinery are located in the narrow space between the IMC and plasma membrane (PM). The subpellicular microtubules (MT) are associated with two proteins. TgSPM1 associates with the full length of the subpellicular microtubules while TgSPM2 is only found along the middle region of these microtubules and is absent from both apical and posterior ends. Two components of the glideosome machinery, TgGAP40 and TgGAP50, are integral membrane proteins within IMC membranes, while other components are acylated (TgGAP45 and TgGAP70) or are soluble proteins (TgMyoA, TgMLC1, actin and aldolase). TgGAP70 is anchored to the plasma membrane in the apical cap region of the pellicle through N-terminal acylations. Current data are consistent with a model whereby immature TgGAP45 is anchored to the lateral region of the pellicle by N-terminal palmitoylation, which prevents premature association of TgMyoA with daughter buds. Once TgGAP45 is correctly located between the plasma membrane and IMC, depalmitoylation of the N-terminus and palmitoylation of the C-terminus redirects TgGAP45 to the IMC.
B) During replication by endodyogeny, some cytoskeletal components are associated with early daughter buds which are defined by IMC. These include the integral membrane proteins TgGAP40 and TgGAP50 as well as the microtubule associated proteins TgSPM1 and TgSPM2 (green box). A number of alveolin family proteins such as TgIMC1 are also found in emerging buds but are not shown in this figure. In order to prevent the premature activation of the glideosome-associated machinery, assembly of the complete motor complex does not occur until daughter parasites emerge from the mother and acquire a mature pellicle structure composed of IMC and plasma membrane. At this time, components such as TgGAP45, TgGAP70, TgMyoA and TgMLC1 associate with the pellicle (red box).

form a functional, pellicle-associated glideosome (Gaskins *et al.*, 2004). The observation that mutation of TgGAP45 N-terminal palmitoylation sites causes premature glideosome assembly at the IMC of daughter buds suggests that N-terminal palmitoylation of GAP45 initially anchors it at the plasma membrane to prevent premature association of TgMyoA with daughter buds (Frenal *et al.*, 2010). Once TgGAP45 is correctly located between the plasma membrane and IMC, depalmitoylation of the N-terminus and palmitoylation of the C-terminus redirects TgGAP45 to the IMC. In addition to acylation, serine phosphorylation of TgGAP45 at residues 163 and 167 inhibits association of the MyoA−MLC1−GAP45 complex with TgGAP50 (Gilk *et al.*, 2009).

13.3.2 Motility, Invasion and Egress

Although the gliding motility displayed by *Toxoplasma* tachyzoites is conserved with many other zoite stage apicomplexans, it is unique to this lineage and is distinct from amoeboid or cilia/flagella-based propulsive mechanisms displayed by other cells (Baum *et al.*, 2006; Carruthers and Boothroyd, 2007; Daher and Soldati-Favre, 2009; Soldati-Favre, 2008). Gliding motility allows zoites to cross biological barriers to seek out the appropriate host cell, to invade that host cell and in some cases to exit (egress) from the host cell upon completion of replication. Studies in *Toxoplasma*, in particular, have demonstrated that gliding motility requires the concerted action of signalling molecules, cytoskeletal components, secreted proteins and a rhomboid protease. While myosins characteristically move along fixed F-actin tracks, TgMyoA myosin is fixed in the IMC and moves short actin filaments from the anterior to the posterior of the tachyzoite in association with the cytoplasmic tail of externally exposed adhesions such as TgMIC2. Gliding requires the release of microneme-secreted adhesins (see Chapter 12). In particular, the short cytoplasmic tail of TgMIC2 interacts with aldolase, which binds to F-actin and this complex is translocated to the tachyzoite posterior by the action of TgMyoA. The peripheral IMC membrane cytoskeleton and associated subpellicular microtubules are critical to gliding motility, because the myosin motor complex is rigidly anchored to the alveolar vesicles and functions in the narrow space that separates the IMC from the plasma membrane. Gliding motility requires the formation of F-actin, which polymerizes in the space between the plasma membrane and the outer membrane of the IMC (Fig. 13.4). Environmental cues that induce release of Ca^{2+} from intracellular stores integrate events to drive microneme secretion, actin polymerization and myosin motor activation in order to induce gliding. The response to Ca^{2+} is mediated by diverse and in several instances unique families of Ca^{2+}-binding proteins (Chapter 10). Established evidence using inhibitors (Kieschnick *et al.*, 2001) and emerging evidence from mass spectrometry (Nebl *et al.*, 2011; Treeck *et al.*, 2011) indicate a role for protein phosphorylation in the signalling processes that regulate these events. Increased Ca^{2+} also induces the TgAKMT lysine methyltransferase to move into the cytoplasm from the parasite apex (Heaslip *et al.*, 2011). Depletion of TgAKMT inhibits tachyzoite motility, suggesting that this enzyme also participates in induction of motility.

The processes associated with *Toxoplasma* gliding motility are also intimately associated with host cell invasion. However, in addition to activation of the TgMyoA myosin motor, induction of actin polymerization and stimulation of microneme secretion, tachyzoites also require conoid extrusion and rhoptry secretion in order to invade host cells (Mondragon and Frixione, 1996; Carey *et al.*, 2004). Conoid extrusion is a reversible Ca^{2+}-mediated process that occurs during gliding motility and at the time of host cell invasion. The cues inducing rhoptry secretion are not well defined, although this process is thought to require prior microneme secretion. Rhoptry neck proteins (RONs) are secreted in order to form the moving junction which serves as an interface between parasite and host cell plasma membranes through which the parasite enters the host cell (Chapter 12). Following successful establishment of a parasitophorous vacuole within a host cell, increased K^+ levels inhibit invasion associated processes such as microneme and rhoptry secretion, myosin motor activity, cortical actin polymerization and conoid extrusion (Moudy *et al.*, 2001; Endo and Yagita, 1990). Motility-associated processes are only reinitiated upon the need to invade new host cells after intracellular replication. In the case that parasite growth triggers a breach in the host cell plasma membrane leading to an efflux of K^+, ionic cues stimulate motility. In other circumstances, parasite growth depletes essential host cell resources and egress requires microneme secretion as well as activated actinomyosin machinery (Hoff and Carruthers, 2002). Studies of invasion and egress have revealed that although gliding and secretion are required for both processes, aspects of the control and execution of these events are specifically tailored to control entry or exit from host cells.

13.3.2.1 Triggering Egress

The process of *Toxoplasma* endodyogeny enables replicating parasites to become motile at nearly any stage in the cell cycle, except during the terminal stages of budding when daughters are emerging from the mother parasite (Gaji *et al.*, 2011). Two fundamentally different events can activate an egress signalling cascade in parasites that are capable of motility. In the first case, parasites exit from a damaged host cell that may be under immune attack while in the second case, exit anticipates the need for additional host resources. In the case that the host cell is damaged (which occurs before the parasite fully exhausts host cell nutrients), egress can be viewed as a survival strategy. Damage to the host cell plasma membrane can be purely mechanical, or can be due to the immune system attacking the infected cell. Both Fas- and perforin-mediated cytotoxic T-cell response mechanisms stimulate egress (Persson *et al.*, 2007). Host cell rupture leads to decreased K^+ and can be mimicked by K^+ ionophores (Moudy *et al.*, 2001; Fruth and Arrizabalaga, 2007). Decreased K^+ triggers the release of Ca^{2+} from intracellular stores. Notably, immune response-mediated egress is increasingly being recognized as a virulence factor and is the dominant egress mechanism in active infections (Persson *et al.*, 2007, 2009; Lambert and Barragan, 2010; Melzer *et al.*, 2008). This pathway is the prevalent route of egress in mouse infections relative to parasite-controlled egress as would occur in standard tissue culture conditions (Tomita *et al.*, 2009). In the absence of host cell damage, tachyzoites replicate until host resources are depleted and parasite waste accumulates. There is some evidence of cues that reflect parasite density, but these are not well understood at the molecular level. For example, the phytohormone abscisic acid (ABA) is produced in the apicoplast and works as a quorum sensor to measure tachyzoite density (Nagamune *et al.*, 2008a). Similarly, *Toxoplasma* synthesizes a family of inactive NTPases that are secreted into the parasitophorous vacuole during tachyzoite replication. It is unclear how the NTPases become activated during infection and whether they act as a sensor or as a facilitator of unmanipulated egress (Silverman *et al.*, 1998). Exogenous activation of the NTPases by treatment with reducing agents such as dithiothreitol (DTT) leads to a rapid depletion of host cell ATP and coincides with initiation of parasite egress (Silverman *et al.*, 1998; Stommel *et al.*, 1997). Both ABA and DTT mediated signalling pathways converge on the release of intracellular Ca^{2+}, as does the K^+-mediated signalling pathway. The Ca^{2+}-mediated processes and proteins involved are described in Chapter 10 and have been reviewed elsewhere as well (Arrizabalaga and Boothroyd, 2004; Billker *et al.*, 2009; Lavine and Arrizabalaga, 2007; Nagamune *et al.*, 2008b; Nagamune and Sibley, 2006). Other critical aspects of motility are regulated by protein kinase G (PKG) activation which acts downstream of Ca^{2+}-mediated signalling pathways (Wiersma *et al.*, 2004).

When parasite egress is inhibited, intracellular tachyzoites continue to replicate and form extremely large vacuoles that eventually mechanically burst out of the host cell. This happens when the microneme-secreted perforin TgPLP1 is genetically deleted, as it is required to form pores in the vacuolar membrane and host cell plasma membrane prior to egress (Kafsack *et al.*, 2009). An identical phenotype is observed at the non-permissive temperature in a parasite line that harbours a temperature sensitive mutation in TgDOC2.1 which regulates exocytic machinery, including microneme secretion (Farrell *et al.*, 2012). Finally, inhibiting host cell calpain proteases, which are thought to degrade the host cytoskeleton to clear an outbound route, also results in an excessively large vacuole phenotype (Chandramohanadas *et al.*, 2009). Since these conditions appear to be downstream of many of the early sensors, such as ABA, downstream defects can block full-fledged activation of egress. These observations suggest that as long as the parasites remain intracellular, replication is default behaviour, even when part of the egress signalling pathway is activated. In essence, a very tight control mechanism prevents premature egress to retain the parasite in a protected intracellular environment.

13.3.2.2 Gliding Motility

Gliding motility requires the transport of apically secreted microneme proteins to the basal end of the parasite where they are released by a rhomboid protease (see Chapter 12). Consequently, the parasite always moves into the apical direction, which is also the end of the parasite where host cell invasion is initiated. However, when parasites are placed on protein coated glass slides they display three distinct modes of motility: circular gliding, helical gliding, and twirling (Hakansson *et al.*, 1999). Circular gliding is conserved across apicomplexan species (Sibley, 2004), and other forms of motility are also observed in related parasites. Notably, parasites never glide in a straight line but follow a defined circular pattern, or half circle after which they flip over to carry out helical gliding. It is generally assumed that the spiral pitch of the subpellicular microtubules defines the diameter of the circle. Twirling, however, is distinct in that the parasite elevates its apical end and spins with the basal end attached to the substrate. It is independent of microneme proteins (Farrell *et al.*, 2012) and appears to be mediated by actin accumulation at the basal end (Wetzel *et al.*, 2003), though the specific mechanism of this action is poorly understood. Since *Plasmodium* sporozoites also twirl at high frequency, twirling does not require a basal complex, which is not present in non-coccidian apicomplexan lineages (Kudryashev *et al.*, 2010; Hegge *et al.*, 2010). A conspicuous limitation of the existing motility assays is that they are performed on an artificial two-dimensional substrate, while apicomplexan parasites normally operate in a three-dimensional tissue environment. Therefore, the relevance of the behaviour described in two-dimensions to *in vivo* motility remains to be seen. Notably, *in vivo* imaging of *Plasmodium* sporozoites has demonstrated that gliding parasites tune the size of the circle that they circumscribe to the diameter of blood vessels in the skin: sporozoites spiral around the exterior of a vessel several times before making their way across the epithelium into the lumen (Amino *et al.*, 2006). Furthermore, motility is not strictly required for invasion by all apicomplexan parasites; notably, *Theileria* sporozoites and merozoites are non-motile and invasion proceeds with the parasite in any orientation (Shaw and Tilney, 1995; Shaw, 2003). However, the innovation of gliding motility was a major breakthrough that permitted parasites to cross biological barriers, greatly expanding their host range and tissue tropism preceding the diversification of Apicomplexa (Leander, 2008).

13.3.2.3 Glideosome Activation

The components of the glideosome are well conserved between *Toxoplasma* and *Plasmodium*. Myosin A (MyoA) is required for invasion of both *Plasmodium* (Siden-Kiamos *et al.*, 2011) and *Toxoplasma* (Meissner *et al.*, 2002) and is anchored in the outer membrane of the IMC by the glideosome. GAP45 and GAP50 anchor MyoA in the membrane (Baum *et al.*, 2008; Frenal *et al.*, 2010; Gaskins *et al.*, 2004; Jones *et al.*, 2006). Phosphorylation of MLC1 (known as Myosin Tail Interacting Protein or MTIP in *Plasmodium*) is critical for activation of motor activity. MTIP phosphorylation is mediated by CDPK1 in *Plasmodium*, at least *in vitro*, which also phosphorylates GAP45 (Green *et al.*, 2006), but it is unclear whether this is the functional orthologue of myosin light chain kinase (MLCK) that regulates MLC status in other organisms (e.g. Stull *et al.*, 2011). TachypleginA, a small molecule inhibitor of *Toxoplasma* tachyzoite motility, modifies TgMLC1 (Heaslip *et al.*, 2010b). Furthermore, MyoA itself is phosphorylated in a Ca^{2+}-dependent fashion, likely mediated by a CDPK, although its identity and whether the modification influences motor activity have yet to be determined (Nebl *et al.*, 2011; Treeck *et al.*, 2011). Recently, a calmodulin-like protein representing the MyoA Essential Light Chain (ELC) was shown to be part of the glideosome as well, suggesting that Ca^{2+}-binding directly mediates motor activity and motility (Nebl *et al.*, 2011).

13.3.2.4 Actin Polymerization

The spatiotemporal control of actin polymerization is critical to activation of gliding motility. Relative to other eukaryotes, apicomplexan genomes contain genes for a limited set of actin-interacting proteins (Baum *et al.*, 2006). Furthermore, differences in the requirement for these proteins throughout the parasite life cycle indicate that they are not equally critical during all cellular programmes. (For recent reviews see Baum *et al.*, 2006, 2008; Gordon and Sibley, 2005; Matuschewski and Schuler, 2008; Sattler *et al.*, 2011; Schuler and Matuschewski, 2006b.) In *Toxoplasma* tachyzoites, the cytoplasmic concentration of G-actin is relatively high (8–10 μM), whereas F-actin is not easily detectable (Dobrowolski *et al.*, 1997a). Actin polymerizes into extremely short (100 nm), unstable filaments (Sahoo *et al.*, 2006; Skillman *et al.*, 2011) which have been observed at the periphery by rapid-freeze electron microscopy (Wetzel *et al.*, 2003). Notably, F-actin is observed in cables or bundles, but not branched networks. This is consistent with the absence of genes for Arp2/3 components, as this complex is required to form actin branches (Gordon and Sibley, 2005). Current observations are consistent with a model whereby filaments are formed transiently and spatiotemporal polymerization dynamics control the directionality and timing of motility. Extrapolating from observations of *Plasmodium* sporozoites, polymerization is likely restricted to the areas of contact between MIC adhesins and substrate or ligands on the host cell (Angrisano *et al.*, 2012; Munter *et al.*, 2009).

The highly dynamic nature of actin polymerization is regulated in large part by its capacity to bind and hydrolyse ATP. G-actin monomers must be loaded with ATP in order to readily assemble into F-actin filaments. Within the context of the microfilament, ATP is hydrolysed to ADP which results in destabilization of the F-actin filament. By having distinct preferences for ATP–actin or ADP–actin, actin binding proteins exert additional regulation of the nucleotide switch and polymerization. In part this is because these proteins modulate ATPase activity or nucleotide exchange rates. Eukaryotes have multiple proteins that bind G-actin to form pools of readily polymerizable monomers, a prerequisite for the fast formation of actin polymers (Sattler *et al.*, 2011). Profilins for example sequester G-actin at locations where polymerization is anticipated by binding to lipids to deliver them to the point of nucleation (Kursula *et al.*, 2008). Unlike mammalian profilin which promotes F-actin assembly, *in vitro* experiments with TgPRF show that it prevents F-actin formation by sequestering G-actin (Skillman *et al.*, 2012) and is essential in tachyzoites (Plattner *et al.*, 2008). The cyclase-associated protein (CAP) is another protein that primarily binds to G-actin (Hliscs *et al.*, 2010). *Arabidopsis* CAP has been shown to promote nucleotide exchange (Chaudhry *et al.*, 2007). Apicomplexans encode a CAP orthologue lacking the typical N-terminal WH2 domain that connects cAMP to RAS signalling (Schuler and Matuschewski, 2006a). *Toxoplasma* CAP localizes to the apical cap in intracellular tachyzoites and rapidly transitions to a cytoplasmic localization in extracellular parasites (Lorestani *et al.*, 2012). It therefore appears to be sequestered at the location where actin polymerization would be first needed when motility is activated, at the apical end, although direct data to support this model are not available yet.

As described above, apicomplexan F-actin is intrinsically unstable. Therefore filament-stabilizing proteins are required for both assembly and sustained filament stability (Skillman *et al.*, 2011). While metazoans usually use the Arp2/3 complex to initiate actin polymerization, in apicomplexans F-actin polymerization is mediated by two formin proteins (Skillman *et al.*, 2012; Daher *et al.*, 2010). A third formin is only found in coccidian lineages and is dispensable for tachyzoite motility (Daher *et al.*, 2012). TgFRM1 and TgFRM2 independently promote actin polymerization (Skillman *et al.*, 2012), although TgFRM1 associates with the plasma membrane and TgFRM2 associates with the IMC membrane (Daher *et al.*, 2010). In both cases the interaction is mediated by electrostatic interactions with membrane proteins or lipid polar heads, suggesting that they may nucleate actin at any time and at any point of contact between the parasite and its substrate. Consistent with other formins, TgFRM1 and TgFRM2 bundle actin filaments (Skillman *et al.*, 2012). Coronin is the only other F-actin bundling protein represented in the *Toxoplasma* genome but has not been extensively studied (Tardieux *et al.*, 1998; Figueroa *et al.*, 2004). There are homologues of both α- and β-subunits of the F-actin capping protein (CP) encoded in the *Toxoplasma* genome, although this protein has not been studied experimentally. CP regulates actin filament growth at the barbed end and a knock-out of the CPβ gene in *P. berghei* impairs ookinete motility, and reduces oocyst numbers in the mosquito (Ganter *et al.*, 2009). The few sporozoites that form cannot migrate to the salivary gland. Although *Plasmodium* fast locomotion is compromised, null sporozoites still exhibit non-productive motility such as bending, flexing and pendulum movements (Ganter *et al.*, 2009). Until parallel studies are performed in *Toxoplasma*, it remains unclear whether CP will be similarly essential for fast actin/myosin dependent gliding in tachyzoites. Lastly, actin depolymerizing factors such as ADF/cofilin family members are required for F-actin filament turnover in many eukaryotes. Although *Toxoplasma* has an ADF homologue that is involved in motility, it does not bind F-actin and functions by sequestering ADP loaded G-actin (Mehta and Sibley, 2010). Moreover, experimental observations and structural features do not support the interaction of TgADF with the phosphoinositide PIP_2, and PIP_2 does not affect the interaction of TgADF with G-actin. This trait distinguishes apicomplexan ADF from vertebrate ADFs, since PIP_2 regulates the affinity of vertebrate ADF for G-actin (Yadav *et al.*, 2011).

13.3.2.5 Energy Requirements

Gliding parasites require energy in the form of ATP which is consumed by myosin A and actin polymerization. Although a knock-out of the gene encoding the glucose transporter TgGT1 influences sustained gliding motility in a defined minimal media lacking glutamine, this effect is explained by the ability of tachyzoites to derive energy from glutaminolysis as well as glycolysis (Blume *et al.*, 2009). Moreover, disruption of oxidative phosphorylation in the mitochondria by atovaquone or cyanide treatment does not result in changes in parasite gliding (Pomel *et al.*, 2008). When tachyzoites egress from host cells, their glycolytic enzymes move from a dispersed cytosolic location to the IMC, a process which starts within minutes after egress but reaches a maximum one hour later with most parasites displaying a peripheral protein localization. This behaviour has been demonstrated for hexokinase (HK), glyceraldehyde-3-phosphate dehydrogenase-1 (GAPDH1), pyruvate kinase-1 (PK1), and lactate dehydrogenase-1 (LDH1). Translocation of aldolase, which directly participates in gliding motility, was not observed in an independent study, perhaps due to differences in tagging, fixation, and detection procedures (Starnes *et al.*, 2009). Translocation of HK, GAPDH1, PK1 and LDH1 to the periphery could be observed with specific antibodies as well as with Myc-tagged enzymes, but not when proteins were tagged with YFP (Pomel *et al.*, 2008). The mechanisms underlying translocation and peripheral anchoring are not known, other than these events are insensitive to disruption of microtubules by oryzalin treatment or inhibition of F-actin formation with cytochalasin D. Evidence in *Plasmodium* suggests that peripheral GAPDH anchoring is mediated by Rab2 through a mechanism comparable to eukaryotic GAPDH membrane anchoring (Daubenberger *et al.*, 2003). It is not clear whether HK, GAPDH1, PK1 and LDH1 are anchored on the cytoplasmic side or on the plasma membrane side of the IMC alveoli. The former scenario is unlikely, it would require these enzymes to cross the IMC in some fashion. In the latter case, if the glycolytic enzymes reside on the cytoplasmic side of the alveoli, ATP would have to cross the IMC in order to be delivered to actin and myosin. There may be uncharacterized ATP transport channels in the IMC that provide for this transfer activity. Finally, there is always the possibility that the translocation of glycolytic enzymes to the periphery is unrelated to providing ATP to myosin.

13.3.2.6 Mechanism of Conoid Extrusion

During gliding motility, the conoid continuously extrudes and retracts, likely in response to fluctuations in cytoplasmic Ca^{2+} levels (Wetzel *et al.*, 2004). A range of pharmacological agents that interfere with Ca^{2+}-signalling and homeostasis also block conoid extrusion (Del Carmen *et al.*, 2009; Mondragon and Frixione, 1996). Since kinase inhibitors also inhibit conoid extrusion, CDPKs may also be required for this process. Inhibition of F-actin (with cytochalasin D) or myosin (with butanedione monoxime) suggests that extrusion requires the actinomyosin machinery. Conoid extrusion is specifically inhibited by a small molecule, named conoidin A, which was identified in a screen for invasion inhibitors (Carey *et al.*, 2004; Haraldsen *et al.*, 2009). Interestingly, conoidin A inhibits invasion without affecting microneme secretion or motility. Similarly, a temperature sensitive mutant of the DOC2.1 gene inhibits microneme secretion without affecting conoid extrusion or motility (Farrell *et al.*, 2012). Collectively, these data demonstrate that conoid extrusion is essential for invasion and is controlled by Ca^{2+}-dependent processes independent of microneme secretion and activation of the actinomyosin motor. However, the processes responsible for conoid extrusion are still poorly understood, and it is equally unclear how conoid extrusion functions in the invasion process.

13.3.2.7 The Role of the Host Cell in the Invasion and Egress Process

Although host cell invasion by *Toxoplasma* has been considered to be independent of the host cell, several recent reports indicate that properties of the host cell cytoskeleton influence invasion and intracellular growth. Changes in the host cell cytoskeleton have been reported at the site of entry (see Chapter 12 for details on the moving junction). The best understood process is the parasite-induced remodelling of host cell cortical actin by the action of a secreted rhoptry protein termed toxofilin (TGGT1_124310) (Delorme-Walker *et al.*, 2012; Lodoen *et al.*, 2010; Poupel *et al.*, 2000). Toxofilin is secreted during invasion and induces host actin depolymerization in the vicinity of the parasite, thereby locally reducing the host cell cortical actin network. This activity facilitates invagination of the plasma membrane at the point of parasite entry in order to enable parasite internalization and vacuole biogenesis (Delorme-Walker *et al.*, 2012). Toxofilin null parasites exhibit delayed invasion kinetics, associated with a slightly narrower aperture of the moving junction. Although F-actin is reduced at the site of tachyzoite entry by the action of toxofilin, host microtubules congregate around the moving junction (Sweeney *et al.*, 2010). Selective disruption of the host microtubules with drugs that do not affect parasite microtubules leads to a reduced efficiency of parasite invasion. This effect only alters the behaviour of parasites that invade rapidly, and does not affect tachyzoites that take longer to invade. Host microtubules may act as a scaffold to resist compression induced by parasite contact with the host cell or operate as a fulcrum in order for gliding parasites to re-orient with apical contact prior to penetrating the host cell. Alternatively, host microtubules may promote efficient formation of the moving junction. It is unknown how host microtubules are recruited to this area, but together with the changes to the host actin network, these results indicate that the parasite manipulates the host to facilitate invasion. Other data also support this conclusion. For example, two recently identified kinase inhibitors act on host cells but interfere with parasite invasion (Kamau *et al.*, 2012). Moreover, parasite egress requires the activity of host cell calpains, which are thought to remodel the host's cytoskeleton and plasma membrane to create an escape tunnel (Chandramohanadas *et al.*, 2009). Since host cell-modifying behaviours are only beginning to be recognized, it is likely that future research will identify other ways that the parasite manipulates the host cell during invasion and egress.

13.3.3 Differentiation to Gametes

Toxoplasma and other apicomplexan parasites typically undergo a transient sexual stage during their life cycle (see Chapters 1 and 2). In preparation for zygote formation, male and female gametocytes differentiate in order to form microgametes and macrogametes (also known as male and female gametes). In some parasites, such as *Plasmodium* spp., completing this portion of the life cycle is an obligatory aspect of transmission to new hosts. Since *Toxoplasma* bradyzoites in tissue cysts are orally infectious, the sexual cycle and consequent formation of oocysts is less critical for parasite transmission to new hosts. In *Toxoplasma*, differentiation to gametocytes (gamete precursors), microgametes and macrogametes only occurs in the intestinal epithelium of cats. Although several studies elegantly describe the morphology of microgametes and macrogametes (Ferguson *et al.*, 1974, 1975; Scholtyseck *et al.*, 1971b), the requirements for differentiation of these forms are not amenable to extensive genetic and biochemical characterization of the *Toxoplasma* cytoskeleton. *Toxoplasma* macrogametes are surrounded by a normal pellicle with numerous micropores (Ferguson *et al.*, 1975; Chobotar and Scholtyseck, 1982; Piekarski *et al.*, 1971). They contain abundant polysaccharide granules and lipid globules earmarked for the post-meiotic assembly of sporozoites within oocysts. Although macrogamete morphology is distinct, it does not imply any particular information about the organization or requirements for the cytoskeleton in this parasite form. Like other apicomplexan male gametes, *Toxoplasma* microgametes have two flagella that originate in the cell apex with a conserved 9+2 axoneme structure (Ferguson *et al.*, 1974). Since there is a large body of literature surrounding flagellar structure and conserved components, we can infer that many axoneme-specific proteins are expressed in this stage (Hodges *et al.*, 2010).

Given the technical barriers that prevent exhaustive study of *Toxoplasma* microgamete and macrogamete stages, it is useful to review information on *Plasmodium* sexual stages, as gamete differentiation can proceed *in vitro*, making it accessible to study with a variety of methods. Mass spectrometry analysis of sorted populations of male and female *Plasmodium* gametocytes indicates that these lineages have distinct protein expression profiles in anticipation of creating cells with distinct sex-specific functions (Khan *et al.*, 2005). Male gametocytes express the greatest proportion of stage-specific proteins, with many of these dedicated to the production of flagella such as PF16, radial spoke proteins and dynein subunits. Validation of the significance of flagellar proteins has been provided by work demonstrating that PF16 is important for motility in *Plasmodium berghei* microgametes (Straschil *et al.*, 2010). While some aspects of microgamete differentiation are likely to be conserved between *Toxoplasma* and *Plasmodium*, others may be distinct. For example, centrioles have not been described in asexual stage *Plasmodium* and ultrastructural studies of microgamete formation describe *de novo* formation of basal bodies. Since asexual stage *Toxoplasma* tachyzoites contain atypical centrioles, it is unclear whether microgamete basal bodies form *de novo* as described for *Plasmodium* or if the atypical centriole found in tachyzoites serves as a template for formation of basal bodies (Fig. 13.2).

13.4 SUMMARY: A STORY OF ADAPTATION, LOSS AND NOVEL COMPONENTS

Research on the *Toxoplasma* cytoskeleton has revealed that this parasite employs a mixture of conserved and novel proteins in order to replicate, move and invade cells. Analysis of the genome databases for *Toxoplasma* and other apicomplexans has identified a reduced subset of the repertoire of defined cytoskeletal proteins used by other eukaryotes (Schuler and Matuschewski, 2006b; Gordon and Sibley, 2005). Moreover, in many cases, apicomplexan homologues of conserved cytoskeletal elements have evolved altered biochemical properties to best serve the particular requirements of these parasites. This is most evident in the case of actin (Skillman *et al.*, 2011; Sahoo *et al.*, 2006) and actin binding proteins such as ADF/cofilin and the *Toxoplasma* formins which have adapted to function in a G-actin-rich context (Mehta and Sibley, 2010, 2011; Poukkula *et al.*, 2011; Yadav *et al.*, 2011; Skillman *et al.*, 2012; Daher *et al.*, 2010, 2012; Daher and Soldati-Favre, 2009). Similarly, in contrast to most if not all other systems where myosins move on fixed actin filaments, apicomplexan myosin A motors remain fixed in place while actin filaments are translocated (Gaskins *et al.*, 2004; Frenal *et al.*, 2010; Johnson *et al.*, 2007). Subpellicular microtubules and a network of alveolin fibres are required to create a rigid IMC structure to immobilize the apicomplexan glideosome complex (Khater *et al.*, 2004; Tremp *et al.*, 2008; Morrissette and Sibley, 2002b). Although it has yet to be demonstrated in apicomplexan organisms, the observation that cytoplasmic dyneins can generate membrane tension (Hook and Vallee, 2006) suggests that microtubule motors may play a role keeping the IMC and associated alveolin network taut against the subpellicular microtubule framework. Apicomplexan parasites likely evolved from an ancestor similar to the predatory flagellated alveolate *Rastrimonas* (*Cryptophagus*) *subtilis* which absorbs its flagella during intracellular growth (Brugerolle, 2002b). This feature may have become more pronounced during the evolution of apicomplexan parasites, ultimately leading to the loss of flagella in all stages other than microgametes and driving the evolution of gliding motility (Leander, 2008). Surprisingly, expression of several flagella-associated components is retained in aflagellate tachyzoites, suggesting that these proteins have been repurposed to function in organelle segregation and formation of the apical complex (Francia *et al.*, 2012; de Leon *et al.*, 2013). Although *Toxoplasma* and *Plasmodium* retain the ability to construct motile flagella in microgametes, bioinformatic analysis indicates that these organisms have a surprisingly reduced complement of conserved centriole/basal body components (Hodges *et al.*, 2010) suggesting that many proteins have been lost from this lineage. In addition, although dynamic regulation of actin polymerization is essential for gliding motility and the ability to invade host cells, the Arp2/3 complex has been lost from this lineage (Gordon and Sibley, 2005). Recent studies on tachyzoite motility identified two unusual proteins (aldolase and GAP50) as having indispensable roles in the glideosome machinery. Although aldolase and other glycolytic enzymes have been shown to associate with the cytoskeleton in metazoan cells (Janmey, 1998), this interaction is critical for linking adhesins to the actin and myosin machinery during gliding motility in apicomplexans (Jewett and Sibley, 2003; Starnes *et al.*, 2006; Starnes *et al.*, 2009). Similarly, GAP50 is essential to anchor myosin A in the IMC for gliding motility (Gaskins *et al.*, 2004; Baum *et al.*, 2006, 2012). Both sequence analysis and structural studies have demonstrated that GAP50 orthologues are closely related to members of the calcineurin-like/purple acid phosphatase superfamily although they lack key enzymatic residues and detectable activity. In these cases, advantageous protein–protein interactions may have been coopted during the evolution of gliding motility. Lastly, other *Toxoplasma* cytoskeletal components (TgGAP45, TgGAP70, TgGAP40, TgTgICMAP1, TgSPM1, TgSPM2) are entirely novel proteins with a distribution that is restricted to all or some apicomplexan organisms (Frenal *et al.*, 2010; Baum *et al.*, 2006; Heaslip *et al.*, 2009; Tran *et al.*, 2012). Given the unusual structures and behaviours of the *Toxoplasma* cytoskeleton, it is likely that we have only begun to identify novel players. Future research is certain to uncover additional examples of cytoskeletal elements with modified biochemistry or novel functions, novel proteins and surprising losses during the evolution of *Toxoplasma* and other apicomplexans. Differences between the *Toxoplasma* cytoskeleton and the cytoskeleton of metazoans represent key targets that can be exploited by future therapeutic agents.

Acknowledgements

We would like to thank our colleagues for useful discussions and/or feedback on this chapter. We thank Chun-Ti Chen, Norikiyo Ueno and Katherine Harker for critically reviewing the manuscript.

References

Agop-Nersesian, C., Naissant, B., Ben Rached, F., Rauch, M., Kretzschmar, A., Thiberge, S., Menard, R., Ferguson, D.J., Meissner, M., Langsley, G., 2009. Rab11A-controlled assembly of the inner membrane complex is required for completion of apicomplexan cytokinesis. PLoS Pathog. 5, e1000270.

Akella, J.S., Wloga, D., Kim, J., Starostina, N.G., Lyons-Abbott, S., Morrissette, N.S., Dougan, S.T., Kipreos, E.T., Gaertig, J., 2010. MEC-17 is an alpha-tubulin acetyltransferase. Nature 467, 218–222.

Allen, M.L., Dobrowolski, J.M., Muller, H., Sibley, L.D., Mansour, T.E., 1997. Cloning and characterization of actin depolymerizing factor from *Toxoplasma gondii*. Mol. Biochem. Parasitol. 88, 43–52.

Amino, R., Thiberge, S., Shorte, S., Frischknecht, F., Menard, R., 2006. Quantitative imaging of *Plasmodium* sporozoites in the mammalian host. C R Biol. 329, 858–862.

Anderson-White, B.R., Ivey, F.D., Cheng, K., Szatanek, T., Lorestani, A., Beckers, C.J., Ferguson, D.J., Sahoo, N., Gubbels, M.J., 2011. A family of intermediate filament-like proteins is sequentially assembled into the cytoskeleton of *Toxoplasma gondii*. Cell Microbiol. 13, 18–31.

Angrisano, F., Riglar, D.T., Sturm, A., Volz, J.C., Delves, M.J., Zuccala, E.S., Turnbull, L., Dekiwadia, C., Olshina, M.A., Marapana, D.S., Wong, W., Mollard, V., Bradin, C.H., Tonkin, C.J., Gunning, P.W., Ralph, S.A., Whitchurch, C.B., Sinden, R.E., Cowman, A.F., McFadden, G.I., Baum, J., 2012. Spatial localisation of actin filaments across developmental stages of the malaria parasite. PLoS One 7, e32188.

Arrizabalaga, G., Boothroyd, J.C., 2004. Role of calcium during *Toxoplasma gondii* invasion and egress. Int. J. Parasitol. 34, 361–368.

Asai, D.J., Wilkes, D.E., 2004. The dynein heavy chain family. J. Eukaryot. Microbiol. 51, 23–29.

Baron, A.T., Salisbury, J.L., 1988. Identification and localization of a novel, cytoskeletal, centrosome-associated protein in PtK2 cells. J. Cell Biol. 107, 2669–2678.

Baum, J., Richard, D., Healer, J., Rug, M., Krnajski, Z., Gilberger, T.W., Green, J.L., Holder, A.A., Cowman, A.F., 2006. A conserved molecular motor drives cell invasion and gliding motility across malaria life cycle stages and other apicomplexan parasites. J. Biol. Chem. 281, 5197–5208.

Baum, J., Tonkin, C.J., Paul, A.S., Rug, M., Smith, B.J., Gould, S.B., Richard, D., Pollard, T.D., Cowman, A.F., 2008. A malaria parasite formin regulates actin polymerization and localizes to the parasite-erythrocyte moving junction during invasion. Cell Host Microbe 3, 188–198.

Beck, J.R., Rodriguez-Fernandez, I.A., Cruz de Leon, J., Huynh, M.H., Carruthers, V.B., Morrissette, N.S., Bradley, P.J., 2010. A novel family of *Toxoplasma* IMC proteins displays a hierarchical organization and functions in coordinating parasite division. PLoS Pathog. 6, e1001094.

Billker, O., Lourido, S., Sibley, L.D., 2009. Calcium-dependent signaling and kinases in apicomplexan parasites. Cell Host Microbe 5, 612–622.

Blume, M., Rodriguez-Contreras, D., Landfear, S., Fleige, T., Soldati-Favre, D., Lucius, R., Gupta, N., 2009. Host-derived glucose and its transporter in the obligate intracellular pathogen *Toxoplasma gondii* are dispensable by glutaminolysis. Proc. Natl. Acad. Sci. U. S. A. 106, 12998–13003.

Bosch, J., Paige, M.H., Vaidya, A.B., Bergman, L.W., Hol, W.G., 2012. Crystal structure of GAP50, the anchor of the invasion machinery in the inner membrane complex of *Plasmodium falciparum*. J. Struct. Biol. 178, 61–73.

Brooks, C.F., Francia, M.E., Gissot, M., Croken, M.M., Kim, K., Striepen, B., 2011. *Toxoplasma gondii* sequesters centromeres to a specific nuclear region throughout the cell cycle. Proc. Natl. Acad. Sci. U. S. A. 108, 3767–3772.

Brugerolle, G., 2002a. *Colpodella vorax*: ultrastructure, predation, life-cycle, mitosis, and phylogenetic relationships. European Journal of Protistology 38, 113–125.

Brugerolle, G., 2002b. *Cryptophagus subtilis*: a new parasite of cryptophytes affiliated with the Perkinsozoa lineage. European Journal of Protistology 37, 379–390.

Bullen, H.E., Tonkin, C.J., O'Donnell, R.A., Tham, W.H., Papenfuss, A.T., Gould, S., Cowman, A.F., Crabb, B.S., Gilson, P.R., 2009. A novel family of Apicomplexan glideosome-associated proteins with an inner membrane-anchoring role. J. Biol. Chem. 284, 25353–25363.

Carey, K.L., Westwood, N.J., Mitchison, T.J., Ward, G.E., 2004. A small-molecule approach to studying invasive mechanisms of *Toxoplasma gondii*. Proc. Natl. Acad. Sci. U. S. A. 101, 7433–7438.

Carruthers, V., Boothroyd, J.C., 2007. Pulling together: an integrated model of *Toxoplasma* cell invasion. Curr. Opin. Microbiol. 10, 83–89.

Cavalier-Smith, T., Chao, E.E., 2004. Protalveolate phylogeny and systematics and the origins of Sporozoa and dinoflagellates (phylum Myzozoa nom. nov.). European Journal of Protistology 40, 185–212.

Chandramohanadas, R., Davis, P.H., Beiting, D.P., Harbut, M.B., Darling, C., Velmourougane, G., Lee, M.Y., Greer, P.A., Roos, D.S., Greenbaum, D.C., 2009. Apicomplexan parasites co-opt host calpains to facilitate their escape from infected cells. Science 324, 794–797.

Chaudhry, F., Guerin, C., Von Witsch, M., Blanchoin, L., Staiger, C.J., 2007. Identification of *Arabidopsis* cyclase-associated protein 1 as the first nucleotide exchange factor for plant actin. Mol. Biol. Cell 18, 3002–3014.

Chobotar, W., Scholtyseck, E., 1982. Ultrastructure. In: Long, P.L. (Ed.), The Biology of the Coccidia. University Park Press, Baltimore.

Cooper, J.A., Sept, D., 2008. New insights into mechanism and regulation of actin capping protein. Int. Rev. Cell Mol. Biol. 267, 183–206.

D'Haese, J., Mehlhorn, H., Peters, W., 1977. Comparative electron microscope study of pellicular structures in coccidia (*Sarcocystis, Besnoitia* and *Eimeria*). Int. J. Parasitol. 7, 505–518.

Daher, W., Klages, N., Carlier, M.F., Soldati-Favre, D., 2012. Molecular characterization of *Toxoplasma gondii* formin 3, an actin nucleator dispensable for tachyzoite growth and motility. Eukaryot. Cell.

Daher, W., Plattner, F., Carlier, M.F., Soldati-Favre, D., 2010. Concerted action of two formins in gliding motility and host cell invasion by *Toxoplasma gondii*. PLoS Pathog. 6, e1001132.

Daher, W., Soldati-Favre, D., 2009. Mechanisms controlling glideosome function in apicomplexans. Curr. Opin. Microbiol. 12, 408–414.

Daubenberger, C.A., Diaz, D., Curcic, M., Mueller, M.S., Spielmann, T., Certa, U., Lipp, J., Pluschke, G., 2003. Identification and characterization of a conserved, stage-specific gene product of *Plasmodium falciparum* recognized by parasite growth inhibitory antibodies. Infect. Immun. 71, 2173–2181.

de Leon, J.C., Scheumann, N., Beatty, W., Beck, J.R., Tran, J.Q., Yau, C., Bradley, P.J., Gull, K., Wickstead, B., Morrissette, N.S. 2013. A SAS-6-Like Protein Suggests that the *Toxoplasma* Conoid Complex Evolved from Flagellar Components. Eukaryot. Cell 12.

Del Carmen, M.G., Mondragon, M., Gonzalez, S., Mondragon, R., 2009. Induction and regulation of conoid extrusion in *Toxoplasma gondii*. Cell Microbiol. 11, 967–982.

Delbac, F., Sanger, A., Neuhaus, E.M., Stratmann, R., Ajioka, J.W., Toursel, C., Herm-Gotz, A., Tomavo, S., Soldati, T., Soldati, D., 2001. *Toxoplasma gondii* myosins B/C: one gene, two tails, two localizations, and a role in parasite division. J. Cell Biol. 155, 613–623.

Delorme-Walker, V., Abrivard, M., Lagal, V., Anderson, K., Perazzi, A., Gonzalez, V., Page, C., Chauvet, J., Ochoa, W., Volkmann, N., Hanein, D., Tardieux, I., 2012. Toxofilin upregulates the host cortical actin cytoskeleton dynamics facilitating *Toxoplasma* invasion. J. Cell Sci.

Desai, A., Walczak, C.E., 2001. Assays for microtubule-destabilizing kinesins. Methods Mol. Biol. 164, 109–121.

Dobrowolski, J.M., Carruthers, V.B., Sibley, L.D., 1997a. Participation of myosin in gliding motility and host cell invasion by *Toxoplasma gondii*. Mol. Microbiol. 26, 163–173.

Dobrowolski, J.M., Niesman, I.R., Sibley, L.D., 1997b. Actin in the parasite *Toxoplasma gondii* is encoded by a single copy gene, ACT1 and exists primarily in a globular form. Cell Motil. Cytoskeleton 37, 253–262.

Dobrowolski, J.M., Sibley, L.D., 1996. *Toxoplasma* invasion of mammalian cells is powered by the actin cytoskeleton of the parasite. Cell 84, 933–939.

Dubremetz, J.F., Elsner, Y.Y., 1979. Ultrastructural study of schizogony of *Eimeria bovis* in cell cultures. J. Protozool. 26, 367–376.

Dubremetz, J.F., Torpier, G., 1978. Freeze fracture study of the pellicle of an eimerian sporozoite (Protozoa, Coccidia). J. Ultrastruct. Res. 62, 94–109.

Dutcher, S.K., 2001. The tubulin fraternity: alpha to eta. Curr. Opin. Cell Biol. 13, 49–54.

Dutcher, S.K., 2004. Dissection of basal body and centriole function in the unicellular green alga *Chlamydomonas reinhardtii*. In: Nigg, E.A. (Ed.), Centrosomes in development and disease. Wiley-VCH.

Endo, T., Yagita, K., 1990. Effect of extracellular ions on motility and cell entry in *Toxoplasma gondii*. J. Protozool. 37, 133–138.

Farrell, A., Thirugnanam, S., Lorestani, A., Dvorin, J.D., Eidell, K.P., Ferguson, D.J., Anderson-White, B.R., Duraisingh, M.T., Marth, G.T., Gubbels, M.J., 2012. A DOC2 protein identified by mutational profiling is essential for apicomplexan parasite exocytosis. Science 335, 218–221.

Ferguson, D.J., Hutchison, W.M., Dunachie, J.F., Siim, J.C., 1974. Ultrastructural study of early stages of asexual multiplication and microgametogony of *Toxoplasma gondii* in the small intestine of the cat. Acta Pathol. Microbiol. Scand. B Microbiol. Immunol. 82, 167–181.

Ferguson, D.J., Hutchison, W.M., Siim, J.C., 1975. The ultrastructural development of the macrogamete and formation of the oocyst wall of *Toxoplasma gondii*. Acta Pathol. Microbiol. Scand. B 83, 491–505.

Ferguson, D.J., Sahoo, N., Pinches, R.A., Bumstead, J.M., Tomley, F.M., Gubbels, M.J., 2008. MORN1 has a conserved role in asexual and sexual development across the apicomplexa. Eukaryot. Cell 7, 698–711.

Figueroa, J.V., Precigout, E., Carcy, B., Gorenflot, A., 2004. Identification of a coronin-like protein in *Babesia* species. Ann. N Y Acad. Sci. 1026, 125–138.

Foth, B.J., Goedecke, M.C., Soldati, D., 2006. New insights into myosin evolution and classification. Proc. Natl. Acad. Sci. U. S. A. 103, 3681–3686.

Francia, M.E., Jordan, C.N., Patel, J.D., Sheiner, L., Demerely, J.L., Fellows, J.D., de Leon, J.C., Morrissette, N.S., Dubremetz, J.F., Striepen, B. 2012. Cell division in apicomplexan parasites is organized by a homologue of the striated rootlet fibre of algal flagella. PLoS Biology 2012.

Frenal, K., Polonais, V., Marq, J.B., Stratmann, R., Limenitakis, J., Soldati-Favre, D., 2010. Functional dissection of the apicomplexan glideosome molecular architecture. Cell Host Microbe 8, 343–357.

Frixione, E., Mondragon, R., Meza, I., 1996. Kinematic analysis of *Toxoplasma gondii* motility. Cell Motil. Cytoskeleton 34, 152–163.

Fruth, I.A., Arrizabalaga, G., 2007. *Toxoplasma gondii*: induction of egress by the potassium ionophore nigericin. Int. J. Parasitol. 37, 1559–1567.

Fujita, M., Sugiura, R., Lu, Y., Xu, L., Xia, Y., Shuntoh, H., Kuno, T., 2002. Genetic interaction between calcineurin and type 2 myosin and their involvement in the regulation of cytokinesis and chloride ion homeostasis in fission yeast. Genetics 161, 971–981.

Fung, C., Beck, J.R., Robertson, S.D., Gubbels, M.J., Bradley, P.J., 2012. *Toxoplasma* ISP4 is a central IMC sub-compartment protein whose localization depends on palmitoylation but not myristoylation. Mol. Biochem. Parasitol. 184, 99–108.

Gaji, R.Y., Behnke, M.S., Lehmann, M.M., White, M.W., Carruthers, V.B., 2011. Cell cycle-dependent, intercellular transmission of *Toxoplasma gondii* is accompanied by marked changes in parasite gene expression. Mol. Microbiol. 79, 192–204.

Galler, S., Litzlbauer, J., Kross, M., Grassberger, H., 2010. The highly efficient holding function of the mollusc 'catch' muscle is not based on decelerated myosin head cross-bridge cycles. Proc. Biol. Sci. 277, 803–808.

Ganter, M., Schuler, H., Matuschewski, K., 2009. Vital role for the *Plasmodium* actin capping protein (CP) beta-subunit in motility of malaria sporozoites. Mol. Microbiol. 74, 1356–1367.

Gaskins, E., Gilk, S., Devore, N., Mann, T., Ward, G., Beckers, C., 2004. Identification of the membrane receptor of a class XIV myosin in *Toxoplasma gondii*. J. Cell Biol. 165, 383–393.

Gilk, S.D., Gaskins, E., Ward, G.E., Beckers, C.J., 2009. GAP45 phosphorylation controls assembly of the *Toxoplasma* myosin XIV complex. Eukaryot. Cell 8, 190–196.

Gilk, S.D., Raviv, Y., Hu, K., Murray, J.M., Beckers, C.J., Ward, G.E., 2006. Identification of PhIL1, a novel cytoskeletal protein of the *Toxoplasma gondii* pellicle, through photosensitized labeling with 5-[125I]iodonaphthalene-1-azide. Eukaryot. Cell 5, 1622–1634.

Goode, B.L., Eck, M.J., 2007. Mechanism and function of formins in the control of actin assembly. Annu. Rev. Biochem. 76, 593–627.

Gordon, J.L., Beatty, W.L., Sibley, L.D., 2008. A novel actin-related protein is associated with daughter cell formation in *Toxoplasma gondii*. Eukaryot. Cell 7, 1500–1512.

Gordon, J.L., Buguliskis, J.S., Buske, P.J., Sibley, L.D., 2010. Actin-like protein 1 (ALP1) is a component of dynamic, high molecular weight complexes in *Toxoplasma gondii*. Cytoskeleton (Hoboken) 67, 23–31.

Gordon, J.L., Sibley, L.D., 2005. Comparative genome analysis reveals a conserved family of actin-like proteins in apicomplexan parasites. BMC Genomics 6, 179.

Gould, S.B., Kraft, L.G., van Dooren, G.G., Goodman, C.D., Ford, K.L., Cassin, A.M., Bacic, A., McFadden, G.I., Waller, R.F., 2011. Ciliate pellicular proteome identifies novel protein families with characteristic repeat motifs that are common to alveolates. Mol. Biol. Evol. 28, 1319–1331.

Gould, S.B., Tham, W.H., Cowman, A.F., McFadden, G.I., Waller, R.F., 2008. Alveolins, a new family of cortical proteins that define the protist infrakingdom Alveolata. Mol. Biol. Evol. 25, 1219–1230.

Green, J.L., Martin, S.R., Fielden, J., Ksagoni, A., Grainger, M., Yim Lim, B.Y., Molloy, J.E., Holder, A.A., 2006. The MTIP-myosin A complex in blood stage malaria parasites. J. Mol. Biol. 355, 933–941.

Gubbels, M.J., Vaishnava, S., Boot, N., Dubremetz, J.F., Striepen, B., 2006. A MORN-repeat protein is a dynamic component of the *Toxoplasma gondii* cell division apparatus. J. Cell Sci. 119, 2236–2245.

Gubbels, M.J., White, M., Szatanek, T., 2008. The cell cycle and *Toxoplasma gondii* cell division: tightly knit or loosely stitched? Int. J. Parasitol. 38, 1343–1358.

Gubbels, M.J., Wieffer, M., Striepen, B., 2004. Fluorescent protein tagging in *Toxoplasma gondii*: identification of a novel inner membrane complex component conserved among Apicomplexa. Mol. Biochem. Parasitol. 137, 99–110.

Hakansson, S., Morisaki, H., Heuser, J., Sibley, L.D., 1999. Time-lapse video microscopy of gliding motility in *Toxoplasma gondii* reveals a novel, biphasic mechanism of cell locomotion. Mol. Biol. Cell 10, 3539–3547.

Hammond, D.M., Roberts, W.L., Youssef, N.N., Danforth, H.D., 1973. Fine Structure of the Intranuclear Spindle Poles in *Eimeria callospermophili* and *E. magna*. The Journal of Parasitology 59, 581–584.

Haraldsen, J.D., Liu, G., Botting, C.H., Walton, J.G.A., Storm, J., Phalen, T.J., Kwok, L.Y., Soldati-Favre, D., Heintz, N.H., Muller, S., Westwood, N.J., Ward, G.E., 2009. Identification of conoidin A as a covalent inhibitor of peroxiredoxin II. Organic & Biomolecular Chemistry Org. Biomol. Chem. 7, 3040–3048.

Harper, J.D., Thuet, J., Lechtreck, K.F., Hardham, A.R., 2009. Proteins related to green algal striated fibre assemblin are present in stramenopiles and alveolates. Protoplasma 236, 97–101.

Hartmann, J., Hu, K., He, C.Y., Pelletier, L., Roos, D.S., Warren, G., 2006. Golgi and centrosome cycles in *Toxoplasma gondii*. Mol. Biochem. Parasitol. 145, 125–127.

Heaslip, A.T., Dzierszinski, F., Stein, B., Hu, K., 2010a. TgMORN1 is a key organizer for the basal complex of *Toxoplasma gondii*. PLoS Pathog. 6, e1000754.

Heaslip, A.T., Ems-McClung, S.C., Hu, K., 2009. TgICMAP1 is a novel microtubule binding protein in *Toxoplasma gondii*. PLoS One 4, e7406.

Heaslip, A.T., Leung, J.M., Carey, K.L., Catti, F., Warshaw, D.M., Westwood, N.J., Ballif, B.A., Ward, G.E., 2010b. A small-molecule inhibitor of *T. gondii* motility induces the posttranslational modification of myosin light chain-1 and inhibits myosin motor activity. PLoS Pathog. 6, e1000720.

Heaslip, A.T., Nishi, M., Stein, B., Hu, K., 2011. The motility of a human parasite, *Toxoplasma gondii*, is regulated by a novel lysine methyltransferase. PLoS Pathog. 7, e1002201.

Hegge, S., Munter, S., Steinbuchel, M., Heiss, K., Engel, U., Matuschewski, K., Frischknecht, F., 2010. Multistep adhesion of *Plasmodium* sporozoites. Faseb J. 24, 2222–2234.

Heintzelman, M.B., 2003. Gliding motility: the molecules behind the motion. Curr. Biol. 13, R57–R59.

Heintzelman, M.B., Schwartzman, J.D., 1999. Characterization of myosin-A and myosin-C: two class XIV unconventional myosins from *Toxoplasma gondii*. Cell Motil. Cytoskeleton 44, 58–67.

Heintzelman, M.B., Schwartzman, J.D., 2001. Myosin diversity in Apicomplexa. J. Parasitol. 87, 429–432.

Herm-Gotz, A., Delbac, F., Weiss, S., Nyitrai, M., Stratmann, R., Tomavo, S., Sibley, L.D., Geeves, M.A., Soldati, D., 2006. Functional and biophysical analyses of the class XIV *Toxoplasma gondii* myosin D. J. Muscle Res. Cell Motil. 27, 139–151.

Herm-Gotz, A., Weiss, S., Stratmann, R., Fujita-Becker, S., Ruff, C., Meyhofer, E., Soldati, T., Manstein, D.J., Geeves, M.A., Soldati, D., 2002. *Toxoplasma gondii* myosin A and its light chain: a fast, single-headed, plus-end-directed motor. Embo J. 21, 2149–2158.

Hettmann, C., Herm, A., Geiter, A., Frank, B., Schwarz, E., Soldati, T., Soldati, D., 2000. A dibasic motif in the tail of a class XIV apicomplexan myosin is an essential determinant of plasma membrane localization. Mol. Biol. Cell 11, 1385–1400.

Hliscs, M., Sattler, J.M., Tempel, W., Artz, J.D., Dong, A., Hui, R., Matuschewski, K., Schuler, H., 2010. Structure and function of a G-actin sequestering protein with a vital role in malaria oocyst development inside the mosquito vector. J. Biol. Chem. 285, 11572–11583.

Hodges, M.E., Scheumann, N., Wickstead, B., Langdale, J.A., Gull, K., 2010. Reconstructing the evolutionary history of the centriole from protein components. J. Cell Sci. 123, 1407–1413.

Hoff, E.F., Carruthers, V.B., 2002. Is *Toxoplasma* egress the first step in invasion? Trends Parasitol. 18, 251–255.

Holloway, S.P., Gerousis, M., Delves, C.J., Sims, P.F., Scaife, J.G., Hyde, J.E., 1990. The tubulin genes of the human malaria parasite *Plasmodium falciparum*, their chromosomal location and sequence analysis of the alpha-tubulin II gene. Mol. Biochem. Parasitol. 43, 257–270.

Holloway, S.P., Sims, P.F., Delves, C.J., Scaife, J.G., Hyde, J.E., 1989. Isolation of alpha-tubulin genes from the human malaria parasite, *Plasmodium falciparum*: sequence analysis of alpha-tubulin. Mol. Microbiol. 3, 1501–1510.

Holzinger, A., 2009. Jasplakinolide: an actin-specific reagent that promotes actin polymerization. Methods Mol. Biol. 586, 71–87.

Hook, P., Vallee, R.B., 2006. The dynein family at a glance. J. Cell Sci. 119, 4369–4371.

Hu, K., 2008. Organizational changes of the daughter basal complex during the parasite replication of *Toxoplasma gondii*. PLoS Pathog. 4, e10.

Hu, K., Johnson, J., Florens, L., Fraunholz, M., Suravajjala, S., Dilullo, C., Yates, J., Roos, D.S., Murray, J.M., 2006. Cytoskeletal components of an invasion machine – the apical complex of *Toxoplasma gondii*. PLoS Pathog. 2, e13.

Hu, K., Mann, T., Striepen, B., Beckers, C.J., Roos, D.S., Murray, J.M., 2002a. Daughter cell assembly in the protozoan parasite *Toxoplasma gondii*. Mol. Biol. Cell 13, 593–606.

Hu, K., Roos, D.S., Murray, J.M., 2002b. A novel polymer of tubulin forms the conoid of *Toxoplasma gondii*. J. Cell Biol. 156, 1039–1050.

Hubberstey, A.V., Mottillo, E.P., 2002. Cyclase-associated proteins: CAPacity for linking signal transduction and actin polymerization. Faseb J. 16, 487–499.

Janmey, P.A., 1998. The cytoskeleton and cell signaling: component localization and mechanical coupling. Physiol. Rev. 78, 763–781.

Jewett, T.J., Sibley, L.D., 2003. Aldolase forms a bridge between cell surface adhesins and the actin cytoskeleton in apicomplexan parasites. Mol. Cell 11, 885–894.

Johnson, T.M., Rajfur, Z., Jacobson, K., Beckers, C.J., 2007. Immobilization of the type XIV myosin complex in *Toxoplasma gondii*. Mol. Biol. Cell 18, 3039–3046.

Jones, M.L., Kitson, E.L., Rayner, J.C., 2006. *Plasmodium falciparum* erythrocyte invasion: a conserved myosin associated complex. Mol. Biochem. Parasitol. 147, 74–84.

Kafsack, B.F., Pena, J.D., Coppens, I., Ravindran, S., Boothroyd, J.C., Carruthers, V.B., 2009. Rapid membrane disruption by a perforin-like protein facilitates parasite exit from host cells. Science 323, 530–533.

Kamau, E.T., Srinivasan, A.R., Brown, M.J., Fair, M.G., Caraher, E.C., Boyle, J.P., 2012. A focused small molecule screen identifies 14 compounds with distinct effects on *Toxoplasma gondii*. Antimicrob. Agents Chemother.

Kao, A.W., Noda, Y., Johnson, J.H., Pessin, J.E., Saltiel, A.R., 1999. Aldolase mediates the association of F-actin with the insulin-responsive glucose transporter GLUT4. J. Biol. Chem. 274, 17742–17747.

Kardon, J.R., Vale, R.D., 2009. Regulators of the cytoplasmic dynein motor. Nat. Rev. Mol. Cell Biol. 10, 854–865.

Keeley, A., Soldati, D., 2004. The glideosome: a molecular machine powering motility and host-cell invasion by Apicomplexa. Trends Cell Biol. 14, 528–532.

Khan, S.M., Franke-Fayard, B., Mair, G.R., Lasonder, E., Janse, C.J., Mann, M., Waters, A.P., 2005. Proteome analysis of separated male and female gametocytes reveals novel sex-specific *Plasmodium* biology. Cell 121, 675–687.

Khater, E.I., Sinden, R.E., Dessens, J.T., 2004. A malaria membrane skeletal protein is essential for normal morphogenesis, motility, and infectivity of sporozoites. J. Cell Biol. 167, 425–432.

Kieschnick, H., Wakefield, T., Narducci, C.A., Beckers, C., 2001. *Toxoplasma gondii* attachment to host cells is regulated by a calmodulin-like domain protein kinase. J. Biol. Chem. 276, 12369–12377.

Kitagawa, D., Vakonakis, I., Olieric, N., Hilbert, M., Keller, D., Olieric, V., Bortfeld, M., Erat, M.C., Fluckiger, I., Gonczy, P., Steinmetz, M.O., 2011. Structural basis of the 9-fold symmetry of centrioles. Cell 144, 364–375.

Kooij, T.W., Franke-Fayard, B., Renz, J., Kroeze, H., van Dooren, M.W., Ramesar, J., Augustijn, K.D., Janse, C.J., Waters, A.P., 2005. *Plasmodium berghei* alpha-tubulin II: a role in both male gamete formation and asexual blood stages. Mol. Biochem. Parasitol. 144, 16–26.

Kovar, D.R., 2006. Molecular details of formin-mediated actin assembly. Curr. Opin. Cell Biol. 18, 11–17.

Kudryashev, M., Lepper, S., Baumeister, W., Cyrklaff, M., Frischknecht, F., 2010. Geometric constrains for detecting short actin filaments by cryogenic electron tomography. PMC Biophys. 3, 6.

Kurokawa, H., Kato, K., Iwanaga, T., Sugi, T., Sudo, A., Kobayashi, K., Gong, H., Takemae, H., Recuenco, F.C., Horimoto, T., Akashi, H., 2011. Identification of *Toxoplasma gondii* cAMP dependent protein kinase and its role in the tachyzoite growth. PLoS One 6, e22492.

Kursula, I., Kursula, P., Ganter, M., Panjikar, S., Matuschewski, K., Schuler, H., 2008. Structural basis for parasite-specific functions of the divergent profilin of *Plasmodium falciparum*. Structure 16, 1638–1648.

Lambert, H., Barragan, A., 2010. Modelling parasite dissemination: host cell subversion and immune evasion by *Toxoplasma gondii*. Cell Microbiol. 12, 292–300.

Lansbergen, G., Akhmanova, A., 2006. Microtubule plus end: a hub of cellular activities. Traffic 7, 499–507.

Lavine, M.D., Arrizabalaga, G., 2007. Invasion and egress by the obligate intracellular parasite *Toxoplasma gondii*: potential targets for the development of new antiparasitic drugs. Curr. Pharm. Des. 13, 641–651.

Leander, B.S., 2008. Marine gregarines: evolutionary prelude to the apicomplexan radiation? Trends Parasitol. 24, 60–67.

Leander, B.S., Keeling, P.J., 2003. Morphostasis in alveolate evolution. Trends in Ecology & Evolution 18, 395–402.

Leander, B.S., Kuvardina, O.N., Aleshin, V.V., Mylnikov, A.P., Keeling, P.J., 2003. Molecular Phylogeny and Surface Morphology of *Colpodella edax* (Alveolata): Insights into the Phagotrophic Ancestry of Apicomplexans. Journal of Eukaryotic Microbiology 50, 334–340.

Lechtreck, K.F., 2003. Striated fibre assemblin in apicomplexan parasites. Mol. Biochem. Parasitol. 128, 95–99.

Lechtreck, K.F., Frins, S., Bilski, J., Teltenkotter, A., Weber, K., Melkonian, M., 1996. The cruciated microtubule-associated fibres of the green alga *Dunaliella bioculata* consist of a 31 kDa SF-assemblin. J. Cell Sci. 109 (Pt 4), 827–835.

Lechtreck, K.F., Melkonian, M., 1991. Striated microtubule-associated fibres: identification of assemblin, a novel 34-kD protein that forms paracrystals of 2 nm filaments *in vitro*. J. Cell Biol. 115, 705–716.

Leidel, S., Delattre, M., Cerutti, L., Baumer, K., Gonczy, P., 2005. SAS-6 defines a protein family required for centrosome duplication in C. elegans and in human cells. Nat. Cell Biol. 7, 115–125.

Lodoen, M.B., Gerke, C., Boothroyd, J.C., 2010. A highly sensitive FRET-based approach reveals secretion of the actin-binding protein toxofilin during *Toxoplasma gondii* infection. Cell Microbiol. 12, 55–66.

Lorestani, A., Ivey, F.D., Thirugnanam, S., Busby, M.A., Marth, G.T., Cheeseman, I.M., Gubbels, M.J., 2012. Targeted proteomic dissection of *Toxoplasma* cytoskeleton sub-compartments using MORN1. Cytoskeleton (Hoboken).

Lorestani, A., Sheiner, L., Yang, K., Robertson, S.D., Sahoo, N., Brooks, C.F., Ferguson, D.J., Striepen, B., Gubbels, M.J., 2010. A *Toxoplasma* MORN1 null mutant undergoes repeated divisions but is defective in basal assembly, apicoplast division and cytokinesis. PLoS One 5, e12302.

Lukasiewicz, K.B., Greenwood, T.M., Negron, V.C., Bruzek, A.K., Salisbury, J.L., Lingle, W.L., 2011. Control of centrin stability by Aurora A. PLoS One 6, e21291.

Lutz, W., Lingle, W.L., McCormick, D., Greenwood, T.M., Salisbury, J.L., 2001. Phosphorylation of centrin during the cell cycle and its role in centriole separation preceding centrosome duplication. J. Biol. Chem. 276, 20774–20780.

Lyons-Abbott, S., Sackett, D.L., Wloga, D., Gaertig, J., Morgan, R.E., Werbovetz, K.A., Morrissette, N.S., 2010. alpha-Tubulin mutations alter oryzalin affinity and microtubule assembly properties to confer dinitroaniline resistance. Eukaryot. Cell 9, 1825–1834.

Ma, C., Li, C., Ganesan, L., Oak, J., Tsai, S., Sept, D., Morrissette, N.S., 2007. Mutations in alpha-tubulin confer dinitroaniline resistance at a cost to microtubule function. Mol. Biol. Cell 18, 4711–4720.

Ma, C., Tran, J., Li, C., Ganesan, L., Wood, D., Morrissette, N., 2008. Secondary mutations correct fitness defects in *Toxoplasma gondii* with dinitroaniline resistance mutations. Genetics 180, 845–856.

Ma, G.Y., Zhang, J.Z., Yin, G.R., Zhang, J.H., Meng, X.L., Zhao, F., 2009. Toxoplasma gondii: proteomic analysis of antigenicity of soluble tachyzoite antigen. Exp. Parasitol. 122, 41–46.

Mahajan, B., Selvapandiyan, A., Gerald, N.J., Majam, V., Zheng, H., Wickramarachchi, T., Tiwari, J., Fujioka, H., Moch, J.K., Kumar, N., Aravind, L., Nakhasi, H.L., Kumar, S., 2008. Centrins, cell cycle regulation proteins in human malaria parasite *Plasmodium falciparum*. J. Biol. Chem. 283, 31871–31883.

Mann, T., Beckers, C., 2001. Characterization of the subpellicular network, a filamentous membrane skeletal component in the parasite *Toxoplasma gondii*. Mol. Biochem. Parasitol. 115, 257–268.

Matuschewski, K., Schuler, H., 2008. Actin/myosin-based gliding motility in apicomplexan parasites. Subcell Biochem. 47, 110–120.

Mehlhorn, H., 1972. Ultrastructural study of developmental stages of *Eimeria maxima* (Sporozoa, Coccidia). II. Fine structure of microgametes. Z. Parasitenkd 40, 151–163.

Mehta, S., Sibley, L.D., 2010. *Toxoplasma gondii* actin depolymerizing factor acts primarily to sequester G-actin. J. Biol. Chem. 285, 6835–6847.

Mehta, S., Sibley, L.D., 2011. Actin depolymerizing factor controls actin turnover and gliding motility in *Toxoplasma gondii*. Mol. Biol. Cell 22, 1290–1299.

Meissner, M., Schluter, D., Soldati, D., 2002. Role of *Toxoplasma gondii* myosin A in powering parasite gliding and host cell invasion. Science 298, 837–840.

Melzer, T., Duffy, A., Weiss, L.M., Halonen, S.K., 2008. The gamma interferon (IFN-gamma)-inducible GTP-binding protein IGTP is necessary for *Toxoplasma* vacuolar disruption and induces parasite egression in IFN-gamma-stimulated astrocytes. Infect. Immun. 76, 4883–4894.

Miki, H., Okada, Y., Hirokawa, N., 2005. Analysis of the kinesin superfamily: insights into structure and function. Trends Cell Biol. 15, 467–476.

Mitra, A., Sept, D., 2006. Binding and interaction of dinitroanilines with apicomplexan and kinetoplastid alpha-tubulin. J. Med. Chem. 49, 5226–5231.

Mondragon, R., Frixione, E., 1996. Ca(2+)-dependence of conoid extrusion in *Toxoplasma gondii* tachyzoites. J. Eukaryot. Microbiol. 43, 120–127.

Morrison, D.K., 2009. The 14-3-3 proteins: integrators of diverse signaling cues that impact cell fate and cancer development. Trends Cell Biol. 19, 16–23.

Morrissette, N.S., Mitra, A., Sept, D., Sibley, L.D., 2004. Dinitroanilines bind alpha-tubulin to disrupt microtubules. Mol. Biol. Cell 15, 1960–1968.

Morrissette, N.S., Murray, J.M., Roos, D.S., 1997. Subpellicular microtubules associate with an intramembranous particle lattice in the protozoan parasite *Toxoplasma gondii*. J. Cell Sci. 110 (Pt 1), 35–42.

Morrissette, N.S., Sept, D., 2008. Dinitroaniline interactions with tubulin: genetic and computational approaches to define the mechanisms of action and resistance. In: Blume, Y.B., Baird, W.V., Yemets, A.I., Breviario, D. (Eds.), The Plant Cytoskeleton: A Key Tool for Agro-Biotechnology. Springer.

Morrissette, N.S., Sibley, L.D., 2002a. Cytoskeleton of apicomplexan parasites. Microbiol. Mol. Biol. Rev. 66, 21–38. Table of contents.

Morrissette, N.S., Sibley, L.D., 2002b. Disruption of microtubules uncouples budding and nuclear division in *Toxoplasma gondii*. J. Cell Sci. 115, 1017–1025.

Moudjou, M., Paintrand, M., Vigues, B., Bornens, M., 1991. A human centrosomal protein is immunologically related to basal body-associated proteins from lower eucaryotes and is involved in the nucleation of microtubules. J. Cell Biol. 115, 129–140.

Moudy, R., Manning, T.J., Beckers, C.J., 2001. The loss of cytoplasmic potassium upon host cell breakdown triggers egress of *Toxoplasma gondii*. J. Biol. Chem. 276, 41492–41501.

Mulder, A.M., Glavis-Bloom, A., Moores, C.A., Wagenbach, M., Carragher, B., Wordeman, L., Milligan, R.A., 2009. A new model for binding of kinesin 13 to curved microtubule protofilaments. J. Cell Biol. 185, 51–57.

Munter, S., Sabass, B., Selhuber-Unkel, C., Kudryashev, M., Hegge, S., Engel, U., Spatz, J.P., Matuschewski, K., Schwarz, U.S., Frischknecht, F., 2009. *Plasmodium* sporozoite motility is modulated by the turnover of discrete adhesion sites. Cell Host Microbe 6, 551–562.

Nagamune, K., Hicks, L.M., Fux, B., Brossier, F., Chini, E.N., Sibley, L.D., 2008a. Abscisic acid controls calcium-dependent egress and development in *Toxoplasma gondii*. Nature 451, 207–210.

Nagamune, K., Moreno, S.N., Chini, E.N., Sibley, L.D., 2008b. Calcium regulation and signaling in apicomplexan parasites. Subcell Biochem. 47, 70–81.

Nagamune, K., Sibley, L.D., 2006. Comparative genomic and phylogenetic analyses of calcium ATPases and calcium-regulated proteins in the apicomplexa. Mol. Biol. Evol. 23, 1613–1627.

Nagel, S.D., Boothroyd, J.C., 1988. The alpha- and beta-tubulins of *Toxoplasma gondii* are encoded by single copy genes containing multiple introns. Mol. Biochem. Parasitol. 29, 261–273.

Nakazawa, Y., Hiraki, M., Kamiya, R., Hirono, M., 2007. SAS-6 is a cartwheel protein that establishes the 9-fold symmetry of the centriole. Curr. Biol. 17, 2169–2174.

Nebl, T., Prieto, J.H., Kapp, E., Smith, B.J., Williams, M.J., Yates 3rd, J.R., Cowman, A.F., Tonkin, C.J., 2011. Quantitative *in vivo* analyses reveal calcium-dependent phosphorylation sites and identifies a novel component of the *Toxoplasma* invasion motor complex. PLoS Pathog. 7, e1002222.

Nichols, B.A., Chiappino, M.L., 1987. Cytoskeleton of *Toxoplasma gondii*. J. Protozool. 34, 217–226.

Nishi, M., Hu, K., Murray, J.M., Roos, D.S., 2008. Organellar dynamics during the cell cycle of *Toxoplasma gondii*. J. Cell Sci. 121, 1559–1568.

O'Connell, C.B., Tyska, M.J., Mooseker, M.S., 2007. Myosin at work: motor adaptations for a variety of cellular functions. Biochim. Biophys. Acta. 1773, 615–630.

Pelster, B., Piekarski, G., 1971. Electron microscopical studies on the microgametogeny of *Toxoplasma gondii*. Z. Parasitenkd 37, 267–277.

Persson, C.M., Lambert, H., Vutova, P.P., Dellacasa-Lindberg, I., Nederby, J., Yagita, H., Ljunggren, H.G., Grandien, A., Barragan, A., Chambers, B.J., 2009. Transmission of *Toxoplasma gondii* from infected dendritic cells to natural killer cells. Infect. Immun. 77, 970–976.

Persson, E.K., Agnarson, A.M., Lambert, H., Hitziger, N., Yagita, H., Chambers, B.J., Barragan, A., Grandien, A., 2007. Death receptor ligation or exposure to perforin trigger rapid egress of the intracellular parasite *Toxoplasma gondii*. J. Immunol. 179, 8357–8365.

Pfister, K.K., Shah, P.R., Hummerich, H., Russ, A., Cotton, J., Annuar, A.A., King, S.M., Fisher, E.M., 2006. Genetic analysis of the cytoplasmic dynein subunit families. PLoS Genet. 2, e1.

Piekarski, G., Pelster, B., Witte, H.M., 1971. Endopolygeny in *Toxoplasma gondii*. Z. Parasitenkd 36, 122–130.

Plattner, F., Yarovinsky, F., Romero, S., Didry, D., Carlier, M.F., Sher, A., Soldati-Favre, D., 2008. *Toxoplasma* profilin is essential for host cell invasion and TLR11-dependent induction of an interleukin-12 response. Cell Host Microbe 3, 77–87.

Polonais, V., Javier Foth, B., Chinthalapudi, K., Marq, J.B., Manstein, D.J., Soldati-Favre, D., Frenal, K., 2011. Unusual anchor of a motor complex (MyoD-MLC2) to the plasma membrane of *Toxoplasma gondii*. Traffic 12, 287–300.

Pomel, S., Luk, F.C., Beckers, C.J., 2008. Host cell egress and invasion induce marked relocations of glycolytic enzymes in *Toxoplasma gondii* tachyzoites. PLoS Pathog. 4, e1000188.

Porchet, E., Torpier, G., 1977. Freeze fracture study of *Toxoplasma* and *Sarcocystis* infective stages (author's transl). Z. Parasitenkd 54, 101–124.

Poukkula, M., Kremneva, E., Serlachius, M., Lappalainen, P., 2011. Actin-depolymerizing factor homology domain: a conserved fold performing diverse roles in cytoskeletal dynamics. Cytoskeleton (Hoboken) 68, 471–490.

Poupel, O., Boleti, H., Axisa, S., Couture-Tosi, E., Tardieux, I., 2000. Toxofilin, a novel actin-binding protein from *Toxoplasma gondii*, sequesters actin monomers and caps actin filaments. Mol. Biol. Cell 11, 355–368.

Poupel, O., Tardieux, I., 1999. *Toxoplasma gondii* motility and host cell invasiveness are drastically impaired by jasplakinolide, a cyclic peptide stabilizing F-actin. Microbes Infect. 1, 653–662.

Preble, A.M., Giddings Jr., T.M., Dutcher, S.K., 2000. Basal bodies and centrioles: their function and structure. Curr. Top Dev. Biol. 49, 207–233.

Rawlings, D.J., Fujioka, H., Fried, M., Keister, D.B., Aikawa, M., Kaslow, D.C., 1992. Alpha-tubulin II is a male-specific protein in *Plasmodium falciparum*. Mol. Biochem. Parasitol. 56, 239–250.

Reed, N.A., Cai, D., Blasius, T.L., Jih, G.T., Meyhofer, E., Gaertig, J., Verhey, K.J., 2006. Microtubule acetylation promotes kinesin-1 binding and transport. Curr. Biol. 16, 2166–2172.

Rees-Channer, R.R., Martin, S.R., Green, J.L., Bowyer, P.W., Grainger, M., Molloy, J.E., Holder, A.A., 2006. Dual acylation of the 45 kDa gliding-associated protein (GAP45) in *Plasmodium falciparum* merozoites. Mol. Biochem. Parasitol. 149, 113–116.

Rieder, C.L., Faruki, S., Khodjakov, A., 2001. The centrosome in vertebrates: more than a microtubule-organizing centre. Trends in Cell Biology 11, 413–419.

Russell, D.G., Burns, R.G., 1984. The polar ring of coccidian sporozoites: a unique microtubule-organizing centre. J. Cell Sci. 65, 193–207.

Sahoo, N., Beatty, W., Heuser, J., Sept, D., Sibley, L.D., 2006. Unusual kinetic and structural properties control rapid assembly and turnover of actin in the parasite *Toxoplasma gondii*. Mol. Biol. Cell 17, 895–906.

Salisbury, J.L., Baron, A., Surek, B., Melkonian, M., 1984. Striated flagellar roots: isolation and partial characterization of a calcium-modulated contractile organelle. J. Cell Biol. 99, 962–970.

Sattler, J.M., Ganter, M., Hliscs, M., Matuschewski, K., Schuler, H., 2011. Actin regulation in the malaria parasite. Eur. J. Cell Biol. 90, 966–971.

Scholtyseck, E., Mehlhorn, H., Haberkorn, A., 1971a. The fine structure of the macrogamete of the mouse coccidium *Eimeria falciformis*. Z. Parasitenkd 37, 44–54.

Scholtyseck, E., Mehlhorn, H., Hammond, D.M., 1971b. Fine structure of macrogametes and oocysts of Coccidia and related organisms. Z. Parasitenkd 37, 1–43.

Scholtyseck, E., Mehlhorn, H., Hammond, D.M., 1972. Electron microscope studies of microgametogenesis in Coccidia and related groups. Z. Parasitenkd 38, 95–131.

Schroer, T.A., Fyrberg, E., Cooper, J.A., Waterston, R.H., Helfman, D., Pollard, T.D., Meyer, D.I., 1994. Actin-related protein nomenclature and classification. J. Cell Biol. 127, 1777–1778.

Schuler, H., Matuschewski, K., 2006a. *Plasmodium* motility: actin not actin' like actin. Trends Parasitol. 22, 146–147.

Schuler, H., Matuschewski, K., 2006b. Regulation of apicomplexan microfilament dynamics by a minimal set of actin-binding proteins. Traffic 7, 1433–1439.

Schwank, S., Sutherland, C.J., Drakeley, C.J., 2010. Promiscuous expression of alpha-tubulin II in maturing male and female *Plasmodium falciparum* gametocytes. PLoS One 5, e14470.

Schwartzman, J.D., 1998. Gliding into the cell: myosins hold the key to invasion by *Toxoplasma gondii*. Trends Microbiol. 6, 98.

Schwartzman, J.D., Pfefferkorn, E.R., 1983. Immunofluorescent localization of myosin at the anterior pole of the coccidian, *Toxoplasma gondii*. J. Protozool. 30, 657–661.

Shaw, M.K., 2003. Cell invasion by *Theileria* sporozoites. Trends Parasitol. 19, 2–6.

Shaw, M.K., Compton, H.L., Roos, D.S., Tilney, L.G., 2000. Microtubules, but not actin filaments, drive daughter cell budding and cell division in *Toxoplasma gondii*. J. Cell Sci. 113 (Pt 7), 1241–1254.

Shaw, M.K., Tilney, L.G., 1995. The entry of *Theileria parva* merozoites into bovine erythrocytes occurs by a process similar to sporozoite invasion of lymphocytes. Parasitology 111, 455–461.

Shaw, M.K., Tilney, L.G., 1999. Induction of an acrosomal process in *Toxoplasma gondii*: visualization of actin filaments in a protozoan parasite. Proc. Natl. Acad. Sci. U. S. A. 96, 9095–9099.

Sheffield, H.G., 1966. Electron Microscope Study of the Proliferative Form of *Besnoitia jellisoni*. The Journal of Parasitology 52, 583–594.

Sheffield, H.G., Melton, M.L., 1968. The fine structure and reproduction of *Toxoplasma gondii*. J. Parasitol. 54, 209–226.

Shimada, H., Koizumi, M., Kuroki, K., Mochizuki, M., Fujimoto, H., Ohta, H., Masuda, T., Takamiya, K., 2004. ARC3, a chloroplast division factor, is a chimera of prokaryotic FtsZ and part of eukaryotic phosphatidylinositol-4-phosphate 5-kinase. Plant Cell Physiol. 45, 960–967.

Sibley, L.D., 2004. Intracellular parasite invasion strategies. Science 304, 248–253.

Sibley, L.D., Hakansson, S., Carruthers, V.B., 1998. Gliding motility: an efficient mechanism for cell penetration. Curr. Biol. 8, R12–R14.

Siden-Kiamos, I., Ganter, M., Kunze, A., Hliscs, M., Steinbuchel, M., Mendoza, J., Sinden, R.E., Louis, C., Matuschewski, K., 2011. Stage-specific depletion of myosin A supports an essential role in motility of malarial ookinetes. Cell Microbiol. 13, 1996–2006.

Silverman, J.A., Qi, H., Riehl, A., Beckers, C., Nakaar, V., Joiner, K.A., 1998. Induced activation of the *Toxoplasma gondii* nucleoside triphosphate hydrolase leads to depletion of host cell ATP levels and rapid exit of intracellular parasites from infected cells. J. Biol. Chem. 273, 12352–12359.

Sinden, R., Talman, A., Marques, S., Wass, M., Sternberg, M., 2010. The flagellum in malarial parasites. Current Opinion in Microbiology: Host-Microbe Interactions: Fungi/Parasites/Viruses 13, 491–500.

Sinden, R.E., 1982. Gametocytogenesis of *Plasmodium falciparum in vitro*: an electron microscopic study. Parasitology 84, 1–11.

Sinden, R.E., 1983. The cell biology of sexual development in *Plasmodium*. Parasitology 86, 7–28.

Sinden, R.E., Smalley, M.E., 1979. Gametocytogenesis of *Plasmodium falciparum in vitro*: the cell-cycle. Parasitology 79, 277–296.

Skillman, K.M., Daher, W., Ma, C.I., Soldati-Favre, D., Sibley, L.D., 2012. *Toxoplasma gondii* profilin acts primarily to sequester G-actin while formins efficiently nucleate actin filament formation *in vitro*. Biochemistry 51, 2486–2495.

Skillman, K.M., Diraviyam, K., Khan, A., Tang, K., Sept, D., Sibley, L.D., 2011. Evolutionarily divergent, unstable filamentous actin is essential for gliding motility in apicomplexan parasites. PLoS Pathog. 7, e1002280.

Soldati-Favre, D., 2008. Molecular dissection of host cell invasion by the apicomplexans: the glideosome. Parasite 15, 197–205.

Soldati, D., Dubremetz, J.F., Lebrun, M., 2001. Microneme proteins: structural and functional requirements to promote adhesion and invasion by the apicomplexan parasite *Toxoplasma gondii*. Int. J. Parasitol. 31, 1293–1302.

Starnes, G.L., Coincon, M., Sygusch, J., Sibley, L.D., 2009. Aldolase is essential for energy production and bridging adhesin-actin cytoskeletal interactions during parasite invasion of host cells. Cell Host Microbe 5, 353–364.

Starnes, G.L., Jewett, T.J., Carruthers, V.B., Sibley, L.D., 2006. Two separate, conserved acidic amino acid domains within the *Toxoplasma gondii* MIC2 cytoplasmic tail are required for parasite survival. J. Biol. Chem. 281, 30745–30754.

Stokkermans, T.J., Schwartzman, J.D., Keenan, K., Morrissette, N.S., Tilney, L.G., Roos, D.S., 1996. Inhibition of *Toxoplasma gondii* replication by dinitroaniline herbicides. Exp. Parasitol. 84, 355–370.

Stommel, E.W., Ely, K.H., Schwartzman, J.D., Kasper, L.H., 1997. *Toxoplasma gondii*: dithiol-induced Ca2+ flux causes egress of parasites from the parasitophorous vacuole. Exp. Parasitol. 87, 88–97.

Straschil, U., Talman, A.M., Ferguson, D.J., Bunting, K.A., Xu, Z., Bailes, E., Sinden, R.E., Holder, A.A., Smith, E.F., Coates, J.C., Rita, T., 2010. The Armadillo repeat protein PF16 is essential for flagellar structure and function in *Plasmodium* male gametes. PLoS One 5, e12901.

Striepen, B., Crawford, M.J., Shaw, M.K., Tilney, L.G., Seeber, F., Roos, D.S., 2000. The plastid of *Toxoplasma gondii* is divided by association with the centrosomes. J. Cell Biol. 151, 1423–1434.

Striepen, B., Jordan, C.N., Reiff, S., van Dooren, G.G., 2007. Building the perfect parasite: cell division in apicomplexa. PLoS Pathog. 3, e78.

Stull, J.T., Kamm, K.E., Vandenboom, R., 2011. Myosin light chain kinase and the role of myosin light chain phosphorylation in skeletal muscle. Arch. Biochem. Biophys. 510, 120–128.

Suvorova, E.S., Lehmann, M.M., Kratzer, S., White, M.W., 2012. Nuclear actin-related protein is required for chromosome segregation in *Toxoplasma gondii*. Mol. Biochem. Parasitol. 181, 7–16.

Swedlow, J.R., Hu, K., Andrews, P.D., Roos, D.S., Murray, J.M., 2002. Measuring tubulin content in *Toxoplasma gondii*: a comparison of laser-scanning confocal and wide-field fluorescence microscopy. Proc. Natl. Acad. Sci. U. S. A. 99, 2014–2019.

Sweeney, K.R., Morrissette, N.S., Lachapelle, S., Blader, I.J., 2010. Host cell invasion by *Toxoplasma gondii* is temporally regulated by the host microtubule cytoskeleton. Eukaryot. Cell 9, 1680–1689.

Takeshima, H., Komazaki, S., Nishi, M., Iino, M., Kangawa, K., 2000. Junctophilins: a novel family of junctional membrane complex proteins. Mol. Cell 6, 11–22.

Tardieux, I., Liu, X., Poupel, O., Parzy, D., Dehoux, P., Langsley, G., 1998. A *Plasmodium falciparum* novel gene encoding a coronin-like protein which associates with actin filaments. FEBS Lett. 441, 251–256.

Tomita, T., Yamada, T., Weiss, L.M., Orlofsky, A., 2009. Externally triggered egress is the major fate of *Toxoplasma gondii* during acute infection. J. Immunol. 183, 6667–6680.

Tran, J.Q., de Leon, J.C., Li, C., Huynh, M.H., Beatty, W., Morrissette, N.S., 2010. RNG1 is a late marker of the apical polar ring in *Toxoplasma gondii*. Cytoskeleton (Hoboken) 67, 586–598.

Tran, J.Q., Li, C., Chyan, A., Chung, L., Morrissette, N.S., 2012. SPM1 Stabilizes Subpellicular Microtubules in *Toxoplasma gondii*. Eukaryot. Cell 11, 206–216.

Treeck, M., Sanders, J.L., Elias, J.E., Boothroyd, J.C., 2011. The phosphoproteomes of *Plasmodium falciparum* and *Toxoplasma gondii* reveal unusual adaptations within and beyond the parasites' boundaries. Cell Host Microbe 10, 410–419.

Tremp, A.Z., Khater, E.I., Dessens, J.T., 2008. IMC1b is a putative membrane skeleton protein involved in cell shape, mechanical strength, motility, and infectivity of malaria ookinetes. J. Biol. Chem. 283, 27604–27611.

Ueno, A., Dautu, G., Saiki, E., Haga, K., Igarashi, M., 2010. *Toxoplasma gondii* deoxyribose phosphate aldolase-like protein (TgDPA) interacts with actin depolymerizing factor (TgADF) to enhance the actin filament dynamics in the bradyzoite stage. Mol. Biochem. Parasitol. 173, 39–42.

van Breugel, M., Hirono, M., Andreeva, A., Yanagisawa, H.A., Yamaguchi, S., Nakazawa, Y., Morgner, N., Petrovich, M., Ebong, I.O., Robinson, C.V., Johnson, C.M., Veprintsev, D., Zuber, B., 2011. Structures of SAS-6 suggest its organization in centrioles. Science 331, 1196–1199.

Verhey, K.J., Hammond, J.W., 2009. Traffic control: regulation of kinesin motors. Nat. Rev. Mol. Cell Biol. 10, 765–777.

Verhey, K.J., Kaul, N., Soppina, V., 2011. Kinesin assembly and movement in cells. Annu. Rev. Biophys. 40, 267–288.

Wetzel, D.M., Chen, L.A., Ruiz, F.A., Moreno, S.N., Sibley, L.D., 2004. Calcium-mediated protein secretion potentiates motility in *Toxoplasma gondii*. J. Cell Sci. 117, 5739–5748.

Wetzel, D.M., Hakansson, S., Hu, K., Roos, D., Sibley, L.D., 2003. Actin filament polymerization regulates gliding motility by apicomplexan parasites. Mol. Biol. Cell 14, 396–406.

Wiersma, H.I., Galuska, S.E., Tomley, F.M., Sibley, L.D., Liberator, P.A., Donald, R.G., 2004. A role for coccidian cGMP-dependent protein kinase in motility and invasion. Int. J. Parasitol. 34, 369–380.

Wloga, D., Gaertig, J., 2010. Post-translational modifications of microtubules. J. Cell Sci. 123, 3447–3455.

Wloga, D., Rogowski, K., Sharma, N., van Dijk, J., Janke, C., Edde, B., Bre, M.H., Levilliers, N., Redeker, V., Duan, J., Gorovsky, M.A., Jerka-Dziadosz, M., Gaertig, J., 2008. Glutamylation on alpha-tubulin is not essential but affects the assembly and functions of a subset of microtubules in *Tetrahymena thermophila*. Eukaryot. Cell 7, 1362–1372.

Xiao, H., El Bissati, K., Verdier-Pinard, P., Burd, B., Zhang, H., Kim, K., Fiser, A., Angeletti, R.H., Weiss, L.M., 2010. Post-translational modifications to *Toxoplasma gondii* alpha- and beta-tubulins include novel C-terminal methylation. J. Proteome Res. 9, 359–372.

Yadav, R., Pathak, P.P., Shukla, V.K., Jain, A., Srivastava, S., Tripathi, S., Krishna Pulavarti, S.V., Mehta, S., Sibley, L.D., Arora, A., 2011. Solution structure and dynamics of ADF from *Toxoplasma gondii*. J. Struct. Biol. 176, 97–111.

Yang, C.H., Kasbek, C., Majumder, S., Yusof, A.M., Fisk, H.A., 2010. Mps1 phosphorylation sites regulate the function of centrin 2 in centriole assembly. Mol. Biol. Cell 21, 4361–4372.

Yarovinsky, F., Sher, A., 2006. Toll-like receptor recognition of *Toxoplasma gondii*. Int. J. Parasitol. 36, 255–259.

CHAPTER

14

Interactions Between *Toxoplasma* Effectors and Host Immune Responses

Barbara A. Butcher, Michael L. Reese†, John C. Boothroyd†, Eric Y. Denkers**

*Department of Microbiology and Immunology, Cornell University College of Veterinary Medicine, Ithaca, New York, USA

†Department of Microbiology and Immunology, Stanford University School of Medicine, Stanford, California, USA

OUTLINE

Toxoplasma gondii, Second Edition
http://dx.doi.org/10.1016/B978-0-12-396481-6.00014-3

14.1 EARLY INDICATIONS THAT *TOXOPLASMA* INTERFERES WITH HOST SIGNALLING

Toxoplasma possesses the machinery to invade virtually any type of warm-blooded vertebrate cell (reviewed elsewhere in this volume). However, *in vivo* studies have found that the parasite preferentially targets macrophage and dendritic lineage cells for infection (Bierly *et al.*, 2008; Chtanova *et al.*, 2008; Courret *et al.*, 2006). These cells are believed to serve as Trojan horses facilitating parasite dissemination, ultimately leading to establishment of latent infection in the central nervous system and skeletal muscle. It seems paradoxical that *Toxoplasma* has adapted to favour infection of macrophages and dendritic cells, since these cells occupy a central position in immunity, serving as microbicidal effectors (in the case of macrophages) and initiators of T lymphocyte-dependent immunity (in the case of dendritic cells).

Within these cellular reservoirs of infection, tachyzoites are sequestered by a parasitophorous vacuole from which they scavenge nutrients and replicate until egress. It is now clear that the parasite actively manipulates host cell biology, in particular intracellular signalling networks of the immune system. As one example, *T. gondii* is known to confer resistance to inducers of apoptosis in infected macrophages (Goebel *et al.*, 2001; Kim and Denkers, 2006; Luder and Gross, 2005; Molestina *et al.*, 2003; Payne *et al.*, 2003). It is also known that the parasite blocks IFN-γ and Toll-like receptor signalling, and this appears due in part to interference with chromatin remodelling (Butcher and Denkers, 2002; Butcher *et al.*, 2001; Denkers *et al.*, 2003; Kim *et al.*, 2007; Lang *et al.*, 2006, 2012; Leng *et al.*, 2009a, b; Luder *et al.*, 1998, 2003). Strikingly, *Toxoplasma* was found to potently phosphorylate signal transducer and activator of transcription (STAT)-3 during macrophage infection, a response that was independent of IL-10 or IL-6 that are major cell autonomous cytokines activating this transcription factor (Butcher *et al.*, 2005b). Yet despite these studies, a clear molecular picture of at least some of the ways that *Toxoplasma* seizes control of the host cell remained elusive until recently.

Toxoplasma is present in humans and domestic animals predominantly as three clonal lineages in Europe and North America. The strain types differ in disease severity and immune responses elicited (Boothroyd and Grigg, 2002; Gavrilescu and Denkers, 2001; Kim *et al.*, 2006; Mordue *et al.*, 2001; Robben *et al.*, 2004). Appreciation of this unusual population distribution, and the disparate effects on the host response, provided the impetus to cross strains and begin to map virulence loci in the parasite genome (Saeij *et al.*, 2006; Taylor *et al.*, 2006). Ultimately, this led to identification of parasite secretory molecules that are injected into host cytoplasm where a subset impacts signalling in the infected cell (Boothroyd and Dubremetz, 2008; Hunter and Sibley, 2012).

14.2 RHOPTRY PROTEIN ROP16

14.2.1 Activation of STAT Signalling

Identification of the *ROP16* locus as a virulence determinant came from quantitative trait locus analysis of F1 progeny from Type II and Type III crosses (Saeij *et al.*, 2006). Moreover, *ROP16* was implicated as a major determinant of differences in the transcriptional responses in fibroblasts infected with Type II versus Type III strains (Saeij *et al.*, 2007). Bioinformatics analysis linked host STAT signalling with the *ROP16* locus, and it was subsequently found that ROP16 expressed by Type I ($ROP16_{I}$) and Type III ($ROP16_{III}$) *Toxoplasma* generates a strong, sustained (at least 20 hours post-infection *in vitro*) activation of STAT3 and STAT6 (Saeij *et al.*, 2007). In contrast, Type II ROP16 ($ROP16_{II}$) elicits an initial activation of these proteins (e.g. 1–2 hours post-infection) but levels quickly return to baseline by 20 hours post-infection. A single ROP16 polymorphism at amino acid position 503 (Leu to Ser) was identified as the switch determining sustained vs. transient STAT3 phosphorylation (Yamamoto *et al.*, 2009).

During invasion, ROP16 (along with a subset of other rhoptry proteins) is injected into the host cell cytoplasm, where it can interact with the JAK/STAT signalling cascade. ROP16 also contains a canonical nuclear localization sequence (NLS), and consistently, it is observed to accumulate in the host cell nucleus. Nevertheless, mutation of the NLS such that ROP16 remains cytoplasmic has no effect on STAT6 (and presumably STAT3) phosphorylation, *in vitro*, at least (Saeij *et al.*, 2007). Because other host genes that are not known to be regulated by STAT3 or STAT6 are also influenced by ROP16, and because this rhoptry protein traffics to the nucleus even though it appears to cytoplasmically activate STAT3 and STAT6, it is possible that other functions of ROP16 await discovery (Melo *et al.*, 2011).

ROP16 was originally identified as a polymorphic serine–threonine kinase, a conundrum since STAT molecules and their upstream JAK proteins are activated through tyrosine phosphorylation. However, *in vitro* kinase assays revealed that recombinant ROP16 is capable of direct tyrosine phosphorylation of STAT3 (Tyr705) and STAT6 (Tyr641) (Ong *et al.*, 2010; Yamamoto *et al.*, 2009). Furthermore, co-immunoprecipitation experiments using HA-tagged ROP16 expressing parasites provided evidence for a physical association of ROP16 with STAT6 in infected cells. Together, these results argue that ROP16 directly tyrosine phosphorylates STAT3 and 6 proteins rather than acting through upstream molecules such as the JAK kinases. An amino acid substitution at position 503 of the ROP16 protein is responsible for differences in STAT phosphorylation mediated by $ROP16_{I/III}$ compared to $ROP16_{II}$ (Yamamoto *et al.*, 2009). Because *bona fide* protein tyrosine kinases are largely absent in protozoa (Manning *et al.*, 2002; Pincus *et al.*, 2008), ROP16 may have originated as a serine–threonine kinase that promiscuously gained tyrosine-phosphorylating enzymatic activity.

Since $ROP16_{II}$ displays impaired STAT3 and STAT6 activation in human and mouse host cells, it has been speculated that polymorphisms in the *rop16* gene may enable STAT activation in other important host species. However, in a study on infection in avians, which represent the most evolutionarily divergent reservoir hosts relative to mice and humans, *rop16* has a similar strain-specific effect on host gene expression (Ong *et al.*, 2011). The implication of this finding is that the differences between hosts that selected for the different alleles of ROP16 may not be the host species *per se* (i.e. its genetic hard-wiring); it might instead have to do with other environmental factors such as the resident microbiota and/or commensurate infections. For example, in a host species that typically harbours a chronic helminth infection, it may benefit *Toxoplasma* to avoid sustained activation of STAT3 or STAT6 that would push the animal further in the Th2 direction and depress inflammation levels to a degree that is not beneficial to the host.

14.2.2 Effects of ROP16-Dependent STAT3 Activation

Unlike other STAT molecules, transcription factor STAT3 has multiple functions outside the immune response, including essential roles in embryogenesis, cell growth, tissue homeostasis, and malignant transformation (Levy and Darnell, 2002). In the immune system, STAT3 paradoxically mediates the pro-inflammatory and anti-inflammatory effects of IL-6 and IL-10, respectively (Alonzi *et al.*, 2001; Lang *et al.*, 2002; Takeda *et al.*, 1999). Expression of suppressor of cytokine synthesis (SOCS)-3 during IL-6 signalling appears to be the critical factor determining whether STAT3 displays pro-inflammatory or anti-inflammatory effects (Croker *et al.*, 2003; Lang *et al.*, 2003; Yasukawa *et al.*, 2003). Deletion of STAT3 in macrophage lineage cells results in mice that develop spontaneous enterocolitis similar to IL-10 knockout mice, implicating an anti-inflammatory function for STAT3 in intestinal homeostasis (Takeda *et al.*, 1999).

The functions of ROP16-mediated STAT3 activation *in vivo* have not been fully explored, but studies on macrophages infected *in vitro* suggest an anti-inflammatory effect that generally mimics IL-10. Initially, it was observed that the suppressive effect of Type I *Toxoplasma* infection on bone marrow-derived macrophages was at least in part dependent on STAT3. Thus, while Type I parasites suppress macrophage IL-12 and TNF-α secretion in response to Toll-like receptor ligands, macrophages lacking STAT3 resist parasite-induced immunosuppression (Butcher *et al.*, 2005b). More recently, it was shown that this suppression is mediated by $ROP16_I$ (Butcher *et al.*, 2011). Furthermore, expression of $ROP16_I$ down-regulates the parasite's own ability to induce IL-12 (Butcher *et al.*, 2011; Saeij *et al.*, 2007). In accord with this result, ectopic expression of $ROP16_I$ in Type II parasites dampens intestinal inflammation that occurs with normal Type II strains (Jensen *et al.*, 2011). These findings are consistent with a view that ROP16 functions in an anti-inflammatory capacity to prevent immunopathology, enabling host survival and permitting establishment of latent infection (Denkers *et al.*, 2012; Melo *et al.*, 2011).

At the same time, it is possible that in other situations $ROP16_I$ might promote parasite persistence. In this regard, IFN-γ-mediated production of nitric oxide by brain astrocytes and microglial cells is suppressed in a Type I ROP16-dependent manner (Butcher *et al.*, 2011). While absence of inducible nitric oxide in $Nos2^{-/-}$ mice has no impact on host survival during acute infection with Type II *Toxoplasma* strains, the animals fail to control chronic infection, suggesting that nitric oxide producing cells in the brain exert a protective function against the parasite (Scharton-Kersten *et al.*, 1997). In this context, ROP16-mediated STAT3 activation could promote parasite persistence.

14.2.3 Effects of ROP16-Mediated STAT6 Activation

The activation of STAT6 by *Toxoplasma* strains expressing $ROP16_I$ endows host macrophages with activation characteristics not previously associated with this microbial pathogen. It is now understood that macrophage activation can take several forms, depending on the cytokine stimuli (Martinez *et al.*, 2009; Murray and Wynn, 2011). At one extreme of what is regarded as a continuum, triggering with proinflammatory cytokines (IFN-γ, TNF-α) alone or with bacterial products such as lipopolysaccharide results in an M1, or classically activated macrophage, characterized by high production of IL-12, IL-23, nitric oxide and reactive oxygen intermediates. As such, M1 macrophages are effective at controlling intracellular microorganisms, but they can also contribute to proinflammatory cytokine pathology. At the other extreme, exposure to IL-4/IL-13, cytokines that signal through STAT6, results in M2, or alternatively activated macrophages. The M2 phenotype is characterized by up-regulation of arginase-1 (Arg-1), resulting in production of polyamines and proline that are involved in wound healing and pathological fibrosis. Alternatively activated macrophages also produce the chitinase-like lectin Ym1, the resistin-like protein FIZZ1, and acidic mammalian chitinase. Other regulatory type macrophage phenotypes also occur as a result of stimulation with IL-10, immune complexes and various combinations of these stimuli (Fleming and Mosser, 2011).

Toxoplasma is widely regarded as a prototypic pro-inflammatory pathogen (Denkers and Gazzinelli, 1998; Munoz *et al.*, 2011). Therefore, it is quite remarkable that we now know that $ROP16_I$, acting via STAT6, induces expression of Arg-1 and other markers of alternative macrophage activation (Butcher *et al.*, 2011; Jensen *et al.*, 2011). It is not completely clear what this means for the parasite and host. Because Arg-1 and iNOS share arginine as a common substrate, it is possible that induction of Arg-1 is used as a means to avoid the microbicidal effects of nitric oxide (El Kasmi *et al.*, 2008; Rutschman *et al.*, 2001). In the mouse intestine, nitric oxide production is reported to contribute to ileitis associated with infection with Type II parasites (Khan *et al.*, 1997). Expression of active ROP16 by Type I or Type III *Toxoplasma* might therefore dampen down this inflammation. On the other hand, iNOS knockout mice are susceptible to reactivation during latent infection, implicating nitric oxide in control of persistent infection (Scharton-Kersten *et al.*, 1997). In this case, ROP16-induced Arg-1 expression could promote host susceptibility by limiting nitric oxide production and the ability to control the parasite.

An alternative perspective is that Arg-1 induction by ROP16 impacts the nutritional needs of the parasite. Here again, two outcomes can be envisioned. Arg-1 catabolizes L-arginine to urea and ornithine, the latter of which is a precursor for amino acid and polyamine synthesis. ROP16-induced Arg-1 could serve to promote parasite replication inasmuch as *Toxoplasma* lacks enzymes required for polyamine synthesis and therefore must scavenge these molecules from the host cell (Cook *et al.*, 2007). In this regard, there is evidence that Type II parasites transgenically expressing $ROP16_I$ display an Arg-1-dependent increase in replication compared to the parental Type II strain (Jensen *et al.*, 2011). Yet, *T. gondii* at the same time is an arginine auxotroph (Fox *et al.*, 2004). Therefore, it might also be expected that ROP16-dependent Arg-1 induction could slow parasite replication. Here again, there is evidence that this is the case during *in vitro* infection, and consistent with this view mice infected with ROP16-deleted Type I *Toxoplasma* display higher parasite burdens than those undergoing infection with wild-type tachyzoites (Butcher *et al.*, 2011). Limiting the pool of intracellular arginine could represent a strategy of slowing replication to facilitate widespread dissemination, or it might reflect the need to avoid parasite overstimulation of the immune system that can result in inflammatory lesions.

14.2.4 Towards an Integrated View of ROP16

How can we understand the biological function of ROP16? This novel parasite kinase tyrosine phosphorylates both STAT3 and STAT6, transcription factors with their own unique transcriptional programmes. STAT3 activation could suppress immunity resulting in decreased inflammatory lesions. At the same time, STAT3 down-regulation of immunity might decrease the ability to control the parasite, causing an increase in *Toxoplasma*-induced lesions (Fig. 14.1). Simultaneous STAT6 induction of Arg-1 could down-regulate nitric oxide production thereby dampening immunopathology. Induction of Arg-1 could also limit arginine that the parasite requires for survival, but it might also promote replication by generating polyamines that the parasite needs (Fig. 14.1). Limited studies so far suggest that $ROP16_I$ is capable of dampening inflammatory ileitis and that its expression slows *in vivo* replication (Butcher *et al.*, 2011; Jensen *et al.*, 2011). Whether these effects are STAT3 or STAT6 mediated is not yet known. Indeed, it seems possible that the dominant function of ROP16 may vary depending on the host or the location within the host.

14.3 DENSE GRANULE PROTEIN GRA15

14.3.1 GRA15 and NFκB Activation

In addition to parasite strain-specific interaction with STAT signalling pathways, a reciprocal

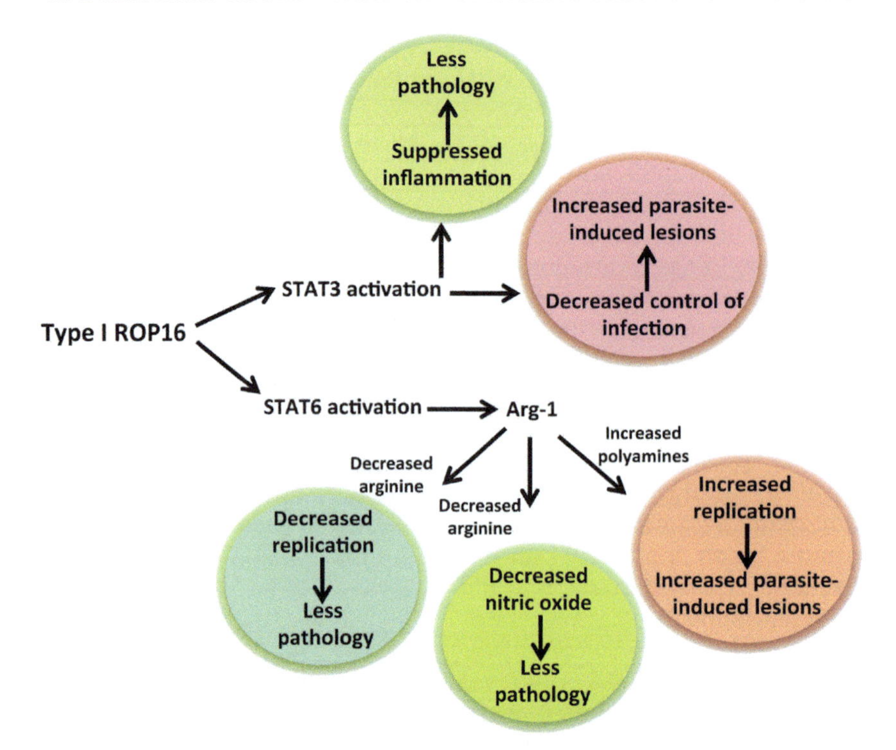

FIGURE 14.1 **Type I and III ROP16 and host inflammation: complexity in action.** The *in vivo* functions of $ROP16_{I/III}$ are likely to be multi-factorial by virtue of its ability to activate both STAT3 and STAT6. Activation of STAT3 can suppress inflammation resulting in less immunopathology during infection, in a manner akin to IL-10 signalling. However, STAT3-mediated immunosuppression could also limit the host's ability to control infection, resulting in an increase in parasite-mediated tissue damage. The activation of STAT6 by $ROP16_{I/III}$ results in an alternatively activated macrophage phenotype that includes expression of arginase-1 (Arg-1). This could have three major effects. Induction of Arg-1 could limit the pool of intracellular arginine that *Toxoplasma* must scavenge from the host cell in order to replicate. This can be expected to result in minimizing parasite-induced lesions during infection. Arg-1-mediated arginine depletion could also limit the supply of arginine that the host cell iNOS enzyme uses to make nitric oxide (NO). This could limit NO-mediated tissue damage associated with intestinal infection. During chronic infection, where NO production is host protective, limiting this anti-microbial molecule could promote host susceptibility. However, Arg-1 induction could also increase the pool of host polyamines that the parasite requires to survive. In this case, ROP16-mediated Arg-1 induction might increase parasite-induced lesions. Note that type II parasites express an active ROP16 that causes an initial activation of STAT3 and STAT6 but this quickly dissipates, returning to baseline, *in vitro*, within a few hours.

strain-specific relationship is observed with activation of NFκB. In this case, infection with Type II tachyzoites triggers nuclear translocation of NFκB p65, and this is associated with higher levels of IL-12 production. In contrast, Type I and III strains elicit low or non-detectable NFκB translocation along with low level IL-12 secretion (Robben *et al.*, 2004). By examining the progeny of Type II × Type III crosses, the *gra15* locus, encoding dense granule protein GRA15, was identified as a determinant of NFκB activation (Rosowski *et al.*, 2011). Thus, while $GRA15_{II}$ can activate NFκB signalling, this is not the case for $GRA15_{I}$ or $GRA15_{III}$.

Using macrophages from gene knockout mice, evidence was obtained that GRA15-mediated NFκB nuclear translocation requires TRAF6 and IKKβ components of the NFκB signalling cascade, but does not involve Toll-like receptor adaptor molecules MyD88 or TRIF. Therefore, rather than acting as a straightforward Toll-like receptor ligand, $GRA15_{II}$ appears to act downstream in a process involving TRAF6 and IKKβ (Rosowski *et al.*, 2011). Since GRA15 is not known as a kinase, how it triggers IκB phosphorylation and subsequent NFκB activation is unclear. However, GRA15 appears to act independently of other *Toxoplasma* factors, inasmuch as its expression in recombinant form in HeLa cells appears sufficient to enable NFκB nuclear translocation (Rosowski *et al.*, 2011). As yet, the subcellular localization of GRA15 in infected cells has not been determined; it is presumed to somehow access the host cytosol and maybe host nucleus but neither has been so far demonstrated.

Interestingly, the GRA15 molecule from Type II parasites possesses in many ways reciprocal biological activity to ROP16 from Type I/III strains. The $ROP16_{I/III}$ kinase polarizes macrophages to an alternatively activated phenotype including increased expression of arginase-1, mannose receptor type C (CD206) and macrophage galactose/N-acetylgalactosamine-specific lectin (mMgl). In contrast $GRA15_{II}$ drives macrophages towards a classically activated phenotype, including up-regulated expression of pro-inflammatory cytokines such as IL-12/23p40 and IL-6, and increased expression of co-stimulatory molecules including CD40 and CD80 (Jensen *et al.*, 2011). In keeping with the view that macrophage polarization represents extremes of a continuum (Mosser and Edwards, 2008), not all M1 associated molecules are up-regulated by infection with GRA15-expressing Type II strains. For example, neither nitric oxide nor TNF-α are produced during these infections (Butcher, Denkers *et al.*, unpublished; Jensen *et al.*, 2011).

14.3.2 Consolidating GRA15 and ROP16 Effects

The observation that Type I/III parasites express a form of ROP16 that activates STAT3/6 and mediates an M2 like macrophage phenotype whereas Type II *Toxoplasma* strains express a GRA15 molecule that activates NFκB and induces M1-like macrophages raises some interesting questions (Fig. 14.2). Chief among these is why different parasite strains induce such profoundly disparate transcriptional programmes in the host cell. One possibility is that allelic polymorphism is maintained because variants of molecules such as ROP16 and GRA15 accommodate different needs of different hosts, as

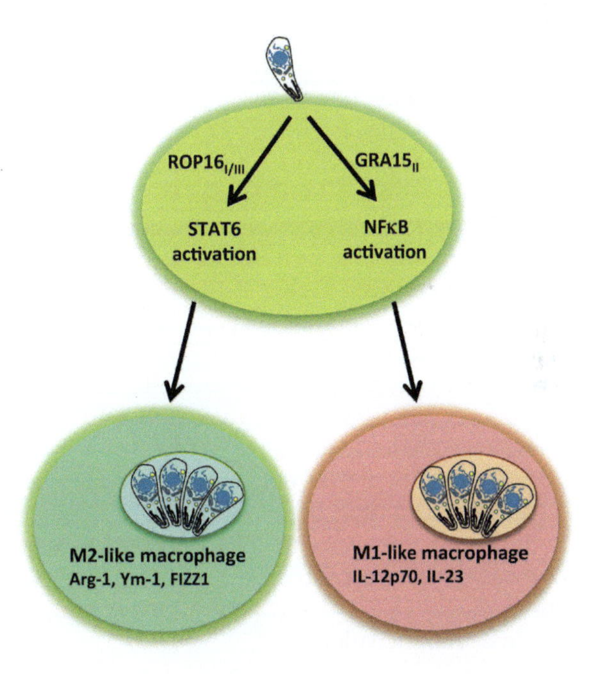

FIGURE 14.2 **Reciprocal effects of $ROP16_{I/III}$ and $GRA15_{II}$ on macrophage activation.** Activation of STAT6 by $ROP16_{I/III}$ drives an alternatively activated macrophage phenotype including up-regulation of Arg-1, Ym-1 and FIZZ-1. Activation of NFκB by $GRA15_{II}$ results in an M1-like macrophage that releases high amounts of IL-12p70 and IL-23. Unlike true classically activated macrophages, they do not express appreciable amounts of nitric oxide or TNF-α.

discussed above. So for example, some hosts (such as C57BL/6 mice) are susceptible to inflammatory damage in the intestine during *Toxoplasma* infection that might be quelled by expression of active $ROP16_{I/III}$ (Jensen *et al.*, 2011). In other cases, avoiding STAT3/6 activation and promoting NFκB signalling could favour development of robust immunity that would favour host survival and establishment of latent infection.

14.4 RHOPTRY PROTEIN ROP18

14.4.1 Identification of ROP18 as a Major Virulence Locus

A third key virulence gene is *ROP18*. This encodes another rhoptry kinase that, like ROP16, is a member of the paralogous 'ROP2' superfamily although ROP18 appears to be a serine/threonine kinase rather than a tyrosine kinase like ROP16. ROP18 was first identified in proteomic analyses of purified rhoptries (initially referred to as ROP2L2) (Bradley *et al.*, 2005) and shown to be an active kinase by expression of recombinant protein (El Hajj *et al.*, 2006). Its specific function and role in virulence, however, only emerged from the same sort of genetic analyses that revealed ROP16's role (Saeij *et al.*, 2006; Taylor *et al.*, 2006). These latter studies used two different crosses, one between a type II and a type III parent (Saeij *et al.*, 2006) and the other between types I and III (Taylor *et al.*, 2006), to reveal the presence of a locus on chromosome VIIa that plays a major role in strain-specific virulence.

14.4.2 Immunity-Related GTPases are a Major Target of ROP18 in Mice

At about the same time, it was found that in infected mouse cells stimulated with IFNγ, the vacuole containing *Toxoplasma* tachyzoites is attacked by a class of GTPases known as 'immunity-related GTPases' (IRGs) (Butcher *et al.*, 2005a; Collazo *et al.*, 2001; Hunn *et al.*, 2008). These proteins, also known as p47 GTPases, are induced upon IFNγ treatment and activate by converting from a GDP- to a GTP-bound form. By an unknown mechanism, they recognize and associate with the parasitophorous vacuole membrane (PVM), where they oligomerize and somehow disrupt the integrity of this membrane (Zhao *et al.*, 2009b). For unknown reasons, this is a lethal event for the parasites within the PV and, thus, an effective host defence. IRGs are found in mice as a family of about 20 closely related proteins, which were initially described due to their strong induction by IFNγ. The IRG family is conserved throughout vertebrates, though it has been lost in some primates, and only a single functional family member, IRGM, remains in humans (Hunn *et al.*, 2011). Human IRGM is not induced by IFNγ and is not known to have the vacuole-attacking properties of the murine versions.

The possibility that IRGs might interface in some way with the polymorphic ROPs came with the finding that different strains of *Toxoplasma* show greater or lesser susceptibility to IRG attack (Zhao *et al.*, 2009a); type I strains are relatively refractory to control by IRGs while types II and III are highly sensitive. This implicated that some combination of polymorphic ROPs might be involved but initial testing of this argued against ROP5, 16 and 18 playing a role (Khaminets *et al.*, 2010). Subsequent work, however, revealed that IRGs are subject to specific phosphorylation by ROP18 and that this leads to their inability to attack the PVM (Fentress *et al.*, 2010; Steinfeldt *et al.*, 2010). Note that the earlier work suggesting that ROP18 was not involved used very high, non-physiologic concentrations of IFNγ that likely overwhelmed the system to the extent that the role of ROP18 was obscured.

Type III strains express virtually no ROP18 and so their susceptibility to IRG attack is easily explained. Type II strains, on the other hand, express substantial amounts of ROP18 but the allele they carry has many polymorphisms relative to type I strains. This was presumed to be the explanation for why the PV within which type II strains grow is efficiently attacked by IRGs. While this might be a factor, subsequent work on another locus, ROP5 (see below), identified what is likely to be the major reason why the ROP18 of type II strains is inefficient at phosphorylating IRGs in infected cells.

14.4.3 ROP18 has at Least One Other Host Target

Protein kinases are rarely specific for just one target substrate and ROP18 appears to be no exception to this rule. To investigate other possible targets, Yamomoto *et al.* used a yeast two-hybrid approach to search for a binding partners of ROP18 with a particular focus on the N-terminal domains, including the N-terminal extension of the kinase domain (Yamamoto *et al.*, 2011). This extension is a structural element that distinguishes the ROP2 superfamily from other kinase families (Labesse *et al.*, 2009; Reese and Boothroyd, 2011). Interestingly, they found that a protein involved in regulating the unfolded protein response in mice, 'Activating Transcription Factor 6 beta' (ATF6β), binds the N-terminal half of ROP18 (including the N-terminal extension of the kinase domain) and this results in proteasomal degradation of ATF6β in a manner dependent on the kinase activity of ROP18. ROP18 can phosphorylate ATF6β *in vitro* and this is presumed to be the mechanism operating *in vivo*, as well.

Importantly, ATF6β knock-out mice are more susceptible to ROP18 knock-out parasites than wild type mice (Yamamoto *et al.*, 2011). This suggests that ATF6β plays an important role in the immune response to *Toxoplasma* and that ROP18, at least in type I strains, blocks ATF6β function, likely through phosphorylation and consequent destabilization. How ATF6β functions in the immune response is not clear but it is known to serve as an important transcription factor mediating the unfolded protein response in mammals. In the experiments of Yamamoto and colleagues, its absence was shown to result in decreased IFNγ production in $CD8^+$ T-cells, rendering the animals susceptible to especially the later stages of an acute infection.

Thus, two targets of ROP18 have now been determined and it is likely that there are others waiting to be identified. These might be key to the immune response or other host functions. The fact that strains exist, like type III strains, that express little if any ROP18 suggests that there is some cost to having this 'loose cannon' firing in some situations, perhaps phosphorylating host or parasite proteins that play a positive role in infection. Hence, it may be that in some non-murine hosts the 'cost' of having abundant and active ROP18 exceeds the 'benefit' of its attack on the immune responses described in detail here. Some of these, like the IRGs, may have distinct roles in different warm-blooded vertebrates.

14.5 RHOPTRY PROTEIN ROP5

14.5.1 ROP5 is a Pseudokinase Member of the ROP2 Family

A somewhat surprising finding has been that a group of tandemly repeated genes known collectively as *ROP5* also play a role in strain-specific differences in virulence (Behnke *et al.*, 2011; Reese *et al.*, 2011). ROP5 proteins are additional members of the ROP2 family but their sequence strongly predicted that they are incapable of catalysing phosphorylation (El Hajj *et al.*, 2007). While difficult to definitively prove such a negative conclusion, this prediction has been experimentally borne out (Reese and Boothroyd, 2011). The structure of one isoform of ROP5 has been solved (Reese and Boothroyd, 2011) and, as expected, it has strong resemblance to that of the previously determined structure for the prototypic member of the family, ROP2 (Labesse *et al.*, 2009). Interestingly, despite its catalytic deficiency, ROP5 still binds ATP, perhaps using this for regulating its other functions, described below.

14.5.2 ROP5 is a Key Virulence Determinant

Genetic mapping of genes that are key to *Toxoplasma* virulence in mouse infections identified a region on chromosome XII as especially important in two experimental crosses: one between a type II and a type III parent (Saeij *et al.*, 2006) and the other between types I and II (Behnke *et al.*, 2011). Upon further study, ROP5 emerged as the responsible locus (Behnke *et al.*, 2011; Reese *et al.*, 2011). Dissection of the mechanism by which this locus impacts virulence was substantially complicated by the fact that ROP5 exists as a tandem repetition of about four to ten genes, depending on the strain. The sequence of the various copies can differ greatly within a strain; some copies appear to be present as identical repeats while others are markedly different from each other. Between strains, there are also marked differences, as expected and as predicted by the genetic analyses. Determining the exact sequence of each copy has been problematic because of this tandem repetition but individual genes have been studied in considerable detail. Remarkably, the polymorphisms between alleles are concentrated within the kinase-like (or 'pseudokinase') domain and form two distinct patches on the surface of the protein structure, suggestive of binding sites for fast-evolving host target proteins (Reese and Boothroyd, 2011).

Deletion of the entire cluster of ROP5 genes has revealed the crucial importance of these genes. A type I Δ*rop5* strain is completely avirulent in mice; even an inoculum of 10^6 parasites causes no obvious pathology (Behnke *et al.*, 2011; Reese *et al.*, 2011), and the parasites are cleared within days (Behnke *et al.*, 2012). As one might expect from highly polymorphic proteins, different alleles of ROP5 were able to potentiate virulence to a different level, suggesting that the variation in ROP5 is due to co-evolution with different host species and/or that the different alleles of ROP5 have evolved distinct functions. It is important to note that the low virulence allele of the *ROP5* locus (the type II allele) has individual copies that are far less divergent from one another than those found in the high virulence alleles in types I and III.

14.5.3 ROP5 Partners with ROP18 to Inhibit the Immunity-Related GTPases

A first function for ROP5 was determined by three independent studies, using genetic (Niedelman *et al.*, 2012), biochemical (Fleckenstein *et al.*, 2012), and immunological (Behnke *et al.*, 2012) techniques. In each of these studies, it was found that only parasites with the virulent allele of ROP5 are able to efficiently escape clearance by the IRG system. This appears to be because ROP18 is unable to efficiently phosphorylate the IRGs loaded on the PVM when ROP5 is not present. There are data that demonstrate a small (three- to 30-fold) increase in ROP18's *in vitro* kinase activity upon addition of recombinantly expressed ROP5 (Behnke *et al.*, 2012; Fleckenstein *et al.*, 2012), although this cannot fully explain the almost complete lack of phosphorylation of IRGs in infected cells. In addition to any interaction ROP5 may have with ROP18 (which appears to be quite weak, as the two proteins are not efficiently co-immunoprecipitated), ROP5 is a competitive inhibitor of IRG oligomerization (Fleckenstein *et al.*, 2012; Niedelman *et al.*, 2012). Fleckenstein *et al.* used NMR to map a rough binding surface and found that ROP5 appears to bind IRG near its GTPase active site, which is thought to be the site that seeds oligomerization on the PVM

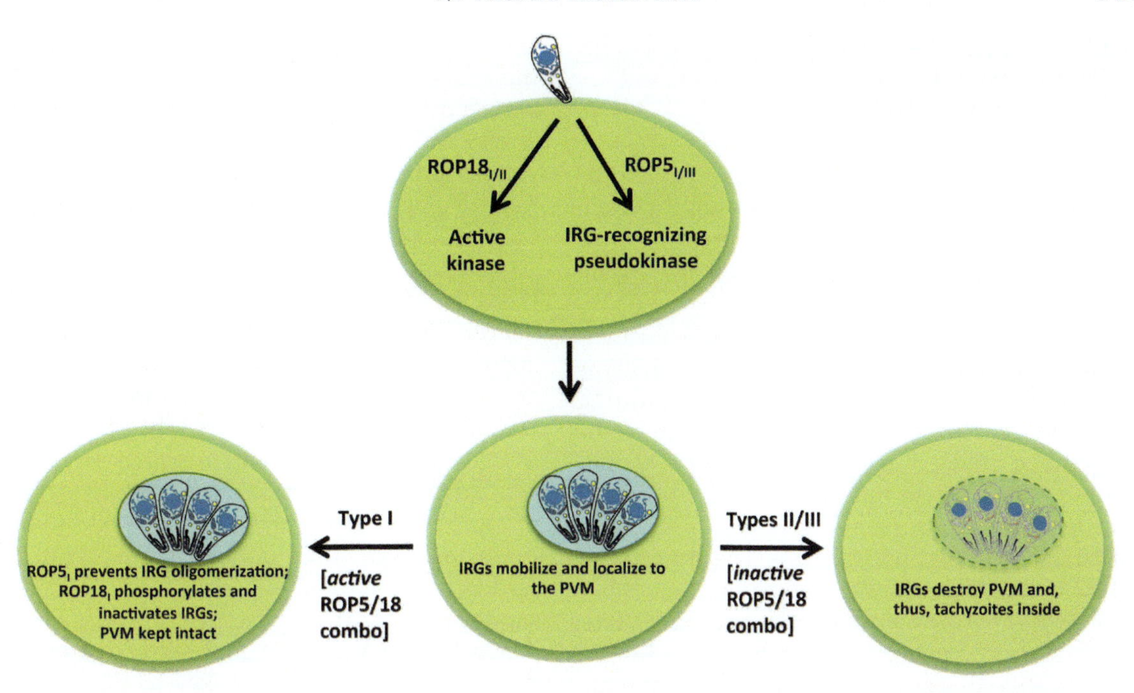

FIGURE 14.3 **ROP18 and ROP5 function together to protect *Toxoplasma* from IRG-mediated host defence.** Parasite strain types I and II express a ROP18 molecule with functional kinase activity, and strain types I and III express a ROP5 pseudokinase molecule that recognizes host IRG proteins. Early during infection, host IRG molecules assemble at the parasitophorous vacuole membrane (PVM). Type I parasites that express an active ROP5/18 combination deactivate IRGs by preventing oligomerization (through the binding of monomeric IRG by ROP5) and by direct, ROP18-mediated IRG phosphorylation. Consequently, the PVM remains intact and tachyzoites proceed through their intracellular life cycle. In contrast, parasite strain types II and III possess an inactive ROP5/18 combination and tachyzoites of these strains are vulnerable to IRG-mediated host defence. This involves IRG assembly at the PVM followed by irreversible damage to its integrity. Tachyzoites are apparently unable to survive in the absence of an intact PVM.

(Pawlowski *et al.*, 2011). This suggests a model where ROP5 is required to hold the IRGs in a monomeric conformation, making them accessible to ROP18 phosphorylation and, thus, permanent inactivation (Fig. 14.3).

14.5.4 ROP5 Likely has Additional Roles that are Independent from ROP18

As important as ROP5's role in inhibiting the IRGs appears to be, it seems likely that, as described for ROP18 above, the ROP5 proteins have multiple targets. This can most clearly be seen from the genetic data. The knockouts for ROP5 do not phenocopy the ROP18 knockout; type I Δ*rop5* parasites are substantially less virulent than are Δ*rop18* parasites (Reese *et al.*, 2011). In addition, the knock-in of the high virulence allele for *ROP5* was able to increase the virulence of an otherwise avirulent strain by 5-logs, and the recipient strain possessed what is essentially a null (type III) allele for *ROP18* (Reese *et al.*, 2011). Thus, while ROP5 partners with ROP18 to disrupt the IRG system, it appears to act independently from ROP18 on other, as yet unknown, targets.

14.6 OTHER PARASITE MOLECULES POSSIBLY INFLUENCING HOST CELL SIGNALLING

It is likely that *Toxoplasma* possesses additional effector molecules that target host transcriptional responses. One of the first rhoptry proteins observed to be injected into the host cell cytoplasm was PP2C-hn, a molecule identified as a type 2C protein phosphatase (Gilbert *et al.*, 2007). Furthermore, PP2C-hn behaves like ROP16 insofar as both accumulate in the host cell nucleus during infection. However, genetic knockout of PP2C-hn resulted in parasites with no observable phenotypic changes in terms of effects on host cell transcriptional responses or fitness and virulence of parasites *in vivo* and *in vitro*. The function of PP2C-hn therefore remains unknown.

Using a comparative genomics approach, Roos and colleagues identified rhoptry protein ROP38 as a putative kinase likely to be important in parasite biology (Peixoto *et al.*, 2010). Type I parasites such as RH express very low levels of ROP38, but Type I parasites transgenically overexpressing Type III ROP38 under the control of the tubulin promoter down-regulate host genes associated with MAPK signalling, apoptosis and proliferation. The sequence of ROP38 is highly conserved between the three archetypal strains indicating that, in this case, evolution has again selected for a difference in quantity of the effector (like ROP18) rather than 'quality' (as for ROP16). Again, it may be that in some hosts, having abundant ROP38 activity is more of a detriment than a benefit, perhaps because the targets for which it evolved are not present or relevant to these latter hosts and phosphorylation of 'unintended' targets results in an unsuitable intracellular environment.

14.7 CONCLUSION

The findings described above reveal that *Toxoplasma* tachyzoites have evolved some remarkable ways to influence the infected host cell. Except for ROP38, however, all of these molecules were identified because they differ between two or more of the three predominant strains of *Toxoplasma* found in Europe and North America; hence they could be genetically mapped through phenotyping the F1 progeny of crosses involving those strains. But these three strains are somewhat closely related, with many loci being represented by the identical or nearly identical allele in all three (Boyle *et al.*, 2006). Such loci cannot be identified by genetic crosses between these strains, no matter how important a role that locus might play. Crosses with other, more distantly related strains are needed and these are almost certain to reveal many other important virulence loci. Similarly, other methods that do not suffer from this limitation (e.g. analysis of mutants generated randomly or through specific targeting of suspect genes) will undoubtedly disclose loci not currently suspected as key to virulence. As detailed elsewhere in this book, the rhoptries and dense granules contain many uncharacterized proteins, most or all of which might have an opportunity to interact with the infected host cell. It is likely that many of these, as well as other ROPs and GRAs yet to be identified, will ultimately be found to have a profound effect on the host–parasite interaction.

The fact that the effectors described here differ so dramatically between strains, either in terms of structure or expression levels, is almost certainly due to differences in the ecological niches within which each strain has evolved. Given the extraordinary host range of *Toxoplasma* in nature, these differences seem likely to include the various species of intermediate hosts that each strain has found itself infecting. As noted above, however, the difference in these hosts need not be at the species level; it could also be in the resident microbiota or the assortment of other infections experienced by those hosts since we know both can have profound influences on the long-term health of the host, especially in terms of immune function.

Essentially all the results described here came from the study of tachyzoites because these are the easiest developmental forms to work with. But *Toxoplasma* has many other intracellular/invasive forms, including the sporozoites that initiate many (maybe even most) infections and the bradyzoites that initiate virtually all the rest. We know almost nothing about what specialized effectors these stages introduce as they launch an infection in the intestinal epithelium of a new host animal. Recent advances in our knowledge of the proteomes and transcriptomes of these stages is providing some tantalizing clues, especially in terms of stage-specific ROPs and GRAs (Buchholz *et al.*, 2011; Fritz *et al.*, 2012a, b; Naguleswaran *et al.*, 2010) (see also a wealth of unpublished data on http://www.ToxoDB.org). And we have yet to gain even the slightest glimpse into the least studied of all the stages of *Toxoplasma*, the schizonts, macrogametocytes and microgametes. These sexual stages can produce a massive infection of the intestinal epithelium of felines. Cats tolerate this extraordinary level of infection with remarkably few symptoms indicating that these sexual stages of *Toxoplasma* must also be engaged in an intense and, in terms of producing minimal disruption, highly productive negotiation with the infected host cells.

Finally, it is worth noting that this chapter has focused entirely on cells that are actually infected. Recently, it has been observed that cells that are not infected can still show evidence of having been injected with rhoptry proteins (Koshy *et al.*, 2010; 2012). *In vitro*, this appears to be the result of abortive invasion where the rhoptries inject, but invasion does not proceed. *In vivo*, it has not been possible to determine what mechanism is operating but it is clear that cells harbouring no parasites have been impacted by rhoptry proteins. Whether these are cells that experienced an abortive invasion or a successful invasion that they then eliminated is not clear but the fact that these cells can massively outnumber the cells that are productively infected (by 50-fold or more) suggests that the impact of ROPs and GRAs in the infected animal may be even greater than we currently imagine (Koshy *et al.*, 2012). Clearly, we have only begun to scratch the surface of the extraordinary dialogue that the many developmental forms of *Toxoplasma* have with their myriad hosts.

References

Alonzi, T., Maritano, D., Gorgoni, B., Rizzuto, G., Libert, C., Poli, V., 2001. Essential role of STAT3 in the control of the acute-phase response as revealed by inducible gene inactivation [correction of activation] in the liver. Mol. Cell Biol. 21, 1621–1632.

Behnke, M.S., Fentress, S.J., Mashayekhi, M., Li, L.X., Taylor, G.A., Sibley, L.D., 2012. The Polymorphic Pseudokinase ROP5 Controls Virulence in Toxoplasma gondii by Regulating the Active Kinase ROP18. PLoS Pathogens 8, e1002992.

Behnke, M.S., Khan, A., Wootton, J.C., Dubey, J.P., Tang, K., Sibley, L.D., 2011. Virulence differences in Toxoplasma mediated by amplification of a family of polymorphic pseudokinases. Proc. Natl. Acad. Sci. U. S. A. 108, 9631–9636.

Bierly, A.L., Shufesky, W.J., Sukhumavasi, W., Morelli, A., Denkers, E.Y., 2008. Dendritic cells expressing plasmacytoid marker PDCA-1 are Trojan horses during Toxoplasma gondii infection. J. Immunol. 181, 8445–8491.

Boothroyd, J.C., Dubremetz, J.F., 2008. Kiss and spit: the dual roles of Toxoplasma rhoptries. Nature reviews. Microbiology 6, 79–88.

Boothroyd, J.C., Grigg, M.E., 2002. Population biology of *Toxoplasma gondii* and its relevance to human infection: do different strains cause different disease? Curr. Opin. Micro. 5, 438–442.

Boyle, J.P., Rajasekar, B., Saeij, J.P., Ajioka, J.W., Berriman, M., Paulsen, I., Roos, D.S., Sibley, L.D., White, M.W., Boothroyd, J.C., 2006. Just one cross appears capable of dramatically altering the population biology of a eukaryotic pathogen like Toxoplasma gondii. Proc. Natl. Acad. Sci. U. S. A. 103, 10514–10519.

Bradley, P.J., Ward, C., Cheng, S.J., Alexander, D.L., Coller, S., Coombs, G.H., Dunn, J.D., Ferguson, D.J., Sanderson, S.J., Wastling, J.M., Boothroyd, J.C., 2005. Proteomic analysis of rhoptry organelles reveals many novel constituents for host-parasite interactions in *Toxoplasma gondii*. The Journal of Biological Chemistry 280, 34245–34258.

Buchholz, K.R., Fritz, H.M., Chen, X., Durbin-Johnson, B., Rocke, D.M., Ferguson, D.J., Conrad, P.A., Boothroyd, J.C., 2011. Identification of tissue cyst wall components by transcriptome analysis of in vivo and in vitro Toxoplasma gondii bradyzoites. Eukaryotic Cell 10, 1637–1647.

Butcher, B.A., Denkers, E.Y., 2002. Mechanism of entry determines ability of *Toxoplasma gondii* to inhibit macrophage proinflammatory cytokine production. Infect. Immun. 70, 5216–5224.

Butcher, B.A., Fox, B.A., Rommereim, L.M., Kim, S.G., Maurer, K.J., Yarovinsky, F., Herbert, D.R., Bzik, D.J., Denkers, E.Y., 2011. Toxoplasma gondii Rhoptry Kinase ROP16 Activates STAT3 and STAT6 Resulting in Cytokine Inhibition and Arginase-1-Dependent Growth Control. PLoS Pathogens 7, e1002236.

Butcher, B.A., Greene, R.I., Henry, S.C., Annecharico, K.L., Weinberg, J.B., Denkers, E.Y., Sher, A., Taylor, G.A., 2005a. p47 GTPases regulate *Toxoplasma gondii* survival in activated macrophages. Infect. Immun. 73, 3278–3286.

Butcher, B.A., Kim, L., Johnson, P.F., Denkers, E.Y., 2001. *Toxoplasma gondii* tachyzoites inhibit proinflammatory cytokine induction in infected macrophages by preventing nuclear translocation of the transcription factor NFkB. J. Immunol. 167, 2193–2201.

Butcher, B.A., Kim, L., Panopoulos, A., Watowich, S.S., Murray, P.J., Denkers, E.Y., 2005b. Cutting Edge: IL-10-independent STAT3 activation by *Toxoplasma gondii* mediates suppression of IL-12 and TNF-alpha in host macrophages. J. Immunol. 174, 3148–3152.

Chtanova, T., Schaeffer, M., Han, S.J., van Dooren, G.G., Nollmann, M., Herzmark, P., Chan, S.W., Satija, H., Camfield, K., Aaron, H., et al., 2008. Dynamics of Neutrophil Migration in Lymph Nodes during Infection. Immunity 29, 487–496.

Collazo, C.M., Yap, G.S., Sempowski, G.D., Lusby, K.C., Tessarollo, L., Vande Woude, G.F., Sher, A., Taylor, G.A., 2001. Inactivation of LRG-47 and IRG-47 reveals a family of interferon-g-inducible genes with essential, pathogen-specific roles in resistance to infection. J. Exp. Med. 194, 181–187.

Cook, T., Roos, D., Morada, M., Zhu, G., Keithly, J.S., Feagin, J.E., Wu, G., Yarlett, N., 2007. Divergent polyamine metabolism in the Apicomplexa. Microbiology 153, 1123–1130.

Courret, N., Darche, S., Sonigo, P., Milon, G., Buzoni-Gatel, D., Tardieux, I., 2006. CD11c and CD11b expressing mouse leukocytes transport single *Toxoplasma gondii* tachyzoites to the brain. Blood 107, 309–316.

Croker, B.A., Krebs, D.L., Zhang, J.-G., Wormald, S., Willson, T.A., Stanley, E.G., Robb, L., Greenhalgh, C.J., Forster, I., Clausen, B.E., et al., 2003. SOCS3 negatively regulates IL-6 signaling in vivo. Nature Immunol. 4, 540–545.

Denkers, E.Y., Bzik, D.J., Fox, B.A., Butcher, B.A., 2012. An inside job: hacking into Janus kinase/signal transducer and activator of transcription signaling cascades by the intracellular protozoan Toxoplasma gondii. Infection and Immunity 80, 476–482.

Denkers, E.Y., Gazzinelli, R.T., 1998. Regulation and function of T cell-mediated immunity during *Toxoplasma gondii* infection. Clin. Microbiol. Rev. 11, 569–588.

Denkers, E.Y., Kim, L., Butcher, B.A., 2003. In the belly of the beast: subversion of macrophage proinflammatory signaling cascades during *Toxoplasma gondii* infection. Cell. Micro. 5, 75–83.

El Hajj, H., Demey, E., Poncet, J., Lebrun, M., Wu, B., Galeotti, N., Fourmaux, M.N., Mercereau-Puijalon, O., Vial, H., Labesse, G., Dubremetz, J.F., 2006. The ROP2 family of Toxoplasma gondii rhoptry proteins: proteomic and genomic characterization and molecular modeling. Proteomics 6, 5773–5784.

El Hajj, H., Lebrun, M., Fourmaux, M.N., Vial, H., Dubremetz, J.F., 2007. Inverted topology of the Toxoplasma gondii ROP5 rhoptry protein provides new insights into the association of the ROP2 protein family with the parasitophorous vacuole membrane. Cellular Microbiology 9, 54–64.

El Kasmi, K.C., Qualls, J.E., Pesce, J.T., Smith, A.M., Thompson, R.W., Henao-Tamayo, M., Basaraba, R.J., Konig, T., Schleicher, U., Koo, M.S., et al., 2008. Toll-like receptor-induced arginase 1 in macrophages thwarts effective immunity against intracellular pathogens. Nat. Immunol. 9, 1399–1406.

Fentress, S.J., Behnke, M.S., Dunay, I.R., Mashayekhi, M., Rommereim, L.M., Fox, B.A., Bzik, D.J., Taylor, G.A., Turk, B.E., Lichti, C.F., et al., 2010. Phosphorylation of Immunity-Related GTPases by a Toxoplasma gondii-Secreted Kinase Promotes Macrophage Survival and Virulence. Cell Host Microbe 8, 484–495.

Fleckenstein, M.C., Reese, M.L., Konen-Waisman, S., Boothroyd, J.C., Howard, J.C., Steinfeldt, T., 2012. A Toxoplasma gondii Pseudokinase Inhibits Host IRG Resistance Proteins. PLoS Biol. 10, e1001358.

Fleming, B.D., Mosser, D.M., 2011. Regulatory macrophages: setting the threshold for therapy. European Journal of Immunology 41, 2498–2502.

Fox, B.A., Gigley, J.P., Bzik, D.J., 2004. Toxoplasma gondii lacks the enzymes required for de novo arginine biosynthesis and arginine starvation triggers cyst formation. Int. J. Parasitol. 34, 323–331.

Fritz, H.M., Bowyer, P.W., Bogyo, M., Conrad, P.A., Boothroyd, J.C., 2012a. Proteomic analysis of fractionated Toxoplasma oocysts reveals clues to their environmental resistance. PloS One 7, e29955.

Fritz, H.M., Buchholz, K.R., Chen, X., Durbin-Johnson, B., Rocke, D.M., Conrad, P.A., Boothroyd, J.C., 2012b. Transcriptomic analysis of toxoplasma development reveals many novel functions and structures specific to sporozoites and oocysts. PloS One 7, e29998.

Gavrilescu, L.C., Denkers, E.Y., 2001. IFN-g overproduction and high level apoptosis are associated with high but not low virulence *Toxoplasma gondii* infection. J. Immunol. 167, 902–909.

Gilbert, L.A., Ravindran, S., Turetzky, J.M., Boothroyd, J.C., Bradley, P.J., 2007. Toxoplasma gondii targets a protein phosphatase 2C to the nuclei of infected host cells. Eukaryotic Cell 6, 73–83.

Goebel, S., Gross, U., Luder, C.G.K., 2001. Inhibition of host cell apoptosis by *Toxoplasma gondii* is accompanied by reduced activation of the caspase cascade and alterations of poly(ADP-ribose) polymerase expression. J. Cell Sci. 114, 3495–3505.

Hunn, J.P., Feng, C.G., Sher, A., Howard, J.C., 2011. The immunity-related GTPases in mammals: a fast-evolving cell-autonomous resistance system against intracellular pathogens. Mammalian Genome: official journal of the International Mammalian Genome Society 22, 43–54.

Hunn, J.P., Koenen-Waisman, S., Papic, N., Schroeder, N., Pawlowski, N., Lange, R., Kaiser, F., Zerrahn, J., Martens, S., Howard, J.C., 2008. Regulatory interactions between IRG resistance GTPases in the cellular response to Toxoplasma gondii. The EMBO Journal 27, 2495–2509.

Hunter, C.A., Sibley, L.D., 2012. Modulation of innate immunity by Toxoplasma gondii virulence effectors. Nature reviews. Microbiology 10, 766–778.

Jensen, K.D., Wang, Y., Wojno, E.D., Shastri, A.J., Hu, K., Cornel, L., Boedec, E., Ong, Y.C., Chien, Y.H., Hunter, C.A., et al., 2011. Toxoplasma polymorphic effectors determine macrophage polarization and intestinal inflammation. Cell Host Microbe 9, 472–483.

Khaminets, A., Hunn, J.P., Konen-Waisman, S., Zhao, Y.O., Preukschat, D., Coers, J., Boyle, J.P., Ong, Y.C., Boothroyd, J.C., Reichmann, G., Howard, J.C., 2010. Coordinated loading of IRG resistance GTPases on to the Toxoplasma gondii parasitophorous vacuole. Cellular Microbiology 12, 939–961.

Khan, I.A., Schwartzman, J.D., Matsuura, T., Kasper, L.H., 1997. A dichotomous role for nitric oxide during acute *Toxoplasma gondii* infection in mice. Proc. Natl. Acad. Sci. U.S.A. 94, 13955–13960.

Kim, L., Butcher, B.A., Lee, C.W., Uematsu, S., Akira, S., Denkers, E.Y., 2006. *Toxoplasma gondii* genotype determines MyD88-dependent signaling in infected macrophages. J. Immunol. 177, 2584–2591.

Kim, L., Denkers, E.Y., 2006. *Toxoplasma gondii* triggers Gi-dependent phosphatidylinositol 3-kinase signaling required for inhibition of host cell apoptosis. J. Cell Sci. 119, 2119–2126.

Kim, S.K., Fouts, A.E., Boothroyd, J.C., 2007. Toxoplasma gondii dysregulates IFN-gamma-inducible gene expression in human fibroblasts: insights from a genome-wide transcriptional profiling. J. Immunol. 178, 5154–5165.

Koshy, A.A., Dietrich, H.K., Christian, D.A., Melehani, J.H., Shastri, A.J., Hunter, C.A., Boothroyd, J.C., 2012. Toxoplasma Co-opts Host Cells It Does Not Invade. PLoS Pathogens 8, e1002825.

Koshy, A.A., Fouts, A.E., Lodoen, M.B., Alkan, O., Blau, H.M., Boothroyd, J.C., 2010. Toxoplasma secreting Cre recombinase for analysis of host-parasite interactions. Nat. Methods 7, 307–309.

Labesse, G., Gelin, M., Bessin, Y., Lebrun, M., Papoin, J., Cerdan, R., Arold, S.T., Dubremetz, J.F., 2009. ROP2 from Toxoplasma gondii: a virulence factor with a protein-kinase fold and no enzymatic activity. Structure 17, 139–146.

Lang, C., Algner, M., Beinert, N., Gross, U., Luder, C.G., 2006. Diverse mechanisms employed by Toxoplasma gondii to inhibit IFN-gamma-induced major histocompatibility complex class II gene expression. Microbes Infect. 8, 1994–2005.

Lang, C., Hildebrandt, A., Brand, F., Opitz, L., Dihazi, H., Luder, C.G., 2012. Impaired chromatin remodelling at STAT1-regulated promoters leads to global unresponsiveness of Toxoplasma gondii-Infected macrophages to IFN-gamma. PLoS Pathogens 8, e1002483.

Lang, R., Patel, D., Morris, J.J., Rutschman, R.L., Murray, P.J., 2002. Shaping gene expression in activated and resting primary macrophages by IL-10. J. Immunol. 169, 2253–2263.

Lang, R., Pauleau, A.-L., Parganas, E., Takahashi, Y., Mages, J., Ihle, J.N., Rutschman, R., Murray, P.J., 2003. SOCS regulates the plasticity of gp130 signaling. Nature Immunol. 4, 546–550.

Leng, J., Butcher, B.A., Denkers, E.Y., 2009a. Dysregulation of macrophage signal transduction by Toxoplasma gondii: Past progress and recent advances. Parasite Immunol. 31, 717–728.

Leng, J., Butcher, B.A., Egan, C.E., Abdallah, D.S., Denkers, E.Y., 2009b. Toxoplasma gondii prevents chromatin remodeling initiated by TLR-triggered macrophage activation. J. Immunol. 182, 489–497.

Levy, D.E., Darnell Jr., J.E., 2002. Stats: transcriptional control and biological impact. Nat. Rev. Mol. Cell Biol. 3, 651–662.

Luder, C.G.K., Algner, M., Lang, C., Bleicher, N., Gross, U., 2003. Reduced expression of the inducible nitric oxide synthase after infection with *Toxoplasma gondii* facilitates parasite replication in activated murine macrophages. Internat. J. Parasitol. 33, 833–844.

Luder, C.G.K., Gross, U., 2005. Apoptosis and its modulation during infection with *Toxoplasma gondii*: molecular mechanisms and role in pathogenesis. Curr. Top. Microbiol. Immunol. 289, 219–238.

Luder, C.G.K., Lang, T., Beurle, B., Gross, U., 1998. Down-regulation of MHC class II molecules and inability to up-regulate class I molecules in murine macrophages after infection with *Toxoplasma gondii*. Clin. Exp. Immunol. 112, 308–316.

Manning, G., Plowman, G.D., Hunter, T., Sudarsanam, S., 2002. Evolution of protein kinase signaling from yeast to man. Trends Biochem. Sci. 27, 514–520.

Martinez, F.O., Helming, L., Gordon, S., 2009. Alternative activation of macrophages: an immunologic functional perspective. Annu. Rev. Immunol. 27, 451–483.

Melo, M.B., Jensen, K.D., Saeij, J.P., 2011. Toxoplasma gondii effectors are master regulators of the inflammatory response. Trends Parasitol. 27, 487–495.

Molestina, R.E., Payne, T.M., Coppens, I., Sinai, A.P., 2003. Activation of NF-kB by *Toxoplasma gondii* correlates with increased expression of antiapoptotic genes and localization of phosphorylated IkB to the parasitophorous vacuole membrane. J. Cell Sci. 116, 4359–4371.

Mordue, D.G., Monroy, F., La Regina, M., Dinarello, C.A., Sibley, L.D., 2001. Acute toxoplasmosis leads to lethal overproduction of Th1 cytokines. J. Immunol. 167, 4574–4584.

Mosser, D.M., Edwards, J.P., 2008. Exploring the full spectrum of macrophage activation. Nature reviews. Immunology 8, 958–969.

Munoz, M., Liesenfeld, O., Heimesaat, M.M., 2011. Immunology of Toxoplasma gondii. Immunol. Rev. 240, 269–285.

Murray, P.J., Wynn, T.A., 2011. Protective and pathogenic functions of macrophage subsets. Nature reviews. Immunology 11, 723–737.

Naguleswaran, A., Elias, E.V., McClintick, J., Edenberg, H.J., Sullivan Jr., W.J., 2010. Toxoplasma gondii lysine acetyltransferase GCN5-A functions in the cellular response to alkaline stress and expression of cyst genes. PLoS Pathogens 6, e1001232.

Niedelman, W., Gold, D.A., Rosowski, E.E., Sprokholt, J.K., Lim, D., Farid Arenas, A., Melo, M.B., Spooner, E., Yaffe, M.B., Saeij, J.P., 2012. The Rhoptry Proteins ROP18 and ROP5 Mediate Toxoplasma gondii Evasion of the Murine, But Not the Human, Interferon-Gamma Response. PLoS Pathogens 8, e1002784.

Ong, Y.C., Boyle, J.P., Boothroyd, J.C., 2011. Strain-dependent host transcriptional responses to toxoplasma infection are largely conserved in mammalian and avian hosts. PloS One 6, e26369.

Ong, Y.C., Reese, M.L., Boothroyd, J.C., 2010. Toxoplasma rhoptry protein 16 (ROP16) subverts host function by direct tyrosine phosphorylation of STAT6. The Journal of Biological Chemistry 285, 28731–28740.

Pawlowski, N., Khaminets, A., Hunn, J.P., Papic, N., Schmidt, A., Uthaiah, R.C., Lange, R., Vopper, G., Martens, S., Wolf, E., Howard, J.C., 2011. The activation mechanism of Irga6, an interferon-inducible GTPase contributing to mouse resistance against Toxoplasma gondii. BMC Biology 9, 7.

Payne, T.M., Molestina, R.E., Sinai, A.P., 2003. Inhibition of caspase activation and a requirement for NF-kB function in the *Toxoplasma gondii*-mediated blockade of host apoptosis. J. Cell Sci. 116, 4345–4358.

Peixoto, L., Chen, F., Harb, O.S., Davis, P.H., Beiting, D.P., Brownback, C.S., Ouloguem, D., Roos, D.S., 2010. Integrative Genomic Approaches Highlight a Family of Parasite-Specific Kinases that Regulate Host Responses. Cell Host Microbe 8, 208–218.

Pincus, D., Letunic, I., Bork, P., Lim, W.A., 2008. Evolution of the phosphotyrosine signaling machinery in premetazoan lineages. Proc. Natl. Acad. Sci. U. S. A. 105, 9680–9684.

Reese, M.L., Boothroyd, J.C., 2011. A conserved non-canonical motif in the pseudoactive site of the ROP5 pseudokinase domain mediates its effect on Toxoplasma virulence. The Journal of Biological Chemistry 286, 29366–29375.

Reese, M.L., Zeiner, G.M., Saeij, J.P., Boothroyd, J.C., Boyle, J.P., 2011. Polymorphic family of injected pseudokinases is paramount in Toxoplasma virulence. Proc. Natl. Acad. Sci. U. S. A. 108, 9625–9630.

Robben, P.M., Mordue, D.G., Truscott, S.M., Takeda, K., Akira, S., Sibley, L.D., 2004. Production of IL-12 by macrophages infected with Toxoplasma gondii depends on the parasite genotype. J. Immunol. 172, 3686–3694.

Rosowski, E.E., Lu, D., Julien, L., Rodda, L., Gaiser, R.A., Jensen, K.D., Saeij, J.P., 2011. Strain-specific activation of the NF-kappaB pathway by GRA15, a novel Toxoplasma gondii dense granule protein. J. Exp. Med. 208, 195–212.

Rutschman, R., Lang, R., Hesse, M., Ihle, J.N., Wynn, T.A., Murray, P.J., 2001. Stat6-dependent substrate depletion regulates nitric oxide production. J. Immunol. 166, 2173–2177.

Saeij, J.P., Boyle, J.P., Coller, S., Taylor, S., Sibley, L.D., Brooke-Powell, E.T., Ajioka, J.W., Boothroyd, J.C., 2006. Polymorphic secreted kinases are key virulence factors in toxoplasmosis. Science 314, 1780–1783.

Saeij, J.P., Coller, S., Boyle, J.P., Jerome, M.E., White, M.W., Boothroyd, J.C., 2007. Toxoplasma co-opts host gene expression by injection of a polymorphic kinase homologue. Nature 445, 324–327.

Scharton-Kersten, T., Yap, G., Magram, J., Sher, A., 1997. Inducible nitric oxide is essential for host control of persistent but not acute infection with the intracellular pathogen *Toxoplasma gondii*. J. Exp. Med. 185, 1–13.

Steinfeldt, T., Konen-Waisman, S., Tong, L., Pawlowski, N., Lamkemeyer, T., Sibley, L.D., Hunn, J.P., Howard, J.C., 2010. Phosphorylation of Mouse Immunity-Related GTPase (IRG) Resistance Proteins Is an Evasion Strategy for Virulent Toxoplasma gondii. PLoS Biol. 8, e1000576.

Takeda, K., Clausen, B.E., Kaisho, T., Tsujimura, T., Terada, N., Forster, I., Akira, S., 1999. Enhanced Th1 activity and development of chronic enterocolitis in mice devoid of stat3 in macrophages and neutrophils. Immunity 10, 39–49.

Taylor, S., Barragan, A., Su, C., Fux, B., Fentress, S.J., Tang, K., Beatty, W.L., Hajj, H.E., Jerome, M., Behnke, M.S., et al., 2006. A secreted serine-threonine kinase determines virulence in the eukaryotic pathogen Toxoplasma gondii. Science 314, 1776–1780.

Yamamoto, M., Ma, J.S., Mueller, C., Kamiyama, N., Saiga, H., Kubo, E., Kimura, T., Okamoto, T., Okuyama, M., Kayama, H., et al., 2011. ATF6{beta} is a host cellular target of the Toxoplasma gondii virulence factor ROP18. J. Exp. Med.

Yamamoto, M., Standley, D.M., Takashima, S., Saiga, H., Okuyama, M., Kayama, H., Kubo, E., Ito, H., Takaura, M., Matsuda, T., et al., 2009. A single polymorphic amino acid on Toxoplasma gondii kinase ROP16 determines the direct and strain-specific activation of Stat3. J. Exp. Med. 206, 2747–2760.

Yasukawa, H., Ohishi, M., Mori, H., Murakami, M., Chinen, T., Aki, D., Hanada, T., Takeda, K., Akira, S., Hoshijima, M., et al., 2003. IL-6 induces an anti-inflammatory response in the absence of SOCS3 in macrophages. Nature Immunol. 4, 551–556.

Zhao, Y., Ferguson, D.J., Wilson, D.C., Howard, J.C., Sibley, L.D., Yap, G.S., 2009a. Virulent Toxoplasma gondii evade immunity-related GTPase-mediated parasite vacuole disruption within primed macrophages. J. Immunol. 182, 3775–3781.

Zhao, Y.O., Khaminets, A., Hunn, J.P., Howard, J.C., 2009b. Disruption of the Toxoplasma gondii parasitophorous vacuole by IFNgamma-inducible immunity-related GTPases (IRG proteins) triggers necrotic cell death. PLoS Pathogens 5, e1000288.

CHAPTER

15

Bradyzoite Development

Laura J. Knoll, Tadakimi Tomita†, Louis M. Weiss†*

*Department of Medical Microbiology and Immunology, University of Wisconsin-Madison, Madison, Wisconsin, USA †Department of Pathology and Medicine, Albert Einstein College of Medicine, Bronx, New York, USA

OUTLINE

15.1 INTRODUCTION

The asexual cycle of *Toxoplasma gondii* can occur in every warm-blooded animal, including humans. The asexual cycle has two developmental stages: a rapidly replicating form called the tachyzoite and a slow growing form called the bradyzoite. The sexual cycle of *T. gondii* occurs only within the feline intestine, after which infected cats excrete between two and 20 million oocysts per day in their faeces. *T. gondii* is acquired orally either by ingestion of

Toxoplasma gondii, Second Edition
http://dx.doi.org/10.1016/B978-0-12-396481-6.00015-5

oocyst-contaminated food or water, or by eating bradyzoite cyst-harbouring meat products. After ingestion, sporozoites or bradyzoites will invade the intestinal epithelium, differentiate into the rapidly growing tachyzoite form and disseminate throughout the body. Epidemiology suggests that the ingestion of bradyzoite cysts in undercooked meat is an important source of *T. gondii* infection for humans (Kimball *et al.*, 1974; McAuley *et al.*, 1994).

In both definitive and intermediate hosts, tachyzoites disseminate throughout the body, and then differentiate into bradyzoites that remain for the lifetime of the host. While tachyzoites can invade virtually any nucleated cell during dissemination, bradyzoite cysts are primarily persistent in the central nervous system and striated muscle of the host. The number of tissue cysts formed in mouse brain appears to be regulated by the class I gene L^d (Brown *et al.*, 1995). More tissue cysts are produced in mice that become mildly ill from infection than in those that become highly symptomatic. Tissue cyst persistence may vary with both the strain of *T. gondii* and the strain of murine host (Ferguson *et al.*, 1994). Mice that are resistant to acute infection can be susceptible to chronic infection with encephalitis suggesting that control of chronic infection is not linked to the loci that control susceptibility to acute infection (Brown and McLeod, 1990; Brown *et al.*, 1994; McLeod *et al.*, 1993, 1989). Bradyzoites contained in cysts are refractory to most chemotherapeutic agents used for treatment of toxoplasmosis and tissue cysts are produced in any animal capable of being infected with *T. gondii.* The persistence and reactivation of bradyzoite forms is a major cause of disease in humans. Most toxoplasmic disease is believed to occur in immunocompromised patients due to recrudescence of a chronic infection of bradyzoites as cellular immune surveillance is lost (Luft and Remington, 1992).

In most individuals acute infection with *T. gondii* is asymptomatic or causes mild symptoms similar to a self-limited mononucleosis-like syndrome. If an immunologically naive, i.e. seronegative, pregnant woman is infected, transmission of this parasite to the foetus can occur with the development of a congenital infection that can result in a fetopathy and a relapsing chorioretinitis (Wong and Remington, 1993; Remington *et al.*, 1995). Although overwhelming disseminated toxoplasmosis has been reported in immunocompromised hosts, e.g. patients with advanced HIV infection, the predilection of this parasite for the central nervous system causing necrotizing encephalitis constitutes its major threat. The incidence of clinically apparent toxoplasmosis has waned with the development of combination antiretroviral treatment (cART). Although less prevalent in AIDS patients, *T. gondii* has been implicated in waterborne outbreaks and deaths in marine wildlife due to water pollution resulting in its listing as a Category B agent (http://www3.niaid.nih.gov/Biodefense/bandc_priority.htm) (Bowie *et al.*, 1997; Miller *et al.*, 2002; Aramini *et al.*, 1999).

The development of *Toxoplasma* encephalitis as well as relapsing chorioretinitis in congenital infection is believed to be due to the transition of bradyzoites in tissue cysts into the active and rapidly replicating tachyzoite stage. It is hypothesized that in chronic toxoplasmosis bradyzoites regularly transform to tachyzoites, but that the tachyzoites are controlled by the immune system. In mice new tissue cysts have been demonstrated to be formed during chronic infection (Gross *et al.*, 1992; McLeod *et al.*, 1991; Ferguson *et al.*, 1994; van der Waaij, 1959). Such a dynamic equilibrium between encysted and replicating forms would lead to recurrent antigenic stimulation, possibly accounting for the life-long persistence of antibody titres found in chronically infected hosts (Ferguson *et al.*, 1989; Frenkel and Escajadillo, 1987). In addition to rodents (mice, hamsters and rats), tissue cyst rupture has been demonstrated to occur in the primate *Aortus lemurinus* (Frenkel and Escajadillo, 1987). Due to its central importance in initiating infection and disease pathogenesis, the biology of the bradyzoite stage has been an active area of research. While progress has been made, bradyzoite to tachyzoite interconversion is not well understood. Many initial seminal observations have been reviewed elsewhere (Weiss and Kim, 2000).

15.2 BRADYZOITE AND TISSUE CYST MORPHOLOGY AND BIOLOGY

Tachyzoites (tachos = fast) refer to the rapidly growing life stage of *T. gondii* that has also been called endozoites or trophozoites. Bradyzoites (brady = slow), also called cystozoites, are the life stage found in the tissue cyst and are believed to replicate slowly. Both stages replicate within a parasitophorous vacuole within the host cell, which is modified by the particular life stage into either a tachyzoite or a bradyzoite (tissue cyst)-specific vacuole. Tissue cysts are intracellular structures in which the bradyzoites divide by endodyogeny, the same replicative mechanism as tachyzoites.

In mature cysts bradyzoites appear to enter a G_0 stage of the cell cycle essentially becoming a non-replicative differentiated organism with a DNA content of 1N (Radke *et al.*, 2003). In mature cysts occasional degenerating bradyzoites are seen (Pavesio *et al.*, 1992); *in vitro* mis-segregation and loss of apicoplasts in mature bradyzoites has been demonstrated (Dzierszinski *et al.*, 2004). Immature tissue cysts may be as small as 5 μm containing only two organisms, suggesting that the commitment to bradyzoite differentiation is an early event in the establishment of a parasitophorous vacuole. Such immature cysts increase in size and thus it is clear that *T. gondii* expressing bradyzoite-specific markers can replicate during the maturation of tissue cyst (van der Waaij, 1959). The size of a tissue cyst is variable, but on average a mature cyst in the brain is spherical and 50 to 70 μm in diameter containing approximately 1000 crescent-shaped 7 × 1.5 μm bradyzoites. In muscle such tissue cysts are more elongated and may be up to 100 μm in length (Dubey *et al.*, 1998).

Tissue cyst size is dependent on cyst age, the host cell parasitized, the strain of *T. gondii* and the cytological method used for measurement. There is a correlation between small cyst size and a decreased oral transmission with some of the exotic strains of *T. gondii* (Fux *et al.*, 2007). Young and old cysts can be distinguished readily by their ultrastructural features (Dubey *et al.*, 1998; Fortier *et al.*, 1996; Sims *et al.*, 1989; Scholytyseck *et al.*, 1974). While tissue cysts can develop in any visceral organ (e.g. lungs, liver and kidneys), they are more common in neural (e.g. brain and eyes) or muscle (e.g. skeletal and cardiac) tissue (Dubey *et al.*, 1998). In the central nervous system cysts have been reported in neurons, astrocytes and microglia (Ferguson *et al.*, 1989; Ferguson and Hutchison, 1987b, 1987a); however, it is not known which is the preferred or predominant cell in which cysts form and whether host cells influence cyst formation. In tissue culture both astrocytes and neurons have been demonstrated to support cyst formation (Halonen *et al.*, 1996, 1998a; Fagard *et al.*, 1999) (Fig. 15.1). Host cells have varying permissiveness for the development of tissue cysts; with cells that are more resistant to bradyzoite formation having elevated levels of lactate, glycolysis and Akt expression (Weilhammer *et al.*, 2012). *In vitro* bradyzoites have been demonstrated to exit a cyst and establish new cysts in the same cell or adjacent cells (Dzierszinski *et al.*, 2004). This observation may account for the observation of clusters of cysts seen *in vivo* in the central nervous system.

The tissue cyst wall or bradyzoite parasitophorous vacuole membrane is elastic, thin (less than 0.5 μm thick), faintly PAS positive and argyrophic, although this depends on the silver staining method used (Sims *et al.*, 1988).

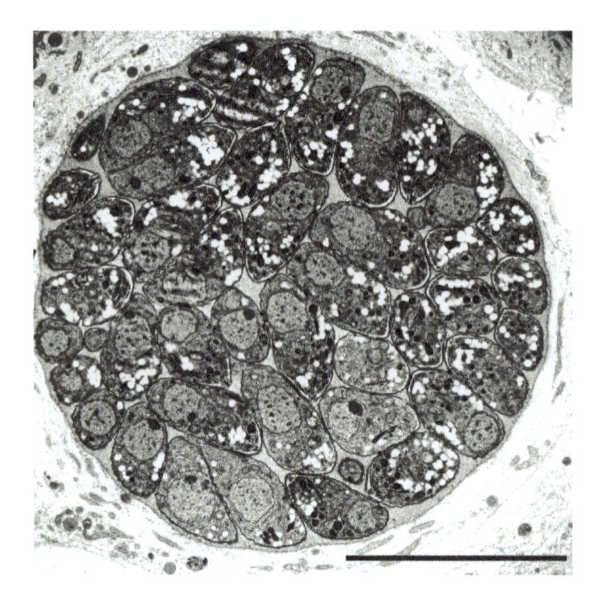

FIGURE 15.1 **Ultrastructure of a *T. gondii* tissue cyst *in vitro*.** This electron micrograph demonstrates multiple bradyzoites within a cyst. The cyst is within a murine astrocyte grown *in vitro* tissue culture. The clear vacuoles are consistent with amylopectin granules seen in bradyzoites. The cyst wall is formed from the parasitophorous vacuole membrane. Bar length is 25 μm. Image courtesy of Dr S. Halonen, Montana State University.

Proteomic data indicate that many cyst wall proteins are, in fact, glycoproteins (L. M. Weiss, unpublished data). In addition, experiments suggest that the proteins in the cyst wall that are glycosylated may be modified by host cell enzymes resulting in host cell-specific glycosylation of the cyst wall (Eller *et al.*, 2012). The tissue cyst wall is phase lucent by phase-contrast microscopy and the vacuole often contains an odd number of club shaped parasites, highlighting that asynchronous division occurs (Weiss *et al.*, 1995; Dzierszinski *et al.*, 2004). Pale blue autofluorescence of the tissue cyst wall can be observed using UV light at 330–385 nm (Lei *et al.*, 2005). The cyst wall appears to be composed of both host and parasite derived materials and lined by granular material, which fills up the space between the bradyzoites, particularly in mature cysts (Sims *et al.*, 1988; Ferguson and Hutchison, 1987a, 1987b). During development in astrocytes, the bradyzoite parasitophorous vacuole is surrounded by a layer of host cell intermediate filaments (glial fibrillary acidic protein) that limits the contact of the vacuole with host ER and mitochondria (Halonen *et al.*, 1998b); however, this material is not incorporated into the cyst wall. Tissue cysts have a specific gravity of 1.056 and have been successfully purified from brain tissue using isopycnic centrifugation in a Percoll gradient (Cornelissen *et al.*, 1981), discontinuous 25% to 30% Percoll gradients (Pettersen, 1988; Blewett *et al.*, 1983) and 20% dextran (Freyre, 1995). (See Chapter 17 for protocol.) In several studies, oral infection with oocysts has been demonstrated to result in more brain cysts than intraperitoneal infections with oocysts, tachyzoites or bradyzoites (Fritz *et al.*, 2012a).

Bradyzoites differ ultrastructurally from tachyzoites in that they have a posteriorly located nucleus, solid rhoptries that often looped back on themselves, numerous micronemes, and polysaccharide (amylopectin) granules (Ferguson and Hutchison, 1987b; Dubey, 1997; Lemgruber *et al.*, 2010). There is at least one bradyzoite rhoptry protein (BRP1) that is absent in tachyzoites; however, knockout of BRP1 did not affect development (Schwarz *et al.*, 2005). Lipid bodies are not seen in bradyzoites, but are numerous in sporozoites and occasionally seen in tachyzoites. The contents of rhoptries in mature bradyzoites are electron dense in contrast to the labyrinthine rhoptries seen in tachyzoites and in immature bradyzoites (Dubey *et al.*, 1998). Bradyzoites stain red with periodic acid-Schiff (PAS) due to the presence of amylopectin granules, whereas tachyzoites are usually PAS negative (Dubey *et al.*, 1998). In general bradyzoites are more slender than tachyzoites. Bradyzoites are more resistant to acid pepsin (1–2 hours' survival in pepsin–HCl) than tachyzoites (10 minutes' survival) (Jacobs and Remington, 1960; Popiel *et al.*, 1996). The time to oocyst shedding, called the prepatent period, in cats following feeding of bradyzoites is shorter (three to seven days) than that following feeding of tachyzoites (over 14 days). The length of the prepatent period is the most sensitive biologic marker of mature functional tissue cysts (Dubey *et al.*, 1997; Dubey, 1997).

15.3 THE DEVELOPMENT OF TISSUE CYSTS AND BRADYZOITES *IN VITRO*

The development of tissue cysts *in vitro* was reported over 40 years ago (Matsubayashi and Akao, 1963; Hogan *et al.*, 1960); however, the morphologic similarity of bradyzoites and tachyzoites by light microscopy made it difficult to study these differentiation events until the development of antibodies to bradyzoite-specific antigens. *In vitro* produced bradyzoite cysts led to oocyte excretion in cats with a prepatent period consistent with that of tissue cysts (Hoff *et al.*, 1977). Using TEM it has been demonstrated that while cyst-like structures were present within three days of infecting host cells in tissue culture, *in vitro* cysts are not mature until six days post infection according to the cat bioassay (Dubey, 1997). Prolonged passage of *T. gondii* or other Apicomplexa *in vitro* may lead to the loss of their ability to differentiate into other stages. For example, prolonged passage of *Besnoitia jellisoni in vitro* leads to a loss of its ability to form tissue cysts in mice and many type II isolates (e.g. PLK) of *T. gondii* cannot form oocysts in cats (Frenkel *et al.*, 1976).

Bradyzoite-specific monoclonal antibodies (see Table 15.1) have greatly facilitated studies of bradyzoite development *in vitro* and the recognition of techniques for the induction of differentiation. Parasite lines that use various GFP constructs (Ds Red, GFP, YFP, etc.) under the control of various stage-specific promoters have also been developed as useful tools to study bradyzoite development (Singh *et al.*, 2002; Ma *et al.*, 2004), this has included lines that express both bradyzoite (usually *BAG1* or *LDH2* promoter driven) and tachyzoite (usually *SAG1* promoter driven)-specific GFP constructs in the same cell (Unno *et al.*, 2009). Using parasites containing chloramphenicol acetyltransferase (CAT) expressed constitutively from the α-tubulin promoter (*TUB1*) and β-galactosidase (βGAL), expressed from a bradyzoite-specific promoter (*BAG1*) one can measure both growth rate and degree of bradyzoite differentiation (Eaton *et al.*, 2005). Studies of the *in vitro* development of bradyzoites using transcriptomic (SAGE and RNAseq) as well as proteomic techniques has confirmed that there are numerous stage-specific genes/proteins in these organisms (data available at http://www.toxodb.org).

In tissue culture studies, it is evident that bradyzoites spontaneously convert to tachyzoites and that tachyzoites spontaneously convert to bradyzoites (Bohne *et al.*, 1993, 1994; De Champs *et al.*, 1997; Jones *et al.*, 1986; Lane *et al.*, 1996; Lindsay *et al.*, 1991, 1993a, 1993b; McHugh *et al.*, 1993; Popiel *et al.*, 1994, 1996; Soete *et al.*, 1993, 1994; Soete and Dubremetz, 1996; Weiss *et al.*, 1994; Parmley *et al.*, 1995). It was noted by Matsubayashi in 1963 and has been confirmed by several groups using bradyzoite-specific monoclonal antibodies that *T. gondii* strains with a slower rate of replication were more likely to develop cysts *in vitro* and that slowing the replication of virulent strains would allow tissue cysts of virulent strains to develop *in vitro* (Matsubayashi and Akao, 1963). Low virulence strains are high cyst forming strains in mice, e.g. type II and III strains such as ME49 or Pru or 76K, and have a higher spontaneous rate of cyst formation in culture than do virulent type I strains such as RH (Soete *et al.*, 1994). Additionally, we have observed that for avirulent *T. gondii* isolates, an increased growth rate and decreased efficiency of cyst production both correlate with prolonged tissue culture passage.

TABLE 15.1 Common Bradyzoite Markers

Name of Antigen	Monoclonal Antibody	Size on Immunoblot (kDa)	Location by IFA	Comments	Cloned ToxodB number
BAG1 (hsp30, BAG5)	7E5 74.1.8	28	cytoplasm	small heat shock protein	TGME49_259020
BSR4 (p36, SRS16C)	T84A12	36	surface	SRS family antigen, also in sporozoites	TGME49_320180
SAG4A (p18, SRS35A)	T83B1	18	surface	SRS family	TGME49_280570
None	DC11	not reactive	surface	surface antigen	No*
p21	T84G11	21	surface	surface antigen	No*
p34	T82C2	34	surface	surface antigen	No*
SRS9 (SRS16B)	murine polyclonal only	43	surface	SRS family	TGME49_320190
MAG1	None	65	matrix	studies indicate also expressed in tachyzoites	TGME49_270240
none	E7B2	29	matrix		No
none	93.2 (Weiss unpublished)	not reactive	matrix		No
none	1.23.29 (Weiss unpublished)	19	matrix		No
CST1	73.18, SalmonE; also recognized by DBA lectin	250 (116)	cyst wall	structural protein	TGME49_064660
CST1?	CC2	115 (bradyzoite) 40 (tachyzoite)	cyst wall	same as CST1?	No
LDH2	polyclonal sera weakly cross reacts to LDH1	35	cytosolic	tachyzoite isoform LDH1	TGME49_291040
ENO1	polyclonal sera to ENO2 and ENO1 do not cross react	48	nuclear and cytosolic	tachyzoite isoform ENO2	TGME49_268860

* *These may be members of the SRS family*

Stress conditions are associated with an induction of bradyzoite development, i.e. there are more bradyzoites under these conditions than would be expected from the simple inhibition of tachyzoite replication. Conditions that induce bradyzoite formation within host cells are temperature stress (43°C (Soete *et al.*, 1994)), pH stress (pH 6.6–6.8 or 8.0–8.2 (Weiss *et al.*, 1995; Soete *et al.*, 1994; Soete and Dubremetz, 1996)), chemical stress from Na arsenite (Soete *et al.*, 1994) and nutrient stress from arginine starvation (Fox *et al.*, 2004). In murine macrophage lines derived from bone marrow, IFNγ increases bradyzoite antigen expression (Bohne *et al.*, 1994). This appears to be due to the production of nitric oxide (NO) as bradyzoite differentiation was inhibited by treating macrophages with an inducible nitric oxide synthase inhibitor N^G-monomethyl-L-arginine (Bohne *et al.*, 1994). Bradyzoite differentiation is also enhanced by sodium nitroprusside (SNP), an exogenous NO donor (Bohne *et al.*, 1994; Kirkman *et al.*, 2001; Weiss *et al.*, 1996). Similarly, both oligomycin, an inhibitor of mitochondrial ATP synthetase function, and antimycin A, an inhibitor of the electron transport of the respiratory chain, increase bradyzoite antigen expression (Bohne *et al.*, 1994; Tomavo and Boothroyd, 1995). Alkaline (pH 8) stress often accompanied by low CO_2 (0.5%) is one of the most commonly used methods to induce bradyzoite differentiation in the laboratory (Fig. 15.2).

The contribution of the host cell to stage conversion remains to be fully elucidated. Two groups have reported that exposure of extracellular tachyzoites to stress conditions (pH 8.1) will result in an increase in bradyzoite differentiation (Weiss *et al.*, 1998; Yahiaoui *et al.*, 1999), consistent with a direct effect of stress on the parasite. Nonetheless, most of the agents that induce differentiation have profound effects upon host cells and it is likely that alterations in host cell signalling also have significant impact upon bradyzoite differentiation. This phenomenon has been demonstrated for compound 1 (Gurnett *et al.*, 2002; Merck Pharmaceuticals Inc.), which requires host cell protein synthesis in order to induce bradyzoite formation suggesting that its effect on differentiation is mediated through a perturbation of the host cell rather than directly on the parasite (Radke *et al.*, 2006). Human cell division autoantigen-1 (CDA1) was identified in this analysis, and small interfering RNA knockdown of this gene demonstrated that CDA1 expression causes the inhibition of parasite replication that leads subsequently to the induction of bradyzoite differentiation (Radke *et al.*, 2006). In addition, the metabolic status of the host cell can affect its ability to support bradyzoite differentiation (Weilhammer *et al.*, 2012). Data also suggest that host cell CD73 generated adenosine can facilitate bradyzoite differentiation in the central nervous system (Mahamed *et al.*, 2012).

When cells are infected by bradyzoites from tissue cysts differentiation to tachyzoites and the appearance of the tachyzoite-specific antigen SAG1 occurs within 15 hours and before cell division had occurred (Soete and Dubremetz, 1996). Vacuoles containing organisms expressing only tachyzoite antigens are clearly evident within 48 hours of infection. When bradyzoite differentiation occurs in cell culture following infection with tachyzoites all of the currently available markers for bradyzoite formation, with the exception of p21 (mAb T84G10), can be detected within 24 hours of infection (Gross *et al.*, 1996; Lane *et al.*, 1996). This detection includes markers of bradyzoite surface antigens as well as those related to cyst wall formation. Conversion between these two stages is a rapid event and the commitment to differentiation may be occurring at the time of or shortly after invasion and formation of the parasitophorous vacuole.

By three days after exposure to conditions that induce bradyzoite development vacuoles are present in tissue culture with the electron microscopic characteristics of cysts; however, reactivity to mAb T84G10 (p21) does not appear until five days. As assessed by the cat bioassay, mature/functional cysts are not formed until at least six days in culture. Additional markers of mature functional cysts are needed to facilitate *in vitro* studies on cyst biology. There may be differences in antigen expression and replication in early bradyzoites and late bradyzoites, but a detailed chronology of marker expression has not yet been developed.

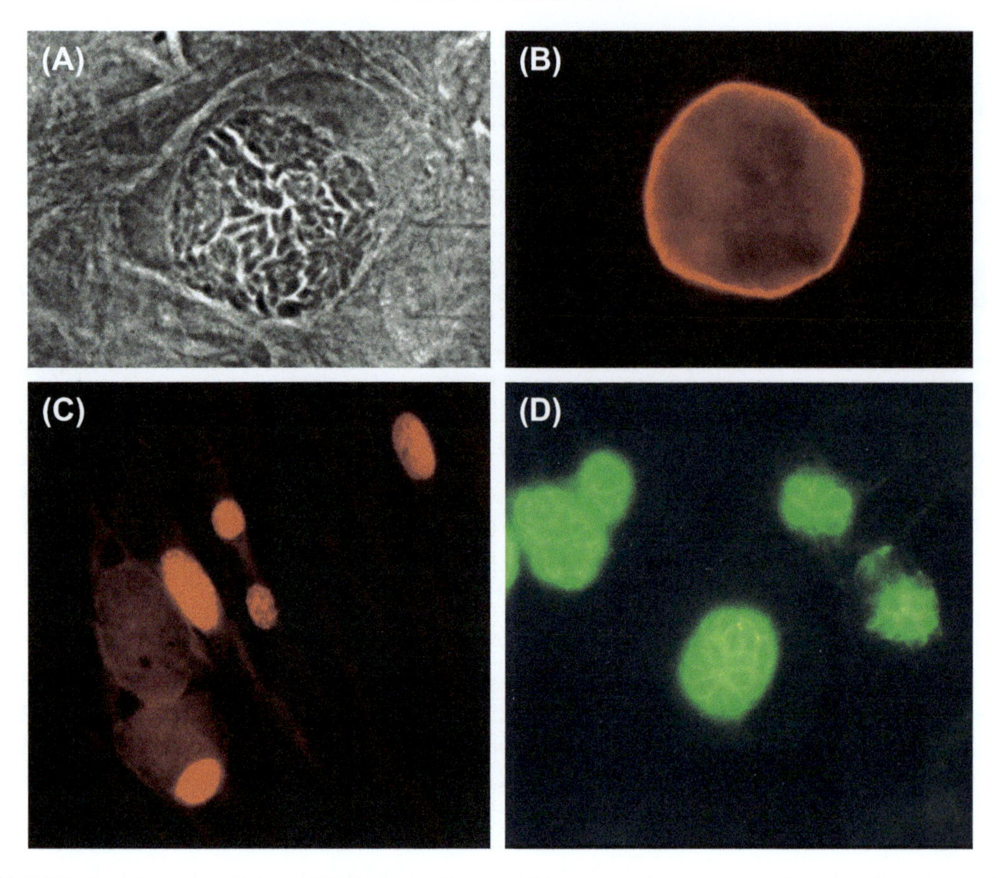

FIGURE 15.2 **Development of *T. gondii* life stages *in vitro*.** Primary murine astrocytes (A, B) or human foreskin fibroblasts (C, D) infected with *T. gondii* (ME49/PLK strain). Human fibroblasts infected with *T. gondii* (ME49/PLK strain) were maintained at pH 8.0 (C) or pH 7.2 (D) for three days.
A) Phase micrograph demonstrating an *in vitro* tissue cyst.
B) IFA using anti-CST1 (mAb 73.18) demonstrating staining of the cyst wall of the bradyzoite parasitophorous vacuole (the corresponding phase microscopy is shown in panel A).
C) IFA using anti-BAG1 (mAb 74.1.8) demonstrating bradyzoite development in several parasitophorous vacuoles. Staining occurs throughout the cytoplasm of the bradyzoites.
D) IFA using anti-SAG1 (mAb DG52) demonstrating tachyzoite development and rosette formation. Staining is localized to the surface of each tachyzoite.

15.4 THE CELL CYCLE AND BRADYZOITE DEVELOPMENT

It is probable that bradyzoite differentiation from tachyzoites is a programmed response related to a slowing of replication and lengthening of the cell cycle (Jerome *et al.*, 1998), similar to the programmed expansion and differentiation reported in other coccidia. The cell cycle in tachyzoites is characterized by major G1 and S phases and a relatively short G2+M. As *T. gondii* replication slows there is an increase in duration of the G1 phase of the cell cycle. It is not known if the checkpoints within this cell cycle differ from those observed in yeast and mammalian cells, but preliminary studies suggest differences (Radke *et al.*, 2001; Khan *et al.*, 2002). It appears that the unique late S/G2 represents a premitotic cell cycle checkpoint at which commitment to bradyzoite differentiation occurs. This is supported by the peak expression of bradyzoite mRNAs in the late mitotic period (Behnke *et al.*, 2010). When *T. gondii* sporozoites from the VEG strain infect human fibroblasts *in vitro* they transform to rapidly dividing tachyzoites with a half-life of 6 hours. After 20 divisions, approximately five days in culture, these tachyzoites shift to a slower growth rate with a half-life of 15 hours (Radke *et al.*, 2001; Khan *et al.*, 2002). Bradyzoite differentiation, as defined by the expression of bradyzoite-specific antigens, occurs spontaneously when the population shifts to a slower growth rate ($t_{½}$ 16 h); but is not seen in the rapidly dividing ($t_{½}$ 6 h) organisms. This finding is consistent with observations that spontaneous bradyzoite differentiation occurs less readily in rapidly dividing strains of *T. gondii* such as RH and that stress conditions that slow growth induce bradyzoite differentiation (Bohne *et al.*, 1994; Jerome *et al.*, 1998; Weiss and Kim, 2000).

Bradyzoites can undergo asynchronous division, resulting in vacuoles with odd numbers of organisms instead of the usual multiples of two seen in tachyzoite vacuoles (Dzierszinski *et al.*, 2004). Bradyzoite differentiation cannot be uncoupled from slowing of the cell cycle and may be a stochastic event that occurs at a specific point in the cell cycle when replication has slowed sufficiently. It appears that conditions that slow the cell cycle result in bradyzoite differentiation and that progression through the cycle is required for full differentiation to occur (Fox and Bzik, 2002; Khan *et al.*, 2002). In support of this idea, it was found that a point mutation in the TgCactin gene causes a temperature-sensitive (ts) cell cycle arrest in G1 (Gubbels *et al.*, 2008) which was confirmed by genome-wide expression profiling of the ts mutant (Szatanek *et al.*, 2012). This genome-wide expression profiling also demonstrated the induction of several extracellular parasite and bradyzoite genes, including AP2 transcription factors associated with extracellular and bradyzoite parasites (Szatanek *et al.*, 2012) To this end, AP2 transcription factors, e.g. TgAP2XI-4 (Gubbels *et al.*, 2008) and TgAP2IX-9 (Radke *et al.*, 2013), appear to be key regulatory factors in bradyzoite development. Interestingly, TgAP2IX-9 appears to function as a repressor of differentiation (Radke *et al.*, 2013).

Conditional knockdown of a ribosomal protein small subunit 13 (RPS13) gene, using a tetracycline regulated rps1 promoter, caused *T. gondii* to arrest in G1, lead to depletion of ribosomes and an increase in BAG1 expression during *in vitro* culture (Hutson *et al.*, 2010). Transcriptional analysis of these parasites demonstrated an early stress response; however, the full repertoire of bradyzoite genes was not seen, suggesting that some progression through G1 is likely needed for full differentiation (Hutson *et al.*, 2010). It has been demonstrated that bradyzoite differentiation is associated with eIF2α phosphorylation (Sullivan *et al.*, 2004; Narasimhan *et al.*, 2008), which dampens global translation initiation favouring the preferential of translation of genes (mRNA) that encode proteins responding to stress conditions including transcription factors (e.g. stage associated AP2s) associated with differentiation (Wek *et al.*, 2006; Sullivan and Jeffers, 2012). It is possible the effect seen with this RPS13 knockdown occurs through a similar common pathway. The rps13 promoter region contains putative AP2 binding regions and pull-down studies suggest a complex of GCN5, AP2 and Swi2/Snf2 ATPases bind to this region of rps13 (Hutson *et al.*, 2010); see Chapter 18 for a detailed discussion on gene regulation in *T. gondii.*

When bradyzoites liberated from tissue cysts are used to infect host cells, the tachyzoite-specific antigen SAG1 is expressed within 15 hours of infection prior to any significant cell division by the infecting bradyzoite (Soete and Dubremetz, 1996). By 48 hours vacuoles containing organisms expressing only tachyzoite antigens are evident. In a similar fashion tachyzoites used for infection with stress exposure (pH 8.1) express bradyzoite antigens at 24 hours and vacuoles containing one or two organisms can be found that have bradyzoite antigens. Conversion between these developmental stages is, therefore, a rapid event and commitment to differentiation may occur at the time of or shortly after invasion.

Although a small proportion of replicating parasites (less than 10%) have 1.8 to 2N DNA content (i.e. are in a G2 premitotic state), parasites that co-express bradyzoite marker BAG1 and tachyzoite marker SAG1 are much more likely (approximately 50% of these parasites) to have a G2 premitotic DNA content (Radke *et al.*, 2003). Interestingly, flow cytometry measurements of DNA content of mature bradyzoites isolated from tissue cysts demonstrate that this stage has a 1N DNA content consistent with their being in a quiescent G_0 state (Radke *et al.*, 2003). Overall, it appears that commitment to bradyzoite differentiation probably occurs at a particular point in the cell cycle and that transit through the cell cycle is required for differentiation. These early 'pre-bradyzoites' can continue to replicate, but at some point in the development and maturation of the tissue cyst the fully mature bradyzoites enter a quiescent G_0 state.

15.5 THE STRESS RESPONSE AND SIGNALLING PATHWAYS FOR BRADYZOITE FORMATION

While differentiation is a programmed response, bradyzoite differentiation is also a stress-related response of *T. gondii* to environmental conditions, e.g. the inflammatory response of the host. Many different classical stress response inducing conditions such as temperature, pH, and mitochondrial inhibitors are associated with bradyzoite development *in vitro*. Bradyzoite differentiation probably shares features common to other stress induced differentiation systems such as glucose starvation and hyphae formation in fungi or spore formation in *Dicyostelium* (Thomason *et al.*, 1999; Soderbom and Loomis, 1998). These systems have demonstrated unique proteins related to specific differentiation structures in each organism as well as the utilization of phylogenetically conserved pathways. Many of these signalling pathways involve cyclic nucleotides and kinases as part of the regulatory system in differentiation. It is also interesting to note that abscisic acid responses which result in the production of cyclic ADP ribose (cADPR), such as those seen in plant stress responses, have been demonstrated in *T. gondii* and that inhibition of abscisic acid synthesis by fluridone triggered bradyzoite differentiation (Nagamune *et al.*, 2008).

Studies of micro-organisms including fungi and other protozoa suggest that differentiation involves conserved signalling pathways, such as cyclic nucleotides, which are also involved in the response to stress or nutrient starvation. The effect of cyclic nucleotides on bradyzoite differentiation has been assessed using non-metabolized analogues of cAMP and cGMP as well as forskolin, which stimulates a short pulse of cAMP. In *Dicyostelium* cAMP in the environment is a trigger for differentiation. Interestingly, it has been demonstrated that extracellular adenosine correlates with the number of cysts of *T. gondii* formed *in vivo* in the brain of mice as well as with the number of cysts formed *in vitro* in murine astrocytes (Mahamed *et al.*, 2012).

Bradyzoite induction in response to cyclic nucleotides was measured using either IFA techniques monitoring bradyzoite markers such as the cyst wall lectin *Dolichos biflorus* (DBA) or a bradyzoite promoter reporter parasite (Kirkman *et al.*, 2001; Eaton *et al.*, 2005). These studies demonstrate that cGMP or forskolin can induce bradyzoite formation. Exposure of extracellular *T. gondii* tachyzoites to conditions such as pH 8.1, forskolin or SNP induces bradyzoite formation and results in a transient 3- to 4-fold elevation in cAMP levels that within 30 minutes returned to the cAMP levels comparable to those seen in control parasites incubated in pH 7.1 media. No reproducible changes in cGMP are observed. Expression of photoactivated adenylate cyclase in either *T. gondii* or the host cell demonstrated multiple roles for parasite-derived cAMP in the biology of this organism, including a role for an increase in cAMP in bradyzoite differentiation (Hartmann *et al.*, 2013).

Most of the effects of cAMP within cells can be attributed to regulation of cAMP-dependent protein kinase A activity (PKA). Effects of cGMP can be attributed to stimulation of a cGMP-dependent kinase (PKG), effects upon phosphodiesterases or other signalling molecules. Several kinases that are potentially involved in differentiation have been cloned. These include *T. gondii* PKA1, PKA2 and PKA3 (K. Kim and L. M. Weiss, unpublished data), a glycogen synthase kinase (GSK-3) homologue (Qin *et al.*, 1998) and a unique apicomplexan PKG (Gurnett *et al.*, 2002; Nare *et al.*, 2002; Donald *et al.*, 2002; Donald and Liberator, 2002). Inhibitors of PKA or apicomplexan PKG inhibit replication and induce differentiation (Eaton *et al.*, 2005; Nare *et al.*, 2002). PKA has effects upon metabolism, gene expression and cell cycle. The dissection of exact signalling cascades is complicated by the presence of more than one PKA isoform and multiple phosphodiesterases that are likely to modulate signalling. Further, inhibitors of PKA or PKG may have off target effects upon other kinases or effects upon host cell signalling. It is likely that there are both inhibitory and stimulatory pathways affected by cyclic nucleotide signalling.

As protein phosphorylation has proven to be a major mechanism of regulation of gene expression and integration and amplification of extracellular signals, the presence of highly conserved signalling molecules suggests that many of the pathways identified in other eukaryotes are likely to be preserved in *T. gondii.* A novel mitogen activated protein kinase (TgMAPK-1) related to p38 (a human MAPK involved in the stress response) is also increased during bradyzoite formation (Brumlik *et al.*, 2004). At least two MAPK orthologues are present in the *T. gondii* genome (http://www.toxodb.org) and one of these MAPK may play a role in the signalling processes involved in bradyzoite differentiation.

In eukaryotes the cellular stress response is associated with phosphorylation of the alpha subunit of eIF2 (eukaryotic initiation factor-2) enhancing the translational expression of bZIP proteins such as GCN4 in yeast and ATF4 in mammals (Hinnebusch and Natarajan, 2002). A novel eIF2 protein kinase, designated TgIF2K-A (*T. gondii* initiation factor-2kinase) has been shown to phosphorylate the *T. gondii* translation initiation factor TgIF2α and that phosphorylation is enhanced by conditions known to induce bradyzoite differentiation (Sullivan *et al.*, 2004). It has been shown that inhibiting TgIF2α dephosphorylation induces bradyzoite development and that TgIF2α is maintained in a phosphorylated state in latent cysts (Narasimhan *et al.*, 2008). *T. gondii* has an additional initiation factor-2kinase (TgIF2K-B) that is localized in the cytoplasm and likely responds to cytoplasmic stresses, whereas TgIF2K-A is localized in the endoplasmic reticulum (Narasimhan *et al.*, 2008).

Epigenetic events, such as histone modification, are probably key factors in the differentiation process resulting in bradyzoite formation. As in other organisms, histone acetylation and methylation patterns correlate with gene activation or repression in *T. gondii* (Saksouk *et al.*, 2005). As bradyzoite differentiation proceeds, gene activation markers are associated with region upstream of transcriptionally regulated bradyzoite genes (Saksouk *et al.*, 2005). Epigenetically regulated changes in gene expression and changes in chromatin modifications typically require transit through S phase, as has been observed for induction of expression of bradyzoite markers (Radke *et al.*, 2003). *T. gondii* has three H2A histones, which are differentially regulated during *in vitro* and *in vivo* bradyzoite conditions (Dalmasso *et al.*, 2009). One of the H2A histone variants, H2AX, is localized to silent areas of the genome during bradyzoite development (Dalmasso *et al.*, 2009). Inhibition of histone deacetylase activity with the compound FR235222 induced bradyzoite differentiation and derepressed stage-specific genes (Bougdour *et al.*, 2009). Derivatives of FR235222 may be useful therapeutics for both acute and chronic infections (Maubon *et al.*, 2010).

T. gondi contains two distinct histone acetyltransferase GCN5 proteins (Hettmann and Soldati, 1999; Sullivan and Smith, 2000). TgGCN5-A is enriched at the upstream regions of transcriptionally controlled bradyzoite genes under alkaline pH (Rodrigues-Pousada *et al.*, 2004). TgGCN5-A knockout parasites do not up-regulate known bradyzoite genes and do not recover as efficiently as wild type from alkaline pH conditions (Rodrigues-Pousada *et al.*, 2004). TgSRCAP (*T. gondii* Snf2-related CBP activator protein) is an SWI2/SNF2 family chromatin remodeller whose expression increases during cyst development (Nallani and Sullivan, 2005). A SRCAP homologue, which in other eukaryotes is involved in the regulation of CREB (cAMP response element binding protein), has been identified in *T. gondii* (Sullivan *et al.*, 2003). Using yeast two hybrid analysis TgSRCAP was found to interact with several proteins that could be transcription regulators including TgLZTR (Nallani and Sullivan, 2005). Insertion into TgRSC8, a *T. gondii* homologue of *Saccharomyces cerevisiae* proteins Rsc8p (remodel the structure of chromatin complex subunit 8) and Swi3p (switch/sucrose nonfermentable (SWI/SNF)), caused parasites to display a bradyzoite development defect (Craver *et al.*, 2010). Further analysis of this TgRSC8 mutant showed that while steady state transcript levels of several known bradyzoite genes were significantly reduced in the mutant, other known bradyzoite genes were not affected, highlighting again the complexity of bradyzoite development (Rooney *et al.*, 2011). Overall, a detailed analysis of these signalling pathways will be required in order to understand the regulatory network triggered during bradyzoite formation.

15.6 HEAT SHOCK PROTEINS

There is a significant body of evidence relating heat shock proteins with differentiation in various phyla (Heikkila, 1993a,b). BAG1 (also known as BAG5) has homology to small heat shock proteins and therefore has also been called hsp30 (Bohne *et al.*, 1995; Parmley *et al.*, 1995). Both *BAG1* mRNA and protein (a 28 kDa cytoplasmic antigen) are up-regulated during bradyzoite formation, suggesting transcriptional regulation of its expression. In fact, BAG1 is one of the most abundant bradyzoite-specific genes found in the *T. gondii* bradyzoite ESTs representing about 3% of all bradyzoite-specific clones. *T. gondii* expressing BAG1 are seen within 24 hours of exposure to pH 8.0 or other stress conditions. BAG1 antibody cross reacts with the corresponding gene in *Neospora caninum* and is a marker for differentiation in this related Apicomplexan (Weiss *et al.*, 1999). The *N. caninum* BAG1 homologue has been cloned and characterized (Kobayashi *et al.*, 2012) .

The carboxyl-terminal region of BAG1 has a small heat shock motif most similar to the small heat shock proteins of plants and near the N-terminus is a synapsin Ia like domain that may be involved in the association of this small heat shock protein with proteins during development. Four other small heat shock proteins are present in *T. gondii,* of these hsp20, hsp21, and hsp29 are expressed in both tachyzoites and bradyzoites and hsp28 is specific for tachyzoites (de Miguel, 2005). None of these other small heat shock proteins are associated with bradyzoite differentiation and all are present as multimeric forms in the cytosol of *T. gondii*. BAG1 appears not to form multimeric forms in *T. gondii* (L. M. Weiss, unpublished data).

A homologue of heat shock protein 70 (hsp70) is induced during both the transition from tachyzoite to bradyzoite and from bradyzoite to tachyzoite (Weiss *et al.*, 1998; Lyons and Johnson, 1995, 1998; Miller *et al.*, 1999; Silva *et al.*, 1998). Induction of hsp70 can be demonstrated at both the protein and RNA level. Quercetin, an inhibitor of hsp synthesis, can suppress hsp70 and decrease the ability of pH shock to induce bradyzoite formation (Weiss *et al.*, 1996, 1998). Extracellular *T. gondii* treated with a one hour exposure to pH 8.1 vs. pH 7.1 expresses a 72 kDa inducible hsp70 (detected with mAb C92F3A-5; Stressgen) (Weiss *et al.*, 1998) and this extracellular treatment induces bradyzoite formation. *T. gondii* infected cultures treated with pH 8.1 show 4-fold induction of the hsp70 levels compared to *T. gondii* grown in pH 7.1 treated cells (Weiss *et al.*, 1998; Weiss *et al.*, 1996), which is similar to the magnitude of the hsp70 response demonstrated in *Trypanosoma cruzi*, *Theileria annulata* and *Plasmodium falciparum* (Shiels *et al.*, 1997). A similar increase in hsp70 is seen with *in vivo* cysts during reactivation in a murine model induced by anti-inteferon-gamma (Silva *et al.*, 1998). The relative level of expression of hsp70 by *T. gondii* is also associated with virulence, and RH strain has four copies of a seven-amino acid repeat unit (GGMPGGM) at the C-terminus of its hsp70 compared with five copies in the ME49 strain (Lyons and Johnson, 1998).

In addition to hsp70, hsp90 mRNA and protein levels also increase during bradyzoite differentiation (Echeverria *et al.*, 2005). Fluorescence microscopy demonstrated that in tachyzoites hsp90 is in the cytosol whereas in mature bradyzoites hsp90 is present in both the nucleus and cytosol. In *T. gondii* mutants that are unable to differentiate, hsp90 is only found in the cytoplasm. Geldanamycin, a benzoquinone ansamycin antibiotic capable of binding and disrupting the function of hsp90, blocks conversion both from the tachyzoite to bradyzoite and the bradyzoite to tachyzoite (Echeverria *et al.*, 2005). Hsp90 forms a complex with p23 (a smHSP) with a nuclear and cytosolic distribution in bradyzoites, but no nuclear localization is seen in tachyzoites (Echeverria *et al.*, 2010). In contrast a complex of Hip–Hsp70–Hsp90 was found in the cytoplasm of both tachyzoites and bradyzoites (Echeverria *et al.*, 2010). DnaK–tetratricopeptide (DnaK–TPR; TGME49_202020) is also associated with bradyzoite differentiation and on yeast two hybrid screening was also found to interact with p23 (Ueno *et al.*, 2011). Three hsp40/DnaJ family members (TgME49_115690, TgME49_010430, and TgME49_023950) were also found to be unregulated in parasites exposed to alkaline stress induced bradyzoite differentiation (Sullivan and Jeffers, 2012).

The expression of reporter genes driven by the hsp70 promoter is responsive to conditions that induce bradyzoite formation (Ma *et al.*, 2004). The pH regulated cis-element of the hsp70 promoter maps to the region -420 through -340 from the initial ATG of the hsp70 gene (Ma *et al.*, 2004). At -650 bp from the initial ATG is the sequence AGAGACG, which has been described as a *cis*-acting element that acts as an enhancer in the transcription of several *T. gondii* genes (Mercier *et al.*, 1996) There are a series of nGAAn repeats -385 from the initial ATG, which have similarity to the heat shock element (HSE) described in other eukaryotes (Morimot *et al.*, 1994). A CCGGGG located next to this HSE is similar to the sp1–hsp70 site in the human hsp70 promoter (Morgan, 1989).

In addition, the hsp70 promoter contains several AGGGG or CCCCT regions which are similar to the core region of the STRE (stress response element) described in many eukaryotic genes (Estruch, 2000). Similar STRE and HSE elements are seen in the promoter region of enolase 1, a bradyzoite-specific isoform of enolase (Kibe *et al.*, 2005). In yeast enolase is also known as hsp48, as it is a stress related heat shock protein (Iida and Yahara, 1985). The STRE-binding activity detected in nuclear extract from stress-induced bradyzoites is significantly higher than that from non-stressed tachyzoites (Kibe *et al.*, 2005). Transcription factors responsible for regulation of hsp70 and enolase 1 have not yet been identified although electromobility shift assays (EMSA) suggest that there are specific proteins that recognize the STRE and HSE elements of these genes (Kibe *et al.*, 2005; Ma *et al.*, 2004). Although there is an area of similarity between the *BAG1* promoter region and that of *hsp70* promoter region, oligonucleotides from this *BAG1* upstream region do not compete in EMSA (Ma *et al.*, 2004).

15.7 TRANSCRIPTIONAL CONTROL OF BRADYZOITE GENES

Expression of most bradyzoite-specific proteins is controlled at the level of transcription and therefore numerous techniques that measure steady state mRNA levels have been used to identify bradyzoite-specific genes. Sequencing of cDNA libraries, called expressed sequence tags (ESTs), serial analysis of gene expression (SAGE), microarray analysis and next generation sequencing projects of bradyzoites and tachyzoites have all been used to identify stage-specific genes, and the data from these methods are organized and available at http://www.toxodb.org. Analysis of ESTs was the first large-scale method used to identify genes that are induced during bradyzoite differentiation; the majority of these genes encode proteins of unknown function (Ajioka, 1998; Ajioka *et al.*, 1998; Manger *et al.*, 1998a). Microarrays have been used to analyse bradyzoite development mutants and to identify groups of genes that are coordinately regulated during bradyzoite differentiation (Manger *et al.*, 1998a; Cleary *et al.*, 2002). We and others have also used RNA-seq (data on RNAseq are available on http://toxodb.org) to examine the bradyzoite transcriptome during *in vitro* stress induced bradyzoite development as well as cysts from chronic infection (L. J. Knoll and L. M. Weiss, unpublished data). These analyses have confirmed the induction of previously described bradyzoite genes, i.e. *BAG1*, *LDH2* and *SAG4A*, and identified other potential bradyzoite-specific genes. In addition, genes not known to be regulated during differentiation were shown to have altered mRNA expression (Cleary *et al.*, 2002). SAGE was used to understand the progression of gene expression during bradyzoite development (Radke *et al.*, 2005). This SAGE data suggests that *T. gondii* undergoes a programmed differentiation response similar to *Plasmodium falciparum* with the coordinate regulation of groups of genes during this developmental programme.

Bradyzoite-specific genes have also been identified by using a subtractive cDNA library approach (Yahiaoui *et al.*, 1999). Sixty-five cDNA clones were analysed from a bradyzoite subtractive cDNA library, of these many were identified that were exclusively or preferentially transcribed in bradyzoites. This included homologues of chaperones (mitochondria heat shock protein 60 and T complex protein 1), nitrogen fixation protein, DNA damage repair protein, KE2 protein, phosphatidylinositol synthase, glucose-6-phosphate isomerase and enolase. Another study has used suppression subtractive hybridization to identify novel bradyzoite-specific genes (Friesen *et al.*, 2008). The data from these expression analysis and genetic studies confirm that there are significant numbers of bradyzoite-specific genes involving many complex pathways.

While it is believed that most of the control for the stage specificity for bradyzoite proteins is at transcription, only one bradyzoite-specific promoter element has so far been characterized. Mapping of cis-acting elements upstream of several bradyzoite genes identified a 6–8 bp sequence that controls bradyzoite gene expression under several bradyzoite induction conditions (Behnke *et al.*, 2008). Gel-shift experiments show that this promoter element is bound by parasite proteins to maintain a 'poised' chromatin state throughout the intermediate host life cycle in low passage strains.

Despite the identification of many bradyzoite specific genes, a promoter and *T. gondii* mutants that are unable to differentiate, a unified model for bradyzoite differentiation has not yet been developed. However, several themes have emerged from the available data: (i) tachyzoites and bradyzoites express related genes encoding structural homologues in a mutually exclusive way; (ii) metabolic genes that are stage specific exist suggesting that each stage is metabolically distinct; (iii) stress related differentiation pathways and stress proteins are associated with these stage transitions; (iv) multiple mechanisms to control bradyzoite development likely exist and may be organized in a hierarchy of gene induction; and (v) chromatin remodelling is an important mechanism used to coordinate bradyzoite development (see Chapter 18 for a detailed discussion of gene regulation).

15.8 CYST WALL AND MATRIX ANTIGENS

The development of the tissue cyst wall and matrix are early events in the process of bradyzoite differentiation. Cyst wall proteins are detected at the same time as other bradyzoite specific antigens such as BAG1 (Gross *et al.*, 1996; Zhang *et al.*, 2001). An important function of the cyst wall and matrix is to protect bradyzoites from harsh environmental conditions such as dehydration. In addition, these structures provide a physical barrier to host immune defences. Much of this may be due to carbohydrates present in the cyst wall. The cyst wall is a modification of the bradyzoite parasitophorous vacuole membrane formed by the parasite which is enclosed in host cell membrane, i.e. tissue cysts are intracellular (Ferguson and Hutchison, 1987b; Scholytyseck *et al.*, 1974). On electron microscopy the membrane of the bradyzoite parasitophorous vacuole has a ruffled appearance and is associated with a precipitation of underlying material creating the cyst wall (Fig. 15.3).

The cyst wall is periodic acid-Schiff (PAS) positive, marks with some silver stains, and binds the lectins *Dolichos biflorus* (DBA) and succinylated-wheat germ agglutinin (S-WGA) suggesting that polysaccharides are present in this structure (Boothroyd *et al.*, 1997; Guimaraes *et al.*, 2003; Sims *et al.*, 1988; Dubey *et al.*, 1998). The binding of these lectins to tissue cysts can be inhibited by competition by their specific sugar haptens; N-acetylgalactosamine (GalNAc) for DBA and N-acetylglucosamine (GlcNAc) for S-WGA (Boothroyd *et al.*, 1997). Using a proteomic approach SUMOylation was also demonstrated on the cyst wall formed *in vitro* under alkaline stress (Braun *et al.*, 2009). Treatment with chitinase disrupts the cyst wall and eliminates S-WGA binding suggesting that chitin or a similar polysaccharide may be present in this structure (Boothroyd *et al.*, 1997). It has been demonstrated that alternatively activated macrophages in CNS detect chitin on the brain cyst wall and it is necessary for the clearance of cysts during chronic infection (Nance *et al.*, 2012) .

Binding of the lectin DBA is a marker of cyst wall formation *in vitro*. CST1 is a 250 kDa protein originally recognized as a 116 kDa antigen using two-dimensional electrophoresis (Weiss *et al.*, 1992; Zhang *et al.*, 2001; Tomita *et al.*, 2013). It is an SRS protein with a large mucin-like domain recognized by DBA as well as the monoclonal antibodies 73.18 and SalmonE (Weiss *et al.*, 1992; Zhang *et al.*, 2001) (Fig. 15.4). Lectin overlay experiments of two-dimensional gels suggest that the lectins DBA and S-WGA recognize different antigens, with S-WGA recognizing a 48 kDa antigen (Zhang *et al.*, 2001). CST1 localizes to the granular material in the cyst wall under the limiting membrane of the bradyzoite

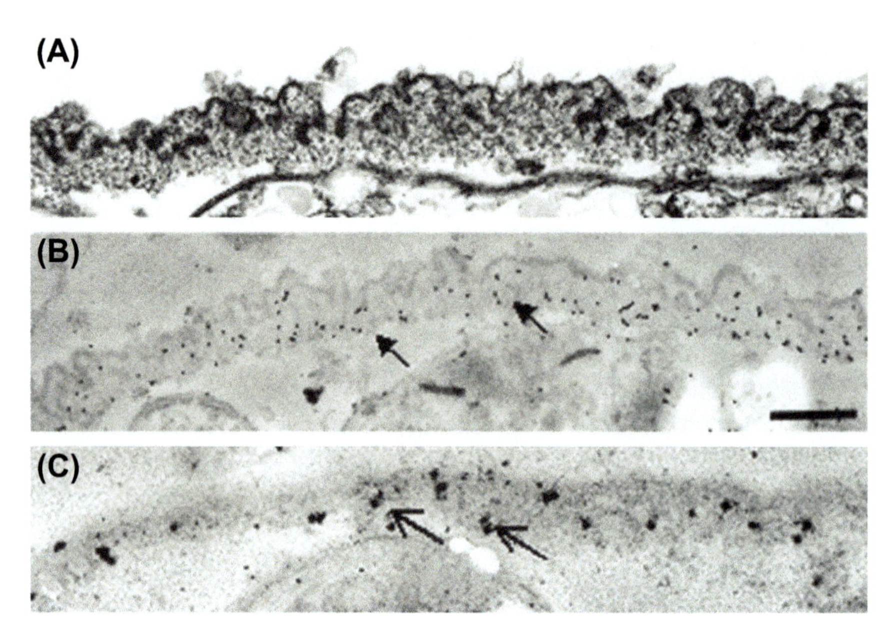

FIGURE 15.3 **Electron microscopy of the cyst wall of *T. gondii* isolated from murine brain.**
A) Transmission electron microscopy of the cyst wall.
B) Immunoelectron microscopy of cyst stained with mAB 73.18 showing labelling of the cyst wall matrix (20 nm gold, arrows).
C) Immunoelectron microscopy of cysts stained with *Dolichos biflorans* lectin demonstrating labelling of the cyst wall matrix (10 nm gold, arrows). Bar is 1 μm.
Reproduced with permission from Zhang et al. (2001).

parasitophorous vacuole (Ferguson, 2004; Zhang *et al.*, 2001). Knockout of *cst1* (Tomita *et al.*, 2013) in the Pru KU80 *T. gondii* background (Fox *et al.*, 2011) eliminates DBA staining of the cyst wall. Such Δ*cst1* parasites are able to form brain cysts in mice, but cysts are much more fragile, rupturing after gentle homogenization of brain. Ultrastructure study of cyst wall by electron microscopy reveals that the lack of CST1 results in the disruption of the cyst wall granular layer. Complementation with full length CST1 rescues the phenotype but the CST1 without the mucin domain did not fully complement the defect, suggesting a role for glycosylation of the mucin domain in cyst wall stability (Tomita *et al.*, 2013).

Protocols for purification of the cyst wall have now been developed and will facilitate identification of component proteins using proteomic approaches (Zhang *et al.*, 2001, 2010). Several glycosyl transferase genes, including a polypeptide N-acetylgalactosaminyltransferase that may be involved in cyst wall formation, have been identified, expressed and characterized in *T. gondii* (Stwora-Wojczyk *et al.*, 2004a; Wojczyk *et al.*, 2003).

Several dense granule proteins (GRA1–8) are known to localize to the parasitophorous vacuolar membrane, the matrix of the vacuole and the tubular structures within the parasitophorous vacuole of *T. gondii* (Ferguson, 2004) (see Fig. 15.5). GRA5 (Lecordier *et al.*, 1993) is found in both tachyzoites and bradyzoites. By immunohistochemistry GRA5 is localized primarily to the cyst wall membrane and not to the granular material under this membrane (Ferguson, 2004; Lane *et al.*, 1996). Less intense staining of the cyst wall membrane is demonstrated by antibodies to GRA1, GRA3 and GRA6 (Ferguson,

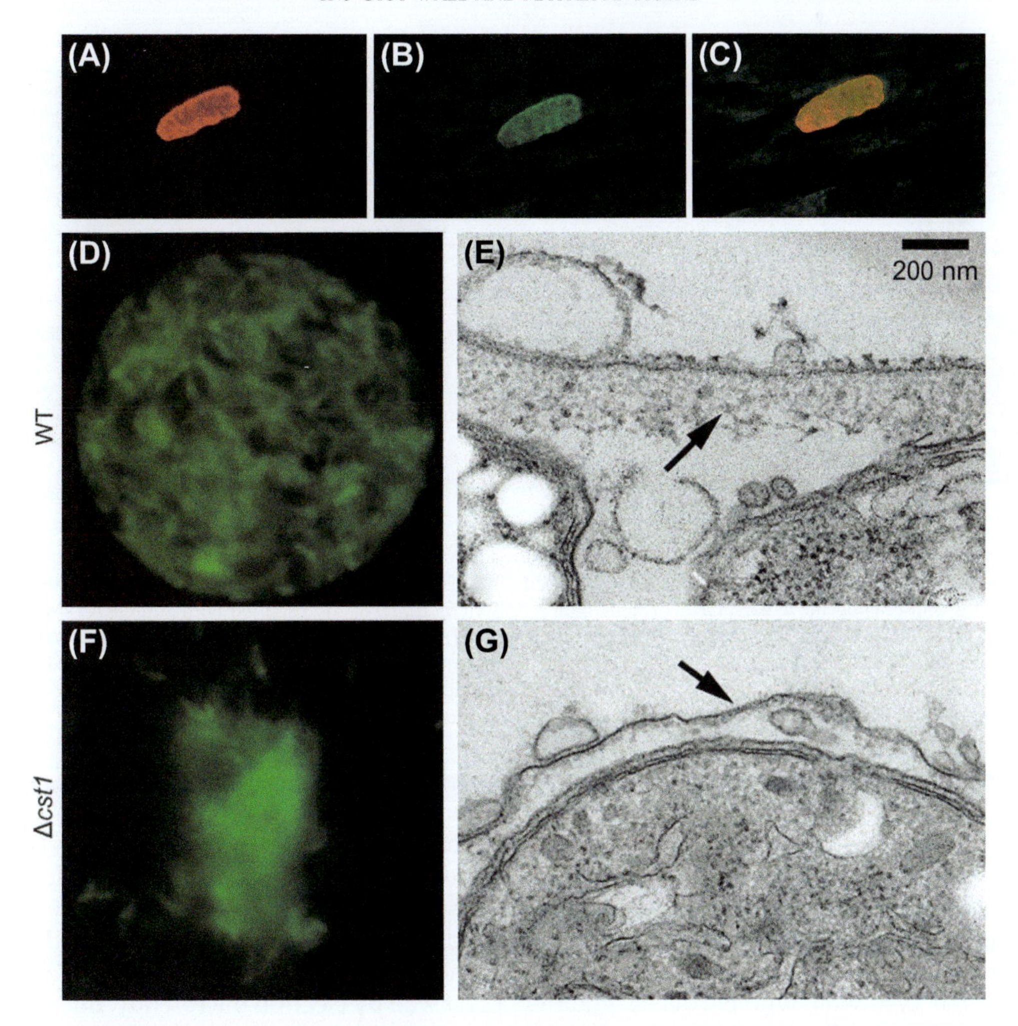

FIGURE 15.4 **Functional characteristic of the cyst wall protein CST1.**
A,B,C) Immunofluorescent assay of HFF infected with EGS strain (Paredes-Santos *et al.*, 2012) of *T. gondii* probed with anti-SalmonE monoclonal antibody (red) DBA lectin (green). Panel A is taken with a rhodamine filter set, Panel B with a fluorescein filter set and Panel C with a filter set that allows simultaneous viewing of red and green fluorescent labels. This confirms that the SalmonE antibody and DBA both label the cyst wall (bradyzoite parasitophorous vacuole membrane). Data indicate that SalmonE recognizes cyst wall protein 1 (CST1) (Tomita *et al.*, 2013).
D, E) Fluorescent microscopy (D) and electron micrograph (E) of purified mouse brain cysts from wild type (WT) PruKU80 *T. gondii.* The PruΔku80 strain expresses GFP using the bradyzoite-specific LDH2 promoter. These cysts do not rupture with the standard purification technique and have a normal appearing cyst wall with amorphous material (see arrow) lining the cyst wall membrane. Bar is 200 nm.
F, G) Fluorescent microscopy (F) and electron micrograph (G) of purified mouse brain cysts from Δ*cst1* PruKU80 *T. gondii.* This demonstrates the 'fragile cyst' phenotype of the *cst1* knock-out and that this knock-out has an abnormal cyst wall morphology on TEM, lacking the amorphous material that is usually present under the cyst wall membrane as indicated by the arrow.
Figures D, E, F, G are adapted with permission from Tomita et al. (2013).

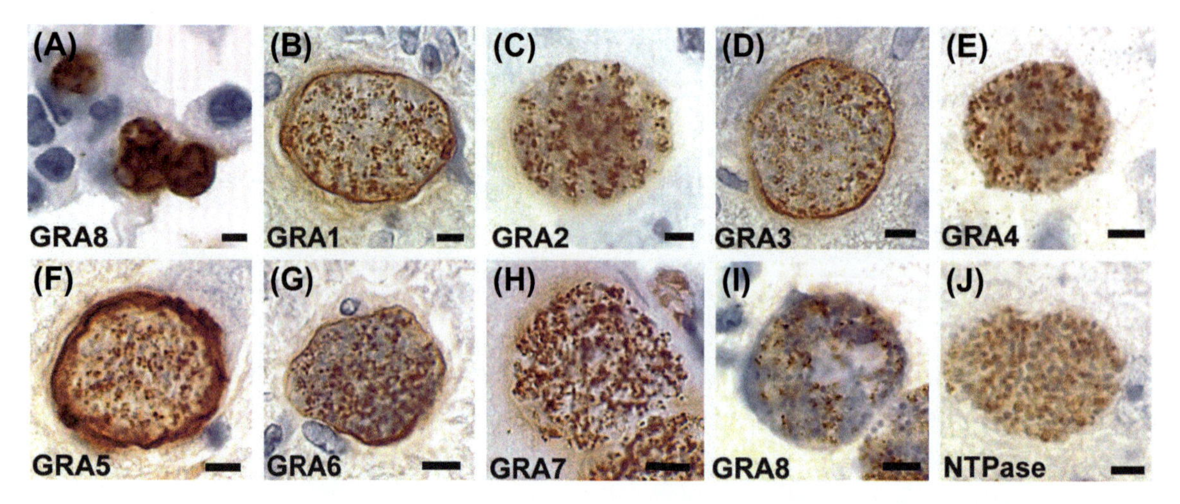

FIGURE 15.5 **Distribution of dense granule proteins (GRA) in bradyzoite parasitophorous vacuoles.** Representative sections from the lung of an acutely infected mouse containing tachyzoites (A) and the brain of a chronically infected mouse containing tissue cysts (B–J). All of the sections were stained for dense granules proteins using the peroxidase technique and corresponding antisera specific to each GRA protein. Bars represent 5 μm. A) Section stained with anti-GRA8 showing strong labelling of the parasitophorous vacuole containing tachyzoites. B–J) Tissue cysts demonstrate relatively uniform staining of the dense granules within the bradyzoites (B–I). There was variable staining of the cyst wall with positive staining for GRA1 (B), GRA3 (D), GRA5 (F), GRA6 (G) and GRA7 (H), but little staining with GRA2 (C), GRA4 (E), GRA8 (I) or NTPase (J). *Courtesy of DJP Ferguson. Reprinted with permission from Ferguson (2004).*

2004; Torpier *et al.*, 1993). Knockouts of either GRA4 or GRA6 and especially the dual GRA4/GRA6 knockout have been shown to have reduced brain cyst burdens in C57BL/6 mice (Fox *et al.*, 2011). Given the localization of most of the dense granule proteins to the parasitophorous vacuole, it seems likely that many components of the cyst wall will be dense granule proteins, perhaps with new carbohydrate modifications or bradyzoite-specific glycoproteins secreted from dense granules. In support of this idea is that rat monoclonal antibody CC2 reacts with a 115 kDa antigen in bradyzoites, but recognizes a 40 kDa protein in tachyzoites (Gross *et al.*, 1995). Deletion of a serine protease inhibitor (TgPI1) that is secreted to the parasitophorous vacuole via the dense granules causes increased frequency of bradyzoite switching and an increased parasite burden during acute infection (Pszenny *et al.*, 2012)

To search for new bradyzoite-specific secreted proteins, transcriptome analysis of *in vitro* and *in vivo* bradyzoites, combined with programmes to identify signal peptides, led to the prediction of over 100 bradyzoite-secreted proteins (Buchholz *et al.*, 2011). The first two proteins identified by this screen, bradyzoite pseudokinase 1 (BPK1) and microneme-adhesive repeat (MAR)-domain containing protein 4 (MCP4), localized to the cyst wall by electron microscopy, but the authors were unable to determine if the proteins originated in the dense granules. Subsequent study showed that deletion of BPK1 did not influence *in vivo* and *in vitro* bradyzoite differentiation but significantly reduced mouse oral infectivity and pepsin-acid resistance (Buchholz *et al.*, 2013). Similar transcriptome analysis of the *T. gondii* sexual stages has identified several putative oocyst wall proteins (Fritz *et al.*, 2012b) .

MAG1 was originally identified as a 65 kDa protein expressed in the cyst matrix that was not expressed in tachyzoites (Parmley *et al.*, 1994). RT-PCR data indicate that mRNA for

MAG1 is present in both tachyzoites and bradyzoites (Parmley *et al.*, 1994). It has now been demonstrated that MAG1 is also expressed in tachyzoites and secreted into the parasitophorous vacuole, albeit less abundantly than in bradyzoites (Ferguson and Parmley, 2002). Antibody to recombinant MAG1 reacts with extracellular material in the cyst matrix and to a lesser extent with the cyst wall, but not with the surface or cytoplasm of bradyzoites. The dense granule proteins GRA1, GRA2, GRA3, GRA5, GRA6 and GRA7 are present in the matrix of both tachyzoites and bradyzoites, but GRA4 and GRA8 appear to be expressed higher in tachyzoites.

The bradyzoite-specific rhoptry protein, BRP1, is secreted into the parasitophorous vacuole on invasion (Schwarz *et al.*, 2005). Knockout of the BRP1 gene did not affect the ability of the parasite to invade or to form cysts or bradyzoites (Schwarz *et al.*, 2005). Monoclonal antibodies E7B2 (Lane *et al.*, 1996), 93.2 (L. M. Weiss, unpublished) and 1.23.29 (L. M. Weiss, unpublished) also recognize matrix antigens, but the corresponding genes have not been identified (see Table 15.1).

15.9 SURFACE ANTIGENS

Most *T. gondii* surface antigens are members of a gene family with similarity to SAG1 (Boothroyd *et al.*, 1998; Manger *et al.*, 1998b; Lekutis *et al.*, 2001; Wasmuth *et al.*, 2012). It is not clear why so many family members exist (with 182 annotated genes), although some are pseudogenes (Wasmuth *et al.*, 2012) because antigenic variation as described for *Trypanosomes* or *Plasmodium* has not been described. All of these SAGs appear to be attached to the plasma membrane by a similar glycolipid anchor. While SAG3 is found in all life stages, several of these surface antigens appear to be stage specific. SAG1 (now SRS29B), SRS1–SRS3, SAG2A and SAG2B are expressed in tachyzoites and SAG2C, SAG2D, SAG4A, SAG5A (SAG 5.1) and SRS9 in bradyzoites (Kim and Boothroyd, 2005; Jung *et al.*, 2004). See Wasmuth *et al.* for details of the proposed renaming and stage expression of the SRS family (Wasmuth *et al.*, 2012). These antigens may be involved in persistence of tissue cysts in their hosts and the relative lack of an immune response to tissue cysts (Kim and Boothroyd, 2005). Both SAG3 (p43) and SAG1 (p30) have been implicated in adhesion to host cells. Disruption of SAG3 leads to 2-fold decreased adhesion of parasites (Dzierszinski *et al.*, 2000), but disruption of SAG1 in RH strain, but not in PLK, results in parasites that are more invasive (Mineo *et al.*, 1993). Either the SAG1 or SAG3 disruption in RH strain results in a decrease in virulence (Dzierszinski *et al.*, 2000). More recently dual disruption of SAG1 (p30, SRS29B) and SAG2 (SRS34A, SAG2A, p22) showed up-regulation of SRS29C (SRS2, p35) and decreased virulence, with overexpression of SRS29C in the virulent RH strain also associated with decreased virulence (Wasmuth *et al.*, 2012).

Bradyzoite-specific recombinant 4 (BSR4) was discovered in a promoter trap to isolate developmentally regulated genes and found to be a surface antigen that interacts with the p36 mAb T84A12 (Knoll and Boothroyd, 1998a). It was later found that SRS9, a highly similar gene immediately downstream of BSR4, is also a bradyzoite-specific surface antigen that reacts with mAb T84A12 (Van *et al.*, 2007). These genes demonstrate a restriction fragment length polymorphism between ME49 (PLK; Type II) and CEP (Type III) strains, which correlates with the lack of mAb T84A12 binding to CEP strain. Antibodies that recognize both BSR4 and SRS9 show that one or both of these proteins are expressed on the merozoite as well as the bradyzoite surface (Ferguson, 2004). While SRS9 is transcriptionally up-regulated in bradyzoites (Cleary *et al.*, 2002), RNA levels of BSR4 are similar in tachyzoites and bradyzoites suggesting that post-transcriptional regulation of BSR4 occurs via some unknown mechanism.

Disruption of BSR4 and SRS9 did not result in a tissue culture bradyzoite development phenotype (Knoll and Boothroyd, 1998a). However, deletion of SRS9 created parasites that were similar to wild type in size and abundance during early chronic infection (three to four weeks), but were not able to persist in the brains of mice (Kim *et al.*, 2007).

Surface antigens similar to SAG1 with a conserved placement of 12 cysteines include SAG1, SAG3, BSR4, and SRS1−4 (SAG-related sequences), SAG5, SAG5.1 and SAG 5.2. A second group of SAG1 members have a less consistent conservation of cysteine spacing. This group includes SAG2A (SAG2 or p22), SAG2B, SAG2C, SAG2D and SRS28 (SporoSAG). SAG2C and SAG2D are only detected on *in vivo* bradyzoites not *in vitro* bradyzoites (Lekutis *et al.*, 2001). Similar to SRS9, parasites deleted in the four-gene surface antigen cluster SAG2CDXY have a chronic infection persistence defect (Saeij *et al.*, 2008). SAG4A (p18) is an 18 kDa protein surface protein transcriptionally regulated during bradyzoite development (Odberg-Ferragut *et al.*, 1996). Crystal structures of SAG1 family members from different life cycle stages, SAG1 (tachyzoite; He *et al.*, 2002), BSR4 (bradyzoite; Crawford *et al.*, 2009) and SRS28 (sporozoite, aka SporoSAG; Crawford *et al.*, 2009), show overall similarly but potentially biologically relevant structural diversity.

A new family of 31 GPI-anchored surface antigens was recently uncovered that are completely unrelated to the SAG1 proteins (Pollard *et al.*, 2008). This new family of proteins was named SAG-unrelated surface antigens (SUSA). Analysis of the single nucleotide polymorphism density shows that the SUSA genes are the most polymorphic within the *T. gondii* genome, highlighting that they are likely under immune pressure. One family member so far, SUSA1, has been shown to be bradyzoite specific (Pollard *et al.*, 2008). Like the SAG1 superfamily, the function(s) of the SUSA family remain to be determined.

15.10 METABOLIC DIFFERENCES BETWEEN BRADYZOITES AND TACHYZOITES

It is probable that the energy metabolism of bradyzoites is different from tachyzoites given the location of bradyzoites in a thick walled vacuole and their slower growth rate. An unusual and unexplained feature of *T. gondii* differentiation is the presence of stage-specific differences in the activity and isoforms of several glycolytic enzymes. It is known that tachyzoites utilize the glycolytic pathway with the production of lactate as their major source of energy. Functional mitochondria with a TCA cycle exist in tachyzoites and are thought to contribute to energy production. While both tachyzoites and bradyzoites utilize the glycolytic pathway for energy, data suggests that bradyzoites lack a functional TCA cycle and respiratory chain (Denton *et al.*, 1996).

Overall, the regulation and activation of glycolysis appear to be different in *T. gondii* than many other eukaryotes (Saito *et al.*, 2002; Maeda *et al.*, 2003; Denton *et al.*, 1996). Lactate dehydrogenase (LDH) and pyruvate kinase activity are higher in bradyzoites than in tachyzoites while PP_i-phosphofructokinase activity is higher in tachyzoites than bradyzoites (Denton *et al.*, 1996). These data suggest that bradyzoite energy metabolism may be dependent on the catabolism of amylopectin, which is present in bradyzoites and essentially absent in tachyzoites, to lactate. The bradyzoite-specific glycolytic isoenzymes are resistant to acidic pH suggesting that bradyzoites are resistant to the acidification resulting from the accumulation of the glycolytic products from amylopectin catabolism to lactate.

Lactate dehydrogenase 2 (LDH2) is a 35 kDa cytoplasmic antigen with a PI of 7.0 that is expressed in bradyzoites but not in tachyzoites (Yang and Parmley 1995, 1997). Its expression appears be transcriptionally regulated because

LDH2 mRNA is detectable by RT–PCR only in bradyzoites. The tachyzoite isoform, LDH1, has also been cloned that is 71.4% identical to LDH2. While the LDH activity of tachyzoite and bradyzoite extracts differs, there are no significant differences in the activity of recombinant enzyme activities of LDH1 or LDH2 (Yang and Parmley, 1995, 1997). Despite its apparent stage specificity, attempts to disrupt *LDH2* have failed (Singh *et al.*, 2002); however, knockdown of LDH2 has been achieved using dsRNA (Al-Anouti *et al.*, 2004). When LDH2 expression was down-regulated bradyzoite differentiation and growth were impaired (Al-Anouti *et al.*, 2004).

Two stage-specific enolases have also been cloned and characterized; consistent with the hypothesis that utilization of the glycolytic pathway is different in tachyzoites compared to bradyzoites (Yahiaoui *et al.*, 1999; Manger *et al.*, 1998a). Enolase catalyses the conversion of 2-phosphoglycerate to phosphoenolpyruvate. In yeast enolase is known to be a stress response protein, i.e. hsp48 (Iida and Yahara, 1985). ENO2, the tachyzoite form, has 3-fold higher specific activity than ENO1, the bradyzoite enzyme, but both have similar Michaelis constants (Km) (Dzierszinski *et al.*, 2001). Surprisingly, polyclonal antisera to each isoform do not cross react despite the similarity of these two isoforms (Dzierszinski *et al.*, 1999, 2001). Both isoforms are found localized to the nucleus in dividing cells, but in late bradyzoites, which are quiescent, ENO1 is cytoplasmic (Ferguson and Parmley, 2002; Ferguson *et al.*, 2002). The significance of these observations is unknown but it is possible that some glycolytic enzymes may have alternate regulatory functions that are not yet fully understood.

An obvious difference between tachyzoites and bradyzoites is the presence of cytosolic granules of amylopectin that are composed of glucose polymers (Guerardel *et al.*, 2005). Structural studies of amylopectin have revealed it to be a plant-like amylopectin with predominantly (α1–4) linkages which is most similar to the semi-crystalline floridean starch accumulated by red algae (Coppin *et al.*, 2005). Amylopectin granules disappear from bradyzoites when they transform into tachyzoites during cell culture (Coppin *et al.*, 2003). While not seen in tachyzoites, amylopectin is present in the sexual cycle in the cat intestine in macrogametes, and persists during oocyst formation and in sporozoites. Merozoites lack amylopectin. Amylopectin is believed to be a carbohydrate store for the bradyzoite or sporozoite during long periods of quiescence and nutrient deprivation. Candidate genes for enzymes involved in amylopectin breakdown and synthesis have been identified (Coppin *et al.*, 2003).

A bradyzoite-specific P type ATPase whose mRNA and protein are preferentially expressed in bradyzoites has been characterized and localizes in punctate pattern to the region of the plasma membrane (Holpert *et al.*, 2001). A second P-type ATPase also may be preferentially expressed in bradyzoites as judged by RT–PCR of steady-state mRNA although mRNA can also be detected in tachyzoites. TgSRCAP is a SWI2/SNF2 whose expression increases during cyst development (Nallani and Sullivan, 2005). By yeast two hybrid, TgSRCAP also binds to several metabolic enzymes (Nallani and Sullivan, 2005), further underscoring that metabolic regulation that is probably different in tachyzoites and bradyzoites.

It is not known if the metabolic changes between bradyzoites and tachyzoites cause differentiation or are a consequence of the differentiation process. One hypothesis is that monitoring for nutrient deprivation, a type of stress response, might serve as the sensor for differentiation in *T. gondii*. However, no sensors that respond to any environmental change have yet been identified in *T. gondii*.

15.11 GENETIC STUDIES ON BRADYZOITE BIOLOGY

Bradyzoite- and tachyzoite-specific promoter regions have been utilized to create reagents for the study of differentiation *in vitro* (Gross, 1996). Reporter genes such as chloramphenicol acetyltransferase (CAT), β-galactosidase and luciferase are useful for mapping promoter activity and the definition of minimal promoter sequences. Studies using these markers have confirmed the stage-specific expression of SAG1 as a tachyzoite marker and LDH2 and BAG1 as bradyzoite makers (Yang and Parmley, 1997; Bohne *et al.*, 1997; Ma *et al.*, 2004; Eaton *et al.*, 2006). Recent studies on the plant-like ApiAP2 transcription factors have defined several that are stage-specific and probably involved in stage specific gene expression (Radke *et al.*, 2013; Walker *et al.*, 2012). Stage-specific reporter constructs have been used in both transient and stable transfection assays, and green fluorescent protein (GFP) reporter parasites have proven useful for the identification of bradyzoite differentiation mutants by FACS (Matrajt *et al.*, 2002; Singh *et al.*, 2002).

One would expect that the knockout of a bradyzoite-specific gene would be feasible as growth should occur in the tachyzoite stage even if bradyzoite development is prevented. This strategy has been applied to the bradyzoite-specific gene BAG1/hsp30 (BAG5) (Bohne *et al.*, 1998; Zhang *et al.*, 1999). A *bag1* knockout was created using HGXPRT as a selectable marker in an HGXPRT^{neg} PLK strain of *T. gondii*. Another *bag1* knockout was performed using CAT as a selectable marker in a clone of PLK strain that had been passaged through mice to ensure it made cysts at the start of the study. Cyst formation *in vitro* and *in vivo* occurred in both knockouts, indicating that BAG1 is not essential for cyst formation. Zhang *et al.* however, found that the number of cysts formed *in vivo* in CD1 mice was reduced in the *bag1* knockouts. Complementation resulted in the production of similar numbers of cysts *in vivo* as the wild type PLK strain (Zhang *et al.*, 1999). When parasites were grown in sodium nitroprusside (SNP) the *bag1* knockout grew faster than PLK. This result may be a difference in transition rate from the rapidly growing tachyzoite to the slowly growing bradyzoite stage in this *bag1* knockout. The decrease in cyst formation is a relatively subtle phenotype, which was not observed when *BAG1* was disrupted in the HGXPRT^{neg} PLK strain background (not passaged in mice prior to the knockout) and cyst formation was tested in highly susceptible C57BL/6 mice (Bohne *et al.*, 1998). The capacity to convert from tachyzoite to bradyzoite is a key feature for *T. gondii* persistence in the host, and thus it is likely that multiple genes with redundant functions are involved in this process. Small heat shock proteins, such as BAG1, have been postulated to act as specialized chaperones for enzymes such as glutathione reductase during differentiation, but the exact function of such small heat shock proteins is unknown. It is possible that the other small heat shock proteins in *T. gondii* can partially compensate for a lack of BAG1.

Promoter trapping has been an effective technique for the identification of genes induced during bradyzoite differentiation. These studies used a promoterless hypoxanthine–xanthine–guanine phosphoribosyltranferase (HGXPRT) gene with 6-thoxanthine (6-TX) or 8-azaguanine (8-AzaH) as negative selection and mycophenolic acid with xanthine (MPA-X) as a positive selection (Bohne *et al.*, 1997; Knoll and Boothroyd, 1998a, 1998b). Selection works by growing transfected parasites at pH 7.0 in the presence of 6-TX which inhibits the growth of all organisms that have the HGXPRT gene on a constitutive or tachyzoite promoter. To confirm bradyzoite specificity this population of organisms is then exposed to pH 8.0 and MPA-X and only parasites that express HGXPRT (i.e. those with a bradyzoite or constitutive promoter in front of the HGXPRT gene) will survive. It should be noted that this approach can be 'leaky', depending on the concentrations of 6-TX and MPA-X used. Nonetheless, when the 6-TX and MPA-X selections are repeated several times one will enrich the population for organisms with HGXPRT under the control of bradyzoite-specific gene promoters. Using this approach additional bradyzoite-specific promoters (Donald and Roos, unpublished; cited in Matrajt *et al.*, 2002) and eight bradyzoite-specific recombinant (BSR) strains were obtained (Knoll and Boothroyd, 1998a).

Additional genetic strategies have been developed to identify mutants unable to undergo bradyzoite differentiation (Matrajt *et al.*, 2002; Singh *et al.*, 2002). Singh *et al.* generated point mutants in a *LDH2-GFP* Prugnaiud (Type II) background, which expresses GFP under the control of the bradyzoite-specific promoter LDH2, to obtain mutants with an altered ability to transform into bradyzoites (Singh *et al.*, 2002). Parasites unable to differentiate were identified by FACS enrichment of GFP-negative parasites when these organisms were exposed to bradyzoite inducing conditions.

Matrajt *et al.* also utilized insertional mutagenesis of an engineered stable line expressing a bradyzoite-specific pT7-HGXPRT cassette in UPRT-deficient RH (Type I) parasites (Matrajt *et al.*, 2002). Earlier studies had demonstrated that RH UPRT disruptants differentiate into bradyzoites under conditions of CO_2 starvation (Bohne and Roos, 1997). The pT7-HGXPRT stable line was obtained by rounds of negative and positive selection, alternating 6-TX (tachyzoite conditions) with MPA-X (bradyzoite conditions) selection. The result was a cell line where HGXPRT was highly regulated by differentiation conditions. Insertional mutagenesis was then performed in this pT7-HGXPRT line using DHFR cassettes that earlier were shown to have a high frequency of non-homologous insertion (Donald and Roos, 1993). An inability to differentiate into bradyzoites was detected by the inability of the disruptant to express HGXPRT.

Both groups were able to demonstrate that these differentiation mutants were unable to make bradyzoites at the same efficiency as the parental strains and both groups demonstrated global defects in expression of previously characterized markers as determined by immunofluorescence and microarray analysis. Due to technical difficulties the exact mutations could not be identified. The insertional mutants identified by Matrajt *et al.* (2002) were similar to the *bag1* knockout (Zhang *et al.*, 1999) and had more rapid growth under bradyzoite inducing conditions than seen in wild-type parasites (Matrajt *et al.*, 2002). Microarray analysis of these mutants identified classes of genes, including a 14-3-3 homologue, a PISTLRE kinase and a probable vacuolar ATPase, whose expression was decreased in the differentiation mutants (Singh *et al.*, 2002; Matrajt *et al.*, 2002). Other genes of interest identified included an AP2 factor (AP2XII-6) and an oocyst wall protein (Lescault *et al.*, 2010). Microarray studies of these mutants also suggest that there was a hierocracy of gene expression during bradyzoite differentiation (Singh *et al.*, 2002), which is consistent with the developmental programme seen in *Plasmodia*. Further microarray analysis of bradyzoite development mutants has also highlighted an extracellular tachyzoite state that is distinct from intracellular tachyzoites and bradyzoites (Lescault *et al.*, 2010).

A FACS strategy, similar to that of Singh *et al.* (2002) and Matrajt *et al.* (2002), combined with insertional mutagenesis has been useful to identify genes necessary for bradyzoite development. Mutant TBD-6 switched to bradyzoites with about half the efficiency of wild-type parasites (Vanchinathan *et al.*, 2005). TBD-6 had an insertion 164 bp upstream of the transcription start site of a gene encoding a zinc finger protein (ZFP1), which causes an up-regulation of the mRNA. The phenotype of decreased bradyzoite development could be replicated by directed integration into the same upstream region. ZFP1 is targeted to the parasite nucleolus by

CCHC motifs, down-regulated in wild type bradyzoites and appears to be essential as multiple attempts yielded no knockouts (Vanchinathan *et al.*, 2005). Mutants TBD-5 and TBD-8 had independent insertions into a pseudouridine synthetase homologue (PUS1) (Anderson *et al.*, 2009). Knockout of *PUS1* had a bradyzoite development phenotype similar to the TBD-8 mutant and could be complemented with the genomic *PUS1* allele. During animal infections, PUS1 deletions had increased mortality during acute infection and higher cyst burdens during chronic infection (Anderson *et al.*, 2009).

Insertional mutagenesis of signature-tag containing parasites was used to isolate mutants defective in the establishment of a chronic infection (Frankel *et al.*, 2007; Knoll *et al.*, 2001). Immunofluorescence microscopy screening of this library for mutant parasites defective in their ability to form an intact cyst wall led to the identification of nine additional bradyzoite development mutants (Craver *et al.*, 2010). In this study, a proteophosphoglycan was found to localize to the parasitophorous vacuole space and enhance cyst wall formation. Two independently identified differentiation deficient mutants (Singh *et al.*, 2002; Frankel *et al.*, 2007) had disruption on a same locus that contains non-coding RNA (*Tg-ncRNA-1*), which encodes a REP-derived small RNA (rdsRNA). Complementation of these mutants with WT Tg-ncRNA-1 locus rescued the differentiation defect (Patil *et al.*, 2012). This suggests that non-coding RNAs probably have a role in the process of differentiation. Overall, these insertional mutagenesis strategies have discovered proteins involved in bradyzoite development that would not have been predicted through other strategies.

15.12 SUMMARY

Investigations into bradyzoite biology and the differentiation of tachyzoites into bradyzoites have been accelerated by the development of *in vitro* techniques to study and produce bradyzoites as well as by the genetic tools that exist for the manipulation of *T. gondii*. The completion of the *Toxoplasma* genome along with development and integration of the expression database with the genome database provide additional important tools for the study of bradyzoite development (http://www.toxodb.org).

Unfortunately, the genetic triggers and sensors for the differentiation response have yet to be identified. Many of the features of this differentiation are reminiscent of epigenetic phenomena described in other systems. Bradyzoite differentiation appears to be coupled with a slowing of the cell cycle. It is likely that many signals can result in appropriate secondary signals that induce bradyzoite formation. The development of bradyzoites appears to be a stress mediated differentiation response that leads to metabolic adaptations. It is abundantly clear that transcription of a whole set of bradyzoite-specific genes occurs during differentiation. These gene products include metabolic enzymes, surface antigens, secretory antigens (including rhoptry proteins) and cyst wall components.

Studies of the mechanism(s) by which development is triggered and coordinated should eventually result in the identification of new therapeutic strategies for the control toxoplasmosis. These advances may ultimately result in the radical cure of infection by eliminating the latent *T. gondii* cyst stage in tissue. Furthermore, genetic strategies that prevent cyst formation may also prove useful in the development of vaccine strains of this pathogenic Apicomplexan.

Acknowledgements

This work was supported by the American Heart Association 0840059N (LJK) and NIH grants AI39454 and AI095094 (LMW), and T32-AI070117 (TT). We thank the members of our laboratories for their dedication and our colleagues in the *Toxoplasma* community for stimulating discussions and sharing unpublished data.

References

Ajioka, J.W., 1998. Toxoplasma gondii: ESTs and gene discovery. Int. J. Parasitol. 28, 1025–1031.

Ajioka, J.W., Boothroyd, J.C., Brunk, B.P., Hehl, A., Hillier, L., Manger, I.D., Marra, M., Overton, G.C., Roos, D.S., Wan, K.L., Waterston, R., Sibley, L.D., 1998. Gene discovery by EST sequencing in Toxoplasma gondii reveals sequences restricted to the Apicomplexa. Genome. Res. 8, 18–28.

Al-Anouti, F., Tomavo, S., Parmley, S., Ananvoranich, S., 2004. The expression of lactate dehydrogenase is important for the cell cycle of Toxoplasma gondii. J. Biol. Chem. 279, 52300–52311.

Anderson, M.Z., Brewer, J., Singh, U., Boothroyd, J.C., 2009. A pseudouridine synthase homologue is critical to cellular differentiation in Toxoplasma gondii. Eukaryot. Cell 8, 398–409.

Aramini, J.J., Stephen, C., Dubey, J.P., Engelstoft, C., Schwantje, H., Ribble, C.S., 1999. Potential contamination of drinking water with Toxoplasma gondii oocysts. Epidemiol. Infect. 122, 305–315.

Behnke, M.S., Radke, J.B., Smith, A.T., Sullivan Jr., W.J., White, M.W., 2008. The transcription of bradyzoite genes in Toxoplasma gondii is controlled by autonomous promoter elements. Mol. Microbiol. 68, 1502–1518.

Behnke, M.S., Wootton, J.C., Lehmann, M.M., Radke, J.B., Lucas, O., Nawas, J., Sibley, L.D., White, M.W., 2010. Coordinated progression through two subtranscriptomes underlies the tachyzoite cycle of Toxoplasma gondii. PLoS One 5. e12354.

Blewett, D.A., Miller, J.K., Harding, J., 1983. Simple technique for the direct isolation of toxoplasma tissue cysts from foetal ovine brain. Vet. Rec. 112, 98–100.

Bohne, W., Gross, U., Ferguson, D.J., Heesemann, J., 1995. Cloning and characterization of a bradyzoite-specifically expressed gene (hsp30/bag1) of Toxoplasma gondii, related to genes encoding small heat-shock proteins of plants. Mol. Microbiol. 16, 1221–1230.

Bohne, W., Heesemann, J., Gross, U., 1993. Induction of bradyzoite-specific Toxoplasma gondii antigens in gamma interferon-treated mouse macrophages. Infect. Immun. 61, 1141–1145.

Bohne, W., Heesemann, J., Gross, U., 1994. Reduced replication of Toxoplasma gondii is necessary for induction of bradyzoite-specific antigens: a possible role for nitric oxide in triggering stage conversion. Infect. Immun. 62, 1761–1767.

Bohne, W., Hunter, C.A., White, M.W., Ferguson, D.J., Gross, U., Roos, D.S., 1998. Targeted disruption of the bradyzoite-specific gene BAG1 does not prevent tissue cyst formation in Toxoplasma gondii. Mol. Biochem. Parasitol. 92, 291–301.

Bohne, W., Roos, D.S., 1997. Stage-specific expression of a selectable marker in Toxoplasma gondii permits selective inhibition of either tachyzoites or bradyzoites. Mol. Biochem. Parasitol. 88, 115–126.

Bohne, W., Wirsing, A., Gross, U., 1997. Bradyzoite-specific gene expression in Toxoplasma gondii requires minimal genomic elements. Mol. Biochem. Parasitol. 85, 89–98.

Boothroyd, J.C., Black, M., Bonnefoy, S., Hehl, A., Knoll, L.J., Manger, I.D., Ortega-Barria, E., Tomavo, S., 1997. Genetic and biochemical analysis of development in Toxoplasma gondii. Philos. Trans. R. Soc. Lond. B. Biol. Sci. 352, 1347–1354.

Boothroyd, J.C., Hehl, A., Knoll, L.J., Manger, I.D., 1998. The surface of Toxoplasma: more and less. Int. J. Parasitol. 28, 3–9.

Bougdour, A., Maubon, D., Baldacci, P., Ortet, P., Bastien, O., Bouillon, A., Barale, J.C., Pelloux, H., Menard, R., Hakimi, M.A., 2009. Drug inhibition of HDAC3 and epigenetic control of differentiation in Apicomplexa parasites. J. Exp. Med. 206, 953–966.

Bowie, W.R., King, A.S., Werker, D.H., Isaac-Renton, J.L., Bell, A., Eng, S.B., Marion, S.A., 1997. Outbreak of toxoplasmosis associated with municipal drinking water. The BC Toxoplasma Investigation Team. Lancet 350, 173–177.

Braun, L., Cannella, D., Pinheiro, A.M., Kieffer, S., Belrhali, H., Garin, J., Hakimi, M.A., 2009. The small ubiquitin-like modifier (SUMO)-conjugating system of Toxoplasma gondii. Int. J. Parasitol. 39, 81–90.

Brown, C.R., David, C.S., Khare, S.J., Mcleod, R., 1994. Effects of human class I transgenes on Toxoplasma gondii cyst formation. J. Immunol. 152, 4537–4541.

Brown, C.R., Hunter, C.A., Estes, R.G., Beckmann, E., Forman, J., David, C., Remington, J.S., Mcleod, R., 1995. Definitive identification of a gene that confers resistance against Toxoplasma cyst burden and encephalitis. Immunology 85, 419–428.

Brown, C.R., Mcleod, R., 1990. Class I MHC genes and CD8+ T cells determine cyst number in Toxoplasma gondii infection. J. Immunol. 145, 3438–3441.

Brumlik, M.J., Wei, S., Finstad, K., Nesbit, J., Hyman, L.E., Lacey, M., Burow, M.E., Curiel, T.J., 2004. Identification of a novel mitogen-activated protein kinase in Toxoplasma gondii. Int. J. Parasitol. 34, 1245–1254.

Buchholz, K.R., Bowyer, P.W., Boothroyd, J.C., 2013. Bradyzoite pseudokinase 1 is crucial for efficient oral infectivity of the Toxoplasma tissue cyst. Eukaryot. Cell.

Buchholz, K.R., Fritz, H.M., Chen, X., Durbin-Johnson, B., Rocke, D.M., Ferguson, D.J., Conrad, P.A., Boothroyd, J.C., 2011. Identification of tissue cyst wall components by transcriptome analysis of in vivo and in vitro Toxoplasma gondii bradyzoites. Eukaryot. Cell 10, 1637–1647.

Cleary, M.D., Singh, U., Blader, I.J., Brewer, J.L., Boothroyd, J.C., 2002. Toxoplasma gondii asexual development: identification of developmentally regulated genes and distinct patterns of gene expression. Eukaryot. Cell 1, 329–340.

Coppin, A., Dzierszinski, F., Legrand, S., Mortuaire, M., Ferguson, D., Tomavo, S., 2003. Developmentally regulated biosynthesis of carbohydrate and storage polysaccharide during differentiation and tissue cyst formation in Toxoplasma gondii. Biochimie 85, 353–361.

Coppin, A., Varre, J.S., Lienard, L., Dauvillee, D., Guerardel, Y., Soyer-Gobillard, M.O., Buleon, A., Ball, S., Tomavo, S., 2005. Evolution of plant-like crystalline storage polysaccharide in the protozoan parasite Toxoplasma gondii argues for a red alga ancestry. J. Mol. Evol. 60, 257–267.

Cornelissen, A.W., Overdulve, J.P., Hoenderboom, J.M., 1981. Separation of Isospora (Toxoplasma) gondii cysts and cystozoites from mouse brain tissue by continuous density-gradient centrifugation. Parasitology 83, 103–108.

Craver, M.P., Rooney, P.J., Knoll, L.J., 2010. Isolation of Toxoplasma gondii development mutants identifies a potential proteophosphoglycan that enhances cyst wall formation. Mol. Biochem. Parasitol. 169, 120–123.

Crawford, J., Grujic, O., Bruic, E., Czjzek, M., Grigg, M.E., Boulanger, M.J., 2009. Structural characterization of the bradyzoite surface antigen (BSR4) from Toxoplasma gondii, a unique addition to the surface antigen glycoprotein 1-related superfamily. J. Biol. Chem. 284, 9192–9198.

Dalmasso, M.C., Onyango, D.O., Naguleswaran, A., Sullivan Jr., W.J., Angel, S.O., 2009. Toxoplasma H2A variants reveal novel insights into nucleosome composition and functions for this histone family. J. Mol. Biol. 392, 33–47.

De Champs, C., Imbert-Bernard, C., Belmeguenai, A., Ricard, J., Pelloux, H., Brambilla, E., Ambroise-Thomas, P., 1997. Toxoplasma gondii: in vivo and in vitro cystogenesis of the virulent RH strain. J. Parasitol. 83, 152–155.

De Miguel, N., Echeverria, P.C., Angel, S.O., 2005. Characterization and stage-specific expression analysis of members of Toxoplasma gondii alpha-crystallin/sHsp family. Eight International Congress on Toxoplasmosis, 85.

Denton, H., Roberts, C.W., Alexander, J., Thong, K.W., Coombs, G.H., 1996. Enzymes of energy metabolism in the bradyzoites and tachyzoites of Toxoplasma gondii. FEMS Microbiol. Lett. 137, 103–108.

Donald, R.G., Allocco, J., Singh, S.B., Nare, B., Salowe, S.P., Wiltsie, J., Liberator, P.A., 2002. Toxoplasma gondii cyclic GMP-dependent kinase: chemotherapeutic targeting of an essential parasite protein kinase. Eukaryot. Cell 1, 317–328.

Donald, R.G., Liberator, P.A., 2002. Molecular characterization of a coccidian parasite cGMP dependent protein kinase. Mol. Biochem. Parasitol. 120, 165–175.

Donald, R.G., Roos, D.S., 1993. Stable molecular transformation of Toxoplasma gondii: a selectable dihydrofolate reductase-thymidylate synthase marker based on drug- resistance mutations in malaria. Proc. Natl. Acad. Sci. U. S. A. 90, 11703–11707.

Dubey, J.P., 1997. Bradyzoite-induced murine toxoplasmosis: stage conversion, pathogenesis, and tissue cyst formation in mice fed bradyzoites of different strains of Toxoplasma gondii [published erratum appears in J. Eukaryot. Microbiol. 1998 May-Jun 45(3), 367]. J. Eukaryot. Microbiol. 44, 592–602.

Dubey, J.P., Lindsay, D.S., Speer, C.A., 1998. Structures of Toxoplasma gondii tachyzoites, bradyzoites, and sporozoites and biology and development of tissue cysts. Clin. Microbiol. Rev. 11, 267–299.

Dubey, J.P., Speer, C.A., Shen, S.K., Kwok, O.C., Blixt, J.A., 1997. Oocyst-induced murine toxoplasmosis: life cycle, pathogenicity, and stage conversion in mice fed Toxoplasma gondii oocysts. J. Parasitol. 83, 870–882.

Dzierszinski, F., Mortuaire, M., Cesbron-Delauw, M.F., Tomavo, S., 2000. Targeted disruption of the glycosylphosphatidylinositol-anchored surface antigen SAG3 gene in Toxoplasma gondii decreases host cell adhesion and drastically reduces virulence in mice. Mol. Microbiol. 37, 574–582.

Dzierszinski, F., Mortuaire, M., Dendouga, N., Popescu, O., Tomavo, S., 2001. Differential expression of two plant-like enolases with distinct enzymatic and antigenic properties during stage conversion of the protozoan parasite Toxoplasma gondii. J. Mol. Biol. 309, 1017–1027.

Dzierszinski, F., Nishi, M., Ouko, L., Roos, D.S., 2004. Dynamics of Toxoplasma gondii differentiation. Eukaryot. Cell. 3, 992–1003.

Dzierszinski, F., Popescu, O., Toursel, C., Slomianny, C., Yahiaoui, B., Tomavo, S., 1999. The protozoan parasite Toxoplasma gondii expresses two functional plant-like glycolytic enzymes. Implications for evolutionary origin of apicomplexans. J. Biol. Chem. 274, 24888–24895.

Eaton, M.S., Weiss, L.M., Kim, K., 2006. Cyclic nucleotide kinases and tachyzoite-bradyzoite transition in Toxoplasma gondii. Int. J. Parasitol. 36 (1), 107–114.

Echeverria, P.C., Figueras, M.J., Vogler, M., Kriehuber, T., De Miguel, N., Deng, B., Dalmasso, M.C., Matthews, D.E., Matrajt, M., Haslbeck, M., Buchner, J., Angel, S.O., 2010. The Hsp90 co-chaperone p23 of Toxoplasma gondii: Identification, functional analysis and dynamic interactome determination. Mol. Biochem. Parasitol. 172, 129–140.

Echeverria, P.C., Matrajt, M., Harb, O.S., Zappia, M.P., Costas, M.A., Roos, D.S., Dubremetz, J.F., Angel, S.O., 2005. Toxoplasma gondii Hsp90 is a potential drug target whose expression and subcellular localization are developmentally regulated. J. Mol. Biol. 350, 723–734.

Estruch, F., 2000. Stress-controlled transcription factors, stress-induced genes and stress tolerance in budding yeast. FEMS Microbiol. Rev. 24, 469–486.

Eller, B., Datta, A., Lynn, B.C., Sinai, A.P. 2012. A case for host involvement in the glycosylation of Toxoplasma tissue cysts. 12th International Workshops on Opportunistic Pathogens August 2012, Tarrytown, NY, Abstract T46, page 64.

Fagard, R., Van Tan, H., Creuzet, C., Pelloux, H., 1999. Differential development of toxoplasma gondii in neural cells. Parasitol. Today 15, 504–507.

Ferguson, D.J., 2004. Use of molecular and ultrastructural markers to evaluate stage conversion of Toxoplasma gondii in both the intermediate and definitive host. Int. J. Parasitol. 34, 347–360.

Ferguson, D.J., Huskinson-Mark, J., Araujo, F.G., Remington, J.S., 1994. A morphological study of chronic cerebral toxoplasmosis in mice: comparison of four different strains of Toxoplasma gondii. Parasitol. Res. 80, 493–501.

Ferguson, D.J., Parmley, S.F., 2002. Toxoplasma gondii MAG1 protein expression. Trends. Parasitol. 18, 482.

Ferguson, D.J., Parmley, S.F., Tomavo, S., 2002. Evidence for nuclear localisation of two stage-specific isoenzymes of enolase in Toxoplasma gondii correlates with active parasite replication. Int. J. Parasitol. 32, 1399–1410.

Ferguson, D.J.P., Huchiso, W.M., Pettersen, E., 1989. Tissue cyst rupture in mice chronically infected with Toxoplasma gondii. An immunocytochemical and ultrastructural study. Parasitol. Res. 73, 599–603.

Ferguson, D.J.P., Hutchison, W.M., 1987a. The host-parasite relationship of Toxoplasma gondii in the brains of chronically infected mice. Virchows Arch. A. 411, 39–43.

Ferguson, D.J.P., Hutchison, W.M., 1987b. An ultrastructural study of the early development and tissue cyst formation of Toxoplasma gondii in the brains of mice. Parasitol. Res. 73, 483–491.

Fortier, B., Coignard-Chatain, C., Soete, M., Dubremetz, J.F., 1996. [Structure and biology of Toxoplasma gondii bradyzoites]. C. R. Seances. Soc. Biol. Fil. 190, 385–394.

Fox, B.A., Bzik, D.J., 2002. De novo pyrimidine biosynthesis is required for virulence of Toxoplasma gondii. Nature 415, 926–929.

Fox, B.A., Falla, A., Rommereim, L.M., Tomita, T., Gigley, J.P., Mercier, C., Cesbron-Delauw, M.F., Weiss, L.M., Bzik, D.J., 2011. Type II Toxoplasma gondii KU80 knockout strains enable functional analysis of genes required for cyst development and latent infection. Eukaryot. Cell 10, 1193–1206.

Fox, B.A., Gigley, J.P., Bzik, D.J., 2004. Toxoplasma gondii lacks the enzymes required for de novo arginine biosynthesis and arginine starvation triggers cyst formation. Int. J. Parasitol. 34, 323–331.

Frankel, M.B., Mordue, D.G., Knoll, L.J., 2007. Discovery of parasite virulence genes reveals a unique regulator of chromosome condensation 1 ortholog critical for efficient nuclear trafficking. Proc. Natl. Acad. Sci. U. S. A. 104, 10181–10186.

Frenkel, J.K., Dubey, J.P., Hoff, R.L., 1976. Loss of stages after continuous passage of Toxoplasma gondii and Besnoitia jellisoni. J. Protozool. 23, 421–424.

Frenkel, J.K., Escajadillo, A., 1987. Cyst rupture as a pathogenic mechanism of toxoplasmic encephalitis. Am. J. Trop. Med. Hyg. 36, 517–522.

Freyre, A., 1995. Separation of toxoplasma cysts from brain tissue and liberation of viable bradyzoites. J. Parasitol. 81, 1008–1010.

Friesen, J., Fleige, T., Gross, U., Bohne, W., 2008. Identification of novel bradyzoite-specific Toxoplasma gondii genes with domains for protein-protein interactions by suppression subtractive hybridization. Mol. Biochem. Parasitol. 157, 228–232.

Fritz, H., Barr, B., Packham, A., Melli, A., Conrad, P.A., 2012a. Methods to produce and safely work with large numbers of Toxoplasma gondii oocysts and bradyzoite cysts. J. Microbiol. Methods 88, 47–52.

Fritz, H.M., Buchholz, K.R., Chen, X., Durbin-Johnson, B., Rocke, D.M., Conrad, P.A., Boothroyd, J.C., 2012b. Transcriptomic analysis of toxoplasma development reveals many novel functions and structures specific to sporozoites and oocysts. PLoS One 7. e29998.

Fux, B., Nawas, J., Khan, A., Gill, D.B., Su, C., Sibley, L.D., 2007. Toxoplasma gondii strains defective in oral transmission are also defective in developmental stage differentiation. Infect. Immun. 75, 2580–2590.

Gross, U., 1996. Toxoplasma gondii. Springer, Berlin.

Gross, U., Bohne, W., Luder, C.G., Lugert, R., Seeber, F., Dittrich, C., Pohl, F., Ferguson, D.J., 1996. Regulation of developmental differentiation in the protozoan parasite Toxoplasma gondii. J. Eukaryot. Microbiol. 43, 114S–116S.

Gross, U., Bohne, W., Windeck, T., Heesemann, J., 1992. New views on the pathogenesis and diagnosis of toxoplasmosis. Immun. Infekt. 20, 151–155.

Gross, U., Bormuth, H., Gaissmaier, C., Dittrich, C., Krenn, V., Bohne, W., Ferguson, D.J., 1995. Monoclonal rat antibodies directed against Toxoplasma gondii suitable for studying tachyzoite-bradyzoite interconversion in vivo. Clin. Diagn. Lab. Immunol. 2, 542–548.

Gubbels, M.J., Lehmann, M., Muthalagi, M., Jerome, M.E., Brooks, C.F., Szatanek, T., Flynn, J., Parrot, B., Radke, J., Striepen, B., White, M.W., 2008. Forward genetic analysis of the apicomplexan cell division cycle in Toxoplasma gondii. PLoS Pathog. 4, e36.

Guerardel, Y., Leleu, D., Coppin, A., Lienard, L., Slomianny, C., Strecker, G., Ball, S., Tomavo, S., 2005. Amylopectin biogenesis and characterization in the protozoan parasite Toxoplasma gondii, the intracellular development of which is restricted in the HepG2 cell line. Microbes Infect. 7, 41–48.

Guimaraes, E.V., De Carvalho, L., Barbosa, H.S., 2003. An alternative technique to reveal polysaccharides in Toxoplasma gondii tissue cysts. Mem. Inst. Oswaldo. Cruz. 98, 915–917.

Gurnett, A.M., Liberator, P.A., Dulski, P.M., Salowe, S.P., Donald, R.G., Anderson, J.W., Wiltsie, J., Diaz, C.A., Harris, G., Chang, B., Darkin-Rattray, S.J., Nare, B., Crumley, T., Blum, P.S., Misura, A.S., Tamas, T., Sardana, M.K., Yuan, J., Biftu, T., Schmatz, D.M., 2002. Purification and molecular characterization of cGMP-dependent protein kinase from Apicomplexan parasites. A novel chemotherapeutic target. J. Biol. Chem. 277, 15913–15922.

Halonen, S.K., Chiu, F., Weiss, L.M., 1998a. Effect of cytokines on growth of Toxoplasma gondii in murine astrocytes. Infect. Immun. 66, 4989–4993.

Halonen, S.K., Lyman, W.D., Chiu, F.C., 1996. Growth and development of Toxoplasma gondii in human neurons and astrocytes. J. Neuropathol. Exp. Neurol. 55, 1150–1156.

Halonen, S.K., Weiss, L.M., Chiu, F.C., 1998b. Association of host cell intermediate filaments with Toxoplasma gondii cysts in murine astrocytes in vitro. Int. J. Parasitol. 28, 815–823.

Hartmann, A., Arroyo-Olarte, R.D., Imkeller, K., Hegemann, P., Lucius, R., Gupta, N., 2013. Optogenetic modulation of an adenylate cyclase in Toxoplasma gondii demonstrates a requirement of the parasite cAMP for host-cell invasion and stage differentiation. J. Biol. Chem. 288, 13705-13717.

He, X.L., Grigg, M.E., Boothroyd, J.C., Garcia, K.C., 2002. Structure of the immunodominant surface antigen from the Toxoplasma gondii SRS superfamily. Nat. Struct. Biol. 9, 606–611.

Heikkila, J.J., 1993a. Heat shock gene expression and development. I. An overview of fungal, plant, and poikilothermic animal developmental systems. Dev. Genet. 14, 1–5.

Heikkila, J.J., 1993b. Heat shock gene expression and development. II. An overview of mammalian and avian developmental systems. Dev. Genet. 14, 87–91.

Hettmann, C., Soldati, D., 1999. Cloning and analysis of a Toxoplasma gondii histone acetyltransferase: a novel chromatin remodelling factor in Apicomplexan parasites. Nucleic Acids Res. 27, 4344–4352.

Hinnebusch, A.G., Natarajan, K., 2002. Gcn4p, a master regulator of gene expression, is controlled at multiple levels by diverse signals of starvation and stress. Eukaryot. Cell 1, 22–32.

Hoff, R.L., Dubey, J.P., Behbehani, A.M., Frenkel, J.K., 1977. Toxoplasma gondii cysts in cell culture: new biologic evidence. J. Parasitol. 63, 1121–1124.

Hogan, M.J., Yoneda, C., Feeney, L., Zweigart, P., Lewis, A., 1960. Morphology and culture of Toxoplasma. Arch. Ophthalmol. 64, 655–667.

Holpert, M., Luder, C.G., Gross, U., Bohne, W., 2001. Bradyzoite-specific expression of a P-type ATPase in Toxoplasma gondii. Mol. Biochem. Parasitol. 112, 293–296.

Hutson, S.L., Mui, E., Kinsley, K., Witola, W.H., Behnke, M.S., El Bissati, K., Muench, S.P., Rohrman, B., Liu, S.R., Wollmann, R., Ogata, Y., Sarkeshik, A., Yates 3rd, J.R., McLeod, R., 2010. T. gondii RP promoters & knockdown reveal molecular pathways associated with proliferation and cell-cycle arrest. PLoS One 5, e14057.

Iida, H., Yahara, I., 1985. Yeast heat shock protein of Mr 48,000 is an isomer of enolase. Nature 315, 688–690.

Jacobs, L., Remington, J., 1960. The resistance of the encysted form of Toxoplasma gondii. J. Parasitol. 46, 11–21.

Jerome, M.E., Radke, J.R., Bohne, W., Roos, D.S., White, M.W., 1998. Toxoplasma gondii bradyzoites form spontaneously during sporozoite-initiated development. Infect. Immun. 66, 4838–4844.

Jones, T.C., Bienz, K.A., Erb, P., 1986. In vitro cultivation of Toxoplasma gondii cysts in astrocytes in the presence of gamma interferon. Infect. Immun. 51, 147–156.

Jung, C., Lee, C.Y., Grigg, M.E., 2004. The SRS superfamily of Toxoplasma surface proteins. Int. J. Parasitol. 34, 285–296.

Khan, F., Tang, J., Qin, C.L., Kim, K., 2002. Cyclin-dependent kinase TPK2 is a critical cell cycle regulator in Toxoplasma gondii. Mol. Microbiol. 45, 321–332.

Kibe, M.K., Coppin, A., Dendouga, N., Oria, G., Meurice, E., Mortuaire, M., Madec, E., Tomavo, S., 2005. Transcriptional regulation of two stage-specifically expressed genes in the protozoan parasite Toxoplasma gondii. Nucleic Acids Res. 33, 1722–1736.

Kim, S.K., Boothroyd, J.C., 2005. Stage-specific expression of surface antigens by Toxoplasma gondii as a mechanism to facilitate parasite persistence. J. Immunol. 174, 8038–8048.

Kim, S.K., Karasov, A., Boothroyd, J.C., 2007. Bradyzoite-specific surface antigen SRS9 plays a role in maintaining Toxoplasma gondii persistence in the brain and in host control of parasite replication in the intestine. Infect. Immun. 75, 1626–1634.

Kimball, A.C., Kean, B.H., Fuchs, F., 1974. Toxoplasmosis: risk variations in New York city obstetric patients. Am. J. Obstet. Gynecol. 119, 208–214.

Kirkman, L.A., Weiss, L.M., Kim, K., 2001. Cyclic nucleotide signalling in Toxoplasma gondii bradyzoite differentiation. Infect. Immun. 69, 148–153.

Knoll, L.J., Boothroyd, J.C., 1998a. Isolation of developmentally regulated genes from Toxoplasma gondii by a gene trap with the positive and negative selectable marker hypoxanthine-xanthine-guanine phosphoribosyltransferase. Mol. Cell. Biol. 18, 807–814.

Knoll, L.J., Boothroyd, J.C., 1998b. Molecular Biology's Lessons about Toxoplasma Development: Stage-Specific Homologs. Parasitol. Today 14, 490–493.

Knoll, L.J., Furie, G.L., Boothroyd, J.C., 2001. Adaptation of signature-tagged mutagenesis for Toxoplasma gondii: a negative screening strategy to isolate genes that are essential in restrictive growth conditions. Mol. Biochem. Parasitol. 116, 11–16.

Kobayashi, T., Narabu, S., Yanai, Y., Hatano, Y., Ito, A., Imai, S., Ike, K., 2012. Gene Cloning and Characterization of the Protein Encoded by the Neospora Caninum Bradyzoite-Specific Antigen Gene Bag1. J. Parasitol.

Lane, A., Soete, M., Dubremetz, J.F., Smith, J.E., 1996. Toxoplasma gondii: appearance of specific markers during the development of tissue cysts in vitro. Parasitol. Res. 82, 340–346.

Lecordier, L., Mercier, C., Torpier, G., Tourvieille, B., Darcy, F., Liu, J.L., Maes, P., Tartar, A., Capron, A., Cesbron-Delauw, M.F., 1993. Molecular structure of a Toxoplasma gondii dense granule antigen (GRA 5) associated with the parasitophorous vacuole membrane. Mol. Biochem. Parasitol. 59, 143–153.

Lei, Y., Davey, M., Ellis, J., 2005. Autofluorescence of Toxoplasma gondii and Neospora caninum cysts in vitro. J. Parasitol. 91, 17–23.

Lekutis, C., Ferguson, D.J., Grigg, M.E., Camps, M., Boothroyd, J.C., 2001. Surface antigens of Toxoplasma gondii: variations on a theme. Int. J. Parasitol. 31, 1285–1292.

Lemgruber, L., Lupetti, P., De Souza, W., Vommaro, R.C., 2010. New details on the fine structure of the rhoptry of Toxoplasma gondii. Microsc. Res. Tech.

Lescault, P.J., Thompson, A.B., Patil, V., Lirussi, D., Burton, A., Margarit, J., Bond, J., Matrajt, M., 2010. Genomic data reveal Toxoplasma gondii differentiation mutants are also impaired with respect to switching into a novel extracellular tachyzoite state. PLoS One 5, e14463.

Lindsay, D.S., Dubey, J.P., Blagburn, B.L., Toivio-Kinnucan, M., 1991. Examination of tissue cyst formation by Toxoplasma gondii in cell cultures using bradyzoites, tachyzoites, and sporozoites. J. Parasitol. 77, 126–132.

Lindsay, D.S., Mitschler, R.R., Toivio-Kinnucan, M.A., Upton, S.J., Dubey, J.P., Blagburn, B.L., 1993a. Association of host cell mitochondria with developing Toxoplasma gondii tissue cysts. Am. J. Vet. Res. 54, 1663–1667.

Lindsay, D.S., Toivio-Kinnucan, M.A., Blagburn, B.L., 1993b. Ultrastructural determination of cystogenesis by various Toxoplasma gondii isolates in cell culture. J. Parasitol. 79, 289–292.

Luft, B.J., Remington, J.S., 1992. Toxoplasmic encephalitis in AIDS. Clin. Infect. Dis. 15, 211–222.

Lyons, R.E., Johnson, A.M., 1995. Heat shock proteins of Toxoplasma gondii. Parasite. Immunol. 17, 353–359.

Lyons, R.E., Johnson, A.M., 1998. Gene sequence and transcription differences in 70 kDa heat shock protein correlate with murine virulence of Toxoplasma gondii. Int. J. Parasitol. 28, 1041–1051.

Ma, Y.F., Zhang, Y., Kim, K., Weiss, L.M., 2004. Identification and characterisation of a regulatory region in the Toxoplasma gondii hsp70 genomic locus. Int. J. Parasitol. 34, 333–346.

Maeda, T., Saito, T., Oguchi, Y., Nakazawa, M., Takeuchi, T., Asai, T., 2003. Expression and characterization of recombinant pyruvate kinase from Toxoplasma gondii tachyzoites. Parasitol. Res. 89, 259–265.

Mahamed, D.A., Mills, J.H., Egan, C.E., Denkers, E.Y., Bynoe, M.S., 2012. CD73-generated adenosine facilitates Toxoplasma gondii differentiation to long-lived tissue cysts in the central nervous system. Proc. Natl. Acad. Sci. U. S. A. 109, 16312–16317.

Manger, I.D., Hehl, A., Parmley, S., Sibley, L.D., Marra, M., Hillier, L., Waterston, R., Boothroyd, J.C., 1998a. Expressed sequence tag analysis of the bradyzoite stage of Toxoplasma gondii: identification of developmentally regulated genes. Infect. Immun. 66, 1632–1637.

Manger, I.D., Hehl, A.B., Boothroyd, J.C., 1998b. The surface of Toxoplasma tachyzoites is dominated by a family of glycosylphosphatidylinositol-anchored antigens related to SAG1. Infect. Immun. 66, 2237–2244.

Matrajt, M., Donald, R.G., Singh, U., Roos, D.S., 2002. Identification and characterization of differentiation mutants in the protozoan parasite Toxoplasma gondii. Mol. Microbiol. 44, 735–747.

Matsubayashi, H., Akao, S., 1963. Morphological studies on the development of the toxoplasma cyst. Am. J. Trop. Med. Hyg. 12, 321–333.

Maubon, D., Bougdour, A., Wong, Y.S., Brenier-Pinchart, M.P., Curt, A., Hakimi, M.A., Pelloux, H., 2010. Activity of the histone deacetylase inhibitor FR235222 on Toxoplasma gondii: inhibition of stage conversion of the parasite cyst form and study of new derivative compounds. Antimicrob. Agents Chemother. 54, 4843–4850.

McAuley, J., Boyer, K.M., Patel, D., Mets, M., Swisher, C., Roizen, N., Wolters, C., Stein, L., Stein, M., Schey, W., 1994. Early and longitudinal evaluations of treated infants and children and untreated historical patients with congenital toxoplasmosis: the Chicago Collaborative Treatment Trial. Clin. Infect. Dis. 18, 38–72.

McHugh, T.D., Gbewonyo, A., Johnson, J.D., Holliman, R.E., Butcher, P.D., 1993. Development of an in vitro model of Toxoplasma gondii cyst formation. FEMS Microbiol. Lett. 114, 325–332.

McLeod, R., Brown, C., Mack, D., 1993. Immunogenetics influence outcome of Toxoplasma gondii infection. Res. Immunol. 144, 61–65.

McLeod, R., Mack, D., Brown, C., 1991. Toxoplasma gondii – new advances in cellular and molecular biology. Exp. Parasitol. 72, 109–121.

McLeod, R., Skamene, E., Brown, C.R., Eisenhauer, P.B., Mack, D.G., 1989. Genetic regulation of early survival and cyst number after peroral Toxoplasma gondii infection of A x B/B x A recombinant inbred and B10 congenic mice. J. Immunol. 143, 3031–3034.

Mercier, C., Lefebvre-Van Hende, S., Garber, G.E., Lecordier, L., Capron, A., Cesbron-Delauw, M.F., 1996. Common cis-acting elements critical for the expression of several genes of Toxoplasma gondii. Mol. Microbiol. 21, 421–428.

Miller, C.M., Smith, N.C., Johnson, A.M., 1999. Cytokines, nitric oxide, heat shock proteins and virulence in Toxoplasma. Parasitol. Today 15, 418–422.

Miller, M.A., Gardner, I.A., Kreuder, C., Paradies, D.M., Worcester, K.R., Jessup, D.A., Dodd, E., Harris, M.D., Ames, J.A., Packham, A.E., Conrad, P.A., 2002. Coastal freshwater runoff is a risk factor for Toxoplasma gondii infection of southern sea otters (Enhydra lutris nereis). Int. J. Parasitol. 32, 997–1006.

Mineo, J.R., Mcleod, R., Mack, D., Smith, J., Khan, I.A., Ely, K.H., Kasper, L.H., 1993. Antibodies to Toxoplasma gondii major surface protein (SAG-1, P30) inhibit infection of host cells and are produced in murine intestine after peroral infection. J. Immunol. 150, 3951–3964.

Morgan, W.D., 1989. Transcription factor Sp1 binds to and activates a human hsp70 gene promoter. Mol. Cell. Biol. 9, 4099–4104.

Morimot, R.I., Tissieres, A., Georgopoulos, C. (Eds.), 1994. The Biology of Heat Shock Proteins and Molecular Chaperones. Cold Spring Harbor Press, New York.

Nagamune, K., Hicks, L.M., Fux, B., Brossier, F., Chini, E.N., Sibley, L.D., 2008. Abscisic acid controls calcium-dependent egress and development in Toxoplasma gondii. Nature 451, 207–210.

Nallani, K.C., Sullivan Jr., W.J., 2005. Identification of proteins interacting with Toxoplasma SRCAP by yeast two-hybrid screening. Parasitol. Res. 95, 236–242.

Nance, J.P., Vannella, K.M., Worth, D., David, C., Carter, D., Noor, S., Hubeau, C., Fitz, L., Lane, T.E., Wynn, T.A., Wilson, E.H., 2012. Chitinase dependent control of protozoan cyst burden in the brain. PLoS Pathog. 8, e1002990.

Narasimhan, J., Joyce, B.R., Naguleswaran, A., Smith, A.T., Livingston, M.R., Dixon, S.E., Coppens, I., Wek, R.C., Sullivan Jr., W.J., 2008. Translation regulation by eukaryotic initiation factor-2 kinases in the development of latent cysts in Toxoplasma gondii. J. Biol. Chem. 283, 16591–16601.

Nare, B., Allocco, J.J., Liberator, P.A., Donald, R.G., 2002. Evaluation of a cyclic GMP-dependent protein kinase inhibitor in treatment of murine toxoplasmosis: gamma interferon is required for efficacy. Antimicrob. Agents Chemother. 46, 300–307.

Odberg-Ferragut, C., Soete, M., Engels, A., Samyn, B., Loyens, A., Van Beeumen, J., Camus, D., Dubremetz, J.F., 1996. Molecular cloning of the Toxoplasma gondii sag4 gene encoding an 18 kDa bradyzoite specific surface protein. Mol. Biochem. Parasitol. 82, 237–244.

Parmley, S.F., Weiss, L.M., Yang, S., 1995. Cloning of a bradyzoite-specific gene of Toxoplasma gondii encoding a cytoplasmic antigen. Mol. Biochem. Parasitol. 73, 253–257.

Parmley, S.F., Yang, S., Harth, G., Sibley, L.D., Sucharczuk, A., Remington, J.S., 1994. Molecular characterization of a 65-kilodalton Toxoplasma gondii antigen expressed abundantly in the matrix of tissue cysts. Mol. Biochem. Parasitol. 66, 283–296.

Patil, V., Lescault, P.J., Lirussi, D., Thompson, A.B., Matrajt, M., 2012. Disruption of the Expression of a Non-Coding RNA Significantly Impairs Cellular Differentiation in Toxoplasma gondii. Int. J. Mol. Sci. 14, 611–624.

Pavesio, C.E., Chiappino, M.L., Setzer, P.Y., Nichols, B.A., 1992. Toxoplasma gondii: differentiation and death of bradyzoites. Parasitol. Res. 78, 1–9.

Pettersen, E.K., 1988. Resistance to avirulent Toxoplasma gondii in normal and vaccinated rats. APMIS 96, 820–824.

Pollard, A.M., Onatolu, K.N., Hiller, L., Haldar, K., Knoll, L.J., 2008. Highly polymorphic family of glycosylphosphatidylinositol-anchored surface antigens with evidence of developmental regulation in Toxoplasma gondii. Infect. Immun. 76, 103–110.

Popiel, I., Gold, M., Choromanski, L., 1994. Tissue cyst formation of Toxoplasma gondii T-263 in cell culture. J. Eukaryot. Microbiol. 41, 17S.

Popiel, I., Gold, M.C., Booth, K.S., 1996. Quantification of Toxoplasma gondii bradyzoites. J. Parasitol. 82, 330–332.

Pszenny, V., Davis, P.H., Zhou, X.W., Hunter, C.A., Carruthers, V.B., Roos, D.S., 2012. Targeted disruption of Toxoplasma gondii serine protease inhibitor 1 increases bradyzoite cyst formation in vitro and parasite tissue burden in mice. Infect. Immun. 80, 1156–1165.

Qin, C.L., Tang, J., Kim, K., 1998. Cloning and in vitro expression of TPK3, a Toxoplasma gondii homologue of shaggy/glycogen synthase kinase-3 kinases. Mol. Biochem. Parasitol. 93, 273–283.

Radke, J.B., Lucas, O., De Silva, E., Ma, Y.F., Sullivan Jr., W.J., Weiss, L.M., Llinas, M., White, M.W., 2013. An ApiAP2 transcription factor is a gatekeeper to development of the Toxoplasma tissue cyst. PNAS, 110, 6871–6876.

Radke, J.R., Behnke, M.S., Mackey, A.J., Radke, J.B., Roos, D.S., White, M.W., 2005. The transcriptome of Toxoplasma gondii. BMC Biol. 3, 26.

Radke, J.R., Donald, R.G., Eibs, A., Jerome, M.E., Behnke, M.S., Liberator, P., White, M.W., 2006. Changes in the expression of human cell division autoantigen-1 influence Toxoplasma gondii growth and development. PLoS Pathog. 2, e105.

Radke, J.R., Guerini, M.N., Jerome, M., White, M.W., 2003. A change in the premitotic period of the cell cycle is associated with bradyzoite differentiation in Toxoplasma gondii. Mol. Biochem. Parasitol. 131, 119–127.

Radke, J.R., Striepen, B., Guerini, M.N., Jerome, M.E., Roos, D.S., White, M.W., 2001. Defining the cell cycle for the tachyzoite stage of Toxoplasma gondii. Mol. Biochem. Parasitol. 115, 165–175.

Remington, J.S., McLeod, R., Desmonts, G., 1995. Toxoplasmosis. In: Remington, J.S., Klein, J.O. (Eds.), Infectious Diseases of the Foetus and Newborn Infant, Fourth ed. W.B. Saunders Company, Philadephia.

Rodrigues-Pousada, C.A., Nevitt, T., Menezes, R., Azevedo, D., Pereira, J., Amaral, C., 2004. Yeast activator proteins and stress response: an overview. FEBS Lett. 567, 80–85.

Rooney, P.J., Neal, L.M., Knoll, L.J., 2011. Involvement of a Toxoplasma gondii chromatin remodeling complex ortholog in developmental regulation. PLoS One 6. e19570.

Saeij, J.P., Arrizabalaga, G., Boothroyd, J.C., 2008. A cluster of four surface antigen genes specifically expressed in bradyzoites, SAG2cdxy, plays an important role in Toxoplasma gondii persistence. Infect. Immun. 76, 2402–2410.

Saito, T., Maeda, T., Nakazawa, M., Takeuchi, T., Nozaki, T., Asai, T., 2002. Characterisation of hexokinase in Toxoplasma gondii tachyzoites. Int. J. Parasitol. 32, 961–967.

Saksouk, N., Bhatti, M.M., Kieffer, S., Smith, A.T., Musset, K., Garin, J., Sullivan Jr., W.J., Cesbron-Delauw, M.F., Hakimi, M.A., 2005. Histone-modifying complexes regulate gene expression pertinent to the differentiation of the protozoan parasite Toxoplasma gondii. Mol. Cell. Biol. 25, 10301–10314.

Scholytyseck, E., Mehlhorn, H., Muller, B.E., 1974. [Fine structure of cyst and cyst wall of Sarcocystis tenella, Besnoitia jellisoni, Frenkelia sp. and Toxoplasma gondii]. J. Protozool. 21, 284–294.

Schwarz, J.A., Fouts, A.E., Cummings, C.A., Ferguson, D.J., Boothroyd, J.C., 2005. A novel rhoptry protein in Toxoplasma gondii bradyzoites and merozoites. Mol. Biochem. Parasitol. 144, 159–166.

Shiels, B., Aslam, N., McKellar, S., Smyth, A., Kinnaird, J., 1997. Modulation of protein synthesis relative to DNA synthesis alters the timing of differentiation in the protozoan parasite Theileria annulata. J. Cell. Sci. 110, 1441–1451.

Silva, N.M., Gazzinelli, R.T., Silva, D.A., Ferro, E.A., Kasper, L.H., Mineo, J.R., 1998. Expression of Toxoplasma gondii-specific heat shock protein 70 during in vivo conversion of bradyzoites to tachyzoites. Infect. Immun. 66, 3959–3963.

Sims, T.A., Hay, J., Talbot, I.C., 1988. Host-parasite relationship in the brains of mice with congenital toxoplasmosis. J. Pathol. 156, 255–261.

Sims, T.A., Hay, J., Talbot, I.C., 1989. An electron microscope and immunohistochemical study of the intracellular location of Toxoplasma tissue cysts within the brains of mice with congenital toxoplasmosis. Br. J. Exp. Pathol. 70, 317–325.

Singh, U., Brewer, J.L., Boothroyd, J.C., 2002. Genetic analysis of tachyzoite to bradyzoite differentiation mutants in Toxoplasma gondii reveals a hierarchy of gene induction. Mol. Microbiol. 44, 721–733.

Soderbom, F., Loomis, W.F., 1998. Cell-cell signalling during Dictyostelium development. Trends Microbiol. 6, 402–406.

Soete, M., Camus, D., Dubremetz, J.F., 1994. Experimental induction of bradyzoite-specific antigen expression and cyst formation by the RH strain of Toxoplasma gondii in vitro. Exp. Parasitol. 78, 361–370.

Soete, M., Dubremetz, J.F., 1996. Toxoplasma gondii: kinetics of stage-specific protein expression during tachyzoite-bradyzoite conversion in vitro. Curr. Top. Microbiol. Immunol. 219, 76–80.

Soete, M., Fortier, B., Camus, D., Dubremetz, J.F., 1993. Toxoplasma gondii: kinetics of bradyzoite-tachyzoite interconversion in vitro. Exp. Parasitol. 76, 259–264.

Stwora-Wojczyk, M., Kissinger, J., Spitalnik, S., Wojczyk, B., 2004a. O-glycosylation in Toxoplasma gondii: identification and analysis of a family of UDP-GalNAc:polypeptide N-acetylgalactosaminyltransferases. Int. J. Parasitol 34 309–322.

Stwora-Wojczyk, M.M., Dzierszinski, F., Roos, D.S., Spitalnik, S.L., Wojczyk, B.S., 2004b. Functional characterization of a novel Toxoplasma gondii glycosyltransferase: UDP-N-acetyl-D-galactosamine:polypeptide N-acetylgalactosaminyltransferase-T3. Arch. Biochem. Biophys. 426, 231–240.

Stwora-Wojczyk, M.M., Kissinger, J.C., Spitalnik, S.L., Wojczyk, B.S., 2004c. O-glycosylation in Toxoplasma gondii: identification and analysis of a family of UDP-GalNAc:polypeptide N-acetylgalactosaminyltransferases. Int. J. Parasitol. 34, 309–322.

Sullivan Jr., W.J., Jeffers, V., 2012. Mechanisms of Toxoplasma gondii persistence and latency. FEMS Microbiol. Rev. 36, 717–733.

Sullivan Jr., W.J., Monroy, M.A., Bohne, W., Nallani, K.C., Chrivia, J., Yaciuk, P., Smith 2nd, C.K., Queener, S.F., 2003. Molecular cloning and characterization of an SRCAP chromatin remodeling homologue in Toxoplasma gondii. Parasitol. Res. 90, 1–8.

Sullivan Jr., W.J., Narasimhan, J., Bhatti, M.M., Wek, R.C., 2004. Parasite-specific eIF2 (eukaryotic initiation factor-2) kinase required for stress-induced translation control. Biochem. J. 380, 523–531.

Sullivan Jr., W.J., Smith 2nd, C.K., 2000. Cloning and characterization of a novel histone acetyltransferase homologue from the protozoan parasite Toxoplasma gondii reveals a distinct GCN5 family member. Gene 242, 193–200.

Szatanek, T., Anderson-White, B.R., Faugno-Fusci, D.M., White, M., Saeij, J.P., Gubbels, M.J., 2012. Cactin is essential for G1 progression in Toxoplasma gondii. Mol. Microbiol. 84, 566–577.

Thomason, P., Traynor, D., Kay, R., 1999. Taking the plunge. Terminal differentiation in Dictyostelium. Trends. Genet. 15, 15–19.

Tomavo, S., Boothroyd, J.C., 1995. Interconnection between organellar functions, development and drug resistance in the protozoan parasite, Toxoplasma gondii. Int. J. Parasitol. 25, 1293–1299.

Tomita, T., Bzik, D.J., Ma, Y.F., Fox, B.A., Markillie, M.L, Taylor, R.C., Kim, K., Weiss, L.M., 2013. The Toxoplasma gondii cyst wall protein CST1 is critical for cyst wall integrity and promotes bradyzoite persistence. Submitted.

Torpier, G., Charif, H., Darcy, F., Liu, J., Darde, M.L., Capron, A., 1993. Toxoplasma gondii: differential location of antigens secreted from encysted bradyzoites. Exp. Parasitol. 77, 13–22.

Ueno, A., Dautu, G., Haga, K., Munyaka, B., Carmen, G., Kobayashi, Y., Igarashi, M., 2011. Toxoplasma gondii: a bradyzoite-specific DnaK-tetratricopeptide repeat (DnaK-TPR) protein interacts with p23 co-chaperone protein. Exp. Parasitol. 127, 795–803.

Unno, A., Suzuki, K., Batanova, T., Cha, S.Y., Jang, H.K., Kitoh, K., Takashima, Y., 2009. Visualization of Toxoplasma gondii stage conversion by expression of stage-specific dual fluorescent proteins. Parasitology 136, 579–588.

Van Der Waaij, D., 1959. Formation, growth and multiplication of Toxoplasma gondii cysts in mouse brain. Trop. Geogr. Med. 11, 345–360.

Van, T.T., Kim, S.K., Camps, M., Boothroyd, J.C., Knoll, L.J., 2007. The BSR4 protein is up-regulated in Toxoplasma gondii bradyzoites, however the dominant surface antigen recognised by the P36 monoclonal antibody is SRS9. Int. J. Parasitol. 37, 877–885.

Vanchinathan, P., Brewer, J.L., Harb, O.S., Boothroyd, J.C., Singh, U., 2005. Disruption of a locus encoding a nucleolar zinc finger protein decreases tachyzoite-to-bradyzoite differentiation in Toxoplasma gondii. Infect. Immun. 73, 6680–6688.

Walker, R., Gissot, M., Croken, M.M., Huot, L., Hot, D., Kim, K., Tomalco, S., 2012. The Toxoplasma nuclear factor TgAP2XI-4 controls bradyzoite gene expression and cyst formation. Mol. Microbiol. 87, 641–655.

Wasmuth, J.D., Pszenny, V., Haile, S., Jansen, E.M., Gast, A.T., Sher, A., Boyle, J.P., Boulanger, M.J., Parkinson, J., Grigg, M.E., 2012. Integrated bioinformatic and targeted deletion analyses of the SRS gene superfamily identify SRS29C as a negative regulator of Toxoplasma virulence. MBio. 3, e00321–12.

Weilhammer, D.R., Iavarone, A.T., Villegas, E.N., Brooks, G.A., Sinai, A.P., Sha, W.C., 2012. Host metabolism regulates growth and differentiation of Toxoplasma gondii. Int. J. Parasitol. 42, 947–959.

Weiss, L.M., Kim, K., 2000. The development and biology of bradyzoites of Toxoplasma gondii. Front. Biosci. 5, D391–D405.

Weiss, L.M., Laplace, D., Takvorian, P., Tanowitz, H.B., Wittner, M., 1996. The association of the stress response and Toxoplasma gondii bradyzoite development. J. Eukaryot. Microbiol. 43, 120S.

Weiss, L.M., Laplace, D., Takvorian, P.M., Cali, A., Tanowitz, H.B., Wittner, M., 1994. Development of bradyzoites of Toxoplasma gondii in vitro. J. Eukaryot. Microbiol. 41, 18S.

Weiss, L.M., Laplace, D., Takvorian, P.M., Tanowitz, H.B., Cali, A., Wittner, M., 1995. A cell culture system for study of the development of Toxoplasma gondii bradyzoites. J. Eukaryot. Microbiol. 42, 150–157.

Weiss, L.M., Laplace, D., Tanowitz, H.B., Wittner, M., 1992. Identification of Toxoplasma gondii bradyzoite-specific monoclonal antibodies. J. Infect. Dis. 166, 213–215.

Weiss, L.M., Ma, Y.F., Halonen, S., McAllister, M.M., Zhang, Y.W., 1999. The in vitro development of Neospora caninum bradyzoites. Int. J. Parasitol. 29, 1713–1723.

Weiss, L.M., Ma, Y.F., Takvorian, P.M., Tanowitz, H.B., Wittner, M., 1998. Bradyzoite development in Toxoplasma gondii and the hsp70 stress response. Infect. Immun. 66, 3295–3302.

Wek, R.C., Jiang, H.Y., Anthony, T.G., 2006. Coping with stress: eIF2 kinases and translational control. Biochem. Soc. Trans. 34, 7–11.

Wojczyk, B.S., Stwora-Wojczyk, M.M., Hagen, F.K., Striepen, B., Hang, H.C., Bertozzi, C.R., Roos, D.S., Spitalnik, S.L., 2003. cDNA Cloning and Expression of UDP-N-acetyl-D-galactosamine: Polypeptide. N-Acetylgalactosaminyltransferase T1 from Toxoplasma gondii. Mol. Biochem. Parasitol. 131, 93–107.

Wong, S.Y., Remington, J.S., 1993. Biology of Toxoplasma gondii. Aids 7, 299–316.

Yahiaoui, B., Dzierszinski, F., Bernigaud, A., Slomianny, C., Camus, D., Tomavo, S., 1999. Isolation and characterization of a subtractive library enriched for developmentally regulated transcripts expressed during encystation of Toxoplasma gondii. Mol. Biochem. Parasitol. 99, 223–235.

Yang, S., Parmley, S.F., 1995. A bradyzoite stage-specifically expressed gene of Toxoplasma gondii encodes a polypeptide homologous to lactate dehydrogenase. Mol. Biochem. Parasitol. 73, 291–294.

Yang, S., Parmley, S.F., 1997. Toxoplasma gondii expresses two distinct lactate dehydrogenase homologous genes during its life cycle in intermediate hosts. Gene 184, 1–12.

Zhang, Y.W., Halonen, S.K., Ma, Y.F., Wittner, M., Weiss, L.M., 2001. Initial characterization of CST1, a Toxoplasma gondii cyst wall glycoprotein. Infect. Immun. 69, 501–507.

Zhang, Y.W., Kim, K., Ma, Y.F., Wittner, M., Tanowitz, H.B., Weiss, L.M., 1999. Disruption of the Toxoplasma gondii bradyzoite-specific gene BAG1 decreases in vivo cyst formation. Mol. Microbiol. 31, 691–701.

Zhang, Y.W., Halonen, S.K., Ma, Y.F., Tanowitz, H.B., Weiss, L.M., , 2010. A purification method for enrichment of the Toxoplasma gondii cyst wall. J Neuroparasitology 1, N101001.

CHAPTER

16

Development and Application of Classical Genetics in *Toxoplasma gondii*

*James W. Ajioka**, *L. David Sibley*†

*Department of Pathology, Cambridge University, Cambridge, England, United Kingdom

†Department of Molecular Microbiology, Washington University School of Medicine, St Louis, Missouri, USA

OUTLINE

Toxoplasma gondii, Second Edition
http://dx.doi.org/10.1016/B978-0-12-396481-6.00016-7

16.1 INTRODUCTION

Apicomplexan parasites have complex life cycles that often involve sexual replication in a single definitive host and asexual replication in a variety of alternative hosts. *Toxoplasma gondii* undergoes its sexual cycle only in cats, yet it also infects a wide range of other vertebrates where it propagates asexually. *T. gondii* exhibits a highly clonal population structure in North America and Europe where it is comprised of three dominant clonal lineages that are currently widespread, yet which only recently originated from a few closely related parental lines. Despite being genetically quite similar, these lineages differ substantially in various biological traits including virulence in laboratory mice and induction of changes in host cell signalling and transcription. The ability to cross strains experimentally has been exploited to develop classical genetics in *T. gondii*. Genetic mapping provides a forward genetic system for analysing complex biological traits including drug resistance, growth, virulence, induction of host gene expression and immune responses. Recent advances have revealed several families of parasite secretory proteins that control important biological differences in virulence and alteration of host cell signalling.

16.2 BIOLOGY OF *TOXOPLASMA*

16.2.1 Life Cycle

Toxoplasma gondii has a typical heteroxenous life cycle, alternating between sexual replication in members of the cat family, which serve as the definitive host, and asexual replication in a wide range of warm-blooded vertebrates that serve as intermediate hosts (Dubey, 1977) (see Chapter 1 in this volume). During the initial infection, *T. gondii* grows rapidly as a haploid form called the tachyzoite, which is capable of invading and replicating within any nucleated cell in its many warm-blooded hosts. In response to stress brought on by the immune response, the parasite converts to a slow growing form called a bradyzoite that is encased in a thick-walled cyst. Ingestion of tissue cysts by cats leads to sexual differentiation within intestinal epithelial cells and the resulting fusion of gametes to form a diploid zygote (Dubey and Frenkel, 1972). Following the development of an impervious wall, oocyst stages are shed in the faeces and they undergo meiosis in the environment to yield eight haploid progeny (Frenkel *et al.*, 1970). Oocysts are long-lived, resistant to environmental conditions and responsible for dissemination due to contamination of food or water (Mead *et al.*, 1999; Dubey, 2004; de Moura *et al.*, 2006).

16.2.2 Defining the Sexual Phase

The sexual cycle of *T. gondii* only takes place in the enterocytes of the small intestine of members of the cat family (*Felidae*) (Dubey, 2010). Factors that restrict development to this host and tissue location are uncertain, but presumably result from co-adaptation of the parasite and host. By comparison, *Hammondia hammondi*, which is closely related to *T. gondii*, also undergoes its sexual cycle in cats, while more distant relatives *Hammondia heydorni* and *Neospora caninum* undergo sexual development only in canines (Dubey, 1977). The relative divergence of *Toxoplasma* and *Neospora* coincides with that of their respective carnivore hosts (Reid *et al.*, 2012), suggesting these parasites speciated following separation of their definitive hosts. Initiation of the sexual cycle in *T. gondii* occurs when bradyzoites infect epithelial cells in the ileum of the cat intestine, although the reasons why this niche is preferred remain unknown. Infection of cats with oocysts or tachyzoites can also lead to oocyst shedding, although with considerable delays (Dubey, 2005, 2006), suggesting that the parasite must first differentiate into bradyzoites prior to commencing the sexual phase. Sexual development proceeds through several rounds of merogony, leading to development of morphologically distinct male and female gametocytes, which fuse to form a zygote (Dubey and Frenkel, 1972). The diploid oocyst is shed in the faeces in an unsporulated state, and it then undergoes meiosis in the environment (Ferguson *et al.*, 1979; Speer *et al.*, 1998). The development of these sexual stages has been extensively studied in histopathology sections by both light and electron microscopy; however, thus far it has not been possible to induce their development *in vitro*.

Following the development of *in vitro* cell culture systems to isolate clones of *T. gondii* tachyzoites (Pfefferkorn and Pfefferkorn, 1976), cats fed with tissue cysts that had developed from a cloned parasite line shed oocysts (Pfefferkorn *et al.*, 1977). Similar findings by another group (Cornelissen and Overdulve, 1985), confirmed that *T. gondii* has the capability to differentiate into both micro- and macrogametocytes starting from a single progenitor parasite cell. A strict genetic basis for gametocyte differentiation (e.g. sex chromosome) can thus be discounted, although the cellular mechanisms that lead to differentiation into male (micro-) or female (macro-) gametocytes are as yet unknown. The developmental programme necessary for completion of transmission through the cat is easily disrupted. Rapid passage of tachyzoites through mice resulted in loss of ability to form oocysts after ~30 passages (likely fewer than 10^3 cell divisions) (Frenkel *et al.*, 1976). This remarkably high rate of loss either implies that any single mutation in large number of essential genes can disrupt the process, or a single target that easily becomes inactivated in an irreversible manner. Alternatively loss of developmental transformation may not be due to mutation, as epigenetic phenomena might result in a similar block in gene expression leading to developmental arrest.

Classic studies by Cornelissen and co-workers examined the ploidy of stages during gametogenesis and meiosis in the sexual cycle by analysing the DNA content of gametes, zygote/oocyst and sporozoites (Cornelissen *et al.*, 1984a; Cornelissen *et al.*, 1984b). These studies concluded that mature macrogametes (presumably fertilized) contain twice the amount of DNA found in all other stages, suggesting that most stages are haploid, and that meiosis takes place during sporogony. The absolute content of DNA in one haploid stage predicted a genome size of ~80 Mbp, which agrees closely with the assembled genome size of 65 Mbp (see below). Morphological studies of sporulation in *T. gondii* indicate that the initial division into two sporocysts is followed by reduction into four haploid sporozoites each (Speer *et al.*, 1998). Presumably the first division represents the meiotic division, although this has not been confirmed experimentally. Classical tetrad analysis using micro dissection and single-cell PCR amplification might resolve this and also provide a rapid means of evaluating the fraction of recombination in future crosses. Other advantages of *T. gondii* as a classical genetic system include: (i) techniques for harvesting and sporulating oocysts (Dubey *et al.*, 1972) allow for infections to be initiated *in vitro* from material collected as the result of naturally or experimentally infected cats and (ii) oocysts can be stored for long periods of time at 4°C (Dubey, 1998), which facilitates the isolation of progeny from experimental crosses as described below.

16.2.3 Population Structure and Major Strain Types

Toxoplasma gondii has an unusual population structure comprised of three clonal lineages, referred to as I, II and III, which are widespread in North America and Europe (Sibley and Ajioka, 2008). Recently a fourth clonal lineage has been described, and strains of this genotype (referred to as haplotype 12) appear to be more common in wild versus domestic animals in North America (Khan *et al.*, 2011). Strains of *T. gondii* are much more diverse in other regions, especially in South America, where distinct non-clonal lineages show much greater evidence of out crossing (Lehmann *et al.*, 2006; Khan *et al.*, 2007; Pena *et al.*, 2008). The global population structure of *T. gondii* has recently been compared using a variety of different typing methods to analyse more than 950 independent isolates grouping them into 16 haplogroups that comprise six major clades (Su *et al.*, 2012). Studies on the population genetic structure of *T. gondii* from different regions are further discussed in Chapter 3 in this volume. Genetic crosses have thus far only been exploited for mapping traits in the three major clonal lineages common to North America and Europe, and hence this chapter will be confined to advances in developing genetic systems to define their unique traits.

The three clonal lineages are themselves quite closely related and derived from a few closely related ancestral strains that underwent limited genetic recombination in the wild (Grigg *et al.*, 2001a; Su *et al.*, 2003). Mapping of strain-specific expressed sequence tags (ESTs) to the whole genome assemblies confirmed that the three clonal lineages arose through a small number of recombination events between separate ancestral strains of type I and III lineages that each crossed with a type II-like parental strain (Boyle *et al.*, 2006). This inheritance pattern reinforces the strongly biallelic nature of DNA polymorphisms in *T. gondii*. The striking observation that virtually all single nucleotide polymorphisms are confined to two alleles may be most simply explained by a severe population bottleneck that reduced the effective population size to $N = 1$. Mutation and recombination during the subsequent population expansion would then give rise to the biallelic pattern because one of the alleles is from the single progenitor ancestor and multiple mutations at any given nucleotide position would be extremely rare (Rosenthal and Ajioka, 2012). The genetic similarity arising from the population ancestry has been crucial in forward genetic approaches for deciphering the molecular basis of major phenotypic differences between the clonal lineages. Despite being very similar genetically, there are major phenotypic differences between the clonal lineages and forward genetic approaches have been very fruitful in deciphering the molecular basis for many of these traits.

16.3 ESTABLISHMENT OF TRANSMISSION GENETICS

16.3.1 Intra-Strain Crosses and Meiosis

Although morphological observations of *T. gondii* infection in the cat provided a framework for the sexual cycle, key mechanistic details such as sexual differentiation/determination and recombination required the establishment of transmission genetics. Early work by Elmer Pfefferkorn and colleagues laid the foundation for *T. gondii* as a genetic model organism by generating necessary tools and carrying out the first genetic crosses. Performing genetic crosses in *T. gondii* is relatively complex and not easily undertaken in most laboratories, in part due to the biological hazard posed by oocysts, which are highly infectious. Oocysts are resistant to chemical treatments and disinfectants (i.e. chlorine, aldehyde fixation, strong acids or bases, UV irradiation). Heat inactivation is the only reliable means of disinfecting surfaces that come in contact with *T. gondii* oocysts (Dubey, 1998). Other stages of *T. gondii* (tachyzoites and bradyzoites) can be used more widely with standard BSL-2 level containment, making these relatively easy to work within the laboratory. Successful completion of a genetic cross requires the production of tissue cysts of two compatible parental strains, obtained by chronically infecting mice, co-feeding of tissue cysts to specific pathogen free cat(s), collection of oocysts shed in the faeces, sporulation, re-initiation of *in vitro* cultures, cloning and phenotyping of progeny (Fig. 16.1). Given the complexity of this process, it is not surprisingly that only a limited number of crosses

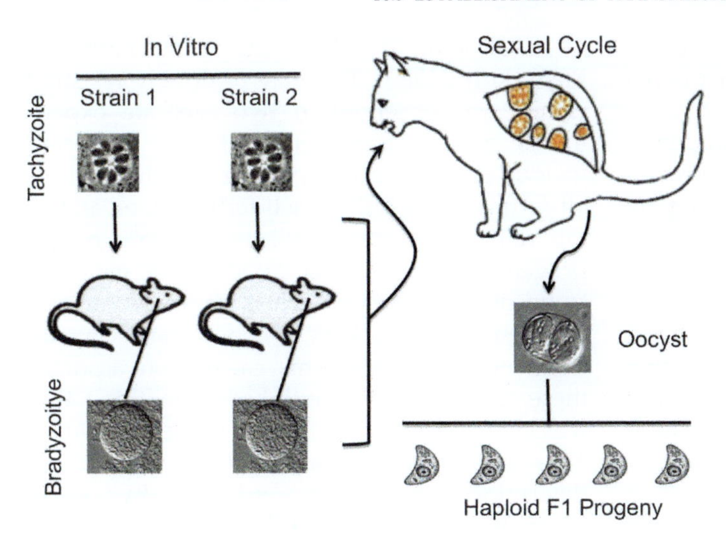

FIGURE 16.1 **Diagram of genetic crosses of *T. gondii*.** Tachyzoites of different genetic types are propagated *in vitro* in host cell monolayers and used to generate drug resistant parental lines. Inoculation of mice leads to chronic infections characterized by bradyzoites found in tissue cysts in the brain. Co-feeding of tissue cysts to cats initiates the sexual cycle, which takes place in enterocytes of the small intestine. Oocysts are shed into the environment where they undergo meiosis. Oocysts are hatched and inoculated onto host cell monolayers to isolate haploid progeny clones that correspond to F1 hybrids. Modified with permission from Sibley and Ajioka (2008).

TABLE 16.1 Summary of Genetic Crosses among Clonal Lineages of *T. gondii*

Parental Strains	Parental Strains	Individual Crosses[a]	Number Unique Progeny	Number of Markers	Frequency of Recombination	Maternal Inheritance
II × III	ME49 B7 × CTG ARA−A^r/SNF^r	S CL	10[b] 9[b]	135[d]	ND	No bias
	PTG FUDR^r/ANO^r × CTG ARA−A^r/SNF^r/ DCL^r	C96	21[b]	135[d]	ND	Type III predominates
I × III	GT−1 FUDR^r × CTG ARA−A^r/SNF^r	C285 C295	11[c] 20[b] 3[d]	175[e]	~100%[g]	Slight type I bias
I × II	GT−1 SYN^r × ME49 B7 FUDR^r	Ct3 Ct7 Ct10	37[b] 3[b]1[c] 4[b]	1,603[f]	~20%−35%[g]	No bias

[a] *Separate crosses were done in parallel cats fed similar mixtures of bradyzoites from the two parental strains for Cl and S clones in the II × III cross (Sibley* et al., *1992), for c285, c295 clones in the I × III cross (Su* et al., *2002) and for the Ct3, Ct7, Ct10 clones in the I × II cross (Behnke* et al., *2011)*
[b] *Clones isolated by drug section*
[c] *Clones selected at random and genotyped by PCR*
[d] *PCR screened for recombination on chromosome VIIa*
[e] *PCR−RFLP markers (Khan* et al., *2005)*
[f] *Affymetrix array hybridizations (Behnke* et al., *2011)*
[g] *Determined by isolating clones and then genotyping with 10 unlinked PCR−RFLP markers*
ND, not determined

have been conducted to date (Table 16.1), yet the results of these experimental crosses have proven highly informative.

As with other genetic model organisms, engineering of drug-resistant lines of *T. gondii* proved useful for tracking recombination in progeny from genetic crosses. These studies exploited the fact that *T. gondii* tachyzoites are haploid and can be maintained indefinitely *in vitro* and cryopreserved, thus aiding in isolation of defined clones (Pfefferkorn, 1988). Alkylating agents such as ethyl-nitrosurea (ENU) proved effective for inducing mutations and isolation of mutant parasite lines (Pfefferkorn and Pfefferkorn, 1979). Among the most useful were lines resistant to 5-fluorodeoxyuridine (FUDR), which is an antimetabolite that is incorporated by the enzyme uracil phosphoribosyl transferase (Pfefferkorn and Pfefferkorn, 1977), and adenosine arabinoside (ARA-A), which is likewise incorporated by adenosine kinase (Pfefferkorn and Pfefferkorn, 1978). The results from these early studies showed that mutants with selectable phenotypes could be easily recovered, opening up the possibility to use genetic crosses to study the genetic basis for phenotypes such as drug resistance.

For the first *T. gondii* cross, two mutant lines resistant to ARA-A and FUDR were separately selected from the parental C strain (a.k.a. CTG, CEP) after ENU mutagenesis (Pfefferkorn and Pfefferkorn, 1980). The CTG strain was isolated from a naturally infected cat in New Hampshire by Elmer Pfefferkorn (Pfefferkorn *et al.*, 1977) and it was later found to be a member of the type III lineage (Howe and Sibley, 1995). All of the oocyst-derived clones recovered from the ARA-A^r line were drug resistant, confirming the haploid nature of the parasite. The ability of this clonal line to produce oocysts also demonstrated the potential to make both micro- and macrogametocytes and to undergo self-fertilization. When roughly equal numbers of ARA-A^r and FUDR^r bradyzoites were fed to a cat, approximately 12% of the clones recovered were resistant to both drugs (Pfefferkorn and Pfefferkorn, 1980). This rate of doubly resistant clones was observed in two separate experiments and is consistent with the expected meiotic recombination yield, given that half of all fertilizations will be self-fertilizations and of the remaining cross-fertilizations; 25% of the progeny from the latter category are expected to be doubly resistant when observed for two independent markers, hence an overall frequency of 12.5%. This calculation assumes these markers to be unlinked, something that was later confirmed by linkage mapping studies. These data argue that mating types are not predetermined, a finding confirmed by others (Cornelissen and Overdulve, 1985), and that parthenogenesis does not contribute substantially to the output of oocysts during experimental crosses. By examining the inheritance of drug resistance in self-crosses versus outcrosses it was also shown that only a haploid model was capable of explaining the inheritance of drug resistance in *T. gondii*.

16.3.2 Genetic Crosses between Different Lineages

The first inter-strain cross was conducted between the CTG line described above and the ME49 strain, originally isolated from a sheep in California (Lunde and Jacobs, 1983) and later shown to be representative of the type II lineage (Howe and Sibley, 1995). Prior to being used in this cross, the ME49 line was passaged through a cat to obtain sporulated oocysts and a clone called B7 was isolated by limiting dilution *in vitro* for use as the parental line. The other parental line used in this cross was a doubly resistant clone of CTG bearing the drug resistance markers ARA-A^r and sinefungin (SNF^r), previously isolated as part of the intra-strain cross among singly drug resistant clones of CTG (Pfefferkorn and Kasper, 1983). Each of the parental strains were used to infect mice, brain homogenates containing tissue cysts were co-fed to two different naive cats, and oocysts were collected after shedding (Fig. 16.1). Nineteen recombinant clones that were singly resistant to either SNF or ARA-A were isolated *in vitro* (Table 16.1). A second cross between type II and III lineages was performed several years later (c. 1996) using a similar strategy (Table 16.1). In this cross, the parental strains were carrying the following drug resistance markers: the type II PTG strain (a separate clone of ME49 (Ware and Kasper, 1987)) was resistant to aprinocide-N-oxide (ANO) and FUDR and the type III strain CTG was resistant to SNF, ARA-A, and diclazuril (DCL) (Khan *et al.*, 2005) (Table 16.1). A total of 21 recombinants from this cross were selected from a pool of parasites that was doubly selected with SNF and ANO and then phenotyped for the other markers (Table 16.1). In both of the crosses between types II and III all of the progeny were isolated based on drug resistance, hence the frequency of outcrossing versus selfing is unknown (Table 16.1).

Expanding on the success of these early interstrain crosses, a genetic cross between the virulent type I strain GT-1 (Dubey, 1980) and the non-virulent type III strain CTG (Pfefferkorn *et al.*, 1977) was conducted to investigate the genetic basis of acute virulence (Su *et al.*, 2002) (Table 16.1). A resistant clone of the GT-1 strain was isolated following chemical mutagenesis and selection with FUDR as described previously (Pfefferkorn and Pfefferkorn, 1977). GT-1 is a highly virulent type 1 strain, originally isolated from a goat (Dubey, 1980), and capable of undergoing the entire life cycle. The GT-1 FUDR^r line was crossed with the CTG strain that was doubly resistant to ARA-A and SNF, described above. Progeny from this cross were initially selected for segregation of these resistance markers to obtain FUDR^R-ARA-A^R and FUDR^R-SNF^R clones (Table 16.1). Subsequent to this, clones were isolated randomly and typed solely by PCR analysis using polymorphic DNA markers (Su *et al.*, 2002). In the absence of any selection, virtually all the clones obtained were recombinants (24 of 25 clones tested), with the exception of a single type III clone (Table 16.1). The very high frequency of outcrossing cannot be explained simply by inefficient gamete formation as each strain readily produced oocysts when fed to cats alone (C. Su and L. D. Sibley, unpublished data). Additionally, examination of the inheritance of apicoplast markers revealed that both parental types were capable for forming microgametes and macrogametes, although there was a slight bias for type I to form macrogametes (C. Su, unpublished data) (Table 16.1).

The final pair-wise combination of crosses between the three clonal lineages was recently completed with the successful crossing of type I and II strains (Behnke *et al.*, 2011). A sinefungin resistant (SNF^r) clone of the type I GT-1 strain was crossed with a FUDR resistant (FUDR^r) B7 clone of ME49. Following co-infection of cats with a mixture of bradyzoites of the GT-1-SYN^r and ME49-FUDR^r strains, recombinant progeny were isolated from sporulated oocysts either by selection with the combination of both drugs, or by random cloning and genotyping using polymorphic DNA markers (Table 16.1). Among the non-drug selected progeny, recombinants were isolated at the expected frequency (i.e. ~20%–35% of randomly selected clones were recombinants) and there was no maternal bias in this cross (Table 16.1, Behnke *et al.*, 2011).

16.3.3 Implications of Selfing vs. Outcrossing for Population Structure

Population genetic studies of *T. gondii* indicate a high level of clonality in North America and Europe, which could be maintained by asexual transmission or by a high proportion of self-fertilization (Sibley and Ajioka, 2008). Occasional hybrids between the clonal types are observed in nature (Su *et al.*, 2012); however, the relative frequency of selfing versus outcrossing is uncertain. In experimental crosses where the frequency of selfing has been monitored it is typically ~50%, with the exception of the I × III cross (Table 16.1). Selfing has been linked to outbreaks of toxoplasmosis spread by a waterborne route due to contamination with oocysts as a single point source (Wendte *et al.*, 2010). Despite having a highly clonal population structure (Dardé *et al.*, 1992; Sibley and Boothroyd, 1992b; Howe and Sibley, 1995; Ajzenberg *et al.*, 2002a), pockets of high genetic diversity also occur in the wild (Ajzenberg *et al.*, 2002b, 2004) suggesting that locally high rates of genetic crossing are important in the population structure of *T. gondii*. Studies in South America have emphasized a much more diverse genetic structure, suggesting that out-crossing is much more frequent (Lehmann *et al.*, 2006; Khan *et al.*, 2007; Pena *et al.*, 2008). The observation that some experimental genetic crosses yield a high proportion of out-crossing (i.e. I × III cross) suggests that this phenomenon contributes to genetic diversity of natural populations, particularly in some regions.

16.4 DEVELOPMENT OF GENETIC MAPPING

16.4.1 Advances in Molecular Genetic Tools

A number of technical advances coincided with the development of classical genetic mapping and these led to synergism in advancing genetic approaches in *T. gondii*. Establishment of linkage groups was aided by the separation of *T. gondii* chromosomes by pulsed-field gel electrophoresis (PFGE) and mapping of DNA probes to these gel-separated bands by Southern blot hybridization (Sibley and Boothroyd, 1992a). Large scale sequencing of ESTs from different strains (Ajioka *et al.*, 1998; Manger *et al.*, 1998; Li *et al.*, 2003), originally undertaken for gene discovery, also allowed identification of single nucleotide polymorphisms (SNPs) that facilitated development of polymorphic genetic markers. Use of PCR-based methods to amplify regions containing polymorphisms that define restriction fragment length polymorphisms (RFLP), allowed higher throughput genotyping (Su *et al.*, 2002). The development of an Affymetrix array for *T. gondii* provided an alternative means for genotyping using single feature polymorphisms (SFP) based on differences in the ESTs as well as a set of previously derived PCR–RFLP genetic markers (Bahl *et al.*, 2010). Completion of whole genome sequences of representatives of the three major lineages *T. gondii* and establishment of a database for housing genomic data (http://www.toxodb.org; Gajria *et al.*, 2007) provided a framework for localizing genes once initial linkages were established. The assembly of the genome went hand-in-hand with the generation of linkage maps, which were used to identify contigs and scaffolds that belonged to individual chromosomes (Khan *et al.*, 2005). Without these efforts, mapping loci responsible for phenotypic differences would not have led to the identification of responsible genes.

16.4.2 Development of Linkage Maps for Forward Genetic Analysis

The generation of genetic linkage maps for *T. gondii* proceeded through several phases. Initially, a set of 64 polymorphic markers that recognized RFLPs detected by Southern blot was used to analyse segregation from a genetic cross between types II and III (Sibley *et al.*, 1992). Segregation of these markers in 19 recombinant progeny led to the first generation linkage map. At this early stage, 11 chromosomes were recognized and the total genetic map distance was less than 150 centimorgans (cM). Although low resolution, this map was sufficient to link several drug resistance markers to specific chromosomes (i.e. SNF, ARA-A) and to establish the basic parameters of meiotic recombination in *T. gondii*. The next advance in genetic mapping in *T. gondii* came with the successful analysis of a cross between the virulent type I (GT-1) and the non-virulent type III (CTG) lineages. The RFLP markers were converted to PCR-based typing and 112 markers were used to generate a linkage map from two parallel crosses (Su *et al.*, 2002). The number of chromosomes remained the same, but the linkage map expanded to ~400 cM. Finally, in phase three, an expanded number of PCR–RFLP based genetic markers (~250 in all) was used to analyse 71 progeny from genetic crosses between two of the three pair-wise groupings (II × III and I × III) (Khan *et al.*, 2005b). MapMaker EXP 3.0 (Lander *et al.*, 1987) was used to generate linkage maps with a LOD score of more than 3.0. The precise order and distance between markers was determined based on physical maps that came from assembly of the whole genome sequence using the linkage map.

The combined genetic linkage map of *T. gondii*, derived by combining the results from II × II and I × III crosses, consists of 14 linkage groups that collectively comprise 590 cM (Khan *et al.*, 2005b). The chromosomes are largely the same as those identified in earlier studies, with the addition of several new groups that had not been resolved by PFGE and/or had been missed due to low density of markers in the earlier studies. The combined genetic map is based on markers that are spaced approximately every 300 kb across the genome. Individual linkage maps for each chromosome and their corresponding markers are shown in Fig. 16.2. As expected, the relative size of each chromosome, based on PFG separation and physical maps from the genome assembly, shows a roughly linear relationship with genetic size in map units (Khan *et al.*, 2005b). The rate of crossover in *T. gondii* averages about 100 Kb/cM although marked differences are seen from a low of 42 Kb/cM on chromosome Ia to a high of ~150 Kb/cM on chromosome VIIb (Khan *et al.*, 2005b). The average crossover frequency for *T. gondii* is substantially lower than for *P. falciparum* (17 Kb/cM) (Su *et al.*, 1999) making it more difficult to fine map genetic traits in *T. gondii*. This combined linkage map was also used to assemble the scaffolds from whole genome sequencing, initially conducted on the same B7 clone of ME49. Scaffolds were assembled and ordered into chromosomes using the linkage maps and based on the position of genetic markers, along with BAC-end and cosmid-end sequences from several large insert libraries (Khan *et al.*, 2005b). More than 95% of the 65 Mb genome was assembled using this strategy, thus providing a framework for genetic mapping and for identify genes within regions of interest.

Separately, a genetic linkage map was generated for the I × II cross based on the segregation of markers among 45 recombinant progeny (Behnke *et al.*, 2011). In this case, the polymorphisms were scored based on SFPs detected following hybridization of gDNA to an Affymetrix array for each of the parental and progeny clones. As the array is based on the genomic sequence of the type II lineage, hybridizations were scored based on perfect match (type II) versus mismatch (type I). Although there are several regions that are non-informative, greater than 85% of the genome is represented in this linkage map. In total, 1603 probes were identified with high confidence genotyping among the parental and all recombinant clones (Table 16.1). On average there are 10 markers for each of the 151 genetic intervals defined by crossovers in this genetic map. Although this map provides a much higher resolution of markers, the precision for mapping is still limited by the number for crossovers, which depends more on the number of progeny than the density of the genetic map.

16.4.3 Limitation of the Current Linkage Maps

The biallelic nature of genetic markers in *T. gondii* presents an unusual situation that complicates assembly of the genetic maps. Because at any given locus, two strains are identical while the third strain is unique, SNP

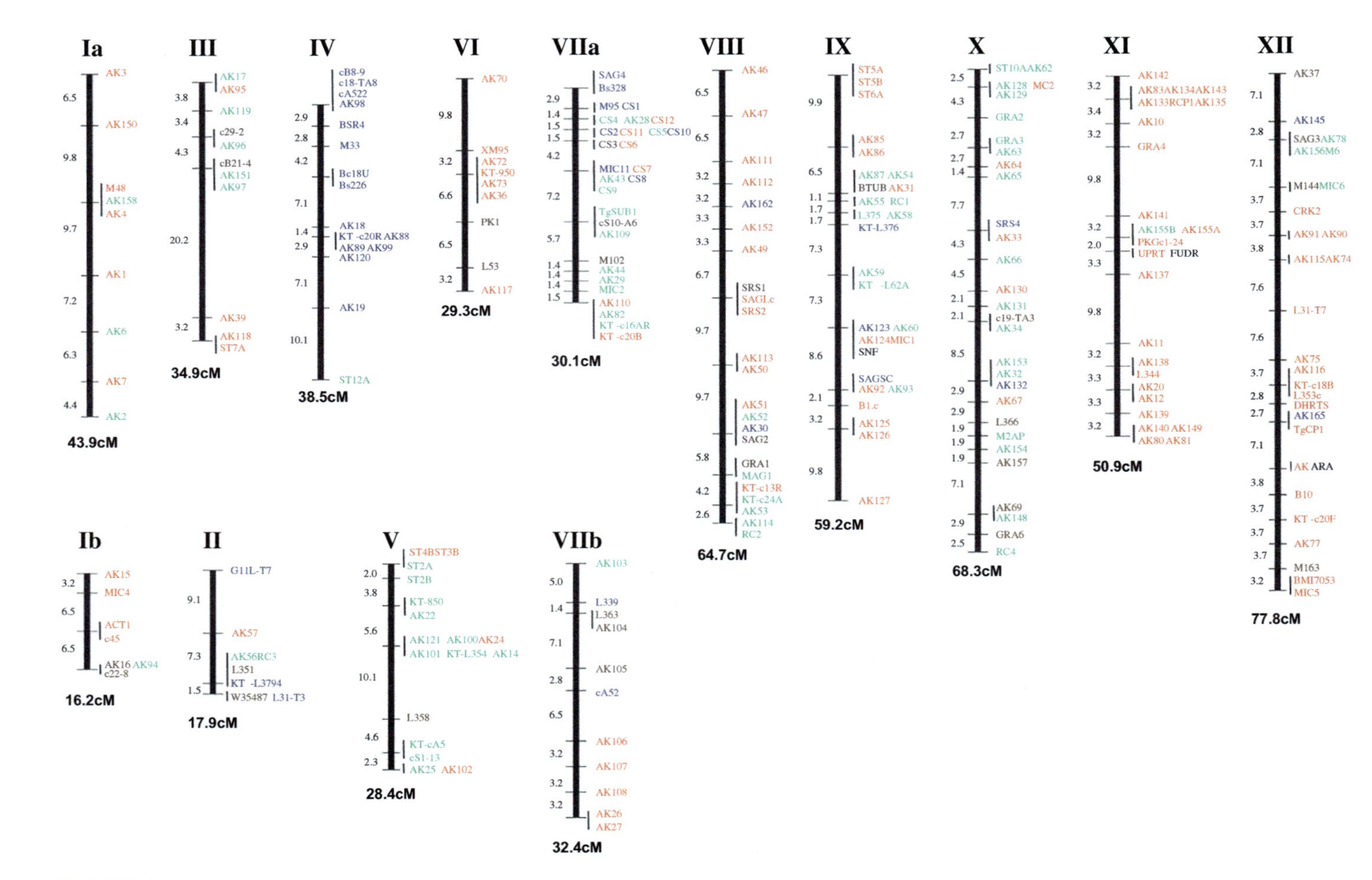

FIGURE 16.2 **Genetic linkage maps for *T. gondii*.** Linkage maps for each of the 14 chromosomes (numbered at the top) were determined by a combination of genetic crosses between types II × III and I × III. Markers are shown to the right of the vertical line, and genetic distances between each nodes are shown to the left in centimorgans. Total genetic distances are shown at the bottom of each chromosome. Colour code: red indicates polymorphism unique to type I, green unique to type II, blue unique to type III, brown indicates multiple polymorphisms. Used with permission from Khan *et al.* (2005b).

markers are not informative in all pair-wise crosses. Strain-specific haplotypes are conserved across large regions of the genome, such that strain-specific SNPs are strongly clustered (Boyle *et al.*, 2006). This leads to a situation where most of the informative markers on some chromosomes (i.e. VI, VIII, XI, XII) are unique to the type I lineage (Khan *et al.*, 2005b). An extreme example of this is chromosome XI where the genetic linkage map is based almost exclusively on progeny from the I × III cross. Hence, this chromosome is effectively invisible in the II × III cross. This situation also leads to linkage disequilibrium between chromosomes VI and VIII, which are strongly linked despite physical evidence that they are separate (Khan *et al.*, 2005b). Interestingly, both *T. gondii* (Khan *et al.*, 2005b) and *P. falciparum* (Su *et al.*, 1999) show a high frequency of closely spaced double-crossover events. These occur at much closer distances than would be predicted by crossover frequency suggesting they are gene conversions rather than reciprocal events. The presence of such events complicates mapping analyses as regions between existing markers might also have undergone exchange, potentially altering important phenotypes without resulting in easily detectable genetic changes. These closely spaced double crossovers appear more frequently in the array based SNP map from the I × II cross (Behnke *et al.*, 2011), reflecting the greater density of markers in this map.

As this book goes to press, ME49, the reference genome strain, has been resequenced using next generation sequencing approaches. This re-sequencing and re-annotation of the genome, done with community input, has been accompanied by the sequencing of prototypic strains representing each of the 16 major lineages of *T. gondii*. RNA-seq transcriptome studies have also been performed on tachyzoites from these strains in addition to RNA-seq of the major replicative forms of ME49 (sporozoite, bradyzoite, tachyzoite). All data are available at NCBI and integrated data will be viewable on http://www.toxodb.org. The cost of sequencing has dropped and technology has improved so that routine genome sequencing with RNA-seq of mutants is feasible, and it is likely that further technological advances will continue apace. Future development of genetic maps based on whole genome sequencing is likely to provide still greater resolution of mapping, along with the ability to verify the nature and extent of these double crossover events.

16.5 MAPPING PHENOTYPIC TRAITS BY CLASSICAL GENETICS

16.5.1 Mapping Drug Resistance

The association between drug resistance and specific genetic markers has been evaluated for several phenotypic markers including FUDR, SNF, and AR-A (Khan *et al.*, 2005b). Resistance to each of these compounds is primarily determined by a single genetic locus. Their association with specific genetic markers was evaluated using quantitative trait locus (QTL) mapping (Lander and Botstein, 1989; Lander and Kruglyak, 1995) to analyse both single-locus effects and secondary loci that influence resistance. QTL analysis showed a single strong association for each of these compounds with no detectable contribution from other loci (Fig. 16.3).

The molecular targets for two of these compounds, FUDR and ARA-A, were previously known and they serve as controls for the precision of mapping by linkage analysis. Resistance to FUDR occurs due to loss of UPRT activity, the gene for which is located in the centre of chromosome XI. Mapping FUDR resistance showed perfect correspondence to this locus (Fig. 16.3). Resistance to ARA-A occurs due to disruption of adenosine kinase activity, the gene for which is located at the end of chromosome XII between markers AK165 and AK163 (Fig. 16.3). Although the target for SNF

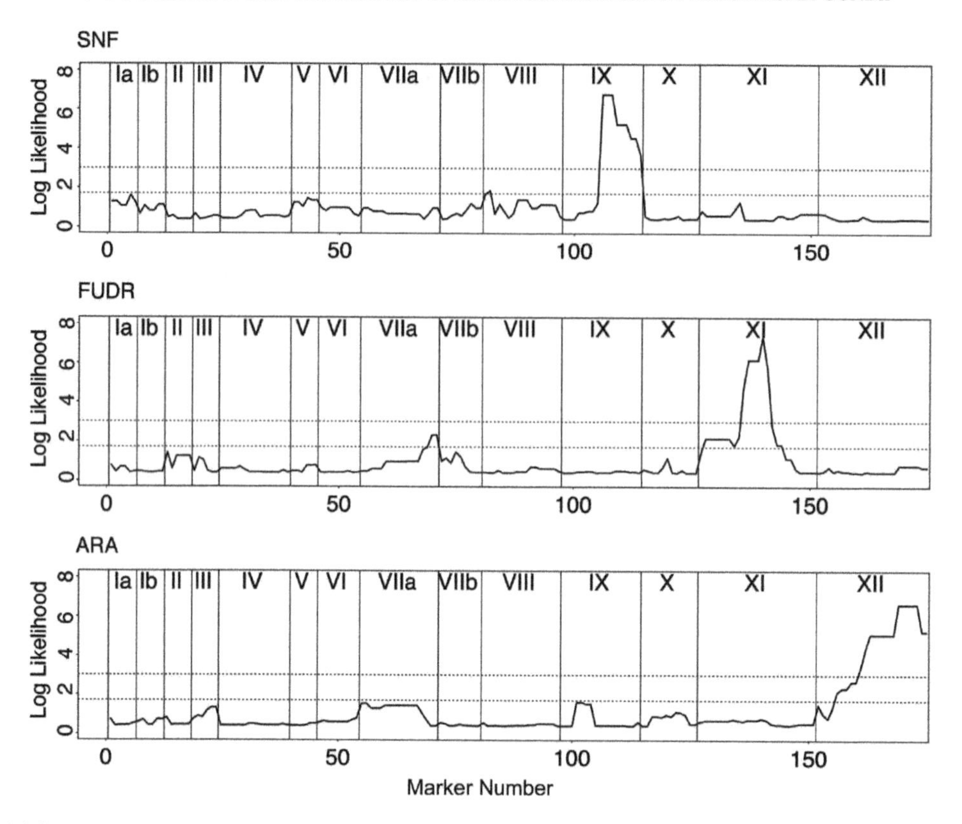

FIGURE 16.3 **Mapping of loci that confer resistance to SNF, FUDR and ARA-A.** Plots indicate the association of resistance based on *in vitro* growth of tachyzoites versus genetic markers aligned along the X-axis. Chromosome numbers are given at the top. A single major peak for resistance is identified for each drug treatment. Dotted horizontal lines indicate significance (lower line) and highly significant (upper line). Used with permission from Khan *et al.* (2005b).

in *T. gondii* is presently unknown, the resistance phenotype mapped to a region on chromosome IX associated with the markers AK123 to MIC1. SNF is known to inhibit methylation reactions (Martin and McMillan, 2002). Preliminary examination of this region of the genome identifies a nearby hypothetical protein TgME49_210310 with a GO (0018024) annotation of histone-lysine N-methyltransferase activity (http://www.ToxoDB.org), although this gene lies just outside the QTL peak.

The relatively large genomic intervals defined by these mapping studies illustrate one of the primary limitations of the linkage mapping in *T. gondii*: the low rate of recombination limits resolution. The map unit (cM) in *T. gondii* is approximately 100 kb (Khan *et al.*, 2005b), predicting that a large number of progeny would have to be evaluated to map a trait to a region this small. Typically a one cM region contains 10–20 genes, many of which have complex splicing of multiple exons. The advent of microarray or Next-Gen sequencing based mapping methods is expected to provide greater resolution for mapping genetic crossovers in individual progeny, thus providing more precise mapping even from a more limited set of progeny. Once a candidate locus has been identified, the completed genome sequence for *T. gondii* (http://www.ToxoDB.org) makes it feasible to readily align markers to specific regions and identify candidate genes, thus allowing application of reverse genetic approaches.

16.5.2 Mapping Quantitative Traits

Many phenotypic traits such as growth and virulence are not governed by a single locus but rather by the combination of genes at different loci. Additionally, other regions of the genome can modulate traits that are strongly influenced by a single locus. Studies by Lander and Botstein (Lander *et al.*, 1987; Lander and Botstein, 1989; Lander and Kruglyak, 1995) provided the foundation for genetic analysis of such complex traits. In particular, these studies demonstrated that interval mapping based on genetic markers can reliably map quantitative loci (QTLs). The power of QTL mapping strategies for analysing complex phenotypes has led to great interest in application of this technology in parasites (Su and Wootton, 2004). QTL analysis not only estimates the contributions of different loci, it provides statistical methods to account for the potential of false-positive associations that occur with large datasets (Lynch and Walsh, 1998). Typically this is expressed as a likelihood ratio that relates the probability that a given association is real versus that it might occur by chance. The outcomes of these calculations are referred to as the log likelihood ratio statistic or log odds ratio (i.e. LOD score) (Lynch and Walsh, 1998). LOD scores of more than three are considered significant, but very strong support requires LOD scores of more than five to six. However, even low LOD scores can provide meaningful clues about loci that may contribute to a trait and this might justify analysis of further progeny in order to confirm or refute the preliminary association.

16.5.3 Genetic Approaches for Defining Virulence Genes

One of the main advantages of classical or forward genetics is that it provides an unbiased analysis of complex phenotypes. In other words, genetic linkage mapping has the potential to identify unknown genes that mediate important, naturally occurring biological phenotypes. The strength of this approach is that it requires no *a priori* assumptions about the underlying molecular basis of the trait, so long as it differs between natural isolates and it can be reliably measured. Strains of *T. gondii* differ substantially in their ability to cause disease in laboratory mice. More specifically, strains of the type I genotype are acutely virulent in the mouse model. Type I strains are uniformly lethal in mice and even at low inocula, infected animals do not survive challenge with tachyzoites given by i.p. inoculation (translates into an effective LD_{100} of a single organism) (Sibley and Boothroyd, 1992a; Howe *et al.*, 1996). Although laboratory passage might affect this trait, the commonly used RH strain was reported to be virulent for mice on primary isolation (Sabin, 1941), and similar high virulence was reported among a collection of type I isolates (Sibley and Boothroyd, 1992). In contrast, type II strains show intermediate virulence (i.e. LD_{50} 10^3–10^5 depending on mouse strain) and type III strains are essentially avirulent (Sibley and Boothroyd, 1992; Howe *et al.*, 1996). Although more susceptible to toxoplasmosis than natural hosts of transmission such as wild deer mice (Frenkel, 1953), rats (Dubey and Frenkel, 1998) or chickens (Dubey, 2008), laboratory mice offer the potential to uncover the molecular basis of differences in acute virulence between natural isolates of *T. gondii.*

16.5.3.1 Mapping Differences in the I × III Cross

Quantitative trait mapping was initially used to evaluate the acute virulence of *T. gondii* in the mouse model using the progeny of a type I (virulent) × III (avirulent) cross (Su *et al.*, 2002). GT-1 has a virulent phenotype typical of type I strains, such that a single viable organism is lethal in outbred laboratory mice, while CTG has an LD_{50} of more than 10^6 in outbred mice. Twenty-six F1 progeny clones were initially tested to establish their virulence phenotype in mice following inoculation with 10^1, 10^2 or 10^3 tachyzoites and virulence was defined based on cumulative mortality. Virulent clones did not give rise to chronic infection (all infected animals died), while other clones showed intermediate or low levels of virulence (Su *et al.*, 2002). Phenotypic analysis of recombinant progeny revealed a range of phenotypes from fully virulent to non-virulent and including intermediate levels not seen in either parental strain (Su *et al.*, 2002). When acute mortality was considered as single gene model, it was mapped to a single QTL in the central region of chromosome VIIa (Su *et al.*, 2002). These results confirmed that virulence was largely heritable rather than epigenetic and that it was primarily controlled by a single locus (Su *et al.*, 2002). This finding was an important conceptual advance, validating use of forward genetics to analyse the molecular basis of virulence. However, at this early stage, the genome was not yet sequenced or assembled, hence it remained only a hypothetical possibility to map and hence identify individual genes that conferred enhanced virulence.

Completion of the genome sequence of the type II strain ME49 and the assembly of the genome using the combined genetic map (Khan *et al.*, 2005) made it possible to map traits to defined genomic intervals and to identify genes that controlled important biological phenotypes. Differences between the virulent type I and avirulent type III lineages were re-examined using an expanded set of 34 unique clones typed with 175 different PCR–RFLP based markers (Taylor *et al.*, 2006). In addition to acute virulence, these strains also differ in their ability to migrate across polarized monolayers and to migrate under agarose *in vitro* (Barragan and Sibley, 2002) and in their intrinsic growth rate (Radke *et al.*, 2001). Differences in migration *in vitro* differ substantially between the type I (high) and type III (low) parental strains (Taylor *et al.*, 2006) (Fig. 16.4A). The migration potential of the progeny varied continuously with some clones having higher migration potential than either parent (Fig. 16.4A). Likewise, differences in acute virulence in the mouse showed a range of different phenotypes from an absence of mortality like the type III parent to uniform lethality like the type I parent (Fig. 16.4B). Analysis of these traits by QTL mapping revealed that migration *in vitro*, transmigration across polarized monolayers, acute mortality (i.e. virulence), and the serum response of surviving mice all mapped to a single peak on chromosome VIIa (Taylor *et al.*, 2006) (Fig. 16.4C, Table 16.2). A minor peak on chromosome Ia was noted for serum response and mortality, but no other peak showed statistical significance (Fig. 16.4C, Table 16.2). The single QTL on chromosome VIIa was predicted to account for ~65% of the variance in virulence between the parental types, leaving open the possibility that other loci contribute smaller effects that were not mapped due to the low number of progeny examined. Nonetheless, finding a single major peak was highly surprising, since it might be expected that a trait like acute virulence would be multigenic and show contributions from many small loci. Such a multi-locus pattern was seen for differences in intrinsic growth rate, which also shared a peak on chromosome VIIa (Taylor *et al.*, 2006) (Fig. 16.4C, Table 16.2).

Finer mapping of the peaks for acute virulence and serum response of surviving animals showed perfect correspondence on chromosome VIIa and addition of new markers within this locus, and several additional recombinant clones with crossovers on chromosome VIIa, allowed finer mapping (Taylor *et al.*, 2006) (Fig. 16.5A). To further refine the choice of candidate genes within this region, RNAs from the parental

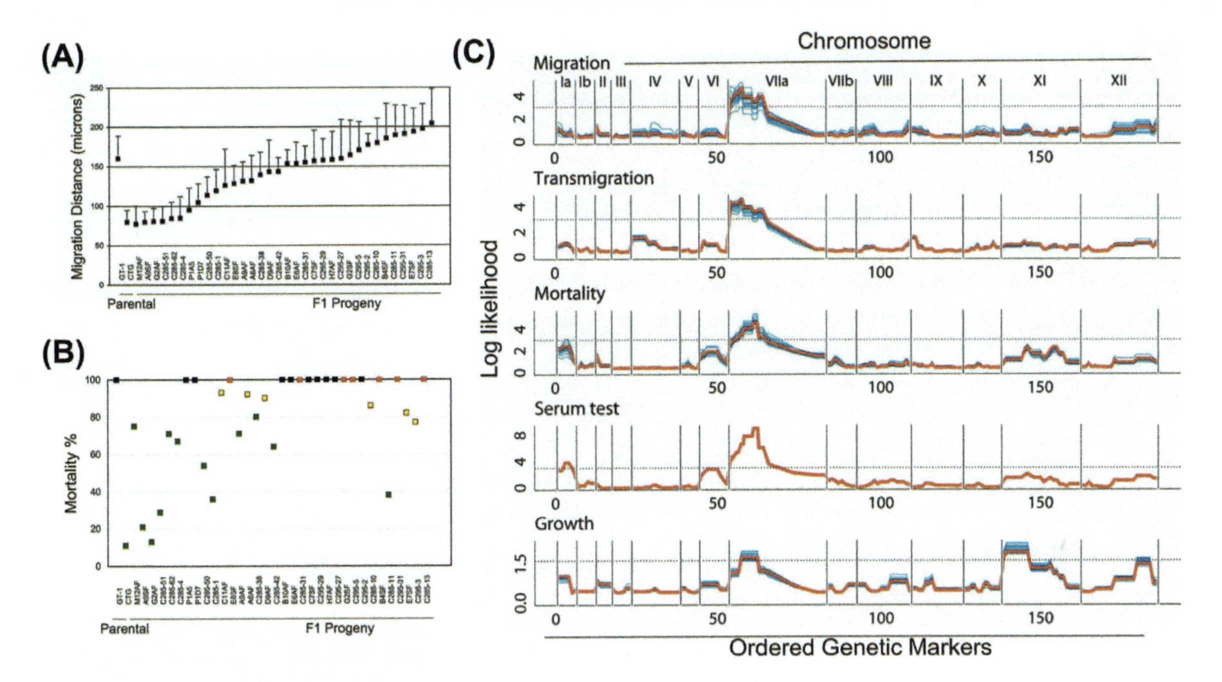

FIGURE 16.4 **Phenotypic analysis of clones from a I × III genetic cross and mapping of genome-wide associations.** A) Differences in migration distance under soft agarose *in vitro*.
B) Differences in acute mortality in outbred CD1 mice.
C) Genome-wide QTL scans for migration, transmigration, mortality, serum responsiveness and growth rate *in vitro*. The first four phenotypes mapped to a single peak on chromosome VIIa, while growth maps to several loci on chromosome VIIa, XI and XII. Genetic markers are ordered along the X-axis and chromosome names are listed at the top. Dotted line indicates significance level, red line corresponds to the log likelihood plot while blue lines represent the 95% confidence interval. Used with permission from Taylor *et al*, (2006).

TABLE 16.2 Summary of Phenotypes Mapped in Genetic Crosses Between Clonal Strains of *T. gondii*

Genetic Cross	Phenotypes	Chromosomes with QTLs	Genes	Function
I × III	Acute virulence	*VIIa, Ia*	*ROP18*	Phosphorylates IRGs, ATF6β
	Growth	*VIIa, XI, XII, Ia*		
	Migration	*VIIa*		
II × III	Acute virulence	*VIIa, VIIb, XII*	*ROP18 ROP5*	Targets IRGs
	Host gene expression	*VIIb*	*ROP16*	Phosphorylates STAT3/6
	IL-12 induction	*X*	*GRA15*	Activates NFκB
I × II	Acute virulence	*XII*	*ROP5*	Activates ROP18 Binds Irga6

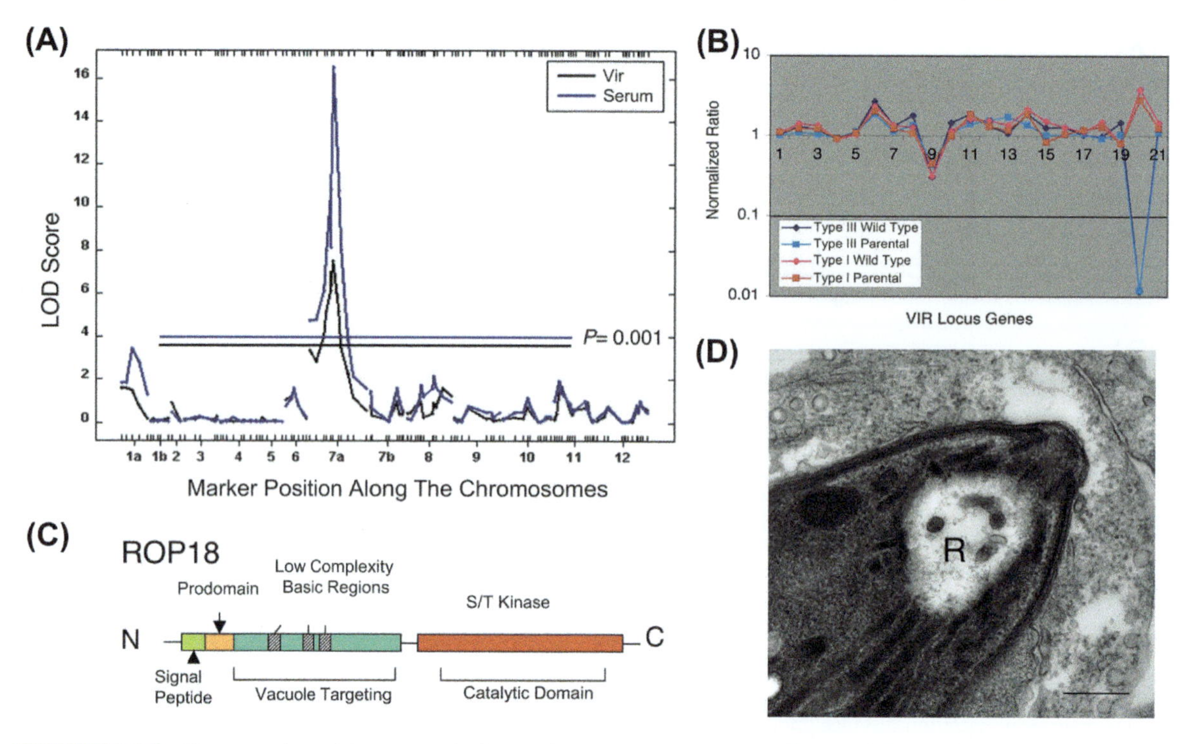

FIGURE 16.5 **Fine mapping of the acute virulence phenotype in the I × III cross.**
A) Plot of acute virulence and serum responsiveness showing the prominent peak on chromosome VIIa, which accounts for more than 65% of the variance in these traits.
B) Microarray showing hybridization to genes in the centre of this region identifies a prominent difference at the right end. This gene, annotated as rhoptry protein 18 (ROP18), is under-expressed by 100-fold in the type III strain.
C) Model of the domain structure of ROP18 revealing a signal peptide pro-piece that is cleaved during maturation, N-terminal low complexity region involved in membrane targeting, and a serine/threonine (S/T) protein kinase domain.
D) Electron micrograph showing discharge of rhoptries (R) during invasion, releasing ROP proteins into the host cell cytoplasm and the lumen of the parasitophorous vacuole. Scale bar is 100 nM. Image provided by Wandy Beatty.

stains were hybridized to an Affymetrix array for *T. gondii* (Bahl *et al.*, 2010), revealing a marked difference in one gene on the right end of the original QTL locus: expression levels for this gene in a type III strain were ~100 fold lower than that in type I (Taylor *et al.*, 2006) (Fig. 16.5B). This differentially expressed gene was annotated as rhoptry protein 18 (ROP18) based on a previous proteomic study of purified rhoptry contents examined by mass spectrometry (Bradley *et al.*, 2005). ROP18 contains a signal peptide and a pro-domain that is processed during secretion, giving rise to a mature protein that consists of an N-terminal amphipathic helical region and a C-terminal serine threonine (S/T) protein kinase domain (Fig. 16.5C). Rhoptries are glandular organelles that discharge their contents into the forming parasitophorous vacuole during invasion (Nichols *et al.*, 1983) (Fig. 16.5D). Previous studies had revealed that ROP proteins are also discharged into the host cell during invasion, leading to the prediction that they might inject effector proteins that would alter host cell signalling (Håkansson *et al.*, 2001). The discovery of ROP18 as a putative virulence factor suggested such a potential role for this secretory protein in modulating host pathways that affect pathogenesis.

Confirmation that ROP18 was indeed responsible for the differences in acute virulence between types I and III relied on reverse genetic approaches, which have also been extensively developed in *T. gondii* (see Chapter 17 in this volume). Taking advantage of the fact that ROP18 gene expression is essentially off in the type III background, transgenic parasites were generated to express the type I ROP18 allele at levels similar to that in the type I strain (Taylor *et al.*, 2006). Restored expression of a kinase-active form of ROP18 in a type III recipient conferred enhanced virulence that nearly matched the type I parental strain (Taylor *et al.*, 2006). However, expression of a mutant form of ROP18, where the catalytic Asp had been changed to Ala, failed to enhance virulence (Taylor *et al.*, 2006). These studies confirmed that ROP18 is largely responsible for the differences in acute virulence between types I and III in the laboratory mouse. The phenotypes of transmigration and enhanced *in vitro* migration also map to a region on chromosome VIIa, however they are not mediated by ROP18, nor is the QTL for growth on chromosome VIIa related to ROP18 (Taylor *et al.*, 2006). Rather these traits are mediated by additional genes found in this relatively broad region of the genome.

Clues to the molecular function of ROP18 came from its cellular location: following injection during host cell invasion, it is targeted to the cytoplasmic surface of the parasitophorous vacuole (Taylor *et al.*, 2006; El Hajj *et al.*, 2007). The N-terminal amphipathic helical region is responsible for tethering ROP18 on the cytoplasmic surface of the parasitophorous vacuole membrane (El Hajj *et al.*, 2007; Reese and Boothroyd, 2009) and this localization is essential to its virulence enhancing potential (Fentress and Sibley, 2012). On the parasitophorous vacuole membrane (PVM), ROP18 targets host immunity related GTPases (IRGs), which are normally recruited to the vacuole causing its vesiculation and destruction (Howard *et al.*, 2011). Destruction by IRGs is strain dependent, and type I strains avoid the IRG pathway while types II and III undergo IRG recruitment, leading to vesiculation and destruction in activated macrophages (Zhao *et al.*, 2009; Khaminets *et al.*, 2010). ROP18 was shown to directly phosphorylate a number of IRGs on key threonine residues in switch region 1, thus presumably inactivating the GTPase domain and preventing oligomerization (Fentress *et al.*, 2010; Steinfeldt *et al.*, 2010) (Table 16.2). In separate studies based on a yeast 2-hybrid screen, it was shown that ROP18 also phosphorylates ATF6β (Table 16.2) an ER stress response transcription factor (Yamamoto *et al.*, 2011). Mice lacking ATF6β show decreased antigen presentation by dendritic cells to $CD8^+$ T cells (Yamamoto *et al.*, 2011), suggesting phosphorylation of this factor by ROP18 might alter adaptive immunity.

16.5.3.2 Mapping Differences in the II × III Cross

In contrast to the very dramatic differences in acute virulence between types I and III, more subtle differences are seen between types II and III (Sibley and Boothroyd, 1992b; Howe *et al.*, 1996). These differences are best appreciated in inbred mice such as BALB/c, where type II shows higher levels of mortality and shorter times to death than challenge with type III parasites (Saeij *et al.*, 2006). Initial studies on the progeny of a II × III cross showed that recombination of genes can lead to higher levels of virulence than is seen in either of the parental strains, albeit still not to the extreme level exhibited by type I strains (Grigg *et al.*, 2001a). Further analysis of the progeny from this cross was based on the time to death following low (100 parasites i.p.) versus high (100,000 parasites i.p.) dose challenge. The largest effect, a locus termed VIR1 located on chromosome XII, was the only QTL to reach high LOD score (i.e. greater than three) (Saeij *et al.*, 2006) (Fig. 16.6A). When VIR1 was treated as a co-variant, it revealed four additional QTLs

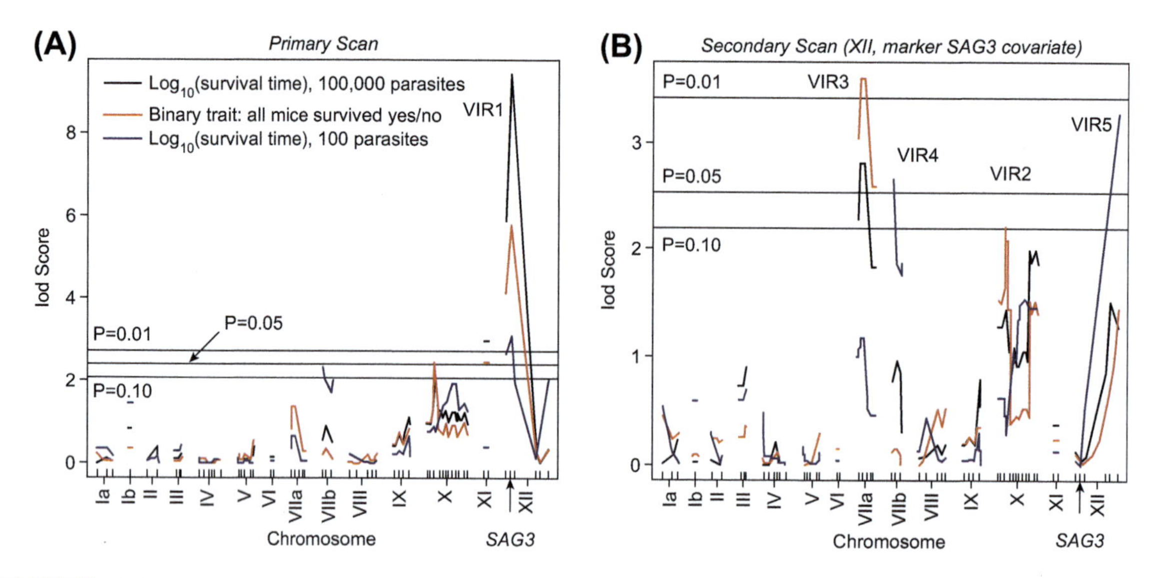

FIGURE 16.6 Genome-wide mapping of acute virulence traits in a genetic cross between types II and III. A) Primary QTL scan reveals a major peak on chromosome XII (VIR1).
B) When this peak is run as a covariate, it revealed several smaller peaks on VIIa (VIR3), VIIb (VIR4), X (VIR2) and a second peak on XII (VIR5). Used with permission from Saeij *et al.* (2006).

that showed moderate to low statistical support (i.e. LOD scores two to four) (Saeij *et al.*, 2006). (Fig. 16.6B). Comparison of other features such as degree of polymorphism, differences in expression level and likelihood of being secreted, and hence interacting with the host, led to ROP18 as the candidate locus for VIR3 on chromosome VIIa (Saeij *et al.*, 2006). Quantitative PCR revealed that ROP18 was under-expressed by more than 1000 fold in type III relative to type II, which expresses similar levels to type I (the fold differences reported here versus above are likely due to sensitivity of microarrays versus qPCR). Generation of transgenic type III parasites expressing the type II ROP18 allele also led to enhanced virulence, confirming the role of this kinase in pathogenesis of *T. gondii* (Saeij *et al.*, 2006). In comparing the two genetic crosses it seems evident that the differences in expression level of ROP18 are largely responsible for the extremely low virulence of type III, since expression of either the type I or type II alleles is sufficient to rescue this phenotype (Table 16.2). Although direct gene swaps of ROP18 in types I and II have not been made, genetic studies below argue that they function similarly in conferring mouse virulence, despite showing a high level of polymorphism between them. The major differences in ROP18 expression have been traced to a difference upstream of the gene where types I and II lack an ancestral region found in type III and the out-group *Neospora caninum* (Khan *et al.*, 2009). Deletion of this upstream region appears to have resulted in increased expression, which is manifest as enhanced virulence in the laboratory mouse (Khan *et al.*, 2009). However, both the low-expression type III allele and the higher expression type II and I alleles show evidence of long-term stability in the wild, suggesting that they are adapted for different niches in nature (Khan *et al.*, 2009).

Using a similar strategy, mapping of the QTL defined by VIR4 on chromosome VIIb led to the identification of ROP16, another polymorphic kinase secreted onto the host cell during invasion (Table 16.2). The type III allele of ROP16 is associated with decreased virulence, and expression of either the type III, or closely related type I, alleles in a type II strain led to slightly reduced mortality following challenge of mice (Saeij *et al.*, 2006). In separate studies, it was shown that ROP16 is targeted to the host cell nucleus and that it drives host gene expression, polarizing toward IL-6 and IL-4 dominated responses while down-modulating IL-12 production (Saeij *et al.*, 2007). ROP16 was subsequently shown to directly phosphorylate STAT3 and STAT6, leading to their prolonged activation and induction of cytokines such as IL-4 and IL-6, while down-regulating IL-12 (Yamamoto *et al.*, 2009; Ong *et al.*, 2010).

Previous studies have also shown that infection with *T. gondii* strains differentially affects nuclear translocation of NFκB and that this leads to much stronger induction of IL-12 by type II strains relative to type I or III (Robben *et al.*, 2004). Further analysis of genes that were differentially altered by infection between types II and III also revealed a network of genes centred on NFκB mediated transcription (Rosowski *et al.*, 2011). Using an assay for nuclear translocation of NFκB, a factor that directly drives expression of IL-12 was mapped using the II × III cross to identify a locus encoding a secretory dense granule protein GRA15 (Rosowski *et al.*, 2011) (Table 16.2). The allele of GRA15 found in type II strains is capable of activating NFκB when expressed in HeLa cells, showing it acts directly (Rosowski *et al.*, 2011). Expression of type II GRA15 in a type I strain also increases NFκB translocation and leads to elevated IL-12 production (Rosowski *et al.*, 2011). Although GRA15 is found on chromosome X, where the VIR2 locus was identified in the II × III cross (Saeij *et al.*, 2006), it does not exactly correspond to this QTL interval. Genetic ablation of GRA15 in the type II strains does not change lethality but does lead to increased parasite growth *in vivo* (Rosowski *et al.*, 2011). Although the precise mechanism of NFκB activation by GRA15 is not known, these findings demonstrate the potential for additional secretory compartments to impact host cell signalling (Table 16.2). Subsequent studies have revealed that ROP16 and GRA15 act together to augment host signalling, driving a polarization of macrophage responses depending on the strain type (Butcher *et al.*, 2011; Jensen *et al.*, 2011) (see also Chapter 14 in this volume).

16.5.3.3 Mapping Differences in the I × II Cross

The final pair-wise comparison came with the completion of a genetic cross between highly virulent type I GT-1 strain and the intermediate virulence type II ME49 strain (Behnke *et al.*, 2011). Analysis of virulence differences among progeny from this cross was again based on mortality in groups of outbred CD1 mice challenged by i.p. inoculation with 100 parasites, a relatively low dose. The parental strain GT-1 results in complete mortality, while ME49 shows no lethality at this dose. The phenotypes of 45 F1 progeny were analysed from this cross, revealing a wide range of virulence in outbred mice (Behnke *et al.*, 2011). QTL analysis of the mortality data revealed a single strong peak on chromosome XII (Table 16.2) (Fig. 16.7A). This QTL was predicted to account for ~90% of the variance in virulence between the parental strains (Behnke *et al.*, 2011). Surprisingly, neither ROP18 nor ROP16 showed QTL peaks, indicating these genes either contribute similarly to acute virulence as in the case for ROP18, or may not contribute substantially to this trait in either strain, as appears to be the case for ROP16 (Fig. 16.7A).

The major virulence QTL on chromosome XII spans ~400 kb and includes 51 genes flanked by several single-feature polymorphism probes that were informative in cross (Fig. 16.7B). To identify the responsible gene with the chromosome XII QTL, differences in copy number were estimated from hybridization of gDNA, while differences in expression level were determined from cDNA hybridization to the *T. gondii* Affymetrix array. Together these approaches identified a cluster of polymorphic genes encoding the pseudokinase

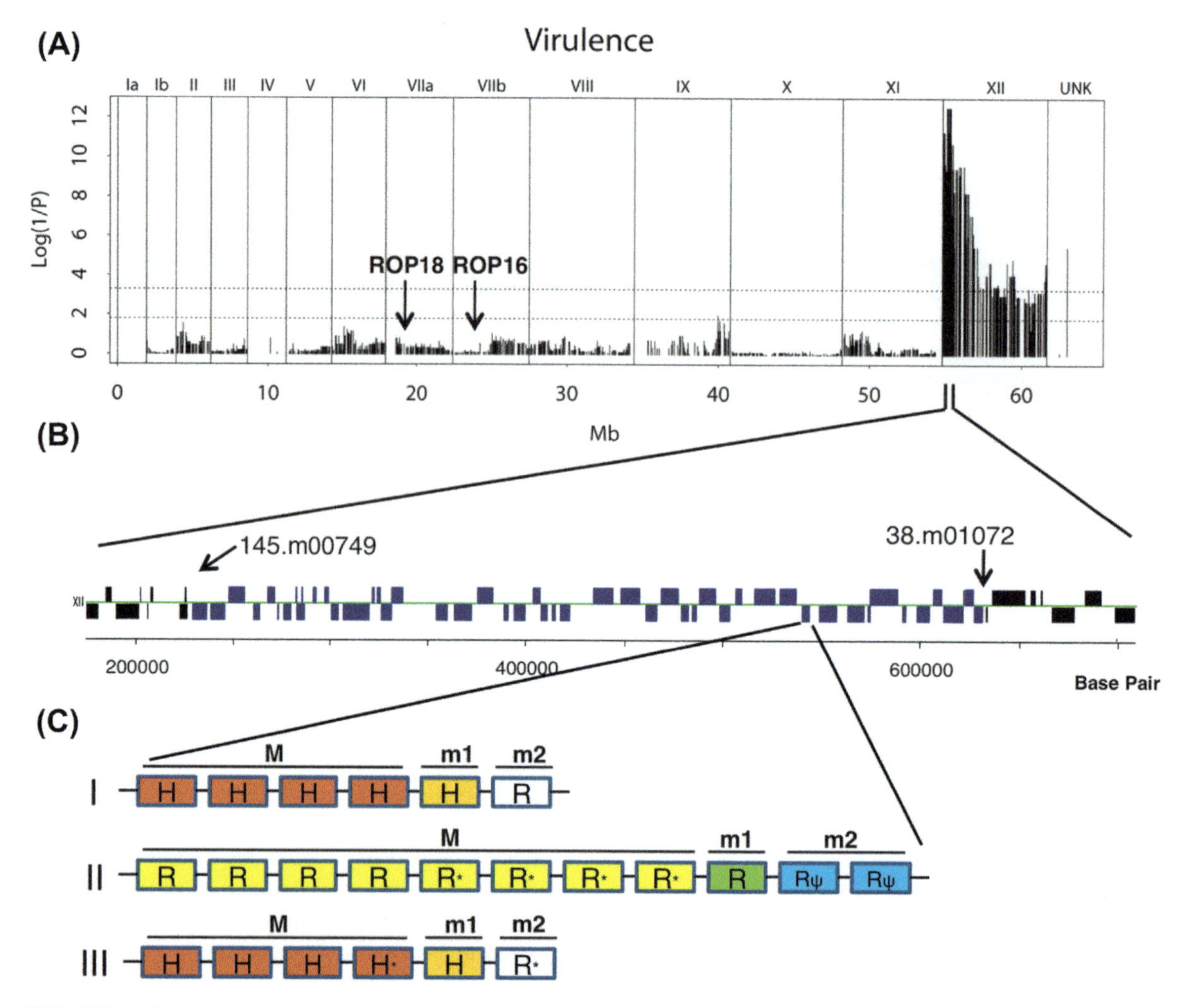

FIGURE 16.7 **Genetic analysis of acute virulence in a cross between type I and II strains.**
A) Genome-wide associate of genetic markers defined by allelic differences detected by hybridization to Affymetrix arrays. A single, strong peak is seen on chromosome XII, while previously mapped genes for *ROP16* and *ROP18* play no role in virulence differences between these strains.
B) Enlargement of the region on the left end of XII reveals ~130 genes lie within this region, including a polymorphic cluster of pseudokinases known as *ROP5* that shows copy number variation.
C) Diagram of the genetic loci encoding ROP5 variants in the three lineages. Types I and III contain almost identical loci consisting of a major allele that contains the catalytic triad KHD (marked H). In contrast, type II strains contain more copies, dominated by a major allele that has a catalytic triad of KRD (marked R). Alleles marked with * represent minor variants. Ψ denotes pseudogene. Used with permission from Behnke *et al*, (2011).

ROP5 as the likely candidate. ROP5 had previously been identified as a pseudokinase related to ROP2 (El Hajj *et al.*, 2006), and to ROP18, although it had no known role in virulence. To confirm the importance of ROP5, deletion of the entire locus in the type I RH strain led to highly attenuated parasites, which were rescued by re-expression of ROP5 from a type I cosmid (Behnke *et al.*, 2011). Reassembly of the genomic locus from types I, II and III revealed two different patterns of gene arrangement (Fig. 16.7C). In types I and III, a cluster of six genes is found, where the dominant alleles are the same and are characterized by an incomplete catalytic triad consisting of KHD (Fig. 16.7C). In contrast, type II contains ~11 genes in tandem (two of which are pseudogenes), where the major allele has a number of differences from type I/III, including having an incomplete catalytic triad consisting of KRD (Fig. 16.7). Consistent with higher copy number, expression levels of ROP5 are actually higher in type II parasites, indicating that unlike ROP18, expression level is unlikely to be the basis for its contribution to virulence.

The functional significance of ROP5 allelic differences is not entirely clear; however, genetic studies demonstrate that the type I allelic cluster is associated with enhanced virulence, while that in type II is associated with lower virulence (Behnke *et al.*, 2012). A similar role for ROP5 was also found in re-analysis of the progeny of the II × III cross, leading to identification of the major peak on chromosome XII as ROP5 (Reese *et al.*, 2011). In this context, the type III allele was again associated with higher levels of virulence, while the type II allele was associated with lower virulence. Hence, types I and III share a virulence enhancing form of ROP5, despite have widely different phenotypes, a result largely due to their differential expression of ROP18. Although ROP5 is catalytically inactive (Reese and Boothroyd, 2011), it plays a major role in virulence, and genetic studies indicate it does so in combination with ROP18 (Behnke *et al.*, 2012). Pseudokinases in other systems have recently been recognized as playing important roles in regulation of active kinases (Boudeau *et al.*, 2006; Zeqiraj and van Aalten, 2010). Subsequent studies have shown that ROP5 and ROP18 act together to disrupt IRG accumulation on the PVM, in part by binding of ROP5 to IRGs, and the ability of ROP5 to directly enhance the enzymatic activity of ROP18 (Behnke *et al.*, 2012; Fleckenstein *et al.*, 2012).

16.5.4 Summary of Differences between Lineages

The natural differences seen in acute virulence in the mouse between the three lineages are likely due to the different assortment of alleles they acquired following the relatively few genetic crosses since their common origin (Grigg and Suzuki, 2003; Su *et al.*, 2003). Comparison of the three pair-wise genetic crosses reveals that the differences in acute virulence are due to the combination of particular alleles of ROP18 and ROP5. Type I strains express a highly active ROP18 allele and a 'virulent' form of ROP5 that enhances virulence. Type II strains have intermediate virulence and contain a ROP5 locus that is genetically 'avirulent', while harbouring a fully functional form of ROP18. Although type II strain parasites do not normally avoid IRG recruitment to the PVM, expression of type II ROP18 in a type III background reduces IRG recruitment, along with enhancing virulence (Niedelman *et al.*, 2012). This finding is consistent with ROP18 working cooperatively with ROP5 to prevent IRG accumulation. Type I and III strains share a major allele of ROP5 that is able to enhance virulence, although this is only fully apparent in the presence of ROP18. In contrast the ROP5 form found in type II is unable to enhance virulence, although the exact mechanistic reason is presently unclear.

It might seem somewhat surprising that so few genes contribute such strong effects to virulence in the mouse model. However, this result may reflect the inbred nature of laboratory mice, which are derived from a narrow genetic stock (Yang *et al.*, 2011) and the relatively inbred state of the three clonal lineage of *T. gondii* (Howe and Sibley, 1995; Su *et al.*, 2003; Boyle *et al.*, 2006). Such limited genetic diversity may magnify the effects of a few genes, which otherwise would contribute less to pathogenesis in either genetically more diverse host or parasite lineages. Collectively the rhoptry kinome includes some 40+ members, about half of which are predicted to be active while the remaining ones are predicted to be pseudokinases (Peixoto *et al.*, 2010). Although current genetic studies have only implicated a few of these, others may influence different aspects of the host-pathogen interaction, including during other life cycle stages and/or in other hosts.

16.5.5 Relevance of the Mouse Model to Other Species

The role of parasite virulence factors in non-rodent hosts is not well understood, in part owing to the fact that most relevant hosts are not experimentally tractable models. However, it can be argued that any evolutionary pressure on traits that influence pathogenesis would only occur in natural hosts, where potential effects on fitness might affect transmission. Hence the interaction between ROP kinases and IRGs has likely been heavily influenced by the role of rodents in natural transmission. IRGs are greatly expanded in the rodents but much more restricted in other mammals and almost entirely absent in humans (Howard *et al.*, 2011). Consequently, it has been argued that ROP5–ROP18 may not contribute to evasion of innate immunity in human cells, although this has only been evaluated in fibroblasts activated *in vitro* with IFNγ (Niedelman *et al.*, 2012). In contrast to IRGs, a second family of GTPases called the p65 or guanylate binding proteins (GBPs) is much more widely expressed including in humans (Shenoy *et al.*, 2008). Recent genetic studies have shown that GBPs are important in innate immunity in the mouse (Kim *et al.*, 2011; Yamamoto *et al.*, 2012), although their role in other hosts including humans remains uncertain.

T. gondii is known to infect a wide range of mammals, many of these such as goats, chickens and pigs are relatively resistant to disease during acute infection, yet commonly become chronically infected. In contrast, other groups show high susceptibility including marsupials and New World monkeys (Dubey, 2010). Rodents show a wide range of susceptibilities: laboratory mice are highly susceptible but deer mice are resistant to type I strains (Frenkel, 1953), and house mice may also harbour increased resistance not found in the lab strains. Rats are relatively resistant to infection (Dubey and Frenkel, 1998) and one factor mediating this has been mapped to a locus called *toxo1* (Cavailles *et al.*, 2006), which contains 1.7 cM region. Among the genes contained in this region is NALP1, a member of the NOD-like receptor (NLR) family, a system of intracellular sensors involved in activating the inflammasome (Rathinam *et al.*, 2012). Recent studies in human susceptibility to congenital toxoplasmosis also report an association with NALP1 (Witola *et al.*, 2011), suggesting they may share a similar pathway for resistance.

Because humans are an accidental host for *T. gondii*, strain differences in pathogenicity seen in humans are not shaped by evolutionary pressure but rather are coincidental. In North America and Europe, human infections are most often caused by type II strains (Howe and Sibley, 1995; Howe *et al.*, 1997; Ajzenberg *et al.*, 2002a,b), yet several studies suggest that type I strains, or strains harbouring alleles typically found in type I strains, are also more pathogenic in humans (Fuentes *et al.*, 2001; Grigg *et al.*, 2001; Khan *et al.*, 2005). Additionally, a number of South American lineages have been associated with severe ocular disease in otherwise healthy individuals (Vallochi *et al.*, 2005; Jones *et al.*, 2006; Khan *et al.*, 2006).

16.6 FUTURE CHALLENGES

16.6.1 Overcoming Current Limitations

Classic genetic analyses in *T. gondii* have been extremely useful for defining basic parameters of recombination, mapping drug resistance loci, assembling the genome, and probing the molecular basis of pathogenesis in laboratory mice. Despite this progress, classical genetics remains a relatively difficult process in *T. gondii* compared to model organisms. Improvements that would accelerate progress include: (i) Development of *in vitro* methods for generation of gametocyte formation. Such techniques have been reported for *Eimeria* (Hofmann and Raether, 1990) and *Plasmodium* (Al-Olayan *et al.*, 2002). Successful completion of the life cycle *in vitro* would remove the single biggest obstacle to performing genetic crosses in *T. gondii.* (ii) Generation of new genetic crosses, especially between the South American lineages, which are mostly virulent in laboratory mice and less virulent North American lineages. (iii) Utilization of Next-Gen sequencing methods for rapid genotyping of recombinant progeny, a technology that might also be adapted to examine allele frequencies in pools hence alleviating the need to isolate clones. (iv) Development of higher throughout assays for monitoring meaningful biological phenotypes including host cell transcriptional differences, growth and effects on host cell signalling.

16.6.2 Expanding Phenotypic Analyses

In addition to the above mentioned phenotypes, there are a number of interesting biological questions that could be probed by classical genetic mapping. Differences in the migration rate of lineages of *T. gondii* have been linked to virulence (Barragan and Sibley, 2002). Although mapped to a broad region of chromosome VIIa, the molecular basis for this trait has not been defined. Differences in growth rates have been described for strains of *T. gondii* (Radke *et al.*, 2001), although it is not known to what extent these contribute to pathogenesis. Differences in the migratory responses of dendritic cells to chemokines (Diana *et al.*, 2004) have been reported to be strain-dependent in the mouse model. The development of intestinal necrosis (Liesenfeld, 2002) and CNS pathology (Suzuki *et al.*, 1989) in the mouse model are also influenced in part by the genotype of the parasite. Future genetic mapping studies may enlighten us about genes that contribute to these unique aspects of parasite biology and better define their roles in pathogenesis.

ACKNOWLEDGEMENTS

We would like to acknowledge the following individuals who have made important contributions to the advances summarized here: Mike Behnke, John Boothroyd, Jon Boyle, J.P. Dubey, Michael Grigg, Asis Khan, Loraine Pfefferkorn, Elmer Pfefferkorn, David Roos, Jeroen Saeij, Chunlei Su, Sonya Taylor, Michael White and John Wootton.

The National Institutes of Health (USA), United States Department of Agriculture (USA), Burroughs Wellcome Fund (USA), American Heart Association (USA), Wellcome Trust (UK), King Abdulla University of Science and Technology (KAUST), and Biotechnology and Biological Sciences Research Council (UK) have supported work in the authors' laboratories.

References

Ajioka, J.A., Boothroyd, J.C., Brunk, B.P., Hehl, A., Hillier, L., Manger, I.D., Overton, G.C., Marra, M., Roos, D., Wan, K.L., Waterston, R.H., Sibley, L.D., 1998. Gene discovery by EST sequencing in *Toxoplasma gondii* reveals sequences restricted to the Apicomplexa. Gen. Res. 8, 18–28.

Ajzenberg, D., Bañuls, A.L., Su, C., Dumètre, A., Demar, M., Carme, B., Dardé, M.L., 2004. Genetic diversity, clonality and sexuality in *Toxoplasma gondii*. Int. J. Parasitol. 34, 1185–1196.

Ajzenberg, D., Bañuls, A.L., Tibayrenc, M., Dardé, M.L., 2002a. Microsatellite analysis of *Toxoplasma gondii* shows considerable polymorphism structured into two main clonal groups. Int. J. Parasitology 32, 27–38.

Ajzenberg, D., Cogné, N., Paris, L., Bessieres, M.H., Thulliez, P., Fillisetti, D., Pelloux, H., Marty, P., Dardé, M.L., 2002b. Genotype of 86 *Toxoplasma gondii* isolates associated with human congenital toxoplasmosis and correlation with clinical findings. J. Infect. Dis. 186, 684–689.

Al-Olayan, E.M., Beetsma, A.L., Butcher, G.A., Sinden, R.E., Hurd, H., 2002. Complete development of mosquito phases of the malaria parasite *in vitro*. Science 295, 677–679.

Bahl, A., Davis, P.H., Behnke, M., Dzierszinski, F., Jagalur, M., Chen, F., Shanmugam, D., White, M.W., Kulp, D., Roos, D.S., 2010. A novel multifunctional oligonucleotide microarray for *Toxoplasma gondii*. BMC Genomics 11, 603.

Barragan, A., Sibley, L.D., 2002. Transepithelial migration of *Toxoplasma gondii* is linked to parasite motility and virulence. J. Exp. Med. 195 (12), 1625–1633.

Behnke, M.S., Fentress, S.J., Mashayekhi, M., Li, L.L., Taylor, G.A., Sibley, L.D, 2012. The polymorphic pseudokinase ROP5 controls virulence in *Toxoplasma gondii* by regulating the active kinase ROP18. PLoS Path. 8, e1002992.

Behnke, M.S., Khan, A., Wootton, J.C., Dubey, J.P., Tang, K., Sibley, L.D., 2011. Virulence differences in *Toxoplasma* mediated by amplification of a family of polymorphic pseuodokinases. Proc Natl Acad Sci (USA) 108, 9631–9636.

Boudeau, J., Miranda-Saavedra, D., Barton, G.J., Alessi, D.R., 2006. Emerging roles of pseudokinases. Trends Cell Biol. 16, 443–452.

Boyle, J.P., Rajasekar, B., Saeij, J.P.J., Ajioka, J.W., Berriman, M., Paulsen, I., Sibley, L.D., White, M., Boothroyd, J.C., 2006. Just one cross appears capable of dramatically altering the population biology of a eukaryotic pathogen like *Toxoplasma gondii*. Proc. Natl. Acad. Sci. (USA) 103, 10514–10519.

Bradley, P.J., Ward, C., Cheng, S.J., Alexander, D.L., Coller, S., Coombs, G.H., Dunn, J.D., Ferguson, D.J., Sanderson, S.J., Wastling, J.M., Boothroyd, J.C., 2005. Proteomic analysis of rhoptry organelles reveals many novel constituents for host-parasite interactions in *T. gondii*. J. Biol. Chem. 280, 34245–34258.

Butcher, B.A., Fox, B.A., Rommereim, L.M., Kim, S.G., Maurer, K.J., Yarovinsky, F., Herbert, D.R., Bzik, D.J., Denkers, E.Y., 2011. Toxoplasma gondii rhoptry kinase ROP16 activates STAT3 and STAT6 resulting in cytokine inhibition and arginase-1-dependent growth control. PLoS Pathog 7 (9), e1002236.

Cavailles, P., Sergent, V., Bisanz, C., Papapietro, O., Colacios, C., Mas, M., Subra, J.F., Lagrange, D., Calise, M., Appolinaire, S., Faraut, T., Druet, P., Saoudi, A., Bessieres, M.H., Pipy, B., Cesbron-Delauw, M.F., Fournie, G.J., 2006. The rat *Toxo1* locus directs toxoplasmosis outcome and controls parasite proliferation and spreading by macrophage-dependent mechanisms. Proc Natl Acad Sci (USA) 103, 744–749.

Cornelissen, A.W.C.A., Overdulve, J.P., 1985. Sex determination and sex differentiation in coccidia: gametogony and oocyst production after monoclonal infection of cats with free-living and intermediate host stages of *Isospora (Toxoplasma) gondii*. Parasitology 90, 35–44.

Cornelissen, A.W.C.A., Overdulve, J.P., Van der Ploeg, M., 1984a. Cytochemical studies on nuclear DNA of four eucoccidian parasites, *Isospora (Toxoplasma) gondii, Eimeria tenella, Sarcocystis cruzi and Plasmodium berghei*. Parasitology 88, 13–25.

Cornelissen, A.W.C.A., Overdulve, J.P., Van der Ploeg, M., 1984b. Determination of nuclear DNA of five eucoccidian parasites, *Isospora (Toxoplasma) gondii, Sarcocystis cruzi, Eimeria tenella, E. acervulina*, and *Plasmodium berghei*, with special reference to gametogenesis and meiosis in *I. (T.) gondii*. Parasitology 88, 531–553.

Dardé, M.L., Bouteille, B., Pestre-Alexandre, M., 1992. Isoenzyme analysis of 35 *Toxoplasma gondii* isolates and the biological and epidemiological implications. Journal of Parasitology 78, 786–794.

de Moura, L., Bahia-Oliveira, L.M., Wada, M.Y., Jones, J.L., Tuboi, S.H., Carmo, E.H., Ramalho, W.M., Camargo, N.J., Trevisan, R., Graca, R.M., da Silva, A.J., Moura, I., Dubey, J.P., Garrett, D.O., 2006. Waterborne toxoplasmosis, Brazil, from field to gene. Emerging Infectious Diseases 12, 326–329.

Diana, J., Persat, F., Staquet, M.J., Assossou, O., Ferrandiz, J., Gariazo, M.J., Peyron, F., Picot, S., Schmitt, D., Vincent, C., 2004. Migration and maturation of human dendritic cells infected with *Toxoplasma gondii* depends on parasite strain type. FEMS Immunol. Med. Microbiol. 42, 321–331.

Dubey, J., 1980. Mouse pathogenicity of *Toxoplasma gondii* isolated from a goat. American Journal of Veterinary Research 41, 427–429.

Dubey, J.P., 1977. *Toxoplasma, Hammondia, Besniotia, Sarcocystis*, and other tissue cyst-forming coccidia of man and animals. Parasitic Protozoa. J. P. Kreier. Academic Press, New York, 101–237.

Dubey, J.P., 1998. *Toxoplasma gondii* oocyst survival under defined temperatures. J. Parasitol. 84, 862–865.

Dubey, J.P., 2004. Toxoplasmosis – a waterborne zoonosis. Vet. Parasitol. 126 (1–2), 57–72.

Dubey, J.P., 2005. Unexpected oocyst shedding by cats fed *Toxoplasma gondii* tachzoites: *in vivo* stage conversion and strain variation. Vet. Parasitol. 133, 289–298.

Dubey, J.P., 2006. Comparative infectivity of oocysts and bradyzoites of *Toxoplasma gondii* for intermediate (mice) and definitive (cats) hosts. Vet. Parasitol. 140, 69–75.

Dubey, J.P., 2008. *Toxoplasma gondii* infections in Chickens (*Gallus domesticus*): Prevalence, clinical disease, diagnosis, and public health significance. Zoonoses and Public Health 57, 60–73.

Dubey, J.P., 2010. Toxoplasmosis of animals and humans. CRC Press, Boca Raton.

Dubey, J.P., Frenkel, J.F., 1972. Cyst-induced toxoplasmosis in cats. Journal of Protozoology 19, 155–177.

Dubey, J.P., Frenkel, J.K., 1998. Toxoplasmosis of rats: a review, with considerations of their value as an animal model and their possible role in epidemiology. Veterinary Parasitology 77, 1–32.

Dubey, J.P., Swan, G.V., Frenkel, J.K., 1972. A simplified method for isolation of *Toxoplasma gondii* from the faeces of cats. Journal of Parasitology 58 (5), 1005–1006.

El Hajj, H., Lebrun, M., Arold, S.T., Vial, H., Labesse, G., Dubremetz, J.F., 2007. ROP18 is a rhoptry kinase controlling the intracellular proliferation of *Toxoplasma gondii*. PLoS Pathogens 3, e14.

El Hajj, H., Lebrun, M., Fourmaux, M.N., Vial, H., Dubremetz, J.F., 2006. Inverted topology of the *Toxoplasma gondii* ROP5 rhoptry protein provides new insights into the association with the parasitophorous vacuole membrane. Cell. Microbiol. 9 (1), 54–64.

Fentress, S.J., Behnke, M.S., Dunay, I.R., Moashayekhi, M., Rommereim, L.M., Fox, B.A., Bzik, D.J., Tayor, G.A., Turk, B.E., Lichti, C.F., Townsend, R.R., Qiu, W., Hui, R., Beatty, W.L., Sibley, L.D., 2010. Phosphorylation of immunity-related GTPases by a parasite secretory kinase promotes macrophage survival and virulence. Cell Host Microbe 16, 484–495.

Fentress, S.J., Sibley, L.D., 2012. An arginine-rich domain of ROP18 is necessary for vacuole targeting and virulence in *Toxopalsma gondii*. Cell. Microbiol. 14, 1921–1933.

Ferguson, D.J.P., Birch-Andersen, A., Siim, J.C., Hutchison, W.M., 1979. Ultrastructural studies of the sporulation of oocysts of *Toxoplasma gondii*. Acta Pathology Microbiology Scandinavia 87, 183–190.

Fleckenstein, M.C., Reese, M.L., Konen-Waisman, S., Boothroyd, J.C., Howard, J.C., Steinfeldt, T., 2012. A *Toxoplasma gondii* Pseudokinase Inhibits Host IRG Resistance Proteins. PLoS Biol. 10 (7), e1001358.

Frenkel, J.K., 1953. Host, strain and treatment variation as factors in the pathogenesis of toxoplasmosis. American Journal of Tropical Medicine and Hygiene 2 (3), 390–415.

Frenkel, J.K., Dubey, J.P., Hoff, R.L., 1976. Loss of stages after continuous passage of *Toxoplasma gondii* and *Besnoitia jellisoni*. Journal of Protozoology 23, 421–424.

Frenkel, J.K., Dubey, J.P., Miller, N.L., 1970. *Toxoplasma gondii* in cats: faecal stages identified as coccidian oocysts. Science 167, 893–896.

Fuentes, I., Rubio, J.M., Ramírez, C., Alvar, J., 2001. Genotypic characterization of *Toxoplasma gondii* strains associated with human toxoplasmosis in Spain: direct analysis from clinical samples. J. Clin. Microbiol. 39, 1566–1570.

Gajria, B., Bahl, A., Brestelli, J., Dommer, J., Fischer, S., Gao, X., Heiges, M., Iodice, J., Kissinger, J.C., M. A.J., Pinney, D.F., Roos, D.S., Stoeckert, C.J., Wang, J., Brunk, B.P., 2007. ToxoDB: an integrated *Toxoplasma gondii* database resource. Nucl. Acids Res. 36, D553–D556.

Grigg, M.E., Bonnefoy, S., Hehl, A.B., Suzuki, Y., Boothroyd, J.C., 2001a. Success and virulence in Toxoplasma as the result of sexual recombination between two distinct ancestries. Science 294, 161–165.

Grigg, M.E., Ganatra, J., Boothroyd, J.C., Margolis, T.P., 2001b. Unusual abundance of atypical strains associated with human ocular toxoplasmosis. J. Infect. Dis. 184, 633–639.

Grigg, M.E., Suzuki, Y., 2003. Sexual recombination and clonal evolution of virulence in *Toxoplasma*. Microbes and Infection 5, 685–690.

Håkansson, S., Charron, A.J., Sibley, L.D., 2001. *Toxoplasma* evacuoles: a two-step process of secretion and fusion forms the parasitophorous vacuole. Embo J. 20 (12), 3132–3144.

Hofmann, J., Raether, W., 1990. Improved techniques for the *in vitro* cultivation of *Eimeria tenella* in primary chick kidney cells. Parasitology Research 76, 479–486.

Howard, J.C., Hunn, J.P., Steinfeldt, T., 2011. The IRG protein-based resistance mechanism in mice and its relation to virulence in *Toxoplasma gondii*. Curr. Opin. Microbiol. 14 (4), 414–421.

Howe, D.K., Honoré, S., Derouin, F., Sibley, L.D., 1997. Determination of genotypes of *Toxoplasma gondii* strains isolated from patients with toxoplasmosis. J. Clin. Microbiol. 35, 1411–1414.

Howe, D.K., Sibley, L.D., 1995. *Toxoplasma gondii* comprises three clonal lineages: correlation of parasite genotype with human disease. J. Infect. Dis. 172, 1561–1566.

Howe, D.K., Summers, B.C., Sibley, L.D., 1996. Acute virulence in mice is associated with markers on chromosome VIII in *Toxoplasma gondii*. Infection and Immunity 64, 5193–5198.

Jensen, K.D., Wang, Y., Wojno, E.D., Shastri, A.J., Hu, K., Cornel, L., Boedec, E., Ong, Y.C., Chien, Y.H., Hunter, C.A., Boothroyd, J.C., Saeij, J.P., 2011. Toxoplasma polymorphic effectors determine macrophage polarization and intestinal inflammation. Cell Host Microbe 9 (6), 472–483.

Jones, L.A., Alexander, J., Roberts, C.W., 2006. Ocular toxoplasmosis: in the storm of the eye. Parasite Immunol. 28, 635–642.

Khaminets, A., Hunn, J.P., Konen-Waisman, S., Zhao, Y.O., Preukschat, D., Coers, J., Boyle, J.P., Ong, Y.C., Boothroyd, J.C., Reichmann, G., Howard, J.C., 2010. Coordinated loading of IRG resistance GTPases on to the *Toxoplasma gondii* parasitophorous vacuole. Cell Microbiol. 12, 939–961.

Khan, A., Dubey, J.P., Su, C., Ajioka, J.W., Rosenthal, B.M., Sibley, L.D., 2011. Genetic analyses of atypical *Toxoplasma gondii* strains reveals a fourth clonal lineage in North America. Int. J. Parasitol. 41 (6), 645–655.

Khan, A., Fux, B., Su, C., Dubey, J.P., Darde, M.L., Ajioka, J.W., Rosenthal, B.M., Sibley, L.D., 2007. Recent transcontinental sweep of *Toxoplasma gondii* driven by a single monomorphic chromosome. Proc. Natl. Acad. Sci. (USA) 104, 14872–14877.

Khan, A., Jordan, C., Muccioli, C., Vallochi, A.L., Rizzo, L.V., Belfort Jr., R., Vitor, R.W., Silveira, C., Sibley, L.D., 2006. Genetic divergence of Toxoplasma gondii strains associated with ocular toxoplasmosis in Brazil. Emerg. Infect. Dis. 12, 942–949.

Khan, A., Su, C., German, M., Storch, G.A., Clifford, D., Sibley, L.D., 2005a. Genotyping of *Toxoplasma gondii* strains from immunocompromised patients reveals high prevalence of type I strains. J. Clin. Micro. 43, 5881–5887.

Khan, A., Taylor, S., Ajioka, J.W., Rosenthal, B.M., Sibley, L.D., 2009. Selection at a single locus leads to widespread expansion of *Toxoplasma gondii* lineages that are virulence in mice. PLoS Genetics 5, e1000404.

Khan, A., Taylor, S., Su, C., Mackey, A.J., Boyle, J., Cole, R.H., Glover, D., Tang, K., Paulsen, I., Berriman, M., Boothroyd, J.C., Pfefferkorn, E.R., Dubey, J.P., Roos, D.S., Ajioka, J.W., Wootton, J.C., Sibley, L.D., 2005b. Composite genome map and recombination parameters derived from three archetypal lineages of *Toxoplasma gondii*. Nucl. Acids Res. 33, 2980–2992.

Kim, B.H., Shenoy, A.R., Kumar, P., Das, R., Tiwari, S., MacMicking, J.D., 2011. A family of IFN-gamma-inducible 65-kD GTPases protects against bacterial infection. Science 332 (6030), 717–721.

Lander, E., Kruglyak, L., 1995. Genetic dissection of complex traits: guidelines for interpreting and reporting linkage results. Nature Genet. 11, 241–247.

Lander, E.S., Botstein, D., 1989. Mapping mendelian factors underlying quantitative traits using RFLP linkage maps. Genetics 121, 185–199.

Lander, E.S., Green, P., Abrahamson, J., Barlow, A., Daly, M.J., Lincoln, S.E., Newburg, L., 1987. MAPMAKER: An interactive computer package for constructing primary genetic linkage maps of experimental and natural populations. Genomics 1, 174–181.

Lehmann, T., Marcet, P.L., Graham, D.H., Dahl, E.R., Dubey, J.P., 2006. Globalization and the population structure of *Toxoplasma gondii*. Proc. Natl. Acad. Sci. (USA) 103, 11423–11428.

Li, L., Brunk, B.P., Kissinger, J.C., Pape, D., Tang, K., Cole, R.H., Martin, J., Wylie, T., Dante, M., Fogarty, S.J., Howe, D.K., Liberator, P.A., Diaz, C., Anderson, J., White, M., Jerome, M.E., Johnson, E.A., Radke, J.A., Stoeckert Jr., C.J., Waterston, R.H., Clifton, S.W., Roos, D.S., Sibley, L.D., 2003. Gene discovery in the Apicomplexa as revealed by EST sequencing and assembly of a comparative gene database. Gen. Res. 13, 443–454.

Liesenfeld, O., 2002. Oral infection of C57BL/6 mice with *Toxoplasma gondii*: a new model of inflammatory bowel disease? J. Infect. Dis. 185, S96–101.

Lunde, M.N., Jacobs, L., 1983. Antigenic differences between endozoites and cystozoites of *Toxoplasma gondii*. Journal of Parasitology 65, 806–808.

Lynch, M., Walsh, B., 1998. Genetics and Analysis of Quantitative Traits. Sinauer Associates, Inc, Sunderland.

Manger, I.D., Adrian, H., Parmley, S., Sibley, L.D., Marra, M., Hillier, L., Waterston, R., Boothroyd, J.C., 1998. Expressed sequence tag analysis of the bradyzoite stage of *Toxoplasma gondii*: Identification of developmentally regulated genes. Infect. Immun. 66 (4), 1632–1637.

Martin, J.L., McMillan, F.M., 2002. SAM (dependent) I AM: the S-adenosylmethionine-dependent methyltransferase fold. Curr. Opin. Struct. Biol. 12, 783–793.

Mead, P.S., Slutsker, L., Dietz, V., McCaig, L.F., Bresee, J.S., Shapiro, C., Griffin, P.M., Tauxe, R.V., 1999. Food-related illness and death in the United States. Emerg. Infect. Dis. 5, 607–625.

Nichols, B.A., Chiappino, M.L., O'Connor, G.R., 1983. Secretion from the rhoptries of *Toxoplasma gondii* during host-cell invasion. J. Ultrastruct. Res. 83, 85–98.

Niedelman, W., Gold, D.A., Rosowski, E.E., Sprokholt, J.K., Lim, D., Farid Arenas, A., Melo, M.B., Spooner, E., Yaffe, M.B., Saeij, J.P., 2012. The Rhoptry Proteins ROP18 and ROP5 Mediate *Toxoplasma gondii* Evasion of the Murine, But Not the Human, Interferon-Gamma Response. PLoS Pathog. 8 (6), e1002784.

Ong, Y.C., Reese, M.L., Boothroyd, J.C., 2010. Toxoplasma rhoptry protein 16 (ROP16) subverts host function by direct tyrosine phosphorylation of STAT6. J. Biol. Chem. 285, 28731–28740.

Peixoto, L., Chen, F., Harb, O.S., Davis, P.H., Beiting, D.P., Brownback, C.S., Ouluguem, D., Roos, D.S., 2010. Integrative genomics approaches highlight a family of parasite-specific kinases that regulate host responses. Cell Host Microbe 8, 208–218.

Pena, H.F., Gennari, S.M., Dubey, J.P., Su, C., 2008. Population structure and mouse-virulence of *Toxoplasma gondii* in Brazil. Int. J. Parasitology 38, 561–569.

Pfefferkorn, E.R., 1988. *Toxoplasma gondii* viewed from a virological perspective. The Biology of Parasitism. P. T. Englund and A. Sher. Alan R, Liss, Inc., New York, 479–501.

Pfefferkorn, E.R., Kasper, L.H., 1983. *Toxoplasma gondii*: Genetic crosses reveal phenotypic suppression of hydroxyurea resistance by fluorodeoxyuridine resistance. Experimental Parasitology 55, 207–218.

Pfefferkorn, E.R., Pfefferkorn, L.C., 1976. *Toxoplasma gondii*: isolation and preliminary characterization of temperature sensitive mutants. Experimental Parasitology 39, 365–376.

Pfefferkorn, E.R., Pfefferkorn, L.C., 1977. *Toxoplasma gondii*: Characterization of a mutant resistant to 5-fluorodeoxyuridine. Experimental Parasitology 42, 44–55.

Pfefferkorn, E.R., Pfefferkorn, L.C., 1978. The biochemical basis for resistance to adenine arabinoside in a mutant of *Toxoplasma gondii*. Journal of Parasitology 64, 486–492.

Pfefferkorn, E.R., Pfefferkorn, L.C., 1979. Quantitative studies of the mutagenesis of *Toxoplasma gondii*. Journal of Parasitology 65, 363–370.

Pfefferkorn, E.R., Pfefferkorn, L.C., Colby, E.D., 1977. Development of gametes and oocysts in cats fed cysts derived from cloned trophozoites of *Toxoplasma gondii*. Journal of Parasitology 63, 158–159.

Pfefferkorn, L.C., Pfefferkorn, E.R., 1980. *Toxoplasma gondii*: Genetic recombination between drug resistant mutants. Experimental Parasitology 50, 305–316.

Radke, J.R., Striepen, B., Guerini, M.N., Jerome, M.E., Roos, D.S., White, M.W., 2001. Defining the cell cycle for the tachyzoite stage of *Toxoplasma gondii*. Molec. Biochem. Parasitol. 115 (2), 165–175.

Rathinam, V.A., Vanaja, S.K., Fitzgerald, K.A., 2012. Regulation of inflammasome signalling. Nat. Immunol. 13 (4), 333.

Reese, M.L., Boothroyd, J.C., 2009. A helical membrane-binding domain targets the Toxoplasma ROP2 family to the parasitophorous vacuole. Traffic 10 (10), 1458–1470.

Reese, M.L., Boothroyd, J.C., 2011. A conserved non-canonical motif in the pseudoactive site of the ROP5 pseudokinase domain mediates its effect on Toxoplasma virulence. J. Biol. Chem. 286 (33), 29366–29375.

Reese, M.L., Zeiner, G.M., Saeij, J.P., Boothroyd, J.C., Boyle, J.P., 2011. Polymorphic family of injected pseudokinases is paramount in Toxoplasma virulence. Proc. Natl. Acad. Sci. U S A 108, 9625–9630.

Reid, A.J., Vermont, S.J., Cotton, J.A., Harris, D., Hill-Cawthorne, G.A., Konen-Waisman, S., Latham, S.M., Mourier, T., Norton, R., Quail, M.A., Sanders, M., Shanmugam, D., Sohal, A., Wasmuth, J.D., Brunk, B., Grigg, M.E., Howard, J.C., Parkinson, J., Roos, D.S., Trees, A.J., Berriman, M., Pain, A., Wastling, J.M., 2012. Comparative genomics of the apicomplexan parasites *Toxoplasma gondii* and *Neospora caninum*: Coccidia differing in host range and transmission strategy. PLoS Pathog. 8 (3), e1002567.

Robben, P.M., Mordue, D.G., Truscott, S.M., Takeda, K., Akira, S., Sibley, L.D., 2004. Production of IL-12 by macrophages infected with *Toxoplasma gondii* depends on the parasite genotype. J. Immunol. 172, 3686–3694.

Rosenthal, B.M., Ajioka, J.W., 2012. Population genetics, diversity and spread of virulence in *Toxoplasma gondii*. In: David Sibley, L., Barbara J., Howlett, Joseph, Heitman (Eds.), Evolution of Virulence in Eukaryotic Microbes. Wiley-Blackwell, Boston, MA. http://dx.doi.org/10.1073/pnas.1203190109. p. 231-245. 1. Proc Natl Acad Sci U S A. 2012 Apr 10;109(15):5844–9.

Rosowski, E.E., Lu, D., Julien, L., Rodda, L., Gaiser, R.A., Jensen, K.D., Saeij, J.P., 2011. Strain-specific activation of the NF-kappaB pathway by GRA15, a novel *Toxoplasma gondii* dense granule protein. J. Exp. Med. 208 (1), 195–212.

Sabin, A.B., 1941. Toxoplasmic encephalitis in children. Journal American Medical Association 116, 801–807.

Saeij, J.P.J., Boyle, J.P., Coller, S., Taylor, S., Sibley, L.D., Brooke-Powell, E.T., Ajioka, J.W., Boothroyd, J.C., 2006. Polymorphic secreted kinases are key virulence factors in toxoplasmosis. Science 314, 1780–1783.

Saeij, J.P.J., Coller, S., Boyle, J.P., Jerome, M.E., White, M.E., Boothroyd, J.C., 2007. *Toxoplasma* co-opts host gene expression by injection of a polymorphic kinase homologue. Nature 445, 324–327.

Shenoy, A.R., Kim, B.H., Choi, H.P., Matsuzawa, T., Tiwari, S., MacMicking, J.D., 2008. Emerging themes in IFN-gamma-induced macrophage immunity by the p47 and p65 GTPase families. Immunobiology 212, 771–784.

Sibley, L.D., Ajioka, J.W., 2008. Population structure of *Toxoplasma gondii*: Clonal expansion driven by infrequent recombination and selective sweeps. Ann. Rev. Microbiol. 62, 329–351.

Sibley, L.D., Boothroyd, J.C., 1992a. Construction of a molecular karyotype for *Toxoplasma gondii*. Molecular and Biochemical Parasitology 51, 291–300.

Sibley, L.D., Boothroyd, J.C., 1992b. Virulent strains of *Toxoplasma gondii* comprise a single clonal lineage. Nature (Lond.) 359, 82–85.

Sibley, L.D., LeBlanc, A.J., Pfefferkorn, E.R., Boothroyd, J.C., 1992. Generation of a restriction fragment length polymorphism linkage map for *Toxoplasma gondii*. Genetics 132, 1003–1015.

Speer, C., Clark, S., Dubey, J., 1998. Ultrastucture of the oocysts, sporocysts, and sporozoites of *Toxoplasma gondii*. Journal of Parasitology 84 (3), 505–512.

Steinfeldt, T., Konen-Waisman, S., Tong, L., Pawlowski, N., Lamkemeyer, T., Sibley, L.D., Hunn, J.P., Howard, J.C., 2010. Phosphorylation of mouse immunity-related GTPase (IRG) resistance proteins is an evasion strategy for virulent *Toxoplasma gondii*. PLoS Biol. 8, e1000576.

Su, C., Evans, D., Cole, R.H., Kissinger, J.C., Ajioka, J.W., Sibley, L.D., 2003. Recent expansion of Toxoplasma through enhanced oral transmission. Science 299, 414–416.

Su, C., Howe, D.K., Dubey, J.P., Ajioka, J.W., Sibley, L.D., 2002. Identification of quantitative trait loci controlling acute virulence in *Toxoplasma gondii*. Proc. Natl. Acad. Sci. (USA) 99, 10753–10758.

Su, C.L., Khan, A., Zhou, P., Majumdar, D., Ajzenberg, D., Dardé, M.L., Zhu, X.Q., Ajioka, J.W., Rosenthal, B., Dubey, J.P., Sibley, L.D., 2012. Globally diverse Toxoplasma gondii isolates comprise six major clades originating from a small number of distinct ancestral lineages. Proc. Natl. Acad. Sci. (USA) 109, 5844–5849.

Su, X., Ferdig, M.T., Huang, Y., Huynh, C.Q., Liu, A., You, J., Wootton, J.C., Wellems, T.E., 1999. A genetic map and recombination parameters of the human malaria parasite *Plasmodium falciparum*. Science 286, 1351–1353.

Su, X.Z., Wootton, J.C., 2004. Genetic mapping in the human malaria parasite *Plasmodium falciparum*. Molec. Microbiol. 53, 1573–1582.

Suzuki, Y., Conley, F.K., Remington, J.S., 1989. Differences in virulence and development of encephalitis during chronic infection vary with the strain of *Toxoplasma gondii*. Journal of Infectious Diseases 159, 790–794.

Taylor, S., Barragan, A., Su, C., Fux, B., Fentress, S.J., Tang, K., Beatty, W.L., Haijj, E.L., Jerome, M., Behnke, M.S., White, M., Wootton, J.C., Sibley, L.D., 2006. A secreted serine-threonine kinase determines virulence in the eukaryotic pathogen *Toxoplasma gondii*. Science 314, 1776–1780.

Vallochi, A.L., Muccioli, C., Martins, M.C., Silveira, C., Belfort Jr., R., Rizzo, L.V., 2005. The genotype of *Toxoplasma gondii* strains causing ocular toxoplasmosis in humans in Brazil. American Journal of Ophthalmology 139, 350–351.

Ware, P.L., Kasper, L.H., 1987. Strain-specific antigens of *Toxoplasma gondii*. Infection and Immunity 55, 778–783.

Wendte, J.M., Miller, M.A., Lambourn, D.M., Magargal, S.L., Jessup, D.A., Grigg, M.E., 2010. Self-mating in the definitive host potentiates clonal outbreaks of the apicomplexan parasites *Sarcocystis neurona* and *Toxoplasma gondii*. PLoS Genet. 6 (12), e1001261.

Witola, W.H., Mui, E., Hargrave, A., Liu, S., Hypolite, M., Montpetit, A., Cavailles, P., Bisanz, C., Cesbron-Delauw, M.F., Fournié, G.J., McLeod, R., 2011. NALP1 influences susceptibility to human congenital toxoplasmosis, proinflammatory cytokine response, and fate of *Toxoplasma gondii*-infected monocytic cells. Infect. Immun. 79, 756–766.

Yamamoto, M., Ma, J.S., Mueller, C., Kamiyama, N., Saiga, H., Kubo, E., Kimura, T., Okamoto, T., Okuyama, M., Kayama, H., Nagamune, K., Takashima, S., Matsuura, Y., Soldati-Favre, D., Takeda, K., 2011. ATF6-beta is a host cellular target of the *Toxoplasma gondii* virulence factor ROP18. J. Exp. Med. 208 (7), 1533–1546.

Yamamoto, M., Okuyama, M., Ma, J.S., Kimura, T., Kamiyama, N., Saiga, H., Ohshima, J., Sasai, M., Kayama, H., Okamoto, T., Huang, D.C.S., Soldati-Favre, D., Horie, K., Takeda, J., Takeda, K., 2012. A cluster of Interferon-gamma-inducible p65 GTPases plays a critical role in host defense against *Toxoplasma gondii*. Immunity 37, 302–313.

Yamamoto, M., Standley, D.M., Takashima, S., Saiga, H., Okuyama, M., Kayama, H., Kubo, E., Ito, H., Takaura, M., Matsuda, T., Soldati-Favre, D., Takeda, K., 2009. A single polymorphic amino acid on *Toxoplasma gondii* kinase ROP16 determines the direct and strain-specific activation of Stat3. J. Exp. Med. 206 (12), 2747–2760.

Yang, H., Wang, J.R., Didion, J.P., Buus, R.J., Bell, T.A., Welsh, C.E., Bonhomme, F., Yu, A.H., Nachman, M.W., Pialek, J., Tucker, P., Boursot, P., McMillan, L., Churchill, G.A., de Villena, F.P., 2011. Subspecific origin and haplotype diversity in the laboratory mouse. Nat. Genet. 43 (7), 648–655.

Zeqiraj, E., van Aalten, D.M., 2010. Pseudokinases-remnants of evolution or key allosteric regulators? Curr. Opin. Struct. Biol. 20 (6), 772–781.

CHAPTER

17

Genetic Manipulation of *Toxoplasma gondii*

Damien Jacot, Markus Meissner†, Lilach Sheiner**, Dominique Soldati-Favre*, Boris Striepen***

***Department of Microbiology and Molecular Medicine, Faculty of Medicine, University of Geneva, Geneva, Switzerland †Wellcome Trust Centre for Molecular Parasitology, Institute of Infection, Immunity and Inflammation, College of Medical, Veterinary and Life Sciences, University of Glasgow, Glasgow, Scotland, United Kingdom**

****Center for Tropical and Emerging Global Diseases, University of Georgia, Athens, Georgia, USA**

OUTLINE

Toxoplasma gondii, Second Edition
http://dx.doi.org/10.1016/B978-0-12-396481-6.00017-9

17.1 INTRODUCTION

The first genetic manipulations applied to *Toxoplasma* were performed by using chemical mutagenesis. These studies were pioneered in the 1970s by the Pfefferkorn laboratory (Pfefferkorn and Pfefferkorn, 1976; Pfefferkorn, 1988) who developed protocols to reproducibly cultivate tachyzoites in a tissue culture system and to mutagenize, select and finally clone parasites by limiting dilution. Based on these protocols, a series of chemically induced mutants were used to map out the parasite's nucleotide biosynthetic pathways. These studies were critical for the establishment of protocols for genetic crosses in the cat (Pfefferkorn and Pfefferkorn, 1980). Crosses can be used to map a given phenotype to a single or multiple genome loci. This classical forward genetic approach has been instrumental to map virulence factors and to analyse *Toxoplasma* population structure and evolution. See Chapters 3 and 16 for further discussion of these topics.

The reverse genetics approach, which introduces foreign DNA into parasites, was achieved using electroporation. Initially the transient

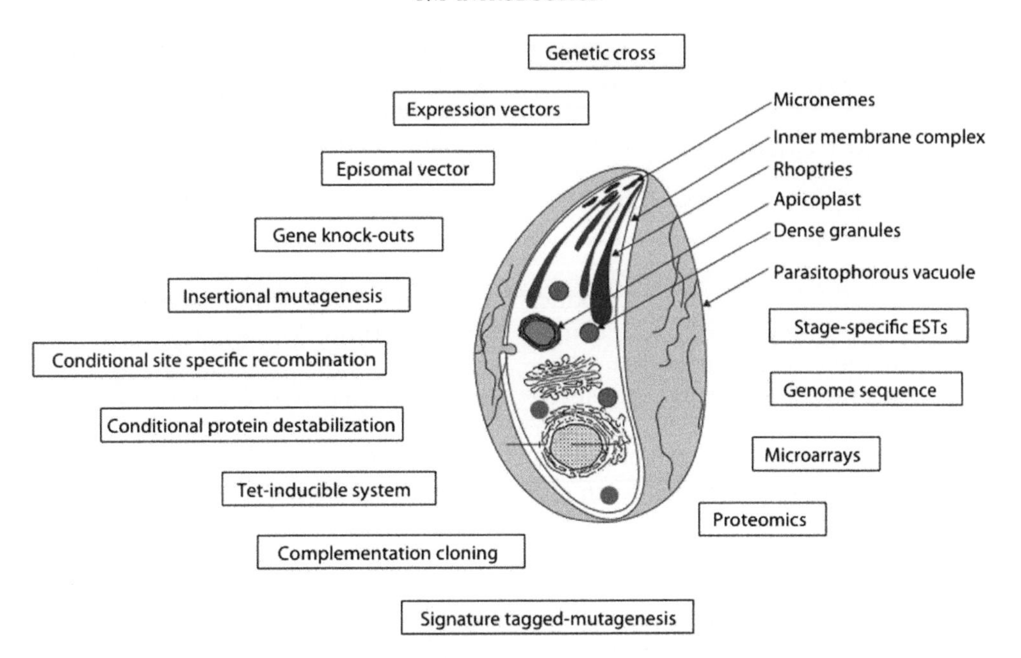

FIGURE 17.1 **Sources of information and manipulation strategies.** Schematic drawing of an intracellular parasite with the subcellular structures and organelles, and the list of the tools currently available for functional analysis. Figure modified from Soldati, D. and Meissner, M. '*Toxoplasma gondii* a model organism for the Apicomplexans' in 'Genomes and the Molecular Cell Biology of Malaria Parasites'. Horizon Press. 5, (2004), 135-167.

transfection of plasmid DNA containing reporter genes flanked by *T. gondii* 5′ and 3′ flanking sequences allowed the expression of reporter genes used for the characterization of the elements controlling transcription. This methodology was rapidly utilized to identify and validate several selectable marker genes, which then opened an avenue for stable transformation and the development of invaluable panoply of tools associated with DNA transfection. A wide range of positive and negative selectable markers have been tailored for homologous recombination leading to allelic replacement and gene knockouts. In addition non-homologous random integration vectors have been designed to express transgenes and as a strategy for random insertional mutagenesis.

The recent completion of the *Toxoplasma* genome sequencing project (http://www.Toxodb.org) and the availability of other apicomplexan genomes for comparison are delivering an unprecedented amount of exciting information. A genome comparison with the closely related parasite *Neosopora caninum* illustrates superbly how rather subtle genetic variation affects the biology of these parasites (Reid *et al.*, 2012). In this new area of post-genomics, the accessibility of *T. gondii* to multiple genetic manipulations approaches and to high throughput studies makes it a very attractive and powerful system to improve our understanding of the basic biology of the apicomplexan parasites. Figure 17.1 summarizes the available sources of information and experimental approaches. There is no limitation to the identification of relevant genes and little or no barrier to experimentally unravel their biological function at a relative large scale.

The purpose of this chapter is to recapitulate and describe the strategies associated with DNA transfection including the most recent acquisitions and to provide a list of the most useful protocols, reagents and strains available to the researchers.

17.2 THE MECHANICS OF MAKING TRANSGENIC PARASITES

17.2.1 Transient Transfection

Successful manipulation of the *Toxoplasma* genome is critically dependent on the efficiency of DNA transfection. Electroporation was and still remains the method of choice to introduce DNA into tachyzoites. Importantly, the combination of this method with media mimicking the cytosolic ion composition of the cells (cytomix) confers the best survival rate (Van den Hoff *et al.*, 1992). The protocol, initially established using a BTX electroporator, led to an efficiency of transient expression that oscillated between 30% and 50% (Soldati and Boothroyd, 1993). The optimal settings chosen on the BTX Electroporator were fixed for the RH strain (Type I, virulent strain) and were slightly modified for the cyst forming strains (ME49 and Prugniaud; Type II strains). It has been frequently observed that the cyst forming strains are less amenable to genetic manipulation probably due to several factors.

To monitor transfection efficiency chloramphenicol acetyl transferase (CAT) and β-galactosidase were originally used as reporter genes and subsequently the β-lactamase, alkaline phosphatase and fire-fly luciferase (LUC). These enzymes are classically used as reporters because their activities can be monitored with great sensitivity and in a quantitative fashion. Additionally, these enzymes are absent in eukaryotic cells, leading to virtually no background activity.

Interestingly, the β-lactamase and alkaline phosphatase exhibit no activity within the parasite probably due to presence of inhibitors and can be exploited to study the secretory pathway and quantify parasite secretion (Chaturvedi *et al.*, 1999; Karsten *et al.*, 1998).

LacZ activity can be measured using a colorimetric assay that transforms yellow chlorophenol red-β-D-galactopyranoside (CPRG) substrate into a red product using an absorbance spectrophotometer at 570 nM (Seeber and Boothroyd, 1996). This colorimetric readout assay can be monitored in live parasites using culture medium without phenol red and in multi-well plates allowing (at a high throughput level) the screening of the efficacy of a drug against the parasite (McFadden *et al.*, 1997).

Faithful expression of a reporter gene requires adequate 5′ and 3′ flanking sequences that are derived from *T. gondii* genes. The flanking sequences must contain the control elements necessary to drive an optimal level of transcription. The monocistronic nature of transcription in *T. gondii* facilitated the identification of promoter elements that are usually in close proximity to the transcription start site. Numerous vectors suitable for transfection are currently available, and as they exhibit a different range of promoter strength and stage specificities, they can be chosen appropriately according to the purpose of the experiment.

A constitutive level of expression can be obtained by using vectors derived from the *TUB1* (α-tubulin), DHFR (dihydrofolate reductase), *ROP1* (rhoptry protein 1), *MIC2* (microneme protein 2), several *GRA* (dense granules proteins) and *HXGPRT* (hypoxanthine–guanine phosphoribosyl transferase) genes. The strength of these and other promoters has not been very systematically compared but the *GRAs* and *MIC2* promoters are the strongest, *TUB1* and *ROP1* promoters are intermediate while *DHFR-TS* is a weak promoter.

Stage-specific expression can be achieved using the 5′-flanking sequences of stage-specific genes and so far no stage-specific regulatory elements have been mapped in the 3′ UTR sequences. Tachyzoite-specific expression is conferred by vectors derived from *SAG1* (surface antigen 1), *ENO1* (enolase 1) and *LDH1* (lactate dehydrogenase 1) genes. In contrast, vectors constructed from *BAG1* (bradyzoite antigen 1), *ENO2* (enolase 2) or *SAG4* genes confer expression in the bradyzoite stage exclusively. Detailed promoter analyses and identification of cis-acting elements have only been undertaken for a limited number of genes (Bohne *et al.*, 1997; Kibe *et al.*, 2005; Mercier *et al.*, 1996; Matrajt *et al.*, 2004; Soldati *et al.*, 1995; Yang and Parmley, 1997). See Chapter 18 for a discussion of regulation of gene expression.

Importantly, a recent study has highlighted the considerable cell cycle dependency of gene expression (Behnke *et al.*, 2010) which stresses the importance of using the endogenous promoter to control the expression of the gene of interest (GOI).

In addition to the promoter elements, sequence features carried on the mRNAs also contribute to the success of transfection. Sequence information derived from the 5′ and 3′ untranslated regions likely affects gene expression, but this level of regulation has not been rigorously investigated to date. The 3′ UTR is an important element as transcription drops to less than 10% when such an element is not included. In *Plasmodium* partial deletion of 3′ UTR regions have been exploited to modulate the level of expression of essential genes offering a way to analyse their function (Thathy and Menard, 2002).

At the start codon, a consensus sequence termed the 'Kozak sequence' is recognized by the ribosome as a favourable sequence to initiate translation. A compilation of abundant expressed genes in *T. gondii* was used to establish a consensus translational initiation sequence *gNCAAa*ATGg, which is similar but not identical to the Kozak sequence found in higher eukaryotes (Seeber, 1997). Several genes including GFP were initially very difficult to express using their native sequence but the lack of expression was solved by the generation of fusions at the N-terminus (Striepen *et al.*, 1998). These observations suggested a significant influence of the N-terminal amino acid sequences in recombinant protein expression. A systematic analysis aiming at the evaluation of the importance of the amino acid following the initiation methionine confirmed the existence of an N-end rule in *T. gondii* (Matrajt *et al.*, 2002b). Amino acids such as Ala, Glu and Asp confer high level of expression of the transgene.

17.2.2 Stable Transformation and Positive and Negative Selectable Markers

Most of the selectable marker genes commonly used for eukaryotic cells are not suitable for selection of stable transformants in *T. gondii* (*T. gondii* is an obligate intracellular parasite). Only drugs selectively affecting the parasite while keeping the host cells intact could be considered. In spite of this restriction, various selection protocols have been developed and are listed in Table 17.1.

Chloramphenicol shows a potent but delayed parasiticidal effect, allowing the use of *E. coli* chloramphenicol acetyl transferase (CAT) not only as reporter enzyme but also as a tight selectable marker gene (Kim *et al.*, 1993). Parasite must complete up to three cycles of host cell lysis (up to seven days) before an effect of the drug is evident. At this point parasites are cloned in 96 well plates for about five days, starting in presence of drug selection.

Another selection strategy based on the resistance to a drug can be achieved by exploiting the protective effect of the *ble* gene product of *Streptoalloteichus* or Tn5 against the DNA breaking activity of phleomycin (Messina *et al.*, 1995; Soldati *et al.*, 1995). Parasites expressing *ble* become resistant to the drug; however, this selection needs to be applied on extracellular parasites to be effective. Phleomycin selection has been used successfully for the random insertion of transgenes (Soldati and Boothroyd, 1995) and to disrupt genes by homologous recombinations (Mercier *et al.*, 1998). As an alternative to drug resistance, stable selection can be achieved by complementation of the naturally occurring tryptophan auxotrophy of *Toxoplasma* by addition of indole to the culture medium (Sibley *et al.*, 1994) following the introduction of the bacterial tryptophan synthase (trpB) gene.

Two genes coding for non-essential nucleotide salvage pathway enzymes have been exploited as negative selectable markers. Loss of uracil phosphoribosyl transferase (*UPRT*)

TABLE 17.1 Selection Strategies, Gene Markers and Conditions

Selectable Marker Genes	Recipient Strain	Drug or Selection Procedure	Concentration Range
CAT *E. coli*	Wild type	Chloramphenicol Drugs treatment during seven days before cloning	20 μM CM
DHFR–TS *T. gondii*	Wild type	Pyrimethamine; treatment during two days before cloning	1 μM PYR
Ble *Streptoalloteichus* or Tn5	Wild type	Phleomycin: two cycles of treatment during 5–10 hours on extracellular parasites	5 μg/ml PHLEO
HXGPRT *T. gondii*	RH hxgprt- ME49 hxgprt- PRU hxgprt-	Positive selection: Mycophenolic acid + xanthine: treatment during three days before cloning Negative selection: 6-Thioxanthine	25 μg/ml MPA 50 μg/ml XAN. 80 μg/ml 6–TX
UPRT *T. gondii*	RH uprt-	Negative selection: 5′-fluo-2′- deoxyuridine	5 μM FUDR
GFP/YFP *Aequorea victoria*	Wild type	FACS	
Essential genes *T. gondii*	TATi-1 conditional KO	Anhydrotetracycline	Max 1 μM ATc
Cre recombinase *Enterobacteria phage P1*	Transgenes flanked by loxP sites (recycling of markers)	Transient transfection with Cre expressing plasmid. Cloning immediately after electroporation	No selection
TK *Herpes simplex*	Wild type	Ganciclovir 24 hours treatment	10 μM GCV
CD *E. coli*	Wild type	5-fluorocytosine	40 μM FLUC

activity confers resistance to the pro-drug 5′-fluo-2′-deoxyuridine (FUDR) (Donald and Roos, 1995) and in absence of hypoxanthine–xanthine–guanine phosphoribosyl transferase (*HXGPRT*) activity, 6-thioxanthine (6-Tx) cannot be converted into an inhibitor of GMP synthase (Donald *et al.*, 1996). In *HXGPRT* deficient mutants, this gene can also be used for positive selection strategies since mycophenolic acid efficiently kills parasites lacking the enzyme.

The frequency of stable transformation fluctuates significantly depending on the type of selectable marker used. The conformation of the transfection plasmid (circular versus linearized by restriction) can also affect transfection efficiency. A much high frequency of stable transformation is achievable using pyrimethamine resistance vectors derived from the parasite's bifunctional dihydrofolate reductase–thymidylate synthase *DHFR–TS*. An artificially mutated *dhfr–ts* gene from *T. gondii* was used to design an expression vector pDHFR*–TSc3 (No. 2854) that confers pyrimethamine resistance (Donald and Roos, 1993). The *DHFR–TS* based selection is unique and shows an exceptional frequency of chromosomal integration of up to 5% (Donald and Roos, 1993). The flanking sequences of the *DHFR–TS* genes are responsible for this unusual property, which can be partially conferred to other selectable marker genes such as the *HXGPRT* if this latter is controlled by the *DHFR–TS* flanking sequences.

Fluorescence activated cell sorting (FACS) is another way to select for stable transformation, when using a fluorescent tag as the marker (Sheiner *et al.*, 2011). To obtain clonal parasite lines stably expressing fluorescent protein, two rounds of fluorescence activated cell sorting and expansion of sorted parasites in culture (Gubbels *et al.*, 2003) is routinely used. Multiple fluorescent proteins can be used and sorted simultaneously; however, an instrument with multiple lasers might be required (see protocol section).

Furthermore, restriction enzyme-mediated integration (REMI) can be used to further enhance the frequency of transformation up to 400-fold (Black *et al.*, 1995) and enables co-transfection of several unselected constructs together with a single selectable marker.

Any of the selectable markers genes listed above can, if needed, be efficiently recycled by the action of the site-specific Cre recombinase. The adaptation of the *cre lox*P system from bacteriophage P1 to *T. gondii* enables the specific *in vivo* excision of any introduced sequence, which was flanked by *lox*P sequences (Brecht *et al.*, 1999).

17.2.3 Homologous Recombination and Random Integration

Unlike the situation in many protozoans, where integration into chromosomes occurs exclusively by homologous recombination and requires only a short segment of homology, homologous recombination is not favoured in *T. gondii*. Vectors lacking long stretches of contiguous genomic DNA typically integrate into chromosomal DNA at random. The high frequency of transformation and random integration throughout the small genome size of haploid *T. gondi* tachyzoites was developed as an efficient strategy to mutagenize the entire genome of *T. gondii* within one single electroporation cuvette (Roos *et al.*, 1997). Such genomic scale tagging allows identification of any gene whose inactivation is not lethal to tachyzoites and for which a suitable functional selection or screen is available (Fig. 17.2). For example, positive/negative selection can be employed for selection schemes for mutants and promoter traps. The *HXGPRT* gene has been exploited to identify genes that are expressed in a stage-specific fashion (Knoll and Boothroyd, 1998). Parasites expressing *HXGPRT* under the control of a bradyzoite-specific promoter were mutagenized by random insertion of a plasmid and subjected to *in vitro* tachyzoite to bradyzoite conversion under 6-thioxanthine selection to isolate mutants deficient in differentiation (Matrajt *et al.*, 2002a). Signature tagged mutagenesis has also been successfully applied to discover virulence genes in forward genetic screens, where growth of insertional mutants has been compared *in vitro* and *in vivo* (Frankel *et al.*, 2007).

Homologous recombination leading to gene replacement is instrumental to study gene function and can be accomplished in *T. gondii* using different strategies (Fig. 17.2). In wild type parasites the efficiency of homologous recombination is favoured if long contiguous stretches of homologous DNA are used to target the locus (Donald and Roos, 1994; Kim, 1993). Previously, the construction of vectors for the homologous removable of genes was a relatively cumbersome approach mainly due to the cloning of long flanking regions. The multisite Gateway® (Invitrogen) recombination technique is an improved strategy to avoid restriction enzyme mediated steps and was efficiently applied to *T. gondii* (Upadhya *et al.*, 2011).

Another restriction enzyme-free system is based on overlap extension polymerase chain reactions where only a few PCR steps are required for the construction of the knockout vector. The 5′ and 3′ flanking regions of the GOI are amplified and fused in a second PCR to either the first or second half of a selection cassette. These two constructs containing an incomplete cassette are then transfected into

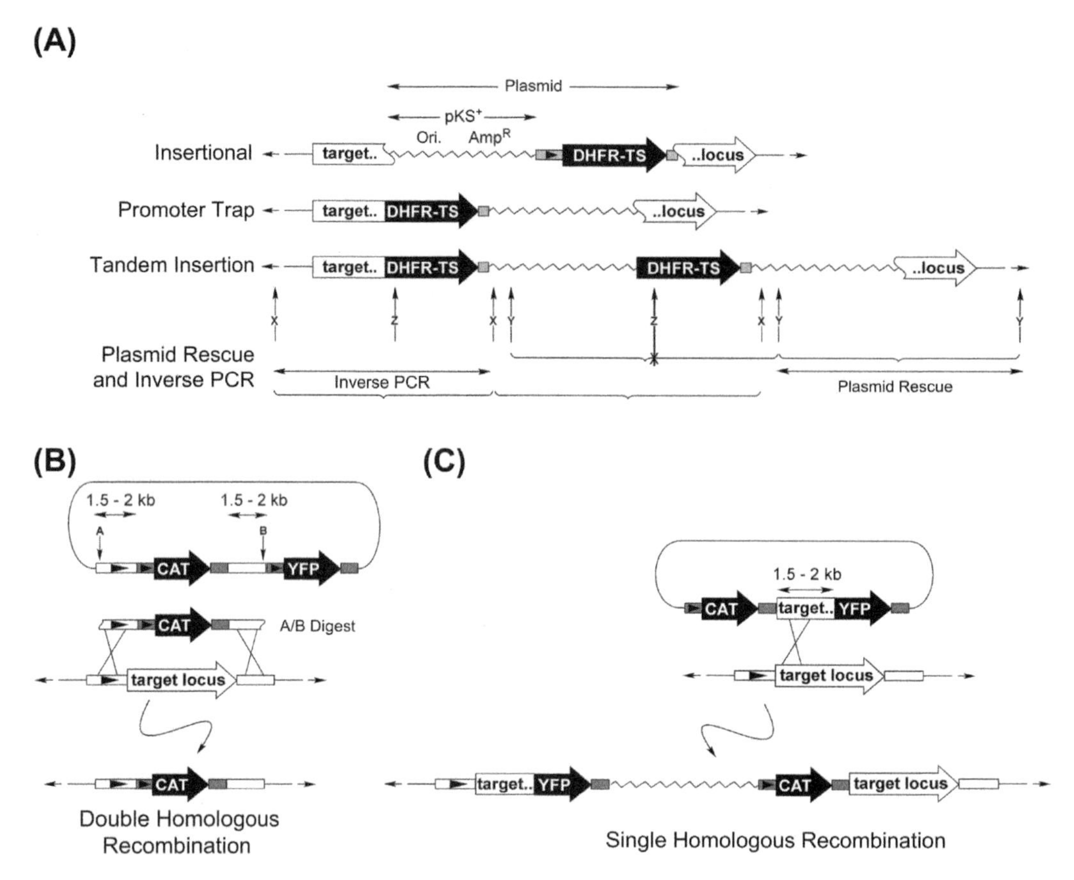

FIGURE 17.2 **Exploiting non-homologous insertion and homologous recombination to manipulate the *T. gondii* genome.**
A) Schematic representation of insertional genomic tagging using a DHFR–TS plasmid (based on Roos *et al.*, 1997). Plasmid DNA is indicated on top, genomic insertions below. For insertional mutagenesis expression of the DHFR–TS pyrimethamine resistance gene is driven by its own promoter, the insertion therefore is not necessarily within the open reading frame but might also act through inactivating a regulatory region (e.g. promoter). In case of promoter trapping DHFR–TS does not carry its own promoter, and expression of the resistance gene depends on insertion close to an active promoter, or as an in-frame fusion into an expressed gene. Tandem insertions can complicate the identification of the tagged locus by plasmid rescue (using restriction enzyme Y) and/or inverse PCR (using restriction enzyme X or Y). However, simultaneously applying restriction enzyme Z cuts the tandem into two fragments incompatible with plasmid rescue or inverse PCR (Sullivan *et al.*, 1999; Roos *et al.*, 1997).
B) Schematic representation of gene knock-out through double homologous recombination. The homologous regions destined for homologous recombination are represented by white boxes. Restriction enzymes A and B are used to generate fully homologous ends. In this case YFP is used as a negative selectable marker to enrich for homologous recombination (YFP is lost and parasites are FACS negative) (Mazumdar *et al.*, 2006).
C) Schematic representation of allelic replacement through single homologous recombination. In this strategy a circular plasmid inserts and tags the locus with a YFP fusion (which can be omitted, or replaced by a shortened ORF to create a functional knock-out). The gene-locus 3′ of the plasmid backbone is functionally inactivated by the lack of a promoter. This figure is taken from 'The Biology of *Toxoplasma gondii*' Manipulating the *Toxoplasma* genome. Gubbels, M-J., Mazumdar, J., van Dooren, G., and Striepen, B. Horizon Press, 2004.

the parasites. In this system, three single independent homologous recombinations are necessary to achieve gene knockout by replacement with a functional (reconstituted) selectable marker gene, i.e. in the 5′, in the 3′ and within the selection cassette (Upadhya *et al.*, 2011).

Additionally, the recent establishment of recombineering approaches using a cosmid library helped to address the need for long regions of homologous DNA (Brooks *et al.*, 2010; Francia *et al.*, 2012). To further enhance this approach a new large insert DNA library was constructed in a copy-control fosmid backbone. Recombineering can thus be performed at single copy, which enhances stability and fidelity, followed by DNA production at higher copy number. The library was arrayed and 100,000 will be end sequenced (Vinayak and Striepen, unpublished). Another way to increase homologous recombination frequencies is by a combination of positive/negative selections (Table 17.1) (Fox *et al.*, 1999, 2001; Mazumdar *et al.*, 2006; Radke and White, 1998).

The efficiency of gene targeting was dramatically enhanced by the isolation of ΔKu80 strains. Ku80 is a component of the non-homologous end joining pathway of DNA repair present in *T. gondii* yet lacking in many other apicomplexans including the *Plasmodium* species. Consequently the usually high frequency of random integration events observed in wild type parasites is almost completely abolished in ΔKu80, resulting in a parasite strain that allows efficient gene replacement and endogenous tagging of genes (Huynh and Carruthers, 2009; Fox *et al.*, 2009).

17.3 USING TRANSGENIC PARASITES TO STUDY THE FUNCTION OF PARASITE GENES

17.3.1 Tagging Subcellular Compartments

Visualizing and following the morphology and behaviour of different subcellular compartments through the cells' life cycle is an essential tool for cell biological analysis. Proteins localizing to almost all organelles of *T. gondii* have been described and a large number of constructs allowing expression of organelle-specific fluorescent proteins is now available (see Fig. 17.3 for examples). Numerous versions of green fluorescent protein (GFP) and related autofluorescent proteins have been successfully expressed in *T. gondii* (Kim *et al.*, 2001; Striepen *et al.*, 1998) and a range of colours is available now for the simultaneous use of multiple markers. Cyan (CFP) and yellow fluorescent protein (YFP) are a suitable pair for double labelling experiments and have been used in *in vivo* microscopic studies of *Toxoplasma* organelle biogenesis (Striepen *et al.*, 2000; Pelletier *et al.*, 2002; Joiner and Roos, 2002). A tandem repeat of the YFP gene yields exceptionally bright fluorescent transgenics that are now widely used to track parasites in tissue culture and in infected animals (Gubbels *et al.*, 2003, 2004, 2005; Egan *et al.*, 2005). Red fluorescent proteins (RFP) further extend the options. DsRed produces brightly fluorescent parasites (Striepen *et al.*, 2001), however, the requirement of tetramerization of this marker can be problematic if the tagged protein is part of a complex or structure. Monomeric variants of RFP (e.g. mRFP (Campbell *et al.*, 2002)) can help overcome these problems but suffer from considerably weaker fluorescence. The newer 'cherry' and 'tomato' variants (Shaner *et al.*, 2004) provide a reasonable compromise and a tandem tomato marker produces exceptionally bright fluorescence when expressed in *T. gondii* (van Dooren *et al.*, 2008).

Fluorescent labelling of organelles can be achieved by simple transient transfection of specific fluorescent proteins. This approach should be used with caution as the transient expression might result in unspecific targeting due to overexpression or inappropriate timing of expression. This is particularly common for proteins that are targeted for the secretory pathway. Alternatively, the respective marker can be stably expressed as a transgene via

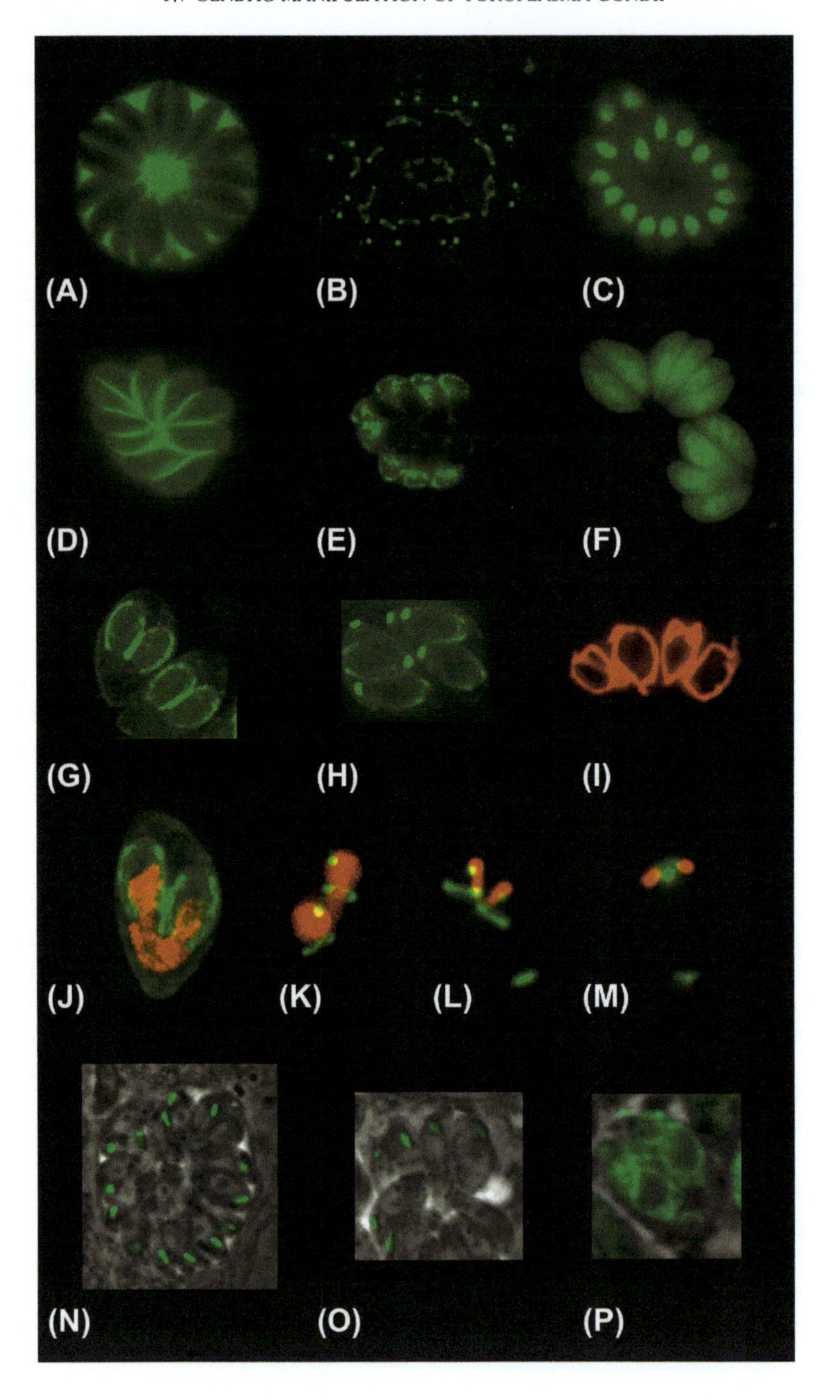
(A)
(B)
(C)
(D)
(E)
(F)
(G)
(H)
(I)
(J)
(K)
(L)
(M)
(N)
(O)
(P)

random integration or via endogenous tagging, using a ΔKu80 strain that allows a high efficiency of endogenous tagging (Huynh and Carruthers, 2009; Fox *et al.*, 2009). Parasites expressing fluorescent proteins can also be analysed and sorted by flow cytometry. Additionally, fluorescent protein expression can be detected using a plate reader. This provides a convenient growth assay for drug screening and genetic selections (Gubbels *et al.*, 2003).

17.3.1.1 Tagging of Parasite Proteins

The cellular localization of a protein of interest is a first important step in order to characterize its function. Specific antibodies raised against subcellular fractions or individual proteins are widely used for this purpose at the light and electron microscopic level. This approach, however, requires the production of antigen, either by purification from the parasite or by recombinant expression and subsequent immunization, which is time-consuming and not always technically feasible. Through transfection experiments, proteins expressed as second copies from a heterologous promoter can be tagged by gene fusion using a generic epitope (for which antibodies are already available) or using fluorescent proteins. However, not all proteins can be studied in this way, as the bulky GFP tag can affect targeting, maturation or function of its fusion partner. In such a case epitope tags can provide an alternative approach. These tags can be inserted internally or placed at the N- and C-terminus. Due to their short length, epitope tags cause limited steric hindrance. Epitope tags require fixation and staining with a specific antibody before visualization. While not suitable for live cell imaging they can be used for subcellular and ultrastructural localization, immunoprecipitation experiments or to monitor protein processing during targeting or maturation. A number of epitope tags have been used successfully in *Toxoplasma* (e.g. cMyc (Delbac *et al.*, 2001), HA (Karsten *et al.*, 1997), FLAG (Sullivan *et al.*, 2005) or Ty-1 (Herm-Gotz *et al.*, 2002)).

It has been frequently observed that the strength, and probably also the timing of expression with respect to the cell cycle, critically influence the outcome of an experiment and can lead to localization artefacts. For example, the overexpression of microneme proteins often results in accumulation in the early compartment of the secretory pathway or mistargeting to the parasitophorous vacuole (Soldati *et al.*, 2001). To

FIGURE 17.3 **Tagging subcellular compartments with fluorescent protein markers in *T. gondii*.** This figure provides examples of single and dual fluorescent protein labelling *T. gondii*; all images were obtained by live cell microscopy.
A) Dense granules and parasitophorous vacuole, P30−GFP (Striepen *et al.*, 1998).
B) Centrocones (outermost dots) and posterior IMC rings of mother (innermost) daughter cells (lines), MORN1−YFP (Gubbels *et al.*, 2006).
C) Nuclei, PCNA−GFP (Radke *et al.*, 2001).
D) Plasma membrane, P30−GFP−GPI (Striepen, unpublished).
E) Micronemes, MIC3−GFP (Striepen *et al.*, 2001).
F) Cytoplasm, YFP−YFP (Gubbels *et al.*, 2003).
G) Inner membrane complex, IMC3−YFP (Gubbels *et al.*, 2004).
H) Microtubules, YFP−TUB (Hu *et al.*, 2002).
I) Mitochondria, HSP60−RFP (G. van Dooren, unpublished).
J) Dividing tachyzoites IMC3−YFP and H2b−mRFP (Hu *et al.*, 2004).
K) Nuclear division and cytokinesis, H2b−mRFP and MORN1−YFP (Gubbels *et al.*, 2006).
L) Apicoplast division, FNR−RFP and MORN1−YFP (Striepen *et al.*, 2000).
M) Golgi division, GRASP−RFP and MORN1−IMC.
N) Apicoplast, ACP−GFP (Waller *et al.*, 1998).
O) Rhoptries, ROP1−GFP (Striepen *et al.*, 1998).
P) Endoplasmatic reticulum, P30−GFP−HDEL (Hager *et al.*, 1999).

overcome this issue, localization of a protein of interest can also be achieved via endogenous tagging in ΔKu80 strain, this provides the advantage of well-matched expression with respect to strength and timing as the native promoter element is used to drive transcription. Several C-terminal tagging constructs have been generated, taking advantages of ligation independent cloning (LIC) (Huynh and Carruthers, 2009). In combination with homologous recombination in a ΔKu80 strain the endogenous protein can be directly tagged (Huynh and Carruthers, 2009) and localized. Similar experiments can be performed using recombineered genomic cosmids and fosmids (Brooks *et al.*, 2010). However, C-terminal tagging can interfere with the function of some proteins preventing the isolation of the tagged strain.

17.3.2 Genetic Analysis of Essential Genes

In order to study the function of essential genes in a haploid organism, tools needed to be developed that allowed the engineering of conditional knockout, knockdown or trans-dominant mutants. Currently, several strategies operating on different levels, such as site-specific recombination, transcriptional control or control of protein stability, have been implemented in *T. gondii*. Each of these technologies has its advantages and disadvantages to be considered for gene-function analysis.

17.3.2.1 *Site-Specific Recombination (SSR)*

The yeast recombinases Cre and Flp recognize DNA sequences (the LOX and FRT sites, respectively) that are short enough for convenient cloning, but long enough to be specific and absent from even large genomes when not deliberately introduced. Both recombinases are highly efficient in excising DNA that lies in between the recognition sites and recombination requires only a minimal amount of recombinase expression. However, in order to generate conditional knockouts, temporal control of Cre is required. This can, in principle, be achieved via transient transfections with a Cre expression construct (Heaslip *et al.*, 2010); however, transfection efficiencies can vary and Cre overexpression is toxic (most likely due to non-specific recombination events (Xiao *et al.*, 2012)). A solution for this problem is provided by conditional Cre-systems, such as ligand controlled Cre-recombinases (Metzger *et al.*, 1995) or dimerisable Cre (DiCre) (Jullien *et al.*, 2003). While fusions of Cre to hormone binding domains have been shown to be constitutively active in *T. gondii* (Brecht *et al.*, 1999), the DiCre-system allows rapid, specific and efficient temporal control of Cre activity (Andenmatten *et al.*, 2013). Here, the Cre recombinase is split into two inactive fragments that are fused to the rapamycin binding proteins FRB and FKBP, respectively. Addition of the ligand rapamycin results in reconstitution of the functional enzyme, and excision of the GOI flanked by LoxP sites.

A clear advantage of the DiCre approach is that the GOI is under the control of its endogenous promoter, ensuring correct timing and levels of gene expression. In addition, future constructs can be easily modified to allow high-content cloning of knockout vectors, comparable to approaches applied in mice (Skarnes *et al.*, 2011). Challenges of the DiCre system include a difficulty to obtain clonal knockout population, since induction of DiCre results in recombination rate between 20% and 90% leading to a mixed population of KO and wild type parasites. Another disadvantage is the irreversibility of the recombination event.

17.3.2.2 *Tetracycline Inducible Systems*

One widely used approach to modulate expression is based on the *E. coli* tetracycline-repressor system, which controls gene expression at the transcriptional level. The original tetracycline-repressor system interferes with transcription and has been optimized and coupled to T7 polymerase to tightly regulate gene expression in *Trypanosoma brucei* (Wirtz *et al.*, 1999). The tet-repressor system has also been developed for other protozoan parasites including *T. gondii* (Meissner *et al.*, 2001). Gene fusion of the tet-repressor (van Poppel *et al.*, 2006) has led to higher transgene expression and tighter regulation.

Although suitable for the expression of toxic genes and dominant negative mutants, this system proved not to be appropriate for the isolation of conditional knockouts in *T. gondii*. Indeed, the necessity to keep the parasites in the presence of drug (anhydrotetracycline, ATc) during a prolonged period in order to maintain the expression of an essential gene led to generation of revertants that lost regulation.

To improve the system, a genetic screen based on random insertion was designed to identify a functional transcriptional activating domain in *T. gondii* and to establish a tetracycline transactivator-based inducible system (Meissner *et al.*, 2002). This screen led to the isolation of two artificial transactivators that were not functional in HeLa cells, illustrating the differences between the transcription machinery in the parasite and its higher eukaryotic hosts. Interestingly, these transactivators corresponding to short stretches of rather hydrophobic amino acids seemed to also be active in *P. falciparum* (Meissner and Soldati, 2005). This system is suitable for the conditional disruption of *Toxoplasma* essential genes with no apparent reversion effect and operates on the parasites in the animal model. A line expressing one of the transactivators (TATi-1) was implemented to functionally analyse numerous genes including TgMyoA (Meissner *et al.*, 2002), TgAMA-1 (Mital *et al.*, 2005), TgMIC2 (Huynh and Carruthers, 2006), TgACP (Mazumdar *et al.*, 2006) and profilin (Plattner *et al.*, 2008), in several pioneering studies.

So far, the tet-inducible system has been relatively laborious, requiring two steps of selection (Fig. 17.4A). The first step is the construction of a stable line expressing an inducible copy of the gene of interest via random integration. The second step is the actual knockout of the target gene (see protocol section for details). More streamlined single step strategies have been established recently (Fig. 17.4B–E). In the first system the native promoter of the gene of interest is either replaced or distanced from the ATG by double homologous recombination with a selectable marker (DHFR) and the tet-inducible promoter (Fig. 17.4B). This can be performed in a Δ*Ku80* TATi-1 expressing line (Sheiner *et al.*, 2011) to favour the homologous recombination event. In the resulting mutant parasite, the GOI is directly controlled by the tet-inducible promoter in its genomic context, and addition of ATc induces knockdown (Sheiner *et al.*, 2011). Analysis of several essential genes was already achieved using this strategy (Francia *et al.*, 2012; Sampels *et al.*, 2012).

An alternative single step approach consists of using vectors that carry the coding sequence of TATi-1 under the control of a tubulin promoter (pTub8) downstream of the 5′ UTR of the GOI. The 3′ recombination results in the replacement of the endogenous promoter by the inducible tet-Operator containing promoter. Simultaneously, a tag can be placed at the N-terminus of the GOI (Fig. 17.4C). To establish C-terminal tagging the plasmid contains the same 5′ UTR, the targeted gene cDNA with the tag under the control of the tet-Operator and a 3′ UTR region for recombination (Fig. 17.4D). In all the strategies, including the two steps, the expression of TATi-1 is under the control of a constitutive promoter, which might not be suitable for genes with a very particular level or timing of expression. To solve this problem the pTub8 promoter in the 'all in one' strategy can be exchanged by the promoter of the GOI, to better mimic the timeline of GOI expression (Fig. 17.4E).

So far the frequency of double homologous recombination was found variable with the 'all in one' approach with frequent integrations only on one side and presumably in more than

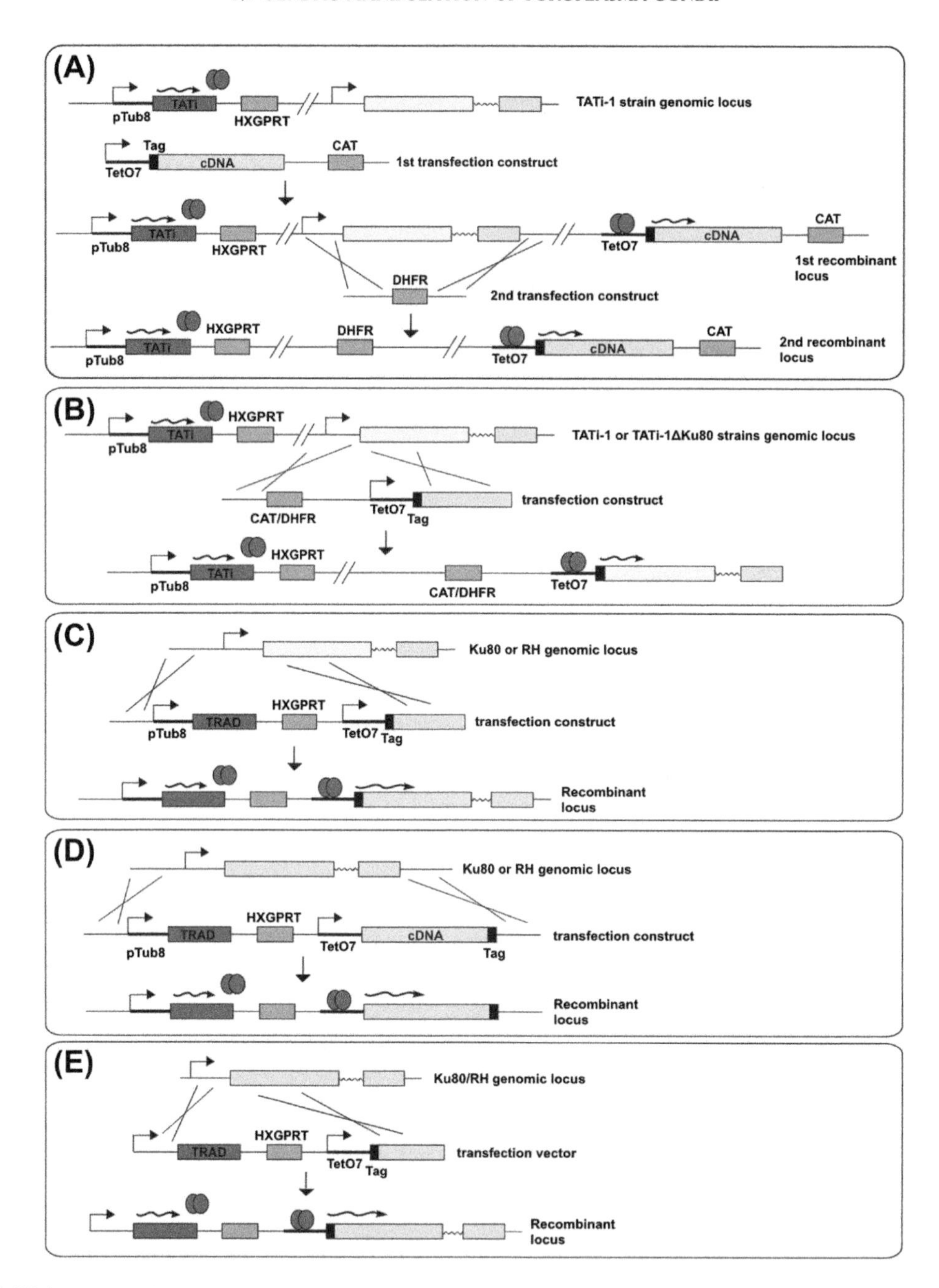

FIGURE 17.4 **Tetracycline inducible systems.**
A) Schematic representation of the 'two steps' process in the TATi-1 expressing strain. The first step consists with the random insertion of the GOI cDNA under the control of the TetO7 promoter. The second step is the knockout of the GOI.
B) The disruption of Ku80 in TATi-1 expressing strain favoured the recovery of double homologous recombination events where the promoter of the GOI is replaced by the TetO7 promoter.
C–D) Alternatively, an 'all in one vector' includes all the different elements for transactivation. This system allows N-term (B) or C-term (D) tagging but also can insert TATi-1 under the control of the GOI promoter.
E) Additionally, the TRADs can be explored to optimize the level of transactivation. GOI, gene of interest; TRAD, transactivator.

one copy. To adjust the expression, more than one copy of TATi-1 might be necessary and several could be integrated. In fact, the TATi-1 strain probably expressed several copies of this transactivator. In consequence, the 'all in one' strategies that result in the integration of one copy of TATi-1 might not be sufficient to drive the expression particularly if the promoter used is weak.

A series of new transactivators have recently been generated to optimize the tet-system for the malaria parasites (Pino *et al.*, 2013). The TATi-2 was previously adapted to *Plasmodium falciparum* and resulted in a successful and tight regulation of transgene expression for multi-copy episomal plasmids (Meissner *et al.*, 2005). In contrast, the level of expression as a single integrated copy dropped dramatically and hence hampered the generation of conditional knock-outs. A group of conserved proteins containing putative Apetala2 DNA-binding domains referred to as the Apicomplexan AP2 (ApiAP2) protein family have been used in a screen in *T. gondii* to identify new transcription activating domains. Four new transactivators (TRADs) have been created that are as or more potent than TATi-1. TRAD4 was exploited in *Plasmodium berghei* to generate knock-downs of essential genes for the intraerythrocytic development of the rodent malaria parasite (Pino *et al.*, 2013). These TRADs could be explored in the 'all in one' strategies to overcome the issue of insufficient transactivation.

The 'two steps' strategy in TATi-1 is a robust but laborious method to produce conditional knockout parasites (Fig. 17.4A). The first step results in different clones of various levels of expression (due to the number of random integrations, integration in highly transcribed loci, etc.) and allows therefore the selection of the most suitable expression. The 'all in one' strategies reduce this process in only one step but the transactivation only takes place in the GOI endogenous locus and in a single copy. The level and the proper timing of transactivation are critical for the establishment of conditional systems and therefore each method has to be considered for different genes.

17.3.2.3 Regulation of Protein Stability

A major limitation of the above conditional systems is their relatively slow response kinetics, as the protein of interest is still present after removal or down-regulation of the respective gene. In the case of stable proteins, it can take up to 96 hours before a phenotype becomes fully evident. Such long incubation times can complicate the interpretation of the observed phenotypes in particular when it comes to distinguishing primary from secondary effects. A more rapid method, based on conditional regulation of protein stability has been developed for mammalian cells (Banaszynski *et al.*, 2006) and successfully adapted to apicomplexans (Herm-Gotz *et al.*, 2007; Armstrong and Goldberg, 2007). This system is based on mutated forms of the FKBP12–rapamycin binding protein that result in its fast degradation. Fusion of this degradation domain (ddFKBP) to a protein of interest results in degradation of the entire protein. Addition of the inducer Shield-1 (a rapamycin analogue) results in rapid stabilization of the protein (Banaszynski *et al.*, 2006). Regulation can be achieved by fusing ddFKBP N- or C-terminally, with N-terminal fusion providing superior performance (Herm-Gotz *et al.*, 2007).

Proteins residing within organelles typically cannot be regulated using ddFKBP, since the protein needs to be accessible to the proteasome, which resides in the cytosol. This system in principle should be suitable to construct conditional mutants by direct endogenous tagging. While direct allelic replacement has been successful in some cases in *P. falciparum* (Dvorin *et al.* 2010; Farrell *et al.*, 2011; Russo *et al.*, 2009), attempts to use this strategy in *T. gondii* failed and resulted in the expression of ddFKBP-tagged proteins that remained stable in the absence of Shield-1.

The ddFKBP-system has been further optimized and new FKBP12 mutants have been tested in the mammalian system (Egeler *et al.*, 2011). In addition, other destabilization domains have been developed, such as DHFR-based systems (DDD) (Iwamoto *et al.*, 2010) that work well in *P. falciparum* (Muralidharan *et al.*, 2011). It will be interesting to see whether these optimized versions can confer better regulation to endogenously tagged proteins in *T. gondii.*

Despite these obstacles the ddFKBP system is very well suited to generate overexpression or trans-dominant mutants (van Dooren *et al.*, 2009; Santos *et al.*, 2011; Daher *et al.*, 2010; Breinich *et al.*, 2009; Agop-Nersesian *et al.*, 2010). The rapid response kinetic of the ddFKBP-system is of particular advantage, when rapid processes are to be analysed, such as components of trafficking systems or signalling cascades.

17.3.3 Insertional Mutagenesis and Promoter Trapping as Tools of Functional Genetic Analysis

Random high frequency integration of a genetic element into the parasite genome can be used to disrupt loci and produce pools of insertional mutants. The integrated sequence can subsequently be exploited to identify the targeted gene with modest effort (Fig. 17.2). The exceptionally high frequency of non-homologous recombination of transgenes in *T. gondii* allows the use of simple plasmid constructs similar to the way transposons are used in other organisms (Donald *et al.*, 1996). Several non-essential genes have been identified using random insertion of a DHFR–TS or HXGPRT element (Sullivan *et al.*, 1999; Donald and Roos, 1995; Chiang *et al.*, 1999; Arrizabalaga *et al.*, 2004). The genomic locus tagged by the insertion can be identified by plasmid rescue or inverse PCR strategies (Roos *et al.*, 1997).

The insertional strategy is not limited to gene disruption but can also be used to trap promoters and genes. Bradyzoite-specific genes (Bohne *et al.*, 1997; Knoll and Boothroyd, 1998) as well as genes controlling differentiation (Matrajt *et al.*, 2002a; Vanchinathan *et al.*, 2005) have been identified using differential HXGPRT selection under culture conditions that favour differentiation to bradyzoites followed by counter-selection under 'tachyzoite' conditions. The trapping of native *T. gondii* transcription factors might also be achievable. For this a recipient strain harbouring a YFP–YFP marker under the control of a tet-regulated promoter would be randomly inserted. The tagging plasmid would harbour a tet-repressor gene lacking a stop codon and 3′ UTR sequences. Translational fusion of this marker with a transcription factor should result in transactivation and hence green fluorescence.

The fact that tachyzoites are haploid precludes the identification of essential genes by insertional mutagenesis. Nevertheless it is possible to generate a library of parasite mutants for essential genes by coupling random insertion to the tet-inducible system (Jammallo *et al.*, 2011).

Signature-tagged mutagenesis is another strategy that has been used to identify essential genes by insertional tagging. In this case screening is performed in a different life-cycle stage or under different growth conditions to permit the identification of 'differentially essential' genes. This approach has been successfully adapted for *Toxoplasma* (Knoll *et al.*, 2001). Wild-type parasite clones are first tagged with unique oligonucleotide insertions (the signature-tag). These clones are then mutagenized (chemical or insertional) followed by another cloning step. Pools of mutants, which are distinguishable by their tag, are subsequently exposed to a selective condition, e.g. infection into an animal. Tagging of genes that are essential in this condition will result in loss of the mutant. 'Missing' mutants are then identified by comparing the tags present in pools before and after selection. Several candidate genes important for parasite persistence in the mouse have been identified using this approach (Craver *et al.*, 2010; Frankel *et al.*, 2007; Payne *et al.*, 2011; Skariah *et al.*, 2012).

17.3.4 Forward Genetic Analysis using Chemical Mutagenesis and Complementation Cloning

Genetic analysis of pathways essential for growth in culture requires conditional mutants. Temperature sensitivity (ts) due to chemically induced point mutations can be exploited to obtain strains that are viable at the permissive temperature and display a mutant phenotype at the restrictive temperature. For *Toxoplasma* heat-sensitive (Pfefferkorn and Pfefferkorn, 1976; Radke *et al.*, 2000; Gubbels *et al.*, 2008) and cold-sensitive (Uyetake *et al.*, 2001) mutants have been isolated. ENU (N-ethyl-N-nitrosourea) induces random point mutations and has been the mutagen of choice in most *T. gondii* studies. Chemical mutagenesis has been successfully used in *T. gondii* to produce mutants with defects in stage differentiation (Singh *et al.*, 2002), invasion and egress (Black *et al.*, 2000; Uyetake *et al.*, 2001) and cell division and cell-cycle progression (Radke *et al.*, 2000; White *et al.*, 2005).

While generating chemical mutants is straightforward, identifying the mutated gene responsible for the phenotype is not. The two avenues most commonly used to accomplish this goal are physical mapping through crosses, and phenotypic complementation by transfection with a wild type DNA library. While crosses are feasible in *T. gondii,* their limited throughput makes them less practical as a general tool for mutant analysis (also the RH strain used as the molecular biology workhorse for *T. gondii* is unable to complete the sexual life cycle). The second approach to identify the gene affected in a given mutant is phenotypic complementation using a library of wild type DNA. This strategy faces two technical challenges: full representation of the genome (or transcriptome) in the complementation library, and efficient recovery of the complementing sequence. Black and colleagues identified a genetic element that maintains stable episomes in *T. gondii* (Black and Boothroyd, 1998) allowing convenient rescue by heat lysis and transformation of bacteria. A library harbouring an episomal maintenance sequence on the backbone successfully complemented the HXGPRT locus in the knockout mutant under mycophenolic acid selection. Analysis of the recovered plasmids, however, suggested that they might undergo extensive recombination, potentially decreasing their stability and usefulness (Black and Boothroyd, 1998).

The second effort to generate a complementation system was built on high frequency integration of library plasmids (Striepen *et al.*, 2002). Mutants are transfected with a plasmid library and subjected to selection. Subsequently complementing DNA sequences (carried as stable chromosomal insets) are rescued into plasmid using an *in vitro* recombination protocol (Invitrogen Gateway system (Hartley *et al.*, 2000)). Rescued library inserts can be shuttled back into a parasite expression plasmid through a second recombination step to confirm their complementation capacity. A cDNA library built on this model successfully complemented the *Toxoplasma* HXGPRT locus at high efficiency (Striepen *et al.*, 2002) and was used to identify a phenotypic suppressor of the *T. gondii* ts cell cycle mutant C9–11 (Radke *et al.*, 2000; White *et al.*, 2005). An analogous library carrying *Cryptosporidium parvum* genomic DNA was used for heterologous complementation resulting in the identification of a *Cryptosporidium* gene encoding the purine salvage enzyme IMPDH (Striepen *et al.*, 2002; Umejiego *et al.*, 2004). Several ts mutants could not be complemented using the cDNA libraries described above (Gubbels, White and Striepen, unpublished). Genes encoding large mRNAs and/or transcribed at low levels are typically underrepresented in cDNA libraries.

To overcome these problems, a large insert (40–50 kb) genomic cosmid library built on a DHFR–TS containing super-cos vector was constructed. This library provides sufficient coverage and transformation efficiency to complement the lack of HXGPRT in every transfection reaction attempted. In addition, this library complemented numerous mutants with a ts cell division defect (Gubbels *et al.*, 2008). Note that the increased power of sequencing technology now also allows sequencing of the complete genome of mutants thus permitting to pinpoint the genetic basis of temperature sensitive defects even in mutants that fail to complement efficiently (Gubbels *et al.*, 2008).

17.4 PERSPECTIVES

T. gondii has proven itself as an excellent experimental model and reverse genetic approaches were key to building a detailed molecular picture of apicomplexan biology. The reverse genetic toolbox has seen constant extension and refinement. The potential and limits of the genetic approaches have been reviewed in light of the biological specificities that differ between *Toxoplasma* and rodent or human malaria parasites (Limenitakis and Soldati-Favre, 2011).

To this joins the tremendous resource in the form of the completed genome sequence, where one can identify a large number of 'candidate' genes of interest by computational screens. With knock-outs and conditional knock-outs becoming easier, one can also target a larger number of candidates. A suitable step forward in the genetic research of *Toxoplasma* biology might be a community effort to establish a phenome project.

Forward genetic approaches have seen considerable progress as well. These approaches could hold the key to mechanistic analysis of phenomena for which the genome does not immediately present an obvious list of candidate genes and proteins. While the tools to complement mutants have improved and may now be at a level to permit robust analysis, the ways to generate and select such mutants still lag behind. Robust screens that reduce a complex cell biological phenomenon to a phenotype that can be easily scored in thousands of mutants with limited effort are needed. The success of visual screens using automated microscopic detection (Carey *et al.*, 2004) points to one avenue to reach this goal. The past decade has seen tremendous progress driven by the ability to transfect and genetically manipulate the parasites. The next decade will require a set of tools with sufficient throughput to take full advantage of the genome sequence.

17.5 THE *TOXOPLASMA* MANIATIS: A SELECTION OF DETAILED PROTOCOLS FOR PARASITE CULTURE, GENETIC MANIPULATION AND PHENOTYPIC CHARACTERIZATION

17.5.1 Propagation of *Toxoplasma* Tachyzoites in Tissue Culture

T. gondii is promiscuous in its choice of host cell and will infect almost any mammalian cell commonly used in tissue culture work. In general large spread-out cells like fibroblasts or Vero cells are most suitable. Infection of these cells results in distinctive rosettes, which makes it easy to monitor parasite development by microscopy. Many laboratories use transformed cell lines like Vero or 3T3, which produce high parasite yields. Immortal lines grow fast, are easy to culture and can be obtained from many sources.

Primary cell lines like human foreskin fibroblasts (HFF) are also widely used. Their strong contact inhibition and slow growth makes them the cell of choice for plaque assays, bradyzoite induction experiments, genetic selections or any experiment in which cultures have to be maintained for longer periods of time. They also provide excellent microscopy for cell biological analysis. The disadvantage of primary lines is that they have to be managed more carefully as they will die at higher passage number due to senescence. A sufficient amount of early passage cells has to be cryopreserved to reinitiate the culture at that point. HTERT cells (BD Biosciences) have emerged as a compromise; these cells are immortal but retain many characteristics of primary fibroblasts. We have found these cells to be equivalent to HFF cells in almost all applications. The protocols below are based on HFF cells but can be used for HTERT cells as well (note the difference in glutamine concentration). Many companies supply reagents for tissue culture, the suppliers mentioned in the following are the ones we have used, products from other sources might work just as well.

17.5.1.1 Maintenance of HFF Cells

- T25 flask tissue cultures typically yield 4–7 10^7 parasites (yields are typically lower for the Type II and III cyst-forming strains). The protocols below are based on this scale. If more material is needed, larger flasks (e.g. T175), roller bottles and cell factories have been used successfully with appropriately scaled protocols.
- Warm media and trypsin solution in a 37°C water bath.
- Aspirate medium from a confluent culture and add 2.5 ml of trypsin solution to the flask (0.25% trypsin and 0.2 g/l EDTA in HBSS, Hyclone, store this solution in smaller 5 ml aliquots at −20°C for convenience). Carefully 'wash' monolayer by tilting flask several times, aspirate most of the solution and leave enough to just cover the cells (~0.5 ml). Incubate at 37°C for 2 minutes. Inspect cells for rounding and detachment using an inverted microscope equipped with phase or interference contrast optics. If cells are still attached after 2 minutes, tap flask with flat hand and/or prolong incubation. HFF are relatively fragile so take care to not over-trypsinize.
- Immediately take up detached cells in a defined volume of Dulbecco's Modified Eagle's Medium (DMEM, Hyclone, if large batches are used medium can be prepared from powder, otherwise use ready-made medium) supplemented with 10% newborn calf serum (NBCS, Hyclone, cosmic calf serum), penicillin and streptomycin (1:200 of a 10,000 unit/ml of antibiotic stock, Hyclone) and glutamine (1:100 of a 200 mM stock in water, note: for HTERT cells do not add glutamine to avoid overgrowing of cultures) and split 1:8 into new flasks. If fungal contaminations are a frequent problem use 1:100 Fungizone (250 μg/ml amphothericin B, Invitrogen). Move to incubators gassed to 5% CO_2 at 37°C. Allow gas exchange by loosening caps. Confluent cultures can be kept for several weeks prior to *T. gondii* infection.

17.5.1.2 Maintenance of Tachyzoites

- Aspirate medium from a confluent HFF culture.
- Add 10 ml of infection medium (DMEM supplemented with 1% foetal calf serum (FCS, Invitrogen. For experiments which require tight control over the small molecule composition use dialysed foetal calf serum), penicillin and streptomycin as above).
- Infect a new flask with culture supernatant of a freshly lysed culture. As a rule of thumb, passing 0.5–1 ml into a T25 culture will result in complete lysis within 2–3 days for RH derived strains. A high inoculum is preferable if parasites are to be used, e.g. in a transfection experiment as the majority of the tachyzoites will egress synchronously resulting in high overall parasite viability. To maintain strains pass smaller number of parasites (e.g. 100 μl of a lysed culture). Transfection efficiency and invasion efficiency are greatly enhanced by using freshly lysed parasites. Host cells should not be over-infected. Ideally every host cell should be infected with one parasite.

17.5.1.3 Cryopreservation of Host Cells and Parasites

- In general, the aim is to freeze slowly and to thaw quickly. Wear a lab coat, face protection and appropriately insulated gloves when handling liquid nitrogen. For best results, have all tubes and reagents prepared and labelled, chill them on ice and work quickly (if you have to freeze many vials at a time divide them into smaller batches).
- Label 2 ml cryo vials (fitted with silicone O-ring, Nalgene) using a pen dispensing ink that resists liquid nitrogen and chill on ice.
- Prepare an isopropanol/water containing freezing container (VWR, using this simple and inexpensive device will result in about 1°C/min cooling in a −80°C freezer, alternatively use a thin-walled foam container to slow cooling).
- Use 'freshly' confluent (T175) HFF cultures for freezing. Trypsinize cells as described above and recover detached cells in DMEM 10% newborn calf serum into a 15 ml sterile centrifuge tube. Pellet cells in a table-top centrifuge at 900 g for 10 minutes at 4°C using a swing bucket rotor.
- Discard the supernatant and re-suspend cells in 1.8 ml chilled DMEM (no serum). Add 1.8 ml freezing medium (25% tissue culture grade DMSO and 20% FBS in DMEM) and mix quickly. Immediately dispense 0.5 ml aliquots into chilled freezing tubes, tightly cap tubes and move into chilled (ice) freezing container and place into a −80°C freezer.
- Thaw one vial the next day to insure that your stocks are viable and move the remaining vials into a liquid nitrogen storage container. Solid bookkeeping to keep track of rack, box and vial position is essential as it is not easy to search for vials in liquid nitrogen stocks.
- Parasites are preserved as extracellular tachyzoites. Pellet a freshly lysed culture (1500 g, 20 minutes, 4°C) and then proceed as described for host cells above. Plan to freeze 2 10^8 per vial which means that you will produce 3 vials from a single T25 culture using 0.8 ml of DMEM and 0.8 ml of freezing medium. Test for viability by thawing before you discontinue the culture of the given line.
- Parasites can also be cryopreserved in host cells at the rosette stage in DMEM with 50% FBS/10% DMSO.
- To thaw HFF cells prepare a flask with medium warmed to 37°C. Remove one vial at a time from liquid nitrogen with thongs and immediately immerse into a beaker filled with water warmed to 37°C gently shaking the vial. Once the medium is thawed, transfer cells to the flask and incubate as described for standard culture. Replace medium after 12 hours.
- To thaw parasites use above procedure and inoculate a confluent T25 culture.

17.5.1.4 Mycoplasma Detection and Removal

- Mycoplasma contamination is a frequent plague of tissue cultures. Heavy infection can affect the growth of host cells, mycoplasma DNA can produce unwanted background in genetic experiments and bacterial contamination is a severe problem for immunological experiments, as mycoplasma derived molecules potently stimulate a variety of immune cells and functions.
- A simple test for contamination can be performed by DNA staining. Culture cells (and/or parasites) for two passages in the absence of antibiotics (which will lead to massive amplification of the bacteria) then transfer to six well plates with coverslips.
- Stain coverslip cultures for bacterial DNA using DAPI using the standard IFA protocol provided below (more sensitive staining can be obtaining by acid/alcohol fixation and Hoechst staining (see Chen, 1977) for a detailed protocol).

- In contaminated cultures you will observe numerous small dots of DNA staining (about the size of the typical apicoplast genome staining) throughout the cytoplasm of the host cell.
- More sensitive PCR (ATCC, Stratagene) or luciferase-based (MycoAlert, Cambrex) assays are also available.
- If you suspect a recent contamination, discard your cultures, thaw fresh vial from liquid nitrogen and retest. Protocols to screen strains obtained from other laboratories should be routine.
- If you have to rescue your particular strain treat with Mycoplasma Removal Agent according to the manufacturer's guidelines (an inhibitor of bacterial gyrase, e.g. MP Pharmaceuticals) for three passages and then retest (this antibiotic is reasonably tolerated by *T. gondii* at the suggested concentration). Other commercial agents kill *T. gondii* and should be screened prior to use as mycoplasma elimination agents.
- Alternatively, passage of the strain through a mouse and re-isolation into tissue culture will remove mycoplasma.

17.5.1.5 Passaging *Toxoplasma* Tachyzoites/Bradyzoite Cysts in Animals

- Tachyzoites of any strain can be maintained by passage in the peritoneal cavities of mice; 10^4 (type I strain, i.e. RH) or 10^5 (type II or III strain, i.e. ME49 or Prugniaud) are injected intraperitoneally into the mouse.
- Replicating *T. gondii* can be harvested from the peritoneal cavity three days later (for type I strains) and five days later (type II or III strain) by peritoneal lavage with 4 ml of sterile saline or PBS.
- This material can be used to serially passage the strain in the peritoneal cavities of mice or to infect tissue culture cells. Murine inflammatory cells (macrophages and neutrophils) will also be seen in this lavage material.
- Passage through mice can be useful to remove microorganisms that have contaminated *T. gondii* tissue culture, provided that they cannot replicate in murine peritoneum. Anecdotal data indicate that periodic murine passage of a *T. gondii* strain passaged continuously in tissue culture helps to maintain the vigour and biologic characteristics of the strain.

17.5.2 Transfection and Stable Transformation Protocols

17.5.2.1 *Transient Transfection*

- Cytomix (120 mM KCl, 0.15 mM CaCl2, 10 mM K2HPO4/KH2PO4 pH 7.6, 25 mM HEPES pH 7.6, 2 mM EGTA, 5 mM MgCl2) can be prepared in larger batches, filter sterilized and stored in aliquots at −20°C or 4°C (Van den Hoof *et al.*, 1992).
- Weigh 12 mg ATP and 15.2 mg glutathione, add to 10 ml of cytomix and sterilize by passing through a 0.22 μM filter.
- Sterilize DNA by ethanol precipitation. Adjust 50 μg of plasmid DNA (typically in ~10 μl and purified using a commercial plasmid purification kit, e.g. Qiagen) to 100 μl with TE (pH 8.0). Add 11 μl 3M NaOAc, and 250 μl ethanol. Precipitate DNA for 5 minutes at −20°C and spin at full speed in a micro-centrifuge.
- Wash the pellet with 1 ml cold 70% ethanol by gently inverting the tube and spin for 2 minutes in a micro-centrifuge.
- Move tubes into the laminar flow hood and discard the ethanol (keep an eye on the pellet).
- Let ethanol evaporate for 5–10 minutes (be careful not to 'over-dry' as it can be hard to redissolve DNA). Re-suspend DNA in 100 μl cytomix.
- Filter parasites from a freshly lysed T175 flask into a 50 ml polypropylene tube and count in a haemocytometer (dilute sample 1:10 in PBS for counting). Pellet parasites at 1500 g,

20 minutes, 4°C and re-suspend in complete cytomix to 3.3 10^7 parasites per ml (if required the parasite concentration can be increased up to eight times).

- Mix 100 μl plasmid DNA and 300 μl parasites in a 2 mm gap electroporation cuvette (genetronix) and electroporate parasites with a single 1.5 kV pulse, a resistance setting of 25 Ω, and a capacitor setting of 25 μF using a BTX ECM 630. If you use a BioRad electroporator set to 1.5 kV, 25 μF and square wave, employing an Amaxa system, use the T-cell solution instead of cytomix and set the electroporation conditions to program U33.
- Transfer parasites immediately into a confluent T25 HFF culture (for selection and biochemical experiments) or onto coverslips for microscopy (see below).
- Expression of the transgene can be detected beginning 8 hours after transfection (depending on the transgene and the sensitivity of the assay employed) and peaks around 36 hours after electroporation. To measure transient transfection efficiency electroporate with a robust and easy to score visual marker (e.g. plasmid tub YFP–YFPsagCAT (Gubbels *et al.*, 2003)). Inoculate coverslips and count total number of vacuoles and number of fluorescent vacuoles for several fields. All three electroporators yield transient efficiencies of 30%–50% 24 hours after electroporation.

17.5.2.2 Selection of Stable Transformants

- **CAT:** Selection for chloramphenicol acetyl transferase (CAT) can start immediately after electroporation in presence of 20 μM chloramphenicol (34 mg/ml stock in ethanol). Since the effect of the drug is delayed, it is important to passage the parasites every two days by inoculating at least 10^6 parasites to keep the pool of parasites as heterogeneous as possible. The minimal amount of plasmid required to generate stable transformants depends on the vector used but 10–50 μg of linearized plasmid will usually yield stable transformants.
- **DHFR-TS:** Electroporate parasites with 50 μg of a plasmid encoding the drug resistant dihydrofolate reductase–thymidylate synthase allele (Donald and Roos, 1993), e.g. plasmid pDHFR*–TScABP (Sullivan *et al.*, 1999). After electroporation culture in the presence of 1 μM pyrimethamine (1 μl of a 10mM stock in ethanol). This plasmid results in the highest frequency of transformation (up to 1%–5%). Be careful handling transgenic strains as pyrimethamine is used in the treatment of human toxoplasmosis.
- **HXGPRT:** This selection requires a hypoxanthine–xanthine–guanine phosphoribosyltransferase null mutant (such mutants are available now for multiple strains, see e.g. Donald *et al.*, 1996 for RH). Twenty-four hours after transfection add 25μg/ml mycophenolic acid (25 mg/ml stock in ethanol) and 50 μg/ml xanthine (50 mg/ml stock in 0.1 N KOH). MPA/xanthine should kill parasites within 2–3 days.
- **BLE:** For phleomycin selection electroporate parasites with an expression vector encoding the resistance marker BLE (Messina *et al.*, 1995) transfer to HFF cells until complete lysis of the host culture occurs (24-48 hours later). The lysed culture is forced three times through a 25-ga needle to ensure that all the parasites are extracellular (see safety section for concerns about needle passing before using this protocol). The suspension of parasites is adjusted to 5 mg/ml of phleomycin (stock solution: 20 mg/ml in water and stored at −20°C) and incubated at 37°C for 10 hours. Parasites are transferred for recovery to

HFF cultures in media containing 5 μg/ml of phleomycin. After a new cycle of lysis the extracellular parasites are treated again in presence of drug for 10 hours and cloned thereafter by limiting dilution in 96-well microtitre plates containing HFF cells in the presence of 5 μg/ml of phleomycin.

17.5.2.3 Restriction Enzyme Mediated Integration (REMI)

Transformation efficiency can be enhanced by adding 50–100 U of BamHI, NotI, or SacII to the cuvette immediately prior to electroporation (these three enzymes have worked in the past; choose one that does not cut an essential part of your plasmid(s)). Note that REMI often results in multi-copy integration of plasmid(s) (Black *et al.*, 1995; Gubbels and Striepen, 2004).

17.5.2.4 Cloning of Transgenic Lines by Limiting Dilution in 96-Well Plates

- Seed tissue culture treated 96-well plates with HFF cells and grow to confluency. Remove medium and add 100 μl DMEM 1% FCS to each well.
- Harvest freshly lysed parasites by filtration and centrifugation as described above.
- Count using a haemocytometer and dilute to 250 parasites per ml.
- Add 100 μl (25 tachyzoites) to each well in the first and seventh vertical column.
- Using a multichannel pippetor perform a serial dilution from left to right transferring 100 μl at each step (mix each well by pipetting up and down three times). Discard medium after you reached column 6 and start over at row 7.
- Incubate for seven days without disturbing the culture.
- Inspect each row from left to right using an inverted microscope and identify wells that contain a single plaque and mark those wells. Expand clonal lines by passage into a T25 flask.

17.5.3 Measuring Parasite Survival and Growth

17.5.3.1 Plaque Assay

- Plaque assays are a reliable way to measure the number of viable and infectious parasites in a sample and are well suited to measure stable transfection efficiency. The following protocol will measure stable transformation using a DHFR–TS resistance plasmid.
- Electroporate tachyzoites as described above using 50 μg of pDHFR*–TScABP (Sullivan *et al.*, 1999). After electroporation, dilute 50 μl of the content of the cuvette into 950 μl cytomix or medium.
- Infect T25 HFF cultures in drug-free medium with 3 μl and 6 μl diluted parasite suspension and two cultures with 6 μl and 60 μl to be cultured in the presence of 1 μM pyrimethamine.
- Incubate for seven days without disturbing the flasks (optimal time may depend on strain used, 2–3 mm plaques are best for scoring, a few extra flasks can be added in a larger experiment to be 'developed' individually to test when the right plaque size is achieved). The period of selection takes longer with type II and III strains.
- To stain the monolayer aspirate the medium, rinse with PBS, fix for 5 minutes with ethanol and stain for 5 minutes with a crystal violet solution (dissolve 12.5 g crystal violet in 125 ml ethanol and mix with 500 ml 1% ammonium oxalate in water).
- Remove crystal violet solution and rinse with PBS.
- Air dry and count the number of plaques.
- This assay can also be used to quantify parasite growth by measuring plaque area. To do this scan stained flasks with a standard

flatbed scanner at 600 dpi and use image analysis for measurements. The area of plaques can be reasonably approximated using an ellipse. Measure the longest and shortest diameters of each plaque and use $\pi ab/4$ to calculate the area.

17.5.3.2 Fluorescence Assay

- This assay will produce dynamic growth curves over the time of the experiment (usually a week).
- Seed tissue culture-treated black 384- or 96-well plates with special optical bottom (Becton, Dickinson and Company) with HFF cells. For larger scale assays an automatic liquid dispenser (e.g. Genetix Q-Fill) will increase throughput and reproducibility.
- Once plates are confluent replace medium with DMEM (without phenol red, Hyclone) 1% FCS and antibiotics as described above.
- Infect each well with 2000 (384 well) or 5000 (96 well) tachyzoites (e.g. the YFP–YFP strain (Gubbels *et al.*, 2003)). Plan to have quadruple wells for each experimental condition (e.g. drug concentration) and include negative (no parasites) and positive controls on each plate. Fill all wells with medium but do not use the outermost wells as they evaporate faster which affects parasite growth.
- Measure fluorescence daily for each well for 5–8 days using a sensitive plate reader (BMG Fluostar, bottom excitation and emission 510/12 and 540/12 nM respectively).
- Plot the results (average of four wells and standard deviation) as percent positive in relation to the untreated positive control in each plate.

17.5.3.3 β-Galactosidase (LacZ) Assay

- This is an endpoint growth assay that can be used in multi-well formats (McFadden *et al.*, 1997), a yellow substrate will be turned into a red product.
- Seed HFF cells into standard tissue culture treated 384 well plates as described above.
- Change medium of confluent cultures to DMEM 1% FCS without phenol red (50 μl/well) and infect with 2000 β-galactosidase expressing tachyzoites (wash parasites in PBS before infection to eliminate phenol red).
- At the desired read-out day (usually 5 days after infection, optimal staining has to be established empirically for each strain and condition) add 4.5 μl chlorophenol red-b-galactopyranoside (CPRG, Boehringer Mannheim, 4.5 mM stock in medium without penol red).
- Develop colour to desired intensity (if you wait too long all wells will turn red, use your negative and positive controls as a guide) and read absorbance at 570 nM. Plot end points as percent positivity as described above.

17.5.3.4 Uracil Incorporation Assay

- In contrast to mammalian cells, *T. gondii* can directly salvage uracil through UPRT. This can be exploited to measure parasite growth as a function of [^{3}H]-uracil incorporation into parasite TCA precipitable nucleic acids (Pfefferkorn and Guyre, 1984; Roos *et al.*, 1994). The advantage of this assay is that it can be used in all strains and does not require a transgene. Recently a 96-well real-time format has been developed for this assay which is described in detail in Nare *et al.* (2002).
- Infect 24 well cultures with parasites and incubate under test conditions (e.g. in presence of a drug).
- Add 5 μCi of [5,6-^{3}H]-uracil (30–60 Ci/mmol) to each well and incubate for 2 hours at 37°C.
- Chill cultures and add an equal volume of ice-cold 0.6 N trichloroacetic acid to the medium of each well and incubate on ice for at least 1 hour.

- Remove TCA and rinse plates under running water overnight (make sure to use a sink designated for radioactivity work).
- Dry plates, add 500 μl of 0.1 N NaOH to each well, incubate for 1 hour and measure radioactivity in half of the sample by liquid scintillation counting. Depending on the scintillation cocktail used neutralization of the base can help to avoid background.

17.5.4 Live-Cell and Indirect Immunofluorescence Microscopy

- Sterilize round 23 mm glass cover slips in 70% ethanol (or autoclave) and transfer to six well plates. Seed cover slips with host cells and culture to confluency. Infect wells with tachyzoites 24–36 hours before microscopic examination.
- To observe parasites expressing fluorescent protein transgenes remove coverslip from dish with sterile forceps, wipe off medium from the bottom side and gently invert onto a microscope glass slide. If longer observation is required (e.g. for time lapse microscopy) use spacer circles (e.g. Secure Seal, Invitrogen) to generate a small reservoir of medium. Alternatively use dishes that have a cover slip bottom (e.g. ΔT3 dishes, Bioptechs).
- To use antibodies to stain cells remove medium, and fix cells in 2 ml of 3% paraformaldehyde in PBS for 10–20 min.
- Remove fixative and permeabilize cells in 2 ml 0.25% Triton X100 (in PBS) for 10 minutes.
- Block in 2 ml 1% w/v BSA in PBS/0.25% Triton X100 for 30 minutes.
- React with primary antibody (diluted 1:100–1:5000 in PBS/BSA/0.25% Triton X100 depending on titre) for 1 hour. This can be done with minimal reagent by inverting the coverslip onto 100 μl drops on parafilm in a moist chamber.
- Place back into six-well dish (cell side up) and wash three times with 3 ml PBS (5 minutes each).
- React with secondary antibody diluted in BSA/PBS for 1 hour.
- Wash four times in 3 ml PBS (5 minutes each). To counter-stain DNA add 2 μl of a 2 mg/ml DAPI stock solution to the first wash.
- Apply a drop of mounting medium to a microscope slide.
- Briefly wash coverslip in dH_2O (to prevent crystal formation after drying) and invert into mounting medium (cells down).
- Some epitopes are sensitive to aldehyde fixation. In that case use 2 ml of methanol for 20 minutes as fixative (methanol will also permeabilize the cells, and no Triton treatment is required). This protocol also works better to stain proteins secreted into the parasitophorous vacuole (these are often washed out by Triton permeabilization). A more elaborate protocol for secreted protein which preserves subcellular structures better than methanol can be found in Lecordier *et al.* (1999).

17.5.5 Cytometry of Parasites and Infected Cells

Toxoplasma tachyzoites can be efficiently sorted using a fluorescence activated cell sorter (FACS) after labelling with specific antibodies to the surface of the parasite (Kim and Boothroyd, 1995; Radke *et al.*, 2004) or based on the expression of autofluorescent protein (Striepen *et al.*, 1998; Gubbels *et al.*, 2004; Gubbels and Striepen, 2004). Parasites expressing fluorescent proteins can also be sorted within their host cells (Gubbels and Striepen, 2004; Gubbels *et al.*, 2005).

- For sorting autofluorescent parasites harvest a freshly lysed culture and filter parasite through a 3 μm polycarbonate filter. Count parasites and take up in sterile PBS at 10^7/ml.
- Use a high-speed sorter equipped with a 488 nM argon laser and the following filter and mirrors (GFP or YFP: DM: 555 SP, F: 530/40 BP; DRFP or Tomato DM: 555 SP, F: 570/40 BP).

Note that for sorting the flow stream is broken into droplets, which carries the potential to produce aerosolized parasites. Extra safety can be provided by an evacuated and HEPA filtered enclosure of the sorting chamber. Discuss biosafety aspects with the FACS facility director and operator.

- For enrichment, sort into tubes preloaded with 0.5 ml of PBS or medium and transfer to a confluent T25 HFF culture. For cloning sort directly into seeded multi-well plates. Using a MOFLO sorter we found three events per well to result in the maximum number of single clones per plate.
- To sort infected cells, inoculate parasites into a confluent HFF culture 1–24 hours prior to sorting.
- Aspirate medium and wash twice with sterile PBS.
- Trypsinize cells as described above and recover in 10 ml DMEM 1% FCS.
- Filter through a 75 μm cell strainer (Becton Dickinson), spin down and re-suspend in 0.5 ml PBS and sort as described above.
- Detail on antibody staining for FACS of tachyzoites is provided in Radke *et al.* (2004).

17.5.6 Disruption of Non-Essential Genes

T. gondii is haploid and non-essential genes can be disrupted by homologous recombination using single or double cross-over. As discussed in detail the main challenge is to overcome the background of non-homologous plasmid insertion. Above we have described and cited several approaches; here we describe a CAT/YFP positive/negative selection for homologous recombination by double cross-over in detail.

- Construct a targeting plasmid that flanks a sagCATsag selectable marker cassette with 1.5–3 kb homologous sequence from the target gene (typically the 5′ and 3′ genomic sequences flanking the actual coding sequence). Introduce a YFP expression cassette 3′ adjacent to the 3′ homologous flanking region. Be sure that your targeting plasmid contains a unique restriction site that will allow you to linearize the construct without cutting into markers or flanking regions (e.g. in the multi-cloning site of the plasmid backbone).
- Test for YFP expression in a transient transfection experiment (~30% of the vacuole should show cytoplasmic fluorescence).
- Transfect with 10, 25 and 50 μg of linearized plasmid and select for stable transformation in the presence of 20 μM chloramphenicol.
- Subject the drug resistant population (typically after 3–4 passages) to FACS (use the non-transfected parent strain and a YFP expressing strain as positive and negative controls). Gate events to be sorted to 'viable' tachyzoites by forward and side scatter and clone non-fluorescent parasites by sorting into confluent 96 well plate cultures.
- Leave plates undisturbed and check for single plaques after seven days and mark clones.
- Suspend parasites by pipetting up and down and transfer 100 μl of each well into a well of a six well plate. Replenish medium in the 96 well plates and keep in the incubator.
- Six well cultures will lyse within 3–4 days. Re-suspend lysed parasites by pipetting and harvest by centrifugation.
- Wash parasites with PBS and pellet again.
- Re-suspend parasites in 500 μl TE, add 1 μl RNAse (10mg/ml), 10 μl 10% SDS and 20 μl proteinase K (10 mg/ml).
- Incubate at 55°C for at least 1 hour (can go overnight).
- Extract twice with 500 μl phenol: chloroform: isoamylalcohol (25:24:1, molecular biology grade), and once with chloroform, always keep the water phase.
- Add 1/10 volume of 3 M NaOAc and 2.5 volumes of ethanol and precipitate DNA for 20 minutes at −20°C.
- Spin for 10 minutes at full speed in a microcentrifuge, wash pellet with 70%

ethanol, spin again, briefly air dry and re-suspend DNA in 50 μl TE.

- Use 5 μl as template in a PCR reaction with primers that will produce different size products for the native and the KO locus (make sure that your primers do not pick up the ectopic mini-gene copy).
- Confirm putative allelic replacements by Southern blot using appropriate probes.

17.5.7 Disruption of Essential Genes

17.5.7.1 Tetracycline Inducible Systems

As detailed above the first approach involves several steps: 1. Introduce an ectopic tet-regulatable copy of the target gene, 2. target the native locus by homologous recombination, and 3. knock-down of the expression of the ectopic copy using ATc treatment. The choice of selectable markers may differ from experiment to experiment (the tet-transactivator line (Meissner *et al.*, 2002) is resistant to mycophenolic acid), this example will use CAT, YFP and DHFR–TS.

- Construct a plasmid for ectopic expression of the target gene, e.g. by replacement of the ACP coding sequence in plasmid ptet07sag4–ACPmyc/DHFR–TS (Mazumdar *et al.*, submitted; Meissner *et al.*, 2002). If you omit the stop codon this should result in an N-terminal translational fusion to a c-myc epitope tag.
- Transfect into the TATi transactivator line (Meissner *et al.*, 2002), select stable transformants in the presence of 1 μM pyrimethamine and clone by limiting dilution.
- Test clones for transgene expression by IFA and Western blot using an anti-myc antibody (mAb 9E10, Roche).
- Choose clones that express the transgene at a similar level as the native gene. Depending on the size of the target gene addition of the tag may result in a noticeable mobility shift on SDS PAGE. In this case an antibody against the target protein can be used to compare both proteins side by side.
- It is critical to identify a tightly regulated clone before proceeding to the KO experiment. Careful characterization of clones will pay off with a clean interpretable phenotype. Test for regulation by culturing parasites in the presence or absence of 1 μg/ml of ATc (0.2 mg/ml stock in ethanol) followed by IFA and Western blot. Note that stable proteins might have to be diluted out by growth. Do your first screen after five days of treatment and then titre the minimal treatment time for complete suppression using your tightest clone.

Target the native locus as described above (using CAT/YFP positive negative selection), establish allelic replacement and analyse regulation of the ectopic copy in confirmed KO clones by IFA and Western blot. To facilitate double homologous recombination, vectors should preferentially be linearized by digestion at both ends of the construct to remove vector sequences. The choice of selectable marker depends on the background strain. A large variety of strains are readily available to the research community but the genetically modified strains significantly differ regarding the set resistance and sensibility to selectable marker genes.

17.5.7.2 Regulation of Protein Stability

The generation of transgenic parasites expressing ddFKBP-fusion proteins can be selected either in the presence (knock-down of an essential gene) or absence (expression of a dominant negative mutant) of Shield-1. Conditional expression is performed with 1 μM Shield-1.

17.5.8 Insertional Mutagenesis and Tag Rescue

- Electroporate tachyzoites by using 50 μg of linearized (e.g. restricted with NotI) plasmid pDHFR*–TScABP. Select for stable transformants in 1 μM pyrimethamine and apply the desired phenotypic screen. Clone mutants by limiting dilution, expand into T25 cultures and isolate genomic DNA as described above.

- Set up parallel 20 μl restriction digests using several restriction enzymes that cut once in your plasmid (e.g. EcoRI, HindIII, XhoI, XbaI for pDHFR*–TScABP see Sullivan *et al.* (1999) for maps and a detailed discussion of enzyme choice). Use 2 μg genomic DNA for each digest and incubate overnight at 37°C.
- Purify DNA from digest using a Qiagen spin column following the manufacturer's protocol and elute in 30 μl elution buffer.
- Mix 5 μl eluate with 2 μl 10 × NEB ligase buffer, 13 μl H_2O and 1 μl T4 DNA ligase and incubate overnight at 16°C.
- Add 1 μl glycogen, 2 μl 3 M NaAc, pH 5.2 and 50 μl ethanol and precipitate DNA for 30 minutes at −20°C.
- Wash pellet with 1 ml 70% ethanol, air dry briefly, and re-suspend pellet in 10 μl H_2O.
- Electroporate 1 μl into 25 μl library efficient electrocompetent bacteria (we found DH12S to result in best recovery).
- Transfer into sterile microcentrifuge tube, add 200 μl LB medium and incubate for 1 hour at 37°C while shaking.
- Plate entire transformation onto an LB agar plate containing suitable antibiotic (in this case ampicillin).
- Tags can also be rescued by inverse PCR. See Sullivan *et al.* (1999) for primer design and a detailed protocol.

17.5.9 Chemical Mutagenesis

- ENU is highly toxic and carcinogenic. Use utmost care with all materials that have come into contact with this chemical. Label tubes and flasks to warn members of your laboratory and dispose contaminated solutions appropriately.
- The mutagenic potency can vary from batch to batch and has to be titrated by plaque assay. Prepare a stock solution (100 mg/ml in DMSO) and store multiple aliquots at −20°C. Perform triplicate plaque assays using 0, 25, 50 and 75 μl of mutagen. Optimal mutagenesis results in 70% parasite killing compared to untreated controls (the protocol below assumes 50 μl as the optimal dose).
- Infect two confluent T25 HFF cultures with 1.2 ml of a freshly lysed culture 24 hours prior to the experiment.
- Replace medium with 10 ml DMEM 0.1% FBS medium.
- Incubate at 37°C for 30 minutes.
- Add 50 l ENU to flask A and 50 μl sterile tissue culture grade DMSO to flask B.
- Treat for 4 hours at 37°C.
- Wash cultures three times with 10 ml cold sterile PBS and discard into a dedicated waste container.
- Add 10 ml PBS, scrape cells with a cell scraper, liberate parasite by two passages through a 25-ga needle (see safety section), and filter through a 3 μm polycarbonate filter.
- Transfer to 50 ml tube, add 40 ml PBS and spin at 1500 × g at 4°C for 20 minutes.
- Re-suspend in 5 ml PBS and count parasites. Proceed to cloning by limiting dilution. It is advisable to control the mutagenesis efficiency of each experiment by plaque assay.

17.5.10 Complementation Cloning using the *Toxoplasma* Genomic Libraries

- Prepare 50 large and 10 small LB–agar Petri dishes (10 μg/ml Kanamycin).
- To titre the ToxoSuperCos library prepare five 1.5 ml Eppendorf tubes with 135 μl LB (no antibiotics), one with 1 ml LB and one empty tube.
- Remove library from the −80°C freezer and keep on ice (work quickly to avoid thawing and immediately refreeze library).

- Scrape a small amount of library (~20 µl) into the empty tube.
- Add 1 µl of thawed scraped bacteria to 1 ml LB (1:10^3 dilution).
- Keep the remainder of the thawed library at 4°C (stable for 1–2 days).
- Prepare a dilution series (10^4–10^8), plate 100 µl of each dilution on pre-warmed small LB–Kan plates, grow overnight at 37°C and count colonies to calculate the number of colony forming units (cfu) per ml.
- To amplify the library DNA pre-warm large LB–Kan plates at 37°C, prepare 10 ml of LB containing 50,000 cfu/ml and plate 200 µl per plate.
- Grow overnight at 37°C (incubate longer if colonies are too small).
- To harvest, add 2 ml of LB to the plate and scrape colonies using a cell scrape, transfer into a 250 ml centrifugation bottle (on ice) and wash with 1 ml of LB. Repeat for each plate and pool.
- Pellet bacteria in a tabletop centrifuge, remove liquid and weigh the pellets (bacteria can be stored at −20°C at this step).
- Purify cosmids using a commercial kit, e.g. Qiagen large construct kit according to the manufacturer's instructions, re-suspend DNA pellet in 150 µl TE per column and store cosmid DNA at 4°C in the dark.
- To complement *T. gondii* mutants perform five independent transfections as described above (8 × 10^7 parasites and 25 µg cosmid DNA per cuvette). Include at least one mock transfection to control for reversion.
- Transfer independently into T175 HFF cultures, incubate overnight at permissive conditions then apply selective pressure.
- For ts mutants plaques can be identified 10–14 days after transfection
- Clone by limiting dilution, prepare genomic DNA and rescue a sequence tag exploiting the Kan marker on the ToxoSuperCos backbone as describe for insertional mutagenesis (use BglII, HindIII and XhoI).
- BLAST rescued sequences against ToxoDB. You should obtain hits to the same genomic region from independent complementations. Check if your candidate region is represented among the sequenced and arrayed cosmids displayed on ToxoDB, acquire these cosmids and test for complementation.

17.5.11 Recombineering Cosmids of *Toxoplasma* Genomic Libraries

- Find a cosmid that covers your gene (using http://www.toxodb.org), and identify the corresponding bacterial clone number (using http://toxomap.wustl.edu/cosmid.html).
- Prepare the cosmid from an overnight 28–30°C culture, confirm by digest and electroporate 100–300 ng into *E. coli* strain EL250 (electroporate in 1 mm gap cuvette at 1.75 kV, 250Ω, and 25 µF).
- Induce the λ phage recombination machinery in a fresh 100 ml culture of EL250 containing cosmid (grown from 2 ml overnight culture at 28–30°C to OD = 0.4) by immediately transferring it to 43°C and shaking 20 minutes at 100 rpm, following 20 minutes cooling in ice-water.
- Use the cooled culture to make competent cells by three consecutive washes in ice cold sterile ddH2O (in 50, 20 and 3 ml, centrifugations at 4000 rpm, 10 minutes at 4°C). Re-suspend the competent pellet in 600 µl 10% sterile glycerol and aliquot 50 µl into ice-cold sterile microfuge tubes for storage at −80°C.
- PCR amplify a modification cassette (see Fig. 17.5 and Table 17.2 below on how to design your desired manipulation) from 0.1 to 50 ng of plasmid template.
- Use 100–300 ng gel purified targeting cassette to electroporate (same as above) to one 50 µl aliquot, rescue in SOC media at 28–30°C for one hour and plate on gentamycin + kanamycin to select for recombineering.

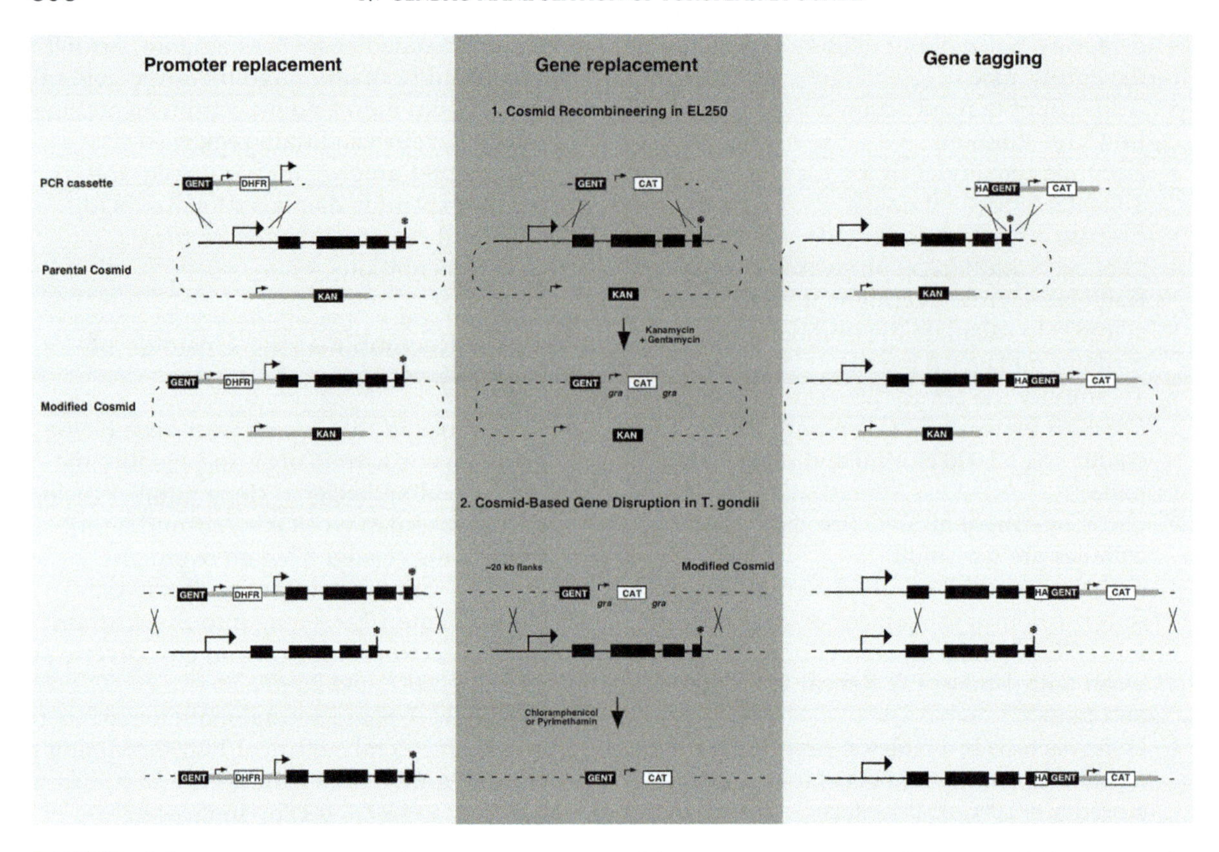

FIGURE 17.5 **Using cosmid recombineering to modify GOI.** Schematic representation of the three available strategies for cosmid modification: promoter replacement, gene replacement and C-terminal gene tagging. The PCR cassette and recombination even into the cosmid are depicted on the top of each panel as Step 1 with the resulting modified cosmid. Step 2 demonstrates the recombination into the genome using each modified cosmid (here shown as linear) and the resulting modified locus.

TABLE 17.2 Primers for PCR Amplification of Cosmid Modification Cassettes

Type of Modification	Toxoplasma Selectable Marker	F Primer	R Primer
Promoter replacement	DHFR	50 bp GOI at the 5′ of the promoter + GAATGGTAACCGACAAACGCGTTC	GCTTTCGTCTGTCTTCAACCAGATCT + 50 bp GOI just upstream of ATG
Gene replacement	BLE/CAT	50 bp GOI upstream of start codon + CCTCGACTACGGCTTCCATTGGCAAC	50 bp GOI downstream from stop codon + ATACGACTCACTATAGGGCGAATTGG
HA tagging	BLE/CAT	50 bp of GOI upstream of stop codon + AGGTACCCGTACGACGTCCCGGACTAC	50 bp GOI downstream from stop codon + ATACGACTCACTATAGGGCGAATTGG

17.5.12 Safety Concerns Working with *T. gondii*

Several aspects of the parasite's biology make working with *T. gondii* relatively safe. In immunocompetent persons the infection usually produces no or only modest symptoms. Depending on the region of the world, 20%–70% of the population is already infected and resistant to reinfection. Lastly, the tachyzoite stage, which is most widely used in experimental work, is not highly infective by aerosol or ingestion. However, *T. gondii* is a human pathogen with the ability to cause severe disease and should be handled with appropriate care (severe lab accidents have occurred in the past).

We summarize a few ground rules in the following (this section does not represent a comprehensive laboratory safety manual).

- Laboratory workers who belong to a specific risk group (active or potential severe immunosuppression, pregnancy) should not work with live parasites.
- Safety procedures should be frequently reviewed with all members of the laboratory.
- Handle parasites in designated biosafety cabinets. Label all work areas, flasks, tubes and waste containers that might harbour infectious material accordingly.
- Wear a lab coat, gloves and goggles. Goggles are especially important for workers who do not wear glasses. An eye splash could potentially deliver a high inoculum of parasites.
- The main route of infection with tachyzoites is direct inoculation by injury or through eye splash. Be extremely careful in all situations that involve sharps. Note that coverslips, microscope slides as well as plastic or glass tubes can break and produce sharp edges. Should you break something, sterilize using 70% ethanol before you attempt clean-up. Needle sticks are the most common source of laboratory infections. The safest approach to minimize such situations is to avoid them. Consider if the use of sharps is really essential to your experiment. If you really have to needle pass infected cells to liberate parasites leave the plastic sheath on the needle and cut off its tip using sturdy scissors several millimetres before the tip of the actual needle. This can help to protect you from accidental sticks and provides extra safety at no additional cost or effort.
- Be especially careful working with strains that encode resistance to drugs commonly used for treatment of humans including pyrimethamine, sulphadiazine, clindamycin and azithromycin.
- Sterilize all materials that were in contact with live parasites (autoclave all plastic tissue culture material, bleach all liquids accumulating in e.g. vacuum bottles and frequently sterilize surfaces by spraying and wiping down with 70% ethanol).
- Have a plan for a potential accident. While the goal is to prevent accidents, they might happen nonetheless. Establish local as well as national contacts to infectious disease specialists who could provide advice for diagnosis and treatment. (Reference laboratories include the Palo Alto Research Foundation (http://www.pamf.org/serology) and the Laboratory of Parasitology and FAO/WHO International Centre for Research and Reference on Toxoplasmosis, Statens Seruminstitut, 2300 Copenhagen S, Denmark).
- Ensure good communication about lab safety and **always** disclose any contamination, accident or inoculation. Inform the head of your laboratory about any accident, even if you feel this was a minor incident.

Acknowledgements

Work in our laboratories is currently funded by grants from the National Institutes of Health (to Boris Striepen) and from the Wellcome Trust, Swiss National Foundation and Howard Hughes Medical Institutes (to Dominique Soldati-Favre). Boris Striepen is a Georgia Research Alliance Distinguished

Investigator. Dominique Soldati-Favre is supported by the Swiss National Foundation (FN3100A0-116722) and Damien Jacot by the Indo-Swiss Joint Research Programme. We thank Louis Weiss for providing protocols for propagation of parasites in mice. We thank all current and former members of our laboratories and many investigators in the field for their protocols and discussion.

References

Agop-Nersesian, C., Egarter, S., Langsley, G., Foth, B.J., Ferguson, D.J., Meissner, M., 2010. Biogenesis of the inner membrane complex is dependent on vesicular transport by the alveolate specific GTPase Rab11B. PLoS Pathog. 6, e1001029.

Armstrong, C.M., Goldberg, D.E., 2007. An FKBP destabilization domain modulates protein levels in Plasmodium falciparum. Nat. Methods 4, 1007–1009.

Andenmatten, N., Engarter, S., Jackson, A.J., Jullien, N., Herman, J.P., Meissner, M., 2013. Conditional genome engineering in *Toxoplasma gondii* uncovers alternative invasion mechanisms. Nat. Methods. In press.

Arrizabalaga, G., Ruiz, F., Moreno, S., Boothroyd, J.C., 2004. Ionophore-resistant mutant of Toxoplasma gondii reveals involvement of a sodium/hydrogen exchanger in calcium regulation. J. Cell Biol. 165, 653–662.

Banaszynski, L.A., Chen, L.C., Maynard-Smith, L.A., Ooi, A.G., Wandless, T.J., 2006. A rapid, reversible, and tunable method to regulate protein function in living cells using synthetic small molecules. Cell 126, 995–1004.

Behnke, M.S., Wootton, J.C., Lehmann, M.M., Radke, J.B., Lucas, O., Nawas, J., Sibley, L.D., White, M.W., 2010. Coordinated progression through two subtranscriptomes underlies the tachyzoite cycle of Toxoplasma gondii. PLoS One 5, e12354.

Black, M., Seeber, F., Soldati, D., Kim, K., Boothroyd, J.C., 1995. Restriction enzyme-mediated intergration elevates transformation frequency and enables co-transfection of *Toxoplasma gondii*. Mol. Biochem. Parasitol. 74, 55–63.

Black, M.W., Arrizabalaga, G., Boothroyd, J.C., 2000. Ionophore-resistant mutants of Toxoplasma gondii reveal host cell permeabilization as an early event in egress. Mol. Cell Biol. 20, 9399–9408.

Black, M.W., Boothroyd, J.C., 1998. Development of a stable episomal shuttle vector for Toxoplasma gondii. J. Biol. Chem. 273, 3972–3979.

Bohne, W., Wirsing, A., Gross, U., 1997. Bradyzoite-specific gene expression in Toxoplasma gondii requires minimal genomic elements. Mol. Biochem. Parasitol. 85, 89–98.

Brecht, S., Erdhart, H., Soete, M., Soldati, D., 1999. Genome engineering of toxoplasma gondii using the site-specific recombinase Cre [In Process Citation]. Gene 234, 239–247.

Breinich, M.S., Ferguson, D.J., Foth, B.J., Van Dooren, G.G., Lebrun, M., Quon, D.V., Striepen, B., Bradley, P.J., Frischknecht, F., Carruthers, V.B., Meissner, M., 2009. A dynamin is required for the biogenesis of secretory organelles in Toxoplasma gondii. Curr. Biol. 19, 277–286.

Brooks, C.F., Johnsen, H., Van Dooren, G.G., Muthalagi, M., Lin, S.S., Bohne, W., Fischer, K., Striepen, B., 2010. The toxoplasma apicoplast phosphate translocator links cytosolic and apicoplast metabolism and is essential for parasite survival. Cell Host Microbe 7, 62–73.

Campbell, R.E., Tour, O., Palmer, A.E., Steinbach, P.A., Baird, G.S., Zacharias, D. A., Tsien, R.Y., 2002. A monomeric red fluorescent protein. Proc. Natl. Acad. Sci. U. S. A. 99, 7877–7882.

Carey, K.L., Westwood, N.J., Mitchison, T.J., Ward, G.E., 2004. A small-molecule approach to studying invasive mechanisms of Toxoplasma gondii. Proc. Natl. Acad. Sci. U. S. A. 101, 7433–7438.

Chaturvedi, S., Qi, H., Coleman, D., Rodriguez, A., Hanson, P.I., Striepen, B., Roos, D.S., Joiner, K.A., 1999. Constitutive calcium-independent release of *Toxoplasma gondii* dense granules occurs through the NSF/SNAP/SNARE/Rab machinery. J. Biol. Chem. 274, 2424–2431.

Chen, T.R., 1977. In situ detection of mycoplasma contamination in cell cultures by fluorescent Hoechst 33258 stain. Exp. Cell Res. 104, 255–262.

Chiang, C.W., Carter, N., Sullivan Jr., W.J., Donald, R.G., Roos, D.S., Naguib, F.N., El Kouni, M.H., Ullman, B., Wilson, C.M., 1999. The adenosine transporter of Toxoplasma gondii. Identification by insertional mutagenesis, cloning, and recombinant expression. J. Biol. Chem. 274, 35255–35261.

Craver, M.P., Rooney, P.J., Knoll, L.J., 2010. Isolation of Toxoplasma gondii development mutants identifies a potential proteophosphoglycan that enhances cyst wall formation. Mol. Biochem. Parasitol. 169, 120–123.

Daher, W., Plattner, F., Carlier, M.F., Soldati-Favre, D., 2010. Concerted action of two formins in gliding motility and host cell invasion by Toxoplasma gondii. PLoS Pathog. 6, e1001132.

Delbac, F., Sanger, A., Neuhaus, E.M., Stratmann, R., Ajioka, J.W., Toursel, C., Herm-Gotz, A., Tomavo, S., Soldati, T., Soldati, D., 2001. Toxoplasma gondii myosins B/C: one gene, two tails, two localizations, and a role in parasite division. J. Cell Biol. 155, 613–623.

Donald, R., Carter, D., Ullman, B., Roos, D.S., 1996. Insertional tagging, cloning, and expression of the Toxoplasma gondii hypoxanthine-xanthine-guanine phosphoribosyltransferase gene. Use as a selectable marker for stable transformation. J. Biol. Chem. 271, 14010–14019.

Donald, R.G., Roos, D.S., 1993. Stable molecular transformation of *Toxoplasma gondii*: a selectable dihydrofolate reductase-thymidylate synthase marker based on drug-resistance mutations in malaria. Proc. Natl. Acad. Sci. U. S. A. 90, 11703–11707.

Donald, R.G., Roos, D.S., 1994. Homologous recombination and gene replacement at the dihydrofolate reductase-thymidylate synthase locus in Toxoplasma gondii. Mol. Biochem. Parasitol. 63, 243–253.

Donald, R.G., Roos, D.S., 1995. Insertional mutagenesis and marker rescue in a protozoan parasite: cloning of the uracil phosphoribosyltransferase locus from Toxoplasma gondii. Proc. Natl. Acad. Sci. U. S. A. 92, 5749–5753.

Dvorin, J.D., Martyn, D.C., Patel, S.D., Grimley, J.S., Collins, C.R., Hopp, C.S., Bright, A.T., Westenberger, S., Winzeler, E., Blackman, M.J., Baker, D.A., Wandless, T.J., Duraisingh, M.T., 2010. A plant-like kinase in Plasmodium falciparum regulates parasite egress from erythrocytes. Science 328, 910–912.

Egan, C.E., Dalton, J.E., Andrew, E.M., Smith, J.E., Gubbels, M.J., Striepen, B., Carding, S.R., 2005. A requirement for the Vgamma1+ subset of peripheral gammadelta T cells in the control of the systemic growth of Toxoplasma gondii and infection-induced pathology. J. Immunol. 175, 8191–8199.

Egeler, E.L., Urner, L.M., Rakhit, R., Liu, C.W., Wandless, T.J., 2011. Ligand-switchable substrates for a ubiquitin-proteasome system. J. Biol. Chem. 286, 31328–31336.

Farrell, A., Thirugnanam, S., Lorestani, A., Dvorin, J.D., Eidell, K.P., Ferguson, D.J., Anderson-White, B.R., Duraisingh, M.T., Marth, G.T., Gubbels, M.J., 2011. A DOC2 protein identified by mutational profiling is essential for apicomplexan parasite exocytosis. Science 335, 218–221.

Fox, B.A., Belperron, A.A., Bzik, D.J., 1999. Stable transformation of Toxoplasma gondii based on a pyrimethamine resistant trifunctional dihydrofolate reductase-cytosine deaminase-thymidylate synthase gene that confers sensitivity to 5-fluorocytosine. Mol. Biochem. Parasitol. 98, 93–103.

Fox, B.A., Belperron, A.A., Bzik, D.J., 2001. Negative selection of herpes simplex virus thymidine kinase in Toxoplasma gondii. Mol. Biochem. Parasitol. 116, 85–88.

Fox, B.A., Ristuccia, J.G., Gigley, J.P., Bzik, D.J., 2009. Efficient gene replacements in Toxoplasma gondii strains deficient for nonhomologous end joining. Eukaryot. Cell 8, 520–529.

Francia, M.E., Jordan, C.N., Patel, J.D., Sheiner, L.L., Demerly, J.L., Fellows, J.D., Cruz De Leon, J., Morrissette, N.S., Dubremetz, J., Striepen, B., 2012. Cell division in apicomplexan parasites is organized by a homolog of the striated rootlet fiber of algal flagella. PLoS Biology 10.

Frankel, M.B., Mordue, D.G., Knoll, L.J., 2007. Discovery of parasite virulence genes reveals a unique regulator of chromosome condensation 1 ortholog critical for efficient nuclear trafficking. Proc. Natl. Acad. Sci. U. S. A. 104, 10181–10186.

Gubbels, M.J., Lehmann, M., Muthalagi, M., Jerome, M.E., Brooks, C.F., Szatanek, T., Flynn, J., Parrot, B., Radke, J., Striepen, B., White, M.W., 2008 Feb. Forward genetic analysis of the apicomplexan cell division cycle in Toxoplasma gondii. PLoS Pathog. 4, e36. http://dx.doi.org/10.1371/journal.ppat.0040036.

Gubbels, M.J., Li, C., Striepen, B., 2003. High-Throughput Growth Assay for Toxoplasma gondii Using Yellow Fluorescent Protein. Antimicrob. Agents Chemother. 47, 309–316.

Gubbels, M.J., Striepen, B., 2004. Studying the cell biology of apicomplexan parasites using fluorescent proteins. Microsc. Microanal. 10, 568–579.

Gubbels, M.J., Striepen, B., Shastri, N., Turkoz, M., Robey, E.A., 2005. Class I major histocompatibility complex presentation of antigens that escape from the parasitophorous vacuole of Toxoplasma gondii. Infect. Immun. 73, 703–711.

Gubbels, M.J., Vaishnava, S., Boot, N., Dubremetz, J.F., Striepen, B., 2006. A MORN-repeat protein is a dynamic component of the Toxoplasma gondii cell division machinary. J. Cell Sci. in press.

Gubbels, M.J., Wieffer, M., Striepen, B., 2004. Fluorescent protein tagging in Toxoplasma gondii: identification of a novel inner membrane complex component conserved among Apicomplexa. Mol. Biochem. Parasitol. 137, 99–110.

Hager, K.M., Striepen, B., Tilney, L.G., Roos, D.S., 1999. The nuclear envelope serves as an intermediary between the ER and golgi complex in the intracellular parasite *Toxoplasma gondii*. J. Cell Sci. 112, 2631–2638.

Hartley, J.L., Temple, G.F., Brasch, M.A., 2000. DNA cloning using in vitro site-specific recombination. Genome Res. 10, 1788–1795.

Heaslip, A.T., Dzierszinski, F., Stein, B., Hu, K., 2010. TgMORN1 is a key organizer for the basal complex of Toxoplasma gondii. PLoS Pathog. 6, e1000754.

Herm-Gotz, A., Agop-Nersesian, C., Munter, S., Grimley, J.S., Wandless, T.J., Frischknecht, F., Meissner, M., 2007. Rapid control of protein level in the apicomplexan Toxoplasma gondii. Nat. Methods 4, 1003–1005.

Herm-Gotz, A., Weiss, S., Stratmann, R., Fujita-Becker, S., Ruff, C., Meyhofer, E., Soldati, T., Manstein, D.J., Geeves, M.A., Soldati, D., 2002. Toxoplasma gondii myosin A and its light chain: a fast, single-headed, plus-end-directed motor. Embo J. 21, 2149–2158.

Hu, K., Roos, D.S., Angel, S.O., Murray, J.M., 2004. Variability and heritability of cell division pathways in Toxoplasma gondii. J. Cell Sci. 117, 5697–5705.

Hu, K., Roos, D.S., Murray, J.M., 2002. A novel polymer of tubulin forms the conoid of Toxoplasma gondii. J. Cell Biol. 156, 1039–1050.

Huynh, M.H., Carruthers, V.B., 2006. Toxoplasma MIC2 is a major determinant of invasion and virulence. PLoS Pathog. 2, e84.

Huynh, M.H., Carruthers, V.B., 2009. Tagging of endogenous genes in a Toxoplasma gondii strain lacking Ku80. Eukaryot. Cell 8, 530–539.

Iwamoto, M., Bjorklund, T., Lundberg, C., Kirik, D., Wandless, T.J., 2010. A general chemical method to regulate protein stability in the mammalian central nervous system. Chem. Biol. 17, 981–988.

Jammallo, L., Eidell, K., Davis, P.H., Dufort, F.J., Cronin, C., Thirugnanam, S., Chiles, T.C., Roos, D.S., Gubbels, M.J., 2011. An insertional trap for conditional gene expression in Toxoplasma gondii: identification of TAF250 as an essential gene. Mol. Biochem. Parasitol. 175, 133–143.

Joiner, K.A., Roos, D.S., 2002. Secretory traffic in the eukaryotic parasite Toxoplasma gondii: less is more. J. Cell Biol. 157, 557–563.

Jullien, N., Sampieri, F., Enjalbert, A., Herman, J.P., 2003. Regulation of Cre recombinase by ligand-induced complementation of inactive fragments. Nucleic Acids Res. 31, e131.

Karsten, V., Qi, H., Beckers, C.J., Joiner, K.A., 1997. Targeting the secretory pathway of Toxoplasma gondii. Methods 13, 103–111.

Karsten, V., Qi, H., Beckers, C.J., Reddy, A., Dubremetz, J.F., Webster, P., Joiner, K.A., 1998. The protozoan parasite Toxoplasma gondii targets proteins to dense granules and the vacuolar space using both conserved and unusual mechanisms. J. Cell Biol. 141, 1323–1333.

Kibe, M.K., Coppin, A., Dendouga, N., Oria, G., Meurice, E., Mortuaire, M., Madec, E., Tomavo, S., 2005. Transcriptional regulation of two stage-specifically expressed genes in the protozoan parasite Toxoplasma gondii. Nucleic Acids Res. 33, 1722–1736.

Kim, K., Soldati, D., Boothroyd, J.C., 1993 Nov 5. Gene replacement in Toxoplasma gondii with chloramphenicol acetyltransferase as selectable marker. Science 262 (5135), 911–914.

Kim, K., Boothroyd, J.C., 1995. Toxoplasma gondii: stable complementation of sag1 (p30) mutants using SAG1 transfection and fluorescence-activated cell sorting. Exp. Parasitol. 80, 46–53.

Kim, K., Eaton, M.S., Schubert, W., Wu, S., Tang, J., 2001. Optimized expression of green fluorescent protein in Toxoplasma gondii using thermostable green fluorescent protein mutants. Mol. Biochem. Parasitol. 113, 309–313.

Kim, K., Soldati, D., Boothroyd, J.C., 1993 Nov. Gene replacement in Toxoplasma gondii with chloramphenicol acetyltransferase as selectable marker. Science 262, 911–914.

Knoll, L.J., Boothroyd, J.C., 1998. Isolation of developmentally regulated genes from Toxoplasma gondii by a gene trap with the positive and negative selectable marker hypoxanthine-xanthine-guanine phosphoribosyltransferase. Mol. Cell Biol. 18, 807–814.

Knoll, L.J., Furie, G.L., Boothroyd, J.C., 2001. Adaptation of signature-tagged mutagenesis for Toxoplasma gondii: a negative screening strategy to isolate genes that are essential in restrictive growth conditions. Mol. Biochem. Parasitol. 116, 11–16.

Lecordier, L., Mercier, C., Sibley, L.D., Cesbron-Delauw, M.F., 1999. Transmembrane insertion of the Toxoplasma gondii GRA5 protein occurs after soluble secretion into the host cell. Mol. Biol. Cell 10, 1277–1287.

Limenitakis, J., Soldati-Favre, D., 2011. Functional genetics in Apicomplexa: potentials and limits. FEBS Lett. 585, 1579–1588.

Matrajt, M., Donald, R.G., Singh, U., Roos, D.S., 2002a. Identification and characterization of differentiation mutants in the protozoan parasite Toxoplasma gondii. Mol. Microbiol. 44, 735–747.

Matrajt, M., Nishi, M., Fraunholz, M., Peter, O., Roos, D.S., 2002b. Amino-terminal control of transgenic protein expression levels in *Toxoplasma gondii*. Mol. Biochem. Parasitol. 120, 285–289.

Matrajt, M., Platt, C.D., Sagar, A.D., Lindsay, A., Moulton, C., Roos, D., 2004. Transcript initiation, polyadenylation, and functional promoter mapping for the dihydrofolate reductase-thymidylate synthase gene of *Toxoplasma gondii*. Molecular and Biochemical Parasitology 137, 229–238.

Mazumdar, J.E.H.W., Masek, K.C.A.H., Striepen, B., 2006. Apicoplast fatty acid synthesis is essential for organelle biogenesis and parasite survival in Toxoplasma gondii. Proc. Natl. Acad. Sci. U. S. A. 103, 13192–13197.

Mcfadden, D.C., Seeber, F., Boothroyd, J.C., 1997. Use of Toxoplasma gondii expressing beta-galactosidase for colorimetric assessment of drug activity in vitro. Antimicrob. Agents Chemother. 41, 1849–1853.

Meissner, M., Brecht, S., Bujard, H., Soldati, D., 2001. Modulation of myosin A expression by a newly established tetracycline repressor-based inducible system in Toxoplasma gondii. Nucleic Acids Res. 29, E115.

Meissner, M., Schluter, D., Soldati, D., 2002. Role of Toxoplasma gondii myosin A in powering parasite gliding and host cell invasion. Science 298, 837–840.

Meissner, M., Soldati, D., 2005. The transcription machinery and the molecular toolbox to control gene expression in Toxoplasma gondii and other protozoan parasites. Microbes Infect. 7, 1376–1384.

Mercier, C., Howe, D.K., Mordue, D., Lingnau, M., Sibley, L.D., 1998. Targeted disruption of the GRA2 locus in toxoplasma gondii decreases acute virulence in mice [In Process Citation]. Infect. Immun. 66, 4176–4182.

Mercier, C., Lefebvre-Van Hende, S., Garber, G.E., Lecordier, L., Capron, A., Cesbron-Delauw, M.F., 1996. Common cis-acting elements critical for the expression of several genes of Toxoplasma gondii. Mol. Microbiol. 21, 421–428.

Messina, M., Niesman, I., Mercier, C., Sibley, L.D., 1995. Stable DNA transformation of Toxoplasma gondii using phleomycin selection. Gene 165, 213–217.

Metzger, D., Clifford, J., Chiba, H., Chambon, P., 1995. Conditional site-specific recombination in mammalian cells using a ligand-dependent chimeric Cre recombinase. Proc. Natl. Acad. Sci. U. S. A. 92, 6991–6995.

Mital, J., Meissner, M., Soldati, D., Ward, G.E., 2005. Conditional expression of Toxoplasma gondii apical membrane antigen-1 (TgAMA1) demonstrates that TgAMA1 plays a critical role in host cell invasion. Mol. Biol. Cell 16, 4341–4349.

Muralidharan, V., Oksman, A., Iwamoto, M., Wandless, T.J., Goldberg, D.E., 2011. Asparagine repeat function in a Plasmodium falciparum protein assessed via a regulatable fluorescent affinity tag. Proc. Natl. Acad. Sci. U. S. A. 108, 4411–4416.

Nare, B., Allocco, J.J., Liberator, P.A., Donald, R.G., 2002. Evaluation of a Cyclic GMP-Dependent Protein Kinase Inhibitor in Treatment of Murine Toxoplasmosis: Gamma Interferon Is Required for Efficacy. Antimicrob. Agents Chemother. 46, 300–307.

Payne, T.M., Payne, A.J., Knoll, L.J., 2011. A Toxoplasma gondii mutant highlights the importance of translational regulation in the apicoplast during animal infection. Mol. Microbiol. 82, 1204–1216.

Pelletier, L., Stern, C.A., Pypaert, M., Sheff, D., Ngo, H.M., Roper, N., He, C.Y., Hu, K., Toomre, D., Coppens, I., Roos, D.S., Joiner, K.A., Warren, G., 2002. Golgi biogenesis in Toxoplasma gondii. Nature 418, 548–552.

Pino, P., Sebastian, S., Kim, E.A., Bush, E., Brochet, M., Volkmann, K., Kozlowski, E., Llinas, M., Billker, O., Soldati-Favre, D., 2013. A tetracycline-repressible transactivator system to study essential genes in malaria parasites. Cell Host Microbe 12, 824–834.

Pfefferkorn, E.R., 1988. Toxoplasma gondii viewed from a virological perspective. In: Englund, P.T., Sher, A. (Eds.), The biology of parasitism. Alan R. Liss. New York.

Pfefferkorn, E.R., Guyre, P.M., 1984. Inhibition of growth of Toxoplasma gondii in cultured fibroblasts by human recombinant gamma interferon. Infection and Immunity 44, 211–216.

Pfefferkorn, E.R., Pfefferkorn, L.C., 1976. Toxoplasma gondii: isolation and preliminary characterization of temperature-sensitive mutants. Exp. Parasitol. 39, 365–376.

Pfefferkorn, L.C., Pfefferkorn, E.R., 1980. Toxoplasma gondii: genetic recombination between drug resistant mutants. Exp. Parasitol. 50, 305–316.

Plattner, F., Yarovinsky, F., Romero, S., Didry, D., Carlier, M.F., Sher, A., Soldati-Favre, D., 2008. Toxoplasma profilin is essential for host cell invasion and TLR11-dependent induction of an interleukin-12 response. Cell Host Microbe 3, 77–87.

Radke, J.R., White, M.W., 1998. A cell-cycle model for the tachyzoite of Toxoplasma gondii using the Herpes simplez virus thymidine kinase. Mol. Biochem. Parsitol. 94, 237–247.

Radke, J.R., Gubbels, M.J., Jerome, M.E., Radke, J.B., Striepen, B., White, M.W., 2004. Identification of a sporozoite-specific member of the *Toxoplasma* SAG-superfamily via genetic complementation. Mol. Microbiol. 52, 93–105.

Radke, J.R., Guerini, M.N., White, M.W., 2000. Toxoplasma gondii: characterization of temperature-sensitive tachyzoite cell cycle mutants. Exp. Parasitol. 96, 168–177.

Radke, J.R., Striepen, B., Guerini, M.N., Jerome, M.E., Roos, D.S., White, M.W., 2001. Defining the cell cycle for the tachyzoite stage of Toxoplasma gondii. Mol. Biochem. Parasitol. 115, 165–175.

Reid, A.J., Vermont, S.J., Cotton, J.A., Harris, D., Hill-Cawthorne, G.A., Konen-Waisman, S., Latham, S.M., Mourier, T., Norton, R., Quail, M.A., Sanders, M., Shanmugam, D., Sohal, A., Wasmuth, J.D., Brunk, B., Grigg, M.E., Howard, J.C., Parkinson, J., Roos, D.S., Trees, A.J., Berriman, M., Pain, A., Wastling, J.M., 2012. Comparative genomics of the apicomplexan parasites Toxoplasma gondii and Neospora caninum: Coccidia differing in host range and transmission strategy. PLoS Pathog. 8, e1002567.

Roos, D.S., Donald, R.G., Morrissette, N.S., Moulton, A.L., 1994. Molecular tools for genetic dissection of the protozoan parasite Toxoplasma gondii. Methods Cell Biol. 45, 27–63.

Roos, D.S., Sullivan, W.J., Striepen, B., Bohne, W., Donald, R.G., 1997. Tagging genes and trapping promoters in Toxoplasma gondii by insertional mutagenesis. Methods 13, 112–122.

Russo, I., Oksman, A., Vaupel, B., Goldberg, D.E., 2009. A calpain unique to alveolates is essential in Plasmodium falciparum and its knockdown reveals an involvement in pre-S-phase development. Proc. Natl. Acad. Sci. U. S. A. 106, 1554–1559.

Sampels, V., Hartmann, A., Dietrich, I., Coppens, I., Sheiner, L., Striepen, B., Herrmann, A., Lucius, R., Gupta, N., 2012. Conditional mutagenesis of a novel choline kinase demonstrates plasticity of phosphatidylcholine biogenesis and gene expression in Toxoplasma gondii. J. Biol. Chem. 287, 16289–16299.

Santos, J.M., Ferguson, D.J., Blackman, M.J., Soldati-Favre, D., 2011. Intramembrane cleavage of AMA1 triggers Toxoplasma to switch from an invasive to a replicative mode. Science 331, 473–477.

Seeber, F., 1997. Consensus sequence of translational initiation sites from *Toxoplasma gondii* genes. Parasitology Research 83, 309–311.

Seeber, F., Boothroyd, J.C., 1996. Escherichia coli beta-galactosidase as an in vitro and in vivo reporter enzyme and stable transfection marker in the intracellular protozoan parasite Toxoplasma gondii. Gene 169, 39–45.

Shaner, N.C., Campbell, R.E., Steinbach, P.A., Giepmans, B.N., Palmer, A.E., Tsien, R.Y., 2004. Improved monomeric red, orange and yellow fluorescent proteins derived from Discosoma sp. red fluorescent protein. Nat. Biotechnol. 22, 1567–1572.

Sheiner, L., Demerly, J.L., Poulsen, N., Beatty, W.L., Lucas, O., Behnke, M.S., White, M.W., Striepen, B., 2011. A systematic screen to discover and analyze apicoplast proteins identifies a conserved and essential protein import factor. PLoS Pathog. 7, e1002392.

Sibley, L.D., Messina, M., Niesman, I.R., 1994. Stable DNA transformation in the obligate intracellular parasite Toxoplasma gondii by complementation of tryptophan auxotrophy. Proc. Natl. Acad. Sci. USA 91, 5508–5512.

Singh, U., Brewer, J.L., Boothroyd, J.C., 2002. Genetic analysis of tachyzoite to bradyzoite differentiation mutants in Toxoplasma gondii reveals a hierarchy of gene induction. Mol. Microbiol. 44, 721–733.

Skariah, S., Bednarczyk, R.B., McIntyre, M.K., Taylor, G.A., Mordue, D.G., 2012. Discovery of a novel Toxoplasma gondii conoid-associated protein important for parasite resistance to reactive nitrogen intermediates. J. Immunol. 188, 3404–3415.

Skarnes, W.C., Rosen, B., West, A.P., Koutsourakis, M., Bushell, W., Iyer, V., Mujica, A.O., Thomas, M., Harrow, J., Cox, T., Jackson, D., Severin, J., Biggs, P., Fu, J., Nefedov, M., De Jong, P.J., Stewart, A.F., Bradley, A., 2011. A conditional knockout resource for the genome-wide study of mouse gene function. Nature 474, 337–342.

Soldati, D., Boothroyd, J.C., 1993. Transient transfection and expression in the obligate intracellular parasite Toxoplasma gondii. Science 260, 349–352.

Soldati, D., Boothroyd, J.C., 1995. A selector of transcription initiation in the protozoan parasite Toxoplasma gondii. Mol. Cell Biol. 15, 87–93.

Soldati, D., Dubremetz, J.F., Lebrun, M., 2001. Microneme proteins: structural and functional requirements to promote adhesion and invasion by the apicomplexan parasite Toxoplasma gondii. Int. J. Parasitol. 31, 1293–1302.

Soldati, D., Kim, K., Kampmeier, J., Dubremetz, J.-F., Boothroyd, J.C., 1995. Complementation of a Toxoplasma gondii ROP1 knock-out mutant using phleomycin selection. Mol. Biochem. Parasitol. 74, 87–97.

Striepen, B., Crawford, M.J., Shaw, M.K., Tilney, L.G., Seeber, F., Roos, D.S., 2000. The plastid of Toxoplasma gondii is divided by association with the centrosomes. J. Cell Biol. 151, 1423–1434.

Striepen, B., He, C.Y., Matrajt, M., Soldati, D., Roos, D.S., 1998. Expression, selection, and organellar targeting of the green fluorescent protein in *Toxoplasma gondii*. Mol. Biochem. Parasitol. 92, 325–338.

Striepen, B., Soldati, D., Garcia-Reguet, N., Dubremetz, J.F., Roos, D.S., 2001. Targeting of soluble proteins to the rhoptries and micronemes in Toxoplasma gondii. Mol. Biochem. Parasitol. 113, 45–54.

Striepen, B., White, M.W., Li, C., Guerini, M.N., Malik, S.B., Logsdon Jr., J.M., Liu, C., Abrahamsen, M.S., 2002. Genetic complementation in apicomplexan parasites. Proc. Natl. Acad. Sci. U. S. A. 99, 6304–6309.

Sullivan Jr., W.J., Chiang, C.W., Wilson, C.M., Naguib, F.N., El Kouni, M.H., Donald, R.G., Roos, D.S., 1999. Insertional tagging of at least two loci associated with resistance to adenine arabinoside in Toxoplasma gondii, and cloning of the adenosine kinase locus. Mol. Biochem. Parasitol. 103, 1–14.

Sullivan Jr., W.J., Dixon, S.E., Li, C., Striepen, B., Queener, S.F., 2005. IMP dehydrogenase from the protozoan parasite Toxoplasma gondii. Antimicrob. Agents Chemother. 49, 2172–2179.

Thathy, V., Menard, R., 2002. Gene targeting in Plasmodium berghei. Methods Mol. Med. 72, 317–331.

Umejiego, N.N., Li, C., Riera, T., Hedstrom, L., Striepen, B., 2004. Cryptosporidium parvum IMP dehydrogenase: identification of functional, structural, and dynamic properties that can be exploited for drug design. J. Biol. Chem. 279, 40320–40327.

Upadhya, R., Kim, K., Hogue-Angeletti, R., Weiss, L.M., 2011. Improved techniques for endogenous epitope tagging and gene deletion in Toxoplasma gondii. J. Microbiol. Methods 85, 103–113.

Uyetake, L., Ortega-Barria, E., Boothroyd, J.C., 2001. Isolation and characterization of a cold-sensitive attachment/invasion mutant of Toxoplasma gondii. Exp. Parasitol. 97, 55–59.

Van Dooren, G.G., Reiff, S.B., Tomova, C., Meissner, M., Humbel, B.M., Striepen, B., 2009. A novel dynamin-related protein has been recruited for apicoplast fission in Toxoplasma gondii. Curr. Biol. 19, 267–276.

Van Dooren, G.G., Tomova, C., Agrawal, S., Humbel, B.M., Striepen, B., 2008. Toxoplasma gondii Tic20 is essential for apicoplast protein import. Proc. Natl. Acad. Sci. U. S. A. 105, 13574–13579.

Van Poppel, N.F., Welagen, J., Duisters, R.F., Vermeulen, A.N., Schaap, D., 2006. Tight control of transcription in Toxoplasma gondii using an alternative tet repressor. Int. J. Parasitol.

Vanchinathan, P., Brewer, J.L., Harb, O.S., Boothroyd, J.C., Singh, U., 2005. Disruption of a locus encoding a nucleolar zinc finger protein decreases tachyzoite-to-bradyzoite differentiation in Toxoplasma gondii. Infect. Immun. 73, 6680–6688.

Waller, R.F., Keeling, P.J., Donald, R.G., Striepen, B., Handman, E., Lang-Unnasch, N., Cowman, A.F., Besra, G.S., Roos, D.S., McFadden, G.I., 1998. Nuclear-encoded proteins target to the plastid in *Toxoplasma gondii* and *Plasmodium falciparum*. Proc. Natl. Acad. Sci. U. S. A. 95, 12352–12357.

White, M.W., Jerome, M.E., Vaishnava, S., Guerini, M., Behnke, M., Striepen, B., 2005. Genetic rescue of a Toxoplasma gondii conditional cell cycle mutant. Mol. Microbiol. 55, 1060–1071.

Wirtz, E., Leal, S., Ochatt, C., Cross, G.A., 1999. A tightly regulated inducible expression system for conditional gene knock-outs and dominant-negative genetics in Trypanosoma brucei. Mol. Biochem. Parasitol. 99, 89–101.

Xiao, Y., Karnati, S., Qian, G., Nenicu, A., Fan, W., Tchatalbachev, S., Holand, A., Hossain, H., Guillou, F., Luers, G.H., Baumgart-Vogt, E., 2012. Cre-mediated stress affects sirtuin expression levels, peroxisome biogenesis and metabolism, antioxidant and proinflammatory signalling pathways. PLoS One 7, e41097.

Yang, S., Parmley, S.F., 1997. Toxoplasma gondii expresses two distinct lactate dehydrogenase homologous genes during its life cycle in intermediate hosts. Gene 184, 1–12.

CHAPTER

18

Epigenetic and Genetic Factors that Regulate Gene Expression in *Toxoplasma gondii*

William J. Sullivan, Jr., Joshua B. Radke†, Kami Kim**, Michael W. White†*

***Department of Pharmacology & Toxicology, and Microbiology & Immunology, Indiana University School of Medicine, Indianapolis, Indiana, USA †Departments of Molecular Medicine and Global Health and the Florida Center for Drug Discovery and Innovation, University of South Florida, Tampa, Florida, USA **Departments of Medicine Pathology, and Microbiology & Immunology, Albert Einstein College of Medicine, Bronx, New York, USA**

OUTLINE

Toxoplasma gondii, *Second Edition*
http://dx.doi.org/10.1016/B978-0-12-396481-6.00018-0

18.1 INTRODUCTION

Toxoplasma gondii is distinct from nearly all other members of the large coccidian family (phylum Apicomplexa) owing to the exceptional range of animals (virtually all warm-blooded animals) that serve as host for its intermediate life cycle. Like other coccidians, *Toxoplasma* completes its definitive life cycle in a single animal host (Dubey *et al.*, 1970); however, the ability of oocysts (shed from the feline definitive host) as well as tissue cysts produced in intermediate hosts to infect either host type (Dubey, 1988) has enabled *Toxoplasma* to increase its host range (Su *et al.*, 2003). Sexual stages in the feline host lead to the development of oocysts that are shed into the environment (Long, 1982) where contamination of soil or water has led to epidemics of human toxoplasmosis (Isaac-Renton *et al.*, 1998; Choi *et al.*, 1997; Bowie *et al.*, 1997; Konishi and Takahashi, 1987; Stray-Pedersen and Lorentzen-Styr, 1980). Moreover, oocysts are the primary source of *Toxoplasma* infections of livestock destined for slaughter and human consumption (Mateus-Pinilla *et al.*, 1999; Andrews *et al.*, 1997).

Given the importance of *Toxoplasma* infections to human populations, understanding developmental mechanisms initiated by sporozoites or bradyzoites leading to tissue cyst formation will be central to ultimately controlling transmission and chronic disease. Studies of *Toxoplasma* primary infections in animals and of sporozoite- and bradyzoite-infected cultures *in vitro* (Dubey and Frenkel, 1976; Dubey, 1998; Jerome *et al.*, 1998; Radke *et al.*, 2003) indicate that development initiated by either the sporozoite or bradyzoite stage is similar and likely the consequence of a unified genetic programme (Radke *et al.*, 2003). Thus, defining the changes in gene expression that accompany this development pathway will be important to understand the underlying mechanisms responsible for toxoplasmosis caused by either route of infection. In the following sections, we review our current understanding of *Toxoplasma* transcription, which undergoes dramatic changes during the parasite intermediate life cycle. Studies have shown that mRNA pools are dynamic and indicate that transcriptional control is a major mechanism employed to regulate developmental transitions in this parasite. It is in this context that we also discuss the evidence that *Toxoplasma* possesses a similar repertoire of epigenetic-based mechanisms to modulate transcription, as observed in other well-studied eukaryotes from yeast to multicellular animals. Finally, we discuss the emerging role of post-transcriptional regulation, which also appears to be active in this parasite.

18.2 TRANSCRIPTION IN *TOXOPLASMA*

Apicomplexan parasites exhibit complicated, multi-stage life cycles that involve a variety of hosts. Coincident with their complex life cycles are wholesale changes in gene expression associated with each developmental stage or host, yet the mechanisms that control gene expression remain elusive.

Early efforts to accelerate gene discovery in *Toxoplasma* led to the sequence for more than 120,000 ESTs from RH and ME49 strain tachyzoites as well as ME49 strain bradyzoites and VEG strain oocysts (Ajioka *et al.*, 1998; Manger *et al.*, 1998; Li *et al.*, 2003, 2004a). A *Toxoplasma* SAGE project and a 10X-whole genome project for the Type II-Me49B7 followed by further genome sequence to provide 5X coverage of GT-1 (Type I) and VEG (Type III) and whole genome microarrays based on the Type II-Me49B7 reference strains were developed and made available to the *Toxoplasma* research community. The most recent release of http://www.toxodb.org features a revised re-sequenced and re-annotated genome (Version 8), incorporating additional data including expression data derived from microarrays and next generation derived RNA-seq, genome-wide chromatin immunoprecipitation studies (ChIP), as well as proteomics studies.

Since the first edition of this book both microarrays and next generation sequencing have been used to develop a comprehensive view of the *T. gondii* transcriptome and permit the characterization of candidate factors that regulate gene expression (Hassan *et al.*, 2012; Minot *et al.*, 2012; Rosowski and Saeij, 2012; Reid *et al.*, 2012; Bahl *et al.*, 2010; Behnke *et al.*, 2010). Much of these data are available on http://www.toxodb.org and more will be available as the results and analysis of the *Toxoplasma* white paper project become available throughout 2013. This community effort resulted in the re-sequencing and re-annotation of the ME49 reference genome, RNA-seq transcriptome studies of ME49, as well as genome sequencing and tachyzoite RNA-seq for the 16 major genetic lineages of *T. gondii* (see Chapter 3 for discussion of the population biology of *T. gondii* and Chapter 16 for an overview of classical genetics). An additional set of representative strains has been sequenced at lower coverage. In addition, many groups have submitted RNA-seq data to http://www.toxodb.org, complementing numerous Affymetrix gene expression datasets already displayed.

18.2.1 The Parasite Transcriptome and Transcriptional Regulation

The first microarray transcriptome studies of the Apicomplexa in *Plasmodium* illustrated that more than 80% of the transcripts were regulated, with most having a peak expression within a single timeframe in the sexual stages or intra-erythrocytic cycle. The proper timing of mRNA accumulation applies not only to genes associated with parasite cell cycle that might be expected to have similar kinetics (Bozdech and Ginsburg, 2005), but also for genes encoding protein components of subcellular structures as the narrow window when these structures form during the division cycle may also require coordinated gene expression (Triglia *et al.*, 2000). The large changes in transcript levels in *Plasmodium* suggested mRNA expression is governed by 'just in time' mechanisms, and the relatively low proportion of constitutive mRNAs in these parasites may reflect this concept (Llinas and DeRisi, 2004). Taken together, these results demonstrate that transcription is a major mechanism controlling gene expression in these parasites and this view is supported by comparison of the changes in the *Plasmodium* transcriptome and proteome that indicates alterations in mRNA levels have a higher correlation to protein changes in this parasite than is observed in yeast or higher eukaryotes (Le Roch *et al.*, 2004).

Initial transcriptome studies in *T. gondii* relied upon cDNA derived from EST projects for apicomplexan protozoa that were subsequently used to construct a limited *Toxoplasma* cDNA microarray that focused on tachyzoite–bradyzoite transitions in cell culture models of bradyzoite differentiation (Cleary *et al.*, 2002) and explored gene expression in mutants that were unable to differentiate (Singh *et al.*, 2002; Matrajt *et al.*, 2002). These studies supported a role for transcriptional mechanisms in determining developmental stage characteristics in *Toxoplasma* and evidence for co-regulation of transcription in this parasite (Singh *et al.*, 2002). In the transcripts of four mutants generated by chemical mutagenesis, and selected against the ability to differentiate, a common set of mRNAs was affected and unable to be induced, while other affected mRNA groups appeared to cluster with two or three mutants suggesting hierarchical gene expression may direct bradyzoite development (Singh *et al.*, 2002). Subsequent SAGE and microarray projects supported the concept of co-regulated transcription and demonstrate that some of the general concepts emerging from the *Plasmodium* studies apply to gene expression in *Toxoplasma*. A community effort led to the fabrication of an Affymetrix gene array that has now become the platform of choice for transcriptome analysis (Bahl *et al.*, 2010). Currently there are numerous published expression datasets from different strains of parasites exposed to different experimental conditions, including bradyzoite inducing conditions (Behnke *et al.*, 2008; Lescault *et al.*, 2010; Buchholz *et al.*, 2011). Most of these datasets are summarized at http://www.toxodb.org and the primary data are usually accessible in the GEO or ArrayExpress databases.

Unlike the high expression of metabolic or structural genes in animal cells, nearly one third of the *Toxoplasma* most abundant mRNA are Apicomplexa-specific genes that have simple genomic structures containing few, if any, introns (Radke *et al.*, 2005). Metabolic or structural genes typically contain introns. Many transcripts encoding proteins of the basal metabolic machinery and subcellular structures appear to be transcribed in *Toxoplasma* only when needed during parasite growth and development (Behnke *et al.*, 2010), consistent with the 'just in time' concept put forth from studies of *Plasmodium* (Llinas and DeRisi, 2004). Overall, development-specific genes (sporozoite, tachyzoite, bradyzoite), genes encoding proteins from biochemical pathways and genes representing mRNA abundance classes are dispersed between all *Toxoplasma* chromosomes. Gene clusters rarely occur, and in those cases mRNA expression patterns do not appear to be strongly influenced by physical proximity. For example, genes encoding enolase 1 and 2 are less than 1500 bp apart on chromosome VIII, yet expressed exclusively in bradyzoite or tachyzoite stages, respectively (Lyons *et al.*, 2002). These observations suggest that local changes in chromatin structure or the recruitment of RNA polymerase to promoters has little influence on nearby genes.

The basal transcriptional complex that controls the expression of protein encoding genes (class II) in most eukaryotes is carried out by RNA polymerase II and its associated general transcription factors (GTF). Comparisons of these transcription factors, as well as the similarities in the three nuclear polymerases (Ranish and Hahn, 1996), have shown that these mechanisms are largely conserved throughout evolution, from the Archaea to mammals. In well-studied unicellular and multicellular eukaryotes, transcription involves a series of co-regulatory complexes that work in concert to control the synthesis of RNA from a particular genomic region. Activating transcription factors (ATF) bind to *cis*-acting promoter element(s) and recruit chromatin remodelling enzymes which relax the chromatin around the *cis*-element-containing region as well as recruit the multi-subunit Mediator complex that contacts the RNA polymerase II pre-initiation complex (PIC) directly (Blazek *et al.*, 2005). The accessibility of the *cis*-element to ATF binding is dependent upon the interaction with these remodelling enzymes, but can also be influenced by other factors such as the chromatin state at the regulatory sequence and the phase of the cell cycle (Fry and Peterson, 2002). In turn, the relaxed chromatin state allows for the formation of the PIC at the core promoter elements through activities contained within the Mediator that facilitate recruitment of RNA polymerase II and the GTFs. Current models of ATFs suggest that activation of RNA polymerase II by these factors occurs indirectly through their recruitment of ATP-dependent chromatin remodelling complexes (Blazek *et al.*, 2005; Li *et al.*, 2004b; Featherstone, 2002).

The analysis of protein encoding genes in the Apicomplexa indicates that conventional RNA polymerases with similarity to other crown eukaryotes are present. Homologues for all known required eukaryotic RNA polymerases have been found in the *Toxoplasma* genome: RNA polymerase I (transcribes ribosomal RNA), RNA polymerase II (transcribes protein encoding transcripts) and RNA polymerase III (transcribes small RNA) (Li *et al.*, 1989, 1991; Fox *et al.*, 1993; Meissner and Soldati, 2005). Thus it appears that *Plasmodium* and *Toxoplasma* possess the conserved eukaryotic machinery whereby RNA polymerase II transcribes protein encoding genes.

The core elements of class II eukaryotic promoters include TATA box, Initiator (Inr), and downstream promoter elements (DPE) that are recognized and bound by several GTFs: TFIIA, TFIIB, TFIID, TFIIE, TFIIF and TFIIH (Blazek *et al.*, 2005; Featherstone, 2002; Ruvalcaba-Salazar *et al.*, 2005; Ranish and Hahn, 1996). The core of the GTF family includes the TATA Binding Protein (TBP), TFIID and RNA polymerase II. Homologues for various subunits for the GTFs (TFIID, TFIIE, TFIIF and TFIIH) and subunits of Mediator have been found in the Apicomplexa, and while GTFs are less conserved in the Apicomplexa, much of the basal transcriptional machinery and chromatin remodelling factors required for cooperative control of gene transcription in eukaryotes are present in these pathogens.

18.2.2 Gene-Specific Cis-Elements

Classical promoter mapping strategies utilizing conventional protein reporters, including chloramphenicol acetyltransferase (CAT), βgalactosidase (βgal), green fluorescent protein (GFP) or firefly/renilla luciferase (luc), have been employed to map regulatory sequences in several promoters. In *Toxoplasma*, deletion studies to identify promoter cis-elements have been reported for various constitutive genes (GRAs and DHFR-TS), tachyzoite-specific genes (SAG1 and enolase 2), and bradyzoite-specific genes (hsp30/BAG1, hsp70, LDH2 and enolase 1) and confirm that promoter elements are primarily located upstream from the translational start site (Nakaar *et al.*, 1998; Kibe *et al.*, 2005; Matrajt *et al.*, 2004; Ma *et al.*, 2004; Roos *et al.*, 1997; Bohne *et al.*, 1997; Mercier *et al.*, 1996; Soldati and Boothroyd, 1995). Promoter elements were observed to be active in either DNA strand, but may have a limited working distance from the transcriptional start or lose their influence when located downstream of the coding region (Soldati and Boothroyd, 1995). The level of detail within these studies varies and minimal sequence elements were determined in only a few studies (Mercier *et al.*, 1996; Matrajt *et al.*, 2004); moreover, no published study has fully resolved the question of functional sufficiency for any putative cis-element. Nonetheless, it is evident that a 27 bp repeat sequence (6X repeat) in the SAG1 promoter is required for function and a sequence element (A/TGAGACG) found in the GRA promoters was demonstrated to be required for basal expression within the context of a 53 bp minimal promoter (Mercier *et al.*, 1996). It is notable that the GAGACG present in the central core of the SAG1 27 bp repeats is also found in regions implicated to contain regulatory cis-elements by deletion analysis of the NTPI/II and DHFR-TS promoters (Nakaar *et al.*, 1998; Matrajt *et al.*, 2004).

Development-specific changes in mRNA levels are a dominant feature of the Apicomplexa transcriptome. Nearly one quarter of the transcripts detected in the *Toxoplasma* SAGE project were observed to be uniquely expressed during parasite development and similar observations have emerged from functional genomics studies of the *Plasmodium* intraerythrocytic cycle (Le Roch *et al.*, 2004; Bozdech and Ginsburg, 2005). Promoters controlling bradyzoite-specific genes, BAG1, hsp70 and LDH2, have now been mapped using alkaline-stress induction at a similar resolution to (Ma *et al.*, 2004; Yang and Parmley 1997; Bohne *et al.*, 1997), while these studies support the role of promoter elements in regulating stress-response in *Toxoplasma*, their resolution is too low to allow for the identification of common cis-elements. The reciprocal regulation of enolase 1 (bradyzoite-specific) and 2 (tachyzoite-specific) is of particular interest given their close proximity in the genome (in an ordered tandem array of enolase 2−1). Repression of enolase 1 expression in tachyzoites appears to require a distal region more than 600 bp from the enolase 1 ATG and these elements are distinct from inductive elements that were mapped closer to the start of transcription (Kibe *et al.*, 2005).

Employing a dual luciferase model, we have mapped bradyzoite-specific cis-elements within a *Toxoplasma* gene encoding a novel NTPase (Brady−NTPase; chromosome X, TGME49_225290) (Behnke *et al.*, 2008). A series of sequential and internal deletions followed by 6 bp substitution mutagenesis have identified a 15 bp cis-element that is responsible for induction of the Brady−NTPase promoter under a variety of drug and stress conditions that co-induce native bradyzoite gene expression. This element lies within the first 500 bp of the Brady−NTPase promoter and 90% of the induction is lost when the element is mutated in the context of the full length promoter fragment (1495 bp). Mutation of this element does not lead to increased expression in the tachyzoite stage indicating that it is a true inductive element.

Approximately 2800 mRNAs have cyclical profiles during parasite division that cluster into two major transcriptional waves; genes with maximum expression in the G1 subtranscriptome encode well-conserved metabolic and biosynthetic functions, while those mRNAs in the S/M subtranscriptome are enriched for genes encoding proteins involved in daughter budding and egress (Behnke, 2010). FIRE (Finding Important Regulatory Elements) analyses of the proximal promoter regions for all cyclical mRNAs were scanned for enrichment of possible DNA regulatory elements. Nine DNA motifs were identified by these analyses (Behnke, 2010) that were distributed in genes with peak transcription spanning the full tachyzoite cell cycle. DNA motifs that were overrepresented in the promoters flanking G1 genes were generally underrepresented in the promoters of S/M genes (and *vice versa*). One of the DNA motifs enriched in G1 promoters (5′-TGCATGC-3′) is identical to the TgTRP2 cis-element required for transcription of ribosomal proteins (Van Poppel *et al.*, 2006; Mullapudi *et al.*, 2009) and is also identical to the 6 bp core DNA binding motif recently determined by PBM for AP2XI-3 (Kim, unpublished; see below). The mRNA for this AP2 also peaks during the G1 period (Behnke *et al.*, 2010).

18.2.3 The Evolution of APETALA-2-Related Proteins

The recent discovery of a class of DNA-binding proteins in the Apicomplexa that are related to the APETALA-2 (AP2) class of plant transcription factors (ApiAP2 proteins) has uncovered an important set of proteins that are likely to have critical roles in parasite gene expression (Balaji *et al.*, 2005). Until recently, AP2 domain-containing proteins were thought to be a plant-specific family of DNA binding proteins (Riechmann and Meyerowitz, 1998; Krizek, 2003). AP2 homologues in non-plant species such as cyanobacteria, ciliates and viruses indicate AP2 DNA-binding domains are widely conserved (Wuitschick *et al.*, 2004; Magnani *et al.*, 2004). Many of these proteins contain a second domain encoding a homing endonuclease function that confers the ability to operate as mobile genetic elements with the capacity to transpose, invade, and self-replicate by exploiting genome repair mechanisms (Chevalier and Stoddard, 2001; Koufopanou *et al.*, 2002). While most homing endonucleases generally have no core cellular function and are eventually discarded, the HO endonuclease in yeast was adapted to trigger mating type interconversion (Raveh *et al.*, 1989) and other proteins in this family function to catalyse intron self-splicing in yeast (Chevalier and Stoddard, 2001). Importantly, homing endonucleases are thought to have expanded through lateral gene transfer and are represented in all biological kingdoms, including mitochondrial and chloroplast genomes in eukaryotes (Chevalier and Stoddard, 2001). Magnani *et al.* (2004) hypothesized that the plant AP2/ERF (ethylene response factor) family of transcription factors arose from the HNH-AP2 family of homing endonucleases present in bacteria or viruses and was incorporated into plant genomes via horizontal gene transfer, or alternatively was acquired indirectly from an endosymbiotic event, likely with a cyanobacterium with an early plant progenitor. Regardless of the source, over time the homing endonuclease function has been lost in plants as the AP2 DNA-binding domain has taken on specific regulatory roles in gene expression associated with plant development and stress response (Magnani *et al.*, 2004; Altschul *et al.*, 2010).

Apicomplexan genomes including *Plasmodium* spp., *Toxoplasma gondii*, *Cryptosporidium parvum* and *Theileria* spp. encode multiple ApiAP2 proteins that may have a similar endosymbiotic origin (Balaji *et al.*, 2005; Altschul *et al.*, 2010). Unlike plant AP2 factors, ApiAP2 proteins have undergone a lineage-specific expansion of a few progenitor genes leading to ApiAp2 factors carrying up to eight AP2 domains (Balaji *et al.*, 2005; Altschul *et al.*, 2010). Interestingly, phylogenetic analysis of ApiAP2 proteins from *Plasmodium* (27 total), *Theileria* (19 total) and *Cryptosporidium* (21 total) suggests the common ancestor of these three protozoa contained nine conserved ApiAP2 proteins, with a higher number of orthologous pairs found in *Plasmodium* spp. and *Theileria* spp. genomes (Balaji *et al.*, 2005). It is notable that with the exception of the ApiAP2 domain, these proteins are otherwise not conserved and there are no other known protein domains (activation, localization, or protein–protein interaction) found in any of the *Plasmodium* spp. or *Toxoplasma* ApiAP2 proteins (Lindner *et al.*, 2010; Altschul *et al.*, 2010), lending further support to lineage-specific expansion of these DNA binding proteins.

18.2.4 ApiAP2 Structure Determination and DNA Binding

The AP2 domain consists of approximately 60 amino acids and was first shown to confer DNA-binding specificity to the AP2/ethylene response element binding family (EREBP) found in plants (Jofuku *et al.*, 1994; Ohme-Takagi and Shinshi, 1995; Balaji *et al.*, 2005). AP2/EREBP proteins represent the second largest class of transcription factors in *Arabidopsis thaliana,* consisting of 145 proteins that are subdivided further into five sub-families based on AP2 domain architecture (Sakuma *et al.*, 2002). DREB (dehydration responsive element binding) and ERF (ERF is identical to EREBP) protein groups constitute the two largest sub-families (56 DREB compared to 65 ERF) and contain one AP2 domain and a conserved WLG motif (Sakuma *et al.*, 2002). While identical in domain architecture, these groups are defined by single amino acid changes within the DNA-binding domain that alter DNA-binding specificity (Sakuma *et al.*, 2002). Proteins that include two AP2 domains (14 total proteins) are further divided into two subfamilies based on the presence or absence of a second non-AP2 DNA binding domain (Sakuma *et al.*, 2002; Magnani *et al.*, 2004).

Alignment of ApiAP2 domains from *Plasmodium*, *Cryptosporidium* and *Theileria* to the structure of the ERF1 from *Arabidopsis* indicates strong conservation of 12 residues that correspond to areas of hydrophobic interactions responsible for the backbone of the DNA binding domain rather than specifying DNA binding (Balaji *et al.*, 2005) (Fig. 18.1). Campbell *et al.* (2010) determined the DNA-binding specificity of 20 of the 27 *P. falciparum* ApiAP2s, revealing an unusual diversity of binding motifs that are either classically palindromic or nucleotide biased, and similar results are seen in *T. gondii* ApiAP2s (Kim, Sullivan, White, Llinas, unpublished). The structural basis of DNA-binding is evident in ApiAP2 factors that share sequence beyond these 12 core hydrophobic residues. For example, orthologues from *P. falciparum* (PF14_0633) and *Cryptosporidium parvum* (cgd2_3490) that share 68% similarity in the AP2 domain show nearly identical DNA-binding specificities, providing evidence that with respect to DNA recognition there is conservation of function across divergent apicomplexan species (De Silva *et al.*, 2008). Whereas the majority of plant AP2 domains cluster tightly around DNA contact residues, ApiAP2 domains exhibit considerable flexibility within the domain, a feature that suggests a greater diversity in DNA-binding specificity

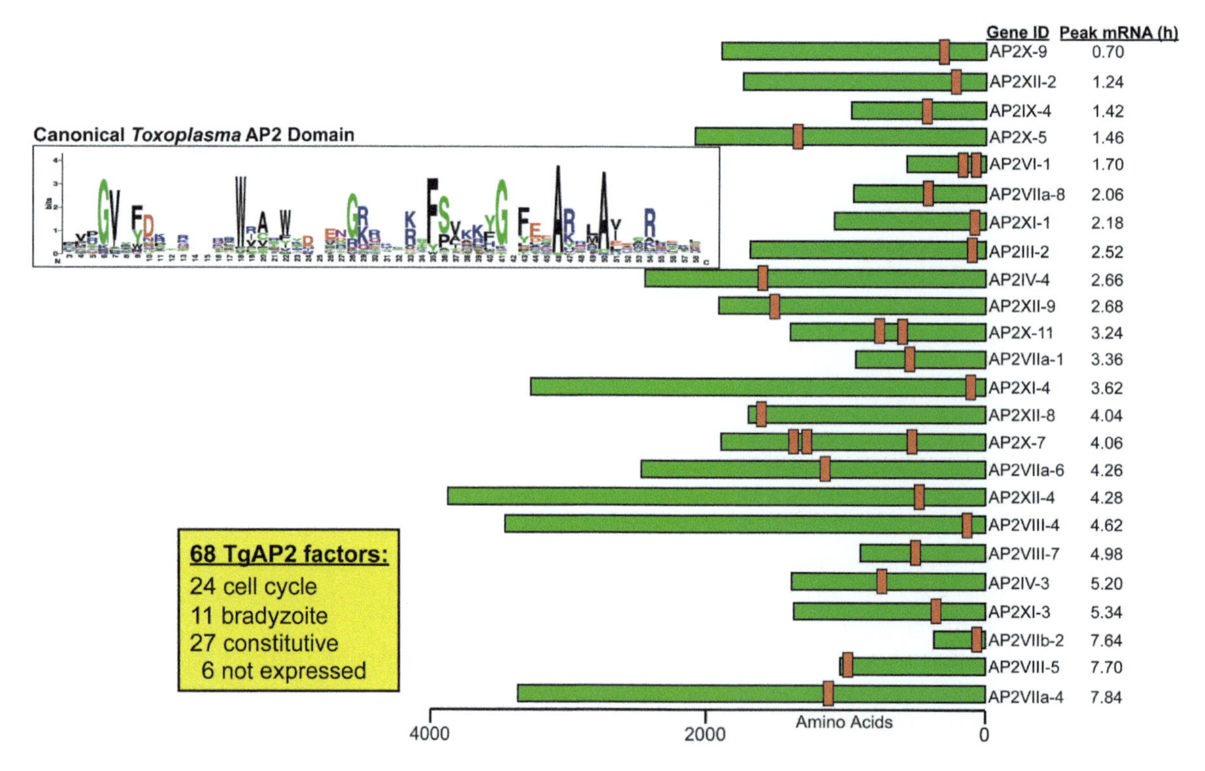

FIGURE 18.1 **The AP2 family of *Toxoplasma gondii*.** There are 68 members of the *T. gondii* ApiAP2 family as determined by a *T. gondii* community annotation effort (White, Sullivan, Kim, Croken and Wootton; see http://www.toxodb.org). A schematic of the consensus DNA binding domain of the *T. gondii* family is shown in the upper left (Altschul *et al.*, 2010). The inferred protein size and AP2 domain location within each TgAP2 is heterogeneous as can be seen in the stick figures representing TgAP2 whose mRNA vary during the cell cycle (right). The peak timing of mRNA of the cell cycle regulated TgAP2 within the cell cycle is indicated, as determined by Affymetrix microarray analysis (Behnke *et al.*, 2010). The mRNA expression of some additional Tg ApiAP2 members suggests that they are expressed in other developmental stages (Behnke *et al.*, 2010).

(Balaji *et al.*, 2005). Supporting this hypothesis, individual PfAP2 domains showed affinity for more than one sequence motif implying a single ApiAP2 factor could indeed regulate multiple gene targets. If ApiAP2 factors operate under a relaxed DNA sequence recognition (De Silva *et al.*, 2008; Campbell *et al.*, 2010), then this feature could account for the smaller repertoire of other transcription factor orthologues encoded in Apicomplexa genomes. Interestingly, the limited domain architecture in plants (one or two domains) is not conserved among the Apicomplexa, where proteins with up to eight AP2 domains are found (Balaji *et al.*, 2005; Altschul *et al.*, 2010). The protein diversity in the ApiAP2 family of proteins is also large. ApiAP2 factors in *P. falciparum* and *T. gondii* range in size from 200 to greater than 4000 amino acids that are highly disordered outside the globular AP2 domain(s) (Campbell *et al.*, 2010; V. Uversky, personal communication; see Fig. 18.1). Proteins of this secondary structure type are thought to undergo conformational changes upon interacting with other proteins (Xue *et al.*, 2012), consistent with transcription factor functions. The high content of disordered proteins in parasitic eukaryotes (Xue *et al.*, 2012) is thought to be a crucial adaptation to distinct environment niches.

Structural determination of representatives of the *Arabidopsis* and *Plasmodium* AP2 families has provided valuable insight into the mechanism by which the AP2 domain binds target DNA sequences. The conserved secondary structure for the monomeric AP2 domain predicts three N-terminal anti-parallel β-sheets and a C-terminal α-helix (Allen *et al.*, 1998; Lindner *et al.*, 2010). The NMR structure of *Arabidopsis* ERF1 GCC box binding domain (GBD is AP2 domain, target sequence = 5′-A/GCCGAC-3′) in complex with DNA revealed a novel interaction. AtEFR1 binds DNA via interaction with the β-sheets at specific locations along the DNA backbone while being supported by the α-helix (Allen *et al.*, 1998). This interaction, as depicted in the structure of the single AP2 domain containing ERF1, is based on 11 highly conserved residues, seven of which target specific interaction with the GCC box (5′-AGCCGCC-3′). These residues are located within the β-sheets that bind to the sugar–phosphate backbone and comprise the framework for specific DNA interaction (Allen *et al.*, 1998). The delineating feature of the DREB sub-family from the ERF sub-family of single domain AP2 proteins is a change in two conserved amino acids: V14 to A, E19 to D respectively. This alters the target DNA-binding sequence (5′-TACCGACAT-3′) illustrating that the diversity of recognition sequence is dictated by a limited number of evolutionarily conserved residues (Sakuma *et al.*, 2002). Interestingly, the dual AP2 domain containing AINTEGUMENTA contains the conserved arginine and tryptophan residues, but mutation of these residues has little to no effect on DNA binding, suggesting dual AP2 domain proteins exhibit a greater complexity in target DNA-binding sequences and each domain utilizes unique residues to facilitate binding (target sequence = 5′-gCAC(A/G)N(A/T)TcCC(a/g)ANG(c/t)-3′) (Krizek, 2003). These results from plants suggest dual AP2 domain proteins are more complex in their DNA binding and this may also apply to ApiAP2 factors. In *Plasmodium*, initial characterization of the dual ApiAP2 protein PFF0200c indicated only one AP2 domain actively bound to a specific 10 mer sequence (De Silva *et al.*, 2008). However, studies of the full-length protein in parasites clearly demonstrate that PfSIP2 (PFF0200c) requires both AP2 domains in order to bind the 16 bp bipartite SPE2 sequence motif (Voss *et al.*, 2003; Flueck *et al.*, 2010). This highlights the limitations associated with using any single approach to determining ApiAP2 function.

Solving the crystal structure of the prototypical ApiAP2 domain from *P. falciparum* (PF14_0633) provided further insight into AP2 function (Lindner *et al.*, 2010). While ApiAP2 domains retain many of the canonical features previously described in the *Arabidopsis thaliana* (Allen *et al.*, 1998), key differences have been identified. In contrast to the *A. thaliana* structure, which acts as a monomer, PfAP2s are thought to dimerize via a domain-swapping mechanism, with the α-helix of one promoter packed against a β-sheet of its partner (Lindner *et al.*, 2010). This model suggests that DNA binding triggers stabilization of the homodimer or that an AP2 binds DNA as a monomer, elucidating a conformational change that then attracts the second monomer to bind (Lindner *et al.*, 2010). In the case of Pf14_0633, dimerization is thought to be critical to combining distal regions of DNA to function in gene-specific transcription of sporozoite stage genes (Lindner *et al.*, 2010). In addition, ApiAP2 may work in concert as proteomics studies of nuclear complexes have often identified more than one ApiAP2 in pull-downs of macromolecular complexes (Flueck *et al.*, 2010; Sullivan, White, Kim, unpublished).

18.2.5 The Function of ApiAP2 Factors

Studies in *Arabidopsis* and other plant species have described major roles for AP2 proteins in a wide variety of developmental and stress responses. These transcription factors systematically regulate a diverse set of plant processes, including meristem, flower and seed development and environmental responses to drought or attack from plant pathogens (Riechmann and Meyerowitz, 1998; Dietz *et al.*, 2010). The number of AP2 domains contained within a plant transcription factor (one or two AP2 domains) predicts function: single AP2 domain containing factors regulate genes associated with pathogenesis and environmental response pathways whereas dual AP2 domain containing proteins regulate genes responsible for plant development (Riechmann and Meyerowitz, 1998). Much less is known concerning the role of ApiAP2 proteins, although the structure–function dichotomy of plant AP2s is not conserved in the Apicomplexa.

The *Toxoplasma* genome encodes 68 ApiAP2 domain-containing genes (Fig. 18.1; for product names see http://www.ToxoDB.org and Altschul *et al.*, 2010), more than twice the number found in *P. falciparum* (27 in total) and other Apicomplexa (Balaji *et al.*, 2005). ApiAP2s in *Toxoplasma* (TgAP2) are expressed during parasite development (Behnke *et al.*, 2010; Buchholz *et al.*, 2011). Roughly a third (24 in total) of TgAP2 genes are cell cycle regulated with mRNA expression profiles that span the tachyzoite division cycle (Behnke *et al.*, 2010). Eleven TgAP2 mRNAs are induced during bradyzoite differentiation, suggesting a role in developmental gene expression. The remaining TgAP2 domain containing proteins are either constitutively expressed (27 in total) or are undetectable in tachyzoite and bradyzoites, indicating possible roles in the sporozoite or oocyst stages of development (six in total) (Behnke *et al.*, 2010).

Of the 68 ApiAP2 domain-containing genes, only about 50 have all structural features predicted to be necessary for DNA binding (Altschul *et al.*, 2010). In most cases, there are no other obvious clues as to protein function, although all studies to date for TgAP2 have shown nuclear localization (Kim, White, Sullivan, unpublished). Yeast two hybrid studies in *Plasmodium* (LaCount *et al.*, 2005) and proteomics studies in *Plasmodium* (Zhang *et al.*, 2011; Flueck *et al.*, 2010), as well as published (Saksouk *et al.*, 2005a; Braun *et al.*, 2010) and unpublished studies in *Toxoplasma* (Kim, White, Sullivan, unpublished), support a role for ApiAP2 in gene regulation.

In the rodent malaria *P. berghei*, it was determined that an ookinete specific AP2 (AP2-O, PF11_0442 orthologue) is critical to mosquito mid-gut invasion. AP2-O was found to directly interact within the proximal promoter regions of 15 genes, including 10 that had previously been defined as ookinete specific or required for ookinete development within the mosquito mid-gut. AP2-O knock-out parasites exhibited normal gametogenesis; however, they lacked the ability to infect mosquitos (Yuda *et al.*, 2009b). A second *P. berghei* AP2, AP2-Sporozoite (AP2-Sp), is a *trans*-acting factor whose cognate binding site is enriched in proximal promoter regions of known sporozoite specific genes, likely interacting with *cis*-elements to promote stage-specific gene expression (Yuda *et al.*, 2009a; Helm *et al.*, 2010). The *P. falciparum* orthologue of this protein, PF14_0633, has a DNA binding domain that shared a conserved DNA binding motif (GCATGC) with both *C. parvum* (De Silva *et al.*, 2008) as well as *T. gondii* orthologues (Kim, unpublished). In contrast to *Plasmodium,* the *T. gondii* gene appears to be essential and to date it has been refractory to disruption. In *T. gondii,* this motif is a cis-acting motif required to drive luciferase activity in reporter constructs (Kim, unpublished). While many of the DNA binding motifs of ApiAP2 appear to be phylogenetically conserved, there is no conclusive evidence as to whether or not these factors have orthologous functions.

Further evidence for transcription factor activity for AP2 proteins comes from a promoter mapping study of a liver stage exclusive promoter. Four repeats of the ApiAP2 PB000252.02.0 (PF11_0404 orthologue) DNA-binding motif were found in the minimal promoter. Interestingly, mutation of a single copy of the binding site within the liver stage promoter increased promoter luciferase gene expression, suggesting PbAP2 PB000252.02.0 has a role as a transcriptional repressor (Helm *et al.*, 2010), which is a common mechanism used by plant AP2 factors to regulate the timing of developmental gene expression (Song *et al.*, 2005; Schmid *et al.*, 2003; Andriankaja *et al.*, 2007). Finally, the Bilker and Soldati groups used the DNA binding domains of ApiAP2 to develop regulated promoters (see description in Chapter 17) that work in both *Plasmodium* species as well as *T. gondii,* further bolstering the hypothesized role of these proteins in transcriptional regulation (Pino *et al.*, 2012).

Genetic studies suggest that *Toxoplasma* AP2 factor AP2IX-9, which is induced by alkaline stress, operates as a suppressor of bradyzoite differentiation through binding to specific bradyzoite gene promoters (Radke *et al.*, 2013). Another AP2 factor, AP2XI-4, is up-regulated in bradyzoites and has been identified as an activator of bradyzoite differentiation (Walker *et al.*, 2013). Thus, it is expected that ApiAP2 factors in *Toxoplasma,* much like plant and *Plasmodium* spp. factors, will have an important role in regulating developmental gene expression in the intermediate and definitive life cycles.

Analysis of ApiAP2 mRNA expression patterns in *P. falciparum* reveals an ordered timing of expression for 22 of 26 ApiAP2 proteins during the intraerythrocytic development cycle (IDC) (Balaji *et al.*, 2005; Campbell *et al.*, 2010). The distribution of expression across the IDC suggests that ApiAP2 factors could be responsible for controlling the dynamic changes in gene expression during apicomplexan replication (Campbell *et al.*, 2010). The enrichment of PfAP2 binding motifs in the promoters of cell cycle transcripts lends support to the idea that the *trans*-acting ApiAP2 regulator of groups of cell cycle genes will likely be co-expressed (Campbell *et al.*, 2010). For example, the putative target genes for Pf13_0235, which includes ribosome function and heat shock genes, all have the same timing in the IDC. Also, enrichment of PfAP2 target sequences that are found in invasion and host cell entry genes may be regulated by Pf10_0075, and transcripts encoding DNA replication factors could be controlled by MAL8P1.153 (Campbell *et al.*, 2010). In each case, the binding motif of the co-expressed PfAP2 was enriched in the promoters of the inclusive mRNA cluster.

Thus far, there is limited experimental evidence available for a role of TgAP2 proteins in parasite replication. However, the sequential profiles of 24 cell cycle regulated TgAP2 factors now provides attractive candidates for an interacting network operating during *Toxoplasma* replication to coordinate a similar cell cycle transcription cascade. Like early work on cell cycle *Plasmodium* ApiAP2s, the study of *Toxoplasma* cell cycle ApiAP2s has focused on determining the DNA-binding specificity for selected proteins and the DNA binding preferences of a substantial fraction of TgAP2 have now been identified using the protein binding array technology previously adapted to *Plasmodium* AP2s (Campbell *et al.*, 2010; De Silva *et al.*, 2008; White, Sullivan, Llinas and Kim, unpublished). These motifs will be useful for genome-wide computational searches for putative regulatory targets within proximal promoters as determined by epigenomics ChIP–chip analysis (Gissot *et al.*, 2007) and inferred transcription start sites (Yamagishi *et al.*, 2010) .

Efforts are also under way to sort essential versus non-essential TgAP2s, which will open the door to determining key regulatory mechanisms critical to cell cycle progression and possible interacting networks of AP2 proteins. While the cell cycle and developmental Tg-ApiAP2s occupy largely unique groups, there are three TgAP2s (AP2VIIa-1, AP2VI-1, and AP2IX-4) that are periodically expressed in the tachyzoite cell cycle and are also elevated in late stage *in vivo* cysts (Behnke *et al.*, 2010; Buchholz *et al.*, 2011). These TgAP2 factors show peak mRNA levels in the S/M phase, which corresponds to a critical point in the tachyzoite cell cycle where the decision to continue replication as a tachyzoite or differentiate into a bradyzoite is made (Radke *et al.*, 2003). It is well established that parasite replication is linked to development, making it possible that these TgAP2s have dual responsibilities in tachyzoite growth and bradyzoite developmental gene expression.

TgAP2s have been identified as proteins that interact with chromatin remodellers HDAC3 (Saksouk *et al.*, 2005a) and GCN5 (Kim and Sullivan, unpublished), as well as complexes implicated in RNA splicing (Kim, Sullivan and White, unpublished). A TgAP2 also was reported to be a component of the macromolecular complex that interacts with TgAgo, an essential component of the RNA induced silencing complex (RISC) that mediates the activity of many small RNAs in gene expression (Braun *et al.*, 2010). Thus TgAP2s are implicated in multiple critical processes in gene regulation.

Studies in *P. falciparum* indicate ApiAP2 proteins may have non-transcriptional roles. Genome-wide interaction studies of PfSIP2 (ChIP–chip), a dual domain PfAP2 protein, found this factor localized to sub-telomeric heterochromatin regions on all chromosomes (Flueck *et al.*, 2010). After proteolytic processing, PfSIP-N exclusively localizes with the SPE2 DNA motif that is present in multiple copies in the promoter regions of upsB *var* genes associated with *var* gene silencing (Flueck *et al.*, 2010; Voss *et al.*, 2003). Over-expression of PfSIP2-N caused no changes in gene expression, lending support to a role in maintaining chromosome end biology (Flueck *et al.*, 2010; Voss *et al.*, 2003). This mechanism is supported by PfSIP2 orthologues that exist in other *Plasmodium* spp. that lack sub-telomeric SPE2 motifs yet maintain the SPE2 motif in internal chromosome regions (Flueck *et al.*, 2010). Taken together, PfSIP2 illustrates a novel function for ApiAP2 proteins in telomeric heterochromatin maintenance and gene silencing. A second PfAP2, Pf11_0091, binds to a motif within the *var* intron that is sufficient to localize DNA to a subnuclear compartment with silenced *var* genes in an actin-specific manner (Zhang *et al.*, 2011). The *var* intron also has promoter activity (Calderwood *et al.*, 2003), and as yet, the role of this ApiAP2 in transcriptional regulation has not been resolved.

Evidence for non-transcriptional roles for *Toxoplasma* AP2 factors are starting to emerge. In genome-wide occupation studies of AP2VI-1 and AP2IV-4, which exhibit maximum expression coinciding with S/M phase, show localization to peri-centromeric regions on all chromosomes (Radke, Kim and White, unpublished). This pattern is reminiscent of the PfSIP2 localization to sub-telomeric heterochromatin and these AP2s could play critical roles in chromosome structure determination or chromatin remodelling.

18.2.6 Other Factors that Regulate Gene Expression

Amongst the other factors that are developmentally regulated are the glycolytic enzymes, including *ENO2* (tachyzoite) and *ENO1* (bradyzoite). Intriguingly these proteins are localized to the nucleus suggesting that they might also play a role in gene regulation in *T. gondii.* Several glycolytic enzymes have been identified as components of transcriptional complexes including LDH and GAPDH in the OCA-S complex (Zheng *et al.*, 2003). As the expression of the ENO genes is also regulated at the mRNA level, the *ENO1* promoter was used as bait to identify nuclear factors that interact with this promoter (Olguin-Lamas *et al.*, 2011). Amongst the 35 nuclear proteins identified, most are hypothetical proteins, but those with inferred function included two Alba family DNA/RNA binding proteins, a potential histone chaperone (NF3), and other proteins predicted to interact with RNA and DNA. NF3 is nucleolar protein whose overexpression alters nucleolar morphology and inhibits parasite virulence. In the bradyzoite stage NF3 is cytosolic. Chromatin immunoprecipitation studies have confirmed that NF3 (Olguin-Lamas *et al.*, 2011) and Eno2 proteins (Tomavo and Kim, submitted) are associated with chromatin.

18.3 EPIGENETICS IN *TOXOPLASMA*

Epigenetic gene regulation refers to heritable changes in gene expression that are not genetically encoded in the DNA sequence of an organism. Among the mechanisms of epigenetic regulation are those affecting accessibility of factors to chromatin such as DNA methylation, histone modification and nucleosome location. Noncoding RNA also affects a myriad of nuclear and cytoplasmic processes that regulate epigenetic gene regulation.

Current models of eukaryotic transcriptional activation implicate a significantly greater number of co-factors than was appreciated more than a decade ago. The simple binding of gene-specific ATFs (activating transcription factors) to local sequences in nucleosomal DNA (Naar *et al.*, 2001) is now recognized to be insufficient to recruit the RNA polymerase II-PIC. ATFs recruit chromatin remodellers to facilitate the assembly of the PIC on core promoter sequences (reviewed in Spector, 2003; Ehrenhofer-Murray, 2004; Li *et al.*, 2004a). These findings bring chromatin dynamics to the forefront of gene expression research, and the discovery that histone proteins can be chemically modified in ways that enhance or inactivate transcription, along with ATPases capable of repositioning nucleosomes, has prompted intensive investigation into how these mechanisms act cooperatively to regulate gene expression. Although the order and assembly of transcriptional factors has not been demonstrated in any apicomplexan parasite, including *Toxoplasma,* the initial forays into understanding transcriptional regulation reveal these parasites possess essential features of the basal transcription machinery as well as a significant collection of chromatin remodelling machinery. Research into chromatin remodelling mechanisms for the purpose of new drug target discovery is an important area of investigation, first illustrated by the HDAC (histone deacetylase) inhibitor, apicidin, which has broad spectrum activity against a variety of apicomplexan parasites including human pathogens *Toxoplasma* and *Plasmodium* and veterinary pathogens from the *Eimeria* genera (Darkin-Rattray *et al.*, 1996).

18.3.1 Chromatin and Chromatin Remodelling

The fundamental building block of chromatin is the histone protein. Four canonical types of histones exist (H2A, H2B, H3 and H4) that form an octamer complexed with DNA (the nucleosome). Histone tails, particularly those of H3 and H4, are subject to a diverse array of covalent modifications that have different consequences on gene transcription (Peterson and Laniel, 2004). Like many other eukaryotes, *Toxoplasma* H3 and H4 are exceptionally well conserved with each residue in the N-terminal tail reported to be susceptible to chemical modification being present (Sullivan *et al.*, 2003; Nardelli *et al.*, 2013). Histone variants, which may be substituted for canonical histones to modulate DNA-driven processes, are also conserved. *Toxoplasma* contains a homologue of variant H3.3 in addition to the canonical H3, the former being associated with genes undergoing transcription in other species (Sullivan *et al.*, 2003). An orthologue of the centromeric H3 variant (CenH3) has also been characterized and localized to an apical subnuclear compartment (Brooks *et al.*, 2011). Beyond the well-conserved H3 and H4 classes, the complement of histone proteins in *Toxoplasma* exhibits a number of unusual features (Dalmasso *et al.*, 2011).

Like yeast, *Toxoplasma* may not possess H1, the extra-nucleosomal 'linker' histone involved in solenoid formation during chromatin condensation, although a small basic protein with homology to the H1 of kinetoplastids (Croken *et al.*, 2012) is present in the genome. Two distinct lineages of H2B are present, including the constitutively expressed TgH2Bv1, a parasite-specific H2B variant, and potential stage-regulated TgH2Ba and TgH2Bb (Dalmasso *et al.*, 2006). The canonical H2A protein is TgH2A1 and both H2AZ and H2AX variants exist (Dalmasso *et al.*, 2009). TgH2A1 and TgH2AX both possess a C-terminal SQ motif, consistent with predicted roles in the DNA damage response. Co-immunoprecipitation experiments are beginning to reveal the composition of *Toxoplasma* nucleosomes (Dalmasso *et al.*, 2009). TgH2AZ dimerizes with TgH2Bv1, but not TgH2AX. TgH2AZ and TgH2Bv1 localize with other acetylated histones to actively transcribed genes (Dalmasso *et al.*, 2009) whereas TgH2AX is present at repressed genes (Dalmasso *et al.*, 2009). TgH2AZ also localizes to gene bodies of developmentally silenced genes (Nardelli and Kim, unpublished). Interestingly, TgH2AX expression increases during bradyzoite conversion, consistent with the increase in repressed genes during the latent stage. These studies are consistent with the hypothesis that TgH2AZ and TgH2Bv1 are involved in transcriptional activation while TgH2AX and TgH2A1 may populate chromatin during stress (Dalmasso *et al.*, 2009).

Histone N-terminal tails are generally rich in positively charged amino acids and interact tightly with negatively charged DNA, facilitating condensation. The assembly of genomic DNA into histone nucleosomes and then into higher order chromatin structure is associated with transcriptional repression and 'silenced' chromatin is thought to be the default mechanism guiding the formation of chromatin following DNA replication (Ehrenhofer-Murray, 2004). Thus, active steps must be taken to alter the normal state of chromatin in order to achieve stable transcriptional activation. A myriad of chemical modifications to histones are now known and are proposed to operate in combinatorial fashion, constituting a 'histone code' that reflects corresponding changes in the local activation (and inactivation) of specific genes (Strahl and Allis, 2000).

The histone code of *T. gondii* has been characterized by mass spectrometry (Fig. 18.2; Nardelli *et al.*, 2013) with *T. gondii* histones possessing novel modifications not previously described in protozoa. Similar to *Plasmodium* (Trelle *et al.*, 2009) and in contrast to the metazoa, *T. gondii* nucleosomes consist primarily of euchromatic acetylated chromatin, consistent with chromatin that is accessible to the transcriptional machinery (Nardelli *et al.*, 2013).

Consistent with the histone code hypothesis is the discovery of protein motifs capable of binding specific histone modifications. Examples include

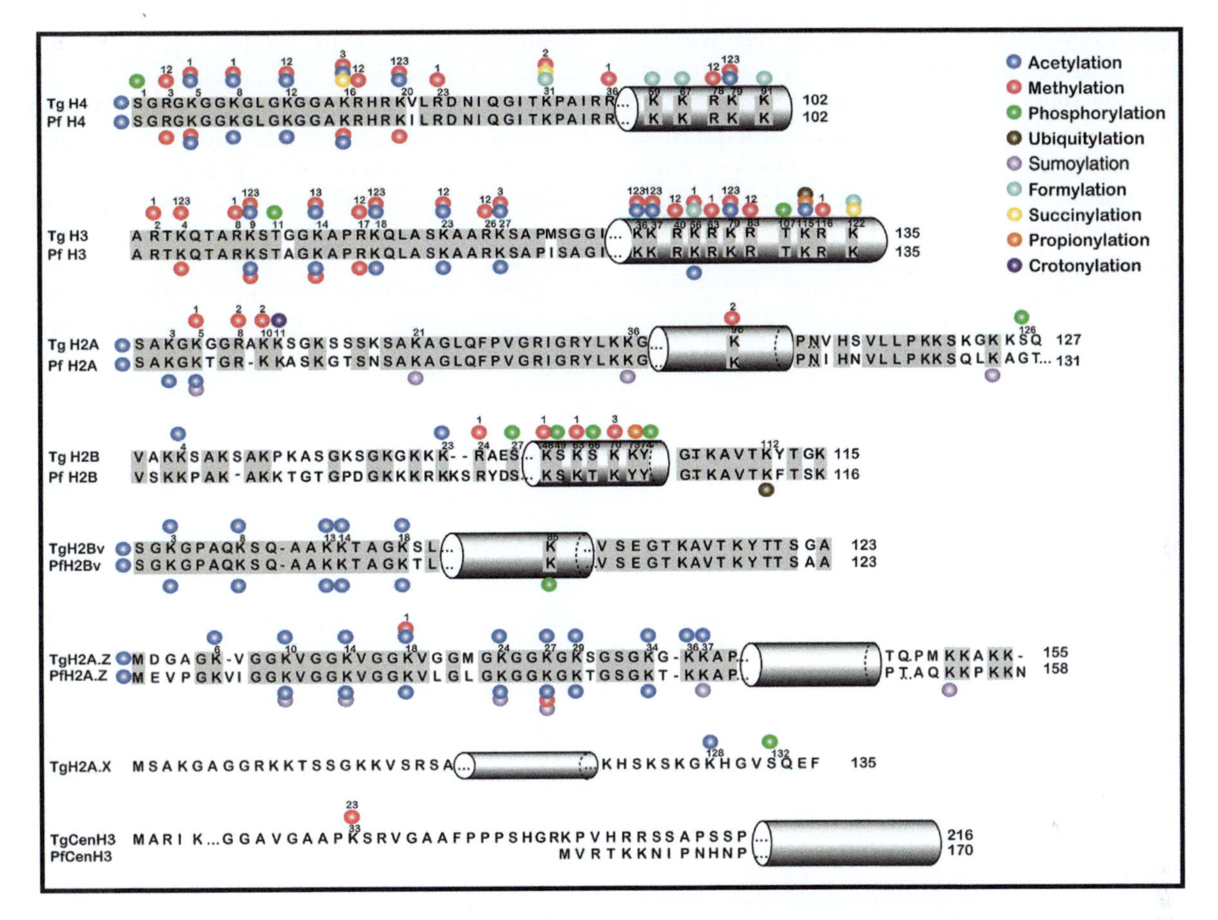

FIGURE 18.2 **The histone code of *Toxoplasma gondii:* a summary of post-translational modifications (PTM) identified on *T. gondii* canonical histones and histone variants by mass spectrometry.** PTM on canonical histones are shown in comparison with *P. falciparum*. Circles in different colours represent the modifications. PTM on canonical and histone variants are shown in comparison with *P. falciparum*. Identical amino acids are represented in grey. Circles above the sequence represent histone PTM in *T. gondii* (Nardelli *et al.*, 2013). Histones depicted are histone 3 (H3), histone 4 (H4), histone H2A (H2A), histone (H2B) and variant histones H2Bv, H2A.Z, H2A.X and CenH3. The grey cylinders indicate the approximate location of the histone globular domain. Numbers above the sequences represent the amino acid position in *T. gondii* while 1x, 2x and 3x above the red circles indicate mono-, di- or tri-methylation respectively. *Figure courtesy of Sheila Nardelli.*

bromodomains that interact with acetylated lysines, chromodomains that bind methylated lysines, and macrodomains that recognize ADP-ribose moieties (Dhalluin *et al.*, 1999; Bannister *et al.*, 2001; Karras *et al.*, 2005). The chromodomain protein TgChromo1 binds H3K9me3 and localizes to centromeres as well as telomeres in *T. gondii* (Gissot *et al.*, 2012), although only centromeres have been demonstrated to be enriched in H3K9me2/3 (Brooks *et al.*, 2011). In the following sections, we present a summary of histone modifications and discuss what is known about their occurrence in *Toxoplasma*.

18.3.2 Mapping the *Toxoplasma* Epigenome

Genome-wide approaches have been used to illuminate the chromatin modifications associated with gene activation as well as define functional regions of the *T. gondii* genome (Fig. 18.3). The major technique used has been chromatin immunoprecipitation, which enriches for the DNA–protein complexes of interest, followed by either hybridization of the enriched DNA to genome-wide arrays (Gissot *et al.*, 2007) or high throughput sequencing (ChIP-seq). Because *T. gondii* histones and histone modifications seen in model organisms are conserved (see Fig. 18.2), commercial antibodies specific for histone modifications can be used for this technique. In other species, direct modification of DNA, primarily cytosine methylation, also affects the accessibility of macromolecular complexes to chromatin, but it appears that *T. gondii* DNA is not cytosine methylated and there currently is no evidence for modification of DNA affecting gene

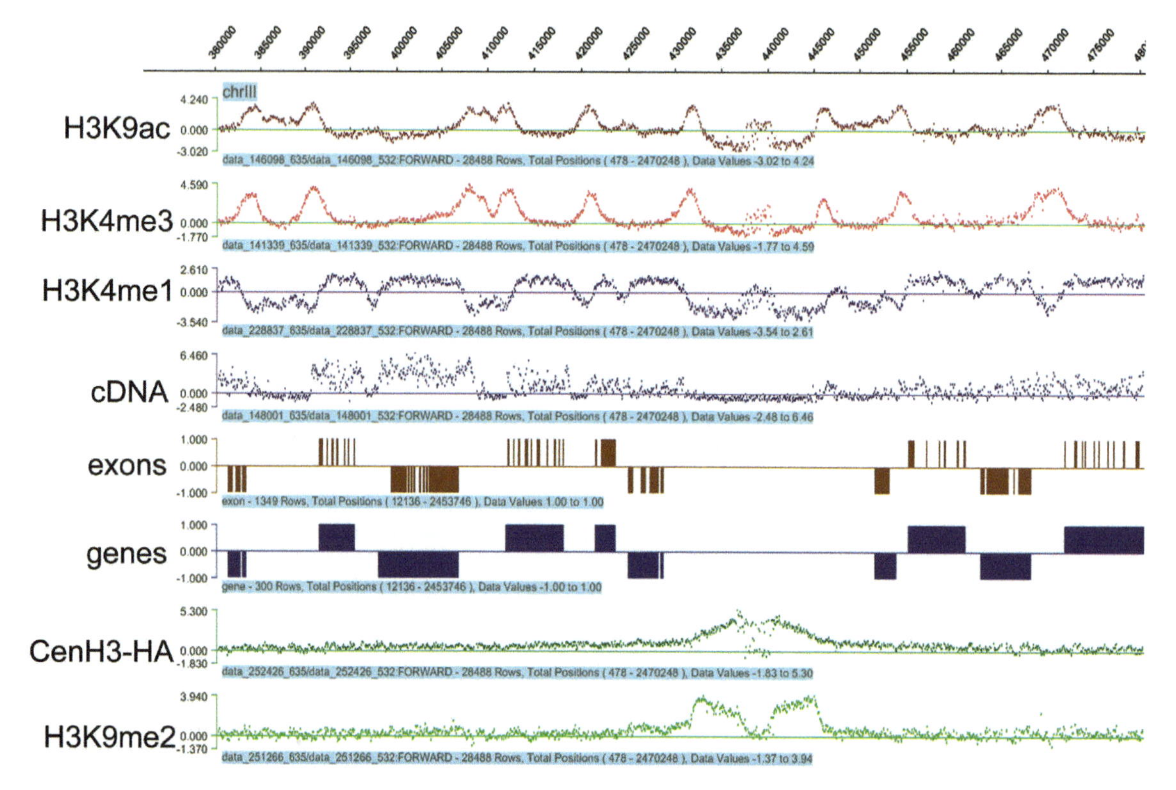

FIGURE 18.3 **Chromatin modifications that define the epigenome of *Toxoplasma gondii*.** Genome-wide chromatin immunoprecipitation microarray hybridization studies (ChIP–chip) define the epigenome of *T. gondii*. Chromatin was harvested from intracellular tachyzoites and chromatin immunoprecipitations were performed with antibodies specific for the indicated histone post-translational modification. Enriched DNA was hybridized to a custom Nimblegen *Toxoplasma* genome tiled microarray and the results of hybridization are shown as log2 ratios of signal to input control DNA. cDNA was harvested in parallel and hybridized to the chip. H3K9ac, H3K4me3 are enriched at sites of active promoters, whereas H3K4me1 co-localizes with gene bodies of actively transcribed genes (cDNA track). Genes and exons are indicated with boxes above the baseline indicating genes predicted to be on the positive strand and boxes below the baseline indicating genes transcribed on the negative strand. The specialized centromeric histone CenH3 marks a gene-poor region of the chromosome and localizes to each chromosomal centromere with H3K9me2.

expression in *T. gondii* or any Apicomplexa (Gissot *et al.*, 2008; reviewed in Croken *et al.*, 2012).

18.3.2.1 Chromatin Signatures in Toxoplasma *Biology*

Through performing ChIP with antibodies to either acetylated H3 or H4, it was demonstrated that relative acetylation versus deacetylation can be correlated with specific gene activation or repression in *Toxoplasma*, respectively (Saksouk *et al.*, 2005b). This was shown in the context of stage conversion to latent bradyzoites, demonstrating that in parasites cultured as tachyzoites, H3 and H4, were acetylated at tachyzoite-specific promoters like SAG1 and SAG2A, while no acetylation is detected at bradyzoite-specific promoters such as LDH2 and BAG1 (Saksouk *et al.*, 2005b). Conversely, in a parasite population induced to enter the bradyzoite pathway, acetylation at tachyzoite promoters was diminished while acetylation at bradyzoite promoters increased. As expected, intergenic regions upstream of constitutively expressed genes were found in the acetylated state in either population *in vitro*. Confirmation of the differential state of histone acetylation that is associated with specific remodelling enzymes was also observed in parasite transgenic lines expressing epitope-tagged TgGCN5-A and TgHDAC3 proteins. Lysine acetyltransferase (KAT) TgGCN5-A was present at tachyzoite promoters in tachyzoites, but absent at bradyzoite promoters, whereas TgHDAC3 was associated with promoters that were down-regulated in each respective developmental stage (Saksouk *et al.*, 2005b).

Studies have also shown that the TgCARM1-mediated methylation of H3R17 is another signature of gene expression in *Toxoplasma*, with the presence of this protein at active genes in either the tachyzoite or bradyzoite stage (Saksouk *et al.*, 2005b). Interestingly, genes marked with methylated H3R17 also displayed enrichment of acetylated H3K18, a potentially synergistic signature for gene activation that relies on crosstalk between acetyl- and methyltransferase complexes. ChIP and mass spectrometry have shown that H3K4 can be mono-, di- or tri-methylated in *Toxoplasma*.

More specifically, tri-methylated H3K4 is enriched at tachyzoite promoters during the tachyzoite stage (Gissot *et al.*, 2007) and becomes enriched at bradyzoite promoters following differentiation, representing another mark of gene activation in this parasite (Saksouk *et al.*, 2005b). More recent genome-wide ChIP studies have established that acetylation of H3K9, H4 and tri-methylation of H3K4 occur at promoters of actively expressed genes (Gissot *et al.*, 2007). Tri-methylation of H3K9 and H4K20 occur at repressed genes in heterochromatic territories (Sautel *et al.*, 2007), with the H3K9me2/3 marks enriched in centromeres (Brooks *et al.*, 2011). Monomethylation of H3K4 is associated with gene bodies of actively transcribed genes (Fig. 18.3). In contrast to *Plasmodium*, *T. gondii* does not encode major antigenic variant gene families whose silencing is associated with deposition of the heterochromatic histone marks H3K9me2/3 (Lopez-Rubio *et al.*, 2007, 2009).

As observed in other species, phosphorylation of TgH2AX has been linked to the parasite DNA damage response. H2AX is phosphorylated in response to double-stranded breaks at its C-terminal SQ(E/D)Φ motif (Φ denoting a hydrophobic residue) (Escargueil *et al.*, 2008). Treatment of *Toxoplasma* with DNA-damaging agents MMS (methyl methanesulphonate) or H_2O_2 led to increased phosphorylation of TgH2AX, as detected by immunoblotting with monoclonal antibody (Vonlaufen *et al.*, 2010; Dalmasso *et al.*, 2009). These studies indicate that phosphorylated TgH2AX can be used as a chromatin biomarker for DNA injury. Phosphorylation of H3S10 has also been reported in *Toxoplasma*, a mark that peaks during mitosis with monomethylation of H4K20 (Sautel *et al.*, 2007). The function of H3S10 phosphorylation has been linked to chromosome condensation in fellow alveolate protozoan *Tetrahymena* (Wei *et al.*, 1998).

18.3.3 Histone Modifying Enzymes

Chromatin remodellers generally fall into two distinct classes: those capable of covalently modifying histones or those that use ATP to reposition nucleosomes (SWI2/SNF2 family ATPases). Both types of chromatin remodelling machinery can be found in *Toxoplasma* and other protozoa as well (Sullivan *et al.*, 2006; Dixon *et al.*, 2010). Apicomplexan histones are subject to a wide variety of covalent modifications that include acetylation, methylation, phosphorylation, ubiquitination, and sumoylation (see Fig. 18.2 for a comparison of *T. gondii* and *Plasmodium falciparum*). In the following sections, we will discuss histone modifying enzymes found in *Toxoplasma* and what we have learned to date about their role in parasite biology.

18.3.3.1 Histone Acetylation

In other eukaryotes, acetylation of lysine residues in the N-terminal histone tails is linked to gene activation. Conversely, the removal of acetyl groups is associated with transcriptional repression. A wide variety of HATs (histone acetyltransferases) and HDACs (histone deacetylases) have been characterized among eukaryotes that control the acetylation status of nucleosomal histones, and hence play an important role in the regulation of gene expression (Sterner and Berger, 2000; Thiagalingam *et al.*, 2003). The *Toxoplasma* genome predicts at least seven HATs and seven HDACs present in the parasite. Recently, it has been proposed that proteins historically referred to as HATs and HDACs be renamed KATs and KDACs (lysine acetyltransferases and deacetylases, respectively) since many of them also act on non-histone substrates (Allis *et al.*, 2007). Given the discovery of widespread lysine acetylation in nonnuclear compartments of *Toxoplasma*, we propose this change in nomenclature be adopted for the parasite (Jeffers and Sullivan, 2012).

Two MYST family KAT proteins (MOZ, Ybf2/Sas3, Sas2, Tip60) exist in *Toxoplasma*, each possessing a chromodomain and the atypical C2HC zinc finger domain upstream of the KAT domain (Smith *et al.*, 2005; Vonlaufen *et al.*, 2010). The predicted proteins, named TgMYST-A (AY578183) and -B (DQ104220), have features consistent with the 'MYST + CHD' subclass, homologous to yeast Esa1, human Tip60, and MOF (Utley and Cote, 2003). Previous studies demonstrate that this type of KAT has a preference for acetylating lysines in H4 and the observation that recombinant TgMYSTs also prefer H4 as substrate in assays using free core histones functionally validates this classification (Smith *et al.*, 2005). Given this similarity, it was not surprising that genes encoding TgMYST-A and -B could not be disrupted by homologous recombination as the Esa1 homologue in yeast is also an essential gene (Smith *et al.*, 1998). TgMYST-A is not amendable to stable over-expression unless the recombinant protein is mutated to nullify its KAT activity, suggesting a delicate balance of TgMYST-A-mediated acetylation exists in *Toxoplasma* (Smith *et al.*, 2005). Despite their histone acetylation abilities, both TgMYST KATs are predominantly cytoplasmic, suggesting they may act on non-histone substrates (Jeffers and Sullivan, 2012).

Over-expression of TgMYST-B is tolerated, but results in a significantly reduced proliferation rate unless enzymatic activity is ablated (Vonlaufen *et al.*, 2010). Interestingly, the delayed replication may be connected to the dramatic resistance to DNA-damage observed for TgMYST-B over-expressing parasites. Consistent with heightened protection from DNA-damaging agents, parasites over-expressing TgMYST-B have increased levels of ataxia telangiectasia mutated (ATM) kinase and phosphorylated H2AX (Vonlaufen *et al.*, 2010). Increased γH2AX leads to cell cycle arrest and a decrease in the number of cells in mitosis, likely explaining why parasites over-expressing TgMYST-B exhibit delayed replication (Vonlaufen *et al.*, 2010; Rios-Doria *et al.*, 2009). The connection between TgMYST-B and the ATM kinase-mediated DNA damage response was further supported when pharmacological inhibitors of ATM kinase or KATs rescued parasites over-expressing TgMYST-B from slowed replication (Vonlaufen *et al.*, 2010).

Two KAT proteins of the GCN5 class also exist in *Toxoplasma* designated TgGCN5-A (AAF29981) and TgGCN5-B (AAW72884), which is a highly unusual arrangement in a lower eukaryote. Aside from the close relative *Neospora caninum*, the presence of two GCN5 KATs in a single cell has not been documented for any other invertebrate. In contrast, mammalian species have two GCN5 KAT enzymes referred to as GCN5 and PCAF (p300/CBP Associating Factor). Deletion of mouse GCN5 is embryonic lethal while the loss of PCAF has no discernible phenotype (Xu *et al.*, 2000; Yamauchi *et al.*, 2000). There is a striking parallel in *Toxoplasma*, in which TgGCN5-A is dispensable in tachyzoites yet TgGCN5-B appears to be essential (Bhatti *et al.*, 2006). The two TgGCN5s differ in other ways as well. GCN5 family members show a strong preference to acetylate H3, particularly lysine 14 (K14). Recombinant TgGCN5-A was found to have an exquisite selectivity to acetylate H3K18 whereas TgGCN5-B was more prototypical and capable of targeting H3K9, H3K14, and H3K18 *in vitro* (Bhatti *et al.*, 2006; Saksouk *et al.*, 2005a). Another difference between the TgGCN5 KATs is their ability to bind with the ADA2 co-activator, for which two homologues have been identified in the *Toxoplasma* (TgADA-A, DQ112184; TgADA2-B, DQ112185). By yeast two-hybrid assay, TgGCN5-B has been shown to interact with either TgADA2 homologue, while TgGCN5-A can only associate with TgADA2-B (Bhatti *et al.*, 2006).

It has been proposed that TgGCN5-B may be required for tachyzoite replication and TgGCN5-A may be required only for specific circumstances, such as the stress response. Indeed, *Saccharomyces cerevisiae* remains viable without GCN5, but is impaired when grown on minimal media (Marcus *et al.*, 1994). Support for this idea was recently obtained in studies of the TgGCN5-A knock-out that showed a significant recovery defect following exposure to alkaline pH stress, a condition commonly used to induce bradyzoite development *in vitro*. Microarray analyses revealed that parasites lacking TgGCN5-A fail to up-regulate ~75% of the genes normally induced during alkaline stress, including bradyzoite-specific induction markers BAG1 and LDH2 (Naguleswaran *et al.*, 2010). While repeated attempts to knock out TgGCN5-B have failed, even in Δ*ku80* parasites, an inducible dominant-negative strategy supports that TgGCN5-B is essential in tachyzoites (Sullivan, unpublished results).

Apicomplexan GCN5 KATs contain an unusual N-terminal extension upstream of the well-conserved catalytic and bromodomains. Curiously, the length and amino acid composition varies greatly among the Apicomplexa, and even among the pair of GCN5s in *Toxoplasma*. Most GCN5s from early eukaryotes do not have appreciable sequence upstream of the KAT domain. In contrast, mammalian GCN5 and PCAF have N-terminal extensions, but they are very similar to each other. The function of the N-terminal extension may be to mediate protein–protein interactions (e.g. the binding of CBP) and/or substrate recognition, as GCN5 lacking the N-terminal extension can only acetylate free histones and not nucleosomal histones (Xu *et al.*, 1998). The N-terminal extensions of TgGCN5-A and -B are required for nuclear localization, but dispensable for enzyme activity on free histones (Bhatti and Sullivan, 2005; Dixon *et al.*, 2011). A six amino acid, basic-rich motif in the N-terminal extension of TgGCN5-A has been mapped as a necessary and sufficient nuclear localization signal (NLS) that interacts with the nuclear chaperone importin alpha

(Bhatti and Sullivan, 2005). The NLS for TgGCN5-B is also rich in basic residues and found in the N-terminal extension, although its sequence is distinct from the NLS of TgGCN5-A (Dixon *et al.*, 2011).

Previous work has also noted that KDAC proteins exist in *Plasmodium* (Joshi *et al.*, 1999; Freitas-Junior *et al.*, 2005) and analysis of *Toxoplasma* genomic sequence indicates there are seven potential KDAC genes with one experimentally characterized to date (TgHDAC3). Recombinant TgHDAC3 exhibits histone deacetylase activity that is inhibited by butyrate, aroyl-pyrrole-hydroxy-amides, and trichostatin A and a native TgHDAC3-containing complex has been purified (TgCRC for CoRepressor Complex) (Saksouk *et al.*, 2005b). The TgCRC contains several protein components that are homologous to subunits found in the human N-CoR and SMRT complexes as well as two large parasite-specific proteins of unknown function (Saksouk *et al.*, 2005b). Pull-down studies suggest HDAC3 may interact with both TgAgo1 (Braun *et al.*, 2010) as well as TgAP2 (Saksouk *et al.*, 2005a; Kim, unpublished), supporting a critical role of TgHDAC3 in gene regulation in *T. gondii*. With the exception of TgSIR2a, the TgKDACs have been refractory to disruption, suggesting that they have essential functions in the biology of *T. gondii* (Kim, unpublished).

18.3.3.2 Histone Methylation

The addition of methyl groups occurs on lysine and arginine residues of histones and can lead to gene activation or silencing (Zhang and Reinberg, 2001). There is an added layer of complexity in methyl-modifications as residues can be mono-, di- or tri-methylated. *Toxoplasma* possesses five **p**rotein a**r**ginine **m**ethyl**t**ransferase (PRMT) homologues, designated TgPRMT1–5. Recombinant TgPRMT1 (AY820756) is capable of methylating H4R3, while TgPRMT4 (referred to as TgCARM1, co-activator associated arginine methyltransferase, AY820755) methylates H3R17 (Saksouk *et al.*, 2005b), which parallels the substrate specificity of their human homologues. The importance of TgPRMT1 for histone methylation is not yet clear, as the phenotype of parasites lacking TgPRMT1 is a cell cycle phenotype that affects daughter cell counting (El Bissati *et al.*, submitted). Human CARM1 has been associated with SWI2/SNF2 ATPases, including the **S**nf2-**R**elated **C**BP **A**ctivator **P**rotein, or SRCAP (Xu *et al.*, 2004; Monroy *et al.*, 2003). Recombinant TgCARM1 incubated with parasite extract enriched for ATP-dependent nucleosome disruption activity indicated that TgCARM1 is likely to interact with a *Toxoplasma* SWI2/SNF2 member (Saksouk *et al.*, 2005b), and an SRCAP SWI2/SNF2 homologue has been characterized in *Toxoplasma* (see section below and Sullivan *et al.*, 2003). Together, these data suggest a conserved connection between CARM1 and SRCAP complexes.

Lysine methyltransferases share a common feature known as the SET (Suv(39)-E(z)-TRX) domain (Dillon *et al.*, 2005). Searching for this domain in the predicted proteins contained in the *Toxoplasma* genome reveals at least 19 candidates and a surprising amount of duplication and divergence within this protein family (Bougdour *et al.*, 2010a). As has been observed with acetyl transferases, novel non-histone substrates for methyltransferases have been described in many systems including *T. gondii*. Commercial methyl lysine antibodies often cross react with cytosolic or cytoskeletal structures in *T. gondii* (Sautel *et al.*, 2007; Xiao *et al.*, 2010) and one SET protein, AKMT or apical lysine methyltransferase, is implicated in methylation of apical cytoskeletal structures (Xiao *et al.*, 2010). In *Plasmodium*, PfSET10, a methylase with H3K4me specificity, has been implicated in maintaining the active *var* gene in a poised state during the cell cycle (Volz *et al.*, 2012). It is likely that some of the TgSET will have analogous functions in lysine methylation of histones, particularly since histone lysine methylation is a common histone post-translational modification

(Fig. 18.2). Although bioinformatics studies have inferred specificity (Bougdour *et al.*, 2010b; Sullivan *et al.*, 2006), in most cases these predictions have not yet been validated experimentally, with the exception of the studies cited below.

ChIP–chip studies localize KMTox (formerly TgSET13) to genes related to antioxidant defences, heat-shock proteins/chaperones, and genes involved in translation and carbohydrate metabolism. KMTox also protects against H_2O_2 exposure, and was found to associate with 2-cys peroxiredoxin-1 (TgPrx1) under oxidative conditions, bolstering the idea that KMTox contributes to the parasite's antioxidant defence system (Sautel *et al.*, 2009). Unlike its monomethylating human counterpart, biochemical and structural modelling analyses show that TgSET8 is capable of mono-, di- and tri-methylation of H4K20 (Sautel *et al.*, 2007). Tachyzoites expressing a mutated version of TgSET8 (F1808Y) that abolishes the monomethylation of H4K20 are not able to progress through the cell cycle, suggesting that monomethyl H4K20 is required for parasite division (Bougdour *et al.*, 2010a). TgSET8 may also play important roles during the latent cyst stage, as suggested by high levels of monomethylated H4K20 in bradyzoites (Sautel *et al.*, 2007).

With regard to the removal of methyl groups, *Toxoplasma* appears to encode seven JmjC (Jumonji) domain demethylating proteins. Only one has both the Jumonji N and C domains characteristic of JARID-like H3K4 and JMJD1-like H3K9 demethylases and preliminary data suggest that this may be a dual function histone demethylase (Kim, unpublished). The other members of this family belong to the JMJD6 family that demethylate H3R2 and H4R3 (Bougdour *et al.*, 2010a). More recent studies have questioned the exclusive role of the JmjC (Jumonji) domain proteins in protein demethylation – with new functions reported in RNA processing (Hong *et al.*, 2010), so further studies will be needed to determine the function of these putative demethylases. *Toxoplasma* also appears to encode homologues of lysine-specific demethylases, but the function of these proteins has not been characterized. This atypical expansion of demethylases in *Toxoplasma* may counter the extensive number of methyltransferases and the expansion of the methyltransferase and demethylase families is an aspect of *T. gondii* biology that differs from *Plasmodium*.

18.3.3.3 Other Histone Covalent Modifications

Considerably less work has been done to dissect the roles of histone phosphorylation, ADP-ribosylation and ubiquitylation in *Toxoplasma*. As mentioned, H3S10 phosphorylation has been reported, and *Toxoplasma* possesses two predicted proteins with strong similarity to histone kinase Snf1. *Toxoplasma* also appears to possess proteins containing PARP and PARG domains, required for the addition or removal (respectively) of ADP-ribose subunits (Dixon *et al.*, 2010). There is also no shortage of ubiquitin-conjugating enzymes in this organism, including Ubc9, which is implicated in gene repression via the sumoylation of H4 (Shiio and Eisenman, 2003). H3 ubiquitination has been confirmed by mass spectrometry (Nardelli, Che *et al.*, submitted and Fig. 18.2). A small ubiquitin-like modifier (SUMO)-conjugating system has been characterized in *Toxoplasma* (Braun *et al.*, 2009), and sumoylation has been reported on *Plasmodium* H2A and H2AZ. While sumoylation of *T. gondii* histones is detectable by immunoblot (Nardelli *et al.*, 2013), the exact modified residues have not yet been mapped. A number of novel histone modifications of unknown function have been reported in the metazoa including propionylation, O-GlcNAcylation and succinylation. These modifications are also present on *T. gondii* histones (Fig. 18.2) and based upon conjectures in other organisms, these histone modifications may provide a mechanism by which changes in metabolism are sensed and can impact epigenetic gene regulation.

In summary, previous reports coupled with bioinformatic analyses of the completed genome demonstrate that *Toxoplasma* is capable of mediating most known histone modifications. The extensive array of chromatin remodelling machinery suggests that histone modifications and epigenetics are likely to be instrumental during progression of the parasite life cycle. These observations underscore the antiquity of epigenetics in the evolution of the eukaryotic cell and indicate that much of this machinery has evolved along parasite-specific trajectories.

18.3.3.4 SWI2/SNF2 ATPases

The second broad class of chromatin remodelling complexes in eukaryotes is comprised of the SWI2/SNF2 DNA-dependent ATPases that have roles in both transcriptional repression as well as activation (Mohrmann and Verrijzer, 2005). While the mechanism of action of these factors is incompletely understood, it is believed the energy of ATP hydrolysis is used to reposition or relocate the nucleosome (Johnson *et al.*, 2005). All members of the SWI2/SNF2 family contain a distinctive ATPase domain consisting of an N-terminal DEXDc portion and a C-terminal HELICc portion. Further classification based on sequence homology and additional structural features leads to four separate types: Snf2 members (contain a bromodomain), ISWI (contain a SANT domain), Mi-2 (contain a chromodomain), and Ino80/SRCAP/p400 (contain a lengthy insert between the DEXDc and HELICc domains). Previous reports have described SWI2/SNF2 factors in Apicomplexa, including an ISWI homologue in *Plasmodium* and an SRCAP homologue in *Toxoplasma* (TgSRCAP, AAL29689), *Cryptosporidium*, and *Plasmodium* (Ji and Arnot, 1997; Sullivan *et al.*, 2003).

Only TgSRCAP has been studied in any great detail. Like human SRCAP, TgSRCAP can function to enhance CREB-mediated transcription in the presence of the HAT CBP *in vitro* (Sullivan *et al.*, 2003). However, a protein with similarity to CBP or CREB has not been found, therefore, its role in *Toxoplasma* remains to be elucidated. To facilitate a better understanding of what TgSRCAP may do, a yeast two-hybrid screen was conducted using the lengthy 'spacer' region separating the DEXDc and HELICc domains as 'bait' (Nallani and Sullivan, 2005). The corresponding region in human SRCAP binds CBP (Johnston *et al.*, 1999). Most of the strongest interacting proteins isolated and confirmed by *in vitro* co-immunoprecipitation are novel parasite-specific proteins having no homologues in other eukaryotes. A few of these are from genes predicted to encode domains suggestive of a role in DNA processes – including transcription. Of particular interest is the first protein described in *Toxoplasma* to contain Kelch repeats and a BTB/POZ domain (Nallani and Sullivan, 2005). POZ domains from several zinc finger proteins have been shown to mediate transcriptional repression and to interact with components of histone deacetylase co-repressor complexes. The gene has subsequently been cloned (DQ174778) and termed TgLZTR since it is most similar to human Leucine-Zipper-like Transcriptional Regulator, a gene deleted in people with DiGeorge syndrome (Kurahashi *et al.*, 1995). Future studies should elucidate the role of TgLZTR and whether it associates with TgSRCAP *in vivo*.

In addition to TgSRCAP, the *Toxoplasma* genome contains at least 17 possible SWI2/SNF2 homologues. Two bear high sequence similarity to Snf2 subclass members, with one harbouring a bromodomain downstream of the ATPase domains. Another SWI2/SNF2 protein has a SANT domain, making it a likely orthologue of ISWI. A second SWI2/SNF2 has strong ISWI homology, but possesses an AT-hook domain instead of a SANT domain. There is also a predicted SWI2/SNF2 family member in *Toxoplasma* that contains a chromodomain, making it a probable Mi-2 orthologue. This protein was identified as part of the TgCRC (Saksouk *et al.*, 2005a). How this extensive family of SWI2/SNF2 ATPases contributes to gene expression in parasites remains an open area for future investigation.

18.3.4 Epigenetic Mechanisms as Drug Targets

It has now been established that chromatin remodelling plays critical roles in various aspects of parasite physiology, prompting discussions about targeting this machinery for novel drug design. Histone acetylation has been validated as a drug target as early as 1996, when it was discovered that a fungal metabolite (now called apicidin) with potent antiprotozoal activity inhibited apicomplexan KDACs (Darkin-Rattray *et al.*, 1996). Given the conservation of histone modifying enzymes in human, and the application of KDAC inhibitors now in cancer, legitimate questions regarding selective toxicity arise. Selective toxicity may be achievable simply because the parasites replicate rapidly, requiring therapeutic levels of drug that have minimal effect on host cells and systems. Second, parasite chromatin remodelling enzymes have significant divergent sequence outside of the catalytic domain that could be targeted selectively by small molecule inhibitors. There are also recent examples that show small divergence within catalytic domains could be sufficient to achieve selective toxicity. Apicomplexan HDAC3 is more susceptible to the KDAC inhibitor FR235222 due to the presence of a unique 2-amino acid insertion in the catalytic domain (Bougdour *et al.*, 2009). In contrast to KDAC inhibitors, surprisingly few specific KAT inhibitors are available. Anacardic acid and curcumin inhibit *Plasmodium* replication and can inhibit PfGCN5 *in vitro*, but these compounds are believed to have many off-target effects (Cui *et al.*, 2007, 2008).

As epigenetic mechanisms contribute to bradyzoite conversion, it may be possible to short-circuit this important pathogenic process with small molecules. Tachyzoites treated with the KDAC inhibitor FR235222 convert into bradyzoites and this compound may have activity against *ex vivo* *Toxoplasma* tissue cysts (Bougdour *et al.*, 2009; Maubon *et al.*, 2010). Genetic studies suggest that pharmacological interference of TgGCN5-A might be able to thwart bradyzoite cyst gene expression (Naguleswaran *et al.*, 2010). Interference with parasite histone methylation may also disrupt control of differentiation, and inhibitors specific for *Plasmodium* methyltransferases have recently been reported to alter expression of variant genes (Malmquist *et al.*, 2012). While the reasons remain unclear, pre-treating tachyzoites with a CARM1 (PRMT4) inhibitor leads to a higher frequency of bradyzoite development upon infecting host cells *in vitro* (Saksouk *et al.*, 2005a).

An important consideration with respect to targeting histone modification enzymes is the recent set of studies demonstrating the abundance of non-histone substrates for these enzymes (Smith and Workman, 2009). Several *Toxoplasma* KATs, namely the MYST KATs, are found predominantly outside of the parasite nucleus. An acetylome has been published for *Toxoplasma* tachyzoites, revealing that lysine acetylation is widespread across proteins of diverse function and location within the parasite (Jeffers and Sullivan, 2012). The activity of KDAC inhibitors, therefore, may not be limited to dysregulation of gene expression, but could exert their antiparasitic effect through inhibition of cytosolic substrate acetylation (Jeffers and Sullivan, 2012). While less extensively studied than lysine acetylation, it appears that methylation of *T. gondii* proteins of diverse function also occurs (Heaslip *et al.*, 2011; Xiao *et al.*, 2010) and inhibitors of these enzymes may also have specific antiparasitic activities.

18.4 POST-TRANSCRIPTIONAL MECHANISMS IN *TOXOPLASMA*

18.4.1 Translational Control

Examples of post-transcriptional mechanisms that regulate the level of specific proteins among protozoa have been described (Rochette *et al.*, 2005; Larreta *et al.*, 2004; Chow and Wirth, 2003; Garcia-Salcedo *et al.*, 2002; Shapira *et al.*, 2001). Transcription in the kinetoplastidae such as *Leishmania* and *Trypanosoma* is polycistronic and post-transcriptional *trans*-splicing mechanisms are required to achieve mature mRNA (Campbell *et al.*, 2003; Shapira *et al.*, 2001). Thus, in *Leishmania* and *Trypanosoma* species, mRNA levels are dictated mostly by post-transcriptional processing and the stability of the mRNA itself (Purdy *et al.*, 2005a, b; Webb *et al.*, 2005b; Cevallos *et al.*, 2005; Webb *et al.*, 2005a; Haile *et al.*, 2003). RNA-binding proteins with demonstrated roles in the regulation of translation and/or RNA stability have been found in *Plasmodium*, including Puf2 (Miao *et al.*, 2010; Muller *et al.*, 2011; Gomes-Santos *et al.*, 2011), the DDX6-class RNA helicase, DOZI (development of zygote inhibited) (Mair *et al.*, 2006), and Alba proteins (Chene *et al.*, 2012; Goyal *et al.*, 2012). Other transcription-associated proteins with known roles in modulating mRNA decay and translation were also found in the genome (Coulson, 2004). The formation of stress-induced RNA granules has been reported for a number of parasite species, including *Toxoplasma* (Cassola, 2011; Lirussi and Matrajt, 2011). Such RNA granules are proposed to be holding areas for translationally regulated mRNAs. As such, indications of post-transcriptional control have been described for protozoal genes with defined roles in differentiation (Vervenne *et al.*, 1994; Dechering *et al.*, 1997; Sullivan *et al.*, 2004; Narasimhan *et al.*, 2008; Miao *et al.*, 2010), mitochondrial RNA processing (Rehkopf *et al.*, 2000), and surface antigens (Lanzer *et al.*, 1993; Levitt *et al.*, 1993; Spano *et al.*, 2002). In *Toxoplasma*, unbalanced ratios of mRNA and protein have been observed for SAG-related *Toxoplasma* surface proteins, designated SAG5A, SAG5B and SAG5C (Spano *et al.*, 2002), and for mRNAs encoding the proliferating cell nuclear antigens, TgPCNA1 and TgPCNA2 (Guerini *et al.*, 2000). TgPCNA1 mRNA was found to be 7-fold higher than that of TgPCNA2, yet TgPCNA1 and -2 on Western blots were expressed at nearly equally levels in all strains examined (Guerini *et al.*, 2000). In the context of stage-specific gene expression in *Toxoplasma*, mRNAs encoding bradyzoite-specific proteins G6-PI and MAG1 can also be detected during the tachyzoite stage, and mRNA encoding tachyzoite-specific proteins like LDH1 can also be detected in bradyzoites (Yang and Parmley, 1997; Dzierszinski *et al.*, 1999; Weiss and Kim, 2000).

These discrepancies in the levels of mRNA and protein indicate that post-transcriptional events occur in apicomplexan parasites, although the mechanisms of translational control are just beginning to be elucidated. Significant advancements have taken place in the past decade particularly in the area of translational control through the phosphorylation of eukaryotic initiation factor-2 alpha subunit, eIF2α. The eIF2 complex governs the rate-limiting step in the initiation of protein synthesis. In response to a wide variety of cellular stresses, eIF2α becomes phosphorylated, leading to a global cessation in translation except for a subset of mRNAs encoding transcription factors that reprogramme the expressed genome to enable cell survival mechanisms (Wek *et al.*, 2006). The eIF2α translational control pathway was first characterized in *Toxoplasma* and linked to stress-induced bradyzoite development; moreover, TgIF2α remains in its phosphorylated state during the latent cyst stage (Sullivan *et al.*, 2004; Narasimhan *et al.*, 2008). A specific inhibitor of TgIF2α dephosphorylation was also found to trigger bradyzoite differentiation, supporting the idea that translational control is a major contributor to the development and maintenance of microbial latency (Joyce *et al.*, 2011; Narasimhan *et al.*, 2008). Studies in *Plasmodium* as well as kinetoplastid parasites lend further support to this model (Zhang *et al.*, 2010; Chow *et al.*, 2011).

Translational control through the phosphorylation of parasite eIF2α has significant roles beyond the modulation of latency and appears to be required even for normal progression through lytic cycles. Allelic replacement of the endogenous TgIF2α gene with a version incapable of being phosphorylated on the regulatory serine residue (Ser-71) exhibits fitness defects *in vitro* and *in vivo* that have been linked to decreased viability following egress from its host cell (Joyce *et al.*, 2010). A similar allelic replacement produces nonviable parasites in *Plasmodium*, demonstrating that PfIF2α phosphorylation is essential during the erythrocytic cycle (Zhang *et al.*, 2012)

Four eIF2α kinases have been identified in the *Toxoplasma* genome, designated TgIF2K-A through -D. TgIF2K-A localizes to the parasite endoplasmic reticulum (ER) and interacts with GRP/BiP in a stress-dependent manner, making it a likely orthologue of PERK, an eIF2α kinase in higher eukaryotes that contributes to the unfolded protein response (UPR) that allows cells to adapt to ER stress (Narasimhan *et al.*, 2008). PERK homologues have also been found to be critical in *Plasmodium* and *Leishmania* (Chow *et al.*, 2011; Zhang *et al.*, 2012). TgIF2K-B appears to be a cytoplasmic eIF2α kinase that is specific to *Toxoplasma* and its function has yet to be resolved. TgIF2K-C and -D are two GCN2-like kinases, which in other species are well-characterized responders to nutritional stress. To date, it has been determined using knock-out strategies that TgIF2K-D is the primary kinase phosphorylating TgIF2α upon egress from the host cell, and its loss phenocopies the non-phosphorylatable TgIF2α mutant (Konrad *et al.*, 2011). A GCN2-like kinase in *Plasmodium* called PfeIK1 has been found to regulate the parasite's response to amino acid starvation (Fennell *et al.*, 2009).

A key area of current investigation is linking translational control to transcriptional control. In other eukaryotes, following eIF2α phosphorylation, a select group of mRNAs is preferentially translated due to the presence of upstream open reading frames in the 5′UTR (untranslated region) (Vattem and Wek, 2004). These messages tend to encode basic-leucine zipper transcription factors such as GCN4/ATF4, which activate genes that facilitate the cellular adaptive response. Such transcription factors are not present in Apicomplexa, which probably use a subset of AP2 factors to regulate transcription. Employing the ability to generate polyribosome profiles in *Toxoplasma* (Narasimhan *et al.*, 2008), it has been shown that several messages encoding AP2 factors are preferentially translated during stress in these parasites (Sullivan, unpublished). Future studies are required to determine if the mechanism of translational control is analogous to other species.

18.4.2 Noncoding and Small RNA

One of the most exciting discoveries over the past two decades has been the role of small RNAs and longer RNAs in regulation of gene expression. *T. gondii* small RNAs have been catalogued (Wang *et al.*, 2012; Braun *et al.*, 2010) and RNA-seq studies have also identified long ncRNA (lncRNA) (Hassan *et al.*, 2012; Kim, unpublished). Many parasitic species use small RNAs to regulate gene expression. Unlike *Plasmodium* species, *T. gondii* encodes the essential components of the RNA induced silencing complex (RISC), including a single Argonaute protein, a Dicer protein (with RNAseIII catalytic domains) and an RNA-dependent RNA polymerase, suggesting that the RNA silencing pathway is fully functional (Braun *et al.*, 2010; Al Riyahi *et al.*, 2006). The single *T. gondii* Argonaute protein is not predicted to have slicer activity and thus the RISC complex is proposed to regulate gene expression by translation repression rather than RNA degradation (Braun *et al.*, 2010). The mechanistic details of the RISC complex activity are not yet completely clear because methylation of TgAgo1 results in recruitment of a staphylococcal endonuclease (Musiyenko *et al.*, 2012) that alters the activity of the RNA slicing complex as well as changing the specificity of the slicing from mismatches to perfect match RNA targets.

Characterization of the small RNAome of *T. gondii* by RNA-seq led to the identification of 14 miRNA families that were unique without significant homology to known miRNA of plants and metazoan, but most were conserved in *Neospora* (Braun *et al.*, 2010; Reid *et al.*, 2012). A second study identified 17 conserved miRNA and 339 novel miRNA (Wang *et al.*, 2012). Many miRNAs are recruited to the Ago complex to affect RNA turnover or translation. Some of the miRNA families of *T. gondii* were associated with polysomes, consistent with a role in translational repression. In addition, several of the miRNA families were differentially expressed in the Type I, II and III lineages and differences were also seen in extracellular versus actively replicating parasites (Wang *et al.*, 2012; Braun *et al.*, 2010). The targets of these miRNA have been predicted, but not yet validated (Braun *et al.*, 2010). A second group of small RNAs matched the repetitive elements REP1–3, mitochondrial-like sequences dispersed throughout the *T. gondii* genome (Braun *et al.*, 2010). The dispersal of these elements has features suggestive of transposon dissemination and one proposed function of the RNAi pathway in *T. gondii* is to prevent expression of these elements (Braun *et al.*, 2010). A third class of repeat associated small RNAs were identified that are proposed to maintain heterochromatin state at satellite DNA within the nucleus (Braun *et al.*, 2010). At present the single Ago protein is proposed to mediate all the potential nuclear and cytoplasmic activities of the *T. gondii* RISC complex. Some of the miRNA families were associated with immunopurified TgAgo, as were proteins associated with RNA processing and the chromatin co-repressor complex (Braun *et al.*, 2010).

Despite reports of successful RNAi in *T. gondii* (Al-Anouti *et al.*, 2003; Al Riyahi *et al.*, 2006), widespread RNAi has not been documented and efforts by several groups to develop RNAi methodology for *T. gondii* gene knock-down have failed (Matrajt, 2010) and reports of successful RNAi has been limited a few select genes (Al-Anouti *et al.*, 2003, 2004; Ananvoranich *et al.*, 2006). These observations suggest that the RNAi pathway has a specialized biological function in *T. gondii*. Consistent with this hypothesis, in the virulent RH strain, conditional disruption of the single Argonaute gene AGO1 has a very modest phenotype (Musiyenko *et al.*, 2012). Future evaluation and functional validation of the catalogued miRNA of *T. gondii* should establish the function and importance of these RNA species.

There is also circumstantial evidence that long non-coding RNA (ncRNA) will have important roles in the biology of *T. gondii.* While the functions of long ncRNA are only now emerging, *Plasmodium falciparum* has several long ncRNA of interest, including a non-coding transcript associated with silenced *var* genes that is transcribed from a promoter located in the conserved *var* intron (Calderwood *et al.*, 2003). In addition, there are long ncRNA associated with telomeric regions of *Plasmodium* that have been hypothesized as important for chromosome stability (Calderwood *et al.*, 2003; Sierra-Miranda *et al.*, 2012; Broadbent *et al.*, 2011). Long ncRNA are associated with developmental regulation in the metazoa (Braun *et al.*, 2010). Several groups have discovered long ncRNA that are upregulated during the process of bradyzoite differentiation (Matrajt, 2010). Some of these long ncRNA are antisense to sense transcripts, raising the hypothesis that long ncRNA may play a role in translational regulation or mRNA stability during the stress response.

Finally, inspection of RNA-seq studies reveals many instances of antisense transcripts that could potentially regulate gene expression of specific genes that are important for the host pathogen interaction (see http://www.toxodb.org for numerous RNA-seq datasets from various strains and developmental conditions). The antisense transcripts are often present in either 5′ or 3′ UTR of genes. At present the significance of these anti-sense RNAs is not known, but these could potentially act as an important regulator of cell cycle progression or developmental transitions.

18.4.3 Other Post-Transcriptional Mechanisms

Another completely unexplored area is the role of RNA processing, RNA trafficking and RNA stability in *T. gondii* gene regulation. While cell cycle transcription is likely to explain the cell cycle periodicity of mRNA, mRNA degradation must also have a prominent role in regulation of steady state RNAs. As in other eukaryotes, *T. gondii* encodes numerous proteins with predicted RNA binding domains. One of these, RRM1, has been demonstrated to have an essential and conserved role in RNA splicing (Suvorova *et al.*, 2013). In other systems, alternative splicing and RNA stability are mechanisms by which gene expression is regulated post-transcriptionally, and these events can be influenced by the metabolic state of the cell. Both chromatin remodellers and TgAP2 appear to interact with RNA binding proteins or the splicing machinery (Kim, White, Sullivan, unpublished) implicating regulation of RNA splicing as another area of potential importance in *T. gondii.* Long ncRNA have also been implicated in the regulation of splicing and RNA trafficking (Guttman and Rinn, 2012).

18.5 CONCLUSIONS AND FUTURE DIRECTIONS

We now have sufficient knowledge of global mRNA expression in *Plasmodium* and *Toxoplasma* to conclude that transcriptional mechanisms play a major role in regulating the developmental programme of these parasites. The observations that co-regulated genes are dispersed across parasite chromosomes, along with the presence of much of the conventional eukaryotic transcriptional machinery in the Apicomplexa genomes including chromatin remodellers, is consistent with growing evidence that promoter structures in these parasites contain *cis*-elements that are regulated by *trans*-acting factors. The discovery and preliminary characterization of plant-like AP2 DNA-binding domains in Apicomplexa has been a major new advance towards completing our picture of transcriptional regulation in these parasites.

ApiAP2 DNA binding domains represent the largest family of putative transcription factors discovered in Apicomplexa parasites. It is early in the study of ApiAP2 factors, yet already there are examples indicating that these factors operate as activators or suppressors of development and stress-responsive gene expression. It is expected that in the next few years, an inventory of sorts will be generated matching specific ApiAP2 factors to the mRNAs they regulate. The new roles for ApiAP2 factors in chromosome structure emerging from studies in *Plasmodium* and *Toxoplasma* are an exciting new development. There is no evidence that in plants AP2s act to maintain or modify chromosome structure, although the majority of these plant AP2s have not been characterized. AP2 proteins are not found in mammals. Thus, AP2 related mechanisms may be a novel invention in the Apicomplexa and a target for therapeutic development.

Recent studies have also made it clear that epigenetic-based gene regulation provides an important contribution to *Toxoplasma* gene expression, with several links to stage-specific gene expression now well established for a variety of different types of histone modifying enzymes (Dixon *et al.*, 2010). In higher eukaryotes, chromatin remodelling machinery and DNA-binding transcription factors work in concert to modulate gene expression, and emerging studies suggest that *Toxoplasma* is no exception. A great deal of work remains, however, in characterizing the large array of chromatin remodelling enzymes in *Toxoplasma* and understanding how these machineries are recruited to target promoters to work coordinately with TgAP2s or other possible DNA-binding factors. Equally important is the characterization of chromatin 'reader' proteins that harbour motifs that can bind specific histone modifications. As initial results with pharmacological agents that interfere with chromatin modifying enzymes have shown promise, epigenetic-based gene regulation remains a high priority for future investigation.

It is also critical to characterize how cellular signals are interpreted by the parasite to result in a reprogramming of gene expression, particularly changes associated with stage conversion. These processes are likely to involve sensing metabolic fluxes and small metabolites, and integrating these changes to alter gene expression. Considerable preliminary data now suggest that mechanisms of translational control and other means of post-transcriptional gene regulation warrant more attention into how they interplay with more conventional components of transcriptional control.

Acknowledgements

Research in our laboratories is supported by the following grants: NIH AI77502 (to WJS), AI087625 (to KK), AI077662 and AI089885 (to MWW) and RC4 AI092801 (to KK, WJS, MWW). We thank Sheila Nardelli for preparation of Fig. 18.2.

References

Ajioka, J.W., Boothroyd, J.C., Brunk, B.P., Hehl, A., Hillier, L., Manger, I.D., Marra, M., Overton, G.C., Roos, D.S., Wan, K.L., Waterston, R., Sibley, L.D., 1998. Gene discovery by EST sequencing in Toxoplasma gondii reveals sequences restricted to the Apicomplexa. Genome Res. 8, 18–28.

Al Riyahi, A., Al-Anouti, F., Al-Rayes, M., Ananvoranich, S., 2006. Single argonaute protein from Toxoplasma gondii is involved in the double-stranded RNA induced gene silencing. Int. J. Parasitol. 36, 1003–1014.

Al-Anouti, F., Quach, T., Ananvoranich, S., 2003. Double-stranded RNA can mediate the suppression of uracil phosphoribosyltransferase expression in Toxoplasma gondii. Biochem. Biophys. Res. Commun. 302, 316–323.

AL-Anouti, F., Tomavo, S., Parmley, S., Ananvoranich, S., 2004. The expression of lactate dehydrogenase is important for the cell cycle of Toxoplasma gondii. J. Biol. Chem. 279, 52300–52311.

Allen, M.D., Yamasaki, K., Ohme-Takagi, M., Tateno, M., Suzuki, M., 1998. A novel mode of DNA recognition by a beta-sheet revealed by the solution structure of the GCC-box binding domain in complex with DNA. EMBO J. 17, 5484–5496.

Allis, C.D., Berger, S.L., Cote, J., Dent, S., Jenuwien, T., Kouzarides, T., Pillus, L., Reinberg, D., Shi, Y., Shiekhattar, R., Shilatifard, A., Workman, J., Zhang, Y., 2007. New nomenclature for chromatin-modifying enzymes. Cell 131, 633–636.

Altschul, S.F., Wootton, J.C., Zaslavsky, E., Yu, Y.K., 2010. The construction and use of log-odds substitution scores for multiple sequence alignment. PLoS Comput. Biol. 6, e1000852.

Ananvoranich, S., Al Rayes, M., Al Riyahi, A., Wang, X., 2006. RNA silencing of glycolysis pathway in Toxoplasma gondii. J. Eukaryot. Microbiol. 53 (Suppl. 1), S162–S163.

Andrews, C.D., Dubey, J.P., Tenter, A.M., Webert, D.W., 1997. Toxoplasma gondii recombinant antigens H4 and H11: use in ELISAs for detection of toxoplasmosis in swine. Vet. Parasitol. 70, 1–11.

Andriankaja, A., Boisson-Dernier, A., Frances, L., Sauviac, L., Jauneau, A., Barker, D.G., De Carvalho-Niebel, F., 2007. AP2-ERF transcription factors mediate Nod factor dependent Mt ENOD11 activation in root hairs via a novel cis-regulatory motif. The Plant Cell 19, 2866–2885.

Bahl, A., Davis, P.H., Behnke, M., Dzierszinski, F., Jagalur, M., Chen, F., Shanmugam, D., White, M.W., Kulp, D., Roos, D.S., 2010. A novel multifunctional oligonucleotide microarray for Toxoplasma gondii. BMC Genomics 11, 603.

Balaji, S., Babu, M.M., Iyer, L.M., Aravind, L., 2005. Discovery of the principal specific transcription factors of Apicomplexa and their implication for the evolution of the AP2-integrase DNA binding domains. Nucleic Acids Res. 33, 3994–4006.

Bannister, A.J., Zegerman, P., Partridge, J.F., Miska, E.A., Thomas, J.O., Allshire, R.C., Kouzarides, T., 2001. Selective recognition of methylated lysine 9 on histone H3 by the HP1 chromo domain. Nature 410, 120–124.

Behnke, M.S., Radke, J.B., Smith, A.T., Sullivan Jr., W.J., White, M.W., 2008. The transcription of bradyzoite genes in Toxoplasma gondii is controlled by autonomous promoter elements. Mol. Microbiol. 68, 1502–1518.

Behnke, M.S., Wootton, J.C., Lehmann, M.M., Radke, J.B., Lucas, O., Nawas, J., Sibley, L.D., White, M.W., 2010. Coordinated progression through two subtranscriptomes underlies the tachyzoite cycle of Toxoplasma gondii. PloS One 5, e12354.

Bhatti, M.M., Livingston, M., Mullapudi, N., Sullivan Jr., W.J., 2006. Pair of unusual GCN5 histone acetyltransferases and ADA2 homologues in the protozoan parasite Toxoplasma gondii. Eukaryot. Cell 5, 62–76.

Bhatti, M.M., Sullivan Jr., W.J., 2005. Histone acetylase GCN5 enters the nucleus via importin-alpha in protozoan parasite Toxoplasma gondii. J. Biol. Chem. 280, 5902–5908.

Blazek, E., Mittler, G., Meisterernst, M., 2005. The mediator of RNA polymerase II. Chromosoma 113, 399–408.

Bohne, W., Wirsing, A., Gross, U., 1997. Bradyzoite-specific gene expression in Toxoplasma gondii requires minimal genomic elements. Mol. Biochem. Parasitol. 85, 89–98.

Bougdour, A., Braun, L., Cannella, D., Hakimi, M.A., 2010a. Chromatin Modifications: Implications in the Regulation of Gene Expression in Toxoplasma gondii. Cell Microbiol. In press.

Bougdour, A., Braun, L., Cannella, D., Hakimi, M.A., 2010b. Chromatin modifications: implications in the regulation of gene expression in Toxoplasma gondii. Cell Microbiol. 12, 413–423.

Bougdour, A., Maubon, D., Baldacci, P., Ortet, P., Bastien, O., Bouillon, A., Barale, J.C., Pelloux, H., Menard, R., Hakimi, M.A., 2009. Drug inhibition of HDAC3 and epigenetic control of differentiation in Apicomplexa parasites. J. Exp. Med. 206, 953–966.

Bowie, W.R., King, A.S., Werker, D.H., Isaac-Renton, J.L., Bell, A., Eng, S.B., Marion, S.A., 1997. Outbreak of toxoplasmosis associated with municipal drinking water. The BC Toxoplasma Investigation Team. Lancet 350, 173–177.

Bozdech, Z., Ginsburg, H., 2005. Data mining of the transcriptome of Plasmodium falciparum: the pentose phosphate pathway and ancillary processes. Malar. J. 4, 17.

Braun, L., Cannella, D., Ortet, P., Barakat, M., Sautel, C.F., Kieffer, S., Garin, J., Bastien, O., Voinnet, O., Hakimi, M.A., 2010. A complex small RNA repertoire is generated by a plant/fungal-like machinery and effected by

a metazoan-like Argonaute in the single-cell human parasite Toxoplasma gondii. PLoS Pathog, 6, e1000920.
Braun, L., Cannella, D., Pinheiro, A.M., Kieffer, S., Belrhali, H., Garin, J., Hakimi, M.A., 2009. The small ubiquitin-like modifier (SUMO)-conjugating system of Toxoplasma gondii. Int. J. Parasitol. 39, 81–90.
Broadbent, K.M., Park, D., Wolf, A.R., Van Tyne, D., Sims, J.S., Ribacke, U., Volkman, S., Duraisingh, M., Wirth, D., Sabeti, P.C., Rinn, J.L., 2011. A global transcriptional analysis of Plasmodium falciparum malaria reveals a novel family of telomere-associated lncRNAs. Genome Biol. 12, R56.
Brooks, C.F., Francia, M.E., Gissot, M., Croken, M.M., Kim, K., Striepen, B., 2011. Toxoplasma gondii sequesters centromeres to a specific nuclear region throughout the cell cycle. Proc. Natl. Acad. Sci. U. S. A. 108, 3767–3772.
Buchholz, K.R., Fritz, H.M., Chen, X., Durbin-Johnson, B., Rocke, D.M., Ferguson, D.J., Conrad, P.A., Boothroyd, J.C., 2011. Identification of tissue cyst wall components by transcriptome analysis of in vivo and in vitro Toxoplasma gondii bradyzoites. Eukaryot. Cell 10, 1637–1647.
Calderwood, M.S., Gannoun-Zaki, L., Wellems, T.E., Deitsch, K.W., 2003. Plasmodium falciparum var genes are regulated by two regions with separate promoters, one upstream of the coding region and a second within the intron. J. Biol. Chem. 278, 34125–34132.
Campbell, D.A., Thomas, S., Sturm, N.R., 2003. Transcription in kinetoplastid protozoa: why be normal? Microbes Infect. 5, 1231–1240.
Campbell, T.L., De Silva, E.K., Olszewski, K.L., Elemento, O., Llinas, M., 2010. Identification and genome-wide prediction of DNA binding specificities for the ApiAP2 family of regulators from the malaria parasite. PLoS Pathog. 6, e1001165.
Cassola, A., 2011. RNA Granules Living a Post-transcriptional Life: the Trypanosomes' Case. Current Chemical Biology 5, 108–117.
Cevallos, A.M., Perez-Escobar, M., Espinosa, N., Herrera, J., Lopez-Villasenor, I., Hernandez, R., 2005. The stabilization of housekeeping transcripts in Trypanosoma cruzi epimastigotes evidences a global regulation of RNA decay during stationary phase. FEMS Microbiol. Lett. 246, 259–264.
Chene, A., Vembar, S.S., Riviere, L., Lopez-Rubio, J.J., Claes, A., Siegel, T.N., Sakamoto, H., Scheidig-Benatar, C., Hernandez-Rivas, R., Scherf, A., 2012. PfAlbas constitute a new eukaryotic DNA/RNA-binding protein family in malaria parasites. Nucleic Acids Res. 40, 3066–3077.
Chevalier, B.S., Stoddard, B.L., 2001. Homing endonucleases: structural and functional insight into the catalysts of intron/intein mobility. Nucleic Acids Res. 29, 3757–3774.
Choi, W.Y., Nam, H.W., Kwak, N.H., Huh, W., Kim, Y.R., Kang, M.W., Cho, S.Y., Dubey, J.P., 1997. Foodborne outbreaks of human toxoplasmosis. J. Infect. Dis. 175, 1280–1282.
Chow, C., Cloutier, S., Dumas, C., Chou, M.N., Papadopoulou, B., 2011. Promastigote to amastigote differentiation of Leishmania is markedly delayed in the absence of PERK eIF2alpha kinase-dependent eIF2alpha phosphorylation. Cellular Microbiology 13, 1059–1077.
Chow, C.S., Wirth, D.F., 2003. Linker scanning mutagenesis of the Plasmodium gallinaceum sexual stage specific gene pgs28 reveals a novel downstream cis-control element. Mol. Biochem. Parasitol. 129, 199–208.
Cleary, M.D., Singh, U., Blader, I.J., Brewer, J.L., Boothroyd, J.C., 2002. Toxoplasma gondii asexual development: identification of developmentally regulated genes and distinct patterns of gene expression. Eukaryot. Cell 1, 329–340.
Coulson, R.M., Hall, N., Ouzounis, C.A., 2004. Comparative genomics of transcriptional control in the human malaria parasite Plasmodium falciparum. Genome Res. 14, 1548–1554.
Croken, M.M., Nardelli, S.C., Kim, K., 2012. Chromatin modifications, epigenetics, and how protozoan parasites regulate their lives. Trends Parasitol. 28, 202–213.
Cui, L., Miao, J., Cui, L., 2007. Cytotoxic effect of curcumin on malaria parasite Plasmodium falciparum: inhibition of histone acetylation and generation of reactive oxygen species. Antimicrob. Agents Chemother. 51, 488–494.
Cui, L., Miao, J., Furuya, T., Fan, Q., Li, X., Rathod, P.K., Su, X.Z., Cui, L., 2008. Histone acetyltransferase inhibitor anacardic acid causes changes in global gene expression during in vitro Plasmodium falciparum development. Eukaryot. Cell 7, 1200–1210.
Dalmasso, M.C., Echeverria, P.C., Zappia, M.P., Hellman, U., Dubremetz, J.F., Angel, S.O., 2006. Toxoplasma gondii has two lineages of histones 2b (H2B) with different expression profiles. Mol. Biochem. Parasitol. 148, 103–107.
Dalmasso, M.C., Onyango, D.O., Naguleswaran, A., Sullivan Jr., W.J., Angel, S.O., 2009. Toxoplasma H2A variants reveal novel insights into nucleosome composition and functions for this histone family. J. Mol. Biol. 392, 33–47.
Dalmasso, M.C., Sullivan Jr., W.J., Angel, S.O., 2011. Canonical and variant histones of protozoan parasites. Frontiers in Bioscience: a journal and virtual library 17, 2086–2105.
Darkin-Rattray, S.J., Gurnett, A.M., Myers, R.W., Dulski, P.M., Crumley, T.M., Allocco, J.J., Cannova, C., Meinke, P.T., Colletti, S.L., Bednarek, M.A., Singh, S.B., Goetz, M.A., Dombrowski, A.W., Polishook, J.D., Schmatz, D.M., 1996. Apicidin: a novel antiprotozoal agent that inhibits parasite histone deacetylase. Proc. Natl. Acad. Sci. U. S. A. 93, 13143–13147.
De Silva, E.K., Gehrke, A.R., Olszewski, K., Leon, I., Chahal, J.S., Bulyk, M.L., Llinas, M., 2008. Specific DNA-binding by apicomplexan AP2 transcription factors. Proc. Natl. Acad. Sci. U. S. A. 105, 8393–8398.
Dechering, K.J., Thompson, J., Dodemont, H.J., Eling, W., Konings, R.N., 1997. Developmentally regulated expression of pfs16, a marker for sexual differentiation of the human malaria parasite Plasmodium falciparum. Mol. Biochem. Parasitol. 89, 235–244.
Dhalluin, C., Carlson, J.E., Zeng, L., He, C., Aggarwal, A.K., Zhou, M.M., 1999. Structure and ligand of a histone acetyltransferase bromodomain. Nature 399, 491–496.
Dietz, K.J., Vogel, M.O., Viehhauser, A., 2010. AP2/EREBP transcription factors are part of gene regulatory networks and integrate metabolic, hormonal and environmental signals in stress acclimation and retrograde signalling. Protoplasma 245, 3–14.
Dillon, S.C., Zhang, X., Trievel, R.C., Cheng, X., 2005. The SET-domain protein superfamily: protein lysine methyltransferases. Genome Biol. 6, 227.
Dixon, S.E., Bhatti, M.M., Uversky, V.N., Dunker, A.K., Sullivan Jr., W.J., 2011. Regions of intrinsic disorder help identify a novel nuclear localization signal in Toxoplasma gondii histone acetyltransferase TgGCN5-B. Mol. Biochem. Parasitol. 175, 192–195.
Dixon, S.E., Stilger, K.L., Elias, E.V., Naguleswaran, A., Sullivan Jr., W.J., 2010. A decade of epigenetic research in Toxoplasma gondii. Mol. Biochem. Parasitol. 173, 1–9.
Dubey, J.P., 1998. Comparative infectivity of Toxoplasma gondii bradyzoites in rats and mice. J. Parasitol. 84, 1279–1282.
Dubey, J.P., Frenkel, J.K., 1976. Feline toxoplasmosis from acutely infected mice and the development of Toxoplasma cysts. J. Protozool. 23, 537–546.
Dubey, J.P., Miller, N.L., Frenkel, J.K., 1970. Toxoplasma gondii life cycle in cats. J. Am. Vet. Med. Assoc. 157, 1767–1770.
Dubey, J.P., Beattie, C.P., 1988. Toxoplasmosis of Animals and Man. CRC Press, Boca Raton.
Dzierszinski, F., Popescu, O., Toursel, C., Slomianny, C., Yahiaoui, B., Tomavo, S., 1999. The protozoan parasite Toxoplasma gondii expresses two functional plant-like glycolytic enzymes. Implications for evolutionary origin of apicomplexans. J. Biol. Chem. 274, 24888–24895.
Ehrenhofer-Murray, A.E., 2004. Chromatin dynamics at DNA replication, transcription and repair. Eur. J. Biochem. 271, 2335–2349.
Escargueil, A.E., Soares, D.G., Salvador, M., Larsen, A.K., Henriques, J.A., 2008. What histone code for DNA repair? Mutat Res. 658, 259–270.
Featherstone, M., 2002. Coactivators in transcription initiation: here are your orders. Curr. Opin. Genet. Dev. 12, 149–155.
Fennell, C., Babbitt, S., Russo, I., Wilkes, J., Ranford-Cartwright, L., Goldberg, D.E., Doerig, C., 2009. PfeIK1, a eukaryotic initiation factor 2alpha kinase of the human malaria parasite Plasmodium falciparum, regulates stress-response to amino-acid starvation. Malar. J. 8, 99.
Flueck, C., Bartfai, R., Niederwieser, I., Witmer, K., Alako, B.T., Moes, S., Bozdech, Z., Jenoe, P., Stunnenberg, H.G., Voss, T.S., 2010. A major role for the Plasmodium falciparum ApiAP2 protein PfSIP2 in chromosome end biology. PLoS Pathog. 6, e1000784.
Fox, B.A., Li, W.B., Tanaka, M., Inselburg, J., Bzik, D.J., 1993. Molecular characterization of the largest subunit of Plasmodium falciparum RNA polymerase I. Mol. Biochem. Parasitol. 61, 37–48.

Freitas-Junior, L.H., Hernandez-Rivas, R., Ralph, S.A., Montiel-Condado, D., Ruvalcaba-Salazar, O.K., Rojas-Meza, A.P., Mancio-Silva, L., Leal-Silvestre, R.J., Gontijo, A.M., Shorte, S., Scherf, A., 2005. Telomeric heterochromatin propagation and histone acetylation control mutually exclusive expression of antigenic variation genes in malaria parasites. Cell 121, 25–36.

Fry, C.J., Peterson, C.L., 2002. Transcription. Unlocking the gates to gene expression. Science 295, 1847–1848.

Garcia-Salcedo, J.A., Nolan, D.P., Gijon, P., Gomez-Rodriguez, J., Pays, E., 2002. A protein kinase specifically associated with proliferative forms of Trypanosoma brucei is functionally related to a yeast kinase involved in the co-ordination of cell shape and division. Mol. Microbiol. 45, 307–319.

Gissot, M., Choi, S.W., Thompson, R.F., Greally, J.M., Kim, K., 2008. Toxoplasma gondii and Cryptosporidium parvum lack detectable DNA cytosine methylation. Eukaryot. Cell 7, 537–540.

Gissot, M., Kelly, K.A., Ajioka, J.W., Greally, J.M., Kim, K., 2007. Epigenomic Modifications Predict Active Promoters and Gene Structure in Toxoplasma gondii. PLoS Pathog. 3, e77.

Gissot, M., Walker, R., Delhaye, S., Huot, L., Hot, D., Tomavo, S., 2012. Toxoplasma gondii chromodomain protein 1 binds to heterochromatin and colocalises with centromeres and telomeres at the nuclear periphery. PLoS One 7, e32671.

Gomes-Santos, C.S., Braks, J., Prudencio, M., Carret, C., Gomes, A.R., Pain, A., Feltwell, T., Khan, S., Waters, A., Janse, C., Mair, G.R., Mota, M.M., 2011. Transition of Plasmodium sporozoites into liver stage-like forms is regulated by the RNA binding protein Pumilio. PLoS Pathog. 7, e1002046.

Goyal, M., Alam, A., Iqbal, M.S., Dey, S., Bindu, S., Pal, C., Banerjee, A., Chakrabarti, S., Bandyopadhyay, U., 2012. Identification and molecular characterization of an Alba-family protein from human malaria parasite Plasmodium falciparum. Nucleic Acids Res. 40, 1174–1190.

Guerini, M.N., Que, X., Reed, S.L., White, M.W., 2000. Two genes encoding unique proliferating-cell-nuclear-antigens are expressed in Toxoplasma gondii. Mol. Biochem. Parasitol. 109, 121–131.

Guttman, M., Rinn, J.L., 2012. Modular regulatory principles of large non-coding RNAs. Nature 482, 339–346.

Haile, S., Estevez, A.M., Clayton, C., 2003. A role for the exosome in the in vivo degradation of unstable mRNAs. RNA 9, 1491–1501.

Hassan, M.A., Melo, M.B., Haas, B., Jensen, K.D., Saeij, J.P., 2012. De novo reconstruction of the Toxoplasma gondii transcriptome improves on the current genome annotation and reveals alternatively spliced transcripts and putative long non-coding RNAs. BMC Genomics 13, 696.

Heaslip, A.T., Nishi, M., Stein, B., Hu, K., 2011. The motility of a human parasite, Toxoplasma gondii, is regulated by a novel lysine methyltransferase. PLoS Pathog. 7, e1002201.

Helm, S., Lehmann, C., Nagel, A., Stanway, R.R., Horstmann, S., Llinas, M., Heussler, V.T., 2010. Identification and characterization of a liver stage-specific promoter region of the malaria parasite Plasmodium. PloS One 5, e13653.

Hong, X., Zang, J., White, J., Wang, C., Pan, C.H., Zhao, R., Murphy, R.C., Dai, S., Henson, P., Kappler, J.W., Hagman, J., Zhang, G., 2010. Interaction of JMJD6 with single-stranded RNA. Proc. Natl. Acad. Sci. U. S. A. 107, 14568–14572.

Isaac-Renton, J., Bowie, W.R., King, A., Irwin, G.S., Ong, C.S., Fung, C.P., Shokeir, M.O., Dubey, J.P., 1998. Detection of Toxoplasma gondii oocysts in drinking water. Appl. Environ. Microbiol. 64, 2278–2280.

Jeffers, V., Sullivan Jr., W.J., 2012. Lysine Acetylation is Widespread on Proteins of Diverse Function and Localization in the Protozoan Parasite Toxoplasma gondii. Eukaryot. Cell 11, 735–742.

Jerome, M.E., Radke, J.R., Bohne, W., Roos, D.S., White, M.W., 1998. Toxoplasma gondii bradyzoites form spontaneously during sporozoite-initiated development. Infect. Immun. 66, 4838–4844.

Ji, D.D., Arnot, D.E., 1997. A Plasmodium falciparum homologue of the ATPase subunit of a multi-protein complex involved in chromatin remodelling for transcription. Mol. Biochem. Parasitol. 88, 151–162.

Jofuku, K.D., Den Boer, B.G., Van Montagu, M., Okamuro, J.K., 1994. Control of Arabidopsis flower and seed development by the homeotic gene APETALA2. The Plant Cell 6, 1211–1225.

Johnson, C.N., Adkins, N.L., Georgel, P., 2005. Chromatin remodelling complexes: ATP-dependent machines in action. Biochem. Cell Biol. 83, 405–417.

Johnston, H., Kneer, J., Chackalaparampil, I., Yaciuk, P., Chrivia, J., 1999. Identification of a novel SNF2/SWI2 protein family member, SRCAP, which interacts with CREB-binding protein. J. Biol. Chem. 274, 16370–16376.

Joshi, M.B., Lin, D.T., Chiang, P.H., Goldman, N.D., Fujioka, H., Aikawa, M., Syin, C., 1999. Molecular cloning and nuclear localization of a histone deacetylase homologue in Plasmodium falciparum. Mol. Biochem. Parasitol. 99, 11–19.

Joyce, B.R., Konrad, C., Wek, R.C., Sullivan Jr., W.J., 2011. Translation control is critical during acute and chronic stages of toxoplasmosis infection. Expert Rev. Anti. Infect. Ther. 9, 1–3.

Joyce, B.R., Queener, S.F., Wek, R.C., Sullivan Jr., W.J., 2010. Phosphorylation of eukaryotic initiation factor-2{alpha} promotes the extracellular survival of obligate intracellular parasite Toxoplasma gondii. Proc. Natl. Acad. Sci. U. S. A. 107, 17200–17205.

Karras, G.I., Kustatscher, G., Buhecha, H.R., Allen, M.D., Pugieux, C., Sait, F., Bycroft, M., Ladurner, A.G., 2005. The macro domain is an ADP-ribose binding module. EMBO J. 24, 1911–1920.

Kibe, M.K., Coppin, A., Dendouga, N., Oria, G., Meurice, E., Mortuaire, M., Madec, E., Tomavo, S., 2005. Transcriptional regulation of two stage-specifically expressed genes in the protozoan parasite Toxoplasma gondii. Nucleic Acids Res. 33, 1722–1736.

Konishi, E., Takahashi, J., 1987. Some epidemiological aspects of Toxoplasma infections in a population of farmers in Japan. Int. J. Epidemiol. 16, 277–281.

Konrad, C., Wek, R.C., Sullivan Jr., W.J., 2011. A GCN2-like eukaryotic initiation factor-2 kinase increases the viability of extracellular Toxoplasma gondii parasites. Eukaryotic. cell 10, 1403–1412.

Koufopanou, V., Goddard, M.R., Burt, A., 2002. Adaptation for horizontal transfer in a homing endonuclease. Mol. Biol. Evol. 19, 239–246.

Krizek, B.A., 2003. Aintegumenta utilizes a mode of DNA recognition distinct from that used by proteins containing a single AP2 domain. Nucleic Acids Res. 31, 1859–1868.

Kurahashi, H., Akagi, K., Inazawa, J., Ohta, T., Niikawa, N., Kayatani, F., Sano, T., Okada, S., Nishisho, I., 1995. Isolation and characterization of a novel gene deleted in DiGeorge syndrome. Hum. Mol. Genet. 4, 541–549.

Lacount, D.J., Vignali, M., Chettier, R., Phansalkar, A., Bell, R., Hesselberth, J.R., Schoenfeld, L.W., Ota, I., Sahasrabudhe, S., Kurschner, C., Fields, S., Hughes, R.E., 2005. A protein interaction network of the malaria parasite Plasmodium falciparum. Nature 438, 103–107.

Lanzer, M., Wertheimer, S.P., De Bruin, D., Ravetch, J.V., 1993. Plasmodium: control of gene expression in malaria parasites. Exp. Parasitol. 77, 121–128.

Larreta, R., Soto, M., Quijada, L., Folgueira, C., Abanades, D.R., Alonso, C., Requena, J.M., 2004. The expression of HSP83 genes in Leishmania infantum is affected by temperature and by stage-differentiation and is regulated at the levels of mRNA stability and translation. BMC Mol. Biol. 5, 3.

Le Roch, K.G., Johnson, J.R., Florens, L., Zhou, Y., Santrosyan, A., Grainger, M., Yan, S.F., Williamson, K.C., Holder, A.A., Carucci, D.J., Yates 3rd, J.R., Winzeler, E.A., 2004. Global analysis of transcript and protein levels across the Plasmodium falciparum life cycle. Genome Res. 14, 2308–2318.

Lescault, P.J., Thompson, A.B., Patil, V., Lirussi, D., Burton, A., Margarit, J., Bond, J., Matrajt, M., 2010. Genomic data reveal Toxoplasma gondii differentiation mutants are also impaired with respect to switching into a novel extracellular tachyzoite state. PLoS One 5, e14463.

Levitt, A., Dimayuga, F.O., Ruvolo, V.R., 1993. Analysis of malarial transcripts using cDNA-directed polymerase chain reaction. J. Parasitol. 79, 653–662.

Li, L., Brunk, B.P., Kissinger, J.C., Pape, D., Tang, K., Cole, R.H., Martin, J., Wylie, T., Dante, M., Fogarty, S.J., Howe, D.K., Liberator, P., Diaz, C., Anderson, J., White, M., Jerome, M.E., Johnson, E.A., Radke, J.A., Stoeckert Jr., C.J., Waterston, R.H., Clifton, S.W., Roos, D.S., Sibley, L.D., 2003. Gene discovery in the apicomplexa as revealed by EST sequencing and assembly of a comparative gene database. Genome Res. 13, 443–454.

Li, L., Crabtree, J., Fischer, S., Pinney, D., Stoeckert Jr., C.J., Sibley, L.D., Roos, D.S., 2004a. ApiEST-DB: analyzing clustered EST data of the apicomplexan parasites. Nucleic Acids Res., 32 Database issue, D326–8.

Li, W.B., Bzik, D.J., Gu, H.M., Tanaka, M., Fox, B.A., Inselburg, J., 1989. An enlarged largest subunit of Plasmodium falciparum RNA polymerase II defines conserved and variable RNA polymerase domains. Nucleic Acids Res. 17, 9621–9636.

Li, W.B., Bzik, D.J., Tanaka, M., Gu, H.M., Fox, B.A., Inselburg, J., 1991. Characterization of the gene encoding the largest subunit of Plasmodium falciparum RNA polymerase III. Mol. Biochem. Parasitol. 46, 229–239.

Li, Y.J., Fu, X.H., Liu, D.P., Liang, C.C., 2004b. Opening the chromatin for transcription. Int. J. Biochem. Cell Biol. 36, 1411–1423.

Lindner, S.E., De Silva, E.K., Keck, J.L., Llinas, M., 2010. Structural determinants of DNA binding by a P. falciparum ApiAP2 transcriptional regulator. Journal of Molecular Biology 395, 558–567.

Lirussi, D., Matrajt, M., 2011. RNA granules present only in extracellular toxoplasma gondii increase parasite viability. International Journal of Biological Sciences 7, 960–967.

Llinas, M., Derisi, J.L., 2004. Pernicious plans revealed: Plasmodium falciparum genome wide expression analysis. Curr. Opin. Microbiol. 7, 382–387.

Long, P.L., 1982. The Biology of the Coccida. University Park Press, Baltimore, Maryland.

Lopez-Rubio, J.J., Gontijo, A.M., Nunes, M.C., Issar, N., Hernandez Rivas, R., Scherf, A., 2007. 5′ flanking region of var genes nucleate histone modification patterns linked to phenotypic inheritance of virulence traits in malaria parasites. Mol. Microbiol. 66, 1296–1305.

Lopez-Rubio, J.J., Mancio-Silva, L., Scherf, A., 2009. Genome-wide analysis of heterochromatin associates clonally variant gene regulation with perinuclear repressive centers in malaria parasites. Cell Host Microbe 5, 179–190.

Lyons, R.E., McLeod, R., Roberts, C.W., 2002. Toxoplasma gondii tachyzoite-bradyzoite interconversion. Trends Parasitol. 18, 198–201.

Ma, Y.F., Zhang, Y., Kim, K., Weiss, L.M., 2004. Identification and characterisation of a regulatory region in the Toxoplasma gondii hsp70 genomic locus. Int. J. Parasitol. 34, 333–346.

Magnani, E., Sjolander, K., Hake, S., 2004. From endonucleases to transcription factors: evolution of the AP2 DNA binding domain in plants. The Plant Cell 16, 2265–2277.

Mair, G.R., Braks, J.A., Garver, L.S., Wiegant, J.C., Hall, N., Dirks, R.W., Khan, S.M., Dimopoulos, G., Janse, C.J., Waters, A.P., 2006. Regulation of sexual development of Plasmodium by translational repression. Science 313, 667–669.

Malmquist, N.A., Moss, T.A., Mecheri, S., Scherf, A., Fuchter, M.J., 2012. Small-molecule histone methyltransferase inhibitors display rapid antimalarial activity against all blood stage forms in Plasmodium falciparum. Proc. Natl. Acad. Sci. U. S. A. 109, 16708–16713.

Manger, I.D., Hehl, A., Parmley, S., Sibley, L.D., Marra, M., Hillier, L., Waterston, R., Boothroyd, J.C., 1998. Expressed sequence tag analysis of the bradyzoite stage of Toxoplasma gondii: identification of developmentally regulated genes. Infect. Immun. 66, 1632–1637.

Marcus, G.A., Silverman, N., Berger, S.L., Horiuchi, J., Guarente, L., 1994. Functional similarity and physical association between GCN5 and ADA2: putative transcriptional adaptors. EMBO J. 13, 4807–4815.

Mateus-Pinilla, N.E., Dubey, J.P., Choromanski, L., Weigel, R.M., 1999. A field trial of the effectiveness of a feline Toxoplasma gondii vaccine in reducing T. gondii exposure for swine. J. Parasitol. 85, 855–860.

Matrajt, M., 2010. Non-coding RNA in apicomplexan parasites. Mol. Biochem. Parasitol. 174, 1–7.

Matrajt, M., Donald, R.G., Singh, U., Roos, D.S., 2002. Identification and characterization of differentiation mutants in the protozoan parasite Toxoplasma gondii. Mol. Microbiol. 44, 735–747.

Matrajt, M., Platt, C.D., Sagar, A.D., Lindsay, A., Moulton, C., Roos, D.S., 2004. Transcript initiation, polyadenylation, and functional promoter mapping for the dihydrofolate reductase-thymidylate synthase gene of Toxoplasma gondii. Mol. Biochem. Parasitol. 137, 229–238.

Maubon, D., Bougdour, A., Wong, Y.S., Brenier-Pinchart, M.P., Curt, A., Hakimi, M.A., Pelloux, H., 2010. Activity of the histone deacetylase inhibitor FR235222 on Toxoplasma gondii: inhibition of stage conversion of the parasite cyst form and study of new derivative compounds. Antimicrob. Agents Chemother. 54, 4843–4850.

Meissner, M., Soldati, D., 2005. The transcription machinery and the molecular toolbox to control gene expression in Toxoplasma gondii and other protozoan parasites. Microbes Infect. 7, 1376–1384.

Mercier, C., Lefebvre-Van Hende, S., Garber, G.E., Lecordier, L., Capron, A., Cesbron-Delauw, M.F., 1996. Common cis-acting elements critical for the expression of several genes of Toxoplasma gondii. Mol. Microbiol. 21, 421–428.

Miao, J., Li, J., Fan, Q., Li, X., Cui, L., 2010. The Puf-family RNA-binding protein PfPuf2 regulates sexual development and sex differentiation in the malaria parasite Plasmodium falciparum. J. Cell Sci. 123, 1039–1049.

Minot, S., Melo, M.B., Li, F., Lu, D., Niedelman, W., Levine, S.S., Saeij, J.P., 2012. Admixture and recombination among Toxoplasma gondii lineages explain global genome diversity. Proc. Natl. Acad. Sci. U. S. A. 109, 13458–13463.

Mohrmann, L., Verrijzer, C.P., 2005. Composition and functional specificity of SWI2/SNF2 class chromatin remodelling complexes. Biochim. Biophys. Acta. 1681, 59–73.

Monroy, M.A., Schott, N.M., Cox, L., Chen, J.D., Ruh, M., Chrivia, J.C., 2003. SNF2-related CBP activator protein (SRCAP) functions as a coactivator of steroid receptor-mediated transcription through synergistic interactions with CARM-1 and GRIP-1. Mol. Endocrinol. 17, 2519–2528.

Mullapudi, N., Joseph, S.J., Kissinger, J.C., 2009. Identification and functional characterization of cis-regulatory elements in the apicomplexan parasite Toxoplasma gondii. Genome Biol. 10, R34.

Muller, K., Matuschewski, K., Silvie, O., 2011. The Puf-family RNA-binding protein Puf2 controls sporozoite conversion to liver stages in the malaria parasite. PLoS One 6, e19860.

Musiyenko, A., Majumdar, T., Andrews, J., Adams, B., Barik, S., 2012. PRMT1 methylates the single Argonaute of Toxoplasma gondii and is important for the recruitment of Tudor nuclease for target RNA cleavage by antisense guide RNA. Cell Microbiol.

Naar, A.M., Lemon, B.D., Tjian, R., 2001. Transcriptional coactivator complexes. Annu. Rev. Biochem. 70, 475–501.

Naguleswaran, A., Elias, E.V., Mcclintick, J., Edenberg, H.J., Sullivan, W.J., 2010. Toxoplasma gondii Lysine Acetyltransferase GCN5-A Functions in the Cellular Response to Alkaline Stress and Expression of Cyst Genes. PLoS Pathog. 6, e1001232.

Nakaar, V., Bermudes, D., Peck, K.R., Joiner, K.A., 1998. Upstream elements required for expression of nucleoside triphosphate hydrolase genes of Toxoplasma gondii. Mol. Biochem. Parasitol. 92, 229–239.

Nallani, K.C., Sullivan Jr., W.J., 2005. Identification of proteins interacting with Toxoplasma SRCAP by yeast two-hybrid screening. Parasitol. Res. 95, 236–242.

Narasimhan, J., Joyce, B.R., Naguleswaran, A., Smith, A.T., Livingston, M.R., Dixon, S.E., Coppens, I., Wek, R.C., Sullivan Jr., W.J., 2008. Translation Regulation by Eukaryotic Initiation Factor-2 Kinases in the Development of Latent Cysts in Toxoplasma gondii. J. Biol. Chem. 283, 16591–16601.

Nardelli, S.C., Che, F.-Y., Silmon De Monerri, N.C., Xiao, H., Nieves, E., Madrid-Aliste, C., Angel, S., Sullivan Jr., W.J., Angeletti, R.H., Kim, K., Weiss, L.M., 2013. The Histone Code of Toxoplasma gondii Comprises Conserved and Unique Post-translational Modifications. submitted.

Ohme-Takagi, M., Shinshi, H., 1995. Ethylene-inducible DNA binding proteins that interact with an ethylene-responsive element. The Plant Cell 7, 173–182.

Olguin-Lamas, A., Madec, E., Hovasse, A., Werkmeister, E., Callebaut, I., Slomianny, C., Delhaye, S., Mouveaux, T., Schaeffer-Reiss, C., Van Dorsselaer, A., Tomavo, S., 2011. A novel Toxoplasma gondii nuclear factor TgNF3 is a dynamic chromatin-associated component, modulator of nucleolar architecture and parasite virulence. PLoS Pathog. 7, e1001328.

Peterson, C.L., Laniel, M.A., 2004. Histones and histone modifications. Curr. Biol. 14, R546–R551.

Pino, P., Sebastian, S., Kim, E.A., Bush, E., Brochet, M., Volkmann, K., Kozlowski, E., Llinas, M., Billker, O., Soldati-Favre, D., 2012. A tetracycline-repressible transactivator system to study essential genes in malaria parasites. Cell Host Microbe 12, 824–834.

Purdy, J.E., Donelson, J.E., Wilson, M.E., 2005a. Leishmania chagasi: the alpha-tubulin intercoding region results in constant levels of mRNA abundance despite protein synthesis inhibition and growth state. Exp. Parasitol. 110, 102–107.

Purdy, J.E., Donelson, J.E., Wilson, M.E., 2005b. Regulation of genes encoding the major surface protease of Leishmania chagasi via mRNA stability. Mol. Biochem. Parasitol. 142, 88–97.

Radke, J.B., Lucas, O., De Silva, E.K., Ma, Y., Sullivan Jr., W.J., Weiss, L.M., Llinas, M., White, M.W., 2013. ApiAP2 transcription factor restricts development of the Toxoplasma tissue cyst. PNAS. 110, 6871–6876.

Radke, J.R., Behnke, M.S., Mackey, A.J., Radke, J.B., Roos, D.S., White, M.W., 2005. The transcriptome of Toxoplasma gondii. BMC Biol. 3, 26.

Radke, J.R., Guerini, M.N., Jerome, M., White, M.W., 2003. A change in the premitotic period of the cell cycle is associated with bradyzoite differentiation in Toxoplasma gondii. Molecular and Biochemical Parasitology 131, 119–127.

Ranish, J.A., Hahn, S., 1996. Transcription: basal factors and activation. Curr. Opin. Genet. Dev. 6, 151–158.

Raveh, D., Hughes, S.H., Shafer, B.K., Strathern, J.N., 1989. Analysis of the HO-cleaved MAT DNA intermediate generated during the mating type switch in the yeast Saccharomyces cerevisiae. Mol. Gen. Genet. 220, 33–42.

Rehkopf, D.H., Gillespie, D.E., Harrell, M.I., Feagin, J.E., 2000. Transcriptional mapping and RNA processing of the Plasmodium falciparum mitochondrial mRNAs. Mol. Biochem. Parasitol. 105, 91–103.

Reid, A.J., Vermont, S.J., Cotton, J.A., Harris, D., Hill-Cawthorne, G.A., Konen-Waisman, S., Latham, S.M., Mourier, T., Norton, R., Quail, M.A., Sanders, M., Shanmugam, D., Sohal, A., Wasmuth, J.D., Brunk, B., Grigg, M.E., Howard, J.C., Parkinson, J., Roos, D.S., Trees, A.J., Berriman, M., Pain, A., Wastling, J.M., 2012. Comparative genomics of the apicomplexan parasites Toxoplasma gondii and Neospora caninum: Coccidia differing in host range and transmission strategy. PLoS Pathog. 8, e1002567.

Riechmann, J.L., Meyerowitz, E.M., 1998. The AP2/EREBP family of plant transcription factors. Biol. Chem. 379, 633–646.

Rios-Doria, J., Velkova, A., Dapic, V., Galan-Caridad, J.M., Dapic, V., Carvalho, M.A., Melendez, J., Monteiro, A.N., 2009. Ectopic expression of histone H2AX mutants reveals a role for its post-translational modifications. Cancer Biol. Ther. 8, 422–434.

Rochette, A., Mcnicoll, F., Girard, J., Breton, M., Leblanc, E., Bergeron, M.G., Papadopoulou, B., 2005. Characterization and developmental gene regulation of a large gene family encoding amastin surface proteins in Leishmania spp. Mol. Biochem. Parasitol. 140, 205–220.

Roos, D.S., Sullivan, W.J., Striepen, B., Bohne, W., Donald, R.G., 1997. Tagging genes and trapping promoters in Toxoplasma gondii by insertional mutagenesis. Methods 13, 112–122.

Rosowski, E.E., Saeij, J.P., 2012. Toxoplasma gondii clonal strains all inhibit STAT1 transcriptional activity but polymorphic effectors differentially modulate IFNgamma induced gene expression and STAT1 phosphorylation. PLoS One 7, e51448.

Ruvalcaba-Salazar, O.K., Del Carmen Ramirez-Estudillo, M., Montiel-Condado, D., Recillas-Targa, F., Vargas, M., Hernandez-Rivas, R., 2005. Recombinant and native Plasmodium falciparum TATA-binding-protein binds to a specific TATA box element in promoter regions. Mol. Biochem. Parasitol. 140, 183–196.

Saksouk, N., Bhatti, M.M., Kieffer, S., Smith, A.T., Musset, K., Garin, J., Sullivan Jr., W.J., Cesbron-Delauw, M.F., Hakimi, M.A., 2005a. Histone-modifying complexes regulate gene expression pertinent to the differentiation of the protozoan parasite Toxoplasma gondii. Mol. Cell Biol. 25, 10301–10314.

Saksouk, N., Bhatti, M.M., Kieffer, S., Smith, A.T., Musset, K., Sullivan Jr., W.J., Cesbron-Delauw, M., Hakimi, M.A., 2005b. Histone modifying complexes regulate gene expression pertinent to the differentiation of protozoan parasite Toxoplasma gondii. Mol. Cell. Biol. In press.

Sakuma, Y., Liu, Q., Dubouzet, J.G., Abe, H., Shinozaki, K., Yamaguchi-Shinozaki, K., 2002. DNA-binding specificity of the ERF/AP2 domain of Arabidopsis DREBs, transcription factors involved in dehydration- and cold-inducible gene expression. Biochem. Biophys. Res. Commun. 290, 998–1009.

Sautel, C.F., Cannella, D., Bastien, O., Kieffer, S., Aldebert, D., Garin, J., Tardieux, I., Belrhali, H., Hakimi, M.A., 2007. SET8-mediated methylations of histone H4 lysine 20 mark silent heterochromatic domains in apicomplexan genomes. Mol. Cell Biol. 27, 5711–5724.

Sautel, C.F., Ortet, P., Saksouk, N., Kieffer, S., Garin, J., Bastien, O., Hakimi, M.A., 2009. The histone methylase KMTox interacts with the redox-sensor peroxiredoxin-1 and targets genes involved in Toxoplasma gondii antioxidant defences. Mol. Microbiol. 71, 212–226.

Schmid, M., Uhlenhaut, N.H., Godard, F., Demar, M., Bressan, R., Weigel, D., Lohmann, J.U., 2003. Dissection of floral induction pathways using global expression analysis. Development 130, 6001–6012.

Shapira, M., Zilka, A., Garlapati, S., Dahan, E., Dahan, I., Yavesky, V., 2001. Post transcriptional control of gene expression in Leishmania. Med. Microbiol. Immunol. (Berl.) 190, 23–26.

Shiio, Y., Eisenman, R.N., 2003. Histone sumoylation is associated with transcriptional repression. Proc. Natl. Acad. Sci. U. S. A. 100, 13225–13230.

Sierra-Miranda, M., Delgadillo, D.M., Mancio-Silva, L., Vargas, M., Villegas-Sepulveda, N., Martinez-Calvillo, S., Scherf, A., Hernandez-Rivas, R., 2012. Two long non-coding RNAs generated from subtelomeric regions accumulate in a novel perinuclear compartment in Plasmodium falciparum. Mol. Biochem. Parasitol. 185, 36–47.

Singh, U., Brewer, J.L., Boothroyd, J.C., 2002. Genetic analysis of tachyzoite to bradyzoite differentiation mutants in Toxoplasma gondii reveals a hierarchy of gene induction. Mol. Microbiol. 44, 721–733.

Smith, A.T., Tucker-Samaras, S.D., Fairlamb, A.H., Sullivan Jr., W.J., 2005. MYST Family Histone Acetyltransferases in the Protozoan Parasite Toxoplasma gondii. Eukaryot. Cell 4, 2057–2065.

Smith, E.R., Eisen, A., Gu, W., Sattah, M., Pannuti, A., Zhou, J., Cook, R.G., Lucchesi, J.C., Allis, C.D., 1998. ESA1 is a histone acetyltransferase that is essential for growth in yeast. Proc. Natl. Acad. Sci. U. S. A. 95, 3561–3565.

Smith, K.T., Workman, J.L., 2009. Introducing the acetylome. Nature Biotechnology 27, 917–919.

Soldati, D., Boothroyd, J.C., 1995. A selector of transcription initiation in the protozoan parasite Toxoplasma gondii. Mol. Cell Biol. 15, 87–93.

Song, C.P., Agarwal, M., Ohta, M., Guo, Y., Halfter, U., Wang, P., Zhu, J.K., 2005. Role of an Arabidopsis AP2/EREBP-type transcriptional repressor in abscisic acid and drought stress responses. The Plant Cell 17, 2384–2396.

Spano, F., Ricci, I., Di Cristina, M., Possenti, A., Tinti, M., Dendouga, N., Tomavo, S., Crisanti, A., 2002. The SAG5 locus of Toxoplasma gondii encodes three novel proteins belonging to the SAG1 family of surface antigens. Int. J. Parasitol. 32, 121–131.

Spector, D.L., 2003. The dynamics of chromosome organization and gene regulation. Annu. Rev. Biochem. 72, 573–608.

Sterner, D.E., Berger, S.L., 2000. Acetylation of histones and transcription-related factors. Microbiol. Mol. Biol. Rev. 64, 435–459.

Strahl, B.D., Allis, C.D., 2000. The language of covalent histone modifications. Nature 403, 41–45.

Stray-Pedersen, B., Lorentzen-Styr, A.M., 1980. Epidemiological aspects of Toxoplasma infections among women in Norway. Acta Obstet Gynecol. Scand. 59, 323–326.

Su, C., Evans, D., Cole, R.H., Kissinger, J.C., Ajioka, J.W., Sibley, L.D., 2003. Recent expansion of Toxoplasma through enhanced oral transmission. Science 299, 414–416.

Sullivan Jr., W.J., Monroy, M.A., Bohne, W., Nallani, K.C., Chrivia, J., Yaciuk, P., Smith 2nd, C.K., Queener, S.F., 2003. Molecular cloning and characterization of an SRCAP chromatin remodelling homologue in Toxoplasma gondii. Parasitol. Res. 90, 1–8.

Sullivan Jr., W.J., Naguleswaran, A., Angel, S.O., 2006. Histones and histone modifications in protozoan parasites. Cell Microbiol. 8, 1850–1861.

Sullivan Jr., W.J., Narasimhan, J., Bhatti, M.M., Wek, R.C., 2004. Parasite-specific eIF2 (eukaryotic initiation factor-2) kinase required for stress-induced translation control. Biochem. J. 380, 523–531.

Suvorova, E., Croken, M., Kratzer, S., Ting, L., Conde De Felipe, M., Balu, B., Markville, L., Weiss, L., Kim, K., White, M., 2013. Discovery of a splicing regulator required for cell cycle progression. PLoS Genetics 9, e1003305.

Thiagalingam, S., Cheng, K.H., Lee, H.J., Mineva, N., Thiagalingam, A., Ponte, J.F., 2003. Histone deacetylases: unique players in shaping the epigenetic histone code. Ann. N. Y. Acad. Sci. 983, 84–100.

Trelle, M.B., Salcedo-Amaya, A.M., Cohen, A.M., Stunnenberg, H.G., Jensen, O.N., 2009. Global histone analysis by mass spectrometry reveals a high content of acetylated lysine residues in the malaria parasite Plasmodium falciparum. J. Proteome. Res. 8, 3439–3450.

Triglia, T., Healer, J., Caruana, S.R., Hodder, A.N., Anders, R.F., Crabb, B.S., Cowman, A.F., 2000. Apical membrane antigen 1 plays a central role in erythrocyte invasion by Plasmodium species. Mol. Microbiol. 38, 706–718.

Utley, R.T., Cote, J., 2003. The MYST family of histone acetyltransferases. Curr. Top Microbiol. Immunol. 274, 203–236.

Van Poppel, N.F., Welagen, J., Vermeulen, A.N., Schaap, D., 2006. The complete set of Toxoplasma gondii ribosomal protein genes contains two conserved promoter elements. Parasitology 133, 19–31.

Vattem, K.M., Wek, R.C., 2004. Reinitiation involving upstream ORFs regulates ATF4 mRNA translation in mammalian cells. Proc. Natl. Acad. Sci. U. S. A. 101, 11269–11274.

Vervenne, R.A., Dirks, R.W., Ramesar, J., Waters, A.P., Janse, C.J., 1994. Differential expression in blood stages of the gene coding for the 21-kilodalton surface protein of ookinetes of Plasmodium berghei as detected by RNA in situ hybridisation. Mol. Biochem. Parasitol. 68, 259–266.

Volz, J.C., Bartfai, R., Petter, M., Langer, C., Josling, G.A., Tsuboi, T., Schwach, F., Baum, J., Rayner, J.C., Stunnenberg, H.G., Duffy, M.F., Cowman, A.F., 2012. PfSET10, a Plasmodium falciparum methyltransferase, maintains the active var gene in a poised state during parasite division. Cell Host Microbe 11, 7–18.

Vonlaufen, N., Naguleswaran, A., Coppens, I., Sullivan Jr., W.J., 2010. MYST family lysine acetyltransferase facilitates ataxia telangiectasia mutated (ATM) kinase-mediated DNA damage response in Toxoplasma gondii. J. Biol. Chem. 285, 11154–11161.

Voss, T.S., Kaestli, M., Vogel, D., Bopp, S., Beck, H.P., 2003. Identification of nuclear proteins that interact differentially with Plasmodium falciparum var gene promoters. Mol. Microbiol. 48, 1593–1607.

Walker, R., Gissot, M., Croken, M.M., Huot, L., Hot, D., Kim, K., Tomavo, S., 2013. The Toxoplasma nuclear factor TgAP2XI-4 controls bradyzoite gene expression and cyst formation. Mol. Microbiol. 87, 641–655.

Wang, J., Liu, X., Jia, B., Lu, H., Peng, S., Piao, X., Hou, N., Cai, P., Yin, J., Jiang, N., Chen, Q., 2012. A comparative study of small RNAs in Toxoplasma gondii of distinct genotypes. Parasit Vectors 5, 186.

Webb, H., Burns, R., Ellis, L., Kimblin, N., Carrington, M., 2005a. Developmentally regulated instability of the GPI-PLC mRNA is dependent on a short-lived protein factor. Nucleic Acids Res. 33, 1503–1512.

Webb, H., Burns, R., Kimblin, N., Ellis, L., Carrington, M., 2005b. A novel strategy to identify the location of necessary and sufficient cis-acting regulatory mRNA elements in trypanosomes. RNA 11, 1108–1116.

Wei, Y., Mizzen, C.A., Cook, R.G., Gorovsky, M.A., Allis, C.D., 1998. Phosphorylation of histone H3 at serine 10 is correlated with chromosome condensation during mitosis and meiosis in Tetrahymena. Proc. Natl. Acad. Sci. U. S. A. 95, 7480–7484.

Weiss, L.M., Kim, K., 2000. The development and biology of bradyzoites of Toxoplasma gondii. Front Biosci. 5, D391–D405.

Wek, R.C., Jiang, H.Y., Anthony, T.G., 2006. Coping with stress: eIF2 kinases and translational control. Biochem. Soc. Trans. 34, 7–11.

Wuitschick, J.D., Lindstrom, P.R., Meyer, A.E., Karrer, K.M., 2004. Homing endonucleases encoded by germ line-limited genes in Tetrahymena thermophila have APETELA2 DNA binding domains. Eukaryot. Cell 3, 685–694.

Xiao, H., El Bissati, K., Verdier-Pinard, P., Burd, B., Zhang, H., Kim, K., Fiser, A., Angeletti, R.H., Weiss, L.M., 2010. Post-translational modifications to Toxoplasma gondii alpha- and beta-tubulins include novel C-terminal methylation. J. Proteome. Res. 9, 359–372.

Xu, W., Cho, H., Kadam, S., Banayo, E.M., Anderson, S., Yates J.R., 3rd, Emerson, B.M., Evans, R.M., 2004. A methylation-mediator complex in hormone signaling. Genes Dev. 18, 144–156.

Xu, W., Edmondson, D.G., Evrard, Y.A., Wakamiya, M., Behringer, R.R., Roth, S.Y., 2000. Loss of Gcn5l2 leads to increased apoptosis and mesodermal defects during mouse development. Nat. Genet. 26, 229–232.

Xu, W., Edmondson, D.G., Roth, S.Y., 1998. Mammalian GCN5 and P/CAF acetyltransferases have homologous amino-terminal domains important for recognition of nucleosomal substrates. Mol. Cell Biol. 18, 5659–5669.

Xue, B., Dunker, A.K., Uversky, V.N., 2012. Orderly order in protein intrinsic disorder distribution: disorder in 3500 proteomes from viruses and the three domains of life. J. Biomol. Struct. Dyn. 30, 137–149.

Yamagishi, J., Wakaguri, H., Ueno, A., Goo, Y.K., Tolba, M., Igarashi, M., Nishikawa, Y., Sugimoto, C., Sugano, S., Suzuki, Y., Watanabe, J., Xuan, X., 2010. High-resolution characterization of Toxoplasma gondii transcriptome with a massive parallel sequencing method. DNA Res. 17, 233–243.

Yamauchi, T., Yamauchi, J., Kuwata, T., Tamura, T., Yamashita, T., Bae, N., Westphal, H., Ozato, K., Nakatani, Y., 2000. Distinct but overlapping roles of histone acetylase PCAF and of the closely related PCAF-B/GCN5 in mouse embryogenesis. Proc. Natl. Acad. Sci. U. S. A. 97, 11303–11306.

Yang, S., Parmley, S.F., 1997. Toxoplasma gondii expresses two distinct lactate dehydrogenase homologous genes during its life cycle in intermediate hosts. Gene 184, 1–12.

Yuda, M., Iwanaga, S., Shigenobu, S., Kato, T., Kaneko, I., 2009a. Transcription Factor AP2–Sp and its Target Genes in Malarial Sporozoites. Mol. Microbiol.

Yuda, M., Iwanaga, S., Shigenobu, S., Mair, G.R., Janse, C.J., Waters, A.P., Kato, T., Kaneko, I., 2009b. Identification of a transcription factor in the mosquito-invasive stage of malaria parasites. Mol. Microbiol. 71, 1402–1414.

Zhang, M., Fennell, C., Ranford-Cartwright, L., Sakthivel, R., Gueirard, P., Meister, S., Caspi, A., Doerig, C., Nussenzweig, R.S., Tuteja, R., Sullivan Jr., W.J., Roos, D.S., Fontoura, B.M., Menard, R., Winzeler, E.A., Nussenzweig, V., 2010. The Plasmodium eukaryotic initiation factor-2alpha kinase IK2 controls the latency of sporozoites in the mosquito salivary glands. J. Exp. Med. 207, 1465–1474.

Zhang, M., Mishra, S., Sakthivel, R., Rojas, M., Ranjan, R., Sullivan Jr., W.J., Fontoura, B.M., Menard, R., Dever, T.E., Nussenzweig, V., 2012. PK4, a eukaryotic initiation factor 2alpha(eIF2alpha) kinase, is essential for the development of the erythrocytic cycle of Plasmodium. Proc. Natl. Acad. Sci. U. S. A. 109, 3956–3961.

Zhang, Q., Huang, Y., Zhang, Y., Fang, X., Claes, A., Duchateau, M., Namane, A., Lopez-Rubio, J.J., Pan, W., Scherf, A., 2011. A critical role of perinuclear filamentous actin in spatial repositioning and mutually exclusive expression of virulence genes in malaria parasites. Cell Host Microbe 10, 451–463.

Zhang, Y., Reinberg, D., 2001. Transcription regulation by histone methylation: interplay between different covalent modifications of the core histone tails. Genes Dev. 15, 2343–2360.

Zheng, L., Roeder, R.G., Luo, Y., 2003. S phase activation of the histone H2B promoter by OCA-S, a coactivator complex that contains GAPDH as a key component. Cell 114, 255–266.

CHAPTER

19

ToxoDB: An Integrated Functional Genomic Resource for *Toxoplasma* and Other Sarcocystidae

Omar S. Harb, David S. Roos
On behalf of the EuPathDB group
Department of Biology, University of Pennsylvania, Philadelphia, Pennsylvania, USA

OUTLINE

19.1 INTRODUCTION

Over the past ten years, ToxoDB (http://www.ToxoDB.org) has expanded in functionality and data content. In addition to including genomes and functional data from *Toxoplasma gondii* strains, ToxoDB now contains genomes and data from additional Sarcocystidae such as *Neospora caninum* and *Eimeria tenella*. From its inception, one of the major goals of ToxoDB has been to provide the global *Toxoplasma* research community with a cutting-edge and useful

Toxoplasma gondii, Second Edition
http://dx.doi.org/10.1016/B978-0-12-396481-6.00019-2

resource that supports daily laboratory research activities. Paramount to its success is accessibility to online data interrogation tools without the need for sophisticated computational skills beyond basic knowledge of Internet browser utility. This chapter summarizes the current status of ToxoDB and provides instructions on basic use and functionality of the database.

ToxoDB belongs to the Eukaryotic Pathogen Database Resources (EuPathDB: http://eupathdb.org) (Aurrecoechea *et al.*, 2013), a National Institutes of Allergy and Infectious Diseases-funded Bioinformatic Resource Center (BRC: http://www.niaid.nih.gov/labsandresources/resources/dmid/brc/Pages/default.aspx) (Greene *et al.*, 2007). The primary mission of ToxoDB is to integrate genomic sequence with genomic-scale experimental data from a variety of sources and provide the global research community with intuitive tools to interrogate this data and facilitate hypothesis driven research. Designed to make bioinformatics easily accessible to the bench scientist, ToxoDB offers over 70 ways to search the underlying data in a user-friendly, graphic web-interface and is updated bimonthly with new data and/or new searches. Each annotated genomic feature (e.g. gene, EST) is represented by an ID, which links to a record page that contains all of the database information for that feature. Users can study and compare the record pages of individual genomic features or they can perform their own *in silico* experiments using the search strategy system to assemble lists of records that share common biological characteristics (Fischer *et al.*, 2011a).

Data submitted to ToxoDB is prioritized and scheduled for release usually three to four releases in advance. Scheduling depends on a number of factors including when data were made available to ToxoDB, effort needed to integrate the data (note that EuPathDB staff concomitantly maintain nine other databases such as PlasmoDB , TriTrypDB, OrthoMCL, etc.) (Aslett *et al.*, 2009; Aurrecoechea *et al.*, 2009; Chen *et al.*, 2006) and advice from the ToxoDB advisory committee. Importantly, scheduling of a dataset for release is not synonymous with actual release. ToxoDB adheres to a strict standard operating procedure *vis-à-vis* data release whose main tenant precludes release of data without prior approval by the data provider (the data release policy is available at http://toxodb.org/EuPathDB_datasubm_SOP.pdf).

Once a dataset is integrated into ToxoDB, the data provider is given access to a password protected quality assurance (QA) site where they can explore their data in the context of all other data in the database. Data providers are asked to confirm proper representation and attribution of the data, and to provide final approval for release. If a dataset is not ready for release (for whatever reason) it is rescheduled for a subsequent version of the database with the QA cycle repeated prior to the next scheduled release date.

The EuPathDB group is organized into teams each of which carries out essential functions needed for the final release of a production database (i.e. the live ToxoDB release available to the community). The teams include database administrators, system administrators, software engineers, user interface developers, data loaders, data developers and community outreach specialists. The organization of the teams and methods of effective communication have evolved over the years to produce a robust mechanism for continuous data integration, software and website innovations and community support. Importantly, the EuPathDB group does not include any students or post-doctoral fellows ensuring that EuPathDB continues to focus on community needs rather than personal research projects (Roos, 2011).

19.2 GENOMES IN TOXODB

ToxoDB includes genome sequence and annotation from *Toxoplasma gondii* (GT1, VEG, ME49 and RH apicoplast), *Neospora caninum* Liverpool (Reid et al., 2012) and *Eimeria tenella* Houghton (Reid *et al.*, unpublished). In addition, the unannotated genome of Gregarina niphandrodes and several *T. gondii* strains (FOU, MAS, RH, RUB, TgCATBR9, TGCKUg2, VAND and P89) have been made available through the sequencing white paper effort conducted at the J. Craig Venter Institute (JCVI). The white paper efforts have also yielded a newly sequenced and annotated version of the *T. gondii* ME49 genome. Gene models have been greatly improved due to the availability of RNA-sequence data to help guide gene model predictions. Curation of genes has also been improved by a community effort to provide expert manual annotation of genes. Updated information regarding genomes in ToxoDB is available from the home page by clicking on the Data Summary link on the left side of the home page (Fig. 19.1).

19.3 DATA CONTENT

In addition to the genomes and annotation described above, ToxoDB contains a variety of functional data types generated by research groups across the globe. Increasingly, data provided to ToxoDB is done so prepublication to enable ample time for integration in the database and to ensure data release concomitantly with publication or prepublication. Detailed information about each data set available in ToxoDB and links to associated queries may be accessed at http://toxodb.org/toxo/getDataSource.do.

Several functional data sets are currently available in ToxoDB including:

- Proteomics data made up of peptide sequences from mass spectrometry experiments that are mapped to translated genes or open reading frames. Genes with peptide matches from the available proteomics experiments may be identified and peptide data are displayed on gene pages and in the genome browser. Experiments are available from multiple

FIGURE 19.1 **The ToxoDB home page.**
A) The banner section contains several useful links including login and registration options and quick ID and text search tools. The banner is present throughout ToxoDB webpages.
B) The left hand section includes expandable menus with news regarding releases and useful links to community related items and educational material, and a link to the data summary table for all EuPathDB sites.
C) The central section contains links to all searches and tools in three categories: searches that return genes, searches that return other data types (SNPs, Isolates, ESTs, SAGE tags, ORFs) and tools (sequence retrieval, BLAST, etc.).

strains, life cycle stages, subcellular fractions as well as phosphopeptide data.

- Expressed sequence tags (ESTs) include sequence data from cloned cDNA libraries retrieved from the GenBank Expressed Sequence Tag database (dbEST) and mapped to the genomes. Genes with EST evidence may be identified based on all or specific EST libraries. In addition, EST alignments are available as a data track in the genome browser. EST libraries are available for *T. gondii, E. tenella, N. caninum* and *G. niphandrodes.*
- Microarray (glass slides or high-density arrays) data in ToxoDB is re-analysed using a standard analysis method to ensure consistency in data representation. Microarray data for individual genes are displayed on gene pages under the expression section as graphs and tables. In addition, mapped probes for each microarray platform can be viewed in the genome browser. Depending on the original microarray experiment, ToxoDB offers searches based on fold induction, percentile expression or similarity of expression. These searches return genes whose expression pattern satisfies the chosen search parameters.
- Serial Analysis of Gene Expression (SAGE) Tags are mapped to the genome and represented in the expression section of gene pages or displayed as a track in the genome browser. Gene with expression evidence based on SAGE tag mapping may be identified. In addition, several searches are available that return SAGE tags records (available under the 'Identify other data types' section of the home page).
- Chromatin immunoprecipitation microarray data (ChIP−chip) for a variety of DNA binding proteins is available and may be displayed in the genome browser. In addition, genes may be identified based on their proximity to ChIP−chip peaks.
- RNA sequence (RNA−seq) reads are mapped to the genome and expression graphs and tables are available on gene pages. Genes may be identified based on their RNA−seq expression profile. In addition, tracks representing depth of coverage and splice junction sites based on intron spanning may be displayed in the genome browser.
- Single Nucleotide Polymorphisms (SNPs) data are available on gene pages or in the genome browser. Genes may be identified based on their SNP characteristics (i.e. synonymous, non-synonymous or non-coding). In addition, several searches return SNPs based on their IDs, gene IDs and genomic location.
- Isolate genotype and meta data (i.e. geographic location, host species, etc.) are integrated from GenBank and several available searches return isolates based on a variety of criteria including isolate ID, taxon/strain, host, isolation source, locus sequence, geographic location, text searches and BLAST similarity. Isolate data based on restriction fragment length polymorphism data are also available.

In addition to the functional data described above, several data types are generated by in house analysis. These include: transmembrane domain and signal peptide predictions, molecular weight and isoelectric point calculations, open reading frame predictions, synteny mapping, splice junction predictions based on RNA−seq data and orthology.

19.4 THE TOXODB HOME PAGE

The ToxoDB home page contains three main sections (Fig. 19.1):

- The top banner section is a constant presence throughout ToxoDB web pages. It contains a clickable ToxoDB logo (which links back to the home page), database release version number and date and Gene ID and text search windows (Fig. 19.1A). In addition, the banner contains many useful help links and links to register, login, 'contact us' form and social media pages (EuPathDB FaceBook, Twitter and YouTube channel). The banner also contains a grey tool bar that provides access to all searches, tools, help and information in ToxoDB via expandable menus.

- The left hand section includes expandable menus including a data summary section which links to a table of all data available in EuPathDB databases, a news section with information regarding data release news and other important community news, a community resources section with links to related sites, community files and upcoming events, education and tutorials which include links to online video tutorials and links to workshop exercises used in previous EuPathDB workshops, and an 'About ToxoDB' section which includes general information regarding usage and citation, EuPathDB publications, scientific working group and advisory team information, funding and website usage statistics (Fig. 19.1B).
- The central section includes links to data searches and tools (Fig. 19.1C). This includes three main categories: The first column includes access to all searches that return genes. The middle column includes access to searches that return other data types such as isolates, single nucleotide polymorphisms, DNA motifs, expressed sequence tags, serial analysis of gene expression tags and open reading frames. The third column includes links to frequently accessed tools such as BLAST, the sequence retrieval tool and the genome bowser.

19.5 THE SEARCH STRATEGY SYSTEM

19.5.1 Running Your First Search

The first step in running a search in ToxoDB is to decide what type of data you are interested in searching. As described above there are several categories of data. For example, if you are interested in searching for genes you would choose one of the searches available under the heading 'Identify Genes By:' on the home page (Fig. 19.2A). Each of the search categories is expandable by clicking on the '+' sign to reveal additional subcategories (Fig. 19.2A, arrow). Selecting one of the subcategories loads a search page revealing search options. In the example in Figure 19.2, a search for genes based on the 'Transmembrane Domain Count' subcategory under 'Protein Features' is selected to reveal search parameters that may be configured by selected or deselecting the check boxes and changing the maximum and minimum transmembrane domain (TM) counts (Fig. 19.2A) (Fig. 19.2B). Once the parameters are selected a search may be run by clicking on the 'Get Answer' button (Fig. 19.2B, rectangle). In the example shown in Figure 19.2, the search would return all genes from all organisms in ToxoDB that contain at least one TM domain.

19.5.2 Understanding and Configuring the Results Page

Once a search is run results are displayed on a results page that is divided into three sections (Fig. 19.3, top to bottom). In the top section is the search strategy graphical system (Fig. 19.3A). Results are displayed graphically in rectangles that contain the number of results and the name of the query. The strategy is interactive and expandable (described below). In the example in Figure 19.3 the strategy is composed of a single step representing the search of all genes from all organisms in ToxoDB containing at least one TM domain (returning 7500 genes). The middle section contains filter tables that allow a user to filter the results of a query based on species and/or strain (Fig. 19.3B). This enables quick toggling between species without the need to rerun the query. Results of filtering are displayed dynamically in the bottom section of the results page. The bottom section displays

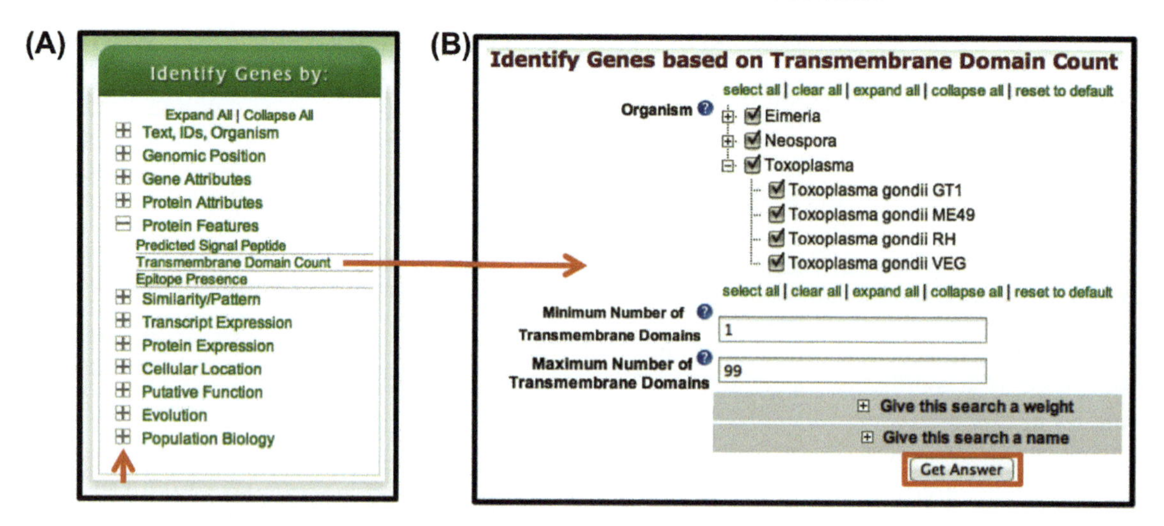

FIGURE 19.2 **Running a first search.**
A) Searches may be selected by clicking on the '+' symbol next to search categories (small arrow) then selecting one of the available searches.
B) A screen shot of the search window that is displayed following clicking on the 'Transmembrane Domain Count' query. Parameters such as organism and the number of transmembrane domains may be modified. Clicking on the 'Get Answer' button runs this query (red box).

the actual results of a query in a dynamic and customizable table (Fig. 19.3C). The table allows you to add more columns including expression graphs (click on the 'Select Columns' button, red box in Fig. 19.3C), remove columns by clicking on the 'x' icon to the right of the column name, move columns (drag and drop), sort items in a column (click on the up or down arrows to the left of the column names), view as many or as few items per page (click on advanced paging), add items to your basket (click on the basket icon on the left of each row, which requires login – a green basket indicates an item is in your basket) and visit hyperlinked items such as gene pages by clicking on the gene ID (first column). In addition, the contents of individual columns may be displayed graphically (histogram or word cloud) by clicking on the graph icon (in the column heading, see red circle in Fig. 19.3C). All results may be downloaded in multiple formats by clicking on the 'Download' link (underlined in Fig. 19.3C). Download formats include Excel formatted tables with any of the available results columns, GFF3, xml, or customized FASTA sequence (Fig. 19.3D insert). Results of a search may also be displayed graphically mapped to chromosomes or scaffolds/contigs (described below).

19.5.3 Building a Multistep Search Strategy

Moving beyond an initial search is achieved by clicking on the 'Add Step' button and growing a search strategy (Fig. 19.3A). Once the 'Add Step' button is selected a popup window offers the user access to all searches in ToxoDB (Fig. 19.4A). Any search may be selected, configured and combined with the results from the previous step in a strategy. In the example shown in Figure 19.4, all genes in ToxoDB with signal peptides are identified and combined with the TM step using a *union* operation (this would define all genes with a signal peptide, a TM domain or both). A search strategy may be extended with as many steps as

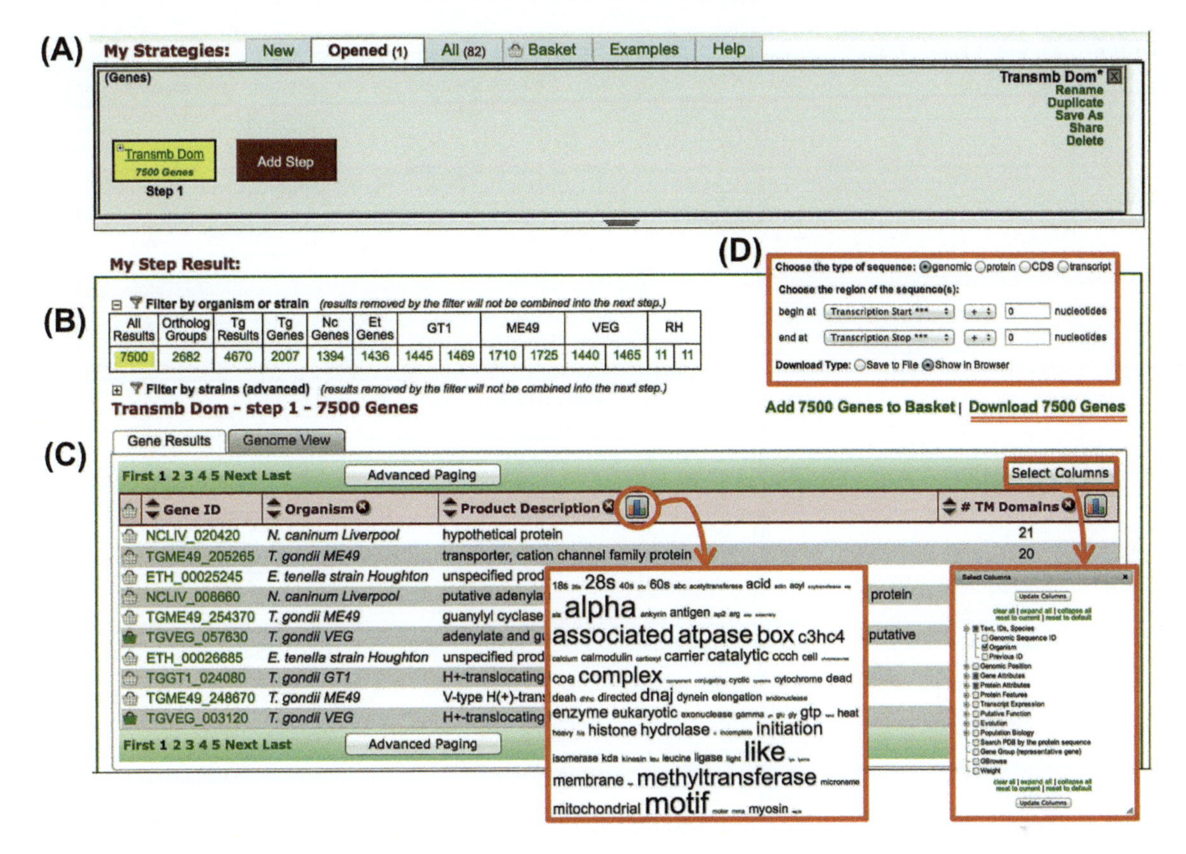

FIGURE 19.3 **Screen shot of the results page.**
A) The search strategy is a graphic display of the results and search workflow. Each search is a step in the strategy; in this example, running a search for genes with transmembrane domains returns 7500 genes.
B) The distribution of genes among species and strains in ToxoDB is displayed in a filter table. Clicking any of the organism specific results will filter the results in the search strategy and in the results table.
C) The results are displayed in a dynamic results table that allows sorting, rearranging, analysing and adding columns. Inserts in red boxes show the results of a word cloud analysis on the product description column (click on the graph icon in red circle) and the list of available columns to load when the 'Select Columns' column is chosen (red box). Individual genes may be added to the basket (green basket icons to the left of gene IDs).
D) Data may be downloaded in multiple formats in addition to tailored sequence data based on user defined coordinates (i.e. for the list of genes retrieve only the 500 nucleotides upstream of the translation start site).

needed. For example, to identify genes from Figure 19.4D that are upregulated in unsporulated and sporulated oocysts compared to tachyzoites, a transcript expression search may be added to the strategy. Figure 19.5 shows the results obtained when adding a search for all genes upregulated by 4-fold in unsporulated and sporulated oocysts compared to tachyzoites based on a microarray experiment (Fritz *et al.*, 2012) (Fig. 19.5A). Note that in this example, the direction of regulation chosen was 'upregulated', the reference samples were the 2, 4, 8 and 21 days and the comparator samples were the unsporulated and 4 and 10 day sporulated samples. In addition, the 4-fold difference between the maximum expression value from the reference samples and the minimum expression value from the comparator samples was

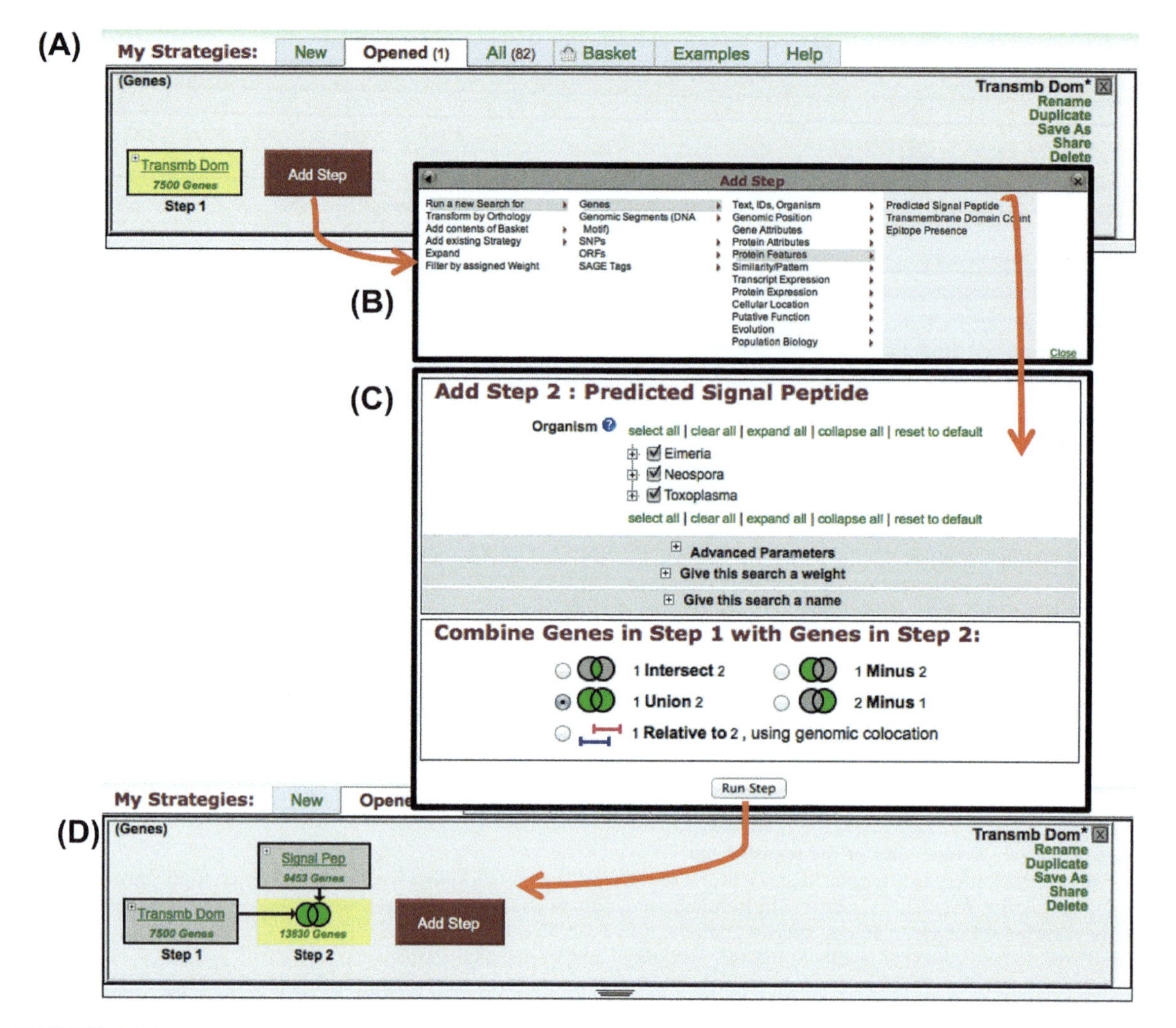

FIGURE 19.4 **Expanding a search strategy.**
A) To combine results from a search with results from another available search, click on the red 'Add Step' button.
B) This displays a pop-up window containing all available searches in ToxoDB.
C) Selecting a new search to run displays a pop-up window with the search specific parameters and options for how to combine the results of the new search with those in the previous one (intersect, union, minus and genomic co-location).
D) Clicking on the 'Add Step' button runs the new search and combines them graphically (in this example, a union operation was applied). Results of the union are shown below the filter table as in Figure 19.3.

selected (Fig. 19.5A). Results of this search were combined with the previous results using the intersect operation (Fig. 19.5B). The results indicate that 144 genes meet these criteria. Note that these genes are only for *T. gondii* ME49, since the *Toxoplasma* array was designed using the original *Toxoplasma* sequence and annotation (Bahl *et al.*, 2010). To define orthologous genes in ToxoDB, a step may be added and the results may be transformed to their orthologues (Fig. 19.5D). This yields a total of 655 genes across all organisms in ToxoDB including possible paralogues within ME49 (Fig. 19.5E). Any step or operation (union, intersect, minus) may be revised at any given time with results of a strategy automatically reflecting the change (Fig. 19.5H). In the example in Figure 19.5E the first step of the strategy may be revised to include proteins with at least 12 TM domain. This is achieved by clicking on the name of the step (Fig. 19.5F) and selecting 'revise' from the popup window, then modifying the search parameters (in this case changing the number of minimum TM domains to 12) (Fig. 19.5G). Strategies themselves can also be renamed, duplicated, saved, shared or deleted (red box in Fig. 19.5E). Clicking on the share link generates a unique URL that can be shared with others. The strategy in Figure 19.5E can be accessed at http://toxodb.org/toxo/im.do?s=650422a412652d0b.

19.5.4 Defining Genes Based on Their Phylogenetic Profile

Orthology in ToxoDB is defined based on protein groupings produced by the OrthoMCL database (Chen *et al.*, 2006; Fischer *et al.*, 2011b; Li *et al.*, 2003). This database includes, in addition to all organisms in EuPathDB resources, 150 organisms representing important branches of the tree of life. Taking advantage of OrthoMCL groups one can define orthologues within ToxoDB as described above in Figure 19.5E or define a phylogenetic profile of genes present in ToxoDB. For example, to determine which genes in Figure 19.5 do not have orthologues in mammals an 'Orthology Phylogenetic Profile' step may be added to the strategy (Fig. 19.6A and B). The parameters for this search (Fig. 19.6C) allow the choice of orthologue presence (clicking once on the grey circles next to entire phyla or organisms; check mark) or orthologue absence (clicking twice on the grey circles next to entire phyla or organisms; red 'x'). Unselected grey circles in Figure 19.6C indicate no preference for orthologue presence or absence. In the example shown in Figure 19.6C genes in ToxoDB with no orthologues in any mammals in OrthoMCL were defined. Adding this step to the strategy results in a total of 460 genes that are conserved in ToxoDB but absent in mammals (Fig. 19.6D).

19.6 GENOMIC COLOCATION

Genomic colocation is a tool in ToxoDB that allows finding results based on defining a relationship between genome-mappable features (i.e. genes, SNPs, DNA motifs, etc.). For instance, this tool can be used to define all genes in *Toxoplasma* that contain a polymorphism in the 10 nucleotides upstream of the translation start site between type I, II and III strains. To do this, two searches need to be combined with the genomic colocation tool as illustrated in Fig. 19.7. In this example the first search is for all SNPs between a type I strain (GT1) and type II (ME49) and type III (VEG) strains (note, the order of searches is important). To define SNPs, the SNPs menu on the home page under 'Identify other data types' is expanded and the 'Genomic location' option is selected. This opens the SNP query that allows the selection of the genomic location of the desired SNPs and the selection of the reference and comparator strains (Fig. 19.7A) (Fig. 19.7B). For this example all chromosomes were selected and the reference chosen was GT1 with ME49 and VEG selected as comparators. Figure 19.7C shows the result of this search which returned 593,111 SNPs distributed across all *Toxoplasma* chromosomes (SNPs are all represented on the reference strain ME49). To define genes that are within 10 nucleotides of SNPs, a step is added to find all genes in ToxoDB (Fig. 19.7D). Note that the only option to combine this search with the SNP search is the genomic colocation option (bottom of Fig. 19.7D). Choosing this option and selecting

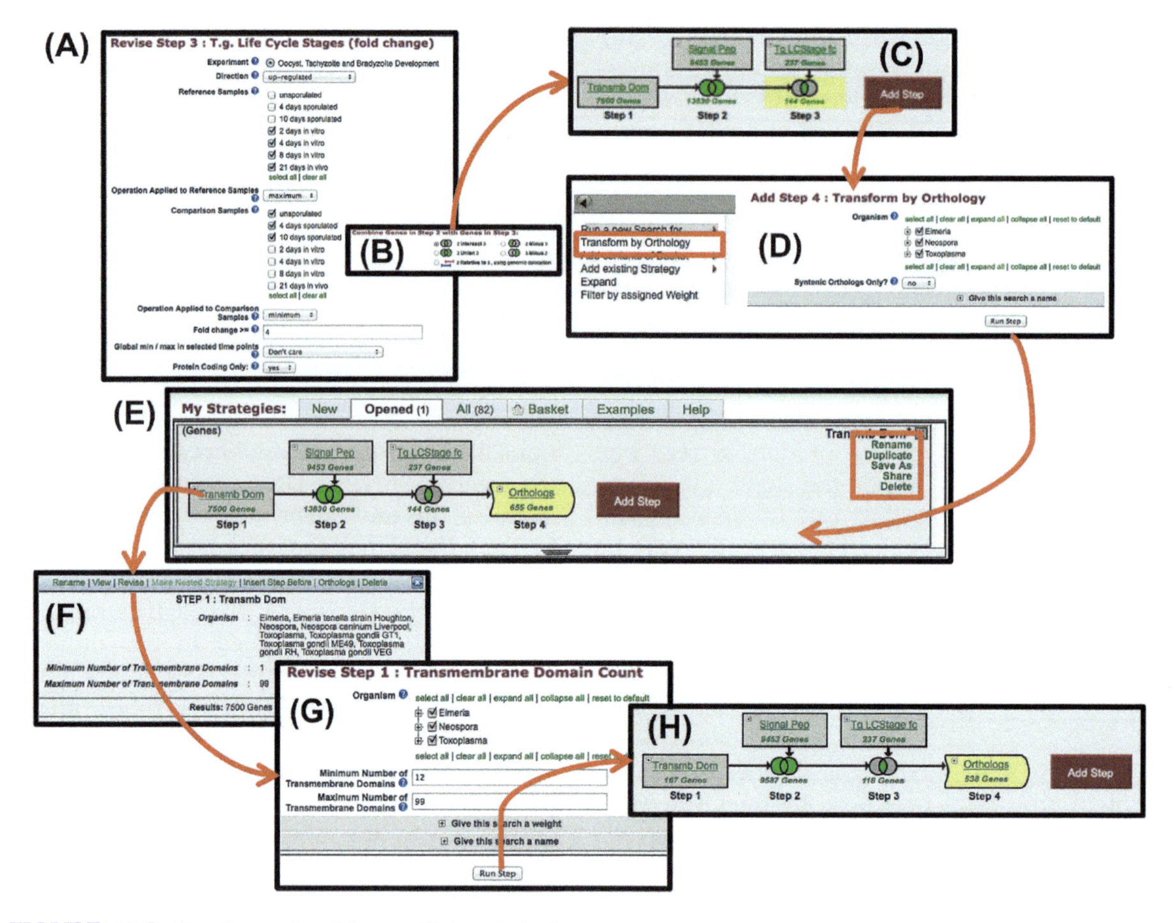

FIGURE 19.5 **Creating and revising multistep strategies.**
A) Following the same logic as in Figure 19.4, additional steps may be added. Shown here is the pop-up window with search parameters for the life cycle stages microarray experiment (Fritz *et al.*, 2012).
B) Results from this search are combined with the previous results using an intersect operation.
C) The strategy grows by an additional step with results updated graphically and tabularly.
D) The microarray Affymetrix chip was created using ME49 sequence (Bahl *et al.*, 2010) so results are only available for this strain. However, to find orthologues of the ME49 genes in all species and strains in ToxoDB, the results may be transformed into orthologues.
E) The strategy is now comprised of four steps showing all genes in ToxoDB that contain transmembrane domains and/or signal peptides and are up-regulated in oocysts and sporozoites.
F) Clicking on the name of the strategy reveals a pop-up window with additional options allowing a step in a strategy to be revised, deleted or expanded into sub-strategies.
G) In this example, the first step is revised to include genes with at least 12 transmembrane domains.
H) Revising the first step results in updating all the subsequent steps in the strategy.

'continue' opens a pop-up window with customizable colocation parameters that include a dynamic logic statement that is updated based on the chosen parameters (Fig. 19.7E).

An example of how the logic statement can be modified is as follows.

Select the type of results of interest. In this case the objective is to return genes, which can be

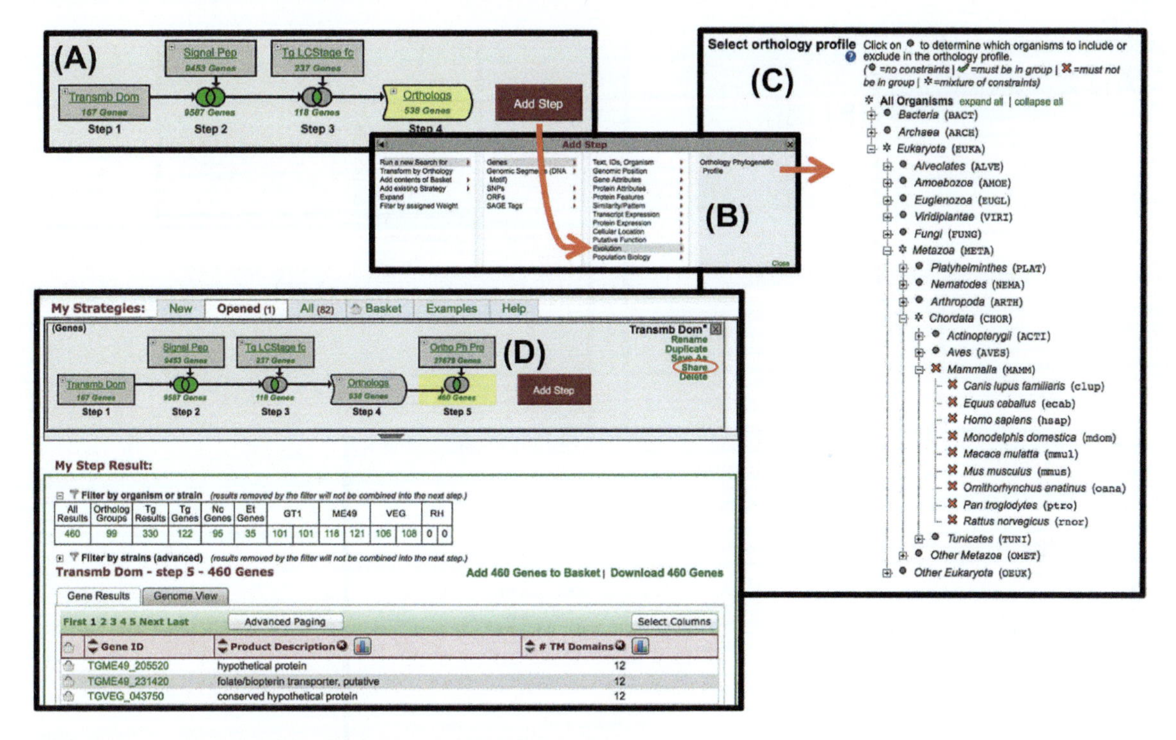

FIGURE 19.6 **Illustration of the Phylogenetic Profile query.**
A) Defining genes based on their phylogenetic profile is achieved by adding a step.
B) Select the 'Orthology Phylogenetic Profile' query under the 'Evolution' category from the pop-up window.
C) Organisms may be included (check mark) or excluded (red 'x') to define the type of genes. In this example only genes without orthologues in mammals are selected.
D) Results of this query are added to the growing strategy revealing 460 genes in ToxoDB that meet all the criteria of this search strategy. Search strategies may be saved and shared with others using unique URLs (generated by clicking on the share link (red circle)): http://toxodb.org/toxo/im.do?s=bcd12f3c24149bfe.

selected from the first drop down menu in the logic statement (Box 1, Fig. 19.7E). Next, the region relative to the genes is customized using the grey highlighted region in the lower right hand side of Figure 19.7E. In this case, the upstream option is chosen and the number of nucleotides is changed to 10. Note that both the graphic and the logic statement are dynamically updated.

Next the type of relationship between the Genes and the SNPs is selected from the drop down menu (Box 2, Fig. 19.7E). The available options include 'overlaps', 'is contained in' or 'contains'. For this query the latter was chosen since the interest is to define an upstream 10 nucleotide region that contains SNPs.

The parameters for the SNP results may now be modified, these include the strand (Box 3, Fig. 19.7E) and the region (lower left side of Fig. 19.7E) – similar to step 2, above. In this case the parameters will not be modified since the SNP may be on either strand and the exact region of the SNP (the single nucleotide polymorphism itself) is desired to be within the upstream 10 nucleotides.

After modifying the parameters the logic statement reads 'Return each gene from step 2 whose upstream region (defined as 10 nucleotides) contains the exact region of a SNP in step 1 on either strand'. Clicking on 'Get Answer' returns all genes that meet the co-location criteria shown

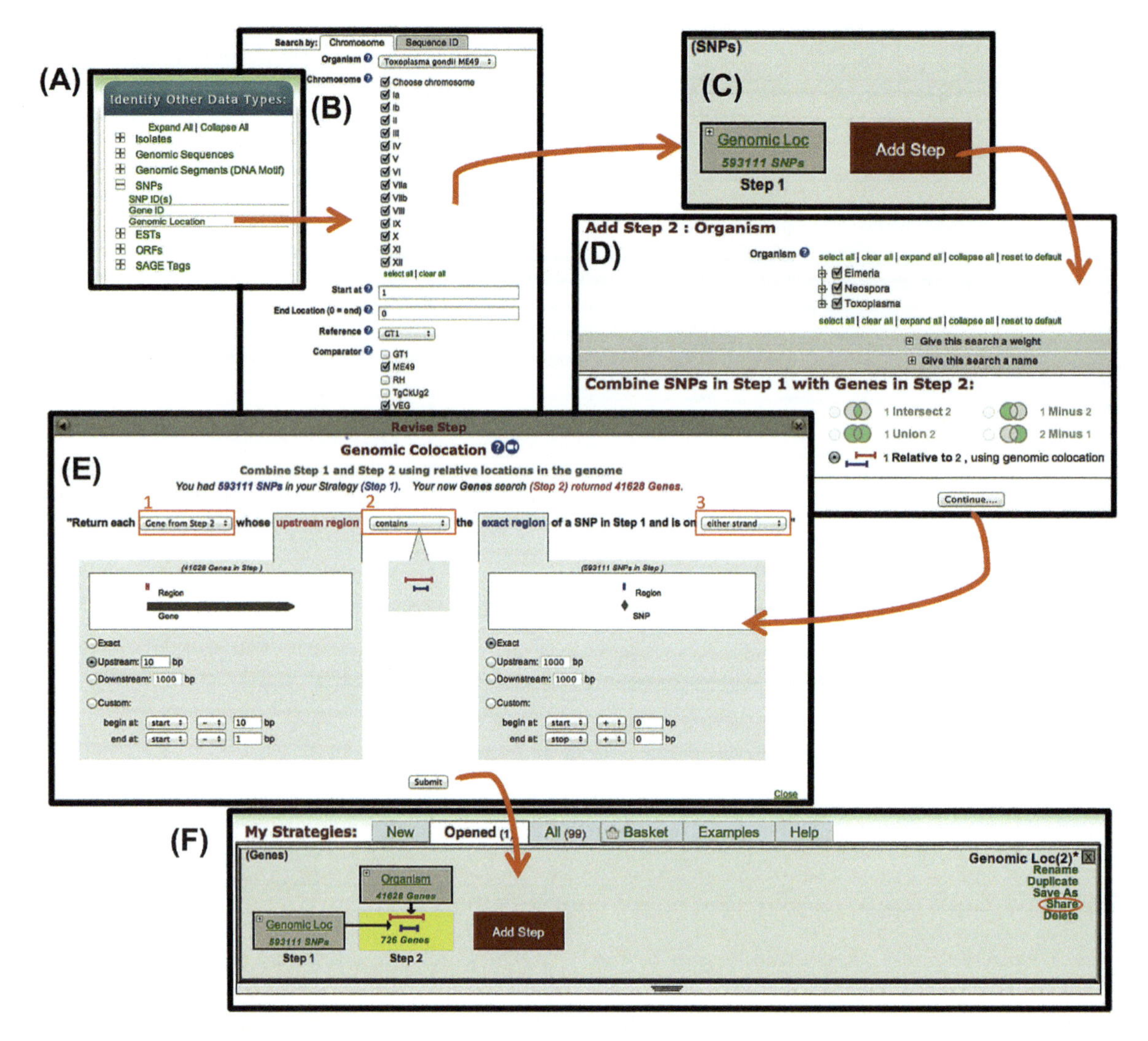

FIGURE 19.7 The Genomic Colocation Tool allows combining any feature that can be mapped to a genomic sequence (i.e. genes, DNA motifs, SNPs, etc.).

A) The SNPs by genomic location query is found under the 'SNPs' category under 'Identify Other Data Types' section (also see Fig. 19.1).

B) The SNPs by genomic location query allows the identification of SNPs between strains and on all or any of the chromosomes. In this example, all chromosomes are selected and ME49 is selected as the reference while GT1 and VEG are selected as comparators.

C) Close to 600,000 SNPs were identified.

D) To define only those SNPs that are proximal to genes, an 'Identify Genes by Organism' is added, organisms of choice selected and the continue button selected.

E) A pop-up window containing a dynamic logic statement allows selection of the relationship between the SNPs and the genes by virtue of their genomic location. In this example, the query is asked to return all genes that contain a SNP in the 10 nucleotides upstream of their starts.

F) Once the parameters of the colocation query are submitted, results are displayed as part of the search strategy. An icon representing colocation is displayed. Any strategy may be shared using a unique URL that can be generated by clicking on the share link (red circle): http://toxodb.org/toxo/im.do?s=8938b673e8b66bab.

in Fig. 19.7F (results include in addition to gene IDs, the number and location of matches).

19.7 THE GENOME BROWSER

ToxoDB includes a genome browser (GBrowse) developed by the Generic Model Organism Database project (Stein *et al.*, 2002). GBrowse allows users to display genomic DNA from genomes in ToxoDB and dynamically decorate regions of interest (which may be as big as an entire chromosome) with additional data tracks. Tracks include a plethora of data such as gene models, proteomics data, transcriptional data (i.e. RNA sequence reads, microarray probes, ChIP–chip and ChIP–seq profiles, etc.), synteny between strains and species in ToxoDB and splice site junctions.

GBrowse has multiple entry points from within ToxoDB web pages, including from the home page (tools section in Fig. 19.1) and from the genomic context section of gene pages. When accessed from the home page or gene pages the annotated genes track is turned on by default (Fig. 19.8). The top section of the GBrowse window includes tabs that allow you to navigate to the 'browser' (this is the section that displays data tracks), the 'select tracks' section allows choosing desired data tracks to display in the 'browser' section, the 'snapshots' section allows saving specific GBrowse views for later access, the 'custom tracks' section allows the loading of user specific tracks (this includes RNA sequence read files) and the 'preferences' section allows setting general preferences for your GBrowse experience (Fig. 19.8A). Below the tabs is the search section (Fig. 19.8B) which enables specifying the exact coordinates of sequence to be viewed (gene IDs may also be entered in the Landmark or Region box) and access to several configuration tools which include a restriction site configuration (choose which restriction sites to view in the browser), design PCR primers (interactive graphical tool for designing primers based on a region of interest) and various track download options. In addition, this section includes the scrolling tool that enables zooming and left/right scrolling. The overview and region sections (Fig. 19.8C) provide a graphical display of the genomic location of the region in view (the overview represents the largest contiguous genomic sequence to which the region of interest belongs, in this case *T. gondii* chromosome VI). The details section (Fig. 19.8D) includes the specific region being viewed and any tracks that have been loaded.

To load tracks in GBrowse click on the 'select tracks' tab (Fig. 19.8A) to reveal available tracks which are distributed into subsections (Fig. 19.8E). Each subsection may be expanded or collapsed by toggling the +/− buttons (arrow in Fig. 19.8E). Desired data tracks can be selected with a checkbox and multiple tracks may be loaded simultaneously. At any point during track selection one may view the loaded tracks by clicking on the 'browser' tab (Fig. 19.8A). The choice of tracks to turn on depends on the question being asked. Examples of GBrowse track combinations are shown in Figure 19.9 with links in URL and QR code formats.

19.8 FUTURE DIRECTIONS

It is expected that the volume and complexity of data sets will continue expanding over the coming years. Datasets such as high throughput sequencing of multiple strains will be integrated to allow searches for population differences through SNP queries. ToxoDB also anticipates incorporating additional genome sequence and annotation from other Sarcocystidae. An expansion of ToxoDB to include compound record pages and metabolic pathways that are fully integrated into the search strategy system is well under way and should be forthcoming in the near future. This would lay the ground work to incorporate metabolomic data and enable queries for compounds and reactions.

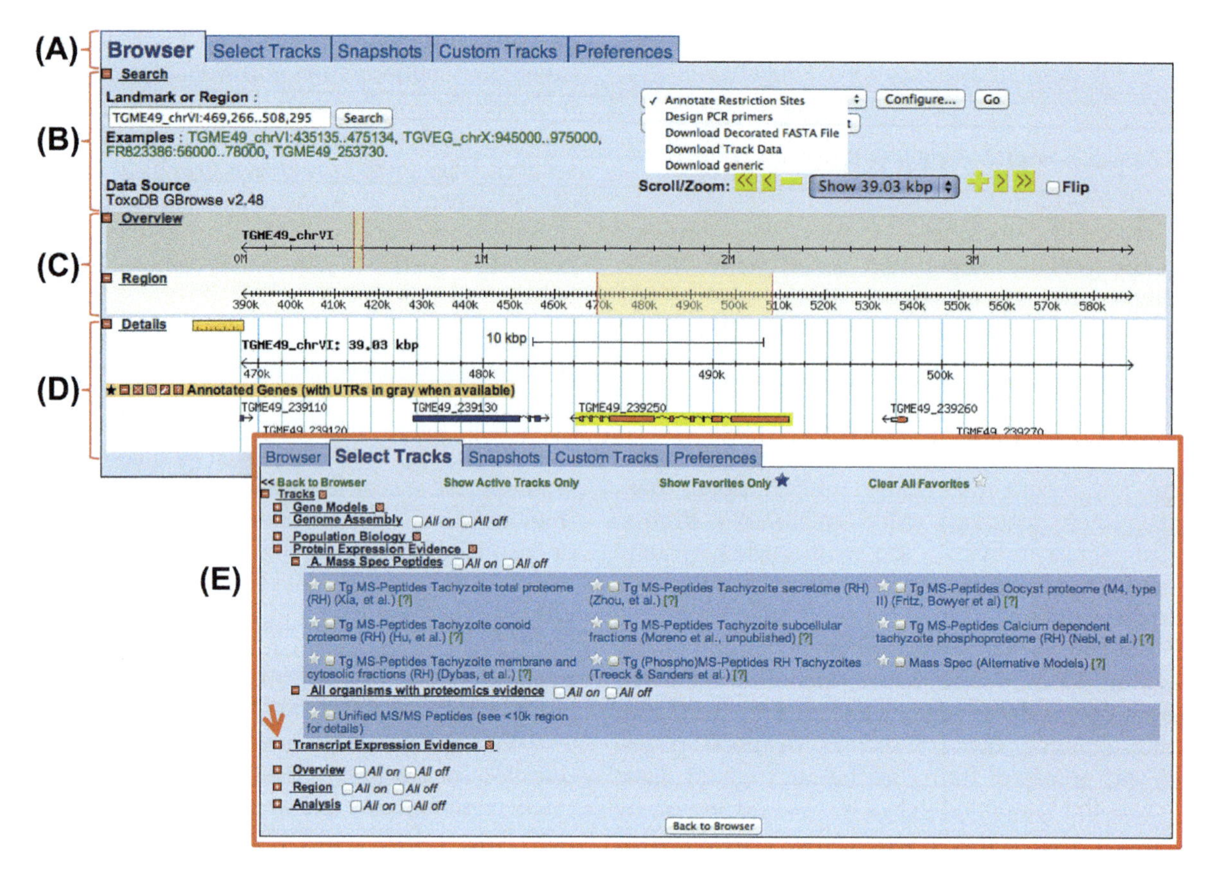

FIGURE 19.8 **The Genome Browser (GBrowse) allows graphical display of data on the genome.**
A) The top section of GBrowse includes tabs that link to a track selection page, snapshots page, custom tracks page and a configuration section.
B) The section below the table includes a 'Landmark or Region' window for enter specific coordinates to visit or gene IDs, a scrolling and zooming button and links to several useful tools (drop down menu on the right hand side). Tools include a PCR primer design tool, restriction site display tool and several data download options.
C) The overview and region sections graphically show the entire reference sequence (i.e. an entire chromosome) and an intermediate size sequence region, respectively.
D) The details section includes a representation of the coordinates being looked at and any data track that has been loaded.
E) A screen shot of the 'select tracks' section showing the various expandable/collapsible sections ('+' /'−', red arrow). In this image the protein expression section is expanded revealing a number of experimental data that may be loaded by selecting the check boxes. Tracks start to load in the browser section as soon as they are selected.

An exciting development for scientists interested in eukaryotic pathogens is current efforts by EuPathDB to make available a host response database that will incorporate host functional datasets in response to infection by parasites such as *Toxoplasma*.

Acknowledgements

EuPathDB is funded with federal funds from the National Institute of Allergy and Infectious Diseases, National Institutes of Health, Department of Health and Human Services, under Contract No. HHSN272200900038C. The TriTrypDB component of EuPathDB is funded by a grant from the Bill

Synteny

View of Chromosome VI of *T.* gondii ME49 with GBrowse tracks "Syntenic Sequences and Genes (shaded by Orthology)" and "Scaffolds" turned on. Both contigs and genes are displayed and synteny is indicated by shading between genes. Gaps in sequence are indicated by red marks in the scaffold track.

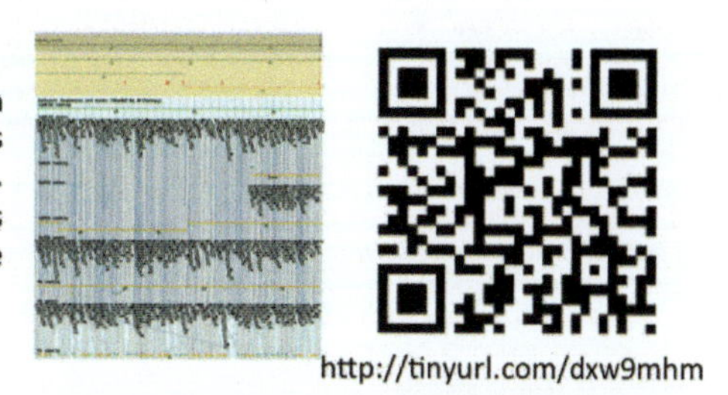

http://tinyurl.com/dxw9mhm

Splice Junctions and RNA-sequence

View of the gene encoding hypoxanthine-xanthine-guanine phosphoribosyl transferase (HXGPRT) with the "Splice Site Junction – Combined and the ME49 strand specific RNAseq data (Gregory et al., unpublished) turned on. In the view, the alternative splicing of HXGPRT is evident and a missing RNAseq data supports the likely existence of another gene model upstream of HXGPRT not in the annotation.

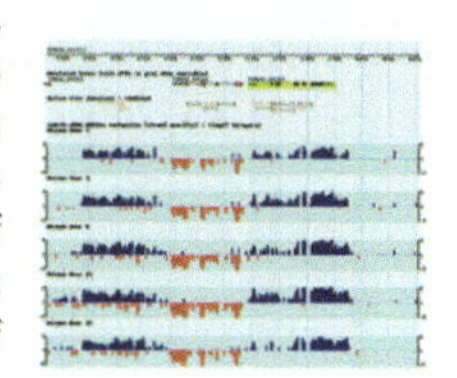

http://tinyurl.com/a8833kv

ChIP-chip and RNA-sequence

View of the gene encoding Rop18 with RNAseq and H3K4Me3 ChIP-chip data tracks from turned on. These data support each other in the observed expression differences between strains.

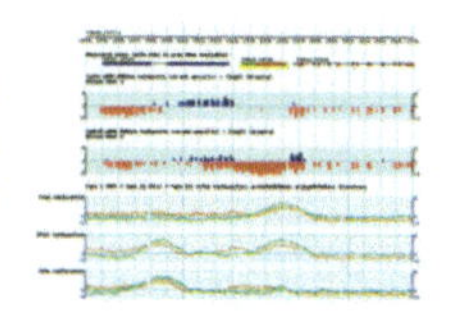

http://tinyurl.com/b8tyh4d

FIGURE 19.9 **Example GBrowse track combinations.** Three track combinations have been generated and bookmarked. To load these GBrowse views in your browser you may use the provided URLs or simply scan the QR code using a smartphone.

and Melinda Gates Foundation with additional support from The Wellcome Trust (Grant No. WT085822MA). EuPathDB wishes to acknowledge all data providers and user comments submitters. The collaborative nature of the *Toxoplasma* research community and their willingness to support data integration and dissemination is admirable.

References

Aslett, M., Aurrecoechea, C., Berriman, M., Brestelli, J., Brunk, B.P., Carrington, M., et al., 2009. TriTrypDB: a functional genomic resource for the Trypanosomatidae. Nucleic Acids Research 38 (Database), D457–D462. http://dx.doi.org/10.1093/nar/gkp851.

Aurrecoechea, C., Barreto, A., Brestelli, J., Brunk, B.P., Cade, S., Doherty, R., et al., 2013. EuPathDB: The Eukaryotic Pathogen database. Nucleic Acids Research 41 (D1), D684–91. http://dx.doi.org/10.1093/nar/gks1113.

Aurrecoechea, C., Brestelli, J., Brunk, B.P., Dommer, J., Fischer, S., Gajria, B., et al., 2009. PlasmoDB: a functional genomic database for malaria parasites. Nucleic Acids Research 37 (Database issue), D539–43. http://dx.doi.org/10.1093/nar/gkn814.

Bahl, A., Davis, P.H., Behnke, M., Dzierszinski, F., Jagalur, M., Chen, F., et al., 2010. A novel multifunctional oligonucleotide microarray for Toxoplasma gondii. BMC Genomics 11, 603. http://dx.doi.org/10.1186/1471-2164-11-603.

Chen, F., Mackey, A.J., Stoeckert, C.J., Roos, D.S., 2006. OrthoMCL-DB: querying a comprehensive multi-species collection of ortholog groups. Nucleic Acids Research 34 (Database issue), D363–8. http://dx.doi.org/10.1093/nar/gkj123.

Fischer, S., Aurrecoechea, C., Brunk, B.P., Gao, X., Harb, O.S., Kraemer, E.T., et al., 2011a. The Strategies WDK: a graphical search interface and web development kit for functional genomics databases. Database: the Journal of Biological Databases and Curation. http://dx.doi.org/10.1093/database/bar027, 2011, bar027.

Fischer, S., Brunk, B.P., Chen, F., Gao, X., Harb, O.S., Iodice, J.B., et al., 2011b. Using OrthoMCL to assign proteins to OrthoMCL-DB groups or to cluster proteomes into new ortholog groups. Curr. Protoc. Bioinformatics, 1–19.

Fritz, H.M., Buchholz, K.R., Chen, X., Durbin-Johnson, B., Rocke, D.M., Conrad, P.A., Boothroyd, J.C., 2012. Transcriptomic analysis of toxoplasma development reveals many novel functions and structures specific to sporozoites and oocysts. PLoS ONE 7 (2), e29998. http://dx.doi.org/10.1371/journal.pone.0029998.

Greene, J.M., Collins, F., Lefkowitz, E.J., Roos, D., Scheuermann, R.H., Sobral, B., et al., 2007. National Institute of Allergy and Infectious Diseases bioinformatics resource centers: new assets for pathogen informatics. Infection and Immunity 75 (7), 3212–3219. http://dx.doi.org/10.1128/IAI.00105-07.

Li, L., Stoeckert, C.J., Roos, D.S., 2003. OrthoMCL: identification of ortholog groups for eukaryotic genomes. Genome Research 13 (9), 2178–2189. http://dx.doi.org/10.1101/gr.1224503.

Reid, A.J., Vermont, S.J., Cotton, J.A., Harris, D., Hill-Cawthorne, G.A., Könen-Waisman, S., et al., 2012. Comparative Genomics of the Apicomplexan Parasites Toxoplasma gondii and Neospora caninum: Coccidia Differing in Host Range and Transmission Strategy. In: Striepen, B. (Ed.), PLoS Pathogens. http://dx.doi.org/10.1371/journal.ppat.1002567.t001, 8(3), e1002567.

Roos, D., 2011. David Roos. Interview by H. Craig Mak. Nature Biotechnology 29 (2), 141–142. http://dx.doi.org/10.1038/nbt.1774.

Stein, L.D., Mungall, C., Shu, S., Caudy, M., Mangone, M., Day, A., et al., 2002. The generic genome browser: a building block for a model organism system database. Genome Research 12 (10), 1599–1610. http://dx.doi.org/10.1101/gr.403602.

CHAPTER

20

Comparative Aspects of Nucleotide and Amino Acid Metabolism in *Toxoplasma gondii* and Other Apicomplexa

*Kshitiz Chaudhary**, *Barbara A. Fox*[†], *David J. Bzik*[†]

*Vaccines Research, Pfizer, Pearl River, New York, USA [†]Department of Microbiology and Immunology, Geisel School of Medicine at Dartmouth, Lebanon, New Hampshire, USA

OUTLINE

Toxoplasma gondii, Second Edition
http://dx.doi.org/10.1016/B978-0-12-396481-6.00020-9

20.1 INTRODUCTION

Nucleotides and amino acids are of fundamental importance in the replication and development of *Toxoplasma gondii* and other apicomplexan parasites. Nucleotides are in particular essential for replication of DNA and transcription of RNA in rapidly dividing stages. DNA synthesis relies on an ample supply of pyrimidine (dCTP, dTTP) and purine (dATP, dGTP) deoxynucleotide 5′-triphosphates, while RNA synthesis utilizes the ribonucleotides ATP, CTP, and GTP, as well as uridine 5′-triphosphate (UTP). Nucleotides are also essential in providing the cellular energy sources (ATP and GTP), and are involved in numerous other metabolic roles. Nucleotides provide precursors of more complex molecules such as folate, serve as nucleotide based enzyme cofactors such as NAD^+ or FAD, serve in regulatory roles as intracellular messengers such as cAMP, and also play roles in controlling metabolic and gene regulation. Nucleotides are either synthesized from small molecules and amino acids, or they are acquired via salvage pathways from preformed host derived nucleobases and nucleosides. The apicomplexan parasites considered in this chapter, *Toxoplasma gondii*, *Plasmodium falciparum* and *Cryptosporidium parvum*, are important pathogens of humans that cause significant morbidity and mortality. Notably, current treatment strategies in human infections caused by *T. gondii* or *P. falciparum* are based on blocking the accumulation of nucleotides. This validated approach to chemotherapy highlights the significance of further research to dissect details of nucleotide metabolism in the Apicomplexa as well as the potential for this research to lead to the development of new treatments.

Apicomplexa are obligate intracellular lower eukaryotic single cell organisms that are only capable of replication when they properly associate with their parasitized host cell. They may exist in a viable form extracellularly for short periods of time but are incapable of even a single cell division. A comparative approach examining three selected apicomplexan pathogens that differ greatly in their interactions with the human host was undertaken to begin to address the possible relationships between the host cell and environment occupied by each parasite and the strategy that each parasite has adopted to ensure the delivery of a sufficient supply of nucleotides and amino acids to support rapid parasite replication. The recent completion of the genome sequences for *T. gondii* (http://toxodb.org), *P. falciparum* (http://plasmodb.org), and *C. parvum* (http://cryptodb.org) enables new predictions about parasite biology, opens new opportunities for development of chemotherapy directed at selected parasite targets, and permits a retrospective comparison of newly identified orthologues of genes involved in nucleotide and amino acid metabolism with previous biochemical, cell biological, and genetic studies of nucleotide and amino acid metabolism.

The study of nucleotide and amino acid metabolism in Apicomplexa is extraordinarily complicated by the obligatory presence of the host cell. Human host cells are much larger than their apicomplexan inhabitants, are themselves rich in nucleotides and amino acids and also possess extensive metabolic capacities to transport, synthesize, interconvert and catabolize nucleotides and amino acids. Nucleotide and amino acid metabolism of the obligate intracellular parasites still largely remains a mystery due to the complex interaction of the parasite and its host cell. While studies have examined the extracellular form of the parasite in regard to the transport, synthesis and interconversion of nucleotides, the isolation of free parasites without contaminating host material is uncertain. Consequently, genetic studies have been a particularly informative approach to examine the phenotype of engineered parasites lacking or gaining a gene involved in nucleotide or amino acid metabolism, or alternatively, by examining the interaction of normal or mutant parasites in a mutant host cell. Most of the genetic studies have been performed in *T. gondii* due to the more rapid and robust genetic models available for manipulating this parasite (Kim and Weiss, 2004). Several important genetic selection models developed for *T. gondii* are also based on parasite nucleotide and amino acid metabolism. Genetic models based on nucleotide metabolism are available in *P. falciparum*, but these models are not yet developed for *C. parvum*.

20.2 PURINES

Purines are crucial to all cells as required components of nucleic acids, a cellular energy source, and cofactors or substrates for specific aspects of cellular metabolism. All aspects of the apicomplexan lifestyle including motility dependent invasion of host cells is powered by nucleotides (Kimata and Tanabe, 1982). Remarkably, parasitic protozoa, with the exception of *Acanthamoeba* (Hassan and Coombs, 1986), lack the ability to synthesize the purine ring *de novo* (Wang, 1984; Berens *et al.*, 1981; Fish *et al.*, 1982; Marr *et al.*, 1978; Miller and Linstead, 1983; Wang and Aldritt, 1983; Wang and Simashkevich, 1981). Since *T. gondii* and other apicomplexan parasites lack the machinery to synthesize the purine ring *de novo*, they rely on essential capture, transport and salvage machinery to steal purines from their hosts for incorporation into the parasite nucleotide pools (Perotto and Keister, 1971; Schwartzman and Pfefferkorn, 1982; Chaudhary *et al.*, 2004; Krug *et al.*, 1989; Ting *et al.*, 2005). The integration of these pathways in Apicomplexa is likely to reflect the specific needs of each parasite, as well as the differing purine resources potentially available in various host cells, tissues and environments inhabited by each parasite. Although incapable of replication extracellularly, even in the richest medium, early investigations in *T. gondii* demonstrated that the parasite could obtain purines and meet all other demands for normal intracellular replication in host cells that are blocked in protein synthesis (Pfefferkorn and Pfefferkorn, 1981), in enucleated host cells that are blocked in host RNA and DNA synthesis (Jones, 1973; Sethi *et al.*, 1973), as well as in host cells that are operating glycolytically due to deficient mitochondrial function (Schwartzman and Pfefferkorn, 1981).

The capture, transport and salvage pathways necessary for purine acquisition have long been viewed as a significant weakness of the parasite that may be targeted for chemotherapy, and significant efforts have been directed towards this objective (Avila and Avila, 1981; Avila *et al.*, 1987; Berens *et al.*, 1984; Fish *et al.*, 1985; Marr and Berens, 1988; Gero *et al.*, 1989, 2003; Pfefferkorn *et al.*, 2001). Current models propose that parasite purine transporters on the parasite plasma membrane are a required mechanism of purine acquisition from the host (Carter *et al.*, 2003; de Koning *et al.*, 2005). For apicomplexan parasites such as *T. gondii* or *P. falciparum*, this model suggests that host cell derived purines either passively or actively accumulate in the parasitophorous vacuole space. Therefore the purines acquired from the host by the intracellular parasite could be derived from some combination of potential resources including *de novo* purine synthesis in the parasitized host cell, existing purine pools within the host cell, host cell purine catabolism or purine transport flux into the parasitized host cell. Few studies have experimentally addressed whether there is any requirement for host cell purine transporters. Significant progress has been made in understanding the purine transport capability of the apicomplexan parasites, as well as the parasite enzymes that facilitate interconversion, salvage and incorporation of host purines into the parasite purine nucleotide pools. The host menu of purine compounds, the purine capture and transport capabilities of the parasite, and the specific activity levels of purine salvage and interconversion enzymes will ultimately determine the potential flux of purines from the host to the parasite.

20.2.1 Capture and Transport

The details of purine transport capacity and purine transporters have been recently elucidated from several protozoans with the majority of work being reported on transporter functions in *Leishmania* spp. and *Trypanosoma* spp. (de Koning *et al.*, 2005; Landfear, 2011; Landfear *et al.*, 2004). Nucleoside and nucleobase transporters fall into two general groups, sodium-dependent concentrative nucleoside transporters (CNT) and non-concentrative equilibrative nucleoside transporters (ENT). Mammals and bacteria possess CNT and ENT transporters while the protozoa only have ENT-type transporters (Landfear *et al.*, 2004; de Koning *et al.*, 2005; Chaudhary, 2005). Other genes for nucleobase transporters in bacteria, fungi and plants are classified as either the plant purine related transporter family, the microbial purine related transporter family, or the nucleobase/ascorbate transporter family (de Koning and Diallinas, 2000). Genes corresponding to these three groups of purine transporters have not been identified among the protozoa.

20.2.1.1 Genome Analysis of Purine Transporters in Apicomplexa

The complete genome sequences of *Toxoplasma gondii*, *Plasmodium falciparum* and *Cryptosporidium parvum* have identified putative ENT-type transporter genes, putative channels and other transporters that may play a role in purine transport. The genome of *C. parvum* revealed remarkable transport capacities and putative purine transporters that await functional characterization (Abrahamsen *et al.*, 2004; Xu *et al.*, 2004). The *P. falciparum* genome encodes a previously characterized purine transporter, PfNT1, as well as three additional putative ENT orthologues (Kirk *et al.*, 2005; Martin *et al.*, 2005; Bahl *et al.*, 2002, 2003; Chaudhary, 2005). Similarly, the *T. gondii* genome reveals TgAT1 and three additional ENT orthologues designated as TgNT1, TgNT2 and TgNT3 (Kissinger *et al.*, 2003; Li *et al.*, 2003; Chaudhary, 2005). Recent work has demonstrated that TgAT1, TgNT1 and TgNT3 are expressed in tachyzoites, bradyzoites and sporulated oocysts, whereas

TgNT2 is not (Chaudhary, 2005). TgAT1 and TgNT3 proteins localized to the plasma membrane of intracellular tachyzoites, while TgNT1 was localized as a punctate labelling of the parasite cytosol within a compartment that did not co-localize with any other marker of a known *T. gondii* organelle (Chaudhary, 2005). Other than *T. gondii* TgAT1 and *P. falciparum* PfNT1, the functional role of each of these recently identified ENT orthologues is currently unknown.

20.2.1.2 An Early Model of Purine Acquisition in T. gondii

The first working model of purine acquisition by *T. gondii* was proposed to be from host ATP, the most abundant host purine present at greater than 4 mM in host cell cytosol (Plagemann, 1986; Traut, 1994; Plagemann *et al.*, 1988). The hypothesis that *T. gondii* used host cell ATP as its purine source was based on data from dual-label experiments showing that extracellular *T. gondii* readily incorporated the nucleoside component of ATP or AMP into nucleic acids, but did not incorporate the phosphate moiety (Schwartzman and Pfefferkorn, 1982). At this time it was also appreciated that the parasitophorous vacuole surrounding the intracellular tachyzoite was covered in a layer of host cell mitochondria, an organelle rich in ATP (Jones *et al.*, 1972). The identification of the parasitophorous vacuole surrounding intracellular tachyzoites as a passive permeation barrier suggested that the high concentration of cytosolic or mitochondrial derived host ATP potentially could permeate into the parasitophorous vacuole space and equilibrate at mM concentrations (Schwab *et al.*, 1994). The discovery of a remarkably abundant nucleoside triphosphate hydrolase (NTPase) activity secreted into the parasitophorous vacuole space suggested host ATP may be hydrolysed to AMP within the vacuolar space (Asai *et al.*, 1995; Sibley *et al.*, 1994; Bermudes *et al.*, 1994). Consequently, it was postulated that *T. gondii* might obtain its purine requirement from the flux created through permeation of ATP into the vacuolar space, conversion of ATP to AMP by vacuolar NTPase activity, and conversion of AMP to adenosine by a putative parasite plasma membrane 5′-ectonucleotidase. Since the pool of host cell adenosine is very small at ~1 μM (Plagemann, 1986; Plagemann *et al.*, 1988; Traut, 1994), utilization of host ATP pools was an attractive model to explain how host adenosine may be concentrated within the parasitophorous vacuole space. The first adenosine transporter characterized for *T. gondii*, TgAT1, was described as a non-concentrative low-affinity (Km ~120 μM) adenosine transport system that likely would require a higher concentration of adenosine than the 1 μM present in host cell cytosol for physiological significance (Schwab *et al.*, 1995). Subsequent work has demonstrated that intracellular tachyzoites as well as the parasitophorous vacuole space, and membrane, have no detectable 5′-ectonucleotidase activity (Ngo *et al.*, 2000). Based on these studies, it appears *T. gondii* has no significant access to intracellular pools of host cell nucleotides. Intriguingly a very recent study showed that CD73 knock-out mice are protected from chronic cyst formation, because the parasites are unable to differentiate to bradyzoites (Mahamed *et al.*, 2012). The investigators concluded that CD73 expression, which is required to convert AMP to adenosine, promotes bradyzoite differentiation in a process that is dependent upon adenosine generation. Thus the role of host purine pools and *T. gondii* survival is not completely resolved.

20.2.1.3 Functional Properties of Purine Transporters in Apicomplexa

Functional studies of purine transporters in *C. parvum* are not yet available. By contrast, several purine transporters have been functionally characterized in *T. gondii*. TgAT1 is a low affinity transporter for adenosine, guanosine and inosine (Chiang *et al.*, 1999; Schwab *et al.*, 1995). Low-affinity transport of adenosine by TgAT1 is blocked by dipyridamole and inhibited by excess inosine, formycin B or hypoxanthine, but not pyrimidines, suggesting a role in transport that is purine selective and broad spectrum.

The subsequent selection and characterization of adenine arabinoside (ara-A) resistant mutants identified mutations at a genetic locus corresponding to a gene, designated as TgAT, and characterized as a putative 11 membrane-spanning region protein of the ENT family (Chiang *et al.*, 1999). Expression of TgAT in *Xenopus laevis* oocytes reconstituted low-affinity adenosine transport function similar to the previously characterized TgAT1 suggesting TgAT is likely to represent the cloned gene of TgAT1 (Chiang *et al.*, 1999; Schwab *et al.*, 1995). Disruption of the TgAT gene caused a complete loss of all adenosine transport function in extracellular tachyzoites (Chiang *et al.*, 1999).

Additional high affinity nucleoside and nucleobase transporters were identified in extracellular tachyzoites of *T. gondii* (De Koning *et al.*, 2003). This study characterized a low-affinity adenosine (Km ~105 μM) and inosine (Km ~134 μM) transporter equivalent to the previously characterized TgAT1 (TgAT). Kinetic evidence revealed the presence of a high-affinity adenosine (Km ~0.49 μM) and inosine (Km ~0.77 μM) transport system that was also a high affinity and broad-spectrum transporter capable of transporting a large number of different purine and pyrimidine nucleosides. This second adenosine transport system was designated as TgAT2 (De Koning *et al.*, 2003). The high affinity and broad specificity of TgAT2 suggested it could be an efficient route for uptake of various therapeutic nucleosides (De Koning *et al.*, 2003). The same work also revealed a third transport system, TgNT1, which is the first purine transporter that has been demonstrated to selectively transport purine nucleobases. Based on competition kinetics, TgNT1 has a high affinity for hypoxanthine (Km ~0.91 μM) as well as guanine and xanthine.

The presence of TgAT2 and TgNT1 on the tachyzoite plasma membrane of extracellular tachyzoites suggests an extraordinary capacity of *T. gondii* to salvage exceedingly low concentrations of purine nucleobases and nucleosides from within the parasitophorous vacuole space in the intracellular environment. This extraordinary capacity to scavenge purines may in part explain the remarkable ability of *T. gondii* to replicate in virtually any mammalian cell type. Yet while the *T. gondii* parasitophorous vacuole membrane is proposed to be a molecular sieve that permits passive permeation of neutral or charged molecules up to 1300 to 1900 daltons between the host cell cytoplasm and the parasitophorous vacuole space (Schwab *et al.*, 1994; Joiner *et al.*, 1994, 1996), experimental studies specifically addressing permeation or concentration of nucleobases, nucleosides, or nucleotides in the *T. gondii* parasitophorous vacuole space have not been reported.

P. falciparum possesses genes encoding four potential ENT transporters (PfNT1, PFNT2, PfNT3 and PfNT4) (Landfear, 2011). Transport properties of PfNT3 are unknown. PfNT2 is proposed to be a transporter for uridine (Downie *et al.*, 2010), although the significance of uridine transport is unclear because *P. falciparum* does not significantly salvage pyrimidine nucleobases or nucleosides (see Section 20.3.2.2). PfNT4 (aka PfENT4) transports adenine, adenosine, and 2′deoxyadenosine with millimolar range affinity, but has no ability to transport hypoxanthine or AMP (Frame *et al.*, 2012). The significance of these transport activities remains unknown, as PfENT4 is not thought to be the major purine transporter.

P. falciparum expresses an 11-membrane-spanning ENT family nucleoside transporter designated PfNT1 (or PfENT1) which is localized to the parasite plasma membrane (Rager *et al.*, 2001; Carter *et al.*, 2000; Parker *et al.*, 2000). PfNT1 is expressed in all life stages and is markedly up-regulated in young trophozoites corresponding with an increased need for purines in replication. PfNT1 expression in *Xenopus laevis* oocytes confers high affinity adenosine transport (Km ~13 μM) and low-affinity inosine transport

(Km ~253 μM). Previous inhibition studies revealed PfNT1 is broad spectrum for a number of purine and pyrimidine nucleosides, and unlike its mammalian host can transport both D- and L-nucleosides. More recent genetic studies suggest PfNT1 is a transporter for purines. Genetic deletion of the gene for PfNT1 established a null strain that could grow if medium was supplemented with 50 μM, or greater concentrations, of hypoxanthine, adenosine, or inosine (El Bissati *et al.*, 2006). Hypoxanthine, adenosine or inosine concentrations below 50 μM could not support parasite replication. PfNT1 transported hypoxanthine, xanthine, guanine and inosine (El Bissati *et al.*, 2008), and either directly transports adenosine and inosine or alternatively adenosine in the host red cell cytosol is converted to inosine by host adenosine deaminase and inosine in the host cell cytosol is converted to hypoxanthine by host purine nucleoside phosphorylase (El Bissati *et al.*, 2006, 2008).

20.2.2 Purine Transport in the Parasitized Host Cell

The *P. falciparum* parasitized erythrocyte exhibits remarkable alterations in transport functions compared with the normal erythrocyte (Kirk *et al.*, 1994; Gero *et al.*, 2003). Nucleosides enter parasitized erythrocytes through at least three parasite-induced permeation pathways including a saturable high-affinity adenosine transport system (Upston and Gero, 1995), through the tubulovesicular membrane (TVM) network connecting the parasitophorous vacuole to the erythrocyte periphery (Lauer *et al.*, 1997), and through a nonsaturable, anion selective-type channel that also has capacity to transport L-nucleosides, amino acids, sugars, and cations (Kirk *et al.*, 1994). *P. falciparum* has a TVM network extending from the parasitophorous vacuole as well as the requisite machinery to enable regulated protein secretion from the parasite cytoplasm to the vacuolar space, the erythrocyte cytoplasm, and to the erythrocyte periphery (Kyes *et al.*, 2001; Lauer *et al.*, 1997).

The ability of parasitized erythrocytes to selectively transport L-nucleosides was recently used to elegantly deliver therapeutic agents to intracellular *P. falciparum* parasites, selectively killing them (Gero *et al.*, 2003). Conjugating 5-fluorouridine to L-adenosine or L-thymidine selectively delivered cytotoxic 5-fluorouridine to intracellular *P. falciparum*, whereas these hybrid molecules showed no transport or cytotoxicity in mammalian cells. It is unknown whether the new transport functions of the parasitized erythrocyte are due to modification or incorporation of new transporters, channels, ducts or TVM near the erythrocyte periphery.

While mammalian cells lack any ability to transport the mammalian adenosine transporter inhibitor 6-thiobenzylthioinosine (NBMPR), *T. gondii* parasitized host cells selectively transports NBMPR (Al Safarjalani *et al.*, 2003). NBMPR is selectively cytotoxic to intracellular *T. gondii* (el Kouni *et al.*, 1999). The activation of NBMPR cytotoxicity is associated with the ability of *T. gondii* derived enzyme extracts, but not host cell derived enzyme extracts, to phosphorylate NBMPR to its nucleoside 5′-monophosphate. Adenosine kinase (AK) deficient *T. gondii* fail to phosphorylate NBMPR to its nucleoside 5′-monophosphate showing AK to be the major pathway to selective incorporation and cytotoxicity (Al Safarjalani *et al.*, 2003; Rais *et al.*, 2005). Similar to *P. falciparum* parasitized erythrocytes, *T. gondii* parasitized host cells also selectively transport non-physiological β-L-enantiomers of purine nucleosides, β-L-adenosine, β-L-deoxyadenosine and β-L-guanosine. Uninfected host cells do not transport NBMPR or the β-L-nucleosides. NBMPR also inhibits the transport function of the host cell nucleoside transporter ENT1 (*es*) (Gupte *et al.*, 2005). Dipyridamole, another inhibitor of nucleoside transport, inhibited transport of NBMPR and β-L-nucleosides into parasitized host cells. Transport of NBMPR and β-L-nucleosides in the parasitized host cell required a functional TgAT1 transporter (Al Safarjalani *et al.*, 2003; Chiang *et al.*, 1999). While these observations explain a requirement for transport into the intracellular tachyzoite from the parasitophorous vacuole space, these studies do not specifically address why the *T. gondii* parasitized host cell selectively transports these compounds. Therefore infection with *T. gondii* confers parasite specific transport mechanisms to the host cell.

The novel transport capacity of parasitized host cells opens a new avenue towards developing chemotherapeutic approaches, as well as addressing other biological modifications of the parasitized host. The novel transport mechanisms specific to the *T. gondii* parasitized host cell may entail numerous possibilities including the recently reported equilibratory high-affinity adenosine transporter (De Koning *et al.*, 2003), a concentrative ion-dependent channel, a tubulovesicular membrane (TVM) system interconnecting the parasitophorous vacuole to the host cell periphery and a duct for transport of macromolecules that bypasses the host cell membrane (Gero *et al.*, 2003). The elegant electrophysiological description of the mechanism of parasitophorous vacuole formation in *T. gondii* suggested that after *T. gondii* invasion and vacuole formation a fission pore remnant is left on the host cell surface (Suss-Toby *et al.*, 1996). Shortly after invasion of the host cell, a protein and membrane rich intravacuolar network derived from electron-dense granules is formed in the parasitophorous vacuole space (Mercier *et al.*, 1998, 2002; Sibley *et al.*, 1995). Three-dimensional imaging of *T. gondii* within recently formed vacuoles revealed fibrous and tubular material that connects the parasite plasma membrane on intracellular tachyzoites within the parasitophorous vacuole to the remnant of the fission pore at the host cell plasma membrane (Schatten and Ris, 2004). Collectively, these observations suggest that the transport of nutrients such as purines to the tachyzoite within the parasitophorous vacuole may be facilitated by additional mechanisms beyond simple diffusion of nutrients within host cell cytosol through proposed pores in the parasitophorous vacuole membrane. Further studies are necessary to assign functional roles to the ENT gene orthologues identified in *T. gondii*, the requirement for host cell transporters, the potential role of the tubulovesicular network in the parasitophorous vacuole space, or to understand other mechanisms that promote permeation of host purines to the intracellular parasite.

20.2.3 Purine Interconversion and Salvage Pathways in Apicomplexa

Early studies in non-replicating extracellular tachyzoites demonstrated that the purine ring precursor, glycine, was poorly incorporated into *T. gondii* nucleic acids (Perotto and Keister, 1971), and by utilizing a mutant host cell deficient in *de novo* purine synthesis, intracellular *T. gondii* was also shown to poorly incorporated glucose and formate precursors of the guanine and adenine ring (Schwartzman and Pfefferkorn, 1982). A third study demonstrated that *T. gondii* could not synthesize the purine nucleotides *de novo* from formate, glycine, or serine (Krug *et al.*, 1989). Collectively these studies clearly demonstrated that *T. gondii* cannot synthesize purines *de novo* and therefore strictly relies on stealing a supply of purines from the host. Early studies illustrated the potential of targeting the purine auxotrophy of apicomplexan parasites to inhibit parasite replication. Replication of *T. gondii* is efficiently inhibited by adenine arabinoside (ara-A) and the establishment of parasite mutants resistant to ara-A demonstrated that a parasite adenosine kinase (AK) activity is required for activation and incorporation of ara-A (Pfefferkorn and Pfefferkorn, 1976, 1978). Most of the reported work on purine auxotrophy in apicomplexan parasites has focused on the machinery responsible for interconverting purines and incorporating host purine nucleobases and nucleosides into the parasite nucleotide pools. These studies are fairly extensive and have included (i) studies of enzyme activities, (ii) gene cloning, expression, kinetic analysis, structure determination, (iii) studies on regulation and cellular localization, (iv) gene knockouts, (v) studies in mutant host cells, and (vi) genome and evolutionary analysis. *T. gondii* has presented the most amenable model for pursuing all of these varied approaches to deciphering purine acquisition in Apicomplexa.

T. gondii possesses significant machinery for purine salvage and this extensive machinery may enable this parasite to survive and replicate in an extensive range of mammalian cell types. By contrast, *C. parvum* possesses the most diminished purine salvage machinery of the apicomplexans, and this diminished capacity is likely to be related to the nutrient rich, but restricted niche of this parasite. The *P. falciparum* purine pathways are only slightly less robust than pathways present in *T. gondii*. Nonetheless, *Plasmodium* spp. is atypical in infecting and replicating within erythrocytes, and this apicomplexan parasite possesses novel adaptations in the purine pathways that are not observed in other parasites.

20.2.3.1 Purine Salvage Pathways in T. gondii

Recent genetic studies in *T. gondii* have complemented biochemical approaches to more clearly define the transport and purine salvage capacity of this parasite. In early studies, [^{3}H] hypoxanthine labelling of *T. gondii* infected Lesch–Nyhan mutant human host cells, deficient in hypoxanthine–guanine phosphoribosyltransferase activity, demonstrated that only intracellular parasites were labelled with no detectable incorporation into host cell nucleic acids (Pfefferkorn and Pfefferkorn, 1977c). Therefore the flux of purines is unidirectional from the host cell to the parasite.

The only comprehensive biochemical investigation of parasite activities involved in salvage, interconversion and incorporation of host purines in *T. gondii* was assessed in viable extracellular tachyzoites (Krug *et al.*, 1989). To some degree, host cell membranes and host purine metabolism enzymes may contaminate preparations of extracellular tachyzoites (Ngo *et al.*, 2000). This study also used high concentrations of radiolabelled purines to maximize transport and incorporation of purines in extracellular tachyzoites, and such high concentrations of purines are unlikely to be available to the intracellular tachyzoite. Extracellular tachyzoites also are not replicating organisms and this model may not accurately reflect the complexity of the purine interactions between intracellular parasites and the host cell.

The purine bases hypoxanthine, xanthine, guanine and adenine were incorporated and indicated the presence of a hypoxanthine–xanthine–guanine phosphoribosyltransferase (HXGPRT) activity as well as an adenine phosphoribosyltransferase (APRT) activity (Krug *et al.*, 1989). Adenine was incorporated one-half as efficiently as hypoxanthine. Guanine was incorporated at 55% and xanthine at 67% of the rate at which hypoxanthine was incorporated (Krug *et al.*, 1989). Subsequent studies have demonstrated *T. gondii* has no APRT gene or activity (Chaudhary *et al.*, 2004). The purine nucleosides adenosine, inosine, guanosine and xanthosine were incorporated into nucleic acids. Adenosine was incorporated more than 12-fold as well as any other purine nucleoside or nucleobase, and suggested a parasite adenosine kinase (AK) to be the major route to AMP. By contrast, hypoxanthine was incorporated at 8.3%, inosine at 8.2%, xanthine at 5.6%, guanine at 4.6%, adenine at 3.9%, guanosine at 2.5% and xanthosine at 0.3% of the rate at which adenosine was incorporated (Krug *et al.*, 1989). Correspondingly, in parasite protein extracts prepared from extracellular tachyzoites AK activity was greater than 15-fold more active than the next most active enzyme.

Guanine, guanosine, xanthine and xanthosine labelled only guanylate nucleotides. Therefore

T. gondii has no pathway from guanylate to adenylate nucleotides (Krug *et al.*, 1989). Adenosine, inosine, and hypoxanthine labelled adenylate and guanylate nucleotides pool at approximately equal ratios (Krug *et al.*, 1989; Pfefferkorn *et al.*, 2001).

Purine nucleoside phosphorylase (PNP) activities were detected only for inosine and guanosine. Deaminase activities were detected for guanine (GUAD), adenine (ADE), adenosine (ADA) and AMP (AMPD) (Krug *et al.*, 1989). While the reported GUAD may be present in *T. gondii*, this activity is not shown on the current model of purine pathways because this activity was low, there is abundant host GUAD that may contaminate tachyzoite preparations, and no gene orthologue is yet detected for a *T. gondii* GUAD (Chaudhary *et al.*, 2004) (Fig. 20.1A). While a *T. gondii* ADA activity was demonstrated in tachyzoites grown in mutant host cell deficient in host adenosine deaminase (Krug *et al.* 1989), the AK pathway is by far the most significant pathway for incorporation of adenosine due to the high specific activity of AK. While no putative gene orthologue for ADA has been yet identified this pathway is present and may be significant, particularly to parasites lacking AK activity (Chaudhary *et al.*, 2004). A recent study reported that adenine was variably but generally poorly incorporated into *T. gondii* nucleic acids during infection of normal host cells, but in host cells that are deficient in APRT activity, adenine incorporation was significantly less than previously reported in Krug *et al.* (1989) (Chaudhary *et al.*, 2004). While a putative gene orthologue for *T. gondii* ADE has been identified, the ADE pathway to hypoxanthine appears to be a minor pathway for this parasite (Chaudhary *et al.*, 2004; Krug *et al.*, 1989). Hypoxanthine is converted to inosine 5′-monophosphate (IMP) by HXGPRT. Once IMP is available AMP can be made in two steps by adenylosuccinate synthetase (ADSS) and adenylosuccinate lyase (ADSL) and GMP can be made in two steps by IMP dehydrogenase (IMPDH) and GMP synthetase (GMPS). Therefore interconversion of nucleotides occurs only in the direction of adenylate to guanylate nucleotides via AMPD (Krug *et al.*, 1989; Pfefferkorn *et al.*, 2001) (Fig. 20.1A).

There is no GMP reductase, and other than AK no other nucleoside kinase or phosphotransferase activities are present (Krug *et al.*, 1989). Therefore *T. gondii* possesses a minimum of 10 enzymes involved in interconversion and salvage of host purines. Gene orthologues have been reported in nine of these enzymes in *T. gondii* (ADE, PNP, AK, HXGPRT, ADSS, ADSL, AMPD, IMPDH and GMPS) (Chaudhary *et al.*, 2004). *T. gondii* can transport and salvage the host nucleosides adenosine, inosine, and guanosine, as well as the host nucleobases adenine, hypoxanthine, xanthine, and guanine (Fig. 20.1A). The parasite can incorporate host purines into the parasite nucleotide pool by two major routes, via AK and HXGPRT.

20.2.3.1.1 ADENOSINE KINASE AND HYPOXANTHINE–XANTHINE–GUANINE PHOSPHORIBOSYLTRANSFERASE

Incorporation of host adenosine into the AMP pool by AK appears to be the most significant purine salvage pathway (Chaudhary *et al.*, 2004; Krug *et al.*, 1989; Ngo *et al.*, 2000). Yet resistance to ara-A due to mutation and disruption of parasite AK was described even in early studies (Pfefferkorn and Pfefferkorn, 1976, 1978). Subsequently, a genome-wide insertional mutagenesis screen was used to select ara-A resistant mutants. One class of isolated mutants was disrupted in the AK gene and activity, demonstrating again that a parasite with disrupted AK function still replicates normally most likely by salvage through the HXGPRT (Sullivan *et al.*, 1999). *T. gondii* AK has been expressed in *E. coli* for biochemical, kinetic, and structural studies that have revealed significant differences between the parasite AK and the mammalian AK which may be exploited for drug design (Schumacher *et al.*, 2000a, b; Darling *et al.*, 1999). Subversive substrates of

T. gondii AK have been shown to be selectively toxic to the parasite (el Kouni *et al.*, 1999; Iltzsch *et al.*, 1995; Pfefferkorn and Pfefferkorn, 1976). Interestingly, the AK gene from *C. parvum* complements AK deficient *T. gondii*. This complementation system may present a useful high-throughput model in which to screen for potential inhibitors of *C. parvum* AK (Striepen *et al.*, 2004).

HXGPRT represents the second major route of incorporation of host purines. Yet, as for AK, disruption of *T. gondii* HXGPRT has no significant effect on replication or viability of tachyzoites (Donald *et al.*, 1996; Pfefferkorn and Borotz, 1994). The unique xanthine phosphoribosyltransferase activity of HXGPRT, which is absent in the human host, may be exploited for drug design using toxic analogues of xanthine. This approach has been validated in studies using 6-thioxanthine, a compound with selective toxicity to *T. gondii* (Pfefferkorn and Borotz, 1994; Pfefferkorn *et al.*, 2001). The crystal structure and enzyme mechanisms of *T. gondii* HXGPRT have been determined and the parasite HXGPRT is a potential drug target (Heroux *et al.*, 1999a, b; Schumacher *et al.*, 1996).

Parasites completely deficient in HXGPRT or AK are viable, suggesting that either pathway can suffice for purine incorporation and parasite replication (Donald *et al.*, 1996; Sullivan *et al.*, 1999). In the *T. gondii* HXGPRT knock-out parasite, host adenosine would be required for parasite replication and would be incorporated into adenylate nucleotides by AK, and then into guanylate nucleotides by AK, AMP deaminase, IMPDH and GMPS (Fig. 20.1A). Conversely, in a *T. gondii* AK knock-out mutant, host guanine, guanosine and xanthine would provide guanylate nucleotides, but no adenylate nucleotides could be formed from these purines. In this mutant host adenine, adenosine, inosine or hypoxanthine could potentially satisfy the parasite's demand for both adenylate and guanylate nucleotides (Fig. 20.1A). All of the potential host purine precursors would funnel through hypoxanthine into the IMP pool, which can go to guanylate nucleotides through IMPDH and GMPS and to adenylate nucleotides through ADSS and ADSL. Therefore, unlike *P. falciparum* and *C. parvum* (discussed below), *T. gondii* possesses a functionally redundant purine salvage pathway (AK and HXGPRT) with the capacity to meet the purine requirement by using an assortment of potential host cell purine nucleobases and nucleosides. This feature of *T. gondii* may help to explain how the parasite is capable of replicating in such a wide variety of host cells and tissues that are likely to present quite varied potential purine resources. The complex purine salvage pathway of *T. gondii* also suggests that this pathway is likely to be subjected to complex regulatory mechanisms (Fig. 20.1A). For instance, parasites that are growth inhibited by treatment with 6-thioxanthine incorporate 4-fold more hypoxanthine and xanthine into nucleic acids that untreated control parasites. This increase in salvage of hypoxanthine and xanthine was not due to any increase in specific activity of HXGPRT, but involves some other aspect of the salvage pathways (Pfefferkorn *et al.*, 2001).

Recent genetic studies have been performed on *T. gondii* AK and HXGPRT that help to clarify previous observations. HXGPRT and AK activities cannot be simultaneously disrupted in *T. gondii*, suggesting that these are the only functional routes to purine nucleotides in the parasite (Chaudhary *et al.*, 2004). Consequently, it is feasible to knock out the parasite AK and HXGPRT activities as long as at least one functional pathway to purine nucleotides is provided in *trans*. A parasite possessing no functional endogenous *T. gondii* genes for AK and HXGPRT is viable when complemented by a functional APRT gene from *Leishmania donovani* (Chaudhary *et al.*, 2004).

This genetic study also demonstrated that a single gene knock-out of parasite AK or HXGPRT has small but detectable defects in fitness as determined by growth rate of tachyzoites. AK deficient parasites exhibit a fitness defect in growth rate of 7.6% per generation,

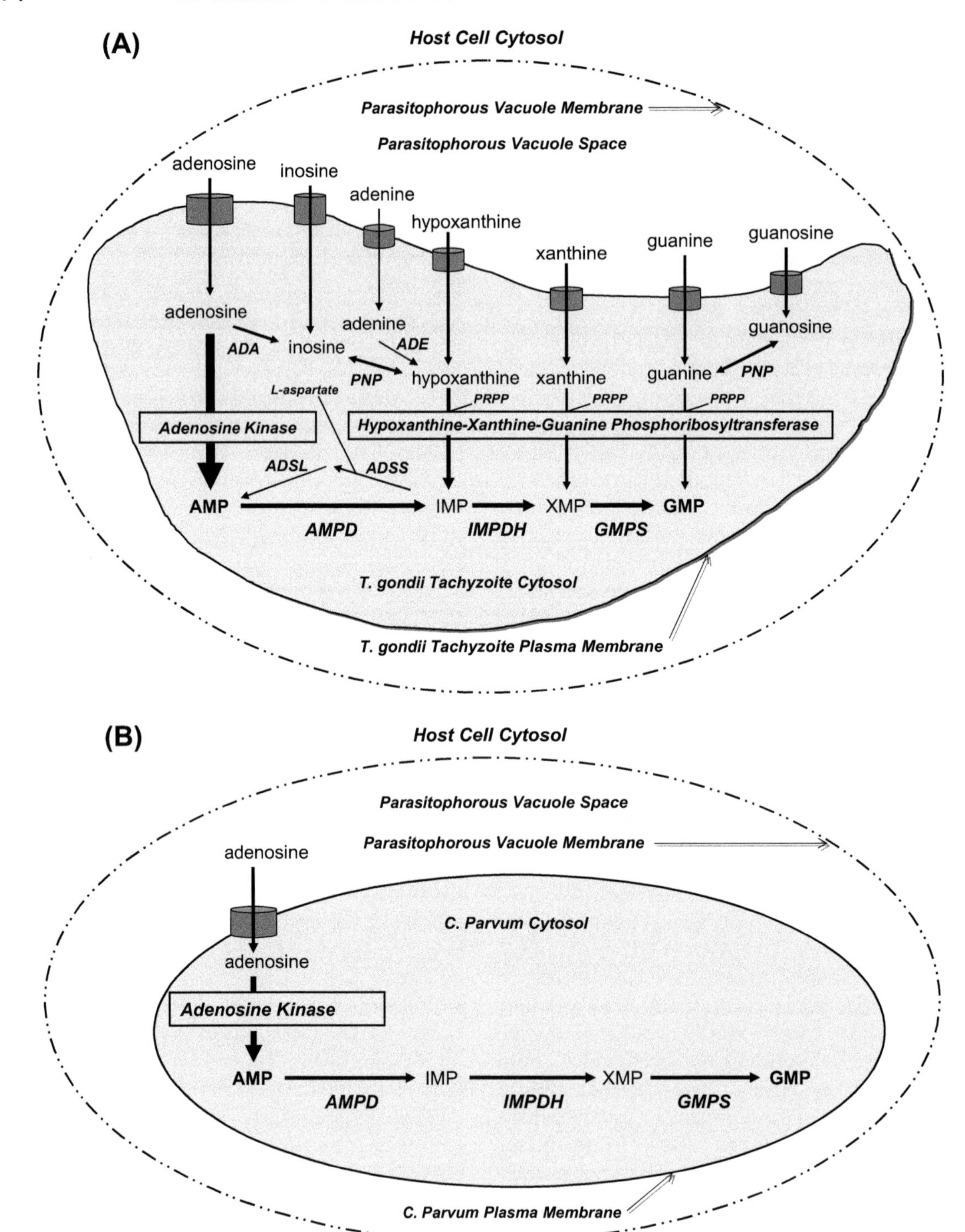
(A)
Host Cell Cytosol
Parasitophorous Vacuole Membrane
Parasitophorous Vacuole Space
adenosine
inosine
adenine
hypoxanthine
xanthine
guanine
guanosine
adenosine
ADA
inosine
adenine
ADE
guanosine
PNP
hypoxanthine
xanthine
guanine
PNP
L-aspartate
PRPP
PRPP
PRPP
Adenosine Kinase
Hypoxanthine-Xanthine-Guanine Phosphoribosyltransferase
ADSL
ADSS
AMP
IMP
XMP
GMP
AMPD
IMPDH
GMPS
T. gondii Tachyzoite Cytosol
T. gondii Tachyzoite Plasma Membrane
(B)
Host Cell Cytosol
Parasitophorous Vacuole Space
Parasitophorous Vacuole Membrane
adenosine
C. Parvum Cytosol
adenosine
Adenosine Kinase
AMP
IMP
XMP
GMP
AMPD
IMPDH
GMPS
C. Parvum Plasma Membrane

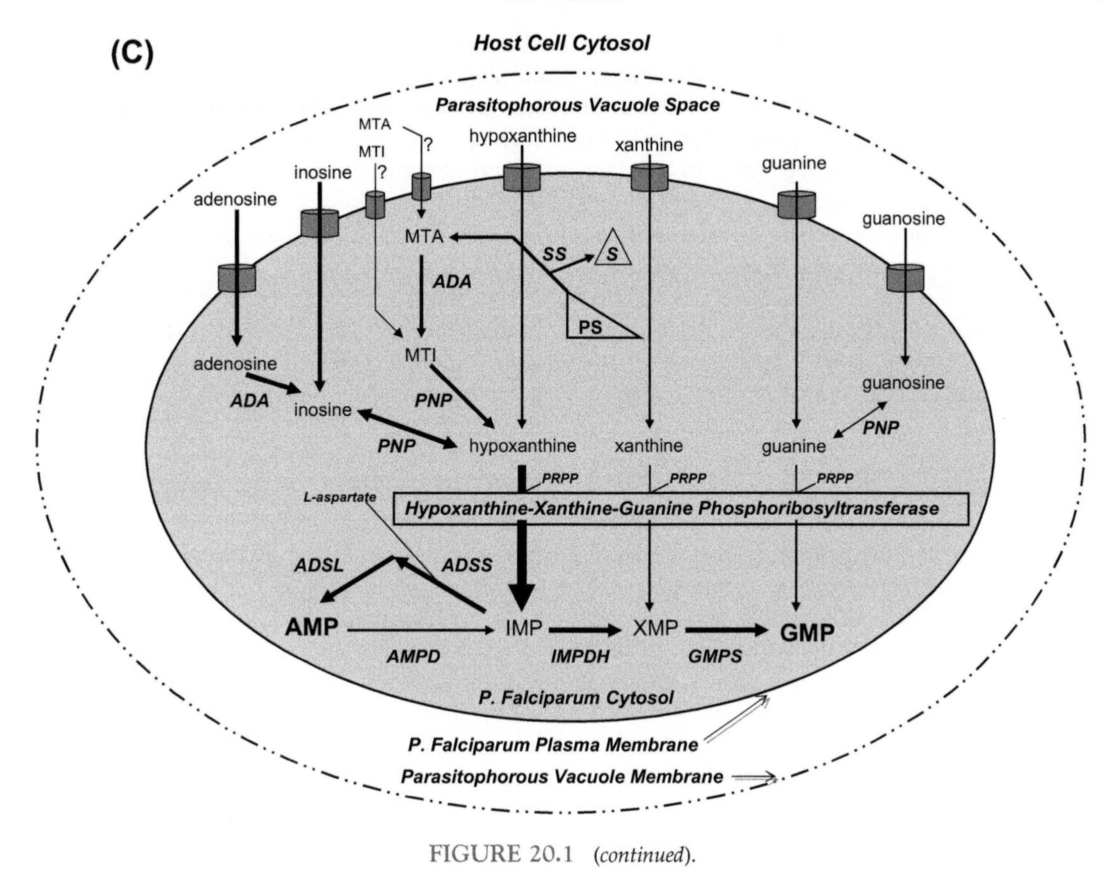

FIGURE 20.1 (*continued*).

FIGURE 20.1 **Model of purine transport and salvage pathways in *T. gondii, P. falciparum* and *C. parvum*.** Model for *Toxoplasma gondii*
A) Model for *Cryptosporidium parvum*
B) Model for *Plasmodium falciparum*
C) The host cell cytosol is shown outside of the parasitophorous vacuole membrane. The parasitophorous vacuole space and the parasite plasma membrane are indicated. Purine transporters are shown as cylinders resting in the parasite plasma membrane. Purine compounds accessible to each parasite are shown inside the parasitophorous vacuole space. Inside the parasite cytosol the enzymic machinery managing purine interconversion and salvage of host purine nucleobases and nucleosides into the adenylate (AMP) and guanylate (GMP) nucleotide pools is shown. Solid lines and arrows depict active pathways. Where information on purine flux is available the weighting of the pathway is emphasized by the weight of the lines and arrows. The weighting of pathways described in this figure reflects the most likely predictions from available data; however, the weighting shown in these diagrams is only hypothetical and the purine flux of host purine to the parasite, as well as interconversion and incorporation within the intracellular parasite, remains to be experimentally tested. Substrates of each enzyme activity are shown on the side of the solid line and the product(s) of each enzyme activity are shown on the arrowhead side. The enzyme activity responsible for each interconversion step is shown in capital italicized text beside the arrowhead line. Adenosine kinase and hypoxanthine–xanthine–guanine phosphoribosyltransferase represent the major pathways for salvage and incorporation into the nucleotide pool and these key activities, if present, are shown enclosed in rectangles. ADA, adenosine deaminase; ADE, adenine deaminase; PNP, purine nucleoside phosphorylase; ADSS, adenylosuccinate synthetase; ADSL, adenylosucccinate lyase; AMPD, AMP deaminase; IMPDH, inosine 5′-monophosphate dehydrogenase; GMPS, GMP synthetase; SS, spermidine synthetase; MTA, 5′-methylthioadenosine; MTI, 5′-methylthioinosine; PRPP, 5-phosphoribosyl-1-pyrophosphate. Enclosed triangles indicate: PS, polyamine synthesis; S, spermidine.

while HXGPRT deficient parasites exhibit a fitness defect in growth rate of 3.7% per generation (Chaudhary *et al.*, 2004). This significant finding suggests that purine acquisition may be rate limiting to parasite growth rate. In the case of the AK knock-out parasite the same host supply of adenosine would be available as in the wild type parent, thus the flux of adenosine, when diverted by lack of AK activity, to inosine to hypoxanthine to IMP, then to guanylate and adenylate nucleotides is not sufficient to fully support the normal parasite growth rate. Therefore parasite transport and incorporation of host adenine, adenosine, inosine, hypoxanthine, guanosine, guanine and xanthine through parasite HXGPRT is insufficient to fully support normal parasite replication (Fig. 20.1A). Similarly, in an HXGPRT knock-out, transport and incorporation of host adenosine through the high activity AK pathway is insufficient to fully support parasite replication suggesting the host supply of adenosine itself is not quite sufficient to fully support the maximum parasite replication rate. Collectively, these observations suggest multiple host purine nucleobases and/or nucleosides and both pathways of incorporation of host purines into the nucleotide pools of *T. gondii* are likely to be required for supporting a maximum replication rate. Considering the very high specific activity of parasite AK (Krug *et al.*, 1989), the bottleneck for purine flux to the replicating intracellular tachyzoite is most likely due to a limited availability of host purines, or a limited transport capacity of the parasitophorous vacuole and intracellular tachyzoite.

20.2.3.1.2 MULTIPLE ISOFORMS AND LOCALIZATION OF HXGPRT

A novel feature of the *T. gondii* purine salvage pathway is the expression of two forms of HXGPRT from a single gene locus by alternatively spliced mRNA (Donald *et al.*, 1996; White *et al.*, 2000). The two isoforms differ by a 49 amino acid segment comprising an extra exon in isoform-II. Both isoforms behave in a kinetically similar manner, although isoform-II is slightly less efficient in recognizing guanine as a substrate (White *et al.*, 2000). The cellular compartmentalization of the two HXGPRT isoforms is different. Isoform-I is cytosolic, while the longer version isoform-II is localized to within the inner membrane complex (IMC) of the tachyzoite (Chaudhary *et al.*, 2005). The 49 amino acid insert at the N-terminus of isoform-II is required for the localization to the IMC. The mechanism of IMC localization was identified to be palmitoylation that occurred at three adjacent cysteine residues within a 49 amino acid insert. Mutation of these three cysteines blocked palmitoylation and localization to the IMC (Chaudhary *et al.*, 2005). The biological basis of functional redundancy of HXGPRT in *T. gondii* is not obvious because both isoforms are functionally competent HXGPRT activities (Donald *et al.*, 1996; Donald and Roos, 1998). It is possible that this functional redundancy in HXGPRT enables *T. gondii* to grow in a wider variety of cell types where purines may be limiting and additional mechanisms may be required for purine transport and salvage. Alternatively, this novel dual localization of HXGPRT may reflect an economy of purine metabolism within the parasite itself, or perhaps some unknown aspect of purine regulation.

20.2.3.1.3 STUDIES ON *T. GONDII* PURINE NUCLEOSIDE PHOSPHORYLASE

The *T. gondii* PNP gene has been cloned and expressed in *E. coli* (Chaudhary *et al.*, 2006). Similar to human and *P. falciparum* PNP, the recombinant *T. gondii* PNP enzyme recognizes inosine and guanosine as good substrates, and adenosine and xanthosine as poor substrates. *T. gondii* PNP, however, is unusual in recognizing deoxynucleosides as poor substrates (Chaudhary, 2005). This unusual substrate property of the parasite PNP suggests that the intracellular parasite may have access to the host pool of guanosine and inosine, but would poorly incorporate host deoxyguanosine or deoxyinosine transported by the parasite. *T. gondii* PNP demonstrated no activity against 5′-methylthioadenosine (MTA) or 5′-methylthioinosine (MTI), and was insensitive to inhibition by methylthio (MT)-immucillin-H (Chaudhary, 2005). Immucillin-H is a strong nM inhibitor of the parasite enzyme *in vitro* (Chaudhary, 2005).

While the replication of wild-type *T. gondii* parasites is completely unaffected by immucillin-H *in vitro*, an AK knock-out mutant of *T. gondii* is inhibited by immucillin-H (Chaudhary, 2005). The growth inhibition of immucillin-H in the AK knock-out background is largely reversed by providing excess hypoxanthine to the *in vitro* culture medium (Chaudhary, 2005). If host adenine and hypoxanthine were being supplied in a sufficient amount to the replicating parasite, this purine supply should confer resistance to immucillin-H even in an AK knock-out mutant (Fig. 20.1A). Therefore based on the growth inhibition of an AK knock-out by immucillin-H, it is possible that the supply of host adenosine and inosine in the intracellular environment exceeds the potential supply of host adenine and hypoxanthine, suggesting that host nucleosides rather than host nucleobases are the more important purine pool (Fig. 20.1A). A caveat to this interpretation of the inhibition of parasite PNP by immucillin-H is the possibility that hypoxanthine may, partly, antagonize inhibition of PNP, or that host PNP is also inhibited by immucillin-H thus reducing host purine pools (Kicska *et al.*, 2002a). Nonetheless, these studies further validate the proposed pathways for purine incorporation and salvage in *T. gondii* (Fig. 20.1A).

20.2.3.1.4 GENETIC SELECTION BASED ON *T. GONDII* HXGPRT

Early studies established the parasite HXGPRT as both a potential drug target and a gene that would be amenable for both positive and negative genetic selection in *T. gondii* (Pfefferkorn and Borotz, 1994). The mechanism of 6-thioxanthine inhibition is based on activation of 6-thioxanthine to 6-thioxanthine 5′-monophosphate by parasite HXGPRT (Pfefferkorn *et al.*, 2001). Unlike mercaptopurine in mammals (Elion, 1989a, b), 6-thioxanthine and its nucleotide product 6-thioxanthine 5′-monophosphate is not a substrate for *T. gondii* GMPS and is not incorporated into nucleic acids. The mechanism of inhibition is parasitostatic and has been suggested to inhibit parasite IMPDH by accumulation of 6-thioxanthine 5′-monophosphate (Pfefferkorn *et al.*, 2001).

T. gondii mutants deficient in HXGPRT are completely resistant to the toxic effects of 6-thioxanthine (Pfefferkorn and Borotz, 1994). Once 6-thioxanthine resistance was selected by knock-out of HXGPRT, parasites with a functional HXGPRT could be positively selected by growth in mycophenolic acid (MPA) with supplements of xanthine or guanine (Pfefferkorn and Borotz, 1994). This selection scheme is based on the ability of MPA to specifically inhibit IMPDH, blocking the conversion of IMP to XMP (Fig. 20.1A). Thus parasites with a non-functional HXGPRT cannot be rescued with xanthine or guanine when IMPDH is inhibited, whereas parasites with a functional HXGPRT will be rescued by xanthine or guanine supplementation of growth medium and this pathway will bypass the inhibition at IMPDH. The biochemical description of this selection strategy also proved that *T. gondii* is perfectly capable of obtaining adenylate nucleotides by the AK pathway, and guanylate nucleotides by HXGPRT salvage of xanthine to XMP, or salvage of guanine to GMP (Pfefferkorn and Borotz, 1994). The identification of the HXGPRT gene enabled a test of this biochemical prediction and resulted in the establishment of a robust genetic selection scheme for positive and negative selection using the selection principles described above (Donald *et al.*, 1996). The HXGPRT selection scheme established the first genetic system for hit and run mutagenesis in *T. gondii* where a stable pseudodiploid could be established during positive selection, then negative selection can be used to force out the HXGPRT gene to create a subtle or major mutation within the gene locus of interest (Donald and Roos, 1998). The HXGPRT genetic model has been extensively used to generate knock out mutants in non-essential parasite genes as well as in genetic studies of parasite development. This genetic system was also adapted for functional cloning studies and has resulted in the initial identification of the ε-proteobacterium-type IMPDH gene from *C. parvum* (Striepen *et al.*, 2002).

20.2.3.2 Purine Salvage Pathways in C. parvum

Experimental work on purine pathways in *C. parvum* has been limited by the lack of robust systems for culturing this parasite, the lack of any genetic models, and the difficulty of obtaining purified parasites for biochemical analysis. The *Cryptosporidium* genomes have given the first detailed insights into the strategy this parasite has adopted to satisfy its appetite for host cell purines (Abrahamsen *et al.*, 2004; Striepen *et al.*, 2004; Xu *et al.*, 2004). In addition to its limited ability to synthesize amino acids and pyrimidines (discussed below), *C. parvum* also possesses a very limited repertoire of purine salvage and interconversion enzymes compared to other apicomplexans (Chaudhary *et al.*, 2004; Striepen *et al.*, 2004). While an early study using crude parasite extracts suggested *C. parvum* expressed HXGPRT and APRT activities (Doyle *et al.*, 1998), *C. parvum* lacks any gene orthologue for APRT or HXGPRT, and the parasite is insensitive to the HXGPRT inhibitor 6-thioxanthine (Pfefferkorn *et al.*, 2001; Striepen *et al.*, 2004; Chaudhary *et al.*, 2004).

Remarkably, adenosine is the only host purine that is of any physiological significance to *C. parvum*, and is supplied by the gut environment or host cells occupied by *C. parvum. C. parvum* (Fig. 20.1B). Similar to *T. gondii* and *P. falciparum*, *C. parvum* possesses an adenosine transporter gene orthologue that is likely to be responsible for transport of host adenosine into the parasite cytosol (Abrahamsen *et al.*, 2004). When host adenosine is available in the parasite cytosol, the parasite can incorporate adenosine into the nucleotide pool by the AK pathway to AMP. Once the parasite has a nucleotide pool of AMP, this pool meets the entire demand for all adenylate and guanylate nucleotides (Abrahamsen *et al.*, 2004). AMP is first deaminated to IMP by AMPD. IMP is then converted to XMP by IMPDH and XMP is converted to GMP by GMPS (Fig. 20.1B). Since there is no reverse pathway from IMP to AMP, the nucleotide flux extends only from AMP to GMP. The *C. parvum* IMPDH gene was acquired by horizontal gene transfer from an ε-proteobacterium (Striepen and Kissinger, 2004; Striepen *et al.*, 2002, 2004). The parasite possesses only these four enzymes to direct the unidirectional flow of transported host adenosine to parasite adenylate and guanylate nucleotides (Fig. 20.1B).

20.2.3.2.1 POTENTIAL DRUG TARGETS IN *C. PARVUM* PURINE SALVAGE PATHWAYS

Unlike *T. gondii*, which has redundant purine salvage pathways, each of the *C. parvum* purine pathway enzymes (AK, AMPD, IMPDH or GMPS) individually represents a potential drug target. The *C. parvum* AK activity has been validated as a drug target in genetic experiments using an AK deficient mutant of *T. gondii*. *C. parvum* AK complements the AK deficiency in *T. gondii*, restoring significant AK activity (Striepen *et al.*, 2004).

IMPDH inhibition using drugs such as ribavirin, mizoribine and MPA has been pursued as a strategy to induce guanylate nucleotide depletion for therapy in cancer, immunosuppressive chemotherapy, and infectious diseases (Pockros *et al.*, 2003; Allison and Eugui, 2000, 2005). The ability of humans and many pathogens to phosphoribosylate guanine to GMP may dilute the potential therapeutic effect of IMPDH inhibition by providing a second route to GMP. The absence of this second route to GMP in *C. parvum* suggests guanylate starvation can be effectively induced via inhibition of

IMPDH or GMPS, as well as AK and AMPD. Ribavirin and MPA were shown in early work to inhibit *C. parvum* development (Woods and Upton, 1998). Recent work has shown *C. parvum* development is inhibited in a dose-dependent manner by these IMPDH inhibitors (Striepen *et al.*, 2004). A screening pipeline for drugs that inhibit *C. parvum* IMPDH has been developed (Sharling *et al.*, 2010). Recently, potent urea inhibitors of IMPDH have been identified (Gorla *et al.*, 2012).

20.2.3.3 *Purine Salvage Pathways in* P. falciparum

The purine salvage and interconversion pathways are more diverse in *P. falciparum* than *C. parvum* and may be slightly less robust than the pathways present in *T. gondii*. Yet, *P. falciparum* (and *Plasmodium* spp.) uniquely possess enzymes of purine metabolism that are highly adapted to the intracellular challenges faced by this parasite. Within the Apicomplexa, *Plasmodium* spp. is unique in surviving and replicating within erythrocytes. The erythrocyte is a highly differentiated cell type that has lost many capabilities normally present in other mammalian cell types and therefore has very limited nucleotide requirements. The human erythrocyte also has no nucleus, does not synthesize DNA, RNA, or protein, and also lacks both purine and pyrimidine *de novo* synthetic pathways.

To fulfil its nucleotide requirements, *P. falciparum* appears to be capable of transporting the purine nucleobases hypoxanthine, xanthine and guanine, as well as the purine nucleosides, adenosine, inosine and guanosine. *P. falciparum* possesses appropriate enzymic machinery for the interconversion and incorporation of any of these six potential purine sources (Chaudhary *et al.*, 2004) (Fig. 20.1C). *P. falciparum* has no reverse pathway from guanylate to adenylate nucleotides, and while the host supplied xanthine, guanine, or guanosine, collectively, could meet the parasite's demand for guanylate nucleotides these precursors could not supply any adenylate nucleotides. Host adenosine, inosine and hypoxanthine, individually or collectively, are therefore likely to be necessary for meeting the parasite's demands for adenylate and guanylate nucleotides. While the parasite can transport additional purine compounds, their physiological relevance, if any, is not clear at this time. For example, adenine is transported by *P. falciparum* yet is not incorporated because the parasite lacks APRT and ADE genes and activities (Chaudhary *et al.*, 2004).

Similar to *T. gondii*, *P. falciparum* can convert adenosine to inosine via ADA, and inosine is then converted to hypoxanthine by parasite PNP (Fig. 20.1C). Similar to human and *T. gondii* PNP, *P. falciparum* PNP does not recognize either xanthosine or adenosine as a good substrate (Kicska *et al.*, 2002a). Hypoxanthine is incorporated into the nucleotide pool as IMP by HXGPRT. Once IMP is available the parasite can meet its entire demand for adenylate and guanylate nucleotides. Notably, *P. falciparum* lacks any detectable AK gene or activity. IMP is converted to AMP in two steps by sequential reactions of ADSS and ADSL. This is the only pathway to AMP. The parasite can balance IMP and AMP pools in the reverse reaction of AMP deamination to IMP (Fig. 20.1C). Since *P. falciparum* has no direct route to AMP from adenosine the most important purine for incorporation into purine nucleotides becomes hypoxanthine. Hypoxanthine is the only purine compound that can completely satisfy the parasite's demand for both adenylate and guanylate nucleotide pools. As also seen in both *T. gondii* and *C. parvum*, IMP is converted to XMP by parasite IMPDH and subsequently to GMP by parasite GMPS (Fig. 20.1). XMP and GMP are also incorporated in *P. falciparum* through the phosphoribosylation of xanthine and guanine, respectively. Thus the parasite ultimately incorporates all purine nucleotides via parasite HXGPRT activity and possesses a minimum of seven distinct purine interconversion or incorporation activities (Fig. 20.1C).

20.2.3.3.1 POTENTIAL DRUG TARGETS IN THE *P. FALCIPARUM* PURINE SALVAGE PATHWAY

The machinery used by *P. falciparum* to satisfy its purine auxotrophy suggests that purine transporters, parasite HXGPRT, certain purine interconversion enzymes or a recently discovered novel purine recycling or salvage pathway may all be amenable targets for drug development. Due to the absence of AK activity in *P. falciparum*, ADA and PNP activities were recognized as potential drug targets in early studies (Daddona *et al.*, 1984, 1986; Webster *et al.*, 1984).

Hypoxanthine could be directly transported by the parasite from the parasitized host cell, or alternatively, hypoxanthine could be supplied from transported inosine and adenosine via ADA and PNP activities. Host erythrocyte hypoxanthine pools may increase during infection from host catabolism of ATP in deteriorating parasitized erythrocytes. Host erythrocyte ATP is catabolized sequentially, via ADP, AMP, adenosine and inosine to produce hypoxanthine via host PNP (Ting *et al.*, 2005). It is known that depletion of hypoxanthine effectively inhibits *P. falciparum* replication *in vitro* (Berman and Human, 1991; Berman *et al.*, 1991), thus the physiological source of hypoxanthine is a critical question to be answered for optimizing strategies to inhibit purine acquisition in *Plasmodium* infections. If the bulk of incorporated hypoxanthine is derived strictly in the parasite cytosol from transported host purines, the parasite ADA and PNP activities may be the optimal drug targets. Complicating this analysis, however, are recent studies that demonstrate *P. falciparum* to be extremely unusual in possessing a pathway from 5′-methylthiopurines to hypoxanthine. If *P. falciparum* can transport 5′-methylthiopurines from the erythrocyte environment the parasite can potentially salvage additional hypoxanthine by this novel pathway (discussed below). Genetic disruption of *P. falciparum* PNP produced a null strain that replicated normally with excess hypoxanthine supplementation, but this strain exhibited a growth defect at low physiological concentrations of hypoxanthine (Madrid *et al.*, 2008). The *P. falciparum* PNP null strain exhibited no PNP activity on inosine or 5′-methylthioinosine (Madrid *et al.*, 2008).

20.2.3.3.2 INHIBITION OF PNP IN *P. FALCIPARUM*

P. falciparum possesses a structurally novel PNP enzyme that is inhibited by immucillin compounds (Kicska *et al.*, 2002a). Several immucillin compounds bind *P. falciparum* PNP with low nanomolar inhibition constants and inhibit enzyme activity. Human PNP is also the target of immucillins, with sub-nanomolar inhibition constants, and currently these compounds are being investigated for therapy in cancer and immunosuppression chemotherapy. Immucillin-H, an immucillin that inhibits both human and *P. falciparum* PNP, induced purine-less death in *P. falciparum* infection (Kicska *et al.*, 2002b). The parasite growth inhibition induced by immucillin-H is reversed by supplementing culture medium with high doses of hypoxanthine, suggesting that the effect of immucillin-H is primarily to block acquisition of essential purines resulting in purine-less death (Kicska *et al.*, 2002b). Crystal structure and energetic mapping studies of *P. falciparum* PNP and immucillin interactions are under investigation to identify more selective inhibitors of the parasite PNP (Lewandowicz *et al.*, 2005; Shi *et al.*, 2004). A more potent immucillin, DADMe-Immucillin-G, which inhibits both *Plasmodium* and human PNP, is able to kill *Plasmodium falciparum* in the Aotus primate model (Cassera *et al.* 2011)

20.2.3.3.3 INHIBITION OF HXGPRT IN *P. FALCIPARUM*

Early studies on inhibitors of *P. falciparum* growth support the parasite HXGPRT activity for drug development (Queen *et al.*, 1990). *P. falciparum* growth *in vitro* is inhibited by purine analogues 6-mercaptopurine, 6-thioguanine and 8-azaguanine. Studies have not been performed to establish whether the mechanism of 6-mercaptopurine inhibition in *P. falciparum* is similar to the mechanism in mammals, which requires activation by phosphoribosylation via host HGPRT, followed by recognition of GMPS to incorporate the toxic analogue into the guanylate nucleotide pool for its toxic incorporation into nucleic acids (Elion, 1989a, b). Early studies characterized the enzymic properties and the gene encoding the *P. falciparum* HXGPRT (Queen *et al.*, 1988; Vasanthakumar *et al.*, 1989, 1990). The crystal structure of the *P. falciparum* HXGPRT revealed unique structural features of the parasite enzyme compared to the structure of the human enzyme (Shi *et al.*, 1999a, b). Additional studies have investigated enzyme mechanisms and features directing substrate specificity and lead transition state inhibitors have been reported (Sarkar *et al.*, 2004; Thomas and Field, 2002; Shi *et al.*, 1999a; Li *et al.*, 1999). Based on the purine incorporation pathways elucidated for *P. falciparum*, selectively blocking incorporation of hypoxanthine by inhibition of HXGPRT has been generally predicted to induce purine-less death in *P. falciparum* (Fig. 20.1C). Screens have been established to identify selective substrates of the *P. falciparum* HXGPRT (Shivashankar *et al.*, 2001). It is also possible that the xanthine phosphoribosyltransferase activity of *P. falciparum* HXGPRT, which is absent in the human host, may also support a mechanism for incorporating toxic xanthine analogues to block parasite replication, analogous to inhibition of *T. gondii* replication by 6-thioxanthine (Pfefferkorn *et al.*, 2001). More recently acyclic immucillin phosphonates (AIPs), and cell permeable AIP prodrugs were developed (Hazleton *et al.*, 2012). Previous immucillin transition state inhibitors of HGXPRT were potent, but unable to cross biological membranes and were ineffective against cultured parasites. The AIP prodrugs are biologically stable inhibitors that block proliferation of cultured parasites and may represent a new strategy to target purine pathways of malaria.

20.2.3.3.4 INHIBITION OF GUANYLATE NUCLEOTIDE POOLS IN *P. FALCIPARUM*

Guanylate nucleotides could arise from guanine supplied by the host, guanine derived from host guanosine by the parasite PNP, and xanthine via HXGPRT. The erythrocyte host of *P. falciparum* lacks a guanosine kinase activity and biochemical evidence suggests that the host guanine and guanosine pools available to *P. falciparum* are limited (Reyes *et al.*, 1982). *P. falciparum* growth *in vitro* is inhibited by psicofuranine, an inhibitor of bacterial GMPS (McConkey, 2000). Psicofuranine inhibition of parasite growth is antagonized by supplying additional guanine to parasite growth medium (McConkey, 2000). These observations provide strong evidence that the primary route to GMP is through GMPS and demonstrates that during *in vitro* cultivation the potential supply of host guanine and guanosine is limited (Fig. 20.1C). Unfortunately, the possible importance of host xanthine in GMP formation is not clarified by these studies. Xanthine can be incorporated into the GMP pool by the sequential reactions of HXGPRT and GMPS. Human erythrocytes are reported to have a concentration of xanthine of $\sim$3.6 μM which may be sufficient for physiological transport and incorporation by HXGPRT and GMPS (Traut, 1994). MPA, a highly specific inhibitor of IMPDH, blocks *P. falciparum* replication at micromolar doses *in vitro* (Queen *et al.*, 1990). While these observations suggest that host xanthine pools are not sufficient to bypass the inhibition at parasite IMPDH, these results cannot be conclusively interpreted because MPA also inhibits host IMPDH and depletes host guanylate nucleotide pools.

20.2.3.3.5 INHIBITION OF ADENYLATE NUCLEOTIDE POOLS IN *P. FALCIPARUM*

The only pathway to AMP is from the IMP nucleotide pool through sequential reactions catalysed by ADSS and ADSL. *P. falciparum* ADSL has been cloned, expressed in *E. coli* for kinetic analysis, and a crystal structure is now available (Eaazhisai *et al.*, 2004; Jayalakshmi *et al.*, 2002). A unique reaction mechanism has been recently described for this essential parasite enzyme (Raman *et al.*, 2004). Targeting either, or both, ADSS and ADSL is predicted to completely deplete *P. falciparum* of essential adenylate nucleotides and inhibit parasite replication (Fig. 20.1C).

20.2.3.4 A Novel Alternative Purine Pathway in Apicomplexa

Of the apicomplexans considered in this chapter, only *P. falciparum* possesses a unique alternative pathway to purines via recycling of 5′-methylthiopurines generated during polyamine metabolism (Ting *et al.*, 2005) (Fig. 20.1C). While earlier studies hinted at the existence of this novel pathway (Sufrin *et al.*, 1995), recent studies have helped to illustrate an important physiological role for the pathway and *in vitro* studies have validated components of the pathway to be drug targets. This is an exciting and emerging area of biology in *P. falciparum* and *Plasmodium* spp. and will require further detailed studies to determine the physiological relevance of these newly described pathways to optimize strategies for drug development. The growth inhibition achieved by treating parasite-infected erythrocytes with immucillins (Kicska *et al.*, 2002a, b) may be multi-faceted because the *P. falciparum* PNP plays a dual role in both conversion of inosine to hypoxanthine, as well as in recycling 5′-methylthioinosine to hypoxanthine (Ting *et al.*, 2005). Therefore, *P. falciparum* has a novel pathway to hypoxanthine. Current evidence suggests this novel pathway plays an important role in parasite metabolism during replication in erythrocytes (Ting *et al.*, 2005).

20.2.3.4.1 THE INTERSECTION OF POLYAMINE AND PURINE METABOLISM IN *P. FALCIPARUM*

Erythrocytes do not synthesize polyamines, and *P. falciparum*, unlike *T. gondii*, must synthesize its own polyamines. Therefore *P. falciparum* replication can be blocked by difluoromethylornithine (DFMO), a mechanism-based inhibitor of ornithine decarboxylase (ODC) involved in conversion of ornithine to putrescine in polyamine synthesis (Muller *et al.*, 2001). The polyamine biosynthesis pathway forms a molecule of 5′-methylthioadenosine (MTA) for each molecule of spermidine, or spermine, that is synthesized. In humans and other organisms the MTA is typically recycled to regenerate both adenine and methionine pools. Surprisingly, the genes associated with purine salvage (APRT) and recycling of MTA into the methionine pool are absent in the *P. falciparum* genome (Gardner *et al.*, 2002a, b; Chaudhary *et al.*, 2004). However, *P. falciparum* ADA was recently discovered to have the novel ability to recognize MTA as a substrate and convert MTA to 5′-methylthioinosine (MTI) (Ting *et al.*, 2005). Additionally, *P. falciparum* PNP was found to have the unique ability to recognize MTI as a substrate and convert MTI to hypoxanthine. This novel pathway was proven to be functional when exogenously supplied MTI was incorporated into *P. falciparum* nucleic acids (Ting *et al.*, 2005). Therefore the unique substrate properties of *P. falciparum* ADA and PNP comprise a novel metabolic pathway from MTA to MTI to hypoxanthine for selective recycling of purines from polyamine biosynthesis (Fig. 20.1C).

Early studies suggested that *P. falciparum* expressed very abundant levels of ADA, PNP and HXGPRT when compared to corresponding activities present in the host erythrocyte, therefore indicating a potentially important secondary role beyond their role in normal purine pathways (Reyes *et al.*, 1982). Although *P. falciparum* grows normally in PNP and ADA deficient host erythrocytes (Daddona *et al.*, 1984, 1986), inhibitors of parasite and host ADA such as coformycin, deoxycoformycin and L-ribosyl analogues of corformycins block *P. falciparum* replication (Daddona *et al.*, 1984; Gero *et al.*, 2003). Selective inhibition of *P. falciparum* PNP, but not host PNP, with MT-immucillin-H blocked parasite replication nearly as efficiently as Immucillin-H (Ting *et al.*, 2005).

This apparent selective inhibition indicates *P. falciparum* PNP is an essential parasite activity. This conclusion is further supported by the significantly decreased ability of MT-immucillin-H to inhibit the PNP null strain of *P. falciparum* (Madrid *et al.*, 2008), mechanistically supporting the parasite PNP as a viable drug target for selective inhibition. It is unlikely that MT-immucillin-H inhibits the polyamine pathway by product inhibition at MTA since *P. falciparum* ADA converts MTA to MTI, and significantly, exogenously supplied hypoxanthine reverses the *in vitro* inhibition achieved by treatment with MT-immucillin-H or immucillin-H (Ting *et al.*, 2005; Kicska *et al.*, 2002b) (Fig. 20.1C). These observations suggest that the host pool (or flux) of hypoxanthine alone is likely to be insufficient to fully support parasite growth during *in vitro* cultivation. Thus the host inosine pool, the host adenosine pool or hypoxanthine recovered in the novel purine recycling pathway is likely to be required for normal replication of *P. falciparum* (Fig. 20.1C). These *in vitro* observations currently validate the *P. falciparum* enzymes ADA and PNP for further drug discovery, as well as further validating enzymes involved in polyamine biosynthesis (Ting *et al.*, 2005; Kicska *et al.*, 2002b). Highly selective inhibitors of *P. falciparum* ADA have been identified (Tyler *et al.*, 2007).

20.2.3.4.2 RECYCLING AND/OR SALVAGE?

A key question remaining to be answered is whether the novel purine recycling pathway in *P. falciparum* plays any direct role in purine salvage. Exogenously supplied MTI is transported by *P. falciparum* and incorporated into parasite nucleic acids (Ting *et al.*, 2005) (Fig. 20.1C). Any supply of host MTA transported by *P. falciparum* could be salvaged to hypoxanthine and incorporated into the IMP nucleotide pool (Fig. 20.1C). MTA may represent a significant source of purines for *P. falciparum* if MTA is abundant in erythrocytes and the parasite has a transporter capable of stealing this molecule from the host cytosol/parasitophorous vacuole space. Alternatively, it is also plausible that *P. falciparum* may either secrete ADA into the erythrocyte cytosol or parasitophorous vacuole space to convert host MTA to MTI or secrete PNP into the erythrocyte cytosol or parasitophorous vacuole space to convert host MTI to hypoxanthine. Both of these strategies could increase the supply of purines to *P. falicparum* because both MTI and hypoxanthine are transported and incorporated into parasite nucleic acids (Ting *et al.*, 2005) (Fig. 20.1C). Therefore studies of the localization of *P. falciparum* ADA and PNP in parasitized erythrocytes, as well as the capabilities of MTI and MTA transport, are necessary to fully resolve the major physiological function of this novel pathway.

20.2.3.5 The Source of Polyamines in Apicomplexa

20.2.3.5.1 PLASMODIUM FALCIPARUM

Host erythrocytes do not synthesize polyamines. By contrast, *P. falciparum* growth *in vitro* is efficiently blocked by the ODC inhibitor DFMO (Muller *et al.*, 2001). ODC catalyses the conversion of ornithine to putrescine in the first step in the synthesis of spermidine and spermine (Tabor and Tabor, 1985). Incorporation of radiolabelled glutamine into the ornithine pool in *P. falciparum* was minor, suggesting that the parasite possesses another major route to acquire or synthesize ornithine (Gafan *et al.*, 2001). In organisms that synthesize polyamines, the other major routes are through enzymes for arginase and ODC, or arginine decarboxylase (conversion of arginine to agmatine) and agmatinase. Arginase directly converts arginine to ornithine, while agmatinase bypasses ornithine entirely by converting agmatine to putrescine (Wu and Morris, 1998). Sensitivity of *P. falciparum* to DFMO suggests that the ODC activity is essential. A *P. falciparum* arginase gene has been cloned and the recombinant enzyme when expressed possessed arginase activity with no detectable agmatinase activity (Muller *et al.*, 2005). As discussed in the pyrimidine and amino acids sections of this chapter all of the apicomplexan parasites are incapable of *de novo* arginine synthesis and therefore must acquire arginine from the host. Arginine taken from the host by *P. falciparum* is utilized in protein synthesis as well as polyamine biosynthesis. The protozoan parasite *Leishmania* spp. also utilizes an arginase activity that is essential for polyamine biosynthesis (Satriano, 2003; Roberts *et al.*, 2004).

P. falciparum has a novel ODC activity that exists as a bifunctional enzyme with S-adenosylmethionine decarboxylase. This bifunctional enzyme enables a balanced synthesis of putrescine from ornithine without involving domain–domain interactions (Birkholtz *et al.*, 2004; Krause *et al.*, 2000; Muller *et al.*, 2000; Wrenger *et al.*, 2001). Putrescine and deoxyadenosylmethionine are combined by spermidine synthase generating a single molecule of spermidine and a molecule of MTA, which is recycled to MTI by ADA, and then to hypoxanthine by PNP (Muller *et al.*, 2000; Chaudhary, 2005; Ting *et al.*, 2005) (Fig. 20.1C). *P. falciparum* also transports host spermidine (Muller *et al.*, 2001). No *P. falciparum* gene orthologue for a spermine synthase has been identified (Chaudhary, 2005).

20.2.3.5.2 *CRYPTOSPORIDIUM PARVUM*

By contrast to *P. falciparum*, *C. parvum* appears to biosynthesize polyamines by the pathway predominantly present in bacteria and plants involving arginine decarboxylase and agmatinase (Yarlett *et al.*, 1996). Based on the extraordinary transport capacities of *C. parvum*, it is also likely that this parasite may also transport putrescine or other polyamines from the rich environment of the human gut. *C. parvum* ADC activity is inhibited by difluoromethylarginine and is unaffected by DFMO (Keithly *et al.*, 1997). Since *C. parvum* lacks any gene orthologue or activity for PNP, this parasite is incapable of recycling 5′-methylthiopurines back into the parasite purine pool (Fig. 20.1B).

20.2.3.5.3 *TOXOPLASMA GONDII*

The origin of polyamines in *T. gondii* is not clearly resolved at this time, although the weight of the current evidence suggests that this parasite is auxotrophic for polyamines and must inhabit host cells that can supply polyamines. The parasite appears to possess transporters capable of transporting ornithine, putrescine and other polyamines (Chaudhary, 2005). *T. gondii* appears to lack ODC activity (Seabra *et al.*, 2004), and growth of tachyzoites is unaffected by DFMO (Chaudhary, 2005). However, the *T. gondii* genome does reveal the presence of a gene orthologue with homology to the ADC/ODC gene family (Kissinger *et al.*, 2003). If functional this gene orthologue would represent a potential ADC based on lack of sensitivity to DFMO. Currently, no gene orthologue can be identified for a member of the related arginase/agmatinase gene family, or any other gene member of the polyamine biosynthetic pathway (Chaudhary, 2005). Collectively, these data suggest that *T. gondii* is incapable of polyamine biosynthesis and most likely relies upon direct transport and salvage of preformed polyamines supplied by the host cell. The absence of any gene orthologue for spermidine synthase or spermine synthase suggests the parasite does not produce MTA. Even in the highly unlikely case that MTA or MTI were present in *T. gondii*, 5′-methylthiopurines could not be recycled into the parasite purine pool. *T. gondii* PNP cannot utilize MTI as a valid substrate (Chaudhary, 2005) (Fig. 20.1A).

20.2.3.5.4 ADAPTATION OF THE POLYAMINE AND PURINE PATHWAYS IN APICOMPLEXA

The selective adaptation of *P. falciparum* ADA and PNP enzymes to recycle 5′-methylthiopurines to hypoxanthine may have arisen in response to the loss of AK activity (Ting *et al.*, 2005). The potential purine flux is reflected by both the demands of the parasite and the purine limitations of the host erythrocyte. Replication of *P. falciparum* occurs by schizogony, where DNA replication and nuclear division are coordinated in a narrow window of time. Schizogony, where many daughter nuclei are rapidly formed, may demand an increased requirement for purine flux into *P. falciparum*. By contrast, *T. gondii* replicates by endodyogeny and creation of daughter parasites every ~7 hours. Endodyogeny, where one daughter nucleus is formed from one mother nucleus, may inherently require a lower flux of purines to sustain replication.

20.3 PYRIMIDINES

Pyrimidines are essential components of nucleic acids and are involved in other aspects of cellular metabolism as well. Most apicomplexan parasites have retained the ability for *de novo* synthesis of the parental pyrimidine molecule, uridine 5′-monophosphate (UMP). The apicomplexan parasites that rely on the *de novo* synthetic pathway are therefore dependent on an appropriate supply of precursor molecules for the biosynthesis of pyrimidines. These precursor molecules include small molecules such as bicarbonate, amino acids and purine nucleotides. Consequently, acquisition of amino acids and satisfying the purine auxotrophy of apicomplexan parasites are necessary for pyrimidine biosynthesis. Most apicomplexans, including *T. gondii* and *P. falciparum*, have functional pathways for *de novo* pyrimidine synthesis. *T. gondii* is unusual in also possessing pyrimidine salvage activities. However, the pyrimidine biosynthetic pathway is essential for replication of both *T. gondii* and *P. falciparum*. Some apicomplexans, such as *C. parvum*, have lost the ability to synthesize UMP by the *de novo* pathway and completely rely on salvage pathways to acquire required pyrimidines from the host. In these apicomplexans acquisition of pyrimidines is completely dependent on transport and salvage of preformed pyrimidine nucleobases and nucleosides from the host. *C. parvum* has uniquely acquired several pyrimidine salvage activities not observed in any other apicomplexan parasite. Relatively little experimental work has been performed on the associated transport pathways essential for pyrimidine biosynthesis or the relationships and likely cross-regulatory talk likely to occur between the pyrimidine and purine pathways. Interestingly, pyrimidine starvation is one of the triggers used to experimentally induce *in vitro* stage differentiation in *T. gondii* from the tachyzoite to the bradyzoite stages (Bohne and Roos, 1997; Roos *et al.*, 1997). Recent studies have clarified the functional organization of pyrimidine metabolism in the apicomplexa and have revealed new strategies to target essential pyrimidine acquisition pathways in apicomplexan parasites.

20.3.1 *De novo* Pyrimidine Synthesis in Apicomplexa

The human host of *T. gondii*, *P. falciparum* and *C. parvum* is capable of significant salvage and biosynthesis of pyrimidines such that either pathway can supply the pyrimidine requirements. Early work in nucleotide metabolism determined that the apicomplexa are generally capable of synthesizing pyrimidines from amino acid precursors glutamine and aspartic acid by the same six-step pathway present in their hosts. While *C. parvum* cannot synthesize pyrimidines *de novo*, *T. gondii* and *P. falciparum* possess the intact pyrimidine biosynthetic pathway (Asai *et al.*, 1983; Hill *et al.*, 1981a, b; O'Sullivan *et al.*, 1981; Schwartzman and Pfefferkorn, 1981; Reyes *et al.*, 1982) (Fig. 20.2). The pyrimidine biosynthetic pathway starts with carbamoyl phosphate

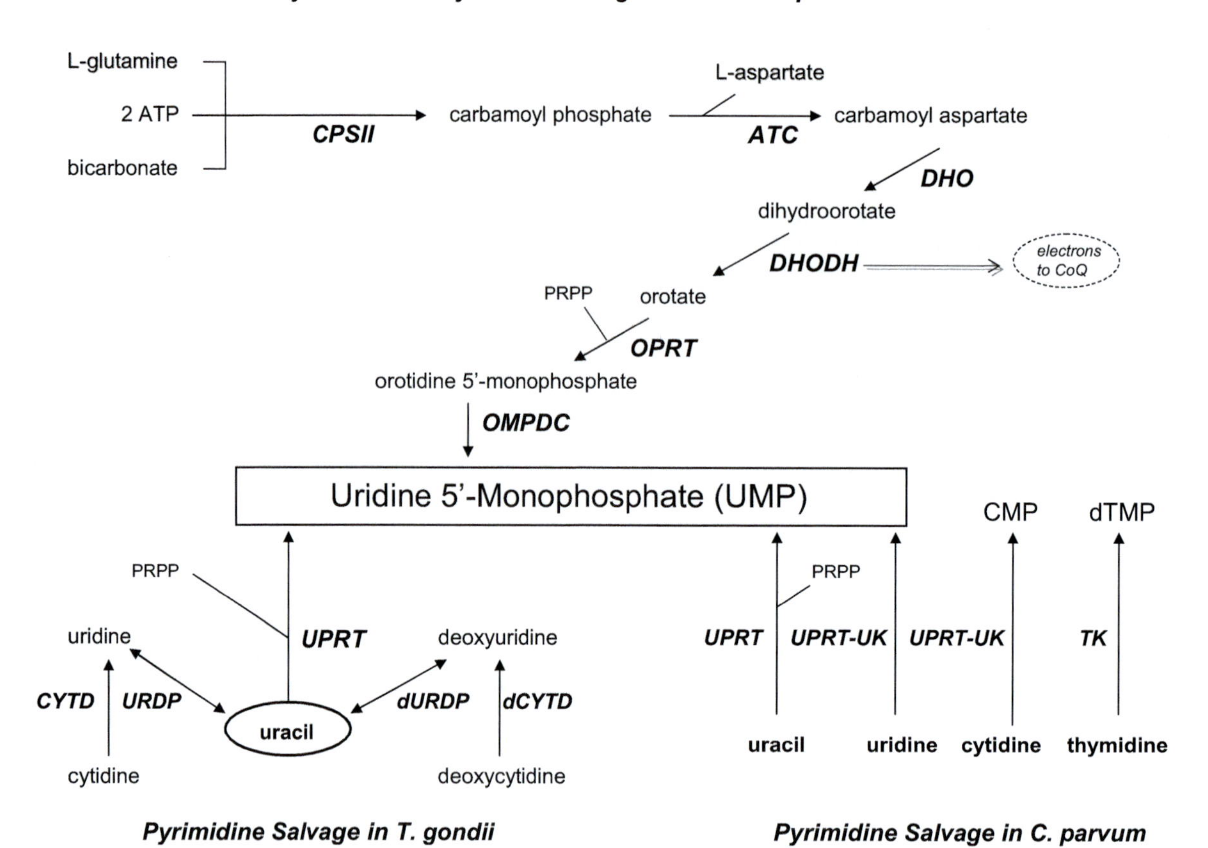

FIGURE 20.2 **Pyrimidine.** Pathways are shown for pyrimidine biosynthesis in *T. gondii* and *P. falciparum* (top left), for pyrimidine salvage in *T. gondii* (bottom left), for pyrimidine salvage in *C. parvum* (bottom right), and for synthesis of CTP from UMP (top right). Solid lines and arrows depict active pathways. Substrates of each enzyme are shown on the side of the solid line and the product(s) of each enzyme activity are shown on the arrowhead side. The enzyme activity responsible for each conversion step is shown in capital italicized text beside the arrowhead line. CPSII, carbamoyl phosphate synthetase II; ATC, aspartate carbamoyltransferase; DHO, dihydroorotase; DHODH, dihydroorotate dehydrogenase; OPRT, orotate phosphoribosyltransferase; OMPDC, orotidine 5′-monophosphate decarboxylase; UPRT, uracil phosphoribosyltransferase; URDP, uridine phosphorylase; dURDP, deoxyuridine phosphorylase; CYTD, cytidine deaminase; dCYTD, deoxycytidine deaminase; UPRT–UK, uracil phosphoribosyltransferase–uridine kinase; TK, thymidine kinase; UMP, uridine 5′-monophosphate; CMP, cytidine 5′-monophosphate; dTMP, thymidine 5′-monophosphate; PRPP, 5′-phosphoribosyl-1-pyrophosphate; CoQ, mitochondrial coenzyme Q.

synthetase II (CPSII) which combines two molecules of ATP, L-glutamine and bicarbonate in a sequence of elegant chemical reactions at multiple active sites to produce a molecule of carbamoyl phosphate (Holden *et al.*, 1999; Kothe *et al.*, 2005). Carbamoyl phosphate is then fused with L-aspartate in the second step by aspartate carbamoyltransferase (ATC) to produce carbamoyl aspartate. Dihydroorotase (DHO) then converts carbamoyl aspartate to dihydroorotate in the third step. In the fourth step of the pathway, dihydroorotate dehydrogenase

(DHODH) creates orotate from dihydrooratate and in doing so also creates electrons to CoQ in the mitochondrion. Orotate is combined with 5-phosphoribosyl-1-diphosphate (PRPP) in the fifth step by orotate phosphoribosyltransferase (OPRT) to produce orotidine-5′-monophosphate. Uridine-5′-monophosphate (UMP), the parent pyrimidine mononucleotide and the precursor of all other pyrimidine nucleotides, is finally produced in the sixth step via the decarboxylation of orotidine-5′-monophosphate (OMP) by orotidine-5′-monophosphate decarboxylase (OMPDC). In all Apicomplexa UMP is phosphorylated to UTP in two sequential steps by UMP kinase and nucleoside diphosphate kinase. UTP is converted to CTP by CTP synthase (CTPS) in a rate-limiting step in all organisms. CTPS is the only known route for *de novo* synthesis of cytidine nucleotides. *T. gondii, P. falciparum* and *C. parvum* all express a CTPS gene (Hendriks *et al.*, 1998; Xu *et al.*, 2004; Abrahamsen *et al.*, 2004; Kissinger *et al.*, 2003). Conversion of ribonucleotides to deoxyribonucleosides occurs at the level of nucleoside diphosphate and is catalysed by a ubiquitous ribonucleotide reductase (RNR).

20.3.1.1 Unique Architecture, Organization and Regulation of CPSII in Apicomplexa

The key regulatory enzyme in the apicomplexan pyrimidine biosynthetic pathway is the first step encoded by CPSII. CPSII catalyses the rate-limiting step and controls the flux through the pathway (Jones, 1980; Evans and Guy, 2004). The architecture of CPSII enzymes from *T. gondii* and *P. falciparum* is unique to the Apicomplexa and other protozoan parasites. *Leishmania* spp., *Trypanosoma* spp., *Babesia bovis, Plasmodium* spp. and *T. gondii* encode a novel glutamine-dependent CPSII activity fused with an N-terminal glutamine amidotransferase (GAT) activity (Aoki *et al.*, 1994; Chansiri and Bagnara, 1995; Flores *et al.*, 1994; Fox and Bzik, 2003; Gao *et al.*, 1998; Gao *et al.*, 1999; Nara *et al.*, 1998). By contrast, mammalian pyrimidine-specific CPSII is contained on the *CAD* gene encoding a multifunctional protein possessing the GAT domain fused via linkers of various lengths in order with CPSII, DHO and ATC (Mori and Tatibani, 1978; Davidson *et al.*, 1993). In *Saccharomyces cerevisiae* a multifunctional protein contains, in order, GAT, CPSII and ATC domains, but is missing the DHO activity found on mammalian *CAD* (Davidson *et al.*, 1993). A different strategy was taken by plants, eubacteria and archebacteria, which express a gene containing the monofunctional CPS, and a separate gene encoding GAT (Jones, 1980; Zhou *et al.*, 2000).

While the *P. falciparum* CPSII gene possesses no introns, the *T. gondii* CPSII is encoded by a single gene with a complex organization of 37 exons and 36 introns specifying a polypeptide of 1687 amino acids (Fox and Bzik, 2003). CPSII from *T. gondii, P. falciparum* (2391 amino acids) and *Babesia bovis* (1645 amino acids) are also much larger polypeptides than is CPSII from other species (Chansiri and Bagnara, 1995; Flores *et al.*, 1994; Fox and Bzik, 2003). Relative to other CPS and GAT domains, the apicomplexan CPSII enzymes have unique sites of insertion within GAT domains, within CPS domains, in the linker region fusing the two CPS halves, as well as within the C-terminal allosteric regulatory domain (Fox and Bzik, 2003).

20.3.1.2 CPSII Activity Required for de Novo Pyrimidine Synthesis is a Validated Drug Target

Conceptually, inhibition of UMP synthesis by the pyrimidine biosynthetic pathway is likely to be a more potent strategy than blocking only thymine nucleotide synthesis and DNA synthesis indirectly through inhibition of folate metabolism (discussed below), because inhibition of pyrimidine biosynthesis will cause starvation for UTP, CTP and TTP nucleotides essential for both RNA and DNA synthesis.

The apicomplexan CPSII enzymes exhibit several differences from the mammalian CPSII that provides a basis for chemotherapy. The novel and large amino acids insertions in *T. gondii* and *P. falciparum* CPSII may provide parasite-specific targets for inhibiting CPSII and *de novo* pyrimidine synthesis in chemotherapy. Targeting ribozymes to a site corresponding to a novel *P. falciparum* CPSII insertion blocked replication of *P. falciparum in vitro* (Flores *et al.*, 1997). A rapid microassay has been developed for *P. falciparum* CPSII activity that may be amenable for high throughput assays (Flores and Stewart, 1998). The GAT activity of *T. gondii* CPSII has also been shown to be a possible target of acivicin (Fox and Bzik, 2003).

Mammalian CPSII is an allosterically regulated enzyme with activity activated by 5-phosphoribosyl-1-pyrophosphate (PRPP), and activity suppressed by UTP (Jones, 1980). *P. falciparum* CPSII is activated by PRPP, and high UTP concentrations cause UTP inhibition (Gero *et al.*, 1984). Remarkably, *T. gondii* CPSII is insensitive to activation by PRPP, but is inhibited by UTP (Asai *et al.*, 1983). The regulatory domain controlling allosteric regulation of CPS activity is contained within the C-terminal ~150 amino acids of the polypeptide (Fresquet *et al.*, 2000; Mora *et al.*, 1999; Evans and Guy, 2004). Both the *P. falciparum* and the *T. gondii* CPSII enzymes possess significant amino acid insertions in the allosteric regulatory domain, as well as divergent amino acid composition (Flores *et al.*, 1994; Fox and Bzik, 2003). The allosteric regulatory domain is the most divergent domain within the entire CPSII suggesting that regulation of CPSII is unique in apicomplexans (Fox and Bzik, 2003). Genetic studies have identified several of the novel CPSII domains to be essential functional domains for enzyme activity or regulation of CPSII (Fox *et al.*, 2009a).

Other enzymes in the *de novo* pathway of *T. gondii* and *P. falciparum* are also under investigation as potential drug targets. *T. gondii* ATC catalyses the second step and is a cytosolic monofunctional enzyme (Asai *et al.*, 1983). Recombinant *T. gondii* ATC product has been produced and characterized (Mejias-Torres and Zimmermann, 2002). The lack of any observed regulation on *T. gondii* ATC further supports the key regulatory role of CPSII in the *de novo* pathway of pyrimidine biosynthesis (Asai *et al.*, 1983; Mejias-Torres and Zimmermann, 2002). *T. gondii* DHO catalyses the third step and is a cytosolic monofunctional enzyme which is unresponsive to any nucleotide (Asai *et al.*, 1983). DHO has been expressed and enzymatically characterized (Robles Lopez *et al.*, 2006). DHODH from *T. gondii* has been cloned and expressed (Sierra Pagan and Zimmermann, 2003). A large number of inhibitors have been recently tested on DHODH and several micromolar inhibitors have been identified (Hortua Triana *et al.*, 2012). Several biosynthetic enzymes from *P. falciparum* including DHO, OPRT and OMPDC have been cloned, expressed and recombinant enzymes characterized (Christopherson *et al.*, 2004; Krungkrai *et al.*, 2004a, b). In *T. gondii* and *P. falciparum* OPRT and OMPDC are present on separate genes (Krungkrai *et al.*, 2004b). Potential inhibitors of apicomplexan CPSII, DHO DHODH, OPRT and OMPDC have been studied (Javaid *et al.*, 1999; Niedzwicki *et al.*, 1984; Krungkrai *et al.*, 2005; Boa *et al.*, 2005; Flores *et al.*, 1997; Seymour *et al.*, 1994; Fox and Bzik, 2003). Recently, highly selective and potent inhibitors of *P. falciparum* DHODH have been identified (Phillips *et al.*, 2008; Phillips and Rathod, 2010).

Genetic inactivation of the pyrimidine biosynthetic pathway by knock-out of the CPSII gene produced a strain of *T. gondii* that is auxotrophic for uracil (Fox and Bzik, 2002) (Fig. 20.2). The *T. gondii* uracil auxotroph replicates at a normal growth rate *in vitro* in the presence of high concentrations of exogenously supplied uracil. By contrast, if uracil is omitted from the *in vitro* culture medium, this mutant will invade host cells normally, but has no detectable growth rate. In murine infections with the *T. gondii* uracil auxotroph, this mutant displayed severe defects in virulence (Fox and Bzik, 2002). Compared with its virulent parental strain RH (Sabin, 1941), the uracil auxotroph was at least 10 million-fold less virulent, or was completely avirulent, suggesting that pyrimidines are not at high enough concentration in the mammalian host to support significant salvage capability. Consequently, the pyrimidine *de novo* synthetic pathway presents a key target for drug development in *T. gondii* and *P. falciparum*. Recently, *T. gondii* OPRT and OMPDC have also been genetically disrupted to establish a severe uracil auxotrophy, suggesting the entire pathway is an excellent target for chemotherapy (Fox and Bzik, 2010; Fox *et al.*, 2009b, 2011). In addition, recent evidence suggests that it is likely that the *de novo* pyrimidine synthesis pathway is also required for sustaining latent infection since disruption of the critical salvage activity mediated by uracil phosphoribosyltransferase (discussed below) has no effect on latent infection (Fox *et al.*, 2011).

20.3.1.3 *Indirect Inhibition of Pyrimidine Biosynthesis*

The pyrimidine biosynthetic pathway of apicomplexans is indirectly inhibited by atovaquone. Atovaquone, a naphthoquinone derivative, in combination with proguanil (Malarone™) is clinically used to treat human malaria infections. Atovaquone is a structural analogue of coenzyme Q (CoQ; ubiquinone) in the mitochondrial electron transport chain. Atovaquone collapses the membrane potential of *Plasmodium* spp. by inhibition of cytochrome *b* in the bc_1 complex (complex III) of the parasite electron transport chain (Srivastava *et al.*, 1999; Korsinczky *et al.*, 2000). Blocking electron flow inhibits mitochondrial membrane associated enzymes such as DHODH that requires electron transfer to CoQ when it oxidizes dihydroorotate to orotate in the fourth step of the pyrimidine biosynthetic pathway (Fig. 20.2). Interestingly, the sole function of mitochondrial electron transport in *P. falciparum* appears to be its role in regeneration of ubiquinone required as the electron acceptor for DHODH to sustain the essential *de novo* pyrimidine synthesis pathway (Painter *et al.*, 2007).

Resistance of *Plasmodium* spp. to atovaquone is associated with specific mutations within the parasite cytochrome *b* (Korsinczky *et al.*, 2000; Srivastava *et al.*, 1999). Atovaquone treatment of *P. falciparum in vitro* causes major accumulations of carbamoyl aspartate and dihydroorotate demonstrating the breakdown in electron flow disrupts DHODH and causes significant substrate accumulation leading to starvation of UMP (Seymour *et al.*, 1994).

Atovaquone is also an approved drug for treatment of acute toxoplasmosis. As with *P. falciparum*, the DHODH of *T. gondii* is most likely associated with mitochondrial membranes. DHODH purifies from *T. gondii* in the particulate fraction of tachyzoites and is inhibited by respiratory chain inhibitors (Asai *et al.*, 1983). The DHODH gene of *T. gondii* is most similar to the family of DHODH enzymes linked to the respiratory chain in mitochondria for their catalytic redox force (Sierra Pagan and Zimmermann, 2003). *T. gondii* mutants resistant to atovaquone can be selected *in vitro* (Pfefferkorn *et al.*, 1993), and these mutants possess mutations within the *T. gondii* cytochrome *b* gene suggesting this function to be the target of atovaquone (McFadden *et al.*, 2000).

20.3.2 Pyrimidine Salvage in Apicomplexa

While *de novo* pyrimidine synthesis is essential for *T. gondii* and *P. falciparum*, *C. parvum* lacks all six enzymes required for synthesis of UMP (Fig. 20.2). Therefore pyrimidine salvage in *C. parvum* presents a key target for drug development.

20.3.2.1 *Salvage of Pyrimidines in* **C. parvum**

The genome sequences of *C. parvum* and *Cryptosporidium hominis* demonstrated that

Cryptosporidium spp. do not retain any of the six enzymes comprising the pyrimidine biosynthetic pathway (Abrahamsen *et al.*, 2004; Striepen *et al.*, 2004; Xu *et al.*, 2004; Puiu *et al.*, 2004) (Fig. 20.2). Thus *C. parvum* is completely dependent on the host for supplying pyrimidines that must be salvaged to meet the pyrimidine demand of the parasite. Remarkably, *C. parvum* has three potentially significant pyrimidine salvage activities expressed by three genes. *C. parvum* encodes a thymidine kinase (TK), a monofunctional uracil phosphoribosyltransferase (UPRT), and a bifunctional polypeptide having UPRT and uridine kinase (UPRT–UK) activities (Striepen *et al.*, 2004). With this repertoire of salvage activities, *C. parvum* can potentially convert thymidine (TK), cytidine and uridine (UK), and uracil (UPRT) to their respective pyrimidine-5′-monophosphate (Fig. 20.2). Biochemical evidence supports the expression of active TK and UK enzymes in *C. parvum* infection based on incorporation of bromodeoxyuridine and cytosine–arabinoside, respectively (Striepen *et al.*, 2004; Woods and Upton, 1998). However, recent expression of active TK indicates that AraT or AraC are not substrates for *C. parvum* TK (Sun *et al.*, 2010). The proposed UPRT gene activity has not yet been experimentally validated by either enzyme assay or demonstrating the incorporation of uracil during *C. parvum* infection. The presence of TK and UK activities in *C. parvum* is highly unusual when compared with their absence in other apicomplexan parasites such as *T. gondii* and *P. falciparum*. Phylogenetic analysis indicates that genes involved in nucleotide metabolism in *C. parvum* are incorporated by horizontal gene transfer from bacterial (TK), algal (UK–UPRT), or protozoan sources (Abrahamsen *et al.*, 2004; Huang *et al.*, 2004; Striepen and Kissinger, 2004; Striepen *et al.*, 2004). *C. parvum* may use the parasitized host cell, the gut environment itself, or both as the source of the pyrimidine nucleobase/nucleoside precursors.

C. parvum growth *in vitro* is inhibited by cytosine–arabinoside, a prodrug that is activated by UK (Woods and Upton, 1998; Pfefferkorn and Pfefferkorn, 1976). The UPRT as well as the bacterial-type TK activity of *C. parvum* potentially could be similarly exploited for selectively targeting incorporation of toxic analogues. *C. parvum* TK has been shown to activate the prodrugs 5-fluorodeoxyuridine and trifluorothymidine and these compounds inhibit *in vitro* replication (Sun *et al.*, 2010). Thus the genome sequence of *C. parvum* has revealed the lack of biosynthetic capability and several newly identified essential salvage activities that may be amenable for drug development. An important goal will be to experimentally determine whether each gene of pyrimidine salvage (UPRT–UK, UPRT or TK) is individually capable of fully supporting parasite replication *in vivo*, or whether multiple activities are required.

20.3.2.2 Salvage of Pyrimidines in P. falciparum

P. falciparum appears to be highly restricted in its ability to salvage pyrimidines marking the *de novo* pyrimidine synthetic pathway a key drug target. Even though salvage is highly restricted, *P. falciparum* is capable of significant transport and accumulation of pyrimidine nucleobases and nucleosides (Lauer *et al.*, 1997). For example, *P. falciparum* is capable of accumulating thymidine, and can incorporate exogenously supplied orotic acid and cytotoxic 5-fluoroorotic acid (Lauer *et al.*, 1997; Rathod *et al.*, 1989). The poor ability of *P. falciparum* to salvage pyrimidines is related to the absence of genes encoding the salvage enzymes UPRT, TK and UK that are required for any incorporation of pyrimidine nucleobases or nucleosides into the nucleotide pool (Gardner *et al.*, 2002a, b; Striepen *et al.*, 2004).

20.3.2.3 Salvage of Pyrimidines in T. gondii

The pyrimidine salvage capabilities of *T. gondii*, *P. falciparum* and *C. parvum* are summarized in Figure 20.2. Unlike *P. falciparum* or *C. parvum*, *T. gondii* has the six-step *de novo* synthetic pathway yet has also retained potentially significant salvage activities. Early labelling studies demonstrated that uracil was well incorporated into intracellular or extracellular tachyzoites but did not label host cell nucleic acids (Pfefferkorn and Pfefferkorn, 1977a). Importantly, these initial studies demonstrated efficient and selective labelling of *T. gondii* nucleic acids by uracil, a method still in use today to assess parasite replication. The authors of this work also made the important biological observation that the host cell has no access to the parasite pyrimidine nucleotide pools.

A parasite mutant was first selected to be resistant to 5-fluorodeoxyuridine (FUDR) and both intracellular (or extracellular) tachyzoites of this mutant were found to be deficient in their ability to incorporate uridine, deoxyuridine and uracil (Pfefferkorn and Pfefferkorn, 1977a, b). Correspondingly, this mutant was also determined to be co-resistant to 5-fluorouracil and 5-fluoruridine. In pioneering work, the basis of the FUDR resistant mutant was identified as a biochemical defect in the parasite UPRT activity (Pfefferkorn, 1978), thus demonstrating that the UPRT activity was nonessential to the tachyzoite stage. Recent genetic evidence also reveals that the UPRT activity is not essential for development or maintenance of latent *T. gondii* infection (Fox *et al.*, 2011). These observations suggest that the *de novo* pyrimidine synthesis pathway is fully capable of supporting parasite replication during the acute and chronic stages of infection. Carbon dioxide starvation was used to suppress *T. gondii de novo* pyrimidine synthesis (blocking step 1, CPSII) and growth of wild-type parasites could be rescued with uracil supplementation, whereas growth of the FUDR-resistant mutant was not rescued. These observations suggested that *T. gondii* is fully capable of growth *in vitro* by salvage of uracil in the absence of a functioning pyrimidine biosynthetic pathway to UMP (Pfefferkorn, 1978).

T. gondii salvage capacities were comprehensively investigated using biochemical measurements of enzyme activities present in protein extracts prepared from isolated tachyzoites. *T. gondii* possesses salvage activities enabling the parasite to salvage a variety of pyrimidines including deoxycytidine, deoxyuridine, cytidine, uridine and uracil (Iltzsch, 1993) (Fig. 20.2). *T. gondii* can also recover pyrimidines arising from degradation of parasite nucleic acids through degradation of dUMP, dCMP, CMP to their corresponding nucleosides by nucleoside 5′-monophosphate phosphorylase (Iltzsch, 1993). The parasite cannot directly obtain phosphorylated nucleotides from the host cell but can transport and incorporate pyrimidine nucleobases and nucleosides (Iltzsch, 1993; Pfefferkorn and Pfefferkorn, 1977a, b). Uracil is the only pyrimidine compound that can be directly incorporated into the pyrimidine pool by conversion to UMP by the major UPRT activity. All other pyrimidine compounds are first catabolized to uracil mediated by parasite activities for uridine phosphorylase, deoxyuridine phosphorylase, cytidine deaminase, and deoxycytidine deaminase (Fig. 20.2). This unique salvage strategy has been described by Dr E. R. Pfefferkorn as a salvage funnel to uracil (Pfefferkorn, 1978). The *T. gondii* genome sequence suggests that the uridine/deoxyuridine/thymidine phosphorylase activities and the cytidine/deoxycytidine deaminase activities are likely to be present on single polypeptides (Iltzsch, 1993; Kissinger *et al.*, 2003).

Although *T. gondii* has retained the ability to interconvert thymine and thymidine, the parasite lacks any TK activity and is incapable of salvaging thymidine. However, these compounds can be incorporated into *T. gondii* which has been transformed with a TK gene derived from herpes simplex virus (Fox *et al.*, 2001). Similarly, *T. gondii* cannot salvage cytosine unless the parasite has been transformed with a bacterial cytosine deaminase (CD) gene (Fox *et al.*, 1999).

20.3.2.3.1 POTENTIAL INHIBITORS OF PYRIMIDINE SALVAGE IN *T. GONDII*

Genetic studies on the pyrimidine salvage pathway have been performed in *T. gondii.* In early mutagenesis and biochemical studies UPRT was demonstrated to be dispensable and therefore UPRT could be used as a target for genetic inactivation (Pfefferkorn, 1978). Integration of plasmid DNA into the UPRT locus was selected using 5-fluorouracil (Donald and Roos, 1995). Mutants selected to be resistant to 5-fluorouracil compounds were defective in UPRT activity. Isolation of plasmid transgenes and surrounding chromosomal sequences demonstrated integration of plasmid in these mutants occurred at the UPRT locus, which was cloned and characterized in the same study. Recent studies disrupted the gene for uridine phosphorylase and this activity was found to be non-essential (Fox and Bzik, 2010; Fox *et al.*, 2011).

Therefore direct inhibition of salvage enzymes in *T. gondii* is unlikely to perturb parasite replication during acute or chronic infection. Nonetheless, parasite enzymes may have unique substrate properties and potential inhibitors of *T. gondii* uridine phosphorylase have been assessed (el Kouni *et al.*, 1996; Iltzsch and Klenk, 1993). UPRT activity expressed by *T. gondii* is largely absent in the mammalian host and this suggests another pathway of drug development. In addition to its natural substrate uracil, *T. gondii* UPRT also recognizes 2,4-dithiouracil and incorporates this analogue into parasite nucleic acids. Interestingly, this property of UPRT has been adapted to enable cell-specific microarray analysis of mRNA synthesis and decay (Cleary *et al.*, 2005). UPRT also recognizes 5′-fluorouracil compounds as well as other uracil analogues that suggest a pathway of drug development based on selective incorporation of toxic analogues by parasite UPRT. This approach has been validated in studies using 5′-fluorouracil, 5′-fluorouridine, FUDR and emimycin (Pfefferkorn, 1978; Pfefferkorn *et al.*, 1989; Pfefferkorn and Pfefferkorn, 1977a, b). 5′-fluorouracil compounds are incorporated by UPRT then ultimately become an inhibitor of thymidylate synthase that prevents synthesis of thymine nucleotides. The crystal structure of *T. gondii* UPRT has been determined and efforts are under way to identify potential analogues that may be selectively incorporated into parasite nucleic acids to block replication of tachyzoites (Iltzsch and Tankersley, 1994; Schumacher *et al.*, 1998, 2002; Iltzsch, 1993; Carter *et al.*, 1997).

20.3.3 Assessment of Pyrimidine Synthesis and Salvage Pathways Related to Parasite Niche

20.3.3.1 P. falciparum *and* C. parvum

The different approaches used by *P. falciparum* and *C. parvum* to acquire their pyrimidines are likely to be related to their particular niche within the human host. *C. parvum* resides in the gut, an environment rich in a nutrient supply, whereas *P. falciparum* resides initially in hepatocytes and subsequently during within the erythrocyte, a host cell that loses its ability to synthesize pyrimidines and is not engaged in the business of transcription or replication. The erythrocyte niche of *P. falciparum* has a low abundance of pyrimidines. *P. falciparum* has compensated over time by abandoning the enzymes, if initially present, that can salvage pyrimidines. *P. falciparum* has retained a robust *de novo* pyrimidine synthetic pathway fully capable of supporting rapid replication.

By contrast, *C. parvum* has no component of the pyrimidine biosynthetic pathway. *C. parvum* as a gut pathogen discovered free food and acquired enzymes via horizontal transfer from other organism to take advantage of pyrimidine nucleobases and nucleosides present in the environment. The *C. parvum* genome reveals in general that this pathogen is rich in genomic bias with a plethora of transporters including putative transporters of nucleobases and nucleosides (Abrahamsen *et al.*, 2004; Puiu *et al.*, 2004).

20.3.3.2 T. gondii

By contrast to *P. falciparum* and *C. parvum*, *T. gondii* can invade and replicate in most human tissues and cells and correspondingly has more significant capabilities in possessing both pyrimidine biosynthesis and salvage pathways. *T. gondii* can rely strictly on the *de novo* pathway and this pathway is required for virulence in mammals (Fox and Bzik, 2002). These observations demonstrate *T. gondii* can readily obtain the amino acid precursors and the necessary ATP from purine salvage to synthesize pyrimidines *de novo* in the wide variety of cells and tissues that this parasite can infect. Surprisingly, *T. gondii* tachyzoites blocked in the pyrimidine biosynthetic pathway can grow strictly on the salvage pathway *in vitro* with uracil supplements (Pfefferkorn, 1978; Fox and Bzik, 2002). Yet complete disruption of the salvage pathway through mutation and disruption of UPRT has no detectable effect on tachyzoite growth or virulence (Donald and Roos, 1995; Pfefferkorn, 1978). These observations raise the puzzling question as to why *T. gondii* has retained the capacity to grow on either pathway, whereas *C. parvum* and *P. falciparum* have not.

20.3.3.2.1 POTENTIAL ROLES OF *T. GONDII* UPRT

It remains a mystery why uracil is so well incorporated into nucleic acids in tachyzoites and why the UPRT activity of the parasite is such a major activity (Pfefferkorn, 1978). The retention of a non-essential gene for the parasite UPRT and other salvage activities suggests that some advantage may be conferred by its expression. *T. gondii* UPRT recognizes uracil as its only natural substrate for pyrimidine compounds that are normally available in the mammalian host (Carter *et al.*, 1997; Pfefferkorn, 1978). It is possible that the expression of UPRT confers some minor advantage to intracellular parasites by enabling the recovery and reincorporation of pyrimidines into the UMP pool that are catabolized from *T. gondii* or host nucleic acids and nucleotides.

The ability of *T. gondii* to grow or survive on uracil alone in the absence of the *de novo* pyrimidine pathway may be essential during another life stage of the parasite other than the tachyzoite stage. Because UPRT is an enzyme that is absent in mammals but commonly found in bacteria and plants it is plausible that uracil may be a nutrient that is present in certain non-mammalian environmental niches. Within these niches the environmentally stable oocyst stage may have some access to uracil. Indirect biochemical evidence suggests *T. gondii* UPRT may interact with the soil environment. The naturally occurring antibiotic emimycin produced by *Streptomyces* spp. (primarily an inhabitant of soil) is an equivalent substrate to uracil for *T. gondii* UPRT (Terao, 1963; Pfefferkorn *et al.*, 1989; Terao *et al.*, 1960). Emimycin is a selective inhibitor of *T. gondii* growth and nucleic acid synthesis and parasite mutants with disrupted UPRT activity are resistant to emimycin. Emymycin is incorporated to emimycin 5′-monophosphate (EMP) by *T. gondii* UPRT and further to EDP and ETP. However, ETP is not well incorporated into nucleic acids and the inhibition of *T. gondii* RNA and DNA synthesis may occur at the level of parasite RNA and DNA polymerases (Pfefferkorn *et al.*, 1989).

It is also possible that the *T. gondii* UPRT activity was retained because it plays an important role in another aspect of cell biology or metabolism. The structure and biochemical properties of UPRT demonstrated that in the absence of substrates or its activator GTP the enzyme behaves as a homodimer composed of two identical subunits. In the presence of GTP, GTP binding stabilizes an active tetrameric structure of UPRT exhibiting high enzyme activity compared to the homodimer (Schumacher *et al.*, 2002). Based on these observations it was suggested that *T. gondii* UPRT is likely to play some role in balancing purine and pyrimidine pools in *T. gondii* (Schumacher *et al.*, 2002). If UPRT is involved in balancing pyrimidine and purine pools this balance may be best achieved in the non-mammalian environment because UPRT cannot substantially contribute to parasite pyrimidine nucleotide pools in infected mammalian cells or animals (Fox and Bzik, 2002).

20.3.4 Folate Pathways and Synthesis of Thymine Nucleotides

For decades, the folate pathway has been a key target for antimicrobial agents directed against *T. gondii* and *P. falciparum*, and is under current investigation for targeting *C. parvum* (Anderson, 2005). Thymine nucleotides are formed during the thymidylate cycle by the methylation of dUMP to produce dTMP in a reaction mediated by thymidylate synthase (TS) in folate metabolism (Fig. 20.3). During conversion of dUMP to dTMP a molecule of tetrahydrofolate is oxidized to dihydrofolate. During the thymidylate cycle, *T. gondii*, *P. falciparum* and *C. parvum* rely on TS to produce dTMP from dUMP, on DHFR for recycling of dihydrofolate to tetrahydrofolate and on serine hydroxymethyltransferase to produce 5,10-methylene-tetrahydrofolate from serine (Fig. 20.3). Therefore thymidine starvation can be induced in *T. gondii* and *P. falciparum*, as well as humans, by reducing the pool of tetrahydrofolate via inhibition of DHFR and recycling of dihydrofolate to tetrahydrofolate. *C. parvum* is unusual in that it also expresses a TK activity that grants this parasite a potentially significant salvage pathway to dTMP (Abrahamsen *et al.*, 2004; Striepen *et al.*, 2004) (Figs 20.2 and 20.3). It is currently unknown whether inhibition of DHFR will inhibit *C. parvum* replication.

20.3.4.1 Biosynthesis of Folates in Apicomplexa

The primary precursors for the *de novo* synthesis of folates are *para*-aminobenzoic acid and guanosine 5′-triphosphate (GTP), a purine nucleotide that protozoan parasites cannot synthesize *de novo*. Humans lack the ability to synthesize folates *de novo* and rely on folate transport from dietary sources. By contrast, several of the apicomplexans possess an endogenous folate biosynthetic pathway that is susceptible to antifolate inhibitors. *T. gondii* and *P. falciparum* possess the full complement of genes encoding all seven enzymic steps in the *de novo* biosynthesis of 7,8-dihydrofolate (Fig. 20.3). By contrast, genome analysis indicates that *C. parvum* lacks any identifiable gene orthologue for folate biosynthetic enzymes and is apparently dependent on folate salvage to obtain 7,8-dihydrofolate for the thymidylate cycle (Abrahamsen *et al.*, 2004; Striepen *et al.*, 2004) (Fig. 20.3). *P. falciparum*, *C. parvum* and *T. gondii* possess gene orthologues for putative folate transporters indicating these parasites have the potential to salvage folates from the host (Striepen *et al.*, 2004; Chaudhary, 2005). Both the salvage pathway to *para*-aminobenzoic acid as well as the *de novo* synthesis of 7,8-dihydrofolate from GTP and *para*-aminobenzoic acid are functional in *P. falciparum* because labelled precursors are incorporated into end products (Wang *et al.*, 2004b). Functional data on folate salvage in *T. gondii* and *C. parvum* are not available. In addition to salvage from the host, a second pathway to *para*-aminobenzoic acid exists as a product of the shikimate pathway via chorismate. While *C. parvum* possesses none of the genes involved in this pathway, *T. gondii* and *P. falciparum* encode several of the enzymes of this pathway (Fig. 20.3) (Roberts *et al.*, 1998; McConkey *et al.*, 2004).

20.3.4.2 Antifolate Chemotherapy and Antifolate Resistance

For several decades antifolates such as pyrimethamine (PYR) and proguanil (PG) that target DHFR and folate recycling and sulpha drugs that target dihydropteroate synthase (DHPS) have been in clinical use for the treatment of *P. falciparum* and *T. gondii* infections (Brooks

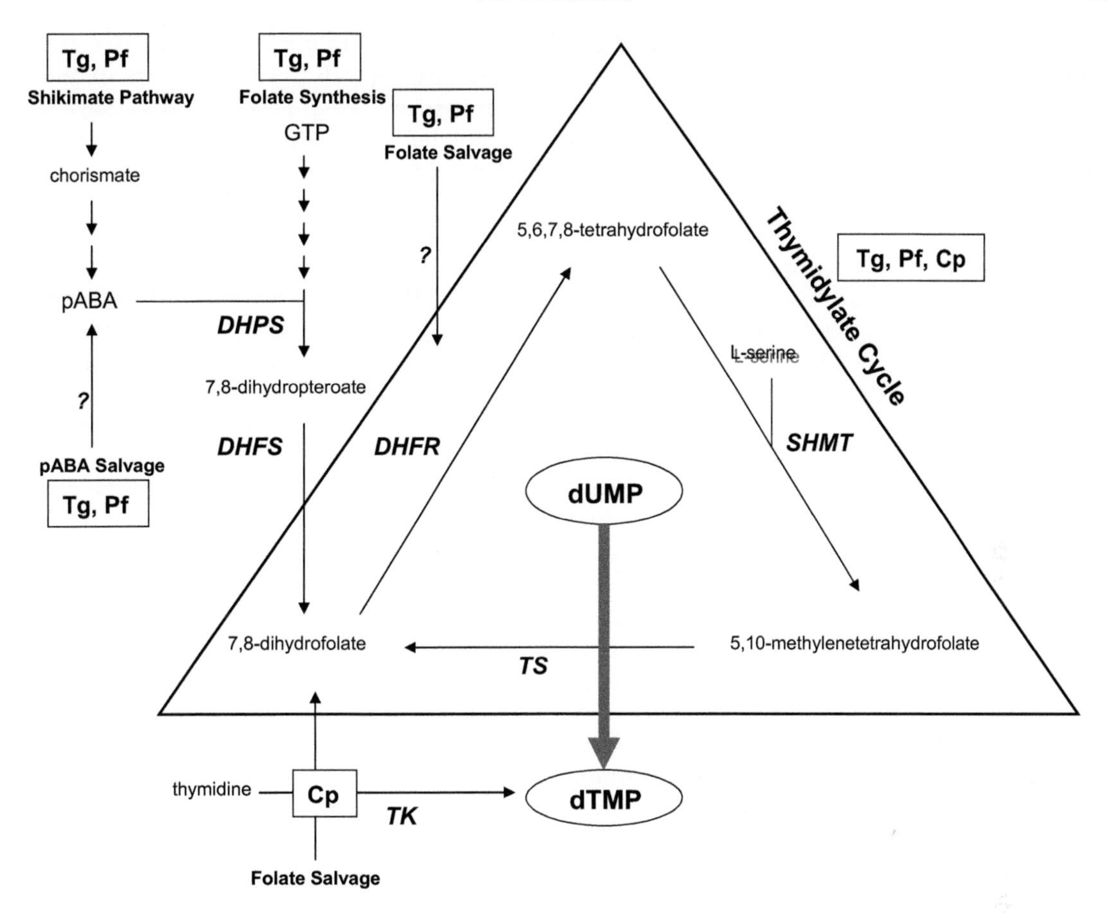

FIGURE 20.3 **Thymine nucleotide synthesis and pathways to folate in Apicomplexa.** Pathways present in *T. gondii* (Tg), *P. falciparum* (Pf) and *C. parvum* (Cp) are indicated in rectangles. The thymidylate cycle is enclosed by a large triangle. Potential synthetic or salvage sources of folate entering the thymidylate cycle are shown. Solid lines and arrows depict active pathways present. Substrates of several enzyme activities are shown at the start of the solid line and the product(s) are shown on the arrowhead side. Several key enzymes are indicated in capital italicized text beside the arrowhead line. All pathways shown appear to be present as indicated; however, the significance of the folate salvage and pABA salvage pathways is unclear in Tg and Pf and is indicated with a question mark. Polyglutamated forms of folate are not shown. dTMP synthesis is shown at the bottom of the thymidylate cycle. Salvage of thymidine in Cp is shown. DHPS, dihydropteroate synthase; DHFS, dihydrofolate synthetase; SHMT, serine hydromethyltransferase; DHFR, dihydrofolate reductase; TS, thymidylate synthetase; TK, thymidine kinase.

et al., 1987; Gregson and Plowe, 2005) (Fig. 20.3). While humans depend upon a monofunctional dihydrofolate reductase (DHFR) for DNA replication, in Apicomplexa the DHFR activity is present on a bifunctional polypeptide with TS activity (DHFR-TS) (Bzik *et al.*, 1987). Sulpha drugs target DHPS, which is a monofunctional enzyme in humans, but is also a bifunctional enzyme in *T. gondii* and *P. falciparum* with hydroxymethylpterin pyrophosphokinase activity (Triglia and Cowman, 1994; Pashley *et al.*, 1997).

Mutations in *P. falciparum* DHPS have been correlated with resistance to sulpha drugs (Wang *et al.*, 2004a; Triglia *et al.*, 1998). Genetic studies have documented that various mutations in DHPS correlate with resistance to sulpha drugs *in vivo* (Triglia *et al.*, 1998).

Resistance to antifolates that target DHFR arose rapidly in *P. falciparum*. Current antifolate therapy is based on synergism achieved in combination therapy of a DHFR inhibitor combined with sulpha drugs (Gregson and Plowe, 2005). The current treatment for *T. gondii* infection employs a similar strategy using PYR and sulphadiazine. While *T. gondii* has not developed resistance to antifolates long-term use of this therapy to treat toxoplasmosis in AIDS patients has proven to be difficult due to significant adverse clinical reactions (Haverkos, 1987; Leport *et al.*, 1988). Antifolates are also under investigation as potential inhibitors of *C. parvum* (Anderson, 2005). However, genome data from *C. parvum* indicates the presence of a TK gene that could bypass inhibition by antifolates if the TK activity is physiologically relevant. The atypical presence of a TK gene in *C. parvum*, as well as the lack of a DHPS gene, the target of sulpha compounds, suggests that directly targeting *C. parvum* DHFR to inhibit replication may be a difficult objective (Striepen *et al.*, 2004).

20.3.4.2.1 ANTIFOLATE RESISTANCE IN *P. FALCIPARUM* AND *C. PARVUM*

In early work, mutation of DHFR in bifunctional DHFR-TS was identified as the basis of resistance to PYR in *P. falciparum* (Inselburg *et al.*, 1987, 1988). In laboratory isolates gene duplication and chromosomal changes have also been reported as a mechanism of resistance to PYR (Inselburg *et al.*, 1987; Tanaka *et al.*, 1990a, b). Early studies of PYR resistance in isolates of *P. falciparum* identified mutations in key amino acid residues (codons 51, 59 and 108) that were associated with high-level resistance to PYR (Cowman *et al.*, 1988; Peterson *et al.*, 1988; Inselburg *et al.*, 1988). The crystal structure of wild type and mutant *P. falciparum* DHFR-TS suggests that these key resistance mutations cause steric interactions with inhibitors or weaken the binding of inhibitors in the active site (Yuvaniyama *et al.*, 2003). Several new antifolates that inhibit current drug resistant alleles of *P. falciparum* DHFR are currently in clinical trials (Anderson, 2005; Gregson and Plowe, 2005).

The crystal structure of *Cryptosporidium hominis* DHFR-TS was determined (O'Neil *et al.*, 2003a, b), and revealed some new insights into the natural resistance of *C. parvum* DHFR to antifolates (Vasquez *et al.*, 1996). Since no current genetic selection model is available for *C. parvum*, the identification of strong inhibitors of *C. parvum* DHFR or TK may be a useful approach to define the most amenable targets for blocking dTMP accumulation and DNA replication in *C. parvum* infection.

20.3.4.2.2 ANTIFOLATE RESISTANCE IN *T. GONDII*

Resistance to pyrimethamine has not been frequently observed in clinical treatment of toxoplasmosis. Even low level resistance to PYR has been difficult to select under drug pressure *in vitro*, but has been demonstrated in a genetic model of direct mutagenesis of a plasmid borne *T. gondii* DHFR-TS followed by transfection of PYR-sensitive *T. gondii* and growth selection under PYR *in vitro* (Reynolds *et al.*, 2001). By modelling *P. falciparum* resistance mutations at codons equivalent to 59 and 108 (Bzik *et al.*, 1987), high-level PYR resistance was obtained in *T. gondii* DHFR-TS (Donald and Roos, 1993; Roos, 1993). Plasmids conferring high-level PYR resistance in *T. gondii* have been useful in genetic dissection of evolution and mechanisms associated with PYR and cycloguanil resistance (Reynolds and Roos, 1998). This approach was adapted to investigate evolutionary fitness of DHFR-TS mutations associated with PYR resistance. This study revealed subtle, but potentially significant, effects on fitness of mutant DHFR-TS enzymes *in vitro* that appear to be associated with the natural appearance, or non-appearance, of these mutants *in vivo* (Fohl and Roos, 2003).

Modelling studies of *T. gondii* DHFR-TS have suggested that the long linker domain connecting DHFR to TS domains in Apicomplexa donates a helix that crosses to the second DHFR domain of the homodimer complex and contacts the outer shell of the DHFR active site (O'Neil *et al.*, 2003a; Belperron *et al.*, 2004). Genetic studies of *T. gondii* DHFR-TS revealed that various mutations within the linker domain inactivated pyrimethamine resistance and enzyme activity *in vitro* and *in vivo* (Belperron *et al.*, 2004).

20.3.5 Genetic Selection Models Based on Enzymes of Pyrimidine Metabolism

High-level PYR resistance based on mutant DHFR-TS was used to establish an early model of positive selection in *T. gondii* (Donald and Roos, 1993; Roos, 1993). The high efficiency of obtaining PYR resistant parasites allowed the refinement of methods for efficiently incorporating plasmid DNA into the parasite (Roos *et al.*, 1994). The isolation of both a genomic DNA version as well as a cDNA version of the PYR resistant DHFR-TS enabled an assessment of general homology requirements for recombination in *T. gondii*. The cDNA version of *T. gondii* DHFR-TS lacks numerous introns contained in the genomic DNA version and randomly incorporates into the genome of the parasite (Donald and Roos, 1993). By contrast, the genomic DNA version was demonstrated to primarily integrate into the homologous DHFR-TS locus (Donald and Roos, 1994; Donald and Roos, 1995). These studies established methods and tools for random insertional mutagenesis, as well as gene replacement approaches in *T. gondii*.

Subsequently the high-level PYR resistant bifunctional DHFR-TS was converted into a trifunctional enzyme by the incorporation of either herpes simplex virus TK, or bacterial cytosine deaminase (CD). The TK and CD enzymes were inserted as in-frame genes into the linker domain of *T. gondii* DHFR-TS. The DHFR-CD-TS plasmid conferred high-level resistance to PYR (positive selection), and stable PYR resistant parasite clones were killed (negative selection) by treatment of parasites with low doses of the normally non-toxic prodrug 5-fluorocytosine (Fox *et al.*, 1999). This study established DHFR-CD-TK as a trifunctional enzyme capable of positive and negative selection in *T. gondii*. Transgenic *T. gondii* parasites expressing CD are killed by 5-fluorocytosine by its conversion to 5-fluorouracil, through salvage by *T. gondii* UPRT to 5-fluorouridine 5′-monophosphate, and conversion to 5-fluoro-dUMP, a suicide inhibitor of TS activity that blocks the accumulation of dTMP and therefore parasite replication (Fox *et al.*, 1999). Similarly, construction of a plasmid encoding the trifunctional enzyme DHFR-TK-TS on a single polypeptide enabled positive selection by high-level resistance to PYR, and negative selection in sub-micromolar doses of ganciclovir (Fox *et al.*, 2001). The DHFR-TK-TS plasmid was used in positive and negative selection experiments to obtain the avirulent uracil auxotroph mutant by targeted knock out of the CPSII gene (Fox and Bzik, 2002). The *T. gondii* CPSII is the most amenable among the apicomplexans for genetic dissection of CPSII activities and regulation. Cloning of the cDNA for *T. gondii* CPSII enables a genetic scheme for positive selection based on complementation of the uracil auxotroph CPSII knock-out mutant. Complementation of CPSII has been demonstrated (Fox and Bzik, 2002; Fox *et al.*, 2009a), as well as complementation of OMPDC (Fox and Bzik, 2010).

20.4 AMINO ACIDS

Amino acids are essential in numerous areas of parasite metabolism and are necessary precursors for nucleotide synthesis. L-aspartate is essential for the synthesis of AMP from IMP in *P. falciparum* and *T. gondii* (Figs 20.1A and 20.1C), and the amino acids L-glutamine and

L-aspartate are essential precursor molecules for *de novo* pyrimidine synthesis of UMP in *T. gondii* and *P. falciparum* (Fig. 20.2). In Apicomplexa, serine is also required for synthesis of dTMP during the thymidylate cycle (Fig. 20.3). This section will briefly review acquisition of amino acids in Apicomplexa. Other reviews of amino acid metabolism in protozoa can be found in previous texts (Gutteridge and Coombs, 1979; North and Lockwood, 1995).

20.4.1 Studies of Amino Acid Auxotrophy in *T. gondii*

Surprisingly, the architecture of CPSII activity in *T. gondii* and *P. falciparum* revealed the presence of a single glutamine-dependent CPSII gene and activity (Flores *et al.*, 1994; Fox and Bzik, 2002, 2003). The existence of a single CPSII in *T. gondii* is highly unusual for a eukaryotic organism. In many prokaryotes, a single CPS polypeptide is typically found and this CPS activity and is responsible for producing carbamoyl phosphate, the precursor molecule for both pyrimidines and arginine. Consequently, disruption of *E. coli* CPS produces a dual pyrimidine and arginine auxotrophy (Beckwith *et al.*, 1962). In many eukaryotes two distinct CPS genes and activities are found, a glutamine-dependent CPSII linked with pyrimidine biosynthesis and a mitochondria associated CPSI dedicated to arginine biosynthesis (Davis, 1986; Makoff and Radford, 1978). The carbamoyl phosphate produced by CPSI in many eukaryotes is sequestered in the mitochondria for immediate conversion to citrulline via ornithine carbamoyltransferase (OCT). Arginine is produced from citrulline in two steps by the sequential actions of argininosuccinate synthetase (AS) and argininosuccinate lyase (AL) (Davis, 1986; Makoff and Radford, 1978).

The availability of a CPSII knock-out mutant in *T. gondii* enabled a functional determination of whether the sole CPSII in Apicomplexa is responsible for pyrimidine and arginine biosynthesis (Fox *et al.*, 2004). This study conclusively demonstrated that the *T. gondii* CPSII and the carbamoyl phosphate product are dedicated to pyrimidine biosynthesis. *T. gondii* is a natural arginine auxotroph. The arginine auxotrophy of *T. gondii* is rescued by supplementing growth media with either arginine or citrulline. Using mutant host cells it was demonstrated that rescue with citrulline was dependent on the presence of host cell AS and AL activities. These experiments demonstrated the functional absence of any arginine biosynthetic enzyme activity in *T. gondii* and this conclusion has been verified by the absence of corresponding gene orthologues in the *T. gondii* genome (Chaudhary and Roos, 2005). Other apicomplexans such as *C. parvum* and *P. falciparum* also are natural arginine auxotrophs (see below). The natural arginine auxotrophy and arginine depletion in *T. gondii* infection have been linked with the differentiation of tachyzoites to bradyzoite containing cysts *in vitro* (Fox *et al.*, 2004). These observations suggest that local depletion of arginine by inducible nitric oxide synthase during the host immune response to *T. gondii* infection may promote signalling or formation of slow growing bradyzoites or cyst development and maintenance (Fox *et al.*, 2004; Grillo and Colombatto, 2004; Wu and Morris, 1998).

Curiously, a second amino acid auxotrophy of *T. gondii* has been previously linked to host immune response and control of infection. Early work demonstrated a natural tryptophan auxotrophy of *T. gondii*, which was also associated with reduced growth rate of tachyzoites during tryptophan depletion elicited by the host immune response to *T. gondii* infection (Pfefferkorn, 1984, 1986a, b; Pfefferkorn and Guyre, 1984; Pfefferkorn, 1986; Khan *et al.*, 1988). In this case, the tryptophan depletion is triggered from downstream responses to the strong induction of interferon gamma during the host immune response to *T. gondii* infection. Collectively, these studies suggest a fundamental link in host immune responses leading to local depletion of amino acids at sites of *T. gondii* infection as a general mechanism to slow parasite growth or trigger differentiation of rapidly replicating tachyzoites into slow growing encysted bradyzoite forms. This tryptophan starvation mechanism restricts availability of tryptophan to *T. gondii* during murine infection (Divanovic *et al.*, 2012).

20.4.2 Genome Studies of Amino Acid Metabolism in Apicomplexa

Relatively little experimental work has been reported on amino acid transport, synthesis, interconversion, or catabolic pathways in the apicomplexa. The genome of *T. gondii*, *P. falciparum* and *C. parvum* has enabled a predictive comparison of the capabilities of these parasites in regard to amino acid metabolism. Humans require nine amino acids in their diet and can synthesize the remaining 11 amino acids required for protein synthesis. Mammals, primarily in liver tissue, can synthesize arginine *de novo* from CPSI, OCT, AS and AL activities and arginine is now considered to be a non-essential amino acid (Grillo and Colombatto, 2004; Wu and Morris, 1998). In addition to arginine, humans can synthesize alanine, serine and glycine (derived from glycolysis), glutamic acid, glutamate, proline, aspartic acid and asparagine (derived from the TCA cycle), and tyrosine and cysteine (derived from essential amino acids phenylalanine and methionine, respectively, from diet).

All of the protozoan and apicomplexan parasites have diminished amino acid biosynthetic capabilities compared with the human host (Chaudhary and Roos, 2005). *C. parvum* as a gut pathogen likely has the most access to host amino acids. This is consistent with *C. parvum* harbouring the most diminished amino acid biosynthetic capability. *C. parvum* is auxotrophic for tyrosine, alanine, serine, glutamic acid, aspartic acid, asparagine, arginine and cysteine and can synthesize only three amino acids (glycine, glutamate and proline). While *C. parvum* has acquired an auxotrophy for serine, it has lost an auxotrophy for the essential amino acid tryptophan that can be metabolized from serine. *C. parvum* may also retain a salvage pathway for tryptophan. *P. falciparum* in the process of digesting host haemoglobin would have reduced needs for amino acid biosynthetic capability. Consistent with this available food source, *P. falciparum* is auxotrophic for tyrosine, alanine, serine, arginine and cysteine, and can synthesize only six amino acids (glycine, glutamic acid, glutamate, proline, aspartic acid and asparagine). Intraerythrocytic parasites can acquire nearly all their required amino acids by the degradation of haemoglobin, but isoleucine, which is not present in haemoglobin, must be acquired exogenously. Isoleucine utilization has been proposed as a potential antimalarial drug target (Itsvan *et al.*, 2011) and erythrocytic stages deprived of this amino acid enter a hibernatory state (Babbitt *et al.*, 2012).

T. gondii has the least diminished amino acid biosynthetic capability and most resembles the human host. Compared with the human host *T. gondii* is auxotrophic only for arginine and cysteine. Collectively, these observations suggest significant adaptations of the different apicomplexan parasites related to their potentially available nutritional amino acid resources.

References

Abrahamsen, M.S., Templeton, T.J., Enomoto, S., Abrahante, J.E., Zhu, G., Lancto, C.A., Deng, M., Liu, C., Widmer, G., Tzipori, S., Buck, G.A., Xu, P., Bankier, A.T., Dear, P.H., Konfortov, B.A., Spriggs, H.F., Iyer, L., Anantharaman, V., Aravind, L., Kapur, V., 2004. Complete genome sequence of the apicomplexan, Cryptosporidium parvum. Science 304, 441–445.

Al Safarjalani, O.N., Naguib, F.N., El Kouni, M.H., 2003. Uptake of nitrobenzylthioinosine and purine beta-L-nucleosides by intracellular Toxoplasma gondii. Antimicrob. Agents Chemother. 47, 3247–3251.

Allison, A.C., Eugui, E.M., 2000. Mycophenolate mofetil and its mechanisms of action. Immunopharmacology 47, 85–118.

Allison, A.C., Eugui, E.M., 2005. Mechanisms of action of mycophenolate mofetil in preventing acute and chronic allograft rejection. Transplantation 80, S181–S190.

Anderson, A.C., 2005. Targeting DHFR in parasitic protozoa. Drug Discovery Today 10, 121–128.

Aoki, T., Shimogawara, R., Ochiai, K., Yamasaki, H., Shimada, J., 1994. Molecular characterization of a carbamoyl-phosphate synthetase II (CPS II) gene from Trypanosoma cruzi. Adv. Exp. Med. Biol. 370, 513–516.

Asai, T., Miura, S., Sibley, L.D., Okabayashi, H., Takeuchi, T., 1995. Biochemical and molecular characterization of nucleoside triphosphate hydrolase isozymes from the parasitic protozoan Toxoplasma gondii. J. Biol. Chem. 270, 11391–11397.

Asai, T., O'Sullivan, W.J., Kobayashi, M., Gero, A.M., Yokogawa, M., Tatibana, M., 1983. Enzymes of the de novo pyrimidine biosynthetic pathway in Toxoplasma gondii. Mol. Biochem. Parasitol. 7, 89–100.

Avila, J.L., Avila, A., 1981. Trypanosoma cruzi: allopurinol in the treatment of mice with experimental acute Chagas disease. Exp. Parasitol. 51, 204–208.

Avila, J.L., Rojas, T., Avila, A., Polegre, M.A., Robins, R.K., 1987. Biological activity of analogs of guanine and guanosine against American Trypanosoma and Leishmania spp. Antimicrob. Agents Chemother. 31, 447–451.

Babbitt, S.E., Altenhofen, L., Cobbold, S.A., Istvan, E.S., Fennell, C., Doerig, C., Llinás, M., Goldberg, D.E., 2012. Plasmodium falciparum responds to amino (acid starvation by entering into a hibernatory state. Proc. Natl. Acad. Sci. 109, E3278–E3287.

Bahl, A., Brunk, B., Coppel, R.L., Crabtree, J., Diskin, S.J., Fraunholz, M.J., Grant, G.R., Gupta, D., Huestis, R.L., Kissinger, J.C., Labo, P., Li, L., Mcweeney, S.K., Milgram, A.J., Roos, D.S., Schug, J., Stoeckert Jr., C.J., 2002. PlasmoDB: the Plasmodium genome resource. An integrated database providing tools for accessing, analyzing and mapping expression and sequence data (both finished and unfinished). Nucleic Acids Res. 30, 87–90.

Bahl, A., Brunk, B., Crabtree, J., Fraunholz, M.J., Gajria, B., Grant, G.R., Ginsburg, H., Gupta, D., Kissinger, J.C., Labo, P., Li, L., Mailman, M.D., Milgram, A.J., Pearson, D.S., Roos, D.S., Schug, J., Stoeckert Jr., C.J., Whetzel, P., 2003. PlasmoDB: the Plasmodium genome resource. A database integrating experimental and computational data. Nucleic Acids Res. 31, 212–215.

Beckwith, J.R., Pardee, A.B., Austrian, R., Jacob, F., 1962. Coordination of the synthesis of the enzymes in the pyrimidine pathway of *Escherichia coli.* J. Mol. Biol. 5, 618–635.

Belperron, A.A., Fox, B.A., O'Neil, R.H., Peaslee, K.A., Horii, T., Anderson, A.C., Bzik, D.J., 2004. Toxoplasma gondii: generation of novel truncation mutations in the linker domain of dihydrofolate reductase-thymidylate synthase. Exp. Parasitol. 106, 179–182.

Berens, R.L., Marr, J.J., Lafon, S.W., Nelson, D.J., 1981. Purine metabolism in Trypanosoma cruzi. Mol. Biochem. Parasitol. 3, 187–196.

Berens, R.L., Marr, J.J., Looker, D.L., Nelson, D.J., Lafon, S.W., 1984. Efficacy of pyrazolopyrimidine ribonucleosides against Trypanosoma cruzi: studies in vitro and in vivo with sensitive and resistant strains. J. Infect. Dis. 150, 602–608.

Berman, P.A., Human, L., 1991. Hypoxanthine depletion induced by xanthine oxidase inhibits malaria parasite growth in vitro. Adv. Exp. Med. Biol. 309A, 165–168.

Berman, P.A., Human, L., Freese, J.A., 1991. Xanthine oxidase inhibits growth of Plasmodium falciparum in human erythrocytes in vitro. J. Clin. Invest 88, 1848–1855.

Bermudes, D., Peck, K.R., Afifi, M.A., Beckers, C.J., Joiner, K.A., 1994. Tandemly repeated genes encode nucleoside triphosphate hydrolase isoforms secreted into the parasitophorous vacuole of Toxoplasma gondii. J. Biol. Chem. 269, 29252–29260.

Birkholtz, L.M., Wrenger, C., Joubert, F., Wells, G.A., Walter, R.D., Louw, A.I., 2004. Parasite-specific inserts in the bifunctional S-adenosylmethionine decarboxylase/ornithine decarboxylase of Plasmodium falciparum modulate catalytic activities and domain interactions. Biochem. J. 377, 439–448.

Boa, A.N., Canavan, S.P., Hirst, P.R., Ramsey, C., Stead, A.M., Mcconkey, G.A., 2005. Synthesis of brequinar analogue inhibitors of malaria parasite dihydroorotate dehydrogenase. Bioorg. Med. Chem. 13, 1945–1967.

Bohne, W., Roos, D.S., 1997. Stage-specific expression of a selectable marker in Toxoplasma gondii permits selective inhibition of either tachyzoites or bradyzoites. Mol. Biochem. Parasitol. 88, 115–126.

Brooks, R.G., Remington, J.S., Luft, B.J., 1987. Drugs used in the treatment of toxoplasmosis. Antimicrob. Agents Annu. 2, 297–306.

Bzik, D.J., Li, W.B., Horii, T., Inselburg, J., 1987. Molecular cloning and sequence analysis of the Plasmodium falciparum dihydrofolate reductase-thymidylate synthase gene. Proc. Natl. Acad. Sci. U. S. A. 84, 8360–8364.

Carter, D., Donald, R.G., Roos, D., Ullman, B., 1997. Expression, purification, and characterization of uracil phosphoribosyltransferase from Toxoplasma gondii. Mol. Biochem. Parasitol. 87, 137–144.

Carter, N.S., Ben Mamoun, C., Liu, W., Silva, E.O., Landfear, S.M., Goldberg, D.E., Ullman, B., 2000. Isolation and functional characterization of the PfNT1 nucleoside transporter gene from Plasmodium falciparum. J. Biol. Chem. 275, 10683–10691.

Carter, N.S., Rager, N., Ullman, B. (Eds.), 2003. Purine and pyrimidine transport and metabolism. Academic Press, London.

Cassera, M.B., Hazleton, K.Z., Merino, E.F., Obaldia 3rd, N., Ho, M.C., Murkin, A.S., Depinto, R., Gutierrez, J.A., Almo, S.C., Evans, G.B., Babu, Y.S., Schramm, V.L., 2011. Plasmodium falciparum parasites are killed by a transition state analogue of purine nucleoside phosphorylase in a primate animal model. PLoS One 6, e26916.

Chansiri, K., Bagnara, A.S., 1995. The structural gene for carbamoyl phosphate synthetase from the protozoan parasite Babesia bovis. Mol. Biochem. Parasitol. 74, 239–243.

Chaudhary, K., 2005. Purine transport and salvage in apicomplexan parasites. Cell and Molcular Biology. Philadelphia. University of Pennsylvania.

Chaudhary, K., Darling, J.A., Fohl, L.M., Sullivan Jr., W.J., Donald, R.G., Pfefferkorn, E.R., Ullman, B., Roos, D.S., 2004. Purine salvage pathways in the apicomplexan parasite Toxoplasma gondii. J. Biol. Chem. 279, 31221–31227.

Chaudhary, K., Donald, R.G., Nishi, M., Carter, D., Ullman, B., Roos, D.S., 2005. Differential localization of alternatively spliced hypoxanthine-xanthine-guanine phosphoribosyltransferase isoforms in Toxoplasma gondii. J. Biol. Chem. 280, 22053–22059.

Chaudhary, K., Roos, D.S., 2005. Protozoan genomics for drug discovery. Nature Biotech. 23, 1089–1091.

Chaudhary, K., Ting, L.M., Kim, K., Roos, D.S., 2006. Toxoplasma gondii purine nucleoside phosphorylase biochemical characterization, inhibitor profiles, and comparison with the Plasmodium falciparum ortholog. J. Biol. Chem. 281, 25652–25658.

Chiang, C.W., Carter, N., Sullivan Jr., W.J., Donald, R.G., Roos, D.S., Naguib, F.N., El Kouni, M.H., Ullman, B., Wilson, C.M., 1999. The adenosine transporter of Toxoplasma gondii. Identification by insertional mutagenesis, cloning, and recombinant expression. J. Biol. Chem. 274, 35255–35261.

Christopherson, R.I., Cinquin, O., Shojaei, M., Kuehn, D., Menz, R.I., 2004. Cloning and expression of malarial pyrimidine enzymes. Nucleosides Nucleotides Nucleic Acids 23, 1459–1465.

Cowman, A.F., Morry, M.J., Biggs, B.A., Cross, G.A., Foote, S.J., 1988. Amino acid changes linked to pyrimethamine resistance in the dihydrofolate reductase-thymidylate synthase gene of Plasmodium falciparum. Proc. Natl. Acad. Sci. U. S. A. 85, 9109–9113.

Daddona, P.E., Wiesmann, W.P., Lambros, C., Kelley, W.N., Webster, H.K., 1984. Human malaria parasite adenosine deaminase. Characterization in host enzyme-deficient erythrocyte culture. J. Biol. Chem. 259, 1472–1475.

Daddona, P.E., Wiesmann, W.P., Milhouse, W., Chern, J.W., Townsend, L.B., Hershfield, M.S., Webster, H.K., 1986. Expression of human malaria parasite purine nucleoside phosphorylase in host enzyme-deficient erythrocyte culture. Enzyme characterization and identification of novel inhibitors. J. Biol. Chem. 261, 11667–11673.

Darling, J.A., Sullivan Jr., W.J., Carter, D., Ullman, B., Roos, D.S., 1999. Recombinant expression, purification, and characterization of Toxoplasma gondii adenosine kinase. Mol. Biochem. Parasitol. 103, 15–23.

Davidson, J.N., Chen, K.C., Jamison, R.S., Musmanno, L.A., Kern, C.B., 1993. The evolutionary history of the first three enzymes in pyrimidine biosynthesis. Bioessays 15, 157–164.

Davis, R.H., 1986. Compartmental and regulatory mechanisms in arginine pathways of *Neurospora crassa* and *Saccharomyces cerevisiae.* Microbiol. Rev. 50, 280–313.

De Koning, H., Diallinas, G., 2000. Nucleobase transporters (review). Mol. Membr. Biol. 17, 75–94.

De Koning, H.P., Al-Salabi, M.I., Cohen, A.M., Coombs, G.H., Wastling, J.M., 2003. Identification and characterisation of high affinity nucleoside and nucleobase transporters in Toxoplasma gondii. Int. J. Parasitol. 33, 821–831.

De Koning, H.P., Bridges, D.J., Burchmore, R.J., 2005. Purine and pyrimidine transport in pathogenic protozoa: from biology to therapy. FEMS Microbiol. Rev. 29, 987–1020.

Divanovic, S., Sawtell, N.M., Trompette, A., Warning, J.I., Dias, A., Cooper, A.M., Yap, G.S., Arditi, M., Shimada, K., Duhadaway, J.B., Prendergast, G.C., Basaraba, R.J., Mellor, A.L., Munn, D.H., Aliberti, J., Karp, C.L., 2012. Opposing biological functions of tryptophan catabolizing enzymes during intracellular infection. J. Infect. Dis. 205, 152–161.

Donald, R.G., Carter, D., Ullman, B., Roos, D.S., 1996. Insertional tagging, cloning, and expression of the Toxoplasma gondii hypoxanthine-xanthine-guanine phosphoribosyltransferase gene. Use as a selectable marker for stable transformation. J. Biol. Chem. 271, 14010–14019.

Donald, R.G., Roos, D.S., 1993. Stable molecular transformation of Toxoplasma gondii: a selectable dihydrofolate reductase-thymidylate synthase marker based on drug-resistance mutations in malaria. Proc. Natl. Acad. Sci. U. S. A. 90, 11703–11707.

Donald, R.G., Roos, D.S., 1994. Homologous recombination and gene replacement at the dihydrofolate reductase-thymidylate synthase locus in Toxoplasma gondii. Mol. Biochem. Parasitol. 63, 243–253.

Donald, R.G., Roos, D.S., 1995. Insertional mutagenesis and marker rescue in a protozoan parasite: cloning of the uracil phosphoribosyltransferase locus from Toxoplasma gondii. Proc. Natl. Acad. Sci. U. S. A. 92, 5749–5753.

Donald, R.G., Roos, D.S., 1998. Gene knock-outs and allelic replacements in Toxoplasma gondii: HXGPRT as a selectable marker for hit-and-run mutagenesis. Mol. Biochem. Parasitol. 91, 295–305.

Downie, M.J., El Bissati, K., Bobenchik, A.M., Nic Lochlainn, L., Amerik, A., Zufferey, R., Kirk, K., Ben Mamoun, C., 2010. PfNT2, a permease of the equilibrative nucleoside transporter family in the endoplasmic reticulum of Plasmodium falciparum. J. Biol. Chem. 285, 20827–20833.

Doyle, P.S., Kanaani, J., Wang, C.C., 1998. Hypoxanthine, guanine, xanthine phosphoribosyltransferase activity in Cryptosporidium parvum. Exp. Parasitol. 89, 9–15.

Eaazhisai, K., Jayalakshmi, R., Gayathri, P., Anand, R.P., Sumathy, K., Balaram, H., Murthy, M.R., 2004. Crystal structure of fully ligated adenylosuccinate synthetase from Plasmodium falciparum. J. Mol. Biol. 335, 1251–1264.

El Bissati, K., Zufferey, R., Witola, W.H., Carter, N.S., Ullman, B., Ben Mamoun, C., 2006. The plasma membrane permease PfNT1 is essential for purine salvage in the human malaria parasite Plasmodium falciparum. Proc. Natl. Acad. Sci. U. S. A. 103, 9286–9291.

El Bissati, K., Downie, M.J., Seong-Kyoun, K., Horowitz, M., Carter, N., Ullman, B., Carter, Ben Mamoun, C., 2008. Genetic evidence for the essential role of PfNT1 in the transport and utilization of xanthine, guanine, guanosine, and adenine by Plasmodium falciparum. Mol. Biochem. Parasitol. 161, 130–139.

El Kouni, M.H., Guarcello, V., Al Safarjalani, O.N., Naguib, F.N., 1999. Metabolism and selective toxicity of 6-nitrobenzylthioinosine in Toxoplasma gondii. Antimicrob. Agents Chemother. 43, 2437–2443.

El Kouni, M.H., Naguib, F.N., Panzica, R.P., Otter, B.A., Chu, S.H., Gosselin, G., Chu, C.K., Schinazi, R.F., Shealy, Y.F., Goudgaon, N., Ozerov, A.A., Ueda, T., Iltzsch, M.H., 1996. Effects of modifications in the pentose moiety and conformational changes on the binding of nucleoside ligands to uridine phosphorylase from Toxoplasma gondii. Biochem. Pharmacol. 51, 1687–1700.

Elion, G.B., 1989a. Nobel Lecture. The purine path to chemotherapy. Biosci. Rep. 9, 509–529.

Elion, G.B., 1989b. The purine path to chemotherapy. Science 244, 41–47.

Evans, D.R., Guy, H.I., 2004. Mammaliam pyrimidine synthesis: Fresh insights into an ancient pathway. J. Biol. Chem. 279, 33035–33038.

Fish, W.R., Marr, J.J., Berens, R.L., 1982. Purine metabolism in Trypanosoma brucei gambiense. Biochim. Biophys. Acta 714, 422–428.

Fish, W.R., Marr, J.J., Berens, R.L., Looker, D.L., Nelson, D.J., Lafon, S.W., Balber, A.E., 1985. Inosine analogs as chemotherapeutic agents for African trypanosomes: metabolism in trypanosomes and efficacy in tissue culture. Antimicrob. Agents Chemother. 27, 33–36.

Flores, M.V., Atkins, D., Wade, D., O'Sullivan, W.J., Stewart, T.S., 1997. Inhibition of Plasmodium falciparum proliferation in vitro by ribozymes. J. Biol. Chem. 272, 16940–16945.

Flores, M.V., O'Sullivan, W.J., Stewart, T.S., 1994. Characterisation of the carbamoyl phosphate synthetase gene from Plasmodium falciparum. Mol. Biochem. Parasitol. 68, 315–318.

Flores, M.V., Stewart, T.S., 1998. Plasmodium falciparum: a microassay for the malarial carbamoyl phosphate synthetase. Exp. Parasitol. 88, 243–245.

Fohl, L.M., Roos, D.S., 2003. Fitness effects of DHFR-TS mutations associated with pyrimethamine resistance in apicomplexan parasites. Mol. Microbiol. 50, 1319–1327.

Fox, B.A., Belperron, A.A., Bzik, D.J., 1999. Stable transformation of Toxoplasma gondii based on a pyrimethamine resistant trifunctional dihydrofolate reductase-cytosine deaminase-thymidylate synthase gene that confers sensitivity to 5-fluorocytosine. Mol. Biochem. Parasitol. 98, 93–103.

Fox, B.A., Belperron, A.A., Bzik, D.J., 2001. Negative selection of herpes simplex virus thymidine kinase in Toxoplasma gondii. Mol. Biochem. Parasitol. 116, 85–88.

Fox, B.A., Bzik, D.J., 2002. De novo pyrimidine biosynthesis is required for virulence of Toxoplasma gondii. Nature 415, 926–929.

Fox, B.A., Bzik, D.J., 2003. Organisation and sequence determination of glutamine-dependent carbamoyl phosphate synthetase II in Toxoplasma gondii. Int. J. Parasitol. 33, 89–96.

Fox, B.A., Bzik, D.J., 2010. Avirulent uracil auxotrophs based on disruption of orotidine-5′-monophosphate decarboxylase elicit protective immunity to Toxoplasma gondii. Infect. Immun. 78, 3744–3752.

Fox, B.A., Falla, A., Rommereim, L.M., Tomita, T., Gigley, J.P., Mercier, C., Cesbron-Delauw, M.F., Weiss, L.M., Bzik, D.J., 2011. Type II Toxoplasma gondii KU80 knockout strains enable functional analysis of genes required for cyst development and latent infection. Eukaryot Cell 10, 1193–1206.

Fox, B.A., Gigley, J.P., Bzik, D.J., 2004. Toxoplasma gondii lacks the enzymes required for de novo arginine biosynthesis and arginine starvation triggers cyst formation. Int. J. Parasitol. 34, 323–331.

Fox, B.A., Ristuccia, J.G., Bzik, D.J., 2009a. Genetic identification of essential indels and domains in carbamoyl phosphate synthetase II of Toxoplasma gondii. Int. J. Parasitol. 39, 533–539.

Fox, B.A., Ristuccia, J.G., Gigley, J.P., Bzik, D.J., 2009b. Efficient gene replacements in Toxoplasma gondii strains deficient for nonhomologous end joining. Eukaryot. Cell 8, 520–529.

Frame, I.J., Merino, E.F., Schramm, V.L., Cassera, M.B., Akabas, M.H., 2012. Malaria parasite type 4 equilibrative nucleoside transporters (ENT4) are purine transporters with distinct substrate specificity. Biochem. J. 446, 179–190.

Fresquet, V., Mora, P., Rochera, L., Ramon-Maiques, S., Rubio, V., Cervera, J., 2000. Site-directed mutagenesis of the regulatory domain of Escherichia coli carbamoyl phosphate synthetase identifies crucial residues for allosteric regulation and for transduction of the regulatory signals. J. Mol. Biol. 299, 979–991.

Gafan, C., Wilson, J., Berger, L.C., Berger, B.J., 2001. Characterization of the ornithine aminotransferase from Plasmodium falciparum. Mol. Biochem. Parasitol. 118, 1–10.

Gao, G., Nara, T., Nakajima-Shimada, J., Aoki, T., 1998. Molecular characterization of a carbamoyl-phosphate synthetase II (CPS II) gene from Leishmania mexicana. Adv. Exp. Med. Biol. 431, 237–240.

Gao, G., Nara, T., Nakajima-Shimada, J., Aoki, T., 1999. Novel organization and sequences of five genes encoding all six enzymes for de novo pyrimidine biosynthesis in Trypanosoma cruzi. J. Mol. Biol. 285, 149–161.

Gardner, M.J., Hall, N., Fung, E., White, O., Berriman, M., Hyman, R.W., Carlton, J.M., Pain, A., Nelson, K.E., Bowman, S., Paulsen, I.T., James, K., Eisen, J.A., Rutherford, K., Salzberg, S.L., Craig, A., Kyes, S., Chan, M.S., Nene, V., Shallom, S.J., Suh, B., Peterson, J., Angiuoli, S., Pertea, M., Allen, J., Selengut, J., Haft, D., Mather, M.W., Vaidya, A.B., Martin, D.M., Fairlamb, A.H., Fraunholz, M.J., Roos, D.S., Ralph, S.A., Mcfadden, G.I., Cummings, L.M., Subramanian, G.M., Mungall, C., Venter, J.C., Carucci, D.J., Hoffman, S.L., Newbold, C., Davis, R.W., Fraser, C.M., Barrell, B., 2002a. Genome sequence of the human malaria parasite Plasmodium falciparum. Nature 419, 498–511.

Gardner, M.J., Shallom, S.J., Carlton, J.M., Salzberg, S.L., Nene, V., Shoaibi, A., Ciecko, A., Lynn, J., Rizzo, M., Weaver, B., Jarrahi, B., Brenner, M., Parvizi, B., Tallon, L., Moazzez, A., Granger, D., Fujii, C., Hansen, C., Pederson, J., Feldblyum, T., Peterson, J., Suh, B., Angiuoli, S., Pertea, M., Allen, J., Selengut, J., White, O., Cummings, L.M., Smith, H.O., Adams, M.D., Venter, J.C., Carucci, D.J., Hoffman, S.L., Fraser, C.M., 2002b. Sequence of Plasmodium falciparum chromosomes 2, 10, 11 and 14. Nature 419, 531–534.

Gero, A.M., Brown, G.V., O'Sullivan, W.J., 1984. Pyrimidine de novo synthesis during the life cycle of the intraerythrocytic stage of Plasmodium falciparum. J. Parasitol. 70, 536–541.

Gero, A.M., Dunn, C.G., Brown, D.M., Pulenthiran, K., Gorovits, E.L., Bakos, T., Weis, A.L., 2003. New malaria chemotherapy developed by utilization of a unique parasite transport system. Curr. Pharm. Des. 9, 867–877.

Gero, A.M., Scott, H.V., O'Sullivan, W.J., Christopherson, R.I., 1989. Antimalarial action of ntrobenzylthioinosine in combination with purine nucleoside antimetabolites. Mol. Biochem. Parasitol. 34, 87–97.

Gorla, S.K., Kavitha, M., Zhang, M., Liu, X., Sharling, L., Gollapalli, D.R., Striepen, B., Hedstrom, L., Cuny, G.D., 2012. Selective and Potent Urea Inhibitors of Cryptosporidium parvum Inosine 5′-Monophosphate Dehydrogenase. J. Med. Chem. 55, 7759–7771.

Gregson, A., Plowe, C.V., 2005. Mechanisms of resistance of malaria parasites to antifolates. Pharmacological Reviews 57, 117–145.

Grillo, M.A., Colombatto, S., 2004. Arginine revisited: Minireview article. Amino Acids 26, 345–351.

Gupte, A., Buolamwini, J.K., Yadav, V., Chu, C.K., Naguib, F.N., El Kouni, M.H., 2005. 6-Benzylthioinosine analogues: Promising anti-toxoplasmic agents as inhibitors of the mammalian nucleoside transporter ENT1 (es). Biochem. Pharmacol. 71, 69–73.

Gutteridge, W.E., Coombs, G.H., 1979. Biochemistry of Parasitic Protozoa. Macmillan, London.

Hassan, H.F., Coombs, G.H., 1986. De novo synthesis of purines by *Acanthamoeba castellani* and *A. astronyxis*. IRCS Med. Sci. 14, 559–560.

Haverkos, H.W., 1987. Assessment of therapy for toxoplasmic encephalitis. Am. J. Med. 82, 907–914.

Hazleton, K.Z., Ho, M.C., Cassera, M.B., Clinch, K., Crump, D.R., Rosario Jr., I., Merino, E.F., Almo, S.C., Tyler, P.C., Schramm, V.L., 2012. Acyclic immucillin phosphonates: second-generation inhibitors of Plasmodium falciparum hypoxanthine-guanine-xanthine phosphoribosyltransferase. Chem. Biol. 19, 721–730.

Hendriks, E.F., O'Sullivan, W.J., Stewart, T.S., 1998. Molecular cloning and characterization of the Plasmodium falciparum cytidine triphosphate synthetase gene. Biochim. Biophys. Acta. 1399, 213–218.

Heroux, A., White, E.L., Ross, L.J., Borhani, D.W., 1999a. Crystal structures of the Toxoplasma gondii hypoxanthine-guanine phosphoribosyltransferase-GMP and -IMP complexes: comparison of purine binding interactions with the XMP complex. Biochemistry 38, 14485–14494.

Heroux, A., White, E.L., Ross, L.J., Davis, R.L., Borhani, D.W., 1999b. Crystal structure of Toxoplasma gondii hypoxanthine-guanine phosphoribosyltransferase with XMP, pyrophosphate, and two Mg(2+) ions bound: insights into the catalytic mechanism. Biochemistry 38, 14495–14506.

Hill, B., Kilsby, J., Mcintosh, R.T., Wrigglesworth, R., Ginger, C.D., 1981a. Pyrimidine biosynthesis in Plasmodium berghei. Int. J. Biochem. 13, 303–310.

Hill, B., Kilsby, J., Rogerson, G.W., Mcintosh, R.T., Ginger, C.D., 1981b. The enzymes of pyrimidine biosynthesis in a range of parasitic protozoa and helminths. Mol. Biochem. Parasitol. 2, 123–134.

Holden, H.M., Thoden, J.B., Raushel, F.M., 1999. Carbamoyl phosphate synthetase: an amazing biochemical odyssey from substrate to product. Cell Mol. Life Sci. 56, 507–522.

Hortua Triana, M.A., Huynh, M.H., Garavito, M.F., Fox, B.A., Bzik, D.J., Carruthers, V.B., Loffler, M., Zimmermann, B.H., 2012. Biochemical and molecular characterization of the pyrimidine biosynthetic enzyme dihydroorotate dehydrogenase from Toxoplasma gondii. Mol. Biochem. Parasitol. 184, 71–81.

Huang, J., Mullapudi, N., Lancto, C.A., Scott, M., Abrahamsen, M.S., Kissinger, J.C., 2004. Phylogenomic evidence supports past endosymbiosis, intracellular and horizontal gene transfer in Cryptosporidium parvum. Genome Biol. 5, R88.

Iltzsch, M.H., 1993. Pyrimidine salvage pathways in Toxoplasma gondii. J. Eukaryot. Microbiol. 40, 24–28.

Iltzsch, M.H., Klenk, E.E., 1993. Structure-activity relationship of nucleobase ligands of uridine phosphorylase from Toxoplasma gondii. Biochem. Pharmacol. 46, 1849–1858.

Iltzsch, M.H., Tankersley, K.O., 1994. Structure-activity relationship of ligands of uracil phosphoribosyltransferase from Toxoplasma gondii. Biochem. Pharmacol. 48, 781–792.

Iltzsch, M.H., Uber, S.S., Tankersley, K.O., El Kouni, M.H., 1995. Structure-activity relationship for the binding of nucleoside ligands to adenosine kinase from Toxoplasma gondii. Biochem. Pharmacol. 49, 1501–1512.

Inselburg, J., Bzik, D.J., Horii, T., 1987. Pyrimethamine resistant Plasmodium falciparum: overproduction of dihydrofolate reductase by a gene duplication. Mol. Biochem. Parasitol. 26, 121–134.

Inselburg, J., Bzik, D.J., Li, W.B., 1988. Plasmodium falciparum: three amino acid changes in the dihydrofolate reductase of a pyrimethamine-resistant mutant. Exp. Parasitol. 67, 361–363.

Javaid, Z.Z., El Kouni, M.H., Iltzsch, M.H., 1999. Pyrimidine nucleobase ligands of orotate phosphoribosyltransferase from Toxoplasma gondii. Biochem. Pharmacol. 58, 1457–1465.

Jayalakshmi, R., Sumathy, K., Balaram, H., 2002. Purification and characterization of recombinant Plasmodium falciparum adenylosuccinate synthetase expressed in Escherichia coli. Protein Expr. Purif. 25, 65–72.

Joiner, K.A., Beckers, C.J., Bermudes, D., Ossorio, P.N., Schwab, J.C., Dubremetz, J.F., 1994. Structure and function of the parasitophorous vacuole membrane surrounding Toxoplasma gondii. Ann. N. Y. Acad. Sci. 730, 1–6.

Joiner, K.A., Bermudes, D., Sinai, A., Qi, H., Polotsky, V., Beckers, C.J., 1996. Structure and function of the Toxoplasma gondii vacuole. Ann. N. Y. Acad. Sci. 797, 1–7.

Jones, M.E., 1980. Pyrimidine nucleotide biosynthesis in animals: genes, enzymes, and regulation of UMP synthesis. Ann. Rev. Biochem. 49, 253–279.

Jones, T.C., 1973. Multiplication of toxoplasmas in enucleate fibroblasts. Proc. Soc. Exp. Biol. Med. 142, 1268–1271.

Jones, T.C., Yeh, S., Hirsch, J.G., 1972. The interaction between Toxoplasma gondii and mammalian cells. I. Mechanism of entry and intracellular fate of the parasite. J. Exp. Med. 136, 1157–1172.

Keithly, J.S., Zhu, G., Upton, S.J., Woods, K.M., Martinez, M.P., Yarlett, N., 1997. Polyamine biosynthesis in Cryptosporidium parvum and its implications for chemotherapy. Mol. Biochem. Parasitol. 88, 35–42.

Khan, I.A., Eckel, M.E., Pfefferkorn, E.R., Kasper, L.H., 1988. Production of gamma interferon by cultured human lymphocytes stimulated with a purified membrane protein (P30) from Toxoplasma gondii. J. Infect. Dis. 157, 979–984.

Kicska, G.A., Tyler, P.C., Evans, G.B., Furneaux, R.H., Kim, K., Schramm, V.L., 2002a. Transition state analogue inhibitors of purine nucleoside phosphorylase from Plasmodium falciparum. J. Biol. Chem. 277, 3219–3225.

Kicska, G.A., Tyler, P.C., Evans, G.B., Furneaux, R.H., Schramm, V.L., Kim, K., 2002b. Purine-less death in Plasmodium falciparum induced by immucillin-H, a transition state analogue of purine nucleoside phosphorylase. J. Biol. Chem. 277, 3226–3231.

Kim, K., Weiss, L.M., 2004. Toxoplasma gondii: the model apicomplexan. Int. J. Parasitol. 34, 423–432.

Kimata, I., Tanabe, K., 1982. Invasion by Toxoplasma gondii of ATP-depleted and ATP-restored chick embryo erythrocytes. J. Gen. Microbiol. 128, 2499–2501.

Kirk, K., Horner, H.A., Elford, B.C., Ellory, J.C., Newbold, C.I., 1994. Transport of diverse substrates into malaria-infected erythrocytes via a pathway showing functional characteristics of a chloride channel. J. Biol. Chem. 269, 3339–3347.

Kirk, K., Martin, R.E., Broer, S., Howitt, S.M., Saliba, K.J., 2005. Plasmodium permeomics: membrane transport proteins in the malaria parasite. Curr. Top. Microbiol. Immunol. 295, 325–356.

Kissinger, J.C., Gajria, B., Li, L., Paulsen, I.T., Roos, D.S., 2003. ToxoDB: accessing the Toxoplasma gondii genome. Nucleic Acids Res. 31, 234–236.

Korsinczky, M., Chen, N., Kotecka, B., Saul, A., Rieckmann, K., Cheng, Q., 2000. Mutations in Plasmodium falciparum cytochrome b that are associated with atovaquone resistance are located at a putative drug-binding site. Antimicrob. Agents Chemother. 44, 2100–2108.

Kothe, M., Purcarea, C., Guy, H.I., Evans, D.R., Powers-Lee, S.G., 2005. Direct demonstration of carbamoyl phosphate formation on the C-terminal domain of carbamoyl phosphate synthetase. Protein Sci. 14, 37–44.

Krause, T., Luersen, K., Wrenger, C., Gilberger, T.W., Muller, S., Walter, R.D., 2000. The ornithine decarboxylase domain of the bifunctional ornithine decarboxylase/S-adenosylmethionine decarboxylase of Plasmodium falciparum: recombinant expression and catalytic properties of two different constructs. Biochem. J. 352 (Pt 2), 287–292.

Krug, E.C., Marr, J.J., Berens, R.L., 1989. Purine metabolism in Toxoplasma gondii. J. Biol. Chem. 264, 10601–10607.

Krungkrai, S.R., Aoki, S., Palacpac, N.M., Sato, D., Mitamura, T., Krungkrai, J., Horii, T., 2004a. Human malaria parasite orotate phosphoribosyltransferase: functional expression, characterization of kinetic reaction mechanism and inhibition profile. Mol. Biochem. Parasitol. 134, 245–255.

Krungkrai, S.R., Delfraino, B.J., Smiley, J.A., Prapunwattana, P., Mitamura, T., Horii, T., Krungkrai, J., 2005. A novel enzyme complex of orotate phosphoribosyltransferase and orotidine 5′-monophosphate decarboxylase in human malaria parasite Plasmodium falciparum: physical association, kinetics, and inhibition characterization. Biochemistry 44, 1643–1652.

Krungkrai, S.R., Prapunwattana, P., Horii, T., Krungkrai, J., 2004b. Orotate phosphoribosyltransferase and orotidine 5′-monophosphate decarboxylase exist as multienzyme complex in human malaria parasite Plasmodium falciparum. Biochem. Biophys. Res. Commun. 318, 1012–1018.

Kyes, S., Horrocks, P., Newbold, C., 2001. Antigenic variation at the infected red cell surface in malaria. Annu. Rev. Microbiol. 55, 673–707.

Landfear, S.M., 2011. Nutrient transport and pathogenesis in selected parasitic protozoa. Eukaryot. Cell 10, 483–493.

Landfear, S.M., Ullman, B., Carter, N.S., Sanchez, M.A., 2004. Nucleoside and nucleobase transporters in parasitic protozoa. Eukaryot. Cell 3, 245–254.

Lauer, S.A., Rathod, P.K., Ghori, N., Haldar, K., 1997. A membrane network for nutrient import in red cells infected with the malaria parasite. Science 276, 1122–1125.

Leport, C., Raffi, F., Katlama, C., Regneir, B., Saimot, A.G., Marche, C., Vedrenne, C., Vilde, J.L., 1988. Treatment of central nervous system toxoplasmosis with pyrimethamine/sulfonamide combination in 35 patients with the acquired immunodeficiency syndrome. Am. J. Med. 84, 94–100.

Lewandowicz, A., Ringia, E.A., Ting, L.M., Kim, K., Tyler, P.C., Evans, G.B., Zubkova, O.V., Mee, S., Painter, G.F., Lenz, D.H., Furneaux, R.H., Schramm, V.L., 2005. Energetic mapping of transition state analogue interactions with human and Plasmodium falciparum purine nucleoside phosphorylases. J. Biol. Chem. 280, 30320–30328.

Li, C.M., Tyler, P.C., Furneaux, R.H., Kicska, G., Xu, Y., Grubmeyer, C., Girvin, M.E., Schramm, V.L., 1999. Transition-state analogs as inhibitors of human and malarial hypoxanthine-guanine phosphoribosyltransferases. Nat. Struct. Biol. 6, 582–587.

Li, L., Brunk, B.P., Kissinger, J.C., Pape, D., Tang, K., Cole, R.H., Martin, J., Wylie, T., Dante, M., Fogarty, S.J., Howe, D.K., Liberator, P., Diaz, C., Anderson, J., White, M., Jerome, M.E., Johnson, E.A., Radke, J.A., Stoeckert Jr., C.J., Waterston, R.H., Clifton, S.W., Roos, D.S., Sibley, L.D., 2003. Gene discovery in the apicomplexa as revealed by EST sequencing and assembly of a comparative gene database. Genome Res. 13, 443–454.

Madrid, D.C., Ting, L.M., Waller, K.L., Schramm, V.L., Kim, K., 2008. Plasmodium falciparum purine nucleoside phosphorylase is critical for viability of malaria parasites. J. Biol. Chem. 283, 35899–35907.

Mahamed, D.A., Mills, J.H., Egan, C.E., Denkers, E.Y., Bynoe, M.S., 2012. CD73-generated adenosine facilitates Toxoplasma gondii differentiation to long-lived tissue cysts in the central nervous system. Proc. Natl. Acad. Sci. U. S. A. 109, 16312–16317.

Makoff, A.J., Radford, A., 1978. Genetics and biochemistry of carbamoyl phosphate biosynthesis and its utilization in the pyrimidine biosynthetic pathway. Microbiol. Rev. 42, 307–328.

Marr, J.J., Berens, R.L., 1988. Hypoxanthine and inosine analogues as chemotherapeutic agents in Chagas' disease. Mem. Inst. Oswaldo Cruz 83 (Suppl. 1), 301–307.

Marr, J.J., Berens, R.L., Nelson, D.J., 1978. Purine metabolism in Leishmania donovani and Leishmania braziliensis. Biochim. Biophys. Acta 544, 360–371.

Martin, R.E., Henry, R.I., Abbey, J.L., Clements, J.D., Kirk, K., 2005. The 'permeome' of the malaria parasite: an overview of the membrane transport proteins of Plasmodium falciparum. Genome Biol. 6, R26.

McConkey, G.A., 2000. Plasmodium falciparum: isolation and characterisation of a gene encoding protozoan GMP synthase. Exp. Parasitol. 94, 23–32.

McConkey, G.A., Pinney, J.W., Westhead, D.R., Plueckhahn, K., Fitzpatrick, T.B., Macheroux, P., Kappes, B., 2004. Annotating the Plasmodium genome and the enigma of the shikimate pathway. Trends Parasitol. 20, 60–65.

McFadden, D.C., Tomavo, S., Berry, E.A., Boothroyd, J.C., 2000. Characterization of cytochrome b from Toxoplasma gondii and Q(o) domain mutations as a mechanism of atovaquone-resistance. Mol. Biochem. Parasitol. 108, 1–12.

Mejias-Torres, I.A., Zimmermann, B.H., 2002. Molecular cloning, recombinant expression and partial characterization of the aspartate transcarbamoylase from Toxoplasma gondii. Mol. Biochem. Parasitol. 119, 191–201.

Mercier, C., Cesbron-Delauw, M.F., Sibley, L.D., 1998. The amphipathic alpha helices of the toxoplasma protein GRA2 mediate post-secretory membrane association. J. Cell Sci. 111 (Pt 15), 2171–2180.

Mercier, C., Dubremetz, J.F., Rauscher, B., Lecordier, L., Sibley, L.D., Cesbron-Delauw, M.F., 2002. Biogenesis of nanotubular network in Toxoplasma parasitophorous vacuole induced by parasite proteins. Mol. Biol. Cell 13, 2397–2409.

Miller, R.L., Linstead, D., 1983. Purine and pyrimidine metabolizing activities in *Trichomonas vaginalis* extracts. Mol. Biochem. Parasitol. 7, 41–51.

Mora, P., Rubio, V., Fresquet, V., Cervera, J., 1999. Localization of the site for the nucleotide effectors of Escherichia coli carbamoyl phosphate synthetase using site-directed mutagenesis. FEBS Lett. 446, 133–136.

Mori, M., Tatibani, M., 1978. Multi-enzyme complex of glutamine-dependent carbamoyl phosphate synthetase with aspartate carbamoyltransferase and dihydroorotase from rat acites-hepatoma cells. Eur. J. Biochem. 86, 381–388.

Muller, I.B., Walter, R.D., Wrenger, C., 2005. Structural metal dependency of the arginase from the human malaria parasite Plasmodium falciparum. Biol. Chem. 386, 117–126.

Muller, S., Coombs, G.H., Walter, R.D., 2001. Targeting polyamines of parasitic protozoa in chemotherapy. Trends Parasitol. 17, 242–249.

Muller, S., Da'dara, A., Luersen, K., Wrenger, C., Das Gupta, R., Madhubala, R., Walter, R.D., 2000. In the human malaria parasite Plasmodium falciparum, polyamines are synthesized by a bifunctional ornithine decarboxylase, S-adenosylmethionine decarboxylase. J. Biol. Chem. 275, 8097–8102.

Nara, T., Gao, G., Yamasaki, H., Nakajima-Shimada, J., Aoki, T., 1998. Carbamoyl-phosphate synthetase II in kinetoplastids. Biochim. Biophys. Acta 1387, 462–468.

Ngo, H.M., Ngo, E.O., Bzik, D.J., Joiner, K.A., 2000. Toxoplasma gondii: are host cell adenosine nucleotides a direct source for purine salvage? Exp. Parasitol. 95, 148–153.

Niedzwicki, J.G., Iltzsch, M.H., El Kouni, M.H., Cha, S., 1984. Structure-activity relationship of pyrimidine base analogs as ligands of orotate phosphoribosyltransferase. Biochem. Pharmacol. 33, 2383–2395.

North, M.J., Lockwood, B.C., 1995. Biochemistry and Molecular Biology of Parasites. Academic Press Inc., San Diego.

O'Neil, R.H., Lilien, R.H., Donald, B.R., Stroud, R.M., Anderson, A.C., 2003a. Phylogenetic classification of protozoa based on the structure of the linker domain in the bifunctional enzyme, dihydrofolate reductase-thymidylate synthase. J. Biol. Chem. 278, 52980–52987.

O'Neil, R.H., Lilien, R.H., Donald, B.R., Stroud, R.M., Anderson, A.C., 2003b. The crystal structure of dihydrofolate reductase-thymidylate synthase from Cryptosporidium hominis reveals a novel architecture for the bifunctional enzyme. J. Eukaryot. Microbiol. 50 (Suppl.), 555–556.

O'Sullivan, W.J., Johnson, A.M., Finney, K.G., Gero, A.M., Hagon, E., Holland, J.W., Smithers, G.W., 1981. Pyrimidine and purine enzymes in Toxoplasma gondii. Aust. J. Exp. Biol. Med. Sci. 59, 763–767.

Painter, H.J., Morrisey, J.M., Mather, M.W., Vaidya, A.B., 2007. Specific role of mitochondrial electron transport in blood-stage Plasmodium falciparum. Nature 446, 88–91.

Parker, M.D., Hyde, R.J., Yao, S.Y., Mcrobert, L., Cass, C.E., Young, J.D., Mcconkey, G.A., Baldwin, S.A., 2000. Identification of a nucleoside/nucleobase transporter from Plasmodium falciparum, a novel target for antimalarial chemotherapy. Biochem. J. 349, 67–75.

Pashley, T.V., Volpe, F., Pudney, M., Hyde, J.E., Sims, P.F., Delves, C.J., 1997. Isolation and molecular characterization of the bifunctional hydroxymethyldihydropterin pyrophosphokinase-dihydropteroate synthase gene from Toxoplasma gondii. Mol. Biochem. Parasitol. 86, 37–47.

Perotto, J., Keister, D.B., 1971. Incorporation of precursors into Toxoplasma DNA. J. Protozool. 18, 470–473.

Peterson, D.S., Walliker, D., Wellems, T.E., 1988. Evidence that a point mutation in dihydrofolate reductase-thymidylate synthase confers resistance to pyrimethamine in falciparum malaria. Proc. Natl. Acad. Sci. U. S. A. 85, 9114–9118.

Pfefferkorn, E.R., 1978. Toxoplasma gondii: the enzymic defect of a mutant resistant to 5-fluorodeoxyuridine. Exp. Parasitol. 44, 26–35.

Pfefferkorn, E.R., 1984. Interferon gamma blocks the growth of Toxoplasma gondii in human fibroblasts by inducing the host cells to degrade tryptophan. Proc. Natl. Acad. Sci. U. S. A. 81, 908–912.

Pfefferkorn, E.R., 1986. Interferon gamma and the growth of Toxoplasma gondii in fibroblasts. Ann. Inst. Pasteur. Microbiol. 137A, 348–352.

Pfefferkorn, E.R., Borotz, S.E., 1994. Toxoplasma gondii: characterization of a mutant resistant to 6-thioxanthine. Exp. Parasitol. 79, 374–382.

Pfefferkorn, E.R., Borotz, S.E., Nothnagel, R.F., 1993. Mutants of Toxoplasma gondii resistant to atovaquone (566C80) or decoquinate. J. Parasitol. 79, 559–564.

Pfefferkorn, E.R., Bzik, D.J., Honsinger, C.P., 2001. Toxoplasma gondii: mechanism of the parasitostatic action of 6-thioxanthine. Exp. Parasitol. 99, 235–243.

Pfefferkorn, E.R., Eckel, M., Rebhun, S., 1986a. Interferon-gamma suppresses the growth of Toxoplasma gondii in human fibroblasts through starvation for tryptophan. Mol. Biochem. Parasitol. 20, 215–224.

Pfefferkorn, E.R., Eckel, M.E., Mcadams, E., 1989. Toxoplasma gondii: the biochemical basis of resistance to emimycin. Exp. Parasitol. 69, 129–139.

Pfefferkorn, E.R., Guyre, P.M., 1984. Inhibition of growth of Toxoplasma gondii in cultured fibroblasts by human recombinant gamma interferon. Infect. Immun. 44, 211–216.

Pfefferkorn, E.R., Pfefferkorn, L.C., 1976. Arabinosyl nucleosides inhibit Toxoplasma gondii and allow the selection of resistant mutants. J. Parasitol. 62, 993–999.

Pfefferkorn, E.R., Pfefferkorn, L.C., 1977a. Specific labelling of intracellular Toxoplasma gondii with uracil. J. Protozool. 24, 449–453.

Pfefferkorn, E.R., Pfefferkorn, L.C., 1977b. Toxoplasma gondii: characterization of a mutant resistant to 5-fluorodeoxyuridine. Exp. Parasitol. 42, 44–55.

Pfefferkorn, E.R., Pfefferkorn, L.C., 1977c. Toxoplasma gondii: specific labelling of nucleic acids of intracellular parasites in Lesch-Nyhan cells. Exp. Parasitol. 41, 95–104.

Pfefferkorn, E.R., Pfefferkorn, L.C., 1978. The biochemical basis for resistance to adenine arabinoside in a mutant of Toxoplasma gondii. J. Parasitol. 64, 486–492.

Pfefferkorn, E.R., Pfefferkorn, L.C., 1981. Toxoplasma gondii: growth in the absence of host cell protein synthesis. Exp. Parasitol. 52, 129–136.

Pfefferkorn, E.R., Rebhun, S., Eckel, M., 1986b. Characterization of an indoleamine 2,3-dioxygenase induced by gamma-interferon in cultured human fibroblasts. J. Interferon. Res. 6, 267–279.

Phillips, M.A., Gujjar, R., Malmquist, N.A., White, J., El Mazouni, F., Baldwin, J., Rathod, P.K., 2008. Triazolopyrimidine-based dihydroorotate dehydrogenase inhibitors with potent and selective activity against the malaria parasite Plasmodium falciparum. J. Med. Chem. 51, 3649–3653.

Phillips, M.A., Rathod, P.K., 2010. Plasmodium dihydroorotate dehydrogenase: a promising target for novel anti-malarial chemotherapy. Infect. Disord. Drug Targets 10, 226–239.

Plagemann, P.G., 1986. Transport and metabolism of adenosine in human erythrocytes: effect of transport inhibitors and regulation by phosphate. J. Cell Physiol. 128, 491–500.

Plagemann, P.G., Wohlhueter, R.M., Woffendin, C., 1988. Nucleoside and nucleobase transport in animal cells. Biochim. Biophys. Acta 947, 405–443.

Pockros, P.J., Reindollar, R., McHutchinson, J., Reddy, R., Wright, T., Boyd, D.G., Wilkes, L.B., 2003. The safety and tolerability of daily infergen plus ribavirin in the treatment of naiive chronic hepatitis C patients. J. Viral. Hepat. 10, 55–60.

Puiu, D., Enomoto, S., Buck, G.A., Abrahamsen, M.S., Kissinger, J.C., 2004. CryptoDB: the Cryptosporidium genome resource. Nucleic Acids Res. 32, D329–D331.

Queen, S.A., Jagt, D.L., Reyes, P., 1990. In vitro susceptibilities of Plasmodium falciparum to compounds which inhibit nucleotide metabolism. Antimicrob. Agents Chemother. 34, 1393–1398.

Queen, S.A., Vander Jagt, D., Reyes, P., 1988. Properties and substrate specificity of a purine phosphoribosyltransferase from the human malaria parasite, Plasmodium falciparum. Mol. Biochem. Parasitol. 30, 123–133.

Rager, N., Mamoun, C.B., Carter, N.S., Goldberg, D.E., Ullman, B., 2001. Localization of the Plasmodium falciparum PfNT1 nucleoside transporter to the parasite plasma membrane. J. Biol. Chem. 276, 41095–41099.

Rais, R.H., Al Safarjalani, O.N., Yadav, V., Guarcello, V., Kirk, M., Chu, C.K., Naguib, F.N., El Kouni, M.H., 2005. 6-Benzylthioinosine analogues as subversive substrate of Toxoplasma gondii adenosine kinase: activities and selective toxicities. Biochem. Pharmacol. 69, 1409–1419.

Raman, J., Mehrotra, S., Anand, R.P., Balaram, H., 2004. Unique kinetic mechanism of Plasmodium falciparum adenylosuccinate synthetase. Mol. Biochem. Parasitol. 138, 1–8.

Rathod, P.K., Khatri, A., Hubbert, T., Milhous, W.K., 1989. Selective activity of 5-fluoroorotic acid against Plasmodium falciparum in vitro. Antimicrob. Agents Chemother. 33, 1090–1094.

Reyes, P., Rathod, P.K., Sanchez, D.J., Mrema, J.E., Rieckmann, K.H., Heidrich, H.G., 1982. Enzymes of purine and pyrimidine metabolism from the human malaria parasite, Plasmodium falciparum. Mol. Biochem. Parasitol. 5, 275–290.

Reynolds, M.G., Oh, J., Roos, D.S., 2001. In vitro generation of novel pyrimethamine resistance mutations in the Toxoplasma gondii dihydrofolate reductase. Antimicrob. Agents Chemother. 45, 1271–1277.

Reynolds, M.G., Roos, D.S., 1998. A biochemical and genetic model for parasite resistance to antifolates. Toxoplasma gondii provides insights into pyrimethamine and cycloguanil resistance in Plasmodium falciparum. J. Biol. Chem. 273, 3461–3469.

Roberts, F., Roberts, C.W., Johnson, J.J., Kyle, D.E., Krell, T., Coggins, J.R., Coombs, G.H., Milhous, W.K., Tzipori, S., Ferguson, D.J., Chakrabarti, D., McLeod, R., 1998. Evidence for the shikimate pathway in apicomplexan parasites. Nature 393, 801–805.

Roberts, S.C., Tancer, M.J., Polinsky, M.R., Gibson, K.M., Heby, O., Ullman, B., 2004. Arginase plays a pivotal role in polyamine precursor metabolism in Leishmania. Characterization of gene deletion mutants. J. Biol. Chem. 279, 23668–23678.

Robles Lopez, S.M., Hortua Triana, M.A., Zimmermann, B.H., 2006. Cloning and preliminary characterization of the dihydroorotase from Toxoplasma gondii. Mol. Biochem. Parasitol. 148, 93–98.

Roos, D.S., 1993. Primary structure of the dihydrofolate reductase-thymidylate synthase gene from Toxoplasma gondii. J. Biol. Chem. 268, 6269–6280.

Roos, D.S., Donald, R.G., Morrissette, N.S., Moulton, A.L., 1994. Molecular tools for genetic dissection of the protozoan parasite Toxoplasma gondii. Methods Cell Biol. 45, 27–63.

Roos, D.S., Sullivan, W.J., Striepen, B., Bohne, W., Donald, R.G., 1997. Tagging genes and trapping promoters in Toxoplasma gondii by insertional mutagenesis. Methods 13, 112–122.

Sabin, A.B., 1941. Toxoplasmic encephalitis in children. J. Am. Med. Assoc. 116, 801–807.

Sarkar, D., Ghosh, I., Datta, S., 2004. Biochemical characterization of Plasmodium falciparum hypoxanthine-guanine-xanthine phosphorybosyltransferase: role of histidine residue in substrate selectivity. Mol. Biochem. Parasitol. 137, 267–276.

Satriano, J., 2003. Agmatine: at the crossroads of the arginine pathways. Ann. N. Y. Acad. Sci. 1009, 34–43.

Schatten, H., Ris, H., 2004. Three-dimensional imaging of *Toxoplasma gondii*-host cell interactions within the parasitophorous vacuole. Microsc. Microanal. 10, 580–585.

Schumacher, M.A., Bashor, C.J., Song, M.H., Otsu, K., Zhu, S., Parry, R.J., Ullman, B., Brennan, R.G., 2002. The structural mechanism of GTP stabilized oligomerization and catalytic activation of the Toxoplasma gondii uracil phosphoribosyltransferase. Proc. Natl. Acad. Sci. U. S. A. 99, 78–83.

Schumacher, M.A., Carter, D., Ross, D.S., Ullman, B., Brennan, R.G., 1996. Crystal structures of Toxoplasma gondii HGXPRTase reveal the catalytic role of a long flexible loop. Nat. Struct. Biol. 3, 881–887.

Schumacher, M.A., Carter, D., Scott, D.M., Roos, D.S., Ullman, B., Brennan, R.G., 1998. Crystal structures of Toxoplasma gondii uracil phosphoribosyltransferase reveal the atomic basis of pyrimidine discrimination and prodrug binding. Embo J. 17, 3219–3232.

Schumacher, M.A., Scott, D.M., Mathews II, Ealick, S.E., Roos, D.S., Ullman, B., Brennan, R.G., 2000a. Crystal structures of Toxoplasma gondii adenosine kinase reveal a novel catalytic mechanism and prodrug binding. J. Mol. Biol. 298, 875–893.

Schumacher, M.A., Scott, D.M., Mathews II, Ealick, S.E., Roos, D.S., Ullman, B., Brennan, R.G., 2000b. Crystal structures of Toxoplasma gondii adenosine kinase reveal a novel catalytic mechanism and prodrug binding. J. Mol. Biol. 296, 549–567.

Schwab, J.C., Afifi Afifi, M., Pizzorno, G., Handschumacher, R.E., Joiner, K.A., 1995. Toxoplasma gondii tachyzoites possess an unusual plasma membrane adenosine transporter. Mol. Biochem. Parasitol. 70, 59–69.

Schwab, J.C., Beckers, C.J., Joiner, K.A., 1994. The parasitophorous vacuole membrane surrounding intracellular Toxoplasma gondii functions as a molecular sieve. Proc. Natl. Acad. Sci. U. S. A. 91, 509–513.

Schwartzman, J.D., Pfefferkorn, E.R., 1981. Pyrimidine synthesis by intracellular Toxoplasma gondii. J. Parasitol. 67, 150–158.

Schwartzman, J.D., Pfefferkorn, E.R., 1982. Toxoplasma gondii: purine synthesis and salvage in mutant host cells and parasites. Exp. Parasitol. 53, 77–86.

Seabra, S.H., Damatta, R.A., De Mello, F.G., De Souza, W., 2004. Endogenous polyamine levels in macrophages is sufficient to support growth of Toxoplasma gondii. J. Parasitol. 90, 455–460.

Sethi, K.K., Pelster, B., Piekarski, G., Brandis, H., 1973. Multiplication of Toxoplasma gondii in enucleated L cells. Nat. New Biol. 243, 255–256.

Seymour, K.K., Lyons, S.D., Phillips, L., Rieckmann, K.H., Christopherson, R.I., 1994. Cytotoxic effects of inhibitors of de novo pyrimidine biosynthesis upon Plasmodium falciparum. Biochemistry 33, 5268–5274.

Sharling, L., Liu, X., Gollapalli, D.R., Maurya, S.K., Hedstrom, L., Striepen, B., 2010. A screening pipeline for antiparasitic agents targeting cryptosporidium inosine monophosphate dehydrogenase. PLoS Negl. Trop. Dis. 4, e794.

Shi, W., Li, C.M., Tyler, P.C., Furneaux, R.H., Cahill, S.M., Girvin, M.E., Grubmeyer, C., Schramm, V.L., Almo, S.C., 1999a. The 2.0 A structure of malarial purine phosphoribosyltransferase in complex with a transition-state analogue inhibitor. Biochemistry 38, 9872–9880.

Shi, W., Li, C.M., Tyler, P.C., Furneaux, R.H., Grubmeyer, C., Schramm, V.L., Almo, S.C., 1999b. The 2.0 A structure of human hypoxanthine-guanine phosphoribosyltransferase in complex with a transition-state analog inhibitor. Nat. Struct. Biol. 6, 588–593.

Shi, W., Ting, L.M., Kicska, G.A., Lewandowicz, A., Tyler, P.C., Evans, G.B., Furneaux, R.H., Kim, K., Almo, S.C., Schramm, V.L., 2004. Plasmodium falciparum purine nucleoside phosphorylase: crystal structures, immucillin inhibitors, and dual catalytic function. J. Biol. Chem. 279, 18103–18106.

Shivashankar, K., Subbayya, I.N., Balaram, H., 2001. Development of a bacterial screen for novel hypoxanthine-guanine phosphoribosyltransferase substrates. J. Mol. Microbiol. Biotechnol. 3, 557–562.

Sibley, L.D., Niesman, I.R., Asai, T., Takeuchi, T., 1994. Toxoplasma gondii: secretion of a potent nucleoside triphosphate hydrolase into the parasitophorous vacuole. Exp. Parasitol. 79, 301–311.

Sibley, L.D., Niesman, I.R., Parmley, S.F., Cesbron-Delauw, M.F., 1995. Regulated secretion of multi-lamellar vesicles leads to formation of a tubulo-vesicular network in host-cell vacuoles occupied by Toxoplasma gondii. J. Cell Sci. 108 (Pt 4), 1669–1677.

Sierra Pagan, M.L., Zimmermann, B.H., 2003. Cloning and expression of the dihydroorotate dehydrogenase from Toxoplasma gondii. Biochim. Biophys. Acta 1637, 178–181.

Srivastava, I.K., Morrisey, J.M., Darrouzet, E., Daldal, F., Vaidya, A.B., 1999. Resistance mutations reveal the atovaquone-binding domain of cytochrome b in malaria parasites. Mol. Microbiol. 33, 704–711.

Striepen, B., Kissinger, J.C., 2004. Genomics meets transgenics in search of the elusive Cryptosporidium drug target. Trends Parasitol. 20, 355–358.

Striepen, B., Pruijssers, A.J., Huang, J., Li, C., Gubbels, M.J., Umejiego, N.N., Hedstrom, L., Kissinger, J.C., 2004. Gene transfer in the evolution of parasite nucleotide biosynthesis. Proc. Natl. Acad. Sci. U. S. A. 101, 3154–3159.

Striepen, B., White, M.W., Li, C., Guerini, M.N., Malik, S.B., Logsdon Jr., J.M., Liu, C., Abrahamsen, M.S., 2002. Genetic complementation in apicomplexan parasites. Proc. Natl. Acad. Sci. U. S. A. 99, 6304–6309.

Sufrin, J.R., Meshnick, S.R., Spiess, A.J., Garofalo-Hannan, J., Pan, X.Q., Bacchi, C.J., 1995. Methionine recycling pathways and antimalarial drug design. Antimicrob. Agents Chemother. 39, 2511–2515.

Sullivan Jr., W.J., Chiang, C.W., Wilson, C.M., Naguib, F.N., El Kouni, M.H., Donald, R.G., Roos, D.S., 1999. Insertional tagging of at least two loci associated with resistance to adenine arabinoside in Toxoplasma gondii, and cloning of the adenosine kinase locus. Mol. Biochem. Parasitol. 103, 1–14.

Sun, X.E., Sharling, L., Muthalagi, M., Mudeppa, D.G., Pankiewicz, K.W., Felczak, K., Rathod, P.K., Mead, J., Striepen, B., Hedstrom, L., 2010. Prodrug activation by Cryptosporidium thymidine kinase. J. Biol. Chem. 285, 15916–15922.

Suss-Toby, E., Zimmerberg, J., Ward, G.E., 1996. Toxoplasma invasion: the parasitophorous vacuole is formed from host cell plasma membrane and pinches off via a fission pore. Proc. Natl. Acad. Sci. U. S. A. 93, 8413–8418.

Tabor, C.W., Tabor, H., 1985. Polyamines in microorganisms. Microbiol. Rev. 49, 81–99.

Tanaka, M., Gu, H.M., Bzik, D.J., Li, W.B., Inselburg, J., 1990a. Mutant dihydrofolate reductase-thymidylate synthase genes in pyrimethamine-resistant Plasmodium falciparum with polymorphic chromosome duplications. Mol. Biochem. Parasitol. 42, 83–91.

Tanaka, M., Gu, H.M., Bzik, D.J., Li, W.B., Inselburg, J.W., 1990b. Dihydrofolate reductase mutations and chromosomal changes associated with pyrimethamine resistance of Plasmodium falciparum. Mol. Biochem. Parasitol. 39, 127–134.

Terao, M., 1963. On A New Antibiotic, Emimycin. II. Studies on the Structure of Emimycin. J. Antibiot. (Tokyo) 16, 182–186.

Terao, M., Karasawa, K., Tanaka, N., Yonehara, H., Umezawa, H., 1960. On a new antibiotic, emimycin. The Journal of Antibiotics 13A, 401–405.

Thomas, A., Field, M.J., 2002. Reaction mechanism of the HGXPRTase from Plasmodium falciparum: a hybrid potential quantum mechanical/molecular mechanical study. J. Am. Chem. Soc. 124, 12432–12438.

Ting, L.M., Shi, W., Lewandowicz, A., Singh, V., Mwakingwe, A., Birck, M.R., Ringia, E.A., Bench, G., Madrid, D.C., Tyler, P.C., Evans, G.B., Furneaux, R.H., Schramm, V.L., Kim, K., 2005. Targeting a novel Plasmodium falciparum purine recycling pathway with specific immucillins. J. Biol. Chem. 280, 9547–9554.

Traut, T.W., 1994. Physiological conc entrations of purines and pyrimidines. Mol. Cell Biochem. 140, 1–22.

Triglia, T., Cowman, A.F., 1994. Primary structure and expression of the dihydropteroate synthetase gene of Plasmodium falciparum. Proc. Natl. Acad. Sci. U. S. A. 91, 7149–7153.

Triglia, T., Wang, P., Sims, P.F., Hyde, J.E., Cowman, A.F., 1998. Allelic exchange at the endogenous genomic locus in Plasmodium falciparum proves the role of dihydropteroate synthase in sulfadoxine-resistant malaria. Embo J. 17, 3807–3815.

Tyler, P.C., Taylor, E.A., Frohlich, R.F., Schramm, V.L., 2007. Synthesis of 5′-methylthio coformycins: specific inhibitors for malarial adenosine deaminase. J. Am. Chem. Soc. 129, 6872–6879.

Upston, J.M., Gero, A.M., 1995. Parasite-induced permeation of nucleosides in Plasmodium falciparum malaria. Biochim. Biophys. Acta 1236, 249–258.

Vasanthakumar, G., Davis Jr., R.L., Sullivan, M.A., Donahue, J.P., 1989. Nucleotide sequence of cDNA clone for hypoxanthine-guanine phosphoribosyltransferase from Plasmodium falciparum. Nucleic Acids Res. 17, 8382.

Vasanthakumar, G., Davis Jr., R.L., Sullivan, M.A., Donahue, J.P., 1990. Cloning and expression in Escherichia coli of a hypoxanthine-guanine phosphoribosyltransferase-encoding cDNA from Plasmodium falciparum. Gene 91, 63–69.

Vasquez, J.R., Gooze, L., Kim, K., Gut, J., Petersen, C., Nelson, R.G., 1996. Potential antifolate resistance determinants and genotypic variation in the bifunctional dihydrofolate reductase-thymidylate synthase gene from human and bovine isolates of Cryptosporidium parvum. Mol. Biochem. Parasitol. 79, 153–165.

Wang, C.C., 1984. Parasite enzymes as potential targets for antiparasitic chemotherapy. J. Med. Chem. 27, 1–9.

Wang, C.C., Aldritt, S., 1983. Purine salvage networks in *Giardia lamblia*. J. Exp. Med. 158, 1703–1712.

Wang, C.C., Simashkevich, P.M., 1981. Purine metabolism in the protozoan parasite *Eimeria tenella*. Proc. Natl. Acad. Sci. USA 78, 6618–6622.

Wang, P., Nirmalan, N., Wang, Q., Sims, P.F., Hyde, J.E., 2004a. Genetic and metabolic analysis of folate salvage in the human malaria parasite Plasmodium falciparum. Mol. Biochem. Parasitol. 135, 77–87.

Wang, P., Wang, Q., Aspinall, T.V., Sims, P.F., Hyde, J.E., 2004b. Transfection studies to explore essential folate metabolism and antifolate drug synergy in the human malaria parasite Plasmodium falciparum. Mol. Microbiol. 51, 1425–1438.

Webster, H.K., Wiesmann, W.P., Pavia, C.S., 1984. Adenosine deaminase in malaria infection: effect of 2′-deoxycoformycin in vivo. Adv. Exp. Med. Biol. 165 Pt A, 225–229.

White, E.L., Ross, L.J., Davis, R.L., Zywno-Van Ginkel, S., Vasanthakumar, G., Borhani, D.W., 2000. The two toxoplasma gondii hypoxanthine-guanine phosphoribosyltransferase isozymes form heterotetramers. J. Biol. Chem. 275, 19218–19223.

Woods, K.M., Upton, S.J., 1998. Efficacy of select antivirals against Cryptosporidium parvum in vitro. FEMS Microbiol. Lett. 168, 59–63.

Wrenger, C., Luersen, K., Krause, T., Muller, S., Walter, R.D., 2001. The Plasmodium falciparum bifunctional ornithine decarboxylase, S-adenosyl-L-methionine decarboxylase, enables a well balanced polyamine synthesis without domain-domain interaction. J. Biol. Chem. 276, 29651–29656.

Wu, G., Morris, S.M., 1998. Arginine metabolism: nitric oxide and beyond. Biochem. J. 336, 1–17.

Xu, P., Widmer, G., Wang, Y., Ozaki, L.S., Alves, J.M., Serrano, M.G., Puiu, D., Manque, P., Akiyoshi, D., Mackey, A.J., Pearson, W.R., Dear, P.H., Bankier, A. T., Peterson, D.L., Abrahamsen, M.S., Kapur, V., Tzipori, S., Buck, G.A., 2004. The genome of Cryptosporidium hominis. Nature 431, 1107–1112.

Yarlett, N., Martinez, M.P., Zhu, G., Keithly, J.S., Woods, K., Upton, S.J., 1996. Cryptosporidium parvum: polyamine biosynthesis from agmatine. J. Eukaryot Microbiol. 43, 73S.

Yuvaniyama, J., Chitnumsub, P., Kamchonwongpaisan, S., Vanichtanankul, J., Sirawaraporn, W., Taylor, P., Walkinshaw, M.D., Yuthavong, Y., 2003. Insights into antifolate resistance from malarial DHFR-TS structures. Nat. Struct. Biol. 10, 357–365.

Zhou, Z., Metcalf, A.E., Lovatt, C.J., Hyman, B.C., 2000. Alfalfa (Medicago sativa) carbamoylphosphate synthetase gene structure records the deep lineage of plants. Gene 243, 105–114.

CHAPTER

21

Toxoplasma gondii Chemical Biology

*Matthew Bogyo**, *Gary Ward*[†]

*Department of Pathology, Stanford University School of Medicine, Stanford, California, USA and [†]Department of Microbiology and Molecular Genetics, University of Vermont College of Medicine, Burlington, Vermont, USA

OUTLINE

Toxoplasma gondii, *second edition*
http://dx.doi.org/10.1016/B978-0-12-396481-6.00021-0

21.1 INTRODUCTION

Our understanding of the biology of *Toxoplasma gondii* has benefited greatly from the availability of an extensive set of molecular genetic tools to manipulate the parasite, including methods for random and targeted integration of exogenous DNA, gene knockouts, conditional gene expression, and chemical and insertional mutagenesis (see Chapter 17). Furthermore, excellent bioinformatics resources for genomic, transcriptomic and proteomic analysis are available (see Chapters 19 and 22). Recent advances in the use of small molecules to study parasites have established chemical biology as another powerful tool in this toolbox, one that is highly complementary to molecular genetics and bioinformatics. 'Chemical biology' is a relatively new term used to describe an experimental approach that shares many techniques and methods with what has traditionally been called pharmacology. However, chemical biology uses small organic molecules not only to induce cellular phenotypes, but also as tools to define and explore biological mechanisms. The emergence of chemical biology as a powerful new experimental approach has been made possible by recent technical advances in organic synthesis, analytical biochemistry, fluorescence microscopy, genomics and proteomics.

In this chapter, we will provide an introduction to the field of chemical biology, with a focus on two strategies: forward and reverse chemical genetics. We will compare different methods for identifying the target(s) of bioactive small molecules of interest, and we will describe approaches to target validation and demonstration of compound specificity. We will then provide a few selected examples of how chemical biology has been used to reveal insights into the complex biology of *T. gondii* and the mechanisms underlying host–parasite interaction. Since chemical biology uses small molecules as the primary experimental tool for mapping biological pathways, the approach necessarily focuses our attention on pathways and targets that are likely to be susceptible to small-molecule perturbation, with potential implications for drug development.

21.2 SMALL MOLECULES AS TOOLS: TO MONITOR OR TO MODULATE?

Small molecules are most frequently used experimentally to *modulate* the function of a protein target and assess the consequences. However, small molecules can also be used to *monitor* the activity of specific protein targets. We present here a brief overview of these two distinct ways of using small molecules, beginning with monitoring. Most of the rest of the chapter is then devoted to the more familiar and widespread use of small molecules as tools to modulate protein function.

21.2.1 Using Small Molecules to Monitor Protein Function

Small-molecule probes that act as tracers to monitor and report a particular enzymatic activity can be used to correlate changes in enzyme expression or activation with the timing of a particular biological process. This kind of information can suggest links between specific proteins and processes, providing a starting point for more targeted validation studies using the powerful genetic tools available in *T. gondii*.

The best examples of small molecules used to monitor enzyme function are a class of reagents called activity based probes (ABPs). These are small molecules that are designed to mimic the substrate of a given enzyme or enzyme class (Fig. 21.1). The ABP binds in the active site of a small fraction of the total pool of target enzyme. Rather than acting as a substrate that is processed and released by the enzyme, it becomes covalently linked in the active site. Because this specific covalent modification is

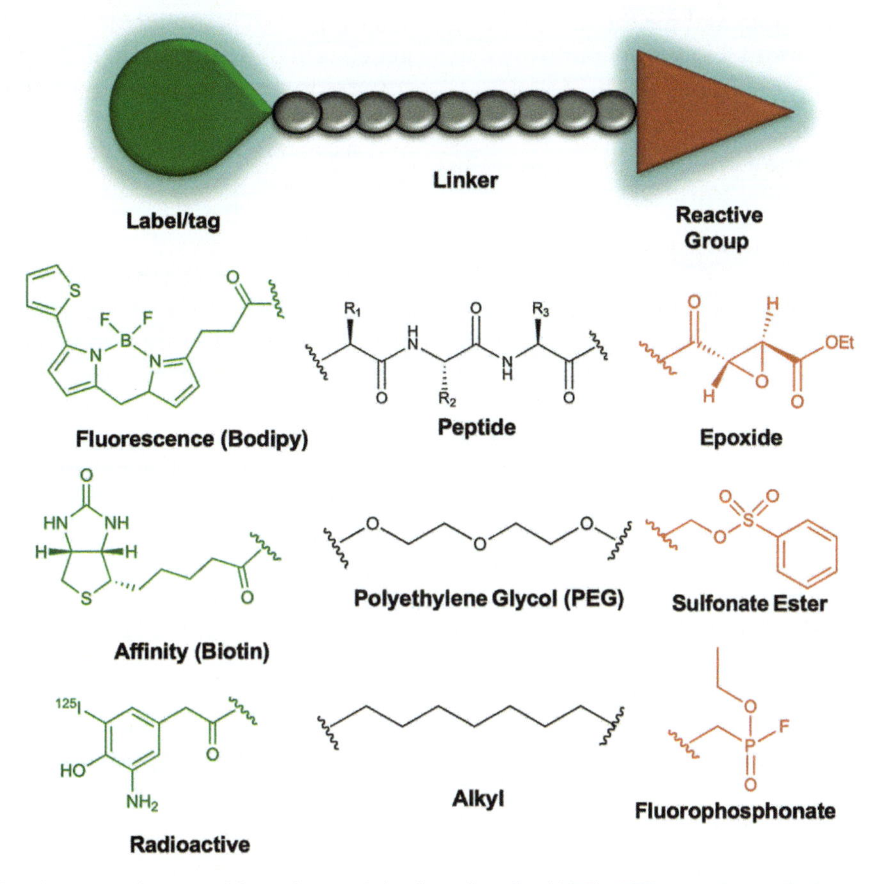

FIGURE 21.1 **The general composition of an activity based probe (ABP).** ABPs contain a reactive group that targets a residue in the enzyme active site, a linker that can function to direct the selectivity of the probe and a tag for visualization or purification of the labelled target protein(s).

irreversible, the ABP-labelled enzyme can be followed using standard analytical methods, such as fluorescence microscopy or SDS–PAGE analysis. Since the binding and modification only occur with active enzymes, the amount of probe bound can be used as an indirect readout of enzymatic activity. This allows dynamic profiling of enzymatic activity with a high degree of temporal control, making the use of ABPs in parasite systems particularly valuable. For example, ABPs have been used to monitor when and how specific proteases are activated during parasite invasion of and egress from host cells, allowing specific functions to be assigned to particular proteases during these processes (Greenbaum *et al.*, 2002; Arastu-Kapur *et al.*, 2008; Chandramohanadas *et al.*, 2009; Larson *et al.*, 2009; Hall *et al.*, 2011; Cong *et al.*, 2012). While there has been rapid growth in the number of validated ABPs reported in the literature over the last decade (for reviews see Evans and Cravatt, 2006; Cravatt *et al.*, 2008), only protease-directed probes have thus far been used in parasite systems, and only two examples of these can be found for *T. gondii* (Larson *et al.*, 2009; Hall *et al.*, 2011). Thus, there remains untapped potential for the use of ABPs to map out important biological pathways in *T. gondii*.

21.2.2 Using Small Molecules to Modulate Protein Function

In the simplest and most familiar application of the use of small molecules to modulate rather than monitor protein function, a compound whose target has been well established in other systems is tested against *T. gondii*. If the compound affects some aspect of parasite biology, the predicted target is implicated in that process. For example, much of what we currently know about the role of actin and myosin in parasite motility and invasion (see Chapter 13) can be traced back to the early observation that these processes are inhibited by the actin inhibitor, cytochalasin D (see Section 21.6.1 for further details). Small-molecule modulators are particularly valuable reagents for studies of essential and/or dynamic processes in *T. gondii*. Genes essential to the haploid parasite can be difficult to disrupt, and while there have been major advances in methods that allow control of gene expression and protein levels in *T. gondii*, these methods generally lack the kind of temporal control that is required to dissect rapidly occurring processes such as parasite invasion, secretion and motility. In contrast, the timing of exposure to a small-molecule modulator can be stringently controlled by the investigator. Therefore, small-molecule modulators of protein function have an important place in the *T. gondii* experimental toolbox. The major limitation of small molecules as modulators is their potential lack of selectivity for a given target protein. As discussed later in this chapter (Section 21.5), there are a number of methods that one can use to define the selectivity of a small-molecule modulator and link its activity to a given target.

The use of selective small molecules, rather than genetic manipulation, to modulate protein function is referred to as 'chemical genetics'. This approach is conceptually similar to classical genetics in that it strives to link a given biological process to a specific gene by inhibiting or otherwise blocking the function of the gene product and assessing the phenotypic consequences. As with classical genetics, chemical genetics can be performed in the forward or reverse direction (Fig. 21.2). In **forward chemical genetics**, small molecules are screened to identify compounds that induce a specific phenotype of interest (e.g. a block in host cell invasion). The resulting hit compounds are then linked to specific gene products using any of a number of different methods (outlined in Sections 21.4.4 and 21.5). In **reverse chemical genetics,** a specific target protein is chosen and small molecules are screened to identify potent modulators of that target. If those molecules are sufficiently specific for the target of interest, they can then be used in intact cells for phenotypic studies that allow assignment of function to the target protein. Both forward (cell-based) and reverse (target-based) chemical genetics have been used to study *T. gondii* biology and will be discussed in more detail below, beginning with the reverse approach.

21.3 REVERSE (TARGET-BASED) CHEMICAL GENETICS

The successful application of reverse chemical genetics requires small molecules with sufficient potency and selectivity for a target that they can be used in cell-based assays. Focusing on a specific target, the approach uses medicinal chemistry, compound screening, modelling and/or structural design methods to build compounds that are then used for detailed studies of the biological function of that target.

In one of the earliest examples of the use of reverse chemical genetics to study parasites, a focused library of cysteine protease inhibitors was used to identify a compound with a high degree of selectivity towards the *Plasmodium*

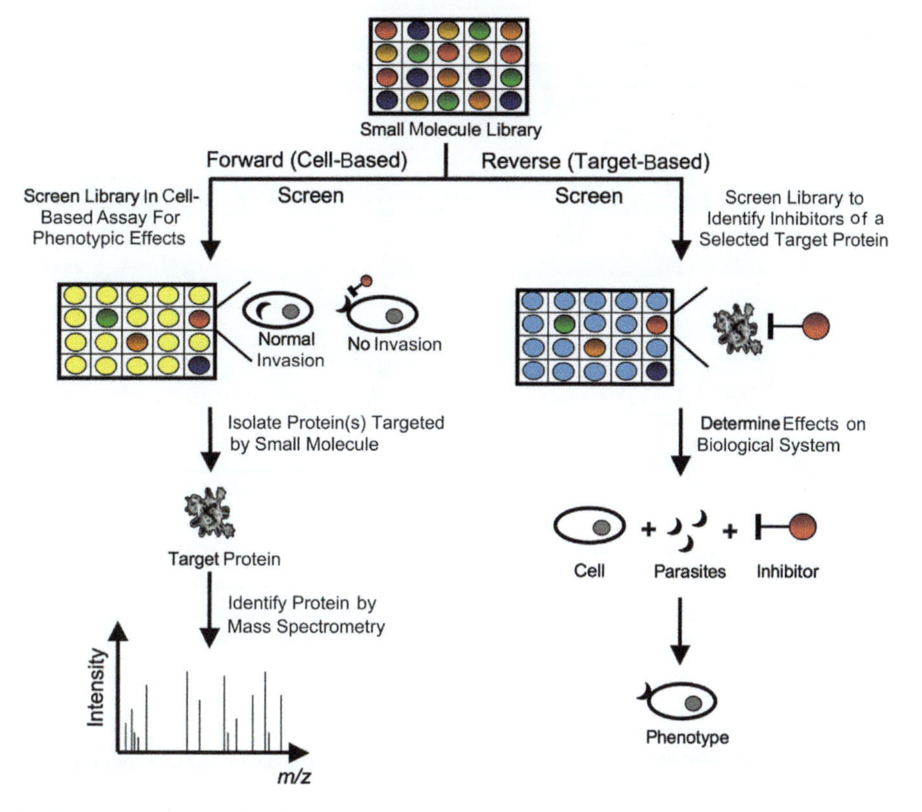

FIGURE 21.2 **Schematic of forward and reverse chemical genetic screens.** In this example, compounds are screened in the forward direction for effects on host cell invasion and in the reverse direction for inhibition of a selected target protein that, ultimately, also proves to be involved in invasion.

falciparum cysteine protease falcipain-1. The resulting selective inhibitor was then used to show that inhibition of this target resulted in a block in host cell invasion (Greenbaum *et al.*, 2002). In a more recent example, a large library of compounds was used to screen for inhibitors of the *P. falciparum* calcium-dependent protein kinase 1 (PfCDPK1). The resulting optimized lead compounds were further elaborated using medicinal chemistry efforts and used to show effects on parasite invasion in both *P. falciparum* and *T. gondii* (Kato *et al.*, 2008; see Section 21.6.5 for details). These initial studies did not assess the selectivity of the compounds in parasites, making it difficult to tell if the observed effects were in fact linked to inhibition of CDPK(s). However, the functional role of TgCDPK1 in invasion was later confirmed using a conditional knock-down system and by engineering highly selective inhibitors (Lourido *et al.*, 2010; Ojo *et al.*, 2010; see Section 21.6.5 for details).

In another example, structural modelling coupled with small-molecule docking was used to identify compounds predicted to disrupt the interaction between the tail of *T. gondii* MyosinA (TgMyoA) and myosin light chain-1 (TgMLC1; Kortagere *et al.*, 2011). These efforts yielded a compound that inhibited *T. gondii* growth with an IC_{50} of 500 nM (Fig. 21.3). Once again, the main caveat of this work was the lack of

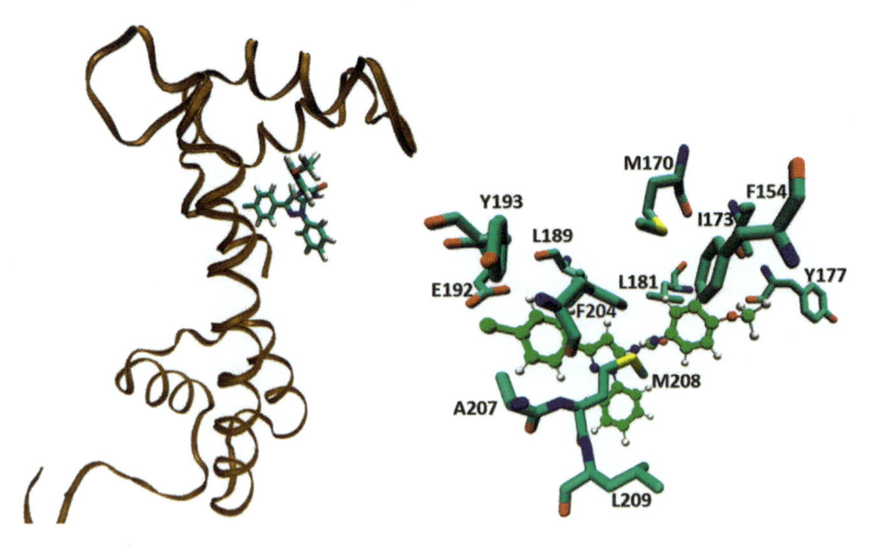

FIGURE 21.3 **Structural model of TgMLC1 (amino acids 84–213) in complex with inhibitor C3-21.** Left: The predicted binding site of C3-21 lies within a deep groove in TgMLC1 (brown ribbon); binding of compound is predicted to directly interfere with TgMyoA binding within this same groove. Right: Detailed image showing the binding pocket of TgMLC1 and key amino acids with which C3-21 interacts. *Image generously provided by Dr. Sandhya Kortagere.*

data confirming that the compound exerts its effects on the parasite as a direct result of blocking the TgMyoA/TgMLC1 interaction.

Powerful *in silico* tools have been developed to identify candidate targets within the genomes of *T. gondii* and other parasites, based on such criteria as essentiality, whether assays for activity have been established, orthology, druggability and the availability of crystal structures or models (Aguero *et al.*, 2008; Crowther *et al.*, 2010). Target selection is rarely the limiting factor; as illustrated by the examples above, the main obstacle to more widespread use of the reverse chemical genetic approach is that once inhibitors of the target of interest are applied to live parasites, proving that the phenotype(s) observed are in fact due to inhibition of that target is challenging. It therefore becomes necessary to use the target validation approaches described in Section 21.5 to definitively connect the observed effect of the compound to its predicted target.

21.4 FORWARD (CELL-BASED) CHEMICAL GENETICS

The application of forward chemical genetics to the study of parasites has been steadily increasing in recent years. Unlike reverse chemical genetic screening, in which a defined target is selected, the forward approach uses unbiased screening to identify compounds that induce a specific phenotype. For a review of cell-based assays available to study *T. gondii* and other protozoan parasites, see Muskavitch *et al.* (2008). The benefit of the cell-based approach is that compounds are selected based on the specific phenotypes they induce or inhibit, with no preconceived ideas about mechanism. The resulting hits can therefore lead to the discovery of unexpected or previously unknown targets and pathways. The major drawback of the approach is that determining the target(s) of a particular hit compound and establishing its mechanism of action can be difficult.

21.4.1 Choice of Compounds to be Screened

Like reverse chemical genetic approaches, which require careful selection of the target to be effective, forward chemical genetic methods require the selection of an optimal library of compounds for screening. Libraries can be large and unbiased, sampling as much 'chemical space' (Eberhardt *et al.*, 2011) as possible, or they can be smaller and more focused, having been pre-selected for their ability to bind to a specific class of target proteins such as protein kinases (e.g. Kamau *et al.*, 2012). Focused libraries that are designed to covalently modify their target proteins can also be useful, facilitating downstream target identification. For example, a library of compounds designed to irreversibly target serine and cysteine proteases was screened in *P. falciparum* cultures to identify compounds that block the process of host cell rupture (Arastu-Kapur *et al.*, 2008). This led to the identification of two compound classes that were used to implicate the serine protease PfSUB1 and the cysteine protease PfDPAP3 as key regulators of parasite egress from the host cell. The same library was later screened in an assay of host cell invasion by *T. gondii* (Hall *et al.*, 2011), resulting in the identification of compounds that inhibited as well as compounds that enhanced host cell invasion. Covalent binding of the identified inhibitory hits allowed direct isolation of the relevant target of one of the leads as the parasite homologue of the human stress response protein DJ-1 (see Section 21.6.4 for details).

21.4.2 Toxic and Promiscuous Inhibitors

If the goal of a forward chemical genetic screening project is to identify probes that can be used to study mechanism, compounds that inhibit *T. gondii* by simply killing either the parasite or the host cell (e.g. through membrane permeabilization) are not of interest. Some compound collections are prefiltered for toxicity, usually to mammalian cells. Alternatively, data may be available from other screens that provide insight into which of the compounds in the collection are toxic. While there are many ways to experimentally measure cell viability in the presence of compounds, the most sensitive and quantitative method in our hands has been luciferase-based measurement of intracellular ATP (Petty *et al.*, 1995; Carey *et al.*, 2004; Hall *et al.*, 2011).

'Promiscuous' inhibitors that self-associate in aqueous solution are a major source of false positive results in high-throughput screens (Feng *et al.*, 2007). These compounds form colloidal aggregates that non-specifically sequester proteins, inhibiting their activity. Fortunately, promiscuous inhibitors can be readily identified in a high-throughput format by measuring their effects on a reference enzyme such as β-lactamase in the presence and absence of non-ionic detergent; the detergent disrupts aggregate formation and reverses non-specific, aggregate-based inhibition (Feng *et al.*, 2005; Feng and Shoichet, 2006).

21.4.3 Host versus Parasite Targets

When a high-throughput screen that involves both parasites and host cells yields a compound of interest, before trying to establish the target(s) of the compound it is important to know which cell the compound is acting upon. In some cases, secondary assays can reveal an effect on either the parasite or the host cell, even if the original screen included both. For example, some compounds first identified as inhibitors of host cell invasion by *T. gondii* were subsequently shown to inhibit motility and/or microneme secretion in isolated parasites (Carey *et al.*, 2004). A more general approach is to preincubate parasites and host cells separately with and without compound, then wash and mix the cells in the four possible combinations

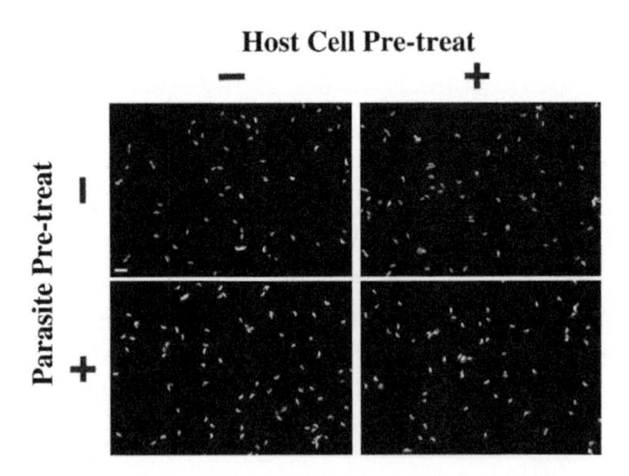

FIGURE 21.4 **Target cell identification.** Parasites and host cells were pre-treated independently with buffer containing DMSO (−) or invasion Inhibitor 10b (+), washed and combined in the four possible combinations for invasion assays. Intracellular parasites are green; extracellular parasites are yellow. With respect to invasion, this compound clearly has a stronger effect on the parasite than the host cell (fewer green parasites in samples where the parasites were pre-treated with compound). Bar is 20 μm. *Reprinted with permission from Carey* et al., *Proc. Natl. Acad. Sci. USA 101: 7433–7438; Copyright 2004 National Academy of Sciences, USA.*

(drug-treated parasites + untreated host cells, untreated parasites + drug-treated host cells, etc.; see Fig. 21.4). Such a 'checkerboard' experiment can reveal whether the compound affects the parasite, the host cell or both (e.g. Carey *et al.*, 2004). It works particularly well with irreversible compounds, but differential sensitivity may be observed even for reversible compounds if the assay time is sufficiently short. Ultimately, the biologically relevant target needs to be identified and validated (Sections 21.4.4 and 21.5) before one can say with certainty which cell the compound is targeting. In some cases, there may be relevant targets in both cells. For example, the trisubstituted pyrrole, Compound 1, inhibits *T. gondii* tachyzoite motility, microneme secretion and invasion through an effect on parasite cGMP-dependent protein kinase (TgPKG; see Section 21.6.6 for further details), but the compound also enhances tachyzoite-to-bradyzoite differentiation, in this case through a host cell pathway that involves cell division autoantigen-1 (CDA-1; Radke *et al.*, 2006).

21.4.4 Target Identification

The central and most challenging aspect of the forward chemical genetic approach is identifying the target(s) of the 'hits' that emerge from screening. Once potential targets have been identified, i.e. proteins to which the compound binds, the specificity and physiological relevance of these targets must be demonstrated (see Section 21.5).

The most direct approach to identifying potential targets is a hypothesis-driven, 'guess-and-test' strategy, i.e. guess what the target might be and test that hypothesis (Ward *et al.*, 2002). Educated guesses about the target can be based on: (i) known genetic mutants that phenocopy the effect of the compound (e.g. Mayer *et al.*, 1999); (ii) the observed biological effect of the compound (e.g. hypothesizing that tachyplegin A, an inhibitor of parasite motility, acts on the myosin motor complex – see Section 21.6.3); (iii) known effects of the compound in other screens or other systems (e.g. identification of *T. gondii* actin as the target of cytochalasin D – see Section 21.6.1); or (iv) comparing chemical scaffold structures to databases of compounds with known mechanisms of action, providing possible targets based on the 'guilt-by-association' method (e.g. Plouffe *et al.*, 2008). Based on the successful track record of hypothesis-driven approaches to target identification, the value of this strategy should not be underestimated.

When no hypothesis is available, more general proteome- or genome-wide approaches are required. Many such approaches have been developed, as summarized in several recent reviews (Lomenick *et al.*, 2011; Cong *et al.*, 2012; Tashiro and Imoto, 2012; Lee and Bogyo, 2013). Historically, the most widely used approaches have been biochemical, affinity-based approaches. In one common permutation, the small molecule is labelled with a non-radioactive (e.g. biotin) or radioactive tag and the tag is used to follow the target during standard biochemical purification (e.g. Gurnett *et al.*, 2002). If necessary, a photoactivatable cross-linking group is added to irreversibly bind the compound to its target(s) during purification. In another common biochemical permutation, the small molecule is linked to a solid phase matrix and compound-binding proteins are isolated by affinity chromatography (e.g. Knockaert *et al.*, 2000, 2002; Deu *et al.*, 2010; Harbut *et al.*, 2011). In any affinity-based approach, specifically and non-specifically bound proteins can be distinguished by conducting parallel experiments with inactive analogues, or by determining whether the interaction can be competed with an excess of active compound. A number of other biochemical/affinity-based methods have been developed, including phage-display biopanning (Jung *et al.*, 2010), drug affinity responsive target stability (DARTS; Lomenick *et al.*, 2009), drug westerns (Tanaka *et al.*, 1999) and protein microarray screening (Huang *et al.*, 2004) but these approaches are not yet as widely used as biochemical purification or affinity chromatography.

As a general rule, the more abundant the binding proteins are and the higher their affinity for compound, the more readily they will be identified by biochemical methods (Cong *et al.*, 2012). A downside to most biochemical approaches is the need to chemically modify the compound of interest, in order to add a tag or to covalently link it to an affinity matrix. This requires not only additional rounds of synthesis, but knowledge of which substituents of the compound can tolerate modification without the compound losing biological activity, i.e. structure–activity relationship (SAR) analysis. For these reasons, screening compounds that already have both a tag for detection/purification and a reactive group to bind covalently to their targets (e.g. activity-based probes; see Section 21.2.1) can greatly facilitate downstream biochemical target identification (Sadaghiani *et al.*, 2007; Fonovic and Bogyo, 2008).

Genetics-based approaches to target identification have been particularly well developed in yeast. These include drug-induced haploinsufficiency profiling, in which a collection of individual yeast strains engineered to lack a single copy of every essential gene in the genome is tested for increased sensitivity to the compound of interest (Giaever *et al.*, 1999, 2004; Parsons *et al.*, 2004). Conversely, libraries of yeast overexpressing individual open reading frames can be tested for decreased sensitivity to the compound (e.g. Hoon *et al.*, 2008). In each of these scenarios, bar-coding of individual strains within the collection (Shoemaker *et al.*, 1996; Winzeler *et al.*, 1999) makes identification of strains with altered sensitivity to the compound straightforward. Yeast made resistant to the compound of interest can also be complemented with bar-coded genome-wide expression libraries to identify gene(s) that restore compound sensitivity (e.g. Ho *et al.*, 2009).

While genome-wide deletion or overexpression libraries are not yet available for *T. gondii*, other promising genetic approaches to target identification are beginning to emerge. Parasites resistant to a particular compound can frequently be recovered from pools of chemically mutagenized parasites (e.g. Pfefferkorn and Borotz, 1994; Dobrowolski and Sibley, 1996; Morrissette *et al.*, 2004). Genomic cosmid-based complementation has been successfully used to identify sequences that rescue temperature-sensitive cell cycle defects (Gubbels *et al.*, 2008), and there is no reason why the same approach could not be used to identify mutations conferring drug resistance. Alternatively, recent developments in high-throughput ('next-generation') sequencing technology have made whole genome sequencing of *T. gondii* a feasible way to identify chemically-induced mutations that generate particular phenotypes (Farrell *et al.*, 2012; Garrison *et al.*, 2012). The generation of drug resistant mutants followed by whole genome sequencing is likely to become a standard approach for small-molecule target identification in *T. gondii* as sequencing costs drop and improved tools for data analysis are developed. A mutagenesis-based approach that requires neither cosmid complementation nor whole genome sequencing is insertional mutagenesis; if resistant parasites can be isolated, identifying the disrupted locus by plasmid rescue and inverse PCR is straightforward (Lescault *et al.*, 2010; Fomovska *et al.*, 2012a). Two caveats to any mutagenesis-based approach are that (i) one must first generate parasites resistant to the small molecule of interest, something that is not always possible (Fomovska *et al.*, 2012b; Q. Tang and GEW, unpublished data) and (ii) the approach will not always identify the direct target of the compound; gene products in pathways upstream or downstream of the actual target of the compound may confer resistance, as can relatively unrelated gene products involved in compound efflux, compound metabolism, etc.

An approach to target identification that circumvents some of these problems is yeast three-hybrid (Y3H) analysis (Licitra and Liu, 1996; Baker *et al.*, 2003; Becker *et al.*, 2004; Chidley *et al.*, 2011). Y3H is based on the more familiar yeast two-hybrid (Y2H) assay to detect protein–protein interactions, but makes transcriptional read-out dependent on the indirect interaction of the bait and prey proteins through a third component, a hybrid small molecule known as a chemical inducer of dimerization (CID; Fig. 21.5). Variations on the method are available for detecting CID-binding membrane proteins and other proteins that cannot translocate into the yeast nucleus (Johnsson and Varshavsky, 1994; Caligiuri *et al.*, 2006; Dirnberger *et al.*, 2006; Chen *et al.*, 2009), and modular methods that streamline CID synthesis have recently been developed (Tran *et al.*, manuscript in preparation). The major downsides to the approach are that gene products that are not represented in the cDNA library or do not fold properly as AD fusions (see Fig. 21.5) will not be detected, and the interaction between

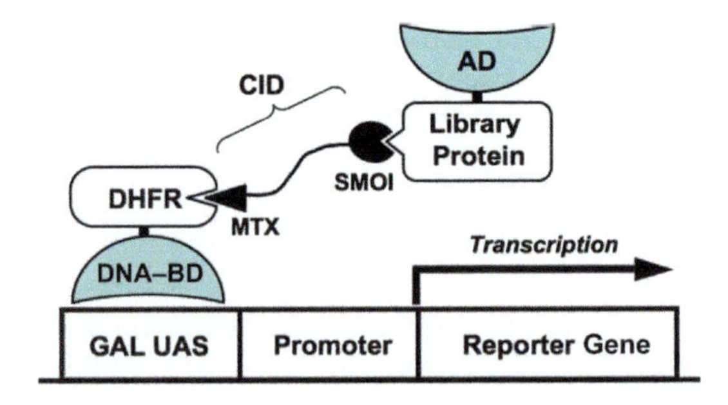

FIGURE 21.5 **Schematic of the Y3H system.** The GAL4 transcription factor is divided into two components (shaded), the DNA-binding domain (DNA-BD), which binds to a GAL4 upstream activator sequence (GAL UAS), and the activation domain (AD). The DNA-BD and AD are expressed as fusions to dihydrofolate reductase (DHFR) and a cDNA library, respectively. Upon addition of a CID, composed of methotrexate (MTX) linked to a small molecule of interest (SMOI), MTX will bind to DHFR and the SMOI binds to its target. This brings the DNA-BD and AD in sufficient proximity to reconstitute the transcription factor, activating the reporter gene in yeast.

the compound and its target(s) in parasites may require more than one parasite protein (e.g. the target can only bind to the compound as part of a protein complex) or a post-translational modification that does not occur in yeast. Despite these limitations, Y3H has recently been used to identify biologically relevant targets of a number of clinically approved drugs (Chidley *et al.*, 2011), and we have used the approach to identify TgBRADIN as a novel target of a small molecule enhancer of *T. gondii* differentiation (Odell *et al.*, submitted). Importantly, Y3H is capable of detecting small-molecule/target interactions with affinities ranging from nanomolar to micromolar (Becker *et al.*, 2004; de Felipe *et al.*, 2004; Chidley *et al.*, 2011).

Transcriptional array profiling represents another approach to target identification with significant potential. As transcriptional profiling becomes more rapid and cost effective, it is becoming increasingly possible to map the effects of many related compounds on parasite gene expression. This allows profiling of compounds with similar or overlapping targets to identify a signature effect on global transcription patterns. Similar patterns can be used to infer that the small molecules function through similar modes of action and may be dependent on the same target. This method has been recently used to map the effects of a range of small molecules on *P. falciparum* growth (Hu *et al.*, 2010; Andrews *et al.*, 2012).

21.5 DEMONSTRATING COMPOUND SPECIFICITY/ SELECTIVITY; TARGET VALIDATION

Once a compound/target pair has been identified by either forward or reverse chemical genetics, it is critical to prove that the phenotype caused by compound treatment is in fact due to that target. In many cases, it is possible to show that a small molecule can physically bind to a particular protein. However, binding of the compound to the protein or even inhibition of that protein's activity may not be the cause of the observed phenotype; more extensive studies are required to validate a given target's role in a particular biological process. The powerful set of molecular genetic tools available to manipulate *T. gondii* can greatly assist in target validation, as described below. Compounds may interact with more than one target in the parasite, but carefully designed validation strategies can determine which of these is of biological relevance. In fact, different targets can be responsible for different biological effects of the same compound (e.g. Donald *et al.*, 2002; Radke *et al.*, 2006). It is therefore important to develop as complete a target profile as possible for any given compound, so that once validation efforts are undertaken, the relevant target is among those being tested.

21.5.1 Structure–Activity Relationships (SAR)

SAR analysis can be useful for linking the binding of a compound to a particular target with the effect(s) of that compound on the parasite. To perform SAR analysis, a set of compounds is generated with structures that are highly related to the original hit, and those with different potencies for the predicted target are identified. In theory, if a compound's phenotypic effects are due to its binding to the predicted target, then changes in potency of the different compounds for that target should result in corresponding changes in potency in the phenotypic assay. For example, a series of analogues of a chloroisocoumarin inhibitor of *P. falciparum* SUB1 serine protease was used to link PfSUB1 to the effects of the compounds on parasite egress from the red blood cell (Arastu-Kapur *et al.*, 2008). The same approach was used to help validate the *T. gondii* protein TgDJ-1 as the target of a small-molecule inhibitor of host cell invasion (Hall *et al.*, 2011; see Section 21.6.4 for details). The approach does not provide direct proof that effects are mediated by a single target, since changes in the compound SAR profiles may result in similar changes in binding to multiple downstream targets. Different compounds in a SAR series may also be taken up by or transported out of the cell differentially. However, when a correlation is observed between SAR profiles and phenotypic potency, it provides evidence of a role for the target in the given biological process.

21.5.2 Competition Profiling with ABPs

In addition to their use for profiling changes in enzyme activity (see Section 21.2.1), ABPs designed to bind to multiple related targets in a given enzyme class can be useful for linking a particular target protein to a given biological

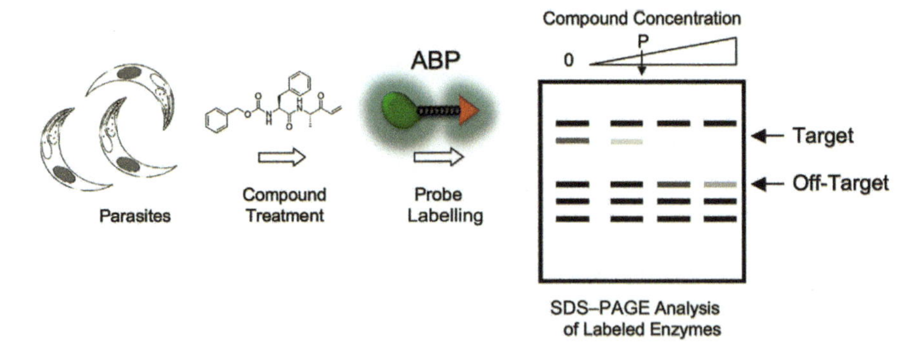

FIGURE 21.6 **Using an ABP to assess selectivity of an inhibitor.** Parasites are treated with a small molecule modulator of interest, which binds to its target protein(s). An ABP with broad specificity is then added to the intact parasite to label residual activity of target enzymes. The concentration of small molecule that saturates target binding will cause complete loss of labelling by the ABP. This is the concentration where the phenotypic effect (P) should be observed. An off target can be identified as a protein whose modification does not correlate with the observed phenotype.

outcome. This is accomplished using a competitive labelling assay in which the small molecule is added to an intact cell or parasite extract and then residual enzymatic activity is measured with a broad spectrum ABP (Fig. 21.6). If the small molecule binds to one of the enzymes targeted by the ABP, it will block the ability of the probe to bind, resulting in a loss of labelling by the ABP. This is a particularly valuable assay since it allows specificity to be assessed without the need to express and purify all of the possible related targets. In addition, competition assays can be performed in intact cells, allowing assessment of overall target selectivity of a small molecule while taking into account parameters such as cell uptake and accumulation. Using this kind of competition profiling, one can more confidently assign a function to a target protein using a small molecule that modulates its function. Competition profiling has been used to study the biological function of *T. gondii* cathepsin L (Larson *et al.*, 2009) and a variety of *P. falciparum* proteases (Greenbaum *et al.*, 2002; Arastu-Kapur *et al.*, 2008; Deu *et al.*, 2010; Harbut *et al.*, 2011). The approach was also used to select the most likely targets of WRR-086, a recently described small-molecule inhibitor of host cell invasion by *T. gondii* (Hall *et al.*, 2011; see Section 21.6.4 for further details).

21.5.3 Molecular Genetic Approaches

21.5.3.1 Gene Deletion

Although RNA interference (RNAi) has proven to be a useful technology for target identification and validation in other parasitic protozoa (e.g. Alsford *et al.*, 2012), robust and reproducible methods for RNA_i in *T. gondii* have been elusive (Kolev *et al.*, 2011). Consequently, the most straightforward approach to target validation in the haploid tachyzoite is to delete the gene encoding the suspected target and determine if the parasite loses some or all of its sensitivity to the compound. As described elsewhere in this volume (see Chapter 17), the generation of gene knock-outs has become routine in *T. gondii*, using parasites lacking the *KU80* gene (Fox *et al.*, 2009, 2011; Huynh and Carruthers, 2009). KU80 is a component of the parasite's non-homologous end joining machinery; in its absence, transfected DNA can be targeted with high efficiency to a specific genetic locus, where it integrates by double crossover homologous recombination.

There are three possible outcomes to a gene knock-out experiment:

(i) *The knock-out shows no change in sensitivity to the compound.* This provides evidence that the target identified is not the biologically relevant target. For example, the small-molecule invasion inhibitor, conoidinA, which inhibits extension of the parasite's conoid (Carey *et al.*, 2004), binds covalently to and inhibits the enzymatic activity of parasite peroxiredoxinII (TgPrxII; Haraldsen *et al.*, 2009). However, wild-type and TgPrxII knock-out parasites proved to be equally sensitive to conoidinA, arguing that TgPrxII is not the biologically relevant target of conoidinA for invasion and conoid extension (Haraldsen *et al.*, 2009). One caveat to this conclusion is that the *T. gondii* genome encodes four putative TgPrx proteins; if the different TgPrx proteins serve redundant functions, disruption of any one TgPrx protein might not be sufficient to confer resistance to the compound. Similarly, if deletion of a potential target gene decreases parasite fitness, compensatory transcriptional or posttranslational changes could develop that confound interpretation of such an experiment.

(ii) *The gene cannot be knocked out.* This is presumably because the gene product is essential for the lytic cycle. In this case one can generate parasites in which target protein levels can be controlled experimentally, either using a tetracycline-regulated promoter or degradation domain-mediated proteolysis (see Chapter 17) and determine if down-regulation of the target protein changes sensitivity of the parasite to the compound.

(iii) *The knock-out is viable, and shows reduced sensitivity to the compound or phenocopies the compound's effects.* This provides strong experimental evidence that the gene product is indeed a biologically relevant target of the compound. Such an approach was used to demonstrate that TgBRADIN is a biologically relevant target of Compound 2 (Odell *et al.*, submitted). Disruption of TgBRADIN affected tachyzoite-to-bradyzoite differentiation but had no apparent effect on the parasite's lytic cycle, facilitating generation of the knock-out in the tachyzoite.

21.5.3.2 Mutations that Block Compound Binding

If point mutations can be identified that block binding of the compound to the suspected target without disrupting the normal function of the protein, then the endogenous copy of the gene can be replaced with a copy containing these mutations and the sensitivity of the resultant parasites to the compound determined. If parasites expressing the mutant protein are less sensitive to the effects of the compound, this provides strong evidence that the target is biologically relevant. Such an approach was used to validate TgDJ-1 as the target of WRR-086 and TgPKG as the target of Compound 1 (see Sections 21.6.4 and 21.6.6). Mutations that block compound binding can be identified by homology modelling and structure-based predictions (e.g. Donald *et al.*, 2002; Hall *et al.*, 2011). Alternatively, if a suspected target interacts with the compound of interest by Y3H (see Section 21.4.4), systematic mutagenesis of the target and analysis of compound binding by Y3H could be used to identify the specific residues involved. Regardless of how the mutation(s) are identified, the functionality of the mutant gene can be readily assessed after allelic replacement of the wild-type gene with the mutant in Δ*ku80* parasites: if the parasites are viable, their sensitivity to the compound can be assessed.

21.5.3.3 Mutations that Confer Resistance to the Compound

As described above (Section 21.4.4), chemical mutagenesis can be used to generate parasites resistant to a compound of interest. The suspected target can then be sequenced in the resistant parasites to determine whether mutations are present. If so, then the mutant alleles can be reintroduced into wild-type parasites, and the sensitivity of the resultant transgenic parasites determined. This approach was used to validate *T. gondii* actin as the target of cytochalasin D and *T. gondii* α-tubulin as the target of oryzalin (see Sections 21.6.1 and 21.6.2 for further details).

21.6 *TOXOPLASMA GONDII* CHEMICAL BIOLOGY: CASE STUDIES

We next describe a few select examples in which chemical biology has provided important insights into the biology of *T. gondii* (see Fig. 21.7 for structures of all compounds discussed).

21.6.1 Cytochalasin D/Actin

In 1976, J. B. Jensen and S. A. Edgar showed that cytochalasin B inhibits host cell invasion by the apicomplexan parasite *Eimeria magna* (Jensen and Edgar, 1976). They used checkerboard analysis (see Section 21.4.3) to show that the compound affected invasion through inhibition of parasite motility. Two years later, Ryning and Remington showed that the more specific actin inhibitor, cytochalasin D (Fig. 21.7), inhibits host cell invasion by *T. gondii,* although they concluded that the effect was likely on host cell actin (Ryning and Remington, 1978). The definitive chemical biology experiments were performed two decades later in an elegant set of experiments in David Sibley's lab (Dobrowolski and Sibley, 1996). First, the authors used cytochalasin D-resistant host cells (Cyt-1 cells; Ohmori and Toyama, 1992) to show that cytochalasin D inhibits invasion through an effect on the parasite, not the host cell. They then chemically mutagenized parasites, selected for cytochalasin D resistance, and isolated three clones that could invade host cells in the presence of cytochalasin D. Using the 'guess-and-test' strategy for target identification (Section 21.4.4), they hypothesized that the mutations in these cytochalasin D-resistant parasites would lie in the single copy actin gene (Tg*ACT1*). Sequencing showed that two of the three resistant clones did in fact have a point mutation that changed a single amino acid (Ala136Gly) in the predicted cytochalasin-binding region of ACT1 (Dobrowolski and Sibley, 1996). To confirm that the cytochalasin D-resistant phenotype was due to this mutation, they reintroduced the mutant allele into wild-type parasites and showed that the transgenic parasites lost their sensitivity to cytochalasin D in invasion and motility assays. Since this seminal paper, we have learned a great deal about the role of actin in the motility and invasion of apicomplexan parasites (see Chapter 13); note that a role for host cell actin during invasion has also recently been proposed (Gonzalez *et al.*, 2009). It should also be noted that the third cytochalasin D-resistant mutant in the Dobrowolski and Sibley study did not result from a mutation in the coding region of Tg*ACT1*, suggesting either an additional target and/or an indirect way to generate cytochalasin D resistance.

21.6.2 Oryzalin/Tubulin

Apicomplexan tubulins are more similar to plant tubulins than animal tubulins (Stokkermans *et al.*, 1996). Consequently, dinitroaniline inhibitors of plant microtubule polymerization (ethalfluralin, oryzalin (Fig. 21.7) and trifluralin) are potent growth inhibitors of *T. gondii* (IC_{50} 100–300 nM) and other apicomplexan parasites

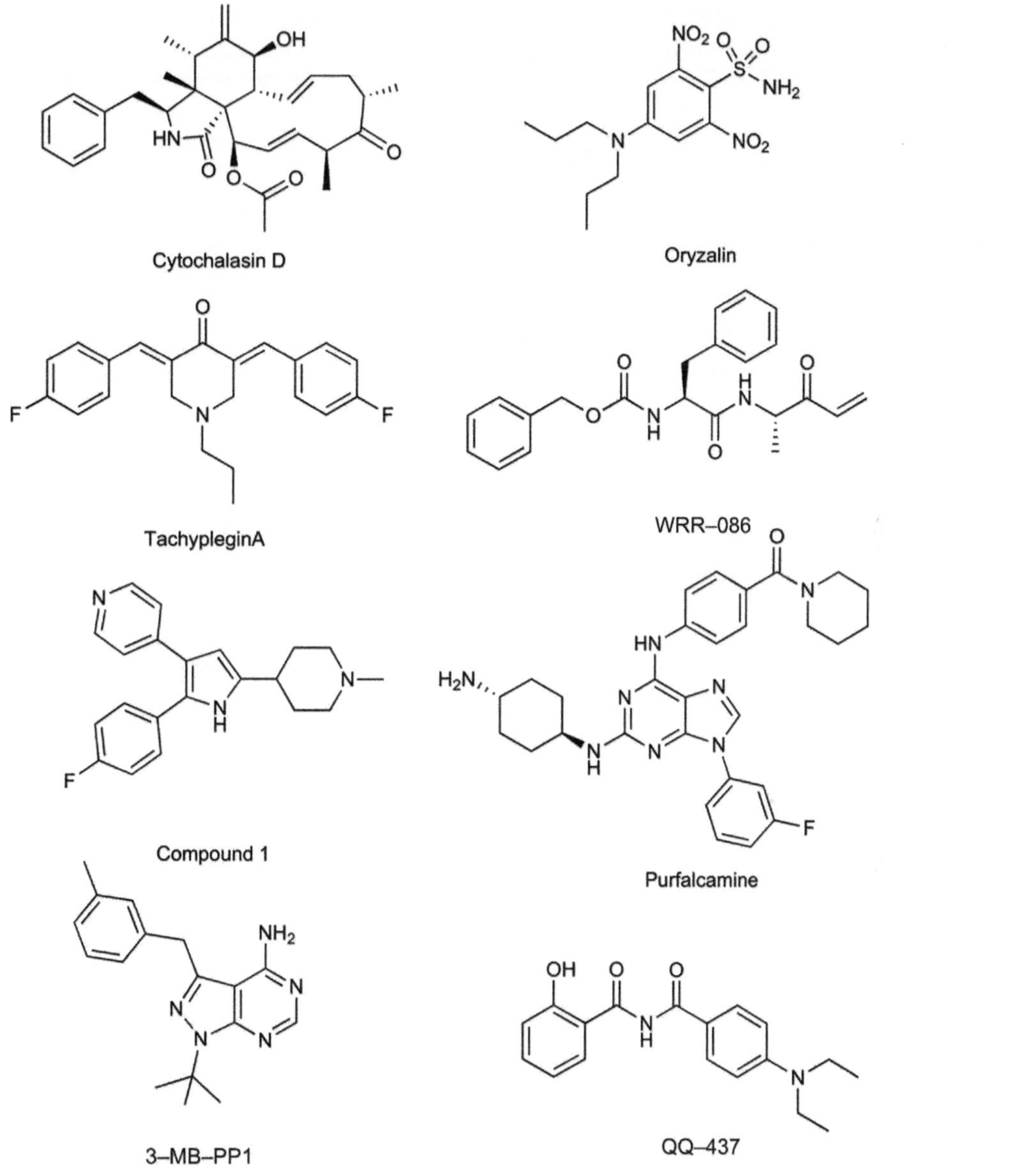

Cytochalasin D

Oryzalin

TachyplegInA

WRR–086

Compound 1

Purfalcamine

3–MB–PP1

QQ–437

FIGURE 21.7 **Structures of compounds used for chemical biological studies in *T. gondii*, as highlighted in Section 21.6, Case Studies.**

(Arrowood *et al.*, 1996; Stokkermans *et al.*, 1996; Dow *et al.*, 2002) but have little, if any, effect on mammalian cells. The dinitroanilines disrupt subpellicular microtubules at the *T. gondii* tachyzoite periphery, resulting in a massive disorganization of the inner membrane complex (Stokkermans *et al.*, 1996; Shaw *et al.*, 2000; Morrissette and Sibley, 2002). Oryzalin-resistant parasites were generated by chemical mutagenesis and found to have point mutations in α1-tubulin. Some of the mutations mapped to regions that coordinate lateral interactions between microtubule protofilaments, while others mapped to a computationally predicted oryzalin binding site, which is not present in mammalian α-tubulin (Morrissette *et al.*, 2004). Allelic replacement of the wild-type tubulin gene in *T. gondii* with the mutant alleles conferred oryzalin resistance (Morrissette *et al.*, 2004; Ma *et al.*, 2007), validating α1-tubulin as the biologically relevant target of oryzalin. Oryzalin shows a differential effect on the subpellicular microtubules and microtubules of the mitotic spindle, disrupting both at 2.5 μM, but only the subpellicular microtubules at 0.5 μM (Morrissette and Sibley, 2002). This differential sensitivity makes oryzalin a useful probe for uncoupling the functions of the different microtubule populations in *T. gondii*, and has revealed new insights into parasite mitotic regulation (Morrissette and Sibley, 2002). The resistance mutations can also be used to study tubulin structure–function relationships. For example, the His28Gln mutation occurs within a loop involved in lateral protofilament interactions and was predicted to stabilize the microtubule; parasites expressing this mutation have a longer, thinner shape and markedly longer subpellicular microtubules (Ma *et al.*, 2007; Fig. 21.8).

21.6.3 TachypleginA/TgMLC1

An automated fluorescence microscopy-based screen of 12,160 compounds identified 24 inhibitors and six enhancers of host cell invasion by *T. gondii* (Carey *et al.*, 2004). Secondary assays revealed different effects of the compounds on parasite motility, microneme secretion and conoid extension, and demonstrated that constitutive and calcium-induced microneme secretion could be pharmacologically uncoupled (Carey *et al.*, 2004). In another example of the guess-and-test approach to target identification, the compounds that inhibited parasite motility were hypothesized to act on the myosin motor complex that powers motility (Heaslip *et al.*, 2010). Immunoprecipitation of the motor complex from untreated and compound-treated parasites demonstrated that treatment with one of the 21 motility inhibitors (tachypleginA; Fig. 21.7) caused ~50% of the TgMLC1 associated with the motor complex to migrate faster on SDS polyacrylamide gels (Fig. 21.9). Isolated motor complexes containing the modified TgMLC1 showed significantly decreased motor activity compared to motor complexes from untreated parasites, likely explaining the effect of tachypleginA on parasite motility (Heaslip *et al.*, 2010). Whether tachypleginA binds directly to TgMLC1 or to another target that indirectly causes the posttranslational modification of TgMLC1 is currently under investigation.

21.6.4 WRR-086/TgDJ-1

A library of small molecules containing reactive electrophiles designed to covalently modify the active site nucleophile of serine and cysteine proteases was screened for effects on *T. gondii* invasion (Hall *et al.*, 2011). The idea was to use compounds likely to target a known enzyme class, providing leads on possible targets (see Section 21.4.4) as well as resulting in their covalent labelling. Thus, hits from the screen could be converted into probes that would allow direct isolation of the corresponding targets using affinity purification methods. The screen identified compound classes that inhibited host cell invasion as well as compounds that enhanced invasion. One of the inhibitors, WRR-086 (Fig. 21.7), was converted into a tagged analogue and used to isolate its bound targets. However, even with direct covalent modification of targets, the resulting list of isolated proteins was extensive. To focus the possible hit list to a smaller number of relevant targets, the authors used a competition approach where parasites were treated with active and inactive analogues of the lead hit (see Section 21.5.2),

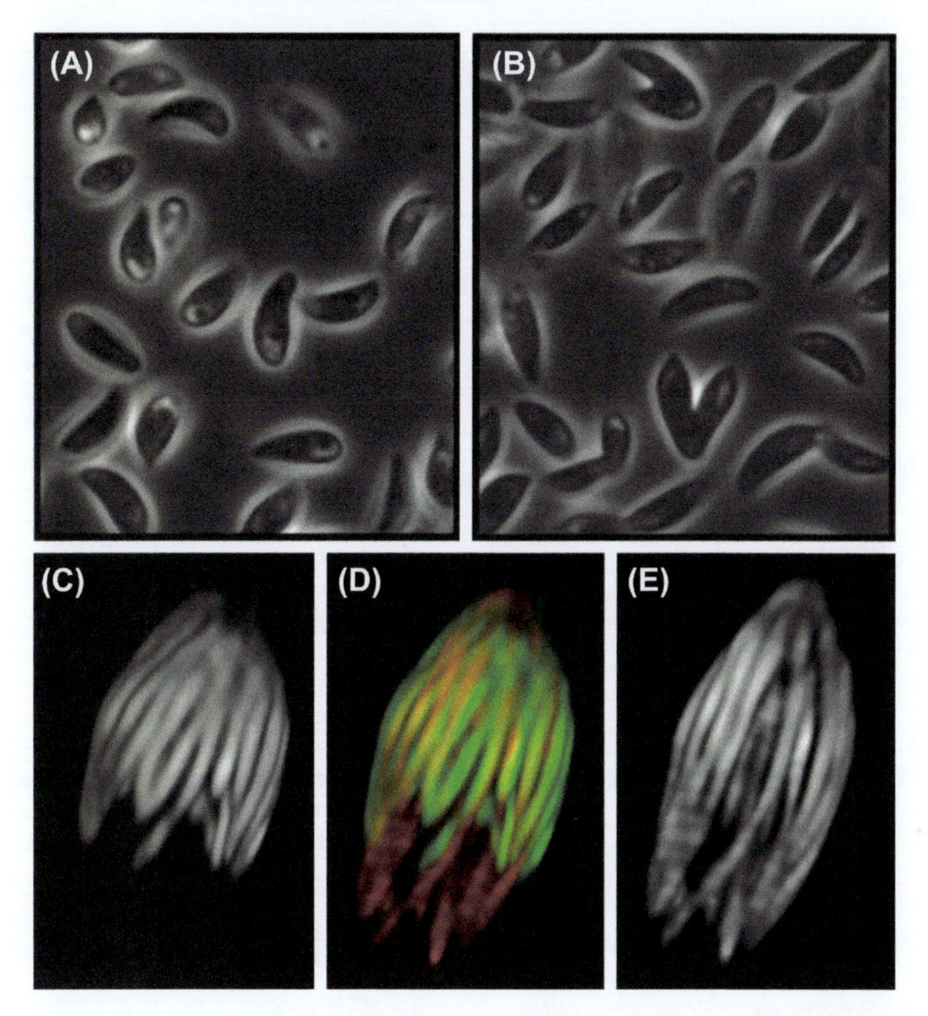

FIGURE 21.8 **Effects of the His28Gln α1-tubulin mutation on parasite morphology and microtubule length.** A, B)Phase-contrast microscopy of extracellular wild-type (A) and His28Gln mutant (B) parasites illustrates that *T. gondii* expressing the His28Gln mutation are thinner and longer and they show an increased frequency of septation defects. C–E) Projections of wild-type (C) and His28Gln mutant (E) parasites indicate that His28Gln mutant parasites have longer microtubules. These three-dimensional projections were created by collecting and deconvolving Z-series images of newly invaded parasites stained with a *T. gondii*-specific tubulin antibody. An overlay of the wild-type parasite projection (green) on the His28Gln mutant parasite projection (red) illustrates the increased subpellicular microtubule length in the mutant parasites (D). *Reprinted from Ma* et al., *Mol. Biol. Cell 18, 4711–4720; Copyright 2007 American Society for Cell Biology.*

reasoning that labelling of the relevant target would be blocked by active, but not inactive analogues. In addition, a protease-cleavable linker was included in the affinity tag to enable specific elution of the chemically modified peptide(s). This served to further confirm the most relevant target protein and allowed direct identification of the exact site of compound labelling. Ultimately, this combination of methods identified a poorly annotated 'intracellular protease' (TGME49_014290; ToxoDB.org), the parasite homologue of the human protein DJ-1, as the

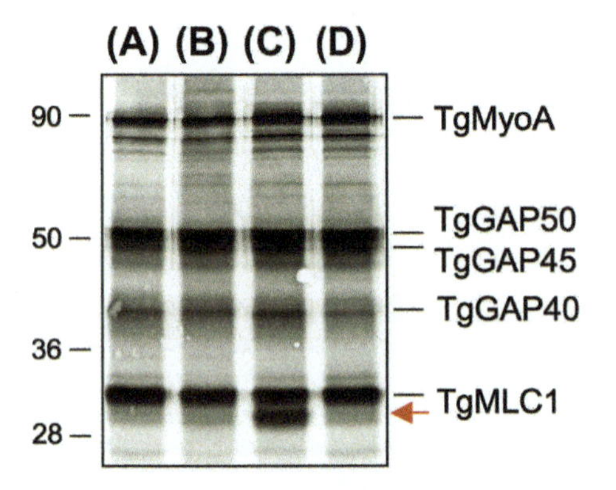

FIGURE 21.9 **Effect of tachypleginA on the electrophoretic mobility of myosin light chain-1 (TgMLC1).** ^{35}S-labelled parasites were treated for 15 minutes in medium containing: A) DMSO; B) 100 μM Inhibitor 22 (Carey *et al.*, 2004); C) 100 μM tachypleginA (see Fig. 21.7); or D) 100 μM Enhancer 5 (Carey *et al.*, 2004). Myosin motor complexes were isolated from the treated parasites by anti-TgGAP45 immunoprecipitation, and the immunoprecipitated proteins were resolved and visualized by SDS–PAGE/autoradiography. Treatment with tachypleginA results in an extra 30 kDa protein associating with the myosin motor complex (red arrow). This protein was subsequently identified as a modified form of TgMLC1. Numbers on the left indicate molecular mass in kDa. *Reprinted with modification from Heaslip* et al., *PLoS Pathog. (2010) 6, e1000720.*

most likely target of the compound (Hall *et al.*, 2011). While both the mammalian and parasite proteins contain a protease domain, one of the predicted key catalytic residues is lacking in *T. gondii*, suggesting that this protein likely lacks protease activity. Because the site of compound modification (Cys127; Fig. 21.10A) had been identified using the cleavable tag, the native Tg*DJ-1* gene was replaced with mutant versions in which Cys127 was mutated to either serine or alanine. The serine mutant parasites showed a partial reduction in sensitivity to WRR-086 in invasion assays (the serine could also be labelled by WRR-086 but to a lesser extent than cysteine) and the alanine mutant parasites showed a complete loss in sensitivity (Fig. 21.10B). This result provided strong genetic proof that TgDJ-1 is the biologically relevant target of WRR-086 and that binding to this target is responsible for the observed inhibition of motility, microneme secretion and invasion. This example combines many of the techniques and methods described in this chapter in a single study that resulted in the identification of a previously unknown regulator of parasite invasion. Studies are currently ongoing to further define the mechanism by which TgDJ-1 regulates invasion.

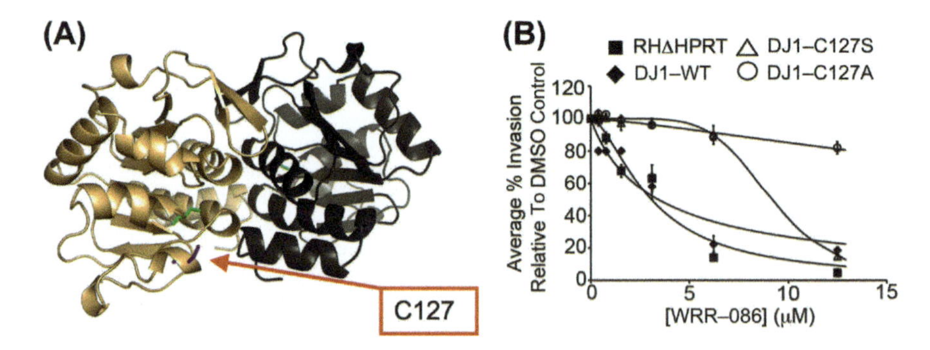

FIGURE 21.10 **TgDJ-1 Is the biologically relevant target of WRR-086.**
A) Structural model of TgDJ-1 based on the human DJ-1 structure. Shown is the cysteine at position 127 that is modified specifically by WRR-086.
B) Effects of replacing the native TgDJ-1 gene with a mutant containing serine (DJ1-C127S) or alanine (DJ1-C127A) at position 127. The loss of the nucleophile in this position prevents WRR-086 binding and eliminates the sensitivity of the parasite to a WRR-086-induced block of host cell invasion.
Part B reprinted with permission from Hall et al., *Proc. Natl. Acad. Sci. USA 108, 10568–10573; Copyright 2011 National Academy of Sciences, USA.*

21.6.5 3-MB-PP1/TgCDPK1, TgCDPK3

A screen of 20,000 compounds against *P. falciparum* calcium-dependent protein kinase-1, PfCDPK1, led to the identification of a class of 2,6,9-substituted purines that inhibited the kinase (Kato *et al.*, 2008). Further optimization yielded a lead compound named purfalcamine (Fig. 21.7). Purfalcamine showed dramatic effects on *P. falciparum* intracellular development, invasion and egress, and PfCDPK1 from parasite extracts bound to purfalcamine affinity resins. Purfalcamine was also shown to block *T. gondii* invasion (Kato *et al.*, 2008), suggesting possible overlapping functional roles for CDPKs in the two parasites, although the data did not formally prove that the effect of purfalcamine on either parasite was due to inhibition of CDPK1 or other CDPK family members.

In 2010, two groups independently used a combination of molecular genetics, chemical biology and structural biology to more precisely define TgCDPK1 function (Lourido *et al.*, 2010; Ojo *et al.*, 2010). Using a molecular genetic approach that enabled experimental manipulation of Tg*CDPK1* expression, the Sibley lab demonstrated a role for the kinase in motility, secretion and invasion (Lourido *et al.*, 2010). Both groups also used chemical genetics to study the function of the kinase by taking advantage of a unique property of TgCDPK1. Protein kinases have a residue in their ATP binding pocket referred to as the 'gatekeeper' residue because this amino acid controls access of small molecule inhibitors to a secondary hydrophobic binding pocket adjacent to the main ATP binding pocket. If a kinase has a large gatekeeper residue then the secondary pocket is not accessible for binding by an inhibitor with a bulky chemical 'bump' on its scaffold (Bishop *et al.*, 2000). Most kinases have either large or moderately sized gatekeeper residues; however, TgCDPK1 is unique in having a glycine in this position. TgCDPK1 can therefore bind and be inhibited by bumped kinase inhibitors with large chemical groups facing this secondary binding pocket (Lourido *et al.*, 2010; Ojo *et al.*, 2010). These compounds are highly selective for TgCDPK1 over other kinases in *T. gondii* and the host. The genetic results were therefore confirmed by showing that specific inhibition of TgCDPK1 using bumped kinase inhibitors such as 3-MB-PP1 (Fig. 21.7) resulted in similar inhibition of parasite motility, secretion and invasion as observed in parasites with down-regulated Tg*CDPK1* expression (Lourido *et al.*, 2010; Ojo *et al.*, 2010). The specificity of the compounds for TgCDPK1 was confirmed by showing that parasites expressing a mutant version of TgCDPK1 containing a large gatekeeper residue (methionine) lost sensitivity to the bumped inhibitors (Lourido *et al.*, 2010; Ojo *et al.*, 2010).

A similar strategy was recently used to study the function of the related protein kinase TgCDPK3 (Lourido *et al.*, 2012). The gatekeeper residue of TgCDPK1 was first mutated to methionine, as described above, rendering the enzyme and the parasites expressing it insensitive to the bumped kinase inhibitors. The gatekeeper residue of TgCDPK3 was then mutated to glycine, rendering it sensitive to bumped kinase inhibitors. Treatment of parasites expressing these two mutations with 3-MB-PP1 demonstrated a role for TgCDPK3 in parasite egress and in microneme secretion (Lourido *et al.*, 2012).

21.6.6 Compound 1/TgPKG

A cell-based screening program at Merck a decade ago identified a 2,3-diarylpyrrole, Compound 1 (Fig. 21.7), which inhibited the *in vitro* growth of *T. gondii* and other coccidian parasites (Gurnett *et al.*, 2002; Wiersma *et al.*, 2004). Compound 1 showed little toxicity to mammalian cells in culture, and protected against a lethal challenge in animal models of *T. gondii* and *Eimeria tenella* infection (Nare *et al.*, 2002; Biftu *et al.*, 2005). Tritiated Compound 1 was used to track compound binding activity through multiple biochemical purification steps, ultimately resulting in the identification of PKG as the Compound 1-binding protein (Gurnett *et al.*, 2002). The compound is a potent inhibitor of recombinant TgPKG ($IC_{50} \sim 1$ nM; Donald and Liberator, 2002). Modelling of the *E. tenella* PKG and TgPKG structures suggested that a threonine residue at the gatekeeper position (T761) plays an important role in the binding of Compound 1 to the kinase. When this residue was mutated to glutamine (T761Q), the kinetic properties of the recombinant enzyme did not change significantly but the IC_{50} of the kinase for Compound 1 increased by more than three orders of magnitude (Donald *et al.*, 2002). To validate TgPKG as a biologically relevant target of Compound 1, the wild-type TgPKG allele was disrupted in parasites stably expressing the T761Q mutant form of the enzyme. Compound 1 inhibits the invasion, microneme secretion and motility of wild-type parasites, and the T761Q mutant showed a significantly reduced sensitivity to the compound in all of these assays (Wiersma *et al.*, 2004). Remarkably, the T761Q mutant parasites were as virulent as wild-type parasites in mice but were completely refractory to treatment with Compound 1 (Donald *et al.*, 2002). These data demonstrated unequivocally that TgPKG is the biologically relevant target for Compound 1 with respect to invasion, microneme secretion, motility and virulence. Compound 1 and the T761Q parasites will be valuable tools for identifying TgPKG substrates involved in these processes.

21.6.7 QQ-437

A screen of 6811 drug-like compounds and subsequent lead optimization yielded two *N*-benzoyl-2-hydroxybenzamide inhibitors of *T. gondii* growth (MP-IV-1 and QQ-437; Fig. 21.7) with IC90 values in the low nanomolar range (Fomovska *et al.*, 2012a). Parasites resistant to MP-IV-1 and QQ-437 were generated by insertional mutagenesis and selection with MP-IV-1. All resistant clones analysed contained a disruption in the gene encoding adaptin AP-3β. In other systems, AP-3β functions in intracellular trafficking between the *trans*-Golgi and endosomal compartments and in biogenesis of the plant lytic vacuole (e.g. see Feraru *et al.*, 2010 and references therein). Consistent with these previous observations, parasites containing the AP-3β mutations and wild-type parasites treated with compound showed similarly disrupted secretory organelles and acidocalcisome/plant-like vacuoles (Fomovska *et al.*, 2012a). Although this work did not establish AP-3β as the direct target for these compounds (i.e. it did not rule out a more indirect role for AP-3β mutation in compound resistance), it did reveal a role for AP-3β in secretory organelle/lytic vacuole biogenesis in *T. gondii*.

21.6.8 Other Examples

The examples above were selected to illustrate different approaches to identifying compounds active against *T. gondii*, as well as different target identification and validation strategies. They are by no means exhaustive. Other examples of the use of chemical biology to study *T. gondii* include: a target-based screen for inhibitors of *T. gondii* nucleotide triphosphatase (Asai *et al.*, 2002); a combined structure- and cell-based screen for inhibitors of parasite secreted kinases, which led to identification of a potent inhibitor of cell cycle progression (Kamau *et al.*, 2011); a cell-based screen of 527 potential kinase inhibitors, which identified 14 compounds that either inhibit or enhance parasite growth (Kamau *et al.*, 2012); a cell-based screen of a series of salicylanides and related compounds that identified several growth inhibitors with low nanomolar potencies (Fomovska *et al.*, 2012b); and a forward chemical genetics project that yielded JCP174, a chloroisocoumarin enhancer of invasion, and identified its target as palmitoyl protein thioesterase-1 (TgPPT1) (Child *et al.*, submitted). For a more extensive list of small molecules that have been shown to affect

T. gondii, see the recent review by S. Kortagere (Kortagere, 2012).

21.7 *TOXOPLASMA GONDII* CHEMICAL BIOLOGY: SUMMARY AND FUTURE PROSPECTS

As the case studies above illustrate, chemical biology has already proven to be a useful approach to studying the biology of *T. gondii*, and is particularly powerful when combined with the sophisticated molecular genetic tools that can be brought to bear on this parasite. To date, the emphasis has been on using small molecules to modulate rather than monitor protein function, but as new protein profiling tools are developed we anticipate their increasing application to studying time-dependent processes in the parasite. Whole genome sequencing represents a major new enabling technology for small-molecule-based studies, and other proteome-, transcriptome- and genome-wide technologies will similarly complement the approach. For example, the availability of protein microarrays and/or genome-wide parasite knock-out collections will greatly facilitate target identification and validation.

In the same way that classical forward genetic screening is unbiased and can lead the investigator in new and unanticipated directions, forward chemical genetic screening and subsequent target identification can reveal completely unexpected insights into parasite biology. Small molecules exhibiting interesting activity in *T. gondii* may also be readily used in cross-species studies; this may prove to be a particularly useful way to study less experimentally tractable, related parasites such as *Cryptosporidium* or *Babesia*.

Although both target- and cell-based screening have been used extensively for drug development in the pharmaceutical industry, the drug development potential of these approaches in the academic setting lies not so much in the discovery of new lead compounds, but rather in the use of the bioactive small molecules that emerge from screening to explore biological mechanisms. This exploration of mechanism will ultimately provide new opportunities for drug development through the discovery of novel targets, new parasite pathways and previously unidentified components of known pathways. This is true for both inhibitors and enhancers of parasite growth and survival, as each can reveal important new insights into parasite mechanisms and potential points of vulnerability.

Based on the success stories described above and the many rapidly emerging technologies that can be brought to bear on compound synthesis, high-throughput screening and target identification and validation, it is our hope that readers of this chapter will begin to think about ways in which the tools of chemical biology might be applied to their own research.

Acknowledgements

The chemical genetics work in our laboratories is currently funded by the National Institutes of Health. We thank past and current lab members for their many contributions to this effort and Gary Ward thanks, in particular, his longstanding collaborator, Nick Westwood. We are grateful to Sandhya Kortagere for generating and giving us permission to publish Figure 21.3 and to Naomi Morrissette for permission to republish Figure 21.8.

References

Aguero, F., Al-Lazikani, B., Aslett, M., Berriman, M., Buckner, F.S., Campbell, R.K., Carmona, S., Carruthers, I.M., Chan, A.W., Chen, F., et al., 2008. Genomic-scale prioritization of drug targets: the TDR Targets database. Nat. Rev. Drug Discov. 7, 900–907.

Alsford, S., Eckert, S., Baker, N., Glover, L., Sanchez-Flores, A., Leung, K.F., Turner, D.J., Field, M.C., Berriman, M., Horn, D., 2012. High-throughput decoding of antitrypanosomal drug efficacy and resistance. Nature 482, 232–236.

Andrews, K.T., Gupta, A.P., Tran, T.N., Fairlie, D.P., Gobert, G.N., Bozdech, Z., 2012. Comparative gene expression profiling of *P. falciparum* malaria parasites exposed to three different histone deacetylase inhibitors. PloS One 7, e31847.

Arastu-Kapur, S., Ponder, E.L., Fonovic, U.P., Yeoh, S., Yuan, F., Fonovic, M., Grainger, M., Phillips, C.I., Powers, J.C., Bogyo, M., 2008. Identification of proteases that regulate erythrocyte rupture by the malaria parasite *Plasmodium falciparum*. Nat. Chem. Biol. 4, 203–213.

Arrowood, M.J., Mead, J.R., Xie, L., You, X., 1996. *In vitro* anti-cryptosporidial activity of dinitroaniline herbicides. FEMS Microbiol. Lett. 136, 245–249.

Asai, T., Takeuchi, T., Diffenderfer, J., Sibley, L.D., 2002. Identification of Small-Molecule Inhibitors of Nucleoside Triphosphate Hydrolase in *Toxoplasma gondii*. Antimicrob. Agents Chemother. 46, 2393–2399.

Baker, K., Sengupta, D., Salazar-Jimenez, G., Cornish, V.W., 2003. An optimized dexamethasone-methotrexate yeast 3-hybrid system for high-throughput screening of small molecule-protein interactions. Anal. Biochem. 315, 134–137.

Becker, F., Murthi, K., Smith, C., Come, J., Costa-Roldan, N., Kaufmann, C., Hanke, U., Degenhart, C., Baumann, S., Wallner, W., et al., 2004. A three-hybrid approach to scanning the proteome for targets of small molecule kinase inhibitors. Chem. Biol. 11, 211–223.

Biftu, T., Feng, D., Ponpipom, M., Girotra, N., Liang, G.B., Qian, X., Bugianesi, R., Simeone, J., Chang, L., Gurnett, A., et al., 2005. Synthesis and SAR of 2,3-diarylpyrrole inhibitors of parasite cGMP-dependent protein kinase as novel anticoccidial agents. Bioorg. Med. Chem. Lett. 15, 3296–3301.

Bishop, A.C., Ubersax, J.A., Petsch, D.T., Matheos, D.P., Gray, N.S., Blethrow, J., Shimizu, E., Tsien, J.Z., Schultz, P.G., Rose, M.D., et al., 2000. A chemical switch for inhibitor-sensitive alleles of any protein kinase. Nature 407, 395–401.

Caligiuri, M., Molz, L., Liu, Q., Kaplan, F., Xu, J.P., Majeti, J.Z., Ramos-Kelsey, R., Murthi, K., Lievens, S., Tavernier, J., et al., 2006. MASPIT: three-hybrid trap for quantitative proteome fingerprinting of small molecule-protein interactions in mammalian cells. Chem. Biol. 13, 711–722.

Carey, K.L., Westwood, N.J., Mitchison, T.J., Ward, G.E., 2004. A small-molecule approach to studying invasive mechanisms of *Toxoplasma gondii*. Proc. Natl. Acad. Sci. USA 101, 7433–7438.

Chandramohanadas, R., Davis, P.H., Beiting, D.P., Harbut, M.B., Darling, C., Velmourougane, G., Lee, M.Y., Greer, P.A., Roos, D.S., Greenbaum, D.C., 2009. Apicomplexan parasites co-opt host calpains to facilitate their escape from infected cells. Science 324, 794–797.

Chen, X., Xie, S., Bhat, S., Kumar, N., Shapiro, T.A., Liu, J.O., 2009. Fumagillin and fumarranol interact with *P. falciparum* methionine aminopeptidase 2 and inhibit malaria parasite growth *in vitro* and *in vivo*. Chem. Biol. 16, 193–202.

Chidley, C., Haruki, H., Pedersen, M.G., Muller, E., Johnsson, K., 2011. A yeast-based screen reveals that sulfasalazine inhibits tetrahydrobiopterin biosynthesis. Nat. Chem. Biol. 7, 375–383.

Cong, F., Cheung, A.K., Huang, S.M., 2012. Chemical genetics-based target identification in drug discovery. Annu. Rev. Pharmacol. Toxicol. 52, 57–78.

Cravatt, B.F., Wright, A.T., Kozarich, J.W., 2008. Activity-based protein profiling: from enzyme chemistry to proteomic chemistry. Annu. Rev. Biochem. 77, 383–414.

Crowther, G.J., Shanmugam, D., Carmona, S.J., Doyle, M.A., Hertz-Fowler, C., Berriman, M., Nwaka, S., Ralph, S.A., Roos, D.S., Van Voorhis, W.C., et al., 2010. Identification of attractive drug targets in neglected-disease pathogens using an *in silico* approach. PLoS Negl. Trop. Dis. 4, e804.

de Felipe, K.S., Carter, B.T., Althoff, E.A., Cornish, V.W., 2004. Correlation between ligand-receptor affinity and the transcription readout in a yeast three-hybrid system. Biochemistry 43, 10353–10363.

Deu, E., Leyva, M.J., Albrow, V.E., Rice, M.J., Ellman, J.A., Bogyo, M., 2010. Functional studies of *Plasmodium falciparum* dipeptidyl aminopeptidase I using small molecule inhibitors and active site probes. Chem. Biol. 17, 808–819.

Dirnberger, D., Unsin, G., Schlenker, S., Reichel, C., 2006. A small-molecule-protein interaction system with split-ubiquitin as sensor. ChemBioChem 7, 936–942.

Dobrowolski, J.M., Sibley, L.D., 1996. *Toxoplasma* invasion of mammalian cells is powered by the actin cytoskeleton of the parasite. Cell 84, 933–939.

Donald, R.G., Allocco, J., Singh, S.B., Nare, B., Salowe, S.P., Wiltsie, J., Liberator, P.A., 2002. *Toxoplasma gondii* cyclic GMP-dependent kinase: chemotherapeutic targeting of an essential parasite protein kinase. Eukaryot. Cell 1, 317–328.

Donald, R.G., Liberator, P.A., 2002. Molecular characterization of a coccidian parasite cGMP dependent protein kinase. Mol. Biochem. Parasitol. 120, 165–175.

Dow, G.S., Armson, A., Boddy, M.R., Itenge, T., McCarthy, D., Parkin, J.E., Thompson, R.C., Reynoldson, J.A., 2002. *Plasmodium*: assessment of the antimalarial potential of trifluralin and related compounds using a rat model of malaria, Rattus norvegicus. ExP. Parasitol. 100, 155–160.

Eberhardt, L., Kumar, K., Waldmann, H., 2011. Exploring and exploiting biologically relevant chemical space. Curr. Drug Targets 12, 1531–1546.

Evans, M.J., Cravatt, B.F., 2006. Mechanism-based profiling of enzyme families. Chem. Rev. 106, 3279–3301.

Farrell, A., Thirugnanam, S., Lorestani, A., Dvorin, J.D., Eidell, K.P., Ferguson, D.J., Anderson-White, B.R., Duraisingh, M.T., Marth, G.T., Gubbels, M.J., 2012. A DOC2 protein identified by mutational profiling is essential for apicomplexan parasite exocytosis. Science 335, 218–221.

Feng, B.Y., Shelat, A., Doman, T.N., Guy, R.K., Shoichet, B.K., 2005. High-throughput assays for promiscuous inhibitors. Nat. Chem. Biol. 1, 146–148.

Feng, B.Y., Shoichet, B.K., 2006. A detergent-based assay for the detection of promiscuous inhibitors. Nat. Protoc. 1, 550–553.

Feng, B.Y., Simeonov, A., Jadhav, A., Babaoglu, K., Inglese, J., Shoichet, B.K., Austin, C.P., 2007. A high-throughput screen for aggregation-based inhibition in a large compound library. J. Med. Chem. 50, 2385–2390.

Feraru, E., Paciorek, T., Feraru, M.I., Zwiewka, M., De Groodt, R., De Rycke, R., Kleine-Vehn, J., Friml, J., 2010. The AP-3 beta adaptin mediates the biogenesis and function of lytic vacuoles in *Arabidopsis*. Plant Cell 22, 2812–2824.

Fomovska, A., Huang, Q., El Bissati, K., Mui, E.J., Witola, W.H., Cheng, G., Zhou, Y., Sommerville, C., Roberts, C.W., Bettis, S., et al., 2012a. Novel N-benzoyl-2-hydroxybenzamide disrupts unique parasite secretory pathway. Antimicrob. Agents Chemother. 56, 2666–2682.

Fomovska, A., Wood, R.D., Mui, E., Dubey, J.P., Ferreira, L.R., Hickman, M.R., Lee, P.J., Leed, S.E., Auschwitz, J.M., Welsh, W.J., et al., 2012b. Salicylanilide Inhibitors of *Toxoplasma gondii*. J. Med. Chem. 55, 8375–8391.

Fonovic, M., Bogyo, M., 2008. Activity-based probes as a tool for functional proteomic analysis of proteases. Expert Rev. Proteomics 5, 721–730.

Fox, B.A., Falla, A., Rommereim, L.M., Tomita, T., Gigley, J.P., Mercier, C., Cesbron-Delauw, M.F., Weiss, L.M., Bzik, D.J., 2011. Type II *Toxoplasma gondii* KU80 knock-out strains enable functional analysis of genes required for cyst development and latent infection. Eukaryot. Cell 10, 1193–1206.

Fox, B.A., Ristuccia, J.G., Gigley, J.P., Bzik, D.J., 2009. Efficient gene replacements in *Toxoplasma gondii* strains deficient for nonhomologous end joining. Eukaryot. Cell 8, 520–529.

Garrison, E., Treeck, M., Ehret, E., Butz, H., Garbuz, T., Oswald, B.P., Settles, M., Boothroyd, J., Arrizabalaga, G., 2012. A Forward Genetic Screen Reveals that Calcium-dependent Protein Kinase 3 Regulates Egress in *Toxoplasma*. PLoS Pathog. 8, e1003049.

Giaever, G., Flaherty, P., Kumm, J., Proctor, M., Nislow, C., Jaramillo, D.F., Chu, A.M., Jordan, M.I., Arkin, A.P., Davis, R.W., 2004. Chemogenomic profiling: identifying the functional interactions of small molecules in yeast. Proc. Natl. Acad. Sci. USA 101, 793–798.

Giaever, G., Shoemaker, D.D., Jones, T.W., Liang, H., Winzeler, E.A., Astromoff, A., Davis, R.W., 1999. Genomic profiling of drug sensitivities via induced haploinsufficiency. Nat. Genet. 21, 278–283.

Gonzalez, V., Combe, A., David, V., Malmquist, N.A., Delorme, V., Leroy, C., Blazquez, S., Menard, R., Tardieux, I., 2009. Host cell entry by apicomplexa parasites requires actin polymerization in the host cell. Cell Host Microbe 5, 259–272.

Greenbaum, D.C., Baruch, A., Grainger, M., Bozdech, Z., Medzihradszky, K.F., Engel, J., DeRisi, J., Holder, A.A., Bogyo, M., 2002. A role for the protease falcipain 1 in host cell invasion by the human malaria parasite. Science 298, 2002–2006.

Gubbels, M.J., Lehmann, M., Muthalagi, M., Jerome, M.E., Brooks, C.F., Szatanek, T., Flynn, J., Parrot, B., Radke, J., Striepen, B., et al., 2008. Forward genetic analysis of the apicomplexan cell division cycle in *Toxoplasma gondii*. PLoS Pathog. 4, e36.

Gurnett, A.M., Liberator, P.A., Dulski, P.M., Salowe, S.P., Donald, R.G.K., Anderson, J.W., Wiltsie, J., Diaz, C.A., Harris, G., Chang, B., et al., 2002. Purification and molecular characterization of cGMP-dependent protein kinase from Apicomplexan parasites: A novel chemotherapeutic target. J. Biol. Chem. 277, 15913–15922.

Hall, C.I., Reese, M.L., Weerapana, E., Child, M.A., Bowyer, P.W., Albrow, V.E., Haraldsen, J.D., Phillips, M.R., Sandoval, E.D., Ward, G.E., et al., 2011. Chemical genetic screen identifies *Toxoplasma* DJ-1 as a regulator of parasite secretion, attachment, and invasion. Proc. Natl. Acad. Sci. USA 108, 10568–10573.

Haraldsen, J.D., Liu, G., Botting, C.H., Walton, J.G., Storm, J., Phalen, T.J., Kwok, L.Y., Soldati-Favre-Favre, D., Heintz, N.H., Muller, S., et al., 2009. Identification of conoidin A as a covalent inhibitor of peroxiredoxin II. Org. Biomol. Chem. 7, 3040–3048.

Harbut, M.B., Velmourougane, G., Dalal, S., Reiss, G., Whisstock, J.C., Onder, O., Brisson, D., McGowan, S., Klemba, M., Greenbaum, D.C., 2011. Bestatin-based chemical biology strategy reveals distinct roles for malaria M1- and M17-family aminopeptidases. Proc. Natl. Acad. Sci. USA 108, E526–534.

Heaslip, A.T., Leung, J.M., Carey, K.L., Catti, F., Warshaw, D.M., Westwood, N.J., Ballif, B.A., Ward, G.E., 2010. A small-molecule inhibitor of *T. gondii* motility induces the posttranslational modification of myosin light chain-1 and inhibits myosin motor activity. PLoS Pathog. 6, e1000720.

Ho, C.H., Magtanong, L., Barker, S.L., Gresham, D., Nishimura, S., Natarajan, P., Koh, J.L., Porter, J., Gray, C.A., Andersen, R.J., et al., 2009. A molecular barcoded yeast ORF library enables mode-of-action analysis of bioactive compounds. Nat. Biotechnol. 27, 369–377.

Hoon, S., Smith, A.M., Wallace, I.M., Suresh, S., Miranda, M., Fung, E., Proctor, M., Shokat, K.M., Zhang, C., Davis, R.W., et al., 2008. An integrated platform of genomic assays reveals small-molecule bioactivities. Nat. Chem. Biol. 4, 498–506.

Hu, G., Cabrera, A., Kono, M., Mok, S., Chaal, B.K., Haase, S., Engelberg, K., Cheemadan, S., Spielmann, T., Preiser, P.R., et al., 2010. Transcriptional profiling of growth perturbations of the human malaria parasite *Plasmodium falciparum*. Nat. Biotechnol. 28, 91–98.

Huang, J., Zhu, H., Haggarty, S.J., Spring, D.R., Hwang, H., Jin, F., Snyder, M., Schreiber, S.L., 2004. Finding new components of the target of rapamycin (TOR) signaling network through chemical genetics and proteome chips. Proc. Natl. Acad. Sci. USA 101, 16594–16599.

Huynh, M.H., Carruthers, V.B., 2009. Tagging of endogenous genes in a *Toxoplasma gondii* strain lacking Ku80. Eukaryot. Cell 8, 530–539.

Jensen, J.B., Edgar, S.A., 1976. Effects of antiphagocytic agents on penetration of *Eimeria magna* sporozoites into cultured cells. J. Parasitol. 62, 203–206.

Johnsson, N., Varshavsky, A., 1994. Split ubiquitin as a sensor of protein interactions *in vivo*. Proc. Natl. Acad. Sci. USA 91, 10340–10344.

Jung, H.J., Shim, J.S., Lee, J., Song, Y.M., Park, K.C., Choi, S.H., Kim, N.D., Yoon, J.H., Mungai, P.T., Schumacker, P.T., et al., 2010. Terpestacin inhibits tumor angiogenesis by targeting UQCRB of mitochondrial complex III and suppressing hypoxia-induced reactive oxygen species production and cellular oxygen sensing. J. Biol. Chem. 285, 11584–11595.

Kamau, E., Meehan, T., Lavine, M.D., Arrizabalaga, G., Mustata Wilson, G., Boyle, J., 2011. A novel benzodioxole-containing inhibitor of *Toxoplasma gondii* growth alters the parasite cell cycle. Antimicrob. Agents Chemother. 55, 5438–5451.

Kamau, E.T., Srinivasan, A.R., Brown, M.J., Fair, M.G., Caraher, E.C., Boyle, J.P., 2012. A focused small molecule screen identifies 14 compounds with distinct effects on *Toxoplasma gondii*. Antimicrob. Agents Chemother. 56, 5581–5590.

Kato, N., Sakata, T., Breton, G., Le Roch, K.G., Nagle, A., Andersen, C., Bursulaya, B., Henson, K., Johnson, J., Kumar, K.A., et al., 2008. Gene expression signatures and small-molecule compounds link a protein kinase to *Plasmodium falciparum* motility. Nat. Chem. Biol. 4, 347–356.

Knockaert, M., Gray, N., Damiens, E., Chang, Y.T., Grellier, P., Grant, K., Fergusson, D., Mottram, J., Soete, M., Dubremetz, J.F., et al., 2000. Intracellular targets of cyclin-dependent kinase inhibitors: identification by affinity chromatography using immobilised inhibitors. Chem. Biol. 7, 411–422.

Knockaert, M., Wieking, K., Schmitt, S., Leost, M., Grant, K.M., Mottram, J.C., Kunick, C., Meijer, L., 2002. Intracellular Targets of Paullones. Identification following affinity purification on immobilized inhibitor. J. Biol. Chem. 277, 25493–25501.

Kolev, N.G., Tschudi, C., Ullu, E., 2011. RNA interference in protozoan parasites: achievements and challenges. Eukaryot. Cell 10, 1156–1163.

Kortagere, S., 2012. Screening for small molecule inhibitors of *Toxoplasma gondii*. Expert Opin. Drug Discov 7, 1193–1206.

Kortagere, S., Mui, E., McLeod, R., Welsh, W.J., 2011. Rapid discovery of inhibitors of *Toxoplasma gondii* using hybrid structure-based computational approach. J. Comput. Aided Mol. Des. 25, 403–411.

Larson, E.T., Parussini, F., Huynh, M.H., Giebel, J.D., Kelley, A.M., Zhang, L., Bogyo, M., Merritt, E.A., Carruthers, V.B., 2009. *Toxoplasma gondii* cathepsin L is the primary target of the invasion-inhibitory compound morpholinurea-leucyl-homophenyl-vinyl sulfone phenyl. J. Biol. Chem. 284, 26839–26850.

Lee, J., Bogyo, M., 2013. Target deconvolution techniques in modern phenotypic profiling. Curr. Opin. Chem. Biol. In Press.

Lescault, P.J., Thompson, A.B., Patil, V., Lirussi, D., Burton, A., Margarit, J., Bond, J., Matrajt, M., 2010. Genomic data reveal *Toxoplasma gondii* differentiation mutants are also impaired with respect to switching into a novel extracellular tachyzoite state. PloS One 5, e14463.

Licitra, E.J., Liu, J.O., 1996. A three-hybrid system for detecting small ligand-protein receptor interactions. Proc. Natl. Acad. Sci. USA 93, 12817–12821.

Lomenick, B., Hao, R., Jonai, N., Chin, R.M., Aghajan, M., Warburton, S., Wang, J., Wu, R.P., Gomez, F., Loo, J.A., et al., 2009. Target identification using drug affinity responsive target stability (DARTS). Proc. Natl. Acad. Sci. USA 106, 21984–21989.

Lomenick, B., Olsen, R.W., Huang, J., 2011. Identification of direct protein targets of small molecules. ACS Chem. Biol. 6, 34–46.

Lourido, S., Shuman, J., Zhang, C., Shokat, K.M., Hui, R., Sibley, L.D., 2010. Calcium-dependent protein kinase 1 is an essential regulator of exocytosis in *Toxoplasma*. Nature 465, 359–362.

Lourido, S., Tang, K., Sibley, L.D., 2012. Distinct signalling pathways control *Toxoplasma* egress and host-cell invasion. EMBO J. 31, 4524–4534.

Ma, C., Li, C., Ganesan, L., Oak, J., Tsai, S., Sept, D., Morrissette, N.S., 2007. Mutations in {alpha}-Tubulin Confer Dinitroaniline Resistance at a Cost to Microtubule Function. Mol. Biol. Cell 18, 4711–4720.

Mayer, T.U., Kapoor, T.M., Haggarty, S.J., King, R.W., Schreiber, S.L., Mitchison, T.J., 1999. Small molecule inhibitor of mitotic spindle bipolarity identified in a phenotype-based screen. Science 286, 971–974.

Morrissette, N.S., Mitra, A., Sept, D., Sibley, L.D., 2004. Dinitroanilines bind alpha-tubulin to disrupt microtubules. Mol. Biol. Cell 15, 1960–1968.

Morrissette, N.S., Sibley, L.D., 2002. Disruption of microtubules uncouples budding and nuclear division in *Toxoplasma gondii*. J. Cell Sci. 115, 1017–1025.

Murphy, R.C., Ojo, K.K., Larson, E.T., Castellanos-Gonzalez, A., Perera, B.G., Keyloun, K.R., Kim, J.E., Bhandari, J.G., Muller, N.R., Verlinde, C.L., et al., 2010. Discovery of Potent and Selective Inhibitors of Calcium-Dependent Protein Kinase 1 (CDPK1) from *C. parvum* and *T. gondii*. ACS Med. Chem. Lett. 1, 331–335.

Muskavitch, M.A., Barteneva, N., Gubbels, M.J., 2008. Chemogenomics and parasitology: small molecules and cell-based assays to study infectious processes. Comb. Chem. High Throughput Screen 11, 624–646.

Nare, B., Allocco, J.J., Liberator, P.A., Donald, R.G., 2002. Evaluation of a cyclic GMP-dependent protein kinase inhibitor in treatment of murine toxoplasmosis: gamma interferon is required for efficacy. Antimicrob. Agents Chemother. 46, 300–307.

Ohmori, H., Toyama, S., 1992. Direct proof that the primary site of action of cytochalasin on cell motility processes is actin. J. Cell Biol. 116, 933–941.

Ojo, K.K., Larson, E.T., Keyloun, K.R., Castaneda, L.J., Derocher, A.E., Inampudi, K.K., Kim, J.E., Arakaki, T.L., Murphy, R.C., Zhang, L., et al., 2010. *Toxoplasma gondii* calcium-dependent protein kinase 1 is a target for selective kinase inhibitors. Nat. Struct. Mol. Biol. 17, 602–607.

Parsons, A.B., Brost, R.L., Ding, H., Li, Z., Zhang, C., Sheikh, B., Brown, G.W., Kane, P.M., Hughes, T.R., Boone, C., 2004. Integration of chemical-genetic and genetic interaction data links bioactive compounds to cellular target pathways. Nat. Biotechnol. 22, 62–69.

Petty, R.D., Sutherland, L.A., Hunter, E.M., Cree, I.A., 1995. Comparison of MTT and ATP-based assays for the measurement of viable cell number. J. Biolumin. Chemilumin. 10, 29–34.

Pfefferkorn, E.R., Borotz, S.E., 1994. *Toxoplasma gondii*: characterization of a mutant resistant to 6-thioxanthine. Exp. Parasitol. 79, 374–382.

Plouffe, D., Brinker, A., McNamara, C., Henson, K., Kato, N., Kuhen, K., Nagle, A., Adrian, F., Matzen, J.T., Anderson, P., et al., 2008. In silico activity profiling reveals the mechanism of action of antimalarials discovered in a high-throughput screen. Proc. Natl. Acad. Sci. USA 105, 9059–9064.

Radke, J.R., Donald, R.G., Eibs, A., Jerome, M.E., Behnke, M.S., Liberator, P., White, M.W., 2006. Changes in the expression of human cell division autoantigen-1 influence *Toxoplasma gondii* growth and development. PLoS Pathog. 2, e105.

Ryning, F.W., Remington, J.S., 1978. Effect of cytochalasin D on *Toxoplasma gondii* cell entry. Infect. Immun. 20, 739–743.

Sadaghiani, A.M., Verhelst, S.H., Bogyo, M., 2007. Tagging and detection strategies for activity-based proteomics. Curr. Opin. Chem. Biol. 11, 20–28.

Shaw, M.K., Compton, H.L., Roos, D.S., Tilney, L.G., 2000. Microtubules, but not actin filaments, drive daughter cell budding and cell division in *Toxoplasma gondii*. J. Cell Sci. 113 (Pt 7), 1241–1254.

Shoemaker, D.D., Lashkari, D.A., Morris, D., Mittmann, M., Davis, R.W., 1996. Quantitative phenotypic analysis of yeast deletion mutants using a highly parallel molecular bar-coding strategy. Nat. Genet. 14, 450–456.

Stokkermans, T.J., Schwartzman, J.D., Keenan, K., Morrissette, N.S., Tilney, L.G., Roos, D.S., 1996. Inhibition of *Toxoplasma gondii* replication by dinitroaniline herbicides. Exp. Parasitol. 84, 355–370.

Tanaka, H., Ohshima, N., Hidaka, H., 1999. Isolation of cDNAs encoding cellular drug-binding proteins using a novel expression cloning procedure: drug-western. Mol. Pharmacol. 55, 356–363.

Tashiro, E., Imoto, M., 2012. Target identification of bioactive compounds. Bioorg. Med. Chem. 20, 1910–1921.

Ward, G.E., Carey, K.L., Westwood, N.J., 2002. Using small molecules to study big questions in cellular microbiology. Cell. Microbiol. 4, 471–482.

Wiersma, H.I., Galuska, S.E., Tomley, F.M., Sibley, L.D., Liberator, P.A., Donald, R.G., 2004. A role for coccidian cGMP-dependent protein kinase in motility and invasion. Int. J. Parasitol. 34, 369–380.

Winzeler, E.A., Shoemaker, D.D., Astromoff, A., Liang, H., Anderson, K., Andre, B., Bangham, R., Benito, R., Boeke, J.D., Bussey, H., et al., 1999. Functional characterization of the *S. cerevisiae* genome by gene deletion and parallel analysis. Science 285, 901–906.

CHAPTER

22

Proteomics of *Toxoplasma gondii*

Jonathan M. Wastling, Dong Xia**

*Department of Infection Biology, Institute of Infection and Global Health, University of Liverpool, Liverpool, England, United Kingdom

OUTLINE

Toxoplasma gondii, second edition
http://dx.doi.org/10.1016/B978-0-12-396481-6.00022-2

22.1 INTRODUCTION

The mechanism of host cell invasion of *Toxoplasma gondii*, the structure and composition of its apical organelles, stage conversion and the evasion of host immune defences have, among others, been important and recurring themes in the study of this parasite. Our understanding of gene expression in the context of these fundamental questions has been transformed by the sequencing of the *T. gondii* genome and the subsequent development of sophisticated genetic tools. Nevertheless, proteins and not genes are ultimately the functional molecules in cells and it is proteins that are central to the fundamental mechanisms of parasitism that we wish to understand. The recent and rapid development of proteomic methods for analysing protein expression in *T. gondii* now represent powerful tools to characterize protein composition and function in this parasite.

To understand the development of proteomic studies in *Toxoplasma*, it is helpful first to look briefly at the development of transcriptional analysis for this parasite. A number of large-scale transcriptional gene expression studies, first with microarrays and serial analysis of gene expression (SAGE) (Blader *et al.*, 2001; Cleary *et al.*, 2002; Matrajt *et al.*, 2002; Radke *et al.*, 2005; Sibley *et al.*, 2002; Singh *et al.*, 2002) and latterly with next-generation sequencing (RNASeq) (Reid *et al.*, 2012) have been undertaken for *T. gondii*. However, transcriptional studies have several disadvantages. First, in common with other eukaryotic models, there is a less than predictable relationship between observed gene transcription and observed protein expression, making the functional interpretation of transcriptional data in *T. gondii* a challenge (Nagaraj *et al.*, 2011; Schwanhausser *et al.*, 2011; Wastling *et al.*, 2009). Next, even in cases of a linear correlation between transcript and protein abundance the activity of a protein is dependent not just on its abundance, but also on its state of activation, often mediated through a post-translational event such as phosphorylation. Finally, the activity of a protein is also dependent on interactions with other proteins, so that identifying and understanding the nature of these interactions is vital in studying protein function. Unfortunately, transcriptome studies provide little help with addressing these questions. The proteome, by definition, contains all the expressed proteins in all their various states of activation and modification. Proteomics or, perhaps more appropriately, functional proteomics thus has the potential to bridge the gap between our knowledge of the genome and the functional biological processes intrinsic to an organism.

Proteomics has now taken its place as one of the pillars of 'systems biology' in an era when a reductionist approach to biological problems is increasingly overshadowed by easy access to massive biological datasets (Wastling *et al.*, 2012). Before the so-called 'post-genomic era', a typical conventional molecular study of *Toxoplasma* might have begun with the identification of a single gene or protein associated with a particular biology event, such as invasion or stage-conversion. This gene would usually be sequenced, cloned and re-expressed to provide localization and other functional data; its gene structure may have been mapped by Southern blotting, its transcriptional expression by Northern blotting and the expression of the protein for which it codes by Western blotting. Consequently, a large amount of detailed information was gathered about a very small number of genes, often based around a single hypothesis. Many excellent studies founded on the above 'hypothesis-driven' approach continue to provide evidence for our present understanding of *T. gondii*; however, the process is relatively slow and does not exploit the vast scope of genomics data in the most efficient way.

Developed in the late 1990s, proteomics is still a relatively young field, although almost since its inception proteomics has been exploited by those interested in the biology of *Toxoplasma*. In this chapter we review the present state of proteomics analysis of *Toxoplasma* and discuss how these data have been used to enhance our understanding of the biology of this parasite. We look at specific examples of how data from proteomics have both confirmed previous hypotheses and revealed unexpected results that would have been difficult to predict without adopting a proteomics approach.

22.2 FUNDAMENTALS OF PROTEOMICS

There are many excellent reviews on proteomics in the literature (Aebersold and Mann, 2003; Bradshaw and Burlingame, 2005; Cox and Mann, 2011; Dhingra *et al.*, 2005; Gorg *et al.*, 2004; Johnson *et al.*, 2004; Lane, 2005; Malmstrom *et al.*, 2011; Morrison *et al.*, 2003; Phillips and Bogyo, 2005; Yates, 2004) and these and others should be referred to for a more thorough introduction to the general topic. However, before dealing specifically with proteomics studies of *Toxoplasma* we will briefly review some of the basic principles of this relatively new area of technology.

Mass spectrometry analysis (MS) of proteins has essentially replaced classical methods for protein micro-sequencing, such as Edman degradation. MS enables the mass measurement of ions derived from molecules, and is capable of forming, separating and detecting molecular ions based on their mass-to-charge ratio (m/z). The development of two ionization methods in the late 1980s made MS routinely available to biological researchers. These were Matrix-Assisted Laser Desorption/Ionization mass spectrometry (MALDI–MS), and Electro-Spray Ionization tandem mass spectrometry (ESI–MS/MS). Proteomics essentially entails the matching of two datasets, one experimental and one virtual, to enable rapid identification of proteins. A general scheme for a proteomics experiment is shown in Figure 22.1. The experimental dataset consists of MS data, which in the simplest form contain the precise mass to charge ratios of peptides obtained from proteins digested with an enzyme such as trypsin. Measured using MALDI–MS, the individual masses of each peptide give rise to a unique peptide mass fingerprint (PMF). Alternatively, peptides can be fragmented by ESI–MS/MS to produce tandem mass spectrometry, or MS/MS, data. Once again a unique fingerprint is obtained, this time for each individual peptide. In both cases, the experimental MS data can then be matched using algorithms to a corresponding virtual peptide mass database. Virtual mass databases are dependent on the availability of a substantial quantity of genome sequence, or experimentally verified expression data, and are derived from translations of all possible open reading frames into protein sequences, or from protein prediction models. Since the availability of an annotated genome sequence for *Toxoplasma*, these virtual or hypothetical protein databases can be readily constructed, thus enabling precise matches with experimental MS data.

22.2.1 Soluble and Insoluble Proteomes

For identification-oriented proteomics experiments, protein samples can be analysed whole or as fractionated sub-proteomes using various approaches such as organelle separation and affinity purification. Depending on the nature of proteins, separation at either the protein level or peptide level is almost always performed to reduce sample complexity prior to MS analysis. Protein separation is typically achieved by techniques such as gel electrophoresis or other forms of chromatography followed by enzyme digestion. Peptide separation is commonly achieved by using liquid chromatography before analysis by tandem mass spectrometry (MS/MS).

In protein separation, gel electrophoresis such as one-dimensional SDS–PAGE (1D SDS–PAGE) and two-dimensional electrophoresis (2-DE) has always played an important role in proteomics experiments. 2-DE is particularly

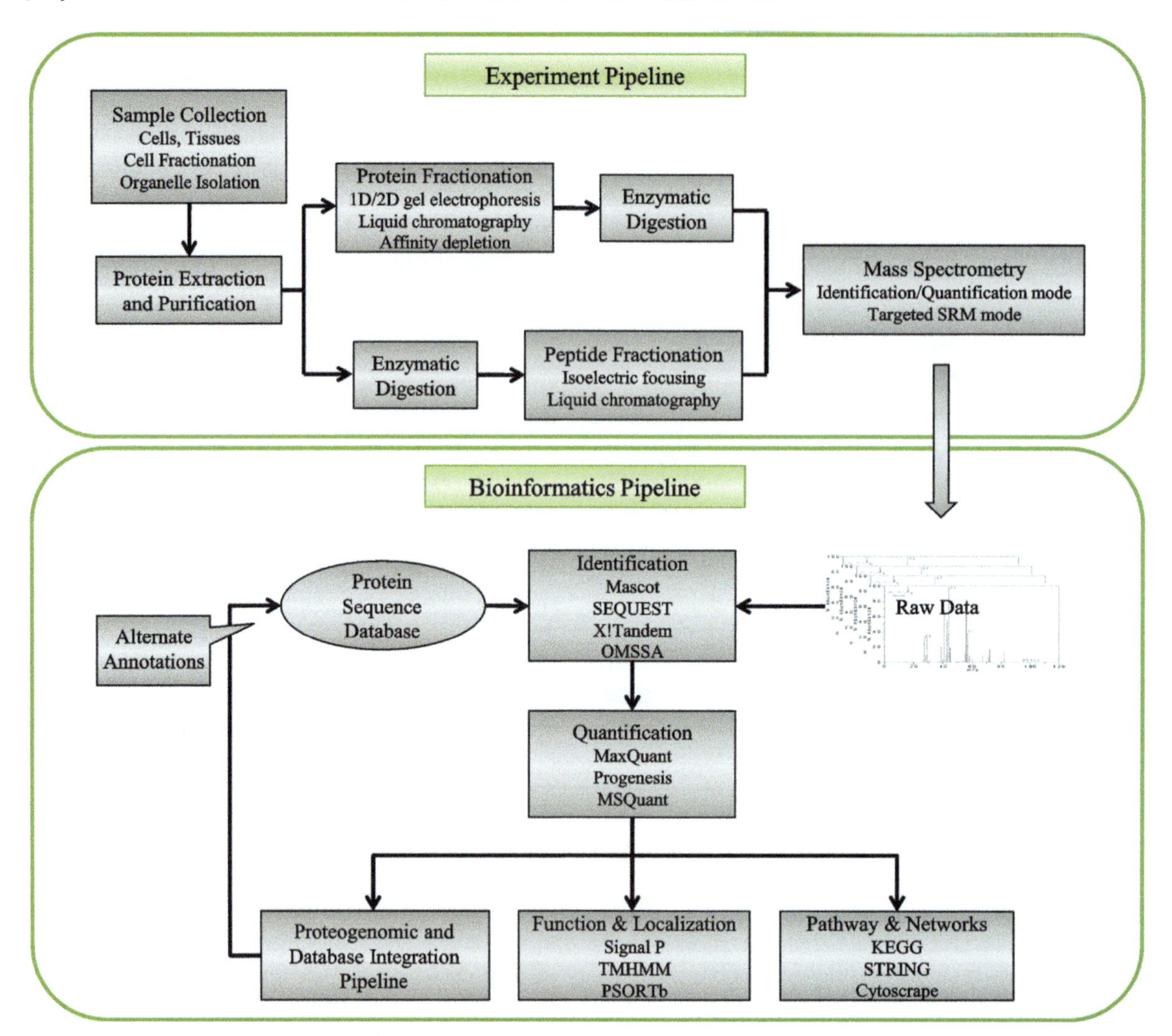

FIGURE 22.1 **A general scheme for a proteomics project.** A typical workflow for an identification and quantification proteomics project is achieved by integrating two pipelines, the experiment pipeline and the bioinformatics pipeline. The experiment pipeline consists of four major steps: sample collection, protein extraction and purification, sample fractionation and digestion (in either protein or peptide space) and mass spectrometry analysis. The raw spectra data collected are then subjected to the bioinformatics pipeline, where protein identifications are acquired based on the comparison between raw data and protein sequence database. Quantification packages are then used to extract and calculate relative or absolute quantifications of the analyte proteins. The biological meaning of the identification and quantification data is inferred by using tools that provide protein function and localization predictions and pathway analysis. Proteogenomic and database integration pipelines are also developed to facilitate data integration with online databases and improve genome annotation using alternate annotations.

suitable for analysis of soluble proteomes and produces high-resolution two-dimensional maps of protein constituents which enable the relative level of steady-state protein expression to be estimated as well as the potential identification of post-translational modifications. Individual protein spots can then be picked from the gel, and the samples containing the proteindigested and analysed by MS to identify each protein spot. Unfortunately, some proteins, notably hydrophobic proteins and those at the extremes of molecular weight, are notoriously difficult to resolve using 2-DE. Very low abundance proteins are also difficult to detect by 2-DE.

An alternative to 2-DE protein separation is to undertake a two-dimensional protein separation using a combination of either 1-D SDS–PAGE followed by nano-liquid chromatography and MS, or two serial liquid chromatographic separations followed by MS. In this latter approach, commonly known as multidimensional protein identification technology (MudPIT), the entire population of proteins in the sample is digested using enzymes such as trypsin, and the peptides are first separated using a strong cation exchanger and second by reverse phase LC. The peptides, which are eluted from the second LC column, are then fed directly into an ESI iontrap mass spectrometer, where they are ionized, mass selected and fragmented to produce MS data. MudPIT has been used partially to resolve the proteomes of a number of protozoan parasites, including those of *Plasmodium falciparum* (Florens *et al.*, 2002; Lasonder *et al.*, 2002), *Trypanosoma cruzi* (Paba *et al.*, 2004), *Cryptosporidium parvum* (Sanderson *et al.*, 2008), as well as the tachyzoites of *T. gondii* (Xia *et al.*, 2008).

At first, MudPIT analysis might seem like the obvious replacement to 2-DE based proteomics as it circumvents many of its inherent limitations and offers a relatively 'hands-off' procedure which can quickly yield large amounts of data. However, the simplicity of MudPIT is at the cost of the loss of information about the proteins which are being analysed. One of the advantageous features of 2-DE protein separation is that it reveals the complexity of the proteome in a way that MudPIT currently cannot. For example, 2-DE proteomic analysis of *T. gondii* (Cohen *et al.*, 2002) reveals that single proteins are often represented by more than one gel spot. These additional spots are not artefacts but represent potentially important functional features of proteins, such as an alternative splicing event, or a post-translational modification like phosphorylation. Moreover, whilst the relative abundance of proteins represented on 2-DE gels can be reproducibly compared using comparative gel approaches (i.e. DIGE).

22.3 WHICH PROTEOME? PROTEOMES AND SUBPROTEOMES OF *TOXOPLASMA GONDII*

The success of proteomic analysis depends on careful experimental design, of which sample preparation is the first and arguably the most important step. The proteome of any cell consists of a heterogeneous mixture of soluble and hydrophobic proteins, some of which may be present in large, stable quantities whilst others are in transitory and almost undetectable amounts. An obvious way to overcome the challenge of complexity is to simplify the mixture of proteins before analysis. This principle has led to development of 'sub-proteomics'. Sub-proteomics relies on the preparation or enrichment of a smaller fraction of the total proteome, a strategy employed, for example, by Zhou and coworkers to analyse the protein repertoire of the excretory–secretory products of *T. gondii* (Zhou *et al.*, 2005). This approach is useful when the study is restricted to the proteins of a particular organelle in which a method for purifying the organelle has been established (Yates *et al.*, 2005). There are, however, several important caveats to a sub-proteomics approach. First, interpretation of the resulting data must take regard of the confidence placed on the preparation method and the degree of enrichment obtained for the organelle or sub-proteome under investigation. Second, the protein composition of organelles may be in a continual flux depending on the biological status of the cell. Third, many molecules in the cytoplasm interact with organellar proteomes and may be associated with these proteins, although not strictly derived from the organelle itself. Thus, with sub-proteomics and organellar proteomics experiments, reproducible sample preparation, immunolocalization to verify the origin of newly discovered proteins, and careful bioinformatics interpretation of proteomics data are important adjuncts to successful studies.

22.4 MASS-SPECTROMETRY ANALYSIS OF *TOXOPLASMA GONDII* PROTEINS

For the purposes of most proteomics analyses of *T. gondii*, the present choice is whether to use MALDI–MS or ESI–MS/MS for protein sequence determination. MALDI–MS and ESI–MS/MS generate different types of data, and the choice of analysis depends on whether PMF or peptide fragmentation data are required. The first protein identifications by MS for *Toxoplasma* were performed before there was substantial genome information for the parasite, and were forced to depend on identification of proteins using the then handful of genes in NCBI and a limited EST database (Cohen *et al.*, 2002; Nischik *et al.*, 2001). Since EST sequences are by nature short, they rarely encode sufficient peptides for unambiguous PMF identification of proteins. However, with the development of a sophisticated, annotated genome for *T. gondii*, it is now possible to search efficiently MALDI–MS data against thousands of *T. gondii* protein sequences, including hypothetical predicted proteins.

The advantage of ESI is that it allows for tandem MS, or MS/MS capability. The key event in MS/MS is that it involves the energetic dissociation of anions to cause peptide fragmentation, and thereby derives structural information relating to the sequence of a peptide. Thus, in contrast to PMF, where only the masses of individual peptides are known, MS/MS peptide fragmentation data give rise to peptide sequence information. For organisms of limited genome data, or where genome data have not been completely annotated, this provides an immediate advantage. It is important to recognize some of the other characteristics of both MALDI–MS and ESI–MS/MS. Easy to use and less expensive, MALDI–MS is generally of a lower resolution than ESI–MS/MS. By contrast, the cost of purchasing and maintaining an ESI–MS/MS instrument can be prohibitively high for many laboratories, and the degree of skill required for successful operation is considerably greater than for MALDI. However, as well as MS/MS capabilities, ESI–MS/MS is excellent for determining peptide modifications, whilst MALDI only has limited capabilities in this area. The ability to determine peptide modifications is critical in what undoubtedly became one of the major areas of proteomics – the characterization of post-translational modifications of proteins.

22.4.1 *Toxoplasma* Protein Databases for MS Searching

The availability of a sophisticated genome database for *T. gondii* is essential to underpin analysis of *Toxoplasma* MS data, since the experimental MS data need to be matched against a suitable protein database. At present, MS data can be searched against theoretical protein sequences contained within the most current version of ToxoDB (http://www.toxodb.org), a database of *Toxoplasma* genomic, transcript and protein prediction models compiled from sequences representing manifold coverage of the *Toxoplasma* genome. A *T. gondii* proteome based on data from ToxoDB is also available at EPIC-DB (http://toro.aecom.yu.edu/cgi-bin/biodefense/main.cgi), along with MS data on *T. gondii*. The latest ToxoDB (version 8.0) contains genome annotations for various *T. gondii* strains such as ME49, RH, VEG and GT1. Translated protein sequence can be directly downloaded from ToxoDB as fasta files and used for MS searching.

Sequence databases for MS searching should not be restricted only to the current version of ToxoDB annotations. An important application of proteomics data is to improve genome annotation, namely proteogenomics (Xia *et al.*, 2008). Since proteomics data represents truly translated and expressed genome sequences, proteogenomics can provide valuable information about the intron–exon boundary, frameshifts and N-terminal methionine excision which are difficult to predict by *in silico* pipelines or transcriptomics studies. A more recent development involves the use of an automated proteogenomic pipeline developed for the integration of mass spectrometry (MS) based proteomics evidence into ToxoDB (Krishna *et al.*, 2011). The pipeline uses MS data for confirming official gene models on the database, but also examines whether there is supporting evidence for alternate annotations at particular loci and for identifying novel genes. The integration of this pipeline to ToxoDB not only assists future genome annotations, but also simplifies the iterative searching process required for each genome annotation update.

22.5 QUANTITATIVE PROTEOMICS

In common with all proteomes, *Toxoplasma* proteomes are highly dynamic and reflect the biological state of the parasites at the time of sampling. Comparing proteomes at different life-stages and under different conditions thus has the potential for revealing something of the nature of the biological response to specific stimuli. For example, we may wish to understand what happens to the proteome when tachyzoites transform to bradyzoites, or what happens when we knock out a gene coding for a protein in a specific biochemical pathway. Such comparisons are useful for understanding not only the basic biology of the parasite, but also how chemotherapeutic compounds may act on the parasite, the basis of drug-resistance and the host response to infection. For these types of analyses, the proteomic method of choice must be capable of some form of quantitative measurement.

It is important to make the distinction between absolute quantitation and relative quantitation in proteomics. Absolute quantification allows data to be compared across several studies including those from different groups and be more easily integrated into transcriptomic and metabolic systems data. The use of absolute quantification in parasite studies is only just beginning to emerge. In the meantime, most of the current 'quantitative proteomics' studies refer to the relative quantitation of proteins – that is, determining the extent to which a protein or group of proteins is up- or down-regulated in different conditions.

Relative quantifications of proteins have been made possible due to important advances in MS instrumentation. These can be categorized into two general approaches: label-based methods and label-free methods. Label-based methods commonly use differential stable isotope labelling through *in vivo* metabolic labelling or *in vitro* chemical modification. In the former approach, intact proteins are labelled *in vivo* through the incorporation of stable isotopes into the media on which cells are growing, for example incorporation of ^{13}C into specific amino acids, such as stable isotope labelling of amino acids in cell culture (SILAC) (Ong *et al.*, 2002). Chemical labelling techniques include protein-level tagging such as with isotope-coded affinity tags (ICAT), applied to cysteine residues (Gygi *et al.*, 1999) and isobaric tag for relative and absolute quantification (iTRAQ), which uses a multiplexed set of isobaric reagents that yield amine-derivatized peptides for quantification (Ross *et al.*, 2004). Fluorescent dye labelling is used in 2-DE fluorescence difference gel electrophoresis (2-DE DIGE) (Unlu *et al.*, 1997). Label-free quantification methods that directly use raw spectral data from parallel MS runs to determine relative protein abundance are increasing in popularity. Spectral counting and intensity-based methods are among the most commonly used approaches. Spectral counting relates protein abundance in a given run to the number of peptide-spectrum matches (PSMs) for each protein. Due to different ionization efficiencies caused by biophysical properties of each peptide, several software packages have been developed to normalize spectral counting, such as APEX (Lu *et al.*, 2007) and emPAI (Ishihama *et al.*, 2005). Intensity based methods align precursor ion spectra of the same peptide from parallel runs according to their retention times (RT); protein quantification is obtained by summation of ion intensities that have been matched to peptides for a given protein. This approach has been implemented by several commercial software packages such as Progenesis LC−MS (NonLinear Dynamics) on all unique peptides and Protein Lynx Global Server (Waters) on the top three unique peptides. Open source packages such as MaxQuant (Cox and Mann, 2008), OpenMS (Sturm *et al.*, 2008) and MSight (Palagi *et al.*, 2005) are also available.

Absolute protein quantification can be achieved using a variation of the intensity based label-free quantification method mentioned, namely intensity based absolute quantification (iBAQ) (Schwanhausser *et al.*, 2011). This method uses the total ion intensity for all of the peptides observed, normalized against the total theoretical number of observable peptides for each protein and then compares these data with a calibration curve produced by a range of spiked protein or peptide internal standards of known concentrations. The approach has been successfully applied to human and mouse cells (Nagaraj *et al.*, 2011; Schwanhausser *et al.*, 2011) and appears to be independent of sequence, is accurate over several logs and is not affected by sample complexities. There is presently no complete absolute quantitative proteome study of *T. gondii*, although a number of large-scale whole proteomic analyses of various strains of tachyzoites and host cell responses as well as other life cycle stages are currently under way (Wastling, unpublished).

Isotope labelling based approaches have also been developed. These are performed by stable isotope dilution, where a reference standard to which a stable isotope has been incorporated is added in known concentration to the sample mixture. The reference peptide shares all the properties of the analyte peptide apart from the mass difference due to the isotopic label. The ratio of the intensities of the analyte and standard ions determined by MS analysis will allow the calculation of the concentration of the analyte. The reference standards can be synthesized chemically individually, e.g. AQUA peptides (Gerber *et al.*, 2003), or expressed from synthetic genes in *E. coli* using stable isotopically enriched media, e.g. QconCAT Absolute quantification (Beynon *et al.*, 2005).

While high-throughput proteomics approaches help generate assumption-free hypotheses, the biological and instrumental resources required to cover a huge diversity of protein species at a broad dynamic range is not always necessary. For a more focused study, targeted proteomics approaches are being developed to tune the MS instrument measurements on specific peptides of proteins of interest. Selected reaction monitoring (SRM) equipped with a triple quadrupole mass spectrometer (QQQ) is typically used in targeted proteomics experiments (Bertsch *et al.*, 2010; Lange *et al.*, 2008). The first quadrupole is used to isolate precursor ions in a narrow mass range and the selected ions are then fragmented in the second quadrupole. The third quadrupole is used specifically to detect a set of fragment ions that is characteristic for the target peptides. This approach has the potential for greater accuracy due to a relatively small sub-set of protein targets and simpler bioinformatics interpretation. Targeted proteomics is particularly useful in achieving a rapid and accurate quantitative profiling of a repeated set of proteins across samples from different conditions.

22.6 APPLICATION OF PROTEOMICS TO THE STUDY OF *TOXOPLASMA GONDII*

In the active, infectious stage of *Toxoplasma*, the tachyzoite has been the main focus of proteomic studies although some data have now been published on the other life cycle stages. The first large-scale proteomic mapping of *T. gondii* tachyzoites was undertaken by Cohen and co-workers, who used a range of immobilized pH gradients and high-resolution 2-DE to separate over 1000 *Toxoplasma* proteins. Mass spectrometry was used to analyse 71 of these proteins by MALDI–MS and MALDI post-source decay analysis (PSD) (Cohen *et al.*, 2002). Notably, this study showed that in many instances several protein spots appeared to be encoded by the same gene, indicating that post-translational modification and/or alternative splicing events may be a common feature of functional gene expression in *T. gondii*. Peptide mass fingerprinting by MALDI–MS enabled unambiguous protein identifications to be made where full gene sequence information was available. However, interpretation of peptide mass fingerprint data using *T. gondii* EST sequences was less reliable. Peptide fragmentation data, acquired by PSD MALDI–MS, proved a more successful strategy for the identification of *Toxoplasma* proteins using the EST database. Post-source decay analyses, like ESI–MS/MS (which has now largely replaced it) yields peptide sequence information and does not rely on matching mass fingerprints to near complete predicted protein sequences. This early experiment demonstrated the viability of proteomic analyses for *T. gondii*, even in the absence of complete genome sequence, as well as the potential value of proteomic data as an aid to annotation of sequence databases in *T. gondii*.

Equipped with the draft genome annotation published on ToxoDB (version 4.0) and improved mass spectrometry instruments, Xia and co-workers published the first multi-platform global proteome analysis of *Toxoplasma* tachyzoites, resulting in the identification of nearly one-third of the entire predicted proteome of *T. gondii* (Xia *et al.*, 2008). Three complementary approaches (1-DE LC-MS/MS, 2-DE LC–MS/MS and MudPIT) were employed to identify 2252 officially annotated genes, with 2477 intron-spanning peptides that confirm correct splice site annotation. This study provided important protein expression evidences for a broad range of proteins, such as proteins predicted to contain signal peptide (10%) and transmembrane domains (18%), as well as a large number of proteins that involved in key metabolic pathways. More interestingly, this study identified 394 non-redundant alternative gene models and ORFs, which suggested required improvement for genome annotation and provided valuable resources for proteogenomic studies of *T. gondii*. The acquired proteomics data have also been compared with transcriptomics data available, such as EST and microarray data, and highlighted important discrepancies between proteome and transcriptome expressions in *T. gondii*. The data have been integrated with other genomics data on ToxoDB for public access and became the first proteomics data hosted on this online database.

Dybas *et al.* examined the genome annotation of *T. gondii* using similar high-throughput proteomics techniques (Dybas *et al.*, 2008). In this study, 1-DE LC–MS/MS was used to identify 2477 gene-coding regions. Using EST and proteomics data, the large scale experimental validation of the hypothetical *T. gondii* genome produced by various gene-prediction algorithms revealed a false negative rate about 31%–42%, where gene prediction missed the experimentally verified coding region. To further examine the quality of genome annotation, predicted gene models were searched against the complete NCBI non-redundant

(NR) database, *Apicomplexa* proteins and human proteins to identify unique and conserved proteins. The results showed that 67% of the predicted *T. gondii* proteome had a homologous sequence in NR and 64% an apicomplexan orthologue. Approximately 52% of *T. gondii* sequences were unique compared with the human genome. A proteomics database was also created (EPIC−DB), which combined current experimental (National Center for Biotechnology information (NCBI)) and predicted *T. gondii* genes. The finding in this study has provided additional proteomics evidence on improvements required for *T. gondii* genome annotation (Madrida-Aliste *et al.*, 2009).

In addition to tachyzoite stage, environmentally resistant oocysts have also been examined by a global proteomics approach. Fritz *et al.* (Fritz *et al.*, 2012) have characterized the proteome of the wall and sporocyst/sporozoite fractions of mature, sporulated oocysts using the 1-DE LC−MS/MS approach. In total, 1021 *T. gondii* proteins were identified in sporocyst/sporozoite fraction and 226 were identified in the oocyst wall fraction. Significantly, 172 proteins identified have not been reported in other *Toxoplasma* proteomic studies, confirming variations of expression profile between different life cycle stages. This study has also identified novel oocyst proteins that are likely involved in conferring environmental resistance including a family of small, tyrosine-rich proteins present in the oocyst wall fractions and late embryogenesis abundant domain-containing (LEA) proteins in the cytosolic fractions.

22.7 PROTEOMICS ANALYSIS OF THE RHOPTRY ORGANELLES OF *TOXOPLASMA GONDII*

The rhoptry organelles are specialized secretory organelles, and have been the focus of considerable attention in an effort to understand the mechanisms by which *Toxoplasma* is able to invade cells. The availability of proteomics methods for *T. gondii* has facilitated the identification of a large number of novel rhoptry proteins and the discovery of a completely new class of proteins which localize to the neck of the organelle (Bradley *et al.*, 2005). This improved understanding of the composition of the rhoptries has led to new insights into their structure and function, setting the groundwork for a more complete understanding of how they contribute to the process of invasion and intracellular survival.

Before the application of proteomics to study *T. gondii* apical components there were only around eight known rhoptry proteins of *Toxoplasma* designated ROP proteins: ROP1, ROP2, ROP4, ROP8 and ROP9 (Beckers *et al.*, 1994, 1997; Ossorio *et al.*, 1992; Reichmann *et al.*, 2002; Sadak *et al.*, 1988), and two rhoptry proteases (a subtilisin-like protein TgSUB2 and a cathepsin B-like protein, Toxopain-1) as well as a sodium−hydrogen exchanger named TgNHE2 (Karasov *et al.*, 2005; Miller *et al.*, 2003; Que *et al.*, 2002). It was anticipated that this handful of proteins did not represent the full complement of rhoptry protein constituents, and with the completion of the genome sequence for *Toxoplasma* it was hoped that bioinformatics analysis might reveal many additional members of the 'rhoptry family' of proteins. This approach unfortunately proved far less successful than was originally anticipated because of an apparent lack of homologous proteins and/or conservation among apicomplexan rhoptry proteins. The paucity of rhoptry sequences in the database meant that it was impossible to determine a rhoptry targeting signal that could be used to identify additional proteins associated with the compartment. However, although bioinformatics approaches proved less successful in the search for additional rhoptry proteins, the development of proteomic methods for *T. gondii* presented a new and unexplored opportunity to characterize rhoptry protein composition.

22.7.1 Preparation of the Rhoptry Sub-Proteome

As with all sub-proteome experiments, the success of the rhoptry proteome project required the development of a reliable method for obtaining a highly purified sub-proteome in quantities sufficient for MS analysis. This was achieved with considerable success by the adaptation of a Percoll gradient method originally devised by Leriche and Dubremetz and used to isolate a fraction enriched for rhoptries (Leriche and Dubremetz, 1991). Analysis of the fraction obtained by the original protocol using 2-DE revealed that it contained substantial proportions of contaminants, mainly dense granules, mitochondria and plastids. To improve on the initial enrichment, this fraction was further subjected to sucrose gradient flotation (Bradley *et al.*, 2005). Subsequent analysis of 1-D SDS–PAGE gels of this fraction by immunoblotting using antibodies to known proteins from the rhoptries, dense granules and mitochondria showed that this new sucrose flotation step was able to enrich for the rhoptries whilst effectively removing the vast majority of dense granules and contaminating mitochondria.

22.7.2 The Hydrophobic Nature of the Rhoptry Contents

The largely hydrophobic nature of rhoptry proteins was determined using sodium carbonate extraction (Bradley *et al.*, 2005) —a procedure that releases proteins that are soluble or associated with membranes through ionic interactions, leaving behind proteins that are anchored in the membranes by transmembrane domains or other strong interactions such as glycosylphosphatidylinositol anchors. Sodium carbonate extraction of partially enriched rhoptry fractions showed the predicted partitioning of known soluble (NTPase) and membrane-associated proteins (ROP2/3/4), but also indicated that the majority of novel rhoptry proteins were likely to be found within the membrane fraction and thus not amenable to analysis by 2-DE (Bradley *et al.*, 2005). Thus, when this enriched rhoptry fraction was subjected to 2-DE the representation of proteins on the gel was poor, with membrane-associated rhoptry proteins such as the ROP2–8 family being poorly resolved because of their hydrophobic nature making them less amenable to solubilization in the isoelectric focusing step of 2-DE.

22.7.3 Proteomic Identification of Rhoptry Proteins

Due to the hydrophobic nature of the rhoptry proteins, an alternative to 2-DE proteome analysis was employed to separate the proteins, using conventional 1-D SDS–PAGE followed by the excision of 51 contiguous gel slices, each of which was subjected to in-gel trypsin digestion and then tandem MS (MS/MS) to obtain peptide fragmentation data suitable for proteomic database searching (a schematic of the complete approach is shown in Fig. 22.2). As expected, known rhoptry proteins were readily identified, including members of the ROP2/4/8 family and ROP9. In addition, 38 previously unidentified candidate novel rhoptry proteins were detected in the fraction. A combination of approaches was used to determine the true localization of the novel proteins identified, including epitope tagging and the production of antibodies against peptides and recombinant proteins.

One of the most exciting findings regarding proteins in the rhoptry proteome was the subgroup of proteins that localize exclusively to the duct-like rhoptry necks. To distinguish these from rhoptry body proteins, called ROPs, rhoptry neck proteins were named RONs (Bradley *et al.*, 2005). As previously noted, four RON proteins (RON1–4) were confirmed by antibody production and localized by IFA to the necks of the organelle. To exclude the possibility that

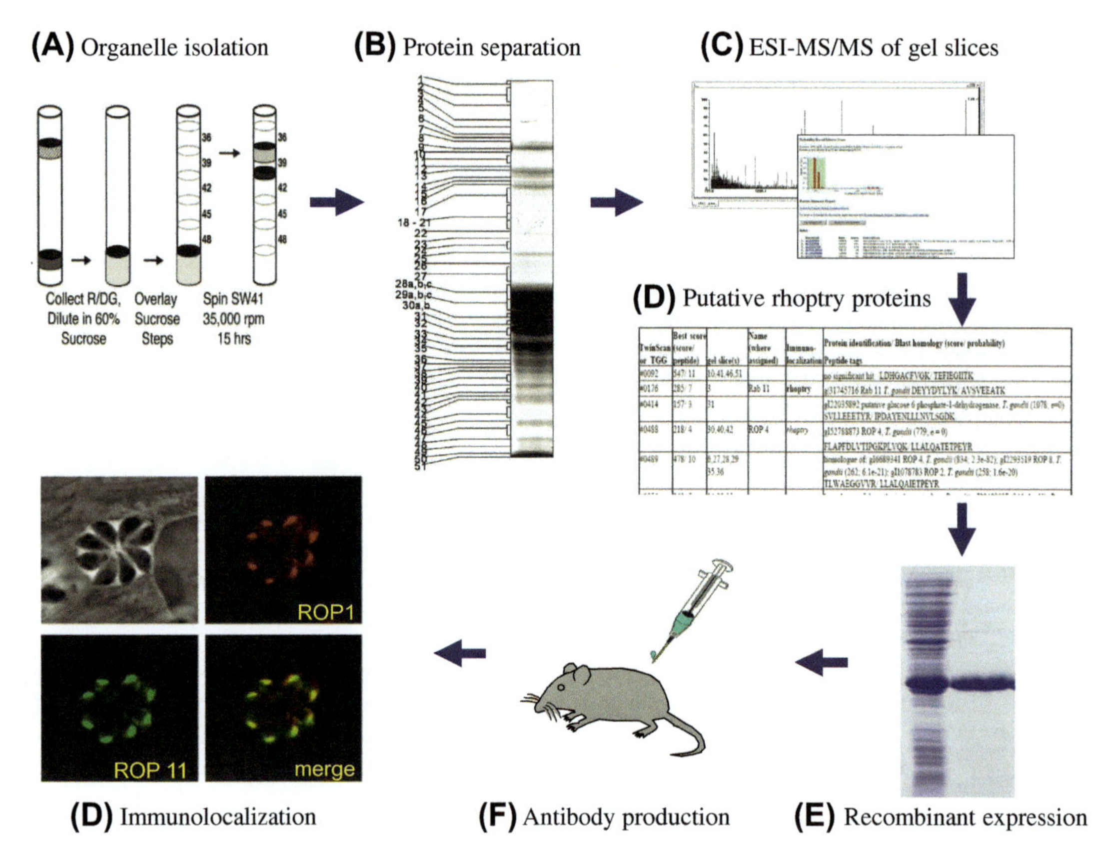

FIGURE 22.2 **Schematic for the proteomic analysis of the rhoptry organelles of *T. gondii*.**
A) Rhoptry organelles were prepared from tachyzoites lysed in isotonic sucrose to maintain intact organelles and the organelles fractionated by Percoll and sucrose gradients (only a single gradient is shown here for simplicity).
B) The purified organelles were then separated by one-dimensional SDS–PAGE and 51 individual bands were excised from the gel and digested.
C) The bands were subjected to LC followed by ESI–MS/MS. MS data were searched against a database containing 881, 411 protein sequences downloaded from ToxoDB to identify the proteins in each gel slice.
D) Thirty-eight novel rhoptry proteins were identified.
E) A sub-set of these were selected and expressed as His6-tag fusion proteins and purified by nickel–agarose chromatography.
F) The purified proteins were injected into mice for polyclonal antibody production.
G) Co-localization of these novel rhoptry proteins was verified by immunofluorescence (Bradley *et al.*, 2005).

these proteins were staining a compartment other than rhoptry necks, localization was confirmed for one of these, RON4, by immuno-electron microscopy (Fig. 22.3). Upon rhoptry release during invasion, the rhoptry neck protein RON4 is localized to the moving junction – a structure that forms the interface between the surface of the parasite and the host cell plasma membrane. RON4 is part of a complex of rhoptry neck proteins that includes RON2 and RON5 (Bradley *et al.*, 2005; Lebrun *et al.*, 2005; Wastling and Bradley, 2007). All three proteins are believed to be present at the moving junction, although localization has only directly been shown for RON4 (Alexander *et al.*, 2005). Proteomic analysis has thus helped to reveal for the

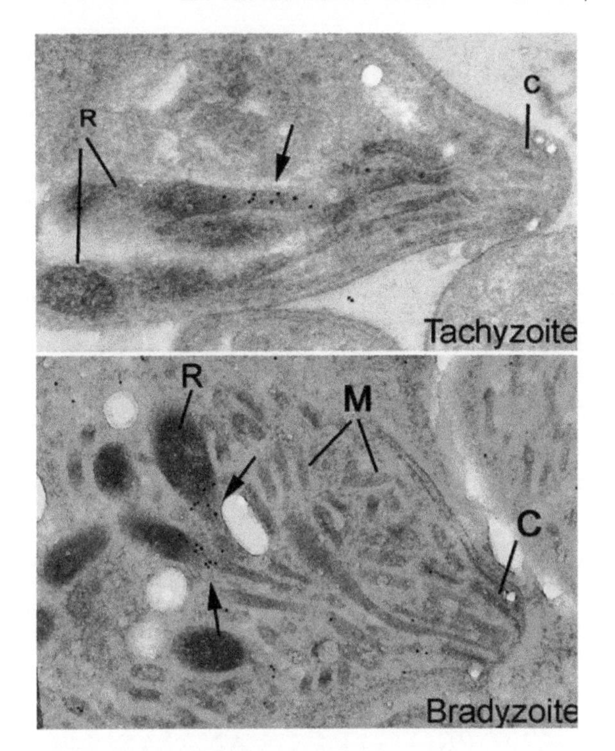

FIGURE 22.3 **Immunolocalization of RON4 to the rhoptry necks of *Toxoplasma gondii*.** Immunoelectron microscopy with anti-RON4 antibodies demonstrates that RON4 is localized to the neck portion (arrows) of the rhoptries (R) and is not present in the bulbous bodies of the organelle. Lines also point to the conoid (C) and the micronemes (M). RON4 appears to be most prominent at the junction of the body and neck portion of the organelle and is present in samples from both tachyzoites and bradyzoites.

first time the presence and participation of rhoptry neck proteins in moving junction formation and strongly suggests the conservation of this structure at the molecular level among *Apicomplexa*.

All *T. gondii* rhoptry proteins studied to date appear to be synthesized as pro-proteins that are then processed to their mature forms. To understand the role of the pro-region in rhoptry protein function, MS analysis has been used to define the processing site of the pro-region of the rhoptry protein ROP1 (Bradley and Boothroyd, 1999). Efforts to determine such processing sites had previously been prevented by blocked N-termini of mature proteins isolated from *T. gondii*, thus preventing analysis by conventional N-terminal amino-acid sequencing. To overcome the problem, an engineered form of ROP1 was designed and MS used to demonstrate that pro-ROP1 is processed to its mature form between the glutamic acid at position 83 and alanine at position 84.

22.8 PROTEOMICS ANALYSIS OF EXCRETORY/SECRETORY PROTEINS OF *TOXOPLASMA GONDII*

The apical complex of *Toxoplasma* is involved in the secretion of compounds that play a key role in host cell invasion and survival during infection. These compounds are derived not just from the rhoptries, but also from the dense granules and the micronemes. Bioinformatics analysis of the *T. gondii* genome reveals more than 800 genes encoding proteins with putative secretory signal peptides (Zhou *et al.*, 2005). Although many of these putative secretory proteins are likely to be associated only with internal organelles, a substantial proportion might be exported externally for interaction with the host. Until recently only a limited number of such secretory proteins had been discovered, and the wider spectrum of these molecules thought to be involved in parasite invasion and survival remained unknown. These proteins, known as excretory/secretory antigen (ESA) proteins, are by nature soluble, and thus ideally suited to mapping and identification by 2-DE. Zhou and co-workers have applied proteomics technology to analyse a large cohort of freely released *Toxoplasma* secretory proteins by using the complementary methodologies of 2-DE and MudPIT (Zhou *et al.*, 2005).

A sub-proteome of ESA proteins can be readily prepared from tachyzoites by separating the parasites from host cells and incubating them in medium containing 1% ethanol before concentration and proteomics analysis. Mapping of these ESA secretory products by 2-DE revealed approximately 100 spots, most of which were successfully identified by protein microsequencing or MALDIMS. Separation of the ESA proteins on 2-DE gels showed clearly that many proteins were present in multiple species, suggesting they are subjected to substantial post-translational modification (PTM), although the nature and significance of these PTMs remains to be elucidated (Zhou *et al.*, 2005). MudPIT proteomics analysis of the same secretory fraction revealed several additional proteins, including novel putative adhesive proteins, proteases and hypothetical secretory proteins similar to products expressed by other related parasites, including *Plasmodium* (Zhou *et al.*, 2005). As with the analysis of rhoptry proteins (Bradley *et al.*, 2005), Zhou and co-workers characterized a subset of these novel ESA proteins by localization experiments, this time by re-expressing them as fusions to yellow fluorescent protein. This screen revealed shared and distinct localizations within the secretory compartments of *T. gondii* tachyzoites, confirming the hypothesis that these proteins were indeed derived from these secretory organelles. This proteomic study has significantly broadened our understanding of the ESA proteins of *Toxoplasma* in a way that would have been difficult to achieve with classical techniques. Interestingly, only one of the 38 novel rhoptry proteins identified by Bradley and co-workers was found in the ESA fraction of Zhou and co-workers. This extremely small overlap between the two sub-proteomes is a demonstration of the purity of the respective sample preparations, and a testament to the importance of good sample preparation when undertaking sub-proteomic studies.

In an elegant experiment by Zhou and co-workers, a variety of proteomics techniques were used to study the proteolytic processing of the *T. gondii* transmembrane micronemal protein MIC2 and its partner M2AP, which form an adhesive complex required for host cell invasion (Zhou *et al.*, 2004). The MIC2/M2AP complex undergoes proteolytic processing on the parasite surface during invasion. In a three-stage proteomics experiment, MALDI intact mass measurements followed by enzymatic digestion and inhibitor profiling were used to demonstrate that the protein M2AP is processed by two proteases: MPP2 and MPP3. Second, differential protein expression patterns detected by 2-DE DIGE were used to define the substrate repertoire of MMP2. Finally, mass spectrometry was used to show that MIC2 is shed by membrane cleavage within its anchor domain (Zhou *et al.*, 2004). This study demonstrates the power and versatility of mass spectrometry analysis to help determine the function of proteins. Notable is the high degree of mass accuracy achievable by MS, which enables intact mass measurements to be made and interpreted in a meaningful way.

22.9 PROTEOMICS ANALYSIS OF MEMBRANE PROTEINS OF *TOXOPLASMA GONDII*

Che and co-workers have reported a comprehensive analysis of the *T. gondii* membrane proteome, where 2241 *T. gondii* proteins with at least one predicted transmembrane segment were identified and grouped into 841 sequentially non-redundant protein clusters, which account for 21.8% of the predicted transmembrane protein clusters in the *T. gondii* genome (Che *et al.* 2011). To maximize the identification of membrane proteins, advanced proteomics strategies were employed in addition to conventional 1-DE LC−MS/MS. In one approach, cell surface proteins were labelled with sulpho-NHS-SS-biotin and affinity purification was carried out by using streptavidin beads. Proteins collected were then identified by 1-DE LC−MS/MS. The other approach has utilized a recently developed gel electrophoresis system called three-layer 'sandwich' gel electrophoresis

(TLSGE) (Liu *et al.*, 2008). TLSGE is capable of concentrating relatively large volumes of protein samples on a small piece of protein gel and removing detergents and salts during electrophoresis. The identification is achieved by MudPIT. A large portion (42%) of the identified membrane proteins in this study was of unknown function, with many proteins being unique to *T. gondii* or to the *Apicomplexa* (Che *et al.*, 2011).

22.10 THE DYNAMIC PROTEOME OF *TOXOPLASMA GONDII*

So far in this chapter we have referred to the proteome of *Toxoplasma* as if it were a relatively stable entity, rather like its genome. All proteomes are in fact highly dynamic, and reflect the biological context of the cell at the time of analysis. Thus proteomics can be used to help understand important facets of the biology of *T. gondii*, such as the changes that occur from one life-stage to another and the analysis of gene knockout phenotypes and virulence factors, as well as for the study of the action of chemotherapeutic agents and the basis of drug resistance.

22.10.1 Proteomics Analysis of Protein Processing in *T. gondii*

Calcium-dependent signal transduction pathways are essential for micronemal release and motility during invasion (Beckers *et al.*, 1997; Blader *et al.*, 2001; Carruthers and Sibley, 1999; Del Carmen *et al.*, 2009) and have been studied using quantitative proteomics. In order to understand the role of these pathways, Nebl and co-workers used a range of proteomics methods to characterize calcium-dependent phosphorylation events during *Toxoplasma* host cell invasion (Nebl *et al.*, 2011). Live *T. gondii* tachyzoites were labelled with ^{32}[P] orthophosphate and stimulated with either ethanol (thought to activate PLC) or the calcium ionophore ionomycin. A 2-DE based screening was conducted to visualize Ca^{2+}-dependent phospho-substrates during *T. gondii* invasion. Digested peptides from ethanol-stimulated parasites were fractionated via hydrophilic interaction chromatography (HILIC) and partitioned *via* titanium dioxide (TiO_2) affinity chromatography to enrich phosphopeptides. MudPIT was used to identify 546 phosphorylation sites on 305 *T. gondii* phospho-proteins, including 10 sites on the actomyosin invasion motor. To further study the invasion motor, a SILAC based approach was carried out to quantitatively monitor the changes in the abundance and phosphorylation of the invasion motor complex upon Ca^{2+} pathway stimulation. Phosphorylation patterns on three of the motor components – GAP45, MLC1 and MyoA – were defined, where calcium-dependent phosphorylation of six residues across GAP45, MLC1 and MyoA is correlated with invasion motor activity. An unknown component of a 15-kDa calmodulin-like protein (TGME49_069440) that likely represents the MyoA Essential Light Chain of the *Toxoplasma* invasion motor was also identified. With its predicted calcium-binding ability, the identification of this motor component suggested that, in addition to phosphorylation, Ca^{2+} binding could directly regulate invasion motor activity.

Protein *N*-glycosylation plays an important role in protein folding, stability and secretion; conventionally it was considered a rare post-translational modification in apicomplexan parasites (Buxton and Innes, 1995; Carey *et al.*, 2004; Cleary *et al.*, 2005; Kimura *et al.*, 1996). Using affinity chromatography followed by mass spectrometry, proteomics analysis carried out by two groups has provided valuable insights into glycoproteome of *T. gondii.* Using MALDI–TOF, Fauquenoy and co-workers identified three major and four minor molecular ions from PNGase F-released glycans of *T. gondii* total protein extract, which suggested that the major *N*-glycans of *T. gondii* have compositions of oligomannosidic ($Man_{5-8}(GlcNAc)_2$) and paucimannosidic ($Man_{3-4}(GlcNAc)_2$) sugars, which are rarely present on mature eukaryotic glycoproteins (Fauquenoy *et al.*, 2008). To further characterize *N*-glycoproteins of *T. gondii*, protein sample was enriched by concanavalin A (ConA) lectin affinity chromatography and identified by LC–MS/MS. A total of 26 proteins were identified to be involved in gliding motility, moving junction and other additional functions implicated in intracellular development. The importance of *N*-glycoproteins was further verified by treating parasites with tunicamycin, an *N*-glycosylation inhibitor, which considerably reduced the motility, host cell invasion and growth of *T. gondii* (Fauquenoy *et al.*, 2008). The work of this study has led to a more detailed study on a component of the gliding motor complex TgGAP50 using a combined glycomic and proteomic approach. An unusual *N*-glycan structure that was rarely found on mature mammalian glycoproteins was identified by MALDI–TOF–MS analysis on PNGase F-digested TgGAP50 peptides (Cravatt and Sorensen, 2000). Using site-directed mutagenesis and affinity purification followed by 1-DE LC–MS/MS, *N*-glycosylation was also proved to be vital for TgGAP50 transport to inner complex membrane and efficient binding to other gliding partners (Cravatt and Sorensen, 2000).

Using lectin affinity chromatography followed by 1-DE LC–MS/MS, Luo and co-workers have identified 132 glycoproteins, including nine microneme proteins, seven dense granule proteins, 15 rhoptry proteins, 17 surface proteins, 19 enzymes, 11 heat shock proteins, 20 other proteins and 32 hypothetical proteins (Luo *et al.*, 2011). A simpler approach that uses trypsin to 'shave' surface proteins off *T. gondii* and tryptically digest the liberated proteins was also used in this study. Peptides acquired were purified by ConA lectin affinity chromatography and the PNGase F-released peptides were analysed by LC–MS/MS. A total of 30 proteins were identified, where all of them had *N*-linked glycoprotein domains and most of them were putative membrane proteins (Luo *et al.*, 2011). Lectin-fluorescence labelling and lectin blotting were also used to confirm the presence of carbohydrates on the surface or cytoplasm of *T. gondii* and the expression of selected hypothetical proteins were demonstrated by PCR (Luo *et al.*, 2011). Together, these glycoproteome studies have provided supporting evidence that glycosylation is not a rare post-translational modification, but occurs in a significant number of the proteins in *T. gondii*.

In addition to phosphorylation and glycosylation, SUMOylation in *T. gondii* was also examined by proteomics approaches. The small ubiquitin-like modifier (SUMO) is covalently linked to a variety of proteins and involved in important processes such as stress response, protein stability and progression through the cell cycle (Delorme *et al.*, 2003). Braun and co-workers have conducted a proteomics study to characterize the SUMO pathway in *T. gondii* (Braun *et al.*, 2009). Proteins extracted from transgenic *T. gondii* strain ectopically expressing HAFlag–TgSUMO were fractionated by FLAG-affinity chromatography. Peptides from the purified SUMOylated proteins were then analysed by 1-DE LC–MS/MS. A total of 120 putative candidate SUMOylated proteins were identified by at least two peptides. These proteins are involved in diverse cellular processes such as host cell invasion, pathogenesis and cyst development during stage conversion from tachyzoite to bradyzoite (Braun *et al.*, 2009).

Other proteomics studies have focused on PTMs of α- and β-tubulins, which form microtubule cytoskeleton that are important for host cell invasion (Morrissette and Sibley, 2002). In an early attempt, Plessmann and co-workers have studied the PTMs of α-tubulin of *T. gondii* (Fischer *et al.*, 1997). In this study, antibodies specific to tubulin PTMs were used to show that α-tubulin of *T. gondii* can be acetylated and detyrosinated. Carboxy-terminal peptides of α-tubulin were further analysed by

MALDI–MS and revealed that the C-terminus of α-tubulin can be truncated by five amino acids and that glutamate 445 can be subjected to polyglutamylation (Fischer *et al.*, 1997). Equipped with the much improved genome annotation of *T. gondii*, Xiao and co-workers have studied various PTMs of *T. gondii* α- and β-tubulins using both the proteomics approach and antibodies (Xiao *et al.*, 2010). Purified cytoskeleton proteins of *T. gondii* tachyzoites were separated by 2-DE and analysed by MALDI–MS or immunoblotting with PTM-specific antibodies. Two β-tubulin isotypes and one α-tubulin isotype were detected and some of the other isotypes that were not identified were thought to be stage specific or expressed in specific sub-cellular structures (Xiao *et al.*, 2010). The PTMs observed in the α-tubulin isotype included acetylation at Lys40, detyrosination, polyglutamylation, methylation and C-terminal truncation of the last five amino acids; whereas polyglutamylation and methylation were the only PTMs observed in β-tubulin. Notably, the identification of methylation, a PTM not previously detected on host cell (human foreskin fibroblasts) tubulins, may represent a specific modification in the *Apicomplexa*, which could provide a new target for therapeutic treatment (Xiao *et al.*, 2010).

22.11 PROTEOMICS AS A TOOL TO DISSECT THE HOST RESPONSE TO INFECTION

To understand the way in which the *T. gondii* proteome interacts with host cells, it is essential to know how the host responds to the parasite during and after invasion of the cell. Proteomics has been used to model changes in the soluble host cell proteome during and after *in vitro* invasion of cells by *T. gondii* using proteomics. In one experiment, conventional 2-DE and DIGE experiments were used to detect several hundred protein changes that occurred after invasion of host cells with *Toxoplasma* (Nelson *et al.*, 2008). After DIGE analysis of parasite-infected cells, differentially expressed proteins were identified by MS analysis and protein functions assigned for each, using protein function assignation tools such as the Bioinformatic Harvester program (http://www.harvester.embl.de) and those at the Human Protein Resource Database (http://www.hprd.org), in order to model pathway changes following parasite invasion. All of the proteins modulated in the study were assigned a functional classification based on a schema put forward by the Munich Information Centre for Protein Sequences (http://mips.gsf.de/projects/funcat). Classes of protein that were modulated included those associated with apoptosis, mitosis, glycolysis, lipid metabolism, nucleoside synthesis and the cytoskeleton. Changes in the host cell proteome were consistent with the arresting of both apoptosis and cell division. In addition, glycolysis was up-regulated, as was the synthesis of cytoskeletal proteins, presumably due to host cell remodelling. Of considerable note was the large number of host mitochondrial proteins that were changed in expression during infection, which accounted for almost one-third of all modulated proteins.

More recently, we have investigated the interactions between *T. gondii* and host cells at both proteome and transcriptome level on simultaneously collected samples. Quantitative label-free proteomics and RNASeq approaches have been carried out to characterize gene expressions of both host cells and various strains and life-cycle stages of *T. gondii.* On the proteomics side, about 1200 host proteins and over 500 *T. gondii* proteins can be quantified from a single batch of samples. This study permits direct comparison of transcript and protein abundance as well as parallel interrogation of host and pathogen responses. Results from statistical and functional bioinformatics analysis have highlighted interesting differences between proteome and transcriptome profiles as well as dynamic changes in host cell responses during infection and to different life-cycle stages of the parasite (Wastling, unpublished). The first batch of these quantitative data has been released to ToxoDB, which will be published in the following database versions.

22.12 DATABASE MANAGEMENT OF *TOXOPLASMA GONDII* PROTEOMICS DATA

With the development of robotics, high-throughput protein separation and MS analysis, even modest proteomics experiments have the capacity to generate vast quantities of data. Database management plays an important role in maximizing the accessibility and handling of proteomics data. Effective database management requires consideration of the following:

1. How to present the data in a summarized, yet informative way.
2. In what form to store raw data so that they can be re-queried at a later date and used by others who might wish to address a different set of biological questions.
3. How to integrate proteomics data with other genomics data.

The minimum information about a proteomics experiment (MIAPE) was developed by the Human Proteome Organization's Proteomics Standards Initiative (HUPO-PSI) to encourage the standardized collection, integration, storage and dissemination of proteomics data (Taylor *et al.*, 2007). This work has been extended to the development of data format standards, which facilitates the conversion from proprietary and open-source formats for downstream bioinformatics interpretation. Under the MIAPE umbrella, data format standards are being developed, such as mzML for capturing mass spectra (Godovac-Zimmermann *et al.*, 2005), PSI−MI for molecular interactions (Kerrien *et al.*, 2007), mzIdentML for peptide and proteins identification (Gastens and Fischer, 2002) and mzQuantML for quantitative proteomics data (Greenbaum *et al.*, 2000).

22.12.1 *T. gondii* Proteomics Data Management and Applications

Proteomics analysis of *T. gondii* was instrumental in providing some of the first data used in developing proteomics standards (Jones *et al.*, 2004), and it remains important that large-scale *Toxoplasma* proteomics experiments and data should attempt to adhere to these developing common formats. Whilst a proteomics experiment in isolation can yield valuable results, there is considerable potential for extending the value of those data to the wider community with the careful use of common standards. Moreover, the principle that raw data should always be stored alongside processed data (as with sequence and microarray data) is also an important one. Whilst raw data are unlikely to be accessed widely by the user-community, its retention preserves the possibility of re-querying the original data at a later time. Similarly, the algorithms for protein identification from MS data are constantly being modified and improved, making it sometimes desirable retrospectively to analyse MS data. In common with microarray experiments, the cost and effort involved in performing experiments justifies careful data storage and archiving, as it is rarely desirable (or possible) to repeat the same work. Large-scale proteomics data for *T. gondii* should thus be viewed as a community resource, with community-wide access, to encourage further exploitation of the data.

For proteomics data to be of benefit to the wider community there must be an easy route of entry for researchers whose primary interest may not be proteomics but for whom proteomics data might be of value. As with genome and expression data, the most obvious portal for this is ToxoDB (http://www.toxodb.org/) (Gajria *et al.*, 2008). Proteomics data for *Toxoplasma* have their greatest value in the context

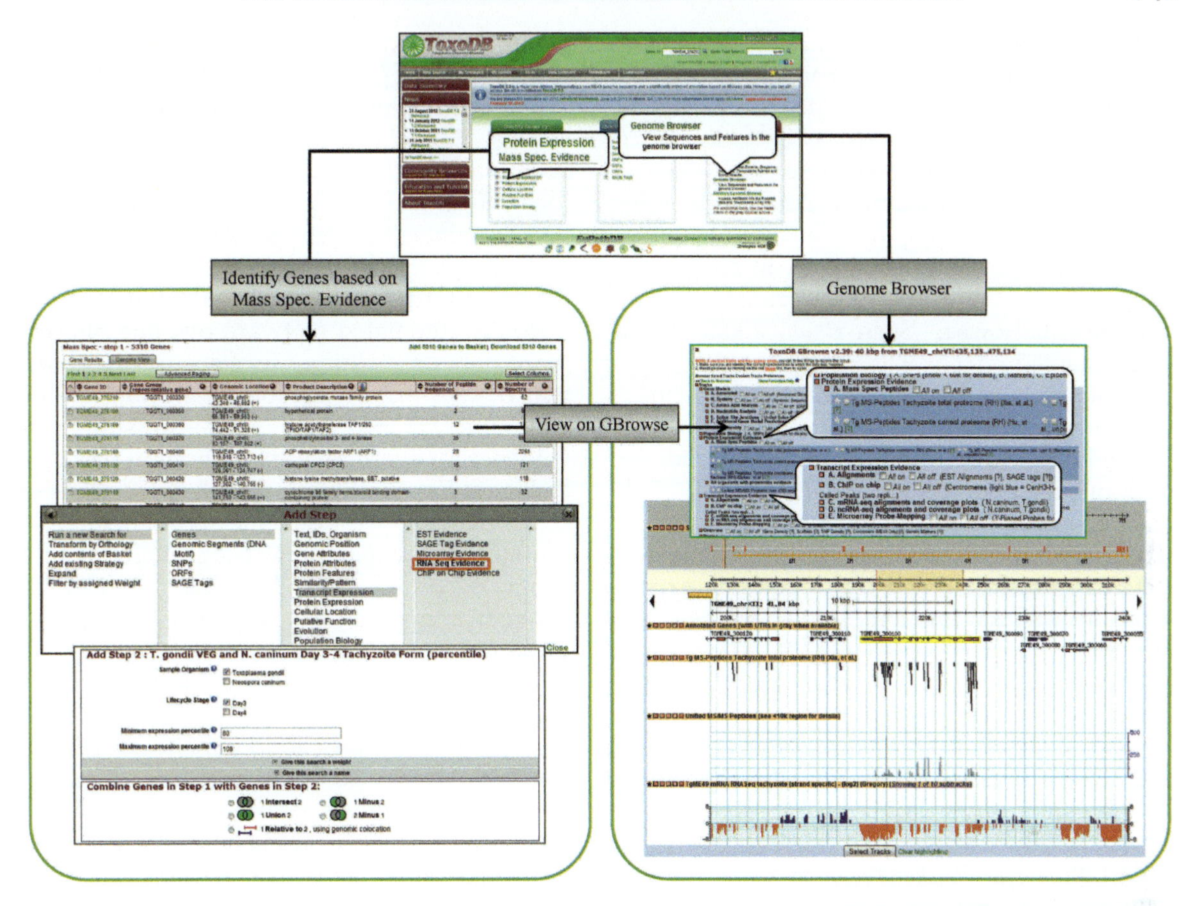

FIGURE 22.4 **Proteomics informatics research on ToxoDB.** Workflows of two important tools for proteomics informatics research are shown. Protein expression evidence can be queried using the 'Identify Genes based on Mass Spec. Evidence' tool according to the experiments and samples. In the result page, columns with various protein properties can be added and sorted. The 'Add Step' tool can be used to compare the results with other genome information. The example shown is the comparison of proteomics results with one of the RNA sequencing datasets (RNA Seq Evidence). 'Genome Browser' is the other useful tool which can be accessed from the ToxoDB front page. Genomic landmark or gene identifiers can be used to define a genomic region of interest, where various genome information (named 'tracks') can be selected, aligned and customized in the browser. Individual proteins from 'Identify Genes based on Mass Spec. Evidence' results can also be viewed directly in Genome Browser from the result page. The example shown is proteomics evidence of gene TGME49_300100 aligned with RNASeq evidence acquired.

of other genome and expression data, where they still have considerable potential for informing gene finding and annotation. Proteomics data are now an established feature of ToxoDB, where protein expression data are integrated with other genomic features of the *T. gondii* genome (Xia *et al.*, 2008). To date, proteomics data from nine individual studies have been hosted on ToxoDB (version 8.0), which provide expression data for 5366 proteins from ME49 strain and 4488 proteins from GT1 strain, representing around two thirds of the predicted proteome. On ToxoDB, proteomics data are shown on individual gene record pages together with other genome and expression data. Proteomics data can also be queried in a batch mode according to the experiments and samples using the 'Identify Genes based on Mass Spec. Evidence' tool. Data analysis can be carried out to interactively compare proteomics data with other genome information, functional annotations and other types of '-omics' data. In addition to these text based tools, the Generic Genome Browser (GBrowse) (Stein *et al.*, 2002) is also implemented on ToxoDB to improve visualization of data mining, where expressed peptides can be viewed in relation to various gene models and other types of '-omics' data in the predicted genomic region of the peptide (Fig. 22.4).

The integration of proteomics data on to ToxoDB has allowed a direct comparison between proteomic and transcriptomic expressions. It has been previously noted in the whole proteomics study carried out by Xia and co-workers that apparent discrepancies have been observed between gene transcription and protein expression (Xia *et al.*, 2008). Wastling and co-workers have carried out a systematic examination of the link between transcription and translation in key apicomplexan parasites *Plasmodium* spp., *T. gondii, C. parvum, Neospora caninum* and *Theileria* spp. (Wastling *et al.*, 2009). Proteomics data with available transcriptomics data such as microarrays, EST collections, serial analysis of gene expression (SAGE) and massively parallel signature sequencing (MPSS) tags have been compared within each species and orthologue analysis has been carried out to compare between species. The study has highlighted significant discrepancies between measures of transcription and protein expression. A large-scale study of *T. gondii* gene expression comparing RNASeq data and quantitative proteomics data is currently under way, which will provide a more in-depth view of gene expression regulations in *T. gondii* in a quantitative fashion (Wastling, unpublished).

As observed in two of the whole proteome studies, proteomics data provide important supporting evidence for gene model annotation as well as suggesting alternative models that can be used to increase the accuracy of the annotations, named proteogenomics (Dybas *et al.*, 2008; Xia *et al.*, 2008). Combining this great idea of improving genome annotation with vast amounts of proteomics data hosted on ToxoDB, an automated proteogenomic pipeline was developed. Krishna and co-workers have developed an automated software pipeline, the ProteoAnnotator (http://proteoannotator.org/), which uses MS data for confirming official gene models, but also examines supporting evidence for alternative annotations at particular loci and for identifying novel genes (Krishna *et al.*, 2011). The pipeline is integrated to EuPathDB (the parent website of ToxoDB) to facilitate proteomics data submission and genome annotation for a number of apicomplexan parasites.

Another proteogenomics project which was inspired by the vast amount of proteomics data has resulted in the development of EPICDB (Madrid-Aliste *et al.*, 2009). EPICDB is a publicly accessible proteomics database that hosts experimentally and computationally generated protein data for *T. gondii* and *C. parvum*. In an effort to reduce the redundancy while retain the unique features from various computationally predicted gene models, the database is structured around clusters of protein sequences. Besides proteomics data, EPICDB also hosts aligned expressed sequence tags (ESTs), open reading frames (ORFs) and antibody experiments, which provide an integrated platform to examine *T. gondii* protein expression against all available protein sequences (Madrid-Aliste *et al.*, 2009).

Proteomics evidence has also been used to assist the metabolic reconstruction of *T. gondii* and other apicomplexan parasites, in which genes are systematically assigned to functions within metabolic pathways and networks. Manually curated biochemical and physiological evidences available in the literature as well as proteomics evidence and gene annotations available on EuPathDB was used and the integrated metabolic pathway database is publically accessible at (Liverpool) Library of Apicomplexan Metabolic Pathways (LAMP) (http://www.llamp.net) (Shanmugasundram *et al.*, 2012). Using this approach, the corresponding genes for several essential processes of the pathways were identified to be absent from the current genome annotation. Integration from EuPathDB gene models to the pathways in LAMP are currently under development which will be available in forthcoming releases.

22.13 CONCLUSION AND PERSPECTIVES

Toxoplasma has been increasingly at the forefront of proteomics and this trend is set to continue with wider accessibility of this rapidly developing technology to researchers. However, major challenges lie ahead if proteomics studies are indeed to revolutionize our understanding of *Toxoplasma* biology. If we are really to use proteomics to establish the link between the genome and its function, then proteomics must move forward to provide more detailed and complete functional data on the proteins that it describes. The key to this lies partly in being better able to profile the activity of proteins, for example, through high-throughput analysis of PTMs and protein–protein interactions. These goals are not trivial. In the case of PTM analysis, for instance, it is not just the presence or absence of PTMs such as phosphorylation that needs to be identified; the degree of phosphorylation and the location of phosphorylation sites for proteins must be determined – all this in the context of an exceptionally transient and labile system. Advances in both relative and absolute protein quantitation also need to be applied to the study of *Toxoplasma* if proteomics data are to be used more effectively with other post-genomic technologies, such as transcriptional analysis and metabolomics in a systems biology approach. Finally, just as mRNA analysis cannot tell us everything about the function of genes, proteins too cannot be understood entirely in isolation. Sugars and lipids are also important cell components, and their large-scale analysis through the emerging fields of glycomics and lipidomics is now a possibility. Proteomics for *Toxoplasma* must develop in the context of these nascent technologies. Perhaps the most certain aspect of the coming years is that the post-genomics picture is going to get more complex before it gets simpler, thus placing bioinformatics at the forefront of all future developments.

Acknowledgements

We are indebted to members of the Wastling laboratory for helpful comments on this manuscript. Jonathan M. Wastling is supported by funding from the BBSRC and the Wellcome Trust.

References

Aebersold, R., Mann, M., 2003. Mass spectrometry-based proteomics. Nature 422 (6928), 198–207. http://dx.doi.org/10.1038/nature01511. nature01511 [pii].

Alexander, D.L., Mital, J., Ward, G.E., Bradley, P., Boothroyd, J.C., 2005. Identification of the moving junction complex of Toxoplasma gondii: a collaboration between distinct secretory organelles. PLoS Pathog. 1 (2), e17. http://dx.doi.org/10.1371/journal.ppat.0010017.

Beckers, C.J., Dubremetz, J.F., Mercereau-Puijalon, O., Joiner, K.A., 1994. The Toxoplasma gondii rhoptry protein ROP 2 is inserted into the parasitophorous vacuole membrane, surrounding the intracellular parasite, and is exposed to the host cell cytoplasm. J. Cell Biol. 127 (4), 947–961.

Beckers, C.J., Wakefield, T., Joiner, K.A., 1997. The expression of Toxoplasma proteins in Neospora caninum and the identification of a gene encoding a novel rhoptry protein. Mol. Biochem. Parasitol. 89 (2), 209–223. doi: S0166-6851(97)00120-5 [pii].

Bertsch, A., Jung, S., Zerck, A., Pfeifer, N., Nahnsen, S., Henneges, C., Nordheim, A., Kohlbacher, O., 2010. Optimal de novo design of MRM experiments for rapid assay development in targeted proteomics. J. Proteome Res. 9 (5), 2696–2704. http://dx.doi.org/10.1021/pr1001803.

Beynon, R.J., Doherty, M.K., Pratt, J.M., Gaskell, S.J., 2005. Multiplexed absolute quantification in proteomics using artificial QCAT proteins of concatenated signature peptides. Nat. Methods 2 (8), 587–589. doi: nmeth774 [pii], 10.1038/nmeth774.

Blader, I.J., Manger, I.D., Boothroyd, J.C., 2001. Microarray analysis reveals previously unknown changes in Toxoplasma gondii-infected human cells. J. Biol. Chem. 276 (26), 24223–24231. http://dx.doi.org/10.1074/jbc.M100951200. M100951200 [pii].

Bradley, P.J., Boothroyd, J.C., 1999. Identification of the pro-mature processing site of Toxoplasma ROP1 by mass spectrometry. Mol. Biochem. Parasitol. 100 (1), 103–109. doi: S0166-6851(99)00035-3 [pii].

Bradley, P.J., Ward, C., Cheng, S.J., Alexander, D.L., Coller, S., Coombs, G.H., Dunn, J.D., Ferguson, D.J., Sanderson, S.J., Wastling, J.M., Boothroyd, J.C., 2005. Proteomic analysis of rhoptry organelles reveals many novel constituents for host-parasite interactions in Toxoplasma gondii. J. Biol. Chem. 280 (40), 34245–34258. doi: M504158200 [pii], 10.1074/jbc.M504158200.

Bradshaw, R.A., Burlingame, A.L., 2005. From proteins to proteomics. IUBMB Life 57 (4-5), 267–272. doi: T45563QK41N42R26 [pii], 10.1080/15216540500091536.

Braun, L., Cannella, D., Pinheiro, A.M., Kieffer, S., Belrhali, H., Garin, J., Hakimi, M.A., 2009. The small ubiquitin-like modifier (SUMO)-conjugating system of Toxoplasma gondii. Int. J. Parasitol. 39 (1), 81–90. doi: S0020-7519(08)00277-4 [pii], 10.1016/j.ijpara.2008.07.009.

Buxton, D., Innes, E.A., 1995. A commercial vaccine for ovine toxoplasmosis. Parasitology 110 Suppl, S11–16.

Carey, K.L., Westwood, N.J., Mitchison, T.J., Ward, G.E., 2004. A small-molecule approach to studying invasive mechanisms of Toxoplasma gondii. Proc. Natl. Acad. Sci. U S A 101 (19), 7433–7438. http://dx.doi.org/10.1073/pnas.0307769101, 0307769101 [pii].

Carruthers, V.B., Sibley, L.D., 1999. Mobilization of intracellular calcium stimulates microneme discharge in Toxoplasma gondii. Mol. Microbiol. 31 (2), 421–428.

Che, F.Y., Madrid-Aliste, C., Burd, B., Zhang, H., Nieves, E., Kim, K., Fiser, A., Angeletti, R.H., Weiss, L.M., 2011. Comprehensive proteomic analysis of membrane proteins in Toxoplasma gondii. Mol. Cell Proteomics 10 (1). M110 000745. doi: M110.000745 [pii], 10.1074/mcp.M110.000745.

Cleary, M.D., Meiering, C.D., Jan, E., Guymon, R., Boothroyd, J.C., 2005. Biosynthetic labeling of RNA with uracil phosphoribosyltransferase allows cell-specific microarray analysis of mRNA synthesis and decay. Nat. Biotechnol. 23 (2), 232–237. doi: nbt1061 [pii], 10.1038/nbt1061.

Cleary, M.D., Singh, U., Blader, I.J., Brewer, J.L., Boothroyd, J.C., 2002. Toxoplasma gondii asexual development: identification of developmentally regulated genes and distinct patterns of gene expression. Eukaryot. Cell 1 (3), 329–340.

Cohen, A.M., Rumpel, K., Coombs, G.H., Wastling, J.M., 2002. Characterisation of global protein expression by two-dimensional electrophoresis and mass spectrometry: proteomics of Toxoplasma gondii. Int. J. Parasitol. 32 (1), 39–51. doi: S0020751901003083 [pii].

Cox, J., Mann, M., 2008. MaxQuant enables high peptide identification rates, individualized p.p.b.-range mass accuracies and proteome-wide protein quantification. Nat. Biotechnol. 26 (12), 1367–1372. doi: nbt.1511 [pii], 10.1038/nbt.1511.

Cox, J., Mann, M., 2011. Quantitative, high-resolution proteomics for data-driven systems biology. Annu. Rev. Biochem. 80, 273–299. http://dx.doi.org/10.1146/annurev-biochem-061308-093216.

Cravatt, B.F., Sorensen, E.J., 2000. Chemical strategies for the global analysis of protein function. Curr. Opin. Chem. Biol. 4 (6), 663–668. doi: S1367-5931(00)00147-2 [pii].

Del Carmen, M.G., Mondragon, M., Gonzalez, S., Mondragon, R., 2009. Induction and regulation of conoid extrusion in Toxoplasma gondii. Cell. Microbiol. 11 (6), 967–982. doi: CMI1304 [pii], 10.1111/j.1462-5822.2009.01304.x.

Delorme, V., Cayla, X., Faure, G., Garcia, A., Tardieux, I., 2003. Actin dynamics is controlled by a casein kinase II and phosphatase 2C interplay on Toxoplasma gondii Toxofilin. Mol. Biol. Cell. 14 (5), 1900–1912. http://dx.doi.org/10.1091/mbc.E02-08-0462. E02-08-0462 [pii].

Dhingra, V., Gupta, M., Andacht, T., Fu, Z.F., 2005. New frontiers in proteomics research: a perspective. Int. J. Pharm. 299 (1-2), 1–18. doi: S0378-5173(05)00226-7 [pii], 10.1016/j.ijpharm.2005.04.010.

Dybas, J.M., Madrid-Aliste, C.J., Che, F.Y., Nieves, E., Rykunov, D., Angeletti, R.H., Weiss, L.M., Kim, K., Fiser, A., 2008. Computational analysis and experimental validation of gene predictions in Toxoplasma gondii. PLoS One 3 (12), e3899. http://dx.doi.org/10.1371/journal.pone.0003899.

Fauquenoy, S., Morelle, W., Hovasse, A., Bednarczyk, A., Slomianny, C., Schaeffer, C., Van Dorsselaer, A., Tomavo, S., 2008. Proteomics and glycomics analyses of N-glycosylated structures involved in Toxoplasma gondii–host cell interactions. Mol. Cell Proteomics 7 (5), 891–910. doi: M700391-MCP200 [pii], 10.1074/mcp.M700391-MCP200.

Fischer, H.G., Nitzgen, B., Reichmann, G., Gross, U., Hadding, U., 1997. Host cells of Toxoplasma gondii encystation in infected primary culture from mouse brain. Parasitol. Res. 83 (7), 637–641.

Florens, L., Washburn, M.P., Raine, J.D., Anthony, R.M., Grainger, M., Haynes, J.D., Moch, J.K., Muster, N., Sacci, J.B., Tabb, D.L., Witney, A.A., Wolters, D., Wu, Y., Gardner, M.J., Holder, A.A., Sinden, R.E., Yates, J.R., Carucci, D.J., 2002. A proteomic view of the Plasmodium falciparum life cycle. Nature 419 (6906), 520–526. http://dx.doi.org/10.1038/nature01107. nature01107 [pii].

Fritz, H.M., Bowyer, P.W., Bogyo, M., Conrad, P.A., Boothroyd, J.C., 2012. Proteomic analysis of fractionated Toxoplasma oocysts reveals clues to their environmental resistance. PLoS One 7 (1), e29955. http://dx.doi.org/10.1371/journal.pone.0029955. PONE-D-11–20508 [pii].

Gajria, B., Bahl, A., Brestelli, J., Dommer, J., Fischer, S., Gao, X., Heiges, M., Iodice, J., Kissinger, J.C., Mackey, A.J., Pinney, D.F., Roos, D.S., Stoeckert, C.J., Jr., Wang, H., Brunk, B.P., 2008. ToxoDB: an integrated Toxoplasma gondii database resource. Nucleic Acids Res. 36 (Database issue), D553–556. doi: gkm981 [pii], 10.1093/nar/gkm981.

Gastens, M.H., Fischer, H.G., 2002. Toxoplasma gondii eukaryotic translation initiation factor 4A associated with tachyzoite virulence is down-regulated in the bradyzoite stage. Int. J. Parasitol. 32 (10), 1225–1234. doi: S0020751902000966 [pii].

Gerber, S.A., Rush, J., Stemman, O., Kirschner, M.W., Gygi, S.P., 2003. Absolute quantification of proteins and phosphoproteins from cell lysates by tandem MS. Proc. Natl. Acad. Sci. U S A 100 (12), 6940–6945. http://dx.doi.org/10.1073/pnas.0832254100, 0832254100 [pii].

Godovac-Zimmermann, J., Kleiner, O., Brown, L.R., Drukier, A.K., 2005. Perspectives in spicing up proteomics with splicing. Proteomics 5 (3), 699–709. http://dx.doi.org/10.1002/pmic.200401051.

Gorg, A., Weiss, W., Dunn, M.J., 2004. Current two-dimensional electrophoresis technology for proteomics. Proteomics 4 (12), 3665–3685. http://dx.doi.org/10.1002/pmic.200401031.

Greenbaum, D., Medzihradszky, K.F., Burlingame, A., Bogyo, M., 2000. Epoxide electrophiles as activity-dependent cysteine protease profiling and discovery tools. Chem. Biol. 7 (8), 569–581. doi: S1074-5521(00)00014-4 [pii].

Gygi, S.P., Rist, B., Gerber, S.A., Turecek, F., Gelb, M.H., Aebersold, R., 1999. Quantitative analysis of complex protein mixtures using isotope-coded affinity tags. Nat. Biotech. 17 (10), 994–999.

Ishihama, Y., Oda, Y., Tabata, T., Sato, T., Nagasu, T., Rappsilber, J., Mann, M., 2005. Exponentially modified protein abundance index (emPAI) for estimation of absolute protein amount in proteomics by the number of sequenced peptides per protein. Mol. Cell Proteomics 4 (9), 1265–1272. doi: M500061-MCP200 [pii], 10.1074/mcp.M500061-MCP200.

Johnson, J.R., Florens, L., Carucci, D.J., Yates J.R., 3rd, 2004. Proteomics in malaria. J. Proteome Res. 3 (2), 296–306.

Jones, A., Hunt, E., Wastling, J.M., Pizarro, A., Stoeckert C.J., Jr., 2004. An object model and database for functional genomics. Bioinformatics 20 (10), 1583–1590. http://dx.doi.org/10.1093/bioinformatics/bth130. bth130 [pii].

Karasov, A.O., Boothroyd, J.C., Arrizabalaga, G., 2005. Identification and disruption of a rhoptry-localized homologue of sodium hydrogen exchangers in Toxoplasma gondii. Int. J. Parasitol. 35 (3), 285–291. doi: S0020-7519(04)00282-6 [pii], 10.1016/j.ijpara.2004.11.015.

Kerrien, S., Orchard, S., Montecchi-Palazzi, L., Aranda, B., Quinn, A.F., Vinod, N., Bader, G.D., Xenarios, I., Wojcik, J., Sherman, D., Tyers, M., Salama, J.J., Moore, S., Ceol, A., Chatr-Aryamontri, A., Oesterheld, M., Stumpflen, V., Salwinski, L., Nerothin, J., Cerami, E., Cusick, M.E., Vidal, M., Gilson, M., Armstrong, J., Woollard, P., Hogue, C., Eisenberg, D., Cesareni, G., Apweiler, R., Hermjakob, H., 2007. Broadening the horizon – level 2.5 of the HUPO-PSI format for molecular interactions. BMC Biol. 5, 44. doi: 1741-7007-5-44[pii], 10.1186/1741-7007-5-44.

Kimura, E.A., Couto, A.S., Peres, V.J., Casal, O.L., Katzin, A.M., 1996. N-linked glycoproteins are related to schizogony of the intraerythrocytic stage in Plasmodium falciparum. J. Biol. Chem. 271 (24), 14452–14461.

Krishna, R., Wastling, J.M., Jones, A.R., 2011. Automated integration of mass spectrometry based proteomics evidence for improvement of gene annotations. In: 8th BSPR-EBI Meeting-From the visible to the hidden proteome. Wellcome Trust Conference Centre, Cambridge, UK.

Lane, C.S., 2005. Mass spectrometry-based proteomics in the life sciences. Cell Mol. Life Sci. 62 (7-8), 848–869. http://dx.doi.org/10.1007/s00018-005-5006-6.

Lange, V., Picotti, P., Domon, B., Aebersold, R., 2008. Selected reaction monitoring for quantitative proteomics: a tutorial. Mol. Syst. Biol. 4, 222. doi: msb200861 [pii], 10.1038/msb.2008.61.

Lasonder, E., Ishihama, Y., Andersen, J.S., Vermunt, A.M., Pain, A., Sauerwein, R.W., Eling, W.M., Hall, N., Waters, A.P., Stunnenberg, H.G., Mann, M., 2002.

Analysis of the Plasmodium falciparum proteome by high-accuracy mass spectrometry. Nature 419 (6906), 537–542. http://dx.doi.org/10.1038/nature01111. nature01111 [pii].

Lebrun, M., Michelin, A., El Hajj, H., Poncet, J., Bradley, P.J., Vial, H., Dubremetz, J.F., 2005. The rhoptry neck protein RON4 re-localizes at the moving junction during Toxoplasma gondii invasion. Cell Microbiol. 7 (12), 1823–1833. doi: CMI646 [pii], 10.1111/j.1462-5822.2005.00646.x.

Leriche, M.A., Dubremetz, J.F., 1991. Characterization of the protein contents of rhoptries and dense granules of Toxoplasma gondii tachyzoites by subcellular fractionation and monoclonal antibodies. Mol. Biochem. Parasitol. 45 (2), 249–259.

Liu, T., Martin, A.M., Sinai, A.P., Lynn, B.C., 2008. Three-layer sandwich gel electrophoresis: a method of salt removal and protein concentration in proteome analysis. J. Proteome Res. 7 (10), 4256–4265. http://dx.doi.org/10.1021/pr800182b.

Lu, P., Vogel, C., Wang, R., Yao, X., Marcotte, E.M., 2007. Absolute protein expression profiling estimates the relative contributions of transcriptional and translational regulation. Nat. Biotechnol. 25 (1), 117–124. doi: nbt1270 [pii], 10.1038/nbt1270.

Luo, Q., Upadhya, R., Zhang, H., Madrid-Aliste, C., Nieves, E., Kim, K., Angeletti, R.H., Weiss, L.M., 2011. Analysis of the glycoproteome of Toxoplasma gondii using lectin affinity chromatography and tandem mass spectrometry. Microbes Infect. 13 (14-15), 1199–1210. doi: S1286-4579(11)00226-7 [pii], 10.1016/j.micinf.2011.08.013.

Madrid-Aliste, C.J., Dybas, J.M., Angeletti, R.H., Weiss, L.M., Kim, K., Simon, I., Fiser, A., 2009. EPIC-DB: a proteomics database for studying Apicomplexan organisms. BMC Genomics 10, 38. doi: 1471-2164-10-38 [pii], 10.1186/1471-2164-10-38.

Malmstrom, L., Malmstrom, J., Aebersold, R., 2011. Quantitative proteomics of microbes: Principles and applications to virulence. Proteomics 11 (15), 2947–2956. http://dx.doi.org/10.1002/pmic.201100088.

Matrajt, M., Donald, R.G., Singh, U., Roos, D.S., 2002. Identification and characterization of differentiation mutants in the protozoan parasite Toxoplasma gondii. Mol. Microbiol. 44 (3), 735–747. doi: 2904 [pii].

Miller, S.A., Thathy, V., Ajioka, J.W., Blackman, M.J., Kim, K., 2003. TgSUB2 is a Toxoplasma gondii rhoptry organelle processing proteinase. Mol. Microbiol. 49 (4), 883–894. doi: 3604 [pii].

Morrison, R.S., Kinoshita, Y., Johnson, M.D., Conrads, T.P., 2003. Proteomics in the postgenomic age. Adv. Protein Chem. 65, 1–23.

Morrissette, N.S., Sibley, L.D., 2002. Cytoskeleton of apicomplexan parasites. Microbiol. Mol. Biol. Rev. 66 (1), 21–38. table of contents.

Nagaraj, N., Wisniewski, J.R., Geiger, T., Cox, J., Kircher, M., Kelso, J., Paabo, S., Mann, M., 2011. Deep proteome and transcriptome mapping of a human cancer cell line. Mol. Syst. Biol. 7, 548. doi: msb201181 [pii], 10.1038/msb.2011.81.

Nebl, T., Prieto, J.H., Kapp, E., Smith, B.J., Williams, M.J., Yates, J.R., Cowman, A.F., Tonkin, C.J., 2011. Quantitative in vivo Analyses Reveal Calcium-dependent Phosphorylation Sites and Identifies a Novel Component of the Toxoplasma Invasion Motor Complex. PLoS Pathogens 7, e1002222. http://dx.doi.org/10.1371/journal.ppat.1002222.

Nelson, M.M., Jones, A.R., Carmen, J.C., Sinai, A.P., Burchmore, R., Wastling, J.M., 2008. Modulation of the host cell proteome by the intracellular apicomplexan parasite Toxoplasma gondii. Infect. Immun. 76 (2), 828–844. doi: IAI.01115-07 [pii], 10.1128/IAI.01115-07.

Nischik, N., Schade, B., Dytnerska, K., Dlugonska, H., Reichmann, G., Fischer, H.G., 2001. Attenuation of mouse-virulent Toxoplasma gondii parasites is associated with a decrease in interleukin-12-inducing tachyzoite activity and reduced expression of actin, catalase and excretory proteins. Microbes Infect. 3 (9), 689–699. doi: S1286-4579(01)01425-3 [pii].

Ong, S.-E., Blagoev, B., Kratchmarova, I., Kristensen, D.B., Steen, H., Pandey, A., Mann, M., 2002. Stable Isotope Labeling by Amino Acids in Cell Culture, SILAC, as a Simple and Accurate Approach to Expression Proteomics. Mol. Cell Proteomics 1 (5), 376–386. http://dx.doi.org/10.1074/mcp.M200025-MCP200.

Ossorio, P.N., Schwartzman, J.D., Boothroyd, J.C., 1992. A Toxoplasma gondii rhoptry protein associated with host cell penetration has unusual charge asymmetry. Mol. Biochem. Parasitol. 50 (1), 1–15. doi: 0166-6851(92)90239-G [pii].

Paba, J., Ricart, C.A., Fontes, W., Santana, J.M., Teixeira, A.R., Marchese, J., Williamson, B., Hunt, T., Karger, B.L., Sousa, M.V., 2004. Proteomic analysis of Trypanosoma cruzi developmental stages using isotope-coded affinity tag reagents. J. Proteome Res. 3 (3), 517–524.

Palagi, P.M., Walther, D., Quadroni, M., Catherinet, S., Burgess, J., Zimmermann-Ivol, C.G., Sanchez, J.C., Binz, P.A., Hochstrasser, D.F., Appel, R.D., 2005. MSight: an image analysis software for liquid chromatography-mass spectrometry. Proteomics 5 (9), 2381–2384. http://dx.doi.org/10.1002/pmic.200401244.

Phillips, C.I., Bogyo, M., 2005. Proteomics meets microbiology: technical advances in the global mapping of protein expression and function. Cell Microbiol. 7 (8), 1061–1076. doi: CMI554 [pii], 10.1111/j.1462-5822.2005.00554.x.

Que, X., Ngo, H., Lawton, J., Gray, M., Liu, Q., Engel, J., Brinen, L., Ghosh, P., Joiner, K.A., Reed, S.L., 2002. The cathepsin B of Toxoplasma gondii, toxopain-1, is critical for parasite invasion and rhoptry protein processing. J. Biol. Chem. 277 (28), 25791–25797. http://dx.doi.org/10.1074/jbc.M202659200. M202659200 [pii].

Radke, J.R., Behnke, M.S., Mackey, A.J., Radke, J.B., Roos, D.S., White, M.W., 2005. The transcriptome of Toxoplasma gondii. BMC Biol. 3, 26. doi: 1741-7007-3-26 [pii], 10.1186/1741-7007-3-26.

Reichmann, G., Dlugonska, H., Fischer, H.G., 2002. Characterization of TgROP9 (p36), a novel rhoptry protein of Toxoplasma gondii tachyzoites identified by T cell clone. Mol. Biochem. Parasitol. 119 (1), 43–54. doi: S0166685101003978 [pii].

Reid, A.J., Vermont, S.J., Cotton, J.A., Harris, D., Hill-Cawthorne, G.A., Konen-Waisman, S., Latham, S.M., Mourier, T., Norton, R., Quail, M.A., Sanders, M., Shanmugam, D., Sohal, A., Wasmuth, J.D., Brunk, B., Grigg, M.E., Howard, J.C., Parkinson, J., Roos, D.S., Trees, A.J., Berriman, M., Pain, A., Wastling, J.M., 2012. Comparative genomics of the apicomplexan parasites Toxoplasma gondii and Neospora caninum: Coccidia differing in host range and transmission strategy. PLoS Pathog. 8 (3), e1002567. http://dx.doi.org/10.1371/journal.ppat.1002567. PPATHOGENS-D-11–02326 [pii].

Ross, P.L., Huang, Y.N., Marchese, J.N., Williamson, B., Parker, K., Hattan, S., Khainovski, N., Pillai, S., Dey, S., Daniels, S., Purkayastha, S., Juhasz, P., Martin, S., Bartlet-Jones, M., He, F., Jacobson, A., Pappin, D.J., 2004. Multiplexed Protein Quantitation in Saccharomyces cerevisiae Using Amine-reactive Isobaric Tagging Reagents. Molecular and Cellular Proteomics 3 (12), 1154–1169. http://dx.doi.org/10.1074/mcp.M400129-MCP200.

Sadak, A., Taghy, Z., Fortier, B., Dubremetz, J.F., 1988. Characterization of a family of rhoptry proteins of Toxoplasma gondii. Mol. Biochem. Parasitol. 29 (2-3), 203–211.

Sanderson, S.J., Xia, D., Prieto, H., Yates, J., Heiges, M., Kissinger, J.C., Bromley, E., Lal, K., Sinden, R.E., Tomley, F., Wastling, J.M., 2008. Determining the protein repertoire of Cryptosporidium parvum sporozoites. Proteomics 8 (7), 1398–1414. http://dx.doi.org/10.1002/pmic.200700804.

Schwanhausser, B., Busse, D., Li, N., Dittmar, G., Schuchhardt, J., Wolf, J., Chen, W., Selbach, M., 2011. Global quantification of mammalian gene expression control. Nature 473 (7347), 337–342. doi: nature10098 [pii], 10.1038/nature10098.

Shanmugasundram, A., Gonzalez-Galarza, F.F., Wastling, J.M., Vasieva, O., Jones, A.R., 2012. Library of Apicomplexan Metabolic Pathways: a manually curated database for metabolic pathways of apicomplexan parasites. Nucleic Acids Res. 41 (D1), D706–713. doi: gks1139 [pii], 10.1093/nar/gks1139.

Sibley, L.D., Mordue, D.G., Su, C., Robben, P.M., Howe, D.K., 2002. Genetic approaches to studying virulence and pathogenesis in Toxoplasma gondii. Philos. Trans. R Soc. Lond. B Biol. Sci. 357 (1417), 81–88. http://dx.doi.org/10.1098/rstb.2001.1017.

Singh, U., Brewer, J.L., Boothroyd, J.C., 2002. Genetic analysis of tachyzoite to bradyzoite differentiation mutants in Toxoplasma gondii reveals a hierarchy of gene induction. Mol. Microbiol. 44 (3), 721–733. doi: 2903 [pii].

Stein, L.D., Mungall, C., Shu, S., Caudy, M., Mangone, M., Day, A., Nickerson, E., Stajich, J.E., Harris, T.W., Arva, A., Lewis, S., 2002. The generic genome browser: a building block for a model organism system database. Genome Res. 12 (10), 1599–1610. http://dx.doi.org/10.1101/gr.403602.

Sturm, M., Bertsch, A., Gropl, C., Hildebrandt, A., Hussong, R., Lange, E., Pfeifer, N., Schulz-Trieglaff, O., Zerck, A., Reinert, K., Kohlbacher, O., 2008.

OpenMS – an open-source software framework for mass spectrometry. BMC Bioinformatics 9, 163. doi: 1471-2105-9-163 [pii], 10.1186/1471-2105-9-163.

Taylor, C.F., Paton, N.W., Lilley, K.S., Binz, P.A., Julian R.K., Jr., Jones, A.R., Zhu, W., Apweiler, R., Aebersold, R., Deutsch, E.W., Dunn, M.J., Heck, A.J., Leitner, A., Macht, M., Mann, M., Martens, L., Neubert, T.A., Patterson, S.D., Ping, P., Seymour, S.L., Souda, P., Tsugita, A., Vandekerckhove, J., Vondriska, T.M., Whitelegge, J.P., Wilkins, M.R., Xenarios, I., Yates J.R., 3rd, Hermjakob, H., 2007. The minimum information about a proteomics experiment (MIAPE). Nat. Biotechnol. 25 (8), 887–893. doi: nbt1329 [pii], 10.1038/nbt1329.

Unlu, M., Morgan, M.E., Minden, J.S., 1997. Difference gel electrophoresis: a single gel method for detecting changes in protein extracts. Electrophoresis 18 (11), 2071–2077. http://dx.doi.org/10.1002/elps.1150181133.

Wastling, J.M., Armstrong, S.D., Krishna, R., Xia, D., 2012. Parasites, proteomes and systems: has Descartes' clock run out of time? Parasitology 139 (9), 1103–1118. doi: S0031182012000716 [pii], 10.1017/S0031182012000716.

Wastling, J.M., Bradley, P.J., 2007. Proteomic analysis of the rhoptry organelles of Toxoplasma gondii. In: Ajioka, J.W., Soldati, D. (Eds.), Toxoplasma: molecular and cellular biology. Horizon Bioscience, Norfolk, England.

Wastling, J.M., Xia, D., Sohal, A., Chaussepied, M., Pain, A., Langsley, G., 2009. Proteomes and transcriptomes of the Apicomplexa – where's the message? Int. J. Parasitol. 39 (2), 135–143. doi: S0020-7519(08)00389-5 [pii], 10.1016/j.ijpara.2008.10.003.

Xia, D., Sanderson, S.J., Jones, A.R., Prieto, J.H., Yates, J.R., Bromley, E., Tomley, F.M., Lal, K., Sinden, R.E., Brunk, B.P., Roos, D.S., Wastling, J.M., 2008. The proteome of Toxoplasma gondii: integration with the genome provides novel insights into gene expression and annotation. Genome Biol. 9 (7), R116. doi: gb-2008-9-7-r116 [pii], 10.1186/gb-2008-9-7-r116.

Xiao, H., El Bissati, K., Verdier-Pinard, P., Burd, B., Zhang, H., Kim, K., Fiser, A., Angeletti, R.H., Weiss, L.M., 2010. Post-translational modifications to *Toxoplasma gondii* alpha- and beta-tubulins include novel C-terminal methylation. J. Proteome Res. 9 (1), 359–372.

Yates J.R., 3rd, 2004. Mass spectral analysis in proteomics. Annu. Rev. Biophys. Biomol. Struct. 33, 297–316. http://dx.doi.org/10.1146/annurev.biophys.33.111502.082538.

Yates J.R., 3rd, Gilchrist, A., Howell, K.E., Bergeron, J.J., 2005. Proteomics of organelles and large cellular structures. Nat. Rev. Mol. Cell Biol. 6 (9), 702–714. http://dx.doi.org/10.1038/nrm1711.

Zhou, X.W., Blackman, M.J., Howell, S.A., Carruthers, V.B., 2004. Proteomic analysis of cleavage events reveals a dynamic two-step mechanism for proteolysis of a key parasite adhesive complex. Mol. Cell Proteomics 3 (6), 565–576. http://dx.doi.org/10.1074/mcp.M300123-MCP200. M300123-MCP200 [pii].

Zhou, X.W., Kafsack, B.F., Cole, R.N., Beckett, P., Shen, R.F., Carruthers, V.B., 2005. The opportunistic pathogen Toxoplasma gondii deploys a diverse legion of invasion and survival proteins. J. Biol. Chem. 280 (40), 34233–34244. doi: M504160200 [pii], 10.1074/jbc.M504160200.

CHAPTER

23

Cerebral Toxoplasmosis: Pathogenesis, Host Resistance and Behavioural Consequences

Yasuhiro Suzuki, Qila Sa*, Eri Ochiai*, Jeremi Mullins*, Robert Yolken†, Sandra K. Halonen***

***Department of Microbiology, Immunology and Molecular Genetics, University of Kentucky College of Medicine, Lexington, Kentucky, USA †Stanley Neurology Laboratory, Johns Hopkins University, Baltimore, Maryland, USA **Department of Microbiology, Montana State University, Bozeman, Montana, USA**

OUTLINE

Toxoplasma gondii, second edition
http://dx.doi.org/10.1016/B978-0-12-396481-6.00023-4

23.1 INTRODUCTION

Toxoplasma gondii is a ubiquitous, obligate intracellular protozoan parasite in humans and animals. Chronic infection with this parasite is likely one of the most common infections of humans, affecting 10% to 25% of the world's population. During the acute stage of infection, tachyzoites quickly proliferate within a variety of nucleated cells and spread throughout host tissues. Interferon-gamma (IFNγ)-dependent cell-mediated immune responses, and humoural immune responses to a lesser extent, control the tachzyoite proliferation (reviewed in Chapters 24 and 25), but the parasite forms cysts in various organs, especially the brain, heart and skeletal muscle, and establishes a chronic infection. Acute acquired infection in immunocompetent individuals is usually unnoticed or causes a benign, self-limiting illness, and results in a chronic infection. Although such chronic infection has been considered 'latent', recent studies indicated a correlation of chronic *T. gondii* infection with cryptogenic epilepsy and schizophrenia. Another important aspect of chronic infection with this parasite is an occurrence of reactivation of the infection in immunocompromised individuals such as those with AIDS, organ transplants or cancer. The reactivation of chronic *T. gondii* infection is initiated by disruption of cysts, followed by proliferation of tachyzoites which causes life-threatening toxoplasmic encephalitis (TE). Since TE occurs almost solely in immunocompromised individuals, it is clear that the immune response to *T. gondii* is crucial to prevent the disease. When an acute acquired infection occurs in pregnant women who have never been infected with the parasite before, the parasite can pass through the placenta and infect the foetus. The brain is the major organ affected by the congenital infection, as well.

Murine models have mainly been used to analyse the mechanisms of the protective immunity to acute acquired infection and development of TE during the later stage of infection. Multiple types of cells are involved as producers of IFNγ in resistance to infection, and a variety of cell types participate as effector cells that become activated by IFNγ to control the parasite. Additionally, multiple cell types also contribute as producers of IL-12, which is required for induction of IFNγ production.

23.2 PRODUCERS OF INTERLEUKIN (IL)-12 REQUIRED FOR IFNγ PRODUCTION

23.2.1 Dendritic Cells

IL-12 is the most important inducer of IFNγ synthesis during the acute stage of infection. Neutralization of IL-12 with anti-IL-12 antibodies results in 100% mortality in mice following infection with an avirulent strain of *T. gondii*, and the mortality is associated with decreased IFNγ production (Gazzinelli *et al.*, 1994b). Dendritic cells were identified to be the source of IL-12 in the spleen in response to *T. gondii* (Reis e Sousa *et al.*, 1997). All IL-12 positive cells in spleens of *T. gondii*-stimulated mice were found in T-cell areas and were $CD8\alpha^{+}$, $CD11c^{+}$, $DEC205^{+}$ dendritic cells. CCR5 expressed on the surface of dendritic cells is responsible for their migration into splenic T-cell areas following stimulation (Reis e Sousa *et al.*, 1997). CCR5 signalling also plays an important role in activation of dendritic cells to produce IL-12 (Aliberti *et al.*, 2000). CCR5-deficient mice are impaired in IL-12 production by dendritic cells and are highly susceptible to *T. gondii* infection. Cyclophilin-18 was identified as the principal molecule derived from the parasite that triggers IL-12 production through CCR5 (Aliberti *et al.*, 2003). More recently, a profilin-like protein of the parasite was found to bind to Toll-like receptor 11 and stimulate IL-12 production by dendritic cells (Yarovinsky *et al.*, 2005). Mice deficient in the gene encoding MyD88, an adaptor molecule essential for most TLR as well as IL-1 and IL-18 signalling, are susceptible to acute infection with an avirulent strain, and their susceptibility is associated with impaired IL-12 responses to the parasite (Scanga *et al.*, 2002). The binding of IL-12 to its receptor leads to the activation of signal transducer and activator of transcription (STAT) 4, and this signalling cascade is crucial for IFNγ producing cells. In agreement with this, STAT4--deficient mice are susceptible to acute infection with *T. gondii*, and their mortality is associated with a defect in the ability to produce IFNγ in response to infection (Cai *et al.*, 2000). It was shown that lack of MyD88 in dendritic cells, but not in macrophages or neutrophils, resulted in high susceptibility to acute infection with *T. gondii* (Hou *et al.*, 2011).

During the chronic stage of infection, cells bearing the dendritic cell markers such as CD11c and 33D1 are located at inflammatory sites in the brains of mice (Fischer *et al.*, 2000b). These brain dendritic cells are mature as indicated by high-level expression of MHC class II molecules, CD40, CD54, CD80 and CD86, and are able to trigger antigen-specific T-cell responses *in vitro*. The dendritic cells were revealed to be the major producers of IL-12 among mononuclear cells isolated from brains of infected animals (Fischer *et al.*, 2000b). GM–CSF is suggested to be important for induction of the dendritic cells in primary brain cell cultures with *T. gondii* (Fischer *et al.*, 2000b). Since IL-12 is important for the maintenance of IFNγ production in T-cells mediating resistance to chronic infection (Yap *et al.*, 2000), dendritic cells in the brains of the infected mice might play a role in maintaining IFNγ production by T-cells in this organ.

Dendritic cells also play a role in disseminating the parasite into the brain. Following oral infection with *T. gondii*, tachyzoites are preferentially detected in $CD11c^{+}$ cells in the lamina propria on day 2 and in the mesenteric lymph nodes from days 3 to 7 of infection (Courret *et al.*, 2006), suggesting that infected $CD11c{+}CD11b^{+/-}$ dendritic cells (DC) are a carrier of the parasite, disseminating the infection from the lamina propria to mesenteric lymph nodes. Tachyzoite-infected dendritic cells exhibit hypermotility *in vitro* (Lambert *et al.*, 2006). When infected DC are injected intravenously into uninfected mice, the parasite can be detected in the brains of the recipients (Lambert *et al.*, 2006; Courret *et al.*, 2006).

23.2.2 Macrophages

Macrophages produce IL-12 in response to tachyzoites or tachyzoite antigens *in vitro* (Gazzinelli *et al.*, 1994b, 1996). *T. gondii* has three predominant genotypes; types I, II and III. Infection of murine macrophages with tachyzoites of a type II (avirulent to mice) strain *in vitro* resulted in a greater production of IL-12 than infection with tachyzoites of a type I (virulent to mice) strain (Robben *et al.*, 2004). Therefore, the lower IL-12 production by macrophages may contribute to the acute virulence of type I parasite to mice. Kim *et al.* (Kim *et al.*, 2006) reported that type I tachyzoites induce IL-12 production through MyD88-independent mechanisms, whereas type II tachyzoites do so by both MyD88-dependent and -independent mechanisms. Since macrophages infiltrate into the brains of mice following infection with *T. gondii* (Suzuki *et al.*, 2005; Wilson *et al.*, 2005), these cells, in addition to dendritic cells, may be an important source of IL-12 in resistance against the parasite in the brain.

As with dendritic cells, monocytes/macrophages appear to play a pathogenic role as well by assisting dissemination of the parasite during the acute stage of infection. Macrophages (Da Gama *et al.*, 2004) have been shown to effectively disseminate tachyzoites into lymph nodes (Da Gama *et al.*, 2004) in mice. In addition, $CD11c^{-}CD11b^{+}$ monocytes are the major cell population that contains tachyzoites in the blood (Courret *et al.*, 2006), suggesting an importance of this cell population in disseminating the infection to various organs, including the brain. In support of this possibility, at one day after an intravenous injection of $CD11b^{+}$ blood cells from infected mice into uninfected animals, the parasite is detectable in mononuclear cells obtained from the brains of the recipient animals (Courret *et al.*, 2006). This is in contrast to an intravenous injection of a small number of free tachyzoites, in which the parasite was not detected one day later. Treatment of infected mice with anti-CD11b mAb at six and seven days after infection resulted in 50% reduction in the number of the parasite in their brains detected at eight days after infection, suggesting an involvement of CD11b integrin in parasite dissemination to the brain.

23.2.3 Neutrophils

Neutrophils rapidly infiltrate into the peritoneal cavity of mice following intraperitoneal infection with *T. gondii*. Approximately 85% of the neutrophils displayed intracellular storage of IL-12 (Bliss *et al.*, 2000). Depletion of neutrophils during the first six days of infection resulted in increased mortality in mice in association with decreased production of IL-12 and IFNγ by splenocytes (Bliss *et al.*, 2001). Rapid infiltration of neutrophils into the site of infection appears to play an important role in induction of the protective Th1-type immune responses against the parasite during the early stage of infection. The production of IL-12 by neutrophils in response to soluble tachyzoite antigens is in strict dependence upon functional MyD88 (Freund *et al.*, 2001). JUNK2 mitogen-associated protein kinase was recently shown to be required for *T. gondii*-induced neutrophil IL-12 production (Aviles *et al.*, 2008). However, it is unknown whether neutrophils are involved in the resistance in the brain to control *T. gondii* during the chronic stage of infection.

23.3 PRODUCERS OF IFNγ

23.3.1 Involvement of 'Innate Immunity'

23.3.1.1 Microglia and Blood-Derived Macrophages

Alpha–beta T-cells are essential to control *T. gondii* in both acute and chronic stages of infection (see Section 23.3.2). However, we found that, in addition to T-cells, IFNγ production by cells other than T-cells is required for prevention of reactivation of *T. gondii* infection (TE) in the brain in chronically infected mice (Kang and Suzuki, 2001). Athymic nude, SCID and IFNγ-deficient mice were infected with *T. gondii* and treated with sulphadiazine to establish chronic infection. After discontinuation of sulphadiazine, each of these animals developed severe TE due to reactivation of the chronic infection. When the animals received adoptive transfer of immune spleen or T-cells before discontinuation of sulphadiazine, infected athymic nude and SCID mice did not develop TE and survived. However, infected IFNγ deficient mice still developed TE even after receiving cell transfers (Kang and Suzuki, 2001). Before cell transfer, IFNγ mRNA was detected in brains of the nude and SCID mice but not in brains of the IFNγ deficient mice. IFNγ mRNA was also detected in brains of infected SCID mice depleted of NK-cells, and such animals did not develop TE after receiving immune T-cells (Kang and Suzuki, 2001). Thus, IFNγ production by non-T-cells, in addition to T-cells, is required for prevention of reactivation of *T. gondii* infection in the brain. The IFNγ producing non-T-cells do not appear to be NK-cells.

In regard to the identity of the non-T-, non-NK-cells that produce IFNγ in the brains of nude and SCID mice chronically infected with *T. gondii*, intracellular staining for IFNγ followed by flow cytometry revealed that approximately 45%–60% of the cells expressing IFNγ in their brains were positive for CD11b or F4/80 (markers for microglia/macrophages) on their surfaces (Suzuki *et al.*, 2005). Smaller portions of the cells were positive for pan-NK marker. Further smaller portions were positive for CD11c (a marker for dendritic cells), and these cells were less than 5% of the IFNγ expressing cells in brains of infected SCID mice. Large amounts of mRNA for IFNγ were detected in $CD11b^+$ cells purified from brains of infected mice, but it was not the case in the cells obtained from uninfected animals. In infected SCID mice depleted of NK-cells by treatment with anti-asialo–GM1 antibody, cells expressing IFNγ in their brains were all positive for CD11b, and the IFNγ producing cells were detected in both $CD45^{low}$ and $CD45^{high}$ populations. These results suggest that $CD11b^+$ $CD45^{low}$ microglia and $CD11b^+$ $CD45^{high}$ blood-derived macrophages are the major non-T-, non-NK-cells which express IFNγ in the brains of mice infected with *T. gondii*. Therefore, it is possible that IFNγ production by microglia and/or macrophages plays an important role in prevention of TE in collaboration with αβ T-cells.

23.3.1.2 Gamma Delta (γδ) T-Cells

During the acute stage of infection with *T. gondii*, increased numbers of T-cells expressing the γδ T-cell receptor have been observed in the spleen and peritoneal cavity of mice and the peripheral blood of humans. Γδ T-cells are cytotoxic to infected target cells and produce IFNγ and TNFα in response to the parasite *in vitro*. Involvement of γδ T-cells in resistance against acute infection with *T. gondii* has been shown in mice. Mice deficient in γδ T-cells due to treatment with anti-TCR γδ mAb (Hisaeda *et al.*, 1995) or lack of the functional TCR δ gene (Kasper *et al.*, 1996) die earlier than control mice during the acute stage of infection.

Γδ T-cells may also be involved in prevention of TE during the late stage of infection, as γδ T-cells are detectable in brains of chronically infected mice and rats. Of interest, the relative percentages of γδ T-cells in lymphocyte preparations isolated from brains of infected mice are significantly higher than in their spleens (Suzuki *et al.*, 1997). This suggests that γδ T-cells preferentially infiltrate into the brain of *T. gondii*-infected mice. In humans, a marked increase in γδ T-cells was observed in the peripheral blood of a patient with CD40L defect (hyper-IgM syndrome) who had developed TE (Leiva *et al.*, 1998). The patient responded well to anti-toxoplasmic chemotherapy and to high dose immunoglobulin replacement therapy. Therefore, γδ T-cells may have contributed to controlling the disease under the chemotherapy although their protective activity is not sufficient by itself to prevent development of TE. Lepage *et al.* (Lepage *et al.*, 1998) suggested a possible role for γδ T-cells to enhance the protective activity of $CD8^+$ αβ T-cells in their studies using adoptive transfer of the lymphocyte populations.

23.3.1.3 NK-cells

NK-cells are an important source of IFNγ in resistance against *T. gondii* during the early stage of infection. Depletion of NK-cells results in early or increased mortality in SCID and wild-type mice. In contrast to the early stage of infection, NK-cells do not appear to be crucial for prevention of TE during the late stage of infection. Depletion of NK-cells in SCID mice, which had received adoptive transfer of immune T-cells, did not abolish resistance of the recipient animals against development of TE (Kang and Suzuki, 2001). In the depleted mice, NK-cells were undetectable by flow cytometry in their brains and spleens.

23.3.2 Importance of 'Acquired Immunity' Involving $CD4^+$ and $CD8^+$αβ T-Cells

It is clear that αβ T-cells are essential for resistance against *T. gondii* since athymic nude and SCID mice, which lack T-cells, succumb to acute infection and their mortality is associated with proliferation of large numbers of tachyzoites in various organs, including the brain. In resistance to acute infection in general, $CD8^+$ T-cells are the major efferent limb of the protective cellular immunity although $CD4^+$ T-cells are also involved. The protective activity of the T-cells is predominantly mediated by IFNγ.

IFNγ also plays a critical role in prevention of TE during the late stage of infection in mice (Suzuki *et al.*, 1989; Gazzinelli *et al.*, 1992). Neutralization of the activity of IFNγ in chronically infected mice by treatment with anti-IFNγ monoclonal antibody (mAb) resulted in severe acute inflammation and development of large areas of necrosis in their brains (Suzuki *et al.*, 1989). In the areas of acute inflammation and necrosis, tachyzoites and *T. gondii* antigens were detected, indicating that such inflammatory responses were caused by proliferation of tachyzoites. A marked increase in numbers of tachyzoites in brains of mice following treatment with anti-IFNγ mAb was also demonstrated by detecting increased amounts of tachyzoite-specific SAG1 and SAG2 mRNA in their brains by the reverse transcriptase–polymerase chain reaction (RT–PCR) (Gazzinelli *et al.*, 1993). Thus, it is clear that IFNγ is critical for prevention of proliferation of tachyzoites in the brains of mice. The same appears to be true in humans, since AIDS patients have an impaired ability to produce IFNγ and they frequently develop TE.

Both $CD4^+$ and $CD8^+$ T-cells infiltrate the brain of mice following infection, and the T-cells are the main source of IFNγ (Schluter *et al.*, 1995; Suzuki *et al.*, 1997; Hunter *et al.*, 1994). $CD8^+$ T-cells are known to be the major mediator of resistance, and this is consistent with evidence that the $H\text{-}2L^d$, a MHC Class I gene, confers resistance to TE in mice (Suzuki *et al.*, 1994; Brown *et al.*, 1995). The protective activity of the T-cells is through their production of IFNγ (Wang *et al.*, 2004). T-cells bearing T-cell receptor Vβ8 are the most abundant population that produces IFNγ in the brains of infected BALB/c mice genetically resistant to TE (Wang *et al.*, 2005), and adoptive transfer of $V\beta8^+$ T-cells alone into infected nude mice prevents the development of TE (Wang *et al.*, 2005; Kang *et al.*, 2003). When immune $V\beta8^+$ T-cells were divided into $CD4^+$ and $CD8^+$ T-cell populations, the $CD8^+$ population conferred much greater resistance to development of TE than did the $CD4^+$ population (Wang *et al.*, 2005). The protective activity of total $V\beta8^+$ T-cells was greater than that of $CD8^+V\beta8^+$ T-cells (Wang *et al.*, 2005). Therefore, the $CD8^+$ population plays a major role in the activity of $V\beta8^+$ immune T-cells against reactivation of infection in the brain although the $CD4^+$ population works additively or synergistically with the $CD8^+$ population. An importance of $CD4^+$ T-cells for optimum IFNγ production by cerebral $CD8^+$ T-cells was also shown in CB6 (BALB/c × C57BL/6) mice infected with *T. gondii* (Chan *et al.*, 2006).

The mechanisms by which $CD4^+$ T-cells enhance IFNγ production and protective activity of $CD8^+$ T-cells are unclear at this moment. One possible mechanism is that IL-4 produced by $CD4^+$ T-cells upregulates IL-12 production of dendritic cells. IL-4 alone or together with IFNγ efficiently enhances the production of bioactive IL-12 in mouse and human DC (Hochrein *et al.*, 2000) (see Section 23.4.3 for additional information). Since IL-12 is important for the maintenance of IFNγ production in T-cells mediating resistance to chronic infection (Yap *et al.*, 2000), it is possible that IL-4 produced by $CD4^+$ T-cells is involved in the activity of this T-cell population to enhance IFNγ production by $CD8^+$ T-cells during the chronic stage of *T. gondii* infection. This possibility is supported by evidence that STAT6, a molecule involved in IL-4 signalling, is important for activation of $CD8^+$ T-cells and their IFNγ production (Jin *et al.*, 2009).

In regard to the protective activity of $CD4^+$ T-cells, it was previously shown that adoptive transfer of $CD4^+$ immune T-cells conferred a partial protection against reactivation of infection in the recipient athymic nude mice whereas the same number of a total population of immune T-cells completely prevented the reactivation of infection (Kang and Suzuki, 2001). Therefore, a large number of $CD4^+$ immune T-cells alone could confer a protection against reactivation of *T. gondii* infection. However, it is unclear how long the protective effect lasts in the absence of $CD8^+$ T-cells. The presence of both T-cell subsets is critical for long-term maintenance of the latency of the chronic infection in the brain.

Interactions of $CD8^+$ T-cells with brain cells during *T. gondii* infection were recently visualized by the use of *T. gondii* transfected to express ovalbumin (OVA) and OT-1 $CD8^+$ T-cells specific to OVA peptide in conjunction with two-photon microscopy of living brain tissue. The study showed that the antigen-specific cerebral $CD8^+$ T-cells make transient contacts with granuloma-like structures containing parasites and with individual $CD11b^+$ antigen-presenting cells (Schaeffer *et al.*, 2009). Another study showed that movement of brain infiltrating OT-1 T-cells is closely associated with an infection-induced reticular system of fibres (Wilson *et al.*, 2009). In the study, seven to 14 days after a transfer of OT-1 cells into infected mice, a reduction in parasite burden in the brains of the recipients occurred (Wilson *et al.*, 2009). However, the parasite burden gradually increased thereafter in association with an increase in PD-1 expression in the transferred OT-1 cells, suggesting that T-cells recruited to the brain during *T. gondii* infection down-regulate their ability to act as effector cells over time.

A number of signalling molecules have been shown to be important for induction and/or maintenance of the protective T-cell responses during infection with *T. gondii.* In T-cell receptor signalling in response to the parasite, mice deficient in two Tec kinases, Rlk and Itk, had increased mortality associated with increased brain cyst numbers and decreased IFNγ production by splenocytes following *in vitro* stimulation with a low dose of *T. gondii* antigens (Schaeffer *et al.*, 1999). Protein kinase C-theta (PKC-θ) is another enzyme involved in the signalling in the T-cell response. Infection of mice deficient in this enzyme ($Pkc\theta^{-/-}$) resulted in impaired production of IFNγ in both $CD4^+$ and $CD8^+$ T-cells, and the animals succumbed to necrotizing TE (Nishanth *et al.*, 2010). The impaired IFNγ production by T-cells in infected $Pkc\theta^{-/-}$ mice is associated with decreased activation of transcription factors including nuclear factor (NF)κB, AP-1, and MAPK pathways. Tumour progression locus 2 (Tpl2), a serine–threonine kinase, and T-cell expression of myeloid differentiation factor 88 (MyD88) are also important for antigen-specific IFNγ production by T-cells after infection (LaRosa *et al*, 2008; Walford *et al*, 2008). However, the mechanisms by which Tpl2 and MyD88 mediate the T-cell response remain to be determined.

In regard to NFκB family transcription factors, RelB, c-Rel, NFκB1 and NFκB2 are all involved in regulating T-cell responses to *T. gondii* infection (Caamano *et al.*, 1999, 2000; Mason *et al.*, 2004; Harris *et al.*, 2010). RelB-deficient ($relB^{-/-}$) mice succumb to death after infection with *T. gondii* and T-cells from $relB^{-/-}$ mice are defective in production of IFNγ when stimulated with CD3 antibody *in vitro* (Caamano *et al*, 1999). Infected NFκB1-deficient mice developed TE associated with a local decrease in the number of $CD8^+$ T-cells and IFNγ production (Harris *et al.*, 2010). A transfer of naive T-cells from the deficient animals to SCID mice conferred less protection against infection than the T-cells from wild-type animals. NFκB2-deficient mice have no defect in their ability to produce IL-12 and IFNγ during the acute stage of infection. However, during the chronic stage of infection, the deficient mice succumbed to TE in association with a reduced capacity of production of IFNγ by splenocytes. Apoptosis of T-cells appears to be involved in the reduced production of this cytokine (Caamano *et al*, 2000). C-Rel-deficient mice also survive the acute phase of infection but develop severe TE associated with decreased numbers of $CD4^+$ T-cells and reduced production of IFNγ in their brains during the later stage of infection (Mason *et al.*, 2004). Bcl-3, a distinct member of the I-κB family, which functions as a positive regulator of nuclear factor NFκB activity, also plays a critical role in mounting a protective Th1 immune response to *T. gondii* (Franzoso *et al.*, 1997). Bcl-3-deficient mice succumb to *T. gondii* infection due to the lack of a protective Th1 response.

Recent studies determined multiple *T. gondii* epitopes recognized by $CD8^+$ T-cells. $H\text{-}2^d$-restricted epitopes of GRA6, GRA4, ROP7, SAG1 and SAG3 of *T. gondii* have recently been identified in mice (Blanchard *et al.*, 2008; Caetano *et al.*, 2006; Frickel *et al.*, 2008; Kirak *et al.*, 2010). The reactivity of $CD8^+$ T-cells to GRA4, GRA6 and ROP7 peaked 2, 4 and 6–8 weeks after infection (Blanchard *et al.*, 2008; Frickel *et al.*, 2008) and these changes would probably reflect changes in antigens available in association with conversion of *T. gondii* from the tachyzoite to the cyst stage during the course of infection. The GRA6 epitope, HF10, has a potent activity to stimulate IFNγ production by $CD8^+$ T-cells obtained from infected mice with $H\text{-}2^d$ haplotype, and an immunization with this epitope peptide prevented mortality of B10.D2 (H-2^d), but not C57BL/6 (H-2^b) mice after challenge infection (Blanchard *et al.*, 2008). Recently, an H-2^b-restricted epitope of tgd057 was also identified (Wilson *et al.*, 2010; Kirak *et al.*, 2010). Tgd057-specific CD8$^+$ T-cells obtained from ES-cloned mice following somatic cell nuclear transfer of individual nuclei from tgd057−tetramer$^+$ CD8$^+$ T-cells into ES cells also mediated a significant protection to lethal challenge infection when transferred into recipient C57BL/6 mice (Kirak *et al.*, 2010).

A *T. gondii* epitope recognized by CD4$^+$ T-cells of C57BL/6 mice has also been identified. This 15-mer epitope AS15 is derived from a *T. gondii* unknown protein CD4Ag28m, and presented by the H-2A^b molecule (Grover *et al.*, 2012). An immunization of C57BL/6 mice with this peptide results in lower parasite burden in the brain of infected mice (Grover *et al.*, 2012).

In humans, HLA-A02, HLA-A03 and HLA-B07 supertype-restricted CD8$^+$ T-cell epitopes have been identified by screening predicted epitope peptides of *T. gondii* antigens for their activity to induce IFNγ production by peripheral mononuclear cells from *T. gondii* seropositive individuals. Identified HLA-A02-restricted epitopes are those of SAG2C, SAG2D, SAG2X, SAG3, GRA6, GRA7, MIC1, MICA2P and SPA. The HLA-A03 supertype-restricted epitopes are those of SAG1, SAG2C, GRA6, GRA7 and SPA and the HLA-B07-restricted epitopes are those of GRA3 and GRA7 (Cong *et al*, 2010, 2011; Tan 2010). In addition, an immunization of transgenic mice expressing these human HLA Class I molecules (HLA-A0201 or HLA-A1101 (an HLA-A03 supertype allele)) with those identified epitope peptides conferred a protection associated with reduced parasite burden against challenge infection (Cong *et al.*, 2010, 2011). Therefore, these epitopes appear to be promising candidates for human vaccine to induce the protective immune responses against *T. gondii* infection.

23.4 OTHER CYTOKINES AND REGULATORY MOLECULES FOR RESISTANCE

23.4.1 TNFα

Murine peritoneal macrophages become activated after treatment with a combination of IFNγ and TNFα *in vitro* and the activated cells inhibit intracellular replication of tachyzoites through generation of NO by inducible NO synthase (NOS2) (Adams *et al.*, 1990b). However, it was demonstrated that TNFα and NOS2 are not essential for controlling acute infection *in vivo* since mice lacking TNF receptor type 1 (R1) and type 2 (R2) and those lacking NOS2 control parasite growth in the peritoneal cavity following intraperitoneal infection (Yap *et al.*, 1998; Deckert-Schluter *et al.*, 1998; Scharton-Kersten *et al.*, 1997b). Thus, the protective mechanism(s) which require IFNγ but do not require TNFα or NOS2 is sufficient for mice to control parasite growth during the acute stage of the infection. Recently, IGTP and LRG-47, members of a new family of IFNγ inducible genes, were shown to be required for the IFNγ-mediated resistance against acute infection with *T. gondii* (see Section 23.6).

In contrast to the acute stage of the infection, mice deficient in TNF R1/R2 or NOS2 succumbed to necrotizing TE during the late stage of the infection (Yap *et al.*, 1998; Deckert-Schluter *et al.*, 1998; Scharton-Kersten *et al.*, 1997b). Their results are consistent with those of earlier studies; treatments of infected wild-type mice with anti-TNFα mAb or aminoguanididine, an NOS2 inhibitor, resulted in development of TE (Gazzinelli *et al.*, 1993; Hayashi *et al.*, 1996a). Thus, TNFα and NOS2 are critical for prevention of proliferation of tachyzoites in the brain. As mentioned earlier in this section, IFNγ plays the central role in resistance of the brain against this parasite (Suzuki *et al.*, 1989; Gazzinelli *et al.*, 1992). Since neutralization of IFNγ or TNFα results in decreased NOS2 expression and development of severe TE (Gazzinelli *et al.*, 1993), activation of NOS2 mediated by INFγ and TNFα appears to play a key role in prevention of TE. Microglia and astrocytes are likely the effector cells involved in this protective mechanism (see Section 23.7). More recently, Collazo *et al.* (Collazo *et al.*, 2002) reported an involvement of IGTP in the IFNγ mediated protection against TE.

In regard to induction of NOS2 in the brain, Yap *et al.* (Yap *et al.*, 1998) reported that NOS2 induction in the brain was unimpaired in infected TNF R1/R2-deficient mice which are susceptible to TE, suggesting that TNF-dependent immune control of *T. gondii* expansion in the brain involves an effector function distinct from NOS2 activation. More recently, Deckert-Schluter *et al.* (Deckert-Schluter *et al.*, 1998) reported that mice lacking TNF R1 but not those lacking TNF R2 developed necrotizing encephalitis following infection and that a remarkable reduction of NOS2 synthesis was observed in the brains of TNF R1-deficient animals as compared with TNF R2-deficient or control animals. They concluded that signalling through TNF R1, but not TNF R2, provides the stimulus required for the induction of NOS2 activation in the brain following infection (Deckert-Schluter *et al.*, 1998). Thus, it appears that there are two pathways to activate NOS2 in the brain of *T. gondii*-infected mice, one TNF-dependent and the other TNF-independent. Since different strains of *T. gondii* were employed in the studies mentioned above (Deckert-Schluter *et al.*, 1998; Yap *et al.*, 1998), the strain of the parasite may be an important factor affecting the activation pathway for iNOS.

In regard to the role of TNFα and NOS2 in prevention of TE, Suzuki *et al.* (Suzuki *et al.*, 2000a) reported that IFNγ deficient mice infected and treated with sulphadiazine developed severe TE after discontinuation of sulphadiazine treatment, although these animals expressed equivalent amounts of mRNA for TNFα and iNOS in their brains when compared to control animals. These results indicate that expression of TNFα and NOS2 in the absence of IFNγ is insufficient for genetic resistance of BALB/c mice against TE.

23.4.2 Lymphotoxin

Lymphotoxin (LT), in addition to TNFα, is the ligand of TNFR1. Schluter *et al.* (Schluter *et al.*, 2003) reported that mice deficient in LT fail to control intracerebral *T. gondii* and succumbed to necrotizing TE following infection. IFNγ expression in their brains was equivalent to those of control mice when the deficient animals had developed TE. Experiments with bone marrow chimera mice showed that hematopoietic cells need to express both LT and TNFα to control *T. gondii* in the brain.

23.4.3 IL-4

$CD4^+$ T-cells are known to be heterogeneous (Tfh, Th1, Th2, Th17 and Treg) with regard to their function and cytokine secretion. Th1 cells preferentially secrete IL-2 and IFNγ whereas Th2 cells preferentially produce IL-4, IL-5, IL-6 and IL-10. IL-4 has been reported to have a dominant effect on determining the pattern of cytokines (Th2-type) produced by $CD4^+$ T-cells upon subsequent antigen stimulation *in vitro*. Since the role of IFNγ is critical for prevention of development of TE as described above in Section 23.3, IL-4 was first expected to play a negative regulatory role in resistance to *T. gondii* infection. Surprisingly, IL-4 deficient ($IL\text{-}4^{-/-}$) mice showed increased mortality compared to control animals (Suzuki *et al.*, 1996; Roberts *et al.*, 1996; Alexander *et al.*, 1998). Therefore, IL-4 plays a protective role in resistance against *T. gondii*. However, the timing of the mortality and development of TE in $IL\text{-}4^{-/-}$ is controversial between the studies. Suzuki *et al.* (Suzuki *et al.*, 1996) reported that $IL\text{-}4^{-/-}$ mice all died during the late stage (from six to 20 weeks) of infection whereas control mice all survived.

The mortality of *IL-4*$^{-/-}$ mice was associated with greater numbers of cysts and areas of acute focal inflammation with tachyzoites in their brains (Suzuki *et al.*, 1996). These results indicate that IL-4 is protective against development of TE by preventing formation of cysts and proliferation of tachyzoites in the brain. In these studies, at eight weeks after infection, spleen cells of wild-type mice produced significantly greater amounts of IFNγ following stimulation *in vitro* with soluble *T. gondii* antigens than those of *IL-4*$^{-/-}$ mice (Suzuki *et al.*, 1996). Therefore, IL-4 plays a role to enhance IFNγ production during the late stage of infection, and the impaired ability of *IL-4*$^{-/-}$ mice to produce IFNγ likely contributes to their susceptibility for development of severe TE (Suzuki *et al.*, 1996). In relation to these findings, it was reported that IL-4 enhances IFNγ production by T-cells which have already been primed (differentiated) whereas it suppresses differentiation of unprimed T-cells to IFNγ producing cells (Noble and Kemeny, 1995). During infection with *T. gondii*, IFNγ production occurs earlier than IL-4 production (Beaman *et al.*, 29–31 January 1993). Thus, it may be that IL-4 does not affect differentiation of unprimed T-cells to IFNγ producing cells following *T. gondii* infection because of the absence (or very low production) of IL-4 in the early stage of the infection whereas it enhances IFNγ production by differentiated T-cells in the late stage of the infection.

In contrast, Roberts *et al.* (Roberts *et al.*, 1996) reported that an increased mortality occurs in *IL-4*$^{-/-}$ mice during the acute stage of infection. However, in the later stage of infection, greater numbers of cysts and more severe pathology in the brain were observed in the control than *IL-4*$^{-/-}$ mice. During the acute stage of infection, which was the time that *IL-4*$^{-/-}$ died, increased IFNγ production was observed in spleen cells of *IL-4*$^{-/-}$ mice when compared to those of control animals. Therefore, the authors suggested that IL-4 plays a protective role in preventing mortality by down-regulating pro-inflammatory cytokine production during the acute stage of infection. Reasons for the different outcomes in the studies described are unclear. However, since genetic backgrounds of mice and the strain of *T. gondii* differ between these two studies, these differences may have contributed to the different outcomes. In support of this possibility, genetic background of mice was shown to affect the outcome of *IL-4*$^{-/-}$ mice following infection (Alexander *et al.*, 1998).

STAT6 is a molecule involved in IL-4 and IL–13 signalling. In agreement with the observations in *IL-4*$^{-/-}$ mice, significantly greater numbers of *T. gondii* cysts were recovered from the brains of STAT6-deficient (*Stat6*$^{-/-}$) than from wild-type mice. CD8^{+} T-cells obtained from the cerebrospinal fluids and spleens of infected wild-type mice produced greater amounts of IFNγ than T-cells from infected *Stat6*$^{-/-}$ animals (Jin *et al.*, 2009). In addition, transfer of CD8^{+} T-cells from wild-type to *Stat6*$^{-/-}$ mice reduced cyst numbers in the brains of recipients. Transfer of splenic adherent cells from wild-type to *Stat6*$^{-/-}$ mice induced activation of CD8^{+} T-cells and decreased brain cyst numbers. Therefore, STAT6 signalling is important in CD8^{+} T-cell activation after *T. gondii* infection, possibly through regulation of the activity of antigen presenting cells. In this regard, it has been shown that IL-4, alone or together with IFNγ, efficiently enhances the production of bioactive IL-12 in mouse and human DCs (Hochein *et al.*, 2000). The lack of this positive regulatory effect of IL-4 on IL-12 production by DCs most likely contributes to the reduced resistance of *IL-4*$^{-/-}$ and *Stat6*$^{-/-}$ mice to *T. gondii* infection described above.

In addition to the regulatory effects of IL-4 on IFNγ production, IL-4 has been shown to have an activity to modify intracellular replication of tachyzoites in murine macrophages (Swierczynski *et al.*, 2000) and human fibroblast cell lines (Chaves *et al.*, 2001). Therefore, IL-4 likely plays important regulatory roles in multiple stages of antimicrobial responses to *T. gondii*. More studies are needed to elucidate the role of IL-4 in the immunopathogenesis of toxoplasmosis.

23.4.4 IL-5

IL-5 enhances expression of IL-2 receptors on B-cells and promotes B-cell proliferation and differentiation (Swain *et al.*, 1988; Takatsu *et al.*, 1988). IL-5-deficient ($IL\text{-}5^{-/-}$) mice showed increased mortality during the late stage of infection and their mortality was associated with greater numbers of cysts and tachyzoites in their brains when compared to infected control mice (Zhang and Denkers, 1999). IL-12 production by spleen cells from infected mice in response to tachyzoite antigens *in vitro* was markedly lower in $IL\text{-}5^{-/-}$ than control animals, and this decrease correlated with a selective loss of B-cells during the culture. Therefore, IL-5 plays a protective role against *T. gondii* and its protective role is related to the production of IL-12.

23.4.5 IL-6

IL-6 is a multifunctional cytokine that regulates various aspects of the immune response, acute-phase reaction and haematopoiesis (Akira *et al.*, 1993) and acts in the nervous system (Hirota *et al.*, 1996). IL-6 mRNA is expressed in brains of mice infected with *T. gondii* (Gazzinelli *et al.*, 1992; Deckert-Schluter *et al.*, 1995; Hunter *et al.*, 1992) and is detected in the CSF of those mice (Schluter *et al.*, 1993). IL-6-deficient ($IL\text{-}6^{-/-}$) mice formed significantly greater numbers of *T. gondii* cysts and areas of inflammation associated with tachyzoites in their brains than control mice (Suzuki *et al.*, 1997; Jebbari *et al.*, 1998). These results indicate that IL-6 is protective against development of TE by preventing formation of cysts and proliferation of tachyzoites in brains of infected mice. In brains of infected $IL\text{-}6^{-/-}$ mice, the amounts of mRNA for IFNγ detected by RT−PCR were significantly less when compared to control mice, whereas the amounts of IL-10 mRNA were greater than in control animals (Suzuki *et al.*, 1997). In addition, lymphocyte preparations isolated from brains of infected $IL\text{-}6^{-/-}$ mice had significantly lower ratios of γδ T-cells and $CD4^{+}$αβ T-cells but higher ratios of $CD8^{+}$αβ T-cells than those of infected wild-type mice (Suzuki *et al.*, 1997). Of interest, no differences were detected in the ratios of these T-cell subsets in spleens between these animals (Suzuki *et al.*, 1997). In another study, serum IFNγ levels were significantly greater in control than $IL\text{-}6^{-/-}$ mice during the early stage of infection (Jebbari *et al.*, 1998). Therefore, the protective activity of IL-6 against development of TE appears to be through its ability to stimulate IFNγ production (systemic in the early stage of infection and in the brain in the later stage) and induce infiltration and accumulation of different T-cell subsets in brains of infected mice. In relation to the protective role of IL-6 against TE, it was reported that human foetal microglia treated with IL-6 inhibits intracellular replication of tachyzoites *in vitro* (Chao *et al.*, 1994a) and that IL-6 acts synergistically with IFNγ to inhibit proliferation of tachyzoites in murine astrocytes (Halonen *et al.*, 1998b).

23.4.6 IL-10

IL-10 is an important negative regulator of inflammatory responses (Kuhn *et al.*, 1993). IL-10-deficient ($IL\text{-}10^{-/-}$) mice all die during the acute stage of infection with *T. gondii* and their mortality is associated with development of severe immunopathology mediated by Th1 immune responses in the liver (Gazzinelli *et al.*, 1996) and intestine (Suzuki *et al.*, 2000b). Therefore, IL-10 is crucial for down-regulating IFNγ mediated immune responses and preventing development of pathology caused by the immune responses. When $IL\text{-}10^{-/-}$ mice were treated with sulphadiazine to control proliferation of *T. gondii* in the early stage of infection, the animals survived the acute stage but developed lethal inflammatory responses in their brains in the later stage of infection (Wilson *et al.*, 2005). The importance of IL-10 dependent signalling for survival of mice during the acute and chronic stages of infection is confirmed by treating infected wild-type animals with antibodies against anti-IL-10 receptor (Jankovic *et al.*, 2007). Therefore, IL-10 plays an important down-regulatory role to prevent immunopathology during the course of infection with *T. gondii*.

Various populations of T-cells, such as Th2 and Tregs, are able to produce IL-10. However, conventional IFNγ secreting T-bet$^+$Foxp3$^-$ Th1 cells were shown to be the main producers of IL-10 in *T. gondii*-infected mice. These IL-10$^+$IFNγ^+CD4$^+$ T-cells possess potent activity to prevent intracellular multiplication of tachyzoites within macrophages, while profoundly suppressing IL-12 production by antigen-presenting cells (Jankovic *et al.*, 2007). In addition, these IL-10-producing T-cells are generated from IL-10$^-$IFNγ^+CD4$^+$ T-cells after re-stimulation with tachyzoite antigens. These results indicate that IL-10 production by CD4$^+$ T-cells after infection with *T. gondii* does not require a distinct regulatory Th subset, but can be generated in Th1 cells as part of the effector response to the parasite, which provides a crucial negative-feedback loop for prevention of pathogenic overstimulation of the Th1 response.

There is more interesting evidence on the existence of a negative-feedback loop of Th1 response in *T. gondii* infection. Tyk2, a member of the Jak family of non-receptor tyrosine kinases, mediates the biological effects of IL-12 and IFN$\alpha\beta$ and promotes IFNγ production by Th1 cells, and *Tyk2*-null mice are susceptible to infection with the parasite. However, Tyk2 is also required for the production of IL-10 by immune CD4$^+$ T-cells after challenge infection of vaccinated mice (Shaw *et al.*, 2006). The Tyk2-dependent production of IL-10 is mediated by IFNγ, indicating negative autoregulation of the Th1 effector response in infection. Therefore, Tyk2 has a dual function mediating induction and suppression of the Th1 effector response, most likely for maximizing pathogen clearance while minimizing immunopathology.

23.4.7 Lipoxin A_4

Lipoxin A_4 (LXA_4), an eicosanoid product generated from arachidonic acid, is an important down-regulator of IL-12 production to prevent pathogenic inflammatory responses in the brain during the chronic stage of *T. gondii* infection. Wild-type mice produced high levels of serum LXA_4 beginning at the onset of chronic infection. 5-Lipoxygenase (5-LO) is an enzyme critical in the generation of LXA_4, and mice deficient in 5-LO (*Alox5*$^{-/-}$) succumbed to infection during the chronic stage displaying a marked inflammation in their brains (Aliberti *et al.*, 2002). The increased mortality in the *Alox5*$^{-/-}$ animals is not due to defective control of the parasite but due to enhanced inflammatory responses associated with elevations of IL-12 and IFNγ, and their mortality is completely prevented by the administration of a stable LXA_4 analogue. Therefore, LXA_4 is important for down-regulation of pro-inflammatory responses during the chronic stage of *T. gondii* infection. Recent studies showed that lipoxins activate two receptors (AhR and LXAR) in dendritic cells, and that this activation triggers expression of suppressor of cytokine signalling (SOCS)-2. SOCS-2-deficient mice succumb to chronic infection with *T. gondii* in association with elevated IL-12 and IFNγ responses and reduced brain cysts (Machado *et al.*, 2006), as observed in *Alox5*$^{-/-}$ animals (Machado *et al.*, 2006).

23.4.8 IL-17 and IL-27

A unique subset of CD4$^+$ T-cells, Th17, produces IL-17, and this T-cell population has been suggested to mediate inflammation in models of autoimmune diseases such as multiple sclerosis and rheumatoid arthritis. In infection with *T. gondii*, IL-17 has been shown to have both protective and pathogenic roles. IL-17-receptor-deficient mice are more susceptible to acute infection with the parasite (Kelly *et al.*, 2005). Their mortality was associated with increased parasite burden in the organs, including the brain, and with a defect in the migration of neutrophils to infected sites during the early stage of infection. Therefore, IL-17-mediated induction of neutrophil migration to infected sites during the initial stage of infection appears to have an important role in resistance against acute infection with *T. gondii*. However, the activity of Th17 cells needs to be appropriately down-regulated by IL-27 to prevent the development of severe inflammatory changes in the brain during the later stage of infection. Chronically infected IL-27-receptor-deficient mice developed severe inflammation in their brains mediated by $CD4^+$ T-cells, and the pathology was associated with a prominent IL-17 response (Stumhofer *et al.*, 2006). In addition, treatment of naive primary T-cells with IL-27 *in vitro* suppressed the development of Th17 cells induced by IL-6 and TGFβ, and the suppressive effect was dependent on the intracellular signalling molecule STAT1 (Stumhofer *et al.*, 2006). Therefore, IL-27 has a crucial role in the prevention of Th17-mediated inflammatory responses in the brain during the chronic stage of *T. gondii* infection.

23.4.9 IL-33

IL-33 is a member of the IL-1 family with the ability to down-regulate IFNγ production by Th1 cells and up-regulate Th2 response. Mice deficient in T1/ST2, a component of the IL-33 receptor, demonstrated increased susceptibility to *T. gondii* infection that correlated with increased pathology and greater parasite burden in the brain (Jones *et al.*, 2010). Real-time PCR analysis of cerebral cytokine levels revealed increased mRNA levels of IFNγ, TNFα and NOS2 in infected T1/ST2-deficient animals. The mechanisms mediated by IL-33 receptor to control *T. gondii* in the brain remain to be determined.

23.5 INVOLVEMENT OF HUMOURAL IMMUNITY IN RESISTANCE

Antibodies are involved in resistance against *T. gondii* although cell-mediated immunity plays the major role as mentioned above. Frenkel and Taylor (Frenkel and Taylor, 1982) examined the effect of depletion of B-cells by treatment with anti-μ antibody on toxoplasmosis in mice infected with a virulent strain and treated with sulphadiazine. They observed mortality associated with pneumonia, myocarditis and/or encephalitis in infected anti-μ-treated mice after discontinuation of sulphadiazine treatment. Administration of antisera to *T. gondii* reduces mortality in these animals. These results suggest that antibody production by B-cells may be important for controlling the latent persistent infection. However, these studies did not provide conclusive information because of the potential side effects of anti-μ antibody treatment on the immune system.

Kang *et al.* (Kang *et al.*, 2000) reported the role of B-cells in resistance to *T. gondii* by using B-cell-deficient (μMT) mice generated by disruption of one of the membrane exons of the μ-chain gene. All B-cell-deficient mice died between three and four weeks after infection whereas no mortality was observed in wild-type mice until eight weeks after infection. At the stage during which μMT animals succumbed to the infection, large numbers of tachyzoites were detected only in their brains. Furthermore, treatment of infected μMT mice with anti-*T. gondii* IgG antibody reduced mortality and prolonged time to death. These results indicate that B-cells play an important role through production of specific antibodies in prevention of TE in mice. In regard to the protective role of antibodies, Johnson and Sayles (Johnson and Sayles, 2002) reported that treatment of CD4-deficient mice with anti-*T. gondii* sera prolonged their survival during the chronic stage of infection.

In regard to antibody production to *T. gondii* in the later stage of infection, our recent study provided evidence indicating that tachyzoite proliferation in the brain of immunocompetent hosts during the chronic stage of infection induces production of IgG antibodies that recognize parasite antigens different from those recognized by the antibodies of infected hosts that do not have the tachyzoite growth (Hester *et al.*, 2012). In this study, two groups of CBA/J mice, which display continuous tachyzoite growth in their brains during the later stage of infection, were infected, and one group received treatment with sulphadiazine to prevent tachyzoite proliferation during the chronic stage of infection. *T. gondii* antigens recognized by the IgG antibodies from these two groups of mice were compared using immunoblotting following separation of tachyzoite antigens by two-dimensional gel electrophoresis. Although several antigens, including the microneme protein MIC2, the cyst matrix protein MAG1, the dense granule proteins GRA4 and GRA7, were commonly recognized by IgG antibodies from both groups of mice, there were multiple antigens recognized mostly by IgG antibodies of only one group of mice, either with or without cerebral tachyzoite growth. The antigens recognized only by or mostly by the antibodies of mice with cerebral tachyzoite growth include MIC6, the rhoptry protein ROP1, GRA2, one heat shock protein HSP90, one (putative) heat shock protein HSP70 and the myosin heavy chain. These results indicate that IgG antibody levels increase only to selected *T. gondii* antigens in association with cerebral tachyzoite proliferation (reactivation of infection) in immunocompetent hosts with chronic infection. Thus, humoural immune responses, in addition to IFNγ dependent cell mediated immune responses, actively respond to cerebral tachzyoite growth, although the roles of these antibodies in controlling the tachyzoite need to be elucidated.

23.6 IFNγ INDUCED EFFECTOR MECHANISMS

Several mechanisms of IFNγ induced anti-*Toxoplasma* activity in various host cells have been described. These anti-*Toxoplasma* effector mechanisms include nitric oxide (NO) production, tryptophan starvation, generation of reactive oxygen intermediates (ROI), iron deprivation, the immunity-related GTPases (IRG), and the p65 guanylate-binding proteins (Gbp). The actions of IFNγ are initiated by the binding of IFNγ with its IFNγ receptor (IFNγ-R) at the cell surface and the initiation of a signalling cascade, involving the JAK family of tyrosine kinases and STAT family of transcription factors. Upon IFNγ stimulation, STAT1 dimerizes and translocates to the nucleus where it binds γ-activated sequence elements in the promoter regions of IFNγ-inducible genes and activates gene transcription. IFNγ has been shown to induce expression of approximately 400–1200 host cell genes in immune effector cells such as macrophages and microglia but also non-professional immune effector cells such as fibroblasts, endothelial cells and astrocytes (Halonen *et al.*, 2006; Moran *et al.*, 2004; Boehm *et al.*, 1997; Rock *et al.*, 2005; MacMicking, 2004). Collectively these IFNγ activated genes regulate the immune response, inducing expression in genes encoding for pro-inflammatory cytokines and components of the MHC-mediated antigen presentation pathway, cell adhesion molecules important in leukocyte activation and trafficking and genes with anti-microbial function (Boehm *et al.*, 1997).

IFNγ-inducible genes, which are of particular importance for resistance to *T. gondii*, include genes involved in generation of NO, tryptophan degradation, reactive oxygen intermediates, the genes encoding for the IRG family of proteins and the p65 GBPs. Studies indicate that the expression of these anti-microbial mechanisms vary among the tissues, with NO synthesis, tryptophan starvation and the genes encoding for the IRG family shown to be of particular importance in controlling *T. gondii* in the brain. The importance of the IFNγ-response genes in protection against *Toxoplasma* is illustrated by studies that have found STAT1-deficient mice ($Stat^{-/-}$), which produce IFNγ but cannot up-regulate anti-microbial effector functions, unable to control parasite replication (Gavrilescu *et al.*, 2004; Lieberman *et al.*, 2004).

23.6.1 Nitric Oxide (NO)-Mediated Mechanism

Nitric oxide is one of the main IFNγ induced anti-toxoplasmacidal mechanisms known to mediate resistance to *T. gondii* in mice (Hayashi *et al.*, 1996b). IFNγ induces synthesis of the enzyme inducible nitric oxide synthase (iNOS) that produces nitric oxide (NO) from L-arginine (Adams *et al.*, 1991). The L-arginine dependent production of NO and subsequent conversion to reactive nitrogen species (RNS) has direct anti-microbial activity and results in parasite killing (James, 1995). *T. gondii* infection also induces apoptosis in non-infected macrophages due to the secretion of NO released by infected cells which may also serve to limit the spread of infection (Nishikawa *et al.*, 2007). While NO-mediated toxoplasmacidal effects in macrophages limit parasite numbers it has not been found to be required for protection against the acute infection in mice, indicating other IFNγ effector mechanisms, such as IRG proteins and likely other mechanisms, are also involved (Hayashi *et al.*, 1996b, c; Scharton-Kersten *et al.*, 1997a).

Conversely, NO has been found to be essential for resistance to the chronic infection in mice (Scharton-Kersten *et al.*, 1997a). IFNγ activated microglia produce NO production that results in parasite killing (Chao *et al.*, 1993a, c, 1994b). Microglia are the resident macrophage cell population in the brain and likely major immune effector cells limiting tachyzoite replication in the brain *via* this NO-mediated mechanism. NO also induces bradyzoite differentiation, indicating NO may also play a role in inducing bradyzoite differentiation and cyst formation in the chronic infection (Bohne *et al.*, 1994).

23.6.2 Tryptophan Starvation Pathway

The IFNγ induced inhibition of *T. gondii via* tryptophan starvation is a mechanism known to operate in many cell types including fibroblasts, epithelial cells, and endothelial cells, in a variety of host species including humans, rat, and mouse, amongst others (Dimier *et al.*, 1992; Nagineni *et al.*, 1996; Daubener *et al.*, 1999; Pfefferkorn, 1984; Pfefferkorn and Guyre, 1984). The mechanism of inhibition is due to induction of the enzyme, idoleamine-2,3-dioxygenase (IDO) resulting in the degradation of tryptophan to kynurenine (Pfefferkorn *et al.*, 1986; Pfefferkorn and Guyre, 1984). Inhibition of parasite growth is due to tryptophan starvation as tryptophan is an essential amino acid that the parasite derives from the host cell.

Evidence indicates that the IDO pathway is important in host resistance to *T. gondii* early in infection and helps control dissemination into different tissues (Silva *et al.*, 2002). For example, IFNγ induction of the tryptophan/IDO pathway significantly inhibits parasite growth in epithelial and endothelial cells indicating this pathway may be important in limiting parasite growth in the intestinal phase of the infection and passage to the foetus in congenital toxoplasmosis (Dimier *et al.*, 1992; Dimier and Bout, 1993, 1996b, 1997, 1998). Likewise, IFNγ induction of the tryptophan/IDO pathway and inhibition of *T. gondii* has been found in human microvascular endothelial cells indicating this pathway may be of importance in restricting entrance of the parasite into the brain. The IFNγ-induced tryptophan starvation pathway has also been found in human pigment epithelial cells (RPE) indicating this pathway may also be of importance in controlling the replication of the parasite in ocular toxoplasmosis (Nagineni *et al.*, 1996).

23.6.3 Immunity-Related GTPase (IRG) Family

The immunity-related GTPases (IRGs) are a family of proteins induced by IFNγ that are important in resistance against a wide variety of intravacuolar bacterial and parasitic pathogens, including *T. gondii* (Taylor *et al.*, 2004, 2007; Zhao *et al.*, 2009b). Of the hundreds of genes increased by IFNγ, the IRG genes are amongst the most abundant. These proteins, formerly called the p47 GTPases, were first described in the 1990s and in the last decade numerous studies have established the role of IRG proteins in resistance to *Toxoplasma* (Hunn *et al.*, 2011; Zhao *et al.*, 2009b). Most of the work has involved the following seven IRG members: Irgm1 (LRG-47), Irgm2 (GTPI), Irgm3 (IGTP), Irga6 (IIGPI), Irgb6 (TGTP), Irgd (IRG-47) and Irgb10. Most of these IRG proteins have been found to be associated with inhibition of *T. gondii in vitro,* and of the four IRG genes that have been knocked out (Irgm1, Irgm3, Irga6 and Irgd), all have been found to significantly increase susceptibility to infection of *T. gondii,* thus establishing the role of IRG proteins in resistance to *T. gondii* in mice.

The IRG proteins are 46–47 kDa GTPases, containing a Ras-like GTP-binding domain (termed G1). The IRG protein family consists of two subfamilies, based upon the nucleotide-binding domain within the G1 GTP-binding domain with one subfamily having a GMS amino acid motif and the other subfamily having a GKS motif. The three IRG members of the GMS subfamily include, Irgm1, Irgm2 and Irgm3 while IRG members, Irga6, Irgb6, Irgd and Irg10 belong to the GKS subfamily. The GMS IRG proteins are regulators of GKS proteins binding to the GKS IRG proteins and maintaining them in the inactivate state *via* a GDP-dependent interaction (Hunn *et al.*, 2008). The IRG genes are present throughout the vertebrate phyla, being present in cephalochordates, amphibians, fish, reptiles and mammals. In mouse, the IRG family is diverse, encoding approximately 23 genes, 21 of which encode proteins (Bekpen *et al.*, 2005). The IRG family, however, seems to have been repeatedly lost during evolution with no IRG genes present in any of the available bird genomes and the number of IRG genes in humans dramatically reduced with only two IRG genes, IRGC and IRGM, present (Bekpen *et al.*, 2009, 2010).

In IFNγ stimulated host cells infected with *Toxoplasma* multiple IRG proteins localize to the *Toxoplasma* parasitophorous vacuole membrane within minutes of invasion, with the parasitophorous vacuolar membrane subsequently becoming vesiculated and finally disrupted, resulting in release of the parasite into the cytosol and degradation of the parasite (Martens *et al.*, 2005; Ling *et al.*, 2006; Melzer *et al.*, 2008). In macrophages infected with *Toxoplasma,* destruction of the *T. gondii* is accompanied by inclusion of the parasite in autophagosomes and subsequent autophagomal delivery to the lysosomes (Ling *et al.*, 2006; Butcher *et al.*, 2005). IRG-mediated vacuolar disruption also occurs in IFNγ-stimulated fibroblasts and astrocytes but the autophagy pathway was not found to be involved (Melzer *et al.*, 2008; Zhao *et al.*, 2009b; Martens *et al.*, 2005). However, mice deficient in the autophagic regulator, atg5, are deficient in their ability to control *T. gondii* replication indicating the autophagic pathway is involved in some way (Konen-Waisman and Howard, 2007). Atg5 has been found to be necessary for delivery of IRG proteins to the PV, although this appears to operate by a mechanism independent of the normal autophagy pathway (Zhao *et al.*, 2008). Finally, in IFNγ-stimulated fibroblasts IRG-mediated PV disruption results in host cell necrosis, following release of the parasite into the host cytoplasm, indicating destruction of the host cell may be part of the IRG mechanism in some cell types (Zhao *et al.*, 2009b).

The IRG mechanism involves a coordinated loading of IRG GTPases on the *Toxoplasma* vacuole with at least six IRG proteins (Irgm2, Irgm3, Irga6, Irgb6, Irgd and Irg10) localizing to the *Toxoplasma* vacuole (Khaminets *et al.*, 2010). The coating of the IRG proteins to the PV occurs within one hour of invasion and is hierarchical with Irgb6 and Irgb10 loading first. Upon infection with *T. gondii,* GKS proteins lose their interaction with GMS proteins and accumulate at the PV membrane (PVM) in the active GTP bound state leading to vesiculation and rupture of the PV (Hunn *et al.*, 2008; Papic *et al.*, 2008). Despite the large amount of information now understood about the molecular and biochemical aspects of IRG-mediated inhibition of *T. gondii*, the mechanisms involved in the vesiculation leading to PV disruption is still not understood. IRG proteins are related to the dynamin-type GTPases known to mediate vesicle formation and deformation of membranes and it has been suggested IRG protein acts in an analogous fashion mediating vesiculation of the PVM, although this has not been demonstrated (Hunn *et al.*, 2011).

The type I strains are resistant to IRG-mediated IFNγ inhibition (Steinfeldt *et al.*, 2010; Howard *et al.*, 2011). This deficiency in IFNγ mediated control is associated with a failure of accumulation of IRG proteins on the PVM (Zhao *et al.*, 2009a). This has been found to be due largely to the polymorphic rhoptry kinase, ROP18, which in type I strains phosphorylates the GKS IRG proteins Irga6, Irgb6 and Irgb10, causing dissociation of IRG from the vacuole and inhibition of PV disruption (Zhao *et al.*, 2009a; Steinfeldt *et al.*, 2010; Fentress *et al.*, 2010). Another rhoptry protein, ROP5, has been found to directly interact with IRG proteins, reducing IRG coating and inactivating IRG proteins (Fleckenstein *et al.*, 2012; Niedelman *et al.*, 2012). ROP5 can interact with IRGs in the absence of ROP18. However, rhoptry proteins, ROP5 and ROP18 while mediating inhibition in IFNγ-activated murine cells, do not affect survival in IFNγ activated human cells (Niedelman *et al.*, 2012). These results suggest that while ROP5 and ROP18 may have evolved to block the IRGs they may not have effects on parasite survival in species that do not have the IRG system, such as humans. Why the IRGs are such a large family of proteins in the murine genome and so reduced in humans, or if functional counterpart(s) exists in humans, is not yet clear.

23.6.4 Guanylate-Binding Proteins (GBPs or p65 GTPases)

Stimulation of immune cells such as macrophages and dendritic cells by IFNγ induces in gene expression of the IRG immunity-related GTPases but also the related the p65 guanylate-binding family proteins (GBPs) (Taylor *et al.*, 2004). The GBPs have been shown to induce antibacterial responses involving phagocytic oxidases, autophagic effectors, and the inflammasome and they have recently also been found to play a critical role in host defence against *T. gondii* (Yamamoto *et al.*, 2012). Mice deficient in a cluster of six GBP family genes (*Gbp1, Gbp2, Gbp3, Gbp5, Gbp7* and *Gbp2ps*) on chromosome 3 (Gbp^{chr3}) are highly susceptible to *T. gondii* infection with increased parasite burden in immune organs. Gbp^{chr3}-deleted macrophages are defective in IFNγ-mediated suppression of *T. gondii* intracellular growth and recruitment of IRG protein, Irgb6, to the parasitophorous vacuole (PV), suggesting GBPs may regulate Irgb6 recruitment to the PV. Another study of the behaviour of GBP1 *in vitro* in murine cells found GBP1 exerts its function in conjunction with GBP family members, GBP2 and GBP5, and that GBP1 recruitment correlates with virulence of the parasite strain (Virreira Winter *et al.*, 2011). *T. gondii* proteins, ROP16, ROP18 and GRA15, were found to be partly responsible for the strain-specific accumulation of GBP1 at the PV. The IFNγ-dependent host factors that regulate this complex process of GBP recruitment to the PV are unknown. However, it appears that along with the IRGs, GBPs are key factors mediating resistance to *T. gondii*. Interestingly, in contrast to the IRGs, the p65 GBPs are represented by 13 genes in the mouse genomes and are similarly represented by seven genes in the human genome indicating this system may also function in human cells (Kresse *et al.*, 2008).

23.6.5 Reactive Oxygen Intermediates (ROI)

IFNγ activated production of reactive oxygen intermediates (ROI) has been demonstrated to induce anti-toxoplasmacidal activity in human macrophages (Murray *et al.*, 1985a, b). The reactive oxygen intermediates generated include superoxide ion (O_2^-) and hydrogen peroxide (H_2O_2). ROI intermediates have also been found to mediate IFNγ inhibition of *T. gondii* in murine dendritic cells but not murine microglia or astrocytes (Jun *et al.*, 1993; Aline *et al.*, 2002a; Halonen and Weiss, 2000b). It has been reported that the parasites are resistant to the oxygen metabolites produced in murine macrophages (Chang and Pechere, 1989) and p47phox-deficient mice (which lack an inducible oxidative burst) can control both the acute and chronic stages of *T. gondii* infection (Scharton-Kersten *et al.*, 1997b). Thus the physiological significance of the ROI pathway still remains unclear, especially in mice.

23.6.6 Iron Deprivation

IFNγ has been shown to inhibit growth of *T. gondii via* iron deprivation in intestinal epithelial cells (Dimier and Bout, 1998; Bout *et al.*, 1999). Cells usually acquire iron *via* the transferrin receptor pathway, in which iron bound to transferrin binds to the transferrin receptor at the cell surface and the receptor–transferrin complex is internalized *via* endocytosis with subsequent acidification of the endocytic vesicle causing the ferric ions to dissociate from transferrin. The iron ions are transported across the vesicle membrane into the cytoplasm and this intracellular pool of iron is available for metabolic processes. *T. gondii* replicates within the parasitophorous vacuole, a membranous compartment within the host cell that does not fuse with the lysosomes and is not contiguous with iron-transferrin as it passes through the endocytic pathway. Rather than acquiring iron from transferrin, experiments with intestinal epithelial cells indicate that IFNγ can inhibit *T. gondii* replication by limiting the availability of intracellular iron for the parasite to scavenge from its host cell. The mechanism by which this occurs is not understood. Inhibition of *T. gondii* growth *via* IFNγ induced iron deprivation has only been shown for enterocytes and may be an important anti-toxoplasmic mechanism on mucosal surfaces as a first line of defence against pathogen invasion. It is not clear if this IFNγ induced iron dependent mechanism also occurs in other cell types and thus has a broader role in the control of *Toxoplasma* in the latent stage of infection and congenital toxoplasmosis.

23.7 IFNγ EFFECTOR CELLS IN THE BRAIN WITH ACTIVITY AGAINST *TOXOPLASMA GONDII*

The establishment of a chronic asymptomatic *T. gondii* infection requires the cytokine IFNγ, as previously discussed, but additional resistance to *T. gondii* requires IFNγ effector cells from haemopoietic and non-haemopoietic compartments (Yap and Sher, 1999; Suzuki *et al.*, 1988). In the brain, these IFNγ effector cells include infiltrating T-cells, dendritic cells and macrophages which stimulate the intracerebral immune responses, while activation of resident CNS IFNγ effector cell populations such as microglia, cerebral microvascular endothelial cells and astrocytes are essential to control the parasite in the brain.

The expression of the IFNγ induced antimicrobial mechanisms varies between phagocytic cells such as macrophages and microglia, and non-phagocytic cells such as endothelial cells and astrocytes (Halonen *et al.*, 2006; Moran *et al.*, 2004; Rock *et al.*, 2005; Kota *et al.*, 2006; Indraccolo *et al.*, 2007; Adams *et al.*, 1990a). IFNγ inhibition in macrophages is primarily mediated *via* a nitric oxide-mediated mechanism while in non-phagocytic cells, IFNγ inhibition is *via* non-NO mediated mechanisms such as IDO/tryptophan degradation, increases in reactive oxygen species, and induction of IRG proteins. The IRG-mediated mechanism is found in macrophages, as well as fibroblasts and astrocytes, and has been shown to be of particular importance in protection against *T. gondii* in mice (Zhao *et al.*, 2009b; Martens *et al.*, 2005). Most recently several members of the related family of IFNγ-inducible GTPases, the guanylate-binding protein (GBP) family, have also been shown to play a role in host defence against *T. gondii* (Yamamoto *et al.*, 2012).

23.7.1 Microglial Cells

Microglia are the resident brain tissue macrophage precursors that rapidly respond to infection in the brain (Nimmerjahn *et al.*, 2005). IFNγ activated microglia produce NO that is responsible for anti-*Toxoplasma* activity (Chao *et al.*, 1993a, c, 1994b). IFNγ activated microglia produce chemokines and cytokines including TNFα, IL-1β, IL-12 and IL-15, and induce expression of MHC I and II, LFA-1 and ICAM-1 indicating IFNγ microglia also regulate infiltration of T-cells into the brain and act as an antigen-presenting cell in the brain (Deckert-Schluter *et al.*, 1999; Schluter *et al.*, 2001). Microglia can also produce IFNγ either in the presence or absence of T-cells (Wang and Suzuki, 2007). IFNγ secretion by microglia may thus play a critical role in the early stage of infection in the brain, prior to the infiltration of T-cells, by stimulating IFNγ effector cells to limit tachyzoite proliferation in the brain and initiating subsequent T-cell immunity. Additionally, IFNγ secretion by microglia may also be important in response to reactivated infections in which T-cell numbers are low.

While IFNγ activated microglia are clearly important effector cells against the parasite in the brain, it has recently been found that unstimulated microglia infected with *T. gondii* exhibit a hypermotility phenotype, similar to infected dendritic cells (Dellacasa-Lindberg *et al.*, 2011). Thus, paradoxically, microglia cells may help disseminate the parasite within the brain. Furthermore, unstimulated microglia infected with *T. gondii* also exhibit increased sensitivity to T-cell mediated killing leading to parasite transfer to effector T-cells. Results of these *in vitro* studies indicate infected microglial might also transfer parasites between different cell populations such as cytotoxic T-cells, thus potentially facilitating dissemination of the parasite within the brain parenchyma in a 'Trojan horse' type mechanism in reactivated infections.

23.7.2 Astrocytes

Astrocytes are the dominant glial cell in the brain and numerous studies indicate they are central to the intracerebral immune response to *T. gondii* in the brain (Wilson and Hunter, 2004b). First, IFNγ activated astrocytes significantly inhibit the growth of tachyzoites of *T. gondii* indicating they can function as an immune effector cells in the brain (Halonen *et al.*, 2001b; Scheidegger *et al.*, 2005). IFNγ inhibition in astrocytes in mice is *via* the IRG-mediated mechanism while inhibition in human astrocytes is *via* nitric oxide or tryptophan degradation (Halonen *et al.*, 1998a, 2001a; Halonen and Weiss, 2000a; Martens *et al.*, 2005; Peterson *et al.*, 1995; Oberdorfer *et al.*, 2003). Additionally, *in vitro* studies indicate IFNγ activated astrocytes can induce bradyzoite differentiation and support cyst formation indicating astrocytes may also foster bradyzoite differentiation/cyst formation in the brain and thus serve to help maintain the chronic infection (Jones *et al.*, 1986b; Halonen *et al.*, 1998c).

In addition to having roles as IFNγ effector cells controlling tachyzoite replication and as host cells to the cysts, astrocytes have other immune functions that are involved in the intracerebral immune response to *T. gondii* (Wilson and Hunter, 2004a). Upon infection with *Toxoplasma* there is a general inflammatory response in the brain characterized by cytokine production and astrocyte activation with astrocytes producing the chemokine's IFNγ inducible protein 10 (IP-10) and MCP-1, which attract activated T-cells and activated macrophages into the site of inflammation in the brain, respectively (Strack *et al.*, 2002; Schluter *et al.*, 1991). Studies in mice indicate astrocytes production of IP-10, precedes infiltration of T-cells indicating astrocytes play a key role in regulation of T-cell trafficking into the brain early in infection, while astrocyte production of MCP-1 may contribute to active recruitment of both macrophages and activated T-cells throughout the chronic infection (Strack *et al.*, 2002; Wilson and Hunter, 2004b). Additionally, IFNγ activated astrocytes are efficient in MHC I presentation and capable of cross-presentation, indicating they may be an important antigen presenting cell for *T. gondii* in the brain, capable of stimulating $CD8^+$ T-cells (Dzierszinski *et al.*, 2007). Finally, recent studies also indicate astrocytes play a crucial regulatory role reducing inflammation and serving to limit sites of parasite replication (Drogemuller *et al.*, 2008), as discussed more fully below.

Collectively, these studies indicate that astrocytes may play a pivotal role in the control of *T. gondii* in the brain, serving to limit replication of the tachyzoites, thus augmenting the more potent anti-microbicidal activity of microglia to clear the parasites from the brain, stimulating bradyzoite differentiation and cyst development while also serving as antigen-presenting cells and key regulators of T-cell trafficking into the brain during *Toxoplasma* infection.

23.7.3 Cerebral Microvascular Endothelial Cells

IFNγ stimulated human brain microvascular endothelial cells have been shown to induce inhibition of *T. gondii via* induction of IDO/tryptophan starvation pathway (Daubener *et al.*, 2001). The anti-parasitic effect was enhanced by TNFα, although TNFα alone had no effect on parasite growth similar to IFNγ inhibition observed in astrocytes. Since one of the first steps in the development of cerebral toxoplasmosis is penetration of the blood–brain barrier, IFNγ induced inhibition in these cells may be important in limiting the number of *T. gondii* parasites that enter the brain *via* a transcellular route. IFNγ also induces expression of VCAM1 on endothelial cells that binds to an integrin on $CD8^+$ T-cells, and thus is of importance for recruitment of T-cells into the brain during the chronic stage of *T. gondii* infection (Wang *et al.*, 2007).

23.7.4 Dendritic Cells

Dendritic cells (DCs) play a crucial role in the initiation of systemic immune responses to acute infection with *T. gondii* as discussed above (see Section 23.2.1). Evidence indicates DCs also play key roles in promoting resistance to cerebral toxoplasmosis. First, recent studies indicate infected DCs exhibit an enhanced migratory response, which facilitates DCs trafficking and may be involved in transporting single tachyzoites into the brain across the blood–brain barrier *via* a Trojan horse-type mechanism (Courret *et al.*, 2006; Unno *et al.*, 2008; Bierly *et al.*, 2008). Additionally, activated DCs are present within the brain during the chronic infection indicating DCs are important in the intracerebral immune response to *Toxoplasma* (Fischer *et al.*, 1993, 2000a). Although the percentage of DCs in the brain is low in

Toxoplasma infection (less than 10%), recent evidence indicates DCs can function as antigen-presenting cells (APCs) capable of processing and presenting antigen to naive T-cells (John *et al.*, 2011). Additionally, both infected and uninfected DCs are able to interact with T-cells, indicating DCs are capable of presenting antigen *via* both direct- and cross-presentation pathways. This study also suggested a constant recruitment and retention of DCs into the CNS during the chronic infection. Real-time imaging studies found DCs are intimately associated with the parasite cysts suggesting a possible innate sensing of the cyst stage by DCs. Finally, in addition to these immune functions, DCs may also function as an IFNγ effector cell as IFNγ stimulation induces anti-toxoplasmacidal activity *via* reactive oxygen mediated mechanism (Aline *et al.*, 2002b).

23.8 THE ROLE OF HOST CELLS HARBOURING *TOXOPLASMA GONDII* IN THE BRAIN

T. gondii persists for the lifetime of the host within cysts located predominantly in the brain. Early in infection the cyst stage can occur in microglia, astrocytes and neurons but cysts persist predominantly in neurons during the chronic infection (Frenkel and Escajadillo, 1987; Sims *et al.*, 1988, 1989a, 1989c; Fagard *et al.*, 1999; Ferguson and Hutchison, 1987b). While neurons serve primarily as the host cell for the cyst stage, infected neurons may also participate in the regulation of the intracerebral immune response and, as neurons are in close proximity to astrocytes and microglia in the brain, infection and development in neurons are also likely affected by these neighbouring cells. Studies do indicate that a complex interplay exists between *T. gondii* cysts in neurons and activated astrocytes and microglial cells, which maintain the chronic *Toxoplasma* infection in the brain and prevent neuropathology.

Requirement of the immune system to maintain the latency of chronic infection is evident from reactivation of infection in immunocompromised patients, such as those with AIDS (Israelski and Remington, 1993; Wong and Remington, 1994). Tissue cysts are the most likely source of reactivation toxoplasmosis in immunocompromised patients. In reactivated infection (i.e. TE) bradyzoite to tachyzoite conversion is thought to occur and the resulting uncontrolled proliferation of the tachyzoite in the brain results in pathology. While the cysts appear to be located primarily in neurons and astrocytes, the tachyzoite may replicate in astrocytes, microglia and possibly neurons. The mechanisms and dynamics of tachyzoite-to-bradyzoite interconversion in the brain are not fully understood and are areas that are important in understanding the persistence of *T. gondii* in the brain in both immunocompetent and immunocompromised individuals and for devising treatments for cerebral toxoplasmosis.

23.8.1 *T. gondii* Infection in Neurons

Neurons are a major host cell for *Toxoplasma* in the brain, supporting both growth of tachyzoite stage and serving as the predominant host cell for the cyst stage in the chronic infection (Ferguson and Hutchison, 1987a; Melzer *et al.*, 2010; Sims *et al.*, 1989b). *In vitro* studies using neuronal cultures indicate neurons are infected less efficiently than glial cells and that growth of tachyzoites is also slower in neurons than glial cells (Halonen *et al.*, 1996; Creuzet *et al.*, 1998; Fischer *et al.*, 1997). The mechanism underlying the lower rates of infection could be due to various parameters such as absence or low level of expression of a cell-surface binding factor required for entry or the mitotic status of the host cell as attachment of *Toxoplasma* has been shown to be host cell-cycle-dependent and to decrease as cells enter the G2–M boundary

(Fagard *et al.*, 1999; Grimwood *et al.*, 1996; Dutta *et al.*, 2000). The slow growth rate in neurons is not well understood but has been suggested to be due to the small cell size. *In vitro* studies using neuronal cultures also indicate stage conversion from the tachyzoite to the bradyzoite stage and cyst formation occurs spontaneously in neurons in the absence of IFNγ or other factors such as NO (Luder *et al.*, 1999a; Fischer *et al.*, 1997). Thus, intrinsic factors of the neuronal host cell environment appear to play a role in inducing cyst formation, partially explaining the predominance of cysts in neurons. It has been suggested that encystation in neurons in the absence of T-cell-derived cytokines may provide a mechanism for a brain-internal spreading of bradyzoites which may help to sustain chronic infection (Fischer *et al.*, 1997). However, neither the phenomenon of brain-internal spreading of bradyzoites in neurons nor the nature of intrinsic neuronal factors favouring cyst formation have been well studied or are well understood.

Neither IFNγ nor TNFα stimulated neurons, however, inhibit replication of *T. gondii* in neurons (Schluter *et al.*, 2001). This is in marked contrast to the inhibition of growth that these cytokines induce in microglia and astrocytes (Chao *et al.*, 1993b; Halonen *et al.*, 1998a). Thus, IFNγ and TNFα while important in controlling *T. gondii* in microglia and astrocytes are not sufficient to control replication of the parasite in neurons. However, the replication of the parasite in neurons is relatively slow and this in itself is indicative of some intrinsic ability of neurons to limit parasite growth. It has been suggested that astrocytes, as the dominant glial cell, intimately involved with neuronal processes and necessary to maintain functional integrity of neurons, may supply additional factors for the induction of inhibition of *Toxoplasma* in neurons, although this has not been well studied.

Evidence indicates infected neurons can, however, participate in the regulation of the intracerebral immune response. For example, similar to infected astrocytes and microglia, neurons induce secretion of cytokines and chemokines in response to infection. Infected neurons secrete the chemokines MIP-1α and MIP-1β that attract inflammatory leukocytes, and the immunosuppressive cytokine, TGFβ (Schluter *et al.*, 2001). Intracellular parasites are commonly surrounded by an infiltrate of T-cells and granulocytes while the bradyzoite containing cysts are usually devoid of inflammatory cells, indicating that the neuronal production of chemokines may be dependent upon the stage of the parasite in neurons. Using synapsin-I (Syn)-Cre gp130$^{-/-}$ mice, which lack gp130, the signal-transducing receptor for the IL-6 family of cytokines in neurons, it has recently been found that loss of gp130 in neurons results in loss of control *T. gondii* infection leading to hyperinflammation, lethal necrotizing TE and a progressive neuronal loss (Handel *et al.*, 2012). IL-6 stimulation of neurons induced gp130-dependent TGF-β1, TGF-β2 and IL-27 production, and reduced death and apoptosis of infected neurons. These findings indicated a protective function of gp130-expressing neurons in chronic toxoplasmosis. Collectively the above results indicate that *T. gondii* infected neurons contribute to the intracerebral immune response through the recruitment of inflammatory cells to the site of infection and *via* down-regulation of the intracerebral immune response resulting in prevention of neuronal loss in chronic toxoplasmosis.

Finally, several recent studies indicate *Toxoplasma* infection affects neuronal functions. For example, an increase in dopamine levels in neurons containing cysts has recently been found (Prandovszky *et al.*, 2011). The parasite contains a tyrosine hydroxylase gene that is able to synthesize L-DOPA, which is expressed only in bradyzoites and it has been proposed that this parasite enzyme is responsible for an increase in dopamine levels in infected neurons (Gaskell *et al.*, 2009). The altered dopamine levels induced by *T. gondii* cysts in the brain could have significant consequences on brain functions, possibly leading to an array of behavioural changes and neurological malfunctions. Another recent study in chronically infected mice found parasite antigens in the host cell cytoplasm, including axons, indicating the parasite might directly interfere with neuronal function and found a decrease in the activity-dependent uptake of the potassium analogue thallium, indicating infected neurons were functionally silenced (Haroon *et al.*, 2012). These studies indicate a functional impairment of cyst harbouring neurons and provide a possible mechanism that may lead to neurological effects and/or behavioural consequences in chronically infected hosts.

23.8.2 Role of Astrocytes in Cerebral Toxoplasmosis

In addition to having a role as IFNγ effector cells controlling tachyzoite replication in the brain, both *in vivo* and *in vitro* studies have found that astrocytes support the cyst stage. For example, cysts in astrocytes have been reported in human brain (Powell *et al.*, 1978) and *in vitro* studies in astrocytes have found to be able to support cyst development (Jones *et al.*, 1986a; Halonen *et al.*, 1998b). Layers of glial intermediate-type filaments (GFAP) surround cysts in astrocytes and this type of GFAP layering of cysts may aid in stabilizing cysts, perhaps preventing cyst rupture (Halonen *et al.*, 1998b; Luder *et al.*, 1999b; Powell *et al.*, 1978). The layers of glial filaments were observed to displace the host cell mitochondria from the surface of the parasitophorous vacuole and thus the glial filament wrapping has also been suggested to play a role in establishing the anaerobic environment helping to facilitate tachyzoite to bradyzoite formation (Halonen *et al.*, 1998b).

Studies indicate astrocytes are also crucial for effective control of inflammation and maintenance of the chronic infection in the brain. The importance of astrocytes in the intracerebral immune response against *Toxoplasma* has been illustrated *in vivo* in mice, in which astrocytes deficient in the expression of their glial fibrillary protein (GFAP$^{-/-}$), showed a reduced capacity to restrict *T. gondii*-associated inflammatory lesions, resulting in increased intracerebral load of the parasite and widespread areas of tachyzoite-induced tissue necrosis (Stenzel *et al.*, 2004). These studies involving GFAP$^{-/-}$ mice indicate a host defence mechanism of astrocytes *via* restricting the pathogenic spread within the CNS. A more recent study in mice in which astrocytes were deficient in gp130 (GFAP-Cre-gp130$^{-/-}$) was similarly found to be unable to contain sites of parasite replication resulting in inflammatory lesions in the brain leading to fatal toxoplasmic encephalitis (Drogemuller *et al.*, 2008). The gp130 effect was not a direct anti-parasitic effect, as GFAP-Cre-gp130$^{-/-}$ astrocytes were capable of IFNγ-stimulated inhibition of parasite growth. Rather the effect appeared to be immunoregulatory as neighbouring uninfected GFAP-Cre-gp130$^{-/-}$ astrocytes become apoptotic. Collectively these studies indicate a crucial role for astrocytes in the immunoregulatory response to *Toxoplasma* in the brain, reducing neuropathology and limiting spread of the parasite in the brain.

Finally, astrocytes have been suggested to play a role in the behavioural effects of the chronic infection in the intermediate host. Activated astrocytes are a hallmark of toxoplasmic encephalitis with activated astrocytes prominent at sites of inflammation and parasite replication (Wilson and Hunter, 2004b). One of the immune effector functions associated with activated astrocytes in humans is induction of IDO, leading to tryptophan degradation, and parasite inhibition. One of the derived products of tryptophan metabolism is kynurenic acid (KYNA), which is released into the extracellular environment and inhibits glutamatergic and nicotinergic neurotransmission. The generation of KYNA in activated astrocytes has been suggested as a mechanism by which *T. gondii* infection possibly can affect the behaviour of the chronically-infected host (Schwarcz and Hunter, 2007). These and other hypotheses are an active area of debate and further discussed in Section 23.13.

23.8.3 Microglia and Astrocyte–Microglia–Neuronal Interactions

Of the three main resident brain cells, microglia appears to function primarily to limit growth of *T. gondii*. IFNγ activated murine microglia produce NO that is responsible for microbiostatic and microbicidal control of parasite growth, as described above in the Section 23.7. Recent studies, however, indicate a more complex interaction between microglia, astrocytes and neurons occurs in the control of *T. gondii* in the brain. For example, while Th1 cytokines are necessary for microglia activation and responsible for protection against *T. gondii* in the brain, products from activated microglia may also cause tissue damage and be detrimental to neuron functions. Nitric oxide (NO), one of the main mediators against *T. gondii* produced by microglia, is also one of the most noxious to the CNS cells. In chronic infections with *T. gondii* in an immunocompetent host in which the parasite persists in the brain, neuronal and tissue damage does not usually occur. Thus, immunomodulary mechanisms are thought to be involved in the prevention of neuronal degeneration and pathological alterations in the CNS during the latent phase of the infection.

Several mechanisms involving down-regulation of NO by microglia have been described. First, *Toxoplasma*-infected IFNγ activated microglia inhibit iNOS expression and a corresponding decrease in NO production (Rozenfeld *et al.*, 2005). Inhibition of iNOS only occurs in infected microglia indicating this inhibition is a parasite-mediated effect. Studies indicate the parasite triggers secretion of TGFβ by infected microglia, with autocrine stimulation of TGFβ–receptor signalling of the infected microglia resulting in a decrease in NO production (Rozenfeld *et al.*, 2005). Additionally, *Toxoplasma* infection induces up-regulation of CD200 on blood vessel endothelial cells and the receptor, CD200R on microglia, the interaction of which down-regulates microglia activation (Deckert *et al.*, 2006). In this study, inhibition of NO production and iNOS expression corresponded with protection of neurite outgrowth, indicating this *T. gondii* triggered mechanism is involved in the neuropreservation observed during immunocompetent host infections. A third mechanism known to down-modulate NO production by IFNγ activated microglia involves release of prostaglandin, PGE2, by infected astrocytes (Rozenfeld *et al.*, 2003). The presence of multiple and independent pathways in the CNS leading to NO inhibition and maintenance of neuronal viability during *T. gondii* infection likely reflect the importance of this neuroprotection mechanism in the brain during chronic toxoplasmosis.

23.9 IMMUNE RESPONSES TO THE CYST STAGE OF *TOXOPLASMA GONDII* IN THE BRAIN

The basis of persistence of chronic infection with *T. gondii* is the tissue cyst, which remains largely quiescent for the life of host, but can reactivate and cause disease. This stage of the parasite is not affected by any of the current drug treatments and has been generally regarded as untouchable. However, our recent study revealed that the immune system can eliminate *T. gondii* cysts from the brains of infected hosts when immune T-cells are transferred into infected immunodeficient animals that have already developed large numbers of the cysts (Suzuki *et al.*, 2010). This T-cell-mediated immune process was associated with accumulation of microglia and macrophages around tissue cysts. Since the accumulated phagocytes penetrate within the cyst, these cells appear to be the main effector cells that destroy the cysts and eliminate them from the brain after initiation of this process by immune T-cells. $CD8^+$ immune T-cells possess a potent activity to remove the cysts. Of interest, the initiation of this process by $CD8^+$ T-cells does not require their production of IFNγ, the major mediator to prevent proliferation of tachyzoites during acute infection, but does require perforin. Perforin is the major molecule that mediates cytolysis of target cells by $CD8^+$ cytotoxic T-cells. Therefore, our results suggest that $CD8^+$ T-cells induce elimination of *T. gondii* cysts through their perforin-mediated cytotoxic activity. In relation to our observation, a study using two-photon microscopy in C57BL/6 mice with ovalbumin-expressing *T. gondii* showed that ovalbumin-specific $CD8^+$ T-cells accumulated in regions containing isolated parasites (tachyzoites) but not around cysts in the brain. C57BL/6 mice are genetically susceptible to chronic infection with the parasite and continuous tachyzoite proliferation occurs in their brain. Therefore, it is possible that the activity of cerebral $CD8^+$ T-cells of these animals is down-regulated in this environment and does not have a potent anti-cyst activity.

Perforin-mediated cytolytic activity by T-cells in resistance against *T. gondii* infection was less appreciated before, when compared to the absolute requirement of IFNγ to control tachyzoite proliferation during the acute stage of infection. Perforin knock-out mice survive acute infection (Wang *et al.*, 2004; Denkers *et al.*, 1997). *In vitro* studies demonstrated that the lysis of tachyzoite-infected cells by cytotoxic $CD8^+$ T-cells results in release of viable parasites (Yamashita *et al.*, 1998). However, the situation of the parasite in the cyst stage seems different. Whereas the cyst resides within an infected cell (Ferguson and Hutchison, 1987c; Ghatak and Zimmerman, 1973) as do tachyzoites, bradyzoites within the cyst are surrounded by a thick cyst wall, which is unique to the cyst. Therefore, cytolysis of cyst-containing cells by perforin-mediated activity of $CD8^+$ T-cells followed by quick accumulation of large numbers of microglia and macrophages would probably provide the parasite little time to escape from the coordinated attack by the T-cells and phagocytes. A previous study (Denkers *et al.*, 1997) reported a three- to four-fold increase in brain cyst numbers in perforin knock-out mice in the later stage of infection. The absence of perforin-dependent anti-cyst activity of $CD8^+$ T-cells in these animals may have contributed to their observation.

23.10 HOST GENES INVOLVED IN REGULATING RESISTANCE

Resistance against *T. gondii* is under genetic control in both acute and chronic stages of infection. Of interest, genes involved in resistance differ between these two stages. Susceptibility of inbred strains of mice to acute infection does not correlate with that to chronic infection (Suzuki *et al.*, 1993). A minimum of five genes are involved in determining survival of mice during the acute stage (McLeod *et al.*, 1989). One of these genes is linked to the major histocompatibility complex (H-2) (McLeod *et al.*, 1989).

During the chronic stage of infection, development of TE in mice is regulated by the gene(s) within the D region of the major histocompatibility complex (H-2) (Brown *et al.*, 1995; Suzuki *et al.*, 1991, 1994). Mice with the *d* haplotype in the D region are resistant to development of TE and those with the *b* or *k* haplotypes are susceptible. Freund *et al.* (Freund *et al.*, 1992) found that polymorphisms in the *Tnf*-gene located in the D region of the H-2 complex correlate with resistance against development of TE and with levels of TNFα mRNA in brains of infected mice. However, more recent studies using deletion mutant mice (Suzuki *et al.*, 1994) and transgenic mice (Brown *et al.*, 1995) demonstrated that the L^d gene in the D region of the H-2 complex, but not the *Tnf*-gene, is important for resistance against development of TE. Resistance of mice to development of TE is observed in association with resistance to formation of *T. gondii* cysts in the brain (Brown *et al.*, 1995; Suzuki *et al.*, 1991, 1994). McLeod *et al.* (McLeod *et al.*, 1989) reported that although the L^d gene has the primary effect on numbers in the brain, the *Bcg* locus on chromosome 1 may also affect it.

In humans, HLA-DQ3 was found to be significantly more frequent in white North American AIDS patients with TE than in the general white population or randomly selected control AIDS patients who had not developed TE (Suzuki *et al.*, 1996). In contrast, the frequency of HLA-DQ1 was lower in TE patients than in healthy controls (Suzuki *et al.*, 1996). Thus, HLA-DQ3 appears to be a genetic marker of susceptibility to development of TE in AIDS patients and DQ1 appears to be a resistance marker. HLA-DQ3 also appears to be a genetic marker of susceptibility to development of cerebral toxoplasmosis in congenitally infected infants since higher frequency of DQ3 was observed in infected infants with hydrocephalus than infected infants without hydrocephalus or normal controls (Mack *et al.*, 1999). The role of the *HLA-DQ3* and *-DQ1* genes in regulation of the susceptibility/resistance of the brain to *T. gondii* infection is supported by the results from a transgenic mouse study (Mack *et al.*, 1999). Expression of the *HLA-DQ1* transgene conferred greater protection against parasite burden and necrosis in brains in mice than did the *HLA-DQ3* transgene (Mack *et al.*, 1999). Expression of the *HLA-B27* and *-Cw3* transgenes had no effects on the parasite burden (Brown *et al.*, 1994). Since the L^d gene in mice and the *HLA-DQ* genes in humans are a part of the MHC which regulate the immune responses, the regulation of the responses by these genes appears to be important to determine the resistance/susceptibility of the hosts to development of TE. Polymorphisms of the genes important for the protective immune responses to *T. gondii*, such as IFNγ, toll-like receptor and Tyk2, have been shown to affect susceptibility to the infection in humans and mice (Shaw *et al.*, 2003; Peixoto-Rangel *et al.*, 2009; Albuquerque *et al.*, 2009).

23.11 IMMUNE EFFECTOR MECHANISMS IN OCULAR TOXOPLASMOSIS

Ocular toxoplasmosis may result from congenital infection or after birth from acquired infections (reviewed in Chapter 5). The primary target of ocular toxoplasmosis is the neural retina, but infection may involve choroids, sclera, optic nerve and retinal pigment epithelial (RPE) cells (Norose *et al.*, 2003b; Delair *et al.*, 2011). The lesions are often necrotic, destroying the neural retina and often the choroid with retinochoroiditis, the most common lesion caused by infection. Ocular toxoplasmosis usually occurs from a reactivation of a latent infection but can also occur from newly acquired infections with *T. gondii* (Holland *et al.*, 1988; Chakroun *et al.*, 1990). Toxoplasmosis also causes retinochoroiditis in AIDS patients (Holland *et al.*, 1988; Chakroun *et al.*, 1990; Vallochi *et al.*, 2008).

The intraocular immune response to *T. gondii* is mediated primarily by $CD4^+$ and $CD8^+$ T-cell responses (Norose *et al.*, 2011). Although the eye is an immune privileged site expressing low levels of MHC class I, infected mice show an increased expression of MHC Class I. Depletion of $CD8^+$ cells results in increased lesion formation and cyst burden in a murine model, indicating $CD8^+$ T-cells can also destroy parasite-infected host cells in an MHC class I restricted manner in the eye (Gazzinelli *et al.*, 1994a; Lyons *et al.*, 2001) . B-cells may also contribute to the immune response to ocular toxoplasmosis by limiting tachyzoite proliferation in the eye (Lu *et al.*, 2004).

IFNγ production has been associated with intraocular inflammation but numerous studies indicate IFNγ is protective against ocular *T. gondii* infection (Norose *et al.*, 2003a; Shen *et al.*, 2001;

Olle *et al.*, 1996). Studies have shown, for example, increased parasite dissemination and load in the retina and other regions of the eye in IFNγ-deficient mice, as compared to wild-type mice irrespective of host genetic background. Control of parasite proliferation in the eye is most likely through IFNγ induced protective mechanisms. In human retinal pigment epithelial cells, an IDO-dependent mechanism was found to be involved in the mechanism of IFNγ induced inhibition while in the murine model of ocular toxoplasmosis, IFNγ induced inhibition was found to be *via* a NO-dependent mechanism (Shen *et al.*, 2001; Nagineni *et al.*, 1996). IFNγ has also been shown to regulate *T. gondii* load and interconversion between the bradyzoite and tachyzoite stages in the murine eye (Norose *et al.*, 2003a). Apoptosis is also an important method used by immunoprivileged sites, including the eye to maintain immune privilege utilizing the Fas/FasL system. Infection with *T. gondii* induces FasL expression on cells of the mouse retina indicating apoptosis is also a mechanism controlling the parasite in the eye (Vallochi *et al.*, 2008).

Clinical presentations of ocular toxoplasmosis vary amongst patients with some patients presenting with only one episode of inflammation while others have multiple recurrences leading to loss of eyesight. Recurrent lesions are usually identified at the borders of the retinochoroidal scars, typically found in clusters that have been attributed to the rupture of cysts within the old lesions (Holland, 2003, 2004). Variation in clinical presentation and severity of disease has been attributed to both host genetic heterogeneity and parasite genotype. A predominance of type I strains has been found to be associated with ocular infection in the US, Poland and Brazil (Grigg et al., 2001; Khan, 2006; Vallochi et al., 2005; Switaj et al., 2006). Ocular toxoplasmosis in southern Brazil is higher than in most other regions in the world and a more severe necrotizing retinochoroiditis is more common (Vallochi *et al.*, 2008). In Brazil and South America, in addition to a predominance of type I strains, divergent parasite genotypes are also present that also appear to be more virulent (Khan *et al.*, 2006). These findings indicate that divergent genotypes likely contribute to severity of clinical outcomes although the underlying mechanisms affecting severity of *T. gondii* infection in the eye remain to be fully resolved.

23.12 IMMUNE EFFECTOR MECHANISMS IN CONGENITAL TOXOPLASMOSIS

Congenital toxoplasmosis is the result of a primary infection with *T. gondii* during pregnancy in which transplacental passage of the tachyzoite stage occurs and infects the foetus (Montoya and Remington, 2008). Foetal involvement is most severe when maternal toxoplasmosis is contracted early in pregnancy leading to spontaneous abortion or severe neurological effects. Conversely, infection in the third trimester is often asymptomatic, with development of chorioretinitis commonly occurring later in life. Congenital transmission occurs almost solely in seronegative women who have acute infection during pregnancy and is not typically seen in women who are seropositive before pregnancy (Montoya and Rosso, 2005; Montoya and Remington, 2008). However, recently exceptions have been found in the case of infections with atypical genotypes, as discussed below. Additionally there have been occasional reports of congenital transmission in women with immune suppression who have reactivation of latent *T. gondii* during pregnancy (Jones *et al.*, 2003).

Studies in humans and murine models that mimic certain key aspects of human congenital toxoplasmosis have contributed to the understanding of congenital toxoplasmosis (Roberts and Alexander, 1992; Menzies *et al.*, 2008). Evidence from these models indicates IFNγ has a dual role, both protecting the foetus and enhancing transmission. For example, increased levels of IFNγ levels can decrease maternal parasitemia and congenital transmission and enhance survival of pregnant mice (Abou-Bacar *et al.*, 2004; Shirahata *et al.*, 1992). Conversely, IFNγ deficient mice have reduced abortion rates although these mice exhibit increased numbers of parasites in their uterus and the placentas as compared with WT mice (Shiono *et al.*, 2007). Studies in human placenta villous explants indicate IFNγ induces parasitized white blood cell adhesion to placental villous cells and IFNγ may thus enhance the probability of vertical transmission (Ortiz-Alegria *et al.*, 2010). Pregnancy has the ability to modulate the immune response and it has been suggested disruption of immunomodulatory mechanisms by *T. gondii* infection could have detrimental effects on pregnancy such as inducing abortion while pregnancy-mediated immunomodulation may also favour parasite replication and congenital transmission (Menzies *et al.*, 2008).

IFNγ induced inhibition of *T. gondii* in umbilical endothelial cells is likely important in the prevention of transplacental transmission of congenital toxoplasmosis and *in vitro* studies investigating IFNγ effector mechanisms using umbilical vein endothelial cells has been studied in several different species. In sheep, treatment of umbilical vein cells with IFNγ blocks the growth of *T. gondii via* a NO and ROI independent mechanism (Dimier and Bout, 1996a). In human umbilical vein endothelial cells, IFNγ induces inhibition of *T. gondii via* a NO, ROI and IDO independent mechanism (Woodman *et al.*, 1991). A study in mice investigating the role of foetus versus maternal IFNγ induced IDO and iNOS expressions following *T. gondii* infection found that IDO production is largely due to foetal expression while iNOS production is largely determined by the maternal phenotype (Pfaff *et al.*, 2008). Placental trophoblast cells which form the border between maternal blood circulation and foetal tissue are also likely important in infection of the foetus. Studies in the BeWo model of human trophoblasts found these cells were refractory to stimulation by IFNγ and TNFα but a lack of polyamine metabolic products *via* cross-regulation by nitric oxide production was found to limit parasite replication (Pfaff *et al.*, 2005). These studies indicate multiple mechanisms may be involved in the host defence of the maternal–foetal interface, although further studies are clearly needed.

Finally, while it has traditionally been thought that transmission of *T. gondii* to the foetus occurs predominantly in women who acquire the infection during gestation, a new concept of congenital toxoplasmosis is emerging due to an understanding of the influence of genotypes of *T. gondii* and virulence (Lindsay and Dubey, 2011). The three major genotypes of *Toxoplasma* differ in their pathogenicity in mice and prevalence in humans although some *Toxoplasma* isolates have been termed atypical because they do not belong to the three major clonal lineages (Su *et al.*, 2006). The type II genotype is responsible for most congenital toxoplasmosis in Europe and the USA, but recently congenital toxoplasmosis in immune mothers infected by an atypical genotype has been found, indicating acquired immunity to the original infecting genotype may not protect against reinfection with an atypical strain (Elbez-Rubinstein *et al.*, 2009). Additionally, congenital toxoplasmosis caused by atypical genotypes results in more severe clinical manifestation of congenital toxoplasmosis than infection caused by typical genotypes (Delhaes *et al.*, 2010; Boughattas *et al.*, 2011). Serological tests to distinguish between typical versus avirulent strains are being developed and may help resolve the underlying mechanisms contributing to severity of congenital toxoplasmosis.

23.13 BEHAVOURIAL CONSEQUENCES OF INFECTION

One important aspect of *Toxoplasma* infection of intermediate hosts is its ability to alter host behaviour. Data regarding the effect of *Toxoplasma* on host behaviour come from two sources. One involves the experimental infection of laboratory animals, generally mice or rats, and the observation of the effects of infection on subsequent behaviour as measured by performance on defined cognitive and behavioural tests. The second set of studies involves the effect of *Toxoplasma* infections in human behaviour. Most of the studies of human behaviour involve the measurement of antibodies to *Toxoplasma* in sets of blood samples from case and control individuals who do or do not have the target behaviour. These studies can be divided into those which involve defined psychiatric diseases or conditions such as schizophrenia or suicide and those which involve the analysis of human personality traits. This section will focus on the effects of chronic *Toxoplasma* on immune competent hosts.

23.13.1 Animal Studies

The first published studies of the effect of persistent *Toxoplasma* infection in learning and memory in mice and rats were reported in 1978 (Piekarski *et al.*, 1978) and in 1979 (Witting, 1979). Since that time there have been multiple studies indicating that infection of mice and rats can result in deficits in learning and memory measured by a variety of standard laboratory measures such as mazes. However, the results of studies on mice and rats have been inconsistent with some studies showing little or no effect on memory or learning (Table 23.1) (Kannan and Pletnikov, 2012). Sources for this variability are likely to be related to differences in experimental methods, timing of infection, the genetic makeup of the infecting *Toxoplasma* organism and the gender and genetic makeup of the host (Kannan *et al.*, 2010).

A novel behavioural effect of *Toxoplasma* on host behaviour involves the apparent ability of *Toxoplasma* to lose its innate fear of felines, as measured by aversion to cat odours (Webster *et al.*, 1994). This 'fatal attraction' (Berdoy *et al.*, 2000) effect of *Toxoplasma* infection on rodents is postulated to have evolutionary significance since it would increase the likelihood of *Toxoplasma* in a secondary host such as a rodent completing its sexual cycle within the cat predator. The plausibility of this process is enhanced by the finding that the fear response is not lost when the smell is that of a non-feline predator such as mink (Lamberton *et al.*, 2008) fitting in with the fact that the transfer of *Toxoplasma* from one intermediate host to another would not result in the completion of the sexual cycle. However, other studies have indicated that mice can have decreased novelty seeking independent of cat behaviour in a way that might result in increased general predation (Hutchison *et al.*, 1980) suggesting that altered behaviour may also play a role in transmission of *Toxoplasma* among intermediate hosts (Afonso *et al.*, 2012). *Toxoplasma* infection has also been reported to alter other behaviours in intermediate hosts which might have evolutional significance, including sexual attractiveness (Dass *et al.*, 2011).

These biological and molecular mechanisms by which *Toxoplasma* modulates host behaviour are not completely understood. Neurophysiological studies indicate that the effect is not mediated by overall changes in general behaviours such as olfaction fear, anxiety, olfaction, or nonaversive learning (Vyas *et al.*, 2007), suggesting a high degree of specificity for some of the effects. However, other studies have documented more generalized effects of *Toxoplasma* infection on neuronal function (Haroon *et al.*, 2012). The neuroanatomical correlates of this specificity are currently under investigation (Vyas and Sapolsky, 2010).

There have also been a number of studies of the molecular basis of the *Toxoplasma* mediated host manipulation. Much of this research has centred on the ability of *Toxoplasma* infection to modulate the levels of dopamine within the host brain, due to the central importance of this neurotransmitter in animal behaviour. In 1985 it was reported that *Toxoplasma* infection can

TABLE 23.1 Learning and Memory in Rodents with *Toxoplasma gondii*

Spatial Learning and Memory			
Test Used	**Rodent Tested**	**Effect**	**Reference**
Labyrinth	Mouse (F), Rat (F)	Impaired	Piekarski *et al.*, 1978
	Mouse (F), Rat (F)	Impaired	Witting, 1979
	Mouse (F)	Impaired	Witting, 1979
	Rat (F)	None	Witting, 1979
Y maze	Mouse (F)	Impaired	Kannan *et al.*, 2010
	Mouse (F)	None	Kannan *et al.*, 2010
	Mouse (F, M)	None	Hay *et al.*, 1984
	Mouse (M)	None	Hutchinson *et al.*, 1980
Radial arm maze	Mouse (F)	Impaired	Hodkova *et al.*, 2007
Morris water maze	Rat (M)	None	Vyas *et al.*, 2007
Object recognition	Mouse (M)	None	Gulinello *et al.*, 2010
Olfactory Based Learning and Memory			
Test Used	**Rodent Tested**	**Effect**	**Reference**
Social transmission of food preference	Mouse (F)	None	Vyas *et al.*, 2007
	Mouse (F)	None	Xiao *et al.*, 2012
	Mouse (M)	Impaired	Xiao *et al.*, 2012
Associative Learning and Memory			
Test Used	**Rodent Tested**	**Effect**	**Reference**
Passive avoidance	Mouse (UK)	Impaired	Wang *et al.*, 2011

(M) male; (F) female; (UK) not stated in paper. Effect: Results related to learning and memory of each type assessed by the listed test. Table adapted from Kannan and Pletnikov, 2012.

result in an increase in dopamine concentrations in the brains of chronically infected mice (Stibbs, 1985), a finding which has been noted in some (Prandovszky *et al.*, 2011), but not all (Goodwin *et al.*, 2012), studies involving rodent brain infection with *Toxoplasma*. The possible role of dopamine is supported by findings indicating that dopamine uptake inhibitors can modulate the effect of *Toxoplasma* on rodent behaviour (Skallova *et al.*, 2006) and that dopamine stimulates the growth of *Toxoplasma* organisms in brain cell cultures (Strobl *et al.*, 2012). The findings that the *Toxoplasma* genome contains DNA capable of encoding two genes homologous to tyrosine hydroxylase, the rate limiting step in the synthesis of dopamine, raises the possibility that *Toxoplasma* organisms may directly manipulate dopamine levels within the brain (Gaskell *et al.*, 2009). However, the fact that immune modulators also change the level of dopamine in the brain (Capuron and Miller, 2011) raises the possibility that some of the alterations in neurotransmitters following *Toxoplasma* infection may be a by-product of the immune responses to that organism.

Persistent *Toxoplasma* infection results in the altered transcription of a large number of genes within the brains of infected mice; the specific genes which are altered varies depending upon the strain of the infecting *Toxoplasma* organism and the sex of the animal host (Xiao *et al.*, 2012). The effect of *Toxoplasma* infection within the brain can affect uninfected as well as infected cells perhaps due to rhoptry effector proteins. This phenomenon may explain the fact that *Toxoplasma* is capable of large scale effects on gene expression in the brain even when the number of cells infected is relatively small (Koshy *et al.*, 2012).

In conclusion, there is a great deal of evidence linking persistent *Toxoplasma* infection to behavioural alterations in experimentally infected mice and rats. The elucidation of the host and parasite biological and molecular mechanisms involved in these effects remains an important goal of research in this area. The investigation of the cognitive and behavioural effects of *Toxoplasma* should also be expanded to include additional species, such as primates, to allow for more relevant models of the effects of *Toxoplasma* infections in humans.

23.13.2 Human Studies

There is a long history of reported psychiatric symptoms in immunocompetent individuals exposed to *Toxoplasma*. The earliest reported cases of psychiatric symptoms were those of laboratory workers accidently infected with *Toxoplasma* in 1951 (Strom, 1951) and 1953 (Sexton *et al.*, 1953). Another early study of 114 individuals who acquired *Toxoplasma* infection between 1940 and 1964 found psychiatric symptoms in 24 (21.1%) of the cases at follow-up (Kramer, 1966). These and other studies led to a number of studies looking at the relationship between serological evidence of *Toxoplasma* infection and schizophrenia. A recent meta-analysis reported on 38 studies which met inclusion criteria and found an odds ratio of approximately 2.5 linking schizophrenia and antibodies to *Toxoplasma* measured by several different methods (Kramer, 1966; Torrey *et al.*, 2012) (Fig. 23.1). Increased levels of antibodies to *Toxoplasma gondii* have also been measured in individuals with bipolar disorder, which shares many features with schizophrenia (Pearce *et al.*, 2012) as well as women with prenatal depression and anxiety (Groer *et al.*, 2011). While most of these studies were retrospective in nature, there have been several prospective studies documenting that the increased level of *Toxoplasma* antibodies can be detected prior to the onset of symptoms (Pedersen *et al.*, 2011). Furthermore, several studies have reported a significant relationship between *Toxoplasma* antibodies in the mother or in the neonate and the development of schizophrenia and psychosis in the offspring (Brown *et al.*, 2005; Mortensen *et al.*, 2007; Blomstrom *et al.*, 2012) although this has not been noted in every population (Buka *et al.*, 2001). One of the variations in terms of the susceptibility of offspring to schizophrenia following maternal exposure may be the serotype of the organism, with the offspring of mothers exposed to type I infections displaying the highest level of risk (Xiao *et al.*, 2009).

Another serious psychiatric behaviour associated with increased levels of exposure to *T. gondii* is suicide and related self-destructive behaviours. Several studies have documented an association between increased levels of antibodies to *Toxoplasma* and increased rates of suicide, suicide attempts, and other forms of self-directed violence with similar odds ratios (Okusaga *et al.*, 2011; Ling *et al.*, 2011; Yagmur *et al.*, 2010; Arling *et al.*, 2009; Zhang *et al.*, 2012). One of the studies was prospective in nature with the *Toxoplasma* exposure being measured several years prior to the suicide or suicide attempt (Pedersen *et al.*, 2012).

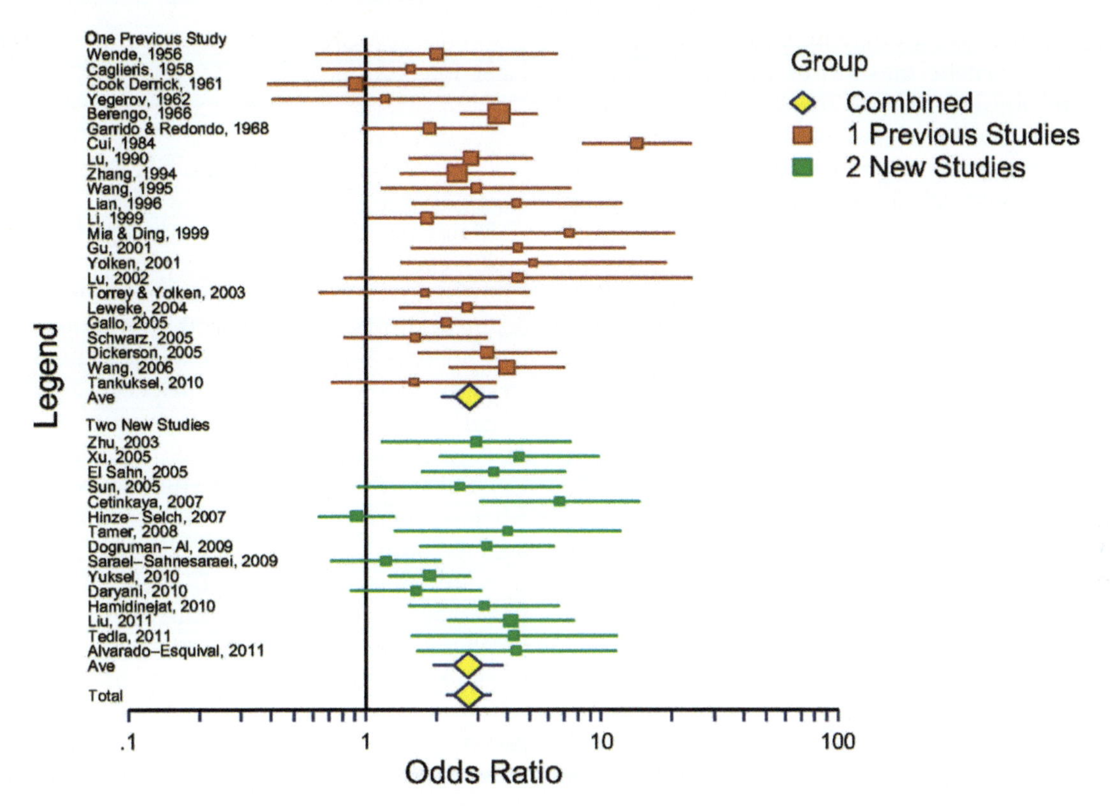

FIGURE 23.1 **plot of 38 studies that address the question of the effect of *Toxoplasma gondii* infection on the risk of schizophrenia in humans.** *Figure reproduced with permission from Figure 1, Schizophrenia Bulletin 2012; 38(3): page 643 (Torrey* et al.*, 2012).*

Increased levels of suicide attempts have been noted in *Toxoplasma* seropositive individuals with schizophrenia, as well as in *Toxoplasma* seropositive individuals without a previous history of a psychiatric disorder (Okusaga *et al.*, 2011). In addition, one study has documented a geographic association between risk of suicide and rates of *Toxoplasma* seropositivity within national populations in Europe (Lester, 2010). The mechanisms responsible for the association of *Toxoplasma* seropositivity and risk of suicide remain to be determined. It will be of interest to determine if the risk of suicide in humans is mechanistically related to the altered level of risk assessment found in *Toxoplasma* infected rodents. Increased levels of antibodies to *Toxoplasma* have also been associated with increased rates of automobile (Flegr *et al.*, 2009; Kocazeybek *et al.*, 2009) and workplace (Alvarado-Esquivel *et al.*, 2012) accidents as well as a range of personality traits (Fekadu *et al.*, 2010).

The association between *Toxoplasma* exposure and altered human behaviour is intriguing in light of the effects of experimental *Toxoplasma* on animal behaviour as noted above. However, it should be noted that there are a number of experimental limitations of the human studies. The most significant is that the human studies rely almost exclusively on the measurement of antibodies to *Toxoplasma* as markers of exposure. The levels measured in cross-sectional or longitudinal studies may thus reflect the robustness and genetic makeup of the host immune system and exposure to other antigens such as intestinally derived antigens (Severance *et al.*, 2012) as well as to previous exposure to *Toxoplasma*. The timing of infection and the antigenic makeup of the infecting strains are additional variables which are difficult to disentangle in serological studies. The level of association between *Toxoplasma* exposure and behavioural alterations is also somewhat variable among populations, with some studies failing to find significant associations. Also, while there are large differences in prevalence of *Toxoplasma* in different geographic regions and among individuals based on geographical, demographic and socio-economic variables, these differences are generally not reflected in rates of schizophrenia and related psychiatric disorders, with the exception of suicide, noted above. For example, most areas of the world which have relatively high rates of *Toxoplasma* exposure, such as parts of Latin America, Eastern and Central Europe, the Middle East and Africa (Pappas *et al.*, 2009), are not known to have particularly high rates of schizophrenia or bipolar disorder. Similarly, the declining rate of *Toxoplasma* noted in young adults in the USA (Jones *et al.*, 2007) and Northern Europe (Hofhuis *et al.*, 2011) does not seem to be associated with a concomitant decrease in rates of schizophrenia (Kirkbride *et al.*, 2012), bipolar disorder (Yutzy *et al.*, 2012), suicide (Hu *et al.*, 2008) or accidents (Karch *et al.*, 2011) in the same populations. The reasons for these discrepancies are not known with certainty but may be related to the effects of age of infection, the serotype and stage of the infecting strain, and the underlying genetic susceptibility of the human host. The availability of better assays to define these factors and to directly detect *Toxoplasma* tissue cysts within the brain and other organs would represent major steps forward in terms of defining the role of *Toxoplasma* in brain disorders in individuals and in populations and in planning appropriate therapeutic and prophylactic interventions.

23.14 CONCLUSIONS

Maintenance of latent infection in the normal host, reactivation of infection in the immunocompromised host and the neuropathologies in congenital infection and in the immunocompetent host are not fully understood although it is clear the immune response (especially IFNγ) is largely responsible for controlling the growth of *T. gondii* in the CNS in all of these conditions. Control of the parasite in the brain involves a balance of these host immune protective mechanisms and immunomodulatory mechanisms that limit neuropathology. In the brain, the relevant IFNγ effector cells include $CD4^+$ and $CD8^+$ T-cells, and dendritic cells, which are involved in the stimulation and maintenance of the intracerebral immune response, while microglia exert potent anti-*Toxoplasma* activity in the brain parenchyma and astrocytes play a crucial immunoregulatory role. Neurons serve as the main host cell for the cyst stage but recent studies indicate neurons also play a crucial immunoprotective role in the brain in chronic infections and that the infection significantly affects neuronal functions. Finally, although the chronic infection has been considered 'latent', increasing evidence indicates neuropsychiatric disorders, such as schizophrenia, and other neuropathologies, such as cryptogenic epilepsy, may be associated with the latent infection. A better understanding of the latent infection and the biology of *T. gondii* is of great importance for a greater understanding of the pathogenesis of cerebral and congenital toxoplasmosis and associated neuropathologies.

Acknowledgements

This work is supported, in part, by National Institues of Health Grants AI078756 (Y.S.), AI095032 (Y.S.), and AI077887 (Y.S.) and a grant from the Stanley Medical Research Insitute 08R-2047 (Y.S.).

References

Abou-Bacar, A., Pfaff, A.W., Georges, S., Letscher-Bru, V., Filisetti, D., Villard, O., Antoni, E., Klein, J.P., Candolfi, E., 2004. Role of NK-cells and gamma interferon in transplacental passage of Toxoplasma gondii in a mouse model of primary infection. Infect. Immun. 72, 1397–1401.

Adams, L.B., Franzblau, S.G., Vavrin, Z., Hibbs Jr., J.B., Krahenbuhl, J.L., 1991. L-arginine-dependent macrophage effector functions inhibit metabolic activity of Mycobacterium leprae. J. Immunol. 147, 1642–1646.

Adams, L.B., Hibbs Jr., J.B., Taintor, R.R., Krahenbuhl, J.L., 1990. Microbiostatic effect of murine-activated macrophages for Toxoplasma gondii. Role for synthesis of inorganic nitrogen oxides from L-arginine. J. Immunol. 144, 2725–2729.

Afonso, C., Paixao, V.B., Costa, R.M., 2012. Chronic Toxoplasma infection modifies the structure and the risk of host behavior. PLoS One 7, e32489.

Akira, S., Taga, T., Kishimoto, T., 1993. Interleukin-6 in biology and medicine. Adv. Immunol. 54, 1–78.

Albuquerque, M.C., Aleixo, A.L., Benchimol, E.I., Leandro, A.C., Das Neves, L.B., Vicente, R.T., Bonecini-Almeida Mda, G., Amendoeira, M.R., 2009. The IFN-gamma +874T/A gene polymorphism is associated with retinochoroiditis toxoplasmosis susceptibility. Mem. Inst. Oswaldo Cruz 104, 451–455.

Alexander, J., Jebbari, H., Bluethmann, H., Brombacher, F., Roberts, C.W., 1998. The role of IL-4 in adult acquired and congenital toxoplasmosis. Int. J. Parasitol. 28, 113–120.

Aliberti, J., Reis E Sousa, C., Schito, M., Hieny, S., Wells, T., Huffnagle, G.B., Sher, A., 2000. CCR5 provides a signal for microbial induced production of IL-12 by CD8 alpha+ dendritic cells. Nat. Immunol. 1, 83–87.

Aliberti, J., Serhan, C., Sher, A., 2002. Parasite-induced lipoxin A4 is an endogenous regulator of IL-12 production and immunopathology in Toxoplasma gondii infection. J. Exp. Med. 196, 1253–1262.

Aliberti, J., Valenzuela, J.G., Carruthers, V.B., Hieny, S., Andersen, J., Charest, H., Reis E Sousa, C., Fairlamb, A., Ribeiro, J.M., Sher, A., 2003. Molecular mimicry of a CCR5 binding-domain in the microbial activation of dendritic cells. Nat. Immunol. 4, 485–490.

Aline, F., Bout, D., Dimier-Poisson, I., 2002. Dendritic cells as effector cells: gamma interferon activation of murine dendritic cells triggers oxygen-dependent inhibition of Toxoplasma gondii replication. Infect. Immun. 70, 2368–2374.

Alvarado-Esquivel, C., Torres-Castorena, A., Liesenfeld, O., Estrada-Martinez, S., Urbina-Alvarez, J.D., 2012. High seroprevalence of Toxoplasma gondii infection in a subset of Mexican patients with work accidents and low socioeconomic status. Parasit. Vectors 5, 13.

Arling, T.A., Yolken, R.H., Lapidus, M., Langenberg, P., Dickerson, F.B., Zimmerman, S.A., Balis, T., Cabassa, J.A., Scrandis, D.A., Tonelli, L.H., Postolache, T.T., 2009. Toxoplasma gondii antibody titers and history of suicide attempts in patients with recurrent mood disorders. J. Nerv. Ment. Dis. 197, 905–908.

Aviles, H., Stiles, J., O'Donnell, P., Orshal, J., Leid, J., Sonnenfeld, G., Monroy, F., 2008. Kinetics of systemic cytokine and brain chemokine gene expression in murine toxoplasma infection. J. Parasitol. 94, 1282–1288.

Beaman, M.H., Wong, S.Y., Abrams, J.S., Remington, J.S., 29–31 January 1993. Splenic IL–4 secretion is not associated with the development of toxoplasmic encephalitis [abstract 81]. Program and abstracts of the 2nd International Congress of Biological Response Modifiers. San Diego.

Bekpen, C., Hunn, J.P., Rohde, C., Parvanova, I., Guethlein, L., Dunn, D.M., Glowalla, E., Leptin, M., Howard, J.C., 2005. The interferon-inducible p47 (IRG) GTPases in vertebrates: loss of the cell autonomous resistance mechanism in the human lineage. Genome Biol. 6, R92.

Bekpen, C., Marques-Bonet, T., Alkan, C., Antonacci, F., Leogrande, M.B., Ventura, M., Kidd, J.M., Siswara, P., Howard, J.C., Eichler, E.E., 2009. Death and resurrection of the human IRGM gene. PLoS Genet. 5, e1000403.

Bekpen, C., Xavier, R.J., Eichler, E.E., 2010. Human IRGM gene 'to be or not to be'. Semin. Immunopathol. 32, 437–444.

Berdoy, M., Webster, J.P., MacDonald, D.W., 2000. Fatal attraction in rats infected with Toxoplasma gondii. Proc. Biol. Sci. 267, 1591–1594.

Bierly, A.L., Shufesky, W.J., Sukhumavasi, W., Morelli, A.E., Denkers, E.Y., 2008. Dendritic cells expressing plasmacytoid marker PDCA-1 are Trojan horses during Toxoplasma gondii infection. J. Immunol. 181, 8485–8491.

Blanchard, N., Gonzalez, F., Schaeffer, M., Joncker, N.T., Cheng, T., Shastri, A.J., Robey, E.A., Shastri, N., 2008. Immunodominant, protective response to the parasite Toxoplasma gondii requires antigen processing in the endoplasmic reticulum. Nat. Immunol. 9, 937–944.

Bliss, S.K., Butcher, B.A., Denkers, E.Y., 2000. Rapid recruitment of neutrophils containing prestored IL-12 during microbial infection. J. Immunol. 165, 4515–4521.

Bliss, S.K., Gavrilescu, L.C., Alcaraz, A., Denkers, E.Y., 2001. Neutrophil depletion during Toxoplasma gondii infection leads to impaired immunity and lethal systemic pathology. Infect. Immun. 69, 4898–4905.

Blomstrom, A., Karlsson, H., Wicks, S., Yang, S., Yolken, R.H., Dalman, C., 2012. Maternal antibodies to infectious agents and risk for non-affective psychoses in the offspring - a matched case-control study. Schizophr. Res. 140, 25–30.

Boehm, U., Klamp, T., Groot, M., Howard, J.C., 1997. Cellular responses to interferon-gamma. Annu. Rev. Immunol. 15, 749–795.

Bohne, W., Heesemann, J., Gross, U., 1994. Reduced replication of Toxoplasma gondii is necessary for induction of bradyzoite-specific antigens: a possible role for nitric oxide in triggering stage conversion. Infect. Immun. 62, 1761–1767.

Boughattas, S., Abdallah, R.B., Siala, E., Aoun, K., Bouratbine, A., 2011. An atypical strain associated with congenital toxoplasmosis in Tunisia. New Microbiol. 34, 413–416.

Bout, D., Moretto, M., Dimier-Poisson, I., Gatel, D.B., 1999. Interaction between Toxoplasma gondii and enterocyte. Immunobiology 201, 225–228.

Brown, A.S., Schaefer, C.A., Quesenberry Jr., C.P., Liu, L., Babulas, V.P., Susser, E.S., 2005. Maternal exposure to toxoplasmosis and risk of schizophrenia in adult offspring. Am. J. Psychiatry 162, 767–773.

Brown, C.R., David, C.S., Khare, S.J., Mcleod, R., 1994. Effects of human class I transgenes on Toxoplasma gondii cyst formation. J. Immunol. 152, 4537–4541.

Brown, C.R., Hunter, C.A., Estes, R.G., Beckmann, E., Forman, J., David, C., Remington, J.S., Mcleod, R., 1995. Definitive identification of a gene that confers resistance against Toxoplasma cyst burden and encephalitis. Immunology 85, 419–428.

Buka, S.L., Tsuang, M.T., Torrey, E.F., Klebanoff, M.A., Bernstein, D., Yolken, R.H., 2001. Maternal infections and subsequent psychosis among offspring. Arch. Gen. Psychiatry 58, 1032–1037.

Butcher, B.A., Greene, R.I., Henry, S.C., Annecharico, K.L., Weinberg, J.B., Denkers, E.Y., Sher, A., Taylor, G.A., 2005. p47 GTPases regulate Toxoplasma gondii survival in activated macrophages. Infect. Immun. 73, 3278–3286.

Caamano, J., Alexander, J., Craig, L., Bravo, R., Hunter, C.A., 1999. The NF-kappa B family member RelB is required for innate and adaptive immunity to Toxoplasma gondii. J. Immunol. 163, 4453–4461.

Caamano, J., Tato, C., Cai, G., Villegas, E.N., Speirs, K., Craig, L., Alexander, J., Hunter, C.A., 2000. Identification of a role for NF-kappa B2 in the regulation of apoptosis and in maintenance of T-cell-mediated immunity to Toxoplasma gondii. J. Immunol. 165, 5720–5728.

Caetano, B.C., Bruna-Romero, O., Fux, B., Mendes, E.A., Penido, M.L., Gazzinelli, R.T., 2006. Vaccination with replication-deficient recombinant adenoviruses encoding the main surface antigens of toxoplasma gondii induces immune response and protection against infection in mice. Hum. Gene. Ther. 17, 415–426.

Capuron, L., Miller, A.H., 2011. Immune system to brain signalling: neuropsychopharmacological implications. Pharmacol. Ther. 130, 226–238.

Chakroun, M., Meyohas, M.C., Pelosse, B., Zazoun, L., Vacherot, B., Derouin, F., Leport, C., 1990. [Ocular toxoplasmosis in AIDS]. Ann. Med. Interne. (Paris) 141, 472–474.

Chan, A., Hummel, V., Weilbach, F.X., Kieseier, B.C., Gold, R., 2006. Phagocytosis of apoptotic inflammatory cells downregulates microglial chemoattractive function and migration of encephalitogenic T-cells. J. Neurosci. Res. 84, 1217–1224.

Chang, H.R., Pechere, J.C., 1989. Macrophage oxidative metabolism and intracellular Toxoplasma gondii. Microb. Pathog. 7, 37–44.

Chao, C.C., Anderson, W.R., Hu, S., Gekker, G., Martella, A., Peterson, P.K., 1993. Activated microglia inhibit multiplication of Toxoplasma gondii *via* a nitric oxide mechanism. Clin. Immunol. Immunopathol. 67, 178–183.

Chao, C.C., Gekker, G., Hu, S., Peterson, P.K., 1994. Human microglial cell defense against Toxoplasma gondii. The role of cytokines. J. Immunol. 152, 1246–1252.

Chao, C.C., Hu, S., Gekker, G., Novick Jr., W.J., Remington, J.S., Peterson, P.K., 1993. Effects of cytokines on multiplication of Toxoplasma gondii in microglial cells. J. Immunol. 150, 3404–3410.

Chaves, A.C., Ceravolo, I.P., Gomes, J.A., Zani, C.L., Romanha, A.J., Gazzinelli, R.T., 2001. IL-4 and IL-13 regulate the induction of indoleamine 2,3-dioxygenase activity and the control of Toxoplasma gondii replication in human fibroblasts activated with IFN-gamma. Eur. J. Immunol. 31, 333–344.

Collazo, C.M., Yap, G.S., Hieny, S., Caspar, P., Feng, C.G., Taylor, G.A., Sher, A., 2002. The function of gamma interferon-inducible GTP-binding protein IGTP in host resistance to *Toxoplasma gondii* is Stat1 dependent and requires expression in both hematopoietic and nonhematopoietic cellular compartments. Infect. Immun. 70, 6933–6939.

Cong, H., Mui, E.J., Witola, W.H., Sidney, J., Alexander, J., Sette, A., Maewal, A., Mcleod, R., 2010. Towards an immunosense vaccine to prevent toxoplasmosis: protective Toxoplasma gondii epitopes restricted by HLA-A*0201. Vaccine 29, 754–762.

Cong, H., Mui, E.J., Witola, W.H., Sidney, J., Alexander, J., Sette, A., Maewal, A., McLeod, R., 2011. Human immunome, bioinformatic analyses using HLA supermotifs and the parasite genome, binding assays, studies of human T-cell responses, and immunization of HLA-A*1101 transgenic mice including novel adjuvants provide a foundation for HLA-A03 restricted CD8+T cell epitope based, adjuvanted vaccine protective against Toxoplasma gondii. Immunome. Res. 6, 12.

Courret, N., Darche, S., Sonigo, P., Milon, G., Buzoni-Gatel, D., Tardieux, I., 2006. CD11c- and CD11b-expressing mouse leukocytes transport single Toxoplasma gondii tachyzoites to the brain. Blood 107, 309–316.

Creuzet, C., Robert, F., Roisin, M.P., Van Tan, H., Benes, C., Dupouy-Camet, J., Fagard, R., 1998. Neurons in primary culture are less efficiently infected by Toxoplasma gondii than glial cells. Parasitol. Res. 84, 25–30.

Da Gama, L.M., Ribeiro-Gomes, F.L., Guimaraes Jr., U., Arnholdt, A.C., 2004. Reduction in adhesiveness to extracellular matrix components, modulation of adhesion molecules and in vivo migration of murine macrophages infected with Toxoplasma gondii. Microbes Infect. 6, 1287–1296.

Dass, S.A., Vasudevan, A., Dutta, D., Soh, L.J., Sapolsky, R.M., Vyas, A., 2011. Protozoan parasite Toxoplasma gondii manipulates mate choice in rats by enhancing attractiveness of males. PLoS One 6, e27229.

Daubener, W., Posdziech, V., Hadding, U., Mackenzie, C.R., 1999. Inducible anti-parasitic effector mechanisms in human uroepithelial cells: tryptophan degradation vs. NO production. Med. Microbiol. Immunol. 187, 143–147.

Daubener, W., Spors, B., Hucke, C., Adam, R., Stins, M., Kim, K.S., Schroten, H., 2001. Restriction of Toxoplasma gondii growth in human brain microvascular endothelial cells by activation of indoleamine 2,3-dioxygenase. Infect. Immun. 69, 6527–6531.

Deckert, M., Sedgwick, J.D., Fischer, E., Schluter, D., 2006. Regulation of microglial cell responses in murine Toxoplasma encephalitis by CD200/CD200 receptor interaction. Acta Neuropathol. 111, 548–558.

Deckert-Schluter, M., Albrecht, S., Hof, H., Wiestler, O.D., Schluter, D., 1995. Dynamics of the intracerebral and splenic cytokine mRNA production in Toxoplasma gondii-resistant and -susceptible congenic strains of mice. Immunology 85, 408–418.

Deckert-Schluter, M., Bluethmann, H., Kaefer, N., Rang, A., Schluter, D., 1999. Interferon-gamma receptor-mediated but not tumor necrosis factor receptor type 1- or type 2-mediated signalling is crucial for the activation of cerebral blood vessel endothelial cells and microglia in murine Toxoplasma encephalitis. Am. J. Pathol. 154, 1549–1561.

Deckert-Schluter, M., Bluethmann, H., Rang, A., Hof, H., Schluter, D., 1998. Crucial role of TNF receptor type 1 (p55), but not of TNF receptor type 2 (p75), in murine toxoplasmosis. J. Immunol. 160, 3427–3436.

Delair, E., Latkany, P., Noble, A.G., Rabiah, P., McLeod, R., Brezin, A., 2011. Clinical manifestations of ocular toxoplasmosis. Ocul. Immunol. Inflamm. 19, 91–102.

Delhaes, L., Ajzenberg, D., Sicot, B., Bourgeot, P., Darde, M.L., Dei-Cas, E., Houfflin-Debarge, V., 2010. Severe congenital toxoplasmosis due to a Toxoplasma gondii strain with an atypical genotype: case report and review. Prenat. Diagn. 30, 902–905.

Dellacasa-Lindberg, I., Fuks, J.M., Arrighi, R.B., Lambert, H., Wallin, R.P., Chambers, B.J., Barragan, A., 2011. Migratory activation of primary cortical microglia upon infection with Toxoplasma gondii. Infect. Immun. 79, 3046–3052.

Denkers, E.Y., Yap, G., Scharton-Kersten, T., Charest, H., Butcher, B.A., Caspar, P., Heiny, S., Sher, A., 1997. Perforin-mediated cytolysis plays a limited role in host resistance to *Toxoplasma gondii*. J. Immunol. 159, 1903–1908.

Dimier, I.H., Bout, D.T., 1993. Rat intestinal epithelial cell line IEC-6 is activated by recombinant interferon-gamma to inhibit replication of the coccidian Toxoplasma gondii. European J. Immunol. 23, 981–983.

Dimier, I.H., Bout, D.T., 1996. Inhibitory effect of interferon-gamma activated ovine umbilical vein endothelial cells on the intracellular replication of Toxoplasma gondii. Vet. Res. 27, 527–534.

Dimier, I.H., Bout, D.T., 1997. Inhibition of Toxoplasma gondii replication in IFN-gamma-activated human intestinal epithelial cells. Immunol. Cell Biol. 75, 511–514.

Dimier, I.H., Bout, D.T., 1998. Interferon-gamma-activated primary enterocytes inhibit Toxoplasma gondii replication: a role for intracellular iron. Immunology 94, 488–495.

Dimier, I.H., Woodman, J.P., Bout, D.T., 1992. [Human endothelial cells activated by interferon gamma, by interleukin-1 and TNF inhibit the replication of Toxoplasma gondii]. Annales de recherches veterinaires. Ann. Rech. Vet. 23, 329–330.

Drogemuller, K., Helmuth, U., Brunn, A., Sakowicz-Burkiewicz, M., Gutmann, D.H., Mueller, W., Deckert, M., Schluter, D., 2008. Astrocyte gp130 expression is critical for the control of Toxoplasma encephalitis. J. Immunol. 181, 2683–2693.

Dutta, C., Grimwood, J., Kasper, L.H., 2000. Attachment of Toxoplasma gondii to a specific membrane fraction of CHO cells. Infect. Immun. 68, 7198–7201.

Dzierszinski, F., Pepper, M., Stumhofer, J.S., Larosa, D.F., Wilson, E.H., Turka, L.A., Halonen, S.K., Hunter, C.A., Roos, D.S., 2007. Presentation of Toxoplasma gondii antigens *via* the endogenous major histocompatibility complex class I pathway in nonprofessional and professional antigen-presenting cells. Infect. Immun. 75, 5200–5209.

Elbez-Rubinstein, A., Ajzenberg, D., Darde, M.L., Cohen, R., Dumetre, A., Yera, H., Gondon, E., Janaud, J.C., Thulliez, P., 2009. Congenital toxoplasmosis and reinfection during pregnancy: case report, strain characterization, experimental model of reinfection, and review. J. Infect. Dis. 199, 280–285.

Fagard, R., Van Tan, H., Creuzet, C., Pelloux, H., 1999. Differential development of Toxoplasma gondii in neural cells. Parasitol. Today 15, 504–507.

Fekadu, A., Shibre, T., Cleare, A.J., 2010. Toxoplasmosis as a cause for behaviour disorders – overview of evidence and mechanisms. Folia Parasitol. 57, 105–113.

Fentress, S.J., Behnke, M.S., Dunay, I.R., Mashayekhi, M., Rommereim, L.M., Fox, B.A., Bzik, D.J., Taylor, G.A., Turk, B.E., Lichti, C.F., Townsend, R.R., Qiu, W., Hui, R., Beatty, W.L., Sibley, L.D., 2010. Phosphorylation of immunity-related GTPases by a Toxoplasma gondii-secreted kinase promotes macrophage survival and virulence. Cell Host Microbe 8, 484–495.

Ferguson, D.J., Hutchison, W.M., 1987a. The host-parasite relationship of Toxoplasma gondii in the brains of chronically infected mice. Virchows Arch. A Pathol. Anat. Histopathol. 411, 39–43.

Ferguson, D.J., Hutchison, W.M., 1987b. An ultrastructural study of the early development and tissue cyst formation of Toxoplasma gondii in the brains of mice. Parasitol. Res. 73, 483–491.

Fischer, H.G., Bielinsky, A.K., Nitzgen, B., Daubener, W., Hadding, U., 1993. Functional dichotomy of mouse microglia developed in vitro: differential effects of macrophage and granulocyte/macrophage colony-stimulating factor on cytokine secretion and antitoxoplasmic activity. J. Neuroimmunol. 45, 193–201.

Fischer, H.G., Bonifas, U., Reichmann, G., 2000. Phenotype and functions of brain dendritic cells emerging during chronic infection of mice with Toxoplasma gondii. J. Immunol. 164, 4826–4834.

Fischer, H.G., Nitzgen, B., Reichmann, G., Gross, U., Hadding, U., 1997. Host cells of Toxoplasma gondii encystation in infected primary culture from mouse brain. Parasitol. Res. 83, 637–641.

Fleckenstein, M.C., Reese, M.L., Konen-Waisman, S., Boothroyd, J.C., Howard, J.C., Steinfeldt, T., 2012. A Toxoplasma gondii Pseudokinase Inhibits Host IRG Resistance Proteins. PLoS Biol. 10, e1001358.

Flegr, J., Klose, J., Novotna, M., Berenreitterova, M., Havlicek, J., 2009. Increased incidence of traffic accidents in Toxoplasma-infected military drivers and protective effect RhD molecule revealed by a large-scale prospective cohort study. BMC Infect. Dis. 9, 72.

Franzoso, G., Carlson, L., Scharton-Kersten, T., Shores, E.W., Epstein, S., Grinberg, A., Tran, T., Shacter, E., Leonardi, A., Anver, M., Love, P., Sher, A., Siebenlist, U., 1997. Critical roles for the Bcl-3 oncoprotein in T-cell-mediated immunity, splenic microarchitecture, and germinal center reactions. Immunity 6, 479–490.

Frenkel, J.K., Escajadillo, A., 1987. Cyst rupture as a pathogenic mechanism of toxoplasmic encephalitis. Am. J. Trop. Med. Hyg. 36, 517–522.

Frenkel, J.K., Taylor, D.W., 1982. Toxoplasmosis in immunoglobulin M-suppressed mice. Infect. Immun. 38, 360–367.

Freund, Y.R., Sgarlato, G., Jacob, C.O., Suzuki, Y., Remington, J.S., 1992. Polymorphisms in the tumor necrosis factor alpha (TNF-alpha) gene correlate with murine resistance to development of toxoplasmic encephalitis and with levels of TNF-alpha mRNA in infected brain tissue. J. Exp. Med. 175, 683–688.

Freund, Y.R., Zaveri, N.T., Javitz, H.S., 2001. In vitro investigation of host resistance to Toxoplasma gondii infection in microglia of BALB/c and CBA/Ca mice. Infect. Immun. 69, 765–772.

Frickel, E.M., Sahoo, N., Hopp, J., Gubbels, M.J., Craver, M.P., Knoll, L.J., Ploegh, H.L., Grotenbreg, G.M., 2008. Parasite stage-specific recognition of endogenous Toxoplasma gondii-derived CD8+ T-cell epitopes. J. Infect. Dis. 198, 1625–1633.

Gaskell, E.A., Smith, J.E., Pinney, J.W., Westhead, D.R., McConkey, G.A., 2009. A unique dual activity amino acid hydroxylase in Toxoplasma gondii. PLoS One 4, e4801.

Gavrilescu, L.C., Butcher, B.A., Del Rio, L., Taylor, G.A., Denkers, E.Y., 2004. STAT1 is essential for antimicrobial effector function but dispensable for gamma interferon production during Toxoplasma gondii infection. Infect. Immun. 72, 1257–1264.

Gazzinelli, R., Xu, Y., Hieny, S., Cheever, A., Sher, A., 1992. Simultaneous depletion of $CD4^{+}$ and $CD8^{+}$ T lymphocytes is required to reactivate chronic infection with *Toxoplasma gondii*. J. Immunol. 149, 175–180.

Gazzinelli, R.T., Brezin, A., Li, Q., Nussenblatt, R.B., Chan, C.C., 1994. Toxoplasma gondii: acquired ocular toxoplasmosis in the murine model, protective role of TNF-alpha and IFN-gamma. Exp. Parasitol. 78, 217–229.

Gazzinelli, R.T., Eltoum, I., Wynn, T.A., Sher, A., 1993. Acute cerebral toxoplasmosis is induced by in vivo neutralization of TNF-alpha and correlates with the down-regulated expression of inducible nitric oxide synthase and other markers of macrophage activation. J. Immunol. 151, 3672–3681.

Gazzinelli, R.T., Wysocka, M., Hayashi, S., Denkers, E.Y., Hieny, S., Caspar, P., Trinchieri, G., Sher, A., 1994. Parasite-induced IL-12 stimulates early IFN-gamma synthesis and resistance during acute infection with Toxoplasma gondii. J. Immunol. 153, 2533–2543.

Gazzinelli, R.T., Wysocka, M., Hieny, S., Scharton-Kersten, T., Cheever, A., Kuhn, R., Muller, W., Trinchieri, G., Sher, A., 1996. In the absence of endogenous IL-10, mice acutely infected with Toxoplasma gondii succumb to a lethal immune response dependent on CD4+ T-cells and accompanied by overproduction of IL-12, IFN-gamma and TNF-alpha. J. Immunol. 157, 798–805.

Ghatak, N.R., Zimmerman, H.M., 1973. Fine structure of Toxoplasma in the human brain. Arch. Pathol. 95, 276–283.

Goodwin, D., Hrubec, T.C., Klein, B.G., Strobl, J.S., Werre, S.R., Han, Q., Zajac, A.M., Lindsay, D.S., 2012. Congenital infection of mice with Toxoplasma gondii induces minimal change in behavior and no change in neurotransmitter concentrations. J. Parasitol. 98, 706–712.

Grimwood, J., Mineo, J.R., Kasper, L.H., 1996. Attachment of Toxoplasma gondii to host cells is host cell cycle dependent. Infect. Immun. 64, 4099–4104.

Grigg, M.E., Ganatra, J., Boothroyd, J.C., Margolis, T.P., 2001. Unusual abundance of atypical strains associated with human ocular toxoplasmosis. J. Infect. Dis 184, 633–639.

Groer, M.W., Yolken, R.H., Xiao, J.C., Beckstead, J.W., Fuchs, D., Mohapatra, S.S., Seyfang, A., Postolache, T.T., 2011. Prenatal depression and anxiety in Toxoplasma gondii-positive women. Am. J. Obstet. Gynecol. 204 (433), e1–e7.

Grover, H.S., Blanchard, N., Gonzalez, F., Chan, S., Robey, E.A., Shastri, N., 2012 Sep. The Toxoplasma gondii peptide AS15 elicits CD4 T cells that can control parasite burden. Infect Immun 80 (9), 3279–3288.

Gulinello, M., Acquarone, M., Kim, J.H., et al., 2010. Acquired infection with Toxoplasma gondii in adult mice results in sensorimotor deficits but normal cognitive behavior despite widespread brain pathology. Microbes Infect. 12, 528–537.

Halonen, S.K., Chiu, F., Weiss, L.M., 1998a Oct. Effect of cytokines on growth of Toxoplasma gondii in murine astrocytes. Infect. Immun. 66, 4989–4993.

Halonen, S.K., Weiss, L.M., Chiu, F.C., 1998b May. Association of host cell intermediate filaments with Toxoplasma gondii cysts in murine astrocytes in vitro. Int J Parasitol 28 (5), 815–823.

Halonen, S.K., Lyman, W.D., Chiu, F.C., 1996. Growth and development of Toxoplasma gondii in human neurons and astrocytes. J. Neuropathol. Exp. Neurol. 55, 1150–1156.

Halonen, S.K., Taylor, G.A., Weiss, L.M., 2001. Gamma interferon-induced inhibition of Toxoplasma gondii in astrocytes is mediated by IGTP. Infect. Immun. 69, 5573–5576.

Halonen, S.K., Weiss, L.M., 2000. Investigation into the mechanism of gamma interferon-mediated inhibition of Toxoplasma gondii in murine astrocytes. Infect. Immun. 68, 3426–3430.

Halonen, S.K., Weiss, L.M., Chiu, F.C., 1998. Association of host cell intermediate filaments with Toxoplasma gondii cysts in murine astrocytes in vitro. Int. J. Parasitol. 28, 815–823.

Halonen, S.K., Woods, T., Mcinnerney, K., Weiss, L.M., 2006. Microarray analysis of IFN-gamma response genes in astrocytes. J. Neuroimmunol. 175, 19–30.

Handel, U., Brunn, A., Drogemuller, K., Muller, W., Deckert, M., Schluter, D., 2012. Neuronal gp130 Expression is Crucial to Prevent Neuronal Loss, Hyperinflammation, and Lethal Course of Murine Toxoplasma Encephalitis. Am. J. Pathol. 181, 163–173.

Haroon, F., Handel, U., Angenstein, F., Goldschmidt, J., Kreutzmann, P., Lison, H., Fischer, K.D., Scheich, H., Wetzel, W., Schluter, D., Budinger, E., 2012. Toxoplasma gondii actively inhibits neuronal function in chronically infected mice. PLoS One 7, e35516.

Harris, T.H., Wilson, E.H., Tait, E.D., Buckley, M., Shapira, S., Caamano, J., Artis, D., Hunter, C.A., 2010. NF-kappaB1 contributes to T-cell-mediated control of Toxoplasma gondii in the CNS. J. Neuroimmunol. 222, 19–28.

Hay, J., Aitken, P.P., Graham, D.I., 1984. Toxoplasma infection and response to novelty in mice. Z. Parasitenkd. 70, 575–588.

Hayashi, S., Chan, C.C., Gazzinelli, R., Roberge, F.G., 1996. Contribution of nitric oxide to the host parasite equilibrium in toxoplasmosis. J. Immunol. 156, 1476–1481.

Hayashi, S., Chan, C.C., Gazzinelli, R.T., Pham, N.T., Cheung, M.K., Roberge, F.G., 1996. Protective role of nitric oxide in ocular toxoplasmosis. Br. J. Ophthalmol. 80, 644–648.

Hester, J., Mullins, J., Sa, Q., Payne, L., Mercier, C., Cesbron-Delauw, M.F., Suzuki, Y., 2012. Toxoplasma gondii Antigens Recognized by IgG Antibodies Differ between Mice with and without Active Proliferation of Tachyzoites in the Brain during the Chronic Stage of Infection. Infect. Immun. 80, 3611–3620.

Hirota, H., Kiyama, H., Kishimoto, T., Taga, T., 1996. Accelerated Nerve Regeneration in Mice by upregulated expression of interleukin (IL) 6 and IL-6 receptor after trauma. J. Exp. Med. 183, 2627–2634.

Hisaeda, H., Nagasawa, H., Maeda, K., Maekawa, Y., Ishikawa, H., Ito, Y., Good, R.A., Himeno, K., 1995. Gamma delta T-cells play an important role in hsp65 expression and in acquiring protective immune responses against infection with Toxoplasma gondii. J. Immunol. 155, 244–251.

Hochrein, H., O'Keeffe, M., Luft, T., Vandenabeele, S., Grumont, R.J., Maraskovsky, E., Shortman, K., 2000. Interleukin (IL)-4 is a major regulatory cytokine governing bioactive IL-12 production by mouse and human dendritic cells. J. Exp. Med. 192, 823–833.

Hodkova, H., Kodym, P., Flegr, J., 2007. Poorer results of mice with latent toxoplasmosis in learning tests: impaired learning processes or the novelty discrimination mechanism? Parasitology 134, 1329–1337.

Hofhuis, A., Van Pelt, W., Van Duynhoven, Y.T., Nijhuis, C.D., Mollema, L., Van Der Klis, F.R., Havelaar, A.H., Kortbeek, L.M., 2011. Decreased prevalence and age-specific risk factors for Toxoplasma gondii IgG antibodies in The Netherlands between 1995/1996 and 2006/2007. Epidemiol. Infect. 139, 530–538.

Holland, G.N., 2003. Ocular toxoplasmosis: a global reassessment. Part I: epidemiology and course of disease. Am J. Ophthalmol. 136, 973–988.

Holland, G.N., 2004. Ocular toxoplasmosis: a global reassessment. Part II: disease manifestations and management. Am J. Ophthalmol. 137, 1–17.

Holland, G.N., Engstrom Jr., R.E., Glasgow, B.J., Berger, B.B., Daniels, S.A., Sidikaro, Y., Harmon, J.A., Fischer, D.H., Boyer, D.S., Rao, N.A., et al., 1988. Ocular toxoplasmosis in patients with the acquired immunodeficiency syndrome. Am. J. Ophthalmol. 106, 653–667.

Hou, B., Benson, A., Kuzmich, L., Defranco, A.L., Yarovinsky, F., 2011. Critical coordination of innate immune defense against Toxoplasma gondii by dendritic cells responding *via* their Toll-like receptors. Proc. Natl. Acad. Sci. U.S.A. 108, 278–283.

Howard, J.C., Hunn, J.P., Steinfeldt, T., 2011. The IRG protein-based resistance mechanism in mice and its relation to virulence in Toxoplasma gondii. Curr. Opin. Microbiol. 14, 414–421.

Hu, G., Wilcox, H.C., Wissow, L., Baker, S.P., 2008. Mid-life suicide: an increasing problem in U.S. Whites, 1999–2005. Am. J. Prev. Med. 35, 589–593.

Hunn, J.P., Feng, C.G., Sher, A., Howard, J.C., 2011. The immunity-related GTPases in mammals: a fast-evolving cell-autonomous resistance system against intracellular pathogens. Mamm. Genome 22, 43–54.

Hunn, J.P., Koenen-Waisman, S., Papic, N., Schroeder, N., Pawlowski, N., Lange, R., Kaiser, F., Zerrahn, J., Martens, S., Howard, J.C., 2008. Regulatory interactions between IRG resistance GTPases in the cellular response to Toxoplasma gondii. EMBO J. 27, 2495–2509.

Hunter, C.A., Litton, M.J., Remington, J.S., Abrams, J.S., 1994. Immunocytochemical detection of cytokines in the lymph nodes and brains of mice resistant or susceptible to toxoplasmic encephalitis. J. Infect. Dis. 170, 939–945.

Hunter, C.A., Roberts, C.W., Alexander, J., 1992. Kinetics of cytokine mRNA production in the brains of mice with progressive toxoplasmic encephalitis. Eur. J. Immunol. 22, 2317–2322.

Hutchison, W.M., Aitken, P.P., Wells, W.P., 1980. Chronic Toxoplasma infections and familiarity-novelty discrimination in the mouse. Ann. Trop. Med. Parasitol. 74, 145–150.

Indraccolo, S., Pfeffer, U., Minuzzo, S., Esposito, G., Roni, V., Mandruzzato, S., Ferrari, N., Anfosso, L., Dell'Eva, R., Noonan, D.M., Chieco-Bianchi, L., Albini, A., Amadori, A., 2007. Identification of genes selectively regulated by IFNs in endothelial cells. J. Immunol. 178, 1122–1135.

Israelski, D.M., Remington, J.S., 1993. Toxoplasmosis in the non-AIDS immunocompromised host. Curr. Clin. Top. Infect. Dis. 13, 322–356.

James, S.L., 1995. Role of nitric oxide in parasitic infections. Microbiol. Rev. 59, 533–547.

Jebbari, H., Roberts, C.W., Ferguson, D.J., Bluethmann, H., Alexander, J., 1998. A protective role for IL-6 during early infection with Toxoplasma gondii. Parasite Immunol. 20, 231–239.

Jin, D., Takamoto, M., Hu, T., Taki, S., Sugane, K., 2009. STAT6 signalling is important in CD8 T-cell activation and defence against Toxoplasma gondii infection in the brain. Immunology 127, 187–195.

John, B., Ricart, B., Tait Wojno, E.D., Harris, T.H., Randall, L.M., Christian, D.A., Gregg, B., De Almeida, D.M., Weninger, W., Hammer, D.A., Hunter, C.A., 2011. Analysis of behavior and trafficking of dendritic cells within the brain during toxoplasmic encephalitis. PLoS Pathog. 7, e1002246.

Johnson, L.L., Sayles, P.C., 2002. Deficient humoral responses underlie susceptibility to Toxoplasma gondii in CD4-deficient mice. Infect. Immun. 70, 185–191.

Jones, J., Lopez, A., Wilson, M., 2003. Congenital toxoplasmosis. Am. Fam. Physician 67, 2131–2138.

Jones, J.L., Kruszon-Moran, D., Sanders-Lewis, K., Wilson, M., 2007. Toxoplasma gondii infection in the United States, 1999–2004, decline from the prior decade. Am. J. Trop. Med. Hyg. 77, 405–410.

Jones, T.C., Bienz, K.A., Erb, P., 1986. In vitro cultivation of Toxoplasma gondii cysts in astrocytes in the presence of gamma interferon. Infect. Immun. 51, 147–156.

Jones, L.A., Roberts, F., Nickdel, M.B., Brombacher, F., McKenzie, A.N., Henriquez, F.L., Alexander, J., Roberts, C.W., 2010. IL-33 receptor (T1/ST2) signalling is necessary to prevent the development of encephalitis in mice infected with Toxoplasma gondii. Eur. J. Immunol 40, 426–436.

Jun, C.D., Kim, S.H., Soh, C.T., Kang, S.S., Chung, H.T., 1993. Nitric oxide mediates the toxoplasmastatic activity of murine microglial cells in vitro. Immunol. Invest. 22, 487–501.

Kang, H., Liesenfeld, O., Remington, J.S., Claflin, J., Wang, X., Suzuki, Y., 2003. TCR V beta 8+ T-cells prevent development of toxoplasmic encephalitis in BALB/c mice genetically resistant to the disease. J. Immunol. 170, 4254–4259.

Kang, H., Remington, J.S., Suzuki, Y., 2000. Decreased resistance of B-cell-deficient mice to infection with Toxoplasma gondii despite unimpaired expression of IFN-gamma, TNF-alpha, and inducible nitric oxide synthase. J. Immunol. 164, 2629–2634.

Kang, H., Suzuki, Y., 2001. Requirement of non-T cells that produce gamma interferon for prevention of reactivation of *Toxoplasma gondii* infection in the brain. Infect. Immun. 69, 2920–2927.

Kannan, G., Moldovan, K., Xiao, J.C., Yolken, R.H., Jones-Brando, L., Pletnikov, M.V., 2010. Toxoplasma gondii strain-dependent effects on mouse behaviour. Folia Parasitol. 57, 151–155.

Kannan, G., Pletnikov, M.V., 2012. Toxoplasma gondii and Cognitive Deficits in Schizophrenia: An Animal Model Perspective. Schizophr. Bull. 38 (6), 1155–1161.

Karch, D.L., Logan, J., Patel, N., 2011. Surveillance for violent deaths – National Violent Death Reporting System, 16 states, 2008. Morb. Mortal. Wkly. Rep. Surveilla. Summa. 60, 1–49.

Kasper, L.H., Matsuura, T., Fonseka, S., Arruda, J., Channon, J.Y., Khan, I.A., 1996. Induction of gammadelta T-cells during acute murine infection with Toxoplasma gondii. J. Immunol. 157, 5521–5527.

Khaminets, A., Hunn, J.P., Konen-Waisman, S., Zhao, Y.O., Preukschat, D., Coers, J., Boyle, J.P., Ong, Y.C., Boothroyd, J.C., Reichmann, G., Howard, J.C., 2010. Coordinated loading of IRG resistance GTPases on to the Toxoplasma gondii parasitophorous vacuole. Cell. Microbiol. 12, 939–961.

Khan, A., Jordan, C., Muccioli, C., Vallochi, A.L., Rizzo, L.V., Belfort Jr., R., Vitor, R.W., Silveira, C., Sibley, L.D., 2006. Genetic divergence of Toxoplasma gondii strains associated with ocular toxoplasmosis. Brazil. Emerging Infect. Dis. 12, 942–949.

Kim, L., Butcher, B.A., Lee, C.W., Uematsu, S., Akira, S., Denkers, E.Y., 2006. Toxoplasma gondii genotype determines MyD88-dependent signalling in infected macrophages. J. Immunol. 177, 2584–2591.

Kirak, O., Frickel, E.M., Grotenbreg, G.M., Suh, H., Jaenisch, R., Ploegh, H.L., 2010. Transnuclear mice with predefined T-cell receptor specificities against Toxoplasma gondii obtained *via* SCNT. Science 328, 243–248.

Kirkbride, J.B., Errazuriz, A., Croudace, T.J., Morgan, C., Jackson, D., Boydell, J., Murray, R.M., Jones, P.B., 2012. Incidence of schizophrenia and other psychoses in England, 1950–2009: a systematic review and meta-analyses. PLoS One 7, e31660.

Kocazeybek, B., Oner, Y.A., Turksoy, R., Babur, C., Cakan, H., Sahip, N., Unal, A., Ozaslan, A., Kilic, S., Saribas, S., Aslan, M., Taylan, A., Koc, S., Dirican, A., Uner, H.B., Oz, V., Ertekin, C., Kucukbasmaci, O., Torun, M.M., 2009. Higher prevalence of toxoplasmosis in victims of traffic accidents suggest increased risk of traffic accident in Toxoplasma-infected inhabitants of Istanbul and its suburbs. Forensic Sci. Int. 187, 103–108.

Konen-Waisman, S., Howard, J.C., 2007. Cell-autonomous immunity to Toxoplasma gondii in mouse and man. Microbes Infect. 9, 1652–1661.

Koshy, A.A., Dietrich, H.K., Christian, D.A., Melehani, J.H., Shastri, A.J., Hunter, C.A., Boothroyd, J.C., 2012. Toxoplasma Co-opts Host Cells It Does Not Invade. PLoS Pathog. 8, e1002825.

Kota, R.S., Rutledge, J.C., Gohil, K., Kumar, A., Enelow, R.I., Ramana, C.V., 2006. Regulation of gene expression in RAW 264.7 macrophage cell line by interferon-gamma. Biochem. Biophys. Res. Commun. 342, 1137–1146.

Kramer, W., 1966. Frontiers of neurological diagnosis in acquired toxoplasmosis. Psychiatr., Neurol., Neurochir. 69, 43–64.

Kresse, A., Konermann, C., Degrandi, D., Beuter-Gunia, C., Wuerthner, J., Pfeffer, K., Beer, S., 2008. Analyses of murine GBP homology clusters based on in silico, in vitro and in vivo studies. BMC Genomics 9, 158.

Kuhn, R., Lohler, J., Rennick, D., Rajewsky, K., Muller, W., 1993. Interleukin-10-deficient mice develop chronic enterocolitis. Cell 75, 263–274.

Lambert, H., Hitziger, N., Dellacasa, I., Svensson, M., Barragan, A., 2006. Induction of dendritic cell migration upon Toxoplasma gondii infection potentiates parasite dissemination. Cell Microbiol. 8, 1611–1623.

Lamberton, P.H., Donnelly, C.A., Webster, J.P., 2008. Specificity of the Toxoplasma gondii-altered behaviour to definitive versus non-definitive host predation risk. Parasitology 135, 1143–1150.

Leiva, L.E., Junprasert, J., Hollenbaugh, D., Sorensen, R.U., 1998. Central nervous system toxoplasmosis with an increased proportion of circulating gamma delta T-cells in a patient with hyper-IgM syndrome. J. Clin. Immunol. 18, 283–290.

Lepage, A.C., Buzoni-Gatel, D., Bout, D.T., Kasper, L.H., 1998. Gut-derived intraepithelial lymphocytes induce long term immunity against Toxoplasma gondii. J. Immunol. 161, 4902–4908.

Lester, D., 2010. Brain parasites and suicide. Psychol. Rep. 107, 424.

Lieberman, L.A., Banica, M., Reiner, S.L., Hunter, C.A., 2004. STAT1 plays a critical role in the regulation of antimicrobial effector mechanisms, but not in the development of Th1-type responses during toxoplasmosis. J. Immunol. 172, 457–463.

Lindsay, D.S., Dubey, J.P., 2011. Toxoplasma gondii: the changing paradigm of congenital toxoplasmosis. Parasitology, 1–3.

Ling, V.J., Lester, D., Mortensen, P.B., Langenberg, P.W., Postolache, T.T., 2011. Toxoplasma gondii seropositivity and suicide rates in women. J. Nerv. Ment. Dis. 199, 440–444.

Ling, Y.M., Shaw, M.H., Ayala, C., Coppens, I., Taylor, G.A., Ferguson, D.J., Yap, G.S., 2006. Vacuolar and plasma membrane stripping and autophagic elimination of Toxoplasma gondii in primed effector macrophages. J. Exp. Med. 203, 2063–2071.

Lu, F., Huang, S., Kasper, L.H., 2004. CD4+ T-cells in the pathogenesis of murine ocular toxoplasmosis. Infect. Immun. 72, 4966–4972.

Luder, C.G., Giraldo-Velasquez, M., Sendtner, M., Gross, U., 1999. Toxoplasma gondii in primary rat CNS cells: differential contribution of neurons, astrocytes, and microglial cells for the intracerebral development and stage differentiation. Exp. Parasitol. 93, 23–32.

Lyons, R.E., Anthony, J.P., Ferguson, D.J., Byrne, N., Alexander, J., Roberts, F., Roberts, C.W., 2001. Immunological studies of chronic ocular toxoplasmosis: up-regulation of major histocompatibility complex class I and transforming growth factor beta and a protective role for interleukin-6. Infect. Immun. 69, 2589–2595.

Machado, F.S., Johndrow, J.E., Esper, L., Dias, A., Bafica, A., Serhan, C.N., Aliberti, J., 2006. Anti-inflammatory actions of lipoxin A4 and aspirin-triggered lipoxin are SOCS-2 dependent. Nat. Med. 12, 330–334.

Mack, D.G., Johnson, J.J., Roberts, F., Roberts, C.W., Estes, R.G., David, C., Grumet, F.C., McLeod, R., 1999. HLA-class II genes modify outcome of Toxoplasma gondii infection. Int. J. Parasitol. 29, 1351–1358.

MacMicking, J.D., 2004. IFN-inducible GTPases and immunity to intracellular pathogens. Trends Immunol. 25, 601–609.

Martens, S., Parvanova, I., Zerrahn, J., Griffiths, G., Schell, G., Reichmann, G., Howard, J.C., 2005. Disruption of Toxoplasma gondii parasitophorous vacuoles by the mouse p47-resistance GTPases. PLoS Pathog. 1, e24.

Mason, N.J., Liou, H.C., Hunter, C.A., 2004. T-cell-intrinsic expression of c-Rel regulates Th1 cell responses essential for resistance to Toxoplasma gondii. J. Immunol. 172, 3704–3711.

McLeod, R., Skamene, E., Brown, C.R., Eisenhauer, P.B., Mack, D.G., 1989. Genetic regulation of early survival and cyst number after peroral Toxoplasma gondii infection of A x B/B x A recombinant inbred and B10 congenic mice. J. Immunol. 143, 3031–3034.

Melzer, T., Duffy, A., Weiss, L.M., Halonen, S.K., 2008. The gamma interferon (IFN-gamma)-inducible GTP-binding protein IGTP is necessary for toxoplasma vacuolar disruption and induces parasite egression in IFN-gamma-stimulated astrocytes. Infect. Immun. 76, 4883–4894.

Melzer, T.C., Cranston, H.J., Weiss, L.M., Halonen, S.K., 2010. Host Cell Preference of Toxoplasma gondii Cysts in Murine Brain: A Confocal Study. J. Neuroparasitol. 1.

Menzies, F.M., Henriquez, F.L., Roberts, C.W., 2008. Immunological control of congenital toxoplasmosis in the murine model. Immunol. Lett. 115, 83–89.

Montoya, J.G., Remington, J.S., 2008. Management of Toxoplasma gondii infection during pregnancy. Clin. Infect. Dis. 47, 554–566.

Montoya, J.G., Rosso, F., 2005. Diagnosis and management of toxoplasmosis. Clin. Perinatol. 32, 705–726.

Moran, L.B., Duke, D.C., Turkheimer, F.E., Banati, R.B., Graeber, M.B., 2004. Towards a transcriptome definition of microglial cells. Neurogenetics 5, 95–108.

Mortensen, P.B., Norgaard-Pedersen, B., Waltoft, B.L., Sorensen, T.L., Hougaard, D., Torrey, E.F., Yolken, R.H., 2007. Toxoplasma gondii as a risk factor for early-onset schizophrenia: analysis of filter paper blood samples obtained at birth. Biol. Psychiatry 61, 688–693.

Murray, H.W., Rubin, B.Y., Carriero, S.M., Harris, A.M., Jaffee, E.A., 1985. Human mononuclear phagocyte antiprotozoal mechanisms: oxygen-dependent vs oxygen-independent activity against intracellular Toxoplasma gondii. J. Immunol. 134, 1982–1988.

Nagineni, C.N., Pardhasaradhi, K., Martins, M.C., Detrick, B., Hooks, J.J., 1996. Mechanisms of interferon-induced inhibition of Toxoplasma gondii replication in human retinal pigment epithelial cells. Infect. Immun. 64, 4188–4196.

Niedelman, W., Gold, D.A., Rosowski, E.E., Sprokholt, J.K., Lim, D., Farid Arenas, A., Melo, M.B., Spooner, E., Yaffe, M.B., Saeij, J.P., 2012. The Rhoptry Proteins ROP18 and ROP5 Mediate Toxoplasma gondii Evasion of the Murine, But Not the Human, Interferon-Gamma Response. PLoS Pathog. 8, e1002784.

Nimmerjahn, A., Kirchhoff, F., Helmchen, F., 2005. Resting microglial cells are highly dynamic surveillants of brain parenchyma in vivo. Science 308, 1314–1318.

Nishanth, G., Sakowicz-Burkiewicz, M., Handel, U., Kliche, S., Wang, X., Naumann, M., Deckert, M., Schluter, D., 2010. Protective Toxoplasma gondii-specific T-cell responses require T-cell-specific expression of protein kinase C-theta. Infect. Immun. 78, 3454–3464.

Nishikawa, Y., Kawase, O., Vielemeyer, O., Suzuki, H., Joiner, K.A., Xuan, X., Nagasawa, H., 2007. Toxoplasma gondii infection induces apoptosis in noninfected macrophages: role of nitric oxide and other soluble factors. Parasite Immunol. 29, 375–385.

Noble, A., Kemeny, D.M., 1995. Interleukin-4 enhances interferon-gamma synthesis but inhibits development of interferon-gamma-producing cells. Immunology 85, 357–363.

Norose, K., Kikumura, A., Luster, A.D., Hunter, C.A., Harris, T.H., 2011. CXCL10 is required to maintain T-cell populations and to control parasite replication during chronic ocular toxoplasmosis. Invest. Ophthalmol. Vis. Sci. 52, 389–398.

Norose, K., Mun, H.S., Aosai, F., Chen, M., Piao, L.X., Kobayashi, M., Iwakura, Y., Yano, A., 2003. IFN-gamma-regulated Toxoplasma gondii distribution and load in the murine eye. Invest. Ophthalmol. Vis. Sci. 44, 4375–4381.

Oberdorfer, C., Adams, O., MacKenzie, C.R., De Groot, C.J., Daubener, W., 2003. Role of IDO activation in anti-microbial defense in human native astrocytes. Adv. Exp. Med. Biol. 527, 15–26.

Okusaga, O., Langenberg, P., Sleemi, A., Vaswani, D., Giegling, I., Hartmann, A.M., Konte, B., Friedl, M., Groer, M.W., Yolken, R.H., Rujescu, D., Postolache, T.T., 2011. Toxoplasma gondii antibody titers and history of suicide attempts in patients with schizophrenia. Schizophr. Res. 133, 150–155.

Olle, P., Bessieres, M.H., Malecaze, F., Seguela, J.P., 1996. The evolution of ocular toxoplasmosis in anti-interferon gamma treated mice. Curr. Eye Res. 15, 701–707.

Ortiz-Alegria, L.B., Caballero-Ortega, H., Canedo-Solares, I., Rico-Torres, C.P., Sahagun-Ruiz, A., Medina-Escutia, M.E., Correa, D., 2010. Congenital toxoplasmosis: candidate host immune genes relevant for vertical transmission and pathogenesis. Genes Immun. 11, 363–373.

Papic, N., Hunn, J.P., Pawlowski, N., Zerrahn, J., Howard, J.C., 2008. Inactive and active states of the interferon-inducible resistance GTPase, Irga6, in vivo. J. Biol. Chem. 283, 32143–32151.

Pappas, G., Roussos, N., Falagas, M.E., 2009. Toxoplasmosis snapshots: global status of Toxoplasma gondii seroprevalence and implications for pregnancy and congenital toxoplasmosis. Int. J. Parasitol. 39, 1385–1394.

Pearce, B.D., Kruszon-Moran, D., Jones, J.L., 2012. The relationship between Toxoplasma gondii infection and mood disorders in the third National Health and Nutrition Survey. Biol. Psychiatry 72, 290–295.

Pedersen, M.G., Mortensen, P.B., Norgaard-Pedersen, B., Postolache, T.T., 2012. Toxoplasma gondii Infection and Self-directed Violence in Mothers. Arch. Gen. Psychiatry, 1–8.

Pedersen, M.G., Stevens, H., Pedersen, C.B., Norgaard-Pedersen, B., Mortensen, P.B., 2011. Toxoplasma infection and later development of schizophrenia in mothers. Am. J. Psychiatry 168, 814–821.

Peixoto-Rangel, A.L., Miller, E.N., Castellucci, L., Jamieson, S.E., Peixe, R.G., Elias Lde, S., Correa-Oliveira, R., Bahia-Oliveira, L.M., Blackwell, J.M., 2009. Candidate gene analysis of ocular toxoplasmosis in Brazil: evidence for a role for toll-like receptor 9 (TLR9). Mem. Inst. Oswaldo Cruz 104, 1187–1190.

Peterson, P.K., Gekker, G., Hu, S., Chao, C.C., 1995. Human astrocytes inhibit intracellular multiplication of Toxoplasma gondii by a nitric oxide-mediated mechanism. J. Infect. Dis. 171, 516–518.

Pfaff, A.W., Mousli, M., Senegas, A., Marcellin, L., Takikawa, O., Klein, J.P., Candolfi, E., 2008. Impact of foetus and mother on IFN-gamma-induced indoleamine 2,3-dioxygenase and inducible nitric oxide synthase expression in murine placenta following Toxoplasma gondii infection. Int. J. Parasitol. 38, 249–258.

Pfaff, A.W., Villard, O., Klein, J.P., Mousli, M., Candolfi, E., 2005. Regulation of Toxoplasma gondii multiplication in BeWo trophoblast cells: cross-regulation of nitric oxide production and polyamine biosynthesis. Int. J. Parasitol. 35, 1569–1576.

Pfefferkorn, E.R., 1984. Interferon gamma blocks the growth of Toxoplasma gondii in human fibroblasts by inducing the host cells to degrade tryptophan. Proc. Natl. Acad. Sci. U.S.A. 81, 908–912.

Pfefferkorn, E.R., Guyre, P.M., 1984. Inhibition of growth of Toxoplasma gondii in cultured fibroblasts by human recombinant gamma interferon. Infect. Immun. 44, 211–216.

Pfefferkorn, E.R., Rebhun, S., Eckel, M., 1986. Characterization of an indoleamine 2,3-dioxygenase induced by gamma-interferon in cultured human fibroblasts. J. Interferon Res. 6, 267–279.

Piekarski, G., Zippelius, H.M., Witting, P.A., 1978. [Effects of a latent Toxoplasma infection on the learning ability in white laboratory rats and mice (author's transl)]. Z. Parasitenkd. 57, 1–15.

Powell, H.C., Gibbs Jr., C.J., Lorenzo, A.M., Lampert, P.W., Gajdusek, D.C., 1978. Toxoplasmosis of the central nervous system in the adult. Electron microscopic observations. Acta. Neuropathol. (Berl.) 41, 211–216.

Prandovszky, E., Gaskell, E., Martin, H., Dubey, J.P., Webster, J.P., McConkey, G.A., 2011. The neurotropic parasite Toxoplasma gondii increases dopamine metabolism. PLoS One 6, e23866.

Reis E Sousa, C., Hieny, S., Scharton-Kersten, T., Jankovic, D., Charest, H., Germain, R.N., Sher, A., 1997. In vivo microbial stimulation induces rapid CD40 ligand-independent production of interleukin 12 by dendritic cells and their redistribution to T-cell areas. J. Exp. Med. 186, 1819–1829.

Robben, P.M., Mordue, D.G., Truscott, S.M., Takeda, K., Akira, S., Sibley, L.D., 2004. Production of IL-12 by macrophages infected with Toxoplasma gondii depends on the parasite genotype. J. Immunol. 172, 3686–3694.

Roberts, C.W., Alexander, J., 1992. Studies on a murine model of congenital toxoplasmosis: vertical disease transmission only occurs in BALB/c mice infected for the first time during pregnancy. Parasitology 104 Pt 1, 19–23.

Roberts, C.W., Ferguson, D.J., Jebbari, H., Satoskar, A., Bluethmann, H., Alexander, J., 1996. Different roles for interleukin-4 during the course of Toxoplasma gondii infection. Infect. Immun. 64, 897–904.

Rock, R.B., Hu, S., Deshpande, A., Munir, S., May, B.J., Baker, C.A., Peterson, P.K., Kapur, V., 2005. Transcriptional response of human microglial cells to interferon-gamma. Genes Immun. 6, 712–719.

Rozenfeld, C., Martinez, R., Figueiredo, R.T., Bozza, M.T., Lima, F.R., Pires, A.L., Silva, P.M., Bonomo, A., Lannes-Vieira, J., De Souza, W., Moura-Neto, V., 2003. Soluble factors released by Toxoplasma gondii-infected astrocytes down-modulate nitric oxide production by gamma interferon-activated microglia and prevent neuronal degeneration. Infect. Immun. 71, 2047–2057.

Rozenfeld, C., Martinez, R., Seabra, S., Sant'Anna, C., Goncalves, J.G., Bozza, M., Moura-Neto, V., De Souza, W., 2005. Toxoplasma gondii prevents neuron degeneration by interferon-gamma-activated microglia in a mechanism involving inhibition of inducible nitric oxide synthase and transforming growth factor-beta1 production by infected microglia. Am. J. Pathol. 167, 1021–1031.

Schaeffer, E.M., Debnath, J., Yap, G., McVicar, D., Liao, X.C., Littman, D.R., Sher, A., Varmus, H.E., Lenardo, M.J., Schwartzberg, P.L., 1999. Requirement for Tec kinases Rlk and Itk in T-cell receptor signalling and immunity. Science 284, 638–641.

Schaeffer, M., Han, S.J., Chtanova, T., Van Dooren, G.G., Herzmark, P., Chen, Y., Roysam, B., Striepen, B., Robey, E.A., 2009. Dynamic imaging of T-cell-parasite interactions in the brains of mice chronically infected with Toxoplasma gondii. J. Immunol. 182, 6379–6393.

Scharton-Kersten, T.M., Yap, G., Magram, J., Sher, A., 1997. Inducible nitric oxide is essential for host control of persistent but not acute infection with the intracellular pathogen Toxoplasma gondii. J. Exp. Med. 185, 1261–1273.

Scheidegger, A., Vonlaufen, N., Naguleswaran, A., Gianinazzi, C., Muller, N., Leib, S.L., Hemphill, A., 2005. Differential effects of interferon-gamma and tumor necrosis factor-alpha on Toxoplasma gondii proliferation in organotypic rat brain slice cultures. J. Parasitol. 91, 307–315.

Schluter, D., Deckert, M., Hof, H., Frei, K., 2001. Toxoplasma gondii infection of neurons induces neuronal cytokine and chemokine production, but gamma interferon- and tumor necrosis factor-stimulated neurons fail to inhibit the invasion and growth of T. gondii. Infect. Immun. 69, 7889–7893.

Schluter, D., Deckert-Schluter, M., Schwendemann, G., Brunner, H., Hof, H., 1993. Expression of major histocompatibility complex class II antigens and levels of interferon-gamma, tumour necrosis factor, and interleukin-6 in cerebrospinal fluid and serum in Toxoplasma gondii-infected SCID and immunocompetent C.B-17 mice. Immunology 78, 430–435.

Schluter, D., Hein, A., Dorries, R., Deckert-Schluter, M., 1995. Different subsets of T-cells in conjunction with natural killer cells, macrophages, and activated microglia participate in the intracerebral immune response to Toxoplasma gondii in athymic nude and immunocompetent rats. Am. J. Pathol. 146, 999–1007.

Schluter, D., Kwok, L.Y., Lutjen, S., Soltek, S., Hoffmann, S., Korner, H., Deckert, M., 2003. Both lymphotoxin-alpha and TNF are crucial for control of Toxoplasma gondii in the central nervous system. J. Immunol. 170, 6172–6182.

Schluter, D., Lohler, J., Deckert, M., Hof, H., Schwendemann, G., 1991. Toxoplasma encephalitis of immunocompetent and nude mice: immunohistochemical characterisation of Toxoplasma antigen, infiltrates and major histocompatibility complex gene products. J. Neuroimmunol. 31, 185–198.

Schwarcz, R., Hunter, C.A., 2007. Toxoplasma gondii and schizophrenia: linkage through astrocyte-derived kynurenic acid? Schizophr. Bull. 33, 652–653.

Severance, E.G., Alaedini, A., Yang, S., Halling, M., Gressitt, K.L., Stallings, C.R., Origoni, A.E., Vaughan, C., Khushalani, S., Leweke, F.M., Dickerson, F.B., Yolken, R.H., 2012. Gastrointestinal inflammation and associated immune activation in schizophrenia. Schizophr. Res. 138, 48–53.

Sexton Jr., R.C., Eyles, D.E., Dillman, R.E., 1953. Adult toxoplasmosis. Am. J. Med. 14, 366–377.

Shaw, M.H., Boyartchuk, V., Wong, S., Karaghiosoff, M., Ragimbeau, J., Pellegrini, S., Muller, M., Dietrich, W.F., Yap, G.S., 2003. A natural mutation in the Tyk2 pseudokinase domain underlies altered susceptibility of B10.Q/J mice to infection and autoimmunity. Proc. Natl. Acad. Sci. U.S.A. 100, 11594–11599.

Shen, D.F., Matteson, D.M., Tuaillon, N., Suedekum, B.K., Buggage, R.R., Chan, C.C., 2001. Involvement of apoptosis and interferon-gamma in murine toxoplasmosis. Invest. Ophthalmol. Vis. Sci. 42, 2031–2036.

Shiono, Y., Mun, H.S., He, N., Nakazaki, Y., Fang, H., Furuya, M., Aosai, F., Yano, A., 2007. Maternal-foetal transmission of Toxoplasma gondii in interferon-gamma deficient pregnant mice. Parasitol. Int. 56, 141–148.

Shirahata, T., Muroya, N., Ohta, C., Goto, H., Nakane, A., 1992. Correlation between increased susceptibility to primary Toxoplasma gondii infection and depressed production of gamma interferon in pregnant mice. Microbiol. Immunol. 36, 81–91.

Silva, N.M., Rodrigues, C.V., Santoro, M.M., Reis, L.F., Alvarez-Leite, J.I., Gazzinelli, R.T., 2002. Expression of indoleamine 2,3-dioxygenase, tryptophan degradation, and kynurenine formation during in vivo infection with Toxoplasma gondii: induction by endogenous gamma interferon and requirement of interferon regulatory factor 1. Infect. Immun. 70, 859–868.

Sims, T.A., Hay, J., Talbot, I.C., 1988. Host-parasite relationship in the brains of mice with congenital toxoplasmosis. J. Pathol. 156, 255–261.

Sims, T.A., Hay, J., Talbot, I.C., 1989a. An electron microscope and immunohistochemical study of the intracellular location of Toxoplasma tissue cysts within the brains of mice with congenital toxoplasmosis. Br. J. Exp. Pathol. 70, 317–325.

Sims, T.A., Hay, J., Talbot, I.C., 1989b. Immunoelectron microscopy for Toxoplasma antigen in association with intact tissue cysts in mouse brain. Ann. Trop. Med. Parasitol. 83, 639–641.

Skallova, A., Kodym, P., Frynta, D., Flegr, J., 2006. The role of dopamine in Toxoplasma-induced behavioural alterations in mice: an ethological and ethopharmacological study. Parasitology 133, 525–535.

Steinfeldt, T., Konen-Waisman, S., Tong, L., Pawlowski, N., Lamkemeyer, T., Sibley, L.D., Hunn, J.P., Howard, J.C., 2010. Phosphorylation of mouse immunity-related GTPase (IRG) resistance proteins is an evasion strategy for virulent Toxoplasma gondii. PLoS Biol. 8, e1000576.

Stenzel, W., Soltek, S., Schluter, D., Deckert, M., 2004. The intermediate filament GFAP is important for the control of experimental murine Staphylococcus aureus-induced brain abscess and Toxoplasma encephalitis. J. Neuropathol. Exp. Neurol. 63, 631–640.

Stibbs, H.H., 1985. Changes in brain concentrations of catecholamines and indoleamines in Toxoplasma gondii infected mice. Ann. Trop. Med. Parasitol. 79, 153–157.

Strack, A., Asensio, V.C., Campbell, I.L., Schluter, D., Deckert, M., 2002. Chemokines are differentially expressed by astrocytes, microglia and inflammatory leukocytes in Toxoplasma encephalitis and critically regulated by interferon-gamma. Acta Neuropathol. 103, 458–468.

Strobl, J.S., Goodwin, D., Rzigalinski, B.A., Lindsay, D.S., 2012. Dopamine Stimulates Propagation of Toxoplasma gondii Tachyzoites in Human Fibroblast and Primary Rat Neonatal Brain Cell Cultures. J. Parasitol.

Strom, J., 1951. Toxoplasmosis due to laboratory infection in two adults. Acta Med. Scand. 139, 244–252.

Su, C., Zhang, X., Dubey, J.P., 2006. Genotyping of Toxoplasma gondii by multilocus PCR-RFLP markers: a high resolution and simple method for identification of parasites. Int. J. Parasitol. 36, 841–848.

Suzuki, Y., Claflin, J., Wang, X., Lengi, A., Kikuchi, T., 2005. Microglia and macrophages as innate producers of interferon-gamma in the brain following infection with Toxoplasma gondii. Int. J. Parasitol. 35, 83–90.

Suzuki, Y., Conley, F.K., Remington, J.S., 1989. Importance of endogenous IFN-gamma for prevention of toxoplasmic encephalitis in mice. J. Immunol. 143, 2045–2050.

Suzuki, Y., Joh, K., Kwon, O.C., Yang, Q., Conley, F.K., Remington, J.S., 1994. MHC class I gene(s) in the D/L region but not the TNF-alpha gene determines development of toxoplasmic encephalitis in mice. J. Immunol. 153, 4649–4654.

Suzuki, Y., Joh, K., Orellana, M.A., Conley, F.K., Remington, J.S., 1991. A gene(s) within the H-2D region determines the development of toxoplasmic encephalitis in mice. Immunology 74, 732–739.

Suzuki, Y., Kang, H., Parmley, S., Lim, S., Park, D., 2000. Induction of tumor necrosis factor-alpha and inducible nitric oxide synthase fails to prevent toxoplasmic encephalitis in the absence of interferon-gamma in genetically resistant BALB/c mice. Microbes Infect. 2, 455–462.

Suzuki, Y., Orellana, M.A., Schreiber, R.D., Remington, J.S., 1988. Interferon-gamma: the major mediator of resistance against Toxoplasma gondii. Science 240, 516–518.

Suzuki, Y., Orellana, M.A., Wong, S.Y., Conley, F.K., Remington, J.S., 1993. Susceptibility to chronic infection with *Toxoplasma gondii* does not correlate with susceptibility to acute infection in mice. Infect. Immun. 61, 2284–2288.

Suzuki, Y., Rani, S., Liesenfeld, O., Kojima, T., Lim, S., Nguyen, T.A., Dalrymple, S.A., Murray, R., Remington, J.S., 1997. Impaired resistance to the development of toxoplasmic encephalitis in interleukin-6-deficient mice. Infect. Immun. 65, 2339–2345.

Suzuki, Y., Sher, A., Yap, G., Park, D., Neyer, L.E., Liesenfeld, O., Fort, M., Kang, H., Gufwoli, E., 2000. IL-10 is required for prevention of necrosis in the small intestine and mortality in both genetically resistant BALB/c and susceptible C57BL/6 mice following peroral infection with Toxoplasma gondii. J. Immunol. 164, 5375–5382.

Suzuki, Y., Wang, X., Jortner, B.S., Payne, L., Ni, Y., Michie, S.A., Xu, B., Kudo, T., Perkins, S., 2010. Removal of Toxoplasma gondii cysts from the brain by perforin-mediated activity of CD8+ T-cells. Am. J. Pathol. 176, 1607–1613.

Suzuki, Y., Yang, Q., Yang, S., Nguyen, N., Lim, S., Liesenfeld, O., Kojima, T., Remington, J.S., 1996. IL-4 is protective against development of toxoplasmic encephalitis. J. Immunol. 157, 2564–2569.

Swain, S.L., Mckenzie, D.T., Dutton, R.W., Tonkonogy, S.L., English, M., 1988. The role of IL-4 and IL5: characterization of a distinct helper T-cell subset that makes IL-4 and IL5 (Th2) and requires priming before induction of lymphokine secretion. Immunol. Rev. 102, 77–105.

Swierczynski, B., Bessieres, M.H., Cassaing, S., Guy, S., Oswald, I., Seguela, J.P., Pipy, B., 2000. Inhibitory activity of anti-interleukin-4 and anti-interleukin-10 antibodies on Toxoplasma gondii proliferation in mouse peritoneal macrophages cocultured with splenocytes from infected mice. Parasitol. Res. 86, 151–157.

Switaj, K., Master, A., Borkowski, P.K., Skrzypczak, M., Wojciechowicz, J., Zaborowski, P., 2006. Association of ocular toxoplasmosis with type I *Toxoplasma gondii* strains: direct genotyping from peripheral blood samples. J. Clin. Microbiol. 44, 4262–4264.

Takatsu, K., Tominaga, A., Harada, N., Mita, S., Matsumoto, M., Takahashi, T., Kikuchi, Y., Yamaguchi, N., 1988. T-cell-replacing factor (TRF)/interleukin 5 (IL-5): molecular and functional properties. Immunol. Rev. 102, 107–135.

Taylor, G.A., 2007. IRG proteins: key mediators of interferon-regulated host resistance to intracellular pathogens. Cell. Microbiol. 9, 1099–1107.

Taylor, G.A., Feng, C.G., Sher, A., 2004. p47 GTPases: regulators of immunity to intracellular pathogens. Nat. Rev. Immunol. 4, 100–109.

Taylor, G.A., Feng, C.G., Sher, A., 2007. Control of IFN-gamma-mediated host resistance to intracellular pathogens by immunity-related GTPases (p47 GTPases). Microbes Infect. 9, 1644–1651.

Torrey, E.F., Bartko, J.J., Yolken, R.H., 2012. Toxoplasma gondii and other risk factors for schizophrenia: an update. Schizophr. Bull. 38, 642–647.

Unno, A., Suzuki, K., Xuan, X., Nishikawa, Y., Kitoh, K., Takashima, Y., 2008. Dissemination of extracellular and intracellular Toxoplasma gondii tachyzoites in the blood flow. Parasitol. Int. 57, 515–518.

Vallochi, A.L., Goldberg, A.C., Falcai, A., Ramasawmy, R., Kalil, J., Silveira, C., Belfort, R., Rizzo, L.V., 2008. Molecular markers of susceptibility to ocular toxoplasmosis, host and guest behaving badly. Clin. Ophthalmol. 2, 837–848.

Vallochi, A.L., Muccioli, C., Martins, M.C., Silveira, C., Belfort, R.Jr., Rizzo, L.V., 2005. The genotype of *Toxoplasma gondii* strains causing ocular toxoplasmosis in humans in Brazil. Am. J. Ophtalmol 139, 350–351.

Virreira Winter, S., Niedelman, W., Jensen, K.D., Rosowski, E.E., Julien, L., Spooner, E., Caradonna, K., Burleigh, B.A., Saeij, J.P., Ploegh, H.L., Frickel, E.M., 2011. Determinants of GBP recruitment to Toxoplasma gondii vacuoles and the parasitic factors that control it. PLoS One 6, e24434.

Vyas, A., Kim, S.K., Giacomini, N., Boothroyd, J.C., Sapolsky, R.M., 2007. Behavioral changes induced by Toxoplasma infection of rodents are highly specific to aversion of cat odors. Proc. Natl. Acad. Sci. U.S.A. 104, 6442–6447.

Vyas, A., Sapolsky, R., 2010. Manipulation of host behaviour by Toxoplasma gondii: what is the minimum a proposed proximate mechanism should explain? Folia Parasitol. 57, 88–94.

Wang, X., Claflin, J., Kang, H., Suzuki, Y., 2005. Importance of CD8(+)Vbeta8(+) T-cells in IFN-gamma-mediated prevention of toxoplasmic encephalitis in genetically resistant BALB/c mice. J. Interferon Cytokine Res. 25, 338–344.

Wang, X., Kang, H., Kikuchi, T., Suzuki, Y., 2004. Gamma interferon production, but not perforin-mediated cytolytic activity, of T-cells is required for prevention of toxoplasmic encephalitis in BALB/c mice genetically resistant to the disease. Infect. Immun. 72, 4432–4438.

Wang, X., Michie, S.A., Xu, B., Suzuki, Y., 2007. Importance of IFN-gamma-mediated expression of endothelial VCAM-1 on recruitment of CD8+ T-cells into the brain during chronic infection with Toxoplasma gondii. J. Interferon Cytokine Res. 27, 329–338.

Wang, X., Suzuki, Y., 2007. Microglia produce IFN-gamma independently from T-cells during acute toxoplasmosis in the brain. J. Interferon Cytokine Res. 27, 599–605.

Wang, T., Liu, M., Gao, X.J., Zhao, Z.J., Chen, X.G., Lun, Z.R., 2011. Toxoplasma gondii: the effects of infection at different stages of pregnancy on the offspring of mice. Exp. Parasitol. 127, 107–112.

Webster, J.P., Brunton, C.F., MacDonald, D.W., 1994. Effect of Toxoplasma gondii upon neophobic behaviour in wild brown rats, Rattus norvegicus. Parasitology 109 (Pt 1), 37–43.

Wilson, D.C., Grotenbreg, G.M., Liu, K., Zhao, Y., Frickel, E.M., Gubbels, M.J., Ploegh, H.L., Yap, G.S., 2010. Differential regulation of effector- and central-memory responses to Toxoplasma gondii Infection by IL-12 revealed by tracking of Tgd057-specific CD8+ T-cells. PLoS Pathog. 6, e1000815.

Wilson, E.H., Harris, T.H., Mrass, P., John, B., Tait, E.D., Wu, G.F., Pepper, M., Wherry, E.J., Dzierzinski, F., Roos, D., Haydon, P.G., Laufer, T.M., Weninger, W., Hunter, C.A., 2009. Behavior of parasite-specific effector CD8+ T-cells in the brain and visualization of a kinesis-associated system of reticular fibers. Immunity 30, 300–311.

Wilson, E.H., Hunter, C.A., 2004. The role of astrocytes in the immunopathogenesis of toxoplasmic encephalitis. Int. J. Parasitol. 34, 543–548.

Wilson, E.H., Wille-Reece, U., Dzierszinski, F., Hunter, C.A., 2005. A critical role for IL-10 in limiting inflammation during toxoplasmic encephalitis. J. Neuroimmunol. 165, 63–74.

Witting, P.A., 1979. Learning capacity and memory of normal and Toxoplasma-infected laboratory rats and mice. Z. Parasitenkd. 61, 29–51.

Wong, S.Y., Remington, J.S., 1994. Toxoplasmosis in the setting of AIDS. In: Broder, S., MerganJr., T.C., Bolognesi, D. (Eds.), Text Book of AIDS Medicine. Williams & Wikins, Baltimore.

Woodman, J.P., Dimier, I.H., Bout, D.T., 1991. Human endothelial cells are activated by IFN-gamma to inhibit Toxoplasma gondii replication. Inhibition is due to a different mechanism from that existing in mouse macrophages and human fibroblasts. J. Immunol. 147, 2019–2023.

Xiao, J., Buka, S.L., Cannon, T.D., Suzuki, Y., Viscidi, R.P., Torrey, E.F., Yolken, R.H., 2009. Serological pattern consistent with infection with type I Toxoplasma gondii in mothers and risk of psychosis among adult offspring. Microbes Infect. 11, 1011–1018.

Xiao, J., Kannan, G., Jones-Brando, L., Brannock, C., Krasnova, I.N., Cadet, J.L., Pletnikov, M., Yolken, R.H., 2012. Sex-specific changes in gene expression and behavior induced by chronic Toxoplasma infection in mice. Neuroscience 206, 39–48.

Yagmur, F., Yazar, S., Temel, H.O., Cavusoglu, M., 2010. May Toxoplasma gondii increase suicide attempt-preliminary results in Turkish subjects? Forensic Sci. Int. 199, 15–17.

Yamamoto, M., Okuyama, M., Ma, J.S., Kimura, T., Kamiyama, N., Saiga, H., Ohshima, J., Sasai, M., Kayama, H., Okamoto, T., Huang, D.C., Soldati-Favre, D., Horie, K., Takeda, J., Takeda, K., 2012. A Cluster of Interferon-gamma-Inducible p65 GTPases Plays a Critical Role in Host Defense against Toxoplasma gondii. Immunity.

Yamashita, K., Yui, K., Ueda, M., Yano, A., 1998. Cytotoxic T-lymphocyte-mediated lysis of *Toxoplasma gondii*-infected target cells does not lead to death of intracellular parasites. Infect. Immun. 66, 4651–4655.

Yap, G., Pesin, M., Sher, A., 2000. Cutting edge: IL-12 is required for the maintenance of IFN-gamma production in T-cells mediating chronic resistance to the intracellular pathogen, Toxoplasma gondii. J. Immunol. 165, 628–631.

Yap, G.S., Scharton-Kersten, T., Charest, H., Sher, A., 1998. Decreased resistance of TNF receptor p55- and p75-deficient mice to chronic toxoplasmosis despite normal activation of inducible nitric oxide synthase in vivo. J. Immunol. 160, 1340–1345.

Yap, G.S., Sher, A., 1999. Effector cells of both nonhemopoietic and hemopoietic origin are required for interferon (IFN)-gamma- and tumor necrosis factor (TNF)-alpha-dependent host resistance to the intracellular pathogen, Toxoplasma gondii. J. Exp. Med. 189, 1083–1092.

Yarovinsky, F., Zhang, D., Andersen, J.F., Bannenberg, G.L., Serhan, C.N., Hayden, M.S., Hieny, S., Sutterwala, F.S., Flavell, R.A., Ghosh, S., Sher, A., 2005. TLR11 activation of dendritic cells by a protozoan profilin-like protein. Science 308, 1626–1629.

Yutzy, S.H., Woofter, C.R., Abbott, C.C., Melhem, I.M., Parish, B.S., 2012. The increasing frequency of mania and bipolar disorder: causes and potential negative impacts. Nerv. Ment. Dis. 200, 380–387.

Zhang, Y., Denkers, E.Y., 1999. Protective role for interleukin-5 during chronic Toxoplasma gondii infection. Infect. Immun. 67, 4383–4392.

Zhang, Y., Traskman-Bendz, L., Janelidze, S., Langenberg, P., Saleh, A., Constantine, N., Okusaga, O., Bay-Richter, C., Brundin, L., Postolache, T.T., 2012. Toxoplasma gondii immunoglobulin G antibodies and nonfatal suicidal self-directed violence. J. Clin. Psychiatry 73, 1069–1076.

Zhao, Y., Ferguson, D.J., Wilson, D.C., Howard, J.C., Sibley, L.D., Yap, G.S., 2009b. Virulent Toxoplasma gondii evade immunity-related GTPase-mediated parasite vacuole disruption within primed macrophages. J. Immunol. 182, 3775–3781.

Zhao, Y.O., Khaminets, A., Hunn, J.P., Howard, J.C., 2009b. Disruption of the Toxoplasma gondii parasitophorous vacuole by IFNgamma-inducible immunity-related GTPases (IRG proteins) triggers necrotic cell death. PLoS Pathog. 5, e1000288.

Zhao, Z., Fux, B., Goodwin, M., Dunay, I.R., Strong, D., Miller, B.C., Cadwell, K., Delgado, M.A., Ponpuak, M., Green, K.G., Schmidt, R.E., Mizushima, N., Deretic, V., Sibley, L.D., Virgin, H.W., 2008. Autophagosome-independent essential function for the autophagy protein Atg5 in cellular immunity to intracellular pathogens. Cell Host Microbe 4, 458–469.

CHAPTER

24

Innate Immunity to *Toxoplasma gondii*

Dana G. Mordue, Christopher A. Hunter*[†]

*Department of Microbiology and Immunology, New York Medical College, Valhalla, New York, USA

[†]Department of Pathobiology, School of Veterinary Medicine, University of Pennsylvania, Philadelphia, Pennsylvania, USA

OUTLINE

24.1 INTRODUCTION

In some instances, typically associated with more virulent strains of parasite, *T. gondii* can cause serious disease in immune competent individuals, but most infections are considered to be largely asymptomatic and resolve with minimal or no pathology. However, the immune system fails to achieve sterile immunity and a stable persistent infection results which can leave individuals at risk for reactivation of disease should they develop defects in cell mediated immunity.

Toxoplasma gondii, second edition
http://dx.doi.org/10.1016/B978-0-12-396481-6.00024-6

In part, because of the public health consequences of this infection there has been a long standing interest in understanding the immunological basis for the ability to control the acute phase of this infection as well as to prevent reactivation. Thus, researchers have elucidated many of the immune mechanisms and cell signalling cascades important for stimulating effective cell mediated immunity to *T. gondii* and for re-establishment of immune homeostasis during and following infection to prevent excessive inflammation. These studies have identified and assigned functions to key players important for innate immunity, dissected mechanisms for antigen presentation and activation of T-cell populations and demonstrated the importance for immune regulation. Consequently, *T. gondii* has been used as a model organism for the study of the immune response to intracellular pathogens. This chapter reviews our current understanding of the innate immune response to *T. gondii* and highlights some of the prominent questions in the field about how *T. gondii* is recognized, controlled by the cytokine IFNγ and recent advances in our understanding of parasite evasion of innate immunity and how these impact on virulence.

24.2 ESTABLISHMENT OF INFECTION

As noted in other chapters, although *T. gondii* can be transmitted through congenital infection, ocular exposure and transplantation, it is most often acquired by ingestion of tissue cysts or oocysts shed by feline species. Following ingestion, parasites invade enterocytes in the small intestine and can penetrate the polarized epithelial barrier by active invasion without causing extensive damage to the epithelium (Barragan and Sibley, 2002). NK- and NKT-cells, epithelial, intraepithelial lymphocytes and enterocytes all contribute to the initial innate immune response to the parasite following mucosal challenge (reviewed in Buzoni-Gatel *et al.*, 2006). These contribute to localized inflammation that typically resolves without extensive pathology. Despite these events, *T. gondii* is able to disseminate, likely as a result of the ability of a few organisms to penetrate across the intestinal barrier, evade mucosal immunity including effector activity of the recruited inflammatory monocytes and spread systemically (Mordue and Sibley, 2003).

While oral infections with *T. gondii* are typically asymptomatic there are conditions where challenge can lead to severe intestinal inflammation. Thus, oral challenge of female C57BL/6 mice with a high dose of cysts or mice deficient in anti-inflammatory mediators such as IL-10 (Suzuki *et al.*, 2000; Liesenfeld *et al.*, 1999) can result in a $CD4^+$ T-cell mediated inflammatory response that damages the intestinal barrier allowing leakage of bacterial components from the lumen leading to further intestinal damage and necrosis (Heimesaat *et al.*, 2006; Liesenfeld *et al.*, 1996). Limited escape of commensal gut bacteria from the lumen, however, may act as an adjuvant to stimulate dendritic cell (DC) production of IL-12 and thereby actually enhance protection against *T. gondii* (Benson *et al.*, 2009). It remains to be seen whether this adjuvant effect by intestinal microflora on the innate system actually stimulates protective *T. gondii* antigen specific T-cells or if the T-cells generated are specific for the commensal bacteria. Once *T. gondii* reaches the lamina propria, the parasite can invade leukocytes or enter the lymphatic and circulatory systems to spread systemically. Early in infection, *T. gondii* can infect DCs and monocytes in the gut and use the $CD11b^+$ monocytes to disseminate to distal sites including the brain (Courret *et al.*, 2006).

24.3 THE CRITICAL IMPORTANCE OF IL-12-DEPENDENT IFNγ PRODUCTION

A series of studies more than 20 years ago revealed that the production of IFNγ by NK- or T-cells is essential for control of acute and chronic *T. gondii* infection and that IFNγ signalling, through the transcription factor STAT1, is critical for activating anti-microbial inducible effectors important for the control of intracellular parasites (Gavrilescu *et al.*, 2004; Lieberman *et al.*, 2004a). Mice that lack IFNγ, IFNγ receptor or STAT1 succumb rapidly to toxoplasmosis due to an inability to control parasite replication (Suzuki *et al.*, 1988; Scharton-Kersten *et al.*, 1996; Yap and Sher, 1999). As noted in other chapters, *T. gondii* can infect all nucleated cells and there is good evidence that optimal control of this challenge is dependent on the ability of IFNγ to activate haematopoietic and non-haematopoietic cells (Lykens *et al.*, 2010; Yap and Sher, 1999). However, macrophages and inflammatory monocytes are regarded as one of the main cell types involved in parasite control. Expression of a dominant negative IFNγ receptor on $CD68^+$ macrophages results in uncontrolled parasite replication during acute infection even in the presence of IFNγ (Dunay *et al.*, 2008, 2010; Robben *et al.*, 2005; Mordue and Sibley, 2003).

During toxoplasmosis the events that lead to the control of infection are dominated by accessory cell production of IL-12 that promotes the ability of NK- and T-cells to make IFNγ (Denkers *et al.*, 1993; Hunter *et al.*, 1994; Scharton-Kersten *et al.*, 1996, 1998). Although IL-12 and IL-23 share a common p40 subunit, IL-12, not IL-23, is the dominant cytokine in resistance to *T. gondii* infection (Lieberman *et al.*, 2004b). Neutrophils, dendritic cells, macrophages and inflammatory monocytes can all make IL-12 p40 in response to *T. gondii*. However, the relevance of each source of IL-12 to control *T. gondii in vivo* remains open to debate and may depend in part on route of infection, parasite dose, whether challenge is with *T. gondii* tachyzoites versus cysts and the tissue type and time examined post-infection. A $CD8\alpha^+$ splenic DC subset has been shown to be critical for production of IL-12 *in vivo* during infection and for priming *T. gondii*-specific $CD8^+$ T-cells (Mashayekhi *et al.*, 2011). Unlike many cell types that produce IL-12 in response to *T. gondii* the $CD8\alpha^+$ DCs do not require initial priming with IFNγ to produce IL-12. Thus, challenge with an avirulent strain of *T. gondii* resulted in a substantial increase in the number of splenic $CD8\alpha^+$ DCs staining positive for IL-12 p40 within three days of infection. Chimeric mice in which only $CD8\alpha^+$ DCs were unable to produce IL-12 died 10 days post-infection with *T. gondii*, implicating these $CD8\alpha^+$ DCs and their production of IL-12 as a major pathway important for IFNγ-mediated control of *T. gondii in vivo* (Mashayekhi *et al.*, 2011).

In addition to the critical importance of IFNγ to activate the anti-microbial effectors required to control *T. gondii* it also impacts on the cellular composition at sites of inflammation. Thus, following challenge with *T. gondii*, at the local site of infection NK-cell-derived IFNγ drives the replacement of resident mononuclear phagocytes with circulating monocytes that differentiate in situ into inflammatory DCs (MoDCs) and F4/80^+ macrophages. In this model, IL-12 p40 production is primarily from non-infected monocyte-derived $CD11b^+$,$CD8\alpha^-$ DCs whose maturation is induced by IFNγ (Goldszmid and Sher, 2012). In a previous study, uninfected inflammatory monocytes ($CD68^+$, $Gr\text{-}1^+$, $F4/80^+$, $CD11b^+$, $CD11c^-$) were shown to be the primary source of IL-12 in response to challenge with tachyzoites (Mordue and Sibley, 2003). These inflammatory monocytes also produce TNFα and suppress parasite replication *in vitro*, in part through production of nitric oxide (Mordue and Sibley, 2003). $CCR2^{-/-}$ and $MCP1^{-/-}$ mice fail to recruit these inflammatory monocytes to primary sites of infection, which results in an inability to control parasite replication even in the presence of normal levels of IFNγ in the peritoneum and systemically. This indicates that the inflammatory monocytes are critical at the site of infection for parasite control but may not be essential for systemic Th1 cytokine production (Robben *et al.*, 2005). Based on the known plasticity in the maturation process of mononuclear phagocytes it seems likely that the inflammatory monocytes are capable of maturing into biologically relevant DC populations.

24.4 PATTERN RECOGNITION RECEPTORS AND IL-12 PRODUCTION

The recognition that the IL-12/IFNγ axis is critical for control of acute *T. gondii* infection (reviewed in Egan *et al.*, 2009) led to basic questions about the receptors involved in the innate recognition of *T. gondii* and the production of IL-12. The importance of Toll-like receptors (TLR) in these events was first shown by the inability of mice deficient in the common TLR-adaptor protein MyD88 to produce IL-12 and control acute toxoplasmosis (Scanga *et al.*, 2002). IL-12 production in response to *T. gondii* is largely MyD88-dependent in dendritic cells, macrophages and neutrophils *in vitro* (Scanga *et al.*, 2002). TLR ligation by pathogen associated molecular patterns results in the recruitment of the adaptor molecule MyD88 followed by IL-1 receptor associated kinases (IRAK) 1 and 4. IRAK forms a complex with TNF-receptor associated factor (TRAF) 6, Ubc13 and Uva1 and the complex functions as an E3 ligase resulting in polyubiquitination of TRAF6 that activates transforming growth factor-β-activated kinase 1 (TAK1). TAK1 triggers both mitogen-activated protein kinase (MAPK) and nuclear factor (NF)-κB signalling pathways leading to the production of inflammatory cytokines including IL-12 (Fig. 24.1). Many parts of this signalling pathway, such as TRAF6, the

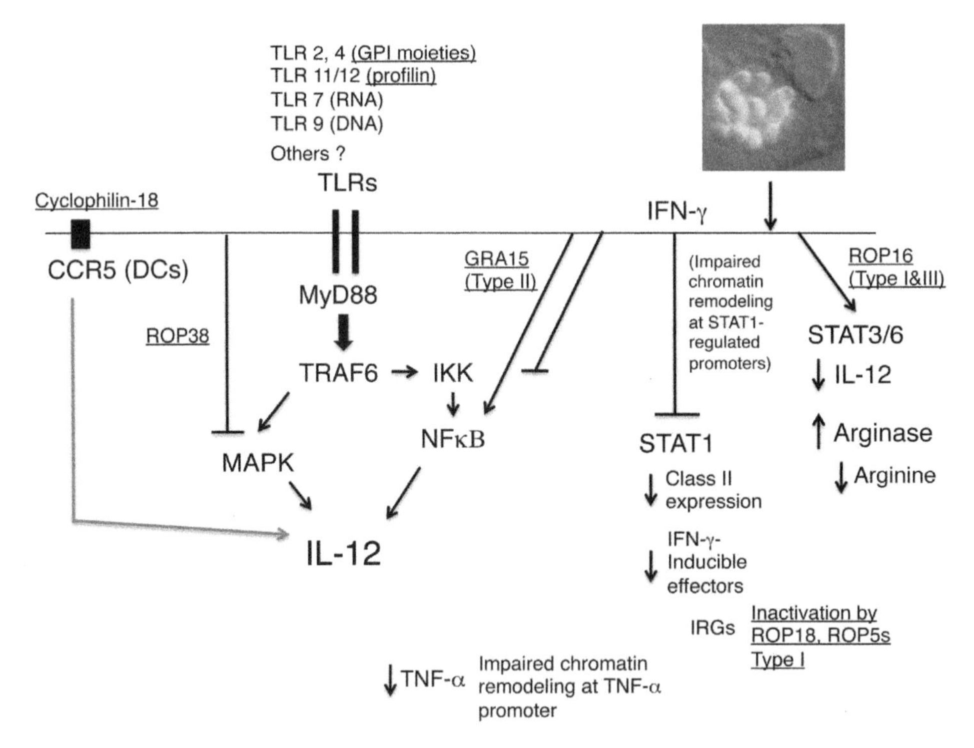

FIGURE 24.1 **Host cell immune signalling cascades induced or modified by *T. gondii*.** Signalling cascades downstream of MyD88 are important for the early production of IL-12 to control *T. gondii* infection. However, *T. gondii* also inhibits activation of its host cell and production of inflammatory cytokines. Parasite products known to effect specific signalling pathways or effector functions are underlined.

NF-κB transcription factor c-Rel, and p38 MAPK have been shown to be relevant to the recognition of *T. gondii* (Mason *et al.*, 2002, 2004; Kim *et al.*, 2005)

In contrast to the inability of MyD88 deficient mice to control *T. gondii* infection, mice deficient in individual TLR receptors including TLR1, 2, 4, 6, 9 or 11 have not demonstrated a marked increase in their susceptibility to acute infection with *T. gondii* (Hitziger *et al.*, 2005; Yarovinsky *et al.*, 2005; Debierre-Grockiego *et al.*, 2007; Melo *et al.*, 2010). Similarly, mice deficient in the endosomal localized TLRs 3, 7 and 9 also do not have an impaired response to *T. gondii*. However, mice deficient in UNC93B1, a transmembrane ER protein important for transport of TLRs from the ER to endosomes (Kim *et al.*, 2008), are as susceptible to acute infection with *T. gondii* as MyD88 deficient mice (Melo *et al.*, 2010; Pifer *et al.*, 2011). This defect is due at least in part to impaired production of IL-12 as exogenous IL-12 restores the ability of UNC93B1 deficient mice to control *T. gondii* infection (Pifer *et al.*, 2011). UNC93B1 was recently discovered to also interact directly with TLR11 and to be critical for DC profilin-dependent IL-12 production through TLR11 (Pifer *et al.*, 2011). However, the inability to control infection with *T. gondii* appears more pronounced in UNC93B1 deficient compared to TLR11 deficient mice implying that UNC93B1 has additional effects on IL-12 production independent of TLR11. Together, these findings suggest that *T. gondii* engages multiple TLRs during infection and that the sum of their parts is more important than any sole contributor. However, it remains possible that MyD88 serves as an adaptor for yet unknown pattern recognition receptors critically important for recognition of *T. gondii* and for IL-12 production. For example, MyD88 is also an adaptor protein for IL-1 and IL-18 receptors. Consequently, it is possible that the protective effect of MyD88 in acute *T. gondii* infection is, in part, mediated by IL-1 and IL-18. However, mice deficient in caspase-1, which is important for IL-1 and IL-18 production, are not more susceptible to acute *T. gondii* infection (Hitziger *et al.*, 2005).

Despite the fact that no single identified TLR is absolutely critical for protection against *T. gondii*, parasite ligands for TLR11 and TLR2 have been identified (reviewed in Pollard *et al.*, 2009). Murine TLR11 recognizes *T. gondii* profilin that stimulates IL-12 production from DCs and contributes to a protective immune response to the parasite (Yarovinsky *et al.*, 2005). Although TLR11 clearly functions in recognition of *T. gondii* and IL-12 production in mice, it does not appear to be important for recognition of *T. gondii* in humans as the full length protein is not expressed due to several stop codons. TLR2 and TLR4, in association with galectin-3, recognize *T. gondii* glycosylphosphatidylinositols resulting in macrophage TNFα production (Debierre-Grockiego *et al.*, 2007, 2010).

MyD88-dependent IL-12 production is critical for early control of *T. gondii*. However, IL-12 production by *T. gondii* and subsequent generation of a protective response is not limited to the MyD88/IL-12/IFNγ pathway. Bone marrow-derived macrophages infected *in vitro* with *T. gondii* can produce IL-12 that is dependent on parasite-induced autophosphorylation of p38 MAPK but independent of MyD88 and independent of CCR5. Both Type I and II *T. gondii* strains induce macrophage p38 MAPK-dependent IL-12 production. However, the contribution of MyD88 is genotype specific; IL-12 production by the RH Type I strain is independent of MyD88 while both MyD88 dependent and independent IL-12 is induced by the ME49 Type II strain (Kim *et al.*, 2006). IL-12 is also produced *in vivo* in the peritoneum of infected MyD88 deficient mice but at lower levels than wild type mice. Experiments with an avirulent temperature sensitive mutant of *T. gondii* indicated that protective immunity could be generated in the absence of MyD88 but not rapidly or robustly enough to protect against acute infection with a non-attenuated strain of the parasite (Sukhumavasi *et al.*, 2008).

MyD88 is also likely important for resistance to *T. gondii* independent of its role in stimulating IL-12 production. Although MyD88$^{-/-}$ mice infected with *T. gondii* succumb rapidly, have decreased IL-12 and fail to control parasites, addition of exogenous IL-12 partially restores the Th1 response but does not protect the mice. Indeed, chimeric mice in which only T-cells lacked expression of MyD88 have increased susceptibility to toxoplasmic encephalitis independent of IL-18 or IL-1, suggesting a role for T-cell TLRs in prolonged resistance to the parasite at least during chronic infection (LaRosa *et al.*, 2008).

While there has been an understandable focus on the role of TLR and downstream signalling pathways such as those mediated by the NF-κB transcription factors, there is also evidence of other host pathways that are activated as a consequence of parasite invasion. G_i protein coupled-receptors (G_iPCR) appear to contribute to macrophage recognition of *T. gondii*, resulting in activation of phosphatidylinositol (PI) 3-kinase-dependent extracellular-regulated kinase 1 and 2 and protein kinase B (Kim and Denkers, 2006). In this case, G_iPCR does not regulate IL-12 production but instead contributes to the ability of *T. gondii* to protect infected cells from apoptosis. A potentially related observation is that parasite invasion of DCs triggers hypermotility which is dependent on host cell G-protein signalling (Lambert *et al.*, 2006) and these events have been linked to parasite dissemination (Collantes-Fernandez *et al.*, 2012; Diana *et al.*, 2004; Lambert *et al.*, 2009). There is also evidence that infection leads to changes in host cell Ca^{2+} stores resulting in activation of PKC, infection induced MAPK activation and IL-12 production (Masek *et al.*, 2006, 2007).

24.5 *TOXOPLASMA GONDII* MODULATION OF HOST CELL SIGNALLING

Given the importance of the MyD88/IL-12/IFNγ axis in limiting the replication of *T. gondii*, and ultimately its levels of transmission, it is not surprising that this organism has evolved numerous mechanisms to modulate these pathways. Thus, *T. gondii* interferes with TLR signalling in part by transiently blocking NF-κB translocation to the host cell nucleus (Butcher *et al.*, 2001; Shapira *et al.*, 2002, 2005). Although the parasite rapidly triggers transient p38 MAPK activation it impairs its subsequent stimulation in response to LPS (Kim *et al.*, 2004), a process that may be mediated by the parasite rhoptry protein ROP38 (Peixoto *et al.*, 2010). Another mechanism for *T. gondii* suppression of signalling cascades and inflammatory cytokine production involves specific modulation of the histone machinery in the host cell to inhibit activation of specific promoters. *T. gondii* prevents chromatin remodelling at the TNFα promoter in infected cells in response to TLR-triggered macrophage activation resulting in inhibition of TNFα production (Leng *et al.*, 2009). Similarly, *T. gondii* inhibits covalent modification of histone H3 at the IL-10 promoter following cross linking of TLR4 and Fc gamma receptor suppressing IL-10 production (Leng and Denkers, 2009). *T. gondii* infection suppresses IFNγ signalling through STAT1 in infected cells by disrupting chromatin remodelling of STAT1 regulated promoters (Lang *et al.*, 2012). As a result, less than 40% of the IFNγ-dependent transcripts induced in naive macrophages by IFNγ are increased in *T. gondii* infected macrophages, resulting in decreased production of IFNγ-dependent antimicrobial mediators (Lang *et al.*, 2012). There are also reports that infected cells cannot up-regulate expression of MHC Class II on the surface and that this would compromise their ability to present antigen (Luder and Gross, 2005; Luder *et al.*, 2003b). In theory, parasite manipulation of specific chromatin remodelling components may be a considerably more efficient approach to globally regulate a cassette of relevant pathways, rather than inhibiting individual genes.

24.6 *TOXOPLASMA GONDII* GENOTYPE-DEPENDENT EFFECTS ON HOST CELL SIGNALLING

A variety of genetic analyses over the last 30 years have identified distinct strains of *T. gondii* that differ in their capacity to modulate host cell signalling cascades and which has been linked to differences in virulence. Typically, the Types I, II and III strains are associated with varying levels of virulence: a single parasite of the Type I strain being able to kill immune competent mice, while the Type III strains are the least virulent. Classic genetic crosses using different parasite strains have led to the identification of secreted rhoptry and dense granule proteins that are important for parasite genotype-dependent effects on host signalling and resistance to IFNγ-induced effectors. The initial observation that infection of macrophages with Type II, but not Types I and III genotypes of *T. gondii* results in substantial translocation of NF-κB and production of IL-12 (Kim *et al.*, 2006; Robben *et al.*, 2004) paved the way for the discovery that the polymorphic protein GRA15 found in Type II strains stimulates NF-κB translocation and IL-12 production independent of MyD88 (Rosowski *et al.*, 2011).

The STAT1 pathway appears to be a major target for *T. gondii* to inhibit, but other members of this family of transcription factors are directly utilized by *T. gondii* to promote its survival. Recent examples include the ability of Types I and III but not Type II parasites to secrete the ROP16 kinase into the host cell cytosol to directly phosphorylate STAT3 and STAT6 and thereby suppress IL-12 (Butcher *et al.*, 2005, 2011; Ong *et al.*, 2010; Saeij *et al.*, 2007). Activation of STAT6 is normally a consequence of IL-4 or IL-13 (reviewed in Denkers *et al.*, 2012; Melo *et al.*, 2011) and in macrophages is associated with an 'alternatively activated' or M2 phenotype implicated in tissue repair (reviewed in Mosser and Edwards, 2008). A major product of M2 macrophages is arginase, which converts arginine to ornithine; a precursor of polyamines. The ROP16 from Type I parasites is a strong inducer of macrophage arginase (Jensen *et al.*, 2011; Butcher *et al.*, 2011). This has led to the proposal that Types I and III strains of *Toxoplasma* are more likely to induce M2 activation of macrophages while Type II genotypes induce macrophages that are 'classically' activated and specialized to kill intracellular organisms. In theory, induction of macrophage arginase could have very diverse outcomes for the parasite; some that favour the parasite while others favour the host immune response (reviewed in Denkers *et al.*, 2012). Since *T. gondii* is a polyamine auxotroph, arginase promotion could enhance parasite replication. However, *T. gondii* must also scavenge the amino acid arginine and increased arginase could limit parasite replication by degrading arginine. This is consistent with the report that Type I *T. gondii* ROP16 null parasites have an *in vivo* growth advantage, suggesting that arginine may become limited during infection perhaps as a consequence of the host immune response (Butcher *et al.*, 2011). ROP16-mediated STAT6 activation and induction of arginase may also decrease the availability of arginine as a substrate for inducible nitric oxide synthase (iNOS) resulting in decreased nitric oxide (NO) and enhanced parasite replication. This is consistent with the finding that ROP16 gene deleted parasites do not persist in activated astrocytes since NO is a known defence mechanism against *T. gondii* in the brain (Butcher *et al.*, 2011; Chao *et al.*, 1993; Peterson *et al.*, 1995; Scharton-Kersten *et al.*, 1997; Schluter *et al.*, 1999). Also, mouse astrocytes and microglial cells produce microbicidal NO following activation with IFNγ and this is suppressed by parasites in a ROP16 dependent manner (Butcher *et al.*, 2011). Recently, high amounts of iNOS and low levels of arginase have been implicated as important for the enhanced resistance of rat peritoneal cells to *T. gondii* compared to mouse peritoneal cells that have low amounts of iNOS and high levels of arginase (Li *et al.*, 2012). Therefore, host and parasite-dependent variables impact the relative ratio of iNOS and arginase in infected macrophages and exert effects on parasite survival. Since *T. gondii* is capable of infecting such a wide range of hosts, *T. gondii* may have exerted a selective pressure on specific host species to evolve mechanisms to combat *T. gondii* infection; different parasite genotypes in response have evolved common and species-specific evasion mechanisms to subvert these defences.

Infection of macrophages with *T. gondii* results in the rapid phosphorylation of STAT3 and this is associated with decreased production of IL-12p40 and TNFα (Butcher *et al.*, 2005). The suppressor of cytokine signalling molecule 3 (SOCS3) is up-regulated downstream of IL-6 and IL-10 mediated activation of STAT3 and this protein has several suppressive activities that include the ability to bind to the gp130 subunit of the IL-6 receptor complex and thereby suppress IL-6 signalling (reviewed in Kubo *et al.*, 2003). SOCS3 is also activated following *T. gondii* invasion of macrophages, independent of parasite genotype, and this led to the idea that it might be involved in the mechanisms used by *T. gondii* to alter macrophage function (Whitmarsh *et al.*, 2011). However, in macrophages that lacked SOCS3 the ability of *T. gondii* to suppress inflammatory cytokines was intact. Unexpectedly, mice engineered to lack SOCS3 in macrophages and neutrophils, when challenged with a Type II strain of *T. gondii*, had reduced production of IL-12, were unable to control parasite replication and succumbed by day 10 post-infection. The basis for this phenotype appears to be due to exaggerated signalling by IL-6 that suppresses IL-12 production and this is supported by the ability of anti-IL-6 antibody or exogenous addition of IL-12 to restore the ability of the mice to control *T. gondii* infection. This indicates that SOCS3 is critical during *T. gondii* infection to limit IL-6 signalling and enable optimal production of IL-12 required to control parasite replication (Whitmarsh *et al.*, 2011).

24.7 CELL AUTONOMOUS IMMUNITY

Although *T. gondii* replication is unrestrained in naive or resident tissue macrophages, stimulation with IFNγ activates STAT1 that induces transcription of host defence pathways critical for control of the parasite. Nearly 2000 human and mouse IFN-stimulated genes have been identified to date and the majority remain uncharacterized (Samarajiwa *et al.*, 2009; Hertzog *et al.*, 2011), suggesting that numerous IFNγ-inducible effectors have yet to be discovered. As mentioned previously, *T. gondii* infection suppresses IFNγ signalling by disrupting chromatin remodelling of STAT1 regulated promoters (Lang *et al.*, 2012). However, even though *T. gondii* modulates the effectiveness of IFNγ and TLR-dependent signalling cascades in infected cells, antimicrobial effectors can still be mobilized to control parasite replication. For example, Type I strains of *T. gondii* inhibit macrophage inducible nitric oxide synthase expression in response to IFNγ and LPS, but the residual nitric oxide produced in infected cells can still impair parasite replication *in vitro* (Luder *et al.*, 2003a).

The relative potency and effectiveness of mechanisms of IFNγ-induced cell autonomous immunity depends on the species and cell specific effectiveness of each mediator, the intracellular niche of the pathogen and the adaptations made by the pathogen to counter antimicrobial effectors. This is a particularly complex evolutionary puzzle for a parasite like *T. gondii* that resides in an intracellular niche that is very distinct from those typically occupied by other organisms found within the phagolysosomal system or those that exist in the host cell cytosol. Thus, many of the antimicrobial mechanisms that function against other intracellular organisms are unlikely to be effective in cells invaded by *T. gondii.*

However, because *T. gondii* is a successful, widespread parasite able to infect any warm-blooded vertebrate or cell type with the potential to kill its host, this organism will have provided a strong selective pressure to evolve appropriate anti-microbial activities. This process may explain why the effectiveness and even presence of antimicrobial mediators differs depending on the host species infected as well as the type of host cell. The aim of this section is to provide a survey of our current knowledge of the many cell-autonomous immune mechanisms involved in the control of *T. gondii.*

24.7.1 IFNγ-Induced Nitrosative and Oxidative Defence

Reactive oxygen (ROS) and nitrogen species (RNS) are potent antimicrobial molecules that are evolutionarily conserved and important cell autonomous defence mechanisms against intracellular pathogens (reviewed in MacMicking, 2012; Fang, 2004; Nathan and Shiloh, 2000). ROS and RNS endow phagocytic cells with a complex array of interacting antimicrobial molecules that act synergistically against microbial pathogens (Fig. 24.2). In mammals, there are three classes of cytokine-inducible oxidoreductases that control ROS and RNS production. NADPH

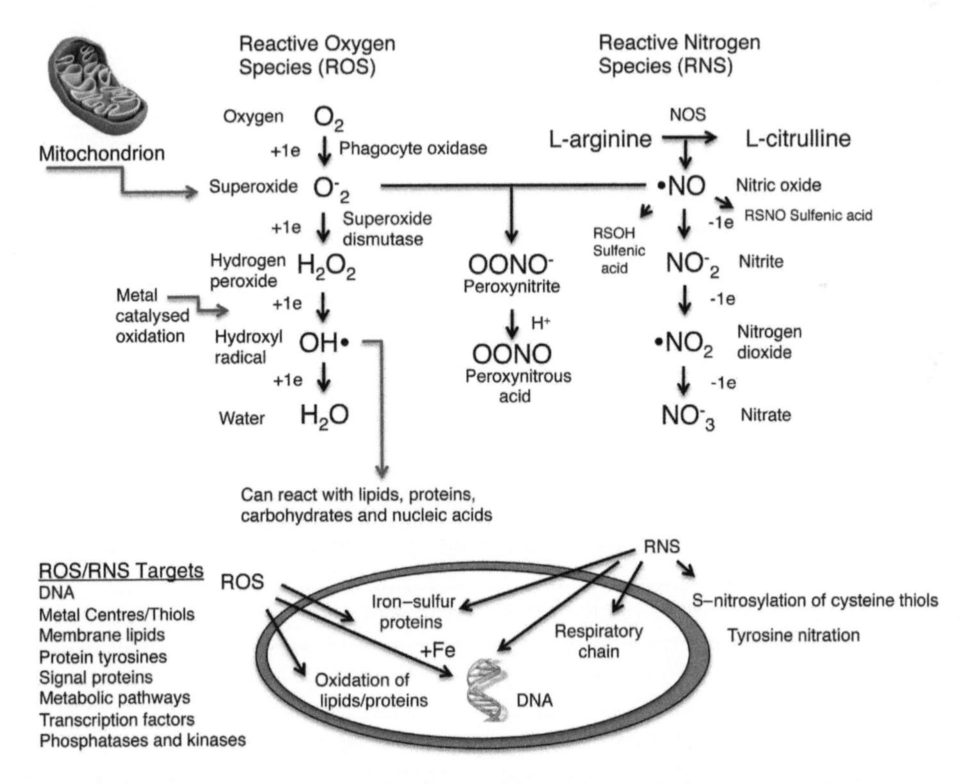

FIGURE 24.2 **Reactive oxygen and nitrogen species and their targets.** Reactive oxygen species can be generated by phagocyte oxidase or leaked from the mitochondrion. Reactive nitrogen species are generated by nitric oxide synthase induced conversion of L-arginine to L-citrulline. Reactive oxygen and nitrogen species act independently and synergistically to create a variety of intermediates toxic to intracellular microbes. *Figure is modified from Fang (2004) and Nathan and Shiloh (2000).*

oxidases (NOX) directly catalyse the production of $O2^-$, dual oxidases produce H_2O_2 and nitric oxide synthases, including iNOS (NOS2), produce nitric oxide (NO) during the metabolism of L-arginine to citrulline. NO interacts with haem, non-haem iron and iron sulphur proteins to disrupt microbial enzymes while NO synergizes with reactive oxygen species to generate additional RNS that interact with protein thiols, tyrosines, lipids and DNA to cause damage and disrupt function. Activation of murine macrophages with IFNγ and a second signal such as TNFα effectively suppress *T. gondii* replication, independent of parasite genotype, *in vitro via* a NO-dependent mechanism (Sibley *et al.*, 1991; Langermans *et al.*, 1992; Zhao *et al.*, 2008. 2009a). NO may be more likely to suppress replication of *T. gondii* rather than kill and degrade it (Zhao *et al.*, 2008) and NO from haematopoietic cells is clearly important for the control of chronic *T. gondii* infection (Khan *et al.*, 1997; Scharton-Kersten *et al.*, 1997; Schluter *et al.*, 1999; Yap and Sher, 1999). Macrophages primed *in vitro* with an avirulent mutant of *T. gondii* suppress the replication of *T. gondii* independent of parasite genotype and through the production of NO (Zhao *et al.*, 2009a). Rats usually develop subclinical toxoplasmosis unlike the more severe disease that occurs in most inbred mouse strains (reviewed in Dubey and Frenkel, 1998). As mentioned previously, rat macrophages produce high levels of NO and low levels of polyamines compared to mice suggesting that NO/RNS are important for protection against *T. gondii in vivo* and effectiveness may depend on relative expression of iNOS versus arginase and the contribution of other antimicrobial mediators (Li *et al.*, 2012) . Consistent with a key role for NO in controlling *T. gondii*, a forward genetics study to identify parasite genes important for survival following macrophage activation uniformly identified genes important for resistance to NO (Mordue *et al.*, 2007; Skariah *et al.*, 2012) and suggests that *T. gondii* has evolved numerous mechanisms to withstand NO/RNS.

ROS include superoxide radicals, hydrogen peroxide and hydroxyl radicals (reviewed in MacMicking, 2012). ROS also synergizes with RNS to produce toxic intermediates including dinitrogen oxides, compound peroxides and nitrosothiol adducts. The NADPH oxidase family of enzymes (NOX1–5) represents the major ROS producers during infection. NOX2 (phagoycyte oxidase) is responsible for the respiratory burst in neutrophils, monocytes, macrophages and eosinophils. However, dual oxidases and non-enzymatic sources of ROS including those from mitochondrial leakage also contribute to host defence. *T. gondii* invasion, in contrast to phagocytosis, fails to trigger production of reactive oxygen intermediates in naive murine resident peritoneal macrophages; however, they are known to exhibit only a weak respiratory burst in the absence of activation. In contrast, NADPH oxidase is stimulated in human monocytes and activated murine macrophages both of which exhibit oxygen-dependent anti-*T. gondii* activity (Wilson *et al.*, 1980; Murray and Cohn, 1979; Murray *et al.*, 1979, 1985). IFNγ-activation of murine dendritic cells also results in oxygen-dependent suppression of *T. gondii* (Aline *et al.*, 2002). In humans, there is evidence that IFNγ primes phagocytes to produce reactive oxygen intermediates including H_2O_2 and leukotrienes to kill *T. gondii* (Murray *et al.*, 1985; Yong *et al.*, 1994). Consistent with a role for ROS in control of *T. gondii*, monocytes from patients with hereditary myeloperoxidase deficiency or chronic granulomatous disease (CGD), characterized by congenital mutations in genes encoding NOX2 subunits, both have a significant defect in their ability to control *Toxoplasma* replication. (Locksley *et al.*, 1982; Murray *et al.*, 1985). Perhaps related to the role of oxygen free radicals in killing microorganisms, *T. gondii* has an intact antioxidant network (Ding *et al.*, 2004; Sautel *et al.*, 2009), components of which are targeted to the parasite mitochondria and apicoplast (Pino *et al.*, 2007). It is likely that this network functions to protect the mitochondrion and apicoplast from oxidative stress during respiration but may also protect against the antimicrobial effects of host reactive oxygen intermediates (Hunn and Howard, 2010).

24.7.2 IFNγ-Inducible Immunity Related GTPases (IRGs)

The IFNγ immunity related GTPases (IRGs) are abundantly represented in small rodents and are present in all orders of mammals. IFNγ-inducible IRGs and p65 guanylate-binding proteins (GBPs) are important intracellular effectors against *T. gondii* (reviewed in Hunn *et al.*, 2011; Howard *et al.*, 2011). Mice have 23 IRG genes while humans only have two; IRGC expressed exclusively in the testis and IRGM that has a truncation in the GTPase binding domain and is no longer regulated by IFNγ (Bekpen *et al.*, 2005). The IRGs have been divided into two groups. Members of the canonical GKS group have a lysine in the G1 motif of the nucleotide binding site whereas members of the GMS group have a methionine in this site and are considered negative regulators of the GKS effectors. Thus, in the absence of the GMS regulators, the GKS effectors spontaneously form GTP-bound aggregates in IFNγ-induced cells. Studies using a Type II genotype of *T. gondii* first elucidated the role of IRGs in control of specific intracellular pathogens (Collazo *et al.*, 2001; Taylor *et al.*, 2000). Mice deficient in IGTP (Irgm3) were acutely susceptible to *T. gondii* during acute infection due to unrestrained parasite replication with a phenotype very similar to that for IFNγ-deficient mice. Importantly, mice displayed normal IL-12, IFNγ and nitric oxide responses yet failed to restrain parasite growth. These findings indicated that IRGs were critical antimicrobial effectors that acted downstream of IFNγ and were important in resistance to acute and chronic phases of this infection. Studies using *in vivo* *T. gondii* primed macrophages demonstrated a critical role for Irgm3 in macrophage clearance of Type II but not Type I genotypes of *T. gondii* (Zhao *et al.*, 2009a). Unexpectedly, the absence of the GMS Irgm1 results in IFNγ-dependent failure of stem cells of the lympho-myeloid system resulting in massive lymphopenia and collapse of T-cell immunity following infection (Feng *et al.*, 2009; Feng *et al.*, 2008). Thus, while most studies on these proteins have focused on their roles in anti-microbial activities, it appears that they also have an immunoregulatory role and this is an aspect of their biology that needs to be explored in more depth.

Since the link of IRGs to control of intracellular pathogens, there has been a concerted effort to define the functions of this family. *T. gondii* invasion by Type II or III parasite genotypes into IFNγ stimulated cells results in trafficking and sequential loading of IRGs onto the PVM over a period of 90 minutes (Martens *et al.*, 2005; Khaminets *et al.*, 2010). This appears to be a process that is hierarchical and cooperative with Irgb6 and possibly Irgb10 being the first to load followed by Irga6 and then Irgd and other IRGs (Khaminets *et al.*, 2010). In macrophages, the Atg5 autophagy protein is important for loading of Irg6a and perhaps other IRGs (Zhao *et al.*, 2008). Mice that lack ATG5 in macrophages and granulocytes are acutely susceptible to infection with *T. gondii* (Zhao *et al.*, 2008). IRG loading ultimately results in disruption of the PVM and exposure of the parasite to the cytosol through yet unknown mechanisms (Zhao *et al.*, 2009b). Observations on the sequential disruption of the PVM in macrophages suggest a mechanism related to extraction of surface area from the PVM by IRG-enforced ruffling and vesiculation resulting in membrane disruption (Howard *et al.*, 2011). Unlike nitric oxide, the effectiveness of IRGs is dependent on parasite genotype; Type II and III genotypes are susceptible while the resistance of the Type I genotype is mediated by ROP18. This parasite kinase is secreted during parasite invasion, traffics to the cytosolic face of the PVM and phosphorylates several effector IRGs on two conserved threonine residues in the conformationally important Switch 1 region of the nucleotide binding region disrupting their enzymatic activity and preventing their loading to the PVM (Steinfeldt *et al.*, 2010; Fentress *et al.*, 2010). The Type I genotype ROP5 locus encodes three isoforms of ROP5 that are each pseudokinases. ROP5 is critical for ROP18 phosphorylation of IRGs (Fleckenstein *et al.*, 2012; Niedelman *et al.*, 2012). The finding that ROP18 and ROP5 mediate *T. gondii* evasion in murine cells, but are not important in IFNγ activated human fibroblast cells, is consistent with the absence of canonical IRGs in humans (Niedelman *et al.*, 2012). This observation illustrates that effective control of *T. gondii* during infection is dependent on both host-specific IFNγ-dependent intracellular immune effectors and parasite genotype-dependent evasion mechanisms. The ability of parasites of different genotypes to express numerous ways to disrupt multiple host responses many of which may be host specific can help us understand why *T. gondii* has been so successful but may also explain the association of different parasite genotypes with different hosts (Howe and Sibley, 1995).

24.7.3 IFNγ-Inducible p65 GTPases

P65 guanylate-binding proteins (GBPs) are another family of IFNγ-inducible GTPases that contribute to immune control of *T. gondii*. There are 13 GBPs identified in the mouse genome and seven in the human genome. Mice deficient in all six GBP genes on chromosome 3 showed significantly increased susceptibility to *T. gondii* (Type II genotype) during the acute stage of infection. Furthermore, IFNγ failed to stimulate GBP^{chr3}-deleted MEFs to suppress *T. gondii* intracellular growth or to recruit the IFNγ-inducible IRG GTPase Irgb6 to the PVM (Yamamoto *et al.*, 2012). Reintroduction of GBP1, GBP5 or GBP7 into GBP^{chr3}-deleted mouse embryonic fibroblasts (MEFs) partially restores IFNγ-dependent clearance of *T. gondii* indicating a specific contribution by these GBPs to parasite clearance. Recruitment of murine GBP1, 2 and 5 to PVs is parasite genotype-dependent. PVs containing Type II or III genotypes accumulate GBP 1, 2 and 5 while the Type I genotype does not (Virreira Winter *et al.*, 2011). The observation that this defect is overcome by adding TNFα and *in vivo* there is a very significant IFNγ-dependent GBP^{chr3}-independent mechanism for control of *T. gondii* highlights that we need additional studies to better understand the cell biology of how IFNγ synergizes with other cytokines to directly promote parasite killing.

24.7.4 IFNγ-Induced Restriction of Nutrients

24.7.4.1 Tryptophan Starvation

Immune-induced nutrient deprivation (nutriprieve) in host cells has recently gained attention as an important cell autonomous defence mechanism (Appelberg, 2006). Indoleamine 2,3-dioxygenases IDO1 and IDO2 are IFNγ-inducible, haem containing oxidoreductases that degrade L-tryptophan to generate N-formlykynurenine; the initial rate-limiting step of the kynurenine pathway. An antimicrobial role for IFNγ-induced IDO mediated by restriction of tryptophan was first shown for restriction of *T. gondii*, a tryptophan auxotroph, in human fibroblasts (Pfefferkorn, 1984; Pfefferkorn *et al.*, 1986). IDO-mediated tryptophan starvation has since been shown to inhibit *T. gondii* in a wide variety of cell types *in vitro* (Daubener *et al.*, 2001; Schwartzman *et al.*, 1990; Dai *et al.*, 1994). Despite the fact that acute toxoplasmosis results in increased expression and activity of IDO, mice deficient in IDO-1 failed to demonstrate increased susceptibility to *T. gondii*. However, with the discovery of IDO-2 it was important to revisit the importance of IDO to control of *T. gondii in vivo*. Treatment of mice with an inhibitor of both IDO-1 and IDO-2 prior to *T. gondii* infection resulted in reduced ability to control parasite replication in the absence of an obvious effect on serum levels of IFNγ, TNFα or IL-10 and succumbed during the chronic stage of infection (Divanovic *et al.*, 2012). In humans, it is likely that IDO also contributes to control of *T. gondii* as IDO expression in human macrophages is often higher than in rodents (Hucke *et al.*, 2004; Samelson-Jones and Yeh, 2006). Overall, a picture is emerging of numerous species-specific differences in the relative contributions of IRGs, reactive nitrogen and oxygen species, IFNγ-inducible GTPases and IDOs to the control of *T. gondii*.

24.7.4.2 Iron Acquisition

Iron is an essential co-factor of many enzymes involved in diverse physiological processes including DNA replication, transcription, respiration and metabolism (reviewed in Hood and Skaar, 2012). Iron acquisition is essential for the majority of pathogens during infection, but free iron within a host is extremely limited. Activation of human monocytes with IFNγ results in decreased expression of the transferrin receptor that is required for transportation of transferrin–iron complexes into the endocytic system. This restricts iron availability and so inhibits replication of the bacteria *Legionella pneumophilia* (Byrd and Horwitz, 1989). However, since *T. gondii* resides in a PV largely segregated from the endocytic system, iron sequestration mechanisms that are effective against intracellular bacteria may not function against *T. gondii*. This is consistent with the observation that excess iron or iron chelators do not influence replication of *T. gondii* in IFNγ activated human monocytes (Murray *et al.*, 1991). At present, a general role for sequestration of host intracellular iron as an immune mechanism for *T. gondii* control appears limited. The exception is in the area of mucosal immunity where the ability of IFNγ to activate enterocytes to inhibit *T. gondii* replication is overcome by addition of excess iron or iron chelators (Dimier and Bout, 1998).

24.7.5 $P2X_7$ ATP

IFNγ induces expression of the purinergic $P2X_7$ receptors on the surface of cells of the macrophage/monocyte lineage (reviewed in Miller *et al.*, 2011). In current models, the binding of ATP, released from damaged cells, binds to $P2X_7$ and opens a cation-specific channel in the membrane resulting in alterations in the ionic environment within the cell that in turn activates a number of intracellular pathways: these include 'inflammasome activation, the stress activated protein kinase pathway stimulating apoptosis, the mitogen-activated protein kinase pathway leading to production of reactive oxygen and nitrogen intermediates, and phospholipase D which enhances phagosome–lysosome fusion' (Miller *et al.*, 2011). Indeed, ATP-activated macrophages killed *T. gondii* concurrently with macrophage apoptosis and independent of nitric oxide (Lees *et al.*, 2010). A role for $P2X_7$ in immunity to *T. gondii* was highlighted when three immune competent individuals who presented with clinical toxoplasmosis were found to have polymorphisms that conferred loss of $P2X_7$ function. Macrophages from these individuals were less able to control *T. gondii* following stimulation with ATP. Similarly, macrophages from $P2X_7R^{-/-}$ mice failed to kill *T. gondii* as effectively as wild type macrophages in response to ATP.

24.7.6 NALP1 Inflammasome

NALP1 is a NOD-like receptor that is important for the activation of the NALP1 inflammasome. Recent studies have identified susceptibility alleles of NALP1 associated with congenital toxoplasmosis. Human monocytes depleted of NALP1 and infected with *T. gondii* were more likely to undergo cell death and less able to control replication of intracellular parasites. *T. gondii* infection also failed to increase expression of IL-1β and IL-18 protein and IL-12 from NALP1 knockdown monocytes. These results suggest that *T. gondii* infection of human monocytes activates the NALP1 inflammasome contributing to control of intracellular parasites (Witola *et al.*, 2011). The refractoriness of the Lewis rat to toxoplasmosis independent of parasite genotype is intrinsic to bone marrow cells and is linked to a genetic locus (q24 region of rat c10 containing 86 putative genes) orthologous to a region in the human genome that contains NALP1 (Witola *et al.*, 2011; Cavailles *et al.*, 2006; Sergent *et al.*, 2005). How *T. gondii* stimulates inflammasome activation from its location within its relatively sequestered PV is currently unknown.

24.7.7 IFNγ-Independent Effector Mechanisms

While IFNγ is considered the major mediator of resistance to *T. gondii*, it does not act in isolation. Rather, IFNγ is most effective when acting in concert with other stimuli and there is genetic evidence in humans that loss of function mutations in the IFNγ receptor may not predispose patients to *T. gondii* (Janssen *et al.*, 2002). One of the most interesting alternative pathways is provided by the CD40/CD154 interaction. CD40, a member of the TNFα superfamily, is expressed on antigen presenting cells as well as on a variety of non-haematopoietic cells. CD40 binds to its ligand CD154 expressed on activated CD4 T-cells; crosslinking of CD40 sends a signal important for regulation of multiple aspects of immunity. The first evidence that the interaction between CD40 and CD154 was important for resistance to *T. gondii* was the finding that patients with X-linked Hyper-IgM syndrome, a congenital condition due to loss of functional CD154, have increased susceptibility to *T. gondii* encephalitis (Leiva *et al.*, 1998; Tsuge *et al.*, 1998). The CD40/CD154 interaction has complex effects on host immunity and is linked to many aspects of cell mediated immunity including the production of IL-12 and class switched B cell responses – all important elements in resistance to *T. gondii*. Mice deficient in CD154 succumbed to toxoplasmic encephalitis four to five weeks after parasite infection due to an inability to control parasite replication in the brain even in the presence of normal production of IFNγ. This discovery along with the finding that IFNγ and soluble CD154 in combination restores the ability of $CD154^{-/-}$ macrophages to suppress intracellular parasites established an important role for CD40/154 in the generation of antimicrobial effector activity against *T. gondii* (Reichmann *et al.*, 2000). Subsequent studies established that activated CD4 T-cells expressing CD154 can crosslink CD40 on the surface of macrophages to induce macrophage inhibition of *T. gondii* infection through a mechanism dependent on macrophage autophagy but independent of IFNγ or nitric oxide (Andrade *et al.*, 2005, 2006). Consistent with a role for CD40-induced autophagy *in vivo*, mice deficient in the autophagy protein Beclin 1 have increased parasite numbers in the brain and eye (Portillo *et al.*, 2010). It is likely that CD40/CD154 is an important contributor to adaptive immunity to *T. gondii* once antigen-specific anti-*T. gondii* T-cells are activated.

24.8 ANTIGEN PRESENTATION

While the focus of this review has been on innate immunity, it is the ability of phagocytic cells to sample and present foreign antigens that is required for the development of adaptive T-cell responses that are critical to control *T. gondii*. Initially, one might assume that the uptake of exogenous parasite antigens leads to the presentation of antigens via the well-described endolysosomal processing pathway in the context of MHC class II. There is also experimental evidence that conventional phagolysosomal processing contributes to antigen processing and presentation via MHC class II that includes *in vitro* studies using opsonized or dead parasites (Goldszmid *et al.*, 2009), the transfer of DC populations pulsed with soluble parasite antigens which can induce protective T-cell responses (Dimier-Poisson *et al.*, 2003), and the identification of uninfected plasmacytoid DC as presenting a class II restricted model antigen from *T. gondii* (Pepper *et al.*, 2008). In addition, parasite debris released by lysis of infected cells, or phagocytosis of infected cells going through programmed cell death provide alternative mechanisms whereby parasite derived material may enter the class II processing pathway (Zhao *et al.*, 2009b). However, following invasion of its host cell, *T. gondii* resides in a unique parasitophorous vacuole that does not fuse with the endolysosomal system and there are reports that it also down-regulates expression of MHC class II (Luder *et al.*, 1998) which would likely impair activation of $CD4^+$ T-cells. Despite these effects, infection with *T. gondii* leads to the activation of DC *in vivo* and deletion of these populations results in increased susceptibility and decreased T-cell activation (Liu *et al.*, 2006a, b; John *et al.*, 2009) and there are data that suggest that strain-specific effects on DC expansion determine the outcome of infection (Tait *et al.*, 2010).

For processing of parasite antigen through MHC class I it is less clear how parasite derived antigens from a PV can access canonical antigen presentation pathways for cytoplasmic antigens. Nevertheless, there is abundant evidence that $CD8^+$ T-cells can recognize and lyse host cells infected with *T. gondii* and that this is TAP-1 and proteasome dependent (Ishii *et al.*, 2006; Tu *et al.*, 2009; Hakim *et al.*, 1991; Subauste *et al.*, 1991; Dzierszinski *et al.*, 2007). Several prominent endogenous antigens that are presented on class I have been characterized and intriguingly these are all secretory proteins including dense granule proteins GRA6 (Blanchard *et al.*, 2008) and GRA4 and the rhoptry protein ROP7 (Frickel *et al.*, 2008). However, given the residence of the parasite in a unique parasitophorous vacuole this raises several questions about how parasite derived antigens can access the host cell cytosol to be processed through the TAP-dependent pathway. In infected cells there is evidence for leakage from the PV (Gubbels *et al.*, 2005) or that fusion of host ER with the PV provides a conduit for antigen into this pathway (Blanchard *et al.*, 2008; Goldszmid *et al.*, 2009). Although this may be sufficient for the ability of effector cells to recognize infected targets, and parasite-specific effector/memory cells have been visualized interacting with infected cells within lymph nodes (Chtanova *et al.*, 2009), whether these events are relevant to the priming of naive $CD8^+$ T-cells *in vivo* is less clear. Imaging studies indicate that uninfected DC are involved in T-cell priming (John *et al.*, 2009), and one implication of these observations is that cross presentation may have a role in priming $CD8^+$ T-cells. Indeed, mice that lack $CD8^+$ DC, a major sub-population involved in cross presentation, are highly susceptible to *T. gondii*, though this has been attributed to their ability to produce IL-12 (Mashayekhi *et al.*, 2011). Regardless, the problem with these scenarios is that in the majority of experimental models it is difficult to rigorously isolate and distinguish the effects of cross presentation versus the role of infected cells in the initiation of the $CD4^+$ and $CD8^+$ T-cell responses.

24.9 CONCLUSION AND PERSPECTIVES

T. gondii is a potent stimulator of IFNγ-dependent cell mediated immunity and inflammatory cytokine production. Yet, *T. gondii* is also adept at limiting activation and inflammatory cytokine production of the cells it infects to create a protective intracellular niche even in the midst of an ongoing innate immune response. These strategies combine to enable parasite dissemination and replication early in infection but prevent uncontrolled parasite replication late in infection that could prove lethal to its host. Early in infection, the parasite encounters and invades naive or resident innate immune cells where it multiplies freely. *T. gondii* readily takes charge of critical signalling pathways in the infected host cell allowing it to some extent to control host cell activation including IFNγ-dependent generation of antimicrobial mediators, production of inflammatory cytokines, and perhaps nutrient acquisition pathways. As infection proceeds the immune system becomes increasingly activated and the majority of the non-infected bystander cells are activated. This makes it increasingly likely that *T. gondii* egress from its relatively safe haven in infected cells will require invasion of cells pre-activated by IFNγ and capable of rapid anti-*T. gondii* activity. The finding that inflammatory macrophages can externally trigger egress of *T. gondii* from infected cells during infection may also be a mechanism to force parasites to invade fully activated bystander host cells where they can be killed (Tomita *et al.*, 2009). In theory, at later stages of infection parasites can be controlled to a large extent by the fact that the only proximal cells left to invade are already activated and primed to kill it.

Studies of the immune response to *T. gondii* have been critical for understanding the development of effective immunity to intracellular pathogens in general as well as understanding how innate immunity directs the generation of antigen-specific cell mediated immunity. The advances in understanding the innate immune response to *T. gondii* have generated new areas of study and open questions about antimicrobial effector mechanisms and the processes that allow antigen presentation. It is clear that the potencies of individual antimicrobial effectors against the parasite are dependent on the cell type infected, the activation stimuli, the presence and effectiveness of antimicrobial mediators within different host species and parasite genotype-dependent molecules. Further studies are needed to understand the relationship between intracellular effectors including IRGs, GBPs, autophagy, reactive nitrogen species and oxidative metabolism in effector activity against *T. gondii*. This includes dissecting the relative potencies of these intracellular mediators in parasite genotype-dependent and independent mechanisms of parasite control and identifying parasite molecules that subvert their activity. Such studies should include the identification of potential vulnerabilities of the parasite to host immunity that could potentially be exploited by targeted therapeutics. On the subject of mucosal immunity an open question is the role commensal bacteria actually play in early stimulation of the innate immune response to *T. gondii* and whether they actually impact the generation of the adaptive immune response to the parasite. This includes the related question as to how *T. gondii* is recognized by and stimulates the innate immune response particularly during human infection. Although MyD88 is critical, the identified parasite and host molecules important for immune stimulation through MyD88 particularly during human infection remain poorly defined.

References

Aline, F., Bout, D., Dimier-Poisson, I., 2002. Dendritic cells as effector cells: gamma interferon activation of murine dendritic cells triggers oxygen-dependent inhibition of *Toxoplasma gondii* replication. Infect. Immun. 70, 2368–2374.

Andrade, R.M., Portillo, J.A., Wessendarp, M., Subauste, C.S., 2005. CD40 signalling in macrophages induces activity against an intracellular pathogen independently of gamma interferon and reactive nitrogen intermediates. Infect. Immun. 73, 3115–3123.

Andrade, R.M., Wessendarp, M., Gubbels, M.J., Striepen, B., Subauste, C.S., 2006. CD40 induces macrophage anti-*Toxoplasma gondii* activity by triggering autophagy-dependent fusion of pathogen-containing vacuoles and lysosomes. J. Clin. Invest. 116, 2366–2377.

Appelberg, R., 2006. Macrophage nutriprive antimicrobial mechanisms. J. Leukoc. Biol. 79, 1117–1128.

Barragan, A., Sibley, L.D., 2002. Transepithelial migration of *Toxoplasma gondii* is linked to parasite motility and virulence. J. Exp. Med. 195, 1625–1633.

Bekpen, C., Hunn, J.P., Rohde, C., Parvanova, I., Guethlein, L., Dunn, D.M., Glowalla, E., Leptin, M., Howard, J.C., 2005. The interferon-inducible p47 (IRG) GTPases in vertebrates: loss of the cell autonomous resistance mechanism in the human lineage. Genome Biol. 6, R92.

Benson, A., Pifer, R., Behrendt, C.L., Hooper, L.V., Yarovinsky, F., 2009. Gut commensal bacteria direct a protective immune response against *Toxoplasma gondii*. Cell Host Microbe 6, 187–196.

Blanchard, N., Gonzalez, F., Schaeffer, M., Joncker, N.T., Cheng, T., Shastri, A.J., Robey, E.A., Shastri, N., 2008. Immunodominant, protective response to the parasite *Toxoplasma gondii* requires antigen processing in the endoplasmic reticulum. Nat. Immunol. 9, 937–944.

Butcher, B.A., Fox, B.A., Rommereim, L.M., Kim, S.G., Maurer, K.J., Yarovinsky, F., Herbert, D.R., Bzik, D.J., Denkers, E.Y., 2011. *Toxoplasma gondii* rhoptry kinase ROP16 activates STAT3 and STAT6 resulting in cytokine inhibition and arginase-1-dependent growth control. PLoS Pathog. 7, e1002236.

Butcher, B.A., Kim, L., Johnson, P.F., Denkers, E.Y., 2001. *Toxoplasma gondii* tachyzoites inhibit proinflammatory cytokine induction in infected macrophages by preventing nuclear translocation of the transcription factor NF-kappa B. J. Immunol. 167, 2193–2201.

Butcher, B.A., Kim, L., Panopoulos, A.D., Watowich, S.S., Murray, P.J., Denkers, E.Y., 2005. IL-10-independent STAT3 activation by *Toxoplasma gondii* mediates suppression of IL-12 and TNF-alpha in host macrophages. J. Immunol. 174, 3148–3152.

Buzoni-Gatel, D., Schulthess, J., Menard, L.C., Kasper, L.H., 2006. Mucosal defences against orally acquired protozoan parasites, emphasis on *Toxoplasma gondii* infections. Cell. Microbiol. 8, 535–544.

Byrd, T.F., Horwitz, M.A., 1989. Interferon gamma-activated human monocytes downregulate transferrin receptors and inhibit the intracellular multiplication of *Legionella pneumophila* by limiting the availability of iron. J. Clin. Invest. 83, 1457–1465.

Cavailles, P., Sergent, V., Bisanz, C., Papapietro, O., Colacios, C., Mas, M., Subra, J.F., Lagrange, D., Calise, M., Appolinaire, S., Faraut, T., Druet, P., Saoudi, A., Bessieres, M.H., Pipy, B., Cesbron-Delauw, M.F., Fournie, G.J., 2006. The rat Toxo1 locus directs toxoplasmosis outcome and controls parasite proliferation and spreading by macrophage-dependent mechanisms. Proc. Natl. Acad. Sci. U.S.A. 103, 744–749.

Chao, C.C., Anderson, W.R., Hu, S., Gekker, G., Martella, A., Peterson, P.K., 1993. Activated microglia inhibit multiplication of *Toxoplasma gondii* via a nitric oxide mechanism. Clin. Immunol. Immunopathol. 67, 178–183.

Chtanova, T., Han, S.J., Schaeffer, M., Van Dooren, G.G., Herzmark, P., Striepen, B., Robey, E.A., 2009. Dynamics of T-cell, antigen-presenting cell, and pathogen interactions during recall responses in the lymph node. Immunity 31, 342–355.

Collantes-Fernandez, E., Arrighi, R.B., Alvarez-Garcia, G., Weidner, J.M., Regidor-Cerrillo, J., Boothroyd, J.C., Ortega-Mora, L.M., Barragan, A., 2012. Infected dendritic cells facilitate systemic dissemination and transplacental passage of the obligate intracellular parasite *Neospora caninum* in mice. PloS One 7, e32123.

Collazo, C.M., Yap, G.S., Sempowski, G.D., Lusby, K.C., Tessarollo, L., Woude, G.F., Sher, A., Taylor, G.A., 2001. Inactivation of LRG-47 and IRG-47 reveals a family of interferon gamma-inducible genes with essential, pathogen-specific roles in resistance to infection. J. Exp. Med. 194, 181–188.

Courret, N., Darche, S., Sonigo, P., Milon, G., Buzoni-Gatel, D., Tardieux, I., 2006. CD11c- and CD11b-expressing mouse leukocytes transport single *Toxoplasma gondii* tachyzoites to the brain. Blood 107, 309–316.

Dai, W., Pan, H., Kwok, O., Dubey, J.P., 1994. Human indoleamine 2,3-dioxygenase inhibits *Toxoplasma gondii* growth in fibroblast cells. J. Interferon Res. 14, 313–317.

Daubener, W., Spors, B., Hucke, C., Adam, R., Stins, M., Kim, K.S., Schroten, H., 2001. Restriction of *Toxoplasma gondii* growth in human brain microvascular endothelial cells by activation of indoleamine 2,3-dioxygenase. Infect. Immun. 69, 6527–6531.

Debierre-Grockiego, F., Campos, M.A., Azzouz, N., Schmidt, J., Bieker, U., Resende, M.G., Mansur, D.S., Weingart, R., Schmidt, R.R., Golenbock, D.T., Gazzinelli, R.T., Schwarz, R.T., 2007. Activation of TLR2 and TLR4 by glycosylphosphatidylinositols derived from *Toxoplasma gondii*. J. Immunol. 179, 1129–1137.

Debierre-Grockiego, F., Niehus, S., Coddeville, B., Elass, E., Poirier, F., Weingart, R., Schmidt, R.R., Mazurier, J., Guerardel, Y., Schwarz, R.T., 2010. Binding of *Toxoplasma gondii* glycosylphosphatidylinositols to galectin-3 is required for their recognition by macrophages. J. Biol. Chem. 285, 32744–32750.

Denkers, E.Y., Bzik, D.J., Fox, B.A., Butcher, B.A., 2012. An Inside Job: Hacking into Janus Kinase/Signal Transducer and Activator of Transcription Signalling Cascades by the Intracellular Protozoan *Toxoplasma gondii*. Infect. Immun. 80, 476–482.

Denkers, E.Y., Gazzinelli, R.T., Martin, D., Sher, A., 1993. Emergence of NK1.1+ cells as effectors of IFN-gamma dependent immunity to *Toxoplasma gondii* in MHC class I-deficient mice. J. Exp. Med. 178, 1465–1472.

Diana, J., Persat, F., Staquet, M.J., Assossou, O., Ferrandiz, J., Gariazzo, M.J., Peyron, F., Picot, S., Schmitt, D., Vincent, C., 2004. Migration and maturation of human dendritic cells infected with *Toxoplasma gondii* depend on parasite strain type. FEMS Immunol. Med. Microbiol. 42, 321–331.

Dimier, I.H., Bout, D.T., 1998. Interferon-gamma-activated primary enterocytes inhibit *Toxoplasma gondii* replication: a role for intracellular iron. Immunology 94, 488–495.

Dimier-Poisson, I., Aline, F., Mevelec, M.N., Beauvillain, C., Buzoni-Gatel, D., Bout, D., 2003. Protective mucosal Th2 immune response against *Toxoplasma gondii* by murine mesenteric lymph node dendritic cells. Infect. Immun. 71, 5254–5265.

Ding, M., Kwok, L.Y., Schluter, D., Clayton, C., Soldati, D., 2004. The antioxidant systems in Toxoplasma gondii and the role of cytosolic catalase in defence against oxidative injury. Mol. Microbiol. 51, 47–61.

Divanovic, S., Sawtell, N.M., Trompette, A., Warning, J.I., Dias, A., Cooper, A.M., Yap, G.S., Arditi, M., Shimada, K., Duhadaway, J.B., Prendergast, G.C., Basaraba, R.J., Mellor, A.L., Munn, D.H., Aliberti, J., Karp, C.L., 2012. Opposing biological functions of tryptophan catabolizing enzymes during intracellular infection. J. Infect. Dis. 205, 152–161.

Dubey, J.P., Frenkel, J.K., 1998. Toxoplasmosis of rats: a review with considerations of their value an an animal model and their possible role in epidemiology. Vet. Parasitol 77, 1–32.

Dunay, I.R., Damatta, R.A., Fux, B., Presti, R., Greco, S., Colonna, M., Sibley, L.D., 2008. Gr1(+) inflammatory monocytes are required for mucosal resistance to the pathogen *Toxoplasma gondii*. Immunity 29, 306–317.

Dunay, I.R., Fuchs, A., Sibley, L.D., 2010. Inflammatory monocytes but not neutrophils are necessary to control infection with *Toxoplasma gondii* in mice. Infect. Immun. 78, 1564–1570.

Dzierszinski, F., Pepper, M., Stumhofer, J.S., Larosa, D.F., Wilson, E.H., Turka, L.A., Halonen, S.K., Hunter, C.A., Roos, D.S., 2007. Presentation of *Toxoplasma gondii* antigens via the endogenous major histocompatibility complex class I pathway in nonprofessional and professional antigen-presenting cells. Infect. Immun. 75, 5200–5209.

Egan, C.E., Sukhumavasi, W., Butcher, B.A., Denkers, E.Y., 2009. Functional aspects of Toll-like receptor/MyD88 signalling during protozoan infection: focus on Toxoplasma gondii. Clin. Exp. Immunol. 156, 17–24.

Fang, F.C., 2004. Antimicrobial reactive oxygen and nitrogen species: concepts and controversies. Nat. Rev. Microbiol. 2, 820–832.

Feng, C.G., Weksberg, D.C., Taylor, G.A., Sher, A., Goodell, M.A., 2008. The p47 GTPase Lrg-47 (Irgm1) links host defense and hematopoietic stem cell proliferation. Cell Stem Cell 2, 83–89.

Feng, C.G., Zheng, L., Lenardo, M.J., Sher, A., 2009. Interferon-inducible immunity-related GTPase Irgm1 regulates IFN gamma-dependent host defense, lymphocyte survival and autophagy. Autophagy 5, 232–234.

Fentress, S.J., Behnke, M.S., Dunay, I.R., Mashayekhi, M., Rommereim, L.M., Fox, B.A., Bzik, D.J., Taylor, G.A., Turk, B.E., Lichti, C.F., Townsend, R.R., Qiu, W., Hui, R., Beatty, W.L., Sibley, L.D., 2010. Phosphorylation of immunity-related GTPases by a *Toxoplasma gondii*-secreted kinase promotes macrophage survival and virulence. Cell Host Microbe 8, 484–495.

Fleckenstein, M.C., Reese, M.L., Konen-Waisman, S., Boothroyd, J.C., Howard, J.C., Steinfeldt, T., 2012. A *Toxoplasma gondii* pseudokinase Inhibits Host IRG Resistance Proteins. PLoS Biol. 10, e1001358.

Frickel, E.M., Sahoo, N., Hopp, J., Gubbels, M.J., Craver, M.P., Knoll, L.J., Ploegh, H.L., Grotenbreg, G.M., 2008. Parasite stage-specific recognition of endogenous *Toxoplasma gondii*-derived CD8+ T-cell epitopes. J. Infect. Dis. 198, 1625–1633.

Gavrilescu, L.C., Butcher, B.A., Del Rio, L., Taylor, G.A., Denkers, E.Y., 2004. STAT1 is essential for antimicrobial effector function but dispensable for gamma interferon production during *Toxoplasma gondii* infection. Infect. Immun. 72, 1257–1264.

Goldszmid, R.S., Coppens, I., Lev, A., Caspar, P., Mellman, I., Sher, A., 2009. Host ER-parasitophorous vacuole interaction provides a route of entry for antigen cross-presentation in *Toxoplasma gondii*-infected dendritic cells. J. Exp. Med.

Goldszmid, R.S., Caspar, P., Rivollier, A., Dzutsez, A., Hieny, S., Kelsall, B., Trinchieri, G., Sher, A., 2012. NK cell-derived interferon-? orchestrates cellular dynamics and the differentiation of monocytes into dendritic cells at the site of infection. Immunity 36, 1047–1059.

Gubbels, M.J., Streipen, B., Shastri, N., Turkoz, M., Robey, E.A., 2005. Class I major histocompatibility complex presentation of antigens that escape from the parasitophorous vacuole of. Toxoplasma gondii. Infect. Immun. 73, 703–711.

Hakim, F.T., Gazzinelli, R.T., Denkers, E., Hieny, S., Shearer, G.M., Sher, A., 1991. CD8+ T-cells from mice vaccinated against *Toxoplasma gondii* are cytotoxic for parasite-infected or antigen-pulsed host cells. J. Immunol. 147, 2310–2316.

Heimesaat, M.M., Bereswill, S., Fischer, A., Fuchs, D., Struck, D., Niebergall, J., Jahn, H.K., Dunay, I.R., Moter, A., Gescher, D.M., Schumann, R.R., Gobel, U.B., Liesenfeld, O., 2006. Gram-negative bacteria aggravate murine small intestinal Th1-type immunopathology following oral infection with Toxoplasma gondii. J. Immunol. 177, 8785–8795.

Hertzog, P., Forster, S., Samarajiwa, S., 2011. Systems biology of interferon responses. J. Interferon Cytokine Res. 31, 5–11.

Hitziger, N., Dellacasa, I., Albiger, B., Barragan, A., 2005. Dissemination of *Toxoplasma gondii* to immunoprivileged organs and role of Toll/interleukin-1 receptor signalling for host resistance assessed by in vivo bioluminescence imaging. Cell. Microbiol. 7, 837–848.

Hood, M.I., Skaar, E.P., 2012. Nutritional immunity: transition metals at the pathogen-host interface. Nat. Rev. Microbiol. 10, 525–537.

Howard, J.C., Hunn, J.P., Steinfeldt, T., 2011. The IRG protein-based resistance mechanism in mice and its relation to virulence in *Toxoplasma gondii*. Curr. Opin. Microbiol. 14, 414–421.

Howe, D.K., Sibley, L.D., 1995. *Toxoplasma gondii* comprises three clonal lineages: correlation of parasite genotype with human disease. J. Infect. Dis. 172, 1561–1566.

Hucke, C., MaCkenzie, C.R., Adjogble, K.D., Takikawa, O., Daubener, W., 2004. Nitric oxide-mediated regulation of gamma interferon-induced bacteriostasis: inhibition and degradation of human indoleamine 2,3-dioxygenase. Infect. Immun. 72, 2723–2730.

Hunn, J.P., Feng, C.G., Sher, A., Howard, J.C., 2011. The immunity-related GTPases in mammals: a fast-evolving cell-autonomous resistance system against intracellular pathogens. Mamm. Genome 22, 43–54.

Hunn, J.P., Howard, J.C., 2010. The mouse resistance protein Irgm1 (LRG-47): a regulator or an effector of pathogen defense? PLoS Pathog. 6, e1001008.

Hunter, C.A., Subauste, C.S., Van Cleave, V.H., Remington, J.S., 1994. Production of gamma interferon by natural killer cells from *Toxoplasma gondii*-infected SCID mice: regulation by interleukin-10, interleukin-12, and tumor necrosis factor alpha. Infect. Immun. 62, 2818–2824.

Ishii, K., Hisaeda, H., Duan, X., Imai, T., Sakai, T., Fehling, H.J., Murata, S., Chiba, T., Tanaka, K., Hamano, S., Sano, M., Yano, A., Himeno, K., 2006. The involvement of immunoproteasomes in induction of MHC class I-restricted immunity targeting *Toxoplasma* SAG1. Microbes Infect. 8, 1045–1053.

Janssen, R., Van Wengen, A., Verhard, E., De Boer, T., Zomerdijk, T., Ottenhoff, T.H., Van Dissel, J.T., 2002. Divergent role for TNF-alpha in IFN-gamma-induced killing of *Toxoplasma gondii* and *Salmonella typhimurium* contributes to selective susceptibility of patients with partial IFN-gamma receptor 1 deficiency. J. Immunol. 169, 3900–3907.

Jensen, K.D., Wang, Y., Wojno, E.D., Shastri, A.J., Hu, K., Cornel, L., Boedec, E., Ong, Y.C., Chien, Y.H., Hunter, C.A., Boothroyd, J.C., Saeij, J.P., 2011. *Toxoplasma* polymorphic effectors determine macrophage polarization and intestinal inflammation. Cell Host Microbe 9, 472–483.

John, B., Harris, T.H., Tait, E.D., Wilson, E.H., Gregg, B., Ng, L.G., Mrass, P., Roos, D.S., Dzierszinski, F., Weninger, W., Hunter, C.A., 2009. Dynamic Imaging of CD8(+) T-cells and dendritic cells during infection with *Toxoplasma gondii*. PLoS Pathog. 5, e1000505.

Khaminets, A., Hunn, J.P., Konen-Waisman, S., Zhao, Y.O., Preukschat, D., Coers, J., Boyle, J.P., Ong, Y.C., Boothroyd, J.C., Reichmann, G., Howard, J.C., 2010. Coordinated loading of IRG resistance GTPases on to the *Toxoplasma gondii* parasitophorous vacuole. Cell. Microbiol. 12, 939–961.

Khan, I.A., Schwartzman, J.D., Matsuura, T., Kasper, L.H., 1997. A dichotomous role for nitric oxide during acute *Toxoplasma gondii* infection in mice. Proc. Natl. Acad. Sci. U.S.A. 94, 13955–13960.

Kim, L., Butcher, B.A., Denkers, E.Y., 2004. *Toxoplasma gondii* interferes with lipopolysaccharide-induced mitogen-activated protein kinase activation by mechanisms distinct from endotoxin tolerance. J. Immunol. 172, 3003–3010.

Kim, L., Butcher, B.A., Lee, C.W., Uematsu, S., Akira, S., Denkers, E.Y., 2006. *Toxoplasma gondii* genotype determines MyD88-dependent signalling in infected macrophages. J. Immunol. 177, 2584–2591.

Kim, L., Del Rio, L., Butcher, B.A., Mogensen, T.H., Paludan, S.R., Flavell, R.A., Denkers, E.Y., 2005. p38 MAPK autophosphorylation drives macrophage IL-12 production during intracellular infection. J. Immunol. 174, 4178–4184.

Kim, L., Denkers, E.Y., 2006. Toxoplasma gondii triggers Gi-dependent PI 3-kinase signalling required for inhibition of host cell apoptosis. J. Cell Sci. 119, 2119–2126.

Kim, Y.M., Brinkmann, M.M., Paquet, M.E., Ploegh, H.L., 2008. UNC93B1 delivers nucleotide-sensing toll-like receptors to endolysosomes. Nature 452, 234–238.

Kubo, M., Hanada, T., Yoshimura, A., 2003. Suppressors of cytokine signalling and immunity. Nat. Immunol. 4, 1169–1176.

Lambert, H., Hitziger, N., Dellacasa, I., Svensson, M., Barragan, A., 2006. Induction of dendritic cell migration upon *Toxoplasma gondii* infection potentiates parasite dissemination. Cell. Microbiol. 8, 1611–1623.

Lambert, H., Vutova, P.P., Adams, W.C., Lore, K., Barragan, A., 2009. The *Toxoplasma gondii*-shuttling function of dendritic cells is linked to the parasite genotype. Infect. Immun. 77, 1679–1688.

Lang, C., Hildebrandt, A., Brand, F., Opitz, L., Dihazi, H., Luder, C.G., 2012. Impaired chromatin remodelling at STAT1-regulated promoters leads to global unresponsiveness of *Toxoplasma gondii*-Infected macrophages to IFN-gamma. PLoS Pathog. 8, e1002483.

Langermans, J.A., Van Der Hulst, M.E., Nibbering, P.H., Hiemstra, P.S., Fransen, L., Van Furth, R., 1992. IFN-gamma-induced L-arginine-dependent toxoplasmastatic activity in murine peritoneal macrophages is mediated by endogenous tumor necrosis factor-alpha. J. Immunol. 148, 568–574.

LaRosa, D.F., Stumhofer, J.S., Gelman, A.E., Rahman, A.H., Taylor, D.K., Hunter, C.A., Turka, L.A., 2008. T-cell expression of MyD88 is required for resistance to *Toxoplasma gondii*. Proc. Natl. Acad. Sci. U.S.A. 105, 3855–3860.

Lees, M.P., Fuller, S.J., Mcleod, R., Boulter, N.R., Miller, C.M., Zakrzewski, A.M., Mui, E.J., Witola, W.H., Coyne, J.J., Hargrave, A.C., Jamieson, S.E., Blackwell, J.M., Wiley, J.S., Smith, N.C., 2010. P2X7 receptor-mediated killing of an intracellular parasite, *Toxoplasma gondii*, by human and murine macrophages. J. Immunol. 184, 7040–7046.

Leiva, L.E., Junprasert, J., Hollenbaugh, D., Sorensen, R.U., 1996. Central nervous system toxoplasmosis with an increased proportion of circulating gamma delta T cells in a patient with hyper-IgM syndrome. J. Clin. Immunol 18, 283–290.

Leiva, L.E., Junprasert, J., Hollenbaugh, D., Sorensen, R.U., 1998 Jul. Central nervous system toxoplasmosis with an increased proportion of circulating gamma delta T cells in a patient with hyper-IgM syndrome. J Clin Immunol 18 (4), 283–290.

Leng, J., Butcher, B.A., Egan, C.E., Abdallah, D.S., Denkers, E.Y., 2009. *Toxoplasma gondii* prevents chromatin remodeling initiated by TLR-triggered macrophage activation. J. Immunol 182, 489–497.

Leng, J., Denkers, E.Y., 2009. *Toxoplasma gondii* inhibits covalent modification of histone H3 at the IL-10 promoter in infected macrophages. PloS One 4, e7589.

Li, Z., Zhao, Z.J., Zhu, X.Q., Ren, Q.S., Nie, F.F., Gao, J.M., Gao, X.J., Yang, T.B., Zhou, W.L., Shen, J.L., Wang, Y., Lu, F.L., Chen, X.G., Hide, G., Ayala, F.J., Lun, Z.R., 2012. Differences in iNOS and arginase expression and activity in the macrophages of rats are responsible for the resistance against *T. gondii* infection. PloS One 7, e35834.

Lieberman, L.A., Banica, M., Reiner, S.L., Hunter, C.A., 2004. STAT1 plays a critical role in the regulation of antimicrobial effector mechanisms, but not in the development of Th1-type responses during toxoplasmosis. J. Immunol. 172, 457–463.

Lieberman, L.A., Cardillo, F., Owyang, A.M., Rennick, D.M., Cua, D.J., Kastelein, R.A., Hunter, C.A., 2004. IL-23 provides a limited mechanism of resistance to acute toxoplasmosis in the absence of IL-12. J. Immunol. 173, 1887–1893.

Liesenfeld, O., Kang, H., Park, D., Nguyen, T.A., Parkhe, C.V., Watanabe, H., Abo, T., Sher, A., Remington, J.S., Suzuki, Y., 1999. TNF-alpha, nitric oxide and IFN-gamma are all critical for development of necrosis in the small intestine and early mortality in genetically susceptible mice infected perorally with. Toxoplasma gondii. Parasite Immunol. 21, 365–376.

Liesenfeld, O., Kosek, J., Remington, J.S., Suzuki, Y., 1996. Association of CD4+ T-cell-dependent, interferon-gamma-mediated necrosis of the small intestine with genetic susceptibility of mice to peroral infection with *Toxoplasma gondii*. J. Exp. Med. 184, 597–607.

Liu, C.H., Fan, Y.T., Dias, A., Esper, L., Corn, R.A., Bafica, A., Machado, F.S., Aliberti, J., 2006. Cutting edge: dendritic cells are essential for *in vivo* IL-12 production and development of resistance against *Toxoplasma gondii* infection in mice. J. Immunol. 177, 31–35.

Locksley, R.M., Wilson, C.B., Klebanoff, S.J., 1982. Role for endogenous and acquired peroxidase in the toxoplasmacidal activity of murine and human mononuclear phagocytes. J. Clin. Invest. 69, 1099–1111.

Luder, C.G., Algner, M., Lang, C., Bleicher, N., Gross, U., 2003. Reduced expression of the inducible nitric oxide synthase after infection with *Toxoplasma gondii* facilitates parasite replication in activated murine macrophages. Int. J. Parasitol. 33, 833–844.

Luder, C.G., Gross, U., 2005. Apoptosis and its modulation during infection with *Toxoplasma gondii*: molecular mechanisms and role in pathogenesis. Current Topics in Microbiol. Immunol. 289, 219–237.

Luder, C.G., Lang, C., Giraldo-Velasquez, M., Algner, M., Gerdes, J., Gross, U., 2003. *Toxoplasma gondii* inhibits MHC class II expression in neural antigen-presenting cells by down-regulating the class II transactivator CIITA. J. Neuroimmunol. 134, 12–24.

Luder, C.G., Lang, T., Beuerle, B., Gross, U., 1998. Down-regulation of MHC class II molecules and inability to up-regulate class I molecules in murine macrophages after infection with. *Toxoplasma gondii*. Clin. Exp. Immunol. 112, 308–316.

Lykens, J.E., Terrell, C.E., Zoller, E.E., Divanovic, S., Trompette, A., Karp, C.L., Aliberti, J., Flick, M.J., Jordan, M.B., 2010. Mice with a selective impairment of IFN-gamma signalling in macrophage lineage cells demonstrate the critical role of IFN-gamma-activated macrophages for the control of protozoan parasitic infections in vivo. J. Immunol. 184, 877–885.

MacMicking, J.D., 2012. Interferon-inducible effector mechanisms in cell-autonomous immunity. Nat. Rev. Immunol. 12, 367–382.

Martens, S., Parvanova, I., Zerrahn, J., Griffiths, G., Schell, G., Reichmann, G., Howard, J.C., 2005. Disruption of *Toxoplasma gondii* parasitophorous vacuoles by the mouse p47-resistance GTPases. PLoS Pathog. 1, e24.

Masek, K.S., Fiore, J., Leitges, M., Yan, S.F., Freedman, B.D., Hunter, C.A., 2006. Host cell Ca2+ and protein kinase C regulate innate recognition of *Toxoplasma gondii*. J. Cell. Sci. 119, 4565–4573.

Masek, K.S., Zhu, P., Freedman, B.D., Hunter, C.A., 2007. *Toxoplasma gondii* induces changes in intracellular calcium in macrophages. Parasitology 134, 1973–1979.

Mashayekhi, M., Sandau, M.M., Dunay, I.R., Frickel, E.M., Khan, A., Goldszmid, R.S., Sher, A., Ploegh, H.L., Murphy, T.L., Sibley, L.D., Murphy, K.M., 2011. CD8alpha(+) dendritic cells are the critical source of interleukin-12 that controls acute infection by *Toxoplasma gondii* tachyzoites. Immunity 35, 249–259.

Mason, N., Aliberti, J., Caamano, J.C., Liou, H.C., Hunter, C.A., 2002. Cutting edge: identification of c-Rel-dependent and -independent pathways of IL-12 production during infectious and inflammatory stimuli. J. Immunol. 168, 2590–2594.

Mason, N.J., Fiore, J., Kobayashi, T., Masek, K.S., Choi, Y., Hunter, C.A., 2004. TRAF6-dependent mitogen-activated protein kinase activation differentially regulates the production of interleukin-12 by macrophages in response to *Toxoplasma gondii*. Infect. Immun. 72, 5662–5667.

Melo, M.B., Jensen, K.D., Saeij, J.P., 2011. *Toxoplasma gondii* effectors are master regulators of the inflammatory response. Trends Parasitol. 27, 487–495.

Melo, M.B., Kasperkovitz, P., Cerny, A., Konen-Waisman, S., Kurt-Jones, E.A., Lien, E., Beutler, B., Howard, J.C., Golenbock, D.T., Gazzinelli, R.T., 2010. UNC93B1 mediates host resistance to infection with *Toxoplasma gondii*. PLoS Pathog. 6.

Miller, C.M., Boulter, N.R., Fuller, S.J., Zakrzewski, A.M., Lees, M.P., Saunders, B.M., Wiley, J.S., Smith, N.C., 2011. The role of the P2X(7) receptor in infectious diseases. PLoS Pathog. 7, e1002212.

Mordue, D.G., Scott-Weathers, C.F., Tobin, C.M., Knoll, L.J., 2007. A patatin-like protein protects *Toxoplasma gondii* from degradation in activated macrophages. Mol. Microbiol. 63, 482–496.

Mordue, D.G., Sibley, L.D., 2003. A novel population of Gr-1+-activated macrophages induced during acute toxoplasmosis. J. Leukoc. Biol. 74, 1015–1025.

Mosser, D.M., Edwards, J.P., 2008. Exploring the full spectrum of macrophage activation. Nat. Rev. Immunol. 8, 958–969.

Murray, H.W., Cohn, Z.A., 1979. Macrophage oxygen-dependent antimicrobial activity. I. Susceptibility of *Toxoplasma gondii* to oxygen intermediates. J. Exp. Med. 150, 938–949.

Murray, H.W., Granger, A.M., Teitelbaum, R.F., 1991. Gamma interferon-activated human macrophages and *Toxoplasma gondii, Chlamydia psittaci*, and *Leishmania donovani*: antimicrobial role of limiting intracellular iron. Infect. Immun. 59, 4684–4686.

Murray, H.W., Juangbhanich, C.W., Nathan, C.F., Cohn, Z.A., 1979. Macrophage oxygen-dependent antimicrobial activity. II. The role of oxygen intermediates. J. Exp. Med. 150, 950–964.

Murray, H.W., Rubin, B.Y., Carriero, S.M., Harris, A.M., Jaffee, E.A., 1985. Human mononuclear phagocyte antiprotozoal mechanisms: oxygen-dependent vs oxygen-independent activity against intracellular Toxoplasma gondii. J. Immunol. 134, 1982–1988.

Nathan, C., Shiloh, M.U., 2000. Reactive oxygen and nitrogen intermediates in the relationship between mammalian hosts and microbial pathogens. Proc. Natl. Acad. Sci. U.S.A. 97, 8841–8848.

Niedelman, W., Gold, D.A., Rosowski, E.E., Sprokholt, J.K., Lim, D., Farid Arenas, A., Melo, M.B., Spooner, E., Yaffe, M.B., Saeij, J.P., 2012. The Rhoptry Proteins ROP18 and ROP5 Mediate *Toxoplasma gondii* evasion of the murine, but not the human, interferon-gamma response. PLoS Pathog. 8, e1002784.

Ong, Y.C., Reese, M.L., Boothroyd, J.C., 2010. *Toxoplasma* rhoptry protein 16 (ROP16) subverts host function by direct tyrosine phosphorylation of STAT6. J. Biol. Chem. 285, 28731–28740.

Peixoto, L., Chen, F., Harb, O.S., Davis, P.H., Beiting, D.P., Brownback, C.S., Ouloguem, D., Roos, D.S., 2010. Integrative genomic approaches highlight a family of parasite-specific kinases that regulate host responses. Cell Host Microbe 8, 208–218.

Pepper, M., Dzierszinski, F., Wilson, E., Tait, E., Fang, Q., Yarovinsky, F., Laufer, T.M., Roos, D., Hunter, C.A., 2008. Plasmacytoid dendritic cells are activated by *Toxoplasma gondii* to present antigen and produce cytokines. J. Immunology 180, 6229–6236.

Peterson, P.K., Gekker, G., Hu, S., Chao, C.C., 1995. Human astrocytes inhibit intracellular multiplication of *Toxoplasma gondii* by a nitric oxide-mediated mechanism. J. Infect. Dis. 171, 516–518.

Pfefferkorn, E.R., 1984. Interferon gamma blocks the growth of *Toxoplasma gondii* in human fibroblasts by inducing the host cells to degrade tryptophan. Proc. Natl. Acad. Sci. U.S.A. 81, 908–912.

Pfefferkorn, E.R., Eckel, M., Rebhun, S., 1986. Interferon-gamma suppresses the growth of *Toxoplasma gondii* in human fibroblasts through starvation for tryptophan. Mol. Biochem. Parasitol. 20, 215–224.

Pifer, R., Benson, A., Sturge, C.R., Yarovinsky, F., 2011. UNC93B1 is essential for TLR11 activation and IL-12-dependent host resistance to *Toxoplasma gondii*. J. Biol. Chem. 286, 3307–3314.

Pino, P., Foth, B.J., Kwok, L.Y., Sheiner, L., Schepers, R., Soldati, T., Soldati-Favre, D., 2007. Dual targeting of antioxidant and metabolic enzymes to the mitochondrion and the apicoplast of *Toxoplasma gondii*. PLoS Pathog. 3, e115.

Pollard, A.M., Knoll, L.J., Mordue, D.G., 2009. The role of specific *Toxoplasma gondii* molecules in manipulation of innate immunity. Trends Parasitol. 25, 491–494.

Portillo, J.A., Okenka, G., Reed, E., Subauste, A., Van Grol, J., Gentil, K., Komatsu, M., Tanaka, K., Landreth, G., Levine, B., Subauste, C.S., 2010. The

CD40-autophagy pathway is needed for host protection despite IFN-Gamma-dependent immunity and CD40 induces autophagy via control of P21 levels. PloS One 5, e14472.

Reichmann, G., Walker, W., Villegas, E.N., Craig, L., Cai, G., Alexander, J., Hunter, C.A., 2000. The CD40/CD40 ligand interaction is required for resistance to toxoplasmic encephalitis. Infect. Immun. 68, 1312–1318.

Robben, P.M., Laregina, M., Kuziel, W.A., Sibley, L.D., 2005. Recruitment of Gr-1+ monocytes is essential for control of acute toxoplasmosis. J. Exp. Med. 201, 1761–1769.

Robben, P.M., Mordue, D.G., Truscott, S.M., Takeda, K., Akira, S., Sibley, L.D., 2004. Production of IL-12 by macrophages infected with *Toxoplasma gondii* depends on the parasite genotype. J. Immunol. 172, 3686–3694.

Rosowski, E.E., Lu, D., Julien, L., Rodda, L., Gaiser, R.A., Jensen, K.D., Saeij, J.P., 2011. Strain-specific activation of the NF-kappaB pathway by GRA15, a novel *Toxoplasma gondii* dense granule protein. J. Exp. Med. 208, 195–212.

Saeij, J.P., Coller, S., Boyle, J.P., Jerome, M.E., White, M.W., Boothroyd, J.C., 2007. *Toxoplasma* co-opts host gene expression by injection of a polymorphic kinase homologue. Nature 445, 324–327.

Samarajiwa, S.A., Forster, S., Auchettl, K., Hertzog, P.J., 2009. INTERFEROME: the database of interferon regulated genes. Nucleic Acids Res. 37, D852–D857.

Samelson-Jones, B.J., Yeh, S.R., 2006. Interactions between nitric oxide and indoleamine 2,3-dioxygenase. Biochemistry 45, 8527–8538.

Sautel, C.F., Ortet, P., Saksouk, N., Kieffer, S., Garin, J., Bastien, O., Hakimi, M.A., 2009. The histone methylase KMTox interacts with the redox-sensor peroxiredoxin-1 and targets genes involved in *Toxoplasma gondii* antioxidant defences. Mol. Microbiol. 71, 212–226.

Scanga, C.A., Aliberti, J., Jankovic, D., Tilloy, F., Bennouna, S., Denkers, E.Y., Medzhitov, R., Sher, A., 2002. Cutting edge: MyD88 is required for resistance to *Toxoplasma gondii* infection and regulates parasite-induced IL-12 production by dendritic cells. J. Immunol. 168, 5997–6001.

Scharton-Kersten, T., Nakajima, H., Yap, G., Sher, A., Leonard, W.J., 1998. Infection of mice lacking the common cytokine receptor gamma-chain (gamma(c)) reveals an unexpected role for CD4+ T lymphocytes in early IFN-gamma-dependent resistance to Toxoplasma gondii. J. Immunol. 160, 2565–2569.

Scharton-Kersten, T.M., Wynn, T.A., Denkers, E.Y., Bala, S., Grunvald, E., Hieny, S., Gazzinelli, R.T., Sher, A., 1996. In the absence of endogenous IFN-gamma, mice develop unimpaired IL-12 responses to *Toxoplasma gondii* while failing to control acute infection. J. Immunol. 157, 4045–4054.

Scharton-Kersten, T.M., Yap, G., Magram, J., Sher, A., 1997. Inducible nitric oxide is essential for host control of persistent but not acute infection with the intracellular pathogen *Toxoplasma gondii*. J. Exp. Med. 185, 1261–1273.

Schluter, D., Deckert-Schluter, M., Lorenz, E., Meyer, T., Rollinghoff, M., Bogdan, C., 1999. Inhibition of inducible nitric oxide synthase exacerbates chronic cerebral toxoplasmosis in *Toxoplasma gondii*-susceptible C57BL/6 mice but does not reactivate the latent disease in *T. gondii*-resistant BALB/c mice. J. Immunol. 162, 3512–3518.

Schwartzman, J.D., Gonias, S.L., Pfefferkorn, E.R., 1990. Murine gamma interferon fails to inhibit *Toxoplasma gondii* growth in murine fibroblasts. Infect. Immun. 58, 833–834.

Sergent, V., Cautain, B., Khalife, J., Deslee, D., Bastien, P., Dao, A., Dubremetz, J.F., Fournie, G.J., Saoudi, A., Cesbron-Delauw, M.F., 2005. Innate refractoriness of the Lewis rat to toxoplasmosis is a dominant trait that is intrinsic to bone marrow-derived cells. Infect. Immun. 73, 6990–6997.

Shapira, S., Harb, O.S., Margarit, J., Matrajt, M., Han, J., Hoffmann, A., Freedman, B., May, M.J., Roos, D.S., Hunter, C.A., 2005. Initiation and termination of NF-kappaB signalling by the intracellular protozoan parasite Toxoplasma gondii. J. Cell Sci. 118, 3501–3508.

Shapira, S., Speirs, K., Gerstein, A., Caamano, J., Hunter, C.A., 2002. Suppression of NF-kappaB activation by infection with. Toxoplasma gondii. J. Infect. Dis. 185 (Suppl 1), S66–S72.

Sibley, L.D., Adams, L.B., Fukutomi, Y., Krahenbuhl, J.L., 1991. Tumor necrosis factor-alpha triggers antitoxoplasmal activity of IFN-gamma primed macrophages. J. Immunol. 147, 2340–2345.

Skariah, S., Bednarczyk, R.B., Mcintyre, M.K., Taylor, G.A., Mordue, D.G., 2012. Discovery of a novel *Toxoplasma gondii* conoid-associated protein important for parasite resistance to reactive nitrogen intermediates. J. Immunol. 188, 3404–3415.

Steinfeldt, T., Konen-Waisman, S., Tong, L., Pawlowski, N., Lamkemeyer, T., Sibley, L.D., Hunn, J.P., Howard, J.C., 2010. Phosphorylation of mouse immunity-related GTPase (IRG) resistance proteins is an evasion strategy for virulent *Toxoplasma gondii*. PLoS Biol. 8, e1000576.

Subauste, C.S., Koniaris, A.H., Remington, J.S., 1991. Murine CD8+ cytotoxic T lymphocytes lyse *Toxoplasma gondii*-infected cells. J. Immunol. 147, 3955–3959.

Subauste, C.S., Wessendarp, M., Sorensen, R.U., Leiva, L.E., 1999. CD40-CD40 ligand interaction is central to cell-mediated immunity against Toxoplasma gondii: patients with hyper IgM syndrome have a defective type 1 immune response that can be restored by soluble CD40 ligand trimer. J. Immunol 162, 6690–6700.

Sukhumavasi, W., Egan, C.E., Warren, A.L., Taylor, G.A., Fox, B.A., Bzik, D.J., Denkers, E.Y., 2008. TLR adaptor MyD88 is essential for pathogen control during oral *Toxoplasma gondii* infection but not adaptive immunity induced by a vaccine strain of the parasite. J. Immunol. 181, 3464–3473.

Suzuki, Y., Orellana, M.A., Schreiber, R.D., Remington, J.S., 1988. Interferon-gamma: the major mediator of resistance against *Toxoplasma gondii*. Science 240, 516–518.

Suzuki, Y., Sher, A., Yap, G., Park, D., Neyer, L.E., Liesenfeld, O., Fort, M., Kang, H., Gufwoli, E., 2000. IL-10 is required for prevention of necrosis in the small intestine and mortality in both genetically resistant BALB/c and susceptible C57BL/6 mice following peroral infection with *Toxoplasma gondii*. J. Immunol. 164, 5375–5382.

Tait, E.D., Jordan, K.A., Dupont, C.D., Harris, T.H., Gregg, B., Wilson, E.H., Pepper, M., Dzierszinski, F., Roos, D.S., Hunter, C.A., 2010. Virulence of *Toxoplasma gondii* is associated with distinct dendritic cell responses and reduced numbers of activated CD8+ T-cells. J. Immunol. 185, 1502–1512.

Taylor, G.A., Collazo, C.M., Yap, G.S., Nguyen, K., Gregorio, T.A., Taylor, L.S., Eagleson, B., Secrest, L., Southon, E.A., Reid, S.W., Tessarollo, L., Bray, M., Mcvicar, D.W., Komschlies, K.L., Young, H.A., Biron, C.A., Sher, A., Vande Woude, G.F., 2000. Pathogen-specific loss of host resistance in mice lacking the IFN-gamma-inducible gene IGTP. Proc. Natl. Acad. Sci. U.S.A. 97, 751–755.

Tomita, T., Yamada, T., Weiss, L.M., Orlofsky, A., 2009. Externally triggered egress is the major fate of *Toxoplasma gondii* during acute infection. J. Immunol. 183, 6667–6680.

Tsuge, I., Matsuoko, H., Nakagawa, A., Kamachi, Y., Aso, K., Negoro, T., Ito, M., Torri, S., Watanabe, K., 1998. Necrotizing toxoplasmic encephalitis in a child with the X-linked hyper-IgM syndrome. Eur. J. Pediatr 157, 735–737.

Tu, L., Moriya, C., Imai, T., Ishida, H., Tetsutani, K., Duan, X., Murata, S., Tanaka, K., Shimokawa, C., Hisaeda, H., Himeno, K., 2009. Critical role for the immunoproteasome subunit LMP7 in the resistance of mice to *Toxoplasma gondii* infection. Eur. J. Immunol. 39, 3385–3394.

Winter, S., Niedelman, W., Jensen, K.D., Rosowski, E.E., Julien, L., Spooner, E., Caradonna, K., Burleigh, B.A., Saeij, J.P., Ploegh, H.L., Frickel, E.M., 2011. Determinants of GBP recruitment to *Toxoplasma gondii* vacuoles and the parasitic factors that control it. PloS One 6, e24434.

Whitmarsh, R.J., Gray, C.M., Gregg, B., Christian, D.A., May, M.J., Murray, P.J., Hunter, C.A., 2011. A critical role for SOCS3 in innate resistance to. Toxoplasma gondii. Cell Host Microbe 10, 224–236.

Wilson, C.B., Tsai, V., Remington, J.S., 1980. Failure to trigger the oxidative metabolic burst by normal macrophages: possible mechanism for survival of intracellular pathogens. J. Exp. Med. 151, 328–346.

Witola, W.H., Mui, E., Hargrave, A., Liu, S., Hypolite, M., Montpetit, A., Cavailles, P., Bisanz, C., Cesbron-Delauw, M.F., Fournie, G.J., Mcleod, R., 2011. NALP1 influences susceptibility to human congenital toxoplasmosis, proinflammatory cytokine response, and fate of *Toxoplasma gondii*-infected monocytic cells. Infect. Immun. 79, 756–766.

Yamamoto, M., Okuyama, M., Ma, J.S., Kimura, T., Kamiyama, N., Saiga, H., Ohshima, J., Sasai, M., Kayama, H., Okamoto, T., Huang, D.C., Soldati-

Favre, D., Horie, K., Takeda, J., Takeda, K., 2012. A Cluster of Interferon-gamma-Inducible p65 GTPases Plays a Critical Role in Host Defense against. Toxoplasma gondii. Immunity.

Yap, G.S., Sher, A., 1999. Effector cells of both nonhemopoietic and hemopoietic origin are required for interferon (IFN)-gamma- and tumor necrosis factor (TNF)-alpha-dependent host resistance to the intracellular pathogen. Toxoplasma gondii. J. Exp. Med. 189, 1083–1092.

Yarovinsky, F., Zhang, D., Andersen, J.F., Bannenberg, G.L., Serhan, C.N., Hayden, M.S., Hieny, S., Sutterwala, F.S., Flavell, R.A., Ghosh, S., Sher, A., 2005. TLR11 activation of dendritic cells by a protozoan profilin-like protein. Science 308, 1626–1629.

Yong, E.C., Chi, E.Y., Henderson Jr., W.R., 1994. *Toxoplasma gondii* alters eicosanoid release by human mononuclear phagocytes: role of leukotrienes in interferon gamma-induced antitoxoplasma activity. J. Exp. Med. 180, 1637–1648.

Zhao, Y., Ferguson, D.J., Wilson, D.C., Howard, J.C., Sibley, L.D., Yap, G.S., 2009a. Virulent *Toxoplasma gondii* evade immunity-related GTPase-mediated parasite vacuole disruption within primed macrophages. J. Immunol. 182, 3775–3781.

Zhao, Y.O., Khaminets, A., Hunn, J.P., Howard, J.C., 2009b. Disruption of the *Toxoplasma gondii* parasitophorous vacuole by IFNgamma-inducible immunity-related GTPases (IRG proteins) triggers necrotic cell death. PLoS Pathog. 5, e1000288.

Zhao, Z., Fux, B., Goodwin, M., Dunay, I.R., Strong, D., Miller, B.C., Cadwell, K., Delgado, M.A., Ponpuak, M., Green, K.G., Schmidt, R.E., Mizushima, N., Deretic, V., Sibley, L.D., Virgin, H.W., 2008. Autophagosome-independent essential function for the autophagy protein Atg5 in cellular immunity to intracellular pathogens. Cell Host Microbe 4, 458–469.

CHAPTER

25

Adaptive Immunity and Genetics of the Host Immune Response

Craig W. Roberts, Sheela Prasad*[†], *Farzana Khaliq**, *Ricardo T. Gazzinelli***, *Imtiaz A. Khan*[‡], *Rima McLeod*[†]

*University of Strathclyde, Glasgow, Scotland, UK [†]The University of Chicago, Chicago, Illinois, USA
**Federal University of Minas Gerais, Minas Gerais, Brazil and University of Massachusetts, Worcester, Massachusetts, USA [‡]Department of Microbiology, Immunology and Tropical Medicine, George Washington University, Washington DC, USA

OUTLINE

Toxoplasma gondii, second edition
http://dx.doi.org/10.1016/B978-0-12-396481-6.00025-8

25.1 INTRODUCTION

Studies of genetics of host and parasite are powerful tools to understand immunity to *T. gondii* and how innate and adaptive immune responses provide key protective and pathogenic effector functions (see Roberts *et al.*, in Weiss and Kim, 2007; Hunter and Sibley, 2012; Benson *et al.*, 2009). The ability of *T. gondii* to infect any cell means that disease can be systemic and affect many organs, but in some cases certain organs including the brain, eye or intestine have the predominant pathology. These individual organs have been studied in detail using animal – most often murine – models of disease. The strong influence of host genetics on this disease process has been evident from early studies and has been exploited to identify immunological functional correlates of protection and pathology (see Weiss and Kim, 2007 and Blackwell, 2008). The selection of genetically different strains of mice with specific susceptibilities to different disease manifestations has allowed a careful dissection of immune responses in various organs and has provided valuable insights into disease manifestations in humans with different disease patterns. Most notably these include toxoplasmic encephalitis (TE). The influence of *T. gondii* genetics on the pathogenesis of both murine and human *T. gondii* infection has become increasingly evident in recent years. In experimental studies, the route and life cycle stage used to initiate infection has been varied according to convenience, scientific rationale or to induce particular disease manifestations and must also be noted when considering studies in the literature.

In the past fifteen years there have been a number of scholarly, comprehensive reviews of cell mediated immunity, newborn immunity, dendritic cells and antigen processing and presentation, and of how each of these are relevant to toxoplasmosis (Aliberti *et al.*, 2003; Denkers *et al.*, 1998, 2003b, 2005; Dowell and McLeod 2000; Gazzinelli *et al.*, 1993, 1996, 2004; Hunter *et al.*, 1996; Lewis and Wilson, 2005; Lüder and Seeber, 2001; McLeod *et al.*, 1995, 1996; Montoya et al., 1997; Pelloux and Ambroise-Thomas, 1996; Ropert and Gazzinelli, 2000; Scott and Hunter, 2002; Taylor *et al.*, 2004; Teixeira *et al.*, 2002; Yap and Sher, 1999a). References included in these reviews are incorporated in the reference list herein for completeness. This chapter does not reiterate the detailed summaries of the literature already presented in these outstanding reviews, but instead, includes relevant references, summary figures and tables from these earlier works, as well as tables created for this chapter that provide overviews and outline the mechanisms involved in the *T. gondii* immune response (Figs 25.1–25.6; Tables 25.1 and 25.2). These tables develop the concepts that the immune response contributes to both protection and pathogenesis of disease, e.g. retinal lesions in both the immune competent and immune compromised person (Fig. 25.7). Innate immunity shapes adaptive immunity needed for protection. We have focused this chapter on work from our own laboratory groups, papers published recently that have been

Processing of Exogenous Antigen, Presentation on MHC Class II and Activation of CD4+ T-lymphocytes

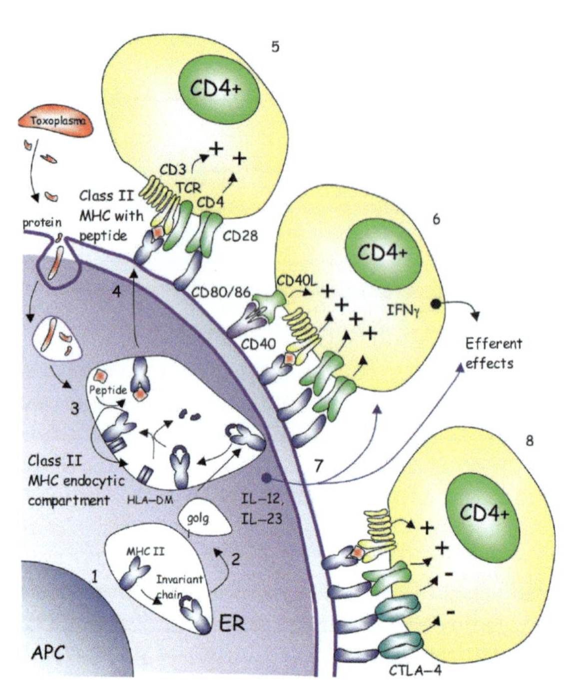

FIGURE 25.1 **Processing of exogenous *T. gondii* antigen, presentation on MHC class II and activation of CD4$^+$ T-lymphocytes.** (1) In the endoplasmic reticulum (ER), Class II MHC (MHC II) molecules are bound by the invariant chain. (2) Export of the class II MHC-invariant chain complex from the ER, through the Golgi and to class II MHC endocytic compartments. (3) Endosomes containing *T. gondii* derived proteins fuse with the class II MHC endocytic compartments. The invariant chain is proteolytically cleaved allowing internalized peptides to bind class II MHC molecules. In humans, the Human Leukocyte Antigen (HLA) DM molecules facilitates peptide loading. (4) Class II MHC molecules carrying peptide cargo are exported to the surface of the Antigen Presenting Cell (APC). (5) Peptides are presented on class II MHC to CD4$^+$ T-lymphocytes which interact *via* their surface α/β-TCR and CD4 molecules. The CD3 complex becomes associated with α/β-TCRs and act as a docking site for tyrosine kinases that transmit activating intracellular signals. CD28 on the surface of T-cells interacts with CD80 and/or CD86 on the APC to provide co-stimulation for the T-cell. (6) CD40 is constitutively expressed on dendritic cells and interacts with CD40 Ligand (CD40L) expressed on the surface of activated T-cells. This increases expression of CD80−86 on dendritic cells and enhances T-cell co-stimulation. (7) Engagement of CD40 activates the dendritic cell. Production of cytokines such as IL-12 by dendritic cells favours the differentiation of CD4$^+$ T-cells into Th1 cells that are characterized by IFNγ production but have low or undetectable IL-4 production. (8) T-cells express CTLA-4 at the later stages of activation. Engagement of this with CD80/86 induces negative signals to the T-cell which terminate their activation. *(Includes concepts from Fig. 5 in McLeod* et al. *(1996) and Fig. 4-1 in Lewis and Wilson (2005), with permission.)*

foundational for and an extension of concepts in the earlier reviews and papers that have recently provided important mechanistic insights into the immune response to *T. gondii* (Abou-Bacar *et al.*, 2004a; Adams *et al.*, 1990; Araujo, 1991; Brown and McLeod, 1995a; Brown *et al.*, 1995a; D'Angelillo *et al.*, 2011; Dannemann *et al.*, 1989; Deckert-Schluter *et al.*, 1994, 1996, 1998b; Denkers *et al.*, 1993a, 1997c; Dobbin *et al.*, 2002; Frenkel and Caldwell, 1975; Gazzinelli *et al.*, 1992a; Handel *et al.*, 2012; Hermes *et al.*, 2008; Kang *et al.*, 2003; Lindberg and Frenkel, 1977; McLeod and Remington, 1979; McLeod *et al.*, 1989a, 1989b; Mennechet *et al.*, 2004; Montoya *et al.*, 1996; Mun *et al.*, 2003, 2005; Murray *et al.*, 1985b; Parker *et al.*, 1991; Parner *et al.*, 1996; Remington and Merigen 1990; Shirahata *et al.*, 1994; Yano *et al.*, 1989).

The innate immune system is also important in driving and dictating the quality of the adaptive immune response Caamano and Hunter, 2002; Chan *et al.*, 2006b; Curiel *et al.*, 1993; Denkers *et al.*, 1997; Gazzinelli *et al.*, 1996a; Kos and Engleman, 1996; Maeckner *et al.*, 2005; Sacks and Sher, 2002; Scharton-Kersten *et al.*, 1996, 1997a, 1997b; Yaro Uinski *et al.*, 2005; Brezin *et al.*, 1994; Brown and McLeod, 1990; Burnett *et al.*, 1998; Biggs *et al.*, 1995; Catterall *et al.*, 1986, 1987; Cavailles *et al.*, 2006a, 2006b; Channon and Kasper, 1996; Couvreur, 1984; Curiel, 1993; Decoster, 1996; Desmonts and Couvreur, 1974; Subauste *et al.*, 1998; McLeod and Dowell, 2000; Eichenwald, 1960; Garweg *et al.*, 2005; Huskinson *et al.*, 1990; Koppe *et al.*, 1974, 1986; Luft *et al.*, 1984a, 1984b; Mack *et al.*, in preparation; McAuley *et al.*, 1994a; McLeod *et al.*, 1980, 1984, 1985, 1988, 1990, 2006; Mets

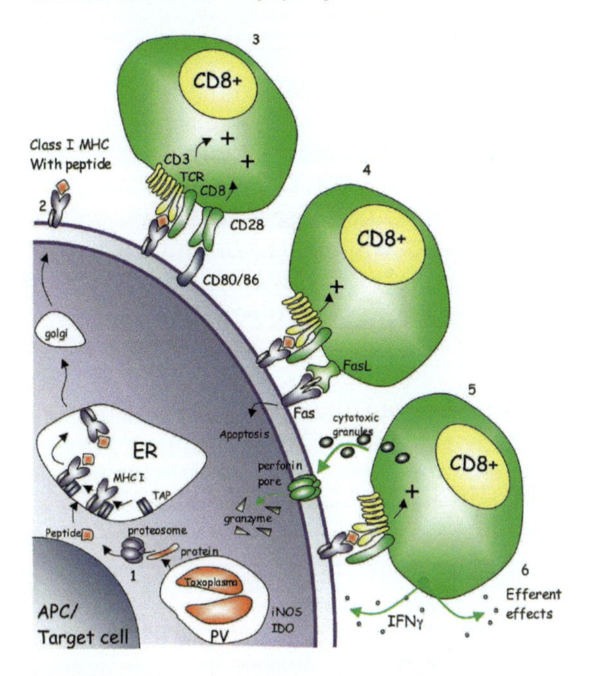

FIGURE 25.2 **Processing of endogenous antigen presentation on MHC class I and activation of CD8$^+$ T-lymphocytes to kill *T. gondii* infected cells.** (1) Proteins escaping or released from the *T. gondii* parasitophorous vacuole are proteolytically cleaved in the proteasome and enter the ER *via* TAP (translocator associated with antigen processing) molecules. These peptides are loaded onto class I MHC (MHC1) molecules in the ER. (2) The class I MHC peptide complex is exported from the ER, through the Golgi and ultimately to the surface of the cell. (3) Peptides presented on class I MHC are recognized by CD8$^+$ T-lymphocytes *via* their surface α/β-TCR and CD8 molecules. The CD3 complex associated with α/β-TCRs acts as a docking site for tyrosine kinases that provide activating intracellular signals. CD28 on the surface of T-cells interacts with CD80 and/or CD86 to provide co-stimulation for T-cells. CD40 is expressed constitutively on dendritic cells and on ligation with CD40 Ligand (CD40L), expressed on the surface of activated cells, increases CD80–86 expression on dendritic cells and enhances T-cell co-stimulation. Engagement of CD40 activates the dendritic cell to produce cytokines such as IL-12 (see Fig. 25.1). Dendritic cells use a process called 'cross-presentation' to transfer proteins taken up as part of necrotic or apoptotic debris into the class I endogenous processing pathway. (4) Cytolytic CD8$^+$ T-cells recognize specific *T. gondii* derived peptides presented by class I MHC molecules. Ligation of FasL on the surface of CD8$^+$ T-cells with Fas on the surface of target cells induces apoptosis of target cells. (5) Following recognition of specific *T. gondii* derived peptides presented by class I MHC molecules, CD8$^+$ T-cells can also kill target cells by release of perforin and granzyme. (6) CD8$^+$ T-cells also release IFNγ which can induce IDO and iNOS in target cells which are effective in killing *T. gondii* tachyzoites. (7) At the later stages of activation, T-cells express CTLA-4 which when engaged with CD80/86 delivers negative signals to the T-cell that favour termination of activation (see Fig. 25.1). *(Includes concepts from Fig. 4-1 in McLeod* et al. *(1996) and Fig. 4-1 in Lewis and Wilson (2005), with permission.)*

et al., 1997; Mitchell *et al.*, 1990; Murray *et al.*, 1979, 1980, 1985a, 1985b; Nathan *et al.*, 1984; Parner *et al.*, 1996; Roberts and Alexander, 1992; Silviera, 2001; Vogel *et al.*, 1997; Wilson *et al.*, 1979, 1980a; Willson and Haas, 1984; Wilson, 1985; Wong *et al.*, 1993; Yano *et al.*, 1989). The following are some more recent reviews: (Iwasaki and Medzhitov, 2010; Abi Abdallah *et al.*, 2011; Denkers *et al.*, 2010, 2012a; Yarnovinsky *et al.*, 2005, 2008; Barragan and Sibley, 2002; Blader and Saeij, 2009; Blanchard and Shastri, 2010; Boothroyd and Dubremetz, 2008; Boothroyd *et al.*, 2012; Brunet *et al.*, 2008; Hall *et al.*, 2012; Garweg and Candolfi, 2009; Ferreira da Silva *et al.*, 2008; Bradley and Sibley, 2007; Feng and Sher, 2010; Dupont and Hunter 2012; Dupont *et al.*, 2012; Egan *et al.*, 2012; Handel *et al.*, 2012; Hermes *et al.*, 2008; Buzoni-Gatel *et al.*, 2008; Begum-Haque *et al.*, 2009; Nance *et al.*, 2012; Mitra *et al.*, 2012; Mirpuri and Yarovinsky, 2012; Melo *et al.*, 2011; Mashayekhi *et al.*, 2011; Lang *et al.*, 2012; Kubo *et al.*, 2003; Koshy *et al.*, 2012; Hunter and Sibley, 2012; Henriquez *et al.*, 2010; Grigg *et al.*, 2001a; Goldszmid *et al.*, 2012; Wiley *et al.*, 2010; Yamamoto *et al.*, 2011; Zeiner and Boothroyd, 2010; Ochoa-Repáraz, 2011; Shapira *et al.*, 2005; Shi *et al.*, 2012; Spear *et al.*, 2006; Suzuki *et al.*, 1994a; 2010; Tait *et al.*, 2010; Unoki *et al.*, 2009; Tussiwand *et al.*, 2012; Vyas, 2013; Webster *et al.*, 1994).

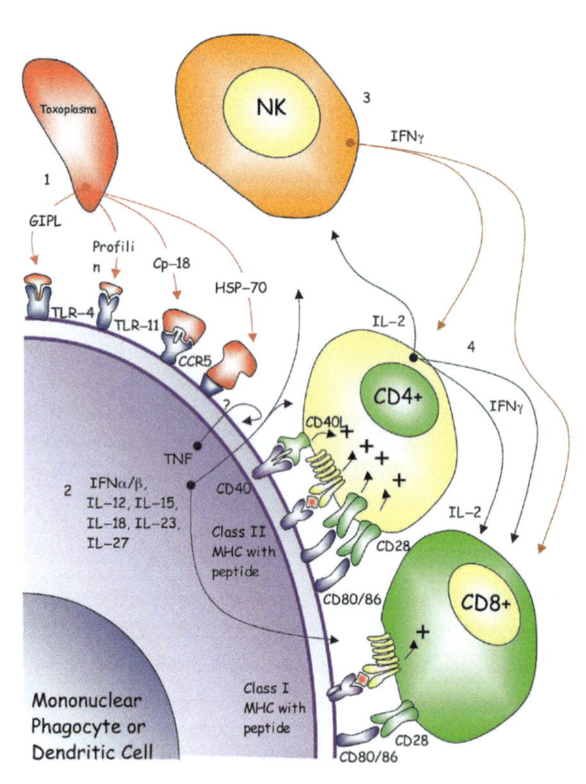

FIGURE 25.3 **Proposed afferent interactions of dendritic cells with *Toxoplasma*, NK-, CD8$^+$ T- and CD4$^+$ T-cells.** (1) *T. gondii* releases a number of molecules that have immunological effects. GIPL bind TLR-4, profilin binds TLR-11, Cp-18 binds CCR5 and HSP70 that induces dendritic cell maturation through an unknown mechanism (not all of these interactions may happen in all mammals and TLR-11, apparently, is not functional in humans). (2) The net effect of these molecules is dendritic cell activation and maturation with the likely production of IFNα/β, IL-12, IL-15, IL-18, IL-23 and IL-27. These mediators act on NK-cells, CD4$^+$ and CD8$^+$ T-cells. (3) NK-cells produce IFNγ that together with IL-12 favours Th1 cell as opposed to Th2 cell maturation. (4) CD4$^+$ T-cell production of IL-2 further activates NK-cells and favours the expansion of cytolytic CD8$^+$ T-cells that can kill *T. gondii* infected cells, by a number of methods (see Fig. 25.2). *(Includes concepts from Fig. 5 in McLeod* et al. *(1996), and Fig. 4-2 in Lewis and Wilson (2005), with permission.)*

25.2 MOUSE GENETIC STUDIES

25.2.1 Mortality

The H-2 complex (MHC) has a role in mediating protection or conferring susceptibility to mice, measured as survival following intraperitoneal infection with *T. gondii* (Williams *et al.*, 1978). In this study, susceptibility was found to be associated with the H-2a and H-2b haplotypes while resistance was associated with the H-2d and H-2k haplotypes. Susceptibility as measured by mortality is a complex trait involving several genes. Jones *et al.* (Jones *et al.*, 1977) recognized that mouse strain influenced brain cyst numbers. To develop a model of acquired infection, the natural history and pathology of infection in a variety of strains of mice and *Toxoplasma* were characterized and patterns of susceptibility based on host genotype and of various parasite isolates were noted (McLeod *et al.*, 1984). The ME49 strain was selected for further studies because it caused little mortality, high cyst burden, and when administered on the eleventh day of gestation led to 100% transmission to the foetus in outbred Swiss–Webster mice. Thus, this model could be used for testing effects of attenuated vaccines (e.g. Ts-4) on congenital transmission (McLeod *et al.*, 1984). These studies provided a foundation for work to characterize genetic susceptibility measured as survival and cyst formation using inbred, and later recombinant congenic, gene deficient and transgenic strains of mice (Brown and McLeod 1990; Brown *et al.*, 1995; Johnson *et al.*, 2002a, 2005; McLeod *et al.*, 1989a, 1989b). The ability of mice to survive oral ME49 strain *T. gondii* infection was found to be under polygenic control and regulated by a minimum of five genes, one of which maps to the H-2 region (McLeod *et al.*, 1989a) (Figs 25.8 and 25.9). An important observation in this work is that there is not a simple association between the development of brain cyst burden and mortality (McLeod *et al.*, 1984, 1989a, 1989b).

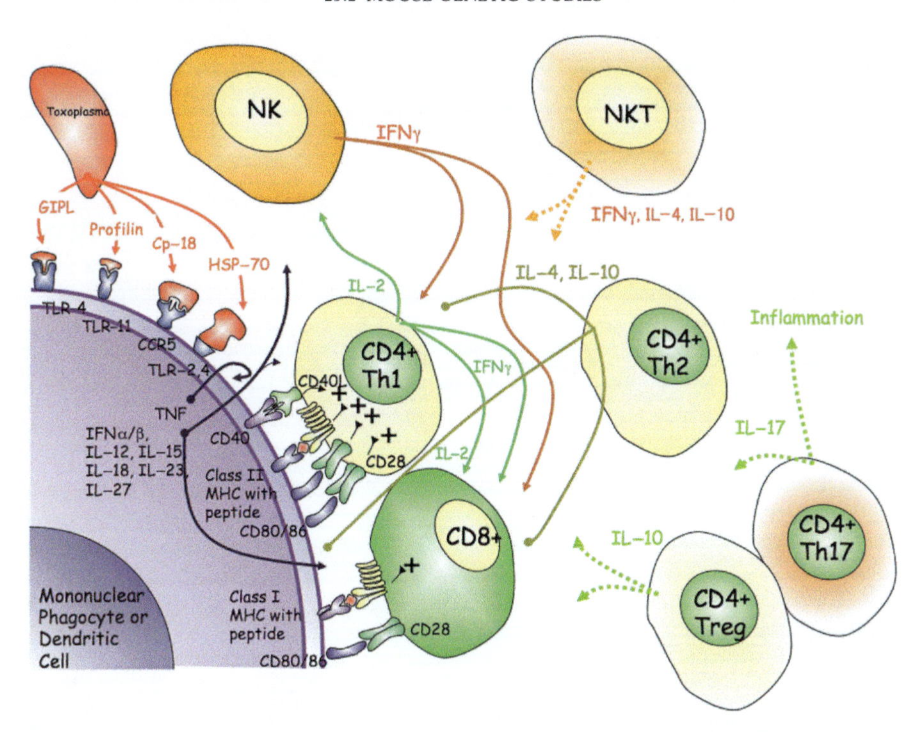

FIGURE 25.4 **Proposed regulatory interactions of dendritic and mononuclear phagocytic cells with *Toxoplasma*, NK-, CD8$^+$ T- and CD4$^+$ T-cells (Th1 And Th2), NKT- and regulatory T-cells.** Note differentiation of CD4 cells into Th1 and Th2 cells and interactions with NKT-cells and regulatory T-cells. NKT-cells recognize glycolipids in the presence of CD1. Th17 cells may modulate/regulate function of other T-cells and IL-17 may induce cell damage. The initial cytokine milieu may subsequently modulate immune response. CD8$^+$ T-cells can also produce IL-2 and make an autocrine loop (Suzuki *et al.*, in progress) *(Includes concepts from Fig. 5 in McLeod* et al.,*(1996), and Fig. 4-2 in Lewis and Wilson (2005), with permission.)*

The route of infection influences mortality in experimental toxoplasmosis (Johnson, 1984; McLeod *et al.*, 1989a), e.g. C57BL/6 mice were found to be susceptible to oral infection, but resistant to intraperitoneal infection (Johnson, 1984; McLeod *et al.*, 1989a), LACA mice exhibited the reverse characteristics (Johnson, 1984), and C57BL/6 mice develop fatal enteritis when infected orally. The characteristics of the inflammatory response, parasite antigens (e.g. SAG1) and genes involved in this inflammatory bowel disease following per-oral infection in C57BL/6 mice has been defined (Alexander *et al.*, 1997, 1998a, 1998b; Kasper *et al.*, 1997, 2001; Liesenfeld *et al.*, 1996, 1999; Nickdel *et al.*, 2004; Suzuki *et al.*, 2000b). Analysis of genetic mapping of survival of AxB and BxA mice following per-oral infection (Johnson *et al.*, 2002a; McLeod *et al.*, 1989b) in the context of further information about the mouse genome and more sophisticated analytical programmes (Collazo *et al.* ,2002; Dai *et al.*, 1994; Fujigaki *et al.*, 2002; Kahn *et al.*, 2002; Kahn *et al.*, in press) has demonstrated that genes or gene regions involved include regions on chromosomes 1, 8, 7 and 11 (Fig. 25.9, Table 25.3; Johnson *et al.*, 2002a).

Mortality following intraperitoneal infection has also been linked to the H-13 locus through examination of recombinant inbred mice originating from BALB/c × C57BL/6J mice. Of the seven recombinant lines obtained, the three most susceptible were found to contain the H-13 allele

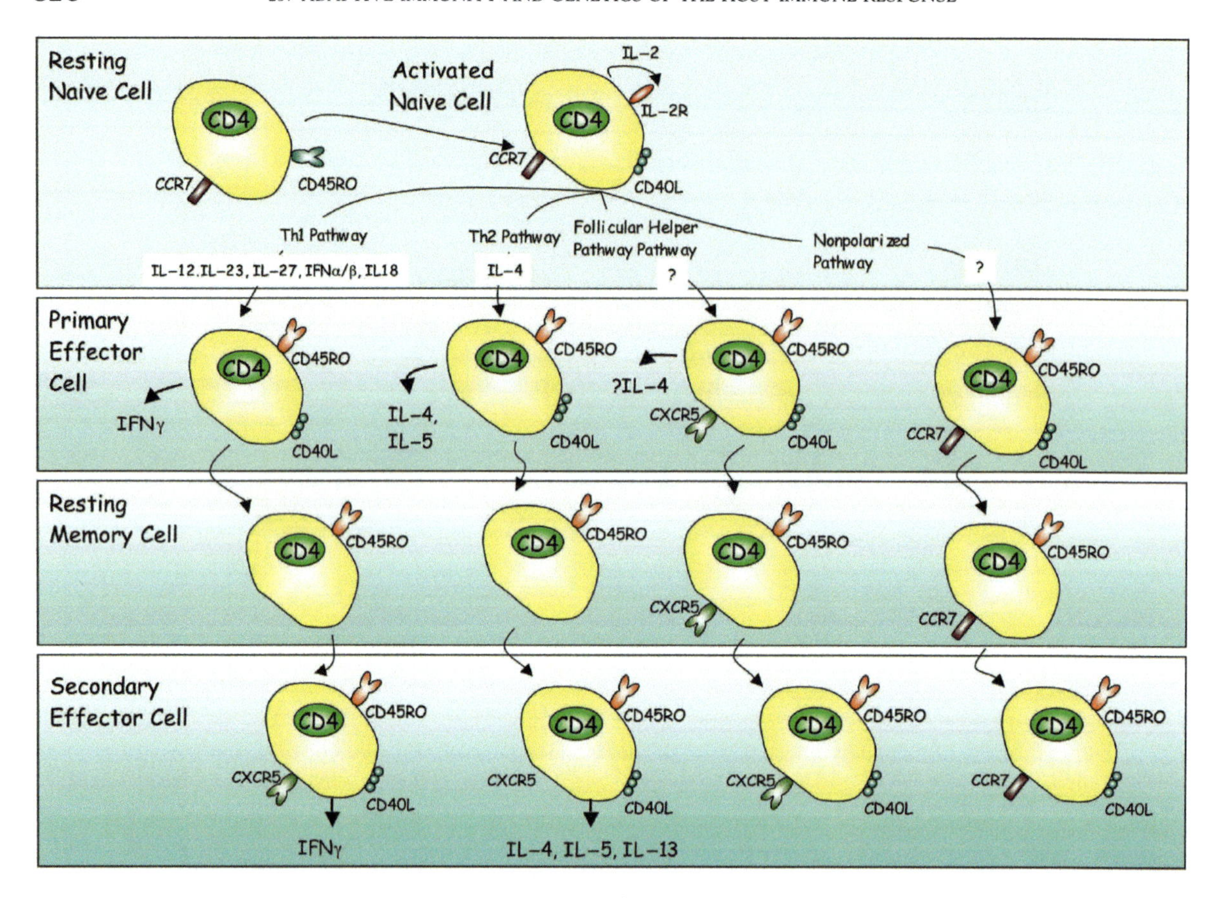

FIGURE 25.5 Differentiation of antigenically naive CD4$^+$ T-cells into Th1, Th2, unpolarized and follicular helper effector and memory T-cells. Differentiation of antigenically naive CD4$^+$ T-cells into Th1, Th2, unpolarized and T-follicular helper effector and memory T-cells. Antigenically naive CD4$^+$ cells express high levels of the CD45RA isoform of the CD45 surface protein tyrosine phosphatase. They are activated by antigen presented by antigen-presenting cells (APCs) to express CD40 ligand and interleukin (IL)-2 and to undergo clonal expansion and differentiation, which is accompanied by expression of the CD45RO isoform and loss of the CD45RA isoform. Most effector cells die by apoptosis, but a small fraction of these cells persist as memory cells which express high levels of CD45RO. Exposure of expanding effector cells to IL−12 family cytokines, IL−18 and interferon (IFN)γ favours their differentiation into Th1 effector cells that secrete IFNγ, whereas exposure to IL−4 favours their differentiation into Th1 effector cells that secrete IL−4, IL−5 and IL−13. Many memory cells are nonpolarized and do not express either Th1 or Th2 cytokines. They may be enriched for cells that continue to express the CCR7 chemokine receptor, which favours their recirculation between Th2 cytokines. They may be enriched for cells that continue to express the CCR7 chemokine receptor, which favours their recirculation between the blood, the lymph nodes and spleen. T-follicular helper cells, which express high levels of CXCR5, move into B-cell follicle areas, where they express CD4O ligand and provide help for B-cell responses. The signals that promote the accumulation of memory T-follicular helper cells and their capacity to produce cytokines are poorly understood. Memory cells rechallenged with antigen undergo rapid clonal expansion into secondary effector cells that mediate the same functions as the initial memory population. Most secondary effector cells eventually die by apoptosis. *(Adapted with permission from Fig. 4-6 in Lewis and Wilson (2005).)*

of the susceptible C57BL/6J parental strain, while the four most resistant strains had the H-13 allele of the resistant BALB/c parent (Williams *et al.*, 1978). The relative importance of the balance of MHC class 2 genes was shown in the work of Johnson *et al.* (2002a) in which the presence of only Ia or Ie led to increased mortality, whereas the presence of both did not lead to this imbalance (see Fig. 25.10A). Mortality is also influenced by gender, age and other factors in addition to genetics (see Fig. 25.10B–G) (Gardner and Remington, 1977; Johnson *et al.*, 2002a).

The work of a number of investigators has demonstrated the importance of both Th1 and Th2 type cytokines in the pathogenesis of *T. gondii* infection (Abou-Bacar *et al.*, 2004b; Araujo *et al.*, 2001; Bermudez *et al.*, 1993; Black *et al.*, 1989; Brown *et al.*, 1995a; Cai *et al.*, 2000; Chang *et al.*, 1990; Channon and Kasper, 1996; Dimier and Bout, 1998; Eisenhouer *et al.*, 1988; Gazzinelli *et al.*, 1991, 1993a, 1993b, 1994a,b,c, 1996a; Hunter *et al.*, 1993, 1994a, 1995b, 1996; Johnson et al., 1997; Lyons *et al.*, 2001; McLeod *et al.*, 1989a; Nathan *et al.*, 1984; Neyer *et al.*, 1997; Nickdel *et al.*, 2001, 2004; Pelloux and Ambroise-Thomas, 1996; Pfefferkorn, 1984; Pfefferkorn *et al.*, 1986; Raymond *et al.*, 1990; Roberts *et al.*, 1996; Roberts F, 1999a; Romagnani, 1996; Rozenfeld et al., 2003; Sher *et al.*, 2003; Shirahata *et al.*, 1992, 1993, 1994; Sibley *et al.*, 1991; Sibley, 2003; Silva *et al.*, 2002; Subauste *et al.*, 1992, 1995, 2002; Suzuki *et al.*, 1988, 1989, 1996a, 2000b; Suzuki and Kobayashi, 1990; Suzuki and Remington, 1990; Suzuki, 1997; Thomas *et al.*, 1993; Wang *et al.*, 2004, 2005). Yap (Yap *et al.*, 2000, 2007) demonstrated that Tyk2 Janus Kinase functions as a nodal regulator of proinflammatory versus anti-inflammatory cytokine balance during toxoplasmosis. They found that Tyk2 Janus Kinase regulates Th1/IL-10 cytokine balance by mediating IL-12 as well as IL-10 receptor signalling and cellular responses. Tyk2 deficient or mutant mice are susceptible to *T. gondii* infection. A dichotomy was also noted in the regulation of interferon-γ (IFNγ) versus IL-10 secretion by $CD4^+$ T-cells primed by *Toxoplasma* vaccination. Primed lymphocytes respond to *in vitro* antigen re-stimulation by immediate synthesis of IFNγ whereas IL-10 secretion requires an *in vivo* reactivation step regulated by Th1 proinflammatory cytokines, i.e. IL-12/Tyk2/IFNγ receptor signalling (Yap *et al.*, 2000, 2007).

The plethora of mice engineered to be deficient in immunologically important genes (Table 25.2) has established or, in some cases, confirmed a role for many immunological components (Alves *et al.*, 2012; Bernardes *et al.*, 2006; Blanchard *et al.*, 2008; Caetano *et al.*, 2011; Cavalcanti *et al.*, 2011; Combe *et al.*, 2006; Debierre-Grockiego *et al.*, 2007a, 2007b; Deckert *et al.*, 2006; Dellacasa-Lindberg *et al.*, 2007; Dunay *et al.*, 2008; Egan *et al.*, 2009a; Egan *et al.*, 2009b; Foureau *et al.*, 2010; Fujigaki *et al.*, 2002; Furuta *et al.*, 2006; Goldszmid *et al.*, 2007; Guiton *et al.*, 2010; Hou *et al.*, 2011; Jebbari *et al.*, 1998; Jensen *et al.*, 2011; Jin *et al.*, 2008; Johnson *et al.*, 2002a, 2003; Jost *et al.*, 2007; Khan *et al.*, 2006; Khan *et al.*, 2001; Koblansky *et al.*, 2012; Lees *et al.*, 2010; Lieberman *et al.*, 2004a, 2004b; Liesenfeld *et al.*, 2011; Lu *et al.*, 2003; Machado *et al.*, 2006a; Machado *et al.*, 2006b; Mahamed *et al.*, 2012; Mashayekhi *et al.*, 2011; Mason *et al.*, 2004; McBerry *et al.*, 2012; Miller *et al.*, 2011a, 2011b; Minns *et al.*, 2006; Mun *et al.*, 2003; Murakami *et al.*, 2012; Noor *et al.*, 2010; Norose *et al.*, 2008; Pawlowski *et al.*, 2007; Poassos *et al.*, 2010; Portillo *et al.*, 2010; Portugal *et al.*, 2008; Reese *et al.*, 2011; Reichmann *et al.*, 1999; Robben *et al.*, 2005; Scharton-Kersten *et al.*, 1998; Schluter *et al.*, 2003; Schulthess *et al.*, 2012; Shaw *et al.*, 2009; Sibley *et al.*, 2010; Stumhofer *et al.*, 2006; Sukhumavasi *et al.*, 2008; Suzuki *et al.*, 1996a; Suzuki, 1997; Tato *et al.*, 2012; Taylor *et al.*, 2004; Terrazas *et al.*, 2010; Thompson *et al.*, 2008; Tu *et al.*, 2009; Tussiwand *et al.*, 2012; Villarino *et al.*, 2003, 2006a; Villegas *et al.*, 2002; Wen *et al.*, 2010; Wiehagen *et al.*, 2012; Willie *et al.*, 2001; Wilson *et al.*, 2005; Yarovinsky *et al.*, 2008; Yarnovinsky *et al.*, 2005, 2008; Zhang and Denkers 1999; Zhao *et al.*, 2008; Zukas *et al.*, 2009). This raises the possibility that naturally occurring polymorphisms in these genes or

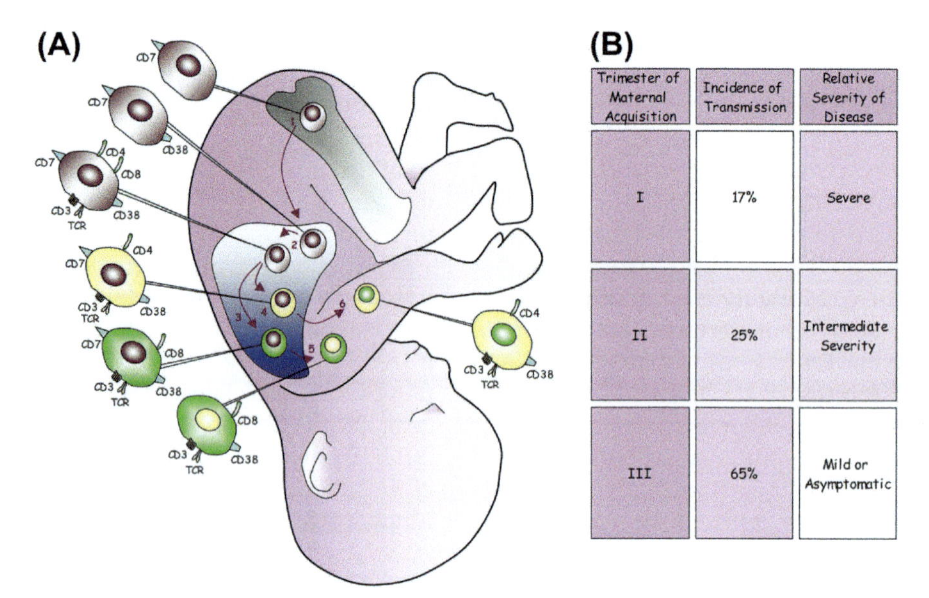

Trimester of Maternal Acquisition	Incidence of Transmission	Relative Severity of Disease
I	17%	Severe
II	25%	Intermediate Severity
III	65%	Mild or Asymptomatic

(C)

	Gestation	Birth	6 months	1 year
NK-cell	6 weeks, present in liver	15% of total lymphocytes Double absolute number present in adults 50% of adult cytolytic activity Overall increase number but decrease function		Full cytolytic activity reached in late infancy
Monocyte/ Macrophage & Dendritic cell	4 weeks, detected in yolk sac then liver and bone marrow	Monocytes number > adults IL–1 equals adult Diminished IL–6 and TNFα Cord blood DC less effective than adult in T-cell support		Chemotaxis < adult for 6–10 years
Neutrophil	Limited storage pool. 14–16 weeks, precursors seen 22–24 weeks, 10% of circulating leukocytes	50–60% of circulating leukocytes 40–45% adhesive capability to endothelium Poor chemotactic ability Normal superoxide anion Decreased hydroxyl radicals		
Eosinophil	18–30 weeks, 10–20% of total granulocytes	Post-natal peak at 3–4 weeks		
T-cell	7 weeks, detected in yolk sac and liver 8 weeks, lymphoid colonization and reduced TCR diversity and TdT enzyme activity 10 weeks, lymphoid tissues, liver and bone marrow 14 weeks, all three thymocytes in proper location 16 weeks, improved TCR diversity and TdT activity	Virgin subset T-cells Normal IL–2 synthesis and IL–2R expression Reduced TNFα, IL–3, IL–4, IL–5, IFNγ and GM-CSF CTL acticity variable Diminished production of IFNγ No DTH skin reaction to antigens and diminished reaction after stimulation	Peak T-cell number	Adult T-cell number, age 4 Peak size of thymus, age 10 Decreased skin reaction persists to age 1
B-cell	8 weeks, maternal IgG crosses placenta (majority in utero) 15 weeks, IgM secreting plasma cells present 17 weeks, some circulating fetal IgG may first be seen 20–30 weeks, IgG and Iga plasma cells first appear	TI-type 1 antigen response T-dependent response primary	TI-type 2 antigen response to encapsulated organismes Maximum IgM at 2–6 months	Circulating IgG is nearly all from infant (nadir at 3–4 months)
CD type	NK CD16 at 6 weeks gestation T-cell thymocytes express CD7 at 7 weeks gestation T-cell thymocytes express CD3, CD7, CD4 and CD8 at 12–14 weeks gestation T-cell thymocytes express CD4 or CD8 > 14 weeks gestation	NK CD56 50% NK CD57 decreased Most CD4$^+$ T-cells express CD45RA$^+$ marker (naïve)	CD45RA$^+$ cells → CD45RO$^+$ with age and antigen stimulation	

promoters may also account for observed difference in murine models and in humans. Parasite genetics has similarly been studied to reveal virulence factors that are secreted into the host cell and take over that cell, and can lead to host and neighbour cell death. Some examples are Type II GRA15 and TRAF6 and NFκB, Type I Rop16 and STAT3 and STAT6, Type III Rop38 and Map p38 kinase, Type I GRA10 and Taf1 B, Rop5 and Rop18 and murine IRGs and ATF6β, among other interactions of ligands with specific host cell receptors or molecules (Yamamoto *et al.*, 2011). Different aspects of newborn immune responses have been characterized (Berman *et al.*, 1978; Hayward *et al.*, 1981; Khalili *et al.*, 1997; Wilson *et al.*, 1986).

25.2.2 Encephalitis

Immune mechanisms that are important in controlling or contributing to encephalitis (Jones, 1978; Brown *et al.*, 1995; Suzuki, 1993; Gazzinelli *et al.*, 1992a, 1993b; Handel *et al.*, 2012; Deckert-Schluter *et al.*, 1994, 1996, 1998a, 1998b, 1999; Lüder *et al.*, 1999, 2003a, 2003b; Suzuki *et al.*, 1989a, 1991, 1993; Suzuki and Kobayashi, 1990; Suzuki and Remington, 1990) and ocular toxoplasmosis (OT) (Gazzinelli *et al.*, 1994b; Roberts and McLeod, 1999; Roberts *et al.*, 2000) have been extensively studied. These have revealed the importance of both host and parasite genetics (Figs 25.9 and 25.11; Table 25.3). MHC genes would appear to play the predominant role, but the NRAMP gene (previously ITY/BCG/Lsh) and loci on chromosomes 1, 4, 17 and 18 have also been implicated in playing an additional role (Blackwell *et al.*, 1994; Johnson *et al.*, 2002a; McLeod *et al.*, 1989a). Careful mapping of inbred, congenic, gene deficient (knock-out) and transgenic strains of mice has shown definitively that the Ld gene, a mouse class 1 MHC gene, plays an important role in protection of certain strains of mice against toxoplasmic encephalitis induced by type II *T. gondii* strains (Johnson *et al.*, 2002b) (Figs 25.10H–L and 25.12–14, Table 25.4). The Ity/BCG/Lsh 1 NRAMP gene exhibits a similar, although minor, effect on tissue cyst production (Fig. 25.11). Mice deficient in immunologically important genes (Table 25.2) have also provided many useful insights and in some cases confirmed a protective or exacerbatory role for many other immunological components in toxoplasmic encephalitis. In other cases the absence of certain genes has resulted in surprising results and polymorphisms in these genes or their promoters is likely to account for the observed differences between murine models.

T. gondii effects on the brain and the immune response are considered in depth in Chapter 23. It has been demonstrated that congenital and

FIGURE 25.6 (A) Putative stages of human αβ T-cell receptor-positive (αβ-TCR$^+$) thymocyte development. Prothymocytes, expressing CD7, are produced in the bone marrow or foetal liver. These enter the thymus *via* vessels at the junction between the thymic cortex and medulla. In the thymus these cells differentiate to more mature αβ-TCR$^+$ thymocytes (defined by their pattern of the αβ-TCR–CD3 complex, CD4, CD8 and CD38). Rearrangement of the TCR and TCRβ chain genes occurs in the outer cortex. Positive selection occurs mainly in the central thymic cortex and involves the interaction of thymic epithelial cells. Negative selection occurs mainly in the medulla and involves the interaction of thymic dendritic cells. Medullary thymocytes emigrate into the circulation and colonize the peripheral lymphoid organs. These T-cells are CD4$^+$ and CD8$^+$ with high levels of the αβ-TCR–CD3 complex. These cells are referred to as 'recent thymic emigrants' (RTEs) and probably lack CD38 surface expression. *(Modified from Fig. 4-3 in Lewis and Wilson (2005), with permission.)*
(B) Transmission and severity of illness in each trimester. In general, as gestation progresses, frequency of maternal to foetal transmission increases and severity of illness in the newborn infant diminishes.
(C) Putative development of human T-cells NK-cells, monocyte/macrophage/dendritic cells, neutrophils, eosinophils and B-cells. *(Adapted from Table 2 in McLeod and Dowell (2000) and Fig. 4-3 in Lewis and Wilson (2005), with permission.)*

TABLE 25.1 Selected Pairs of Surface Molecules Involved in T-Cell-Antigen-Presenting Cell (APC) Interactions

T-Cell Surface Molecule	T-Cell Distribution	Corresponding Ligand(s) on APCs	APC Distribution
CD2	Most T-cells; higher on memory cells, lower on adult naïve and neonatal T-cells	LFA-3 (CD58), CD59	Leukocytes
CD4	Subset of αβ T-cells with predominantly helper activity	MHC class II β chain	Dendritic cells, Mϕ, B-cells, others (see text)
CD5	All T-cells	CD72	B-cells, Mϕ
CD8	Subset of αβ T-cells with predominantly cytotoxic activity	MHC glass I heavy chain	Ubiquitous
LFA-1 (CD1 la/CD18)	All T-cells; higher on memory cells, lower on adult naïve and neonatal T-cells	ICAM-1 (CD54) ICAM-2 (CD102) ICAM-3 (CD50)	Leukocytes (ICAM-3 > ICAM-1,-2) and endothelium (ICAM-1, ICAM-2); most ICAM-1 expression requires activation
CD28	Most $CD4^+$ T-cells, subset of $CD8^+$ T-cells	CD80 (B7-1) CD86 (B7-2)	Dendritic cells, Mϕ, activated B-cells
ICOS	Effector and memory T-cells; not on resting naive cells	B7RP-1 (B7h)	B-cells, Mϕ, dendritic cells, endothelial cells
VLA-4 (CD49d/CD29)	All T-cells; higher on memory cells, lower on adult naïve and neonatal T-cells	VCAM-1 (CD106)	Activated or inflamed endothelium (increased by TNF, IL-1, IL-4)
ICAM-1 (CD54)	All T-cells; higher on memory cells, lower on adult virgin and neonatal T-cells	LFA-1 (CD11a/CD18)	Leukocytes
TLA-4 (CD152)	Activated T-cells	CD80 CD86	Dendritic cells, Mϕ, activated B-cells, activated T-cells
CD40 ligand (CD154)	Activated $CD4^+$ T-cells; lower on neonatal $CD4^+$ T-cells	CD40	Dendritic cells, Mϕ, B-cells, thymic epithelial cells
PD-1	Activated $CD4^+$ and $CD8^+$ T-cells	PD-L1, PD-L2	Dendritic cells, Mϕ, B-cells, regulatory T-cells

CTLA-4, cytotoxic T lymphocyte antigen-4; ICAM, intercellular adhesion molecule; ICOS, inducibe co-stimulator; IL, interleukin; LFA, leukocyte function antigen; Mϕ, mononuclear phagocytes; MHC, major histocompatibility complex; PD, programmed death [molecule]; VCAM, vascular all adhesion molecule; VLA-4, very late antigen-4.

Adapted with permission from Lewis and Wilson (2005).

TABLE 25.2A The Effect of Interleukin or Interleukin Receptor Gene Deficiency on Murine *T. gondii* Infection

Gene Deficiency	Mouse Background	Parasite Strain	Route of Infection	Effect on Parasite	Effect on Pathology	Effect on Survival	Immunological Effects	References
IL–2	C57BL/6	Me49	i/p	No difference in parasite number in brain	None reported	Reduced	Reduced IL–12p40 and IFNγ	Villegas *et al.* (2002)
IL–4	C57BL/6	Beverley	Oral	No difference in intestine	Decreased necrosis in intestine	Enhanced	Increased plasma IFNγ and IL–12 Increased IL–10 transcripts in intestine	Nickdel *et al.* (2004)
IL–4	129/J	Me49	Oral	Increased tachyzoites and cysts in brain	Increased areas of focal inflammation	Reduced	No difference in transcripts for IFNγ, TNFα, IL–6 and IL–10 Reduced IFNγ production by antigen stimulated splenocytes	Suzuki *et al.* (1996a, 1996b)
IL–4	129/ SvxC57BL/6	Beverley	Oral	Reduced cysts in brain	Decreased encephalitis Decreased necrotic lesions	Reduced	Increased IFNγ (day 7) Reduced IL–10 (day 28)	Roberts *et al.* (1996)
IL–5	C57BL/6	Me49	i/p	Increased parasite number in brain	Not reported	Reduced	Reduced splenocyte IL–12 production	Zhang *et al.* (1999)
IL–5	C57BL/6	Beverley	Oral	No difference in intestine	Decreased necrosis in intestine	Enhanced	Increased plasma IFNγ and IL–12 Reduced eosinophilia	Nickdel *et al.* (2001)
IL–6	129/Sv x C57BL/6	Me49	i/p	Increased % of infected peritoneal cells	Increased necrosis in brain	Not reported	Reduced transcripts for IFNγ and increased transcripts for IL–10 in brain	Suzuki (1997)

(*Continued*)

TABLE 25.2A The Effect of Interleukin or Interleukin Receptor Gene Deficiency on Murine *T. gondii* Infection (*cont'd*)

Gene Deficiency	Mouse Background	Parasite Strain	Route of Infection	Effect on Parasite	Effect on Pathology	Effect on Survival	Immunological Effects	References
IL−6	129/SvJ	Beverley	Oral	Increased cyst number and uncontrolled tachyzoites replication in brain	Increased brain pathology	Reduced	Reduced serum IL−6 levels	Jebbari *et al.* (1998)
IL−6	129/SVJ	Beverley	i/p	Increased tachyzoite number	Increased inflammation in the eye	Not reported	Decreased ocular IL−1 mRNA transcripts Increased ocular TNFα mRNA transcripts	Lyons *et al.* (2001)
IL−6	C57BL/6	Me49	i/p	Not reported	Not reported	Not reported	Increased serum levels of IL−17 Decrease in frequency of IL−17-producing $CD3^+$ and $NK1.1^+$ cells	Poassos *et al.* (2010)
IL−10	C57BL/6	Me49	Oral	Not reported	Increased necrosis in ileum	Reduced	Increased IFNγ mRNA transcripts in	Suzuki *et al.* (2000a)
IL−10	BALB/c	Me49	Oral	Not reported	Focal necrosis in ileum	Reduced	Not reported	Suzuki *et al.* (2000b)
IL−10	C57BL/6	Me49	i/p	None observed in brain	Increase inflammation and necrosis in brain	Reduced	Increased $CD4^+$ T-cells in brain Partial depletion of $CD4^+$ T-cells decreased inflammation and increased survival	Wilson *et al.* (2005)
IL−10	C57BL/6	Me49	Oral	Reduced number (brain)	Increased liver pathology	Increased	Increased serum IL−12(p70), IL−12p(40) and IFNγ	Gazzinelli *et al.* (1996b)

IL–10	C57BL/6	Me49	i/p	Increased numbers (intestine and liver)	Not reported	Reduced	Not reported	Wille *et al.* (2004)
IL–10	BALB/c	RH	i/p	Not reported	Not examined	None	No difference in serum IL–12p40 or IFNγ levels	Wille (2001)
IL–10	C57BL/6	RH	Intraocular	None reported	Increased necrosis and inflammation in the eye	Not reported	Increased serum IFNγ	Lu *et al.* (2003)
IL–10	BALB/c	RH	Intraocular	None reported	Increased necrosis and inflammation in the eye	Not reported	Increased serum IFNγ	Lu *et al.* (2003)
p40 (IL–12/ IL–23)	C57BL/6	Ts-4	i/p	Not reported	Not reported	Reduced	Surviving mice have increased resistance to increased doses of ts-4 or infection with 76K strain of *T. gondii*	Ely *et al.* (1999)
p40 (IL–12/ IL–23)	BALB/c	Ts-4	i/p	Not reported	Not reported	Reduced	Reduced serum IFNγ	Lieberman *et al.* (2004)
p35 (IL–12)	BALB/c	Ts-4	i/p	Increased numbers (peritoneal exudate)	Not reported	Reduced	Reduced serum IFNγ	Lieberman *et al.* (2004)
p40 (IL–12/ IL–23)	C57BL/6	Me49	i/p	Increased number in brain and lung	Not reported	Reduced	Not reported	Villarino *et al.* (2003)

(*Continued*)

TABLE 25.2A The Effect of Interleukin or Interleukin Receptor Gene Deficiency on Murine *T. gondii* Infection (*cont'd*)

Gene Deficiency	Mouse Background	Parasite Strain	Route of Infection	Effect on Parasite	Effect on Pathology	Effect on Survival	Immunological Effects	References
p40 (IL−12/ IL−23)	C57BL/6	Me49	i/p	Not reported	Not reported	Reduced	Exogenous IL−12 restores resistance but reactivation occurs on withdrawal	Yap *et al.* (2000)
p40 (IL−12/ IL−23)	C57BL/6	Me49	i/p	Increased % of infected peritoneal cells	Not reported	Reduced	Not examined	Scharton-Kersten *et al.* (1997)
p40 (IL−12/ IL−23)	C57BL/6	Me49	i/p	Increased numbers (peritoneal exudate)	Not reported	Reduced	None reported	Lieberman *et al.* (2004)
p35 (IL−12)	C57BL/6	Me49	i/p	None	Not reported	Reduced	None reported	Lieberman *et al.* (2004)
IL−15	C57BL/6	Me49	i/p	Not reported	None	None	Similar levels of IFNγ production and T-cell activation	Lieberman *et al.* (2004)
IL−15	C57BL/6	Me49	Oral	No difference in intestine or MLN	Reduced intestinal pathology	Reduced	Reduced IL1b, TNFα and IL−6 by LP-cells	Schulthess *et al.* (2012)
IL−15	C57BL/6	76K	i/p Oral	Not reported Increased parasite levels in gut, spleen liver and brain	Not reported Reduced intestinal and liver pathology	Increased	Not reported Reduced IFNγ producing cells in MLN and spleen Reduced IFNγ transcripts in gut	Combe *et al.* (2006)
IL−17RA	C57BL/6	76K	Oral	Reduced cysts in brain	Reduced brain and acute intestinal inflammation Reduced systemic imflammtion	Increased	Increased Th1 cytokine response in particular IFNγ and IL−2 levels	Guiton *et al.* (2010)

IL–18	C57BL/6	Me49	Oral	No difference in lung Increased number in liver Reduced number in intestine	Reduced necrosis in intestine and liver	Increased	Reduced serum IFNγ Reduced intestinal IFNγ production	Vossenkamper *et al.* (2004)
p19 (IL–23)	C57BL/6	Me49	i/p	None	Not reported	None	No difference in serum IFNγ or IL–12 levels	Lieberman *et al.* (2004)
WSX-1 (IL–27Ra)	C57BL/6	Me49	i/p	Increased numbers (peritoneal exudate)	Increased inflammation and necrosis in liver and lungs	Reduced	Increased serum IL–12 and IFNγ Increased T-cell activation	Villarino *et al.* (2003)
WSX-1 (IL–27Ra)	C57BL/6	Me49	i/p	Not reported	Not reported	Reduced	Increased IL–2 production	Villarino *et al.* (2006)
WSX–1 (IL–27Ra)	C57BL/6	Me49	i/p	None in brain	Increased brain pathology	Reduced	Increased IL–17 production	Stumhofer *et al.* (2006)
IL–33	BALB/c	Beverley	Oral	Increased parasite levels in the brain	Increased brain pathology	Reduced	Increased IFNγ, iNOS and TNFα transcript levels in the brain	Jones *et al.* (2010)
GP130 (neurons)	C57BL/6	DX	i/p	Increased number of cysts and tachyzoites in brain	Increased necrosis in brain	Reduced	Increased IL–17 and IFNγ producing CD4 and CD8 T-cells, reduce intracerebral TGFβ and IL–127	Handel *et al.* (2012)

TABLE 25.2B The Effect of TLR and Other Innate Sensing Receptor Gene Deficiency on Murine *T. gondii* Infection

Gene Deficiency	Mouse Background	Parasite Strain	Route of Infection	Effect on Parasite	Effect on Pathology	Effect on Survival	Immunological Effects	References
TLR1	129/Ola x C57BL/6	PTGluc	i/p	None	None	None	None examined	Hitziger *et al.* (2005)
TLR2	129/Ola x C57BL/6	PTGluc	i/p	None	None	None	None examined	Hitziger *et al.* (2005)
TLR2	C57BL/6	Fukaya	Oral	None	Damaged renal function Glomerular and extracellular matrix swelling, advancing glomerular tissue proliferation, thickened Bowman's capsules and vacuolization of tubules	Not reported	None reported	Kudo *et al.* (2004)
TLR2	C57BL/6	Fukaya	Increased inflammation in lungs	No difference in brain or lung	Increased pathology in lungs	Reduced	Reduced PEC production of IL−12 and IFNγ Increased PEC production of IL−4 and IL−10 Reduced iNOS and IDO expression and NO production	Mun *et al.* (2003)
TLR2	C57BL/6	Me49	Oral	No difference in cyst numbers in brain	Not reported	Reduced	No difference in splenocyte IFNγ, IL−4, IL−10, IL−6 or IL−12 production	Furuta *et al.* (2006)
	C57BL/6	Me49	i/p	No difference in cyst numbers in brain		Reduced	Not reported	

TLR2	C57BL/6	Me49	i/p	No difference in cyst numbers in brain	Not reported	None	Increased splenocyte production of IFNγ, similar levels of IL–12	Debierre-Grockiego *et al.*, 2007a
TLR2/4	C57BL/6	Me49	i/p	Increased cyst numbers in brain	Not reported	None	Increased splenocyte production of IFNγ, similar levels of IL–12	Debierre-Grockiego *et al.*, 2007a
TLR4	129/Ola x C57BL/6	PTGluc	i/p	None	None	None	None examined	Hitziger *et al.* (2005)
TLR4	C57BL/6	Fukaya	Oral	Increased numbers (kidney)	Damaged renal function	Not reported	None reported	Kudo *et al.* (2004)
TLR4	C57BL/6	Fukaya	None	No difference in brain or lung	Not reported	None	Not reported	Mun *et al.* (2003)
TLR4	C57BL/6	Me49	i/p	No difference in cyst numbers in brain	Not reported	None	Similar levels of splenocyte production of IFNγ and IL–12	Debierre-Grockiego *et al.*, 2007a
TLR4	C57BL/6	Me49	Oral	Increased cyst numbers in brain	Not reported	Reduced	Reduced splenocyte IFNγ, IL–6 and IL–12 production, no difference in IL–4, IL–10 production	Furuta *et al.* (2006)
	C57BL/6	Me49	i/p	No difference in cyst numbers in brain		Reduced	Not reported	
TLR6	129/Ola x C57BL/6	PTGluc	i/p	None	None	None	None examined	Hitziger *et al.* (2005)
TLR9	129/Ola x C57BL/6	PTGluc	i/p	Increased dissemination	None	None	None examined	Hitziger *et al.* (2005)
TLR9	C57BL/6	76K	Oral	Not reported	Not reported	Not reported	Failure to modulate Crp mRNA expression	Foureau *et al.* (2010)

(*Continued*)

TABLE 25.2B The Effect of TLR and Other Innate Sensing Receptor Gene Deficiency on Murine *T. gondii* Infection (*cont'd*)

Gene Deficiency	Mouse Background	Parasite Strain	Route of Infection	Effect on Parasite	Effect on Pathology	Effect on Survival	Immunological Effects	References
TLR9	C57BL/6	76K	Oral	Increased parasite in intestine	Reduced intestinal pathology	Increased	Reduced intestinal TNFα and IFNγ production	Minns *et al.* (2006)
TLR11	C57BL/6	Me49	i/p	Not reported	Not reported	Reduced	Failure to produce significant amount of IL−12 Decreased numbers of $CD4^+$ and $CD8^+$ T-cells	Yarovinsky *et al.* (2008)
TLR11	C57BL/6	Me49	i/p	Not reported	Fat necrosis in peritoneal serosa and acute pancreatitis	Not reported	TLR11-independent induction of IL−12 and IFNγ	Yarovinsky *et al.* (2008)
		RH-YFP	i/p	Altered parasite distribution to WT but no difference in levels detected at pancreas				
TLR12	129/Ola x C57BL/6	PTGluc	i/p	Increased dissemination	None	None	None examined	Koblansky *et al.* (2012)
NOD2	C57BL/6	Me49	i/p	None	None	None	None	Caetano *et al.* (2011)
NOD2	C57BL/6	Me49	i/p	Increased parasite burden in peritoneal cavity	Not reported	Reduced	Lower IFN γ levels	Shaw *et al.* (2009)
Gal3	C57BL/6	RH	i/p	Not reported	Not reported	None	Reduced ROS	Alves *et al.* (2012)
Gal3	C57BL/6	Me49	Oral	Increased parasite numbers in the brain	Reduced inflammation in liver, intestine and brain	None	Increased spenic DC production of IL−12	Bernardes *et al.* (2006)
$P2X_7$	C57BL/6	Me49	i/p	Increased parasite numbers in liver	Increased liver pathology	Reduced	Similar IFNγ levels, increased nitric oxide levels and delayed IL−10 production	Miller *et al.* (2011)

TABLE 25.2C The Effect of Cell Signaling and Transcription Factor Gene Deficiency on Murine *T. gondii* Infection

Gene Deficiency	Mouse Background	Parasite Strain	Route of Infection	Effect on Parasite	Effect on Pathology	Effect on Survival	Immunological Effects	References
MyD88	129/Ola x C57BL/6	PTGluc	i/p	None	None examined	Reduced	None examined	Hitziger *et al.* (2005)
MyD88	C57BL/6	Fukaya	Increased inflammation in lungs	No difference in brain or lung	Increased pathology in lungs	Reduced	Reduce PEC production of IL−12 and IFNγ Increased PEC production of IL−4 and IL−10 Reduced iNOS and IDO expression and NO production	Mun *et al.* (2003)
MyD88	129/Ola xC57BL/6	Me49	i/p	Increased parasite numbers in liver, lung, heart, spleen and brain	None reported	Reduced	Reduced plasma IL−12p40 and IFNγ	Scanga *et al.* (2002)
MyD88	C57BL/6	Me49	i/p	Increased parasite numbers in PEC	None reported	Reduced	None reported	Hou *et al.* 2011
MyD88 (dendritic cells)	C57BL/6	Me49	i/p	Not reported	None reported	Reduced	Reduced IL−12 production and delayed IFNγ production	Hou *et al.* 2011
MyD88 (macrophages and neutrophils)	C57BL/6	Me49	i/p	Increased parasite numbers in PEC	None reported	Reduced	None reported	Hou *et al.* 2011

(*Continued*)

TABLE 25.2C The Effect of Cell Signaling and Transcription Factor Gene Deficiency on Murine *T. gondii* Infection (*cont'd*)

Gene Deficiency	Mouse Background	Parasite Strain	Route of Infection	Effect on Parasite	Effect on Pathology	Effect on Survival	Immunological Effects	References
MyD88	C57BL/6	Me49	Oral	Not reported	No significant pathology but inflammatory infiltrates observed in liver		Reduction in early neutrophil recruitment Similar macrophage, dentritic, B- and T-cell levels	Sukhumavasi *et al.* (2008)
			i/p	Not reported		Reduced Rapidly reduced	Reduced IL–12/23p40 levels	
MyD88	C57BL/6	Me49	i/p	No difference in cyst numbers in brain	Not reported	Reduced	Not reported	Debierre-Grockiego *et al.*, 2007a
c-Rel	C57BL/6	Me49	Oral	Increased numbers (peritoneal macrophages)	More severe encephalitis	Reduced	Decreased T-cell activation, proliferation and IFNγ production	Mason *et al.* (2004a, 2004b)
c-Rel	C57BL/6	RH (OVA)	i/p	Not reported	Not reported	Reduced	Decreased CD8 T-cell responses to immunization	Jordon *et al.* (2010b)
STAT1	129Sv/Ev	Me49	i/p	Increased numbers (peritoneal exudate)	Not reported	Reduced	Reduced iNOS, IGTP and LRG-47. Failure to upregulate IL–12Rβ2 Reduced IFNγ producing $CD4^+$ and $CD8^+$ T-cells Increased IL–4, Arg1, Ym1 and Fizz1	Gavrilescu *et al.* (2004)

STAT1	SvEv	Me49	i/p	Not reported	Not reported	Reduced	Failure to upregulate MHC, reduced NO production reduced IFN-inducible GTPase proteins	Lieberman *et al.* (2004b)
STAT1	129/Sv/Ev	Me49	i/p	Not reported	Not reported	Reduced	Not reported	Collazo *et al.* (2002)
STAT4	129/Sv/ C57BL/6	Me49	i/p	Not reported	Not reported	Reduced	Reduced plasma IFNγ	Cai *et al.* (2000)
STAT6	C57BL/6	Fukaya	Oral	Increased cyst burden on the brain	No significant inflammation observed in brain	Not reported	Reduced IFNγ in cerebrospinal fluid Reduced activated $CD8^+$ T-cells	Jin *et al.* (2009)
NFκB(2)	C57BL/6	Me49	i/p	Increased percent infected PEC	Increased toxoplasmic encephalitis	Reduced	Reduced IFNγ production and loss of $CD4^+$ and $CD8^+$ T-cells Increased expression of Fas and apoptosis in spleen	Caamano *et al.* (2000)
NFκB/p52	C57BL/6	Me49	i/p	Increased percent of infected PEC	Not reported	Reduced	Not reported	Franzoso *et al.* (1998)
NFκB1	BALB/c	Me49	i/p	Increase in parasite burden in brain and liver	Increase in inflammation and necrosis of the brain	Increased	Reduced levels of $CD8^+$ T-cells in the brain	Harris *et al.* (2010)

(*Continued*)

TABLE 25.2C The Effect of Cell Signaling and Transcription Factor Gene Deficiency on Murine *T. gondii* Infection (*cont'd*)

Gene Deficiency	Mouse Background	Parasite Strain	Route of Infection	Effect on Parasite	Effect on Pathology	Effect on Survival	Immunological Effects	References
Rel-B	C57BL/6	Me49	i/p	No difference in brain cyst burdens	Necrotizing myocardiatis and interstitial pneumonia in moribund animals	Reduced	Reduced IFNγ levels and NK-cell function	Caamano *et al.* (1999)
ICSBP	C57BL/6	Me49	i/p	Increased cysts in brain Increased numbers of tachyzoites in brain	Not reported	Reduced	Reduced serum IFNγ and IL−12	Scharton-Kersten *et al.* (1997)
LRG-47	C57BL/6 x 129SvJ	Me49	i/p	Increased cysts in brain	Not examined	Reduced	Modest increase in serum IFNγ and IL−12p40	Collazo *et al.* (2001)
IRG-47	C57BL/6 x 129SvJ	Me49	i/p	None	Not examined	Reduced	None reported	Collazo *et al.* (2001)
IRF	C57BL/6	PLK	i/p	Not determined	Not determined	Reduced	Increased splenocyte IL−10 mRNA transcripts Similar splencoyte IL−2, IL−12 and IFNγ mRNA transcripts	Khan *et al.* (1996)
IRF	129	Me49	i/p	Increased parasite numbers in liver, lung, and spleen	Not reported	Reduced	Reduced IDO and iNOS expression in lungs	Silva *et al.* (2002)
IRF-8	C57BL/6	Me49	Oral	Increased parasite burden	Not reported	Reduced	Not reported	Jost *et al.* (2007)

IGTP	C57BL/6x129Sv	Me49	i/p	Increased percent of infected peritoneal exudates cells	Not reported	Reduced	Cannot be made resistant by bone marrow donation from WT mice	Collazo *et al.* (2002)
IGTP	C57BL/6 x 129Sv	Me49	i/p	Not reported	Not reported	Reduced	Increased plasma IFNγ and IL−12 Increased iNOS transcripts in liver	Taylor *et al.* (2004)
FoxP4 (CD4 cells only)	C57BL/6	Me49	i/p	No difference in brain cyst burden	No difference	None	Reduced splenocyte IFNγ	Wiehagen *et al.* (2012)
Socs2	C57BL/6	Me49	i/p	Not reported	Increased monocyte and neutrophil infiltration to peritoneal cavity	Reduced	Increased serum IL−12p40 and IFNγ, increased peritoneal CCL2	Machado *et al.* (2006a, 2006b)
Socs3 (Neutrophil and macrophages only)	C57BL/6	Pru	i/p	Increased parasite numbers in peritoneal exudates cells, liver	Increased necrosis and inflammation in liver and lungs	Reduced	Reduced IL−12	Whitmarsh *et al.* (2011)
Batf3	C57BL/6	Pru (expressing luciferase)	i/p Oral	Increased parasites by IVIS Increased parasites by IVIS	Not reported	Reduced	Reduced IL−12 and IFNγ CD8 T-cell priming reduced	Mashayekhi *et al.* (2011)

(*Continued*)

TABLE 25.2C The Effect of Cell Signaling and Transcription Factor Gene Deficiency on Murine *T. gondii* Infection (*cont'd*)

Gene Deficiency	Mouse Background	Parasite Strain	Route of Infection	Effect on Parasite	Effect on Pathology	Effect on Survival	Immunological Effects	References
IRAK4	C57BL/6 129Sv	Me49	Oral	Increased parasite numbers in brain, liver and spleen	Increased inflammation in brain	Reduced	Reduced IL−12 production	Bela *et al.* (2012)
Batf2	C57BL/6	Pru	Not specified	No difference	Not reported	Reduced	Reduced percent of CS45.2 cells and reduced numbers of lung-resident CD1031CD11b2 dendritic cells and CD1032CD11b2	Tussiwand *et al.* (2012)
AhR	C57BL/6	Me49	i/p	Reduced number of brain cysts	Greater liver damage	Reduced	Reduced IL−10 and increased TNFα in sera Overproduction of IL−12 and IFNγ in PECs	Sanchez *et al.* (2010)

TABLE 25.2D The Effect of IFNγ, TNFα, LTα and Their Induced Effectors or Receptor Gene Deficiency on Murine *T. gondii* Infection

Gene Deficiency	Mouse Background	Parasite Strain	Route of Infection	Effect on Parasite	Effect on Pathology	Effect on Survival	Immunological Effects	References
IFNγ	C57BL/6 BALB/c	Fukaya strain cysts	Oral	Increased bradyzoite numbers in the eye Increased cyst burden in the eye	Increased inflammation in the eye	Not reported	None reported	Norose *et al.* (2003)
IFNγ	C57BL/6 BALB/c	Fukaya strain cysts	Oral	Dramatic increase in tachyzoites observed in peripheral blood	Not reported	Reduced	Significant increase in neutrophils	Norose *et al.* (2008)
IFNγ	C57BL/6	Me49	i/p	Not reported	Not reported	Reduced	Not reported	Collazo *et al.* (2002)
IFNγ	C57BL/6	Me49	i/p	Increased parasite numbers in brain	Not reported	Reduced	None reported	Silva *et al.* (2002)
IFNγ	C57BL/6	Me49	i/p	Not reported	Not reported	Reduced	Reduced IDO and iNOS expression in lungs	Silva *et al.* (2002)
IFNγ	C57BL/6	Fukaya	i/p	Increase parasite numbers in brain, spleen, gut and liver	Not reported	Not reported	Reduced IDO expression in lung and brain	Fujigaki *et al.* (2002)
IFNγ	C57BL/6	Me49	i/p	Increased cyst burden in brain	Not reported	Reduced	Not examined	Scharton-Kersten *et al.* (1997b)
IFNγ	BALB/c	Fukaya	Oral	Increased numbers (brain, liver, lung, gut and spleen)	Not reported	Reduced	Not examined	Norose *et al.* (2008)

(*Continued*)

TABLE 25.2D The Effect of IFNγ, TNFα, LTα and Their Induced Effectors or Receptor Gene Deficiency on Murine *T. gondii* Infection (*cont'd*)

Gene Deficiency	Mouse Background	Parasite Strain	Route of Infection	Effect on Parasite	Effect on Pathology	Effect on Survival	Immunological Effects	References
IFNγ	C57BL/6	Fukaya	Oral	Increased numbers (brain, liver, lung, gut and spleen)	Not reported	Reduced	Not examined	Norose *et al.* (2008)
IFNγ	BALB/c	Me49	i/p	Increased percent of infected peritoneal cells	Not reported	Reduced	Similar levels of serum IL−12	Scharton-Kersten *et al.* (1996a)
IFNγ	C57BL/6	Fukaya	Oral	Increased numbers (kidney)	No change to renal function	Not reported	None reported	Kudo *et al.* (2004)
IFNγ	BALB/c	Me49	Oral	None (brain)	No change to renal function	Not reported	None reported	Wang *et al.* (2004)
IFNγ	BALB/c	Me49	Oral	Not reported	Not reported	Not reported	Increased transcripts for CXCL10, CCL2, CCL3, CCL5 Reduced transcripts for CXCL10, CXCL9 and CCL5	Wen *et al.* (2010)
TNFα	C57BL/6	DX	Oral	Increased parasite numbers in brain	Necrosis in brain Extracellular parasites	Reduced	Reduced intracerebral splenic IFNγ production	Schluter *et al.* (2003)
LTα	C57BL/6	DX	Oral	Increased parasite numbers in brain	Necrosis in brain Extracellular parasites	Reduced	Reduced intracerebral splenic IFNγ production	Schluter *et al.* (2003)
TNF/LTα	C57BL/6	DX	Oral	No difference in parasite numbers in liver	Necrosis in brain Extracellular parasites	Reduced	Reduced intracerebral splenic IFNγ production	Schluter *et al.* (2003)
iNOS	C57BL/6	Me49	i/p	Increased parasite numbers in liver, lung, heart and spleen	Not reported	Reduced	None reported	Silva *et al.* (2002)

iNOS	C57BL/6	Me49	i/p	Not reported	Not reported	Reduced	Not reported	Collazo *et al.* (2002)
iNOS	C57BL/6	Me49	i/p	Increased parasite numbers in lung and brain	Not reported	Reduced	None reported	Silva *et al.* (2002)
iNOS	C57BL/6 x 129	76K	Oral	Increased parasites in brain and liver	Reduced intestinal and liver pathology	Increased (moderately but parasite dose dependent)	Reduced IL−10 and IFNγ mRNA transcripts in spleens Similar levels of serum IFNγ and TNFα	Khan *et al.* (1997)
iNOS	C57BL/6x129	Me49	i/p	No difference in percent infected peritoneal cells Increased brain cyst burdens	Increased necrosis in brain	Reduced	IFNγ, IL−12 neutrophil dependent, but iNOS independent mechanism responsible for early survival	Scharton-Kersten *et al.* (1997a)
IFN−γR	SvEv	Me49	i/p	Not reported	Not reported	Reduced	Not reported	Lieberman *et al.* 2004b
IFNα/β/R	SvEv	Me49	i/p	None	Not reported	None	Not reported	Lieberman *et al.* 2004b
IFNγR	129/Sv	DX	Oral	Increased parasite numbers in intestine, lymphatic tissue liver and spleen	Necrotizing hepatitis	Reduced	Reduced expression of MHC II on macropgages Decreased TNFα iNOS and IL−1β transcripts	Deckert-Schluter *et al.* (1996)

(*Continued*)

TABLE 25.2D The Effect of IFNγ, TNFα, LTα and Their Induced Effectors or Receptor Gene Deficiency on Murine *T. gondii* Infection (*cont'd*)

Gene Deficiency	Mouse Background	Parasite Strain	Route of Infection	Effect on Parasite	Effect on Pathology	Effect on Survival	Immunological Effects	References
IFNγR	129/Sv	DX	Oral	Increased parasite numbers in brain	Encephalitis present, but no comparison with WT control mice reported	Not reported	Failure to upregulate ICAM-1, MHC class I and II in brain endothelial cells Failure to upregulate LFA-I, ICAM-1, MHC class I and II in microglial cells Reduced TNFα expression in brain	Deckert-Schluter *et al.* (1999)
IFNγR	C57BL/6	CTGluc	i/p	Increased parasite burden	Not reported	Reduced	Brain has high levels of infiltrating leukocytes	Lindberg and Frenkel (2007)
IFNγR	C57BL/6	RH (luciferase expressing)	i/p	Increased parasite numbers by IVIS in peritoneum and brain	Increased in brain	Not reported	None reported	Dellacasa-Lindberg *et al.* (2007)
TNFRp55	C57BL/6	Me49	i/p	Increased parasite numbers in lung, heart, spleen and brain	None reported	Reduced	None reported	Silva *et al.* (2002)

TNFRp55	129/Sv x C57BL/6	DX	Oral	Increased parasite numbers in brain	Encephalitis present, but no comparison with WT control mice reported	Not reported	No difference in ability to upregulate ICAM-1, MHC class I and II in brain endothelial cells No difference in ability to upregulate LFA-I, ICAM-1, MHC class I and II in microglial cells Reduced TNFα expression in brain	Deckert-Schluter *et al.* (1999)
TNFRp55	C57BL/6	Me49	i/p	Increased parasite numbers in liver, lung, and spleen	Not reported	Reduced	None reported	Silva *et al.* (2002)
TNFRp55	129Sv x C57BL/6	DX	i/p	Increased parasites in brain and lung	Increased severity of brain pathology including encephalitis and necrosis Pneumonia	Reduced	Reduced levels of mRNA transcripts for iNOS in brain	Deckert-Schluter *et al.* (1998a)
TNFRp75	129Sv x C57BL/6	DX	i/p	No difference	Not reported	No difference	Similar levels of mRNA transcripts for iNOS in brain	Deckert-Schluter *et al.* (1998a)

(*Continued*)

TABLE 25.2D The Effect of IFNγ, TNFα, LTα and Their Induced Effectors or Receptor Gene Deficiency on Murine *T. gondii* Infection (*cont'd*)

Gene Deficiency	Mouse Background	Parasite Strain	Route of Infection	Effect on Parasite	Effect on Pathology	Effect on Survival	Immunological Effects	References
TNFRp55/Rp75	129Sv x C57BL/6	DX	i/p	Increased parasites in brain and lung	Increased severity of brain pathology including encephalitis and necrosis Pneumonia	Reduced	Reduced levels of mRNA transcripts for iNOS in brain	Deckert-Schluter *et al.* (1998a)
TNFRp55/Rp75	C57BL/6 x 129	Me49	i/p	Increased cyst numbers	Increased severity of TE including necrosis	Reduced	iNOS and IFNγ induction in brain and peritoneal cavity unimpaired	Yap *et al.* (1998a)
TNFRp55/Rp75	129/Sv x C57BL/6	DX	Oral	Not reported	Encephalitis present, but no comparison with WT control mice reported	Not reported	No difference in ability to upregulate ICAM-1, MHC class I and II in brain endothelial cells No difference in ability to upregulate LFA-I, ICAM-1, MHC class I and II in microglial cells Reduced TNFα expression in brain	Deckert-Schluter *et al.* (1999)
Irga6	C57BL/6 129/SvJ	Me49	Oral or i/p	Increased brain cyst numbers	Increased inflammation in brain	Reduced	Reduced IFNγ and TNFα in brain	Liesenfeld *et al.* (2011)
IDO	C57BL/6	Fukaya	i/p	Reduced tachyzoites levels in lungs	Not reported	None	Reduced IL–6 and IFNγ in lungs	Murakami *et al.* (2012)

TABLE 25.2E The Effect of CD Gene Deficiency on Murine *T. gondii* Infection

Gene Deficiency	Mouse Background	Parasite Strain	Route of Infection	Effect on Parasite	Effect on Pathology	Effect on Survival	Immunological Effects	References
CD1d	BALB/c	Me49	Oral	Increased numbers	Not reported	None	None reported	Smiley *et al.* (2005)
CD1d	C57BL/6	Me49	Oral	Increased numbers	Increased intestinal pathology Increased weight loss	Reduced	Increased serum IFNγ	Smiley *et al.* (2005)
CD2	C57BL/6	Me49	Oral	Reduced parasite numbers in the intestine	Reduced intestinal pathology	Increased	Reduced IFNγ and IL–6 production by MLN cells and increased IFNγ production by splenocytes	Pawlowski *et al.* (2007)
CD4	C57BL/6	RH	Intraocular	Increased numbers of tachyzoites	Decreased necrosis and inflammation eye	Not reported	No increase in IFNγ or TNFα as seen in WT	Lu *et al.* (2004)
CD4	C57BL/6	76K	Oral	Increased cysts (brain)	Reduced pathology in liver and gut Increased pathology in brain	Increased	Reduced IFNγ expression in liver, lung, gut and spleen	Casciotti *et al.* (2002)
CD4	C57BL/6	Me49	Oral	Increased cysts (brain)	Not reported	Reduced	Reduced splenocyte production of IFNγ	Johnson and Sayles (2002)
CD8	C57BL/6	RH	Intraocular	Increased numbers of tachyzoites in the eye	Increased necrosis and inflammation in the eye	Not reported	Similar levels of IFNγ or TNFα as seen in WT	Lu *et al.* (2004)
CD14	C57BL/6	Me49	i/p	No difference in cyst numbers in brain	Not reported	None	Similar levels of splenocyte production of IFNγ and IL–12	Debierre-Grockiego et al., 2007a

(Continued)

TABLE 25.2E The Effect of CD Gene Deficiency on Murine *T. gondii* Infection (*cont'd*)

Gene Deficiency	Mouse Background	Parasite Strain	Route of Infection	Effect on Parasite	Effect on Pathology	Effect on Survival	Immunological Effects	References
CD28	C57BL/6	Me49	i/p	No difference in brain cyst burdens	Reduced severity of encephalitis	Increased	Reduced IFNγ levels Reduced $CD4^+$ T-cells in brain	Reichmann *et al.* (2000)
CD28	C57BL/6	Me49	i/p	Increased tachyzoites in peritoneal exudate	No difference in brain	None	Chronically infected mice have defective memory response and are susceptible to challenge with the RH strain of *T. gondii*	Villegas *et al.* (2002)
CD40	C57BL/6	76K	Oral	Increased parasite numbers in spleen No difference in intestine	Reduced intestinal pathology	Increased	Reduced transcripts for IL−1β, IL−18 IL−6 and IFNγ, MIP1β, MIP2, MCP3, MCP1, IP10 and RANTES in ileum Reduced serum IL−12 (p70)	Li, *et al.* (2002)
CD40	C57BL/6	Me49	Not reported	Increased parasite burden in eyes	Disruption of retinal architecture	Reduced	Increased IFNγ, TNFα and iNOS	Portillo *et al.* (2010)
CD73	C57BL/6	Me49	Oral	Reduced cysts in brain	Similar brain inflammation	Increased	Similar IFNγ and iNOS transcript levels	Mahamed *et al.* (2012)
CD154	C57BL/6	76K	Oral	Increased parasite numbers in spleen No difference in intestine	Reduced intestinal pathology	Increased	Reduced transcripts for IL−1β, IL−18 IL−6 and IFNγ, MIP1β, MIP2, MCP3, MCP1, IP10 and RANTES in ileum Reduced serum IL−12 (p70)	Li *et al.* (2002)
CD200	C57BL/6	DX	i/p	Reduced parasite numbers in brain	Not reported	Increased	Increased microglial activation and brain TNFα and iNOS levels	Deckert *et al.* (2006)

TABLE 25.2F The Effect of Chemokine Receptor Gene Deficiency on Murine *T. gondii* Infection

Gene Deficiency	Mouse Background	Parasite Strain	Route of Infection	Effect on Parasite	Effect on Pathology	Effect on Survival	Immunological Effects	References
CCR1	C57BL/6	76K	Oral	Not reported	Increased necrotic lesions in liver Reduced necrotic lesions in intestine	Reduced	Reduced PMNs influx into blood and liver	Khan *et al.* (2001)
CCR2	C57BL/6J	PTG	i/p		Not examined	Reduced	Reduced recruitment of Gr-1$^+$ monocytes	Robben *et al.* (2005)
CCR2	C57BL/6	Pru (luciferase expressing)	Oral	Increased parasite burden	Ileitis and inflammation of the spleen	Reduced	Deficient in inflammatory monocyte recruitment to the site of infection	Dunay et al., (2010)
CCR2	C57BL/6	Me49	Oral	Not reported	Resistant to damage to the intestinal mucosa	Reduced	Recruitment of CD130$^+$ T-cells into mucosal sites	Egan *et al.* (2009)
CCR2	C57BL/6	Me49	i/p	Increased parasite burden in the ileum	Extensive imflammation of ileum	Reduced	Similar to slightly higher levels of IL−12 and IFNγ Deficient in inflammatory monocyte recruitment to the site of infection	Dunay *et al.* (2008)
CCR5	B6129F2/J	Not specified	Oral	Increased parasite numbers in intestine	Increased inflammation in intestine	Not reported	CD8$^+$ lymphocytes have reduced ability to migrate *in vitro*	Luangsay *et al.* (2003)

(*Continued*)

TABLE 25.2F The Effect of Chemokine Receptor Gene Deficiency on Murine *T. gondii* Infection (*cont'd*)

Gene Deficiency	Mouse Background	Parasite Strain	Route of Infection	Effect on Parasite	Effect on Pathology	Effect on Survival	Immunological Effects	References
CCR5	C57BL/6 C57BL/6x129	76K 76K	Oral Oral	Increased parasites in liver, similar parasite numbers in brain spleen, lung and gut Increased parasites in liver, spleen, lung and gut, similar parasite numbers in the brain	Not quantified Not quantified	Increased Reduced	Reduced NK-cell infiltration into tissues Reduced NK-cell infiltration into tissues	Khan *et al.* (2006)
CCR7	C57BL/6	Pru	i/p	Increased parasite burden	Increased inflammation and necrosis	Reduced	Defective recruitment of effector cells Reduced proportion of inflammatory monocytes ant site of infection	Noor *et al.* (2010)
CXCR2	BALB/c	Me49	i/p	Increased tachyzoites in peritoneal exudate	Not reported	Not reported	Reduced serum IFNγ Defective neutorphil recruitment	Del Rio *et al.* (2001a)
CXCR2	BALB/c	RH	i/p	Increased tachyzoites in peritoneal exudates cells	Not reported	Not reported	Defective neutorphil recruitment	Del Rio *et al.* (2001a)

TABLE 25.2G Antigen Processing Components on Murine *T. gondii* Infection

Gene Deficiency	Mouse Background	Parasite Strain	Route of Infection	Effect on Parasite	Effect on Pathology	Effect on Survival	Immunological Effects	References
LMP2	C57BL/6	PLK	Not reported	Not reported	Not reported	Reduced	Impaired development of functional $CD8^+$ T-cells	Tu *et al.* (2009)
LMP7	C57BL/6	PLK	Not reported	Not reported	Not reported	None	Not reported	Tu *et al.* (2009)
TAP1	C57BL/6	Me49	i/p	Increased parasite burden in lung, brain and heart	Not reported	Reduced	Reduced IFNγ producing $CD4^+$ cells at infection site and spleen Similar levels of IL–12 Defective NK-cells response	Goldszmid *et al.* (2007)
ERAAP	C57BL/6	Pru-GFP	i/p	Increased parasite burden in brain, spleen and liver	Not reported	Reduced	Not reported	Blanchard *et al.* (2008)

TABLE 25.2H The Effect of Various Gene Deficiencies on Murine *T. gondii* Infection

Gene Deficiency	Mouse Background	Parasite Strain	Route of Infection	Effect on Parasite	Effect on Pathology	Effect on Survival	Immunological Effects	References
γ(c)	C57BL/6	Me49	i/p	Increased cysts in brain	Not reported	Reduced	Similar IFNγ and IL−12 production measured in serum and produced by spleen cells and peritoneal cells Reduced NK-cell activity to YAC-1 cells	Scharton-Kersten *et al.* (1998)
TCR−Vγ1	C57BL/6	RH	Oral	None	No difference in ileitis and liver pathology	None reported	Increase macrophage and granulocyte infiltration of liver Increased serum TNFα and IFNγ	Egan *et al.* (2005)
NKT (Jα28)	C57BL/6	Me49	Oral	Not reported	Increased weight loss	Reduced	None reported	Smiley *et al.* (2005)
NKT (Jα28)	C57BL/6	76K	Oral	Increased parasite numbers in intestine	Reduced severity intestinal pathology	Reduced	Reduced IFNγ production in the intestine Increased number of FoxP3 lymphocytes	Ronet *et al.* (2005)
MCP	C57BL/6J	PTG	i/p	None	Increased neuropathology	Reduced	Reduced recruitment of Gr−1+ monocytes	Robben *et al.* (2005)
MCP1	C57BL/6	Me49	i/p	Increased parasite burden in the ileum	Extensive inflammation of ileum	Reduced	Similar to slightly higher levels of IL-12 and IFNγ Deficient in inflammatory monocyte recruitment to the site of infection	Dunay *et al.* (2008)

ICE	C57BL/6	PTGluc	i/p		None	None	None examined	Hitziger *et al.* (2005)
Perforin		Me49	Oral	Not reported	None (brain)	None	Increased splenocyte IFNγ production	Wang *et al.* (2004)
Fibrinogen	C57BL/6	Me49	Oral	Not reported	Haemorrhagic foci in liver	Reduced	No difference in IFNγ, iNOS, IL−10 or TNFα transcripts or IFNγ protein levels or nitric oxide levels in liver	Johnson *et al.* (2003)
Fibrinogen-like protein 2 (Fgl2, fibroleukin)	C57BL/6	Me49	Oral	None in peritoneal exudate	None	None	None	Hancock *et al.* (2004)
Ucp2	C57BL/6	Me49	i/p	Reduced parasites in brain	None reported	Enhanced	Not reported	Arsenijevic *et al.* (2000)
SLAM-associated protein (SAP)	C57BL/6 or 129Sv/Ev/Tac	Me49	i/p	Not reported	Not reported	None	Increased IFNγ production	Czar *et al.* (2001)
A1α	C57BL/6/ 129Sv/SLJ	RH	i/p	Increased SAG2 expression in lung and brain	Not reported	Enhanced	Reduced peritonitis	Orlofsky *et al.* (2002)
β2m	C57BL/ 6 x129	RH	Subcutaneous	Increased numbers of tachyzoites	Not determined	None	Expansion of spleen NK1.1	Denkers *et al.* (1993)
B-cell	C57BL/6	Me49	Oral	Increased numbers of cysts and tachyzoites in brain	Necrosis in brain of muMT mice only	Reduced	Similar levels of IFNγ, IL−10 and iNOS, but increased TNFα mRNA transcripts in brains	Kang *et al.* (2000)

(*Continued*)

TABLE 25.2H The Effect of Various Gene Deficiencies on Murine *T. gondii* Infection (*cont'd*)

Gene Deficiency	Mouse Background	Parasite Strain	Route of Infection	Effect on Parasite	Effect on Pathology	Effect on Survival	Immunological Effects	References
B-cell	C57BL/6	RH	Intraocular	Increased dissemination of parasites in multiple ocular tissues	Increased necrosis and inflammation in the eye	Not reported	No increases in serum Ig levels	Lu *et al.* (2004)
TFF2	C57BL/6	Me49	Oral	Reduced intestinal parasite numbers	Reduced intestinal pathology	Enhanced	Increased IL-10 transcripts in jejunum	McBerry *et al.* (2012)
PILRβ	C57BL/6	Me49	i/p Oral	Reduced cyst burdens in brain	Reduced inflammation in the brain Reduced intestinal pathology	Enhanced Enhanced	Increased splenocyte production of IFN γ, TNFα and IL-12 (i.p. infection) Increased systemic IL-27p28, but decreased IFNγ (i.p. infection)	Tato *et al.* (2012)
MIF	BALB/c	Me49	Oral	Increased cyst numbers in brain and liver	Not reported	Increased	Reduced IL-12 production and reduced maturation of MLNDCs	Terrazas *et al.* (2010)
MIF	C57BL/6	Me49	Oral	Increased intestinal parasite burdens	Reduced ileal inflammation and liver pathology	Increased	Reduced TNF1, IL-12, IFNγ and IL-23 and increased IL-22 in ileal mucosa Increased IL-4, IL10 and TGFβ at ileal mucosa	Cavalcanti *et al.* (2011)

Mmp	C57Bl/6	76K	Oral	Increased parasite burden in small intestine	Not reported	Increased	Not reported	Foureau *et al.* (2010)
12/15-LOX	Alox15 mice (C57BL/6)	Me49	i/p	Increased parasite burden in chronic not acute infection	Not reported	Reduced	IL−12 production by macrophages not dendritic cells Regulates Th1 cytokine production during chronic but not acute infection	Middleton *et al.*, 2009
ATG5 (macrophages, monocytes and neutrophils)	Not reported	Pru-FLUC	Not reported	Increased parasite burden in spleen and MLN	Not reported	Reduced	IFNγ at similar levels in serum	Zhao *et al.* (2008)
CAT2	C57BL/6	Me49	i/p	Not reported	Not reported	Reduced	Th1 immunity compromised	Thompson *et al.* (2008)
LDLr	C57BL/6	Me49	i/p	Reduced cyst burden in brain and CNS	Artherosclerotic lesions reduced in aorta	Not reported	Not reported	Portugal *et al.* (2008)

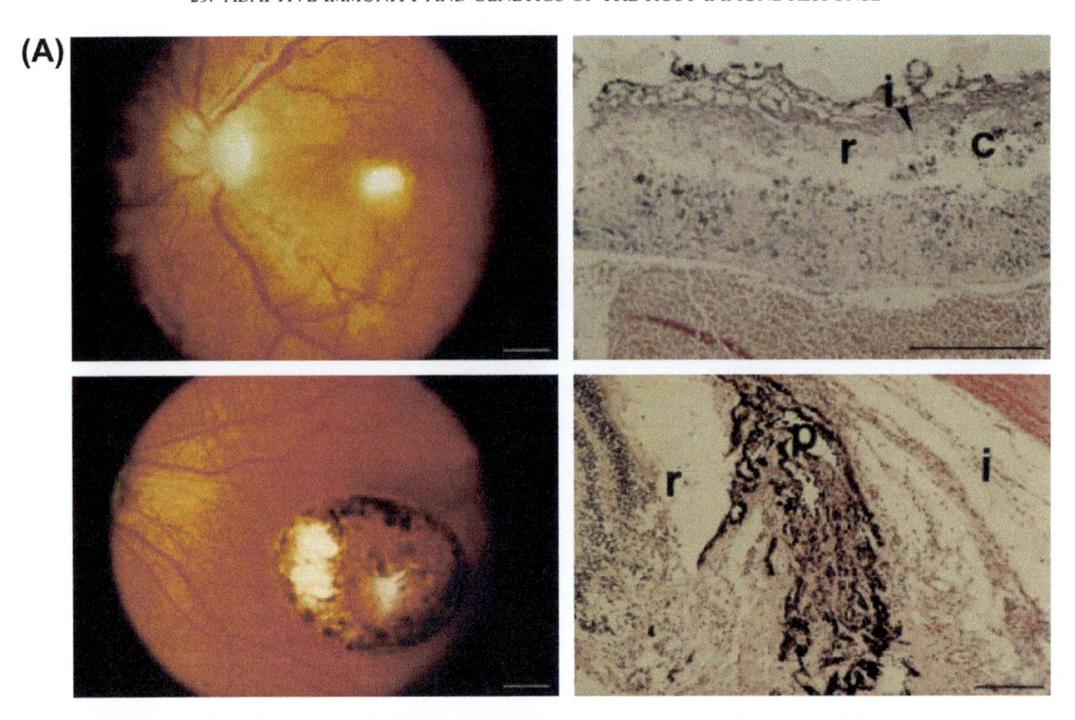

FIGURE 25.7 **(A) Toxoplasmic retinochoroiditis in immunocompetent patients.** *Upper left.* Active toxoplasmic retinochoroiditis in a patient with acute acquired infection (provided by Dr Jack S. Remington, Stanford University). Scale bar is 1.5 μm. *Upper right.* Acute lesion in a five day old infant born prematurely. There is complete necrosis in all layers of the retina (r) with numerous inflammatory cells (i) and focal calcification (c). Scale bar is 100 μm. *Lower left.* Quiescent toxoplasmic retinochoroidal scar in a congenitally infected patient. Scale bar is 1.5 μm. *Lower right.* Retinochoroidal scar in a two year old child. The hyperpigmentation at the edge of a retinochoroidal scar seen with ophthalmoscopy as in *lower left* is the result of disruption and proliferation of the retinal pigment epithelium (p). The retina adjacent to this also shows disruption of the normal architecture (r). Scattered chronic inflammatory cells persist in the lesion (i). Scale bar is 100 μm. *(Adapted from Roberts F. and McLeod R. (1999). Pathogenesis of Toxoplasmic Retinochoroiditis. Parasitol. Today. 15(2), 51-57, with permission.)*
B) Toxoplasmic retinochoroiditis in immunocompromised patients. *Upper left.* Fundus photograph with red-free light showing irregular margins and presence of multiple satellite lesions. *Upper right.* Multiple cysts of *T. gondii* in necrotic retina (haematoxylin–eosin, ×440). *Upper centre.* Mononuclear inflammation within the optic nerve, with several *T. gondii* cysts on the left. *Lower centre.* Two young cysts (YC) within a host retinal cell. The cell on the left has many microtubules (mt) and Nissl body-like structures (open arrows), suggesting a neuroretinal cell. *(Adapted with permission from Yeo J.H., Jakobiec F.A., Iwamoto T., Richard G., Kreissig I. (1983). Opportunistic toxoplasmic retinochoroiditis following chemotherapy for systemic lymphoma. A light and electron microscopic study. Ophthalmol. 90(8), 885-98.)* *Lower left.* Fundus appearance of right eye in an AIDS patient with bilateral miliary toxoplasmic retinitis, showing multiple small, round, white, inflammatory retinal lesions, as well as haemorrhage and oedema in the macula. *(Adapted with permission from Berger B.B., Egwuagu C.E., Freeman W.R., Wiley C.A. (1993). Miliary toxoplasmic retinitis in acquired immunodeficiency syndrome. Arch. Ophthalmol. 111(3), 373-6.)* *Lower right.* Fundus appearance of left eye from an 8 year old child treated with systemic corticosteroids, showing active retinal inflammatory lesion adjacent to an old retinochoroidal scar along the inferonasal vascular arcade. *(Adapted with permission from Morhun P.J., Weisz J.M., Elias S.J., Holland G.N. (1996). Recurrent ocular toxoplasmosis in patients treated with systemic corticosteroids. Retina. 16(5), 383-7.)* *Bottom.* Histologic section from an adult receiving long-term corticosteroid therapy, showing focal zones of inner retinal necrosis adjacent to the vessels. Arrows showing cysts and released organisms lie at the interface of intact and necrotic retina. V, vitreous cavity; L, blood vessel lumen (haematoxylin–eosin, ×700). *(Adapted from Nicholson D.H., Wolchok E.B. (1976). Ocular toxoplasmosis in an adult receiving long-term corticosteroid therapy. Arch. Ophthalmol. 94(2), 248-54, with permission.)*

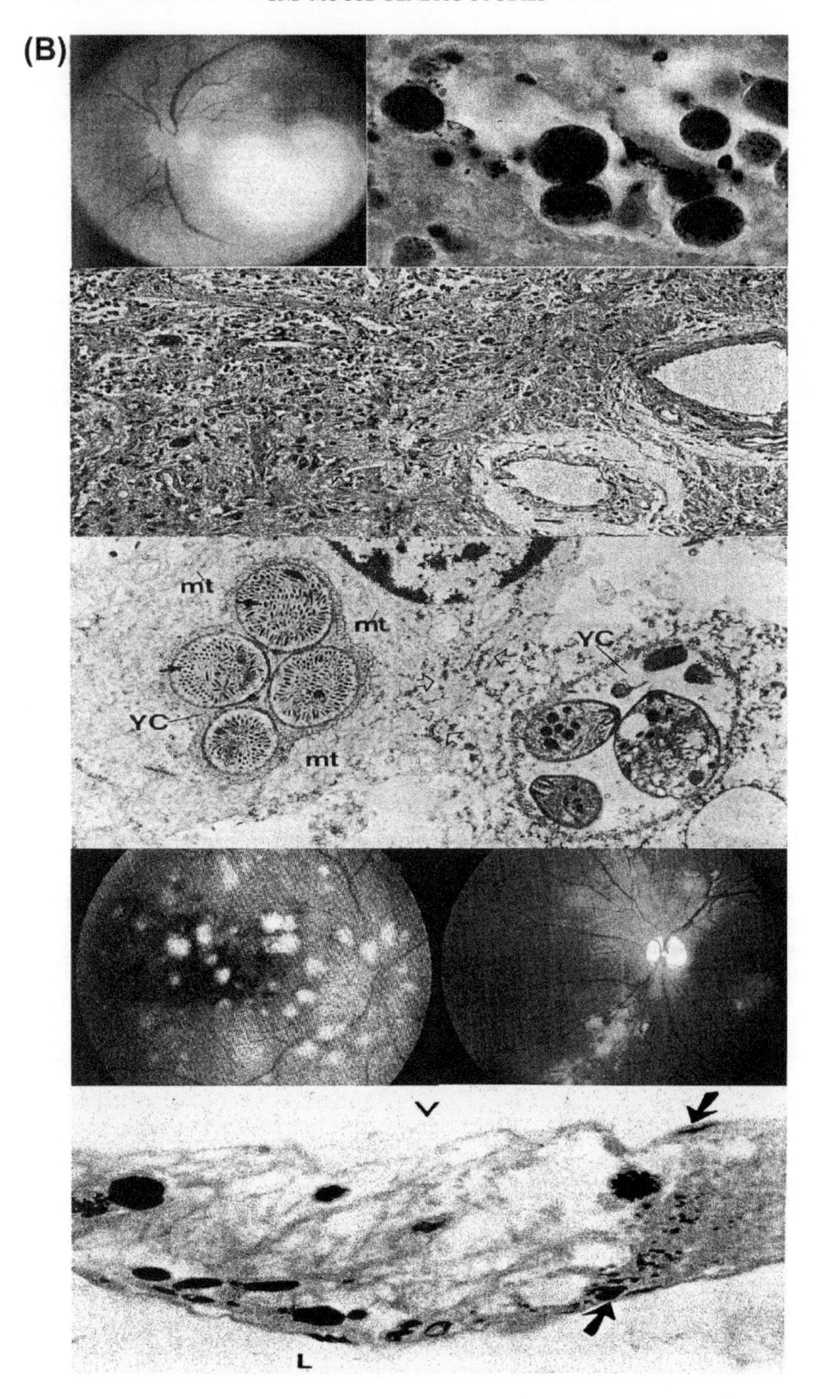

FIGURE 25.7 (*continued*).

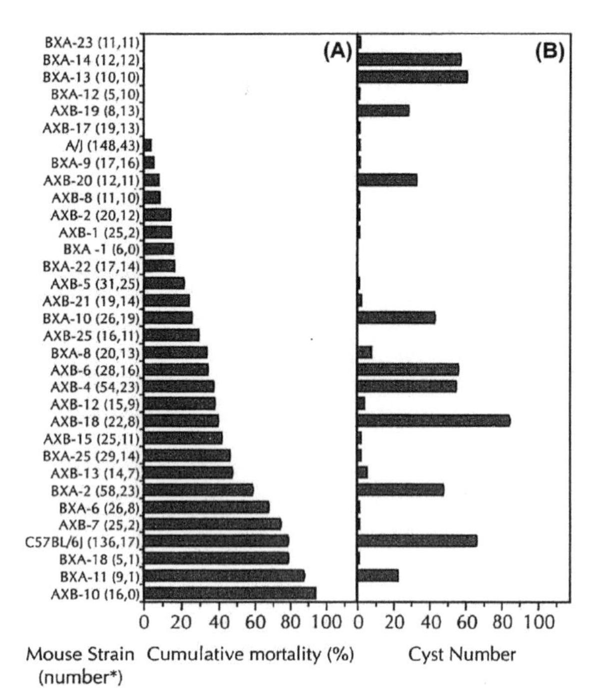

FIGURE 25.8A, B **Survival and brain cyst burden in AXB/BXA recombinant inbred strains of mice.** Cumulative mortality
(A) and cyst number
(B) 30 days after the peroral infection of AXB and BXA recombinant inbred strains of mice.
*The first number within parentheses represents the number of mice studied for mortality, the second number within parentheses represents the number of mice studied for brain cyst number. In this study, genetics of two traits, survival and brain cyst number after per-oral *T. gondii* infection, were studied by using recombinant inbred strains of mice derived from resistant A/J (indicated by the letter A) and susceptible C57BL/6J (indicated by the letter B) progenitors, F1 progeny of crosses between A/J and C57BL/6J mice, and congenic mice (B10 background). The continuous variation in the percentage survival indicated that control of this trait involved multiple genes. Analysis of strain distribution pattern of survival AXB/BXA recombinant mice indicated that survival is regulated by a minimum of five genes. One of these genes appears to be linked to the H-2 complex, another is related to an as yet unmapped gene controlling resistance to *Ectromelia* virus and another in the *Wnt1* locus. The large versus low magnitude phenotypes indicated that cyst formation is regulated by one or only a few genes. Associations of defined traits with resistance or susceptibility to *Toxoplasma* cyst formation were also analysed. Cyst number is regulated by a locus on chromosome 17 within 0–4 cM of the H-2 complex ($p = 0.001$). Mice with the H-2a haplotype are resistant and those with the H-2b haplotype are susceptible. This analysis also indicated that the *Bcg* (*Nramp*) locus on chromosome 1 may effect cyst number (map distance is 12 cM, $p = 0.05$). Resistance to cyst formation is a dominant trait. *(Adapted from McLeod* et al. *(1989b), with permission.)*

adult acquisition of infection by mice leads to distinct patterns, much like disease patterns in humans (Deckert-Schluter *et al.*, 1994). Specifically, in NMRI mice following prenatal infection with low dose DK strain *T. gondii*, newborn mice had foci of necrosis, intracerebral calcifications and ventriculitis, resembling human congenital toxoplasmosis. Inflammatory responses included macrophages, granulocytes and astrocytes (Deckert-Schluter, *et al.*, 1994). In chronic toxoplasmic encephalitis recruitment of T-cells stopped and apoptosis of $CD4^+$ and $CD8^+$ T-cells occurred; nonetheless, intracerebral T-cells that had already been recruited persisted (Schluter *et al.*, 1997; Deckert-Schluter *et al.*, 1999). Suzuki has noted that the immune responses to different strains of *T. gondii* in the brains of mice are not always the same during acute and chronic infection (Suzuki *et al.*, 1993b) and different cell types may participate in earlier and later immune responses. Suzuki identified Vβ8 $CD8^+$ L^d restricted T-cells that are *T. gondii* antigen specific in the brains of chronically infected, genetically resistant BALB/c mice and these cells can transfer protection to naive mice (Suzuki *et al.*, 2005).

In a study of chemokines in C57BL/6 (susceptible to toxoplasmic encephalitis) adult mice, CR G-2/IP10, MuMIG, RANTES, MCP-1, MIP1α and MIP1β reached higher maximum levels earlier, when compared with BALB/c (resistant to toxoplasmic encephalitis) mice. In both murine models astrocytes and microglia produced CRG-2/IP10 and MCP or RANTES and MuMiG respectively and leukocytes transcribed CRG-2/IP10, MCP-1 and RANTES (Strack *et al.*, 2002). Genetic factors exerted a strong impact on intracerebral chemokines. In a separate study, chemokines were differentially expressed by

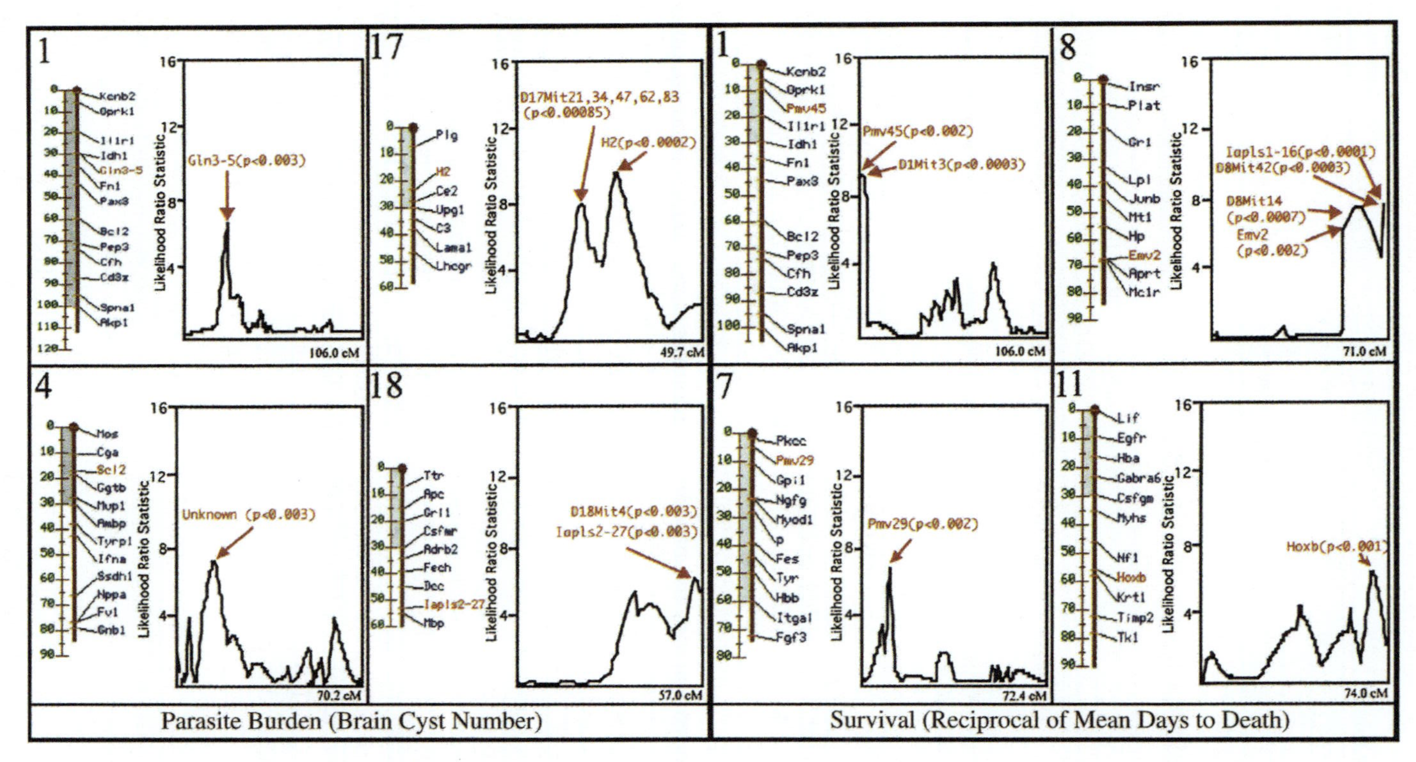

FIGURE 25.9 **Qualitative trait analysis (likelihood analysis) of genes influencing parasite burden and survival of AXB/BXA mice per-orally infected with *T. gondii* (Me49 strain).** Chromosome number (upper left). Peaks with accompanying red line and typography indicate loci with high association with outcome (i.e. survival) following per-oral infection. Significance of associations indicated with red typography is $p < 0.003$. Candidate loci for susceptibility or resistance and their chromosome locations are indicated in Table 25.1 along with possible human orthologues. *(Adapted from Johnson* et al., *(2002), with permission.)*

TABLE 25.3 Gene Regions Associated with Resistance to *Toxoplasma gondii*

T. gondii Parasite Burden	Survival after *T. gondii* Infection
1: *Slcllal* @ 39.2 cM [*SLCl IAl* @ 2q35][a]	**1**: *ILIO* @ 69.9 cM [*ILIO* @ 1q31−32] and Tgfβ2 @ 101.5 cM [*TGFβ2* @ lq41][b]
4: *Sc12* @ 17.2 cM [*JAK* @ 9p−9p]	**7**: *NA* [*NA*]
17: *H-2-Ld, la* @ 23.0 cM [*HLA* @ 6p21.3]	**8**: *NA* [*NA*]
18: *NA* [*NA*]	**11**: *Sc11 Syntenic* [IL4, IRFIlL3, CSF2, IL5, *IL9* @ 5q23−32] and [*NOS2, SCYA1−5* @ 17q12]

[a] *Chromosome number is shown in bold with candidate genes involved in resistance to other pathogens at approximately the same locations. Distance in centimorgans from the centromere is indicated. Shown inside brackets are human orthologues of these candidate genes with their locations indicated by cyto-band.*
[b] *There are other associations of genes at these approximate loci (see Fig. 25.6), but these are less significant,* $P > 0.001$.
Adapted with permission from Johnson et al. *(2002).*

astrocytes (GRG-2/IP10, MCP-1), microglia (RANTES) and inflammatory leucocytes, and were critically regulated by IFNγ. IFNγ-deficient mice did not produce CRG-2/IP10, MuMiG, RANTES and expressed reduced amounts of MIP1α, MIP1β and MCP-1 mRNA diminishing the recruitment of leukocytes across the blood−brain barrier. These investigators stated that 'T Cells are the single source of IFNγ gamma in toxoplasmic encephalitis and thus lead to parasites in brain parenchyma' (Strack *et al.*, 2002). Using per-oral infection of B6C (H-2 (BxD)) mice with *T. gondii* expressing β-galactosidase and monitoring CD8 T-cells with MHC class I tetramer staining, Kwok *et al.* (Kwok *et al.*, 2003) found that in primary infection only tachyzoites induced $CD8^+$ T-cells, in secondary $CD8^+$ T-cells, but after that time, numbers of these cells diminished in spleens, but T-cells remained at high levels in brain. These cells produced IFNγ and were cytolytic.

Inactivation of the VCAM gene in VCAM (flox/flox Mx Cre) mice resulted in lack of induction of VCAM-1 on cerebral blood vessel endothelial cells (but not choroid plexus epithelial cells or ependyma) and resistance to *T. gondii* was abolished in conjunction with diminished B-cell response, *T. gondii* specific intracerebral T-cells and microglial activation; however, leukocytes continued to home across cerebral blood vessels (Deckert *et al.*, 2003). LTα and TNFα were essential for the control of intracerebral toxoplasmosis, including that caused by the *T. gondii* temperature sensitive 4 (ts4) mutant (Schluter *et al.*, 2003). Antioxidant systems in *T. gondii* and cytoplasmic catalase in protection against oxidative injury were also studied (Ding *et al.*, 2004). Using CD45 congenic and chimera mice, there was microglial cytokine production (TNFα, ILIβ, IL-10, IL-15 in normal brain; IL-12 p 40, iNOS, increased IL-1β and TNFα, continuous IL-10, IL-15 and induction of MHC class I and II molecules, and ICAM, and LFA-1). Depletion of $CD4^+$ and $CD8^+$ T-cells showed that cytokine expression was regulated by $CD8^+$ T-cells and expression of cell surface molecules was less dependent on T-cells. T-lymphocytes regulated microglia in brain (Schluter *et al.*, 2001). In a separate study, interleukin 10 downregulated intracerebral immune response in chronic *T. gondii* encephalitis and may contribute to parasite persistence in brain (Deckert-Schluter *et al.*, 1995, 1999; Schluter *et al.*, 1991, 1997). In another study, IFNγ receptor mediated signalling, but not TNF receptor type I or 2 mediated signalling, was crucial for the activation of cerebral blood vessel endothelial cells and microglia in murine *Toxoplasma* encephalitis. This was demonstrated in mice without IFNγ receptors in which expression of ICAM, LFA-1 and MHC class I and II antigens but not ILIα, IL-10, IL-12, p 40 or IL-15 was independent of IFNγ receptor signalling (Deckert-Schluter *et al.*, 1996).

Signalling in certain strains of mice through TNFR1, but not TNFR2, induces protective nitric oxide (Deckert-Schluter *et al.*, 1998). Interestingly, the role of iNOS and NO in containing toxoplasmic encephalitis appears to depend on host genetics. In mice with the C57BL6 background, NOS and NO are essential for protection, but this is not the case in resistant BALB/c mice, 30 days after infection (Schluter *et al.*, 1999). NO also plays an immunoregulatory role by inhibiting the proliferation of spleen cells early in infection (Hayashi *et al.*, 1996b). The role of iNOS in some neurodegenerative diseases (Kroncke *et al.*, 1998) suggests that chronic *T. gondii* infection in the brain might also be controlled in an iNOS/NO-independent manner for at least some humans, similar to the BALB/c mouse.

There is expression of certain cell-surface molecules induced by *T. gondii* by both the central nervous system and immune cells. There are novel molecules expressed by neural and haematopoietic lineage cells (Deckert-Schluter *et al.*, 1998): specifically, heat stable antigen (HSA, CD24, nectadrin) and GL7 are haematolymphoid differentiation antigens that play a role in antigen presentation, cell adhesion, signal transduction and activation in normal brain ependymal cells, choroid plexus macrophages and some blood vessel endothelial cells. HSA can be detected in brain parenchymal cells by immunohistochemistry and GL7 can be detected in choroid plexus epithelium. Toxoplasmic encephalitis did not modify this GL7 expression, but in acute and chronic toxoplasmic encephalitis HSA and GL7 were strongly induced in resident brain cells and activated astrocytes were the predominant HSA^{+} and $GL7^{+}$ cell type. HSA^{+} microglia were present, but were a small fraction of the total microglia and increased only a limited amount in toxoplasmic encephalitis. HSA and GL7 may have anatomically and functionally diverse immunological and non-immunological roles (Blackwell *et al.*, 1993, 1994; Blackwell 1998; Brown and McLeod 1990; Brown *et al.*, 1995; Johnson, 1984; McLeod *et al.*, 1989a, 1989b, 1995; 1996; Mack *et al.*, in preparation; Morley *et al.*, 2005; Montoya *et al.*, 1997; Mordue *et al.*, 2001; Ottenhoff *et al.*, 2002; Trowsdale and Parham, 2004; Vilches and Parham, 2002). Two very important recent observations are those of the Suzuki and Deckert-Schluter groups. Suzuki *et al.* (2011) and Handel *et al.* (2012) recently noted that perforin, but not IFNγ, was needed to protect against toxoplasmic encephalitis. Handel *et al.* (2012) noted that modulation of gp130 (part of the IL-6 receptor) modulated presence of toxoplasmic encephalitis in mice.

25.3 STUDIES OF LEWIS AND FISCHER RATS

Genetic mapping studies (Cavailles *et al.*, 2006a, 2006b) with immunologic correlates have extended an initial fascinating observation that the Lewis rat strain shows remarkable resistance to infection with *T. gondii* (Gross *et al.*, 1993). This resistance is probably due to an initial innate gastrointestinal or other very early immune response in the intestine before the parasite can even disseminate (Fig. 25.4). This refractoriness to infection is intrinsic to bone marrow derived cells and is dependent on IFNγ. Using a genome-wide search with F2 progeny of susceptible BN and resistant Lewis rats, resistance was found to be controlled by a single locus on chromosome 10 designated *'Toxo'*. This effect was found to be independent of the background genes. Using rats from congenic sublines characterized by genomic recombination with *Toxo1* the interval has been reduced to a 1.8 cM region homologous to human 17p13. *In vitro* functional studies demonstrated

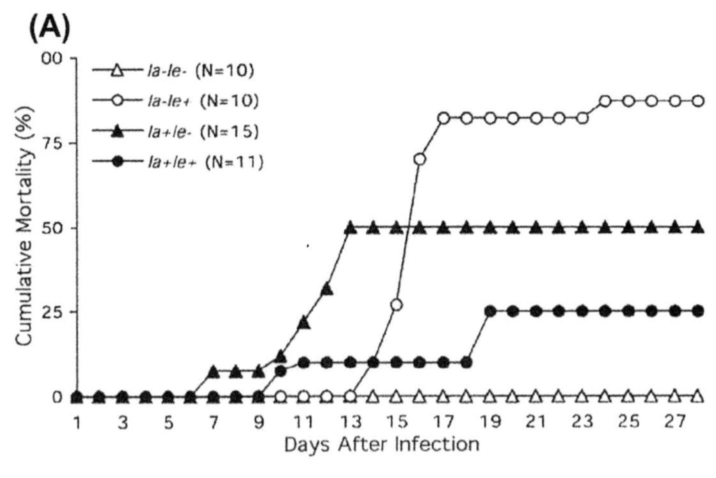
(A)
Ia-Ie- (N=10)
Ia-Ie+ (N=10)
Ia+Ie- (N=15)
Ia+Ie+ (N=11)
Cumulative Mortality (%)
Days After Infection

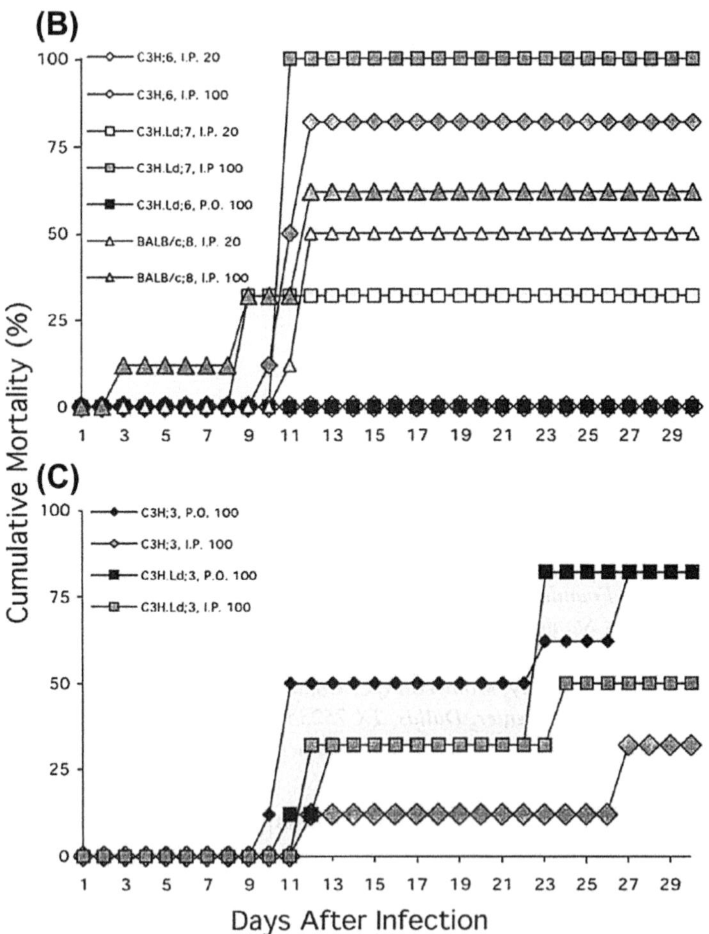
(B)
C3H;6, I.P. 20
C3H,6, I.P. 100
C3H.Ld;7, I.P. 20
C3H.Ld;7, I.P 100
C3H.Ld;6, P.O. 100
BALB/c;8, I.P. 20
BALB/c;8, I.P. 100
(C)
C3H;3, P.O. 100
C3H;3, I.P. 100
C3H.Ld;3, P.O. 100
C3H.Ld;3, I.P. 100
Cumulative Mortality (%)
Days After Infection

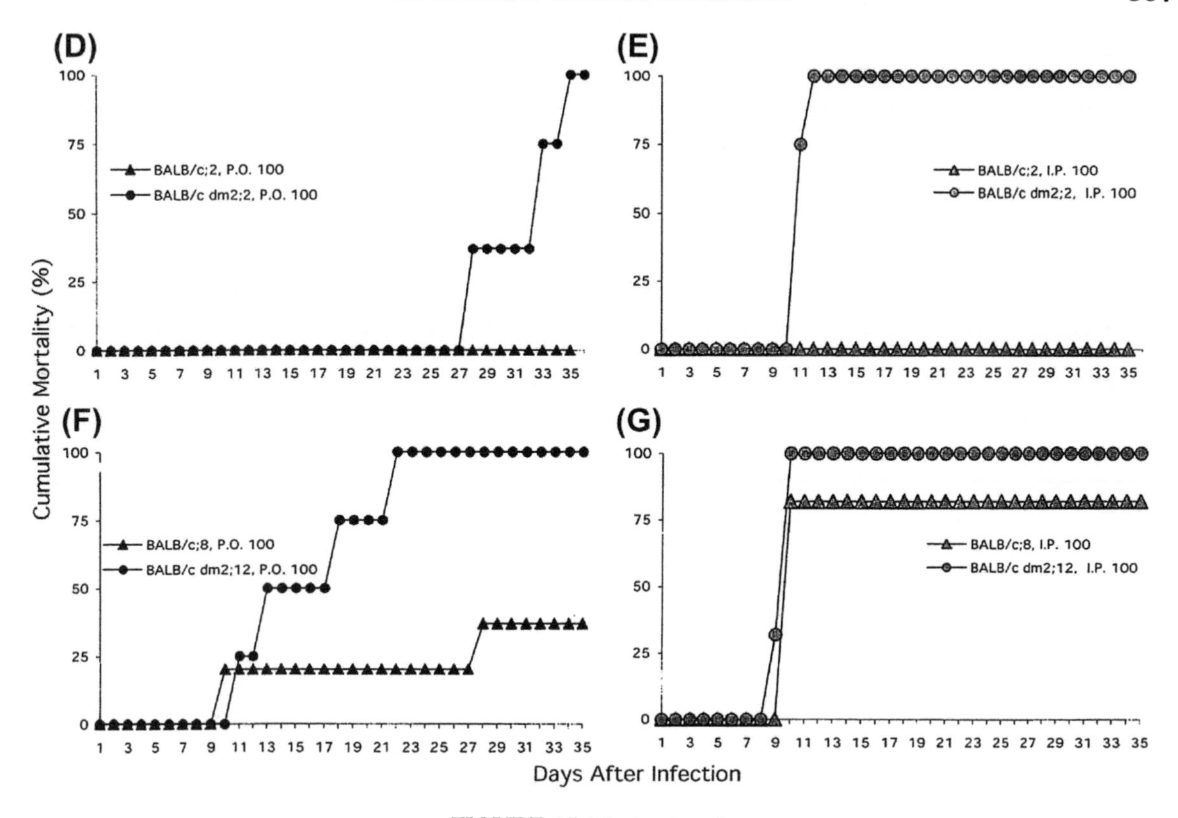

FIGURE 25.10 (*continued*).

FIGURE 25.10 **(A) Survival following per-oral infection of mice without murine class II genes, with murine class II transgenes and wild-type controls.** Symbols: square, *Ia–Ie–*; diamond, *Ia–Ie*$^{+}$; circle, *Ia*$^{+}$*Ie–*; triangle, *Ia*$^{+}$*Ie*$^{+}$. All differences in survival between *Ia–Ie–* mice and mice with only one of these genes (*Ia* wild type or *Ie* transgene) were significant ($p < 0.05$) in this experiment. In the replicate experiment, the trends were the same, although not all differences achieved statistical significance. Survival of the *Ia–Ie–* and the *Ia*$^{+}$*Ie*$^{+}$ mice was not significantly different ($p > 0.05$). (*Adapted from Johnson et al. (2002), with permission.*)

(B, C) Lack of protective effect of the L^{D} gene on survival with differing doses or routes of *T. gondii*. (B) L^{d} does not increase or significantly decrease survival with high or low inocula ($p > 0.05$). (C) L^{d} does not increase or significantly decrease survival in per-oral or parenteral infection ($p > 0.05$). Age in months is indicated after strain of mouse. Po, per-oral; i.p., intraperitoneal. *(Adapted from Johnson et al. (2002), with permission.)*

(D, E, F, G) Influence of Dm^{2} mutation on survival of young and old BALB/c mice. The Dm^{2} mutation diminishes survival of younger and older mice following per-oral infection (D,F). The Dm^{2} mutation diminishes survival in young mice following parenteral infxection (E,G) and older mice of both the wild-type BALB/c and with the Dm^{2} mutation are more susceptible than younger mice ($p < 0.05$) (F). Age in months is indicated after the strain of mouse. (*Adapted from Johnson* et al. *(2002), with permission.*)

(H, I, J, K) Influence of route and age on survival of C3H.L^{d} mice. Increase in age (months of age indicated after strain of mouse) only slightly increases susceptibility to per-oral infection ($p > 0.05$) (H, J), but markedly increases susceptibility to parental infection ($p < 0.05$) (I, K). L^{d} does not significantly increase or decrease survival. (*Adapted from Johnson et al. (2002), with permission.)*

(L) Influence of gender on survival of Dm^{2} mice. Female mice of the Dm^{2} strain are markedly more susceptible than males ($p < 0.001$). *(Adapted from Johnson et al. (2002), with permission.)*

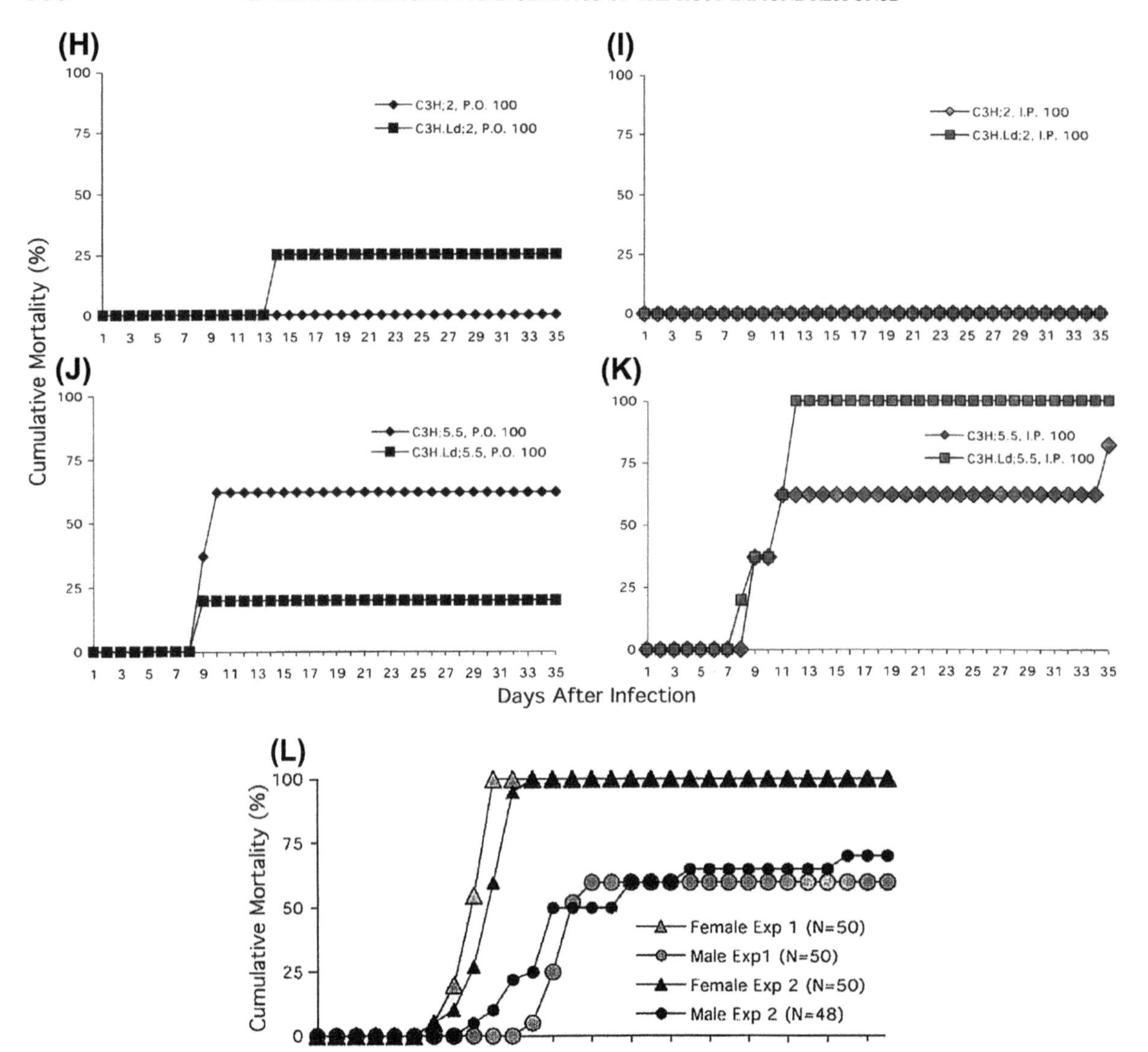

FIGURE 25.10 *(continued)*.

that this gene controls the ability of *T. gondii* to proliferate within macrophages. Using this functional correlate of resistance, the interval was further reduced in congenic sublines to a 0.3 cM region of 1.8 Mb that contains 26 identified rat genes and should provide very useful insights about key protective mechanisms in this infection (Cavailles *et al.*, 2006b) (Fig. 25.15A). Additional studies have shown the importance of macrophages, a cell death phenotype and mapping of the region with congenic rats is ongoing (Sergent *et al.*, 2005; Cavailles *et al.*, 2006a, 2006b). NALP1 in the

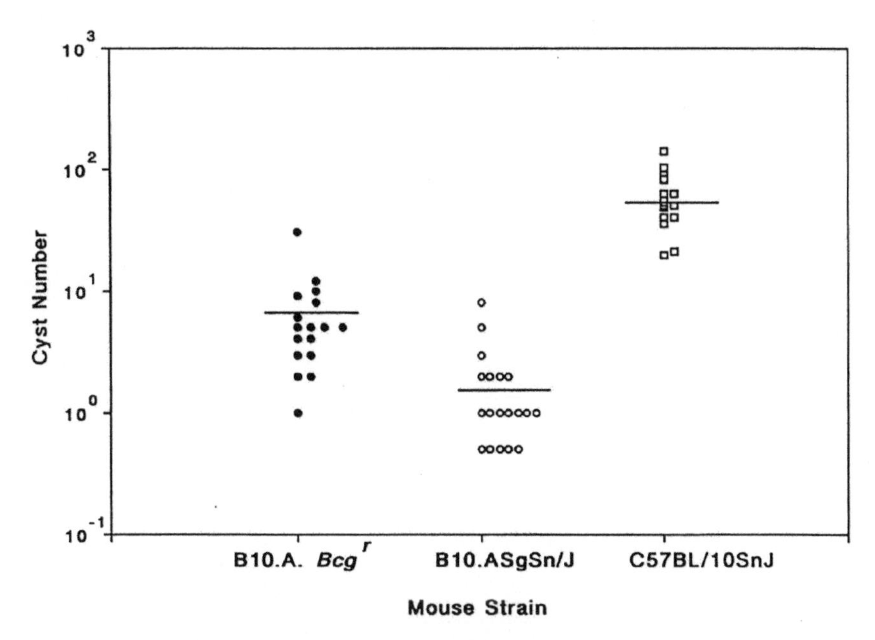

FIGURE 25.11 **Studies with congenic mice demonstrate a major influence of the H-2 complex and minor influence of the *Bcg (Nramp)* locus on the number of brain cysts following per-oral infection.** Number of cysts (per 10 μL) in brains of B10 congenic mice 30 days after per-oral infection with *T. gondii.* Circles represent mice that have the H-2a haplotype. *Solid symbols* represent mice that are *Bcg* resistant and *open symbols* represent mice that are *Bcg* susceptible. Data are from two replicate experiments with similar results. Differences between H-2a and H-2b mice were highly significant regardless of their *Bcg* type ($p < 0.001$). The smaller differences between B10.A.*Bcg*^{r} and B10.AsgSn/J mice also were significant ($p < 0.01$). In these experiments, control A/J mice had low cyst numbers and C57BL/6J mice had high cyst numbers, as in all other experiments. *(Adapted from McLeod* et al. *(1989b), with permission.)*

human syntenic region also has susceptibility alleles and is associated with cell and parasite death in MonoMac6 cells and modulates both caspase cleavage and cytokine production (Witola *et al.*, 2011).

25.4 STUDIES IN HUMANS CONCERNING GENES THAT CONFER RESISTANCE OR SUSCEPTIBILITY AND THE USE OF MURINE MODELS WITH HUMAN TRANSGENES

25.4.1 Studies to 2006

Early studies that aimed to determine whether HLA haplotypes conferred susceptibility to eye disease demonstrated no associations (Fig. 25.15 B–C) (Fig. 25.16). More recent results have found that there are associations between HLA haplotypes and the severity of toxoplasmic encephalitis in AIDS patients and the manifestations of disease in congenitally infected children (Suzuki *et al.*, 1996a; Mack *et al.*, 1999). The importance of the HLA DQ3 gene in determining the susceptibility of patients with AIDS to toxoplasmic encephalitis has been demonstrated by Suzuki (Suzuki *et al.*, 1996b) (Table 25.4). In addition, the HLA DQ3 gene has also been demonstrated to be an indicator of the susceptibility of congenitally infected children developing hydrocephalous (Mack *et al.*, 1999) (Fig. 25.16). The observation that there is concordance in monozygotic twins, but discordance of dizygotic twins for manifestations of congenital toxoplasmosis (Couvreur *et al.*, 1976; Farquhar *et al.*, 1950;

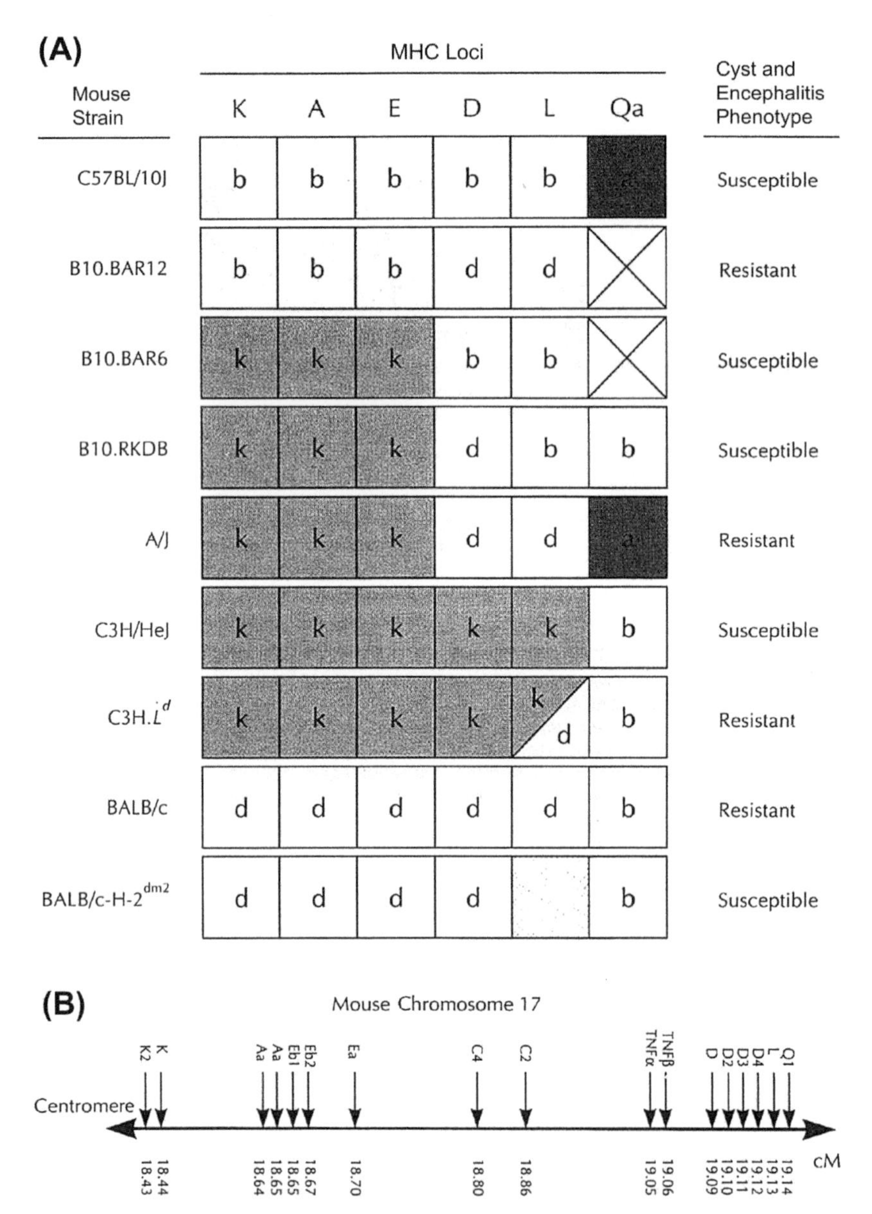

FIGURE 25.12 **Mapping studies which indicate that resistance to brain arasite burden and encephalitis are regulated by the L^d gene.**

(A) MHC haplotypes of mouse strains used to determine the controlling locus for cyst formation following per-oral *T. gondii* infection.

(B) Schematic diagram indicating the location of these MHC loci on mouse chromosome 17. Control of resistance to cyst burden following per-oral infection with *T. gondii* had been mapped previously to a region of mouse chromosome 17 of approximately 140 kb (McLeod *et al.*, 1989b; Brown and McLeod, 1990). This region is contiguous with and contains the class 1 gene L^d. Resistance to development of toxoplasmic encephalitis had also been reported to be controlled by genes in this region of H-2 by Suzuki *et al.* (1991). *TNFα*, *D* and *L* genes as well as unidentified genes are in this region. Studies were

Wiswell *et al.*, 1984; Murphy *et al.*, 1952; Remington *et al.*, 2005) emphasizes the importance of the genetic background of a foetus in determining the susceptibility to congenital toxoplasmosis. In these studies, infants with congenital toxoplasmosis and hydrocephalous had a statistically significant greater frequency of having the DQ3 allele. The observation that there were fewer than predicted children who were homozygous for the DQ3 allele suggested that homozygosity for this allele might lead to increased loss of foetuses infected with *T. gondii* (i.e. perhaps from spontaneous abortion) (Mack *et al.*, 1999) (Fig. 25.16 A-C; Tables 25.5 and 25.6). Furthermore, taking advantage of the fact that human MHC class 2 genes like DQ3 can function in mice, studies were performed that compared the susceptibility of mice expressing HLA DQ3 or HLA DQ1 to cerebral toxoplasmosis. Mice expressing the DQ3 gene were found to develop significantly more severe encephalitis than those expressing DQ1 (Fig. 25.16). This indicates that this type of murine model could be used to further validate the association of certain human genes with susceptibility to diseases, such as that which occurred in this work (Mack *et al.*, 1999) (Fig. 25.17; Tables 25.5 and 25.6).

Other studies (Brown and McLeod, 1990) have also demonstrated that human MHC class 1 transgenes can function in mice (Fig. 25.18). Although these mice are yet to be used in models in which pathogenesis and protection in *T. gondii* infection are characterized, they are likely to provide valuable insights into the role these genes in humans and the *T. gondii* peptide HLA molecules are capable of presenting. Polymorphisms in a number of other genes have also recently emerged as being important in determining disease. Thus, mutations in the gene for CD40 which is a ligand for B71 has been demonstrated to be associated with increased susceptibility of patients (Subauste *et al.*, 1998). There have been reports of a calcium channel gene upstream of NFκB as being important in susceptibility to toxoplasmosis as well as other intracellular infections such as tuberculosis and salmonella infections (Boulter *et al.*, 2005).

There are clearly patients who have differing manifestations of their *Toxoplasma* infections: some asymptomatic, some with adenopathy of varying duration, some with a flu-like illness, some with chronic fatigue and some with eye disease. Small numbers of apparently immunologically normal individuals have damage to certain organ systems, e.g. encephalitis (Townsend *et al.*, 1978; Couvreur and Desmonts, 1977; Ajzenberg and Darde, 1996, 2002; Carme *et al.*, 2002), pericarditis or myocarditis (Remington *et al.*, 2005). Some of this variability may be due to parasite clonal type or inoculum or form of the parasite acquired, but almost certainly some is due to differences in genetic susceptibility.

Other genetic mutations, knock-outs or alleles conferring susceptibility are shown in Table 25.2 (Arsenijevic *et al.*, 2000; Caamano *et al.*, 1999, 2000; Czar *et al.*, 2001; Collazo, 2002; Casciotti *et al.*, 2002; Del Rio, 2001a; Fujigaki *et al.*, 2002; Deckert-Schluter *et al.*, 1996, 1998, 1999; Denkers,

performed to identify the gene(s) in the 140 kb region that confers resistance to cysts and encephalitis (Brown *et al.*, 1995). In this study relative resistance to *T. gondii* organisms and cyst burden in brain, and toxoplasmic encephalitis 30 days following per-oral *T. gondii* infection were correlated with presence of the L^d gene in inbred, recombinant, mutant and C3H.L^d transgenic mice. Mice that were resistant to cysts and encephalitis had little detectable brain cytokine mRNA expression, whereas mice that were susceptible had elevated levels of mRNA for a wide range of cytokines, consistent with their greater amounts of inflammation. This work definitively demonstrates that and L^d-restricted response decreases the number of organisms and cysts within the brain and thereby limits toxoplasmic encephalitis and levels of IFNγ, TNFα, IL−2, IL−6, IL−10, TGFβ, IL−1α, IL−1β and macrophage inhibitory protein mRNA in the brain 30 days after per-oral infection. *Boxed X* indicates that the haplotype at this locus has not yet been determined. *Hatched square* indicates deletion of the L^d gene. *(Adapted from Brown* et al. *(1995), with permission.)*

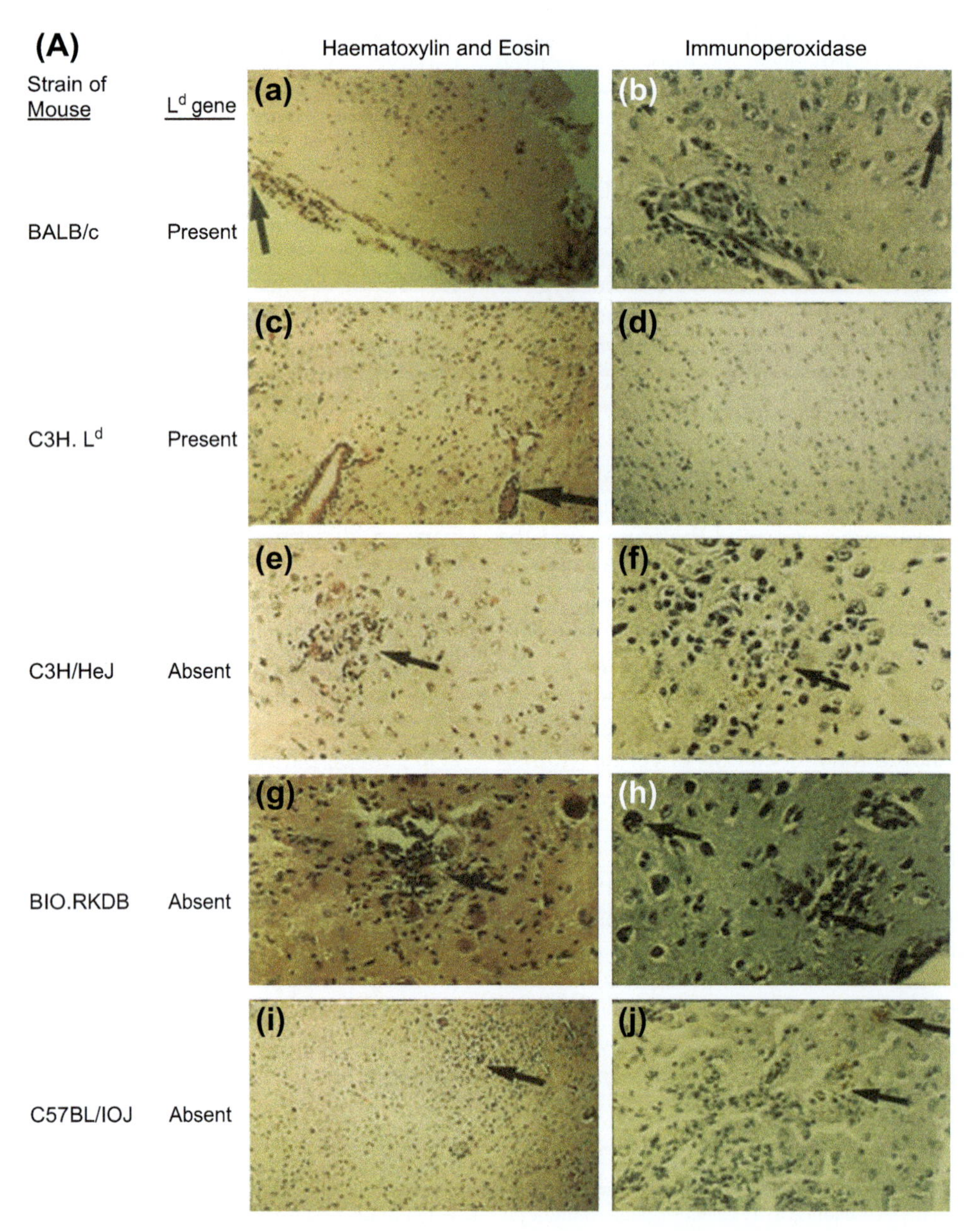

FIGURE 25.13 **(A) Representative examples of brains of resistant and susceptible strains of mice 30 days after per-oral infection.** Resistant mice with the L^d gene had minimal inflammation and parasite burden, and susceptible mice with the L^d gene had greater inflammation and parasite burden. Brains of resistant BALB/c (a, b) and C3H.L^d (c, d) mice. Brains of susceptible C3H/HeJ (d, e), B10.RKDB (f, g) and C57BL/10J (h, i) mice. In (a) and (c), HandE-stained sections (×250) from the resistant mice demonstrated only mild meningeal and perivascular inflammation (marked by arrows). These figures are representative of mild inflammation graded as ≤2 by the pathologist. In (b) and (d), immunoperoxicase-stained sections representative of resistant mice did not demonstrate presence of *T. gondii* tachyzoites in either the meninges or in association with vessels. Arrow in (b) marks a *T. gondii* cyst, seen only rarely in sections from resistant mice. In (e), (g) and

1993; Egan et al., 2005; Ely et al., 1999; Franzoso *et al.*, 1998; Gavrilescu *et al.*, 2004; Gazzinelli *et al.*, 1996b; Gazzinelli *et al.*, 2004; Hancock, 2004; Johnson *et al.*, 2003; Jebbari *et al.*, 1998; Khan, 1996b; Khan *et al.*, 1997, 2001. Kang, 2000; Kobayashi *et al.*, 1999; Li 2002; Kudo *et al.*, 2004; Kelly 2005; Lieberman 2004; Lu *et al.*, 2003, 2004; Lyons *et al.*, 2003; Mason *et al.*, 2004a, 2004b; Mun *et al.*, 2003; Nickdel *et al.*, 2004, 2001; Norose, 2011, 2003; Orlofsky *et al.*, 2002; Roberts *et al.*, 1996; Ronet *et al.*, 2005; Reichman *et al.*, 1999; Robben *et al.*, 2005; Smiley *et al.*, 2005; Suzuki *et al.*, 1996a, 1996b, 2000; Suzuki, 1997; Scharton-Kersten 1996b; Scharton-Kersten *et al.*, 1997b, 1998; Silva *et al.*, 2002; Scanga *et al.*, 2002; Taylor *et al.*, 2004; Schluter *et al.*, 2003; Villarino *et al.*, 2003; Vossenkamper, 2004; Villegas *et al.*, 2002; Wang *et al.*, 2004; Wille *et al.*, 2001; Wilson *et al.*, 2005; Yap, 1998a; Yap *et al.*, 2000; Zhang, 1999).

Earlier studies concerning parasite genetics, innate and adaptive immunity, B cells and antibodies, and *T. gondii* infections have provided a foundation for recent studies (Ajzenberg *et al.*, 2009; Aosai *et al.*, 2002; Araujo 1992; Bala *et al.*, 1994; Bruning *et al.*, 1997; Butcher *et al.*, 2002; Buzoni-Gatel *et al.*, 1999; Ceravolo *et al.*, 1999; Dellacasa-Lindberg *et al.*, 2007; Denkers *et al.*, 1993c; Denkers *et al.*, 1994; Desmonts *et al.*, 1980; Dobrowolski *et al.*, 1996; Dubey *et al.*, 1981, 1988, 1997; Dunay *et al.*, 2010; Dutton *et al.*, 1986; Dzierszinski *et al.*, 2000; Eisenhauer *et al.*, 1988; Ellis Neyer *et al.*, 1998; European Collaborative Study and Research Network on Congenital Toxoplasmosis, 1996; Farquhar *et al.*, 1950; Fazaelia *et al.*, 2000; Ferreira *et al.*, 2004; Fiorentino *et al.*, 1991; Frenkel, 1988; Frenkel *et al.*, 1982, 1955; Freund *et al.*, 1992; Furtado *et al.*, 1992; Gardner *et al.*, 1978a; Gavrilescu *et al.*, 2003; Gazzinelli *et al.*, 1993c; Gazzinelli *et al.*, 2004; Gross *et al.*, 1994; Handman *et al.*, 1980; Hauser *et al.*, 1981; He *et al.*, 2002; H of H *et al.*, 1976; Hunter *et al.*, 1995c; Israelski *et al.*, 1989; Jensen *et al.*, 2011; Johnson *et al.*, 1979; Johnson *et al.*, 2002; Jones

(i), HandE-stained sections (×250) from the susceptible mice demonstrated substantial meningeal and perivascular inflammation and, in contrast to the cyst-resistant strains, there was also substantial parenchymal inflammation (each marked by arrows). These figures with larger amounts of inflammation are represented by those graded as ≥3 by the pathologist. In (f), (h) and (j), immunoperoxidase stains of representative brain sections from susceptible mice demonstrated many extracellular *T. gondii* tachyzoites or bradyzoites (marked by arrows) within foci of inflammatory cells. The morphologically distinct cyst (e.g. in the upper right-hand corner of b and g) were also readily demonstrated both with and without accompanying inflammatory cell infiltrates. Results with BALB/c-H-2dm2 mice were similar (data not shown). The polyclonal anti-*Toxoplasma* antisera used for immunoperoxidase staining recognizes both tachyzoites and bradyzoites. *(Adapted from Brown et al. (1995), with permission.)*

(B) Representative examples of CD4$^+$ and CD8$^+$ T-cells in brains of resistant BALB/c and C3H.L^d (A–D) and C3H/Hej, BALB/c-H-2dm2 and C57 BL L0/J (E–I) mice that were per-orally infected 30 days earlier. CD4$^+$ and DC8$^+$ T-cells were demonstrated by immunoperoxidase stain. (a) BALB/c mouse, CD4$^+$ cells. Note that there was only a small number of CD4$^+$ cells (arrow). These cells were also present around a blood vessel and in the meninges (data not shown). (b) BALB/c mouse, CD8$^+$ cells. Note that there were also occasional CD8$^+$ cells (arrow). (c) C3.L^d CD8$^+$ cells. Arrow indicates CD8$^+$ cell. (e) C3H/HeJ mouse, CD4$^+$cells. Arrow demonstrates substantial numbers of CD4$^+$ cells in the brain parenchyma. (f) C3h/HeJ mouse, CD8$^+$ cells. Arrows demonstrate CD8$^+$ cells in the wall of a blood vessel in the brain parenchyma. (g) BALB/c-H-2dm2 mouse, CD4$^+$ cells. Large numbers of CD4$^+$ T-cells were present. (h) BALB/c-H-2dm2 mouse, CD8$^+$cells. Very small numbers of CD8$^+$ T-cells were present. (i) C57BL/10 mouse, CD4$^+$ cells. Large numbers of CD4$^+$ cells were present in clusters in the brain parenchyma (arrow). (j) C57BL/10 mouse, CD8$^+$cells. Large numbers of CD8$^+$ lymphocytes were present in brain parenchyma of control mice that had not been infected with *T. gondii* (data not shown). *(Adapted from Brown* et al. *(1995), with permission.)*

(C) TGFβ and actin MrRNA in brains of resistant and susceptible mice and demonstration of semiquantitative grading. Ethidium bromide-stained gels, which demonstrate reverse transcriptase PCR products from brains of resistant and susceptible mice. Neither further semiquantification of cytokine messenger RNA nor study of additional cytokines because of the conclusive all-or-none results. In uninfected controls, cytokine message was absent (data not shown). *(From Brown* et al. *(1995), with permission.)*

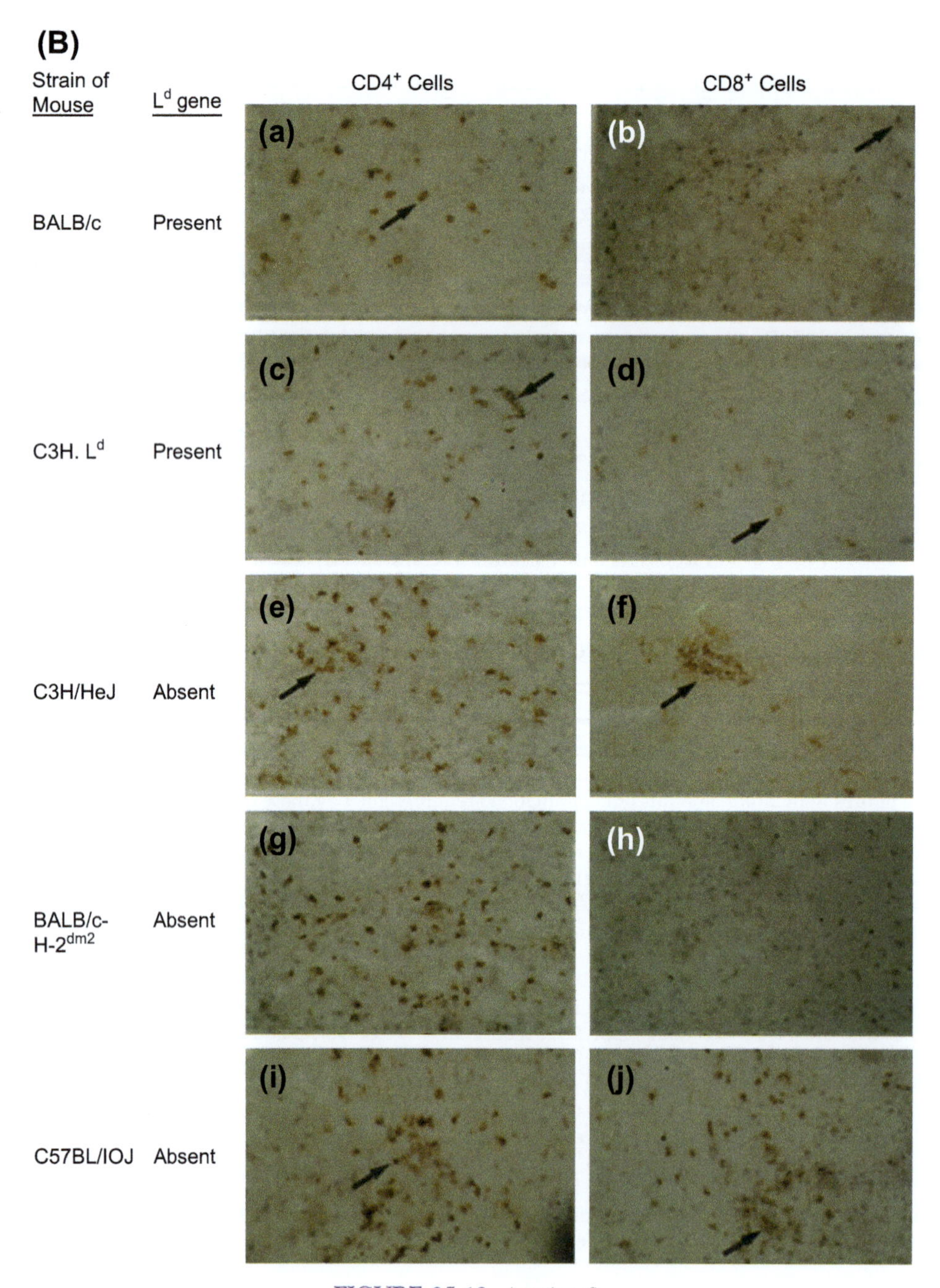

FIGURE 25.13 (*continued*).

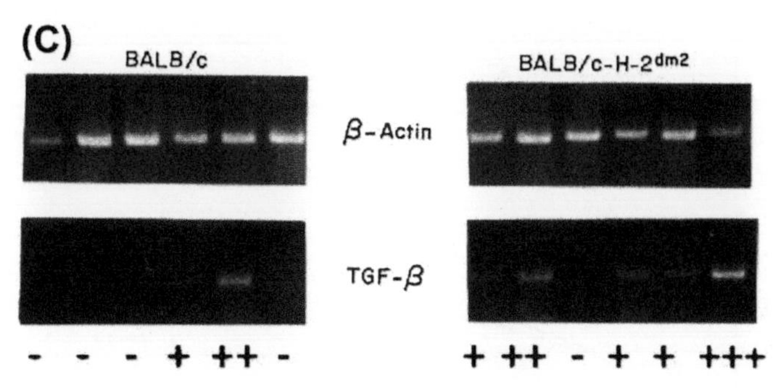

FIGURE 25.13 (*continued*).

et al., 1986; Kang *et al.*, 2001; Kasper *et al.*, 2004; Khan *et al.*, 1991, 1999, 2002; Kim *et al.*, 2004; Kobayashi and Susaki, 1987; Krahenbuhl *et al.*, 1980; Kwon *et al.*, 1992; Langermans *et al.*, 1992; Liesenfeld *et al.*, 1997, 2002, 2004; Locksley *et al.*, 1983; Lu *et al.*, 2005; Luangsay *et al.*, 2003; Makioka *et al.*, 1991; Manger *et al.*, 1998; McLeod and Mack, 1986; Mennechet *et al.*, 2002; Mévélec et al., 2010; Minamidani *et al.*, 1996; Mordue *et al.*, 1997; Nash *et al.*, 1998; Ohtake *et al.*, 1992; Oliveira *et al.*, 2000; Olle *et al.*, 1994; Payne *et al.*, 2003; Pfaff *et al.*, 2005; Phillip *et al.*, 2012; Potasman *et al.*, 1986; Purner *et al.*, 1996; Rachinel *et al.*, 2004; Rizzo *et al.*, 1998; Roberts *et al.*, 2001a; Rodgers *et al.*, 2005; Saavedra and Herion, 1993; Saeij *et al.*, 2005b; Schluter *et al.*, 1998; Schwartzman, 1987; Seder *et al.*, 1993; Shen *et al.*, 2001; Sibley and Howe, 1996; Sinai *et al.*, 2004; Soete *et al.*, 1994; Su *et al.*, 2003; Subauste *et al.*, 2001; Suzuki *et al.*, 1987; Suzuki *et al.*, 1993; Suzuki *et al.*, 1994a, 1994b; Suzuki *et al.*, 1995; Vallochi *et al.*, 2005; Vaudaux *et al.*, 2010; Ware and Kasper, 1987; Watanabe *et al.*, 1993; Weiss and Kim, 2000; Bahia-Oliveira *et al.*, 2003; Berger *et al.*, 1993; Couvreur and Thulliez, 1996; Daffos *et al.*, 1988; Edelson and Unanue, 2000; Fardeau *et al.*, 2002; Foulon *et al.*, 1990; Gilbert *et al.*, 2000; Gormley *et al.*, 1999; Hennequin, 1997; Herb *et al.*, 1977; Hitziger *et al.*, 2005; Hoerni *et al.*, 1978; Hogan, 1961; Hogan *et al.*, 1958; Hohlfeld *et al.*, 1989; Holland 2004; Howe *et al.*, 1997; Labalette *et al.*, 2001; Lucet *et al.*, 1993; Luft *et al.*, 1983; McLeod *et al.*, 1992; Mead *et al.*, 1999; Navia *et al.*, 1986; Peacock *et al.*, 1995; Peyron *et al.*, 2003; Rizzo *et al.*, 1998; Roberts *et al.*, 2001b; Rothova, 2003; Sibalic *et al.*, 1990; Silveira *et al.*, 1988; Vaudaux *et al.*, 2010; Wallon *et al.*, 2004; Zenner *et al.*, 1993; Bendelac *et al.*, 1995; George and Schroeder, 1992; Graca *et al.*, 2000; Hancock *et al.*, 2004; Hayashi *et al.*, 2003; Heyes *et al.*, 1992; Kappler *et al.*, 1988; Ma *et al.*, 1996; Mackay, 2001; Maggi *et al.*, 1994; Masson and Tschopp, 1987; Montecino-Rodriguez *et al.*, 2006; Moore *et al.*, 1993; Morrell, 1995; Murphy *et al.*, 1995; Nossal, 1994; Oswald *et al.*, 1992; Reis e Sousa *et al.*, 1997; Robey *et al.*, 1992; Romagnani, 1996; Scollay *et al.*, 1984; Sher *et al.*, 1993; Shirasawa *et al.*, 1993; Steffen *et al.*, 1996; Terpenning and Bradley, 1991; Tham *et al.*, 2003; Thilaganathan *et al.*, 1994; Unanue, 1996; von Boehmer *et al.*, 1989; Wang and Hakanson, 1995; Yamamoto *et al.*, 2003.

25.4.2 Studies from 2006 to 2012

25.4.2.1 Introduction

Recent studies have provided information about ligands, signalling pathways of innate immunity and how they shape the mechanisms of adaptive immunity that are stimulated and the effector mechanisms which result. These studies indicate that strain-specific virulence factors modulate IFNγ response genes and cytokine responses which shape adaptive

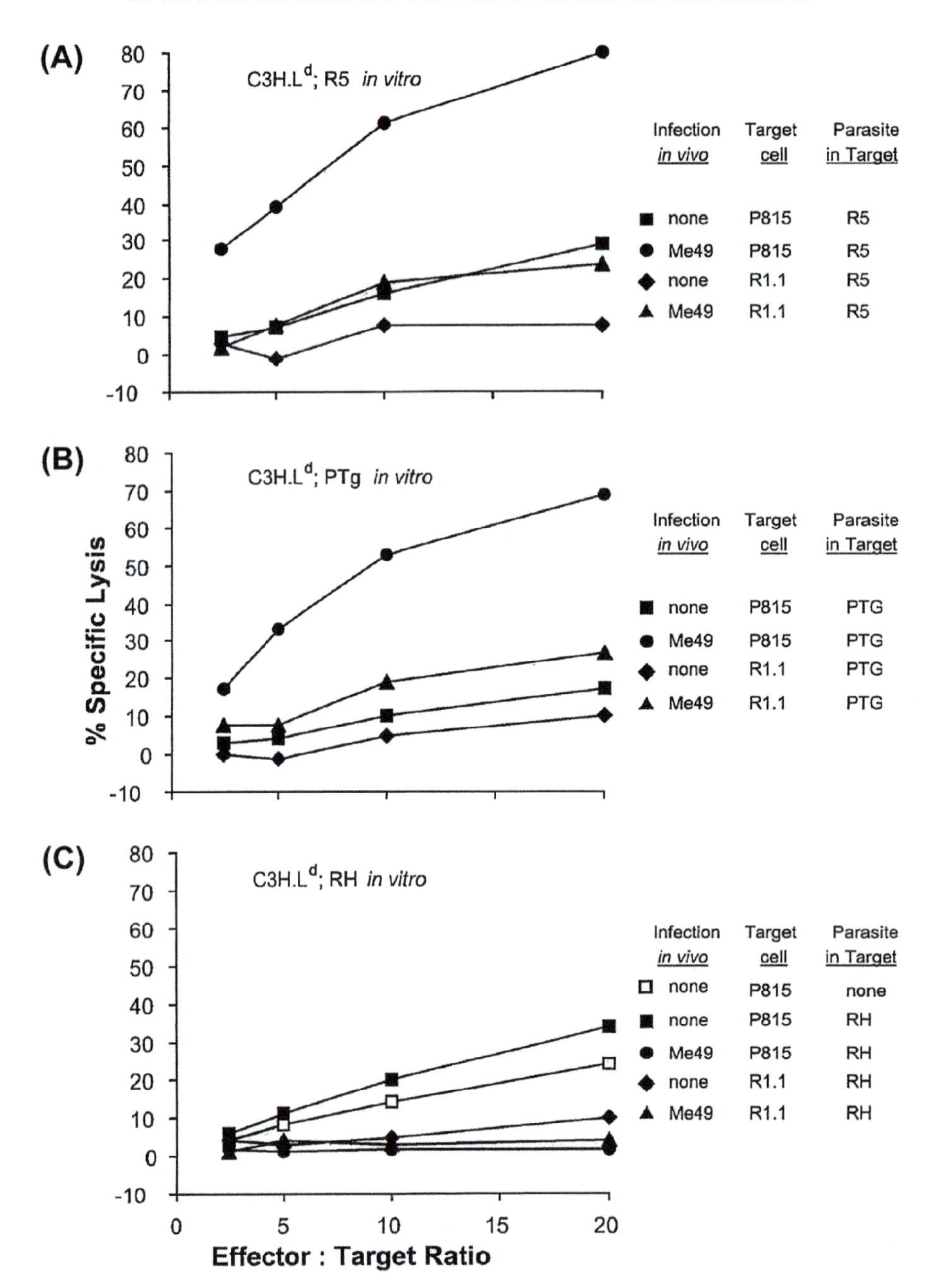

FIGURE 25.14 **Splenocytes from *T. gondii*-infected C3H.L^d mice exhibit L^d specific lysis when cultured *in vitro* with either of two type II strains, but not a type I strain of *T. gondii*.**
(A, B) Splenocytes were harvested from uninfected C3H.L^d mice or from mice infected 12 days earlier with cysts of the Me-49 strain *T. gondii* and stimulated for six days *in vitro* with R5
(A) or PTg (B) organisms attenuated by gamma irradiation. Effectors were tested against uninfected targets (data not shown, as lysis was <10% at all E:T-cell ratios) and against R5 strain (A) and PTg strain (B) *T. gondii*-infected P815(H−2d_ and

TABLE 25.4 Ab to L[d] Abrogates Cytolytic Activity of Me49-Infected C3H. L[d] Effectors

	P815 Cells				Ri.1 Cells			
Treatment of Cultures	**U[a]**	**I[b]**	**% lysis[c]**	**% reduction[a]**	**U[a]**	**I[b]**	**% lysis[c]**	**% reduction[d]**
No Ab	10	72	62		0	34	34	
Isotype control	20	72	52	16	1	36	35	0
Antibody to L[d]	43	43	31	50	11	45	34	0

[a] *Percent lysis of uninfected cells.*
[b] *Percent lysis of R5-infected cells.*
[c] *Percent lysis of infected cells minus percent lysis of uninfected cells.*
[d] *Percent lysis of Ab-treated cells/percent lysis of untreated ceils (100%). Spontaneous lysis of Ab-treated cells was 18% or lower for all target cells, with the exception of 41% for infected RI.1 cells.*
Adapted with permission from Johnson et al. *(2002).*

immunity. Pattern recognition receptors and the inflammasome play a significant role in shaping the adaptive immune response. There have been a number of outstanding recent reviews (e.g. Ishikawa and Medzhitov, 2010) discussing the influence of innate or adaptive immune response. Studies of human genes that are key in protection have been done, including those noting persons with mutations in genes or with treatments who have developed disseminated toxoplasmosis. Utilizing transmission disequilibrium testing for candidate genes has been another powerful approach. Such detection of susceptible alleles can indicate that a gene may be important in resistance even if the SNP is not the functional variant. Then, studies of putative mechanisms with either studies of human cells or murine models with knock-out mice have been informative. This is outlined below and shown schematically in Figures 25.20–25.29 (Abi Abdallah and Denkers, 2012; Abi Abdallah *et al.*, 2011, 2011; Afonso *et al.*, 2012; Akira *et al.*, 1993; Albuquerque *et al.*, 2009; Ali *et al.*, 2005; Aline *et al.*, 2002a, 2002b; Alvarado-Esquivel *et al.*, 2012; Alves *et al.*, 2012; Andrade *et al.*, 2005; Andrade *et al.*, 2006; Arling *et al.*, 2009; Arsenijevic *et al.*, 2007; Aviles *et al.*, 2008; Avunduk *et al.*, 2007; Baird *et al.*, 2013; Beaman *et al.*, 1993; Barragan and Sibley, 2002; Beauvillain *et al.*, 2007; Begum-Haque *et al.*, 2009; Bekpen *et al.*, 2005; Bekpen *et al.*, 2010; Benson *et al.*, 2009; Benson *et al.*, 2012; Berdoy *et al.*, 2000; Bernardes *et al.*, 2006; Beug *et al.*, 2012; Bhadra *et al.*, 2011; Bierly *et al.*, 2008; Blader and Saeij, 2009; Blanchard and Shastri, 2010; Blanchard *et al.*, 2008; Bliss *et al.*, 2000; Blomstrom *et al.*, 2012; Boothroyd and Dubremetz, 2008; Bottova *et al.*, 2009; Boyle *et al.*, 2008; Boyle *et al.*, 2007; Bradley and Sibley, 2007; Brandão *et al.*, 2009; Brenier-Pinchart *et al.*, 2006; Brown *et al.*, 2005; Brown *et al.*, 1994; Brown *et al.*, 1995; Brunet *et al.*, 2008; Butcher *et al.*, 2011; Butcher *et al.*, 2005; Buzoni-Gatel *et al.*, 2008; Buzoni-Gatel *et al.*, 2006; Byrd and Horwitz, 1989; Caetano *et al.*, 2011; Capuron and Miller, 2011; Cavailles *et al.*, 2006a; Cavailles *et al.*, 2006b; Cavalcanti *et al.*, 2011; Cesbron-Delauw *et al.*, 2008;

Rl.1(H-2k) target cells. Spontaneous lysis of all target cells was 23% or less, with the exception of 40% for R5-infected Rl.1 target cells Rl.1 target cells were 38%–55% infected. P815 target cells were 65% infected.
(C) Splenocytes were harvested as described in A and B from mice infected 11 days and stimulated with the UV-attenuated RH strain of *T. gondii*. Effectors were tested against uninfected (open box: data not shown when lysis was <10% at all E:T-cell ratios) and RH-infected (black box) P815 and Rl.1 target cells. Spontaneous lysis of targets was 14% or less for all target cells. Target cells were 50%–55% infected. *(Adapted with permission from Johnson* et al. *(2002) with permission.)*

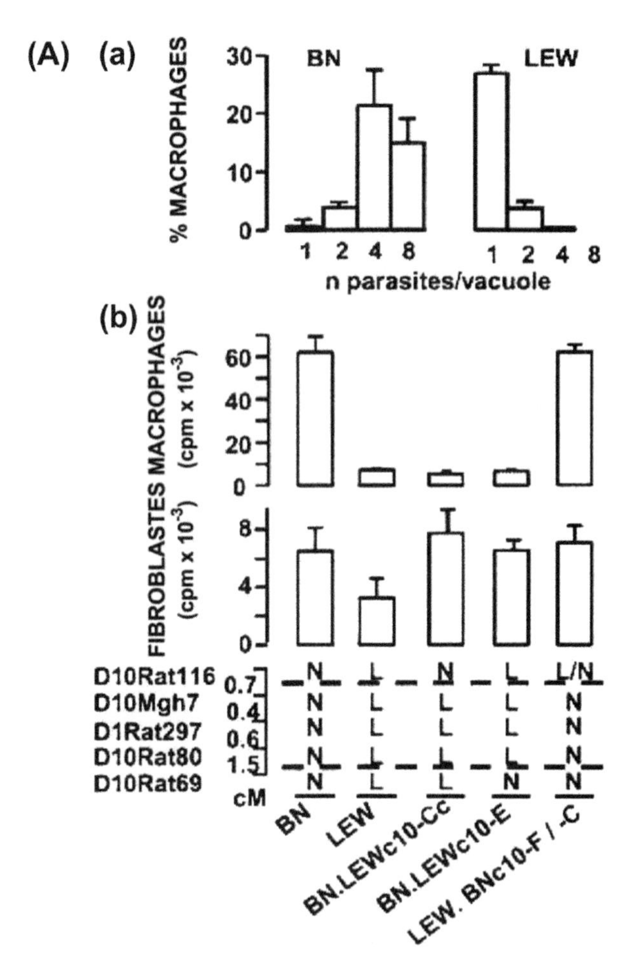

FIGURE 25.15 **(A)** ***Toxo 1*** **controls the proliferation of** ***T. gondii*** **within macrophages.** (a) BN or LEW macrophages were mixed with *T. gondii* for 1 hour, washed and cultured for 20 hours. The figure represents the repartition of infected macrophages according to the number of parasites per parasitophorous vacuole. The columns and the bars show the mean result and the standard deviation of three independent experiments. (b) The intracellular growth of *T. gondii* on macrophage and fibroblast monolayers from BN, LEW and congenic (BN.LEWc10-E, BN.LEWc10-CC,LEW.BNc10F, LEW.BNc10-C) was measured by monitoring tritiated uracil incorporation into *T. gondii* RNA. From the two different LEW.BNc10 lines of the same BN genotype at *Toxo 1*, the -F line was used for macrophage studies and the -C line was used for fibroblast studies. The columns and the bars show the mean result and the standard deviation of triplicates in one rat. These results are representative of two (fibroblasts) and three (macrophages) independent experiments. As a whole, studies on macrophages were performed on six BN, four LEW and four rats of each congenic line with similar results. Dotted lines indicate the limits of *Toxo 1* (boundary markers: D10Rat 116 and D10Rat80), N, homozygous BN; L, homozygous LEW. *(Adapted from Cavailles (2006), with permission.)*

(B) Ocular histopathology in congenital toxoplasmosis. (a, top) A well-demarcated area of retinal necrosis (n) at the posterior pole in the eye of a 22 weeks' gestation foetus (haematoxylin–eosin, original magnification ×250). (a, bottom) The edge of a large retinochoroidal scar from the eye of a 2 year old child. The scar is well demarcated with tubuloacinar proliferation of the retinal pigment epithelium (rpe) at the edge of the scar. The centre of the scar is devoid of retina (haematoxylin–eosin, original magnification ×250). (b, top) Eye from a 32 weeks' gestation foetus showing a large, hyperpigmented scar with a white rim, in the superotemporal region of the eye (arrow). (b, bottom) The retina from the edge of the scar shows disorganization with formation of Flexner–Wintersteiner rosettes (arrows) (haematoxylin–eosin, original magnification ×400).

Chan, A. *et al.*, 2006; Chan, C. *et al.*, 2006; Chandramohanadas *et al.*, 2009; Chang and Pechere, 1989; Chao *et al.*, 1994; Chao *et al.*, 1993; Charles *et al.*, 2010; Charles *et al.*, 2007; Chou *et al.*, 2012; Chtanova *et al.*, 2009; Clark *et al.*, 2011; Combe *et al.*, 2006; Cong *et al.*, 2012; Cong *et al.*, 2010; Cong *et al.*, 2011; Cong *et al.*, 2008; Correa *et al.*, 2010; Courret *et al.*, 2006; Craven *et al.*, 2012; Crawford *et al.*, 2009; Creuzet *et al.*, 1998; Da Gama *et al.*, 2004; de Alencar *et al.*, 2009; D'Angelillo *et al.*, 2011; Dalton *et al.*, 2006; Dass *et al.*, 2011; Daubener *et al.*, 2001; Daubener *et al.*, 1999; Debierre-Grockiego *et al.*, 2012; Debierre-Grockiego *et al.*, 2003; Debierre-Grockiego *et al.*, 2007a; Debierre-Grockiego *et al.*, 2007b; Debierre-Grockiego *et al.*, 2009; Debierre-Grockiego *et al.*, 2010; Deckert *et al.*, 2006; Dellacasa-Lindberg *et al.*, 2011; Dellacasa-Lindberg *et al.*, 2007; Denkers *et al.*, 2012; Denkers *et al.*, 2011a; Denkers *et al.*, 2009; Denkers *et al.*, 2008; Denkers, 2010; Diana *et al.*, 2004; Dimier and Bout, 1997; Dimier and Bout, 1998; Ding *et al.*, 2004; Divanovic *et al.*, 2012; Doğ *et al.*, 2006; Drogemuller *et al.*, 2008; Dunay *et al.*, 2008; Dunn *et al.*, 2008; Dupont *et al.*, 2012; Dupont and Hunter, 2012; Dutra *et al.*, 2012; Dzierszinski *et al.*, 2007; Egan *et al.*, 2012; Egan *et al.*, 2008; Egan *et al.*, 2009a; Egan *et al.*, 2009b; Egan *et al.*, 2011; El Kasmi *et al.*, 2008; Fagard *et al.*, 1999; Fang *et al.*, 2008; Fehervari *et al.*, 2006; Feng *et al.*, 2008a; Feng *et al.*, 2009; Feng and Sher, 2010; Feng *et al.*, 2008b; Fentress *et al.*, 2010; Ferguson and Hutchison, 1987a; Ferguson and Hutchison, 1987b; Ferreira da Silva *et al.*, 2008; Fleckenstein *et al.*, 2012; Foureau *et al.*, 2010; Frankel *et al.*, 2007; Frenkel and Escajadillo, 1987; Frickel *et al.*, 2008; Ferreira da Silva *et al.*, 2009; Fox and Bzik, 2010; Fujigaki *et al.*, 2002; Furuta, *et al.*, 2006; Gaddi and Yap, 2007; Gais *et al.*, 2008; Garweg and Candolfi, 2009; Gaskell *et al.*, 2009; Gazzinelli *et al.*, 1994a; Gazzinelli *et al.*, 1994b; Ghatak and Zimmerman, 1973; Gigley

(c, left) Retina from a 5 day old infant eye showing retinal detachment with an exudate (e) between retina and choroids. The inner retinal layer is oedematous and inflamed (haematoxylin–eosin, original magnification ×100). (c, right) Retina from the eye of a 22–23 weeks' gestation foetus showing gliosis (g) of the inner retinal layers (haematoxylin–eosin, original magnification ×250). (d) Eye from a 23–23 weeks' gestation foetus showing a moderate inflammatory infiltrate (i) within the primary vitreous and surrounding the hyaloid artery (ha) (haematoxylin–eosin, original magnification ×20). (e) Optic nerve from the eye of a 23 weeks' gestation foetus with congenital toxoplasmosis. The nerve architecture is disrupted with an inflammatory cell infiltrate (haematoxylin–eosin, original magnification ×100). *(Adapted from Roberts* et al. *(2001), with permission.)*

(C) Inflammatory cells and *Toxoplasma gondii* organisms present in ocular toxoplasmosis. (a) through (d) represent the same discrete ocular lesion from a 21 weeks' gestation foetus with a peripapillary lesion. (a) Disruption of the retinal pigment epithelium (RPE) with choroidal congestion and inflammation (haematoxylin–eosin, original magnification ×400). Immunohistochemical staining for T-cells, CD3 (b) and T-cell subset. CD4 (c), shows numerous positive lymphocytes within the choroids (arrows). (d) In this case, CD68-positive macrophages are numerous within the choroids underlying the area of RPE disruption (arrows). No staining was identified in the negative control or in sections stained with anti-CD8 (not shown). (e) and (f) demonstrate staining for *T. gondii*. (e, left) Retina from a 5 day old infant showing a collection of intracellular *T. gondii* within the retina (arrow) (haematoxylin–eosin), original magnification ×100). Note the small blood vessel (v). The inset shows a high-power view of these organisms (haematoxylin–eosin, original magnification ×400). (e, right) Retina from the 5 day old infant eye showing immunohistochemical staining for *T. gondii* antigen. Note the same small blood vessel (v) also identified in (e, left). Many extracellular *T. gondii* organisms are identified (arrows). In addition, the inset shows the presence of organisms in a perivascular location (L43 stain, original magnification ×100 and ×250). (f, left) Gliotic retina in an eye from a 22 weeks' gestation foetus showing extracellular organisms scattered throughout the retinal layers (arrows) (polyclonal antibody, original magnification ×250). (f, right) Disrupted retina and necrotic debris in an eye from a 23 weeks' gestation foetus. Numerous extracellular *T. gondii* organisms (arrows) are present within the necrotic debris (L43 stain, original magnification ×400). The red staining product allows distinction from melanin pigment granules (pg) of disrupted RPE. *(Adapted with permission from Roberts, F.* et al. *(1999), with permission.)*

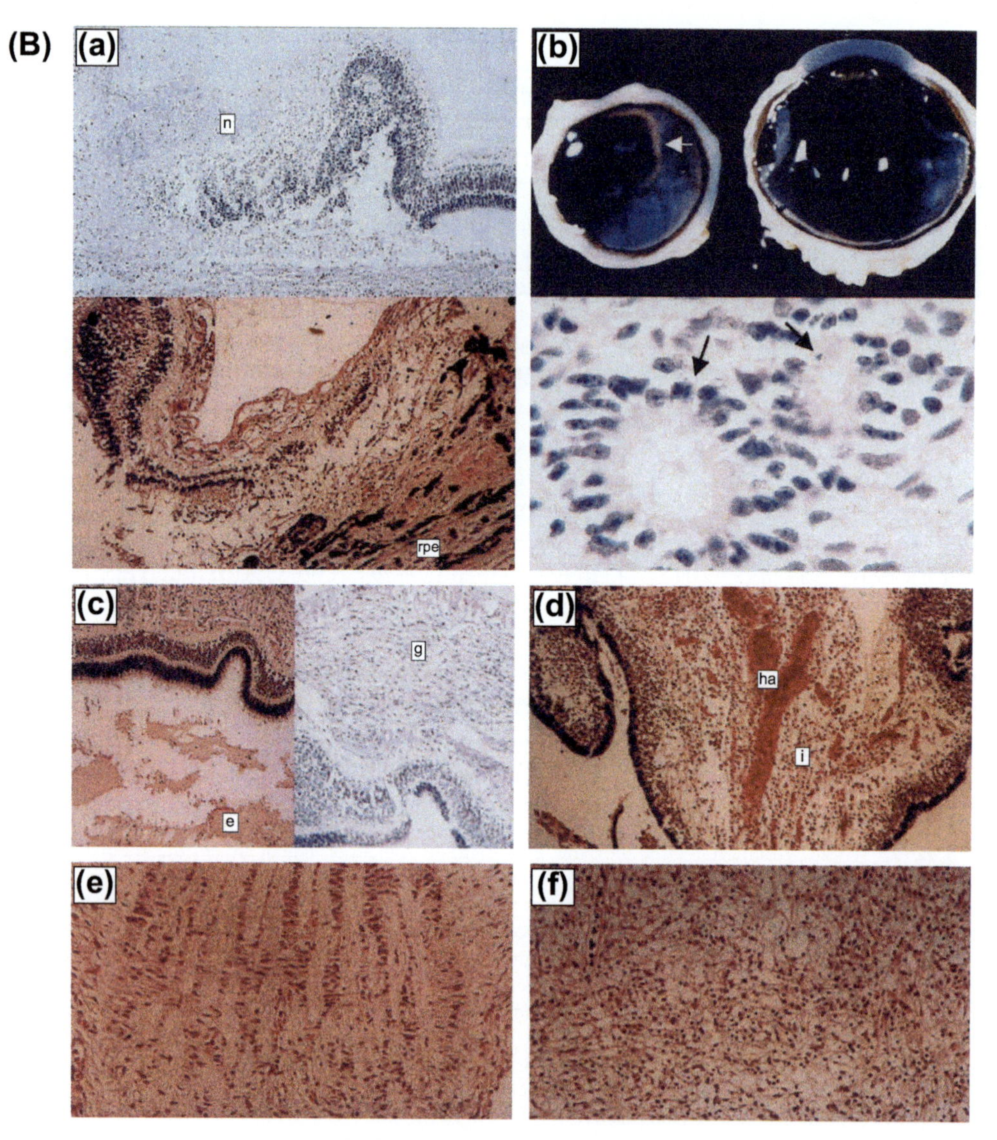

FIGURE 25.15 *(continued)*.

et al., 2009a; Gigley *et al.*, 2009b; Gilbert *et al.*, 2007; Gilbert *et al.*, 2008; Glatman and Zaretsky, 2012; Goodwin *et al.*, 2012; Goldszmid *et al.*, 2009; Goldszmid and Sher 2010; Goldszmid *et al.*, 2007; Goldszmid *et al.*, 2012; Gregg *et al.*, 2011; Grimwood *et al.*, 1996; Groer *et al.*, 2011a; Gröer *et al.*, 2011b; Grover *et al.*, 2012; Guan *et al.*, 2007; Guiton *et al.*, 2010; Guglietta *et al.*, 2007; Hall *et al.*, 2012; Halonen *et al.*, 1996, 1998a, 1999, 2001, 2006; Haroon *et al.*, 2012; Harris *et al.*, 2012; Harris *et al.*, 2010; Hassan *et al.*, 2012; Hauser *et al.*, 1982; Hauser and Tsai, 1986; Hayashi *et al.*, 1996a; Hayashi *et al.*, 1996b; Heimesaat *et al.*, 2006, 2007; Henriquez *et al.*, 2010; Henry *et al.*, 2009; Hermes *et al.*, 2008; Hertzog *et al.*, 2011; Hippe *et al.*, 2008, 2009a, 2009b; Hirota *et al.*,

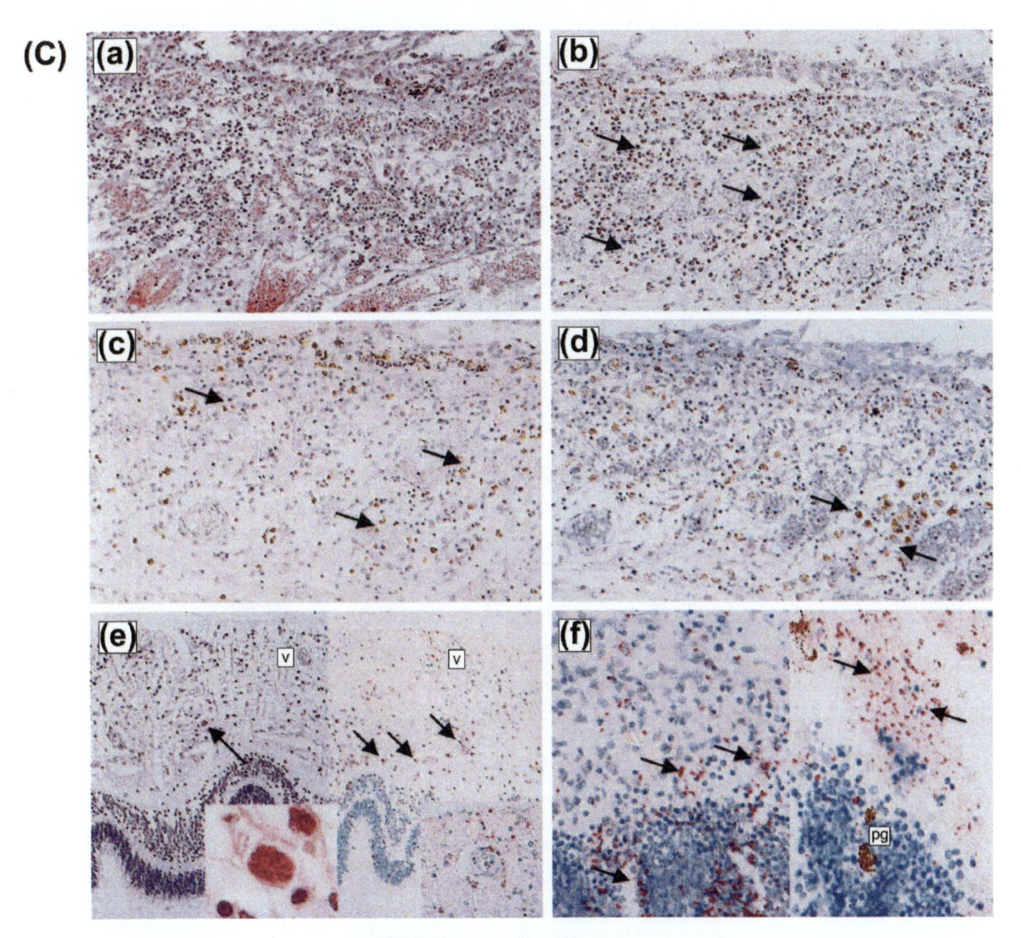

FIGURE 25.15 *(continued)*.

1996; Honore *et al.*, 2000; Hortua Triana *et al.*, 2012; Hoshi *et al.*, 2010; Hou *et al.*, 2011; House *et al.*, 2011; Howard *et al.*, 2011; Hunter and Sibley, 2012; Hutchison *et al.*, 1980; Indraccolo *et al.*, 2007; Ishii *et al.*, 2006; James, 1995; Jamieson *et al.*, 2008; Jamieson *et al.*, 2010; Jankovic *et al.*, 2007; Janssen *et al.*, 2002; Jebbari *et al.*, 1998; Jensen *et al.*, 2011; Jin *et al.*, 2008; Jin *et al.*, 2009; John *et al.*, 2009; John *et al.*, 2011; Johnson *et al.*, 2003; Johnson *et al.*, 2002a; Jones *et al.*, 2011; Jones *et al.*, 2010; Jordan *et al.*, 2009a; Jordan and Hunter, 2010; Jordan *et al.*, 2009b; Jordan *et al.*, 2010b; Jost *et al.*, 2007; Kafsack *et al.*, 2009; Kannan *et al.*, 2010; Khan *et al.*, 2006; Khan *et al.*, 2001; Kikumura *et al.*, 2012; Kim *et al.*, 2006a; Kim *et al.*, 2005; Kim, S.K. *et al.*, 2008; Kim, Karasov and Boothroyd, 2007; Kim and Denkers, 2006b; Kim, Y. *et al.*, 2008; Kirak *et al.*, 2010; Koblansky *et al.*, 2012; Konen–Waisman *et al.*, 2007; Koshy *et al.*, 2012; Koshy *et al.*, 2010; Kubo *et al.*, 2003; Lahmar *et al.*, 2010; LaRosa *et al.*, 2008; Lambert *et al.*, 2006; Lambert *et al.*, 2009; Lamberton *et al.*, 2008; Lang *et al.*, 2006, 2007, 2012; Lee *et al.*, 2007; Leiva *et al.*, 1998; Leng and Denkers 2009; Leng *et al.*, 2009; Lepage *et al.*, 1998; Lees *et al.*, 2010; Lieberman *et al.*, 2004a; Lieberman *et al.*, 2004b;

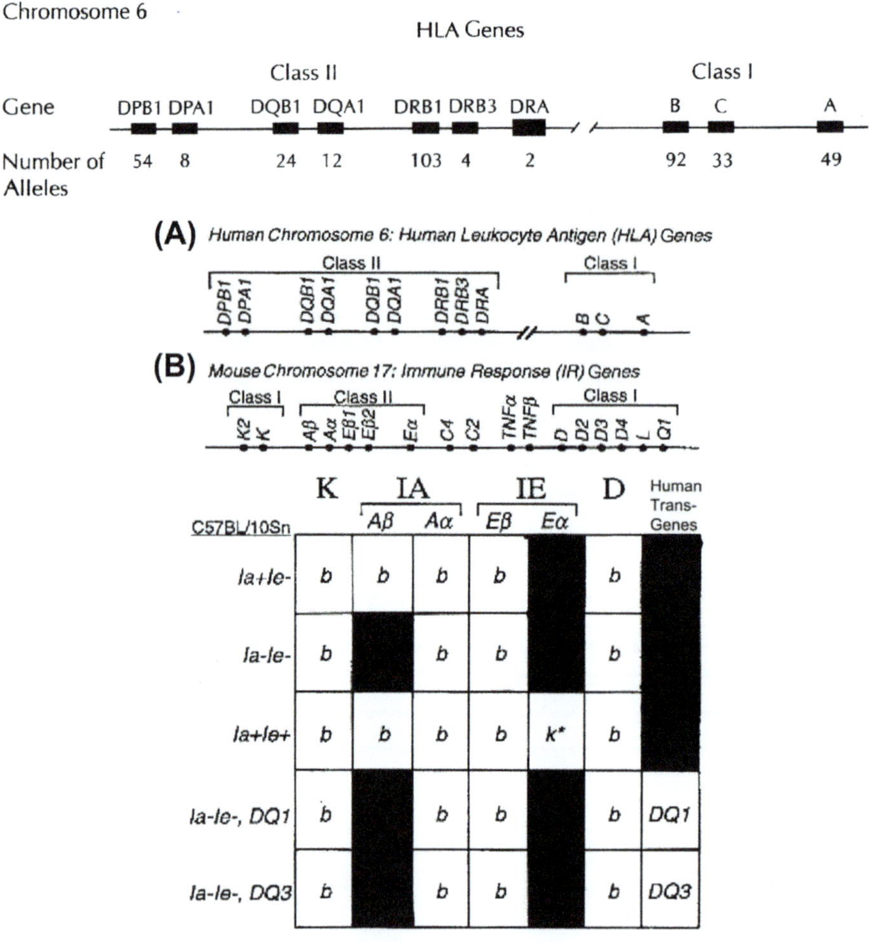

FIGURE 25.16 **Class II HLA gene and susceptibility to toxoplasmosis.**
(A) Map of location of polymorphic HLA class I and class II genes in the human MHC on chromosome 6. The number of known alleles of each gene, whether identified serologically or by sequence analysis, is indicated.
(B) Human and murine class II genes. (a) Schematic representation of the relationships of human and murine MHC class II genes. (b) Diagrams of expression of murine and human class II molecules on cells from mice used in the studies described herein. *The Ea b gene is non-functional. A functional Ea k gene was inserted to produce E^+ mice. Shaded area equals not present.
(C) Parasite burden and histopathology in *T. gondii* infected mice with differing class II phenotypes. (a, b) Numbers of cysts in wet mounts of brains from wild type, knock-out and transgenic mice. (c) Comparison of magnitude of parasite burden in histopathologic preparations of tissue. In panel f, antibody was polyclonal rabbit anti-*Toxoplasma* antibody produced by immunizing rabbits with tachyzoites of the C56 strain followed by tachyzoites of the RH strain of *T. gondii.* (d) Comparison of magnitude of necrosis. (e) Comparison of magnitude of parenchymal inflammatory activity. (f) Clusters of *T. gondii* cysts (arrow). (g) Necrosis (arrow) and parenchymal inflammation in the brain of an *Ia– Ie–* strain mouse. (h) Severe meningitis (arrow) in the brain of a DQ3 strain mouse. (i) Severe perivascular inflammation (arrow) in the brain of a DQ3 strain mouse. *(Adapted from Mack* et al. *(1999), with permission.)*

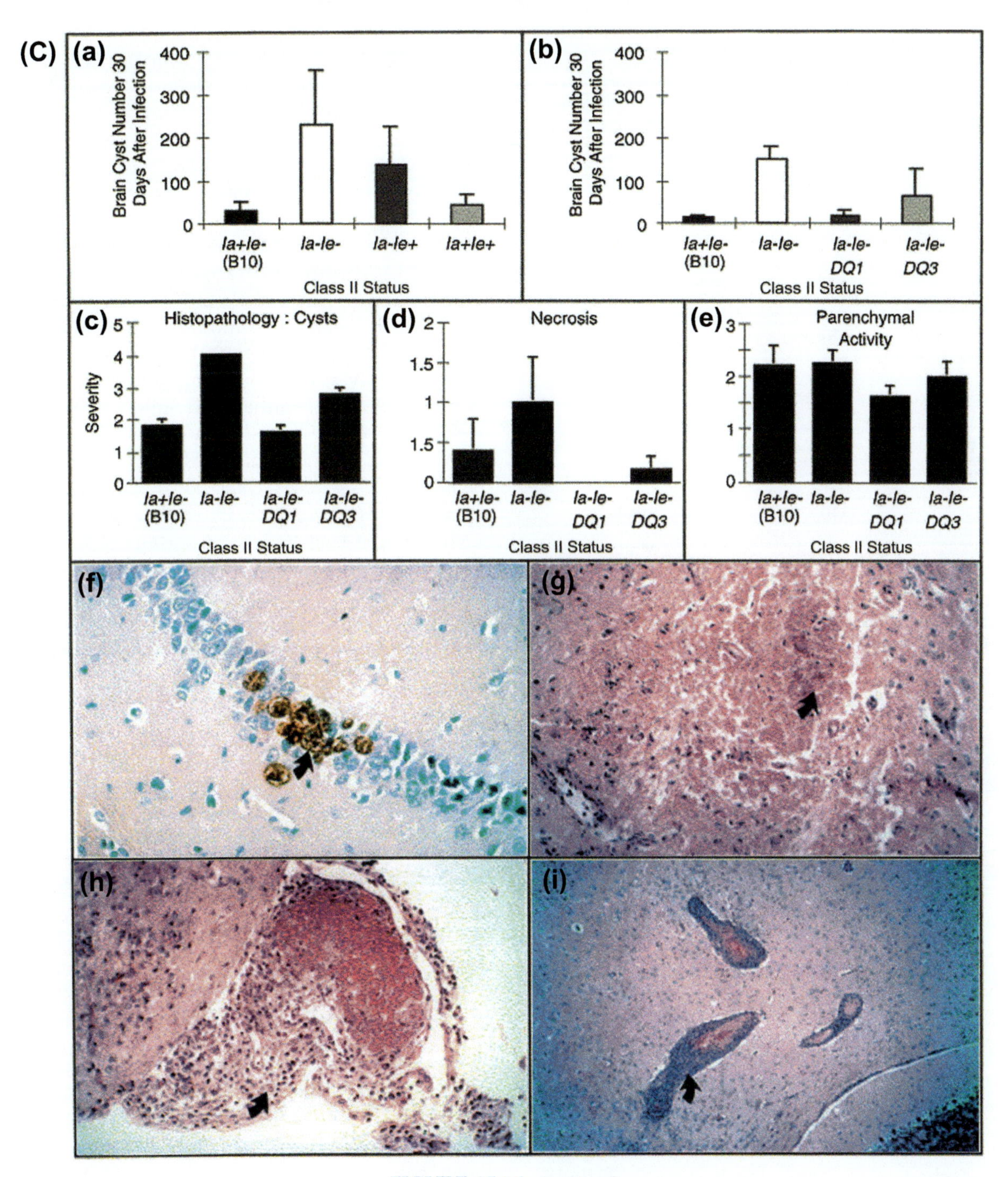

FIGURE 25.16 (*continued*).

TABLE 25.5A Findings in 20 AIDS Patients with Toxoplasmic Encephalitis

Patient No.	CD4 Cell Count	No. of Lesions on Brain CT or MRI
1	30	Multiple
2	20	Multiple
3	90	Multiple
4	43	Multiple
5	35	Multiple
6	20	Multiple
7	NA	Single*
8	41	Multiple
9	10	Single†
10	71	Multiple
11	10	Single
12	20	Multiple
13	113	Multiple*
14	45	Multiple*
15	60	Multiple
16	60	Multiple‡
17	50	Multiple*
18	4	Single
19	260	Single
20	34	Multiple

* *Brain biopsy revealed* Toxoplasma gondii.
† *Basal ganglia lesion.*
‡ *Cerebrospinal fluid positive for* T. gondii *DNA by polymerase chain reaction.*
NOTE. All patients were positive for *Toxoplasma* IgG antibodies, and all patients except nos. 13 (NA, not available) and 16 (died) had positive response to specific therapy. CT, computed tomography; MRI, magnetic resonance imaging.

Liesenfeld *et al.*, 2011; Lim *et al.*, 2012; Ling *et al.*, 2006; Liu *et al.*, 2006; Locksley *et al.*, 1982; Lu *et al.*, 2003; Lu *et al.*, 2009; Lüder *et al.*, 2009, 2010; Lykens *et al.*, 2010; Machado *et al.*, 2006a; Machado *et al.*, 2006b; Machado *et al.*, 2010; MacMicking, 2012; Mahamed *et al.*, 2012; Makino *et al.*, 2011; Marshall and Denkers, 1998; Marshall *et al.*, 1999; Masek *et al.*, 2007; Mashayekhi *et al.*, 2011; Mason *et al.*, 2004; McBerry *et al.*, 2012; McGeachy *et al.*, 2009; McLeod *et al.*, 2012; Melo *et al.*, 2010; Melzer *et al.*, 2010; Melzer *et al.*, 2008; Menard *et al.*, 2007; Mendes *et al.*, 2011; Menzies *et al.*, 2008; Mévélec *et al.*, 2010; Meylan *et al.*, 2008; Middleton *et al.*, 2009; Miller *et al.*, 2011a; Miller *et al.*, 2011b; Minns *et al.*, 2006; Minot *et al.*, 2012; Mirpuri and Yarovinsky, 2012; Mitchell *et al.*, 2006; Mitra *et al.*, 2012; Montfort *et al.*, 2009; Mun *et al.*, 2003; Muñoz *et al.*, 2009; Mordue *et al.*, 2007; Morgado *et al.*, 2011; Murakami *et al.*, 2012; Murray *et al.*, 1991; Nagineni *et al.*, 1996; Nance *et al.*, 2012; Nathan and Shiloh, 2000; Niedelman *et al.*, 2012; Nishanth *et al.*, 2010; Nishikawa *et al.*, 2007; Nishikawa *et al.*, 2011; Noor *et al.*, 2010; Norose *et al.*, 2011; Norose *et al.*, 2003; Norose *et al.*, 2008; Oberdorfer *et al.*, 2003; Ochoa-Repáraz *et al.*, 2011; Okusaga *et al.*, 2011; Oldenhove *et al.*, 2009; Ong *et al.*, 2010; Ortiz-Alegria *et al.*, 2010; Osborn *et al.*, 2010; Passos *et al.*, 2010; Pawlowski *et al.*, 2007; Peixoto–Rangel *et al.*, 2009; Peixoto *et al.*, 2010; Pepper *et al.*, 2008; Perona-Wright *et al.*, 2009; Peterson *et al.*, 1995; Pfaff *et al.*, 2008a; Pfaff *et al.*, 2008a; Pfaff *et al.*, 2007; Pfaff *et al.*, 2008b; Pfefferkorn and Guyre, 1984; Phelps *et al.*, 2008; Pifer *et al.*, 2011; Pino *et al.*, 2007; Plattner *et al.*, 2008; Ploix *et al.*, 2011; Poassos *et al.*, 2010; Pollard *et al.*, 2009a; Pollard *et al.*, 2009b; Powell *et al.*, 1978; Portillo *et al.*, 2012; Portillo *et al.*, 2010; Portugal *et al.*, 2008; Prandovszky *et al.*, 2011; Rajapakse *et al.*, 2007; Ramón *et al.*, 2010; Ravindran and Boothroyd, 2008; Reese *et al.*, 2011; Reichmann *et al.*, 1999; Reichmann *et al.*, 2000; Resende *et al.*, 2008; Rincon, 2012; Robben *et al.*, 2005; Robbins *et al.*, 2012; Rock *et al.*, 2005; Romano *et al.*, 2012a; Romano *et al.*, 2012b; Rosowski and Saeij, 2012; Rosowski *et al.*, 2011; Rozenfeld *et al.*, 2005; Saeij *et al.*, 2008; Saeij *et al.*, 2007; Samarajiwa *et al.*, 2009; Sanchez *et al.*, 2010; Salek-Ardakani and Croft, 2009; Sauer *et al.*, 2009; Sautel *et al.*, 2009; Schaeffer *et al.*, 1999; Schaeffer *et al.*, 2009; Scharton-Kersten *et al.*, 1998; Schluter *et al.*,

TABLE 25.5B Frequencies of HLA-DQ Antigens in White North American AIDS Patients with Toxoplasmic Encephalitis (TE) and Controls

	% Phenotype Frequency						
		Healthy Controls				TE-negative Local AIDS Controls	
			From the Literature				
HLA Antigen	TE Patients ($n = 20$)	Local ($n = 136$)	Serotyped* ($n = 232$)	DNA Typed[†] ($n = 167$)	All Healthy Controls ($n = 535$)	Random ($n = 15$)	*Toxoplasma* Seropositive[‡] ($n = 8$)
DQ1	40.0[§]	64.7	68.1	65.9	66.5[§]	60.0	75.0
DQ2	30.0	38.2	40.5	38.9	39.6	26.7	25.0
DQ3	85.0[‖]	52.2	48.7	55.7	51.8[‖]	40.0	62.5
DQ4	5.0	11.0	4.7	5.4	6.5	20.0	12.5

* *11th International Histocompatibility Workshop*

[‡] *Includes one patient from local TE-negative random group.*

Significant differences (TE patients versus combined healthy controls): [§]$P = 0.028$ and corrected P (P_c) $= 0.108$, [‖]$P = 0.007$ and $P_c = 0.028$.

Adapted with permission from Suzuki.

TABLE 25.6A Association of the *DQ3* Gene with Presence of Hydrocephalus in Infants with Congenital Toxoplasmosis and their Mothers

		Children with Toxoplasmosis		Mothers of Children with Toxoplasmosis	
	USA Population	Without Hydrocephalus	With Hydrocephalus	Without Hydrocephalus	With Hydrocephalus
Number in group	232[a]	45	23	41	21
Gene frequency	0.487[a]	0.444	0.783[b]	0.537	0.762[c]

[a] *Literature source is [13].*

[b] *Differences between* DQ3 *gene frequencies of infants with and without hydrocephalus ($P < 0.02$) and infants with hydrocephalus and the USA population ($P < 0.02$) were significant.*

[c] *The difference between* DQ3 *gene frequencies of mothers of infants with and without hydrocephalus were not significant ($P > 0.05$) but the difference between mothers of children with hydrocephalus and the USA population was significant ($P < 0.03$).*

Adapted with permission from Mack et al. *(1999a).*

TABLE 25.6B Fewer than Expected DQ3 +/DQ3 + (Homozygous) Children with Hydrocephalus

Group	*n*	Observed DQ3+/+	Observed DQ3+/−	Observed DQ3−/−	Expected DQ3+/+	Expected DQ3+/−	Expected DQ3−/−	Chi Square Total
Children with hydrocephalus	23	2	16	5	4.35	11.30	7.35	3.97[a]
Mothers of children with hydrocephalus	21	3	13	5	4.30	10.40	6.30	1.31
Children without hydrocephalus	45	2	18	25	2.69	16.62	25.69	0.31
Mothers of children without hydrocephalus	41	4	18	19	4.12	17.76	19.12	0.008

[a] *Note: Among infants with hydrocephalus, there were fewer than expected DQ3+/+, i.e. DQ3 homozygotes ($\chi2 = 3.97$, $P < 0.04$), whereas differences for other groups were not significant ($P > 0.05$).*

TABLE 25.7 Immune Function in the Foetus and Newborn Infant

	Gw	Foetus	Newborn
Ag presentation in the foetus and neonate			
MHC molecule expression in the Foetus and Neonate	12	Expression of MHC class I and MHC class II molecules by foetal tissues. All APCs, including mononuclear phagocytes, B-cells and DCs are present.	MHC class I expression lower than on adult cells. The amount of MHC class II expressed by neonatal monocytes or B-cells is similar to or greater than expressed by adult cells. Most of neonatal monocytes lack HLA−DR surface expression.
Circulating neonatal dendritic cells			Most DCs found in the tissues, small numbers in the blood, including immature DC1 and pre-DC2 0.5% of circulating mononuclear cells. Predominance of pre-DC2s.
Foetal Tissue Dendritic Cells	16	Epidermal Langerhans cells and dermal DCs in foetal skin. Immature DC2 lineage.	Expression of MHC class II (HLA−DR) and co-stimulatory molecules on neonatal and adult blood
	19−21	Cells with the features of pre-DC2 cells in foetal lymph nodes.	DC1 and pre-DC2 is similar. Diminished ability of neonatal DCs to produce type I IFN and IL−12.
Neonate Derived Dendritic Cells			Neonate dendritic cells from cord blood express less HLA−DR, CD1α and co-stimulatory molecules (CD40 and CD80), they have also decreased all stimulatory activity for T-cells. Limitations in T_H1 immunity, such as delayed-type hypersensitivity skin reactions and antigen-specific $CD4^+$ T-cell IFNγ. APC function of monocytes and B-cells from human neonates appears to be intact. Functional pre-immune T-cell receptor repertoire is fully formed at birth.
Neonatal T-cell			IL−12 production by circulating neonatal T-cells and by adult T-cells is similar.
Proliferation and IL−2 Production			Decreased responsiveness of neonatal T-cells that are generally antigenically naive.
T-Cell Development and Function in a Foetus and Neonate	6−8	Foetal liver contains CD34+ lymphoid cells and appear to include prothymocytes.	The production of most other cytokines and their mRNAs by unfractionated neonatal T-cells or the CD4+ T-cell subset is reduced in response to different stimuli (e.g. anti-

			CD3 mAb mitogen). For most cytokines (IL−3, IL−4, IL−5, IL−6, IL−10, IL−13, IFNγ, GM−CSF) this is an important reduction, for a few, such as TNFα, the reduction is modest. The low capacity of neonatal T-cells to produce IFNγ and IFN4 is due to an almost complete absence of IFNγ and IL−4 mRNA-expressing cells. Neonatal naive $CD8^+$ T-cells produce substantially more IL−13 than adult cells.
	8–9	Initial colonization of the foetal thymus by prothymocytes followed by expression of proteins that are characteristic of T-lineage cells, including CD4, CD8, CD38 and the αβ-TCR−CD3 complex	
	12	Pattern of expression of other proteins such as CD2, CD5, CD38 and the CD45 isoforms matches that in the postnatal thymus. Clear separation between the thymic cortex and medulla.	
	14	Major human thymocyte subsets present.	
	17	ICOS co-stimulatory molecule expressed by thymus	
	18−23	Foetal thymocytes express chemokine receptors CXCR4 and CCR5 (co-receptors for entry of the HIV-1). Medullary CDs in the foetus express high levels of CD80.	
Foetal and Neonatal T-cells Receptors		Thymic involution begins at the end of the third trimester.	
	8–9	Generation of the αβ-TCRs.	
	11−13	Increased diversity of the use of D and J segments in the rearrangement of the TCRβ chain gene in the thymus.	
Ontogeny of T-cell Surface Phenotype	13−15	The CDR3 region of the TCRβ chain transcripts is reduced in length and sequence and increases by the second trimester.	
	12−13	Circulation T-cells detectable.	
	14	$CD4^+$ and $CD8^+$ cells are found in the foetal liver and spleen and $CD4^+$ T-cells are detectable in lymph nodes. The percentage of the T-cells in the foetal circulation gradually increases during the	

(Continued)

TABLE 25.7 Immune Function in the Foetus and Newborn Infant (*cont'd*)

	Gw	Foetus	Newborn
		2nd and 3rd trimesters through 6 months of age, then gradually declines. The ratio of $CD4^+$ to $CD8^+$ T-cells in the circulation is relatively high during foetal life (about 3.5) and gradually declines with age. All peripheral foetal and neonatal T-cells express the CD38 molecule.	
	14–20	A significant fraction of T-cells in the foetal spleen lack CD38 expression.	
Foetal T-cell Function		Circulating T-cells in the 2nd and 3rd trimester foetus and term and preterm neonate predominantly express a $CD45RA^{hi}$–$CD45RA^{lo}$ surface phenotype. A substantial proportion of T-cells in the second trimester foetal spleen are $CD45RA^{hi}$–$CD45RA^{lo}$, a T-cell population that is absent from the spleen of young infants. Foetal $CD45RA^{hi}$ express high levels of CD25 and proliferate with IL-12, however they are not fully functional. Their αβ-TCR repertoire is diverse.	
Foetal and Neonatal T-cell	15–16	Mucosal T-cells with the capacity to secrete substantial amounts of IFNγ after stimulation with anti-CD3 with exogenous IFNα are present in the foetal intestine.	
Expression of TNF ligand family members	19–31	Substantial proportion of circulating foetal T-cells express CD40 ligand *in vitro* in response to polyclonal activation.	Decreased CD40 ligand production by neonatal cells. T-cells have decreased Fas ligand expression after anti-CD28 mAb stimulation compared with adults.
Neonatal T-cell and Co-stimulation and Anergy			Neonatal T-cells have a greater tendency to become anergic, particularly under conditions in which production of inflammatory mediators or co-stimulation (e.g. by CD40, CD80 or CD 86 on the APC) may be limited.

Foetal and Neonatal T-cell Chemokine Receptor Expression		Neonatal naive T-cells lack CCR1 surface expression, they do not increase CXCR3 expression, and they do not decrease CCR7 expression, after activation by anti-CD3 and anti-CD28 mAbs. The CCR7 expressed on neonatal T-cells is functional and mediates chemotaxis of these cells in response to CCL19 and CCL21. Neonatal T-cells can increase their surface expression of CCR5 by treatment with IL-2. Neonatal naive $CD4^+$ T-cells also have the capacity to acquire expression of chemokines characteristic of T_H1 or T_H2 effectors following their differentiation in presence of IL–12 and anti-IL–4 for T_H1 and IL–4 and anti-IL-12 for T_H2.
Foetal and Neonatal T-Cell-Mediated Cytotoxicity	Capacity to generate a functional $CD8^+$ T-cell effector population *in vivo* during chronic stimulation.	Neonatal T-cells are moderately less effective than adult T-cells as cytotoxic effector cells.
Neonatal T-cell Apoptosis		Circulating mononuclear cells from cord blood, including naive $CD4^+$ T-cells, are more prone than those from adults to undergo spontaneous apoptosis *in vitro*. Fas levels are undetectable on neonatal lymphocytes, including $CD4^+$ and $CD8^+$ T-cells. Increased tendency of neonatal naive CD4 and unfractioned T-cells to undergo apoptosis. Expression of lower ratio of Bcl-2 to Bax protein compared to adults. Treatment of neonatal naive $CD4^+$ T-cells with IL–7 can block spontaneous apoptosis. The circulating levels of soluble Fas, TNF and p55 TNFR increase in the first days of life. Neonatal mononuclear cells are also more prone than adults to undergo apoptosis after engagement of MHC class I achieved by mAb treatment.
Regulatory T-cells of the Neonate		About 5% of circulating $CD4^+$ T-cells in neonates but also infants and young children express high levels of CD45RA and CD25 and contain CD45RO transcripts.
Natural Killer T-cells in the Neonate		NKT-cells represent less than 1% of circulating T-cells in the neonatal circulation. Neonatal NKT-cells are similar to adult NKT-cells in having a memory/effector-like cell surface phenotype, including expression of CD25, the CD45RO isoform, and the low level of expression of

(*Continued*)

TABLE 25.7 Immune Function in the Foetus and Newborn Infant (*cont'd*)

	Gw	Foetus	Newborn
			L-selectin. Neonatal NKT-cells produce only limited amounts of IL−4 or INFγ on primary stimulation (functional immaturity). Neonatal $CD56^+$ T-cells express less perforin than do adult cells. Because $CD56^{+\ T\text{-cell}}$ population is highly enriched in CD1d-restricted NKT-cells, NKT-cell cytotoxicity probably is limited at birth.
T-Cell Reactivity to Environmental Antigens	20	Foetal T-cells become primed to environmental and dietary protein allergens as a result of maternal exposure and transfer to the foetus. Regulatory (IL−10-dominant) response. Production of IFNγ 100-fold higher than IL−4, however the ratio is still reduced compared to adults. Protein allergen-specific T-cell proliferation detected at birth is more common when allergen exposure occurs in the first or second trimester rather that in the third trimester.	
Foetal T-cell Sensitization to Maternally Administered Vaccines and Maternally Derived Antigens		Foetal T-cell sensitization can occur in cases of antigen exposure due to chronic infection of the mother with parasites or viruses. Foetal exposure to parasitic antigens without infection can down-regulate subsequent postnatal T_H1 responses to unrelated antigens.	
Maternal Transfer of T-cell Immunity to the Foetus		Maternal-to-foetal transfer of leukocytes occurs, however their number in the foetus is very low ($< 0.1\%$).	Neonatal T-cell responses as a result of transfer of maternal immunity should remain suspect unless the T-cell population is identified and antigen specificity and MHC restriction are demonstrated.
T-Cell Response to Congenital Infection $CD4^+$ T-cells		Reduced $CD4^+$ T-cells responses, particularly at the first- or second-trimester infections, are not absolute and dual T_H1- and T_H2-type immune responses can develop after some congenital infections or foetal exposure to pathogenic antigens from the mother. They may be the result of antigen-specific	Pathogen-specific T-cell proliferative responses and cytokine responses (IL−2 and IFNγ) in infants and children with congenital infection are markedly decreased or absent compared with postnatal infection. T-cells from children with *Toxoplasma* infection retain the ability to respond to alloantigen and mitogen.

		unresponsiveness (anergy, deletion or the failure of the $CD4^+$ T-cell to be initially activated by antigen).	
$CD8^+$ T-cells		$CD8^+$ T-cell responses to congenital infections appear to be relatively robust. Congenital infection with *Toxoplasma* or viruses during the second and the third trimesters may result in appearance of $CD45RO^{hi}$ memory cells and an inverse ratio of $CD4^+$ to $CD8^+$ cells, findings that also suggest that the foetal $CD8^+$ T-cells are activated and expanded in response to serious infection.	The alterations in CD45RO expression by T-cells in congenital infection may persist at least through early infancy.
B-cells and Igs Ontogenesis of B-cells and Igs B-cell maturation and Pre-immune selection Development of Igs	8 10 8–11 13 16 18–22 22	Pre-B-cells detected in the foetal liver and omentum. B-cells express surface IgM but not IgD. Those IgM+IgD– B-cells are transitory stage and express CD21 surface molecule. Transcripts for IgA and IgG present in the foetal liver. Pre-B-cells detected in the bone marrow which becomes the predominant site in the mid-gestation. Foetal bone marrow B-cells expressing Igs of all heavy chain isotypes are detectable. Pro-B cells and Pre-B cells in smaller numbers in the liver, lung and kidney. The proportion of B-cells in spleen, blood and bone marrow is similar to that in adult, further increase of B-cells concentration in 3rd trimester. • B-1 cells predominate during early foetal development, high frequency of CD5 expression (>40% in mid-gestation) indicates predominance of B-1a subset. • CXCL12 (SDF-1) is a critical for B-cells expressing CXCR4 chemokine receptor	• Pre-B-cells are developed solely in bone marrow. • Declining concentration of B-cells until adulthood. • High levels of CD34+CD38– progenitor cells in circulation which are capable to differentiate into B-cells in bone marrow of children and adults. • B-cells have increased surface levels of IgM compared to the adult. • High but gradually decreasing level of $CD5^+$ B-cells. • B-1 is the major source of the low amounts of the low source of circulating natural IgM at birth.

(Continued)

TABLE 25.7 Immune Function in the Foetus and Newborn Infant (*cont'd*)

Gw	Foetus	Newborn
	• The first recognized pre-B cells contain cytoplasmic IgM heavy chains but no light chains or sIg. • In foetal bone marrow light chain rearrangement can occur in the absence of productive heavy chain rearrangement. • Pre-immune immunoglobulin repertoire is limited and significantly shaped by self-antigens. • There are differences in the usage of particular heave chains D and J segments between 1st, 2nd and 3rd trimesters. Overrepresentations of certain fragments flex. DH7-27.	
B-cell surface genotype	• CDR3 region of the immunoglobulin heavy chain is relatively short at the beginning of the 3rd trimester and increases until birth. • Most foetal bone marrow and spleen B-cells express CD10. • Absence of significant differences in MHC class II expression by circulating foetal B-cells and in adults.	• Small numbers of CD10 are found at birth and their decline during infancy. • Increased expression of CD38 on neonatal B-cells. • Lower levels of chemokine receptor CCR7 than in adults. • Reduced CD21 and CD32 expression on neonatal B-cells. • Neonatal CD5 cells have reduced expression of: CD11a, CD44, CD54 (ICAM-1) and L-selectin. • Circulating neonatal cells have lower levels of MHC class II and inability to use intracellular calcium after engagement of MHC class II by mAb. • B-cells (90% B-1a) express CD28 (typically found on T-cells) and more SD27 and CD80 than adults.
Response to T-Dependent Antigens		The capacity of the neonate to respond to T-dependent antigens is well established at birth and is modestly reduced in comparison with the response in the adult.

Response to T-Independent Antigens			Antibody production by human neonatal B-cells to TI type antigen *in vitro* is modestly reduced. The response to TI type II antigens is the last to appear chronologically. Decreased expression of CD21 on neonatal B-cells has been proposed as possible mechanism for limitations in TI type II response in a neonate.
Specific Antibody Response by the Foetus to Maternal		Antibody response by the human foetus may occur following maternal immunization during the third trimester.	Human neonatal B-cells demonstrate a marked decrease in CD22 expression following engagement of IgM, a stimulus to mimic a TI type II antigen.
Immunization and Congenital Infection		Specific antibody may be present at birth to agents of intrauterine infection. T-cell dependent isotype switching and immunoglobulin production occurs during foetal life, at least for certain pathogens.	
Maternally derived IgG		The mechanism of the transmission of maternal IgG to the foetus depends on the recognition of maternal IgG through its Fc domain. Maternal IgG and intracellular receptor FcRn expression can be detected in placental syncytiotrophoblasts during the first trimester but transport does not occur.	
	17	IgG detectable in the foetus.	
	30	Half of the serum concentration at term.	
	38	IgG concentration equal that of the mother or higher.	
Inhibition of Neonatal Antibody Responses by Maternal Antibodies		Maternal antibody may inhibit the production by the foetus or newborn of antibodies of the same specificity. The inhibition varies with the maternal antibody titre and with the type and amount of antigen.	

(*Continued*)

TABLE 25.7 Immune Function in the Foetus and Newborn Infant (*cont'd*)

	Gw	Foetus	Newborn
Foetal and Neonatal T-cell-Dependent Immunoglobulin Production	12 20–30	IgG and IgM synthesis in foetal organ cultures. IgG and IgA secreting plasma cells detectable. Pre-B cells have capacity for isotype switching during foetal ontogeny.	Neonatal plasma cells can differentiate into IgM-secreting plasma cells as efficiently as adult cells. Production of IgM, IgG and IgE by neonatal activated B-cells is similar to that in antigenically naive B-cells.
Immunoglobulin Synthesis by the Foetus and Neonate		Passively transmitted maternal IgG are the source of all subclasses detected in a foetus and neonate.	
IgG		Maternal IgG may inhibit certain postnatal antibody responses by binding to FcγRII receptors and by rapidly clearing or masking potential antigens.	Some of the human neonatal IgM is monomeric and therefore non-functional. Elevated IgM concentrations in cord blood suggest possible intrauterine infections, but many congenitally infected infants have normal IgM cord blood levels.
IgM		Does not cross the placenta. Natural IgM, which is not the result of a B-cell response to foreign antigens, plays an important role in innate defence against infection, allowing time for the initiation of antigen-specific B-cell response; it also enhances antigen-specific B-cell responses.	
IgA			At birth, the frequency of IgA_1 and IgA_2 bearing cells is equivalent.
IgD		IgA does not cross the placenta.	Functions of IgM and IgD are largely redundant.
IgE	11	Synthesis detectable.	
NK-cells NK development and surface phenotype	6 2nd trimester 3rd trimester	Foetal liver is the main site of increasing gradually foetal NK-cells production, probably due to CD34+CD38+Lin− cells development. NK are present in the greatest numbers in mid-gestation. From late gestation onwards the bone marrow is a site of NK production.	• The number of NK-cells is the same or greater than in adult. • 50% of NK-cells at birth are CD56 cells.

	• The development of NK-cells precedes that of αβ-thymic independence. • Commitment to NK lineage is made first then acquisition of inhibitory receptors. • IL−15 critical for directing some lymphocyte precursors into NK lineage and survival of mature NK-cells. • Most foetal liver NK-cells express the CD3-ε and CD3-δ components associated with TCR and CD16 surface expression.	
NK mediated cytotoxicity	• All foetal and neonatal NK-cells lack expression of CD57 and have reduced expression of CD2 and CD56 about 30−50%. • Foetal NK-cells commonly express CD28. • Cytolytic function of NK-cells increases progressively in foetal life and reaches approximately 50% at term of those in adult life.	• Preterm neonates have reduced cytolytic NK function compared to those of the term. • Cytolytic activity depends on the target − pathogen related differences. • Neonatal NK-cells express on their surface IL−2 and IFNγ in numbers equal or greater than those of adult NK-cells. • Neonatal NK-cells are less responsive that in adults to activation by the combination of IL−12 and IL−15 by induction of CD69 surface expression.
Cytokine production		• IL−12 induced production of INF-γ by neonatal mononuclear cells (mostly NK-cells) may be reduced than in adults. • They produce however more INF-γ after stimulation with combination of IL−12 and IL−18. • Neonatal lymphokine-activated killer cells (LAK) have normal capacity to be primed by exogenous cytokines.
Congenital infection and NK-cells	• Congenital viral or *Toxoplasma* infection may increase the number of circulating NK-cells. • NK-cells number remains elevated until birth and is accompanied by decreased NK expression of CD45RO isoform of the CD 45 tyrosine phosphatase.	• Neonatal CD56-NK population is phenotypically and functionally immature NK-cells that give rise to mature CD56+ population.

(*Continued*)

TABLE 25.7 Immune Function in the Foetus and Newborn Infant (*cont'd*)

	Gw	Foetus	Newborn
Phagocytes Neutrophils in the Foetus and Neonate	Yolk sac 14–16 Mid-gestation	Neutrophil precursors detected. In the liver, spleen and bone marrow they appear later than macrophage precursors. Mature neutrophils. The numbers of post-mitotic neutrophils in the foetal liver and bone marrow are markedly lower than in term newborns and adults. Neutrophils constitute less than 10% of circulation leucocytes rising to values of 50% to 60% at term. • The number of circulating neutrophil precursors (CFU–GMs) is 10- to 20-fold higher in a foetus and neonate than in the adult. • Mononuclear cells and monocytes from mid-gestation foetuses and premature neonates generally produce less than in adults.	• Bone marrow contains abundance of neutrophil precursors. • Maximal rate of CFU–GMs. • Within hours of birth the number of CFU–GMs increases sharply in term and preterm neonates. 15% of immature neutrophils in healthy neonate.
Migration to Sites of Infection		• Granulocyte–CSF and granulocyte monocyte GM–CSF after stimulation *in vitro*.	
Eosinophilic Granulocytes	18–30	• The most critical deficiency in phagocyte defences in the foetus and premature neonate is their limited ability to accelerate neurtophil production in response to infection.	
Mononuclear Phagocytes in the Foetus and Neonate			
Humoral Mediators of Inflammation and Opsonization			

Complement	6–14	• Little if any maternal complement is transferred to the foetus.	• Substantial interindividual variability in term neonates but alternative pathway components are more often decreased than are the classic pathway activity. Decreased abundance of some terminal components. Preterm infants have more often both pathways components decreased.
C-reactive protein Mannan-Binding-Lectin (MBL) Fibronectin Opsonic activity Chemotactic Factor Generation		• Does not cross placenta.	• Small for gestational age neonates have the values similar to healthy neonates. • Serum levels in healthy neonates are higher that in adults. • Concentration of MBL in preterm neonates is approximately 50% lower than in term neonates. • Plasma fibronectin concentrations are low in neonates, particularly in the premature; they are further reduced in infections; they reach normal ranges in one year of age. • Efficiency with which neonatal sera opsonize organisms is quite variable but they are less able to opsonize organisms in the absence of antibody. • Sera from term neonates generate less chemotactic activity than adult sera and this diminished activity reflects a defect in complement activation rather than antibody.

Concepts adapted from Lewis and Wilson (2005) with permission.
Gw, gestational week at which response appears.

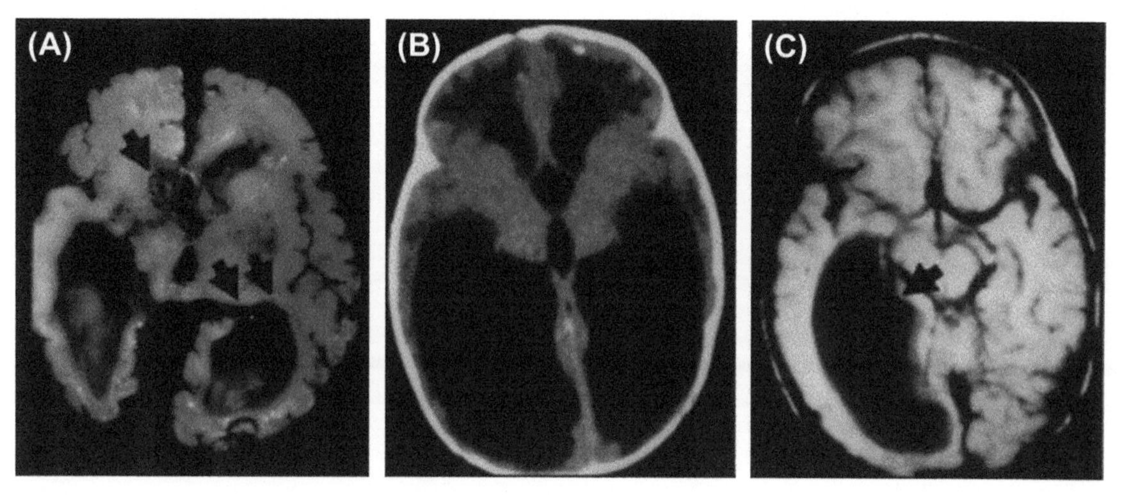

FIGURE 25.17 **Hydrocephalus in congenital toxoplasmosis.**
(A) Pathologic specimen which demonstrates the periaqueductal and periventricular necrosis and inflammation (single arrow) which has obstructed the aqueduct of Sylvius and caused ventricular dilatation (double arrow). *(Photograph kindly provided by J.S. Remington, Stanford, California, USA.)*
(B) Computed tomography of the brain of an infant with hydrocephalus with bilateral ventricular dilation.
(C) Magnetic resonance image of the brain of an infant with unilateral hydrocephalus (arrow) secondary to obstruction of the foramen of Monro. *(Adapted from Mack* et al. *(in preparation), with permission.)*

2001; Schluter *et al.*, 1995; Schluter *et al.*, 2003; Schulthess *et al.*, 2008; Schulthess *et al.*, 2012; Schwartz and Hunter 2007; Senegas *et al.*, 2009; Severance *et al.*, 2012; Sergent *et al.*, 2005; Shapira *et al.*, 2005; Shaw *et al.*, 2006; Shaw, *et al.*, 2009; Shi *et al.*, 2012; Shiono *et al.*, 2007; Sibley and Ajioka, 2008; Sibley *et al.*, 2010; Silva *et al.*, 2010; Silva *et al.*, 2009; Silver *et al.*, 2011; Sims *et al.*, 1989; Skallova *et al.*, 2006; Skariah and Mordue, 2012; Skariah *et al.*, 2012; Spear *et al.*, 2006; Starnes *et al.*, 2009; Stenzel *et al.*, 2004; Stibbs, 1985; Strack *et al.*, 2002; Strobl *et al.*, 2012; Stumhofer *et al.*, 2006; Stutz *et al.*, 2012; Subauste 2009a; Subauste 2009b; Subauste 2009c; Subauste *et al.*, 1991; Subauste *et al.*, 2002; Sukhumavasi *et al.*, 2008; Sukhumavasi *et al.*, 2010; Sukhumavasi *et al.*, 2007; Sun *et al.*, 2007; Suzuki *et al.*, 1994a; Suzuki *et al.*, 2005; Suzuki, 1997; Suzuki and Remington, 1988; Suzuki and Remington, 1990; Suzuki *et al.*, 1996b; Suzuki *et al.*, 1989; Suzuki *et al.*, 1994b; Suzuki *et al.*, 2010; Tait *et al.*, 2010; Tait and Hunter, 2009; Takács *et al.*, 2012; Tan *et al.*, 2010; Tato *et al.*, 2012; Taylor *et al.*, 2004; Terrazas *et al.*, 2010; Thompson *et al.*, 2008; Tomita *et al.*, 2009; Torrey *et al.*, 2012; Tu *et al.*, 2009; Tussiwand *et al.*, 2012; Unoki *et al.*, 2009; Vallochi *et al.*, 2008; Van *et al.*, 2007; Vaudaux *et al.*, 2010; Villard *et al.*, 2012; Villarino *et al.*, 2003; Villarino *et al.*, 2006; Villegas *et al.*, 2002;

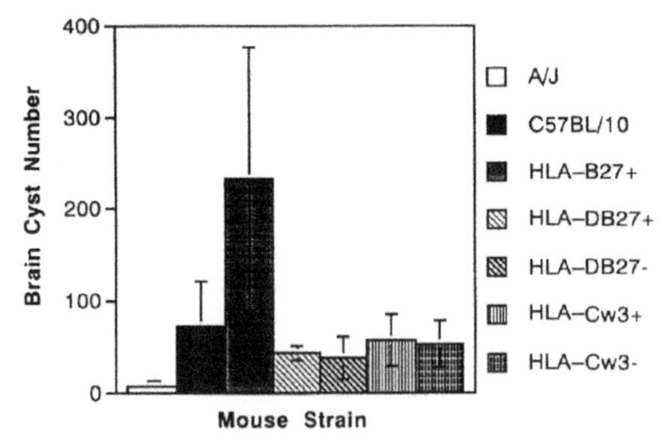

FIGURE 25.18 **Effect of human MHC transgenes on parasite burden in brain.** Cyst numbers in brains of control and HLA-transgenic mice. Presence (+) or absence (−) of transgene expression in littermates is indicated. *(Adapted from Brown* et al. *(1994), with permission.)*

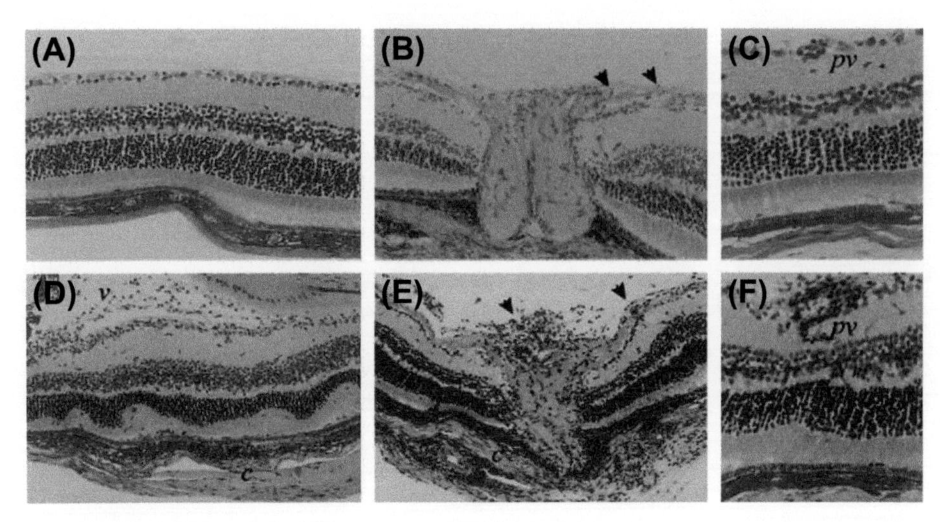

FIGURE 25.19 **C57BL6 mice develop toxoplasmic retinochoroiditis that can be exacerbated by inhibition of NO production by L-NAME treatment.**
(A) In infected mice not treated with L-NAME, the inflammatory infiltrate is focal and in areas where the retina appears normal.
(B) In other areas, there is a mild inflammatory cell infiltrate (arrows), confined to the inner retinal layers and optic nerve head.
(C) In these mice there is only mild perivascular cuffing (pv).
(D, E) In L-NAME treated mice, inflammatory infiltrate is severe and diffuse.
(F) There are numerous cells in the vitreous (v) and expanding choroid as well as within the inner retinal layers and optic nerve head (arrows). Perivascular cuffing by inflammatory cells is more severe in L-NAME treated mice than in untreated mice. *(Adapted from Roberts* et al. *(2000), with permission.)*

Virreira *et al.*, 2011; Vutova *et al.*, 2007; Vyas *et al.*, 2007a; Vyas *et al.*, 2007b; Vyas and Sapolsky, 2010; Vyas, 2013; Wang *et al.*, 2009a; Wang *et al.*, 2010a; Washino *et al.*, 2012; Wang *et al.*, 2007; Wang and Suzuki, 2007; Wasmuth *et al.*, 2012; Watford *et al.*, 2008; Webster *et al.*, 1994; Wen *et al.*, 2010; Wen *et al.*, 2010; Wendte *et al.*, 2011; Whitmarsh *et al.*, 2011; Whitmarsh and Hunter, 2008; Wiehagen *et al.*, 2012; Willie *et al.*, 2001; Wilson *et al.*, 2009; Wilson, D.C. *et al.*, 2010; Wilson, M.S. *et al.*, 2010; Witting, 1979; Wiley *et al.*, 2010; Wilson *et al.*, 2006; Wilson *et al.*, 2005; Wilson and Hunter, 2004; Wilson *et al.*, 2008; Witola *et al.*, 2011; Woodman *et al.*, 1991; Yamamoto *et al.*, 2012; Yamamoto *et al.*, 2011; Yamashita *et al.*, 1998; Yamauchi *et al.*, 2007; Yap *et al.*, 2007; Yap *et al.*, 2000; Yap and Sher, 1999b; Yarovinsky *et al.*, 2008; Yarovinsky *et al.*, 2005; Yarnovinsky *et al.*, 2005, 2008; Yong *et al.*, 1994; Young *et al.*, 2012; Xiao *et al.*, 2009; Xiao *et al.*, 2012; Zakrzewski, 2011; Zeiner and Boothroyd, 2010b; Zhang *et al.*, 2007; Zhang and Denkers, 1999; Zhao *et al.*, 2008; Zhao *et al.*, 2009b; Zhou *et al.*, 2012; Zhao *et al.*, 2009a; Zhao *et al.*, 2007; Zukas *et al.*, 2009).

25.4.2.2 TLR2/TLR4 and Signalling Pathways

Toll-like receptors (TLRs) recognize microbial components that are distinct from the host self-molecules. Stimulus of TLRs activates transcription factor NFκB and leads to production of cytokines. TLRs use MyD88 to cause translocation of NFκB to the nucleus (Fig. 25.21). Recent excellent reviews (Denkers, 2010; Yarnovinsky *et al.*, 2005, 2008) summarize these interactions as they apply to a variety of pathogens and polymorphisms in TLRs that have been shown to influence outcomes of a wide variety of human infections and other diseases. The proinflammatory and regulatory cytokines produced shape the adaptive

(A)

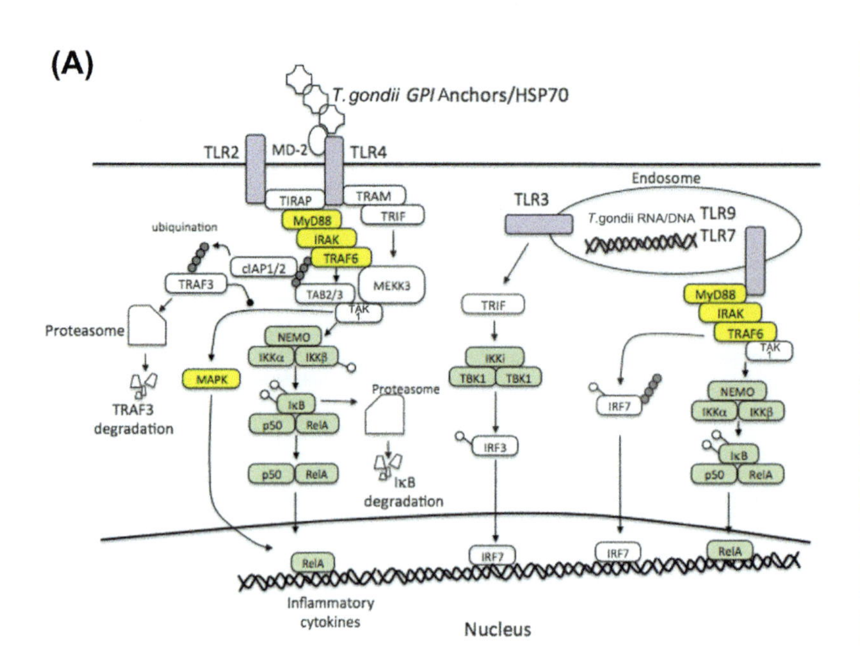

TLR4 and TLR2
T. gondii GPI anchors are ligands for human **TLR2**. **TLR9** SNPs associated with susceptibility to eye disease in Brazil, congenital toxoplasmosis in North America (manuscript in preparation); RNA, DNA, and CPGs stimulate **TLR7** and **TLR9** respectively. **TLR 3, 7, and 9** UNC knockout mice are susceptible.
Mice deficient in TLR4 and TLR2 have increased cyst burden.

IRAK4
SNPs in human IRAK4 are associated with susceptibility to congenital toxoplasmosis
Mice deficient in IRAK4 have reduced survival and increased parasite numbers.
MyD88
Mice deficient in MyD88 and ICE have reduced survival
TRAF
T. gondii GRA15 has been shown to interfere with TRAF6
MAPK
T. gondii ROP38 has been demonstrated interfere with p38 MAPK

NFκB
Mice deficient in components of NFκB are known to have reduced survival. *T.gondii* GRA15 II modulates NFκB translocation to the nucleus (also please refer to the figure for P2X7R)

(B)

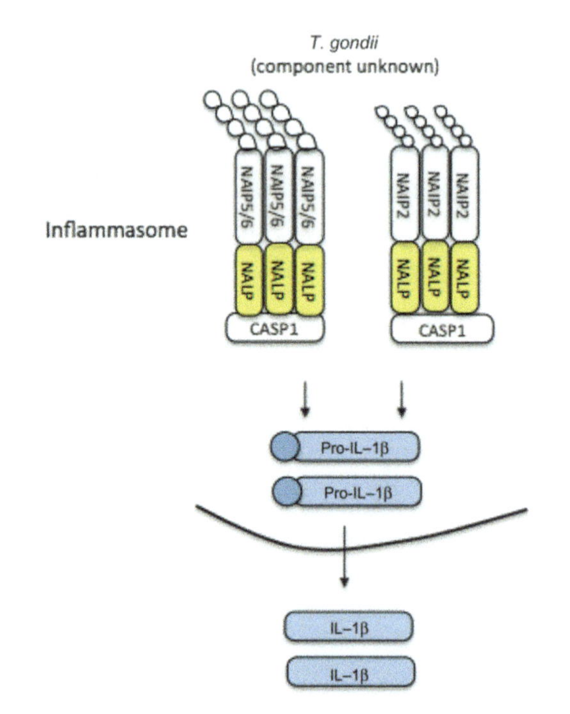

NALP1
The *NALP1* allele in humans is associated with susceptibility to congenital toxoplasmosis. *NALP1* is involved in the cell death phenotype following infection of a human monocytic cell line.

immune response. For example, they may lead to production of Th17 (TgFβ + IL-6), Foxβ regulatory T-cells, or T-cells that produce harmful IL-17 (IL-6 and other proinflammatory cytokines not balanced by downmodulatory cytokines). Stimulus of TLRs also induces production of IL-12 and IFNγ by human monocytes and mouse macrophages. Glycosylphosphatidylinositol-anchored proteins (GPIs) expressed on the surface of *T. gondii* tachyzoites have been demonstrated by Debierre-Grockiego and Schwartz (2012) to play a role in inducing NFκB activation in human monocytes. The ability of *T. gondii* whole GPIs or their fragments to promote NFκB translocation and activate macrophages to produce proinflammatory cytokines such as TNFα also prompted the study of their ability to activate macrophages from wild type and TLR2/4^{-}, MyD88 deficient mice (Debierre-Grockiego *et al.*, 2007).

In order to further characterize whether TLR4 and TLR2 contribute to susceptibility to toxoplasmosis, peritoneal macrophages from mice of the TLR$2^{-/-}$, TLR$4^{-/-}$, TLR2/$4^{-/-}$, CD$14^{-/-}$ and MyD$88^{-/-}$ strains on a C57BL/6 background were obtained and stimulated with various molecules of *T. gondii*. UV-killed *S. aureus and E. coli* LPS were used as controls for TLR2 and TLR4 dependent TNFα production. The peritoneal macrophages were incubated for 24 hours in the presence of six different GPIs (I–VI) extracted from *T. gondii* RH strain tachyzoites before the supernatants were assayed for TNFα using a sandwich ELISA. No difference in TNFα production was found between macrophages of wild-type and TLR$2^{-/-}$ or TLR$4^{-/-}$ mice in response to whole, highly purified *T. gondii* GPIs. However, significantly reduced amounts of TNFα were found in the supernatant of MyD$88^{-/-}$ macrophage cultures. No production of TNFα was seen in response to GPIs in macrophages of mice deficient in both TLR2 and TLR4 (Fig. 25.21A) (Debierre-Grockiego *et al.*, 2007).

The results obtained from *in vitro* activation by *T. gondii* GPIs prompted investigation into the significance of TLR2 and TLR4 during *in vivo* infection. C57BL/6, TLR$2^{-/-}$, TLR$4^{-/-}$, TLR2/$4^{-/-}$, CD$14^{-/-}$, and MyD$88^{-/-}$ mouse strains on a C57BL/6 background were infected intraperitoneally with 10 cysts of the ME-49 strain of *T. gondii*. Mouse mortality was assessed daily. After 40 days of infection, surviving mice were euthanized and brains were harvested in order for the number of brain cysts to be counted. Cysts were not obtained from MyD$88^{-/-}$ mice. Although the survival rate of TLR$2^{-/-}$/TLR$4^{-/-}$ mice reached 100% at day 40 post-infection, the number of brain cysts in these mice was higher than the other compared strains (Debierre-Grockiego *et al.*, 2012). Cytokine production from spleen cells from the uninfected controls and infected C57BL/6, TLR$2^{-/-}$, TLR$4^{-/-}$, TLR2/$4^{-/-}$, CD$14^{-/-}$ and MyD$88^{-/-}$ mice was also measured after stimulation with F_3, an antigenic preparation enriched in

FIGURE 25.20 **Signalling pathways *T. gondii* infection.**
(A) TLR signalling activates MAPK and NFκB pathways.
(B) NAIP regulates (NOD)-like receptor signalling. The NALP inflammasome requires NAIP for formation. NAIP binds to NALP, and this complex activates caspase-1. This activation leads to an inflammatory response *via* IL−1B and IL−18, which leads to cell death.
(C) NOD1/2 activates the MAPK and NFκB pathways.
(D) TNFα activates the classical NFκB pathway.
(E) The CD40 ligand or tumour necrosis factor-like weak (TWEAK) inducer of apoptosis activates the alternative NFκB pathway.
(F) Diagram showing metabolic regulation of M2 macrophages and Th17 cells. Diagram (a) demonstrates the role of M2 macrophages in processes such as tissue repair and antiparasitic effects. Diagram (b) shows the effects of HIF-1α, which induces the key transcription factor RORγt for IL−17 and promotes degradation of the T_{reg} transcription factor Foxp3. *(A−E adapted with permission from Akira* et al. *(2010). F adapted with permission from O'Neill and Hardie (2013).)*

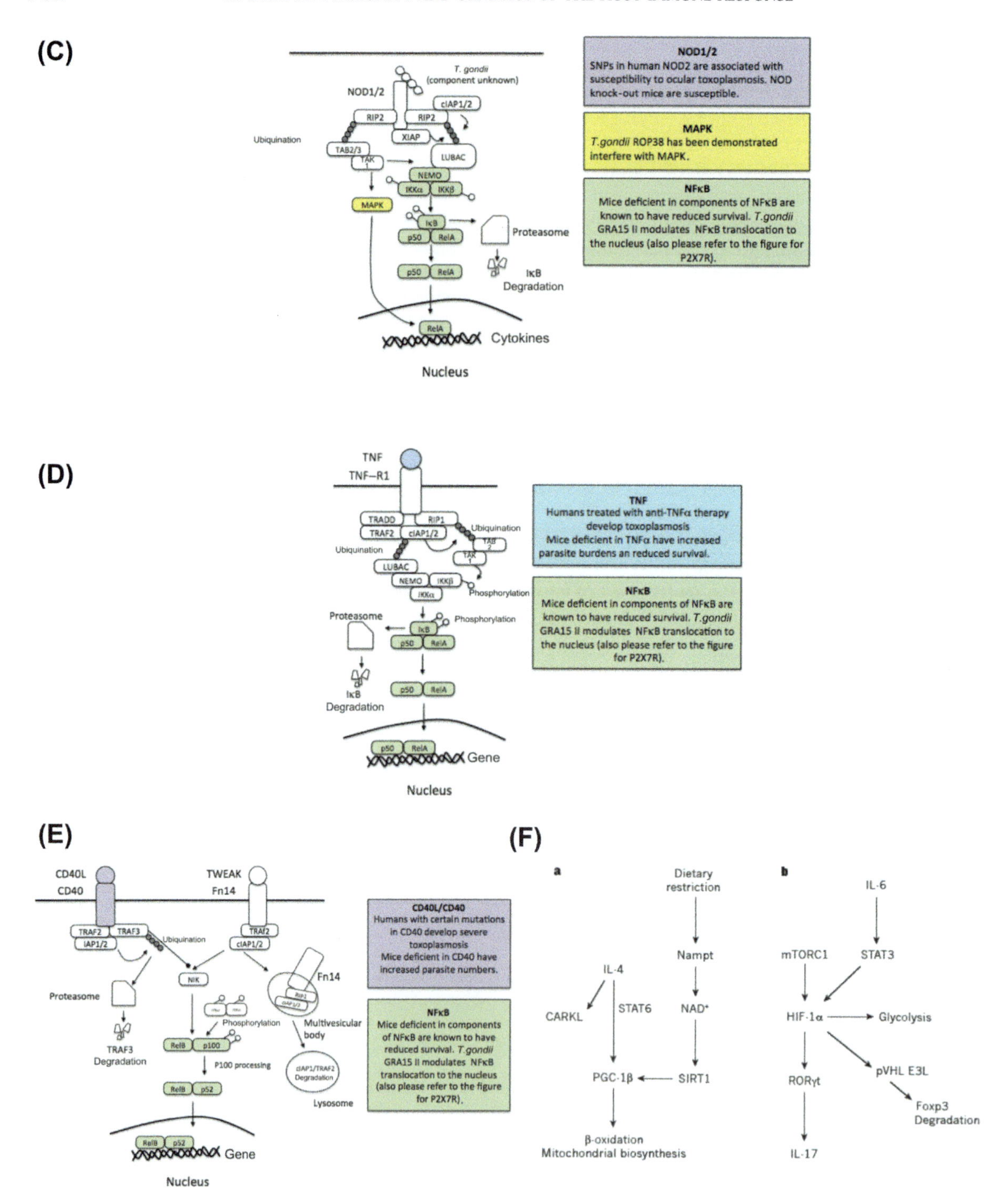

FIGURE 25.20 *(continued)*.

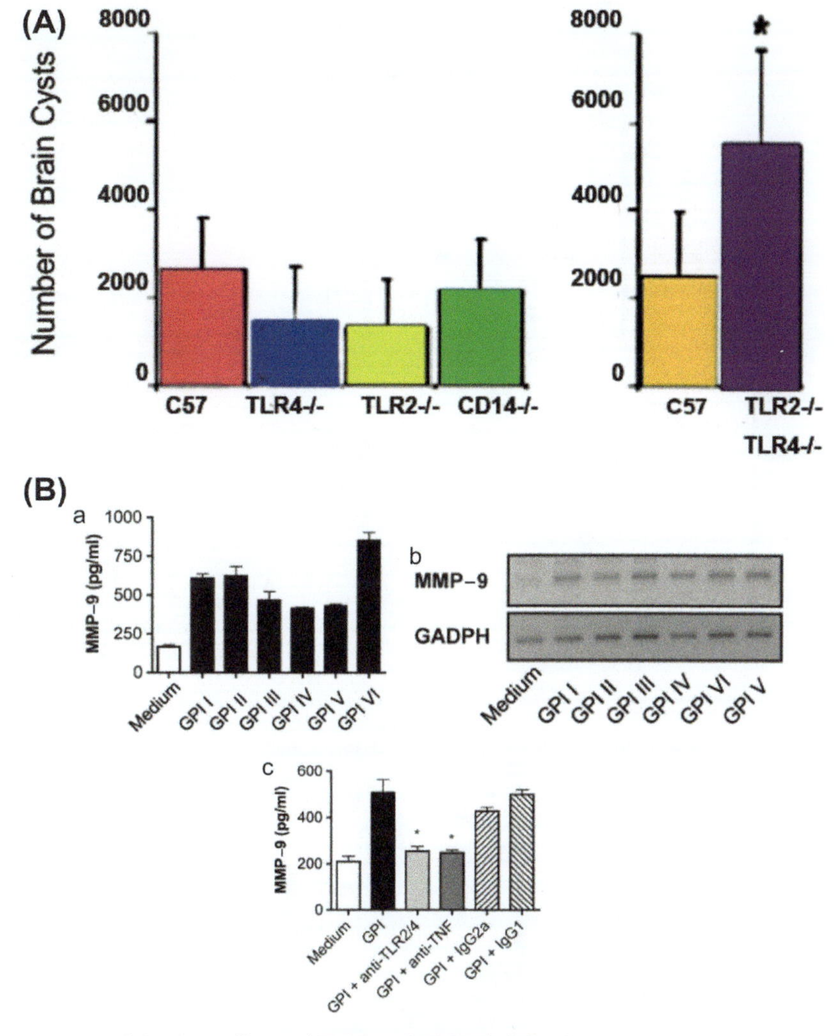

FIGURE 25.21 ***In vivo*** **and** ***in vitro*** **effects of TLR2 and TLR4 deficiencies.**
(A) Number of brain cysts is increased in TLR2$^{-/-}$/TLR4$^{-/-}$ double knock-out mice but not in the single TLR4$^{-/-}$/TLR2$^{-/-}$ knock-out mice.
(B) (a) THP-1-derived macrophages incubated for 24 hours with *T. gondii* GPIs (each at 2 μm) have increased MMP-9 production compared to macrophages incubated with medium alone. (b) Expression of *mmp*-9 evaluated by RT–PCR is also elicited by GPIs. (c) THP-1 derived macrophages pre-incubated with neutralizing antibodies against human TLR2 and TLR4, or neutralizing antibody for human TNFα, or with medium demonstrate TLR2 and TLR4 and TNF contribute to production of MMP-9.
(C) Nucleic Acid Sensing TLRs (TLR7 and TLR9) are important for IL–12, p70, TNFα and IL–1β production and NFKβ production particularly when primed with IFNγ. *T. gondii* RNA and DNA including CPGs from *T. gondii* identified, activate NFkβ transfected HEK cells and human PBMCs (peripheral blood mononuclear cells), but recombinant profilin does not. *(A is adapted with permission from Debierre-Grockiego* et al. *(2007). B is adapted with permission from Niehus* et al. *(2012). C is adapted with permission from Andrade* et al. *(2012).)*

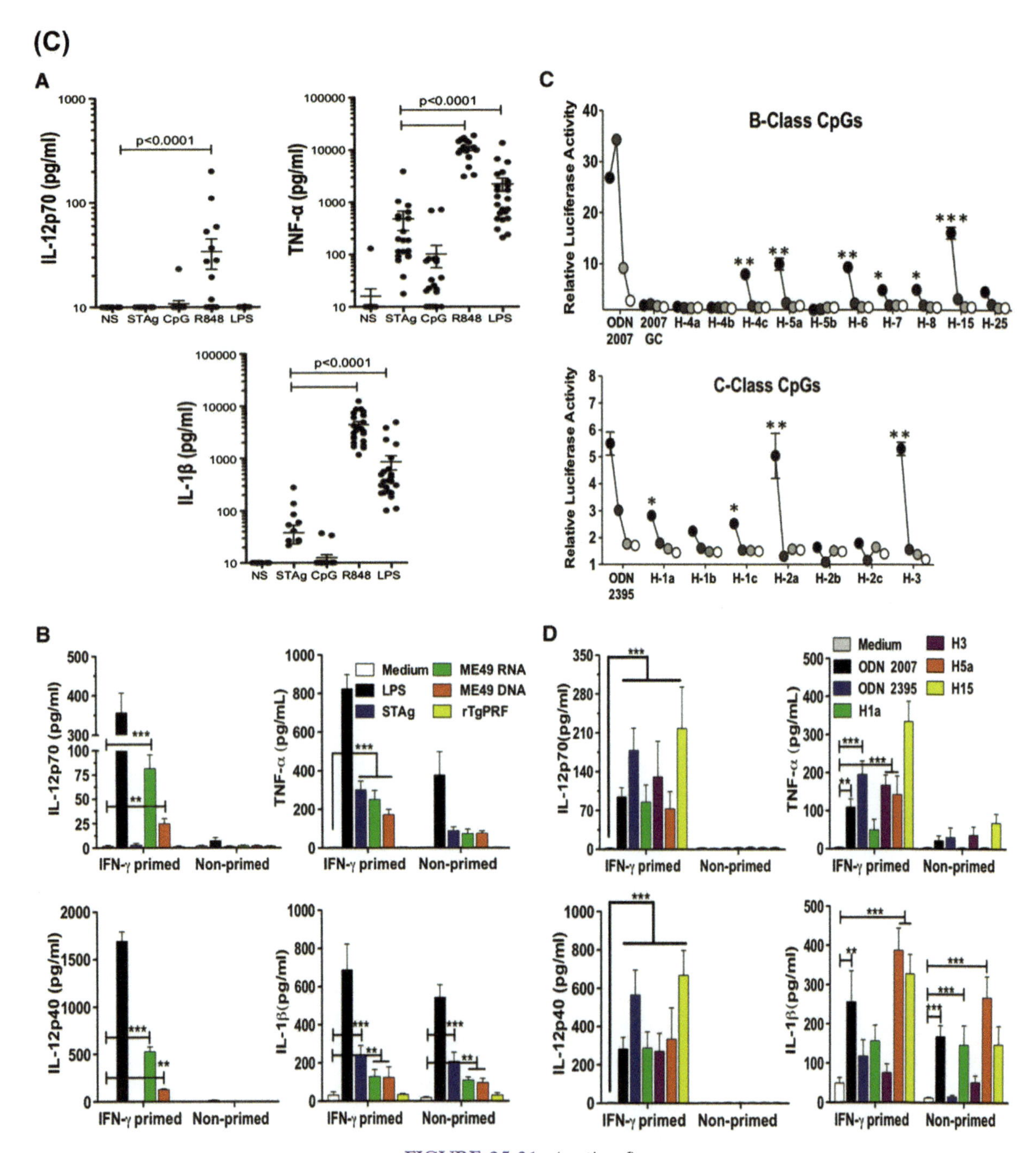

FIGURE 25.21 *(continued)*.

GPI-anchored surface proteins from *T. gondii* tachyzoites. F_3 stimulated splenocytes from ME-49 *T. gondii* infected TLR2/4$^{-/-}$ mice produced higher levels of IFNγ and IL-12 (Debierre-Grockiego *et al.*, 2012). The *in vivo* and *in vitro* results of this study together indicate that activation of both TLR2 and TLR4 by *T. gondii* GPIs is necessary for TNFα production by host-cell macrophages in the mouse model. TLR2/4$^{-/-}$ mice are more susceptible to infection by *T. gondii* and present with a larger number of cysts in the brain than TLR2$^{-/-}$, TLR4$^{-/-}$, and CD14$^{-/-}$ mice (Debierre-Grockiego *et al.*, 2012). Additional studies that characterize the TLR2 and TLR4 ligands and their downstream signalling and cytokine production have been completed (Hargrave *et al.*, manuscript in preparation 2013).

25.4.2.3 TLR9

Hargrave (manuscript in preparation, 2013), using NCCCTS congenital toxoplasmosis data, and Peixieto-Rangel (2009), using Brazilian ocular toxoplasmosis data, have demonstrated TLR9 susceptibility alleles. Andrade *et al.* (2006) found 3d mice with a mutation in UNC93B1, a chaperone for the Nucleic Acid Sensing (NAS) Toll-like receptors TLR3, TLR7 and TLR9, are highly susceptible to *T. gondii* infection. In contrast, none of the single or triple NAS–TLR-deficient mice had the 3d susceptible phenotype. TLR11 and TLR12 are required to work as heterodimers for murine cell sensing and responding to *T. gondii* profilin, while parasite DNA and RNA activate innate immune responses via TLR3, TLR7 and TLR9. Triple TLR7/TLR9/TLR11-deficient mice were highly susceptible to *T. gondii.* Humans do not have functional *TLR11* and *TLR12* genes. Human cells did not produce proinflammatory cytokines (IL-12, TNFα) in response to recombinant *T. gondii* profilin (Fig. 25.21C). Parasite RNA, DNA, and CPGs all activate NFκβ and elicit IL-12, TNFα, and IL-1β in HEK TLR transfected cells and with IFNγ priming in human monocytes (Fig. 25.21C). Collectively, these findings support a critical role for NAS–TLRs acting together in protection against toxoplasmosis in humans (Andrade *et al.*, 2006) and also support that there is potentially RNA and DNA sensing through other pathways as well.

25.4.3.4 NALP1

NALP1 was noted to be the most telomeric gene within the TOXO1 region. This is a 30 gene region that was associated with marked resistance of Lewis strain rats and recombinant congenic (RCS) rats based on crosses between the Lewis (resistant) and a susceptible strain background against *Toxoplasma* infection. At the time it was selected as a candidate gene within this region that had 11 candidate cell death genes because it was the most telomeric and it was thought that if the gene were syntenic with the rat gene in protection against toxoplasmosis for humans it might also help to narrow the 30 gene region. Thus, whether NALP1 might be a (or the) susceptibility/resistance determinant in human congenital infection was tested as a candidate gene. This was performed using SNPs in this gene, and Transmission Disequilibrium Testing (TDT) with DNA samples from the North American cohort of families with a child with congenital toxoplasmosis (McLeod *et al.*, 1990, 2006, 2012; McAuley *et al.*, 1994). DNA from 149 case-parent samples from the National Collaborative Chicago-Based Congenital Toxoplasmosis Study (NCCCTS) was genotyped at 23 single-nucleotide polymorphism tags distributed throughout the NALP1 gene (Witola *et al.*, 2011) (Fig. 25.22A). There are susceptibility/resistance alleles of NALP1 for congenital toxoplasmosis ($p < 0.03$).

NALP1 is a member of the NOD-like receptor family of proteins. NOD-like receptors (NLRs) function to modulate the expression of proinflammatory cytokines IL-1β and IL-18. NALP1 consists of a C-terminal leucine-rich repeat domain,

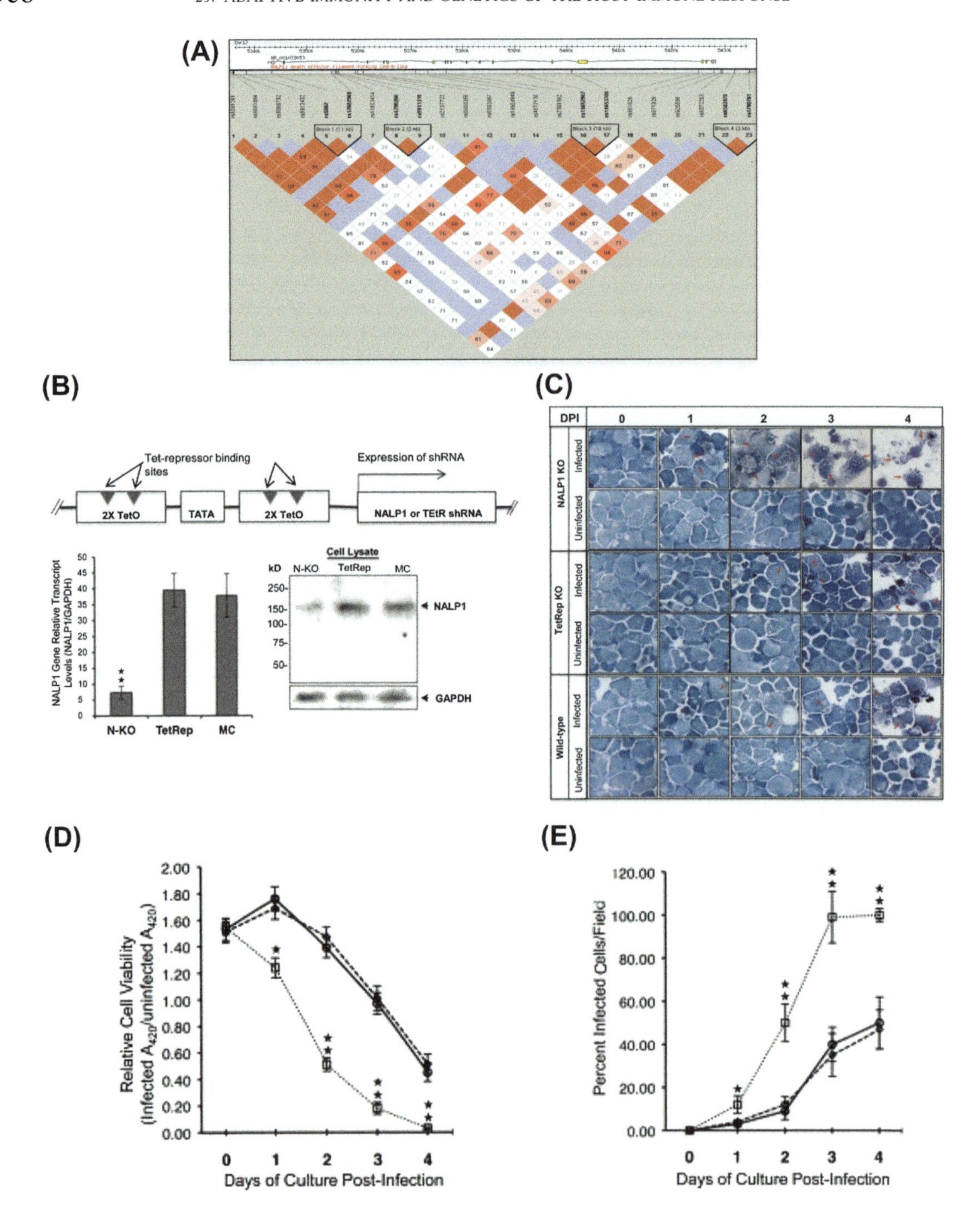
(A)
(B)
Tet-repressor binding sites
Expression of shRNA
2X TetO
TATA
2X TetO
NALP1 or TEtR shRNA
NALP1 Gene Relative Transcript Levels (NALP1/GAPDH)
N-KO
TetRep
MC
Cell Lysate
kD
N-KO
TetRep
MC
NALP1
GAPDH
(C)
DPI
NALP1 KO
TetRep KO
Wild-type
Infected
Uninfected
(D)
Relative Cell Viability (Infected A_{420}/uninfected A_{420})
Days of Culture Post-Infection
(E)
Percent Infected Cells/Field
Days of Culture Post-Infection

a central nucleotide-binding oligomerization domain, and a variable N-terminal protein–protein interaction domain (Witola *et al.*, 2011). The N-terminal domain is a caspase recruitment domain that mediates protein–protein interactions for downstream signalling. NALP1 assembles in an inflammasome complex containing NALP1, the apoptosis associated protein containing the C-terminal caspase activation and recruitment domain, caspase-1 and caspase-5 and NTPs. The minimum requirements are NALP1, caspase-1 and NTP (Bruey *et al.*, 2007). During cellular stress or infection, this inflammasome complex is activated and this activation results in cell death (Witola *et al.*, 2011). In order to better understand the mechanism of NALP1 in susceptibility to human congenital toxoplasmosis, an *in vitro* culture-adapted human monocytic cell line called MonoMac6 (MM6) that was either uninfected or infected with *T. gondii* for various lengths of time was studied. Stable lentivirus knockdown transfected cells of this line were created for these studies. To study the effect of NALP1 gene knockdown on NALP1 gene expression and NALP1 protein levels, IL-1ß, IL-12 and IL-18 levels, the viability of human monocytic cell cultures and appearance of *T. gondii* in those cultures were studied when wild-type MonoMac6 cells, and those modified to knockdown NALP1 or TetR gene, were cultured in the presence or absence of *T. gondii*. Both wild-type MonoMac6 cells and those modified with knockdown of the NALP1 or TetR genes were studied at different times (–, 1, 2, 3 and 4 days) (Witola *et al.*, 2011). NALP1 knock-down in human monocytic cells increases infection by *T. gondii* and reduces the increase in proinflammatory cytokines I IL-1β, IL-18 and IL-12

FIGURE 25.22 **NALP1 SNPs and *T. gondii* viability in MonoMac6 cells with wild-type or knock down of *NALP1*.**
(A) Analysis of NALP1 SNPs. The upper diagram shows positions of genotyped SNPs relative to the intron–exon structure of the gene and the lower diagram shows the L^d plots generated in Haploview, using NALP1 gene data from the North American patient cohort. Bright red and pink diamonds in boxes depict SNPs with high probabilities of being associated with susceptibility to congenital toxoplasmosis.
(B) The diagram of the promoter (top) shows that there is constitutive expression of the NALP1 or TetR shRNA fragment in the absence of a Tet repressor protein. Graph in bottom left shows results of quantitative real-time PCR analysis of NALP1 transcript levels in MonoMac6 cells expressing the NALP1 gene shRNA (N–KO) or the tetracycline repressor gene shRNA (TetRep) or in wild-type MonoMac6 cells (MC) after 72 hours of culture. Reduction in NALP1 transcript levels in N–KO cells was significant ($p < 0.001$). The figure on the lower right shows a Western blot of NALP1 protein expression in MC, TetRep, and N–KO cells after 72 hours of culture using anti-NALP1 antibody.
(C) Effect of NALP1 gene knockdown on viability of monocytic cells and presence of *T. gondii* shown *via* Giemsa-stained preparations. Red arrows indicate cells infected with parasites. Monocytic cells modified for knock-down of NALP1 (NALP1 KO) or tetracycline repressor (TetRep KO) and wild-type cells are shown at various days post-infection (DPI).
(D) Effect of NALP1 knock-down on viability of MonoMac6 during infection with RH *T. gondii*. Parts D, E and F depict wild-type MonoMac6 cells (solid lines with circles), NALP1 knockdown cells (dotted lines with squares) and tetracycline repressor knockdown cells (dashed lines with rhombuses). Relative cell viability decreased significantly in MonoMac6 cells with NALP1 knocked out.
(E) *T. gondii* RH strain viability remained high in MonoMac6 cells with NALP1 knock-down after one day post-infection.
(F) The mean number of *T. gondii* RH strain parasites per vacuole remained high post-infection in MonoMac6 cells with NALP1 knock-down.
(G) Western blot shows NALP1 knock-down on expression of IL–1β in MonoMac6 cells infected with or without *T. gondii*. Mature IL–1β was shown as a ~17-kDa band and pro-IL–1β was shown as an ~35-kDa band.
(H) Effect of NALP1 gene knock-down on expression of IL–1β.
(I) Effect of NALP1 gene knock-down on expression of IL–18.
(J) Effect of NALP1 gene knock-down on expression of TNFα.
(K) Effect of NALP1 gene knock-down on expression of IL–12. *(A–K are adapted with permission from Witola* et al. *(2011).)*

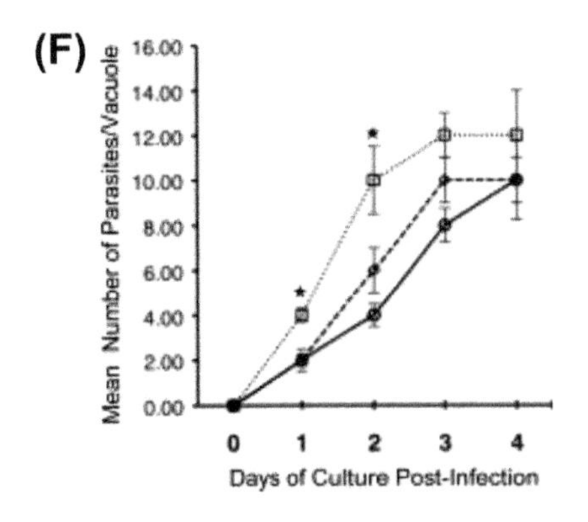

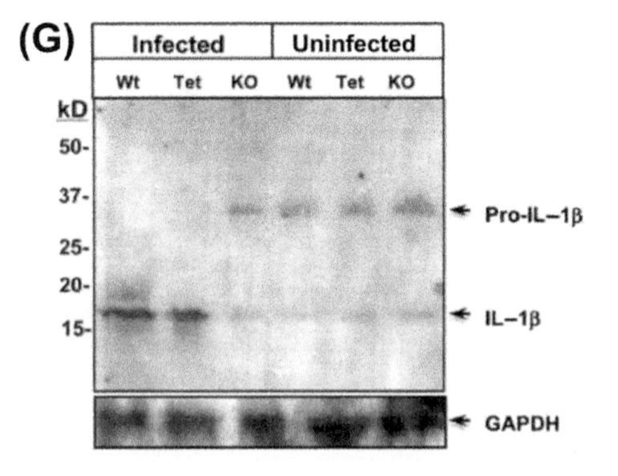

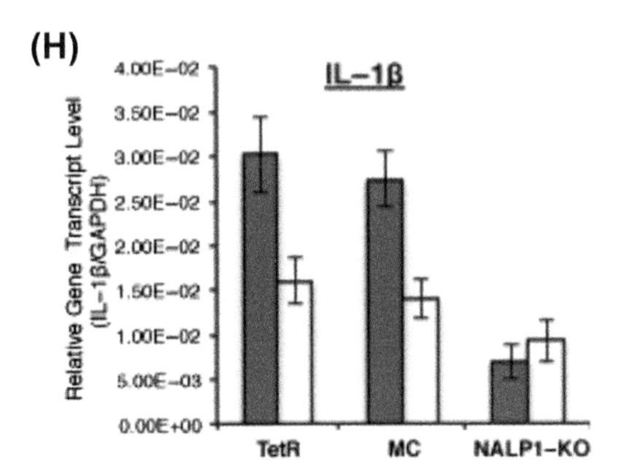

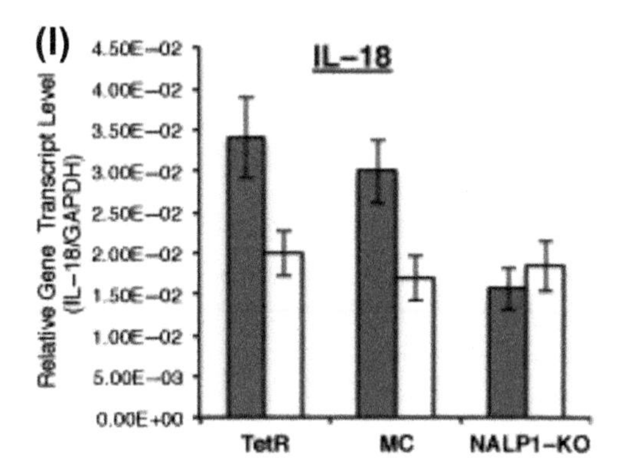

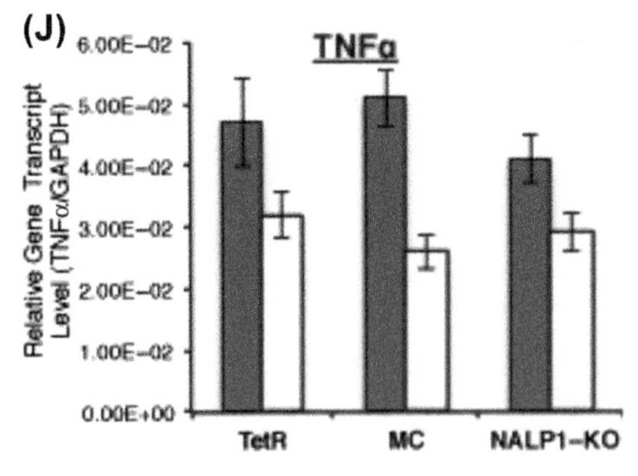

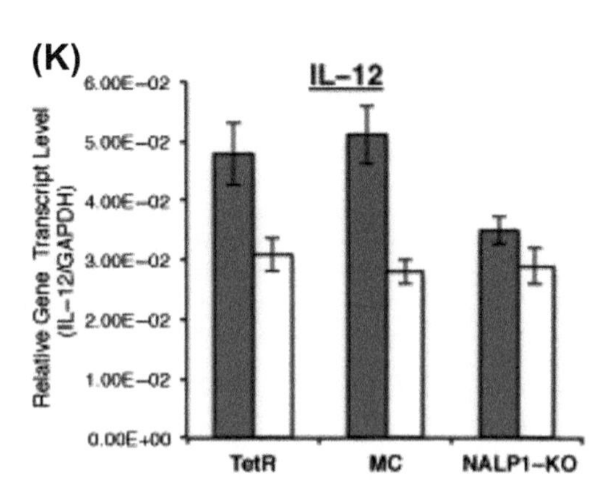

FIGURE 25.22 (*continued*).

and cleavage of pro-IL-1β by caspase-1 (Fig. 25.22). These studies demonstrated that human NALP1 has alleles associated with susceptibility to human congenital toxoplasmosis. Human NALP1 influences proinflammatory cytokine response elicited by *T. gondii* and fate of *T. gondii*-infected monocytic cells (Witola *et al.*, 2011) (Fig. 25.22B).

25.4.3.5 NOD2 and IL-17A

NOD2 (nucleotide-binding oligomerization domain containing 2) is an intracellular pattern-recognition receptor, which recognizes bacterial peptidoglycans and is expressed in neutrophils, monocytes, macrophages and dendritic cells. NOD2 knock-out mice have increased susceptibility to *T. gondii*. In mice, *T. gondii* infection activated NOD2, triggering translocation of NFκB, inducing the production of proinflammatory mediators and thereby activating antigen-specific $CD4^+$ T-cell responses (Dutra *et al.*, 2012). In order to determine whether NOD2 contributes to susceptibility to ocular toxoplasmosis in humans, DNA was extracted from PBMC of 301 individuals from cohorts from Vale do Jequitinhonha in the Minas Gerais state (the MG cohort) and Campos dos Goytacazes in the Rio de Janeiro state (the RJ cohort) in Brazil. SNPs in NOD2 were genotyped for all individuals and allelic associations were assessed (Fig. 25.23A). Phenotypes of POT (presumed ocular toxoplasmosis) and AOT (active ocular toxoplasmosis) were noted for each individual from the MG and RJ cohorts. Association was found between the *NOD2* SNP rs3135499 with toxoplasmic retinal scars and retinochoroiditis, perhaps due to strong linkage disequilibrium between this SNP and the true variant in the *NOD2* gene. This variant may influence cytokine production by T-lymphocytes (i.e. Fig. 25.23B) (Dutra *et al.*, 2012).

IL-17A, a cytokine stimulated by activation of NOD2, has been implicated in the inflammatory process and it was hypothesized that it might influence or reflect subsequent development of ocular toxoplasmosis. To further investigate the role of NOD2 in the production of IL-17A, levels of the cytokine were measured in the supernatant of peripheral blood mononuclear cells (PBMCs) from individuals who were seronegative, those who had probable ocular toxoplasmosis (POT), and asymptomatic retinal scars that morphologically were areas of pigment with contiguous smaller areas of pigmentation. The PBMCs were cultured for five days in the presence of soluble tachyzoites antigens (STAg) from the RH strain of *T. gondii*. PBMCs from a subset of persons with POT produced higher levels of IL-17A compared to those from persons in the NL group (Fig. 25.23C) (Dutra *et al.*, 2012). Following this, production of IL-17A within the different genotypes for three tag-SNPs in the *NOD2* gene was analysed. No significant difference in production of IL-17A was seen among the individuals with the SNPs, rs2076753 and rs2111235, not associated with ocular disease. However, higher production of IL-17A was found in a subset of the persons with the rs3135499 SNP (Fig. 25.23D). These results suggest that SNPs at the *NOD2* gene may contribute to increased production of IL-17A in some seropositive individuals with POT. Thus, Dutra *et al.* speculate that an association between NOD2 and increased production of IL-17A may contribute to the development of ocular toxoplasmosis via a deleterious inflammatory process (Dutra *et al.*, 2012). In other studies, TGFβ and IL-6 present together lead to T-cell production of excessive amounts of IL-17, which damages tissue. The IL-6 receptor component of gp130 was essential for *T. gondii* encephalitis (Handel *et al.*, 2012).

25.4.3.6 CD36

CD36 is a macrophage pattern receptor for which there is some intriguing data from murine models concerning *T. gondii* infection (Hermes *et al.*, 2008) and that interacts with malaria GPIs.

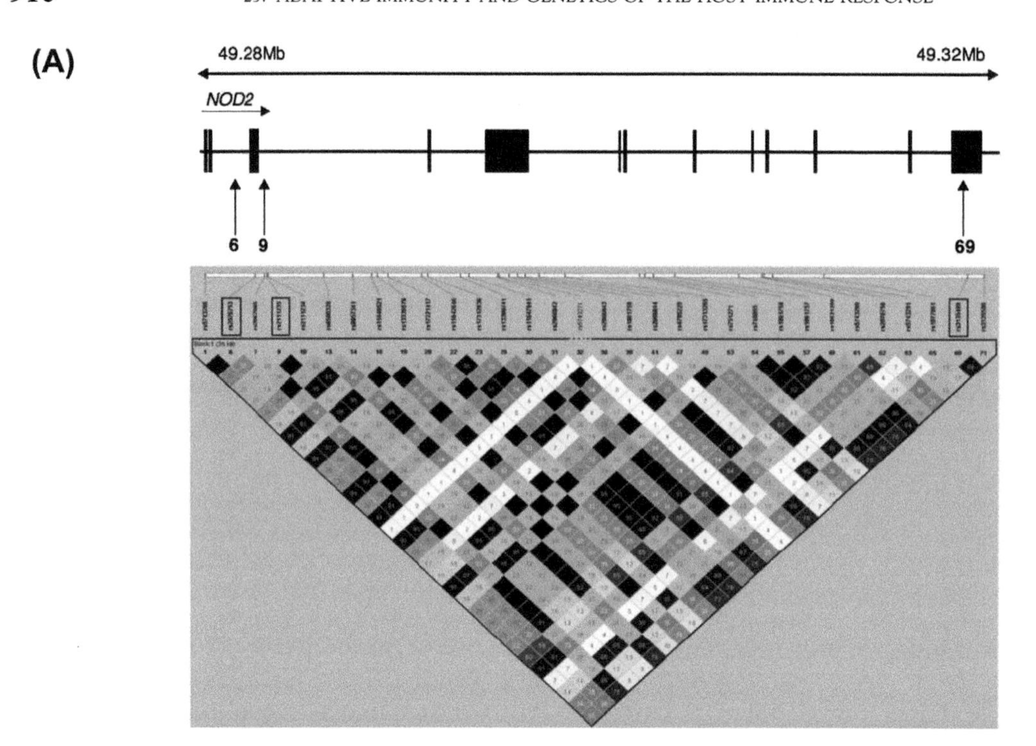

(B)

SNP Characteristic				FBAT Parameter[b]							
Name	Location	Physical Position[a]	MAF	Genotype	Genotype Frequency	No. of Families[c]	S	E(S)	Var(S)	Z^d	P^e
NOD2rs3135499	3′ UTR	50766127	0.45	CC	0.211	19	3.00	7.17	4.09	−2.06	**0.039**
				CA	0.487	53	34.00	28.38	13. 9	1.55	0.121
				AA	0.302	44	18.00	19.45	10.25	−0.45	0.651
NOD2rs2111235	Intron 1	50733969	0.32	GG	0.488	40	17.00	17.73	9.53	−0.24	0.814
				GA	0.381	46	27.00	23.55	11.67	1.01	0.312
				AA	0.131	19	5.00	7.73	4.42	−1.30	0.194
NOD2rs2076753	Intron 2	50733374	0.24	TT	0.044	4	2.00	1.50	0.875	0.535	0.593
				TG	0.383	26	11.00	12.20	6.174	−0.483	0.629
				GG	0.573	26	14.00	13.30	6.049	0.285	0.776

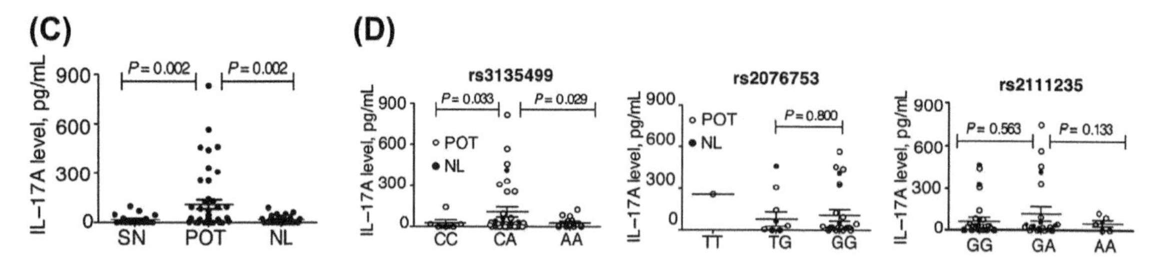

FIGURE 25.23 **Associations between NOD2, IL−17A and ocular toxoplasmosis in the Minas Gerais cohort from Brazil.** (A) Haploview analysis for measures of linkage disequilibrium between NOD2 SNPs in unrelated family founders in Brazil. 6 indicates the SNP rs2076753, 9 indicates the SNP rs2111235, and 69 indicates the SNP rs3135499. (*Adapted from Dutra* et al. *(2012).*)

Hermes *et al.* (2008) noted that transcription of CD36 in murine brain is increased significantly during chronic *T. gondii* infection in murine brain (Hermes *et al.*, 2008) (Fig. 25.30). GPIs from *P. falciparum* activate TLRs and then activate MAPK and NFκB signalling pathways and CD36, thereby increasing cerebral pathology. Rosiglitazone, a proliferator-activated receptor [PPARγ] agonist, decreased inflammation and reduced parasitemia in a CD36 dependent manner in a *P. chaubaudi* hyperparasitaemia model. It will be of interest to determine whether human CD36 interacts with *T. gondii* GPIs, GIPLs or the phosphethanolamine ceramide found in lipidomics study of type 1 *T. gondii*, or retinylidene phosphethanolamine, and whether rosiglitazone modulates the chronic inflammatory process that leads to neuronal cell loss caused by *T. gondii* in mouse brain.

25.4.2.7 P2X7R

The P2X7 receptor (P2X7R) is inserted into macrophage, neuronal, and other cell membranes and is increased by IFNγ (Fig. 25.24A). Ligation of P2X7R by ATP kills host cells, leading to killing of intracellular pathogens (Lees *et al.*, 2010) (Fig. 25.24B). ATP levels are high in the cytoplasm of these cells, and ATP is released into the extracellular environment when cells are damaged. P2X7R stimulation with extracellular ATP opens a cation-specific ion channel that allows the influx of Ca^{++} and Na^{+} and efflux of K^{+} into the cell (Fig. 25.24D). Prolonged exposure to ATP creates a pore in the cell membrane that increases intracellular Ca^{++} concentration and allows passage of larger molecules (Fig. 25.24D). This pore formation causes cell lysis via activation of NLRP3 and NFκB, phospholipase D, release of reactive oxygen and nitrogen intermediates, and stimulation of caspases (Zakrzewski, 2011).

In order to determine whether P2X7R contributes to susceptibility to congenital toxoplasmosis, P2X7R SNPs were genotyped from a cohort of 149 children with congenital toxoplasmosis and their available parents from the NCCCTS and 68 individuals with ocular toxoplasmosis from Campos dos Goyatacazes in Brazil (Jamieson *et al.*, 2010) (Fig 25.24E, F). Using transmission disequilibrium testing (TDT), susceptibility alleles of P2X7R were identified (Jamieson *et al.*, 2010; Lees *et al.*, 2010).

To further determine relevance of this susceptibility allele to illness due to the parasite and possible mechanisms involved in susceptibility and resistance mediated by this gene, assays were performed to characterize P2X7R in human monocytes. Macrophage P2X7R function was measured by treatment of human monocyte-derived macrophages with ATP, followed by quantitation of ethidium bromide flux through P2X7R generated pores using flow cytometry (Lees *et al.*, 2010). First, it was found that three persons from the Nepean hospital in Australia with severe toxoplasmosis had reduced or absent P2X7R function (Fig. 25.24G) (Lees *et al.*, 2010): One was heterozygous for 1513A>C, the most common loss-of-function polymorphism. The second was heterozygous for 946G>A and 1096C>G and the third was heterozygous for 946G>A and 1513A>C loss-of-function SNPs. Then, persons in the NCCCTS with a loss-of-function polymorphism in P2X7R

(B) Results of a TDT test between NOD2 SNPs and POT in samples from Brazilian patients. A significant association was found between SNP rs3135499 and POT ($p = 0.039$).
(C) Production of IL−17A by PBMCs from individuals who were seronegative (SN), with POT or asymptomatic (NL).
(D) Production of IL−17A plotted against three possible genotypes for the SNP rs3135499 shown on the far left. There is evidence of an association between this SNP and ocular toxoplasmosis ($p = 0.039$). The second and third figures show production of IL−17A plotted against three possible genotypes for SNPs rs2076753 and rs2111235 respectively, neither of which was associated with ocular disease. *(A−D are adapted with permission from Dutra* et al. *(2012).)*

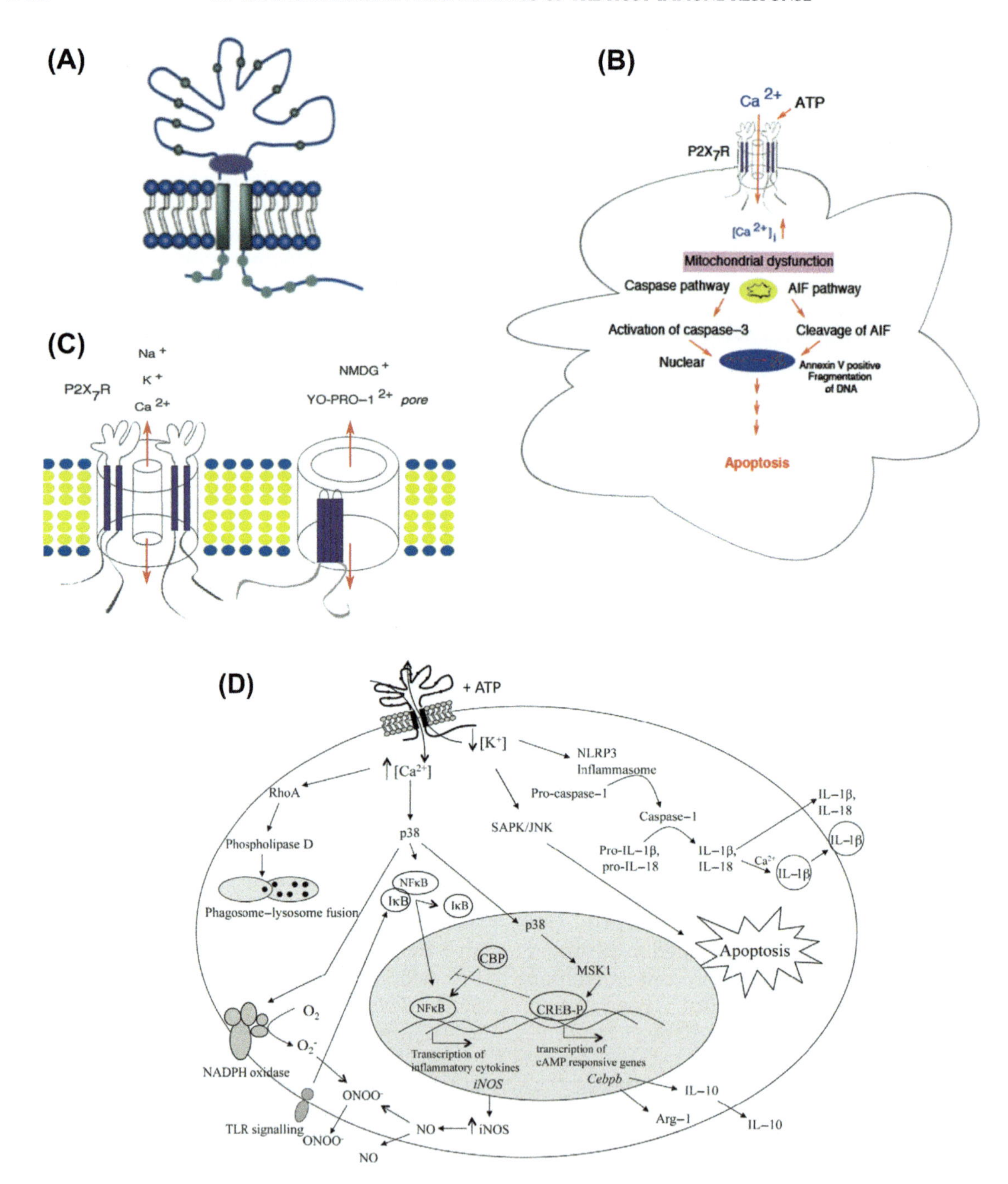
(A)
(B)
Ca 2+
ATP
P2X7R
[Ca 2+]i
Mitochondrial dysfunction
Caspase pathway
AIF pathway
Activation of caspase-3
Cleavage of AIF
Nuclear
Annexin V positive
Fragmentation
of DNA
Apoptosis
(C)
P2X7R
Na +
K +
Ca 2+
NMDG +
YO-PRO-1 2+ pore
(D)
+ ATP
[K+]
[Ca2+]
NLRP3
Inflammasome
RhoA
Pro-caspase-1
IL-1β,
IL-18
Caspase-1
SAPK/JNK
p38
Phospholipase D
Pro-IL-1β,
pro-IL-18
IL-1β,
IL-18
Ca2+
IL-1β
IL-1β
NFκB
IκB
IκB
Phagosome-lysosome fusion
p38
Apoptosis
CBP
MSK1
O2
NFκB
CREB-P
O2-
Transcription of
inflammatory cytokines
iNOS
transcription of
cAMP responsive genes
Cebpb
NADPH oxidase
IL-10
ONOO-
TLR signalling
ONOO-
NO
iNOS
Arg-1
IL-10
NO

(the 1514A>C polymorphism), in addition to those with wild-type P2RX7 who had been identified as above, were studied (Lees *et al.*, 2010). ATP treatment of monocyte-derived macrophages cultured from persons homozygous for the 1513A>C loss-of-function polymorphism did not significantly reduce the number of intracellular *T. gondii* tachyzoites, whereas ATP treatment of monocyte-derived macrophages from persons with wild-type, with no loss-of-function polymorphisms, reduced the number of parasites (Lees *et al.*, 2010) (Fig. 25.24K).

To prove that P2X7R played a role in toxoplasmosis a murine model of toxoplasmosis was studied. P2X7R function of macrophages from BALB/c mice (i.e. with fully functional P2X7R), C57BL/6J mice (i.e. with a polymorphism with variable effects on P2X7R function), and P2X7R$^{-/-}$ mice on a C57BL/6J background then were studied. ATP treatment of macrophages from BALB/c and C57BL/6J reduced viability of *T. gondii*, while ATP treatment of P2X7R$^{-/-}$ murine macrophages did not produce a loss of *T. gondii* viability (Lees *et al.*, 2010) (Fig. 25.24L). There was a similar effect on parasite burden following challenge with type 1 and type 2 parasites *in vivo* (Lees *et al.*, 2010). Death of the parasite was a function of host cell death and was not due to NO.

Thus, P2X7R plays an important role in immune response to *T. gondii* in humans. Loss of function of the P2X7R gene inhibits ATP-dependent killing of *T. gondii* tachyzoites, while ATP-dependent killing contributes to

FIGURE 25.24 **Structure and function of the P2X7 receptor.**

(A) Structure of P2X7 receptor. Sub-units of the receptor assemble as homo- or hetero-trimers. Each sub-unit is formed by two transmembrane domains, an extracellular loop with 10 cysteine residues, and intracellular N- and C-terminal domains in the cytosolic domain. Regions of loss-of-function alleles are noted. *(Adapted from Zakrzewski (2011).)*

(B) Cascade for P2X7R mediated cell death via caspase dependent apoptosis. *(Adapted from Zakrzewski* (2011), *with permission.)*

(C) Intracellular signalling pathways in immune cells stimulated by P2X7R activation. The efflux of K^+ from the cell stimulates an inflammasome, which results in the activation of caspase-1 and secretion of IL−1β and IL−18. The efflux of K^+ and the influx of Na^+ also activates stress-activated protein kinase (SAPK)/c-Jun N-terminal kinases (JNK) pathway, inducing apoptosis. *(Adapted from Miller* et al. *(2011), with permission.)*

(D) Na^+ and Ca^{++} influx and K^+ efflux from a channel formed by two sub-units of the P2X7R and the formation of pore permeable to larger cations such as $NMDG^+$ and YO−PRO-1^+.

(E) Haploview analysis for D′ and r^2 pairwise measures of LD between *P2X7R* SNPs in unrelated family founders for NCCCTS. *(Adapted from Jamieson (2010), with permission.)*

(F) Haploview analysis for D′ and r^2 pairwise measures of LD between *P2X7R* SNPs in unrelated family founders for Brazil. *(Adapted from Jamieson (2010), with permission.)*

(G) Presence of one or more *P2X7R* SNPs in all three persons with severe toxoplasmosis from the Nepean hospital in Australia. *(Adapted from Lees* et al. *(2010), with permission.)*

(H) Monocyte-derived macrophages from S1−S3 from the Nepean Hospital and from subjects' wild-type for loss-of-function P2X7R SNPs were treated with 1 mM ATP. Function of P2X7R pores was quantified with ethidium bromide flux and flow cytometry. P2X7R function of monocyte-derived macrophages from S1−S3 was decreased compared with three control subjects. *(Adapted from Lees (2009), with permission.)*

(I) *T. gondii* RH strain in cells with functional P2X7 receptors and treated with ATP. Wild-type monocyte-derived macrophages from P2X7 receptor genotypes from the NCCCTS and monocyte derived macrophages with the non-loss-of-function P2X7 polymorphism from the NCCCTS.

(J) Intracellular *T. gondii* tachyzoites numbers dependent on ATP-treated monocyte derived macrophages with wild type P2X7R SNPs, non-loss-of-function P2X7R SNPs, and loss-of-function P2X7R SNPs.

(K, L) Knock-out of the P2X7R gene affects the ability of mouse macrophages to control *T. gondii*. *T. gondii* viability was reduced in ATP-treated macrophages from BALB/c and C57BL/6J mice compared with untreated controls. *T. gondii* viability was not significantly reduced after addition of ATP to macrophages from P2X7R−/− mice. *(I−L are adapted from Lees (2009) diss.)*

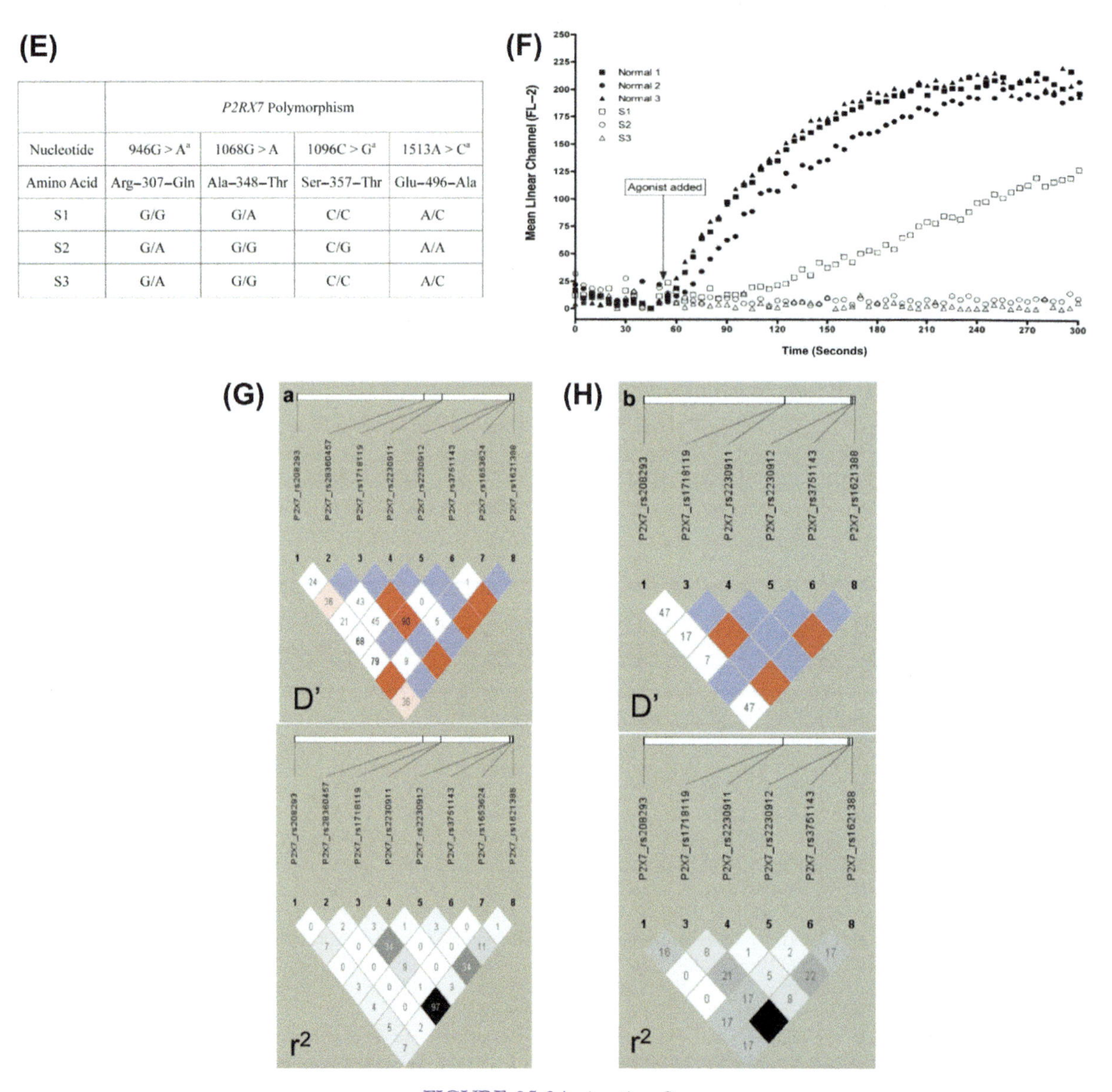

(E)

	P2RX7 Polymorphism			
Nucleotide	946G > A[a]	1068G > A	1096C > G[a]	1513A > C[a]
Amino Acid	Arg–307–Gln	Ala–348–Thr	Ser–357–Thr	Glu–496–Ala
S1	G/G	G/A	C/C	A/C
S2	G/A	G/G	C/G	A/A
S3	G/A	G/G	C/C	A/C

FIGURE 25.24 (*continued*).

protection when there are wild-type P2X7R or polymorphisms that do not cause loss of function in the receptors. P2X7R is proven to be causal in protection against *T. gondii* in mouse models (Lees *et al.*, 2010). Lees *et al.* also recently summarized and discussed P2X7R and some remaining questions about its role in toxoplasmosis in humans.

25.4.2.8 *IRAK4*

IL-1R-associated kinase 4 (IRAK4) plays an important role in downstream signalling from

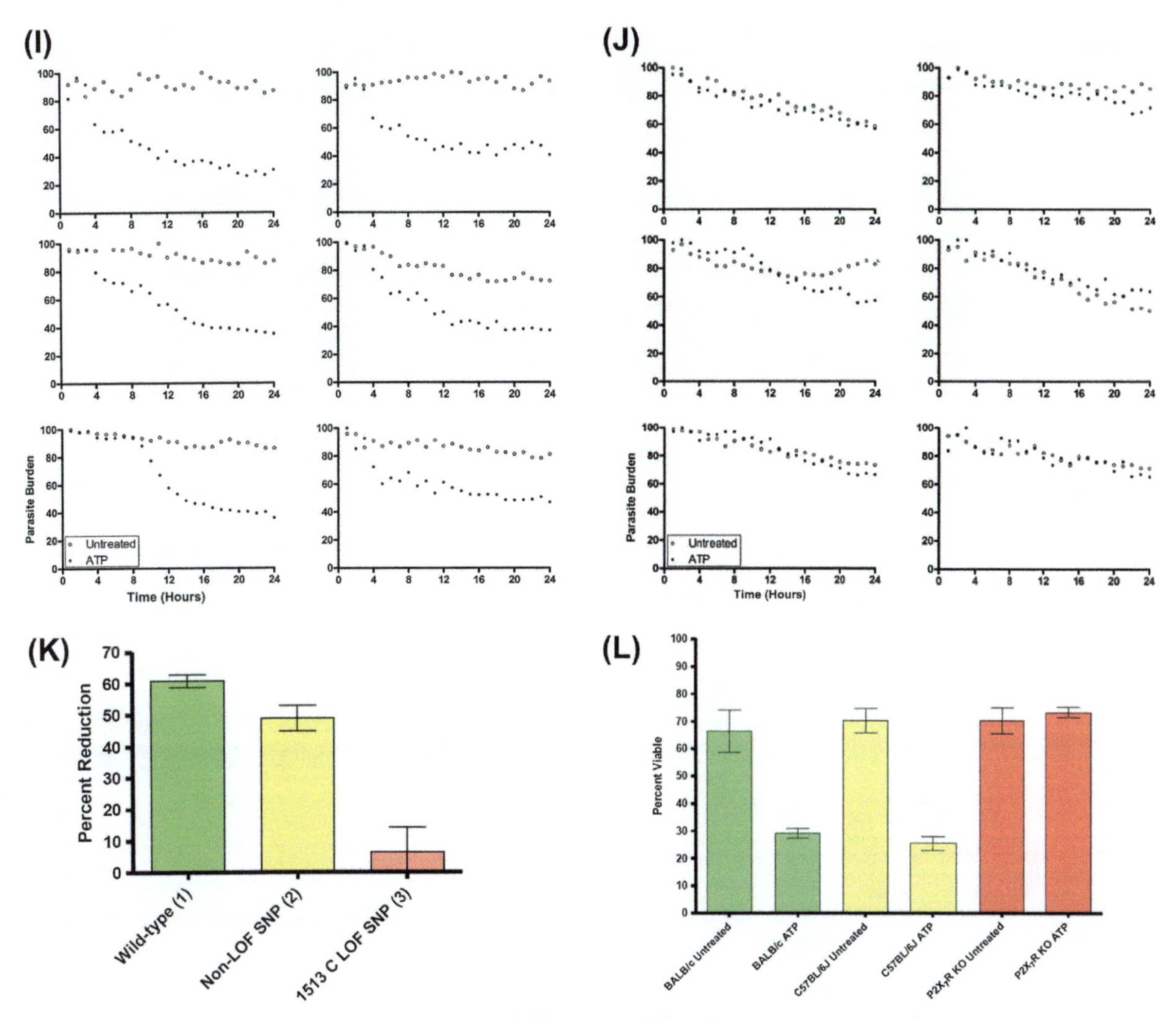

FIGURE 25.24 *(continued)*.

cell membrane TLRs, producing inflammatory mediators. Myeloid cells produce IL-12 during infection with *T. gondii* after Toll-like receptors (TLRs) are activated. TLRs associate with the adapter protein MyD88, which recruits IRAK4 to the TLRs. Deficiency in IRAK4 leads to impaired activation of the transcription factors NFκB, AP-1 and cJUN, and reduced production of inflammatory cytokines induced by signalling through TLRs (Bela *et al.*, 2012). In order to determine whether there was an association between allelic variants of the *IRAK4* gene and susceptibility or resistance to human congenital toxoplasmosis, peripheral blood mononuclear cells from 124 case-parent trios from the NCCCTS were placed in a buffer for DNA extraction. DNA from all persons studied was genotyped for 7 tag-SNPs, selected from the HapMap project, from the *IRAK4* gene. Allelic association analysis was done for 124 children and their

(A)

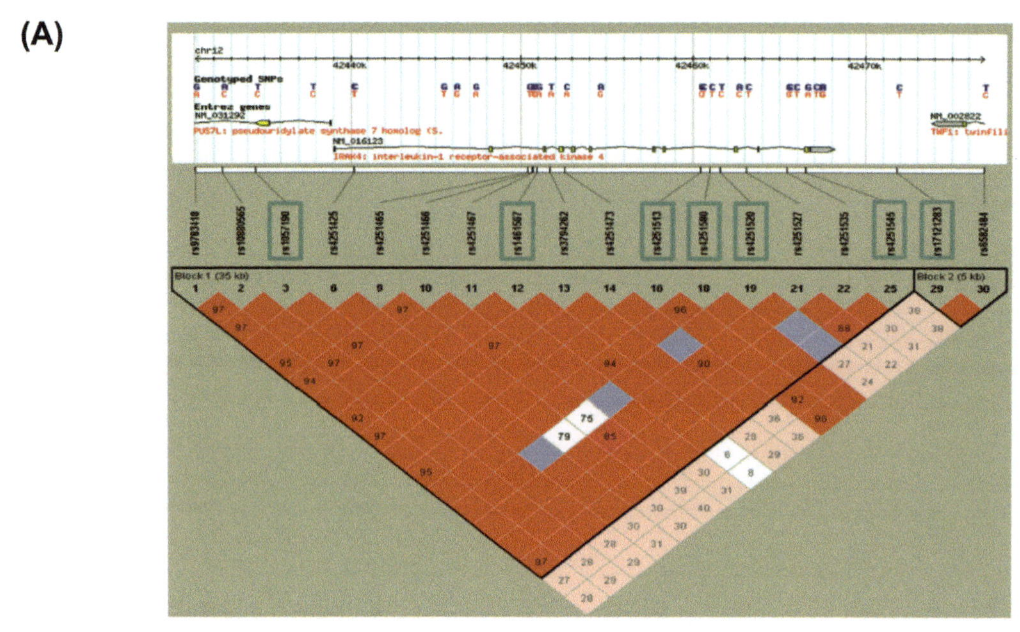

(B) (C)

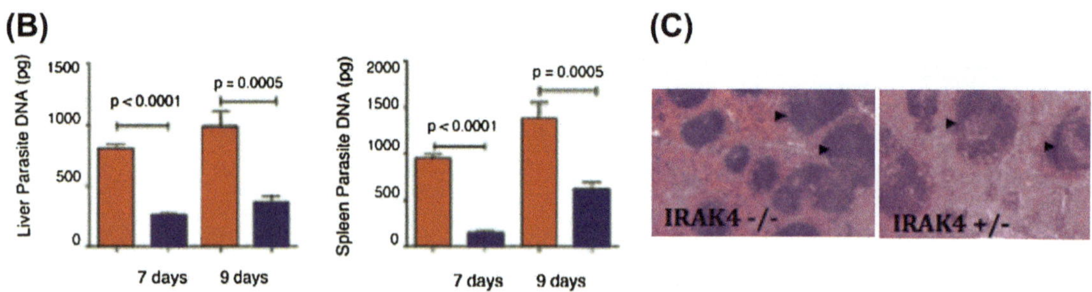
Liver Parasite DNA (pg)
p < 0.0001
p = 0.0005
7 days
9 days
Spleen Parasite DNA (pg)
p < 0.0001
p = 0.0005
7 days
9 days
IRAK4 -/-
IRAK4 +/-

(D)

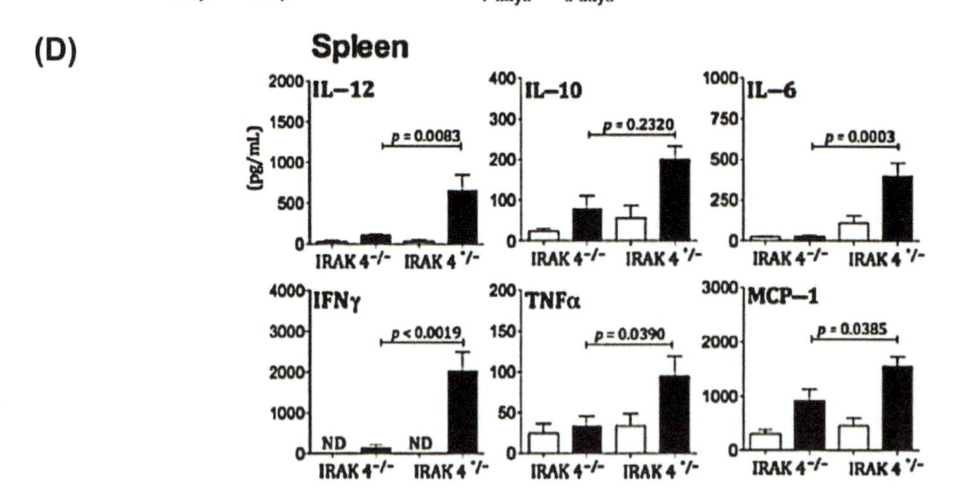
Spleen
IL–12
p = 0.0083
IL–10
p = 0.2320
IL–6
p = 0.0003
IFNγ
p < 0.0019
ND
TNFα
p = 0.0390
MCP–1
p = 0.0385
(pg/mL)
IRAK 4-/-
IRAK 4+/-

parents in the cohort with symptoms in the eye and/or brain due to toxoplasmosis. The two *IRAK4* SNPs associated most strongly with congenital toxoplasmosis ($p < 0.023$ and $p < 0.045$) were rs14161567 and rs4251513 (Fig. 25.25) (Bela *et al.*, 2012).

In order to better understand mechanism(s) whereby *IRAK4* susceptibility/resistance alleles played a role during *T. gondii* infection, *IRAK4*$^{+/-}$ mice with C57BL6 and S129 genetic backgrounds were used as heterozygote breeders in order to generate both homozygous and heterozygous mice. IRAK4$^{-/-}$, IRAK4$^{+/-}$, and wild type mice between the ages of eight and 12 weeks were orally infected with 20 *T. gondii* ME49 cysts contained in 0.1 mL of brain homogenates. They were then monitored for mortality and morbidity for 30 days or euthanized after infection at days 7, 9 or 20–30 days for cyst counts. Several methods, including RT–PCR, histopathology, isolation of splenic leukocytes, cytokine measurements and flow cytometry were used to assess the role of IRAK4 after *T. gondii* infection in mice. *T. gondii* DNA levels were higher in both the spleen and liver of *IRAK4*$^{-/-}$ mice than in the heterozygous *IRAK4*$^{+/-}$ mice, indicating that control of tachyzoite replication is dependent on functional IRAK4 (Bela *et al.*, 2012) (Fig. 25.25B). The area of follicles and germinal centres in spleens from infected *IRAK4*$^{-/-}$ mice were relatively smaller than those in *IRAK4*$^{+/-}$ mice, indicating that lymphoid cells were less active in the homozygous mice than in the heterozygous mice during *T. gondii* infection (Bela *et al.*, 2012) (Fig. 25.25C). Additionally, supernatants of splenocyte cultures and serum samples from *IRAK4*$^{-/-}$ and *IRAK4*$^{+/-}$ mice were assessed for cytokine levels at seven days post-infection. IFNγ, TNFα, IL-10, IL-6 and MCP-1 levels were determined using the BD Cytometric Bead Array Mouse Inflammation kit. Production of IL-12, IFNγ and TNFα by cells from *IRAK4*$^{-/-}$ mice infected with the *T. gondii* ME49 strain was impaired compared to production from cells of the homozygotes. The phenotypes of splenocyte subpopulations were then analysed using cell suspensions stained with monoclonal antibodies and analysed for surface markers expressed by dendritic cells, macrophages, CD4^{+} T-, CD8^{+} T-, B- and NK-cells. The frequency of monocytes and macrophages from mouse cells that produced IL-12 or TNFα when stimulated by STAg was lower in *IRAK4*$^{-/-}$ mice than in *IRAK4*$^{+/-}$ mice, indicating a failure on activation of dendritic cells from homozygous mice infected with *T. gondii* (Bela *et al.*, 2012) (Fig. 25.25D).

Thus, IRAK4 has a significant role in host immune response to *T. gondii* in both humans with the mechanism confirmed in mouse models. The SNPs in the human *IRAK4* gene found to associate with congenital toxoplasmosis indicate that IRAK4 is critical in mediating resistance to *T. gondii* infection in humans. Loss of function of the *IRAK4* gene in the mouse model led to

FIGURE 25.25 **Allelic association of *IRAK4* SNPs in humans and host cell response to *T. gondii* infection in IRAK4$^{-/-}$ And IRAK4$^{+/-}$ mice.**
(A) Haploview analysis depicting association of SNPs in the *IRAK4* gene with human congenital toxoplasmosis. Two *IRAK4* SNPs associated most strongly with congenital toxoplasmosis are rs14161567 ($p < 0.023$) and rs4251513 ($p < 0.045$), respectively.
(B) Quantitation of parasite DNA in the liver and spleen at seven and nine days post-infection in IRAK4$^{-/-}$ (grey bars) and IRAK4$^{+/-}$ (black bars) mice.
(C) Structures of follicles and germinal centres in spleens of IRAK4$^{-/-}$ and IRAK4$^{+/-}$ mice infected with *T. gondii*. Black arrows indicate activated and non-activated germinal centres. Germinal centres are less active in IRAK4$^{-/-}$ than in IRAK4$^{+/-}$ spleens.
(D) Impaired cytokine response in spleens of infected and uninfected IRAK4$^{-/-}$ mice as compared with infected and uninfected IRAK4$^{+/-}$ mice. *(A–D adapted with permission from Bela* et al. *(2012).)*

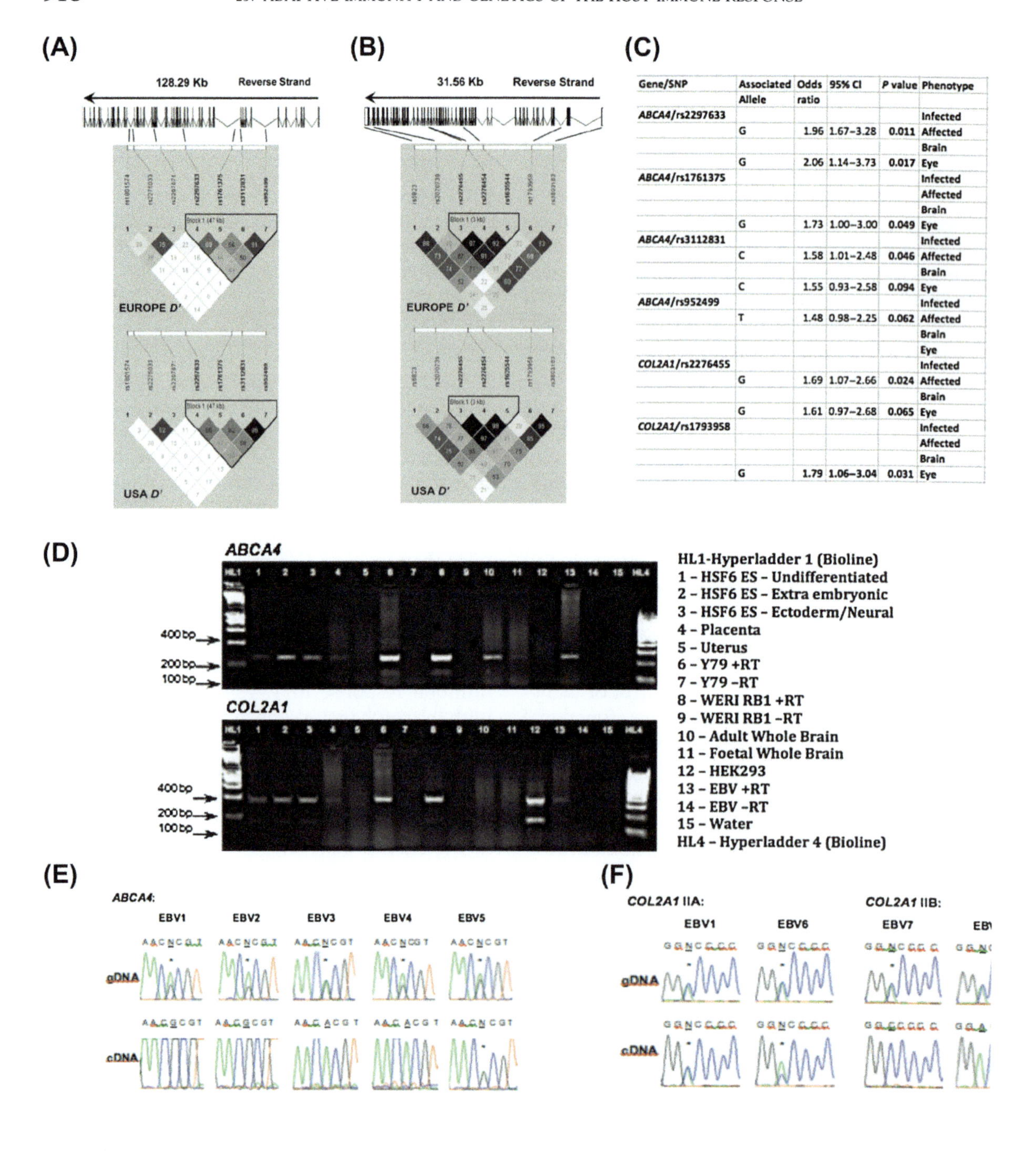

Gene/SNP	Associated Allele	Odds ratio	95% CI	*P* value	Phenotype
ABCA4/rs2297633					Infected
	G	1.96	1.67–3.28	0.011	Affected
					Brain
	G	2.06	1.14–3.73	0.017	Eye
ABCA4/rs1761375					Infected
					Affected
					Brain
	G	1.73	1.00–3.00	0.049	Eye
ABCA4/rs3112831					Infected
	C	1.58	1.01–2.48	0.046	Affected
					Brain
	C	1.55	0.93–2.58	0.094	Eye
ABCA4/rs952499					Infected
	T	1.48	0.98–2.25	0.062	Affected
					Brain
					Eye
COL2A1/rs2276455					Infected
	G	1.69	1.07–2.66	0.024	Affected
					Brain
	G	1.61	0.97–2.68	0.065	Eye
COL2A1/rs1793958					Infected
					Affected
					Brain
	G	1.79	1.06–3.04	0.031	Eye

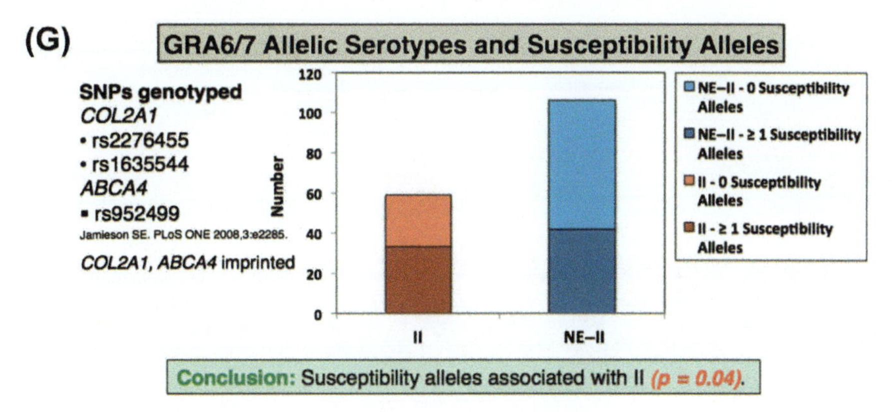

FIGURE 25.26 (*continued*).

increased parasitism in the spleen, liver and brain, as well as impaired production of proinflammatory cytokines IL-12, TNFα and IFNγ in mice due to disruption in signalling triggered by TLRs exposed to components from *T. gondii*.

25.4.2.9 *COL2A1* and *ABCA4*

Two genes were selected as candidate genes to study because they had been associated with the genetic causes of eye disease and/or hydrocephalus. Transmission disequilibrium testing of candidate SNPs in these genes were tested to determine whether there was an association with congenital toxoplasmosis or eye or brain disease due to *T. gondii* congenital infection. When there was a suggestion of parent-of-origin effect, definitive studies indicating imprinting (epigenetic effects) were noted. *COL2A1* is a gene encoding type II collagen, found in the vitreous humour, cornea, lens, ciliary body, retina and sclera of the eye. Mutations cause Stargart's disease, which is a congenital lattice degeneration of the retina, which leads to loss of sight. The *ABCA4* gene encodes an ATP-binding cassette transporter subfamily. ABC4r is a transporter of the toxic oxidized lipid retinylidene phosphethanolamine

FIGURE 25.26 **Allelic associations of *COL2A1* and *ABCA4*.**
(A) Gene structure and linkage disequilibrium for *ABCA4* in the EMSCOT (Europe) and NCCCTS (USA) cohorts. Haplotype blocks within each gene are outlined within the black triangles. Black to grey to white shading (high to low) indicates degree of confidence in the estimate of linkage disequilibrium between markers. Association at *ABCA4* is due to SNPs rs2997633 ($p = 0.011$) and rs1761375 ($p = 0.049$).
(B) Gene structure and linkage disequilibrium for *COL2A1*. Haplotype blocks within each gene are outlined within the black triangles. Black to grey to white shading (high to low) indicates degree of confidence in the estimate of linkage disequilibrium between markers. At *COL2A*, the SNPs rs2276455, rs1635544 and rs3803183 all add significant main effects.
(C) Effect of alleles at *ABCA4* and *COL2A1* on clinical outcomes in children in the EMSCOT cohort infected *in utero*.
(D) Figure shows transcripts of *ABCA4* and *COL2A1* expressed in human embryonic stem cell lines.
(E) Panel shows lines EBV1 to EBV4 appear homozygous (i.e. monoallelic) in cDNA specific for the exon 10 containing isoform and EBV lines that are heterozygous for *ABCA4* SNP rs3112831 in the genomic DNA. Line EBV5 is heterozygous, which indicates that mono-allelic silencing is polymorphic.
(F) Panel shows monoallelic expression for *COL2A1* SNP rs3737548 in PCR products specific for isoform IIB in EBV cell lines, but not for isoform IIA.
(G) Figure shows GRA6/7 allelic serotypes and susceptibility alleles. Susceptibility alleles associated with II have been compared ($p = 0.04$). (*A–F adapted with permission from Jamieson* et al. *(2008). G is adapted from McLeod* et al. *(2012).*)

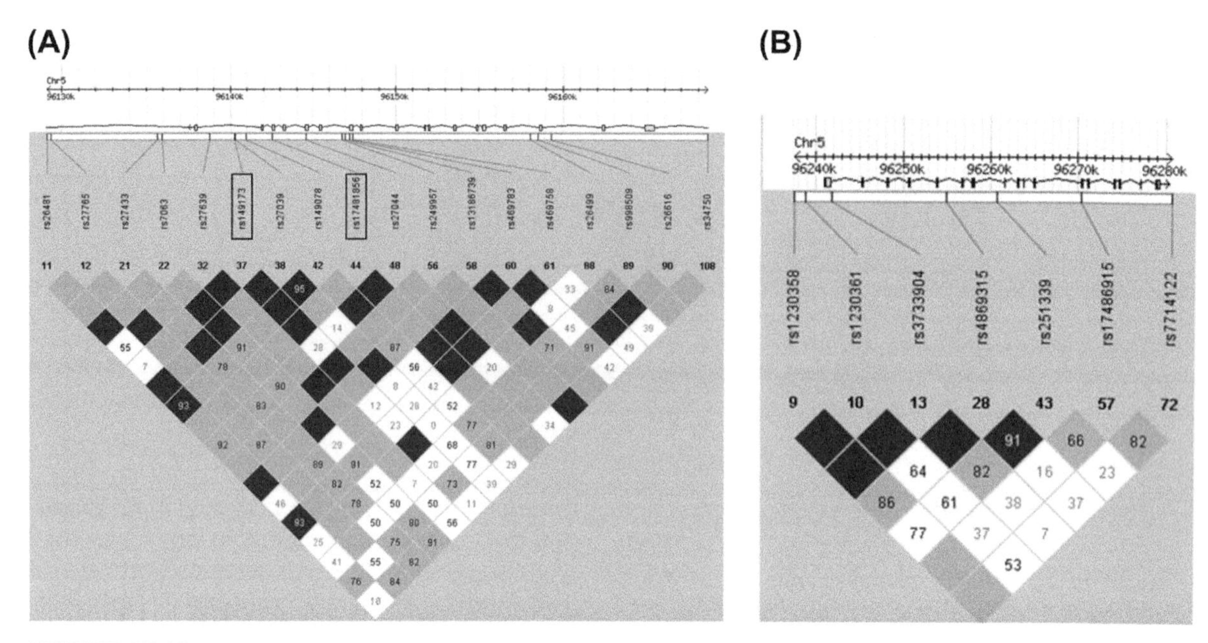

FIGURE 25.27 **Genotyping of *ERAP1* and *ERAP2* SNPs.**
(A) Gene structure and Haploview analysis for human *ERAP1* from the NCCCTS cohort. Upper diagrams show positions of SNPs relative to intron/exon structure of each gene. Black to grey to white shading indicates degree of confidence in estimate of linkage disequilibrium between markers (high to low). Two SNPs of *ERAP1* were found to associate with congenital toxoplasmosis: rs149173 ($p = 0.0077$) and rs17481856 ($p = 0.0253$) are outlined in black rectangles.
(B) Gene structure and Haploview analysis for human *ERAP2* from the NCCCTS cohort. *(A and B are adapted with permission from Tan* et al. *(2010).)*

out of rod cells in the retina. It also is expressed along the aqueduct of Sylvius, the location in the foetal brain where an inflammatory process that congenital toxoplasmosis elicits causes hydrocephalus. ABC4r is also associated with juvenile onset retinal dystrophies, adult macular degeneration and congenital aqueductal obstruction due to mutations in this gene. To determine whether *ABCA4* and *COL2A1* contribute to clinical outcomes from congenital toxoplasmosis patients and their families from the NCCCTS were studied (Jamieson *et al.*, 2008) and then, as a replicate cohort to confirm associations, inheritance patterns for alleles of both genes were studied in individuals from the European Multicentre Cohort Study on Congenital Toxoplasmosis (EMSCOT).

DNA was obtained from 124 children with confirmed congenital toxoplasmosis in the NCCCTS cohort and their mothers. Sixty-two percent of these affected children had brain calcifications and retinal lesions, 14% had retinal lesions only, 7% had brain calcifications only, and 17% were without these clinical findings. The use of case-parent trios and transmission disequilibrium testing controlled for ethnic diversity within this cohort. Individuals in this study were also genotyped for seven SNPs at *ABCA4* (rs1801574, rs2275033, rs2297671, rs2297633, rs176375, rs3112831, rs952499) and *COL2A1* (rs6823, rs2070739, rs2276455, rs2276454, rs1635544, rs1793958, and rs3803183). Linkage disequilibrium between markers was determined and graphed using the Haploview software (Fig 25.26A, B). From this cohort, significant allelic associations were found for one SNP at *ABCA4* (rs952499) and five SNPs at *COL2A1* (rs6823, rs2007039, rs2276454, rs1635544 and rs3803183). SNPs rs2276455, rs2276454 and rs1635544 for *COL2A1* showed improved significance when the analysis was specific for the eye lesion phenotype (with or without brain lesions) (Fig. 25.26C). DNA from 457 mother–child pairs from the EMSCOT cohort was obtained from buccal swabs from each individual and genotyped for seven SNPs at *ABCA4* (rs1801574, rs2275033, rs2297671, rs2297633, rs176375, rs3112831, rs952499) and *COL2A1* (rs6823, rs2070739, rs2276455, rs2276454, rs1635544, rs1793958 and rs3803183) (COL2A1/ABCA4 A) (Jamieson *et al.*, 2008). Linkage disequilibrium between markers was determined and graphed using the Haploview software (COL2A1/ABCA4 B, COL2A1/ABCA4 C). Of the children infected with *T. gondii*, 34% had clinical signs of congenital toxoplasmosis, 67% had ocular lesions, 57% had brain lesions and 24% had both brain and ocular lesions. From this cohort, allelic associations were found for two SNPs at *ABCA4* (rs2997633 and rs3112831) and one SNP at *COL2A1* (rs2276455) when children with retinal disease, brain disease or both were compared with unaffected children (infected but without disease) with *T. gondii* (Jamieson *et al.*, 2008) (Fig. 25.26G).

In both the NCCCTS and EMSCOT cohorts, linkage disequilibrium patterns across both loci showed that all SNPs contributing effects in the associations in each cohort fell within the same haplotype blocks within the *ABCA4* and *COL2A1* genes (Jamieson *et al.*, 2008). Evidence for these genetic associations was confirmed in two cohorts (Jamieson *et al.*, 2008).

A parent-of-origin association was noted for both *COL2A1* and *ABCA4*. That susceptibility alleles of both genes demonstrate a parent-of-origin effect indicates/suggests that the parasite secretes a product that modifies expression of these genes during embryonic human life, and perhaps later also. This suggests that the parasite may alter methylation of host genes during embryogenesis. Parent-of-origin effects indicate modification of transcription or translation of these genes has an effect on whether they are susceptibility alleles or not. They occur in gene regions known to be imprinted.

Genetic association between ocular disease in congenital toxoplasmosis and SNPs of *COL2A1* suggest that differences in collagen expression in the retina and vitreous humour may influence inflammatory pattern responses such as fibrosis during parasite replication in the eye. *ABCA4* alleles or mutations are associated with both macular degeneration and hydrocephalus due to aqueductal obstruction. Like VCAM and ICAM, *ABCA4* is expressed along the aqueduct of Sylvius.

In a study in which associations of these two genes with type II or non-type II infections were studied, the predominance of the associations with *COL2A* and *ABCA4* were due to persons with type 2 infections. This is consonant with the replication in the European (predominantly type 2 infections) and USA–North American (mixed different types of parasites in North America with ~1/3 being type 2 parasites) cohorts

Thus, alleles of *COL2A1* and *ABCA4* associated with eye and brain disease in persons congenitally affected by congenital toxoplasmosis are imprinted and more often associated with genetic type parasite infections. Both genes are associated with eye disease in congenital toxoplasmosis and *ABC4r* alleles also are associated with brain disease.

25.4.2.9 ERAAP

The peptide repertoire presented by MHC class I molecules to $CD8^+$ T-cells is determined by endoplasmic reticulum-associated aminopeptidase (ERAAP). ERAAP is a protease that trims precursor peptides in the endoplasmic reticulum lumen to lengths of 8–10 amino acids before they are loaded onto MHC class I molecules (Tan *et al.*, 2010). *T. gondii* parasites reside in a vacuole in the host cell separated from the host cell cytoplasm. Thus, peptides from proteins secreted by *T. gondii* must be able to enter into an

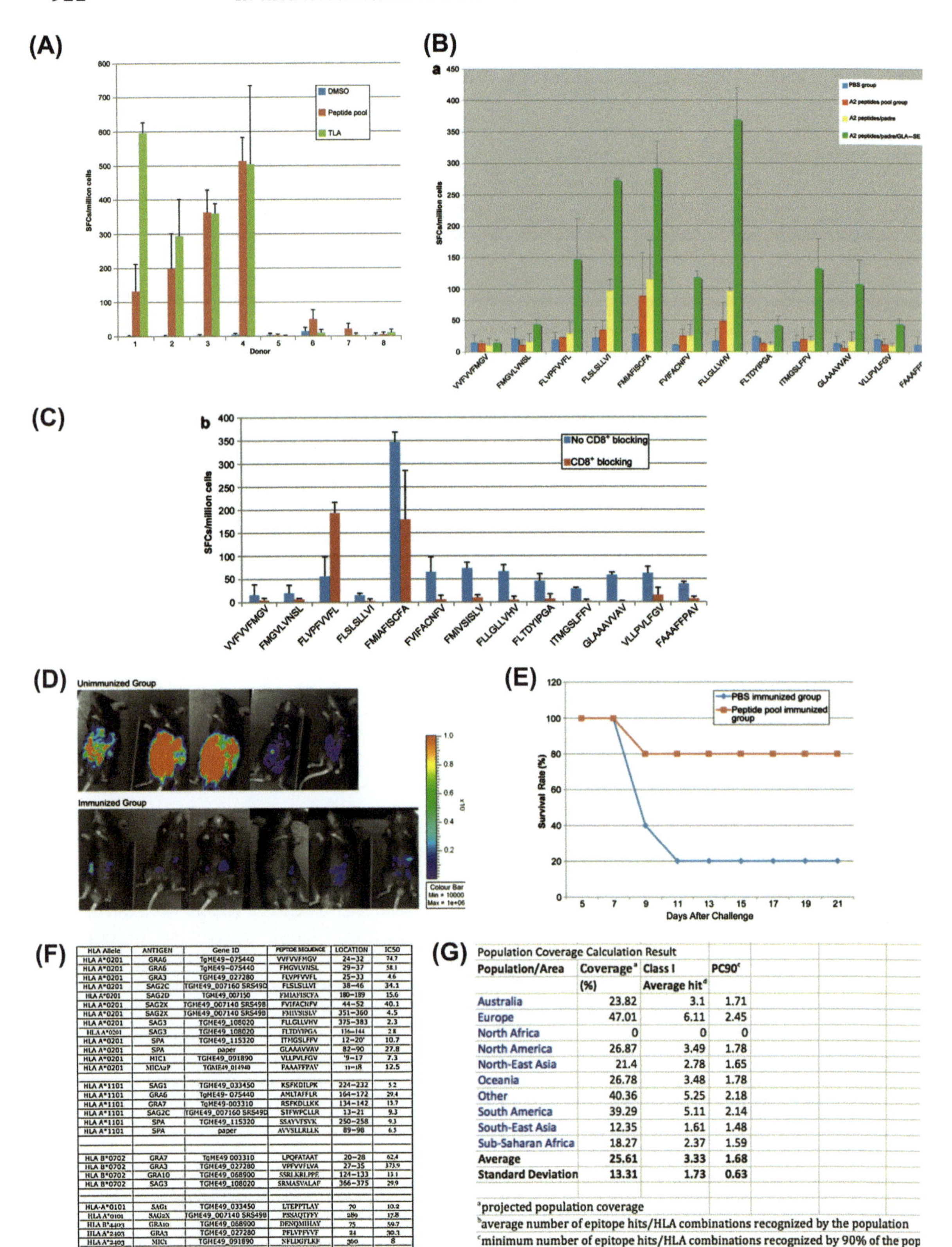

HLA Allele	ANTIGEN	Gene ID	PEPTIDE SEQUENCE	LOCATION	IC50
HLA A*0201	GRA6	TgME49–075440	VVFVVFMGV	24–32	74.7
HLA A*0201	GRA6	TgME49–075440	FMGVLVNSL	29–37	58.1
HLA A*0201	GRA3	TGME49_027280	FLVPFVVFL	25–33	4.6
HLA A*0201	SAG2C	TGME49_007160 SRS49D	FLSLSLLVI	38–46	34.1
HLA A*0201	SAG2D	TGME49_007150	FMIAFISCFA	180–189	15.6
HLA A*0201	SAG2X	TGME49_007140 SRS49B	FVIFACNFV	44–52	40.1
HLA A*0201	SAG2X	TGME49_007140 SRS49B	FMIVSISLV	351–360	4.5
HLA A*0201	SAG3	TGME49_108020	FLLGLLVHV	375–383	2.3
HLA A*0201	SAG3	TGME49_108020	FLTDYIPGA	136–144	2.8
HLA A*0201	SPA	TGME49_115320	ITMGSLFFV	12–20'	10.7
HLA A*0201	SPA	paper	GLAAAVVAV	82–90	27.8
HLA A*0201	MIC1	TGME49_091890	VLLPVLFGV	'9–17	7.3
HLA A*0201	MICA2P	TGME49_014940	FAAAFFPAV	11–18	12.5
HLA A*1101	SAG1	TGME49_033450	KSFKDILPK	224–232	5.2
HLA A*1101	GRA6	TgME49- 075440	AMLTAFFLR	164–172	29.4
HLA A*1101	GRA7	TgME49-003310	RSFKDLLKK	134–142	13.7
HLA A*1101	SAG2C	TGME49_007160 SRS49D	STFWPCLLR	13–21	9.3
HLA A*1101	SPA	TGME49_115320	SSAYVTSVK	250–258	9.3
HLA A*1101	SPA	paper	AVVSLLRLLK	89–98	6.5
HLA B*0702	GRA7	TgME49 003310	LPQFATAAT	20–28	62.4
HLA B*0702	GRA3	TGME49_027280	VPFVVFLVA	27–35	373.9
HLA B*0702	GRA10	TGME49_068900	SSRLKRLPPE	124–133	13.1
HLA B*0702	SAG3	TGME49_108020	SRMASVALAF	366–375	29.9
HLA-A*0101	SAG1	TGME49_033450	LTEPPTLAY	70	10.2
HLA A*0101	SAG2X	TGME49_007140 SRS49B	PSSAQTFFY	289	17.8
HLA B*4403	GRA10	TGME49_068900	DENQMIHAY	75	59.7
HLA A*2403	GRA3	TGME49_027280	PFLVPFVVF	24	30.3
HLA A*2403	MIC1	TGME49_091890	NFLDGFLKF	360	8
HLA A*2403	SRS9	TGME49_120190	KFVDILPNF	286	43.5

Population Coverage Calculation Result

Population/Area	Coverage[a] (%)	Class I Average hit[b]	PC90[c]
Australia	23.82	3.1	1.71
Europe	47.01	6.11	2.45
North Africa	0	0	0
North America	26.87	3.49	1.78
North-East Asia	21.4	2.78	1.65
Oceania	26.78	3.48	1.78
Other	40.36	5.25	2.18
South America	39.29	5.11	2.14
South-East Asia	12.35	1.61	1.48
Sub-Saharan Africa	18.27	2.37	1.59
Average	**25.61**	**3.33**	**1.68**
Standard Deviation	**13.31**	**1.73**	**0.63**

[a]projected population coverage
[b]average number of epitope hits/HLA combinations recognized by the population
[c]minimum number of epitope hits/HLA combinations recognized by 90% of the pop

MHC class I pathway, which requires ERAAP to prepare peptides for loading onto MHC molecules (Tan *et al.*, 2010). In order to determine whether the ERAAP genes *ERAP1* and *ERAP2* contribute to susceptibility to congenital toxoplasmosis, *ERAP1* SNPs were genotyped from case-parent trios from the NCCCTS. DNA was obtained from 149 case-parent samples and genotyped at 18 SNPs in *ERAP1* and 7 SNPs in *ERAP2*. Allelic association analysis was done using a TDT for 124 infected children in the NCCCTS cohort with clinical findings in the eye and/or brain (Fig 25.27A, B) (Tan *et al.*, 2010). Two SNPs in *ERAP1* (rs149173 and rs17481856) associated with congenital toxoplasmosis, although none of the SNPs in *ERAP2* associated with the disease. These associates are likely responsible for a change in the protein sequence or regulation of ERAP1 expression *in vivo* (Tan *et al.*, 2010).

Thus, ERAAP plays an important role in host immunity to *T. gondii*, as it is necessary for MHC class I molecules to effectively present parasite-produced peptides to $CD8^+$ T-cells and elicit a protective immune response.

25.4.2.10 TAP

Dogu *et al.* (2006), from Ankara, Turkey, describe two siblings (a girl – the older child – and a boy) with a newly described mutation in the peptide transporter (TAP). This transporter functions in HLA antigen processing. The boy had granulomatous skin lesions and recurrent sino-pulmonary infections, known to be associated with this deficiency and severe pulmonary toxoplasmosis. His infections were treated successfully.

25.4.2.11 HLA Class I

Earlier studies concerning HLA class I and class II molecule associations with human congenital toxoplasmosis and studies of mechanisms utilized HLA class II, DQ 1 and 3 and HLA class I transgenic mice (see above). That HLA class I molecules that bind *T. gondii* peptides can elicit protective immune responses also is shown by studies performed recently to develop an HLA class I dependent vaccine. In these studies, the fact that HLA supermotif molecules that bind peptides with specific motifs into the binding pocket of HLA A2, All01 or B7 molecules, which are present collectively in ~90% of the human population, is utilized. In this approach a predictive immunosense algorithm is used to identify such peptides in *T. gondii* molecules that are secreted and known to reach the host cell cytoplasm or known to elicit protective immune responses or antibody. That they empirically bind the specific HLA class I molecule is then tested (Fig. 25.28F). Then, the peptides are tested for their ability to elicit $CD8^+$ T-cell production of IFNγ by human peripheral blood T-cells. Then, the ability of these peptides to elicit IFNγ production by $CD8^+$ T-cells from

FIGURE 25.28 **HLA bound peptides. HLA is key in resistance and susceptibility.**
(A) Frequency of SFCs using an ELISpot assay for IFNγ production of an A02 peptide pool identified with bioinformatics and A2 binding assays tested in HLA-A02 restricted human cells. Donors 1–4 were infected, donors 5–8 were uninfected.
(B) $CD8^+$ T-cell responses in HLA-A*0201 mice immunized with PBS, peptide pool, peptide pool with PADRE and peptide pool with PADRE in GLA-SE. ACD8 eliminates IFNγ.
(C) $CD8^+$ T-cell responses in HLA-A*0201 mice to a peptide pool with PADRE and GLA-SE. Frequency of SFCs is demonstrated.
(D) When challenged with 10,000 Pru(Fluc)– *T. gondii* luciferase expressing parasites, HLA-A02 transgenic mice immunized with adjuvant and peptide pool were protected compared to control mice.
(E) Figure shows survival of HLA-A02 transgenic mice after challenge with *T. gondii* Type II parasites.
(F) Peptide candidates derived from GRA10, GRA15, SAG2C, SAG2D, SAG2X, SAG3, SRS9, BSR4, SRS, MIC1, MICA2P were used to screen $CD8^+$ T-cells.
(G) Class I coverage by population. *(A–G adapted with permission from Cong* et al. *(2011).)*

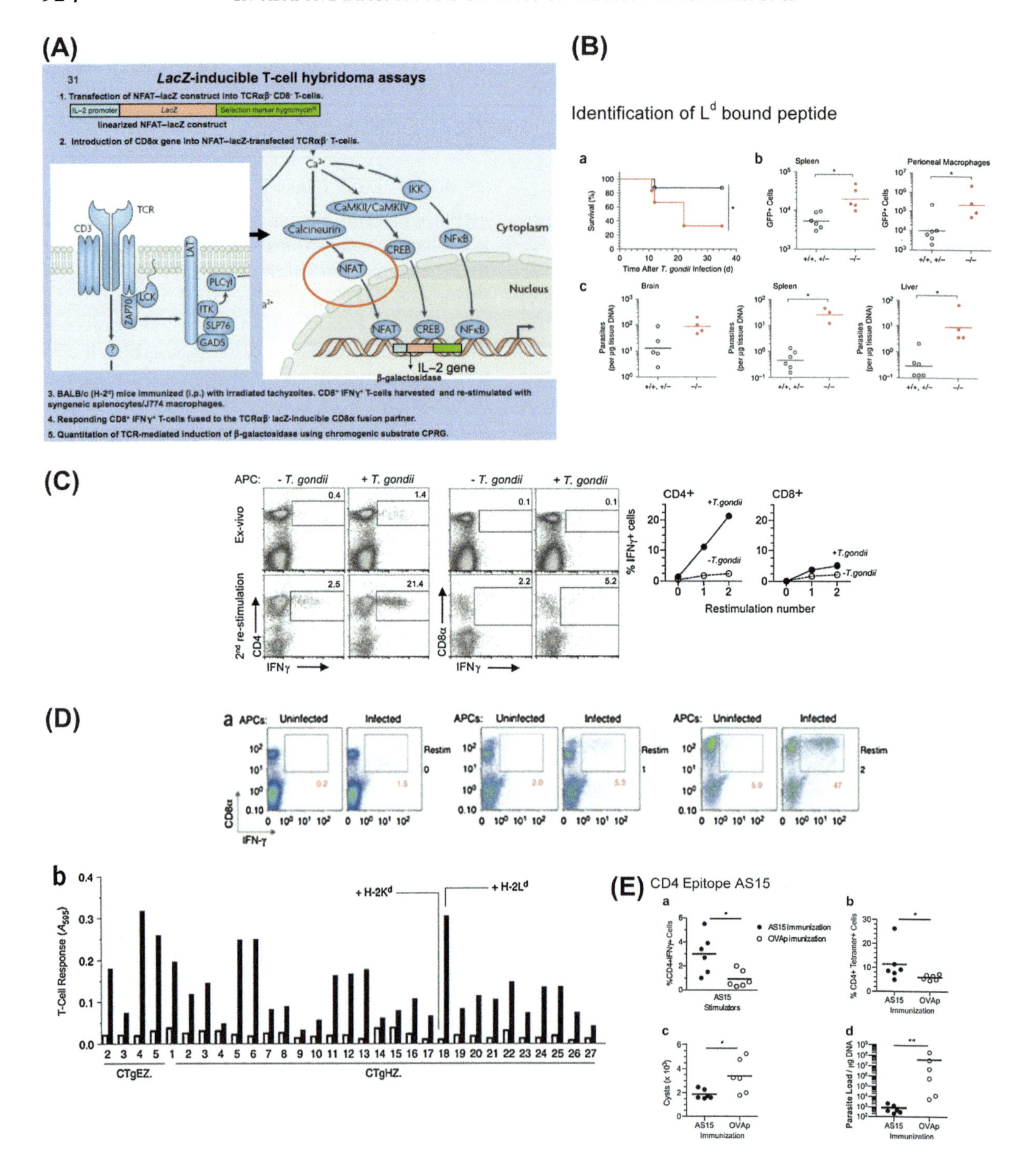
(A)
LacZ-inducible T-cell hybridoma assays
1. Transfection of NFAT–lacZ construct into TCRαβ⁻ CD8⁻ T-cells.
linearized NFAT–lacZ construct
2. Introduction of CD8α gene into NFAT–lacZ-transfected TCRαβ⁻ T-cells.
TCR
CD3
LAT
Calcineurin
CaMKII/CaMKIV
IKK
NFAT
CREB
NFκB
Cytoplasm
Nucleus
IL-2 gene
β-galactosidase
3. BALB/c (H-2ᵈ) mice immunized (i.p.) with irradiated tachyzoites. CD8⁺ IFNγ⁺ T-cells harvested and re-stimulated with syngeneic splenocytes/J774 macrophages.
4. Responding CD8⁺ IFNγ⁺ T-cells fused to the TCRαβ⁻ lacZ-inducible CD8α fusion partner.
5. Quantitation of TCR-mediated induction of β-galactosidase using chromogenic substrate CPRG.
(B)
Identification of Lᵈ bound peptide
Time After T. gondii Infection (d)
Spleen
Perioneal Macrophages
Brain
Liver
(C)
APC: - T. gondii + T. gondii
Ex-vivo
2nd re-stimulation
IFNγ
CD4+
CD8+
% IFNγ+ cells
Restimulation number
(D)
APCs: Uninfected Infected
Restim
b
T-Cell Response (A595)
+ H-2Kᵈ
+ H-2Lᵈ
CTgEZ.
CTgHZ.
(E) CD4 Epitope AS15
AS15 immunization
OVAp imunization
AS15 Stimulators
Immunization
Cysts (x 10³)
Parasite Load / μg DNA

HLA transgenic mouse splenocytes after they have been administered to mice with a universal CD4 epitope called PADRE to elicit $CD4^+$ T-cell help, along with adjuvants, is determined (Fig. 25.28B). Then, the ability of this immunization to elicit protection in HLA transgenic mice is determined (Fig. 25.28E). Adjuvants so far have included lipopeptides and a TLR ligand adjuvant called GLA-SE (IDRI) or GLA-SE plus Pam Cys or other adjuvants in DNA vaccine vectors. This approach has demonstrated that HLA-A2, -A1101 and -B7 supermotif bound peptides, as well as the L^d bound peptide recognized by BALB/c mouse $CD8^+$ T-cells, can elicit protective immune responses in conjunction with presence of $CD8^+$ T-cells that recognize these peptides and produce interferon-γ (Cong *et al.*, 2011, 2012; Tan *et al.*, 2010). Frequency of HLA class I haplotypes vary depending on the ethnicity of the population and the area of the world (Fig. 25.28G). For example, ~50% of persons of European descent have HLA-A2 supermotif haplotypes, whereas virtually none of those in North Africa, and substantially fewer of those in East Asia have HLA-A2 haplotypes. These studies prove the importance of both MHC class I and class II molecules in protection against parasite burden and death due to *T. gondii* in mice and that human $CD8^+$ T-cells recognize and produce IFNγ in response to these peptides. Whether additional or more robustly protective epitopes can be identified remains to be determined. Similarly, whether more robust adjuvants can result in protection which is as good as that elicited by live vaccines (Gigley *et al.*, 2009; Hutson *et al.*, 2010; McLeod *et al.*, 1984; Baird *et al.*, 2013) remains to be determined. However, these studies confirm that HLA classes I and II and adaptive immunity

FIGURE 25.29 **An immunodominant L^d peptide confers protection.**
(A) Diagram showing creation of *LacZ*-inducible T-cell hybridomas created by Tze Guan Tan.
(B) Panel (a) shows a Kaplan Meier survival curve of B10.D2 mice which were challenged intraperitoneally with 500 *T. gondii* tachyzoites of the Prugneaud strain ($p = 0.047$). Panel (b) shows flow cytometry of GFP^+ infected splenocytes ($p = 0.0043$) and peritoneal macrophages ($p = 0.038$). $^{+/+}$ indicates mice homozygous for ERAAP, $^{+/-}$ indicates mice heterozygous for ERAAP, and $^{-/-}$ indicates ERAAP-deficient mice. Panel (c) shows semiquantitative PCR of parasite burden in the brain ($p > 0.05$), spleen ($p = 0.012$), and liver ($p = 0.013$) of mice infected with Prugneaud tachyzoites. $^{+/+}$ indicates mice homozygous for ERAAP, $^{+/-}$ indicates mice heterozygous for ERAAP, and $^{-/-}$ indicates ERAAP-deficient mice. These results show that in the absence of ERAAP, the ability to control parasite replication is impaired.
(C) AS15 immunization of C57BL/6 mice shows a weak $CD8^+$ T-cell response but strong $CD4^+$ T-cell response. *T. gondii*-specific T-cell responses were measured by staining for IFNγ. Expansion of $CD4^+$ T-cells is shown on the left, expansion of $CD8^+$ T-cells is shown in the middle, and expansion over a time period including two *in vitro* re-stimulations is shown on the right.
(D) All *T. gondii*-specific $CD8^+$ T-cell hybridomas are restricted to the L^d MHC I molecule. Figure (a) shows *T. gondii*-specific $CD8^+$ T-cells expanded in *in vitro* cultures and produced IFNγ as measured by flow cytometry. Spleen cells from *T. gondii* immunized BALB/c mice were analysed *ex vivo* one week after challenge (top panels) or after one or two *in vitro* re-stimulations with infected J774 macrophages (middle and bottom panels). (b) T-cell hybridomas were generated by fusing the responding cells from two independent groups of mice to a *lacZ*-inducible fusion partner. The lacZ response of a panel of T-cell hybridomas after overnight co-culture with *T. gondii* infected L^d or K^d expressing L-cells was measured using a chromogenic substrate.
(E) C57BL/6 mice were immunized with LPS-activated bone marrow derived cells (BMDCs) pulsed with AS15 or control peptide (OVAp). Mice were infected with 1×10^4 *T. gondii* tachyzoites seven days after immunization. Panel (A) shows IFNγ response by brain $CD4^+$ T-cells measured by flow cytometry. Panel (B) shows MHC class II I-A^b-AS15 tetramer staining on brain leukocytes, which were also co-stained with CD4 antibody. Panel (C) shows the number of cysts in the brain, detected with a fluorescent lectin stain. Panel (D) shows the *T. gondii* load in the brain, measured using semiquantitative PCR on gDNA. Panels (C) and (D) show that immunization with AS15 peptide leads to $CD4^+$ immune response and lowers cyst burden and parasite load in the brains of infected mice. *(A is prepared by T. Tan, R. McLeod* et al. *(2009). B and D are adapted with permission from Blanchard* et al. *(2008). C and E are adapted with permission from Grover* et al. *(2012).)*

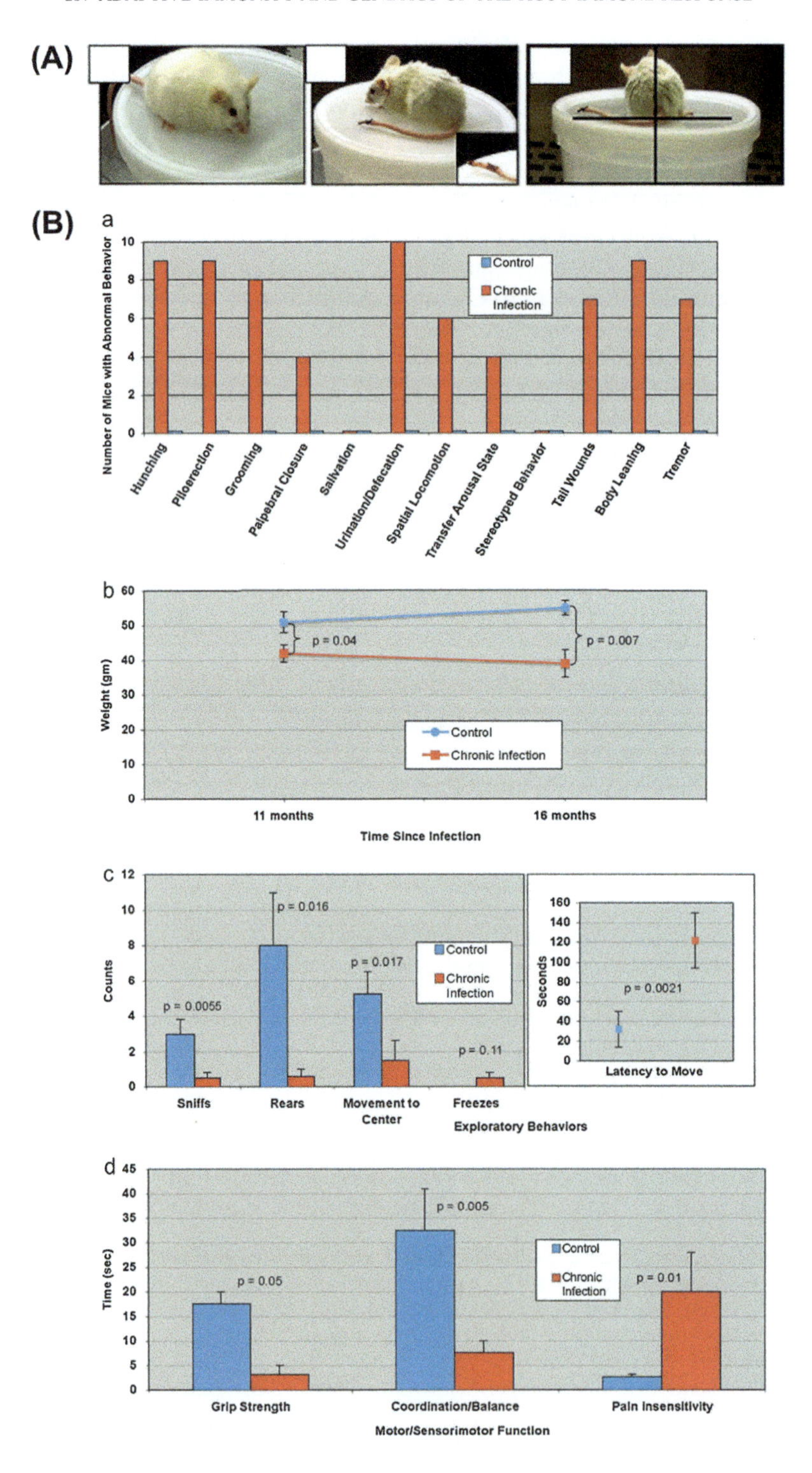
(A)
(B)
a
Number of Mice with Abnormal Behavior
Control
Chronic Infection
Hunching
Piloerection
Grooming
Palpebral Closure
Salivation
Urination/Defecation
Spatial Locomotion
Transfer Arousal State
Stereotyped Behavior
Tail Wounds
Body Leaning
Tremor
b
Weight (gm)
p = 0.04
p = 0.007
Control
Chronic Infection
11 months
16 months
Time Since Infection
c
Counts
p = 0.0055
p = 0.016
p = 0.017
p = 0.11
Control
Chronic Infection
Sniffs
Rears
Movement to Center
Freezes
Exploratory Behaviors
Seconds
p = 0.0021
Latency to Move
d
Time (sec)
p = 0.05
p = 0.005
p = 0.01
Control
Chronic Infection
Grip Strength
Coordination/Balance
Pain Insensitivity
Motor/Sensorimotor Function

stimulated by adjuvants to prime this immune response comprise part of a protective immune response against parasite burden due to *T. gondii* following infection. That the HLA molecules confer different amounts of genetically mediated protection and can be used in a murine model to characterize protection against oocysts dependent on HLA classes I and II molecules also has been demonstrated recently (Dubey *et al.*, 2011; Gigley *et al.*, 2013, in preparation).

Consistent with these studies, HLADQ B0402, DRB but not DQ3 (as was found for the NCCCTS, and for USA persons with AIDS studied by Suzuki and Kobayashi (1990)) were found to have susceptibility alleles in a South American study (Habbeger de Sorrentino *et al.*, 2005).

Other recent studies in mice further confirm that HLA molecules (classes I and II) are key in protection. Blanchard and Shastri studied the MHC class I L^d epitope as did Frickel *et al.* (2008) (Fig. 25.29). A GRA6 C terminus peptide (called HF10) provided a protective immunodominant epitope (Blanchard *et al.*, 2010) and GRA4 and ROP7 also provided L^d bound epitopes (Frickel *et al.*, 2008). Grover *et al.* defined a partially protective $CD4^+$ T-cell eliciting MHC class II restricted epitope for C57Bl6/J mice (Grover *et al.*, 2012). Dendritic, T-cell and T-cell target cell interactions have been studied elegantly by Robey *et al.* (2005), Hunter *et al.* (2010) and Wilson *et al.* (2012) have defined the role of chitinase and Suzuki *et al.* (2005) the role of perforin dependent protective $CD8^+$ T-cell interactions with cysts in brain. $CD8^+$ T-cell clones can produce IL2 and stimulate IFNγ in an autocine loop (Sa *et al.*, 2013).

25.4.2.12 CD40–CD40–L(154) Interactions

CD40 (a TNFα superfamily member) binds to CD154 on activated $CD4^+$ T-cells. In X-linked

FIGURE 25.30 **Chronic infection.**
(A) Relative to uninfected mice, infected mice presented with a decrease in autonomic nervous system function, which was measured by increased levels of urination and defecation. The mouse on the left is an uninfected eleven month old control. The mouse in the middle is an eleven month old infected female mouse ten months after infection with *T. gondii*. The mouse on the far right is a chronically infected mouse from the same colony. The infected mice show symptoms such as poor grooming, piloerection, tail wounding, lower weight, grip strength, coordination and balance.
(B) SPF mice chronically infected with *T. gondii* show neurologic findings or abnormal behaviour. The weight of chronically infected mice at eleven and sixteen months of age is significantly lower than the weight of control mice. Infected mice also show less exploratory behaviours and lower motor and sensorimotor function.
(C) MRI scans of uninfected and infected mice. Ventricular dilatation and some asymmetry were present in the brain parenchyma of infected mice, compared to uninfected mice.
(D) Perivascular inflammatory infiltrates are seen adjacent to the hippocampus in infected mice, while none are present in uninfected control mice.
(E) The brain from a mouse that is chronically infected with *T. gondii* shows perivascular inflammation, a cluster of microglia and isolated cysts in (a-c). A BALB/c mouse that is more resistant has less prominent perivascular cuffing (black circle) in (d). Figures (e) and (f) show more perivascular inflammation for a mouse with mild lateral ventricular dilatation in an MRI.
(F) Figure shows $CD4^+$, $CD8^+$, T-lymphocytes and microglia present in nodules and perivascular spaces in the frontal cortex and activated microglia in the brain of a mouse infected with *T. gondii*.
(G) Perivascular staining of plasmacytoid B-cells secreting antibody around blood vessels.
(H) Tissue cysts in mouse neurons during the formation of synapses (black arrows). CW, cyst wall; Br, bradyzoites; HC, host cell; M, inflammatory cells (prepared by D. Ferguson).
(I) Expression of ABCA4, VCAM1 and ICAM5 in hippocampal, perihippocampal and periventricular areas of infected mice (from Allen Atlas).
(J) *T. gondii* infection results in inflammation, neuronal cell loss, behavioural abnormalities, and other neurologic deficits in mice. Figure summarizes molecules, cells, and processes influenced by infection with the parasite.
(K) Figure shows the genes that are up-regulated in transcription from the brains of mice that have been infected with *T. gondii* for more than one year. *(A–K are adapted with permission from Hermes* et al. *(2008).)*

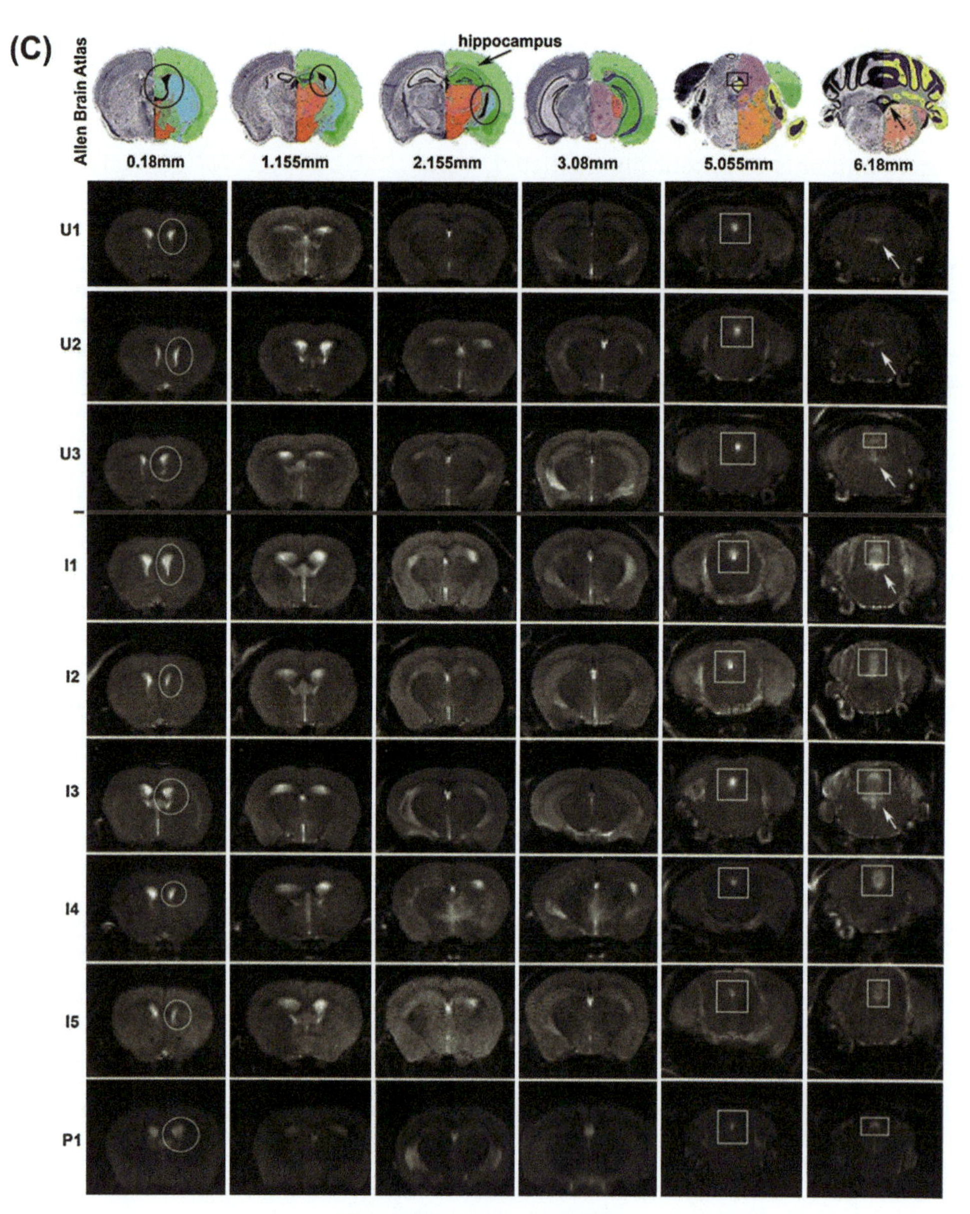

FIGURE 25.30 (*continued*)

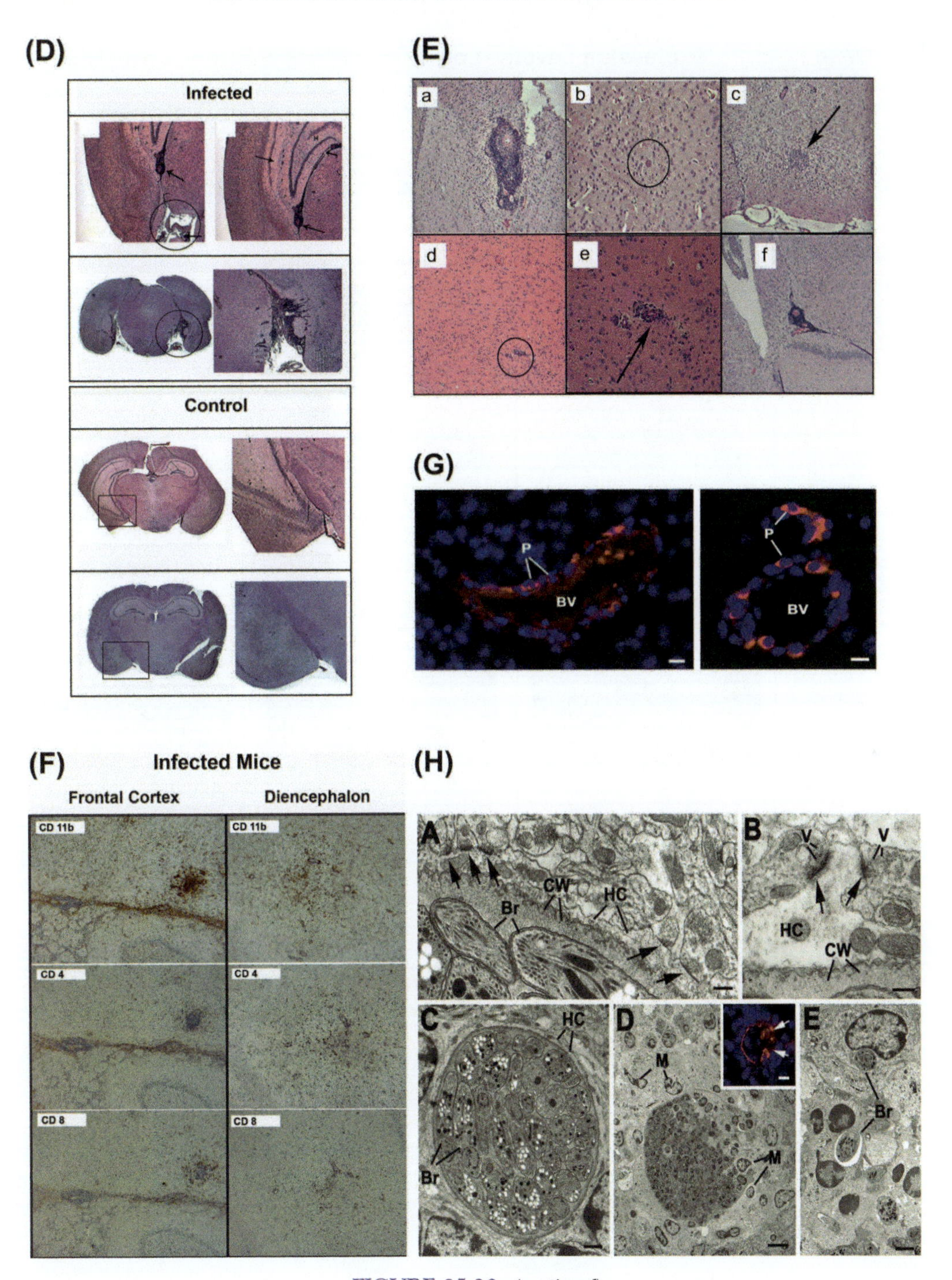

FIGURE 25.30 (*continued*)

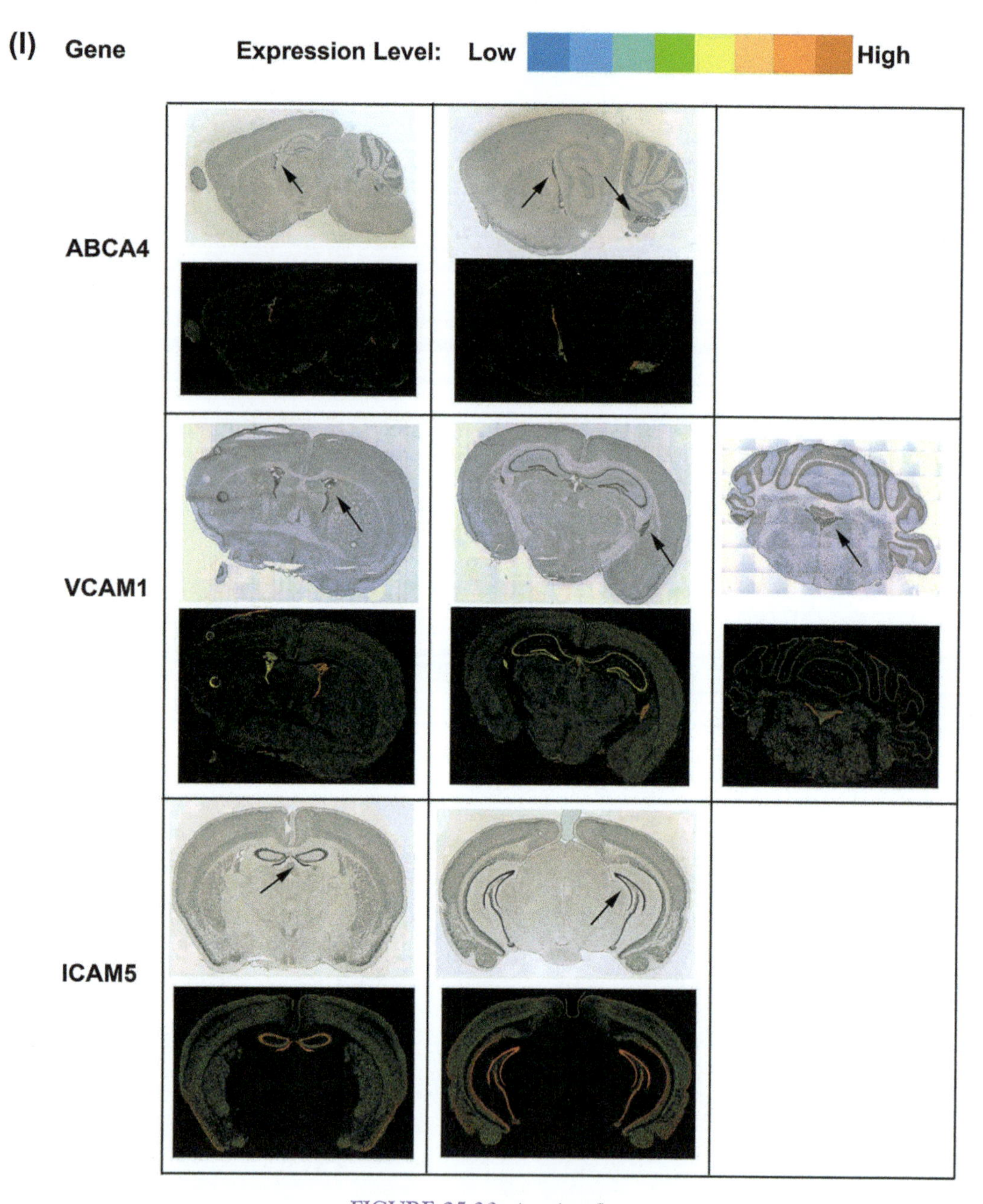

FIGURE 25.30 *(continued)*

(J)

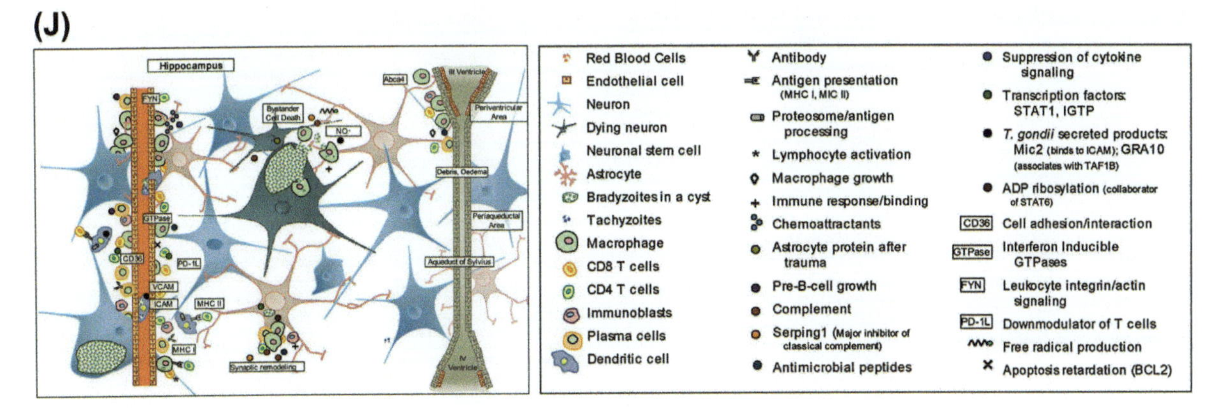

FIGURE 25.30 (*continued*).

hyper IGM syndrome which is due to a loss of CD154 there is increased susceptibility to toxoplasmic encephalitis (Reichmann *et al.*, 2000). CD40 or CD154 deficiency is known to diminish production of IL-12 and class switched B-cells. This also affects autophagy which is dependent on Beclin 1. Experiments in mice confirm the importance of this pathway. HyperIgM autophagy is associated with mutations in CD40 ligand (Subauste *et al.*, 2001). The mechanism for this defect in causing increased susceptibility to toxoplasmosis has also been recently defined (Subauste *et al.*, 2001).

25.4.2.13 *TNF*

Antibody to TNF is used therapeutically for treatment of rheumatoid arthritis. In that setting its use has been associated with the development of toxoplasmosis. This is consistent with the roles TNFα plays in protection against toxoplasmosis. This is consistent with the need for TNFα coincident with IFNγ for macrophage activation. It is consistent with the need for TNF in ligation of TLR4 with galectin. It may be part of the second signal for dendritic cells to prepare for antigen presentation. Studies by Deckert-Schluter *et al.* (2006) and Hunter and Sibley (2010) show the importance of TNFβ-lymphotoxin, which is essential for protection of mice.

25.4.2.14 *IL-10 and TGFβ*

The down-modulatory cytokine IL-10 also was found to have a promoter allele that was associated with risk for toxoplasmosis (Cordeiro *et al.*, 2008). Buffolano found that GRA1 led to TGFβ production associated with monocyte cell death (Buffolano *et al.*, 2005; D'Angelillo *et al.*, 2011) as well as production of IFNγ in approximately 50% of those persons whose lymphocytes were tested. Further, Minns (2004) found in a mouse model that a novel terpenoid induces TGFβ production by intraepithelial lymphocytes that prevent ileitis.

25.4.2.15 *IFNγ and Interferon-γ Receptor*

Mice without INFγ are not protected during primary infection, highlighting the critical importance of this cytokine in protection. It is of interest that perforin seems to be the key molecule for protection of murine brain. Menard found that B-cells amplify IFNγ production by T-cells *via* a TNFα-mediated mechanism which further elucidates how IFNγ and TNFα function together to protect against toxoplasmosis (Menard *et al.*, 2007).

(K) Genes Upregulated in the Brains of Mice Infected with *T. gondii* for a Year

Function	Gene* [Number/ID]+
Antibody	Ig heavy chain V regions [116], Ig kappa V regions [119], Ig lambda 1 germline V region (Igl-V1, 1810027O01Rik, IgL) [V00811], Ig alpha chain (IGHV1S44) [M19402], Rearranged IgA-chain gene, V region (IGHV1S46) [M20774], IgG-1 gene, D-J-C region: 3' exon for secreted form (Igh-4, IgG1) [J00453], Ig germline D-J-C region alpha gene and secreted tail (Igh-VJ558, AI893585, MGC118142, MGC6727, Igh-A (1g2), Vh186.2/Jh2) [J00475], Ig gamma2a-b(c57bl/6 allele) c gene and secreted tail (IGHG2C) [J00479], Immunoglobulin joining chain (Igj, 9530090F24Rik, AI323815, Jch) [NM_152839]
Antigen Presentation	MHC-I Similar to H-2 MHC-I antigen, D-37 alpha chain precursor (C920025E04Rik) [AK083387], MHC-I Transporter 2, ATP-binding cassette, sub-family B (MDR/TAP) (Tap2, ABC18, AI462429, APT2, Abcb3, Ham-2, Ham2, MTP2, PSF2, RING11, Tap-2, Y1, jas), [NM_011530], MHC-II [22], MHC-III (D17H6S56E-3, C6orf27, G7c, NG37) [NM_138582], Beta-2 microglobulin (B2m, Ly-m11, beta2-m) [NM_009735]
Complement	Complement component 4B (C4b, C4, Ss) [NM_009780], Complement component 1, r subcomponent (C1r, AI132558, C1rb) [NM_023143] Complement component C1SB (C1sb) [NM_173864,] Complement component 1, q subcomponent, alpha polypeptide (C1qa, AI255395, C1q) [NM_007572], Complement component 1, q subcomponent, C chain (C1qc, AI385742, C1qg, Ciqc) [NM_007574]
Major inhibitor of classical complement	Serine (or cysteine) peptidase inhibitor, clade G, member 1 (Serping1, C1INH, C1nh) [NM_009776], Serine (or cysteine) peptidase inhibitor, clade A, member 3N (Serpina3n, Spi2-2, Spi2.2, Spi2/eb.4) GeneCards: Although its physiological function is unclear, it can inhibit neutrophil cathepsin G and mast cell chymase. [NM_009252]
GTPase	Interferon inducible GTPase 1 (Iigp1, 2900074L10Rik, AI046432, AW111922, Iigp) [AK013785] Interferon inducible GTPase 1 (AW111922, Iigp1) [NM_021792], Interferon inducible GTPase 2 (Iigp2, RP24-499A8.1, AI481100, GTPI, (GC102455) [NM_019440], Interferon gamma inducible protein 47 (Ifi47, IRG-47, Iigp4, Iipg4, Olfr56) [NM_008330], Interferon gamma induced GTPase (Igtp, RP24-499A8.4, AW558444) [NM_018738], GTPase, IMAP family member 4 (Ian1, Gimap4, AU019574, E430007K16Rik, IMAP4, MGC11734, mIAN1), transcript
Transcription	Signal transducer and activator of transcription (Stat1, 2010005J02Rik, AA408197), GeneCards: Mediates signaling by interferons. Results in induction of a cellular antiviral state. [NM_009283 BC057690], Transcription factor MafB (v-maf musculoaponeurotic fibrosarcoma oncogene homolog B).(Mafb, Kreisler, Krml, kr), GeneCards: Plays a pivotal role in regulating lineage-specific hematopoiesis by repressing ETS1-mediated transcription of erythroid-specific genes in myeloid cells (By similarity). [NM_010658], Down regulator of transcription 1/Differentially regulated in lymphoid organs and differentiation (Dr1d) [AF043513]
Suppression of cytokine signaling	Suppressor of cytokine signaling 1 (Socs1, Cish1, Cish7, JAB, SOCS-1, SSI-1), GeneCards: SOCS1 is involved in negative regulation of cytokines that signal through the JAK/STAT3 pathway. Appears to be a major regulator of signaling by interleukin 6 (IL6) and leukemia inhibitory factor (LIF). Regulates interferon-gamma mediated sensory neuron survival (By similarity). [NM_009896]
Proteosome/antigen processing	Proteosome (prosome, macropain) subunit, beta type 8 (large multifunctional peptidase 7) (Psmb8, Lmp-7, Lmp7) GeneCards: This subunit is involved in antigen processing to generate class I binding peptides. Stimulated by interferon gamma, involved in the degradation of cytoplasmic antigens for MHC class I antigen presentation pathways. [NM_010724] Proteosome (prosome, macropain) subunit, beta type 9 (large multifunctional peptidase 2) (Psmb9, Lmp-2, Lmp2) GeneCards: stimulated by interferon gamma, involved in the degradation of cytoplasmic antigens for MHC class I antigen presentation pathways. [NM_013585], Proteasome (prosome, macropain) 28 subunit, alpha (Psme1, AW413925, MGC113815, PA28a), GeneCards: Implicated in immunoproteasome assembly and required for efficient antigen processing. The PA28 activator complex enhances the generation of class I binding peptides by altering the cleavage pattern of the proteasome. Induction by interferon gamma. [NM_011189]
Antimicrobial	Secretory leukocyte peptidase inhibitor (Slpi) Acid-stable proteinase inhibitor with strong affinities for trypsin, chymotrypsin, elastase, and cathepsin G. Secretory leukoprotease inhibitor, involved in antineutrophil elastase protection at inflammatory sites. [NM_011414] Defensin related cryptdin 17 (Defcr17, AU014719, Cryp17) GeneCards: Acid-stable proteinase inhibitor with strong affinities for trypsin, chymotrypsin, elastase, and cathepsin G. Secretory leukoprotease inhibitor, involved in antineutrophil elastase protection at inflammatory sites [S73391]
Pre-B-cell growth	Bone marrow stromal cell antigen 2 (Bst2, 2310015I10Rik, C87040, DAMP-1) GeneCards: May be involved in pre-B-cell growth. [NM_198095]
T lymphocyte proliferation	CD274 antigen/Programmed cell death 1 ligand 1 precursor (Cd274, B7-H1, PD-L1, Pdcd1l1, Pdcd1lg1) Genecards: Involved in costimulatory signal, essential for T lymphocyte proliferation and production of IL10 and IFNG, in an IL2-dependent and a PDCD1-independent manner. Interaction with PDCD1 inhibits T-cell proliferation and cytokine production. Up-regulated on T and B cells, dendritic cells, keratinocytes and monocytes after LPS and IFNG activation. Up-regulated in B cells activated by surface Ig cross-linking. [NM_021893]

FIGURE 25.30 (*continued*).

Lymphocyte activation	Lymphocyte-activation gene 3 (Lag3, CD223, LAG-3, Ly66), GeneCards: Involved in lymphocyte activation. Binds to HLA class-II antigens. Highly homologous to CD4, expressed exclusively in activated T and NK, lymphocytes, major MHC class 2 ligand potentially involved in the regulation of immune response. [NM_008479]
Macrophage growth	Guanylate nucleotide binding protein 2 (Gbp2), Uniprot: Interferon-induced Gbp 2 [NM_010260], Guanylate nucleotide binding protein 4 (Gbp4, AW228655, Gbp3) [NM_018734], Macrophage activation 2/guanylate-binding protein 4 homolog (Mpa2, AW228052, Gbp4, KIAA4245, Mag-2, Mpa-2, mKIAA4245) [NM_008620], Macrophage activation 2 like (MPA2l, AI595338) [NM_194336], Guanylate nucleotide binding protein 5 (Gbp5, 5330409J06Rik) Uniprot: Interferon-induced Gbp5 [NM_153564], RIKEN cDNA 5830443L24 gene/guanylate binding protein 8(5830443L24Rik, mGBP8) [NM_029509], cDNA sequence BC057170/guanylate binding protein like (BC057170, E430029F06) [NM_172777]
Immune response/binding	Interferon-induced protein with tetratricopeptide repeats 3 (Ifit3, Ifi49, MGC107331) [NM_010501] GO:0000004 biological process unknown, GO:0005488 binding, GO:0005554 molecular function unknown, GO:0006955 immune response, GO:0008372 cellular component unknown
Astrocyte protein after trauma	Glial fibrillary acidic protein (Gfap, AI836096), GeneCards: Class-III intermediate filament, cell-specific marker; during development of CNS distinguishes astrocytes from other glial cells. Almost exclusively expressed in astrocytes; interacts with S100A1., S100A1 home page: S100A1 and S100B can be isolated as a complex from bovine brain; mixture of S100A1 and S100B inhibited the assembly of tubulin into, microtubules [NM_010277]
Chemo-attractants	Chemokine (C-C motif) ligand 5 (Ccl5, MuRantes, RANTES, SISd, Scya5, TCP228) GeneCards: Chemoattractant for blood monocytes, memory T helper cells and eosinophils. [NM_013653] Chemokine (C-C motif) ligand 8/monocyte chemoattractant protein-2 precursor (Ccl8, RP23-446K18.1, 1810063B20Rik, AB023418, HC14, MCP-2, Mcp2, Scya8), GeneCards: Interferon gamma induced chemotaxis Attracts monocytes, lymphocytes, basophils, eosinophils [NM_021443]
Cell adhesion/interaction	CD36 antigen/fatty acid translocase (Cd36, FAT, GPIV, Scarb3), May function as a cell adhesion molecule. GeneCards: Directly mediates cytoadherence of Plasmodium falciparum parasitized erythrocytes. Mediates free, radical production in cerebral ischemia. Publication (Khoury, J Exp Med 2003, 197(12) 1657–1666): CD36, a major pattern recognition receptor, mediates microglial and macrophage response to beta-amyloid, and imply that CD36 plays a key role in the proinflammatory events associated with Alzheimer's disease. [NM_007643], EGF-like module containing, mucin-like, hormone receptor-like sequence 1/cell surface glycoprotein F4/80/lymphocyte antigen 71 (Emr1, D7A5-7, EGF-TM7, F4/80, Gpf480, Ly71, TM7LN3), GeneCards: Could be involved in cell-cell interactions. [NM_010130], Lectin, galactoside-binding, soluble, 3 binding protein/cyclophilin, associated. protein (Lgals3bp, 90K, CyCAP, MAC-2BP, Ppicap), GeneCards: Promotes integrin-mediated cell adhesion. May stimulate host defense against viruses and tumor cells. [NM_011150]
Leukocyte integrin/actin signaling	FYN binding protein (Fyb, ADAP, B630013F22Rik), GeneCards: Acts as an adapter protein of the FYN and SH2-domain-containing leukocyte protein-76 (SLP76) signaling cascades in T cells. Modulates expression of interleukin-2 (IL-2). [NM_011815]
Free radical production	Cytochrome b-245, beta polypeptide (Cybb, C88302, Cgd, Nox2, gp91<phox>, gp91phox) GeneCards: Critical component of the membrane-bound oxidase of phagocytes that generates superoxide. [NM_007807]
Apoptosis retardation	B-cell leukemia/lymphoma 2 related protein A1c(Bcl2a1, A1-c c), GeneCards: Retards apoptosis induced by IL-3 deprivation. May function in response of hemopoietic cells to external signals and in maintaining endothelial, survival during infection (By similarity). [NM_007535]
ADP ribosylation	Poly (ADP-ribose) polymerase family, member 14/collaborator of STAT6 (Parp14, 1600029O10Rik, BC021340, KIAA1268, MGC29390, mKIAA1268), [BC021340]
Unknown	[XM_207778]

*Gene names and symbols from Entrez Gene [87] and function information from GeneCards [88]
†Either the number of genes in this category or the accession number
Function of all genes significantly upregulated in the brains of mice infected with T. gondii for a year (p-value after adjustment for multiple testing < 0.01). No genes were significantly downregulated. Gene names and symbols were retrieved from Entrez Gene [87] using the accession number and function information from GeneCards [88]. Each distinct gene symbol is included in the column headed Gene, except for the functions Antibody and Antibody presentation, where there were too many similar accession numbers. The Number/ID is in brackets contains the accession number where there were < 4 for a function or else a count of the accession numbers for that function.

FIGURE 25.30 (*continued*).

25.4.2.16 *Studies of Parasite Genetics Elucidate Critical Host–Parasite Interactions and the Parasite Molecules (Virulence Factors) which are Responsible for these Interactions*

A number of recent, elegant studies have defined parasite proteins and miRs that modulate host gene expression of proteins and in some instances immune response. Some of these are shown in Figure 25.20. Interactions of parasite with the host and especially the host's immune response have been studied extensively (Abi Abdallah and Denkers, 2012; Abi Abdallah *et al.*, 2011; Abi Abdallah *et al.*, 2011; Afonso *et al.*, 2012; Akira *et al.*, 1993; Albuquerque *et al.*, 2009; Ali *et al.*, 2005; Aline *et al.*, 2002a, 2002b; Alvarado-Esquivel *et al.*, 2012; Alves *et al.*, 2012; Andrade *et al.*, 2005; Andrade *et al.*, 2006; Arling *et al.*, 2009; Arsenijevic *et al.*, 2007; Aviles *et al.*, 2008; Avunduk *et al.*, 2007; Baird *et al.*, 2013; Beaman *et al.*, 1993; Barragan and Sibley, 2002; Beauvillain *et al.*, 2007; Begum-Haque *et al.*, 2009; Bekpen *et al.*, 2005; Bekpen *et al.*, 2010; Benson *et al.*, 2009; Benson *et al.*, 2012; Berdoy *et al.*, 2000; Bernardes *et al.*, 2006; Beug *et al.*, 2012; Bhadra *et al.*, 2011; Bierly *et al.*, 2008; Blader and Saeij, 2009; Blanchard and Shastri, 2010; Blanchard *et al.*, 2008; Bliss *et al.*, 2000; Blomstrom *et al.*, 2012; Boothroyd and Dubremetz, 2008; Bottova *et al.*, 2009; Boyle *et al.*, 2008, 2007; Bradley and Sibley, 2007; Brandão *et al.*, 2009; Brenier-Pinchart *et al.*, 2006; Brown *et al.*, 2005; Brown *et al.*, 1994; Brown *et al.*, 1995; Brunet *et al.*, 2008; Butcher *et al.*, 2011, 2005; Buzoni-Gatel *et al.*, 2006, 2008; Caetano *et al.*, 2011; Capuron and Miller, 2011; Cavailles *et al.*, 2006a, 2006b; Cavalcanti *et al.*, 2011; Cesbron-Delauw *et al.*, 2008; Chan, A. *et al.*, 2006; Chan, C. *et al.*, 2006; Chandramohanadas *et al.*, 2009; Chang and Pechere, 1989; Chao *et al.*, 1993, 1994; Charles *et al.*, 2007, 2010; Chou *et al.*, 2012; Chtanova *et al.*, 2009; Clark *et al.*, 2011; Combe *et al.*, 2006; Cong *et al.*, 2012; Cong *et al.*, 2008, 2010, 2011; Correa *et al.*, 2010; Courret *et al.*, 2006; Craven *et al.*, 2012; Crawford *et al.*, 2009; Creuzet *et al.*, 1998; Da Gama *et al.*, 2004; de Alencar *et al.*, 2009; D'Angelillo *et al.*, 2011; Dalton *et al.*, 2006; Dass *et al.*, 2011; Daubener *et al.*, 2001; Daubener *et al.*, 1999; Debierre-Grockiego *et al.*, 2003, 2007a, 2007b, 2009, 2010, 2012; Deckert *et al.*, 2006; Dellacasa-Lindberg *et al.*, 2007, 2011; Denkers *et al.*, 2008, 2009, 2010, 2011, 2012a; Diana *et al.*, 2004; Dimier and Bout, 1997, 1998; Ding *et al.*, 2004; Divanovic *et al.*, 2012; Dogu *et al.*, 2006; Drogemuller *et al.*, 2008; Dunay *et al.*, 2008; Dunn *et al.*, 2008; Dupont *et al.*, 2012; Dupont and Hunter, 2012; Dutra *et al.*, 2012; Dzierszinski *et al.*, 2007; Egan *et al.*, 2012; Egan *et al.*, 2008, 2009a, 2009b, 2011; El Kasmi *et al.*, 2008; Fagard *et al.*, 1999; Fang *et al.*, 2008; Fehervari *et al.*, 2006; Feng *et al.*, 2008a, 2008b, 2009; Feng and Sher, 2010; Fentress *et al.*, 2010; Ferguson and Hutchison, 1987a, 1987b; Ferreira da Silva *et al.*, 2008; Fleckenstein *et al.*, 2012; Foureau *et al.*, 2010; Frankel *et al.*, 2007; Frenkel and Escajadillo, 1987; Frickel *et al.*, 2008; Ferreira da Silva *et al.*, 2009; Fox and Bzik, 2010; Fujigaki *et al.*, 2002; Furuta *et al.*, 2006; Gaddi and Yap 2007; Gais *et al.*, 2008; Garweg and Candolfi, 2009; Gaskell *et al.*, 2009; Gazzinelli *et al.*, 1994a, 1994b; Ghatak and Zimmerman, 1973; Gigley *et al.*, 2009a, 2009b; Gilbert *et al.*, 2007, 2008; Glatman and Zaretsky, 2012; Goodwin *et al.*, 2012; Goldszmid *et al.*, 2007, 2009, 2012; Goldszmid and Sher, 2010; Gregg *et al.*, 2011; Grimwood *et al.*, 1996; Groer *et al.*, 2011a; Gröer *et al.*, 2011b, 2012; Guan *et al.*, 2007; Guiton *et al.*, 2010; Guglietta *et al.*, 2007; Hall *et al.*, 2012; Halonen *et al.*, 1996, 1998a, 1998b, 2001, 2006; Haroon *et al.*, 2012; Harris *et al.*, 2010, 2012; Hassan *et al.*, 2012; Hauser *et al.*, 1982; Hauser and Tsai, 1986; Hayashi *et al.*, 1996a, 1996b; Heimesaat *et al.*, 2006, 2007; Henriquez *et al.*, 2010; Henry *et al.*, 2009; Hermes *et al.*, 2008; Hertzog *et al.*, 2011; Hippe *et al.*, 2008, 2009a, 2009b; Hirota *et al.*, 1996; Honore *et al.*, 2000; Hortua Triana *et al.*, 2012; Hoshi *et al.*, 2010; Hou *et al.*, 2011; House *et al.*, 2011; Howard *et al.*, 2011; Hunter and Sibley, 2012; Hutchison *et al.*, 1980; Indraccolo *et al.*, 2007; Ishii *et al.*, 2006; James 1995; Jamieson *et al.*, 2008, 2010; Jankovic *et al.*, 2007; Janssen *et al.*, 2002; Jebbari *et al.*, 1998; Jensen *et al.*, 2011; Jin *et al.*, 2008, 2009; John *et al.*, 2009, 2011; Johnson and Sayles, 2002; Johnson *et al.*, 2003; Jones *et al.*, 2010, 2011; Jordan *et al.*, 2009a, 2009b, 2010a, 2010b; Jost *et al.*, 2007; Kafsack *et al.*, 2009; Kannan *et al.*, 2010; Khan *et al.*, 2006; Khan *et al.*, 2001; Kikumura *et al.*, 2012; Kim *et al.*, 2005, 2006a, 2008; Kim and Denkers, 2006b; Kim, Karasov and Boothroyd, 2007; Kim, Y. *et al.*, 2008; Kirak *et al.*, 2010; Koblansky *et al.*, 2012; Konen-Waisman and Howard, 2007; Koshy *et al.*, 2010, 2012; Kubo *et al.*, 2003; Lahmar *et al.*, 2010; LaRosa *et al.*, 2008; Lambert *et al.*, 2006; Lambert *et al.*, 2009; Lamberton *et al.*, 2008; Lang *et al.*, 2006, 2007, 2012; Lee *et al.*, 2007; Leiva *et al.*, 1998; Leng and Denkers, 2009; Leng *et al.*, 2009; Lepage *et al.*, 1998; Lees *et al.*, 2010; Lieberman *et al.*, 2004a, 2004b; Liesenfeld *et al.*, 2011; Lim *et al.*, 2012; Ling *et al.*, 2006; Liu *et al.*, 2006; Locksley *et al.*, 1982; Lu *et al.*, 2003, 2009;

Lüder *et al.*, 2009, 2010; Lykens *et al.*, 2010; Machado *et al.*, 2006a, 2006b, 2010; MacMicking, 2012; Mahamed *et al.*, 2012; Makino *et al.*, 2011; Marshall and Denkers 1998; Marshall *et al.*, 1999; Masek *et al.*, 2007; Mashayekhi *et al.*, 2011; Mason *et al.*, 2004; McBerry *et al.*, 2012; McGeachy *et al.*, 2009; McLeod *et al.*, 2012; Melo *et al.*, 2010; Melzer *et al.*, 2008, 2010; Menard *et al.*, 2007; Mendes *et al.*, 2011; Menzies *et al.*, 2008; Mévélec *et al.*, 2010; Meylan *et al.*, 2008; Middleton *et al.*, 2009; Miller *et al.*, 2011a, 2011b; Minns *et al.*, 2006; Minot *et al.*, 2012; Mirpuri and Yarovinsky, 2012; Mitchell *et al.*, 2006; Mitra *et al.*, 2012; Montfort *et al.*, 2009; Mun *et al.*, 2003; Muñoz *et al.*, 2009; Mordue *et al.*, 2007; Morgado *et al.*, 2011; Murakami *et al.*, 2012; Murray *et al.*, 1991; Nagineni *et al.*, 1996; Nance *et al.*, 2012; Nathan and Shiloh, 2000; Niedelman *et al.*, 2012; Nishanth *et al.*, 2010; Nishikawa *et al.*, 2007; Nishikawa *et al.*, 2011; Noor *et al.*, 2010; Norose *et al.*, 2011; Norose *et al.*, 2003, 2008; Oberdorfer *et al.*, 2003; Ochoa-Repáraz *et al.*, 2011; Okusaga *et al.*, 2011; Oldenhove *et al.*, 2009; Ong *et al.*, 2010; Ortiz-Alegria *et al.*, 2010; Osborn *et al.*, 2010; Passos *et al.*, 2010; Pawlowski *et al.*, 2007; Peixoto-Rangel *et al.*, 2009; Peixoto *et al.*, 2010; Pepper *et al.*, 2008; Perona-Wright *et al.*, 2009; Peterson *et al.*, 1995; Pfaff *et al.*, 2008a; Pfaff *et al.*, 2007, 2008a, 2008b; Pfefferkorn and Guyre, 1984; Phelps *et al.*, 2008; Pifer *et al.*, 2011; Pino *et al.*, 2007; Plattner *et al.*, 2008; Ploix *et al.*, 2011; Poassos *et al.*, 2010; Pollard *et al.*, 2009a, 2009b; Powell *et al.*, 1978; Portillo *et al.*, 2010, 2012; Portugal *et al.*, 2008; Prandovszky *et al.*, 2011; Rajapakse *et al.*, 2007; Ramón *et al.*, 2010; Ravindran and Boothroyd 2008; Reese *et al.*, 2011; Reichmann *et al.*, 1999, 2000; Resende *et al.*, 2008; Rincon, 2012; Robben *et al.*, 2005; Robbins *et al.*, 2012; Rock *et al.*, 2005; Romano *et al.*, 2012a, 2012b; Rosowski and Saeij, 2012; Rosowski *et al.*, 2011; Rozenfeld *et al.*, 2005; Saeij *et al.*, 2007, 2008; Samarajiwa *et al.*, 2009; Sanchez *et al.*, 2010; Salek-Ardakani and Croft, 2009; Sauer *et al.*, 2009; Sautel *et al.*, 2009; Schaeffer *et al.*, 1999, 2009; Scharton-Kersten *et al.*, 1998; Schluter *et al.*, 1995, 2001, 2003; Schulthess *et al.*, 2008, 2012; Schwartz and Hunter, 2007; Senegas *et al.*, 2009; Severance *et al.*, 2012; Sergent *et al.*, 2005; Shapira *et al.*, 2005; Shaw *et al.*, 2006; Shaw *et al.*, 2009; Shi *et al.*, 2012; Shiono *et al.*, 2007; Sibley and Ajioka, 2008; Sibley *et al.*, 2010; Silva *et al.*, 2009, 2010; Silver *et al.*, 2011; Sims *et al.*, 1989; Skallova *et al.*, 2006; Skariah and Mordue, 2012; Skariah *et al.*, 2012; Spear *et al.*, 2006; Starnes *et al.*, 2009; Stenzel *et al.*, 2004; Stibbs, 1985; Strack *et al.*, 2002; Strobl *et al.*, 2012; Stumhofer *et al.*, 2006; Stutz *et al.*, 2012; Subauste, 2009a, 2009b, 2009c; Subauste *et al.*, 1991, 2002; Sukhumavasi *et al.*, 2007, 2008, 2010; Sun *et al.*, 2007; Suzuki *et al.*, 1989, 1994a, 1994b, 1996b, 2005, 2010; Suzukui, 1997; Suzuki and Remington 1988, 1990; Tait *et al.*, 2010; Tait and Hunter, 2009; Takács *et al.*, 2012; Tan *et al.*, 2010; Tato *et al.*, 2012; Taylor *et al.*, 2004; Terrazas *et al.*, 2010; Thompson *et al.*, 2008; Tomita *et al.*, 2009; Torrey *et al.*, 2012; Tu *et al.*, 2009; Tussiwand *et al.*, 2012; Unoki *et al.*, 2009; Vallochi *et al.*, 2008; Van *et al.*, 2007; Vaudaux *et al.*, 2010; Villard *et al.*, 2012; Villarino *et al.*, 2003, 2006; Villegas *et al.*, 2002; Virreira *et al.*, 2011; Vutova *et al.*, 2007; Vyas *et al.*, 2007a, 2007b; Vyas and Sapolsky, 2010; Vyas, 2013; Wang *et al.*, 2007, 2009a, 2010a; Wang and Suzuki, 2007; Washino *et al.*, 2012; Wasmuth *et al.*, 2012; Watford *et al.*, 2008; Webster *et al.*, 1994; Wen *et al.*, 2010; Wendte *et al.*, 2011; Whitmarsh and Hunter, 2008; Whitmarsh *et al.*, 2011; Wiehagen *et al.*, 2012; Wiley *et al.*, 2010; Willie *et al.*, 2001; Wilson, E.H. *et al.*, 2009; Wilson, D.C. *et al.*, 2010; Wilson, M.S. *et al.*, 2010; Wilson and Hunter, 2004; Wilson *et al.*, 2005, 2008, 2006; Witola *et al.*, 2011; Witting, 1979; Woodman *et al.*, 1991; Yamamoto *et al.*, 2011, 2012; Yamashita *et al.*, 1998; Yamauchi *et al.*, 2007; Yap and Sher, 1999b, 2007; Yap *et al.*, 2000; Yarovinsky *et al.*, 2005, 2008, 2010; Yong *et al.*, 1994; Young *et al.*, 2012; Xiao *et al.*, 2009, 2012b; Zakrzewski, 2011; Zeiner and Boothroyd, 2010; Zhang *et al.*, 2007; Zhang and Denkers, 1999; Zhao *et al.*, 2007, 2008, 2009a, 2009b; Zhou *et al.*, 2012; Zukas *et al.*, 2009).

There are some studies which show interactions of parasite molecules with certain human cells. These include: type 1 GRA1 and TgFβ-monocytes (Beghetto *et al.*, 2012); type 1 GRA10 Taf1b-fibroblasts (Ahn *et al.*, 2010); type 2 GRA15 NFκB IkB–HDACs and ubiquitin E3 ligase human fibroblasts, mouse macrophages (Brunet *et al.*,2008; Candolfi *et al.*, 2008); type 1 Rop16 STAT3 and 6-fibroblasts, mouse macrophages (Boothroyd *et al.*, 2011); type 1 Rop5/18 murine GTPases (Reese *et al.*, 2010; Sibley and Ajioka, 2008), human AP-1 ATFβ (Yamamoto *et al.*, 2012); types 1, 2, 3, Rop16 Stat 1–mouse macrophages, human fibroblasts (Saeij *et al.*, 2008; Hunter and Sibley, 2012; Denkers, 2003a, 2003b; Lüder *et al.*, 2003a, 2003b); type 3 Rop38 human monocytes p38 Map kinase (Peixoto *et al.*, 2010); mouse brain transcriptome PD-1 L, SOCS3, GFAP, CD36 (Hermes *et al.*, 2008); adult neuronal tumour cells (Xiao *et al.*, 2009); HIF 1α and human fibroblasts (Blader *et al.*, 2001); type 1 miR 17∼92 and 106b∼25, Mycland fibroblasts and E2F3 (Zeiner and Boothroyd, 2010). *T. gondii* proteins known to elicit lymphocyte blastogenesis or antibodies in humans have been described by a variety of research groups. These include, among others, SAG1, GRA1, GRA4, GRA6, GRA7, MAG1 complete. There is polarization of macrophage phenotypes which are altered by different types of parasites: Type I, Rop16 (M2 associated with fibrosis, TGFβ), Type II GRA 15 (M1 proinflammatory). This type of macrophage polarization also is associated in separate studies with different metabolic phenotypes. That the metabolism of inflammation is limited by AMPK and pseudo-starvation was described recently (O'Neill and Hardie, 2013) (Fig. 25.20).

25.4.2.17 *T. gondii* Strains Associated with Premature Births and Severity of Manifestation of Congenital Toxoplasmosis at Birth

In recent work by McLeod *et al.* (2012) persons with congenital toxoplasmosis and their mothers in North America in the NCCCTS were studied. An ELISA which identifies antibodies to peptides of GRA6 and 7 allelic variants allows differentiation of persons infected with genetic type 2 and non-type 2 parasite strains by testing serum. This identifies antibodies to parasites that elicit antibody to either type 2 or non-type 2 parasites. Presence of antibodies to both type 2 and non- type 2 GRA6/7 were present in a limited number of persons where parasite isolates and serum were both available. In this case infection was with type 4 (haplogroup 12) parasites. Studies are ongoing to identify other peptides useful to separate parasite genetic types further through such serotyping (McLeod *et al.*, 2012). Multiple genetically distinct *T. gondii* strain types have been found throughout the world. In France, where research has been done to establish which strains are most common, type II strains predominate. In this study of persons in the NCCCTS, type II parasites were distinguished from all other strains, which were collectively termed not-exclusively-type II strains (or NE-II). Either type II or NE-II infections were present in 183 of the mother–child pairs in the NCCCTS. Statistical analysis revealed that NE-II parasites were more likely to be associated with premature birth, and infants infected with these strains were more likely to have severe manifestations of disease than infants infected by type II parasites. For example, severe eye damage was seen at birth in 67% of NE-II cases (59 out of 88), while such eye damage was present in only 39% of type II cases (18 out of 46). Association is not absolute and mild, moderate or severe disease can result, regardless of the infecting strain. When clinical histories of those children in the long-term study who had been diagnosed with congenital toxoplasmosis during gestation and whose mothers had received treatment with medicines prior to giving birth, the association between N-II and severe disease at birth was not present. This study also demonstrated that outcomes were equally good following postnatal treatment for type II and N-II parasites, although not all outcomes are favourable for all children. This suggested that both treatment *in utero* and/or postnatally was effective against type II and N-II parasites.

In this work of McLeod *et al.* (2012), for the first time, it was determined systematically that congenitally infected people and their mothers in the USA have European-type strains. However, there also is a predominance of other types, and McLeod *et al.* (2012) determined that N-II strains in the USA and Canada would have more severe disease symptoms associated with them at birth for congenitally infected persons who were not treated *in utero*, although this was not absolute. Unlike in France, where type 2 is the most common strain detected, McLeod *et al.* (2012) found that NE-II parasites predominated (61%) in the USA over the three-decade span of the national collaborative study. NE-II parasites were more common than type 2 along the Gulf Coast, the Pacific coast and in Hawaii. NE-II strains were also more common among lower-income and rural populations. Similar to the effect of prenatal treatment on outcomes at birth, postnatal treatment appeared to be efficacious for type 2 and NE-II strains for infections that were either type 2 or NE-II. Whether this will affect influences of parasite genetics on subsequent immune response during or after treatment is not known.

This study also suggested that there is an interaction of parasite and host genetics as shown in Figure 25.26, where susceptibility alleles for *COL2A* and *ABC4r* were more often noted in persons who had serologic evidence of infection with type II parasites. Whether this might be associated with a more robust proinflammatory immune response or differential processing and presentation of a specific immunodominant parasite peptide and a different consequent host response remains to be determined. Since *T. gondii* epigenetically influences these genes, the mechanism for the parent-of-origin effect observed also remains to be determined. Figure 25.20 summarizes some of the interactions, including mechanisms of sensing by TLRs, NLRs and NAIPs, regulatory molecules, downstream signalling and effector mechanisms.

25.5 INFLUENCE OF PARASITE STRAIN ON IMMUNE RESPONSE AND DISEASE IN MURINE MODELS

Earlier studies of parasite strain and immune response are summarized here. Earlier, *T. gondii* strains were divided into three main lineages (types I, II and III) based on various genetic markers (Sibley and Boothroyd, 1992a; Howe and Sibley, 1995), although considerably greater complexity has been established recently. Earlier studies in mice showed that infection with each of the three lineages of *T. gondii* results in different outcomes: type I strains are highly virulent, whereas types II and III strains are relatively avirulent in mice (Kaufman *et al.*, 1958, 1959; Sibley and Boothroyd, 1992a; Howe and Sibley, 1995). Type I strains differ genetically by 1% or less from type II and type III strains (Sibley and Boothroyd, 1992a); however, the main determinants that dramatically affect the virulence of different *T. gondii* strains in the host and pathogenesis of toxoplasmosis are only partially understood (Saeij *et al.*, 2005b; Robben *et al.*, 2004; Mordue *et al.*, 2001; Miller *et al.*, 1999a, 1999b, 2000). Type II strains of *T. gondii* earlier were reported to be dominant in the USA and are frequently isolated from AIDS patients with toxoplasmic encephalitis (Howe *et al.*, 1997). Interestingly, other studies suggest that type I and type I/III or recombinant strains are involved in the development of ocular disease (Grigg *et al.*, 2001a; Kahn and Sibley, 2004). In one study, an atypical strain parasite was identified as responsible for a toxoplasmosis epidemic outbreak in North America (Victoria, Canada) associated with a high rate of development of acquired ocular disease (Burnett *et al.*, 1998).

In contrast with the frequencies of clonal types of parasites observed in the USA and much of Europe (Howe and Sibley, 1995; Howe *et al.*,

1997), in a study of strains in the Minas Gerais region of Brazil, there were mixed genotypes, with typical alleles of types I, II or III at almost all loci assessed, with a clear dominance of alleles characteristics of type I strains, followed by a high frequency of type III strains and a very low frequency of alleles from type II strains (Ferreira *et al.*, 2005). Others have reported on atypical isolates in South America and Africa (Sibley, 2003; Howe and Sibley, 1995; Darde *et al.*, 1992; Ajzenberg *et al.*, 2002, 2004; Silveira *et al.*, 2001; Glasner *et al.*, 1992; Khan *et al.*, 2005; Fazaelia *et al.*, 2000; Neto *et al.*, 2000; da Silva *et al.*, 2005; Dubey *et al.*, 2002, 2003a, 2003b, 2003c, 2004, 2005a, 2005b). The fact that the Brazilian *T. gondii* strains are more closely related to the type I lineage is noteworthy. These strains were studied in mice and 85% were highly virulent, virulent or of intermediate virulence and only 15% of the strains were avirulent (Ferreira *et al.*, 2005). These findings contrast with studies performed in Europe, where most strains are avirulent of type II or type III (Howe and Sibley, 1995; Howe *et al.*, 1997). Although speculative at this point, it is plausible that the high frequency of virulent strains closely related to type I lineage may be in part responsible for the high frequency of severe, acquired, congenital ocular toxoplasmosis commonly found in Minas Gerais, Brazil, although host factors and size of inoculum may also play a role. There are a considerable number of genetic polymorphisms of parasites in Brazil (Portela *et al.*, 2004; Glasner *et al.*, 1992; Silveira *et al.*, 2001).

Clonal type I and type II *T. gondii* strains have been shown to stimulate host cells to produce proinflammatory cytokines differently. For example, the RH strain, the prototype strain of the Type I group, induces a robust proinflammatory response during the very early stages of infection. This immune response appears to be in part responsible for the pathology and lethality observed during the first week of infection (Mordue *et al.*, 2001). In addition, type II strains and not type I strains were shown to elicit the production of IL-12 by host macrophages. IL-12 has a critical role in inducing IFNγ and host resistance to infection with *T. gondii* (Robben *et al.*, 2004). Similar findings have been described by Suzuki *et al.* (2005). Thus, the ability to induce IL-12 may contribute to the control of parasite replication and relative avirulence of type II strains, as opposed to type I strains. The avirulent type I/III strains showed a similar pattern of induction of cytokines, such as IL-12 and IFNγ, as compared to the ME49 strain, a Type II strain (Fux *et al.*, 2003). There have been suggestions that a clonal type of parasite may be associated with magnitude of eye disease, IgA antibody to GPILs or other disease manifested during infection in humans (Portela *et al.*, 2004; Grigg *et al.*, 2001a). In addition, Dobbin *et al.* (2002) found that HSP70 of *T. gondii* inhibits iNOS expression, NO production and NFκB activity which can have effects on cytokine production, but that these effects occur only with the virulent RH and ENT strains and not with the avirulent ME49 and e-strains.

As previously noted for a type II strain (ME49), IL-12/IFNγ axis and iNOS were defined as main determinants of resistance during the acute infection with the Brazilian strains. Different from the type II strain of *T. gondii* (ME49), per-oral infection with the Type I/III strains led to only a mild inflammatory infiltrate and no major lesions in the intestine of C57BL/6 mice. In addition, BALB/c (resistant to ME49) and C57BL/6 (susceptible to ME49) mice were shown, respectively, to be more susceptible and resistant to cyst formation and toxoplasmic encephalitis, when infected with type I/III strains. Studies with congenic BALB/c strain mice that had MHC haplotype 'b' and C57BL/6 containing MHC haplotype 'd' showed that whereas the 'd' haplotype was a critical element conferring host resistance to parasites of the ME49 strain, the genetic background (and not MHC haplotype) of C57BL/6 mice was responsible for resistance of these mice to recombinant type I/III strains isolated in the Minas Gerais region of Brazil. These results indicate that MHC haplotype 'b' is a major determinant of susceptibility to cyst formation and toxoplasmic encephalitis induced during infection with type II, but not with type I/III strains of *T. gondii* (Fux *et al.*, 2003).

For mice with the same genetic background, different clonal types of parasites may not elicit the same immune response. For example, in C3H L^d mice, $CD8^+$ cytolytic T-cells could lyse target cells that were infected with the homologous type 2 strain with which the mice were per-orally infected, but not MHC matched target cells infected with a heterologous clonal type (Johnson *et al.*, 2002b) (Fig. 25.19).

25.6 GENERAL ASPECTS OF IMMUNITY

25.6.1 Relationship of Adaptive Immunity to the Innate Immune System

Studies in mice demonstrate the important role of cytokines such as IL-12, TNFα and IFNγ and reactive nitrogen intermediates (RNI) as mediators of host resistance to *T. gondii* infection (Denkers and Gazzinelli, 1998). Animals deficient in IL-12, IFNγ and inducible nitric oxide synthase (NOS2), treated with neutralizing antibodies, anti-cytokines or specific inhibitors of NOS2 (on a C57B16 background) are highly susceptible to infection with *T. gondii* (Suzuki *et al.*, 1989; Gazzinelli *et al.*, 1991, 1992, 1993; 1994; Hayashi *et al.*, 1996a, 1996b; Scharton-Kersten, 1996a, 1996b, 1997a). Pathology associated with excessive immune stimulation of Th1 responses and high levels of IFNγ has been also demonstrated during acute infection with *T. gondii* in a mouse model (Liesenfeld, 1996). Production of IL-10 is stimulated during infection with *T. gondii* and is involved in regulating production of IL-12, IFNγ and TNFα and its absence is associated with exacerbated pathology (Gazzinelli, 1996b; Neyer *et al.*, 1997). Acquired immunity to *T. gondii* is associated with a Th1-type response (Gazzinelli *et al.*, 1991, 1992a, 1994a). During chronic toxoplasmosis, neutralization of IFNγ, TNFα or inhibition of NOS2 results in the reactivation of disease and the development of encephalitis as well as uveitis (Gazzinelli *et al.*, 1992a, 1993b, 1994a; Hayashi *et al.*, 1996a, 1996b). The relative importance of some of these factors, including TNFα and NOS2 appears to be dependent on the host genetic background (Johnson, 1992b; Rozenfeld *et al.*, 2003). Among other factors, specific MHC haplotypes are known to be important determinants of host resistance and susceptibility to early infection in mice (McLeod *et al.*, 1989a, 1989b; Brown and McLeod, 1990; Suzuki *et al.*, 1991a; Blackwell, 1998; Johnson *et al.*, 2002b) and appear to influence the relative contribution of some of these factors in different mouse strains. Consistently, both $CD4^+$ T- and $CD8^+$ T-lymphocytes are important components in host resistance to this parasite (Gazzinelli *et al.*, 1991c; Suzuki *et al.*, 1991a, 1991b, 1991c). Thus, one of the important contributions of innate immunity and of the Toll-like receptors (TLRs) is in shaping protective T-cell mediated (adaptive immunity) immunity. Some of these interactions between components of the innate and adaptive immune systems are shown in Figure 25.5.

In other systems major advances have been made in the assignment of individual TLRs to specific components of bacterial cell walls (LPS, lipoproteins and peptidoglycan) that have potent proinflammatory activity. Characterization of TLR ligands is only beginning to be performed for protozoan parasites. Several studies have identified that glycosylphosphatidylinositol (GPI) anchors (or fragments from GPI anchors) from *P. falciparum, L. major, T. gondii, T. brucei* and *T. cruzi* act as molecules that activate cells of both lymphocytic and monocytic lineages. Structural requirements for the bioactivity of protozoan-derived GPI anchors have been partially defined. Minor structural modifications of these glycolipids markedly affect their biological activity, including the induction of proinflammatory cytokines, chemokines and apoptosis. Importantly, native GPI anchors purified from the tachyzoite stage of *T. gondii* as well as synthetic fragments mimicking the native GPI anchors were shown to promote NFκB activation and stimulate the synthesis of TNFα by a murine macrophage cell line (Debierre-Grockiego *et al.*, 2003). Heat shock protein and other partially purified protein preparations from *T. gondii* tachyzoites were shown to activate TLR4 and TLR2, respectively. More recently a profilin-like protein from *T. gondii* (PFTG) was also shown to activate TLR11 in murine cells as is discussed below under dendritic cells and antigen processing and presentation (Yarovinsky *et al.*, 2005). In humans, the TLR11 gene has a stop codon in the middle and humans express a truncated form of TLR11 without the functional domain. Nevertheless, live tachyzoites have been shown to elicit IL-12 production by human dendritic cells suggesting that other TLR agonists may be acting to activate TLRs in humans. Recently, mouse TLR12 was also found to recognize profilin and to confer partial protection in mice (Koblansky *et al.*, 2012). TLR12 functions more robustly with TLR11 in mice. TLR12 also is a pseudogene in humans (Roach *et al.*, 2005).

The strongest data suggesting that TLRs are important in the host resistance and pathogenesis of protozoan infections, often acting together, come from microbial challenge studies in MyD88 or TLR or ICE deficient mice. Since MyD88 signals from other protein receptors ICE knock-out mice are also informative (Denkers, 2010). MyD88 is involved in the activation of all of the known TLRs with the exception of TLR3. In addition to TLRs, MyD88 has only been described as essential for IL-1 and IL-18 receptor functions (Adachi *et al.*, 1998). Neither IL-1 nor IL-18 has been described as critical in host resistance to *T. gondii* infection (Graefe *et al.*, 2003); hence an impaired response of MyD88 deficient mice to a pathogen provides reasonable evidence that a member of the TLR family is engaged. Indeed, MyD88 deficient mice infected with *T. gondii* had diminished production of IL-12 and IFNγ, enhanced parasitemia and accelerated mortality (Campos, 2004). Thus, TLRs appear to be involved in early resistance to infection with *T. gondii*. The identification of a single TLR involved in the *in vivo* host responses to protozoan parasites has been a more difficult task. Enhanced susceptibility to infection in mice deficient in a specific TLR, such as TLR2 and TLR4, is highly dependent on experimental conditions including parasite strain, size and route of inoculum. In most cases, TLR deficiency has no major impact on resistance to protozoan infections (Adachi *et al.*, 2001; Mun *et al.*, 2003; Campos, 2004; Kropf *et al.*, 2004; Oliveira *et al.*, 2004). On the other hand mice lacking TLR11, the counterpart receptor for PFTG, showed a significant phenotype, with a great increase on cyst numbers. However, no published information concerning the role for TLRs in human toxoplasmosis is available. As mentioned above, in all humans tested to date, the TLR11 gene has a stop codon and does not appear to function (Zhang *et al.*, 2004). Thus, it is likely that alternative TLR(s) or another cognate receptor from innate immunity will be important in human toxoplasmosis.

Since MYD88 is not essential for live attenuated vaccines to elicit protection (Denkers *et al.*, 2010) and since all TLRs signal via MYD88, other than TLR3, this suggests that other PAMPs and DAMPs may be critical. Nucleic acid sensing, roles of NALP1, P2X7R, NLRP3, NOD2, TLRs, HLA in humans, and polarization of macrophages in mice, have been discussed above.

25.6.2 Antigen Presentation, and Processing and MHC Class I and Class II Molecules

Schematic diagrams of antigen presentation to the MHC class I and MHC class II pathways, the effector functions elicited and the cytokines produced, and some of the evidence for the importance of these mechanisms are summarized in Figures 25.1–25.10 and Tables 25.1 and 25.2. CD8α dendritic cell production of IL-12 is essential for generation of adaptive immunity. Several *T. gondii* proteins (e.g. SAG1, GRA2, GRA6, GRA7, BAG1, etc.) have elicited antibodies, providing evidence for Class II MHC processing and presentation and $CD4^+$ T-cell help (e.g. Kasper and Buzoni-Gatel, 1998; Couper *et al.*, 2003; Letscher-Bru *et al.*, 2003). The presence of CD4 and CD8 cytolytic and IFNγ producing T-cells have provided evidence that there is MHC processing either through conventional MHC class I processing or via bystander uptake of antigen which enters an MHC class I pathway (Fig. 25.18). Pores in the vacuolar membrane (Joiner *et al.*, 1990; Schwab *et al.*, 1994) and secretion from the parasite into the vacuole (Gubbels *et al.*, 2005) contribute *T. gondii* proteins (peptides) which can enter a class I MHC pathway.

As described above, primary infection by *T. gondii* stimulates an effective protective immune response through production of type 1 cytokines and continuous cell-mediated immunity. After innate resistance and IFNγ-dependent mechanisms have controlled tachyzoite replication, both $CD4^+$ and $CD8^+$ T-cells are essential to maintain bradyzoites in a latent stage. Control of acute and chronic phases of *T. gondii* infection requires both $CD4^+$ and $CD8^+$ T-cells. There are only limited data about antigens that elicit cell mediated immunity (Khan *et al.*, 1988a, 1988b, 1994; Yano *et al.*, 1999) and endogenous immunodominant epitopes have not yet been characterized.

One approach to studying the generation of antigen-specific T-cell responses during toxoplasmosis has been to introduce model antigens for which the TCR specificity is known (Kumar and Tarleton, 2001; Pope *et al.*, 2001; Yrlid *et al.*, 2001, Deckert and Soldati, 2000). Many genetic tools and approaches have been developed to study *T. gondii* biology (Boothroyd *et al.*, 1997; Roos *et al.*, 1994). These genetic approaches are currently being utilized to study antigen presentation through both MHC-I and MHC-II, and naive and effector or memory $CD4^+$ and $CD8^+$ T-cells responses in the context of acute and chronic infection by *T. gondii* (Deckert *et al.*, 2004; Dzierszinski Gordon Conference, *Toxoplasma* meeting, Corsica, 2005; Dzierzinski, personal communication, 2005). Dzierszinski *et al.* used a transgenic system of ovalbumin-expressing *T. gondii* parasites (Pepper *et al.*, 2008). This system was shown to be appropriate to study $CD4^+$ T-cell responses *in vivo* (Pepper *et al.*, 2008). It has been used to better understand antigen presentation by professional or non-professional antigen presenting cells through MHC class I and/or MHC class II molecules, as well as $CD8^+$ T-cell responses, *in vitro* and *in vivo* by Pepper *et al.* (2008). Taking advantage of another model antigen, Eα:I–A(b) complex-red fluorescent protein fusion (EαRFP) (Barlow *et al.*, 1998; Itano *et al.*, 2003), a second *T. gondii* transgenic system was generated to track antigen presenting cells and follow MHC class II antigen presentation *in vivo* (Hunter and Sibley, 2010). In order to better characterize antigen presentation and T-cell responses during toxoplasmosis, transgenic parasites were also generated expressing the model antigens under the control of different promoters, which vary in strength and stage specificity and in multiple parasite strain backgrounds. A variety of assays and genetic screens are also currently being developed to determine *T. gondii* and host molecules implicated in parasite–host interactions (Dzierszinski *et al.*, 2007).

Transgenic ovalbumin expressing parasites have also been utilized to study the antigen presentation proteins that enter the parasite secretory pathway (Gubbels *et al.*, 2005). These antigens were found to enter the classical class I MHC presentation pathway (Gubbels *et al.*, 2005). Presentation of ovalbumin secreted by the parasite required the peptide transporter TAP and occurred primarily in actively infected cells, rather than through bystander cells. Dendritic cells are a major target of *T. gondii* infection and account for much of the antigen presentation by the spleen (Robey *et al.*, 2005). Further, Cre protein secreted by *T. gondii* is capable of mediating recombination in the nucleus of the host cell. Thus, using this heterologous system, it has been shown that proteins can escape from *Toxoplasma* into the parasitophorous vacuole and then enter a class I MHC presentation pathway and be recognized by CD8 T-cells (Gubbels *et al.*, 2005).

Aliberti, Jankovic and Sher reviewed dendritic cell function in *T. gondii* infection (Aliberti *et al.*, 2003). Dendritic cells play a major role in the initiation of IL-12 driven host resistance. IL-12 synthesis by dendritic cells is carefully regulated to avoid overproduction. Dendritic cells play a critical role in determining the highly polarized T-helper 1 type response triggered by *T gondii*. Dendritic cell function is initiated by *T. gondii* and parasite-primed dendritic cells drive the Th1 effector choice. There is endogenous regulation of dendritic cell IL-12 production during *T. gondii* infection. IL-12 is the primary cytokine initiating IFNγ synthesis by NK- and T-lymphocytes during *T. gondii* infection and is essential for survival following *T. gondii* infection. Injection of live tachyzoites or STAg (soluble *T. gondii* antigen) led to dendritic cells clustering in T-cell areas producing significant levels of IL-12 (also *in vitro*). The dendritic cells that respond to STAg are of the CD8 $\alpha\beta$ T-cell subset. The response is rapid (peaks in 3–6 hours), returning to baseline by 24 hours and then is actively turned off for up to a week. Stimulation of mouse TLR 11 is responsible (Plattner *et al.*, 2008). TLR 11 activation of dendritic cells by a profilin-like protein is responsible for stimulation of dendritic cell IL-12 production by a MyD88 dependent manner (Plattner *et al.*, 2008). CCR5, which is activated by parasite cyclophilin-18, is MyD88 independent.

Lipoxin A4 is an arachidonic acid metabolite generated by a 5-lipoxygenase (LO)-dependent pathway. LXA4 inhibits dendritic cell IL-12 production (Aliberti *et al.*, 2002a; Mandal *et al.*, 2005; Karp *et al.*, 2004; Lewis *et al.*, 1990) by inhibiting CCR5 expression. LXA4 synthesis is stimulated by STAg. Acute infection of mice leads to LXA4 production that reaches a plateau at 15 days and persists during the chronic phase of infection (Aliberti *et al.*, 2002b). Mice deficient in LXA4 have more IL-12/IFNγ synthesis, reduced parasite cyst numbers in brain, but die 38 days after infection probably from uncontrolled proinflammatory responses. In 5-LO deficient mice LXA4 administration prevented mortality. In 5-LO deficient mice exogenous IL-10 prolonged survival, however a relapse of infection in the CNS occurred in surviving mice. LXA4 protects mice against cytokine-mediated pathology primarily during the chronic phase of infection and appears to affect CD11 cells. The down-modulatory effect of LXA4 is associated with an induction of SOCS-2 (suppressor of cytokine signalling). IL-10 has a down-modulatory effect associated with the induction of SOCS-1 and SOCS-3. *T. gondii* can generate LXA4 by using host lipid precursors. Extracts of the parasite have 15-lipoxygenase activity (Bannenberg *et al.*, 2004) and 15-LO from plants induces LXA4 when inoculated into mice.

In other models IL-12 appears to serve as a bridge between innate and adaptive immunity by promoting the development of Th1 effector cells (Blanco *et al.*, 2000; Uematsu *et al.*, 2002). It has been shown to be critical for generation of an adaptive immune response in wild type and vaccine induced immune responses. IL-10 appears to have a major down-modulatory effect on IFNγ mediated effector function when it is at a maximum during the acute phase of infection. IL-15 is needed for optimal priming of CD4$^+$ T-cell response (Combe *et al.*, 2006).

In considering antigen presentation and processing by MHC molecules, work has demonstrated that *T. gondii* may subvert the immune response by down-modulating MHC molecule expression as well as iNOS and NO production (Lüder *et al.*, 1998, 2001, 2003a, 2003b; Rozenfeld *et al.*, 2003). This effect of *T. gondii* on MHC expression had varied depending on the cell type and experimental model system under study (Bohne *et al.*, 1994; Denkers and Butcher, 2005; Goebel, 2001; Lüder *et al.*, 2001, 2003a, 2003b;

Lüder and Seeber, 2001; Schluter *et al.*, 1991, 1993). This subversion of the immune system may contribute to *T. gondii* cysts that persist in the central nervous system. PD1 L, $CD8^+$ T-cells and cyst rupture related to chitinase (Wilson *et al.*, 2012) may also play a role in cyst rupture and containment and persistence.

The elegant studies which have helped to decipher the importance of sensors of pathogens and danger and their downstream signalling molecules and how the parasite interferes with this in a strain specific manner are also summarized in chapters in this book on innate immunity, *T. gondii* and the brain and parasite genetics. Relevant references in the past six years are included in those chapters and are only mentioned briefly here and cited here and in our tables and figures (Abi Abdallah and Denkers, 2012; Abi Abdallah *et al.*, 2011; Afonso *et al.*, 2012; Akira *et al.*, 1993; Albuquerque *et al.*, 2009; Ali *et al.*, 2005; Aline *et al.*, 2002a, 2002b; Alvarado-Esquivel *et al.*, 2012; Alves *et al.*, 2012; Andrade *et al.*, 2005, 2006; Arling *et al.*, 2009; Arsenijevic *et al.*, 2007; Aviles *et al.*, 2008; Avunduk, *et al.*, 2007; Baird *et al.*, 2013; Beaman, *et al.*, 1993; Barragan and Sibley, 2002; Beauvillain *et al.*, 2007; Begum-Haque *et al.*, 2009; Bekpen *et al.*, 2005, 2010; Benson *et al.*, 2009, 2012; Berdoy *et al.*, 2000; Bernardes *et al.*, 2006; Beug *et al.*, 2012; Bhadra *et al.*, 2011; Bierly *et al.*, 2008; Blader and Saeij, 2009; Blanchard and Shastri, 2010; Blanchard *et al.*, 2008; Bliss *et al.*, 2000; Blomstrom *et al.*, 2012; Boothroyd and Dubremetz, 2008; Bottova *et al.*, 2009; Boyle *et al.*, 2007, 2008; Bradley and Sibley, 2007; Brandão *et al.*, 2009; Brenier-Pinchart *et al.*, 2006; Brown *et al.*, 1994, 1995, 2005; Brunet *et al.*, 2008; Butcher *et al.*, 2005, 2011; Buzoni-Gatel *et al.*, 2006, 2008; Byrd and Horwitz, 1989; Caetano *et al.*, 2011; Capuron and Miller, 2011; Cavailles *et al.*, 2006a, 2006b; Cavalcanti *et al.*, 2011; Cesbron-Delauw *et al.*, 2008; Chan, A. *et al.*, 2006; Chan, C. *et al.*, 2006; Chandramohanadas *et al.*, 2009; Chang and Pechere, 1989; Chao *et al.*, 1994, 1993; Charles *et al.*, 2007, 2010; Chou *et al.*, 2012; Chtanova *et al.*, 2009; Clark *et al.*, 2011; Combe *et al.*, 2006; Cong *et al.*, 2008, 2010, 2011, 2012; Correa *et al.*, 2010; Courret *et al.*, 2006; Craven *et al.*, 2012; Crawford *et al.*, 2009; Creuzet *et al.*, 1998; Da Gama *et al.*, 2004; de Alencar *et al.*, 2009; D'Angelillo *et al.*, 2011; Dalton *et al.*, 2006; Dass *et al.*, 2011; Daubener *et al.*, 1999, 2001; Debierre-Grockiego *et al.*, 2003, 2007a, 2007b, 2009, 2010, 2012; Deckert *et al.*, 2006; Dellacasa-Lindberg *et al.*, 2007, 2011; Denkers, 2010; Denkers *et al.*, 2008, 2009, 2011, 2012; Diana *et al.*, 2004; Dimier and Bout, 1997, 1998; Ding *et al.*, 2004; Divanovic *et al.*, 2012; Dogu *et al.*, 2006; Drogemuller *et al.*, 2008; Dunay *et al.*, 2008; Dunn *et al.*, 2008; Dupont *et al.*, 2012; Dupont and Hunter, 2012; Dutra *et al.*, 2012; Dzierszinski *et al.*, 2007; Egan *et al.*, 2008, 2009a, 2009b, 2011, 2012; El Kasmi *et al.*, 2008; Fagard *et al.*, 1999; Fang *et al.*, 2008; Fehervari *et al.*, 2006; Feng *et al.*, 2008a, 2008b, 2009; Feng and Sher, 2010; Fentress *et al.*, 2010; Ferguson and Hutchison, 1987a, 1987b; Ferreira da Silva *et al.*, 2008; Fleckenstein *et al.*, 2012; Foureau *et al.*, 2010; Frankel *et al.*, 2007; Frenkel and Escajadillo, 1987; Frickel *et al.*, 2008; Fox and Bzik, 2010; Fujigaki *et al.*, 2002; Furuta, *et al.*, 2006; Gaddi and Yap 2007; Gais *et al.*, 2008; Garweg and Candolfi, 2009; Gaskell *et al.*, 2009; Gazzinelli *et al.*, 1994a, 1994b; Ghatak and Zimmerman, 1973; Gigley *et al.*, 2009a, 2009b; Gilbert *et al.*, 2007, 2008; Glatman and Zaretsky, 2012; Goodwin *et al.*, 2012; Goldszmid *et al.*, 2009, 2007, 2012; Goldszmid and Sher, 2010; Gregg *et al.*, 2011; Grimwood *et al.*, 1996; Grover *et al.*, 2011a, 2011b, 2012; Guan *et al.*, 2007; Guiton *et al.*, 2010; Guglietta *et al.*, 2007; Hall *et al.*, 2012; Halonen *et al.*, 1996, 1998a, 1998b, 2001, 2006; Haroon *et al.*, 2012; Harris *et al.*, 2010, 2012; Hassan *et al.*, 2012; Hauser *et al.*, 1982; Hauser and Tsai, 1986; Hayashi *et al.*, 1996a, 1996b; Heimesaat *et al.*, 2006, 2007; Henriquez *et al.*, 2010; Henry *et al.*, 2009; Hermes *et al.*, 2008; Hertzog *et al.*, 2011; Hippe *et al.*, 2008, 2009a, 2009b; Hirota *et al.*, 1996; Honore *et al.*, 2000; Hortua Triana *et al.*, 2012; Hoshi *et al.*,

2010; Hou *et al.*, 2011; House *et al.*, 2011; Howard *et al.*, 2011; Hunter and Sibley, 2012; Hutchison *et al.*, 1980; Indraccolo *et al.*, 2007; Ishii *et al.*, 2006; James, 1995; Jamieson *et al.*, 2008, 2010; Jankovic *et al.*, 2007; Janssen *et al.*, 2002; Jebbari *et al.*, 1998; Jensen *et al.*, 2011; Jin *et al.*, 2008, 2009; John *et al.*, 2009, 2011; Johnson *et al.*, 2002a, 2003; Jones *et al.*, 2011; Jones *et al.*, 2010; Jordan *et al.*, 2009a, 2009b, 2010a, 2010b; Jost *et al.*, 2007; Kafsack *et al.*, 2009; Kannan *et al.*, 2010; Khan *et al.*, 2001, 2006; Kikumura *et al.*, 2012; Kim *et al.*, 2005, 2006a, 2008; Kim and Denkers, 2006b; Kim, Karasov and Boothroyd, 2007; Kim, Y. *et al.*, 2008; Kirak *et al.*, 2010; Koblansky *et al.*, 2012; Konen-Waisman and Howard *et al.*, 2011; Koshy *et al.*, 2010, 2012; Kubo *et al.*, 2003; Lahmar *et al.*, 2010; LaRosa *et al.*, 2008; Lambert *et al.*, 2006, 2009; Lamberton *et al.*, 2008; Lang *et al.*, 2006, 2007, 2012; Lee *et al.*, 2007; Leiva *et al.*, 1998; Leng and Denkers, 2009; Leng *et al.*, 2009; Lepage *et al.*, 1998; Lees *et al.*, 2010; Lieberman *et al.*, 2004a, 2004b; Liesenfeld *et al.*, 2011; Lim *et al.*, 2012; Ling *et al.*, 2006; Liu *et al.*, 2006; Locksley *et al.*, 1982; Lu *et al.*, 2003, 2009; Lüder *et al.*, 2009, 2010; Lykens *et al.*, 2010; Machado *et al.*, 2006a, 2006b, 2010; MacMicking, 2012; Mahamed *et al.*, 2012; Makino *et al.*, 2011; Marshall and Denkers, 1998; Marshall *et al.*, 1999; Masek *et al.*, 2007; Mashayekhi *et al.*, 2011; Mason *et al.*, 2004; McBerry *et al.*, 2012; McGeachy *et al.*, 2009; McLeod *et al.*, 2012; Melo *et al.*, 2010; Melzer *et al.*, 2008, 2010; Menard *et al.*, 2007; Mendes *et al.*, 2011; Menzies *et al.*, 2008; Mévélec *et al.*, 2010; Meylan *et al.*, 2008; Middleton *et al.*, 2009; Miller *et al.*, 2011a, 2011b; Minns *et al.*, 2006; Minot *et al.*, 2012; Mirpuri and Yarovinsky, 2012; Mitchell *et al.*, 2006; Mitra *et al.*, 2012; Montfort *et al.*, 2009; Mun *et al.*, 2003; Muñoz *et al.*, 2009; Mordue *et al.*, 2007; Morgado *et al.*, 2011; Murakami *et al.*, 2012; Murray *et al.*, 1991; Nagineni *et al.*, 1996; Nance *et al.*, 2012; Nathan and Shiloh, 2000; Niedelman *et al.*, 2012; Nishanth *et al.*, 2010; Nishikawa *et al.*, 2007, 2011; Noor *et al.*, 2010; Norose *et al.*, 2003, 2008, 2011; Oberdorfer *et al.*, 2003; Ochoa-Repáraz *et al.*, 2011; Okusaga *et al.*, 2011; Oldenhove *et al.*, 2009; Ong *et al.*, 2010; Ortiz-Alegria *et al.*, 2010; Osborn *et al.*, 2010; Passos *et al.*, 2010; Pawlowski *et al.*, 2007; Peixoto-Rangel *et al.*, 2009; Peixoto *et al.*, 2010; Pepper *et al.*, 2008; Perona-Wright *et al.*, 2009; Peterson *et al.*, 1995; Pfaff *et al.*, 2008a; Pfaff *et al.*, 2007, 2008a, 2008b; Pfefferkorn and Guyre, 1984; Phelps *et al.*, 2008; Pifer *et al.*, 2011; Pino *et al.*, 2007; Plattner *et al.*, 2008; Ploix *et al.*, 2011; Poassos *et al.*, 2010; Pollard *et al.*, 2009a, 2009b; Powell *et al.*, 1978; Portillo *et al.*, 2010, 2012; Portugal *et al.*, 2008; Prandovszky *et al.*, 2011; Rajapakse *et al.*, 2007; Ramón *et al.*, 2010; Ravindran and Boothroyd 2008; Reese *et al.*, 2011; Reichmann *et al.*, 1999, 2000; Resende *et al.*, 2008; Rincon 2012; Robben *et al.*, 2005; Robbins *et al.*, 2012; Rock *et al.*, 2005; Romano *et al.*, 2012a, 2012b; Rosowski and Saeij, 2012; Rosowski *et al.*, 2011; Rozenfeld *et al.*, 2005; Saeij *et al.*, 2007, 2008; Samarajiwa *et al.*, 2009; Sanchez *et al.*, 2010; Salek-Ardakani and Croft, 2009; Sauer *et al.*, 2009; Sautel *et al.*, 2009; Schaeffer *et al.*, 1999, 2009; Scharton-Kersten *et al.*, 1998; Schluter *et al.*, 1995, 2001, 2003; Schulthess *et al.*, 2008, 2012; Schwartz and Hunter, 2007; Senegas *et al.*, 2009; Severance *et al.*, 2012; Sergent *et al.*, 2005; Shapira *et al.*, 2005; Shaw *et al.*, 2006, 2009; Shi *et al.*, 2012; Shiono *et al.*, 2007; Sibley and Ajioka, 2008; Sibley *et al.*, 2010; Silva *et al.*, 2009, 2010; Silver *et al.*, 2011; Sims *et al.*, 1989; Skallova *et al.*, 2006; Skariah and Mordue, 2012; Skariah *et al.*, 2012; Spear *et al.*, 2006; Starnes *et al.*, 2009; Stenzel *et al.*, 2004; Stibbs, 1985; Strack *et al.*, 2002; Strobl *et al.*, 2012; Stumhofer *et al.*, 2006; Stutz *et al.*, 2012; Subauste, 2009a, 2009b, 2009c; Subauste *et al.*, 1991, 2002; Sukhumavasi *et al.*, 2008, 2010; Sukhumavasi *et al.*, 2007; Sun *et al.*, 2007; Suzuki *et al.*, 1989, 1994a, 1994b, 1996a, 1996b, 2005, 2010; Suzukui, 1997; Suzuki and Remington, 1988, 1990; Tait, *et al.*, 2010; Tait and Hunter, 2009; Takács *et al.*, 2012; Tan *et al.*, 2010; Tato *et al.*, 2012; Taylor *et al.*, 2004; Terrazas *et al.*, 2010; Thompson *et al.*, 2008; Tomita *et al.*, 2009; Torrey *et al.*, 2012; Tu *et al.*,

2009; Tussiwand *et al.*, 2012; Unoki *et al.*, 2009; Vallochi *et al.*, 2008; Van *et al.*, 2007; Vaudaux *et al.*, 2010; Villard *et al.*, 2012; Villarino *et al.*, 2003, 2006; Villegas *et al.*, 2002; Virreira *et al.*, 2011; Vutova *et al.*, 2007; Vyas *et al.*, 2007a, 2007b; Vyas and Sapolsky, 2010a; Vyas, 2013; Wang *et al.*, 2009b, 2010; Washino *et al.*, 2012; Wang *et al.*, 2007; Wang and Suzuki, 2007; Wasmuth *et al.*, 2012; Watford *et al.*, 2008; Webster *et al.*, 1994; Wen *et al.*, 2010; Wendte *et al.*, 2011; Whitmarsh and Hunter, 2008; Whitmarsh *et al.*, 2011; Wiehagen *et al.*, 2012; Willie *et al.*, 2001; Wilson *et al.*, 2006; Wilson *et al.*, 2005; Wilson and Hunter, 2004; Wilson *et al.*, 2008; Wilson, E.H. *et al.*, 2009; Wilson, D.C. *et al.*, 2010; Wilson, M.S. *et al.*, 2010; Witting, 1979; Wiley *et al.*, 2010; Witola *et al.*, 2011; Woodman *et al.*, 1991; Yamamoto *et al.*, 2012; Yamamoto *et al.*, 2011; Yamashita *et al.*, 1998; Yamauchi *et al.*, 2007; Yap *et al.*, 2000, 2007; Yarovinsky *et al.*, 2005, 2008, 2010; Yong *et al.*, 1994; Young *et al.*, 2012; Xiao *et al.*, 2009, 2012; Zakrzewski 2011; Zeiner and Boothroyd, 2010b; Zhang *et al.*, 2007; Zhang and Denkers, 1999; Zhao *et al.*, 2007, 2008, 2009a, 2009b; Zhou *et al.*, 2012; Zukas *et al.*, 2009)

Recently, the regulatory role of NAIPs in other systems in regulating this type of innate sensing and signalling of DNA and other pathogen and self-molecules has been defined. It will be of interest to determine if, as expected, this will also be relevant in human cells during T. gondii infections. In the milieu of the cytokines and chemokines that this initial interaction, which also may involve parasite ligands with host danger signals, CD8α dendritic cells present antigen and are critical for IL-12 production. Type I and III Rop16 act on STAT3 and STAT6, Type II GRA15 acts on NFκB and Rop38 acts on p3MAP kinase. Rop16 and GRA15 appear in murine macrophages and in fibroblasts to polarize effector functions to arginase fibrotic type reactions (Rop16) or proinflammatory (GRA15) phenotypes (Saeij, 2011). Other cell types such as macrophages also contribute. Dendritic cell expansion is influenced in a strain specific manner (Tait *et al.*, 2010). TCR signalling is depending on Tec kinases RLk, TK and PKCθ (Nishanth *et al.*, 2010). Tyk2 is critical. IFNγ and TNFα and TNFβ in mice and in some cells studied *in vitro* alter macrophage killing mechanisms. IFNγ responsive genes are key. IRG GTPases which are so critical to killing mechanisms and survival of mice are not present in humans, but humans do have functional GBP (p65 GTPases), IFNγ inducible genes that may subserve a similar function (Krese, 2008).

A delicate balance exists between proinflammatory and down-modulatory cytokines, and cytokines stimulated by innate immune mechanisms influence outcomes. Work with IL-6 and the IL-6 receptor and signalling pathway that uses p130 demonstrates this (Hunter *et al.*, 1994; Handel *et al.*, 2012). IL-10, TGFβ, Tyk 2 (Shaw *et al.*, 2006), SOCs 2 and 3 play a role in protection. Excess harmful IL-17, usually primed by IL-6 plus TGFβ, appears in some cases to be pathogenic (Dutra *et al.*, 2012). Th17 regulatory T-cells stimulated by IL-23, IL-27 and IL-33 contribute to protection (Shaw *et al.*, 2006).

25.6.3 Effector Cells

25.6.3.1 Importance of CD8$^+$ T-Cell Response and CD4$^+$ T-Cell Help

Early cell transfer studies (Frenkel, 1967; Frenkel and Caldwell, 1975; Brinkman *et al.*, 1986) that investigated illness due to *T. gondii* infection in HIV patients who had low CD4$^+$ T-cell counts (Remington *et al.*, 2005) and the increased susceptibility of nude and SCID mice (Lindberg and Frenkel, 1977) compared with immunocompetent mice (Beaman *et al.*, 1994; Johnson, L., 1992; Gazzinelli *et al.*, 1993a, 1994a; Hunter *et al.*, 1994), first indicated the importance of CD4$^+$ T-cells in protection against toxoplasmosis. CD4$^+$ and CD8$^+$ T-cells both are important in limiting chronic CNS infection and CD4$^+$ T-cells are needed for long-term protection. Acute infection CD8$^+$ T-cells are important and CD4$^+$ T-cells play a smaller role. CD4$^+$ and CD8$^+$

T-cells both produce IFNγ in mice. Perforin and granzyme are more important in protection against chronic infections in mice and are produced by $CD8^+$ cells in mice. In humans $CD4^+$ and $CD8^+$ T-cells that produce IFNγ and $CD4^+$ and $CD8^+$ T-cells that are cytolytic for infected cells have been described (Curiel *et al.*, 1993; Montoya *et al.*, 1996). Interestingly, Vβ7 appears to be used preferentially by these cells (Montoya *et al.*, 1996). Lymphokine activated killer cells have also been described (Subauste *et al.*, 1992). Th1 cytokines produced by effector and memory T-cells that have expanded from naive precursors are essential to prevent reactivation. CD45RAhi CD45 R0lo and αβ T-cells from naive donors proliferate in response to *T. gondii* antigens (Subauste et al., 1998). This depends on the increased expression of CD80, CD86 and IL-12 by monocytes which requires CD40 in antigen processing cells (APCs) and CD40 ligand on activated T-cells (Subauste *et al.*, 1998).

$CD8^+$ T-cells play an important role in protective immunity against *T. gondii* infection (Kahn *et al.*, 1988, 1990; Suzuki and Remington 1988; Gazzinelli *et al.*, 1991; Brown and McLeod, 1990). This immunity is, in part, a result of production of high titre IFNγ by this T-cell population (Gazzinelli *et al.*, 1991; Kahn *et al.*, 1991). Studies in mice using mutant (ts-4) parasite and *in vivo* depletion of $CD8^+$ T-cells indicate that immune mediated resistance is dependent upon $CD8^+$ T-cell lymphocytes and IFNγ (Suzuki and Remington, 1990). IFNγ also has been reported to be important for regulating $CD8^+$ T-cell response against the parasite. Mice deficient for the p40 chain of IL-12 heterodimer exhibit increased susceptibility and severely depressed $CD8^+$ T-cell immunity to infection (Ely *et al.*, 1999); however, treatment of these mutant animals with exogenous IFNγ restores their $CD8^+$ T-cell response and ability to withstand *Toxoplasma* infection.

The specific parasite antigens responsible for induction of $CD8^+$ T-cell responses have not been fully described. SAG1 has been reported to evoke a strong $CD8^+$ T-cell response in immunized mice (Kahn *et al.*, 1991) and induce high titre IFNγ production and antigen-specific cytotoxic $CD8^+$ T-cells that are directly parasiticidal to extracellular parasites (Kahn *et al.*, 1991). In combination with adjuvants (e.g. QuIL-A or liposomes) SAG1 is able to induce nearly 100% protection against both acute (Kahn *et al.*, 1991; Bulow and Boothroyd, 1991) and chronic infection (measured as number of brain cysts in survivors) (Kasper *et al.*, 1992). T-cell depletion studies followed by adoptive transfer into naive mice support high IFNγ titre producing $CD8^+$ T-cells as the primary mediators of this protection. A $CD8^+$ T-cell clone generated against SAG1 was able to induce almost 100% protection against *T. gondii* challenge in naive recipients (Kahn *et al.*, 1994). A SAG1 specific $CD4^+$ T-cell clone failed to transfer significant protection in naive animals in spite of its ability to produce quantities of IFNγ similar to the $CD8^+$ SAG1 specific T-cell clone. ROP2 has also been found to stimulate IFN production (Herion *et al.*, 1993; Saavedra and Hérion 1991, 1996) and secreted proteins (ESA) have also been noted to induce protection (Sharma *et al.*, 1999; Cesbron-Delauw and Darcy 1992, Mercier *et al.*, 1998).

A role for $CD4^+$ T-cells in the induction of $CD8^+$ T-cell immunity against the parasite has been demonstrated (Casciotti *et al.*, 2002). *T. gondii* infection induces a normal antigen-specific $CD8^+$ T-cell immune response in $CD4^{-/-}$ mice. The frequency of the antigen-specific $CD8^+$ T-cell population in both $CD4^{-/-}$ and wild type mice was measured at various time points after infection (e.g. on days 30, 90 and 180), by precursor cytolytic T-lymphocyte (pCTL) assay. There was no significant difference in the generation of number of antigen-specific $CD8^+$ T-cell response between $CD4^{-/-}$ and wild type animals on days 30 and 90 after infection (Fig. 25.31). However, the mutant mice were not able to sustain $CD8^+$ T-cell immunity and at 180 days after infection, the $CD8^+$ T-cell

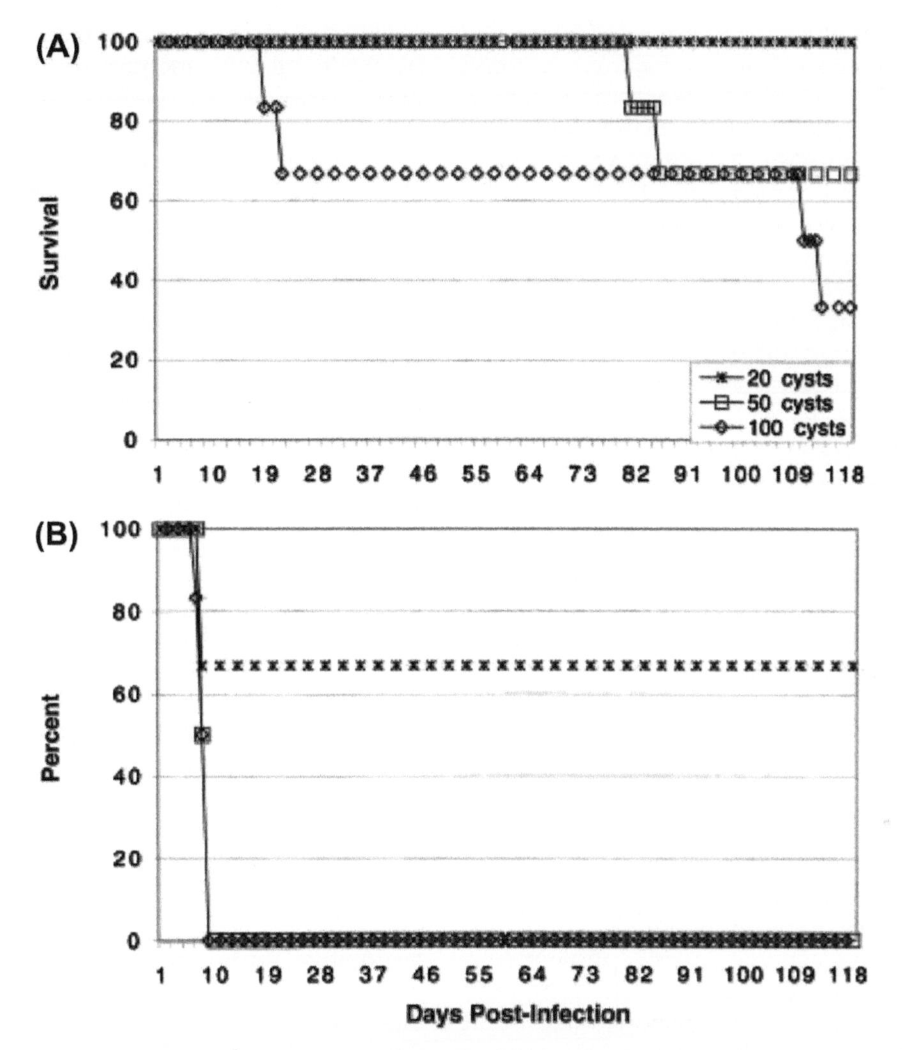

FIGURE 25.31 **Survival of CD4$^{-/-}$ (A) and wild-type (B) mice infected with different doses of *T. gondii*.** Female CD4$^{-/-}$ mice and parental C57BL/6 mice that were five to six weeks old were challenged per-orally with 20, 50 or 100 cysts of *T. gondii* 76K. Survival of animals was monitored daily. There were six animals per group, and the experiment was performed twice with similar results. The data shown are from one of the two experiments. *(Adapted with permission from Casciotti* et al. *(2002).)*

response in the gene deficient mice was depressed, as determined by pCTL assay (Fig. 25.31). This study demonstrated that although CD8^{+} T-cell immunity can be induced in the absence of conventional CD4^{+} T-cells, it cannot be maintained without such cells. In a subsequent study using CD4$^{-/-}$ mice *T. gondii*-infected mice exhibited an extended NK-cell response, which was mediated by continued IL-12 secretion. This prolonged NK-cell response was critical for priming parasite-specific CD8^{+} T-cell immunity. Depletion of NK-cells inhibited the generation of CD8^{+} T-cell immunity in CD4$^{-/-}$ mice. Similarly neutralization of IL-12 reduced NK-cell numbers in infected animals and to the down-regulation of CD8^{+} T-cell immunity against *T. gondii*. Adoptive transfer of NK-cells into the IL-12-depleted animals restored their $CD8^+$ T-cell immune response, and animals exhibited reduced mortality. NK-cell IFNγ was essential for cytotoxic T-lymphocyte priming. This report demonstrated that NK-cells can play an important role in induction of primary $CD8^+$ T-cell immunity, in the absence of $CD4^+$ T-cells, against an intracellular infection. These observations have therapeutic implications for immunocompromised individuals, including those with human immunodeficiency virus infection. In other systems T-bet and eomesodermin induce enhanced expression of CD122, the receptor that specifies IL-15 responsiveness (Intlekofer *et al.*, 2005).

Interestingly, there is also a proliferative response of T-cells from unexposed individuals to high doses of *T. gondii* antigens (Subauste *et al.*, 1998; McLeod *et al.*, unpublished data), but this does not appear to be similar in mechanism to the superantigen response observed by Denkers *et al.* (1996) in mice.

25.6.3.2 Chronic Infection

In earlier studies (Hermes *et al.*, 2008) brain transcriptomes from chronically infected mice had increases in mrRNA for PD1 ligand. Although $CD8^+$ T-cell immunity is critical for control of toxoplasmosis, in susceptible mouse strains it does not ensure their long-term survival. Studies have demonstrated that during chronic toxoplasmosis $CD8^+$ T-cells undergo progressive functional exhaustion, poor recall response and elevated apoptosis (Bhadra *et al.*, 2011). This dysfunction leads to reactivation of latent infection and subsequent death of infected host. Concomitant with loss of polyfunctional T-cell response, $CD8^+$ T-cells exhibit high PD1 (Programmed death-1) expression, a known inhibitory receptor which plays a central role in regulating T-cell exhaustion (Simone *et al.*, 2009). PD1 expression was found in the polyfunctional memory $CD8^+$ T-cell population, which are critical for maintaining the chronicity of infection (Bhadra *et al.*, 2011). Blockade of PD1 interaction with its ligand PDL1 prevented reactivation and host mortality (Bhadra *et al.*, 2011). It will be interesting to determine if PD1 is a sole inhibitory molecule up-regulated on the $CD8^+$ T-cells from chronically infected mice or others like TIM-3 or LAG-3 (Ngiow *et al.*, 2011; Norde *et al.*, 2012) are also involved. Nevertheless, $CD8^+$ T-cell exhaustion is an important factor which determines if the infection is reactivated or remains in a chronic state and therapeutic interventions for its blockade need to be considered. Whether there will be regulation by T-bet and Eomes (Virgin *et al.*, 2009; Paley and Wherry, 2010), as occurs with LCMV where $CD8^+$ T-cell exhaustion was first defined, remains to be determined.

Recent advances in our understanding of the immune response to *T. gondii* in the brain using brain 2 photon microscopy include: killing mechanisms with perforin resulting in lysis of cysts (Suzuki *et al.*, 2010), gp130 driving a harmful immune response (Wilson *et al.*, 2012; Handel *et al.*, 2012) and chitinase being involved in cyst rupture of targets by T-lymphocytes (Wilson *et al.*, 2012; Robey *et al.*, 2005; Hunter and Sibley, 2012). Studies demonstrating conventional antigen processing also have been reported beginning with antigen presenting $CD8^+$ dendritic cells that both presented antigen and produced IL-12 that stimulates IFNγ (Dzierszinski *et al.*, 2007).

25.6.3.3 Macrophages and Microbicidal Mechanisms

In vitro experiments have shown a crucial role for both IFNγ and TNFα in the induction of inducible nitric oxide synthase (iNOS). iNOS degrades L-arginine into citruline and reactive nitrogen intermediates (RNI) that have a potent microbicidal activity in murine macrophages infected with *T. gondii* tachyzoites (Murray *et al.*, 1979, 1980, 1985a, 1985b; Nathan *et al.*, 1984). Intermediate RNI levels, while not killing the tachyzoites, appear to reduce parasite multiplication and favour stage conversion of *T. gondii.* Consistently, *in vivo* IFNγ deficient mice and TNFα deficient mice, or wild type mice treated with antiIFNγ or antiTNFα express lower levels of iNOS, when infected with *T. gondii.* RNI appears to be only one of the various effector mechanisms induced by IFNγ that are involved in the control of tachyzoite replication. Certain strains of mice (e.g. C57BL/6 background mice) treated with iNOS inhibitors (Roberts *et al.*, 2000; Deckert-Schluter *et al.*, 1998; Gazzinelli *et al.*, 1993) or deficient in iNOS will succumb to infection only at three to four weeks post-infection, due to large number of cysts in the brain and intense toxoplasmic encephalitis. Mice deficient in iNOS function are relatively more resistant than the IFNγ deficient mice, which die of a systemic infection in the first two weeks after *T. gondii* infection (Gazzinelli *et al.*, 1993). In mice iNOS is needed for control of chronic infection and iNOS is dependent on IFNγ and TNFα (Sibley, 1991; Adams *et al.*, 1990). It is not clear the role iNOS plays in differing cell types (Nathan *et al.*, 1984; Murray *et al.*, 1985a, 1985b; Wilson and Remington, 1980b; Wilson *et al.*, 1980).

Alternative IFNγ inducible mechanisms that are involved in *T. gondii* control have been identified. Indoleamine 2,3-dioxygenase (INDO) is an enzyme that catalyses the initial rate-limiting step of tryptophan (Trp) catabolism to *N*-formylkynurenine and kynurenine (Pfefferkorn, 1984; Pfefferkorn *et al.*, 1986). Different human cells express INDO upon stimulation of IFNγ; and INDO expression is regulated by IL-4 and IL-13 (Chaves *et al.*, 2001). The restriction of the available essential amino acid, tryptophan, due to degradation by INDO leads to the control of various intracellular pathogens including *T. gondii.* The expression of INDO mRNA and activity is induced in mice infected with *T. gondii* in an IFNγ and IRF-1 dependent manner (Silva *et al.*, 2002). However, the *in vivo* role of INDO in the control of *T. gondii* replication remains to be demonstrated. Another mechanism involved in the control of *T. gondii in vivo* involves a family of GTPases of 47 kDa (p47-GTPase) that are induced by IFNγ (Collazo *et al.*, 2001; Butcher *et al.*, 2005). Mice deficient in either IGTP or Lrg47, two distinct members of the p47-GTPase family, are highly susceptible to *T. gondii* infection as they fail to control *T. gondii* (ME49 strain) multiplication and succumb to infection at nine to 11 days post-infection. Importantly, both IGTP and Lrg47 deficient mice showed a robust IL-12 and IFNγ production, as well as regular expression of iNOS gene. Similarly, IGTP expression was normal in the Lrg47 deficient mice and *vice versa*. Importantly, IGTP is required for both haematopoietic and non-haematopoietic cell resistance to *T. gondii* (Collazo *et al.*, 2002). Together, these findings suggest that the members of the p47-GTPase family are more important than iNOS for resistance during the early stages of infection with *T. gondii* in mice. Interestingly, IGTP is not essential for IFNγ induced resistance in the chronic stage of infection, when RNI was shown to have a more pronounced role.

25.6.3.4 Mononuclear Phagocytes

IFNγ produced by T-lymphocytes stimulates macrophage microbicidal mechanisms including NO and iron-related mechanisms (Adams *et al.*, 1990). In other cell types, IFNγ induced tryptophan starvation results in reductions of intracellular parasite growth (Pfefferkorn, 1984; Pfefferkorn *et al.*, 1986). CCL2 and CCL3 (MIPIα) attract mononuclear phagocytes to an inflammatory focus (Denkers and Butcher, 2005; Lewis and Wilson, 2005). Infection of human monocytes with *T. gondii* induces CD80 and CD86 (Subauste *et al.*, 1998) and may enhance T-cell IFNγ early and Th1 differentiation (Catterall *et al.*, 1986, 1987). Human monocytes have greater innate antimicrobial effect against *T. gondii* than do human monocyte-derived macrophages or tissue macrophages (McLeod *et al.*, 1980, 1983, 1990, 1993, 1994; McLeod and Estes, 1982; McLeod and Remington, 1977; Anderson and Remington,

1974). Oxidative and non-oxidative mechanisms are involved (Wilson *et al.*, 1979, 1980a, 1980b; Willson and Haas, 1984; Murray *et al.*, 1979, 1980, 1985a, 1985b; Nathan *et al.*, 1984). Human alveolar macrophages and peritoneal macrophages have also been studied and non-oxidative killing mechanisms were found (Catterall *et al.*, 1986). HIV reduces killing of *T. gondii* by human monocytes. Mouse macrophages require TNFα in addition to IFNγ for enhanced microbicidal capacity (Sibley *et al.*, 1991). IFNγ also stimulates non-oxidative mechanisms (Wilson *et al.*, 1979, 1980a, 1980b; Willson and Haas, 1984; Lewis and Wilson, 2005). Indoleamine 2,3-dioxygenase which degrades tryptophan deprives *T. gondii* of this essential nutrient and is one such non-oxidative mechanism found to be present in human fibroblasts (Pfefferkorn, 1984; Pfefferkorn *et al.*, 1986). Lewis and Wilson reviewed findings concerning newborn monocytes (Lewis and Wilson, 2005): There appears to be reduced chemotaxis of newborn monocytes. MHC class II-mediated antigen presentation by monocytes appears to be intact. Blood monocytes from neonates are normal in number and similar to adults in microbicidal capacity and are activated by IFNγ. There is modestly lower production of cytokines and MyD88.

T. gondii (differently for different strains) (Saeij *et al.*, 2005b) commandeers the host's NFκB-I-κB signalling pathway (Adams *et al.*, 1990; Arsenijevic *et al.*, 2000; Butcher *et al.*, 2001; Bellman *et al.*, 1996; Dobbin *et al.*, 2002; Feinstein *et al.*, 1996; Goebel *et al.*, 2001; Kim *et al.*, 1997; Langemans *et al.*, 1992; Lyons and Johnson, 1995; Mason, 2004a, 2004b; Miller *et al.*, 1999a, 1999b, 2000; Molestina *et al.*, 2003, 2005; Saeij *et al.*, 2005b; Shapira *et al.*, 2004; Scharton-Kersten, 1997a; Torre *et al.*, 1999; Zhang *et al.*, 1994). *T. gondii* can also block host cell apoptosis (Curiel *et al.*, 1993; Brown *et al.*, 1995). Antibodies (serum and secretory) may opsonize and interfere with or specify a different pathway for parasite entry into host cells (Mineo *et al.*, 1993; Chardes *et al.*, 1990; Joiner *et al.*,1993). Heparin binding proteins and lectins may play a role in initial interactions of *T. gondii* with somatic cells (Ortega and Boothroyd, 1999). Activated macrophages also down modulate lymphocyte blastogenesis at least in part by NO production (Wing and Remington, 1977; Gazzinelli *et al.*, 1996; Candolfi *et al.*, 1994) and partly by induction of apoptosis (reviewed in Denkers *et al.*, 1998). *T. gondii* also alters leukotriene release by macrophages (Locksley *et al.*, 1985).

25.6.3.5 Gamma Delta (γδ) T-Cells

Gamma delta T-cells are induced early in murine and human *T. gondii* infections (Kasper *et al.*, 1996; Scalise *et al.*, 1992; Sayles *et al.*, 1995; Subauste *et al.*, 1995; Hara *et al.*, 1996, DePaoli *et al.*, 1992). Vγ2Vδ2CR CD45RAlo CD45hi (activated *in vivo*) T-cells are present in children with postnatally acutely acquired toxoplasmosis. They secrete cytokines (IL-2, IFNγ and TNFα) and proliferate and lyse *T. gondii* infected cells in an MHC unrestricted manner. IFNγ producing γδ T-cells have conferred protection, but not in another study against bradyzoites (Egan *et al.*, 2005; Subauste *et al.*, 1995; Hara *et al.*, 1996; De Paoli *et al.*, 1992). A 65 kDa heat shock protein that is induced in macrophages by γδ T-cells contributes to protection against *T. gondii* (Himeno and Hisaeda, 1996; Hisaeda *et al.*, 1995; Nagawasa *et al.*, 1992, 1994).

25.6.3.6 Chemokines and Chemokine Receptors

Chemokines (chemotactic cytokines) are small heparin-binding proteins that couple the detection of pathogens with infiltration of tissues with neutrophils and monocytes. They guide circulating white blood cells to areas of inflammation and injury, thus participating in adaptive immunity and pathogenesis. There are approximately 50 human chemokines which are considered to be in four families based on their structure and function. CC chemokines have their initial two cysteines adjacent and include CCL2 (MCP1) which attracts monocytes, dendritic cells, basophiles and memory T-cells with the receptor CCR2; CCL3 (MIP1α) which attracts macrophages with the receptor CCR5; CCL4 (MIP1β) which attracts T-cells and monocytes with the receptor CCR5; and CCL5 (RANTES) which attracts T-cells and monocytes with the receptors CCR1 or CCR5. A second family is CXC that have an amino acid between the canonical adjacent cysteines. These attract polymorphonuclear leukocytes, activate monocytes recruiting them to vascular lesions and mediate interactions of APCs and T-cells. The CX3C has a mucin-like stalk with transmembrane and cytoplasmic regions. CX3 is cleaved from the cell membrane by TNFα converting enzyme. The fourth family contains XCL1. Chemokines bind to 7-transmembrane domain-6-protein-coupled receptors activating signalling cascades and alter movement of cell action. CCR7 and its ligands CCL19 and CCL21 link innate and adaptive immunity and organize T-cell zones in T-cells (Denkers, 1998).

25.6.3.7 Other Cell Types

Platelets (Yong *et al.*, 1991), eosinophils (Capron *et al.*, 1997), lymphokine activate killer cells (Subauste *et al.*, 1992), polymorphonuclear leukocytes (Bliss *et al.*, 2001; Butcher *et al.*, 2001; Del Rio *et al.*, 2001b; Denkers *et al.*, 2005; Marshall and Denkers, 1998) and mast cells have also been noted to have microbicidal activity for or play a role in the immune response to *T. gondii*.

25.6.3.8 Cytokines

Cytokines have been discussed throughout the sections above and are summarized in Table 25.2. IFNγ, IL-12, IL-18 and TNFα produced by NK-cells, dendritic cells and monocytes directly limit acute infection and lead to the differentiation of antigen specific $CD4^+$ T-cells into Th1 effector T-cells that produce more IFNγ. This IFNγ and IL-12, IL-15 and IL-23 produced by non-T-cells sustain the Th1 response. Dendritic cells infected with *T. gondii* produce little IL-12 but engagement of CD40 with CD40 ligand markedly up-regulates IL-12 production. Cytokines may lead to evasion of initial activation of protective immunity or curtail a potentially lethal immune response (e.g. Neyer *et al.*, 1997). For example, in the absence of IL-10, the inflammatory process is lethal. Presence of IL-4 early in infection inhibits protective Th1 responses, but later inhibits *Toxoplasma* replication within the CNS. Thus, the temporal sequence of cytokine production is important. A balance of Th1 and Th2 cytokines is critical in a protective immune response to *T. gondii* (see Table 25.2). Antibody to TNFα and steroids were reported to be associated with reactivation of human CNS toxoplasmosis (Young and Maguire, 2005). The studies which define the stimulation of IL-27, IL-23 and IL-33 and the down-modulatory effect on IL-6 along with TGFβ were noted.

25.6.3.9 Prostaglandins

As discussed under dendritic cells there are prostaglandins which promote an inflammatory response and those that down-modulate the inflammatory response. Prostaglandins appear to contribute significantly to the outcome of *T. gondii* infection. In other systems, PGA1 supresses NFκB and Cox2 gene expression, which is critical for proinflammatory prostaglandins (Mandal *et al.*, 2005).

25.6.3.10 Regulatory T-Cells

Regulatory T-cells are very important in a variety of infections and autoimmune and neoplastic disease states in mice. Their roles in human *Toxoplasma* infections have not yet been defined.

25.6.3.11 NK-Cells and KIR

NK-cells are important early after infection in mice (Hunter *et al.*, 1997; Dannemann *et al.*, 1989; Hauser *et al.*, 1982, 1986; Hughes *et al.*, 1988; Kasper *et al.*, 1982; Kasper and Ware, 1985). NK-cells also have been found in humans. KIRs (killer-cell immunoglobulin-like receptors) are modulating NK-cell receptors that interact with T-cells with certain MHC haplotypes (Stanley and Luzio, 1998; Parham, 2005). They are important in a variety of different infections, but they have not been studied in human toxoplasmosis. In other systems, CPG stimulate TLR9 and thus NK-cells. With loss of NKG2D receptor (also known as KLrk1) and its adaptors Dap 10, DA12, cytotoxicity is lost and dendritic cell-like APC capacity is gained with upregulation of MHC II and co-stimulatory molecules. These cells kill target cells and present antigen linking innate and adaptive immunity. Finally, ordered crosstalk occurs between NK, NKT and DC. The relevance of these findings and KIR receptors (Parham, 2005, Rajagopalan and Long, 2005; Bryceson *et al.*, 2005) to the function of these cells in adaptive immunity in *Toxoplasma* infection and toxoplasmosis remains to be determined.

25.6.3.12 NKT-Cells

NKT-cells are a minor subset of T-lymphocytes. They share receptors with NK-cells and T-cells. Murine NKT-cells use a semi-variant Vα14-Jα281TCR B chain paired with Vβ8-7 or -2 TCR B chain with NK-cell receptors NKR-P1, Ly-49 and NK 1.1 in C57B1/6 mice. They are located primarily in the liver, spleen, thymus and bone marrow and also are in the gastrointestinal tract. They recognize hydrophobic lipid antigens presented by CD1d, and CD1d and invariant TCRβ chain are essential for their development. NKT-cells have been found to provide help for $CD8^+$ T-cell function by producing IL-2 (Denkers *et al.*, 1997a). They produce large amounts of IL-4 and IFNγ when activated, are cytotoxic, help in antibody production, regulate Th1/Th2 differentiation and participate in parasite clearance by shifting the cytokine profile toward a Th1 pattern. They contribute to an immunopathologic gastrointestinal response when the Th1 response remains uncontrolled (Ronet *et al.*, 2005). SAG1 has elicited this gastrointestinal immunopathology (Buzoni-Gatel *et al.*, 2001). The *T. gondii* ligand for NKT-cells remains to be defined. Mice with $CD4^+$ T-cells but without CD8αβ T-cells and NKT-cells cannot develop a vaccine-induced protective immune response (Denkers and Sher, 1997).

25.6.3.13 Placenta

The interaction of the foetal and the maternal side of the placenta, and adhesion molecules that affect that interaction have also been characterized recently. There are differences in the functioning of immune cells on the foetal and maternal sides of the placenta (Robert-Gangneux *et al.*, 2010; Candolfi *et al.*, 1994).

25.6.3.14 Neuronal Cells

Neuronal cells are discussed at length in Chapter 23. To study effects of chronic brain infection with *T. gondii*, Hermes *et al.* (2008) infected specific pathogen free, outbred mice with *T. gondii* of the ME49 strain. Behavioural and neurologic signs were assessed when mice were 1.5, 7 and 12 months of age or older (Fig. 25.30). Infected mice exhibited 'illness behaviours', including: abnormal body position, hunching, piloerection, tremor, lacrimation, palpebral closure and abnormal grooming. Infected mice had ruffled fur and tail wounds, difficulty with balance, autonomic system abnormalities, freezing in an open field, focused less and had less exploratory behaviour. Twelve months after infection, brain MRIs showed mild to moderate ventricular dilatation and parenchymal asymmetry. Brain weight was less. Host genome microarrays revealed effects of infection on host cell protein processing (ubiquitin ligase), synapse remodelling and inflammation causing neuronal damage. In histopathologic sections there were prominent perivascular, leptomeningeal inflammatory cells contiguous to the hippocampus and the aqueduct of Sylvius in infected mice. These mice also had clusters of $CD4^+$ and $CD8^+$ T-lymphocytes and microglia present in inflammatory nodules in the brain parenchyma and, in perivascular spaces, especially in the frontal cortex and diencephalon. Neurologic function and brain histopathologic abnormalities were found to correlate with lower brain weight and inflammation. Electron micrographs of the brain from chronically infected mice demonstrated cysts in neurons that connected with other neurons with synapses, bradyzoites egressing from ruptured cysts and in monocytic cells contiguous to cysts and cysts without contiguous inflammation, as had been described earlier (Ferguson *et al.*, 1997). Although sulphadiazine treatment for four months given to the chronically infected mice did not modify inflammation and perivascular cuffing in the brain, it reduced the number of *T. gondii* cysts. Mice with knock-outs of *IL-4, IL-6, IL-13* and *NRAMP* genes showed the same histopathology as those without the gene knock-outs, indicating that NRAMP and these cytokines were not necessary to cause the perivascular cuffing and parenchymal abnormalities observed in the non-knock-out mice. BALB/c (L^d) mice also had the perihippocampal, perivascular and leptomeningeal findings, although they were less pronounced.

25.6.3.15 Neutrophils and Neutrophil NETs

Neutrophils can function as antigen presenting cells, as cells which store IL-12, and are involved in the innate immune response to *T. gondii* (see Chapter 24). DNA in neutrophil NETs interacts with DNA sensors (Denkers *et al.*, 2012).

25.7 IMMUNOLOGICAL CONTROL IN ANIMAL MODELS

25.7.1 Systemic Disease and Harmful Effects of Immune Response

Toxoplasma infection rapidly overcomes hosts with impaired T-cell function and diminished ability to produce Type 1 cytokines (i.e. IL-12, IFNγ and TNFα). The parasite is itself a strong stimulus of this type of immunity during acute toxoplasmosis, reflecting the need to keep the host alive during the early stages of infection. However, in many instances this early Type 1 cytokine response is so intense that it can result in pathologic changes associated with acute toxoplasmosis. Therefore, regulatory mechanisms, in particular the production of IL-10, are critical to control tissue damage and lethality due to excessive Type 1 cytokine production during the early stages of infection with *T. gondii.*

The most frequent clinical symptoms associated with acute toxoplasmosis are lymphadenopathy and fever, which occur simultaneously with parasite-induced activation of the immune system and concurrent release of proinflammatory cytokines. However, it may be difficult to differentiate the pathologic changes caused by the parasite by direct tissue destruction from systemic damage caused by parasite-induced cytokines. A study suggested that the pathology and lethality caused by the highly virulent RH strain is caused by the excessive production of proinflammatory cytokines, especially IL-1β (Mordue *et al.*, 2001). Further, pre-treatment with β-galactosamine, a model to study mechanisms induced by low doses of endotoxin, makes the injection of tachyzoite extracts lethal. This work indicates that granulocytes contribute to the inflammatory pathologic changes triggered by *Toxoplasma* infection (Marshall and Denkers, 1998).

During oral infection, the susceptible C57BL/6 mouse strain develops a severe intestinal inflammatory reaction, characterized by necrosis of the villi and mucosal cells of the small intestine (e.g. Liesenfeld *et al.*, 1999; Buzoni-Gatel *et al.*, 1997, 2001). The pathologic changes are largely reversed by administration of antiCD4 or anti-IFNγ mAb, indicating the critical role of Th1 lymphocytes in the intestinal lesions during per-oral infection of this strain of mouse with *T. gondii.* However, it has also been noted that IL-4 and IL-5 deficient mice do not develop lethal necrosis in their intestines (e.g. Nickdel *et al.*, 2001). In addition, in a model of ocular infection, parasites and T-cells both contribute to the pathologic process (Hu *et al.*, 1999a, 1999b; Lu *et al.*, 2003). *T. gondii* also can make atherosclerotic lesions in large vessels in a mouse model worse (Portugal *et al.*, 2004).

IL-10 is a cytokine that was first identified by its ability to inhibit IFNγ production by Th1 lymphocytes (Gazzinelli *et al.*, 1993a), mainly through the inhibition of IL-12 production by accessory/antigen presenting cells. During acute toxoplasmosis in mice, the immunoregulatory role of IL-10 appears to be critical to avoid excessive production of IFNγ by Th1 cells and systemic pathology. IL-10 deficient mice on a BL6 background have enhanced susceptibility to infection, with all such deficient mice dying by day 14 post-infection. Mortality of these mice is delayed by depletion of $CD4^+$ T-cells, indicating that these cells have a role in promoting early death. Increased mortality is associated with elevated levels of IL-12 and IFNγ in sera from infected IL-10 deficient mice, relative to those from uninfected controls or infected wild type mice. In addition, the co-stimulatory molecules CD28/B7 and CD40/CD40 ligand have been shown to play a critical role in the excessive Th1 lymphocyte stimulation and pathology observed during acute infection with *Toxoplasma* in IL-10 deficient mice (Gazzinelli *et al.*, 1996b). Importantly, parasite burden is similar or diminished in tissues of IL-10 deficient mice. Further, BALB/c mice are resistant to the intestinal pathology observed in the C57BL/6 mice per-orally infected with the ME49 strain of *T. gondii.* Nevertheless, IL-10 deficient mice on a BALB/c background become susceptible and die of the Th1 lymphocyte mediated intestinal pathology, when they are per-orally infected with *T. gondii* (Gazzinelli *et al.*, 1996b). Taken together, these studies indicate the critical role of IL-10 as a physiological regulator of production of Type 1 cytokines and systemic disease during acute toxoplasmosis (Gazzinelli *et al.*, 1996b).

25.7.2 Toxoplasmic Encephalitis

In contrast to acute disease, most of the pathologic findings associated with chronic toxoplasmosis have been thought to be caused by lack of appropriate T-cell immunity rather than an excessively vigorous response in humans. In general, toxoplasmic encephalitis positively correlates with the number of cysts in the central nervous system of mice experimentally infected with *T. gondii.* Immunogenetic studies in mice have demonstrated a major role for MHC loci, more specifically the L^d gene, in the development of toxoplasmic encephalitis (ME49 strain) implicating T-cell involvement in resistance of cyst formation in the central nervous system (Suzuki *et al.*, 1991; Brown *et al.*, 1995b). These studies suggest the role of $CD8^+$ T-cell responses restricted by this MHC class I molecule as a critical element in protective immunity to toxoplasmic encephalitis (Brown *et al.*, 1995). MHC class II-restricted $CD4^+$ T-cells appear to be equally important, as animals depleted of $CD4^+$ T-cells (Brown *et al.*, 1995a; Gazzinelli *et al.*, 1991, 1992a; McLeod *et al.*, 1995, 1996) as well as mice carrying mutations in the class II αβ locus also display increased cyst numbers in their central nervous system.

Infection with *T. gondii* induces a strong and polarized Th1 response characterized by the production of high levels of IFNγ in addition of IL-2 and TNFα. The levels of IL-4, IL-13 and IL-5 produced by $CD4^+$ T-cells from infected mice are rather small, and their role in host resistance and control of toxoplasmic encephalitis is more limited. $CD8^+$ T-cells are also an important source of IFNγ and their function appears to be, in part, dependent on IL-2 produced by $CD4^+$ T-cells. IFNγ produced by a specific subset of $CD8^+$ T-lymphocytes that express Vαβ chain in the TCR appears to have a primary role in preventing toxoplasmic encephalitis in BALB/c mice. The most critical cytokine involved appears to be IFNγ. Mice treated with anti-IFNγ develop severe TE characterized by a dramatic increase in cyst numbers and the appearance of tachyzoites in brain. This leads to death in 100% of mice seven to 10 days after the beginning of treatment that neutralizes endogenous IFNγ. Once the chronic infection is established, simultaneous treatment with anti-CD4 and anti-CD8 monoclonal antibodies is necessary to promote increased numbers of cysts, severe encephalitis and lethality. Thus, both $CD4^+$ T- and $CD8^+$ T-cells contribute IFNγ which controls parasite replication in the brain. Other cytokines, e.g. IL-6 and TNFα, are also produced by T-cells and contribute to host resistance to toxoplasmic encephalitis. C57BL/6 mice treated with neutralizing anti-TNFα mAbs also display a dramatic increase in cyst numbers, severe encephalitis and die. Similarly, mice lacking the functional receptor(s) for TNFα have an increased number of parasites in their brains and increased severity of TE (Gazzinelli *et al.*, 1993b). Additionally, mice deficient in IL-6 also have an increase in cyst numbers, areas of necrosis and tachyzoite-associated inflammation in the central nervous system.

The effector mechanisms elicited by these cytokines that are involved in host resistance to cyst formation, release of tachyzoite replication and development of toxoplasmic encephalitis are not completely understood. Nevertheless, it is clear that in some strains of mice, expression with inducible nitric oxide synthase (iNOS) and production of reactive nitrogen intermediates (RNI) are involved in this process. Mice treated with specific inhibitors of iNOS and those lacking the functional iNOS gene develop a large number of cysts in the brain and develop severe toxoplasmic encephalitis and succumb to infection rapidly. In contrast perforin is not important in the control of parasite replication and development of toxoplasmic encephalitis in mice (Denkers *et al.*, 1997a).

There is both an active inflammatory process (Hermes *et al.*, 2008) and bystander cell and infected cell death due to *T. gondii* infection in the murine brain (Boothroyd *et al.*, 2012). There are behavioural and neurologic correlates for this (Hermes *et al.*, 2008; Vyas, 2007a, 2007b; Webster *et al.*, 1994; Hay *et al.*, 1981) some of which may appear to be associated with dopamine (Webster *et al.*, 1994). Especially noteworthy are the importance in mice of Vβ8 TCR cells and an IFNγ independent perforin dependent mechanism of cyst lysis and protection of the brain (Suzuki *et al.*, 2012). Another key concept is that there are specific proteins which elicit antibodies during protective versus harmful immune responses in the brain (Suzuki *et al.*, 2012); further, that gp130, part of the IL-6 receptor, is key for protection from a chronic inflammatory process in the murine brain (Handel *et al.*, 2012). In some murine models there is localization of this process and neurobehavioural correlates that are specific to the amygdala (Vyas, 2013), whereas in other reports (Hermes *et al.*, 2008) the process is more widespread. The relevance of the studies of seroprevalence of *T. gondii* antibodies with a variety of neurologic and behavioural diseases, in which dopamine and tryptophan metabolites play a role, and also phenotypic personality traits is summarized in Chapter 23.

25.7.3 Intestinal Immunity and Toxoplasmic Enteritis

The gut represents the first line of defence for most naturally acquired *T. gondii* infections (see Chapter 24), but is often bypassed in experimental infections that use intraperitoneal or subcutaneous routes. Infection by the oral route can be initiated by *T. gondii* cysts or oocysts. These are susceptible to acid pepsin digestion in the stomach of the host and release bradyzoites or sporozoites, respectively, which are pepsin resistant. The bradyzoites or sporozoites are assumed to invade the gut epithelial cells and ultimately disseminate through the blood and lymphatic systems. Immunity in the gut could, in theory, prevent systemic infection and is therefore worthy of study for the purpose of vaccine design alone.

Mice fall into two categories regarding susceptibility to orally induced *T. gondii* infection. While most mouse strains examined to date can tolerate an orally induced infection, at least one, the C57BL/6 strain, is exquisitely susceptible. C57BL/6 mice infected orally with either the ME49 (Liesenfeld *et al.*, 1996) or Beverley strains of *T. gondii* (Nickdel *et al.*, 2001) develop severe intestinal necrosis that normally results in death within eight to 10 days post-infection. The mechanism of this intestinal pathology in C57BL/6 mice is complex and would appear to involve aspects of the immune response normally associated with Th1 and Th2 cells. Initial studies demonstrated that necrosis and death are dependent on $CD4^+$ T-cells as their depletion at strategic time points could reduce necrosis and promote survival. Furthermore, neutralization of IFNγ has similar effects implying that Th1 cells were responsible for the production of IFNγ and intestinal necrosis (Liesenfeld *et al.*, 1996). IL-10 gene deficient C57BL/6 mice have increased susceptibility to oral infection than wild type mice, and pathology is reduced by neutralizing IL-12. IL-10 gene deficient BALB/c mice were found to be susceptible to intestinal necrosis. Pathology in the intestine of these mice could be prevented and the time to death prolonged by neutralization of IFNγ (Suzuki *et al.*, 2000b).

In contrast with these studies implying a detrimental role for Th1 cells and their products, other studies have implicated Th2 cells and their products in the disease process. IL-5 gene deficient C57BL6 mice developed significantly less severe intestinal pathology than wild type control mice following oral inoculation of *T. gondii* (Beverley strain) (Nickdel *et al.*, 2001). A similar decrease in susceptibility was evident in IL-4 deficient mice compared with wild type mice infected with *T. gondii* (Beverley strain) (Nickdel *et al.*, 2004). Administration of L-NAME to these mice increased susceptibility, indicating that susceptibility was not associated with excess NO production (Nickdel *et al.*, 2004). IL-4 deficient mice had increased levels of transcripts for IL-10 compared with wild type mice, consistent with a protective role for this cytokine at this anatomical site (Nickdel *et al.*, 2004).

Studying the immunological events in mice strains that are largely resistant or strains that are susceptible to intestinal necrosis has provided valuable information on what constitutes a protective or pathological response in the intestine. Intraepithelial cells (IEL) from mice infected 11 days previously are able to transfer immunity to naive syngeneic animals of the CBA, BALB/c or C57BL/6 strain. As C57BL/6 mice are susceptible to intestinal necrosis and BALB/c are resistant, these data would indicate that IELs are important in a protective immune response and as regulators of the pathological immune response that can occur in C57BL/6 mice (Buzoni-Gatel *et al.*, 1997). Isolated $CD8\alpha\beta^+$ IELs could again confer protection, but this was abrogated by co-administration of neutralizing IFNγ. The anti-*T. gondii* action of IFNγ in the intestine would appear to be independent of arginine and not due to tryptophan starvation, but dependent on iron (Dimier and Bout, 1998). IELs were found to produce IL-10 and TGFβ, both of which are known to inhibit many aspects of inflammation. Adoptive transfer of primed IELs from IL-10 deficient mice was able to prevent pathology, indicating that a molecule other than IL-10 was responsible for protection (Mennechet *et al.*, 2004). However, neutralization of TGFβ can ablate the protective effects of IELs indicating that this molecule is critical to their protective effects (Buzoni-Gatel *et al.*, 2001). Buzoni-Gatel *et al.* also found that $TLR9^{-/-}$ mice had diminished antimicrobial peptide from Paneth cells and less intestinal pathology. Yarovinsky *et al.* have also demonstrated that some of the intestinal pathology is secondary to TLR stimulated by the intestine's bacterial microbiome (Yarovinsky *et al.*, 2008).

Overall, these observations indicate that the adaptive immune response plays an important role in control of parasite replication and in prevention of immune pathology at this anatomical site. T-cell derived IFNγ is not only important in parasite killing but also contributes to pathology. However, there is also a clear role for the Th2 associated cytokines IL-4 and IL-5 as mice deficient in these cytokines are resistant to intestinal necrosis.

25.7.4 Ocular Toxoplasmosis

In addition to congenital toxoplasmosis, there is strong evidence that acquired toxoplasmosis can also be a relatively common cause of uveitis (Burnett *et al.*, 1998; Silveira *et al.*, 2001; Glasner *et al.*, 1992; Couvreur and Thulliez, 1996; Boch-Dreissen 2000) (see Chapter 5). Pregnancy has been reported to be associated with reactivation of *T. gondii* infection in humans (Garweg *et al.*, 2005), but how often this occurs has not been defined. In a mouse model of ocular toxoplasmosis, based mainly in studies of mice with acquired infection with *T. gondii*, both $CD4^+$ T- and $CD8^+$ T-cells are observed in the inflammatory process in the eye and are associated with the presence of the parasite (Roberts, F. *et al.*, 2000; Khan *et al.*, 1997). Similar to toxoplasmic encephalitis, IFNγ, TNFα, IL-6 and iNOS appear to be critical elements in the control of parasite replication in the eye. Specifically, neutralization of endogenous IFNγ or TNFα as well as inhibition of iNOS with specific inhibitors results in enhancement of parasite replication in the eye tissue and development of severe (Gazzinelli *et al.*, 1994a, 1994b, 1994c; Hayashi *et al.*, 1996b; Roberts *et al.*, 2000; Hu *et al.*, 1999a, 1999b; Lu *et al.*, 2003). Consistently, IFNγ mRNA was expressed in high levels in the eyes of animals infected with *T. gondii* and IFNγ deficient mice were highly susceptible to toxoplasmic uveitis. IL-6 deficient mice also developed more severe toxoplasmic uveitis. Employing a particular model of ocular toxoplasmosis in which ocular infection developed after mice were immunized with temperature sensitive mutant *Toxoplasma* and then received an intraocular inoculation of RH tachyzoites, IL-10 deficient mice were more susceptible and IL-10 transgenic mice more resistant to toxoplasmic uveitis (Lu *et al.*, 2003). Similarly, Fas—FasL interaction and apoptosis of T-lymphocytes were important in control of cellular infiltrates and intensity of eye inflammation elicited by *T. gondii* antigens (Hu *et al.*, 1999b). Parasite proliferation and $CD4^+$ and $CD8^+$ T-cells were important in pathogenesis and the latter important in protection as well (Hu *et al.*, 1999a; Lu *et al.*, 2003).

Whether the mechanisms for congenital and acquired ocular disease in humans infected with *T. gondii* are similar or not remains to be defined. Nevertheless, congenital infection with *T. gondii* is associated with decreased responsiveness of T-cells to parasite antigens (McLeod *et al.*, 1990) and episodes of recrudescence of the parasite in the eye. It is likely that congenital ocular toxoplasmosis is the result of a less effective immune response against the parasite. The critical role of IFNγ, TNFα, IL-6 and iNOS in control of parasite replication in the eye in mice may be relevant to the pathogenesis of congenital ocular toxoplasmosis in humans.

25.7.5 Influence of Co-Infection with Other Pathogens

In addition to inducing an effective parasite-specific T-cell mediated immune response, *T. gondii* infection elicits strong cell-mediated immunity and host resistance to non-related pathogens and tumour cells (Ruskin and Remington 1968; Hibbs *et al.*, 1971). These 'nonspecific' immune responses are thought to be dependent on activation of innate immunity, which plays an important role in the development of pathogen specific Th1 lymphocytes. The Th1 lymphocytes will then mediate activation of cellular effector mechanisms involved in the control of unrelated pathogens. More specifically, *T. gondii* infection protects the host against pathology elicited during infections with *Schistosoma mansoni* and *Leishmania major* and the murine leukaemia viruses (Mahmoud *et al.*, 1976, 1977; Gazzinelli *et al.*, 1992b; Santiago *et al.*, 1999) in which the pathogenesis is mediated by Th2 lymphocytes. In addition, infection with *T. gondii* protects the host against *Listeria monocytogenes* and certain tumours, where host resistance is largely mediated by T-cell mediated immunity. HIV interferes with macrophage lulling of *T. gondii* (Biggs *et al.*, 1995), and also alters Th1 cytokines and IL-12 but not monokine production (Gazzinelli *et al.*, 1995).

Timing of infection with *T. gondii* appears to be critical in terms of specific effects on infections with other parasites. In the case of infection with *L. major*, acute but not chronic infection with *T. gondii* promoted resistance to lesion development and parasite growth, by promoting a *Leishmania* specific Th1 response in the susceptible BALB/c mice. In the case of *Schistosoma mansoni*, chronic infection with *T. gondii* resulted in resistance to the trematode infection with a remarkable modulatory effect on granulomas, probably through the control of Th2 responses in co-infected animals. In contrast, acute *Toxoplasma* infection associated with *S. mansoni* infection resulted in more intense hepatic lesions and lethality that was associated with high TNFα levels (Mahmoud *et al.*, 1976, 1977). This effect was dependent on endogenous IL-12 levels (Araujo *et al.*, 2001; Mahmoud *et al.*, 1976, 1977). The interaction of *T. gondii* and *Nippostongylus braziliensis* was evaluated in mice, where *T. gondii* infection largely inhibited the Th2 responses (i.e. IL-4 and IL-5) elicited by this intestinal worm. In contrast, there was no major change in Th1 responses (i.e. IFNγ) elicited by *T. gondii* in co-infected mice. No change in the course of diseases caused by either of these pathogens was observed. In summary, these studies suggest that infection with *T. gondii* in mice is highly efficient in suppressing Th2 responses elicited by other parasites (i.e. *L. major*, *S. mansoni* and *N. braziliensis*). However, the opposite is not true, since Th1 responses elicited by *T. gondii* are not controlled when mice are co-infected with infectious agents that preferentially stimulate Th2 responses.

25.7.6 Summary and Generalized Scheme

Figures 25.1 to 25.6 summarize interactions of innate and adaptive immunity, antigen presentation to CD8$^+$ T-cells and CD4$^+$ T-cells, effector mechanisms elicited and the initiation and regulation of effector functions.

25.8 IMMUNOLOGICAL CONTROL IN HUMANS

25.8.1 Immunologically Normal Older Children and Adults

S. E. Anderson first described lymphocyte blastogenesis in response to *Toxoplasma* antigens in individuals with chronic infection, and occasional patients who had little or no response early on after infection (1976). In patients with a prolonged lymphadenopathy syndrome, there was an elevation in suppressor (Leu-2 positive) T-cells. In patients with asymptomatic lymphadenopathy, a decrease in the number of T-helper cells or an increase in the number of suppressor cells or both were noted. Patients who were completely asymptomatic with only serological evidence of acute infections did not have abnormalities in their T-cell populations (Luft *et al.*, 1984a). Others have reported similar subset modifications (De Waele, 1985; Luft *et al.*, 1984b). Vγ2Vδ2 T-cells are increased in the circulation of children who have acquired toxoplasmosis postnatally (Hara *et al.*, 1996). These are predominantly CD45RAlo−CD45ROhi suggesting that they are activated cells. Lysis by such cells is not MHC restricted. Γδ T-cells can lyse infected targets when obtained from individuals without prior infection. Other mechanisms important in human cells are discussed above.

25.8.2 Immunocompromised Older Children and Adults

This is reviewed in Chapter 4. One interesting report of recrudescent toxoplasmosis in conjunction with steroid treatment augmented by antibody to TNFα was noted (Young and Maguire, 2005) illustrating the importance of this cytokine in chronic infection. Disease due to immunological reconstitution in patients with AIDS is another described problem (Petrof and McLeod, 2002).

25.8.3 Protecting the Foetus by Immunity in the Mother and Implications for Vaccine Development

Mothers who are chronically infected with *T. gondii* do not appear to transmit *T. gondii* infection to their foetus if they are immunologically normal once an established immune response has developed. This finding makes it likely that it would be possible to elicit a protective immune response that would allow a woman to be immunized before becoming pregnant. This approach has been used for vaccines for animals (Buxton and Innes, 1995) and in experimental murine models (McLeod *et al.*, 1988; Roberts *et al.*, 1994).

25.8.4 The Foetus and Infant

Many components of the immune system of the newborn infant function less robustly than in the older child. This has been reviewed by Dowell and McLeod (2000), Lewis and Wilson (2005) and others (Arnold *et al.*, 2005; Barakonyi *et al.*, 1999; Blackman *et al.*, 1990; Marodi *et al.*, 1994; Marchant *et al.*, 2003) and is summarized in Figure 25.6 and Table 25.8. In newborn infants infected congenitally with *T. gondii*, there have been studies describing percentages of lymphocyte subpopulations (Hohlfeld *et al.*, 1990; McLeod *et al.*, 1985a, 1985b, 1990; Lecolier *et al.*, 1989) and their immune responses to *T. gondii* antigens. Wilson and Remington (1979) initially described a robust blastogenic response to *Toxoplasma* lysate antigens. In a larger series McLeod *et al.* (1985, 1990) found that the responses were not present or not robust in most of the children in a larger study in the newborn period. This was especially true of infants who had severe disease and were infected early in gestation and those infected late in gestation. By one year of age most children developed some lymphocyte blastogenic responses to *T. gondii* antigens, once medicines were discontinued, but the responses were often less than those of their mothers. This was the case for prolonged periods of time, and a similar diminished response was also noted in a Brazilian cohort of congenitally infected children. Fatoohi *et al.* (2003) noted response of most congenitally infected children to soluble *Toxoplasma* antigen with IL-2 production and CD25 expression. This occurred in about 7% of apparently uninfected children as well (Fatoohi *et al.*, 2003). Many of these children had been treated *in utero* and had much milder disease than those children in the USA cohort (Fatoohi *et al.*, 2003). Differences in results in these studies may be due to differences in methodology or to the major differences in the clinical status of the patients studied. Rizzo *et al.* (2000) also described lower lymphocyte blastogenic responses in children they believed to be congenitally infected relative to those they believed to have acute acquired infection. Foetuses and infants with congenital toxoplasmosis produce IFNγ (Thulliez *et al.*, 1994; Lewis and Wilson, 2005).

In an interesting study of γδ T-cell function in infants with congenital toxoplasmosis, some infants had increased expansion of Vγ2Vδ2 bearing γδ T-cells expressing a CD45RAlo–DC45R0hi surface phenotype. These cells had a poor proliferative response when cultured with peripheral blood mononuclear cells infected

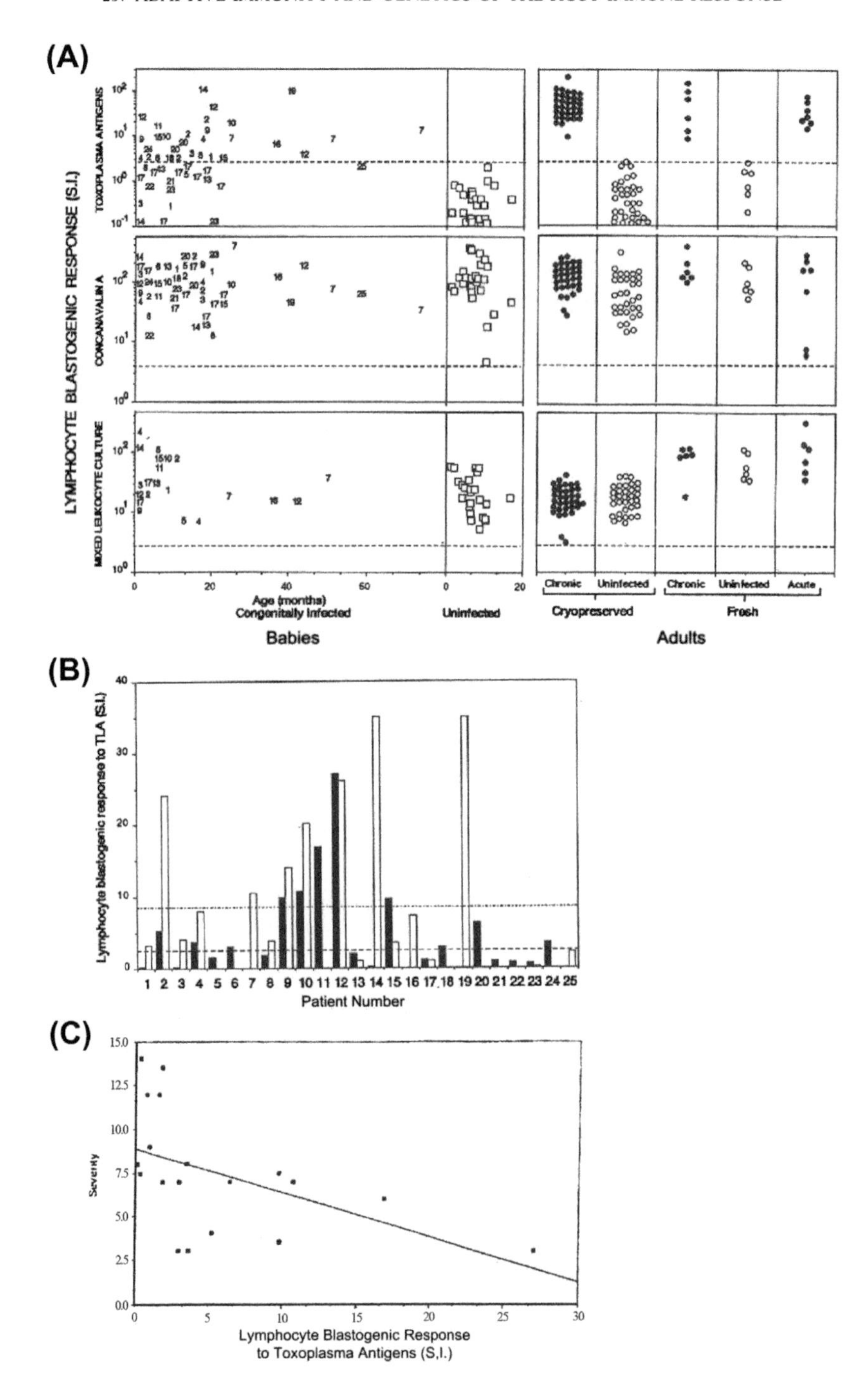
(A)
LYMPHOCYTE BLASTOGENIC RESPONSE (S.I.)
TOXOPLASMA ANTIGENS
CONCANAVALIN A
MIXED LEUKOCYTE CULTURE
Age (months)
Congenitally Infected
Uninfected
Babies
Chronic
Uninfected
Chronic
Uninfected
Acute
Cryopreserved
Fresh
Adults
(B)
Lymphocyte blastogenic response to TLA (S.I.)
Patient Number
(C)
Severity
Lymphocyte Blastogenic Response
to Toxoplasma Antigens (S,I.)

with live unirradiated or irradiated *T. gondii* tachyzoites.

25.8.5 Foetal and Neonatal Defences

Foetal and neonatal defences that have been described include monocyte derived and placental macrophages activated by IFNγ that kill *T. gondii* (Wilson and Remington, 1979). T-cell numbers and the repertoire for antigen recognition are limited in the first half of gestation (Lewis and Wilson, 2005). Sequellae occur regardless of the time in gestation that infection is acquired (Koppe *et al.*, 1986; Wilson *et al.*, 1980; Eichenwald, 1960; Remington *et al.*, 2005), suggesting that protective mechanisms are immature even late in gestation.

Congenitally infected infants generally have less lymphocyte blastogenesis, IL-2 and IFNγ production in response to *T. gondii* (Fig. 25.32) (McLeod *et al.*, 1990). Fatoohi *et al.* (2003) found $CD25^+$ CD4 T-cells response measured as IL-2 production in response to a soluble *T. gondii* antigen preparation (STAg) suggesting that perhaps these might be regulatory T-cells. The children studied by Fatoohi *et al.* (2003) were often born to mothers treated in gestation with mild or no manifestations of infection, whereas McLeod (1990) studied children with severe as well as milder manifestations. The most profound deficit in the McLeod (1990) series was present in children with the most severe disease (Fig. 25.32). No data are available about $CD8^+$ T-cell responses in congenitally infected newborns. Some of the possible mechanisms to explain absent or less robust lymphocyte blastogenesis in response to *T. gondii* antigens in congenitally infected infants include: (i) limitations of antigen presentation; (ii) down-modulation of responding T-cells, e.g. by regulatory T-cells; (iii) sequestration of responding T-cells; (iv) deletion of responder T-cells; or (v) immaturity of responding T-cells or co-stimulatory molecules that are not present, different or in different amounts. No antibodies specific to *T. gondii* were detectable in ~50% of infected newborn infants. Newborns lack DTH, have IgG similar to their mother and have little or no circulating IgM or IgA (Lewis and Wilson, 2005).

25.8.6 Pregnancy and Recrudescent Eye Disease

Recrudescent eye disease occurs in some pregnant women in successive pregnancies, but the pathogenesis of this process is not understood.

25.8.7 Chronic Infection

Congenital toxoplasmosis in humans is a recrudescent, recurring disease (Desmonts and Couvreur, 1984; Koppe *et al.*, 1974, 1982, 1986; Koppe and Rothova, 1989; Binquet *et al.*, 2003). Rarely, chronically infected older children or adults with postnatally acquired infection have had unusual, severe or protracted neurologic signs, some with persistent CSF local antibody production (Couvreur and Thulliez, 1996; Ajzenberg *et al.*, 2002, 2004; Bossi *et al.*, 1998; Carme, 2000; Darde, 1996; Darde *et al.*, 1998;

FIGURE 25.32 **Lymphocyte blastogenic responses in infants with congenital toxoplasmosis. *Toxoplasma* lysate antigens (top), concanavalin A (centre) and mixed leukocyte culture (bottom).** (A) The dashed lines demarcate positive and negative responses. (B) Lymphocyte response of all study children to *Toxoplasma* lysate antigens. Infants older than 13.5 months are represented by solid bars and those children older than 15 months by open bars. The horizontal dashed line at S.I. = 2.5 demarcates positive and negative responses. The horizontal dashed and dotted line at S.I. = 8 indicates the lowest responses of lymphocytes from infected adults. (C) Correlation of lymphocyte blastogenic response to *Toxoplasma* lysate antigen and severity score. Correlation between diminished transformation and severity was significant ($p = 0.002$). Patient numbers are those used in publications from the US National Collaborative Study Group (NCCTS). *(Adapted with permission from McLeod* et al. *(1990).)*

Desmonts and Couvreur, 1984). Other than these isolated cases, the consequences of chronic *T. gondii* infection in the brains of two billion people worldwide remain to be determined. In humans, remote chronic infection is not transmitted to the foetus unless there is immune compromise in the mother (Remington *et al.*, 2005), but in mice such vertical transmission sometimes does occur (Beverley, 1959).

25.8.8 Guillain Barre Syndrome

There have been isolated cases of Guillain Barre syndrome in individuals with postnatally acquired toxoplasmosis (Ajzenberg *et al.*, 2002, 2004; Couvreur and Thulliez, 1996), including a case due to an atypical strain of *T. gondii* in a patient in Guyana (Ajzenberg and Darde *et al.*, 2004; Carme *et al.*, 2002). The pathogenic mechanism(s) were not determined in these cases.

25.8.9 Recurrent Signs and Symptoms in Congenitally Infected Persons and those with Acute Acquired Infection and Eye Disease

The pathogenesis of this clinically important problem is unknown. It appears that recrudescence is more frequent in adolescence and slightly more frequent around the age that children enter secondary school (McLeod *et al.*, in preparation; Binquet *et al.*, 2003; Hogan *et al.*, 1956, 1957, 1964; Holland *et al.*, 1988, 1999, 2003). Persons affected with congenital toxoplasmosis appear to have recrudescent disease most frequently at school age entry, puberty, during pregnancy and with significant stress (Phan *et al.*, 2008a, 2008b; Binquet *et al.*, 2003). Mechanisms that lead to this recrudescence are poorly understood. Possible explanations may be related to effects of sex hormones and corticosteroids. Sex hormones increase during puberty. Serum levels of steroid androgens increase in both males and females, which may influence immune responses. For example, Kutlu *et al.* (2003) investigated effects of age-related changes of serum dehydroepiandrosterone sulphate (DHEAS) and androstenediol (AED) concentrations on tuberculin response to BCG vaccination before and during puberty. Response to BCG at puberty was higher than in those who were pre-pubertal without differences between sexes. Higher levels of DHEAS and AED in the pubertal group correlated with increased tuberculin response. If this were to influence a harmful or too robust immune response it could be reflected as greater or more clinically obvious recurrences, but this is only a speculation. How oestrogen might influence increase in recurrences of toxoplasmosis in puberty is not known. Oestrogens do play a role in immunity, which could contribute, but again this is speculation. For example, the oestrogen receptors ERα and ERβ are expressed in antigen presenting cells, macrophages, dendritic cells and monocytes. Levels of TNFα and IL-1β are highest during the luteal phase of the female reproductive cycle (Cunningham *et al.*, 2011).

Glucocorticoids have a paradoxical role in the expression of inflammatory genes. They interfere with proinflammatory signalling cascades and inhibit the binding of transcription factors to DNA by competing for the binding of co-activators such as CREB-binding protein, p300, and GRIP. In addition, they inhibit activity of NFκB and AP-1 responsive genes (Bellavance *et al.*, 2012). Inhibition occurs when glucocorticoids bind to glucocorticoid receptors, which bind to DNA sequences called glucocorticoid receptor response elements. NFκB and AP-1 responsive genes contain negative glucocorticoid receptor response elements, which glucocorticoids bind to and down-regulate the immune and inflammatory response (Bellavance *et al.*, 2012). Isoforms of glucocorticoid receptors can be induced by proinflammatory pathways. Bellavance *et al.* (2012) noted that stress and high levels of GCs increase inflammation in mouse brains. These types of stress induced GCs could compromise the capacity of neurons to tolerate parasite challenges, and could promote neuronal death by increasing extracellular levels of glutamate (an excitatory neurotransmitter).

Thymic involution may also play a role in immune response to *T. gondii* infection in humans. Thymic atrophy is observed during immune responses to certain micro-organisms including *T. gondii* in experimental animals, and thereby may reduce thymic functions and prevent protective immune responses. The thymus involutes during puberty. The simultaneous decline in growth hormone (GH) and increase in sex steroid production with age has been proposed as an explanation for thymic involution. The exogenous administration of androgens or oestrogens results in thymic atrophy, suggesting that increases in sex steroids at puberty contribute to thymic involution. Blocking GH/IGF-I production and sex steroids did not delay thymic involution, which is not consistent with the speculation that thymic involution at puberty is due to sex hormones and increase in endogenous glucocorticoids. Actual mechanisms remain to be determined.

25.8.10 Repeated Infections with Different Clonal Types of Parasites

In mice, a first infection protects against a second infection (McLeod *et al.*, 1988). The relevance of this observation and the natural history of repeated *T. gondii* infections with the same or different clonal types of parasites is not known. However, a seropositive recipient of a transplanted heart from an acutely infected individual develops IgM specific antibody and an increase in *T. gondii* specific IgG antibody, but does not become clinically ill and, in contrast, a seronegative recipient may develop life-threatening infection when transplanted from a seropositive donor (Ryning *et al.*, 1979). This suggests that, in humans, there is probably protection from an initial infection against subsequent infections.

25.9 INFLUENCE OF CO-INFECTION WITH OTHER PATHOGENS

The effect of HIV infection on depletion of $CD4^+$ T-lymphocytes and the results of this immunosuppression predisposing to *T. gondii* encephalitis and disseminated infection are commonly recognized. Cytomegalovirus (CMV) modified the pattern of intracellular replication of *T. gondii* (Ghatak *et al.* in Remington *et al.*, 2005) and such conjoint congenital infections have been reported. Little is known about the influences of these two pathogens on the host when acquired together. Coppens *et al.* (2012) have emphasized the importance of co-infection with other pathogens in considering the competition for nutrients and host cell structures (Coppens *et al.*, 2012). Knoll has considered the effect of *T. gondii* proteins on outcomes of influenza and *Staphylococcal* infections (Knoll, personal communication). Bzik has demonstrated that an attenuated *T. gondii* can adjuvant $CD8^+$ T-cells in their protective effect on both melanoma and ovarian tumours, but not on Lewis lung cell carcinoma in murine models (Bzik *et al.*, 2012).

25.10 PREGNANCY AND CONGENITAL DISEASE

During pregnancy there are number of immunological changes that facilitate implantation of the placenta and survival of the foetus (Morrel, 1995). It is perhaps not surprising that susceptibility to a number of pathogens is noted during pregnancy. Modulation of the immune response is mediated by changes in a number of sex and pregnancy associated hormones including oestrogen, testosterone and progesterone, all of which are raised during pregnancy compared with normal physiological levels. Most notably, progesterone, which cycles between 1.8 ± 0.34 and 43 ± 13 mmol/litre in a non-pregnant woman is raised incrementally as pregnancy proceeds to over 500 nmol/litre in the late stages of pregnancy. This hormonally induced regulation of immune cell function means that in spite of the presence of macrophages, NK-cells, mast cells, eosinophils, neutrophils and T-cells in the decidua, the trophoblast is not rejected (reviewed in Roberts *et al.*, 1996a, 2001a).

Local T-cell function is tightly regulated in fetoplacental tissues. In mice, Treg cells have been demonstrated to accumulate in the uterus in preparation for implantation and expand greatly as pregnancy progresses (Kallikourdis and Betz, 2007). In humans, a failure to increase circulating Treg numbers during pregnancy has been associated with a poor prognosis and IL-17^{+} T-cell accumulation in the decidua has been associated with abortion (reviewed in Lee *et al.*, 2012). Furthermore, a Th2 phenotype is favoured with the production of IL-4, IL-5 and IL-10 (Lin *et al.*, 1993; Reghupathy, 1997; Piccinni *et al.*, 2000). Notably disruption of this by administration of Th1 associated cytokines such as IFNγ or IL-2 can mediate abortion in pregnant mice (Lin *et al.*, 1993; Reghupathy, 1997). The actions of these cytokines are most probably mediated downstream by TNFα and/or reactive nitrogen intermediates (RNI), as administration of LPS, for example, can also induce similar effects (Gendron, 1990). The importance of progesterone in polarizing the developing T-cell response towards Th2 has been demonstrated and this hormone has even been demonstrated to induce IL-4 and IL-5 production by established Th1 clones (Piccinni *et al.*, 1995). Progesterone and IL-4 have been shown to induce LIF (leukaemia inhibitory factor), which is necessary for embryo implantation (Piccinni *et al.*, 2000, 2001). The expression of CCR5 and CXCR4 is down-regulated on activated T-cells in the presence of progesterone (Vassiliadou *et al.*, 1999).

Progesterone also modulates NK-cell function in a number of different ways. For example, progesterone induces lymphocyte production of PIBF (pregnancy induced blocking factor) which binds NK-cells and inhibits their degranulation (Faust *et al.*, 1999). PIBF is produced by lymphocytes during normal pregnancy and its production positively correlates with successful conception (Check *et al.*, 1997). The gene has recently been cloned from humans and the expressed recombinant protein demonstrated to down-regulate NK-cell function (Polgar *et al.*, 2003). Administration of PIBF to mice has also been demonstrated to prevent abortion in mice injected with the progesterone antagonist, RU486 (Polgar *et al.*, 2003). Progesterone has also been shown to be responsible for the expansion of γ1.4δ1 T-cells that produce IL-10 throughout pregnancy and inhibit NK-cell function and IL-12 production (Barakonyi *et al.*, 1999; Polgar *et al.*, 1999).

The need for tight regulation of inflammatory products and a local Th2 response for successful pregnancy is opposed to what is required for control of *T. gondii* infection. Conversely, a strong immune response to *T. gondii* involving NK stimulation and Th1 cell activation is incompatible with successful pregnancy.

25.10.1 Pregnancy Increases Susceptibility to *T. gondii* Infection

Mice have increased susceptibility to *T. gondii* when infected during pregnancy (Shirahata *et al.*, 1992). In keeping with the ability of pregnancy to polarize the immune response towards Th2, pregnant mice infected with *T. gondii* produce less IFNγ than similarly infected non-pregnant female mice (Shirahata *et al.*, 1992). This reduced ability to produce IFNγ correlates with increased mortality, but can be ablated by administration of exogenous IFNγ or IL-2 (Shirahata *et al.*, 1993).

Early observations suggested a chronic *T. gondii* infection in humans presented no risk of congenital transmission. While this is generally true, there are now a number of cases reported in the literature that demonstrate congenital transmission from a chronic infection (Vogel *et al.*, 1993; Kodjikian *et al.*, 2004); this has often been due to immune suppression in the mother (e.g. HIV infection or high dose steroids) (Bachmeyer *et al.*, 2005; Remington *et al.*, 2005; Vogel *et al.*, 2000). This implies that there has been a degree of reactivation of disease. Indeed, a recent study that followed pregnancy in 18 females with chronic *T. gondii* infection, seven cases of reactivation in the ocular site were observed (Garweg *et al.*, 2005). The incidence of disease reactivation is likely to be higher as it may go unobserved in other anatomical sites. This has implications for the health of the pregnant mother and the foetus. Similarly, it was largely believed that sheep were only at risk of congenital transmission if infected during pregnancy. Furthermore, chronic infection was thought to prevent congenital transmission if the sheep were exposed to a further infection. However, a study found consistently high levels of congenital transmission in pedigree flocks of sheep, implying that this is not always the case (Morley *et al.*, 2005).

In mice, congenital transmission of *T. gondii* through successive generations of mice was reported as early as 1959 (Beverley, 1959), but has since been demonstrated to be mouse strain dependent (Roberts and Alexander, 1992). Thus chronically infected BALB/c mice do not transmit disease to their offspring even if rechallenged with *T. gondii* during pregnancy (Roberts and Alexander, 1992). This demonstrates that the ability of mice to prevent congenital transmission has a genetic component. The BALB/c model has been used by a number of workers to determine the functional correlates of this protection both for prevention of congenital transmission following infection during pregnancy or in pregnant mice with chronic infections (Roberts *et al.*, 1994: Abou-Bacar *et al.*, 2004a, 2004b). $RAG^{-/-}$ mice have reduced levels of congenital transmission compared with BALB/c mice when infected during pregnancy (Abou-Bacar *et al.*, 2004a). This indicated an important role for innate immunity, which was confirmed by depletion of NK-cells in $RAG^{-/-}$ mice. Interestingly, similar NK-cell depletion experiments had no effect on the rate of transmission in BALB/c mice, suggesting that T- or B-cells play a role even at these early stages in immunocompetent mice (Abou-Bacar *et al.*, 2004a). A surprising finding is that neutralizing IFNγ increases the rate of congenital transmission, which would suggest that transmission is enhanced by an inflammatory response (Abou-Bacar *et al.*, 2004a). A role for $CD8^+$ cells and IFNγ has been demonstrated in prevention of congenital transmission from chronically infected mice. In contrast, depletion of $CD4^+$ cells did not cause disease reactivation or disease transmission (Abou-Bacar *et al.*, 2004b). The role of IL-4 in prevention of congenital transmission is complex and is likely to be mouse strain dependent. A study using BALB/c IL-4 deficient mice found these to have reduced rates of congenital transmission compared with wild type mice (Thouvenin *et al.*, 1997), whereas a separate study using B6/129 mice found no difference in congenital transmission rates between IL-4 deficient mice and wild types (Alexander *et al.*, 1998).

The role of various components of the immune response in preventing congenital disease is difficult to study in humans. In keeping with the studies that found no role for $CD4^+$ cells in preventing congenital transmission from chronically infected mice (Abou-Bacar *et al.*, 2004b) rates of congenital *T. gondii* infection in HIV infected mothers are low (European Collaborative Study and Referral Network on Congenital Toxoplasmosis, 1996).

25.10.2 *T. gondii* Infections can have an Adverse Effect on Pregnancy

In apparently healthy, pregnant women who become infected with *T. gondii* during pregnancy, the rate of transmission and consequence for the foetus varies by trimester. Thus infection during the first trimester has a relatively higher risk of spontaneous abortion, although still appears relatively small compared with the second or third trimester. In contrast, the likelihood of congenital infection increases by the trimester in which infection occurs: first trimester, 25%; second trimester, 54%; and third trimester, 65% (for a review see McCabe *et al.*, 1987). In spite of the relatively short pregnancy in mice, a number of similarities have been observed in the BALB/c model. Infection during the first trimester normally results in abortion or foetal resorption, whereas infection in the second trimester normally results in approximately 50% of the pups becoming infected (Roberts and Alexander, 1992). The ability of *T. gondii* to induce a Th1 response may be sufficient to induce abortion during the first trimester when pregnancy induced Th2 bias is low. In contrast, by trimester three, when pregnancy has induced a strong Th2 polarization, this may counteract the *T. gondii* induced Th1 response and prevent abortion. As a consequence, parasite survival may be favoured in the third semester and account for the increased rates of transmission. More recently, *T. gondii* infection has been shown to disrupt the balance of Treg and Th17 cells in mice, raising the possibility that this might also contribute to abortion (Zhang *et al.*, 2012).

25.10.3 Immune Response in Congenital Infection is Abrogated

Congenitally infected humans develop a reduced immune response to *T. gondii* compared with that observed in adults. Specifically, lymphocytes from congenitally infected infants have been found to have poorer blastogenic responses to *T. gondii* antigen extracts than those of recently or chronically infected adults (McLeod *et al.*, 1985b). Poor blastogenic responses correlated with more severe disease symptoms in congenitally infected infants. Furthermore, IFNγ and IL-2 production was impaired in congenitally infected infants (McLeod *et al.*, 1990). A further study found that both $\alpha\beta$ T-cells and $\gamma\delta$ T-cells are anergized in congenitally infected infants less than one year of age. However, whereas $\gamma\delta$ cells were no longer anergic to *T. gondii* antigen after one year, and produced IFNγ, $\alpha\beta$ cells were still unresponsive to *T. gondii* antigen (Hara *et al.*, 1996). A degree of impaired immune function to *T. gondii* is likely to be present in all those congenitally infected for the remainder of their lives. Recurrent ocular disease is common in congenitally infected people. Although it is now recognized that not all cases of ocular disease are due to congenital infection, recurrent ocular disease in people with adult acquired *T. gondii* disease is considerably rarer.

25.11 SUMMARY AND CONCLUSIONS

Adaptive immunity and the genes that specify this response and the innate immune response and their interplay are critical to the outcome of *T. gondii* infection. Understanding these immune responses and other genetic factors that influence outcome are central to understanding pathogenesis and protection in this infection, and will provide the foundation for development of protective preparations. This infection causes significant morbidity, costs for care and loss of productivity and suffering (Roberts and Frenkel, 1990). Understanding pathogenesis and protective mechanisms in this infection will lead to strategies to prevent this disease.

Acknowledgements

This work was supported by NIH R01 AI 27530, NIH R01 43228 and the Research to Prevent Blindness Foundation. We also thank the Rooney Alden, Pritzker, Taub, Mussillami, Samuel, Powers, Engel and Mann Cornwell families for their support.

References

Abi Abdallah, D.S., Denkers, E.Y., 2012. Neutrophils cast extracellular traps in response to protozoan parasites. Front Immunol. 3, 382. http://dx.doi.org/10.3389/fimmu.2012.00382.

Abi Abdallah, D.S., Lin, C., Ball, C.J., King, M.R., Duhamel, G.E., Denkers, E.Y., 2012 Feb. *Toxoplasma gondii* triggers release of human and mouse neutrophil extracellular traps. Infect. Immun. 80 (2), 768–777. http://dx.doi.org/10.1128/IAI.05730-11.

Abi Abdallah, D.S., Egan, C.E., Butcher, B.A., Denkers, E.Y., 2011 May. Mouse neutrophils are professional antigen-presenting cells programmed to instruct Th1 and Th17 T-cell differentiation. Int. Immunol. 23 (5), 317–326. http://dx.doi.org/10.1093/intimm/dxr007.

Abou-Bacar, A., Pfaff, A.W., Georges, S., Letscher-Bru, V., Filisetti, D., VIllard, O., Antoni, E., Klein, J.P., Candolfi, E., 2004a. Role of NK-cells and gamma IFN in transplacental passage of *Toxoplasma gondii* in a mouse model of primary infection. Infect. Immun. 72, 1397–1401.

Abou-Bacar, A., Pfaff, A.W., Letscher-Bru, V., Filisetti, D., Rajapakse, R., Antoni, E., Villard, O., Klein, J.P., Candolfi, E., 2004b. Role of gamma interferon and T-cells in congenital Toxoplasma transmission. Parasite Immunol. 26, 315–318.

Adachi, O., Kawai, T., Takeda, K., Matsumoto, M., Tsutsui, H., Sakagami, M., Nakanishi, K., Akira, S., 1998. Targeted disruption of the MyD88 gene results in loss of IL-1- and IL-18-mediated function. Immunity 9, 143–150.

Adachi, K., Tsutsui, H., Kashiwamura, S., Seki, E., Nakano, H., Takeuchi, O., Takeda, K., Okumura, K., Van Kaer, L., Okamura, H., Akira, S., Nakanishi, K., 2001. Plasmodium berghei infection in mice induces liver injury by an IL-12- and toll-like receptor/myeloid differentiation factor 88-dependent mechanism. J. Immunol. 167, 5928–5934.

Adams, L.B., Hibbs Jr., J.B., Taintor, R.R., et al., 1990. Microbiostatic effect of murine-activated macrophages for *Toxoplasma gondii*. Role for synthesis of inorganic nitrogen oxides from L-arginine. J. Immunol. 144, 2725–2729.

Afonso, C., Paixao, V.B., Costa, R.M., 2012. Chronic Toxoplasma infection modifies the structure and the risk of host behaviour. PLoS One 7, e32489.

Ajzenberg, D., Cogné, N., Paris, L., Bessieres, M.H., Thulliez, P., Fillisetti, D., Pelloux, H., Marty, P., Dardé, M.L., 2002. Genotype of 86 Toxoplasmosis gondii isolates associated with human congenital toxoplasmosis and correlation with clinical findings. J. Infect. Dis. 186, 684–689.

Ajzenberg, D., Bañuls, A.L., Su, C., Dumétre, A., Demar, M., Carme, B., Dardé, M.L., 2004. Genetic diversity, clonality and sexuality in *Toxoplasma gondii*. Int. J. Parasitol. 34, 1185–1196.

Ajzenberg, D., Dumetre, A., Darde, M.L., 2005. Multiplex PCR for typing strains of *Toxoplasma gondii*. J. Clin. Microbiol. 4, 1940–1943.

Akira, S., Taga, T., Kishimoto, T., 1993. Interleukin-6 in biology and medicine. Adv. Immunol. 54, 1–78.

Albuquerque, M.C., Aleixo, A.L., Benchimol, E.I., Leandro, A.C., Das Neves, L.B., Vicente, R.T., Bonecini-Almeida Mda, G., Amendoeira, M.R., 2009. The IFNgamma +874T/A gene polymorphism is associated with retinochoroiditis toxoplasmosis susceptibility. Mem. Inst. Oswaldo Cruz 104, 451–455.

Alexander, J., Jebbari, H., Bluethman, H., Brombacher, F., Roberts, C.W., 1998a. The role of IL-4 in adult acquired and congenital toxoplasmosis. Int. J. Parasitol. 28, 113–120.

Alexander, J., et al., 1997. Mechanisms of innate resistance to *Toxoplasma gondii* infection. Philos. Trans. R. Soc. Lond. B. Biol. Sci. 352 (1359), 1355–1359.

Alexander, J., Hunter, C.A., 1998b. Immunoregulation during toxoplasmosis. Chem. Immunol. 70, 81–102.

Ali, K., Middleton, M., Puré, E., Rader, D.J., 2005 Oct 28. Apolipoprotein E suppresses the type I inflammatory response in vivo. Circ. Res. 97 (9), 922–927. Epub 2005 Sep 22.

Aliberti, J., Valenzuela, J.G., Carruthers, V.B., Hieny, S., Andersen, J., Charest, H., Reis e Sousa, C., Fairlamb, A., Ribeiro, J.M., Sher, A., 2003. Molecular mimicry of a CCR5 binding-domain in the microbial activation of dendritic cells. Nat. Immunol. 4 (5), 485–490.

Aliberti, J., Hieny, S., Reis e Sousa, C., Serhan, C.N., Sher, A., 2002a. Lipoxin-mediated inhibition of IL-12 production by DCs: a mechanism for regulation of microbial immunity. Nat. Immunol. (1), 76–82.

Aliberti, J., Serhan, C., Sher, A., 2002b. Parasite-induced lipoxin A4 is an endogenous regulator of IL-12 production and immunopathology in *Toxoplasma gondii* infection. J. Exp. Med. 9, 1253–1262.

Aline, F., Bout, D., Dimier-Poisson, I., 2002a. Dendritic cells as effector cells: gamma interferon activation of murine dendritic cells triggers oxygen-dependent inhibition of *Toxoplasma gondii* replication. Infect. Immun. 70, 2368–2374.

Aline, F., Bout, D., Dimier-Poisson, I., 2002b. Dendritic cells as effector cells: gamma interferon activation of murine dendritic cells triggers oxygen-dependent inhibition of *Toxoplasma gondii* replication. Infect. Immun. 70, 2368–2374.

Alvarado-Esquivel, C., Torres-Castorena, A., Liesenfeld, O., Estrada-Martinez, S., Urbina-Alvarez, J.D., 2012. High seroprevalence of *Toxoplasma gondii* infection in a subset of Mexican patients with work accidents and low socioeconomic status. Parasit. Vectors 5, 13.

Alves, C.M., Silva, D.A., Azzolini, A.E., Marzocchi-Machado, C.M., Lucisano-Valim, Y.M., Roque-Barreira, M.C., Mineo, J.R., 2012 Sep 14. Galectin-3 is essential for reactive oxygen species production by peritoneal neutrophils from mice infected with a virulent strain of *Toxoplasma gondii*. Parasitology, 1–10. [Epub ahead of print].

Amichay, D., Gazzinelli, R., Karupiah, G., Moench, T., Sher, A., Farber, J., 1996. The genes for the chemokines MuMig and Crg-2 are induced in protozoan and viral infections in response ot IFNgamma with patterns of tissue expression that suggest nonredundant roles in vivo. J. Immunol. 157, 4511–4520.

Anderson, S.E., Remington, J.S., 1974. Effect of normal and activated human macrophages on *Toxoplasma gondii*. J. Exp. Med. 139, 1154–1174.

Anderson, S.E., Bautista, S., Remington, J.S., 1976. Induction of resistance to *Toxoplasma gondii* in human macrophages by soluble lymphocyte products. J. Immunol. (2), 381–387.

Andrade, R.M., Portillo, J.A., Wessendarp, M., Subauste, C.S., 2005. CD40 signalling in macrophages induces activity against an intracellular pathogen independently of gamma interferon and reactive nitrogen intermediates. Infect. Immun. 73, 3115–3123.

Andrade, R.M., Wessendarp, M., Gubbels, M.J., Striepen, B., Subauste, C.S., 2006. CD40 induces macrophage anti-*Toxoplasma gondii* activity by triggering autophagy-dependent fusion of pathogen-containing vacuoles and lysosomes. J. Clin. Invest. 116, 2366–2377.

Aosai, F., Yang, T., Ueda, M., Yano, A., 1994. Isolation of naturally processed peptides from a Toxoplasma gondii-infected human B lymphoma cell line that are recognized by cytotoxic T-lymphocytes. J. Parasitol. 80, 260–266.

Aosai, F., Chen, M., Kang, H.K., Mun, H.S., Norose, K., Piao, L.X., Kobayashi, M., Takeuchi, O., Akira, S., Yano, A., 2002. Toxoplasma gondii-derived heat shock protein HSP70 functions as a B-cell mitogen. Cell Stress Chaperones 7, 357–364.

Araujo, F.G., 1991. Depletion of L3T4$^+$ (CD4$^+$) T-lymphocytes prevents development of resistance to *Toxoplasma gondii* in mice. Infect. Immun. 59, 1614–1619.

Araujo, F.G., 1992. Depletion of CD4$^+$ T-cells but not inhibition of the protective activity of IFNgamma prevents cure of toxoplasmosis mediated by drug therapy in mice. J. Immunol. 149, 3003–3007.

Araujo, M.I., Bliss, S.K., Suzuki, Y., Alcaraz, A., Denkers, E.Y., Pearce, E.J., 2001. Interleukin-12 promotes pathologic liver changes and death in mice coinfected with Schistosoma mansoni and *Toxoplasma gondii*. Infect. Immun. 69, 1454–1462.

Arling, T.A., Yolken, R.H., Lapidus, M., Langenberg, P., Dickerson, F.B., Zimmerman, S.A., Balis, T., Cabassa, J.A., Scrandis, D.A., Tonelli, L.H., Postolache, T.T., 2009. *Toxoplasma gondii* antibody titres and history of suicide attempts in patients with recurrent mood disorders. J. Nerv. Ment. Dis. 197, 905–908.

Arnold, B., Schuler, T., Hammerling, G., 2005. Control of peripheral T-lymphocyte tolerance in neonates and adults. Trends Immunol. 8, 406–411.

Arsenijevic, D., Onuma, H., Pecqueur, C., Raimbault, S., Manning, B.S., Miroux, B., Couplan, E., Alves-Guerra, M.C., Goubern, M., Surwit, R., Bouillaud, F., Richard, D., Collins, S., Ricquier, D., 2000. Disruption of the uncoupling protein-2 gene in mice reveals a role in immunity and reactive oxygen species production. Nat. Genet. 4, 435–439.

Arsenijevic, D., Clavel, S., Sanchis, D., Plamondon, J., Huang, Q., Ricquier, D., Rouger, L., Richard, D., 2007 May. Induction of Ucp2 expression in brain phagocytes and neurons following murine toxoplasmosis: an essential role of IFNgamma and an association with negative energy balance. J. Neuroimmunol. 186 (1–2), 121–132. Epub 2007 Apr 30.

Aviles, H., Stiles, J., O'donnell, P., Orshal, J., Leid, J., Sonnenfeld, G., Monroy, F., 2008. Kinetics of systemic cytokine and brain chemokine gene expression in murine toxoplasma infection. J. Parasitol. 94, 1282–1288.

Avunduk, A.M., Avunduk, M.C., Baltaci, A.K., Moğulkoç, R., 2007. Effect of melatonin and zinc on the immune response in experimental Toxoplasma retinochoroiditis. Ophthalmologica 6, 421–425.

Bachmeyer, C., Mouchnino, G., Thulliez, P., Blum, L., 2005. Congenital toxoplasmosis from an HIV-infected woman as a result of reactivation. J. Infect. (2), e55–e57.

Bahia-Oliveira, L.M., Jones, J.L., Azevedo-Silva, J., Alves, C.C., Orefice, F., Addiss, D.G., 2003. Highly endemic, waterborne toxoplasmosis in north Rio de Janeiro state, Brazil. Emerg. Infect. Dis. 9, 55–62.

Baird, J.R., Byrne, K.T., Lizotte, P.H., Toraya-Brown, S., Scarlett, U.K., Alexander, M.P., Sheen, M.R., Fox, B.A., Bzik, D.J., Bosenberg, M., Mullins, D.W., Turk, M.J., Fiering, S., 2013 Jan 1. Immune-Mediated Regression of Established B16F10 Melanoma by Intratumoural Injection of Attenuated *Toxoplasma gondii* Protects against Rechallenge. J. Immunol. 190 (1), 469–478. http://dx.doi.org/10.4049/jimmunol.1201209.

Bala, S., Englund, G., Kovacs, J., Wahl, L., Martin, M., Sher, A., Gazzinelli, R., 1994. *Toxoplasma gondii* soluble products induce cytokine secretion oby macrophages and potentiate in vitro replication of a monotropic strain of HIV. J. Eukaryot. Microbiol. 41, 7S.

Barakonyi, A., Polgar, B., Szekeres-Bartho, J., 1999. The role of gamma/delta T-cell receptor-positive cells in pregnancy: part II. Am. J. Reprod. Immunol. 42, 83–87.

Barragan, A., Sibley, L.D., 2002. Transepithelial migration of *Toxoplasma gondii* is linked to parasite motility and virulence. The Journal of Experimental Medicine 195, 1625–1633.

Beaman, M.H., Wong, S.Y., Abrams, J.S., Remington, J.S., January 1993. Splenic IL-4 secretion is not associated with the development of toxoplasmic encephalitis [abstract 81]. Program and abstracts of the 2nd International Congress of Biological Response Modifiers, San Diego, 29–31.

Beaman, M.H., Araujo, F.G., Remington, J.S., 1994. Protective reconstitution of the SCID mouse against reactivation of toxoplasmic encephalitis. J. Infect. Dis. 169 (2), 375–383.

Beauvillain, C., Ruiz, S., Guiton, R., Bout, D., Dimier-Poisson, I., 2007 Nov–Dec. A vaccine based on exosomes secreted by a dendritic cell line confers protection against T. gondii infection in syngeneic and allogeneic mice. Microbes Infect. 9 (14–15), 1614–1622.

Begum-Haque, S., Haque, A., Kasper, L.H., 2009 Nov. Apoptosis in *Toxoplasma gondii* activated T-cells: the role of IFNgamma in enhanced alteration of Bcl-2 expression and mitochondrial membrane potential. Microb Pathog. 47 (5), 281–288. http://dx.doi.org/10.1016/j.micpath.2009.09.004.

Bekpen, C., Hunn, J.P., Rohde, C., Parvanova, I., Guethlein, L., Dunn, D.M., Glowalla, E., Leptin, M., Howard, J.C., 2005. The interferon-inducible p47 (IRG) GTPases in vertebrates: loss of the cell autonomous resistance mechanism in the human lineage. Genome Biology 6, R92.

Bekpen, C., Xavier, R.J., Eichler, E.E., 2010. Human IRGM gene 'to be or not to be'. Semin. Immunopathol. 32, 437–444.

Béla, S.R., Dutra, M.S., Mui, E., Montpetit, A., Oliveira, F.S., Oliveira, S.C., Arantes, R.M., Antonelli, L.R., McLeod, R., Gazzinelli, R.T., 2012. Impaired innate immunity in mice deficient in interleukin-1 receptor-associated kinase 4 leads to defective type 1 T cell responses, B cell expansion, and enhanced susceptibility to infection with Toxoplasma gondii. Infect Immun. 80 (12), 4298–4308.

Bellmann, K., Jaattela, M., Wissing, D., Burkart, V., Kolb, H., 1996. Heat shock protein hsp70 overexpression confers resistance against nitric oxide. FEBS Lett. 391 (1-2), 185–188.

Bendelac, A., 1995. Positive selection of mouse NK1$^+$ T-cells by CD1-expressing cortical thymocytes. J. Exp. Med. 182 (6), 2091–2096.

Bendelac, A., Lantz, O., Quimby, M.E., Yewdell, J.W., Bennink, J.R., Brutkiewicz, R.R., 1995. CD1 recognition by mouse NK1$^+$ T-lymphocytes. Science 268 (5212), 863–865.

Benson, A., Pifer, R., Behrendt, C.L., Hooper, L.V., Yarovinsky, F., 2009. Gut commensal bacteria direct a protective immune response against *Toxoplasma gondii.* Cell Host and Microbe 6, 187–196.

Benson, A., Murray, S., Divakar, P., Burnaevskiy, N., Pifer, R., Forman, J., Yarovinsky, F., 2012 Jan 15. Microbial infection-induced expansion of effector T-cells overcomes the suppressive effects of regulatory T-cells via an IL-2 deprivation mechanism. J. Immunol. 188 (2), 800–810. http://dx.doi.org/10.4049/jimmunol.1100769.

Berdoy, M., Webster, J.P., Macdonald, D.W., 2000. Fatal attraction in rats infected with *Toxoplasma gondii.* Proc. Biol. Sci. 267, 1591–1594.

Berger, B.B., Egwuagu, C.E., Freeman, W.R., Wiley, C.A., 1993. Miliary toxoplasmic retinitis in acquired immunodeficiency syndrome. Arch. Ophthalmol. (3), 373–376.

Berman, J.D., Johnson, W.D., 1978. Monocyte function in human neonates. Infect. Immun. 19, 898–902.

Bermudez, L.E., Covaro, G., Remington, J., 1993. Infection of murine macrophages with *Toxoplasma gondii* is associated with release of transforming growth factor beta and down-regulation of expression of tumour necrosis factor receptors. Infect. Immun. 61 (10), 4126–4130.

Bernardes, E.S., Silva, N.M., Ruas, L.P., Mineo, J.R., Loyola, A.M., Hsu, D.K., Liu, F.T., Chammas, R., Roque-Barreira, M.C., 2006. *Toxoplasma gondii* infection reveals a novel regulatory role for galectin-3 in the interface of innate and adaptive immunity. Am. J. Pathol. (6), 1910–1920.

Beug, Shawn T., Cheung, Herman H., LaCasse, Eric C., Korneluk, Robert G., 2012. Modulation of immune signalling by inhibitors of apoptosis. Trends Immunol. 33, 535–545.

Beverley, J.K., 1959. Congenital transmission of toxoplasmosis through successive generations of mice. Nature 183, 1348–1349.

Bhadra, R., Gigley, J.P., Khan, I.A., 2011 Nov 1. Cutting edge: CD40-CD40 ligand pathway plays a critical CD8-intrinsic and -extrinsic role during rescue of exhausted CD8 T-cells. J. Immunol. 187 (9), 4421–4425. http://dx.doi.org/10.4049/jimmunol.1102319. Epub 2011 Sep 26.

Bierly, A.L., Shufesky, W.J., Sukhumavasi, W., Morelli, A.E., Denkers, E.Y., 2008. Dendritic cells expressing plasmacytoid marker PDCA-1 are Trojan horses during *Toxoplasma gondii* infection. J. Immunol. 181 (12), 8485–8491.

Biggs, B., Hewish, M., Kent, S., et al., 1995. HIV-1 infection of human macrophages impairs phagocytosis and killing of *Toxoplasma gondii.* J. Immunol. 154, 6132–6139.

Binquet, C., Wallon, M., Quantin, C., Kodjikian, L., Garweg, J., et al., 2003. Prognostic factors for the long-term development of ocular lesions in 327 children with congenital toxoplasmosis. Epidemiol. Infect. 131, 1157–1168.

Black, C.M., et al., 1989. Effect of recombinant tumour necrosis factor on acute infection in mice with *Toxoplasma gondii* or Trypanosoma cruzi. Immunology 68 (4), 570–574.

Blackman, M., Kappler, J., Marrack, P., 1990. The role of the T-cell receptor in positive and negative selection of developing T-cells. Science 248 (4961), 1335–1341.

Blackwell, J.M., Roberts, C.W., Alexander, J., 1993. Influence of genes within the MHC on mortality and brain cyst development in mice infected with *Toxoplasma gondii*: kinetics of immune regulation in BALB H-2 congenic mice. Parasite Immunol. 15, 317–324.

Blackwell, J.M., Roberts, C.S., Roach, T.I., Alexander, J., 1994. Influence of macrophage resistance gene Lsh/S/Ity/Bcg (candidate NRamp on *Toxoplasma gondii* infection in mice). Clin. Exp. Immunol. 97, 107–112.

Blackwell, J.M., 1998. Genetics of host resistance and susceptibility to intramacrophage pathogens: a study of multicase families of tuberculosis, leprosy and leishmaniasis in north-eastern Brazil. Int. J. Parasitol. 28 (1), 21–28.

Blader, I.J., Manger, I.D., Boothroyd, J.C., 2001. Microarray analysis reveals previously unknown changes in Toxoplasma gondii-infected human cells. J. Biol. Chem. 276, 24223–24231.

Blader, I.J., Saeij, J.P., 2009. Communication between *Toxoplasma gondii* and its host: impact on parasite growth, development, immune evasion, and virulence. APMIS 117 (5–6), 458–476.

Blanchard, N., Shastri, N., 2010. Topological journey of parasite-derived antigens for presentation by MHC class I molecules. Trends Immunol. 31 (11), 414–421.

Blanchard, N., Gonzalez, F., Schaeffer, M., Joncker, N.T., Cheng, T., Shastri, A.J., Robey, E.A., Shastri, N., 2008. Immunodominant, protective response to the parasite *Toxoplasma gondii* requires antigen processing in the endoplasmic reticulum. Nat. Immunol. 9 (8), 937–944.

Bliss, S.K., Gavrilescu, L.C., Alcaraz, A., Denkers, E.Y., 2001. Neutrophil depletion during *Toxoplasma gondii* infection leads to impaired immunity and lethal systemic pathology. Infect. Immun. 69 (8), 4898–4905.

Bliss, S.K., Butcher, B.A., Denkers, E.Y., 2000. Rapid recruitment of neutrophils containing prestored IL-12 during microbial infection. J. Immunol. 165, 4515–4521.

Blomstrom, A., Karlsson, H., Wicks, S., Yang, S., Yolken, R.H., Dalman, C., 2012. Maternal antibodies to infectious agents and risk for non-affective psychoses in the offspring-a matched case-control study. Schizophr. Res. 140, 25–30.

Bogdan, C., Rollinghoff, M., 1999. How do protozoan parasites survive inside macrophages? Parasitol. Today 15 (1), 22–28.

Bohne, W., Heesemann, J., Gross, U., 1994. Reduced replication of *Toxoplasma gondii* is necessary for induction of bradyzoite-specific antigens: a possible role for nitric oxide in triggering stage conversion. Infect. Immun. 62, 1761–1767.

Boothroyd, J.C., Dubremetz, J.F., 2008. Kiss and spit: the dual roles of Toxoplasma rhoptries. Nat. Rev. Microbiol. 6 (1), 79–88.

Bossi, P., Caumes, E., Paris, L., Darde, M.L., Bricaire, F., 1998. Toxoplasma gondii-associated Guillain-Barre syndrome in an immunocompetent patient. J. Clin. Microbiol. 36 (12), 3724–3725.

Bottino, C., Castriconi, R., Moretta, L., Moretta, A., 2005. Cellular ligands of reactivating NK receptors. Trends in Immun. 26.

Bottova, I., Hehl, A.B., Stefanić, S., Fabriàs, G., Casas, J., Schraner, E., Pieters, J., Sonda, S., 2009 Jun 26. Host cell P-glycoprotein is essential for cholesterol uptake and replication of *Toxoplasma gondii*. J. Biol. Chem. 284 (26), 17438–17448. http://dx.doi.org/10.1074/jbc.M809420200. Epub 2009 Apr 22.

Boyle, J.P., Saeij, J.P., Harada, S.Y., Ajioka, J.W., Boothroyd, J.C., 2008. Expression quantitative trait locus mapping of toxoplasma genes reveals multiple mechanisms for strain-specific differences in gene expression. Eukaryot. Cell 7 (8), 1403–1414. http://dx.doi.org/10.1128/EC.00073-08, 2008 Aug.

Boyle, J.P., Saeij, J.P., Boothroyd, J.C., 2007 Jul. *Toxoplasma gondii*: inconsistent dissemination patterns following oral infection in mice. Exp. Parasitol. 116 (3), 302–305.

Bradley, P.J., Sibley, L.D., 2007. Rhoptries: an arsenal of secreted virulence factors. Curr. Opin. Microbiol. 10 (6), 582–587.

Brandão, G.P., Melo, M.N., Gazzinelli, R.T., Caetano, B.C., Ferreira, A.M., Silva, L.A., Vitor, R.W., 2009. Mem. Inst. Oswaldo Cruz. 104 (2), 241–245.

Brenier-Pinchart, M.P., Villena, I., Mercier, C., Durand, F., Simon, J., Cesbron-Delauw, M.F., Pelloux, H., 2006 Jan. The Toxoplasma surface protein SAG1 triggers efficient in vitro secretion of chemokine ligand 2 (CCL2) from human fibroblasts. Microbes Infect. 8 (1), 254–261.

Brezin, A.P., Kasner, L., Thulliez, P., Li, Q., Daffos, F., Nussenblatt, R.B., Chan, C.C., 1994. Ocular toxoplasmosis in the foetus. Immunohistochemistry analysis and DNA amplification. Retina 14 (1), 19–26.

Brinkman, V., Sharma, S., Remington, J.S., 1986. Different regulation of the L3T4-T cell subset by B-cells in different mouse strains bearing the H-2 haplotype. J. Immunol. 137, 2991–2997.

Brown, C.R., McLeod, R., 1990. Class I MHC genes and $CD8^+$ T-cells determine cyst number in *Toxoplasma gondii* infection. J. Immunol. 145, 3438–3441.

Brown, C.R., McLeod, R., 1995a. Fate of an intracellular parasite during lysis of its host cell by $CD8^+$ lymphocytes. Keystone meeting, Keystone, CO.

Brown, C.R., Hunter, C.A., Estes, R.G., Beckmann, E., Forman, J., David, C., Remington, J.S., McLeod, R., 1995a. Definitive identification of a gene that confers resistance against Toxoplasma cyst burden and encephalitis. Immunology 85, 419–428.

Brown, A.S., Schaefer, C.A., Quesenberry Jr., C.P., Liu, L., Babulas, V.P., Susser, E.S. , 2005. Maternal exposure to toxoplasmosis and risk of schizophrenia in adult offspring. The American Journal of Psychiatry 162, 767–773.

Brown, C.R., David, C.S., Khare, S.J., McLeod, R., 1994. Effects of human class I transgenes on *Toxoplasma gondii* cyst formation. J. Immunol. 152, 4537–4541.

Brown, C.R., Hunter, C.A., Estes, R.G., Beckmann, E., Forman, J., David, C., Remington, J.S., McLeod, R., 1995. Definitive identification of a gene that confers resistance against Toxoplasma cyst burden and encephalitis. Immunology 85, 419–428.

Brunet, J., Pfaff, A.W., Abidi, A., Unoki, M., Nakamura, Y., Guinard, M., Klein, J.P., Candolfi, E., Mousli, M., 2008. *Toxoplasma gondii* exploits UHRF1 and induces host cell cycle arrest at G2 to enable its proliferation. Cell Microbiol. 10, 908–920.

Bruning, T., Daiminger, A., Enders, G., 1997. Diagnostic value of CD45RO expression on circulating T-lymphocytes of foetuses and newborn infants with pre-, peri- or early post-natal infections. Clin. Exp. Immunol. 107, 306–311.

Bryceson, Y.T., Foster, J.A., Kuppusamy, S.P., Henkenham, M., Long, E.O., 2005. Expression of a killer cell receptor-like gene in plastic regions of the central nervous system. J. Neuroimmunol. 161, 177–172.

Buffolano, W., Beghetto, E., Del Pezzo, M., Spadoni, A., Di Cristina, M., Petersen, E., Gargano, N., 2005. Use of recombinant antigens for early postnatal diagnosis of congenital toxoplasmosis. J. Clin. Microbiol. 43, 5916–5924.

Bulow, R., Boothroyd, J.C., 1991. Protection of mice from fatal *Toxoplasma gondii* infection by immunization with p30 antigen in liposomes. J. Immunol. 157, 979.

Burnett, A.J., Shortt, S.G., Isaac-Renton, J., King, A., Werker, D., Bowie, W.R., 1998. Multiple cases of acquired toxoplasmosis retinitis presenting in an outbreak. Ophthalmology 105, 1032–1037.

Butcher, B., Kim, L., Johnson, P., Denkers, E., 2001. *Toxoplasma gondii* tachyzoites inhibit proinflammatory cytokine induction in infected macrophages by preventing nuclear translocation of the transcription factor NF-kB. J. Immunol. 167, 2193.

Butcher, B., Denkers, E., 2002. Mechanism of entry determines the ability of *Toxoplasma gondii* to inhibit macrophage proinflammatory cytokine production. Infect. Immun. 70, 5216–5224.

Butcher, B.A., Fox, B.A., Rommereim, L.M., Kim, S.G., Maurer, K.J., Yarovinsky, F., Herbert, D.R., Bzik, D.J., Denkers, E.Y., 2011 Sep. *Toxoplasma gondii* rhoptry kinase ROP16 activates STAT3 and STAT6 resulting in cytokine inhibition and arginase-1-dependent growth control. PLoS Pathog. 7 (9), e1002236. http://dx.doi.org/10.1371/journal.ppat.1002236.

Butcher, B.A., Kim, L., Panopoulos, A.D., Watowich, S.S., Murray, P.J., Denkers, E.Y., 2005. IL-10-independent STAT3 activation by *Toxoplasma gondii* mediates suppression of IL-12 and TNF-alpha in host macrophages. J. Immunol 174, 3148–3152.

Buxton, D., Innes, E., 1995. A commercial vaccine for ovine toxoplasmosis. Parasitology 110 (Suppl 6), S11–S16.

Buzoni-Gatel, D., Lepage, A.C., Dimier-Posson, I.H., Bout, D.T., Kasper, L.H., 1997. Adoptive transfer of gut intraepithelial lymphocytes protects against murine infection with *Toxoplasma gondii*. J. Immunol. 158, 5883–5889.

Buzoni-Gatel, D., Debbabi, H., Moretto, M., Dimier-Poisson, I.H., Lepage, A.C., Bout, D.T., Kasper, L.H., 1999. Intraepithelial lymphocytes traffic to the intestine and enhance resistance to *Toxoplasma gondii* oral infection. J. Immunol. 162 (10), 5846–5852.

Buzoni-Gatel, D., Debbabi, H., Mennechet, F.J., Martin, V., Lepage, A.C., Schwartzman, J.D., Kasper, L.H., 2001. Murine ileitis after intracellular parasite infection is controlled by TGF-beta-producing intraepithelial lymphocytes. Gastroenterology 129, 914–924.

Buzoni-Gatel, D., Dubremetz, J.F., Werts, C., 2008 Feb. Molecular cross talk between *Toxoplasma gondii* and the host immune system. Med. Sci. (Paris) 24 (2), 191–196. http://dx.doi.org/10.1051/medsci/2008242191.

Buzoni-Gatel, D., Schulthess, J., Menard, L.C., Kasper, L.H., 2006. Mucosal defences against orally acquired protozoan parasites, emphasis on *Toxoplasma gondii* infections. Cell. Microbiol. 8, 535–544.

Byrd, T.F., Horwitz, M.A., 1989. Interferon gamma-activated human monocytes down-regulate transferrin receptors and inhibit the intracellular multiplication of *Legionella pneumophila* by limiting the availability of iron. J. Clin. Invest. 83, 1457–1465.

Caamano, J., Tato, C., Cai, G., Villegas, E.N., Speirs, K., Craig, L., Alexander, J., Hunter, C.A., 2000 Nov 15. Identification of a role for NF-kappa B2 in the regulation of apoptosis and in maintenance of T-cell-mediated immunity to *Toxoplasma gondii*. J. Immunol. 165 (10), 5720–5728.

Caamano, J., Hunter, C.A., 2002. NF-kappaB family of transcription factors: central regulators of innate and adaptive immune functions. Clin. Microbiol. Rev. 15, 414–429.

Caamano, J., Alexander, J., Craig, L., Bravo, R., Hunter, C.A., 1999 Oct 15. The NF-kappa B family member RelB is required for innate and adaptive immunity to *Toxoplasma gondii*. J. Immunol. 163 (8), 4453–4461.

Caetano, B.C., Biswas, A., Lima Jr., D.S., Benevides, L., Mineo, T.W., Horta, C.V., Lee, K.H., Silva, J.S., Gazzinelli, R.T., Zamboni, D.S., Kobayashi, K.S., 2011 Dec. Intrinsic expression of Nod2 in $CD4^+$ T-lymphocytes is not necessary for the development of cell-mediated immunity and host resistance to *Toxoplasma gondii*. Eur. J. Immunol. 41 (12), 3627–3631. http://dx.doi.org/10.1002/eji.201141876.

Cai, G., Kastelein, R., Hunter, C.A., 2000. Interleukin-18 (IL-18) enhances innate IL-12 mediated resistance to *Toxoplasma gondii*. Infect. Immun. 68, 6932–6938.

Candolfi, E., Hunter, A., Remington, J.S., 1994. Mitogen- and antigen-specific proliferation of T-cells in murine toxoplasmosis is inhibited by reactive nitrogen intermediates. Infect. Immun. 62, 1995–2001.

Canessa, A., Pistonia, V., Roncella, S., Merli, A., Melioli, G., Terragna, A., Ferrarini, 1988. An in vitro model for Toxoplasma infection in man. Interaction between CD4_ monoclonal T-cells and macrophages results in killing of trophozoites. J. Immunol. 140, 3580–3588.

Capuron, L., Miller, A.H., 2011. Immune system to brain signalling: neuropsychopharmacological implications. Pharmacol. Ther. 130, 226–238.

Capron, M., et al., 1997. Differentiation of eosinophils from cord blood cell precursors: kinetics of Fc epsilon RI and Fc epsilon RII expression. Int. Arch. Allergy Immunol. 113 (1–3), 48–50.

Carme, B., Bissuel, F., Ajzenberg, D., Bouyne, R., Aznar, C., 2002. Severe acquired toxoplasmosis in immunocompetent adult patients in French Guiana. J. Clin. Microbio. 40 (11), 4037–4044.

Casciotti, L., Ely, K.H., Williams, M.E., Khan, I.A., 2002. $CD8(^+)$-T-cell immunity against *Toxoplasma gondii* can be induced but not maintained in mice lacking conventional $CD4(^+)$ T-cells. Infect. Immun. 2, 434–443.

Catterall, J., Black, C., Leventhal, J.P., et al., 1987. Nonoxidative microbicidal activity in normal human alveolar and peritoneal macrophages. Infect. Immun. 55, 1635–1640.

Catterall, J., Sharma, S., Remington, J., 1986. Oxygen-independent killing by alveolar macrophages. J. Exp. Med. 163, 1113–1131.

Cavailles, P., Bisanz, C., Papapietro, O., Colacios, C., Sergent, V., Pipy, B., Saoudi, A., Cesbron-Delauw, M.F., Fournie, G.J., 2006a. The rat Toxo1 locus control the outcome of the toxoplasmic infection according to a mendelian model. Med. Sci. 8–9, 679–680.

Cavailles, P., Sergent, V., Bisanz, C., Papapietro, O., Colacios, C., Mas, M., Subra, J.F., Lagrance, D., Calise, M., Appolinaire, S., Faraut, T., Druet, P., Saoudi, A., Bessieres, M.H., Pipy, B., Cesbron-Delauw, M.F., Fournie, G.J., 2006b. The rat Toxo1 locus directs toxoplasmosis outcome and controls parasite proliferation and spreading by macrophage-dependent mechanisms. Proc. Natl. Acad. Sci. 103, 744–749.

Cavalcanti, M.G., Mesquita, J.S., Madi, K., Feijó, D.F., Assunção-Miranda, I., Souza, H.S., Bozza, M.T., 2011. MIF participates in Toxoplasma gondii-induced pathology following oral infection. PLoS One 6 (9), e25259. http://dx.doi.org/10.1371/journal.pone.0025259. Epub 2011 Sep 22.

Cesbron-Delauw, M.F., Gendrin, C., Travier, L., Ruffiot, P., Mercier, C., 2008 May. Apicomplexa in mammalian cells: trafficking to the parasitophorous vacuole. Traffic 9 (5), 657–664. http://dx.doi.org/10.1111/j.1600-0854.2008.00728.x.

Ceravolo, I.P., Chaves, A.C., Bonjardim, C.A., Sibley, D., Romanha, A.J., Gazzinelli, R.T., 1999. Replication of *Toxoplasma gondii*, but not Trypanosoma cruzi, is regulated in human fibroblasts activated with gamma interferon: requirement of a functional JAK/STAT pathway. Infect. Immun. 67, 2233–2240.

Chan, A., Hummel, V., Weilbach, F.X., Kieseier, B.C., Gold, R., 2006a. Phagocytosis of apoptotic inflammatory cells down-regulates microglial chemoattractive function and migration of encephalitogenic T-cells. J. Neurosci. Res. 84, 1217–1224.

Chan, C., Crafton, E., Fan, H., Flook, J., Yoshimura, K., Skarica, M., Brockstedt, D., Dubensky, T., Stins, M., Lanier, L., Pardoll, D., Housseau, F., 2006b. Interferon-producing killer dendritic cells provide a link between innate and adaptive immunity. Nat. Med. 12 (2), 207–213.

Chandramohanadas, R., Davis, P.H., Beiting, D.P., Harbut, M.B., Darling, C., Velmourougane, G., Lee, M.Y., Greer, P.A., Roos, D.S., Greenbaum, D.C., 2009 May 8. Apicomplexan parasites co-opt host calpains to facilitate their escape from infected cells. Science 324 (5928), 794–797. http://dx.doi.org/10.1126/science.1171085. Epub 2009 Apr 2.

Chang, H.R., Grau, G.E., Pechere, J.C., 1990. Role of TNF and IL-1 in infections with *Toxoplasma gondii*. Immunology 69 (1), 33–37.

Chang, H.R., Pechere, J.C., 1989. Macrophage oxidative metabolism and intracellular *Toxoplasma gondii*. Microb. Pathog. 7, 37–44.

Channon, J.Y., Kasper, L.H., 1996. Toxoplasma gondii-induced immune suppression by human peripheral blood monocytes: role of gamma interferon. Infect. Immun. 64 (4), 1181–1189.

Channon, J.Y., Seguin, R.M., Kasper, L.H., 2000. Differential infectivity and division of *Toxoplasma gondii* in human peripheral blood leukocytes. Infect. Immun. 68 (8), 4822–4826.

Chao, C.C., Gekker, G., Hu, S., Peterson, P.K., 1994. Human microglial cell defence against *Toxoplasma gondii*. The role of cytokines. J. Immunol. 152, 1246–1252.

Chao, C.C., Anderson, W.R., Hu, S., Gekker, G., Martella, A., Peterson, P.K., 1993. Activated microglia inhibit multiplication of *Toxoplasma gondii* via a nitric oxide mechanism. Clin. Immunol. Immunopathol. 67, 178–183.

Chardes, T., Buzoni Gatel, D., Lepage, A., Bernard, F., Bout, D., 1994. *Toxoplasma gondii* oral infection induces specific cytotoxic CD8alpha/$beta^+$ Thy-1^+ gut intraepithelial lymphocytes, lytic for parasite-infected enterocytes. J. Immunol. 153, 4596–4603.

Chardes, T., Bout, D., 1993. Mucosal immune response in toxoplasmosis. Res. Immunol. 144 (1), 57–60.

Chardes, T., Bourguin, I., Mevelec, M., Dubremetz, J.F., Bout, D., 1990. Antibody response to *Toxoplasma gondii* in the sera, intestinal secretions, and milk from orally infected mice and characterization of target antigens. Infect. Immune. 58, 1240–1246.

Charles, E., Joshi, S., Ash, J.D., Fox, B.A., Farris, A.D., Bzik, D.J., Lang, M.L., Blader, I.J., 2010. CD4 T-cell suppression by cells from Toxoplasma gondii-infected retinas is mediated by surface protein PD-L1. Infect. Immun. 78 (8), 3484–3492.

Charles, E., Callegan, M.C., Blader, I.J., 2007. The SAG1 *Toxoplasma gondii* surface protein is not required for acute ocular toxoplasmosis in mice. Infect. Immun. 75 (4), 2079–2083.

Chaves, A.C., Ceravolo, I.P., Gomes, J.A., Zani, C.L., Romanha, A.J., Gazzinelli, R.T., 2001. IL-4 and IL-13 regulate the induction of indoleamine 2,3-dioxygenase activity and the control of *Toxoplasma gondii* replication in human fibroblasts activated with IFNgamma. Eur. J. Immunol. 31, 333–344.

Check, J.H., Arwitiz, M., Gross, J., Szekeres-Bartho, J., Wu, C.H., 1997. Evidence that the expression of progesterone-induced blocking factor by maternal T-lymphocytes is positively correlated with conception. Am. J. Reprod. Immunol. 38, 6–8.

Chon, S.Y., Hassanain, H.H., Gupta, S.L., 1996. Cooperative role of interferon regulatory factor 1 and p91 (STAT1) response elements in interferon-gamma-inducible expression of human indoleamine 2,3-dioxygenase gene. J. Biol. Chem. 271 (29), 17247–17252.

Chou, D.B., Sworder, B., Bouladoux, N., Roy, C.N., Uchida, A.M., Grigg, M., Robey, P.G., Belkaid, Y., 2012 Jul. Stromal-derived IL-6 alters the balance of myeloerythroid progenitors during *Toxoplasma gondii* infection. J. Leukoc. Biol. 92 (1), 123–131. http://dx.doi.org/10.1189/jlb.1011527.

Chtanova, T., Han, S.J., Schaeffer, M., Van Dooren, G.G., Herzmark, P., Striepen, B., Robey, E.A., 2009. Dynamics of T-cell, antigen-presenting cell, and pathogen interactions during recall responses in the lymph node. Immunity 31, 342–355.

Clark, R.T., Nance, J.P., Noor, S., Wilson, E.H., 2011. T-cell production of matrix metalloproteinases and inhibition of parasite clearance by TIMP-1 during chronic Toxoplasma infection in the brain. ASN Neuro. (1), e00049.

Cobbold, S.P., Castejon, R., Adams, E., Zelenika, D., Graca, L., Humm, S., Waldmann, H., 2004. Induction of $FoxP3^+$ regulatory T-cells in the periphery of T-cell receptor transgenic mice tolerized to transplants. J. Immunol. 172, 6003–6010.

Collazo, C.M., Yap, G.S., Sempowski, G.D., Lusby, K.C., Tessarollo, L., Woude, G.F., Sher, A., Taylor, G.A., 2001. Inactivation of LRG-47 and IRG-47 reveals a family of interferon gamma-inducible genes with essential, pathogen-specific roles in resistance to infection. J. Exp. Med. 194, 181–188.

Collazo, C.M., Yap, G.S., Hieny, S., Caspar, P., Feng, C.G., Taylor, G.A., Sher, A., 2002. The function of gamma interferon-inducible GTP-binding protein

IGTP in host resistance to *Toxoplasma gondii* is Stat1 dependent and requires expression in both hematopoietic and nonhematopoietic cellular compartments. Infect. Immun. 70, 6933–6939.

Combe, C.L., Moretto, M.M., Schwartzman, J.D., Gigley, J.P., Bzik, D.J., Khan, I.A., 2006 Apr 25. Source Lack of IL-15 results in the suboptimal priming of CD4$^+$ T-cell response against an intracellular parasite. Proc. Natl. Acad. Sci. U S A 103 (17), 6635–6640.

Cong, H., Mui, E.J., Witola, W.H., Sidney, J., Alexander, J., Sette, A., Maewal, A., El Bissati, K., Zhou, Y., Suzuki, Y., Lee, D., Woods, S., Sommerville, C., Henriquez, FL., Roberts, CW., McLeod, R., 2012 Jan. *Toxoplasma gondii* HLA-B*0702-restricted GRA7(20-28) peptide with adjuvants and a universal helper T-cell epitope elicits CD8($^+$) T-cells producing interferon-γ and reduces parasite burden in HLA-B*0702 mice. Hum. Immunol. 73 (1), 1–10. http://dx.doi.org/10.1016/j.humimm.2011.10.006.

Cong, H., Mui, E.J., Witola, W.H., Sidney, J., Alexander, J., Sette, A., Maewal, A., McLeod, R., 2010 Dec 3. Human immunome, bioinformatic analyses using HLA supermotifs and the parasite genome, binding assays, studies of human T-cell responses, and immunization of HLA-A*1101 transgenic mice including novel adjuvants provide a foundation for HLA-A03 restricted CD8$^+$T cell epitope based, adjuvanted vaccine protective against *Toxoplasma gondii*. Immunome Res. 6, 12. http://dx.doi.org/10.1186/1745-7580-6-12.

Cong, H., Mui, E.J., Witola, W.H., Sidney, J., Alexander, J., Sette, A., Maewal, A., McLeod, R., 2011 Jan 17. Towards an immunosense vaccine to prevent toxoplasmosis: protective *Toxoplasma gondii* epitopes restricted by HLA-A*0201. Vaccine 29 (4), 754–762. http://dx.doi.org/10.1016/j.vaccine.2010.11.015.

Cong, H., Gu, Q.M., Yin, H.E., Wang, J.W., Zhao, Q.L., Zhou, H.Y., Li, Y., Zhang, J.Q., 2008 Jul 23. Multi-epitope DNA vaccine linked to the A2/B subunit of cholera toxin protect mice against *Toxoplasma gondii*. Vaccine 26 (31), 3913–3921. http://dx.doi.org/10.1016/j.vaccine.2008.04.046.

Cordeiro, C.A., Moreira, P.R., Andrade, M.S., et al., 2008. Interleukin-10 gene polymorphism (-1082G/A) is associated with toxoplasmic retinochoroiditis. Invest. Ophthalmol. Vis. Sci. 49, 1979–1982.

Corrêa, G., Marques da Silva, C., de Abreu Moreira-Souza, A.C., Vommaro, R.C., Coutinho-Silva, R., 2010 Jun. Activation of the P2X(7) receptor triggers the elimination of *Toxoplasma gondii* tachyzoites from infected macrophages. Microbes Infect. 12 (6), 497–504. http://dx.doi.org/10.1016/j.micinf.2010.03.004.

Couper, K., Nielsen, H., Petersen, E., et al., 2003. DNA vaccination with the immunodominant tachyzoite surface antigen (SAG1) protects against adult acquired *Toxoplasma gondii* infection but does not prevent maternofoetal transmission. Vaccine 21, 2813–2820.

Couper, K.N., Roberts, C.W., et al., 2005. Toxoplasma gondii-specific immunoglobulin M limits parasite dissemination by preventing host cell invasion. Infect. Immun. 73 (12), 8060–8068.

Courret, N., Darche, S., Sonigo, P., Milon, G., Buzoni-Gatel, D., Tardieux, I., 2006. CD11c- and CD11b-expressing mouse leukocytes transport single *Toxoplasma gondii* tachyzoites to the brain. Blood 107, 309–316.

Couvreur, J., et al., 1984. Study of a homogeneous series of 210 cases of congenital toxoplasmosisin infants aged 0 to 11 months detected prospectively. Ann. Pediatr. 31, 815–819.

Couvreur, J., Desmonts, G., Girre, J.Y., 1976. Congenital toxoplasmosis in twins: a series of 14 pairs of twins: absence of infection in one twin in two pairs. J. Pediatr. 89 (2), 235–240.

Couvreur, J., Thulliez, P., 1996. Toxoplasmose acquise a localization oculaire ou neurologique. Presse. Med. 25 (9), 438–442.

Couvreur, J., Thulliez, P., Daffos, F., et al., 1993. In utero treatment of toxoplasmic fetopathy with the combination pyrimethamine-sulfadiazine. Foetal Diagn. Ther. 8, 45–50.

Couvreur, J., Thulliez, T., Daffos, F., Aufrant, C., Bompard, Y., et al., 1991. [6 cases of toxoplasmosis in twins]. Ann. Pediatr. (Paris) 38 (2), 63–68.

Craven, M., Egan, C.E., Dowd, S.E., McDonough, S.P., Dogan, B., Denkers, E.Y., Bowman, D., Scherl, E.J., Simpson, K.W., 2012. Inflammation drives dysbiosis and bacterial invasion in murine models of ileal Crohn's disease. PLoS One 7 (7), e41594. http://dx.doi.org/10.1371/journal.pone.0041594.

Crawford, J., Grujic, O., Bruic, E., Czjzek, M., Grigg, M.E., Boulanger, M.J., 2009 Apr 3. Structural characterization of the bradyzoite surface antigen (BSR4) from *Toxoplasma gondii*, a unique addition to the surface antigen glycoprotein 1-related superfamily. J. Biol. Chem. 284 (14), 9192–9198. http://dx.doi.org/10.1074/jbc.M808714200.

Creuzet, C., Robert, F., Roisin, M.P., Van Tan, H., Benes, C., Dupouy-Camet, J., Fagard, R., 1998. Neurons in primary culture are less efficiently infected by *Toxoplasma gondii* than glial cells. Parasitology Research 84, 25–30.

Curiel, T., Krug, E., Purner, M., et al., 1993. Cloned human CD4$^+$ cytotoxic T-lymphocytes specific for *Toxoplasma gondii* lyse tachyzoite-infected target cells. J. Immunol. 151, 2024–2031.

Czar, M.J., Kersh, E.N., Mijares, L.A., Lanier, G., Lewis, J., Yap, G., Chen, A., Sher, A., Duckett, C.S., Ahmed, R., Schwartzberg, P.L., 2001 Jun 19. Altered lymphocyte responses and cytokine production in mice deficient in the X-linked lymphoproliferative disease gene SH2D1A/DSHP/SAP. Proc. Natl. Acad. Sci. U S A 98 (13), 7449–7454.

Da Gama, L.M., Ribeiro-Gomes, F.L., Guimaraes Jr., U., Arnholdt, A.C., 2004. Reduction in adhesiveness to extracellular matrix components, modulation of adhesion molecules and in vivo migration of murine macrophages infected with *Toxoplasma gondii*. Microbes Infect. 6, 1287–1296.

Dai, W., Pan, H., Kwok, O., et al., 1994. Human indoleamine 2,3-dioxygenase inhibits *Toxoplasma gondii* growth in fibroblast cells. J. Interferon. Res. 14, 313–317.

de Alencar, B.C., Persechini, P.M., Haolla, F.A., de Oliveira, G., Silverio, J.C., Lannes-Vieira, J., Machado, A.V., Gazzinelli, R.T., Bruna-Romero, O., Rodrigues, M.M., 2009 Oct. Perforin and gamma interferon expression are required for CD4$^+$ and CD8$^+$ T-cell-dependent protective immunity against a human parasite, Trypanosoma cruzi, elicited by heterologous plasmid DNA prime-recombinant adenovirus 5 boost vaccination. Infect. Immun. 77 (10), 4383–4395. http://dx.doi.org/10.1128/IAI.01459-08.

D'Angelillo, A., De Luna, E., Romano, S., Bisogni, R., Buffolano, W., Gargano, N., Del Porto, P., Del Vecchio, L., Petersen, E., Romano, M.F., 2011 Jun. *Toxoplasma gondii* Dense Granule Antigen 1 stimulates apoptosis of monocytes through autocrine TGF-β signalling. Apoptosis 16 (6), 551–562. http://dx.doi.org/10.1007/s10495-011-0586-0.

Daffos, F., Forestier, F., Capella-Pavlovsky, M., Thulliez, P., Aufrant, C., et al., 1988. Prenatal management of 746 pregnancies at risk for congenital toxoplasmosis. N. Engl. J. Med. 318, 271–275.

Dalton, J.E., Cruickshank, S.M., Egan, C.E., Mears, R., Newton, D.J., Andrew, E.M., Lawrence, B., Howell, G., Else, K.J., Gubbels, M.J., Striepen, B., Smith, J.E., White, S.J., Carding, S.R., 2006. Intraepithelial gammadelta$^+$ lymphocytes maintain the integrity of intestinal epithelial tight junctions in response to infection. Gastroenterology (3), 818–829.

Dannemann, B.R., et al., 1989. Assessment of human natural killer and lymphokine-activated killer cell cytotoxicity against *Toxoplasma gondii* trophozoites and brain cysts. J. Immunol. 143 (8), 2684–2691.

Dardé, M.L., Bouteille, B., Pestre-Alexandre, M., 1992. Isoenzyme analysis of 35 *Toxoplasma gondii* isolates and the biological epidemiological implications. J. Parasitol. 78, 786–794.

Darde, M.L., 1996. Biodiversity in *Toxoplasma gondii*. Curr. Top. Microbiol. Immunol. 219, 27–41.

Darde, M.L., Villena, I., Pinon, J.M., Beguinot, I., 1998. Severe toxoplasmosis caused by a *Toxoplasma gondii* strain with a new isoenzyme type acquired in French Guyana. J. Clin. Microbiol. 36 (1), 324.

Dass, S.A., Vasudevan, A., Dutta, D., Soh, L.J., Sapolsky, R.M., Vyas, A., 2011. Protozoan parasite *Toxoplasma gondii* manipulates mate choice in rats by enhancing attractiveness of males. PLoS One 6 (11), e27229. http://dx.doi.org/10.1371/journal.pone.0027229.

Daubener, W., Spors, B., Hucke, C., Adam, R., Stins, M., Kim, K.S., Schroten, H., 2001. Restriction of *Toxoplasma gondii* growth in human brain microvascular endothelial cells by activation of indoleamine 2,3-dioxygenase. Infect. Immun. 69, 6527–6531.

Daubener, W., Posdziech, V., Hadding, U., Mackenzie, C.R., 1999. Inducible antiparasitic effector mechanisms in human uroepithelial cells: tryptophan degradation vs. NO production. Med. Microbiol. Immunol. 187, 143–147.

Debierre-Grockiego, F., Campos, M.A., Azzouz, N., Schmidt, J., et al., 2012. Activation of TLR2 and TLR4 by glycosylphosphatidylinositols derived from *Toxoplasma gondii*. The Journal of Immunology 179, 1129–1137.

Debierre-Grockiego, F., Azzouz, N., Schmidt, J., Dubremetz, J.F., Geyer, H., Geyer, R., Weingart, R., Schmidt, R.R., Schwarz, R.T., 2003. Roles of glycosylphosphatidylinositols of *Toxoplasma gondii*. Induction of tumour necrosis factor-alpha production in macrophages. J. Biol. Chem. 278, 32987–32993.

Debierre-Grockiego, F., Campos, M.A., Azzouz, N., Schmidt, J., Bieker, U., Resende, M.G., Mansur, D.S., Weingart, R., Schmidt, R.R., Golenbock, D.T., Gazzinelli, R.T., Schwarz, R.T., 2007a. Activation of TLR2 and TLR4 by glycosylphosphatidylinositols derived from *Toxoplasma gondii*. J. Immunol. 179 (2), 1129–1137.

Debierre-Grockiego, F., Hippe, D., Schwarz, R.T., Lüder, C.G., 2007b. *Toxoplasma gondii* glycosylphosphatidylinositols are not involved in T. gondii-induced host cell survival. Apoptosis 4, 781–790.

Debierre-Grockiego, F., Molitor, N., Schwarz, R.T., Lüder, C.G., 2009. *Toxoplasma gondii* glycosylphosphatidylinositols up-regulate major histocompatibility complex (MHC) molecule expression on primary murine macrophages. Innate. Immun. (1), 25–32.

Debierre-Grockiego, F., Niehus, S., Coddeville, B., Elass, E., Poirier, F., Weingart, R., Schmidt, R.R., Mazurier, J., Guérardel, Y., Schwarz, R.T., 2010 Oct 22. Binding of *Toxoplasma gondii* glycosylphosphatidylinositols to galectin-3 is required for their recognition by macrophages. J. Biol. Chem. 285 (43), 32744–32750. http://dx.doi.org/10.1074/jbc.M110.137588. Epub 2010 Aug 20.

Deckert-Schluter, M., et al., 1994. Activation of the innate immune system in murine congenital Toxoplasma encephalitis. J. Neuroimmunol. 53 (1), 47–51.

Deckert-Schluter, M., Albrecht, S., Hof, H., Wiestler, O.D., Schluter, D., 1995. Dynamics of the intracerebral and splenic cytokine mRNA production in Toxoplasma gondii-resistant and -susceptible congenic strains of mice. Immunology 85, 408.

Deckert-Schluter, M., Schluter, D., Hof, H., Wiestler, O.D., Lassmann, H., 1994. Differential expression of ICAM-1, VCAM-1 and their ligands LFA-1, Mac-1, CD43, VLA-4, and MHC class II antigens in murine Toxoplasma encephalitis: a light microscopic and ultrastructural immunohistochemical study. J. Neuropathol. Exp. Neurol. 53 (5), 457–468.

Deckert-Schluter, M., Rang, A., Weiner, D., Huang, S., Wiestler, O.D., Hof, H., Schluter, D., 1996. Interferon-gamma receptor-deficiency renders mice highly susceptible to toxoplasmosis by decreased macrophage activation. Lab. Invest. 75, 827–841.

Deckert-Schluter, M., Bluethmann, H., Kaefer, N., Rang, A., Schluter, D., 1999. Interferon-gamma receptor-mediated but not tumour necrosis factor receptor type 1- or type 2-mediated signalling is crucial for the activation of cerebral blood vessel endothelial cells and microglia in murine Toxoplasma encephalitis. Am. J. Pathol. 154 (5), 1549–1561. May.

Deckert-Schluter, M., Bluethmann, H., Rang, A., Hof, H., Schluter, D., 1998a. Crucial role of TNF receptor type 1 (p55), but not of TNF receptor type 2 (p75), in murine toxoplasmosis. J. Immunol. 160, 3427–3436.

Deckert-Schluter, M., Buck, C., Schluter, D., 1998b. Kinetics and differential expression of heat-stable antigen and GL7 in the normal and Toxoplasma gondii-infected murine brain. Acta Neuropathol. 98, 97–106.

Deckert, M., Sedgwick, J.D., Fischer, E., Schluter, D., 2006. Regulation of microglial cell responses in murine Toxoplasma encephalitis by CD200/CD200 receptor interaction. Acta Neuropathol. 111, 548–558.

Decoster, A., 1996. Detection of IgA anti-P30 SAG1 antibodies in acquired and congenital toxoplasmosis. Curr. Top. Microbiol. Immunol. 219, 199–207.

Dellacasa-Lindberg, I., Fuks, J.M., Arrighi, R.B., Lambert, H., Wallin, R.P., Chambers, B.J., Barragan, A., 2011. Migratory activation of primary cortical microglia upon infection with *Toxoplasma gondii*. Infect. Immun. 79, 3046–3052.

Dellacasa-Lindberg, I., Hitziger, N., Barragan, A., 2007 Sep. Localized recrudescence of Toxoplasma infections in the central nervous system of immunocompromised mice assessed by in vivo bioluminescence imaging. Microbes Infect. 9 (11), 1291–1298.

Del Rio, L., Butcher, B.A., Bennouna, S., Hieny, S., Sher, A., Denkers, E.Y., 2004. *Toxoplasma gondii* triggers myeloid differentiation factor 88-dependent IL-12 and chemokine ligand 2 (monocyte chemoattractant protein 1) responses using distinct parasite molecules and host receptors. J. Immunol. 172, 6954–6960.

Del Rio, L., Bennouna, S., Salinas, J., Denkers, E.Y., 2001a. CXCR2 deficiency confers impaired neutrophil recruitment and increased susceptibility during *Toxoplasma gondii* infection. J. Immunol. 167 (11), 6503–6509.

Del Rio, L., Bennouna, S., Salinas, J., Denkers, E.Y., 2001a. CXCR2 deficiency confers impaired neutrophil recruitment and increased susceptibility during *Toxoplasma gondii* infection. J. Immunol. 167 (11), 6503–6509.

De Paoli, P., Basaglia, G., Gennari, D., et al., 1992. Phenotypic profile and functional characteristics of human gamma and delta T-cells during acute toxoplasmosis. J. Clin. Microbiol. 30, 729–731.

Denkers, E.Y., et al., 1993a. Emergence of NK1.1$^+$ cells as effectors of IFNgamma dependent immunity to *Toxoplasma gondii* in MHC class I-deficient mice. J. Exp. Med. 178 (5), 1465–1472.

Denkers, E.Y., Sher, A., Gazzinelli, 1993c. CD8$^+$ T-cell interactions with toxoplasma gondii: implications for processing of antigen for class I-restricted recognition. Res. Immunol. 144, 51–56.

Denkers, E.Y., Del Rio, L., Bennouna, S., 2003a. Neutrophil production of IL-12 and other cytokines during microbial infection. Chem. Immunol. Allergy 83, 95–114.

Denkers, E.Y., Gazzinelli, R., Hieny, S., Caspar, P., Sher, A., 1993b. Bone marrow macrophages process exogenous toxoplasma gondii peptides for recognition by parasite-specific cytolytic T-lymphocytes. J. Immunol. 150, 517–526.

Denkers, E.Y., Caspar, P., Sher, A., 1994. *Toxoplasma gondii* possesses a superantigen activity that selectively expands murine T-cell receptor Vbeta5-bearing CD8$^+$ lymphocytes. J. Exp. Med. 180, 985–995.

Denkers, E.Y., Scharton-Kersten, T., Barbieri, S., Caspar, P., Sher, A., 1996. A role for Cd4_NK1.1$^+$ T-lymphocytes as MHC class II independent helper cells in the generation of CD8$^+$ effector function against intracellular infection. J. Exp. Med. 184, 131–139.

Denkers, E.Y., 1996. A *Toxoplasma gondii* Superantigen: Biological Effects and Implications for the Host-Parasite Interaction 12 (9), 362–366.

Denkers, E.Y., Scharton-Kersten, Gazzinelli R., Yap, G., Charest, H., Sher, A., 1997c. Cell-mediated immunity to *Toxoplasma gondii*: redundant and required mechanisms as revealed by studies in gene knock-out mice. In: Kaufmann, S.H.E. (Ed.), Medical intelligence unit: host response to intracellular pathogens. R.G. Landes Co, Austin, Tex, pp. 167–181.

Denkers, E.Y., Yap, G., Scharton-Kersten, T., Charest, H., Butcher, B.A., Caspar, P., Heiny, S., Sher, A., 1997a. Perforin-mediated cytolysis plays a limited role in host resistance to *Toxoplasma gondii*. J. Immunol. 159, 1903–1908.

Denkers, E., Sher, A., 1997b. Role of natural killer and NK1_ T-cells in regulating cell-mediated immunity during *Toxoplasma gondii* infection. Biochem. Soc. Trans. 25, 699–703.

Denkers, E.Y., Gazzinelli, R.T., 1998. Regulation and function of T-cell-mediated immunity during *Toxoplasma gondii* infection. Clin. Microbiol. Rev. 11, 569–588.

Denkers, E.Y., Kim, L., Butcher, B.A., 2003b. In the belly of the beast: subversion of macrophage proinflammatory signalling cascades during *Toxoplasma gondii* infection. Cell Microbiol. 5, 75–83.

Denkers, E.Y., Butcher, B.A., 2005. Sabotage and exploitation in macrophages parasitized by intracellular protozoans. Trends Parasitol. 21 (1), 35–41.

Denkers, E.Y., Schneider, A.G., Cohen, S.B., Butcher, B.A., 2012a. Phagocyte responses to protozoan infection and how *Toxoplasma gondii* meets the challenge. PLoS Pathog. 8 (8), e1002794. http://dx.doi.org/10.1371/journal.ppat.1002794.

Denkers, E.Y., Bzik, D.J., Fox, B.A., Butcher, B.A., 2012b. An inside job: hacking into Janus kinase/signal transducer and activator of transcription signalling cascades by the intracellular protozoan *Toxoplasma gondii*. Infect. Immun. 80 (2), 476–482. http://dx.doi.org/10.1128/IAI.05974-11.

Denkers, E.Y., 2009 Aug 20. A gut feeling for microbes: getting it going between a parasite and its host. Cell Host Microbe. 6 (2), 104–106. http://dx.doi.org/10.1016/j.chom.2009.07.009.

Denkers, E.Y., Striepen, B., Cell Host Microbe, 2008. Deploying parasite profilin on a mission of invasion and danger 2, 61–63. http://dx.doi.org/10.1016/j.chom.2008.01.003.

Denkers, E.Y., 2010. Toll-like receptor initiated host defence against *Toxoplasma gondii*. J. Biomed. Biotechnol. http://dx.doi.org/2010:737125. doi: 10.1155/2010/737125.

De Paoli, Basaglia, G., Gennari, D., Crovatto, M., Modolo, M., Santini, G., 1992. Phenotypic profile and functional characteristics of human gamma and delta T-cells during murine acute toxoplasmosis. J. Clin. Microbiol. 30, 729–731.

Desmonts, G., Couvreur, J., 1984. Natural history of congenital toxoplasmosis. Ann. Pediatr. (Paris) 31 (10), 799–802.

Desmonts, G., Couvreur, J., 1974. Congenital toxoplasmosis: a prospective study of 378 pregnancies. N. Engl. J. Med. 290, 1110–1116.

Desmonts, G., 1966. Definitive serological diagnosis of ocular toxoplasmosis. Arch. Ophthalmol. 76, 839–851.

Desmonts, G., Naot, Y., Remington, J.S., 1981. Immunoglobulin M-immunosorbent agglutination assay for diagnosis of infectious diseases: diagnosis of acute congenital and acuired Toxoplasma infections. J. Clin. Microbiol. 14, 486–491.

Desmonts, G., Remington, J.S., 1980. Direct agglutination test for diagnosis of Toxoplasma infection: method for increasing sensitivity and specificity. J. Clin. Microbiol. 11, 562–568.

De Waele, M., 1985. Immunoregulatory T-cells in acute toxoplasmosis. J. Infect. Dis. 152 (2), 424.

Diana, J., Persat, F., Staquet, M.J., Assossou, O., Ferrandiz, J., Gariazzo, M.J., Peyron, F., Picot, S., Schmitt, D., Vincent, C., 2004. Migration and maturation of human dendritic cells infected with *Toxoplasma gondii* depend on parasite strain type. FEMS Immunol. Medical Microbiol. 42, 321–331.

Dimier, I.H., Bout, D.T., 1998. Interferon-gamma-activated primary enterocytes inhibit *Toxoplasma gondii* replication: a role for intracellular iron. Immunology 94, 488–495.

Dimier, I.H., Bout, D.T., 1997. Inhibition of *Toxoplasma gondii* replication in IFNgamma-activated human intestinal epithelial cells. Immunol. Cell Biol. 75, 511–514.

Ding, M., Kwok, L.Y., Schluter, D., Clayton, C., Soldati, D., 2004. The antioxidant systems in *Toxoplasma gondii* and the role of cytosolic catalase in defence against oxidative injury. Mol. Microbiol. 51, 47–61.

Divanovic, S., Sawtell, N.M., Trompette, A., Warning, J.I., Dias, A., Cooper, A.M., Yap, G.S., Arditi, M., Shimada, K., Duhadaway, J.B., Prendergast, G.C., Basaraba, R.J., Mellor, A.L., Munn, D.H., Aliberti, J., Karp, C.L., 2012. Opposing biological functions of tryptophan catabolizing enzymes during intracellular infection. J. Infect. Dis. 205, 152–161.

Dobbin, C.A., Smith, N.C., Johnson, A.M., 2002. Heat shock protein 70 is a potential virulence Factor in Murine Toxoplasma Infection via Immunomodulation of Host NF-KB and Nitric Oxide. J. Immunol. 169, 958–965.

Dobrowolski, J.M., Sibley, L.D., 1996. Toxoplasma invasion of mammalian cells is powered by the actin cytoskeleton of the parasite. Cell 84 (6), 933–939.

Doğu, F., Ikincioğullari, A., Fricker, D., Bozdoğan, G., Aytekin, C., Ileri, M., Teziç, T., Babacan, E., De La Salle, H., 2006. A novel mutation for TAP deficiency and its possible association with Toxoplasmosis. Parasitol. Int. 3, 219–222.

Drogemuller, K., Helmuth, U., Brunn, A., Sakowicz-Burkiewicz, M., Gutmann, D.H., Mueller, W., Deckert, M., Schluter, D., 2008. Astrocyte gp130 expression is critical for the control of Toxoplasma encephalitis. J. Immunol. 181, 2683–2693.

Dubey, J.P., Sharma, S., Juranek, D.D., Sulzer, A.J., Teutsch, S.M., 1981. Characterization of *Toxoplasma gondii* isolates from an outbreak of toxoplasmosis in Atlanta, Georgia. Am. J. Vet. Res. 42 (6), 1007–1010.

Dubey, J.P., Beattie, C.P., 1988. Toxoplasmosis of animals and man. CRC Press, Boca Raton.

Dubey, J.P., Graham, D.H., Blackston, C.R., Lehmann, T., Gennari, S.M., Ragozo, A.M.A., Nishi, S.M., K.SS., Kwok, O.C.H., Hill, D.E., Thulliez, P., 2002. Biological and genetic characterization of *Toxoplasma gondii* isolates from chickens (Gallus domesticus) from São Paulo, Brazil: unexpected findings. Int. J. Parasitol. 32, 99–105.

Dubey, J.P., Graham, D.H., da Silva, D.S., Lehmann, T., Bahia-Oliveira, L.M., 2003a. *Toxoplasma gondii* isolates of free-ranging chickens from Rio de Janeiro, Brazil: mouse mortality, genotype, and oocyst shedding by cats. J. Parasitol. 89, 851–853.

Dubey, J.P., Venturini, M.C., Venturini, L., Piscopo, M., Graham, D.H., Dahl, E., Sreekuman, C., Vianna, M.C., Lehmann, T., 2003b. Isolation and genotyping of *Toxoplasma gondii* from free-ranging chickens from Argentina. J. Parasitol. 89, 1063–1064.

Dubey, J.P., Karhemere, S., Dahl, E., Sreekumar, C., Diabate, A., Dabire, K.R., Vianna, M.C., Kwok, M.C., Lehmann, T., 2005a. First biological and genetic characterization of *Toxoplasma gondii* isolates from chickens from Africa (Democratic Republic of Congo, Mali, Burkina Faso, and Kenya). J. Parasitol. 91, 69–72.

Dubey, J.P., Gomez-Marin, J.E., Bedoya, A., Lora, F., Vianna, M.C., Hill, D., Kwok, O.C., Shen, S.K., Marcet, P.L., Lehmann, T., 2005b. Genetic and biologic characteristics of *Toxoplasma gondii* isolates in free-ranging chickens from Africa (Democratic Republic of Congo, Mali, Burkina Faso, and Kenya). J. Parasitol. 91, 69–72.

Dubey, J.P., Graham, D.H., Dahl, E., Hilali, M., El-Ghaysh, A., Streekumar, C., Kwok, O.C.H., Shen, S.K., Lehmann, T., 2003c. Isolation and molecular characteristics of *Toxoplasma gondii* from chickens and ducks from Egypt. Vet. Parasitol. 114, 89–95.

Dubey, J.P., Navarro, I.T., Sreekumar, C., Dahl, E., Freire, R.L., Kawabata, H.H., Vianna, M.C., Kwok, O.C.H., Shen, S.K., Thulliez, P., Lehmann, T., 2004. *Toxoplasma gondii* infections in cats from Paraná, Brazil: seroprevalence, tissue distribution, and biologic and genetic characterization of isolates. J. Parasitol. 90, 721–726.

Dubey, J.P., 1997. Bradyzoite-induced murine toxoplasmosis: stage conversion, pathogenesis, and tissue cyst formation in mice fed bradyzoites of different strains of *Toxoplasma gondii*. J. Eukaryot. Microbiol. 44 (6), 592–602.

Dubey, J.P., Speer, C.A., Shen, S.K., Kwok, O.C., Blixt, J.A., 1997. Oocyst-induced murine toxoplasmosis: life cycle, pathogenicity, and stage conversion in mice fed *Toxoplasma gondii* oocysts. J. Parasitol. 83 (5), 870–882.

Dunay, I.R., Sibley, L.D., 2010 Aug. Monocytes mediate mucosal immunity to *Toxoplasma gondii*. Curr. Opin. Immunol. 22 (4), 461–466. http://dx.doi.org/10.1016/j.coi.2010.04.008.

Dunay, I.R., Fuchs, A., Sibley, L.D., 2010 Apr. Inflammatory monocytes but not neutrophils are necessary to control infection with *Toxoplasma gondii* in mice. Infect. Immun. 78 (4), 1564–1570. http://dx.doi.org/10.1128/IAI.00472-09.

Dunay, I.R., Damatta, R.A., Fux, B., Presti, R., Greco, S., Colonna, M., Sibley, L.D., 2008. Gr1($^+$) inflammatory monocytes are required for mucosal resistance to the pathogen *Toxoplasma gondii*. Immunity 2, 306–317.

Dunn, J.D., Ravindran, S., Kim, S.K., Boothroyd, J.C., 2008. The *Toxoplasma gondii* dense granule protein GRA7 is phosphorylated upon invasion and forms an unexpected association with the rhoptry proteins ROP2 and ROP4. Infect. Immun. 76 (12), 5853–5861. http://dx.doi.org/10.1128/IAI.01667-07.

Dupont, C.D., Christian, D.A., Hunter, C.A., 2012. Immune response and immunopathology during toxoplasmosis. Semin. Immunopathol. 34 (6), 793–813. http://dx.doi.org/10.1007/s00281-012-0339-3.

Dupont, C.D., Hunter, C.A., 2012. Guanylate-binding proteins: niche recruiters for antimicrobial effectors. Immunity 37 (2), 191–193. http://dx.doi.org/10.1016/j.immuni.2012.08.005.

Dutra, Miriam S., Bela, Samatha R., Peixoto-Rangel, Alba, et al., 2012. Association of a NOD2 Gene Polymorphism and T-helper 17 cells with Presumed Ocular Toxoplasmosis. J. Infect. Dis. 207, 152–163.

Dutton, G.N., 1989. Toxoplasmic retinochorioditis. A historical review and current concepts. Ann. Acad. Med. Singap. 18 (2), 216–220.

Dutton, G.N., McMenamin, P.G., et al., 1986. 'The ultrastructural pathology of congenital murine toxoplasmic retinochoroiditis. Part II: The morphology of the inflammatory changes'. Exp. Eye Res. 43 (4), 545–560.

Dzierszinski, F., Mortuaire, M., Cesbron-Delauw, M.F., Tomavo, S., 2000. Targeted disruption of the glycosylphosphatidylinositol-anchored surface antigen SAG3 gene in *Toxoplasma gondii* decreases host cell adhesion and drastically reduces virulence in mice. Mol. Microbiol. 37 (3), 574–582.

Dzierszinski, F., Pepper, M., Stumhofer, J.S., LaRosa, D.F., Wilson, E.H., Turka, L.A., Halonen, S.K., Hunter, C.A., Roos, D.S., 2007 Nov. Presentation of *Toxoplasma gondii* antigens via the endogenous major histocompatibility complex class I pathway in nonprofessional and professional antigen-presenting cells. Infect. Immun. 75 (11), 5200–5209.

Edelson, B., Unanue, E., 2000. Immunity to Listeria infection. Curr. Opin. Immunol. 12, 425–431.

Egan, C.E., Dalton, J.E., Andrew, E.M., Smith, J.E., Gubbels, M.J., Striepen, B., Carding, S.R., 2005 Dec 15. A requirement for the Vgamma1$^+$ subset of peripheral gammadelta T-cells in the control of the systemic growth of *Toxoplasma gondii* and infection-induced pathology. J. Immunol. 175 (12), 8191–8199.

Egan, C.E., Cohen, S.B., Denkers, E.Y., 2012 Aug. Insights into inflammatory bowel disease using *Toxoplasma gondii* as an infectious trigger. Immunol. Cell Biol. 90 (7), 668–675. http://dx.doi.org/10.1038/icb.2011.93.

Egan, C.E., Sukhumavasi, W., Bierly, A.L., Denkers, E.Y., 2008. Understanding the multiple functions of Gr-1($^+$) cell subpopulations during microbial infection. Immunol. Res. 40 (1), 35–48. http://dx.doi.org/10.1007/s12026-007-0061-8.

Egan, C.E., Sukhumavasi, W., Butcher, B.A., Denkers, E.Y., 2009a. Functional aspects of Toll-like receptor/MyD88 signalling during protozoan infection: focus on *Toxoplasma gondii*. Clin. Exp. Immunol. 156, 17–24.

Egan, C.E., Craven, M.D., Leng, J., Mack, M., Simpson, K.W., Denkers, E.Y., 2009b. CCR2-dependent intraepithelial lymphocytes mediate inflammatory gut pathology during *Toxoplasma gondii* infection. Mucosal Immunol. 2 (6), 527–535. http://dx.doi.org/10.1038/mi.2009.105. Epub 2009 Sep 9.

Egan, C.E., Maurer, K.J., Cohen, S.B., Mack, M., Simpson, K.W., Denkers, E.Y., 2011 Nov. Synergy between intraepithelial lymphocytes and lamina propria T-cells drives intestinal inflammation during infection. Mucosal Immunol. 4 (6), 658–670. http://dx.doi.org/10.1038/mi.2011.31.

Eisenhauer, P., Mack, D., McLeod, R., 1988. Prevention of peroral and congenital acquisition of *Toxoplasma gondii* by antibody and activated macrophages. Infect. Immun. 56, 83–87.

El Kasmi, K.C., Qualls, J.E., Pesce, J.T., Smith, A.M., Thompson, R.W., Henao-Tamayo, M., Basaraba, R.J., König, T., Schleicher, U., Koo, M.S., Kaplan, G., Fitzgerald, K.A., Tuomanen, E.I., Orme, I.M., Kanneganti, T.D., Bogdan, C., Wynn, T.A., Murray, P.J., 2008 Dec. Toll-like receptor-induced arginase 1 in macrophages thwarts effective immunity against intracellular pathogens. Nat. Immunol. 9 (12), 1399–1406. http://dx.doi.org/10.1038/ni.1671.

Ellis Neyer, L., Remington, J.S., Suzuki, Y., 1998. Mesenteric lymph node T-cells but not splenic T-cells maintain their proliferative response to Concanavalin A following peroral infection with *Toxoplasma gondii*. Parasite Immunol. 20, 573–581.

Ely, K.H., Kasper, L.H., Khan, I.A., 1999. Augmentation of the $CD8^+$ T-cell response by IFNgamma in IL-12-deficient mice during *Toxoplasma gondii* infection. J. Immunol. 162 (9), 544. May 1.

European Collaborative Study and Research Network on Congenital Toxoplasmosis. 1996. Low incidence of congenital toxoplasmosis in children born to women infected with human immunodeficiency virus. Eur. J. Obstet Gynecol. Reprod. Biol. 68, 93–96.

Fagard, R., Van Tan, H., Creuzet, C., Pelloux, H., 1999. Differential development of *Toxoplasma gondii* in neural cells. Parasitology Today 15, 504–507.

Fang, H., Mun, H.S., Kikumura, A., Sayama, Y., Norose, K., Yano, A., Aosai, F., 2008 Jul. Toxoplasma gondii-derived heat shock protein 70 induces lethal anaphylactic reaction through activation of cytosolic phospholipase A2 and platelet-activating factor via Toll-like receptor 4/myeloid differentiation factor 88. Microbiol. Immunol. 52 (7), 366–374. http://dx.doi.org/10.1111/j.1348-0421.2008.00047.x.

Fardeau, C., Romand, S., Rao, N., Cassoux, N., Bettembourg, O., 2002. Diagnosis of toxoplasmic retinohoroiditis with atypical clinical features. Am. J. Ophthalmol. 134 (2), 196–203.

Farquhar, H.G., et al., 1950. Congenital toxoplasmosis: Report of two cases in twins. Lancet 2, 562–564.

Fatoohi, A., Cozon, G., Wallon, M., et al., 2003. Cellular immunity to *Toxoplasma gondii* in congenitally infected newborns and immunocompetent infected hosts. Eur. J. Clin. Microbiol. Infect. Dis. 22, 181–184.

Faust, Z., Laskarin, G., Rukavina, D., Szekeres-Bartho, J., 1999. Progesterone-induced blocking factor inhibits degranulation of natural killer cells. Am. J. Reprod. Immunol. 42, 71–75.

Fazaelia, A., Carter, P.E., Dardé, M.L., Pennington, T.H., 2000. Molecular typing of *Toxoplasma gondii* strains by GRA6 gene sequence analysis. Int. J. Parasitol. 30, 637–642.

Fehervari, Z., Yamaguchi, T., Sakaguchi, S., 2006. The dichotomous role of IL-2: tolerance versus immunity. Trends in Immunol. 27, 109–111.

Feng, C.G., Weksberg, D.C., Taylor, G.A., Sher, A., Goodell, M.A., 2008a. The p47 GTPase Lrg-47 (Irgm1) links host defence and hematopoietic stem cell proliferation. Cell Stem Cell 2, 83–89.

Feng, C.G., Zheng, L., Lenardo, M.J., Sher, A., 2009. Interferon-inducible immunity-related GTPase Irgm1 regulates IFN gamma-dependent host defence, lymphocyte survival and autophagy. Autophagy 5, 232–234.

Feng, C.G., Sher, A., 2010. Parasites paralyze cellular host defence system to promote virulence. Cell Host Microbe 8 (6), 463–464.

Feng, C.G., Zheng, L., Jankovic, D., Báfica, A., Cannons, J.L., Watford, W.T., Chaussabel, D., Hieny, S., Caspar, P., Schwartzberg, P.L., Lenardo, M.J., Sher, A., 2008b. The immunity-related GTPase Irgm1 promotes the expansion of activated $CD4^+$ T-cell populations by preventing interferon-gamma-induced cell death. Nat. Immunol. 9 (11), 1279–1287. http://dx.doi.org/10.1038/ni.1653.

Fentress, S.J., Behnke, M.S., Dunay, I.R., Mashayekhi, M., Rommereim, L.M., Fox, B.A., Bzik, D.J., Taylor, G.A., Turk, B.E., Lichti, C.F., Townsend, R.R., Qiu, W., Hui, R., Beatty, W.L., Sibley, L.D., 2010 Dec 16. Phosphorylation of immunity-related GTPases by a *Toxoplasma gondii*-secreted kinase promotes macrophage survival and virulence. Cell Host Microbe 8 (6), 484–495. http://dx.doi.org/10.1016/j.chom.2010.11.005.

Ferreira, A.M., A, Vitor, R.W., Carneiro, ACAV, Brandão, G.P., Melo, M.N., 2004. Genetic variability of Brazilian *Toxoplasma gondii* strains detected by random amplified polymorphic DNA-polymerase chain reaction (RAPD-PCR) and simple sequence repeat anchored-PCR (SSR-PCR). Infect. Genet. Evol. 4, 131–142.

Ferreira, A.M., Vitor, R.W., Gazzinelli, R.T., Melo, M.N., 2005. Genetic analysis of natural recombinant Brazilian *Toxoplasma gondii* strains by multilocus PCR-RFLP. Infect. Genet. Evol. (in press).

Ferreira da Silva Mda, F., Barbosa, H.S., Gross, U., Lüder, C.G., 2008 Aug. Stress-related and spontaneous stage differentiation of *Toxoplasma gondii*. Mol. Biosyst. 4 (8), 824–834. http://dx.doi.org/10.1039/b800520f. Epub 2008.

Ferreira-da-Silva Mda, F., Takács, A.C., Barbosa, H.S., Gross, U., Lüder, C.G., 2009. Primary skeletal muscle cells trigger spontaneous *Toxoplasma gondii* tachyzoite-to-bradyzoite conversion at higher rates than fibroblasts. Int. J. Med. Microbiol. 5, 381–388.

Fiorentino, D.F., Zlotnik, A., Mosmann, T.R., Howard, M., O'Garra, A., 1991. IL-10 inhibits cytokine production by activated macrophages. J. Immunol. 147, 3815–3822.

Fleckenstein, M.C., Reese, M.L., Könen-Waisman, S., Boothroyd, J.C., Howard, J.C., Steinfeldt, T., 2012. A *Toxoplasma gondii* pseudokinase inhibits host IRG resistance proteins. PLoS Biol. 10 (7), e1001358. http://dx.doi.org/10.1371/journal.pbio.1001358.

Foulon, W., et al., 1990. Detection of congenital toxoplasmosis by chorionic villus sampling and early amniocentesis. Am. J. Obstet. Gynecol. 163 (5 Pt 1), 1511–1513.

Foureau, D.M., Mielcarz, D.W., Menard, L.C., Schulthess, J., Werts, C., Vasseur, V., Ryffel, B., Kasper, L.H., Buzoni-Gatel, D., 2010. TLR9-dependent induction of intestinal alpha-defensins by *Toxoplasma gondii*. J. Immunol. 184 (12), 7022–7029. http://dx.doi.org/10.4049/jimmunol.0901642.

Frankel, M.B., Mordue, D.G., Knoll, L.J., 2007. Discovery of parasite virulence genes reveals a unique regulator of chromosome condensation 1 ortholog critical for efficient nuclear trafficking. Proc. Natl. Acad. Sci. U S A 104 (24), 10181–10186.

Franzoso, G., Carlson, L., Poljak, L., Shores, E.W., Epstein, S., Leonardi, A., Grinberg, A., Tran, T., Scharton-Kersten, T., Anver, M., Love, P., Brown, K., Siebenlist, U., 1998. Mice deficient in nuclear factor (NF)-kappa B/p52 present with defects inhumoral responses, germinal center reactions, and splenic microarchitecture. J. Exp. Med. 187 (2), 147–159. Jan 19.

Frenkel, J.K., 1967. Adoptive immunity to intracellular infection. J. Immunol. 98, 1309–1319.

Frenkel, J.K., 1988. Pathophysiology of toxoplasmosis. Parasitol. Today 4, 273–278.

Frenkel, J.K., Caldwell, S.A., 1975. Specific immunity and nonspecific resistance to infection: listeria, protozoa, and viruses in mice and hamsters. J. Infect. Dis. 131, 201–209.

Frenkel, J., Taylor, D., 1982. Toxoplasmosis in immunoglobulin M-suppressed mice. Infect. Immun. 38, 360–367.

Frenkel, J.K., 1990. Toxoplasmosis in human beings. JAVMA 196, 240–248.

Frenkel, J.K., Amare, M., Larsen, W., 1978. Immune competence in a patient with Hodgkin's disease and relapsing toxoplasmosis. Infection 6 (2), 84–91.

Frenkel, J.K., 1955. Ocular lesions in hamsters with chronic Toxoplasma and Besnoita infection. Am. J. Ophthalmol. 39, 203–225.

Frenkel, J.K., Escajadillo, A., 1987. Cyst rupture as a pathogenic mechanism of toxoplasmic encephalitis. Am. J. Trop. Med. Hyg. 36, 517–522.

Freund, Y., Sgarlato, G., Jacob, C.O., Suzuki, Y., Remington, J.S., 1992. Polymorphisms in the tumour necrosis factor alpha (TNF-a) gene correlate with murine resistance to development of toxoplasmic encephalitis and with levels of TNF-a mRNA in infected brain tissue. J. Exp. Med. 175, 683–688.

Frickel, E.M., Sahoo, N., Hopp, J., Gubbels, M.J., Craver, M.P., Knoll, L.J., Ploegh, H.L., Grotenbreg, G.M., 2008. Parasite stage-specific recognition of endogenous Toxoplasma gondii-derived CD8$^+$ T-cell epitopes. J. Infect. Dis. 198, 1625–1633.

Fujigaki, S., Saito, K., Takemura, M., Maekawa, N., Yamada, Y., Wada, H., Seishima, M., 2002. L-tryptophan-L-kynurenine pathway metabolism accelerated by *Toxoplasma gondii* infection is abolished in gamma interferon-gene-deficient mice: cross-regulation between inducible nitric oxide synthase and indoleamine-2,3-dioxygenase. Infect. Immun. 70, 779–786.

Ferguson, D.J., Hutchison, W.M., 1987a. The host-parasite relationship of *Toxoplasma gondii* in the brains of chronically infected mice. Virchows Arch. A Pathol. Anat. Histopathol. 411 (1), 39–43.

Ferguson, D.J., Hutchison, W.M., 1987b. An ultrastructural study of the early development and tissue cyst formation of *Toxoplasma gondii* in the brains of mice. Parasitol. Res. 73 (6), 483–491.

Fox, B.A., Bzik, D.J., 2010 Sep. Avirulent uracil auxotrophs based on disruption of orotidine-5′-monophosphate decarboxylase elicit protective immunity to *Toxoplasma gondii*. Infect. Immun. 78 (9), 3744–3752. http://dx.doi.org/10.1128/IAI.00287-10.

Furtado, G.C., Cao, Y., Joiner, K.A., 1992. Laminin on *Toxoplasma gondii* mediates parasite binding to the beta 1 integrin receptor alpha 6 beta 1 on human foreskin fibroblasts and Chinese hamster ovary cells. Infect. Immun. 60 (11), 4925–4931.

Furuta, T., Kikuchi, T., Akira, S., Watanabe, N., Yoshikawa, Y., 2006 Dec. Roles of the small intestine for induction of toll-like receptor 4-mediated innate resistance in naturally acquired murine toxoplasmosis. Int. Immunol. 18 (12), 1655–1662. Epub 2006 Oct 11.

Fux, B., Rodrigues, C.V., Portela, R.W., Silva, N.M., Su, C., Sibley, D., Vitor, R.W., Gazzinelli, R.T., 2003. Role of cytokines and major histocompatibility complex restriction in mouse resistance to infection with a natural recombinant strain (type I–III) of *Toxoplasma gondii*. Infect. Immun. 71, 6392–6401.

Gaddi, P.J., Yap, G.S., 2007. Cytokine regulation of immunopathology in toxoplasmosis. Immunol. Cell Biol. 2, 155–159.

Gais, A., Beinert, N., Gross, U., Lüder, C.G., 2008 Apr. Transient inhibition of poly(ADP-ribose) polymerase expression and activity by *Toxoplasma gondii* is dispensable for parasite-mediated blockade of host cell apoptosis and intracellular parasite replication. Microbes Infect. 10 (4), 358–366. http://dx.doi.org/10.1016/j.micinf.2007.12.010.

Gardner, I.D., Remington, J.S., 1977. Age-related decline in the resistance of mice to infection with intracellular fpathogens. Infect. Immun. 16, 593–598.

Gardner, I.D., Remington, J.S., 1978b. Aging and the immune response. II. Lymphocyte responsiveness and macrophage activation in Toxoplasma gondii-infected mice. J. Immunol. 120 (3), 944–949.

Gardner, I.D., Remington, J.S., 1978a. Aging and the immune response. I. Antibody formation and chronic infection in Toxoplasma gondii-infected mice. J. Immunol. 120 (3), 939–943.

Garweg, J.G., Scherrer, J., wallon, M., Kodjikian, L., Peyron, F., 2005. Reactivation of ocular toxoplasmosis during pregnancy. BJOG 112, 241–242.

Garweg, J.G., Candolfi, E., 2009. Immunopathology in ocular toxoplasmosis: facts and clues. Mem. Inst. Oswaldo Cruz 2, 211–220.

Gaskell, E.A., Smith, J.E., Pinney, J.W., Westhead, D.R., Mcconkey, G.A., 2009. A unique dual activity amino acid hydroxylase in *Toxoplasma gondii*. PLoS One 4. e4801.

Gavrilescu, L.C., Denkers, E.Y., 2003. Interleukin-12 p40- and Fas ligand-dependent apoptotic pathways involving STAT-1 phosphorylation are triggered during infection with a virulent strain of *Toxoplasma gondii*. Infect. Immun. 71 (5), 2577–2583.

Gavrilescu, L.C., Butcher, B.A., Del Rio, L., Taylor, G.A., Denkers, E.Y., 2004. STAT1 is essential for antimicrobial effector function but dispensable for gamma interferon production during *Toxoplasma gondii* infection. Infect. Immun. 72 (3), 1257–1264. Mar.

Gazzinelli, R.T., Hakim, F.T., Hieny, S., Shearer, G.M., Sher, A., 1991. Synergistic role of CD4$^+$ and CD8$^+$ T-lymphocytes in IFNgamma production and protective immunity induced by an attenuated *Toxoplasma gondii* vaccine. J. Immunol. 146, 286–292.

Gazzinelli, R., Xu, Y., Hieny, S., Cheever, A., Sher, A., 1992a. Simultaneous depletion of CD4$^+$ and CD8$^+$ T-lymphocytes is required to reactivate chronic infection with *Toxoplasma gondii*. J. Immunol. 149, 175–180.

Gazzinelli, R., Oswald, I., Hieny, S., James, S., Sher, A., 1992b. IL-10 inhibits parasite killing and nitrogen oxide prodution by IFNgamma activated macrophages. J. Immunol. 22, 2501–2506.

Gazzinelli, R.T., Hartley, J.W., Fredrickson, T.N., Chattopadhyay, S.K., Sher, A., Morse 3rd, H.C., 1992c. Opportunistic infections and retrovirus-induced immunodeficiency: studies of acute and chronic infections with *Toxoplasma gondii* in mice infected with LP-BM5 murine leukaemia viruses. Infect. Immun. 60, 4394–4401.

Gazzinelli, R.T., Hieny, S., Wynn, T.A., Wolf, S., Sher, A., 1993a. Interleukin 12 is required for the T-lymphocyte-independent induction of interferon gamma by an intracellular parasite and induces resistance in T-cell-deficient hosts. Proc. Natl. Acad. Sci. U S A 90, 6115–6119.

Gazzinelli, R.T., Eltoum, I., Wynn, T.A., Sher, A., 1993b. Acute cerebral toxoplasmosis is induced by in vivo neutralization of TNF-alpha and correlates with the down-regulated expression of inducible nitric oxide synthase and other markers of macrophage activation. J. Immunol. 151, 3672–3681.

Gazzinelli, R.T., Denkers, E.Y., Sher, A., 1993c. Host resistance to *Toxoplasma gondii*: model for studying the selective induction of cell-mediated immunity by intracellular parasites. Infect. Agents Dis. 2 (3), 139–149.

Gazzinelli, R.T., Wysocka, M., Hayashi, S., Denkers, E.Y., Hieny, S., Caspar, P., Trinchieri, G., Sher, A., 1994a. Parasite-induced IL-12 stimulates early IFN-gamma synthesis and resistance during acute infection with *Toxoplasma gondii*. J. Immunol. 153, 2533–2543.

Gazzinelli, R.T., Brezin, A., Li, Q., Nussenblatt, R.B., Chan, C.C., 1994b. *Toxoplasma gondii*: acquired ocular toxoplasmosis in the murine model, protective role of TNF-alpha and IFNgamma. Experimental Parasitology 78, 217–229.

Gazzinelli, R.T., Giese, N.A., Morse 3rd, H.C., 1994c. In vivo treatment with interleukin 12 protects mice from immune abnormalities observed during murine acquired immunodeficiency syndrome (MAIDS). J. Exp. Med. 180, 2199–2208.

Gazzinelli, R., Bala, S., Stevens, R., Baseler, M., Wahl, L., Kovacs, J., Sher, A., 1995. HIV infection suppresses Type 1 lymphokine and IL-12 responses to *Toxoplasma gondii* but fails to inhibit the synthesis of other parasite-induced monokines. J. Immunol. 155, 1565–1574.

Gazzinelli, R.T., Amichay, D., Sharton-Kersten, et al., 1996a. Role of macrophage-derived cytokines in the induction and regulation of cell-mediated immunity to Toxoplasma gondi. Curr. Top Microbiol. Immunol. 219, 155–163.

Gazzinelli, R.T., Wysocka, M., Hieny, S., Scharton-Kersten, T., Cheever, A., Kuhn, R., Muller, W., Trinchieri, G., Sher, A., 1996b. In the absence of endogenous IL-10, mice acutely infected with *Toxoplasma gondii* succumb to a lethal immune response dependent on CD4$^+$ T-cells and accompanied by overproduction of IL-12, IFNgamma and TNF-alpha. J. Immunol. 157, 798–805.

Gazzinelli, R.T., Ropert, C., Campos, M.A., 2004. Role of the Toll/interleukin-1 receptor signalling pathway in host resistance and pathogenesis during infection with protozoan parasites. Immunol. Rev. 201, 9–25.

George, J.F.J.R., Schroeder Jr., H.W., 1992. Developmental regulation of D beta reading frame and junctional diversity in T-cell receptor-beta transcripts from human thymus. J. Immunol. 148 (4), 1230–1239.

Gendron, R.L., 1990. Lipopolysaccharide induced foetal resorption in mice is associated with intrauterine production of Tumour necrosis factor-alpha. J. Reprod. Fertil. 90, 395–402.

Ghatak, N.R., Zimmerman, H.M., 1973. Fine structure of Toxoplasma in the human brain. Arch. Pathol. 95, 276–283.

Gigley, J.P., Fox, B.A., Bzik, D.J., 2009a. Long-term immunity to lethal acute or chronic type II *Toxoplasma gondii* infection is effectively induced in genetically susceptible C57BL/6 mice by immunization with an attenuated type I vaccine strain. Infect. Immun. 77 (12), 5380–5388. http://dx.doi.org/10.1128/IAI.00649-09.

Gigley, J.P., Fox, B.A., Bzik, D.J., 2009b. Cell-mediated immunity to *Toxoplasma gondii* develops primarily by local Th1 host immune responses in the absence of parasite replication. J. Immunol. 182 (2), 1069–1078.

Gigley, J.P., 2013 in preparation.

Gilbert, L.A., Ravindran, S., Turetzky, J.M., Boothroyd, J.C., Bradley, P.J., 2007 Jan. *Toxoplasma gondii* targets a protein phosphatase 2C to the nuclei of infected host cells. Eukaryot. Cell. 6 (1), 73–83.

Gilbert, R.E., Dunn, D.T., Lightman, S., et al., 1999. Incidence of symptomatic Toxoplasma eye disease: aetiology and public health implications. Epidemiol. Infect. 123, 283–289.

Gilbert, R.E., Stanford, M.R., 2000. Is ocular toxoplasmosis caused by prenatal or postnatal infection? Br. J. Ophthalmol. 84, 224–226.

Gilbert, R.E., Freeman, K., Lago, E.G., Bahia-Oliveira, L.M., Tan, H.K., Wallon, M., Buffolano, W., Stanford, M.R., Petersen, E., 2008 Aug 13. European Multicentre Study on Congenital Toxoplasmosis (EMSCOT). Ocular sequelae of congenital toxoplasmosis in Brazil compared with Europe. PLoS Negl. Trop. Dis. 2 (8), e277. http://dx.doi.org/10.1371/journal.pntd.0000277.

Glasner, P.D., Silveira, C., Kruszon-Moran, D., Martins, M.C., Burnier Junior, M., Silveira, S., Camargo, M.E., Nussenblatt, R.B., Kaslow, R.A., Belfort Junior, R., 1992. An unusually high prevalence of ocular toxoplasmosis in southern Brazil. Am. J. Ophthalmol. 114, 136–144.

Glatman Zaretsky, A., Silver, J.S., Siwicki, M., Durham, A., Ware, C.F., Hunter, C.A., 2012. Infection with *Toxoplasma gondii* alters lymphotoxin expression associated with changes in splenic architecture. Infect. Immun. 80 (10), 3602–3610. http://dx.doi.org/10.1128/IAI.00333-12.

Goodwin, D., Hrubec, T.C., Klein, B.G., Strobl, J.S., Werre, S.R., Han, Q., Zajac, A.M., Lindsay, D.S., 2012. Congenital infection of mice with *Toxoplasma gondii* induces minimal change in behaviour and no change in neurotransmitter concentrations. J. Parasitol. 98, 706–712.

Goebel, S., Gross, U., Luder, C., 2001. Inhibition of host cell apoptosis by *Toxoplasma gondii* is accompanied by reduced activation of the caspase cascade and alterations of poly (ADP-ribose) polymerase expression. J. Cell Science 114, 3495–3505.

Goldszmid, R.S., Coppens, I., Lev, A., Caspar, P., Mellman, I., Sher, A., 2009 Feb 16. Host ER-parasitophorous vacuole interaction provides a route of entry for antigen cross-presentation in *Toxoplasma gondii*-infected dendritic cells. J. Exp. Med. 206 (2), 399–410. http://dx.doi.org/10.1084/jem.20082108.

Goldszmid, R.S., Sher, A., 2010 Feb. Processing and presentation of antigens derived from intracellular protozoan parasites. Curr. Opin. Immunol. 22 (1), 118–123. http://dx.doi.org/10.1016/j.coi.2010.01.017.

Goldszmid, R.S., Bafica, A., Jankovic, D., Feng, C.G., Caspar, P., Winkler-Pickett, R., Trinchieri, G., Sher, A., 2007. TAP-1 indirectly regulates $CD4^+$ T-cell priming in *Toxoplasma gondii* infection by controlling NK-cell IFN-gamma production. J. Exp. Med. 204 (11), 2591–2602.

Goldszmid, R.S., Caspar, P., Rivollier, A., White, S., Dzutsev, A., Hieny, S., Kelsall, B., Trinchieri, G., Sher, A., 2012. NK-cell-derived interferon-γ orchestrates cellular dynamics and the differentiation of monocytes into dendritic cells at the site of infection. Immunity 36 (6), 1047–1059.

Gormley, P.D., Pavesio, C.E., et al., 1999. Retinochoroiditis is induced by oral administration of *Toxoplasma gondii* cysts in the hamster model. Exp. Eye Res. 68 (6), 657–661.

Graca, L., Honey, K., Adams, E., Cobbold, S.P., Waldmann, H., 2000. Cutting edge: anti-CD154 therapeutic antibodies induce infectious transplantation tolerance. J. Immunol. 165, 4783–4786.

Graefe, S.E., Jacobs, T., Gaworski, I., Klauenberg, U., Steeg, C., Fleischer, B., 2003. Interleukin-12 but not interleukin-18 is required for immunity to Trypanosoma cruzi in mice. Microbes Infect. 5, 833–839.

Gregg, B., Dzierszinski, F., Tait, E., Jordan, K.A., Hunter, C.A., Roos, D.S., 2011. Subcellular antigen location influences T-cell activation during acute infection with *Toxoplasma gondii*. PLoS One 6 (7), e22936. http://dx.doi.org/10.1371/journal.pone.0022936.

Grigg, M.E., Ganatra, J., Boothroyd, J.C., Margolis, T.P., 2001a. Unusual abundance of atypical strains associated with human ocular toxoplasmosis. J. Infect. Dis. 184, 633–639.

Grigg, M.E., Bonnefoy, S., Hehl, A.B., Boothroyd, J.C., Suzuki, Y., 2001b. Success and virulence in Toxoplasma as the results of sexual recombination between two distinct ancectries. Science 294, 161–165.

Grimwood, J., Mineo, J.R., Kasper, L.H., 1996. Attachment of *Toxoplasma gondii* to host cells is host cell cycle dependent. Infect. Immun. 64, 4099–4104.

Groer, M.W., Yolken, R.H., Xiao, J.C., Beckstead, J.W., Fuchs, D., Mohapatra, S.S., Seyfang, A., Postolache, T.T., 2011a. Prenatal depression and anxiety in Toxoplasma gondii-positive women. Am. J. Obstet. Gynecol. 204 (433), e1–e7.

Groër, M.W., Yolken, R.H., Xiao, J.C., Beckstead, J.W., Fuchs, D., Mohapatra, S.S., Seyfang, A., Postolache, T.T., 2011b. Prenatal depression and anxiety in *Toxoplasma gondii*-positive women. Am. J. Obstet Gynecol. (5), 433 e1–7.

Gross, U., Bohne, W., Schroder, J., Roos, T., Heesemann, J., 1993. Comparison of a commercial enzyme immunoassay and an immunoblot technique for detection of immunoglobulin A antibodies to *Toxoplasma gondii*. Eur. J. Clin. Microbiol. Infect. Dis. (8), 636–639.

Gross, U., Bohne, W., 1994. *Toxoplasma gondii*: strain- and host cell-dependent induction of stage differentiation. J. Eukaryot. Microbiol. 41 (5), 10S–11S.

Grover, H.S., Blanchard, N., Gonzalez, F., Chan, S., Robey, E.A., Shastri, N., 2012. The *Toxoplasma gondii* peptide AS15 elicits CD4 T-cells that can control parasite burden. Infect. Immun. 80 (9), 3279–3288.

Guan, H., Moretto, M., Bzik, D.J., Gigley, J., Khan, I.A., 2007. NK-cells enhance dendritic cell response against parasite antigens via NKG2D pathway. J. Immunol. 179 (1), 590–596, 2007.

Gubbels, M.J., Striepen, B., Shastri, N., Turkoz, M., Robey, E.A., 2005. Class I major histocompatibility complex presentation of antigens that escape from the parasitophorous vacuole of *Toxoplasma gondii*. Infect Immun. 73 (2), 703–11.

Guiton, R., Vasseur, V., Charron, S., Arias, M.T., Van Langendonck, N., Buzoni-Gatel, D., Ryffel, B., Dimier-Poisson, I., 2010. Interleukin 17 receptor signalling is deleterious during *Toxoplasma gondii* infection in susceptible BL6 mice. J. Infect. Dis. 3, 427–435 http://dx.doi.org/10.1086/653738.

Guglietta, S., Beghetto, E., Spadoni, A., Buffolano, W., Del Porto, P., Gargano, N., 2007. Age-dependent impairment of functional helper T-cell responses to immunodominant epitopes of *Toxoplasma gondii* antigens in congenitally infected individuals. Microbes Infect. 9 (2), 127–133.

Habbeger de Sorrentino, A., Lopez, R., Motta, P., et al., 2005. HLA class II involvement in HIV-associated toxoplasmic encephalitis development. Clin. Immunol. 115, 133–137.

Hakim, F., Gazzinelli, R., Denkers, E., Hieny, S., Shearer, G., Sher, A., 1991. $CD8^+$ T-cells from mice vaccinated against *Toxoplasma gondii* are cytotoxic for parasite-infected or antigen-pulsed host cells. J. Immunol. 147, 2310–2316.

Hall, A.O., Beiting, D.P., Tato, C., John, B., Oldenhove, G., Lombana, C.G., Pritchard, G.H., Silver, J.S., Bouladoux, N., Stumhofer, J.S., Harris, T.H., Grainger, J., Wojno, E.D., Wagage, S., Roos, D.S., Scott, P., Turka, L.A., Cherry, S., Reiner, S.L., Cua, D., Belkaid, Y., Elloso, M.M., Hunter, C.A., 2012 Sep 21. The cytokines interleukin 27 and interferon-γ promote distinct Treg cell populations required to limit infection-induced pathology. Immunity 37 (3), 511–523. http://dx.doi.org/10.1016/j.immuni.2012.06.014.

Halonen, S.K., Lyman, W.D., Chiu, F.C., 1996. Growth and development of *Toxoplasma gondii* in human neurons and astrocytes. J. Neuropathol. Exp. Neurol. 55 (11), 1150–1156.

Halonen, S.K., Chiu, F., Weiss, L.M., 1998a. Effect of cytokines on growth of *Toxoplasma gondii* in murine astrocytes. Infect. Immun. 66, 4989–4993.

Halonen, S.K., Taylor, G.A., Weiss, L.M., 2001. Gamma interferon-induced inhibition of *Toxoplasma gondii* in astrocytes is mediated by IGTP. Infect. Immun. 69, 5573–5576.

Halonen, S.K., Weiss, L.M., Chiu, F.C., 1998b. Association of host cell intermediate filaments with *Toxoplasma gondii* cysts in murine astrocytes in vitro. Int. J. Parasitol. 28, 815–823.

Halonen, S.K., Chiu, F.C., Weiss, L.M., 1999 Sep-Oct. Effect of cytokines and quercetin on Toxoplasma gondii cyst induction in murine astrocytes. J Eukaryot Microbiol. 46 (5), 83S–84S.

Halonen, S.K., Woods, T., McInnerney, K., Weiss, L.M., 2006. Microarray analysis of IFNgamma response genes in astrocytes. Journal of Neuroimmunology 175, 19–30.

Handman, E., Remington, J., 1980. Antibody response to Toxoplasma antigens in mice infected with strais of different virulence. Infect. Immun. 40, 215–225.

Handel, U., Brunn, A., Drogemuller, K., Muller, W., Decker, M., Schluter, D., 2012. Neuronal gp130 expression is crucial to prevent neuronal loss, hyperinflammation, and lethal course of murine Toxoplasma encephalitis. Am. J. Pathol. 181 (1), 163–173.

Hancock, W.W., Szaba, F.M., Berggren, K.N., Parent, M.A., Mullarky, I.K., Pearl, J., Cooper, A.M., Ely, K.H., Woodland, D.L., Kim, I.J., Blackman, M.A., Johnson, L.L., Smiley, S.T., 2004 Mar 2. Intact type 1 immunity and immune-associated coagulative responses in mice lacking IFN gamma-inducible fibrinogen-like protein 2. Proc. Natl. Acad. Sci. U S A 101 (9), 3005–3010.

Haque, S., Hanna, S., Gharbi, S., Franck, J., Dumon, H., Haque, A., 1999. Infection of mice by a *Toxoplasma gondii* isolate from an AIDS patient: virulence and activation of hosts' immune responses are independent of parasite genotype. Parasite Immunol. 21 (12), 649–657.

Haroon, F., Handel, U., Angenstein, F., Goldschmidt, J., Kreutzmann, P., Lison, H., Fischer, K.D., Scheich, H., Wetzel, W., Schluter, D., Budinger, E., 2012. *Toxoplasma gondii* actively inhibits neuronal function in chronically infected mice. PLoS One 7, e35516.

Hara, T., Ohashi, S., Yamashita, Y., Abe, T., Hisaeda, H., Himeno, K., Good, R.A., Takeshita, K., 1996. Human V delta 2^+ gamma delta T-cell tolerance to foreign antigens of *Toxoplasma gondii*. Proc. Natl. Acad. Sci. U S A 93, 5136–5140.

Harris, T.H., Banigan, E.J., Christian, D.A., Konradt, C., Tait Wojno, E.D., Norose, K., Wilson, E.H., John, B., Weninger, W., Luster, A.D., Liu, A.J., Hunter, C.A., 2012. Generalized Lévy walks and the role of chemokines in migration of effector $CD8^+$ T-cells. Nature 486 (7404), 545–548. http://dx.doi.org/10.1038/nature11098.

Harris, T.H., Wilson, E.H., Tait, E.D., Buckley, M., Shapira, S., Caamano, J., Artis, D., Hunter, C.A., 2010 May. NF-kappaB1 contributes to T-cell-mediated control of *Toxoplasma gondii* in the CNS. J. Neuroimmunol. 222 (1–2), 19–28. http://dx.doi.org/10.1016/j.jneuroim.2009.12.009.

Hassan, M.A., Melo, M.B., Haas, B., Jensen, K.D., Saeij, J.P., 2012 Dec 12. De novo reconstruction of the *Toxoplasma gondii* transcriptome improves on the current genome annotation and reveals alternatively spliced transcripts and putative long non-coding RNAs. BMC Genomics 13 (1), 696.

Hauser Jr., W.E., Remington, J.S., 1981. Effect of monoclonal antibodies on phagocytosis and killing of *Toxoplasma gondii* by normal macrophages. Infect. Immun. 32 (2), 637–640.

Hauser, W., Sharma, S., Remington, J., 1982. Natural killer cells induced by acute and chronic Toxoplasma infection. Cell. Immunol. 69, 330–346.

Hauser, W., Tsai, V., 1986. Acute Toxoplasma infection of mice induces spleen NK-cells that are cytotoxic for t. gondii in vitro. J. Immunol. 136, 313–319.

Hay, J., Hutchison, W.M., et al., 1981. Cataract in mice congenitally infected with *Toxoplasma gondii*. Ann. Trop. Med. Parasitol. 75 (4), 455–457.

Hayashi, N., Kimura, H., Morishima, T., et al., 2003. Flow cytometric analysis of cytomegalovirus-specific cell-mediated immunity in the congenital infection. J. Med. Virol. 71, 251–258.

Hayashi, S., Chan, C.C., Gazzinelli, R., Roberge, F.G., 1996a. Contribution of nitric oxide to the host parasite equilibrium in toxoplasmosis. J. Immunol. 156, 1476–1481.

Hayashi, S., Chan, C.C., Gazzinelli, R.T., Pham, N.T., Cheung, M.K., Roberge, F. G., 1996b. Protective role of nitric oxide in ocular toxoplasmosis. Br. J. Ophthalmol. 80, 644–648.

Hayward, A.R., Kurnick, J., 1981. Newborn T-cell suppression: early appearance, maintenance in culture, and lack of growth factor suppression. J. Immunol. 126 (1), 50–53.

He, X.L., Grigg, M.E., Boothroyd, J.C., Garcia, K.C., 2002. Structure of the immunodominant surface antigen from the *Toxoplasma gondii* SRS superfamily. Nat. Struct. Biol. 9 (8), 606–611.

Heimesaat, M.M., Bereswill, S., Fischer, A., Fuchs, D., Struck, D., Niebergall, J., Jahn, H.K., Dunay, I.R., Moter, A., Gescher, D.M., Schumann, R.R., Gobel, U.B., Liesenfeld, O., 2006. Gram-negative bacteria aggravate murine small intestinal Th1-type immunopathology following oral infection with *Toxoplasma gondii*. J. Immunol. 177, 8785–8795.

Heimesaat, M.M., Fischer, A., Jahn, H.K., Niebergall, J., Freudenberg, M., Blaut, M., Liesenfeld, O., Schumann, R.R., Göbel, U.B., Bereswill, S., 2007 Jul. Exacerbation of murine ileitis by Toll-like receptor 4 mediated sensing of lipopolysaccharide from commensal Escherichia coli. Gut 56 (7), 941–948. Epub 2007 Jan 25.

Hennequin, C., 1997. Congenital toxoplasmosis acquired from an immune woman. Pediatr. Infect. Dis. J. 16, 75–76.

Henriquez, F.L., Woods, S., Cong, H., McLeod, R., Roberts, C.W., 2010 Nov. Immunogenetics of *Toxoplasma gondii* informs vaccine design. Trends Parasitol. 26 (11), 550–555. http://dx.doi.org/10.1016/j.pt.2010.06.004.

Henry, S.C., Daniell, X.G., Burroughs, A.R., Indaram, M., Howell, D.N., Coers, J., Starnbach, M.N., Hunn, J.P., Howard, J.C., Feng, C.G., Sher, A., Taylor, G.A., 2009 May. Balance of Irgm protein activities determines IFNgamma-induced host defence. J. Leukoc. Biol. 85 (5), 877–885. http://dx.doi.org/10.1189/jlb.1008599. Epub 2009 Jan 27.

Herb, H.M., Jontofosohn, R., Loffler, H.D., et al., 1977. Toxoplasmosis after renal transplantation. Clin. Nephrol. 8, 329–332.

Herion, P., Hernandez Pando, R., Dubremetz, J.F., Saavedra, R., 1993. Subcellular localization of the 54-kDa antigen of toxoplasma gondii. J. Parasitol. 79, 216–222.

Hermes, G., Ajioka, James W., Kelly, Kystyna A., Mui, Ernest, et al., 2008. Neurological and behavioural abnormalities, ventricular dilatation, altered cellular functions, inflammation, and neuronal injury in brains of mice due to common, persistent, parasitic infection. J. Neuroinflammation 5, 48. ePub October 23.

Hertzog, P., Forster, S., Samarajiwa, S., 2011. Systems biology of interferon responses. Journal of Interferon and Cytokine Research: the official journal of the International Society for Interferon and Cytokine Research 31, 5–11.

Heyes, M., Saito, K., Crowley, J., 1992. Quinolinic acid and kynurenine pathway in inflammatory and non-inflammatory neurological disease. Brain 115, 1249–1273.

Hibbs Jr., J.B., Lambert Jr., L.H., Remington, J.S., 1971. Resistance to murine tumours conferred by chronic infection with intracellular protozoa, *Toxoplasma gondii* and Besnoitia jellisoni. J. Infect. Dis. 124, 587–592.

Hippe, D., Weber, A., Zhou, L., Chang, D.C., Häcker, G., Lüder, C.G., 2009a. *Toxoplasma gondii* infection confers resistance against BimS-induced apoptosis by preventing the activation and mitochondrial targeting of pro-apoptotic Bax. J. Cell Sci. 122 (Pt 19), 3511–3521.

Hippe, D., Gais, A., Gross, U., Lüder, C.G., 2009b. Modulation of caspase activation by *Toxoplasma gondii*. Methods Mol. Biol. 470, 275–288.

Hippe, D., Lytovchenko, O., Schmitz, I., Lüder, C.G., 2008 Jul. Fas/CD95-mediated apoptosis of type II cells is blocked by *Toxoplasma gondii* primarily via interference with the mitochondrial amplification loop. Infect. Immun. 76 (7), 2905–2912. http://dx.doi.org/10.1128/IAI.01546-07.

Himeno, K., Hisaeda, H., 1996. Contribution of 65-kDa heat shock protein induced by gamma and delta T-cells to protection against *Toxoplasma gondii* infection. Immunol. Res. 15 (3), 258–264.

Hirota, H., Kiyama, H., Kishimoto, T., Taga, T., 1996. Accelerated Nerve Regeneration in Mice by up-regulated expression of interleukin (IL) 6 and IL-6 receptor after trauma. J. Exp. Med. 183, 2627–2634.

Hisaeda, H., Nagasawa, H., Maeda, K., et al., 1995. Gamma delta T-cells play an important role in hsp65 expression and in acquiring protective immune responses against infection with *Toxoplasma gondii*. J. Immunol 155, 244–251.

Hitziger, N., Dellacasa, I., Albiger, B., Barragan, A., 2005. Dissemination of *Toxoplasma gondii* to immunoprivileged organs and role of Toll/interleukin-1 receptor signalling for host resistance assessed by in vivo bioluminescence imaging. Cell Microbiol. 7 (6), 837–848. Jun.

H of H, Emmerling, P., Hoehne, K., Seeliger, H.R.P., 1976. Infection of congenitally athymic (nude) mice with *Toxoplasma gondii*. Ann. Microbiol. (Inst. Pasteur) 127b, 503–507.

Hoerni, B., Vallat, M., Durand, M., Pesme, D., 1978. Ocular toxoplasmosis and Hodgkin's disease: report of two cases. Arch. Ophthalmol. 96 (1), 62–63.

Hogan, M.J., 1961. Ocular toxoplasmosis in adult patients. Surv. Ophthalmol. 6, 835–851.

Hogan, M.J., Kimura, S.J., O'Connor, G.R., 1964. Ocular toxoplasmosis. Arch. Ophthalmol. 72, 592–600.

Hogan, M.J., Zweigart, P.A., et al., 1956. Persitence of *Toxoplasma gondii* in ocular, tissues: I. Am. J. Ophthalmol. 42, 84–89.

Hogan, M.J., Zweigart, P.A., et al., 1957. Experimental ocular toxoplasmosis. Am. J. Ophthalmol. 43, 291–292.

Hogan, M.J., Zweigart, P.A., et al., 1958. Experimental ocular toxoplasmosis: II. Arch. Ophthalmol. 60, 448–449.

Hohlfeld, P., Daffos, F., Thulliez, P., Aufrant, C., Couvreur, J., et al., 1989. Foetal toxoplasmosis: outcome of pregnancy and infant follow-up after in utero treatment. J. Pediatr. 115 (5), 765–769.

Hohlfeld, P.F., Marion, S., Thulliez, P., Marcon, P., Daffos, F., 1990. *Toxoplasma gondii* infection during pregnancy: T lymphocyte subpopulations in mothers and foetuses. Pedatr. Infect. Dis. J. 9, 878–881.

Holland, G.N., 1999. Reconsidering the pathogenesis of ocular toxoplasmosis. Am. J. Ophthalmol. 128, 502–505.

Holland, G.N., 2003. Ocular toxoplasmosis: A global reassessment. Part I; Epidemiology and course of disease. Am. J. Ophthalmol. 136, 973–988.

Holland, G.N., 2004. Ocular Toxoplasmosis: A global reassessment. Part II: Disease manifestations and management. Am. J. Ophthalmol. 137 (1), 1–17.

Holland, G.N., Engstrome, R.E., Glasgow, B.J., Berger, B.B., Daniels, S.A., Sidikaro, Y., 1988. Ocular toxoplasmosis in patients with the acquired immunodeficiency syndrome. Am. J. Ophthalmol. 106, 653–667.

Honore, S., Couvelard, A., Garin, Y.J., Bedel, C., Henin, D., et al., 2000. [Genotyping of *Toxoplasma gondii* strains from immmunocompromised patients.]. Pathol. Biol (Paris) 48, 541–547.

Hortua Triana, M.A., Huynh, M.H., Garavito, M.F., Fox, B.A., Bzik, D.J., Carruthers, V.B., Löffler, M., Zimmermann, B.H., 2012 Aug. Biochemical and molecular characterization of the pyrimidine biosynthetic enzyme dihydroorotate dehydrogenase from *Toxoplasma gondii*. Mol. Biochem. Parasitol. 184 (2), 71–81. http://dx.doi.org/10.1016/j.molbiopara.2012.04.009.

Hoshi, M., Saito, K., Hara, A., Taguchi, A., Ohtaki, H., Tanaka, R., Fujigaki, H., Osawa, Y., Takemura, M., Matsunami, H., Ito, H., Seishima, M., 2010 Sep 15. The absence of IDO up-regulates type I IFN production, resulting in suppression of viral replication in the retrovirus-infected mouse. J. Immunol. 185 (6), 3305–3312. http://dx.doi.org/10.4049/jimmunol.0901150.

Hou, B., Benson, A., Kuzmich, L., Defranco, A.L., Yarovinsky, F., 2011. Critical coordination of innate immune defence against *Toxoplasma gondii* by dendritic cells responding via their Toll-like receptors. Proceedings of the National Academy of Sciences of the USA of America 108, 278–283.

House, P.K., Vyas, A., Sapolsky, R., 2011. Predator cat odors activate sexual arousal pathways in brains of *Toxoplasma gondii* infected rats. PLoS One 6 (8), e23277. http://dx.doi.org/10.1371/journal.pone.0023277.

Howard, J.C., Hunn, J.P., Steinfeldt, T., 2011. The IRG protein-based resistance mechanism in mice and its relation to virulence in *Toxoplasma gondii*. Curr. Opin. Microbiol. 14, 414–421.

Howe, D.K., Sibley, L.D., 1995. *Toxoplasma gondii* comprises three clonal lineages: correlation of parasite genotype with human disease. J. Infect. Dis. 172, 1561–1566.

Howe, D.K., Honore, S., Derouin, F., Sibley, L.D., 1997. Determination of genotypes of *Toxoplasma gondii* strains isolated from patients with toxoplasmosis. J. Clin. Microbiol. 35, 1411–1414.

Hu, M.S., Schwartzman, J.D., Lepage, A.C., Khan, I.A., Kasper, L.H., 1999a. Experimental ocular toxoplasmosis induced in naive and preinfected mice by intracameral inoculation. Ocul. Immunol. Inflamm. 7, 17–26.

Hu, M.S., Schwartzman, J.D., Yeaman, G.R., Collins, J., Seguin, R., Khan, I.A., Kasper, L.H., 1999b. Fas-FasL interaction involved in pathogenesis of ocular toxoplasmosis in mice. Infect. Immun. 67, 928–935.

Hughes, H.P., et al., 1988. absence of a role for natural killer cells in the control of acute infection by *Toxoplasma gondii* oocysts. Clin. Exp. Immunol. 72 (3), 394–399.

Hunter, C.A., Abrams, J.S., Beaman, M.H., Remington, J.S., 1993. Cytokine mRNA in the central nervous system of SCID mice infected with *Toxoplasma gondii*: importance of T-cell-independent regulation of resistance to T. gondii. Infect. Immun. 61, 4038–4044.

Hunter, C.A., Sibley, L.D., 2012. Modulation of innate immunity by *Toxoplasma gondii* virulence effectors. Nat. Rev. Microbiol. 10 (11), 766–778. http://dx.doi.org/10.1038/nrmicro2858.

Hunter, C.A., Subauste, C.S., Van Cleave, V.H., Remington, J.S., 1994a. Production of gamma interferon by natural killer cells from Toxoplasma gondii-infected SCID mice: regulation by interleukin-10, interleukin-12, and tumour necrosis factor alpha. Infect. Immun. 62, 2818–2824.

Hunter, C.A., Bermudez, L., Beernink, H., Waegell, W., Remington, J., 1995a. Transforming growth factor-beta inhibits interleukin-12-induced production of interferon-gamma by natural killer cells: a role for transforming growth factor-beta in regulation of T-cell-independent regulation of resistance to T. gondii. Infect. Immun. 61, 4038–4044.

Hunter, C., Roberts, C.W., Alexander, J., 1992. Detection of cytokine mRNA in the brains of mice with toxoplasmic encephalitis. Parasite Immunol. 14, 405–413.

Hunter, C., Candolfi, E., Subauste, C., Van Cleave, V., Remington, J., 1995b. Studies on the role of IL-12 in murine toxoplasmosis. Immunology 84, 16–21.

Hunter, C., Chizzonite, R., Remington, J., 1995c. IL-1beta is required for IL-12 to induce production of IFNgamma by NK-cells. J. Immunol. 155, 4347–4354.

Hunter, C.A., Litton, M.J., Remington, J.S., Abrams, J.S., 1994b. Immunocyotchemical detection of cytokines in the lymph nodes and brains of mice resistant or susceptible to toxoplasmic encephalitis. Immunochemical JID 170, 939–945.

Hunter, C.A., Suzuki, Y., Subauste, C.S., 1996. Cells and cytokines in resistance to *Toxoplasma gondii*. Curr. Top Microbiol., 113–125.

Hunter, C.A., Ellis-Neyer, L., Gabriel, K.E., Kennedy, M.K., Grabstein, K.H., Linsley, P.S., Remington, J.S., 1997. The role of the CD28/B7 interaction in the regulation of NK-cell responses during infection with *Toxoplasma gondii*. J. Immunol. 158, 2285–2293.

Hunter, C.A., Sibley, L.D., 2012. Modulation of innate immunity by *Toxoplasma gondii* virulence effectors. Nat. Rev. Microbiol. 10 (11), 766–778. http://dx.doi.org/10.1038/nrmicro2858.

Huskinson, J., Thulliez, P., Remington, 1990. Toxoplasma antigens recognized by human immunoglobulin A antibodies. J. Clin. Microbiol. 28, 2632–2636.

Hutchison, W.M., Aitken, P.P., Wells, W.P., 1980. Chronic Toxoplasma infections and familiarity-novelty discrimination in the mouse. Ann. Trop. Med. Parasitol. 74, 145–150.

Indraccolo, S., Pfeffer, U., Minuzzo, S., Esposito, G., Roni, V., Mandruzzato, S., Ferrari, N., Anfosso, L., Dell'eva, R., Noonan, D.M., Chieco-Bianchi, L., Albini, A., Amadori, A., 2007. Identification of genes selectively regulated by IFNs in endothelial cells. Journal of Immunology 178, 1122–1135.

Israelski, D., Araujo, F., Conley, F., Suzuki, Y., Sharma, S., Remington, J., 1989. Treatment with anti-L3T4 (CD4) monoclonal antibody reduces the inflammatory response in toxoplasmic encephalitis. J. Immunol. 142, 954–958.

Ishii, K., Hisaeda, H., Duan, X., Imai, T., Sakai, T., Fehling, H.J., Murata, S., Chiba, T., Tanaka, K., Hamano, S., Sano, M., Yano, A., Himeno, K., 2006. The involvement of immunoproteasomes in induction of MHC class I-restricted immunity targeting *Toxoplasma* SAG1. Microbes Infect. 8, 1045–1053.

James, S.L., 1995. Role of nitric oxide in parasitic infections. Microbiol. Rev. 59, 533–547.

Jamieson, Sarra, E., de Roubaix, Lee-Anne, Cortina-Borja, Mario, et al., 2008. Genetic and epigenetic factors at *COL2A1* and *ABCA4* influence clinical outcome in congenital toxoplasmosis. PLoS One 3, e2285.

Jamieson, Sarra E., Peixoto-Rangel, Alba L., Hargrave, Aubrey C., et al., 2010. Evidence for associations between the purinergic receptor P2X7 (P2RX7) and toxoplasmosis. Genes Immun. 11, 374–383.

Jankovic, D., Kullberg, M.C., Feng, C.G., Goldszmid, R.S., Collazo, C.M., Wilson, M., Wynn, T.A., Kamanaka, M., Flavell, R.A., Sher, A., 2007. Conventional T-bet(+)Foxp3(-) Th1 cells are the major source of host-protective regulatory IL-10 during intracellular protozoan infection. J. Exp. Med. 2, 273–283.

Janssen, R., Van Wengen, A., Verhard, E., De Boer, T., Zomerdijk, T., Ottenhoff, T.H., Van Dissel, J.T., 2002. Divergent role for TNF-alpha in IFNgamma-induced killing of *Toxoplasma gondii* and *Salmonella typhimurium* contributes to selective susceptibility of patients with partial IFNgamma receptor 1 deficiency. J. Immunol. 169, 3900–3907.

Jebbari, H., Roberts, C.W., Ferguson, D.J., Bluethmann, H., Alexander, J., 1998. A protective role for IL-6 during early infection with *Toxoplasma gondii*. Parasite Immunol. 20, 231–239.

Jensen, K.D., Wang, Y., Wojno, E.D., Shastri, A.J., Hu, K., Cornel, L., Boedec, E., Ong, Y.C., Chien, Y.H., Hunter, C.A., Boothroyd, J.C., Saeij, J.P., 2011. Toxoplasma polymorphic effectors determine macrophage polarization and intestinal inflammation. Cell Host Microbe 9 (6), 472–483. http://dx.doi.org/10.1016/j.chom.2011.04.015.

Jin, D., Takamoto, M., Hu, T., Taki, S., Sugane, K., 2009. STAT6 signalling is important in CD8 T-cell activation and defence against *Toxoplasma gondii* infection in the brain. Immunology 127, 187–195.

John, B., Harris, T.H., Tait, E.D., Wilson, E.H., Gregg, B., Ng, L.G., Mrass, P., Roos, D.S., Dzierszinski, F., Weninger, W., Hunter, C.A., 2009 Jul. Dynamic Imaging of CD8(+) T-cells and dendritic cells during infection with *Toxoplasma gondii*. PLoS Pathog. 5 (7), e1000505. http://dx.doi.org/10.1371/journal.ppat.1000505.

John, B., Ricart, B., Tait Wojno, E.D., Harris, T.H., Randall, L.M., Christian, D.A., Gregg, B., De Almeida, D.M., Weninger, W., Hammer, D.A., Hunter, C.A.,

2011. Analysis of behaviour and trafficking of dendritic cells within the brain during toxoplasmic encephalitis. PLoS Pathog. 7(9):e1002246. PLoS Pathog. 7 (9), e1002246. http://dx.doi.org/10.1371/journal.ppat.1002246.

Johnson, LL, Berggren, KN, Szaba, FM, Chen, W, Smiley, ST, 2003. Fibrin-mediated protection against infection-stimulated immunopathology. J. Exp. Med. 197 (6), 801–806.

Johnson, A.M., McDonald, P.J., Neo, Sh.H., 1979. Kinetics of the growth of *Toxoplasma gondii* (RH strain) in mice. Int. J. Parasitol. 9, 55–56.

Johnson, A.M., 1984. Strain-dependent, route of challenge-dependent, murine susceptibility to toxoplasmosis. Z. Parasitenkd 70 (3), 303–309, 1984.

Johnson, L.L., 1992a. A protective role for endogenous tumour necrosis factor in *Toxoplasma gondii* infection. Infect. Immun. 60, 1979–1983.

Johnson, L.L., 1992b. SCID mouse models of acute and relapsing chronic *Toxoplasma gondii* infections. Infect. Immun. 60 (9), 3719–3724.

Johnson, L.L., Sayles, P.C., 2002. Deficient humoral responses underlie susceptibility to *Toxoplasma gondii* in CD4-deficient mice. Infect. Immun. 70, 185–191.

Johnson, J., Suzuki, Y., Mack, D., Mui, E., Estes, R., David, C., Skamene, E., Forman, J., McLeod, R., 2002a. Genetic analysis of influences on survival following *Toxoplasma gondii* infection. Int. J. Parasitol. 32, 179–185.

Johnson, J., Roberts, C., Pope, C., Roberts, R., Kirisits, M., Estes, R., Mui, E., Krieger, T., Brown, C., Forman, McLeod R., 2002b. Vitro Correlates of Ld-Restricted Resistance to Toxoplasmic Encephalitis and their Critical Dependence on Parasite Strain. J. Immunol. 169, 966–973.

Johnson, M.W., Greven, C.M., Jaffe, G.J., et al., 1997. Atypical, severe toxoplasmic retinochoroiditis in elderly patients. Ophthalmology 104, 48–57.

Joiner, K., Fuhrman, S., Miettinen, H., et al., 1990. Toxoplasma gondi: fusion competence of parasitophorous vacuoles in Fc receptor transfected fibroblasts. Science 249, 641–646.

Joiner, K.A., Dubremetz, J.F., 1993. *Toxoplasma gondii*: a protozoan for the nineties. Infect. Immun. 61 (4), 1169–1172.

Jones, Simon, A., Scheller, Jurgen, Rose-John, Stefan, 2011. Therapeutic strategies for the clinical blockade of IL-6/gp130 signalling. J. Clin. Invest. 121, 3375–3383.

Jones, T.C., Bienz, K.A., Erb, P., 1986. In vitro cultivation of *Toxoplasma gondii* cysts in astrocytes in the presence of gamma interferon. Infect. Immun. 51 (1), 147–156.

Jones, T.C., Masur, H., Len, L., Fu, T.L., 1977. Lymphocyte-macrophage interaction during control of intracellular parasitism. Am. J. Trop. Med. Hyg. 26, 187–193.

Jones, L.A., Roberts, F., Nickdel, M.B., Brombacher, F., McKenzie, A.N., Henriquez, F.L., Alexander, J., Roberts, C.W., 2010. IL-33 receptor (T1/ST2) signalling is necessary to prevent the development of encephalitis in mice infected with *Toxoplasma gondii*. Eur. J. Immunol. (2), 426–436 http://dx.doi.org/10.1002/eji.200939705.

Jordan, K.A., Wilson, E.H., Tait, E.D., Fox, B.A., Roos, D.S., Bzik, D.J., Dzierszinski, F., Hunter, C.A., 2009a. Kinetics and phenotype of vaccine-induced $CD8^+$ T-cell responses to *Toxoplasma gondii*. Infect. Immun. 77 (9), 3894–3901. http://dx.doi.org/10.1128/IAI.00024-09.

Jordan, K.A., Hunter, C.A., 2010. Regulation of $CD8^+$ T-cell responses to infection with parasitic protozoa. Exp. Parasitol. 126 (3), 318–325. http://dx.doi.org/10.1016/j.exppara.2010.05.008. Epub 2010 May 21.

Jordan, K.A., Wilson, E.H., Tait, E.D., Fox, B.A., Roos, D.S., Bzik, D.J., Dzierszinski, F., Hunter, C.A., 2009b. Kinetics and phenotype of vaccine-induced $CD8^+$ T-cell responses to *Toxoplasma gondii*. Infect. Immun. 77 (9), 3894–3901. http://dx.doi.org/10.1128/IAI.00024-09.

Jordan, K.A., Dupont, C.D., Tait, E.D., Liou, H.C., Hunter, C.A., 2010b. Role of the NF-κB transcription factor c-Rel in the generation of $CD8^+$ T-cell responses to *Toxoplasma gondii*. Int. Immunol. 11, 851–861 http://dx.doi.org/10.1093/intimm/dxq439.

Jost, C., Reiter-Owona, I., Liesenfeld, O., 2007 Nov. The timing of sulfadiazine therapy impacts the reactivation of latent Toxoplasma infection in IRF-8−/− mice. Parasitol. Res. 101 (6), 1603–1609. Epub 2007 Sep 12.

Kafsack, B.F., Pena, J.D., Coppens, I., Ravindran, S., Boothroyd, J.C., Carruthers, V.B., 2009. Rapid membrane disruption by a perforin-like protein facilitates parasite exit from host cells. Science 323 (5913), 530–533. http://dx.doi.org/10.1126/science.1165740.

Kang, H., Remington, J.S., et al., 2000. Decreased resistance of B-cell-deficient mice to infection with *Toxoplasma gondii* despite unimpaired expression of IFNgamma, TNF-alpha, and inducible nitric oxide synthase. J. Immunol. 164 (5), 2629–2634.

Kang, H., Suzuki, Y., 2001. Requirement of non-T cells that produce interferon-gamma for prevention of reactivation of *Toxoplasma gondii* infection in the brain. Infect. Immun. 69, 2920–2927.

Kang, H., Liesenfeld, O., Remington, J.S., Claflin, J., Wang, X., Suzuki, Y., 2003. TCR V beta 8^+ T-cells prevent development of toxoplasmic encephalitis in BALB/c mice genetically resistant to the disease. J. Immunol. 170, 4254–4259.

Kannan, G., Moldovan, K., Xiao, J.C., Yolken, R.H., Jones-Brando, L., Pletnikov, M.V., 2010. *Toxoplasma gondii* strain-dependent effects on mouse behaviour. Folia Parasitol. (Praha) (2), 151–155.

Kappler, J.W., et al., 1988. Self-tolerance eliminates T-cells specific for Mls-modified products of the major histocompatibility complex. Nature 332 (6159), 35–40.

Karp, C.L., Flick, L.M., Park, K.W., Softic, S., Greer, T.M., Keledjian, R., Yang, R., Uddin, J., Guggino, W.B., Atabani, S.F., Belkaid, Y., Xu, Y., Whitsett, J.A., Accurso, F.J., Wills-Karp, M., Petasis, N.A., 2004. Defective lipoxin-mediated anti-inflammatory activity in the cystic fibrosis airway. Nat. Immunol. 5 (4), 388–392.

Kasper, L., Matsuura, T., Fonseka, S., et al., 1996. Induction of gamma delta T-cells during acute murine infection with *Toxoplasma gondii*. J. Immunol. 157, 5521–5527.

Kasper, L.H., kahn, I.A., Ely, R., Buelow, R., Boothroyd, J.C., 1992. Antigen-specific (p30) mouse CD8+ cells are cytotoxic against Toxoplasma gondii-infected peritoneal macrophages. J. Immunol. 148, 1493–1498.

Kasper, L.H., 1989. Identification of stage-specific antigens of *Toxoplasma gondii*. Infect. Immun. 57 (3), 668–672.

Kasper, L., Courret, N., Darche, S., Luangsay, S., Mennechet, F., Minns, L., Rachinel, N., Ronet, C., Buzoni-Gatel, D., 2004. *Toxoplasma gondii* and mucosal immunity. Int. J. Parasitol. 34 (3), 401–409.

Kaufman, H.E., et al., 1958. Toxoplasmosis: the nature of virulence. Am. J. Ophthalmol. 46, 255–260.

Kaufman, H.E., et al., 1959. Strain differences of *Toxoplasma gondii*. J. Parasitol. 45, 189–190.

Khalili, H., DR, Chang, M.Y., 1997. The defective antigen-presenting activity of murine foetal macrophage cell lines. Immunology 4, 487–493.

Khan, A., Su, C., German, M., Storch, G.A., Clifford, D., Sibley, L.D., 2005. Genotyping of *Toxoplasma gondii* strains from immunocompromised patients reveals high prevalence of type I strains. J. Clin. Micro. 43, 5881–5887.

Khan, I.A., Smith, K.A., Kasper, L.H., 1988a. Induction of antigen-specific parasiticidal cytotoxic T-cell splenocytes by a major membrane protein (P30) of *Toxoplasma gondii*. J. Immunol. 10, 3600–3605.

Khan, I., Eckel, M., Pfefferkorn, E., Kasper, L.H., 1988b. Production of gamma interferon by cultured human lymphocytes stimulated with a purified membrane protein (P30) from *Toxoplasma gondii*. J. Inf. Dis. 157 (5), 979–984.

Khan, I.A., Ely, K.H., Kasper, L.H., 1991. A purified parasite antigen (p30) mediates $CD8^+$ T-cell immunity against fatal *Toxoplasma gondii* infection in mice. J. Immunol. 157, 3501–3506.

Khan, I.A., Ely, K.H., Kasper, L.H., 1994. Antigen-specific $CD8^+$ T-cell clone protects against acute *Toxoplasma gondii* infection in mice. J. Immunol. 152, 1856–1860.

Khan, I.A., Matsuura, S., Fonseka, S., Kaasper, L.H., 1996a. Production of nitric oxide (NO) is not essential for protection against acute *Toxoplasma gondii* infection in IRF mice. J. Immunol. 156 (646), 43.

Khan, I.A., Matsuura, T., Kasper, L.H., 1996b. Activation-mediated $CD4^+$ T-cell unresponsiveness during acute *Toxoplasma gondii* infection in mice. Int. Immunol. 8, 887–896.

Khan, I.A., Schwartzman, J.D., Matsuura, T., Kasper, L.H., 1997. A dichotomous role for nitric oxide during acute *Toxoplasma gondii* infection in mice. Proc. Natl. Acad. Sci. USA 94, 13955–13960.

Khan, I.A., Casciotti, L., 1999. IL-15 prolongs the duration of $CD8^+$ T-cell mediated immunity in mice infected with a vaccine strain of *Toxoplasma gondii*. J. Immunol. 163, 4503–4509.

Khan, I.A., Murphy, P.M., Casciotti, L., Schwartzman, J.D., Collins, J., Gao, J.L., Yeaman, G.R., 2001. Mice lacking the chemokine receptor CCR1 show increased susceptibility to *Toxoplasma gondii* infection. J. Immunol. 166 (3), 1930–1937.

Khan, I.A., Moretto, M., Wei, X.Q., et al., 2002. Treatment with soluble interkleukin-15Ralpha exacerbates intracellular parasitic infection by blocking the development of memory CD8$^+$T cell response. J. Exp. Med. 195, 1463–1470.

Khan, I.A., Thomas, S.Y., Moretto, M.M., Lee, F.S., Islam, S.A., Combe, C., Schwartzman, J.D., Luster, A.D., 2006 Jun. CCR5 is essential for NK-cell trafficking and host survival following *Toxoplasma gondii* infection. PLoS Pathog. 2 (6), e49. Epub 2006 Jun 9.

Kikumura, A., Ishikawa, T., Norose, K., 2012 Sep. Kinetic analysis of cytokines, chemokines, chemokine receptors and adhesion molecules in murine ocular toxoplasmosis. Br. J. Ophthalmol. 96 (9), 1259–1267. http://dx.doi.org/10.1136/bjophthalmol-2012-301490. Epub 2012 Jul 11.

Kim, Y.M., de Vera, M.E., Watkins, S.C., Billiar, T.R., 1997. Nitric oxide protects cultured rat hepatocytes from tumour necrosis factor-alpha-induced apoptosis by inducing heat shock protein 70 expression. J. Biol. Chem. 272 (2), 1402–1411.

Kim, L., Butcher, B.A., Denkers, E.Y., 2004. *Toxoplasma gondii* interferes with lipopolysaccharide-induced mitogen-activated protein kinase activation by mechanisms distinct from endotoxin tolerance. J. Immunol. 172 (5), 3003–3010.

Kim, L., Butcher, B.A., Lee, C.W., Uematsu, S., Akira, S., Denkers, E.Y., 2006a. *Toxoplasma gondii* genotype determines MyD88-dependent signalling in infected macrophages. J. Immunol. 177, 2584–2591.

Kim, L., Del Rio, L., Butcher, B.A., Mogensen, T.H., Paludan, S.R., Flavell, R.A., Denkers, E.Y., 2005. p38 MAPK autophosphorylation drives macrophage IL-12 production during intracellular infection. J. Immunol. 174, 4178–4184.

Kim, S.K., Fouts, A.E., Boothroyd, J.C., 2008. *Toxoplasma gondii* dysregulates IFNgamma-inducible gene expression in human fibroblasts: insights from a genome-wide transcriptional profiling. J. Immunol. 178 (8), 5154–5165.

Kim, S.K., Karasov, A., Boothroyd, J.C., 2007 Apr. Bradyzoite-specific surface antigen SRS9 plays a role in maintaining *Toxoplasma gondii* persistence in the brain and in host control of parasite replication in the intestine. Infect. Immun. 75 (4), 1626–1634.

Kim, L., Denkers, E.Y., 2006b. *Toxoplasma gondii* triggers Gi-dependent PI 3-kinase signalling required for inhibition of host cell apoptosis. J. Cell Sci. 119, 2119–2126.

Kim, Y.M., Brinkmann, M.M., Paquet, M.E., Ploegh, H.L., 2008. UNC93B1 delivers nucleotide-sensing toll-like receptors to endolysosomes. Nature 452, 234–238.

Kirak, O., Frickel, E.M., Grotenbreg, G.M., Suh, H., Jaenisch, R., Ploegh, H.L., 2010. Transnuclear mice with predefined T-cell receptor specificities against *Toxoplasma gondii* obtained via SCNT. Science 328, 243–248.

Koblansky, A.A., Jankovic, D., Oh, H., Hieny, S., Sungnak, W., Mathur, R., Hayden, M.S., Akira, S., Sher, A., Ghosh, S., 2012 Dec 12. Recognition of Profilin by Toll-like Receptor 12 Is Critical for Host Resistance to *Toxoplasma gondii*. Immunity. doi:pii: S1074–7613(12)00515-8. 10.1016/j.immuni.2012.09.016.

Kobayashi, A., Suzuki, Y., 1987. Suppression of antibody responses by Toxoplasma infection in mice. Zentral Bakteriol. Mikrobiol. Hyg. 264, 312–318.

Kodjikian, L., Hoigne, I., Adam, O., Jacquier, P., Aebi-Ochsner, C., Aerbi, C., Garweg, J.G., 2004. Vertical transmission of toxoplasmosis from a chronically infected immunocompetent woman. Pediatr. Infect. Dis. J. 23, 272–274.

Konen-Waisman, S., Howard, J.C., 2007. Cell-autonomous immunity to *Toxoplasma gondii* in mouse and man. Microbes Infect. 9, 1652–1661.

Koppe, et al., 1974. Toxoplasma and pregnancy, with a long term follow-up of the children. Eur. J. Obstet Gynecol. Reprod. Biol. 413, 101–110.

Koppe, J.G., Kloosterman, G.J., 1982. Congenital toxoplasmosis: long-term follow-up. Padiatr. Padol. 17 (2), 171–179.

Koppe, J.G., Loewer-Sieger, D.H., de Roever-Bonnet, H., 1986. Results of 20-year follow-up of congenital toxoplasmosis. Lancet 1 (8475), 254–256.

Koppe, J.G., Rothova, A., 1989. Congenital toxoplasmosis: a long-term follow-up of 20 years. Int. Ophthalmol. 13, 387–390.

Kos, F.J., Engleman, E.G., 1996. Immune regulation: a critical link between NK-cells and CTLs. Immunol. Today 17 (4), 174–176.

Koshy, A.A., Dietrich, H.K., Christian, D.A., Melehani, J.H., Shastri, A.J., Hunter, C.A., Boothroyd, J.C., 2012. Toxoplasma co-opts host cells it does not invade. PLoS Pathog. 8 (7), e1002825. http://dx.doi.org/10.1371/journal.ppat.1002825.

Koshy, A.A., Fouts, A.E., Lodoen, M.B., Alkan, O., Blau, H.M., Boothroyd, J.C., 2010. Toxoplasma secreting Cre recombinase for analysis of host-parasite interactions. Nat. Methods 7 (4), 307–309. http://dx.doi.org/10.1038/nmeth.1438.

Krahenbuhl, J.L., Remington, J.S., McLeod, R., 1980. Cytotoxic and Microbicidal properties of macrophages. In: Phagocytes, Mononuclear (Ed.), Functional Aspects Parts I and II. R Van Furth. Martinus Midjhoff Publishers, The Hague, pp. 1631–1653.

Kroncke, K., Fehsel, K., Kolb-Bachyofer, V.K., 1998. Clin. Exp. Immunol.

Kropf, P., Freudenberg, N., Kalis, C., Modolell, M., Herath, S., Galanos, C., Freudenberg, M., Muller, I., 2004. Infection of C57BL/10ScCr and C57BL/10ScNCr mice with Leishmania major reveals a role for Toll-like receptor 4 in the control of parasite replication. J. Leukoc Biol. 76, 48–57.

Kubo, M., Hanada, T., Yoshimura, A., 2003. Suppressors of cytokine signalling and immunity. Nature Immunol. 4, 1169–1176.

Kudo, M., Aosai, F., Mun, H.S., Norose, K., Akira, S., Iwakura, Y., Yano, A., 2004. The role of IFN-gamma and Toll-like receptors in nephropathy induced by Toxoplasma gondii infection. Microbiol. Immunol. 48 (8), 617–628.

Kwon, O.C., Takahara, M., Katayama, T., Suzuki, Y., 1992. Fine structural and immuno-histological changes in prolactin cells in the anterior pituitaries during infection with *Toxoplasma gondii* in mice. Jikeikai Med. J. 39, 73–81.

Labalette, P., Delhaes, L., Margaron, F., Fortier, B., Rouland, J.F., 2001. Ocular toxoplasmosis after the fifth decade. Am. J. Ophthalmol. 133 (4), 506–515.

Lahmar, I., Guinard, M., Sauer, A., Marcellin, L., Abdelrahman, T., Roux, M., Mousli, M., Moussa, A., Babba, H., Pfaff, A.W., Candolfi, E., 2010 Feb. Murine neonatal infection provides an efficient model for congenital ocular toxoplasmosis. Exp. Parasitol. 124 (2), 190–196. http://dx.doi.org/10.1016/j.exppara.2009.09.010.

LaRosa, D.F., Stumhofer, J.S., Gelman, A.E., Rahman, A.H., Taylor, D.K., Hunter, C.A., Turka, L.A., 2008 Mar 11. T cell expression of MyD88 is required for resistance to *Toxoplasma gondii*. Proc. Natl. Acad. Sci. U S A 105 (10), 3855–3860. http://dx.doi.org/10.1073/pnas.0706663105.

Lambert, H., Hitziger, N., Dellacasa, I., Svensson, M., Barragan, A., 2006. Induction of dendritic cell migration upon *Toxoplasma gondii* infection potentiates parasite dissemination. Cell Microbiol. 8, 1611–1623.

Lambert, H., Vutova, P.P., Adams, W.C., Lore, K., Barragan, A., 2009. The *Toxoplasma gondii*-shuttling function of dendritic cells is linked to the parasite genotype. Infect. Immun. 77, 1679–1688.

Lamberton, P.H., Donnelly, C.A., Webster, J.P., 2008. Specificity of the *Toxoplasma gondii*-altered behaviour to definitive versus non-definitive host predation risk. Parasitology 135, 1143–1150.

Lang, C., Hildebrandt, A., Brand, F., Opitz, L., Dihazi, H., Lüder, C.G., 2012 Jan. Impaired chromatin remodelling at STAT1-regulated promoters leads to global unresponsiveness of *Toxoplasma gondii*-Infected macrophages to IFNγ. PLoS Pathog. 8 (1), e1002483 http://dx.doi.org/10.1371/journal.ppat.1002483.

Lang, C., Gross, U., Lüder, C.G., 2007. Subversion of innate and adaptive immune responses by *Toxoplasma gondii*. Parasitol. Res. 2, 191–203. Epub 2006 Oct 6.

Lang, C., Algner, M., Beinert, N., Gross, U., Lüder, C.G., 2006 Jul. Diverse mechanisms employed by *Toxoplasma gondii* to inhibit IFNgamma-induced major histocompatibility complex class II gene expression. Microbes Infect. 8 (8), 1994–2005. Epub 2006 Jun 12.

Langermans, J.A., Van der Hulst, M.E., Nibbering, P.H., Hiemstra, P.S., Fransen, L., Van Furth, R., 1992. IFNgamma-induced L-arginine-dependent toxoplasmastatic activity in murine peritoneal macrophages is mediated by endogenous tumour necrosis factor-alpha. J. Immunol. 148, 568–574.

Lecolier, B., et al., 1989. T-cell subpopulations of foetuses infected by *Toxoplasma gondii* [letter]. Eur. J. Clin. Microbiol. Infect. Dis. 8 (6), 572–573.

Lee, C.W., Sukhumavasi, W., Denkers, E.Y., 2007 Dec. Phosphoinositide-3-kinase-dependent, MyD88-independent induction of CC-type chemokines characterizes the macrophage response to *Toxoplasma gondii* strains with high virulence. Infect. Immun. 75 (12), 5788–5797.

Lee, Y.F., Chen, S.J., Chung, Y.M., Liu, J.H., Wong, W.W., 2000. Diffuse toxoplasmic retinochoroiditis as the initial manifestation of acquired immunodeficiency syndrome. J. Formos. Med. Assoc. 99 (3), 219–223.

Lees, Michael P., Fuller, Stephen J., McLeod, Rima, et al., 2010. P2X7 receptor-mediated killing of an intracellular parasite, *Toxoplasma gondii*, by human and murine macrophages. J. Immunol. 184, 7040–7046.

Leiva, L.E., Junprasert, J., Hollenbaugh, D., Sorensen, R.U., 1998. Central nervous system toxoplasmosis with an increased proportion of circulating gamma delta T-cells in a patient with hyper-IgM syndrome. J. Clin. Immunol. 18, 283–290.

Leng, J., Denkers, E.Y., 2009. *Toxoplasma gondii* inhibits covalent modification of histone H3 at the IL-10 promoter in infected macrophages. PLoS One 4 (10), e7589. http://dx.doi.org/10.1371/journal.pone.0007589.

Leng, J., Butcher, B.A., Egan, C.E., Abi Abdallah, D.S., Denkers, E.Y., 2009. *Toxoplasma gondii* prevents chromatin remodeling initiated by TLR-triggered macrophage activation. J. Immunol. 182 (1), 489–497.

Lepage, A.C., Buzoni-Gatel, D., Bout, D.T., Kasper, L.H., 1998. Gut-derived intraepithelial lymphocytes induce long term immunity against *Toxoplasma gondii*. J. Immunol. 161, 4902–4908.

Letscher-Bru, V., Pfaff, A., Abou-Bacar, A., et al., 2003. Vaccination with *Toxoplasma gondii* SAG1 protein is protective against congenital toxoplasmosis in BALB/c mice but not in CBA/J mice. Infect. Immun. 71, 6615–6619.

Lewis, D.B., Wilson, C.B., 2005. Developmental Immunology and Role of Host Defences in Foetal and Neonatal susceptibility to Infections. In: Remington, J.S., Klein, J., Baber, C., Wilson, C.B. (Eds.), Infectious Disease of the Foetus and Newborn Infant, Chapter 4, pp. 88–210.

Lewis, R.A., Austen, K.F., Soberman, R.J., 1990. Leukotrienes and other products of the 5-lipoxygenase pathway. Biochemistry and relation to pathobiology in human diseases. N. Engl. J. Med. 323 (19), 645–655.

Li, W., Buzoni-Gatel, D., Debbabi, H., Hu, M.S., Mennechet, F.J., Durell, B.G., Noelle, R.J., Kasper, L.H., 2002 Mar. CD40/CD154 ligation is required for the development of acute ileitis following oral infection with an intracellular pathogen in mice. Gastroenterology 122 (3), 762–.773

Lieberman, L.A., Banica, M., Reiner, S.L., Hunter, C.A., 2004a. STAT1 plays a critical role in the regulation of antimicrobial effector mechanisms, but not in the development of Th1-type responses during toxoplasmosis. J. Immunol. 172, 457–463.

Lieberman, L.A., Hunter, C.A., 2002. The role of cytokines and their signalling pathways in the regulation of immunity to *Toxoplasma gondii*. Int. Rev. Immunol. 21 (4-5), 373–403.

Lieberman, L.A., Cardillo, F., Owyang, A.M., Rennick, D.M., Cua, D.J., Kastelein, R.A., Hunter, C.A., 2004. IL-23 provides a limited mechanism of resistance to acute toxoplasmosis in the absence of IL-12. J. Immunol. 173 (3), 1887–1893.

Liesenfeld, O., Kosek, J., Remington, J.S., Suzuki, Y., 1996. Association of $CD4^+$ T-cell-dependent, interferon-gamma-mediated necrosis of the small intestine with genetic susceptibility of mice to peroral infection with *Toxoplasma gondii*. J. Exp. Med. 184, 597–607.

Liesenfeld, O., Kosek, J.C., Suzuki, Y., 1997. Gamma interferon induces Fas-dependent apoptosis of Peyer's patch T-cells in mice following peroral infection with *Toxoplasma gondii*. Infect. Immun. 65, 4682–4689.

Liesenfeld, O., Kang, H., Park., D., Nguyen, T.A., Parkhe, C.V., Watanabe, H., Abo, T., Sher, A., Remington, J.S., Suzuki, Y., 1999. TNF-alpha, nitric oxide and IFNgamma are all critical for development of necrosis in the small intestine and early mortality in genetically susceptible mice infected perorally with *Toxoplasma gondii*. Parasite Immunol. 21, 365–376.

Liesenfeld, O., 2002. Oral infection of C57BL/6 mice with *Toxoplasma gondii*: a new model of inflammatory bowel disease. J. Infect. Dis. 185 (Suppl 1), S96–S101.

Liesenfeld, O., Nguyen, T.A., Pharke, C., Suzuki, Y., 2001. Importance of gender and sex hormones in regulation of susceptibility of the small intestine to peroral infection with *Toxoplasma gondii* tissue cysts. J. Parasitol. 87 (6), 1491–1493.

Liesenfeld, O., Dunay, I.R., Erb, K.J., 2004. Infection with *Toxoplasma gondii* reduces established and developing Th2 responses induced by Nippostrongylus brasiliensis infection. Infect. Immun. 72, 3812–3822.

Liesenfeld, O., Parvanova, I., Zerrahn, J., Han, S.J., Heinrich, F., Muñoz, M., Kaiser, F., Aebischer, T., Buch, T., Waisman, A., Reichmann, G., Utermöhlen, O., von Stebut, E., von Loewenich, F.D., Bogdan, C., Specht, S., Saeftel, M., Hoerauf, A., Mota, M.M., Könen-Waisman, S., Kaufmann, S.H., Howard, J.C., 2011. The IFNγ-inducible GTPase, Irga6, protects mice against *Toxoplasma gondii* but not against Plasmodium berghei and some other intracellular pathogens. PLoS One 6 (6), e20568. http://dx.doi.org/10.1371/journal.pone.0020568. Epub 2011 Jun 17.

Lim, A., Kumar, V., Hari Dass, S.A., Vyas, A., 2012. *Toxoplasma gondii* infection enhances testicular steroidogenesis in rats. Mol. Ecol. 1, 102–110.

Lin, H., Mosmann, T.R., Guilbert, L., Tuntipopipat, S., Wegmann, T.G., 1993. Synthesis of T helper 2-type cytokines at the maternal foetal interface. J. Immunol. 151, 4562–4573.

Lindberg, R.E., Frenkel, J.K., 1977. Toxoplasmosis in nude mice. J. Parasitol. 63 (2), 219–221.

Ling, Y.M., Shaw, M.H., Ayala, C., Coppens, I., Taylor, G.A., Ferguson, D.J., Yap, G.S., 2006. Vacuolar and plasma membrane stripping and autophagic elimination of *Toxoplasma gondii* in primed effector macrophages. J. Exp. Med. 203, 2063–2071.

Liu, C.H., Fan, Y.T., Dias, A., Esper, L., Corn, R.A., Bafica, A., Machado, F.S., Aliberti, J., 2006. Cutting edge: dendritic cells are essential for in vivo IL-12 production and development of resistance against *Toxoplasma gondii* infection in mice. J. Immunol. 177, 31–35.

Locksley, et al., 1983. Oxygen-dependent microbicidal systems of phagocytes against intracellular protazoal. J. Cell Biochem. 22, 173–180.

Locksley, et al., 1985. Alteration of leukotriene release by macrophages ingesting *Toxoplasma gondii*. Proc. Natl. Acad. Sci. USA 82, 6922–6927.

Locksley, R.M., Wilson, C.B., Klebanoff, S.J., 1982. Role for endogenous and acquired peroxidase in the toxoplasmacidal activity of murine and human mononuclear phagocytes. J. Clin. Invest. 69, 1099–1111.

Lu, F., Huang, S., Kasper, L.H., 2003. Interleukin-10 and pathogenesis of murine ocular toxoplasmosis. Infect. Immun. 71, 7159–7163.

Lu, F., Huang, S., et al., 2004. $CD4^+$ T-cells in the pathogenesis of murine ocular toxoplasmosis. Infect. Immun. 72 (9), 4966–4972.

Lu, F., Huang, S., et al., 2005. Experimental ocular toxoplasmosis in genetically susceptible and resistant mice. Infect. Immun. 73 (8), 5160–5165.

Lu, F., Huang, S., Kasper, L.H., 2009 Jan 22. The temperature-sensitive mutants of *Toxoplasma gondii* and ocular toxoplasmosis. Vaccine 27 (4), 573–580. http://dx.doi.org/10.1016/j.vaccine.2008.10.090. Epub 2008 Nov 19.

Luangsay, S., Kasper, L.H., Rachinel, N., Minns, L.A., Mennechet, F.J., Vandewalle, A., Buzoni-Gatel, D., 2003. CCR5 mediates specific migration of Toxoplasma gondii-primed CD8 lymphocytes to inflammatory intestinal epithelial cells. Gastroenterology 125 (2), 491–500.

Lucet, J., et al., 1993. Septic shock due to toxoplasmosis in patients with the human immunodeficiency virus. Chest 104, 1054–1058.

Lüder, C.H., et al., 1998. Down-regulation of MHC class II molecules and inability to up-regulate class I molecules in murine macrophges after infection with *Toxoplasma gondii*. Clin. Exp. Immunol. 112 (2), 308–316.

Lüder, C.H., et al., 2003a. Reduced expression of the inducible nitric oxide synthase after infection with *Toxoplasma gondii* facilitates parasite replication in activated murine macrophages. Int. J. Parasitol. 33, 833–844.

Lüder, C.H., et al., 2003b. *Toxoplasma gondii* inhibits MHC class II expression in neural antigen-presenting cells by down-regulating the class II transactivator CIITA. J. Neuroimmunol. 134, 12–24.

Lüder, C., Seeber, F., 2001. *Toxoplasma gondii* and MHC-restricted antigen presentation: on degradation, transport and modulation. Int. J. Parasitol. 31(12), 1355–1369.

Lüder, C., Walter, W., Beuerle, B., Maeurer, U., Gross, M., 2001. *Toxoplasma gondii* down-regulates MHC class II gene expression and antigen presentation by murine macrophages via interference with nuclear translocation of STAT1alpha. Eur. J. Immunol. 31, 1475–1484.

Lüder, C., Giraldo-Velasquez, M., Sendtner, M., Gross, U., 1999. *Toxoplasma gondii* in Primary Rat CNS Cells: Differential Contribution of Neurons, Astrocytes, and Microglial Cells for the Intracerebral Development and Stage Differentiation. Exp. Parasitol. 93, 23–32.

Lüder, C.G., Campos-Salinas, J., Gonzalez-Rey, E., van Zandbergen, G., 2010. Impact of protozoan cell death on parasite-host interactions and pathogenesis. Parasit. Vectors 3, 116.

Lüder, C.G., Stanway, R.R., Chaussepied, M., Langsley, G., Heussler, V.T., 2009. Intracellular survival of apicomplexan parasites and host cell modification. Int. J. Parasitol. 39, 163–173. Epub 2008 Oct 25. Review.

Luft, B., Kansas, G., Engleman, E., Remington, J., 1984a. Functional and quantitative alterations in T lymphocyte subpopulations in acute toxoplasmosis. J. Infect. Dis. 150, 761–767.

Luft, B., Brooks, R., Conley, F., et al., 1984b. Toxoplasmic encephalitis in patients with acquired immune deficiency syndrome. JAMA 252, 913–917.

Luft, B.J., Naot, Y., arujo, F.G., Stinson, E.B., Remington, J.S., 1983. Primary and reactivated toxoplasmosis infection in patients with cardiac transplants. Ann. Intern. Med. 99, 27–31.

Lykens, J.E., Terrell, C.E., Zoller, E.E., Divanovic, S., Trompette, A., Karp, C.L., Aliberti, J., Flick, M.J., Jordan, M.B., 2010. Mice with a selective impairment of IFNgamma signalling in macrophage lineage cells demonstrate the critical role of IFNgamma-activated macrophages for the control of protozoan parasitic infections in vivo. J. Immunol. 184, 877–885.

Lyons, R.E., Anthony, J.P., Ferguson, D.J., Byrne, N., Alexander, J., Roberts, F., Roberts, C.W., 2001. Immunological studies of chronic ocular toxoplasmosis: up-regulation of major histocompatibility complex class I and transforming growth factor beta and a protective role for interleukin-6. Infect. Immun. 69, 2589–2595.

Lyons, R.E., Johnson, A.M., 1995. Heat shock proteins of *Toxoplasma gondii*. Parasite Immunol. 17 (7), 353–359.

Lyons, R.E., Johnson, A.M., 1998. Gene sequence and transcription differences in 70 kDa heat shock protein correlate with murine virulence of *Toxoplasma gondii*. Int. J. Parasitol. 28 (7), 1041–1051.

Ma, X., et al., 1996. The interleukin-12 p40 promoter is primed by interferon gamma in moncytic cells. J. Exp. Med. 183, 147–157.

Machado, F.S., Johndrow, J.E., Esper, L., Dias, A., Bafica, A., Serhan, C.N., Aliberti, J., 2006a. Anti-inflammatory actions of lipoxin A4 and aspirin-triggered lipoxin are SOCS-2 dependent. Nat. Med. 12, 330–334.

Machado, F.S., Johndrow, J.E., Esper, L., Dias, A., Bafica, A., Serhan, C.N., Aliberti, J., 2006b. Anti-inflammatory actions of lipoxin A4 and aspirin-triggered lipoxin are SOCS-2 dependent. Nat. Med. 12 (3), 330–334. Epub 2006 Jan 15.

Machado, A.V., Caetano, B.C., Barbosa, R.P., Salgado, A.P., Rabelo, R.H., Garcia, C.C., Bruna-Romero, O., Escriou, N., Gazzinelli, R.T., 2010 Apr 19. Prime and boost immunization with influenza and adenovirus encoding the *Toxoplasma gondii* surface antigen 2 (SAG2) induces strong protective immunity. Vaccine 28 (18), 3247–3256. http://dx.doi.org/10.1016/j.vaccine.2010.02.003.

Mack, D.M., Holfels, E., McLeod, R., in preparation. Immune responses in human congenital toxoplasmosis.

Mack, D.G., Johnson, J.J., Roberts, F., Roberts, C.W., Estes, R.G., David, C., Grumet, F.C., McLeod, R., 1999. HLA-class II genes modify outcome of *Toxoplasma gondii* infection. Int. J. Parasitol. 29, 1351–1358.

Mackay, C.R., 2001. Chemokines: immunology's high impact factors. Nat. Immunol. 2 (2), 95–101.

MacMicking, J.D., 2012. Interferon-inducible effector mechanisms in cell-autonomous immunity. Nature reviews. Immunology 12, 367–382.

Maeckner, H., et al., 2005. TWEAK Attenuates the Transition from Innate to Adaptive Immunity. Cell 123, 931–944.

Mahamed, D.A., Mills, J.H., Egan, C.E., Denkers, E.Y., Bynoe, M.S., 2012 Oct 2. CD73-generated adenosine facilitates *Toxoplasma gondii* differentiation to long-lived tissue cysts in the central nervous system. Proc. Natl. Acad. Sci. U S A 109 (40), 16312–16317. http://dx.doi.org/10.1073/pnas.1205589109. Epub 2012 Sep 17.

Mahmoud, A.A., Warren, K.S., Strickland, G.T., 1976. Acquired resistance to infection with Schistosoma mansoni induced by *Toxoplasma gondii*. Nature 263, 56–57.

Mahmoud, A.A., Strickland, G.T., Warren, K.S., 1977. Toxoplasmosis and the host-parasite relationship in murine infection with schistosomi mansoni. J. Infect. Dis. 135, 408–413.

Maggi, E., et al., 1994. Ability of HIV to promote a Th1 to Th0 shift and to replicate preferentially in Th2 and Th0 cells. Science 265, 244–248.

Makino, M., Uemura, N., Moroda, M., Kikumura, A., Piao, L.X., Mohamed, R.M., Aosai, F., 2011 Feb 24. Innate immunity in DNA vaccine with Toxoplasma gondii-heat shock protein 70 gene that induces DC activation and Th1 polarization. Vaccine 29 (10), 1899–1905. http://dx.doi.org/10.1016/j.vaccine.2010.12.118. Epub 2011 Jan 12.

Makioka, A., Suzuki, Y., Kobayashi, A., 1991. Recognition of tachyzoite and bradyzoite antigens of *Toxoplasma gondii* by infected hosts. Infect. Immun. 59, 2763–2766.

Mandal, A., Zhang, Z., Kim, S., Tsai, P., Mukherjee, 2005. Yin-yang: balancing act of prostaglandins with opposing functions to regulate inflammation. J. Immunol. 175 (10), 6271–6273.

Manger, I., Hehl, A., Boothroyd, J., 1998. The surface of Toxoplasma tachyzoites is dominated by a family of glycosylphosphatidylinositol-anchored antigens related to SAG1. Infect. Immun. 66, 2237–2244.

Marchant, A., Appay, V., Van Der Sande, M., et al., 2003. Mature $CD8^+$ T lymphocyte response to viral infection during foetal life. J. Clin. Invest. 111, 1747–1755.

Marodi, L., Kaposzta, R., Campbell De, et al., 1994. Candidacidal mechanisms in the human neonate. Impaired IFNgamma activation of macrophages in newborn infants. J. Immunol. 153, 5643–5649.

Marshall, A.J., Denkers, E.Y., 1998. *Toxoplasma gondii* triggers granulocyte-dependent cytokine-mediated lethal shock in D-galactosamine-sensitized mice. Infect. Immun. 66, 1325–1333.

Marshall, A.J., Brunet, L.R., van Gessel, Y., Alcaraz, A., Bliss, S.K., Pearce, E.J., Denkers, E.Y., 1999. *Toxoplasma gondii* and Schistosoma mansoni synergize to promote hepatocyte dysfunction associated with high levels of plasma TNF-alpha and early death in C57BL/6 mice. J. Immunol. 163, 2089–2097.

Masek, K.S., Zhu, P., Freedman, B.D., Hunter, C.A., 2007. *Toxoplasma gondii* induces changes in intracellular calcium in macrophages. Parasitology 134, 1973–1979.

Mashayekhi, M., Sandau, M.M., Dunay, I.R., Frickel, E.M., Khan, A., Goldszmid, R. S., Sher, A., Ploegh, H.L., Murphy, T.L., Sibley, L.D., Murphy, K.M., 2011 Aug 26. $CD8\alpha(^+)$ dendritic cells are the critical source of interleukin-12 that controls acute infection by *Toxoplasma gondii* tachyzoites. Immunity 35 (2), 249–259. http://dx.doi.org/10.1016/j.immuni.2011.08.008.

Mason, N.J., Artis, D., Hunter, C.A., 2004a. New lessons from old pathogens: what parasitic infections have taught us about the role of nuclear factor-kappaB in the regulation of immunity. Immunol. Rev. 201, 48–56.

Mason, N.J., Fiore, J., Kobayashi, T., Masek, K.S., Choi, Y., Hunter, C.A., 2004b. TRAF6-dependent mitogen-activated protein kinase activation differentially regulates the production of interleukin-12 by macrophages in response to Toxoplasma gondii. Infection Immun. 72, 5662–5667.

Masson, D., Tschopp, J., 1987. A family of serine esterases in lytic granules of cytolytic T-lymphocytes. Cell 49 (5), 679–685.

McAuley, J., Boyer, K.M., Patel, D., Mets, M., Swisher, C., Roizen, N., Wolters, C., Stein, L., Stein, M., Schey, W., et al., 1994. Early and longitudinal evaluations of treated infants and children and untreated historical patients with congenital toxoplasmosis: the Chicago Collaborative Treatment Trial. Clin. Infect. Dis. 18 (1), 38–72.

McBerry, C., Egan, C.E., Rani, R., Yang, Y., Wu, D., Boespflug, N., Boon, L., Butcher, B., Mirpuri, J., Hogan, S.P., Denkers, E.Y., Aliberti, J., Herbert, D.R., 2012 Sep 15. Trefoil factor 2 negatively regulates type 1 immunity against *Toxoplasma gondii*. J. Immunol. 189 (6), 3078–3084. http://dx.doi.org/10.4049/jimmunol.1103374.

McGeachy, M.J., Chen, Y., Tato, C.M., Laurence, A., Joyce-Shaikh, B., Blumenschein, W.M., McClanahan, T.K., O'Shea, J.J., Cua, D.J., 2009 Mar. The interleukin 23 receptor is essential for the terminal differentiation of interleukin 17-producing effector T helper cells in vivo. Nat. Immunol. 10 (3), 314–324. http://dx.doi.org/10.1038/ni, 1698. Epub 2009 Feb 1.

McGwire, B.S., Young, J.D., 2005. Infliximab and reactivation of cerebral toxoplamosis. NEJM 353 (14), 1531.

McLeod, R., Remington, J.S., 1977. A new method for evaluation of intracellular inhibition of multiplication or killing by mononuclear phagocytes. J. Immunol. 119, 1894.

McLeod, R., Remington, J.S., 1979. A method to evaluate the capacity of monocytes and macrophages to inhibit multiplication of an intracellular pathogen. J. Immunol. Methods 27, 19–29.

McLeod, R., Bensch, K.G., Smith, S.M., Remington, J.S., 1980. Effects of human peripheral blood monocytes, monocyte-derived macrophages and spleen mononuclear cells on *Toxoplasma gondii*. Cell Immunol. 54, 330–350.

McLeod, R., Estes, R., 1982. Microbicidal activity of peripheral blood monocytes from patients with Hodgkin's disease for *Toxoplasma gondii*. JID 146, 565.

McLeod, R., Estes, R., McLeod, E.G., Mack, D., 1983. Effects of human alveolar macrophages and peripheral blood monocytes on *Toxoplasma gondii*. JID 147, 957.

McLeod, R., Mack, D., McLeod, E., Estes, R., Campbell, E., 1985a. Alveolar macrophage function and inflammatory stimuli in smokers with and without COPD. Amer Rev. Resp. Dis. 131, 377–384.

McLeod, R., Mack, D., 1986. Secretory IgA specific for *Toxoplasma gondii*. J. Immunol. 136, 2640–2643.

McLeod, et al., 1988. Subcutaneous and intestinal vaccination with tachyzoites of *Toxoplasma gondii* and acquisition of immunity to peroral and congenital toxoplasma challenge. J. Immunol. 140 (5), 1632–1637.

McLeod, R., et al., 1990. Phenotypes and functions of lymphocytes in congenital toxoplasmosis. J. Lab. Clin. Med. 116 (5), 623–635.

McLeod, R., Beem, M.O., Estes, R.G., 1985b. Lymphocyte anergy specific to *Toxoplasma gondii* antigens in a baby with congenital toxoplasmosis. J. Clin. Lab Immunol. 17, 149–153.

McLeod, R., Cohen, H., Estes, R., 1984. Immune response of mice to ingested *Toxoplasma gondii*: a model of Toxoplasma infection acquired by ingestion. J. Inf. Disease 149 (2), 234–244.

McLeod, R., Dowell, M., 2000. Basic Immunology: the foetus and the newborn. In: Ambroise-Thomas, P., Peterson, E. (Eds.), Congenital Toxoplasmosis: Scientific Background, Clinical Management and Control. Springer, New York, pp. 37–68.

McLeod, R., Eisenhauer, P., Mack, D., Filice, G., Spitalny, G., 1989a. Immune responses associated with early survival after peroral infection with *Toxoplasma gondii*. J. Immunol. 142, 3247–3255.

McLeod, R., Skamene, E., Brown, C.R., Eisenhauer, P.B., Mack, D.G., 1989b. Genetic regulation of early survival and cyst number after peroral *Toxoplasma gondii* infection of A x B/B x A recombinant inbred and B10 congenic mice. J. Immunol. 143, 3031–3034.

McLeod, R., Mack, D., Foss, R., Boyer, K., Withers, S., Levin, S., Hubbell, J., 1992. Levels of pyrimethamine in sera and cerebrospinal and ventricular fluids from infants treated for congenital toxoplasmosis. Antimicrob. Agents Chemother. 36, 1040–1048.

McLeod, R., Bushman, E., Arbuckle, D., Skamene, E., 1995. Immunogenetics in the analysis of resistance to intracellular pathogens. Curr. Opin. Immunol. 7, 539–545.

McLeod, R.L., Johnson, J.J., Estes, E., Mack, D., 1996. *Toxoplasma gondii* immunogenetics in pathogenesis of and protection against toxoplasmosis. In: Gross, U. (Ed.), Current Topics in Microbiology and Immunology. Springer, Heidelberg, pp. 95–112.

McLeod, R., Boyer, K.M., Lee, D., Mui, E., Wroblewski, K., Karrison, T., Noble, A.G., Withers, S., Swisher, C.N., Heydemann, P.T., Sautter, M., Babiarz, J., Rabiah, P., Meier, P., Grigg, M.E., 2012 Jun. Toxoplasmosis Study Group. Prematurity and severity are associated with *Toxoplasma gondii* alleles (NCCCTS, 1981-2009). Clin. Infect. Dis. 54 (11), 1595–1605. http://dx.doi.org/10.1093/cid/cis258.

Mead, P.S., Slutsker, L., Dietz, V., McCaig, L.F., Bresee, J.S., Shapiro, C., Griffin, P.M., Tauxe, R.V., 1999. Food-related illness and death in the USA. Emerg. Infect. Dis. 5, 607–625.

Melo, M.B., Kasperkovitz, P., Cerny, A., Könen-Waisman, S., Kurt-Jones, E.A., Lien, E., Beutler, B., Howard, J.C., Golenbock, D.T., Gazzinelli, R.T., 2010 Aug 26. UNC93B1 mediates host resistance to infection with *Toxoplasma gondii*. PLoS Pathog. 6 (8), e1001071. http://dx.doi.org/10.1371/journal.ppat.1001071.

Melo, M.B., Jensen, K.D., Saeij, J.P., 2011 Nov. *Toxoplasma gondii* effectors are master regulators of the inflammatory response. Trends Parasitol. 27 (11), 487–495. http://dx.doi.org/10.1016/j.pt.2011.08.001.

Melzer, T.C., Cranston, H.J., Weiss, L.M., Halonen, S.K., 2010. Host Cell Preference of *Toxoplasma gondii* Cysts in Murine Brain: A Confocal Study. Journal of Neuroparasitology. 1 (1), 1–12.

Melzer, T., Duffy, A., Weiss, L.M., Halonen, S.K., 2008 Nov. The gamma interferon (IFNgamma)-inducible GTP-binding protein IGTP is necessary for toxoplasma vacuolar disruption and induces parasite egression in IFNgamma-stimulated astrocytes. Infect. Immun. 76 (11), 4883–4894. http://dx.doi.org/10.1128/IAI.01288-07.

Menard, L.C., Minns, L.A., Darche, S., Mielcarz, D.W., Foureau, D.M., Roos, D., Dzierszinski, F., Kasper, L.H., Buzoni-Gatel, D., 2007. B-cells amplify IFNgamma production by T-cells via a TNF-alpha-mediated mechanism. J. Immunol. 179 (7), 4857–4866.

Mendes, E.A., Caetano, B.C., Penido, M.L., Bruna-Romero, O., Gazzinelli, R.T., 2011 Jun 15. MyD88-dependent protective immunity elicited by adenovirus 5 expressing the surface antigen 1 from *Toxoplasma gondii* is mediated by CD8(+) T-lymphocytes. Vaccine 29 (27), 4476–4484. http://dx.doi.org/10.1016/j.vaccine.2011.04.044.

Mennechet, F.J., Hasper, L.H., Rachinel, N., Minns, L.A., Luangsay, S., Vandewalle, A., Buzoni-Gatel, D., 2004. Intestinal intraepithelial lymphocytes prevent pathogen-driven inflammation and regulate the Smad/T-bet pathway of lamina propria CD4+ T-cells. Eur. J. Immunol. 34, 1059–1067.

Mennechet, F.J., Kasper, L.H., Rachinel, N., Li, W., Vandewalle, A., Buzoni-Gatel, D., 2002. Lamina propria CD4+ T-lymphocytes synergize with murine intestinal epithelial cells to enhance proinflammatory response against an intracellular pathogen. J. Immunol. 168 (6), 2988–2996.

Menzies, F.M., Henriquez, F.L., Roberts, C.W., 2008. Immunological control of congenital toxoplasmosis in the murine model. Immunology Letters 115, 83–89.

Mévélec, M.N., Ducournau, C., Bassuny Ismael, A., Olivier, M., Sèche, E., Lebrun, M., Bout, D, Dimier-Poisson, I., 2010 Jul-Aug. Mic1-3 Knock-out *Toxoplasma gondii* is a good candidate for a vaccine against *T. gondii*-induced abortion in sheep. Vet. Res. 41 (4), 49. http://dx.doi.org/10.1051/vetres/2010021.

Mercier, C., et al., 1998. Targeted disruption of the GRA2 locus in toxoplasma gondii decreases host cell adhesion and drastically reduces virulence in mice. Infect. Immun. 66, 4176–4182.

Mets, M.B., Holfels, E., Boyer, K.M., Swisher, C.N., Roizen, N., et al., 1997. Eye manifestations of congenital toxoplasmosis. Am. J. Ophthalmol. 123 (1), 1–16.

Meylan, F., Davidson, T.S., Kahle, E., Kinder, M., Acharya, K., Jankovic, D., Bundoc, V., Hodges, M., Shevach, E.M., Keane-Myers, A., Wang, E.C., Siegel, R.M., 2008 Jul 18. The TNF-family receptor DR3 is essential for diverse T-cell-mediated inflammatory diseases. Immunity 29 (1), 79–89. http://dx.doi.org/10.1016/j.immuni.2008.04.021. Epub 2008 Jun 19.

Middleton, M.K., Zukas, A.M., Rubinstein, T., Kinder, M., Wilson, E.H., Zhu, P., Blair, I.A., Hunter, C.A., Puré, E., 2009. 12/15-lipoxygenase-dependent myeloid production of interleukin-12 is essential for resistance to chronic toxoplasmosis. Infect. Immun. 12, 5690–5700.

Miller, C., Smith, N., Johnson, A.M., 1999a. Cytokines, nitric oxide, heat shock proteins and virulence in Toxoplasma. Parasitol. Today 15, 418.

Miller, C.M.D., et al., 1999b. Immunogenicity of toxoplasma gondii heat shock protein 70. J. Protozool. Res. 9, 113.

Miller, C.M.D., et al., 2000. the production of a 70 kDa heat shock protein by toxoplasma gondii RH strain in immunocompromised mice. Int. J. Parasitol. 30, 1476.

Miller C., M., Zakrzewski, A.M., Ikin, R.J., Boulter, N.R., Katrib, M., Lees, M.P., Fuller, S.J., Wiley, J.S., Smith, N.C., 2011a. Dysregulation of the inflammatory response to the parasite, *Toxoplasma gondii*, in P2X7 receptor-deficient mice. Int. J. Parasitol. 41, 301–308.

Miller, C.M., Boulter, Nicola R., Fuller, Stephen J., Zakrzewski, Alana M., Lees, Michael P., et al., 2011b. The role of the P2X7 receptor in infectious diseases. PLoS Pathog. 11, 1–6.

Minamidani, M., Tanaka, J., Suzuki, Y., 1996. Pathomechanism of cerebral hypoplasia in experimental toxoplasmosis in murine foetuses. Early Human Develop. 44, 37–50.

Mineo, J.R., McLeod, R., Mack, D.M., et al., 1993. Antibodies to *Toxoplasma gondii* major surface protein SAG1, P30 inhibit infection of host cells and are produced in murine intestine after peroral infection. J. Immunol. 150, 3951–3964.

Minns, L.A., Menard, L.C., Foureau, D.M., Darche, S., Ronet, C., Mielcarz, D.W., Buzoni-Gatel, D., Kasper, L.H., 2006. TLR9 is required for the gut-associated lymphoid tissue response following oral infection of *Toxoplasma gondii*. J. Immunol. 176 (12), 7589–7597.

Minns, A., Buzoni-Gatel, D., Ely, K.H., Rachinel, N., Luangsay, S., Kasper, L.H., 2004. A novel triterpenoid induces transforming growth factor beta production by intraepithelial lymphocytes to prevent ileitis. Gastroenterology 127, 119–126.

Minot, S., Melo, M.B., Li, F., Lu, D., Niedelman, W., Levine, S.S., Saeij, J.P., 2012 Aug 14. Admixture and recombination among *Toxoplasma gondii* lineages explain global genome diversity. Proc. Natl. Acad. Sci. U S A 109 (33), 13458–13463. http://dx.doi.org/10.1073/pnas.1117047109.

Mirpuri, J., Yarovinsky, F., 2012. IL-6 signalling SOCS critical for IL-12 host response to *Toxoplasma gondii*. Future Microbiol. 1, 13–16. http://dx.doi.org/10.2217/fmb.11.147.

Mitchell, C.D., Erlich, S.S., Mastrucci, M.T., Hutto, S.C., Parks, W.P., Scott, G.B., 1990. Congenital toxoplasmosis occurring in infants perinatally infected with human immunodeficiency virus 1. Pediatr. Infect. Dis. J. 9 (7), 512–518.

Mitchell, S.M., Zajac, A.M., Kennedy, T., Davis, W., Dubey, J.P., Lindsay, D.S., 2006. Prevention of recrudescent toxoplasmic encephalitis using ponazuril in an immunodeficient mouse model. J. Eukaryot. Microbiol. 53 (Suppl 1), S164–S165.

Mitra, R., Sapolsky, R.M., Vyas, A., 2012 Oct 25. *Toxoplasma gondii* infection induces dendritic retraction in basolateral amygdala accompanied by reduced corticosterone secretion. Dis. Model Mech. 6 (2), 516–20. http://dx.doi.org/10.1242/dmm.009928. Epub 2012 Oct 25.

Molestina, R., Payne, T., Coppens, I., Sinai, A., 2003. Activation of NF-kB by *Toxoplasma gondii* correlates with increased expression of antiapoptotic genes and localization of phosphorylated IkB to the parasitophorous vacuole membrane. J. Cell Sci. 116, 4359–4371.

Molestina, R.E., Sinai, A.P., 2005. Detection of a novel parasite kinase activity at the *Toxoplasma gondii* parasitophorous vacuole membrane capable of phosphorylating host IkappaBalpha. Cell Microbiol. 7 (3), 351–362.

Montecino-Rodriguez, E., Leathers, H., Dorshkind, K., 2006. Identification of a B-1 B-cell-specified progenitor. Nat. Immunol. 7 (3), 293–301.

Moore, K.W., O'Garra, A., de Waal Malefyt, R., Vieira, P., Mosmann, T.R., 1993. Interleukin-10. Annu. Rev. Immunol. 11, 165–190.

Montfort, A., de Badts, B., Douin-Echinard, V., Martin, P.G., Iacovoni, J., Nevoit, C., Therville, N., Garcia, V., Bertrand, M.A., Bessières, M.H., Trombe, M.C., Levade, T., Benoist, H., Ségui, B., 2009 Oct 15. FAN stimulates TNF(alpha)-induced gene expression, leukocyte recruitment, and humoral response. J. Immunol. 183 (8), 5369–5378. http://dx.doi.org/10.4049/jimmunol.0803384. Epub 2009 Sep 28.

Montoya, J.G., et al., 1996. Human $CD4^+$ and $CD8^+$ T-lymphocytes are both cytotoxic to Toxoplasma gondii-infected cells. Infect. Immun. 64 (1), 176–181.

Montoya, G., Remington, J.S., 1997. Toxoplasmosis of the Central Nervous System. In: Peterson, R. (Ed.), In Defence of the brain. Blackwell Science, Malden, pp. 163–188.

Mordue, D.G., Scott-Weathers, C.F., Tobin, C.M., Knoll, L.J., 2007. A patatin-like protein protects *Toxoplasma gondii* from degradation in activated macrophages. Mol. Microbiol. 63, 482–496.

Mordue, D.G., Monroy, F., La Regina, M., Dinarello, C.A., Sibley, L.D., 2001. Acute toxoplasmosis leads to lethal overproduction of Th1 cytokines. J. Immunol. 167, 4574–4584.

Mordue, D.G., Desai, N., Dustin, M., Sibley, L.D., 1999a. Invasion by *Toxoplasma gondii* establishes a moving junction that selectively excludes host cell plasma membrane proteins on the basis of their membrane anchoring. J. Exp. Med. 190 (12), 1783–1792.

Mordue, D.G., Hakansson, S., Niesman, I., Sibley, L.D., 1999b. *Toxoplasma gondii* resides in a vacuole that avoids fusion with host cell endocytic and exocytic vesicular trafficking pathways. Exp. Parasitol. 92 (2), 87–99.

Mordue, D.G., et al., 1997. Intracellular fate of vacuoles containing *Toxoplasma gondii* is determined att ehtime of formation and depends upon the mechanism of entry. J. Immunol. 159, 4452–4459.

Morgado, P., Ong, Y.C., Boothroyd, J.C., Lodoen, M.B., 2011. *Toxoplasma gondii* induces B7-2 expression through activation of JNK signal transduction. Infect. Immun. 79 (11), 4401–4412. http://dx.doi.org/10.1128/IAI.05562-11.

Morley, E.K., Williams, R.H., Hughes, J.M., Terry, R.S., Duncanson, P., Smith, J.E., Hide, G., 2005. Significant familial differences in the frequency of abortion and *Toxoplasma gondii* infection within a flock of Charollais sheep. Parasitology 131, 181–185.

Morrell, V., 1995. Zeroing in on how hormones affect the immune system. Science 269, 773–775.

Mun, H.S., Aosai, F., Norose, K., Chen, M., Piao, L.X., Takeuchi, O., Akira, S., Ishikura, H., Yano, A., 2003. TLR2 as an essential molecule for protective immunity against *Toxoplasma gondii* infection. Int. Immunol. 15, 1081–1087.

Mun, H.S., Aosai, F., Norose, K., Piao, L.X., Fang, H., Akira, S., Yano, A., 2005. Toll-like receptor 4 mediates tolerance in macrophages stimulated with Toxoplasma gondii-derived heat shock protein 70. Infect. Immun. 8, 4634–4642.

Muñoz, M., Heimesaat, M.M., Danker, K., Struck, D., Lohmann, U., Plickert, R., Bereswill, S., Fischer, A., Dunay, I.R., Wolk, K., Loddenkemper, C., Krell, H. W., Libert, C., Lund, L.R., Frey, O., Hölscher, C., Iwakura, Y., Ghilardi, N., Ouyang, W., Kamradt, T., Sabat, R., Liesenfeld, O., 2009 Dec 21. Interleukin (IL)-23 mediates Toxoplasma gondii-induced immunopathology in the gut via matrixmetalloproteinase-2 and IL-22 but independent of IL-17. J. Exp. Med. 206 (13), 3047–3059. http://dx.doi.org/10.1084/jem.20090900. Epub 2009 Dec 7.

Murakami, Y., Hoshi, M., Hara, A., Takemura, M., Arioka, Y., Yamamoto, Y., Matsunami, H., Funato, T., Seishima, M., Saito, K., 2012 Aug. Inhibition of increased indoleamine 2,3-dioxygenase activity attenuates *Toxoplasma gondii* replication in the lung during acute infection. Cytokine 59 (2), 245–251. http://dx.doi.org/10.1016/j.cyto.2012.04.022. Epub 2012 May 18.

Murphy, T., et al., 1995. Regulation of interleukin-12 p40 expression through an NF-kB half-site. Mol. Cell Biol. 15, 5258–5267.

Murphy, W.F., Flannery, J.L., 1952. Congenital toxoplasmosis in triplets. J. Pediatr. 105, 59–61.

Murray, H.W., Spitalny, G.L., Nathan, C.F., 1985a. Activation of mouse peritoneal macrophages in vitro and in vivo by interferon-gamma. J. Immunol. 134, 1619–1622.

Murray, H.W., Nathan, C.F., Cohn, Z.A., 1980. Macrophage oxygen-dependent antimicrobial activity. IV. Role of endogenous scavengers of oxygen intermediates. J. Exp. Med. 152, 1610–1624.

Murray, H.W., Juangbhanich, C.W., Nathan, C.F., Cohn, Z.A., 1979. Macrophage oxygen-dependent antimicrobial activity. II. The role of oxygen intermediates. J. Exp. Med. 150, 950–964.

Murray, H.W., Rubin, B.Y., Carriero, S.M., et al., 1985b. Human mononuclear phagocyte antiprotozoal mechanisms: oxygen-dependent vs oxygen-independent activity against intracellular *Toxoplasma gondii*. Immunol. 134, 1982–1988.

Murray, H.W., Granger, A.M., Teitelbaum, R.F., 1991. Gamma interferon-activated human macrophages and *Toxoplasma gondii, Chlamydia psittaci*, and *Leishmania donovani*: antimicrobial role of limiting intracellular iron. Infect. Immun. 59, 4684–4686.

Nagasawa, H., et al., 1994. Gamma delta T-cells play a crucial role in the expression of 65000 MW heat-shock protein in mice immunized with Toxoplasma antigen. Immunology 83, 347–352.

Nagasawa, H., et al., 1992. Induction of a heat shock protein closely correlates with protection against *Toxoplasma gondii* infection. Proc. Natl. Acad. Sci. USA 89, 3155–3158.

Nagineni, C.N., Pardhasaradhi, K., Martins, M.C., Detrick, B., Hooks, J.J., 1996. Mechanisms of interferon-induced inhibition of *Toxoplasma gondii* replication in human retinal pigment epithelial cells. Infect. Immun. 64, 4188–4196.

Nance, J.P., Vannella, K.M., Worth, D., David, C., Carter, D., Noor, S., Hubeau, C., Fitz, L., Lane, T.E., Wynn, T.A., Wilson, E.H., 2012 Nov. Chitinase dependent control of protozoan cyst burden in the brain. PLoS Pathog. 8 (11), e1002990. http://dx.doi.org/10.1371/journal.ppat.1002990.

Nathan, C.F., Prendergast, T.J., Wiebe, M.E., Stanley, E.R., Platzer, E., Remold, H.G., Welte, K., Rubin, B.Y., Murray, H.W., 1984. Activation of human macrophages. Comparison of other cytokines with interferon-gamma. J. Exp. Med. 160, 600–605.

Nathan, C., Shiloh, M.U., 2000. Reactive oxygen and nitrogen intermediates in the relationship between mammalian hosts and microbial pathogens. Proceedings of the National Academy of Sciences of the USA of America 97, 8841–8848.

Nash, P., Purner, M., Leon, R., Clarke, P., Duke, R., Curiel, T., 1998. Toxoplasma gondii-infected cells are resistant to multiple inducers of apoptosis. J. Immunol. 160, 1824–1830.

Navia, B.A., et al., 1986. Cerebral toxoplasmosis complicating the acquired immune deficiency syndrom:clinical and neuropathological findings in 27 patients. Ann. Neurol. 19, 224–238.

Neto, E.C., Anele, E., Rubim, R., Brites, A., Schulte, J., Becker, D., Tuuminen, T., 2000. High prevalence of congenital toxoplasmosis in Brazil estimated in a 3-year prospective neonatal screening study. Int. J. Epidemiol. 29, 941–947.

Neyer, L.E., Grunig, G., Fort, M., Remington, J.S., Rennick, D., Hunter, C.A., 1997. Role of interleukin-10 in regulation of T-cell-dependent and T-cell-independent mechanisms of resistance to *Toxoplasma gondii*. Infect. Immun. 65, 1675–1682.

Ngiow, S.F., von Scheidt, B., Akiba, H., Yagita, H., Teng, M.W., Smyth, M.J., et al., 2011. Cancer Res Anti-TIM3 antibody promotes T-cell IFNγ-mediated antitumour immunity and suppresses established tumours. Cancer Res. 71, 3450–3451.

Nguyen, T.D., Bigaignon, G., Markine-Goriaynoff, D., Heremans, H., Nguyen, T.N., Warnier, G., Delmee, M., Warny, M., Wolf, S.F., Uyttenhove, C., Van Snick, J., Coutelier, J.P., 2003. Virulent *Toxoplasma gondii* strain RH promotes T-cell-independent overproduction of proinflammatory cytokines IL-12 and gamma-interferon. J. Med. Microbiol. 52 (Pt 10), 869–876.

Nicholson, D.H., Wolchok, E.B., 1976. Ocular toxoplasmosis in an adult receiving long-term corticosteroid therapy. Arch. Ophthalmol. (2), 248–254.

Nickdel, M.B., Roberts, F., Brombacher, F., Alexander, J., Roberts, C.W., 2001. A counter-protective role for IL-5 during acute *Toxoplasma gondii* infection. Infect. Immun. 69, 1044–1052.

Nickdel, M.B., Lyons, R.E., Roberts, F., Brombacher, F., Hunter, C.A., Alexander, J., Roberts, C.W., 2004. Intestinal pathology during acute toxoplasmosis is IL-4 dependent and unrelated to parasite burden. Parasite Immunol. 26, 75–82.

Niedelman, W., Gold, D.A., Rosowski, E.E., Sprokholt, J.K., Lim, D., Farid Arenas, A., Melo, MB., Spooner, E., Yaffe, MB., Saeij, JP., 2012. The rhoptry proteins ROP18 and ROP5 mediate *Toxoplasma gondii* evasion of the murine, but not the human, interferon-gamma response. PLoS Pathog. 8 (6), e1002784. http://dx.doi.org/10.1371/journal.ppat.1002784.

Nishanth, G., Sakowicz-Burkiewicz, M., Handel, U., Kliche, S., Wang, X., Naumann, M., Deckert, M., Schluter, D., 2010. Protective Toxoplasma gondii-specific T-cell responses require T-cell-specific expression of protein kinase C-theta. Infect. Immun. 78, 3454–3464.

Nishikawa, Y., Kawase, O., Vielemeyer, O., Suzuki, H., Joiner, K.A., Xuan, X., Nagasawa, H., 2007. *Toxoplasma gondii* infection induces apoptosis in noninfected macrophages: role of nitric oxide and other soluble factors. Parasite Immunol. 29, 375–385.

Nishikawa, Y., Ibrahim, H.M., Kameyama, K., Shiga, I., Hiasa, J., Xuan, X., 2011 May. Host cholesterol synthesis contributes to growth of intracellular *Toxoplasma gondii* in macrophages. J. Vet. Med. Sci. 73 (5), 633–639. Epub 2010 Dec 24.

Noor, S., Habashy, A.S., Nance, J.P., Clark, R.T., Nemati, K., Carson, M.J., Wilson, E.H., 2010. CCR7-dependent immunity during acute *Toxoplasma gondii* infection. Infect. Immun. 5, 2257–2263.

Norde, W.J., Hobo, W., van der Voort, R., Dolstra, H., et al., 2012. Coinhibitory molecules in hematologic malignancies: targets for therapeutic intervention. Blood 120, 728–736.

Norose, K., Kikumura, A., Luster, A.D., Hunter, C.A., Harris, T.H., 2011. CXCL10 is required to maintain T-cell populations and to control parasite replication during chronic ocular toxoplasmosis. Invest. Ophthalmol. Vis. Sci. 52, 389–398.

Norose, K., Mun, H.S., Aosai, F., Chen, M., Piao, L.X., Kobayashi, M., Iwakura, Y., Yano, A., 2003. IFNgamma-regulated *Toxoplasma gondii* distribution and load in the murine eye. Invest. Ophthalmol. Vis. Sci. 44, 4375–4381.

Norose, K., Naoi, K., Fang, H., Yano, A., 2008 Dec. In vivo study of toxoplasmic parasitemia using interferon-gamma-deficient mice: absolute cell number of leukocytes, parasite load and cell susceptibility. Parasitol. Int. 57 (4), 447–453. http://dx.doi.org/10.1016/j.parint.2008.05.007. Epub 2008 Jun 8.

Nossal, G.J., 1994. Negative seletion of lymphocytes. Cell 76 (2), 229–239.

Oberdorfer, C., Adams, O., MacKenzie, C.R., De Groot, C.J., Daubener, W., 2003. Role of Ido activation in anti-microbial defence in human native astrocytes. Adv. Exp. Med. Biol. 527, 15–26.

Ochoa-Repáraz, J., Mielcarz, D.W., Begum-Haque, S., Kasper, L.H., 2011. Gut, bugs, and brain: role of commensal bacteria in the control of central nervous system disease. Ann. Neurol. (2), 240–247.

O'Connor, G.R., Frenkel, J.K., 1976. Editorials: Dangers of steroid treatment in toxoplasmosis. Periocular injections and systemic therapy. Arch. Ophthalmol. 94, 213.

Ohtake, R., Suzuki, Y., Kwon, O.C., 1992. Folliculo-stellate cell activation in the anterior pituitary in mice during Toxoplasma infection. Jikeikai Med. J. 39, 11–19.

Okusaga, O., Langenberg, P., Sleemi, A., Vaswani, D., Giegling, I., Hartmann, A.M., Konte, B., Friedl, M., Groer, M.W., Yolken, R.H., Rujescu, D., Postolache, T.T., 2011. *Toxoplasma gondii* antibody titres and history of suicide attempts in patients with schizophrenia. Schizophr Res. 133 (1–3), 150–155.

Oldenhove, G., Bouladoux, N., Wohlfert, E.A., Hall, J.A., Chou, D., Dos Santos, L., O'Brien, S., Blank, R., Lamb, E., Natarajan, S., Kastenmayer, R., Hunter, C., Grigg, M.E., Belkaid, Y., 2009 Nov 20. Decrease of $Foxp3^+$ Treg cell number and acquisition of effector cell phenotype during lethal infection. Immunity 31 (5), 772–786. http://dx.doi.org/10.1016/j.immuni.2009.10.001.

Oliveira, A.C., Peixoto, J.R., de Arruda, L.B., Campos, M.A., Gazzinelli, R.T., Golenbock, D.T., Akira, S., Previato, J.O., Mendonca-Previato, L., Nobrega, A., Bellio, M., 2004. Expression of functional TLR4 confers proinflammatory responsiveness to Trypanosoma cruzi glycoinositolphospholipids and higher resistance to infection with T. cruzi. J. Immunol. 173, 5688–5696.

Oliveira, M.A., Santiago, H.C., Lisboa, C.R., Ceravollo, I.P., Trinchieri, G., Gazzinelli, R.T., Vieira, L.Q., 2000. Leishmania sp: comparative study with *Toxoplasma gondii* and Trypanosoma cruzi in their ability to initialize IL-12 and IFNgamma synthesis. Exp. Parasitol. 95 (2), 96–105.

Olle, P., Bessieres, M.H., et al., 1994. Experimental murine toxoplasmic retinochoroiditis. J. Eukaryot. Microbiol. 41 (5), 16S.

O'Neill, Luke A., Hardie, Grahame D., 2013. Metabolism of inflammation limited by AMPK and pseudo-starvation. Nature 493, 346–355.

Ong, Y.C., Reese, M.L., Boothroyd, J.C., 2010. *Toxoplasma* rhoptry protein 16 (ROP16) subverts host function by direct tyrosine phosphorylation of STAT6. J. Biol. Chem. 285, 28731–28740.

Orellana, M.A., Suzuki, Y., Araujo, F., Remington, J.S., 1991. Role of beta interferon in resistance to *Toxoplasma gondii* infection. Infect. Immun. 59, 3287–3290.

Orlofsky, A, Weiss, LM, Kawachi, N, Prystowsky, MB, 2002. Deficiency in the anti-apoptotic protein A1-a results in a diminished acute inflammatory response. J Immunol 168 (4), 1840–1846.

Ortiz-Alegria, L.B., Caballero-Ortega, H., Canedo-Solares, I., Rico-Torres, C.P., Sahagun-Ruiz, A., Medina-Escutia, M.E., Correa, D., 2010. Congenital toxoplasmosis: candidate host immune genes relevant for vertical transmission and pathogenesis. Genes Immun. 11, 363–373.

Osborn, S.L., Diehl, G., Han, S.J., Xue, L., Kurd, N., Hsieh, K., Cado, D., Robey, E.A., Winoto, A., 2010 Jul 20. Fas-associated death domain (FADD) is a negative regulator of T-cell receptor-mediated necroptosis. Proc. Natl. Acad. Sci. U S A 107 (29), 13034–13039. http://dx.doi.org/10.1073/pnas.1005997107. Epub 2010 Jul 6.

Oswald, I.P., et al., 1992. IL-10 synergizes with IL-4 and transforming growth factor beta to inhibit macrophage cytoxic activity. J. Immunol. 150, 3578–3582.

Ottenhoff, T., Verreck, F., Lichtenauer-Kaligis, E., Hoeve, M., Sanal, O., van Diesel, J., 2002. Genetics, cytokines and human infectious disease: lessons from weakly pathogenic mycobacteria and salmonellae. Nature Genet. 32, 97–105.

Paley, M., Wherry, E.J., 2010. TCF-1 flips the switch on Eomes. Immunity 33, 145–147.

Parham, P., 2005. MHC class I molecules and KIR in human history, health and survival. Nature Rev. Immunol. 5, 201–214.

Parker, S.J., Roberts, C.W., Alexander, 1991. $CD8^+$ T-cells are the major lymphocyte subpopulation involved in the protective immune response to toxoplasma gondii in mice. Clin. Exp. Immunol. 84 (2), 207–212.

Passos, S.T., Silver, J.S., O'Hara, A.C., Sehy, D., Stumhofer, J.S., Hunter, C.A., 2010. IL-6 promotes NK-cell production of IL-17 during toxoplasmosis. J. Immunol. 184 (4), 1776–1783. http://dx.doi.org/10.4049/jimmunol.0901843.

Pawlowski, N.N., Struck, D., Grollich, K., Kuhl, A.A., Zeitz, M., Liesenfeld, O., Hoffmann, J.C., 2007. CD2 deficiency partially prevents small bowel inflammation and improves parasite control in murine *Toxoplasma gondii* infection. World J. Gastroenterol. 13 (31), 4207–4213.

Payne, T., Molestina, R., Sinai, A., 2003. inhibition of caspase activation and a requirement for NF-kB function in the *Toxoplasma gondii*-mediated blockade of host apoptosis. J. Cell Sci. 116, 4245–4358.

Peacock, J.E., Greven, C.M., Cruz, J.M., Hurd, D.D., 1995. Reactivation toxoplasmic retinochoroiditis in patients undergoing bone marrow transplantation: is there a role for chemoprophylaxis. Bone Marrow Transplant 15 (6), 983–987.

Peixoto-Rangel, A.L., Miller, E.N., Castellucci, L., Jamieson, S.E., Peixe, R.G., Elias Lde, S., Correa-Oliveira, R., Bahia-Oliveira, L.M., Blackwell, J.M., 2009. Candidate gene analysis of ocular toxoplasmosis in Brazil: evidence for a role for toll-like receptor 9 (TLR9). Mem. Inst. Oswaldo Cruz 104 (8), 1187–1190.

Peixoto, L., Chen, F., Harb, O.S., Davis, P.H., Beiting, D.P., Brownback, C.S., Ouloguem, D., Roos, D.S., 2010. Integrative genomic approaches highlight a family of parasite-specific kinases that regulate host responses. Cell Host Microbe 8, 208–218.

Pelloux, H., Ambroise-Thomas, 1996. Cytokine production by human cells after *Toxoplasma gondii* infection. Curr. Top Microbiol. Immunol. 219, 155–163.

Pepper, M., Dzierszinski, F., Wilson, E., Tait, E., Fang, Q., Yarovinsky, F., Laufer, T.M., Roos, D., Hunter, C.A., 2008. Plasmacytoid dendritic cells are activated by *Toxoplasma gondii* to present antigen and produce cytokines. J. Immunol. 180 (9), 6229–6236.

Perona-Wright, G., Mohrs, K., Szaba, F.M., Kummer, L.W., Madan, R., Karp, C.L., Johnson, L.L., Smiley, S.T., Mohrs, M., 2009 Dec 17. Systemic but not local infections elicit immunosuppressive IL-10 production by natural killer cells. Cell Host Microbe 6 (6), 503–512. http://dx.doi.org/10.1016/j.chom.2009.11.003.

Peterson, P.K., Gekker, G., Hu, S., Chao, C.C., 1995. Human astrocytes inhibit intracellular multiplication of *Toxoplasma gondii* by a nitric oxide-mediated mechanism. J. Infect. Dis. 171, 516–518.

Peyron, F., Ateba, A., Wallon, M., Kodjikian, L., Binquet, C., et al., 2003. Congenital toxoplasmosis in twins: a report of fourteen consecutive cases and a comparison with published data. Pediatr. Infect. Dis. J. 22, 695–701.

Pfaff, A.W., Candolfi, E., 2008 Jun. New insights in toxoplasmosis immunology during pregnancy. Perspective for vaccine prevention. Parassitologia 50 (1-2), 55–58.

Pfaff, A.W., Georges, S., Candolfi, E., 2008a. Different effect of *Toxoplasma gondii* infection on adhesion capacity of fibroblasts and monocytes. Parasite Immunol. 30 (9), 487–490. http://dx.doi.org/10.1111/j.1365-3024.2008.01047.x.

Pfaff, A.W., Abou-Bacar, A., Letscher-Bru, V., Villard, O., Senegas, A., Mousli, M., Candolfi, E., 2007. Cellular and molecular physiopathology of congenital toxoplasmosis: the dual role of IFNgamma. Parasitology 134 (Pt 13), 1895–1902.

Pfaff, A.W., Villard, O., Klein, J.P., Mousli, M., Candolfi, E., 2005. Regulation of *Toxoplasma gondii* multiplication in BeWo trophoblast cells: cross-regulation of nitric oxide production and polyamine biosynthesis. Int. J. Parasitol. 35, 1569–1576.

Pfaff, A.W., Mousli, M., Sénégas, A., Marcellin, L., Takikawa, O., Klein, J.P., Candolfi, E., 2008b. Impact of foetus and mother on IFNgamma-induced indoleamine 2,3-dioxygenase and inducible nitric oxide synthase expression in murine placenta following *Toxoplasma gondii* infection. Int. J. Parasitol. 38 (2), 249–258.

Pfefferkorn, E.R., 1984. Interferon gamma blocks the growth of *Toxoplasma gondii* in human fibroblasts by inducing the host cells to degrade tryptophan. Proc. Natl. Acad. Sci. U S A 81, 908–912.

Pfefferkorn, E.R., Eckel, M., Rebhun, S., 1986. Interferon-gamma suppresses the growth of *Toxoplasma gondii* in human fibroblasts through starvation for tryptophan. Mol. Biochem. Parasitol. 20, 215–224.

Pfefferkorn, E.R., Guyre, P.M., 1984. Inhibition of growth of *Toxoplasma gondii* in cultured fibroblasts by human recombinant gamma interferon. Infect. Immun. 44, 211–216.

Phan, L., Kasza, K., Jalbrzikowski, J., et al., 2008a. Longitudinal study of new eye lesions in treated congenital toxoplasmosis. Ophthalmology 115.

Phan, L., Kasza, K., Jalbrzikowski, J., et al., 2008b. Longitudinal study of new eye lesions in children with toxoplasmosis who were not treated during the first year of life. Am. J. Ophthalmol. 146.

Phelps, E.D., Sweeney, K.R., Blader, I.J., 2008. *Toxoplasma gondii* rhoptry discharge correlates with activation of the early growth response 2 host cell transcription factor. Infect. Immun. 76 (10), 4703–4712. http://dx.doi.org/10.1128/IAI.01447-07.

Phillip, J., Nance, J.P., Vannella, K.M., Worth, D., David, C., Carter, D., Noor, S., Hubeau, C., Fitz, L., Lane, T.E., Wynn, T.A., Wilson, E.H., 2012. Chitinase dependent control of protozoan cyst burden in the brain. PLOS Pathogens 8 (11), e1002990.

Piccinni, M.P., Maggi, E., Romagnani, S., 2000. Role of hormone controlled T-cell cytokines in the maintenance of pregnancy. Biochem. Soc. Trans. 28, 212–215.

Piccinni, M.P., Giudizi, M.G., Biagiotti, R., Beloni, R., Giannarini, L., Sampognaro, S., Parronchi, P., Manetti, R., Annunziato, F., Livi, C., Romagnani, S., Maggi, E., 1995. Progesterone favours the development of human T-helper cells producing Th2-type cytokines and promotes both IL-4 production and membane CD30 expression in established Th1 cell clones. J. Immunol. 155, 128–133.

Piccinni, M.P., Scaletti, C., Mavilia, C., Lazzeri, E., Romagnani, P., Natali, I., Pellegrini, S., Livi, C., Romagnani, S., Maggi, E., 2001. Production of IL-4 and leukaemia inhibitory factor by T-cells of the cumulus oophorus: a favourable microenvironment for pre-implantation embryo development. Eur. J. Immunol. 31, 2431–2437.

Pifer, R., Benson, A., Sturge, C.R., Yarovinsky, F., 2011. UNC93B1 is essential for TLR11 activation and IL-12-dependent host resistance to *Toxoplasma gondii*. J. Biol. Chem. 286, 3307–3314.

Pino, P., Foth, B.J., Kwok, L.Y., Sheiner, L., Schepers, R., Soldati, T., Soldati-Favre, D., 2007. Dual targeting of antioxidant and metabolic enzymes to the mitochondrion and the apicoplast of *Toxoplasma gondii*. PLoS Pathogens 3, e115.

Plattner, F., Yarovinsky, F., Romero, S., Didry, D., Carlier, M.F., Sher, A., Soldati-Favre, D., 2008. Toxoplasma profilin is essential for host cell invasion and TLR11-dependent induction of an interleukin-12 response. Cell Host Microbe. 3 (2), 77–87.

Ploix, C.C., Noor, S., Crane, J., Masek, K., Carter, W., Lo, D.D., Wilson, E.H., Carson, M.J., 2011 Jul. CNS-derived CCL21 is both sufficient to drive homeostatic $CD4^+$ T-cell proliferation and necessary for efficient $CD4^+$ T-cell migration into the CNS parenchyma following *Toxoplasma gondii* infection. Brain Behav. Immun. 25 (5), 883–896. http://dx.doi.org/10.1016/j.bbi.2010.09.014.

Polgar, B., Kispal, G., Lachman, M., Parr, C., Nagy, E., Csere, P., Miko, E., Szereday, L., Sarga, P., Szekeres-Barthos, J., 2003. Molecular cloning and immunologic characterization of a novel cDNA coding for progesterone-induced blocking factor. J. Immunol. 171, 5956–5963.

Polgar, B., Barakonyi, A., Xynos, I., Szekeres-Bartho, J., 1999. The role of gamma/delta T-cell receptor positive cells in pregnancy. Am. J. Reprod. Immunol. 41, 239–244.

Pollard, A.M., Knoll, L.J., Mordue, D.G., 2009a. The role of specific *Toxoplasma gondii* molecules in manipulation of innate immunity. Trends Parasitol. 11, 491–494.

Pollard, A.M., Skariah, S., Mordue, D.G., Knoll, L.J., 2009b. A transmembrane domain-containing surface protein from *Toxoplasma gondii* augments replication in activated immune cells and establishment of a chronic infection. Infect. Immun. 77 (9), 3731–3739.

Powell, H.C., Gibbs Jr., C.J., Lorenzo, A.M., Lampert, P.W., Gajdusek, D.C., 1978. Toxoplasmosis of the central nervous system in the adult. Electron microscopic observations. Acta Neuropathol. (Berl.) 41, 211–216.

Portela, R.W., Bethony, J., Costa, M.I., Gazzinelli, A., Vitor, R.W., Hermeto, F.M., Correa-Oliveira, R., Gazzinelli, R.T., 2004. A multihousehold study reveals a positive correlation between age, severity of ocular toxoplasmosis, and levels of glycoinositolphospholipid-specific. immunoglobulin A. J. Infect. Dis. 190 (1), 175–183.

Portillo, J.A., Feliciano, L.M., Okenka, G., Heinzel, F., Subauste, M.C., Subauste, C. S., 2012 Feb. CD40 and tumour necrosis factor-α co-operate to up-regulate inducible nitric oxide synthase expression in macrophages. Immunology 135 (2), 140–150. http://dx.doi.org/10.1111/j.1365-2567.2011.03519.x.

Portillo, J.A., Okenka, G., Reed, E., Subauste, A., Van Grol, J., Gentil, K., Komatsu, M., Tanaka, K., Landreth, G., Levine, B., Subauste, C.S., 2010. The CD40-autophagy pathway is needed for host protection despite IFNΓ-dependent immunity and CD40 induces autophagy via control of P21 levels. PLoS One (12), e14472.

Portugal, L.R., Fernandes, L.R., Cesar, G.C., Santiago, H.C., Oliveira, D.R., Silva, N. M., Silva, A.A., Lannes-Vieira, J., Arantes, R.M., Gazzinelli, R.T., Alvarez-Leite, J.I., 2004. Infection with *Toxoplasma gondii* increases atherosclerotic lesion in ApoE-deficient mice. Infect. Immun. 72, 3571–3576.

Portugal, L.R., Fernandes, L.R., Pietra Pedroso, V.S., Santiago, H.C., Gazzinelli, R.T., Alvarez-Leite, J.I., 2008 Mar. Influence of low-density lipoprotein (LDL) receptor on lipid composition, inflammation and parasitism during *Toxoplasma gondii* infection. Microbes Infect. 10 (3), 276–284. http://dx.doi.org/10.1016/j.micinf.2007.12.001.

Potasman, I., Araujo, F.G., Desmonts, G., Remington, J.S., 1986. Analysis of *Toxoplasma gondii* antigens recognized by human sera obtained before and after acute infection. J. Infect. Dis. 154 (4), 650–657.

Prandovszky, E., Gaskell, E., Martin, H., Dubey, J.P., Webster, J.P., McConkey, G. A., 2011. The neurotropic parasite *Toxoplasma gondii* increases dopamine metabolism. PLoS One 6, e23866.

Purner, M.B., et al., 1996. CD4-mediated and CD8-mediated cytotoxic and proliferative immune responses to *Toxoplasma gondii* in seropositive humans. infect. Immune 64 (10), 4330–4338.

Rachinel, N., Buzoni-Gatel, Putta C., Mennechet, Luangsay S., Minns, L., Grigg, M., Tomavo, S., Boothroyd, J., Kasper, L., 2004. The induction of acute ileitis by a single microbial antigen of *Toxoplasma gondii*. J. Imunol. 173, 2725–2735.

Rajagopalan, S., Long, E.D., 2005. Understanding how combinations of HLA and KIR genes influence disease. J. Exp. Med. 201, 1025–1029.

Rajapakse, R., Uring-Lambert, B., Andarawewa, K.L., Rajapakse, R.P., Abou-Bacar, A., Marcellin, L., Candolfi, E., 2007 Mar. 1,25(OH)2D3 inhibits in vitro and in vivo intracellular growth of apicomplexan parasite *Toxoplasma gondii*. J. Steroid Biochem. Mol. Biol. 103 (3-5), 811–814.

Raymond, J., et al., 1990. Presence of gamma interferon in human acute and congenital toxoplasmosis [published erratum appears in J Clin Microbiol 1990 28(12):2853]. J. Clin. Microbiol. 28 (6), 1434–1437.

Ravindran, S., Boothroyd, J.C., 2008. Secretion of proteins into host cells by Apicomplexan parasites. Traffic. 2008 May 9 (5), 647–656. http://dx.doi.org/10.1111/j.1600-0854.2008.00723.x.

Reese, M.L., Zeiner, G.M., Saeij, J.P., Boothroyd, J.C., Boyle, J.P., 2011 Jun 7. Polymorphic family of injected pseudokinases is paramount in Toxoplasma virulence. Proc. Natl. Acad. Sci. U S A 108 (23), 9625–9630. http://dx.doi.org/10.1073/pnas.1015980108.

Reichmann, G., Villegas, E.N., Craig, L., Peach, R., Hunter, C.A., 1999. The CD28/B7 interaction is not required for resistance to *Toxoplasma gondii* in the brain but contributes to the development of immunopathology. J. Immunol 163 (6), 3354–3362.

Reichmann, G., Walker, W., Villegas, E.N., Craig, L., Cai, G., Alexander, J., Hunter, C.A., 2000. The CD40/CD40 ligand interaction is required for resistance to toxoplasmic encephalitis. Infect. Immun. 68, 1312–1318.

Reis e Sousa, et al., 1997. In vivo microbial stimulation induces rapid CD40L-independent production of IL-12 by dendritic cells and their re-distribution to T-cell areas. J. Exp. Med. 186, 1819–1829.

Remington, J.S., McLeod, R., Thulliez, P., Desmonts, G., 2005. Toxoplasmosis. In: Remington, J.S., Klein, J.O., Baker, C., Wilson, C.B. (Eds.), Infectious diseases of the foetus and the newborn infant. W.B. Saunders, Philadelphia, U.S.A. chapter 31.

Resende, M.G., Fux, B., Caetano, B.C., Mendes, E.A., Silva, N.M., Ferreira, A.M., Melo, M.N., Vitor, R.W., Gazzinelli, R.T., 2008. The role of MHC haplotypes H2d/H2b in mouse resistance/susceptibility to cyst formation is influenced by the lineage of infective *Toxoplasma gondii* strain. An Acad. Bras. Cienc. 1, 85–99.

Rincon, Mercedes, 2012. Interleukin-6: from an inflammatory marker to a target for inflammatory diseases. Trends Immunol. 33, 571–577.

Rizzo, L.V., et al., 1998. Patients with acquired toxoplasmosis can be discriminated from patients with congenital toxoplasmosis based on their T-cell activity but not their T-cell receptor (Vbeta family) repertoire.

Roach, J., Glusman, G., Rowen, L., Kaur, A., Purcell, M., Smith, K., Hood, L., Aderem, A., 2005. The evolution of vertebrate Toll-like receptors. PNAS 102, 9577–9582.

Robben, P.M., Laregina, M., Kuziel, W.A., Sibley, L.D., 2005. Recruitment of Gr-1^+ monocytes is essential for control of acute toxoplasmosis. J. Exp. Med. 201, 1761–1769.

Robben, P.M., Mordue, D.G., Truscott, S.M., Takeda, K., Akira, S., Sibley, L.D., 2004. Production of IL-12 by macrophages infected with *Toxoplasma gondii* depends on the parasite genotype. J. Immunol. 172, 3686–3694.

Robbins, J.R., Zeldovich, V.B., Poukchanski, A., Boothroyd, J.C., Bakardjiev, A.I., 2012. Tissue barriers of the human placenta to infection with *Toxoplasma gondii*. Infect. Immun. 80 (1), 418–428. http://dx.doi.org/10.1128/IAI.05899-11.

Robert-Gangneux, F., Dupretz, P., Yvenou, C., Quinio, D., Poulain, P., Guiguen, C., Gangneux, J.P., 2010. Clinical relevance of placenta examination for the diagnosis of congenital toxoplasmosis. Pediatr. Infect. Dis. J. 29, 33–38.

Roberts, F., McLeod, R., 1999. Pathogenesis of Toxoplasmic Retinochoroiditis. Parasitol. Today 15 (2), 51–57.

Roberts, C.W., Satoskar, A., Alexander, J., 1996. Sex-steroids, pregnancy-associated hormones and immunity to parasites. Parasitology Today 12, 382–388.

Roberts, C.W., Ferguson, D.J.P., Jebbari, H., Satoskar, A., Bluethmann, H., Alexander, J., 1996. Different roles for IL4 during the course of *Toxoplasma gondii* infection. Infect. Immun. 64, 897–904.

Roberts, C.W., Walker, W., Alexander, J., 2001a. Sex-associated hormones and immunity to protozoan parasites. Clinical Microbiology Reviews 14, 476–488.

Roberts, C.W., Alexander, J., 1992. Studies on a murine model of congenital toxoplasmosis: vertical disease transmission only occurs in BALB/c mice infected for the first time during pregnancy. Parasitology 104, 19–23.

Roberts, C.W., Brewer, J.M., Alexander, J., 1994. Congenital toxoplasmosis in the BALB/c mouse: prevention of vertical disease transmission and foetal death by vaccination. Vaccine 12, 1389–1394.

Roberts, F., Roberts, C.W., Ferguson, D.J., McLeod, R., 2000. Inhibition of nitric oxide production exacerbates chronic ocular toxoplasmosis. Parasite Immunol. 22, 1–5.

Roberts, F., McLeod, R., 1999. Pathogenesis of Toxoplasmic Retinochoroiditis. Parasitol. Today 15 (2), 51–57.

Roberts, F., Mets, M.B., et al., 2001b. Histopathological features of ocular toxoplasmosis in the foetus and infant. Arch. Ophthalmol. 119, 51–58.

Roberts, T., Frenkel, J.K., 1990. Estimating losses and other preventable costs caused by congenital toxoplasmosis in people in the USA. J. Am. Med. Assoc. 2, 249–256.

Robey, E.A., et al., 1992. The level of CD8 expression can determine the outcome of thymic selection. Cell 69 (7), 1089–1096.

Robey, E., et al., 2005. The alphabeta versus gammadelta T cell fate decision: when less is more. Immunity. (5), 533–534.

Rock, R.B., Hu, S., Deshpande, A., Munir, S., May, B.J., Baker, C.A., Peterson, P.K., Kapur, V., 2005. Transcriptional response of human microglial cells to interferon-gamma. Genes Immun. 6, 712–719.

Rodgers, J., Wang, X., Wen, X., Dunford, B., Miller, R., Suzuki, Y., 2005. Strains of *Toxoplasma gondii* used for tachyzoite antigens to stimulate spleen cells of infected mice in vitro affect cytokine responses of the cells in the culture. Parasitol. Res. 97, 332–335.

Romagnani, S., 1996. Th1 and Th2 in human diseases. Clin. Immunol. 80 (3 Pt 1), 225–235.

Romano, J.D., de Beaumont, C., Carrasco, J.A., Ehrenman, K., Bavoil, P.M., Coppens, I., 2012a. Fierce competition between Toxoplasma and Chlamydia for host cell structures in dually infected cells. Eukaryot. Cell.

Romano, J.D., de Beaumont, C., Carrasco, J.A., Ehrenman, K., Bavoil, P.M., Coppens, I., 2012b. A novel co-infection model with Toxoplasma and Chlamydia trachomatis highlights the importance of host cell manipulation for nutrient scavenging. Cell Microbiol. http://dx.doi.org/10.1111/cmi.12060. [Epub ahead of print].

Ronet, C., Darche, S., Leite de Moraes, M., Miyake, S., Yamamura, T., Louis, J.A., Kasper, L.H., Buzoni-Gatel, D., 2005. NKT-cells are critical for the initiation of an inflammatory bowel response against *Toxoplasma gondii*. J. Immunol. 175 (2), 899–908.

Ropert, C., Gazzinelli, R.T., 2000. Signalling of immune system cells by glycosylphosphatidylinositol (GPI) anchor and related structures derived from parasitic protozoa. Curr. Opin. Microbiol. 3, 395–403.

Rosowski, E.E., Saeij, J.P., 2012. *Toxoplasma gondii* Clonal Strains All Inhibit STAT1 Transcriptional Activity but Polymorphic Effectors Differentially Modulate IFNγ Induced Gene Expression and STAT1 Phosphorylation. PLoS One 7 (12), e51448. http://dx.doi.org/10.1371/journal.pone.0051448.

Rosowski, E.E., Lu, D., Julien, L., Rodda, L., Gaiser, R.A., Jensen, K.D., Saeij, J.P., 2011 Jan 17. Strain-specific activation of the NF-kappaB pathway by GRA15, a novel *Toxoplasma gondii* dense granule protein. J. Exp. Med. 208 (1), 195–212. http://dx.doi.org/10.1084/jem.20100717.

Rothova, A., 2003. Ocular manifestations of toxoplasmosis. Curr. Opin. Ophthalmol. 14, 384–388.

Rozenfeld, C., Martinez, R., Figueiredo, R.T., Bozza, M.T., Lima, F.R., Pires, A. L., Silva, P.M., Bonomo, A., Lannes-Vieira, J., De Souza, W., Moura-Neto, V., 2003. Soluble factors released by Toxoplasma gondii-infected astrocytes down-modulate nitric oxide production by gamma interferon-activated microglia and prevent neuronal degeneration. Infect. Immun. 71, 2047–2057.

Rozenfeld, C., Martinez, R., Seabra, S., Sant'anna, C., Goncalves, J.G., Bozza, M., Moura-Neto, V., De Souza, W., 2005. *Toxoplasma gondii* prevents neuron degeneration by interferon-gamma-activated microglia in a mechanism involving inhibition of inducible nitric oxide synthase and transforming growth factor-beta1 production by infected microglia. Am. J. Pathol. 167, 1021–1031.

Ryning, F.W., McLeod, R., Maddox, J.C., Hunt, S., Remington, J.S., 1979. Probable transmission fo *Toxoplasma gondii* by organ transplantation. Ann. Intern. Med. 90, 47–49.

Sa, Q., Woodward, J., Suzuki, Y., 2013. IL-2 Produced by CD8+ Immune T-cells Can Augment Their IFNγ Production Independently from Their Proliferation in the Secondary Response to an Intracellular Pathogen. J. Immunol. [Epub ahead of print].

Saavedra, R., et al., 1996. Epptiopes recognized by human T-lymphocytes in the ROP2 protien antigen of *Toxoplasma gondii*. Infect. Immunol. 64, 897–904.

Saavedra, R., Herion, P., 1993. Human T-cell clones as tools for the characterization of the cell-mediated immune response to *Toxoplasma gondii*. Res. Immunol. 144 (1), 48–51.

Sacks, D., Sher, A., 2002. Evasion of innate immunity by parasitic protozoa. Nat. Immunol. 3, 1041–1047.

Saeij, J.P., Boyle, J.P., Boothroyd, J.C., 2005a. Differences among the three major strains of *Toxoplasma gondii* and their specific interactions with the infected host. Trends Parasitol. 21 (10), 476–478.

Saeij, J.P., Boyle, J.P., Grigg, M.E., Arrizabalaga, G., Boothroyd, J.C., 2005b. Bioluminescence imaging of *Toxoplasma gondii* infection in living mice reveals dramatic differences between strains. Infect. Immun. 73 (2), 695–702.

Saeij, J.P., Arrizabalaga, G., Boothroyd, J.C., 2008. A cluster of four surface antigen genes specifically expressed in bradyzoites, SAG2CDXY, plays an important role in *Toxoplasma gondii* persistence. Infect. Immun. 76 (6), 2402–2410. http://dx.doi.org/10.1128/IAI.01494-07, 2008 Jun.

Saeij, J.P., Coller, S., Boyle, J.P., Jerome, M.E., White, M.W., Boothroyd, J.C., 2007. *Toxoplasma* co-opts host gene expression by injection of a polymorphic kinase homologue. Nature 445, 324–327.

Santiago, H.C., Oliveira, M.A., Bambirra, E.A., Faria, A.M., Afonso, L.C., Vieira, L. Q., Gazzinelli, R.T., 1999. Coinfection with *Toxoplasma gondii* inhibits antigen-specific Th2 immune responses, tissue inflammation, and parasitism in BALB/c mice infected with Leishmania major. Infect. Immun. 67, 4939–4944.

Samarajiwa, S.A., Forster, S., Auchettl, K., Hertzog, P.J., 2009. INTERFEROME: the database of interferon regulated genes. Nucleic Acids Research 37, D852–D857.

Sanchez, Y., Rosado Jde, D., Vega, L., Elizondo, G., Estrada-Muñiz, E., Saavedra, R., Juárez, I., Rodríguez-Sosa, M., 2010. The unexpected role for the aryl hydrocarbon receptor on susceptibility to experimental toxoplasmosis. J. Biomed. Biotechnol. 2010, 505694. http://dx.doi.org/10.1155/2010/505694. Epub 2010 Jan 11.

Salek-Ardakani, S., Croft, M., 2009 Dec. T cells need Nod too? Nat. Immunol. 10 (12), 1231–1233 http://dx.doi.org/10.1038/ni1209-1231.

Sauer, A., Lahmar, I., Scholler, M., Villard, O., Speeg-Schatz, C., Brunet, J., Pfaff, A., Bourcier, T., Candolfi, E., 2009. Development of murine models of ocular toxoplasmosis and preliminary results of ocular inflammatory transcriptome. J. Fr. Ophtalmol. 32, 742–749.

Sautel, C.F., Ortet, P., Saksouk, N., Kieffer, S., Garin, J., Bastien, O., Hakimi, M.A., 2009. The histone methylase KMTox interacts with the redox-sensor peroxiredoxin-1 and targets genes involved in *Toxoplasma gondii* antioxidant defences. Mol. Microbiol. 71, 212–226.

Sayles, P., Rakhmilevich, A., Johnson, L., 1995. Gamma delta T-cells and acute primary *Toxoplasma gondii* infection in mice. J. Infect. Dis. 171, 249–252.

Scalise, F., Gerli, R., Castellucci, G., et al., 1992. Lymphocytes bearing the gamma delta T-cell receptor in acute toxoplasmosis. Immunology 76, 668–670.

Scanga, C.A., Aliberti, J., Jankovic, D., Tilloy, F., Bennouna, S., Denkers, E.Y., Medzhitov, R., Sher, A., 2002. Cutting edge: MyD88 is required for resistance to *Toxoplasma gondii* infection and regulates parasite-induced IL-12 production by dendritic cells. J. Immunol. 168 (12), 5997–6001.

Schaeffer, E.M., Debnath, J., Yap, G., McVicar, D., Liao, X.C., Littman, D.R., Sher, A., Varmus, H.E., Lenardo, M.J., Schwartzberg, P.L., 1999. Requirement for Tec kinases Rlk and Itk in T-cell receptor signalling and immunity. Science 284, 638–641.

Schaeffer, M., Han, S.J., Chtanova, T., Van Dooren, G.G., Herzmark, P., Chen, Y., Roysam, B., Striepen, B., Robey, E.A., 2009. Dynamic imaging of T-cell-parasite interactions in the brains of mice chronically infected with *Toxoplasma gondii*. J. Immunol. 182, 6379–6393.

Scharton-Kersten, T.M., Wynn, T.A., Denkers, E.Y., Bala, S., Grunvald, E., Hieny, S., Gazzinelli, R.T., Sher, A., 1996a. In the absence of endogenous IFNgamma, mice develop unimpaired IL-12 responses to *Toxoplasma gondii* while failing to control acute infection. J. Immunol. 157, 4045–4054.

Scharton-Kersten, T.M., Yap, G., Magram, J., Sher, A., 1997a. Inducible nitric oxide is essential for host control of persistent but not acute infection with the intracellular pathogen *Toxoplasma gondii*. J. Exp. Med. 185, 1261–1273.

Scharton-Kersten, T., et al., 1997b. Interferon consensus sequence binding protein-deficient mice display impaired resistance to intracellular infection due to a primary defect in interleukin 12 p40 induction. J. Exp. Med. 186 (9), 1523–1534.

Scharton-Kersten, T., Caspar, P., Sher, A., Denkers, E.Y., 1996b. *Toxoplasma gondii*: evidence for interleukin-12-dependent and -independent pathways of interferon-gamma production induced by an attenuated parasite strain. Exp. Parasitol. 84 (2), 102–114.

Scharton-Kersten, T., Nakajima, H., Yap, G., Sher, A., Leonard, W.J., 1998. Infection of mice lacking the common cytokine receptor gamma-chain (gamma(c)) reveals an unexpected role for $CD4^+$ T-lymphocytes in early IFNgamma-dependent resistance to *Toxoplasma gondii*. J. Immunol. 160, 2565–2569.

Schluter, D., Lohler, J., Deckert, M., Hof, H., Schwendemann, G., 1991. Toxoplasma encephalitis of immunocompetent and nude mice: immunohistochemical characterisation of Toxoplasma antigen, infiltrates and major histocompatibility complex gene products. J. Neuroimmunol. 31, 185–198.

Schluter, D., Deckert-Schluter, M., Schwendemann, G., Brunner, H., Hof, H., 1993. Expression of major histocompatibility complex class II antigens and levels of interferon-gamma, tumour necrosis factor, and interleukin-6 in cerebrospinal fluid and serum in *Toxoplasma gondii*-infected SCID and immunocompetent C.B-17 mice. Immunology 78 (3), 430–435.

Schluter, D., Kaefer, N., Hof, H., Wiestler, O.D., Deckert-Schluter, M., 1997. Expression pattern and cellular origin of cytokines in the normal and Toxoplasma gondii-infected murine brain. Am. J. Pathol. 150, 1021–1035.

Schluter, D., Bertsch, D., Frei, K., Hubers, S.B., Wiestler, O.D., Hof, H., Fontana, A., Deckert-Schluter, M., 1998. Interferon-gamma antagonizes transforming growth factor-beta2-mediated immunosuppression in murine Toxoplasma encephalitis. J. Neuroimmunol. 81, 38–48.

Schluter, D., Deckert-Schluter, M., et al., 1999. Inhibition of inducible nitric oxide synthase exacerbates chronic cerebral toxoplasmosis in Toxoplasma gondii-susceptible C57BL/6 mice but does not reactivate the latent disease in T. gondii-resistant BALB/c mice. J. Immunol. 162 (6), 3512–3518.

Schluter, D., Deckert, M., Hof, H., Frei, K., 2001. *Toxoplasma gondii* infection of neurons induces neuronal cytokine and chemokine production, but gamma interferon- and tumour necrosis factor-stimulated neurons fail to inhibit the invasion and growth of T. gondii. Infect. Immun. 69, 7889–7893.

Schluter, D., Hein, A., Dorries, R., Deckert-Schluter, M., 1995. Different subsets of T-cells in conjunction with natural killer cells, macrophages, and activated microglia participate in the intracerebral immune response to *Toxoplasma gondii* in athymic nude and immunocompetent rats. Am. J. Pathol. 146, 999–1007.

Schluter, D., Kwok, L.Y., Lutjen, S., Soltek, S., Hoffmann, S., Korner, H., Deckert, M., 2003. Both lymphotoxin-alpha and TNF are crucial for control of *Toxoplasma gondii* in the central nervous system. J. Immunol. 170, 6172–6182.

Schulthess, J., Fourreau, D., Darche, S., Meresse, B., Kasper, L., Cerf-Bensussan, N., Buzoni-Gatel, D., 2008. Mucosal immunity in *Toxoplasma gondii* infection. Parasite 15 (3), 389–395.

Schulthess, J., Meresse, B., Ramiro-Puig, E., Montcuquet, N., Darche, S., Bègue, B., Ruemmele, F., Combadière, C., Di Santo, J.P., Buzoni-Gatel, D., 2012 Jul 27. Cerf-Bensussan N. Interleukin-15-dependent NKp46^{+} innate lymphoid cells control intestinal inflammation by recruiting inflammatory monocytes. Immunity 37 (1), 108–121. http://dx.doi.org/10.1016/j.immuni.2012.05.013.

Schwab, J.C., Beckers, C.J., Joiner, K.A., 1994. The parasitophorous vacuole membrane surrounding intracellular *Toxoplasma gondiii* functions as a molecular sieve. Proc. Natl. Acad. Sci. USA 91, 509–513.

Schwartzman, J.D., 1987. Quantitative comparison of infection of neural cell and fibroblast monolayers by two strains of *Toxoplasma gondii*. Proc. Soc. Exp. Biol. Med. 186 (1), 75–78.

Schwarcz, R., Hunter, C.A., 2007 May. *Toxoplasma gondii* and schizophrenia: linkage through astrocyte-derived kynurenic acid? Schizophr. Bull. 33 (3), 652–653.

Scollay, R., Bartlett, P., Shortman, K., 1984. T-cell development in the adult murine thymus: changes in the expression of the surface antigens Ly2, L3T4 and B2A2 during development from early precursor cells to emigrants. Immunol. Rev. 82, 79–103.

Scott, P., Hunter, C.A., 2002. Dendritic cells and immunity to leishmaniasis and toxoplasmosis. Curr. Opin. Immunol. 14, 466–470.

Seder, R.A., Gazzinelli, R., Sher, A., Paul, W.E., 1993. Interleukin 12 acts directly on CD4^{+} T-cells to enhance priming for interferon gamma production and diminishes interleukin 4 inhibition of such priming. Proc. Natl. Acad. Sci. U S A 90 (21), 10188–10192.

Senegas, A., Villard, O., Neuville, A., Marcellin, L., Pfaff, A.W., Steinmetz, T., Mousli, M., Klein, J.P., Candolfi, E., 2009 Mar. Toxoplasma gondii-induced foetal resorption in mice involves interferon-gamma-induced apoptosis and spiral artery dilation at the maternofoetal interface. Int. J. Parasitol. 39 (4), 481–487. http://dx.doi.org/10.1016/j.ijpara.2008.08.009.

Severance, E.G., Kannan, G., Gressitt, K.L., Xiao, J., Alaedini, A., Pletnikov, M.V., Yolken, R.H., 2012. Anti-Gluten Immune Response following *Toxoplasma gondii* Infection in Mice. PLoS One 7 (11), e50991.

Sergent, V., Cautain, B., Khalife, J., Deslee, D., Bastien, P., Dao, Al, Dubremetz, J.F., Fournie, G.J., Saoudi, A., Cesbron-Delauw, M.F., 2005. Innate refractoriness of the Lewis rat to toxoplasmosis is a dominant trait that is intrinsic to bone marrow-derived cells. Infect. Immun. 73, 6990–6997.

Shapira, S., Harb, O., Caamano, J., Hunter, A., 2004. The NF-kB signalling pathway: immune evasion and immunoregulation during toxoplasmosis. Int. J. Parasitol. 34 (3), 393–400.

Shapira, S., Harb, O.S., Margarit, J., Matrajt, M., Han, J., Hoffmann, A., Freedman, B., May, M.J., Roos, D.S., Hunter, C.A., 2005. Initiation and termination of NF-kappaB signalling by the intracellular protozoan parasite *Toxoplasma gondii*. J. Cell Sci. 118, 3501–3508.

Shaw, M.H., Freeman, G.J., Scott, M.F., Fox, B.A., Bzik, D.J., Belkaid, Y., Yap, G.S., 2006. Tyk2 negatively regulates adaptive Th1 immunity by mediating IL-10 signalling and promoting IFNgamma-dependent IL-10 reactivation. J. Immunol. 12, 7263–7271.

Shaw, M.H., Reimer, T., Sánchez-Valdepeñas, C., Warner, N., Kim, Y.G., Fresno, M., Nuñez, G., 2009 Dec. T cell-intrinsic role of Nod2 in promoting type 1 immunity to *Toxoplasma gondii*. Nat. Immunol. 10 (12), 1267–1274. http://dx.doi.org/10.1038/ni, 1816. Epub 2009 Nov 1.

Shen, D.F., Matteson, D.M., et al., 2001. Involvement of apoptosis and interferon-gamma in murine toxoplasmosis. Invest. Ophthalmol. Vis. Sci. 42 (9), 2031–2036.

Sher, A., Collazo, C., Scanga, C., et al., 2003. Induction and regulation of IL-12-dependent host resistance to *Toxoplasma gondii*. Immunol. Res. 27, 521–528.

Sher, A., Coffman, R.L., 1992. Regulation of immunity to parasites by T-cells and T-cell-derived cytokines. Annu. Rev. Immunol. 10, 385–409.

Sher, A., Fiorentino, D., Caspar, P., Pearce, E., Mosmann, T., 1991. Production of IL-10 by CD4^{+} T-lymphocytes correlates with down-regulation of Th1 cytokine synthesis in helminth infection. J. Immunol. 147 (8), 2713–2716.

Sher, A., Gazzinelli, R.T., Oswald, I.P., Clerici, M., Kullberg, M., Pearce, E.J., Berzofsky, J.A., Mosmann, T.R., James, S.L., Morse 3rd, H.C., 1992. Role of T-cell derived cytokines in the down-regulation of immune responses in parasitic and retroviral infection. Immunol. Rev. 127, 183–204.

Sher, A., Oswald, I.P., Wynn, T.A., Williams, M.E., Eltoum, I., Cheever, A.W., James, S.L., 1993. Regulatory and immunopathological roles of IL4 in experimental schistosomiasis. Res. Immunol. (8), 643–648.

Shi, C.S., Shenderov, K., Huang, N.N., Kabat, J., Abu-Asab, M., Fitzgerald, K.A., Sher, A., Kehrl, J.H., 2012. Activation of autophagy by inflammatory signals limits IL-1β production by targeting ubiquitinated inflammasomes for destruction. Nat. Immunol. 13 (3), 255–263.

Shiono, Y., Mun, H.S., He, N., Nakazaki, Y., Fang, H., Furuya, M., Aosai, F., Yano, A., 2007. Maternal-foetal transmission of *Toxoplasma gondii* in interferon-gamma deficient pregnant mice. Parasitol. Int. 56, 141–148.

Shirahata, T., Muroya, N., Ohta, C., Goto, H., Nakane, A., 1992. Correlation between increased susceptibility to primary *Toxoplasma gondii* infection and depressed production of gamma interferon in pregnant mice. Microbiol. Immunol. 36, 81–91.

Shirahata, T., Muroya, N., Ohta, C., Goto, H., Nakane, A., 1993. Enhancement by recombinant human interleukin 2 of host resistance to *Toxoplasma gondii* infection in pregnant mice. Microbiol. Immunol. 37, 583–590.

Shirahata, T., Yamashita, T., Ohta, C., et al., 1994. CD8^{+} T-lymphocytes are the major cell population involved in the early gamma interferon response and resistance to acute primary *Toxoplasma gondii* infection in mice. Microbiol. Immunol. 38, 789–796.

Shirasawa, T., Akashi, T., Sakamoto, K., Takahashi, H., Maruyama, N., Hirokawa, K., 1993. Gene expression of CD24 core peptide molecule in developing brain and developing non-neural tissues. Dev. Dyn. 198 (1), 1–13.

Sibalic, D., Djurkovic, O., Bobic, B., 1990. Onset of ocular complications in congenital toxoplasmosis assoicated with immunoglobulin M antiboides ot *Toxoplasma gondii*. Eur. J. Clin. Microbiol. Infect. Dis. 9, 671–674.

Sibley, L.D., Adams, L.B., Fukutomi, Y., Krahenbuhl, J.L., 1991. Tumor necrosis factor-alpha triggers antitoxoplasmal activity of IFNgamma primed macrophages. J. Immunol. 147, 2340–2345.

Sibley, L.D., Howe, D.K., 1996. Genetic basis of pathogenicity in toxoplasmosis. Curr. Top Microbiol. Immunol. 219, 3–15.

Sibley, L.D., Boothroyd, J.C., 1992a. Virulent strains of *Toxoplasma gondii* comprise a single clonal lineage. Nature 359, 82–85.

Sibley, L.D., Boothroyd, J.C., 1992b. Construction of a molecular karyotype for *Toxoplasma gondii*. Mol. Biochem. Parasitol. 51 (2), 291–300.

Sibley, L.D., 2003a. Recent origins among ancient parasites. Vet. Parasitol. 115, 185–195.

Sibley, L.D., 2003b. *Toxoplasma gondii*: perfecting an intracellular life style. Traffic 4, 581–586.

Sibley, L.D., Ajioka, J.W., 2008. Population structure of *Toxoplasma gondii*: clonal expansion driven by infrequent recombination and selective sweeps. Annu. Rev. Microbiol. 62, 329–351.

Silva, N.M., Rodrigues, C.V., Santoro, M.M., Reis, L.F., Alvarez-Leite, J.I., Gazzinelli, R.T., 2002. Expression of indoleamine 2,3-dioxygenase, tryptophan degradation, and kynurenine formation during in vivo infection with *Toxoplasma gondii*: induction by endogenous gamma interferon and requirement of interferon regulatory factor 1. Infect. Immun. 70, 859–868.

Silva, N.M., Manzan, R.M., Carneiro, W.P., Milanezi, C.M., Silva, J.S., Ferro, E.A., Mineo, J.R., 2010 Oct. *Toxoplasma gondii*: the severity of toxoplasmic encephalitis in C57BL/6 mice is associated with increased ALCAM and VCAM-1 expression in the central nervous system and higher blood-brain barrier permeability. Exp. Parasitol. 126 (2), 167–177. http://dx.doi.org/10.1016/j.exppara.2010.04.019.

Silva, N.M., Vieira, J.C., Carneiro, C.M., Tafuri, W.L., 2009 Sep. *Toxoplasma gondii*: the role of IFNgamma, TNFRp55 and iNOS in inflammatory changes during infection. Exp. Parasitol. 123 (1), 65–72. http://dx.doi.org/10.1016/j.exppara.2009.05.011.

da Silva, A.V., Pezerico, S.B., de Lima, V.Y., d'Arc Moretti, L., Pinheiro, J.P., Tanaka, E.M., Ribeiro, M.G., Langoni, H., 2005. Genotyping of *Toxoplasma gondii* strains isolated from dogs with neurological signs. Vet. Parasit. 127, 23–27.

Silveira, C., Belfort Jr., R., Burnier Jr., M., Nussenblatt, R., 1988. Acquired toxoplasmic infection as the cause of toxoplasmic retinochoiditis in families. Am. J. Ophthalmol. 106 (3), 362–364.

Silveira, C., Belfort Jr., R., Muccioli, C., Abreu, M.T., Martins, M.C., Victora, C., Nussenblatt, R.B., Holland, G.N., 2001. A follow-up study of *Toxoplasma gondii* infection in southern Brazil. Am. J. Ophthalmol. 131, 351–354.

Silver, J.S., Stumhofer, J.S., Passos, S., Ernst, M., Hunter, C.A., 2011 Jul 1. IL-6 mediates the susceptibility of glycoprotein 130 hypermorphs to *Toxoplasma gondii*. J. Immunol. 187 (1), 350–360. http://dx.doi.org/10.4049/jimmunol.1004144.

Sims, T.A., Hay, J., Talbot, I.C., 1989. Immunoelectron microscopy for Toxoplasma antigen in association with intact tissue cysts in mouse brain. Ann. Trop. Med. Parasitol. 83, 639–641.

Sinai, A.P., Payne, T.M., Carmen, J.C., Hardi, L., Watson, S.J., Molestina, R.E., 2004. Mechanisms underlying the manipulation of host apoptotic pathways by *Toxoplasma gondii*. Int. J. Parasitol. 34 (3), 381–391.

Skallova, A., Kodym, P., Frynta, D., Flegr, J., 2006. The role of dopamine in Toxoplasma-induced behavioural alterations in mice: an ethological and ethopharmacological study. Parasitology 133, 525–535.

Skariah, S., Mordue, D.G., 2012. Identification of *Toxoplasma gondii* genes responsive to the host immune response during in vivo infection. PLoS One. 10:e46621.

Skariah, S., Bednarczyk, R.B., Mcintyre, M.K., Taylor, G.A., Mordue, D.G., 2012. Discovery of a novel *Toxoplasma gondii* conoid-associated protein important for parasite resistance to reactive nitrogen intermediates. J. Immunol. 188, 3404–3415.

Smiley, ST, Lanthier, PA, Couper, KN, Szaba, FM, Boyson, JE, Chen, W, Johnson, LL, 2005. Exacerbated susceptibility to infection-stimulated immunopathology in CD1d-deficient mice. J Immunol 174 (12), 7904–7911.

Soete, M., Camus, D., Dubremetz, J.F., 1994. Experimental induction of bradyzoite-specific antigen expression and cyst formation by the RH strain of *Toxoplasma gondii* in vitro. Exp. Parasitol. 78 (4), 361–370.

Spear, W., Chan, D., Coppens, I., Johnson, R.S., Giaccia, A., Blader, I.J., 2006. The host cell transcription factor hypoxia-inducible factor 1 is required for *Toxoplasma gondii* growth and survival at physiological oxygen levels. Cell Microbiol. 8 (2), 339–352.

Starnes, G.L., Coincon, M., Sygusch, J., Sibley, L.D., 2009 Apr 23. Aldolase is essential for energy production and bridging adhesin-actin cytoskeletal interactions during parasite invasion of host cells. Cell Host Microbe 5 (4), 353–364. http://dx.doi.org/10.1016/j.chom.2009.03.005.

Stanley, K., Luzio, P., 1998. Perforin. A family of killer proteins [news]. Nature 334 (6182), 475–476.

Stenzel, W., Soltek, S., Schluter, D., Deckert, M., 2004. The intermediate filament GFAP is important for the control of experimental murine Staphylococcus aureus-induced brain abscess and Toxoplasma encephalitis. J. Neuropathol. Exp. Neurol. 63, 631–640.

Stibbs, H.H., 1985. Changes in brain concentrations of catecholamines and indoleamines in *Toxoplasma gondii* infected mice. Ann. Trop. Med. Parasitol. 79, 153–157.

Strack, A., Asensio, V.C., Campbell, I.L., Schluter, D., Deckert, M., 2002. Chemokines are differentially expressed by astrocytes, microglia and inflammatory leukocytes in Toxoplasma encephalitis and critically regulated by interferon-gamma. Acta Neuropathologica 103, 458–468.

Strobl, J.S., Goodwin, D., Rzigalinski, B.A., Lindsay, D.S., 2012. Dopamine Stimulates Propagation of Toxoplasma Gondii Tachyzoites in Human Fibroblast and Primary Rat Neonatal Brain Cell Cultures. J. Parasitol. 98, 1296–1299.

Steffen, B.J., Breier, G., Butcher, E.C., Schulz, M., Engelhardt, B., 1996. ICAM-1, VCAM-1, and MAdCAM-1 are expressed on choroid plexus epithelium but not endothelium and mediate binding of lymphocytes in vitro. Am. J. Pathol. 148 (6), 1819–1838.

Stumhofer, J.S., Laurence, A., Wilson, E.H., Huang, E., Tato, C.M., Johnson, L.M., Villarino, A.V., Huang, Q., Yoshimura, A., Sehy, D., Saris, C.J., O'Shea, J.J., Hennighausen, L., Ernst, M., Hunter, C.A., 2006 Sep. Interleukin 27 negatively regulates the development of interleukin 17-producing T helper cells during chronic inflammation of the central nervous system. Nat. Immunol. 7 (9), 937–945.

Stutz, A., Kessler, H., Kaschel, M.E., Meissner, M., Dalpke, A.H., 2012 Jan. Cell invasion and strain dependent induction of suppressor of cytokine signalling-1 by *Toxoplasma gondii*. Immunobiology 217 (1), 28–36. http://dx.doi.org/10.1016/j.imbio.2011.08.008. Epub 2011 Aug 27.

Su, C., Evans, D., Cole, R.H., Kissinger, J.C., Ajioka, J.W., Sibley, L.D., 2003. Recent expansion of Toxoplasma through enhanced oral transmission. Science 299, 414–416.

Subauste, C.S., et al., 1998. Alpha beta T-cell response to *Toxoplasma gondii* in previously unexposed individuals. J. Immunol. 160 (7), 3403–3411.

Subauste, C., Chung, J., Do, D., et al., 1995. Preferential activation and expansion of human peripheral blood gamma delta T-cells in response to *Toxoplasma gondii* in vitro and their cytokine production and cytotoxic activity against T. gondii-infected cells. J. Clin. Invest 96, 610–619.

Subauste, C.S., Dawson, L., Remington, J.S., 1992. Human lymphokine-activated killer cells are cytotoxic against cells infected with Toxopwlasma gondii. J. Exp. Med. 176 (6), 1511–1519.

Subauste, C.S., de Waal Malefyt, R., Fuh, F., 1998. Role of CD80 (B7.1) and CD86 (B7.2) in the immune response to an intracellular pathogen. J. Immunol. 160 (4), 1831–1840.

Subauste, C.S., Wessendarp, M., Smulian, A.G., et al., 2001. Role of CD40 ligand signalling in defective type 1 cytokine response in human immunodeficiency virus infection. J. Infect. Dis. 183, 1722–1731.

Subauste, C.S., 2002. CD154 and type-1 cytokine response: from hyper IgM syndrome to human immunodeficiency virus infection. J. Infect. Dis. 185 (Suppl 1), S83–S89.

Subauste, C.S., 2009a. Autophagy in immunity against *Toxoplasma gondii*. Curr. Top. Microbiol. Immunol. 335, 251–265.

Subauste, C.S., 2009b. Autophagy as an antimicrobial strategy. Expert Rev. Anti. Infect. Ther. 6, 743–752.

Subauste, C.S., 2009c. CD40 and the immune response to parasitic infections. Semin. Immunol. 21 (5), 273–282. http://dx.doi.org/10.1016/j.smim.2009.06.003. Epub 2009 Jul 18.

Subauste, C.S., Koniaris, A.H., Remington, J.S., 1991. Murine $CD8^+$ cytotoxic T-lymphocytes lyse *Toxoplasma gondii*-infected cells. Journal of Immunology 147, 3955–3959.

Subauste, C.S., Wessendarp, M., 2006. CD40 restrains in vivo growth of *Toxoplasma gondii* independently of gamma interferon. Infect. Immun. 74 (3), 1573–1579.

Sukhumavasi, W., Egan, C.E., Warren, A.L., Taylor, G.A., Fox, B.A., Bzik, D. J., Denkers, E.Y., 2008. TLR adaptor MyD88 is essential for pathogen control during oral *toxoplasma gondii* infection but not adaptive immunity induced by a vaccine strain of the parasite. J. Immunol. 181 (5), 3464–3473.

Sukhumavasi, W., Warren, A.L., Del Rio, L., Denkers, E.Y., 2010 Nov. Absence of mitogen-activated protein kinase family member c-Jun N-terminal kinase-2 enhances resistance to *Toxoplasma gondii*. Exp. Parasitol. 126 (3), 415–420. http://dx.doi.org/10.1016/j.exppara.2010.01.019.

Sukhumavasi, W., Egan, C.E., Denkers, E.Y., 2007. Mouse neutrophils require JNK2 MAPK for *Toxoplasma gondii*-induced IL-12p40 and CCL2/MCP-1 release. J. Immunol. 179 (6), 3570–3577.

Sun, G.D., Kobayashi, T., Abe, M., Tada, N., Adachi, H., Shiota, A., Totsuka, Y., Hino, O., 2007 Aug 17. The endoplasmic reticulum stress-inducible protein Niban regulates eIF2alpha and S6K1/4E-BP1 phosphorylation. Biochem. Biophys. Res. Commun. 360 (1), 181–187. Epub 2007 Jun 12.

Suzuki, Y., Kobayashi, A., 1987. Presence of high concentrations of circulating Toxoplasma antigens during acute Toxoplasma infection in athymic nude mice. Infect. Immun. 55, 1017–1018.

Suzuki, Y., Joh, K., Kobayashi, A., 1987. Macrophage-mediated suppression of immune responses in Toxoplasma-infected mice. III. Suppression of antibody responses to parasite itself. Cell Immunol. 110, 218–225.

Suzuki, Y., Orellana, M.A., Schreiber, R.D., Remington, J.S., 1988. Interferon-gamma: the major mediator of resistance against *Toxoplasma gondii*. Science 240, 516–518.

Suzuki, Y., Remington, J.S., 1988. Dual regulation of resistance against *Toxoplasma gondii* infection by Lyt-2^+ and Lyt-1^+, L3T4^+ T-cells in mice. J. Immunol. 140 (11), 3943–3946.

Suzuki, Y., Conley, F.K., Remington, J.S., 1989. Importance of endogenous IFNgamma for prevention of toxoplasmic encephalitis in mice. J. Immunol. 143, 2045–2050.

Suzuki, Y., Kobayashi, A., 1990. Induction of tolerance to *Toxoplasma gondii* in newborn mice by maternal antibody. Parasitol. Res. 76, 424–427.

Suzuki, Y., Remington, J.S., 1990. The effect of anti-IFNgamma antibody on the protective effect of Lyt- 2^+ immune T-cells against toxoplasmosis in mice. J. Immunol. 144 (5), 1954–1956.

Suzuki, Y., Joh, K., Kobayashi, A., 1991a. Tumor necrosis factor-independent protective effect of recombinant IFNg against acute toxoplasmosis in T-cell-deficient mice. J. Immunol. 147, 2728–2733.

Suzuki, Y., Bunazawa, M., Kobayashi, A., 1991b. Toxoplasma antigens recognized by immunoglobulin M and G antibodies during acute and chronic infections in humans. Jpn. J. Parasitol. 40, 446–450.

Suzuki, Y., Joh, K., Orellana, M.A., Conley, F.K., Remington, J.S., 1991c. A gene(s) within the H-2D region determines the development of toxoplasmic encephalitis in mice. Immunology 74, 732–739.

Suzuki, Y., Kobayashi, A., Nishizawa, T., Inagawa, H., Morikawa, A., Soma, G.I., Mizuno, D., 1992. Homeostasis as regulated by activated macrophage. VI. Protective effect of LPSw (a lipopolysaccharide from wheat flour) against acute infection with *Toxoplasma gondii* in mice. Chem. Pharm. Bull. 40, 1266–1267.

Suzuki, Y., Kobayashi, A., Ohtomo, H., 1993a. Presence of interferon-gamma mediated resistance against *Toxoplasma gondii* in T-cell-deficient mice. Jpn. J. Parasitol. 42, 507–510.

Suzuki, Y., Orellana, M.A., Conley, F.K., Remington, J.S., 1993b. Susceptibility to chronic infection with *Toxoplasma gondii* does not correlate with susceptibility to acute infection in mice. Infect. Immun. 61, 2284–2288.

Suzuki, Y., Remington, J.S., 1993. Toxoplasmic encephalitis in AIDS patients and experimental models for study of the disease and its treatment. Res. Immunol. 144, 66–67.

Suzuki, Y., Yang, Q., Conley, F.K., Abrams, J.S., Remington, J.S., 1994a. Antibody against interleukin-6 reduces inflammation and numbers of cysts in brains of mice with toxoplasmic encephalitis. Infect. Immun. 62, 2773–2778.

Suzuki, Y., Joh, K., 1994. Effect of the strain of *Toxoplasma gondii* on the development of toxoplasmic encephalitis in mice treated with antibody to interferon-gamma. Parasitol. Res. 80 (2), 125–130.

Suzuki, Y., Joh, K., Kwon, O.C., Yang, Q., Conley, F.K., Remington, J.S., 1994b. MHC class I gene(s) in the D/L region but not the TNF-alpha gene determines development of toxoplasmic encephalitis in mice. J. Immunol. 153 (10), 4649–4654.

Suzuki, Y., Yang, Q., Remington, J.S., 1995. Genetic resistance against acute toxoplasmosis depends on the strain of *Toxoplasma gondii*. J. Parasitol. 81, 1032–1034.

Suzuki, Y., Yang, Q., Yang, S., Nguyen, N., Lim, S., Liesenfeld, O., Kojima, T., Remington, J.S., 1996a. IL-4 is protective against development of toxoplasmic encephalitis. J. Immunol. 157, 2564–2569.

Suzuki, Y., Wong, S.-Y., Grumet, F.C., Fessel, J., Montoya, J.G., Zolopa, A.R., Portmore, A., Schumacher-perdreau, F., Schrappe, M., Koppen, S., Ruf, B., Brown, B.W., Remington, J.S., 1996b. Evidence for genetic regulation of susceptibility to toxoplasmic encephalitis in AIDS patients. J. Infect. Dis. 173, 265–268.

Suzuki, Y., 1997. Cells and cytokines in host defence of the central nervous sytem. In: Peterson, R. (Ed.), In Defence of the Brain. Blackwell Science, Malden, pp. 56–73. l.

Suzuki, Y., Rani, S., Liesenfeld, O., Kojima, T., Lim, S., Nguyen, T.A., Dalrymple, S.A., Murray, R., Remington, J.S., 1997. Impaired resistance to the development of toxoplasmic encephalitis in interleukin-6-deficient mice. Infect. Immun. 65 (6), 2339–2345.

Suzuki, Y., Kang, H., Parmley, S., Lim, S., Park, D., 2000a. Induction of tumour necrosis factor-α and inducible nitric oxide synthase fails to prevent toxoplasmic encephalitis in the absence of interferon-g in genetically resistant BALB/c mice. Microbes Infect. 2, 455–462.

Suzuki, Y., Sher, A., Yap, G., Park, D., Neyer, L.E., Liesenfeld, O., Fort, M., Kang, H., Gufwoli, E., 2000b. IL-10 is required for prevention of necrosis in the small intestine and mortality in both genetically resistant BALB/c and susceptible C57BL/6 mice following peroral infection with *Toxoplasma gondii*. J. Immunol. 164, 5375–5382.

Suzuki, Y., Claflin, J., Wang, X., Lengi, A., Kikuchi, T., 2005. Microglia and macrophages as innate producers of interferon-gamma in the brain following infection with *Toxoplasma gondii*. Int. J. Parasitol. 35, 83–90.

Suzuki, Y., Wang, X., Jortner, B.S., Payne, L., Ni, Y., Michie, S.A., Xu, B., Kudo, T., Perkins, S., 2010. Removal of *Toxoplasma gondii* cysts from the brain by perforin-mediated activity of CD8^+ T-cells. Am. J. Pathol. 176, 1607–1613.

Tait, E.D., Jordan, K.A., Dupont, C.D., Harris, T.H., Gregg, B., Wilson, E.H., Pepper, M., Dzierszinski, F., Roos, D.S., Hunter, C.A., 2010 Aug 1. Virulence of *Toxoplasma gondii* is associated with distinct dendritic cell responses and reduced numbers of activated CD8^+ T-cells. J. Immunol. 185 (3), 1502–1512. http://dx.doi.org/10.4049/jimmunol.0903450.

Tait, E.D., Hunter, C.A., 2009. Advances in understanding immunity to *Toxoplasma gondii*. Mem. Inst. Oswaldo Cruz (2), 201–210.

Takács, A.C., Swierzy, I.J., Lüder, C.G., 2012. Interferon-γ restricts *Toxoplasma gondii* development in murine skeletal muscle cells via nitric oxide production and immunity-related GTPases. PLoS One 7 (9), e45440.

Tan, Guan, Tze, Mui, Ernest, Cong, Hua, Witola, William, Montpetit, Alexandre, et al., 2010. Identification of *T. gondii* epitopes, adjuvants, and host genetic factors that influence protection of mice and humans. Vaccine 28, 3977–3989.

Tato, C.M., Joyce-Shaikh, B., Banerjee, A., Chen, Y., Sathe, M., Ewald, S.E., Liu, M.R., Gorman, D., McClanahan, T.K., Phillips, J.H., Heyworth, P.G., Cua, D.J., 2012. The myeloid receptor PILRβ mediates the balance of inflammatory responses through regulation of IL-27 production. PLoS One 7 (3), e31680. http://dx.doi.org/10.1371/journal.pone.0031680. Epub 2012 Mar 27.

Taylor, G.A., Feng, C.G., Sher, A., 2004. p47 GTPases: regulators of immunity to intracellular pathogens. Nat. Rev. Immunol. 4, 100–109.

Teixeira, M.M., Almeida, I.C., Gazzinelli, R.T., 2002. Introduction: innate recognition of bacteria and protozoan parasites. Microbes Infect. 4, 883–886.

Terpenning, M.S., Bradley, S.F., 1991. Why aging leads to increased susceptibility to infection. Geriatrics 48, 77–80.

Terrazas, C.A., Juarez, I., Terrazas, L.I., Saavedra, R., Calleja, E.A., Rodriguez-Sosa, M., 2010 Nov. *Toxoplasma gondii*: impaired maturation and proinflammatory response of dendritic cells in MIF-deficient mice favours susceptibility to infection. Exp. Parasitol. 126 (3), 348–358. http://dx.doi.org/10.1016/j.exppara.2010.03.009. Epub 2010 Mar 21.

Teutsch, S.M., Juranek, D.D., Sulzer, A., Dubey, J.P., Sikes, R.K., 1979. Epidemic toxoplasmosis associated with infected cats. N. Engl. J. Med. 300 (13), 695–699.

Tham, C., Lin, F., Rao, T., Yu, N., Webb, M., 2003. Microglial activation state and lysophospholipid acid receptor expression. Int. J. Dev. Neurosci. 21 (8), 431–443.

Thilaganathan, B., Abbas, A., Nicolaides, K.H., 1993. Foetal blood natural killer cells in human pregnancy. Foetal Diagn. Ther. 8, 149–153.

Thilaganathan, B., Carroll, S.G., Plachouras, N., et al., 1994. Foetal immunological and haematological changes in intrauterine infection. Br. J. Obstet Gynaecol. 101, 418–421.

Thomas, S., Garrity, L., Brandt, C., et al., 1993. IFNgamma-mediated antimicrobial response. Indoleamine 2,3-dioxygenase-deficient mutant host cells no longer inhibit intracellular Chlamydia spp. or Toxoplasma growth. J. Immunol. 150, 5529–5534.

Thompson, R.W., Pesce, J.T., Ramalingam, T., Wilson, M.S., White, S., Cheever, A.W., Ricklefs, S.M., Porcella, S.F., Li, L., Ellies, L.G., Wynn, T.A., 2008 Mar 14. Cationic amino acid transporter-2 regulates immunity by modulating arginase activity. PLoS Pathog. 4 (3), e1000023. http://dx.doi.org/10.1371/journal.ppat.1000023.

Thouvenin, M., Candolfi, E., Villard, O., Klein, J.P., Kien, T., 1997. Immune response in a murine model of congenital toxoplasmosis: increased susceptibility of pregnant mice and transplacental passage of *Toxoplasma gondii* are type 2-dependent. Parassitologia 39, 279–283.

Tomita, T., Yamada, T., Weiss, L.M., Orlofsky, A., 2009. Externally triggered egress is the major fate of *Toxoplasma gondii* during acute infection. J. Immunol. 183, 6667–6680.

Torre, D., Zeroli, C., Ferrario, G., Pugliese, A., Speranza, F., Orani, A., Casari, S., Bassi, P., Poggio, A., Carosi, G.P., Fiori, G.P., 1999. Levels of nitric oxide, gamma interferon and interleukin-12 in AIDS patients with toxoplasmic encephalitis. Infection 27 (3), 218–220.

Torrey, E.F., Bartko, J.J., Yolken, R.H., 2012. *Toxoplasma gondii* and other risk factors for schizophrenia: an update. Schizophrenia Bulletin 38, 642–647.

Trowsdale, J., Parham, P., 2004. Mini-review: defence strategies and immunity-related genes. Eur. J. Immunol. 34, 7–17.

Tu, L., Moriya, C., Imai, T., Ishida, H., Tetsutani, K., Duan, X., Murata, S., Tanaka, K., Shimokawa, C., Hisaeda, H., Himeno, K., 2009. Critical role for the immunoproteasome subunit LMP7 in the resistance of mice to *Toxoplasma gondii* infection. Eur. J. Immunol. 39, 3385–3394.

Tussiwand, R., Lee, W.L., Murphy, T.L., Mashayekhi, M., Wumesh, K.C., Albring, J.C., Satpathy, A.T., Rotondo, J.A., Edelson, B.T., Kretzer, N.M., Wu, X., Weiss, L.A., Glasmacher, E., Li, P., Liao, W., Behnke, M., Lam, S.S., Aurthur, C.T., Leonard, W.J., Singh, H., Stallings, C.L., Sibley, L.D., Schreiber, R.D., Murphy, K.M., 2012 Oct 25. Compensatory dendritic cell development mediated by BATF-IRF interactions. Nature 490 (7421), 502–507. http://dx.doi.org/10.1038/nature11531.

Unanue, E.R., 1996. Macrophages, NK-cells and neutrophils in the cytokine loop of Listeria resistance. Res. Immunol. 147, 499–505.

Unoki, M., Brunet, J., Mousli, M., 2009 Nov 15. Drug discovery targeting epigenetic codes: the great potential of UHRF1, which links DNA methylation and histone modifications, as a drug target in cancers and toxoplasmosis. Biochem. Pharmacol. 78 (10), 1279–1288. http://dx.doi.org/10.1016/j.bcp.2009.05.035.

Vainisi, S.J., Campbell, L.H., 1969. Ocular toxoplasmosis in cats. J. Am. Vet. Med. Assoc. 154, 141–152.

Vallochi, A.L., Muccioli, C., Martins, M.C., Silveira, C., Belfort Jr., R., Rizzo, L.V., 2005. The genotype of *Toxoplasma gondii* strains causing ocular toxoplasmosis in humans in Brazil. Amer. J. Ophthalmol. 139, 350–351.

Vallochi, A.L., Goldberg, A.C., Falcai, A., Ramasawmy, R., Kalil, J., Silveira, C., Belfort, R., Rizzo, L.V., 2008. Molecular markers of susceptibility to ocular toxoplasmosis, host and guest behaving badly. Clin. Ophthalmol. 2, 837–848.

Van, T.T., Kim, S.K., Camps, M., Boothroyd, J.C., Knoll, L.J., 2007 Jul. The BSR4 protein is up-regulated in *Toxoplasma gondii* bradyzoites, however the dominant surface antigen recognised by the P36 monoclonal antibody is SRS9. Int. J. Parasitol. 37 (8-9), 877–885.

Vaudaux, J.D., Muccioli, C., James, E.R., Silveira, C., Magargal, S.L., Jung, C., Dubey, J.P., Jones, J.L., Doymaz, M.Z., Bruckner, D.A., Belfort Jr., R., Holland, G.N., Grigg, M.E., 2010. Identification of an atypical strain of *toxoplasma gondii* as the cause of a waterborne outbreak of toxoplasmosis in Santa Isabel do Ivai. Brazil. J. Infect. Dis. 202 (8), 1226–1233. http://dx.doi.org/10.1086/656397.

Vassiliadou, M., Tucker, L., Anderson, D.J., 1999. Progesterone-induced inhibition of chemokine receptor expression on peripheral blood mononuclear cells correlates with reduced HIV-1 infectability in vitro. J. Immunol. 162, 7510–7518.

Vilches, C., Parham, P., 2002. KIR: diverse, rapidly evolving receptors of innate and adaptive immunity. Annu. Rev. Immunol. 20, 217–251.

Villard, O., Breit, L., Cimon, B., Franck, J., Fricker-Hidalgo, H., Godineau, N., Houze, S., Paris, L., Pelloux, H., Villena, I., Candolfi, E., 2012. The French National Reference Center for Toxoplasmosis Network. Comparison of four commercially available avidity tests for Toxoplasma -specific IgG antibodies. Clin. Vaccine Immunol. 20 (2), 197–204.

Villarino, A., Hibbert, L., Lieberman, L., Wilson, E., Mak, T., Yoshida, H., Kastelein, R.A., Saris, C., Hunter, C.A., 2003. The IL-27R (WSX-1) is required to suppress T-cell hyperactivity during infection. Immunity 19 (5), 645–655.

Villarino, A.V., Stumhofer, J.S., Saris, C.J., Kastelein, R.A., de Sauvage, F.J., Hunter, C.A., 2006. IL-27 limits IL-2 production during Th1 differentiation. J. Immunol. 176 (1), 237–247, 2006 Jan 1.

Villegas, E.N., Lieberman, L.A., Mason, N., Blass, S.L., Zediak, V.P., Peach, R., Horan, T., Yoshinaga, S., Hunter, C.A., 2002. A role for inducible costimulator protein in the CD28-independent mechanism of resistance to *Toxoplasma gondii*. J. Immunol 169 (2), 937–943.

Virgin, H., Wherry, E.J., Ahmed, R., 2009. Redefining chronic viral infection. Cell 138, 30–50.

Virreira Winter, S., Niedelman, W., Jensen, K.D., Rosowski, E.E., Julien, L., Spooner, E., Caradonna, K., Burleigh, B.A., Saeij, J.P., Ploegh, H.L., Frickel, E.M., 2011. Determinants of GBP recruitment to *Toxoplasma gondii* vacuoles and the parasitic factors that control it. PLoS One 6, e24434.

Vogel, N., Kirisits, M., Michael, E., Bach, H., Hostetter, M., Boyer, K., Simpson, R., Holfels, E., Hopkins, J., Mack, D., Mets, M.B., Swisher, C.N., Roizen, N., Stein, L., Stein, M., Withers, S., Mui, E., Egwuagu, C., Remington, J., Dorfman, R., McLeod, R., 1996. Congenital toxoplasmosis transmitted from an immunologically competent mother infected before conception. Clin. Infect. Dis. 23, 1055–1060.

von Boehmer, H., The, H.S., Kisielow, P., 1989. The thymus selects the useful, neglects the useless and destroys the harmful. Immunol. Today 10 (2), 57–61.

Vutova, P., Wirth, M., Hippe, D., Gross, U., Schulze-Osthoff, K., Schmitz, I., Lüder, C.G., 2007 Jun. *Toxoplasma gondii* inhibits Fas/CD95-triggered cell death by inducing aberrant processing and degradation of caspase 8. Cell Microbiol. 9 (6), 1556–1570.

Vyas, A., Kim, S.K., Giacomini, N., Boothroyd, J.C., Sapolsky, R.M., 2007a. Behavioural changes induced by Toxoplasma infection of rodents are highly specific to aversion of cat odors. Proc. Natl. Acad. Sci. U S A 104 (15), 6442–6447.

Vyas, A., Kim, S.K., Sapolsky, R.M., 2007b. The effects of toxoplasma infection on rodent behaviour are dependent on dose of the stimulus. Neuroscience 148 (2), 342–348.

Vyas, A., Sapolsky, R., 2010. Manipulation of host behaviour by *Toxoplasma gondii*: what is the minimum a proposed proximate mechanism should explain? Folia Parasitol. (Praha) 57 (2), 88–94.

Vyas, A., 2013. Parasite-augmented mate choice and reduction in innate fear in rats infected by *Toxoplasma gondii*. J. Exp. Biol. 216 (Pt 1), 120–126.

Wallon, M., Kodjikian, L., Binquet, C., Garweg, J., Fleury, J., et al., 2004. Long-term ocular prognosis in 327 children with congenital toxoplasmosis. Pediatrics 113, 1567–1572.

Wang, H., He, S., Yao, Y., Cong, H., Zhao, H., Li, T., Zhu, X.Q., 2009a. *Toxoplasma gondii*: protective effect of an intranasal SAG1 and MIC4 DNA vaccine in mice. Exp. Parasitol. 122 (3), 226–232. http://dx.doi.org/10.1016/j.exppara.2009.04.002.

Wang, X., Kang, H., Kikuchi, T., Suzuki, Y., 2004. Gamma interferon production, but not perforin-mediated cytolytic activity, of T-cells is required for prevention of toxoplasmic encephalitis in BALB/c mice genetically resistant to the disease. Infect. Immun. 72, 4432–4438.

Wang, X., Claflin, J., Kang, H., Suzuki, Y., 2005. Importance of CD8($^+$)Vbeta8($^+$) T-cells in IFNgamma-mediated prevention of toxoplasmic encephalitis in genetically resistant BALB/c mice. J. Interferon Cytokine Res. 25, 338–344.

Wang, Y., Weiss, L.M., Orlofsky, A., 2010 May 14. Coordinate control of host centrosome position, organelle distribution, and migratory response by *Toxoplasma gondii* via host mTORC2. J. Biol. Chem. 285 (20), 15611–15618. http://dx.doi.org/10.1074/jbc.M109.095778. Epub 2010 Mar 17.

Wang, Z.Y., Hakanson, R., 1995. Role of nitric oxide (NO) in ocular inflammation. Br. J. Pharmacol. 116 (5), 2447–2450.

Wang, X., Michie, S.A., Xu, B., Suzuki, Y., 2007. Importance of IFNgamma-mediated expression of endothelial VCAM-1 on recruitment of $CD8^+$ T-cells into the brain during chronic infection with *Toxoplasma gondii*. J. Interferon. Cytokine Res. 27, 329–338.

Wang, X., Suzuki, Y., 2007. Microglia produce IFNgamma independently from T-cells during acute toxoplasmosis in the brain. J. Interferon. Cytokine Res. 27, 599–605.

Wang, Y., Weiss, L.M., Orlofsky, A., 2009b. Host cell autophagy is induced by *Toxoplasma gondii* and contributes to parasite growth. J. Biol. Chem. 284 (3), 1694–1701. http://dx.doi.org/10.1074/jbc.M807890200. Epub 2008 Nov 21.

Washino, T., Moroda, M., Iwakura, Y., Aosai, F., 2012 Apr. *Toxoplasma gondii* infection inhibits Th17-mediated spontaneous development of arthritis in interleukin-1 receptor antagonist-deficient mice. Infect. Immun. 80 (4), 1437–1444. http://dx.doi.org/10.1128/IAI.05680-11. Epub 2012 Jan 30.

Ware, P.L., Kasper, L.H., 1987. Strain-specific antigens of *Toxoplasma gondii*. Infect. Immun. 55 (3), 778–783.

Wasmuth, J.D., Pszenny, V., Haile, S., Jansen, E.M., Gast, A.T., Sher, A., Boyle, J.P., Boulanger, M.J., Parkinson, J., Grigg, M.E., 2012. Integrated Bioinformatic and Targeted Deletion Analyses of the SRS Gene Superfamily Identify SRS29C as a Negative Regulator of *Toxoplasma* Virulence. MBio. 3 (6). doi:pii: e00321–12. 10.1128/mBio.00321-12.

Watanabe, H., Suzuki, Y., Makino, M., Fujiwara, M., 1993. *Toxoplasma gondii*: Induction of toxoplasmic encephalitis in mice with chronic infection by inoculation of a murine leukaemia virus inducing immunodeficiency. Exp. Parasitol. 76, 39–45.

Watford, W.T., Hissong, B.D., Durant, L.R., Yamane, H., Muul, L.M., Kanno, Y., Tato, C.M., Ramos, H.L., Berger, A.E., Mielke, L., Pesu, M., Solomon, B., Frucht, D.M., Paul, W.E., Sher, A., Jankovic, D., Tsichlis, P.N., O'Shea, J.J., 2008. Tpl2 kinase regulates T-cell interferon-gamma production and host resistance to *Toxoplasma gondii*. J. Exp. Med. 205 (12), 2803–2812.

Webster, J.P., Brunton, C.F., Macdonald, D.W., 1994. Effect of *Toxoplasma gondii* upon neophobic behaviour in wild brown rats, Rattus norvegicus. Parasitology 109 (Pt 1), 37–43.

Wegman, T.G., et al., 1993. bidirectional cytokine interactions in the maternal-foetal relationship: is successful pregnancy a TH2 phenomenon? [see comments]. Immunol. Today 14 (7), 353–356.

Wen, X., Kudo, T., Payne, L., Wang, X., Rodgers, L., Suzuki, Y., 2010. Predominant interferon-γ-mediated expression of CXCL9, CXCL10, and CCL5 proteins in the brain during chronic infection with *Toxoplasma gondii* in BALB/c mice resistant to development of toxoplasmic encephalitis. J. Interferon Cytokine Res. 30 (9), 653–660. http://dx.doi.org/10.1089/jir.2009.0119.

Wendte, J.M., Gibson, A.K., Grigg, M.E., 2011 Nov 24. Population genetics of *Toxoplasma gondii*: new perspectives from parasite genotypes in wildlife. Vet. Parasitol. 182 (1), 96–111. http://dx.doi.org/10.1016/j.vetpar.2011.07.018.

Weiss, L.M., Laplace, D., Takvorian, P.M., Tanowitz, H.B., Cali, A., Wittner, M., 1995. A cell culture system for study of the development of *Toxoplasma gondii* bradyzoites. J. Eukaryot. Microbiol. 42 (2), 150–157.

Weiss, L., Kim, K., 2000. The development and biology of bradyzoites of *Toxoplasma gondii*. Front. Biosci. 5, 391.

Wille, U., Villegas, E.N., Striepen, B., Roos, D.S., Hunter, C.A., 2001. Interleukin-10 does not contribute to the pathogenesis of a virulent strain of Toxoplasma gondii. Parasite Immunol. 23 (6), 291–296.

Wille, U., Nishi, M., Lieberman, L., Wilson, E.H., Roos, D.S., Hunter, C.A., 2004. IL-10 is not required to prevent immune hyperactivity during memory responses to Toxoplasma gondii. Parasite Immunol. 26 (5), 229–236.

Whitmarsh, R.J., Gray, C.M., Gregg, B., Christian, D.A., May, M.J., Murray, P.J., Hunter, C.A., 2011. A critical role for SOCS3 in innate resistance to *Toxoplasma gondii*. Cell Host Microbe 10 (3), 224–236. http://dx.doi.org/10.1016/j.chom.2011.07.009.

Whitmarsh, R.J., Hunter, C.A., 2008. Digest this! A role for autophagy in controlling pathogens. Cell Host Microbe (5), 413–414. http://dx.doi.org/10.1016/j.chom.2008.10.008.

Wiehagen, K.R., Corbo-Rodgers, E., Li, S., Staub, E.S., Hunter, C.A., Morrisey, E.E., Maltzman, J.S., 2012. Foxp4 is dispensable for T-cell development, but required for robust recall responses. PLoS One 7 (8), e42273. http://dx.doi.org/10.1371/journal.pone.0042273.

Wiley, M., Sweeney, K.R., Chan, D.A., Brown, K.M., McMurtrey, C., Howard, E.W., Giaccia, A.J., Blader, I.J., 2010. *Toxoplasma gondii* activates hypoxia-inducible factor (HIF) by stabilizing the HIF-1alpha subunit via type I activin-like receptor kinase receptor signalling. J. Biol. Chem. 285 (35), 26852–26860.

Williams, D.M., Grumet, F.C., Remington, 1978. Genetic control of murine resistance to *Toxoplasma gondii*. Infect. Immun. 19, 416–420.

Wilson, E.H., Harris, T.H., Mrass, P., John, B., Tait, E.D., Wu, G.F., Pepper, M., Wherry, E.J., Dzierzinski, F., Roos, D., Haydon, P.G., Laufer, T.M., Weninger, W., Hunter, C.A., 2009 Feb 20. Behaviour of parasite-specific effector $CD8^+$ T-cells in the brain and visualization of a kinesis-associated system of reticular fibers. Immunity 30 (2), 300–311. http://dx.doi.org/10.1016/j.immuni.2008.12.013.

Wilson, D.C., Grotenbreg, G.M., Liu, K., Zhao, Y., Frickel, E.M., Gubbels, M.J., Ploegh, H.L., Yap, G.S., 2010a. Differential regulation of effector- and central-memory responses to *Toxoplasma gondii* Infection by IL-12 revealed by tracking of Tgd057-specific $CD8^+$ T-cells. PLoS Pathog. 6, e1000815.

Wilson, M.S., Feng, C.G., Barber, D.L., Yarovinsky, F., Cheever, A.W., Sher, A., Grigg, M., Collins, M., Fouser, L., Wynn, T.A., 2010b. Redundant and pathogenic roles for IL-22 in mycobacterial, protozoan, and helminth infections. J. Immunol. 184 (8), 4378–4390. http://dx.doi.org/10.4049/jimmunol.0903416. Epub 2010 Mar 10.

Wilson, C.B., Remington, J.S., 1979. Activity of human blood leukocytes against *Toxoplasma gondii*. J. Infect. Dis. 140, 890–895.

Wilson, C.B., Tsai, V., Remington, J.S., 1980a. Failure to trigger the oxidative metabolic burst by normal macrophages: possible mechanism for survival of intracellular pathogens. J. Exp. Med. 151, 328–346.

Wilson, C.B., Remington, J.S., Stagno, S., Reynolds, D.W., 1980b. Development of adverse sequelae in children born with subclinical congenital Toxoplasma infection. Pediatrics 66 (5), 767–774.

Wilson, C.B., Haas, J.E., 1984. Cellular defences against *Toxoplasma gondii* in newborns. J. Clin. Invest. 73 (6), 1606–1616.

Wilson, C.B., et al., 1986. Decreased production of interferon-gamma by human neonatal cells. Intrinsic and regulatory deficiencies. J. Clin. Invest. 77 (3), 860–867.

Wilson, C.B., 1985. Congenital nonbacterial infections: diagnosis, treatment and prevention. Perinatol. Neonatal. 9, 9.

Wilson, E.H., Zaph, C., Mohrs, M., Welcher, A., Siu, J., Artis, D., Hunter, C.A., 2006. B7RP-1-ICOS interactions are required for optimal infection-induced expansion of $CD4^+$ Th1 and Th2 responses. J. Immunol. 177 (4), 2365–2372.

Wilson, E.H., Wille-Reece, U., Dzierszinski, F., Hunter, C.A., 2005. A critical role for IL-10 in limiting inflammation during toxoplasmic encephalitis. J. Neuroimmunol. (1–2), 63–74.

Wilson, E.H., Hunter, C.A., 2004. The role of astrocytes in the immunopathogenesis of toxoplasmic encephalitis. Int. J. Parasitol. 34 (5), 543–548.

Wilson, D.C., Matthews, S., Yap, G.S., 2008. IL-12 signalling drives $CD8^+$ T-cell IFNgamma production and differentiation of $KLRG1^+$ effector subpopulations during *Toxoplasma gondii* Infection. J. Immunol. 180 (9), 5935–5945.

Wiswell, T.E., Fajardo, J.E., Bass, J.W., Brien, J.H., Forstein, S.H., 1984. Congenital toxoplasmosis in triplets. J. Pediatr. 105 (1), 59–61.

Witola, William, H., Mui, Ernest, Hargrave, Aubrey, et al., 2011. NALP1 influences susceptibility to human congenital toxoplasmosis, proinflammatory cytokine response, and fate of Toxoplasma gondii-infected monocytic cells. Infect. Immun. 79, 756–766.

Witting, P.A., 1979. Learning capacity and memory of normal and Toxoplasma-infected laboratory rats and mice. Zeitschrift fur Parasitenkunde 61, 29–51.

Wong, S., Hajdu, M., Ramirez, R., et al., 1993. Role of specific immunoglobulin E in diagnosis of acute Toxoplasma infection and toxoplasmosis. J. Clin. Microbiol. 31, 2952–2959.

Woodman, J.P., Dimier, I.H., Bout, D.T., 1991. Human endothelial cells are activated by IFNgamma to inhibit *Toxoplasma gondii* replication. Inhibition is due to a different mechanism from that existing in mouse macrophages and human fibroblasts. J. Immunol. 147, 2019–2023.

Yamamoto, J.H., Vallochi, A.L., Silveira, C., Filho, J.K., Nussenblatt, R.B., Cunha-Neto, E., Gazzinelli, R.T., Belfort Jr., R., Rizzo, L.V., 2000. Discrimination between patients with acquired toxoplasmosis and congenital toxoplasmosis on the basis of the immune response to parasite antigens. J. Infect. Dis. 181, 2018–2022.

Yamamoto, M., Sato, S., Hemmi, H., Hoshino, K., Kaisho, T., Sanjo, H., Takeuchi, O., Sugiyama, M., Okabe, M., Takeda, K., Akira, S., 2003. Role of adaptor TRIF in the MyD88-independent toll-like receptor signalling pathway. Science 301, 640–643.

Yamamoto, M., Okuyama, M., Ma, J.S., Kimura, T., Kamiyama, N., Saiga, H., Ohshima, J., Sasai, M., Kayama, H., Okamoto, T., Huang, D.C., Soldati-Favre, D., Horie, K., Takeda, J., Takeda, K., 2012. A Cluster of Interferon-gamma-Inducible p65 GTPases Plays a Critical Role in Host Defence against *Toxoplasma gondii*. Immunity 37 (2), 302–13. Epub 2012 Jul 12.

Yamamoto, M., Ma, J.S., Mueller, C., Kamiyama, N., Saiga, H., Kubo, E., Kimura, T., Okamoto, T., Okuyama, M., Kayama, H., Nagamune, K., Takashima, S., Matsuura, Y., Soldati-Favre, D., Takeda, K., 2011 Jul 4. ATF6beta is a host cellular target of the *Toxoplasma gondii* virulence factor ROP18. J. Exp. Med. 208 (7), 1533–1546. http://dx.doi.org/10.1084/jem.20101660. Epub 2011 Jun 13.

Yamashita, K., Yui, K., Ueda, M., Yano, A., 1998. Cytotoxic T-lymphocyte-mediated lysis of *Toxoplasma gondii*-infected target cells does not lead to death of intracellular parasites. Infect. Immun. 66, 4651–4655.

Yamauchi, L.M., Aliberti, J.C., Baruffi, M.D., Portela, R.W., Rossi, M.A., Gazzinelli, R.T., Mineo, J.R., Silva, J.S., 2007 Jan. The binding of CCL2 to the surface of *Trypanosoma cruzi* induces chemo-attraction and morphogenesis. Microbes Infect. 9 (1), 111–118.

Yano, A., et al., 1989. Antigen presentation by *Toxoplasma gondii* infected cells to $CD4^+$ proliferative T-cells and $CD8^+$ cytotoxic cells. J. Parasitol. 75 (3), 411–416.

Yap, G.S., Sher, A., 1999a. Cell-mediated immunity to *Toxoplasma gondii*: initiation, regulation and effector function. Immunobiology 201, 240–247.

Yap, G.S., Scharton-Kersten, T., Charest, H., Sher, A., 1998a. Decreased resistance of TNF receptor p55- and p75-deficient mice to chronic toxoplasmosis despite normal activation of inducible nitric oxide synthase in vivo. J. Immunol. 160 (3), 1340–1345.

Yap, G.S., Scharton-Kersten, T., Ferguson, D.J.P., Howe, D., Suzuki, Y., Sher, A., 1998b. Partially protective vaccination permits the development of latency in a normally virulent strain of *Toxoplasma gondii*. Infect. Immun. 66, 4382–4388.

Yap, G.S., Ling, Y., Zhao, Y., 2007. Autophagic elimination of intracellular parasites: convergent induction by IFNgamma and CD40 ligation? Autophagy (2), 163–165.

Yap, G., Pesin, M., Sher, A., 2000. Cutting edge: IL-12 is required for the maintenance of IFNgamma production in T-cells mediating chronic resistance to the intracellular pathogen, *Toxoplasma gondii*. J. Immunol. 165, 628–631.

Yap, G.S., Sher, A., 1999b. Effector cells of both nonhemopoietic and hemopoietic origin are required for interferon (IFN)-gamma- and tumour necrosis factor (TNF)-alpha-dependent host resistance to the intracellular pathogen, *Toxoplasma gondii*. The Journal of Experimental Medicine 189, 1083–1092.

Yarovinsky, F., Zhang, D., Andersen, J.F., Bannenberg, G.L., Serhan, C.N., Hayden, M.S., Hieny, S., Sutterwala, F.S., Flavell, R.A., Ghosh, S., Sher, A., 2005. TLR11 activation of dendritic cells by a protozoan profilin-like protein. Science 308, 1626–1629.

Yarovinsky, F., Hieny, S., Sher, A., 2008. Recognition of *Toxoplasma gondii* by TLR11 prevents parasite-induced immunopathology. J. Immunol. 181 (12), 8478–8484.

Yeo, J.H., Jakobiec, F.A., Iwamoto, T., Richard, G., Kreissig, I., 1983. Opportunistic toxoplasmic retinochoroiditis following chemotherapy for systemic lymphoma. A light and electron microscopic study. Ophthalmology (8), 885–889.

Yong, R.Y., Henderson, R.A., Krissansen, G.W., Leung, E., Watson, J.D., Dholakia, J.N., 1994a. The delta-subunit of murine guanine nucleotide exchange factor eIF-2B. Characterization of cDNAs predicts isoforms differing at the amino-terminal end. J. Biol. Chem. 269 (48), 30517–30523.

Yong, E.C., Chi, E.Y., Fritsche, T.R., Henderson Jr., W.R., 1991. Human platelet-mediated cytotoxicity against *Toxoplasma gondii*: role of thromboxane. J. Exp. Med. 173 (1), 65–78.

Yong, E.C., Chi, E.Y., Henderson Jr., W.R., 1994b. *Toxoplasma gondii* alters eicosanoid release by human mononuclear phagocytes: role of leukotrienes in interferon gamma-induced antitoxoplasma activity. J. Exp. Med. 180, 1637–1648.

Young, A., Linehan, E., Hams, E., O'Hara Hall, A.C., McClurg, A., Johnston, J.A., Hunter, C.A., Fallon, P.G., Fitzgerald, DC., 2012 Sep 1. Cutting edge: suppression of GM-CSF expression in murine and human T-cells by IL-27. J. Immunol. 189 (5), 2079–2083. http://dx.doi.org/10.4049/jimmunol.1200131.

Xiao, J., Buka, S.L., Cannon, T.D., Suzuki, Y., Viscidi, R.P., Torrey, E.F., Yolken, R.H., 2009. Serological pattern consistent with infection with type I *Toxoplasma gondii* in mothers and risk of psychosis among adult offspring. Microbes Infect. 11, 1011–1018.

Xiao, J., Kannan, G., Jones-Brando, L., Brannock, C., Krasnova, I.N., Cadet, J.L., Pletnikov, M., Yolken, R.H., 2012. Sex-specific changes in gene expression and behaviour induced by chronic Toxoplasma infection in mice. Neuroscience 206, 39–48.

Zakrzewski, Alana, 2011. The role of the purinergic receptor, P2X7, in the intestinal inflammatory response to the parasite *Toxoplasma gondii*. PLoS Pathogens 7 (11), 1–7.

Zeiner, G.M., Boothroyd, J.C., 2010. Use of two novel approaches to discriminate between closely related host microRNAs that are manipulated by *Toxoplasma gondii* during infection. RNA 16 (6), 1268–1274. http://dx.doi.org/10.1261/rna.2069310.

Zenner, L., Darcy, F., Cesbron-Delauw, M.F., Capron, A., 1993. Rat model of congenital toxoplasmosis: rate of transmission of three *Toxoplasma gondii* strains to foetuses and protective effect of a chronic infection. Infect. Immun. 61 (1), 360–363.

Zhang, J., Dawson, V.L., Dawson, T.M., Snyder, S.H., 1994. Nitric oxide activation of poly(ADP-ribose) synthetase in neurotoxicity. Science 263 (5147), 687–689.

Zhang, J., He, S., Jiang, H., Yang, T., Cong, H., Zhou, H., Zhang, J., Gu, Q., Li, Y., Zhao, Q., 2007 Jul. Evaluation of the immune response induced by multi-antigenic DNA vaccine encoding SAG1 and ROP2 of *Toxoplasma gondii* and the adjuvant properties of murine interleukin-12 plasmid in BALB/c mice. Parasitol. Res. 101 (2), 331–338.

Zhang, D., Zhang, G., Hayden, M.S., Greenblatt, M.B., Bussey, C., Flavell, R.A., Ghosh, S., 2004. A toll-like receptor that prevents infection by uropathogenic bacteria. Science 303, 1522–1526.

Zhang, Y., Denkers, E.Y., 1999. Protective role for interleukin-5 during chronic Toxoplasma gondii infection. Infect. Immun. 67, 4383–4392.

Zhao, Z., Fux, B., Goodwin, M., Dunay, I.R., Strong, D., Miller, B.C., Cadwell, K., Delgado, M.A., Ponpuak, M., Green, K.G., Schmidt, R.E., Mizushima, N., Deretic, V., Sibley, L.D., Virgin, H.W., 2008. Autophagosome-independent essential function for the autophagy protein Atg5 in cellular immunity to intracellular pathogens. Cell Host Microbe 5, 458–469.

Zhao, Y.O., Khaminets, A., Hunn, J.P., Howard, J.C., 2009b. Disruption of the *Toxoplasma gondii* parasitophorous vacuole by IFNgamma-inducible immunity-related GTPases (IRG proteins) triggers necrotic cell death. PLoS Pathogens 5, e1000288.

Zhao, Y., Ferguson, D.J., Wilson, D.C., Howard, J.C., Sibley, L.D., Yap, G.S., 2009a. Virulent *Toxoplasma gondii* evade immunity-related GTPase-mediated parasite vacuole disruption within primed macrophages. J. Immunol. 182 (6), 3775–3781.

Zhao, Y., Wilson, D., Matthews, S., Yap, G.S., 2007. Rapid elimination of *Toxoplasma gondii* by gamma interferon-primed mouse macrophages is independent of CD40 signalling. Infect. Immun. 75 (10), 4799–4803.

Zhou, H., Min, J., Zhao, Q., Gu, Q., Cong, H., Li, Y., He, S., 2012 Feb 27. Protective immune response against *Toxoplasma gondii* elicited by a recombinant DNA vaccine with a novel genetic adjuvant. Vaccine 30 (10), 1800–1806. http://dx.doi.org/10.1016/j.vaccine.2012.01.004.

CHAPTER

26

Vaccination against Toxoplasmosis: Current Status and Future Prospects

Craig W. Roberts, Rima McLeod[†], Fiona L. Henriquez**, James Alexander**

*University of Strathclyde, Glasgow, Scotland, United Kingdom [†]The University of Chicago, Chicago, Illinois, USA **Institute of Biomedical and Environmental Health Research, School of Science, University of the West of Scotland, Paisley, Scotland, United Kingdom

OUTLINE

Toxoplasma gondii, second edition
http://dx.doi.org/10.1016/B978-0-12-396481-6.00026-X

26.1 INTRODUCTION

Toxoplasma gondii is a zoonotic disease and as such a successful vaccine would have both beneficial medical and veterinary impacts. An effective vaccine for use in humans, while serving, in the first instance, to reduce mortality and morbidity associated with infection, would also have economic benefits as it would reduce the financial burden of lifelong care needed for those with severe chronic disease. The ideal vaccine for veterinary use would have the dual advantages of increasing livestock productivity while reducing the public health risk associated with eating meat.

26.2 SCOPE OF PROBLEM AND POTENTIAL BENEFITS OF VACCINATION

26.2.1 Toxoplasmosis in Animals and the Potential Benefits of Vaccination

T. gondii is a widely disseminated parasite, which is capable of infecting all warm blooded vertebrates to different degrees of severity, depending on the species infected. Prior to a discussion on the benefits of vaccination, we shall define the problems associated with toxoplasmosis in different economically important livestock animals such as sheep, goats, pigs, cattle and chickens and in the definitive host species, the cat.

26.2.1.1 Toxoplasmosis in Sheep and Goats

The main problem associated with *T. gondii* infections in sheep and goats is foetal death. In countries with a high prevalence of *T. gondii*, such as the UK and Spain, the parasite is responsible for 25% of all abortions (Buxton, 1998; Pereira-Bueno *et al.*, 2004). Sheep become infected with oocysts derived from contaminated feed, pasture or water. Since most sheep and goats are kept outdoors (with the exception of milk goats) they are all at risk. In non-pregnant sheep a *T. gondii* infection normally goes undetected and is associated with mild flu-like symptoms, which coincide with the presence of tachyzoites in the circulation. The induction of protective immunity results in reduced parasitaemia and conversion of tachyzoites to bradyzoites, followed by a persistent lifelong infection with tissue cysts. Most evidence indicates that after a primary infection, sheep become immune to reinfection and this immunity prevents foetal infections during subsequent pregnancies (reviewed in Buxton and Innes, 1995; Munday, 1972; Frenkel, 1990; McColgan *et al.*, 1988). However, a study by Williams *et al.* (2005) would suggest that there is a high rate (54%) of transplacental transmission of *Toxoplasma* in sheep, which was proposed to be due to reactivation of chronic infections. These data conflicted with older reports (Hartley, 1966; Munday, 1972) and led to much discussion, which quickly resulted in another independent study that could not confirm the results from Williams (Rodger *et al.*, 2005). Thus, although pregnant, immune ewes may incidentally transmit *Toxoplasma* to their offspring, it seems to remain more the exception than the rule.

Sheep and goats that contract a primary infection during gestation have a high risk (greater than 80%) that tachyzoites will infect the placenta and traverse to the foetus, leading to resorption, abortion or stillbirth (Buxton, 1998). The clinical outcome depends on when during the 145 day gestation period a pregnant ewe becomes infected. Infection early in pregnancy is likely to cause foetal resorption, whereas infection late in pregnancy (after day 120) usually results in the birth of an apparently normal lamb, which may in fact be infected. Such infected newly born lambs usually have developed a protective immune response to *T. gondii*, therefore few congenital defects are observed (Buxton, 1998). Infection in midterm (days 50–120) will cause foetal death, mummification and abortion; the time from infection to abortion being about 40 days. *T. gondii* specific antibodies may be detected in the foetal circulation 30 days after initial infection of the ewe (reviewed in Buxton, 1998).

The impact and prevalence of *T. gondii* infection in sheep and goats in the USA has recently been reviewed in detail (Hill and Dubey, 2012). In one study cited, approximately 27% of lambs butchered and on sale to the public harboured *T. gondii* infection. Types II and III lineages predominated in this sample, although some atypical and mixed infections were recorded. In a separate study, approximately 53% of goats were found to be seropositive for *T. gondii*. Approximately 26% of commercially available goats' meat (heart) was found to contain viable *T. gondii* and Types II and III lineages predominated with some atypical and mixed infections noted (Hill and Dubey, 2012). These recent insights, specifically the identification of atypical lineages, highlight the potential risk of consuming these meats and the potential benefits of vaccination.

26.2.1.2 Toxoplasmosis in Pigs

Foetal *T. gondii* infections in pigs can lead to abortion and stillbirth, similar to sheep (Dubey and Urban, 1990). Transplacental infection in pigs is less common than postnatal infections (reviewed in Dubey, 1986, 2009). In particular, young nursing pigs are susceptible to toxoplasmosis, showing fever, coughing, weakness and wasting. In adults, toxoplasmosis is mostly sub-clinical, but infection is persistent with tissue cysts being present in many different tissues (reviewed in Dubey, 1986, 2009). As such, the contamination of pork meat with tissue cysts for human consumption defines the major problem in pigs.

The prevalence of *T. gondii* positive pigs is complicated by the different farm facilities and farm managements that are used and, furthermore, depends on the age of the animal. There is an extremely high incidence of toxoplasmosis in pigs reared outdoors. In the late 1960s, when pigs were kept outdoors, 75% of pigs were infected with *T. gondii* (Tenter *et al.*, 2000). The introduction of indoor farming facilities has dramatically reduced infection rates to as low as 1% (Davies *et al.*, 1998). In Argentina seroprevalence was only 4% in indoor reared pigs, whereas in some farms outdoor reared sows were 100% positive (Venturini *et al.*, 2004). A study in the Netherlands showed that pigs reared indoors were completely free of *T. gondii* infections while 3% of animal welfare-friendly reared pigs were seropositive for *T. gondii* (Kijlstra *et al.*, 2004). Thus, *T. gondii* infections, at present, are mainly an issue for the minority of pigs that are reared outdoors. Currently within the EU there is broad concern about animal welfare and an increasing trend to purchase animal welfare-friendly products (Kyprianou, 2005), which will undoubtedly lead to more free ranging pigs within the EU and is likely to coincide with a re-emergence of infected pigs and pork meat. Indeed, to highlight the likelihood of this happening a recent study in the USA has demonstrated for the first time a high prevalence of *T. gondii* in 'organic' pigs with Type II and Type III clonal lineages being identified (Dubey *et al.*, 2012). Throughout Europe and the USA, Type II and III strains predominate (Dubey, 2009), although in China and South America distinct genotypes may be the norm (Bezerra *et al.*, 2012; Zhou *et al.*, 2010).

26.2.1.3 Toxoplasmosis in Cattle

Cattle do not get clinically ill from *T. gondii* infection and the only substantive concern is whether or not beef can be infective to consumers. Various surveys could not identify cattle that had been infected with *T. gondii* in the field, and controlled infections with *T. gondii* displayed only transient infections, which were quickly eliminated (Dubey, 1990). Experimental infection followed by feeding of edible tissue to mice and cats demonstrated that *T. gondii* could not be isolated in significant quantities from these tissues; mice fed with homogenized organs remained negative and cats shed oocysts after feeding on heart and tongue (Dubey *et al.*, 1993). A recent study has indicated that although infrequent congenital infection can occur following natural infection (Costa *et al.*, 2011). Abortions in cattle reported as being due to *T. gondii* in the past were probably a result of *Neospora caninum* infection, which was first recognized in 1989 (Dubey and Lindsay, 1996). Overall, beef are only transiently *T. gondii* positive generally, and this was thought to pose only a small risk for human health. A quantitative risk assessment study from the Netherlands to determine the relative contribution of sheep, pork and cattle to human infection, taking into account prevalence, meat processing techniques and consumption, demonstrated that, even with a low prevalence of infection in cattle, consumption of beef remained not only an important source of infection for humans but probably the major source (Opsteegh *et al.*, 2011).

26.2.1.4 Toxoplasmosis in Chickens

Chickens can be infected with *T. gondii*, which results in the development of tissue cysts in multiple organs and generation of specific antibodies, but not clinical toxoplasmosis (Dubey *et al.*, 1993). The parasite is present in free ranging chickens often at high levels and in the USA plays a major role in the epidemiology of the disease in the rural environment (Hill and Dubey, 2012). However, chicken meat is not considered a risk factor for humans, because most chickens are reared indoors and poultry products are usually frozen for storage that suffices to kill the parasites, as well as being thoroughly cooked to avoid infections with other pathogens.

26.2.1.5 Toxoplasmosis in Cats

Cats, as the definitive hosts, normally become infected by ingestion of tissue cysts and, although adult cats (apart from a few isolated reports (Henriksen *et al.*, 1994; Dubey, 1995)) usually remain clinically healthy, kittens sometimes die of acute toxoplasmosis (Dubey and Frenkel, 1972). A large proportion of all cats are seropositive, as evidenced in a recent study from Ohio, where 48% of analysed cats were seropositive (Dubey *et al.*, 2002). Cats that become infected shed around 20 million oocysts over a short period of about two weeks, before the development of a strong protective immune response which limits further oocyst shedding (Dubey, 1995). However, some immune cats can re-shed oocysts if they are challenged after a long period (six years) with a heterologous strain (Dubey, 1995). The shedding of highly infectious oocysts, which remain viable for more than one year, into the environment by cats poses a major risk for both humans and livestock. Oocysts can contaminate food and thereby create an effective route of infection for both humans and livestock. Faeces from cats and wild felids can contaminate drinking water, thereby causing outbreaks of toxoplasmosis in humans, as shown in Panama, the USA and Brazil (Benenson *et al.*, 1982; Bowie *et al.*, 1997; Bahia-Oliveira *et al.*, 2003). Finally, in contrast to contact with cat litter, petting of cats that had recently shed oocysts was shown to pose a minor risk of infection for humans (Dubey, 1995).

26.2.2 Benefits of Animal Vaccination

In the case of the domestic cat, successful vaccination would limit oocyst shedding and ultimately reduce the incidence of toxoplasmosis generally. Otherwise, vaccination should improve livestock productivity by reducing foetal damage and hopefully also the incidence of human disease by limiting contamination of meat products with tissue cysts.

As discussed above, clinical toxoplasmosis resulting in foetal damage occurs mostly in sheep and goats. For these animals a commercial live tachyzoite vaccine already exists, Toxovax (derived from the S48 'incomplete' strain). Toxovax does efficiently reduce foetal deaths and as a non-persistent strain has a good safety record (further discussed in Section 26.3.2.1). It is unknown if Toxovax also aids in the reduction of contaminated lamb meat with tissue cysts and future vaccine studies on sheep and goats are needed to address this question.

While the vast majority of pigs are reared indoors, toxoplasmosis in this species remains a minor problem and the market for a vaccine is perhaps too limited to be of commercial importance. However, there is an increasing public perception in the developed world that 'organic' farming and 'free range' produce is superior to that intensely reared indoors, both in terms of quality and public health, not to mention improved animal welfare considerations. Consequently, free ranging pigs are becoming more common with the subsequent increasing human health risk associated with an undoubted increase in the incidence of toxoplasmosis in pigs. Therefore, vaccination to protect both young nursing pigs, as well as to block cyst formation, is likely to become necessary for a free ranging population that will increase in the future.

Since cattle display no clinical signs of toxoplasmosis and as *Toxoplasma* infections do not persist in these animals, it was thought previously that they did not require vaccination. However, the recent quantitative risk assessment study undertaken in the Netherlands suggests that when all parameters, such as consumption, processing and incidence, are taken into account beef was found to be the likeliest conduit in infecting humans (Opsteegh *et al.*, 2011). Consequently, given the relative resistance of bovines to natural infection, they may offer an excellent target for successful vaccination that would have real impact on the incidence in humans. Chickens, also; although they can have persistent infection with tissue cysts, they do not require a vaccine, as they display no clinical manifestations and chicken meat is processed in such a way that tissue cysts are unlikely to survive.

Finally, as the definitive hosts, cats are a major risk responsible for contaminating food, pastures and drinking water with oocysts. Cats are a particular risk around farms and vaccinating farm cats with an experimental live bradyzoite vaccine (T-263) not only demonstrated it was possible to neutralize oocyst shedding, but within a few years local mice were found to be *Toxoplasma* seronegative and the incidence of seropositivity in finishing pigs had decreased (Mateus-Pinilla *et al.*, 1999). This trial clearly demonstrates the feasibility and the potential benefits of vaccinating cats. Reduced oocyst shedding from farm cats would not only limit the incidence of toxoplasmosis in domestic livestock, but also in humans *via* contaminated food and water (Bahia-Oliveira *et al.*, 2003). Similarly, vaccinating household cats would also reduce the likelihood of oocyst initiated infection to humans and, in particular, to pregnant women.

In conclusion, a veterinary vaccine against toxoplasmosis already exists for sheep and goats that limits the incidence of abortion. The question that remains to be addressed is whether it also limits cyst burdens. Obviously, an ideal vaccine would also have this outcome. If new veterinary *Toxoplasma* vaccines are developed, successful vaccination of cats would be expected to have the biggest impact on reducing infection in both livestock and humans.

26.2.3 Toxoplasmosis in Humans and the Potential Benefits of Vaccination

Essentially all humans (approximately 6.5 billion) are at risk of *T. gondii* infection and all could arguably benefit from vaccination against this parasite. Current treatments are inadequate as they only control the proliferative tachyzoite stage of the life cycle, but do not eliminate the cyst stages associated with chronic infection (reviewed in Roberts *et al.*, 2002). There are a number of groups where the consequences of infection could be particularly severe and where the potential of vaccination would be great. *T. gondii* is a major cause of congenital disease with potentially severe sequelae. Pregnant women are more likely to acquire the infection (Avelino *et al.*, 2004; Gilbert and Gras, 2003). Consequently, vaccination of women before they reach childbearing age may be a reasonable strategy to reduce or eliminate this risk. The benefits of such a programme would vary according to country and clearly would have greatest impact where the incidence of congenital infection is highest. For example, the incidence of congenital toxoplasmosis is estimated to be 10 per 1000 births in Paris, France (Desmonts and Couvreur, 1974), 0.5 per 1000 births in the UK (Williams *et al.*, 1981) and one to 10 per 10,000 births in the USA (Lopez *et al.*, 2000). The financial cost of such a programme would be considerable, but would be offset by a reduction in the cost of caring for those congenitally infected. In the USA, the estimated total medical costs and loss of productivity as a consequence of human toxoplasmosis, excluding AIDS patients, is approximately $3 billion per annum and an annual loss of 11,000 quality adjusted years (Batz *et al.*, 2012; Hoffmann *et al.*, 2012).

At one time it was a widely held assumption that ocular toxoplasmosis only occurred in the immune competent following congenital infection, but recent evidence has found that it occurs in a significant number of adult acquired infections (reviewed in Roberts and McLeod, 1999). It has been estimated to occur in 2% to 3% of adult acquired infections (Perkins, 1973) and 49 cases were identified over a 13 year study in France (Couvreur and Thulliez, 1996). The incidence of adult acquired ocular disease varies considerably by geographical region. In certain populations in Brazil, seropositivity was found to be over 80% in adults over 25 years of age. The incidence of ocular disease within this population was as high as 14%, although the contribution of congenital infections to this figure is difficult to estimate (Petersen *et al.*, 2001). Certain atypical strains of *T. gondii* have increased association with ocular disease and may account for geographical differences in incidence of ocular disease (Grigg *et al.*, 2001). Consequently the whole human population is at risk of adult acquired ocular disease and there is a good argument for vaccinating the entire population. This may be especially important in regions where atypical strains are more abundant.

The vast majority of congenitally infected children appear asymptomatic at birth. However, in one USA based study, around 20% of apparently asymptomatic children had ocular involvement at birth. Ocular disease had risen to over 80% in the subjects of this study by adolescence (reviewed in Roberts and McLeod, 1999). It has been suggested that therapeutic vaccination in childhood may be useful in reducing the incidence of ocular disease amongst asymptomatic, congenitally infected individuals (Wilson *et al.*, 1980; Koppe *et al.*, 1986). Such a vaccine would have to overcome the mechanism that prevents this patient group from developing solid immunity that is normally exhibited by those who have adult acquired infection. Although immunological tolerance could play a role, the precise mechanism is currently unknown. Therefore, rational design of such a vaccine may prove challenging.

Other possible groups that might significantly benefit from a therapeutic vaccine are those with active disease. This would include those with active adult acquired disease or the immunosuppressed. There would be inherent problems in vaccinating people with ongoing disease as the immune response to the vaccine may be influenced in a detrimental manner by the natural infection. Successful vaccination of immunosuppressed people as an alternative to antimicrobial therapy would also prove challenging due to the very fact that they have poorly functioning immune systems.

26.3 CURRENT STATUS OF VACCINES FOR INTERMEDIATE HOSTS

26.3.1 Vaccination Using Extracts or Killed Parasites

The earliest studies to test the vaccine potential of killed or crude antigen extracts against toxoplasmosis were carried out in 1956 (Cutchins and Warren, 1956; Jacobs, 1956). A summary of more recent studies carried out since the early 1970s is listed in Table 26.1. The vaccine potential of whole fixed tachyzoites has been examined, as well as whole tachyzoite lysates, soluble fractions, particulate fractions, excretory/secretory products, detergent extracts, cysts, soluble cyst fractions and crude whole rhoptry extracts. Numerous adjuvants have also been employed as part of the vaccine formulation including Freund's Complete (FCA) and Freund's Incomplete Adjuvant (FIA), lipid vesicles, ISCOMs (Immuno Stimulating Complexes), BCG, cholera toxin, PLG microspheres, exosomes derived from an infected DC cell line, and CpG for vaccination. Guinea pigs, mice, rats, sheep and pigs have all been used as models and challenge infections have utilized both virulent and avirulent tachyzoites, cysts and bradyzoites.

Overall, while some vaccines increased survival following challenge infection (Krahenbuhl *et al.*, 1972; Eissa *et al.*, 2012), others did little (Waldeland and Frenkel, 1983; Saavedra *et al.*, 2004). The fact that the immune system tends to recognize life cycle stage-specific antigens (Kasper, 1989) in some respect may account for some discrepancies. In addition, although not a hard and fast rule, lipid vesicles such as liposomes, non-ionic surfactant vesicles and ISCOMs may be better adjuvants than FCA and FIA; excretory/secretory antigens may induce stronger protection than tachyzoite lysates, while the addition of cyst antigens to tachyzoite preparations has improved efficacy (Elsaid *et al.*, 2001). Again, while some vaccines reduced cyst burden following challenge (Alexander *et al.*, 1996; El-Malky *et al.*, 2005), others failed to do so (McLeod *et al.*, 1985; Lunden *et al.*, 1993); however, in murine and rodent models of congenital toxoplasmosis vaccination with crude tachyzoite lysate (Roberts *et al.*, 1994; Elsaid *et al.*, 2001; Beauvillain *et al.*, 2009), tachyzoite and cyst lysate (Elsaid *et al.*, 2001) or excretory secretory factors (Zenner *et al.*, 1999) proved extremely effective at limiting both maternal—foetal transmission and foetal death. Nevertheless, killed whole tachyzoites alone (Beverley *et al.*, 1971) or in FIA (Wilkins and O'Connell, 1992) could not protect sheep from aborting.

More recently, tachyzoite sonicates were formulated in QuilA-containing vesicles 'ISCOMs' and used to vaccinate pregnant sheep. One vaccination given four weeks prior to mating was followed by two injections in the first 10 weeks of gestation. A challenge was given at day 91 of pregnancy using oocysts of the M1 strain (Buxton *et al.*, 1989). Ewes were not protected against the acute phase of the infection as was apparent from their febrile response comparable to non-vaccinated controls. However, fewer abortions were induced in the vaccinated ewes and the gestational time was comparable to non-infected controls; probably due to low numbers these differences were not significant. Foetal infection could not be prevented using this vaccine. Using a similar vaccine formulation, vaccination of pigs with rhoptry proteins in ISCOMs had little effect on the febrile response following oral oocyst infection, although the cyst burden was slightly reduced (Garcia *et al.*, 2005).

A similar approach was applied by Stanley *et al.* (2004) using tachyzoite extracts formulated

TABLE 26.1 Killed and Crude Antigen Vaccine Studies and Their Outcomes in Animal Models

Reference	Antigen	Adjuvants or Carrier	Route of Vac.	Animal Model	Immunology	Challenge	Survival	Parasite Burden	Other
(Krahenbuhl *et al.*, 1972)	Formalin fixed tachyzoites, total lysate, soluble and particulate fractions	w/wo FIA or FCA	s.c. and i.p.	Swiss–Webster mice	Antibodies	C56 strain tachyzoites (i.p.)	++		
(Araujo and Remington, 1974)	Tachyzoite soluble and particulate fractions and RNA	w/wo FIA	i.p.	Swiss–Webster mice		C56 tachyzoites (i.p.)	++		
(Beverley *et al.*, 1971)	Tachyzoite lysate		s.c.	Sheep	Antibodies	Cysts (s.c.)			++Foetal death–maternal foetal transmission
(Waldeland and Frenkel, 1983)	Tachyzoite lysate	FCA Liposomes Fatty acid anhydrides	s.c.	Outbred CF-1 mice	Antibodies	Tachyzoites bradyzoites M-7741 strain (s.c.)	+/++ FCA > liposomes		
(McLeod *et al.*, 1985)	Tachyzoite lysate	Liposomes	i.m. Oral	Swiss–Webster mice	Antibodies	Me49 cysts (oral)		–	
(Duquesne *et al.*, 1990)	Tachyzoite excretory/secretory	FCA	s.c.	Fischer/nude rats	Antibodies Lymphoproliferation	RH tachyzoites (i.p.)	++		
(Overnes *et al.*, 1991)	Detergent extract Plasma membrane tachyzoite	ISCOM	s.c.	Outbred white mice	Antibodies Lymphoproliferation	M-7441 tachyzoites (s.c.)	+/-		

(Lunden *et al.*, 1993)	Detergent extract tachyzoites	ISCOM	s.c.	Swiss–Webster mice	Antibodies DTH	C56 tachyzoites (i.p.)	+	–	
						C56 cysts (oral)	+	–	
						Me49 oocysts (oral)	++	–	
(Roberts *et al.*, 1994)	Soluble tachyzoites	w/wo non-ionic surfactant vesicles (NISV)	s.c.	BALB/c	Antibodies Lymphoproliferation IFNγ	Beverley cysts (oral)			+++ Foetal death Ag w NISV –– Foetal death Ag wo NISV ++ Maternal-foetal transmission Ag w NISV
(Alexander *et al.*, 1996)	Cysts + tachyzoite lysate	FCA/FIA	s.c.	BALB/K		Beverley cysts (oral)			– Tachyzoites ++ Cysts +++ Tachyzoite cysts
(Zenner *et al.*, 1999)	Excretory/secretory	FIA	s.c.	Fischer rats	Antibodies	76K cysts (oral)			++ maternalfoetal transmission
(Elsaid *et al.*, 1999)	Soluble tachyzoite Soluble cysts Soluble cysts plus tachyzoite	Liposomes FCA	s.c.	Swiss mice	Antibodies	P strain (oral)		++ w liposomes –FCA	+ w liposomes –FCA
(Elsaid *et al.*, 2001)	Soluble tachyzoite Soluble cysts Soluble cysts plus tachyzoite	Liposomes FCA	s.c.	BALB/c	Antibodies Lymphoproliferation	P strain cysts (oral)			++ Foetal Death + maternal foetal transmission Ag plus liposomes –Ag plus FCA

(*Continued*)

TABLE 26.1 Killed and Crude Antigen Vaccine Studies and Their Outcomes in Animal Models (*cont'd*)

Reference	Antigen	Adjuvants or Carrier	Route of Vac.	Animal Model	Immunology	Challenge	Survival	Parasite Burden	Other
(Daryani *et al.*, 2003)	Tachyzoite excretory/ secretory, lysate	FCA/FIA	s.c.	BALB/c mice	DTH Lymphopro liferation	RH tachyzoites (s.c.)		++/+ ESA > TCA	
(Garcia *et al.*, 2005)	Rhoptry proteins	ISCOM	s.c.	Pigs	Antibodies	VEG oocysts (oral)			++− Febrile response
(Beauvillain *et al.*, 2007)	Tachyzoite lysate	SRDC derived Exosomes	s.c.	CBA/J; C57B L/6	Antibodies IFN-γ, IL-4, IL-5, IL-10, IL-2	76 K cysts Oral			+++ CBA/J + C57BL/6
(Hedhli *et al.*, 2009)	Tag	Eimeria profilin g-like antigen	i.p.	CBA/J	Antibodies, IFN-γ, IL-10, IL-2	70 76k cysts		62% less cyst burden	
(Beauvillain *et al.*, 2009)	Tachyzoite lysate	SRDC derived Exosomes	s.c.	Pregnant CBA/J	Antibodies, IFN-γ, IL-4, IL-5, IL-10, IL-2	76K cysts Oral	Pups +++		
(Eissa *et al.*, 2012)	Autoclaved Tachyzoite lysate	BCG	i.d.	Swiss albino	Splenic $CD8^+$	RH i.p.	++		

Key for Survival: −− decreased survival: − no difference in survival; + moderate (≤50% increased survival); ++ significant (≥50% increased survival); +++ highly significant (≥90% increased survival), compared with control groups.

Key for Parasite Burden: −− increased parasite burden: − no difference in parasite burden; + moderate (≤50% decrease in parasite burden); ++significant (≥50% decrease in parasite burden); +++ highly significant (≥90% decrease in parasite burden), compared with control groups.

in cholera-toxin containing microspheres (Stanley *et al.*, 2004). Vaccination of sheep was performed intranasally and induced mucosal and serum IgA. Since the oral route is the natural route of infection in sheep, mucosal immunity may contribute to reducing the numbers of infectious parasites as was shown previously in mice (Bourguin *et al.*, 1993), and reviewed in Kasper *et al.* (2004). Stanley and co-workers vaccinated non-pregnant sheep three times, but the effect on the acute infection measured as a febrile response was only marginal. Antigenic differences between sporozoites, the stages invading the mucosal lining and the later developing tachyzoites, which cause the febrile response, may account for these observations.

Thus, while vaccines based on killed or lysed tachyzoite antigens could induce protective immunity in mice and limit maternal–foetal transmission, results from a more practical model employing outbred sheep were less promising and could not compare to the effects induced by the incomplete S48 strain. However, relatively few studies have been documented in sheep and further improvement could be achieved by applying different immunization schedules or better adjuvants. The success of the killed *Neospora* vaccine which protects against abortion in cattle warrants further studies using antigen preparations to enhance protective immunity against *Toxoplasma* induced abortion in sheep.

Although no studies are documented on vaccination of humans, it is known that chronically infected women can protect their foetus from congenital infection. From the studies described above it is clear that sterile immunity cannot be achieved using killed or sub-unit vaccines in any of the infection models used. Since tachyzoites are very efficient in reaching the foetus and since prevention of congenital infection is a prime target for vaccination of humans, a killed vaccine will probably not be sufficiently efficacious.

26.3.2 Vaccination Using Live, Attenuated Parasites

The ubiquitous RH strain is a type I strain and infection with only a few tachyzoites is sufficient to kill a mouse. This strain is, however, significantly less pathogenic in other animals, such as pigs. Moreover, inoculated tachyzoites did not cause a persistent infection in these animals, making this strain useful as a vaccine strain. Pigs, which can harbour a huge cyst burden, are considered a major source of infection for humans. Vaccination of pigs using the RH strain protected against challenge with a persistent strain and reduced the presence of tissue cysts in the meat. When challenge occurred at 220 days post-vaccination all vaccinated pigs were negative for tissue cysts in a mouse bio-assay, whereas all control challenged animals were positive (Dubey *et al.*, 1994). Attenuated lines were derived from this RH strain by chemical mutation using N-methyl-N′-nitro-N-nitrosoguanidine. Temperature-sensitive mutants were isolated, of which the TS-4 strain demonstrated the most favourable phenotype of retarded growth at 37°C while maintaining immunogenicity in mammals (Pfefferkorn and Pfefferkorn, 1976). Mice vaccinated with tachyzoites of TS-4 were protected against a lethal challenge infection with RH. While vaccination could also reduce congenital transmission in a pregnancy, model tissue cyst formation was not prevented after vaccination with TS-4 and challenge with an avirulent strain (McLeod *et al.*, 1988).

Frenkel *et al.* (1991) used similar mutation methodology and selected a strain deficient in the coccidial cycle of the parasite (the sexual replication in the intestine of the cat) (Frenkel *et al.*, 1991). In contrast to RH and TS-4, this T-263 strain did produce tissue cysts in the intermediate host. It lost, however, the capacity to form oocysts in cats that had ingested tissue cysts. The deficiency was associated with infertile microgametes. Vaccinating young cats with T-263 bradyzoites could prevent oocyst shedding in 80% of recipients when challenged with an oocyst-forming strain of *T. gondii*. Such effects could not be induced by vaccination with live tachyzoites only (Freyre *et al.*, 1993). This data confirms that bradyzoites and tachyzoites carry different antigens and immunity differs between intermediate and definitive hosts relating to the interfaces where the infection is occurring. Although commercialization of the T-263 strain as a vaccine for cats was considered, a product was never released.

Attenuation, while preserving immunogenicity, can also be achieved by gamma-irradiation as has been shown for many different organisms, including parasites, such as in a commercial lungworm vaccine (McKeand, 2000). Dosages of less than 1000 Gy resulted in tachyzoites that could invade cells, and although they could not replicate they were still able to induce cell mediated immune responses and some protection (Seah and Hucal, 1975). However, the effectiveness of different irradiation doses differed from study to study (summarized in Dubey, 1996). Oocysts of the VEG strain failed to induce persistent infections, when irradiated with doses higher than 200 Gy (Dubey, 1996). Oocysts irradiated with 200 Gy could induce partial protection in mice against oral challenge as measured by extended survival and fewer brain cysts. This was once again confirmed in a more recent experiment (Hiramoto *et al.*, 2002). Tachyzoites (10^7) irradiated with 200 Gy extended the survival time of mice by roughly four days when challenged with 1000 RH tachyzoites. More importantly, it reduced tissue cyst development in the brain by more than 10-fold when mice were challenged orally with 25 ME-49 tissue cysts. Since irradiated tachyzoites do not persist and are able to reduce the number of tissue cysts, such an approach may have implications in the design of a therapeutic vaccine.

In addition to inducing an attenuated phenotype by chemical mutation or irradiation, serial passage of a type II strain has also been shown to evoke changes that lead to an attenuated phenotype. This methodology was used to develop the S48 strain, which is now commercially applied as an effective vaccine against abortion in sheep (Buxton, 1993). This strain was originally isolated from the cotyledon of an aborted lamb and passaged twice weekly thereafter through mice for more than 30 years. The strain consequently lost its capacity to form tissue cysts in any animal challenged, but kept its immunogenic characteristics. S48 is described in greater detail below.

In conclusion, a *T. gondii* infection is generally able to induce a status of lifelong protection, which would be ideal if the primary infection did not have the potential to damage neural tissues or interfere with the outcome of pregnancy. Attenuated vaccines may have the potential to induce similar protection to natural primary infection in the absence of such pathology. The preferred attenuation should be stable and, thus, irradiation is not a particularly good option. Selection of sufficiently attenuated strains after mutagenic actions may have the potential to deliver stable vaccine strains. However, the use of such strains in immune compromised animals or humans should be restricted. New reverse genetic techniques will allow specific genes to be deleted such that the reversion to virulence can be excluded (discussed below).

26.3.2.1 A Commercial Vaccine (S48) Against Toxoplasmosis in Sheep and Goats

T. gondii normally causes disease when infection occurs for the first time while the animal is in gestation, allowing the parasite to invade the non-immune foetus. However, an animal that is already immune prior to gestation is normally sufficiently protected to prevent foetal infection (Frenkel, 1990; McColgan *et al.*, 1988). Indeed, while a natural infection with *T. gondii* in sheep induced protective immunity (McColgan *et al.*, 1988), vaccination with inactivated *Toxoplasma* tachyzoites could not protect pregnant sheep from infection (Beverley, 1971; Wilkins *et al.*, 1987). This illustrated that vaccination with live

T. gondii, but perhaps not a killed preparation, prior to gestation, could prevent subsequent foetal death. Accordingly, the 'incomplete' S48 strain was developed into a safe, live vaccine (Wilkins *et al.*, 1988). S48 was isolated in 1956 from an aborted lamb in New Zealand. Since this time pre-dates the era of cryopreservation and tissue culturing, it was maintained by passaging the tachyzoites through mice twice weekly for a period of 30 years. During this time S48 lost its pathogenicity for sheep, as well as its ability to form tissue cysts (Wilkins and O'Connell, 1992). Not only has S48 lost its ability to form tissue cysts, it also does not generate oocysts in cats (Intervet, unpublished observations). The safety of the strain was proven by its absence in any animal from four weeks post-vaccination (Buxton, 1993). In 1988 it was introduced as Toxovax, a live vaccine, in New Zealand. Subsequently, it was registered for use in the UK and Ireland in 1992 and is now sold by Intervet as Ovilis™ Toxovax in many European countries.

Ovilis™ Toxovax is a live tissue culture grown vaccine that is supplied as a frozen product and is distributed to end users at 4°C. The vaccine is applied either intramuscularly or subcutaneously no later than three weeks prior to mating. To demonstrate the efficacy of the vaccine, the Moredun Research Institute together with Intervet performed a series of vaccination-challenge experiments (Buxton *et al.*, 1991; Buxton and Innes, 1995). Using 2000 sporulated oocysts of the M3 strain they induced a severe infection in pregnant S48-vaccinated and control sheep. While less than 18% of the lambs from control sheep were born alive and viable, 80% of lambs survived in the vaccinated group and their weights were comparable with lambs from non-challenged ewes. Toxovax induced an IgG antibody response, although this response waned by 20 months post-vaccination, in contrast to persistent infections that maintain high antibody titres in the blood throughout life. S48 vaccination also induced specific $CD4^+$ and $CD8^+$ T-cells that produced IFNγ as documented for normal live tachyzoite infections in sheep (Buxton and Innes, 1995). Further studies demonstrated that animals were still immune after 18 months post-vaccination despite their antibody levels having waned, highlighting the ability of S48 to induce a potent adaptive type I cell mediated response (Buxton, 1993). Field trials in the UK have shown the economic profitability of vaccination against *T. gondii* induced abortion in sheep (Bos and Smith, 1993). Vaccination with S48 has also shown moderate efficacy in goats (Chartier and Mallereau, 2001).

26.3.3 Vaccination Using Gene Deletion Attenuated Parasites

Increased knowledge of *T. gondii* at the molecular level, greatly facilitated by the completion of the genome project and in combination with rapid progress in the development of genetic tools to manipulate the parasite, has generated opportunities to create new attenuated vaccines. Such targeted approaches have been used to create parasites with incomplete life cycle stages with reduced proliferative capacity or with reduced virulence. Mutant parasites have been generated either with an irreversible gene deletion or more recently by conditionally inhibiting or activating expression of an essential gene. Genetically 'crippled' parasites have been analysed *in vitro* and *in vivo* to characterize their mutant phenotypes and to determine whether a gene is possibly redundant. At present with the exception of a Mic1-3 deletion mutant (Mevelec *et al.*, 2010), the infectivity of all *Toxoplasma* mutants that have been generated by reverse genetics has only been analysed in mice and not in larger animals. Whether vaccine potential or its absence demonstrated in mice can be translated into success or failure in larger animals is a matter of some conjecture. For example, while both the RH strain and the incomplete S48 strain are highly lethal to mice, RH is not persistent in pigs and induces protective immunity upon challenge with oocysts (Dubey *et al.*, 1994). Similarly, the incomplete S48 strain does not persist in any animal so far examined, and is commercially used as a live vaccine. Thus, the potential of a live mutant parasite cannot be determined in mice only, but will ultimately have to be established in a larger animal.

In recent years, various *Toxoplasma* gene deletion mutants have been generated. Although the objective was usually to gain further insight into the function of a particular gene rather than to generate a vaccine, it is likely that any biologically important gene will contribute either to the fitness and/or virulence of the parasite and consequently the vaccine potential of these mutants has been of significant interest (see Table 26.2).

The first genetically engineered mutant to be extensively studied was the RH strain with a disruption of carbamoyl phosphate synthetase II (CPSII) (Fox and Bzik, 2002). CPSII is the first

TABLE 26.2 Toxoplasma Mutant, Deficient and Conditional Deficient Strains and Induced Immunity in Mice

Reference	Targeted Gene or Mutant Name	Parental Strain	Dosage	Animal Model	Survival and/or Protective Effect
(McLeod *et al.*, 1988)	Ts-4	RH			Reduced mortality against parenteral M7741 and per-oral Me49
(Lindsay *et al.*, 1993)	Ts-4	RH	5×10^5 tachyzoites	Pigs	Less severe disease
(Pinckney *et al.*, 1994)	Ts-4	RH	3×10^5 tachyzoites	Pigs	Less severe disease
(Fox and Bzik, 2002)	CPSII	RH	Up to 10^7 tachyzoites	BALB/c	No proliferation, all survive, protects against lethal challenge
(Rachinel *et al.* 2004)	SAG1	RH	10^4 tachyzoites*	C57BL/6	30% reduced mortality
(Dzierszinski *et al.*, 2000)	SAG3	RH	20 tachyzoites	BALB/c	85% reduced mortality
(Soldati *et al.*, 1995)	ROP1	RH	50 tachyzoites	Swiss	Lethal
(Mercier *et al.*, 1998)	GRA2	RH	10 tachyzoites	Swiss (CD1)	50% reduced mortality
(Bohne *et al.*, 1998)	BAG1	PLK**	10^4 tachyzoites	C57BL/6	No reduced mortality
(Zhang *et al.*, 1999)	*BAG1*	PLK**	10^5 tachyzoites	Swiss (CD1)	5-fold reduced cyst burden
(Moire *et al.*, 2009)	MIC1−3	RH	Up to 10^6 tachyzoites	CBA/J	Reduced cyst burden
(Lu *et al.*, 2005)	Ts-4	RH	10^5 tachyzoites	C57BL/6, CBA/JBALB/c	Protects against ocular toxoplasmosis
(Lu *et al.*, 2009)	Ts-4	RH	2×10^4 tachyzoites	C57BL/6, CBA/JBALB/c	Protects against ocular toxoplasmosis
(Fox and Bzik *et al.*, 2010)	ΔOMPDC	RH	10^5–10^7 tachyzoites	C57BL/6	Reduced mortality ananinst RH challenge
(Hutson *et al.*, 2010)	ΔRPS13	RH	10^5 tachyzoites	SW mice	100% reduced mortality
(Zorgi *et al.*, 2011)	Irradiated	RH	VEG, Me49, P strain	C57BL/6 BALB/c	Reduced mortality ananinst Me49 challenge

*) *applied via surgical injection in the intestines.*

**) *PLK is a clonal derivative of ME49.*

C57BL/6 is a susceptible mouse, BALB/c is a relative resistant mouse and CD1 is an outbred, relative resistant line.

enzyme in the metabolic pathway for *de novo* pyrimidine synthesis (generating the building blocks for RNA and DNA) and disrupting this enzyme made *Toxoplasma* dependent on externally supplied uracil, which it can salvage. Disruption of CPSII thus created a uracil auxotroph, which only grew *in vitro* when host cells were supplemented with uracil. In the absence of uracil, CPSII knock-out parasites invaded host cells normally but failed to replicate. No growth was observed (without added uracil) *in vitro* and *in vivo*. Injection of mice with CPSII knock-out parasites did not kill BALB/c mice, and mice infected with CPSII knock-out parasites 40 days previously were resistant to a lethal challenge with 200 pfu of RH strain tachyzoites.

Surface antigens (SAGs) of *Toxoplasma* have also been targeted for deletion. SAGs are thought to be involved in host cell attachment and the activation of a host immune response. The major tachyzoite surface antigen is SAG1 and two types of SAG1 mutants have been generated. One was made by chemical mutagenesis and the other was recently genetically engineered (ΔSAG1). Both attach to, enter and proliferate at approximately normal rates within host cells (Mineo and Kasper, 1994; Kasper and Khan, 1993; Rachinel *et al.*, 2004). ΔSAG1 tachyzoites were lethal in susceptible C57BL/6 mice (although survival was slightly prolonged compared with wild-type infected mice), but deletion of SAG1 prevented an acute ileitis when tachyzoites were directly injected into the intestine.

SAG3 deletion mutants showed more pronounced effects than ΔSAG1 tachyzoites; these had significantly reduced adherence to host cells *in vitro* and mortality was reduced by 80% upon infection in BALB/c mice compared with wild-type organisms (Dzierszinski *et al.*, 2000).

Recently, a genomic cluster containing four bradyzoite specific SAGs (SAG2c, SAG2d, SAG2x and SAG2y) was deleted in one knock-out. Deleting these four SAGs (ΔSAG2cdxy) yielded viable tachyzoites that could still differentiate into bradyzoites *in vitro*. In contrast, preliminary studies *in vivo* showed that nine out of 10 ΔSAG2cdxy infected mice were negative for brain cysts when assayed three months after infection (J. Saeij and J. Boothroyd, personal communication). Consequently, ΔSAG2cdxy appears promising as a vaccine candidate because although it is still infective and can transform into bradyzoites, it persists poorly if at all.

Two secretory vesicle proteins, ROP1 and GRA2, gene deletion mutants, have also been generated. Disruption of either ROP1 or GRA2 resulted in no difference in growth rates or host cell invasiveness *in vitro*, although ΔGRA2 was less virulent in mice (Soldati *et al.*, 1995; Mercier *et al.*, 1998).

Disruption of BAG1, a bradyzoite specific heat shock protein, was thought to interfere with the formation or viability of tissue cysts. However, in one study (Bohne *et al.*, 1998) disruption of BAG1 had no effect on tissue cyst formation, while in a second study (Zhang *et al.*, 1999) disruption of BAG1 could only reduce the number of tissue cysts in mouse brains by roughly five-fold. In the latter study, the lethal dose with ΔBAG1 did increase from 2×10^6 to 5×10^7 compared with the parental PLK strain (a clonal line derived from ME49). Nevertheless, tissue cysts were still being formed and were completely normal, proving that BAG1 is not essential for bradyzoites. The authors suggested that BAG1 homologous genes may exist in *Toxoplasma*, generating some redundancy.

Recently a Mic1−3 deficient RH strain of *T. gondii* was developed. This strain is capable of conferring a degree of protection in mice against congenital infection. Fewer mice in vaccinated groups were found to be infected than in control groups and those infected had fewer cysts in their brains compared with those born to unvaccinated control groups (Ismael *et al.*, 2006). Similarly, this vaccine was found to confer a degree of protection against congenital transmission when the pregnant ewes were challenged with oocysts (Mévélec, 2010).

Attenuated parasites can also be generated by targeting expression of essential genes. Since deletion of such genes will immediately result in non-viable parasites, targeting the expression of essential genes should occur in a conditional way. A number of systems have been developed that achieve this goal (Meissner *et al.*, 2001, 2002b; van Poppel *et al.*, 2006; Hutson *et al.*, 2010). A number of conditional knock-out parasites have been generated and one could envisage a vaccine application using these conditional knock-outs. A conditional *T. gondii* knock-out for the small subunit ribosomal protein 13 (Hutson *et al.*, 2010) has been generated using the tetracycline repressor system (Hutson *et al.*, 2010). In the absence of anhydrotetracycline the conditional knock-out parasite exits the cell cycle in G1, but is arrested in the G0 phase. Although able to persist a sustained time *in vitro* in this stage, it appears to be cleared by the immune system of mice (Hutson *et al.*, 2010). Notably, mice vaccinated with these parasites are protected against challenge with either type 1 or type 2 strains of *T. gondii.*

In summary, reverse genetic techniques have enabled the creation of mutant attenuated *Toxoplasma* strains with vaccine potential. The uracil auxotrophic mutant generated by disruption of CPSII and the RPS13 conditional knock-out strains are particularly promising. It will be particularly important to demonstrate the safety of the mutants, before such genetically modified organisms (GMO) can be tested and used in the field.

26.3.4 Vaccination Using Viral Vectors

Viral delivery of immunogenic antigens has been tested widely against various cancers and infectious diseases including malaria, although only a few studies have used this technique to express *Toxoplasma* antigens Table 26.5. Multiple viral vectors are available that are considered safe and are being tested in animals and humans. Poxviruses, adenoviruses and herpesviruses are mostly used as vectors. Some poxviruses such as vaccinia can be a minor human pathogen whereas others, such as fowlpox and MVA (modified vaccinia Ankara) cannot replicate in mammalian cells and are non-pathogenic (Paoletti, 1996). Likewise, replication defective adenoviruses, such as Ad5, are being used (Graham *et al.*, 1977). Due to their safety, replication defective viruses may also be considered for a human *Toxoplasma* vaccine.

If we consider the use of viral vectors as vaccine carriers for parasitic diseases, then most work has focused on malaria. For example, vaccination of mice with a *Plasmodium yoelii* circumsporozoite protein, expressed by either a replication defective Ad5 adenovirus or a combination of Ad5 with a vaccinia, induced sterile and long lasting immunity (Rodrigues *et al.*, 1997; Bruna-Romero *et al.*, 2001). In another study, the multi-epitope vaccine fused to TRAP (ME−TRAP) induced sterile immunity in some human subjects (Webster *et al.*, 2005). ME−TRAP was delivered with a prime and boost combination of a recombinant fowlpox vector followed by an MVA vector. These examples illustrate the potential of such vectors.

A few infectious viral vaccine vectors have been tested against *T. gondii* including a feline herpesvirus (FHV1) tested in cats (Mishima *et al.*, 2002). FHV1 was selected as carrier because it spreads amongst cats contagiously, thereby hopefully disseminating the vaccine to neighbouring domestic and stray cats. An attenuated strain was generated by deleting thymidine kinase and inserting *Toxoplasma* ROP2. FHV−ROP2 induced specific antibodies and, upon bradyzoite challenge of cats, the number of brain tissue cysts was reduced. However, it did not reduce oocyst secretion, an essential prerequisite of a cat vaccine. In another study, the vaccine potential of ROP2 was tested with MVA in mice (Roque-Resendiz *et al.*, 2004). High dosages of MVA−ROP2 did induce specific antibody titres and delayed the time of death (by two days), but could not protect mice against a challenge with 300 RH tachyzoites. Vaccination with MVA−ROP2 also failed to reduce the formation of brain cysts if mice were orally challenged with 20 ME49 cysts. Gazzinelli recently reported preliminary studies with three recombinant adenoviruses, expressing SAG1, SAG2 or SAG3 (Gazzinelli *et al.*, 2005; Caetano, 2006). Immunization of BALB/c mice with these viruses induced both antibodies and IFNγ responses. However, upon challenge with a lethal dose of RH parasites these mice were not protected. Conversely, challenge with a P-BR strain did show a reduction of tissue cysts in the brain. Further studies have utilized baculovirus (Fang *et al.*, 2009; 2012), vaccinia (Zhang *et al.*, 2007a), pseudorabies (Liu *et al.* 2008; Shang *et al.*, 2009; Nie *et al.*, 2011) and influenza (Machado *et al.*, 2010) as expression systems and additional antigens examined include AMA1 (Yu *et al.*, 2012), GRA4 (Zhang *et al.*, 2007) and MIC3 (Nie *et al.*, 2011; Fang *et al.*, 2012). Generally vaccines incorporating some antigens are more effective than others (Mendes *et al.*, 2011) and those expressing more than one antigen are more effective than those expressing a single antigen (Qu *et al.*, 2009; Nie *et al.*, 2011; Fang *et al.*, 2012).

Thus, recombinant viruses do have potential as vaccine vectors, but undoubtedly the right mix of *Toxoplasma* antigens expressed and the vector system utilized to yield the highest level of protective immunity have yet to be found (see Table 26.5).

26.3.5 Vaccination Using Bacterial Vectors

Live bacterial vaccine vectors have been extensively used to deliver and express heterologous vaccine antigens to protect against cancer and various infectious agents, including AIDS (reviewed in Drabner and Guzman, 2001). Live bacterial vaccines have the advantage that they can express multiple antigens, are easily mass produced, can be orally or intranasally applied and induce strong immune responses. However, relatively few studies have tested whether heterologous expression of parasitic antigens with bacterial vaccine vector strains can lead to protective immunity Table 26.5.

Invasive bacteria such as *Salmonella, Listeria, Yersinia, Shigella and Mycobacterium bovis BCG* have been used as vaccine vectors, capable of mounting potent humoral and cellular immune responses. Since these are pathogenic bacteria they were attenuated to generate suitable non-pathogenic vaccine strains. Many attenuated strains have been reported that are non-pathogenic and have limited proliferative capacity *in vivo*. Attenuation can, however, lead to reduced immune stimulation. Moreover, overexpression of heterologous genes can result in a rapid selection for low or non-expressers. To circumvent this potential obstacle, different approaches have been used to obtain stable expression with bacterial vaccine strains for *in vivo* use. For example, inducible promoters were used, such as the *Salmonella* nirB promoter, which becomes activated *in vivo* under anaerobic conditions (Chatfield *et al.*, 1992). Alternatively, a mixed population approach was tested whereby expressing bacteria are constantly derived from non-expressing carrier cells (Yan and Meyer, 1996). Finally, bacterial vaccine strains have been successfully used to deliver eukaryotic expression plasmids. In one convincing example, eukaryotic expression plasmids containing *Listeria* antigens were successfully delivered with *Salmonella typhimurium*, protecting mice from a lethal challenge with *Listeria monocytogenes* (Darji *et al.*, 1997).

The earliest use of a bacterial vector as a vaccine against *T. gondii* was oral immunization with a live attenuated *Salmonella typhimurium* vaccine strain (Cong *et al.*, 2005). SAG1 and SAG2 were delivered with a eukaryotic expression plasmid which also contained cholera toxin sub-units A2 and B. Cholera toxin (CT) is known to have an adjuvant effect and indeed the addition of CT sub-units A2 and B induced a strong cellular immune response, as measured by induced specific IgG2A titres, splenocyte proliferation and IFNγ production. Upon challenge with 1000 RH tachyzoites the vaccinated mice survived longer and 40% of the mice survived the whole trial period. There have been three further studies using bacterial vectors delivered orally in the mouse model, two using attenuated *Salmonella* incorporating either SAG1 (Qu *et al.*, 2008) or SAG1 and MIC3 alone or in combination (Qu *et al.*, 2009) and one using BCG expressing ROP2 (Wang *et al.*, 2007). While all formulations delivered a level of protection, vaccines comprising more than one antigen were generally more effective (Qu *et al.*, 2008) and were improved by incorporating an adjuvant (Cong *et al.*, 2005).

In conclusion, vaccination with live bacterial vectors can induce both strong humoral and cellular immunity and as they are delivered orally should also induce protection at the mucosal level. However, as all studies to date have utilized i.p. challenge with RH and their ability to protect against natural infection with cyst forming lineages remains to be ascertained (see Table 26.5).

26.3.6 DNA Vaccines

It is the general consensus of opinion that a type 1 response, particularly associated with $CD8^+$ T-cells producing IFNγ, is the major mediator of immunity against *T. gondii* infection. Nevertheless, numerous vaccine and immunological studies have also demonstrated that a broad spectrum of immune response requiring elements of type 2 immunity with antibodies provides the best overall protection against infection. Consequently, as DNA vaccines are known to induce $CD8^+$ T-cell responses in addition to broad spectrum immunity, there has been in recent years a substantial effort to determine their effectiveness against toxoplasmosis. While the majority of studies have concentrated on SAG1 (Table 26.3), the vaccine potential of numerous other molecules GRA1, GRA2, GRA4, GRA6, GRA7, ROP1, ROP2, ROP16, ROP 18, HSP70, HSP30, MIC1, MIC2, MIC3, MIC4, MIC6, MIC8, M2AP, MAG1, AMA1, ROM1, BAG1, IMP1, Perforin-like-1 EFG and binding lectin domains have also been investigated, either alone or in combination, with varying degrees of success. Vaccination with SAG1 has been found to be particularly effective at limiting mortality against both virulent and avirulent challenge (Nielsen *et al.*, 1999; Angus *et al.*, 2000; Chen *et al.*, 2002; Couper *et al.*, 2003; Liu *et al.*, 2010a). In addition, the effectiveness of SAG1 DNA vaccines was generally enhanced by utilizing cocktail vaccines comprising other antigens such as ROP2 (Fachado *et al.*, 2003c), GRA4 (Mevelec *et al.*, 2005), MIC2 and MIC3 (Rosenburg *et al.*, 2009) and GRA2 (Zhou *et al.*, 2012). The incorporation of adjuvants into the vaccines, particularly pGM-CSF (Desolme *et al.*, 2000; Ismael *et al.*, 2003; Mevelec *et al.*, 2005), cholera toxin (Wang *et al.*, 2009) and pIL-18 (Liu *et al.*, 2010b; Yan *et al.*, 2012b), also enhanced general efficacy irrespective of the antigen under investigation. Surprisingly, while a number of studies have indicated the potential of pIL-12 as adjuvant (Xue *et al.*, 2008a, 2008b; Cui *et al.*, 2008) a more recent study suggested an adverse effect for this cytokine (Khosroshahi *et al.*, 2012) with regard to protection. These studies generally used similar antigens, SAG1 and ROP2, the BALB/c mouse model and RH challenge and consequently it is difficult to draw a conclusion. However, the general consensus would favour multi-epitoped, adjuvanted vaccines being more efficacious than those comprising single antigens or non-adjuvanted preparations. Consequently, while a SAG1 DNA vaccine failed to limit maternal–foetal transmission (Couper *et al.*, 2003) a SAG1/GRA4 vaccine adjuvanted with pGM-CSF did increase pup survival if dams were infected during pregnancy, but again without preventing vertical transmission (Mevelec *et al.*, 2005). However, the former study utilized BALB/c mice while the latter used Swiss OF1 mice, which could also provide an explanation for relative success and failure. Indeed, it has been shown that the degree of protection afforded by a particular DNA vaccine can be dependent on the animal model used and

TABLE 26.3 DNA Vaccine Studies and Their Outcomes in Animal Models

Reference	Antigen (Route)	Adjuvant or Carrier	Route of Vac.	Animal Model	Immunology	Challenge (Route)	Survival	Parasite Burden
(Nielsen *et al.*, 1999)	SAG1		i.m.	BALB/c	Antibodies $CD8^+$ T-cells	RH (i.p.)	+++	
(Angus *et al.*, 2000)	SAG1		i.m.	C57BL/6 Rats	Antibodies, splenocyte IFN-γ and IL-2 production Antibodies	Me49 cysts (oral) VEG Oocysts (oral)	+++	++ ++
(Vercammen *et al.*, 2000)	GRA1		i.m.	C57BL/6 BALB/c C3H	Antibodies T-cell proliferation IFN-γ	IPB-G or 76K cysts (oral)	++ C3H – BALB/c – C57 BL/6	++ C3H – BALB/c
(Desolme *et al.*, 2000)	GRA4	w/wo pGM-CSFw/wo pIL-12	i.m.	C57BL/6	Antibodies Splenocyte proliferation IFN-γ IL-2, IL-10	76K cysts (oral)	++ GRA4 and GRA4 and pGM–CSF	
(Leyva *et al.*, 2001)	ROP2		i.m.	BALB/c C57BL/6 CBA/J	Antibodies	RH (s.c.)	+ BALB/c – C57BL/6 – CBA/J	
(Chen *et al.*, 2001)	ROP1		i.m.	BALB/c	Antibodies			
(Chen *et al.*, 2002)	SAG1	pIL-2	i.m.	BALB/c	Antibodies IFN-γ	RH	+	
(Chen *et al.*, 2003)	SAG1	Liposomes	i.m.	BALB/c	Antibodies IFN-γ, IL-2			
(Couper *et al.*, 2003)*1	SAG1		i.m.	BALB/c	Antibodies IFN-γ	Beverley cysts (oral)	++	+++
(Bivas-Benita *et al.*, 2003)	GRA1	Chitosan microparticles	oral i.m.	C3H/HeN	Antibodies			
(Mohamed *et al.*, 2003)	HSP70 HSP30 SAG1		i.d. i.m. i.p.	C57BL/6 BALB/c	IFN-γ	Fukaya cysts (oral)		++/+ HSP70 > HSP30 and SAG1 i.d. > i.m. and i.p.

(*Continued*)

TABLE 26.3 DNA Vaccine Studies and Their Outcomes in Animal Models (*cont'd*)

Reference	Antigen (Route)	Adjuvant or Carrier	Route of Vac.	Animal Model	Immunology	Challenge (Route)	Survival	Parasite Burden
(Fachado *et al.*, 2003a)	SAG1 ROP2 SAG1and ROP2		i.m.	BALB/c	Antibodies, T-cell proliferation IFNγ	RH (i.p.)	++ SAG1 and ROP2	
(Fachado *et al.*, 2003b)	Genomic library		i.m.	BALB/c	Antibodies, T-cell proliferation, $CD4^+$ and $CD8^+$ activation IFN-γ	RH (i.p.)	++	
(Ismael *et al.*, 2003)	MIC3	± pGM-CSF	i.m.	CBA/J	Antibodies, lymphocyte proliferation IFN-γ, IL-2	76K cysts (oral)		++ pMIC3 and pGM-CSF > MIC3
(Scorza *et al.*, 2003)	GRA1		i.m.	C3H/HeN	Antibodies, CD4 Cytolytic CD8 IFN-γ	IPB-G cysts (i.p.)	++	++
(Martin *et al.*, 2004)	GRA4		i.m.	C3H	Antibodies	Me49 cysts (oral)		++
(Mévélec *et al.*, 2005)*2	SAG1 GRA4 SAG1and GRA4	pGM-CSF	i.m.	C57BL/6 Swiss OF1 *2	Antibodies, splenocyte proliferation IFN-γ	76K cysts (oral)	+++/++ SAG1 and GRA4 and pGM-SCF > SAG1 and GRA4	++ SAG1 and GRA4 and pGM-CSF
(Beghetto *et al.*, 2005)	MIC 1,2,3,4, M2AP, AMA1		i.m.	BALB/c	Antibodies	SS1119 cysts (oral)	++	
(Mévélec *et al.*, 2005)	SAG1mut or GRA4 or SAG1 mut+ GRA4	Cardiotoxin Plasmid containing GMCSF	i.m.	C57BL/6 for acute Swiss OF1 for chronic and congenital	Antibodies, splenocyte, IL-4, IL-10, IFN-γ	76K cysts Oral	+++	+
(Dimier-Poisson *et al.*, 2006)	RNA	?	i.n	C57BL/6	Antibodies, lymphocyte proliferation	76K cysts Oral	++	+++

(Nielsen *et al.*, 2006)	MAG1, BAG1		i.m.	C3H/HeN	Antibodies	SSI 119 cysts Oral		+
(Zhang *et al.*, 2007b)	ROP2 + SAG1, ROP2, SAG1	w/wo pIL-12	i.m.	BALB/c	Antibodies, lymphocyte proliferation IFN-γ IL-12, IL-4	RH IP	− ROP2 − SAG1 + ROP2 + SAG1 ++ ROP2 + SAG1 + pIL-12	
(Dautu *et al.*, 2007)	MIC2, M2AP, AMA1, BAG1	Gold particles	Gene gun into abdomen	BALB/c C57BL/6	Antibodies, IL-4, IFN-γ, splenocyte proliferation	Beverley cysts Oral	+ BALB/c − C57BL/6 + C57BL/6 (AMA1)	
(Jongert *et al.*, 2007)	GRA1, GRA7, ROP2,		i.m.	C3H/HeN	Antibodies, IFN-γ	76K cysts Oral		+
(Jongert *et al.*, 2008a)	MIC2, MIC3, SAG1,	DNA only of protein primed boost GERBU	i.m.	ID	Antibodies, IL-2, IL-10, IFN-γ, splenocyte proliferation		+++ protein − protein + DNA − protein primer boost	
(Jongert *et al.*, 2008b)	GRA1, GRA7		i.m.	ID	Antibodies, IFN-γ, PMBCs	IPB-G cysts IP		
(Cui *et al.*, 2008)	SAG1 + SAG2 + ROP2	w/wo pIL-12	i.m.	BALB/c	Antibodies, IFN-γ, IL-4, IL-12, splenocyte proliferation	RH	− SAG2 + ROP2 + SAG1+ SAG2 ++ ROP2 + SAG1 + pIL-12	
(Cong *et al.*, 2008)	Fragments from SAG1, GRA1, GRA4, ROP2	Cholera toxin A2/B plasmid	i.m.	BALB/c	Antibodies, splenocyte proliferation, CTLs, IL-2, IL-4, IL-5	RH IP		++
(Xue *et al.*, 2008a)	SAG1, ROP2	Cholera toxin A2/B plasmid (pCTA2/B) or pIL-12	i.m.	BALBc	Antibodies IFN-γ, IL-4, IL-12, splenocyte proliferation	RH IP	+ SAG1 + ROP2 ++ SAG1 + ROP2 pCTA2/B +++ SAG1 + ROP2 pIL-12	+++

(*Continued*)

TABLE 26.3 DNA Vaccine Studies and Their Outcomes in Animal Models (*cont'd*)

Reference	Antigen (Route)	Adjuvant or Carrier	Route of Vac.	Animal Model	Immunology	Challenge (Route)	Survival	Parasite Burden
(*et al.*, 2008b)	SAG1 + SAG2 + GRA2	pIL-12	i.m.	BALB/c	Antibodies, IFN-γ, IL-4, IL-12, splenocyte proliferation	RH IP	+ SAG1 + ROP2 + GRA2 ++ ROP2 +SAG1 +GRA2+ pIL-12	
(Liu *et al.*, 2009)	Multiepitope vaccine SAG1 GRA1 GRA4 GRA2	w/wo CpG	i.m.	BALB/c C57BL/6	Antibodies, IFN-γ, IL-10, splenocyte proliferation	RH i.p.	BALB/c +++ TLA control ++ pME/CpG C57BL/6 ++ pME − pME/CpG	
(Wang *et al.*, 2009)	SAG1, MIC4, SAG1 + MIC4	Cholera toxin A2/B plasmid (pCTA2/B)	i.n,	BALB/c	Antibodies, IFN-γ, IL-4, IL-12, splenocyte proliferation	RH i.p.	++ SAG1 + MIC4 + pCTA2/B + SAG1 + MIC4	
(Rosenburg *et al.*, 2009)	Multiepitope vaccine MIC2, MIC3, SAG1	CpG	i.m.	BALB/c	Antibodies, lymphocyte proliferation, IFN-γ, IL-2	Prugniaud, Trosseau cysts Oral	+	+++
(Ismael *et al.*, 2009)	MIC3 complete EFG domains Lectin domains	Cardiotoxin Plasmid containing GMCSF	i.m.	CBA/J	Antibodies, adoptive transfer, splenocyte proliferation, IFN-γ, IL-2, IL-4, IL-10,	76K cysts Oral		MIC3 + EFG + Lectin +
(Chen *et al.*, 2009)	GRA4	Cardiotoxin w/woliposome	i.m.	BALB/c C57BL/6	Antibodies, IFN-γ, IL-2, IL-4, splenocyte proliferation	RH Me 49 cysts IP	+++ w ++ w/o in BALB/c + for both in C57BL/6	+ for all
(Fang *et al.*, 2009)	MIC3 w/wo suicidal gene		i.m.	BALB/c	Antibodies, splenocyte proliferation, IFN-γ, IL-4	RH IP	++ w ++ w/o	

(Peng *et al.*, 2009)	MIC6		i.m.	Kunming	Antibodies, splenocyte proliferation IFN-γ, IL-4	RH IP	++	++
(Xiang *et al.*, 2009)	MIC3		footpad	Kunming	Antibodies, T-cell subsets	RH IP	++	
(Hiszczynsk-Sawicka *et al.*, 2010a)	GRA7	Liposomes Emulsigen Emulsigen D	i.m. (neck dorsal)	Coopworth ewes	Antibodies, IFN-γ			
(Liu *et al.*, 2010a)	SAG1	w/wo IL-18	i.m.	C3H/HeN	Antibodies, lymphocyte proliferation, IFN-γ, IL-2, IL-4, IL-10,	RH IP	++ with IL-18 + wo IL-18	
(Kikumura *et al.*, 2010)	HSP70	Gold particles	Gene gun into abdomen	C57BL/6	IFN-γ, effector cells	Fukaya cysts Oral	+++	
(Hiszczynsk-Sawicka *et al.*, 2010b)	MAG1	IL-6, liposomes	i.m. (neck dorsal)	Coopworth ewes	Antibodies, IFN-γ			
(Liu *et al.*, 2010b)	MIC8		i.m.	Kunming	Antibodies, lymphocyte proliferation, IFN-γ, IL-2, IL-4, IL-10,	RH IP	++	
(Yao *et al.*, 2010)	SAG1, MIC8		i.m.		Antibodies, IFN-γ, IL-4, T-cell proliferation	RH IP	++	
(Li *et al.*, 2010)	SAG1, ROP1	pGM-CSF or CpG	i.m.	Coopworth ewes	Antibodies, IFN-γ			
(Laguía-Becher *et al.*, 2010)	SAG1	Tabcco leaf extract containing optimally expressed SaG1. Boost is rSAG1	Oral, s.c.	C3H/HeN C57nL/6	Antibodies DTH, IFN-γ	Me49 cysts Oral	–	–
(Hoseinian Khosroshahi *et al.*, 2011)	SAG1, ROP2	Freund's complete and incomplete adjuvant	i.m.	BALB/c	Antibodies, splenocyte proliferation, IL-2, TNF-α, IFN-γ	RH IP	+	

(*Continued*)

TABLE 26.3 DNA Vaccine Studies and Their Outcomes in Animal Models (*cont'd*)

Reference	Antigen (Route)	Adjuvant or Carrier	Route of Vac.	Animal Model	Immunology	Challenge (Route)	Survival	Parasite Burden
(Yuan *et al.*, 2011a)	ROP16		i.m.	Kunming	Antibodies, IL-2, IFN-γ, IL-4, IL-10, CTL activity, splenocyte proliferation	RH IP	++	
(Makino *et al.*, 2011)	HSP70	Gold particle	Gene gun into abdomen	C57BL/6	IFN-γ, IL-4, IL-17, T-cell polarization and DC IL-12	Fukaya cysts Oral		+
(Hiszczyńsk-Sawicka *et al.*, 2011a)	GRA1, GRA4, GRA6, GRA7	CpG, liposomes	i.m. (neck dorsal)	Coopworth ewes	Antibodies, IFN-γ			
(Rashid *et al.*, 2011)	RON4	DNA (pGMSF) and protein (choleratoxin)	i.m.	CBA/J	Antibodies, IFN-γ, IL-2, IL-4, IL-10 Splenocyte proliferation			
(Yan *et al.*, 2011)	Perforin-like I	IL-18	i.m.	Kunming	Antibodies, IFN-γ, IL-2, IL-4, IL-10, splenocyte proliferation	RH i.p.	PLP-1 alone ++ PLP-1 w IL-18 +++	
(Li *et al.*, 2011)	ROP2-SAG1	Freund's incomplete	i.m.	BALB/c	Antibodies, lymphocyte proliferation, IFN-γ		Comparison with rROP2-SAG1 and DNA boost to protein vaccination	
(Hiszczyńsk-Sawicka *et al.*, 2011b)	ROP1	CD154 CpG ODN, liposomes	i.m. (neck dorsal)	Coopworth ewes	Antibodies, IFN-γ			
(Yuan *et al.*, 2011b)	ROP18		i.m.	Kunming	Antibodies, IL-2, IFN-γ, IL-4, IL-10, CTL activity, splenocyte proliferation	RH IP	++	
(LI *et al.*, 2011)	ROM1		i.m.	BALB/c	Antibodies, IL-2, IFN-γ, IL-4, IL-10, %CD4+/CD8+, splenocyte proliferation	RH IP	+++	

(Sun *et al.*, 2011)	GRA6	1% LMS	i.m.	BALB/c Kunming	Antibodies, splenocyte proliferation	RH IP	+++ with LMS in BALB/c ++wo LMS – in Kunming	
(Zhou *et al.*, 2012)	SAG1, GRA2	PreS2 (HepB)	i.m.	BALB/c	Antibodies, lymphocyte proliferation, IFN-γ, IL-4, IL-10	RH IP	GRA2 + SAG1 + GRA2 – SAG1 ++ GRA2 – SAG1 – PreS2 +++	
(Cui *et al.*, 2012)	IMP-1		i.m.	BALB/c	Antibodies, IL-2, IFN-γ, IL-4, IL-10, splenocyte proliferation	RH IP	++	
(Hoseinian-Khosroshahi *et al.*, 2011)	SAG1, ROP2	IL-12 ALUM	i.m.	BALB/c	Antibodies, IFN-γ, IL-4	RH IP	++ wo adjuvant + w IL-12 +ALUM	
(Yan *et al.*, 2012b)	Perforin-like I	IL-18	i.m.	Kunming	Antibodies, IFN-γ, IL-2, IL-4, IL-10, splenocyte proliferation	Prugniaud Oral	++	+
(Quan *et al.*, 2012)	GRA7, ROP1	w/wo IL-12			Antibodies, IFN-γ, IL-10, TNF-α, T-cell proliferation		++	+
(Min *et al.*, 2012)	GRA7	FCA, IFA	i.m.	BAL B/c	Antibodies, IFN-γ,	RH	++ DNA prime, protein boost	+
(Hiszczynsk-Sawicka *et al.*, 2012)	MIC3	Liposomes	i.m. (neck dorsal)	Coopworth ewes	Antibodies, IFN-γ			

*1 *Vaccination did not prevent congenital toxoplasmosis in BALB/c mice*

*2 *Vaccination did not limit congenital transmission in SWISS OF1 mice but increased survival*

Key for Survival: –– decreased survival; – no difference in survival; + moderate (≤50% increased survival); ++ significant (≥50% increased survival); +++ highly significant (≥90% increased survival), compared with control groups.

Key for Parasite Burden: –– increased parasite burden; – no difference in parasite burden; + moderate (≤50% decrease in parasite burden); ++significant (≥50% decrease in parasite burden); +++ highly significant (≥90% decrease in parasite burden), compared with control groups whereas GRA1 protected C3H mice both with regard to survival and cyst burden it did not protect BALB/c or C57BL/6 mice (Vercammen *et al.*, 2000). Conversely, ROP2 improved survival of BALB/c mice following infection with RH tachyzoites but not survival of C57BL/6 or CBA/J mice (Leyva *et al.*, 2001). Similarly, GRA6 protected BALB/c mice but not Kunmings against IP challenge with RH strain (Sun *et al.*, 2011). In addition to adjuvanting the vaccines, efficacy may also be enhanced by changing the route of vaccination as in one study in which intradermal inoculation proved more effective than the intramuscular or intraperitoneal routes (Mohamed *et al.*, 2003). Many of the studies to date utilizing nucleic acid vaccines can be viewed as incomplete as they merely measure immune responses after vaccination; this is particularly true of numerous recent studies utilizing Coopworth ewes. While many studies utilizing rodent models do measure survival after challenge infection, relatively few then monitor parasite burdens and to date there are only two studies that investigate maternal–foetal transmission.

The normal portal of entry of *T. gondii* is via the gut mucosa. Furthermore, intraepithelial IFNγ producing $CD8^+$ T-cells cytolytic for parasitized enterocytes have been shown to be generated following infection (Chardes and Bout, 1993; Chardes *et al.*, 1994) while IgA may protect mucosal surfaces from parasite invasion (Mineo *et al.*, 1993). Unfortunately, conventional parenterally administered vaccines do not generally induce mucosal immune responses and the most successful method to induce this type of response in addition to systemic immunity has been to administer vaccines orally (Gallichan and Rosenthal, 1996). Current evidence would suggest that entrapment of such vaccines in lipid vesicles would enhance the immune response generated by both protecting the DNA from degradation and targeting the DNA directly to APC (Gregoriadis *et al.*, 2002). Furthermore, it has been demonstrated that DNA vaccines can influence both mucosal and systemic immunity by the oral route if suitably encapsulated (Chen and Langer, 1998). However, only one study to date has tested this route using chitosan microparticle entrapped pGRA1 (Bivas-Benita *et al.*, 2003). Unfortunately, mucosal immunity was not measured and no challenge infections were undertaken in this study. Nevertheless, a few recent studies that utilized the intranasal route have highlighted the potential of specifically targeting the mucosal immune system (Dimier-Poisson *et al.*, 2006; Wang *et al.*, 2009). C57BL/6 mice receiving three intranasal doses of tachyzoite mRNA developed systemic and mucosal humoral immunity as well as systemic and mucosal cell mediated immunity. Furthermore, survival rates were significantly improved and a partial reduction in brain cyst burdens noted following normally lethal or sublethal oral challenge with brain cysts of the 76K strain (Dimier-Poisson *et al.*, 2006). Cholera toxin (pCTA2/B) may improve the efficacy of vaccination by the oral route (Wang *et al.*, 2009)

While overall DNA vaccine formulations based on the above antigens have resulted in reduced mortality, in those studies where brain cyst burdens have also been quantified only a reduction has been achieved with no sterile immunity observed. What was thought significant initially was that the antigens utilized tended to be either tachyzoite specific, e.g. SAG1, or shared, e.g. GRA1, 2 and 4, while none were primarily bradyzoite specific. Similar incomplete protection has also resulted when attenuated or 'crude' tachyzoite antigen preparations have constituted the vaccine (Roberts *et al.*, 1994; Buxton, 1993). What could be of significance in this respect is that serological studies in *T. gondii* infected humans (Kasper, 1989; Zhang *et al.*, 1995; Lunden *et al.*, 1993), as well as mice, have indicated that recognition is generally of immunodominant stage specific antigens with little recognition of shared antigens. This would imply a requirement for additional antigens to those expressed in tachyzoites for complete protection following vaccination which can be concluded from some experimental studies (Alexander *et al.*, 1996; Freyre *et al.*, 1993). To date, the bradyzoite antigens BAG1 (Nielsen *et al.*, 2006; Dautu *et al.*, 2007) and MAG1 (Nielsen *et al.*, 2006; Hiszczyńska-Sawicka *et al.*, 2010b) have been utilized in DNA vaccine studies with some limited success in reducing mortality (Dautu *et al.*, 2007) and reducing cyst burdens (Nielsen *et al.*, 2006). However, these antigens were not combined with those primarily associated with tachyzoite stages. Thus, the ideal vaccine to induce complete immunity against *T. gondii* may be one that would promote protection against more than one life cycle stage, induce mucosal as well as systemic immunity and, in addition, a strong CD8 response. A DNA vaccine delivered by the oral or other appropriate mucosal route would offer a rational solution.

26.3.7 Sub-Unit Vaccines

While various crude antigen preparations have been tested for their vaccine potential against toxoplasmosis for almost 50 years, studies on the efficacy of purified sub-units or recombinant products is relatively new, with the earliest investigations taking place around 20 and 10 years ago, respectively (Table 26.4). Sub-unit vaccines have the advantage that specific immunogenic antigens are presented without adding antigens that are at best irrelevant and at worst capable of inducing febrile or disease exacerbating immune responses. In addition, antigens that induce little immunity in the context of the parasite can be boosted to be more immunogenic if applied in a non-natural context. Thus, sub-unit vaccines are focused in their immune objectives and safe but tend to lack immunological potency and, therefore, require formulation with appropriate adjuvants to enhance their effectiveness. Typically, SAG1, the immunodominant tachyzoite life cycle specific surface antigen, has been the most extensively studied product both as a purified sub-unit and as a recombinant antigen (Table 26.4). Additional vaccine candidates that have been investigated include ROP1, ROP2, ROP4, GRA1, GRA2, GRA4, GRA5, GRA6, HF10 (a GRA6 derived decapeptide), GRA7, $GRA7_{20\text{-}28}$ peptide, MIC1, MIC4, HSP70, SAG2, SAG3, SRS1, TgPI-1, Actin Depolymerization Factor, TCP, NTPasesII, P54 and P24 as well as uncharacterized antigens recognized by monoclonal antibodies such as F3G3 (Brinkmann *et al.*, 1993) or SDS–PAGE purified proteins associated with cell invasion (Azzouz *et al.*, 2012). Collectively, these have been studied both individually and as 'cocktail' vaccines administered by a variety of routes and using a variety of adjuvants. Recombinant sub-unit vaccines have also utilized live infectious expression vectors such as BCG, *Salmonella*, feline herpes virus and vaccinia virus (Supply *et al.*, 1999; Mishima *et al.*, 2002; Roque-Resendiz *et al.*, 2004; Cong *et al.*, 2005). In addition to tachyzoite life cycle stage specific antigens, some success has been achieved with antigens such as MAG1 known to have shared expression in both tachyzoite and bradyzoite stages (Parmley *et al.*, 2002).

The choice of adjuvant can be crucial and profoundly influence vaccine efficacy. Thus, a purified sub-unit vaccin which is protective as measured by vaccine survival (Bulow and Boothroyd, 1991; Khan *et al.*, 1991) and cyst burden following infection when entrapped in liposomes or formulated with saponin Quil A is disease exacerbatory if adjuvanted with FCA (Kasper *et al.*, 1985). Adjuvants so far utilized for subcutaneous administration include: FCA, FIA, liposomes, saponin QuilA, ISCOMs, SBAS1, Monophosphoryl lipid A, lipopeptide, PADRE, GLA-SE and IL-12; for intraperitoneal administration: FCA, FIA, liposomes, VetL-10, Lactobacillus, FMA and aluminium hydroxide (ALUM); for intramuscular administration: ALUM, FIA and CpG; for oral administration: cholera toxin, chitosan microparticles; for intranasal administration: salbutamol, cholera toxin and enterotoxin.

Overall, SAG1, adjuvant withstanding, has induced comparatively good protection in terms of decreasing mortality (e.g. Bulow and Boothroyd, 1991), cyst burden (Khan *et al.*,

TABLE 26.4 Sub-Unit Vaccine Studies and Their Outcomes in Animal Models

Reference	Antigen	Adjuvant or Carrier	Route of Vac.	Animal Model	Immunology	Challenge (Route)	Survival	Parasite Burden	Other
(Kasper *et al.*, 1985)	SAG1	FCA	i.p./s.c.	BALB/c CD1 mice	Antibodies	tachyzoites C strain (i.p.)	−−	−−	
(Khan *et al.*, 1991)	SAG1	Saponin Q and A	s.c.	Outbred A/J mice CD1 mice C57BL/6 mice	$CD8^+$ cytolytic T-cells, IFN-γ, IL-2	P strain (Me49) cysts (i.p.)	+++	+++	
(Bulow and Boothroyd, 1991)	SAG1	Liposomes	i.p.	Female Swiss–Webster mice	Antibodies	tachyzoites C strain (i.p.)	+++/++ SAG1 liposomes > SAG1		
(Duquesne *et al.*, 1991)	P24	Vaccine	s.c. or i.p.	Fischer/ nude rat	T-cells	tachyzoites (i.p.)	++		
(Darcy *et al.*, 1992)	SAG1 monomeric peptide SAG1 multiple antigenic peptide	IFA	s.c.	Mice OF1 Fischer/ nude rat	T-cells Antibodies	76K cysts (oral) RH tachyzoites (i.p.)	− SAG1 MP ++ SAG1 MAP + SAG1 MAP		
(Brinkmann *et al.*, 1993)	F3G3 antigen 2G11-Ag 1E11	IFA	s.c. and i.p.	Outbred Swiss–Webster mice	Antibodies $CD4^+$ T-cells, IL-2	C56 tachyzoites (i.p.) Me49 cysts (i.p.)	+++ F3G3 Antigen		
(Debard *et al.*, 1996)	SAG1	Cholera toxin	i.n.	CBA/J mice	Antibodies IgG and IgA T cells IL-2, IL-5	76K cysts (oral)		++	
(Velge-Roussel *et al.*, 1997)	SAG2	FCA/IFA	s.c.	C57BL/6 CBA/J	Antibodies T-cells	76K cysts (oral)		− +	

(Lunden *et al.*, 1997)	SAG2	ISCOM GST	s.c.	Swiss–Webster	Antibodies	Oocysts Me49 cysts C56 (oral)	−		
(Mevelec *et al.*, 1998)	GRA4	Cholera toxin GST	Oral	C57BL/6	Antibodies IgA IgG	76K cysts (oral)	+	++	
(Letscher-Bru *et al.*, 1998)	SAG1	IL-12	s.c.	CBA/J	Antibodies IFN-γ	PRU cysts (oral)		+	
(Petersen *et al.*, 1998)	SAG1	ALUM	i.m.	Outbred NMRI	Antibodies	RH tachyzoites (i.p.) SS1119 cysts (s.c.)	+	−	
(Fermin *et al.*, 1999)	SAG1	Salbutamol	i.n.	CBA	T-cell proliferation	Cysts		++	
(Mun *et al.*, 1999)	HSP70 HSP 70/ bag1			C57B/6 BALB/c		Fukaya strain cysts (oral) RH tachyzoites (i.p.)	+ HSP70/ bag1 −− HSP70	+ HSP70/ bag1 −− HSP70	
(Supply *et al.*, 1999)	GRA1	BCG	i.p., s.c, i.v.	OF1 Outbred mice Sheep	Antibodies Lymphocytes IFN-γ	Virulent oocysts (oral)	−		+ temp
(Aosai *et al.*, 1999)	SAG1	RNAs (lymphoma)				RH tachyzoites (i.p.)			
(Bonenfant *et al.*, 2001)	SAG1	Cholera toxin Heat-labile enterotoxin	i.n.	CBA/J	IgG and IgA Lymphocytes IFN-γ IL-2	76K cysts (oral)		++	
(Haumont *et al.*, 2000)	SAG1	SBAS1	s.c.	Dunkin Harley Guinea pigs	Antibodies	C57 tachyzoites (i.d.)			++ congenital transmission
(Mishima *et al.*, 2001)	SAG1 SAG2 SAG3 SRS1 P54	FCA/FIA	i.p.	BALB/c	Antibodies	Beverley bradyzoites (i.p)	+ SAG2 + SRS1 + P54		

(Continued)

TABLE 26.4 Sub-Unit Vaccine Studies and Their Outcomes in Animal Models (*cont'd*)

Reference	Antigen	Adjuvant or Carrier	Route of Vac.	Animal Model	Immunology	Challenge (Route)	Survival	Parasite Burden	Other
(Parmley *et al.*, 2002)	MAG-1	Quil A/GST	s.c.	Swiss–Webster mice	Antibodies	Me49 cysts (oral)	+	++	+ Inflammation
(Mishima *et al.*, 2002)	ROP2	Feline herpes virus type 1		Cats	Antibodies			+	
(Letscher-Bru *et al.*, 2003)	SAG1		s.c.	BALB/c CBA/J	Antibodies, IFN-γ, IL-10 Antibodies, IL-10	Me49 cysts (oral)			++ BALB/c −− CBA/J congenital transmission
(Bivas-Benita *et al.*, 2003)	GRA1	Chitosan microparticles	Oral	C3H/HeN	Antibodies				
(Martin *et al.*, 2004)	GRA4 ROP2 GRA4 + ROP2	ALUM	i.m.	C57BL/6 C3H	Antibodies, lymphocytes IFN-γ, IL-4	Me49 cysts (oral)		+/++	
(Yang *et al.*, 2004)	SAG1/2	Vet L-10	i.p.	BALB/c	Antibodies, lymphocytes IFN-γ, IL-4	RH tachyzoites (s.c.)	+++		
(Roque-Resendiz *et al.*, 2004)	ROP2	MVA vaccine			Antibodies	RH tachyzoites (i.p.)	+++		
(Cong *et al.*, 2005)	SAG1/ SAG2	*Salmonella typhimurium* Cholera toxin A2/B	Oral	BALB/c	Antibodies Lympho-proliferation IFN-γ, IL-4	RH tachyzoites (i.p.)	+++/++ SAG1–2 and CT A2/B > SAG1–2		
(Garcia *et al.*, 2005)	Crude rhoptry proteins	ISCOM (immunostimulating complexes)	s.c.	Pigs	Antibodies IgG	VEG oocysts Oral		+++ (through mouse bioasssay)	No clinical signs

(Echeverria *et al.*, 2006)	rROP2	LiHSP83	Footpad	C57BL/6 C3H BALB/c	Antibodies Splenocyte proliferation, IFN-γ, IL-4	Me49 cysts Oral	C57BL/6 − C3H BALB/c −	C57BL/6 + C3H +	
(Siachoque *et al.*, 2006)	SAG1 (17 peptides)	BSA	i.v.	C3H	Antibodies	RH i.p.	++ 4 out 17 peptides		All 4 peptides in C-terminus
(Lourenço *et al.*, 2006)	MIC1, MIC4	FCA	s.c.	C57BL/6	Antibodies IgG subclasses, IFN-γ, IL-4, IL-2, IL-10, IL-12p40	Me49 cysts Oral	+++	++	+
(Cuppari *et al.*, 2008)	TgPI-1	ALUM	i.m.	C3H/HeN	Antibodies, IFN-γ, IL-10	Me49 cysts Oral	++		
(Igarashi *et al.*, 2008)	ROP2, GRA5, GRA7	Not specified	i.n.	BALB/c	IgA, IgG	VEG cysts Oral		++	
(Golkar *et al.*, 2007)	GRA2, GRA6	Mono-phosphoryl lipid A	s.c.	CBA/J	Antibodies	Pruβgal cysts i.p.	++	+	
(Blanchard *et al.*, 2008)	HF10 (GRA6)	Peptide-loaded BMDCs	footpad	B10.D2 C57BL/6	IFN-γ CD8 responses	PruGFP Oral	+++ B10.D2 − C57BL/6	− B10.D2	
(Martinez-Gomez *et al.*, 2009)	TCP	*Lactobacillus casei* and FMA	i.p.	NIH	IgM	Me49 cysts Oral	Not specified	+	
(Cong *et al.*, 2010)	Peptide SAG1, GRA6, GRA7	Lipopeptide, PADRE	s.c.	HLA-A1101/Kb	IFN-γ CD8 responses	PruFLUC i.p.	+	+	
(Laguia-Becher *et al.*, 2010)	SAG1	Tobacco leaf extract containing optimally expressed SAG1. Boost is rSAG1	Oral, s.c.	C3H/HeN C57BL/6	Antibodies DTH, IFN-γ	Me49 cysts Oral	Not specified	+	−

(*Continued*)

TABLE 26.4 Sub-Unit Vaccine Studies and Their Outcomes in Animal Models (*cont'd*)

Reference	Antigen	Adjuvant or Carrier	Route of Vac.	Animal Model	Immunology	Challenge (Route)	Survival	Parasite Burden	Other
(Tan *et al.*, 2010)	GRA6 HF10	Lipopeptide, PADRE	s.c.	BALB/c	IFN-γ CD8 responses	PruFLUC i.p.	+	+	
(Cong *et al.*, 2011)	Peptides from BSR4, GRA15, GRA10, SAG2C, SAG2D, SAG2X, SAG3, SRS9, SPA, MIC1, MIC4, MIC6, MIC8, MICA2P	PADRE GLA-SE	s.c.	HLA-A0201/Kb	IFN-γ CD8 responses	PruFLUC i.p.	++ For different pools of peptide	+ For different pools of peptide	
(Dziadek *et al.*, 2011)	ROP2, GRA4, SAG1, ROP4	Incomplete Freund's adjuvant	s.c.	C3H/HeN C57BL/6	Proliferation, IFN-γ, IL-2, antibodies	DX cysts i.p.	Not specified	+ ROP2, ROP4, SAG1 ++ ROP2, GRA4, SAG1 ++ ROP2, ROP4, GRA4	
(Lau *et al.*, 2011)	SAG1+SAG2		s.c.	BALB/c	IFN-γ	RH i.p.	++		
(Tan *et al.*, 2011)	NTPases II	ALUM	i.m.	BALB/c	Antibodies Proliferation, IFN-γ, IL-4, IL-2, IL-10	RH i.p.	+	+	

(Wang *et al.*, 2011)	GRA1, GRA4, SAG1	Freund's incomplete adjuvant	i.m.	BALB/c Kunming	Proliferation, IFN-γ, IL-4, IL-2, IL-10,	GJS tachyzoites i.p.	+	++	
(Sánchez *et al.*, 2011)	ROP2, GRA4	CpG	i.m.	C3H/HeN	Antibodies, IFN-γ, IL-4, IL-10	Me49 cysts Oral	Not specified	+	
(Azzouz *et al.*, 2012)	Proteins involved in invasion	ALUM	i.p.	BALB/c	Antibodies, IFN-γ, IL-4, IL-2, IL-10, IL-12, IL-6	RH i.p.	Not specified	Not specified	
(Huang *et al.*, 2012)	Actin depoly-merization factor	Not specified	i.m.	BALB/c		RH i.p.	+	++	
(Cong *et al.*, 2012)	$GRA7_{20\text{-}28}$ peptide	PADRE, GLA-SE	s.c	HLA-B0702	IFN-γ, T-cell proliferation	Pru FLUC i.p.	Not specified	+	
(Da Cunha*et al.*, 2012)	Rhoptry proteins	Quil-A	i.n.	pigs	antibodies	VEG oocysts	Not specified	Not specified	Reduced burden
(Grover *et al.*, 2012)	AS15	BMDCs	f.p.	C57BL/6	IFN-γ, T cell proliferation	Pru tachyzoites i.p.	++	++	

Key for Survival: −− decreased survival; − no difference in survival; + moderate (≤50% increased survival); ++ significant (≥50% increased survival); +++ highly significant (≥90% increased survival), compared with control groups

Key for Parasite Burden: −− increased parasite burden; − no difference in parasite burden; + moderate (≤ 50% decrease in parasite burden); ++significant (≥50% decrease in parasite burden); +++ highly significant (≥90% decrease in parasite burden), compared with control groups

TABLE 26.5 Studies on *Toxoplasma* Nucleic Acid Incorporated into Live Infectious Vectors

Reference	Antigen	Adjuvant/ Carrier	Route Vaccination	Animal Model	Immunology	Challenge	Survival	Parasite Burden
(Cong *et al.*, 2005)	SAG1 + SAG2	CTA2B/ Attenuated Salmonella	i.g. via oral gavage	BALB/c	Antibodies, splenocyte proliferation, CTLs, IFN-γ, IL-2	RH IP	+++ w CTA2B ++ wo CTA2B	
(Caetano *et al.*, 2006)	SAG1, SAG2, SAG3	None/ Adenovirus	s.c.	BALB/c	Antibodies, IFN-γ, IL-4	RH IP P-Br cysts Oral	–	++
(Zhang *et al.*, 2007a)	GRA4	Gold Particles/ Vaccinia	Gene gun in shaved ventral	C57BL/6	Antibodies, splenocyte proliferation IFN-γ, IL-4	PLK/GFP IP	+++	+
(Wang *et al.*, 2007)	ROP2	None/BCG	s.c.	BALB/c	Antibodies, splenocyte proliferation IFN-γ, IL-2,	RH IP	+	
(Gatkowska *et al.*, 2008)	SAG1, GRA1, MAG	*E.coli*	s.c. i.p.	C3H/HeJ	Antibodies	Dx cysts	–	
(Qu *et al.*, 2008)	SAG1	None/ Attenuated Salmonella	Oral	ICR	Antibodies, splenocyte proliferation, IFN-γ, IL-4	RH IP	++	
(Fang *et al.*, 2009)	SAG1	None/ Baculovirus	i.m.	BALBc	Antibodies, splenocyte proliferation, IFN-γ, IL-4, IL-10	RH IP	++	
(Machado *et al.*, 2010)	SAG2	None/ Influenza (boost by Adenovirus)	i.n. boost i.n. or s.c.	BALB/c	Antibodies, IFN-γ, IL-2	P-Br cysts Oral	+++	
(Zhang *et al.*, 2010)	SAG1	*Neospora caninum*	i.p.		Antibodies, IFN-γ	Beverley, 500 bradyzoites	++	

(Mendes *et al.*, 2011)	SAG1, SAG2, SAG3	None/ Adenovirus	s.c.	C57BL/6	Antibodies, T-cell proliferation IFN-γ, IL-4, IL-10	Me49 cysts Oral	+++ SAG1 + SAG2, SAG3	+SAG1 ++ SAG2, SAG3
(Liu *et al.*, 2008)	SAG1	None/ Pseudorabies	i.m.	BALB/c	Antibodies, IFN-γ, IL-2, IL-4, IL-10, CTLs	RH IP	++	
(Shang *et al.*, 2009)	SAG1	None/ Pseudorabies (boost with pVAXSAG1)	i.m.	BALB/c	Antibodies, IFN-γ, IL-2, IL-4, IL-10, CTLs, splenocyte proliferation	RH IP	+	
(Nie *et al.*, 2011)	SAG1, MIC3 SAG1+ MIC3	None/ Pseudorabies	i.m.	BALB/c	Antibodies, T-cell proliferation, IFN-γ, IL-2, IL-10	PRV Ea strain IP	++ MIC3 + SAG1 SAG1 + MIC ++	
(Fang *et al.*, 2012)	SAG1, MIC3 SAG1+ MIC3	None/ Baculovirus	i.m.	BALB/c	Antibodies, splenocyte proliferation, IFN-γ, IL-4, IL-10	RH IP	SAG1+ MIC3 ++	
(Yu *et al.*, 2012)	AMA1	Gold particles/ Adenovirus boost with plasmid	Gene gun into shaved abdomen	C57BL/6	Antibodies, splenocyte proliferation, IFN-γ, IL-4	PLK-GFP IP	+/++	++

– no protection; + moderate (≤50%); ++ significant (≥50%); +++ highly significant (≥90%) protection

1991) and limiting maternal–foetal transmission (Letscher-Bru *et al.*, 2003). Variation in outcomes following vaccination while, in part, dependent on adjuvant employed and, undoubtedly, route of administration is often also dependent on strain and life cycle stage of the parasite used and route of challenge infection. The animal model (species/strain) utilized is also crucial to success. Thus, subcutaneous SAG1 vaccination protected guinea pigs (Haumont *et al.*, 2000) and BALB/c mice (Letscher-Bru *et al.*, 2003) against maternal–foetal transmission but failed to protect CBA/J mice (Letscher-Bru *et al.*, 2003). Subcutaneous vaccination of Fischer rats with GRA2 and GRA5 adjuvanted with FIA also inhibited maternal–foetal transmission (Zenner *et al.*, 1999), but whether these sub-units would also be effective in other murine models or elsewhere requires further investigation. Interestingly, vaccination with the Th2 inducing adjuvant ALUM, incorporating SAG1 (Petersen *et al.*, 1998) or various combinations of ROP2 and GRA4 (Martin *et al.*, 2004), TgPI-1 (Cuppari *et al.*, 2008) or NTPase II (Tan *et al.*, 2011), was able to promote survival of BALB/c mice infected with virulent parasites (Petersen *et al.*, 1998; Tan *et al.*, 2011) along with reduced parasite burdens (Tan *et al.*, 2011) and increase survival in C3H mice (Cuppari *et al.*, 2008) and reduce parasite burdens of C57BL/6 and C3H mice infected with Me49 cysts (Martin *et al.*, 2004; Cuppari *et al.*, 2008). In these studies mixed Type 1/Type 2 responses were induced although the bias was antigen dependent: SAG1, ROP2 and TgPI-1 induced primarily a Th2 response and GRA4 and NTPase II primarily a Type 1 response. ALUM (as one of a few licensed human adjuvants) has been used in humans for over 60 years and although a type 2 adjuvant if it is formulated with type 1 inducers such as CpG (Stacey and Blackwell, 1999) or IL-12 (Pollock *et al.*, 2003) a strong type 1 response can also be induced. In this context *Toxoplasma* immunogens such as cyclophilin (Aliberti *et al.*, 2003) may be ideal co-stimulators in an ALUM formulated vaccine to enhance a potent type 1 response.

It has also been suggested that multivalent vaccines may be more successful than those comprising a single antigen (Martin *et al.*, 2004). Using various combinations of SAG1, ROP2, ROP4 and GRA4 adjuvanted with FIA this appears to have been confirmed by Dziadek *et al.* (2011). Nevertheless, a recent study suggested that a combined ROP2 + GRA4 vaccine adjuvanted with CpG was not as effective as either antigen individually (Sánchez *et al.*, 2011). However, in addition to other experimental differences, the former study utilized an i.p. challenge infection while in the latter study infection was by the oral route. More importantly, targeting more than one life cycle stage immunodominant antigen may further enhance vaccine efficacy. Bradyzoite specific sub-unit antigens or antigens shared between tachyzoites could have potential for vaccination. In this respect, MAG1, originally reported as a bradyzoite specific antigen, was later demonstrated to be secreted from tachyzoites stages (Parmley *et al.*, 2002). Significantly, vaccination with MAG1 adjuvanted with QuilA not only reduced the cyst burden in the brain but also reduced inflammation. As SAG1 is also effective by this route using QuilA, the outcome of vaccination with a combined SAG1/MAG1 vaccine would be extremely interesting. Indeed, as bradyzoites initiate infection in the intestine the outcome of mucosal immunization with this combination linked to cholera toxin, for example, would be intriguing.

26.3.8 Genomics and Immunosense

The sequencing of the three major lineages and a number of atypical strains will soon provide an almost complete supragenome (data available at http://www.toxodb.org). This, with the appropriate predictive algorithms, should, in theory, allow the identification of *T. gondii*, MHC1 binding peptides for any host species. Recent work has taken this 'immunosense' approach to epitope identification as previously successfully used for other vaccines. This approach predicts HLA binding nonamer peptides from pathogen proteins predicted or proven to have vaccine potential. Importantly, these algorithms can be used for multiple HLA class I supermotif haplotypes, HLA-B*0702, HLA-A03 and HLA-A*0201, at least one of which has been found to be present in ~80%–90% of the human population. These peptides are then tested *in vitro* for their ability to elicit IFNγ production from PBLs isolated from humans of the matching HLA with chronic *T. gondii* infection. Peptides that are strong inducers of IFNγ are then tested in HLA transgenic mice for their ability to elicit CD8 IFNγ production and protection. Thus far, using this approach a number of candidate peptides capable of eliciting IFNγ from human PBMCs expressing one of the HLA-B*0702, HLA-A03 and HLA-A*0201 supertypes and capable of protecting mice against challenge with a type II strain of *T. gondii* have been identified. In spite of the use of diverse adjuvanting systems, including conjugation to lipopeptides, or administration with a new TLR-4 binding adjuvant derived from *Salmonella* produced by IDRI/IDC, protection was impressive but not equivalent to that observed following vaccination with the RPS13 conditional knock-out parasites (Cong *et al.*, 2010, 2011, 2012). The ability of a single MHC1 binding peptide to protect mice from challenge has been demonstrated as a proof of principle with the GRA6 HF10 peptide in BALB/c mice, albeit using non-translational *in vitro* peptide pulsed dendritic cells as the delivery mechanism (Blanchard *et al.*, 2008).

26.4 THE RODENT AS A MODEL TO STUDY CONGENITAL DISEASE AND VACCINATION

The success of an attenuated vaccine for preventing *Toxoplasma* induced abortion in sheep is reported in detail above. However, experimental studies are extremely expensive and comparatively difficult to perform and coordinate in domestic animals. Consequently, much effort has been expended in elucidating the immunology of congenital toxoplasmosis in the rodent model and exploring the efficacy of prophylactic therapies. Very early studies on the mouse model indicated that vertical transmission through successive generations was the normal situation in mice (Beverley, 1959) unlike either humans (Cook, 1990) or ovids (Beverley and Watson, 1971) where only a primary infection during pregnancy resulted in congenital infection. These differences would suggest that mice would make poor substitutes for studying immune prophylaxis of human or ovine congenital disease. More recently, vertical disease transmission has also been observed to occur occasionally in humans in the apparent absence of reinfection (Vogel *et al.*, 1996; Kodjikian *et al.*, 2004). Conversely, the authors and others have now demonstrated that vertical disease transmission in mice in the absence of primary infection during pregnancy is, in fact, mouse strain dependent and that the BALB/c mouse is an excellent model of the human disease (Roberts and Alexander, 1992; Roberts *et al.*, 1994; Alexander *et al.*, 1998; Thouvenin *et al.*, 1997; Elsaid *et al.*, 2001; Couper *et al.*, 2003; Letscher-Bru *et al.*, 2003; Abou-Bacar *et al.*, 2004a, 2004b). BALB/c dams previously infected with and recovered from cyst forming strains of *T. gondii,* unlike mice infected for the first time during pregnancy, produced healthy litters even if infected for a second time by the oral route during pregnancy. Similarly, congenital transmission is significantly inhibited following oral challenge with cysts in Swiss outbred OF1 mice dams previously vaccinated with Mic1–3 deficient parasites (Ismael *et al.*, 2006). Thus the mouse, if the appropriate strain is utilized, provides an appropriate model to study the immunology of vertical disease transmission and design appropriate vaccine strategies.

The physiological/immunological environment associated with pregnancy, which favours a type 2 immune response (reviewed in Roberts *et al.*, 1996, 2001), probably promotes a permissive environment for vertical transmission to occur. Indeed, a type 2 response bias in the murine placenta has been demonstrated to facilitate successful implantation, maintenance of early pregnancy and suppression of inflammation: a switch towards a type 1 response can result in foetal death (Krishnan *et al.*, 1996a, 1996b). It is well recorded that humans, as well as mice, develop more severe primary infections during pregnancy (Luft and Remington, 1982; Shirahata *et al.*, 1992, 1993). This is associated with reduced IFNγ levels (Shirahata *et al.*, 1992) and administration of recombinant IFNγ promotes the resistance of pregnant mice against toxoplasmosis. Similarly, the Th1 cytokine IL-2 also promotes resistance against lethal challenge in pregnant mice (Shirahata *et al.*, 1993). Furthermore, IL-4 deficient mice that should have a type 1 bias are more resistant to *T. gondii* infection during gestation than their wild-type counterparts (Alexander *et al.*, 1998; Thouvenin *et al.*, 1997). In addition, although vertical disease transmission from infected dams to pups was not impaired in IL-4 deficient B6/129 (Alexander *et al.*, 1998), there was a 50% decrease in congenital infections in BALB/c IL-4 deficient mice compared with wild-type controls (Thouvenin *et al.*, 1997). Neutralization of IFNγ and depletion of $CD8^+$ T-cells increased the incidence of maternal–foetal transmission in BALB/c mice (Abou-Bacar *et al.*, 2004a) although studies using $RAG2^{-/-}$ mice which lack B- and T-cells also indicated a protective role for NK-cells, although in this study, paradoxically, neutralizing IFNγ inhibited transmission to the foetus in this model (Abou-Bacar *et al.*, 2004b).

As previously infected BALB/c dams are able to totally prevent vertical disease transmission to their pups even when re-infected during pregnancy this model offers a gold standard for testing putative vaccines. Consequently, vaccination of dams with crude soluble tachyzoite antigens, with or without cyst antigens, entrapped in lipid vesicles prior to pregnancy has been demonstrated to limit vertical disease transmission and pup mortality following infection with cysts orally on day 11 of pregnancy associated with enhanced IFNγ production (Roberts *et al.*, 1994; Elsaid *et al.*, 2001). Indeed, a recombinant SAG1 vaccine, although not a SAG1 DNA vaccine (Couper *et al.*, 2003), has been demonstrated to be sufficient to inhibit vertical transmission in BALB/c mice associated with $CD8^+$ T-cells and IFNγ production (Letscher-Bru *et al.*, 2003). While rSAG1, though adjuvanted with a type 1 promoter, also was protective in a congenital guinea pig infection model (Haumont *et al.*, 2000), rSAG1 actually promoted vertical disease transmission in CBA/J mice (Letscher-Bru *et al.*, 2003). Thus the success of a vaccine may be mouse strain or species dependent but how the vaccine is adjuvanted is also crucial to success. For example while lipid vesicle entrapped STAg (Roberts *et al.*, 1994) successfully vaccinated against maternal–foetal transmission, unadjuvanted STAg and FCA adjuvanted STAg increased rates of foetal death, abortion and vertical transmission. FCA, reputably the gold standard type 1 adjuvant, had previously been shown to be counterprotective in a SAG1 vaccine (Kasper *et al.*, 1985), promoting increased death and parasite burdens following challenge infection. However, while vesicular adjuvants such as lipid vesicles and ISCOMS have been demonstrated to induce $CD8^+$ T-cell responses (Debrick *et al.*, 1991; Zhou *et al.*, 1992) there is little evidence that emulsion systems such as FCA are capable of inducing a similar response (Roberts *et al.*, 1994).

These observations, however, do not explain why rSAG1, but not a SAG1 DNA vaccine, induces sufficient protective immunity to prevent maternal–foetal transmission in BALB/c mice as both vaccines induce or should induce $CD8^+$ T-cell responses and IFNγ production (Couper *et al.*, 2003; Letscher-Bru *et al.*, 2003). This could perhaps be a result of the utilization of the different parasite strains, Beverley (Couper *et al.*, 2003) and Me49 (Letscher-Bru *et al.*, 2003), used for challenge although both these are type 2 strains. The protection afforded by SAG1 in the guinea pig study required inclusion of a type 1 inducing immune response adjuvant (Haumont *et al.*, 2000), SABS1, and the challenge involved a type 3 strain which is less virulent than type 2 strains and not often associated with the human disease. However, using the Beverley strain, Roberts *et al.* (1994) did demonstrate significant protection against maternal–foetal transmission with a vaccine comprising a cocktail of soluble tachyzoite antigens suggesting that an approach utilizing more than one vaccine candidate may be a more effective one (Roberts *et al.*, 1994). This has been confirmed by Mevelec *et al.* (2005) who found that a combined SAG1, GRA4 DNA vaccine adjuvanted with plasmid GM–CSF was more effective than vaccines expressing single antigens (Mevelec *et al.*, 2005). Consequently, vaccination of outbred Swiss OF1 dams with this 'cocktail' reduced parasite induced foetal death during pregnancy although it did not limit maternal–foetal transfer. Overall, these reports demonstrate that a type 1 response induced by an appropriately adjuvanted multivalent vaccine would provide the most effective protective immunity against murine congenital toxoplasmosis. The protective activity of vaccines may also be enhanced firstly by introducing bradyzoite antigens to the cocktail (Elsaid *et al.*, 2001) and secondly by inducing mucosal, as well as systemic, responses. Thus, McLeod *et al.* (1988) found that intraintestinal immunization of mice with the temperature sensitive mutant TS-4, but not subcutaneous immunization, was effective in reducing the incidence of congenital disease (McLeod *et al.*, 1988).

The need to immunize using different life cycle stage specific antigens of *T. gondii* has been recently highlighted in the rat model of congenital transmission (Freyre *et al.*, 2006). Previous studies using Fischer (Zenner *et al.*, 1993) and Wistar and Holtzman (Paulino and Vitor, 1999) rats had indicated that rats recovered from a primary infection with *T. gondii* and produced healthy non-infected pups even if dams were re-infected during the gestation period. However, Freyre and colleagues (2006), using Sprague–Dawley rats, have demonstrated that immunization with RH tachyzoites induced only low rates of protection against cyst or oocyst challenge. Furthermore, immunization with cysts provided incomplete protection against oocyst challenge even if the same parasite strain was used. Indeed, complete protection was only demonstrated in cyst immunized rats challenged with cysts of the same strain and complete protection was rarely achieved following challenge with different *Toxoplasma* strains. Conversely, Zenner *et al.* (1999) found that full protection against vertical transmission was irrespective of which *Toxoplasma* strain was used for immunization and which for challenge (Zenner *et al.*, 1999). Differences between these two studies perhaps reflect the different rat strains used in each. However, in the latter study, vaccinating dams with a vaccine comprising excretory/secretory tachyzoite antigens did significantly protect against congenital infection, highlighting the potential importance of the rat model. The hamster may provide a further model to study the effectiveness of vaccination, as it has recently been shown that previous infection almost completely protects against congenital toxoplasmosis initiated by inoculation with cysts or oocysts (Freyre *et al.*, 2012).

26.5 REVIEW OF VACCINES FOR DEFINITIVE HOST (CATS)

Whether or not a cat vaccine is required is a matter of debate. Although cats are the definitive host for *Toxoplasma* they seldom develop disease. A future cat vaccine should therefore not be aimed at protecting the cat from illness, but should prevent oocyst shedding, thereby reducing oocyst contamination of the environment and risk to livestock and/or humans. Cats shed high numbers of oocysts after a first infection and most (but not all) cats remain immune afterwards (Dubey, 1995). Kittens frequently become infected soon after weaning when they start eating prey. Vaccinations should thus be applied as early as possible in kittens of outdoor-roaming cats. Ideally, domestic and stray cats should be vaccinated if one would like to prevent oocyst contamination of food, water and soil. It is clear that this can only be accomplished if nationwide vaccination campaigns are organized, including the vaccination of stray cats. Vaccination of stray cats could be accomplished by a similar approach as with rabies. Foxes were orally immunized by distributing vaccine baits using a vaccinia–rabies glycoprotein recombinant virus, which successfully controlled the disease in Europe and the USA (Pastoret, 2002). Clearly, such a radical vaccination approach for *Toxoplasma* would be most beneficial to areas with high incidences of toxoplasmosis, being mainly countries with poor hygiene conditions, as exemplified in Brazil, where it was shown that contaminated drinking water caused a high incidence of human toxoplasmosis (Bahia-Oliveira *et al.*, 2003). Although postnatal acquired infections are mostly asymptomatic, it has been demonstrated to cause ocular disease (Vallochi *et al.*, 2002; Burnett *et al.*, 1998), making a strong case to control *Toxoplasma*. Apart from extensive nationwide vaccination campaigns, one can also envisage a cat vaccine for domestic cat owners to prevent or reduce the incidence of *Toxoplasma* infections during pregnancy, which may lead to congenitally infected babies or abortion.

A cat vaccine should thus prevent oocyst shedding, which can be induced in cats upon infection with each of the three infectious forms of *Toxoplasma*, being tachyzoites, bradyzoites or sporozoites (Dubey, 1998). The natural and most efficient route of infection of cats is *via* tissue cysts that are present in prey. Released bradyzoites can generate tachyzoites, but also directly initiate the enteroepithelial life cycle (Dubey and Frenkel, 1972). A vaccine should be focused on inhibiting this enteroepithelial cycle to prevent the formation of oocysts and should not be limited to tachyzoites. Currently, no enteroepithelial antigens have been defined that are specific to schizonts, gametocytes or zygotes. A few *Toxoplasma* antigens (including HXGPRT, HSP70 and a 14-3-3 homologue) have been cloned from enteroepithelial stages, but these are not specific to these stages (Koyama *et al.*, 2000). No antigens from enteroepithelial stages have currently been tested as vaccines candidates. However, it may be questioned if a sub-unit vaccine will induce strong enough immunity to prevent or dramatically reduce oocyst shedding. *Eimeria* can serve as an example, where various sub-unit vaccines against coccidiosis in chickens were tested and were shown to reduce oocyst shedding by only 50% (Vermeulen, 1998). This demonstrates the difficulty of making an effective sub-unit vaccine to significantly reduce oocyst shedding. *Eimeria* is a related coccidial parasite with only enteroepithelial stages, and currently all commercial *Eimeria* vaccines contain live sporulated oocysts. These vaccines can reduce oocyst shedding by more than 90% and frequently even 100% (Vermeulen *et al.*, 2001). It is therefore most likely that a live attenuated *Toxoplasma* vaccine will be required to prevent oocyst shedding.

The T-263 mutant *Toxoplasma* strain has been extensively tested as a live vaccine for cats (Frenkel *et al.*, 1991). The T-263 strain is unable to produce oocysts in cats (due to a defect in the sexual stages), but can be propagated as tachyzoites and bradyzoites. Oral vaccination with T-263 bradyzoites prevented oocyst shedding in 84% of kittens following a single dose and was 100% effective following a secondary vaccination (Freyre *et al.*, 1993). Importantly, live tachyzoites from the T-263 strain could not protect cats from oocyst shedding upon challenge (Freyre *et al.*, 1993), indicating that live tachyzoite vaccines are not an option for cats. The effectiveness of the T-263 strain as a cat vaccine was subsequently tested in a field trial on eight commercial swine farms in the USA (Mateus-Pinilla *et al.*, 1999). Over a three year period, during which time cats around the farms were trapped and vaccinated, the number of oocyst shedding cats was reduced and the number of seropositive pigs was reduced. Trapped mice were also found to be seronegative for *Toxoplasma*. Despite this success, T-263 has never been commercialized. A serious drawback for such a vaccine is the production and shelf life. Bradyzoites or tissue cysts are most efficiently produced *in vivo* and cannot be frozen without considerable loss of viability. On the other hand, it is unknown how long tissue cysts remain viable at room temperature and/or at 4°C. Currently, no alternatives to T-263 are available, since attenuated strains such as S48 do not develop into bradyzoites and the TS4 mutant may be too limited to develop *in vivo* (apart from the safety issue for TS4). Interestingly, a T-263 vaccine may be easily integrated into bait for wild animals, by simply infecting bait animals prior to their deployment. As a precaution it should be realized that the use of bait with live *Toxoplasma* parasites can be lethal to some highly sensitive animal species, such as New World monkeys and Australian marsupials (Hill *et al.*, 2005).

More recent studies have used crude rhoptry preparations administered either through the nasal or rectal route with QuilA as an adjuvant (Garcia *et al.*, 2007; Zulpo *et al.*, 2012). Intranasal immunization was able to prevent oocyst shedding in two out of three cats challenged with *T. gondii* (VEG strain) oocysts (Garcia *et al.*, 2007). In a subsequent study that challenged with Me49 cysts, intranasal administration with a similar preparation was found to be superior to intrarectal administration and was also demonstrated to reduce the prepatent period (Zulpo *et al.*, 2012).

In conclusion, the impact of reduced oocyst burden on the risk of transmission to animals and humans justifies the vaccination of the feline definitive host. Although such a target is technically achievable, this would require large investments from authorities and health organizations making vaccination of cats compulsory. It is as yet not expected that this would be given such priority. Until then, hygiene measures remain the only tools to reduce the risk of *Toxoplasma* infection.

26.6 FUTURE STRATEGIES TO DESIGN NEW VACCINES FOR COCCIDIAL PARASITES IN GENERAL AND *TOXOPLASMA GONDII* IN PARTICULAR

It is generally believed that adult acquired infection of humans results in lifelong immunity to *T. gondii*. The evidence would suggest that immunity prevents reactivation of disease from the bradyzoite stage and reinfection by other strains of *T. gondii*. Furthermore, immunity normally prevents congenital transmission from chronically infected individuals even if those individuals are re-exposed to infection. This would suggest that a vaccine is feasible and this should be achieved by mimicking the immune response that occurs during a natural infection.

Immunity to *T. gondii* is complex and involves many facets of the immune system. The innate immune system is important in containing parasite proliferation during the early stages of infection and drives the adaptive immune response (Denkers and Gazzinelli, 1998). $CD4^+$ T-lymphocytes play an important role in shaping the immune responses and provide IL-2 for the development of $CD8^+$ T lymphocytes which produce IFNγ and would appear to be the effectors of long term immunity. Although $CD8^+$

T-lymphocytes from experimentally vaccinated mice are capable of killing *T. gondii* infected cells in an MHC restricted, perforin dependent manner, IFNγ would appear to be the major effector mechanism of long term immunity *in vivo* (Denkers *et al.*, 1997; Wang *et al.*, 2004). Thus, if a vaccine is to mimic natural immunity, it should comprise the proteins or peptides thereof that are capable of being presented on MHC class I. These should be administered in such a manner that facilitates MHC class I processing and development of $CD8^+$ T-lymphocytes. In addition to the immunogen, an appropriate adjuvant is likely to be required. Live attenuated parasites would appear to fulfil many of these requirements. Assuming they maintain their fitness to invade cells, they effectively deliver the complete set of proteins to the class II and class I processing pathways as occurs during the course of natural infection. In addition, *T. gondii* has a number of endogenous adjuvants including cyclophilin 18, which has the ability to bind CCR-5 receptor and elicit IL-12 production (Aliberti *et al.*, 2003), profilin which is a ligand for murine TLR-11 (not present in all mammals) (Yarovinsky *et al.*, 2005) and HSP70 which induces dendritic cell maturation (Kang *et al.*, 2004). Sub-unit vaccines are likely to be critically dependent on adjuvant for effective delivery to the endogenous class I processing pathway. DNA vaccination or the use of viral vectors may represent alternative means of achieving this end.

Not surprisingly, live attenuated vaccines that mimic a natural infection have been extremely successful in inducing protection in murine models of infection. The use of live attenuated parasites for human and coccidial veterinary vaccines is still the best option if solid immunity is required. However, a number of concerns would have to be overcome for a live attenuated *Toxoplasma* vaccine. These would include the potential of the parasite to revert to a virulent phenotype, the ability of these parasites to infect cells of the central nervous system, the possibility of persistence and the potential to cause congenital infection or abortion. However, a live attenuated vaccine (Toxovax) has been used in livestock for some time and has been successful in limiting abortion in sheep. The nature of the defect in this parasite that prevents it from completing a full life cycle is unknown. Consequently, the appeal of a rationally engineered, highly attenuated vaccine with multiple deleted genes conferring multiple auxotrophy or multiple developmental disabilities is obvious.

The completion of the *T. gondii* genome has provided the amino acid sequence for all potentially immunogenic components of this pathogen. As algorithms are refined it shall be possible to make predictions of which predicted peptides are likely to have MHC class I and class II epitopes not only for humans, but also for other species. These peptide predictions will also need to take into account the polymorphisms in MHC molecules as outlined in Section 26.3.8 for humans and the various animal species that it may be desirable to vaccinate. To enable this information to be fully exploited for a sub-unit vaccine, adjuvants will be of key importance.

Alternatively, genome information can be used in combination with mass spectrometry to identify relevant sets of proteins (Mann *et al.*, 2001). For example, surface antigens from different *Toxoplasma* life cycle stages could be isolated, analysed with MALDI-TOF and compared to peptide masses from a database for identification. This may yield interesting new vaccine candidates, in particular from bradyzoite stages and gastrointestinal stages.

Another approach to identify novel vaccine antigens would be to use the full theoretical set of genes from *Toxoplasma* to select a subgroup of promising candidates and have all of them expressed and assessed for their protective capacity. Such a major task has been undertaken with the genome data from Meningococcus (serogroup B), from which 570 surface antigens were identified and tested, resulting in seven new vaccine candidates (Pizza *et al.*, 2000). Since *T. gondii* has a larger genome with 6927 genes currently annotated (http://www.toxodb.org

(02/06)) this approach cannot be performed by a single group but would require a joint effort by multiple teams.

An ideal vaccine would be able to protect against all strains of *T. gondii.* This in theory should not be challenging as *T. gondii* is remarkably clonal with most strains examined, falling into one of five types (Type I–V). However, a number of naturally occurring recombinant strains have been identified in the USA and a number of exotic strains isolated in South America. Natural infection has generally been thought to protect from a secondary infection. However, there is now some evidence that in at least some circumstances secondary infections can occur with separate strains in mice (Araujo *et al.*, 1997; Dao *et al.*, 2001). In addition, it has been demonstrated in a murine system that cytotoxic $CD8^+$ lymphocytes can be parasite strain specific (Johnson *et al.*, 2002). This raises the possibility that a vaccine might not be protective against all strains of *T. gondii.*

Most infections in humans are initiated by bradyzoites or sporozoites, but these stages only persist for a short length of time before giving rise to the tachyzoite form. The tachyzoite multiplies extensively for around 14 days before transforming into bradyzoites that reside inside cyst structures. Each of these stages have stage specifically expressed surface proteins (e.g. Kim and Boothroyd, 2005; Lyons *et al.*, 2002) and secreted proteins (e.g. Reichmann *et al.*, 2002; Meissner *et al.*, 2002a). There is a clear advantage in targeting a vaccine against sporozoite or bradyzoite antigens as this might prevent infection entirely. However, such a strategy may necessitate two components, one targeting sporozoites and one targeting bradyzoites. Most vaccine studies have to date used tachyzoite derived fractions (and in some cases specific components), which, in theory, could be protective in infections initiated by either of the infective stages. However, the ability of the tachyzoite to differentiate into the bradyzoite form with a different antigenic profile provides a means of escaping an immune response as it develops against the tachyzoite stage. Targeting the bradyzoite stage in a vaccine may prevent this means of escape and result in sterile immunity. Such a strategy might not be necessary for a vaccine aimed at preventing congenital transmission as the tachyzoite stage would appear to be responsible for transplacental transmission. Notably, the live attenuated vaccine (S48) used in sheep to prevent congenital transmission consists of tachyzoites. Thus, the choice of antigens may be affected by the intended use of the vaccine, but in reality a vaccine may require multiple components from multiple life cycle stages.

To summarize, there is currently only a commercially available *Toxoplasma* vaccine for sheep and goats, but a vaccine for pigs, cats and humans is still lacking. In recent years, tremendous progress has been made not only in cell and molecular biology of *T. gondii,* but also in characterizing the immunobiology of the parasite in host species. This information allied to current biochemical, molecular and immunological progress must make us optimistic about the likelihood of developing new, safe and successful vaccines for both clinical and veterinary medicine.

References

Abou-Bacar, A., Pfaff, A.W., Georges, S., Letscher-Bru, V., Filisetti, D., Villard, O., Antoni, E., Klein, J.P., Candolfi, E., 2004a. Role of NK cells and gamma interferon in transplacental passage of *Toxoplasma gondii* in a mouse model of primary infection. Infect. Immun. 72, 1397–1401.

Abou-Bacar, A., Pfaff, A.W., Letscher-Bru, V., Filisetti, D., Rajapakse, R., Antoni, E., Villard, O., Klein, J.P., Candolfi, E., 2004b. Role of gamma interferon and T-cells in congenital *Toxoplasma* transmission. Parasite Immunol. 26, 315–318.

Alexander, J., Jebbari, H., Bluethmann, H., Brombacher, F., Roberts, C.W., 1998. The role of IL-4 in adult acquired and congenital toxoplasmosis. Int. J. Parasitol. 28, 113–120.

Alexander, J., Jebbari, H., Bluethmann, H., Satoskar, A., Roberts, C.W., 1996. Immunological control of *Toxoplasma gondii* and appropriate vaccine design. Curr. Top. Microbiol. Immunol. 219, 183–195.

Aliberti, J., Valenzuela, J.G., Carruthers, V.B., Hieny, S., Andersen, J., Charest, H., Reis E Sousa, C., Fairlamb, A., Ribeiro, J.M., Sher, A., 2003. Molecular mimicry of a CCR5 binding-domain in the microbial activation of dendritic cells. Nat. Immunol. 4, 485–490.

Aosai, F., Mun, H.S., Norose, K., Chen, M., Hata, H., Kobayashi, M., Kiuchi, M., Stauss, H.J., Yano, A., 1999. Protective immunity induced by vaccination with SAG1 gene-transfected cells against *Toxoplasma gondii*-infection in mice. Microbiol. Immunol. 43, 87–91.

Angus, C.W., Klivington-Evans, D., Dubey, J.P., Kovacs, J.A., 2000. Immunization with a DNA plasmid encoding the SAG1 (P30) protein of *Toxoplasma gondii* is immunogenic and protective in rodents. J. Infect. Dis. 181, 317–324.

Araujo, F., Slifer, T., Kim, S., 1997. Chronic infection with *Toxoplasma gondii* does not prevent acute disease or colonization of the brain with tissue cysts following reinfection with different strains of the parasite. J. Parasitol. 83, 521–522.

Araujo, F.G., Remington, J.S., 1974. Protection against *Toxoplasma gondii* in mice immunized with *Toxoplasma* cell fractions, RNA and synthetic polyribonucleotides. Immunology 27, 711–721.

Avelino, M.M., Campos Jr., D., Parada, J.B., Castro, A.M., 2004. Risk factors for Toxoplasma gondii infection in women of childbearing age. Braz. J. Infect. Dis. 8, 164–174.

Azzouz, S., Maache, M.O., Suna, A., Lawton, P., Pétavy, A.F., 2012. *Toxoplasma gondii*: identification and immune response against a group of proteins involved in cellular invasion. Exp. Parasitol. 1, 63–68.

Bahia-Oliveira, L.M., Jones, J.L., Azevedo-Silva, J., Alves, C.C., Orefice, F., Addiss, D.G., 2003. Highly endemic, waterborne toxoplasmosis in north Rio de Janeiro state. Brazil. Emerg. Infect. Dis. 9, 55.

Batz, M.B., Hoffmann, S., Morris Jr., J.G., 2012. Ranking the disease burden of 14 pathogens in food sources in the united states using attribution data from outbreak investigations and expert elicitation. J. Food Prot. 75, 1278–1291.

Beauvillain, C., Juste, M.O., Dion, S., Pierre, J., Dimier-Poisson, I., 2009. Exosomes are an effective vaccine against congenital toxoplasmosis in mice. Vaccine 27, 1750–1757.

Beauvillain, C., Ruiz, S., Guiton, R., Bout, D., Dimier-Poisson, I., 2007. A vaccine based on exosomes secreted by a dendritic cell line confers protection against *T. gondii* infection in syngeneic and allogeneic mice. Microbes Infect. 14-15, 1614–1622.

Beghetto, E., Nielsen, H.V., Del Porto, P., Buffolano, W., Guglietta, S., Felici, F., Petersen, E., Gargano, N., 2005. A combination of antigenic regions of *Toxoplasma gondii* microneme proteins induces protective immunity against oral infection with parasite cysts. J. Infect. Dis. 191, 637–645.

Benenson, M.W., Takafuji, E.T., Lemon, S.M., Greenup, R.L., Sulzer, A.J., 1982. Oocyst-transmitted toxoplasmosis associated with ingestion of contaminated water. N. Engl. J. Med. 307, 666.

Beverley, J.K., 1959. Congenital transmission of toxoplasmosis through successive generations of mice. Nature 183, 1348–1349.

Beverley, J.K., 1971. [Clinical aspects, pathogenesis and prevention of animal toxoplasmosis (with special reference to toxoplasmosis abortion in sheep)]. Monatsh. Veterinarmed. 26, 893.

Beverley, J.K., Archer, J.F., Watson, W.A., Fawcett, A.R., 1971. Trial of a killed vaccine in the prevention of ovine abortion due to toxoplasmosis. Br. Vet. J. 127, 529–535.

Beverley, J.K., Watson, W.A., 1971. Prevention of experimental and of naturally occurring ovine abortion due to toxoplasmosis. Vet. Rec. 88, 39–41.

Bezerra, R.A., Carvalho, F.S., Guimarães, L.A., Rocha, D.S., Maciel, B.M., Wenceslau, A.A., Lopes, C.W., Albuquerque, G.R., 2012. Genetic characterization of *Toxoplasma gondii* isolates from pigs intended for human consumption in Brazil. Vet. Parasitol. 189, 153–161.

Bivas-Benita, M., Laloup, M., Versteyhe, S., Dewit, J., De Braekeleer, J., Jongert, E., Borchard, G., 2003. Generation of *Toxoplasma gondii* GRA1 protein and DNA vaccine loaded chitosan particles: preparation, characterization, and preliminary *in vivo* studies. Int. J. Pharm. 266, 17–27.

Blanchard, N., Gonzalez, F., Schaeffer, M., Joncker, N.T., Cheng, T., Shastri, A.J., Robey, E.A., Shastri, N., 2008. Immunodominant, protective response to the parasite *Toxoplasma gondii* requires antigen processing in the endoplasmic reticulum. Nat. Immunol. 8, 937–944.

Bonenfant, C., Dimier-Poisson, I., Velge-Roussel, F., Buzoni-Gatel, D., Del Giudice, G., Rappuoli, R., Bout, D., 2001. Intranasal immunization with SAG1 and nontoxic mutant heat-labile enterotoxins protects mice against *Toxoplasma gondii*. Infect. Immun. 69, 1605–1612.

Bohne, W., Hunter, C.A., White, M.W., Ferguson, D.J., Gross, U., Roos, D.S., 1998. Targeted disruption of the bradyzoite-specific gene BAG1 does not prevent tissue cyst formation in *Toxoplasma gondii*. Mol. Biochem. Parasitol. 92, 291.

Bos, H.J., Smith, J.E., 1993. Development of a live vaccine against ovine toxoplasmosis. In the series analytic: Toxoplasmosis. Proceedings of a workshop held June 28–July 2, 1992, Fontevraud, France.

Bourguin, I., Chardes, T., Bout, D., 1993. Oral immunization with *Toxoplasma gondii* antigens in association with cholera toxin induces enhanced protective and cell-mediated immunity in C57BL/6 mice. Infect. Immun. 61, 2082.

Bowie, W.R., King, A.S., Werker, D.H., Isaac-Renton, J.L., Bell, A., Eng, S.B., Marion, S.A., 1997. Outbreak of toxoplasmosis associated with municipal drinking water. The BC Toxoplasma Investigation Team. Lancet 350, 173.

Brinkmann, V., Remington, J.S., Sharma, S.D., 1993. Vaccination of mice with the protective F3G3 antigen of Toxoplasma gondii activates CD4+ but not CD8+ T-cells and induces *Toxoplasma specific* IgG antibody. Mol. Immunol. 30, 353–358.

Bruna-Romero, O., Gonzalez-Aseguinolaza, G., Hafalla, J.C., Tsuji, M., Nussenzweig, R.S., 2001. Complete, long-lasting protection against malaria of mice primed and boosted with two distinct viral vectors expressing the same plasmodial antigen. Proc. Natl. Acad. Sci. U.S.A. 98, 11491.

Bulow, R., Boothroyd, J.C., 1991. Protection of mice from fatal *Toxoplasma gondii* infection by immunization with p30 antigen in liposomes. J. Immunol. 147, 3496–3500.

Burnett, A.J., Shortt, S.G., Isaac-Renton, J., King, A., Werker, D., Bowie, W.R., 1998. Multiple cases of acquired toxoplasmosis retinitis presenting in an outbreak. Ophthalmology 105, 1032.

Buxton, D., 1993. Toxoplasmosis: the first commercial vaccine. Parasitol. Today 9, 335.

Buxton, D., 1998. Protozoan infections (*Toxoplasma gondii*, *Neospora caninum* and *Sarcocystis* spp.) in sheep and goats: recent advances. Vet. Res. 29, 289.

Buxton, D., Innes, E.A., 1995. A commercial vaccine for ovine toxoplasmosis. Parasitology 110 (Suppl), S11.

Buxton, D., Thomson, K., Maley, S., Wright, S., Bos, H.J., 1991. Vaccination of sheep with a live incomplete strain (S48) of *Toxoplasma gondii* and their immunity to challenge when pregnant. Vet. Rec. 129, 89.

Buxton, D., Uggla, A., Lovgren, K., Thomson, K., Lunden, A., Morein, B., Blewett, D.A., 1989. Trial of a novel experimental *Toxoplasma* iscom vaccine in pregnant sheep. Br. Vet. J. 145, 451.

Caetano, B.C., Bruña-Romero, O., Fux, B., Mendes, E.A., Penido, M.L., Gazzinelli, R.T., 2006. Vaccination with replication-deficient recombinant adenoviruses encoding the main surface antigens of *Toxoplasma gondii* induces immune response and protection against infection in mice. Hum. Gene. Ther. 17, 415–426.

Chardes, T., Bout, D., 1993. Mucosal immune response in toxoplasmosis. Res. Immunol. 144, 57–60.

Chardes, T., Buzoni-Gatel, D., Lepage, A., Bernard, F., Bout, D., 1994. *Toxoplasma gondii* oral infection induces specific cytotoxic CD8 alpha/beta+ Thy-1+ gut intraepithelial lymphocytes, lytic for parasite-infected enterocytes. J. Immunol. 153, 4596–4603.

Chartier, C., Mallereau, M.P., 2001. Vaccinal efficacy of *Toxoplasma gondii* S48 strain tested in an experimental trial in goats. Ann. MÇd. VÇt. 145, 202.

Chatfield, S.N., Charles, I.G., Makoff, A.J., Oxer, M.D., Dougan, G., Pickard, D., Slater, D., Fairweather, N.F., 1992. Use of the nirB promoter to direct the stable expression of heterologous antigens in *Salmonella* oral vaccine strains: development of a single-dose oral tetanus vaccine. Biotechnology (N.Y.) 10, 888.

Chen, G., Chen, H., Guo, H., Zheng, H., 2002. Protective effect of DNA-mediated immunization with a combination of SAG1 and IL-2 gene adjuvant against infection of *Toxoplasma gondii* in mice. Chin. Med. J. (Engl.) 115, 1448–1452.

Chen, H., Langer, R., 1998. Oral particulate delivery: status and future trends. Adv. Drug Deliv. Rev. 34, 339–350.

Chen, H., Chen, G., Zheng, H., Guo, H., 2003. Induction of immune responses in mice by vaccination with Liposome-entrapped DNA complexes encoding *Toxoplasma gondii* SAG1 and ROP1 genes. Chin. Med. J. 116, 1561–1566.

Chen, G., Guo, H., Lu, F., Zheng, H., 2001. Construction of a recombinant plasmid harbouring the rhoptry protein 1 gene of *Toxoplasma gondii* and preliminary observations on DNA immunity. Chin. Med. J. 114, 837–840.

Chen, R., Lu, S.H., Tong, Q.B., Lou, D., Shi, D.Y., Jia, B.B., Huang, G.P., Wang, J.F., 2009. Protective effect of DNA-mediated immunization with liposome-encapsulated GRA4 against infection of *Toxoplasma gondii*. J. Zhejiang Univ. Sci. B. 7, 512–521.

Cong, H., Gu, Q.M., Yin, H.E., Wang, J.W., Zhao, Q.L., Zhou, H.Y., Li, Y., Zhang, J.Q., 2008. Multi-epitope DNA vaccine linked to the A2/B subunit of cholera toxin protect mice against Toxoplasma gondii. Vaccine 26 (31), 3913–21.

Cong, H., Mui, E.J., Witola, W.H., Sidney, J., Alexander, J., Sette, A., Maewal, A., El Bissati, K., Zhou, Y., Suzuki, Y., Lee, D., Woods, S., Sommerville, C., Henriquez, F.L., Roberts, C.W., McLeod, R., 2012. *Toxoplasma gondii* HLA-B*0702-restricted GRA7(20-28) peptide with adjuvants and a universal helper T-cell epitope elicits CD8(+) T-cells producing interferon-γ and reduces parasite burden in HLA-B*0702 mice. Hum. Immunol. 73, 1–10.

Cong, H., Mui, E.J., Witola, W.H., Sidney, J., Alexander, J., Sette, A., Maewal, A., McLeod, R., 2011. Towards an immunosense vaccine to prevent toxoplasmosis: protective *Toxoplasma gondii* epitopes restricted by HLA-A*0201. Vaccine 29, 754–762.

Cong, H., Mui, E.J., Witola, W.H., Sidney, J., Alexander, J., Sette, A., Maewal, A., Mcleod, R., 2010. Human immunome, bioinformatic analyses using HLA supermotifs and the parasite genome, binding assays, studies of human T-cell responses, and immunization of HLA-A*1101 transgenic mice including novel adjuvants provide a foundation for HLA-A03 restricted CD8+T cell epitope based, adjuvanted vaccine protective against *Toxoplasma gondii*. Immunome. Res. 6, 12.

Cong, H., Gu, Q.M., Jiang, Y., He, S.Y., Zhou, H.Y., Yang, T.T., Li, Y., Zhao, Q.L., 2005. Oral immunization with a live recombinant attenuated *Salmonella typhimurium* protects mice against *Toxoplasma gondii*. Parasite Immunol. 27, 29–35.

Cook, G.C., 1990. *Toxoplasma gondii* infection: a potential danger to the unborn foetus and AIDS sufferer. Q. J. Med. 74, 3–19.

Costa, G.H., Da Costa, A.J., Lopes, W.D., Bresciani, K.D., Dos Santos, T.R., Esper, C.R., Santana, A.E., 2011. *Toxoplasma gondii*: infection natural congenital in cattle and an experimental inoculation of gestating cows with oocysts. Exp. Parasitol. 127, 277–281.

Couper, K.N., Nielsen, H.V., Petersen, E., Roberts, F., Roberts, C.W., Alexander, J., 2003. DNA vaccination with the immunodominant tachyzoite surface antigen (SAG-1) protects against adult acquired *Toxoplasma gondii* infection but does not prevent maternal–foetal transmission. Vaccine 21, 2813–2820.

Couvreur, J., Thulliez, P., 1996. Acquired toxoplasmosis of ocular or neurologic site: 49 cases. Presse. Med. 25, 438–442.

Cutchins, E.C., Warren, J., 1956. Immunity patterns in the guinea pig following *Toxoplasma* infection and vaccination with killed *Toxoplasma*. Am. J. Trop. Med. Hyg. 5, 197–209.

Cui, X., Lei, T., Yang, D., Hao, P., Li, B., Liu, Q., 2012. *Toxoplasma gondii* immune mapped protein-1 (TgIMP1) is a novel vaccine candidate against toxoplasmosis. Vaccine 30, 2282–2287.

Cui, Y.L., He, S.Y., Xue, M.F., Zhang, J., Wang, H.X., Yao, Y., 2008. Protective effect of a multiantigenic DNA vaccine against *Toxoplasma gondii* with co-delivery of IL-12 in mice. Parasite Immunol. 30, 309–313.

Cuppari, A.F., Sanchez, V., Ledesma, B., Frank, F.M., Goldman, A., Angel, S.O., Martin, V., 2008. *Toxoplasma gondii* protease inhibitor-1 (TgPI-1) is a novel vaccine candidate against toxoplasmosis. Vaccine 26, 5040–5045.

Da Cunha, I.A., Zulpo, D.L., Bogado, A.L., De Barros, L.D., Taroda, A., Igarashi, M., Navarro, I.T., Garcia, J.L., 2012. Humoral and cellular immune responses in pigs immunized intranasally with crude rhoptry proteins of *Toxoplasma gondii* plus QuIL-A. Vet. Parasitol. 186, 216–221.

Daryani, A., Hosseini, A.Z., Dalimi, A., 2003. Immune responses against excreted/secreted antigens of *Toxoplasma gondii* tachyzoites in the murine model. Vet Parasitol 113 (2), 123–134.

Dimier-Poisson, I., Aline, F., Bout, D., Mévélec, M.N., 2006. Induction of protective immunity against toxoplasmosis in mice by immunization with *Toxoplasma gondii* RNA. Vaccine 24, 1705–1709.

Dziadek, B., Gatkowska, J., Grzybowski, M., Dziadek, J., Dzitko, K., Dlugonska, H., 2012. *Toxoplasma gondii*: the vaccine potential of three trivalent antigen-cocktails composed of recombinant ROP2, ROP4, GRA4 and SAG1 proteins against chronic toxoplasmosis in BALB/c mice. Exp. Parasitol. 131, 133–138.

Dziadek, B., Gatkowska, J., Brzostek, A., Dziadek, J., Dzitko, K., Grzybowski, M., Dlugonska, H., 2011. Evaluation of three recombinant multi-antigenic vaccines composed of surface and secretory antigens of *Toxoplasma gondii* in murine models of experimental toxoplasmosis. Vaccine 29, 821–830.

Dziadek, B., Gatkowska, J., Brzostek, A., Dziadek, J., Dzitko, K., Dlugonska, H., 2009. *Toxoplasma gondii*: the immunogenic and protective efficacy of recombinant ROP2 and ROP4 rhoptry proteins in murine experimental toxoplasmosis. Exp. Parasitol. 123, 81–89.

Dao, A., Fortier, B., Soete, M., Plenat, F., Dubremetz, J.F., 2001. Successful reinfection of chronically infected mice by a different *Toxoplasma gondii* genotype. Int. J. Parasitol. 31, 63–65.

Dautu, G., Munyaka, B., Carmen, G., Zhang, G., Omata, Y., Xuenan, X., Igarashi, M., 2007. *Toxoplasma gondii*: DNA vaccination with genes encoding antigens MIC2, M2ap, AMA1 and BAG1 and evaluation of their immunogenic potential. Exp. Parasitol. Jul. 116 (3), 273–282.

Darcy, F., Maes, P., Gras-Masse, H., Auriault, C., Bossus, M., Deslee, D., Godard, I., Cesbron, M.F., Tartar, A., Capron, A., 1992. Protection of mice and nude rats against toxoplasmosis by a multiple antigenic peptide construction derived from *Toxoplasma gondii* P30 antigen. J. Immunol. 149, 3636–3641.

Darji, A., Guzman, C.A., Gerstel, B., Wachholz, P., Timmis, K.N., Wehland, J., Chakraborty, T., Weiss, S., 1997. Oral somatic transgene vaccination using attenuated *S. typhimurium*. Cell 91, 765.

Davies, P.R., Morrow, W.E., Deen, J., Gamble, H.R., Patton, S., 1998. Seroprevalence of *Toxoplasma gondii* and *Trichinella spiralis* in finishing swine raised in different production systems in North Carolina, USA. Prev. Vet. Med. 36, 67.

Debard, N., Buzoni-Gatel, D., Bout, D., 1996. Intranasal immunization with SAG1 protein of *Toxoplasma gondii* in association with cholera toxin dramatically reduces development of cerebral cysts after oral infection. Infect. Immun. 64, 2158–2166.

Debrick, J.E., Campbell, P.A., Staerz, U.D., 1991. Macrophages as accessory cells for class I MHC-restricted immune responses. J. Immunol. 147, 2846–2851.

Denkers, E.Y., Gazzinelli, R.T., 1998. Regulation and function of T-cell-mediated immunity during *Toxoplasma gondii* infection. Clin. Microbiol. Rev. 11, 569–588.

Denkers, E.Y., Yap, G., Scharton-Kersten, T., Charest, H., Butcher, B.A., Caspar, P., Heiny, S., Sher, A., 1997. Perforin-mediated cytolysis plays a limited role in host resistance to *Toxoplasma gondii*. J. Immunol. 159, 1903–1908.

Desmonts, G., Couvreur, J., 1974. Toxoplasmosis in pregnancy and its transmission to the foetus. Bull. N. Y. Acad. Med. 50, 146–159.

Desolme, B., Mevelec, M.N., Buzoni-Gatel, D., Bout, D., 2000. Induction of protective immunity against toxoplasmosis in mice by DNA immunization with a plasmid encoding *Toxoplasma gondii* GRA4 gene. Vaccine 18, 2512–2521.

Dimier-Poisson, I., Aline, F., Bout, D., Mevelec, M.N., 2006. Induction of protective immunity against toxoplasmosis in mice by immunization with *Toxoplasma gondii* RNA. Vaccine 18, 1705–1709.

Drabner, B., Guzman, C.A., 2001. Elicitation of predictable immune responses by using live bacterial vectors. Biomol. Eng. 17, 75.

Dubey, J.P., 1986. A review of toxoplasmosis in pigs. Vet. Parasitol. 19, 181.

Dubey, J.P., 2009. Toxoplasmosis in pigs – The last 20 years. Vet. Parasitol. 164, 89.

Dubey, J.P., 1990a. Status of toxoplasmosis in cattle in the United States. J. Am. Vet. Med. Assoc. 196, 257.

Dubey, J.P., 1995. Duration of immunity to shedding of *Toxoplasma gondii* oocysts by cats. J. Parasitol. 81, 410.

Dubey, J.P., 1996. Strategies to reduce transmission of *Toxoplasma gondii* to animals and humans. Vet. Parasitol. 64, 65.

Dubey, J.P., 1998. Advances in the life cycle of *Toxoplasma gondii*. Int. J. Parasitol. 28, 1019.

Dubey, J.P., Baker, D.G., Davis, S.W., Urban Jr., J.F., Shen, S.K., 1994. Persistence of immunity to toxoplasmosis in pigs vaccinated with a nonpersistent strain of *Toxoplasma gondii*. Am. J. Vet. Res. 55, 982.

Dubey, J.P., Frenkel, J.K., 1972. Cyst-induced toxoplasmosis in cats. J. Protozool. 19, 155.

Dubey, J.P., Lindsay, D.S., Saville, W.J., Reed, S.M., Granstrom, D.E., Speer, C.A., 2001. A review of *Sarcocystis neurona* and equine protozoal myeloencephalitis (EPM). Vet. Parasitol. 95, 89.

Dubey, J.P., Ruff, M.D., Camargo, M.E., Shen, S.K., Wilkins, G.L., Kwok, O.C., Thulliez, P., 1993. Serologic and parasitologic responses of domestic chickens after oral inoculation with *Toxoplasma gondii* oocysts. Am. J. Vet. Res. 54, 1668.

Dubey, J.P., Saville, W.J., Stanek, J.F., Reed, S.M., 2002. Prevalence of *Toxoplasma gondii* antibodies in domestic cats from rural Ohio. J. Parasitol. 88, 802.

Dubey, J.P., Urban Jr., J.F., 1990. Diagnosis of transplacentally induced toxoplasmosis in pigs. Am. J. Vet. Res. 51, 1295–1299.

Dubey, J.P., Hill, D.E., Rozeboom, D.W., Rajendran, C., Choudhary, S., Ferreira, L.R., Kwok, O.C., Su, C., 2012. High prevalence and genotypes of Toxoplasma gondii isolated from organic pigs in northern USA. Vet. Parasitol. 2012 (188), 14–28.

Duquesne, V., Auriault, C., Darcy, F., Decavel, J.P., Capron, A., 1990. Protection of nude rats against *Toxoplasma* infection by excreted-secreted antigen-specific helper T-cells. Infect. Immun. 58, 2120–2126.

Duquesne, V., Auriault, C., Gras-Masse, H., Boutillon, C., Darcy, F., Cesbron-Delauw, M.F., Tartar, A., Capron, A., 1991. Identification of T-cell epitopes within a 23-kD antigen (P24) of *Toxoplasma gondii*. Clin. Exp. Immunol. 84, 527–534.

Dzierszinski, F., Mortuaire, M., Cesbron-Delauw, M.F., Tomavo, S., 2000. Targeted disruption of the glycosylphosphatidylinositol-anchored surface antigen SAG3 gene in *Toxoplasma gondii* decreases host cell adhesion and drastically reduces virulence in mice. Mol. Microbiol. 37, 574.

Echeverria, P.C., De Miguel, N., Costas, M., Angel, S.O., 2006. Potent antigen-specific immunity to *Toxoplasma gondii* in adjuvant-free vaccination system using Rop2-*Leishmania infantum* Hsp83 fusion protein. Vaccine 24, 4102–4110.

Eissa, M., El-Azzouni, M.Z., Mady, R.F., Fathy, F.M., Baddour, N.M., 2012. Initial characterization of an autoclaved *Toxoplasma* vaccine in mice. Exp. Parasitol. 131, 310–316.

El-Malky, M., Shaohong, L., Kumagai, T., Yabu, Y., Noureldin, M.S., Saudy, N., Maruyama, H., Ohta, N., 2005. Protective effect of vaccination with *Toxoplasma* lysate antigen and CpG as an adjuvant against Toxoplasma gondii in susceptible C57BL/6 mice. Microbiol. Immunol. 49, 639–646.

Elsaid, M.M., Vitor, R.W., Frézard, F.J., Martins, M.S., 1999. Protection against toxoplasmosis in mice immunized with different antigens of *Toxoplasma gondii* incorporated into liposomes. Mem. Inst. Oswaldo Cruz. 94, 485–490.

Elsaid, M.M., Martins, M.S., Frezard, F., Braga, E.M., Vitor, R.W., 2001. Vertical toxoplasmosis in a murine model. Protection after immunization with antigens of Toxoplasma gondii incorporated into liposomes. Mem. Inst. Oswaldo Cruz. 96, 99.

Fachado, A., Rodriguez, A., Angel, S.O., Pinto, D.C., Vila, I., Acosta, A., Amendoeira, R.R., Lannes-Vieira, J., 2003. Protective effect of a naked DNA vaccine cocktail against lethal toxoplasmosis in mice. Vaccine 21, 1327–1335.

Fachado, A., Rodriguez, A., Molina, J., Silverio, J.C., Marino, A.P., Pinto, L.M., Angel, S.O., Infante, J.F., Traub-Cseko, Y., Amendoeira, R.R., Lannes-Vieira, J., 2003. Long-term protective immune response elicited by vaccination with an expression genomic library of *Toxoplasma gondii*. Infect. Immun. 71, 5407–5411.

Fachado, A., Rodriguez, A., Angel, S.O., Pinto, D.C., Vila, I., Acosta, A., Amendoeira, R.R., Lannes-Vieira, J., 2003b. Protective effect of a naked DNA vaccine cocktail against lethal toxoplasmosis in mice. Vaccine 21, 1327–1333.

Fang, R., Feng, H., Hu, M., Khan, M.K., Wang, L., Zhou, Y., Zhao, J., 2012. Evaluation of immune responses induced by SAG1 and MIC3 vaccine cocktails against *Toxoplasma gondii*. Vet. Parasitol. 187, 140–146.

Fang, R., Feng, H., Nie, H., Wang, L., Tu, P., Song, Q., Zhou, Y., Zhao, J., 2010. Construction and immunogenicity of pseudotype baculovirus expressing *Toxoplasma gondii* SAG1 protein in BALB/c mice model. Vaccine 28, 1803–1807.

Fang, R., Nie, H., Wang, Z., Tu, P., Zhou, D., Wang, L., He, L., Zhou, Y., Zhao, J., 2009. Protective immune response in BALB/c mice induced by a suicidal DNA vaccine of the MIC3 gene of *Toxoplasma gondii*. Vet. Parasitol. 164, 134–140.

Fermin, Z., Bout, D., Ricciardi-Castagnoli, P., Hoebeke, J., 1999. Salbutamol as an adjuvant for nasal vaccination. Vaccine 17, 1936–1941.

Fox, B.A., Bzik, D.J., 2002. De novo pyrimidine biosynthesis is required for virulence of *Toxoplasma gondii*. Nature 415, 926.

Fox, B.A., Bzik, D.J., 2010. Avirulent uracil auxotrophs based on disruption of orotidine-5′-monophosphate decarboxylase elicit protective immunity to *Toxoplasma gondii*. Infect. Immun. 9, 3744–3752.

Frenkel, J.K., 1990. Transmission of toxoplasmosis and the role of immunity in limiting transmission and illness. J. Am. Vet. Med. Assoc. 196, 233.

Frenkel, J.K., Pfefferkorn, E.R., Smith, D.D., Fishback, J.L., 1991. Prospective vaccine prepared from a new mutant of *Toxoplasma gondii* for use in cats. Am. J. Vet. Res. 52, 759.

Freyre, A., Choromanski, L., Fishback, J.L., Popiel, I., 1993. Immunization of cats with tissue cysts, bradyzoites, and tachyzoites of the T-263 strain of *Toxoplasma gondii*. J. Parasitol. 79, 716.

Freyre, A., Falcon, J., Mendez, J., Rodriguez, A., Correa, L., Gonzalez, M., 2006. *Toxoplasma gondii*: partial cross-protection among several strains of the parasite against congenital transmission in a rat model. Exp. Parasitol. 112, 8–12.

Freyre, A., Araujo, F.A., Fialho, C.G., Bigatti, L.E., Falcón, J.D., 2012. Protection in a hamster model of congenital toxoplasmosis. Vet. Parasitol. 183, 359–363.

Gallichan, W.S., Rosenthal, K.L., 1996. Long-lived cytotoxic T lymphocyte memory in mucosal tissues after mucosal but not systemic immunization. J. Exp. Med. 184, 1879–1890.

Garcia, J.L., Gennari, S.M., Navarro, I.T., Machado, R.Z., Sinhorini, I.L., Freire, R.L., Marana, E.R., Tsutsui, V., Contente, A.P., Begale, L.P., 2005. Partial protection against tissue cysts formation in pigs vaccinated with crude rhoptry proteins of *Toxoplasma gondii*. Vet. Parasitol. 129, 209–217.

Garcia, J.L., Navarro, I.T., Biazzono, L., Freire, R.L., Da Silva Guimarães Junior, J., Cryssafidis, A.L., Bugni, F.M., Da Cunha, I.A., Hamada, F.N., Dias, R.C., 2007. Protective activity against oocyst shedding in cats vaccinated with crude rhoptry proteins of the *Toxoplasma gondii* by the intranasal route. Vet. Parasitol. 145, 197–206.

Gatkowska, J., Gasior, A., Kur, J., Dlugonska, H., 2008. *Toxoplasma gondii*: chimeric Dr fimbriae as a recombinant vaccine against toxoplasmosis. Exp. Parasitol. 118, 266–270.

Gazzinelli, R.T., Caetano, B., Fux, B., Garcia, M., Machado, A., Ferreira, A.M., Vitor, R.W.A., Melo, M.N., Bruna-Romero, O., 2005. Protective immune responses involved in host resistance to Brazilian isolates of *Toxoplasma gondii*: implications on vaccine development employing recombinant viral vectors. Proceedings of the IXth International Coccidiosis Conference, 84–90.

Gilbert, R., Gras, L., 2003. Effect of timing and type of treatment on the risk of mother to child transmission of *Toxoplasma gondii*. Bjog 110, 112–120.

Golkar, M., Shokrgozar, M.A., Rafati, S., Musset, K., Assmar, M., Sadaie, R., Cesbron-Delauw, M.F., Mercier, C., 2007. Evaluation of protective effect of recombinant dense granule antigens GRA2 and GRA6 formulated in monophosphoryl lipid A (MPL) adjuvant against *Toxoplasma* chronic infection in mice. Vaccine 25, 4301–4311.

Graham, F.L., Smiley, J., Russell, W.C., Nairn, R., 1977. Characteristics of a human cell line transformed by DNA from human adenovirus type 5. J. Gen. Virol. 36, 59.

Gregoriadis, G., Bacon, A., Caparros-Wanderley, W., McCormack, B., 2002. A role for liposomes in genetic vaccination. Vaccine 20 (Suppl. 5), B1–B9.

Grigg, M.E., Ganatra, J., Boothroyd, J.C., Margolis, T.P., 2001. Unusual abundance of atypical strains associated with human ocular toxoplasmosis. J. Infect. Dis. 184, 633–639.

Grover, H.S., Blanchard, N., Gonzalez, F., Chan, S., Robey, E.A., Shastri, N., 2012. The *Toxoplasma gondii* peptide AS15 elicits CD4 T-cells that can control parasite burden. Infect. Immun. 9, 3279–3288.

Hartley, W.J., 1966. Some investigations into the epidemiology of ovine toxoplasmosis. N. Z. Vet. J. 14, 106–107.

Haumont, M., Delhaye, L., Garcia, L., Jurado, M., Mazzu, P., Daminet, V., Verlant, V., Bollen, A., Biemans, R., Jacquet, A., 2000. Protective immunity against congenital toxoplasmosis with recombinant SAG1 protein in a guinea pig model. Infect. Immun. 68, 4948–4953.

Hedhli, D., Dimier-Poisson, I., Judge, J.W., Rosenberg, B., Mévélec, M.N., 2009. Protective immunity against *Toxoplasma* challenge in mice by coadministration of *T. gondii* antigens and Eimeria profilin-like protein as an adjuvant. Vaccine 27, 2274–2281.

Henriksen, P., Dietz, H.H., Henriksen, S.A., 1994. Fatal toxoplasmosis in five cats. Vet. Parasitol. 55, 15.

Hill, D.E., Chirukandoth, S., Dubey, J.P., 2005. Biology and epidemiology of Toxoplasma gondii in man and animals. Anim. Health Res. Rev. 6, 41.

Hill, D.E., Dubey, J.P., 2012. *Toxoplasma gondii* prevalence in farm animals in the United States. Int. J. Parasitol. (in press).

Hiramoto, R.M., Galisteo, A.J., Do, N.N., De Jr., A.H.F., 2002. 200 Gy sterilised *Toxoplasma gondii* tachyzoites maintain metabolic functions and mammalian cell invasion, eliciting cellular immunity and cytokine response similar to natural infection in mice. Vaccine 20, 2072.

Hiszczyńska-Sawicka, E., Li, H., Boy, U., Xu, J., Akhtar, M., Holec-Gasior, L., Kur, J., Bickerstaffe, R., Stankiewicz, M., 2012. Induction of immune responses in sheep by vaccination with liposome-entrapped DNA complexes encoding *Toxoplasma gondii* MIC3 gene. Pol. J. Vet. Sci. 15, 3–9.

Hiszczyńska-Sawicka, E., Oledzka, G., Holec-Gasior, L., Li, H., Xu, J.B., Sedcole, R., Kur, J., Bickerstaffe, R., Stankiewicz, M., 2011a. Evaluation of immune responses in sheep induced by DNA immunization with genes encoding GRA1, GRA4, GRA6 and GRA7 antigens of *Toxoplasma gondii*. Vet. Parasitol. 177, 281–289.

Hiszczyńska-Sawicka, E., Li, H., Xu, J.B., Holec-Gasior, L., Kur, J., Sedcole, R., Bickerstaffe, R., Stankiewicz, M., 2011b. Modulation of immune response to *Toxoplasma gondii* in sheep by immunization with a DNA vaccine encoding ROP1 antigen as a fusion protein with ovine CD154. Vet. Parasitol. 183, 72–78.

Hiszczyńska-Sawicka, E., Akhtar, M., Kay, G.W., Holec-Gasior, L., Bickerstaffe, R., Kur, J., Stankiewicz, M., 2010a. The immune responses of sheep after DNA immunization with, *Toxoplasma gondii* MAG1 antigen-with and without co-expression of ovine interleukin 6. Vet. Immunol. Immunopathol. 136, 324–329.

Hiszczyńska-Sawicka, E., Li, H., Xu, J.B., Oledzka, G., Kur, J., Bickerstaffe, R., Stankiewicz, M., 2010b. Comparison of immune response in sheep immunized with DNA vaccine encoding *Toxoplasma gondii* GRA7 antigen in different adjuvant formulations. Exp. Parasitol. 124, 365–372.

Hoffmann, S., Batz, M.B., Morris Jr., J.G., 2012. Annual cost of illness and quality-adjusted life year losses in the United States due to 14 foodborne pathogens. J. Food Prot. 75, 1292–1302.

Hoseinian-Khosroshahi, K., Ghaffarifar, F., D'souza, S., Sharifi, Z., Dalimi, A., 2011. Evaluation of the immune response induced by DNA vaccine cocktail expressing complete SAG1 and ROP2 genes against toxoplasmosis. Vaccine 29, 778–783.

Huang, X., Li, J., Zhang, G., Gong, P., Yang, J., Zhang, X., 2012. *Toxoplasma gondii*: protective immunity against toxoplasmosis with recombinant actin depolymerizing factor protein in BALB/c mice. Exp. Parasitol. 130, 218–222.

Hutson, S.L., Mui, E., Kinsley, K., Witola, W.H., Behnke, M.S., El Bissati, K., Muench, S.P., Rohrman, B., Liu, S.R., Wollmann, R., Ogata, Y., Sarkeshik, A., Yates Jr, 3rd, McLeod, R., 2010. *T. gondii* RP promoters and knockdown reveal molecular pathways associated with proliferation and cell-cycle arrest. PLoS One 5, e14057.

Igarashi, M., Zulpo, D.L., Cunha, I.A., Barros, L.D., Pereira, V.F., Taroda, A., Navarro, I.T., Vidotto, O., Vidotto, M.C., Jenkins, M.C., Garcia, J.L., 2010. *Toxoplasma gondii*: humoral and cellular immune response of BALB/c mice immunized via intranasal route with rTgROP2. Rev. Bras. Parasitol. Vet. 19, 210–216.

Igarashi, M., Kano, F., Tamekuni, K., Machado, R.Z., Navarro, I.T., Vidotto, O., Vidotto, M.C., Garcia, J.L., 2008. *Toxoplasma gondii*: evaluation of an intranasal vaccine using recombinant proteins against brain cyst formation in BALB/c mice. Exp. Parasitol. 118, 386–392.

Ismael, A.B., Sekkai, D., Collin, C., Bout, D., Mevelec, M.N., 2003. The MIC3 gene of *Toxoplasma gondii* is a novel potent vaccine candidate against toxoplasmosis. Infect. Immun. 71, 6222–6228.

Ismael, A.B., Hedhli, D., Cérède, O., Lebrun, M., Dimier-Poisson, I., Mévélec, M.N., 2009. Further analysis of protection induced by the MIC3 DNA vaccine against *T. gondii*: CD4 and CD8 T-cells are the major effectors of the MIC3 DNA vaccine-induced protection, both Lectin-like and EGF-like domains of MIC3 conferred protection. Vaccine 27, 2959–2966.

Ismael, A.B., Dimier-Poisson, I., Lebrun, M., Dubremetz, J.F., Bout, D., Mevelec, M.N., 2006. Mic1-3 knock-out of *Toxoplasma gondii* is a successful vaccine against chronic and congenital toxoplasmosis in mice. J. Infect. Dis. 194, 1176–1183.

Jacobs, L., 1956. Propogation, morphology and biology of *Toxoplasma*. Annal of the New York Academy of Science 64, 154–159.

Johnson, J.J., Roberts, C.W., Pope, C., Roberts, F., Kirisits, M.J., Estes, R., Mui, E., Krieger, T., Brown, C.R., Forman, J., McLeod, R., 2002. *In vitro* correlates of Ld-restricted resistance to toxoplasmic encephalitis and their critical dependence on parasite strain. J. Immunol. 169, 966–973.

Jongert, E., Verhelst, D., Abady, M., Petersen, E., Gargano, N., 2008a. Protective Th1 immune responses against chronic toxoplasmosis induced by a protein-protein vaccine combination but not by its DNA-protein counterpart. Vaccine 26, 5289–5295.

Jongert, E., Melkebeek, V., De Craeye, S., Dewit, J., Verhelst, D., Cox, E., 2008b. An enhanced GRA1-GRA7 cocktail DNA vaccine primes anti-*Toxoplasma* immune responses in pigs. Vaccine 26, 1025–1031.

Jongert, E., De Craeye, S., Dewit, J., Huygen, K., 2007. GRA7 provides protective immunity in cocktail DNA vaccines against *Toxoplasma gondii*. Parasite Immunol. 29, 445–453.

Kang, H.K., Lee, H.Y., Lee, Y.N., Jo, E.J., Kim, J.I., Aosai, F., Yano, A., Kwak, J.Y., Bae, Y.S., 2004. *Toxoplasma gondii*-derived heat shock protein 70 stimulates the maturation of human monocyte-derived dendritic cells. Biochem. Biophys. Res. Commun. 322, 899–904.

Kasper, L., Courret, N., Darche, S., Luangsay, S., Mennechet, F., Minns, L., Rachinel, N., Ronet, C., Buzoni-Gatel, D., 2004. *Toxoplasma gondii* and mucosal immunity. Int. J. Parasitol. 34, 401.

Kasper, L.H., 1989. Identification of stage-specific antigens of *Toxoplasma gondii*. Infect. Immun. 57, 668–672.

Kasper, L.H., Currie, K.M., Bradley, M.S., 1985. An unexpected response to vaccination with a purified major membrane tachyzoite antigen (P30) of *Toxoplasma gondii*. J. Immunol. 134, 3426–3431.

Kasper, L.H., Khan, I.A., 1993. Role of P30 in host immunity and pathogenesis of T. gondii infection. Res. Immunol. 144, 45.

Khan, I.A., Ely, K.H., Kasper, L.H., 1991. A purified parasite antigen (p30) mediates CD8+ T-cell immunity against fatal *Toxoplasma gondii* infection in mice. J. Immunol. 147, 3501.

Khosroshahi, K.H., Ghaffarifar, F., Sharifi, Z., D'Souza, S., Dalimi, A., Hassan, Z.M., Khoshzaban, F., 2012. Comparing the effect of IL-12 genetic adjuvant and alum non-genetic adjuvant on the efficiency of the cocktail DNA vaccine containing plasmids encoding SAG-1 and ROP-2 of *Toxoplasma gondii*. Parasitol. Res. 111, 403–411.

Kijlstra, A., Eissen, O.A., Cornelissen, J., Munniksma, K., Eijck, I., Kortbeek, T., 2004. *Toxoplasma gondii* infection in animal-friendly pig production systems. Invest Ophthalmol. Vis. Sci. 45, 3165.

Kim, S.K., Boothroyd, J.C., 2005. Stage-specific expression of surface antigens by *Toxoplasma gondii* as a mechanism to facilitate parasite persistence. J. Immunol. 174, 8038–8048.

Kikumura, A., Fang, H., Mun, H.S., Uemura, N., Makino, M., Sayama, Y., Norose, K., Aosai, F., 2010. Protective immunity against lethal anaphylactic reaction in *Toxoplasma gondii*-infected mice by DNA vaccination with *T. gondii*-derived heat shock protein 70 gene. Parasitol. Int. 59, 105–111.

Kodjikian, L., Hoigne, I., Adam, O., Jacquier, P., Aebi-Ochsner, C., Aebi, C., Garweg, J.G., 2004. Vertical transmission of toxoplasmosis from a chronically infected immunocompetent woman. Pediatr. Infect. Dis. J. 23, 272–274.

Koppe, J.G., Loewer-Sieger, D.H., De Roever-Bonnet, H., 1986. Results of 20-year follow-up of congenital toxoplasmosis. Lancet 1, 254–256.

Koyama, T., Shimada, S., Ohsawa, T., Omata, Y., Xuan, X., Inoue, N., Mikami, T., Saito, A., 2000. Antigens expressed in feline enteroepithelial-stages parasites of *Toxoplasma gondii*. J. Vet. Med. Sci. 62, 1089.

Krahenbuhl, J.L., Ruskin, J., Remington, J.S., 1972. The use of killed vaccines in immunization against an intracellular parasite: *Toxoplasma gondii*. J. Immunol. 108, 425.

Krishnan, L., Guilbert, L.J., Russell, A.S., Wegmann, T.G., Mosmann, T.R., Belosevic, M., 1996a. Pregnancy impairs resistance of C57BL/6 mice to *Leishmania major* infection and causes decreased antigen-specific IFNgamma response and increased production of T helper 2 cytokines. J. Immunol. 156, 644–652.

Krishnan, L., Guilbert, L.J., Wegmann, T.G., Belosevic, M., Mosmann, T.R., 1996b. T helper 1 response against *Leishmania major* in pregnant C57BL/6 mice increases implantation failure and foetal resorptions. Correlation with increased IFNgamma and TNF and reduced IL-10 production by placental cells. J. Immunol. 156, 653–662.

Kyprianou, M., 2005. Speech to the Animal Welfare Intergroup of the European Parliament. SPEECH/05/335. European Commission.

Laguía-Becher, M., Martín, V., Kraemer, M., Corigliano, M., Yacono, M.L., Goldman, A., Clemente, M., 2010. Effect of codon optimization and subcellular targeting on *Toxoplasma gondii* antigen SAG1 expression in tobacco leaves to use in subcutaneous and oral immunization in mice. BMC Biotechnol. 10, 52.

Lau, Y.L., Thiruvengadam, G., Lee, W.W., Fong, M.Y., 2011. Immunogenic characterization of the chimeric surface antigen 1 and 2 (SAG1/2) of *Toxoplasma gondii* expressed in the yeast *Pichia pastoris*. Parasitol. Res. 109, 871–878.

Letscher-Bru, V., Pfaff, A.W., Abou-Bacar, A., Filisetti, D., Antoni, E., Villard, O., Klein, J.P., Candolfi, E., 2003. Vaccination with *Toxoplasma gondii* SAG-1 protein is protective against congenital toxoplasmosis in BALB/c mice but not in CBA/J mice. Infect. Immun. 71, 6615–6619.

Leyva, R., Herion, P., Saavedra, R., 2001. Genetic immunization with plasmid DNA coding for the ROP2 protein of *Toxoplasma gondii*. Parasitol. Res. 87, 70–79.

Li, B., Oledzka, G., McFarlane, R.G., Spellerberg, M.B., Smith, S.M., Gelder, F.B., Kur, J., Stankiewicz, M., 2010. Immunological response of sheep to injections of plasmids encoding *Toxoplasma gondii* SAG1 and ROP1 genes. Parasite Immunol. 32, 671–683.

Li, J., Han, Q., Gong, P., Yang, T., Ren, B., Li, S., Zhang, X., 2012. *Toxoplasma gondii* rhomboid protein 1 (TgROM1) is a potential vaccine candidate against toxoplasmosis. Vet. Parasitol. 184, 154–160.

Li, W.S., Chen, Q.X., Ye, J.X., Xie, Z.X., Chen, J., Zhang, L.F., 2011. Comparative evaluation of immunization with recombinant protein and plasmid DNA vaccines of fusion antigen ROP2 and SAG1 from *Toxoplasma gondii* in mice: cellular and humoral immune responses. Parasitol. Res. 109, 637–644.

Lindsay, D.S., Blagburn, B.L., Dubey, J.P., 1993. Safety and results of challenge of weaned pigs given a temperature-sensitive mutant of *Toxoplasma gondii*. J. Parasitol. 79, 71–76.

Liu, M.M., Yuan, Z.G., Peng, G.H., Zhou, D.H., He, X.H., Yan, C., Yin, C.C., He, Y., Lin, R.Q., Song, H.Q., Zhu, X.Q., 2010a. *Toxoplasma gondii* microneme protein 8 (MIC8) is a potential vaccine candidate against toxoplasmosis. Parasitol. Res. 106, 1079–1084.

Liu, Q., Shang, L., Jin, H., Wei, F., Zhu, X.Q., Gao, H., 2010b. The protective effect of a *Toxoplasma gondii* SAG1 plasmid DNA vaccine in mice is enhanced with IL-18. Res. Vet. Sci. 89, 93–97.

Liu, Q., Gao, S., Jiang, L., Shang, L., Men, J., Wang, Z., Zhai, Y., Xia, Z., Hu, R., Zhang, X., Zhu, X.Q., 2008. A recombinant pseudorabies virus expressing TgSAG1 protects against challenge with the virulent *Toxoplasma gondii* RH strain and pseudorabies in BALB/c mice. Microbes Infect. 10, 1355–1362.

Liu, S., Shi, L., Cheng, Y.B., Fan, G.X., Ren, H.X., Yuan, Y.K., 2009. Evaluation of protective effect of multi-epitope DNA vaccine encoding six antigen segments of *Toxoplasma gondii* in mice. Parasitol. Res. 105, 267–274.

Lopez, A., Dietz, V.J., Wilson, M., Navin, T.R., Jones, J.L., 2000. Preventing congenital toxoplasmosis. MMWR Recomm. Rep. 49, 59–68.

Lourenço, E.V., Bernardes, E.S., Silva, N.M., Mineo, J.R., Panunto-Castelo, A., Roque-Barreira, M.C., 2006. Immunization with MIC1 and MIC4 induces protective immunity against *Toxoplasma gondii*. Microbes Infect. 8, 1244–1251.

Lu, F., Huang, S., Kasper, L.H., 2009. The temperature-sensitive mutants of *Toxoplasma gondii* and ocular toxoplasmosis. Vaccine 27, 573–580.

Lu, F., Huang, S., Hu, M.S., Kasper, L.H., 2005. Experimental ocular toxoplasmosis in genetically susceptible and resistant mice. Infect. Immun. 73, 5160–5165.

Luft, B.J., Remington, J.S., 1982. Effect of pregnancy on resistance to *Listeria monocytogenes* and *Toxoplasma gondii* infections in mice. Infect. Immun. 38, 1164–1171.

Lunden, A., Lovgren, K., Uggla, A., Araujo, F.G., 1993. Immune responses and resistance to *Toxoplasma gondii* in mice immunized with antigens of the parasite incorporated into immunostimulating complexes. Infect. Immun. 61, 2639–2643.

Lunden, A., Parmley, S.F., Bengtsson, K.L., Araujo, F.G., 1997. Use of a recombinant antigen, SAG2, expressed as a glutathione-S-transferase fusion protein to immunize mice against *Toxoplasma gondii*. Parasitol. Res. 83, 6–9.

Lyons, R.E., McLeod, R., Roberts, C.W., 2002. *Toxoplasma gondii* tachyzoite-bradyzoite interconversion. Trends Parasitol. 18, 198–201.

Machado, A.V., Caetano, B.C., Barbosa, R.P., Salgado, A.P., Rabelo, R.H., Garcia, C.C., Bruna-Romero, O., Escriou, N., Gazzinelli, R.T., 2010. Prime and boost immunization with influenza and adenovirus encoding the *Toxoplasma gondii* surface antigen 2 (SAG2) induces strong protective immunity. Vaccine 28, 3247–3256.

Mann, M., Hendrickson, R.C., Pandey, A., 2001. Analysis of proteins and proteomes by mass spectrometry. Annu. Rev. Biochem. 70, 437–473.

Makino, M., Uemura, N., Moroda, M., Kikumura, A., Piao, L.X., Mohamed, R.M., Aosai, F., 2011. Innate immunity in DNA vaccine with *Toxoplasma gondii*-heat shock protein 70 gene that induces DC activation and Th1 polarization. Vaccine 29, 1899–1905.

Mateus-Pinilla, N.E., Dubey, J.P., Choromanski, L., Weigel, R.M., 1999. A field trial of the effectiveness of a feline *Toxoplasma gondii* vaccine in reducing *T. gondii* exposure for swine. J. Parasitol. 85, 855.

Martin, V., Supanitsky, A., Echeverria, P.C., Litwin, S., Tanos, T., De Roodt, A.R., Guarnera, E.A., Angel, S.O., 2004. Recombinant GRA4 or ROP2 protein combined with alum or the gra4 gene provides partial protection in chronic murine models of toxoplasmosis. Clin. Diagn. Lab. Immunol. 11, 704–710.

Martínez-Gómez, F., García-González, L.F., Mondragón-Flores, R., Bautista-Garfias, C.R., 2009. Protection against *Toxoplasma gondii* brain cyst formation in mice immunized with *Toxoplasma gondii* cytoskeleton proteins and *Lactobacillus casei* as adjuvant. Vet. Parasitol. 160, 311–315.

McColgan, C., Buxton, D., Blewett, D.A., 1988. Titration of *Toxoplasma gondii* oocysts in non-pregnant sheep and the effects of subsequent challenge during pregnancy. Vet. Rec. 123, 467.

McKeand, J.B., 2000. Vaccine development and diagnostics of Dictyocaulus viviparus. Parasitology 120 (Suppl), S17–S23.

McLeod, R., Estes, R.G., Mack, D.G., 1985. Effects of adjuvants and *Toxoplasma gondii* antigens on immune response and outcome of peroral *T. gondii* challenge. Trans. R. Soc. Trop. Med. Hyg. 79, 800–804.

McLeod, R., Frenkel, J.K., Estes, R.G., Mack, D.G., Eisenhauer, P.B., Gibori, G., 1988. Subcutaneous and intestinal vaccination with tachyzoites of *Toxoplasma gondii* and acquisition of immunity to peroral and congenital toxoplasma challenge. J. Immunol. 140, 1632.

Mendes, E.A., Caetano, B.C., Penido, M.L., Bruna-Romero, O., Gazzinelli, R.T., 2011. MyD88-dependent protective immunity elicited by adenovirus 5 expressing the surface antigen 1 from *Toxoplasma gondii* is mediated by CD8(+) T lymphocytes. Vaccine 29, 4476–4484.

Mévélec, M.N., Bout, D., Desolme, B., Marchand, H., Magné, R., Bruneel, O., Buzoni-Gatel, D., 2005. Evaluation of protective effect of DNA vaccination with genes encoding antigens GRA4 and SAG1 associated with GM-CSF plasmid, against acute, chronical and congenital toxoplasmosis in mice. Vaccine 23, 4489–4499.

Mévélec, M.N., Ducournau, C., Bassuny Ismael, A., Olivier, M., Sèche, E., Lebrun, M., Bout, D., Dimier-Poisson, I., 2010. Mic1-3 Knock-out Toxoplasma gondii is a good candidate for a vaccine against *T. gondii*-induced abortion in sheep. Vet. Res. 41 (4), 49.

, M.N., Buzoni-Gatel, D., Bourguin, I., Chardes, T., Dubremetz, J.F., Bout, D., 1998. Mapping of B epitopes in GRA4, a dense granule antigen of *Toxoplasma gondii* and protection studies using recombinant proteins administered by the oral route. Parasite Immunol. 20, 183–195.

Meissner, M., Brecht, S., Bujard, H., Soldati, D., 2001. Modulation of myosin A expression by a newly established tetracycline repressor-based inducible system in *Toxoplasma gondii*. Nucleic. Acids Res. 29, E115.

Meissner, M., Reiss, M., Viebig, N., Carruthers, V.B., Toursel, C., Tomavo, S., Ajioka, J.W., Soldati, D., 2002a. A family of transmembrane microneme proteins of *Toxoplasma gondii* contain EGF-like domains and function as escorters. J. Cell Sci. 115, 563–574.

Meissner, M., Schluter, D., Soldati, D., 2002b. Role of *Toxoplasma gondii* myosin A in powering parasite gliding and host cell invasion. Science 298, 837.

Mercier, C., Howe, D.K., Mordue, D., Lingnau, M., Sibley, L.D., 1998. Targeted disruption of the GRA2 locus in *Toxoplasma gondii* decreases acute virulence in mice. Infect. Immun. 66, 4176.

Min, J., Qu, D., Li, C., Song, X., Zhao, Q., Li, X.A., Yang, Y., Liu, Q., He, S., Zhou, H., 2012. Enhancement of protective immune responses induced by *Toxoplasma gondii* dense granule antigen 7 (GRA7) against toxoplasmosis in mice using a prime-boost vaccination strategy. Vaccine 30, 5631–5636.

Mineo, J.R., Kasper, L.H., 1994. Attachment of *Toxoplasma gondii* to host cells involves major surface protein, SAG-1 (P30). Exp. Parasitol. 79, 11.

Mineo, J.R., McLeod, R., Mack, D., Smith, J., Khan, I.A., Ely, K.H., Kasper, L.H., 1993. Antibodies to *Toxoplasma gondii* major surface protein (SAG-1, P30) inhibit infection of host cells and are produced in murine intestine after peroral infection. J. Immunol. 150, 3951–3964.

Mishima, M., Xuan, X., Yokoyama, N., Igarashi, I., Fujisaki, K., Nagasawa, H., Mikami, T., 2002. Recombinant feline herpesvirus type 1 expressing *Toxoplasma gondi* ROP2 antigen inducible protective immunity in cats. Parasitol. Res. 88, 144.

Mishima, M., Xuan, X., Shioda, A., Omata, Y., Fujisaki, K., Nagasawa, H., Mikami, T., 2001. Modified protection against *Toxoplasma gondii* lethal infection and brain cyst formation by vaccination with SAG2 and SRS1. J. Vet. Med. Sci. 63, 433–438.

Mital, J., Meissner, M., Soldati, D., Ward, G.E., 2005. Conditional expression of *Toxoplasma gondii* apical membrane antigen-1 (TgAMA1) demonstrates that TgAMA1 plays a critical role in host cell invasion. Mol. Biol. Cell 16, 4341.

Mohamed, R.M., Aosai, F., Chen, M., Mun, H.S., Norose, K., Belal, U.S., Piao, L.X., Yano, A., 2003. Induction of protective immunity by DNA vaccination with *Toxoplasma gondii* HSP70, HSP30 and SAG1 genes. Vaccine 21, 2852–2861.

Moiré, N., Dion, S., Lebrun, M., Dubremetz, J.F., Dimier-Poisson, I., 2009. Mic1-3KO tachyzoite a live attenuated vaccine candidate against toxoplasmosis derived from a type I strain shows features of type II strain. Exp. Parasitol. 23, 111–117.

Mun, H.S., Aosai, F., Yano, A., 1999. Role of *Toxoplasma gondii* HSP70 and *Toxoplasma gondii* HSP30/bag1 in antibody formation and prophylactic immunity in mice experimentally infected with *Toxoplasma gondii*. Microbiol. Immunol. 43, 471–479.

Munday, B.L., 1972. Transmission of *Toxoplasma* infection from chronically infected ewes to their lambs. Br. Vet. J. 128, lxxi.

Nie, H., Fang, R., Xiong, B.Q., Wang, L.X., Hu, M., Zhou, Y.Q., Zhao, J.L., 2011. Immunogenicity and protective efficacy of two recombinant pseudorabies viruses expressing *Toxoplasma gondii* SAG1 and MIC3 proteins. Vet. Parasitol. 2011 (181), 215–221.

Nielsen, H.V., Di Cristina, M., Beghetto, E., Spadoni, A., Petersen, E., Gargano, N., 2006. *Toxoplasma gondii*: DNA vaccination with bradyzoite antigens induces protective immunity in mice against oral infection with parasite cysts. Exp. Parasitol. 12, 274–279.

Nielsen, H.V., Lauemoller, S.L., Christiansen, L., Buus, S., Fomsgaard, A., Petersen, E., 1999. Complete protection against lethal *Toxoplasma gondii* infection in mice immunized with a plasmid encoding the SAG1 gene. Infect. Immun. 67, 6358–6363.

Opsteegh, M., Prickaerts, S., Frankena, K., Evers, E.G., 2011. A quantitative microbial risk assessment for meatborne *Toxoplasma gondii* infection in The Netherlands. Int. J. Food Microbiol. 150, 103–114.

Overnes, G., Nesse, L.L., Waldeland, H., Lovgren, K., Gudding, R., 1991. Immune response after immunization with an experimental *Toxoplasma gondii* ISCOM vaccine. Vaccine 9, 25–28.

Paoletti, E., 1996. Applications of pox virus vectors to vaccination: an update. Proc. Natl. Acad. Sci. U.S.A. 93, 11349.

Parmley, S., Slifer, T., Araujo, F., 2002. Protective effects of immunization with a recombinant cyst antigen in mouse models of infection with *Toxoplasma gondii* tissue cysts. J. Infect. Dis. 185 (Suppl. 1), S90–S95.

Pastoret, P.P., 2002. Rabies. Virus Res. 82, 61.

Paulino, J.P., Vitor, R.W., 1999. Experimental congenital toxoplasmosis in Wistar and Holtzman rats. Parasite 6, 63–66.

Peng, G.H., Yuan, Z.G., Zhou, D.H., He, X.H., Liu, M.M., Yan, C., Yin, C.C., He, Y., Lin, R.Q., Zhu, X.Q., 2009. *Toxoplasma gondii* microneme protein 6 (MIC6) is a potential vaccine candidate against toxoplasmosis in mice. Vaccine 27, 6570–6574.

Pereira-Bueno, J., Quintanilla-Gozalo, A., Perez-Perez, V., Alvarez-Garcia, G., Collantes-Fernandez, E., Ortega-Mora, L.M., 2004. Evaluation of ovine abortion associated with *Toxoplasma gondii* in Spain by different diagnostic techniques. Vet. Parasitol. 121, 33.

Perkins, E.S., 1973. Ocular toxoplasmosis. Br. J. Ophthalmol. 57, 1–17.

Petersen, E., Nielsen, H.V., Christiansen, L., Spenter, J., 1998. Immunization with E. coli produced recombinant *T. gondii* SAG1 with alum as adjuvant protect mice against lethal infection with *Toxoplasma gondii*. Vaccine 16, 1283–1289.

Petersen, E., Pollak, A., Reiter-Owona, I., 2001. Recent trends in research on congenital toxoplasmosis. Int. J. Parasitol. 31, 115–144.

Pfefferkorn, E.R., Pfefferkorn, L.C., 1976. *Toxoplasma gondii*: isolation and preliminary characterization of temperature-sensitive mutants. Exp. Parasitol. 39, 365.

Pinckney, R.D., Lindsay, D.S., Blagburn, B.L., Boosinger, T.R., Mclaughlin, S.A., Dubey, J.P., 1994. Evaluation of the safety and efficacy of vaccination of nursing pigs with living tachyzoites of two strains of *Toxoplasma gondii*. J. Parasitol. 80, 438–448.

Pizza, M., Scarlato, V., Masignani, V., Giuliani, M.M., Arico, B., Comanducci, M., Jennings, G.T., Baldi, L., Bartolini, E., Capecchi, B., Galeotti, C.L., Luzzi, E., Manetti, R., Marchetti, E., Mora, M., Nuti, S., Ratti, G., Santini, L., Savino, S., Scarselli, M., Storni, E., Zuo, P., Broeker, M., Hundt, E., Knapp, B., Blair, E., Mason, T., Tettelin, H., Hood, D.W., Jeffries, A.C., Saunders, N.J., Granoff, D.M., Venter, J.C., Moxon, E.R., Grandi, G., Rappuoli, R., 2000. Identification of vaccine candidates against serogroup B meningococcus by whole-genome sequencing. Science 287, 1816–1820.

Pollock, K.G., McNeil, K.S., Mottram, J.C., Lyons, R.E., Brewer, J.M., Scott, P., Coombs, G.H., Alexander, J., 2003. The *Leishmania mexicana* cysteine protease, CPB2.8, induces potent Th2 responses. J. Immunol. 170, 1746–1753.

Qu, D., Yu, H., Wang, S., Cai, W., Du, A., 2009. Induction of protective immunity by multiantigenic DNA vaccine delivered in attenuated *Salmonella typhimurium* against *Toxoplasma gondii* infection in mice. Vet. Parasitol. 166, 220–227.

Qu, D., Wang, S., Cai, W., Du, A., 2008. Protective effect of a DNA vaccine delivered inattenuated *Salmonella typhimurium* against *Toxoplasma gondii* infection in mice. Vaccine 26, 4541–4548.

Quan, J.H., Chu, J.Q., Ismail, H.A., Zhou, W., Jo, E.K., Cha, G.H., Lee, Y.H., 2012. Induction of protective immune responses by a multiantigenic DNA vaccine encoding GRA7 and ROP1 of *Toxoplasma gondii*. Clin. Vaccine Immunol. 19, 666–674.

Rachinel, N., Buzoni-Gatel, D., Dutta, C., Mennechet, F.J., Luangsay, S., Minns, L.A., Grigg, M.E., Tomavo, S., Boothroyd, J.C., Kasper, L.H., 2004. The induction of acute ileitis by a single microbial antigen of *Toxoplasma gondii*. J. Immunol. 173, 2725.

Rashid, I., Hedhli, D., Moiré, N., Pierre, J., Debierre-Grockiego, F., Dimier-Poisson, I., Mévélec, M.N., 2011. Immunological responses induced by a DNA vaccine expressing RON4 and by immunogenic recombinant protein RON4 failed to protect mice against chronic toxoplasmosis. Vaccine 29, 8838–8846.

Reichmann, G., Dlugonska, H., Fischer, H.G., 2002. Characterization of TgROP9 (p36), a novel rhoptry protein of *Toxoplasma gondii* tachyzoites identified by T-cell clone. Mol. Biochem. Parasitol. 119, 43–54.

Roberts, C.W., Alexander, J., 1992. Studies on a murine model of congenital toxoplasmosis: vertical disease transmission only occurs in BALB/c mice infected for the first time during pregnancy. Parasitology 104 (Pt 1), 19.

Roberts, C.W., Brewer, J.M., Alexander, J., 1994. Congenital toxoplasmosis in the BALB/c mouse: prevention of vertical disease transmission and foetal death by vaccination. Vaccine 12, 1389.

Roberts, C.W., Roberts, F., Lyons, R.E., Kirisits, M.J., Mui, E.J., Finnerty, J., Johnson, J.J., Ferguson, D.J., Coggins, J.R., Krell, T., Coombs, G.H., Milhous, W.K., Kyle, D.E., Tzipori, S., Barnwell, J., Dame, J.B., Carlton, J., McLeod, R., 2002. The shikimate pathway and its branches in apicomplexan parasites. J. Infect. Dis. 185 (Suppl. 1), S25–S36.

Roberts, C.W., Satoskar, A., Alexander, J., 1996. Sex steroids, pregnancy-associated hormones and immunity to parasitic infection. Parasitol. Today 12, 382–388.

Roberts, C.W., Walker, W., Alexander, J., 2001. Sex-associated hormones and immunity to protozoan parasites. Clin. Microbiol. Rev. 14, 476–488.

Roberts, F., McLeod, R., 1999. Pathogenesis of toxoplasmic retinochoroiditis. Parasitol. Today 15, 51–57.

Rodger, S.M., Maley, S.W., Wright, S.E., MacKellar, A., Sales, J., Buxton, D., 2005. Toxoplasmosis in sheep; the possibility of endogenous transplacental transmission. Wiadomosci. Parazytologiczne 51, s15–s18.

Rodrigues, E.G., Zavala, F., Eichinger, D., Wilson, J.M., Tsuji, M., 1997. Single immunizing dose of recombinant adenovirus efficiently induces CD8+ T-cell-mediated protective immunity against malaria. J. Immunol. 158, 1268.

Roque-Resendiz, J.L., Rosales, R., Herion, P., 2004. MVA ROP2 vaccinia virus recombinant as a vaccine candidate for toxoplasmosis. Parasitology 128, 397.

Rosenberg, C., De Craeye, S., Jongert, E., Gargano, N., Beghetto, E., Del Porto, P., Vorup-Jensen, T., Petersen, E., 2009. Induction of partial protection against infection with *Toxoplasma gondii* genotype II by DNA vaccination with recombinant chimeric tachyzoite antigens. Vaccine 27, 2489–2498.

Saavedra, R., Leyva, R., Tenorio, E.P., Haces, M.L., Rodriguez-Sosa, M., Terrazas, L.I., Herion, P., 2004. CpG-containing ODN has a limited role in the protection against *Toxoplasma gondii*. Parasite Immunol. 26, 67–73.

Sánchez, V.R., Pitkowski, M.N., Fernández Cuppari, A.V., Rodríguez, F.M., Fenoy, I.M., Frank, F.M., Goldman, A., Corral, R.S., Martin, V., 2011. Combination of CpG-oligodeoxynucleotides with recombinant ROP2 or GRA4 proteins induces protective immunity against *Toxoplasma gondii* infection. Exp. Parasitol. 128 (4), 448–453.

Scorza, T., D'Souza, S., Laloup, M., Dewit, J., De Braekeleer, J., Verschueren, H., Vercammen, M., Huygen, K., Jongert, E., 2003. A GRA1 DNA vaccine primes cytolytic CD8(+) T-cells to control acute *Toxoplasma gondii* infection. Infect. Immun. 71, 309–316.

Seah, S.K., Hucal, G., 1975. The use of irradiated vaccine in immunization against experimental murine toxoplasmosis. Can. J. Microbiol. 21, 1379.

Shang, L., Liu, Q., Liu, W., Men, J., Gao, S., Jiang, L., Wang, Z., Zhai, Y., Jin, H., Lian, H., Chen, C., Xia, Z., Yuan, Z., Zhu, X.Q., 2009. Protection in mice immunized with a heterologous prime-boost regime using DNA and recombinant pseudorabies expressing TgSAG1 against *Toxoplasma gondii* challenge. Vaccine 27, 2741–2745.

Shirahata, T., Muroya, N., Ohta, C., Goto, H., Nakane, A., 1992. Correlation between increased susceptibility to primary *Toxoplasma gondii* infection and depressed production of gamma interferon in pregnant mice. Microbiol. Immunol. 36, 81–91.

Shirahata, T., Muroya, N., Ohta, C., Goto, H., Nakane, A., 1993. Enhancement by recombinant human interleukin 2 of host resistance to *Toxoplasma gondii* infection in pregnant mice. Microbiol. Immunol. 37, 583–590.

Siachoque, H., Guzman, F., Burgos, J., Patarroyo, M.E., Gomez Marin, J.E., 2006. *Toxoplasma gondii*: immunogenicity and protection by P30 peptides in a murine model. Exp. Parasitol. 114, 62–65.

Soldati, D., Kim, K., Kampmeier, J., Dubremetz, J.F., Boothroyd, J.C., 1995. Complementation of a *Toxoplasma gondii* ROP1 knock-out mutant using phleomycin selection. Mol. Biochem. Parasitol. 74, 87.

Stacey, K.J., Blackwell, J.M., 1999. Immunostimulatory DNA as an adjuvant in vaccination against *Leishmania major*. Infect. Immun. 67, 3719–3726.

Stanley, A.C., Buxton, D., Innes, E.A., Huntley, J.F., 2004. Intranasal immunization with *Toxoplasma gondii* tachyzoite antigen encapsulated into PLG microspheres induces humoral and cell-mediated immunity in sheep. Vaccine 22, 3929.

Sun, X.M., Zou, J., A A, E.S., Yan, W.C., Liu, X.Y., Suo, X., Wang, H., Chen, Q.J., 2011. DNA vaccination with a gene encoding *Toxoplasma gondii* GRA6 induces partial protection against toxoplasmosis in BALB/c mice. Parasit. Vectors 9 (4), 213.

Supply, P., Sutton, P., Coughlan, S.N., Bilo, K., Saman, E., Trees, A.J., Cesbron Delauw, M.F., Locht, C., 1999. Immunogenicity of recombinant BCG producing the GRA1 antigen from *Toxoplasma gondii*. Vaccine 17, 705–714.

Tan, F., Hu, X., Luo, F.J., Pan, C.W., Chen, X.G., 2011. Induction of protective Th1 immune responses in mice by vaccination with recombinant *Toxoplasma gondii* nucleoside triphosphate hydrolase-II. Vaccine 29, 2742–2748.

Tan, T.G., Mui, E., Cong, H., Witola, W.H., Montpetit, A., Muench, S.P., Sidney, J., Alexander, J., Sette, A., Grigg, M.E., Maewal, A., Mcleod, R., 2010. Identification of *T. gondii* epitopes, adjuvants, and host genetic factors that influence protection of mice and humans. Vaccine 28, 3977–3989.

Tenter, A.M., Heckeroth, A.R., Weiss, L.M., 2000. *Toxoplasma gondii*: from animals to humans. Int. J. Parasitol. 30, 1217.

Thouvenin, M., Candolfi, E., Villard, O., Klein, J.P., Kien, T., 1997. Immune response in a murine model of congenital toxoplasmosis: increased susceptibility of pregnant mice and transplacental passage of *Toxoplasma gondii* are type 2-dependent. Parassitologia 39, 279–283.

Vallochi, A.L., Nakamura, M.V., Schlesinger, D., Martins, M.C., Silveira, C., Belfort Jr., R., Rizzo, L.V., 2002. Ocular toxoplasmosis: more than just what meets the eye. Scand. J. Immunol. 55, 324.

Van Poppel, N.F.J., Welagen, J., Duisters, R.F.J.J., Vermeulen, A.N., Schaap, D., 2006. Tight control of transcription in *Toxoplasma gondii* using an alternative tet repressor. Int. J. Parasitol. 36, 443–452.

Velge-Roussel, F., Moretto, M., Buzoni-Gatel, D., Dimier-Poisson, I., Ferrer, M., Hoebeke, J., Bout, D., 1997. Differences in immunological response to a *T. gondii* protein (SAG1) derived peptide between two strains of mice: effect on protection in *T. gondii* infection. Mol. Immunol. 34, 1045–1053.

Venturini, M.C., Bacigalupe, D., Venturini, L., Rambeaud, M., Basso, W., Unzaga, J.M., Perfumo, C.J., 2004. Seroprevalence of *Toxoplasma gondii* in sows from slaughterhouses and in pigs from an indoor and an outdoor farm in Argentina. Vet. Parasitol. 124, 161.

Vercammen, M., Scorza, T., Huygen, K., De Braekeleer, J., Diet, R., Jacobs, D., Saman, E., Verschueren, H., 2000. DNA vaccination with genes encoding *Toxoplasma gondii* antigens GRA1, GRA7, and ROP2 induces partially protective immunity against lethal challenge in mice. Infect. Immun. 68, 38–45.

Vermeulen, A.N., 1998. Progress in recombinant vaccine development against coccidiosis. A review and prospects into the next millennium. Int. J. Parasitol. 28, 1121.

Vermeulen, A.N., Schaap, D.C., Schetters, T.P., 2001. Control of coccidiosis in chickens by vaccination. Vet. Parasitol. 100, 13.

Vogel, N., Kirisits, M., Michael, E., Bach, H., Hostetter, M., Boyer, K., Simpson, R., Holfels, E., Hopkins, J., Mack, D., Mets, M.B., Swisher, C.N., Patel, D., Roizen, N., Stein, L., Stein, M., Withers, S., Mui, E., Egwuagu, C., Remington, J., Dorfman, R., McLeod, R., 1996. Congenital toxoplasmosis transmitted from an immunologically competent mother infected before conception. Clin. Infect. Dis. 23, 1055–1060.

Waldeland, H., Frenkel, J.K., 1983. Live and killed vaccines against toxoplasmosis in mice. J. Parasitol. 69, 60.

Wang, X., Kang, H., Kikuchi, T., Suzuki, Y., 2004. Gamma interferon production, but not perforin-mediated cytolytic activity, of T-cells is required for prevention of toxoplasmic encephalitis in BALB/c mice genetically resistant to the disease. Infect. Immun. 72, 4432–4438.

Wang, H., He, S., Yao, Y., Cong, H., Zhao, H., Li, T., Zhu, X.Q., 2009. *Toxoplasma gondii*: protective effect of an intranasal SAG1 and MIC4 DNA vaccine in mice. Exp. Parasitol. 122, 226–232.

Wang, H., Liu, Q., Liu, K., Zhong, W., Gao, S., Jiang, L., An, N., 2007. Immune response induced by recombinant *Mycobacterium bovis* BCG expressing ROP2 gene of *Toxoplasma gondii*. Parasitol. Int. 56, 263–268.

Wang, Y., Wang, M., Wang, G., Pang, A., Fu, B., Yin, H., Zhang, D., 2011. Increased survival time in mice vaccinated with a branched lysine multiple antigenic peptide containing B- and T-cell epitopes from *T. gondii* antigens. Vaccine 29, 8619–8623.

Webster, D.P., Dunachie, S., Vuola, J.M., Berthoud, T., Keating, S., Laidlaw, S.M., McConkey, S.J., Poulton, I., Andrews, L., Andersen, R.F., Bejon, P., Butcher, G., Sinden, R., Skinner, M.A., Gilbert, S.C., Hill, A.V., 2005. Enhanced T-cell-mediated protection against malaria in human challenges by using the recombinant poxviruses FP9 and modified vaccinia virus Ankara. Proc. Natl. Acad. Sci. U.S.A. 102, 4836.

Wilkins, M.F., O'Connell, E., 1992. Vaccination of sheep against *Toxoplasma* abortion. Surveillance 19, 20.

Wilkins, M.F., O'Connell, E., Te Punga, W.A., 1987. Toxoplasmosis in sheep. I. Effect of a killed vaccine on lambing losses caused by experimental challenge with *Toxoplasma gondii*. N. Z. Vet. J. 35, 31.

Wilkins, M.F., O'Connell, E., Te Punga, W.A., 1988. Toxoplasmosis in sheep III. Further evaluation of the ability of a live *Toxoplasma gondii* vaccine to prevent lamb losses and reduce congenital infection following experimental oral challenge. N. Z. Vet. J. 36, 86.

Williams, K.A., Scott, J.M., MacFarlane, D.E., Williamson, J.M., Elias-Jones, T.F., Williams, H., 1981. Congenital toxoplasmosis: a prospective survey in the West of Scotland. J. Infect. 3, 219–229.

Williams, R.H., Morley, E.K., Hughes, J.M., Duncanson, P., Terry, R.S., Smith, J.E., Hide, G., 2005. High levels of congenital transmission of *Toxoplasma gondii* in longitudinal and cross-sectional studies on sheep farms provides evidence of vertical transmission in ovine hosts. Parasitology 130, 301–307.

Wilson, W.B., Sharpe, J.A., Deck, J.H., 1980. Cerebral blindness and oculomotor nerve palsies in toxoplasmosis. Am. J. Ophthalmol. 89, 714–718.

Xiang, W., Qiong, Z., Li-Peng, L., Kui, T., Jian-Wu, G., Heng-Ping, S., 2009. The location of invasion-related protein MIC3 of *Toxoplasma gondii* and protective effect of its DNA vaccine in mice. Vet. Parasitol. 166, 1–7.

Xue, M., He, S., Zhang, J., Cui, Y., Yao, Y., Wang, H., 2008a. Comparison of cholera toxin A2/B and murine interleukin-12 as adjuvants of *Toxoplasma* multi-antigenic SAG1-ROP2 DNA vaccine. Exp. Parasitol. 119, 352–357.

Xue, M., He, S., Cui, Y., Yao, Y., Wang, H., 2008b. Evaluation of the immune response elicited by multi-antigenic DNA vaccine expressing SAG1, ROP2 and GRA2 against *Toxoplasma gondii*. Parasitol. Int. 57, 424–429.

Yan, Z.X., Meyer, T.F., 1996. Mixed population approach for vaccination with live recombinant Salmonella strains. J. Biotechnol. 44, 197.

Yan, H., Yan, H., Tao, Y., Chen, H., Li, G., Gong, W., Jiao, H., Tian, F., Ji, M., 2012a. Application and expression of *Toxoplasma* gondii surface antigen 2 (SAG2) and rhoptry protein 2 (ROP2) from recombinant *Escherichia coli* strain. Trans. R. Soc. Trop. Med. Hyg. 106, 356–362.

Yan, H.K., Yuan, Z.G., Song, H.Q., Petersen, E., Zhou, Y., Ren, D., Zhou, D.H., Li, H. X., Lin, R.Q., Yang, G.L., Zhu, X.Q., 2012b. Vaccination with a DNA vaccine coding for perforin-like protein 1 and MIC6 induces significant protective immunity against *Toxoplasma gondii*. Clin. Vaccine. Immunol. 19, 684–689.

Yan, H.K., Yuan, Z.G., Petersen, E., Zhang, X.X., Zhou, D.H., Liu, Q., He, Y., Lin, R.Q., Xu, M.J., Chen, X.L., Zhong, X.L., Zhu, X.Q., 2011. *Toxoplasma gondii*: protective immunity against experimental toxoplasmosis induced by a DNA vaccine encoding the perforin-like protein 1. Exp. Parasitol. 128, 38–43.

Yang, C.D., Chang, G.N., Chao, D., 2004. Protective immunity against *Toxoplasma gondii* in mice induced by a chimeric protein rSAG1/2. Parasitol. Res. 92, 58–64.

Yao, Y., He, S.Y., Wang, H.X., Zhou, H.Y., Zhao, H., Li, T., Xue, M.F., Zhu, X.Q., 2010. Protective immunity induced in mice by multiantigenic DNA vaccine with genes encoding SAG1 and MIC8 of *Toxoplasma gondii*. Zhongguo Ji Sheng Chong Xue Yu Ji Sheng Chong Bing Za Zhi. 28, 81–88.

Yarovinsky, F., Zhang, D., Andersen, J.F., Bannenberg, G.L., Serhan, C.N., Hayden, M.S., Hieny, S., Sutterwala, F.S., Flavell, R.A., Ghosh, S., Sher, A., 2005. TLR11 activation of dendritic cells by a protozoan profilin-like protein. Science 308, 1626–1629.

Yu, L., Yamagishi, J., Zhang, S., Jin, C., Aboge, G.O., Zhang, H., Zhang, G., Tanaka, T., Fujisaki, K., Nishikawa, Y., Xuan, X., 2012. Protective effect of a prime-boost strategy with plasmid DNA followed by recombinant adenovirus expressing TgAMA1 as vaccines against *Toxoplasma gondii* infection in mice. Parasitol. Int. 61, 481–486.

Yuan, Z.G., Zhang, X.X., Lin, R.Q., Petersen, E., He, S., Yu, M., He, X.H., Zhou, D. H., He, Y., Li, H.X., Liao, M., Zhu, X.Q., 2011a. Protective effect against toxoplasmosis in mice induced by DNA immunization with gene encoding *Toxoplasma gondii* ROP18. Vaccine 29, 6614–6619.

Yuan, Z.G., Zhang, X.X., He, X.H., Petersen, E., Zhou, D.H., He, Y., Lin, R.Q., Li, X.Z., Chen, X.L., Shi, X.R., Zhong, X.L., Zhang, B., Zhu, X.Q., 2011b. Protective immunity induced by *Toxoplasma gondii* rhoptry protein 16 against toxoplasmosis in mice. Clin. Vaccine Immunol. 18, 119–124.

Zenner, L., Darcy, F., Cesbron-Delauw, M.F., Capron, A., 1993. Rat model of congenital toxoplasmosis: rate of transmission of three *Toxoplasma gondii* strains to foetuses and protective effect of a chronic infection. Infect. Immun. 61, 360–363.

Zenner, L., Estaquier, J., Darcy, F., Maes, P., Capron, A., Cesbron-Delauw, M.F., 1999. Protective immunity in the rat model of congenital toxoplasmosis and the potential of excreted-secreted antigens as vaccine components. Parasite Immunol. 21, 261.

Zhang, Y.W., Fraser, A., Balfour, A.H., Wreghitt, T.G., Gray, J.J., Smith, J.E., 1995. Serological reactivity against cyst and tachyzoite antigens of *Toxoplasma gondii* determined by FAST-ELISA. J. Clin. Pathol. 48, 908–911.

Zhang, Y.W., Kim, K., Ma, Y.F., Wittner, M., Tanowitz, H.B., Weiss, L.M., 1999. Disruption of the *Toxoplasma gondii* bradyzoite-specific gene BAG1 decreases *in vivo* cyst formation. Mol. Microbiol. 31, 691.

Zhang, G., Huang, X., Boldbaatar, D., Battur, B., Battsetseg, B., Zhang, H., Yu, L., Li, Y., Luo, Y., Cao, S., Goo, Y.K., Yamagishi, J., Zhou, J., Zhang, S., Suzuki, H., Igarashi, I., Mikami, T., Nishikawa, Y., Xuan, X., 2010. Construction of *Neospora caninum* stably expressing TgSAG1 and evaluation of its protective effects against *Toxoplasma gondii* infection in mice. Vaccine 28, 7243–7247.

Zhang, G., Huong, V.T., Battur, B., Zhou, J., Zhang, H., Liao, M., Kawase, O., Lee, E.G., Dautu, G., Igarashi, M., Nishikawa, Y., Xuan, X., 2007a. A heterologous prime-boost vaccination regime using DNA and a vaccinia virus, both expressing GRA4, induced protective immunity against *Toxoplasma gondii* infection in mice. Parasitology 134, 1339–1346.

Zhang, J., He, S., Jiang, H., Yang, T., Cong, H., Zhou, H., Zhang, J., Gu, Q., Li, Y., Zhao, Q., 2007b. Evaluation of the immune response induced by multi-antigenic DNA vaccine encoding SAG1 and ROP2 of *Toxoplasma gondii* and the adjuvant properties of murine interleukin-12 plasmid in BALB/c mice. Parasitol. Res. 101, 331–338.

Zhou, X., Berg, L., Motal, U.M., Jondal, M., 1992. *In vivo* primary induction of virus-specific CTL by immunization with 9-mer synthetic peptides. J. Immunol. Methods 153, 193–200.

Zhou, D.H., Liang, R., Yin, C.C., Zhao, F.R., Yuan, Z.G., Lin, R.Q., Song, H.Q., Zhu, X.Q., 2010. Seroprevalence of *Toxoplasma gondii* in pigs from southern China. J. Parasitol. 96, 673–674.

Zhou, H., Min, J., Zhao, Q., Gu, Q., Cong, H., Li, Y., He, S., 2012. Protective immune response against *Toxoplasma gondii* elicited by a recombinant DNA vaccine with a novel genetic adjuvant. Vaccine 30, 1800–1806.

Zorgi, N.E., Costa, A., Galisteo Jr., A.J., Do Nascimento, N., De Andrade Jr., H.F., 2011. Humoral responses and immune protection in mice immunized with irradiated *T. gondii* tachyzoites and challenged with three genetically distinct strains of *T. gondii*. Immunol. Lett. 138, 187–196.

Zulpo, D.L., Headley, S.A., Biazzono, L., Da Cunha, I.A., Igarashi, M., De Barros, L.D., Taroda, A., Cardim, S.T., Bogado, A.L., Navarro, I.T., Garcia, J.L., 2012. Oocyst shedding in cats vaccinated by the nasal and rectal routes with crude rhoptry proteins of *Toxoplasma gondii*. Exp. Parasitol. 131, 223–230.

Epilogue

This revised edition was assembled to address the numerous advances in the *Toxoplasma* field in the past five years and provide a single volume that continues to represent the breadth of the research efforts focusing upon *Toxoplasma gondii.*

As the 'omics' era has unfolded, we are faced with the challenges of organizing, understanding and integrating large amounts of new data so that we can form and test new hypotheses about the biology of *Toxoplasma gondii.* These data will build upon the large base of biological and genetic data about the organism, and now must also integrate similar data reflecting similar advances in studying human responses to infectious diseases. Many investigators are now applying new approaches and technologies to address important aspects of the host–parasite interaction. These include studies on innate and acquired immunity, studies that reflect the growing appreciation of the influence of the host microbiome on the immune response to pathogens, as well as studies that reflect our deepening understanding of how extensively *T. gondii* remodels the transcriptome, proteome and metabolome of its host. The role of the parasite in neuropsychiatric disease and immune modulation reflects our appreciation of the potential effects of chronic infection with toxoplasmosis.

We look forward to many more insights into the biology of *Toxoplasma gondii.*

LMW and KK
Bronx, New York, USA, 2013

Index

Note: Page numbers followed by "f" and "t" indicate figures and tables, respectively.

A

D

F

N

O

P

Q

R

S

T

U

CPI Antony Rowe
Chippenham, UK
2017-12-12 21:27